Local and Nonlocal Micromechanics of Heterogeneous Materials

Valeriy A. Buryachenko

Local and Nonlocal Micromechanics of Heterogeneous Materials

Springer

Valeriy A. Buryachenko
Micromechanics and Composites LLC
Cincinnati, OH, USA

ISBN 978-3-030-81783-1 ISBN 978-3-030-81784-8 (eBook)
https://doi.org/10.1007/978-3-030-81784-8

This Springer imprint is published by the registered company Springer Nature Switzerland AG
The registered company address is: Gewerbestrasse 11, 6330 Cham, Switzerland

Elena, Andrey, Nina, Alexander, Irina
for their love and support over the years

Preface

Materials that are either manufactured or occur in nature and used both in industry and in our daily lives (metals, rocks, wood, soil, suspensions, and biological tissue) are very seldom homogeneous and have complicated internal structures. Although the combination of two or more constituents to produce materials with controlled distinguish properties has been exploited since at least ancient civilization, modern composite materials were developed only a few decades ago and have found intensive application in contemporary life and in all branches of industry. The establishment of a link between the structure and properties in order to understand which kind of structure provides the necessary properties is an objective of "micromechanics," which exploits information about microtopology and properties of constituents of the heterogeneous medium for the development of mathematical models predicting the macroproperties. The problem of micromechanical modeling of the mechanical properties of engineering materials is today a crucial part of the design process, and sample testing is usually performed only during the final stage for validation of the "virtual" design. An accuracy of the classical "trial and error" testing method of the new materials and constructions is no longer be affordable in modern industry and science.

Owing to wide applications of composite materials, their modeling has been developed very intensively over recent decades, as reflected in the numerous papers and books only partially presented in the reference section of this book. A variety of materials and approaches appearing in apparently different contexts and among different scientific disciplines (solid mechanics, geophysics, solid-state physics, hydromechanics, biomechanics, chemical technology, etc.) do not allow the opportunity to investigate adequately the whole field in a single book. In light of this, I was challenged with a natural question as to why it was necessary to write another book and what is the difference between this book and the ones published earlier. In parallel with this book, there are a few fundamental books combining readily applicable results useful for material scientists with a significant contribution to progress in theoretical research itself. However, a fundamental difference of this book is a systematic analysis of statistical distributions of local microfields rather than only effective properties based on the average field concentrator factors inside the phases.

The uniqueness of this book consists of the development and expressive representation of statistical methods quantitatively describing random structures which are most adopted for the subsequent evaluation of a wide variety of macroscopic transport, electromagnetic, and elastic properties of heterogeneous media. The popular methods in micromechanics are essentially one-particle ones that are invariant with respect to statistical second- and higher-order quantities examining the association of one particle relative to other particles. This book expressively reflects the explosive progress of modern micromechanics resulting from the development of image analyses and computer-simulation methods on one hand and improved materials processing on the other hand, since processing controls the prescribed microstructure. With the appearance of new experimental techniques, it is now possible to study the microtopology of disordered materials much more deeply to understand their properties. Modern techniques are also available to design materials with morphological properties that are suitable for planned applications. This progress in micromechanics is based on methods allowing for statistical mechanics of a multiparticle system considering n-point correlation functions and direct multiparticle interaction of inclusions, and the book presents a universally rigorous scheme for both analyses of microstructures and prediction of macroscopic properties which leaves room for correction of their individual elements if improved methods are utilized for the analysis of these individual elements. The book successfully combines advanced numerical methods for the analysis of a finite number of interacting inhomogeneities in either the bounded or unbounded domain with analytical methods.

It should be mentioned that there are two coupled classes of micromechanical problems for which averaging is critical. Averaging is usually suitable for predicting effective elastic properties. However, failure and elastoplastic deformations will depend on specific details of the local stress fields when fluctuation is important. In the framework of computational micromechanics in such a case, one must check the specific observable stress fields for many large system realizations of the microstructure with the use of an extremal statistical technique. A more effective approach used in analytical mechanics is the estimation of the statistical moments of the stress field at the interface between the matrix and inhomogeneities. The inherent characteristics of this interface are critical to understanding the failure mechanisms usually localized at the interface. In the framework of a unique scheme of the proposed multiparticle effective field method (MEFM), we attempted to analyze the wide class of statical and dynamical, local and nonlocal, linear and nonlinear micromechanical problems of composite materials with deterministic (periodic and nonperiodic), random (statistically homogeneous and inhomogeneous, so-called graded), and mixed (periodic structures with random imperfections) structures in bounded and unbounded domains, containing coated or uncoated inclusions of any shape and orientation and subjected to coupled or uncoupled, homogeneous or inhomogeneous external fields of different physical natures.

I do not pretend to cover the whole field of micromechanics in this book (restricted by my interests); there are many other topics in micromechanics, of much industrial and scientific importance, that are either treated schematically or only mentioned. In particular, the homogenization theory of periodic structures, the geometrically nonlinear problems, flow in porous media, viscoelasticity problems, and cross-property relations are not considered at all in this book while stochastic geometry, variation methods, propagation of waves in composites, and multiscale discrete modeling are not treated in depth they deserve and sometimes were only mentioned. Interested readers are referred to the references cited in the appropriate sections to achieve a deeper understanding of these topics. This book finalizes my research in micromechanics that began with the papers published in 1986. It was written piece by piece at different times and in different countries facing new challenging problems in micromechanics. The book is suitable as a reference for researchers in different disciplines (applied mathematicians; physicists; geophysicists; material scientists; and electrical, chemical, civil, and mechanical engineers) working in micromechanics of heterogeneous media and providing a rigorous interdisciplinary treatment through experimental investigation. The book is also appropriate as a textbook for an advanced graduate course.

I gratefully acknowledge the financial support for 30 years of research in micromechanics of heterogeneous materials by the National Research Council, the Air Force Office of Scientific Research (AFOSR), the Office of Naval Research of the United States, Regione Autonoma della Sardegna of Italy, Austrian Society for the Promotion of Scientific Research, Max-Planck-Society of Germany, as well as by the Ministry of Higher Education, National Academy of Science, and the Ministry of Machine Construction of USSR and Russia. This book would not have been possible without this support. The excellent cooperation with the team at Springer is gratefully acknowledged; I owe special thanks to Charles B. Glaser, Michael Luby, Brian Halm, Hemalatha Velarasu, and Kala Palanisamy for their careful and patient editing of this book.

Also, thank you ASME for providing permission to use Figs. 12. 9, 12.10 originally published in [273]; Begell House Inc. for permission to use Figs. 9.8, 9.9, 17.1, 17.2, 18.17-18.22, 19.11-19.13 originally published in [213, 222, 229, 233]; Elsevier for allowing permission to use Figs. 3.1, 4.3, 4.4, 7.2, 8.1–8.3, 8.13, 8.20–8.22, 9.2–9.7, 9.10, 9.11, 9.12, 10.1–10.18, 12.1, 12.2, 12.6, 12.7, 13.3, 14.1–14.6, 16.4, 18.11–18.14, and Tables 18.1, 19.6 originally published in [182, 185, 204, 206, 208, 209, 215, 220, 221, 228, 230, 232, 235, 265, 270, 274]; Kluwer for permission to use Fig. 13.11 originally published in [265]; SAGE Publication for permission to use Figs. 11.1–11.4, 13.4–13.6, 17.1–17.6, 19.1–19.5, 19.7–19.10 and Tables 13.1–13.3 originally published in [216, 223, 224, 227, 253]; Springer Nature for providing permission to use Figs. 8.4–8.7, 8.12, 8.13, 8.14, 8.16–8.19, 9.1, 9.13, 11.5–11.8, 12.3–12.5, 12.8, 13.1, 13.2, 13.8–13.10, 18.1–18.4 originally published in [200, 203, 219, 225, 241, 242, 269, 271]; Tech Science Press for permission to use Figs. 5.1–5.10, 8.8–8.11 originally published in [207, 237].

My scientific interests go back to my student days in Moscow State University, where I had the good fortune of being nurtured by brilliant teachers and scientists such as V. I. Arnold, G. I. Barenblatt, A. A. Ilyushin, A. N. Kolmogorov, S. P. Novikov, A. N. Shiryaev, and my first research adviser O. A. Oleinik. In later years I met most of the key players in the development of the subject, many of whom have had a profound influence on my own work. I would like to acknowledge M. Brun, G. J. Dvorak, J. Fish, T. Jackson, E. Kröner, V. I. Kushch, A. M. Lipanov, N. J. Pagano, V. Z. Parton, F. G. Rammerstorfer, T. D. Shermergor, S. A. Silling, S. Torquato, G. Z. Voyiadjis, and G. J. Weng for the helpful and inspiring exchanges of ideas, fantastic collaboration, stimulating discussions, encouraging remarks, and thoughtful criticism I have received over the years. Last, but certainly not least, my deepest gratitude goes to my wife Elena and our son Andrey for their love, devotion, care, tolerance, and understanding that the most beneficial scientific insights often come at time that were destined to be spent with the family. I am pleased to dedicate this book to them.

I am aware of the fact that the book may contain controversial statements, too personal or one-sided arguments, inaccuracies, and typographical errors. Any remarks and comments regarding the book will be fully appreciated (buryach@yahoo.com).

Cincinnati, OH, USA
October 2021

Valeriy A. Buryachenko

Contents

1 Introduction 1
1.1 Classification of Composites and Nanocomposites 1
1.1.1 Geometrical Classification of Composite Materials (CM) 2
1.1.2 Classification of Mechanical Properties of CM Constituents 5
1.1.3 Classification of CM Manufacturing 8
1.2 Effective Material and Field Characteristics 9
1.3 Homogenization of Random Structure CM 13
1.4 Overview of the Book 16

2 Foundations of Solid Mechanics 19
2.1 Elements of Tensor Analysis 19
2.2 The Theory of Strains and Stresses 23
2.3 Basic Equations of Solid Mechanics 27
2.3.1 Conservation Laws, Boundary Conditions, and Constitutive Equation 27
2.3.2 The Equations of Linear Elasticity 33
2.3.3 Extremum Principles of Elastostatic 34
2.4 Basic Equations of Thermoelasticity and Electroelasticity 36
2.4.1 Thermoelasticity Equations 36
2.4.2 Electroelastic Equations 40
2.4.3 Matrix Representation of Some Symmetric Tensors ... 43
2.5 Symmetry of Elastic Properties 46
2.6 Basic Equations of Thermoelastoplastic Deformations 54
2.6.1 Incremental Theory of Plasticity 54
2.6.2 Deformation Theory of Plasticity 57

3 Green's Functions, Eshelby, and Related Tensors 59
3.1 Static Green's Function 59
3.2 The Second Derivative of Green's Function and Related Problems 63
3.2.1 The Second Derivative of Green's Function 63
3.2.2 The Tensors Related to the Green's Function 66
3.3 Dynamic Green's and Related Functions 67
3.4 Inhomogeneity in an Elastic Medium 72
3.4.1 General Case of Inhomogeneity in an Elastic Medium 72
3.4.2 Interface Boundary Operators 75
3.5 Ellipsoidal Inhomogeneity in the Elastic Medium 78
3.6 Eshelby Tensor 81
3.6.1 Tensor Representation of Eshelby Tensor 81
3.6.2 Eshelby and Related Tensors in a Special Basis 86
3.7 Coated Ellipsoidal Inclusion and Imperfect Interface Models 90
3.7.1 Elastic Moduli of Fictitious Ellipsoidal Homogeneous Inclusion 91
3.7.2 Ellipsoidal Heterogeneity with Thin Coating 94
3.7.3 Imperfect Interface 96
3.8 Related Problems for Ellipsoidal Inhomogeneity in an Infinite Medium 100
3.8.1 Conductivity Problem 100
3.8.2 Scattering of Elastic Waves by Ellipsoidal Inclusion in a Homogeneous Medium 105
3.8.3 Piezoelectric Problem 109

4 Multiscale Analysis of the Multiple Interacting Inclusion Problem: Finite Number of Interacting Inclusions 111
4.1 Description of Numerical Approaches Used for Analyses of Multiple Interacting Heterogeneities 111
4.2 Basic Equations for Multiple Heterogeneities and Numerical Solution for One Inclusion 116
4.2.1 Basic Equations for Multiple Heterogeneities 116
4.2.2 The Heterogeneity v_i Inside an Imaginary Ellipsoid v_i^0 119
4.2.3 Estimation of Concentration Factor Tensors by FEA 122
4.3 Boundary Integral Equation Methods 125
4.3.1 Formulation of Boundary Integral Equations 126
4.3.2 Interface Polarization Tensors and Perturbators for a Single Inclusion Inside Macrodomain 129
4.3.3 Method of Fundamental Solutions (MFS) 134
4.3.4 Matrix Representation of the MFS 138

4.4 Volume Integral Equation Method ... 142
4.4.1 Regularized Representation of Integral Equations ... 142
4.4.2 The Iteration Method ... 145
4.4.3 Initial Approximation for Interacting Inclusions in an Infinite Medium ... 147
4.4.4 First-Order and Subsequent Approximations ... 151
4.4.5 Operator Representations of Solutions for One and Two Inclusions ... 153
4.4.6 Approximate Representations of Some Tensors and Operators ... 156
4.4.7 Concluding Remarks ... 158

5 Statistical Description of Composite Materials ... 159
5.1 Basic Terminology and Properties of Random Variables and Random Point Fields ... 160
5.1.1 Random Variables ... 160
5.1.2 Random Point Fields ... 164
5.1.3 Basic Descriptors of Random Point Fields ... 167
5.2 Some Random Point Field Distributions ... 171
5.2.1 Poisson Distribution ... 171
5.2.2 Statistically Homogeneous Clustered Point Fields ... 174
5.2.3 Inhomogeneous Poisson Fields ... 176
5.2.4 Gibbs Point Fields ... 178
5.3 Ensemble Averaging of Random Structures ... 181
5.3.1 Ensemble Distribution Functions ... 182
5.3.2 Statistical Averages of Functions ... 188
5.3.3 Statistical Description of Indicator Functions ... 191
5.3.4 Geometrical Description and Averaging of Doubly and Triply Periodic Structures ... 195
5.3.5 Representations of ODF ... 197
5.4 Numerical Simulation of Particle Random Composites ... 199
5.4.1 Materials and Image Analysis Procedures ... 203
5.4.2 Hard-Core Model ... 204
5.4.3 Hard-Core Shaking Model (HCSM) ... 205
5.4.4 Collective Rearrangement Model (CRM) ... 207
5.4.5 Numerical Results for Unimodal Distribution ... 208
5.4.6 Numerical Results for Bimodal Distribution (Application to Solid Propellants) ... 213

6 Effective Properties and Energy Methods in Thermoelasticity and Thermoelectroelasticity of Composites ... 219
6.1 Effective Thermoelastic Properties ... 219
6.1.1 Hill's Condition and Representative Volume Element ... 219

6.1.2 Effective Elastic Moduli 223
6.1.3 Apparent Elastic Moduli for Finite-Size RVE 226
6.1.4 Overall Thermoelastic Properties 229
6.2 Effective Energy Functions 233
6.3 Some General Exact Results 239
6.3.1 Two-Phase Composites 239
6.3.2 Polycrystals Composed of Transversally Isotropic Crystals 247
6.3.3 Exact Representation for First and Second Moments of Stresses Averaged over the Phase Volumes 250
6.4 Variational Principle of Hashin and Shtrikman 255
6.5 Hill's Bounds of Effective Elastic Moduli 260
6.6 Some General Representations for Thermoelectroelasticity of Composites 265
6.6.1 Generalized Hill's Conditions and Effective Properties 266
6.6.2 Effective Energy Functions 269
6.6.3 Two-Phase Composites 270
6.6.4 Discontinuities of Generalized Fields at the Interface Between Components 271
6.6.5 Phase-Averaged First and Second Moments of the Field Σ 272

7 General Integral Equations of Micromechanics of Composite Materials 277
7.1 General Integral Equations for Matrix Composites of Any Structure 279
7.2 Random Structure Composites 282
7.2.1 General Integral Equation for Random Structure Composites 282
7.2.2 Some Particular Cases 288
7.2.3 Comparison with Related Equations 289
7.3 Effective Field Hypothesis and Asymptotic Representation of the GIE (7.19) and (7.20) 295
7.3.1 Effective Field Hypothesis 295
7.3.2 About Renormalized Terms in Eqs. (7.19) and (7.32) 297
7.3.3 Model Example 298
7.4 Periodical Structure Composites 300
7.4.1 Short Review of Analyses of Periodical Structure Composites 300
7.4.2 Doubly and Triply Periodical Structure Composites 302

7.4.3 Triply Periodic Particulate Matrix Composites with Imperfect Unit Cells ... 305
7.5 Concluding Remarks ... 308

8 Multiparticle Effective Field and Related Methods in Micromechanics of Random Structure Composites ... 311
8.1 Definitions of Effective Fields and Effective Field Hypotheses ... 312
8.1.1 Approximate Effective Field Hypothesis ... 312
8.1.2 Closing Effective Field Hypothesis ... 314
8.1.3 Effective Field Hypothesis and Composites with One Sort of Inhomogeneities ... 316
8.2 Multiparticle Effective Field Method ... 319
8.2.1 Average Stresses in the Components ... 319
8.2.2 Effective Properties of the Composite ... 321
8.2.3 Some Related Multiparticle Methods ... 323
8.3 One-Particle Approximation of the MEFM and Mori-Tanaka Approach ... 325
8.3.1 One-Particle ("Quasi-Crystalline") Approximation of MEFM ... 325
8.3.2 Mori-Tanaka Approach ... 330
8.3.3 Effective Properties Estimated via the EFM and MTM at $\mathbf{Q}_i \equiv \mathbf{Q}_i^0$ ... 334
8.3.4 Some Methods Related to the One-Particle Approximation of the MEFM ... 336
8.3.5 Some Analytical Representations for Effective Moduli ... 338
8.4 Hashin–Shtrikman Bounds Estimations ... 342
8.4.1 The Scheme by Willis for Obtaining Hashin–Shtrikman Bounds ... 342
8.4.2 EFH and Hashin–Shtrikman Bounds for CMs with Noncanonical Inhomogeneous Inclusions ... 346
8.4.3 Interface Boundary Integral Representations ... 349
8.4.4 Numerical Results ... 351
8.4.5 Some Representations Related with Hashin–Shtrikman Bounds ... 354
8.5 Related Perturbation Methods ... 358
8.5.1 Combined MEFM–Perturbation Method ... 358
8.5.2 Perturbation Method for Small Concentrations of Inclusions ... 360
8.5.3 Perturbation Method for Weakly Inhomogeneous Media ... 361
8.6 Effective Medium Methods ... 362
8.6.1 Application to Composite Materials ... 362
8.6.2 Analysis of Polycrystal Materials ... 366

8.7 Differential Methods 367
8.8 Numerical Results 369
8.8.1 Composites with Spheroidal Inhomogeneities 369
8.8.2 Composites with Polifractional Inhomogeneities (Solid Propellants) 374
8.8.3 Nanocomposites 380
8.8.4 Some Prospective Problems in Modeling of Nanocomposites 388
8.9 Critical Analysis of Maxwell Homogenization Schemes 392
8.9.1 Classical Maxwell Scheme and Its Comparison with EFM and MTM 392
8.9.2 Deficiencies of Maxwell Scheme 395
8.9.3 Generalized Maxwell Schemes Taking Inclusion Interactions into Account 398

9 Basic Representations of New Background of Analytical Micromechanics 403
9.1 Stress and Strain Distributions for One Heterogeneity Inside Infinite Matrix 404
9.2 Some Classical Hypotheses and Approaches Based on the Classical GIE (7.32) and (7.33) 407
9.2.1 Effective Field Hypothesis H1 and Some Perturbator Representations 407
9.2.2 The Hypotheses H2 and H3 and Some Classical Approaches 409
9.3 Estimation of Both the Average Fields and Effective Properties Based on the New GIE (7.19) and (7.20) 412
9.3.1 General Case of Arbitrary Number of Components 412
9.3.2 Some Particular Cases 414
9.3.3 Iteration Method and Abandonment of the Hypothesis H1a 417
9.3.4 Elastically Homogeneous Medium with Random Residual Stresses 418
9.3.5 Comparisons with Some Exact Results 420
9.4 Qualitative Analysis of the New GIE, Some Classical Hypotheses, and Propositions 422
9.4.1 Analysis of Proposition 9.2.1 and Hypothesis H1 422
9.4.2 Analysis of Proposition 9.2.2 and Hypotheses H1b and H3 424
9.5 Numerical Results 426
9.5.1 Finite Element Analysis in the Framework of the Hypothesis H1a 426
9.5.2 Iteration Method (Abandonment of the Hypothesis H1a) 435

9.6 Hashin–Shtrikman Bounds of Effective Moduli Based on the New GIE 439
9.7 Concluding Remarks 442

10 Inhomogeneity of the First and Second Statistical Moments of Stresses Inside the Heterogeneities of CMs 445
10.1 Effective Moduli and the First Statistical Moments Inside Inclusions 447
10.1.1 General Representations 447
10.1.2 Closing Assumptions 448
10.1.3 Initial Approximations Based on New Eq. (7.20) 450
10.1.4 Initial Approximations Based on Classical Eq. (7.33) 453
10.1.5 Some Related Integral Equations 454
10.2 Integral Equations for Second Moments of Stresses 456
10.2.1 Second Moment of Stresses Inside the Heterogeneities 456
10.2.2 Stress Fluctuations Inside the Matrix 460
10.2.3 Second Moment of Stresses at the Interface 461
10.3 Second Moments of Stresses Inside the Components 462
10.3.1 Initial Approximation 462
10.3.2 High-Order Approximation 464
10.4 Elastically Homogeneous Composites with Randomly Distributed Residual Microstresses 466
10.5 Related Method of Estimations of the Second Moments of Stresses 467
10.6 Numerical Results 468
10.6.1 Effective Moduli and the First Statistical Moments of Stresses Inside the Inclusions 468
10.6.2 Stress Fluctuations Inside the Inclusions 474
10.7 Conclusion 479

11 Interface Integral Technique in Thermoelasticity of Random Structure Matrix CMs 483
11.1 Interface Polarization Tensors and Effective Properties 483
11.2 Interface Integral Technique 485
11.2.1 General Integral Equation for Microinhomogeneous Infinite Medium 485
11.2.2 Some Classical Hypotheses and Approaches 488
11.2.3 Solution of GIEss 490
11.2.4 Iteration Method in Terms of the Interface Perturbator Factor 493
11.2.5 Numerical Results 495

11.3 General Integral Equations in Micromechanics of CMs with Imperfectly Bonded Interfaces ... 502
11.3.1 Stress and Strain Distributions for One Heterogeneity Inside Macrodomain w ... 503
11.3.2 General Integral Equations ... 506
11.3.3 Simple Example of Residual Stresses ... 509
11.4 General Boundary Integral Equations of Stokes Flow Inporous Media ... 510
11.4.1 Stokes Equations and Fundamental Solutions ... 511
11.4.2 Single Rigid Inclusion in a Macrodomain ... 514
11.4.3 General Integral Equation for Microinhomogeneous Infinite Medium ... 514
11.4.4 Known GIEs as the Approximations of the New Exact GIEs ... 518

12 Nonlinear Problems in Micromechanics of CMs ... 521
12.1 Effective Limiting Surface ... 522
12.1.1 Limiting Surface for Bulk Stresses ... 522
12.1.2 Limiting Surface for Interface Stresses ... 525
12.1.3 Effective Fracture Surface for an Isolated Crack ... 527
12.1.4 Numerical Results for the Effective Limiting Surfaces ... 530
12.1.5 Possible Generalizations ... 535
12.2 Nonlinear Composites ... 537
12.2.1 Linearization Schemes ... 538
12.2.2 Modified Linearization Scheme ... 541
12.2.3 Deformation Plasticity Theory of Composite Materials ... 543
12.2.4 Power-Law Creep ... 547
12.2.5 Elastic–Plastic Behavior of Elastically Homogeneous Materials with Random Residual Microstresses ... 550
12.2.6 A Local Theory of Elastoplastic Deformations of Metal Matrix Composites ... 556
12.3 Concluding Remarks ... 564

13 Nonlocal Effects in Micromechanics of Locally Elastic CMs ... 569
13.1 The Nonlocal Integral Equation ... 574
13.2 Methods for the Solution of the Nonlocal Integral Equations ... 576
13.2.1 Direct Quadrature Method ... 576
13.2.2 The Iteration Method ... 576
13.2.3 The Fourier Transform Method for Statistically Homogeneous Media ... 578

13.3 Average Stresses in the Components and Effective Properties for Statistically Homogeneous Media 580
13.3.1 Differential Representations 580
13.3.2 The Reduction of Integral Overall Constitutive Equations to Differential Ones 581
13.3.3 "Quasi–Crystalline" Approximation 583
13.3.4 Approximations in the Framework of the Hypothesis H1b 584
13.3.5 Numerical Analysis of Nonlocal Effects for Statistically Homogeneous Composites 585
13.4 The Effective Properties of Statistically Inhomogeneous Media 589
13.4.1 Local Effective Properties of FGMs 589
13.4.2 Numerical Results of Estimation of Effective Properties of FGMs 592
13.4.3 Perturbation Method 597
13.4.4 Combined MEFM–Perturbation Method 598
13.5 Random Structure Matrix Composites in a Half-Space 599
13.6 Concluding Remarks 604

14 Micromechanics of Strongly Nonlocal Strain-Type and Weakly Nonlocal CMs 607
14.1 Nonlocal Constitutive Equation 610
14.2 Effective Elastic Moduli 615
14.2.1 Influence Functions and Effective Stiffness and Compliance 615
14.2.2 Energetic Definition of Effective Elasticity 619
14.3 Effective Eigenfields end Overall Elastic and Stored Energies 620
14.3.1 Field Decomposition and Effective Eigenstrain 620
14.3.2 Transformation Influence Functions 624
14.3.3 Overall Elastic Energy and Stored Energy 626
14.4 Perturbation Method for Estimation of Statistical Moments of Fields 629
14.4.1 Stress Fluctuations 629
14.4.2 Local Fluctuation of Stresses 633
14.4.3 Average Stresses in the Constituents 633
14.5 General Integral Equations and Effective Fields 635
14.6 One Heterogeneity Inside Infinite Matrix 637
14.7 Effective Elastic Properties 641
14.8 Numerical Results 644
14.9 Strain Gradient Theories 649
14.9.1 First Strain Gradient Medium 649
14.9.2 Homogenization of the First Strain Gradient CMs 654
14.9.3 GIE for the First Strain Gradient CMs and Effective Properties 657

14.9.4 Strain Gradient Plasticity Model of CMs and Polycrystals ... 659
14.10 Stress-Gradient Theories ... 661
14.10.1 Stress-Gradient Elasticity Model ... 661
14.10.2 Homogenization of Stress-Gradient CMs ... 663
14.10.3 GIE of Micromechanics of Stress-Gradient CMs and Effective Moduli ... 664
14.11 Short Review of Higher-Order Micromechanical Modeling of CMs and Concluding Remarks ... 666

15 Subsequent Generalizations of Theory and Related Problems ... 671
15.1 Basic Constitutive Equations ... 671
15.2 General Integral Equations (GIEs) for CMs ... 676
15.2.1 General Integral Equation in Terms of the Green's Functions ... 676
15.2.2 One Inclusion in the Matrix ... 680
15.2.3 Operator Form of General Integral Equations ... 683
15.2.4 Some Particular Cases and Approximate GIEs ... 686
15.2.5 Composites of Periodical Structure ... 688
15.3 General Representations for Solution ... 689
15.3.1 Two Heterogeneities in the Infinite Matrix ... 689
15.3.2 General Representations for the Effective Fields ... 692
15.4 Some Particular Cases of the Perturbators and Corresponding Tensors for Some Constitutive Laws ... 692
15.4.1 Closing Assumption ... 693
15.4.2 Effective Field Hypothesis **H1a** and Initial Approximation ... 695
15.4.3 Some Additional Hypotheses and Initial Approximations ... 698
15.4.4 Approximate Representation of Some Tensors and Operators. Some Asymptotic Representations ... 701
15.5 Second Moment of Fields ... 704
15.5.1 General Representations of Second Moment of Fields ... 704
15.5.2 Field Fluctuations Inside the Matrix ... 707
15.6 Particular Cases ... 708
15.6.1 Thermoelasticity of CMs ... 708
15.6.2 Thermoconductivity of CMs ... 712
15.6.3 Coupled Problems in Micromechanics of CMs ... 718
15.6.4 Wave Propagation in CMs ... 721

16 Bond-Based Peridynamics ... 725
16.1 Basic Formulation ... 726
16.1.1 Peridynamic Equilibrium Equation ... 726
16.1.2 Elasticity ... 729
16.1.3 Peridynamic Stress ... 731

16.2 Boundary Volumetric Conditions 734
16.3 Linearized Microelastic Model 736
16.4 Homogeneous Deformations 738
16.5 Damage and Failure Models 741
16.6 Thermo-mechanical Peridynamics 744

17 Background of Peridynamic Micromechanics 747
17.1 Decomposition of Material Properties and Some Averages 749
17.2 General Representations for Effective Moduli 752
17.2.1 Admissible Fields and Generalized Hill's Equality 752
17.2.2 Effective Elastic Moduli 755
17.2.3 Energetic Definition of Effective Elastic Moduli 757
17.3 Thermoperidynamic Composites 759
17.3.1 Basic Equations of Thermoperidynamics 759
17.3.2 Effective Eigenfields 760
17.3.3 Transformation Influence Functions 762
17.4 Effective Thermoelastic Properties for the Matrix Composites 764
17.4.1 Effective Moduli and Effective Eigenstrains 764
17.4.2 Violation of the Generalized Levin's Representation 768
17.4.3 Interface Integral Representations of Effective Properties 771
17.5 Constitutive Model and Strain Energy 773
17.6 Minimum Principles of Linear Peridynamics 776
17.6.1 Some Properties of Strain Energy 776
17.6.2 Minimum Principles 778
17.6.3 Upper and Lower Bounds for the Strain Energy 780
17.7 Generalized Hill's Bounds 780
17.7.1 Effective Elastic Moduli 781
17.7.2 Generalized Hill's Bounds on the Effective Stiffness and Compliance 783
17.7.3 Numerical Examples 786
17.8 General Integral Equations 789
17.8.1 General Integral Equation in the Operator Form 789
17.8.2 Infinite Coupled System of General Integral Equations 792

18 Peridynamic Micromechanics of Random Structure Composites 795
18.1 Modeling of One Inclusion in the Infinite Matrix 796
18.1.1 Truncation Method 797
18.1.2 Periodization Method 800
18.1.3 Iterative Methods 801
18.1.4 Green Function Method 805
18.1.5 Numerical Results for 1D Case 806

18.2 Basic Hypotheses of Peridynamic Micromechanics 809
18.3 Micromechanical Modeling 810
18.3.1 Dilute Approximation Method (DAM) 811
18.3.2 Mori-Tanaka Method (MTM) 812
18.3.3 Effective Field Method (Abandonment of the Hypotheses H1b and H3) 815
18.3.4 Effective Field Method (Abandonment of the Hypotheses H1a, H1b, and H3) 817
18.3.5 The Scheme of Estimation of the Effective Moduli 818
18.3.6 Nonlocal Effective Operator 821
18.3.7 Numerical Schemes and Results for 1D Case 822
18.4 Effective Deformation of Peridynamic Random Structure Bar Subjected to Inhomogeneous Body Force 833
18.4.1 Locally Elastic Solution for the Infinite Bar with Inclusion 833
18.4.2 Single Inclusion in the Infinite Homogeneous Medium: Decomposition and Truncation Methods 836
18.4.3 Micromechanical Modeling 839
18.4.4 Numerical Results 840
18.5 The Effective Displacements and Stresses for CMs with Prescribed Body Force 845
18.6 Concluding Remarks 846

19 Computational Homogenization in Linear Peridynamic Micromechanics of Periodic Structure CMs 849
19.1 Periodic and Volumetric Periodic Boundary Conditions 850
19.1.1 Locally Elastic Unit Cell with Periodic Boundary Condition 851
19.1.2 Volumetric Boundary Conditions in Peridynamic Micromechanics 852
19.2 General Representations for Effective Moduli 855
19.2.1 Generalization of Hill's Equality 855
19.2.2 Effective Elastic Moduli 857
19.3 Decomposition of the Material and Field Parameters and Effective Modulus Representations 859
19.4 Numerical Schemes 862
19.4.1 Bounded Domain Subjected to the Volumetric Displacement Homogeneous Boundary Condition 862
19.4.2 Direct Approach for the Infinite Periodic Medium 862
19.4.3 Decomposition Approach for the Infinite Periodic Medium 865
19.4.4 Estimation of Effective Moduli 867
19.4.5 Choice of the Unit Cell 868

19.5 Numerical Schemes for 1D Case ... 870
19.5.1 Locally Elastic Heterogeneous Bar of Periodic Structure ... 870
19.5.2 Peridynamic Heterogeneous Bar of Periodic Structure ... 871
19.5.3 Numerical Examples ... 876
19.6 Effective Moduli of Peridynamic Medium with Periodically Distributed Damage ... 880
19.6.1 Decomposition of the Material and Field Parameters ... 881
19.6.2 Numerical Results ... 882
19.7 Volumetric Bloch-Floquet Periodic Boundary Condition ... 888
19.7.1 Discretized Peridynamic Equation in Frequency Domain ... 889
19.7.2 Bloch-Floquet Condition for Locally Elastic CMs ... 890
19.7.3 Volumetric Bloch-Floquet Periodic Boundary Condition ... 891
19.8 Scheme of Data-Driven Learning of Nonlocal Models ... 894
19.9 Concluding Remarks ... 898

20 Conclusion ... 901

21 Appendix ... 905
21.1 Parametric Representation of Rotation Matrix ... 905
21.2 Second- and Fourth-Order Tensors of Special Structures ... 907
21.2.1 E-Basis ... 907
21.2.2 P-Basis ... 909
21.2.3 B-Basis ... 912
21.3 Analytical Representation of Some Tensors ... 914
21.3.1 Exterior-Point Eshelby Tensor ... 914
21.3.2 Some Tensors Describing Fluctuations of Residual Stresses ... 917

References ... 919

Index ... 989

Author Biography

Valeriy A. Buryachenko received his MS degree in Mathematics from M.V. Lomonosov Moscow State University, Russia, in 1976; his PhD degree in Material Science from the Chemical Production Engineering in Moscow, Russia, in 1982; and his DSc degree in Mechanical Engng (Mechanics of Solids) from SP Timoshenko Institute of Mechanics (NAS of Ukraine, Kiev) in 1993. He is the author of 160 papers in micromechanics and the books "Micromechanics of Heterogeneous Materials," Springer, NY (2007) and "Local and Nonlocal Micromechanics of Heterogeneous Materials," Springer, NY (2021). His main achievement is a creation of a new (second) background of analytical micromechanics of composites (the first background based on the effective field hypothesis was proposed by Poisson, Faraday, Mossotti, Clausius, Lorenz, and Maxwell, 1824–1880). It offers great opportunities for a fundamental jump in multiscale and multi-physics modeling of random heterogeneous media with drastically improved accuracy of local field estimations.

He is a President of a small consulting company Micromechanics & Composites LLC. Prior this, he worked in the leading Research centers and universities of USA, Italy, Austria, and Germany. Prior this, he worked as a Senior Res. Scientist in the Research institutes and a Full Professor of Math in the Moscow Univ. of Chemical Engineering Industry, Moscow, Russia

Chapter 1
Introduction

Abstract To introduce readers to the subject of composite (or heterogeneous) materials, it is necessary to begin with the main notions and definitions which are presented with the understanding that any definition and classification scheme is neither comprehensive nor complete. This chapter outlines a general classification of CM based on the geometry of structural elements, their mechanical properties, and their manufacture. In particular, the geometrical classification is defined by connectivity of structural elements (matrix, skeletal, and polycrystals structures), its shape (particles, fibers, flakes and hybrid ones), and arrangement (random, periodic, graded, and clustered). The mechanical properties of constituents correspond to polymer matrix composites (PMCs), metal matrix composites (MMCs), ceramic matrix composites (CMCs), carbon-carbon (CC), and hybrids. The replacement of a governing integral equation with rapidly oscillating coefficients by an equation with either constant or smoothly varying "overall" coefficients is a main problem of micromechanics shortly described in this chapter with classification of basic groups of methods. Overview of the book is also presented.

1.1 Classification of Composites and Nanocomposites

The engineering level of composite material definition is given in ASTM D 3878-95c: "Composite material. A substance consisting of two or more materials, insoluble in one another, which are combined to form a useful engineering material possessing certain properties not possessed by the constituents." This definition is usually added by the notion of length scale when the characteristic sizes of structural elements (the number of which is statistically large) are much smaller than characteristic sizes of composite medium but much larger than the molecular dimensions with the result that the mechanical state of structural elements is described by the equations of continuous mechanics.

The materials used for structural applications are very seldom homogeneous, and because of this, the problem of modeling the mechanical properties of engineering materials has to be investigated in the framework of heterogeneous materials. The chief purpose of introduction of fillers into composite materials (CM) is to

V. A. Buryachenko, *Local and Nonlocal Micromechanics of Heterogeneous Materials*, https://doi.org/10.1007/978-3-030-81784-8_1

create materials with a prescribed set of physico-mechanical properties, which can be controlled, for example, by varying the type of base matrix and fillers and its particle size distribution, shape, and arrangement (see for details, e.g., [64, 334, 437, 553, 1559, 1565]). Thanks to considerable advances in CM research, their use in a broad range of industries—machine building, construction, aerospace technology, etc.—has become extensive. Development of an assortment of CMs, and decisions concerning their possible scope of applicability and service conditions involve research into a very wide spectrum of problems: (1) choice of the matrix, fillers, functional adjuvants, and design of the optimal microtopology of CM for creating materials with prescribed properties; (2) processing of a particular CM into constructions, with optimization of the process parameters to provide the best trade-off between the properties of each of the constituents of the CM; and (3) prediction of the serviceability of a construction under prescribed service conditions.

By now, an almost incomprehensible amount of data on the properties and processing technology of the individual constituents and CM has been amassed. However, the diversity of different materials with different geometrical and mechanical properties of constituents is such that it is useful to have a classification of these materials to choose the appropriate modeling tool. A number of classification schemes are known that are distinguished by the criteria used for these classifications: geometry of structural elements, their mechanical properties, and their manufacture. The description of CM encountered in materials science can be useful to mechanical engineers if one avoids excessive intractable details and considers at first the geometrical characteristics and then the mechanical properties of constituent phases. Such a classified description should be favorable to a qualitative understanding of the relation between microstructure and properties on one hand and of the relation between process parameters and microstructure on the other.

This section outlines a general classification of CM based on the geometry of structural elements, their mechanical properties, and their manufacture.

1.1.1 Geometrical Classification of Composite Materials (CM)

Geometrical microstructure of composites can be classified according to the following features: (1) nature of structural element connectivity; (2) the characteristic relative sizes and shapes of structural elements; and (3) arrangement of structural elements.

The nature of structural element connectivity defines three groups of composite materials depending on the existence of a phase "containing" the others. To the first group, one can assign the composite with one continuous over the volume constituent called the matrix reinforced by the isolated inhomogeneities of the different shapes. Such composites are called the matrix ones. The second group of composites forms the skeletal (or mutually penetrated or multipercolated) composites where at least two phases make a monolithic frame so they can be seen as containing the other. The laminate composites as well as open cellular structures such as some metallic foam or porous silicon [551] are particular cases of percolated structures. A more complicated structure is intrinsic to entangled materials such as cellulose-reinforced polymers [553], metallic wool, and glass wool. These materials have both a network connecting them to the penetrated structure and also an incomplete network with only one connection to the network tree. It should be mentioned that variation of the constituent concentration can lead to changeover from one connectivity group to another. For example, closed cell foams [551] can be considered as porous metal matrix composites, while the open cell foams have a percolated type of structure. The third group of composites is characterized by a structure in which there is no contact among the elements of each constituent and the different phases are contiguous but none can be seen as containing the other. For example, polycrystalline single-phase metal containing the structural elements of the same anisotropic crystals with different orientations of crystallographic axes pertain to the third group of composite materials, which in this case are called the heterogeneous materials. The grains fill the whole space, and no element with one orientation contains the other element with another orientation.

The lack or existence of a continuous matrix in CM is an essential geometrical feature, but the aspect ratio that may characterize the phases is also important. The matrix composites can be classified by the characteristic relative sizes and shapes of structural elements, as continuously reinforced (typically plies and continuous tapes and fibers), short fiber (or chopped whiskers) reinforced (aspect ratio noticeably larger than 1), particle reinforced (aspect ratio around 1), flake or platelet reinforced (aspect ratio smaller than 1), and hybrid reinforcement. If the ratio of characteristic sizes of the reinforcement and the composite materials is taken as a dimension of the composite structure, then the particle, fiber, and laminated composites are called zero-dimensional, one-dimensional, and three-dimensional ones, respectively. Polyreinforced (simple and hybrid) CMs contain reinforcements from different materials. In the simple polyreinforced CM, the reinforcements from different materials have the same dimensions. In the hybrid CM, the reinforcements from different materials have different dimensions (e.g., fibers and particles). The ultimate objective of hybridization is to create and combine an engineered proportion of material components for a range of applications, including the following hybrid systems: ternary alloys, hybrid particulate composites, glass fiber-containing toughened plastics, liquid crystalline polymer hybrid composites, and nanocomposites (for references see [1639]).

The composites can be further structured for identification by noting the arrangement of structural elements. A distinction is made between the periodic (or deterministic) and random structure composites. For example, for woven fiber-reinforced composites, one can indicate the following structures: periodic, cylindrical, braided, orthogonal interlock, and angle interlock structures. Such a perfectly regular structure, of course, does not exist in actual cases, although the periodic modeling can be quite useful, as it provides rigorous estimations with a priori prescribed accuracy for various material properties. The random structure composites, in turn, should be subdivided into statistically homogeneous structures that are invariant under a shift of the space and into statistically inhomogeneous (or functionally graded, FG) composites in which the arrangement (e.g., the concentration and orientation) of elements is position-dependent. Statistically homogeneous composites are divided on statistically isotropic one which structure is invariant with respect to the rotation and statistically anisotropic with anisotropy assessing from directional measurements, e.g., variation of the chord size distribution with their orientation. A particular sort of FG composites is the clustered (or aggregated) composites in which the areas with high and small concentration of reinforcements can be separated. In a polymeric matrix, the filler (even if present in minimum quantities) is always more or less agglomerated. The ability of the filler to form the structural network is largely responsible for its reinforcing action [940]. On the other hand, the structuralization in melt increases the viscosity of the material and hampers its processability. Microscopic studies have confirmed the existence of two types of primary structures in CMs, i.e., filler aggregates with particles bound together firmly enough, and agglomerates—systems of weakly interrelated aggregates. For a fixed filler content, the viscosity and moduli of solid CM based on the same matrix with agglomerates is always higher than that of the "well-dispersed" sample (for references see, e.g., [437, 940]). However, the mentioned behavior of increasing the effective stiffness with the agglomeration of the spherical particles changes drastically with variation of the inclusion shape. So, well-known experimental data point to decrease of effective elastic properties with the growth of clustering (when the aligned clusters contain the parallel nanoplates with a small aspect ratio $\alpha \sim 0.001$), with the maximum effective stiffness of silicate nanocomposites corresponding to the completely exfoliated structures [1374]. At the same time, the strength of CM is usually lower in materials with agglomerates of spherical particles than in CM with uniformly distributed filler. Clustered composites in the simplest case have so-called ideal cluster structure that contains either finite or infinite, deterministic or random ellipsoidal domains called particle clouds distributed in the composite matrix. In so doing the concentration of particles is a piecewise and homogeneous one within the areas of ellipsoidal clouds and composite matrix. An alternative case of clustered structures is a dual scale one existing, e.g., in Al7%Si alloy [363, 553], where the elementary unit of the microstructure is a core of almost pure aluminum, surrounding by a thick layer that can be considered as CM consisting of aluminum matrix reinforced by silicon particles. The ratio of linear sizes of the core and the layer as well as the ratio of the volume fraction of Si particles to the Al matrix in the layer are defined by both the composition of the alloy and technological features.

As an example of composites combining random and periodic structure (also called periodic structure with random imperfections), we can indicate the initially periodic structure with randomly distributed damage as well as composites with small random perturbation of inclusion location with respect to the periodic grid due to manufacturing errors. From another aspect, the laminated composites usually classified at a mesoscale level as either periodic or deterministic structures can be considered at the microscale level as the FG composites with the statistically inhomogeneous packing of fibers in each ply. Nanocomposites are randomly reinforced by the nanofibers which in turn have a periodic or deterministic structure.

This leads to the notion of coupling of different scales. The growing recognition that the properties of materials involve different scales expressively reflects the explosive progress of modern nano-and micromechanics resulting from the development of image analysis and computer-simulation methods on one hand and both the advanced experimental technique (e.g., computed X-ray tomography and electron microscopy) and improved materials processing on the other hand, since processing controls the prescribed structure. One recalls that at the upper level, the collective results of events or processes taking place at the lower level are registered and are measured on a finer scale. Although in material modeling, one can distinguish four levels of simulation, electronic structure, atomistic, microscale, and macroscale, micromechanics of CMs is usually concentrated just on the micro-macro-connection of the multiscale concept when the continuum mechanics technique is appropriate.

1.1.2 Classification of Mechanical Properties of CM Constituents

The composites also can be categorized with respect to the mechanical properties of constituents according to the following features: polymer matrix composites (PMCs), metal matrix composites (MMCs), ceramic matrix composites (CMCs), carbon-carbon (CC), and hybrids. The polymer resins fall into two groups: thermosets and thermoplastics.

The cured thermoset matrix has polymer chains that have become irreversibly crosslinked in a three-dimensional frame like a space truss. A cured thermoset matrix has the following mechanical properties: elasticity modulus 2.5–10.0 GPa, tensile strength 30–120 MPa, compressive strength 80–250 MPa, and density 1200–1400 kg/m^3. Among the most often used thermoset resins are polyester, vinylester, epoxy, phenolic, polyurethane, and polybutadiene. Thermoplastic resins are processed at high temperatures and solidify on cooling. That is, polymer crosslinking does not occur, and thermoplastic resins remain plastic and can be reheated and reshaped. The solidified thermoplastic resins have high tensile strength (50–100 MPa), high stiffness (1.4–4.2 GPa), and high density 1100–1700 kg/m^3. The most widely used thermoset resins are polyethylene, polystyrene, polypropylene, acrilonitride-butadiene styrene, acetate, polycarbonate, polyvinyl chloride, polysulfone, polyphenylene sulfide, and nylon.

MMCs have several advantages [1472, 1495, 1496]: high strength, elastic moduli, fracture toughness, and impact properties; high electrical and thermal conductivity; low sensitivity to temperature change or thermal shock; high surface durability; and low sensitivity to surface flaws.

Reinforcement of the matrix composites is accomplished by the fillers carrying the majority of the loading. Conventional fillers are separated into three categories: reinforcing, active, and inert. The reinforcing class includes mainly fiber composites. Disperse fillers may also perform the reinforcing function, and then they are called active. The basic properties of commonly used fibers are as follows: continuous glass fibers have a diameter $3\text{–}20 \cdot 10^{-6}$ m, tensile strength 2–6 GPa, stiffness 50–100 GPa, and 2400–2600 kg/m^3. They are used in fabrication of composite elements in the form of elementary yarns, rovings, and cloth. The advantage of glass fibers is in their high tensile strength and compressive strength, low cost of the materials and manufacturing, and good compatibility with polymeric matrix. The disadvantages are associated with their low stiffness and heat stability. The carbon fibers have the following characteristics: 1.5–7.0 GPa tensile strength, 150–800 GPa elasticity modulus, 1500–2000 kg/m^3 density. They have higher stiffness in comparison with glass fibers and are characterized by a negative expansion coefficient. Their mechanical properties do not deteriorate with temperature increases up to 450 °C, a fact that allows the use of carbon fibers in composites with both polymeric and metal matrices. Boron fibers have 2–4 GPa tensile strength, 370–430 GPa modulus of elasticity, and 2500–2700 kg/m^3 density. The main advantage of boron fibers is their high stiffness and compression strength, but they are rather expensive, more brittle, and less processable than glass fibers. Boron fibers are used with both polymeric and metal matrices in the form of elementary or hybrid threads which are bundles of parallel continuous fibers wound with an auxiliary glass thread. Organic fibers (e.g., Kevlar) with different polymer compositions and method of molding can be obtained with 2–4 GPa tensile strength, 70–150 GPa modulus of elasticity, 1410–1450 kg/m^3 density. Organic fibers preserve their initial characteristics up to 180 °C. They are characterized by high tensile strength and stiffness in comparison with glass fibers, but under compression, composites with organic fibers have substantially lower properties than fiberglass composites.

It should be emphasized that the interfaces are not always as perfect as sometimes assumed in micromechanical society; they may have different tension and shear strength as well as different constitutive equations describing their mechanical properties. The strength must be higher in CMs with fillers featuring the absolute adhesion to the matrix than that in systems with little or no adhesion. The relative elongation and specific impact strength must, on the contrary, go up with increasing adhesion. In so doing, the modulus of CMs is significantly less sensitive to the adhesion properties of the interphase. The most promising method of modification of CM is by pretreating the filler or adding to the matrix specific depressants or modifiers with the aim of creating chemical bonds at the interphase. It has been convincingly demonstrated that on the filler-polymer boundary, there exist adsorbed layers of polymer characterized by a density greater than that of the unfilled matrix. Fixation of macromolecules on the filler surface, their orientation in the direction coplanar or normal to the surface, and condensation result in “hard” interphases. Alternatively, selective sorption by the filler sorption of one of the matrix low molecular components may lead to plasticization of the boundary layers and appearance of “soft” interphases [437]. The role of interfaces increases with reduction of a reinforcement size.

Fiber Nanotube Composites Carbon fibers are currently considered to be most promising. The recent discovery of carbon nanotubes (CNTs) by Iijima [682] has gained ever-broaded interest due to providing unique properties generated by their structural perfection, small size, low density, high strength, and excellent electronic properties (see, e.g., comprehensive reviews in [1254, 1503, 1511]), depending on the diameter and chirality [295, 1216, 1432, 1681]. Indeed, the longitudinal Young's modulus of CNTs falls between 0.4 and 4.15 TPa, and a tensile strength approaching 100 GPa. Carbon nanotubes occur in two distinct forms, single-walled nanotubes (SWNT), which are composed of a graphite sheet rolled into a cylinder, and multi-walled nanotubes (MWNT), which consist of multiple concentric graphite cylinders. Compared with multi-walled nanotubes, single-walled nanotubes are expensive and difficult to obtain and clean, but they have been of great interest owing to their expected novel electronic, mechanical, and gas adsorption properties [429]. Nanofibers can have a number of different internal structures, wherein graphene layers are arranged as concentric cylinders, nested truncated cones, segmented structures, or stacked coins (see, e.g., [1019, 1558]). External morphologies include kinked and branched structures and diameter variation. A high aspect ratio of CNT and extraordinary mechanical properties (strength and flexibility) make them ideal reinforcing fibers in nanocomposites, which can produce advanced nanocomposites with improved stiffness and strength [1510, 1516]. Tensile test on composites show that 1 wt% nanotube additions result in 40 and 30% increase in elastic modulus and strength, respectively, indicating significant load transfer across the nanotube—matrix interface. An additional significant impact of CNT on the effective properties of nanocomposites is based on the fact that CNTs provide very high interfacial area if embedded in a matrix, and assemblies with unique architectures might be constructed by interconnected CNTs. It is emphasized in [1194] that the planar graphite surface structure of these nanotubes provides relatively few sites suitable for facile chemical modification. It is believed that graphitic CNTs are perhaps more attractive additives for fabricating polymer/carbon nanofiber composite materials of enhanced mechanical and physical properties due to their unusual atomic structure. Depending on the metal catalyst and thermal syntheses conditions, CNTs can be grown with widths ranging from 5 to 1000 nm and lengths ranging from 5 to 10 μm. The design and fabrication of these materials are performed on the nanometer scale with the ultimate goal to obtain highly desirable macroscopic properties.

Silicate Clay Nanocomposites Polymeric composites reinforced with clay crystals of nanometer scale recently attracted tremendous attention in material engineering. A particular class of nanocomposites is clay nanocomposites (see Chap. 8 and [498, 1549]. In general, layered silicate-reinforced presenting many challenges for mechanics are conditionally divided into three types (see [286, 540, 676]): conventional composites where the layered silicate is presented as the original aggregate state of the clay particles (tactoids) with no intercalation of the matrix material into the layered silicate; intercalated nanocomposites (modeled in Chap. 8) where the matrix material is inserted in the form of thin (a few nm) layers into the space between the parallel silicate layers of extremely small thickness ($\sim$1 nm);

and completely exfoliated or delaminated nanocomposites where the individual silicate platelets of 1 nm thickness are randomly dispersed in a continuous polymer matrix. Single clay layers were proposed to be an ideal reinforcing agent due to their extremely small aspect ratio ($\alpha = 0.001$–0.01), the nanometer filler thickness being comparable to the scale of the polymer chain structure, as well as to their high stiffness (200–400 GPa). A doubling of the tensile modulus and strength is achieved, and the heat distortion temperature increases by 100 °C for nylon-layered silicate nanocomposites containing as little as 2% vol. nanoplates [890, 1677]. However, exfoliation of layered minerals is seriously hampered by the fact that sheet-like materials exhibit a strong tendency to agglomeration due to their large contact surfaces. Generally, exfoliated nanocomposites demonstrate better properties than intercalated ones at the same nanoplate concentration. Vaia and Giannelis [1550] emphasized that the presence of many chains at interfaces means that much of the polymer is really "interphase-like" instead of having bulk-like properties that modify the thermodynamics of polymer chain confirmations and kinetics of chain motion. Since an interface limits the confirmations that polymer molecules can adopt, the properties of polymer in this interface are fundamentally different from that of polymer far removed from the interface. Consequently, almost all matrixes in CMs with a few volume percent of nanofiller can be considered as nanoscopically confined interfacial polymer. Thus, many mechanical, physical, and chemical factors could potentially formulate the properties of nanocomposites, and a better understanding of their relative contribution is needed.

1.1.3 Classification of CM Manufacturing

Classification of composites with respect to the manufacturing route divides them into "artificial" (i.e., man-made) and "natural" (i.e., in situ) composites. "Artificial" composites include long and short fiber composites, particulate and plate composites, interpenetrating multiphase composites, cellular and foam solids, and concrete. The manufacturing features can lead to significant variation of the microtopology and mechanical properties of CM. For example, depending on the thermomechanical treatment, the grains of a polycrystalline material can be isotropic spheroid (after full recrystallization and grain growth) or flake (after rolling) or elongated "cigars" (after extrusion). Among the "natural" composites, one can mention polycrystals, soil, sandstone, ice, wood, bone, cell aggregates, and the Earth's crust. Because the mechanical properties and the microtopology of composites are assumed to be known, their classification based on the manufacturing features will not be considered (see for details [1559, 1565]).

1.2 Effective Material and Field Characteristics

A wide class of phenomena for microinhomogeneous media are described by linear partial differential equations with random rapidly oscillating coefficients that can be reduced to linear integral equations by the use of a Green's function for a selected problem for a homogeneous comparison medium in bounded or infinite domains. The replacement of a governing integral equation with rapidly oscillating coefficients by an equation with either constant or smoothly varying "overall" coefficients is a main problem of micromechanics.

In classical elasticity, under assumption of geometrical and physical linearity, the relations between the local stress $\boldsymbol{\sigma}(\mathbf{x})$ and strain $\boldsymbol{\varepsilon}(\mathbf{x})$ tensors of microinhomogeneous media are established by Hooke's law: $\boldsymbol{\sigma}(\mathbf{x}) = \mathbf{L}(\mathbf{x})\boldsymbol{\varepsilon}(\mathbf{x})$, $\boldsymbol{\varepsilon}(\mathbf{x}) = \mathbf{M}(\mathbf{x})\boldsymbol{\sigma}(\mathbf{x})$, where $\mathbf{L}(\mathbf{x})$ and $\mathbf{M}(\mathbf{x}) = \mathbf{L}^{-1}$ are the known local stiffness and compliance fourth-order tensors. Effective tensors of elastic properties $\mathbf{L}^*$ and $\mathbf{M}^*$ of statistically homogeneous medium are determined as the proportionality factors $\langle\boldsymbol{\sigma}(\mathbf{x})\rangle = \mathbf{L}^*\langle\boldsymbol{\varepsilon}(\mathbf{x})\rangle$, $\langle\boldsymbol{\varepsilon}(\mathbf{x})\rangle = \mathbf{M}^*\langle\boldsymbol{\sigma}(\mathbf{x})\rangle$ between the macroscopic stresses $\langle\boldsymbol{\sigma}(\mathbf{x})\rangle$ and strains $\langle\boldsymbol{\varepsilon}(\mathbf{x})\rangle$ averaged over the volume $\bar{w}$ of a large sample w containing statistically large numbers of inhomogeneities. Thus, at the macroscale significantly exceeding the length scale of the material inhomogeneity (microscale), the behavior of a microinhomogeneous medium can be described by that of a suitably chosen equivalent homogeneous medium with the macroscopic stresses and strains slowly varying according to the macroscopic geometry and boundary conditions. In so doing, the local stress and strains, and other values, can be split into slow and fast (marked by primes) constituents that lead to the average constitutive equations presented in the following form: $\langle\boldsymbol{\sigma}\rangle = \langle\mathbf{L}\rangle\langle\boldsymbol{\varepsilon}\rangle + \langle\mathbf{L}'\boldsymbol{\varepsilon}'\rangle$, $\langle\boldsymbol{\varepsilon}\rangle = \langle\mathbf{M}\rangle\langle\boldsymbol{\varepsilon}\rangle + \langle\mathbf{M}'\boldsymbol{\sigma}'\rangle$, where $\mathbf{f}' = \mathbf{f} - \langle\mathbf{f}\rangle$ $[\mathbf{f} = \boldsymbol{\varepsilon}(\mathbf{x}), \boldsymbol{\sigma}(\mathbf{x}), \mathbf{L}(\mathbf{x}), \mathbf{M}(\mathbf{x})]$. Obviously, fluctuating parts of stresses $\boldsymbol{\sigma}'$ and strains $\boldsymbol{\varepsilon}'$ can be expressed through the random influence functions:

$$\boldsymbol{\varepsilon}'(\mathbf{x}) = \mathbf{A}^*(\mathbf{x})\langle\boldsymbol{\varepsilon}\rangle, \quad \boldsymbol{\sigma}'(\mathbf{x}) = \mathbf{B}^*(\mathbf{x})\langle\boldsymbol{\sigma}\rangle, \tag{1.1}$$

which exist (but are generally unknown) due to linearity of the problem. Substitution of the micro-macrolinkages (1.1) into the right-hand sides of the averaged constitutive equations leads to representations for the effective stiffness and compliance:

$$\mathbf{L}^* = \langle\mathbf{L}\rangle + \langle\mathbf{L}'\mathbf{A}^*\rangle, \quad \mathbf{M}^* = \langle\mathbf{M}\rangle + \langle\mathbf{M}'\mathbf{B}^*\rangle. \tag{1.2}$$

Thus, the problem of effective properties estimation is reduced to finding $\mathbf{A}^*$ (or $\mathbf{B}^*$), which is turns out some functional involving all points $\mathbf{y}$ surrounding the considered point $\mathbf{x}$. The mentioned problem of many bodies appearing in many sections of astronomy, mechanics, and physics can be generally solved only approximately.

The problem formulated in the preceding, which is central to our understanding of micromechanics, looks very trivial. However, this illusion of triviality disappears if we recall the first steps of research in this field. Voight [1566] proposed in 1889 to consider the average of the elastic moduli of the crystallites in a polycrystal as effective elastic moduli of the aggregate $\mathbf{L}^* = \langle \mathbf{L} \rangle$. Much later in 1929, Reuss [1290] proposed an analogous operation for obtaining the effective compliance $\mathbf{M}^* = \langle \mathbf{M} \rangle$. Much later still, Hill [639] proved in 1952 that Voight and Reuss estimations are upper and lower bounds of the true effective elastic moduli of the polycrystals: $\langle \mathbf{M} \rangle^{-1} \leq \mathbf{L}^* \leq \langle \mathbf{L} \rangle$.

The random values involved in the estimation of effective properties (1.2) can be classified as two sets: the values consisting of the material tensors (e.g., $\mathbf{L}, \mathbf{M}$) and some functions of these tensors, and the values formed by the field tensors (e.g., $\boldsymbol{\varepsilon}, \boldsymbol{\sigma}$). The distribution parameters of random material tensors can be evaluated through the known properties of constitutive phases and their random topological structure. The distributions of the random field tensors are expressed via the random material tensors by the use of the appropriate equations of elasticity theory. The simplest characteristic of a random variable is a one-point probability density whose first statistical moments defined the mean and dispersion. For example, the Voight and Reuss estimations are based on the first one-point distribution functions such as the phase volume fractions. Significantly more detailed information is contained in the multipoint probability densities incorporating all points of the medium. So, a knowledge of at least two-point distributions of the material tensors is necessary for estimation of the corrections introduced the terms $\langle \mathbf{L}'\mathbf{A}^* \rangle$ and $\langle \mathbf{M}'\mathbf{B}^* \rangle$ into the effective properties representations (1.2). The progress in micromechanics of random structure composites is based on the methods of statistical mechanics of a multiparticle system considering n-point correlation functions and direct multiparticle interactions of inhomogeneities.

For the sake of brevity, we intentionally sacrificed rigorousness for simplification of a presentation of basic concepts of micromechanics. For example, the average can be introduced as either the volume or ensemble averages $\{\mathbf{f}\} = (\bar{w})^{-1} \int_w \mathbf{f}(\mathbf{x})\, d\mathbf{x}$, $\langle \mathbf{f} \rangle = n^{-1} \sum_{i=1}^n \mathbf{f}_i(\mathbf{x})$, respectively, where $\mathbf{f}_i$ is an i's realization of a random process in the point $\mathbf{x}$. However, we will demonstrate that the size of RVE depends on the problem being considered and the method used for solution of this problem, and, because of this, RVE is simply a notation convenience reflecting a nature of nonlocal interaction between microelements that interact without touching of elements as in the case of local interaction. Probably the first demonstration of the nonlocality principle was the discovery of the Newtonian law of gravitation, which implies the existence of interaction between bodies remote from one another. A similar law for electric forces was demonstrated by Coulomb a century later.

The nonlocal nature of the effective constitutive law is manifested most conspicuously in the case of the eventual abandonment of the hypothesis of statistically homogeneous fields leading to a nonlocal coupling between statistical averages of the stress and strain tensors when the statistical average stress is given by an integral of the field quantity weighted by some tensorial kernel, i.e., the *nonlocal* effective elastic operator $\boldsymbol{\mathcal{L}}^*$: $\langle \boldsymbol{\sigma} \rangle(\mathbf{x}) = \int \boldsymbol{\mathcal{L}}^*(\mathbf{x}, \mathbf{y}) \langle \boldsymbol{\varepsilon} \rangle(\mathbf{y})\, d\mathbf{y}$. Perhaps the most challenging issue is how micromechanics can contribute to the understanding of the bridging mechanism between the coupled scales, which is described by the nonlocal constitutive equations involving the parameters of a relevant effective nonlocal operator with the microstructure. Considering dispersed media, this approach makes intuitive sense since the stress at any point will depend on the arrangement of the surrounding inclusions. Therefore the value of the statistical average field will locally depend on its value at the other points in its vicinity. This approach is especially appropriate if the inclusion number density varies over distances that are comparable to the particle size.

It should be mentioned that a multiscale hierarchy of composite materials is typically investigated via the use of the RVE concept when parameters of larger scale models are measured or calculated on a smaller scale. The RVE concept serving as a bridge between different scales is rigorously justified if a field scale (field inhomogeneity) infinitely exceeds a material scale (material inhomogeneity). However, even in this case in a large class of problems, the length scales cannot be separated in this way; the coupling between them is strongly "bidirectional." It takes place, for example, when the microscopic structure (e.g., the plastic strain and damage accumulation) develops due to some macroscopic forces. The coupling phenomenon is manifested even more visually in the case of comparable scales of the material scale of analyzed structures and the inhomogeneity scale of internal stresses, which of necessity yields a nonlocal character of the constitutive law in the area of coupling of the mentioned scales, and, therefore, the popular basic assumption acting as a bridge between adjusted scales can be considered as an approximation of the real nonlocal constitutive law.

We need to emphasize the emphasized an importance of analysis of statistical distributions of local microfields rather than only effective properties, which in actuality is a secondary problem deriving from estimation of the average field concentrator factors inside the phases. Averaging is usually suitable for predicting effective elastic properties. However, failure and elastoplastic deformations will depend on specific details of the local stress fields when fluctuation is important. A determination of failure mechanism usually localized at the interfaces is essentially important for the ceramic and metal matrix composites that have relatively weak interfaces and multiple interphase regions. Plastic deformations usually emanating in the vicinity of the interphases lead to loss of the homogeneity of the mechanical properties of the constituents when the local properties of the phases become position dependent. In the framework of computational micromechanics in such a case, one must check the specific observable stress fields for many large system realizations of the microstructure with the use of an extremal statistical technique. A more effective approach used in analytical mechanics is the estimation of the statistical moments of the stress field at the interface between the matrix and inhomogeneities. The inherent characteristics of this interface are critical to understanding the failure mechanisms usually localized at the interface.

The widespread accessibility of a unique level of computing power provides promotes the argument that modeling and simulation, properly formulated and performed, can be a credible partner to the traditional approaches of theory and experiments. Micromechanical modeling and simulation of random structures are becoming more and more ambitious as a result of advances in modern computer software and hardware (see [193]). On one hand, some models have been developed with the goal of minimizing empirical elements and assumptions. Specifically, several examples are available in the area of nonlocal and nonlinear problems such as localization of the plastic strains and failure based on the estimations of local stress concentration factors. In many cases, the resolution of microscopic phenomena has led to improved accuracy and offers the possibility of solving previously intractable problems. On the other hand, there are ambitions to attack increasingly large systems. Such methods, usually referred to as computational micromechanics, are based on the wide exploitation of Monte Carlo simulation with forthcoming numerical analysis for each random realization of multiparticle interactions of microinhomogeneities. However, at the present capacities of computer hardware and software, they are practical only for realizations containing no more than a few thousand inhomogeneities. In the case of periodic structures, the asymptotic unit cell homogenization method coupled with the finite element method (FEM) has been justified to be the most successful tool especially for the establishment of a link between the unit cell microstructure with periodic boundary conditions and overall properties of the heterogeneous media. Such a perfectly periodic problem, of course, does not exist in actual cases, although periodic modeling can be quite useful, since it provides rigorous estimations with a priori prescribed accuracy for various material properties. However, if the microscopic crack is considered, it should lead to the unrealistic assumption that the unit cell model with a crack is repeated. The case of a single microcrack (imperfection) in a composite material of periodic structure was investigated in the framework of the scheme of "matching rules" coupling the stresses in the outer and inner regions of the cell containing the imperfection (see [475, 482, 731, 732, 1476]). Implementation of the mentioned scheme accompanied by the Monte Carlo simulation has prohibitive computation costs for the periodic structures with random field of imperfections. Because of this, combining opportunities of computational micromechanics with basic assumptions of analytical micromechanics is very promising and considered in this book. The starting point of any average scheme is the choice of the elastic properties of the comparison medium that usually coincides with the homogeneous matrix. This allows one to restrict the estimation of stress fields to the fields just inside the inclusions. The choice of the comparison medium in the form of the perfect periodic structure has several benefits, which are considered in this book. These corrections incorporate specially adopted versions of the finite element analysis (FEA) method of analysis of multiple interacting inclusion problems into the modified versions of the MEFM combining a well-established theory of homogenization of periodic media with the methodology of averaging of random structures in the infinite domains.

Another barrier limiting Monte Carlo simulation appears in the analysis of statistically inhomogeneous random structures. Statistically homogeneous media are typically reduced to the averaging over a set of pseudorandom generations based on the determination of the "exact" values of the effective moduli of a periodic unit cell containing a few tens of particles. Trivialness of this problem is explained by the a priori known fact that the effective elastic moduli is a constant relating the average stresses and average strains. The case of the statistically inhomogeneous random structure is essentially more complicated because the effective elastic properties are now described by a nonlocal integral operator, which is a two-point one with a priori unknown structure. However, combining the MEFM with the numerical solution for a finite number of inclusions (considered in this book) presents no difficulties in the analysis of the indicated problem. In parallel with computational micromechanics mentioned above, classical or analytical micromechanics are also widely used and schematically considered in the next section.

1.3 Homogenization of Random Structure CM

For random structure composites, in general, the problem of estimation of effective properties cannot be solved exactly and is usually treated in an approximate manner. According to Willis [1628], the numerous methods in micromechanics can be classified into four broad categories: perturbation methods, self-consistent methods of truncation of a hierarchy, variational methods, and the model methods, among which there are no rigorous boundaries. Perturbation methods are effectively utilized only for the analysis of weakly inhomogeneous media and are based on the theory of a small parameter having controlled accuracy [106, 107, 966]. Variational methods generate the bounds of effective properties via substitution of approximate fields into the strict energy bounds obtained by the volume average of the energy density with the help of lower-order correlation functions (see the fundamental works [1075, 1239, 1520, 1625, 1627] and references therein). The variational methods represent the most rigorous trend of micromechanics of statistically homogeneous media; however, in the case of either statistically inhomogeneous structures or strong inhomogeneity, these bounds are not instructive. The model methods based on the replacement of the real random structure of composites by a specific one (see, e.g., [336]) have intuitive physical interpretations; but, majority of model methods are self-contained: usually the modification of either geometrical structure of composites or constitutive equations generates a need for creation of a new model with a priori unknown prediction accuracy. The variational and model replacement methods lead to some exact results; however, the analysis of these methods is considered schematically in this book.

We will consider in detail the self-consistent methods based on some approximate and closing assumptions for truncating an infinite system of integral equations involved and its approximate solution. Solution methods considered include a multiparticle effective field method (MEFM), effective medium method, Mori-Tanaka method, differential methods, and some others. The following also tend to concentrate on methods and concepts and their possible generalizations and connections of different methods rather than explicit results. For linear problems, the estimation methods of effective properties (compliance, thermal expansion, stored energy) as well as first and second statistical moments in the components for the general case of nonhomogeneity of the thermoelastic inclusion properties are analyzed. Both statistically homogeneous and statistically inhomogeneous, e.g., functionally graded, composites with a homogeneous matrix are considered, and the dependence of the effective properties (both local and nonlocal) on the concentration of the inclusions is presented. Both the Fourier transform method and iteration method are analyzed. It is likely that all of the methods discussed for linear elastic static problems can be generalized and in many cases are generalized to the dynamic case as well as to coupled phenomena such as thermoelectroelasticity.

Consideration of composites with nonlinear constitutive behavior, including plasticity and creep, requires an acceptance of some additional assumptions. The popular hypothesis of the mean field method takes into account only average stresses in the components. However, the statistical higher moments of stresses are extremely useful for understanding the evaluation of nonlinear phenomena. So, for estimation of nonlinear functions described the nonlinear process (the strength surface in fatigue, the yield function in plasticity, the stress potential in power-law creep, the energy release rate in fracture), one can use the statistical second moment of stresses in the phase. The hypothesis of homogeneity of plastic strains in the components and its generalization for plastic strain concentrations in thin layers around the inclusions are also considered.

It should be mentioned that many micromechanical methods, including self-consistent ones, use the hypothesis (called hypothesis H1) according to which each inclusion is located inside a homogeneous so-called effective field. Exploiting this hypothesis allows for replacement of the real inclusions with, perhaps, an arbitrary shape and structure by the fictitious ellipsoidal homogeneous inclusions with the same average thermoelastic response (see for details Chap. 8). No restrictions are imposed on the microtopology of the microstructure and the shape of inclusions as well as on the inhomogeneity of the stress field inside the inclusions. It will be justified that the most general the MEFM includes as the particular cases the well-known methods of mechanics of strongly heterogeneous media (such as the effective medium and the mean field methods). The MEFM is based on the theory of functions of random variables and Green's functions. Within this method a hierarchy of statistical moment equations for conditional averages of the fields estimated in the inclusions is derived. The hierarchy is established by introducing the notion of an effective field. In this way the interaction of different inclusions is taken directly into account in the framework of the hypothesis of the effective field. The main advantage of the MEFM proposed is a universal calculation scheme of general integral equations, which leaves room for corrections of their individual elements if either the initial integral equations and boundary conditions are changed or more improved methods are utilized for the analysis of these individual elements. In particular, the hypothesis H1 is merely a zero-order approximation of binary interacting inclusions that results in a significant shortcoming of the MEFM. This substantial obstacle will be overcome in the upgraded version of the MEFM in light of the generalized schemes based on the numerical solution of the problem for two inclusions in the infinite media, subjected to the effective field evaluated from forthcoming self-consistent estimations. The mentioned self-consistent methods are usually based on such fundamental notions as the Green's function and Eshelby tensor. However, a combination of the general anisotropy of the matrix and the general shape of randomly located inclusions with continuously variable anisotropic properties poses a significant barrier to the classical approaches based on either analytical or numerical representations for the Eshelby tensor for inclusions. Because of this, the combination of computational micromechanics with analytical micromechanics seems very promising because it allows for exploring the most powerful features of both mentioned groups of methods. The so-called numerical MEFM presents a universally rigorous scheme of both analyses of the microstructures and prediction of macroscopic properties and the statistical distributions of the local stress fields. One of the main advantages of the MEFM is in an efficient calculation of the general integral equations, which allows for hierarchical improvement of the method. The progress of the modern mathematical materials science demands the subsequent development of the MEFM providing a wide implementation of advanced numerical methods. It is possible to get a numerical solution in the form of a table in which each line among a few hundred lines corresponds to an accurate solution for two inclusions in an infinite medium subjected to unit loading. Such a table is be used subsequently for the estimation of local and nonlocal effective properties of statistically homogeneous and inhomogeneous composites in both the bounded and unbounded domains with a wide range of both inclusion concentration and elastic mismatch.

Lastly, we should recognize that the properties of materials involve different scales reflects the explosive character of progress in modern nano- and micromechanics resulting from the development of image analyses and computer-simulation methods on one hand and advanced experimental technique (such as X-ray tomography and electron microscopy) and improved materials processing (prescribed structure controlled by processing) on the other hand. At the coarse level, the collective results of events or processes taking place at the fine level are accounted for. In material modeling, one can distinguish four levels of simulation: electronic structure, atomistic, microscale, and macroscale. Although this book concentrates on the micro-macro-connection of the multiscale concept within the framework of continuum mechanics, we also consider some elements of molecular modeling and bridging nanomechanics to micromechanics in nanocomposites (NCs).

Unlike many other fields in material science, the evolution of NCs to its anticipated level of importance to material engineering significantly depends on the contributions from modeling and simulation. Computational approaches, based on the Newton's equations of motion at the atomistic level [referred to as the molecular dynamics (MD) approach], are currently limited to the nanoscale (where all dimensions are at the nanoscale) and cannot deal with the micro-length scales (which are traditionally analyzed by a continuum mechanics approach). These studies rely on fitting of atomistic simulation results to determine the important elastic parameters such as elastic moduli that implicitly assumes a local nature of constitutive law of continuum mechanics at the nanoscale based in turn on the assumption that a field scale (internal stress inhomogeneity) infinitely exceeds a material scale (molecular inhomogeneity). This popular basic assumption acting as a bridge between nano- and micromechanics is questionable due to comparable scales of the material and field inhomogeneities that with necessity yields nonlocal character of the constitutive law in the area of coupling the mentioned scales. The numerical technique proposed in this book is particularly well suited for a fast and accurate prediction on the effective elastic properties in the random nanocomposites where other existing methods may not be reliable. This goal is accomplished by combining an advanced MD simulation just for one nanoinclusion in the matrix with the methodology of averaging of random functionally graded structures recently developed.

1.4 Overview of the Book

The contents give a good of what this publication is about. Briefly stated, the first chapter discusses the classification of composites and nanocomposites, motivation for its studying, and outlines homogenization techniques of random structure composites from various viewpoints. Chapter 2 is a brief version of a solid mechanics course particularly tailored to provide background for the rest of this book. The key tools of random structure composite materials (CMs) such as Green's function, Eshelby, and related tensors are considered in Chap. 3 for the problems of static and dynamic elasticity, conductivity, and thermoelectroelasticity. Analytical and numerical solutions for one and a few ellipsoidal and nonellipsoidal heterogeneities in an infinite matrix and in a half space are considered in Chaps. 3 and 4.

Chapter 5 describes basic concepts and recent advances in quantitative characterization and numerical simulation of the microstructure of random structure composites. General representations of both the effective properties and effective energy functions are presented through the local stress and strain concentration factors in Chap. 6. The principal formulae of thermoelastostatics of statistically homogeneous composites satisfying the ergodicity condition adapted to our intended application to composites are discussed; minimum energy principles are derived that lead to variational bounds on all the effective properties in terms of trial fields. A new general integral equations (GIE) of micromechanics of CMs with deterministic (periodic and nonperiodic), random (statistically homogeneous and inhomogeneous, so-called graded), and mixed (periodic structures with random imperfections) structures are derived in Chap. 7, where also presented a connection of the new exact GIE with the known approximate GIEs. Chapter 8 describes the multiparticle effective field method (MEFM) (put forward and developed by the author) which is based on the theory of functions of random variables and Green's functions. Within this method a hierarchy of statistical moment equations for conditional averages of the stresses in the inclusions is derived. The hierarchy is established by introducing the notion of an effective field. In this way the interaction of different inclusions is taken directly into account. One observes in Chap. 8 that the MEFM includes as particular cases the well-known methods of mechanics of strongly heterogeneous media (such as the effective medium and the mean field methods and some others). In Chap. 9 we consider the solutions of the new GIEs by the use of acceptance of different combinations of the classical hypotheses **H1a**, **H1b**, **H2**, and **vH3**. A larger difference between the use of the new and old GIEs was obtained for composites reinforced by inclusions demonstrating essentially inhomogeneous stress distribution inside isolated inclusions even in the framework of the EFH. Inhomogeneity of the first and second statistical moments of stresses inside the heterogeneities (even for the spherical homogeneous inclusions) of CMs are estimated in Interface integral technique in thermoelasticity of CMs is proposed in Chap. 11 by the use of introduction of both the interface version of the GIE and interface polarization tensors; nimerical results are obtained by the method of fundamental solutions. Although the effective behavior of the composite is traditionally the main focus of micromechanics, it is also essential to supply insight into the statistical description of the local strains and stresses (considered in Chaps. 6 and 10 by both the perturbation method and the method of integral equations) which are extremely useful for understanding the evolution of nonlinear phenomena (analyzed in Chap. 12) such as plasticity, creep, and damage. Nonlocal effects in statistically homogeneous and inhomogeneous random structure composites are analyzed in Chap. 13 by the MEFM which also has qualitative benefits following immediately from the consideration of multiparticle interactions. From such considerations, it can be concluded that the final relations for effective properties described by nonlocal integral operators depend explicitly not only on the local concentration of the inclusions but also on at least binary correlation functions of the inclusions.

Micromechanics of CM with nonlocal properties of constituents is considered in Chap. 14: strongly nonlocal strain type model and weakly nonlocal theories covering higher-grade models, higher-order models, and micromorphic models. In Chap. 15, we generalize the approaches of Chaps. 7–11 and 14 from the point of view both the unified operator form representations and coupling of physical phenomena; as the particular case, one considers the problem of thermoelasticity, conductivity, thermoelectroelasticity, and wave propagation in CMs. Basic representations for bond-based peridynamics for homogeneous materials are described in Chap. 16. Background of peridynamic micromechanics of CMs are presented in Chap. 17: the GIE proposed for locally elastic CMs are generalized to their peridynamic counterpart, one expresses the effective thermoelastic properties through the influence functions, and both the micropolarization tensor and interface polarization tensors are defined. The admissible displacement and force fields are introduced. The theorem of work and energy, Betti's reciprocal theorem, and the theorem of virtual work are proved. Principles of minimum of both potential energy and complimentary energy are generalized. The strain energy bounds are estimated for both the displacement and force homogeneous volumetric boundary conditions. Generalized Hill's bounds on the effective elastic moduli of peridynamic random structure composites are obtained. The results of Chaps. 6 and 9 for locally elastic CMs, are generalized to the peridynamic counterpart in Chap. 18. So, a solution (called basic problem) for one inclusion inside the infinite matrix subjected to some effective field is substituted into the one or another micromechanical scheme (e.g., Mori-Tanaka approach of Effective field method). Effective deformation of peridynamic random structure bar subjected to inhomogeneous body-force is estimated. Computational homogenization the technique for peridynamic periodic structure CMs is presented in Chap. 19. One introduces the new volumetric periodic boundary conditions. Decomposition of the material and field parameters, and effective modulus representations are performed for periodic media. Volumetric Bloch-Floquet periodic boundary conditions are introduced. Introduction of volumetric periodic boundary conditions opens wide opportunities for a straightforward generalization of well-developed local theory of computational homogenization to their peridynamic counterparts including different physical phenomena (e.g., state-based models, conductivity, viscosity, thermoelastoplasticity, damage accumulation, wave propagation) for CMs with different periodic structure. A short sketch of limitations and ideas as well as possible generalizations of some basic concepts of micromechanics is presented in the Conclusion.

Chapter 2
Foundations of Solid Mechanics

Abstract A major problem in continuum mechanics consists of predicting the deformations and internal forces arising in a material body subjected to a given set of external fields such as the forces of pure mechanical nature, temperature, and electromagnetic fields. This chapter summarizes some of important definitions, relations, and methods commonly employed for the analysis of reformable media although, of course, many aspects of mechanical behavior are left unaccounted for in this approach and referring the unsatisfied reader to monographs in which a level of higher generality is adopted. However, it is hoped that sufficient details are given so that the reader can understand remainder chapters without constant reference to other books.

As is customary in continuum mechanics studies, material properties and fields are expressed in tensor form in this book. The general technique of tensor analysis is given, e.g., in [146, 1426, 1482] which present the tensor theory not only as an autonomous mathematical discipline but also as a preparation for theories of continuum mechanics. The special applications of tensors are described in the books [989, 1535]. The books [42, 515] provide an introduction to the theories of linear elasticity and nonlinear elasticity. More comprehensive treatments of nonlinear and anisotropic elasticity can be found, e.g., in the books [569, 585, 989, 1153, 1378, 1409, 1500, 1515]. The coupling effects of mechanical, temperature, and electromagnetic fields are considered in [1033, 1188, 1349]. The general reviews of phenomenological plasticity with an extensive list of references are given in [726, 977, 1017].

2.1 Elements of Tensor Analysis

This section provides a remainder of basic notions of tensors of a certain rank which components transform in a particular way under the transformation of coordinates $(\mathbf{e}_1, \mathbf{e}_2, \mathbf{e}_3) \rightarrow (\mathbf{e}'_1, \mathbf{e}'_2, \mathbf{e}'_3)$.

V. A. Buryachenko, *Local and Nonlocal Micromechanics of Heterogeneous Materials*, https://doi.org/10.1007/978-3-030-81784-8_2

We concern with the rectangular Cartesian coordinate systems and require that the orthonormalized basis is right-handedness $\mathbf{e}_1 \cdot (\mathbf{e}_2 \times \mathbf{e}_3) = 1$ and $\mathbf{e}_i \cdot \mathbf{e}_i = \delta_{ij}$, where a scalar multiplication $\mathbf{e}_i \cdot \mathbf{e}_j$ denotes the projection of the ith basic vector on the jth one (or vice-versa) and

$$\delta_{ij} = \begin{cases} 1 & \text{if } i = j, \\ 0 & \text{otherwise} \end{cases} \tag{2.1}$$

is the Kronecker delta. The vector (cross) product of two basic vectors $\mathbf{e}_i \times \mathbf{e}_j$ can be defined by the permutation tensor ϵ (summation over repeated Latin indices is implied)

$$\mathbf{e}_i \times \mathbf{e}_j = \epsilon_{ijk}\mathbf{e}_k, \; \epsilon_{ijk} = \begin{cases} +1 & \text{if } ijk \text{ is an even permutation of 123,} \\ -1 & \text{if } ijk \text{ is an odd permutation of 123,} \\ 0 & \text{if any two indices are identical.} \end{cases} \tag{2.2}$$

The vector product of the vectors **a**, **b** and the triple scalar product of the vectors **a**, **b**, **c** can consequently be written as

$$\mathbf{a} \times \mathbf{b} = \epsilon_{ijk} a_i b_j \mathbf{e}_k, \quad (\mathbf{a} \times \mathbf{b}) \cdot \mathbf{c} = \epsilon_{ijk} a_i b_j c_k = \begin{vmatrix} a_1 & b_1 & c_1 \\ a_2 & b_2 & c_2 \\ a_3 & b_3 & c_3 \end{vmatrix}. \tag{2.3}$$

The expansion of another orthonormalized basis $\mathbf{e}'$ ($\mathbf{e}'_i \cdot \mathbf{e}'_j = \delta_{ij}$) with the same origin of coordinates into the old one $\mathbf{e}'_{i'} = g_{i'j}\mathbf{e}_j$ is defined by coefficients $g_{i'j}$ called a *cosine matrix*, because each element $g_{i'j}$ is the cosine of the angle between two corresponding axes

$$g_{i'j} = \mathbf{e}'_i \cdot \mathbf{e}_j = \cos\alpha_{i'j}. \tag{2.4}$$

The most frequently used methods of parametric representation of orthogonal transformation of Cartesian coordinates $\mathbf{e}'$ and $\mathbf{e}$ are presented in Sect. 21.1. An arbitrary vector $\mathbf{x}$ is completely defined by its magnitude (modulus) and direction in the space

$$\mathbf{x} = \sum x_i \mathbf{e}_i = x_i \mathbf{e}_i, \; x_i = \mathbf{x} \cdot \mathbf{e}_i = |\mathbf{x}| \cos(\mathbf{x}, \mathbf{e}_i). \tag{2.5}$$

The orthogonal matrix $\mathbf{g} = \|g_{ij}\|, \; (i, j = 1, \ldots, 3)$ called the rotation matrix links the coordinates $\mathbf{x}'$ of an arbitrary point in the reference orthonormalized frame $\mathbf{e}'$ ($\mathbf{e}'_i \cdot \mathbf{e}'_j = \delta_{ij}$) with the coordinates $\mathbf{x}$ in the crystal orthonormalized coordinate system $\mathbf{e}$ ($\mathbf{e}_i \cdot \mathbf{e}_j = \delta_{ij}$) with the same origin of coordinates (see for details Sect. 21.1)

$$\mathbf{x}' = \mathbf{g}\mathbf{x}. \tag{2.6}$$

The definitions of tensors of order 0 (scalar) ${}^0\mathbf{T}(\mathbf{e}_1, \mathbf{e}_2, \mathbf{e}_3) \equiv {}^0\mathbf{T}'(\mathbf{e}_1', \mathbf{e}_2', \mathbf{e}_3')$ and 1 [contravariant vector (2.6)] can be easily generalized to contravariant tensors of order n

$$^n\mathbf{T} =^n T_{i_1 i_2 \ldots i_n} \mathbf{e}_{i_1} \otimes \ldots \mathbf{e}_{in} = {}^nT'_{i_1 i_2 \ldots i_n} \mathbf{e}'_{i_1} \otimes \ldots \mathbf{e}'_{in}, \tag{2.7}$$

where the 3^n tensor components in the prime and unprime coordinates are linked by the following transformation

$${}^nT'_{j_1 j_2 \cdots j_n} = g_{j_1 i_1} \cdots g_{j_n i_n} {}^nT_{i_1 i_2 \ldots i_n}. \tag{2.8}$$

A tensor ${}^{n+m}\mathbf{T}$ of rank $n+m$, contravariant of rank n and covariant of rank m, is an object with the induced component transformation

$${}^{n+m}T'^{j_1 j_2 \ldots j_n}_{j_1' j_2' \ldots j_m'} = g_{j_1 i_1} \cdots g_{j_n i_n} g_{i_1' j_1'} \cdots g_{i_m' j_m'} {}^{n+m}T^{i_1 i_2 \ldots i_n}_{i_1' i_2 \ldots i_m'}. \tag{2.9}$$

It is possible to distinguish four operations with the tensors. The first one joins the summation and multiplication over the number of the tensors $T_{i_1,\ldots,i_n}$ and $S_{i_1,\ldots,i_n}$ of the same order

$$\lambda T_{i_1,\ldots,i_n} + \beta S_{i_1,\ldots,i_n} = R_{i_1,\ldots,i_n} \Leftrightarrow \lambda \mathbf{T} + \beta \mathbf{S} = \mathbf{R}, \tag{2.10}$$

where λ and β are scalars. The tensor (or diadic) production of two tensors of orders n and m is a tensor of order $n+m$, defined as

$$\mathbf{T} \otimes \mathbf{S} = (T_{i_1,\ldots,i_n} S_{j_1,\ldots,j_m}) \mathbf{e}_{i_1} \ldots \mathbf{e}_{i_n} \mathbf{e}_{j_1} \ldots \mathbf{e}_{j_m}. \tag{2.11}$$

In so doing, in general $\mathbf{T}$ and $\mathbf{S}$ are not commutative, $\mathbf{T} \otimes \mathbf{S} \neq \mathbf{S} \otimes \mathbf{T}$. The third operation called convolution (or contraction) is the action with one tensor reducing his order on two in the framework of the Einstein summation convention, for example

$$T_{i_1,\ldots,i_n,k,k,j_1,\ldots,j_m} = \mathbf{S}_{i_1,\ldots,i_n,j_1,\ldots,j_m}. \tag{2.12}$$

The fourth operation produces the index permutation leads to a new tensor of the same order but with different index order. For example, one can generates only one new tensor (called transposed one) of the second order: $\mathbf{S} = \mathbf{T}^\top \Leftrightarrow S_{ij} = T_{ji}$. In so doing, a second-order tensor $\mathbf{T}$ is called symmetric if $\mathbf{T}^\top = \mathbf{T}$, and it is called antisymmetric if $\mathbf{T} = -\mathbf{T}^\top$.

By straightforward check, we are able to ascertain that the new objects formed by mentioned operations are really the tensors. These operations can be arranged in the different combinations. For example, the combination of the diadic production and convolution denoting by a dot leads to the following tensors

$$\mathbf{R} = \mathbf{T} \cdot \mathbf{S} \Leftrightarrow R_{ij} = T_{ik} S_{kj}, \tag{2.13}$$

$$\lambda = \mathbf{R} \cdot \cdot \mathbf{T} \Leftrightarrow \lambda = R_{ij} T_{ji}, \tag{2.14}$$

$$\lambda = \mathbf{R} : \mathbf{T} \Leftrightarrow \lambda = R_{ij} T_{ij}, \tag{2.15}$$

where multiple dots are appropriate for complex constructions.

In particular, the simplest tensors ${}^2\mathbf{T}$ of order 2 are the unit tensor $\boldsymbol{\delta}$ (1.1) and the rotation tensor $\mathbf{g}$ (2.4). They hold the property of *orthogonality*: ${}^2\mathbf{T} \cdot {}^2\mathbf{T} = \boldsymbol{\delta}$ and *isotropy*: ${}^2\mathbf{T}' = {}^2\mathbf{T}$. The tensor $\boldsymbol{\delta}$ transform a vector $\mathbf{a}$ into itself while $\mathbf{g}$ transforms the unit base vector $\mathbf{e}_i$ to the unit vector $\mathbf{e}'_i$ of the primed coordinate system: $\mathbf{e}'_i = \mathbf{g} \cdot \mathbf{e}_i$. The simplest tensor of order 4 is the unit tensor $\mathbf{I}$ with the components $I_{ijkl} = \frac{1}{2}(\delta_{ik}\delta_{jl} + \delta_{il}\delta_{jk})$ which can be decomposed into the *bulk* and *deviatoric* parts

$$\mathbf{I} = \mathbf{N}_1 + \mathbf{N}_2, \ \mathbf{N}_1 \equiv \frac{1}{3}\boldsymbol{\delta} \otimes \boldsymbol{\delta}, \ \ \mathbf{N}_2 = \mathbf{I} - \mathbf{N}_1 \tag{2.16}$$

which have the property of orthogonality

$$\mathbf{N}_1 : \mathbf{N}_1 = \mathbf{N}_1, \ \mathbf{N}_2 : \mathbf{N}_2 = \mathbf{N}_2, \ \mathbf{N}_1 : \mathbf{N}_2 = \mathbf{0}. \tag{2.17}$$

Differentiation denoting by the *del operator* $\partial_i \equiv \partial/\partial x_i$ of a tensor with the suitably smooth components forms other tensors

$$\nabla \cdot \mathbf{T} \equiv \mathrm{Div}\mathbf{T} = (\partial_j T_{ji_1,\ldots,i_{n-1}})(\mathbf{x})\mathbf{e}_{i_1} \otimes \mathbf{e}_{i_2} \ldots \mathbf{e}_{i_{n-1}}, \tag{2.18}$$

$$\nabla \otimes \mathbf{T} \equiv \mathrm{Grad}\mathbf{T} = (\partial_{i_1} T_{i_2,\ldots,i_{n+1}})(\mathbf{x})\mathbf{e}_{i_1} \otimes \mathbf{e}_{i_2} \ldots \mathbf{e}_{i_{n+1}}, \tag{2.19}$$

$$\nabla \times \mathbf{T} \equiv \mathrm{Curl}\mathbf{T} = [\epsilon_{i_1 jk}(\partial_j T_{k,i_2,\ldots,i_n})](\mathbf{x})\mathbf{e}_{i_1} \otimes \mathbf{e}_{i_2} \ldots \mathbf{e}_{i_{n+1}}, \tag{2.20}$$

which are called *divergence, gradient*, and *curl* of the tensor field $\mathbf{T} = {}^n\mathbf{T}$ and have the orders $n-1, n+1$, and n, respectively. We will also use the other differential operators of the first- and second-order tensors,

$$\mathrm{Def}\mathbf{a} \equiv \frac{1}{2}(\nabla \otimes \mathbf{a} + \mathbf{a} \otimes \nabla) = \frac{1}{2}(a_{i,j} + a_{j,i})\mathbf{e}_i \otimes \mathbf{e}_j, \tag{2.21}$$

$$\mathrm{Inc}\,\mathbf{b} \equiv \epsilon_{ijk}\epsilon_{lmn} b_{jn,km} \mathbf{e}_i \otimes \mathbf{e}_j \tag{2.22}$$

called the operator of *deformation* and *incompatibility*, respectively.

Analogously to differentiation, the integration of a tensor field $\mathbf{T}$ on both the domain w and their boundary ∂w can be defined in terms of the integration of its components. The *Gauss theorem* relates the action of the ∇ operator on the tensor field $\mathbf{T}$ in domain w to its flux across the boundary ∂w, under suitable smoothness of both the tensor **T** and the domain boundary ∂w with the unit vector of an outward normal $\mathbf{n}(\mathbf{s}) \perp \partial w$ ($\mathbf{s} \in \partial w$)

$$\int_w \nabla * \mathbf{T}\, d\mathbf{x} = \int_{\partial w} \mathbf{n} * \mathbf{T}\, ds, \tag{2.23}$$

where $*$ can stand for " $\cdot$ ",$\otimes$, or $\times$. Let $\partial w'$ be a portion of an oriented surface with unit outward normal **n** and with the surface closed edge C. Then for the tensor field **T** continuously differentiable in $\partial w'$ and continuous in C, the generalized Stokes theorem can be presented in the form

$$\int_{\partial w'} (\mathbf{n} \times \nabla) * \mathbf{T}\, ds = \oint_C \mathbf{T} * d\mathbf{l}. \tag{2.24}$$

Equation (2.23) is also generalized for the volume w and surface ∂w, excluding the points of a discontinuity surface τ with unit normal **n** which may be sweeping the body. In a similar manner, the surface $\partial w'$ with edge C in Eq. (2.24) is accompanied by a discontinuity line γ that may be sweeping the surface. Thus (see, e.g., [446]),

$$\int_{w\setminus\tau} \nabla \cdot \mathbf{T}\, d\mathbf{x} = \int_{\partial w\setminus\tau} \mathbf{T} \cdot \mathbf{n}\, ds - \int_\tau [\mathbf{T}] \cdot \mathbf{n}\, ds, \tag{2.25}$$

$$\int_{\partial w'\setminus\gamma} (\mathbf{n} \times \nabla) \cdot \mathbf{T}\, ds = \oint_{C\setminus\gamma} \mathbf{T} \cdot d\mathbf{l} - \oint_\gamma [\mathbf{T}]\, d\mathbf{l}, \tag{2.26}$$

where the brackets $[\cdot]$ denote the jump of its enclosure across the discontinuity surface τ or the discontinuity line γ and $w \setminus \tau$, $\partial w \setminus \tau$, and $\partial w' \setminus \gamma$, $C \setminus \gamma$ exclude points of τ and γ, respectively.

2.2 The Theory of Strains and Stresses

We consider a body occupying the regions w_0 and w, with boundaries w_0 and ∂w, in some fixed reference configuration and in the current one, respectively. Suppose an arbitrary material point $P^0 \in w_0$ at time $t = 0$ has the spatial coordinates **x** in the orthonormalized coordinate system $\mathbf{e}_i$ ($i = 1, 2, 3$). The point P^0 moves during the time t of deformations to the point $P \in w$ with coordinates $\boldsymbol{\xi}(t)$ in the same coordinate system. The relations between the coordinates $\boldsymbol{\xi}$ and **x** are defined by the smooth one-to-one invertible functions

$$\boldsymbol{\xi} = \boldsymbol{\xi}(\mathbf{x}, t), \;\; (\xi_i = \xi_i(x_1, x_2, x_3, t)) \;\text{ and }\; \mathbf{x} = \mathbf{x}(\boldsymbol{\xi}, t), \;\; (x_i = x_i(\xi_1, \xi_2, \xi_3, t)), \tag{2.27}$$

which are provided by an assumption that the Jacobian

$$J = \det\left\| \frac{\partial(\xi_1, \xi_2, \xi_3)}{\partial(x_1, x_2, x_3)} \right\| = \begin{vmatrix} \partial\xi_1/\partial x_1 & \partial\xi_1/\partial x_2 & \partial\xi_1/\partial x_3 \\ \partial\xi_2/\partial x_1 & \partial\xi_2/\partial x_2 & \partial\xi_2/\partial x_3 \\ \partial\xi_3/\partial x_1 & \partial\xi_3/\partial x_2 & \partial\xi_3/\partial x_3 \end{vmatrix} \tag{2.28}$$

exists at each point of the configuration w and that $J > 0$ meaning that the material cannot penetrate itself and that material element of non-zero volume cannot be compressed to a point or expanded to infinite volume during the motion. The independent variables $\mathbf{x}$ and $\boldsymbol{\xi}$ in Eq. (2.27) are referred to as Lagrange and Euler variables, respectively.

We define the strain tensor (specifying the change between the points in the deformable body) by introducing the displacement $\mathbf{u} = \boldsymbol{\xi} - \mathbf{x}, \quad u_i = \xi_i - x_i$ as well as the tensors of the deformation gradient $\mathbf{F}$, material displacement gradient $\mathbf{H}$, and spatial displacement gradient ($J \equiv \det\mathbf{F}(\mathbf{x}) > 0$)

$$\frac{\partial\boldsymbol{\xi}}{\partial\mathbf{x}} = \mathbf{F}(\mathbf{x}), \quad \frac{\partial\mathbf{u}(\mathbf{x})}{\partial\mathbf{x}} = \mathbf{H}(\mathbf{x}) \equiv \mathbf{F}(\mathbf{x}) - \mathbf{I}, \quad \frac{\partial\mathbf{u}(\boldsymbol{\xi})}{\partial\boldsymbol{\xi}} = \mathbf{h}(\boldsymbol{\xi}), \;\; \mathbf{Fh} = \mathbf{H}, \tag{2.29}$$

respectively. We will reproduce the correspondences between some elements of the regions w_0 and w in the reference and current configurations, respectively. The equation $d\boldsymbol{\xi} = \mathbf{F}\,d\mathbf{x}$ describes how an infinitesimal line element $d\mathbf{x} \in w_0$ transforms linearly into the element $d\boldsymbol{\xi} \in w$. The areas ds and dS of the surface elements of ∂w and ∂w_0 with the (positive) unit normals $\mathbf{n}$ and $\mathbf{N}$, respectively, are connected by the *Nanson's formula*

$$\mathbf{n}ds = J\mathbf{BN}dS, \;\; \mathbf{B} = (\mathbf{F}^{-1})^\top. \tag{2.30}$$

The volume dv in the deformed configuration is related with the volume dV in the reference configuration by the equation

$$dv = JdV. \tag{2.31}$$

The ratio of current $|\boldsymbol{\xi}|$ to reference $|d\mathbf{x}|$ lengths of a line element that was in the direction $\mathbf{M}$ in the reference configuration defines the *stretch*:

$$\lambda(\mathbf{M}) \equiv \frac{|d\boldsymbol{\xi}|}{|d\mathbf{x}|} = |\mathbf{FM}| = \;[\mathbf{M} \cdot (\mathbf{F}^\top\mathbf{FM})]^{1/2} \tag{2.32}$$

in the direction $\mathbf{M}$ at $\mathbf{x}$. The quality $\lambda(\mathbf{M}) - 1$ is called the extension ratio in the direction $\mathbf{M}$, while $|d\boldsymbol{\xi}| - |d\mathbf{x}|$ is the extension.

The deformed states of the body in the vicinity of P^0 are defined the strain tensor specified by the displacement $\mathbf{u}$. Working with the material $\mathbf{x}$ and spatial $\boldsymbol{\xi}$

variables, we can define the strain tensors in Lagrange's (Green's tensor) and Euler's (Almansi's tensor) representations

$$\boldsymbol{\varepsilon} = \frac{1}{2}(\mathbf{H}^\top + \mathbf{H} + \mathbf{H}^\top \mathbf{H}),\ \varepsilon_{ij}(\mathbf{x}) = \frac{1}{2}\left(\frac{\partial u_i}{\partial x_j} + \frac{\partial u_j}{\partial x_i} + \frac{\partial u_k}{\partial x_i}\frac{\partial u_k}{\partial x_j}\right), \quad (2.33)$$

$$\tilde{\boldsymbol{\varepsilon}} = \frac{1}{2}(\mathbf{h}^\top + \mathbf{h} + \mathbf{h}^\top \mathbf{h}),\ \tilde{\varepsilon}_{ij}(\mathbf{x}) = \frac{1}{2}\left(\frac{\partial u_i}{\partial \xi_j} + \frac{\partial u_j}{\partial \xi_i} - \frac{\partial u_k}{\partial \xi_i}\frac{\partial u_k}{\partial \xi_j}\right), \quad (2.34)$$

Hereafter the values referred to the variables $\boldsymbol{\xi}$ are marked by the symbol $\tilde{}$. In many practical applications, it is possible to neglect the products of derivatives in Eqs. (2.33) and (2.34). Then the coincidence of the tensors of Green and Almansi yields the well-known infinitesimal strain expression

$$\boldsymbol{\varepsilon} = \mathrm{Def}\mathbf{u},\ \varepsilon_{ij}(\mathbf{x}) = \frac{1}{2}\left(\frac{\partial u_i}{\partial x_j} + \frac{\partial u_j}{\partial x_i}\right). \quad (2.35)$$

If displacements $\mathbf{u}$ are given, the components of the strain tensors (2.33)–(2.35) are easily calculated by proper differentiation of these quantities. However, if six component of a strain tensor are given, they must be interrelated to serve as an integrability conditions ensuring a compatible set of the three independent displacement components. For the linear infinitesimal strain (2.35), these six interrelations, called *Saint-Venan't compatibility equations*, produce only three independent relations between the six components of the strain

$$\mathrm{Inc}\boldsymbol{\varepsilon} = \mathbf{0},\quad \epsilon_{ilm}\epsilon_{jpq}\frac{\partial^2 \varepsilon_{mn}}{\partial x_l \partial x_q} = 0. \quad (2.36)$$

For an arbitrary symmetric tensor of the second order (including Green and Almansi strain tensors), the characteristic equation $\det\|\varepsilon_{ij} - \varepsilon\delta_{ij}\| = 0$, defining a polynomial of the third degree in ε

$$\varepsilon^3 - I_\varepsilon \varepsilon^2 + II_\varepsilon \varepsilon - III_\varepsilon = 0 \quad (2.37)$$

has three real roots $\varepsilon_1, \varepsilon_2, \varepsilon_3$ called *principal values*. The coefficients $I_\varepsilon, II_\varepsilon, III_\varepsilon$ of Eq. (2.37) are scalars called *principal basic invariants* and are given by

$$I_\varepsilon = \mathrm{tr}\boldsymbol{\varepsilon} \equiv \varepsilon_{ii} = \varepsilon_1 + \varepsilon_2 + \varepsilon_3, \quad (2.38)$$

$$II_\varepsilon = \frac{1}{2}\epsilon_{ijk}\epsilon_{ilm}\varepsilon_{jl}\varepsilon_{km} \equiv \frac{1}{2}(I_\varepsilon^2 - \boldsymbol{\varepsilon} : \boldsymbol{\varepsilon}) = \varepsilon_{11}\varepsilon_{22} + \varepsilon_{22}\varepsilon_{33} + \varepsilon_{33}\varepsilon_{11}$$
$$-\varepsilon_{12}^2 - \varepsilon_{13}^2 - \varepsilon_{23}^3 = \varepsilon_1\varepsilon_2 + \varepsilon_2\varepsilon_3 + \varepsilon_1\varepsilon_3, \quad (2.39)$$

$$III_\varepsilon = \det\boldsymbol{\varepsilon} \equiv \frac{1}{6}(2\varepsilon_{ij}\varepsilon_{jk}\varepsilon_{ki} - 3I_\varepsilon\boldsymbol{\varepsilon} : \boldsymbol{\varepsilon} + I_\varepsilon^3) = \varepsilon_1\varepsilon_2\varepsilon_3. \quad (2.40)$$

To each principal value ε_i, the corresponding principal direction $\mathbf{m}_i$ is defined by the equation (i not summed) $\boldsymbol{\varepsilon} \cdot \mathbf{m}_i - \varepsilon_i \mathbf{m}_i = \mathbf{0}$. The tensor $\boldsymbol{\varepsilon}$ can be expressed in diagonal form in the principal triad $(\mathbf{m}_1, \mathbf{m}_2, \mathbf{m}_3)$

$$\boldsymbol{\varepsilon} = \sum_{i=1}^{3} \varepsilon_i \mathbf{m}_i \otimes \mathbf{m}_i, \tag{2.41}$$

which is orthogonal if $\varepsilon_1 \neq \varepsilon_2 \neq \varepsilon_3$: $\mathbf{m}_i \cdot \mathbf{m}_j = \delta_{ij}$. In such a case, the components of the tensor $\boldsymbol{\varepsilon}$ in the principle triad have the form (no sum on i) $\varepsilon_{ij} = \varepsilon_i \delta_{ij}$.

In parallel with the principle basic invariants (2.38)–(2.40), one can introduce other invariants such as *basic algebraic invariants*

$$A_1 = \varepsilon_{ii} = \varepsilon_1 + \varepsilon_2 + \varepsilon_3, \tag{2.42}$$

$$A_2 = \varepsilon_{ij}\varepsilon_{ji} = \varepsilon_1^2 + \varepsilon_2^2 + \varepsilon_3^2, \tag{2.43}$$

$$A_3 = \varepsilon_{ij}\varepsilon_{jk}\varepsilon_{ki} = \varepsilon_1^3 + \varepsilon_2^3 + \varepsilon_3^3 \tag{2.44}$$

as well as the invariant system specified by the intensity e_i of the tensor deviator $\boldsymbol{\varepsilon}'$ and its phase ψ

$$e = (\varepsilon_1 + \varepsilon_2 + \varepsilon_3)/3, \tag{2.45}$$

$$e_i = \frac{1}{3}\sqrt{(\varepsilon_1 - \varepsilon_2)^2 + (\varepsilon_2 - \varepsilon_3)^2 + (\varepsilon_1 - \varepsilon_3)^2}, \tag{2.46}$$

$$\frac{1}{\sqrt{2}} \cos 3\psi = e_i^{-3}(\varepsilon_1 - e)(\varepsilon_2 - e)(\varepsilon_3 - e). \tag{2.47}$$

The following relationships between the invariants hold true

$$I_\varepsilon = A_1,\ 2II_\varepsilon = a_1^2 - A_2,\ III_\varepsilon = (2A_3 - 3A_2A_1 + A_1^3)/6, \tag{2.48}$$

$$e = \frac{1}{3}A_1,\ e_i = \frac{1}{3}\sqrt{3A_2 - A_1^2},$$

$$\cos 3\psi = \sqrt{2}(9A_3 - 9A_1A_2 + 2A_1^3)(3A_2 - A_1^2)^{3/2}. \tag{2.49}$$

Now we will consider some definitions and notations of stress analysis. Let an element of the internal surface with an area ds have a unit normal vector $\mathbf{n}$. The total force $d\mathbf{f}$ acting on this element of surface defines the surface traction $\mathbf{t}^{(n)} = d\mathbf{f}/ds$, where the superscript (n) designates the normal to the surface element. The second-order Cauchy (or Eulerian) stress tensor $\mathbf{T}$ is formally related to the stress traction by the symmetric tensor equation

$$\mathbf{t}^{(n)} = \mathbf{n} \cdot \mathbf{T},\ t_j^{(n)} = n_i T_{ij}. \tag{2.50}$$

The tensor **T** is symmetric if and only if balance of moment of momentum holds. Then the stress state in the point $\boldsymbol{\xi}$ can be described either by six components T_{ij} of the tensor **T** or by three invariants $I_T, II_T, and III_T$ called the *principal stresses* [introduced analogously to Eqs. (2.38)–(2.40), $\mathbf{T} = \boldsymbol{\sigma}$]

$$I_\sigma = \text{tr}\boldsymbol{\sigma} \equiv \sigma_{ii} = \sigma_1 + \sigma_2 + \sigma_3, \tag{2.51}$$

$$II_\sigma = \frac{1}{2}\epsilon_{ijk}\epsilon_{ilm}\sigma_{jl}\sigma_{km} \equiv \frac{1}{2}(I_\sigma^2 - \boldsymbol{\sigma} : \boldsymbol{\sigma}) = \sigma_{11}\sigma_{22} + \sigma_{22}\sigma_{33} + \sigma_{33}\sigma_{11}$$
$$-\sigma_{12}^2 - \sigma_{13}^2 - \sigma_{23}^3 = \sigma_1\sigma_2 + \sigma_2\sigma_3 + \sigma_1\sigma_3, \tag{2.52}$$

$$III_\sigma = \det\boldsymbol{\sigma} \equiv \frac{1}{6}(2\sigma_{ij}\sigma_{jk}\sigma_{ki} - 3I_\sigma\boldsymbol{\sigma} : \boldsymbol{\sigma} + I_\sigma^3) = \sigma_1\sigma_2\sigma_3 \tag{2.53}$$

and by the Eulerian angles specifying the principle triad.

The Cauchy stress tensor defined by Eq. (2.50) refers to the deformed body in the spatial coordinate system. In a similar manner, we can introduce a *nominal stress tensor* (that is transposed of the first Piola-Kirchhoff stress tensor) associated with an original element of undeformed area

$$\mathbf{T}^0 = J\mathbf{F}^{-1}\mathbf{T}, \quad T_{ij}^0 = J\frac{\partial x_i}{\partial \xi_k}T_{kj}, \tag{2.54}$$

where the Jacobian J is written as a determinant (2.28). The first Piola–Kirchhoff stress tensor is a two-point and, in general, nonsymmetric tensor. For elimination of these two disadvantages, one introduces the symmetric *second Piola-Kirchhoff stress tensor*

$$\boldsymbol{\sigma} = \mathbf{T}^0\mathbf{B} = J\mathbf{B}^\top\mathbf{T}\mathbf{B}, \quad \sigma_{ij} = T_{im}^0\frac{\partial x_j}{\partial \xi_m} = J\frac{\partial x_i}{\partial \xi_m}T_{mn}\frac{\partial x_j}{\partial \xi_n}, \tag{2.55}$$

which may be inverted

$$T_{ij} = J^{-1}\frac{\partial \xi_i}{\partial x_m}\sigma_{mn}\frac{\partial \xi_i}{\partial x_n}, \quad T_{im}^0 = \sigma_{ij}\frac{\partial \xi_m}{\partial x_j}. \tag{2.56}$$

2.3 Basic Equations of Solid Mechanics

2.3.1 Conservation Laws, Boundary Conditions, and Constitutive Equation

There are four conservation laws in continuum mechanics: (1) mass conservation, (2) conservation of linear momentum, (3) conservation of angular momentum, and (4) conservation of energy. We will reproduce the first, second, and fourth conservation laws.

The *balance equation of linear momentum* in spatial coordinates in terms of the Cauchy stress tensor **T** is

$$\frac{\partial T_{ij}}{\partial \xi_i} + \rho b_i = \rho\left(\frac{\partial v_i}{\partial t} + v_i \frac{\partial v_j}{\partial \xi_i}\right), \tag{2.57}$$

where **b** is the body force per unit of mass acting upon the volume element; ρ is the mass density at time t related to the initial mass density ρ_0 by the *equation of continuity* (the first conservation law):

$$\rho = J^{-1}\rho_0 \tag{2.58}$$

according to Eq. (2.31). v_i is a velocity of a particle occupying the coordinate ξ_i at some time t found by taking the time derivative and holding the material coordinates constant: $v_i = \partial \xi_i / \partial t$.

Since the boundary conditions in elastic problems are most easily expressed in material coordinates, we shall reproduce an equivalent equation of motion in Lagrangian coordinates in terms of the first and second Piola–Kirchhiff stress tensors:

$$\frac{\partial T^0_{im}}{\partial x_i} + \rho_0 b_{0m} = \rho_0 \frac{d^2 u_m(\mathbf{x})}{dt^2}, \tag{2.59}$$

$$\frac{\partial}{\partial x_i}\left[\sigma_{ij}\left(\delta_{jm} + \frac{\partial u_m(\mathbf{x})}{\partial x_j}\right)\right] + \rho_0 b_{0m} = \rho_0 \frac{d^2 u_m(\mathbf{x})}{dt^2}, \tag{2.60}$$

respectively, where the time derivative is actually a simple time derivative.

The basic *energy conservation equation* (or *first principle of thermodynamic*) equates the rate of change of kinetic K and internal energy E to the rate at which surface and body forces do mechanical work P and the rate at which nonmechanical energy Q is transferred (per unit volume)

$$\dot{K} + \dot{E} = \dot{Q} + P, \tag{2.61}$$

where

$$K = \frac{1}{2\bar{w}} \int_w \rho v_i v_i \, dv, \quad E = \frac{1}{\bar{w}} \int_w \rho e \, dv, \tag{2.62}$$

$$Q = -\frac{1}{\bar{w}} \int_{\partial w} q_k n_k \, ds + \frac{1}{\bar{w}} \int_w h \, dv = \frac{1}{\bar{w}} \int_w (-\nabla \cdot \mathbf{q} + h) \, dv, \tag{2.63}$$

$$P = \frac{1}{\bar{w}} \int_w b_i v_i \, dv + \frac{1}{\bar{w}} \int_{\partial w} T_{ij} n_j v_i \, ds, \tag{2.64}$$

where e is a specific internal energy per unit mass, $\mathbf{q}$ is a nonmechanical energy flux tensor defining the rate $q_k n_k ds$ of any nonmechanical energy transmitted outward through the surface element ds with outward directed normal $\mathbf{n}$ and h is heat created per unit volume in the body. In Eq. (2.61) K, E, Q, and P are global quantities represented as integrals over the volume and its surface (2.62)–(2.64), but they also have local analogies (for which we will use the same notations) defined pointwise within the body. Equation (2.61) can be transformed into the differential form by the use of the equation of continuity (2.58) and the equations of motion

$$\rho\frac{de}{dt} = T_{ij}d_{ij} - \frac{\partial q_j}{\partial \xi_j} + \rho h,\ \ d_{ij} = \frac{1}{2}\Big(\frac{\partial v_j}{\partial \xi_i} + \frac{\partial v_i}{\partial \xi_j}\Big), \tag{2.65}$$

where $\mathbf{d}$, called the *rate strain tensor*, is the symmetric part of the tensor $v_{i,j}$ and $d/dt = \partial/\partial t + v_i \partial/\partial \xi_i$ is a *material derivative*.

The *second law of thermodynamics* states the existence of the total differential for reversible processes

$$dS = \frac{dQ}{T}, \tag{2.66}$$

where S is the entropy regarded as a measure of energy dissipation with respect to the absolute temperature T and

$$\dot{S} + \int_{\partial w} \Big(\frac{\mathbf{q}}{T}\Big) \cdot \mathbf{n}\, ds - \int_w \frac{h}{T}\, dv \geq 0. \tag{2.67}$$

The last expression is referred to as the second law of thermodynamics, known as the Clausius–Duhem inequality, which can be rewritten as

$$\int_w \Big[\dot{s} + \nabla \cdot \Big(\frac{\mathbf{q}}{T}\Big) - \frac{h}{T}\Big] div \geq 0, \tag{2.68}$$

or

$$T\dot{s} + \nabla \cdot \mathbf{q} - \frac{1}{T}\mathbf{q} \cdot \nabla T - h \geq 0, \tag{2.69}$$

which is the local form of the Clausius–Duhem inequality. Here and henceforth, s is the specific entropy density, and the term "specific" will mean "per unit volume." Without loss of generality, it is assumed mes $w = 1$.

If the right-hand side of motion equations (2.57), (2.59), (2.60) are zero, the resulting equations are called the equilibrium equation. One way to maintain a body in equilibrium is to apply suitable boundary conditions. The typical boundary conditions involves the prescribed displacements $\mathbf{u}$ specifying on part of the boundary $\partial w^0_{\mathbf{u}} \subset \partial w^0$ and the stress vector on the remained, ∂w^0_σ, so that

$$\mathbf{u}(\mathbf{x}) = \mathbf{u}^\partial(\mathbf{x}), \quad \text{on } \partial w^0_{\mathbf{u}}, \tag{2.70}$$

$$\mathbf{T}^{0\top} \cdot \mathbf{N} = \mathbf{t}^{0(\mathbf{N})}(\mathbf{F}, \mathbf{x}) \quad \text{on } \partial w^0_\sigma, \tag{2.71}$$

where $\partial w^0_{\mathbf{u}} \cap \partial w^0_\sigma = \partial w^0$, $\partial w^0_{\mathbf{u}} \cup \partial w^0_\sigma = 0$. A general *configuration-dependent loading* (2.70) if reduced to a *dead-load traction* in the surface traction introduced by (2.70) is independent of $\mathbf{F}$.

The aforementioned equations of conservation laws are in general insufficient to determine the body motion produced by given boundary conditions and body forces. They need to be accompanied by a *constitutive equation* characterizing the constitution of the body. It is assumed that there are no stresses and no strains in the initial (virgin) strain of the body. For *Cauchy elastic materials*, it is described by a symmetric single-valued *response function* $\mathbf{G}$: $\mathbf{T} = \mathbf{G}(\mathbf{F})$, which does not depend on the path of deformation. In a special case of Cauchy elasticity called hyperelasticity, there exists a *specific strain-energy function* $w = w(\mathbf{F})$ defined on the space of deformation gradients

$$\mathbf{T}^0 = \frac{\partial w}{\partial \mathbf{F}}, \quad \mathbf{T} = J^{-1}\mathbf{F}\frac{\partial w}{\partial \mathbf{F}}. \tag{2.72}$$

Equations (2.72) are the constitutive equations for finite-deformation elasticity. The function w defined per unit volume in w^0 represents the work done per unit volume at $\mathbf{x} \in w^0$ in changing the deformation gradient from $\mathbf{I}$ to $\mathbf{F}$. The function w can be considered as depending on either $\boldsymbol{\varepsilon}$ (2.33) or H_{ij} (2.29). Then we get the representations

$$\sigma_{ij} = \frac{\partial w(\boldsymbol{\varepsilon})}{\varepsilon_{ij}}, \quad T^0_{ij} = \frac{\partial w(\mathbf{H})}{H_{ij}} \tag{2.73}$$

relating the second Piola–Kirchhoff stress tensor $\boldsymbol{\sigma}$ to the Green deformation tensor $\boldsymbol{\varepsilon}$ and the nominal stress tensor $\mathbf{T}^0$ to the material displacement gradient $\mathbf{H}$, respectively.

A material is said to be *isotropic* if there is no orientation effect in the material. In such a case, the strain energy can be presented either as a symmetric function of the principal stretches or as a function of three independent invariants, such as (2.38)–(2.40) or (2.42)–(2.44). For an incompressible material $J = III_\varepsilon \equiv 1$ holds, and the function w depends on only two independent invariants. In particular, the *Mooney-Rivlind* form of strain energy

$$w = C_1(I_\varepsilon - 3) + C_2(II_\varepsilon - 3), \ (C_1, C_2 = \text{const.}) \tag{2.74}$$

is reduced to the *neo-Hookean* form

$$w = C_1(I_\varepsilon - 3). \tag{2.75}$$

Using Eq. (2.73$_1$) and the forms of the strain-energy functions (2.74) and (2.75), the stress tensor take the forms $\boldsymbol{\sigma} = -p\mathbf{I} + 2C_1\mathbf{F}\mathbf{F}^\top - 2C_2(\mathbf{F}\mathbf{F}^\top)^{-1}$ and $\boldsymbol{\sigma} = -p\mathbf{I} + 2C_1\mathbf{F}\mathbf{F}^\top$ for Mooney-Rivlin and neo-Hookean materials, respectively; here p is an arbitrary scalar. Many other forms of the strain energy have been analyzed for both incompressible and compressible nonlinear rubberlike solids in [57, 1151]. Now we will consider the strain-energy functions of physically nonlinear elasticity defined on the space of infinitesimal strain tensors (2.35). A general quadratic form of the function w composed by the algebraic invariants (2.42) and (2.43) is

$$w = a_1 A_1^2 + a_2 A_2, \tag{2.76}$$

leading to the known constitutive equation (2.73$_1$) for the isotropic medium:

$$\boldsymbol{\sigma} = \mathbf{L}\boldsymbol{\varepsilon}, \ \ \mathbf{L} = 3\lambda\mathbf{N}_1 + 2\mu\mathbf{I} \tag{2.77}$$

by the use of two, λ and μ, Lame's elastic constants. A more general special form of specific strain energy depending on three basic algebraic invariants (2.42)–(2.44) is the *Murnaghan* form, defined by

$$w = \frac{1}{2}\lambda A_1^2 + \mu A_2 + \frac{a}{3}A_1^3 + bA_1A_2 + \frac{c}{3}A_3, \tag{2.78}$$

where λ and μ are the Lame's elastic constants of the second order, and a, b, c are the elastic constants of the third order. Equations (2.73$_1$) and (2.78) yield the following tensor constitutive equations:

$$\boldsymbol{\sigma} = \mathbf{L}\boldsymbol{\varepsilon} + \mathbf{L}^{(3)}\boldsymbol{\varepsilon} \otimes \boldsymbol{\varepsilon}, \tag{2.79}$$

$$L_{ijkl} = \lambda\delta_{ij}\delta_{kl} + \mu(\delta_{ik}\delta_{jl} + \delta_{il}\delta_{jk}), \tag{2.80}$$

$$L^{(3)}_{ijklmn} = a\delta_{ij}\delta_{mn}\delta_{kl} + b(\delta_{ij}I_{mnkl} + \delta_{mn}I_{ijkl} + \delta_{kl}I_{mnij}) + cJ_{ijmnkl}, \tag{2.81}$$

$$J_{ijmnkl} = \frac{1}{2}(I_{ipkl}I_{pqmn} + I_{ipmn}I_{pjkl}). \tag{2.82}$$

The Kauderer potential depending on two basic algebraic invariants (2.42)–(2.43) and three constant ($\lambda,\ \mu,\ \gamma$) is known

$$w = \frac{1}{2}\lambda A_1^2 + \mu A_2 + \frac{1}{3}\gamma\left(A_2 - \frac{1}{3}A_1\right)^2. \tag{2.83}$$

The strain energy (2.83) is associated with constitutive equations

$$\sigma_{ij} = \lambda\varepsilon_{kk}\delta_{ij} + 2\mu\varepsilon_{ij} + 2\gamma\varepsilon_{\rm eq}\left(\varepsilon_{ij} - \frac{1}{3}\delta_{ij}\varepsilon_{kk}\right), \tag{2.84}$$

$$\varepsilon_{ij} = \frac{1}{9k}\sigma_{kk}\delta_{ij} + \frac{1}{2\mu}\left(1 - \frac{\gamma}{9\mu^3}\tau^2\right)s_{ij}, \tag{2.85}$$

where $k \equiv \lambda + 2\mu/3$ is the bulk modulus and τ and $\varepsilon_{\rm eq}$ are the stress and strain intensities

$$\tau = \left(\frac{3}{2}\mathbf{s} : \mathbf{s}\right)^{1/2}, \quad \varepsilon_{\rm eq} = \left(\frac{2}{3}\mathbf{e} : \mathbf{e}\right)^{1/2}, \tag{2.86}$$

respectively, where $\mathbf{s} = \mathbf{N}_2\boldsymbol{\sigma}$ and $\mathbf{e} = \mathbf{N}_2\boldsymbol{\varepsilon}$. The constitutive equation (2.84) can be presented in more compact form

$$\boldsymbol{\sigma} = \mathbf{L}\boldsymbol{\varepsilon} + \boldsymbol{\Psi}(\boldsymbol{\varepsilon}),\ \mathbf{L} = 3k\mathbf{N}_1 + 2\mu\mathbf{N}_2,\ \boldsymbol{\Psi} = 2\gamma\varepsilon_{\rm eq}^2\mathbf{N}_2, \tag{2.87}$$

where the tensor $\boldsymbol{\Psi}$ depending on the effective strains $\varepsilon_{\rm eq}$ describes the nonlinear material properties.

Also used is an asymptotic expansion of the specific strain energy in a Taylor series about the state of zero strain and stress as

$$w = \frac{1}{2}L^{(2)}_{ijkl}\varepsilon_{ij}\varepsilon_{kl} + \frac{1}{6}L^{(3)}_{ijklmn}\varepsilon_{ij}\varepsilon_{kl}\varepsilon_{mn} + \cdots \tag{2.88}$$

assuming $w(\mathbf{0}) = 0$. One can get from Eq. (2.88) that

$$\mathbf{L}^{(n)} = \frac{\partial^{(n)}w}{\partial\boldsymbol{\varepsilon}^{(n)}} \tag{2.89}$$

where $n = 2, 3, \ldots$ and $\boldsymbol{\varepsilon}^{(n)} = \boldsymbol{\varepsilon} \otimes \ldots \otimes \boldsymbol{\varepsilon}$ is n-multiple tensor production of the tensor $\boldsymbol{\varepsilon}$. The equality (2.89) yields the following symmetry properties of the elasticity tensors $\mathbf{L}^{(n)}$ of the $2n$ order with respect to the index pairs

$$L^{(2)}_{ijkl} = L^{(2)}_{jikl} = L^{(2)}_{ijlk} = L^{(2)}_{lkij}, \tag{2.90}$$

$$L^{(3)}_{ijklpq} = L^{(3)}_{jiklpq} = L^{(3)}_{ijlkpq} = L^{(3)}_{ijklqp} = L^{(3)}_{klijpq} = L^{(3)}_{pqklij} = L^{(3)}_{ijpqkl}, \tag{2.91}$$

and so forth. In so doing, the "pair symmetry" described by two first equalities of Eq. (2.90) and corresponding to an interchange of the indices of the first pair and interchange of the indices of the last pair are followed from the symmetry of both the $\boldsymbol{\sigma}$ and $\boldsymbol{\varepsilon}$ tensors. The "diagonal symmetry" corresponding to an interchange of the first pair of indices with the second one is described by the third equality of

Eq. (2.90) and defined by Eq. (2.88). Fourth-order tensors with pair and diagonal-symmetry are referred as "full-symmetric" tensors.

2.3.2 *The Equations of Linear Elasticity*

We will reproduce the reduction of the elasticity theory of finite deformation to a linear theory, which is a special case of small deformations superposed on a finite deformations with the special values $\mathbf{b}_0 = \mathbf{0}$, $\mathbf{F}_0 = \mathbf{I}$, $\mathbf{T}_0^0 = \mathbf{0}$. Then all stress tensors coincide and can be recognized as a Cauchy stress. All strain tensors likewise reduce to the infinitesimal strain tensor (2.35). Then the equalities

$$\boldsymbol{\sigma} = \mathbf{T} = \mathbf{T}^0, \quad \boldsymbol{\varepsilon} = \tilde{\boldsymbol{\varepsilon}}, \quad J = 1 + \varepsilon_{ii} \tag{2.92}$$

simplify the basic equations of elasticity theory

$$\nabla\boldsymbol{\sigma} + \rho\mathbf{b} = \rho\ddot{\mathbf{u}}, \quad \frac{\partial\sigma_{ij}}{\partial x_j} + \rho b_i = \rho\frac{\partial^2 u_i}{\partial t^2}, \tag{2.93}$$

$$\boldsymbol{\sigma} = \mathbf{L}\boldsymbol{\varepsilon}, \quad \sigma_{ij} = L_{ijkl}\varepsilon_{kl}, \tag{2.94}$$

$$\boldsymbol{\varepsilon} = \mathrm{Def}\mathbf{u}, \quad \varepsilon_{ij} = \frac{1}{2}\Big(\frac{\partial u_i}{\partial x_j} + \frac{\partial u_j}{\partial x_i}\Big). \tag{2.95}$$

The mixed boundary conditions on ∂w with the unit outward normal $\mathbf{n}^{\partial w}$ will be considered

$$\mathbf{u}(\mathbf{x}) = \mathbf{u}^{\partial w}(\mathbf{x}), \quad \mathbf{x} \in \partial w_{\mathbf{u}}, \tag{2.96}$$

$$\boldsymbol{\sigma}(\mathbf{x}) \cdot \mathbf{n}^{\partial w}(\mathbf{x}) = \mathbf{t}^{\partial w}(\mathbf{x}), \quad \mathbf{x} \in \partial w_{\sigma}, \tag{2.97}$$

where $\partial w_{\mathbf{u}}$ and ∂w_{σ} are prescribed displacement and traction non-intersected boundary conditions such that $\partial w_{\mathbf{u}} \cup \partial w_{\sigma} = \partial w$; $\mathbf{u}^{\partial w}(\mathbf{x})$ and $\mathbf{t}^{\partial w}(\mathbf{x})$ are prescribed the displacement on $\partial w_{\mathbf{u}}$ and traction on ∂w_{σ}, respectively; mixed boundary conditions, such as in the case of elastic supports, are also possible. When ∂w_{σ} is empty, the mixed boundary conditions (2.96) and (2.97) reduce to the *displacement problem* or the *first* boundary-value problem. If $\partial w_{\mathbf{u}}$ is empty, the boundary conditions (2.96), (2.97) becomes the *traction problem* or the *second* boundary-value problem. As usual we shall distinguish the interior from the exterior problem according to whether the body occupies the interior or the exterior domain with respect to w.

Of special practical interest are the homogeneous boundary conditions

$$\mathbf{u}^{\partial w}(\mathbf{x}) = \boldsymbol{\varepsilon}^{\partial w}\mathbf{x},\ \boldsymbol{\varepsilon}^{\partial w}(\mathbf{x}) \equiv \text{const.},\ \mathbf{x} \in \partial w, \tag{2.98}$$

$$\boldsymbol{\sigma}(\mathbf{x}) \cdot \mathbf{n}^{\partial w}(\mathbf{x}) = \boldsymbol{\sigma}^{\partial w}(\mathbf{x}) \cdot \mathbf{n}^{\partial w}(\mathbf{x}),\ \boldsymbol{\sigma}^{\partial w}(\mathbf{x}) = \text{const.},\ \mathbf{x} \in \partial w, \tag{2.99}$$

where $\boldsymbol{\varepsilon}^{\partial w}(\mathbf{x}) = \frac{1}{2}\left[\nabla \otimes \mathbf{u}^{\partial w}(\mathbf{x}) + (\nabla \otimes \mathbf{u}^{\partial w}(\mathbf{x}))^{\top}\right]$, $\mathbf{x} \in \partial w_{\mathbf{u}}$.

2.3.3 Extremum Principles of Elastostatic

The principle of minimum potential energy and the principle of minimum complementary energy completely characterize the solution of the mixed boundary problem (2.93)–(2.97) of elastostatic ($\ddot{\mathbf{u}} \equiv \mathbf{0}$ in Eq. (2.93)). For definition of these energy functions, we introduce the notion of admissible fields. A sufficiently smooth (of class C^1) displacement field $\mathbf{u}$ with the displacement boundary conditions

$$\mathcal{A}^k(\mathbf{u}) = \{\mathbf{u}|\mathbf{u}(\mathbf{x}) \in C^1(w),\ \mathbf{u}(\mathbf{y}) = \mathbf{u}^{\partial w}(\mathbf{y}),\ \mathbf{y} \in \partial w_{\mathbf{u}}\} \tag{2.100}$$

is called a *kinematically admissible displacement field.* A kinematically admissible state is an ordered array $s = \{\mathbf{u}, \boldsymbol{\varepsilon}, \boldsymbol{\sigma}\}$ (forming the set $\mathcal{A}^k(s)$) of kinematically admissible displacement field $\mathbf{u} \in \mathcal{A}^k(\mathbf{u})$ (2.100) generating the symmetric fields $\boldsymbol{\varepsilon}$ and $\boldsymbol{\sigma}$ according to Eqs. (2.95) and (2.94), respectively. While $\boldsymbol{\varepsilon}(\mathbf{u})$ is compatible, $\boldsymbol{\sigma}(\mathbf{u})$ may not necessarily satisfy the equilibrium equations (2.93).

A sufficiently smooth (of class C^1) stress field satisfying the equilibrium equation (2.93) and the traction boundary condition (2.97)

$$\mathcal{A}^s(\boldsymbol{\sigma}) = \{\boldsymbol{\sigma}|\boldsymbol{\sigma}(\mathbf{x}) \in C^1(w),\ \nabla \cdot \boldsymbol{\sigma} + \rho\mathbf{b} = \mathbf{0},\ \boldsymbol{\sigma}(\mathbf{y}) \cdot \mathbf{n}^{\partial w}(\mathbf{y}) = \mathbf{t}^{\partial w}(\mathbf{y}),\ \mathbf{y} \in \partial w_{\boldsymbol{\sigma}}\} \tag{2.101}$$

is referred to as a *statically admissible stress field.* The ordered array $s^s = \{\mathbf{u}, \boldsymbol{\varepsilon}, \boldsymbol{\sigma}\}$ (forming the set $\mathcal{A}^s(s)$) is called a statically admissible state if $\boldsymbol{\sigma} \in \mathcal{A}^s(\boldsymbol{\sigma})$ (2.101) generates the symmetric tensor $\boldsymbol{\varepsilon}$ according to the constitutive equation (2.94) and the displacement field $\mathbf{u}$ is related to $\boldsymbol{\varepsilon}$ by the use of Eq. (2.95). In so doing, $\boldsymbol{\varepsilon}(\boldsymbol{\sigma})$ is not necessarily compatible.

We also define the total strain energy $W(\boldsymbol{\varepsilon})$ and the total stress energy $W^c(\boldsymbol{\sigma})$ by

$$W(\boldsymbol{\varepsilon}) = \frac{1}{2}\int_w \boldsymbol{\varepsilon} : \mathbf{L} : \boldsymbol{\varepsilon}\, d\mathbf{x},\quad W^c(\boldsymbol{\sigma}) = \frac{1}{2}\int_w \boldsymbol{\sigma} : \mathbf{M} : \boldsymbol{\sigma}\, d\mathbf{x}, \tag{2.102}$$

respectively. Let $\delta\mathbf{u} = \delta\mathbf{u}(\mathbf{x})$ be a virtual, or imaginative, infinitesimal displacement field associated with a virtual strain field from the current state $\delta\boldsymbol{\varepsilon} = \text{Def}(\delta\mathbf{u})$. The components of the virtual displacement vector $\delta\mathbf{u}$ vanish at $\partial w_{\mathbf{u}}$: $\delta\mathbf{u}(\mathbf{y}) \equiv \mathbf{0}$ at

$\mathbf{y} \in \partial w_{\mathbf{u}}$. It can be shown from the equations of linear and angular momentum that it holds the *principle of virtual work* (PVW)

$$\int_w \boldsymbol{\sigma} : \delta\boldsymbol{\varepsilon}\, d\mathbf{x} = \int_{\partial w} \mathbf{t}^{\partial w_\sigma} \cdot \delta\mathbf{u}\, ds + \int_w \rho\mathbf{b} \cdot \delta\mathbf{u}\, d\mathbf{x}, \tag{2.103}$$

which is a starting point in developing minimum principles.

As a consequence, one can obtain a work–energy relation (theorem of work expended) involving the admissible stress field $\boldsymbol{\sigma}$ and the admissible displacement field $\mathbf{u}$ [$\boldsymbol{\varepsilon}(\mathbf{u})$ is the related strain field] corresponding to the given external forces $\rho\mathbf{b}(\mathbf{x})$ and $\mathbf{t}^{\partial w_\sigma}(\mathbf{y})$ ($\mathbf{x} \in w,\ \mathbf{y} \in \partial w_\sigma$):

$$\int_w \boldsymbol{\sigma} : \boldsymbol{\varepsilon}\, d\mathbf{x} = \int_{\partial w} \mathbf{t}^{\partial w_\sigma} \cdot \mathbf{u}\, ds + \int_w \rho\mathbf{b} \cdot \mathbf{u}\, d\mathbf{x}, \tag{2.104}$$

which looks similar to the first law of thermodynamics (2.61), although there is no thermodynamic expression in (2.104). We note that $\boldsymbol{\sigma}$ and $\boldsymbol{\varepsilon}$ need not be connected by a specific stress-strain relation.

We will present the classical extremum principles.

Principle of Minimum of Potential Energy Let Π, called the *potential energy* (corresponding to the given external forces $\rho\mathbf{b}(\mathbf{x})$ and $\mathbf{t}^{\partial w_\sigma}(\mathbf{y})$, $\mathbf{x} \in w,\ \mathbf{y} \in \partial w_\sigma$), be the function defined on the set of kinematically admissible states $\tilde{s}^k = \{\tilde{\mathbf{u}}, \tilde{\boldsymbol{\varepsilon}}, \tilde{\boldsymbol{\sigma}}\} \in \mathcal{A}^k(\tilde{s})$ by the equation

$$\Pi(\tilde{s}^k) = W(\tilde{\boldsymbol{\varepsilon}}) - \int_w \rho\mathbf{b} \cdot \tilde{\mathbf{u}}\, d\mathbf{x} - \int_{\partial w_\sigma} \mathbf{t}^{\partial w_\sigma} \cdot \tilde{\mathbf{u}}\, ds. \tag{2.105}$$

Then the actual displacement field $\mathbf{u}$ with corresponding admissible state s^k renders the potential energy Π an absolute minimum

$$\Pi(s^k) \le \Pi(\tilde{s}^k). \tag{2.106}$$

Principle of Minimum of Complementary Energy Let Π^c called the *complementary energy* (corresponding to the given boundary displacement $\mathbf{u}^{\partial w_{\mathbf{u}}}(\mathbf{y}) \in \partial w_{\mathbf{u}}$) be the function defined on the set of statically admissible states $\tilde{s}^s = \{\tilde{\mathbf{u}}, \tilde{\boldsymbol{\varepsilon}}, \tilde{\boldsymbol{\sigma}}\} \in \mathcal{A}^s(\tilde{s})$ by the equation

$$\Pi^c(\tilde{s}^s) = W^c(\tilde{\boldsymbol{\sigma}}) - \int_{\partial w_{\mathbf{u}}} \tilde{\mathbf{t}}_{\mathbf{n}} \cdot \mathbf{u}^{\partial w_{\mathbf{u}}}\, ds,\ \tilde{\mathbf{t}}_{\mathbf{n}} \equiv \tilde{\boldsymbol{\sigma}} \cdot \mathbf{n}^{\partial w}. \tag{2.107}$$

Then the actual stress field $\boldsymbol{\sigma}$ with corresponding admissible state s^s renders the complementary energy Π^c an absolute minimum

$$\Pi^c(s^s) \le \Pi^c(\tilde{s}^s). \tag{2.108}$$

2.4 Basic Equations of Thermoelasticity and Electroelasticity

2.4.1 *Thermoelasticity Equations*

In classical thermodynamics, we are concerned with the small neighborhood of thermodynamic equilibrium. We also consider the infinitesimal deformations (2.92). In such a case the first law of thermodynamics (2.65) can be recast in the form (hereafter a unit volume is considered)

$$dE = dQ + \sigma_{ij} d\varepsilon_{ij}, \tag{2.109}$$

To derive the constitutive relations of thermoelasticity theory, we transform (2.109) for the reversible processes (2.66) to the thermodynamic relation

$$dE = \sigma_{ij}\, d\varepsilon_{ij} + T\, dS \tag{2.110}$$

equating the total differential of the internal energy (2.62_2) and the sum of the increment of the deformation work and the amount of the heat introduced into the considered volume; here T is the absolute temperature and S is the entropy density (2.68). We see that E is a state function of $\boldsymbol{\varepsilon}$ and S for the reversible adiabatic (isentropic, S =const.) processes, and according to the ordinary rules of differentiation, we get

$$\sigma_{ij} = \left.\frac{\partial E}{\partial \varepsilon_{ij}}\right|_S, \quad T = \left.\frac{\partial E}{\partial S}\right|_\varepsilon. \tag{2.111}$$

On the other hand, if the process is isothermal (T =const.), the dependence of the stresses $\boldsymbol{\sigma}$ on the strains $\boldsymbol{\varepsilon}$ is found by introducing the *Helmholtz free-energy function*:

$$F = E - TS, \quad dF = \sigma_{ij}\, d\varepsilon_{ij} - S\, dT \tag{2.112}$$

with the strains and temperature as its independent variables. The value $-TS$ is the irreversible heat energy due to entropy as related to temperature, with the negative sign indicating that the compressive reaction results from thermal expansion in a restrained body. Then, from Eqs. (2.110) and (2.112), we get

$$\sigma_{ij} = \left.\frac{\partial F}{\partial \varepsilon_{ij}}\right|_T, \quad S = -\left.\frac{\partial F}{\partial T}\right|_\varepsilon. \tag{2.113}$$

Thus, the strain energy density w can be identified with the internal energy E in an isentropic process (S $\equiv$const.) and the free energy F in an isothermal process ($T \equiv$const.). The temperature is determined by Eq. (2.111_2) when using the potential E and the entropy is determined by Eq. (2.113) when using the potential F.

For small deformations and small temperature changes $\theta = T - T^0$ ($\theta/T_0 \ll 1$), $F(\boldsymbol{\varepsilon}, T)$ can be expanded in a power series of its arguments in the neighborhood of the virgin state ($\boldsymbol{\varepsilon} = \mathbf{0},\ \theta = 0$)

$$F(\boldsymbol{\varepsilon}, T) = \frac{1}{2} L_{ijkl}\varepsilon_{ij}\varepsilon_{kl} + \alpha_{ij}^T \varepsilon_{ij}\theta - \frac{C_\varepsilon}{2T_0}\theta^2, \tag{2.114}$$

where one introduces the notations

$$L_{ijkl} = \frac{\partial^2 F(\mathbf{0}, T_0)}{\partial\varepsilon_{ij}\partial\varepsilon_{kl}}, \quad \alpha_{ij}^T = \frac{\partial^2 F(\mathbf{0}, T_0)}{\partial\varepsilon_{ij}\partial T}, \quad \frac{C_\epsilon}{T_0} = -\frac{\partial^2 F(\mathbf{0}, T_0)}{\partial T^2} \tag{2.115}$$

and C_ε denotes the specific heat at constant strain, which can be also determined as

$$C_\epsilon = \left.\frac{dQ}{dT}\right|_\epsilon = \left.\frac{\partial E}{\partial T}\right|_\epsilon = T\left.\frac{\partial S}{\partial T}\right|_\epsilon. \tag{2.116}$$

The negative sign for the last term on the right-hand side of Eq. (2.114) indicates that a temperature rise leads to a compressive reaction on the restricted body. This term signifies the thermal energy due to temperature, while the second term $\boldsymbol{\alpha}^T\boldsymbol{\varepsilon}\theta$ describes the coupling effect between temperature and mechanical deformation. The difference between the isentropic moduli $\tilde{\mathbf{L}}$ and isothermic moduli $\mathbf{L}$:

$$\tilde{\mathbf{L}} - \mathbf{L} = \frac{T_0}{\rho C_\epsilon}\boldsymbol{\alpha}^T \otimes \boldsymbol{\alpha}^T, \quad \tilde{\mathbf{L}} = \left.\frac{\partial^2 E}{\partial\boldsymbol{\varepsilon}\otimes\partial\boldsymbol{\varepsilon}}\right|_S, \quad \mathbf{L} = \left.\frac{\partial^2 F}{\partial\boldsymbol{\varepsilon}\otimes\partial\boldsymbol{\varepsilon}}\right|_T$$

is of as order of 1% or less for metals and ceramics.

To determine the entropy as a function of $\boldsymbol{\varepsilon}$ and T, consider the total differential of the function $S(\boldsymbol{\varepsilon}, T)$

$$dS = \left.\frac{\partial S}{\partial\boldsymbol{\varepsilon}}\right|_T : d\boldsymbol{\varepsilon} + \left.\frac{\partial S}{\partial T}\right|_\epsilon dT. \tag{2.117}$$

Taking Eqs. (2.113) and (2.116) into account and that

$$\left.\frac{\partial\boldsymbol{\sigma}}{\partial T}\right|_\varepsilon = \left.\frac{\partial^2 F}{\partial T\partial\boldsymbol{\varepsilon}}\right|_\varepsilon = \boldsymbol{\alpha}^T, \tag{2.118}$$

equation (2.117) becomes

$$dS = -\boldsymbol{\alpha}^T : d\boldsymbol{\varepsilon} + \frac{C_\varepsilon}{T}dT, \tag{2.119}$$

which after integration under the virgin state conditions $\boldsymbol{\varepsilon} = \mathbf{0}$ and $T = T_0$ gives, for small temperature change ($\theta / T_0 \ll 1$),

$$S = -\boldsymbol{\alpha}^T : \boldsymbol{\varepsilon} + \frac{C_\varepsilon}{T_0}\theta. \tag{2.120}$$

The Gibbs thermodynamic potential per unit volume is defined for the case where the stress and temperature are chosen as the independent state variables

$$G_e = F - TS, \quad dG_e = -\varepsilon_{ij}\, d\sigma_{ij} - S\, dT. \tag{2.121}$$

Since dG_e is a total differential (as dE and dF), it follows that

$$\left.\frac{\partial G_e}{\partial T}\right|_\sigma = -S, \quad \left.\frac{\partial G_e}{\partial \sigma_{ij}}\right|_T = -\varepsilon_{ij}. \tag{2.122}$$

Thus, $-G_e$ can be identified with the complementary energy function W^c, which has the property that $\partial W^c / \partial \sigma_{ij} = \varepsilon_{ij}$. In the neighborhood of the virgin state ($\theta / T_0 \ll 1$, $\mathbf{M} :: (\boldsymbol{\sigma} \otimes \boldsymbol{\sigma}) \ll 1$; $\mathbf{M}$ is the compliance tensor), we have a power series expansion

$$G_e(\boldsymbol{\sigma}, T) = -\frac{1}{2} M_{ijkl}\sigma_{ij}\sigma_{kl} - \beta^T_{ij}\sigma_{ij}\theta + \frac{1}{2} C_\sigma \frac{\theta^2}{T_0} \tag{2.123}$$

depending on the coefficient of thermal expansion $\boldsymbol{\beta}^T$ and the specific heat at constant stress C_σ

$$\boldsymbol{\beta}^T \equiv \left.\frac{\partial \boldsymbol{\varepsilon}}{\partial T}\right|_\sigma = -\mathbf{M}\boldsymbol{\alpha}^T, \quad C_\sigma = \left.\frac{dQ}{dT}\right|_\sigma, \tag{2.124}$$

where C_σ is related to C_ϵ by the equation

$$C_\sigma = C_\varepsilon + L_{ijkl}\boldsymbol{\beta}^T_{ij}\beta^T_{kl}T_0. \tag{2.125}$$

Analogously to the tensor $\mathbf{L}$ (2.90), the compliance tensor $\mathbf{M}$ is also full symmetric.

Substitution of the representation (2.114) into Eq. (2.113$_1$) yields the general *Duhamel–Neumann form of Hook's law* for an anisotropic body in the form of invertible relations

$$\boldsymbol{\sigma} = \mathbf{L}\boldsymbol{\varepsilon} + \boldsymbol{\alpha}, \quad \boldsymbol{\varepsilon} = \mathbf{M}\boldsymbol{\sigma} + \boldsymbol{\beta} \tag{2.126}$$

where $\boldsymbol{\alpha} \equiv \boldsymbol{\alpha}^T\theta$, $\boldsymbol{\beta}^T = \boldsymbol{\beta}/\theta$ is the coefficient of thermal expansion related to $\boldsymbol{\alpha}$ by $\boldsymbol{\alpha} = -\mathbf{L}\boldsymbol{\beta}$, $\alpha_{ij} = -L_{ijkl}\beta_{kl}$; see Eq. (2.124). $\boldsymbol{\beta}(\mathbf{x})$ and $\boldsymbol{\alpha}(\mathbf{x}) \equiv -\mathbf{L}(\mathbf{x})\boldsymbol{\beta}(\mathbf{x})$ are second-order tensors of local eigenstrains and eigenstresses, respectively (frequently called transformation fields), which may arise by thermal expansion, phase transformation, twinning, and other changes of shape or volume of the material. Substitution of the constitutive equation (2.126$_1$) into the motion equation (2.60) at small deformations (2.92) leads to the Navier equation

$$\nabla(\mathbf{L}\nabla\mathbf{u}) + \rho\mathbf{b} = \rho\ddot{\mathbf{u}} - \boldsymbol{\alpha}^T\nabla\theta, \quad \frac{\partial}{\partial x_j}\left(L_{ijkl}\frac{\partial u_k}{\partial x_l}\right) + \rho b_i = \rho\frac{\partial^2 u_j}{\partial t^2} - \alpha^T_{ij}\frac{\partial\theta}{x_j}, \tag{2.127}$$

which should be considered simultaneously with the generalized heat condition equation:

$$\nabla(\boldsymbol{\lambda}\nabla\theta) - C_\varepsilon\dot{\theta} + T_0\boldsymbol{\alpha}^T : \dot{\boldsymbol{\varepsilon}} = -h, \quad \frac{\partial}{\partial x_i}\left(\lambda_{ij}\frac{\partial}{\partial x_j}\theta\right) - C_\varepsilon\frac{\partial\theta}{\partial t} + T_0\alpha^T_{ij}\frac{\partial\varepsilon_{ij}}{\partial t} = -h, \tag{2.128}$$

where h is the heat source function. The coefficient $\boldsymbol{\alpha}^T$ couples the equations of the dynamic thermoelastic problem (2.127) and (2.128). Dropping the term $T_0\boldsymbol{\alpha}^T : \dot{\boldsymbol{\varepsilon}}$ in the left-hand side of Eq. (2.128$_1$), we obtain the uncoupled equations of dynamic thermoelasticity. Further neglecting the inertial term $\rho\ddot{\mathbf{u}}$ in Eq. (2.127) leads to a system of equations of the quasistatic thermoelastic problem. The system (2.127) and (2.128) should be accompanied by appropriate boundary and initial conditions. The initial conditions specify the temperature, displacements, and velocities at all points $\mathbf{x} \in w$ within w: $\theta(\mathbf{x}, 0) = \theta_0(\mathbf{x})$, $\mathbf{u}(\mathbf{x}, 0) = \mathbf{u}_0(\mathbf{x})$, $\dot{\mathbf{u}}(\mathbf{x}, 0) = \mathbf{v}_0(\mathbf{x})$ with the prescribed functions $\theta_0, \mathbf{u}_0, \mathbf{v}_0$ on w. The most widely used boundary condition for the temperature function θ is expressed by

$$\boldsymbol{\lambda} : (\nabla\theta(\mathbf{x}, t) \otimes \mathbf{n}) + \alpha_{\partial w}[\theta(\mathbf{x}, t) - \theta_{\partial w}(\mathbf{x}, t)] = q_0(\mathbf{x}, t) \tag{2.129}$$

with $\theta_{\partial w}$, q_0 as known functions of the time and point $\mathbf{x} \in \partial w$. If $q_0 \equiv 0$, Eq. (2.129) describes heat exchange of the third kind called Newton's law. The conditions $\alpha_{\partial w} = \infty$ and $q_0 \equiv 0$ reduce Eq. (2.129) to the boundary conditions of the first kind. We obtain the second kind of boundary conditions if $\alpha_{\partial w} \equiv 0$. The mechanical boundary conditions (2.96) and (2.97) should be added to the thermal boundary conditions (2.129). In so doing, the traction $\mathbf{t}(\mathbf{x}) = \boldsymbol{\sigma}(\mathbf{x})\mathbf{n}(\mathbf{x})$ acting on any surface with the normal $\mathbf{n}(\mathbf{x})$ through the point $\mathbf{x}$ can be represented in terms of displacements:

$$\mathbf{t}(\mathbf{x}) = \hat{\mathbf{t}}(\mathbf{n}, \nabla)\mathbf{u}(\mathbf{x}) + \boldsymbol{\alpha}\mathbf{n}, \quad \hat{t}_{ik}(\mathbf{n}, \nabla) \equiv L_{ijkl}n_j(\mathbf{x})\partial/\partial x_l, \tag{2.130}$$

where $\hat{t}_{ik}(\mathbf{n}, \nabla)$ is the so-called stress operator.

We will consider the uncoupled quasistatic thermoelasticity theory [$\rho\ddot{\mathbf{u}} \equiv \mathbf{0}$ in Eq. (2.127) and $T_0\boldsymbol{\alpha}^T : \dot{\boldsymbol{\varepsilon}} \equiv \mathbf{0}$ in Eq. (2.128)], where the temperature field is determined by Eq. (2.128) with no influence of the latent heat due to the change of strain. It takes place according to the so-called body-force analogy [285] asserting that

$\{\mathbf{u}, \boldsymbol{\varepsilon}, \boldsymbol{\sigma}\}$ is a solution of the mixed problem of thermoelastostatics corresponding to the external loading conditions $(\mathbf{b}, \mathbf{t}^{\partial w_\sigma}, \mathbf{u}^{\partial w_\mathbf{u}}, \theta)$ on $w \cup \partial w_\sigma \cup \partial w_\mathbf{u}$ if and only if $\{\mathbf{u}, \boldsymbol{\varepsilon}, \boldsymbol{\sigma} - \boldsymbol{\alpha}\}$ is a solution of the mixed problem of elastostatic corresponding to the external loading $(\mathbf{b} + 1/\rho\nabla\boldsymbol{\alpha}, \mathbf{t}^{\partial w_\sigma} - \boldsymbol{\alpha} \cdot \mathbf{n}, \mathbf{u}^{\partial w_\mathbf{u}})$ on $w \cup \partial w_\sigma \cup \partial w_\mathbf{u}$.

To present the extremum principles of linear thermoelastostatic, determine the total strain energy $W(\boldsymbol{\varepsilon})$ and total stress energy $W^c(\boldsymbol{\sigma})$

$$W(\boldsymbol{\varepsilon}) = \int_w w(\boldsymbol{\varepsilon})\,d\mathbf{x}, \quad w(\boldsymbol{\varepsilon}) = \frac{1}{2}(\boldsymbol{\varepsilon} - \boldsymbol{\beta}) : \mathbf{L} : (\boldsymbol{\varepsilon} - \boldsymbol{\beta}),$$

$$W^c(\boldsymbol{\sigma}) = \int_w w^c(\boldsymbol{\sigma})\,d\mathbf{x}, \quad w^c(\boldsymbol{\sigma}) = \frac{1}{2}(\boldsymbol{\sigma} - \boldsymbol{\alpha}) : \mathbf{M} : (\boldsymbol{\sigma} - \boldsymbol{\alpha}) \qquad (2.131)$$

yielding Eqs. (2.102) at $\boldsymbol{\alpha} = \boldsymbol{\beta} \equiv \mathbf{0}$. $w(\boldsymbol{\varepsilon})$ and $w^c(\boldsymbol{\sigma})$ are potential functions for determining the stress and strain according to the equations

$$\sigma_{ij} = \frac{\partial w(\boldsymbol{\varepsilon})}{\partial \varepsilon_{ij}}, \quad \varepsilon_{ij} = \frac{\partial w^c(\boldsymbol{\sigma})}{\partial \sigma_{ij}}, \qquad (2.132)$$

respectively, which are valid for both elasticity and thermoelasticity; in linear theory $w(\boldsymbol{\varepsilon})$ and $w^c(\boldsymbol{\sigma})$ are equal: $w = w^c$. Positiveness of w and w^c (at $\boldsymbol{\varepsilon}, \boldsymbol{\sigma} \neq \mathbf{0}$) leads to the positive definiteness of $\mathbf{L}$ and $\mathbf{M}$, respectively (i.e. $\boldsymbol{\varepsilon} : \mathbf{L} : \boldsymbol{\varepsilon} > 0$ for $\forall \boldsymbol{\varepsilon} \neq \mathbf{0}$).

The classical extremum principles (2.106) and (2.108) for pure mechanical loading can be generalized for the thermostatic case in the following manner. We determine the potential Π and complementary Π^c energies:

$$\Pi(\tilde{s}^k) = W(\tilde{\boldsymbol{\varepsilon}}) - \int_w \rho\mathbf{b} \cdot \tilde{\mathbf{u}}\,d\mathbf{x} - \int_{\partial w_\sigma} \mathbf{t}^{\partial w_\sigma} \cdot \tilde{\mathbf{u}}\,ds,$$

$$\Pi^c(\tilde{s}^s) = W^c(\tilde{\boldsymbol{\sigma}}) - \int_{\partial w_\mathbf{u}} \tilde{\mathbf{t}}_\mathbf{n} \cdot \mathbf{u}^{\partial w_\mathbf{u}}\,ds, \quad \tilde{\mathbf{t}}_\mathbf{n} \equiv \tilde{\boldsymbol{\sigma}} \cdot \mathbf{n}^{\partial w}. \qquad (2.133)$$

defined on the sets of kinematically $\tilde{s}^k = \{\tilde{\mathbf{u}}, \tilde{\boldsymbol{\varepsilon}}, \tilde{\boldsymbol{\sigma}}\} \in \mathcal{A}^k(\tilde{s})$ (2.100) and statically $\tilde{s}^s = \{\tilde{\mathbf{u}}, \tilde{\boldsymbol{\varepsilon}}, \tilde{\boldsymbol{\sigma}}\} \in \mathcal{A}^s(\tilde{s})$ (2.101) admissible states, respectively. In so doing one assumes that Eqs. (2.94) is replaced by Eq. (2.126_1) to include the $\tilde{\boldsymbol{\sigma}}$ and $\tilde{\boldsymbol{\varepsilon}}$ generated by $\tilde{\mathbf{u}} \in \mathcal{A}^k(\tilde{\mathbf{u}})$ and by $\tilde{\boldsymbol{\sigma}} \in \mathcal{A}^s(\tilde{\boldsymbol{\sigma}})$, respectively. Then the actual state $s = \{\mathbf{u}, \boldsymbol{\varepsilon}, \boldsymbol{\sigma}\}$ renders the potential energy Π and complementary energy Π^c an absolute minimum

$$\Pi(s) \leq \Pi(\tilde{s}^k), \quad \Pi^c(s) \leq \Pi^c(\tilde{s}^s). \qquad (2.134)$$

2.4.2 *Electroelastic Equations*

Another example of the theory of coupled fields is associated with electroelasticity and deals with phenomena caused by interactions between the elastic, electric, and thermal fields. The relations between the mechanical, electrical, and thermal properties can be demonstrated from equilibrium thermodynamics described by the potential *Gibbs function*

$$G_e = E - \mathbf{E} \cdot \mathbf{D} - ST, \tag{2.135}$$

whose differential form

$$dG_e = \sigma_{ij} d\varepsilon_{ij} - D_i dE_i - S\, dT \tag{2.136}$$

yields the relations

$$\sigma_{ij} = \left.\frac{\partial G_e}{\partial \varepsilon_{ij}}\right|_{\mathbf{E},T}, \quad D_m = -\left.\frac{\partial G_e}{\partial E_m}\right|_{\varepsilon,T}, \quad S = -\left.\frac{\partial G_e}{\partial T}\right|_{\varepsilon,\mathbf{E}}. \tag{2.137}$$

Differentiating Eqs. (2.137) gives

$$\left.\frac{\partial \sigma_{ij}}{\partial E_m}\right|_{\varepsilon,T} = -\left.\frac{\partial D_m}{\partial \varepsilon}\right|_{\mathbf{E},T}, \quad \left.\frac{\partial \sigma_{ij}}{\partial T}\right|_{\varepsilon,\mathbf{E}} = -\left.\frac{\partial S}{\partial \varepsilon_{ij}}\right|_{\mathbf{E},T}, \quad \left.\frac{\partial D_m}{\partial T}\right|_{\varepsilon,\mathbf{E}} = \left.\frac{\partial S}{\partial E_m}\right|_{\varepsilon,T}. \tag{2.138}$$

It was assumed that the strains $\boldsymbol{\varepsilon}$, the electric field intensity $\mathbf{E}$, and the temperature $T = T_0 + \theta$ are considered the independent variables, and the dependent variables will be the stresses $\boldsymbol{\sigma}$, the vector of induction $\mathbf{D}$, and the entropy S. G_e is a state function of $\boldsymbol{\varepsilon}$, $\mathbf{E}$, and T, and the ordinary rules of differentiation yield the explicit representation of $d\boldsymbol{\sigma}$, $d\mathbf{D}$, and dS as the functions of $d\boldsymbol{\varepsilon}$, $d\mathbf{E}$, and dT

$$\begin{aligned}
d\sigma_{ij} &= \left.\frac{\partial \sigma_{ij}}{\varepsilon_{kl}}\right|_{\mathbf{E},T} d\varepsilon_{kl} + \left.\frac{\partial \sigma_{ij}}{E_m}\right|_{\varepsilon,T} dE_m + \left.\frac{\partial \sigma_{ij}}{\partial T}\right|_{\mathbf{E},\varepsilon} dT, \\
dD_m &= \left.\frac{\partial D_m}{\varepsilon_{kl}}\right|_{\mathbf{E},T} d\varepsilon_{kl} + \left.\frac{\partial D_m}{E_k}\right|_{\varepsilon,T} dE_k + \left.\frac{\partial D_m}{\partial T}\right|_{\varepsilon,\mathbf{E}} dT, \\
dS &= \left.\frac{\partial S}{\varepsilon_{kl}}\right|_{\mathbf{E},T} d\varepsilon_{kl} + \left.\frac{\partial S}{E_m}\right|_{\varepsilon,T} dE_m + \left.\frac{\partial S}{\partial T}\right|_{\varepsilon\mathbf{E}} dT,
\end{aligned} \tag{2.139}$$

For small variations of $\boldsymbol{\varepsilon}$, $\mathbf{E}$, and T in the neighborhood of virgin state ($\boldsymbol{\varepsilon}, \mathbf{E}, \theta = \mathbf{0}$), the partial derivatives can be considered as constants and Eqs. (2.139) becomes integrable. This allows us to obtain the constitutive relations of a deformable piezoelectric medium (see, e.g., [1033, 1188])

$$\begin{aligned}
\sigma_{ij} &= L_{ijkl}\varepsilon_{kl} + e_{kij}E_k + \alpha^T_{ij}\theta, \\
D_i &= e_{ijk}\varepsilon_{jk} - k_{ij}E_j + p_i\theta, \quad dS = \alpha_{ij}\varepsilon_{ij} + p_i E_i + c_\varepsilon \theta / T_0,
\end{aligned} \tag{2.140}$$

where the coefficients of the system (2.140) denote the particular derivatives:

$$L_{ijkl} = \left.\frac{\partial \sigma_{ij}}{\partial \varepsilon_{kl}}\right|_{\mathbf{E},T}, \quad e_{ijk} = \left.\frac{\partial \sigma_{ij}}{\partial E_k}\right|_{\boldsymbol{\varepsilon},T},$$
$$p_i = \left.\frac{\partial D_i}{\partial T}\right|_{\varepsilon,\mathbf{E}}, \quad k_{ij} = \left.\frac{\partial D_i}{\partial E_j}\right|_{\varepsilon,T},$$
$$\alpha^T_{ij} = \left.\frac{\partial \sigma_{ij}}{\partial T}\right|_{\varepsilon,\mathbf{E}}, \quad \frac{c_\varepsilon}{T_0} = \left.\frac{\partial S}{\partial T}\right|_{\varepsilon,\mathbf{E}}. \tag{2.141}$$

The constitutive Eqs. (2.140_1) and (2.140_2) can be inverted

$$\varepsilon_{ij} = M_{ijkl}\sigma_{kl} + d_{kij}D_k + \beta^T_{ij}\theta, \tag{2.142}$$
$$E_i = d_{ijk}\sigma_{jk} - b_{ij}D_j + q_i\theta. \tag{2.143}$$

The coefficients of Eqs. (2.140) and (2.142), (2.143) are denoted as follows: **L** and **M** are the elastic and compliance tensors, **k** and **b** are the tensors of dielectric permeability and impermeability, **e** and **d** are the piezoelectric moduli, $\boldsymbol{\alpha}^T$ and $\boldsymbol{\beta}^T$ are the coefficients of thermoelastic stress and expansion, **p** and **q** are the pyroelectric coefficients, and c_ε is a specific heat per unit volume at constant strain. To obtain a symmetric matrix of coefficients, we replace the electric field **E** by $-\mathbf{E}$. Hereinafter in Sect. 2.4.3, the Latin indexes range from 1 to 3, and the Greek ones range from 1 to 4. We will assume that the electric and elastic fields are fully coupled, but temperature enters the problem only as a parameter through the constitutive equation. The tensor coefficients in the above equations are related as $e_{kij} = d_{kmn}L_{mnij}$, $d_{kij} = e_{kmn}M_{mnij}$, $M_{ijmn}L_{mnkl} = I_{ijkl}$. We will use the matrix notation instead of the tensor one, as it is accepted in the theory of elasticity. For notational convenience the elastic and electric variable will be treated on equal footing, and with this in mind we recast the local linear constitutive relations of thermoelectroelasticity for this material in the notation introduced in [70]. For this purpose, we introduce the matrices of generalized stresses $\boldsymbol{\Sigma}$ and strains $\boldsymbol{\mathcal{E}}$ $(i, j = 1, 2, 3)$

$$\boldsymbol{\Sigma} = \begin{pmatrix} \boldsymbol{\sigma} \\ \mathbf{D} \end{pmatrix}, \quad \Sigma_{ij} = \sigma_{ij}, \quad \Sigma_{i4} = D_i, \tag{2.144}$$

$$\boldsymbol{\mathcal{E}} = \begin{pmatrix} \boldsymbol{\varepsilon} \\ \mathbf{E} \end{pmatrix}, \quad w_{ij} = \varepsilon_{ij}, \quad w_{4i} = E_i, \tag{2.145}$$

which are interrelated through the matrices of generalized elastic coefficients $\mathbb{L}_{\alpha\beta\gamma\delta}$, compliance $\mathbb{M}_{\alpha\beta\gamma\delta}$, and generalized coefficients of thermal expansion $\Lambda_{\alpha\beta}$

$$\boldsymbol{\Sigma} = \mathbb{L}(\boldsymbol{\mathcal{E}} - \boldsymbol{\Lambda}), \quad \boldsymbol{\mathcal{E}} = \mathbb{M}\boldsymbol{\Sigma} + \boldsymbol{\Lambda}, \tag{2.146}$$

where the generalized coefficients can be presented in matrix form

$$\mathbb{L} = \begin{pmatrix} \mathbf{L} & \mathbf{e}^\top \\ \mathbf{e} & -\mathbf{k} \end{pmatrix}, \quad \mathbb{M} = \begin{pmatrix} \mathbf{M} & \mathbf{d}^\top \\ \mathbf{d} & -\mathbf{b} \end{pmatrix}, \quad \boldsymbol{\Lambda} = \begin{pmatrix} \boldsymbol{\beta}^T\theta \\ \mathbf{q}\theta \end{pmatrix} \tag{2.147}$$

as well as in component form

$$\mathbb{L}_{ijkl} = L_{ijkl}, \quad \mathbb{L}_{4ikl} = \mathbb{L}_{i4kl} = e_{ikl}, \tag{2.148}$$

$$\mathbb{L}_{ij4l} = \mathbb{L}_{ijl4} = e_{lij}, \quad \mathbb{L}_{44kl} = \mathbb{L}_{kl44} = -k_{kl}, \tag{2.149}$$

$$\mathbb{M}_{ijkl} = M_{ijkl}, \quad \mathbb{M}_{4ikl} = \mathbb{M}_{i4kl} = d_{ikl}, \tag{2.150}$$

$$\mathbb{M}_{ij4l} = \mathbb{M}_{ijl4} = d_{lij}, \quad \mathbb{M}_{44kl} = \mathbb{M}_{kl44} = -b_{kl}, \tag{2.151}$$

$$\Lambda_{ij} = \beta_{ij}\theta, \quad \Lambda_{4i} = \Lambda_{i4} = q_i\theta \tag{2.152}$$

The generalized static equation and the Cauchy conditions of small generalized deformations are expressed in the form

$$\nabla_\alpha \Sigma_{\alpha\beta} = f_\beta, \quad \nabla_4 \equiv 0, \tag{2.153}$$

$$w_{\alpha\beta} = (\nabla_\alpha U_\beta + \nabla_\beta U_\alpha)/2, \tag{2.154}$$

where $U = (u_1, u_2, u_3, -\phi)^\top$, ϕ is the electrostatic potential, and f_β is a density of generalized body forces. Equations (2.153) and (2.154) should be complemented by the boundary conditions of the first or the second type. Except for notations, these equations coincide with the equations of linear thermoelasticity (2.126). Because of this, the theory of piezoelectric composite materials (PCMs) retraces at a particular instant the path of development of the theory of microinhomogeneous elastic media, exhibiting substantial progress. In light of the analogy mentioned in our brief survey, we will not consider in detail the PCM; one may refer instead to the appropriate scheme of thermoelastic composites.

2.4.3 Matrix Representation of Some Symmetric Tensors

Many material (e.g., elastic moduli (2.90)) and field (e.g., the stresses (2.50)) tensors have internal symmetry with respect to pairs of indices. Although all calculations with these tensors can be done by common methods of tensor analysis (see Sect. 2.2), it is sometimes convenient to introduce a contracted notation significantly simplifying the calculations. It is possible to form six different pairs ij for permuted indices i, j $(i, j = 1, 2, 3)$. Then the pair of these Latin indices ij can be substituted by a single Greek index α ranging from 1 to 6 according to the Voight notation

$$\begin{array}{lcccccc} \text{tensor notation } ij = & 11 & 22 & 33 & 23\ 32 & 13\ 31 & 12\ 21 \\ \text{matrix notation } \alpha = & 1 & 2 & 3 & 4 & 5 & 6 \end{array}, \tag{2.155}$$

which can be presented in a formula form (no sum on i, j)

$$\alpha = i\delta_{ij} + (9 - i - j)(1 - \delta_{ij}). \tag{2.156}$$

To avoid confusion, it should be mentioned that the prescribed reduction rule $ij \leftrightarrow \alpha$ between the Latin and Greek indices is also used for Latin indices $ij \leftrightarrow k$ $(i, j = 1, 2, 3,\ k = 1, \ldots, 6)$ where the correspondence between the pair indices and one index is provided by the scheme (2.155). We will also use this traditional form of an accordance.

We accept the convention that for the tensors with the prescribed aforementioned symmetry, Greek indices α, β, and γ range from 1 to 6 and are connected with the pairs of Latin indices ij, kl, and others by relation (2.155). In such a case, the tensor representations of the thermoelastic constitutive equations (2.126) can be presented in matrix notations

$$\sigma_\gamma = L_{\gamma\delta}\varepsilon_\delta + \alpha_\gamma, \quad \varepsilon_\gamma = M_{\gamma\delta}\sigma_\delta + \beta_\gamma. \tag{2.157}$$

It is also required that the tensor representations of widely used scalar productions $\mathbf{L} : \mathbf{M}$ and $\boldsymbol{\sigma} : \boldsymbol{\varepsilon}$ have an appropriate vector form $L_{\lambda\mu}M_{\mu\nu} = \delta_{\lambda\nu}$, $\sigma_{ij}\varepsilon_{ij} = \sigma_\lambda\varepsilon_\lambda$ with the standard assumption of summation over the repeated indices. However, it is conceivable to show that if the quantities corresponding to the different tensors are formed by one and the same rule (such as $\sigma_{ij} = \sigma_\lambda$), the equalities either (2.156) or (2.157) cannot be fulfilled. The mentioned inconsistency can be avoided if the following matrix representations of tensors are used $(i, j, k, l = 1, 2, 3;\ \lambda, \mu = 1, \ldots, 6)$

$$\varepsilon_{ij} = \frac{\varepsilon_\lambda}{2 - \delta_{ij}}, \quad \sigma_{ij} = \sigma_\lambda, \tag{2.158}$$

$$L_{ijkl} = L_{\lambda\mu}, \quad M_{ijkl} = \frac{M_{\lambda\mu}}{(2 - \delta_{ij})(2 - \delta_{kl})}. \tag{2.159}$$

In so doing, the transition to Greek indices for all symmetric second-rank tensors could be done in the unique manner $\alpha_{ij} = (2 - \delta_{ij})^{-1/2}\alpha_\lambda$.

Since the components of the stress and strain tensors are functions of the orientation of the reference system, the elastic coefficients in Eqs. (2.126_1) (and, analogously, in Eq. (2.126_2)) are also tensor functions of this orientation and they can be presented in a new coordinate system:

$$\mathbf{L}' = \mathbf{g}\mathbf{g}\mathbf{L}\mathbf{g}^\top\mathbf{g}^\top, \ L'_{ijkl} = q_{im}g_{jn}g_{ko}g_{lp}L_{mnop} \tag{2.160}$$

with the relevant transformation for the strain tensors:

$$\boldsymbol{\varepsilon}' = \mathbf{g}\mathbf{g}\boldsymbol{\varepsilon}, \quad \varepsilon'_{ij} = g_{ik}g_{jl}\varepsilon_{kl}. \tag{2.161}$$

These tensors as well as the thermal expansion and thermoelasticity coefficients are transformed using Voight reduced components as

$$\mathbf{L}' = \mathbf{q}^L \mathbf{L} (\mathbf{q}^L)^\top, \quad \mathbf{M}' = \mathbf{q}^R \mathbf{M} (\mathbf{q}^R)^\top, \tag{2.162}$$

$$\boldsymbol{\sigma}' = \mathbf{q}^L \boldsymbol{\sigma}, \quad \boldsymbol{\varepsilon}' = \mathbf{q}^R \boldsymbol{\varepsilon}, \quad \boldsymbol{\alpha}^{T'} = \mathbf{q}^L \boldsymbol{\alpha}^T, \quad \boldsymbol{\beta}^{T'} = \mathbf{q}^R \boldsymbol{\beta}^T. \tag{2.163}$$

where the fourth-order tensors $\mathbf{q}^L$ and $\mathbf{q}^R = \left[(\mathbf{q}^L)^{-1}\right]^\top$ presented in the form of (6×6) transformation matrices:

$$\mathbf{q}^L = \begin{pmatrix} \mathbf{K}_1 & 2\mathbf{K}_2 \\ \mathbf{K}_3 & \mathbf{K}_4 \end{pmatrix}, \quad \mathbf{q}^R = \begin{pmatrix} \mathbf{K}_1 & \mathbf{K}_2 \\ 2\mathbf{K}_3 & \mathbf{K}_4 \end{pmatrix}, \tag{2.164}$$

are combined by four matrices (3×3):

$$\mathbf{K}_1 = \begin{pmatrix} g_{11}^2 & g_{12}^2 & g_{13}^2 \\ g_{21}^2 & g_{22}^2 & g_{23}^2 \\ g_{31}^2 & g_{32}^2 & g_{33}^2 \end{pmatrix}, \quad \mathbf{K}_2 = \begin{pmatrix} g_{12}g_{13} & g_{13}g_{11} & g_{11}g_{12} \\ g_{22}g_{23} & g_{23}g_{21} & g_{21}g_{22} \\ g_{32}g_{33} & g_{33}g_{31} & g_{31}g_{32} \end{pmatrix},$$

$$\mathbf{K}_3 = \begin{pmatrix} g_{21}g_{31} & g_{22}g_{32} & g_{23}g_{33} \\ g_{31}g_{11} & g_{32}g_{121} & g_{33}g_{13} \\ g_{11}g_{21} & g_{12}g_{22} & g_{13}g_{23} \end{pmatrix},$$

$$\mathbf{K}_4 = \begin{pmatrix} g_{22}g_{33} + g_{23}g_{32} & g_{23}g_{31} + g_{21}g_{33} & g_{21}g_{32} + g_{22}g_{31} \\ g_{32}g_{13} + g_{33}g_{12} & g_{33}g_{11} + g_{31}g_{13} & g_{31}g_{12} + g_{32}g_{11} \\ g_{12}g_{23} + g_{13}g_{22} & g_{13}g_{21} + g_{11}g_{23} & g_{11}g_{22} + g_{12}g_{21} \end{pmatrix}. \tag{2.165}$$

The elements of the matrices $\mathbf{q}^L$ and $\mathbf{q}^R$ can be represented by means of the formulae

$$\frac{1}{2}(g_{i'k}g_{j'l} + g_{i'l}g_{j'k}) = \frac{q^R_{\lambda'\mu}}{2 - \delta_{i'j'}} = \frac{q^L_{\lambda'\mu}}{2 - \delta_{kl}}, \tag{2.166}$$

where the index λ' corresponds to the pair $i'j'$ and μ to the pair kl. The equations describing the piezoelectric effect (2.142) and (2.143) can be presented in the matrix form

$$\varepsilon_\gamma = M_{\gamma\delta}\sigma_\delta + d_{\gamma i}E_i + \lambda_\gamma, \quad D_i = e_{i\gamma}\varepsilon_\gamma - k_{ij}E_j + p_i\theta, \tag{2.167}$$

which leads to the following recalculation rules for the coefficients ($jk \leftrightarrow \gamma$): $d_{ijk} = (2 - \delta_{jk})^{-1} d_{i\gamma}$, $e_{ijk} = e_{i\gamma}$ with Latin indices ranging from 1 to 3. The transformation formulae for quantities $d_{i\gamma}$ and $e_{i\gamma}$ in the old $\mathbf{e}_k$ and new $\mathbf{e}'_k$ $(k = 1, 2, 3)$ coordinate systems with the cosine matrix g_{ij} (2.4) are

$$d'_{i\gamma} = g_{ij}q^R_{\gamma\delta}d_{j\delta}, \quad d_{j\delta} = g^\top_{ji}q^L_{\delta\gamma}d'_{i\gamma}, \quad e'_{i\gamma} = g_{ij}q^L_{\gamma\delta}e_{j\delta}, \quad e_{j\delta} = g^\top_{ji}q^R_{\delta\gamma}e'_{i\gamma}. \tag{2.168}$$

The recalculation rules and the transformation formulae can be obtained by the same method for other coefficients characterizing piezoelectric properties of crystals (for details see [1409]).

2.5 Symmetry of Elastic Properties

Under an orthogonal transformation (2.6), the elastic stiffnesses L'_{ijkl} referred to the $\mathbf{e}'_i$ coordinate system are described by Eq. (2.160). When $\mathbf{L}' = \mathbf{L}$, i.e.,

$$L_{ijkl} = g_{ip}g_{jq}g_{kr}g_{st}L_{pqrt}, \tag{2.169}$$

the material is said to possess a *symmetry* with respect to $\mathbf{g}$, which is called a symmetry transformation. If $\mathbf{g}$ is a symmetry transformation, $\mathbf{g}^{-1} = \mathbf{g}^\top$ is also a symmetry transformation, and if $\mathbf{g}$ and $\mathbf{q}$ are two symmetry transformations then $\mathbf{gq}$ is also a symmetry transformation. The set of all symmetry transformations at $\mathbf{x} \in w$ forms a symmetry group $S_\mathbf{x}$ with the unit transformation $\boldsymbol{\delta}$. The group $S_\mathbf{x}$ always contains the two-element subgroup $\{\boldsymbol{\delta}, -\boldsymbol{\delta}\}$. We describe by $\mathbf{g}(\omega\mathbf{n})$ (21.1.2) a right-handed rotation by the angle ω, $0 < \omega < \pi$, about an axis oriented in the direction of unit vector $\mathbf{n}$. The transformation $\mathcal{R}(\mathbf{n}) = -\mathbf{g}(\pi\mathbf{n})$ is called a reflection in the plane $P(\mathbf{n})$ with the normal unit vector $\mathbf{n}$. A unit vector $\mathbf{n}$ is called an axis of symmetry at $\mathbf{x}$ if $\mathbf{gn} = \mathbf{n}$ for some $\mathbf{g} \in S_\mathbf{x}$, with $\mathbf{g} \neq \boldsymbol{\delta}$. A plane $P(\mathbf{n}_1, \mathbf{n}_2)$, spanned by two mutually orthogonal unit vectors $\mathbf{n}_1$ and $\mathbf{n}_2$, is called a plane of symmetry at $\mathbf{x}$ if $\mathbf{gn}_1 = \mathbf{n}_1$ and $\mathbf{gn}_2 = \mathbf{n}_2$, for some $\mathbf{g} \in S_\mathbf{x}$, with $\mathbf{g} \neq \boldsymbol{\delta}$. The condition symmetry (2.169) introduces a restriction on the elastic moduli $\mathbf{L}$. A *triclinic* material has a minimum symmetry group $S_\mathbf{x}$ formed by the two-element group $\{\boldsymbol{\delta}, -\boldsymbol{\delta}\}$. According to Eq. (2.90), the number of independent elastic coefficients for the general anisotropic linearly elastic triclinic materials is reduced from $3^4 = 81$ to 21, which is the number of terms on the main diagonal of the matrix $\mathbf{L}$ and the five subdiagonals above it: $6 + 5 + 4 + 3 + 2 + 1 = 21$. This number is reduced when the symmetry of the crystal class of medium is accounted for. A material is called *monoclinic* if for any $\mathbf{x} \in w$ the symmetry grope $S_\mathbf{x}$ is formed by the transformations $\pm\boldsymbol{\delta}, \pm\mathbf{g}(\pi, \mathbf{n})$ for any $\mathbf{x} \in w$. An *orthotropic* has the symmetry group $S_\mathbf{x}$ formed for any $\mathbf{x} \in w$ by the transformations $\pm\boldsymbol{\delta}, \pm\mathbf{g}(\pi, \mathbf{n}_1), \pm\mathbf{g}(\pi, \mathbf{n}_2), \pm\mathbf{g}(\pi, \mathbf{n}_3)$, where $\mathbf{n}_1, \mathbf{n}_2, \mathbf{n}_3$ are three mutually orthogonal unit vectors. A material is called *transversally isotropic* with respect to the direction $\mathbf{n}$ if its symmetry group $S_\mathbf{x}$ $(\forall \mathbf{x} \in w)$ consists of the transformations $\pm\boldsymbol{\delta}$ and $\pm\mathbf{g}(\omega, \mathbf{n})$ with $0 < \omega < \pi$.

Depending on the number of rotations and/or reflection symmetry a crystal possesses, the 6×6 matrix $\mathbf{L}$ can be represented by one of the eight groups summarized in [1142]. If the matrix $\mathbf{L}'$ referred to a different coordinate system $\mathbf{e}'_i$ is desired, the transformation (2.160) is used.

Let us consider polycrystalline aggregates of an orthorhombic system. Then for the classes of symmetry with nine independent elastic constants, the tensors of the elastic moduli and compliance of a crystallite in the crystallographic system of coordinates can be represented in the form (see, e.g., [1378]):

$$L_{ijkl} = \sum_n \Big[\lambda_n \delta_{in}\delta_{jn}\delta_{kn}\delta_{ln} + \mu_n(\delta_{in}\delta_{jn}\delta_{kl} + \delta_{ij}\delta_{kn}\delta_{ln})$$

$$+\nu_n(\delta_{in}\delta_{jk}\delta_{ln} + \delta_{jn}\delta_{ik}\delta_{ln} + \delta_{in}\delta_{jl}\delta_{kn} + \delta_{jn}\delta_{il}\delta_{kn})\Big], \quad (2.170)$$

$$M_{ijkl} = \sum_n \Big[p_n \delta_{in}\delta_{jn}\delta_{kn}\delta_{ln} + q_n(\delta_{in}\delta_{jn}\delta_{kl} + \delta_{ij}\delta_{kn}\delta_{ln})$$

$$+r_n(\delta_{in}\delta_{jk}\delta_{ln} + \delta_{jn}\delta_{ik}\delta_{ln} + \delta_{in}\delta_{ji}\delta_{kn} + \delta_{jn}\delta_{il}\delta_{kn})\Big], \quad (2.171)$$

where the elastic constants λ_n, μ_n, ν_n and $p_n,\ q_n,\ r_n$ $(n = 1, 2, 3)$ are defined by the matrix representation of the elastic moduli L_{mn} and stiffness M_{ij} in the crystallographic system:

$$\begin{aligned}
\lambda_1 &= L_{11} + L_{23} + 2L_{44} - (L_{12} + L_{13} + 2L_{55} + 2L_{66}),\\
\lambda_2 &= L_{22} + L_{13} + 2L_{55} - (L_{12} + L_{23} + 2L_{44} + 2L_{66}),\\
\lambda_3 &= L_{33} + L_{12} + 2L_{66} - (L_{13} + L_{23} + 2L_{44} + 2L_{55}),\\
2\mu_1 &= L_{12} + L_{13} - L_{23},\quad 2\nu_1 = L_{55} + L_{66} - L_{44},\\
2\mu_2 &= L_{12} + L_{23} - L_{13},\quad 2\nu_2 = L_{44} + L_{66} - L_{55},\\
2\mu_3 &= L_{13} + L_{23} - L_{12},\quad 2\nu_3 = L_{44} + L_{55} - L_{66},\\
p_1 &= M_{11} + M_{23} + M_{44}/2 - (2M_{12} + 2M_{13} + M_{55} + M_{66})/2,\\
p_2 &= M_{22} + M_{13} + M_{55}/2 - (2M_{12} + 2M_{23} + M_{44} + M_{66})/2,\\
p_3 &= M_{33} + M_{12} + M_{66}/2 - (2M_{13} + 2M_{23} + M_{44} + M_{55})/2,\\
2q_1 &= M_{12} + M_{13} - M_{23},\quad 8r_1 = M_{55} + M_{66} - M_{44},\\
2q_2 &= M_{12} + M_{23} - M_{13},\quad 8r_2 = M_{44} + M_{66} - M_{55},\\
2q_3 &= M_{13} + M_{23} - M_{12},\quad 8r_3 = M_{44} + M_{55} - M_{66},
\end{aligned} \quad (2.172)$$

In particular, for tetragonal crystals, the relations $L_{11} = L_{22}$, $L_{13} = L_{23}$, and $L_{44} = L_{55}$ reduce the number of unknown elastic constants from nine to six

$$\begin{aligned}
\lambda_1 &= \lambda_2 = L_{11} - (L_{12} + L_{66}),\\
\lambda_3 &= L_{33} + L_{12} + 2L_{66} - 2(L_{13} + 2L_{44}),\\
2\mu_1 &= 2\mu_2 = L_{12},\quad 2\mu_3 = 2L_{13} - L_{12},\\
2\nu_1 &= 2\nu_2 = L_{66},\quad 2\nu_3 = 2L_{44} - L_{66}.
\end{aligned} \quad (2.173)$$

For hexagonal crystals, the addition condition $2L_{66} = L_{11} - L_{12}$ takes place substitution of which into Eq. (2.173) leads to

$$\begin{aligned}
&\lambda_1 = \lambda_2 = 0, \quad \lambda_3 = L_{11} + L_{33} - 2(L_{13} + 2L_{44}), \\
&2\mu_1 = 2\mu_2 = L_{12}, \quad 2\mu_3 = 2L_{13} - L_{12}, \\
&4\nu_1 = 4\nu_2 = L_{11} - L_{12}, \quad 4\nu_3 = 4L_{44} + L_{12} - L_{11}.
\end{aligned} \tag{2.174}$$

For cubic symmetry, substitution of the conditions $L_{22} = L_{33}$, $L_{12} = L_{13}$, $L_{44} = L_{66}$ into Eq. (2.173) gives

$$\begin{aligned}
&\lambda_1 = \lambda_2 = \lambda_3 = L_{11} - L_{12} - 2L_{44}, \\
&\mu_1 = \mu_2 = \mu_3 = L_{12}/2, \quad \nu_1 = \nu_2 = \nu_3 = L_{44}/2.
\end{aligned} \tag{2.175}$$

At last, a material is called elastically isotropic if its elastic properties are independent of direction. In the case of cubic crystal symmetry, this is realized when $L_{44} = (L_{11} - L_{12})/2$. The deviation of the cubic lattice from an isotropic one can be quantified by the so-called *Zener anisotropy ratio*

$$Z = \frac{2L_{44}}{L_{11} - L_{12}}. \tag{2.176}$$

Substitution of an additional condition $2L_{44} = L_{11} - L_{12}$ for the isotropic medium into Eq. (2.175) yields

$$\begin{aligned}
&\lambda_1 = \lambda_2 = \lambda_3 = 0, \quad \mu_1 = \mu_2 = \mu_3 = L_{12}/2, \\
&\nu_1 = \nu_2 = \nu_3 = (L_{11} - L_{12})/2.
\end{aligned} \tag{2.177}$$

The representation of elastic moduli in the reference coordinate system (Ox', Oy', Oz') can be obtained by the use of rotation $\mathbf{g} \in O^3$ from the tensors **L** (2.170) and **M** (2.171) according to Eqs. (2.160) and (2.161), respectively. The final representation is simplified in exploiting of the identities $g_{ij}\delta_{jn} = g_{in}$ and $g_{ij}g_{kj} = g_{ik}$

$$\begin{aligned}
L'_{ijkl} = \sum_n \Big[&\lambda_n g_{in} g_{jn} g_{kn} g_{ln} + \mu_n (g_{in} g_{jn} \delta_{kl} + \delta_{ij} g_{kn} g_{ln}) \\
&+ \nu_n (g_{in} \delta_{jk} g_{ln} + g_{jn} \delta_{ik} g_{ln} + g_{in} \delta_{jl} g_{kn} + g_{jn} \delta_{il} g_{kn}) \Big],
\end{aligned} \tag{2.178}$$

$$\begin{aligned}
M'_{ijkl} = \sum_n \Big[&p_n g_{in} g_{jn} g_{kn} g_{ln} + q_n (g_{in} g_{jn} \delta_{kl} + \delta_{ij} g_{kn} g_{ln}) \\
&+ r_n (g_{in} \delta_{jk} g_{ln} + g_{jn} \delta_{ik} g_{ln} + g_{in} \delta_{jl} g_{kn} + g_{jn} \delta_{il} g_{kn}) \Big].
\end{aligned} \tag{2.179}$$

where the elastic constants λ_n, μ_n, ν_n, and p_n, q_n, r_n $(n = 1, 2, 3)$ are defined by the matrix representations L_{ij} and M_{ij} in the crystallographic coordinate system (2.172).

In particular, for the *tetragonal* crystals (2.173),

$$\begin{aligned} L'_{ijkl} &= L_{12}\delta_{ij}\delta_{kl} + L_{66}(\delta_{ik}\delta_{jl} + \delta_{il}\delta_{jk}) + (L_{11} - L_{12} - 2L_{66})\sum_{ijkl} \\ &+(L_{33} - L_{11} + 2L_{12} - 2L_{13} + 4L_{66} - 4L_{44})T_{ijkl} \\ &+2(L_{44} - L_{66})(T_{(ik}\delta_{jl)} + T_{(il}\delta_{jk)}) + 2(L_{13} - L_{12})T_{(ij}\delta_{kl)}, \end{aligned} \tag{2.180}$$

for *hexagonal* crystals (2.174),

$$\begin{aligned} L'_{ijkl} &= L_{12}\delta_{ij}\delta_{kl} + (L_{11} - L_{12})(\delta_{ik}\delta_{jl} + \delta_{il}\delta_{jk}) \\ &+(L_{33} + L_{11} - 2L_{13} - 4L_{44})T_{ijkl} + 2(L_{13} - L_{12})T_{(ij}\delta_{kl)} \\ &+(L_{12} - L_{11} + 2L_{44})(T_{(ik}\delta_{jl)} + T_{(il}\delta_{jk)}), \end{aligned} \tag{2.181}$$

for *cubic* crystals (2.175),

$$L'_{ijkl} = L_{12}\delta_{ij}\delta_{kl} + L_{44}(\delta_{ik}\delta_{jl} + \delta_{il}\delta_{jk}) + (L_{11} - L_{12} - 2L_{44})\sum_{ijkl}, \tag{2.182}$$

and for the *isotropic* material (2.177),

$$L'_{ijkl} = L_{12}\delta_{ij}\delta_{kl} + L_{44}(\delta_{ik}\delta_{jl} + \delta_{il}\delta_{jk}). \tag{2.183}$$

Here one introduces the notations

$$\sum_{ijkl} = \sum_{n=1}^{3} g_{in}g_{jn}g_{kn}g_{ln}, \quad T_{ij} = g_{i3}g_{j3}, \quad T_{ijkl} = g_{i3}g_{j3}g_{k3}g_{l3}, \tag{2.184}$$

where the tensors T_{ijkl} and $\sum_{ijkl}$ show the properties

$$T_{iikl} = T_{kl}, \quad T_{ijkl}T_{klpq} = T_{ijpq}, \quad \sum_{iikl} = \delta_{kl}, \quad \sum_{ijkl}\sum_{klpq} = \sum_{ijpq}. \tag{2.185}$$

The representation (2.182) for the *cubic* crystals can be also presented in more detailed form:

$$\begin{aligned} L'_{11} &= L_{11} - 2L(g_{11}^2g_{12}^2 + g_{12}^2g_{13}^2 + g_{13}^2g_{11}^2), \\ L'_{12} &= L_{12} + L(g_{11}^2g_{21}^2 + g_{12}^2g_{22}^2 + g_{13}^2g_{23}^2), \\ L'_{44} &= L_{44} + L(g_{12}^2g_{13}^2 + g_{22}^2g_{32}^2 + g_{23}^2g_{33}^2), \quad L \equiv L_{11} - L_{12} - 2L_{44}. \end{aligned} \tag{2.186}$$

The inverse transformations from the coefficients $L_{\gamma\nu}$ to $M_{\gamma\nu}$ can be carried out by direct inversion of the matrix of known coefficients because $L_{\gamma\delta}M_{\delta\nu} = \delta_{\gamma\nu}$. However, in some cases, the simple relations between the coefficients $L_{\gamma\nu}$ and $M_{\gamma\nu}$ are useful. They are presented below for some symmetry groups described, at most, by seven independent parameters:

Trigonal Materials

$$\begin{aligned} &M_{11} + M_{12} = L_{33}/l,\ M_{11} - M_{12} = L_{44}/l',\ M_{13} = -L_{13}/l, \\ &\quad M_{14} = -L_{14}/l',\ \ M_{33} = (L_{11} + L_{12})/l,\ M_{44} = (L_{11} - L_{12})/l', \\ &\quad\quad l \equiv L_{33}(L_{11} + L_{12}) - 2L_{13}^2,\ \ l' \equiv L_{44}(L_{11} - L_{12}) - 2L_{14}^2. \end{aligned} \tag{2.187}$$

Hexagonal Materials

$$\begin{aligned} &M_{11} + M_{12} = L_{33}/l,\ M_{11} - M_{12} = (L_{11} - L_{12})^{-1},\ M_{13} = -L_{13}/l, \\ &\quad M_{33} = (L_{11} + L_{12})/l,\ M_{44} = L_{44}^{-1}, \\ &\quad\quad l \equiv L_{33}(L_{11} + L_{12}) - 2L_{13}^2. \end{aligned} \tag{2.188}$$

Cubic Materials

$$\begin{aligned} M_{11} &= (L_{11} + L_{12})(L_{11} - L_{12})^{-1}(L_{11} + 2L_{12})^{-1}, \\ M_{12} &= -L_{12}(L_{11} - L_{12})^{-1}(L_{11} + 2L_{12})^{-1}, \\ M_{44} &= L_{44}^{-1}. \end{aligned} \tag{2.189}$$

In particular, the elastic constants in the isotropic matrix ($\mathbf{A} = \mathbf{L}, \mathbf{M};\ A_{ij} = L_{ij}, M_{ij}$):

$$\mathbf{A} = \begin{pmatrix} A_{11} & A_{12} & A_{12} & 0 & 0 & 0 \\ A_{12} & A_{11} & A_{12} & 0 & 0 & 0 \\ A_{12} & A_{12} & A_{11} & 0 & 0 & 0 \\ 0 & 0 & 0 & A_{44} & 0 & 0 \\ 0 & 0 & 0 & 0 & A_{44} & 0 \\ 0 & 0 & 0 & 0 & 0 & A_{44} \end{pmatrix}, \tag{2.190}$$

(where $L_{44} = (L_{11} - L_{12})/2,\ M_{44} = 2(M_{11} - M_{12})$) are usually written in the engineering notations:

$$L_{11} = \frac{E}{1+\nu}\frac{1-\nu}{1-2\nu},\ \ L_{12} = \frac{E}{1+\nu}\frac{\nu}{1-2\nu},\ \ L_{44} = \frac{E}{2(1+\nu)}, \tag{2.191}$$

$$M_{11} = \frac{1}{E},\ M_{12} = -\frac{\nu}{E},\ M_{44} = \frac{2(1+\nu)}{E}, \tag{2.192}$$

where two independent constants *Young's modulus* E and *Poisson's ratio* ν can be also used in the tensor representations ($\mathbf{N}_1 = \delta \otimes \delta/3$)

$$\mathbf{L} = \frac{3\nu E}{(1+\nu)(1-2\nu)}\mathbf{N}_1 + \frac{E}{1+\nu}\mathbf{I}, \quad \mathbf{M} = -\frac{3\nu}{E}\mathbf{N}_1 + \frac{1+\nu}{E}\mathbf{I}. \tag{2.193}$$

Another popular representations of the elastic isotropic properties using *Lame's constants* λ and μ can be formed in both the matrix

$$L_{11} = \lambda + \mu, \; L_{12} = \lambda, \; L_{66} = \mu, \tag{2.194}$$

and tensor

$$\mathbf{L} = 3\lambda\mathbf{N}_1 + 2\mu\mathbf{I}, \quad \mathbf{M} = 3s\mathbf{N}_1 + 2q\mathbf{I}, \tag{2.195}$$

notations where the compliances s and q are expressed in terms of elastic constants:

$$s = \frac{1}{3(3\lambda + 2\mu)}, \quad q = \frac{1}{4\mu}. \tag{2.196}$$

Due to the decomposition of the unit tensor $\mathbf{I}$ into the orthogonal bulk $\mathbf{N}_1$ and deviatoric $\mathbf{N}_2$ tensors (2.16), the isotropic tensors $\mathbf{L}$ and $\mathbf{M}$ can also be decomposed:

$$\mathbf{L} = (3k, 2\mu) \equiv 3k\mathbf{N}_1 + 2\mu\mathbf{N}_2, \quad \mathbf{M} = (3p, 2q). \tag{2.197}$$

where k and μ, called the bulk and shear moduli, are related to the appropriate compliance coefficients p, q by means of the "multiplication table" for the products between the elementary idempotent tensors $\mathbf{N}_1$ and $\mathbf{N}_2$ (2.17). Then the inverse tensor $\mathbf{L}_1^{-1} \equiv \mathbf{M}$, and the product $\mathbf{L}_1 : \mathbf{L}_2$ of such tensors, $\mathbf{L}_1 = (3k_1, 2\mu_1)$ and $\mathbf{L}_2 = (3k_2, 2\mu_2)$, can be estimated by the compact formulae

$$\mathbf{L}_1^{-1} = (3p, 2q) = \left(\frac{1}{3k}, \frac{1}{2\mu}\right), \quad \mathbf{L}_1 : \mathbf{L}_2 = (9k_1k_2, 4\mu_1\mu_2). \tag{2.198}$$

Table 2.1 shows the relationships between the various elastic constants of isotropic materials.

The bulk and shear components of the isotropic tensors $\mathbf{L}$ and $\mathbf{M}$ are found through a suitable contraction of the indices of $\mathbf{L}$ and $\mathbf{M}$ in the following relations

$$3k = \frac{1}{3}L_{iikk}, \quad 2\mu = \frac{1}{5}\left(L_{ikik} - \frac{1}{3}L_{iikk}\right), \tag{2.199}$$

$$3p = \frac{1}{3}M_{iikk}, \quad 2q = \frac{1}{5}\left(M_{ikik} - \frac{1}{3}M_{iikk}\right), \tag{2.200}$$

which can be used at the estimation of average moduli of isotropic polycrystals.

Table 2.1 Relationships between elastic moduli

Constant	$\lambda,\ \mu$	$K,\ \mu$	$\mu,\ \nu$	$E,\ \nu$	$E,\ \mu$
λ	λ	$K-\frac{2}{3}\mu$	$\frac{2\mu\nu}{1-2\nu}$	$\frac{\nu E}{(1+\nu)(1-3\nu)}$	$\frac{\mu(E-2\mu)}{3\mu-E}$
μ	μ	μ	μ	$\frac{E}{2(1+\nu)}$	μ
K	$\lambda+\frac{2}{3}\mu$	K	$\frac{2\mu(1+\nu)}{3(1-2\nu)}$	$\frac{E}{3(1-2\nu)}$	$\frac{E\mu}{3(3\mu-E)}$
E	$\frac{(3\lambda+2\mu)\mu}{\lambda+\mu}$	$\frac{9K\mu}{3K+\mu}$	$2(1+\nu)\mu$	E	E
ν	$\frac{\lambda}{2(\lambda+\mu)}$	$\frac{3K-2\mu}{6K+2\mu}$	ν	ν	$\frac{E}{2\mu}-1$

Transversely isotropic materials have five independent elastic parameters. If the plane of isotropy coincides with the x_1, x_2-plane, the elastic and compliance matrices ($\mathbf{A}=\mathbf{L}, \mathbf{M};\ A_{ij}=L_{ij}, M_{ij}$):

$$\mathbf{A}=\begin{pmatrix} A_{11} & A_{12} & A_{13} & 0 & 0 & 0\\ A_{12} & A_{11} & A_{13} & 0 & 0 & 0\\ A_{13} & A_{13} & A_{33} & 0 & 0 & 0\\ 0 & 0 & 0 & A_{44} & 0 & 0\\ 0 & 0 & 0 & 0 & A_{44} & 0\\ 0 & 0 & 0 & 0 & 0 & A_{55} \end{pmatrix}, \tag{2.201}$$

(where $L_{55}=(L_{11}-L_{12})/2,\ M_{55}=2(M_{11}-M_{12})$) can be presented in engineering notations

$$L_{11}=l\Big(\frac{1}{EE_3}-\frac{\nu_3^2}{E_3^2}\Big),\quad L_{12}=l\Big(\frac{\nu}{EE_3}+\frac{\nu_3^2}{E_3^2}\Big),\quad L_{13}=l\frac{(1+\nu)\nu_3}{EE_3},$$
$$L_{33}=l\frac{1-\nu^2}{E^2},\quad L_{44}=\mu_3,\ l=\frac{E^2E_3^2}{(1+\nu)[(1-\nu)E_3-2\nu_3^2E]}, \tag{2.202}$$

$$M_{11}=\frac{1}{E},\quad M_{12}=-\frac{\nu}{E},\quad M_{13}=-\frac{\nu_3}{E_3},\quad M_{33}=\frac{1}{E_3},\quad M_{44}=\frac{1}{\mu_3}. \tag{2.203}$$

Here the Young modulus and the Poisson ratios in the x_1, x_2-plane are the same, say, $E_1=E_2=E$ and $\nu_{12}=\nu_{21}=\nu$, respectively. The corresponding shear modulus $\mu_{12}=\mu_{21}=\mu$ is formed as $\mu=E/2(1+\nu)$. Young modulus, Poisson ratio, and the shear modulus associated with the x_3-direction and a direction in the x_1x_2-plane are denoted by, say, E_3, $\nu_{13}=\nu_{23}=\nu_3$, and $\mu_{13}=\mu_{23}=\mu_3$, respectively.

For the *orthotropic* case, the elastic and compliance matrices are defined by nine independent elastic parameters and can be presented in a coordinate system coincident with the material symmetry directions as ($\mathbf{A} = \mathbf{L}, \mathbf{M};\ A_{ij} = L_{ij}, M_{ij}$)

$$\mathbf{A} = \begin{pmatrix} A_{11} & A_{12} & A_{13} & 0 & 0 & 0 \\ A_{12} & A_{22} & A_{23} & 0 & 0 & 0 \\ A_{13} & A_{23} & A_{33} & 0 & 0 & 0 \\ 0 & 0 & 0 & A_{44} & 0 & 0 \\ 0 & 0 & 0 & 0 & A_{55} & 0 \\ 0 & 0 & 0 & 0 & 0 & A_{66} \end{pmatrix}, \tag{2.204}$$

where, for example, the compliance matrix in engineering terminology becomes

$$\mathbf{M} = \begin{pmatrix} 1/E_1 & -\nu_{21}/E_2 & -\nu_{31}/E_3 & 0 & 0 & 0 \\ -\nu_{12}/E_1 & 1/E_2 & -\nu_{32}/E_3 & 0 & 0 & 0 \\ -\nu_{13}/E_1 & -\nu_{23}/E_2 & 1/E_3 & 0 & 0 & 0 \\ 0 & 0 & 0 & 1/(2\mu_{23}) & 0 & 0 \\ 0 & 0 & 0 & 0 & 1/(2\mu_{13}) & 0 \\ 0 & 0 & 0 & 0 & 0 & 1/(2\mu_{12}) \end{pmatrix}. \tag{2.205}$$

We will now consider a unique representation of the constitutive laws and elastic moduli for a d-dimensional linear isotropic material with bulk modulus $k_{[d]}$, shear modulus $\mu_{[d]}$, Young's modulus $E_{[d]}$, and Poisson's ratio $\nu_{[d]}$ in the space dimensionality d ($i, j, k, l = 1, \dots, d;\ d = 2, 3$) (see, e.g., [1523])

$$\sigma_{ij} = \left(k_{[d]} - \frac{2}{d}\mu_{[d]}\right)\varepsilon_{kk}\delta_{ij} + 2\mu_{[d]}\varepsilon_{ij}, \quad \varepsilon_{ij} = -\frac{\nu_{[d]}}{E_{[d]}}\sigma_{kk}\delta_{ij} + \frac{(1+\nu_{[d]})}{E_{[d]}}\sigma_{ij} \tag{2.206}$$

and

$$\mathbf{L} = dk_{[d]}\mathbf{N}_1 + 2\mu_{[d]}\mathbf{N}_2, \quad \mathbf{M} = \frac{1+\nu_{[d]}(1-d)}{E_{[d]}}\mathbf{N}_1 + \frac{1+\nu_{[d]}}{E_{[d]}}\mathbf{N}_2, \tag{2.207}$$

where the projection tensors

$$\mathbf{N}_{1|ijkl} = \frac{1}{d}\delta_{ij}\delta_{kl}, \quad \mathbf{N}_{2|ijkl} = \frac{1}{2}(\delta_{ik}\delta_{jl} + \delta_{il}\delta_{jk}) - \frac{1}{d}\delta_{ij}\delta_{kl} \tag{2.208}$$

satisfy the normality conditions (2.17) allowing the inverse of relation (2.207_1)

$$\mathbf{M} = \frac{1}{dk_{[d]}}\mathbf{N}_1 + \frac{1}{2\mu_{[d]}}\mathbf{N}_2. \tag{2.209}$$

Comparison of representations (2.207$_2$) and (2.209) leads to the interrelations

$$k_{[d]} = \frac{E_{[d]}}{d[1+\nu_{[d]}(1-d)]}, \quad \mu_{[d]} = \frac{E_{[d]}}{2(1+\nu_{[d]})},$$
$$\frac{1}{E_{[d]}} = \frac{1}{d^2 k_{[d]}} + \frac{d-1}{2d\mu_{[d]}}, \quad \nu_{[d]} = \frac{dk_{[d]} - 2\mu_{[d]}}{d(d-1)k_{[d]} + 2\mu_{[d]}}. \tag{2.210}$$

Passages to the limits of the positive moduli $k_{[d]}/\mu_{[d]} \to \infty$ and $\mu_{[d]}/k_{[d]} \to \infty$ in the expression (2.210$_4$) yield the upper and lower limits of $\nu_{[d]}$: $-1 \le \nu_{[d]} \le (d-1)^{-1}$, respectively.

Finally, we will consider either *plane-strain* ($\varepsilon_{11} = \varepsilon_{12} = \varepsilon_{13} = 0$ in Eq. (2.206) with $d = 3$) or *plane-stress* ($\sigma_{11} = \sigma_{12} = \sigma_{13} = 0$ in Eq. (2.206) with $d = 3$) elasticity. If we compare these simplified three-dimensional expressions to relations (2.206) with $d = 2$, we obtained the interrelations

$$E_{[2]} = \frac{E_{[3]}}{(1-\nu_{[3]})(1+\nu_{[3]})}, \quad \nu_{[2]} = \frac{\nu_{[3]}}{1-\nu_{[3]}}, \quad k_{[2]} = k_{[3]} + \frac{1}{3}\mu_{[3]}, \quad \mu_{[2]} = \mu_{[3]}, \tag{2.211}$$

for plane-strain elasticity and

$$E_{[2]} = E_{[3]}, \quad \nu_{[2]} = \nu_{[3]}, \quad k_{[2]} = \frac{9k_{[3]}\mu_{[3]}}{3k_{[3]} + 4\mu_{[3]}}, \quad \mu_{[2]} = \mu_{[3]} \tag{2.212}$$

for plane-stress elasticity, respectively.

2.6 Basic Equations of Thermoelastoplastic Deformations

2.6.1 *Incremental Theory of Plasticity*

We assume that the thermomechanical properties of the composite medium with, generally speaking, anisotropic components are described by the theory of small elastic–plastic strains under arbitrarily varying external loading. Additive decomposition of the increments of total strain tensor ε is assumed

$$d\boldsymbol{\varepsilon} = d\boldsymbol{\varepsilon}^e + d\boldsymbol{\varepsilon}^t + d\boldsymbol{\varepsilon}^p, \tag{2.213}$$

with the increments of the elastic strains $d\boldsymbol{\varepsilon}^e$, the so-called transformation strains $d\boldsymbol{\varepsilon}^t$ including thermal strains and plastic strains $d\boldsymbol{\varepsilon}^p$.

The stress increment $d\boldsymbol{\sigma}$ relates to the elastic part of the strain increment $d\boldsymbol{\varepsilon}$ by the elasticity relation given in the form

$$d\boldsymbol{\sigma}(\mathbf{x}) = \mathbf{L}(\mathbf{x})\, d\boldsymbol{\varepsilon}^e(\mathbf{x}), \tag{2.214}$$

where $\mathbf{L}(\mathbf{x})$ is the fourth-order elasticity tensor (2.88).

The transformation strain increment $d\boldsymbol{\varepsilon}^t$ (2.213) may consist of contributions of different physical origins. For example, if only thermal effects are considered, $\boldsymbol{\varepsilon}^t = \boldsymbol{\beta}^T\theta$, where $\boldsymbol{\beta}^T$ is the tensor of linear thermal expansion coefficients and θ is the temperature change from the reference value to the current temperature.

In the six-dimensional stress space, consider a yield surface $f(\mathbf{x}, \ldots) = 0$ bonding the region of plastic deformation; here dots stand for temperature and for internal variables characterizing material hardening. The behavior is elastic if $f < 0$ or if $f = 0$ and $[(\partial f/\partial \boldsymbol{\sigma}) : d\boldsymbol{\sigma} + (\partial f/\partial \theta)d\theta] \leq 0$ (for the elastic unloading and neutral loading); elastic-plastic deformations take place under active loading, when $f = 0$ and $[(\partial f/\partial \boldsymbol{\sigma}) : d\boldsymbol{\sigma} + (\partial f/\partial \theta)d\theta] > 0$. In the general case, the yield surface can depend on a variety of tensor $\mathbf{a}^p$ and scalar hardening parameters. The hardening tensor parameter, in particular, may simply coincide with the plastic-strain tensor. Regarding the constitutive equations for the elastic-plastic materials we use the so-called J_2-flow theory with combined isotropic-kinematic hardening. The von Mises form of the yield surface is given by

$$f \equiv \tau - F(\gamma, \theta) = 0, \qquad F(0, \theta) = \tau_0(\theta) \tag{2.215}$$

in terms of von Mises effective stress, τ, and the accumulated effective plastic strain increment, γ, respectively, as defined by

$$\tau = \left(\frac{3}{2} s^a_{ij} s^a_{ij}\right)^{1/2}, \quad d\gamma = \left(\frac{2}{3} d\varepsilon^p_{ij}\, d\varepsilon^p_{ij}\right)^{1/2}, \quad \mathbf{s}^a = \mathbf{N}_2(\boldsymbol{\sigma} - \mathbf{a}^p). \tag{2.216}$$

Here τ_0 is the initial yield stress, and F is a nonlinear function describing the hardening effect and temperature θ dependence, for example, as in the modified Ludwik equation

$$F(\gamma, \theta) = \tau_0(\theta) + h(\theta)\gamma^{n(\theta)}, \tag{2.217}$$

where h and n are hardening parameters and $\mathbf{s}^a$ is the active stress deviator; $\mathbf{a}^p$ is a symmetric second-order tensor corresponding to the "back-stresses" defining the location of the center of the yield surface in the deviatoric stress space. For evaluation of the back stress tensor $\mathbf{a}^p$, we use Ziegler's assumption

$$d\mathbf{a}^p = d\gamma A \mathbf{s}^a, \qquad A \equiv A(\gamma), \tag{2.218}$$

or Prager's rule

$$d\mathbf{a}^p = B d\boldsymbol{\varepsilon}^p, \quad B \equiv B(\gamma), \tag{2.219}$$

respectively, with $\mathbf{a}^p = 0$ if $\gamma = 0$. In the case $\mathbf{a}^p \equiv \mathbf{0}$, $h \neq 0$, Eqs. (2.215) and (2.216) are reduced to the isotropic hardening, the case $h \equiv 0$, $\mathbf{a}^p \neq \mathbf{0}$ correspondents to the kinematic hardening. Though the von Mises yield criterion (2.215) is assumed in this study, modifications of the present method for introducing other yield criteria and hardening laws are possible.

Drucker's postulate states that the work of additional stresses is positive over the whole cycle. Then for any stress state $\boldsymbol{\sigma}^0$ within or on the convex yield surface the local maximum principle $(\boldsymbol{\sigma} - \boldsymbol{\sigma}^0) : d\boldsymbol{\varepsilon}^p \geq 0$. Drucker's postulate leads to the necessity of the associated law of plastic flow when the smooth yield function f is taken as plastic potential function from which the incremental plastic strain can be derived as

$$d\boldsymbol{\varepsilon}^p = d\lambda \frac{\partial f}{\partial \boldsymbol{\sigma}}, \quad \text{for } f = 0, \ \frac{\partial f}{\partial \boldsymbol{\sigma}} : d\boldsymbol{\sigma} > 0, \tag{2.220}$$

where the plastic flow is fixed in the direction along the normal to the yield surface, while its magnitude $d\lambda$, called a proportionality factor, is undetermined. For consistency during plastic deformation:

$$df \equiv \frac{\partial f}{\partial \boldsymbol{\sigma}} : d\boldsymbol{\sigma} + \frac{\partial f}{\partial \mathbf{a}^p} : d\mathbf{a}^p + \frac{\partial f}{\partial \gamma} d\gamma + \frac{\partial f}{\partial \theta} d\theta = 0. \tag{2.221}$$

Substitution of (2.216), (2.218), (2.219) in (2.221) leads to the determination of $d\lambda$ by

$$d\lambda = G \left(\frac{\partial f}{\partial \boldsymbol{\sigma}} : d\boldsymbol{\sigma} + \frac{\partial f}{\partial \theta} d\theta \right), \tag{2.222}$$

where the proportionality parameter G is defined by the explicit form of hardening rule (2.217)–(2.219). For example, for isotropic-kinematic hardening, (2.217), (2.218), we have

$$G = -\left(\frac{\partial f}{\partial \gamma} + A(\gamma) \frac{\partial f}{\partial \mathbf{a}^p} : \mathbf{s}^a \right)^{-1} \left(\frac{2}{3} \frac{\partial f}{\partial \boldsymbol{\sigma}} : \frac{\partial f}{\partial \boldsymbol{\sigma}} \right)^{-1/2}. \tag{2.223}$$

The associated flow law (2.220) requires the smoothness of the yield surface with a defined normal to the surface. However, the plastic flow corresponding to edges or vertexes of the yield surface $f = 0$ originated by the intersections of smooth surfaces $f_\alpha = 0$ ($\alpha = 1, \dots$) can be estimated from superposition principle by Koiter [787]:

$$d\boldsymbol{\varepsilon}^p = \sum_{\alpha=1} d\lambda_\alpha \frac{\partial f_\alpha}{\partial \boldsymbol{\sigma}}, \tag{2.224}$$

$$f_\alpha = 0, \quad \frac{\partial f_\alpha}{\partial \boldsymbol{\sigma}} : d\boldsymbol{\sigma} \geq 0. \tag{2.225}$$

A state of stress on the yield surface is described by a value of zero of one or more yield functions, all other yield functions being negative. In so doing, the nonzero proportionality factors $d\lambda_\alpha$ correspond only to the yield surfaces satisfying the conditions (2.225).

2.6.2 *Deformation Theory of Plasticity*

It is assumed that the rheological properties of isotropic media are described by the theory of small elastoplastic strains under monotonic, proportional loading when the ratio of the stress components $\sigma_{11} : \sigma_{22} : \sigma_{33} : \sigma_{13} : \sigma_{23} : \sigma_{12}$ is held constant at all time. Specifically, the total strain $\boldsymbol{\varepsilon}$ is written as the sum of elastic $\boldsymbol{\varepsilon}^e$ and plastic $\boldsymbol{\varepsilon}^p$ contributions

$$\boldsymbol{\varepsilon} = \boldsymbol{\varepsilon}^e + \boldsymbol{\varepsilon}^p, \tag{2.226}$$

where the mean stress and mean strain are linearly related: $\sigma_{ii} = 3k\varepsilon_{ii}$. The local equation for the elastic material state, which relates the stress tensor $\boldsymbol{\sigma}(\mathbf{x})$ and the elastic strain tensor $\boldsymbol{\varepsilon}^e(\mathbf{x})$, is given in the form (2.214) where $\mathbf{L}(\mathbf{x})$ is an isotropic fourth-order elasticity tensor described by the Young's modulus E and the Poisson's ratio ν (2.193).

The relation between flow stress and plastic strains is represented by the expression

$$\tau = \tau_0 + f(\varepsilon^p_{\text{eq}}), \quad f(0) = 0 \tag{2.227}$$

in terms of the von Mises effective stress τ and the effective plastic strain ϵ^p_{eq}, defined as

$$\tau = \left(\frac{3}{2} s_{ij} s_{ij}\right)^{1/2}, \quad \varepsilon^p_{\text{eq}} = \left(\frac{2}{3} \varepsilon^p_{ij} \varepsilon^p_{ij}\right)^{1/2}, \quad \mathbf{s} \equiv \mathbf{N}_2 \sigma. \tag{2.228}$$

In Eq. (2.227), τ_0 is the initial yield stress and f is a nonlinear function describing the material's hardening behavior, for example,

$$f(\epsilon_{\text{eq}}^{p}) = h(\epsilon_{\text{eq}}^{p})^{n}, \tag{2.229}$$

where h and n are the strength coefficient and the work-hardening exponent, respectively. The material remains elastic, i.e., $\boldsymbol{\varepsilon}^{p} \equiv \mathbf{0}$, when $\tau < \tau_0$. In addition, Hencky's flow rule

$$\varepsilon_{ij}^{p} = \frac{3\varepsilon_{\text{eq}}^{p}}{2\tau}\sigma_{ij} \tag{2.230}$$

is adopted. A more general relation given in Prager's form is

$$\varepsilon_{ij}^{p} = \alpha s_{ij} + \beta t_{ij}, \tag{2.231}$$

where $t_{ij} = s_{ik}s_{kj} - 2/3J_2\delta_{ij}$ is the deviation of the square of the stress deviation, and for an isotropic materials, α and β are the functions of the invariants $J_2 = s_{ij}s_{ij}/2$ and $J_3 = s_{ij}s_{jk}s_{ki}/3$.

Based on relation (2.226), one may determine the secant modulus as

$$E^{s} = \left[\frac{1}{E} + \frac{\epsilon_{\text{eq}}^{p}}{\tau_0 + f(\epsilon_{\text{eq}}^{p})}\right]^{-1}. \tag{2.232}$$

Due to the plastic incompressibility, the secant bulk modulus k^{s} is equal to k, and we may accordingly define the secant Poisson's ratio and shear modulus

$$\nu^{s} = \frac{1}{2} - \left(\frac{1}{2} - \nu\right)\frac{E^{s}}{E}, \quad \mu^{s} = \frac{E^{s}}{2(1+\nu^{s})}. \tag{2.233}$$

Both here and below the superscript, ^{s} indicates the calculation of the parameter under consideration with the help of the secant modulus E^{s}. Therefore, under a monotonic, proportional loading, the plastic state can be described by a single secant modulus, say E^{s}, and the other two elastic constants. For these loading the incremental flow rule can be integrated to yield the deformation rule.

Chapter 3
Green's Functions, Eshelby, and Related Tensors

Abstract The key tools of random structure composite materials (CMs) such as Green's function and related tensors are considered in this chapter for the problems of static and dynamic elasticity, conductivity, and thermoelectroelasticity. Inhomogeneity in an elastic medium is considered for thermoelastic problem with the interface boundary operators. One presents the theorem of polynomial conservation (the p-property) and Eshelby tensors and related tensors in a special basis for ellipsoidal inclusion. Coated ellipsoidal inclusion and imperfect interface models with estimation of elastic moduli of fictitious ellipsoidal homogeneous inclusion are considered. Related problems (conductivity, scattering of elastic waves, and piezoelectric problem) for ellipsoidal inhomogeneity in an infinite medium are also analyzed.

3.1 Static Green's Function

Let $f_i(\mathbf{x})$ be the distribution density of volume forces with finite support in an infinite medium with an elastic modulus L_{ijkl}. Then the elastic fields of small displacements are found from the equilibrium equation:

$$\boldsymbol{\mathcal{L}}\mathbf{u} = -\mathbf{f}, \quad (\boldsymbol{\mathcal{L}})_{ik} \equiv \nabla_j L_{ijkl} \nabla_l, \tag{3.1}$$

where $\mathbf{u}(\mathbf{x})|_{|\mathbf{x}|\to\infty} = \mathbf{0}$. The Green's tensor for the displacement $G_{ij}(\mathbf{x})$ in the infinite medium is defined from the relation ($\mathbf{x}, \mathbf{y} \in R^d,\ d = 2, 3$)

$$u_k = G_{ki} * f_i \equiv \int G_{ki}(\mathbf{x} - \mathbf{y}) f_i(\mathbf{y})\, d\mathbf{y}. \tag{3.2}$$

For the unique definition of the Green's tensor, one specifies the boundary conditions vanishing at infinity. From the equilibrium equation (3.1) and Eq. (3.2), one obtains an equation for the estimation of the Green's function:

$$L_{ijkl} G_{lm,jk}(\mathbf{x}) = -\delta_{im}\delta(\mathbf{x}), \tag{3.3}$$

V. A. Buryachenko, *Local and Nonlocal Micromechanics of Heterogeneous Materials*, https://doi.org/10.1007/978-3-030-81784-8_3

where $\delta(\mathbf{x})$ is the Dirac delta function equal to zero in the whole space except the point $\mathbf{x} = \mathbf{0}$ and having the property $\int \mathbf{f}(\mathbf{x})\delta(\mathbf{x})\, d\mathbf{x} = \mathbf{f}(\mathbf{0})$ if the integration area contains the point $\mathbf{x} = \mathbf{0}$. Equation (3.2) defines the physical meaning of the Green's tensor component $G_{lm}(\mathbf{x} - \mathbf{y})$ as the component of the displacement vector $u_i(\mathbf{x})$ arising from the unit force directed along the axis x_m and applied at the point $\mathbf{y}$.

Estimation of the Green's tensor $\mathbf{G}$ will be carried out by the use of the Fourier transform

$$\mathbf{g}(\mathbf{k}) = F(\mathbf{g}) = \int \mathbf{g}(\mathbf{x})e^{-i\mathbf{k}\cdot\mathbf{x}}\, d\mathbf{x},$$

$$\mathbf{g}(\mathbf{x}) = F^{-1}(\widetilde{\mathbf{g}}) = (2\pi)^{-d} \int \mathbf{g}(\mathbf{k})e^{i\mathbf{k}\cdot\mathbf{x}}\, d\mathbf{k}, \tag{3.4}$$

provided, of course, that the integrals on the right-hand sides of the equations are convergent; here $\xi \cdot \mathbf{x} = \xi_1 x_1 + \cdots + \xi_d x_d$, $i = \sqrt{-1}$, and for the tensorial function $\mathbf{g}(\mathbf{x})$ and its Fourier transform $\mathbf{g}(\mathbf{k})$ one uses the same notations with the replacement of an argument $\mathbf{x}$ over $\mathbf{k}$. Performing the Fourier transform of Eq. (3.3) yielding $L_{ijkl}k_jk_lG_{kn}(\mathbf{k}) = \delta_{ij}$, to say, the problem of estimation of the Fourier transform of the Green's tensor is formally reduced to the solution of a linear algebraic system

$$K_{ik}G_{kn}(\mathbf{k}) = \delta_{in}, \quad K_{ik} = L_{ijkl}k_jk_l. \tag{3.5}$$

Then $G_{kn}(\mathbf{k}) = K^*_{kn}/\det\{\mathbf{K}\}$, where $\mathbf{K}^*$ is a cofactor of an element K_{kn} in a symmetric matrix

$$\mathbf{K}(\mathbf{k}) = \begin{pmatrix} K_{11}(\mathbf{k}) & K_{12}(\mathbf{k}) & K_{13}(\mathbf{k}) \\ K_{21}(\mathbf{k}) & K_{22}(\mathbf{k}) & K_{23}(\mathbf{k}) \\ K_{31}(\mathbf{k}) & K_{32}(\mathbf{k}) & K_{33}(\mathbf{k}) \end{pmatrix}, \tag{3.6}$$

and $\det\{\mathbf{K}\}$ is a determinant of the Christoffel matrix $\|K(\mathbf{k})\|$ which is positive define. The matrices $\|\mathbf{K}(\mathbf{k})\|$ and $\|\mathbf{K}^*(\mathbf{k})\|$ are symmetric due to the symmetry of the elastic modulus tensor. The following representations for $\det\{\mathbf{K}\}$ and for $\|\mathbf{K}^*(\mathbf{k})\|$ take place

$$\begin{aligned} \det\{\mathbf{K}\} &= \epsilon_{mnl}K_{m1}(\mathbf{k})K_{n2}(\mathbf{k})K_{l3}(\mathbf{k}), \\ K^*_{ij}(\mathbf{k}) &= \frac{1}{2}\epsilon_{ijl}\epsilon_{jmn}K_{lm}(\mathbf{k})K_{ln}(\mathbf{k}) = K_{im}K_{mj} - K_{mm}K_{ij} \\ &\quad + (\epsilon_{mn1}K_{m2}(\mathbf{k})K_{n2}(\mathbf{k}) + \epsilon_{mn3}K_{m1}(\mathbf{k})K_{n2}(\mathbf{k}))\delta_{ij}. \end{aligned} \tag{3.7}$$

Then, passing on from $\mathbf{k}$-representation to the $\mathbf{x}$-representation of the Green's tensor leads to

$$G_{ij}(\mathbf{x}) = (2\pi)^{-3} \int K^*_{ij}(\mathbf{k})(\det\{\mathbf{K}\})^{-1}e^{i\mathbf{k}\mathbf{x}}\, d\mathbf{k}. \tag{3.8}$$

As can be seen from (3.5)–(3.7), $K_{ij}(\mathbf{k})$ and $\det\{K(\mathbf{k})\}$ are the homogeneous polynomials of the forth and six order, respectively. Then Eq. (3.8) can be presented in the form of the surface integral over the surface of a unit sphere S_1 in the **k**-space

$$G_{ij}(\mathbf{x}) = (8\pi^2)^{-1}\int_{S_1} \delta(\overline{\mathbf{k}}\overline{\mathbf{x}})K^*_{ij}(\overline{\mathbf{k}})(\det\{\mathbf{K}(\overline{\mathbf{k}})\})^{-1}\, ds(\overline{\mathbf{k}}), \tag{3.9}$$

where $\overline{\mathbf{x}} \equiv \mathbf{x}/|\mathbf{x}|$, $\overline{\mathbf{k}} = \mathbf{k}/|\mathbf{k}|$. The representation (3.9), in turn, is reduced to the counter integral over the curve of the intersection $\gamma_1 = \{\boldsymbol{\xi} \in R^3 | \|\boldsymbol{\xi}\| = 1,\ \boldsymbol{\xi}\cdot\mathbf{x} = 0\}$ of the sphere S_1 with the plate orthogonal to the vector $\mathbf{x}$ and passing through the coordinate center

$$G_{ij}(\mathbf{x}) = \frac{1}{8\pi^2|\mathbf{x}|}\oint_{\gamma_1} K_{ij}(\overline{\mathbf{k}})(\det\{\mathbf{K}(\overline{\mathbf{k}})\})^{-1}\, d\gamma_1(\overline{\mathbf{k}}). \tag{3.10}$$

The integrand in Eq. (3.10) has no singularities, and, therefore, the tensorial Green's function can be estimated by a straightforward numerical quadrature, e.g., Romberg integration, that was performed by Barnett [68], who also demonstrated that corresponding contour integral formulae for derivatives of **G** can be derived in a similar manner.

Construction of the Green's function in an analytical form is reduced to finding of the roots the algebraic equation of the six orders with the coefficients depending on L_{ijkl}. The roots of this equation are found in the explicit form only for the particular cases of the isotropic materials and the media with a hexagonal symmetry.

Namely, one obtains in an isotropic case from Eqs. (3.6) and (3.7)

$$\det\{\mathbf{K}(\mathbf{k})\} = \mu^2(\lambda + 2\mu)|\mathbf{k}|^6,$$
$$K_{ij}(\mathbf{k}) = \frac{1}{(2\pi)^3\mu}\Big[\delta_{ij}\int \frac{1}{|\mathbf{k}|^2}e^{i\mathbf{kx}}\, d\mathbf{k} - \frac{\lambda+\mu}{\lambda+2\mu}\int \frac{k_i k_j}{|\mathbf{k}|^4}e^{i\mathbf{kx}}\, d\mathbf{k}\Big]. \tag{3.11}$$

Estimation of integrals in Eq. (3.11) can be carried out by means of differentiation over x_i and x_j of the identity

$$\pi^{-2}\int \mathbf{k}^{-4}e^{i\mathbf{kx}}\, d\mathbf{k} = -|\mathbf{x}|,$$

leading to the relations

$$|\mathbf{x}|_{,ij} = \frac{1}{\pi^2}\int \frac{k_i k_j}{\mathbf{k}^4}e^{i\mathbf{kx}}\, d\mathbf{k}, \quad |\mathbf{x}|_{,ii} = \frac{1}{\pi^2}\int \frac{1}{\mathbf{k}^2}e^{i\mathbf{kx}}\, d\mathbf{k}. \tag{3.12}$$

This allows us to obtain the final representation for the Green's function

$$G_{ij}(\mathbf{x}) = \frac{1}{8\pi\mu}\Big[\delta_{ij}|\mathbf{x}|_{,mm} - \varkappa|\mathbf{x}|_{,ij}\Big] = \frac{1}{4\pi\mu|\mathbf{x}|}[(2-\varkappa)\delta_{ij} + \varkappa n_i n_j], \quad \varkappa = \frac{\lambda+\mu}{\lambda+2\mu}, \tag{3.13}$$

where $n_i = |\mathbf{x}|_{,i} = x_i/|\mathbf{x}|$.

For completeness, we reproduce the compact representation of the Green's function valid for both 2D and 3D cases (see, e.g., [59, 302, 1521])

$$G_{ij}(|\mathbf{x}|) = \frac{1}{4\omega_d(1-\bar{\nu})\mu}\Big[(3-4\bar{\nu})\gamma(|\mathbf{x}|)\delta_{ij} + \frac{1}{|\mathbf{x}|^{d-2}}n_in_j\Big], \tag{3.14}$$

where

$$\gamma(|\mathbf{x}|) = \begin{cases} -\ln|\mathbf{x}|, & d=2, \\ |\mathbf{x}|^{-1}, & d=3, \end{cases}$$

$$\bar{\nu} = \begin{cases} \nu, & \text{for 3D and plane strain,} \\ \frac{\nu}{1+\nu}, & \text{for plane stress problems,} \end{cases} \tag{3.15}$$

and ω_d is the surface area of the unit sphere in R^d.

Construction of the Green's function for the transversally isotropic medium is more complicated. Kröner [811] has obtained a close representation of the Green's function for the case $(L_{1111}L_{3333})^{1/2} - L_{1133} - 2L_{2323} \neq 0$. This problem was analyzed [935, 1619] by the perturbation and Fourier methods, which are generally limited to "small" anisotropy. They found that as $x_1 + x_2 \to 0$ (x_3 is the symmetry axis of a medium), the items in the representation for $G_{ij}(\mathbf{x})$ tend to infinity although the total value of $G_{ij}(\mathbf{x})$ is finite. This drawback is absent in the representation obtained in [1180] by the potential method for an arbitrary transversally isotropic media; the compact form of these formulae was proposed in [754]. An alternative to analytic expansions is direct numerical estimation of the contour integral (3.10) which has an obvious drawback for the problems of micromechanics due to the necessity of calculating the integrand at many points, with a nontrivial amount of work at each evaluation. To avoid this difficulty, one [1634] has evaluated the Green's function and its derivatives by interpolation of precalculated tables. Gray et al. [564] proposed a hybrid analytical–numerical perturbation approach based on an expansion of $\mathbf{G}$ in which the "expansion point" (zero-order term) is an isotropic Green's function. The higher-order contributions are expressed as contour integrals of matrix products and evaluated with a symbolic manipulation program. For moderately anisotropic materials, employing the first few terms in the series provides an accurate solution and a fast computational algorithm, which, however, is not competitive with the method [1634] for strongly anisotropic solids. The basic idea of splitting off an isotropic part has been also used by Schclar [1340], who treated the anisotropic part of the Green's function as a generalized body force.

3.2 The Second Derivative of Green's Function and Related Problems

3.2.1 *The Second Derivative of Green's Function*

It should be mentioned that in micromechanics, the second derivative of the Green's tensor $\nabla\nabla\mathbf{G}$ is more likely to be used than the Green's function $\mathbf{G}$. The first gradient of the Green's function $\nabla\mathbf{G}(\mathbf{x}-\mathbf{y})$ found by formal differentiation are the homogeneous function of degree $1-d$ and locally integrable everywhere. At the same time, the second gradient $\nabla\nabla\mathbf{G}$ is considered the generalized functions and for the construction of the integral regularized representation of the generalized function of the type of derivatives of a homogeneous regular function one usually considers a scheme proposed in [537] (see also [836]). Then the components of the function representation of an even homogeneous distribution $\mathbf{U}(\mathbf{x}) = \nabla\nabla\mathbf{G}$ of the order $-d$ is defined in the form of a sum of the formal $\mathbf{U}^{(f)}(\mathbf{n})$ and singular $\mathbf{U}^{(s)}$ components depending on the shape of the integration domain in the improper integral involved (which is chosen for definiteness sake as a ball, see, e.g., [173, 1378, 1521]) at the regularization of the generalized function $\mathbf{U}(\mathbf{x})$, though itself, $\mathbf{U}(\mathbf{x})$ does not depend on this region shape

$$(G_{ij,kl},\phi) = (G^{(s)}_{ij,kl},\phi) + (G^{(f)}_{ij,kl},\phi), \tag{3.16}$$

$$(G^{(s)}_{ij,kl},\phi) = \int_v G_{ij,kl}(\mathbf{x})\,d\mathbf{x}\phi(\mathbf{0}) = \int_S G_{ij,k}(\mathbf{x})\,ds_m\cdot\phi(\mathbf{0}),$$

$$(G^{(f)}_{ij,kl},\phi) = \int_v G_{ij,kl}(\mathbf{x})[\phi(\mathbf{x})-\phi(\mathbf{0})]\,d\mathbf{x} + \int_{R^3\setminus v} G_{ij,kl}(\mathbf{x})\phi(\mathbf{x})\,d\mathbf{x}, \tag{3.17}$$

where $\phi(\mathbf{x})$ is an arbitrary twice differentiable finite function and V and S are the volume and the surface of integration domain containing an origin of coordinates where the power singularity is located. The integral of the formal constituent of the Green's function converges in the singular point because $\phi(\mathbf{x})-\phi(\mathbf{0})\to 0$ as $\mathbf{x}\to\mathbf{0}$. The surface integral of the singular constituent also converges because the singular point is not located at the integration domain.

Let us obtain the decomposition (3.17) for the ellipsoidal domain v described by the equation

$$\mathbf{x}\mathbf{a}^{-2}\mathbf{x} = x_i(\mathbf{a}^{-2})_{ij}x_j = 1 \tag{3.18}$$

where $\mathbf{a}$ is the second-order symmetric tensor with $\det\mathbf{a} > 0$. According to the regularization (3.17), the distribution

$$U_{kijl}(\mathbf{x}) = [\partial_k\partial_l G_{ij}(\mathbf{x})]_{(ki)(jl)} \tag{3.19}$$

can be presented in the form

$$\mathbf{U}(\mathbf{x}) = \mathbf{U}^{(f)}(\mathbf{x}) - \mathbf{P}\delta(\mathbf{x}), \tag{3.20}$$

where $\mathbf{U}^{(f)}(\mathbf{x}) = \nabla\nabla\mathbf{G}^{(f)}$ is a regular function and $\mathbf{P}$ is a constant tensor

$$\mathbf{P} = -\int_S \nabla\mathbf{G}(\mathbf{s})\mathbf{n}(\mathbf{s})\, ds, \tag{3.21}$$

where $\mathbf{n}(\mathbf{s})$ is an outer normal in the point $\mathbf{s} \in S$ to the surface S of the domain v. Transforming Eq. (3.21) by application of the Gauss theorem (2.23) yields

$$\mathbf{P} = \int \nabla\mathbf{G}(\mathbf{x})\nabla V(\mathbf{x})\, d\mathbf{x}, \tag{3.22}$$

where $V(\mathbf{x})$ is a characteristic function of the domain v. The relation (3.22) can be recast in the form

$$\mathbf{P} = (2\pi)^{-3}\int \mathbf{k}\mathbf{G}(\mathbf{k})\mathbf{k}V(\mathbf{k})\, d\mathbf{k} \equiv -(2\pi)^{-3}\int \mathbf{U}(\mathbf{k})V(\mathbf{k})\, d\mathbf{k}, \tag{3.23}$$

where the Parserval equality was exploited. The affine transform $\mathbf{x} = \mathbf{a}\mathbf{y}$ transforms the ellipsoidal domain v into a unit sphere with a characteristic function $V_1(\mathbf{y}) = V(\mathbf{a}\mathbf{y})$ with a Fourier transform $V_1(\mathbf{a}\mathbf{k}) = (\det\mathbf{a})^{-1}V(\mathbf{k})$. Transition of the integration over the unit sphere in the Fourier space transforms Eq. (3.23) to the form

$$\mathbf{P} = -(2\pi)^{-3}\int \mathbf{U}(\mathbf{a}^{-1}\mathbf{k})V_1(\mathbf{k})\, d\mathbf{k}. \tag{3.24}$$

Integration in Eq. (3.24) over the variable $\mathbf{k} = |\mathbf{k}|\bar{\mathbf{k}}$ is carried out over the surface of unit sphere $\bar{\mathbf{k}}^2 = 1$ and over the scalar $|\mathbf{k}|$. Obviously, $\mathbf{U}(\mathbf{a}^{-1}\mathbf{k}) = \mathbf{U}(\mathbf{a}^{-1}\bar{\mathbf{k}})$ and $V_1(\mathbf{k}) = V_1(|\mathbf{k}|)$ as far as $\mathbf{U}(\mathbf{k})$ is a homogeneous function of zero order, and $V_1(\mathbf{k})$ is a spherically symmetric, and, therefore,

$$\mathbf{P} = -\frac{1}{4\pi}\int \mathbf{U}(\mathbf{a}^{-1}\bar{\mathbf{k}})\, d\bar{\mathbf{k}}, \tag{3.25}$$

where the integration is performed over the unit sphere.

It might be well to point out that the constant tensor $\mathbf{P}$ (3.21) depends on the shape (but not on the size) of the ellipsoid v because $\mathbf{U}(\mathbf{k})$ is a homogeneous function of zero order. Then the integral over v in the definition of the formal constituent (3.17) does not depend on the size of v and vanishes at the shrinking of the ellipsoid v into the point $\mathbf{x} = \mathbf{0}$. Thus, one can define

$$(\mathbf{U}^{(f)}, \phi) = \det\{\mathbf{a}\} \fint \mathbf{U}(\mathbf{a}\mathbf{x})\phi(\mathbf{a}\mathbf{x})\, d\mathbf{x}, \tag{3.26}$$

where the integral in the Cauchy principle sense exists due to the existence of $(\mathbf{U}^{(f)}, \phi)$.

In the particular case of an isotropic medium, we formally note the first and the second derivatives of the Green's function (3.13) in the form

$$G_{ij,k}(\mathbf{x}) = \frac{1}{8\pi\mu}(\delta_{ij}|\mathbf{x}|_{,ppk} - \varkappa|\mathbf{x}|_{,ijk}),$$
$$G_{ij,kl}(\mathbf{x}) = \frac{1}{8\pi\mu}(\delta_{ij}|\mathbf{x}|_{,ppkl} - \varkappa|\mathbf{x}|_{,ijkl}), \tag{3.27}$$

where the derivatives from $|\mathbf{x}|$ until the third order can be directly calculated as follows:

$$|\mathbf{x}|_{,i} = n_i \equiv \frac{x_i}{|\mathbf{x}|}, \quad |\mathbf{x}|_{,ij} = n_{i,j} = \frac{1}{|\mathbf{x}|}(\delta_{ij} - n_{ij}), \quad |\mathbf{x}|_{,pp} = \frac{2}{|\mathbf{x}|},$$
$$|\mathbf{x}|_{,ijk} = n_{i,jk} = \frac{1}{|\mathbf{x}|^2}(3n_{ijk} - n_i\delta_{jk} - n_k\delta_{ik} - n_l\delta_{ik}), \quad |\mathbf{x}|_{,ppk} = -\frac{2}{|\mathbf{x}|^2}n_k, \tag{3.28}$$

The finding of the higher-order derivatives is related to the exploiting the regularization (3.16): $|\mathbf{x}|_{,ijkl} = |\mathbf{x}|^{(s)}_{,ijkl} + |\mathbf{x}|^{(f)}_{,ijkl}$. Then for the spherical domain in the regularization (3.16), we get

$$|\mathbf{x}|^{(s)}_{,ijkl} = \frac{-8\pi}{15}\delta(|\mathbf{x}|)\delta_{ijkl}, \quad |\mathbf{x}|^{(s)}_{,ppkl} = \frac{-8\pi}{3}\delta(|\mathbf{x}|)\delta_{kl}, \quad |\mathbf{x}|^{(s)}_{,ppqq} = -8\pi\delta(|\mathbf{x}|),$$
$$|\mathbf{x}|^{(s)}_{,ijkl} = n_{i,jkl} = \frac{1}{|\mathbf{x}|^3}\Big[3(n_{ij}\delta_{kl} + n_{ik}\delta_{jl} + n_{il}\delta_{jk} + n_{kl}\delta_{ij} + n_{jl}\delta_{ik} + n_{jk}\delta_{il})$$
$$-\delta_{ijkl} - 15n_{ijkl}\Big], \quad |\mathbf{x}|^{(f)}_{,ppkl} = \frac{2}{|\mathbf{x}|^3}(3n_{kl} - \delta_{kl}), \quad |\mathbf{x}|^{(f)}_{,ppii} = 0. \tag{3.29}$$

Calculation of the singular constituent $|\mathbf{x}|^{(s)}_{ijkl}$ is carried out exploiting the transfer to a spherical coordinate system

$$(|\mathbf{x}|^{(s)}_{,ijkl}, \phi) = \int |\mathbf{x}|^{(s)}_{,ijk} n_l |\mathbf{x}|^2 d\omega\phi(\mathbf{0}).$$

The $\phi(\mathbf{0})$ in the singular constituent of derivatives is stipulated by the known relation $(\delta, \phi) = \phi(\mathbf{0})$.

The determined values of derivatives (3.29) allow us to represent the formulae (3.27) in the closed form

$$G_{ij,k} = -\frac{1}{8\pi\mu|\mathbf{x}|^2}[3\varkappa n_{ijk} + (2-\varkappa)n_k\delta_{ij} - \varkappa(n_i\delta_{jk} + n_j\delta_{ik})],$$
$$G^{(s)}_{ij,kl} = -\frac{1}{3\mu}\delta(|\mathbf{x}|)\Big[\delta_{ij}\delta_{kl} - \frac{\varkappa}{5}\delta_{ijkl}\Big],$$

$$G^{(f)}_{ij,kl} = \frac{1}{8\pi\mu|\mathbf{x}|^3}\Big\{(2-\varkappa)\delta_{ij}(3n_{kl}-\delta_{kl}) + \varkappa[2I_{ijkl} + 15n_{ijkl} \\ -3(n_{ij}\delta_{kl} + n_{ik}\delta_{jl} + n_{il}\delta_{jk} + n_{jl}\delta_{ik} + n_{jk}\delta_{il})]\Big\} \quad (3.30)$$

For completeness, we reproduce the compact representations of these tensors valid for both 2D and 3D cases of the isotropic media

$$G_{ij,k}(\mathbf{x}) = \frac{1}{4\omega_d(1-\bar{\nu})\mu|\mathbf{x}|^{d-1}}\Big[(4\bar{\nu}-3) + \delta_{ij}n_k + \delta_{ik}n_j \\ = \delta_{jk}n_i + (d-4)n_i n_j n_k\Big], \quad (3.31)$$

$$\mathbf{U}^{(s)}(\mathbf{x}) = \frac{1}{(d+2)[dk^{(0)} + 2(d-1)\mu^{(0)}]}\Big[\frac{k^{(0)}+2\mu^{(0)}}{\mu^{(0)}}d\mathbf{E}^{(1)} - \frac{\alpha}{d}\mathbf{E}^{(2)}\Big]\delta(\mathbf{x}), \quad (3.32)$$

$$\mathbf{U}^{(f)} = \sum_{p=1}^{6} b^U_p \mathbf{E}^{(p)}, \quad (3.33)$$

where the values b^U_p $(p = 1, 2, \ldots, 6)$ are defined by the equations

$$b^U_1 = \frac{-2d}{2\omega_d[dk^{(0)} + 2(d-1)\mu^{(0)}]|\mathbf{x}|^d}, \quad b^U_2 = \frac{\alpha}{2\omega_d[dk^{(0)} + 2(d-1)\mu^{(0)}]|\mathbf{x}|^d}, \quad (3.34)$$

$$b^U_3 = b^U_4 = \frac{-d\alpha}{2\omega_d[dk^{(0)} + 2(d-1)\mu^{(0)}]|\mathbf{x}|^d}, \quad (3.35)$$

$$b^U_5 = \frac{-2d(d-2)\alpha}{2\omega_d[dk^{(0)} + 2(d-1)\mu^{(0)}]|\mathbf{x}|^d}, \quad b^U_6 = \frac{d(d+2)\alpha}{2\omega_d[dk^{(0)} + 2(d-1)\mu^{(0)}]|\mathbf{x}|^d}, \quad (3.36)$$

with the parameters ω_d and $\bar{\nu}$ as defined in Eqs. (3.14), $\alpha = dk^{(0)}/\mu^{(0)} + (d-2)$ is a dimensionless parameter, and the tensors $\mathbf{E}^{(p)}$ are the basic tensors of the special structure that depend on the direction $\mathbf{n}$ (see Sect. 21.2).

3.2.2 *The Tensors Related to the Green's Function*

Analogously for the distribution (called the fundamental stress tensor)

$$\boldsymbol{\Gamma}(\mathbf{x}) = -\mathbf{L}(\mathbf{I}\delta(\mathbf{x}) + \mathbf{U}(\mathbf{x})\mathbf{L}), \quad \Gamma_{ijkl}(\mathbf{x}) = -L_{ijkl}\delta(\mathbf{x}) - L_{ijmn}U_{mnpq}(\mathbf{x})L_{pqkl} \quad (3.37)$$

we can define the regular and singular constituents

$$\boldsymbol{\Gamma}(\mathbf{x}) = \boldsymbol{\Gamma}^{(f)}(\mathbf{x}) + \boldsymbol{\Gamma}^{\delta}(\mathbf{x}) = \boldsymbol{\Gamma}^{(f)}(\mathbf{x}) - \mathbf{Q}\delta(\mathbf{x}), \\ (\boldsymbol{\Gamma}^{(f)}, \phi) = \det\{\mathbf{a}\} \fint \boldsymbol{\Gamma}(\mathbf{ax})\phi(\mathbf{ax})\, d\mathbf{x}, \quad \mathbf{Q} = -\frac{1}{4\pi}\int \boldsymbol{\Gamma}(\mathbf{a}^{-1}\bar{\mathbf{k}})\, d\bar{\mathbf{k}}, \quad (3.38)$$

and the integration in Eq. (3.26) is carried out over the unit sphere $\bar{\mathbf{k}}^2 = 1$, and $\mathbf{Q} = \mathbf{L} - \mathbf{LPL}$.

In parallel with the tensor of "fundamental displacement" **G**, we need the associated tensor of the the fundamental stresses, tractions (called also Kupradze tensor), and strains

$$\mathbf{D}(\mathbf{x}-\mathbf{s}) = \mathbf{L}\nabla\mathbf{G}(\mathbf{x}-\mathbf{s}), \quad \mathbf{T}(\mathbf{x},\mathbf{s}) = -\mathbf{n}^S(\mathbf{x})\mathbf{D}(\mathbf{x}-\mathbf{s}),$$
$$\boldsymbol{\mathcal{E}}(\mathbf{x}-\mathbf{s}) = \nabla\mathbf{G}(\mathbf{x}-\mathbf{s}), \tag{3.39}$$

and $\mathbf{S}(\mathbf{x},\mathbf{s}) = \mathbf{L}\nabla\mathbf{T}^\top(\mathbf{x},\mathbf{s})$. In particular, for isotropic medium, we reproduce the compact representation of the mentioned Green functions valid for both 2D and 3D cases (see also, e.g., [59, 1523]) at $\mathbf{x} \neq \mathbf{s}$

$$T_{ij}(\mathbf{x},\mathbf{s}) = -2\mu C\big[\big((1-2\bar\nu)\delta_{ij} + dn_in_j\big)n_kn_k^S(\mathbf{x}) + (1-2\bar\nu)(n_in_j^S(\mathbf{x}) - n_jn_i^S(\mathbf{x}))\big], \tag{3.40}$$
$$D_{ijk}(r) = -2\mu C[(1-2\bar\nu)(\delta_{ik}n_j + \delta_{jk}n_i - \delta_{ij}n_k) + dn_in_jn_k], \tag{3.41}$$
$$\mathcal{E}_{ijk}(r) = -C[(3-4\bar\nu)\delta_{ij}n_k - \delta_{ik}n_j - \delta_{jk}n_i + dn_in_jn_k], \tag{3.42}$$
$$S_{ijk}(\mathbf{x},\mathbf{s}) = 4\mu^2C/r\Big\{d\big[(1-2\bar\nu)\delta_{ij}n_k + \bar\nu(\delta_{ik}n_j + \delta_{jk}n_i) - (d+2)n_in_jn_k\big] \times n_mn_m^S(\mathbf{x}) - (1-4\bar\nu)\delta_{ij}n_k^S(\mathbf{x}) + d\bar\nu n_k[n_in_j^S(\mathbf{x}) + n_jn_i^S(\mathbf{x})] + (1-2\bar\nu)[dn_in_jn_k^S(\mathbf{x}) + \delta_{ik}n_j^S(\mathbf{x}) + \delta_{jk}n_i^S(\mathbf{x})]\Big\}, \tag{3.43}$$

respectively, where $\gamma(r)$ and $\bar\nu$ are defined by Eq. (3.15), $C = [4\omega_d\mu(1-\bar\nu)r^{d-1}]^{-1}$, ω_d is the surface area of the unit sphere in R^d, and the tensors $\mathbf{T}(\mathbf{x},\mathbf{s})$ and $\mathbf{S}(\mathbf{x},\mathbf{s})$ are defined at the surface $\mathbf{s} \in S$ with the unit outward normal $\mathbf{n}^S = (n_1^S, \ldots, n_d^S)^\top$ at $\mathbf{s} \in S$, $r \equiv |\mathbf{x}-\mathbf{s}|$, $r_{,i} = r_i/r = n_i$.

3.3 Dynamic Green's and Related Functions

By analogy with the static Green's tensor (3.3), we introduce its generalization to dynamic problems. For the differential equation of motion

$$\boldsymbol{\mathcal{L}}\mathbf{u} = -\mathbf{f}, \quad (\boldsymbol{\mathcal{L}})_{ik}(\mathbf{x},t) \equiv \nabla_jL_{ijkl}\nabla_l - \delta_{ik}\rho\frac{\partial^2}{\partial t^2}, \tag{3.44}$$

where $\mathbf{u}(\mathbf{x})|_{|\mathbf{x}|\to\infty} = \mathbf{0}$ and the Green's tensor for the displacement $G_{ij}(\mathbf{x}, t)$ in the infinite medium with density ρ is defined from the relation $(\mathbf{x}, \mathbf{y} \in R^d,\ d = 2, 3)$:

$$u_k = G_{ki} * f_i \equiv \iint_{-\infty}^{\infty} G_{ki}(\mathbf{x} - \mathbf{y}, t - t') f_i(\mathbf{y})\, dt' \tag{3.45}$$

assuming vanishing of $\mathbf{G}(\mathbf{x}, t)$ at the infinity $|\mathbf{x}| \to \infty$. The differential equation for $\mathbf{G}$ is produced by the action of the operator $(\boldsymbol{\mathcal{L}})_{ik}$ on Eq. (3.45):

$$\boldsymbol{\mathcal{L}}\mathbf{G}(\mathbf{x}, t) = -\boldsymbol{\delta}\delta(\mathbf{x})\delta(t), \ \ \mathcal{L}_{ik}G_{kj}(\mathbf{x}, t) = -\delta_{ij}\delta(\mathbf{x})\delta(t), \ \ (t \geq 0), \tag{3.46}$$

and $\mathbf{G}(\mathbf{x}, t) = \mathbf{0},\ G_{kj}(\mathbf{x}, t) = 0,\ \ (t < 0)$. Thus, the kj-component of the symmetric Green's tensor $G_{kj}(\mathbf{x}, t)$ defines the displacement component in the x_k-direction at point $\mathbf{x}$ and time t, generated by a unit impulsive force applied in the x_j direction at point $\mathbf{x} = \mathbf{0}$ and time $t = 0$. Addition of the second equation (3.46_2) is stipulated by a causality principle, according to which the response in the point $\mathbf{x}$ should arise only after applying the unit impulse in the origin.

The construction of the dynamic Green's tensor is carried out by the use of the Fourier transforms over the coordinates and time (for details, see [1104])

$$\mathbf{G}(\mathbf{k}, \omega) = \iint_{-\infty}^{\infty} \mathbf{G}(\mathbf{x}, t) e^{i(\mathbf{k}\cdot\mathbf{x} - \omega t)}\, d\mathbf{x}\, dt,$$

$$\mathbf{G}(\mathbf{x}, t) = \frac{1}{(2\pi)^{-d}} \iint_{-\infty}^{\infty} \mathbf{G}(\mathbf{k}, \omega) e^{-i(\mathbf{k}\cdot\mathbf{x} - \omega t)}\, d\mathbf{k}\, d\omega.$$

Performing the Fourier transform of Eq. (3.46) yields the algebraic equation

$$L_{ijkl}k_j k_l G_{kn}(\mathbf{k}, \omega) - \rho\omega^2 G_{kn}(\mathbf{k}, \omega)\delta_{ik} = \delta_{in}, \tag{3.47}$$

whence it follows that

$$G_{ij}(\mathbf{k}, \omega) = K^*_{ij}(\mathbf{k}, \omega)/|\mathbf{K}(\mathbf{k}, \omega)|, \tag{3.48}$$

where $\mathbf{K}^*$ is a cofactor of an element K_{kn} in a symmetric matrix

$$\mathbf{K}(\mathbf{k}, \omega) = \begin{pmatrix} K_{11}(\mathbf{k}) - \rho\omega^2 & K_{12}(\mathbf{k}) & K_{13}(\mathbf{k}) \\ K_{21}(\mathbf{k}) & K_{22}(\mathbf{k}) - \rho\omega^2 & K_{23}(\mathbf{k}) \\ K_{31}(\mathbf{k}) & K_{32}(\mathbf{k}) & K_{33}(\mathbf{k}) - \rho\omega^2 \end{pmatrix}, \tag{3.49}$$

with the following representation of the tensor $\mathbf{K}^*$ and the determinant $|\mathbf{K}(\mathbf{k}, \omega)|$

$$K^*_{ij} = K_{im}K_{mj} + (\rho\omega^2)K_{ij} + [(\rho\omega^2)^2 - \rho\omega^2 A + B]\delta_{ij},$$

$$|\mathbf{K}(\mathbf{k}, \omega)| = -(\rho\omega^2)^3 + (\rho\omega^2)^2 A - \rho\omega^2 B + C, \tag{3.50}$$

where

$$A = K_{mm}, \quad C = \epsilon_{mnl} K_{m1} K_{n2} K_{l3},$$
$$B = \epsilon_{mn1} K_{m2} K_{n3} + \epsilon_{mn2} K_{m3} K_{n1} + \epsilon_{mn3} K_{m1} K_{n2}.$$

By virtue of the fact that the equation $|\mathbf{K}(\mathbf{k}, \omega)| = 0$ contains the degrees of $\rho\omega^2$, it has the roots $\rho\omega_1^2$, $\rho\omega_2^2$, $\rho\omega_3^2$. Then removing the parentheses in the representation $|\mathbf{K}(\mathbf{k}, \omega)| = \rho^3(\omega_1^2 - \omega^2)(\omega_2^2 - \omega^2)(\omega_3^2 - \omega^2)$ and comparison of the relation obtained with Eq. (3.50$_1$) leads to $A = \rho(\omega_1^2 + \omega_2^2 + \omega_3^2)$, $C = \rho^3\omega_1^2\omega_2^2\omega_3^2$, and $B = \rho^2(\omega_1^2\omega_2^2 + \omega_2^2\omega_3^2 + \omega_1^2\omega_3^2)$. Then Eq. (3.48) can be presented in the form

$$G_{ij}(\mathbf{k}, \omega) = \sum_{\nu=1}^{3} \frac{{}^{\nu}\phi_i {}^{\nu}\phi_j}{\rho(\omega_\nu^2 - \omega^2)}, \tag{3.51}$$

where

$${}^1\phi_i {}^1\phi_j = \frac{(K_{im} - \rho\omega_2^2\delta_{im})(K_{mj} - \rho\omega_3^2\delta_{mj})}{\rho^2(\omega_1^2 - \omega_2^2)(\omega_1^2 - \omega_3^2)} \tag{3.52}$$

and ${}^2\phi_i$, ${}^2\phi_j$, ${}^3\phi_i$, ${}^3\phi_j$ are obtained by cyclic permutation of indexes $(1, 2, 3)$. ${}^{\nu}\phi_i$ and $\rho\omega_\nu^2$ are the eigenvectors and eigenvalues, respectively, of the matrix $\|\mathbf{K}(\mathbf{k}, \omega)\|$ (3.49):

$$(K_{ij}(\mathbf{k}, \omega) - \rho\omega_\nu^2\delta_{ij}){}^{\nu}\phi_j = 0, \quad \sum_{\nu=1}^{3} {}^{\nu}\phi_i {}^{\nu}\phi_j = \delta_{ij}. \tag{3.53}$$

By use of the inverse Fourier transform, one can find from Eq. (3.51)

$$G_{ij}(\mathbf{x}, t) = \frac{1}{(4\pi)^4\rho} \int_{-\infty}^{\infty} \sum_{\nu=1}^{3} \frac{\exp(-i\omega t)}{\omega_\nu^2 - \omega^2}\, d\omega \int_0^{\infty} |\mathbf{k}|^2 d|\mathbf{k}|$$
$$\times \int_{S_1} {}^{\nu}\phi_i(\bar{\mathbf{k}}){}^{\nu}\phi_j(\bar{\mathbf{k}}) \exp(i\mathbf{k} \cdot \mathbf{x})\, d\mathbf{s}(\bar{\mathbf{k}}), \tag{3.54}$$

where S_1 is the surface of the unit sphere $S_1 = \{\mathbf{k} | |\bar{\mathbf{k}}| = 1\}$, $\bar{\mathbf{k}} \equiv \mathbf{k}/|\mathbf{k}|$, and it is taken into account that ${}^{\nu}\phi_i {}^{\nu}\phi_j$ are the homogeneous functions of $\mathbf{k}$ of degree 0.

We present only the final relation for the integral (3.54) obtained by the use of the residue theory:

$$G_{ij}(\mathbf{x}, t) = \frac{-H(t)}{8\pi^2\rho} \frac{\partial}{\partial t} \sum_{\nu=1}^{3} \int_{S_1} \delta(\bar{\omega}_\nu t - \bar{\mathbf{k}} \cdot \mathbf{x}){}^{\nu}\phi_i(\bar{\mathbf{k}}){}^{\nu}\phi_j(\bar{\mathbf{k}})\bar{\omega}_\nu^{-2}(\bar{\mathbf{k}})\, d\mathbf{s}(\bar{\mathbf{k}}), \tag{3.55}$$

where $\bar{\omega}_\nu = \omega_\nu/|\mathbf{k}|$, and $H(t)$ is the Heaviside step function.

Let us consider the case of an isotropic medium when

$$
\begin{aligned}
K_{ij}(\mathbf{k}) &= (\lambda+\mu)k_i k_j - \mu\delta_{ij}|\mathbf{k}|^2,\\
|\mathbf{K}| &= (\mu|\mathbf{k}|^2 - \rho\omega^2)[(\lambda+2\mu)|\mathbf{k}|^2 - \rho\omega^2],\\
\rho\omega_1^2 &= (\lambda+2\mu)|\mathbf{k}|^2, \quad \rho\omega_2^2 = \rho\omega_3^2 = \mu|\mathbf{k}|^2,
\end{aligned}
\tag{3.56}
$$

and we get from (3.52) and (3.53):

$$
{}^1\phi_i{}^1\phi_j = k_i k_j/|\mathbf{k}|^2, \quad {}^2\phi_i{}^2\phi_j + {}^3\phi_i{}^3\phi_j = \frac{\delta_{ij}|\mathbf{k}|^2 - k_i k_j}{|\mathbf{k}|^2}. \tag{3.57}
$$

The representation (3.55) can be simplified after substitution of Eqs. (3.57):

$$
\begin{aligned}
G_{ij}(\mathbf{x},t) = \frac{-H(t)}{8\pi^2\rho}\frac{\partial}{\partial t}\bigg\{&\int_{S_1}\delta(c_{\rm L}t - \bar{\mathbf{k}}\cdot\mathbf{x})\bar{k}_i\bar{k}_j c_{\rm L}^{-2}\,d\mathbf{s}(\bar{\mathbf{k}})\\
&+\int_{S_1}\delta(c_{\rm T}t - \bar{\mathbf{k}}\cdot\mathbf{x})(\delta_{ij} - \bar{k}_i\bar{k}_j)c_{\rm T}^{-2}\,d\mathbf{s}(\bar{\mathbf{k}})\bigg\},
\end{aligned}
\tag{3.58}
$$

where one introduces the notations

$$
c_{\rm L} = \sqrt{(\lambda+2\mu)/\rho}, \quad c_{\rm T} = \sqrt{\mu/\rho}, \tag{3.59}
$$

and uses the integrals

$$
\int_{S_1}\delta(c_{\rm L}t - \bar{\mathbf{k}}\cdot\mathbf{x})\bar{k}_i\bar{k}_j ds(\bar{\mathbf{k}}) = |\mathbf{x}|^{-1}H(|\mathbf{x}| - c_{\rm L}t)\int_0^{2\pi}\bar{k}_i\bar{k}_j\,d\varphi.
$$

When $c_{\rm L}t - \bar{\mathbf{k}}\cdot\mathbf{x} = 0$, we have

$$
\begin{aligned}
\bar{k}_i = c_{\rm L}t\bar{x}_i/|\mathbf{x}| + \eta_i, \quad &\int_0^{2\pi}\eta_i\eta_j\,d\varphi = \pi(\delta_{ij} - \bar{x}_i\bar{x}_j)(1 - c_{\rm L}^2t^2|\mathbf{x}|^{-2}),\\
\int_0^{2\pi}\bar{\xi}_i\bar{\xi}_j\,d\varphi = 2\pi(c_{\rm L}t/|\mathbf{x}|)^2\bar{x}_i\bar{x}_j &+ \pi(\delta_{ij} - \bar{x}_i\bar{x}_j)(1 - c_{\rm L}^2t^2|\mathbf{x}|^{-2}),
\end{aligned}
\tag{3.60}
$$

and finally

$$
\begin{aligned}
G_{ij}(\mathbf{x},t) = \frac{H(t)}{4\pi\rho|\mathbf{x}|}\Big[&\delta(|\mathbf{x}| - c_{\rm L}t)\bar{x}_i\bar{x}_j c_{\rm L}^{-1} + \delta(|\mathbf{x}| - c_{\rm T}t)(\delta_{ij} - \bar{x}_i\bar{x}_j)c_{\rm T}^{-1}\\
&+t|\mathbf{x}|^{-2}(\delta_{ij} - 3\bar{x}_i\bar{x}_j)\{H(|\mathbf{x}| - c_{\rm L}t) - H(|\mathbf{x}| - c_{\rm T}t)\}\Big].
\end{aligned}
\tag{3.61}
$$

As can be seen, c_{L} and c_{T} introduced by the equalities (3.59) have the meaning of the longitudinal and transverse wave speeds, respectively. The summand with a cofactor $\delta(|\mathbf{x}|-c_{\mathrm{T}}t)$ (3.61) describes the impulses where the displacements are perpendicular to the direction of its propagation, and the summand with the cofactor $\delta(\mathbf{x}|-c_{\mathrm{L}}t)$ corresponds to the impulses with the transversal direction of displacement. The last summand in Eq. (3.61) describes the perturbation arising between the arrival of impulses produced the longitudinal and transverse displacements.

From Eqs. (3.48) and (3.49), the following relationships between the dynamic and static tensors in the Fourier space $\mathbf{K}(\mathbf{k},0) = \mathbf{K}(\mathbf{k})$, $|\mathbf{K}(\mathbf{k},0)| = |\mathbf{K}(\mathbf{k})|$ follow, and therefore, according to (3.48) and (3.55), $\mathbf{G}(\mathbf{k},0) = \mathbf{G}(\mathbf{k})$ leading to the connection between the static and dynamic Green's tensors

$$\mathbf{G}(\mathbf{x}) = \int_{-\infty}^{\infty} \mathbf{G}(\mathbf{x},t)\,dt. \tag{3.62}$$

We will demonstrate employing (3.62) that for the isotropic medium Eq. (3.61) immediately yields the representation (3.13) for the static Green's tensor. Actually, integration of Eq. (3.61) over time leads to

$$G_{ij}(\mathbf{x}) = \frac{1}{8\pi\rho|\mathbf{x}|}\left[\frac{2}{c_{\mathrm{T}}^2}\delta_{ij} + \left(\frac{1}{c_{\mathrm{L}}^2} - \frac{1}{c_{\mathrm{T}}^2}\right)(\delta_{ij} - \bar{x}_i\bar{x}_j)\right] \tag{3.63}$$

coinciding with formula (3.13) according to Eqs. (3.59).

Now we will consider the widely occurring case of dynamic process such as the wave or steady-state process when the displacements in the medium are described by the relation $\mathbf{u}(\mathbf{x},t) = \mathbf{u}(\mathbf{x})\exp(-i\omega t)$ where $\mathbf{u}(\mathbf{x})$ is the amplitude satisfying to the Helmholtz equation

$$\mathcal{L}^{\omega}(\mathbf{x})\mathbf{u}(\mathbf{x}) = -\mathbf{f}(\mathbf{x}),\ \mathcal{L}^{\omega}_{ik} = \delta_{ik}\rho\omega^2 + \nabla_j L_{ijkl}\nabla_l. \tag{3.64}$$

If a unit body force applied at the origin of coordinates in the x_m-direction with time dependence $\exp(-i\omega t)$, then the ith component of a displacement at $\mathbf{x}$ can be presented in the form $g_{im}(\mathbf{x})\exp(i\omega t)$ where $g_{im}(\mathbf{x})$, called the wave Green's tensor, satisfies the motion equation

$$L_{ijkl}g_{km,lj}(\mathbf{x}) + \rho\omega^2 g_{jm}(\mathbf{x})\delta_{ij} + \delta_{im}\delta(\mathbf{x}) = 0. \tag{3.65}$$

To find the solution of Eq. (3.65), we will apply to the obtained value $\mathbf{G}(\mathbf{k},\omega)$ (3.51) the inverse Fourier transform only over the coordinates

$$g_{ij}(\mathbf{x}) = (2\pi)^{-3}\int_{\infty}^{\infty} G_{ij}(\mathbf{k},\omega)\exp(i\mathbf{k}\cdot\mathbf{x})\,d\mathbf{k}. \tag{3.66}$$

For the isotropic medium, we have from Eq. (3.61):

$$G_{ij}(\mathbf{k},\omega) = \frac{k_i k_j}{|\mathbf{k}|^2(\lambda+2\mu)(|\mathbf{k}|^2-k_{\rm L})} + \frac{\delta_{ij}|\mathbf{k}|^2 - k_i k_j}{|\mathbf{k}|\mu(|\mathbf{k}|^2-k_{\rm T})}, \tag{3.67}$$

where $k_{\rm L} = \omega/c_{\rm L}$ and $k_{\rm T} = \omega/c_{\rm T}$ are the wave numbers. Substitution of Eq. (3.67) into (3.46) leads to the final representation for the time-reduced wave Green's tensor for the isotropic medium:

$$g_{ij}(\mathbf{x}) = \frac{1}{4\pi\rho\omega^2|\mathbf{x}|}\left\{k_{\rm T}^2\delta_{ij}e^{ik_{\rm T}|\mathbf{x}|} - |\mathbf{x}|\frac{\partial^2}{\partial x_i\partial x_j}\left[\frac{\exp(ik_{\rm L}|\mathbf{x}|)}{|\mathbf{x}|} - \frac{\exp(ik_{\rm T}|\mathbf{x}|)}{|\mathbf{x}|}\right]\right\}. \tag{3.68}$$

3.4 Inhomogeneity in an Elastic Medium

3.4.1 General Case of Inhomogeneity in an Elastic Medium

Let an inhomogeneity v with an arbitrary shape and characteristic function V be included in an infinite homogeneous matrix. Stresses and strains are related to each other via the constitutive equations $\boldsymbol{\sigma}(\mathbf{x}) = \mathbf{L}(\mathbf{x})\boldsymbol{\varepsilon}(\mathbf{x}) + \boldsymbol{\alpha}(\mathbf{x})$ or $\boldsymbol{\varepsilon}(\mathbf{x}) = \mathbf{M}(\mathbf{x})\boldsymbol{\sigma}(\mathbf{x}) + \boldsymbol{\beta}(\mathbf{x})$ (2.126). All tensors of material properties $\mathbf{g}$ (e.g., $\mathbf{g} = \mathbf{L}, \mathbf{M}, \boldsymbol{\alpha}, \boldsymbol{\beta}$) are decomposed as

$$\mathbf{g} \equiv \mathbf{g}^{(0)} + \mathbf{g}_1(\mathbf{x}) = \mathbf{g}^{(0)} + \mathbf{g}_1(\mathbf{x})V(\mathbf{x}), \tag{3.69}$$

where $\mathbf{g}^{(0)}(\mathbf{x}) \equiv \mathbf{g}^{(0)}$ =const in the matrix ($\mathbf{x} \in v^{(0)}$). For the homogeneous inclusion, the material properties are described by the step functions

$$\mathbf{g}(\mathbf{x}) = \begin{cases} \mathbf{g}^{(0)}, & \mathbf{x} \in v^{(0)}, \\ \mathbf{g}^{(0)} + \mathbf{g}_1, & \mathbf{x} \in v. \end{cases} \tag{3.70}$$

It is assumed that $\mathbf{L}(\mathbf{x})$ and $\mathbf{L}^{(0)}$ are positive definite. To hold this property in the cases of a cavity and rigid inclusions, the latter are be modeled by the use of limiting passing ($\mathbf{x} \in v$): $\mathbf{L}_1(\mathbf{x}) \to -\mathbf{L}^{(0)}$, $\mathbf{M}_1(\mathbf{x}) \to \mathbf{0}$, for a cavity and $\mathbf{L}_1(\mathbf{x}) \to \infty$, $\mathbf{M}_1(\mathbf{x}) \to -\mathbf{M}^{(0)}$, for a rigid inclusions.

The initial equilibrium equation for the displacement $\mathbf{u}(\mathbf{x})$ has form

$$\mathcal{L}\mathbf{u} + \nabla\boldsymbol{\alpha} \equiv (\mathcal{L}^{(0)} + \mathcal{L}_1)\mathbf{u} + \nabla\boldsymbol{\alpha} = -\mathbf{f},$$

$$\nabla_j[\mathbf{L}_{ijkl}(\mathbf{x})\nabla_l u_k(\mathbf{x}) + \alpha_{lk}(\mathbf{x})] = -f_i(\mathbf{x}), \tag{3.71}$$

where $\mathcal{L}^{(0)} = \nabla\mathbf{L}^{(0)}\nabla,\ \mathcal{L}_1 = \nabla\mathbf{L}_1\nabla$ and the density of external forces is assumed to be a generalized function, decreasing sufficiently rapidly at infinity. The body force tensor $\mathbf{f}$ can be generated, e.g., by either gravitational loads or a centrifugal load. If $\mathbf{f}$ has no singularities of the type of the single or double layer assumed hereafter on the boundary S of the inhomogeneity v, Eq. (3.71) automatically provides conjunction conditions at the boundary $\mathbf{x} \in S$

$$\mathbf{u}^-(\mathbf{x}^-) = \mathbf{u}^+(\mathbf{x}^+),\ \ \mathbf{n}\boldsymbol{\sigma}^-(\mathbf{x}^-) = \mathbf{n}\boldsymbol{\sigma}^+(\mathbf{x}^+),\ \ \mathbf{x} \in S, \tag{3.72}$$

where $\mathbf{g}^-$ and $\mathbf{g}^+$ ($\mathbf{g} = \mathbf{u}, \boldsymbol{\sigma}$) are the limiting values of the function $\mathbf{g}$ outside and inside, respectively, near the boundary S

$$\mathbf{g}^-(\mathbf{x}^-) = \lim_{\mathbf{y}} \to \mathbf{x}\mathbf{g}(\mathbf{y}),\ \ \ \mathbf{g}^+(\mathbf{x}^+) = \lim_{\mathbf{z}} \to \mathbf{x}\mathbf{g}(\mathbf{z}), \tag{3.73}$$

where $\mathbf{n}\perp S$ is an outward normal vector on S at the point $\mathbf{x} \in S$, and $\mathbf{y} \in v^{(0)},\ \mathbf{z} \in v$.

Let $\mathbf{u}^0(\mathbf{x})$ be a displacement field that would be present in a homogeneous medium ($\mathbf{L}_1(\mathbf{x}) = \boldsymbol{\beta}_1(\mathbf{x}) \equiv \mathbf{0}$) under the action of the body force $\mathbf{f}^0$:

$$\mathcal{L}\mathbf{u}^0 = -\mathbf{f}^0. \tag{3.74}$$

This field tends to zero at infinity and can be uniquely presented in the form

$$\mathbf{u}^0 = \mathbf{G} * \mathbf{f}^0,\ u_i^0(\mathbf{x}) = \int G_{ij}(\mathbf{x}-\mathbf{y}) f_j^0(\mathbf{x})\, d\mathbf{x}, \tag{3.75}$$

where $\mathbf{G}$ (3.2) is the Green's function for the displacement.

For the inhomogeneous displacement field $\mathbf{u}_1(\mathbf{x}) \equiv \mathbf{u}(\mathbf{x}) - \mathbf{u}^0(\mathbf{x})$ produced by the inhomogeneity v, we get from Eqs. (3.71) and (3.74) ($\mathbf{u}_1(\mathbf{x}) \to \mathbf{0},\ $ as $\mathbf{x} \to \infty$)

$$\begin{aligned} \mathcal{L}^{(0)}\mathbf{u}_1 &= -\mathcal{L}_1\mathbf{u} - \nabla\boldsymbol{\alpha}_1 - \mathbf{f}_1, \\ \nabla_j L^{(0)}_{ijkl}\nabla_l u_{1k}(\mathbf{x}) &= -\nabla_j L_{1ijkl}(\mathbf{x})\nabla_l u_k(\mathbf{x}) - \nabla_j \alpha_{1ij}(\mathbf{x}) - \mathbf{f}_{1i}. \end{aligned} \tag{3.76}$$

In so doing, according to the potential theory, the solutions of Eqs. (3.74) and (3.76) exist and are unique, and $\boldsymbol{\varepsilon}^1(\mathbf{x}) = \nabla\mathbf{u}^1(\mathbf{x})$ has at the surface S of the inhomogeneity a jump providing a continuity of the normal constituent of stresses $\boldsymbol{\sigma}(\mathbf{x})$. The right-hand side of Eq. (3.76) is formally equivalent to the presence of a fictitious body force that allows us to reduce Eq. (3.76) to an integral equation by the use of the Green's tensor $\mathbf{G}(\mathbf{x})$ (3.2) for the displacement

$$\begin{aligned} \mathbf{u}_1 &= \mathbf{G} * \nabla(\mathbf{L}_1\boldsymbol{\varepsilon} + \boldsymbol{\alpha}_1) + \mathbf{G} * \mathbf{f}_1, \\ u_{1k} &= \int G_{ki}(\mathbf{x}-\mathbf{y})\nabla_j[L_{1ijml}\varepsilon_{lm}(\mathbf{y}) + \alpha_{1ij}(\mathbf{x})]d\mathbf{y} + \int G_{ki}(\mathbf{x}-\mathbf{y}) f_{1i}(\mathbf{y})d\mathbf{y}. \end{aligned} \tag{3.77}$$

Symmetrized derivation of Eq. (3.77) with subsequent integration of the first integral obtained by parts leads to the equations for strains:

$$\boldsymbol{\varepsilon}(\mathbf{x}) = \boldsymbol{\varepsilon}^0(\mathbf{x}) + \int \mathbf{U}(\mathbf{x}-\mathbf{y})[\mathbf{L}_1(\mathbf{y})\boldsymbol{\varepsilon}(\mathbf{y}) + \boldsymbol{\alpha}_1(\mathbf{y})]\, d\mathbf{y} + \int \nabla\mathbf{G}(\mathbf{x}-\mathbf{y})\mathbf{f}_1(\mathbf{y})\, d\mathbf{y}, \tag{3.78}$$

where it was used that $\nabla_{\mathbf{y}} = -\nabla_{\mathbf{x}}$. Equation (3.78) can be recast in terms of stresses

$$\boldsymbol{\sigma}(\mathbf{x}) = \boldsymbol{\sigma}^0(\mathbf{x}) + \int \boldsymbol{\Gamma}(\mathbf{x}-\mathbf{y})[\mathbf{M}_1(\mathbf{y})\boldsymbol{\sigma}(\mathbf{y}) + \boldsymbol{\beta}_1(\mathbf{y})]\, d\mathbf{y} + \int \mathbf{L}^{(0)}\nabla\mathbf{G}(\mathbf{x}-\mathbf{y})\mathbf{f}_1(\mathbf{y})\, d\mathbf{y}, \tag{3.79}$$

by the use of identities $\mathbf{L}_1(\boldsymbol{\varepsilon}-\boldsymbol{\beta}) = -\mathbf{L}^{(0)}\mathbf{M}_1\boldsymbol{\sigma}$ and $\boldsymbol{\varepsilon} = [\mathbf{M}^{(0)}\boldsymbol{\sigma} + \boldsymbol{\beta}^{(0)}] + [\mathbf{M}_1\boldsymbol{\sigma} + \boldsymbol{\beta}_1]$; here the convolution kernels $\mathbf{U}(\mathbf{x}-\mathbf{y})$ and $\boldsymbol{\Gamma}(\mathbf{x}-\mathbf{y})$ are found via Eqs. (3.19) and (3.37), respectively.

Equations (3.78) and (3.79) can be represented in the operator form:

$$\boldsymbol{\varepsilon} = \boldsymbol{\varepsilon}^0 + \mathbf{U}(\mathbf{L}_1\boldsymbol{\varepsilon} + \boldsymbol{\alpha}_1)V, \quad \boldsymbol{\sigma} = \boldsymbol{\sigma}^0 + \boldsymbol{\Gamma}(\mathbf{M}_1\boldsymbol{\sigma} + \boldsymbol{\beta}_1)V, \tag{3.80}$$

where it was assumed that $\mathbf{f}_1 \equiv \mathbf{0}$ and the operators $\mathbf{U}$ and $\boldsymbol{\Gamma}$ have the convolution kernels $\mathbf{U}(\mathbf{x}-\mathbf{y})$ (3.19) and $\boldsymbol{\Gamma}(\mathbf{x}-\mathbf{y})$ (3.37).

In particular, for a cavity and a rigid inclusion, Eqs. (3.80_1) and (3.80_2) are reduced to

$$\boldsymbol{\varepsilon} = \boldsymbol{\varepsilon}^0 - \mathbf{U}(\mathbf{L}^{(0)}\boldsymbol{\varepsilon} + \boldsymbol{\alpha}^{(0)})V, \quad \boldsymbol{\sigma} = \boldsymbol{\sigma}^0 - \boldsymbol{\Gamma}(\mathbf{M}^{(0)}\boldsymbol{\sigma} - \boldsymbol{\beta}_1)V, \tag{3.81}$$

respectively.

Let us consider a related problem for a gas-saturated pore at the gas pressure p^0 ($\mathbf{L}_1(\mathbf{x}) \to -\mathbf{L}^{(0)}$, $\mathbf{x} \in v$;. Substituting the constitutive equation in the pore

$$\boldsymbol{\sigma}(\mathbf{x}) = (1 - V(\mathbf{x}))\mathbf{L}^{(0)}\boldsymbol{\varepsilon}(\mathbf{x}) - \mathbf{p}(\mathbf{x}), \tag{3.82}$$

where $p_{ij}(\mathbf{x}) = p^0\delta_{ij}V(\mathbf{x})$, into the equilibrium equation leads to the differential equation for the displacement analogous to Eq. (3.76) ($\mathbf{u}_1(\mathbf{x}) \to \mathbf{0}$, as $\mathbf{x} \to \infty$):

$$\mathcal{L}^{(0)}\mathbf{u}_1 = \mathcal{L}^{(0)}(V\mathbf{u}) + \nabla\mathbf{p} - \mathbf{f}_1,$$
$$\nabla_j L^{(0)}_{ijkl}\nabla_l u_{1k}(\mathbf{x}) = \nabla_j L^{(0)}_{ijkl}[\nabla_l u_k(\mathbf{x})V(\mathbf{x})] + \nabla_j p_{ij}(\mathbf{x}) - f_{1i}, \tag{3.83}$$

which can be transformed to the integral one similar to obtaining Eq. (3.78)

$$\boldsymbol{\varepsilon}(\mathbf{x}) = \boldsymbol{\varepsilon}^0(\mathbf{x}) - \int \mathbf{U}(\mathbf{x}-\mathbf{y})[\mathbf{L}^{(0)}(\mathbf{y})\boldsymbol{\varepsilon}(\mathbf{y}) + \mathbf{p}(\mathbf{y})]V(\mathbf{y})\,d\mathbf{y}$$
$$+ \int \nabla\mathbf{G}(\mathbf{x}-\mathbf{y})\mathbf{f}_1(\mathbf{y})\,d\mathbf{y}, \qquad (3.84)$$

It should be noted that Eqs. (3.78) and (3.84) coincide with $\boldsymbol{\beta}^{(0)} = \mathbf{0},\ \mathbf{L}_1(\mathbf{x}) \to -\mathbf{L}^{(0)},\ \mathbf{L}(\mathbf{x})\boldsymbol{\beta}_1(\mathbf{x}) = \mathbf{p}(\mathbf{x})$ taken at $\mathbf{x} \in v$. Therefore, one needs only to investigate the Eq. (3.78).

It should be mentioned that the integrands in Eqs. (3.78), (3.79), and (3.84) have the bounded support $V(\mathbf{y})$. Therefore, the problem of the estimation of the fields $\boldsymbol{\sigma}(\mathbf{x})$, $\boldsymbol{\varepsilon}(\mathbf{x})$ is solved inside and outside the inclusion v if the fields $\boldsymbol{\sigma}(\mathbf{y})$ and $\boldsymbol{\varepsilon}(\mathbf{y})$ $(\mathbf{y} \in v)$ are found inside the domain v.

3.4.2 *Interface Boundary Operators*

Let us turn to the analysis of the fields $\boldsymbol{\sigma}(\mathbf{x})$ and $\boldsymbol{\varepsilon}(\mathbf{x})$ in the matrix in the vicinity of the sufficiently smooth inclusion surface S. It happened that the relations for the jump of the stresses and strains at the interface between two media with homogeneous elastic moduli $\mathbf{L}^-$ and $\mathbf{L}^+$ and stress-free strains $\boldsymbol{\beta}^-$ and $\boldsymbol{\beta}^+$, respectively, can be explicitly found without a concrete representation of the Green's tensor $\mathbf{G}(\mathbf{x})$ (3.3). Ideal boundary conditions of continuity of displacement and of the normal stress components are assumed

$$u_i^-(\mathbf{x}^-) = u_i^+(\mathbf{x}^+),\ \ n_i\sigma_{ij}^-(\mathbf{x}^-) = n_i\sigma_{ij}^+(\mathbf{x}^+), \qquad (3.85)$$

at the interface with the unit outward tensor $\mathbf{n}\perp S$ in the point $\mathbf{x} \in S$. Here and below the symbols $^+$ and $^-$ relate to the different boundary sides.

According to [644] (see also [836]), one defines the projective operators $\boldsymbol{\eta},\ \boldsymbol{\nu}$ and $\mathbf{E},\ \mathbf{F}$ of the second order and fourth order

$$\eta_{kl} \equiv n_k n_l, \qquad \nu_{kl} \equiv \delta_{kl} - \eta_{kl},$$
$$F_{klmn} \equiv (\nu_{km}\nu_{ln} + \nu_{ln}\nu_{km})/2, \qquad E_{klmn} \equiv I_{klmn} - F_{klmn} \qquad (3.86)$$

prescribed at the surface S with the outward normal vector $\mathbf{n}$ at the considered surface point; the tensors F_{ijkl} and E_{ijkl} are symmetric with respect to the replacement $i \leftrightarrow j,\ k \leftrightarrow l,\ ij \leftrightarrow kl$.

By testing, we obtain immediately "orthogonal" properties of operators (3.86)

$$\boldsymbol{\eta\eta} = \boldsymbol{\eta}, \quad \boldsymbol{\nu\nu} = \boldsymbol{\nu}, \quad \boldsymbol{\nu\eta} = \mathbf{0},\ \boldsymbol{\eta} + \boldsymbol{\nu} = \boldsymbol{\delta},\ \mathbf{E} + \mathbf{F} = \mathbf{I},$$
$$\mathbf{FF} = \mathbf{F},\ \mathbf{EE} = \mathbf{E}, \quad \mathbf{E}\boldsymbol{\nu} = 0, \quad \mathbf{F}\boldsymbol{\nu} = 0, \quad \mathbf{FE} = 0. \qquad (3.87)$$

Orthogonal tensors **E** and **F** allow us to decompose an arbitrary tensor of the second-order t_{ij} on two constituents:

$$t_{ij} = \tau_{ij} + \theta_{ij}, \quad \tau_{ij} = E_{ijkl}t_{kl}, \quad \theta_{ij} = F_{ijkl}t_{kl}, \tag{3.88}$$

which can be also presented in the form

$$\tau_{ij} = [n_i(\delta_{jk} - n_j n_k) + n_j(\delta_{ik} - n_i n_k) + n_i n_j n_k]t_{kl}n_l,$$
$$\theta_{ij} = (\delta_{ik} - n_i n_k)(\delta_{jl} - n_j n_l)t_{kl}.$$

The decompositions (3.88) are most readily illustrated in the local coordinate system with the origin at the point $\mathbf{x} \in S$ when for the outward normal takes the representation $\mathbf{n} = (0, 0, 1)^\top$. Then the tensorial operator **E** conserves the tensor components t_{ij} containing the index 3 and vanishes all others. On the contrary, operator **F** conserves components t_{ij}, which do not contain the index 3, and vanishes the components containing index 3. Therefore, the tensors τ_{ij} and θ_{ij} can be named the normal and tangential constituents of the tensor t_{ij}, respectively, with respect to the point $\mathbf{x} \in S$ being considered.

Perfect contact (3.85) between two materials means

$$\mathbf{E}\boldsymbol{\sigma}^+ = \mathbf{E}\boldsymbol{\sigma}^-, \quad \mathbf{F}\boldsymbol{\varepsilon}^+ = \mathbf{F}\boldsymbol{\varepsilon}^-. \tag{3.89}$$

Further, the surface tensors are defined by

$$\mathbf{L}^{\pm}(\mathbf{n}) = \mathbf{L}^{\pm}\eta, \ \mathbf{G}^{\pm}(\mathbf{n}) = [\mathbf{L}^{\pm}(n)]^{-1}, \quad \boldsymbol{\Gamma}^{\pm}(\mathbf{n}) = \mathbf{L}^{\pm} - \mathbf{L}^{\pm}\mathbf{U}^{\pm}(\mathbf{n})\mathbf{L}^{\pm},$$
$$U(n)^{\pm}_{klmn} = [n_k G^{\pm}(n)_{lm} n_n]_{(kl)(mn)}. \tag{3.90}$$

The second-order tensor $L_{ij}(\mathbf{n})$ is symmetric and positive definite due to the positiveness of elastic energy, and therefore the tensor $L_{ij}(\mathbf{n})$ has an inverse tensor $G_{ij}(\mathbf{n})$, which is also symmetric and positive definite. By straightforward testing, one can obtain

$$\mathbf{FU} = \mathbf{UF} = \mathbf{0}, \quad \mathbf{E\Gamma} = \mathbf{\Gamma E} = \mathbf{0},$$
$$\mathbf{EU} = \mathbf{UE} = \mathrm{U}, \quad \mathbf{E\Gamma} = \mathbf{\Gamma E} = \mathbf{0}. \tag{3.91}$$

It should be mentioned that the tensors $\boldsymbol{\Gamma}(\mathbf{n})$ and $\mathbf{U}(\mathbf{n})$ are obviously related to the Green's tensors $\boldsymbol{\Gamma}(\mathbf{x})$ (3.37) and $\mathbf{U}(\mathbf{x})$ (3.19). Indeed, let, for example, $\boldsymbol{\Gamma}(\mathbf{k})$ be a Fourier transform of the tensor $\boldsymbol{\Gamma}(\mathbf{x})$. Then $\boldsymbol{\Gamma}(\mathbf{n})$ coincides with the value of $\boldsymbol{\Gamma}(\mathbf{k})$ at the surface of a unit sphere $|\mathbf{k}| = 1$ at $\mathbf{n} = \mathbf{k}/|\mathbf{k}|$. The surface tensors $\boldsymbol{\Gamma}(\mathbf{n})$ and $\mathbf{U}(\mathbf{n})$ are found by the use of the Fourier transforms $\boldsymbol{\Gamma}(\mathbf{k})$ and $\mathbf{U}(\mathbf{k})$ of the involved Green's tensors rather than their original. The tensors $\boldsymbol{\Gamma}(\mathbf{n})$ and $\mathbf{U}(\mathbf{n})$ are always finite and depend only on the normal $\mathbf{n}$ in the point being considered rather than on the size and the shape on the inhomogeneity v.

Estimating the components of the tensors **ELE** and **FMF** in terms of the projective operators leads to

$$(\mathbf{ELE})\mathbf{U} = \mathbf{E} = \mathbf{U}(\mathbf{ELE}), \quad (\mathbf{FMG})\boldsymbol{\Gamma} = \mathbf{F} = \boldsymbol{\Gamma}(\mathbf{FMF}), \tag{3.92}$$

or, in other words:

$$\mathbf{U}(\mathbf{n}) = (\mathbf{ELE})^{-1}\mathbf{E}, \quad \boldsymbol{\Gamma}(\mathbf{n}) = (\mathbf{FMF})^{-1}\mathbf{F}. \tag{3.93}$$

Let us come back to the estimation of the jump of stresses and strains at the interface S. Exploiting the contact conditions (3.89) leads to representation of the constitutive equation in the vicinity of the interface in the form

$$\boldsymbol{\sigma}^- = \mathbf{L}^-(\mathbf{F}\boldsymbol{\varepsilon}^+ + \mathbf{E}\boldsymbol{\varepsilon}^- - \boldsymbol{\beta}^-), \quad \boldsymbol{\varepsilon}^- = \mathbf{M}^-(\mathbf{E}\boldsymbol{\sigma}^+ + \mathbf{F}\boldsymbol{\sigma}^-) + \boldsymbol{\beta}^-. \tag{3.94}$$

Multiplying Eqs. (3.94_1) and (3.94_2) over the tensors **E** and **F**, respectively, yields

$$\mathbf{E}\boldsymbol{\sigma}^+ = \mathbf{EL}^-(\mathbf{F}\boldsymbol{\varepsilon}^+ + \mathbf{E}\boldsymbol{\varepsilon}^- - \boldsymbol{\beta}^-), \quad \mathbf{F}\boldsymbol{\varepsilon}^+ = \mathbf{FM}^-(\mathbf{E}\boldsymbol{\sigma}^+ + \mathbf{F}\boldsymbol{\sigma}^-) + \mathbf{F}\boldsymbol{\beta}^-, \tag{3.95}$$

from which, with inclusion of Eqs. (3.93), it follows

$$\boldsymbol{\varepsilon}^- = [\mathbf{I} + \mathbf{U}^-(\mathbf{n})(\mathbf{L}^+ - \mathbf{L}^-)]\boldsymbol{\varepsilon}^+ + \mathbf{U}^-(\mathbf{n})(\boldsymbol{\alpha}^+ - \boldsymbol{\alpha}^-), \tag{3.96}$$

$$\boldsymbol{\sigma}^- = [\mathbf{I} + \boldsymbol{\Gamma}^-(\mathbf{n})(\mathbf{M}^+ - \mathbf{M}^-)]\boldsymbol{\sigma}^+ + \boldsymbol{\Gamma}^-(\mathbf{n})(\boldsymbol{\beta}^+ - \boldsymbol{\beta}^-). \tag{3.97}$$

The replacement $+ \leftrightarrow -$ in Eqs. (3.96) and (3.101) leads to the representations

$$\boldsymbol{\varepsilon}^+ = [\mathbf{I} + \mathbf{U}^+(\mathbf{n})(\mathbf{L}^- - \mathbf{L}^+)]\boldsymbol{\varepsilon}^- + \mathbf{U}^+(\mathbf{n})(\boldsymbol{\alpha}^- - \boldsymbol{\alpha}^+), \tag{3.98}$$

$$\boldsymbol{\sigma}^+ = [\mathbf{I} + \boldsymbol{\Gamma}^+(\mathbf{n})(\mathbf{M}^- - \mathbf{M}^+)]\boldsymbol{\sigma}^- + \boldsymbol{\Gamma}^+(\mathbf{n})(\boldsymbol{\beta}^- - \boldsymbol{\beta}^+). \tag{3.99}$$

Substitution of Eqs. (3.96), (3.97) into the right-hand side of Eqs. (3.98), (3.99), respectively, leads to

$$\mathbf{U}^-(\mathbf{n}) - \mathbf{U}^+(\mathbf{n}) = \mathbf{U}^-(\mathbf{n})(\mathbf{L}^+ - \mathbf{L}^-)\mathbf{U}^+(\mathbf{n}), \tag{3.100}$$

$$\boldsymbol{\Gamma}^-(\mathbf{n}) - \boldsymbol{\Gamma}^+(\mathbf{n}) = \boldsymbol{\Gamma}^-(\mathbf{n})(\mathbf{M}^+ - \mathbf{M}^-)\boldsymbol{\Gamma}^+(\mathbf{n}). \tag{3.101}$$

Let us denote $\mathbf{A}^-(\mathbf{n}) = [\mathbf{I} + \mathbf{U}^-(\mathbf{n})(\mathbf{L}^+ - \mathbf{L}^-)]$, $\mathbf{B}^-(\mathbf{n}) = [\mathbf{I} + \boldsymbol{\Gamma}^-(\mathbf{n})(\mathbf{M}^+ - \mathbf{M}^-)]$. Then from the expressions (3.96) and (3.97), we get the correlations between the stress and strain fields in the vicinity of the interface

$$\boldsymbol{\varepsilon}^-(\mathbf{n}) = \mathbf{A}^-(\mathbf{n})\boldsymbol{\sigma}^+(\mathbf{x}) + (\mathbf{A}^-(\mathbf{n}) - \mathbf{I})(\mathbf{L}^+ - \mathbf{L}^-)^{-1}(\boldsymbol{\alpha}^+ - \boldsymbol{\alpha}^-), \tag{3.102}$$

$$\boldsymbol{\sigma}^-(\mathbf{n}) = \mathbf{B}^-(\mathbf{n})\boldsymbol{\sigma}^+(\mathbf{x}) + (\mathbf{B}^-(\mathbf{n}) - \mathbf{I})(\mathbf{M}^+ - \mathbf{M}^-)^{-1}(\boldsymbol{\beta}^+ - \boldsymbol{\beta}^-), \tag{3.103}$$

which are also valid at the replacement $+ \leftrightarrow -$.

In particular for an isotropic medium with the elastic modulus $\mathbf{L} = (3k, 2\mu)$ (2.197), an inversion of the matrix $\mathbf{L}(\mathbf{n})$ may be simplified, and we obtain

$$L_{kl}(\mathbf{n}) = \mu\delta_{kl} + \left(k + \frac{\mu}{3}\right) n_k n_l, \; G_{kl}(\mathbf{n}) = \mu^{-1}\left(\delta_{kl} - \frac{2k+\mu}{3k+4\mu} n_k n_l\right),$$
$$U_{klmn}(\mathbf{n}) = \frac{1}{2\mu}\left(E_{klmn} - \frac{3k-2\mu}{3k+4\mu} n_k n_l n_m n_n\right),$$
$$\Gamma_{klmn}(\mathbf{n}) = 2\mu\Big[F_{klmn} + \frac{3k-2\mu}{3k+4\mu}(\delta_{kl} - n_k n_l)(\delta_{mn} - n_m n_n)\Big]. \tag{3.104}$$

From Eqs. (3.101) and (3.104), one can obtain the formulae for the strain jump $[\boldsymbol{\varepsilon}] \equiv \boldsymbol{\varepsilon}^+ - \boldsymbol{\varepsilon}^-$ at the interface S between joint isotropic media in the local coordinate system when the normal vector has the coordinate $\mathbf{n} = (0, 0, 1)^\top$:

$$[\varepsilon_{11}] = [\varepsilon_{12}] = [\varepsilon_{22}] = 0, \; [\varepsilon_{13}] = -\frac{\mu^+ - \mu^-}{\mu^-}\varepsilon^+_{13}, \; [\varepsilon_{23}] = -\frac{\mu^+ - \mu^-}{\mu^-}\varepsilon^+_{23},$$
$$[\varepsilon_{33}] = -\frac{1}{\lambda^- + 2\mu^-}\Big\{(\lambda^+ - \lambda^-)(\varepsilon^+_{11} + \varepsilon^+_{22} + \varepsilon^+_{33}) + 2(\mu^+ - \mu^-)\varepsilon^+_{33}\Big\}, \tag{3.105}$$

where $\lambda^\pm = k^\pm - \frac{2}{3}\mu^\pm$.

3.5 Ellipsoidal Inhomogeneity in the Elastic Medium

Let us consider a particular case of the ellipsoidal homogeneous domain v described by Eq. (3.18) in the Cartesian coordinate system connected to the semi-axes a_1, a_2, a_3 when $\mathbf{a} = \mathrm{diag}(a_1, a_2, a_3)$ (3.18). If the elastic modulus is homogeneous inside the matrix and inclusion of an arbitrary anisotropy: $\mathbf{L}(\mathbf{x}) \equiv \mathbf{L}^{(0)} + \mathbf{L}_1(\mathbf{x})$, $\mathbf{L}_1(\mathbf{x}) = \mathbf{L}_1^+ V(\mathbf{x})$, $\mathbf{L}_1^+ \equiv$ const., then the solutions of some important problems can be found explicitly that is based on the following theorem of polynomial conservation (the p-property) [838] (see also [836]).

Theorem *If the external field $\boldsymbol{\varepsilon}^0(\mathbf{x})$ acting on the inhomogeneity $\mathbf{x} \in v$ is a polynomial of degree m, then the field $\boldsymbol{\varepsilon}(\mathbf{x})$ generated inside v is also polynomial of the same order m.*

The most important case of the homogeneous external loading $\boldsymbol{\sigma}^0(\mathbf{x})$ =const. was proved by Eshelby [449] for the isotropic matrix and in [779, 1577] for the anisotropic matrix (see also [505, 1186]); the cases of the polynomial fields were considered in [450] and in [35, 838] for the isotropic and anisotropic materials, respectively.

Let us consider the consequence of this theorem when $\boldsymbol{\varepsilon}^0$, $\boldsymbol{\sigma}^0(\mathbf{x})$ =const. and, therefore, the produced fields $\boldsymbol{\varepsilon}^+(\mathbf{x})$ and $\boldsymbol{\sigma}^+(\mathbf{x})$ are also homogeneous. By this means it is necessary to define the action of generalized functions $\mathbf{U}(\mathbf{x})$ and $\boldsymbol{\Gamma}(\mathbf{x})$ in the operator Eqs. (3.80_1) and (3.80_2) on the finite functions $\boldsymbol{\varepsilon}^+ V(\mathbf{x})$ and $\boldsymbol{\sigma}^+ V(\mathbf{x})$. According to the regularizations (3.26) and (3.38), the action of the formal constituents $\mathbf{U}^{(f)}$ and $\boldsymbol{\Gamma}^{(f)}$ of tensors $\mathbf{U}$ and $\boldsymbol{\Gamma}$, respectively, on the mentioned functions equal zero, and the effect of the operators $\mathbf{U}$ and $\boldsymbol{\Gamma}$ equivalent to the multiplication on the constant tensors:

$$\mathbf{P} = -\frac{1}{4\pi}\int \mathbf{U}(a^{-1}\bar{\mathbf{k}})\, d\bar{\mathbf{k}}, \quad \mathbf{Q} = -\frac{1}{4\pi}\int \boldsymbol{\Gamma}(a^{-1}\bar{\mathbf{k}})\, d\bar{\mathbf{k}}, \tag{3.106}$$

where integration is performed over the surface of unit sphere $\bar{\mathbf{k}}^2 = 1$ and the regularization domain has a shape of the ellipsoid v. A relationship follows between the tensors $\mathbf{P}$ and $\mathbf{Q}$: $\mathbf{Q} = \mathbf{L}^{(0)} - \mathbf{L}^{(0)}\mathbf{P}\mathbf{L}^{(0)}$, derived from definition (3.37). From Eqs. (3.80_1), (3.80_2) and (3.106_1), (3.106_2), it follows the linear algebraic equations relating the generated fields inside the inclusion with the remote fields ($\boldsymbol{\beta}_1 \equiv \mathbf{0}$):

$$(\mathbf{I} + \mathbf{P}\mathbf{L}_1)\boldsymbol{\varepsilon}^+ = \boldsymbol{\varepsilon}^0, \quad (\mathbf{I} + \mathbf{Q}\mathbf{M}_1)\boldsymbol{\sigma}^+ = \boldsymbol{\sigma}^0. \tag{3.107}$$

Let us represent another form of the tensors $\mathbf{P}$ (3.106_1) based on the homogeneity of the function $\mathbf{U}(\mathbf{k})$:

$$\int \mathbf{U}(\mathbf{a}^{-1}\bar{\mathbf{k}})d\bar{\mathbf{k}} = \int \mathbf{U}(\mathbf{n})d\bar{\mathbf{k}} = \det \mathbf{a} \int \mathbf{U}(\mathbf{n})\rho^3(\mathbf{n})\, d\mathbf{n}, \tag{3.108}$$

$$\mathbf{n} = \frac{\mathbf{a}^{-1}\bar{\mathbf{k}}}{|\mathbf{a}^{-1}\bar{\mathbf{k}}|}, \quad \rho(\mathbf{n}) = \frac{1}{\sqrt{\mathbf{n}\mathbf{a}^2\mathbf{n}}}. \tag{3.109}$$

The point $\mathbf{x} \in S = \partial v$ belonging to the surface S described by the equation $\mathbf{x}\mathbf{a}^{-2}\mathbf{x} = 1$ (3.18) is related with the normal $\mathbf{n} \perp S$ in this point by the relations

$$\mathbf{x} = \frac{\mathbf{a}^2\mathbf{n}}{\sqrt{\mathbf{n}\mathbf{a}^2\mathbf{n}}}, \quad \mathbf{n} = \frac{\mathbf{a}^{-2}\mathbf{n}}{\sqrt{\mathbf{x}\mathbf{a}^{-4}\mathbf{x}}} \tag{3.110}$$

establishing one-to-one correspondence between the points $\mathbf{x}$ on the surface S and the points $\mathbf{n}$ on the unit sphere $S_1 = \{\mathbf{n} |\ \mathbf{n}^2 = 1\}$. Then the averages of an arbitrary function f over the ellipsoid v and the sphere S_1

$$\langle f(\mathbf{x})\rangle = \frac{1}{4\pi \det \mathbf{a}}\int_v \frac{f(\mathbf{x})}{\rho(\mathbf{x})}\, d\mathbf{x}, \quad \text{and} \quad \langle f(\mathbf{n})\rangle = \frac{\det \mathbf{a}}{4\pi}\int_{S_1} f(\mathbf{n})\rho^3(\mathbf{n})\, d\mathbf{n}, \tag{3.111}$$

respectively, coincide. Then from Eqs. (3.108)–(3.111), we get the representations for the tensors $\mathbf{P}$ and, analogously, for $\mathbf{Q}$

$$\mathbf{P} = -\langle \mathbf{U}(\mathbf{n})\rangle, \quad \mathbf{Q} = -\langle \boldsymbol{\Gamma}(\mathbf{n})\rangle \tag{3.112}$$

defined by the averages over the unit sphere of the tensors $\mathbf{U}(\mathbf{n})$ and $\boldsymbol{\Gamma}(\mathbf{n})$ describing the jumps of the strains and stresses on the interface S (3.96)–(3.99).

Let us consider the case of the linear external field $\varepsilon^0_{ij}(\mathbf{x}) = b_{ijk}x_k$ ($\mathbf{x} \in v$) when, according to the p-property theorem, $\varepsilon_{ij}(\mathbf{x}) = d_{ijk}x_k$, ($\mathbf{x} \in v$). The tensors b_{ijk} and d_{ijk}, symmetrical over the first pair of indexes, are related by the linear ratio which can be presented in the form:

$$b_{i_1i_2i_3} = 3\Lambda_{i_1i_2i_3j_1j_2j_3}d_{j_1j_2j}(\mathbf{a}^2)_{jj_3}. \tag{3.113}$$

To obtain $\boldsymbol{\Lambda}$, let us multiply Eq. (3.112$_1$) by $\mathbf{x}$ and average the result over v. Then, analogously to the case of the homogeneous field (3.108)–(3.113), we get

$$\boldsymbol{\Lambda} = \langle \rho(\mathbf{n})\mathbf{n}(\mathbf{I} - \mathbf{U}(\mathbf{n})\mathbf{L}_1)\mathbf{n}\rho(\mathbf{n})\rangle. \tag{3.114}$$

In conclusion, we consider one more practical problem for the ellipsoidal inhomogeneity with eigenstrain ($\boldsymbol{\beta}_1(\mathbf{x}) = \text{const.} \neq \mathbf{0}$) subjected to the homogeneous remote fields. From Eqs. (3.80$_1$), (3.80$_2$) and (3.106$_1$), (3.106$_2$), follow the linear algebraic equations related to the generated fields inside the inclusion with the remote fields

$$\boldsymbol{\varepsilon}^+ = \mathbf{A}\boldsymbol{\varepsilon}^0 + \mathbf{C}^\epsilon, \quad \boldsymbol{\sigma}^+ = \mathbf{B}\boldsymbol{\sigma}^0 + \mathbf{C}, \tag{3.115}$$

with obvious correlations between the tensors involved into Eqs. (3.115)

$$\mathbf{A} = (\mathbf{I} + \mathbf{P}\mathbf{L}_1)^{-1} = \mathbf{M}^{(i)}\mathbf{B}(\mathbf{M}^{(0)})^{-1}, \tag{3.116}$$

$$\mathbf{B} = (\mathbf{I} + \mathbf{Q}\mathbf{M}_1)^{-1} = \mathbf{L}^{(i)}\mathbf{A}(\mathbf{L}^{(0)})^{-1}, \tag{3.117}$$

$$\mathbf{C}^\epsilon = -\mathbf{A}\mathbf{P}\boldsymbol{\alpha}_1, \quad \mathbf{C} = -\mathbf{B}\mathbf{Q}\boldsymbol{\beta}_1, \tag{3.118}$$

$$\mathbf{M}_1 = -\mathbf{M}^{(i)}\mathbf{L}_1\mathbf{M}^{(0)}, \quad \boldsymbol{\beta}_1 = -\mathbf{M}^{(i)}\boldsymbol{\alpha}^{(i)} + \mathbf{M}^{(0)}\boldsymbol{\alpha}^{(0)}, \tag{3.119}$$

$$\mathbf{L}_1 = -\mathbf{L}^{(0)}\mathbf{M}_1\mathbf{L}^{(i)}, \quad \boldsymbol{\alpha}_1 = -\mathbf{L}^{(i)}\boldsymbol{\beta}^{(i)} + \mathbf{L}^{(0)}\boldsymbol{\alpha}^{(0)}, \tag{3.120}$$

$$\mathbf{Q} = \mathbf{L}^{(0)} - \mathbf{L}^{(0)}\mathbf{P}\mathbf{L}^{(0)}, \tag{3.121}$$

$$\mathbf{P} = \mathbf{M}^{(0)} - \mathbf{M}^{(0)}\mathbf{Q}\mathbf{M}^{(0)}, \tag{3.122}$$

where the tensors **P**, **Q** and **A**, **B** are called Hill's tensors [640], (see also [505, 1186, 1575, 1576]), and Wu's tensors [1642], respectively.

Let an ellipsoidal inclusion v_i with the homogeneous compliance $\mathbf{M}^+$ be located in an infinite homogeneous matrix with compliance $\mathbf{M}^-$ and loaded by the homogeneous stress $\boldsymbol{\sigma}^0$. Then, according to Eshelby's theorem (with $\boldsymbol{\beta} \equiv 0$), we have the representations for the stresses

$$\boldsymbol{\sigma}^+ = \boldsymbol{\sigma}^0 - \mathbf{Q}(\mathbf{M}^+ - \mathbf{M}^-)\boldsymbol{\sigma}^+, \tag{3.123}$$

$$\boldsymbol{\sigma}(\mathbf{x}) = \boldsymbol{\sigma}^0 + \overline{v}_i\mathbf{T}_i(\mathbf{x}_i - \mathbf{x})(\mathbf{M}^+ - \mathbf{M}^-)\boldsymbol{\sigma}^+, \tag{3.124}$$

and the strains

$$\boldsymbol{\varepsilon}^+ = \boldsymbol{\varepsilon}^0 - \mathbf{P}(\mathbf{L}^+ - \mathbf{L}^-)\boldsymbol{\varepsilon}^+, \tag{3.125}$$

$$\boldsymbol{\varepsilon}(\mathbf{x}) = \boldsymbol{\varepsilon}^0 + \overline{v}_i \mathbf{T}_i^\epsilon(\mathbf{x}_i - \mathbf{x})(\mathbf{L}^+ - \mathbf{L}^-)\boldsymbol{\varepsilon}^+, \tag{3.126}$$

where the tensors $\mathbf{T}_i^\epsilon(\mathbf{x} - \mathbf{x})$ and $\mathbf{T}_i(\mathbf{x}_i - \mathbf{x})$ are defined by the relations

$$\mathbf{T}_i^\epsilon(\mathbf{x}_i - \mathbf{x}) = \frac{1}{\bar{v}_i}\int V_i(\mathbf{y})\mathbf{U}(\mathbf{y} - \mathbf{x})\,d\mathbf{y},$$

$$\mathbf{T}_i(\mathbf{x}_i - \mathbf{x}) = \frac{1}{\bar{v}_i}\int V_i(\mathbf{y})\boldsymbol{\Gamma}(\mathbf{y} - \mathbf{x})\,d\mathbf{y} \tag{3.127}$$

for the points $\mathbf{x} \notin v_i$.

If $\mathbf{x}$ is a limiting point $\mathbf{x} = \mathbf{x}^-$ in the matrix near the ellipsoidal surface ∂v_i, then, substituting the relations (3.96), (3.97) and (3.126), (3.124) into Eq. (3.125), (3.123), respectively, we obtain

$$\overline{v}_i \mathbf{T}_i^\epsilon(\mathbf{x}_i - \mathbf{x}^-) = \mathbf{U}(\mathbf{n})^- - \mathbf{P}, \quad \overline{v}_i \mathbf{T}_i(\mathbf{x}_i - \mathbf{x}^-) = \boldsymbol{\Gamma}(\mathbf{n})^- - \mathbf{Q}. \tag{3.128}$$

3.6 Eshelby Tensor

3.6.1 Tensor Representation of Eshelby Tensor

The Eqs. (3.115) define the strains and stresses inside ellipsoidal inclusion in terms of the remote fields in the explicit form utilizing the Eshelby's solution in the form obtained in [838] (see also [836]). However, for completeness, we will reproduce another more descriptive method of the *equivalent inclusions* proposed by Eshelby [449, 450], which has played a key role in the micromechanics of composites. Namely, it is assumed that the perturbation of a stress field, introduced by the inhomogeneity v with elastic moduli $\mathbf{L}^{(1)}(\mathbf{x})$ $(\mathbf{x} \in v)$ into the applied remote stress, can be modeled by the fictitious eigenstrains $\boldsymbol{\varepsilon}^*$ localized in the area v with elastic modulus $\mathbf{L}^{(0)}$ called inclusion v. The choice of $\boldsymbol{\varepsilon}^*$ called equivalent eigenstrains is defined from the condition of equivalence of a stress states of the media with the inhomogeneity and inclusion

$$\mathbf{L}^{(1)}(\mathbf{x})\boldsymbol{\varepsilon}(\mathbf{x}) = \mathbf{L}^{(0)}(\boldsymbol{\varepsilon}(\mathbf{x}) - \boldsymbol{\varepsilon}^*(\mathbf{x})). \tag{3.129}$$

For ellipsoidal inhomogeneity subjected to the homogeneous remote stress $\boldsymbol{\sigma}^0 = \mathbf{L}^{(0)}\boldsymbol{\varepsilon}^0$, the homogeneity of the fields $\boldsymbol{\varepsilon}(\mathbf{x})$ and $\boldsymbol{\varepsilon}^*(\mathbf{x})$ $(\mathbf{x} \in v)$ occurs, and

$$\boldsymbol{\varepsilon} = \boldsymbol{\varepsilon}^0 + \mathbf{S}\boldsymbol{\varepsilon}^*, \tag{3.130}$$

where the well-known Eshelby tensor

$$\mathbf{S} = -\int \mathbf{U}(\mathbf{x}-\mathbf{y})V(\mathbf{y})\,d\mathbf{y}\mathbf{L}^{(0)} \equiv \text{const.}, \quad \mathbf{x} \in v \tag{3.131}$$

is related to Hill's tensors $\mathbf{P}$ (3.106$_1$) and $\mathbf{Q}$ (3.106$_2$) by the equations

$$\mathbf{S} = \mathbf{P}\mathbf{L}^{(0)}, \quad \mathbf{S} = \mathbf{I} - \mathbf{M}^{(0)}\mathbf{Q}. \tag{3.132}$$

Substitution of (3.130) into (3.129) yields the equation

$$\mathbf{L}^{(1)}(\mathbf{x})(\boldsymbol{\varepsilon}^0 + \mathbf{S}\boldsymbol{\varepsilon}^*) = \mathbf{L}^{(0)}(\boldsymbol{\varepsilon}^0 + \mathbf{S}\boldsymbol{\varepsilon}^* - \boldsymbol{\varepsilon}^*), \tag{3.133}$$

from which we get the representation for both the equivalent eigenstrain and the strain inside the inhomogeneity

$$\begin{aligned} \boldsymbol{\varepsilon}^*(\mathbf{x}) &= -(\mathbf{L}_1^{(1)}\mathbf{S} + \mathbf{L}^{(0)})^{-1}\mathbf{L}_1^{(1)}\boldsymbol{\varepsilon}^0, \\ \boldsymbol{\varepsilon}(\mathbf{x}) &= (\mathbf{I} + \mathbf{S}\mathbf{M}^{(0)}\mathbf{L}_1^{(1)})^{-1}\boldsymbol{\varepsilon}^0, \end{aligned} \tag{3.134}$$

respectively.

The method of the equivalent inclusions is also appropriate in the case when the inhomogeneity v is subjected to the stress-free strains $\boldsymbol{\beta}_1^{(1)}(\mathbf{x})$ ($\mathbf{x} \in v$, $\boldsymbol{\beta}^{(0)} \equiv \mathbf{0}$). Then the inhomogeneity with the properties $\mathbf{L}^{(1)}$ and $\boldsymbol{\beta}_1^{(1)}$ is modeled as an inclusion with the moduli $\mathbf{L}^{(0)}$ and with the eigenstrain $\boldsymbol{\beta}_1^{(1)}$ subjected to the fictitious eigenstrain $\boldsymbol{\varepsilon}^*(\mathbf{x}) = \boldsymbol{\varepsilon}^* V(\mathbf{x})$. Equivalence of stress states of the inhomogeneity and the inclusions means that

$$\mathbf{L}^{(1)}(\mathbf{x})(\boldsymbol{\varepsilon}(\mathbf{x}) - \boldsymbol{\beta}_1^{(1)}) = \mathbf{L}^{(0)}(\boldsymbol{\varepsilon}(\mathbf{x}) - \boldsymbol{\beta}_1^{(1)} - \boldsymbol{\varepsilon}^*(\mathbf{x})). \tag{3.135}$$

If remote stress $\boldsymbol{\sigma}^0$ and the eigenstrain $\boldsymbol{\beta}^{(1)}$ are homogeneous then Eq. (3.135) is valid if $\boldsymbol{\varepsilon}$ is a solution to the problem of the inclusion with the eigenstrain. According to Eq. (3.130), we have a representation for the strain inside the inclusion

$$\boldsymbol{\varepsilon} = \boldsymbol{\varepsilon}^0 + \mathbf{S}(\boldsymbol{\beta}_1^{(1)} + \boldsymbol{\varepsilon}^*). \tag{3.136}$$

Substitution of (3.136) into (3.135) yields

$$\mathbf{L}^{(1)}(\mathbf{x})\Big[\boldsymbol{\varepsilon}^0 + \mathbf{S}(\boldsymbol{\beta}_1^{(1)} + \boldsymbol{\varepsilon}^*) - \boldsymbol{\beta}_1^{(1)}\Big] = \mathbf{L}^{(0)}\Big[\boldsymbol{\varepsilon}^0 + (\mathbf{S} - \mathbf{I})(\boldsymbol{\beta}_1^{(1)} + \boldsymbol{\varepsilon}^*)\Big] \tag{3.137}$$

from which we can find the equivalent eigenstrain $\boldsymbol{\varepsilon}^*$ and the strains inside the inhomogeneity ($\mathbf{x} \in v$):

$$\boldsymbol{\varepsilon}^*(\mathbf{x}) = -(\mathbf{L}_1^{(1)}\mathbf{S} + \mathbf{L}^{(0)})^{-1}[\mathbf{L}_1^{(1)}\boldsymbol{\varepsilon}^0 - \mathbf{L}^{(0)}\boldsymbol{\beta}_1^{(1)}] - \boldsymbol{\beta}_1^{(1)}, \tag{3.138}$$

$$\boldsymbol{\varepsilon}(\mathbf{x}) = (\mathbf{I} + \mathbf{S}\mathbf{M}^{(0)}\mathbf{L}_1^{(1)})^{-1}[\boldsymbol{\varepsilon}^0 + \mathbf{S}\mathbf{M}^{(0)}\mathbf{L}^{(1)}\boldsymbol{\beta}^{(1)}]. \tag{3.139}$$

The method of equivalent inclusions (3.129)–(3.139) can be easily recast in terms of stresses by replacement of the tensors $\mathbf{L}$, $\boldsymbol{\beta}$, $\boldsymbol{\varepsilon}$, $\boldsymbol{\varepsilon}^0$, $\boldsymbol{\varepsilon}^*$ with the tensors $\mathbf{M}$, $\boldsymbol{\alpha}$, $\boldsymbol{\sigma}$, $\boldsymbol{\sigma}^0$, $\boldsymbol{\sigma}^*$ where $\boldsymbol{\sigma}^*$ is an equivalent eigenstress accompanying the eigenstress $\boldsymbol{\alpha}$ in a similar manner as the equivalent eigenstrain $\boldsymbol{\varepsilon}^*$ accompanying the eigenstrain $\boldsymbol{\beta}$.

The Eshelby tensor depending only on the elastic properties of the matrix is symmetric with respect to the indexes inside the first and the second pairs $S_{ijkl} = S_{jikl} = S_{ijlk}$; however, it is not symmetric, in general, with respect to the exchange of the index pairs $S_{ijkl} \neq S_{klij}$. Moreover, for the isotropic medium, the Eshelby tensor depends only on Poisson ratio which we will see from an explicit representation. To accomplish this, we at first estimate

$$S_{imn} = -\int G_{ij,k}(\mathbf{x}-\mathbf{y})V(\mathbf{y})\, d\mathbf{y} L^{(0)}_{jkmn}, \tag{3.140}$$

where the Green's found $G_{ij}(\mathbf{x})$ (3.13) is known for the isotropic matrix. Substituting of Eq. (3.13) into Eq. (3.140) yields

$$S_{imn} = -\frac{1}{8\pi(1-\nu^{(0)})}\int g_{imn}(\mathbf{l})V(\mathbf{y})\, d\mathbf{y}, \tag{3.141}$$

where the unit vector $\mathbf{l} = (\mathbf{x}-\mathbf{y})/|\mathbf{x}-\mathbf{y}|$ and one introduces

$$g_{imn}(\mathbf{l}) = (1-2\nu^{(0)})(\delta_{im}l_n + \delta_{in}l_m - \delta_{mn}l_i) + 3l_il_ml_n. \tag{3.142}$$

For the internal point $\mathbf{x}$ (3.141) of the ellipsoid v, the volume element $d\mathbf{y}$ can be presented in the form $d\mathbf{y} = r^2 dr d\omega$, where $r = |\mathbf{x}-\mathbf{y}|$ and $d\omega$ is the differential element of the surface of the unit sphere S_1 with the center in the point $\mathbf{x}$. After integration with respect to r in (3.141), we get

$$S_{imn} = -\frac{1}{8\pi(1-\nu^{(0)})}\int_{S_1} r(\mathbf{l}) g_{imn}(\mathbf{l}) d\omega, \tag{3.143}$$

where the positive root of the equation

$$(x_1 + rl_1)^2/a_1^2 + (x_2 + rl_2)/a_2^2 + (x_3 + rl_3)^2/a_3^2 = 1$$

equals $r(\mathbf{l}) = -f/h + (f^2/h^2 + e/h)^{1/2}$, where

$$h = \sum_{i=1}^{3}(l_i/a_i)^2,\ f = \sum_{i=1}^{3} l_i x_i/a_i^2,\ e = 1 - \sum_{i=1}^{3}(x_i/a_i)^2.$$

After substitution of $r(\mathbf{l})$ into Eq. (3.143), $(f^2/h^2 + e/h)^{1/2}$ which is even with respect to l, vanishes because $g_{imn}(\mathbf{l})$ is odd in $\mathbf{l}$. Then

$$S_{imn} = \frac{x_k}{8\pi(1-\nu^{(0)})}\oint_{S_1}\frac{\lambda_k g_{imn}}{h}d\omega, \tag{3.144}$$

where one introduces the vector $\lambda_k = l_k/a_k^2$ (do not sum over k, $k = 1, 2, 3$). Deriving Eq. (3.144) leads to representation of Eshelby tensor

$$S_{ijmn} = \frac{1}{16\pi(1-\nu^{(0)})}\int_{S_1}\frac{\lambda_i g_{jmn} + \lambda_j g_{imn}}{h}d\omega. \tag{3.145}$$

The Eshelby tensor for the 2D case can be obtained by passage to the limit in Eq. (3.145) at $a_1 \to \infty$. It was obtained [722] the representation for 2D Eshelby tensor for the plane stresses by the use of a straightforward following the scheme for the 3D case

$$S_{ijmn} = \frac{1}{8\pi(1-\nu^{(0)})}\int_0^{2\pi}\frac{\lambda_i g_{jmn} - \lambda_j g_{imn}}{h}d\omega \tag{3.146}$$

where the parameters g_{imn}, λ_i, h are also determined by the formulae (3.142), and (3.144) where in such a case the indexes i, m, n take the values 1 and 2.

The 3D Eshelby tensor (3.131), (3.145) does not depend on $\mathbf{x}$ as confirmed in the theorem of polynomial conservation. The surface integrals (3.145) are reduced to simple ones

$$I_1 = \int_{S_1}\frac{l_1^2}{a_1^2 h}d\omega = 2\pi a_1 a_2 a_3\int_0^{\infty}\frac{1}{(a_1^2+s)\Delta(s)}ds,$$
$$I_{11} = \int_{S_1}\frac{l_1^4}{a_1^4 h}d\omega = 2\pi a_1 a_2 a_3\int_0^{\infty}\frac{1}{(a_1^2+s)^2\Delta(s)}ds,$$
$$I_{12} = 3\int_{S_1}\frac{l_1^2 l_2^2}{a_1^2 a_2^2 h}d\omega = 2\pi a_1 a_2 a_3\int_0^{\infty}\frac{1}{(a_1^2+s)(a_2^2+s)\Delta(s)}ds, \tag{3.147}$$

where $\Delta(s) = (a_1^2+s)^{1/2}(a_2^2+s)^{1/2}(a_3^2+s)^{1/2}$ and the other integrals I_i and I_{ij} are found by the simultaneous cyclic permutations of indexes $i, j = (1, 2, 3)$. In turn, the integrals (3.119) can be reduced to the standard elliptic integrals. Indeed, the I_i

and I_{ij} integrals are given by the formulae assuming $a_1 > a_2 > a_3$:

$$I_1 = \frac{4\pi a_1 a_2 a_3}{(a_1^2 - a_2^2)(a_1^2 - a_3^2)^{1/2}}[\mathrm{F}(\theta, \mathrm{k}) - \mathrm{E}(\theta, \mathrm{k}],$$

$$I_3 = \frac{4\pi a_1 a_2 a_3}{(a_2^2 - a_3^2)(a_1^2 - a_3^2)^{1/2}}\Big[\frac{a_2(a_1^2 - a_3^2)^{1/2}}{a_1 a_3} - \mathrm{E}(\theta, \mathrm{k})\Big], \tag{3.148}$$

and by the equations connecting the tensors I_i and I_{ij}:

$$I_1 + I_2 + I_3 = 4\pi, \quad I_{12} = (I_2 - I_1)/(a_1^2 - a_2^2)$$

$$3I_{11} + I_{12} + I_{13} = 4\pi/a_1^2, \quad 3a_1^2 I_{11} + a_2^2 I_{12} + a_3^2 I_{13} = 3I_1, \tag{3.149}$$

where F and E are the elliptic integrals of the first and the second kind

$$\mathrm{F}(\theta, \mathrm{k}) = \int_0^\theta \frac{dw}{(1 - \mathrm{k}^2 \sin^2 w)^{1/2}}, \quad \mathrm{E}(\theta, \mathrm{k}) = \int_0^\theta (1 - \mathrm{k}^2 \sin^2 w)^{1/2}\, dw,$$

$$\theta = \arcsin\left(\frac{a_1^2 - a_3^2}{a_1^2}\right)^{1/2}, \quad \mathrm{k} = \left(\frac{a_1^2 - a_2^2}{a_1^2 - a_3^2}\right)^{1/2}. \tag{3.150}$$

The values I_1, I_3 (3.148) and Eqs. (3.149) accompanied by the pertinent equations obtained by the cyclic permutation of indexes $(1, 2, 3)$ enable the relations for I_i and I_{ij} to be found in terms of I_1 and I_3.

Thus, if the unbounded medium is an isotropically elastic and the semi-axes a_i coincide with the coordinate axes x_i $(i = 1, 2, 3)$ of a rectangular Cartesian coordinate system, then the Eshelby tensor **S** has the symmetry of an ellipsoid and is defined by the nine nonzero components:

$$S_{1111} = \frac{1}{8\pi(1 - \nu^{(0)})}[3a_1^2 I_{11} + (1 - 2\nu^{(0)})I_1],$$

$$S_{1122} = \frac{1}{8\pi(1 - \nu^{(0)})}[a_2^2 I_{12} - (1 - 2\nu^{(0)})I_1],$$

$$S_{1212} = \frac{1}{16\pi(1 - \nu^{(0)})}[(a_1^2 + a_2^2)I_{12} + (1 - 2\nu^{(0)})(I_1 + I_2)] \tag{3.151}$$

with the remaining nonzero six components obtained by the cyclic index permutation $(1, 2, 3)$.

Eshelby tensor for the particular cases has the representations

1. Sphere ($a_1 = a_2 = a_3 = a$)

$$S_{ijkl} = \frac{5\nu^{(0)} - 1}{15(1 - \nu^{(0)})}\delta_{ij}\delta_{kl} + \frac{4 - 5\nu^{(0)}}{15(1 - \nu^{(0)})}(\delta_{ik}\delta_{jl} + \delta_{il}\delta_{jk}). \tag{3.152}$$

2. Elliptic cylinder ($a_3 \to \infty$)

$$
\begin{aligned}
S_{1111} &= \frac{1}{2(1-\nu^{(0)})}\Big[\frac{a_2^2+2a_1a_2}{(a_1+a_2)^2} + (1-2\nu^{(0)})\frac{a_2}{a_1+a_2}\Big],\\
S_{2222} &= \frac{1}{2(1-\nu^{(0)})}\Big[\frac{a_1^2+2a_1a_2}{(a_1+a_2)^2} + (1-2\nu^{(0)})\frac{a_1}{a_1+a_2}\Big],\\
S_{1122} &= \frac{1}{2(1-\nu^{(0)})}\Big[\frac{a_2^2}{(a_1+a_2)^2} - (1-2\nu^{(0)})\frac{a_2}{a_1+a_2}\Big],\\
S_{2233} &= \frac{1}{2(1-\nu^{(0)})}\frac{2\nu^{(0)}a_1}{a_1+a_2}, \quad S_{3333} = 0,\\
S_{2211} &= \frac{1}{2(1-\nu^{(0)})}\Big[\frac{a_1^2}{(a_1+a_2)^2} - (1-2\nu^{(0)})\frac{a_1}{a_1+a_2}\Big],\\
S_{1133} &= \frac{1}{2(1-\nu^{(0)})}\Big[\frac{2\nu^{(0)}a_2}{a_1+a_2}\Big], \quad S_{3311} = S_{3322} = 0,\\
S_{1212} &= \frac{1}{4(1-\nu^{(0)})}\Big[\frac{a_1^2+a_2^2}{(a_1+a_2)^2} + 1 - 2\nu^{(0)}\Big],\\
S_{2323} &= \frac{a_1}{2(a_1+a_2)}, \quad S_{3131} = \frac{a_2}{2(a_1+a_2)}.
\end{aligned} \tag{3.153}
$$

3. Penny-shape inclusion ($a_1 = a_2 \gg a_3$)

$$
\begin{aligned}
S_{1111} &= S_{2222} = \frac{\pi(13-8\nu^{(0)})}{32(1-\nu^{(0)})}\frac{a_3}{a_1}, \quad S_{3333} = 1 - \frac{\pi(1-2\nu^{(0)})}{4(1-\nu^{(0)})}\frac{a_3}{a_1},\\
S_{1122} &= S_{2211} = \frac{\pi(8\nu^{(0)}-1)}{32(1-\nu^{(0)})}\frac{a_3}{a_1}, \quad S_{1133} = S_{2233} = \frac{\pi(2\nu^{(0)}-1)}{8(1-\nu^{(0)})}\frac{a_3}{a_1},\\
S_{3311} &= S_{3322} = \frac{\nu^{(0)}}{1-\nu^{(0)}}\Big[1 - \frac{\pi(4\nu^{(0)}+1)}{8\nu^{(0)}}\frac{a_3}{a_1}\Big],\\
S_{1212} &= \frac{\pi(7-8\nu^{(0)})}{32(1-\nu^{(0)})}\frac{a_3}{a_1}, \quad S_{3131} = S_{2323} = \frac{1}{2}\Big[1 + \frac{\pi(\nu^{(0)}-2)}{4(1-\nu^{(0)})}\frac{a_3}{a_1}\Big].
\end{aligned} \tag{3.154}
$$

In the case of the transversally isotropic media, the scheme of the Eshelby tensor estimation is analogous to Eqs. (3.141)–(3.148) and was performed in [1637].

3.6.2 *Eshelby and Related Tensors in a Special Basis*

We will consider the representations of the Eshelby and some related tensors in a special **P**-basis (21.2.7) (for details, see [746, 836, 1364]. Eshelby tensor for spheroidal inclusion (with the semi-axes $a_1 = a_2 = a$, a_3, aspect ratio $\gamma = a_3/a$, and a symmetry axis $\mathbf{n}$ forming the $\mathbf{P}$-basis) $\mathbf{S} = (s_1, \ldots, s_6)$ has the components

$$
\begin{aligned}
s_1 &= \frac{1}{2(1-\nu^{(0)})} f_0 + f_1, \quad s_2 = \frac{3-4\nu^{(0)}}{2(1-\nu^{(0)})} f_0 + f_1, \\
s_3 &= \frac{\nu^{(0)}}{1-\nu^{(0)}} f_0 - 2f_1, \quad s_4 = \frac{\nu^{(0)}}{1-\nu^{(0)}} (1-2f_0) - 2f_1, \\
s_5 &= 2(1 - f_0 - 4f_1), \quad s_6 = 1 - 2f_0 + 4f_1,
\end{aligned} \tag{3.155}
$$

where

$$
f_0 = \frac{\gamma^2(1-g)}{2(\gamma^2-1)}, \quad f_1 = \frac{\varkappa^{(0)}\gamma^2}{4(\gamma^2-1)^2}[(2\gamma^2+1)g - 3], \quad \varkappa^{(0)} = \frac{1}{2(1-\nu^{(0)})},
$$

$$
g(\gamma) = \begin{cases} \frac{1}{\gamma\sqrt{1-\gamma^2}} \arctan \frac{\sqrt{1-\gamma^2}}{\gamma}, & \text{oblate shape } (\gamma < 1), \\ \frac{1}{2\gamma\sqrt{\gamma^2-1}} \ln \frac{\gamma+\sqrt{\gamma^2-1}}{\gamma-\sqrt{\gamma^2-1}}, & \text{prolate shape } (\gamma > 1). \end{cases} \tag{3.156}
$$

The components of the tensors $\mathbf{P} = (p_1, \ldots, p_6)$ and $\mathbf{Q} = (q_1, \ldots, q_6)$ are

$$
\begin{aligned}
p_1 &= \frac{1}{2\mu^{(0)}}[(1-\varkappa^{(0)})f_0 + f_1], \quad p_2 = \frac{1}{2\mu^{(0)}}[(2-\varkappa^{(0)})f_0 + f_1], \\
p_3 &= p_4 = -\frac{f_1}{\mu^{(0)}}, \quad p_5 = \frac{1 - f_0 - 4f_1}{\mu^{(0)}}, \quad p_6 = \frac{(1-\varkappa^{(0)})(1-2f_0) + 2f_1}{\mu^{(0)}}, \\
q_1 &= \mu^{(0)}[4\varkappa^{(0)} - 1 - 2(3\varkappa^{(0)} - 1)f_0 - 2f_1], \\
q_2 &= 2\mu^{(0)}[1 - (2-\varkappa^{(0)})f_0 - f_1], \quad q_3 = g_4 = 2\mu^{(0)}[(2\varkappa^{(0)} - 1)f_0 + 2f_1], \\
q_5 &= 4\mu^{(0)}[f_0 + 4f_1], \; q_6 = 4\mu^{(0)}[(1+2\varkappa^{(0)})f_0 - 2f_1].
\end{aligned} \tag{3.157}
$$

If $\gamma \ll 1$ (a crack), we can expand the functions f_0 and f_1 in a series over a parameter γ, and retaining terms with the first order of γ reduces the components of the tensor $\mathbf{Q}$ to the following:

$$
\begin{aligned}
q_1 &= \mu^{(0)}(4\varkappa^{(0)} - 1) - \pi\mu^{(0)}\gamma(7\varkappa^{(0)} - 2)/4, \\
q_2 &= 2\mu^{(0)} - \pi\mu^{(0)}\gamma(4-\varkappa^{(0)})/4, \quad q_3 = g_4 = \pi\mu^{(0)}\gamma(3\varkappa^{(0)} - 1)/2, \\
q_5 &= \pi\mu^{(0)}\gamma(1+2\varkappa^{(0)}), \quad q_6 = \pi\mu^{(0)}\gamma(1+\varkappa^{(0)}).
\end{aligned} \tag{3.158}
$$

In another limiting case $\gamma \gg 1$ (a needle), the representations for the tensors $\mathbf{P}$ and $\mathbf{Q}$ also have the form (3.157) where in the functions f_0 and f_1, the main terms of a serial expansion over the small parameter γ^{-1} need to be conserved: $f_0 = (\gamma^2 + 1 - \ln 2\gamma)/(2\gamma^2)$, $f_1 = -\varkappa^{(0)}(3 - 2\ln 2\gamma)/(4\gamma^2)$.

The polarization concentration tensors

$$\mathbf{R}^{v\sigma} = \mathbf{M}_1(\mathbf{I}+\mathbf{Q}\mathbf{M}_1)^{-1}, \quad \mathbf{R}^{v\epsilon} = \mathbf{L}_1(\mathbf{I}+\mathbf{P}\mathbf{L}_1)^{-1} \tag{3.159}$$

in a special **P**-basis have the components $\mathbf{R}^{v\sigma} = (r_1^\sigma, \ldots, r_6^\sigma)$, $\mathbf{R}^{v\epsilon} = (r_1^\epsilon, \ldots, r_6^\epsilon)$

$$\begin{aligned}
r_1^\sigma &= \frac{1}{2\Delta_1}\Big[\delta k + \frac{4}{3}\delta\mu + q_6\Big],\ r_2^\sigma = \frac{1}{2\delta\mu + q_2},\ r_5^\sigma = \frac{4}{4\delta\mu + q_5},\\
r_3^\sigma &= r_4^\sigma = -\frac{1}{\Delta_1}\Big[\delta k - \frac{2}{3}\delta\mu + q_3\Big],\ r_6^\sigma = \frac{2}{\Delta_1}\Big[\delta k + \frac{1}{3}\delta\mu + q_1\Big],\\
r_1^\epsilon &= \frac{1}{2\Delta_2}\Big[\frac{\lambda_1+\mu_1}{\mu_1(3\lambda_1+2\mu_1)} + p_6\Big],\ r_2^\epsilon = \frac{2\mu_1}{1+2p_2\mu_1},\\
r_3^\epsilon &= r_4^\epsilon = -\frac{1}{\Delta_2}\Big[-\frac{\lambda_1}{2\mu_1(3\lambda_1+2\mu_1)} + p_1\Big],\ r_5^\epsilon = \frac{4\mu_1}{1+\mu_1 p_5},\\
r_6^\epsilon &= \frac{1}{\Delta_2}\Big[\frac{\lambda_1+2\mu_1}{2\mu_1(3\lambda_1+2\mu_1)} + 2p_1\Big],
\end{aligned} \tag{3.160}$$

where $\lambda_1 = \lambda^{(1)} - \lambda^{(0)}$, $\mu_1 = \mu^{(1)} - \mu^{(0)}$, and

$$\begin{aligned}
\Delta_1 &= 2\Big[3\delta\mu\delta k + \delta k(q_1+q_6-2q_3) + \frac{\delta\mu}{3}(4q_1+q_6+4q_3) + q_1q_6 - q_3^2\Big],\\
\Delta_2 &= \frac{1}{2\mu_1(3\lambda_1+2\mu_1)}[1+(\lambda_1+2\mu_1)p_6 + 4(\mu_1+\lambda_1)p_1 + 4\lambda_1 p_3]\\
&\quad +2p_1p_6 - 2p_3^2,\\
\delta k &= k^{(1)}k^{(0)}/(k^{(0)}-k^{(1)}), \quad \delta\mu = \mu^{(1)}\mu^{(0)}/(\mu^{(0)}-\mu^{(1)}),
\end{aligned} \tag{3.161}$$

For the particular case of a spherical inclusion, all abovementioned tensors are isotropic $\mathbf{g} = (3g^k, 2g^\mu)$ ($\mathbf{g} = \mathbf{S}, \mathbf{P}, \mathbf{Q}, \mathbf{R}^{v\sigma}, \mathbf{R}^{v\epsilon}$, see Eq. (2.197)) and can be represented in a compact form

$$\mathbf{S} = \left(\frac{3k^{(0)}}{3k^{(0)}+4\mu^{(0)}}, \frac{6(k^{(0)}+2\mu^{(0)})}{5(3k^{(0)}+4\mu^{(0)})}\right), \tag{3.162}$$

$$\mathbf{P} = \left(\frac{1}{3k^{(0)}+4\mu^{(0)}}, \frac{3(k^{(0)}+2\mu^{(0)})}{5(3k^{(0)}+4\mu^{(0)})\mu^{(0)}}\right), \tag{3.163}$$

$$\mathbf{Q} = \left(\frac{12k^{0)}\mu^{(0)}}{3k^{(0)}+4\mu^{(0)}}, \frac{2(9k^{(0)}+8\mu^{(0)})\mu^{(0)}}{5(3k^{(0)}+4\mu^{(0)})}\right), \tag{3.164}$$

$$\mathbf{A} = \left(\frac{k^{(0)}+k_0^*}{k^{(1)}+k_0^*}, \frac{\mu^{(0)}+\mu_0^*}{\mu^{(1)}+\mu_0^*}\right), \tag{3.165}$$

$$\mathbf{B} = \left(\frac{k^{(0)} + k_0^* \, k^{(1)}}{k^{(1)} + k_0^* \, k^{(0)}}, \ \frac{\mu^{(0)} + \mu_0^* \, \mu^{(1)}}{\mu^{(1)} + \mu_0^* \, \mu^{(0)}} \right), \tag{3.166}$$

$$\mathbf{R}^{\upsilon\epsilon} = \left(3(k^{(1)} - k^{(0)}) \frac{k^{(0)} + k_0^*}{k^{(1)} + k_0^*}, \ 2(\mu^{(1)} - \mu^{(0)}) \frac{\mu^{(0)} + \mu_0^*}{\mu^{(1)} + \mu_0^*} \right), \tag{3.167}$$

$$\mathbf{R}^{\upsilon\sigma} = -\left(\frac{k^{(1)} - k^{(0)}}{3(k^{(0)})^2} \frac{k^{(0)} + k_0^*}{k^{(1)} + k_0^*}, \ \frac{\mu^{(1)} - \mu^{(0)}}{2(\mu^{(0)})^2} \frac{\mu^{(0)} + \mu_0^*}{\mu^{(1)} + \mu_0^*} \right), \tag{3.168}$$

where

$$k_0^* = \frac{4}{3}\mu^{(0)}, \ \mu_0^* = \frac{3}{2}\left(\frac{1}{\mu^{(0)}} + \frac{10}{9k^{(0)} + 8\mu^{(0)}} \right)^{-1}. \tag{3.169}$$

It should be mentioned that the representation of Eshelby and related tensors (such as **P**, **Q**, **A**, **B**) are also known in the Walpole's tensor **B**-basis (21.2.17) (for details see [1256, 1579, 1625]). For example, for an inclusion oriented in direction $\mathbf{e}_1$, Walpole's components of **S** are given by

$$\mathbf{S} = (S_{2222} + S_{2233}, S_{1111}, 2S_{2323}, 2S_{1212}, S_{1122}, S_{2211})^W \tag{3.170}$$

where the components S_{ijkl} are presented by Eqs. (3.151)–(3.154). Eshelby tensor for the inclusion oriented in the direction **n** can be found using the same components (3.170) and Walpole's basis relative to the direction **n** (21.2.17) and (21.2.20). Analogous representations of the Eshelby and related tensors in the basis $\mathbf{E}^{(\alpha)}$ (21.2.1) with an altered numeration were used in [894].

Let both materials of the matrix $v^{(0)}$ and inclusion $v^{(1)}$ be transversally isotropic with the common axis of isotropy directed along the fiber's axis **n** and presented in the special **P**-basis (21.2.7) ($i = 0, 1$):

$$\mathbf{L}^{(i)} = (k^{(i)}, 2m^{(i)}, l^{(i)}, l^{(i)}, 4\mu^{(i)}, n^{(i)}) \tag{3.171}$$

with the parameters related with the tensor components of the tensor $\mathbf{L}^{(i)}$ by Eqs. (21.2.9) and (21.12). The Eshelby tensor $\mathbf{S}(\mathbf{L}^{(0)})$ for an ellipsoidal inclusion in the transversally isotropic medium was obtained in [1637], while the **P**-basis representations of the nonzero components of the related tensors $\mathbf{P} = (p_1, \ldots, p_6)$ and $\mathbf{A} = (a_1, \ldots, a_6)$ were proposed in [746] for an infinite cylindrical fiber

$$p_1 = \frac{1}{4m^{(0)}(k^{(0)} + m^{(0)})}, \ p_2 = \frac{k^{(0)} + 2m^{(0)}}{4m^{(0)}(k^{(0)} + m^{(0)})}, \ p_5 = \frac{1}{2\mu^{(0)}}, \tag{3.172}$$

$$a_1 = \frac{1}{2}\left(1 + \frac{k_1^{(1)}}{k^{(0)} + m^{(0)}}\right)^{-1}, \ a_2 = \left[1 + \frac{m_1^{(1)}(k^{(0)} + 2m^{(0)})}{2m^{(0)}(k^{(0)} + m^{(0)})}\right]^{-1}, \tag{3.173}$$

$$a_3 = -\frac{l_1^{(1)}}{2(m^{(0)} + k^{(1)})}, \quad a_5 = \frac{2\mu^{(0)}}{\mu^{(0)} + \mu^{(1)}}, \quad a_6 = 1, \tag{3.174}$$

where $f_1^{(1)} = f^{(1)} - f^{(0)}, \quad f = (k, m, l)$.

3.7 Coated Ellipsoidal Inclusion and Imperfect Interface Models

More detailed consideration of the mechanical behavior of composite materials requires analysis of the interface between the reinforcement and the matrix. Interfacial bonding conditions play an important role that control key properties of the composite materials (CM) such as stiffness, strength, and fracture behavior. In the context of classical elasticity, the interface between constituent phases are assumed to be perfect that means continuity of both the displacement and traction vector across interface. However, in many cases imperfect interfacial bonding may exist in CMs. These situations of imperfect interfaces include the following cases: unilateral contact of two adjacent parts; weak interfacial layer due to imperfect bonding between the two phases; and inter-diffusion and/or chemical interaction zones (with properties varying through the thickness and/or along the surface) at the interface between the two phases. Significance of interface effects are drastically increased for nanocomposites when surface to volume ratio becomes appreciable and the coupled deformation-dependent interfacial energy cannot be negligible with respect to the bulk energy. In the literature, three kinds of model are often used to simulate the properties of interface regions in CM. The first kind of model can be referred to as interface models in which displacement discontinuities are assumed to exist at an interface. Interface models include the dislocation-like model, free sliding model, linear-spring interface model, and its nonlinear generalization such as cohesive zone model, where interfaces are assumed to be comprised of zero thickness nonlinear springs with a specific traction-displacement law. The second kind of model (coherent interface model) is characterized by traction discontinuities at the interface where additional interface stresses appear due to the different coordination numbers of the surface/interface atoms, different bond lengths, angles, and a different charge distribution This interface is assumed to be "attached" to both the heterogeneity and matrix without any discontinuity of displacement across it. The third kind of model is the interphase model which describes the interface region as a layer, called an interphase, between the inhomogeneity and matrix [757, 1505]. The elastic moduli of the interphase are different from those of the matrix and inhomogeneity and can be homogeneous or variable. Perfect bonding is generally assumed to prevail at both the matrix/interphase and interphase/inhomogeneity interfaces. We will consider the third kind of models in Sect. 3.7.2, whereas the interface models of the first and second kinds will be considered in Sect. 3.7.3.

We shortly consider the basic results of micromechanics for one heterogeneity in the infinite homogeneous matrix (see for references [1107, 1691]). Classical works dealing with three-phase solids with spherical or cylindrical coated inclusions are discussed, e.g., in [607, 617]. More general cases of mechanical loading, including location dependent transformation were considered in [338, 986] (see references in [99]). The effect of variation of elastic properties with radial distance from the fiber's boundary in continuously reinforced fiber composites [690, 1505] was analyzed. The above studies are restricted to isotropic materials and spherical or cylindrical reinforcement shapes. Micata and Taya [1059] applied Boussinesq–Sadowsky stress-functions when calculating the stress field for two confocal prolate spheroids embedded in an infinite body.

The thin-layer hypothesis appeared as a principal step in the investigation of coated inclusions, because it allows the use of the well-developed Eshelby [450] theory and Hill [644] interface operators for the general case of anisotropy of the materials being in contact. In his pioneering paper in this direction, Walpole [1578] assumed that the stress and strain components inside the inclusion coincide with those already determined before the coating was introduced. Afterwards this assumption was replaced by the hypothesis of homogeneity of the stress state inside the core (inclusion), and thermoelastic problems were considered in [293, 328, 627, 1257]. In most of the papers, homogeneous thermoelastic properties were assumed for the coating (a case of the multi-coated inclusion was analyzed in a similar manner, see [942]). In the present work, we relax these restrictive assumptions. The case of inhomogeneity of elastic mismatch properties in the coating is a typical situation the production of the coated inclusions and due to thermal and plastic deformations of the matrix near the inclusion. Even in situations in which for a specific reference system (connected to the unit normal and tangential vectors of the surface of the inclusion) homogeneous stress and strain fields may be assumed, the introduction of a global coordinate system requires consideration of inhomogeneous fields in the coating. We consider in the next Subsection (following [270], see also [263, 269]) thermoelastic solution for an ellipsoidal heterogeneity subjected to remote homogeneous loading with thin inhomogeneous coatings of the inclusions along the surface.

3.7.1 Elastic Moduli of Fictitious Ellipsoidal Homogeneous Inclusion

At first, let us consider a single heterogeneity (no restrictions are imposed on the shape and microstructure of the heterogeneity) v_i in an infinite medium subjected to the homogeneous fields (called the *homogeneous effective fields* in subsequent manipulations) $\overline{\boldsymbol{\sigma}}(\mathbf{x}) \equiv \overline{\boldsymbol{\sigma}} =$ const. and $\overline{\boldsymbol{\varepsilon}}(\mathbf{x}) \equiv \overline{\boldsymbol{\varepsilon}} = \mathbf{M}^{(0)}\overline{\boldsymbol{\sigma}} =$ const. are assumed to be known. Then, in view of the linearity of the problem, there exist fourth- and second-rank tensors $\mathbf{A}_i(\mathbf{x}), \mathbf{R}_i^{\epsilon}(\mathbf{x}),\ \mathbf{B}_i(\mathbf{x}),\ \mathbf{R}_i(\mathbf{x})$ and $\mathbf{C}_i(\mathbf{x}),\ \mathbf{F}_i^{\epsilon}(\mathbf{x}),\ \mathbf{C}_i(\mathbf{x}),\ \mathbf{F}_i(\mathbf{x})$, such that (with no sum on i, $\mathbf{x} \in R^d$)

$$\begin{aligned}
&\boldsymbol{\varepsilon}(\mathbf{x}) = \mathbf{A}_i(\mathbf{x})\overline{\boldsymbol{\varepsilon}} + \mathbf{C}_i^{\epsilon}(\mathbf{x}), \quad \overline{v}_i\boldsymbol{\tau}(\mathbf{x}) = \mathbf{R}_i^{\epsilon}(\mathbf{x})\overline{\boldsymbol{\varepsilon}} + \mathbf{F}_i^{\epsilon}(\mathbf{x}),\\
&\boldsymbol{\sigma}(\mathbf{x}) = \mathbf{B}_i(\mathbf{x})\overline{\boldsymbol{\sigma}} + \mathbf{C}_i(\mathbf{x}), \quad \overline{v}_i\boldsymbol{\eta}(\mathbf{x}) = \mathbf{R}_i(\mathbf{x})\overline{\boldsymbol{\sigma}} + \mathbf{F}_i(\mathbf{x}),\\
&\mathbf{R}_i^{\epsilon}(\mathbf{x}) = \bar{v}_i\mathbf{L}_1(\mathbf{x})\mathbf{A}_i(\mathbf{x}), \quad \mathbf{F}_i^{\epsilon}(\mathbf{x}) = \bar{v}_i[\mathbf{L}_1(\mathbf{x})\mathbf{C}_i^{\epsilon}(\mathbf{x}) + \boldsymbol{\alpha}_1(\mathbf{x})],\\
&\mathbf{R}_i(\mathbf{x}) = \bar{v}_i\mathbf{M}_1(\mathbf{x})\mathbf{B}_i(\mathbf{x}), \quad \mathbf{F}_i(\mathbf{x}) = \bar{v}_i[\mathbf{M}_1(\mathbf{x})\mathbf{C}_i(\mathbf{x}) + \boldsymbol{\beta}_1(\mathbf{x})],
\end{aligned} \tag{3.175}$$

where we define the stress and strain polarization tensors (which are simply a notation convenience)

$$\boldsymbol{\tau}(\mathbf{x}) = \mathbf{L}_1(\mathbf{x})\boldsymbol{\varepsilon}(\mathbf{x}) + \boldsymbol{\alpha}_1(\mathbf{x}), \ \boldsymbol{\eta}(\mathbf{x}) = \mathbf{M}_1(\mathbf{x})\boldsymbol{\sigma}(\mathbf{x}) + \boldsymbol{\beta}_1(\mathbf{x}), \ \boldsymbol{\eta}(\mathbf{x}) = -\mathbf{M}^{(0)}\boldsymbol{\tau}(\mathbf{x}), \tag{3.176}$$

respectively, vanishing in the matrix $\boldsymbol{\tau}(\mathbf{y}) \equiv \boldsymbol{\eta}(\mathbf{y}) \equiv \mathbf{0}$ ($\mathbf{y} \in v^{(0)}$). In the general case, the estimation of the tensors $\mathbf{A}_i(\mathbf{x}),\ \mathbf{B}_i(\mathbf{x}),\ \mathbf{C}_i^{\epsilon}(\mathbf{x}),\ \mathbf{C}_i(\mathbf{x})$ is a particular problem of the transformation field analysis method [425]. which can be realized by different numerical methods, for example, by finite element analysis (for details see Sect. 4.1).

We determine an arbitrary imaginary ellipsoid v_i^0 (with the smallest possible volume) with the indicator function $V_i^0(\mathbf{x})$ inside which the real inclusion $v_i \subset v_i^0$ is to be inscribed. Averaging of Eq. (3.80_2) over the volume v_i^0 and exploiting Eshelby theorem [449] leads to the algebraic equation (the case of strains can be considered in a similar manner)

$$\langle\boldsymbol{\sigma}\rangle_i^0 = \overline{\boldsymbol{\sigma}} - \frac{\bar{v}_i}{v_i^0}\mathbf{Q}_i^0\langle\boldsymbol{\eta}\rangle_i, \tag{3.177}$$

where $\mathbf{Q}_i^0(\mathbf{x}) = \mathbf{Q}_i^0 \equiv -\bar{v}_i^0\langle\boldsymbol{\Gamma}(\mathbf{x}-\mathbf{y})\rangle_i^0 = \text{const.}$ $(\mathbf{x} \in v_i \subset v_i^0,\ \mathbf{y} \in v_i^0)$ is associated with the well-known Eshelby tensor by $\mathbf{S}_i = \mathbf{I} - \mathbf{M}^{(0)}\mathbf{Q}_i$. Hereafter

$$\mathbf{f}_i \equiv \langle\mathbf{f}(\mathbf{y})\rangle_i = \bar{v}_i^{-1}\int \mathbf{f}(\mathbf{y})V_i(\mathbf{y})\,d\mathbf{y}, \ \ \mathbf{f}_i^0 \equiv \langle\mathbf{f}(\mathbf{y})\rangle_i^0 = (\bar{v}_i^0)^{-1}\int \mathbf{f}(\mathbf{y})V_i^0(\mathbf{y})\,d\mathbf{y} \tag{3.178}$$

denotes averaging of some tensor $\mathbf{f}(\mathbf{y})$ (e.g., $\mathbf{f}(\mathbf{x}) = \mathbf{A}(\mathbf{x}),\ \mathbf{B}(\mathbf{x}),\ \mathbf{C}^{\epsilon}(\mathbf{x}), \mathbf{C}(\mathbf{x})$) over the volume of the regions $\mathbf{y} \in v_i \subset v_i^0$ and $\mathbf{y} \in v_i^0$, respectively; in so doing $\mathbf{x} \in v_i$ is fixed. In deriving Eq. (3.178), it was essentially exploited that the correlation hole v_i^0 is ellipsoidal and, therefore $\mathbf{Q}_i^0(\mathbf{x}) = \mathbf{Q}_i^0 \equiv$const., and that $\boldsymbol{\eta}(\mathbf{x}) \equiv \mathbf{0}$ at $\mathbf{x} \in v_i^0 \setminus v_i$.

Substitution of Eq. (3.177) into the right-hand side of Eq. (3.175_3) averaged over the volume v_i^0 leads to the equation

$$\mathbf{B}_i^0\overline{\boldsymbol{\sigma}} + \mathbf{C}_i^0 = \overline{\boldsymbol{\sigma}}_i - \frac{\bar{v}_i}{\bar{v}_i^0}\mathbf{Q}_i^0\langle\boldsymbol{\eta}\rangle_i, \tag{3.179}$$

yielding the following representation for the average strain polarization tensors

$$\bar{v}_i\langle \boldsymbol{\eta}\rangle_i = \bar{v}_i^0(\mathbf{Q}_i^0)^{-1}[\mathbf{I}-\mathbf{B}_i^0]\overline{\boldsymbol{\sigma}}_i - \bar{v}_i^0(\mathbf{Q}_i^0)^{-1}\mathbf{C}_i^0, \tag{3.180}$$

and the relations between the averaged tensors (3.175) (no sum on i)

$$\mathbf{R}_i^0 = \bar{v}_i^0(\mathbf{Q}_i^0)^{-1}(\mathbf{I}-\mathbf{B}_i^0),\quad \mathbf{F}_i^0 = -\bar{v}_i^0(\mathbf{Q}_i^0)^{-1}\mathbf{C}_i^0, \tag{3.181}$$

$$\mathbf{B}_i^0 = \mathbf{I}-(\bar{v}_i^0)^{-1}\mathbf{Q}_i^0\mathbf{R}_i^0,\quad \mathbf{C}_i^0 = -(\bar{v}_i^0)^{-1}\mathbf{Q}_i^0\mathbf{F}_i^0. \tag{3.182}$$

We consider a particular case of the homogeneous ellipsoidal heterogeneity $v_i \equiv v_i^0$ $(k,l=1,\ldots,d)$

$$\begin{aligned}
&(\mathbf{x}-\mathbf{x}_i)\mathbf{a}^{-2}(\mathbf{x}-\mathbf{x}_i) = (x_l-x_i)(\mathbf{a}^{-2})_{kl}(x_l-x_i) \le 1,\\
&\mathbf{L}_1(\mathbf{x}) \;=\; \mathbf{L}_1^{(i)} = \text{const},\ \boldsymbol{\alpha}_1(\mathbf{x}) = \boldsymbol{\alpha}_1^{(i)} = \text{const}\quad \text{at } \mathbf{x}\in v_i,\\
&\mathbf{M}_1(\mathbf{x}) \;=\; \mathbf{M}_1^{(i)} = \text{const},\ \boldsymbol{\beta}_1(\mathbf{x}) = \boldsymbol{\beta}_1^{(i)} = \text{const}\quad \text{at } \mathbf{x}\in v_i,
\end{aligned} \tag{3.183}$$

where $\mathbf{a}$ is the second-order symmetric tensor with $\det\mathbf{a} > 0$. Then the tensors $\mathbf{A}_i$, $\mathbf{B}_i$, $\mathbf{C}_i^{\epsilon}$ and $\mathbf{C}_i$ (3.116)–(3.118) as well as the tensors

$$\begin{aligned}
&\mathbf{R}_i^{\epsilon} = \bar{v}_i\mathbf{L}_1^{(i)}\mathbf{A}_i,\ \ \mathbf{F}_i^{\epsilon} = \bar{v}_i(\mathbf{I}+\mathbf{L}_1^{(i)}\mathbf{P}_i)^{-1}\boldsymbol{\alpha}_1^{(i)},\ \ \mathbf{C}_i^{\epsilon} = -\mathbf{A}_i\mathbf{P}_i\boldsymbol{\alpha}_1^{(i)},\\
&\mathbf{R}_i = \bar{v}_i\mathbf{M}_1^{(i)}\mathbf{B}_i,\ \ \mathbf{F}_i = \bar{v}_i(\mathbf{I}+\mathbf{M}_1^{(i)}\mathbf{Q}_i)^{-1}\boldsymbol{\beta}_1^{(i)},\ \ \mathbf{C}_i = -\mathbf{B}_i\mathbf{Q}_i\boldsymbol{\beta}_1^{(i)}
\end{aligned} \tag{3.184}$$

are expressed through the Eshelby tensor.

By comparing the relations (3.175) and (3.184), we see that the average thermoelastic responses (i.e., the tensors $\mathbf{A}_i$, $\mathbf{B}_i$, $\mathbf{C}_i^{\epsilon}$, $\mathbf{C}_i$ and $\mathbf{R}_i^{\epsilon}$, $\mathbf{F}_i^{\epsilon}$, $\mathbf{R}_i$, $\mathbf{F}_i$) of heterogeneity v_i of both any shape and microstructure are the same as the responses of some fictitious ellipsoidal homogeneous inclusion with thermoelastic parameters

$$\begin{aligned}
\mathbf{L}_1^{f(i)} &= (\mathbf{P}_i^0)^{-1}((\mathbf{A}_i^0)^{-1}-\mathbf{I}),\quad \boldsymbol{\alpha}_1^{f(i)} = -(\mathbf{P}_i^0)^{-1}(\mathbf{A}_i^0)^{-1}\mathbf{C}_i^{\epsilon 0},\\
\mathbf{M}_1^{f(i)} &= (\mathbf{Q}_i^0)^{-1}((\mathbf{B}_i^0)^{-1}-\mathbf{I}),\quad \boldsymbol{\beta}_1^{f(i)} = -(\mathbf{Q}_i^0)^{-1}(\mathbf{B}_i^0)^{-1}\mathbf{C}_i^0,
\end{aligned} \tag{3.185}$$

$$\begin{aligned}
\mathbf{L}_1^{f(i)} &= (\mathbf{I}\bar{v}_i-\mathbf{R}_i^{\epsilon 0}\mathbf{P}_i^0)^{-1}\mathbf{R}_i^{\epsilon 0},\quad \boldsymbol{\alpha}_1^{f(i)} = \bar{v}_i^{-1}(\mathbf{L}_1^{f(i)}\mathbf{P}_i^0+\mathbf{I})\mathbf{F}_i^{\epsilon 0},\\
\mathbf{M}_1^{f(i)} &= (\mathbf{I}\bar{v}_i-\mathbf{R}_i^0\mathbf{Q}_i^0)^{-1}\mathbf{R}_i^0,\quad \boldsymbol{\beta}_1^{f(i)} = \bar{v}_i^{-1}(\mathbf{M}_1^{f(i)}\mathbf{Q}_i^0+\mathbf{I})\mathbf{F}_i^0.
\end{aligned} \tag{3.186}$$

The parameters (3.185) and (3.186) of fictitious ellipsoidal inclusions are simply a notational convenience. No restrictions are imposed on the microtopology of the inclusions that have any shape and arbitrarily varying (continuously or discontinuously) thermoelastic properties as well as on the inhomogeneity of the stresses and strains (e.g., singularity) and displacements (e.g., jump as in debonding) in the inclusions. The assumption of an ellipsoidal shape of the region v_i^0 was used only in order to obtain analytical representation (3.116)–(3.122) which will be used as the ingredients in the different micromechanical models (see Chap. 8). Moreover, the similar scheme of fictitious ellipsoidal inclusions is realized in [1683] (see also [378, 906, 1169]) for partially debonded inclusions.

3.7.2 Ellipsoidal Heterogeneity with Thin Coating

Now an analytical method for estimating the tensors $\mathbf{B}(\mathbf{x})$ and $\mathbf{C}(\mathbf{x})$, see (3.175), is considered for the example of a single ellipsoidal inclusion with a thin coating in an infinite matrix loaded by a constant macroscopic stress $\overline{\boldsymbol{\sigma}}$. Let the coated inclusion v_1 consist of an ellipsoidal core $v^i \subset v_1$ with a characteristic function $V^i(\mathbf{x})$ and thermoelastic parameters $\mathbf{M}^i$, $\boldsymbol{\beta}^i \equiv$ const, and a thin coating $v^c \equiv v_1 \setminus v^i$ with a characteristic function $V^c \equiv V_1 - V^i$ and thermoelastic inhomogeneous properties $\mathbf{M}^c(\mathbf{x})$, $\boldsymbol{\beta}^c(\mathbf{x}) \neq$ const (see Fig. 3.1). In the considered case of a single inclusion the origin of the coordinate system is chosen to be the center of the inclusion $\mathbf{x}^i = \mathbf{0}$ and the coordinate axes coincide with the axes of the inclusions. In addition to the decomposition (3.69), we define the jump of the material properties $\mathbf{f}$ ($\mathbf{f} = \mathbf{M}, \boldsymbol{\beta}$) across the boundary s^i between the core and the coating as $\mathbf{f}_2(\mathbf{x}) \equiv \mathbf{f}^i - \mathbf{f}^c(\mathbf{x})$. For the single coated inclusion, Eq. (3.79) yields

$$\boldsymbol{\sigma}(\mathbf{x}) = \overline{\boldsymbol{\sigma}} + \int \left[V^i(\mathbf{y}) + V^c(\mathbf{y})\right] \boldsymbol{\Gamma}(\mathbf{x}-\mathbf{y}) \left[\mathbf{M}_1(\mathbf{y})\boldsymbol{\sigma}(\mathbf{y}) + \boldsymbol{\beta}_1(\mathbf{y})\right] d\mathbf{y}. \quad (3.187)$$

In analogy to [293, 328, 1262, 1263], we find an approximative solution of Eq. (3.187) under the approximative assumption of an homogeneous stress state in the core

$$\boldsymbol{\sigma}(\mathbf{x}) \equiv \boldsymbol{\sigma}^i = \text{const}, \quad \mathbf{x} \in v^i \subset v_1. \quad (3.188)$$

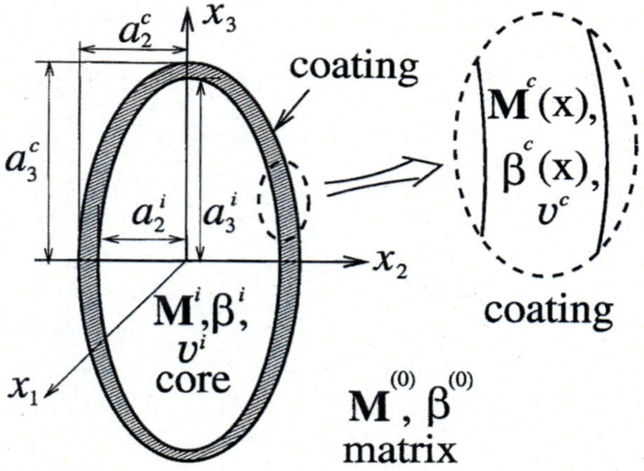

Fig. 3.1 Single coated inclusion

and the thin-layer hypothesis, which means that the characteristic function $V^c(\mathbf{y})$, can be replaced by a surface δ-function (with weighting function ρ) at the outer surface S^i_- of the boundary $S^i = S^i_- \cup S^i_+$ and the volume integral of the continuous function $\mathbf{g(y)}$, $\mathbf{y} \in v^c$ is equal to a surface integral over outer surface S^i_- [537]:

$$\int V^c(\mathbf{y})\mathbf{g}(\mathbf{y})\,d\mathbf{y} = \int_{S^i_-} \mathbf{g}(\mathbf{s})\rho\,ds, \tag{3.189}$$

where some continuous function $\mathbf{g(s)} \equiv \lim \mathbf{g(y)}$, $(\mathbf{y} \to \mathbf{s},\ \mathbf{y} \in v^c,\ \mathbf{s} \in S^i_-)$ is integrated over the surface S^i_-. In the particular case considered hereafter, the weighting function ρ for a domain v^c bounded by two ellipsoidal surfaces with the same center and with identically oriented semi-axes a_j and a^c_j $(j = 1, 2, 3)$, respectively, is estimated in [328] by $(\mathbf{y} \equiv (y_1, y_2, y_3)^\top \in s^i_-)$:

$$\rho(\mathbf{y}) = \left(\frac{y_1^2}{a_1^4} + \frac{y_2^2}{a_2^4} + \frac{y_3^2}{a_3^4}\right)^{-1/2} \sum_{j=1}^{3} \frac{(a^c_j - a_j)}{a_j}\frac{y_j^2}{a_j^2}. \tag{3.190}$$

Under these assumptions the integral equation (3.187) is, after averaging over the domain v^i, reduced to

$$\boldsymbol{\sigma}^i = \overline{\boldsymbol{\sigma}} - \mathbf{Q}^i\boldsymbol{\eta}^i + \int S^i_-(\mathbf{s})\mathbf{T}^i(\mathbf{x}^i - \mathbf{s})\left[\mathbf{M}^c_1(\mathbf{s})\boldsymbol{\sigma}(\mathbf{s}) + \boldsymbol{\beta}^c_1(\mathbf{s})\right]\rho\,ds, \tag{3.191}$$

where $\boldsymbol{\eta}^i \equiv \mathbf{M}^i_1\boldsymbol{\sigma}^i + \boldsymbol{\beta}^i_i$. Here and in the following the upper index i for the tensors $\mathbf{Q}^i$, $\mathbf{B}^i$, $\mathbf{T}^i(\mathbf{x}^i - \mathbf{s})$ stands for the calculation of these tensors for the core v^i by the use of the formulae (3.177), (3.117), (3.127$_2$), respectively. Obviously, discarding the integral term in (3.191) leads to the Eshelby solution.

Taking the properties of the interface operator $\boldsymbol{\Gamma}(\mathbf{n})$ (3.97) and (3.127$_2$) into account leads to

$$\boldsymbol{\sigma}^c(\mathbf{s}) \equiv \boldsymbol{\sigma}(\mathbf{s}) = \boldsymbol{\sigma}^i + \boldsymbol{\Gamma}(\mathbf{n}, \mathbf{M}^c)\left[\mathbf{M}^i_2(\mathbf{s})\boldsymbol{\sigma}^i + \boldsymbol{\beta}^i_2(\mathbf{s})\right],$$
$$\overline{v}^i\mathbf{T}^i(\mathbf{x}^i - \mathbf{s}) = \boldsymbol{\Gamma}(\mathbf{n}, \mathbf{M}^{(0)}) - \mathbf{Q}^i, \tag{3.192}$$

where $\mathbf{n}$ is a unit outward normal vector on ∂v^i in the point $\mathbf{s}$. Here both interface operators $\boldsymbol{\Gamma}(\mathbf{n}, \mathbf{M}^{(0)})$ and $\boldsymbol{\Gamma}(\mathbf{n}, \mathbf{M}^c)$ are defined by formula (3.127$_2$) applied with the compliances $\mathbf{M}^{(0)}$ and $\mathbf{M}^c$, respectively; $\mathbf{M}^i_2(\mathbf{s}) \equiv \mathbf{M}^i - \mathbf{M}^c(\mathbf{s})$, $\boldsymbol{\beta}^i_2(\mathbf{s}) \equiv \boldsymbol{\beta}^i - \boldsymbol{\beta}^c(\mathbf{s})$.

Thus, the Eqs. (3.191) and (3.192) represent the closed system for estimation of the stresses $\boldsymbol{\sigma}^i$ and $\boldsymbol{\sigma}(\mathbf{s})$ $(\mathbf{s} \in S^i_-$. Therefore, the stress concentration tensors $\mathbf{B}(\mathbf{x})$, $\mathbf{C}(\mathbf{x})$ in Eq. (3.175) are found to be $\mathbf{B}(\mathbf{x})$, $\mathbf{C}(\mathbf{x})$ $=$ const at $\mathbf{x} \in v^i$ and $\mathbf{B}(\mathbf{x})$, $\mathbf{C}(\mathbf{x}) \neq$ const at $\mathbf{x} \in v^c$. After that the tensors $\mathbf{R}$ and $\mathbf{F}$ are defined by Eq. (3.175) and the thermoelastic properties of the fictitious homogeneous inclusions, $\mathbf{M}_1^{f(i)}$ and $\boldsymbol{\beta}_1^{f(i)}$, are evaluated by the relations (3.185) and (3.186). Interested readers are referred to [200] and [270] (see also [263, 269]) for details.

Clearly, the thin-layer hypothesis can be rejected if the elastic properties of the coating and the matrix are the same: $\mathbf{M}^c(\mathbf{x}) \equiv \mathbf{M}^{(0)}$, $\mathbf{x} \in v^c$. Then under assumption (3.188) the Eq. (3.187) can be solved immediately, leading to

$$\boldsymbol{\sigma}^i = \mathbf{B}^i \left[\overline{\boldsymbol{\sigma}} - \mathbf{Q}^i \boldsymbol{\beta}_1^i + \int V^c(\mathbf{y}) \mathbf{T}^i(\mathbf{x}^i - \mathbf{y}) \boldsymbol{\beta}_1^c(\mathbf{y})\, d\mathbf{y} \right], \tag{3.193}$$

where $\mathbf{B}^i = (\mathbf{I} + \mathbf{Q}^i \mathbf{M}_1^i)^{-1}$. In particular, let $\boldsymbol{\beta}^i = \boldsymbol{\beta}^{(0)} = \mathbf{0}$, and $\boldsymbol{\beta}^c$ has a special form with some physical meaning represented by

$$\boldsymbol{\beta}^c = \gamma \boldsymbol{\delta} + (\chi - \gamma) \mathbf{n} \otimes \mathbf{n}, \tag{3.194}$$

where $\mathbf{n}$ is the unit outward normal vector on ∂v^i. It can easily be shown that γ and χ are the transformation parameters of the coating in the tangential and normal directions, respectively. If $\gamma = \chi$ this constitutive characterization of the coating corresponds to the particular case of an isotropical thermal expansion, considered by a number of authors (see, e.g., [293, 627]). We will analyze a less trivial case $\gamma \neq \chi$ allowing the existence of a prestress in the coating. This situation is typical for the case of production of coated inclusions separately from the matrix. Moreover, under purely thermal deformations (e.g., $\overline{\boldsymbol{\sigma}} = \mathbf{0}$), the plastic strains of the matrix near the inclusions also have the form (3.194) (see [272]). Thus, Eq. (3.193) is reduced to the following equation

$$\boldsymbol{\sigma}^i = \mathbf{B}^i \left[\boldsymbol{\sigma}^0 + (\chi - \gamma) \int V^c(\mathbf{y}) \mathbf{T}^i(\mathbf{x}^i - \mathbf{y}) (\mathbf{n} \otimes \mathbf{n})\, d\mathbf{y} \right]. \tag{3.195}$$

The numerical results obtained from finite element analysis (presented in [263]) differ from the analytical solution (3.195) by not more than 1%. Thin-layer hypothesis (3.189) provides an acceptable exactness for a not too thick coating, let's say for $h = (a^c - a^i)/a^i < 0.2$.

3.7.3 *Imperfect Interface*

In this Subsection, the second and third kinds of imperfection models are considered. If the phases are perfectly bonded, the displacements and the traction components are continuous across the interphase boundaries, i.e.,

$$[[\boldsymbol{\sigma}]] \cdot \mathbf{n} = \mathbf{0}, \quad [[\mathbf{u}]] = \mathbf{0} \tag{3.196}$$

on the interface boundary $\Gamma = \cup \Gamma_i$ $(i = 1, \ldots)$ (assumed to be sufficiently smooth) where $\mathbf{n}$ is the outward normal vector on Γ from v to $v^{(0)}$ and $[[(.)]]$ =(out)−(in) is the jump operator. The traction $\mathbf{t}(\mathbf{x}) = \boldsymbol{\sigma}(\mathbf{x})\mathbf{n}(\mathbf{x})$ acting on any plane with the normal $\mathbf{n}(\mathbf{x})$ through the point $\mathbf{x}$ can be represented in terms of displacements $\mathbf{t}(\mathbf{x}) = \hat{\mathbf{t}}(\mathbf{n}, \nabla)\mathbf{u}(\mathbf{x}) + \boldsymbol{\alpha}(\mathbf{x})\mathbf{n}$, where $\hat{t}_{ik}(\mathbf{n}, \nabla) = L_{ijkl}n_j(\mathbf{x})\partial/\partial x_l$ is the conormal derivative operator.

The assumption of perfect interfaces (2.6) is not appropriate in a lot of cases. In the first kind of imperfect interface models, the displacement components are discontinuous $[[\mathbf{u}(\mathbf{s})]] \neq \mathbf{0}$ at the surface $\mathbf{s} \in \Gamma_i^u \subset \Gamma_i$, and the strain (2.3_1) contains a delta function at the interface Γ_i^u (with zero thickness) due to differentiation of discontinuous function $\mathbf{u}(\mathbf{s})$ on Γ_i^u

$$\boldsymbol{\varepsilon}(\mathbf{x}) = \nabla \overset{S}{\otimes} \mathbf{u}(\mathbf{x}) + [[\mathbf{u}(\mathbf{x})]] \overset{S}{\otimes} \mathbf{n}_i(\mathbf{x})\delta_{\Gamma_i^u}, \tag{3.197}$$

where $\mathbf{n}_i(\mathbf{s})$ is the normal vector on Γ_i^u (assumed to be sufficiently smoothed) defined by the equation $\Gamma_i^u(\mathbf{x}) = 0$, and $\delta_{\Gamma_i^u}$ is the Dirac distribution on Γ_i^u. Hear we used the properties of the distributional derivatives of $V_i(\mathbf{x})$ (see for details, e.g., p. 127 in [657])

$$\nabla V_i(\mathbf{x}) = -\mathbf{n}(\mathbf{x})\delta_{\Gamma_i} \tag{3.198}$$

with the Dirac distribution δ_{Γ_i} on Γ_i defined by

$$(\delta_{\Gamma_i}, \phi)_{R^d} \equiv \int_{R^d} \delta_{\Gamma_i}(\mathbf{x})\phi(\mathbf{x})d\mathbf{x} := \int_{\Gamma_i} \phi(\mathbf{s})ds \tag{3.199}$$

for any test function $\phi \in C_0^\infty(R^d)$.

The normal $\mathbf{n}_i(\mathbf{s})$ is typically presented in the form of two normals $\mathbf{n}_i(\mathbf{s}) = \mathbf{n}_i^+(\mathbf{s})$ and $\mathbf{n}_i^-(\mathbf{s}) \equiv -\mathbf{n}_i^+(\mathbf{s})$ at the surfaces Γ_i^+ and Γ_i^- $(\Gamma_i = \Gamma_i^+ \cup \Gamma_i^-)$, respectively, and the last item in Eq. (3.197) is conventionally represented as $[[\mathbf{u}(\mathbf{x})]] \otimes \mathbf{n}_i(\mathbf{x})\delta_{\Gamma_i} = [\mathbf{u}(\mathbf{s}^+) \otimes \mathbf{n}_i^+(\mathbf{s}^+) + \mathbf{u}(\mathbf{s}^-) \otimes \mathbf{n}_i^-(\mathbf{s}^-)]\delta_{\Gamma_i}$, where $\mathbf{s} = \mathbf{s}^\pm$ and $\mathbf{s}^+ \in \Gamma_i^+$, $\mathbf{s}^- \in \Gamma_i^-$ are placed from the side of the inhomogeneity v_i (Γ_i^+) and the matrix $v^{(0)}$ (Γ_i^-), respectively. For later use, we introduce the two orthogonal complementary projection operators of the second order $\mathbf{P}^n = \mathbf{n}\otimes\mathbf{n}$ and $\mathbf{P}^s = \boldsymbol{\delta} - \mathbf{P}^n$. Geometrically, $P_{kl}^n(\mathbf{x})$ describes the projection along the direction $n_k(\mathbf{x})$ normal to $\mathbf{x} \in \Gamma_i$ at point $\mathbf{x}$, whereas $P_{kl}^s(\mathbf{x})$ characterizes the projection onto the plane tangent to Γ_i at $\mathbf{x}$.

In particular, for the perfect interface the last term in Eq. (3.197) is omitted since the jump of displacement is zero. In the case of a microcrack $\Gamma_i = \Gamma_i^+ \cup \Gamma_i^-$, we have zero traction at the crack surface $\boldsymbol{\sigma}(\mathbf{s}^\pm) \cdot \mathbf{n}_i(\mathbf{s}^\pm) \equiv \mathbf{0}$ $(\mathbf{s}^\pm \in \Gamma_i^\pm)$. Intermediate cases between the ideal interface and microcrack are described by the different interactions of the discontinuity surfaces Γ_i^+ and Γ_i^-. The simplest case of the first kind of model associates the material displacement discontinuities discussed above with Somigliana [1428] dislocations describing the displacements given by $\mathbf{b}(\mathbf{s})$ at the interface $\mathbf{s} \in \Gamma$. Asaro [34] proved that for the ellipsoidal transformed region and polynomial $\mathbf{b}(\mathbf{s})$ in the spatial variables, the internal strain within the region caused by the constraint of the matrix is also polynomial. Grain boundary sliding in polycrystalline and granular materials can be observed even at room temperature, and the authors [693, 1105, 1106], among others solved the problem for ellipsoidal inclusion with a frictionless sliding condition at the interface, whereas the thermoelastic problem for a functionally graded coated inhomogeneity with sliding interfaces was considered in [623]. In *free sliding model* mentioned, the normal traction and displacement are assumed to be continuous across an interface, and the shear traction is assumed to be zero

$$\mathbf{n} \cdot [[\boldsymbol{\sigma}]] \cdot \mathbf{n} = 0, \quad [[\mathbf{u}]] \cdot \mathbf{n} = 0, \quad \mathbf{P}^s \cdot \boldsymbol{\sigma} = \mathbf{0}. \tag{3.200}$$

Numerous researchers (see, e.g., [424, 614–616, 668, 1688]) have analyzed a more realistic model, so-called linear spring imperfect interface model (LSM), where the resistance of the interface is accommodated by assuming that the discontinuous components of displacement are proportional to the corresponding traction field at the interface

$$[[\boldsymbol{\sigma}]] \cdot \mathbf{n} = \mathbf{0}, \quad [[\mathbf{u}]] = \mathbf{a} \cdot \boldsymbol{\sigma} \cdot \mathbf{n} + \boldsymbol{\beta}^{\text{int}}, \tag{3.201}$$

where $\mathbf{a}$ is a second-order tensor, $\mathbf{a} = a_n\mathbf{n} \otimes \mathbf{n} + a_s\mathbf{s} \otimes \mathbf{s} + a_t\mathbf{t} \otimes \mathbf{t}$, a_n, a_s, a_t represent the interface compliance parameters in the normal and tangential directions, respectively, and $\mathbf{s}$ and $\mathbf{t}$ represent two orthogonal unit vectors in the tangent plane of the interface, while the vector $\boldsymbol{\beta}^{\text{int}} = \beta_n^{\text{int}}\mathbf{n} + \beta_s^{\text{int}}\mathbf{s} + \beta_t^{\text{int}}\mathbf{t}$ (see [400]), and β_n^{int}, β_s^{int}, β_t^{int} represent the interface parameters in the normal $\mathbf{n}$ and tangential $\mathbf{s}, \mathbf{t}$ directions, respectively. Equations (3.201) can be decomposed in the tangential and the normal directions of the interface (at $a_s = a_t$, $\beta_s^{\text{int}} = \beta_t^{\text{int}}$)

$$[[\boldsymbol{\sigma}]] \cdot \mathbf{n} = \mathbf{0}, \quad \mathbf{P}^s \cdot [[\mathbf{u}]] = a_s\mathbf{t}^s + \mathbf{P}^s \cdot \boldsymbol{\beta}^{\text{int}}, \quad \mathbf{P}^n \cdot [[\mathbf{u}]] = a_n\mathbf{t}^n + \mathbf{P}^n \cdot \boldsymbol{\beta}^{\text{int}}, \tag{3.202}$$

where $\mathbf{t}^s \equiv \mathbf{P}^s \cdot \boldsymbol{\sigma} \cdot \mathbf{n}$ and $\mathbf{t}^n \equiv \mathbf{P}^n \cdot \boldsymbol{\sigma} \cdot \mathbf{n}$ represents the shear and the normal traction at the interface, respectively. Equations (3.195) imply that the interfacial traction remains continuous (3.195_1) across the interface, while both the tangential (3.195_2) and the normal (3.195_3) displacements undergo a jump defined by the parameters a_s and a_n, respectively. For example, the case where $a_s = 0$ and $a_n = 0$ implies vanishing interface displacement jumps and therefore a perfect bonding interface, while case $a_n = 0$ and $a_s \neq 0$, only considers interfacial sliding taking place with normal contact remaining intact. Furthermore, the case where $a_n = 0$ and $a_s \to \infty$ represents the free sliding interface (3.200). Finite positive values of the interface parameters define an imperfect interface. It should be mentioned that the popular solution of the problem for an ellipsoidal inclusion with the linear spring imperfect interface (see, e.g., [86, 328, 378, 905, 906, 1169]) is based, in fact, on the thin-layer hypothesis (3.189) and (3.190). Then analog of the surface integral in Eq. (3.191) is

$$\int_{S^i_-} \mathbf{T}^{\epsilon i}(\mathbf{x}_i - \mathbf{s})\mathbf{L}^{(0)}[[\mathbf{u}(\mathbf{s})]] \otimes \mathbf{n}(\mathbf{s})\, ds, \tag{3.203}$$

where $\mathbf{T}^{\epsilon i}(\mathbf{x}_i - \mathbf{s})$ and $[[\mathbf{u}]](\mathbf{s})$ are defined by Eq. (3.127) and (3.201), respectively. For the sake of simplicity (see, e.g., [86, 378, 1671]), the integral in (3.203) is calculated by separate integration of integrants $\mathbf{T}^{\epsilon i}(\mathbf{x}_i - \mathbf{s})$ and $[[\mathbf{u}(\mathbf{s})]] \otimes \mathbf{n}(\mathbf{s})$ over the surface $\mathbf{S}^i_-$. However, the integral in Eq. (3.203) is regular one with known analytical nonsingular representation of the kernel $\mathbf{T}^{\epsilon i}(\mathbf{x}_i - \mathbf{s})$ (3.127) and can be estimated by a standard quadrature method.

The linear interfacial condition (3.201) was generalized to a *cohesive zone model* (CZM), where the traction vector assumed to be continuous is a nonlinear function of the displacement jump (bilinear, trapezoidal, exponential, and polynomial cohesive laws); linear approximation of this function at the origin leads to the spring-layer model. The cohesive model originated by Barenblatt [65] in fracture mechanics (see also [1118, 1162]) has received wide development in micromechanics of CM (see, e.g., [1169, 1483, 1484, 1544]).

The second kind of interface model (called *Gurtin–Murdoch interface stress model* (ISM) or coherent interface model, see, e.g., [587, 588, 679, 1242] where additional references can be found) can be considered as dual with respect to the linear spring-layer model, specifying that the displacement vector is continuous

$$[[\mathbf{u}(\mathbf{s})]] = \mathbf{0}, \tag{3.204}$$

across $\mathbf{s} \in \Gamma_i^\sigma \subset \Gamma_i$ where the tangential or surface projection of the strain field $\boldsymbol{\varepsilon}^s \equiv \mathbf{P}^s\boldsymbol{\varepsilon}\mathbf{P}^s$ is also continuous $[[\boldsymbol{\varepsilon}^s]] = 0$ $(\mathbf{s} \in \Gamma_i^\sigma)$, whereas the stress $\boldsymbol{\sigma}$ and strain $\boldsymbol{\varepsilon}$ fields are in general discontinuous across Γ_i^σ. The concept of surface or interface elasticity is introduced which is excluded in the classical elasticity theory and governed by 2D constitutive equation at the interface Γ_i^σ

$$\boldsymbol{\sigma}^s = \tau_0\mathbf{I}^s + 2(\mu^s - \tau_0)\boldsymbol{\varepsilon}^s + (\lambda^s + \tau_0)\mathrm{Tr}(\boldsymbol{\varepsilon}^s)\mathbf{I}^s, \tag{3.205}$$

where the considered isotropic interface is characterized by surface Lamè constants λ^s, μ^s, and surface tension τ_0; $\mathbf{I}^s$ and Tr represent the 2×2 unit tensor and the trace, respectively. Additional surface stress appears due to the difference of the equilibrium lattice spacing in the surface and the bulk that manifested for solids with large interface surface-to-volume ratio as for materials containing nanoparticles and nanovoids (see for references, e.g., the review articles [1006, 1585]). Generalization of the ISMI was proposed by Steigmann and Ogden [1440], who introduced the resistance of the surface to both stretching and bending implying that the surface energy includes both the surface membrane strain tensor and the surface curvature tensor. The coupling of the bulk and surface tensors is performed through the static equilibrium at the interface Γ_i^σ governed by the generalized Young-Laplace equation

$$[[\sigma]] \cdot \mathbf{n} + \nabla^s \cdot \boldsymbol{\sigma}^s = \mathbf{0}, \tag{3.206}$$

where the action of the surface gradient ∇^s on a vector $\mathbf{v}$ is defined through the usual 3D gradient ∇ operator: $\nabla^s \mathbf{v} \equiv \nabla \mathbf{v} \mathbf{P}^s$. In the case of absent surface terms denoted by the upper index ^{s}, Eq. (3.206) reduces to the usual traction continuity equation of classical elasticity (3.196$_1$).

Recently the coherent model was applied for analysis of nanoscale structures including the stretching, bending and torsion of rods [1064, 1375]. The classical Eshelby formalism was extended for inclusion/inhomogeneity problems to the nanoscale by taking the linear coherent model into account in [314, 315, 401, 631, 1368, 1373], see also [473, 474]. Kushch et al. [854] have obtained a complete and accurate solution for the problem of multiple interacting spherical inhomogeneities with coherent interface by the use a vectorial spherical harmonics-based analytical technique developed. The presence of interface stresses results in a size-dependent of the elastic response of nanocomposites, consistent with some experimental observations (see, e.g., [1638]).

It is interesting that both two dual basic models of the imperfect interface (spring-layer and coherent ones) were at the beginning proposed as phenomenological ones and then derived rigorously as a limiting asymptotic approach of thin interphase between two anisotropic solid (see for references and details [95, 1264]; see also [424, 614, 615, 668, 691, 979, 1104, 1262, 1383]) Namely, when a linearly elastic interphase between two linearly elastic bodies is very thin, the linear spring-layer model or the linear coherent model can be considered as a limiting case of the interphase elastic properties which are much softer or stiffer, respectively, than each of the bodies (otherwise, an imperfect interface—intermediate—model is more appropriate). The contact conditions considered can be described by the limited cases (3.201) and (3.206) where only either the displacement jump or traction jump is presented. However, a generalization of these conditions to a formalism which is able to take into account both previous schemes at the same time is of strong practical importance (see, e.g., [296, 473, 474, 578, 579, 1193]).

3.8 Related Problems for Ellipsoidal Inhomogeneity in an Infinite Medium

3.8.1 *Conductivity Problem*

It is well-known that the equations describing the steady-state conditions of such processes as heat and mass transfer, electric conductivity and permittivity, and filtration of a Newtonian liquid in undeformable cracked-porous media (see, e.g., [74, 415, 599, 881, 1384, 1389]) are mathematically equivalent. For reasons of this mathematical analogy, we consider the basic equations of steady-state transfer process:

$$\nabla \mathbf{q} = 0, \quad \mathbf{q} = -\boldsymbol{\kappa}\nabla T, \tag{3.207}$$

where T is a potential (e.g., temperature), $\boldsymbol{\kappa}$ is a conductivity tensor, $\mathbf{q}$ is a flux vector, and the first Eq. (3.207) is a transfer equation (Fourier, Ficks, Ohm, Darcy, etc.). Equations (3.207) are simpler than Eq. (3.1) that simplifies its analysis.

For materials admitting an energy density function, the conductivity tensor $\boldsymbol{\kappa}$ must be symmetric: $\varkappa_{ij} = \varkappa_{ji}$. This symmetry arises also from Onsager's (see, e.g., [1142]) reciprocity theorem for irreversible processes, in which fluxes are linearly related to intensities. For symmetry with respect to one plane (monoclinic symmetry), say the $x_1 - x_2$ plane, the tensor $\boldsymbol{\kappa}$ has four independent components:

$$\boldsymbol{\kappa} = \begin{pmatrix} \varkappa_{11} & \varkappa_{12} & 0 \\ \varkappa_{12} & \varkappa_{22} & 0 \\ 0 & 0 & \varkappa_{33} \end{pmatrix}. \tag{3.208}$$

For symmetry with respect to three orthogonal plates (orthogonal symmetry), the tensor $\boldsymbol{\kappa}$ has three independent components $\boldsymbol{\kappa} = \text{diag}(\varkappa_{11}, \varkappa_{22}, \varkappa_{33}) \equiv \text{diag}(\varkappa_1, \varkappa_2, \varkappa_3)$ referred to its principal axes. For symmetry with respect to a 90° rotation about one axis (transversely isotropic symmetry), say the x_3-axis, we have two independent components: $\boldsymbol{\kappa} = \text{diag}(\varkappa_1, \varkappa_1, \varkappa_3)$. Finally, for symmetry with respect to any rotations (isotropic symmetry), the conductivity tensor is defined by the scalar $\varkappa_0$: $\varkappa_{ij} = \varkappa_0\delta_{ij}$. All numbers $\varkappa_1, \varkappa_2$, and $\varkappa_3$ are always found to be positive that provides the positive definiteness of the tensor $\boldsymbol{\kappa}$.

Analogous to the elasticity problem (3.3), we define the Green's function as

$$\varkappa_{ij} G_{,ij}(\mathbf{x}) = -\delta(\mathbf{x}), \tag{3.209}$$

where $\delta(\mathbf{x} - \mathbf{y})$ is the delta function. For media with both arbitrary anisotropic and isotropic ($\boldsymbol{\kappa} = \varkappa_0\boldsymbol{\delta}$) properties, Green's functions G have the representation (see, e.g., [833, 879])

$$G(\mathbf{x}) = \frac{1}{4\pi r^k(\mathbf{x})}, \quad G(\mathbf{x}) = \frac{1}{4\pi \varkappa_0 |\mathbf{x}|}, \tag{3.210}$$

respectively, where $r^k(\mathbf{x}) \equiv (|\boldsymbol{\kappa}| x_i k_{ij} x_j)^{1/2}$, $|\boldsymbol{\kappa}| \equiv \det\boldsymbol{\kappa}$, $\mathbf{k} = \boldsymbol{\kappa}^{-1}$. In a case of orthotropic media in coordinates coinciding with the principal axes of $\varkappa_{ij}$, Eq. (3.209) is reduced to $\varkappa_1 G_{,11}(\mathbf{x}) + \varkappa_2 G_{,22}(\mathbf{x}) + \varkappa_3 G_{,33}(\mathbf{x}) = \delta(\mathbf{x})$, which can be solved [318] as $G(\mathbf{x}) = (4\pi)^{-1}(\varkappa_2\varkappa_3 x_1^2 + \varkappa_1\varkappa_3 x_2^2 + \varkappa_1\varkappa_2 x_3^2)^{-1/2}$, by the use of the coordinate transformation $x_i = \sqrt{\varkappa_i}\xi_i$ (no sum over i).

Following [318], we estimate local fields $T(\mathbf{x})$ and $\mathbf{q}(\mathbf{x})$ in an ellipsoidal inclusion v with conductivity of the matrix $\boldsymbol{\kappa}(\mathbf{x}) \equiv \boldsymbol{\kappa}^{(0)}$ and polynomial eigen-intensity

$$\beta_i^*(\mathbf{x}) = \beta_i + \beta_{ij}x_i x_j + \beta_{ijk}x_i x_j x_k + \cdots , \tag{3.211}$$

($\beta_{i\ldots}$ =const., $\mathbf{x} \in v$) prescribed in the coordinate system coinciding with the principal axes of the ellipsoid v. Using the Green's function technique analogously to the elastic problem (3.2), we can get the representations for the local field in the inclusion and matrix

$$T(\mathbf{x}) = \frac{1}{4\pi\varkappa_0}\frac{\partial}{\partial x_i}\int_v \frac{\beta_i^*(\mathbf{y})}{|\mathbf{x}-\mathbf{y}|}\,d\mathbf{y},\quad q_j = \frac{1}{4\pi\varkappa_0}\frac{\partial^2}{\partial x_i \partial x_j}\int_v \frac{\beta_i^*(\mathbf{y})}{|\mathbf{x}-\mathbf{y}|}\,d\mathbf{y}, \tag{3.212}$$

where we consider the isotropic matrix $\boldsymbol{\kappa}(\mathbf{x}) \equiv \varkappa_0\boldsymbol{\delta}$ ($\mathbf{x} \in R^3$) just for simplicity. Substituting (3.211) into (3.212) yields

$$T(\mathbf{x}) = \frac{1}{4\pi\varkappa_0}\frac{\partial}{\partial x_i}[\beta_i\phi(\mathbf{x}) + \beta_{ir}\phi_r(\mathbf{x}) + \beta_{irs}\phi_{rs}(\mathbf{x}) + \cdots], \tag{3.213}$$

$$q_j(\mathbf{x}) = \frac{1}{4\pi\varkappa_0}\frac{\partial^2}{\partial x_i \partial x_j}[\beta_i\phi(\mathbf{x}) + \beta_{ir}\phi_r(\mathbf{x}) + \beta_{irs}\phi_{rs}(\mathbf{x}) + \cdots], \tag{3.214}$$

where $\phi(\mathbf{x}) = \int_v |\mathbf{x}-\mathbf{y}|^{-1}\,d\mathbf{y}$, $\phi_{rs\ldots t}(\mathbf{x}) = \int_v y_r y_s \ldots y_t |\mathbf{x}-\mathbf{y}|^{-1}\,d\mathbf{y}$, and the Newtonian potential ϕ for the domain v is a solution of the equation $\Delta\phi(\mathbf{x}) = -V(\mathbf{x})$.

It turns out that the theorem of polynomial conservation (the p-property) (see Sect. 3.5) holds also for the conductivity problem. Moreover, according to Ferrers and Dyson (see, e.g., [1104]) ϕ and ϕ_i can be written in terms of elliptic integrals:

$$\phi = \frac{1}{2}[I(\lambda) - x_n x_n I_N(\lambda)],\quad \phi_i = \frac{1}{2}a_I x_i[I_i(\lambda) - x_n x_n I_{Ni}(\lambda)],\ldots \tag{3.215}$$

where

$$I(\lambda) = 2\pi a_1 a_2 a_3 \int_\lambda^\infty \frac{1}{\Delta(s)}ds,\quad I_i(\lambda) = 2\pi a_1 a_2 a_3 \int_\lambda^\infty \frac{1}{(a_i^2+s)\Delta(s)}ds,$$

$$I_{ij}(\lambda) = 2\pi a_1 a_2 a_3 \int_\lambda^\infty \frac{1}{(a_i^2+s)(a_j^2+s)\Delta(s)}ds, \tag{3.216}$$

$\Delta(s) = [(a_1^2+s)(a_2^2+s)(a_3^2+s)]^{1/2}$, and the constant λ is zero for $\mathbf{x} \in v$ and is the largest positive root of $x_1^2(a_1^2+\lambda)^{-1} + x_2^2(a_2^2+\lambda)^{-1} + x_3^2(a_3^2+\lambda)^{-1} = 1$ for $\mathbf{x} \notin v$. One follows Mura's [1104] tensorial index notation when uppercase indices always take on the same numbers as the corresponding lowercase ones but are not summed up, e.g., $x_n x_n I_N = x_1^2 I_1 + x_2^2 I_2 + x_3^2 I_3$, $x_i I_i = x_1 I_1$, if $i = 1$.

Turning back to Eqs. (3.213) and (3.214), we can recast the fields $T(\mathbf{x})$ and $\mathbf{q}(\mathbf{x})$ in the form

$$T(\mathbf{x}) = a_i(\mathbf{x})\beta_i + a_{ij}(\mathbf{x})\beta_{ij} + a_{ijk}\beta_{ijk} + \cdots,$$
$$q_r(\mathbf{x}) = b_{ri}(\mathbf{x})\beta_i + b_{rij}(\mathbf{x})\beta_{ij} + b_{rijk}\beta_{rijk} + \cdots, \tag{3.217}$$

where

$$a_i(\mathbf{x}) = \frac{1}{4\pi}\phi_i,\ a_{ij}(\mathbf{x}) = \frac{1}{4\pi}\phi_{j,i},\ a_{ijk}(\mathbf{x}) = \frac{1}{4\pi}\phi_{jk,i},\ldots$$
$$b_{ri}(\mathbf{x}) = \frac{1}{4\pi}\phi_{,ri},\ b_{rij}(\mathbf{x}) = \frac{1}{4\pi}\phi_{j,ri},\ b_{rijk}(\mathbf{x}) = \frac{1}{4\pi}\phi_{jk,ri},\ldots. \tag{3.218}$$

The explicit representations of the integrals $I(\lambda)$, $I_{i\ldots}(\lambda)$ for spheroidal inclusions can be found in [318]. It is interesting that for a uniform eigenfield $\beta_i^*(\mathbf{x}) \equiv \beta_i$, the interior intensity $\mathbf{q}(\mathbf{x})$ ($\mathbf{x} \in v$) (see Eqs. (3.217) and (3.218)) involving the second derivative of the potential $\phi(\mathbf{x})$ is spatially uniform because $\phi(\mathbf{x})$ is simply a quadratic function of $\mathbf{x} \in v$ (see Eq. (3.215_1)).

In particular, for the homogeneous ellipsoidal inhomogeneity with the conductivity $\boldsymbol{\kappa}$ inside remote homogeneous field ∇T^0, the local field ∇T inside the inhomogeneity is also homogeneous:

$$\nabla T = \left[\boldsymbol{\delta} + \mathbf{S}(\boldsymbol{\kappa}^{(0)})^{-1}\boldsymbol{\kappa}_1\right]^{-1}\nabla T^0, \tag{3.219}$$

where $\mathbf{S}$ is a symmetric *depolarization* tensor, which for the isotropic matrix and in the principal axes frame has diagonal structure

$$\mathbf{S} = \mathrm{diag}[M_1, M_2, (1 - M_1 - M_2)] \tag{3.220}$$

with the factors M_i expressed as the classical elliptic integrals [879, 1457]

$$M_i = \frac{1}{2}a_1a_2a_3\int_0^\infty \frac{1}{(a_i^2 + s)\Delta(s)}\,ds. \tag{3.221}$$

For our calculations, we present results for a spheroidal inclusions with the semi-axes $a_1 = a_2 = a$ and a_3 when

$$M_1 = M_2 = M = \frac{(1-g)\gamma^2}{2(\gamma^2 - 1)} \tag{3.222}$$

where $\gamma = a_3/a$ and g is defined by Eq. (3.156). The factors M_i can also be simply evaluated for an elliptical cylinder with $a_3 = \infty$: $M_1 = a_2/(a_1 + a_2)$, $M_2 = a_1/(a_1 + a_2)$. In particular, for a spherical inclusion ($\gamma = 1$) $M = 1/3$, for a cylinder ($\gamma = \infty$) $M = 1/2$, and for a disk-shaped inclusion ($\gamma = 0$) $M = 0$.

We assumed ideal contact between the phases when the temperature and the normal component of the heat flux are continuous at the interphase of the phases. In reality, however, perfect bonding does not exist. The availability of an interphase between the inhomogeneities and the surrounding matrix has also attracted the attention of researchers. In a simplest case the interface is considered as an additional homogeneous isotropic phase. So, one [626] (the case of n-layered isotropic spherical inclusion is considered in [637]) solves the problem for two-layered spheroids consisting of isotropic materials in an isotropic matrix in spherical coordinates using Legendre polynomials. The thickness of the interphase may be comparable to the size of the heterogeneities for some cases, for example, in thick interphases deliberately produced for various purposes. The usual scheme of work on heterogeneous media is based on modeling an inclusion surrounded by an inhomogeneous interphase by mapping it onto an effective homogeneous particle of identical size in order to predict the effective moduli. The problem is simplified in the case of a thin interphase when a boundary value problem for the inclusion-interphase-matrix configuration can be replaced by a simpler problem involving the inclusion and matrix only, plus certain matching conditions simulating the interphase. So, Benveniste and Miloh [101] considered the case of a spherical inclusion with as imperfect thermal contact at a constituent interface, when the boundary conditions on the surface of the inclusion have the form

$$\varkappa_0^{(0)}\frac{\partial T^-}{\partial \mathbf{n}} = \varkappa_0^{(1)}\frac{\partial T^+}{\partial \mathbf{n}} = \beta(T^- - T^+), \tag{3.223}$$

where $\mathbf{n}$ is the outer normal of the inclusion and T^+ and T^- are the limiting values of the potential inside and outside the inclusion, near the surface of the heterogeneity. The case $\beta \to 0$ corresponds to "adiabatic" boundary conditions and $\beta \to \infty$ corresponds to "isothermal" boundary conditions. The boundary condition (3.223) is mathematically equivalent to the surface of a homogeneous inclusion having an infinitesimally thin layer with thickness $\delta \ll a$ and conduction coefficient $\varkappa_0^{\rm int} = \beta/\delta$. Transforming the results [101], one [252] estimated the average response of such a sphere in the terms of the tensor $\mathbf{R}^t = R_0^t\boldsymbol{\delta}$ defined as $\bar{v}_1\langle \boldsymbol{\varkappa}_1\nabla T\rangle \equiv \mathbf{R}^t\nabla T^0$ analogously to the tensor $\mathbf{R}^\varepsilon$ (3.160):

$$R_0^t = 3\bar{v}_1\varkappa_0^{(0)}\Big[\beta a\varkappa_0^{(1)} + (\varkappa_0^{(0)} + \varkappa_0^{(1)})\varkappa_0^{(0)}\Big]\Big[\beta a(\varkappa_0^{(1)} + 3\varkappa_0^{(0)}) + (2\varkappa_0^{(1)} + \varkappa_0^{(0)})\varkappa_0^{(0)}\Big]^{-1}. \tag{3.224}$$

The error that arises when the interphase of finite thickness is replaced by the boundary condition (3.223) was evaluated in [626].

One [1012, 1526] treats nonideal contact of a spherical inclusion of the radius a in two limiting case of both the large and small conductivity $\boldsymbol{\kappa}^{\rm int} = \varkappa_0^{\rm int}\boldsymbol{\delta}$ of the interphase with a small thickness $\delta \ll a$ (the cases of an ellipsoidal heterogeneity with these nonideal contacts were analyzed in [101, 1070]). In the first case of the high-conductive interface with $\varkappa_0^{\rm int} \to \infty$ being of the order $1/\delta$, the limit

$$C = \frac{1}{ak_0^{(0)}} \lim_{1/\delta, \varkappa_0^{\rm int} \to \infty} \delta \varkappa_0^{\rm int} \tag{3.225}$$

remains finite. For the boundary conditions of "superconducting" interface (3.225), the temperature field is continuous at the interface, but the heat flux suffers a jump. In the elastic context Eq. (3.225) corresponds to a rigid "flake" in the matrix when the displacement is continuous, but the normal stress component has a jump on the flake.

The second case of the low-conductive (LC), or Kapitza interface model corresponds to $\varkappa_0^{\rm int} \to 0$, being of the order δ, as $\delta \to 0$ the limit

$$R = \frac{\varkappa_0^{(1)}}{a} \lim_{\delta, \varkappa_0^{\rm int} \to 0} \frac{\delta}{\varkappa_0^{\rm int}} \tag{3.226}$$

remains finite. For "resisting" boundary conditions (3.226), the heat flux remains continuous, but the temperature suffers a jump on the interface. In the elastic context, the resisting condition (3.226) corresponds to a crack where the displacement has a jump, but the normal stress component is continuous on the interface.

The solutions [1012, 1526] (see also [144, 1265]) for the sphere with the boundary conditions (3.225) and (3.226) make it possible to get the representations for the tensor $\mathbf{R}^t = R_0^t \boldsymbol{\delta}$:

$$R_0^t = \frac{3\bar{v}_1 \varkappa_0^{(0)} (\varkappa_0^{(1)} - k^{(0)})}{2k^{(0)} + 3\varkappa_0^{(1)}} + \frac{6\bar{v}_1 (\varkappa_0^{(0)})^2 C}{\varkappa_0^{(1)} + 2(1+C)\varkappa_0^{(0)}},$$
$$R_0^t = \frac{3\bar{v}_1 \varkappa_0^{(0)} (\varkappa_0^{(1)} - k^{(0)})}{2k^{(0)} + 3\varkappa_0^{(1)}} - \frac{3\bar{v}_1 (\varkappa_0^{(0)})^2 R}{\varkappa_0^{(1)} + 2(1+R)\varkappa_0^{(0)}}, \tag{3.227}$$

for the conditions (3.225) and (3.226), respectively.

3.8.2 *Scattering of Elastic Waves by Ellipsoidal Inclusion in a Homogeneous Medium*

A fundamental task of nondestructive methods of testing the structure of composites is the establishment of the relation between the elastic-wave propagation microstructural parameters of the material [509, 510]. The structural and material study of the material begins by solving the problem of a single ellipsoidal inclusion in an infinite matrix. For an inclusion of simple shape in an isotropic matrix, the T-matrix approximation was employed [792, 1557], which relates the coefficients in the series expansion, in vector spherical wave functions, of the incident and scattered waves. However, for an anisotropic matrix, [581, 1547, 1622] demonstrated an advantage of an integral equation method, starting from a representation of the scattered field as an integral involving a Green's function for the matrix analogously to the static limit (3.78). In the longwave region (lengths of the propagating waves are more than the characteristic size of inclusions), the first approximation, called the Born (or Rayleigh-Gaus) approximation [582], is obtained by substituting the incident field into the integrand of the appropriate wave analog of Eq. (3.78). In the isotropic case, the method of equivalent inclusions was used in [511] to obtain a generalization of Eshelby's theorem to dynamic problems for moderate wave-number values. Similar results have been obtained for an anisotropic material by means of series expansion, in a small vibration frequency, of the fundamental solution of the wave operator for a homogeneous medium [745, 1622]. For the case in which the mechanical properties of the phases differ slightly, the perturbation method has been used in [1641]. In this section, we will present the problem of the scattering of elastic waves by a homogeneous ellipsoidal inclusion in the longwave region following [256, 743, 745, 1622].

Consider now a homogeneous medium with the elastic moduli $\mathbf{L}^{(0)}$ and density $\rho^{(0)}$ containing a homogeneous ellipsoidal inclusion v with the moduli $\mathbf{L}^{(1)}$ and density $\rho^{(1)}$. The differential equation of motion (3.44) can be presented in the form

$$\boldsymbol{\mathcal{L}}^{(0)}\mathbf{u} = -\boldsymbol{\mathcal{L}}_1\mathbf{u} - \mathbf{f}, \tag{3.228}$$

where $\boldsymbol{\mathcal{L}}^{(0)} = \nabla\mathbf{L}^{(0)}\nabla - \boldsymbol{\delta}\rho^{(0)}\partial^2/\partial t^2$, $\boldsymbol{\mathcal{L}}_1 = \nabla\mathbf{L}_1(\mathbf{x})\nabla - \boldsymbol{\delta}\rho_1(\mathbf{x})\partial^2/\partial t^2$ and the parameters $\mathbf{L}_1(\mathbf{x})$ and $\rho_1(\mathbf{x})$ are nonzero only over the volume v. Equation (3.228) will now be applied to the problem of the scattering of a plane wave by the inclusion v when an incident plane wave with angular frequency ω and amplitude $\mathbf{u}^0$ is expressed by $\mathbf{u}^0(\mathbf{x}, t) = \mathbf{u}^0(\mathbf{x})\exp(-i\omega t)$. A steady-state solution $\mathbf{u}(\mathbf{x}, t)$ depending upon t through a factor $\exp(-i\omega t)$ satisfies the equation for the wave amplitude $\mathbf{u}(\mathbf{x})$ (see (3.64)) $\boldsymbol{\mathcal{L}}^{\omega(0)}\mathbf{u} = -\boldsymbol{\mathcal{L}}_1^{\omega}\mathbf{u} - \mathbf{f}$, which can be reduced analogously to (3.77) to the following integral equation:

$$\mathbf{u}(\mathbf{x}) = \mathbf{u}^{(0)}(\mathbf{x}) + \int \mathbf{g}(\mathbf{x}-\mathbf{y})\nabla\boldsymbol{\tau}(\mathbf{y})\,d\mathbf{y} + \int \mathbf{g}(\mathbf{x}-\mathbf{y})\boldsymbol{\pi}(\mathbf{y})\,d\mathbf{y}, \tag{3.229}$$

Here we used the time-reduced Green's function (3.66) and an assumption $\mathbf{f}_1(\mathbf{x}) \equiv \mathbf{0}$; the introduced stress polarization tensor $\boldsymbol{\tau}$ and momentum polarization $\boldsymbol{\pi}$

$$\boldsymbol{\tau}(\mathbf{x}) = \mathbf{L}_1(\mathbf{x})\boldsymbol{\varepsilon}(\mathbf{x}), \quad \boldsymbol{\pi}(\mathbf{x}) = \omega^2\rho_1(\mathbf{x})\mathbf{u}(\mathbf{x}) \tag{3.230}$$

vanish in the matrix $R^d \setminus v$. Transforming the first integral by parts and taking the equality $\nabla_{\mathbf{y}} = -\nabla_{\mathbf{x}}$ into account, we obtain

$$\mathbf{u}(\mathbf{x}) = \mathbf{u}^{(0)}(\mathbf{x}) + \int \nabla\mathbf{g}(\mathbf{x}-\mathbf{y})\boldsymbol{\tau}(\mathbf{y})\,d\mathbf{y} + \int \mathbf{g}(\mathbf{x}-\mathbf{y})\boldsymbol{\pi}(\mathbf{y})\,d\mathbf{y},$$

$$u_i(\mathbf{x}) = u_i^{(0)}(\mathbf{x}) + \int \nabla_j g_{ik}(\mathbf{x}-\mathbf{y})\tau_{jk}(\mathbf{y})\,d\mathbf{y} + \int g_{ij}(\mathbf{x}-\mathbf{y})\pi_j(\mathbf{y})\,d\mathbf{y} \tag{3.231}$$

Differentiating both sides of (3.210), we obtain the equation for the strain tensor $\varepsilon_{ij}(\mathbf{x}) = \nabla_{(i}u_{j)}(\mathbf{x})$

$$\boldsymbol{\varepsilon}(\mathbf{x}) = \boldsymbol{\varepsilon}^0(\mathbf{x}) + \int \mathbf{K}^{\omega}(\mathbf{x}-\mathbf{y})\boldsymbol{\tau}(\mathbf{y})\, d\mathbf{y} + \int \nabla\mathbf{g}(\mathbf{x}-\mathbf{y})\boldsymbol{\pi}(\mathbf{y})\, d\mathbf{y},$$

$$\varepsilon_{ij}(\mathbf{x}) = \varepsilon_{ij}^{(0)}(\mathbf{x}) + \int K_{ijkl}^{\omega}(\mathbf{x}-\mathbf{y})\tau_{kl}(\mathbf{y})d\mathbf{y} + \int \nabla_{(i}g_{j)k}(\mathbf{x}-\mathbf{y})\pi_k(\mathbf{y})d\mathbf{y}, \quad (3.232)$$

where $K_{ijkl}^{\omega}(\mathbf{x}) = \nabla_{i)}\nabla_{(j}g_{k)(l}(\mathbf{x})$.

To obtain tractable results, we approximately solve Eqs. (3.231) and (3.232) (following [743, 746]) in the longwave limit when the length of the incident wave significantly exceeds the characteristic size of v. In such a case, keeping only essential terms in the real and imaginary parts of the expansion of $\mathbf{g}$ in ω:

$$\mathbf{g}(\mathbf{x}) = \mathbf{g}_0(\mathbf{x}) + i\omega\mathbf{g}_1(\mathbf{x}) - i\omega^3|\mathbf{x}|^2\mathbf{g}_3(\mathbf{n}), \quad (3.233)$$

where for the arbitrary anisotropic medium, we have the following series expansion with respect to a small parameter ω

$$\mathbf{g}(\mathbf{x}) = \mathbf{g}_0(\mathbf{x}) + \frac{1}{|\mathbf{x}|}\sum_{k=1}^{\infty}\frac{(i\omega|\mathbf{x}|)^k}{(k-1)!}\mathbf{g}_k(\mathbf{n}), \quad \mathbf{g}_0(\mathbf{x}) = \frac{1}{8\pi^2|\mathbf{x}|}\int_{|\xi|=1}\mathbf{L}^{-1}(\boldsymbol{\xi})\delta(\mathbf{n}\boldsymbol{\xi})\, ds_{\xi},$$

$$\mathbf{g}_k(\mathbf{x}) = \frac{(\rho^{(0)})^{k/2}}{16\pi^2}\int_{|\xi|=1}\mathbf{L}^{-(k+2)/2}(\boldsymbol{\xi})|\mathbf{n}\boldsymbol{\xi}|^{k-1}\, ds_{\xi}, \quad L_{ik}(\boldsymbol{\xi}) = L_{ijkl}^{(0)}\xi_j\xi_l, \quad (3.234)$$

where $\mathbf{n} = \mathbf{x}/|\mathbf{x}|$, $\boldsymbol{\xi}$ is a vector on the unit sphere $|\boldsymbol{\xi}| = 1$ and $\mathbf{g}_0$ is the static limit of the operator $\mathbf{g}$ obeying Eq. (3.3): $\mathbf{g}_0(\mathbf{x}) = \mathbf{G}(\mathbf{x})$.

The expansion (3.233) leads to the following expression for the kernel $\mathbf{K}^{\omega}(\mathbf{x})$:

$$\mathbf{K}^{\omega}(\mathbf{x}) = \mathbf{K}_0(\mathbf{x}) + i\omega^3\mathbf{K}_3, \quad K_{3|ijkl} = -\frac{(\rho^{(0)})^{3/2}}{16\pi^2}\int_{|\xi|=1}\xi_{(i}L_{j)(k}^{-5/2}(\boldsymbol{\xi})\xi_{l)}\, ds_{\xi}, \quad (3.235)$$

where $\mathbf{K}_0(\mathbf{x}) = \mathbf{U}(\mathbf{x})$ is a static limit (3.19) of the tensor $\mathbf{K}^{\omega}(\mathbf{x})$ at $\omega = 0$, and $\mathbf{K}_3$ is a constant tensor. In a similar manner, we will find the solution of Eqs. (3.230) and (3.232) in the form of the expansion in ω. We will consider a linear approximation of the incident field

$$\mathbf{u}^0(\mathbf{x}) = \mathbf{u}^0(\mathbf{0}) + \boldsymbol{\varepsilon}^0(\mathbf{0})\mathbf{x}, \quad \mathbf{x} \in v \quad (3.236)$$

with respect to $\mathbf{x}$ in the neighborhood of the center of gravity in the inclusion v. Substituting (3.236) and the series expansion of $\mathbf{u}$ and $\boldsymbol{\varepsilon}$ in ω in Eqs. (3.230) and (3.232) and equating the terms of the same order with respect to ω, we will obtain a linear dependence for the fields $\mathbf{u}(\mathbf{x})$ and $\boldsymbol{\varepsilon}(\mathbf{x})$ inside the ellipsoid v:

$$\mathbf{u}(\mathbf{x}) = \boldsymbol{\lambda}\mathbf{u}^0(\mathbf{0}) + \boldsymbol{\Lambda}(\omega, \mathbf{a})[\mathbf{H}(\mathbf{x}) \otimes \boldsymbol{\varepsilon}^0(\mathbf{0})], \quad \boldsymbol{\varepsilon}(\mathbf{x}) = \boldsymbol{\Lambda}(\omega, \mathbf{a})\boldsymbol{\varepsilon}^0(\mathbf{0}),$$
$$\boldsymbol{\lambda} = \boldsymbol{\delta} + i\omega^3\rho_1\bar{v}\mathbf{g}_1, \; \boldsymbol{\Lambda}(\omega, \mathbf{a}) = \mathbf{A} + i\omega^3\bar{v}\mathbf{A}\mathbf{K}_3\mathbf{L}_1^{(1)}\mathbf{A}, \tag{3.237}$$

where the terms of order ω^2 in the real parts and of order ω^5 in the imaginary ones are discarded and $\mathbf{H}(\mathbf{x}) = \mathbf{x}$ in the coordinate system with origin at the center of the inclusion v with the volume $\bar{v}$. The tensor $\mathbf{A} = \mathbf{A}(\mathbf{a})$ determined by Eq. (3.116) for the static limit depends on the shape and orientation of the ellipsoid v, while the constant tensors $\mathbf{g}_1$ and $\mathbf{K}_3$ do not depend on the shape and orientation of v.

Substitution of (3.237) into the integrands of Eqs. (3.230) and (3.232) leads to the representation of the wave fields outside the inclusion $\mathbf{x} \notin v$. From (3.237) we obtain, for the ellipsoid v with center $\mathbf{x}_1 = \mathbf{0}$, an analytical representation for the displacement amplitude ($\mathbf{x} \notin v$):

$$\mathbf{u}(\mathbf{x}) = \mathbf{u}^{(0)}(\mathbf{x}) + \mathbf{G}^{\omega}(\mathbf{x})\bar{v}\omega^2\rho_1\boldsymbol{\lambda}\mathbf{u}^0(\mathbf{0}) + \nabla\mathbf{G}^{\omega}(\mathbf{x})\bar{v}\mathbf{L}_1\boldsymbol{\Lambda}\boldsymbol{\varepsilon}^0(\mathbf{0}),$$
$$u_i(\mathbf{x}) = u_i^0 + G^{\omega}_{im}(\mathbf{x})\bar{v}\omega^2\rho_1\lambda_{mk}\mathbf{u}_k^0(\mathbf{0}) + G^{\omega}_{im,k}(\mathbf{x})\bar{v}L_{1|mkpq}\Lambda_{pqln}u^0_{l,n}(\mathbf{0}) \tag{3.238}$$

and, in the general case of anisotropy of the matrix, we can assume a point approximation $\mathbf{G}^{\omega}(\mathbf{x}) = \mathbf{g}(\mathbf{x})$. A more accutrate representation has been obtained for an isotropic matrix in [1430]:

$$\mathbf{G}^{\omega}(\mathbf{x}) = \mathbf{g}^{\mathrm{T}}I^{\mathrm{T}} + \mathbf{g}^{\mathrm{L}}I^{\mathrm{L}},$$
$$\mathbf{g}^{\mathrm{T}} = \frac{1}{4\pi\rho^{(0)}\omega^2}\Big(k_{\mathrm{T}}\frac{\exp(ik_{\mathrm{T}}|\mathbf{x}|)}{|\mathbf{x}|}\boldsymbol{\delta} + \nabla\nabla\frac{\exp(ik_{\mathrm{T}}|\mathbf{x}|)}{|\mathbf{x}|}\Big),$$
$$\mathbf{g}^{\mathrm{L}} = \frac{1}{4\pi\rho^{(0)}\omega^2}\nabla\nabla\frac{\exp(ik_{\mathrm{L}}|\mathbf{x}|)}{|\mathbf{x}|}, \quad I^{\alpha} = \bar{v}^{-1}\int e^{ik_{\alpha}\mathbf{x}\cdot\xi}dv_{\xi}, \tag{3.239}$$

where $k_\alpha = \omega/c_\alpha$ and c_α (3.45) ($\alpha = L, T$) are the wave numbers and propagation velocities of longitudial and transverse waves in a matrix. As was shown in [511], approximation (3.238) and (3.239) yields an error of not more than 5% in estimating $\mathbf{u}(\mathbf{x})$, $\mathbf{x} \notin v$ for $ka^1 \le 2$. For the isotropic medium (3.56) with Lamé moduli $\lambda^{(0)}$ and $\mu^{(0)}$, we have

$$(\rho^{(0)})^{k/2}L_{ij}^{-k}(\boldsymbol{\xi}) = \frac{1}{c_{\mathrm{T}}^{2k}}(\delta_{ij} - \xi_i\xi_j) + \frac{1}{c_{\mathrm{L}}^{2k}}\xi_i\xi_j, \tag{3.240}$$

allowing to obtain the representations

$$\mathbf{g}_k(\mathbf{n}) = \frac{c_{\mathrm{T}}^{-(k+2)}}{4\pi\rho^{(0)}k(k+2)}[(1 + k + \eta^{k+2})\boldsymbol{\delta} + (k-1)(\eta^{k+2} - 1)\mathbf{n} \otimes \mathbf{n}]. \tag{3.241}$$

In particular, the isotropic tensors $\mathbf{g}_1 = g_1\boldsymbol{\delta}$ and $\mathbf{K}_3 = (3K_3^k, 2K_3^\mu)$ are defined by the expressions

$$g_1 = \frac{2+\eta^3}{12\pi\rho^{(0)}k_{\rm T}^3}, \quad 3K_3^k = -\frac{\eta^5}{36\pi\rho^{(0)}k_{\rm T}^5}, \quad 3K_3^\mu = -\frac{3+2\eta^5}{60\pi\rho^{(0)}k_{\rm T}^5}. \tag{3.242}$$

For a spherical inclusion, the tensor $\mathbf{A}$ is also isotropic and expressed by Eq. (3.165), and $I^\alpha = [3/(k_\alpha a)^2][\sin(k_\alpha a) - k_\alpha a\cos(k_\alpha a)]$, $(\alpha = \text{L,T})$.

Further development of the approaches to the scattering of elastic waves by ellipsoidal inclusions deserve attention. A wave generalization of the theorem of polynomial conservation was obtained in [464]; the limiting cases of the cracks, discs, or fibers were considered in [744, 1622]. Binary interaction of inclusions is analyzed in [925] taking the polynomial dependence on coordinates of the displacement field inside the inclusions. For isotropic materials the problems of elastic wave propagation in the cylindrical and spherical inclusions coated by any number of homogeneous layers of constant thickness were considered [780, 1140] (see also [282, 1274]). Scattering of anti-plane shear waves by a single piezoelectric cylindrical inhomogeneity partially bonded to an unbounded piezomagnetic matrix are obtained in [395] by means of the wave function expansion method. Recently a comprehensive database of the T-matrix method since 2017 through 2019 has been published [1080] (where the references on previous databases can be found).

3.8.3 *Piezoelectric Problem*

Analogously to the linear elastic composite materials, we will consider the Eshelby tensor for a piezoelectric medium that play a central role in the characterization of piezoelectric composites. A solution of a single inclusion in an infinite piezoelectric solid has been analyzed in [91, 408, 410, 411, 671, 919, 1057, 1582]. Following [919], we will reproduce the relevant schematic representations of the Eshelby and related tensors.

Consider an infinite piezoelectric medium R^d containing an ellipsoidal inclusion v with piecewise constant parameters $\mathbf{g}(\mathbf{x}) = \mathbf{g}^{(0)} + \mathbf{g}_1(\mathbf{x})$ $(\mathbf{g} = \mathbb{L}, \mathbb{M}, \boldsymbol{\Lambda})$ (2.146): $\mathbf{g}^{(0)} \equiv$const., $\mathbf{g}_1(\mathbf{x}) =$const. if $\mathbf{x} \in v$ and $\mathbf{g}_1(\mathbf{x}) = \mathbf{0}$ otherwise; $v = \{\mathbf{x}\,|\,x_i(a^{-2})_{ij}x_j \le 1\}$, $a_{ij} = a_i\delta_{ij}$ (no sum on i). Analogously to Eq. (3.3), we define the Green's function under isothermal conditions $\nabla\mathbb{L}^{(0)}\nabla\mathbb{G}(\mathbf{x}) = -\delta(\mathbf{x})\text{diag}(\boldsymbol{\delta}, -1)$, which can be calculated through the integral over a unit sphere in a Fourier space:

$$\mathbb{G}(\mathbf{x}) = \frac{1}{8\pi^2}\int_{|\zeta|=1} \mathbb{G}(\boldsymbol{\zeta})\delta(\boldsymbol{\zeta}\cdot\mathbf{x})\,ds(\boldsymbol{\zeta}), \quad \mathbb{G}(\boldsymbol{\zeta}) = \begin{pmatrix} G_{ij}(\boldsymbol{\zeta}) & \gamma_i^\top(\boldsymbol{\zeta}) \\ -\gamma_i(\boldsymbol{\zeta}) & g(\boldsymbol{\zeta}) \end{pmatrix},$$
$$G_{ij} = \left(L_{ij} - \lambda^{-1}H_i h_j\right)^{-1}, \ \gamma_j = \lambda^{-1}h_i G_{ij}, \ g = -(\lambda + h_i L_{ij}^{-1}H_j)^{-1},$$
$$L_{ij}(\boldsymbol{\zeta}) = L^{(0)}_{ijkl}\zeta_k\zeta_l, \ H_i(\boldsymbol{\zeta}) = e^{(0)}_{kil}\zeta_k\zeta_l, \ h_j(\boldsymbol{\zeta}) = e^{(0)}_{jkl}\zeta_k\zeta_l, \ \lambda(\boldsymbol{\zeta}) = k^{(0)}_{ij}\zeta_i\zeta_j, \quad (3.243)$$

where $|\boldsymbol{\zeta}| = 1$ is a unit sphere centered at $\boldsymbol{\zeta} = \mathbf{0}$. Then the Eshelby tensor $\mathbb{S} = \mathbb{PL}$ is expressed through the tensor

$$\mathbb{P} = -\int \mathcal{D}\mathbb{G}(\mathbf{x}-\mathbf{y})\mathcal{D}V(\mathbf{y})\,d\mathbf{y} = \frac{|\det a|}{4\pi}\int_{|\zeta|=1} \rho^{-3}(\boldsymbol{\zeta})\mathbb{U}(\boldsymbol{\zeta})\,ds(\boldsymbol{\zeta}), \qquad (3.244)$$

where $\mathbb{U}(\boldsymbol{\zeta}) = \boldsymbol{\zeta}\mathbb{G}(\boldsymbol{\zeta})\boldsymbol{\zeta}$, $\rho(\boldsymbol{\zeta}) = \sqrt{\zeta_i(a^2)_{ij}\zeta_j}$, and $\mathcal{D} =$ diag(def, grad). Thus, the generalized homogeneous field inside the homogeneous ellipsoidal inclusion $\boldsymbol{\mathcal{E}}(\mathbf{x})$ $(\mathbf{x} \in v)$ (2.145) is expressed via a homogeneous remote filed $\boldsymbol{\mathcal{E}}^0$ by the equation

$$\boldsymbol{\mathcal{E}}(\mathbf{x}) = \mathbb{A}(\boldsymbol{\mathcal{E}}^0 - \mathbb{P}\boldsymbol{\Lambda}_1), \ \mathbb{A} = (\mathbb{I} + \mathbb{PL}_1)^{-1}, \qquad (3.245)$$

where $\mathbb{I} = \text{diag}(\mathbf{I}, \boldsymbol{\delta})$, $\theta \equiv$const.

Dunn [408] has obtained the explicit representations of Eshelby tensor for elliptical cylindrical inclusion and thin-disc inclusion embedded in a transversaly isotropic matrix. Explicit results for the piezoelectric Eshelby tensors are obtained in [1057] for a spheroidal inclusion aligned along the axis of the anisotropy in a transversely isotropic piezoelectric material. For a spheroidal inclusion in a transversely isotropic piezoelectric matrix, the coupled fields inside the inclusion are expressed in [1582, 1583] (see also [918]) in terms of a system of the linear algebraic equations containing only some single integral. Shermergor and Yakovlev [1379] considered the surface integral representation of the Green's function $\mathcal{G}$ and proved the property of polynomial conservativity of an integral operator with the kernel $\mathcal{DGD}$ for the ellipsoidal homogeneous inclusion. Shermergor and Yakovlev [1379] suggested generalization of the interface operator by Hill [644] for piezoelectric contacted materials (at $\boldsymbol{\Lambda} \equiv \mathbf{0}$), while Buryachenko and Shermergor [280] (see also Sect. 6.6.4) analyzed a similar problem at $\boldsymbol{\Lambda} \not\equiv \mathbf{0}$ by other more compact method. Because of the awkwardness of the obtained Eshelby tensors, interested readers are referred to the cited publications.

Chapter 4
Multiscale Analysis of the Multiple Interacting Inclusion Problem: Finite Number of Interacting Inclusions

Abstract Semi-analytical and numerical solutions for one and a few ellipsoidal and nonellipsoidal heterogeneities in an infinite matrix are considered in this chapter. These solutions are analyzed by the finite element analysis (FEA), boundary integral equation (BIE) methods (in most details the method of fundamental solution, MFS), and volume integral equation (VIE) method. One describes a series of advantages and disadvantages of these methods in one or another particular problems, and it is crucial for the analyst to be aware of their range of applications. Operator representations of solutions for one and two inclusions are presented for subsequent exploitation in the rest of the book.

4.1 Description of Numerical Approaches Used for Analyses of Multiple Interacting Heterogeneities

Elastic analysis of local fields in heterogeneous solids containing multiple inclusions subjected to remote loading is of considerable interest in many engineering disciplines. Several techniques have been proposed for computing multiparticle interactions in an infinite medium. In the two-dimensional case, the method of Kolosov–Muskhelishvili's complex potentials [653, 1110, 1332] is now well recognized as a powerful and easy-to-use tool for solving a variety of the plane many-inclusion linear elastic problems with an accurate account for the inclusion-inclusion interaction. Combining the body force method with complex stress function theory, a complex integral equation method was proposed in [403, 651] (see also [1083, 1129]) to study interaction problems between holes and other defects in an infinite or semi-infinite elastic plane. The conformal mapping technique was generalized in [1306] for the analysis of elastic fields in two joined half-planes with an inclusion of an arbitrary shape. Sherman [1377] constructed an elegant complex variable method based on the Kolosov–Muskhelishvili potentials leading to a singular integral equation (see also [1129]) for a complex-valued density. An effective numerical algorithm utilizing this method [1670] was developed, and Greengard and Helsing [571] combined Sherman's method with the fast multipole method and adaptive quadrature technique; the methods mentioned

V. A. Buryachenko, *Local and Nonlocal Micromechanics of Heterogeneous Materials*, https://doi.org/10.1007/978-3-030-81784-8_4

can be implemented in two-dimensional piecewise inhomogeneous media with isotropic constituents. Application of complex potentials to solve some particular problems for the interacting elliptic inclusions was considered in [795] in the framework of the point-to-point collocation scheme. Kaloerov [734] has applied least-squares fitting to determine the unknown series expansion coefficients. The authors [862, 863] (see also [239, 852, 857]) proposed the solution of interacting elliptical inclusions with no restrictions on their number, size, aspect ratio, elastic properties, and arrangement. By exact satisfaction of all the interface conditions, a primary boundary-value problem stated on a complicated heterogeneous domain has been reduced to an ordinary well-posed set of linear algebraic equations. A properly chosen form of potentials provides a remarkably simple form of a solution and thus an efficient computational algorithm.

However, in three dimensions, there are no analogous powerful methods to analyze the multiparticle interaction problems. Thus, Sternberg and Sadowsky [1447] analyzed the axisymmetric problem for two spherical cavities, whereas the case of spherical inclusions was investigated by Chen and Acrivos [304] by the use of the Boussinesq–Papkovich stress functions and multipole expansion technique in which the solutions are expanded into a series of spherical harmonics with respect to the centers of spheres. In the multipole expansion method [1094], the Fredholm integral equation of the second kind was approximated using Taylor's expansion with the simulation of a neighboring inclusion by a singular source of the polarization strain located at the center. Moschovidis and Mura [1096] used Taylor's expansion twice: for the transformation field inside the inclusion being considered as well as for the strain fields induced by the neighbors. In a similar integral equation, Chen and Young [301] (see also [708, 903]) used a Taylor series expansion of the constrained displacement about the origin of inclusions; the solution of this equation reduced to the determination of the biharmonic potential functions and their derivatives for an isotropic system and for the inclusion morphology under consideration.

In recent years, the multiparticle interaction problem was considered via several techniques. McPedran and Movchan [1047] developed a method of multipole expansions for an analysis of interacting long cylinders for the two-dimensional case, while the same method for the case of elastic interactions of non-rigid spherical particles with imperfect interfaces was developed by Sangani and Mo [1327]. The essence of the method [558] analyzed N interacting spheres is the representation of the matrix displacement vector as a sum of general solutions for a single-particle problem. The latter is expressed by a series of partial vectorial solutions of Navier's equation in a spherical basis. Satisfaction of interfacial boundary conditions leads to an infinite linear algebraic system with a normal determinant. A transversely isotropic solid with a single inclusion subjected to the polynomial external loading was considered Podil'chuk [1223, 1224] by the use of separation of variables in the properly introduced curvilinear coordinates. The last two methods were essentially developed and generalized by Kushch [847] (see also [849, 856, 859, 861]) to transversely isotropic phases and ellipsoidal inhomogeneities with no restrictions on their number, size, aspect ratio, elastic properties, and arrangement. Application of this approach to a polymer nanocomposite of clustered structure was considered in [242] (see also Chap. 8).

Of course, a realistic model of the problem cannot be solved by analytical methods in the general case of inclusion shape and its coating structure. Powerful numerical methods, such as finite element method (FEM) (see [281, 529, 546, 589, 800, 1459, 1463, 1536, 1694]) and boundary integral equations (BIE) method (see [5, 320, 432, 685, 960, 1179]) are useful for the solution of these problems. However, the FEM (including both the generalized [1459] and extended [1463] versions incorporating special functions which reflect the local character of the solution into the construction of the approximation) requires discretization of the full domain containing the inclusions and both a triangulation of microinhomogeneous structures and efficient solution methods may still be a different task.

The BIE methods (see, e.g., [59, 156, 657, 1101, 1189, 1442, 1548]) are well established and most favorable ones when solving homogeneous differential equations of elasticity especially when considering exterior boundary-value problems. The BEMs are ideally suited for macroscale analysis with the well-known advantages of reduction of dimension by one (because only a boundary discretization is needed) and can easily handle complex geometries along with general loading conditions (see, e.g., [458]). The efficiency of the BEMs is significantly improved with the fast multipole method (FMM, [573]) that allows one to reduce both the CPU time and memory requirement in FMM accelerated BEM from $O(N^2)$ (when an iterative solver is applied) to $O(N)$ and increase the number N of unknowns to 10 millions (see, e.g., [957, 1128, 1680]). When dealing with inhomogeneous partial differential equations, a volume potential requires some volume triangulation of the whole domain and the main advantage of BEMs is lost. Similar difficulties for BEMs appear at the analysis of random structure composites because the impact of surrounding heterogeneities on the fixed one is similar to the action of some auxiliary body forces (or equivalent stresses). Moreover, if the inclusion has a complicated coating structure with continuously inhomogeneous layers [309]), the boundary integral equations (BIEs) become extremely cumbersome in dealing with the problem.

Despite the FEM's and BEE methods's popularity, there are many problems (e.g., meshing surface in 3D, computing of singular integrals) where it is desirable to improve the efficiency of traditional methods. In the alternative (proposed by Trefftz), the trial functions must satisfy the governing equation exactly and the error in the satisfaction of the boundary conditions is minimized. As a consequence, it is interesting to develop some alternative so-called meshless methods (see [461]) for an overview and historical background on the subject) of the local boundary integral equation, boundary knot method, boundary collocation method, non-dimensional dynamic influence functions method, and method of fundamental solutions (MFS proposed by Kupradze and Aleksidze [841]) belong to the BIE methods for solving boundary-value problems, which can be recognized as a discrete type of the indirect boundary element method with a concentrated source instead of distribution (see for references [43, 82, 319, 557]). In this study, we will bring the MFS systematically for multiple inclusions inside infinite medium subjected to inhomogeneous loading.

The MFS belonging to the BIE technique shares the same advantages as the BEM over domain discretization methods. Moreover, the MFS has certain attractive advantages over the BEM, since it is truly meshless (which requires neither domain nor boundary mesh structure), is simple to program, and is able to take into account sharp changes in geometry. The MFS and related methods over the last years have found extensive application in computing solutions to a broad range of problems [303, 455].

To avoid using the fundamental solution for materials with continuously varying anisotropy, the volume integral equation (VIE) method was proposed in [899, 900, 1004, 1319] (see also [896]) for the analysis of the stress field in an isotropic space containing a finite number of anisotropic inclusions of an arbitrary shape. The hybrid boundary-domain element method [386, 897] (which we will call also the hybrid BIE and VIE method) is applied to the discretization of the integral equations thus obtained, in which the boundary is divided into surface elements while the inner domain is divided into volume elements. The hybrid BIE and VIE methods considered in this chapter enable one to restrict discretization to the inclusions only (in contrast to the FEM), and an inhomogeneous structure of inclusions presents no problem in the framework of the same numerical scheme (compared to the standard BIE method). A version of the hybrid method proposed by Dong and Bonnet [385] involves the fundamental solution for an isotropic half-plane medium. If the number of interacting inclusions n is small enough, the standard numerical method for the solution of the Fredholm integral equation of the second kind can be used. Increasing n to 10^4 leads to the necessity of employing the iterative solution strategies with usually a small number of required iterations (or a combination of the analytical and numerical methods proposed in this article) reducing the operation count from $O(n^3)$ to $O(n^2)$ [1204]. In so doing at each iteration, the fast multipole method (FMM) [571, 573], for solving the many-body electrostatics problem can be extremely effective which reduces the computing cost from $O(n^2)$ to $O(n)$. Very recently some versions of FMM were generalized for the solution of 2D [571, 1196], and 3D (see [514] where additional references can be found) problems of linear elasticity. Since FMM is usually combined with the BIE method, which employs Green's functions for each isotropic homogeneous constituent, it cannot be easily extended to the composite materials containing the inclusions with anisotropic and especially with either spherically or cylindrically anisotropic anisotropic [309, 828] and nonlinear properties, although the VIE method used in this chapter requires no corrections for these cases if the matrix is isotropic.

Additional fundamental difficulties appear in the analysis of micro-macro problems when micro-inclusions and their spacing have a length scale that is a few orders of magnitude smaller than the length scales of the macroscopic problem (geometric, loading, and boundary conditions); related problems are the boundary-value problem for a solid containing inclusions leading to the so-called edge effect and inclusions in bimaterials consisting of two semi-infinite solids (see [1107] for references). In the consideration of a spherical (or circular) inclusion in an elastic half-space, one Jasiuk et al. [692, 904] used an infinite series of harmonic functions, which corresponds to an infinite number of eigenstrains; the method is restricted by both regular shape of inclusions and isotropic properties of constituents. A similar problem for a circular tunnel in an isotropic half-space was solved by Verruijt [1562, 1563] via the method of Kolosov–Muskhelishvili's complex potentials. In an alternative approach for the analysis of an inclusion in transversely isotropic bimaterials, Yu et al. [1665] use the fundamental solution for transversely isotropic bimaterials and the equivalent inclusion method by Eshelby [450]; a system of singular integral equations obtained can be solved numerically. Particular cases of spherical and finite cylindrical inclusions in an isotropic bimaterial are investigated in [1580] in detail. A finite number of plies with cylindrical inclusions in a finite body were analyzed by Pagano and Rybicki [1175, 1285, 1312] by the finite element method; they discovered a fundamental role of an edge effect leading to the redistribution of local stresses in a boundary layer region compared to the stress estimations in remote cells. The authors [1176, 1270] proposed a combination of the effective medium and micromechanical theories leading to reasonably accurate interface stresses. Several authors [1276, 1280, 1343] have proposed a hybrid method connecting a finite element discretization for the near field and a boundary element discretization for a far-field as well as a hierarchical coupling of standard FE meshes on the micro and the macro scales.

A hybrid method based on the combination of the VIE and BIE methods is proposed by Buryachenko [200] for the micro–macro solution of elastostatic 2D and 3D *multiscale problems* in bounded or unbounded solids containing interacting multiple inclusions of essentially different scale. The hybrid micro–macro formulation allows decomposition of the complete multiscale problem into two associated subproblems, one residing entirely at the micro-level and the other at the macro-level. The efficiency of the standard iterative scheme of the BIE and VIE methods for the singular integral equation involved is enhanced by the use of a modification in the spirit of a subtraction technique as well as by a judicious choice of the initial analytical approximation for interacting inclusions (micro-level) in an unbounded medium subjected to inhomogeneous loading.

The judicious choice of the initial analytical approximation is based on the decomposition of the original problem for n macro and micro inclusions into n uncoupled problems for a single inclusion v_i $(i = 1, \ldots, n)$ subjected to a complicated so-called effective field $\overline{\boldsymbol{\sigma}}_i(\mathbf{x})$. The approximate solution of the n-particle interaction problem obtained in the framework of the homogeneity hypothesis of effective fields in the vicinity of micro inclusions was chosen as a zero-order approximation for effective fields in the surrounding inclusions v_j $(j = 1, \ldots, n;\ j \neq i)$. This step of our approach coincides with the previous scheme [188, 260, 270] proposed for coated inclusions and generalizes the approaches [116, 712, 722, 776, 1299, 1632] for homogeneous ellipsoidal inclusions as well as the approach [836] for the point approximation of interacting inclusions $\mathbf{T}_{ij}(\mathbf{x}_i - \mathbf{x}_j) = \boldsymbol{\Gamma}(\mathbf{x}_i - \mathbf{x}_j)$ (for references see also [494]). To obtain the first-order approximation, the inclusion v_i is loaded by inhomogeneous effective stress $\overline{\boldsymbol{\sigma}}_i(\mathbf{x})$ obtained as a superposition of the remote stress and the disturbances caused by the other inclusions loaded by the known homogeneous fields $\overline{\boldsymbol{\sigma}}_j$ $(j = 1, \ldots, n;\ j \neq i)$. This second step is motivated by an assumption made by Kachanov [729] in an analysis of interacting cracks who has used an analytical solution for a single crack loaded by the effective field $\overline{\sigma}_i(\mathbf{x})$. The numerical solution of singular integral equations (SIEs) involved is determined by the BIE and VIE methods with the evaluation of the appropriate integrals with a Gauss quadrature formula. In doing so, the singular terms in the Gauss formulae are handled in the spirit of a subtraction technique [323] that eliminates the difficulties connected with the use of singular finite elements in the VIE considered in [897, 899, 900]. The next-order approximations are constructed by the iteration method. The accuracy and efficiency of the method are examined through comparison with results obtained from FE methods.

4.2 Basic Equations for Multiple Heterogeneities and Numerical Solution for One Inclusion

4.2.1 *Basic Equations for Multiple Heterogeneities*

Let a linear elastic body occupy an open-bounded domain $w \subset R^d$ with a smooth boundary Γ and with a characteristic function W and space dimensionality d ($d = 2$ and $d = 3$ for 2D and 3D problems, respectively). The domain w contains a homogeneous matrix $v^{(0)}$ and a finite set $X = (v_i)$ of inclusions v_i with characteristic functions V_i and is bounded by the closed smooth surfaces Γ_i $(i = 1, 2, \ldots, n)$; $\Gamma \cap X = \emptyset$. For the sake of definiteness, in the 2D case, we will consider a Plane–strain problem. At first no restrictions are imposed on the elastic symmetry[1] of the phases or on the geometry of the inclusions. Stresses and strains are related to each other via the constitutive equations $\boldsymbol{\sigma}(\mathbf{x}) = \mathbf{L}(\mathbf{x})\boldsymbol{\varepsilon}(\mathbf{x}) + \boldsymbol{\alpha}(\mathbf{x})$ or $\boldsymbol{\varepsilon}(\mathbf{x}) = \mathbf{M}(\mathbf{x})\boldsymbol{\sigma}(\mathbf{x}) + \boldsymbol{\beta}(\mathbf{x})$ (2.126). In particular, for isotropic constituents the local stiffness tensor $\mathbf{L}(\mathbf{x}) = (dk(\mathbf{x}), 2\mu(\mathbf{x}))$ (2.207) ($d = 2, 3 = R^d$) is given in terms of the local bulk modulus $k(\mathbf{x})$ and the local shear modulus $\mu(\mathbf{x})$, and the local eigenstrain $\boldsymbol{\beta}(\mathbf{x}) = \beta_0(\mathbf{x})\boldsymbol{\delta}$ is given in terms of the bulk component $\beta_0(\mathbf{x})$. All tensors $\mathbf{f}$ ($\mathbf{f} = \mathbf{L}, \mathbf{M}, \boldsymbol{\alpha}, \boldsymbol{\beta}$) of material properties are decomposed as $\mathbf{f} \equiv \mathbf{f}^{(0)} + \mathbf{f}_1(\mathbf{x})$; see Eq. (3.69), where $\mathbf{f}_1(\mathbf{x}) = \mathbf{f}_1^{(i)}(\mathbf{x})$ is an inhomogeneous function inside the inclusions v_i $(i = 1, 2, \ldots, n)$. The upper index of the material properties tensor in parentheses shows the number of the respective component. The subscript 1 denotes a jump of the corresponding quantity (e.g., of the material tensor). We assume that the phases are perfectly bonded [see Eq. (3.85)], so that the displacements and the traction components are continuous across the interphase boundaries. The traction $\mathbf{t}(\mathbf{x}) = \boldsymbol{\sigma}(\mathbf{x})\mathbf{n}(\mathbf{x})$ acting on any plane with the normal $\mathbf{n}(\mathbf{x})$ through the point $\mathbf{x}$ can be represented in terms of displacements; see Eq. (2.130). The boundary conditions

[1] It is known that for 2D problems the plane-strain state is possible only for material symmetry no lower than orthotropic (see, e.g., [910]) that will be assumed hereafter in 2D case.

at the interface boundary will be considered together with the mixed boundary conditions on Γ with the unit outward normal $\mathbf{n}^\Gamma$

$$\mathbf{u}(\mathbf{x}) = \mathbf{u}^\Gamma(\mathbf{x}), \quad \mathbf{x} \in \Gamma_u,$$
$$\boldsymbol{\sigma}(\mathbf{x})\mathbf{n}^\Gamma(\mathbf{x}) = \mathbf{t}^\Gamma(\mathbf{x}), \quad \mathbf{x} \in \Gamma_t, \tag{4.1}$$

where Γ_u and Γ_t are prescribed displacement and traction nonintersected boundary conditions such that $\Gamma_u \cup \Gamma_t = \Gamma$, $\Gamma_u \cap \Gamma_t = \emptyset$. $\mathbf{u}^\Gamma(\mathbf{x})$ and $\mathbf{t}^\Gamma(\mathbf{x})$ are prescribed displacement on Γ_u and traction on Γ_t, respectively; mixed boundary conditions, such as in the case of elastic supports, are possible. As usual we shall distinguish the interior from the exterior problem according to whether the body occupies the interior or the exterior domain with respect to Γ.

The introduction of jumps of material properties allows one to define the stress $\boldsymbol{\tau}$ and strain $\boldsymbol{\eta}$ polarization tensors by two equivalent ways ($\mathbf{x} \in R^d$) (see also Eqs. (3.175) and (3.176))

$$\boldsymbol{\tau}(\mathbf{x}) = \mathbf{L}_1(\mathbf{x})\boldsymbol{\varepsilon}(\mathbf{x}) + \boldsymbol{\alpha}(\mathbf{x}), \quad \boldsymbol{\eta}(\mathbf{x}) = \mathbf{M}_1(\mathbf{x})\boldsymbol{\sigma}(\mathbf{x}) + \boldsymbol{\beta}_1(\mathbf{x}), \quad \boldsymbol{\tau} = -\mathbf{L}^{(0)}\boldsymbol{\eta}, \tag{4.2}$$
$$\boldsymbol{\tau}(\mathbf{x}) = \boldsymbol{\sigma}(\mathbf{x}) - [\mathbf{L}^{(0)}\boldsymbol{\varepsilon}(\mathbf{x}) + \boldsymbol{\alpha}^{(0)}], \quad \boldsymbol{\eta}(\mathbf{x}) = \boldsymbol{\varepsilon}(\mathbf{x}) - [\mathbf{M}^{(0)}\boldsymbol{\sigma}(\mathbf{x}) + \boldsymbol{\beta}^{(0)}], \tag{4.3}$$

which are simply a notational convenience and vanish inside the matrix $\boldsymbol{\tau}(\mathbf{x}) \equiv \boldsymbol{\eta}(\mathbf{x}) \equiv \mathbf{0}$ ($\mathbf{x} \in v^{(0)}$). It is interesting that for the elasticity with $\boldsymbol{\alpha}(\mathbf{x}) \equiv \mathbf{0}$, the popular polarization tensors (4.2) (see, e.g., [1625]) are attributed to Hashin and Shtrikman [618] and Hill [641] who introduced, in fact, these notions in the equivalent form (4.3).

We now summarize the principal formulae of thermostatics in a decomposition form that is appropriate to our intended application [59, 302, 1189]:

$$\mathbf{u}(\mathbf{x}) = \mathbf{u}^0(\mathbf{x}) + \mathbf{u}^1(\mathbf{x}), \quad \mathbf{u}^1(\mathbf{x}) = \int_w \nabla\mathbf{G}(\mathbf{x}-\mathbf{y})\boldsymbol{\tau}(\mathbf{y})\, d\mathbf{y} \tag{4.4}$$

$$\mathbf{u}^0(\mathbf{x}) = \int_\Gamma \Big[\mathbf{G}(\mathbf{x}-\mathbf{s})\underline{\mathbf{t}}(\mathbf{s}) - \mathbf{T}^\top(\mathbf{x},\mathbf{s})\mathbf{u}(\mathbf{s})\Big]\, ds, \tag{4.5}$$

where $\underline{\mathbf{t}} \equiv \mathbf{t} - \mathbf{n}^{\Gamma\top}\boldsymbol{\alpha}^{(0)}$ is the modified boundary traction and the tensor of "fundamental displacement" $\mathbf{G}$ (3.3) is the infinite-homogeneous-body Green's function (vanishing at infinity $|\mathbf{x}| \to \infty$) of the Navier equation (3.1) with an elastic modulus tensor $\mathbf{L}^{(0)}$. The tensor of the "fundamental traction" $\mathbf{T}$ associated with the tensor of "fundamental displacement" $\mathbf{G}$ is given by Eq. (3.40). The field $\mathbf{u}^1(\mathbf{x})$ is a perturbation induced by the inclusions, while the function $\mathbf{u}^0(\mathbf{x})$ is the displacement field which would exist in the domain with homogeneous properties $\mathbf{L}^{(0)}$ (with $\mathbf{L}_1(\mathbf{x}), \boldsymbol{\beta}_1(\mathbf{x}) \equiv \mathbf{0}$) and appropriate boundary conditions (4.1).

Equations (4.4) and (4.5) are the well-known Somigliana identity written for a fictitious "body-force" which can be rewritten as well in terms of stresses:

$$\boldsymbol{\sigma}(\mathbf{x}) = \boldsymbol{\sigma}^0(\mathbf{x}) + \boldsymbol{\sigma}^1(\mathbf{x}), \tag{4.6}$$

$$\boldsymbol{\sigma}^0(\mathbf{x}) = \int_{\Gamma} \left[\mathbf{L}^{(0)} \nabla \mathbf{G}(\mathbf{x}-\mathbf{s})\underline{\mathbf{t}}(\mathbf{s}) - \mathbf{L}^{(0)} \nabla \mathbf{T}(\mathbf{x},\mathbf{s})\mathbf{u}(\mathbf{s}) \right] ds + \boldsymbol{\alpha}^{(0)}., \tag{4.7}$$

$$\boldsymbol{\sigma}^1(\mathbf{x}) = \int_{w} \boldsymbol{\Gamma}(\mathbf{x}-\mathbf{y})\boldsymbol{\eta}(\mathbf{y})\, d\mathbf{y} \tag{4.8}$$

The integral operator kernel, $\boldsymbol{\Gamma}(\mathbf{x}-\mathbf{y}) \equiv -\mathbf{L}^{(0)}\ [\mathbf{I}\delta(\mathbf{x}-\mathbf{y}) + \mathbf{U}(\mathbf{x}-\mathbf{y})\mathbf{L}^{(0)}]$, called the Green's stress tensor, is defined by the Green's tensor; see Eq. (3.37). The stress field $\boldsymbol{\sigma}^0(\mathbf{x})$ (4.8) is derived from the field $\mathbf{u}^0(\mathbf{x})$: $\boldsymbol{\sigma}^0(\mathbf{x}) = \mathbf{L}^{(0)}\boldsymbol{\varepsilon}^0(\mathbf{x}) + \boldsymbol{\alpha}^{(0)}$, $\boldsymbol{\varepsilon}^0 = [\nabla \otimes \mathbf{u}^0 + (\nabla \otimes \mathbf{u}^0)^\top]/2$. The decomposition of the displacement field $\mathbf{u}(\mathbf{x})$ ($\mathbf{x} \in w$) (4.4) (or stress field $\boldsymbol{\sigma}(\mathbf{x})$ (4.6)) from the fields induced by the boundary conditions and by the fictitious "body force" was carried out for clarity of the forthcoming presentation in the spirit of the *superposition techniques*.

In particular, for the conditions (4.1_2) (with $\Gamma_t \equiv \Gamma$) the right-hand side of Eq. (4.8) can be considered as a continuation of $\boldsymbol{\sigma}(\mathbf{x})$, $\mathbf{x} \in \Gamma_t$ (4.1_2), into w being the stress field that the boundary condition (4.1_2) would generate in the homogeneous domain w with homogeneous moduli $\mathbf{L}^{(0)}$ (with $\mathbf{L}_1, \boldsymbol{\beta}_1(\mathbf{x}) \equiv \mathbf{0}$). On the other hand, if $\mathbf{u}^0(\mathbf{x})$ (or $\boldsymbol{\sigma}^0(\mathbf{x})$), $\mathbf{x} \in w$, is known, then Eqs. (4.5) (or (4.7)) can be considered as the equations for the field $\mathbf{u}(\mathbf{x})$ (or $\boldsymbol{\sigma}(\mathbf{x})$) induced by the inclusions in an infinite medium loaded by the prescribed field $\mathbf{u}^0(\mathbf{x})$ (or $\boldsymbol{\sigma}^0(\mathbf{x})$). Such decomposition permits one to use separately the known methods of analyses of problems both in an infinite medium and the boundary-value problems that will be shown below.

It should be mentioned that the integrand terms $\boldsymbol{\tau}(\mathbf{y})$ and $\boldsymbol{\eta}(\mathbf{y})$ in Eqs. (4.4) and (4.8), respectively, are nonzero within the inclusions only, since $\mathbf{M}_1,\ \boldsymbol{\beta}_1 \equiv \mathbf{0}$, outside the inclusions. This permits one to restrict consideration of the fields $\boldsymbol{\sigma}^0(\mathbf{x})$, $\boldsymbol{\eta}(\mathbf{x})$ in Eqs. (4.6) and (4.8) to the analysis of the stress state only inside inclusions.

The main advantage of (4.4) and (4.6) is the possibility of making systematic approximations by the use of standard methods of solutions for Fredholm volume integral equations of the second kind in a full space $w = R^d$. The accompanying surface integral equations (4.5) and (4.7) can be treated via the BIE method which is ideally suited for macro scale analysis and can handle complex geometries along with general boundary conditions. Thus the decomposition of the initial equation into associated equations containing the volume and surface integrals permits the use of a hybrid method based on the combination of well-known analytical and numerical methods.

We will consider now a thermoelastic problem for a single inhomogeneity v_i in the infinite homogeneous matrix subjected to the effective fields $\overline{\boldsymbol{\varepsilon}}_i(\mathbf{x})$ and $\overline{\boldsymbol{\sigma}}_i(\mathbf{x})$ coinciding with $\boldsymbol{\varepsilon}^0(\mathbf{x})$ and $\boldsymbol{\sigma}^0(\mathbf{x})$ (4.7), respectively, at $w = R^d$. For this purpose, we will at first consider the general approach to the solution of this problem and then apply FEA for a concrete solution.

Let the heterogeneity v_i inside the infinite matrix be subjected to a *homogeneous* field $\overline{\boldsymbol{\sigma}}(\mathbf{x}_i)$ (no sum on i)

$$\boldsymbol{\varepsilon}(\mathbf{x}) = \overline{\boldsymbol{\varepsilon}}(\mathbf{x}) + \int \mathbf{U}(\mathbf{x}-\mathbf{y})V_i(\mathbf{y})\boldsymbol{\tau}(\mathbf{y})\,d\mathbf{y},$$

$$\boldsymbol{\sigma}(\mathbf{x}) = \overline{\boldsymbol{\sigma}}(\mathbf{x}) + \int \boldsymbol{\Gamma}(\mathbf{x}-\mathbf{y})V_i(\mathbf{y})\boldsymbol{\eta}(\mathbf{y})\,d\mathbf{y}, \tag{4.9}$$

where $\overline{\boldsymbol{\varepsilon}} = \mathbf{M}^{(0)}\overline{\boldsymbol{\sigma}}$. Then in view of the linearity of the problem, the solutions of Eqs. (4.9) can be presented by Eqs. (3.175) which are expressed through the Eshelby tensor (see Eqs. (3.184)) for the ellipsoidal homogeneous inclusion (3.183).

4.2.2 The Heterogeneity v_i Inside an Imaginary Ellipsoid v_i^0

We determine an arbitrary imaginary ellipsoid v_i^0 (with the smallest possible volume) with the characteristic function $V_i^0(\mathbf{x})$ inside which the real inclusion $v_i \subset v_i^0$ is to be inscribed. The results of this section are valid for an arbitrary ellipsoid v_i^0, but in subsequent presentations (e.g., in Chap. 8), v_i^0 is assumed to be connected with statistical distribution of surrounding inclusions in a composite material. Averaging of Eq. (4.9) over the volume v_i^0 and exploiting of the Eshelby theorem [449] leads to the algebraic equation $\langle\boldsymbol{\sigma}\rangle_i^0 = \overline{\boldsymbol{\sigma}}_i - \bar{v}_i(v_i^0)^{-1}\mathbf{Q}_i^0\langle\boldsymbol{\eta}\rangle_i$ (3.177), where $\mathbf{Q}_i^0(\mathbf{x}) = \mathbf{Q}_i^0 =$ const. (3.121) for the ellipsoidal domain v_i^0 at $\mathbf{x} \in v_i^0$, and the averaging operations $\langle\cdot\rangle_i$ and $\langle\cdot\rangle_i^0$ are defined by Eq. (3.178).

Substitution of Eq. (3.177) into the right-hand side of Eq. (3.175$_3$) averaged over the volume v_i^0 leads to the equation $\mathbf{B}_i^0\overline{\boldsymbol{\sigma}}_i + \mathbf{C}_i^0 = \overline{\boldsymbol{\sigma}}_i - \bar{v}_i(\bar{v}_i^0)^{-1}\mathbf{Q}_i^0\langle\boldsymbol{\eta}\rangle_i$ (3.179) yielding the following representation for the average strain polarization tensors

$$\bar{v}_i\langle\boldsymbol{\eta}\rangle_i = \bar{v}_i^0(\mathbf{Q}_i^0)^{-1}[\mathbf{I}-\mathbf{B}_i^0]\overline{\boldsymbol{\sigma}}_i - \bar{v}_i^0(\mathbf{Q}_i^0)^{-1}\mathbf{C}_i^0,$$

$$\bar{v}_i\langle\boldsymbol{\eta}\rangle_i = \bar{v}_i^0(\mathbf{Q}_i^0)^{-1}[(\mathbf{B}_i^0)^{-1}-\mathbf{I}]\langle\boldsymbol{\sigma}\rangle_i^0 - \bar{v}_i^0(\mathbf{Q}_i^0)^{-1}(\mathbf{B}_i^0)^{-1}\mathbf{C}_i^0. \tag{4.10}$$

The estimations $\mathbf{f}_i$ and $\mathbf{f}_i^0$ of the tensors $\mathbf{f}(\mathbf{x})$ averaged over the volumes of the inclusion v_i and over the imaginary ellipsoid v_i^0, respectively [e.g., $\mathbf{f}(\mathbf{x}) = \mathbf{B}_i(\mathbf{x}),\ \mathbf{C}_i(\mathbf{x}), \mathbf{R}_i(\mathbf{x}),\ \mathbf{F}_i(\mathbf{x})$] are carried out in a straightforward manner; in so doing $\mathbf{g}_i = \bar{v}_i^0\bar{v}_i^{-1}\mathbf{g}_i^0$ ($\mathbf{g}_i(\mathbf{x}) = \mathbf{R}_i(\mathbf{x}),\ \mathbf{F}_i(\mathbf{x})$) because $\mathbf{M}_1^{(i)}(\mathbf{x}) \equiv \mathbf{0}$ as $\mathbf{x} \in v_i^0 \setminus v_i$. In particular, in the case of nonellipsoidal v_i^0, the tensor $\mathbf{Q}_i^0 \neq$ const. and Eqs. (3.179) and (4.10$_1$) should be recast ($\boldsymbol{\beta}_1 \equiv \mathbf{0}$):

$$\langle\boldsymbol{\sigma}\rangle_i^0 = \overline{\boldsymbol{\sigma}} - \frac{1}{v_i^0}\int \mathbf{Q}_i^0(\mathbf{x})\boldsymbol{\eta}(\mathbf{x})V_i(\mathbf{x})\,d\mathbf{x}, \quad \langle\mathbf{Q}_i^0\boldsymbol{\eta}\rangle_i = \frac{\bar{v}_i^0}{\bar{v}_i}[\mathbf{I}-\mathbf{B}_i^0]\overline{\boldsymbol{\sigma}}. \tag{4.11}$$

A homogeneous effective field $\overline{\sigma}_i$ can be produced not only by an applied remote stress field but also by some fictitious loading by a homogeneous eigenstrain concentrated in an arbitrary ellipsoidal domain enclosing v_i. In particular, the ellipsoidal correlation hole v_i^0 can be chosen as such a domain with a fictitious homogeneous eigenstrain $V_i^0(\mathbf{x})\boldsymbol{\beta}^{\rm f}$ ($\boldsymbol{\beta}^{\rm f} = \text{const}$). Then the problem (4.9) is reduced to

$$\boldsymbol{\sigma}(\mathbf{x}) - \int V_i(\mathbf{y})\boldsymbol{\Gamma}(\mathbf{x}-\mathbf{y})\mathbf{M}_1(\mathbf{y})\boldsymbol{\sigma}(\mathbf{y})\,d\mathbf{y} = \int V_i(\mathbf{y})\boldsymbol{\Gamma}(\mathbf{x}-\mathbf{y})\mathbf{M}_1(\mathbf{y})\boldsymbol{\beta}_1(\mathbf{y})\,d\mathbf{y} - \mathbf{Q}_i^0\boldsymbol{\beta}^{\rm f}. \tag{4.12}$$

The solution of Eq. (4.12) can be formally presented in an operator form ($\mathbf{x} \in v_i$):

$$\boldsymbol{\sigma}(\mathbf{x}) = \int\int V_i(\mathbf{z})\boldsymbol{\mathcal{B}}(\mathbf{x}-\mathbf{y})\boldsymbol{\Gamma}(\mathbf{y}-\mathbf{z})\boldsymbol{\beta}_1(\mathbf{z})\,d\mathbf{z}\,d\mathbf{y} - \mathbf{B}_i(\mathbf{x})\mathbf{Q}_i^0\boldsymbol{\beta}^{\rm f}, \tag{4.13}$$

where it was taken into account that the action of the operator $\boldsymbol{\mathcal{B}} = (\mathbf{I} - \boldsymbol{\Gamma}\mathbf{M}_1)^{-1}V_i$ with the kernel $\boldsymbol{\mathcal{B}}(\mathbf{x})$ at the constant is reduced to the multiplication on a tensor $\mathbf{B}_i(\mathbf{x}) \equiv \int \boldsymbol{\mathcal{B}}(\mathbf{x}-\mathbf{y})\,d\mathbf{y}$. Equation (4.13) allows one to get a representation for the strain polarization tensor

$$\begin{aligned}\boldsymbol{\eta}(\mathbf{x}) = &\int\int V_i(\mathbf{z})\mathbf{M}_1(\mathbf{x})\boldsymbol{\mathcal{B}}(\mathbf{x}-\mathbf{y})\boldsymbol{\Gamma}(\mathbf{y}-\mathbf{z})\boldsymbol{\beta}_1(\mathbf{z})\,d\mathbf{z}\,d\mathbf{y}\\ &-\mathbf{M}_1(\mathbf{x})\mathbf{B}_i(\mathbf{x})\mathbf{Q}_i^0\boldsymbol{\beta}^{\rm f} + \boldsymbol{\beta}_1(\mathbf{x}).\end{aligned} \tag{4.14}$$

Comparing Eqs. (3.1757.8) and (4.13), (4.14) we can see that the tensors $\mathbf{B}_i(\mathbf{x})$, $\mathbf{C}_i(\mathbf{x})$, $\mathbf{R}_i(\mathbf{x})$, $\mathbf{F}_i(\mathbf{x})$ (as well as their averages $\mathbf{B}_i$, $\mathbf{C}_i$, $\mathbf{R}_i$, $\mathbf{F}_i$) are expressed in terms of the operator $\boldsymbol{\mathcal{B}}$. It should be mentioned that our current goal is an establishment of relationships between the tensors $\mathbf{B}_i(\mathbf{x})$, $\mathbf{C}_i(\mathbf{x})$, $\mathbf{R}_i(\mathbf{x})$, $\mathbf{F}_i(\mathbf{x})$ and some another tensors (which will be used in the forthcoming estimations of effective properties of composite materials) with the numerical solutions of different model problems obtained via the popular numerical methods (such as FEA). A formal introduction of the integral operator $\boldsymbol{\mathcal{B}}$ is intended only for ascertainment of these relationships rather than for the numerical solution of Eq. (4.9) via the volume integral equation method (see, e.g., Sect. 4.3). Because of this, a concrete construction of the integral operator $\boldsymbol{\mathcal{B}}$ is not in our current interest.

In the general case, the estimation of the tensors $\mathbf{B}(\mathbf{x})$, $\mathbf{C}(\mathbf{x})$, $\mathbf{Q}_i^0$ is a particular problem of the transformation field analysis methods [425] which can be realized by different numerical methods, for example, by finite element analysis (see also [1131]).

Indeed, let the inclusion v_i be subjected to homogeneous remote stress $\overline{\boldsymbol{\sigma}} = \text{const.}$ with a single nonzero component $\overline{\sigma}_j = 1$; otherwise $\overline{\sigma}_k \equiv 0$ ($j, k = 1, \ldots, 3d-3$; $k \neq j$), and $\boldsymbol{\beta}_1 \equiv \mathbf{0}$. We assume that the stress field $\boldsymbol{\sigma}(\mathbf{x})$ ($\mathbf{x} \in v_i$)

inside the inclusion v_i is estimated by some numerical method. Then the tensor $\mathbf{B}_i(\mathbf{x})$ is represented explicitly over the known stress field $\boldsymbol{\sigma}(\mathbf{x})$:

$$B_{i|jm}(\mathbf{x}) = \sigma_m(\mathbf{x}) \quad \text{for } \overline{\sigma}_j = 1,\ \overline{\sigma}_k \equiv 0\ (j \neq k), \tag{4.15}$$

where $j, k, m = 1, \ldots, 3d - 3$ and $\mathbf{x} \in v_i$, $(i = 1, \ldots, n)$.

Furthermore, instead of the real eigenstrain $\boldsymbol{\beta}_1(\mathbf{x})$, let the inclusion be subjected to the fictitious eigenstrain $\boldsymbol{\beta}^{\rm f}$ such that $\beta^{\rm f}_j = \beta_{1|j}(\mathbf{x})$ $(\mathbf{x} \in v_i)$ $(j = 1, \ldots, d)$; otherwise $\beta^{\rm f}_k \equiv 0$ $(k = 1, \ldots, 3d - 3;\ k \neq j)$ and $\overline{\boldsymbol{\sigma}} \equiv \mathbf{0}$. Then, we obtain the explicit representation of the tensor $\mathbf{C}_i(\mathbf{x})$ over the known stress field $\boldsymbol{\sigma}(\mathbf{x})$ inside the inclusion v_i:

$$C_{i|jm}(\mathbf{x}) = \sigma_m(\mathbf{x}) \quad \text{for } \overline{\boldsymbol{\sigma}} \equiv \mathbf{0},\ \beta^{\rm f}_j(\mathbf{x}) = \beta_{1|j}(\mathbf{x}),\ \beta^{\rm f}_k = 0\ (j \neq k), \tag{4.16}$$

where $j = 1, \ldots, d$; $k, m = 1, \ldots, 3d - 3$ and $\mathbf{x} \in v_i$, $i = 1, \ldots, n$.

At last, for the domain v^0_i of any shape, the tensor $\mathbf{Q}^0_i(\mathbf{x}) \equiv -\overline{v}^0_i \langle \boldsymbol{\Gamma}(\mathbf{x} - \mathbf{y}) \rangle^0_i(\mathbf{x})$ $(\mathbf{x} \in v_i \subset v^0_i,\ \mathbf{y} \in v^0_i)$ can be found for $\overline{\boldsymbol{\sigma}} = \boldsymbol{\beta}_1 \equiv \mathbf{0}$ from estimations of the stresses inside the inclusion $\mathbf{x} \in v^0_i$ [$\boldsymbol{\sigma}(\mathbf{x}) = -\mathbf{Q}^0 \boldsymbol{\beta}^{\rm f}$; see Eqs. (3.175$_{3,4}$) and (3.115)] subjected to homogeneous fictitious eigenstrain $V^0_i(\mathbf{x})\boldsymbol{\beta}^{\rm f}$ ($\boldsymbol{\beta}^{\rm f} =$ const.):

$$\mathbf{Q}^0_{i|mj}(\mathbf{x}) = \sigma_m(\mathbf{x}) \quad \text{for } \beta^{\rm f}_j = 1,\ \beta^{\rm f}_k = 0\ (j \neq k), \tag{4.17}$$

where $j = 1, k, m = 1, \ldots, 3d - 3$ and $\mathbf{x} \in v^0_i$. In particular, for an ellipsoidal domain v^0_i usually considered in the subsequent discussion, the tensor $\mathbf{Q}^0_i(\mathbf{x})$ is a constant: $\mathbf{Q}^0_i(\mathbf{x}) = \mathbf{Q}^0_i \equiv$ const.

The constant tensor $\mathbf{Q}^0_i$ can also be estimated from the pure elastic problem (4.15) for ellipsoidal domain v^0_i. Indeed, the replacement of the modulus $\mathbf{L}(\mathbf{x})$ $(\mathbf{x} \in v^0_i)$ by the infinite modulus $\mathbf{L}(\mathbf{x}) \equiv \infty$ (a rigid inclusion) and estimation of the tensor $\mathbf{B}^{\rm rigid}_i$ analogously to Eq. (4.15) leads to the evaluation

$$\mathbf{Q}^0_i = -[(\mathbf{B}^{\rm rigid}_i)^{-1} - \mathbf{I}]\mathbf{L}^{(0)}. \tag{4.18}$$

For the ellipsoidal domain, v^0_i Eq. (4.18) is more convenient for the estimation of the tensor $\mathbf{Q}^0_i$ than Eq. (4.17) because it already uses what is considered a pure elastic case (4.15), which is necessary anyway. However, for a nonellipsoidal domain v^0_i the possibility of the estimation of the inhomogeneous tensor $\mathbf{Q}^0_i(\mathbf{x})$ from the pure elastic problem ($\boldsymbol{\beta}_1 \equiv \mathbf{0},\ \overline{\boldsymbol{\sigma}} \neq \mathbf{0}$) is questionable.

Thus, the tensor $\mathbf{Q}^0_i$ involved in the eigenstrain problem was estimated from the pure elastic problem (4.18) as well as from the eigenstrain problem (4.17). It turns out that the stress concentration factor $\mathbf{B}_i(\mathbf{x})$ (3.175$_3$) can also be found from a fictitious eigenstrain problem rather than only from the pure elastic one (4.15). For this purpose, we will consider a fictitious eigenstrain loading $V^0_i(\mathbf{x})\boldsymbol{\beta}^{\rm f}$ ($\boldsymbol{\beta}^{\rm f} =$ const.)

accompanied by the conditions $\overline{\boldsymbol{\sigma}}_i = 0, \boldsymbol{\beta}_1 \equiv \mathbf{0}$. Let the applied eigenstrain $\boldsymbol{\beta}^{\mathrm{f}}$ have a single nonzero component $\beta^{\mathrm{f}}_j = 1$ ($j = 1, \ldots, d$); otherwise $\beta^{\mathrm{f}}_k \equiv 0$ ($k = 1, \ldots, 3d-3; k \neq j$). Then we form the square matrix (6×6) $\boldsymbol{\Sigma}(\mathbf{x})$ with a j-tuple ($j = 1, \ldots, 6$) composed of the six components of the stress vector $\boldsymbol{\sigma}(\mathbf{x})$ generated by the eigenstrain $\boldsymbol{\beta}^{\mathrm{f}}$ with only one nonzero component $\beta^{\mathrm{f}}_j = 1$. Then we obtain the explicit representation of the tensor $\mathbf{B}(\mathbf{x})$ through the known matrices (6×6) of both the stress fields $\boldsymbol{\sigma}^M(\mathbf{x})$ and $\mathbf{Q}^0_i$ ($\mathbf{x} \in v^0_i$):

$$\mathbf{B}_i(\mathbf{x}) = \boldsymbol{\Sigma}(\mathbf{x})(\mathbf{Q}^0_i)^{-1}. \tag{4.19}$$

The tensor $\mathbf{B}_i(\mathbf{x})$ (and $\mathbf{R}_i(\mathbf{x}) = \bar{v}_i \mathbf{M}_1(\mathbf{x})\mathbf{B}_i(\mathbf{x})$) does not depend on both the shape and size of the domain v^0_i of the auxiliary fictitious problem although the tensor $\mathbf{Q}^0_i$ explicitly enters into the right-hand side of Eq. (4.19).

4.2.3 *Estimation of Concentration Factor Tensors by FEA*

It was mentioned in the preceding text that estimation of the tensors $\mathbf{B}(\mathbf{x})$, $\mathbf{C}(\mathbf{x})$, and $\boldsymbol{Q}^0_i$ can in general be realized by numerical methods only: in the present study, FEA will be utilized. Namely, following the Bubnov–Galerkin approach, we multiply the equilibrium equation with the shape function N_m (associated with nodes with number m of the finite element (FE) mesh, $m = 1, 2, \ldots, K$, K being a total number of nodes in a given FE model) and then integrate it over the volume V. It gives us a "weak" form of the equilibrium equation:

$$\int_V \nabla \cdot \boldsymbol{\sigma}\, N_m\, dv + \int_V \mathbf{f} N_m\, dv = 0, \tag{4.20}$$

where $\mathbf{f}$ is the body force vector. Using the well-known identities $\nabla \cdot (\sigma\, N_m) = \nabla N_m \cdot \boldsymbol{\sigma} + N_m \nabla \cdot \boldsymbol{\sigma}$, and

$$\int_V \nabla \cdot \boldsymbol{\sigma}\, dv = \int_S \mathbf{n} \cdot \boldsymbol{\sigma}\, ds, \tag{4.21}$$

one obtains from (4.20) $\int_V \boldsymbol{\sigma} \cdot \nabla N_m\, dv = \int_V \mathbf{f} N_m\, dv + \int_S N_m \mathbf{n} \cdot \boldsymbol{\sigma}\, ds$. Here, $\mathbf{n}$ is the unit vector, normal to the surface S. After we substitute the constitutive and equilibrium equations into (4.21), we obtain the equation

$$\frac{1}{2}\int_V [\mathbf{L}(\nabla \mathbf{u} + (\nabla \mathbf{u})^{\top})] \nabla N_m\, dv = \int_V \mathbf{f} N_m\, dv - \int_V \boldsymbol{\alpha} \cdot \nabla N_m\, dv + \int_S N_m \mathbf{n} \cdot \boldsymbol{\sigma}\, ds. \tag{4.22}$$

The last term in (4.22) is nonzero in cases where the stress condition $\boldsymbol{\sigma} \cdot \mathbf{n} = \mathbf{T}$ is prescribed at the outer boundary.

In accordance with the FEA scheme, the fields **u**, **p**, and $\boldsymbol{\alpha}$ on each finite element are approximated as

$$W = W_p N_p, \quad p = 1, 2, \ldots, K^e, \ (W = (\mathbf{u}, \mathbf{f}, \boldsymbol{\alpha})), \tag{4.23}$$

where K^e is the node number of elements numbered $e, e = 1, 2, \ldots, M, M$ is a total number of FEs. By substituting the expressions (4.23) in (4.22), we come to the equations for the nodal values of the displacement vector $\mathbf{u}(\mathbf{r}_k, t) = \mathbf{U}_k(t) \equiv \mathbf{U}_k$:

$$\frac{1}{2}\sum_{e=1}^{M} \int_{V^e} [\mathbf{L}(\nabla(\mathbf{U}_k N_k) + (\nabla(\mathbf{U}_k N_k))^\top)] \nabla N_m \, dv$$
$$= \sum_{e=1}^{M} \Big[\int_{V^e} f_k N_k N_m \, dv - \int_{V^e} \alpha_k N_k \cdot \nabla N_m \, dv + \int_{S^e} N_k N_m \mathbf{n} \cdot \boldsymbol{\sigma}_k \, ds \Big]. \tag{4.24}$$

After calculating the volume and surface integrals in the local linear sets (4.24) for each element, we apply the assembling procedure to obtain a global set of linear algebraic equations with the unknown vector {**U**} of nodal displacements: [**A**] {**U**} = {**R**}. Here, the vector {**R**} is constructed from the right-hand parts of the local equations (4.24), and [**A**] is $K \times K$ is the symmetric and positively defined matrix, which enables use of the efficient iterative solvers. Specifically, convergence of the Jacobi conjugate gradient solver utilized by us was found to be strongly dependent on the aspect α^r ratio of inclusion. So, for very high values of α^r (say, 1000), the tolerance level of order 10^{-10} must be taken to get the fully convergent solution.

We are coming now to the concrete numerical analysis of the elastic problems for one nonellipsoidal inhomogeneity in an infinite matrix (see for details [200, 281]). The 3D problem for one cylindrical inclusion with the smooth ends will be considered in Chap. 8 and [241] as an application to the modeling of nanocomposites. Here we will consider the 2D plane strain problem for the noncircular inclusion. Let the elastic properties of the isotropic components be $E^{(0)} = 1$ GPa, $E^{(1)} = 100$ GPa, $\nu^{(0)} = \nu^{(1)} = 0.45$. Figure 4.1a is a schematic representation of the quarter of the inclusion boundary in the first quadrant described by the curve

$$\begin{cases} (x-l)^2 + (y-l)^2 = (r-l)^2, & \text{for } x > l \text{ and } y > l, \\ |x+y| + |x-y| = 2r, & \text{for } x < l \text{ or } y < l, \end{cases} \tag{4.25}$$

which reduces to a circle and a square in the limiting cases $l = 0$ and $l = r$, respectively. In parallel with the inclusion (4.25), we will consider the inclusion with the same shape but with the different orientation obtained by the rotation of

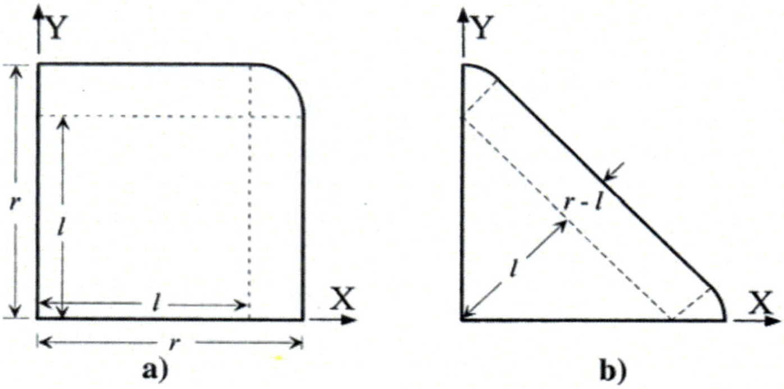

Fig. 4.1 Schematic of a noncircular inclusion characterized by the shape parameter, l/r: (**a**) initial configuration, (**b**) configuration rotated by $\pi/4$ about the Z-axis

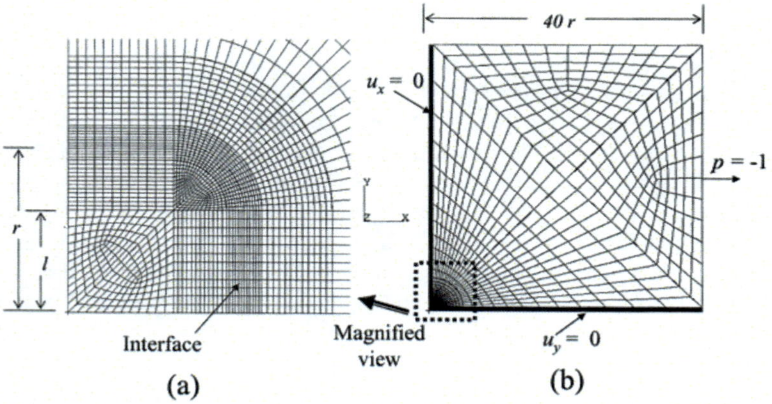

Fig. 4.2 (**a**) Magnified view and (**b**) complete view of the FE mesh utilized for solution of a single inclusion in an infinite matrix subjected to normal loading at infinity

the inclusion (4.25) on the angle $\pi/4$ as shown in Fig. 4.1b, where the quarter of the inclusion in the first quadrant is described by the equation

$$\begin{cases} x^2 + (y - l\sqrt{2})^2 = (r - l)^2, & \text{for } y > (r + l)/\sqrt{2}, \\ (x - l\sqrt{2})^2 + y^2 = (r - l)^2, & \text{for } x > (r + l)/\sqrt{2}, \\ |x| + |y| = \sqrt{2}r, & \text{for } |x| < (r + l)/\sqrt{2} \text{ and } |y| < (r + l)/\sqrt{2}, \end{cases}$$

The commercial finite element code, ABAQUS [1], was used for the FE modeling with the mesh designed for the noncircular inclusion shape shown in Fig. 4.2b along with an expanded view of the mesh inside and near the inclusion boundary in Fig. 4.2a. The inclusion and matrix materials were assumed to be isotropic, linear elastic, and modeled with plane strain second-order isoparametric (CPE8) elements, while the interface was modeled as perfectly bonded. The infinite dimensions of the matrix were approximated with a length of 40 inclusion diameters: $R = 40r$. Increasing the matrix dimensions further did not significantly change the results. To capture the rapidly changing stresses near the interface, elements with a radial span of 0.0083 inclusion diameters were used near the interface. Symmetry conditions were implemented (i.e., $u_y = 0$ along $y = 0$ and $u_x = 0$ along $x = 0$, as shown in Fig. 4.2b) so that only one quadrant of the problem required meshing. Typical meshes contained a total of 3200 elements (approx. 20,000 d.o.f.s) with 1000 elements in the inclusion and 2200 in the matrix. Normal loading of the matrix boundary in the X- and Y-directions was implemented by applying unit pressure on the corresponding surfaces (e.g., $\sigma_{ij} = \delta_{1i}\delta_{1j}$ and $\sigma_{ij} = \delta_{2i}\delta_{2j}$), such as is illustrated in Fig. 4.2b for $\sigma_{ij} = \delta_{1i}\delta_{j1}$.

Unlike the case of pure normal loading where only a quarter of the problem required meshing, the full problem had to be considered for the shear loading case. The complete view of the FE model used to consider pure shear in the X-direction of the coordinate system (4.25) is shown in [281]. Typical meshes for the full problem contained a total of 8600 elements (approx. 52,000 d.o.f.s) with 2520 elements in the inclusion and 6080 in the matrix.

We assume that for all inclusions (4.25), the tensors $\mathbf{Q}_i^0$ are identical and coincide with the tensor $\mathbf{Q}^0$ for the circular inclusion which was estimated both numerically by the use of FEA and analytically by means of the Eshelby tensor. The error of the numerical solution is not more than 0.15%, where the ratio $R/r = 40$ is considered to represent an infinite medium. It was demonstrated that the stress inside the inclusion approaches a constant value for mesh with the inclusion and matrix divisions equal 24 and 72, respectively, along the X-axis.

4.3 Boundary Integral Equation Methods

Obtaining analytical solutions is not feasible in general even for a finite number of interacting particles, so various numerical methods have been developed, mostly based on the finite element analysis (FEA, see, e.g., [478, 1700]) and boundary integral equations (BIE, see, e.g., [59, 156, 1101]) technique. A distinct advantage of the BIE compared to the FEA is that the BIEs require meshing only the boundary surface of computational domain as opposed to the entire 3D domain for FEA. In the BIE singular forces distributed, e.g., on the surface of particles, depend on the external field, thus yielding an integral equation for the singularity strengths. The most popular formulation is the direct one where the variables of interest are the surface displacement and traction fields that leads to a set of ill-conditioned linear systems on discretization. The indirect formulation (e.g., the completed double layer boundary element method, see [458, 947]) results in a set of integral equations of the second kind and, therefore, is numerically well-behaved. Although this formulation is more complicated than the direct method, however, it leads to an unconditionally convergent iterative scheme.

Despite the FEM's and BEM's popularity, there are many problems (e.g., meshing surface in 3D, computing of singular integrals) where it is desirable to improve the efficiency of traditional methods. In the alternative (proposed by Trefftz), the trial functions must satisfy the governing equation exactly and the error in the satisfaction of the boundary conditions is minimized. As a consequence, it is interesting to develop some alternative so-called meshless methods (see [461] for an overview and historical background on the subject) of the local boundary integral equation, boundary knot method, boundary collocation method, non-dimensional dynamic influence functions method, and method of fundamental solutions (MFS) belong to the boundary integral equations method for solving boundary-value problems, which can be recognized as a discrete type of the indirect boundary element method with a concentrated source instead of distribution (see for references [43, 82, 319, 557]). In this study, we will bring the MFS systematically for multiple inclusions inside infinite medium subjected to inhomogeneous loading. The MFS belonging to the BIE technique shares the same advantages as the BEM over domain discretization methods. Moreover, the MFS has certain attractive advantages over the BEM, since it is truly meshless (which requires neither domain nor boundary mesh structure), simple to program, and is able to take into account sharp changes in geometry. The MFS and related methods over the last years have found extensive application in computing solutions to a broad range of problems [303, 455].

4.3.1 Formulation of Boundary Integral Equations

We consider a single inclusion v inside the infinite homogeneous medium. Without loss of generality, it is assumed that $\boldsymbol{\alpha}^{(0)} = \boldsymbol{\beta}^{(0)} \equiv \mathbf{0}$, and thereby $\boldsymbol{\alpha}_1(\mathbf{x}) = \boldsymbol{\alpha}^{(1)} V(\mathbf{x})$ and $\boldsymbol{\beta}_1(\mathbf{x}) = \boldsymbol{\beta}^{(1)} V(\mathbf{x})$. For notational convenience, in parallel with Eqs. (3.70), we consider the equivalent local boundary-value problems

$$\nabla \mathbf{L}^{(k)} \nabla \mathbf{u}_k(\mathbf{x}) = -\nabla \boldsymbol{\alpha}^{(k)}(\mathbf{x}) \tag{4.26}$$

for the connected subdomains $\mathbf{x} \in v^{(k)}$ $(k = 0, 1)$ with the natural coupling conditions (called transmission conditions) across all local coupling boundaries $\mathbf{s} \in \Gamma_{km}$ $(m = 1 - k)$

$$\mathbf{u}_k(\mathbf{s}) = \mathbf{u}_m(\mathbf{s}), \;\; \mathbf{t}_k(\mathbf{s}) + \mathbf{t}_m(\mathbf{s}) = \mathbf{0} \;\; \text{for } \mathbf{s} \in \Gamma_{km}. \tag{4.27}$$

Here $\mathbf{u}_k$ is the restriction of $\mathbf{u}$ to v^k and $\mathbf{n}_k$ is the outward normal to $v^{(k)}$ $(k, m = 0, 1)$. The traction $\mathbf{t}(\mathbf{s})$ acting on any plane with the normal outward normal vector

$\mathbf{n}(\mathbf{s})$ on Γ_i from v_i through the point $\mathbf{s} \in \Gamma_i$ can be represented in terms of displacements $\mathbf{t}(\mathbf{s}) = \hat{\mathbf{t}}(\mathbf{n}, \nabla)\mathbf{u}(\mathbf{s})$, where

$$\hat{t}_{ik}(\mathbf{n}, \nabla) = L^{(k)}_{ijkl} n_j(\mathbf{x})\partial/\partial x_l \tag{4.28}$$

is the conormal derivative operator.

According to the essence of the boundary decomposition method reducing the solution of the original boundary-value problem (3.46) to the solutions of local boundary problems (4.26) and (4.27), we can use the different discretization techniques for solution of Eqs. (4.26) and (4.27) (see for details [33, 34, 656, 1529]). The boundary decomposition method assume that each heterogeneity v_k can be considered as embedded into the infinite homogeneous matrix and subjected to the effective fields (formally considered as the remoted fields)

$$\overline{\mathbf{u}}(\mathbf{x}) \quad \text{and} \quad \overline{\boldsymbol{\sigma}}(\mathbf{x}), \quad (\mathbf{x} \in v_k) \tag{4.29}$$

(no restrictions are imposed on the inhomogeneity of these random effective fields (4.29)).

Let us consider the well-known integral representation formulae of the displacement and stress fields for the thermoelastostatic problem inside and outside the heterogeneity $v_k \in v^{(1)}$ subjected to the remoted fields $\overline{\mathbf{u}}(\mathbf{x})$, $\overline{\boldsymbol{\sigma}}(\mathbf{x})$ (see, e.g., [59, 1189]):

$$(1-e)\mathbf{u}(\mathbf{x}) = e\overline{\mathbf{u}}(\mathbf{x}) + \int_{\Gamma_k} \Big[\mathbf{G}^{(m)}(\mathbf{x}-\mathbf{s})\underline{\mathbf{t}}(\mathbf{s}) - \mathbf{T}^{(m)\top}(\mathbf{x}, \mathbf{s})\mathbf{u}(\mathbf{s})\Big] ds, \tag{4.30}$$

$$(1-e)\boldsymbol{\sigma}(\mathbf{x}) = e\overline{\boldsymbol{\sigma}}(\mathbf{x}) + \int_{\Gamma_k} \Big[\mathbf{L}^{(m)}\nabla\mathbf{G}^{(m)}(\mathbf{x}-\mathbf{s})\underline{\mathbf{t}}(\mathbf{s}) - \mathbf{S}^{(m)}(\mathbf{x}, \mathbf{s})\mathbf{u}(\mathbf{s})\Big] ds, \tag{4.31}$$

where without loss of generality $m = 0, 1$; $e \equiv e(\mathbf{x}) = 0$ and $e \equiv e(\mathbf{x}) = 1$ for the internal ($\mathbf{x} \in v_k \in v^{(1)}$, $m = 1$) and external ($\mathbf{x} \notin v_k$, $m = 0$) problem, respectively; $\underline{\mathbf{t}}(\mathbf{s}) \equiv \mathbf{t}_m(\mathbf{s}) - \boldsymbol{\alpha}^{(m)}\mathbf{n}_m(\mathbf{s})$ (no sum on m) is the modified traction on the interface Γ_m with the outer normal $\mathbf{n}_m(\mathbf{s})$ ($\mathbf{s} \in \Gamma_m$); the Green functions $\mathbf{T}$, $\nabla\mathbf{G}$, $\mathbf{S}$ are defined by Eqs. (3.67).

Taking in (4.26) and (4.26) the limit $\mathbf{x} \to \Gamma_k$ leads the well-known boundary integral equations (BIEs) between the Cauchy data $\mathbf{u}$ and $\boldsymbol{\sigma}$ on $\mathbf{x} \in \Gamma_k$

$$\mathbf{k}(\mathbf{x})\mathbf{u}(\mathbf{x}) = e\overline{\mathbf{u}}(\mathbf{x}) + \int_{\Gamma_k} \Big[\mathbf{G}^{(m)}(\mathbf{x}-\mathbf{s})\underline{\mathbf{t}}(\mathbf{s}) - \mathbf{T}^{(m)\top}(\mathbf{x}, \mathbf{s})\mathbf{u}(\mathbf{s})\Big] ds, \tag{4.32}$$

$$\mathbf{k}(\mathbf{x})\boldsymbol{\sigma}(\mathbf{x}) = e\overline{\boldsymbol{\sigma}}(\mathbf{x}) + \int_{\Gamma_k} \Big[\mathbf{L}^{(m)}\nabla\mathbf{G}^{(m)}(\mathbf{x}-\mathbf{s})\underline{\mathbf{t}}(\mathbf{s}) - \mathbf{S}^{(m)}(\mathbf{x}, \mathbf{s})\mathbf{u}(\mathbf{s})\Big] ds, \tag{4.33}$$

for the external ($e = 1$, $\mathbf{x} \in \Gamma_k^- = \Gamma_{0k}$, $m = 0$) and internal ($e = 0$, $\mathbf{x} \in \Gamma_k^+ = \Gamma_{k0}$, $m = 1$) problems, respectively; the free term coefficient (see, e.g.,

[354]) $\mathbf{k}(\mathbf{x}) = (1 - e(\mathbf{x}))\boldsymbol{\delta}$ for $\mathbf{x} \in R^d \setminus \Gamma_k$ and $\mathbf{k}(\mathbf{x}) = \boldsymbol{\delta}/2$ for $\mathbf{x}$ belonging to the smooth boundary Γ_k at point $\mathbf{x} \in \Gamma_k$. Hereafter the superscripts $^-$ and $^+$ denote the limiting values of parameters involved in the outside and inside of the inclusion v_i, respectively, near the inclusion boundary $\mathbf{x} \in \Gamma_k$.

The BIEs (4.30) and (4.31) deal with strongly-singular and hypersingular integrals. Significant efforts of researchers were centered on the the subject of how to accurately compute the singular integrals analytically or numerically. The BIE representations are referred to as the nonregularized ones, because of singularity of integral kernels as the source point $\mathbf{x}$ approaches the boundary. The boundary integral representation of displacement field (4.32) contains the weakly singular kernel (logarithmic or r^{-1} for $d = 2, 3$, respectively, $r = |\boldsymbol{x} - \mathbf{y}|$ is the distance between the source $\mathbf{y}$ and field $\mathbf{x}$ points) and strongly singular kernel (r^{1-d}). For the stresses field (4.33), the order of singularities in the kernels formed from differentiation of the kernel (5.3) increase by one, that are the kernels of the first and second integrands of the boundary integral representation for stresses (5.4) have a strong singularity $O(r^{1-d})$ and hypersingularity $O(r^{-d})$, respectively.

The weak singularities in the surface integrals can be removed by using a transformation of the integration variable, and the integrals being considered are integrable in the ordinary sense even if $\mathbf{x} \to \boldsymbol{\zeta} \in \Gamma$ (see [1111]). The strongly singular integrals are considered in the Cauchy principal value (CPV) sense, while the hypersingular integrals are considered in the Hadamard finite part sense (see for references and details [306, 1061, 1708]) when the integrals involved are estimated analytically at the elements of the mesh containing the singularities, for example, for boundary densities approximated within finite-size elements polynomially (see, e.g., [1410]). However, the analyses of the BIE formulations have revealed that the BIEs can be regularized by using the subtraction technique when the operation of the Taylor series expansion of the densities about the singular point, subtraction, and adding back the singular terms as well as the use of some fundamental solution identities leads to elimination of singularities, which are not integrable in the ordinary sense before discretization. (see, e.g., [1411, 1485]). So Liu and Rudolphi [961] (see also [354, 956]) proposed the operational approach generalizing the simple solution approach to obtain some identities for the fundamental solutions exploited in so-called physical regularization ($\mathbf{x} \in w \cup \Gamma$)

$$\mathbf{k}(\mathbf{x}) = -\int_{\Gamma} \mathbf{T}(\mathbf{x}, \mathbf{s})ds, \quad \mathbf{0} = \int_{\Gamma} \nabla_{\mathbf{x}} \mathbf{T}(\mathbf{x}, \mathbf{s})ds,$$

$$\mathbf{k}(\mathbf{x})\boldsymbol{\delta} = \mathbf{L}(\mathbf{x}) \int_{\Gamma} \nabla_{\mathbf{x}} \mathbf{G}(\mathbf{x} - \mathbf{s}) \otimes \mathbf{n}(\mathbf{s})ds - \int_{\Gamma} \nabla_{\mathbf{x}} \mathbf{T}(\mathbf{x}, \mathbf{s}) \otimes (\mathbf{s} - \mathbf{x})ds,$$

$$\mathbf{0} = \int_{\Gamma} \mathbf{T}(\mathbf{x}, \mathbf{s}) \otimes (\mathbf{s} - \mathbf{x})ds - \mathbf{L}(\mathbf{x}) \int_{\Gamma} \mathbf{G}(\mathbf{x} - \mathbf{s}) \otimes \mathbf{n}(\mathbf{s})ds, \tag{4.34}$$

Both the strongly (Cauchy type) and weakly singular integrals existing in the conventional BIEs are removed from the BIE formulations by the use of two-term subtraction technique. An alternative method for estimation of the Cauchy data

$[\mathbf{u}(\mathbf{x}), \boldsymbol{\sigma}(\mathbf{x})]$ for $\mathbf{x} \in \Gamma_k$ is based on the solution of the conventional BIE (4.32) presented for both the matrix and inclusion in the matrix forms by the substitution method (see, e.g., [387, 1584, 1657]). The scheme of this method can be recast in the theoretical framework of well-developed boundary operators (see for details [657, 1442]).

However, in more details we will only consider in this section the Methods of Fundamental Solutions (MFS) belonging to the BIE technique and having some advantages mentioned in Sect. 4.1.

4.3.2 *Interface Polarization Tensors and Perturbators for a Single Inclusion Inside Macrodomain*

The concept of the polarization tensors (4.2) and (4.3) is an important ingradient of the effective moduli theory based on the estimations of the polarization tensors $\boldsymbol{\tau}$ (4.2$_1$) and $\boldsymbol{\eta}$ (4.2$_2$) averaged over the inclusion volume v. However, the representations $\boldsymbol{\tau}$ (4.3$_1$) and $\boldsymbol{\eta}$ (4.3$_2$) allows one to use only surface averages of some transformations of these tensors. So, we introduce the new notions of the stress $\boldsymbol{\tau}^s(\mathbf{s})$ and strain $\boldsymbol{\eta}^s(\mathbf{s})$ *interface polarization tensors*

$$\boldsymbol{\tau}^s(\mathbf{s}) = \mathbf{t}(\mathbf{s}) \overset{s}{\otimes} \mathbf{s} - \mathbf{L}^{(0)}[\mathbf{u}(\mathbf{s}) \otimes \mathbf{n}(\mathbf{s})] - [\boldsymbol{\alpha}^{(0)}\mathbf{n}(\mathbf{s})] \overset{s}{\otimes} \mathbf{s},$$
$$\boldsymbol{\eta}^s(\mathbf{s}) = \mathbf{u}(\mathbf{s}) \overset{s}{\otimes} \mathbf{n}(\mathbf{s}) - \mathbf{M}^{(0)}[\mathbf{s} \otimes \mathbf{t}(\mathbf{s})] - [\boldsymbol{\beta}^{(0)}\mathbf{n}(\mathbf{s})] \overset{s}{\otimes} \mathbf{s}, \tag{4.35}$$

defined at the inclusion surface $\mathbf{s} \in \Gamma$.

For thermoelasticity, and elasticity with a body force, the conventional boundary element method (BEM) involves a domain integral. In general, the body force term is approximated by some special functions that are amenable to a treatment of the domain integral to a boundary one (see for details and references [321, 1191]) with subsequent implementation of a superposition principle for solution of the corresponding inhomogeneous differential equation. However, for the particular case of the pseudo-body force (4.26) with a step function type of eigenstrain, the domain integral is originally presented in the form of the boundary one over the interface Γ_i. In such a case, the thermoelasticity integral equation can be derived using the body force analogy

$$\mathcal{L}\mathbf{u} \equiv \nabla \mathbf{L} \nabla \mathbf{u} = -\mathbf{b}, \tag{4.36}$$

where the body force is concentrated at the interface Γ_i

$$\mathbf{b}(\mathbf{x}) = -\boldsymbol{\alpha}_1 \mathbf{n}(\mathbf{s})\delta_{\Gamma_i}. \tag{4.37}$$

To solve the inhomogeneous differential equation (4.36) with the pseudo-body force (4.37), we utilize a split of the solution into two constituents

$$\mathbf{u} = \mathbf{u}^p + \mathbf{u}^c, \ \boldsymbol{\sigma} = \boldsymbol{\sigma}^p + \boldsymbol{\sigma}^c \tag{4.38}$$

where the particular solution $\mathbf{u}^p$ satisfies

$$\mathcal{L}\mathbf{u}^p = \boldsymbol{\alpha}_1 \mathbf{n}\delta_{\Gamma_i}, \tag{4.39}$$

whereas the complementary solution $\mathbf{u}^c$, free from the body force, satisfies the homogeneous equation

$$\mathcal{L}\mathbf{u}^c = \mathbf{0}. \tag{4.40}$$

The original prescribed $\hat{\mathbf{u}}(\mathbf{s})$, $\hat{\mathbf{t}}(\mathbf{s})$ ($\mathbf{s} \in \Gamma_{km}$) boundary conditions (4.27) are modified for the complimentary solution as

$$\mathbf{u}^c(\mathbf{s}) = \hat{\mathbf{u}}(\mathbf{s}) - \mathbf{u}^p(\mathbf{s}), \quad \mathbf{t}^c(\mathbf{s}) = \hat{\mathbf{t}}(\mathbf{s}) - [\boldsymbol{\alpha}_1\mathbf{n}(\mathbf{s}) + \mathbf{t}^p(\mathbf{s})], \tag{4.41}$$

where $\mathbf{u}^p(\mathbf{s})$ and $\mathbf{t}^p(\mathbf{s})$ are defined from the solution (4.39). On applying the superposition principle (4.38), we can determine the unknown boundary displacement and traction

$$\mathbf{u}(\mathbf{s}) = \mathbf{u}^c(\mathbf{s}) + \mathbf{u}^p(\mathbf{s}), \quad \mathbf{t}(\mathbf{s}) = \mathbf{t}^c(\mathbf{s}) + [\boldsymbol{\alpha}_1\mathbf{n}(\mathbf{s}) + \mathbf{t}^p(\mathbf{s})], \tag{4.42}$$

while the mechanical fields of the inhomogeneous problem may be recovered from (4.38) in the space

$$\mathbf{u}(\mathbf{x}) = V_i(\mathbf{x})\mathbf{u}^c(\mathbf{s}) + \mathbf{u}^p(\mathbf{s}), \quad \boldsymbol{\sigma}(\mathbf{x}) = V_i(\mathbf{x})\boldsymbol{\sigma}^c(\mathbf{x}) + \boldsymbol{\alpha}_1(\mathbf{x}) + \boldsymbol{\sigma}^p(\mathbf{x}). \tag{4.43}$$

Let us assume that the domain w contains one heterogeneity $v_k \subset w$. We define the *perturbators* $\mathcal{L}_k^{\vartheta}(\mathbf{x} - \mathbf{x}_k, \boldsymbol{\zeta})$ and $\mathcal{L}_k^{\chi}(\mathbf{x} - \mathbf{x}_k, \boldsymbol{\chi})$ as the perturbations of the corresponding fields introduced by the heterogeneity v_k subjected to the effective fields $\overline{\boldsymbol{\vartheta}}(\mathbf{x})$ and $\overline{\boldsymbol{\chi}}(\mathbf{x})$, respectively,

$$\mathcal{L}_k^{\vartheta}(\mathbf{x} - \mathbf{x}_k, \boldsymbol{\zeta}) = \mathcal{L}_k^{\vartheta}(\mathbf{x} - \mathbf{x}_k, \overline{\boldsymbol{\vartheta}}) \equiv \boldsymbol{\vartheta}(\mathbf{x}) - \overline{\boldsymbol{\vartheta}}(\mathbf{x}), \tag{4.44}$$

$$\mathcal{L}_k^{\chi}(\mathbf{x} - \mathbf{x}_k, \boldsymbol{\chi}) = \mathcal{L}_k^{\chi}(\mathbf{x} - \mathbf{x}_k, \overline{\boldsymbol{\chi}}) \equiv \boldsymbol{\chi}(\mathbf{x}) - \overline{\boldsymbol{\chi}}(\mathbf{x}). \tag{4.45}$$

Hereafter for contraction of calculations, we introduce the substitutions

$$(\boldsymbol{\varepsilon}, \boldsymbol{\sigma}) \leftrightarrow \boldsymbol{\vartheta}, \quad (\boldsymbol{\tau}, \boldsymbol{\eta}) \leftrightarrow \boldsymbol{\zeta}, \tag{4.46}$$

indicating that for each duplet $\boldsymbol{\vartheta} = \boldsymbol{\varepsilon}$, $\boldsymbol{\zeta} = \boldsymbol{\tau}$ and $\boldsymbol{\vartheta} = \boldsymbol{\sigma}$, $\boldsymbol{\zeta} = \boldsymbol{\eta}$, Eq. (4.46) and the subsequent equations with the variables $\boldsymbol{\vartheta}$ and $\boldsymbol{\zeta}$ are reduced to the corresponding equations for the strain field $\boldsymbol{\varepsilon}$ and stress $\boldsymbol{\sigma}$. In a similar manner, one introduce a duplet

$$(\mathbf{u}, \mathbf{t}) \leftrightarrow \boldsymbol{\chi}, \tag{4.47}$$

which is used in Eq. (4.45), however, for both variables $\mathbf{u}$ and $\mathbf{t}$ simultaneously. In so doing, the perturbators $\mathcal{L}_k^{\vartheta}(\mathbf{x}-\mathbf{x}_k, \boldsymbol{\zeta})$ and $\mathcal{L}_k^{\chi}(\mathbf{x}-\mathbf{x}_k, \boldsymbol{\chi})$ are defined by the volume $\boldsymbol{\zeta}(\mathbf{x})$ ($\mathbf{x} \in v_i$) and the boundary $\boldsymbol{\chi}(\mathbf{s})$ ($\mathbf{s} \in \Gamma_i$) values of the corresponding fields. Strictly speaking, the perturbators $\mathcal{L}_k^{\vartheta}(\mathbf{x}-\mathbf{x}_k, \boldsymbol{\zeta})$ (4.44$_1$) and $\mathcal{L}_k^{\chi}(\mathbf{x}-\mathbf{x}_k, \boldsymbol{\chi})$ (4.45$_1$) are just the notations of some problem which is destined to be solved while the perturbators $\mathcal{L}_k^{\vartheta}(\mathbf{x}-\mathbf{x}_k, \overline{\boldsymbol{\vartheta}})$ (4.44$_2$) and $\mathcal{L}_k^{\chi}(\mathbf{x}-\mathbf{x}_k, \overline{\boldsymbol{\chi}})$ (4.45$_2$) are the solutions of this problem.

Let us consider the well-known integral representation formulae of the strains and the stresses (4.9) ($\mathbf{x} \in R^d$, $\mathbf{y} \in v_k$). In such a case, the perturbators $\mathcal{L}^{\epsilon}(\mathbf{x}-\mathbf{x}_k, \boldsymbol{\tau})$ and $\mathcal{L}^{\sigma}(\mathbf{x}-\mathbf{x}_k, \boldsymbol{\eta})$ can be expressed through the Green's functions

$$\mathcal{L}^{\epsilon}(\mathbf{x}-\mathbf{x}_k, \boldsymbol{\tau}) = \int \mathbf{U}^{(0)}(\mathbf{x}-\mathbf{y})\boldsymbol{\tau}(\mathbf{y})d\mathbf{y},$$

$$\mathcal{L}^{\sigma}(\mathbf{x}-\mathbf{x}_k, \boldsymbol{\eta}) = \int \boldsymbol{\Gamma}^{(0)}(\mathbf{x}-\mathbf{y})\boldsymbol{\eta}(\mathbf{y})d\mathbf{y}. \tag{4.48}$$

The surface perturbators $\mathcal{L}^{\chi}(\mathbf{x}-\mathbf{x}_k, \boldsymbol{\chi})$ can be found in a similar manner as the volume perturbators $\mathcal{L}^{\vartheta}(\mathbf{x}-\mathbf{x}_k, \boldsymbol{\zeta})$ were found from Eqs. (4.48). Namely, we introduce the single- and double -layer operators $\mathcal{G}$, $\mathcal{T}$, $\mathcal{D}$, $\mathcal{S}$ with the kernels $\mathbf{G}(\mathbf{x}-\mathbf{s})$, $\mathbf{T}^{\top}(\mathbf{x},\mathbf{s})$, $\mathbf{D}(\mathbf{x}-\mathbf{s})$, and $\mathbf{S}(\mathbf{x},\mathbf{s})$, respectively ($\mathbf{x} \in R^d$)

$$(\mathcal{G}\mathbf{t})(\mathbf{x}) = \int_{\Gamma_k} \mathbf{G}(\mathbf{x},\mathbf{s})\mathbf{t}(\mathbf{s})ds, \quad (\mathcal{T}\mathbf{u})(\mathbf{x}) = \int_{\Gamma_k} \mathbf{T}^{\top}(\mathbf{x},\mathbf{s})\mathbf{u}(\mathbf{s})ds,$$

$$(\mathcal{D}\mathbf{t})(\mathbf{x}) = \int_{\Gamma_k} \mathbf{D}(\mathbf{x},\mathbf{s})\mathbf{t}(\mathbf{s})ds, \quad (\mathcal{S}\mathbf{u})(\mathbf{x}) = \int_{\Gamma_k} \mathbf{S}^{\top}(\mathbf{x},\mathbf{s})\mathbf{u}(\mathbf{s})ds.$$

Then Eqs. (4.32) and (4.33) can be presented in the matrix operator form

$$\begin{pmatrix} \mathbf{k}\mathbf{u} \\ \mathbf{k}\boldsymbol{\sigma} \end{pmatrix} - \begin{pmatrix} e\overline{\mathbf{u}} \\ e\overline{\boldsymbol{\sigma}} \end{pmatrix} = \begin{pmatrix} -\mathcal{T}^{(m)} & \mathcal{G}^{(m)} \\ -\mathcal{S}^{(m)} & \mathcal{D}^{(m)} \end{pmatrix} \begin{pmatrix} \mathbf{u} \\ \mathbf{t} \end{pmatrix}$$

that leads to the obvious matrix representation for the perturbator $\mathcal{L}_i^{\chi}(\mathbf{x}-\mathbf{x}_i, \boldsymbol{\chi})$. For example, for the external problem

$$\mathcal{L}_i^{\chi}(\mathbf{x}-\mathbf{x}_i, \boldsymbol{\chi}) = \begin{pmatrix} -\mathcal{T}^{(0)} & \mathcal{G}^{(0)} \\ -\mathbf{n}\mathcal{S}^{(0)} & \mathbf{n}\mathcal{D}^{(0)} \end{pmatrix} \begin{pmatrix} \mathbf{u} \\ \mathbf{t} \end{pmatrix}. \tag{4.49}$$

In a similar manner, a direct method for estimation of the Cauchy data $[\mathbf{u}(\mathbf{x}), \boldsymbol{\sigma}(\mathbf{x})]$ for $\mathbf{x} \in \Gamma_k$ is based on the solution of the conventional BIE (4.32) presented for both the matrix and inclusion in the matrix forms by the substitution method (see, e.g., [387, 1584, 1657]). The scheme of this method can be recast in the theoretical framework of well-developed boundary operators (see for details [657, 1441, 1442]). Namely, sending in (4.32) $\mathbf{x} \to \Gamma_k$ yields the following operator form of the well-known conventional BIE (4.32) ($\mathbf{s} \in \Gamma_k$)

$$(\mathcal{G}^{(0)}\mathbf{t}_0)(\mathbf{s}) + \overline{\mathbf{u}}(\mathbf{s}) = (\mathbf{k} + \mathcal{T}^{(0)})\mathbf{u}_0(\mathbf{s}), \quad (\mathcal{G}^{(k)}\mathbf{t}_k)(\mathbf{s}) = (\mathbf{k} + \mathcal{T}^{(k)})\mathbf{u}_k(\mathbf{s}). \tag{4.50}$$

From the BIE (4.50_2), the elliptic invertible operator $\mathcal{G}$ defines the Dirichlet–Neumann map

$$\mathbf{t}_k(\mathbf{s}) = (\widetilde{\mathcal{S}}^{(k)}\mathbf{u}_0)(\mathbf{s}), \tag{4.51}$$

where the Steklov–Poincaré operator $(\widetilde{\mathcal{S}}^{(k)}\mathbf{u}_0)(\mathbf{s}) = (\mathcal{G}^{(k)})^{-1}(\mathbf{k} + \mathcal{T}^{(k)})\mathbf{u}_0(\mathbf{s})$ is self-adjoint with a known alternating symmetric approximation which is well suited for a symmetric coupling of finite and boundary elements (see [724]). Substituting of Eq. (4.50) into the traction coupling condition (4.27_2) and Eq. (4.50_2) yields the closed BIE for the boundary displacement ($\mathbf{s} \in \Gamma_k$) $(\mathbf{k} + \mathcal{G}^{(0)}\widetilde{\mathcal{S}}^{(k)} + \mathcal{T}^{(0)})\mathbf{u}_0(\mathbf{s}) = \overline{\mathbf{u}}(\mathbf{s})$ with the solution

$$\mathbf{u}_0(\mathbf{s}) = (\mathcal{A}^u_k\overline{\mathbf{u}})(\mathbf{s}), \tag{4.52}$$

where $\mathcal{A}^u_k = (\mathbf{k} + \mathcal{G}^{(0)}\widetilde{\mathcal{S}}^{(k)} + \mathcal{T}^{(0)})^{-1}$. The fields $\mathbf{u}(\mathbf{x})$ and $\boldsymbol{\sigma}(\mathbf{x})$ (and, therefore, the estimation of the boundary inclusion perturbator $\mathcal{L}^{\chi}(\mathbf{x} - \mathbf{x}_k, \chi)$) in the whole region $\mathbf{x} \in w$ can be found by substitution of the solutions $\mathbf{u}_0(\mathbf{s})$ (4.52) and $\mathbf{t}_k(\mathbf{s})$ (4.51) ($\mathbf{s} \in \Gamma_k$) into Eqs. (4.32) and (4.33). By this means the boundary integral operator $\mathcal{A}_k$ describing a displacement perturbation introduced by the heterogeneity v_k ($\forall \mathbf{x} \in w$)

$$\mathbf{u}(\mathbf{x}) - \overline{\mathbf{u}}(\mathbf{x}) \equiv (\mathcal{A}_k\overline{\mathbf{u}})(\mathbf{x}) \equiv \int \mathcal{A}_k(\mathbf{x}, \mathbf{s})\overline{\mathbf{u}}(\mathbf{s})ds$$

and acting on $\overline{\mathbf{u}}(\mathbf{s})$ ($\mathbf{s} \in \Gamma_k$) is found. In particular, for the external problem ($\mathbf{x} \in w \setminus v_k$)

$$(\mathcal{A}_k\overline{\mathbf{u}})(\mathbf{x}) = (\mathcal{T}^{(0)}\mathcal{A}^u_k\overline{\mathbf{u}})(\mathbf{x}) + (\mathcal{G}^{(0)}\mathcal{S}^{(k)}\mathcal{A}^u_k\overline{\mathbf{u}})(\mathbf{x}).$$

Thus, the boundary inclusion perturbator $\mathcal{L}^{\chi}(\mathbf{x} - \mathbf{x}_k, \overline{\chi}) = (\mathcal{A}_k\overline{\mathbf{u}})(\mathbf{x})$ ($\chi = \mathbf{u}$) is found. Construction of the perturbator $\mathcal{L}^{\chi}_i(\mathbf{x} - \mathbf{x}_i, \chi)$ by the MFS will be performed in Sect. 4.3.3.

Analytical representations for the operators $\mathcal{L}^{\vartheta}_k(\mathbf{x} - \mathbf{x}_k, \overline{\boldsymbol{\vartheta}})$ and $\mathcal{L}^{\chi}_k(\mathbf{x} - \mathbf{x}_k, \overline{\chi})$ are only known for the particular canonical (i.e., ellipsoidal) shapes of particle and specific effective fields (see, e.g., [532, 1186, 1244]). In the case of a homogeneous effective fields ($\mathbf{x} \in v_i$)

$$\overline{\boldsymbol{\vartheta}}(\mathbf{x}) \equiv \text{const.} \tag{4.53}$$

the operators $\boldsymbol{\mathcal{L}}_k^{\vartheta}(\mathbf{x}-\mathbf{x}_k,\boldsymbol{\zeta})$ and $\boldsymbol{\mathcal{L}}_k^{\vartheta}(\mathbf{x}-\mathbf{x}_k,\overline{\boldsymbol{\vartheta}})$ are reduced to the tensorial multiplication

$$\boldsymbol{\mathcal{L}}_k^{\vartheta}(\mathbf{x}-\mathbf{x}_k,\boldsymbol{\zeta}) = \boldsymbol{\mathcal{L}}_k^{\zeta}(\mathbf{x}-\mathbf{x}_k)\langle\boldsymbol{\zeta}\rangle_i, \quad \boldsymbol{\mathcal{L}}_k^{\vartheta}(\mathbf{x}-\mathbf{x}_k,\overline{\boldsymbol{\vartheta}}) = \boldsymbol{\mathcal{L}}_k^{\vartheta}(\mathbf{x}-\mathbf{x}_k)\overline{\boldsymbol{\vartheta}}. \tag{4.54}$$

for the ellipsoidal homogeneous inclusion v_k ($\mathbf{f} = \mathbf{L}, \mathbf{M}, \boldsymbol{\alpha}, \boldsymbol{\beta}$, no sum on i)

$$\mathbf{f}(\mathbf{x}) \equiv \text{const.}\ (\mathbf{x} \in v_k), \quad v_k = \{\mathbf{x} \mid x_m(a^{-2})_{ml}x_l \le 1\}, \quad a_{lm} = a_l\delta_{lm}. \tag{4.55}$$

In particular, for $\boldsymbol{\vartheta} = \boldsymbol{\sigma},\ \boldsymbol{\zeta} = \boldsymbol{\eta}$ (the case $\boldsymbol{\vartheta} = \boldsymbol{\varepsilon},\ \boldsymbol{\zeta} = \boldsymbol{\tau}$ can be considered in a similar manner)

$$\boldsymbol{\mathcal{L}}_k^{\zeta}(\mathbf{x}-\mathbf{x}_k) = \mathbf{T}_k^{\sigma}(\mathbf{x}-\mathbf{x}_k), \quad \boldsymbol{\mathcal{L}}_k^{\vartheta}(\mathbf{x}-\mathbf{x}_k) = \mathbf{T}_k^{\sigma}(\mathbf{x}-\mathbf{x}_k)\mathbf{R}_k,$$

where the tensors $\mathbf{T}_k^{\sigma}$ and $\mathbf{R}_k$ are defined by Eqs. (3.127$_2$) and (3.184$_3$), respectively.

For the inhomogeneous loading $\overline{\boldsymbol{\vartheta}}(\mathbf{x}) \not\equiv$const. and noncanonical shape (or inhomogeneous structure) of the inclusion, the operator concentration factors for the values $\boldsymbol{\sigma}(\mathbf{x})$ and $\boldsymbol{\eta}(\mathbf{x})$ ($\mathbf{x} \in R^d$) can be expressed through the perturbator $\boldsymbol{\mathcal{L}}^{\vartheta}(\mathbf{x}-\mathbf{x}_k,\overline{\boldsymbol{\vartheta}})$ ($\boldsymbol{\vartheta} = \boldsymbol{\sigma},\ \boldsymbol{\zeta} = \boldsymbol{\eta}$)

$$\boldsymbol{\sigma}(\mathbf{x}) = \boldsymbol{\mathcal{B}}_k^{\sigma}(\overline{\boldsymbol{\sigma}})(\mathbf{x}) + \mathbf{C}_k(\mathbf{x}), \quad \bar{v}_k\boldsymbol{\eta}(\mathbf{x}) = \boldsymbol{\mathcal{R}}_k^{\sigma}(\overline{\boldsymbol{\sigma}})(\mathbf{x}) + \mathbf{F}_k(\mathbf{x}), \tag{4.56}$$

where

$$\boldsymbol{\mathcal{B}}_k^{\sigma}(\overline{\boldsymbol{\sigma}})(\mathbf{x}) = \boldsymbol{\mathcal{L}}^{\vartheta}(\mathbf{x}-\mathbf{x}_k,\overline{\boldsymbol{\vartheta}}) + \mathbf{I}\overline{\boldsymbol{\sigma}}(\mathbf{x}), \tag{4.57}$$

$$\boldsymbol{\mathcal{R}}_k^{\sigma}(\overline{\boldsymbol{\sigma}})(\mathbf{x}) = \bar{v}_k\mathbf{M}_1^{(k)}\boldsymbol{\mathcal{B}}_k^{\sigma}(\overline{\boldsymbol{\sigma}})(\mathbf{x}), \tag{4.58}$$

$$\mathbf{F}_k(\mathbf{x}) = \bar{v}_k[\mathbf{M}_1^{(k)}\mathbf{C}_k(\mathbf{x}) + \boldsymbol{\beta}_1(\mathbf{x})]. \tag{4.59}$$

The concentration operators for the surface variables $\boldsymbol{\chi}(\mathbf{s})$ and $\boldsymbol{\eta}^s(\mathbf{s})$ (4.35) can be expressed in a similar manner

$$\boldsymbol{\chi}(\mathbf{s}) = \boldsymbol{\mathcal{B}}_k^{\chi}(\overline{\boldsymbol{\chi}})(\mathbf{s}) + \mathbf{C}_k^{\chi}(\mathbf{s}), \quad \bar{v}_k\boldsymbol{\eta}^s(\mathbf{x}) = \boldsymbol{\mathcal{R}}_k^{\chi}(\overline{\boldsymbol{\chi}})(\mathbf{s}) + \mathbf{F}_k^{\chi}(\mathbf{s}), \tag{4.60}$$

where $\boldsymbol{\mathcal{B}}_k^{\chi}(\overline{\boldsymbol{\chi}})(\mathbf{s}) = \boldsymbol{\mathcal{L}}^{\chi}(\mathbf{s}-\mathbf{x}_k,\overline{\boldsymbol{\chi}}) + \mathbf{I}\overline{\boldsymbol{\chi}}(\mathbf{s})$. It should be mentioned that even for the homogeneous volume effective fields (4.53), the surface effective fields $\overline{\boldsymbol{\chi}}(\mathbf{s})$ ($\mathbf{s} \in v_k$) are inhomogeneous ones, e.g., for the boundary condition $\overline{\boldsymbol{\sigma}} \equiv \boldsymbol{\sigma}^{\Gamma}$ =const.

$$\overline{\boldsymbol{\chi}}(\mathbf{s}) = (\mathbf{M}^{(0)}\overline{\boldsymbol{\sigma}}(\mathbf{s}-\mathbf{x}_k), \overline{\boldsymbol{\sigma}}\mathbf{n}^{\Gamma_k}(\mathbf{s}))^{\top} \tag{4.61}$$

and, therefore, the operators in the right-hand sides of Eqs. (4.60) are not reduced to the tensorial convolution with the surface effective field $\overline{\boldsymbol{\chi}}(\mathbf{s})$ as in their counterpart (4.53) for the volume effective field $\overline{\boldsymbol{\sigma}}$. However, homogeneity of the volume field $\overline{\boldsymbol{\sigma}}$ in Eq. (4.61) allows one to get another version of splitting of the operators $\mathcal{L}^{\chi}(\mathbf{s}-\mathbf{x}_k,\overline{\boldsymbol{\chi}})$, $\mathcal{B}_k^{\chi}(\overline{\boldsymbol{\chi}})(\mathbf{s})$, and $\mathcal{R}_k^{\chi}(\overline{\boldsymbol{\chi}})(\mathbf{s})$

$$\mathcal{L}^{\chi}(\mathbf{s}-\mathbf{x}_k,\overline{\boldsymbol{\chi}}) = \mathbf{L}^{\chi\sigma}(\mathbf{s}-\mathbf{x}_k)\overline{\boldsymbol{\sigma}} + \mathbf{C}^{\chi}(\mathbf{s}-\mathbf{x}_k), \tag{4.62}$$

$$\mathcal{B}_k^{\chi}(\overline{\boldsymbol{\chi}})(\mathbf{s}) = \mathbf{B}_k^{\chi\sigma}(\mathbf{s})\overline{\boldsymbol{\sigma}}, \quad \mathcal{R}_k^{\chi}(\overline{\boldsymbol{\chi}})(\mathbf{s}) = \mathbf{R}_k^{\chi\sigma}(\mathbf{s})\overline{\boldsymbol{\sigma}}. \tag{4.63}$$

Estimation of the perturbators $\mathcal{L}_k^{\vartheta}(\mathbf{x}-\mathbf{x}_k,\overline{\boldsymbol{\vartheta}})$ (4.44) and $\mathcal{L}_k^{\chi}(\mathbf{x}-\mathbf{x}_k,\overline{\boldsymbol{\chi}})$ (4.45) can be performed by any available numerical method, such as the volume integral equation (VIE), boundary element method (BEM), FEM, hybrid FEM-BEM, multipole expansion method, complex potential method, and other (see for references [200, 544, 849, 1353]). Each method has advantages and disadvantages, and it is crucial for the analyst to be aware of their range of applications. In particular, the VIE method for the ideal contact conditions enables one to restrict discretization to the inclusions only (in contrast to the FEA), and an inhomogeneous structure of inclusions (see, e.g., [313, 695, 1662]) presents no problem in the framework of the same numerical scheme (compared to the standard BEM). The VIE method (see [203]) has well-developed routines for the solution of integral equations (e.g., the iteration method and the quadrature schemes) and allows to analyze arbitrary inhomogeneous fields $\boldsymbol{\varepsilon}^0(\mathbf{x})$ and $\boldsymbol{\sigma}^0(\mathbf{x})$. However, the VIE method is quite time-consuming and no optimized commercial softwares exist for its application. From other side, the FEM is supported by well-developed commercial softwares (see, e.g., [1]) and gives strong advantages in term of CPU-time. The FEA is especially effective for estimations of perturbators at the constant fields $\boldsymbol{\varepsilon}^0, \boldsymbol{\sigma}^0$ =const. (see for references e.g. [211]). In the current section, one will express the perturbators $\mathcal{L}_k^{\vartheta}(\mathbf{x}-\mathbf{x}_k,\boldsymbol{\zeta})$ (4.44) and $\mathcal{L}_k^{\chi}(\mathbf{x}-\mathbf{x}_k,\boldsymbol{\chi})$ (4.45) by the use of MFS which have the advantages described in Sect. 4.1.

4.3.3 Method of Fundamental Solutions (MFS)

The basic idea for the formulation of the MFS proposed by Kupradze and Aleksidze [841] is based on approximation of the solution by a finite linear combination of fundamental solutions (FS, assumed to be explicitly defined) with respect to source points which are positioned outside the solution domain. The unknown coefficients in the mentioned linear approximation can be determined by matching the boundary conditions in a variety of ways, the simplest (and most popular) being by collocation on a set of physical boundary points. In contrast to domain discretization methods such as the finite element (FEM), the MFS is a boundary method which means that only the boundary of the solution domain needs to be considered. Moreover, the MFS is a boundary-type meshless method which does not need a boundary element mesh, either for purposes of interpolation of the trial and test functions of the solution variables, or for the integration, and only a set of source points is required for the discretization of the problem being analyzed. However, unlike the BEM, no potentially troublesome integration is required in the MFS due to the placement of source points outside the solution domain when the singular integrals are avoided. At last MFS is adaptive in the sense that it can take into account sharp changes in the solution and in the geometry of the domain and can easily handle complex boundary conditions. Despite the mentioned advantages, a few disadvantages are that the positioning of the source points are preassigned and also the resulting system of algebraic equations is ill-conditioned that leads to oscillation of the convergence curve of the numerical solution when a large number of source points are used. Optimization of source points allocation substantially reduces computational time while some regularization methods, such as the damped singular value decomposition, the truncated singular value decomposition, or the Tikhonov regularization, can be used to mitigate the ill-conditioned effect (see [937]).

The boundary integral equations (4.32) and (4.33) can be regarded as a continuous form of the discrete MFS which approximates the solution within a given elastic domain v_i as a finite linear combination of fundamental solutions generated by a set of virtual sources positioned outside the domain (to avoid singularities). Namely, let the domains v_i and $\hat{v}_i^{(1)}$ be open subsets in R^d separated by the smooth boundary Γ_i. (see, e.g., [1417]). We say that $\hat{v}_i^{(1)}$ with the smooth pseudo-boundary $\hat{\Gamma}_i^{(1)} := \partial\hat{v}_i^{(1)}$ embraces v_i if $(v_i \cup \Gamma_i) \subset \hat{v}_i^{(1)}$. In a similar manner, one can define a pseudo-boundary $\hat{\Gamma}_i^{(0)} := \partial\hat{v}_i^{(0)}$ of an open subset $\hat{v}_i^{(0)}$ embracing the open domain $R^d\backslash(v_i \cup \Gamma_i)$. So, the gray regions in Fig. 4.3a, b are the inclusion v_i with the boundary $\Gamma_i = \partial v_i$. The $\hat{\Gamma}_i^{(1)}$ and $\hat{\Gamma}_i^{(0)}$ are constructed from $\Gamma_i = \partial v_i$ by a homothety transformation (4.70) (Fig. 4.3a) and by a normal direction choice for the source points (4.72) (Fig. 4.3b) which will be considered later.

At first, we consider the problem (4.39) for one inclusion in the infinite matrix loaded by both the effective field $(\overline{\mathbf{u}}(\mathbf{s}), \overline{\mathbf{t}}(\mathbf{s}))$ and the delta function type body force (4.37). We place m points of the admissible singularities (or sources)

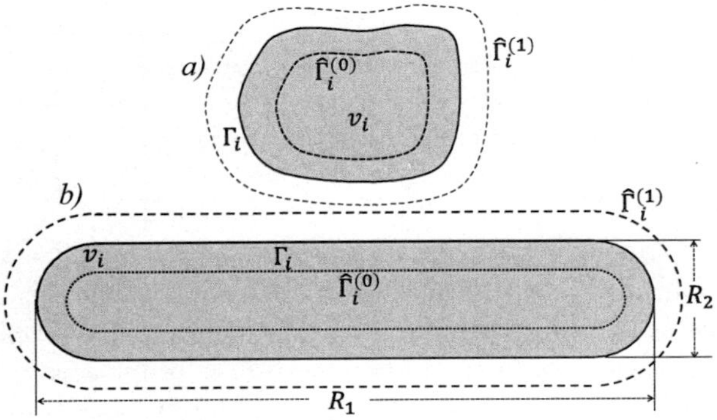

Fig. 4.3 The boundary $\Gamma_i = \partial v_i$ and the pseudo-boundaries $\hat{\Gamma}_i^{(0)}$ and $\hat{\Gamma}_i^{(1)}$ constructed by Eqs. (4.70) (4.93a) and (4.72) (4.93b), which will be considered later

$\{\mathbf{y}_s^k\}_{s=1}^m$ corresponding to the external ($k = 0$) and internal ($k = 1$) problems on the prescribed pseudo-boundary surfaces $\hat{\Gamma}_i^{(0)}$ and $\hat{\Gamma}_i^{(1)}$, respectively. A slight modification of [27] (see also [1018]) results shows that it is possible to approach the boundary conditions for the displacement approximated by linear combinations of the corresponding fundamental solutions. Let the effective fields $\overline{\mathbf{u}}(\mathbf{x}_j)$ and $\overline{\mathbf{t}}(\mathbf{x}_j)$ be prescribed in the collocation points $\{\mathbf{x}_j\}_{j=1}^m \in \Gamma_i$. Then the strength of FSs of the particular solution (4.39) $\mathbf{f}_i^{p(k)} = (\mathbf{f}_{is|1}^{p(k)}, \ldots, \mathbf{f}_{is|d}^{p(k)})^\top$ ($s = 1, \ldots, m$) hold a $dm \times dm$ linear system of equations ($j = 1, \ldots, d,\ l = 1, \ldots, m$)

$$\sum_{s=1}^{m}\sum_{q=1}^{d} f_{isq}^{p(1)} G_{jq}(\mathbf{r}_{ls}^1) = \sum_{s=1}^{m}\sum_{q=1}^{d} f_{isq}^{p(0)} G_{jq}(\mathbf{r}_{ls}^0) + \overline{u}_j(\mathbf{x}_l), \tag{4.64}$$

where $\mathbf{r}_{ls}^k = (\mathbf{x}_l - \mathbf{y}_s^k)$ and $|\mathbf{r}_{ls}^k|$ is the distance between the collocation point (the field point), $\mathbf{x}_l$, and the location of the FSs (the source point), $\mathbf{y}_s^k \in \hat{\Gamma}_i^{(k)}$ ($k = 0, 1$). In a similar manner, the traction boundary conditions (4.27_2) can be obtained directly by taking summation of the traction FSs with the same strength $\mathbf{f}_i^{p(k)} = (\mathbf{f}_{is|1}^{p(k)}, \ldots, \mathbf{f}_{is|d}^{p(k)})^\top$ (4.64) over the source points $\mathbf{y}_s^k \in \hat{\Gamma}_i^{(k)}$ outside the boundary Γ_i ($j, q = 1, \ldots, d;\ s = 1, \ldots, m;\ k = 0, 1$):

$$\sum_{s=1}^{m}\sum_{q=1}^{d} f_{isq}^{p(1)} T_{jq}(\mathbf{r}_{ls}^1) = \sum_{s=1}^{m}\sum_{q=1}^{d} f_{isq}^{p(0)} T_{jq}(\mathbf{r}_{ls}^0) + \overline{t}_j(\mathbf{x}_l) + t_j^\alpha(\mathbf{x}_l), \tag{4.65}$$

where

$$\overline{\mathbf{t}}(\mathbf{x}_l) = \overline{\boldsymbol{\sigma}}(\mathbf{x}_l)\mathbf{n}^{\Gamma_i}(\mathbf{x}_l) \quad \text{and} \quad \mathbf{t}^\alpha(\mathbf{x}_l) = -\boldsymbol{\alpha}_1\mathbf{n}^{\Gamma_i}(\mathbf{x}_l), \tag{4.66}$$

are the tractions produced by the effective field (4.29) the pseudo-body force (4.37), respectively. We assumed in the system (4.64) and (4.65) that the numbers of both the field points $\{\mathbf{x}_l\}_{l=1}^m$ and source ones $\{\mathbf{y}_s^k\}_{s=1}^m$ ($k = 0, 1$) coincide. In such a case, the strength of the FSs, $\mathbf{f}_i^p$, is determined uniquely, provided that the corresponding coefficient matrix is non-singular. If $\overline{\mathbf{u}}(\mathbf{x})$ is prescribed in more points $\mathbf{x}_1, \ldots, \mathbf{x}_n \in \Gamma_i$, with $n > S$, the system (4.64) and (4.65) is over-determined, and the strength can be chosen by the discrete least-squares method (see [27, 1417]) which is not considered in more details in the current book.

Once the strength of the FS, $\mathbf{f}_i^{p(k)}$, is found by solving the system (4.64) and (4.65) at $\{\mathbf{x}_l\}_{l=1}^S \in \Gamma_i$, the displacement vector in both the inclusion v_i and matrix $R^d \backslash v_i$ is then calculated straightforwardly by using the left-right side and right-hand side, respectively, of Eq. (4.65) with values of replaced radius–vectors $\mathbf{x}_l \to \mathbf{x}_s$ corresponding to location of a point $\mathbf{x}_s$. All other field variables can be obtained directly by taking summation of the same strength $\mathbf{f}_i^{p(k)}$ (4.65) and (4.64) of the corresponding FSs over the source points $\mathbf{y}_s^k \in \hat{\Gamma}_i^{(k)}$ ($k = 0, 1$) outside the boundary Γ_i, e.g. the displacement, traction, and stress induced by S FSs can be estimated by the equations $(j, l = 1, \dots, d)$:

$$u_j(\mathbf{x}_r) = \sum_{s=1}^{S}\sum_{q=1}^{d} f_{isq}^{p(k)} G_{jq}^{(k)}(\mathbf{r}_{rs}^k) + \overline{u}_j(\mathbf{x}_r)\delta_{k0}, \tag{4.67}$$

$$t_j(\mathbf{x}_r) = \sum_{s=1}^{S}\sum_{q=1}^{d} f_{isq}^{p(k)} T_{jq}^{(k)}(\mathbf{r}_{rs}^k) + \overline{t}_j(\mathbf{x}_r)\delta_{k0}, \tag{4.68}$$

$$\sigma_{jl}(\mathbf{x}_r) = \sum_{s=1}^{S}\sum_{q=1}^{d} f_{isq}^{p(k)} D_{jlq}^{(k)}(\mathbf{r}_{rs}^k) + \overline{\sigma}_{jl}(\mathbf{x}_r)\delta_{k0} \tag{4.69}$$

for $\mathbf{x}_s \in v_i$ $(k = 1)$ and $\mathbf{x}_s \in (R^d \backslash v_i)$ $(k = 0)$, respectively.

A specific feature of the MFS is some freedom in choosing the source points with respect to the accuracy of the numerical solution. Young et al. [1663] (see also [27]) estimated an excellent accuracy of the MFS for the positions of the source points $\mathbf{y}$ in computational domain through the field points $\mathbf{x}_l$ $(l = 1, \dots, m)$ by a homothety transformation:

$$\mathbf{y}_l = \mathbf{x}_i + b(\mathbf{x}_l - \mathbf{x}_i), \tag{4.70}$$

where $\mathbf{x}_i$ is a homothetic center coinciding with the particle center. The pseudo-boundaries $\hat{\Gamma}_i^{(0)}$ and $\hat{\Gamma}_i^{(1)}$ in Fig. 4.3a are obtained from the boundary $\hat{\Gamma}_i$ of the gray inclusion v_i by the homothety transformation (4.70).

However, for the thin inclusions with the thickness $2h$ (e.g., for the tri-axial ellipsoid with distinct semi-axis lengths $a_1 > a_2 \gg a_3$, $h^a = a_3$), the distances $D(\mathbf{y}_l, \Gamma_i)$ [defining the matrix coefficients $G_{jq}^{(k)}(\mathbf{r}_{lp}^1)$ (4.64) and $T_{jq}^{(k)}(\mathbf{r}_{lp}^1)$ (4.65)] essentially differ one from another point for the different $\mathbf{x}_l$. Here $D(\mathbf{y}_l, \Gamma_i)$ denotes the Euclidean distance between $\mathbf{y}_l$ and the boundary of v_i or, in other words, the radius of the largest inscribed sphere (or disk) with center $\mathbf{y}_l$. Because of this, for the thin inclusions the following allocations of the source points $(k = 0, 1)$

$$\mathbf{y}_l^k = \mathbf{x}_l + (2k - 1)D_k\mathbf{n}^{\Gamma_i}(\mathbf{x}_l) \tag{4.71}$$

with $D = D(\mathbf{y}_l, \Gamma_i) = 10\text{–}30\%h^a$ is preferable.

The algorithm of the normal direction choice for the source points (4.71) proposed in [26] was accomplished by so called global choice for the coefficient D_k $(k = 0, 1)$ recently presented in [25] for convex and concave regions. However, more detail consideration of optimization of FS's allocation substantially reducing computational time for the elastic problem solution is beyond the scope of the current study. A simplest choice of a constant D_k is a non-optimal one that works in most cases. In such a case, the pseudo-boundaries $\hat{\Gamma}^{(1)}$ and $\hat{\Gamma}^{(0)}$ can be presented as the boundaries of Minkowski addition and subtraction

$$\hat{v}^{(1)} = v_i \oplus b(\mathbf{0}, D_0), \quad \hat{v}^{(0)} = v_i \ominus b(\mathbf{0}, D_1), \tag{4.72}$$

respectively, of the domain v_i, where $b(\mathbf{0}, D_k)$ $(k = 0, 1)$ stands the ball of radius D_k centered at the origin $\mathbf{0}$. As an example, a grey cross-section of a spherocylinder (or a capsule) with the aspect ratio $\alpha = R_2/R_1$ is presented in Fig. 4.3b. The pseudo-boundaries $\hat{\Gamma}_i^{(0)}$ and $\hat{\Gamma}_i^{(1)}$ in Fig. 4.3b are obtained from the boundary $\hat{\Gamma}_i$ of the grey inclusion v_i by the normal direction choice for the source points (4.71) with $D_1 = R_2/2$ and $D_0 = R_2/2$, respectively. In so doing, the aspect ratios of the domains with the pseudo-boundaries $\hat{\Gamma}_i^{(1)}$ and $\hat{\Gamma}_i^{(0)}$ are $(2\alpha + 1)/3$ and $(2\alpha - 1)$ while the aspect ratios of the corresponding domains obtained by the homothety transform (4.71) do not change and equal α. Alves [25] (see also [26, 1060]) analyzed optimization of more general stresslet allocation techniques for convex and concave regions. However, more detail consideration of optimization of FS's allocation substantially reducing computational time for the elastic problem solution is beyond the scope of the current study.

4.3.4 *Matrix Representation of the MFS*

Construction and solution of the linear algebraic system (4.64) is straightforward (see, e.g., [752, 1663]). However, we consider it in more details for the 2D domain v_i. The singularities $Y_s^0 = (y_{s|1}^0, y_{s|2}^0)$ and $Y_s^1 = (y_{s|1}^1, y_{s|2}^1)$ $(s = 1, \ldots, m)$ are fixed on the boundaries $\hat{\Gamma}_i^{(0)}$ of $\hat{v}_i^{(0)}$ and $\hat{\Gamma}_i^{(1)}$ of $\hat{v}_i^{(1)}$, respectively. A set of m collocation points $X_s = (x_{s|1}, x_{s|2})$ $(s = 1, \ldots, m)$ is chosen on Γ_i. Then the system (4.64) can be presented in the matrix form

$$\mathbf{A}\mathbf{f}^p = \begin{pmatrix} \mathbf{A}_{11} & \mathbf{A}_{1,2} & \ldots & \mathbf{A}_{1,m} \\ \mathbf{A}_{2,1} & \mathbf{A}_{2,2} & \ldots & \mathbf{A}_{2,m} \\ \vdots & \vdots & \ddots & \vdots \\ \mathbf{A}_{m,1} & \mathbf{A}_{m,2} & \ldots & \mathbf{A}_{m,m} \end{pmatrix} \begin{pmatrix} \mathbf{f}_1^p \\ \mathbf{f}_2^p \\ \vdots \\ \mathbf{f}_m^p \end{pmatrix} = \begin{pmatrix} \mathbf{g}_1^p \\ \mathbf{g}_2^p \\ \vdots \\ \mathbf{g}_m^p \end{pmatrix} = \mathbf{g}^p, \tag{4.73}$$

where the $4m \times 4m$ matrix $\mathbf{A}$ with the submatrices

$$\mathbf{A}_{\alpha,\beta} = \begin{pmatrix} \mathbf{A}_{\alpha,\beta}^{1,1} & \mathbf{A}_{\alpha,\beta}^{1,2} \\ \mathbf{A}_{\alpha,\beta}^{2,1} & \mathbf{A}_{\alpha,\beta}^{2,2} \end{pmatrix} \tag{4.74}$$

are defined through the submatrices

$$\mathbf{A}^{1,\delta+1}_{\alpha,\beta} = (-1)^{\delta+1}\mathbf{G}^{(\delta)}(|X_\alpha - Y^\delta_\beta|) \in R^{2\times 2}, \tag{4.75}$$

$$\mathbf{A}^{2,\delta+1}_{\alpha,\beta} = (-1)^{\delta+1}\mathbf{T}^{(\delta)}(|X_\alpha - Y^\delta_\beta|) \in R^{2\times 2}, \tag{4.76}$$

and

$$\mathbf{f}^p_s = (\mathbf{f}^{p(0)}_s, \mathbf{f}^{p(1)}_s)^\top = (f^{p(0)}_{s|1}, f^{p(0)}_{s|2}, f^{p(1)}_{s|1}, f^{p(1)}_{s|2})^\top \in R^4, \tag{4.77}$$

$$\mathbf{g}^p_s = (\mathbf{g}^{p(0)}_s, \mathbf{g}^{p(1)}_s)^\top = (g^{p(0)}_{s|1}, g^{p(0)}_{s|2}, g^{p(1)}_{s|1}, g^{p(1)}_{s|2})^\top \in R^4, \tag{4.78}$$

where $f^{p(\delta)}_{s|j} = f^{p(\delta)}_{s|j}(Y^\delta_s)$, $g^{p(\delta)}_{s|j} = \overline{u}_{s|j}\delta_{0\delta} + \overline{t}_{s|j}\delta_{1\delta}$, and $j = 1,2$; $s, \alpha, \beta = 1, \ldots, m$; $\delta = 0, 1$. It is observed for $d = 2, 3$ that accurate numerical results could be obtained when the source points for the pseudo-boundary $\hat{\Gamma}^{(0)}_i$ of the submerged stresslets are located off the boundary at about 70–90% (4.70) of the distance from the center (see for details [1060, 1663]).

Thus, for a set of m collocation points $X_s = (x_{s|1}, x_{s|2})$ $(s = 1, \ldots, m)$ chosen on Γ_i, the system (4.64) and (4.65) can be presented in the matrix form

$$\mathbf{A}\mathbf{f}^p = \mathbf{g}^p, \tag{4.79}$$

where the elements of $2dm \times 2dm$ matrix $\mathbf{A}$ are expressed through the fundamental displacements $\mathbf{G}$ and tractions $\mathbf{T}$ while $\mathbf{f}^p_s\,(\mathbf{f}^{p(0)}_s, \mathbf{f}^{p(1)}_s)^\top$, $\mathbf{g}^p_s\,(\mathbf{g}^{p(0)}_s, \mathbf{g}^{p(1)}_s)^\top$, $f^{p(\delta)}_{s|j} = f^{p(\delta)}_{s|j}(Y^\delta_s)$, and $g^{p(\delta)}_{s|j} = \overline{u}_{s|j}\delta_{0\delta} + [\overline{t}_{s|j} + t^\alpha_{s|j}]\delta_{1\delta}$ $(j = 1, 2;\ s = 1, \ldots, m;\ \delta = 0, 1)$ is defined by the effective fields of displacements $\overline{\mathbf{u}}$ and tractions $\overline{\mathbf{t}}$ in the collocation points X_s.

However, the MFS coefficient matrix $\mathbf{A}$ (4.79) of the system (4.64) and (4.65) is often severely ill-conditioned and the system of linear algebraic equations (4.64) and (4.65) cannot be solved by direct methods, e.g., the least-squares method producing a highly unstable solution which increases dramatically as the number of boundary collocation points S increases. Several regularization procedures have been proposed to remedy the instability and accuracy loss in the solution of such ill-conditioned matrix equations (see, e.g., [603, 937]). However, we only consider the Tikhonov regularization (TR) method (see, e.g., [1507]), since it is simple, noniterative, and it provides an explicit solution. Namely, the TR procedure consists of seeking a solution which minimizes a criterion made up of the sum of two summands. The first one is a possibly weighted least square term, while the second one is a quadratic penalty term on the solution, that is

$$\min\{||\mathbf{A}\mathbf{f} - \mathbf{g}||^2_2 + \lambda||\mathbf{I}^{\rm dm}\mathbf{f}||^2_2\}, \tag{4.80}$$

where $||(\cdot)||_2$ denotes the Euclidean norm while for the zero-order TR method, $\mathbf{I}^{\rm dm} = {\rm diag}(1, \ldots, 1) \in R^{dm\times dm}$ is the identity matrix.

Formally, the TR procedure provides an explicit solution of Eq. (4.73)

$$\mathbf{f}^p = \boldsymbol{\mathcal{A}}\mathbf{g}^p, \quad \boldsymbol{\mathcal{A}} := (\mathbf{A}^\top \mathbf{A} + \lambda \mathbf{I})^{-1}\mathbf{A}^\top, \tag{4.81}$$

where the optimal value λ is chosen according to the method described in, e.g., [603, 1008], and the partitioned matrix $\boldsymbol{\mathcal{A}}$ can be expressed in the form (4.73) with the blocks 4×4 $\boldsymbol{\mathcal{A}}_{\alpha,\beta}$ presented as

$$\boldsymbol{\mathcal{A}}_{\alpha,\beta} = (\boldsymbol{\mathcal{A}}^u_{\alpha,\beta}\mathbf{0}) + (\mathbf{0}\boldsymbol{\mathcal{A}}^t_{\alpha,\beta}), \tag{4.82}$$

where the 4×2 submatrices $\boldsymbol{\mathcal{A}}^u_{\alpha,\beta}$ and $\boldsymbol{\mathcal{A}}^t_{\alpha,\beta}$, in its turn, can be expressed through the blocks $\boldsymbol{\mathcal{A}}^{1,1}_{\alpha,\beta}$, $\boldsymbol{\mathcal{A}}^{2,1}_{\alpha,\beta}$ and $\boldsymbol{\mathcal{A}}^{1,2}_{\alpha,\beta}$, $\boldsymbol{\mathcal{A}}^{2,2}_{\alpha,\beta}$, respectively, in much the same manner as (4.74).

The found solution $\mathbf{f}^p$ allows one to present all field variables in an arbitrary point $\mathbf{x} \in R^d$ in a symbolic matrix notations ($k = 0, 1$)

$$\mathbf{u}^p(\mathbf{x}) - \overline{\mathbf{u}}(\mathbf{x})\delta_{k0} = \mathbf{G}^{(k)}(\boldsymbol{\mathcal{A}}^u\overline{\mathbf{u}}^p(\mathbf{x}) + \overline{\boldsymbol{\mathcal{A}}}^t\overline{\mathbf{t}}^p(\mathbf{x})), \tag{4.83}$$

$$\mathbf{t}^p(\mathbf{x}) - \overline{\mathbf{t}}(\mathbf{x})\delta_{k0} = \mathbf{T}^{(k)}(\boldsymbol{\mathcal{A}}^u\overline{\mathbf{u}}^p(\mathbf{x}) + \overline{\boldsymbol{\mathcal{A}}}^t\overline{\mathbf{t}}^p(\mathbf{x})), \tag{4.84}$$

$$\boldsymbol{\sigma}^p(\mathbf{x}) - \overline{\boldsymbol{\sigma}}(\mathbf{x})\delta_{k0} = \mathbf{D}^{(k)}(\boldsymbol{\mathcal{A}}^u\overline{\mathbf{u}}^p(\mathbf{x}) + \overline{\boldsymbol{\mathcal{A}}}^t\overline{\mathbf{t}}^p(\mathbf{x})), \tag{4.85}$$

which can be considered as a representation of the boundary inclusion perturbations $\boldsymbol{\mathcal{L}}^\chi_i(\mathbf{x} - \mathbf{x}_i, \overline{\boldsymbol{\chi}})$ defined at the pseudo-boundaries either $\hat{\Gamma}^{(1)}_i$ or $\hat{\Gamma}^{(0)}_i$ at $k = 1$ or $k = 0$, respectively. Here the blocks $4 \times 2\, \overline{\boldsymbol{\mathcal{A}}}^t_{\alpha,\beta} = (\boldsymbol{\mathcal{A}}^{1,2}_{\alpha,\beta} - \delta_{\alpha,\beta}\boldsymbol{\delta}, \boldsymbol{\mathcal{A}}^{2,2}_{\alpha,\beta} - \delta_{\alpha,\beta}\boldsymbol{\delta})^\top$ of the partitioned matrix $\overline{\boldsymbol{\mathcal{A}}}^t$ are defined through the blocks $\boldsymbol{\mathcal{A}}^{1,2}_{\alpha,\beta}$, $\boldsymbol{\mathcal{A}}^{2,2}_{\alpha,\beta}$, and $\boldsymbol{\delta}$ is a unite 2×2 matrix.

Thus, the problem (4.39) for the particular solution is solved in the matrix form (4.60). The complimentary solution of Eq. (4.40) with the boundary conditions (4.41) can be found in a similar manner

$$\mathbf{f}^c = \boldsymbol{\mathcal{A}}\mathbf{g}^c, \tag{4.86}$$

where $\mathbf{g}^c = (\mathbf{g}^c_1, \ldots, \mathbf{g}^c_m)^\top$ and $\mathbf{g}^c_s = (\mathbf{g}^{c(0)}_s, \mathbf{g}^{c(1)}_s)^\top = (g^{c(0)}_{s|1}, g^{c(0)}_{s|2}, g^{c(1)}_{s|1}, g^{c(1)}_{s|2})^\top \in R^4$, $g^{c(\delta)}_{s|j} u^p_{s|j}\delta_{0\delta} + [t^p_{s|j} - t^\alpha_{s|j}]\delta_{1\delta}$, $j = 1, 2$; $s = 1, \ldots, m$; $\delta = 0, 1$.

Substitution of both the particular (4.83)–(4.85) and complimentary (4.81) solutions into Eq. (4.43) finalizes the solution of the thermoelastic problem (4.26) and (4.27) presented in the form ($\mathbf{x} \in R^d$, $\mathbf{s} \in \Gamma_i$)

$$\mathbf{u}(\mathbf{x}) - \overline{\mathbf{u}}(\mathbf{x}) = \boldsymbol{\mathcal{L}}^u_i(\mathbf{x} - \mathbf{x}_i, \overline{\boldsymbol{\chi}}), \tag{4.87}$$

$$\boldsymbol{\sigma}(\mathbf{x}) - \overline{\boldsymbol{\sigma}}(\mathbf{x})) = \boldsymbol{\mathcal{L}}^\sigma_i(\mathbf{x} - \mathbf{x}_i, \overline{\boldsymbol{\chi}}), \tag{4.88}$$

$$\mathbf{t}(\mathbf{s}) - \overline{\mathbf{t}}(\mathbf{s})) = \boldsymbol{\mathcal{L}}^t_i(\mathbf{s} - \mathbf{x}_i, \overline{\boldsymbol{\chi}}), \tag{4.89}$$

where the perturbators $\mathcal{L}_i^u(\mathbf{x}-\mathbf{x}_i,\overline{\chi})$, $\mathcal{L}_i^\sigma(\mathbf{x}-\mathbf{x}_i,\overline{\chi})$, and $\mathcal{L}_i^t(\mathbf{s}-\mathbf{x}_i,\overline{\chi})$ are the linear functions of $\boldsymbol{\alpha}_1$ which are estimated by two step applications of the MFSs at the solution of the problems (4.79) and (4.86). It is interesting that $\mathcal{L}_i^\sigma(\mathbf{x}-\mathbf{x}_i,\overline{\chi})$ is exactly expressed through the volume integral (4.48$_2$).

Algorithm of the solution as described above can be summarized as follows. Estimate the strength of FSs $\mathbf{f}^p$ from Eq. (4.81) with the known vector $\mathbf{g}^p$ (4.79) prescribed at the interface Γ_i. From Eqs. (4.83) and (4.84), determine a particular solution $(\mathbf{u}^p,\mathbf{t}^p)^\top$ of the non-homogeneous equilibrium equations (4.39). Estimate the right-hand side $\mathbf{g}^c$ of Eq. (4.39) through the found particular solution $(\mathbf{u}^p,\mathbf{t}^p)^\top$ and the pseudo-traction $\mathbf{t}^\alpha$ (4.66). Find the strength of FSs $\mathbf{f}^c$ (4.86) as well as the complimentary solution $(\mathbf{u}^c,\mathbf{t}^c)^\top$ by the use of the modified Eqs. (4.83) and (4.85) with a replacement $p\to c$ of the index p. Find the mechanical fields $\mathbf{u}(\mathbf{x})$ and $\boldsymbol{\sigma}(\mathbf{x})$ in the space (4.43).

It should be mentioned that the MFS was widely used for analyses of the thermoelastic problems in the bounded domains with continuous temperature distribution (see, e.g., [752, 1009, 1010, 1537]). Here, we aim to provide, apparently for the first time, a comprehensive study of the application of the MFS to the thermoelastic problem for the unbounded homogeneous domain with an isolated heterogeneity with a step function type of eigenstrains. This problem is a basic one in micromechanics of random structure composites.

We start our estimation from evaluation of the stress perturbator $\mathcal{L}_j^\chi(\mathbf{s}-\mathbf{x}_j,\overline{\chi})$ (4.62) for one heterogeneity in an infinite matrix (see for details [215, 217–219, 224]). In order to estimate the approximations introduced by discretizations implemented in the MFS and for the evaluation of the solution (4.79), we compare numerical results with a reference analytical solution for a single circle inclusion inside the infinite matrix in a plane strain problem subjected to the uniform remote loading $\overline{\boldsymbol{\sigma}}$ =const., and the isotropic constituents with the Young's moduli $E^{(1)}=100$, $E^{(0)}=1$ and Poisson ratios $\nu^{(1)}=0.49$, $\nu^{(0)}=0.45$ (see also [215]). In the following experiments, we consider uniformly spaced collocation points on Γ_i and the same number of source points on $\hat{\Gamma}_i^{(1)}$ and $\hat{\Gamma}_i^{(0)}$ when the source sets prescribed by Eqs. (4.70) and (4.71) coincide. We consider a uniform remote tension $\overline{\sigma}_{ij}\equiv\delta_{i1}\delta_{j1}$const. generating a homogeneous stress distribution inside circle inclusion $\boldsymbol{\sigma}(\mathbf{x})=\mathbf{B}_i(\mathbf{x})\overline{\boldsymbol{\sigma}}$ ($\mathbf{B}_i(\mathbf{x})\equiv$ const., $\mathbf{x}\in v_i$) which is analytically estimated by means of the Eshelby tensor. The value $B_{i|1111}(\mathbf{x})\equiv 1.4587$ analytically evaluated is compared with their numerical values $B_{i|1111}(\mathbf{x})$ estimated by both the direct (least-squares) method of solution of Eq. (4.79) and the pseudo-inverse TR procedure (4.81) in the vicinity of the inclusion boundary $B_{i|1111}(\mathbf{s})\equiv\lim B_{i|1111}(\mathbf{x})$ ($\mathbf{x}\to\mathbf{s}$, $\mathbf{x}\in v_i$, $\mathbf{s}\in\Gamma_i$) In Fig. 4.4 the curves 1 and 2 are estimated by the direct method and the TR procedure, respectively, for $S=30$, $D=0.6$ (4.71), and $\lambda=10^{-5}$. The errors corresponding to the curves 1 and 2 equal 6.5% and 0.23%, respectively. Increasing of the number of source points $S=60$ leads to increasing of accuracy of the TR procedure (4.81) (curve 4 with the error $6.8\cdot10^{-4}$%) and to osculation of the direct solution (4.73) (curve 3, the error 35%). Subsequent increasing of the number of source points $S=200$

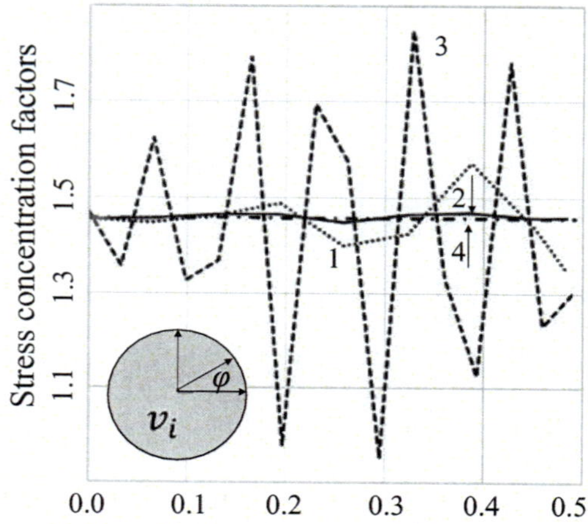

Fig. 4.4 Stress concentration factor $B_{i|1111}(\mathbf{s})$ vs a polar angle φ/π

with $D = 0.1$ provide the error $1.6 \cdot 10^{-5}\%$ of the stress distribution $B_{i|1111}(\mathbf{s})$ ($\mathbf{s} \in \Gamma_i$) (4.81) while the approximative osculating solution (4.73) takes the negative values ($B_{i|1111}(\mathbf{s}) < 0$. at some $\mathbf{x} \in \Gamma_i$) and loses a physical meaning.

Buryachenko [215] also demonstrated an advantage of the allocation scheme (4.71) with respect to the set (4.70) for the prolate elliptical heterogeneities with an aspect ratio 10. Numerical examples for thermoelastic problems for the noncanonical shape of inclusions are considered in Chap. 10 (see also [217–219, 224]).

4.4 Volume Integral Equation Method

4.4.1 Regularized Representation of Integral Equations

At first, we will consider Eq. (4.6) assuming that the stress field $\boldsymbol{\sigma}^0(\mathbf{x})$ ($\mathbf{x} \in v_i$, $i = 1, \ldots, n$) is known. A known quadrature method for obtaining an approximate solution of Eq. (4.8) is to evaluate the volume integrals with a Gauss quadrature formula. Then the corresponding equations at the Gauss points will contain a singular term that results when the field point $\mathbf{x}$ and source point $\mathbf{y}$ are coincident: $\mathbf{x} = \mathbf{y}$ (4.8). The difficulties with the troublesome singularities can be avoided if rearrangement of Eq. (4.8) is performed in the spirit of a subtraction technique used in the modified quadrature method [370]:

$$\boldsymbol{\eta}(\mathbf{x}) = \mathbf{M}_1(\mathbf{x})\boldsymbol{\sigma}^0(\mathbf{x}) + \boldsymbol{\beta}_1(\mathbf{x}) + \mathbf{M}_1(\mathbf{x}) \int V_i^0(\mathbf{y})\boldsymbol{\Gamma}(\mathbf{x}-\mathbf{y})\, d\mathbf{y}\boldsymbol{\eta}(\mathbf{x})$$

$$+\mathbf{M}_1(\mathbf{x}) \int \boldsymbol{\Gamma}(\mathbf{x}-\mathbf{y})\big[\boldsymbol{\eta}(\mathbf{y}) - V_i^0(\mathbf{y})\boldsymbol{\eta}(\mathbf{x})\big]\, d\mathbf{y}, \quad \mathbf{x} \in v_i, \qquad (4.90)$$

where v_i, perhaps, is not an ellipsoid. The equation is valid for any domain $\mathbf{x} \in v_i^0$ with the characteristic function $V_i^0(\mathbf{x})$. We assume that the first integral in (4.90) is easy computable: $\bar{v}_i^0 \langle \mathbf{\Gamma}(\mathbf{x}-\mathbf{y})\rangle_i^0 = -\mathbf{Q}_i^0(\mathbf{x})$ (3.178). In such a case the problem is reduced to computing the second integral in (4.90) which is not a singular as will be shown later. In a general case of the inclusion shape v_i the function $\mathbf{Q}_i^0(\mathbf{x})$ can be found numerically, e.g., by finite element analysis (FEA; see, e.g., Sect. 4.3). For ellipsoidal domain $v_i^0 \supset v_i$, the first integral on the right-hand side of (4.90) is known and is associated with the well-known Eshelby tensor by Eq. (3.132): $\mathbf{S}_i^0 = \mathbf{I} - \mathbf{M}^{(0)}\mathbf{Q}_i^0$. The assumption of an ellipsoidal shape of the domain v_i^0, was used only in order to obtain an analytical representation of the integral (3.178).

This is because the tensor $\langle \mathbf{\Gamma}(\mathbf{x}-\mathbf{y})\rangle_i^0$ is homogeneous for $\mathbf{x} \in v_i$, $\mathbf{y} \in v_i^0$ for an ellipsoid. For nonellipsoidal inclusions v^{n-e}, one could assume that in some parts of the region $v_i^c \subset v_i$ the properties $\mathbf{M}_1(\mathbf{x}) \equiv \mathbf{0}$, $\beta_1(\mathbf{x}) \equiv \mathbf{0}$, i.e., it is sufficient to replace a real nonellipsoidal inclusion $v^{n-e} = v_i \setminus v_i^c$ with an fictitious ellipsoid (with smallest possible volume) and call it the inclusion v_i with a "coating" v_i^c. In so doing, at the estimation of the second integral in (4.90) we keep in mind that $\int \mathbf{\Gamma}(\mathbf{x}-\mathbf{y})(V_i(\mathbf{y}) - V_i^0(\mathbf{y}))\, d\mathbf{y} = \mathbf{Q}_i^0 - \mathbf{Q}_i(\mathbf{x})$ $(\mathbf{x} \in v_i)$. With the nonessential restriction on the shape of the inclusion v_i mentioned above, we can consider without loss of generality an ellipsoidal inclusion $v_i^0 = v_i$; in so doing $\mathbf{M}_1(\mathbf{y}) \equiv \mathbf{0}$, $\beta_1(\mathbf{y}) \equiv \mathbf{0}$ at $\mathbf{y} \in v_i^c \subset v_i$. Then Eq. (4.90) can be rewritten in compact form:

$$\boldsymbol{\eta}(\mathbf{x}) = \boldsymbol{\eta}_i^0(\mathbf{x}) + \int \mathbf{K}(\mathbf{x},\mathbf{y})\big[\boldsymbol{\eta}(\mathbf{y}) - \boldsymbol{\eta}(\mathbf{x})\big]\, d\mathbf{y}, \quad \mathbf{x} \in v_i, \tag{4.91}$$

where $\boldsymbol{\eta}_i^0(\mathbf{x}) = \mathbf{E}_i(\mathbf{x})\boldsymbol{\sigma}^0(\mathbf{x}) + \mathbf{H}_i(\mathbf{x})$, $(\mathbf{x} \in v_i)$ is called the external strain polarization tensor in the inclusion v_i, and (no sum on i):

$$\begin{aligned} &\mathbf{K}(\mathbf{x},\mathbf{y}) = \mathbf{E}_i(\mathbf{x})\mathbf{\Gamma}(\mathbf{x}-\mathbf{y})V_i(\mathbf{y}), \quad \mathbf{E}_i(\mathbf{x}) = \mathbf{M}_1(\mathbf{x})[\mathbf{I} + \mathbf{Q}_i^0\mathbf{M}_1(\mathbf{x})]^{-1}, \\ &\mathbf{H}_i(\mathbf{x}) = [\mathbf{I} + \mathbf{M}_1(\mathbf{x})\mathbf{Q}_i^0]^{-1}\boldsymbol{\beta}_1(\mathbf{x}). \end{aligned} \tag{4.92}$$

Equation (4.91) is in any case less singular than the original form (4.7) and (4.8), since $\boldsymbol{\eta}(\mathbf{y}) - \boldsymbol{\eta}(\mathbf{x}) = \mathbf{0}$ at the singular point $\mathbf{y} = \mathbf{x}$. Moreover, if $\boldsymbol{\eta}(\mathbf{y}) \in C^{0,\alpha}$ at $\mathbf{y} = \mathbf{x} \in v_i$ (Hölder continuity, i.e., $|\boldsymbol{\eta}(\mathbf{y}) - \boldsymbol{\eta}(\mathbf{x})| < K|\mathbf{y}-\mathbf{x}|^\alpha$, $0 < K$, $0 \le \alpha \le 1$) then the subtracted form (4.91) is regular, and the improper volume integral (4.91) over the region $v_i \setminus \omega$ absolutely converges and is independent of the shape of an excluded contracting region ω. We recast the integral equation obtained in [190] in a form (4.91) and (4.92) that is appropriate to our intended application. One has justified, by doing so, the applicability of a standard quadrature rule for the numerical estimation of the integral (4.91). Only for the homogeneous ellipsoidal inclusion (3.183) with $v_i^c \equiv \emptyset$, Eq. (4.91) is reduced to the equation $(\mathbf{x} \in v_i)$ $\boldsymbol{\sigma}(\mathbf{x}) = \boldsymbol{\sigma}^0(\mathbf{x}) - \mathbf{Q}_i\mathbf{M}_1^{(i)}\boldsymbol{\sigma}(\mathbf{x}) + \int \mathbf{\Gamma}(\mathbf{x}-\mathbf{y})\mathbf{M}_1(\mathbf{y})\big[\boldsymbol{\sigma}(\mathbf{y}) - \boldsymbol{\sigma}(\mathbf{x})\big]\, d\mathbf{y}$ (when $\mathbf{Q}_i = \mathbf{Q}_i^0$) obtained in [323].

We will formulate the popular Effective Field Hypothesis (**H**1) which is widely used in micromechanics of random structure composites (see Chap. 7 for details). However, this hypothesis is also exploited in this section for obtaining of an approximate solution of multiple interacting inclusion problems. The EFH is usually formulated in the following form of a combination of two hypotheses:

Hypothesis (H1a) *Each heterogeneity v_i is located in the effective fields*

$$\overline{\boldsymbol{\varepsilon}}_i(\mathbf{y}) \equiv \overline{\boldsymbol{\varepsilon}}(\mathbf{x}_i), \quad \overline{\boldsymbol{\sigma}}_i(\mathbf{y}) \equiv \overline{\boldsymbol{\sigma}}(\mathbf{x}_i) \ (\mathbf{y} \in v_i) \tag{4.93}$$

which are homogeneous over the inclusion v_i.

Hypothesis (H1b) *The perturbation introduced by the heterogeneity v_i at the point $\mathbf{y} \in R^d$ is defined by the relation*

$$\begin{aligned}
\int \mathbf{U}(\mathbf{y}-\mathbf{x})V_i(\mathbf{x})\boldsymbol{\tau}(\mathbf{x})d\mathbf{x} &= \bar{v}_i \mathbf{T}_i^{\epsilon}(\mathbf{y}-\mathbf{x}_i)\boldsymbol{\tau}_{(i)}, \\
\mathbf{T}_i^{\epsilon}(\mathbf{y}-\mathbf{x}_i) &:= \frac{1}{\bar{v}_i}\int \mathbf{U}(\mathbf{y}-\mathbf{x})V_i(\mathbf{x})\,d\mathbf{x}, \\
\int \boldsymbol{\Gamma}(\mathbf{y}-\mathbf{x})V_i(\mathbf{x})\boldsymbol{\eta}(\mathbf{x})d\mathbf{x} &= \bar{v}_i \mathbf{T}_i^{\sigma}(\mathbf{y}-\mathbf{x}_i)\boldsymbol{\eta}_{(i)}, \\
\mathbf{T}_i^{\sigma}(\mathbf{y}-\mathbf{x}_i) &:= \frac{1}{\bar{v}_i}\int \boldsymbol{\Gamma}(\mathbf{y}-\mathbf{x})V_i(\mathbf{x})\,d\mathbf{x}.
\end{aligned} \tag{4.94}$$

and $\mathbf{T}^{\varepsilon}_{ji}(\mathbf{x}_j-\mathbf{x}_i) = \langle \mathbf{T}^{\varepsilon}_i(\mathbf{y}-\mathbf{x}_i)\rangle_j$, $\mathbf{T}^{\sigma}_{ji}(\mathbf{x}_j-\mathbf{x}_i) = \langle \mathbf{T}^{\sigma}_i(\mathbf{y}-\mathbf{x}_i)\rangle_j$ are defined by the average of over some inclusion $\mathbf{y} \in v_j$ (either $v_j \neq v_i$ or $v_j = v_i$).

For the ellipsoidal inclusions, the tensors ($\mathbf{y} \in v_i, \ \ \mathbf{x} \in v_j$)

$$\begin{aligned}
\mathbf{T}_i^{\varepsilon}(\mathbf{x}-\mathbf{x}_i) &:= \langle \mathbf{U}(\mathbf{x}-\mathbf{y})\rangle_i = \begin{cases} -(\overline{v}_i)^{-1}\mathbf{P}_i(\mathbf{x}) & \text{for } \mathbf{x} \in v_i, \\ (\overline{v}_i)^{-1}\int \mathbf{U}(\mathbf{x}-\mathbf{y})V_i(\mathbf{y})d\mathbf{y} & \text{for } \mathbf{x} \notin v_i, \end{cases} \\
\mathbf{T}_i^{\sigma}(\mathbf{x}-\mathbf{x}_i) &:= \langle \boldsymbol{\Gamma}(\mathbf{x}-\mathbf{y})\rangle_i = \begin{cases} -(\overline{v}_i)^{-1}\mathbf{Q}_i(\mathbf{x}) & \text{for } \mathbf{x} \in v_i, \\ (\overline{v}_i)^{-1}\int \boldsymbol{\Gamma}(\mathbf{x}-\mathbf{y})V_i(\mathbf{y})d\mathbf{y} & \text{for } \mathbf{x} \notin v_i, \end{cases}
\end{aligned} \tag{4.95}$$

have analytical representations for ellipsoidal inclusions in an isotropic matrix; the tensors $\mathbf{P}_i = \mathbf{P}_i(\mathbf{x}) \equiv -\int \mathbf{U}(\mathbf{x}-\mathbf{y})V_i(\mathbf{y})d\mathbf{y} \equiv \mathbf{S}_i\mathbf{M}^{(0)} \equiv$const. and $\mathbf{Q}_i = \mathbf{Q}_i(\mathbf{x}) \equiv -\int \boldsymbol{\Gamma}(\mathbf{x}-\mathbf{y})V_i(\mathbf{y})d\mathbf{y} \equiv \mathbf{L}^{(0)} - \mathbf{L}^{(0)}\mathbf{P}_i\mathbf{L}^{(0)} \equiv$const. (for $\forall \mathbf{x} \in v_i$) are defined through the Eshelby [449] tensor $\mathbf{S}_i$. The tensor $\mathbf{T}_{ji}$ (originated in [1632]) has the analytical representation for d-dimensional ($d = 2, 3$) spherical inclusions in an isotropic matrix presented in Sect. 21.3.1 (see for details [200]). For two identical inclusions the matrix $\mathbf{T}_{ij}$ is symmetric, and $\mathbf{T}_j(\mathbf{x}-\mathbf{x}_j) = \lim \mathbf{T}_{ij}(\mathbf{x}_i-\mathbf{x}_j)$ as $\overline{v}_i \to 0$ and $\mathbf{x} \in v_i$, the case of ellipsoidal inclusions of different sizes and orientations is analyzed in [501] and [503] (see also [716]).

4.4.2 The Iteration Method

Thus we obtain a regular representation of the integral equation (4.3) in (4.91) and can use known methods for their solution. Numerical methods of solving integral equations are based in the first place on the possibility of evaluating the integrals appearing in the equations, no matter which method of solution is used: whether it is the method of successive approximations (when the whole integrand is known at each stage) or the mechanical quadrature method (when the unknown function is assumed to be constant or varying in a certain manner within small regions, which enables one to pass to an integral of a known expression). These approximation methods for integral equations lead to full linear systems. Application of the method of mechanical quadratures is very popular [897, 899, 900] although it requires solving linear systems of very high order even for fairly smooth surfaces of inclusions and a smoothly varying load. Only if the number of unknowns is reasonably small, these equations can be solved by direct methods such as Gaussian elimination. But, in general, a satisfactory accuracy of the approximate solution to the integral equation will require a comparatively large number of unknowns, in particular for integral equations in more than one dimension. Therefore iterative methods for the resulting linear systems will be preferable: first, their implementation requires the retention of only the last iteration and the sum of the preceding ones in the memory of a computer, and second, since there is a proof of convergence (in the case of its exact implementation) and the approximate determination of a finite sum of the series reduces to a finite number of quadratures, it may be stated that the numerical algorithm leads to a solution with given accuracy for a sufficiently fine discretization.

Therefore, we solve Eq. (4.91) by the method of successive approximations, which is also called the Neumann series method. With this aim in view we rewrite Eq. (4.91) in symbolic form

$$\eta = \eta^0 + \mathcal{K}\eta, \tag{4.96}$$

where $\big(\mathcal{K}\eta\big)(\mathbf{x}) = \int \mathcal{K}(\mathbf{x},\mathbf{y})\eta(\mathbf{y})\, d\mathbf{y}$ defines the integral operator $\mathcal{K}$ with the kernel formally represented as $\mathcal{K}(\mathbf{x},\mathbf{y}) = \mathbf{K}(\mathbf{x},\mathbf{y}) - \delta(\mathbf{x}-\mathbf{y})\int V_i(\mathbf{z})\mathbf{K}(\mathbf{x},\mathbf{z})\, d\mathbf{z}$. We formally write the solution of Eq. (4.96) as

$$\eta = \mathcal{L}\eta^0, \tag{4.97}$$

where the inverse operator $\mathcal{L} = (\mathbf{I} - \mathcal{K})^{-1}$ will be constructed by the iteration method based on the recursion formula

$$\eta^{[k+1]} = \eta^0 + \mathcal{K}\eta^{[k]} \tag{4.98}$$

to construct a sequence of functions $\left\{\eta^{[k]}\right\}$ that can be treated as an approximation of the solution of Eq. (4.96). We have used point Jacobi (called also Richardson and point total-step) iterative scheme for ease of calculations. The details of the real iteration method used for the solution of Eq. (4.96) will be presented in Sect. 4.4.4. Usually the driving term of this equation is used as an initial approximation:

$$\boldsymbol{\eta}^{[0]}(\mathbf{x}) = \boldsymbol{\eta}^0(\mathbf{x}), \tag{4.99}$$

which is exact for a homogeneous ellipsoidal inclusion subjected to remote homogeneous stress field $\boldsymbol{\sigma}^0(\mathbf{x}) \equiv \boldsymbol{\sigma}^0 = \text{const}$. It suggests the Neumann series form for the solution $\boldsymbol{\eta}$ of (4.91) $\boldsymbol{\eta} = \boldsymbol{\mathcal{L}}\boldsymbol{\eta}^0$, $\boldsymbol{\mathcal{L}} \equiv \sum_{k=0}^{\infty} \boldsymbol{\mathcal{K}}^k$, where $\boldsymbol{\mathcal{L}}$ is the inverse integral operator with the kernel $\boldsymbol{\mathcal{L}}(\mathbf{x}, \mathbf{y})$, and the power $\boldsymbol{\mathcal{K}}^k$ is defined recursively by the condition $\boldsymbol{\mathcal{K}}^0 = \mathbf{I}$, $\boldsymbol{\mathcal{K}}^1 = \boldsymbol{\mathcal{K}}$ and the kernel of $\boldsymbol{\mathcal{K}}^k$ is $\boldsymbol{\mathcal{K}}_k(\mathbf{x}, \mathbf{y}) = \int \boldsymbol{\mathcal{K}}(\mathbf{x}, \mathbf{z})\boldsymbol{\mathcal{K}}_{k-1}(\mathbf{z}, \mathbf{y})\, d\mathbf{z}$ [1217]. The sequence $\{\boldsymbol{\eta}^{[k]}\}$ (4.98) with arbitrary continuous $\boldsymbol{\eta}^{[0]}(\mathbf{x})$ converges to a unique solution $\boldsymbol{\eta}$ if the norm of the integral operator $\boldsymbol{\mathcal{K}}$ turns out to be small "enough" (less than 1) $\|\boldsymbol{\mathcal{K}}\|_{\infty, v_i} \equiv \max_{\mathbf{x}\in v_i} \int |\boldsymbol{\mathcal{K}}(\mathbf{x}, \mathbf{y})|\, d\mathbf{y} < 1$ and the problem is reduced to computation of the integrals involved, the density of which is given. In effect the iteration method (4.98) transforms the integral equation problem (4.98) into a linear algebra problem in any case.

The iteration method (4.98) has two known drawbacks. First, the Neumann series ensures the existence of solutions to integral equations of the second kind only for sufficiently small kernels (4.95), and second, in general, it cannot be summed in closed form. Of course Eq. (4.95) can be solved directly by the quadrature method even if this condition of smallness is not valid. However, strongly inhomogeneous problems may lead to much larger numbers of quadrature points, making iteration potentially worthwhile. Moreover, increasing the problem dimensionality (from 2D to 3D) raises the number of nodes to the dimensional power and the situation changes radically. As will be shown in the concrete examples, only a few iterations of Eq. (4.98) are necessary; these iterations converge very much faster than a direct inversion of the operator $\mathbf{I} - \boldsymbol{\mathcal{K}}$ by the quadrature method (a comparative analysis of the inversion of the matrices by the iteration and direct methods is presented in [370]).

The convergence of successive approximations in elasticity theory has been proved in [887] for the exact evaluation of boundary integrals at each iteration [1343]. However, in the concrete examples of high matrix-inclusion elastic contrast and some others, these standard popular iterative schemes may diverge or converge very slowly (i.e., the iteration scheme (4.98) does not work in general) so that an implementation of the improved algorithm proposed in this chapter becomes more complicated. In such a case, Mikhlin [1061] and Perlin [1201] (for references see also [1189, 1670]). proved that the convergence of the scheme (4.98) is provided by their modification $\boldsymbol{\eta}^{[k+1]} = [\boldsymbol{\eta}^0 + \boldsymbol{\eta}^{[k]} + \boldsymbol{\mathcal{K}}\boldsymbol{\eta}^{[k]}]/2$. Although the convergence of this method was rigorously proved in [1061] for the elastic problems of an arbitrary dimensions, its effectiveness in [254] was demonstrated only for 2D problems.

4.4.3 *Initial Approximation for Interacting Inclusions in an Infinite Medium*

4.4.3.1 A Single Inclusion Subjected to Inhomogeneous Remote Stress

At first, we have considered in Sect. 4.3 a case of a single ellipsoidal inclusion v_i in an infinite medium subjected to the homogeneous field $\boldsymbol{\sigma}^0(\mathbf{x}) \equiv \boldsymbol{\sigma}^0 = \text{const.}$ which is assumed to be known. The relationships between the averaged tensors $\mathbf{B}_i$, $\mathbf{C}_i$ and $\mathbf{R}_i$, $\mathbf{R}_i$ describing the thermoelastic solution (3.175) were established.

We now turn our attention to the analysis of the inhomogeneous field $\boldsymbol{\sigma}^0(\mathbf{x})$, $\mathbf{x} \in w$. Solving Eq. (4.97) with the initial approximation (4.99) leads to the solution (4.98) by the series of integral operators recurrently coupled. The solution (4.97) obtained inside the region v_i for strain polarization tensor $\boldsymbol{\eta}(\mathbf{x})$ ($\mathbf{x} \in v_i$) can be rewritten for stress field inside the the inclusion v_i (if $\mathbf{M}_1(\mathbf{x}) \neq \mathbf{0}$) $\boldsymbol{\sigma}(\mathbf{x}) = \mathbf{M}_1^{-1}(\mathbf{x})[\boldsymbol{\eta}(\mathbf{x}) - \boldsymbol{\beta}_1(\mathbf{x})]$, or, taking Eqs. (4.7) and (4.8) into account, the alternative representation for the stress field has the form $\boldsymbol{\sigma}(\mathbf{x}) = \boldsymbol{\sigma}^0(\mathbf{x}) + \int\int \boldsymbol{\Gamma}(\mathbf{x},\mathbf{z})\mathcal{L}_i(\mathbf{z},\mathbf{y})\boldsymbol{\eta}^0(\mathbf{y})\,d\mathbf{y}\,d\mathbf{z}$. for both inside and outside the inclusion v_i.

For the homogeneous external loading $\boldsymbol{\sigma}^0(\mathbf{x}) = \boldsymbol{\sigma}^0 = \text{const.}$ the integral operator $\mathcal{L}_i$ (4.97) reduces the tensors (3.175) to ($\mathbf{x} \in v_i$):

$$\mathbf{R}_i(\mathbf{x}) = \bar{v}_i\big(\mathcal{L}_i\mathbf{E}_i\big)(\mathbf{x}), \quad \bar{v}_i\mathbf{F}_i(\mathbf{x}) = \big(\mathcal{L}_i\mathbf{H}_i\big)(\mathbf{x}), \tag{4.100}$$

If $\mathbf{M}_1(\mathbf{x}) \neq \mathbf{0}$ ($\mathbf{x} \in v_i$) then Eqs. ($3.175_{7,8}$) and (4.100) lead to the representation of tensors $\mathbf{B}_i(\mathbf{x})$ and0 $\mathbf{C}_i(\mathbf{x})$ ($\mathbf{x} \in v_i$):

$$\mathbf{B}_i(\mathbf{x}) = \mathbf{M}_1^{-1}(\mathbf{x})(\mathcal{L}_i\mathbf{E}_i)(\mathbf{x}), \quad \mathbf{C}_i(\mathbf{x}) = \mathbf{M}_1^{-1}(\mathbf{x})\Big[(\mathcal{L}_i\mathbf{H}_i)(\mathbf{x}) - \boldsymbol{\beta}_1(\mathbf{x})\Big]. \tag{4.101}$$

To put this another way, Eqs. (4.100) and (4.101) give the explicit representation of tensors $\mathbf{R}_i(\mathbf{x})$, $\mathbf{F}_i(\mathbf{x})$ and $\mathbf{B}_i(\mathbf{x})$, $\mathbf{C}_i(\mathbf{x})$, respectively, obtained by the use of VIE method for the homogeneous $\boldsymbol{\sigma}^0$.

Moreover, taking Eqs. (4.10) and (4.100) into account, we find that for the homogeneous ellipsoidal inclusion v_i the operator $\mathcal{L}_i$ is an identity operator

$$\mathbf{R}_i = \mathcal{L}_i\mathbf{R}_i, \quad \mathbf{F}_i = \mathcal{L}_i\mathbf{F}_i. \tag{4.102}$$

4.4.3.2 Effective Fields and the General Scheme of the Solution for n Inclusions in an Infinite Medium

Let us consider a certain set $X = (v_i)$ of inclusions v_i with characteristic functions V_i ($i = 1, \ldots, n$). To solve the problem through a superposition technique, one replaces the original problem by the superposition of n problems where each involves an infinite space with a single inclusion at the designated location. Then the stress field in the medium is defined from the general integral equations (4.9) and

$$\overline{\boldsymbol{\sigma}}(\mathbf{x}) = \boldsymbol{\sigma}^0(\mathbf{x}) + \sum_{j\neq i} \int \boldsymbol{\Gamma}(\mathbf{x}-\mathbf{y})\boldsymbol{\eta}(\mathbf{y})V_j(\mathbf{y})\,d\mathbf{y}, \tag{4.103}$$

where we defined the effective field $\overline{\boldsymbol{\sigma}}(\mathbf{x})$ ($\mathbf{x} \in v_i, \ldots, v_n$) as a stress field in which the chosen inclusion v_i is embedded. This effective field is a function of all the other positions of the surrounding inhomogeneities and is a superposition of the remote stress and the disturbances (unknown) caused by the other inclusions of the considered set. No restrictions on the homogeneity of the field $\overline{\boldsymbol{\sigma}}(\mathbf{x})$ were used. The field $\overline{\boldsymbol{\sigma}}(\mathbf{x})$ defined by Eq. (4.103) in the region v_i is nothing more than a notational convenience that may be interpreted as the remote field of the inclusion v_i (analysis of this concept will be considered in Chap. 8). In this field $\overline{\boldsymbol{\sigma}}(\mathbf{x})$, the inclusion v_i behaves as an isolated one, and, therefore, the strain polarization tensor $\boldsymbol{\eta}(\mathbf{x}) = \boldsymbol{\eta}_i(\mathbf{x})$ inside the inclusion v_i can be found in much the same way as a solution of Eq. (4.95):

$$\boldsymbol{\eta}_i = \boldsymbol{\mathcal{L}}_i \overline{\boldsymbol{\eta}}_i, \tag{4.104}$$

where we have introduced the effective strain polarization tensor $\bar{v}_i \overline{\boldsymbol{\eta}}_i = \mathbf{E}_i(\mathbf{x})\overline{\boldsymbol{\sigma}}(\mathbf{x}) + \mathbf{H}_i(\mathbf{x})$.

Substituting Eq. (4.104) into (4.103) leads to

$$\overline{\boldsymbol{\eta}}_i = \boldsymbol{\eta}_i^0 + \sum_{j\neq i} \boldsymbol{\mathcal{L}}_{ij}^{\text{out}} \overline{\boldsymbol{\eta}}_j, \tag{4.105}$$

where the kernel of the "outer" integral operator $\boldsymbol{\mathcal{L}}_{ij}^{\text{out}}$ is $\boldsymbol{\mathcal{L}}_{ij}^{\text{out}}(\mathbf{x},\mathbf{y}) = \mathbf{E}_i(\mathbf{x}) \int \boldsymbol{\Gamma}(\mathbf{x}-\mathbf{z})\boldsymbol{\mathcal{L}}_j(\mathbf{z},\mathbf{y})V_j(\mathbf{z})\,d\mathbf{z}$ where $\mathbf{y},\ \mathbf{z} \in v_j,\ \mathbf{x} \in v_i;\ i \neq j$.

In principle, the system of Eqs. (4.105) can be analyzed directly, using different numerical methods, but multiparticle problems that involve many inhomogeneities require a great deal of computer time. As mentioned for a general case of a single inhomogeneity, the iteration schemes are usually more efficient. We now define a sequence of approximations $\overline{\boldsymbol{\eta}}_i^{[(k]}$ ($k \to \infty$) to $\overline{\boldsymbol{\eta}}_i$ as follows:

$$\overline{\boldsymbol{\eta}}_i^{[k+1]} = \boldsymbol{\eta}_i^0 + \sum_{j\neq i} \boldsymbol{\mathcal{L}}_{ij}^{\text{out}} \overline{\boldsymbol{\eta}}_j^{[k]}, \quad (i = 1, \ldots, n;\ k = 0, 1, \ldots). \tag{4.106}$$

The performance of the iterative scheme proposed can be improved by choice of the initial approximation $\overline{\boldsymbol{\eta}}_i^{[0]}$ of the solution of Eq. (4.106) proposed below.

4.4.3.3 Initial Approximation for *n* Inclusions in an Infinite Medium

In the traditional scheme of iteration methods, the driving term of the integral equation involved is taken as the initial approximation

$$\boldsymbol{\eta}_i^{[0]}(\mathbf{x}) = \boldsymbol{\eta}_i^0(\mathbf{x}), \quad (i = 1, \dots, n), \tag{4.107}$$

or, expressing in terms of the interaction problem, the iterative procedure is initiated by assuming that in each inclusion the effective strain polarization tensor is induced by the remote field as if the neighbors are absent. As will be shown in the numerical examples, the choice of the initial approximation on the basis of an approximate analytical solution leads to a significant increase of the accuracy in the first steps of an iterative procedure.

With this in mind, let us rewrite Eq. (4.103) in the spirit of a subtraction technique ($\mathbf{x} \in v_i$):

$$\overline{\boldsymbol{\sigma}}(\mathbf{x}) = \boldsymbol{\sigma}^0(\mathbf{x}) + \sum_{j \neq i} \mathbf{T}_j(\mathbf{x}-\mathbf{x}_j)\boldsymbol{\xi}_j + \sum_{j \neq i} \int \boldsymbol{\Gamma}(\mathbf{x}-\mathbf{y})\big[\boldsymbol{\eta}(\mathbf{y}) - \boldsymbol{\xi}_j\big] V_j(\mathbf{y})\, d\mathbf{y}. \tag{4.108}$$

Equation (4.108) is valid for any constant tensors $\boldsymbol{\xi}_j$ $(i, j = 1, \dots, n;\ j \neq i)$. For definitiveness, we shall use $\bar{v}_j \boldsymbol{\xi}_j = \mathbf{R}_j \langle \overline{\boldsymbol{\sigma}} \rangle_j + \mathbf{F}_j$ $(j \neq i)$ that can be considered as an approximation of strain polarization tensors in the inclusions v_j in the spirit of the Saint–Venant principle for constant tensors:

$$\boldsymbol{\eta}(\mathbf{y}) \equiv \boldsymbol{\xi}_j, \quad (\mathbf{y} \in v_j), \tag{4.109}$$

which can be formulated in an alternative form (see effective field hypothesis (4.93) and (4.94)):

$$\int \boldsymbol{\Gamma}(\mathbf{x}-\mathbf{y})\boldsymbol{\eta}(\mathbf{y}) V_j(\mathbf{y})\, d\mathbf{y} = \mathbf{T}_j^{\sigma}(\mathbf{x}-\mathbf{x}_j)(\mathbf{R}_j \langle \overline{\boldsymbol{\sigma}} \rangle_j + \mathbf{F}_j), \quad (\mathbf{x} \notin v_j), \tag{4.110}$$

where the tensors $\mathbf{T}_j^{\sigma}(\mathbf{x}-\mathbf{x}_j)$ are defined by Eq. (4.95).

In the framework of the hypothesis (4.110), a zero-order approximation of the average effective stresses can be estimated from Eqs. (4.108) and (4.109) $\langle \overline{\boldsymbol{\sigma}} \rangle_i^{[0]} = \langle \boldsymbol{\sigma}^0 \rangle_i + \sum_{j \neq i} \mathbf{T}_{ij}(\mathbf{x}-\mathbf{x}_j)\big[\mathbf{R}_j \langle \overline{\boldsymbol{\sigma}} \rangle_j^{[0]} + \mathbf{F}_j\big]$ leading to explicit representation

$$\bar{v}_i \boldsymbol{\eta}_i^{[0]} = \mathbf{R}_i \langle \overline{\boldsymbol{\sigma}} \rangle_i^{[0]} + \mathbf{F}_i = \sum_{j=1}^{n} \mathbf{Z}_{ij}\big[\mathbf{R}_j \langle \boldsymbol{\sigma}^0 \rangle_j + \mathbf{F}_j\big], \tag{4.111}$$

where the matrix $\mathbf{Z}^{-1}$ has elements $(\mathbf{Z}^{-1})_{ij}$ $(i, j = 1, \dots, n)$ (the elements represent fourth-order tensors; no sum $i,\ j$)

$$(\mathbf{Z}^{-1})_{ij} = \mathbf{I}\delta_{ij} - (1 - \delta_{ij})\mathbf{R}_j \mathbf{T}_{ij}(\mathbf{x}_i - \mathbf{x}_j). \tag{4.112}$$

The representation for the strain polarization tensor (4.111) can be rewritten as

$$\bar{v}_i\boldsymbol{\eta}_i^{[0]} = \bar{v}_i\boldsymbol{\eta}_i^0 + \mathbf{R}_i\sum_{j\neq i}^{n}\mathbf{T}_{ij}(\mathbf{x}_i-\mathbf{x}_j)\sum_{k=1}^{n}\mathbf{Z}_{jk}\boldsymbol{\xi}_k^0\bar{v}_k, \tag{4.113}$$

where $\bar{v}_i\boldsymbol{\xi}_i^0 = \mathbf{R}_i\langle\boldsymbol{\sigma}^0\rangle_i + \mathbf{F}_i$.

By this means, we have estimated the average strain polarization tensors (4.111) obtained in the framework of the homogeneity hypothesis of effective fields inside the inclusions: $\overline{\boldsymbol{\sigma}}(\mathbf{x}) = \langle\overline{\boldsymbol{\sigma}}\rangle_j$ $(\mathbf{x}\in v_j)$. Substituting (4.111) in (4.9$_2$) leads to the inhomogeneous strain polarization tensors obtained in the framework of the same homogeneity hypothesis of effective fields (4.110):

$$\bar{v}_i\boldsymbol{\eta}_i^{[0]}(\mathbf{x}) = \mathbf{R}_i(\mathbf{x})\mathbf{R}_i^{-1}\Big(\sum_{j=1}^{n}\mathbf{Z}_{ij}\boldsymbol{\xi}_j^0\bar{v}_j - \mathbf{F}_i\Big) + \mathbf{F}_i(\mathbf{x}). \tag{4.114}$$

The initial approximations (4.113) and (4.114) will be used at the analysis of interacting inclusions of commensurable sizes; otherwise at the analysis of a macroinclusion, we will use the initial approximation (4.107).

We turn our attention to a discussion of different simplified assumptions for obtaining zero-order approximations (4.111) which are known for the analysis of two interacting homogeneous ellipsoidal inclusions.

The so-called point approximation of interacting inclusions [836] is exact for infinitely spaced heterogeneities

$$\mathbf{T}_{ij}(\mathbf{x}-\mathbf{x}_j) = \mathbf{T}_j(\mathbf{x}_i-\mathbf{x}_j) = \boldsymbol{\Gamma}(\mathbf{x}_i-\mathbf{x}_j), \tag{4.115}$$

and, therefore, $(\mathbf{Z}^{-1})_{ij} = \mathbf{I}\delta_{ij} - (1-\delta_{ij})\mathbf{R}_j\boldsymbol{\Gamma}(\mathbf{x}_i-\mathbf{x}_j)$.

The *approximation of widely space inclusions* (see, e.g., [728, 729]) is defined by a quite restrictive geometrical inequality

$$\varepsilon_r = \text{(inclusion sizes)/(spacing between inclusions)} \ll 1, \tag{4.116}$$

leading to the solution

$$\bar{v}_i\boldsymbol{\eta}_i^{[0]} = \bar{v}_i\boldsymbol{\eta}_i^0 + \mathbf{R}_i\sum_{j\neq i}^{n}\boldsymbol{\Gamma}(\mathbf{x}_i-\mathbf{x}_j)\boldsymbol{\xi}_j^0\bar{v}_j \tag{4.117}$$

Other simplifications can be obtained as a truncation of the expansion (4.114). So, the iteration solution of (4.111) was used in [243] in the analysis of two interacting homogeneous ellipsoidal inclusions, whereas Willis and Acton [1632] have considered two first iterations of this scheme for two spherical homogeneous inclusions; their approximation can be rephrased in our notations as

$$\mathbf{Z}_{ij} = \mathbf{I}\delta_{ij} + (1-\delta_{ij})\mathbf{R}_j\mathbf{T}_{ij}(\mathbf{x}_i - \mathbf{x}_j),$$
$$\bar{v}_i\boldsymbol{\eta}_i^{[0]} = \bar{v}_i\boldsymbol{\eta}_i^0 + \big[\mathbf{I} + \mathbf{R}_j\mathbf{T}_{ij}(\mathbf{x}_i - \mathbf{x}_j)\big]\mathbf{R}_j\mathbf{T}_{ij}(\mathbf{x}_i - \mathbf{x}_j)\boldsymbol{\xi}_j^0\bar{v}_j. \qquad (4.118)$$

The right-hand side of Eq. (4.118$_2$) is a sum of terms of order $O(\varepsilon_r^\alpha)$ (if $\varepsilon_r \ll 1$), where the nonnegative integer parameter α varies from 0 to $2d+4$. Then the *asymptotic approximation* can be obtained by truncation of terms having the order less than $O(\varepsilon_r^{d+2})$:

$$\bar{v}_i\boldsymbol{\eta}_i^{[0]} = \bar{v}_i\boldsymbol{\eta}_i^0 + \big[\mathbf{I} + \mathbf{R}_j\boldsymbol{\Gamma}(\mathbf{x}_i - \mathbf{x}_j)\big]\mathbf{R}_j\mathbf{T}_{ij}(\mathbf{x}_i - \mathbf{x}_j)\boldsymbol{\xi}_j^0\bar{v}_j. \qquad (4.119)$$

For two identical spherical homogeneous inclusions subjected to homogeneous external loading, Eq. (4.119) was obtained by much lengthier means in [712] (3D case) and in [722] (2D case).

At last, the simplest initial approximation (4.107) in the framework of effective field hypothesis (4.109) follows from Eq. (4.118$_2$) by neglecting of the second summand in the square brackets in the right-landside of Eq. (4.118$_2$). All related simplified methods mentioned above yield only a moderate reduction of computer costs and can be derived in the framework of the method proposed (4.111).

4.4.4 First-Order and Subsequent Approximations

To obtain the first-order approximation of strain polarization tensors $\overline{\boldsymbol{\eta}}_i^{[1]}(\mathbf{x})$ ($\mathbf{x} \in v_i$, $i = 1, \ldots, n$) we should substitute (4.112) (or (4.107)) into (4.106) which leads to the necessity of evaluation of integrals in the right-landside of Eq. (4.106) if $\overline{\boldsymbol{\eta}}_i[0](\mathbf{x}) \neq$ const. For example, choosing the simplest initial approximation (4.107) leads to the first-order approximation

$$\overline{\boldsymbol{\eta}}_i^{[1]}(\mathbf{x}) = \boldsymbol{\eta}_i^0(\mathbf{x}) + \mathbf{E}_i(\mathbf{x})\sum_{j\neq i}^{n}\mathbf{T}_j(\mathbf{x}-\mathbf{x}_j)\boldsymbol{\xi}_j^0\bar{v}_j. \qquad (4.120)$$

Substituting zero-order approximation (4.111) considering interacting effects between inclusions into (4.106) leads to another representation of the first-order approximation

$$\overline{\boldsymbol{\eta}}_{i(1)}(\mathbf{x}) = \boldsymbol{\eta}_i^0(\mathbf{x}) + \mathbf{E}_i(\mathbf{x})\sum_{j\neq i}^{n}\mathbf{T}_j(\mathbf{x}-\mathbf{x}_j)\sum_{k=1}^{n}\mathbf{Z}_{jk}\boldsymbol{\xi}_k^0\bar{v}_k. \qquad (4.121)$$

Substituting (4.121) (or (4.120)) into (4.106) yields the first-order approximation for the strain polarization tensor inside the inclusion v_i

$$\boldsymbol{\eta}_i^{[1]}(\mathbf{x}) = \big(\boldsymbol{\mathcal{L}}_i\overline{\boldsymbol{\eta}}_i^{[1]}\big)(\mathbf{x}). \tag{4.122}$$

It should be mentioned that the derivation of the relation (4.121) is based on the hypothesis equivalent to the assumption proposed in [728, 729] for interacting cracks if (and only if) the inclusions are the homogeneous and have an ellipsoidal shape, and subject to a remote homogeneous stress field. This assumption can be formulated in an alternative form. The initial problem for n inhomogeneities can be represented by n problems for an isolated single inclusion inside the infinite medium subject to a remote stress field induced by the remaining inhomogeneities and the real remote stress. The perturbation induced by remaining inclusions to the remote field of a reference inclusion is defined by average equivalent strain polarization tensors. In light of the analogy mentioned, the first-order approximation (4.120) is called *approximation of weak interaction* which can be interpreted as a truncation of "feedbacks": transmission of the $\boldsymbol{\sigma}^0$-induced effective fields from the jth inclusion on the ith inclusion is not followed by feedback.

The iterative solution strategy with the simplest initial approximation (4.120) for the homogeneous ellipsoidal inclusions subjected to mechanical loading ($\boldsymbol{\beta}_1 \equiv \mathbf{0}$) was considered in [198] where this approximation is called as *a simple method of analysis*. The relations (4.121) and (4.122) generalize a similar equation proposed in [1299] under the analysis of two spherical cavities subjected to homogeneous remote stress $\boldsymbol{\sigma}^0$. After evaluation of the effective field $\overline{\boldsymbol{\sigma}}_i^{[1]}$ (4.121) at the surface of cavities, Rodin and Hwang [1299] have solved the problem for a single isolated inclusion by the finite element analysis using the basic idea of Eshelby. However, the field $\overline{\boldsymbol{\eta}}_i^{[1]}(\mathbf{x})$ (and, therefore $\overline{\boldsymbol{\sigma}}_i^{[1]}(\mathbf{x})$) is defined inside the domain $\mathbf{x} \in v_i$ $(i \neq j,\ j = 1, \ldots, n)$, and the use of the finite element analysis becomes too involved as the inclusions themselves become inhomogeneous. At the same time, there is no change in the basic formulation for the method proposed (4.121) and (4.122) for the cavities and the inhomogeneous inclusions.

Obtaining the next approximations is carried out in the framework of a standard schema of the iteration method:

$$\overline{\boldsymbol{\eta}}_i^{[k+1]}(\mathbf{x}) = \boldsymbol{\eta}_i^{(0)}(\mathbf{x}) + \sum_{j\neq i}(\boldsymbol{\mathcal{L}}_{ij}^{\rm out}\overline{\boldsymbol{\eta}}_j^{[k]})(\mathbf{x}), \quad \boldsymbol{\eta}_i^{[k+1]}(\mathbf{x}) = (\boldsymbol{\mathcal{L}}_i\overline{\boldsymbol{\eta}}_i^{[k+1]})(\mathbf{x}), \tag{4.123}$$

where $\mathbf{x} \in v_i,\ i = 1, \ldots, n$.

In order to check the accuracy of the VIE (see for details [190]), the influence of the interaction between two circular inclusions v_i and v_j (e.g., fibers) with the same radius a in an infinite matrix on the stress field in the inclusions is studied. The method proposed (4.123) (fifth iteration), the VIE method [899, 900], as well as FEA produce almost identical results. One also considered a pair of noncanonical inclusion described by equation $|x_1|^\alpha + |x_2|^\alpha = 1$ in the Cartesian coordinate system. At the estimation of zero-order approximation (4.114), the average concentration of the strain polarization tensor $\mathbf{R}$ (4.10) is calculated by the use of the scheme (4.15) and a concept of a fictitious ellipsoidal homogeneous inclusion (3.185). The first-order approximation (4.121) and (4.122) provides an acceptable accuracy if the distance between the inclusion centers $|\mathbf{x}_i - \mathbf{x}_j| \geq 2.4a$.

4.4.5 Operator Representations of Solutions for One and Two Inclusions

The solution (4.97) obtained inside the region v_i for strain polarization tensor $\boldsymbol{\eta}(\mathbf{x})$ ($\mathbf{x} \in v_i$) can be rewritten in terms of the stress and effective stress inside the inclusion v_i

$$\boldsymbol{\sigma}(\mathbf{x}) = \mathcal{B}_i(\overline{\boldsymbol{\sigma}})(\mathbf{x}) + \mathbf{C}_i(\mathbf{x}), \quad \bar{v}_i\boldsymbol{\eta}(\mathbf{x}) = \mathcal{R}_i(\overline{\boldsymbol{\sigma}})(\mathbf{x}) + \mathbf{F}_i(\mathbf{x}), \quad \mathbf{x} \in v_i, \tag{4.124}$$

where the linear operators $\mathcal{B}_i$ and $\mathcal{R}_i$ and the tensors $\mathbf{C}_i(\mathbf{x})$ and $\mathbf{F}_i(\mathbf{x})$ have the explicit representations (if $\mathbf{M}_1(\mathbf{x}) \neq \mathbf{0}$, with no sum on i):

$$\mathcal{R}_i(\overline{\boldsymbol{\sigma}})(\mathbf{x}) = \bar{v}_i\mathcal{L}_i\big(\mathbf{E}_i\overline{\boldsymbol{\sigma}}\big)(\mathbf{x}), \quad \mathcal{B}_i(\overline{\boldsymbol{\sigma}})(\mathbf{x}) = \mathbf{M}_1^{-1}(\mathbf{x})\mathcal{L}_i(\mathbf{E}_i\overline{\boldsymbol{\sigma}})(\mathbf{x}), \tag{4.125}$$

$$\mathbf{F}_i(\mathbf{x}) = \bar{v}_i\big(\mathcal{L}_i\mathbf{H}_i\big)(\mathbf{x}), \quad \mathbf{C}_i(\mathbf{x}) = \mathbf{M}_1^{-1}(\mathbf{x})\Big[(\mathcal{L}_i\mathbf{H}_i)(\mathbf{x}) - \boldsymbol{\beta}_1(\mathbf{x})\Big]. \tag{4.126}$$

The solution (4.124_1) obtained inside the region v_i can be extended on the whole space $\mathbf{x} \in R^d$ by the use of the definition

$$\boldsymbol{\sigma}(\mathbf{x}) = \overline{\boldsymbol{\sigma}}(\mathbf{x}) + \mathcal{B}_i(\overline{\boldsymbol{\sigma}})(\mathbf{x}) + \mathbf{C}_i(\mathbf{x}), \ \mathbf{x} \in R^d \setminus v_i, \tag{4.127}$$

where $\mathcal{B}_i(\overline{\boldsymbol{\sigma}})(\mathbf{x}) = \bar{v}_i^{-1}\int\int \boldsymbol{\Gamma}(\mathbf{x}-\mathbf{y})\mathcal{R}_i(\mathbf{y},\mathbf{z})\overline{\boldsymbol{\sigma}}(\mathbf{z})V_i(\mathbf{y})V_i(\mathbf{z})d\mathbf{y}d\mathbf{z}$, $\mathbf{C}_i(\mathbf{x}) = \bar{v}_i^{-1} \times \int \boldsymbol{\Gamma}(\mathbf{x}-\mathbf{y})\mathbf{F}_i(\mathbf{y})V_i(\mathbf{y})d\mathbf{y}$, describes the perturbations introduced by the heterogeneity v_i into the point $\mathbf{x} \notin v_i$; $\mathcal{R}_i(\mathbf{y},\mathbf{z})$ is a kernel of the operator $\mathcal{R}_i(\overline{\boldsymbol{\sigma}})(\mathbf{x}) = \int \mathcal{R}_i(\mathbf{x},\mathbf{z})\overline{\boldsymbol{\sigma}}(\mathbf{z})V_i(\mathbf{z})d\mathbf{z}$.

For the homogeneous external loading $\overline{\boldsymbol{\sigma}}(\mathbf{x}) = \overline{\boldsymbol{\sigma}} =$ const. the integral operators $\mathcal{R}_i$ and $\mathcal{B}_i$ (4.126) are reduced to the tensors

$$\mathcal{R}_i \equiv \mathbf{R}_i(\mathbf{x}) = \bar{v}_i\big(\mathcal{L}_i\mathbf{E}_i\big)(\mathbf{x}), \quad \mathcal{B}_i \equiv \mathbf{B}_i(\mathbf{x}) = \bar{v}_i^{-1}\mathbf{M}_1^{-1}(\mathbf{x})\mathbf{R}_i(\mathbf{x}) \ \ (\mathbf{x} \in v_i). \tag{4.128}$$

For the homogeneous ellipsoidal domain v_i with the properties $\mathbf{M}_1(\mathbf{x}) = \mathbf{M}_1^{(i)} =$ const., $\boldsymbol{\beta}_1(\mathbf{x}) = \boldsymbol{\beta}_1^{(i)}$ =const. ($\mathbf{x} \in v_i$), all tensors (4.124)–(4.126) are constant and represented through the Eshelby [449] tensor $\mathbf{S}_i = \mathbf{S}_i^0$ (see Chap. 3) whereas the operator $\mathcal{L}_i$ is an identity operator $\mathcal{L}_i \equiv \mathbf{I}$.

Let us assume that two inclusions v_i and v_j $(i = 1, 2;\quad j = 3 - i)$ are placed in an infinite homogeneous matrix subjected to the inhomogeneous stress field $\widetilde{\boldsymbol{\sigma}}_{i,j}(\mathbf{x})$ $(\mathbf{x} \in R^d)$. For $\mathbf{x} \in v_i$, Eqs. (4.96) can be recast in the form

$$\boldsymbol{\eta}(\mathbf{x}) = \overline{\boldsymbol{\eta}}_i(\mathbf{x}) + \int \mathcal{K}_i(\mathbf{x}, \mathbf{y})V_i(\mathbf{y})\boldsymbol{\eta}(\mathbf{y})d\mathbf{y}, \tag{4.129}$$

$$\overline{\boldsymbol{\eta}}_i(\mathbf{x}) = \widetilde{\boldsymbol{\eta}}_{i,j}(\mathbf{x}) + \int \mathcal{K}_j(\mathbf{x}, \mathbf{y})V_j(\mathbf{y})\boldsymbol{\eta}(\mathbf{y})d\mathbf{y}, \tag{4.130}$$

where $\overline{\boldsymbol{\eta}}_i$ and $\widetilde{\boldsymbol{\eta}}_{i,j}(\mathbf{x}) = \mathbf{E}_i(\mathbf{x})\widetilde{\boldsymbol{\sigma}}_{i,j}(\mathbf{x}) + \mathbf{H}_i(\mathbf{x})$, $(\mathbf{x} \in v_i)$ are two kinds of the effective strain polarization tensor introduced analogously to the tensor $\boldsymbol{\eta}^0$ (4.97).

Let the integral operator $\mathcal{L}_i$ (4.97) expressing the strain polarization tensor inside the inclusion $\boldsymbol{\eta}(\mathbf{x})$ though the effective strain polarization tensor $\overline{\boldsymbol{\eta}}(\mathbf{x})$ be known. The solution $\mathcal{L}_i$ (4.97) of the volume integral equation (4.96) can be found by the direct quadrature method (called also the Nystrom method), by the iteration method as well as by the Fourier transform method (see, e.g., [199]). It should be mentioned that each of the mentioned methods has a series of advantages and disadvantages, and it is crucial for the analyst to be aware of their range of application. However, the knowledge of the operator $\mathcal{L}_i$ at the current stage is not necessary and the assumption of the existence of such an operator is enough. At first no restrictions are imposed on the method of estimation of the operator $\mathcal{L}_i$. On the effective stress tensor $\widetilde{\boldsymbol{\sigma}}_{i,j}(\mathbf{x})$, according to Eq. (4.129$_2$), the linear operator $\mathcal{L}_i$ acts as

$$\bar{v}_i \mathcal{L}_i \widetilde{\boldsymbol{\eta}}(\mathbf{x}) = \mathcal{R}_i(\widetilde{\boldsymbol{\sigma}}_{i,j})(\mathbf{x}) + \mathbf{F}_i(\mathbf{x}), \quad (\mathbf{x} \in v_i). \tag{4.131}$$

Acting on Eq. (4.130) by the operator $\mathcal{L}_i$ and taking Eqs. (4.131) into account transforms Eq. (4.130) into the following $(\mathbf{x} \in v_i)$

$$\begin{aligned}\bar{v}_i \mathcal{L}_i \int \mathcal{K}(\mathbf{y}, \mathbf{z})\boldsymbol{\eta}(\mathbf{z})V_j(\mathbf{z})d\mathbf{z} &= \bar{v}_i \boldsymbol{\eta}(\mathbf{x}) - \mathcal{R}_i(\widetilde{\boldsymbol{\sigma}}_{i,j})(\mathbf{x}) - \mathbf{F}_i(\mathbf{x}) \\ &\equiv \mathcal{R}_{i,j}(\widetilde{\boldsymbol{\sigma}}_{i,j})(\mathbf{x}) + \mathbf{F}_{i,j}(\mathbf{x}),\end{aligned} \tag{4.132}$$

where the introduced operator $\mathcal{R}_{i,j}$ and the tensor $\mathbf{F}_{i,j}(\mathbf{x})$ can be found by any numerical method (e.g., such as BIE and FEA, complex potential method and others) analogously to the scheme described in Sect. 4.4.2. It should be mentioned that the operators $\mathcal{R}_i$ (4.131) and $\mathcal{R}_{i,j}$ (4.132) act on the effective fields $\widetilde{\boldsymbol{\sigma}}_{i,j}(\mathbf{x})$ at $\mathbf{x} \in v_i$ and $\mathbf{x} \in v_i, v_j$, respectively, and the kernel of the operator $\mathcal{R}_{i,j}$ can be decomposed $(K = I, J)$:

$$\mathcal{R}_{i,j}(\mathbf{x}, \mathbf{y}) = \mathcal{R}^I_{i,j}(\mathbf{x}, \mathbf{y}) + \mathcal{R}^J_{i,j}(\mathbf{x}, \mathbf{y}), \quad \mathcal{R}^K_{i,j}(\mathbf{x}, \mathbf{y}) = \mathcal{R}_{i,j}(\mathbf{x}, \mathbf{y})V_k(\mathbf{y}), \tag{4.133}$$

where one follows Mura's [1104] tensorial indicia notation; i.e., repeated lowercase indices are summed up from 1 to d, while uppercase indices always take on the same numbers as the corresponding lowercase ones but are not summed up.

The stresses inside the heterogeneity v_i can be estimated from the second Eq. (4.132)

$$\boldsymbol{\sigma}(\mathbf{x}) - \mathcal{B}_i(\widetilde{\boldsymbol{\sigma}}_{i,j})(\mathbf{x}) - \mathbf{C}_i(\mathbf{x}) = \mathcal{B}_{i,j}(\widetilde{\boldsymbol{\sigma}}_{i,j})(\mathbf{x}) + \mathbf{C}_{i,j}(\mathbf{x}), \tag{4.134}$$

where

$$\mathcal{B}_{i,j}(\widetilde{\boldsymbol{\sigma}}_{i,j})(\mathbf{x}) \equiv \bar{v}_i^{-1}\mathbf{M}_1^{-1}(\mathbf{x})\mathcal{R}_{i,j}(\widetilde{\boldsymbol{\sigma}}_{i,j})(\mathbf{x}), \;\; \mathbf{C}_{i,j}(\mathbf{x}) \equiv \bar{v}_i^{-1}\mathbf{M}_1^{-1}(\mathbf{x})\mathbf{F}_{i,j}(\mathbf{x}) \tag{4.135}$$

are defined on the domain $\mathbf{x} \in v_i$.

Let us replace the inhomogeneity v_j by the fictitious inclusion with the matrix elastic properties and the fictitious eigenstrain

$$\mathbf{L}(\mathbf{y}) \equiv \mathbf{L}^{(0)}, \quad \bar{v}_j \boldsymbol{\beta}_1^{\text{fict}}(\mathbf{y}) = \mathcal{R}_j(\widetilde{\boldsymbol{\sigma}}_{i,j})(\mathbf{y}) + \mathbf{F}_j(\mathbf{y}), \quad (\mathbf{y} \in v_j) \tag{4.136}$$

so that the constitutive equation in the inclusion v_j is $\boldsymbol{\sigma}(\mathbf{y}) = \mathbf{L}^{(0)}\boldsymbol{\varepsilon}(\mathbf{y}) + \boldsymbol{\beta}_1^{\text{fict}}(\mathbf{y})$. Then the strain polarization tensor $\boldsymbol{\eta}(\mathbf{y}) = \boldsymbol{\beta}_1^{\text{fict}}(\mathbf{y})$ $(\mathbf{y} \in v_j)$ can be assumed to be known and Eq. (4.132) can be reduced to

$$\begin{aligned}\bar{v}_i \mathcal{L}_i \int \mathcal{K}_j(\mathbf{y}, \mathbf{z})\boldsymbol{\beta}_1^{\text{fict}}(\mathbf{z})V_j(\mathbf{z})d\mathbf{z} &= \bar{v}_i\boldsymbol{\eta}(\mathbf{x}) - \mathcal{R}_i(\widetilde{\boldsymbol{\sigma}}_{i,j})(\mathbf{x}) - \mathbf{F}_i(\mathbf{x}) \\ &\equiv \mathcal{R}^{\infty}_{i,j}(\widetilde{\boldsymbol{\sigma}}_{i,j})(\mathbf{x}) + \mathbf{F}^{\infty}_{i,j}(\mathbf{x}),\end{aligned} \tag{4.137}$$

defining the new concentration factors $\mathcal{R}^{\infty}_{i,j}(\mathbf{x})$ and $\mathbf{F}^{\infty}_{i,j}(\mathbf{x})$ $(\mathbf{x} \in v_i)$ which can be estimated analogously to the tensors $\mathcal{R}_{i,j}(\mathbf{x})$ and $\mathbf{F}_{i,j}(\mathbf{x})$ $(\mathbf{x} \in v_i)$ (4.131).

Thus, the tensors $\mathcal{R}_i(\mathbf{x}, \mathbf{y})$ and $\mathbf{F}_i(\mathbf{x})$ $(\mathbf{x}, \mathbf{y} \in v_i)$ describe the strain polarization tensor $\boldsymbol{\eta}_i(\mathbf{x})$ into the isolated inclusion v_i in the infinite matrix subjected to the loading $\widetilde{\boldsymbol{\sigma}}_i(\mathbf{x})$ $(\mathbf{x} \in v_i)$. In so doing, the tensors $\mathcal{R}_{i,j}(\widetilde{\boldsymbol{\sigma}})(\mathbf{x})$ and $\mathbf{F}_{i,j}(\mathbf{x})$ $(\mathbf{x} \in v_i)$ define the perturbation of the strain polarization tensor $\boldsymbol{\eta}_i(\mathbf{x})$ introduced by the placement of the inclusion v_j interacting with the inclusion v_i. The tensors $\mathcal{R}^{\infty}_{i,j}(\widetilde{\boldsymbol{\sigma}})(\mathbf{x})$ and $\mathbf{F}^{\infty}_{i,j}(\mathbf{x})$ describe another perturbation of the strain polarization tensor $\boldsymbol{\eta}_i(\mathbf{x})$ at $\mathbf{x} \in v_i$ produced by the inclusion v_j interacting just with the external loading $\widetilde{\boldsymbol{\sigma}}_j(\mathbf{y})$ $(\mathbf{y} \in v_j)$ but not with the inclusion v_i. In so doing, the tensors $\mathcal{R}^{\infty}_{i,j}(\widetilde{\boldsymbol{\sigma}})(\mathbf{x})$ and $\mathbf{F}^{\infty}_{i,j}(\mathbf{x})$ are defined on the whole space $\mathbf{x} \in R^d$ while $\mathcal{R}_{i,j}(\widetilde{\boldsymbol{\sigma}})(\mathbf{x})$ and $\mathbf{F}_{i,j}$ are defined only on $\mathbf{x} \in R^d \setminus v_j$. Comparison of Eqs. (4.137) defining the tensors $\mathcal{R}^{\infty}_{i,j}(\widetilde{\boldsymbol{\sigma}})(\mathbf{x})$ and $\mathbf{F}^{\infty}_{i,j}(\mathbf{x})$ with the iteration scheme (4.97) leads to a conclusion that the tensors $\mathcal{R}^{\infty}_{i,j}(\widetilde{\boldsymbol{\sigma}})(\mathbf{x})$ and $\mathbf{F}^{\infty}_{i,j}(\mathbf{x})$ present the first-order approximation of the solution (4.132_1) by the iteration method (4.97) and are, in fact, the first-order approximation of the tensors $\mathcal{R}_{i,j}(\widetilde{\boldsymbol{\sigma}})(\mathbf{x})$ and $\mathbf{F}_{i,j}(\mathbf{x})$, respectively, describing the solution of Eq. (4.132_2).

In a similar manner, we can define $\mathcal{T}^{\sigma}_{i,j}(\mathbf{x}, \mathbf{y})$, $\mathbf{T}^{\beta}_{i,j}(\mathbf{x})$ and $\mathcal{T}^{\sigma\infty}_{j}(\mathbf{x}, \mathbf{y})$, $\mathbf{T}^{\beta\infty}_{j}(\mathbf{x})$ describing the perturbation of the effective field $\overline{\boldsymbol{\sigma}}_i(\mathbf{x}) - \widetilde{\boldsymbol{\sigma}}_{i,j}(\mathbf{x})$ introduced by the heterogeneity v_j and fictitious inclusion with the elastic modulus $\mathbf{L}^{(0)}$ and eigenstrain $\boldsymbol{\beta}_1^{\text{fict}}(\mathbf{y})$ $(\mathbf{y} \in v_j,\ \mathbf{x} \in v_i,\ \mathbf{z} \in R^d)$ (4.136)

$$\overline{\boldsymbol{\sigma}}_i(\mathbf{x}) - \widetilde{\boldsymbol{\sigma}}_{i,j}(\mathbf{x}) = \mathcal{T}^{\sigma}_{i,j}(\widetilde{\boldsymbol{\sigma}}_{i,j})(\mathbf{x}) + \mathbf{T}^{\beta}_{i,j}(\mathbf{x}), \tag{4.138}$$

$$\overline{\boldsymbol{\sigma}}_i(\mathbf{z}) - \widetilde{\boldsymbol{\sigma}}_{i,j}(\mathbf{z}) = \mathcal{T}^{\sigma\infty}_{j}(\widetilde{\boldsymbol{\sigma}}_{i,j})(\mathbf{z}) + \mathbf{T}^{\beta\infty}_{j}(\mathbf{z}), \tag{4.139}$$

where the operators $\mathcal{T}^{\sigma}_{i,j}$ and $\mathcal{T}^{\sigma\infty}_{j}$ are reduced to the tensors $\mathbf{T}^{\sigma}_{i,j}(\mathbf{x}) = \langle \mathcal{T}^{\sigma}_{i,j}(\mathbf{x},\mathbf{y})\rangle_{(j)}$ and $\mathbf{T}^{\sigma\infty}_{j}(\mathbf{z}) = \langle \mathcal{T}^{\sigma\infty}_{j}(\mathbf{z},\mathbf{y})\rangle_{(j)}$ $(\mathbf{y} \in v_j,\ \mathbf{x} \in v_i,\ \mathbf{z} \in R^d)$, respectively, on the constant $\widetilde{\boldsymbol{\sigma}}_{i,j}(\mathbf{y})$ =const. For ascertaining the connections between $\mathcal{R}_{i,j}(\mathbf{x})$, $\mathbf{F}_{i,j}(\mathbf{x})$ and $\mathcal{T}^{\sigma}_{i,j}(\mathbf{x})$, $\mathbf{T}^{\beta}_{i,j}(\mathbf{x})$, respectively, we recast Eq. (4.138) in terms of the strain polarization tensor

$$\overline{\boldsymbol{\eta}}_i(\mathbf{x}) = \widetilde{\boldsymbol{\eta}}_{i,j} + \mathbf{E}_i(\mathbf{x})\mathcal{T}^{\sigma}_{i,j}(\widetilde{\boldsymbol{\sigma}}_{i,j})(\mathbf{x}) + \mathbf{E}_i(\mathbf{x})\mathcal{T}^{\beta}_{i,j}(\mathbf{x}) \tag{4.140}$$

Acting with the operator $\mathcal{L}_i$ on Eq. (4.139) and comparison of the obtained equation with Eq. (4.134) yields the following relationships $(\mathbf{x} \in v_i)$

$$\mathcal{R}_{i,j} = \mathcal{L}_i\mathbf{E}_i\mathcal{T}^{\sigma}_{i,j},\quad \mathbf{F}_{i,j}(\mathbf{x}) = (\mathcal{L}_i\mathbf{E}_i\mathbf{T}^{\beta}_{i,j})(\mathbf{x}). \tag{4.141}$$

4.4.6 *Approximate Representations of Some Tensors and Operators*

Effective field hypothesis **H1** means that $\mathbf{T}^{\sigma\infty}_{j}(\mathbf{x}) = \mathbf{T}_j(\mathbf{x}-\mathbf{x}_j)\mathbf{R}_j$, $\mathbf{T}^{\beta\infty}_{j}(\mathbf{x}) = \mathbf{T}_j(\mathbf{x}-\mathbf{x}_j)\mathbf{F}_j$, where hereafter the tensors $\mathbf{T}_j(\mathbf{x}-\mathbf{x}_j) \equiv \langle \boldsymbol{\Gamma}(\mathbf{x}-\mathbf{y})\rangle_{(j)}$, $\mathbf{T}_{i,j}(\mathbf{x}_i-\mathbf{x}_j) \equiv \langle \mathbf{T}_j(\mathbf{x}-\mathbf{x}_j)\rangle_{(i)}$ are defined through the external Eshelby tensor $\mathbf{S}_j(\mathbf{x}-\mathbf{x}_j) \equiv -\bar{v}_j\langle \mathbf{U}(\mathbf{x}-\mathbf{y})\rangle_{(j)}\mathbf{L}^{(0)}$, $(\mathbf{y} \in v_j,\ \mathbf{x} \in R^d)$: $\mathbf{T}_j(\mathbf{x}) = \mathbf{L}^{(0)}\big[\mathbf{S}_j(\mathbf{x}-\mathbf{x}_j) - \mathbf{I}V_j(\mathbf{x})\big]$. The acceptance of the effective field hypothesis **H1** for both interacting inclusions v_i and v_i reduces the integral Eqs. (4.132) and (4.137) describing the solution for the binary interacting inclusions to the algebraic ones. This allows us to significantly simplify the estimation of the tensors $\mathbf{R}_{i,k}(\mathbf{x})$, $\mathbf{F}_{i,k}(\mathbf{x})$ (4.4.5), and $\mathbf{R}^{\infty}_{i,k}(\mathbf{x})$, $\mathbf{F}^{\infty}_{i,k}(\mathbf{x})$ (4.4.8) $\mathbf{R}^{\infty}_{i,k}(\mathbf{x}) = \mathbf{R}_i(\mathbf{x})\mathbf{T}_{i,j}(\mathbf{x}_i-\mathbf{x}_j)\mathbf{R}_j$, $\mathbf{F}^{\infty}_{i,k}(\mathbf{x}) = \mathbf{R}_i(\mathbf{x})\mathbf{T}_{i,j}(\mathbf{x}_i-\mathbf{x}_j)\mathbf{F}_j$.

The tensors $\mathbf{R}_{i,k}(\mathbf{x})$ and $\mathbf{F}_{i,k}(\mathbf{x})$ can also easily be estimated in the framework of the effective field hypothesis (4.93) an (4.94) from the matrix representation of the solution of Eq. (4.132) $(\mathbf{x} \in v_i$, no sum on $i, j)$

$$\overline{v}_i\boldsymbol{\eta}_i(\mathbf{x}) = \mathbf{F}_i(\mathbf{x}) - \mathbf{R}_i(\mathbf{x})\mathbf{R}_i^{-1}\mathbf{F}_i + \mathbf{R}_i(\mathbf{x})\mathbf{R}_i^{-1}\sum_{k=1}^{2}\mathbf{Z}^{2k}_{ij}[\mathbf{R}_k\widetilde{\boldsymbol{\sigma}}_{i,j} + \mathbf{F}_k], \tag{4.142}$$

where for simplicity $i = 1, 2, \ \ j = 3 - i$ are taken and the matrix $\mathbf{Z}^{-1}$ has the elements

$$(\mathbf{Z}^{-1})_{ij} = \mathbf{I}\delta_{ij} - (1 - \delta_{ij})\mathbf{R}_j \mathbf{T}_{i,j}(\mathbf{x}_i - \mathbf{x}_j), \tag{4.143}$$

while $\mathbf{Z}^{21}_{ij}$ and $\mathbf{Z}^{22}_{ij}$ are nondiagonal and diagonal elements, respectively, of the matrix $\mathbf{Z}_{ij} = (\mathbf{Z}^{lm}_{ij})$ $(l, m = 1, 2)$. Substitution of the solution (4.142) into Eq. (4.134) leads to the following desired representation ($i = 1, 2; \ j = 3 - i$; no sum on i, j)

$$\mathbf{R}^I_{i,j}(\mathbf{x}) = \mathbf{R}_i(\mathbf{x})\mathbf{R}_i^{-1}(\mathbf{Z}^{22}_{ij} - \mathbf{I})\mathbf{R}_i, \ \ \mathbf{R}^J_{i,j}(\mathbf{x}) = \mathbf{R}_i(\mathbf{x})\mathbf{R}_i^{-1}\mathbf{Z}^{21}_{ij}\mathbf{R}_j, \tag{4.144}$$

$$\mathbf{F}_{i,j}(\mathbf{x}) = \mathbf{R}_i(\mathbf{x})\mathbf{R}_i^{-1}[(\mathbf{Z}^{22}_{ij} - I)\mathbf{F}_i + \mathbf{Z}^{21}_{ij}\mathbf{F}_j], \tag{4.145}$$

which can be used for construction of the tensors ($\mathbf{x} \in v_i$)

$$\mathbf{T}^{I\sigma}_{i,j}(\mathbf{x}) = \mathbf{T}_j(\mathbf{x})\mathbf{Z}^{21}_{ji}\mathbf{R}_i, \ \ \mathbf{T}^{J\sigma}_{i,j}(\mathbf{x}) = \mathbf{T}_j(\mathbf{x})\mathbf{Z}^{22}_{ji}\mathbf{R}_j, \ \ \mathbf{T}^{\beta}_{i,j}(\mathbf{x}) = \mathbf{T}_j(\mathbf{x})\mathbf{F}_{j,i}. \tag{4.146}$$

The representations for the effective parameters (at the hypotheses **H1b** and some additional hypotheses) by the use of the tensors $\mathbf{R}^I_{i,j}, \mathbf{R}^J_{i,j}, \mathbf{F}_{i,j}$ are considered in Chap. 8. Detailed constructions of the operators $\mathcal{T}^{\sigma\infty}_{i,j}(\langle\overline{\boldsymbol{\sigma}}_j\rangle)(\mathbf{x})$ and $\mathcal{R}_i(\langle\overline{\boldsymbol{\sigma}}_i\rangle)(\mathbf{y})$ ($\mathbf{x} \in R^d, \mathbf{y} \in v_i$) are presented in Chap. 9 by the iteration method. We will use approximate construction of the operator $\mathcal{T}^{\sigma}_{i,j}(\langle\widetilde{\boldsymbol{\sigma}}_{i,j}\rangle)(\mathbf{x})$. In particular, for $|\mathbf{x}_i - \mathbf{x}_j| \geq 2(a_i + a_j)$, we will use the first order approximation ($\mathbf{x} \in v_i$)

$$\mathcal{T}^{\sigma}_{i,j}(\langle\widetilde{\boldsymbol{\sigma}}_{i,j}\rangle)(\mathbf{x}) = \mathbf{T}^{I\sigma}_{i,j}(\mathbf{x})\langle\overline{\boldsymbol{\sigma}}_i\rangle + \mathbf{T}^{J\sigma}_{i,j}(\mathbf{x})\langle\overline{\boldsymbol{\sigma}}_j\rangle, \tag{4.147}$$

which accuracy was estimated in [190] (see also Sect. 4.4.4). In the right-hand side of Eq. (4.147), we use the average effective fields which can be considered as a zero order approximation leading to the first-order approximation of the effective stress field in the left-hand side of Eq. (4.147). For $|\mathbf{x}_i - \mathbf{x}_j| < 2(a_i + a_j)$, we will take the inhomogeneities of $\langle\overline{\boldsymbol{\sigma}}_i\rangle(\mathbf{x})$ and $\langle\overline{\boldsymbol{\sigma}}_j\rangle(\mathbf{y})$ ($\mathbf{x} \in v_i, \ \mathbf{y} \in v_j$) into account

$$\mathcal{T}^{\sigma}_{i,j}(\langle\widetilde{\boldsymbol{\sigma}}_{i,j}\rangle)(\mathbf{x}) = \int \boldsymbol{\Gamma}(\mathbf{x} - \mathbf{y})\mathbf{Z}^{21}_{ji}\mathbf{R}_i\langle\overline{\boldsymbol{\sigma}}_i\rangle(\mathbf{y})V_i(\mathbf{y})d\mathbf{y}$$
$$+ \int \boldsymbol{\Gamma}(\mathbf{x} - \mathbf{y})\mathbf{Z}^{22}_{jj}\mathbf{R}_j\langle\overline{\boldsymbol{\sigma}}_j\rangle(\mathbf{y})V_j(\mathbf{y})d\mathbf{y}. \tag{4.148}$$

The operator $\mathcal{T}^{\sigma}_{i,j}(\langle\widetilde{\boldsymbol{\sigma}}_{i,j}\rangle)(\mathbf{x})$ can be constructed by the iteration method with any desirable accuracy analogously to the operators for one particle $\mathcal{T}^{\sigma\infty}_{i,j}(\langle\overline{\boldsymbol{\sigma}}_j\rangle)(\mathbf{x})$ and $\mathcal{R}^{\infty}_i(\langle\overline{\boldsymbol{\sigma}}_i\rangle)(\mathbf{y})$ ($\mathbf{x} \in R^d, \mathbf{y} \in v_i$) (see for details [203]). However, the first-order approximation (4.147) and (4.148) provides a reasonable level of accuracy.

4.4.7 Concluding Remarks

It should be mentioned that the VIE and BIE methods have a series of advantages and disadvantages, and it is crucial for the analyst to be aware of their range of applications. So, the high efficiency of the method proposed is realized only if the accuracy of the first-order approximation (see [198, 200]) is sufficient for a researcher and the VIE method is more efficient in the case being considered than other methods. However, for example, for composite materials with isotropic homogeneous inclusions and a multiscale analysis (such as finite number of inclusions in a half space), the BIE method is more effective than the VIE method because it has a well-known advantage of reduction of dimension by one. Because of this the hybrid BIE and VIE method supplied by an iterative solution strategy presented in [254] can be very effective for the anisotropic inclusions. On the other hand, the use of a reference isotropic fundamental solution with some residual volume integral [1200, 1340] eliminates the main well-known advantage of the BIE method of reduction of dimension by one. There are strong grounds for believing that the harnessing of the analytical solution for a single ellipsoidal homogeneous inclusion in either a bimaterial or half-space [904, 1665] will have culminated in a refinement of the first-order approximation considered in Sect. 4.4.4. If both the macro- and microinclusions are isotropic, then the use of BIE for both the macro- and microinclusions is more effective. In so doing efficiency of the BIE method is provided by use of an asymptotically adapted micromesh supplied by the iteration method. So, the estimation of stress distribution in the problem for 102 holes with essentially different sizes with the factor 10^7 was in [254] (with comparison with the classical solution [1078, 1202, 1497]) at laptop PC, whereas Oden and Zohdi [1148] (see also [1412]) exploited a multiprocessor supercomputer thousands of times more powerful for the analysis of the stress field in a 3D body containing inclusions differing by the factor 20 from a size of the body.

Furthermore, the regular shape of the macroinclusion v_1 in Sect. 4.4 was considered in order to make it easier to obtain the analytical representation of the initial approximation $\boldsymbol{\sigma}^{0[0]}(\mathbf{x})$ induced by the macroinclusion in the absence of small inclusions. In the case of nonregular shape of the inclusion v_1, we can use the BIE and/or VIE methods. Then we can analyze a nonsmooth surface Γ_1 in the neighborhood of small inclusions v_i $(i = 2, \ldots, n)$, in particular this part of Γ_1 can have the shape of either a quarter or an eighth space. In addition, the case of a few macroinclusions can be analyzed in the framework of the same scheme as well. So, we can consider two anisotropic macroinclusions with their smooth boundaries close together. In so doing, the small inclusion can be located in a layer between two macroinclusions that have parallel boundaries. In all cases mentioned above, the anisotropy of both the micro- and macroinclusions creates no complications, and we do not need to know the Green functions either for a half or quarter of the space or for the layer as used in the standard BIE method.

Chapter 5
Statistical Description of Composite Materials

Abstract The quantitative description of the microtopology of heterogeneous media, such as composite materials, porous and cracked solids, suspensions, and amorphous materials, is crucial in the prediction of overall mechanical and physical properties of these materials. In composite materials, internal structure refers to the size and shape of inhomogeneities, defects, and crystallites (grains); to the distribution of their orientations (textures); and to spatial correlations between these mechanical, geometrical, and crystallographic features. In this chapter the basic concepts and hypotheses associated with quantification of microstructure morphology of CM are introduced in a quite general form applicable to a wide class of both random and deterministic structures. Finally, the methods of stochastic simulation of real random structures of fiber composites and particle ones are considered.

After many years of comprehensive study by direct measurements and empirical relations that is extremely laborious, the structures of microinhomogeneous materials are still not completely understood; see [8, 174, 700, 1249, 1523, 1524]. Apart from obvious goal of quantitative analysis of the real morphology of random media, this research has been fueled by the fact that these statistical descriptors arise in rigorous bounds of diverse physical quantities [700, 1523]. The observable details of internal structure are entirely dependent upon the length scales that are accessible to the microscopic technique. Physical observation and simulation of internal structures are tractable at many different length scales. A central challenge in material science is to distinguish those aspects of internal structure that form the material properties. Experimentally, the complete description of local stress-strain states is not accessible at the grain level, and the main descriptor is the geometrical one representing statistical quantitative information concerning the orientational and spatial correlation between the structure elements. In the last decades, increasing attention has been paid to the combination of microstructure quantification with stochastic or deterministic numerical simulation of mechanical properties of microinhomogeneous media.

V. A. Buryachenko, *Local and Nonlocal Micromechanics of Heterogeneous Materials*, https://doi.org/10.1007/978-3-030-81784-8_5

5.1 Basic Terminology and Properties of Random Variables and Random Point Fields

5.1.1 Random Variables

The basic notion in probability theory is a *probability space* $(\Omega, \mathcal{F}, \mathcal{P})$ comprising three quantities: a sample space Ω, a collection $\mathcal{F}$ of certain subsets of Ω called a σ-algebra of events, and a probability measure $\mathcal{P}$. These quantities are defined as following. In the simplest case, Ω consists of a denumerable collection of elementary events $\omega_1, \omega_2, \ldots$, while nondenumerated *sample spaces* are more common, e.g., $\Omega = \{\omega | 0 \le \omega \le 1\}$. The σ*-algebra* of events $\mathcal{F}$ is a particular collection of events posing the properties: $\mathcal{F}$ is not empty; if $\mathcal{A} \subset \mathcal{F}$, then its compliment is $\mathcal{A}^c = \{\omega | \omega \in \Omega, \omega \notin \mathcal{A}\}$; if $\mathcal{A}_1, \mathcal{A}_2, \ldots \subset \mathcal{F}$, then its union is $\cup_i \mathcal{A}_i = \{\omega | \omega \in \mathcal{A}_i$ for some $i\}$. The *probability measure* $\mathcal{P}$ is a real-valued function mapping each set in $\mathcal{F}$ into the unit interval $[0, 1]$: $0 \le \mathcal{P}(\mathcal{A}) \le 1$ for every $\mathcal{A} \subset \mathcal{F}$; $\mathcal{P}(\Omega) = 1$; if $\mathcal{A}_1, \mathcal{A}_2, \ldots$ are disjoint sets in $\mathcal{P}$, then $\mathcal{P}(\cup_i \mathcal{A}_i) = \sum_i \mathcal{P}(\mathcal{A}_i)$.

We specify a real *random variable* (r.v.) $X(\omega)$ (called also *structure function*) defined on $(\Omega, \mathcal{F}, \mathcal{P})$ as a mapping from Ω to $\mathcal{R} \equiv \mathcal{R}^1$. This mapping is such that the set $\{\omega | X(\omega) \le x\}$ is in $\mathcal{F}$ for every choice of $x \in \mathcal{R}$, and probability of this set is well defined for $X(\omega)$ $F(x) = \Pr(\{\omega | X(\omega) \le x\})$ and called the *probability distribution*, which has the properties: $\Pr(a \le X < b) = F(b) - F(a)$; $F(x_1) \le F(x_2)$, if $x_1 \le x_2$; $\lim_{x \to +\infty} F(x) = 1$; $\lim_{x \to -\infty} F(x) = 0$. If there is a function $f(x) \ge 0$, called the *probability density function* for $X(\omega)$, such that $\Pr(X \in \mathcal{B}) = \int_{\mathcal{B}} f(x)\, dx$ hold for an arbitrary set $\mathcal{B} \subset R$ for which probability $\mathcal{P}(\{\omega | X(\omega) \in \mathcal{B}\})$ is defined, then the random variable $X(\omega)$ is called continuous. The notion of probability density $f(x)$ can also be introduced for a discrete r.v. by setting $f(x) = \sum_{i=1}^n p_i \delta(x - x_i)$, in which x_i denote the possible values of the r.v., p_i are their corresponding probabilities $p_i = \Pr(X = x_i)$ and $\delta(x)$ is the δ-function.

The extension to more than one random variable defined on the same probability space is straightforward. For example, the *joint distribution function* $F(\mathbf{x}) \equiv F_{\mathbf{X}}(\mathbf{x})$ $(\mathbf{x} = [x_1, \ldots, x_n]^\top)$ of a r.v. $\mathbf{X}(\omega) = [X_1(\omega), \ldots, X_n(\omega)]^\top$ is defined by

$$F(\mathbf{x}) \equiv F_{X_1,\ldots,X_n}(x_1, \ldots, x_n) = \Pr(\{\omega | \mathbf{X}(\omega) \le \mathbf{x}\}), \tag{5.1}$$

for all $\mathbf{x} \in R^n$, here the vector inequality $\mathbf{X}(\omega) \le \mathbf{x}$ means $X_i(\omega) \le x_i$ for $i = 1, \ldots, n$. For a system of continuous random variables $\mathbf{X}(\omega)$, there can exist a probability density $f(\mathbf{x}) \equiv f_{\mathbf{X}}(\mathbf{x}) \ge 0$ related with $F(\mathbf{x})$ by the formulae

$$f(\mathbf{x}) = \frac{\partial^n F(x_1, \ldots, x_n)}{\partial x_1 \ldots \partial x_n}, \quad F(\mathbf{x}) = \int_{-\infty}^{x_1} \cdots \int_{-\infty}^{x_n} f(y_1, \ldots, y_n)\, dy_1 \ldots dy_n. \tag{5.2}$$

According to the consistency theorem, the functions $F(\mathbf{x})$ and $f(\mathbf{x})$ must satisfy the compatibility conditions $F(x_1, \ldots, x_m) = F(x_1, \ldots, x_m, \infty, \ldots, \infty)$

and $F_{X_{i_1},\ldots,X_{i_n}}(x_{i_1},\ldots,x_{i_n}) = F_{X_1,\ldots,X_n}(x_1,\ldots,x_n)$ for both any permutation $\{i_1, i_2, \ldots, i_n\}$ of $\{1, 2, \ldots, n\}$ and $\forall m < n$. In particular, the probability density should satisfy the condition

$$f(x_1,\ldots,x_m) = \int_{-\infty}^{\infty} \cdots \int_{-\infty}^{\infty} f(x_1,\ldots,x_n)\, dx_{m+1}\ldots dx_n. \tag{5.3}$$

If $\mathbf{Y}(\omega) = \mathbf{g}(\mathbf{X}(\omega))$ is a random variable defined by $\mathbf{X}(\omega)$ and the function $\mathbf{g}$ mapping R^n into R^m is integrable over Ω with respect to $\mathcal{P}$, we can define the mathematical expectation, which for the continuous r.v. has the form

$$E(\mathbf{g}(\mathbf{X})) = \int_{R^n} \mathbf{g}(\mathbf{x})\, dF_{\mathbf{X}}(\mathbf{x}). \tag{5.4}$$

In particular, let us consider

$$\mathbf{g}(\mathbf{x}) = \prod_{i=1}^{n} x_i^{\nu_i}, \tag{5.5}$$

where ν_i, $(i = 1, \ldots, n)$ are nonnegative integers. Then $\alpha_{\nu_1,\ldots,\nu_n} = E(X_1^{\nu_1}\ldots X_n^{\nu_n})$, if it exists, is called the moment of order ν_1 in $X_1, \ldots,$ and ν_n in X_n. In the particular case of one r.v. X_k, we can define the mean, variance, and moment of qth order of X_k $(i_l = \delta_{lk},\ j_l = 2\delta_{lk},\ p_l = q\delta_{lk},\ l = 1, \ldots, n)$:

$$E(X_k) \equiv \bar{x}_k = \alpha_{i_1,\ldots,i_n}, \quad \mathrm{var}(X_k) \equiv \sigma_{kk} = \alpha_{j_1,\ldots,j_n} - \alpha^2_{i_1,\ldots,i_n}, \quad m_q = \alpha_{p_1,\ldots,p_n}, \tag{5.6}$$

as well as the central qth moment of X_k:

$$\mu_q = E[X_k - E(X_k)]^q. \tag{5.7}$$

Another special case takes place when $\nu_i = \nu_j = 1 - \delta_{ij}$ and $\nu_k = 0$ for $k = 1, \ldots, n,\ k \neq i,\ k \neq j$. Then the value

$$\sigma_{ij} = E[(X_i - EX_i)(X_j - EX_j)] \equiv E(X_iX_j) - E(X_i)E(X_j), \tag{5.8}$$

if it exists, is called the *covariance* between X_i and X_j $(i, j = 1, \ldots, n,\ i \neq j$. The real symmetric, positive semidefinite matrix (σ_{ij}) $(i, k = 1, \ldots, n)$ is called *variance–covariance* (or *dispersion*) matrix of the random vector $\mathbf{X}$. The random variables $X_1, \ldots, X_n$ are said to be uncorrelated if the nondiagonal elements of the covariance matrix are zero.

The probability for the product of n events $\mathcal{A}_1, \ldots, \mathcal{A}_n$ can be estimated through the conditional probabilities $\mathcal{P}(A_m | \prod_{k=1}^{m-1} \mathcal{A}_k)$:

$$\mathcal{P}\left(\prod_{k=1}^{n} \mathcal{A}_k\right) = \mathcal{P}(\mathcal{A}_1)\mathcal{P}(\mathcal{A}_2|\mathcal{A}_1)\mathcal{P}(\mathcal{A}_3|\mathcal{A}_1\mathcal{A}_2) \cdot \ldots \cdot \mathcal{P}\left(\mathcal{A}_n | \prod_{k=1}^{n-1} \mathcal{A}_k\right). \tag{5.9}$$

In so doing, an appropriate formula for the probability density of a system can be expressed in terms of the conditional densities according to the chain rule

$$f(x_1, x_2, \ldots, x_n) = f_1(x_1) f_2(x_2|x_1) f_3(x_3|x_1, x_2) \ldots f_n(x_n|x_1, x_2, \ldots, x_n). \tag{5.10}$$

as well in terms of the conditional probability density of the subsystem of the random variables $(X_1, \ldots, X_k)$, estimated under the assumption that the remaining random variables $X_{k+1}, \ldots, X_n$ assume certain values

$$f(x_1, \ldots, x_n) = f(x_1, \ldots, x_k|x_{k+1}, \ldots, x_n) f_{k+1,\ldots,n}(x_{k+1}, \ldots, x_n). \tag{5.11}$$

If for any m, where $m = 2, 3, \ldots, n$, and any k_j $(j = 1, 2, \ldots, n)$, $1 \leq k_1 < k_2 < \ldots < k_m \leq n$, Eq. (5.9) can be simplified $\mathcal{P}(\prod_{k=1}^{n} \mathcal{A}_k) = \prod_{k=1}^{n} \mathcal{P}(\mathcal{A}_k)$, then the events $\mathcal{A}_1, \ldots, \mathcal{A}_n$ are said to be independent, and the next properties occur for them: $F(x_1, \ldots, x_n) = F(x_1) \cdot \ldots \cdot F(x_n)$, $f(x_1, \ldots, x_n) = f(x_1) \cdot \ldots \cdot f(x_n)$. If the conditional distribution $f(x_1, x_2, \ldots, x_k|x_{k+1}, x_{k+2}, \ldots, x_n)$ (5.11) does not coincide with the "unconditional" distribution $f(x_1, x_2, \ldots, x_k)$ (5.3), then there is a statistical dependence between the sets of random variables $X_1, X_2, \ldots, X_k$ and $X_{k+1}, \ldots, X_n$. The measure of statistical dependence between random variables vanishing if they are statistically independent is described by multiple correlations of various orders. The covariance $\mathbf{K}(X_1, X_2) = \sigma_{ij}$ of two random variables X_1 and X_2 is defined by Eq. (5.8) vanishing if X_1 and X_2 are independent, when we have $E(X_1X_2) = E(X_1)E(X_2)$. A triple correlation $\mathbf{K}(X_1, X_2, X_3)$ describes statistical dependence on all three random variables X_1, X_2, X_3, vanishes when at least one of the variables is independent of the others, and can be taken in the form of

$$\mathbf{K}(X_1, X_2, X_3) = E(X_1X_2X_3) - \bar{x}_1\sigma_{23} - \bar{x}_2\sigma_{13} - \bar{x}_3\sigma_{12} + 2\bar{x}_1\bar{x}_2\bar{x}_3. \tag{5.12}$$

In a similar manner, the formulae for higher-order correlations can be obtained.

The concrete probability distributions described analytically by distribution function $F(x)$ can take a very different form. We reproduce in Table 5.1 the elementary properties of some distribution functions (discrete and continuous) which have proved to be extremely useful in micromechanics: the Poisson, the binominal, the uniform, the normal (or Gaussian), the lognormal, and the gamma distributions. In particular, the probability that an event occurs m times in n independent trial, in which the probability of occurrence of the event is $0 \leq p \leq 1$, is given by the binominal distribution presented in the second line of Table 5.1.

Table 5.1 Some distribution functions

Distr.	Poss. values	Density	Mean	Variance
Poisson	$0<\lambda,\ n=0,.$	$\frac{1}{n!}\lambda^n e^{-\lambda}$	λ	λ
Binomin.	$m=0,1,..,n$	$C_n^m p^m q^{n-m}$	np	npq
Uniform	$a<b-$ real	$(b-a)^{-1}\chi_{[a,b]}(x)$	$(a+b)/2$	$(b-a)^2/12$
Normal	$0<\sigma, m-$real	$(2\pi\sigma)^{-1/2}\exp\big[-\frac{(x-m)^2}{2\sigma^2}\big]$	m	σ^2
Lognorm.	$0<\xi$	$\frac{1}{\xi x\sqrt{2\pi}}\exp\big[-\frac{(\ln x-m)^2}{2\xi^2}\big]$	$\exp(m+\frac{\xi}{2})$	$m_X^2[\exp\xi^2-1]$
Gamma	$0<a,p$	$a^p[\Gamma(p)]^{-1}x^{p-1}\exp(-ax)$	p/a	pa^{-2}

The normal r.v. is completely characterized by its mean m and variance σ^2. Since the normal distribution is symmetric with respect to its mean values m, all odd central moments vanish ($\mu_{2k+1} = 0$), and, moreover, $\mu_{2k} = (2k-1)!!\sigma^{2k}$ ($k = 1, 2, \ldots$). The lognormal r.v. X is defined by the parameters m and ξ, which are the mean and variance of the normal random variable $\ln X$, respectively. The mean $m_X = \exp(m + \xi/2)$ and variance σ_X^2 of a lognormal distribution are given in terms of m and ξ. A probability density $f(x)$ ($0 \le x$) of a gamma distribution $G(a, p)$ is defined by the parameters $a, p > 0$ and the gamma function $\Gamma(p)$.

As an example of the system of random variables, we will consider the probability density of a system of n normal random variable $X_1, \ldots, X_n$:

$$f(x_1,\ldots,x_n) = \frac{1}{\sqrt{(2\pi)^n \det(\sigma)}} \exp\Big(-\frac{1}{2}\sum_{i,j=1}^{n}(x_i-\bar{x}_i)\sigma_{ij}^{(-1)}(x_j-\bar{x}_j)\Big), \quad (5.13)$$

where $\det(\sigma)$ is the determinant of the covariance matrix (5.8) and $\sigma_{ij}^{(-1)}$ are the elements of the inverse covariance matrix (5.8): $\sigma_{ij}^{(-1)} = A_{ij}\det(\sigma)^{-1}$, A_{ij} is the cofactor of the element σ_{ij} (5.8). For the normal distribution (5.13), the higher correlations vanish, and the moments of third, fourth, ... orders can be expressed by $\bar{x}_i, \sigma_{ij}$

$$E(X_k X_l X_m) = \bar{x}_k\bar{x}_l\bar{x}_m + \bar{x}_k\sigma_{lm} + \bar{x}_l\sigma_{km} + \bar{x}_m\sigma_{kl}, \quad (5.14)$$

$$\begin{aligned} E(X_k X_l X_m X_n) &= \bar{x}_k\bar{x}_l\bar{x}_m\bar{x}_n + \sigma_{kl}\sigma_{mn} + \sigma_{km}\sigma_{ln} + \sigma_{kn}\sigma_{lm} + \bar{x}_k\bar{x}_l\sigma_{nm} \\ &= \bar{x}_k\bar{x}_m\sigma_{ln} + \bar{x}_k\bar{x}_n\sigma_{lm} + \bar{x}_m\bar{x}_n\sigma_{kl} + \bar{x}_l\bar{x}_n\sigma_{km} + \bar{x}_l\bar{x}_m\sigma_{kn}. \quad (5.15) \end{aligned}$$

In particular, one obtains $M(X_k^3) = (\bar{x}_k)^3 + 3\bar{x}_k\sigma_{kk}$, and $M(X_k^4) = (\bar{x}_k)^4 + 3(\sigma_{kk})^2 + 6(\bar{x}_k)^3\sigma_{kk}$. Many important models are based on the normal distribution. The reason for this popularity originates in the central limit theorem of probability theory, which contends that, under some general conditions, the sum of independent random values is approximately normal distributed if their number tends to infinity.

5.1.2 Random Point Fields

Let each $\omega \in \Omega$ correspond to a realization of random medium that occupies some domain w called the window of observation in d-dimensional Euclidean space, i.e., $w \subset \mathcal{R}^d$. The *random point field* (called also *random process*) studied is denoted by Φ assuming that Φ is a random field with a finite number of points in any bounded area and with the property that different points are not coincide (simple point process). The random set of all points with the coordinates $\mathbf{x}_1, \ldots, \mathbf{x}_n$ is denoted as $\Phi = \{\mathbf{x}_i\}$ and $\mathbf{x} \in \Phi$; Φ_B denotes the restriction of Φ to a given area $B \subset \mathcal{R}^d$. Note that physicists speak of a "field," while statisticians use the term "process" even if no time dependence is considered. In so doing, we will use a parameterization by space ($\mathbf{x}$) rather than (t) coordinate owing to our concern with spatially random microstructure. The goal of statistical researches leading to so-called nonparametric estimators for different statistical parameters is to obtain *unbiased* estimators, which are free of systematic errors. It should be mentioned that the term "random" is sometimes used in a restrictive intuitive sense to mean uniformly distributed points. However, we use the term "random" in its more general sense of "non-deterministic" and not in the sense of specific distribution. Furthermore, since the object with which our processes are concerned are geometrical, we are dealing with topics in geometric probability and, in particularly, in point processes [1031, 1453].

For a given set $\mathcal{A} \subset \mathcal{R}^d$, $\Phi(\mathcal{A})$ is the random number points of Φ containing in $\mathcal{A}$. An observed sample of Φ_w in the observation window w is denoted by ϕ_w and $\phi(w)$ is the observed number of points in w. $N = \Phi(w)$ denotes the random number of points $\mathbf{x}_i$ falling in w; $\hat{N}(w)$ is the observed number of points in w. A random point field is a random variable in the sense of probability theory. We will also introduce the notations for some sets and set operations. So, $\chi_{\mathcal{A}}(\mathbf{x})$ stands the *indicator function* of $\mathcal{A}$:

$$\chi_{\mathcal{A}}(\mathbf{x}) = \begin{cases} 1 & (\mathbf{x} \in \mathcal{A}), \\ 0 & (\mathbf{x} \notin \mathcal{A}). \end{cases} \tag{5.16}$$

Translation $\mathcal{A}_{\mathbf{x}}$ is the set $\mathcal{A}$ shifted by the vector $\mathbf{x}$:

$$\mathcal{A}_{\mathbf{x}} = \mathcal{A} + \mathbf{x} = \{\mathbf{y} + \mathbf{x} : \mathbf{y} \in \mathcal{A}\}. \tag{5.17}$$

Let $\mathbf{g}$ be a rotation matrix (see Sect. 21.1) rotating the initial coordinate system around o. If $\mathbf{x} \in \mathcal{A}$ is a point of $\mathcal{R}^d$ then $\mathbf{gx}$ is the rotated point, and $\mathbf{g}\mathcal{A}$ is a rotated point field $\mathbf{g}\mathcal{A} = \{\mathbf{y} : \mathbf{y} = \mathbf{gx},\ \mathbf{x} \in \mathcal{A}\}$. The symbols $\oplus$ and $\ominus$ denote the *Minkowski addition* and *subtraction* ($\mathcal{A}^c := R^d \setminus \mathcal{A}$):

$$\mathcal{A} \oplus \mathcal{B} = \cup_{\mathbf{x}\in\mathcal{A},\mathbf{y}\in\mathcal{B}}\{\mathbf{x} + \mathbf{y}\} = \cup_{\mathbf{y}\in\mathcal{B}}\mathcal{A}_{\mathbf{y}} = \cup_{\mathbf{x}\in\mathcal{A}}\mathcal{B}_{\mathbf{x}}, \tag{5.18}$$

$$\mathcal{A} \ominus \mathcal{B} = \cap_{\mathbf{y}\in\mathcal{B}}\mathcal{A}_{\mathbf{y}} = (\mathcal{A}^c \oplus \mathcal{B})^c. \tag{5.19}$$

For the special important case of a ball $b(o, r)$ of radius r centered at the origin o, the sets $\mathcal{A}_r = \mathcal{A} \oplus b(o, r)$ and $\mathcal{A}^r = \mathcal{A} \ominus b(o, r)$ are known as the *outer parallel* (or *delation*) set of $\mathcal{A}$ and the *inner parallel* (or *erosion*) set of $\mathcal{A}$, respectively.

A quantitative description of the composite microstructure is based on the consideration of the inclusion centers statistically described by the multiparticle probability densities $\varphi_m(\mathbf{x}_1, \ldots, \mathbf{x}_m)$ that give the probability $\varphi_m(\mathbf{x}_1, \ldots, \mathbf{x}_m)$ $d\mathbf{x}_1 \ldots d\mathbf{x}_m$ to find an inclusion center in the vicinities $d\mathbf{x}_1 \ldots d\mathbf{x}_m$ of the point $\mathbf{x}^m = (\mathbf{x}_1, \ldots, \mathbf{x}_m)$. The φ_m are the most basic descriptors that characterize the structure of a multiparticle system and have been well studied in the statistical mechanics of liquids [604]. In particular, $\varphi_1 = n$, where n is the number density of inclusions v_i connected with the volume concentration (or fraction) $c = n\bar{v}_i$, $\bar{v}_i \equiv \text{mes} v_i$.

The point field Φ defined by the point coordinates can be generalized to the *marked process* denoted by the same letter $\Phi = \{(\mathbf{x}_i, \mathbf{m}_i)\}$ if each point $\mathbf{x}_i$ of the point process $\{\mathbf{x}_i\}$ has a *mark* $\mathbf{m}_i$ of the mark space $\{\mathbf{m}_i\} = \mathcal{M}$ distinguishing it to a great extent. The most popular marks used in micromechanics of CM are qualitative marks to distinguish different phases; quantitative marks to characterize the object such as, e.g., the orientation, shape, and size. Of course, any marked point process can be interpreted as a nonmarked point process in more general space, namely, in the product space $\mathcal{R}^d \times \mathcal{M}$. However, the points and marks are usually treated separately. For example, the translation of the marked point field $\mathcal{A} = \{(\mathbf{x}_i, \mathbf{m}_i)\}$ becomes $\mathcal{A}_\mathbf{x} = \{(\mathbf{x}_i + \mathbf{x}, \mathbf{m}_i)\}$.

Although the point pattern analyzed by statisticians is usually finite, the stochastic analysis is simplified by the assumption of an infinity of random field with subsequent accompanying assumptions, namely, those of homogeneity and isotropy. A point field $\Phi = \{\mathbf{x}^m\}$ is called *statistically homogeneous* (or *stationary*) if Φ and translated field $\Phi_\mathbf{x}$ (5.17) have the same distribution for $\forall \mathbf{x} \in \mathcal{R}^d$

$$\varphi_m(\mathbf{x}_1 + \mathbf{x}, \ldots, \mathbf{x}_m + \mathbf{x}) = \varphi_m(\mathbf{x}^m), \ m = 1, 2, \ldots; \ \forall \mathbf{x} \in \mathcal{R}^d. \tag{5.20}$$

Sometimes the properties (5.20) are replaced by the hypothesis that the mean point numbers are translation-invariant $E\Phi(\mathcal{A}) = E\Phi(\mathcal{A}_\mathbf{x})$ for any compact set $\mathcal{A}$ and all $\mathbf{x} \in \mathcal{R}^d$. For the homogeneous point field Φ, the last property makes it possible to define the *intensity*:

$$\lambda = E\Phi(\mathcal{A})/\bar{\mathcal{A}}, \ (\bar{\mathcal{A}} = \text{mes}\mathcal{A}) \tag{5.21}$$

where the position, size, and shape of $\mathcal{A}$ are irrelevant. The intercity λ noted also n is a mean number of points in the unit cube $[0, 1]^d \subset \mathcal{R}^d$. The intensity measure $E\Phi(\mathcal{A})$ plays a similar role in the theory of point fields as the mean does for random variables.

It should be mentioned that mathematical notion of "homogeneous pattern" is wider of the sense of this term used by many physicists and material scientists. For them it means "uniform," which is not the same for, e.g., clustered fields scattered similarly on the whole space (these fields are homogeneous but not uniform). In a general case, the rigorous deviation of the notions homogeneity and uniformity is very difficult and impossible if only one pattern in a bounded observation window w is given. If more than one sample (or a few parts of a sample) is given which can be considered as independent, then at least some aspects of homogeneity can be tested in the framework of the theory of testing of statistical hypotheses.

The isotropy of the point field Φ is defined in a similar manner. The point process Φ is called *isotropic* if

$$\Pr(\Phi \subset \mathcal{F}) = \Pr(\mathbf{g}\Phi \subset \mathcal{F}) \tag{5.22}$$

for any rotation $\mathbf{g}$ of the set Φ around the origin O. The point field which is both homogeneous and isotropic has the same distribution for the rotation around arbitrary points ($\neq O$) and is called motion-invariant. A bonded part Φ_w of Φ is assumed to be derived via a brick from a larger ensemble of random field Φ that is stationary and isotropic.

A stronger property is *ergodicity* ensuring an equality of the spatial average estimated over one a sufficiently large sample w and the mean value over the random fluctuations. Implicitly often supposed ergodic hypothesis enables us the replace ensemble average with volume averaging, assuming that observation window tends to infinity. A necessary condition of ergodicity is $\lim_{w\uparrow R^d} \Phi(w)/\bar{w} = \lambda$, where λ is the intensity and $w \uparrow \mathcal{R}^d$ is means that w contains a sphere of radius r limiting to infinity. A sufficient condition for ergodicity is the some "mixing" property which should be hold for any subsets of a homogeneous point field Φ. In particular, if $\mathcal{A}$ means that the sphere $b(O, r)$ does not contain a point and $\mathcal{B}$ means that the same sphere contains two points, then the fulfillment of the mixing condition means

$$\Pr(\Phi(b(O,r)) = 0,\ \Phi(b(O,r)) = 2) \to \Pr\big(\Phi(b(O,r)) = 0\big)\Pr\big(\Phi(b(O,r)) = 2\big).$$

A popular example of a nonergodic field is the lattice point fields with the points located at the nodes of a periodic grid, which are randomly positioned in the space (see other examples in [1031, 1210, 1453], where some philosophical problems of the ergodic hypothesis are discussed).

Thus, we summarized some basic ideas, notations, and properties of random point processes [351, 375, 789, 1295, 1297, 1452, 1454, 1523]. We usually will consider statistically homogeneous (or stationary) and isotropic ergodic random field Φ, keeping in mind that the stationarity, isotropy, and ergodicity can never be tested statistically in their full generality. Stationarity means invariance under arbitrary translation, and isotropy means invariant under arbitrary rotation. The ergodicity ensured that one sample (one point pattern) is sufficient for obtaining statistically secure results, assuming the convergence of results obtained for infinitely expanding observation window w.

5.1.3 Basic Descriptors of Random Point Fields

The packed random structure can be characterized by several parameters, such as packing density, coordination number, radial distribution function, particle cage, inter-particle spacing, and others. The various methods of estimating the effective properties of a composite material use a knowledge of statistical geometrical information about the microstructure. In order to incorporate the spatial arrangement of components in a micromechanical simulation, it is essential to quantitatively characterize the random structure of the composite. The most important factors characterizing the microstructure of composite material are the shape, volume fraction, and arrangement (random or regular) of the components that permit the calculation of bounds of effective moduli.

We will indicate three basic classes of descriptors and their estimators for unmarked point fields: (1) the intensity (mean number of points); (2) two distance distribution functions (for the distance analysis); and (3) several second-order quantities (for the descriptions based on the counting and counting-distance analysis).

The common unbiased estimator of the intensity (5.21) $\hat{n} = \phi(w)/\overline{w}$ ($\overline{w} = \text{mes} w$), which is also consistent (i.e., accuracy increases with increasing window area). The intensity is a sample statistical parameter used in the estimation of effective moduli.

The interaction effects generating the local stresses produced by inclusions are highly sensitive to their locations. In so doing, the nonlinear processes such as fracture and fatigue are far more sensitive to local stresses and, therefore, to local variations of microstructure then other mechanical phenomena. Because of this, in the deformation and fracture processes, another important statistical parameter of the location of inclusions is a *nearest neighbor distribution* (NND) function $D(r)$ [1248] defined by the density $d^{nd}(r)$ such that $d^{nd}(r)dr$ equals to the probability that there is no other inclusion centroid in a circle of radius r with the center $\mathbf{x}_i$, and there is at least one inclusion centroid in the ring of radius r and $(r + dr)$. In other words, the NND function describes a probability of the shortest of the distances from an arbitrary chosen point $\mathbf{x}_n \in \Phi$ to all other points of Φ:

$$D(r) = \Pr(\text{distance from } \mathbf{x}_n \text{ to nearest point of } \Phi \text{ is less then } r), \ (r \geq 0). \tag{5.23}$$

Many problems in point field theory require the consideration of an arbitrary "typical" point of point field Φ. The term "typical point" was used in [1454] as a synonym of "arbitrarily chosen point," where "arbitrarily" means that a selection scheme is used in which every point has the same chance to be selected. The typical point concept is a simplified notion of the Palm distribution referring to individual points in a point process (see for details [683]).

$D(r)$ is in fact a conditional probability function described by the Palm probability distribution P^o, which can be interpreted as being the probability distribution of point field events given under the condition that a point $\mathbf{x}_n \in \Phi$ called "a typical point" is observed at a specific location in the origin O (or any arbitrary chosen point of $\mathcal{R}^d$) $D(r) = \Pr^o(\Phi(b^0(O,r)) > 1)$, where $b^o(O,r)$ denotes the open ball of radius r centered at the origin O. Palm distribution theory makes precise the notion of a typical point that is heuristically clear. The estimation of $D(r)$ is based on the points $\mathbf{x}_i \in \Phi_w$ $(i = 1,\ldots,N)$ whose distance ρ_i to the nearest neighbor in w is smaller than r and whose shortest distance ρ_i^∂ to the boundary ∂w is larger than r. Repley [1295] proposed the following unbiased estimator for $D(r)$ taking the edge effect into account

$$\hat{D}(r) = \frac{\#\{\rho_i \le r \wedge \rho_i^\partial > r\}}{\#\{\rho_i^\partial > r\}}, \tag{5.24}$$

where $\#(\cdot)$ is the counting function which tallies the number of points in the specified set. In the estimator (5.24), only the points are analyzed for which the nearest neighbor distance ρ_i is smaller than the smallest distance ρ_i^∂ to the boundary ∂w. The Eq. (5.24) can be presented in a formalized form

$$\hat{D}(r) = \sum_{\mathbf{x}_i\in\Phi} \chi_{(O,r)}(\rho_i)\chi_{w\ominus b(O,\rho_i^\partial)}(\mathbf{x}_i)\Big[\sum_{\mathbf{x}_i\in\Phi} \chi_{w\ominus b(O,\rho_i^\partial)}(\mathbf{x}_i)\Big]^{-1}, \tag{5.25}$$

where the eroded window $w \ominus b(O,\rho_i^\partial)$ is defined by the Minkowski subtraction (5.19). The average neighbor distance $\langle D\rangle$ is found by the formula $\langle D\rangle = \int_0^\infty rD(r)\,dr$. The nearest neighbor and higher-order neighbor distributions can be estimated experimentally by measuring the frequency of occurrence of different distances of different order neighbors of the inclusions for a large number of inclusions. It should be mentioned that basing conclusions upon mean values or distance functions, or second-order quantities only, is not enough in general because none of these quantities can, in general, determine uniquely the distribution of point fields. Therefore, a combined use of intensities, a distance distribution, and second-order quantity is desirable.

An important step forward to a more refined description and analysis of the point arrangement is the use of *second-order quantities* such as the pair-correlation function, the radial distribution function, and the second-order product density. The widely used informative function which describes the point distribution is the *second-order intensity function* $K(r)$ (also called Ripley's [1295] K function) defined as the number of further points expected to be located within a distance r of an arbitrary point divided by the number of points per unit area n ([375, 1295]; see also [1249, 1394]). This notion looks like the definition of the nearest-neighbor distance distribution function $D(r)$ (5.23), but we are now interested in numbers of points rather than in distances. Heuristically speaking, $K(r)$ equals the mean number of points that have the distance smaller than $r \ge 0$ from the "typical point," which itself is excluded:

$$\hat{K}(r) = \frac{1}{N} \sum_{i=1}^{N} \Phi(b(\mathbf{x}_i, r) \setminus \{\mathbf{x}_i\}), \ (N = \Phi(w)). \tag{5.26}$$

Due to the assumed homogeneity of Φ, we may translate the typical point $\mathbf{x}_i$ to the original O and then again use the Palm probability distribution P^o, introduced in the definition of $D(r)$: $K(r) = n^{-1} E_{P^o}(\Phi(b(O, r)) - 1)$, where the mean is taken with respect to the Palm probability P^o. The unbiased estimator (5.26) has the disadvantage that it cannot be completely obtained from the observation window w because parts of the spheres $b(\mathbf{x}_i, r)$ may be outside w. Since points lying outside the observation window w are unobserved, the latter depends strongly on the shape and the size of w. It is our aim to simulate a typical realization that also includes interaction with the structure element outside of the window while avoiding systematic errors or biases in the estimation procedure. The effect of the edge of the domain w becomes increasingly dominant as the dimension increases. A number of special edge corrections are known. A naive way proposed in *the minus-sampling* method is to consider $w^* = w \ominus b(O, r)$ within domain w and allow measurements from an object in w^* to an object in w. To estimate $K(r)$, points $\mathbf{x}_i$ are used for which the sphere $b(\mathbf{x}_i, r)$ lies completely inside w

$$\widehat{K}(r) = \sum_{\mathbf{x}_i \in w \ominus b(O,r)} \frac{\Phi(b(\mathbf{x}_i, r) \setminus \{\mathbf{x}_i\})}{\text{mes}(w \ominus b(O, r))}, \ (r \geq 0) \tag{5.27}$$

Although the effective sample size is then the number of points in w, the method, of course, leads to a large loss of information because only the inner points in $w \ominus b(O, r)$ are sphere centers. A much better idea of edge correction of the estimator for $K(r)$ was suggested in [1295]:

$$\hat{K}(r) = \frac{\bar{w}}{\hat{N}^2} \sum_{i=1}^{\hat{N}} \sum_{j \neq i}^{\hat{N}} w_{ij}^{-1} I_{ij}(r_{ij}), \tag{5.28}$$

where $\hat{N}$ is the number of points $\mathbf{x}_k$ $(k = 1, \ldots \hat{N})$ in the field of observation w with the area $\bar{w}$, $I_{ij}(r_{ij})$ is the indicator function equals 1 if $r_{ij} \leq r$ and zero otherwise where r_{ij} is a distance between the points $\mathbf{x}_i$ and $\mathbf{x}_j$. w_{ij} is the ratio of the circumference contained within w to the whole circumference with radius r_{ij}. For circles intersecting the boundary ∂w, the function w_{ij} compensating for the boundedness of w is less than one and has an explicit formula when the field of observation w is rectangular [375]; the formulae and algorithm for the 3D case were proposed in [789]. The function $\hat{K}(r)$ (5.28) is obtained by averaging over all inclusions at each value of r. Equation (5.28) is an approximately unbiased estimator, which is free of systematic errors, for sufficiently small r because $N/\bar{w}$ is a slightly biased estimator for n. One [683] considered different improvements for the quality of estimations of $K(r)$ and n. In 2D, Diggle [375] recommended an upper limit of r equal to half the length of the diagonal of a square sampling region.

The plots of K-functions may difficult to interpret because their usual power behavior of order r^d for large r. In such a case, there are some clear advantages in using the *L-functions* instead of working directly with K-functions: $L(r) = (K(r)/\omega_d)^{1/d}$, where

$$\omega_d = \frac{\pi^{d/2}}{\Gamma(1+d/2)} \tag{5.29}$$

is the value of the unit sphere in $\mathcal{R}^d$ and Γ is a gamma function. For example, for $d = 1, 2$ and 3, ω_d equals 2, π, and $4\pi/3$, respectively. The estimators for $L(r)$ can be obtained through the estimators for $K(r)$ (5.26)–(5.28).

Some related further second-order quantities, based on the derivatives of the K-function, is known for the analysis of motion-invariant point fields. The *radial distribution function* (RDF) $g(r)$ (called also the pair correlation function) which plays a role similar to the variance in a classical analysis of random variables is defined as the radial distribution of the average number of sphere centers per unit area in a spherical shell. The RDF can be estimated from second-order intensity function as

$$g(r) = \frac{1}{d\omega_d r^{d-1}} \frac{dK(r)}{dr}. \tag{5.30}$$

The K-function is seldom used in the study of packing, and the RDF $g(r)$ is used instead despite the fact that the estimation of $g(r)$ is more delicate due to numerical evaluation of the derivative (5.30). The RDF is related to the derivative of $K(r)$ (4) and is therefore more sensitive to changes in the spatial order than is the function $K(r)$. For a completely random point process (i.e., a homogeneous Poisson process) $K(r) = \omega_d r^d$, $g(r) \equiv 1$. Values of the RDF $g(r)$ larger than 1 indicate that the interposing distances around r appear more frequently compared to those in a completely random point process (typically there is clustering), whereas values of g(r) smaller than 1 indicate that the corresponding distances are rare, that there is mutual inhibition. RDF takes all values from zero to large values at $g(r \to a^+) = g(a^+)$ ($a^+ = \lim_{\epsilon\to 0} a + \epsilon$, $\epsilon > 0$), and the second neighbor shell at the large c becomes split exhibiting a shoulder. For $r \to \infty$ it tends to 1, and the rate of convergence reflects a degree of randomness of the structure considered in a comparison with a completely random Poisson field. The radius r_0 such that $g(r) \approx 1$ at $r > r_0$ determines the range of geometrical disorder where a reference heterogeneity does not distinguish individual neighbors and "fills" the surrounding as a continuum.

Up to now, we only considered unimodal distribution of identical heterogeneities. The introduction of mark correlation function (see [533, 1249, 1307, 1454]) may facilitate the characterization of more complex aspect of random structure. One attaches to each point with a position $\mathbf{x}_i$ of the center set a typical feature which may correspond to the size, shape, and type of inclusions served as "point identifiers." In our case, the "points" are the particle locations, while "marks" describe the "particle type" with reference to the particle radii $a^{(1)}$ and $a^{(2)}$. Then Eq. (5.28) for the toroidal edge correction (when $w_{ij} \equiv 1$, see Sect. 5.4.1) can be generalized to the partial K_{km} functions

$$\hat{K}_{km}(r) = \frac{\bar{w}}{\hat{N}^{(k)}\hat{N}^{(m)}} \sum_{i=1}^{N^{(k)}} \sum_{j, j \neq i}^{N^{(m)}} I_{ij}(r_{ij}), \tag{5.31}$$

where the indexes i and j correspond to the particles $v^{(k)}$ and $v^{(m)}$ $(k, m = 1, 2)$, respectively. The interpretation of the partial RDF g_{km} is similar to that of $g(r)$ $(k, m = 1, 2)$

$$g_{km}(r) = \frac{1}{d\omega_d r^{d-1}} \frac{d\hat{K}_{km}(r)}{dr}, \tag{5.32}$$

In accordance with the definition, $g_{km}(r) = g_{mk}(r)$ is symmetric with respect to $k \leftrightarrow m$ for $k, m = 1, 2$ and $\forall r \geq a^{(k)} + a^{(m)}$ (see, e.g., [683]). Introduction of partial descriptors (5.31) and (5.32) enables one to describe more accurately the random structures than their mass-weighted version of the spatial repartitions proposed by [521] for analysis of solid propellants.

5.2 Some Random Point Field Distributions

5.2.1 *Poisson Distribution*

Various models have been proposed for the generation of center coordinates of randomly packed spheres. If the sphere radii are small enough, then their centroid coordinates will be described by the stationary (or homogeneous) Poisson point process for which: for any bounded region w, the number of points of X falling in w follows the Poisson one-dimensional and m-dimensional distributions

$$P(\Phi(w) = k) = \frac{(n\bar{w})^k}{k!} \exp(-n\bar{w}), \quad k = 0, 1, \ldots \tag{5.33}$$

and for nonoverlapping sets $w_1, \ldots, w_m$, the numbers $\Phi(w_1), \ldots, \Phi(w_m)$ are independent random variables

$$P(\Phi(w_1) = k_1, \ldots, \Phi(w_m) = k_m) = \frac{(n\bar{w}_1)^k}{k_1!} \cdots \frac{(n\bar{w}_m)^k}{k_m!} \exp\left(-n \sum_{i=1}^{m} \bar{w}_i\right), \tag{5.34}$$

($k_1, \ldots, k_m \geq 0$). These properties imply stationarity and isotropy because of the translation and rotation invariance of volume. A further implication is that the point positions are independent and uniformly distributed within w, when the points are placed in the area where each possible location for a point is equally likely to be chosen, and the location of each point is independent of the location of any other point. The uniform (binomial) point process of a given number $\Phi(w) = N$ of points $\mathbf{x}_1, \ldots, \mathbf{x}_N$, independent and uniformly distributed in w, if for nonoverlapping bounded $w_1, \ldots, w_N \subset w$, $P(\mathbf{x}_1 \in w_1, \ldots, \mathbf{x}_N \in w_N) = (\bar{w}_1 \ldots \bar{w}_N)/(\bar{w})^N$, with the binomial distribution of the number of points in any bounded region $w_j \subset w$

$$P(\Phi(w_j) = k) = C_N^k (p(w_j))^k (1 - p(w_j))^{N-k}, \ k = 0, 1, \ldots, N, \qquad (5.35)$$

where $p(w_j) = \bar{w}_j/\bar{w}$, $E\Phi(w_j) = Np(w_j)$, and $VarN(w_j) = Np(w_j)(1 - p(w_j))$.

A stationary Poisson point process may serve as a reference model for complete spatial randomness and can be easily simulated with a computer. For example, $N(w)$ points are generated with uniform random position in a region $w = [0, L_1] \times \ldots \times [0, L_d]$ as a sequence of $d \cdot N(w)$ independent random numbers $x_1, \ldots, x_{d \cdot N(w)}$ uniformly distributed on [0, 1] and generated by a random number generator. The coordinates of the ith ($i = 1, N(w)$) point are then $\mathbf{x}_i = (L_1 x_{d \cdot i - d+1}, \ldots, L_d x_{d \cdot i})^\top$. Although the hypothesis of a Poisson set of centers for non-overlapping spheres is not fulfilled for finite sphere radii, it can often be used as a useful approximate description of the observed structures [1295]. We recall that for Poisson distribution

$$H_s(r) = D(r) = 1 - e^{-n\omega_d r^d}, \quad K(r) = \omega_d r^d, \quad L(r) = r, \quad g(r) = 1, \qquad (5.36)$$

where the first equality (5.36) means that the point field Φ seen from the position of an arbitrary chosen point $\mathbf{x}$ has the same distribution as the point field with the point $\mathbf{x} \in \Phi$. The expectation mean $\langle D \rangle$ and variation $E(s^2)$ of nearest neighbor distances are estimated as ($d = 2$, see [1433]): $\langle D \rangle = 0.4n^{-1/2}$, $E(s^2) = (4 - \pi)/(2\pi n)$. For a Poisson point process, an equivalency $H_s(r) \equiv D(r)$ provides the basis for a test of whether or not a given point pattern is consistent with Poisson process. The Poisson distribution depends only on one parameter n, which has the meaning of the mean number of points per unit volume. A common unbiased estimator $\hat{n} = \phi(w)/\overline{w}$ can be used for calculation of all interesting quantities and distributions mentioned above.

Many important mathematical models of random point fields are constructed from Poisson fields. For infinitely extended stationary field, the following are known: the (1) Boolean model, (2) non-overlapping sphere model, and (3) the system of edges or faces of the Poisson Voronoi tessellation [1451, 1454].

The most famous Boolean model is generated from the Poisson field of primary (or *germ*) points $\mathbf{x}_1, \mathbf{x}_2, \ldots$ of intensity n. The *grains* form a sequence of independent identically distributed compact sets K_i (e.g., the spheres). The union of all these grains shifted to the germs is the Boolean model $\Xi = \bigcup_{i=1}^{\infty}(K_i + \mathbf{x}_i)$. Usually the grains K_i are convex, although the non-convex grains are also important (the case where K_i are a finite point set of a Poisson field considered in the next section). The parameters of a Boolean model are the intensity n of the germ and parameters characterizing the grains K_i which can be estimated from the observation of appropriate geometric functionals of Ξ [1087]. The generalization of the prescribed Poisson germ-grain model is the case where the primary grains are "tubes" around either the straight lines or planes uniformly distributed in space modeled as an isometric Poisson line field or Poisson plane field. The mentioned Boolean models do not coincide with real microstructures because any overlapping is in conflict with fundamental formation mechanism. However, the "complete randomness" of Boolean models often fits well to low-density approximations of material structures.

Space Dirichlet tessellations subdividing an Euclidian space into n-dimensional bounded convex polytopes (polygons in 2D case) are widely used to characterize the spatial distribution, size, and shape of a filled phase [546, 547]. One provides a natural and unique approach [547] for defining a particle's neighbors and neighborhood. At first, a Poisson process of germ points is taken and the particle centers are placed at these germ points (the last problem is a nonoverlapping model considered later). The Dirichlet tessellation of the two-dimensional domain w yields a network of convex Voronoi polygons containing one germ point (inclusion) with the center $\mathbf{x}_i$ $(i = 1, \ldots, n)$ each, at most. The cells of this tessellation consist of those points that are closer to a germ point than to all other germ points. The interior of the Voronoi cell associated with the point $\mathbf{x}_i$ is the region $w_i = \{\mathbf{x} \in w : |\mathbf{x} - \mathbf{x}_i| < |\mathbf{x} - \mathbf{x}_j|, \ \forall j \neq i\}$ that is the neighborhood of $\mathbf{x}_i$. The tessellation is constructed by plotting lines to the centers of all nearby particles and then constructing perpendicular bisecting planes to those lines. Green and Simpson [570] have proposed the algorithm generating Voronoi polygons for n points by computing in $O(n \log n)$ time by tracing boundary adjustment, as a new polygon is fitted into a previously generated set. Ghosh and coworkers (for references see [544, 548]) have developed a material-based Voronoi cell method for directly treating multiple-phase Voronoi polygons as elements in a finite element model for elastic and thermoelastoplastic problems; they suggested a modification to the standard tessellation procedure to preclude the situation where neighboring fibers are substantially different in size and are closely spaced, which may result in polygons that do not completely envelope their corresponding fibers and may instead "cut" though the fibers. Since each Voronoi cell contains a single particle surrounded by the matrix, the Dirichlet tessellation [1154] can be used for a description of a statistical structure of composites in the form of the frequency distribution, for all cells, of the ratio in particle-to-cell volume that is also a measure of particle clustering in microstructure [119]. The Dirichlet tessellation is usually constructed from a set of points. Another partitions called skeletization by influence zones (SKIZs) [503] is built from a continuous and uniform dilatation procedure, starting from the inclusion surface, up to total erosion of the matrix. The SKIZ partitions (in contrast to Voronoi one) ensure that each cell completely envelopes its inclusion, which makes it possible to define a local concentration of inclusions.

In a nonoverlapping spherical model, we will again take a Poisson field Φ of germ point and start a uniform radial growth of grain in each of the points. The grain growth stops when its diameter is the distance to the nearest neighbor in the point field. For this model, called the Stienen model, the diameter of the typical sphere has the same distribution as the nearest neighbor distance $D(r)$ of the Poisson point field (5.36) that leads to the representations for the mean volume of a typical sphere $\bar{v}_i = 2^{-d}/n$ and the volume concentration $\langle V \rangle = 2^{-d}$. The case of greater volume concentrations of nonoverlapping spheres with non-Poisson field of particle centers will be defined operationally in Sect. 5.4.

5.2.2 *Statistically Homogeneous Clustered Point Fields*

Many important mathematical models of random point fields are constructed from Poisson fields. The simplest generalization, called a *mixed Poisson field*, is generated by a Poisson field where the intensity $n(\mathbf{x})$ is a constant for each sample, but $n(\mathbf{x})$ varies from sample to sample. The basis of more usable *Neyman-Scott fields* is generated by the points (called "parent points") of a homogeneous Poisson field of constant intensity n_p. The set of "daughter points" forming the *cluster field* is independently scattered around each parent point according to the same density function (see, e.g., [1249]). Because a cluster center is not usually a point in the process, there is no loss of generality in accepting the cluster center as the center of symmetry. For the sake of simplicity, assume that the center of representative isotropic cluster is placed at the origin O. Then the coordinates of the ith daughter point are described by a random vector $\mathbf{x}_i$ with the same probability density function $d(\mathbf{x})$. In the simplest case of *Matern cluster field* (called also ideal cluster field), the number k of points has a Poisson distribution (5.33) with parameter μ. Then the positions $\mathbf{x}_i$ are uniformly distributed in the disc $b(O, R)$ of radius R centered at O

$$d(\mathbf{x}) = (\pi R^2)^{-1}\chi_{b(O,R)}(\mathbf{x}), \quad \text{or} \quad d(r) = 2rR^{-2}\chi_{[0,R]}(r). \tag{5.37}$$

The parent points of a stationary Neyman-Scott field can be easily generated by a Poisson field (see Sect. 5.3.1) of intensity n_p in a delation $w \oplus b(O, R)$ that avoids the edge effect. Then the number of daughter points is determined for each cluster generated in the vicinity of each parent point (in the case of the Matern clusters a Poisson random number of the intensity μ is generated). The coordinates of each daughter point are simulated with a random position in a polar coordinate system $[0, R] \times [0, 2\pi]$ when the radial coordinate has the distribution $d(r)$ (5.37_2) and the angle coordinate is uniformly distributed on $[0, 2\pi]$. The pair correlation function $g(r)$ of cluster fields decreases for increasing r with the range characterization less than or equal to $2R$; it is possible that $g(0) = \infty$.

Matern cluster fields are most usable and for them some formulae are derived easily [1454]. So, the intensity n of the field is $n = n_p\mu$. The pair distribution function satisfies $g(r) = 1 + (2\pi n_p r)^{-1} f(r)$, where

$$f(r) = \begin{cases} \frac{4r}{\pi R^2}\Big[\cos^{-1}(\frac{r}{2R}) - \frac{r}{2R}\sqrt{1-\frac{r^2}{4R^2}}\Big], & (0 \le r \le 2R), \\ 0, & (r > 2R). \end{cases} \tag{5.38}$$

In particular, $g(0) = 1 + (\pi n_p R^2)^{-1}$. The K-function has the form

$$K(r) = \pi r^2 + \frac{1}{n_p} \times \begin{cases} 2 + \pi^{-1}[(8z^2 - 4)\cos^{-1} z - 2\sin^{-1} z \\ +4z\sqrt{(1-z^2)^3} - 6z\sqrt{1-z^2}] & (r \le 2R), \\ 1, & (r > 2R), \end{cases} \tag{5.39}$$

where $z = r/2R$. We considered the simplest case of Matern cluster fields. The applicability of one or another cluster shape is defined by the physical problem being considered [157]. If the problem of recrystallization or residual stresses, it is reasonable that the the important clusters have a convex shape. If one is concerned by the damage properties of dielectric breakdown, a long chain of broken or conducting particles forming a “percolating network” can be more deleterious than the same number of particles arranged in a convex cluster; for chain clusters not only the local density of particles but also their arrangement are relevant, demonstrated in a limiting case of the “weakest link.” Due to the numerous possible cluster configurations and limited prediction models, one usually considers a degenerate case of cluster materials that, nevertheless, has a wide applications. Namely, one analyzes so-called ideal cluster materials [1494] that contain either finite or infinite, deterministic, or random ellipsoidal domains called particle clouds distributed in the composed matrix. In so doing, the concentration of particles is piecewise and homogeneous within the areas of ellipsoidal clouds and composed matrix. In particular, in Chap. 12, we will consider a single particle cloud with the shape of a thick ply located in an infinite matrix with zero concentration of particles. The ideal cluster configuration may be created by distributing cloud centers deterministically or randomly and placing offspring points randomly and uniformly around parent points within a predetermined area (see, e.g., [1352, 1434]). For a description of the cluster arrangement, the notion of “cluster radius” [774] is used for the presentation of a cluster as a particle of the second generation with the radius $R^{\rm clust} = \sum_i^N \sum_{j\neq i}^N |r_{ij}|/N(N-1)$ where i, j are particle numbers; r_{ij} is the interparticle distance; and N is a total number of particles in the cluster. Usually the cluster covers 80–90% of the total volume of the sphere with $R^{\rm clust}$ radius.

5.2.3 Inhomogeneous Poisson Fields

For researchers the statistical description of statistically homogeneous structures is a fascinating subject studied in many papers and books. In contrast, the analogous analysis for statistically inhomogeneous media such as clustered and functionally graded materials pose annoying problems. In such a case, the ergodicity fails, and ensemble and volume averages do not coincide. The degenerate case of this material is a random matrix composite bounded in some directions; in addition, the composite media for the inclusions are located in a region bounded in some directions, although unrestrictedness of the domain of inclusion locations does not preclude statistical inhomogeneity. For example, any laminated composite materials randomly reinforced with aligned fibers in each ply are a statistically inhomogeneous (or functionally graded) material (see Chap. 12 and Fig. 12.2). There are just a few theoretical papers which study structures with a gradient or that clustered objects [468, 596, 632, 1267].

An important sort of random point field constructed from a homogeneous Poisson field is an inhomogeneous Poisson field with deterministic differences in the point density. Instead of the intensity n (5.33), an intensity measure Λ on R^d is introduced, so that the measure $\Lambda(\mathcal{B})$ of any measurable set $\mathcal{B}$ has a sense of the number points in $\mathcal{B}$. It is assumed that this field has the same independent property of the random variables $\Phi(w_1), \ldots \Phi(w_n)$ (if $w_i \cap w_j = \emptyset$, for $\forall i \neq j$) as the homogeneous Poisson field. However, the property (5.33) is modified:

$$P(\Phi(w) = k) = \frac{(\Lambda(w))^k}{k!} e^{-\Lambda(w)}. \quad k = 0, 1, \ldots \tag{5.40}$$

There is often a density $n(\mathbf{x})$ called the *density function* of Λ so that $\Lambda(w) = \int_w n(\mathbf{x})\, d\mathbf{x}$, and $\Lambda(w)$ is the mean number of points in w. In the homogeneous case $n(\mathbf{x}) \equiv n =$ const., an inhomogeneous Poisson field is reduced to a homogeneous Poisson field with intensity n (5.33). However, even if $n(\mathbf{x}) \neq$ const., it has a similar intuitive interpretation as a probability $n(\mathbf{x})\, d\mathbf{x}$ to detect a point of Φ in an infinitesimally small disc of center $\mathbf{x}$ and area $d\mathbf{x}$. In a similar manner, the probability of finding a point in each of two infinitesimally small discs centered at $\mathbf{x}_1$ and $\mathbf{x}_2$ with area $d\mathbf{x}_1$ and $d\mathbf{x}_2$ is $n(\mathbf{x}_1)n(\mathbf{x}_2)\, d\mathbf{x}_1\, d\mathbf{x}_2$. In a similar manner, one can define systems on a finite region $\mathcal{B}$ by choosing the intensity function to be a restriction of n to $\mathcal{B}$: $n(\mathbf{x})\chi_{\mathcal{B}}(\mathbf{x})$. We note that the formulae for the number distribution as in the homogeneous case (see Sect. 5.3.1) holds if $n\bar{w}$ (5.33) is replaced by $\Lambda(w)$. The same is true for $P(\Phi(w) = 0) = e^{-\Lambda(w)}$.

The elegant method of simulation of an inhomogeneous Poisson field with intensity function $n(\mathbf{x})$ is based on the following thinning procedure of a homogeneous Poisson field. Let the intensity $n(\mathbf{x})$ be bounded $n(\mathbf{x}) \leq n^*$ and the function $p(\mathbf{x}) = n(\mathbf{x})/n^*$ is the so-called thinning function. This means that at first a homogeneous field of intensity n^* is generated in the observation window w (see Sect. 5.3.1). Then for every point, $0 \leq p(\mathbf{x}) \leq 1$ is estimated and compared with a random number $z \in [0, 1]$ independently generated. If $z > p(\mathbf{x})$, the point $\mathbf{x}$ is eliminated; otherwise it is a point of a newly generated inhomogeneous Poisson field. The probability density $n(\mathbf{x})$ as well as all another microstructural functions for the inhomogeneous fields will depend on the absolute positions of their arguments. The statistical descriptions for particulate, statistically inhomogeneous two-phase random media were proposed [596, 1267] (see also [468]) by the use of the theory of a general Poisson process [1454] and for some simulated fully penetrable (Poisson distribution) spheres the canonical n-point microstructure function, the nearest-neighbor functions, and the linear-path function were estimated that, unlike the homogeneous case, will depend on their absolute positions.

In the graded structures, there is a naturally given direction (e.g., the z-axis of the Cartesian coordinate system) such that the structure shows a trend-like variability in this direction, while in the planes perpendicular to the z-axis, the random field is the homogeneous and isotropic. Many materials consist of layers and have a gradient in their interior or near to their boundaries owing to structural inhomogeneities. For statistical description of the microstructure of ceramic matrix composites containing fiber-rich and fiber-poor regions that have a laminated structure, Yang et al. [1655] have proposed so-called the strip model. The key parameters of the strip model are the number densities of fiber in the fiber-poor n^p and fiber-rich n^r regions, the widths of the fiber-poor l^p and fiber-rich l^r regions, and the global number density of the fibers $n \equiv (n^p l^p + n^r l^r)/(l^p + l^r)$. Thus, the number density in a strip model is described by a piecewise constant function. Generalization of this model is based on the introduction of a variable number of densities $n(z)$ defined as the average number of disc centers per unit in a thin ply parallel to the plate $z = 0$.

The most important problem of inhomogeneous Poisson fields is the inverse one of estimation of the intensity function $n(\mathbf{x})$ from the observed data $\mathbf{x}_i$. The determination of the unknown parameters in an analytical representation of an intensity function can be done using the maximum likelihood method. This approach was described in [596, 1454], for graded structures that are inhomogeneous along a particular gradient direction but homogeneous perpendicular to that direction. Usually one analyzes so-called ideal cluster materials that contain either a finite or infinite, deterministic, or random ellipsoidal domains called particle clouds distributed in the composite matrix. In so doing, the concentration of particles is a piecewise constant and a homogeneous one within the areas of ellipsoidal clouds and composite matrix [891, 1367]. The most used descriptor for the clustered and graded materials is a volume concentration of inclusions [358, 577, 1081, 1289, 1471], which is not enough for the characterization of the micromorphology of fillers, simply because one can present other morphologies with the same descriptor. However, just taking into account binary interacting inclusions effects directly depending on the radial distribution function allows to detect some fundamentally new nonlocal effects for graded materials (see Chap. 12). In light of this, the further development of a statistical quantitative description for so many prospective clustered and graded materials (see, e.g., [468, 1267, 1367, 1504, 1655]) is of profound importance in both practically and theoretically sense.

5.2.4 Gibbs Point Fields

For the Poisson fields, the location of a point is independent of the location of any other point. The Gibbs fields studied in both bounded regions and in the whole space are more regular and take the interactions of particles into account (statistically isotropic fields of unmarked spherical particles with the radius a are considered). This approach starts with a potential function $\Phi^{\rm p}(\mathbf{x}^N)$ defining the probability of a configuration $\mathcal{C}^N = \{\mathbf{x}^N\}$, $[\mathbf{x}^N = (\mathbf{x}_1, \ldots, \mathbf{x}_N)]$ with fixed particle number N in a bounded box w [604, 1453, 1523, 1701]. In the absence of an external field, we can assume that the potential function contains only pairwise additive components

$$\Phi^{\rm p}(\mathbf{x}^N) = \sum_{i<j}^{N} \phi^{\rm p}(|\mathbf{x}_{ij}|) \tag{5.41}$$

because higher-order potentials are negligibly small compared to two-particle terms; $\mathbf{x}_{ij} = \mathbf{x}_i - \mathbf{x}_j$. In the considered case with fixed particle number N and pair potential $\Phi^{\rm p}$ (5.41), the joint distribution of the N particles is the Boltzmann-Gibbs canonical distribution

$$\varphi(\mathbf{x}^N) = \frac{1}{Z_N} \exp\Big[-\beta \sum_{i=1}^{N} \sum_{j=i+1}^{N} \phi^{\rm p}(|\mathbf{x}_{ij}|)\Big], \tag{5.42}$$

where $\beta = 1/(kT)$ is a reciprocal temperature T, k is Boltzmann's constant, and the normalizing constant Z_N called configurational partitional function ensures that $\varphi(\mathbf{x}^N)$ is a density function $Z_N = \int \exp(-\beta\Phi^{\rm p}(\mathbf{x}^N))\, d\mathbf{x}^N$, where $\int(\cdot)\, d\mathbf{x}^N$ denotes integration over w^N. Due to the difficulties of computation of Z_N, Ogata and Tanemura [1149, 1150] proposed an approximate method of the second cluster approximation

$$Z_N = \overline{w}^n (1 - a\overline{w}^{-1})^{n(n-1)/2}, \tag{5.43}$$

where $a = d\omega_d \int_0^\infty \big(1 - e^{-\beta\Phi^{\rm p}(r)}\big) r^{d-1}\, dr$ and $d\omega_d$ is the surface area of the unit ball in d-space (5.29). The positive and negative values of $\phi^{\rm p}(r)$ lead to "repulsion" and "attraction" of particles, so that the interparticle distances r appear rarely and frequently, respectively.

By considering different $\Phi^{\rm p}$, one gets a wide variety of distributions from (5.42). For example, for a system of noninteracting inclusions, we have $\Phi^{\rm p}(\mathbf{x}^N) \equiv 0$ for $\mathbf{x}^N \in w$, and Eq. (5.42) defines a uniform distribution of N particles in the box w. If $\phi^{\rm p}(r) = \infty$ for $r < 2a$, then interpoint distances smaller than $2a$ are impossible. If additionally $\phi^{\rm p}(r) = 0$ at $r > 2a$, then any interaction between the particle is absent and the potential function

$$\phi^{\mathrm{p}}(r) = \begin{cases} \infty, & (r \le 2a), \\ 0, & (r > 2a), \end{cases} \tag{5.44}$$

forms a Poisson hard core field (or nonoverlapping sphere model considered in Sect. 5.3.1) that is ergodic and the most popular in micromechanics. The important particular cases of the more complicated potential functions are

$$\phi^{\mathrm{p}}(r) = \begin{cases} \infty, & (r \le \sigma), \\ -\epsilon, & (\sigma < r \le 2a), \\ 0, & (r > 2a), \end{cases} \quad \phi^{\mathrm{p}}(r) = 4\epsilon\Big[\Big(\frac{\sigma}{r}\Big)^{12} - \Big(\frac{\sigma}{r}\Big)^{6}\Big], \tag{5.45}$$

both of which incorporate both repulsive and attractive interactions of particles. Here the parameter ϵ is the attractive well depth in the minimum in $\phi^{\mathrm{p}}(r)$. In Eq. (5.45_1) the parameter σ is a hard-core distance estimated by the minimum interpoint distance in the pattern while $2a$ is a radius of interaction. The parameter ϵ describes a jump at $r = 2a$ of the radial distribution function (5.30) of corresponding Gibbs field $\lim_{r\uparrow 2a} g(r) = e^{\epsilon} \lim_{r\downarrow 2a} g(r)$. Ogata and Tanemura [1150] describe maximum-likelihood estimation of parameters for these models in case of sparseness by the use of Eqs. (5.42) and (5.43). The square-well potential (5.45_1) at $\epsilon = 0$ is reduced to the cherry-pit potential in which each sphere has a core of diameter σ that is impenetrable to other cores. The surrounding shells, however, can penetrate other shells and cores that can be appropriately modified to model polymerization-filled composites [437]. A limiting case of the square-well potential (5.45_1) reducing attractive interactions to a delta function at interphase is called the sticky hard-sphere potential, which provides a modeling of aggregation processes. Lennard-Jones potential (5.45_2), widely used in the theories of classical simple liquids and colloidal dispersions [604, 1311], vanishes $\phi^{\mathrm{p}}(r) = 0$ at r equals the collision diameter σ.

It is rather difficult to estimate the characteristics of the Gibbs fields (e.g., $g(r)$ and $D(r)$) in terms of the intensity n and potential function $\phi^{\mathrm{p}}(r)$ determining the character of the point distribution (hard core, cluster, etc.). For the solution of this problem, Ornstein and Zernike [1160] proposed an integral equation linking dimensionless parameters such as the total correlation function $h(\mathbf{x}) = g_2(\mathbf{x}) - 1$ and the direct correlation function $c(\mathbf{x})$:

$$h(\mathbf{x}_{12}) = c(\mathbf{x}_{12}) + n(c * h)(\mathbf{x}_{12}) \tag{5.46}$$

where $c(\mathbf{x}_{12})$ describes a direct short-range effect of particle at $\mathbf{x}_1$ on one at $\mathbf{x}_2$ and the convolution integral depends on the indirect effects weighted by the intensity and averaged over all possible $\mathbf{x}_3$. This integral can be presented in 2D case as

$$(c * h)(r) = \int_0^{2\pi} \int_0^{\infty} c(s)h((r^2 + s^2 - 2rs\cos\xi)^{1/2})\, ds\, d\xi.$$

Considering that Eq. (5.46) is a convolution integral equation with constant coefficients, the standard method of its solution is using the Fourier transform method to transform the integral equation (5.46) into the division problem of solving the multiplicative algebraic equation

$$\tilde{h}(\mathbf{k}) = \tilde{c}(\mathbf{k}) + n\tilde{c}(\mathbf{k})\tilde{h}(\mathbf{k}) = \frac{\tilde{c}(\mathbf{k})}{1 - n\tilde{c}(\mathbf{k})}, \tag{5.47}$$

where the Fourier transformation $\tilde{\mathbf{f}}(\mathbf{k})$ of a function $\mathbf{f}(\mathbf{x})$ and its inverse are defined by the formulae (3.4).

According to Eq. (5.42), we can put the initial low-density approximation of $c(\mathbf{x}_{12})$ as the Mayer f-function

$$c^{(0)}(\mathbf{x}_{12}) \approx f(\mathbf{x}_{12}) = \exp(-\beta\Phi^{\mathrm{p}}(\mathbf{x}_{12})) - 1 \tag{5.48}$$

closing Eq. (5.46). The next step resides in modification of Eq. (5.48) for better fit of the solution to the results of computer simulation. An appropriate *Percus–Yevick* (PY) approximation [1198] contains an additional factor in Eq. (5.47)

$$c^{(0)}(\mathbf{x}_{12}) \approx \Big[\exp(-\beta\Phi^{\mathrm{p}}(\mathbf{x}_{12})) - 1\Big] \exp(\beta\Phi^{\mathrm{p}}(\mathbf{x}_{12}))g_2(\mathbf{x}_{12}) \tag{5.49}$$

so that $c(\mathbf{x}_{12})$ is zero wherever the potential vanishes.

The numerical method of solution of Eq. (5.46) is based on the solution of Eq. (5.47). The Fourier transform $\tilde{c}(\mathbf{k})$ of an initial approximation $c^{(0)}(\mathbf{x}_{12})$ is calculated and inserted in Eq. (5.47), and an inverse transform leads to an initial approximation $h^{(0)}(\mathbf{x}_{12})$. After that, the appropriate relation between $f(r)$ and $c(r)$ is used to obtain of the first iteration $c^{(1)}(r)$. An obvious iteration scheme continues until convergence is achieved. However, for the hard-core field of 3D spheres, the PY approximation enables one to solve Eq. (5.46) analytically by the Laplace transform method [1614]: $c(\rho) = -a_1 - 6ca_2\rho - \frac{1}{2}ca_1\rho^3$ at $(\rho \le 1)$, and $c(\rho) = 0$ at $(\rho > 1)$, where $a_1 = (1 + 2c)^2(1 - c)^{-4}$, $a_2 = -(1 + c/2)^2(1 - c)^{-4}$, and $\rho = r/(2a)$, $c = n\bar{v}_i$ is a volume concentration of inclusions. The solution $h(r)$ (and $g_2(r)$) can be found by Fourier transformation of $c(r)$, substituting $\tilde{c}(\mathbf{k})$ into (5.47), and inverting the last representation (5.47)

$$h(r) = \frac{1}{2\pi r} \int_0^{\infty} \frac{\tilde{c}(k)k\sin kr}{1 - n\tilde{c}(k)}\, dk, \tag{5.50}$$

where

$$\tilde{c}(k) = -\frac{4\pi}{q^3}\Big\{a_1[\sin q - q\cos q] + \frac{6\pi c a_2}{q}[2q\sin q + (2-q^2)\cos q - 2]$$
$$+\frac{ca_1}{2q^3}[4q(q^2-6)\sin q - (24 - 12q^2 + q^4)\cos q + 24]\Big\}, \tag{5.51}$$

and $q = 2ak$. Unfortunately, the mentioned problem was not solved analytically for the hard-disk system. A semiempirical modification of the PY estimation of $g_2(r)$ proposed in [1513, 1561] provides a good fit of the theory to the numerical data. The equation of the direct correlation function of hard d-spheres was obtained [77] allowing a unified treatment of both the odd and even space dimensions d. In the particular case $d = 2$, they obtained

$$c(\rho) = -\frac{\partial}{\partial c}[cz((c)]\chi_{[0,1]}(\rho)\Big\{1 - b^2c + \frac{2}{\pi}b^2c\Big[\cos^{-1}\frac{\rho}{b} - \frac{\rho}{b^2}\sqrt{b^2-\rho^2}\Big]\Big\}, \tag{5.52}$$

where $b = b(c)$ is obtained by solving the equation

$$\frac{2}{\pi}\Big[b^2(b^2-4)\sin^{-1}\frac{1}{b} - (b^2+2)\sqrt{b^2-1}\Big] = \frac{1}{c^2}\Big\{1 - 4c - \Big[\frac{\partial}{\partial c}[cz(c)]\Big]^{-1}\Big\} \tag{5.53}$$

with subsequent calculation of the direct and inverse Fourier transform for the functions $\tilde{c}(k)$ and $g_2(r)$, respectively. Baus and Colot [77] analyzed different sophisticated representations for z; the most simple of them is $z(c) = (1 + 0.125c^2)(1-c)^{-2}$.

5.3 Ensemble Averaging of Random Structures

The method of ensemble averaging is very popular in micromechanics [31, 73, 74, 103, 170, 508, 634, 646, 818, 821, 1247, 1523] and will be summarized below in a form selected for the subsequent presentation. The advantage with respect to volume-averaging is complete generality, which does not require the separation of scalars needed in volume averaging the existence of "an elementary macrovolume dV ...the characteristic linear dimensions of which are many times greater than the nonuniformities a but at the same time much less than the characteristic macrodimensions L of the problem" [1126]. For example, Cemlins [288] explicitly demonstrated a highly oscillatory of the volume fraction defined on the basis of volume averaging unless the size of averaging volume is several interparticle distances. However, in the context of ensemble averaging, these spurious oscillations are easily removed by the use of the very natural definition of volume fraction as the probability that the point $\mathbf{x}$ is occupied by the phase of interest. Working with ensemble averages (rather than with volume averages) is convenient since ensemble averaging commutes with differentiating and integrating for statistically homogeneous media (although this requires the validity of the ergodic hypothesis, see Sect. 5.1.2). The ensemble averaging produces the results which have a clear meaning in terms of the probability densities, and there is no necessity to require the presence of a great number of particles in physically small volumes of a composite as in the case for volume averaging. Ensemble averaging can be used in a unified way as a basis for the formulation of a variety of approximations of the closure problem in which the unknown variables are expressed in a computable form. Ensemble averaging can be applied both to random and periodic structure composites because an exclusive case of strictly ordered periodic systems can be included in the analysis by allowing for the ensemble distribution function to be expressed in terms of proper delta-functions. The advantage of ensemble averaging is most pronounced in the analysis of statistically inhomogeneous heterogeneous media (bounded or infinite) subjected to essentially inhomogeneous loading by the fields of the traction, temperature, and body forces. Each distinguishing feature of either the composite structure or the parameters of loading mentioned above leads to the loss of the statistical homogeneity of stress fields estimated that makes the use of the volume averaging technique impossible for the purpose of evaluating effective properties. However, a feasible way of generalizing all results and conclusions, obtained for statistically homogeneous composites by the use of ensemble average, to the more complex situation of nonlocal problems is quite straightforward, so that there is no need to devise special methods in order to treat statistically inhomogeneous media.

In order to incorporate the spatial arrangement of components in a micromechanical simulation, it is essential to quantitatively characterize the random structure of the composite. The most important factors characterizing the microstructure of composite material are the shape, volume fraction, and arrangement (random or regular) of the components that permit the calculation of bounds of effective moduli. There are two possible ways of describing the distribution (or configuration $\mathcal{F}$) of constituents in the composite. In the first way, the statistical moments of the probability densities of vectors describing locations, shapes, and orientations of different inhomogeneities in the medium are given. The second way is based on specification of which constituent each point of medium belongs and prescribes the multipoint moments of characteristic function of the region occupied by the inclusions.

5.3.1 Ensemble Distribution Functions

Let a linear elastic infinite body occupying full space R^d contain an open-bounded domain $w \subset R^d$ (window of observation) with a boundary Γ and with an indicator function W and space dimensionality d ($d = 2$ and $d = 3$ for 2D and 3D problems, respectively). The domain w contains a homogeneous matrix $v^{(0)}$ with indicator function $V^{(0)}(\mathbf{x})$ and a random finite set $\Omega = (v_i)$ $(i = 1, \ldots, N(w))$ of inclusions v_i with centers $\mathbf{x}_i$ and with indicator functions $V_i(\mathbf{y})$ equal to one in $\mathbf{y} \in v_i$ and zero otherwise and bounded by the surfaces Γ_i; $v = \cup v_i$, $v^{(0)} = w \setminus v$ and all domains v_i, $v^{(0)}$ are open $(i = 1, 2, \ldots, N)$. It is assumed that the representative mesodomain w contains a statistically large number of cranes v_i and that the grain can be grouped into components (phases) $v^{(k)}$ $(k = 1, 2, \ldots, N)$ with identical mechanical and geometrical properties (such as the shape, size, orientation, and microstructure of inclusions). Members of the ensemble are symbolized by discernible sets of the vectors $\mathbf{x}_i$ acceptable from a physical point of view. In a more general case, when the inhomogeneities are of the same but arbitrary form and of different size, it is necessary to incorporate into the analysis a set of unit orientation vectors $\mathbf{g}$ locked inside each particle and to introduce likewise a scale of particles $\mathbf{a}$ as the new independent ensemble variables. Thus, the random point field Φ defined by the points $\{\mathbf{x}_i\}$ is determined now by a triplet $\{(\mathbf{x}_i, \mathbf{g}_i, \mathbf{a}_i)\}$. In so doing, the set $\{(\mathbf{g}_i, \mathbf{a}_i)\}$ can be considered as a mark $\{\mathbf{m}_i\} = \{(\mathbf{g}_i, \mathbf{a}_i)\}$ of the marked process $\Phi = \{(\mathbf{x}_i, \mathbf{m}_i)\}$; the mark $\mathbf{m}_i$ can also contain some other information about the particles, e.g., their material "properties," particularly elastic moduli and stresses [320, 548, 1249, 1250]. We will use the term *configuration* $\mathcal{C}^N$ to indicate the set of values of numbers of quantities $\{(\mathbf{x}_i, \mathbf{m}_i)\}$ sufficient to specify uniquely the state of a system. The configuration $\mathcal{C}^N$ is completely described by a detailed distribution function, $\varphi(t, \mathbf{x}^N, \mathbf{m}^N)$, for N particles with the centers $\mathbf{x}^N = (\mathbf{x}_i, \ldots, \mathbf{x}_N)$ and marks $\mathbf{m}^N = (\mathbf{m}_1, \ldots, \mathbf{m}_N)$. Thus, $\varphi(t, \mathcal{C}^N)d\mathcal{C}^N \equiv \varphi(t, \mathbf{x}^N, \mathbf{m}^N)\, d\mathbf{x}^N\, d\mathbf{m}^N$ represent the probability of finding, at time t, the system in a configuration in which the first inclusion is within $d\mathbf{x}_1$, $d\mathbf{m}_1$ of $(\mathbf{x}_1, \mathbf{m}_1)$, etc. The time dependence can describe evolving microstructure that may arise as a result of, e.g., phase transitions and plastic deformations. However, we will usually assume that the microstructures are static or can be considered as static, and the function $\varphi(\mathbf{x}^N, \mathbf{m}^N)$ will be taken to be independent of time. The center locations $\mathbf{x}^N$ are arbitrary within w, except for the restriction $V_i(\mathbf{x})V_j(\mathbf{x}) = 0$ and $V_i(\mathbf{x})V^{(0)}(\mathbf{x}) = 0$ for any numbers $i \neq j$ and $\forall \mathbf{x} \in w$ that result from the mutual conditions of nonoverlapping of inclusions and of impenetrability of the boundaries. The distribution functions $\varphi(\mathcal{C}^N)$ are regarded as the probability density in which to find in the domains $v_1, \ldots, v_N$ the inhomogeneities with the centers $\mathbf{x}^N$ and marks $\mathbf{m}^N$. The normalized condition for $\varphi(\mathcal{C}^N)$ reads in the proper phase space as

$$\int \varphi(\mathcal{C}^N)d\mathcal{C}^N = 1. \tag{5.54}$$

The information contained in the marked random density function $\varphi(\mathcal{C}^N)$ can be too detailed and one has to use a reduced probability density of lower order which can be obtained by integrating $\varphi(\mathcal{C}^N)$ over the positions of $m < n$ inhomogeneous leading to less detailed distribution functions for the remaining $n - m$ particles. In particular, the one- and two-particle distribution functions for one and two fixed particles can be obtained from Eq. (5.3) $(i, j, k = 1, \ldots, N)$

$$\varphi(\mathbf{x}_i, \mathbf{m}_i) = \int \ldots \int \varphi(\mathcal{C}^N) \prod_{k \neq i} d\mathbf{x}_k \, d\mathbf{m}_k, \tag{5.55}$$

$$\varphi(\mathbf{x}_i, \mathbf{x}_j, \mathbf{m}_i, \mathbf{m}_j) = \int \ldots \int \varphi(\mathcal{C}^N) \prod_{k \neq i \neq j} d\mathbf{x}_k \, d\mathbf{m}_k, \tag{5.56}$$

The first equation (5.55) gives the unconditional probability density of the center of a particle with mark $\mathbf{m}_i$ being placed at $\mathbf{x}_i$. The Eq. (5.56) gives the probability density of an event in which simultaneously the centers of two particles with marks $\mathbf{m}_i$ and $\mathbf{m}_j$ are at $\mathbf{x}_i$ and $\mathbf{x}_j$, respectively. It follows from these definitions and from Eq. (5.55) that all $\varphi(\mathbf{x}^M, \mathbf{m}^M)$ $(M = 1, \ldots, N)$ are normalized.

In addition to probability densities $\varphi(\mathbf{x}^M, \mathbf{m}^M)$ $(M = 1, \ldots, N)$, we will follow to the definition (5.11) and introduce various conditional probability density configurations in which some particles are fixed. For example, the densities $\varphi(\mathbf{x}^{N-1}, \mathbf{m}^{N-1}|\mathbf{x}_i, \mathbf{m}_i)$ and $\varphi(\mathbf{x}^{N-2}, \mathbf{m}^{N-2}|\mathbf{x}_i, \mathbf{x}_j, \mathbf{m}_i, \mathbf{m}_j)$ corresponding to the configurations where the inclusions v_i as well as v_i and v_j are fixed, respectively, $\{\mathbf{x}^{N-1}, \mathbf{m}^{N-1}\}$ and $\{\mathbf{x}^{N-2}, \mathbf{m}^{N-2}\}$ are the hypersurfaces of $\{\mathbf{x}^N, \mathbf{m}^N\}$ which play the role of the phase space for all other $N-1$ and $N-2$ inclusions. For example, $\varphi(\mathbf{x}^{N-1}, \mathbf{m}^{N-1}|\mathbf{x}_i, \mathbf{m}_i)$ is the probability of a configuration $\mathcal{C}^{N-1}$ being found in the range $d\mathcal{C}^{N-1}$ about $\mathcal{C}^{N-1}$ where a particle with the mark $\mathbf{m}_i$ at the point $\mathbf{x}_i$ is fixed. These conditional probability densities are defined analogously to Eq. (5.11):

$$\varphi(\mathbf{x}^{N-1}, \mathbf{m}^{N-1}|\mathbf{x}_i, \mathbf{m}_i) = \varphi(\mathbf{x}^N, \mathbf{m}^m)/\varphi(\mathbf{x}_i, \mathbf{m}_i), \tag{5.57}$$

$$\varphi(\mathbf{x}^{N-2}, \mathbf{m}^{N-2}|\mathbf{x}_i, \mathbf{x}_j, \mathbf{m}_i, \mathbf{m}_j) = \varphi(\mathbf{x}^{N-1}, \mathbf{m}^{N-1}|\mathbf{x}_i, \mathbf{m}_i)/\varphi(\mathbf{x}_i, \mathbf{x}_j, \mathbf{m}_i, \mathbf{m}_j). \tag{5.58}$$

Extension of the chain rule used in Eqs. (5.57) and (5.58) leads to the definition of conditional distributions $\varphi(\mathcal{C}^{N-K}|\mathcal{C}^K)$ $(0 < K < N)$, $\mathcal{C}^K$ and $\mathcal{C}^{N-K}$ are the hypersurfaces of $\mathcal{C}^N$, $\mathcal{C}^K \cap \mathcal{C}^{N-K} = \emptyset$. Let us introduce a conditional probability density ($\mathcal{C}^{N-K} = \mathcal{C}^{N-K-1} \cup \{\mathbf{x}_m, \mathbf{m}_m\}$):

$$\varphi(\mathbf{x}_m, \mathbf{m}_m|\mathcal{C}^K) = \int \varphi(\mathcal{C}^{N-K}|\mathcal{C}^K) d\mathcal{C}^{N-K-1},$$

which is a probability density to find the m-th inclusion with the center $\mathbf{x}_m$ and the mark $\mathbf{m}_m$ in the domain v_m with fixed inclusions $v_1, \ldots, v_K$ with the centers $\mathbf{x}_1, \ldots, \mathbf{x}_K$ and the marks $\mathbf{m}_1, \ldots, \mathbf{m}_K$. The notation $\varphi(\mathbf{x}_m, \mathbf{m}_m|; \mathcal{C}^K)$ denotes the case $\mathbf{x}_m \neq \mathbf{x}_1, \ldots, \mathbf{x}_K$. The condition of nonoverlapping leads to an obvious restriction of the function $\varphi(\mathbf{x}_m, \mathbf{m}_m|; \mathcal{C}^K)$ which would be quite complicated in general, especially in the case of spatial correlation of location $\mathbf{x}_i$ and $\mathbf{m}_i$ of inclusions.

We will consider the limiting behavior and some simplifying assumption for this function. For an "overlapping inclusion" model, $\varphi(\mathcal{C}^N) = \prod_{i=1}^N \varphi(\mathbf{x}_i, \mathbf{m}_i)$. A more realistic probability density must account for the mutual exclusiveness of the inclusions. Of course, $\varphi(\mathbf{x}_m, \mathbf{m}_m|; \mathcal{C}^K) = 0$ for values of $\mathbf{x}_m$ lying inside the "excluded volumes" (or "correlation holes") $\cup v_{0i}$ $(i = 1, \ldots, K)$, where $v_{0i} \supset v_i$ with characteristic functions V_{0i} (since inclusions cannot overlap). It will be usually assumed that there is no long-range order in the medium and that the state probabilities of particles whose separation is large compared with a particle size are independent: $\varphi(\mathbf{x}_m, \mathbf{m}_m|; \mathcal{C}^K) \to \varphi(\mathbf{x}_m, \mathbf{m}_m)$ at $|\mathbf{x}_i - \mathbf{x}_m| \to \infty, \quad i = 1, \ldots, n$. The basic simplifying assumption uses these limiting behaviors as an approximation of the distribution function

$$\varphi(\mathbf{x}_m, \mathbf{m}_m|; \mathcal{C}^K) = \sum_{i=1}^{K}(1 - \chi_{v_{0i}}(\mathbf{x}_m))\varphi(\mathbf{x}_m, \mathbf{m}_m), \ (l = 1, \ldots, K) \tag{5.59}$$

which is called a well-stirred approximation and is in fact a low-density limit for multiparticle distribution function. It means that the inclusions are forbidden only to overlap; otherwise their locations are statistically independent. The next assumption is the absence of a spatial correlation between location $\mathbf{x}^N$ and mark $\mathbf{m}^N$ of inclusions, i.e., spatial location and mark are statistically independent

$$\varphi(\mathcal{C}^N) = \varphi_{\mathbf{x}}(\mathbf{x}^N)\varphi_{\mathbf{m}}(\mathbf{m}^N), \quad \varphi_{\mathbf{m}}(\mathbf{m}^N) = \varphi_{\mathbf{m}}(\mathbf{m}_1) \cdot \ldots \cdot \varphi_{\mathbf{m}}(\mathbf{m}_N).$$

It means that the marks of different inclusions are also statistically independent, i.e., the mark distribution functions are also factorized.

For statistically homogeneous media, $\varphi(\mathcal{C}^N)$ is translationary invariant $\varphi(\mathcal{C}^N) = \varphi(\mathbf{x}_{12}, \ldots, \mathbf{x}_{1n}, \mathbf{m}^N)$, where $\mathbf{x}_{ij} = \mathbf{x}_i - \mathbf{x}_j$. In particular, for unmarked fields, we can define the number density $\varphi(\mathbf{x}_1) \equiv \lim_{\bar{w} \to \infty} N/\bar{w}$ and the n-particle correlation function $g_N(\mathbf{x}^N) = n^{-N}\varphi(\mathcal{C}^N)$, which tends to unity for the system with no long-range order and $|\mathbf{x}_{ij}| \to \infty$ $(1 \le i, j \le N, \ i \ne j)$.

We have considered probability densities treating the particles as distinguishable with states $\mathbf{x}_i, \mathbf{m}_i$ labeled by the index i $(i = 1, \ldots, N)$. If N particles in the region w are identical, we should not consider their states as different from each other, and, because of this, we need to introduce suitable corrections to the normalization condition (5.54) to account for the indistinguishability of the states. The number of placement events of distinguishable and identical particles differs by the permutation number and so we have the normalized relations

$$\frac{1}{N!}\int \varphi(\mathcal{C}^N)\, d\mathcal{C}^N = 1, \quad \varphi(\mathcal{C}^K) = \frac{1}{(N-K)!}\int \varphi(\mathcal{C}^N)\, d\mathcal{C}^{N-K},$$

$$\int \varphi(\mathcal{C}^K) d\mathcal{C}^K = \frac{N!}{(N-K)!}.$$

In particular, the last equation can be used to obtain of the normalized condition for the one-particle probability density $\varphi(\mathbf{x}_1, \mathbf{m}_1)$: $\int \varphi(\mathbf{x}_1, \mathbf{m}_1)\, d\mathbf{x}_1\, d\mathbf{m}_1 = N$ justifying the following definition of particle number density $n(\mathbf{x}_1) = \int \varphi(\mathbf{x}_1, \mathbf{m}_1)\, d\mathbf{m}_1$ as the average number of particles per unit volume.

For unimodal distributions, the widely used two-point density $\varphi(r) = \varphi(\mathbf{x}_1, \mathbf{x}_2)$ ($r = |\mathbf{x}_1 - \mathbf{x}_2|$) is expressed in terms of *radial distribution function* $g(r)$ (RDF) (5.30) (called also the pair correlation function, see [1454]) as

$$\varphi(r) = n^2 g(r), \tag{5.60}$$

which is a function of the inter-point distance r. Hence, $ng(r)\bar{s}_d dr$ is a probability to find a point whose center lies in an infinitesimal spherical shell $s_d \subset R^d$ of inner radius r with a surface $\bar{s}_d$ ($\bar{s}_d = 2\pi r, 4\pi r^2$ for $d = 2, 3$, respectively) thickness dr and centered in a specific particle. For the probability density $\varphi_{2|km}(\mathbf{x}_1, \mathbf{x}_2)d\mathbf{x}_1 d\mathbf{x}_2$ of having one of particle type k and the other of type m, we can define $g_{km}(r)$ analogously to $g(r)$ ($k, m = 1, 2$)

$$\varphi_{2|km}(\mathbf{x}_1, \mathbf{x}_2) = n^{(k)} n^{(m)} g_{km}(r), \tag{5.61}$$

where $n^{(k)}$ is the intensity of particles $v^{(k)}$, the mean number k-points per unit volume, $n = n^{(1)} + n^{(2)}$. In accordance with the definition, $\varphi_{2|km}$ is symmetric with respect to $k \leftrightarrow m$ that implies the symmetry $g_{km}(r) = g_{mk}(r)$ for $k, m = 1, 2$ and $\forall r \geq a^{(k)} + a^{(m)}$ (see e.g., [683]).

In parallel with the conditional probability densities $\varphi(\mathbf{x}_m, \mathbf{m}_m|\mathcal{C}^K)$ and $\varphi(\mathbf{x}_m, \mathbf{m}_m|; \mathcal{C}^K)$ of the marked field Φ, other equivalent notations are used. Namely, for the description of the random structures, let us introduce at first the multipoint probability densities $\varphi(v_1, \mathbf{x}_1, \ldots, v_n, \mathbf{x}_n)$ that give the probability $\varphi(v_1, \mathbf{x}_1, \ldots, v_n, \mathbf{x}_n)\ d\mathbf{x}_1 \ldots d\mathbf{x}_n$ to find an inclusion center in the vicinity $\mathbf{x}_1 \ldots \mathbf{x}_n$ of the points $\mathbf{x}_1, \ldots, \mathbf{x}_n$, respectively. A conditional probability density $\varphi(v_i, \mathbf{x}_i|v_1, \mathbf{x}_1, \ldots, v_n, \mathbf{x}_n)$ is a probability density to find the ith inclusion with the center $\mathbf{x}_i$ in the domain v_i with fixed inclusions $v_1, \ldots, v_n$ with the centers $\mathbf{x}_1, \ldots, \mathbf{x}_n$. The notation $\varphi(v_i, \mathbf{x}_i|; v_1, \mathbf{x}_1, \ldots, v_n, \mathbf{x}_n)$ denotes the case $\mathbf{x}_i \neq \mathbf{x}_1, \ldots, \mathbf{x}_n$. The notation $\varphi(v_i, \mathbf{x}_i|; v_1, \mathbf{x}_1, \ldots, v_n, \mathbf{x}_n; v^{(0)}, \mathbf{x}_0)$ is considered under the condition that the inclusions $v_1, \ldots, v_n$ are located in the points $\mathbf{x}_1, \ldots, \mathbf{x}_n$, whereas $\mathbf{x}_0 \in v^{(0)}$ is the matrix position vector. In a general case of statistically inhomogeneous media with the homogeneous matrix (e.g., for so-called FGM), the conditional probability density is not invariant with respect to translation, that is, the microstructure functions depend their absolute positions [1267]. In particular, a random medium is called statistically homogeneous in a narrow sense if its multipoint statistical moments of any order are shift-invariant functions of spatial variables. Thus,

$$\varphi(v_i, \mathbf{x}_i|v_1, \mathbf{x}_1, \ldots, v_n, \mathbf{x}_n) \neq \varphi(v_i, \mathbf{x}_i|v_1, \mathbf{x}_1 + \mathbf{x}, \ldots, v_n, \mathbf{x}_n + \mathbf{x}), \exists \mathbf{x}, n,$$
$$\varphi(v_i, \mathbf{x}_i|v_1, \mathbf{x}_1, \ldots, v_n, \mathbf{x}_n) = \varphi(v_i, \mathbf{x}_i|v_1, \mathbf{x}_1 + \mathbf{x}, \ldots, v_n, \mathbf{x}_n + \mathbf{x}), \forall \mathbf{x}, n \tag{5.62}$$

for statistical inhomogeneous and homogeneous structures, respectively.

Of course, $\varphi(v_i, \mathbf{x}_i|; v_1, \mathbf{x}_1, \ldots, v_n, \mathbf{x}_n) = 0$ (since heterogeneities cannot overlap) for values of $\mathbf{x}_i$ lying inside some area $\cup v^0_{mi}$ ($m = 1, \ldots, n$) called "excluded volumes", where $v_{mi}^0 \supset v_m$ with indicator function V_{mi}^0 is the "excluded volumes" of $\mathbf{x}_i$ with respect to v_m (it is usually assumed that $v_{mi}^0 \equiv v_m^0$) and $\varphi(v_i, \mathbf{x}_i|; v_1, \mathbf{x}_1, \ldots, v_n, \mathbf{x}_n) \to \varphi(v_i, \mathbf{x}_i)$ as $|\mathbf{x}_i - \mathbf{x}_m| \to \infty$, $m = 1, \ldots, n$ (since no long-range order is assumed). $\varphi(v_i, \mathbf{x})$ is a number density, $n^{(k)} = n^{(k)}(\mathbf{x})$ of component $v^{(k)} \ni v_i$ at the point $\mathbf{x}$ and $c^{(k)} = c^{(k)}(\mathbf{x})$ is the concentration, i.e., volume fraction, of the component $v_i \in v^{(k)}$ at the point $\mathbf{x}$: $c^{(k)}(\mathbf{x}) = \langle V^{(k)}\rangle(\mathbf{x}) = \overline{v}_i n^{(k)}(\mathbf{x})$, $\overline{v}_i = \text{mes} v_i$ $(k = 1, 2, \ldots, N;\ i = 1, 2, \ldots)$, $\quad c^{(0)}(\mathbf{x}) = 1 - \langle V\rangle(\mathbf{x})$. If the system is isotropic (5.22), its two-particle distribution function $\varphi(v_m\mathbf{x}_m, v_i, \mathbf{x}_i)$ depends only on $r = |\mathbf{x}_m - \mathbf{x}_i|$; in this case written as $g(r)$ and called the radial distribution function (RDF) (5.61). In addition to the average $\langle(.)\rangle(\mathbf{x})$, the notation $\langle(.)|v_1, \mathbf{x}_1; \ldots; v_n, \mathbf{x}_n\rangle(\mathbf{x})$ will be used for the conditional average taken for the ensemble of a statistically inhomogeneous set $X = (v_i)$ at the point $\mathbf{x}$, on the condition that there are heterogeneities at the points $\mathbf{x}_1, \ldots, \mathbf{x}_n$ and $\mathbf{x}_i \neq \mathbf{x}_j$ if $i \neq j$ $(i, j = 1, \ldots, n)$. The notations $\langle(.)|; v_1, \mathbf{x}_1; \ldots; v_n, \mathbf{x}_n\rangle(\mathbf{x})$ are used for the additional condition $\mathbf{x} \notin v_1, \ldots, v_n$. We will distinguish macro-coordinate $\mathbf{x}$ used above (with "resolution" equal to Λ) and micro-coordinate $\mathbf{z} \in v_i \subset v^{(k)}$ So, the notation $\langle(\cdot)\rangle_i(\mathbf{x}, \mathbf{z})$ at $\mathbf{z} \in v_i \subset v^{(k)}$ means the average over an ensemble realization of surrounding heterogeneities (but not over the volume v_i of a particular inhomogeneity, in contrast to $\langle(\cdot)\rangle_{(i)}$) at the fixed v_i. The subdomains $v^{(k)}$ $(k = 1, \ldots, N)$ are called floating subdomains if they do not touch the boundary ∂w. In such a case, the body w is considered as one cut out from an infinite random medium and the inclusions v_i intersected with the boundary ∂w are replaced by the matrix material.

The well-stirred approximation (5.59) can be presented in these notations as $(v_k \subset v^{(K)})$:

$$\varphi(v_k, \mathbf{x}_k|; v_i, \mathbf{x}_i, v_j, \mathbf{x}_j) = (1 - V_i^0(\mathbf{x}_k))(1 - V_j^0(\mathbf{x}_k))n^{(K)},$$
$$\varphi(v_k, \mathbf{x}_k|; v_i, \mathbf{x}_i) = (1 - V_i^0(\mathbf{x}_k))n^{(K)}. \tag{5.63}$$

For both spherical and circle inclusions, the RDF $\varphi(v_k, \mathbf{x}_k|; v_i, \mathbf{x}_i)$ is well analyzed theoretically, that is, in agreement with both the results of Monte-Carlo simulations and experimental data. Various approximations have been proposed with varying degrees of success [255, 1523]. Much less investigated is the three-particle distribution function: $g_3(\mathbf{x}_1, \mathbf{x}_2, \mathbf{x}_3) = \varphi(v_1, \mathbf{x}_1, v_2, \mathbf{x}_2, v_3, \mathbf{x}_3)/(n^{(1)})^3$ [121, 560, 1444] because it depends on two or more variables $g_3(\mathbf{x}_1, \mathbf{x}_2, \mathbf{x}_3) = g_3(r_1, r_2, r_3)$ $(r_1 = |\mathbf{x}_1 - \mathbf{x}_2|$, $r_2 = |\mathbf{x}_1 - \mathbf{x}_3|$, $r_3 = |\mathbf{x}_2 - \mathbf{x}_3|)$, while the two-particle functions depend only on the radial coordinate. The earliest approximation for the three-particle functions of a homogeneous isotropic system was the superposition approximation of Kirkwood [782]:

$$g_3(r_1, r_2, r_3) = g(r_1)g(r_2)g(r_3),$$
$$\varphi(v_k, \mathbf{x}_k|; v_i, \mathbf{x}_i, v_j, \mathbf{x}_j) = \varphi(v_k, \mathbf{x}_k|; v_i, \mathbf{x}_i)\varphi(v_k, \mathbf{x}_k|; v_j, \mathbf{x}_j), \tag{5.64}$$

which overestimated the probability to find the inclusion v_k between the inclusions v_i and v_j for the small distance $|\mathbf{x}_i - \mathbf{x}_j|$. The Kirkwood superposition approximation (5.64) turns out to describe the triplet structure in an inadequate way, and because of this, Bildstein and Kahl [121] have strongly advised not to use it. Approximation (5.64) is rigorous only in the limiting case $|\mathbf{x}_i - \mathbf{x}_j| \to \infty$. Another approximation of triplet-structure

$$\varphi(v_k, \mathbf{x}_k|; v_i, \mathbf{x}_i, v_j, \mathbf{x}_j) = (1 - V_i^0(\mathbf{x}_k))(1 - V_j^0(\mathbf{x}_k)) \times \max_{R^d \setminus (v_i^0 \cup v_j^0)} [\varphi(v_k, \mathbf{x}_k|; v_i, \mathbf{x}_i)\varphi(v_k, \mathbf{x}_k|; v_j, \mathbf{x}_j)]$$

is also valid only in the limiting case $|\mathbf{x}_i - \mathbf{x}_j| \to \infty$. We will also use the notation $V(\mathbf{y}|v_1, \mathbf{x}_i; \ldots; v_n, \mathbf{x}_n)$ denoting a random characteristic function of inclusions under the condition that the inclusions $v_1 \ldots \neq v_n$ are located in the domains with the centers $\mathbf{x}_1, \ldots, \mathbf{x}_n$. In particular, the terms with $\mathbf{y} \in v_1$ and $\mathbf{y} \in v_2$ may be isolated in the characteristic function $V(\mathbf{y}|v_1, \mathbf{x}_1; v_2, \mathbf{x}_2)$ with the help of the equality

$$V(\mathbf{y}|v_1, \mathbf{x}_1; v_2, \mathbf{x}_2) = V_1(\mathbf{y}) + V_2(\mathbf{y}) + V(\mathbf{y}|; v_1, \mathbf{x}_1; v_2, \mathbf{x}_2), \quad (5.65)$$

which can be also presented in terms of the conditional probability density:

$$\varphi(v_q, \mathbf{x}_q|v_1, \mathbf{x}_1; v_2, \mathbf{x}_2) = \delta(\mathbf{x}_q - \mathbf{x}_1) + \delta(\mathbf{x}_q - \mathbf{x}_2) + \varphi(v_q, \mathbf{x}_q|; v_1, \mathbf{x}_1; v_2, \mathbf{x}_2). \quad (5.66)$$

5.3.2 Statistical Averages of Functions

Let $\mathbf{g}(\mathbf{x}|\mathcal{C}^N)$ be an arbitrary local physical function of coordinates $\mathbf{x} \in w$ which is continuous inside each constituent, but may have discontinuities on the inclusion surfaces. This function depends in addition on the arrangement of the configuration $\mathcal{C}^N$. The ensemble average conditional average of the function $\mathbf{g}(\mathbf{x}|\mathcal{C}^N)$ over the ensemble can be defined as

$$\langle \mathbf{g} \rangle (\mathbf{x}) = \int \mathbf{g}(\mathbf{x}|\mathcal{C}^N)\varphi(\mathcal{C}^N)\, d\mathcal{C}^N, \quad \langle \mathbf{g}|\mathcal{C}^K \rangle (\mathbf{x}) = \int \mathbf{g}(\mathbf{x}|\mathcal{C}^N)\varphi(\mathcal{C}^{N-K}|\mathcal{C}^K)\, d\mathcal{C}^{N-K}, \quad (5.67)$$

respectively. Equation (5.67_2) is calculated by integrating over all physically permissible values of $(\mathbf{x}_i, \mathbf{m}_i)$, $i = 1, N - K$, compatible with the condition that the centers of the inclusions with marks $\mathbf{m}_1, \ldots, \mathbf{m}_K$ lie at the points $\mathbf{x}_1, \ldots, \mathbf{x}_K$. It follows from Eqs. (5.67)

$$\langle \mathbf{g} \rangle(\mathbf{x}) = \int \mathbf{g}(\mathbf{x}|\mathbf{x}_i, \mathbf{m}_i)\varphi(\mathbf{x}_i, \mathbf{m}_i)\, d\mathbf{x}_i\, d\mathbf{m}_i,$$

$$\langle \mathbf{g}|\mathbf{x}_i, \mathbf{m}_i \rangle(\mathbf{x}) = \int \mathbf{g}(\mathbf{x}|\mathbf{x}_i, \mathbf{x}_j, \mathbf{m}_i, \mathbf{m}_j)\varphi(\mathbf{x}_i, \mathbf{x}_j, \mathbf{m}_i, \mathbf{m}_j)\, d\mathbf{x}_j\, d\mathbf{m}_j, \ldots .$$

Analogously, the multipoint correlation function of the function $\mathbf{g}(\mathbf{x}|\mathcal{C}^N)$ is defined as

$$\langle \mathbf{g}(\mathbf{x}_1), \ldots, \mathbf{g}(\mathbf{x}_k) \rangle = \int \mathbf{g}(\mathbf{x}_1|\mathcal{C}^N) \cdot \ldots \cdot \mathbf{g}(\mathbf{x}_k|\mathcal{C}^N)\varphi(\mathcal{C}^N)\, d\mathcal{C}^N.$$

We will also introduce the multipoint "phase" statistical averages over the constituent $\mathbf{x} \in v^{(i)}$ by means of the following formulae:

$$\langle \mathbf{g} \rangle^{(i)}(\mathbf{x}_1, \ldots, \mathbf{x}_k) = \left\langle \prod_{i=1}^{k} \chi_{v^{(i)}}(\mathbf{x}_i) \right\rangle \int \prod_{i=1}^{k} \chi_{v^{(i)}}(\mathbf{x}_i)\mathbf{g}(\mathbf{x}|\mathcal{C}^N)\varphi(\mathcal{C}^N)\, d\mathcal{C}^N, \qquad (5.68)$$

where $\chi_{v^{(i)}}(\mathbf{x})$ is an indicator function of the constituent $v^{(i)}$.

In analysis of statistically homogeneous media (5.20) we shall use an ergodicity hypothesis that will be assumed when the spatial average estimated over one sufficiently large sample (with realization $V^{*(i)}(\mathbf{x})$ $(i = 1, \ldots, N)$ and corresponding function $\mathbf{g}^*(\mathbf{x})$) and statistical mean $\langle \mathbf{g} \rangle(\mathbf{x})$ coincide for both the whole volume and the individual constituent:

$$\langle \mathbf{g} \rangle = \{\mathbf{g}\} := \lim_{w \uparrow R^d} \frac{1}{\overline{w}} \int_w \mathbf{g}^*(\mathbf{x})\, d\mathbf{x}, \quad \langle \mathbf{g} \rangle^{(i)} = \{\mathbf{g}\}^{(i)} := \lim_{w \uparrow R^d} \frac{1}{\overline{v^{(i)}}} \int_w \mathbf{g}^*(\mathbf{x}) V^{*(i)}\, d\mathbf{x}, \qquad (5.69)$$

where $\sum V^{(k)} = \sum V_i \equiv V$, $k = 1, 2, \ldots, N$; $i = 1, 2, \ldots$. $V^{(k)}$ is the characteristic functions of $v^{(k)}$. The bar appearing above the region represents its measure, e.g., $\overline{v} \equiv \mathrm{mes}\, v$. $V^{(k)}$ is the characteristic function of $v^{(k)}$. The average over an individual inclusion $v_i \subset v^{(k)}$ $(i = 1, 2, \ldots)$ is $\langle (.) \rangle_i = \langle (.) \rangle^{(k)}$. In the case of statistically homogeneous fields, the averages $\langle \mathbf{g} \rangle$ and $\langle \mathbf{g} \rangle^{(i)}$ (5.69) do not depend on $\mathbf{x}$. However, for statistically inhomogeneous fields, the statistical averages $\langle \mathbf{g} \rangle(\mathbf{x})$ (5.67_1) and $\langle \mathbf{g} \rangle^{(i)}(\mathbf{x})$ (5.68) depend on $\mathbf{x}$ and do not coincide with the spatial averages (5.69).

In particular, in the case under treatment, the structural function is described by the field $\Omega \subset w$ of inhomogeneities with indicator functions $V_i(\mathbf{x})$ $(i = 1, 2, \ldots)$ and can be presented by the marked random density function [836, 1014, 1456]

$$\mathcal{C}^N(\mathbf{x}, \mathbf{m}) = \sum_{\mathbf{x}_i \in \Omega} \delta(\mathbf{x} - \mathbf{x}_i)\delta(\mathbf{m} - \mathbf{m}_i). \qquad (5.70)$$

The standard method of estimation of the statistical average consists of exploiting the ergodic property (5.69) with subsequent estimation of the spatial average. For example, for a simplified unmarked field (5.70), we find that the statistical average

$$\langle \mathcal{C}^N(\mathbf{x})\rangle = \lim_{w\uparrow R^d} \frac{1}{\bar{w}} \int_w \sum_{\mathbf{x}_i\in\Omega} \delta(\mathbf{x}-\mathbf{x}_i)\, d\mathbf{x} = \lim_{w\uparrow R^d} \frac{\hat{N}}{\bar{w}} = n \tag{5.71}$$

defines the intensity of the field Ω; here $\hat{N} = \phi(w)$ is the number of points $\mathbf{x}_i$ in the field of observation w. In a similar manner, by the use of the phase average of the marked field with the absence of correlations between location $\mathbf{x}$ and mark $\mathbf{m}$ of inclusions (5.59), we get

$$\langle \mathcal{C}^N(\mathbf{x},\mathbf{m})\rangle = n\varphi_{\mathbf{m}}(\mathbf{m}),$$
$$\langle \mathcal{C}^N(\mathbf{x}_1,\mathbf{m}_1)\mathcal{C}^N(\mathbf{x}_2,\mathbf{m}_2)\rangle = n\varphi_{\mathbf{m}}(\mathbf{m}_1)\delta(\mathbf{x}_{12})\delta(\mathbf{m}_{12}) + \varphi_{\mathbf{x}}(\mathbf{x}_{12})\varphi_{\mathbf{m}}(\mathbf{m}_1)\varphi_{\mathbf{m}}(\mathbf{m}_2),$$

where for brevity, we denote $\mathbf{x}_{12} = \mathbf{x}_1 - \mathbf{x}_2$, $\mathbf{m}_{12} = \mathbf{m}_1 - \mathbf{m}_2$, and $\varphi_{\mathbf{x}}(\mathbf{x}_1,\mathbf{x}_2) = \varphi_{\mathbf{x}}(\mathbf{x}_{12})$ for statistically homogeneous fields. We can define $\varphi(v_i) = n\varphi_{\mathbf{m}}(\mathbf{m}_k)$ ($v_i \in v^{(k)}$) as a numberical density $n^{(k)}$ of the inclusions of the kth phase $v^{(k)}$.

It should be mentioned that the equality

$$\langle \mathbf{g}\rangle = \sum_{q=1}^{N} c^{(q)}\langle \mathbf{g}\rangle_q \tag{5.72}$$

(e.g., $\mathbf{g} = \boldsymbol{\varepsilon}, \boldsymbol{\sigma}, \boldsymbol{\tau}, \boldsymbol{\eta}$), is fulfilled only for statistically homogeneous media subjected to the homogeneous boundary conditions; here the summation in the right-hand side is performed over the volume of the representative inclusions $v_q \in v^{(q)}$ ($q = 1,\ldots,N$). If any of these conditions are broken, then Buryacheko [200] proposed to consider two sorts of conditional averages. At first, the conditional statistical average in the inclusion phase $\langle \mathbf{g}\rangle^{(q)}(\mathbf{x})$ (at the condition that the point with the macro-coordinate $\mathbf{x}$ is located in the inclusion phase $\mathbf{x}\in v^{(q)}$) can be found as $\langle \mathbf{g}\rangle^{(q)}(\mathbf{x}) = \langle V^{(q)}\rangle^{-1}(\mathbf{x})\langle \mathbf{g}V^{(q)}\rangle(\mathbf{x})$ where the averages $\langle(.)\rangle(\mathbf{x})$ are defined by Eq. (5.72). Usually, it is simpler to estimate the conditional averages of these tensors in the concrete point with the micro-coordinate $\mathbf{z}$ of the fixed inclusion $\mathbf{z}\in v_q$: $\langle \mathbf{g}|v_q,\mathbf{x}_q\rangle(\mathbf{x},\mathbf{z}) \equiv \langle \mathbf{g}\rangle_q(\mathbf{x},\mathbf{z})$. Although in a general case

$$\langle \mathbf{g}\rangle(\mathbf{x}) \equiv \sum_{q=1}^{N} c^{(q)}(\mathbf{x})\langle \mathbf{g}\rangle^{(q)}(\mathbf{x}) \neq \sum_{q=1}^{N} c^{(q)}(\mathbf{x})\langle\langle \mathbf{g}|v_q,\mathbf{x}_q\rangle(\mathbf{x},\mathbf{z})\rangle_{(q)}, \tag{5.73}$$

where $\mathbf{z}\in v_q \subset v^{(q)}$, it can be easy to establish a straightforward relation between these averages for the ellipsoidal inclusions v_q with the semi-axes $\mathbf{a}_q = (a_q^1,\ldots,a_q^d)^\top$. Indeed, at first we built some auxiliary set $v_q^1(\mathbf{x})$ with the boundary

$\partial v_q^1(\mathbf{x})$ formed by the centers of translated ellipsoids $v_q(\mathbf{0})$ around the fixed point $\mathbf{x}$. We construct $v_q^1(\mathbf{x})$ as a limit $v_{kq}^0 \to v_q^1(\mathbf{x})$ if a fixed ellipsoid v_k is shrinking to the point $\mathbf{x}$. Then we can get a relation between the mentioned averages [$\mathbf{x} = (x_1, \ldots, x_d)^\top$]:

$$c^{(q)}(\mathbf{x})\langle \mathbf{g}\rangle^{(q)}(\mathbf{x}) = \int_{v_q^1(\mathbf{x})} n^{(q)}(\mathbf{y})\langle \mathbf{g}|v_q, \mathbf{y}\rangle(\mathbf{x}, \mathbf{y}-\mathbf{x})\, d\mathbf{y}. \tag{5.74}$$

Formula (5.74) is valid for any material inhomogeneity of inclusions of any concentration in the macrodomain w of any shape (if $v_q^1(\mathbf{x}) \subset w$). In this relation (5.74), the total probability is expressed in term of conditional probabilities over portioned probability space. Obviously, the general Eq. (5.74) is reduced to Eq. (5.72) for both the statistically homogeneous media subjected to homogeneous boundary conditions and statistically homogeneous fields $\mathbf{g}$ (e.g., $\mathbf{g} = \boldsymbol{\sigma}, \boldsymbol{\varepsilon}$). However, in a general case $\mathbf{g}(v_q, \mathbf{y})(\mathbf{x}) \equiv \mathbf{f}(\mathbf{x}, \mathbf{y})\mathbf{g}^1(v_q, \mathbf{y})$ [$\mathbf{g}^1(v_q, \mathbf{y})$ is a statistically homogeneous field and $\mathbf{f}(\mathbf{x}, \mathbf{y})$ is a function of $\mathbf{x}, \mathbf{y}$], Eq. (5.74) is not reduced to Eq. (5.72). It should be mentioned that for statistically homogeneous fields $\langle \mathbf{g}\rangle^{(q)}(\mathbf{x}) \equiv \langle \mathbf{g}\rangle^{(q)}$ =const. and $v_q^1(\mathbf{x}) \equiv v_q^1$ =const., while in general $\langle \mathbf{g}\rangle_q(\mathbf{x}, \mathbf{z}) \equiv \langle \mathbf{g}\rangle_q(\mathbf{z}) \not\equiv$const. at the micro-coordinate $\mathbf{z} \in v_q$. Because of this, no confusion will arise below in notations of the average $\langle \mathbf{g}\rangle^{(q)}$ and conditional average $\langle \mathbf{g}\rangle_q(\mathbf{z})$ (at the fixed inclusion v_q) which are functions of macro-coordinate (with resolution equal to Λ) and micro-coordinate $\mathbf{z} \in v_q$ (used in the case of fixed inclusion v_q), respectively.

5.3.3 Statistical Description of Indicator Functions

Since the overall properties of composite materials are sensitive to the details of the microstructure, the geometrical basis for modeling actual microstructures is needed. We considered the first method of random structure composite description. The second popular statistical description of microinhomogeneous media is based on the specification as to which constituent each point of medium belongs, and on estimations of expectations of products of the indicator function $V^{(i)}(\mathbf{y})$, assuming that the role of the matrix is assigned to phase "0." In a simple case the bounding of effective properties uses the multipoint statistic reducing to only one point probability density (volume fraction) $S_1^{(i)}(\mathbf{y}_1)$ defined as ($i = 1, 2, \ldots, N$): $S_1^{(i)}(\mathbf{y}_1) = \langle V_i(\mathbf{y}_1)\rangle \equiv c^{(i)}$, where the angle brackets $\langle(\cdot)\rangle$ denote an ensemble average, and a substitution of the one-point correlation function by its volume average (i.e., *volume concentration* of the ith phase) is provided by an ergodic assumption. Improved bounds on a variety of different effective properties have been derived in terms of n-point probability densities

$$S_n^{(i)}(\mathbf{y}^n) = \langle V_i(\mathbf{y}_1), \ldots V_i(\mathbf{y}_n)\rangle, \tag{5.75}$$

which are symmetric in their arguments, i.e., the probability of simultaneously finding n points in a specified geometrical arrangement $\mathbf{y}^n \equiv \mathbf{y}_1, \ldots, \mathbf{y}_n$ in one of the phases. For example, the one-point correlation function is the probability that any point lies in any particular. The two-point probability function is the probability that both two points $\mathbf{y}_1$ and $\mathbf{y}_2$ lie in the same phase. These functions provide a method for experimentally determining the low-order multipoint moments for real two-phase media, as considered in [348], where Corson used an unautomated painstaking procedure to compute three-point functions from micrographs of cross-sections of composites.

Digital image analysis is now available for estimation of descriptors of the spatial arrangement of microstructural features observed in a transverse cross section of the material based on the *stereological* technique [113, 1452] allowing us to measure the geometry of such two-dimensional structures more rapidly, more precisely, and more comprehensively than ever before. Classical stereology investigates the spatial structures by planar sections and statistically analyzes the visible structures from a range of microscopic techniques such as transmission electron microscopy and scanning tunneling electron microscopy (see for references [1524]). For example, local stereology [698] shows how mean particle volumes can be estimated by length measurement. In general, some important spatial characteristics cannot be estimated stereologically, and statistical methods based on new microscopic techniques (e.g., computed X-ray tomography, magnetic resonance imaging, and confocal microscopy [753, 789, 1156, 1524]) using three-dimensional measurement were proposed. All these imaging methods are nondestructive, leaving the sample intact and unaltered. For example, by moving the local plane of a confocal scanning laser microscope up or down through the specimen, a stack of serial sections, called a brick, is obtained, from which a 3D image may be produced without the need to use physical sections.

Since it is possible to determine multipoint probabilities only up to some relatively low order, Kröner [820] proposed classifying materials as uniform "of grade k" if probabilities involving up to k points were known to be translation-invariant. It is also known that the conventional bounds on effective properties are given in terms of other types of statistical quantities such as point/q-particle functions, surface-surface correlation functions, nearest-neighbor distribution function, linear-path function, two-point cluster function, chord-length distribution function, as well as the generalized n-point distribution function for the system of identical spheres (see for details [1523]). However, although higher-order correlation functions ($N > 3$) are obtained on theoretical grounds, this is not a very practicable approach.

We will consider some analytical representations of second-order characteristics for two-phase media. For statistically homogeneous media, the functions (also called autocorrelation functions) $S_2^{(i)}(\mathbf{y}_1, \mathbf{y}_2) \equiv S_2^{(i)}(\mathbf{y})$ $(i = 0, 1)$ provide a measure of how the end points of a vector $\mathbf{y} = \mathbf{y}_1 - \mathbf{y}_2$ fall in phase $v^{(i)}$. For statistically isotropic media, one point can be taken at the origin and the functions $S_2^{(i)}(\mathbf{y}) \equiv S_2^{(i)}(|\mathbf{y}|)$ are invariant with respect to the direction of the vector $\mathbf{y}$ and reach their maximum value of $c^{(i)}$ at $|\mathbf{y}| = 0$ with nonmonotonic decay to their asymptotic value of $(c^{(i)})^2$ in the case of *no long-range order* in the medium being assumed. For statistically homogeneous media, it is convenience to define the autocovariance of phase $v^{(i)}$

$$R^{(i)}(\mathbf{y}) = \langle V^{(i)\prime}(\mathbf{x})V^{(i)\prime}(\mathbf{x}+\mathbf{y})\rangle = S_2^{(i)}(\mathbf{y}) - (c^{(i)})^2 \tag{5.76}$$

taking the limiting values $R^{(i)}(0) = c^{(0)}c^{(1)}$ and $R^{(i)}(\infty) = 0$; here $V^{(i)\prime}(\mathbf{x}) = V^{(i)}(\mathbf{x}) - c^{(i)}$ is the fluctuating part of the random field $V^{(i)}(\mathbf{x})$, and $\langle (V^{(i)\prime})^2(\mathbf{0})\rangle = c^{(0)}c^{(1)}$. It can be easily verified that $R^{(0)}(\mathbf{y}) = R^{(1)}(\mathbf{y})$, and, therefore $S_2^{(1)}(\mathbf{y}) = S_2^{(0)}(\mathbf{y}) + (c^{(1)})^2 - (c^{(0)})^2$. The two-point inclusion probability function $S_2^{(1)}(\mathbf{y}_1, \mathbf{y}_2)$ is connected with the two-point correlation even function $\gamma_2(r)$ ($r = |\mathbf{y}_1 - \mathbf{y}_2|$) for statistically uniform two-phase materials consisting of an isotropic distribution of phases

$$\gamma_2(r) = \langle (V^{(i)\prime})^2(\mathbf{0})\rangle^{-1}\langle V^{(i)\prime}(\mathbf{0})V^{(i)}(\mathbf{x})\rangle = \frac{1}{c^{(1)}c^{(0)}}[S_2^{(1)}(r) - (c^{(1)})^2], \tag{5.77}$$

which has the limiting values $\gamma_2(0) = 1$ and $\gamma_2(\infty) = 0$. For future use, we will represent some properties of the two-point probabilities $S_2^{(ij)}(\mathbf{y}_1, \mathbf{y}_2)$:

$$\begin{aligned} S_2^{(0)}(\mathbf{y}_1, \mathbf{y}_2) + S_2^{(01)}(\mathbf{y}_1, \mathbf{y}_2) &= S_1^{(0)}(\mathbf{y}_1) = S_2^{(0)}(\mathbf{y}_2, \mathbf{y}_1) + S_2^{(01)}(\mathbf{y}_2, \mathbf{y}_1), \\ S_2^{(01)}(\mathbf{y}_2, \mathbf{y}_1) + S_2^{(1)}(\mathbf{y}_2, \mathbf{y}_1) &= S_1^{(1)}(\mathbf{y}_1) = S_2^{(10)}(\mathbf{y}_1, \mathbf{y}_2) + S_2^{(1)}(\mathbf{y}_1, \mathbf{y}_2), \end{aligned} \tag{5.78}$$

where $S_1^{(0)}(\mathbf{y}) + S_1^{(1)}(\mathbf{y}) = 1$. For statistically homogeneous media

$$\begin{aligned} S_2^{(0)} &= (c^{(0)})^2 + c^{(0)}c^{(1)}g_2(\mathbf{y}), \ S_2^{(01)} = c^{(0)}c^{(1)}(1 - g_2(\mathbf{y})), \\ S_2^{(1)} &= (c^{(1)})^2 + c^{(0)}c^{(1)}g_2(\mathbf{y}), \end{aligned} \tag{5.79}$$

where $g_2(\mathbf{y})$ is a function of $r = |\mathbf{y}|$ only if the distribution of phases is isotropic. It should be mentioned that this statement does not refer to the elastic symmetries of the phases so that the composite could still be elastically anisotropic.

Sets of expansions are known that yield systematic bounds for material and transport properties [1523] in terms of the $S_n^{(i)}$ and related functions for statistically homogeneous media. The results can then be re-expressed in terms of $\varphi(\mathbf{x}_1, \ldots, \mathbf{x}_n)$ if the equation $S_n^{(i)}(\mathbf{x}^n) \sim \varphi(\mathbf{x}^n)$ is known. In various applications it is necessary to interconnect these two kinds of statistical descriptions. For example, the functions $\varphi(\mathbf{x}_1, \mathbf{x}_2)$ and $\varphi(\mathbf{x}_1, \mathbf{x}_2, \mathbf{x}_3)$ are usually assumed to be known, and the integrals involving the two- and three-point moments $S_2^{(1)}(\mathbf{x}_1, \mathbf{x}_2)$ and $S_3^{(1)}(\mathbf{x}_1, \mathbf{x}_2, \mathbf{x}_3)$ should be estimated in bounds on the effective properties of particular composites. Torquato and Stell [1527] have related the n-point matrix probability function $S_N^{(0)}$ to multidimensional integrals over the infinite set of n-particle probability densities $\varphi(\mathbf{x}_1, \ldots, \mathbf{x}_n)$ of impenetrable spheres of unit radius:

$$S_n^{(0)}(\mathbf{x}_1,\ldots,\mathbf{x}_n) = 1 + \sum_{s=1}^{n} \frac{(-1)^s}{s!} \int \varphi(\mathbf{x}_{n+1},\ldots,\mathbf{x}_{n+s})$$
$$\times \prod_{j=n+1}^{n+s} \left\{1 - \prod_{i=1}^{n}[1 - \chi_{b(0,1)}(\mathbf{x}_{ij})]\right\} d\mathbf{x}_{n+j}. \qquad (5.80)$$

In the framework of the simplest "two-point" level, Markov and Willis [1014] demonstrated a simple interconnection expressing $S_2^{(1)}(\mathbf{y}^2)$ for two-phase media as a simple one-tuple integral containing the radial distribution function $g(r)$ (5.30). They obtained the representation for the two-point correlation function $\gamma_2(r)$ (5.77) $\gamma_2(r) = \gamma_2^{\rm ws}(r) + \tilde{\gamma}_2(r)$. For the well-stirred approximation (5.63), the constituent $\tilde{\gamma}_2(r) \equiv 0$ vanishes and the representation (5.77) is reduced to the formula ($\rho = r/a$)

$$\gamma_2(r) = \gamma_2^{\rm ws}(r) \equiv \begin{cases} 1 - \frac{3\rho}{4(1-c)} + \frac{(1+3c)\rho^3}{16(1-c)} - \frac{9c\rho^4}{160(1-c)} + \frac{c\rho^6}{2240(1-c)} & (0 \le \rho \le 2), \\ \frac{c}{1-c}\frac{(\rho-4)^2(36-34\rho-16\rho^2-\rho^3)}{2240\rho} & (2 \le \rho \le 4), \\ 0 & (4 \le \rho). \end{cases} \qquad (5.81)$$

obtained in [1527]. For an arbitrary $g(r)$,

$$\tilde{\gamma}_2(r) = \frac{3c}{160\rho(1-c)} \begin{cases} \int_2^{\rho+2} f(\rho-\tau)\tau h(\tau)\, d\tau, \\ \int_2^{\rho+2} f(\rho-\tau)\tau h(\tau)\, d\tau + \int_2^{\rho} f(\rho-\tau)\tau h(\tau)\, d\tau, \\ \int_{\rho-2}^{\rho+2} f(\rho-\tau)\tau h(\tau)\, d\tau + \int_{\rho}^{\rho+2} f(\rho-\tau)\tau h(\tau)\, d\tau, \end{cases} \qquad (5.82)$$

at $0 \le \rho \le 2$, $2 \le \rho \le 4$, and $4 \le \rho$, respectively; here $h(r) = g_2(r) - 1$ and $f(t) = (2+t)^3(4-6t+t^2)$. Thus, the two-point correlation $\gamma_2(r)$ is represented as a simple one-tuple integral containing the radial distribution function of statistically isotropic dispersion of nonoverlapping spheres. Markov and Willis [1014] mentioned that due to the well-known positive-definiteness of the two-point correlation function $\gamma_2(r)$ for any statistically homogeneous medium, the integral

$$\int_0^{\infty} r^2 \gamma_2^{\rm ws}(r) dr = \frac{(1-8c)}{3(1-c)} a^3 \qquad (5.83)$$

(which can obviously be estimated through an elementary integration of Eq. (5.81)) should be non-negative which is violated at $c > 1/8$.

Generalization of Eqs. (5.77)–(5.80) was obtained in [1266], where the model of impenetrable ellipsoidal aligned particles within correlation spherical holes v_i^0 ("security" spheres) was used to derive $S_2^{(0)}(\mathbf{x}_1, \mathbf{x}_2)$. This assumption is more restrictive than simply requiring the particles to be nonoverlapping, but it made possible the use of well-known statistical models for spherical inclusions (5.77)–(5.80); Quintanilla [1266] used the Fourier transform of the density function and the Percus–Yevick–Wertheim approximation for the radial distribution function. Monetto and Drugan [1089] obtained an explicit analytical result for the two-point correlation function. In the framework of the same assumption of spherical shape v_i^0 and aligned inclusions, they separated the effects of inclusion shape and their spatial distribution and incorporated the Verlet-Weis improvement to the standard Percus-Yevick-Wertheim approximation.

5.3.4 Geometrical Description and Averaging of Doubly and Triply Periodic Structures

It is assumed that the representative mesodomain w contains a statistically large number of particles $v_i \subset v^{(1)}$ $(i = 1, 2, \ldots)$ with identical shape, orientation, and mechanical properties (i.e., the unmarked point field is analyzed). We now consider a composite medium with particle centers distributed at the nodes of some spatial lattice Λ. Suppose $\mathbf{e}_i$ $(i = 1, 2, 3)$ are linearly independent vectors, so that we can represent any node $\mathbf{m} \in \Lambda$:

$$\mathbf{x_m} = f_1(m_1)\mathbf{e}_1 + f_2(m_2)\mathbf{e}_2 + f_3(m_3)\mathbf{e}_3, \tag{5.84}$$

where $\mathbf{m} = (m_1, m_2, m_3)$ are integer-valued coordinates of the node $\mathbf{m}$ in the basis $\mathbf{e}_i$ which are equal in modulus to $|\mathbf{e}_i|$, and $f_i(m_i) - f_i(m_i+1) \not\equiv$ const. $(i = 1, 2, 3)$.

For triply periodic structures with linear-independent vectors of the principal period of Λ determining a unit cell Ω of volume $\bar{\Omega} = |\mathbf{e}_1 \cdot (\mathbf{e}_2 \otimes e_3)|$, we can represent any node $\mathbf{m} \in \Lambda$ in the form

$$\mathbf{x_m} = m_1\mathbf{e}_1 + m_2\mathbf{e}_2 + m_3\mathbf{e}_3. \tag{5.85}$$

If, for example, the basis $\mathbf{e}_i$ is orthonormal, and the coefficients $\mathbf{m} = (m_1, m_2, m_3)$ are the integer set Z^3, independent of one other; Λ defines a simple cubic (SC) packing; in the case where the coefficients m_i $(i = 1, 2, 3)$ are either all even or odd, we have a body-centered cubic structure (BCC); a cubic face-centered structure (FCC) is obtained in the case where the coefficients m_i are either all even or two are odd, while the third is even. The method of assigning the lattice Λ is also possible where several nodes are located within the limits of a cell, and the coefficients m_i are the integer set Z^3, independent of one another [868].

For doubly periodical structures

$$\mathbf{x_m} = m_1\mathbf{e}_1 + m_2\mathbf{e}_2 + f_3(m_3)\mathbf{e}_3, \tag{5.86}$$

where $f_3(m_3) - f_3(m_3+1) \neq$ const. In the plane $f(m_3) =$ const. the composite is reinforced by periodic arrays Λ_{m_3} of inclusions in the direction of the $\mathbf{e}_1$ axis and the $\mathbf{e}_2$ axis. The type of lattice Λ_{m_3} is defined by the law governing the variation in the coefficients m_i $(i = 1, 2)$ and also by the magnitude and orientation of the vectors $\mathbf{e}_i$ $(i = 1, 2)$. In the functionally graded direction $\mathbf{e}_3$, the inclusion spacing between adjacent arrays may vary ($f_3(m_3) - f_3(m_3+1) \not\equiv$ const.). For a doubly periodic array of inclusions in a finite ply containing $2m^l + 1$ layers of inclusions, we have $f(m_3) \equiv 0$ at $|m_3| > m^l$; in the more general case of doubly periodic structures $f(m_3) \not\equiv 0$ at $m_3 \to \pm\infty$. To make the exposition more clear, we will assume that the basis $\mathbf{e}_i$ is an orthogonal one and the axes $\mathbf{e}_i$ $(i = 1, 2, 3)$ are directed along axes of the global Cartesian coordinate system (these assumptions are not obligatory).

The statistical description of random structures presented above can also be used for periodic structures by allowing the ensemble distribution function to be expressed in terms of proper delta-functions, when, e.g., the configuration $\mathcal{C}^N$ (5.85) has a distribution function $\varphi(\mathbf{x}) = \sum_{i=1}^{N} \delta(\mathbf{x} - \mathbf{x}_i)$ (see [1454]).

The composite material is constructed using the building blocks or cells: $w = \cup\Omega_{\mathbf{m}}$ containing the inclusions $v_{\mathbf{m}} \subset \Omega_{\mathbf{m}}$. Hereafter the notation $\mathbf{f}^{\Omega}(\mathbf{x})$ will be used for the average of the function $\mathbf{f}$ over the cell $\mathbf{x} \in \Omega_i$ with the center $\mathbf{x}_i^{\Omega} \in \Omega_i$:

$$\mathbf{f}^{\Omega}(\mathbf{x}) = \mathbf{f}^{\Omega}(\mathbf{x}_i^{\Omega}) \equiv n(\mathbf{x}) \int_{\Omega_i} \mathbf{f}(\mathbf{y})\, d\mathbf{y}, \quad \mathbf{x} \in \Omega_i, \tag{5.87}$$

$n(\mathbf{x}) \equiv 1/\overline{\Omega}_i$ is the number density of inclusions in the cell Ω_i.

Let $\mathcal{V}_{\mathbf{x}}$ be a "moving averaging" cell (or moving-window [563]) with the center $\mathbf{x}$ and characteristic size $a_{\mathcal{V}} = \sqrt[3]{\overline{\mathcal{V}}}$, and let for the sake of definiteness $\boldsymbol{\xi}$ be a random vector uniformly distributed on $\mathcal{V}_{\mathbf{x}}$ whose value at $\mathbf{z} \in \mathcal{V}_{\mathbf{x}}$ is $\varphi_{\boldsymbol{\xi}}(\mathbf{z}) = 1/\overline{\mathcal{V}}_{\mathbf{x}}$ and $\varphi_{\boldsymbol{\xi}}(\mathbf{z}) \equiv 0$ otherwise. Then we can define the average of the function $\mathbf{f}$ with respect to translations of the vector $\boldsymbol{\xi}$:

$$\langle \mathbf{f} \rangle_{\mathbf{x}}(\mathbf{x} - \mathbf{y}) = \frac{1}{\overline{\mathcal{V}}_{\mathbf{x}}} \int_{\mathcal{V}_{\mathbf{x}}} \mathbf{f}(\mathbf{z} - \mathbf{y})\, d\mathbf{z}, \quad \mathbf{x} \in \Omega_i. \tag{5.88}$$

Among other things, "moving averaging" cell $\mathcal{V}_{\mathbf{x}}$ can be obtained by translation of a cell Ω_i and can vary in size and shape during motion from point to point. Clearly, contracting the cell $\mathcal{V}_{\mathbf{x}}$ to the point $\mathbf{x}$ occurs in passing to the limit $\langle \mathbf{f} \rangle_{\mathbf{x}}(\mathbf{x} - \mathbf{y}) \to \mathbf{f}(\mathbf{x} - \mathbf{y})$. To make the exposition more clear we will assume that $\mathcal{V}_{\mathbf{x}}$ results from Ω_i by translation of the vector $\mathbf{x} - \mathbf{x}_i^{\Omega}$; it can be seen, however, that this assumption is not mandatory.

5.3.5 Representations of ODF

To detail the properties of the general density function $\varphi(v_m \mathbf{x}_m|; \mathbf{x}_1, \dots, \mathbf{x}_n)$, one uses different notions. In the simplest case, the one-point function $\varphi(v_m, \mathbf{x})$ depends only on the orientation but not on the position $\mathbf{x}$ of a volume element in the sample. So, the orientation distribution function (ODF) f specifying the volume fraction of crystals with an orientation placed between $\mathbf{g}$ and $\mathbf{g}+d\mathbf{g}$ is $dV/V \equiv f(\mathbf{g})\, d\mathbf{g}$, where $d\mathbf{g}$ is the invariant orientation probability measure (Haar measure) in SO(3) [536] of dV which is the total of all volume elements of the sample with the orientation $\mathbf{g}$ and V is the total sample volume. Arguably, the most representative is volume-fraction distribution (including as a particular case the orientation distribution function) describing the one-point statistics of elements in the structure. The term fundamental zone of crystal orientation expressed by the symbol $SO(3)/G$ refers to the set of all physically distinct orientations of the local crystal that can occur in nature. In the absence of any texture, all rotations of the Euler-angle parametrization belong to the fundamental zone of crystal orientation. If the shape and the size of grains are not altered, then $\varphi(v_m) = f$. The ODF describes the probability density for the occurrence of specified crystallographic orientation in a composite material; it contains no information about positional or morphological measures of the microstructure such as the shape and the size of inclusions or grains. However, some other recent studies have combined the morphological and orientational measures of structure to a limited degree. The misorientation distribution function (MDF) gives the probability density for the occurrence of specified intercrystalline misorientation between adjacent grains in the polycrystals. The MDF is closely connected with structure of intercrystalline boundaries and is not predicted from the ODF. Owing to a significant influence across intercrystalline boundaries which occurs in the consideration of many material properties, microstructural measures containing a positional dependence in conjunction with the orientation are widely used, and the two-point conditional probability function $\varphi(v_j|\mathbf{x}_j; \mathbf{x}_i)$ and two-point orientation coherence (or correlation) function (OCF) [7] are exploited. This function gives the probability density for the simultaneous occurrence of crystallites of specified orientation $\mathbf{g}_i$ and $\mathbf{g}_j$ with the centers in the points $\mathbf{x}_i$ and $\mathbf{x}_j$ independently located in a specified measurement volume separated by a vector $\mathbf{x}_j - \mathbf{x}_i$, and contains morphological information such as inclusion (or crystalline) size and shape as a function of orientation. The OCF is reduced to the ODF upon integration over the domain of one set of orientation variables; the OCF is reduced to $\varphi(v_j|\mathbf{x}_j; \mathbf{x}_i)$ for the fixed shape of the grain and their location centers $\mathbf{x}_j$ and $\mathbf{x}_i$. For the remote centers $|\mathbf{x}_i - \mathbf{x}_j| \to \infty$, the orientation coherence relationship is absent, and the OCF equals the product of the ODF $f(\mathbf{x}_i)f(\mathbf{x}_j)$.

The ODF is used as a "weight function" in the estimation of the average of some function $\mathbf{F}$ over the group O_3:

$$\langle \mathbf{F} \rangle^{\mathbf{g}} = \int \mathbf{F}(\mathbf{g}) f(\mathbf{g})\, d\mathbf{g} \tag{5.89}$$

The ODF f is chosen such that for the one-phase polycrystals (when $\langle\cdot\rangle^{\mathbf{g}} \equiv \langle\cdot\rangle$), the integral of the function $\mathbf{F} = \mathbf{I}$ is unity, i.e.,

$$\int f(\mathbf{g})\, d\mathbf{g} = 1. \tag{5.90}$$

The invariant measure for the Euler angle parameters (ϕ_1, θ, ϕ_2) (21.1.4) is $d\mathbf{g} = \sin\theta\, d\phi_1\, d\theta\, d\phi_2$. For the axis-angle parametrization (21.1.2), the invariant measure is [1092]: $d\mathbf{g} = \left[\pi(1+\rho^2)/\rho\right]^{-2} d\rho\, d\mathbf{n}$, where $\rho = |\tan(\omega/2)|$ is the magnitude of the Rodrigues vector $\boldsymbol{\rho}$ (see Sect. 21.1) and $d\mathbf{n}$ is the invariant measure of an infinitesimal patch lying on the unit sphere S^2 of directions $\mathbf{n} = \sin\theta\cos\psi\mathbf{e}_1 + \sin\theta\sin\psi\mathbf{e}_2 + \cos\theta\mathbf{e}_3$ (see Sect. 21.1); for the spherical coordinate representation $d\mathbf{n} = \sin\theta\, d\theta\, d\psi$.

For the fibers, the ODF should also comply with the physical condition of periodicity, because one end of the fiber is indistinguishable from the other $f(\theta, \phi_1, \phi_2) = f(\pi - \theta, \phi_1 + \pi, \phi_2)$ The ODF is a complete and unambiguous description of the fiber orientation state. In particular, for uniformly distributed fiber orientation with $f \equiv$ const. and the Euler angles parametric representation for the rotation group, the condition (5.90) leads to the equality

$$f\, d\mathbf{g} = (8\pi^2)^{-1}\sin(\theta)\, d\theta\, d\phi_1\, d\phi_2, \tag{5.91}$$

which may be termed as a completely random distribution. Let the preferable direction be the Ox_1-axis

$$f\, d\mathbf{g} = (2\pi)^{-1}\delta(\theta - \pi/2)\delta(\phi_1 - 0)\, d\theta\, d\phi_1\, d\phi_2, \tag{5.92}$$

which represents the situation in which all fibers are aligned and parallel to the Ox_1-axis, and hence fibers are free to rotate only with respect to the angle ϕ_2. The ODF

$$f\, d\mathbf{g} = (2\pi)^{-1}\delta(\theta - \pi/2) d\theta d\phi_1 d\phi_2, \tag{5.93}$$

represents the case in which the fibers are uniformly lying on the planes parallel to the plane Ox_1x_2 and is hence a two-dimensional in plane distribution. The ODF (5.92), (5.93), and (5.91) will be termed as the $1D$, $2D$, and $3D$, respectively, uniformly random distributions.

The description using three angles (ϕ_1, θ, ϕ_2) is obviously more general than is necessary for fibers with circular cross sections and transversely isotropic symmetry about the fiber axis. In the last case, there is no ϕ_2 dependence to the orientation distribution. In such a case, the axis-angle parametrization for the rotation matrix $\mathbf{g} \in SO(3)$ (21.1.2) is more convenient. Owing to the symmetry of fibers, $\mathbf{n}(\theta, \psi) = \mathbf{n}(\pi - \theta, \pi - \psi)$. One usually assumes that the fiber orientations are uniformly distributed with respect to ψ, which is usually the case for injection-molded components. An independence of ODF $f(\theta, \psi)$ on ψ implies transverse isotropy, with Ox_3 being the symmetry axis. With this assumption, a two-parameter exponential function was proposed [1001, 1643] to describe the fiber orientation in the injection-molded specimens

$$f(\theta) = \left[\int_{\theta_{\min}}^{\theta_{\max}} (\sin\theta)^{2p-1}(\cos\theta)^{2q-1}\, d\theta\right]^{-1} (\sin\theta)^{2p-1}(\cos\theta)^{2q-1}, \tag{5.94}$$

where $\theta_{\min}$ and $\theta_{\max}$ are the upper and lower limits of the angle θ: $0 \le \theta_{\min} \le \theta \le \theta_{\max} \le \pi/2$ and the parameters $p \ge 0.5$ and $q \ge 0.5$. The ODF (5.94) continuously link the extreme cases of fiber orientations from aligned to uniform random orientation with the most probable fiber orientation angle $\theta_{mod} = \tan^{-1}\{[(2p-1)/(2q-1)]^{1/2}\}$. Among one-parameter axisymmetric ODFs, one can indicate the function $f(\theta) = a\exp\left[-\frac{1}{2}(\theta/s)^2\right]$, where a is a normalized parameter found from the condition (5.90) and s is the standard deviation in a meridional section which can vary from $s \to 0$ (aligned fibers) to $s \to \infty$ (uniform random orientation).

In the general case, the ODF is determined from the complete set of single-orientation measurements called pole figures and is represented as a series expansion (see for details [174]):

$$f(\mathbf{g}) = \sum_{l=0}^{\infty} \sum_{m=-l}^{l} \sum_{n=-l}^{l} C_l^{mn} T_l^{mn}(\mathbf{g}) \tag{5.95}$$

in the generalized spherical harmonics T_l^{mn} that form an orthogonal functional system

$$\int T_{l'}^{m'n'}(\mathbf{g}) T_l^{mn}(\mathbf{g})\, d\mathbf{g} = (2l+1)^{-1}\delta_{ll'}\delta_{mm'}\delta_{nn'}. \tag{5.96}$$

The representation of the ODF in the form of an expansion in a series of generalized spherical harmonics is useful at the analytical estimations of orientation averages $\langle \mathbf{F} \rangle^{\mathbf{g}}$ (5.89).

5.4 Numerical Simulation of Particle Random Composites

Influence of microstructure on mechanical properties of microinhomogeneous media drastically increases for highly packed particulate media such as concentrated suspensions (see, e.g., [333]), concrete (see [28, 528]), dental restorative composite materials (see [939, 1486, 1636]), and solid propellants. Heterogeneous solid propellants (HSPs) commonly used in aerospace propulsion are likely to play an important role for as long as rockets are built. Fundamentally, they consist of multimodal spherical particles of solid oxidizer with extremely high volume fractions (>0.9), typically ammonium perchlorate, dispersed in a rubbery fuel binder fuel such as hydroxi-terminated-polybutadiene (HTPB) or polybutadiene acrylonitrile (PBAN). Designers of rockets are concerned with a number of propellant-related issues, including the burning rate and the thermal and mechanical properties of the propellants, and, for metallized propellants, the behavior of the metal particles at the surface, including agglomeration (see [1003, 1025, 1027, 1028, 1599, 1646]).

The prediction of effective mechanical properties of polymer-bonded explosives provides a new challenge. These particulate composites contain extremely high volume fractions of explosive particles (>0.9) in a polymer binder typically exhibiting viscoelastic behavior (see, e.g., [61]). In explosives, particle sizes range from a few μm for oxidizer particles in the "dirty binder" to 10s of μm for the metal flakes and 100s of μm for the largest oxidizer crystals. The oxidizer (HMX) particle shapes are more polyhedral than spherical (see [1437]). Large dynamic deformation during processing and shock loading can fracture the microstructure and microcracks can create additional internal surfaces or porosity that greatly influences explosive initiation sensitivity (see, e.g., [50]). In many important cases of practical interest (e.g., solid propellants), the shape of the heterogeneities can be assumed as a spherical. Computer simulation of topologically disordered structures by the random packing of hard spherical particles in $2D$ and $3D$ cases has a long history originated in the theory of liquids. This problem is closely connected with the known fundamental problem of statistical physics—description of the behavior of the particle system with interaction potential of hard spheres (see, e.g., [124]). Random packing of spheres has been studied very extensively due to their technological importance, and their capability of modeling the simple liquid, concentrated suspensions, amorphous and powder materials.

Computer simulation of packing problems can be classified into three groups of methods: molecular dynamic, Monte Carlo, and dense random packing (see for references [67, 502]). Much progress in the theory of the dense random packing was approached by the use of two kinds of methods: the sequential generation models and the collective rearrangement models. In the sequential model with the fixed radii of particles proposed in [85] (see also the modified algorithms [149, 699, 750, 970]), so-called cluster growth model, a particle being added to the surface of particle cluster which grows outwards is placed sequentially to the point closest to the original such that the new particle established contact with three existing spheres in the cluster. The phenomenological character of the construction algorithm moved the particle cluster from the initial term containing three particles leads to the inhomogeneous and anisotropic inclusion fields with different densities that was demonstrated in [149]. Moreover, the configurations generated do not demonstrate the characteristic split second peak in the radial distribution function observed in experimental packing.

In the second type of sequential generation model, called the model of "rigid sphere free fall into a virtual box," one particle is dropped vertically each time from the random point onto the surface of an existing particle cluster growing upwards [289, 517, 751, 788, 1133, 1134]. The different densities were approached by introducing a phenomenological parameter limiting the number of rotations of each fallen sphere until it becomes a permanent part of the structure. Effects of boundaries of the virtual box are eliminated by introducing conventional cyclic boundary conditions. The algorithms described belong to the class of static methods where the particles are placed at a given time step and cannot thereafter move. For contrast, dynamic methods assume the reorganization of whole packing due to either short- or long-range interactions between particles.

In the collective rearrangement model, N points randomly distributed in a virtual box are assigned both radii and random motion vectors. Each sphere is moved until there are no overlaps. Then the radii are increased and the process is repeated until any further increase in radii or any displacement of the spheres create overlaps that cannot be eliminated (the different versions of this method can be found, e.g., in [67, 341, 502, 629, 784, 785, 974, 975, 1154, 1704]). Classical molecular dynamics (see, e.g., [502]) can also be considered as a collective rearrangement model (CRM, see Sect. 5.4.4), where particle size is fixed. More recently, numerical simulations were performed to realize homogeneous and isotropic packing of spheres by various methods, for instance, by assuming hypothetical spheres having dual structure whose inner diameter defines the true density and the outer one a nominal density [707]. An alternative approach (eliminating the boundary effect of the virtual box) is based on the use of spherical boundary conditions instead of periodic ones [597, 1517]. There one simulates hard discs (more exactly a circular cap which can be visualized as a contact lens on the surface of an eyeball) on the surface of the ordinary three-dimensional sphere and hard spheres on the "surface" of a four-dimensional hyposphere. The advantage of this procedure is that there is no preferential direction, and it is impossible to pack particles into perfect regular periodic configurations.

The a priori prediction of the maximum packing fraction, c_m, for a system of particles is still an open question, despite much study. Packing densities for close lattice packing of monomodal particles are $\pi/\sqrt{12} \approx 0.9069$ (triangular) in the case of disks packing into the plane and $\pi/\sqrt{18} \approx 0.7405$ (fcc or hcp) in the case of spheres packed into R^3. However, a well-mixed random packing does not self-assemble into one of the periodic arrangements but instead forms a so-called random close-packed (RCP) arrangement which is not well-defined, protocol dependent (in both the numerical simulation and experimental sense, see [41, 112, 327, 1042, 1528]) and can vary in the range $0.59 \div 0.64$ for 3D unimodal spheres. Here the low value corresponds to the loose packs (LRP), while the upper value is a characteristic of so-called dense packs (DRP). A shaking process makes it possible to unlock the spheres from the jamming configuration and allows them to find the most homogeneous and mixed arrangement. So, [1139] demonstrated by an electromagnetic vibration exciter that under vibrations the bead packing evolves from an initial, low-density configuration towards higher density. The measured densities are extrapolated to eliminate finite-size effect that enables one to estimate the DRP's density $c_m = 0.6366 \pm 0.004$ (Stoyan [1451] proposed elegant hypothesis for $c_m = 2/\pi \approx 0.63662$). Various algorithms have been devised to simulate reordering due to shaking or vibration of dense packing (see, e.g., [66, 907]) which reduces the volume concentration of the high density jam configuration.

For multimodal systems, small spheres may be placed into the spaces between packed large spheres. For the infinitely different sizes of packed spheres, the maximum $c_m = 0.637 \cdot (2 - 0.637) = 0.868$. The volume fraction ratio ξ between large and small spheres is also important, with maximum packing obtained at about $\xi = 0.60 \div 0.75$ large particles [1002, 1003, 1370]. So, for the packs with size ratio $\lambda = 6.5 : 1$, one [1003] simulated the packing fractions of 0.687 and 0.731 for the coarse-to-fine volume ratio $\xi = 0.90$ and $\xi = 0.75$, respectively. Trimodal, multimodal, and polydisperse systems can reach even higher packing fractions [1707]). However, packing configurations containing a wide range of inclusion concentrations have been investigated less and invite further investigation.

It should be mentioned that the most experimentally investigated cases of the RDF correspond to the close packing (from the loose packs $c = 0.59$ to dense packs $c = 0.64$), see [40, 41, 127, 552, 842, 1314]. The volume fractions $0 < c < 0.6$ are less well understood, and, moreover, their investigations are usually limited by suspensions with the different effects of interparticle interactions (see, e.g., [1392]) that makes difficult their application to the solid composites. Estimation of the RDF from numerical simulations of random structures with $0 < c < 0.6$ is still less explored (see the use of the hard core model for simulation in [1351]).

The RDF $g(r)$ is well investigated only for identical spherical ($3D$ and $2D$ cases) inclusions with a radius a. Two alternative analytical RDFs of inclusion will be examined for 3D case

$$g(r) = 1 - \chi_{[0,2a)}(r), \tag{5.97}$$

$$g(r) = (1 - \chi_{[0,2a)}(r)) \left[1 + g_c^W(c) g_r^W(r)\right], \tag{5.98}$$

where $\chi_{[0,2a)}(r)$ is the indicator function (5.16) of $[0, 2a)$, $r \equiv |\mathbf{x}_i - \mathbf{x}_q|$ is the distance between the nonintersecting fixed inclusions v_i and moving one v_q, c is the volume fraction of particles of the radius a, $g_r^W(r) = 1$ at $r = 2a^+$, and $g_c^W(c) \equiv g(2a^+) - 1$. The formula (5.97) describes a well-stirred approximation (differing from the RDF for a Poisson distribution by the availability of "excluded volume" with the center $\mathbf{x}_i$ where $g(|\mathbf{x}_i - \mathbf{x}_q|) \equiv 0$), while Eq. (5.98), proposed by Willis [1621], takes into account a neighboring order in the distribution of the inclusions (see also [1445]):

$$g_c^W(c) = g_{c0}^W(c) \equiv 0.8006 \left(\frac{8c - 1}{c}\right), \quad g_r^W(r) = \cos\left(\frac{\pi r}{a}\right) e^{2(2-r/a)}. \tag{5.99}$$

The restriction is that $c \geqslant 1/8$, when $g_{c0}^W(c) \geqslant 0$ (12_1). For avoiding of this restriction, one [739] proposed a modification of the formula (5.99_1) when $g_c^W(c)+1$ coincide with the first peak of the popular Percus–Yevick [1198] approximation (see [1337, 1513]): $g_c^W(c) = g_c^{PY}(c) \equiv (1+c/2)/(1-c^2)^{-1}$. Equation (5.98) (following [739]) with the representations $g_c^W(c)$ and $g_r^W(r)$

$$g(r) = [1 - \chi_{[0,2a)}(r)]\Big[1 + \Big(\frac{2+c}{2(1-c)^2} - 1\Big)\cos\Big(\frac{\pi r}{a}\Big)e^{2(2-r/a)}\Big] \tag{5.100}$$

is called the Willis approximation. The representation for the 2D case of identical circle system [604, 1525] is also known

$$g(r) = [1 - \chi_{[0,2a)}(r)]\Big\{1 + \frac{4c}{\pi}\Big[\pi - 2\sin^{-1}(\frac{r}{4a}) - \frac{r}{2a}\sqrt{1 - \frac{r^2}{16a^2}}\Big]\chi_{[0,4a]}(r)\Big\}. \tag{5.101}$$

In this section, we will follow [237, 255].

5.4.1 *Materials and Image Analysis Procedures*

A digital image processing technique is considered for measurement of centroid coordinates of fibers with subsequent estimation of statistical parameters and functions describing the stochastic structure of a real fiber composite. The fiber volume $c^{(1)}$ and spatial distribution (e.g., $S_1^{(i)}(\mathbf{y})$, $S_2(i)(\mathbf{y}^2)$, $i = 0, 1$) are determined from the image analyzer that is capable of calculating the fiber volume and the coordinate of each fiber [1416, 1673]. To determine $S_1^{(i)}$, a simple Monte-Carlo like simulation can be utilized for a statistically homogeneous medium $S_1^{(i)} = \hat{N}^{(i)}/\hat{N}$, where $\hat{N}$ is a total number of random throws into a sample w and $\hat{N}^{(i)}$ denotes the number of successful hits into the phase $v^{(i)}$. Some sort of the naive *minus-samping* method (5.27) can be employed to determine $S_2^{(r)}(\mathbf{y}^2)$ $(r = 0, 1)$ from a digitized micrograph imagined as a discretization of the characteristic function $V^{(r)}(\mathbf{y})$, usually presented in terms of a $M \cdot N$ bitmap. Denoting the value of $V^{(r)}$ for a pixel located in the ith row and j-column as $V^{(r)}(i, j)$ allows writing the first two statistical descriptors as

$$S_1^{(r)} = \frac{1}{MN}\sum_{i=0}^{M-1}\sum_{j=0}^{N-1} V^{(r)}(i, j),$$

$$S_2^{(r,s)}(m, n) = \frac{1}{(i_M - i_m)(j_N - j_n)}\sum_{i=i_m}^{i_M-1}\sum_{j=j_n}^{j_N-1} V^{(r)}(i, j)V^{(s)}(i + m, j + n), \tag{5.102}$$

where $i_m = \max(0, -m)$, $i_M = \min(M, M - m)$, $j_n = \max(0, -n)$, $j_N = \min(N, N - n)$; $r, s = 0, 1$.

Digital images of transverse sections through the fiber composite material were obtained by digitizing high-resolution optical micrographs, using standard methods [1367, 1434, 1504, 1655]. Care was taken to maximize the contrast between the fibers and matrix in the original micrograph. One [255] described in detail the analysis of the 1024×1024 digital images using a commercially available desktop software package (Adobe Photoshop 5.5) in conjunction with a plug-in Image Processing Tool Kit (IPTK; see [1309]). Once the image is reduced to an array of black pixels corresponding to the centroid positions of each fiber profile, it is a simple matter to extract the coordinates, using the IPTK software or any other suitable computer code. The coordinates are output as integer pairs in the range [1, 1024] and are subsequently renormalized for input into the microstructural models.

Since random packed structures are strongly dependent on the method of their generation, we will consider a few popular algorithms and their combinations and will compare the statistical parameters of configurations generated by each different method.

5.4.2 *Hard-Core Model*

The extension of the Poisson distribution process as a static model is the generation of random assemblies of n nonoverlapping discs by the *Hard-Core Model* (HCM) (called also random sequential adsorption model and simple sequential inhibition) [463, 647]: discs with radius a are placed one by one with the center positions $X = (\mathbf{x}_1, \ldots, \mathbf{x}_N)$ being distributed randomly and uniformly over the set of all points in a rectangular region ω of size $[-0.5, 0.5] \times [-0.5, 0.5]$. Although the distribution of discs depends heavily on the shape and size of w, usually researchers assume a homogeneous structure in the whole space which is observed only in w.

To avoid this discrepancy, one supposes the periodic boundary conditions, that is, ω and X are periodically replicated in all directions. (see for details [519, 1443]). In such a case, an alternative *toroidal edge correction* (see [1296]) is often used in which each 2D rectangular box w can be regarded as a torus, so that points on opposite edges are considered to be closed (a 3D case is analyzed in a similar manner). Then w can be considered to be part of a grid of identical boxes, forming a border around the pattern inside w. If a moving particle leaves the central simulation box, then one of its image particles enters the central box from the opposite direction. Each of the image particles in the neighboring boxes (the numbers of these neighboring boxes equal 8 and 26 in 2D and 3D case, respectively) moves in exactly the same way. Precisely this method will be explored hereafter in this chapter for the elimination of the boundary effect. So, we start with a cube $w = \{[-1, 1]^d\}$ with the center $\mathbf{0} \in R^d$ and its periodically located neighbors labeled by the multiplet of integer numbers $\boldsymbol{\alpha} = (\alpha_1, \ldots, \alpha_d) \in Z^+$ where $\alpha_i = 0, \pm 1$ $(i = 1, \ldots, d)$. N random points $\mathbf{x}_i$ $(i = 1, \ldots, N)$ in the central cube w are periodically reflected into the neighboring cubes $w^{\boldsymbol{\alpha}}$. Distances are then measured taking into account the minimum image convention according to which the real distance between any two particles is given by the shortest distance of any of their images in the surrounding periodic cubes $w^{\boldsymbol{\alpha}}$ (see [1297]): $|\mathbf{x}_i - \mathbf{x}_j| = \min_{\boldsymbol{\alpha}} |\mathbf{x}_i - \mathbf{x}_j^{\boldsymbol{\alpha}}| = \min_{\boldsymbol{\alpha}} |\mathbf{x}_i - \mathbf{x}_j - 2(\alpha_1 + \ldots + \alpha_d)|$

where $\mathbf{x}_j^\alpha \in w^\alpha$ is a periodic mapping of $\mathbf{x}_j \in w$ into w^α. Toroidal edge correction allows us to analyze all N particles in the central box w without using the minus-sampling method because even for the points $\mathbf{x}_i$ placed near the boundary of w all surrounding points are visible. An alternative approach (eliminating the boundary effect of the virtual box w) is based on the use of spherical boundary conditions instead of periodic ones (see, e.g., [597, 1517]). There one simulates hard disks (more exactly a circular cap which can be visualized as a contact lens on the surface of an eyeball) on the surface of the ordinary three-dimensional sphere and hard spheres on the "surface" of a four-dimensional hypersphere. The advantage of this procedure is that there is no preferential direction, and it is impossible to pack particles into perfect regular periodic configurations. The considered both toroidal and spherical edge corrections hold for simulated structures rather than for the real observation window w obtained by visual cutting out from a large sample. In the last case, either the edge correction (2) or minus-sampling method should be used.

If the new disc in the HCM does not overlap already deposited discs, its position is fixed and does not move anymore; otherwise, it is rejected and another random center position is generated. The process is finished when either a preassigned packing fraction is achieved or when no more particles can be added (jamming limit) which occurs at a volume fraction $c \approx 0.55$ (2D case) or $c = 0.38$ (3D case) [969, 1492]. The mathematically formalized descriptions of other versions of the HCM using either a birth-and-death process with vanishing death rate or the Matern's thinning rule can be found in [1453].

The HCM provides a more realistic reference model than a Poisson point process, in which arbitrary small distances between points are allowed. The advantage of protocol independence in the HCM is sacrificed in the case of generation of binary or polydisperse structures [364, 630, 1408]. Indeed, after an unacceptable trial one can keep the previous choice of inclusion radius or choose to replace it. The real situation then lies between two limiting cases of jamming limits, the packing density generated by first placing the larger particles and then the smaller particles will be higher than the opposite case, where the small particles are placed before the big ones. The geometrical blocking effect and the process irreversibility leads to packing configurations that are essentially different from the corresponding equilibrium configuration [1617]. To prevent just this kind of low-density jamming in the following model, we will shake the discs.

5.4.3 Hard-Core Shaking Model (HCSM)

Hard-core shaking model (HCSM) is a type of HCM that generates an increasing number of inclusions in a virtual box w accompanied by a shaking process (also called Monte-Carlo simulation and local-update algorithms [66, 985], see also [502]), i.e., giving each disc a small random displacement independent of its neighbors' positions. This makes it possible to unlock the discs from the jamming configuration (which takes place for HCM at $c \approx 0.55$) and allows them to find the most homogeneous and mixed arrangement.

For speeding up the calculations, it is necessary to carry out the shaking process only within a local shake up window [255] rather than within a whole domain w [320]. The random local shaking of a particle $v_i \subset v^{(k)}$ $(k = 1, 2)$ is established in a shake up window $w_{\rm sh}^{(k)} = \{\mathbf{x} : \ |\mathbf{x} - \mathbf{x}_i| < R_{\rm sh}^{(k)} a^{(k)}\}$ whereby the selected particle center $\mathbf{x}_i$ is randomly moved to a new position $\mathbf{x}_i'$ uniformly distributed in the window $w_{\rm sh}^{(k)}$ with the normalized radius $R_{\rm sh}^{(k)}$ =const. $(k = 1, 2)$ depending on the size of spherical particle in the multimodal system. The displacement $\mathbf{x}_i' - \mathbf{x}_i$ can be chosen as a randomly oriented vector within a sphere with radius $R_{\rm sh}^{(k)} a^{(k)}$ which permits control over the efficiency of the simulation. If the particle does not overlap with any other inclusions the shaking is accepted, otherwise the trial shaking is repeated until the number of attempted trial shakings exceed some limit $N_{\rm sh}^{(k)}$. A valid shaking scheme must be ergodic. This means that there is a path in phase space from every state to every other state, via a succession of trial moves.

Another way to speed up the calculations is to check the collision partners v_j $(j = j_1^i, \ldots, j_{n_t}^i)$ of v_i only in some restricted neighborhood of $\mathbf{x}_i$ called testing window $w_{\rm t}^{(km)} = \{\mathbf{x} : \ |\mathbf{x} - \mathbf{x}_i| < r_{\rm t}^{(km)}\}$ $(r_{\rm t}^{(km)} = (1 + R_{\rm sh}^{(k)}) a^{(k)} + a^{(m)} =$ const., $v_i \subset v^{(k)}$, $v_j \subset v^{(m)})$ which was introduced for unimodal particles in 2D case in [255]. Thus for bimodal packs of particles $v^{(1)}$ and $v^{(2)}$, we have four different sorts of testing windows $w_{\rm t}^{(km)}$ $(k, m = 1, 2)$. A similar concept of a near-neighbor list (see [384]) can be considered as a generalization of testing window to the case of nonspherical particles. In this model, the neighbors of a particle (near-neighbor list, $\mathcal{N}(i)$) are considered to be the set of particles whose centers lie within some maximum distance called the testing window $w_{\rm t}^i$ of that particle rather than the "geometric neighbors" which can be determined by the more computationally expensive Delaunay tessellation (see, e.g., [1157]). One [384] demonstrated advantage of the near-neighbor list method with respect to a traditional cell method (see, e.g., [20]) used for speeding up for the neighbor search. One wants to have the testing window $w_{\rm t}^{(km)}$ as large as possible so that there is more room for the particle $v_i \subset v^{(k)}$ to move without leaving of $w_{\rm t}^{(km)}$. However, for the larger $w_{\rm t}^{(km)}$, the more neighbors there will be to examine. The optimal balance, as determined by the choice of $w_{\rm t}^{(km)}$ $(k, m = 1, 2)$, is chosen numerically at the testing of concrete packs.

Only the near neighbor set of inclusions v_j $(j = j_1^i, \ldots, j_{n_{\rm t}}^i$; $n_{\rm t}$ depends on $k, m)$ which are located in the testing window $w_{\rm t}^{(km)}$ centered at the point $\mathbf{x}_i$ $(v_i \subset v^{(k)}$, $v_j \subset v^{(m)}$, $k, m = 1, 2)$ are checked for overlap, which also reduces the computer time. One determines the optimal size for the shake up window $R_{\rm sh}^{(k)} a^{(k)}$ which provides the minimum average number of trial shaking attempts. This size of shake up window provides the fastest stabilization of statistical parameter estimations. However, the values of these parameters do not depend on the size of the shaking window and are, furthermore, protocol independent (in contrast with the known methods of random close packing simulations, see, e.g.,

[1528]). The shaking process passes through all inclusion (so called global shaking) with re-estimation of the neighbor inclusions in the testing window $w_t^{(km)}$ after each local shaking of the inclusion v_i. In so doing, the increasing number of shaking has a twofold effect. On the one hand, the system becomes more homogeneous and well-mixed And, on the other hand, the stochastic fluctuations of statistical parameters (such as $g_{km}(r)$) estimated by averaging of these structures tend to diminish.

5.4.4 *Collective Rearrangement Model (CRM)*

Collective rearrangement model (CRM) was previously proposed in [975] for the packs of unimodal spheres and generalized in [784] and [1003] to polydisperse packs (application to heterogeneous solid propellants). Just for completeness, we will briefly reproduce the last version of the CRM accompanied by the shaking procedure which is adapted to our goal to generate the most homogeneous and highly mixed configurations.

Let N random points $\mathbf{x}_i$ $(i = 1, \ldots, N)$ be located in a central box, periodically reflected into neighboring boxes. Assign initial velocities $\mathbf{v}_i = (v_{i1}, \ldots, v_{id})$ to each random point, with velocity components that are independently distributed at random between -1 and $+1$ and to the uniformly growing inclusions $v_i \subset v^{(k)}$ $(k = 1, 2)$ with the radii $a^{(k)}(t) = a_0^{(k)}t$. The centers of the inclusions move according to the equations $d\mathbf{x}/dt = \mathbf{v}_i$ with a discontinuous change of the vectors $\mathbf{v}_i$ at the moment the particle exits through a face of a central box as well as during collisions with other inclusions. The collision time is obtained from the condition that the separation distance is the current diameter, that for the inclusions $v_i \subset v^{(k)}$ and $v_j \subset v^{(m)}$ $(k, m = 1, 2)$ is

$$|\mathbf{x}_i + \mathbf{v}_i\Delta t - \mathbf{x}_j - \mathbf{v}_j\Delta t| = (a_0^{(k)}t + a_0^{(m)}t + a_0^{(k)}\Delta t + a_0^{(m)}\Delta t). \tag{5.103}$$

In the case of the collision, the smaller positive root Δt of Eq. (5.103) defines the collision time τ_{ij}. If the collision takes place between the inclusions v_i and v_j from central and neighboring squares (e.g., identified by $(-1, 1, 0)$), respectively, then $\mathbf{x}_j$ in Eq. (5.103) should be replaced by $\mathbf{x}_j + (-1, 1, 0)$ with a subsequent estimation of a collision time τ_{ij}. The exit time τ_i^Γ is estimated as the smallest positive time for exiting of the inclusion v_i being considered through one of the sides of the central box. The determination of $\Delta t_* = \min(\tau_{ij}, \tau_i^\Gamma)$ allows us to estimate the new inclusion radii $a_0(t + \Delta t_*)$ and identify either the colliding (v_{i*} and v_{j*}) or exiting (v_{i*}) inclusions for which the new velocities and locations are re-estimated. For all other inclusions, the position vectors are updated: $\mathbf{x}_i \rightarrow \mathbf{x}_i + \mathbf{v}_i\Delta t$. For these inclusions $\Delta t = \Delta t_*$ and velocities $\mathbf{v}_i$ remain unchanged. In addition, the collision and exit times are corrected $\tau_{ij} \rightarrow \tau_{ij} - \Delta t_*$ and $\tau_i^\Gamma \rightarrow \tau_i^\Gamma - \Delta t_*$, respectively. An optimized cell method with analysis of only collisions between particles in neighboring cells is used for reduction of the computation time. The details of the re-estimation of velocities for the inclusions v_{i*} and v_{j*} based on the conservation laws of the momentum and energy can be found in [784].

At the next step, the inclusion shaking is carried out with introduction of the testing window $w_{\rm t}^{(km)}$ allows one to improve an efficiency of the shaking simulation if the displacement $\mathbf{x}_i' - \mathbf{x}_i$ is defined by possible collision of $v_i \subset v^{(k)}$ with $v_j \subset w_{\rm t}^{(km)}$. Namely, let $\mathbf{x}_i' - \mathbf{x}_i = \mathbf{u}_i^l \Delta t^l$ ($|\mathbf{v}_i^l| = 1$) be a displacement of $\mathbf{x}_i$ in l's trying of $\mathbf{x}_i$ random shaking where $\Delta t^l = \max_j \Delta t_j^l$, $v_j \subset w_{\rm t}^{(km)}$ and Δt_j^l is obtained from a particular case of Eq. (5.103) ($j \in \mathcal{N}(i)$; $l = 1, \ldots, N_{\rm sh}^{(k)}$) $|\mathbf{x}_i + \mathbf{u}_i^l \Delta t_j^l - \mathbf{x}_j| = a^{(k)} + a^{(m)}$, where the smaller positive root Δt_j^l defines the collision time (or, by other words, the amplitude of l's shaking) of inclusions v_i and v_j. Found Δt^l with the corresponding moving directions $\mathbf{u}_i^l$ allows one to find both the random amplitude of shaking uniformly distributed on the interval $[0, \tau]$ ($\tau = \min(\Delta t, R_{\rm sh}^{(k)})$, $\Delta t = \max_l \Delta t^l$) with the corresponding direction $\mathbf{u}_i^l$ of v_i's moving. Comparison of $\Delta t > R_{\rm sh}^{(k)}$ (or $<$) makes it possible to increase (or decrease, respectively) the size $R_{\rm sh}^{(k)}$ of the shaking window. Moreover, because of the ideally parallel nature of the problem, the second goal of shaking related with diminishing of stochastic fluctuations of statistical parameters estimated can be very effectively reached by using of parallel computing at the standard instruction-level parallelism that is performed in the computer codes developed.

5.4.5 *Numerical Results for Unimodal Distribution*

The analytical Willis RDF (5.100) and Percus-Yervik (P-Y, [1198]) approximation (see numerical results for P-Y approximation in [1513]) are compared with the numerical simulations by the CRM and SCRM (the number of spheres is $n = 2000$) for $c = 0.575$ in Figs. 5.1 and 5.2 for the ranges of a normalized coordinates $2 < R \equiv r/a < 2.2$ and $2.5 < R < 4.5$, respectively. As it is well known, the values $g_c^{PY}(c) = g_c^W(c)$ are underestimations of the corresponding first picks at $r = 2a^+$ obtained by both the CRM and SCRM. It should be mentioned that the Monte Carlo shaking process is both the most structurally influential and, computationally, the most intensive part of the whole packing process. So, packing by the CRM for $n = 2000$ takes a few seconds for a PC with a 2.4 GHz processor while one global shaking requires 5 min and stabilization (rather than elimination of a statistical noise of descriptors evaluated) of statistical parameter estimations is reached after 10 global shakings. Thus, the shaking process is very time-consuming and sensitive to a few competitive variables, such as $R_{\rm sh}^{(1)}$ and $N_{\rm sh}^{(1)}$, which are optimized in [237]. The radial step Δr in the numerical differential (5.30) is also optimized; one demonstrated of the method proposed with respect to [40] (see also [1395]).

As can be seen in Fig. 5.2, the SCRM leads to the smoothing of the RDFs $g^{SCRM}(r, c)$. But the smoothing of the RDFs (see the curve 1 in Fig. 5.2) could be reached significantly cheaper in the framework of the initial version of the CRM by

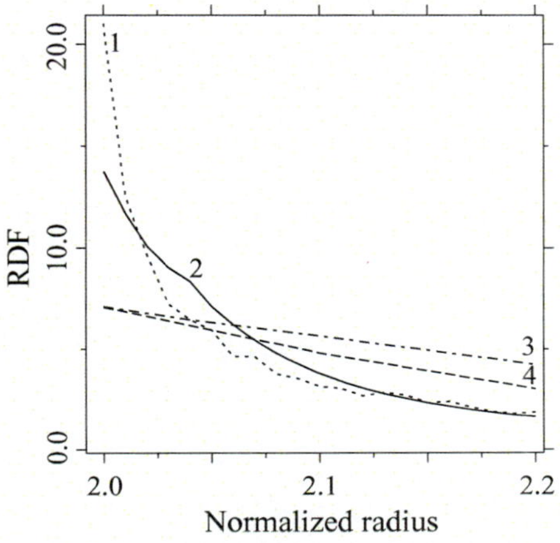

Fig. 5.1 The RDFs $g(r)$ vs r/a at $c = 0.575$ estimated by CRM (1), SCRM (2), Eq. (5.100) (3), P-Y (4)

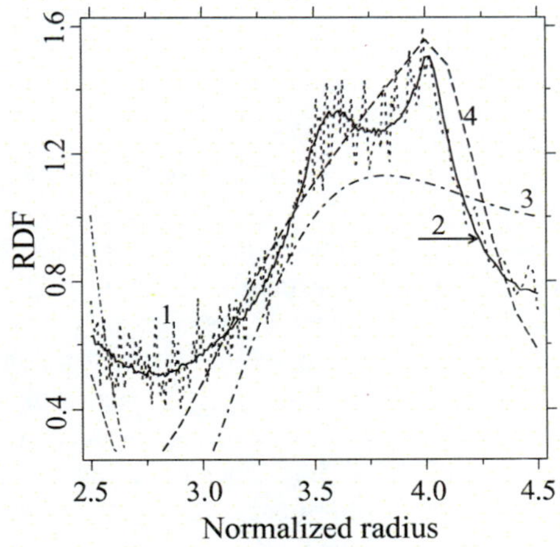

Fig. 5.2 The RDFs $g(r)$ vs r/a at $c = 0.575$ estimated by CRM (1), SCRM (2), Eq. (5.100) (3), P-Y (4)

increasing of either N or the number of realizations X. Nevertheless, any smoothed RDF obtained by averaging of RDFs of configurations generated by the CRM don't remove the birthmarks of a fundamental shortcoming of the CRM such as protocol-dependent and too strong local arrangement in the nearest zone ($r = 2a^+$). In doing so, a fundamental advantage of the SCRM is, first of all, a destruction of dense "locked" local configurations (which are protocol dependent) leading to the most homogeneous and mixed structures with stabilization of estimations of $g(r)$ (see [1388]). It is especially pronounced in Fig. 5.1 where SCRM demonstrates a reduction of $g(2a^+)$ estimated by the CRM of 43% (destruction of the nearest zone arrangement which is responsible for the local jamming).

The RDFs estimated from the CRM and SCRM simulations in Figs. 5.1 and 5.2 have a well-known form (see, e.g., [965, 1003]) with the main maxima appearing at well-defined positions, Namely, at $r = 2a, 2\sqrt{3}a, 4a$ (the first, second, and third

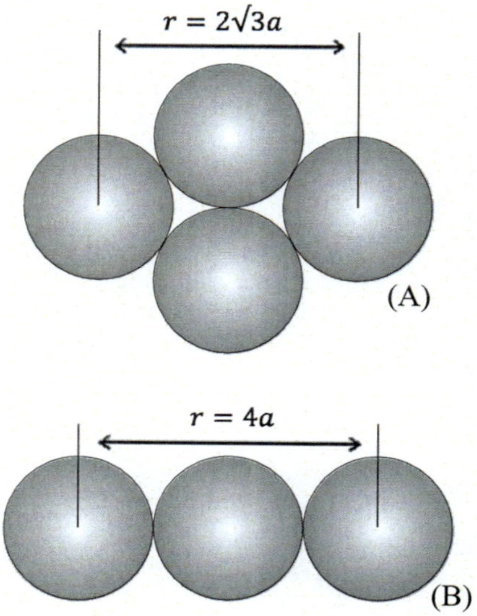

Fig. 5.3 The RDFs $g(r)$ vs r/a at $c = 0.575$ estimated by CRM (1), SCRM (2), Eq. (5.100) (3), P-Y (4)

peaks, respectively), and so on (which can be also presented in a nondimensional form $R = 2, 2\sqrt{3}, 4$) and then $g(r)$ continues to fluctuate with decreasing amplitude. The smoothing of the RDFs obtained from the SCRM simulation was performed by the averaging over $P = 50$ realizations of paralleled simulations. These maxima represent the distances between the centers of a reference sphere to its first, second, third neighbors, etc. The first very pronounced peak (discontinuity) at the hard-core distance $r = 2a$ is a result of a direct contact of spheres which is a most frequent nearest-neighbor distance. The second peak at $r = 2\sqrt{3}a \approx 3.464a$ (see Figs. 5.2 and 5.3a) corresponding to a Face-Centered Cubic (FCC) lattice and appearing at the high packing fraction is completely lost at the low volume fraction of particles. The peak at $r = 4a$ (growing faster with growing c than the second one) sorts well with the placement of three touching spheres in a row (see Fig. 5.3b). The gap distance $r = 2.75a \div 2.85a$ corresponding to the empty space between the first and second neighbors of typical particle is also clearly observable. The third peak at $r = 4a$ is well predicted by the Percus–Yevick [1198] approximation which does not expect the second peak at $r = 2\sqrt{3}a$ for any c. Willis approximation (5.100) (curves 3 in Figs. 5.1 and 5.2) is in agreement with simulation observation (curves 1 and 2) only qualitatively. Experimentally detected for high inclusion concentrations (see [40, 41, 284, 1109]), the split peak demonstrate the presence of large clusters with close locally periodic inclusion packing [1541]. Arrangement of inclusions presenting in Fig. 5.3a, b the fixed left inclusion and defining the second and third, respectively, peaks can be denoted in the forms $v_i - (v_i + v_i) - v_i$ and $v_i - (v_i) - v_i$, respectively. Here the inclusions placed within the parentheses represent the first shell of nearest moving particles.

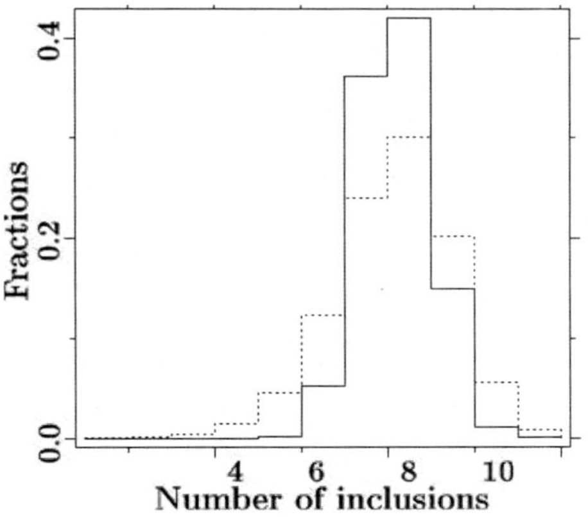

Fig. 5.4 $p(n_{\rm h})$ of testing window: generated by the HCSM (solid curve), experimental data (dotted curve)

It should be mentioned that the number of points in a sphere of radius r about a randomly chosen point $\mathbf{x}_i$ is a random parameter. Estimation of its expectation defining K-function leads to loss of statistical information which can be presented in more details if one uses a histogram form. Namely, we evaluate the histograms of the average number $n_{\rm h}$ of inclusions in the spherical testing window $w_{\rm h}^k = \{\mathbf{x} : |\mathbf{x} - \mathbf{x}_i| < R_{\rm h}^{(k)} a^{(k)}\}$ (i.e., the fractions $p(n_{\rm t})$ of the histogram windows containing the numbers $n_{\rm h}$ of inclusions) as a measure of inhomogeneity of an inclusion distribution inside $w_{\rm h}^k$. Obviously, the descriptor $p(n_{\rm h})$ is a more sensitive measure of the local statistical homogeneity of the configuration analyzed than the $K(r)$ is.

In Fig. 5.4, the histograms of distributions of fibers in the testing window $w_{\rm h}^k = \{\mathbf{x} : |\mathbf{x} - \mathbf{x}_i| < R_{\rm h}^{(k)} a^{(k)}\}$ with the size $R_{\rm h}^{(1)} = 3.1$ are presented for the averaging over the configurations simulated by the HCSM as well as the averaging over the 100 experimental samples containing 1800 fibers each (2D case, see for details [200, 255]). As can be seen, the HCSM ($n = 3700$, $R_{\rm h}^{(1)} = 3.1$) generates significantly more homogeneous and mixed arrangements than the real slightly clustered structures: the fractions $p(n_t)$ of testing windows containing both the small and large numbers n_t of inclusions decrease. Thus, the shaking procedure with the modified HCM leads to a more homogeneous and well-mixed arrangement than HCM that is more effective than combination of the various Metropolis-Hastings algorithms (see [383]) to obtain a simulation algorithm with good mixing properties.

It is important to be sure that the packs simulated by proposed virtual methods have the same statistics as the real material packs (manufactured at other physico-mechanical and chemical conditions, see [19]). Aste et al. [40] and [41] have measured the RDFs for a number of very large monodisperse packs (~150,000 spheres) using X-ray tomography. Our packs use 35,000 spheres in a periodic cube and are generated using the T algorithm (see for details [1003]) with packing fractions of 59.25%; the experimental packing fractions are 59.3%. As can be seen in Figs. 5.5 and 5.6, the curves simulated by the SCRM ($n = 2000$; $P = 80$)

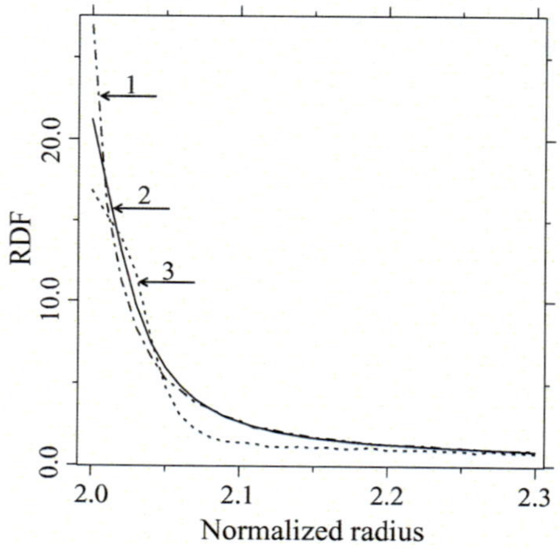

Fig. 5.5 The RDFs $g(r)$ vs r/a at $c = 0.593$ estimated by CRM (1), SCRM (2), Aste's data (3)

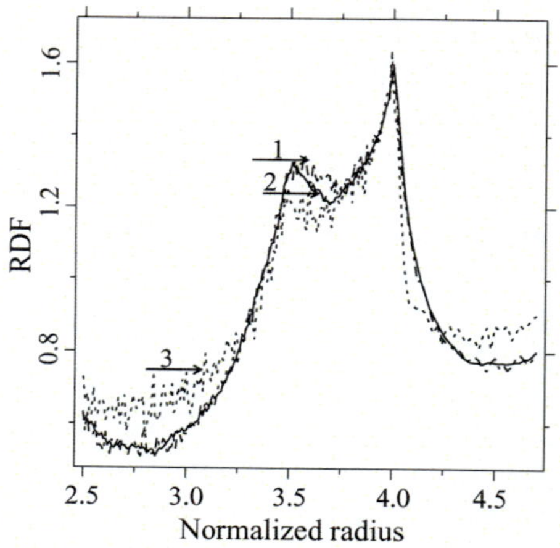

Fig. 5.6 The RDFs $g(r)$ vs r/a at $c = 0.593$ estimated by CRM (1), SCRM (2), Aste's data (3)

is better correlated with experimental data [40] and [41] (see also [1395]) than CRM's simulation. In doing so, it was proved in [237] that estimations of $g(r)$ from experimental data are biased because the estimations of the inclusion size are also biased. Both experiments and simulations suggest a transition between random and ordered configurations in the vicinity of a density $c \approx 0.63$ (3D) [237] and $c \approx 0.8$ (2D) [1281]. The packing density increases much more slowly beyond these points.

5.4.6 Numerical Results for Bimodal Distribution (Application to Solid Propellants)

For polydisperse structures, the counterpart to the RDF is less investigated than for monodisperse case and described by the partial RDF $g_{ij}(r)$ which is the normalized (by the condition $g_{ij}(\infty) = 1$) number density of inclusions $v^{(j)}$ of type j that are within shells r to $r + \Delta r$ centered on the inclusions $v^{(i)}$ of type i. $g_{ij}(r)$ were estimated by the modified CRM [1003] for the size ratio λ of spherical inclusions more than 6.5. As it was expected, for the large inclusions, $g_{11}(r)$ is stochastically very noisy, although trends are apparent. Because of this, as we have already noted, incorporation of the shaking procedure in the proposed algorithm is very effective for smoothing of $g_{ij}(r)$.

It is clear that we have to measure the distances between the particles $v_i \subset v^{(k)}$ ($a_i = a^{(k)}$) and $v_j \subset v^{(m)}$ ($a_j = a^{(m)}$) as the excess distances with respect to contact, that is $\Delta r^{\rm con} = r - (a_i + a_j)$; rather than the distances r themselves. Then the term $\chi_{[0,2a)}(r)$ in Eq. (5.100) should be replaced by $\chi_{[0,a_i+a_j)}(r)$. The decoupling approximation (see [71, 797, 965]) uses the same scaling parameter for the distances in all partial RDFs g_{km} $(k, m = 1, 2)$ of the mixture. We will exploit the partial RDFs g_{km} as the functions of a scaling parameter $R_{km} = (r - a^{(k)})/a^{(m)} + 1$ proposed in [237]. In a general case, the peaks are a consequence of the bidisperse morphology, but some of them can be explained within the monodisperse framework. For example, the coordinates of the first peak of g_{km} and g_{mk} for $k \neq m$ coincide in both arguments $R_{km} = 2$ and $r = a^{(k)} + a^{(m)}$, while the second and third peaks depends on additional information. The nature of the third peak at $R_{km} = 4$ (or $r = a^{(k)} + 3a^{(m)}$) described within the monodisperse framework reflects the placement of the touching inclusions $v_i - (v_j) - v_j$ ($v_i \subset v^{(k)}$, $v_j \subset v^{(m)}$) in a raw (see Fig. 5.3b). However, a specific bidisperse morphology can be manifested by a lot of additional specific configurations. So, it can be generated by the configuration $v_i - (v_i) - v_j$ of touching inclusions in a raw that produces a new peak at $r = 3a^{(k)} + a^{(m)}$ or, in normalized coordinates, at $R_{km} = 2(1 + \lambda_{km})$ ($\lambda_{km} = a^{(k)}/a^{(m)}$). Analogously, the related peaks of $g_{mk}(r)$ corresponding to the configurations $v_j - (v_i) - v_i$ and $v_j - (v_j) - v_i$ are placed at $r = a^{(k)} + 3a^{(m)}$ and $r = 3a^{(m)} + a^{(k)}$, respectively, that matches with the peak coordinates of $g_{km}(r)$. Thus, the third peak ($R = 4$) of unimodal pack has two possible counterparts of their bidisperse pack with $R_{km} = 4$ and $R_{km} = 2(1+\lambda_{km})$ coinciding only at $\lambda_{km} = 1$. In a similar manner, the second peak of unimodal pack with a symbolic configuration $v_i - (v_i + v_i) - v_i$ (see Fig. 5.3a) have three counterparts for the fixed inclusion $v_i \subset v^{(k)}$: $v_i - (v_i + v_i) - v_i$, $v_i - (v_j + v_j) - v_j$, and $v_i - (v_i + v_j) - v_j$. For example, the configurations $v_i - (v_i + v_i) - v_j$ and $v_i - (v_j + v_j) - v_j$ corresponding to the peaks $R_{km} = (\sqrt{3} - 1)\lambda_{km} + \sqrt{1 + 2\lambda_{km}} + 1$ and $R_{km} = \sqrt{3} - \lambda_{km} + \sqrt{\lambda_{km}^2 + 2\lambda_{km}} + 1$, respectively. Of course for other the volume fractions c, size ratio λ_{km}, and volume fraction ratio $\xi_{km} = c^{(k)}/c^{(m)}$ $(k \neq m)$ being considered, one obtains different results for the peak locations (see, e.g., [965]) when the different "monodisperse" and "bidisperse" peaks are pronounced in varying degrees. So, visualization of peaks in unimodal packs is only managed by a single parameter c while the peaks of bimodal packs are managed by three parameters c, λ_{km}, and ξ_{km} $(k \neq m)$.

Due to the absence of the analytical representations of $g_{ij}(r)$ $(i, j = 1, 2)$ for polydisperse structures, it is possible to construct $g_{ij}(r)$ within monodisperse considered framework (5.97) and (5.100) in the following manner ($R_{ij} = (r - a_i)/a_j + 1$):

$$g_{ij}(r) = 1 - \chi_{[0,2)}(R_{ij}), \tag{5.104}$$

$$g_{ij}(r) = [1 - \chi_{[0,2)}(R_{ij})]\left[1 + [g^c_{ij}(\xi, c) - 1]\cos(\pi R_{ij})e^{2(2-R_{ij})}\right], \tag{5.105}$$

where $c = c^{(1)} + c^{(2)}$. The RDFs contact values $g^c_{ij}(\xi, c)$ were obtained in [44] from Ornstein-Zernike equation using Percus–Yevick closure of the pair and triplet levels. In particular, the P-Y result obtained for the bulk mixture at the pair level is

$$g^c_{ij}(a_{ij}) = [a_j g_{ii}(a_{ii}) + a_i g_{jj}(a_{jj}]/a_{ij},\ g^c_{ii}(a_{ii}) = [1-\zeta_3 + 3a_i\zeta_2](1-\zeta_3)^2, \tag{5.106}$$

where $a_{ij} = a_i + a_j$ is a mutual diameter, and $\zeta_\alpha = (\pi 2^{\alpha-1}/3)\sum_i a_i^\alpha$. Taking the limit as $c^{(2)} \to 0$, so that $\zeta_3 = c^{(1)}$, one has

$$g^c_{12}(a_{12}) = \left[1 + c^{(1)}\frac{(2\lambda_{21} - 1)}{(\lambda_{21} + 1)}\right]\frac{1}{(1 - c^{(1)})^2}, \tag{5.107}$$

which reproduces the respective limit $g^c_{12}(a_{12}) = (1 + c^{(1)}/2)/(1 - c^{(1)})^{-2}$ corresponding to the usual P-Y value for unimodal pack $g_c^{PY}(c)$ (5.100).

The paper [333] is dedicated to the experimental measurement of effective Newtonian suspensions $\eta^*/\eta^{(0)}$ of spherical bimodal particles with the size ratio $\lambda \equiv \lambda_{21} = a^{(2)}/a^{(1)} = 0.477, 0.313,\ 0.138$, and the volume fraction ratio $\xi \equiv \xi_{21} = c^{(2)}/c^{(1)} = 0.333$. Because of this one considers the partial RDF's $g_{ij}(R_{ij})$ (no summation over $i, j = 1, 2$) simulation for bimodal packs with $\lambda = 0.313$ and $c = 0.6$. The typical first peak at $R_{ij|1} = 2$ of the $g_{ij}(R_{ij})$ presented for the RDF $g_{11}(R_{11})$ in Fig. 5.7 is analogous to the first peak for the unimodal pack (see Fig. 5.1); shaking parallel procedure SCRM yields a reduction of the first CRM's peak on 25%. The curves $g_{ij}(R_{ij})$ (see the curves 2 in Figs. 5.9 and 5.10) behave at $R_{ij} > 2.2$ less trivial than $g(R)$ (see Fig. 5.2). So the second peak $R_{11|2} = 2.64$ holds a clearly defined bidisperse nature and corresponds to the configuration $v_1 - (v_2) - v_1$: $R_{11|2} = 2.64 \simeq 1 + 2\lambda + 1$. The "monodisperse" peaks $R_{11|3} = 3.464$ and $R_{11|4} = 4.0$ correspond to the peaks $R = 2\sqrt{3} \simeq 3.462$ and $R = 4$ of the unimodal pack in Fig. 5.2. The pronounced peak in Fig. 5.8 at $R_{11|5} = 4.54$ corresponds to the case when the shell either $-(v_2 + v_2)-$ or $-(v_2)-$ is wedged in between the three spheres in a row $v_1 - (v_1) - v_1$ that moves the peak $R_{11|4} = 4.0$ on either $2(\sqrt{1 + 2\lambda} - 1) \simeq 0.55$ (see Fig. 5.3a) or $2\lambda \simeq 0.62$

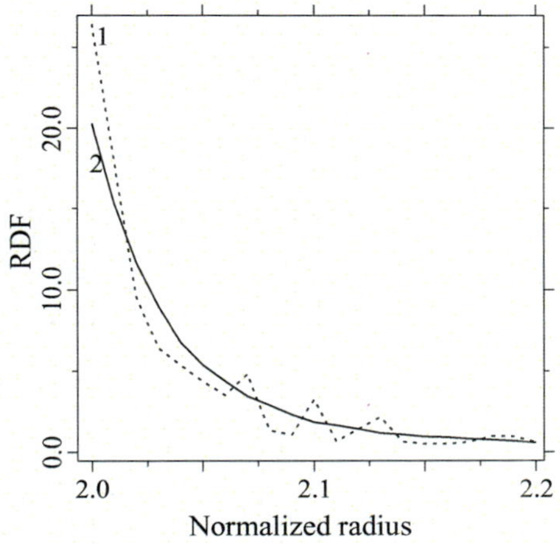

Fig. 5.7 The RDFs $g_{11}(R_{11})$ vs R_{11} at $c = 0.6$ estimated by CRM (1), SCRM (2)

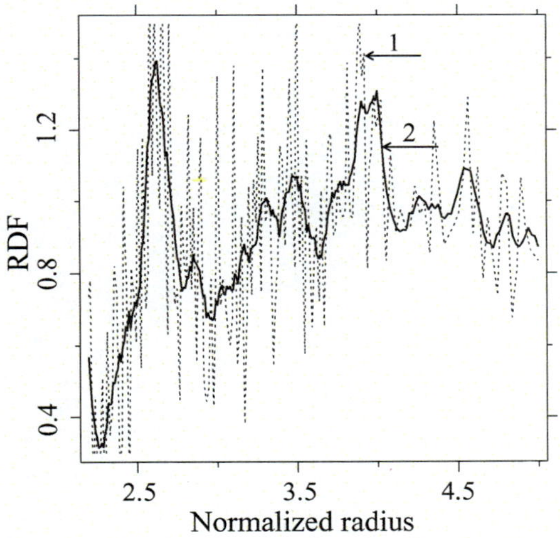

Fig. 5.8 The RDFs $g_{11}(R_{11})$ vs R_{11} at $c = 0.6$ estimated by CRM (1), SCRM (2)

(see Fig. 5.3b), respectively, that is matched up to the average displacement 0.59. It is interesting that the first gap distance $R = 2.8 \div 2.9$ in unimodal pack (see Fig. 5.1) corresponds to the empty space between the first and the second neighbors of typical particle v_1 and symbolically denoted by the form $v_1 - (v_1 + v_1) \smile$ at $R = 2 + \sqrt{3}/2 \simeq 2.87$. This gap is manifested in $g(R)$ but not in $g_{12}(R_{12})$ and $g_{22}(R_{22})$ where the first gap distance is formed by a bidisperse configuration $v_1 - (v_1) \smile$ and $v_2 - (v_1) \smile$ at $R_{12} = 1/\lambda \simeq 3.2$ and $R_{22} = 1/\lambda \simeq 3.2$ (observed as 3.3 in Figs. 5.9 and 5.10), respectively, rather than by a configuration $v_1 - (v_2 + v_2) \smile$ and $v_2 - (v_2 + v_2) \smile$ at R_{12}, $R_{22} = 2.87$. The distance R_{12}, $R_{22} = 1/\lambda \simeq 3.2$ corresponds to the middle surface of the shell $-(v_1)-$ surrounding a typical fixed point v_1 (or v_2), occupied by the centers of moving particles v_1 and inaccessible to the centers v_2.

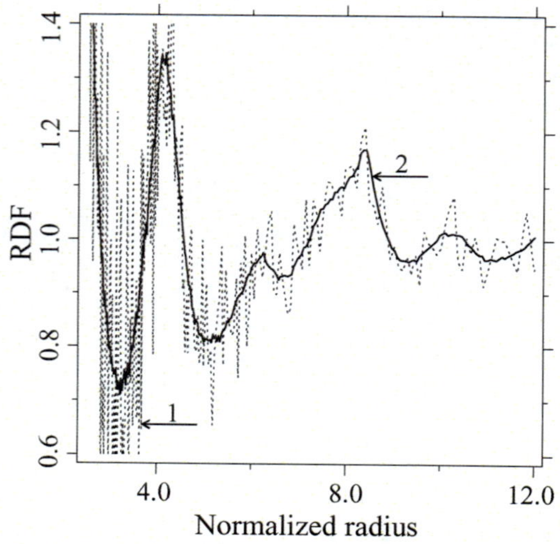

Fig. 5.9 The RDFs $g_{12}(R_{12})$ vs R_{12} at $c = 0.6$ estimated by CRM (1), SCRM (2)

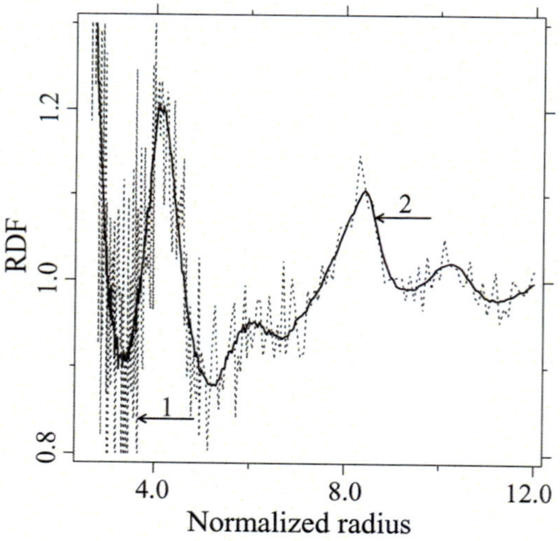

Fig. 5.10 The RDFs $g_{22}(R_{22})$ vs R_{22} at $c = 0.6$ estimated by CRM (1), SCRM (2)

Moreover, a similar gap at the distance $R_{11} = 2+\lambda \simeq 2.3$ (see Fig. 5.8) of $g_{11}(R_{11})$ holds the same bidisperse nature and is produced by the shell $-(v_2)-$ surrounding a typical point v_1 and unattainable to the centers of moving particles v_1. The next (after $R_{12} = 4$) most pronounced peak of $g_{12}(R_{12})$ in Fig. 5.8 is analogous to the peak $R_{12} = 4$ with the configuration $v_1 - (v_2) - v_2$ where a middle sphere $-(v_2)-$ is replaced by a sphere $-(v_1)-$ in the configuration $v_1 - (v_1) - v_2$ that leads to displacement of the peak $R_{12} = 4$ on $2/\lambda - 2$ to the right to $R_{12} = 2 + 2/\lambda \simeq 8.3$. In a similar manner, the peak of $g_{22}(R_{22})$ in Fig. 5.10 at $R_{22} = 2 + 2/\lambda \simeq 8.3$ produced by the configuration $v_2 - (v_1) - v_2$ is clearly defined.

The RDFs curves in Figs. 5.7, 5.8, 5.9, and 5.10 obtained for $N = 2000$ were smoothed by averaging over $P = 80$ realizations of the paralleled SCRM's generations. It was also demonstrated a sensitivity of results with respect to the sample size N and the number of the global shaking M. It should be mentioned that one of the basic challenges of the rocket industry is the optimization of size ratios of packs which are usually performed experimentally. Only at the present time this problem can be solved numerically. The problem is reduced to a minimization problem $f(c^{(1)}, \ldots, c^{(N)}) \to \min(\max)$ at the constraint $c^{(1)} + \ldots + c^{(N)} \equiv c = c_{\max}$. The objective function $f(c^{(1)}, \ldots, c^{(N)})$ can be described as, e.g., either a jamming limit, the values of the RDFs $g_{ij}(a^{(i)} + a^{(j)})$ $(i, j = 1, \ldots, N)$ estimated by the parallelized SCRM, or the effective properties and stress concentrator factors of local stresses estimated in [207] (see also Sect. 8.5).

Chapter 6
Effective Properties and Energy Methods in Thermoelasticity and Thermoelectroelasticity of Composites

Abstract We now summarize the principal formulae of thermoelastostatics of statistically homogeneous composites satisfying the ergodicity condition in a form that is appropriate for our intended application to composites. General representations of both effective properties and effective energy functions are presented through the local stress and strain concentrator factors. Some general exact results for both two-phase composites and polycrystals composed of transversally isotropic crystals are presented. We will consider the variational methods representing the most rigorous trend of micromechanics which generate the bounds of effective properties by the substitution of approximate fields into the strict energy bounds obtained by the volume average of the energy density with the help of lower-order correlation functions.

6.1 Effective Thermoelastic Properties

6.1.1 Hill's Condition and Representative Volume Element

We will consider the local basic equations of thermoelastostatic composites

$$\nabla\boldsymbol{\sigma}(\mathbf{x}) + \mathbf{f}(\mathbf{x}) = \mathbf{0}, \tag{6.1}$$

$$\boldsymbol{\sigma}(\mathbf{x}) = \mathbf{L}(\mathbf{x})\boldsymbol{\varepsilon}(\mathbf{x}) + \boldsymbol{\alpha}(\mathbf{x}), \quad \text{or} \quad \boldsymbol{\varepsilon}(\mathbf{x}) = \mathbf{M}(\mathbf{x})\boldsymbol{\sigma}(\mathbf{x}) + \boldsymbol{\beta}(\mathbf{x}), \tag{6.2}$$

$$\boldsymbol{\varepsilon}(\mathbf{x}) = \text{Def}\mathbf{u}(\mathbf{x}), \quad \text{Inc}\boldsymbol{\varepsilon}(\mathbf{x}) = \mathbf{0}, \tag{6.3}$$

accompanied by one of two homogeneous boundary conditions (also called the kinematic uniform boundary conditions (KUBC) and static uniform boundary conditions (SUBC), respectively)

$$\mathbf{u}(\mathbf{x}) = \mathbf{u}^{\partial w}(\mathbf{x}) = \boldsymbol{\varepsilon}^{\partial w} \cdot \mathbf{x}, \ \boldsymbol{\varepsilon}^{\partial w} \equiv \text{const.}, \ \mathbf{x} \in \partial w, \tag{6.4}$$

$$\boldsymbol{\sigma}(\mathbf{x}) \cdot \mathbf{n}^{\partial w}(\mathbf{x}) = \mathbf{t}^{\partial w}(\mathbf{x}) \equiv \boldsymbol{\sigma}^{\partial w} \cdot \mathbf{n}^{\partial w}(\mathbf{x}), \ \ \boldsymbol{\sigma}^{\partial w} = \text{const.}, \ \mathbf{x} \in \partial w, \tag{6.5}$$

V. A. Buryachenko, *Local and Nonlocal Micromechanics of Heterogeneous Materials*, https://doi.org/10.1007/978-3-030-81784-8_6

where $\boldsymbol{\varepsilon}^{\partial w}(\mathbf{x}) = \frac{1}{2}[\nabla \otimes \mathbf{u}^{\partial w}(\mathbf{x}) + (\nabla \otimes \mathbf{u}^{\partial w}(\mathbf{x}))^\top]$, $\mathbf{x} \in \partial w$, and both $\boldsymbol{\varepsilon}^{\partial w}$ and $\boldsymbol{\sigma}^{\partial w}$ are symmetric constant given tensors. In the case described by Eq. (6.4), the area w is subjected to the prescribed surface displacements of the kind that would produce uniform strain $\boldsymbol{\varepsilon}^{\partial w}$ in a homogeneous material. The surface constraints in the case (6.5) are provided by the surface tractions of the kind that would produce uniform internal stress in a homogeneous material. We assumed that only one of the following two representations of ∂w takes place: $\partial w = \partial w_{\mathbf{u}}$ ($\partial w_\sigma \equiv 0$) or $\partial w = \partial w_\sigma$ ($\partial w_{\mathbf{u}} \equiv 0$).

We consider statistically homogeneous microinhomogeneous medium (see detailed statistical description in Sects. 5.3.1 and 5.3.2) subjected to remote boundary conditions either KUBC (6.4) or SUBC (6.5) with accepted ergodic hypothesis which enables us to replace ensemble average with volume averaging, assuming that observation window tends to infinity (5.69). In partiqular, we define the *mean value* (or volume average) of an integrable tensor field $\mathbf{g}$ (e.g., $\mathbf{g} = \boldsymbol{\varepsilon}, \boldsymbol{\sigma}$) given on the macrodomain $w \subset R^d$ $(d = 2, 3)$ as $\{\mathbf{g}\} = (\bar{w})^{-1}\int_w \mathbf{g}(\mathbf{x})\, d\mathbf{x}$, where $\bar{w} = \text{mes} w$. Then, according to the work–energy (2.104) and Cauchy (6.3_1) relations, we can express the volume averages of stress and strain by integrals over the respective sources and boundary conditions

$$\{\sigma_{ij}\} = \frac{1}{\bar{w}}\Big[\int_w f_{(i}(\mathbf{x})x_{j)}\, d\mathbf{x} + \int_{\partial w} t_{(i}(\mathbf{x})x_{j)}\, d\mathbf{s}\Big],$$

$$\{\varepsilon_{ij}\} = \frac{1}{\bar{w}}\int_{\partial w} u_{(i}(\mathbf{x})n_{j)}(\mathbf{x})\, ds. \tag{6.6}$$

The relation (6.6_1) yields an interesting conclusion that the average value of residual stresses ($\mathbf{f}, \mathbf{t} \equiv \mathbf{0}$, $\boldsymbol{\beta}_1(\mathbf{x}) \not\equiv \mathbf{0}$) vanishes. In the particular form of the homogeneous boundary conditions (6.4), Eq. (6.6_2) with $\nabla\mathbf{x} = \mathbf{x}\nabla = \boldsymbol{\delta}$ leads to the conclusion that for any kinematically admissible strain field corresponding to $\boldsymbol{\varepsilon}^{\partial w} \equiv$ const., the mean value $\{\boldsymbol{\varepsilon}\}$ of $\boldsymbol{\varepsilon}$ is just $\boldsymbol{\varepsilon}^{\partial w}$

$$\{\boldsymbol{\varepsilon}\} = \text{const}. \tag{6.7}$$

In a similar manner, the homogeneous traction boundary condition (6.5) incorporated into Eq. (6.6_1) (at $\mathbf{f} \equiv \mathbf{0}$) leads to the equality for the statically admissible stress field corresponding to $\boldsymbol{\sigma}^{\partial w}(\mathbf{x})$

$$\{\boldsymbol{\sigma}\} = \boldsymbol{\sigma}^{\partial w} = \text{const}. \tag{6.8}$$

Equalities (6.7) and (6.8) are called the *average-strain theorem* and *average-stress theorem*, respectively, which allows us to recast the homogeneous boundary conditions (6.4) or (6.5) as ($\mathbf{x} \in \partial w$)

$$\mathbf{u}(\mathbf{x})|_{\mathbf{x}\in\partial w} = \{\boldsymbol{\varepsilon}\}\cdot\mathbf{x}, \quad \text{or} \ \boldsymbol{\sigma}(\mathbf{x})\cdot\mathbf{n}^{\partial w}(\mathbf{x})|_{\mathbf{x}\in\partial w} = \{\boldsymbol{\sigma}\}\cdot\mathbf{n}^{\partial w}(\mathbf{x}), \tag{6.9}$$

respectively.

Now we will consider the Hill [640] condition (called also *Hill–Mandel* [1005] condition; see for references and generalization [645, 1318]) which enables one to split a volume average of an energy-like representation into a product of volume averages of stress and strain fields

$$\{\boldsymbol{\sigma} : \boldsymbol{\varepsilon}\} = \{\boldsymbol{\sigma}\} : \{\boldsymbol{\varepsilon}\} \tag{6.10}$$

for any admissible $\mathbf{u} \in \mathcal{A}^k(\mathbf{u})$ (2.100) ($\boldsymbol{\varepsilon} = \text{Def}\mathbf{u}$) and equilibrium $\boldsymbol{\sigma} \in \mathcal{A}^s(\boldsymbol{\sigma})$ ($\mathbf{f} \equiv \mathbf{0}$) (2.101) compatible with the homogeneous boundary conditions (6.4) and (6.5), respectively.

Indeed, let us consider the next equation

$$\begin{aligned}
&\frac{1}{\bar{w}}\int_{\partial w}(u_i - x_j\{\epsilon_{ij}\})(\sigma_{ik}n_k - \{\sigma_{ik}\}n_k)d\mathbf{s},\\
&\quad = \frac{1}{\bar{w}}\int_{\partial w}(u_i\sigma_{ik}n_k - u_i n_k\{\sigma_{ik}\} - \sigma_{ik}n_k x_j\{\epsilon_{ij}\} + x_j n_k\{\epsilon_{ij}\}\{\sigma_{ik}\})d\mathbf{s},
\end{aligned} \tag{6.11}$$

where the integrand in the right-hand side of the equation is obtained by expanding the integrand in the left-hand side. Substitution of the following obvious equalities

$$\begin{aligned}
&\int_{\partial w} u_i\sigma_{ik}n_k d\mathbf{s} = \int_w u_{i,j}\sigma_{ij}d\mathbf{x} = \bar{w}\{\epsilon_{ij}\sigma_{ij}\},\\
&\int_{\partial w} u_i n_k\sigma_{ik}d\mathbf{s} = \{\sigma_{ik}\}\int_w u_{i,k}d\mathbf{x} = \bar{w}\{\epsilon_{ik}\}\{\sigma_{ik}\},\\
&\int_{\partial w} \sigma_{ik}n_k x_j d\mathbf{s} = \int_w \sigma_{ik}\delta_{jk}d\mathbf{x} = \bar{w}\{\sigma_{ij}\}, \int_{\partial w} x_j n_k d\mathbf{s} = \int_w \delta_{i,k}\, d\mathbf{x} = \bar{w}\delta_{jk}
\end{aligned}$$

into the right-hand side of Eq. (6.11) proves the *Hill's lemma*

$$\{\sigma_{ij}\epsilon_{i}\} - \{\sigma_{ij}\}\{\epsilon_{ij}\} = \frac{1}{\bar{w}}\int_{\partial w}(u_i - x_j\{\epsilon_{ij}\})(\sigma_{ik}n_k - \{\sigma_{ik}\}n_k)d\mathbf{s}, \tag{6.12}$$

which, in its turn, proves the Hill's condition (6.10) for prescribed homogeneous boundary conditions either (6.9_1) or (6.9_2). We note that $\boldsymbol{\sigma}$ and $\boldsymbol{\varepsilon}$ need not be connected by a specific stress–strain relation.

We will show that for statistically homogeneous composites satisfying the ergodicity condition, the Hill's condition (6.10) exploiting the volume average operations over the volume containing statistically large number of realizations of inclusions can be recast in terms of statistical averages. At the prescribed assumptions, the operations of averaging over the macrovolume and the ensemble averaging are commutative, and the next chain of equalities takes place

$$\langle \boldsymbol{\sigma} : \boldsymbol{\varepsilon}\rangle = \{\langle \boldsymbol{\sigma} : \boldsymbol{\varepsilon}\rangle\} = \langle\{\boldsymbol{\sigma} : \boldsymbol{\varepsilon}\}\rangle = \frac{1}{\bar{w}}\left\langle\int_{\partial w} t_i u_i\, d\mathbf{s}\right\rangle = \frac{1}{\bar{w}}\int_{\partial w}\langle t_i u_i\rangle\, d\mathbf{s}. \tag{6.13}$$

A splitting of the displacement field $\mathbf{u}(\mathbf{x}) = \delta\mathbf{u}(\mathbf{x}) + \langle\mathbf{u}(\mathbf{x})\rangle$ transforms the last integral (6.13) to the form

$$\frac{1}{\bar{w}}\int_{\partial w}\langle t_i u_i\rangle\, d\mathbf{s} = \frac{1}{\bar{w}}\int_{\partial w}\langle t_i\rangle\langle u_i\rangle\, d\mathbf{s} + \frac{1}{\bar{w}}\left\langle\int_{\partial w_\sigma} t_i\delta u_i\, d\mathbf{s}\right\rangle + \frac{1}{\bar{w}}\left\langle\int_{\partial w_\mathbf{u}} t_i\delta u_i\, d\mathbf{s}\right\rangle, \tag{6.14}$$

where the last two integrals vanish for prescribed deterministic boundary conditions because the operations $\{\cdot\}$ can be replaced by $\langle\cdot\rangle$ owing to the ergodicity condition. Thus, Eqs. (6.13) and (6.14) lead to the equation

$$\langle\boldsymbol{\sigma} : \boldsymbol{\varepsilon}\rangle = \frac{1}{\partial\bar{w}}\int_{\partial w}\{\langle\mathbf{t}\rangle : \langle\mathbf{u}\rangle\}\, ds, \tag{6.15}$$

which can be transformed to the desired statistical form of the Hill's condition

$$\langle\boldsymbol{\sigma} : \boldsymbol{\varepsilon}\rangle = \langle\boldsymbol{\sigma}\rangle : \langle\boldsymbol{\varepsilon}\rangle, \tag{6.16}$$

by the use of substitution of the relations

$$\langle\boldsymbol{\varepsilon}\rangle = \text{Def}\langle\mathbf{u}\rangle, \;\; \langle\mathbf{t}(\mathbf{x})\rangle = \langle\boldsymbol{\sigma}\rangle\cdot\mathbf{n}(\mathbf{x}), \;\; \text{Div}\langle\boldsymbol{\sigma}\rangle = \mathbf{0}$$

into the right-hand side of Eq. (6.15) and subsequent application of the Gauss theorem (2.23).

Thus, for statistically homogeneous composites satisfying the ergodicity condition, the Hill's condition can be presented in two equivalent forms involving either the volume averages (6.10) or statistical averages (6.16). The Hill's condition holds for any compatible strain field $\boldsymbol{\varepsilon}(\mathbf{x})$ and equilibrium stress field $\boldsymbol{\sigma}(\mathbf{x})$ not necessarily related to each other by a specific stress–strain relation. The result (6.16) does not demand statistical independence of stress and strains fields because it is only a scalar condition.

The basic subject of the theory of statistically homogeneous linear elastic composites is the characterization of the *overall* or *equivalent* elastic behavior of composites. It means that one is asking not for the detailed stress–strain state in each point but only for the (volume or statistical) averages of the stress state. This involves finding the *overall* or *effective* moduli as the moduli of some fictitious homogeneous medium with the same average stress–strain state as for the microinhomogeneous one being considered. It means that there exists a *representative volume element* Ω (RVE) of the body under consideration that [640] "(a) is structurally entirely typical of the whole mixture on average, and (b) contains a sufficient number of inclusions for the apparent overall moduli to be effectively independent of the surface values of traction and displacement, so long as these values are 'macroscopically uniform'" (see Eqs. (6.4) and (6.5)). These apparent or *effective* moduli fluctuate about a mean value with a wavelength smaller than the dimension l of the sample, which is, however, much larger than the characteristic size of the body inhomogeneity d and much smaller than the characteristic dimension L of the body itself

$$d \ll l \ll L, \tag{6.17}$$

which is called *scale separation hypothesis*. The effect of such fluctuations becomes insignificant within a few wavelengths from the boundary of Ω. The contribution of this boundary layer effect to any volume averages becomes negligible if the RVE becomes large enough (for details see [398, 1166]).

6.1.2 Effective Elastic Moduli

Thus, a notion of RVE is critical to the concrete transition from micro- to macro levels. This transition involves two main problems: (a) to define field and material macrovariables of an effective constitutive equation and (b) to determine how the macrovariables are related to the microvariables of the local constitutive equations as well as to microtopology of a composite material. The term RVE was originally proposed in the analysis of statistically homogeneous linear elastic media ($\mathbf{f}, \boldsymbol{\beta} \equiv \mathbf{0}$) when the answer to the problem (a) is obvious. Indeed, the primary microfields $\boldsymbol{\sigma}(\mathbf{x})$ and $\boldsymbol{\varepsilon}(\mathbf{x})$ correspond to the field macrovariables $\{\boldsymbol{\sigma}\}$ and $\{\boldsymbol{\varepsilon}\}$ dependent only on the homogeneous boundary conditions either KUBC (6.4) or SUBC (6.5) and related by a material effective tensor of either the elasticity or compliance

$$\{\boldsymbol{\sigma}\} = \mathbf{L}^*\{\boldsymbol{\varepsilon}\}, \ \text{ or } \ \{\boldsymbol{\varepsilon}\} = \mathbf{M}^*\{\boldsymbol{\sigma}\}, \tag{6.18}$$

respectively, which are understood as a series of $d(d+1)/2$ calculations (with linearly independent prescribed macroscopic strains or stresses). The problem (b) establishing the constitutive linkage between the suitable chosen micro- and macrofield is significantly more complicated. This problem is reduced to the estimation of the tensor influence functions $\mathbf{A}^*(\mathbf{x})$ and $\mathbf{B}^*(\mathbf{x})$ for the displacement (6.4) and traction (6.5) homogeneous boundary conditions

$$\boldsymbol{\varepsilon}(\mathbf{x}) = \mathbf{A}^*(\mathbf{x})\{\boldsymbol{\varepsilon}\}, \quad \boldsymbol{\sigma}(\mathbf{x}) = \mathbf{B}^*(\mathbf{x})\{\boldsymbol{\sigma}\}, \tag{6.19}$$

respectively. Due to linearity of the problem, these influencing functions exist (but are generally unknown) and have a fundamental role for obtaining the macroscopic elastic response, since they act as weighting functions in averages presenting the overall elastic moduli. Indeed, substitution of the micro–macro linkages (6.20) into the right-hand sides of the averaged constitutive equations $\{\boldsymbol{\sigma}(\mathbf{x})\} = \{\mathbf{L}(\mathbf{x})\boldsymbol{\varepsilon}(\mathbf{x})\}$, and $\{\boldsymbol{\varepsilon}(\mathbf{x})\} = \{\mathbf{M}(\mathbf{x})\boldsymbol{\sigma}(\mathbf{x})\}$ leads to the representations

$$\{\boldsymbol{\sigma}(\mathbf{x})\} = \{\mathbf{L}(\mathbf{x})\mathbf{A}^*(\mathbf{x})\}\{\boldsymbol{\varepsilon}(\mathbf{x})\}, \quad \{\boldsymbol{\varepsilon}(\mathbf{x})\} = \{\mathbf{M}(\mathbf{x})\mathbf{B}^*(\mathbf{x})\}\{\boldsymbol{\sigma}(\mathbf{x})\}, \tag{6.20}$$

respectively. Subsequent substitutions of Eqs. (6.18$_1$) and (6.18$_2$) into the left-hand sides of Eqs. (6.20$_1$) and (6.20$_2$) yield estimations of the effective stiffness and compliance as well as the symmetry conditions for the influence functions

$$\mathbf{L}^* = \{\mathbf{L}(\mathbf{x})\mathbf{A}^*(\mathbf{x})\}, \quad \mathbf{M}^* = \{\mathbf{M}(\mathbf{x})\mathbf{B}^*(\mathbf{x})\},$$
$$\{\mathbf{A}^*(\mathbf{x})\} = \{\mathbf{B}^*(\mathbf{x})\} = \mathbf{I}', \quad I'_{ijkl} = \delta_{ik}\delta_{jl} \tag{6.21}$$

respectively, due to a symmetry of prescribed fields either $\boldsymbol{\varepsilon}^{\partial w}(\mathbf{x}) = \text{const.}$ or $\boldsymbol{\sigma}^{\partial w}(\mathbf{x}) = \text{const.}$ at $\mathbf{x} \in \partial w$, respectively. Thus, the influence functions $\mathbf{A}^*(\mathbf{x})$ and $\mathbf{B}^*(\mathbf{x})$ act as weighting factors in averages giving the effective stiffness (6.21$_1$) and compliance (6.21$_2$). We will consider the properties of these functions in more detail. From the symmetry of tensors $\{\boldsymbol{\varepsilon}\}$ and $\{\boldsymbol{\sigma}\}$, it follows that the influence tensors A^*_{ijkl} and B^*_{ijkl} are symmetric inside the index pairs ($\mathbf{C} = \mathbf{A}^*,\ \mathbf{B}^*$): $C_{ijkl} = C_{jikl} = C_{ijlk}$. Averaging of Eq. (6.20) leads to the symmetry conditions (6.21$_3$) for the average influence functions $\{\mathbf{A}^*\}$ and $\{\mathbf{B}^*\}$.

From (6.21), and from the symmetry properties of the microscopic values $\mathbf{L}(\mathbf{x})$, $\mathbf{M}(\mathbf{x})$ and $\mathbf{A}^*(\mathbf{x})$, $\mathbf{B}^*(\mathbf{x})$, the symmetry of macroproperties $\mathbf{L}^*$ and $\mathbf{M}^*$ inside the index pairs follows. The symmetry of $\mathbf{L}^*$ and $\mathbf{M}^*$ between the index pairs (such as, e.g., $L^*_{ijkl} = L^*_{klij}$) can be obtained from energetic analysis. Indeed, for the homogeneous displacement boundary conditions (6.4), substitution of Eqs. (6.4) (at $\boldsymbol{\beta} \equiv \mathbf{0}$) and (6.20$_1$) into Hill's condition (6.10) leads to

$$\{\boldsymbol{\varepsilon} : \boldsymbol{\sigma}\} = \Big\{(\mathbf{A}^*\{\boldsymbol{\varepsilon}\}) : (\mathbf{L}\mathbf{A}^*\{\boldsymbol{\varepsilon}\})\Big\} = \Big\{(\{\boldsymbol{\varepsilon}\}\mathbf{A}^{*\top}) : (\mathbf{L}\mathbf{A}^*\{\boldsymbol{\varepsilon}\})\Big\}, \tag{6.22}$$

where $\mathbf{A}^{*\top}$ denotes the transposition of the tensor $\mathbf{A}^*$. Since the volume average $\{\boldsymbol{\varepsilon}\}$ is a constant tensor, Eq. (6.22) can be recast as

$$\{\boldsymbol{\varepsilon} : \boldsymbol{\sigma}\} = \{\boldsymbol{\varepsilon}\} : \{\mathbf{A}^{*\top}\mathbf{L}\mathbf{A}^*\}\{\boldsymbol{\varepsilon}\}, \tag{6.23}$$

and compared with the result of substitution of the effective constitutive equation (6.18$_1$) into Hill's condition (6.10)

$$\{\boldsymbol{\varepsilon} : \boldsymbol{\sigma}\} = \{\boldsymbol{\varepsilon}\} : \mathbf{L}^*\{\boldsymbol{\varepsilon}\}, \tag{6.24}$$

which can be considered as an *energetic definition* of the effective elasticity $\mathbf{L}^*$. The two definitions (6.18) and (6.24) are not in general equivalent to one another for arbitrary inhomogeneous media, and the extent to which they agree in their estimated for $\mathbf{L}^*$ provides a partial check on the validity of the "overall modulus" concept.

Since Eqs. (6.23) and (6.24) coincide for any constant symmetric field $\{\boldsymbol{\varepsilon}\}$, the effective module $\mathbf{L}^*$ can be expressed as

$$\mathbf{L}^* = \{\mathbf{A}^{*\top}\mathbf{L}\mathbf{A}^*\}. \tag{6.25}$$

Owing to the symmetry properties of $\mathbf{L}$, Eq. (6.25) yields the symmetry of $\mathbf{L}^*$: $\mathbf{L}^{*\top} = \mathbf{L}^*$. In a similar manner, for homogeneous traction boundary conditions (6.5), we get an analogous expression of symmetric effective compliance

$$\mathbf{M}^* = \{\mathbf{B}^{*\top}\mathbf{M}\mathbf{B}^*\}, \quad \mathbf{M}^{*\top} = \mathbf{M}^*. \tag{6.26}$$

Furthermore, from the positive definiteness of local elastic properties $\mathbf{L}(\mathbf{x})$ and $\mathbf{M}(\mathbf{x})$, and Hill's condition (6.24), we can obtain the positive definiteness of macroscopic properties $\{\boldsymbol{\varepsilon}\} : \mathbf{L}^*\{\boldsymbol{\varepsilon}\} \geq 0$, $\{\boldsymbol{\sigma}\} : \mathbf{M}^*\{\boldsymbol{\sigma}\} \geq 0$, and $\{\boldsymbol{\varepsilon}\} : \mathbf{L}^*\{\boldsymbol{\varepsilon}\} = 0$, $\{\boldsymbol{\sigma}\} : \mathbf{M}^*\{\boldsymbol{\sigma}\} = 0$ if and only if $\{\boldsymbol{\varepsilon}\} = \mathbf{0}$ or $\{\boldsymbol{\sigma}\} = \mathbf{0}$, respectively. Then the effective tensors $\mathbf{L}^*$ and $\mathbf{M}^*$ are invertible, and the inverse tensors $(\mathbf{L}^*)^{-1}$ and $(\mathbf{M}^*)^{-1}$ exist.

First, we will analyze the homogeneous displacement boundary condition (6.4) corresponding to $\{\boldsymbol{\varepsilon}\}$, which can be presented as a function of $\{\boldsymbol{\sigma}\}$: $\{\boldsymbol{\varepsilon}\} = (\mathbf{L}^*)^{-1}\{\boldsymbol{\sigma}\}$. Then from the definition of the influence function (6.20_1) as well as from the local relation $\boldsymbol{\sigma} = \mathbf{L}(\mathbf{x})\boldsymbol{\varepsilon}$, we can find the symmetric stress influence function $\hat{\mathbf{B}}^*(\mathbf{x})$

$$\boldsymbol{\sigma}(\mathbf{x}) = \hat{\mathbf{B}}^*(\mathbf{x})\{\boldsymbol{\sigma}\}, \quad \hat{\mathbf{B}}^*(\mathbf{x}) = \mathbf{L}(\mathbf{x})\mathbf{A}^*(\mathbf{x})(\mathbf{M}^*)^{-1}, \tag{6.27}$$

satisfying the averaging condition $\{\hat{\mathbf{B}}^*(\mathbf{x})\} = \mathbf{I}$. In a similar manner, for the homogeneous traction boundary condition (6.5) corresponding to $\{\boldsymbol{\sigma}\}$, we can get the symmetric strain influence function $\hat{\mathbf{A}}^*(\mathbf{x})$

$$\boldsymbol{\varepsilon}(\mathbf{x}) = \hat{\mathbf{A}}^*(\mathbf{x})\{\boldsymbol{\varepsilon}\}, \quad \hat{\mathbf{A}}^*(\mathbf{x}) = \mathbf{M}(\mathbf{x})\mathbf{B}^*(\mathbf{x})(\mathbf{M}^*)^{-1}, \tag{6.28}$$

with averaging condition $\{\hat{\mathbf{A}}^*(\mathbf{x})\} = \mathbf{I}$.

Thus, we defined two groups of influence functions $\mathbf{A}^*(\mathbf{x})$, $\mathbf{B}^*(\mathbf{x})$ and $\hat{\mathbf{A}}^*(\mathbf{x})$, $\hat{\mathbf{B}}^*(\mathbf{x})$ as well as two effective parameters $\mathbf{L}^*$ and $\mathbf{M}^*$. Up to now we did not establish any connections between the influence tensor functions $\mathbf{A}^*(\mathbf{x})$ and $\hat{\mathbf{A}}^*(\mathbf{x})$ as well as between $\mathbf{B}^*(\mathbf{x})$ and $\hat{\mathbf{B}}^*(\mathbf{x})$ appearing in the different homogeneous boundary conditions (6.4) and (6.5). We also did not define a connection between $\mathbf{L}^*$ and $\mathbf{M}^*$. However, the main goal of micromechanical research is to replace a large microinhomogeneous sample (RVE) by an equivalent homogeneous medium. This replacement makes sense if and only if the two effective parameters $\mathbf{L}^*$ and $\mathbf{M}^*$, obtained at two different homogeneous boundary conditions, are reciprocally inverse: $\mathbf{L} = (\mathbf{M}^*)^{-1}$. Therefore, we need to install some additional supplementary conditions providing the equivalence of effective constitutive equations obtained using two different homogeneous boundary conditions. It turns out that such conditions can be formulated in the form of so-called Hill's weak assumption (see for details and analyses of other conditions [353]):

If $\{\mathbf{u}, \boldsymbol{\varepsilon}, \boldsymbol{\sigma}\}$ is the solution of the first homogeneous boundary-value problem (6.4) corresponding to a prescribed symmetric $\{\boldsymbol{\varepsilon}\}$, if $\{\mathbf{u}_1, \boldsymbol{\varepsilon}_1, \boldsymbol{\sigma}_1\}$ is the solution of the second homogeneous boundary-value problem (6.5) corresponding to a prescribed symmetric $\{\boldsymbol{\sigma}_1\}$ and if $\{\boldsymbol{\varepsilon}_1\} = \{\boldsymbol{\varepsilon}\}$, then $\{\boldsymbol{\sigma}\} = \{\boldsymbol{\sigma}_1\}$ and, moreover,

$$\mathbf{A}^*(\mathbf{x}) \equiv \hat{\mathbf{A}}^*(\mathbf{x}), \quad \mathbf{B}^*(\mathbf{x}) \equiv \hat{\mathbf{B}}^*(\mathbf{x}), \tag{6.29}$$

$$\mathbf{L}^* = (\mathbf{M}^*)^{-1}. \tag{6.30}$$

Thus, Hill's weak assumption means the compatibility of the influence tensor functions (6.29) as well as the equivalence of the dual effective material laws (6.30). We will assume in a subsequent account that Hill's weak assumption is fulfilled.

6.1.3 Apparent Elastic Moduli for Finite-Size RVE

We recall that the theory presented above is based on the conception of a RVE for which Hill's condition (6.10) is valid only asymptotically with a tolerance of order d/l: $\{\boldsymbol{\sigma} : \boldsymbol{\varepsilon}\} = \{\boldsymbol{\sigma}\} : \{\boldsymbol{\varepsilon}\} + O(d/l)$ just in the case of the homogeneous boundary conditions, either KUBC (6.4) or SUBC (6.5). If a microstructural domain w does not contain sufficient number of inclusions, its overall responses under the boundary conditions KUBC (6.4) or SUBC (6.5) will differ. The homogenized properties determined in this way are called *apparent* ($\mathbf{L}^{\rm A}$ and $\mathbf{M}^{\rm A}$), a term introduced by Huet [674] (see for reference and details [1167, 1168]). For comparison of the apparent elastic moduli, we also consider the periodic boundary conditions (PBC) for periodic structure composites forming by a representative unite cell Ω_{00} (see for details Sect. 5.1.1). The displacement field over the entire volume w takes the form $\mathbf{u} = \{\boldsymbol{\varepsilon}\} \cdot \mathbf{x} + \mathbf{v}$, where the fluctuation $\mathbf{v}$ with zero average $\{\mathbf{v}\}^{\Omega} = \mathbf{0}$ (5.87) is periodic. It takes the same values at two homologous points on opposite faces of Ω_{00}. The traction vector $\boldsymbol{\sigma} \cdot \mathbf{n}$ is skew periodic and takes opposite values at two homologous points on opposite faces of Ω_{00}.

Strictly speaking, the response of the material must be independent of the choice of boundary conditions KUBC (6.4) or SUBC (6.5) imposed on the RVE. In practice, however, the mentioned response is estimated from direct numerical simulations of microstructural volume elements (MVEs) extracted, for example, from microcomputer tomography (micro-CT) or scanning electron microscopy (SEM) images of composite samples and typically consist of several hundreds or thousands of inclusions [430]. Then the apparent properties obtained from the MVEs smaller than the true RVE have the following bounding inequalities [674] (see also [703]):

$$(\mathbf{M}^{\rm A}_{\rm SUBC})^{-1} \leq \mathbf{L}^* \leq \mathbf{L}^{\rm A}_{\rm KUBC}, \quad (\mathbf{M}^{\rm A}_{\rm SUBC})^{-1} \leq \mathbf{L}^{\rm A}_{\rm PER} \leq \mathbf{L}^{\rm A}_{\rm KUBC} \tag{6.31}$$

with respect to the effective moduli $\mathbf{L}^*$ and the periodic estimation $\mathbf{L}^{\rm A}_{\rm PER}$; here the inequalities are understood in the sense of quadratic forms. However, the inequalities (6.31) do not mean that the results $\mathbf{L}^{\rm A}_{\rm PER}$ obtained under periodic boundary conditions are always the closest ones to the effective moduli $\mathbf{L}^*$ as clearly stated by Terada et al. [1501] "there is no guarantee that periodic boundary conditions are the best among a class of possible boundary conditions. Nonetheless, the periodic boundary conditions provide the reasonable estimates on the effective moduli in the sense that they are always bounded by the other." The mean values $\mathbf{L}^{\rm A}_{\rm KUBC} - (\mathbf{M}^{\rm A}_{\rm SUBC})^{-1}$ and a given confidence interval demonstrate how these properties converge with respect to the MVE-size for different statistical properties of the random microstructure. There is ample numerical evidence that periodic boundary conditions (and their modifications, see [1329]) are efficient (in the sense of rapid convergence of apparent modulus estimations when the MVE-size is increased) even when the microstructure is non-periodic (which is the most common situation). Kanit et al. [749] (see also [1098]) have shown that this size depends on the investigated five parameters: morphological or physical property, the contrast in the properties of the constituents, their volume fractions, the relative precision for the estimation of the effective property, and the number of realizations of the microstructure. At the example of the critical size of the RVE for discontinuous fiber composites with increasing fiber length and volume fraction, it was established in [605] that it is computationally more efficient to study fewer large microsamples rather than many small ones. Comparative analysis of computational effectiveness of a single microstructural realizations with increasing material volume sizes and several microstructural realizations at the same material volume size was performed for particle reinforced random composites in [520] where generally equivalent results were demonstrated for two methods. A few approaches reducing the size of RVE estimated were proposed. In particular, in the use of self-consistent window method [1498], the authors demonstrated that, given a sufficiently thick embedding frame (with the effective properties), the convergence of different boundary conditions to the effective value exhibits a much faster trend compared to the case that boundary conditions are directly applied to the external boundary of the MVE. In practice, self-consistency is satisfied within a few iterations to very good accuracy, as demonstrated [1498]. The comparison of spherical and cubical RVE with different boundary conditions in [345] concludes that the smaller surface-to-volume ratio associated with a spherical RVE yields less influence of the boundary resulting in a better convergence to the effective material behavior as the size of the RVE increases.

Determining of the RVE's size is related with the notion of *statistically equivalent representative volume element* (SERVE) based on image-based (data-driven) modeling for obtaining detailed three-dimensional data sets from imaging methods, for example, microcomputer tomography (micro-CT), and constructing optimal computational domains. It is important to identify statistically equivalent RVEs or SERVEs for meaningful simulation of microscopic regions MVE. The SERVE is identified as the smallest, statistically equivalent region of the microstructure (although the five parameters indicated in [749] (see above) will also define the size of SERVEs). These statistical equivalence is confirmed by minimizing a least-square functional comparing statistical measures describing the microstructure morphology of the real microstructure and the SERVE. Various statistical descriptors (see Chap. 5) have been used to estimate the volume of the MVE that needs to be sampled for estimating the SERVE size. In particular, one uses the n-point correlation functions $S_n^{(1)}(\mathbf{y}^n)$ (5.75) with subsequent exploitation in the self-consistent method [893] of effective moduli estimation of simulated medium (see also Sect. 8.8.2).

Conventionally, the SERVE boundary-value problem is solved by prescribing the boundary conditions either KUBC (6.4) or SUBC (6.5) on the boundary of the SERVE that usually leads to a large dispersion in Eq. (6.31) and to significantly larger than optimal SERVE domains, needed for converged properties. An attempt to eliminate the mentioned boundary layer effects was for the first time initiated, apparently, in the self-consistent window method [1498] (considered above). Other way for correction of the boundary condition is incorporation of the ellipsoidal SERVE into the infinite matrix and using the theory of functional graded materials (see Sect. 13.4). So, Buryachenko and Ramerstorfer [267, 271] estimated the anisotropic effective moduli as a proportionality factor between statistically average stresses and statistically average strain along a cross section of the spherical cloud (or random cluster $|\mathbf{x}| \leq a_w$) with constant concentration of rigid spherical inclusions. By the use of the local approximation of the MEFM (see Chap. 8 and Sect. 13.4), it was demonstrated that in the interior of a large cloud, sufficiently far away from its boundary, $\mathbf{L}^*(\mathbf{x})$ coincides with the isotropic effective moduli $\mathbf{L}^*$ for the statistically homogeneous medium. Near the boundary of the cloud, the tensors of the effective moduli $\mathbf{L}^*(\mathbf{x})$ are transversally isotropic and vary significantly within the boundary layer $a_w - 4a \leq |\mathbf{x}| \leq a_w$ (nonlocal boundary layer effect). Thus, the effective moduli can be obtained after "removing" of the boundary layers of the cloud; this procedure is conceptually similar to the minus-sampling method (5.27) of statistical analysis. Of course, the mentioned ellipsoidal cluster can be directly extracted from the image (e.g., micro-CT) of composite sample MVE. In such a case, the effective moduli can be estimated as a proportionality factor between the volume average (over the cloud with "removed" boundary layer) strain and the volume average stress, or, that is the same, as a fictitious effective ellipsoidal inclusion; see Eqs. (3.185) and (3.186) (similar representations can be obtained for the effective stiffness). Estimation of the stress–strain fields can be estimated by different methods which are most adapted for analysis of the inclusions inside the infinite matrix (see for references Chap. 4). The mentioned straightforward scheme does not require intermedium steps such as estimations statistical descriptors (see Sect. 5.1.3) and subsequent statistical simulation (see Sect. 5.4); see for references [1366] with review of the elastic and conductivity problems.

Other ways for correction of the boundary conditions either (6.4) or (6.5) are related with attempts to estimate an influence of the inclusions in the regions $v_i \subset w^{\rm MVE} \setminus w^{\rm SERVE}$ exterior to the SERVE. Systematic researches in this new direction of micromechanics were performed by the different methods by Ghosh and Kubair [545, 827] and by Kanaun [740] (see also [741]). For shortening, we reproduce the schemes of these results in the terms of the current book. So, in the exterior statistics method [545, 827] for interaction of finite number of the circle inclusions inside the infinite matrix, one uses the general integral equation (GIE) (7.40), where the regularization (7.33) is not required. Binary interaction of inclusions is performed in the framework of the EFH (4.111). Integration over the domain $w^{\rm MVE} \setminus w^{\rm SERVE}$ is performed by the use of 2-point correlation functions $S_2^{(1)}(\mathbf{y}^n)$ (5.75) (rather than RDF $\varphi(v_j, \mathbf{x}_j|; v_i, \mathbf{x}_i)$ as in Eq. (8.25)). It should be mentioned that the integral in Eq. (7.30) is not absolutely convergent, and because of this, the integration over the domain $w^{\rm MVE} \setminus w^{\rm SERVE}$ depends on both the size and the shape of this domain; these difficulties can be easily avoided by using Eq. (7.19) instead of Eq. (7.30) (see for details Sect. 7.2.3) with subsequent consideration of the binary interactions of inclusions (8.24) exploiting the RDF $\varphi(v_j, \mathbf{x}_j|; v_i, \mathbf{x}_i)$ [advantage of $\varphi(v_j, \mathbf{x}_j|; v_i, \mathbf{x}_i)$ with respect to $S_2^{(1)}(\mathbf{y}^n)$ (5.75) was described in Sect. 8.8.2]. Kanaun [740] also initially used Eq. (7.30) with fixed inclusions v_i $(i = 1, \ldots, n)$ inside the ellipsoidal domain $v_i \subset w^{\rm SERVE}$ that is equivalent to the definition of the effective field $\widetilde{\boldsymbol{\sigma}}_{1,\ldots,n}(\mathbf{x})$ acting on these inclusions by Eq. (7.23) with the term $\langle \boldsymbol{\Gamma}(\mathbf{x}-\mathbf{y})\boldsymbol{\eta}\rangle(\mathbf{y}) \equiv \mathbf{0}$. Subsequent manipulations we reproduce at the example of more accurate approximation GIE (7.33) in the framework of the reasonable assumptions [740] of both a homogeneity of the statistical average field $\langle \widetilde{\boldsymbol{\sigma}}_{1,\ldots,n}\rangle(\mathbf{x})$ =const. $(\mathbf{x} \in w^{\rm SERVE})$ and $\langle \boldsymbol{\Gamma}(\mathbf{x}-\mathbf{y})\boldsymbol{\eta}\rangle(\mathbf{y})V(\mathbf{y}|; v_1, \mathbf{x}_1; \ldots; v_n, \mathbf{x}_n)(\mathbf{y}) = \langle \boldsymbol{\Gamma}(\mathbf{x}-\mathbf{y})\boldsymbol{\eta}\rangle(\mathbf{y})$ at $\mathbf{y} \in R^d \setminus w^{\rm SERVE}$. Of course, the last assumption can be relaxed taking into account far-field approximation$\langle \boldsymbol{\Gamma}(\mathbf{x}-\mathbf{y})\boldsymbol{\eta}\rangle(\mathbf{y})V(\mathbf{y}|; v_1, \mathbf{x}_1; \ldots; v_n, \mathbf{x}_n)(\mathbf{y}) - \langle \boldsymbol{\Gamma}(\mathbf{x}-\mathbf{y})\boldsymbol{\eta}\rangle(\mathbf{y}) = \sum_i \langle \mathbf{T}_{iq}(\mathbf{x}-\mathbf{y}_q)\mathbf{R}_q\mathbf{T}_{iq}(\mathbf{x}-\mathbf{y}_q)\rangle_{(i)}\mathbf{R}_i\langle\overline{\boldsymbol{\sigma}}\rangle_i$ $(i = 1, \ldots, n)$ for estimation of influence of inclusions v_i inside the domain $v_i \subset w^{\rm SERVE}$ on the inclusions in the exterior domain $v_q \subset w^{\rm MVE} \setminus w^{\rm SERVE}$. Self-consistency for estimation of the effective fields $\langle\overline{\boldsymbol{\sigma}}\rangle_i$ is satisfied within a few iterations of subsequent solutions for accurate analysis of n interacting inclusions in domain $w^{\rm SERVE}$ and their approximate impacts on the exterior inclusions $v_q \subset w^{\rm MVE} \setminus w^{\rm SERVE}$ (that leads to absolutely convergent integrals). Of course, the next improvement of both methods [545, 740, 827] can be performed by using both the exact GIE (7.20) instead of the classical GIE (7.33) and the operator form of a solution for the binary interacting inclusions (see Sect. 4.4.5).

6.1.4 Overall Thermoelastic Properties

In addition to purely mechanical loading $(\mathbf{f}, \boldsymbol{\beta} \equiv \mathbf{0})$, an eigenstress field $\boldsymbol{\alpha}(\mathbf{x})$ and an eigenstrain field $\boldsymbol{\beta}(\mathbf{x})$ may exist in w. They cause an effective eigenstress $\boldsymbol{\alpha}^*$ or eigenstrain $\boldsymbol{\beta}^*$ involved in the overall constitutive relation

$$\langle\boldsymbol{\sigma}\rangle = \mathbf{L}^*\langle\boldsymbol{\varepsilon}\rangle + \boldsymbol{\alpha}^*, \quad \langle\boldsymbol{\varepsilon}\rangle = \mathbf{M}^*\langle\boldsymbol{\sigma}\rangle + \boldsymbol{\beta}^*, \tag{6.32}$$

where $\mathbf{L}^* = (\mathbf{M}^*)^{-1}$ and $\mathbf{M}^*$ are the tensor of effective stiffness and compliance $\boldsymbol{\beta}^*$ and $\boldsymbol{\alpha}^* \equiv -\mathbf{L}^*\boldsymbol{\beta}^*$ are second-order tensors of macroscopic eigenstrains and eigenstresses, respectively. Without loss of generality, we can assume that the macrodomain (or RVE) w is divided into subvolumes or local volumes $w^{(r)}$ $(w = \cup w^{(r)},\ w^{(r)} \cap w^{(s)} = \emptyset,\ r, s = 0, 1, \ldots, N,\ r \neq s)$ with constant parameters $\mathbf{g}(\mathbf{x}) = \mathbf{g}^{(r)} \equiv$ const. at $\mathbf{x} \in w^{(r)}$ $(\mathbf{g} = \mathbf{L}, \mathbf{M}, \boldsymbol{\alpha}, \boldsymbol{\beta})$. Then the local constitutive equations can be presented in the form

$$\boldsymbol{\sigma}(\mathbf{x}) = \mathbf{L}^{(r)}\boldsymbol{\varepsilon}(\mathbf{x}) + \boldsymbol{\alpha}^{(r)}, \quad \boldsymbol{\varepsilon}(\mathbf{x}) = \mathbf{M}^{(r)}\boldsymbol{\varepsilon}(\mathbf{x}) + \boldsymbol{\beta}^{(r)}, \tag{6.33}$$

while the extensions of local mechanical concentrator factors (6.20) have the following form $(\mathbf{x} \in w_s$, see for details [417]):

$$\boldsymbol{\varepsilon}(\mathbf{x}) = \mathbf{A}^*(\mathbf{x})\langle\boldsymbol{\varepsilon}\rangle + \overline{\mathbf{D}}^*(\mathbf{x}), \quad \overline{\mathbf{D}}^*(\mathbf{x}) \equiv \sum_{r=0}^{N} \mathbf{D}_r^*(\mathbf{x})\boldsymbol{\beta}^{(r)}, \tag{6.34}$$

$$\boldsymbol{\sigma}(\mathbf{x}) = \mathbf{B}^*(\mathbf{x})\langle\boldsymbol{\sigma}\rangle + \overline{\mathbf{F}}^*(\mathbf{x}), \quad \overline{\mathbf{F}}^*(\mathbf{x}) \equiv \sum_{r=0}^{N} \mathbf{F}_r(\mathbf{x})\boldsymbol{\alpha}^{(r)}, \tag{6.35}$$

where $\mathbf{D}_r^*(\mathbf{x})$ and $\mathbf{F}_r^*(\mathbf{x})$ are called the eigenstrain influence function and the eigenstress influence function, respectively. Without loss of generality, we consider a matrix composite where $w^{(0)}$ is the matrix.

For the estimation of effective transformation $\boldsymbol{\alpha}^*$ and $\boldsymbol{\beta}^*$, we decompose the overall field as

$$\begin{aligned} \boldsymbol{\sigma}(\mathbf{x}) &= \boldsymbol{\sigma}^I(\mathbf{x}) + \boldsymbol{\sigma}^{II}(\mathbf{x}), \quad \boldsymbol{\varepsilon}(\mathbf{x}) = \boldsymbol{\varepsilon}^I(\mathbf{x}) + \boldsymbol{\varepsilon}^{II}(\mathbf{x}), \\ \boldsymbol{\tau}(\mathbf{x}) &= \boldsymbol{\tau}^I(\mathbf{x}) + \boldsymbol{\tau}^{II}(\mathbf{x}), \quad \boldsymbol{\eta}(\mathbf{x}) = \boldsymbol{\eta}^I(\mathbf{x}) + \boldsymbol{\eta}^{II}(\mathbf{x}), \end{aligned} \tag{6.36}$$

with the sources

$$\boldsymbol{\alpha}^I(\mathbf{x}) = \boldsymbol{\beta}^I(\mathbf{x}) = 0, \quad \text{and} \quad \boldsymbol{\alpha}^{II}(\mathbf{x}) = \boldsymbol{\alpha}(\mathbf{x}), \quad \boldsymbol{\beta}^{II}(\mathbf{x}) = \boldsymbol{\beta}(\mathbf{x}). \tag{6.37}$$

Hereafter, the polarization tensors $\boldsymbol{\tau}$ and $\boldsymbol{\eta}$ are defined by Eqs. (4.2) and (4.3).

At first, the *second homogeneous (traction) boundary-value problem* (6.5) is considered

$$\mathbf{t}^I(\mathbf{x}) = \mathbf{t}^{\partial w_\sigma}(\mathbf{x}) = \langle\boldsymbol{\sigma}\rangle \cdot \mathbf{n}(\mathbf{x}), \quad \mathbf{t}^{II}(\mathbf{x}) = \mathbf{0}, \tag{6.38}$$

$\mathbf{x} \in \partial w_\sigma = \partial w$. The field I is solely created at $\boldsymbol{\beta}(\mathbf{x}) \equiv \mathbf{0}$ by the applied traction compatible with a homogeneous stress field $\langle\boldsymbol{\sigma}\rangle$. The field II is generated by the random stress-free strains $\boldsymbol{\beta}(\mathbf{x})$ in the absence of boundary traction. Because of this, we may call $\boldsymbol{\sigma}^I$ the load stress field and $\boldsymbol{\sigma}^{II}$ the residual stress field; knowledge of residual stresses is important for strength and fracture analyses or for phase transformation predictions (see, e.g., [52, 1334]). Consequently, the constitutive equations decompose into two kinds of equations at the local

$$\boldsymbol{\sigma}^I = \mathbf{L}\boldsymbol{\varepsilon}^I, \text{ or } \boldsymbol{\varepsilon}^I = \mathbf{M}\boldsymbol{\sigma}^I, \tag{6.39}$$

$$\boldsymbol{\sigma}^{II} = \mathbf{L}\boldsymbol{\varepsilon}^{II} + \boldsymbol{\alpha}, \text{ or } \boldsymbol{\varepsilon}^{II} = \mathbf{M}\boldsymbol{\sigma}^{II} + \boldsymbol{\beta} \tag{6.40}$$

and the macroscopic

$$\langle\boldsymbol{\sigma}^I\rangle = \mathbf{L}^*\langle\boldsymbol{\varepsilon}^I\rangle, \text{ or } \langle\boldsymbol{\varepsilon}^I\rangle = \mathbf{M}^*\langle\boldsymbol{\sigma}^I\rangle, \tag{6.41}$$

$$\langle\boldsymbol{\sigma}^{II}\rangle = \mathbf{L}^*\langle\boldsymbol{\varepsilon}^{II}\rangle + \boldsymbol{\alpha}^*, \text{ or } \langle\boldsymbol{\varepsilon}^{II}\rangle = \mathbf{M}^*\langle\boldsymbol{\sigma}^{II}\rangle + \boldsymbol{\beta}^*, \tag{6.42}$$

levels, respectively, while the overall constitutive relations are defined by Eq. (6.32).

Taking into account the definitions of the two fields I and II and the boundary conditions, we find the following tensor equations for the average fields:

$$\langle\boldsymbol{\sigma}^I\rangle = \langle\boldsymbol{\sigma}\rangle, \quad \langle\boldsymbol{\varepsilon}^I\rangle = \mathbf{M}^*\langle\boldsymbol{\sigma}\rangle, \quad \langle\boldsymbol{\sigma}^{II}\rangle = \mathbf{0}, \quad \langle\boldsymbol{\varepsilon}^{II}\rangle = \boldsymbol{\beta}^*. \tag{6.43}$$

Since $\boldsymbol{\sigma}^I$ and $\boldsymbol{\sigma}^{II}$ are statically admissible stress fields, while $\boldsymbol{\varepsilon}^I$ and $\boldsymbol{\varepsilon}^{II}$ are kinematically admissible strain fields, then, according to Hill's condition, we obtain

$$\langle\boldsymbol{\sigma}^I\boldsymbol{\varepsilon}^I\rangle = \langle\boldsymbol{\sigma}\rangle\mathbf{M}^*\langle\boldsymbol{\sigma}\rangle, \quad \langle\boldsymbol{\varepsilon}^I\boldsymbol{\sigma}^{II}\rangle = 0, \quad \langle\boldsymbol{\varepsilon}^{II}\boldsymbol{\sigma}^I\rangle = \boldsymbol{\beta}^*\langle\boldsymbol{\sigma}\rangle, \quad \langle\boldsymbol{\varepsilon}^{II}\rangle\langle\boldsymbol{\sigma}^{II}\rangle = 0. \tag{6.44}$$

We can recast Eq. (6.44_3) by the use of Eqs. (6.32) and (6.40)

$$\boldsymbol{\beta}^*\langle\boldsymbol{\sigma}\rangle = \langle\boldsymbol{\varepsilon}^{II}\mathbf{L}\boldsymbol{\varepsilon}^I\rangle = \langle(\boldsymbol{\varepsilon}^{II} - \boldsymbol{\beta})\mathbf{L}\boldsymbol{\varepsilon}^I\rangle + \langle\boldsymbol{\beta}\mathbf{L}\boldsymbol{\varepsilon}^I\rangle = \langle\boldsymbol{\varepsilon}^I\boldsymbol{\sigma}^{II}\rangle + \langle\boldsymbol{\beta}\boldsymbol{\sigma}^I\rangle. \tag{6.45}$$

Substitution of Eqs. (6.44_2) and (6.20_2) into Eq. (6.45) yields the representation of the effective eigenstrain through the stress concentrator factors of the pure mechanical loading problem (6.20_2)

$$\boldsymbol{\beta}^* = \langle\mathbf{B}^{*\top}\boldsymbol{\beta}\rangle. \tag{6.46}$$

Thus, the effective residual strain $\boldsymbol{\beta}^*$ is defined by the averaging of eigenstrains $\boldsymbol{\beta}(\mathbf{x})$, with the weight $\mathbf{B}^{*\top}(\mathbf{x})$ depending only on the elastic properties of the medium.

However, another form of Eq. (6.46) is widely encountered in practice. To obtain it, we average the strain polarization tensor $\boldsymbol{\eta}$ (4.3_2) and (6.37)

$$\mathbf{M}^* = \mathbf{M}^{(0)} + \mathbf{R}^{\sigma *}, \quad \mathbf{R}^{\sigma *}(\mathbf{x})\langle\boldsymbol{\sigma}\rangle = \boldsymbol{\eta}^I(\mathbf{x}), \tag{6.47}$$

$$\boldsymbol{\beta}^* = \boldsymbol{\beta}^{(0)} + \mathbf{R}^{\beta *}, \quad \mathbf{R}^{\beta *}(\mathbf{x}) = \boldsymbol{\eta}^{II}(\mathbf{x}), \tag{6.48}$$

where the representations (6.47) and (6.48) were obtained for the case of the matrix homogeneity ($\mathbf{M}(\mathbf{x}) = \mathbf{M}^{(0)} \equiv$ const., $\boldsymbol{\beta}\mathbf{x}) = \boldsymbol{\beta}^{(0)} \equiv$ const., $\mathbf{x} \in v^{(0)}$) by the use of equalities (6.43_1), (6.43_2) and (6.43_3), (6.43_4), respectively.

For the special case of piecewise local uniform eigenstrains and elastic moduli, Eqs. (6.46) and (6.48) can be evaluated in each local volume $w^{(r)}$ and the results added together to provide the following expressions:

$$\boldsymbol{\beta}^* = \boldsymbol{\beta}^{(0)} + \sum_{r=1}^{N} c^{(r)}\langle\mathbf{B}^{*\top}\rangle_r\boldsymbol{\beta}_1^{(r)}, \tag{6.49}$$

$$\boldsymbol{\beta}^* = \langle\boldsymbol{\beta}\rangle - \sum_{r=1}^{N}\sum_{s=1}^{N} c^{(r)}\mathbf{M}_1^{(r)}\langle\mathbf{F}_s^*\rangle_r\mathbf{L}^{(s)}\boldsymbol{\beta}_1^{(s)}. \tag{6.50}$$

In a similar manner, the *first homogeneous boundary-value problem* (6.4) can be analyzed when the following decomposition of the boundary displacement takes place:

$$\mathbf{u}^I(\mathbf{x}) = \boldsymbol{\varepsilon}^{\partial w_\mathbf{u}} \cdot \mathbf{x}, \quad \boldsymbol{\varepsilon}^{\partial w_\mathbf{u}} \equiv \text{const.}, \quad \mathbf{u}^{II}(\mathbf{x}) \equiv \mathbf{0}, \quad \mathbf{x} \in \partial w_\mathbf{u} \equiv \partial w, \tag{6.51}$$

which leads to the average fields replacing the averages (6.43) and (6.44)

$$\langle \boldsymbol{\varepsilon}^{I} \rangle = \langle \boldsymbol{\varepsilon} \rangle, \ \ \langle \boldsymbol{\sigma}^{I} \rangle = \mathbf{L}^{*} \langle \boldsymbol{\varepsilon} \rangle, \ \ \langle \boldsymbol{\varepsilon}^{II} \rangle = \mathbf{0}, \ \ \langle \boldsymbol{\sigma}^{II} \rangle = \boldsymbol{\alpha}^{*}, \tag{6.52}$$

$$\langle \boldsymbol{\sigma}^{I} \boldsymbol{\varepsilon}^{I} \rangle = \langle \boldsymbol{\varepsilon} \rangle \mathbf{L}^{*} \langle \boldsymbol{\varepsilon} \rangle, \ \ \langle \boldsymbol{\varepsilon}^{I} \boldsymbol{\sigma}^{II} \rangle = \boldsymbol{\alpha}^{*} \langle \boldsymbol{\varepsilon} \rangle, \ \ \langle \boldsymbol{\varepsilon}^{II} \boldsymbol{\sigma}^{I} \rangle = 0, \ \ \langle \boldsymbol{\varepsilon}^{II} \rangle \langle \boldsymbol{\sigma}^{II} \rangle = 0. \tag{6.53}$$

Consider the loading case $\langle \boldsymbol{\varepsilon} \rangle \equiv \mathbf{0}$, and note from (6.32) there is $\boldsymbol{\alpha}^{*} = \langle \mathbf{L}(\mathbf{x}) \boldsymbol{\varepsilon}(\mathbf{x}) + \boldsymbol{\alpha}(\mathbf{x}) \rangle$. Then we will get the representations of the effective eigenstress analogous to the representations for the effective eigenstrain (6.46) and (6.48)

$$\boldsymbol{\alpha}^{*} = \langle \mathbf{A}^{*\top} \boldsymbol{\alpha} \rangle, \quad \boldsymbol{\alpha}^{*} = \langle \boldsymbol{\alpha} \rangle + \langle \mathbf{L}_1 \boldsymbol{\varepsilon}^{II} \rangle, \tag{6.54}$$

Analogs of the representations $\mathbf{M}^{*}$ (6.47) and $\boldsymbol{\beta}^{*}$ (6.48) are

$$\begin{aligned} \mathbf{L}^{*} &= \mathbf{L}^{(0)} + \mathbf{R}^{\epsilon *}, \quad \mathbf{R}^{\epsilon *}(\mathbf{x}) \langle \boldsymbol{\varepsilon} \rangle = \boldsymbol{\tau}^{I}(\mathbf{x}), \\ \boldsymbol{\alpha}^{*} &= \boldsymbol{\alpha}^{(0)} + \mathbf{R}^{\alpha *}, \quad \mathbf{R}^{\alpha *}(\mathbf{x}) = \tau^{II}(\mathbf{x}), \end{aligned} \tag{6.55}$$

The representations (6.54) can be recast for piecewise uniform local properties (such as $\boldsymbol{\alpha}$, $\mathbf{L}$) as

$$\boldsymbol{\alpha}^{*} = \boldsymbol{\alpha}^{(0)} + \sum_{r=1}^{N} c^{(r)} \langle \mathbf{A}^{*\top} \rangle_r \boldsymbol{\alpha}_1^{(r)}, \tag{6.56}$$

$$\boldsymbol{\alpha}^{*} = \langle \boldsymbol{\alpha} \rangle - \sum_{r=1}^{N} \sum_{s=1}^{N} c^{(r)} \mathbf{L}_1^{(r)} \langle \mathbf{D}_s^{*} \rangle_r \mathbf{M}^{(s)} \boldsymbol{\alpha}_1^{(s)}. \tag{6.57}$$

The consistency of the results (6.56), (6.57), and (6.49), (6.50), respectively, are assured as follows. Substitution of Eq. (6.56) into (6.57) at $\boldsymbol{\alpha}^{(r)} = \delta_{rp} \boldsymbol{\alpha}^{(p)}$ ($\boldsymbol{\alpha}^{(p)} \neq \mathbf{0}$) leads to ($s = 0, 1, \ldots, N$)

$$\langle \mathbf{A}^{*\top} \rangle_s = \mathbf{I} - (c^{(s)})^{-1} \sum_{r=1}^{N} c^{(r)} \mathbf{L}^{(r)} \langle \mathbf{D}_s \rangle_r \mathbf{M}^{(s)}. \tag{6.58}$$

In a similar manner, we can get from (6.49) and (6.50)

$$\langle \mathbf{B}^{*\top} \rangle_s = \mathbf{I} - (c^{(s)})^{-1} \sum_{r=0}^{N} c^{(r)} \mathbf{M}^{(r)} \langle \mathbf{F}_s \rangle_r \mathbf{L}^{(s)}. \tag{6.59}$$

Summation over s of Eqs. (6.58) and (6.59) then gives another pair of consistency conditions for $\langle \mathbf{D}_s \rangle_s$ and $\langle \mathbf{F}_s \rangle_s$

$$\sum_{s=0}^{N} \sum_{r=0}^{N} c^{(r)} \mathbf{L}^{(r)} \langle \mathbf{D}_s \rangle_r \mathbf{M}^{(s)} = 0, \quad \sum_{s=0}^{N} \sum_{r=0}^{N} c^{(r)} \mathbf{M}^{(r)} \langle \mathbf{F}_s \rangle_r \mathbf{L}^{(s)} = 0. \tag{6.60}$$

6.2 Effective Energy Functions

We will show that the effective energy functions can be expressed in terms of only the average values of the state variables and the composite effective properties. For this purpose, the average energy function densities are presented in terms of the constitutive properties

$$W = \frac{1}{2\bar{w}} \int_w \boldsymbol{\sigma} : (\boldsymbol{\varepsilon} - \boldsymbol{\beta})\, d\mathbf{x}, \tag{6.61}$$

$$\Pi = W - \frac{1}{\bar{w}} \int_w \rho \mathbf{b} \cdot \mathbf{u}\, d\mathbf{x} - \frac{1}{\bar{w}} \int_{\partial w_\sigma} \mathbf{t}^{\partial w_\sigma} \cdot \mathbf{u}\, ds, \tag{6.62}$$

$$\langle F(\boldsymbol{\varepsilon}, T) \rangle = \frac{1}{\bar{w}} \int_w \Big[\frac{1}{2} L_{ijkl} \varepsilon_{ij} \varepsilon_{kl} + \alpha^T_{ij} \varepsilon_{ij} \theta - C_\varepsilon \frac{\theta^2}{2T_0} \Big]\, d\mathbf{x}, \tag{6.63}$$

$$\langle G_e(\boldsymbol{\sigma}, T) \rangle = \frac{1}{\bar{w}} \int_w \Big[-\frac{1}{2} M_{ijkl} \sigma_{ij} \sigma_{kl} - \beta^T_{ij} \sigma_{ij} \theta + C_\sigma \frac{\theta^2}{2T_0} \Big]\, d\mathbf{x}, \tag{6.64}$$

where at the estimation of the effective strain energy W (2.131), potential energy Π (2.133_1), Helmholtz free energy (2.114), and Gibbs potential (2.123) for the domain devoid of non-integrable singularities in the mechanical or physical variables, the simple averaging procedure is applicable. Hereafter we assume without loss of generality that $\bar{w} = 1$.

At first we will establish some relations for both the potential energy and strain energy for the composite materials. We can consider a strain energy function W presented, according to the principle of virtual work [PVW, see (2.103)], in the form

$$W = \frac{1}{2} \int_{\partial w_\sigma} \mathbf{t}^{\partial w_\sigma} \cdot \mathbf{u}\, ds + \frac{1}{2} \int_w \rho \mathbf{b} \cdot \mathbf{u}\, d\mathbf{x} - \frac{1}{2} \int_w \boldsymbol{\sigma} : \boldsymbol{\beta}\, d\mathbf{x}. \tag{6.65}$$

The expression for the potential energy (6.62) can also be recast taking the principle of virtual work into account

$$\begin{aligned} \Pi &= \int_w \Big[\frac{1}{2} \boldsymbol{\sigma} : (\boldsymbol{\varepsilon} - \boldsymbol{\beta}) - \rho \mathbf{b} \cdot \mathbf{u} \Big]\, d\mathbf{x} - \int_{\partial w_\sigma} \mathbf{t}^{\partial w_\sigma} \cdot \mathbf{u}\, ds \\ &= -\frac{1}{2} \Big[\int_w (\boldsymbol{\sigma} : \boldsymbol{\beta} + \rho \mathbf{b} \cdot \mathbf{u})\, d\mathbf{x} + \int_w \mathbf{t} \cdot \mathbf{u}\, ds \Big] + \int_{\partial w_\mathbf{u}} \mathbf{t} \cdot \mathbf{u}\, ds \\ &= -\frac{1}{2} \int_w \boldsymbol{\sigma} \cdot (\boldsymbol{\varepsilon} + \boldsymbol{\beta})\, d\mathbf{x} + \int_{\partial w_\mathbf{u}} \mathbf{t} \cdot \mathbf{u}\, ds. \end{aligned} \tag{6.66}$$

It is interesting to measure the change of potential energy due to the change of elastic constants from $\mathbf{L}^0$ to $\mathbf{L}^1$ at the fixed sources ($\mathbf{u}^{\partial w}, \mathbf{t}^{\partial w}, \mathbf{b}, \boldsymbol{\beta}$). Let $\mathbf{f}^0$ and $\mathbf{f}^1$ ($\mathbf{f} = \boldsymbol{\sigma}, \boldsymbol{\varepsilon}, \mathbf{u}$) be the fields related with the material tensors $\mathbf{L}^0$ and $\mathbf{L}^1$, respectively.

Then we get from (6.66)

$$
\begin{aligned}
2(\Pi^1 - \Pi^0) &= \int_{\partial w_{\mathbf{u}}} (\mathbf{t}^1 - \mathbf{t}^0) \cdot \mathbf{u}\, ds - \int_{\partial w_\sigma} \mathbf{t} \cdot (\mathbf{u}^1 - \mathbf{u}^0)\, ds \\
&= \int_w \Big[\rho \mathbf{b} \cdot (\mathbf{u}^1 - \mathbf{u}^0) + (\boldsymbol{\sigma}^1 - \boldsymbol{\sigma}^0) : \boldsymbol{\beta}\Big]\, d\mathbf{x} \\
&= \int_{\partial w} (\mathbf{t}^1 \cdot \mathbf{u}^0 - \mathbf{t}^0 \cdot \mathbf{u}^1)\, ds - \int_w \Big[\rho \mathbf{b} \cdot (\mathbf{u}^1 - \mathbf{u}^0) + (\boldsymbol{\sigma}^1 - \boldsymbol{\sigma}^0) : \boldsymbol{\beta}\Big]\, d\mathbf{x} \\
&= \int_w (\boldsymbol{\sigma}^1 : \boldsymbol{\varepsilon}^0 - \boldsymbol{\sigma}^0 : \boldsymbol{\varepsilon}^1)\, d\mathbf{x} - \int_w (\boldsymbol{\sigma}^1 - \boldsymbol{\sigma}^0) : \boldsymbol{\beta}\, d\mathbf{x} \\
&= \int_w (\boldsymbol{\varepsilon}^1 - \boldsymbol{\beta}):(\mathbf{L}^1 - \mathbf{L}^0):(\boldsymbol{\varepsilon}^0 - \boldsymbol{\beta})\, d\mathbf{x} = \int_w \boldsymbol{\sigma}^1 : (\mathbf{M}^0 - \mathbf{M}^1) : \boldsymbol{\sigma}^0\, d\mathbf{x},
\end{aligned}
\tag{6.67}
$$

where the PVW was again used to obtain the third equality. In particular, the case $\mathbf{L}^0 \equiv$ const. and $\mathbf{L}^1(\mathbf{x}) \equiv \mathbf{L}^0$ at $\mathbf{x} \in v^{(0)}$, where $v^{(0)} \subset w$ is a simply connected domain called the matrix, models the change of potential energy produced by an "introduction" of inclusions $v^{(1)} \equiv w \setminus v^{(0)}$ with characteristic function $V(\mathbf{x})$ into the matrix

$$
2(\Pi^1 - \Pi^0) = \int_w (\boldsymbol{\varepsilon}^1 - \boldsymbol{\beta}):(\mathbf{L}^1 - \mathbf{L}^0):(\boldsymbol{\varepsilon}^0 - \boldsymbol{\beta}) V(\mathbf{x})\, d\mathbf{x} = \int_w \boldsymbol{\sigma}^1 : (\mathbf{M}^0 - \mathbf{M}^1) : \boldsymbol{\sigma}^0 V(\mathbf{x})\, d\mathbf{x},
\tag{6.68}
$$

where, in contrast to (6.67), the integration is carried out only over the inclusion volume $v^{(1)}$ rather than over the entire domain w.

Further simplification of Eq. (6.62) can be performed for two particular cases of mixed boundary conditions (2.96) and (2.97):

1. Fixed displacement described by the conditions

$$
\mathbf{u}(\mathbf{x}) = \mathbf{u}^{\partial w}(\mathbf{x}), \quad \mathbf{x} \in \partial w_{\mathbf{u}}; \quad \mathbf{b} \equiv \mathbf{0}, \quad \mathbf{t}(\mathbf{y}) = \mathbf{0}, \quad \mathbf{y} \in w_\sigma \quad \text{or} \quad w_\sigma = \emptyset.
\tag{6.69}
$$

2. Fixed traction described by the conditions

$$
\boldsymbol{\sigma}(\mathbf{x}) \cdot \mathbf{n}^{\partial w}(\mathbf{x}) = \mathbf{t}^{\partial w}(\mathbf{x}), \quad \mathbf{x} \in \partial w_\sigma, \quad \boldsymbol{\beta} \equiv \mathbf{0}, \quad \mathbf{u}(\mathbf{y}) = \mathbf{0}, \quad \mathbf{y} \in w_{\mathbf{u}} \quad \text{or} \quad w_{\mathbf{u}} = \emptyset.
\tag{6.70}
$$

In particular, the boundary conditions (6.69) and (6.70) at $w_\sigma = \emptyset$ and $w_{\mathbf{u}} = \emptyset$ yield the displacement and traction boundary conditions, respectively. Moreover, the boundary conditions (6.69) and (6.70) lead to the homogeneous boundary conditions (2.98) and (2.99), respectively, at some obvious additional constraints.

Then for the fixed displacement boundary conditions (6.69), Eqs. (6.65) and (6.67) lead to

$$\Pi = W = \frac{1}{2}\int_{\partial w_{\mathbf{u}}} \mathbf{t}\cdot\mathbf{u}^{\partial w_{\mathbf{u}}}\, ds - \frac{1}{2}\int_{w} \boldsymbol{\sigma} : \boldsymbol{\beta}\, d\mathbf{x} = \frac{1}{2}\int_{w} \boldsymbol{\sigma} : (\boldsymbol{\varepsilon}^{\partial w_{\mathbf{u}}} - \boldsymbol{\beta})\, d\mathbf{x}, \tag{6.71}$$

where $\boldsymbol{\varepsilon}^{\partial w_{\mathbf{u}}}$ is an arbitrary strain field compatible with the given displacement $\mathbf{u}^{\partial w}(\mathbf{x})$ on $\partial w_{\mathbf{u}}$. For the fixed traction boundary conditions (6.70), Eqs. (6.65) and (6.67) yield

$$\Pi = -W = -\frac{1}{2}\int_{\partial w_{\sigma}} \mathbf{t}^{\partial w_{\sigma}}\cdot\mathbf{u}\, ds - \frac{1}{2}\int_{w} \rho\mathbf{b}\cdot\mathbf{u}\, d\mathbf{x} = -\frac{1}{2}\int_{w} \boldsymbol{\sigma}^{\partial w_{\sigma}} : \boldsymbol{\varepsilon}\, d\mathbf{x}, \tag{6.72}$$

where $\boldsymbol{\sigma}^{\partial w_{\sigma}}$ is an arbitrary equilibrium stress field compatible with the given traction $\mathbf{t}^{\partial w_{\sigma}}$ on ∂w_{σ}. Thus, the potential energy Π coincides with elastic energy W to within sign for both boundary conditions (6.69) and (6.70).

In the elastic case ($\boldsymbol{\beta} \equiv \mathbf{0}$), applying Gauss's divergence theorem to Eq. (17.8) reduces the volume integrals over the inclusions $v^{(1)}$ to the surface integrals over their surfaces

$$\Pi^1 - \Pi^0 = \int_{\partial v^{(1)}} \left[\boldsymbol{\sigma}^1 : (\mathbf{n}\otimes\mathbf{u}^0) - \boldsymbol{\sigma}^0 : (\mathbf{n}\otimes\mathbf{u}^1)\right] ds \tag{6.73}$$

where $\mathbf{n}$ is an outward unit normal to the inclusion surfaces $\partial v^{(1)}$. The relation (6.73) recast in the terms of the elastic energy (6.61) according to Eqs. (6.71) and (6.72)

$$W^1 - W^0 = \pm\int_{\partial v^{(1)}} \left[\boldsymbol{\sigma}^1 : (\mathbf{n}\otimes\mathbf{u}^0) - \boldsymbol{\sigma}^0 : (\mathbf{n}\otimes\mathbf{u}^1)\right] ds \tag{6.74}$$

bears the name of Eshelby [448] formula (see also [336]) reducing the usual volume integrations for calculating strain energy to a particular type of surface integrations. Here signs + and − correspond to the boundary conditions (6.69) and (6.70), respectively. The relations obtained in the current section have been concerned with deterministic location of inclusions in the macrodomain w that can be considered as a particular realization of a random inclusion field. For the homogeneous boundary conditions (6.4) and (6.5), due to the assumed statistical homogeneity of the random function involved and realizability of the ergodicity hypothesis, both the volume and surface averages can be replaced by the appropriate statistical averages. However, some interesting results can be obtained directly by the use of a standard technique of random variables that was already explored in previous sections. So, we now turn to the estimation of contributions introduced by the load stress field $\boldsymbol{\sigma}^I$ and residual stress field $\boldsymbol{\sigma}^{II}$ into the elastic energy W for the homogeneous boundary conditions (6.38) and (6.51) at $\mathbf{b} \equiv \mathbf{0}$. We will also use the decomposition of both stress and strain fields (6.36) for the estimation of the average elastic energy (6.61):

$$W \equiv \frac{1}{2}\langle(\boldsymbol{\varepsilon}-\boldsymbol{\beta})\boldsymbol{\sigma}\rangle = \frac{1}{2}\langle\boldsymbol{\varepsilon}^I\boldsymbol{\sigma}^I\rangle + \frac{1}{2}\langle\boldsymbol{\varepsilon}^I\boldsymbol{\sigma}^{II}\rangle + \frac{1}{2}\langle(\boldsymbol{\varepsilon}^{II}-\boldsymbol{\beta})\boldsymbol{\sigma}^I\rangle$$
$$+\frac{1}{2}\langle(\boldsymbol{\varepsilon}^{II}-\boldsymbol{\beta})\boldsymbol{\sigma}^{II}\rangle = W^I + W^{II} + W^{\mathrm{Int}}, \quad W^I = \frac{1}{2}\langle\boldsymbol{\varepsilon}^I\boldsymbol{\sigma}^I\rangle, \tag{6.75}$$

$$W^{II} = \frac{1}{2}\langle(\boldsymbol{\varepsilon}^{II}-\boldsymbol{\beta})\boldsymbol{\sigma}^{II}\rangle, \quad W^{\mathrm{Int}} = \frac{1}{2}\langle\boldsymbol{\varepsilon}^I\boldsymbol{\sigma}^{II}\rangle + \frac{1}{2}\langle(\boldsymbol{\varepsilon}^{II}-\boldsymbol{\beta})\boldsymbol{\sigma}^I\rangle, \tag{6.76}$$

where the superscripts $I,\ II$, and Int stand for energetic values caused by the fields $\boldsymbol{\sigma}^I$ and $\boldsymbol{\sigma}^{II}$ as well as by the interaction of $\boldsymbol{\sigma}^I$ and $\boldsymbol{\sigma}^{II}$, respectively. We will also introduce the average value of the strain energy density stored in the residual stress field (shortly called stored energy)

$$W^* \equiv \frac{1}{2}\langle\boldsymbol{\sigma} : (\boldsymbol{\varepsilon}-\boldsymbol{\beta})\rangle = -\frac{1}{2}\langle\boldsymbol{\sigma} : \boldsymbol{\beta}\rangle, \tag{6.77}$$

where the Hill's condition (6.16) was used together with the condition $\langle\boldsymbol{\sigma}\rangle = \mathbf{0}$ being assumed. As we will see, the stored energy W^* depends only on $\mathbf{L}(\mathbf{x})$ and $\boldsymbol{\beta}(\mathbf{s})$ and is thus as effective constant of a composite material in parallel with $\mathbf{L}^*$ and $\boldsymbol{\beta}^*$. For the *traction boundary conditions* (6.38), according to the average representations (6.43), (6.44), and (6.46), we can write

$$W^I = \frac{1}{2}\langle\boldsymbol{\sigma}^I\rangle\langle\boldsymbol{\varepsilon}^I\rangle, \quad W^{II} = -\frac{1}{2}\langle\boldsymbol{\beta}\boldsymbol{\sigma}^{II}\rangle, \quad W^{\mathrm{Int}} = 0. \tag{6.78}$$

Thus, the overall strain energy density W is the sum of the strain energy density W^I due to the applied load and W^{II} stored by the transformation stress field

$$W = W^I + W^{II}, \quad W^I = \frac{1}{2}\langle\boldsymbol{\sigma}\rangle\mathbf{M}^*\langle\boldsymbol{\sigma}\rangle, \quad W^{II} = -\frac{1}{2}\langle\boldsymbol{\beta}\boldsymbol{\sigma}^{II}\rangle. \tag{6.79}$$

From Eqs. (6.44_2) and (6.45), the energy W^{II} is just the stored energy (6.77) $W^{II} = W^* = -0.5\langle\boldsymbol{\beta} : \boldsymbol{\sigma}^{II}\rangle$. The straightforward substitution of the decompositions (6.77) and (6.78) into the representation for the potential energy density (6.62) at $\mathbf{b} \equiv \mathbf{0}$ and applying the divergence theorem to the last integral in Eq. (6.62) leads to the decomposition of the potential energy at the traction boundary conditions

$$\Pi = \Pi^I + \Pi^{II} + \Pi^{\mathrm{Int}}, \tag{6.80}$$

$$\Pi^I = W^I - \int_{\partial w} \mathbf{t}^{\partial w} \cdot \langle\mathbf{u}^I\rangle\, ds = -W^I, \tag{6.81}$$

$$\Pi^{II} = W^{II}, \quad \Pi^{\mathrm{Int}} = -\boldsymbol{\beta}^* : \langle\boldsymbol{\sigma}^I\rangle. \tag{6.82}$$

Thus, the strain energy W and potential energy Π can be expressed through the average stresses and the effective parameters

$$W=\frac{1}{2}\langle\boldsymbol{\sigma}\rangle\mathbf{M}^*\langle\boldsymbol{\sigma}\rangle+W^*,\quad \Pi=-\frac{1}{2}\langle\boldsymbol{\sigma}\rangle\mathbf{M}^*\langle\boldsymbol{\sigma}\rangle-\boldsymbol{\beta}^*\langle\boldsymbol{\sigma}\rangle+W^*. \tag{6.83}$$

For the *displacement boundary conditions* (6.51), the decomposition of the strain energy W presented in the form

$$\begin{aligned}W\equiv\frac{1}{2}\langle(\boldsymbol{\sigma}-\boldsymbol{\alpha})\boldsymbol{\varepsilon}\rangle&=\frac{1}{2}\langle\boldsymbol{\sigma}^I\boldsymbol{\varepsilon}^I\rangle+\frac{1}{2}\langle\boldsymbol{\sigma}^I\boldsymbol{\varepsilon}^{II}\rangle+\frac{1}{2}\langle(\boldsymbol{\sigma}^{II}-\boldsymbol{\alpha})\boldsymbol{\varepsilon}^I\rangle\\&+\frac{1}{2}\langle(\boldsymbol{\sigma}^{II}-\boldsymbol{\alpha})\boldsymbol{\varepsilon}^{II}\rangle=W^I+W^{II}+W^{\mathrm{Int}},\quad W^I=\frac{1}{2}\langle\boldsymbol{\sigma}^I\rangle\langle\boldsymbol{\varepsilon}^I\rangle,\\W^{II}=\frac{1}{2}\langle(\boldsymbol{\sigma}^{II}-\boldsymbol{\alpha})\boldsymbol{\varepsilon}^{II}\rangle,&\quad W^{\mathrm{Int}}\equiv\frac{1}{2}\langle\boldsymbol{\sigma}^I\boldsymbol{\varepsilon}^{II}\rangle+\frac{1}{2}\langle(\boldsymbol{\sigma}^{II}-\boldsymbol{\alpha})\boldsymbol{\varepsilon}^I\rangle\end{aligned} \tag{6.84}$$

can be simplified due to the averages (6.52), (6.53), and (6.54_1). Therefore, the decompositions (6.78) and (6.80) should be replaced by

$$W^I=\Pi^I=\frac{1}{2}\langle\boldsymbol{\varepsilon}\rangle\mathbf{L}^*\langle\boldsymbol{\varepsilon}\rangle,\quad W^{II}=\Pi^{II}=-\frac{1}{2}\langle\boldsymbol{\alpha}\boldsymbol{\varepsilon}^{II}\rangle,\quad W^{\mathrm{Int}}=\Pi^{\mathrm{Int}}=\boldsymbol{\alpha}^*\langle\boldsymbol{\varepsilon}\rangle. \tag{6.85}$$

Therefore, the total strain energy and potential energy density become

$$W=\Pi=\frac{1}{2}\langle\boldsymbol{\varepsilon}\rangle\mathbf{L}^*\langle\boldsymbol{\varepsilon}\rangle+W^*. \tag{6.86}$$

Thus, the average energy densities (6.61) and (6.62) can be expressed by the effective thermoelastic constants $\mathbf{L}^*$, $\boldsymbol{\beta}^*$, and W^* for the homogeneous boundary conditions (6.4) and (6.5).

At the estimation of the effective entropy, the simple averaging procedure is also applicable: $S^*=\langle S\rangle$. With the help of (2.120), the average entropy for the composite medium at the displacement boundary conditions (6.51) is found as

$$S^*=-\frac{1}{\theta}\boldsymbol{\alpha}^*\langle\boldsymbol{\varepsilon}\rangle+\frac{C^*_\varepsilon}{T_0}\theta,\quad \boldsymbol{\alpha}^*=\langle\mathbf{A}^{*\top}\boldsymbol{\alpha}\rangle, \tag{6.87}$$

$$\frac{C^*_\varepsilon}{T_0}\theta=\frac{\langle C_\epsilon\rangle}{T_0}\theta-\frac{1}{\theta}\sum_{s,r=1}^{N}c^{(s)}\langle\boldsymbol{\alpha}^\top\mathbf{D}_r\rangle_s\boldsymbol{\beta}^{(r)}. \tag{6.88}$$

The integrals (6.63) and (6.64) can be evaluated term by term [1303]. However, following [883], $\langle F\rangle$ can be easily calculated by adopting Hill's condition (6.10) and taking advantage of the equation $F=(\boldsymbol{\sigma}:\boldsymbol{\varepsilon}-S\theta)/2$ and combining the results with Eqs. (6.32), (6.34), and (6.88)

$$\langle F\rangle=\frac{1}{2}\langle\boldsymbol{\sigma}\rangle\langle\boldsymbol{\varepsilon}\rangle-\frac{1}{2}\langle S\rangle\theta=\frac{1}{2}\langle\boldsymbol{\varepsilon}\rangle\mathbf{L}^*\langle\boldsymbol{\varepsilon}\rangle+\boldsymbol{\alpha}^*\langle\boldsymbol{\varepsilon}\rangle-\frac{C^*_\varepsilon}{2T_0}\theta^2. \tag{6.89}$$

In the case of the traction boundary conditions (6.38), following the same procedure as above yields

$$S^* = \frac{1}{\theta}\boldsymbol{\beta}^*\langle\boldsymbol{\sigma}\rangle + \frac{C^*_\sigma}{T_0}\theta, \quad \boldsymbol{\beta}^* = \langle\mathbf{B}^{*\top}\boldsymbol{\beta}\rangle, \tag{6.90}$$

$$\frac{C^*_\sigma}{T_0}\theta = \frac{\langle C_\sigma\rangle}{T_0}\theta - \frac{1}{\theta}\sum_{s,r=1}^{N} c^{(s)}\langle\boldsymbol{\beta}^\top\mathbf{F}_r\rangle_s\boldsymbol{\alpha}^{(r)}, \tag{6.91}$$

$$\langle F\rangle = \frac{1}{2}\langle\boldsymbol{\sigma}\rangle\mathbf{M}^*\langle\boldsymbol{\sigma}\rangle + \boldsymbol{\beta}^*\langle\boldsymbol{\sigma}\rangle - \frac{C^*_\sigma}{2T_0}\theta^2. \tag{6.92}$$

Since (6.89) and (6.92) should be the same when the constant traction boundary conditions $\langle\boldsymbol{\sigma}\rangle$ (6.38) correspond with the displacement boundary conditions (6.51) according to Eq. (6.32), it leads to $C^*_\sigma = C^*_\varepsilon + \boldsymbol{\alpha}^{T*}\mathbf{L}^*\boldsymbol{\alpha}^{T*}T_0$ and as expected

$$\frac{C^*_\sigma}{T_0}\theta = \frac{\langle C_\sigma\rangle}{T_0}\theta - \frac{1}{\theta}\sum_{s,r=1}^{N} c^{(s)}\langle\boldsymbol{\beta}^\top\mathbf{F}_r\rangle_s\boldsymbol{\alpha}^{(r)}. \tag{6.93}$$

Moreover, from Eqs. (6.34) and (6.88) and Eqs. (6.35) and (6.93) at the boundary conditions (6.51) and (6.38), we obtain

$$2W^{II} = -\langle\boldsymbol{\alpha}\boldsymbol{\varepsilon}^{II}\rangle = (C^*_\varepsilon - \langle C_\varepsilon\rangle)\frac{\theta^2}{T_0}, \quad 2W^{II} = -\langle\boldsymbol{\beta}\boldsymbol{\sigma}^{II}\rangle = (C^*_\sigma - \langle C_\sigma\rangle)\frac{\theta^2}{T_0}, \tag{6.94}$$

respectively.

In a similar manner, the average of the Gibbs thermodynamic potential per unit volume $\langle G_e\rangle$ (6.64) can be treated as

$$\langle G_e\rangle = -\frac{1}{2}\langle\boldsymbol{\sigma}\rangle\mathbf{M}^*\langle\boldsymbol{\sigma}\rangle - \boldsymbol{\beta}^*\langle\boldsymbol{\sigma}\rangle + C^*_\sigma\frac{\theta^2}{2T_0}. \tag{6.95}$$

In so doing, the representations (6.89) and (6.95) can be written in other terms of temperature and either average stresses or average strains by the use of the effective relations (6.32). Thus, the average energy functions (6.89), (6.92), and (6.95) for the statistically homogeneous composite medium subjected to the homogeneous boundary conditions either (6.4) or (6.5) are the same as for a homogeneous continuum with average values of the state variables replacing the local values and with the effective composite properties replacing the homogeneous properties. The stored energies W^{II} ($6.94_{1,2}$), like $\boldsymbol{\alpha}^*$ and $\boldsymbol{\beta}^*$, depend only on the fields $\mathbf{L}(\mathbf{x})$, $\mathbf{M}(\mathbf{x})$ and $\boldsymbol{\alpha}(\mathbf{x})$, $\boldsymbol{\beta}(\mathbf{x})$ and act like effective material constants, which can be observed macroscopically. The stored energies W^{II} can be determined experimentally by measuring the specific heats of the composite according to Eq. (6.94). The tensors of effective transformations $\boldsymbol{\alpha}^* = \langle\boldsymbol{\sigma}^{II}\rangle$ (6.52_4) and $\boldsymbol{\beta}^* = \langle\boldsymbol{\varepsilon}^{II}\rangle$ (6.43_4) for the displacement (6.51) and traction (6.38) boundary conditions can also be defined though averages (6.46) and (6.54_1) with the weighting functions in the form of mechanical concentrator factors of stresses $\mathbf{B}^*(\mathbf{x})$ (6.35) and strains $\mathbf{A}^*(\mathbf{x})$ (6.34), respectively. It should be mentioned that the effective thermoelastic constants $\mathbf{M}^*, \boldsymbol{\beta}^*, W^*$ play a key role in micromechanics because of their close relationship to the internal stress fluctuations (see Chap. 13).

6.3 Some General Exact Results

6.3.1 *Two-Phase Composites*

The binary composites are characterized by piecewise constant constitutive equations (6.33) at $N = 1$ when the shape, orientation, and mechanical properties in a sample-fixed coordinate system of each element of the phases are identical. The feature that distinguishes binary composites from all other composites is that the stress concentrator factors (6.34) and (6.35) can be expressed directly in terms of the mechanical properties of both the composite and all constituents.

Indeed, substitution of Eq. (6.35) into Eqs. ($6.43_{1,2}$) leads to

$$c^{(0)}\langle\mathbf{B}^*\rangle_0 + c^{(1)}\langle\mathbf{B}^*\rangle_1 = \mathbf{I}, \quad c^{(0)}\mathbf{M}^{(0)}\langle\mathbf{B}^*\rangle_0 + c^{(1)}\mathbf{M}^{(1)}\langle\mathbf{B}^*\rangle_1 = \mathbf{M}^*, \tag{6.96}$$

which can unequally be solved for $\langle\mathbf{B}^*\rangle_i$, $(i = 0, 1)$ provided that $(\mathbf{M}^{(1)} - \mathbf{M}^{(0)})^{-1}$ exists

$$\begin{aligned}\langle\mathbf{B}^*\rangle_0 &= (c^{(0)})^{-1}(\mathbf{M}^{(0)} - \mathbf{M}^{(1)})^{-1}(\mathbf{M}^* - \mathbf{M}^{(1)}),\\ \langle\mathbf{B}^*\rangle_1 &= (c^{(1)})^{-1}(\mathbf{M}^{(1)} - \mathbf{M}^{(0)})^{-1}(\mathbf{M}^* - \mathbf{M}^{(0)}).\end{aligned} \tag{6.97}$$

Analogous substitution of Eq. (6.34) into Eqs. ($6.52_{1,2}$) leads to the estimation of average strain concentrator factors

$$\begin{aligned}\langle\mathbf{A}^*\rangle_0 &= (c^{(0)})^{-1}(\mathbf{L}^{(0)} - \mathbf{L}^{(1)})^{-1}(\mathbf{L}^* - \mathbf{L}^{(1)}),\\ \langle\mathbf{A}^*\rangle_1 &= (c^{(1)})^{-1}(\mathbf{L}^{(1)} - \mathbf{L}^{(0)})^{-1}(\mathbf{L}^* - \mathbf{L}^{(0)}).\end{aligned} \tag{6.98}$$

In a similar manner, the transformation concentrator factors can be estimated. So, substitution of Eq. (6.35) and (6.40_2) into Eq. ($6.43_{3,4}$) yields

$$c^{(0)}\langle\overline{\mathbf{F}}^*\rangle_0 + c^{(1)}\langle\overline{\mathbf{F}}^*\rangle_1 = \mathbf{0}, \quad \langle\boldsymbol{\beta}\rangle + c^{(0)}\mathbf{M}^{(0)}\langle\overline{\mathbf{F}}^*\rangle_0 + c^{(1)}\mathbf{M}^{(1)}\langle\overline{\mathbf{F}}^*\rangle_1 = \boldsymbol{\beta}^*. \tag{6.99}$$

Hence

$$\begin{aligned}\langle\overline{\mathbf{F}}^*\rangle_0 &= (c^{(0)})^{-1}(\mathbf{M}^{(0)} - \mathbf{M}^{(1)})^{-1}(\boldsymbol{\beta}^* - \langle\boldsymbol{\beta}\rangle),\\ \langle\overline{\mathbf{F}}^*\rangle_1 &= (c^{(1)})^{-1}(\mathbf{M}^{(1)} - \mathbf{M}^{(0)})^{-1}(\boldsymbol{\beta}^* - \langle\boldsymbol{\beta}\rangle).\end{aligned} \tag{6.100}$$

Hereinafter $\langle \mathbf{g} \rangle = c^{(0)}\mathbf{g}^{(0)} + c^{(1)}\mathbf{g}^{(1)}$, $(\mathbf{g} = \boldsymbol{\alpha}, \boldsymbol{\beta}, \mathbf{L}, \mathbf{M})$. Analogous treatment of Eqs. (6.34) and (6.52$_{3,4}$) allows one to obtain the representations of the transformation strain concentrator factors

$$\langle \overline{\mathbf{D}}^* \rangle_0 = (c^{(0)})^{-1}(\mathbf{L}^{(0)} - \mathbf{L}^{(1)})^{-1}(\boldsymbol{\alpha}^* - \langle \boldsymbol{\alpha} \rangle),$$
$$\langle \overline{\mathbf{D}}^* \rangle_1 = (c^{(1)})^{-1}(\mathbf{L}^{(1)} - \mathbf{L}^{(0)})^{-1}(\boldsymbol{\alpha}^* - \langle \boldsymbol{\alpha} \rangle). \tag{6.101}$$

Thus, the average strain concentrator factors $\langle \mathbf{A}^* \rangle_i$, $\langle \overline{\mathbf{D}}^* \rangle_i$ and $\langle \mathbf{B}^* \rangle_i$, $\langle \overline{\mathbf{F}}^* \rangle_i$ are completely determined by the effective properties $\mathbf{L}^*$, $\boldsymbol{\alpha}^*$ and $\mathbf{M}^*$, $\boldsymbol{\beta}^*$, respectively, assumed to be known. Moreover, it can be shown that $\boldsymbol{\alpha}^*$ and $\boldsymbol{\beta}^*$ are the unique functions of $\mathbf{L}^*$ and $\mathbf{M}^*$, respectively. Indeed, introducing Eqs. (6.98) and (6.97) into Eqs. (6.54$_1$) and (6.46), respectively, leads to the following symmetric connections:

$$\boldsymbol{\alpha}^* = -(\mathbf{L}^* - \mathbf{L}^{(1)})[\mathbf{L}_1^{(1)}]^{-1}\boldsymbol{\alpha}^{(0)} + (\mathbf{L}^* - \mathbf{L}^{(0)})[\mathbf{L}_1^{(1)}]^{-1}\boldsymbol{\alpha}^{(1)}, \tag{6.102}$$

$$\boldsymbol{\beta}^* = -(\mathbf{M}^* - \mathbf{M}^{(1)})[\mathbf{M}_1^{(1)}]^{-1}\boldsymbol{\beta}^{(0)} + (\mathbf{M}^* - \mathbf{M}^{(0)})[\mathbf{M}_1^{(1)}]^{-1}\boldsymbol{\beta}^{(1)} \tag{6.103}$$

as well as to other representations of the effective eigenstresses and eigenstrains

$$\boldsymbol{\alpha}^* = \boldsymbol{\alpha}^{(0)} + (\mathbf{L}^* - \mathbf{L}^{(0)})(\mathbf{L}^{(1)} - \mathbf{L}^{(0)})^{-1}(\boldsymbol{\alpha}^{(1)} - \boldsymbol{\alpha}^{(0)}), \tag{6.104}$$

$$\boldsymbol{\alpha}^* = \langle \boldsymbol{\alpha} \rangle + (\mathbf{L}^* - \langle \mathbf{L} \rangle)(\mathbf{L}^{(1)} - \mathbf{L}^{(0)})^{-1}(\boldsymbol{\alpha}^{(1)} - \boldsymbol{\alpha}^{(0)}), \tag{6.105}$$

$$\boldsymbol{\beta}^* = \boldsymbol{\beta}^{(0)} + (\mathbf{M}^* - \mathbf{M}^{(0)})(\mathbf{M}^{(1)} - \mathbf{M}^{(0)})^{-1}(\boldsymbol{\beta}^{(1)} - \boldsymbol{\beta}^{(0)}), \tag{6.106}$$

$$\boldsymbol{\beta}^* = \langle \boldsymbol{\beta} \rangle + (\mathbf{M}^* - \langle \mathbf{M} \rangle)(\mathbf{M}^{(1)} - \mathbf{M}^{(0)})^{-1}(\boldsymbol{\beta}^{(1)} - \boldsymbol{\beta}^{(0)}). \tag{6.107}$$

The stored energy W^{II} can also be expressed in terms of effective moduli by the use of substitution of Eqs. (6.35), (6.100), and (6.103) into (6.76$_2$)

$$W^{II} = \frac{1}{2}\boldsymbol{\beta}_1^{(1)}[\mathbf{M}_1^{(1)}]^{-1}(\langle \mathbf{M} \rangle - \mathbf{M}^*)[\mathbf{M}_1^{(1)}]^{-1}\boldsymbol{\beta}_1^{(1)}. \tag{6.108}$$

with an alternative expression of W^{II} (6.79$_3$) in terms of the eigenstress $\boldsymbol{\alpha}$ and elastic moduli $\mathbf{L}$. Substitution of Eqs. (6.98), (6.102) and (6.97), (6.103) into Eqs. (6.101) and (6.100), respectively, leads to a unique expression of the transformation concentration functions averaged in the phases through the mechanical concentrator factors and local thermoelastic properties ($i = 0, 1$)

$$\langle \overline{\mathbf{D}}^* \rangle_i = (\mathbf{I} - \langle \mathbf{A}^* \rangle_i)(\mathbf{L}^{(i)} - \mathbf{L}^{(1-i)})^{-1}(\boldsymbol{\alpha}^{(1-i)} - \boldsymbol{\alpha}^{(i)}), \tag{6.109}$$

$$\langle \overline{\mathbf{F}}^* \rangle_i = (\mathbf{I} - \langle \mathbf{B}^* \rangle_i)(\mathbf{M}^{(i)} - \mathbf{M}^{(1-i)})^{-1}(\boldsymbol{\beta}^{(1-i)} - \boldsymbol{\beta}^{(i)}). \tag{6.110}$$

All the results for both the concentration factors and effective transformation fields hold independently in phase topology. The only assumptions are statistical homogeneity and the existence of just two phases. Information about phase topology is completely provided by the effective properties $\mathbf{L}^*$ or $\mathbf{M}^*$.

As an example we will consider the special case of a composite with isotropic phases (2.197): $\mathbf{L}^{(i)} = (3k^{(i)}, 2\mu^{(i)})$, $\boldsymbol{\beta}^{T(i)} = \beta^{T(i)}\boldsymbol{\delta}$. Then the effective eigenstrain (6.102) is

$$\beta_{ij}^{T*} = \langle \beta^T \rangle \delta_{ij} + \frac{(\beta^{T(1)} - \beta^{T(0)})k^{(0)}k^{(1)}}{(k^{(0)} - k^{(1)})} \left[3M^*_{kkij} - \left\langle \frac{1}{k} \right\rangle \delta_{ij} \right]. \tag{6.111}$$

For transversally isotropic composites of two isotropic phase, Eq. (6.111) can be written as

$$\beta_a^{T*} = \langle \beta^T \rangle + \frac{(\beta^{T(1)} - \beta^{T(0)})k^{(0)}k^{(1)}}{(k^{(0)} - k^{(1)})} \left[\frac{3(1-\nu_a^*)}{E_a^*} - \left\langle \frac{1}{k} \right\rangle \right],$$

$$\beta_t^{T*} = \langle \beta^T \rangle + \frac{(\beta^{T(1)} - \beta^{T(0)})k^{(0)}k^{(1)}}{(k^{(0)} - k^{(1)})} \left[\frac{3}{2k_t^*} - \frac{3\nu_a^*(1-2\nu_a^*)}{E_a^*} - \left\langle \frac{1}{k} \right\rangle \right], \tag{6.112}$$

where β_a^{T*} and β_t^{T*} are the effective axial and transversal thermal expansion coefficients, E_a^* and ν_a^* are the effective axial Young's modulus and Poisson's ratio, and k_t^* is the effective plane strain transverse bulk modulus.

If both the composite and its constituents are isotropic, Eq. (6.111) simplifies to the formula of Levin [912]

$$\beta^{T*} = \langle \beta^T \rangle + \frac{(\beta^{T(1)} - \beta^{T(0)})k^{(0)}k^{(1)}}{(k^{(0)} - k^{(1)})} \left[\frac{1}{k^*} - \left\langle \frac{1}{k} \right\rangle \right] \tag{6.113}$$

with the stress concentrator factors presented in the form (2.197) for the isotropic tensors ($i = 0, 1$)

$$\langle \mathbf{B}^* \rangle_i = \left(\frac{k^{(i)}}{c^{(i)}k^*} \frac{k^{(1-i)} - k^*}{k^{(1-i)} - k^{(i)}}, \frac{\mu^{(i)}}{c^{(i)}\mu^*} \frac{\mu^{(1-i)} - \mu^*}{\mu^{(1-i)} - \mu^{(i)}} \right), \tag{6.114}$$

$$\langle \overline{\mathbf{F}}^* \rangle_i = \frac{3(-1)^i}{c^{(i)}} \frac{k^{(0)}k^{(1)}}{k^{(1)} - k^{(0)}} (\beta^* - \langle \beta \rangle)\boldsymbol{\delta}. \tag{6.115}$$

Thus, average eigenstress concentrator factors $\langle \overline{\mathbf{F}}^* \rangle_i$ ($i = 0, 1$) are purely isotropic, and this holds independently of phase geometry. The stored energy (6.108) is given by

$$W^{II} = \frac{9}{2} \frac{k^{(0)}k^{(1)}}{k^{(1)} - k^{(0)}} (\beta^* - \langle \beta \rangle)(\beta^{(1)} - \beta^{(0)}). \tag{6.116}$$

The results of this section have been derived and recovered by several authors [96, 417, 418, 804, 808, 883, 912, 1304, 1338].

Equations (6.113)–(6.116) show that average phase stresses, effective thermal expansion, and stored energy are monotonic functions of the effective module k^* (concerning (6.114), this holds for the isotropic part; the deviatoric stresses are monotonic functions of μ^*). Therefore, the bounds on the effective moduli lead to bounds on the average phase stresses, effective expansion, and stored energy.

It is interesting that the direct expressions of the transformation concentrator factors through the mechanical ones (6.109), (6.110) were obtained for their averages over the phases. However, Dvorak [417] (see also [96, 425]) has generalized these equalities to the local level by a proposed *uniform field theory*.

Namely, let the homogeneous traction boundary conditions (6.5) be prescribed. The derivation of the local eigenstress concentrator factor $\overline{\mathbf{F}}^*(\mathbf{x})$ from the mechanical one $\mathbf{B}^*(\mathbf{x})$ will be performed via the following scheme. At first, superpose an auxiliary overall stress $\tilde{\boldsymbol{\sigma}}$, so that the total overall stress is given by $\hat{\boldsymbol{\sigma}} = \boldsymbol{\sigma}^{\partial w} + \tilde{\boldsymbol{\sigma}}$, and that makes both the local stress and strain uniform in w. The goal is to find the overall stress $\hat{\boldsymbol{\sigma}}$ providing the compatibility of strain fields in the phases

$$\hat{\boldsymbol{\varepsilon}} = \mathbf{M}^{(0)}\hat{\boldsymbol{\sigma}} + \boldsymbol{\beta}^{(0)} = \mathbf{M}^{(1)}\hat{\boldsymbol{\sigma}} + \boldsymbol{\beta}^{(1)}, \tag{6.117}$$

$$\hat{\boldsymbol{\sigma}} = (\mathbf{M}^{(1)} - \mathbf{M}^{(0)})^{-1}(\boldsymbol{\beta}^{(0)} - \boldsymbol{\beta}^{(1)}). \tag{6.118}$$

At the last step of the decomposition procedure, we apply the overall traction $-\hat{\boldsymbol{\sigma}}\cdot\mathbf{n}$ to ∂w to cancel the traction introduced at the previous step. Then we can find from Eq. (6.34) that the eigenstrain problem in two-phase composites can be converted into a solution of a mechanical loading problem ($\mathbf{x} \in w^{(i)},\ i = 0, 1$)

$$\overline{\mathbf{F}}^*(\mathbf{x}) = (\mathbf{I} - \mathbf{B}^*(\mathbf{x}))(\mathbf{M}^{(i)} - \mathbf{M}^{(1-i)})^{-1}(\boldsymbol{\beta}^{(1-i)} - \boldsymbol{\beta}^{(i)}). \tag{6.119}$$

at the local level which yields the analogous problem at the level of phase averages (6.110) after the averaging of Eq. (6.119) over the phase i.

In a similar manner, a displacement boundary condition (6.4) associated with the homogeneous strain tensor $\boldsymbol{\varepsilon}^{\partial\mathcal{E}}$ may be considered. Then the actual $\boldsymbol{\varepsilon}^{\partial\mathcal{E}}$ is superposed by an auxiliary overall strain $\tilde{\boldsymbol{\varepsilon}}$ generating the overall total strain $\hat{\boldsymbol{\varepsilon}} = \boldsymbol{\sigma}^{\partial\mathcal{E}} + \tilde{\boldsymbol{\varepsilon}}$ and both the local uniform stresses and strains in w

$$\hat{\boldsymbol{\sigma}} = \mathbf{L}^{(0)}\hat{\boldsymbol{\varepsilon}} + \boldsymbol{\alpha}^{(0)} = \mathbf{L}^{(1)}\hat{\boldsymbol{\varepsilon}} + \boldsymbol{\alpha}^{(1)}, \tag{6.120}$$

$$\hat{\boldsymbol{\varepsilon}} = (\mathbf{L}^{(1)} - \mathbf{L}^{(0)})^{-1}(\boldsymbol{\alpha}^{(0)} - \boldsymbol{\alpha}^{(1)}). \tag{6.121}$$

The last step is application of the boundary conditions $\mathbf{u}(\mathbf{x}) = -\hat{\boldsymbol{\varepsilon}}\cdot\mathbf{x}$ that leads to the expression of eigenstrain concentrator factor in Eq. (6.34) by the equation ($\mathbf{x} \in w^{(i)},\ i = 0, 1$)

$$\overline{\mathbf{D}}^*(\mathbf{x}) = (\mathbf{I} - \mathbf{A}^*(\mathbf{x}))(\mathbf{L}^{(i)} - \mathbf{L}^{(1-i)})^{-1}(\boldsymbol{\alpha}^{(1-i)} - \boldsymbol{\alpha}^{(i)}). \tag{6.122}$$

Thus, for any two-phase medium, there exists a unique pair of uniform overall strain $\hat{\boldsymbol{\varepsilon}}$ and stress $\hat{\boldsymbol{\sigma}}$ (see Eqs. (6.117), (6.118) and (6.120), (6.121)), which, in superposition with transformation fields, make the total local field uniform in the entire volume. Applying the homogeneous boundary conditions associated with either the field $\hat{\boldsymbol{\varepsilon}}$ or $\hat{\boldsymbol{\sigma}}$ leads to the unique explicit expressions of local transformation concentrator factors through the mechanical concentration factors (6.119) and (6.122). The obtained representations are exact ones and hold independently on phase topology. The only assumptions are statistical homogeneity and the existence of just two phases with the ideal contact conditions between them.

Moreover, in two-phase media with isotropic constituents and slipping interfaces, analogous exact relationships are found between mechanical and thermal stress or strain fields in [87, 97, 417, 424] where the emphasis was on evaluation of general thermomechanical connections rather than the formulation of micromechanical models. Following the aforementioned works, we will consider *imperfectly bonded interfaces* allowing nonvanishing relative displacements to exist together with nonzero traction. Let an orthonormalized basis $(\mathbf{n}, \mathbf{q}, \mathbf{s})$ with unit normal $\mathbf{n} \perp \partial w_{01}$ be considered at each point $\mathbf{x}$ on the interface $\mathbf{x} \in \partial w^{(01)}$ between the phases. At any generic point $\mathbf{x}$ of the interphase, let us define the decomposition of both the displacement $\mathbf{u} = (\mathbf{u}_n, \mathbf{u}_q, \mathbf{u}_s)^\top$ and traction $\mathbf{t} = (\mathbf{t}_n, \mathbf{t}_q, \mathbf{t}_s)^\top$ vectors in the basis $(\mathbf{n}, \mathbf{q}, \mathbf{s})$; additional indexes $\mathbf{g}^{(ij)}$ ($\mathbf{g} = \mathbf{n}, \mathbf{u}, \mathbf{t}, \partial w$; $i, j = 0, 1$) denote a direction from phase i to phase j. The interphase $\partial w^{(01)}$ is modeled by the following equations:

$$[\mathbf{u}_n]_{\partial w^{(01)}} = \mathbf{0}, \quad [\mathbf{t}_n]_{\partial w^{(01)}} = \mathbf{0}, \tag{6.123}$$

$$[\mathbf{u}_\alpha]_{\partial w^{(01)}} = M^{\partial}_{\alpha\beta} \mathbf{t}_\beta, \tag{6.124}$$

where $[\mathbf{g}_n]_{\partial w^{(ij)}} = \mathbf{g}^{(j)}(\mathbf{y}) - \mathbf{g}^{(i)}(\mathbf{z})$ is a jump of the function $\mathbf{g} = \mathbf{u}, \mathbf{t}$ at the transition from phase i to phase j ($i, j = 0, 1$; $i \neq j$; $\mathbf{y} \in w^{(j)}$, $\mathbf{z} \in w^{(i)}$; $\mathbf{y}, \mathbf{z} \to \mathbf{x} \in \partial w^{(ij)}$). $M^{\partial}_{\alpha\beta}$ ($\alpha, \beta = q, s$) are the "compliance" of the interface layer modeling in the limiting cases both the ideal contact ($M^{\partial}_{\alpha\beta} \equiv 0$) and sliding ($M^{\partial}_{\alpha\beta} \equiv \infty$).

Let the field components of the local decompositions $\boldsymbol{\varepsilon}(\mathbf{x}) = \boldsymbol{\varepsilon}^I(\mathbf{x}) + \boldsymbol{\varepsilon}^{II}(\mathbf{x})$ and $\boldsymbol{\sigma}(\mathbf{x}) = \boldsymbol{\sigma}^I(\mathbf{x}) + \boldsymbol{\sigma}^{II}(\mathbf{x})$ satisfy the equations for the stress–strain concentrator factors in the phases

$$\boldsymbol{\varepsilon}^I(\mathbf{x}) = \mathbf{A}^*(\mathbf{x})\langle\boldsymbol{\varepsilon}\rangle, \quad \boldsymbol{\sigma}^I(\mathbf{x}) = \mathbf{L}(\mathbf{x})\mathbf{A}^*(\mathbf{x})\langle\boldsymbol{\varepsilon}\rangle, \tag{6.125}$$

$$\boldsymbol{\varepsilon}^{II}(\mathbf{x}) = \overline{\mathbf{D}}^*(\mathbf{x}), \quad \boldsymbol{\sigma}^{II}(\mathbf{x}) = \mathbf{L}(\mathbf{x})\overline{\mathbf{D}}^*(\mathbf{x}) + \boldsymbol{\alpha}^*(\mathbf{x}), \tag{6.126}$$

$$\boldsymbol{\sigma}^I(\mathbf{x}) = \mathbf{B}^*(\mathbf{x})\langle\boldsymbol{\sigma}\rangle, \quad \boldsymbol{\varepsilon}^I(\mathbf{x}) = \mathbf{M}(\mathbf{x})\mathbf{B}^*(\mathbf{x})\langle\boldsymbol{\sigma}\rangle, \tag{6.127}$$

$$\boldsymbol{\sigma}^{II}(\mathbf{x}) = \overline{\mathbf{F}}^*(\mathbf{x}), \quad \boldsymbol{\varepsilon}^{II}(\mathbf{x}) = \mathbf{M}(\mathbf{x})\overline{\mathbf{F}}^*(\mathbf{x}) + \boldsymbol{\beta}^*(\mathbf{x}), \tag{6.128}$$

where Eqs. (6.125), (6.126) and (6.127), (6.128) correspond to the homogeneous boundary conditions (6.4) and (6.5), respectively. Furthermore, at the interphase between the phases, the following equations for the displacement jumps $\mathbf{u} \equiv \mathbf{u}^I + \mathbf{u}^{II}$ should be added:

$$[\mathbf{u}^I(\mathbf{x})]_{\partial w^{01}} = \mathbf{d}^I(\mathbf{x})\langle\boldsymbol{\varepsilon}\rangle, \quad [\mathbf{u}^{II}(\mathbf{x})]_{\partial w^{01}} = \mathbf{d}^{II}(\mathbf{x}), \tag{6.129}$$

$$[\mathbf{u}^I(\mathbf{x})]_{\partial w^{01}} = \mathbf{f}^I(\mathbf{x})\langle\boldsymbol{\sigma}\rangle, \quad [\mathbf{u}^{II}(\mathbf{x})]_{\partial w^{01}} = \mathbf{f}^{II}(\mathbf{x}), \tag{6.130}$$

where Eqs. (6.129) and (6.130) are consistent with the boundary conditions (6.4) and (6.5), respectively, and the local displacement fields (6.129) and (6.130) hold the contact conditions (6.123), (6.124).

We will show that in two-phase systems with isotropic phases and imperfectly bonded interface (6.123) and (6.124), the transformation concentrator factors can be explicitly expressed through the mechanical ones, i.e., the two pairs of transformation tensors $\overline{\mathbf{D}}^*$, $\mathbf{d}^{II}$ and $\overline{\mathbf{F}}^*$, $\mathbf{f}^{II}$ are explicitly expressed through the tensor pairs of pure mechanical loading $\mathbf{A}^*$, $\mathbf{d}^I$ and $\mathbf{B}^*$, $\mathbf{f}^I$, respectively. To accomplish this, we will modify the scheme (6.117)–(6.121) previously proposed for the ideal contact between the phases.

First let the homogeneous traction boundary conditions (6.38) be considered. For imperfectly bonded interface (6.123) and (6.124), Eqs. (6.117)–(6.119) are also valid and define the homogeneous field $\hat{\boldsymbol{\sigma}}$ leading to coinciding of the strain distributions in the phases (6.117). This field $\hat{\boldsymbol{\sigma}}$ provides the continuation of displacement fields in the interface vicinity because the field $\hat{\boldsymbol{\sigma}}$ (6.118) in the microinhomogeneous media with the isotropic constituents is isotropic, the shear stresses are absent at the interface $\partial w^{(01)}$, and the composite responds as if interfaces were perfectly bounded. Zero tractions at the external interface ∂w are provided by applying of the traction at the boundary ∂w: $\mathbf{t}(\mathbf{y}) = -\hat{\boldsymbol{\sigma}} \cdot \mathbf{n}$, $\mathbf{y} \in \partial w$, which produces the local fields inside the medium $\mathbf{x} \in w$

$$\boldsymbol{\sigma}^I(\mathbf{x}) = -\mathbf{B}^*(\mathbf{x})\hat{\boldsymbol{\sigma}}, \quad [\mathbf{u}^I(\mathbf{x})]_{\partial w^{(01)}} = -\mathbf{f}^I(\mathbf{x})\hat{\boldsymbol{\sigma}}. \tag{6.131}$$

Substitution of Eq. (6.131) into the equality (6.35) yields ($\mathbf{x} \in w^{(i)}$, $i = 0, 1$)

$$\boldsymbol{\sigma}^{II}(\mathbf{x}) = (\mathbf{I} - \mathbf{B}^*(\mathbf{x}))(\mathbf{M}^{(i)} - \mathbf{M}^{(1-i)})^{-1}(\boldsymbol{\beta}^{(1-i)} - \boldsymbol{\beta}^{(i)}), \tag{6.132}$$

$$[\mathbf{u}^{II}(\mathbf{x})]_{\partial w^{(01)}} = -\mathbf{f}^I(\mathbf{x})(\mathbf{M}^{(i)} - \mathbf{M}^{(1-i)})^{-1}(\boldsymbol{\beta}^{(1-i)} - \boldsymbol{\beta}^{(i)}), \tag{6.133}$$

from which the concentrator factors can be found ($\mathbf{x} \in w^{(i)}$, $i = 0, 1$)

$$\overline{\mathbf{F}}^*(\mathbf{x}) = (\mathbf{I} - \mathbf{B}^*(\mathbf{x}))(\mathbf{M}^{(i)} - \mathbf{M}^{(1-i)})^{-1}(\boldsymbol{\beta}^{(1-i)} - \boldsymbol{\beta}^{(i)}), \tag{6.134}$$

$$\mathbf{f}^{II}(\mathbf{x}) = -\mathbf{f}^I(\mathbf{x})(\mathbf{M}^{(i)} - \mathbf{M}^{(1-i)})^{-1}(\boldsymbol{\beta}^{(1-i)} - \boldsymbol{\beta}^{(i)}). \tag{6.135}$$

In a similar manner, for the displacement boundary conditions associated with the homogeneous tensor $\boldsymbol{\varepsilon}^{\partial w_{\mathbf{u}}}$ (6.54), a connection between the local strain concentrator factors can be found ($\mathbf{x} \in w^{(i)}$, $i = 0, 1$):

$$\overline{\mathbf{D}}^*(\mathbf{x}) = (\mathbf{I} - \mathbf{A}^*(\mathbf{x}))(\mathbf{L}^{(i)} - \mathbf{L}^{(1-i)})^{-1}(\boldsymbol{\alpha}^{(1-i)} - \boldsymbol{\alpha}^{(i)}), \tag{6.136}$$

$$\mathbf{d}^{II}(\mathbf{x}) = -\mathbf{d}^I(\mathbf{x})(\mathbf{L}^{(i)} - \mathbf{L}^{(1-i)})^{-1}(\boldsymbol{\alpha}^{(1-i)} - \boldsymbol{\alpha}^{(i)}). \tag{6.137}$$

It is notable that the equalities (6.134) and (6.136) coincide with the analogous expression for the ideal interface (6.119) and (6.122), respectively, with the distinguished values of the tensors $\mathbf{A}^*(\mathbf{x}), \mathbf{B}^*(\mathbf{x})$. At the obtaining Eqs. (6.134)–(6.137) the isotropy condition of constituents was essentially explored because in the case of their anisotropy, both the displacement jumps and shear components of interface tractions do not vanish, which makes establishment of the interconnections of the concentration tensors (6.134)–(6.137) difficult.

We now turn our attention to the specification of the link between the overall stresses and strains of microinhomogeneous media. We will show that the effective constitutive equation has the form (6.32) and find the dependences of the effective parameters $\mathbf{L}^*$, $\mathbf{M}^*$, $\boldsymbol{\alpha}^*$, $\boldsymbol{\beta}^*$ on the average concentrator tensors (6.125) and (6.127). We start by writing the average strain in phase r $(r = 0, \ldots, N)$

$$\langle \varepsilon_{ij} \rangle^{(r)} = \frac{1}{2\bar{w}^{(r)}} \int_{\partial w} (u_i n_j + u_j n_i)\, ds + \frac{1}{2\bar{w}^{(r)}} \sum_{s=1}^{N} \int_{\partial w^{(rs)}} (u_i n_j + u_j n_i)\, ds, \tag{6.138}$$

where Gauss's divergence theorem (2.23) has been applied to the phase r with the outward normal $\mathbf{n}$ to the bounding surfaces ∂w and $\partial w^{(rs)}$. Here there are not any limitation on the interface boundary conditions (e.g., (6.123) and (6.124)), and the surface of cracks or cavities that are internal to phase r is denoted by $\partial w^{(rr)}$. Multiplying Eq. (6.138) by $c^{(r)} \equiv \bar{w}^{(r)}/\bar{w}$, and summing over r, results in

$$\langle \boldsymbol{\varepsilon} \rangle = \sum_{r=0}^{N} c^{(r)} \langle \boldsymbol{\varepsilon} \rangle^{(r)} - \sum_{r=0}^{N} \sum_{s=0}^{N} \mathbf{J}^{(rs)}, \quad \mathbf{J}^{(rs)} \equiv \frac{1}{2\bar{w}} \int_{\partial w^{(rs)}} (u_i n_j + u_j n_i)\, ds, \tag{6.139}$$

where Gauss's divergence theorem has again been applied to the total domain w bounded by the surface ∂w. Equation (6.139$_2$) can also be recast in terms of displacement jumps at the interfaces $\partial w^{(rs)}$

$$(\mathbf{J}^{(rs)} + \mathbf{J}^{(sr)})_{ij} = \frac{1}{2\bar{w}} \int_{\partial w^{(rs)}} ([u_i] n_j + [u_j] n_i)\, ds \tag{6.140}$$

taking into account that $\mathbf{n}^{(rs)} = -\mathbf{n}^{(sr)}$, $[\mathbf{u}]_{\partial w^{(rs)}} = \mathbf{u}^{(s)}(\mathbf{y}) - \mathbf{u}^{(r)}(\mathbf{z})$ $(r, s = 0, 1, \ldots N;\ r \neq n;\ \mathbf{y} \in w^{(s)},\ \mathbf{z} \in w^{(r)};\ \mathbf{y}, \mathbf{z} \to \mathbf{x} \in \partial w^{(rs)})$ and that the internal interphase $\partial w^{(rr)}$ of phase r can be decomposed as $\partial w^{(rr)} = \partial w^{+(rr)} + \partial w^{-(rr)}$, where $\partial w^{+(rr)}$ and $\partial w^{-(rr)}$ denote the upper and lower surfaces of the interphase $\partial w^{(rr)}$. The interface tensors $\mathbf{J}^{(rs)} + \mathbf{J}^{(sr)} = \tilde{\mathbf{d}}^{I(rs)} \langle \boldsymbol{\varepsilon} \rangle + \tilde{\mathbf{d}}^{II(rs)}$ can be decomposed with respect to both the strain concentrator factor $\mathbf{d}^I$ (6.129$_1$) and eigenstrain concentrator factor $\mathbf{d}^{II}$ (6.129$_2$)

$$\tilde{\mathbf{d}}^{I(rs)} = \frac{1}{\bar{w}} \int_{\partial w^{(rs)}} d^I_{(ikl}(\mathbf{x}) n_{j)}(\mathbf{x})\, d\mathbf{x}, \quad \tilde{\mathbf{d}}^{II(rs)} = \frac{1}{\bar{w}} \int_{\partial w^{(rs)}} d^{II}_{(ikl}(\mathbf{x}) n_{j)}(\mathbf{x})\, d\mathbf{x}, \tag{6.141}$$

where the concentration factors $\tilde{\mathbf{d}}^{I(sr)}$ and $\tilde{\mathbf{d}}^{II(sr)}$ are symmetric with respect to the index interchange $r \leftrightarrow s$. Once the concentrator factors $\langle\mathbf{A}^*\rangle_r$, $\langle\overline{\mathbf{D}}^*\rangle_r$, $\tilde{\mathbf{d}}^{I(rs)}$, $\tilde{\mathbf{d}}^{II(rs)}$ $(r, s = 0, \ldots, N)$ are known, we can obtain the effective properties

$$\mathbf{L}^* = \mathbf{L}^{(0)} + \sum_{r=1}^{N} c^{(r)}\mathbf{L}_1^{(r)}\langle\mathbf{A}^*\rangle_r + \mathbf{L}^{(0)}\sum_{r=0}^{N}\sum_{s=r}^{N}\tilde{\mathbf{d}}^{I(rs)}, \tag{6.142}$$

$$\boldsymbol{\alpha}^* = \boldsymbol{\alpha}^{(0)} + \sum_{r=1}^{N} c^{(r)}\mathbf{L}_1^{(r)}\langle\overline{\mathbf{D}}^*\rangle_r + \mathbf{L}^{(0)}\sum_{r=0}^{N}\sum_{s=r}^{N}\tilde{\mathbf{d}}^{II(rs)} \tag{6.143}$$

with similar representations for $\mathbf{M}^*$ and $\boldsymbol{\beta}^*$. By the use of the theorem of virtual work, Benveniste and Dvorak [97] have proved a generalization of effective eigenvector representations (6.54_1) and (6.49) also [as Eqs. (6.142) and (6.143)] for the case of multiphase materials with anisotropic phases and imperfect interfaces of the type described by (6.123) and (6.124)

$$\boldsymbol{\alpha}^* = \boldsymbol{\alpha}^{(0)} + \sum_{r=0}^{N}\langle\mathbf{A}^{*\top}\rangle_r\boldsymbol{\alpha}_1^{(r)}, \quad \boldsymbol{\beta}^* = \boldsymbol{\beta}^{(0)} + \sum_{r=1}^{N}\langle\mathbf{B}^{*\top}\rangle_r\boldsymbol{\beta}_1^{(r)}. \tag{6.144}$$

Initially, we will consider N phase composite material with zero boundary conditions (6.51) $\boldsymbol{\varepsilon}^{\partial w_{\mathbf{u}}} \equiv \mathbf{0}$, when the left-hand site of Eq. (6.139_1) vanishes $\langle\boldsymbol{\varepsilon}\rangle \equiv \mathbf{0}$, and the average stresses can be expressed as

$$\langle\boldsymbol{\sigma}\rangle = \sum_{r=0}^{N} c^{(r)}(\mathbf{L}^{(r)}\langle\boldsymbol{\varepsilon}\rangle^{(r)} + \boldsymbol{\alpha}^{(r)}) = \boldsymbol{\alpha}^*. \tag{6.145}$$

In subsequent presentations, we will analyze the case of two-phase composites $(r = 0, 1)$, when the closed analogy with the calculations for the perfect interphase (6.96)–(6.103) takes place. Expressing the average phase strain $\langle\boldsymbol{\varepsilon}\rangle^{(1)}$ from Eq. (6.139_1), and substituting into Eq. (6.145), leads to

$$\boldsymbol{\alpha}^* = \langle\boldsymbol{\alpha}\rangle + c^{(1)}\mathbf{L}_1^{(1)}\langle\overline{\mathbf{D}}^*\rangle_1 + c^{(1)}\mathbf{L}^{(0)}\tilde{\mathbf{d}}^I[\mathbf{L}_1^{(1)}]^{-1}\boldsymbol{\alpha}_1^{(1)} \tag{6.146}$$

where the tensor $\langle\overline{\mathbf{D}}^*\rangle_1$ is the average of $\overline{\mathbf{D}}^*(\mathbf{x})$ (6.34) defined by Eq. (6.109), and the last item in the right-hand side of Eq. (6.146) is obtained by substituting (6.129_2), (6.137) into (6.139_2) with a subsequent notation $\tilde{\mathbf{d}}^I = \sum_{r=0}^{1}\sum_{s=r}^{1}\tilde{\mathbf{d}}^{I(rs)}$. Hence, due to Eq. (6.109), Eq. (6.146) can be presented in terms of the constituent phase properties and the mechanical concentrator factors $\langle\mathbf{A}^*\rangle_1$ and $\tilde{\mathbf{d}}^{I(01)}$

$$\boldsymbol{\alpha}^* = \langle\boldsymbol{\alpha}\rangle - c^{(1)}\mathbf{L}_1^{(1)}(\mathbf{I} - \langle\mathbf{A}^*\rangle_1)[\mathbf{L}_1^{(1)}]^{-1}\boldsymbol{\alpha}_1^{(1)} + c^{(1)}\mathbf{L}^{(0)}\tilde{\mathbf{d}}^I[\mathbf{L}_1^{(1)}]^{-1}\boldsymbol{\alpha}_1^{(1)}. \tag{6.147}$$

The representation (6.147) in turn can be simplified by expressing $\langle \mathbf{A}^* \rangle_1$ through the effective elastic modulus (6.142)

$$\mathbf{L}^* = \mathbf{L}^{(0)} + c^{(1)}\mathbf{L}_1^{(1)}\langle \mathbf{A}^* \rangle_1 + c^{(1)}\mathbf{L}^{(0)}\tilde{\mathbf{d}}^I \tag{6.148}$$

and substituting the result into Eq. (6.147), which leads to the formula

$$\boldsymbol{\alpha}^* = \boldsymbol{\alpha}^{(0)} + (\mathbf{L}^* - \mathbf{L}^{(0)})[\mathbf{L}_1^{(1)}]^{-1}\boldsymbol{\alpha}_1^{(1)}, \tag{6.149}$$

which formally coincides with Eq. (6.104) for composites with perfect interface where at this point the tensor $\mathbf{L}^*$ is defined by Eq. (6.148) rather than by Eq. (6.21_1).

In a similar manner, we can obtain for the zero traction boundary conditions (6.38) $\mathbf{t}^{\partial w_\sigma}(\mathbf{x}) = \mathbf{0}$, $\mathbf{x} \in \partial w$ equations analogous to Eqs. (6.149) and (6.148)

$$\boldsymbol{\beta}^* = \boldsymbol{\beta}^{(0)} + (\mathbf{M}^* - \mathbf{M}^{(0)})[\mathbf{M}_1^{(1)}]^{-1}\boldsymbol{\beta}_1^{(1)}, \tag{6.150}$$

$$\mathbf{M}^* = \mathbf{M}^{(0)} + c^{(1)}\mathbf{M}_1^{(1)}\langle \mathbf{B}^* \rangle_1 + c^{(1)}\mathbf{M}^{(0)}\tilde{\mathbf{f}}^I, \tag{6.151}$$

respectively, where $\tilde{f}_{ijkl}^I = \frac{1}{w}\int_{\partial w^{(01)}} f_{(ikl}^I(\mathbf{x})n_{j)}(\mathbf{x})\, d\mathbf{x}$.

Recall that the effective eigenstrain $\boldsymbol{\beta}^*$ (6.150) and eigenstress $\boldsymbol{\alpha}^*$ (6.149) are obtained for composites with two isotropic phases and slipping interface. For the case of anisotropic constituents, Benveniste and Dvorak [97] have proposed the counterparts of these representations

$$\boldsymbol{\alpha}^* = \boldsymbol{\alpha}^{(0)} + (\mathbf{L}^* - \mathbf{L}^{(0)})[\mathbf{L}_1^{(1)}]^{-1}\boldsymbol{\alpha}_1^{(1)} + c^{(1)}(\tilde{\mathbf{d}}^I)^\top\left\{\boldsymbol{\alpha}^{(0)} - \mathbf{L}^{(0)}[\mathbf{L}_1^{(1)}]^{-1}\boldsymbol{\alpha}_1^{(1)}\right\}, \tag{6.152}$$

$$\boldsymbol{\beta}^* = \boldsymbol{\beta}^{(0)} + (\mathbf{M}^* - \mathbf{M}^{(0)})[\mathbf{M}_1^{(1)}]^{-1}\boldsymbol{\beta}_1^{(1)} + c^{(1)}(\tilde{\mathbf{f}}^I)^\top[\mathbf{M}_1^{(1)}]^{-1}\boldsymbol{\beta}_1^{(1)}, \tag{6.153}$$

where the last terms in Eqs. (6.152) and (6.153) vanish in the special cases of isotropic phases.

6.3.2 *Polycrystals Composed of Transversally Isotropic Crystals*

It turns out that a similar exact relationship between the thermal expansion coefficient and the bulk modulus holds for statistically isotropic polycrystalline aggregates (PA) composed of crystals of hexagonal, tetragonal, or trigonal symmetry. We will consider one-phase polycrystals composed of one kind of identical crystals with 3D uniform random orientation of their crystallographic axes described by the orientation distribution function $f(\mathbf{g})$ (ODF) (5.91) with respect to a fixed set of axes $\mathbf{x}_i$. Then the components of the material tensors, e.g, $\mathbf{M}$ and $\boldsymbol{\beta}$ referred to $\mathbf{x}_i$ $(i = 1, 2, 3)$, are denoted $\mathbf{M}(\mathbf{g})$ and $\boldsymbol{\beta}(\mathbf{g})$, respectively. Consequently, the polycrystal is a microinhomogeneous medium composed of anisotropic crystals of different orientation $\mathbf{g}$. For statistically homogeneous media being considered, $f(\mathbf{g})$ depends only on the orientation $\mathbf{g}$ but not on the position $\mathbf{x}$ of a crystal in the sample, and the statistical average $\langle(\cdot)\rangle$ (5.67_1) and orientation average $\langle(\cdot)\rangle^{\mathbf{g}}$ (5.89) coincide: $\langle(\cdot)\rangle = \langle(\cdot)\rangle^{\mathbf{g}}$. It follows that, for such a PA, the general representations for effective properties of microinhomogeneous media (6.21_2), (6.29), and (6.46) can be recast in the form

$$\mathbf{M}^* = \langle \mathbf{M}(\mathbf{g})\mathbf{B}^*(\mathbf{g})\rangle, \quad \mathbf{I} = \langle \mathbf{B}^*(\mathbf{g})\rangle, \quad \boldsymbol{\beta}^* = \langle \mathbf{B}^{*\top}(\mathbf{g})\boldsymbol{\beta}(\mathbf{g})\rangle. \tag{6.154}$$

The tensors $\mathbf{M}^*$ and $\boldsymbol{\beta}^*$ are assumed to be isotropic $\mathbf{M}^* = (1/3k^*, 1/2\mu^*)$ (2.198_1) and $\boldsymbol{\beta}^* = \beta^*\boldsymbol{\delta}$, respectively, where their components can be found by contraction of all free indices in Eq. (6.154) and analogously in Eq. (2.199)

$$1/k^* = \langle M_{rrkl}(\mathbf{g})B^*_{klii}(\mathbf{g})\rangle, \quad 3 = \langle B^*_{kkii}(\mathbf{g})\rangle, \quad 3\beta^* = \langle \beta_{kl}(\mathbf{g})B^*_{klii}(\mathbf{g})\rangle. \tag{6.155}$$

Since, moreover, each term in angle brackets $\langle(\cdot)\rangle$ is unaltering in the different coordinate systems, Eq. (6.155) may be written in the crystallographic axes of a crystal

$$1/k^* = M^c_{rrkl}B^c_{klii}, \quad 3 = B^c_{kkii}, \quad 3\beta^* = \beta^c_{kl}B^c_{klii}, \tag{6.156}$$

where the indexes c denote crystallographic axes.

We will consider a particular case of the traction boundary condition (6.5) restricted to a pure isotropic applied load

$$\boldsymbol{\sigma}^{\partial w}_{ij}(\mathbf{x}) = \sigma^0\delta_{ij} \tag{6.157}$$

which allows us to interpret the tensor B^c_{klii} as a stress concentrator factor corresponding to the loading (6.157)

$$\sigma^c_{kl} = B^c_{klii}\sigma^0. \tag{6.158}$$

Furthermore, we will consider all those crystal symmetries for which the second-order tensors β_{ij}, M_{ijmm}, M_{imjm}, and $M_{ijkl}\beta_{kl}$ show transversal isotropy where $\mathbf{x}_3$ is the preferred, axial direction. This includes the hexagonal, tetragonal, and trigonal symmetries [611]. For such an ideal statistically isotropic PA subjected to isotropic loading (6.157), there is no distinction between the statistical averages of the stresses, strains, and eigenstrains in the crystals of the specific direction which must show the same transversal symmetry as the crystals

$$\begin{aligned}&\boldsymbol{\sigma}(\mathbf{g}) = \boldsymbol{\sigma}^c = \mathrm{diag}(\sigma_1, \sigma_2, \sigma_3), \quad \boldsymbol{\varepsilon}(\mathbf{g}) = \boldsymbol{\varepsilon}^c = \mathrm{diag}(\varepsilon_1, \varepsilon_2, \varepsilon_3),\\ &\boldsymbol{\beta}(\mathbf{g}) = \boldsymbol{\beta}^c = \mathrm{diag}(\beta_1, \beta_2, \beta_3);\end{aligned} \tag{6.159}$$

for the present we do not distinguish between fields I and II. The relevant components of the elastic compliance tensor $\mathbf{M}^c$ (and, analogously, the elastic stiffness $\mathbf{L}^c$) are

$$M^c_{1111} = M_{2222},\ M^c_{3333},\ M^c_{1122},\ M^c_{1133} = M^c_{2233},\ M^c_{1212},\ M^c_{1313} = M^c_{2323} \tag{6.160}$$

(the other components vanish or do not enter into the subsequent analysis). Thus, all M_{rrij} with $i \neq j$ vanish, and the only surviving groups of $\mathbf{B}$ components are $B_{11ii} = B_{22ii}$ and B_{33ii}. Then, due to the absence of orientation dependence, Eq. (6.156) is reduced to

$$1/k^* = 2M^c_{rr11}B^c_{11ii} + M^c_{rr33}B^c_{33ii}, \tag{6.161}$$

$$3\beta^* = 2\beta_1 B^c_{11ii} + \beta_3 B^c_{33ii},\quad 3 = 2B^c_{11ii} + B^c_{33ii}. \tag{6.162}$$

which allows us to find the components B^c_{11ii}, B^c_{33ii} (and, therefore, σ^I_1 and σ^I_3 (6.158), (6.159)), as well as β^*

$$\sigma^I_1 = \frac{1}{2}\frac{1/k^* - 3M_3}{M_1 - M_3}\sigma^0,\quad \sigma^I_3 = \frac{3M_1 - 1/k^*}{M_1 - M_3}\sigma^0, \tag{6.163}$$

$$\beta^* = \beta^0 + \frac{\beta_3 - \beta_1}{3(M_3 - M_1)}(1/k^* - 2M_1 - M_3), \tag{6.164}$$

where $\beta^0 = (2\beta_1 + \beta_3)/3$ and

$$M_1 = M^c_{rr11} = M_{11} + M_{12} + M_{13} = \frac{L_{33} - L_{13}}{L_{33}(L_{11} + L_{12}) - 2L^2_{13}}, \tag{6.165}$$

$$M_3 = M^c_{rr33} = 2M_{13} + M_{33} = \frac{L_{11} + L_{12} - 2L_{13}}{L_{33}(L_{11} + L_{12}) - 2L^2_{13}}. \tag{6.166}$$

A more general representation of Eqs. (6.161) and (6.162) was proposed in [1347] where the case of a statistically homogeneous, but not isotropic, PA was analyzed.

In a similar manner, the components of the residual stresses $\boldsymbol{\sigma}^{II} = \text{diag}(\sigma^{II}_1, \sigma^{II}_1, \sigma^{II}_3)$ can be

$$\sigma^{II}_1 = \frac{3}{2}\frac{\beta^* - \beta^0}{M_1 - M_3},\quad \sigma^{II}_3 = -2\sigma^{II}_1, \tag{6.167}$$

which demonstrates that the statistical average residual stresses in the crystals are purely deviatoric ones because the isotropic part vanishes. Finally, the expression for the stored energy W^{II} can be obtained from (6.77$_2$) and (6.167)

$$W^{II} = \frac{3}{2}\frac{(\beta^* - \beta^0)(\beta_3 - \beta_1)}{M_1 - M_3}. \tag{6.168}$$

The expressions (6.164) and (6.168) are reminiscent of the corresponding expressions for two-phase materials (6.113) and (6.116), respectively.

Again all effective properties are monotonic functions of the effective bulk modulus k^*, and, therefore, the bounds on k^* lead to the bounds on effective eigenstrain, stored energy, as well as stresses. The basic results of this section were obtained in [611] (see also [804]) where the limiting case of elastic isotropy $(M_1 = M_3,\ \mathbf{M}^c = (3k, 2\mu))$ was also analyzed by the use of a passage to the limit $M_1 - M_3 \to \mathbf{0}$ in Eqs. (6.164), (6.167), and (6.168)

$$\beta^* = \beta^0, \quad W^{II} = \frac{2\mu}{15}\frac{9k + 8\mu}{3k + 4\mu}(\beta_3 - \beta_1)^2, \tag{6.169}$$

$$\sigma_1^{II} = \frac{2\mu}{15}\frac{9k + 8\mu}{3k + 4\mu}(\beta_3 - \beta_1), \quad \sigma_3^{II} = -2\sigma_1^{II}. \tag{6.170}$$

The exact relations (6.169) and (6.170) hold for arbitrary crystal shape. They can also be used as an approximation for PA demonstrating large anisotropic thermal deformation of crystals but weak anisotropy of elastic constants.

6.3.3 Exact Representation for First and Second Moments of Stresses Averaged over the Phase Volumes

In this section, we derive exact representations for all components for second-order moments of the stress and strain fields averaged over the volume of a component by utilizing the functional dependence of effective properties on the thermoelastic constants of the component being considered [110, 132, 133, 280, 808, 1187].

Just for concreteness, we will consider homogeneous traction boundary conditions (6.38) and decompose the overall field as $\boldsymbol{\sigma}(\mathbf{x}) = \boldsymbol{\sigma}^I(\mathbf{x}) + \boldsymbol{\sigma}^{II}(\mathbf{x})$, with the sources (6.37) and (6.38) $\boldsymbol{\beta}^I(\mathbf{x}) = \mathbf{0},\ \ \mathbf{t}^I(\mathbf{y}) = \mathbf{t}^{\partial w_\sigma}(\mathbf{y})$ and $\boldsymbol{\beta}^{II}(\mathbf{x}) = \boldsymbol{\beta}(\mathbf{x}),\ \ \mathbf{t}^{II}(\mathbf{y}) = \mathbf{0}$ for the field $\boldsymbol{\sigma}^I$ and $\boldsymbol{\sigma}^{II}$, respectively, where $\mathbf{t}^{\partial w_\sigma}$ $(\mathbf{x} \in w,\ \ \mathbf{y} \in \partial w_\sigma)$ is a prescribed traction in the boundary conditions (6.38). We now summarize the results of Sect. 6.2 in a form appropriated to our anticipated purpose

$$\langle \boldsymbol{\sigma}^I \rangle = \langle \boldsymbol{\sigma} \rangle, \quad \langle \boldsymbol{\sigma}^{II} \rangle = \mathbf{0}, \tag{6.171}$$

$$\langle \boldsymbol{\sigma}^I \mathbf{M} \boldsymbol{\sigma}^I \rangle = \langle \boldsymbol{\sigma} \rangle \mathbf{M}^* \langle \boldsymbol{\sigma} \rangle, \quad \langle \boldsymbol{\sigma}^{II} \mathbf{M} \boldsymbol{\sigma}^{II} \rangle = 2W^*, \quad \langle \boldsymbol{\sigma}^I \mathbf{M} \boldsymbol{\sigma}^{II} \rangle = 0, \tag{6.172}$$

$$\langle \boldsymbol{\sigma}^I \boldsymbol{\beta} \rangle = \langle \boldsymbol{\sigma} \rangle \boldsymbol{\beta}^*, \quad \langle \boldsymbol{\sigma}^{II} \boldsymbol{\beta} \rangle = -2W^*, \tag{6.173}$$

where the right-hand side of the mentioned scalar equations only depends on the effective constant $\mathbf{M}^*,\ \boldsymbol{\beta}^*,\ W^*$ assumed to be known in the following.

For fixed boundary conditions (6.38), we first consider a small variation of the local compliance $\delta\mathbf{M}(\mathbf{x})$ as compared to $\mathbf{M}(\mathbf{x})$: $\mathbf{M}(\mathbf{x}) \to \mathbf{M}(\mathbf{x}) + \delta\mathbf{M}(\mathbf{x})$ implying corresponding variations of the stress and strain fields: $\boldsymbol{\sigma}(\mathbf{x}) \to \boldsymbol{\sigma}(\mathbf{x}) + \delta\boldsymbol{\sigma}(\mathbf{x})$ and $\boldsymbol{\varepsilon}(\mathbf{x}) \to \boldsymbol{\varepsilon}(\mathbf{x}) + \delta\boldsymbol{\varepsilon}(\mathbf{x})$, respectively. Since both $\boldsymbol{\sigma}(\mathbf{x})$ and $\boldsymbol{\sigma}(\mathbf{x}) + \delta\boldsymbol{\sigma}(\mathbf{x})$ satisfy the same boundary condition, we obtain from (6.171) $\langle\delta\boldsymbol{\sigma}(\mathbf{x})\rangle = \mathbf{0}$. Moreover, $\delta\boldsymbol{\sigma}(\mathbf{x})$ and $\delta\boldsymbol{\varepsilon}(\mathbf{x})$ accomplish the requirements of the Hill's condition (6.16) leading to the scale equalities

$$\langle\delta\boldsymbol{\sigma}\,\boldsymbol{\varepsilon}\rangle = \langle\delta\boldsymbol{\sigma}\rangle\langle\boldsymbol{\varepsilon}\rangle = 0, \quad \langle\boldsymbol{\sigma}\,\delta\boldsymbol{\varepsilon}\rangle = \langle\boldsymbol{\sigma}\rangle\langle\delta\boldsymbol{\varepsilon}\rangle, \tag{6.174}$$

which are fulfilled for both $\boldsymbol{\varepsilon}^{I}$ and $\boldsymbol{\varepsilon}^{II}$ combined with both fields $\boldsymbol{\sigma}^{I}$ and $\boldsymbol{\sigma}^{II}$.

Varying (6.172_1) in an accompany with Eqs. (6.174_1) and (6.39) leads to

$$\langle\boldsymbol{\sigma}\rangle\delta\mathbf{M}^*\langle\boldsymbol{\sigma}\rangle = \langle\boldsymbol{\sigma}^{I}\delta\mathbf{M}\boldsymbol{\sigma}^{I}\rangle + 2\langle\boldsymbol{\sigma}^{I}\mathbf{M}\delta\boldsymbol{\sigma}^{I}\rangle = \langle\boldsymbol{\sigma}^{I}\delta\mathbf{M}\boldsymbol{\sigma}^{I}\rangle. \tag{6.175}$$

In a similar manner, variation of Eq. (6.172_2) by varying $\mathbf{M}(\mathbf{x})$ at the fixed $\boldsymbol{\beta}(\mathbf{x})$ yields

$$2\delta W^* = \langle\boldsymbol{\sigma}^{II}\delta\mathbf{M}\boldsymbol{\sigma}^{II}\rangle + 2\langle\boldsymbol{\sigma}^{II}\mathbf{M}\delta\boldsymbol{\sigma}^{II}\rangle = \langle\boldsymbol{\sigma}^{II}\delta\mathbf{M}\boldsymbol{\sigma}^{II}\rangle + 2\langle\boldsymbol{\varepsilon}^{II}\delta\boldsymbol{\sigma}^{II}\rangle - 2\langle\boldsymbol{\beta}\delta\boldsymbol{\sigma}^{II}\rangle \tag{6.176}$$

where the second term in the right-hand side of Eq. (6.176) vanishes in virtue of (6.174_1). Combining Eqs. (6.173_2) and (6.176) leads to

$$\langle\boldsymbol{\sigma}^{II}\delta\mathbf{M}\boldsymbol{\sigma}^{II}\rangle = -2\delta W^*. \tag{6.177}$$

Finally, variation of (6.172_3) with varying $\mathbf{M}(\mathbf{x})$ and fixed $\boldsymbol{\beta}(\mathbf{x})$ gives

$$\begin{aligned} 0 = \langle\boldsymbol{\sigma}^{I}\delta\mathbf{M}\boldsymbol{\sigma}^{II}\rangle + \langle\delta\boldsymbol{\sigma}^{I}\mathbf{M}\boldsymbol{\sigma}^{II}\rangle + \langle\boldsymbol{\sigma}^{I}\mathbf{M}\delta\boldsymbol{\sigma}^{II}\rangle = \langle\boldsymbol{\sigma}^{I}\delta\mathbf{M}\boldsymbol{\sigma}^{II}\rangle \\ +\langle\delta\boldsymbol{\sigma}^{I}\boldsymbol{\varepsilon}^{II}\rangle - \langle\delta\boldsymbol{\sigma}^{I}\boldsymbol{\beta}\rangle + \langle\boldsymbol{\sigma}^{I}\delta\boldsymbol{\sigma}^{II}\rangle = \langle\boldsymbol{\sigma}^{I}\delta\mathbf{M}\boldsymbol{\sigma}^{II}\rangle - \langle\boldsymbol{\sigma}\rangle\delta\boldsymbol{\beta}^*, \end{aligned} \tag{6.178}$$

where the last equality was obtained utilizing Eqs. (6.173_1), (6.174_1), and (6.174_2).

Here, one may use a special variation of the compliance tensor ($k = 0, 1, \ldots$) :

$$\delta\mathbf{M}(\mathbf{x}) \equiv \delta\mathbf{M}^{(k)}V^{(k)}(\mathbf{x}), \ \ \delta\mathbf{M}^{(k)} = \text{const}, \ \ \mathbf{x} \in v^{(k)}. \tag{6.179}$$

In this case $\delta\mathbf{M}$ may be taken outside the averaging sign over the volume of component $v^{(k)}$ ($k = 0, 1, \ldots, N$). Furthermore, we have

$$\delta\mathbf{M}^* = \frac{\partial\mathbf{M}^*}{\partial\mathbf{M}^{(k)}}\delta\mathbf{M}^{(k)}, \quad \delta W^* = \frac{\partial W^*}{\partial\mathbf{M}^{(k)}}\delta\mathbf{M}^{(k)}, \quad \delta\boldsymbol{\beta}^* = \frac{\partial\boldsymbol{\beta}^*}{\partial\mathbf{M}^{(k)}}\delta\mathbf{M}^{(k)} \tag{6.180}$$

which together with Eqs. (6.175), (6.177), (6.178), and (6.179) leads to the final expressions for the second moments of stresses

$$\langle\boldsymbol{\sigma}^{I}\otimes\boldsymbol{\sigma}^{I}\rangle^{(k)} = \frac{1}{c^{(k)}}\frac{\partial\mathbf{M}^*}{\partial\mathbf{M}^{(k)}}\langle\boldsymbol{\sigma}\rangle\otimes\langle\boldsymbol{\sigma}\rangle, \tag{6.181}$$

$$\langle\boldsymbol{\sigma}^{II}\otimes\boldsymbol{\sigma}^{II}\rangle^{(k)} = -\frac{2}{c^{(k)}}\frac{\partial W^*}{\partial\mathbf{M}^{(k)}}, \quad \langle\boldsymbol{\sigma}^{I}\otimes\boldsymbol{\sigma}^{II}\rangle^{(k)} = \frac{2}{c^{(k)}}\frac{\partial\boldsymbol{\beta}^*}{\partial\mathbf{M}^{(k)}}\langle\boldsymbol{\sigma}\rangle \tag{6.182}$$

or in index form

$$\langle\sigma^I_{pq}\sigma^I_{rs}\rangle^{(k)} = \frac{1}{c^{(k)}}\frac{\partial M^*_{ijmn}}{\partial M^{(k)}_{pqrs}}\langle\sigma_{ij}\rangle\langle\sigma_{mn}\rangle,$$

$$\langle\sigma^{II}_{pq}\sigma^{II}_{rs}\rangle^{(k)} = -\frac{2}{c^{(k)}}\frac{\partial W^*}{\partial M^{(k)}_{pqrs}}, \quad \langle\sigma^{I}_{pq}\sigma^{II}_{rs}\rangle^{(k)} = \frac{2}{c^{(k)}}\frac{\partial\beta^*_{ij}}{\partial M^{(k)}_{pqrs}}\langle\sigma_{ij}\rangle. \quad (6.183)$$

Summation of Eqs. (6.181) and (6.182) yields the representation for the second moment of the overall stresses $\boldsymbol{\sigma} = \boldsymbol{\sigma}^I + \boldsymbol{\sigma}^{II}$ (see [280])

$$\langle\boldsymbol{\sigma}\otimes\boldsymbol{\sigma}\rangle^{(k)} = \frac{1}{c^{(k)}}\frac{\partial\mathbf{M}^*}{\partial\mathbf{M}^{(k)}}\langle\boldsymbol{\sigma}\rangle\otimes\langle\boldsymbol{\sigma}\rangle - \frac{2}{c^{(k)}}\frac{\partial W^*}{\partial\mathbf{M}^{(k)}} + \frac{2}{c^{(k)}}\frac{\partial\boldsymbol{\beta}^*}{\partial\mathbf{M}^{(k)}}\langle\boldsymbol{\sigma}\rangle, \quad (6.184)$$

where the partial derivatives are calculated under the assumption of fixed transformation fields $\boldsymbol{\beta}(\mathbf{x}),\ \mathbf{x}\in w$. Equation (6.184) is equivalently expressed through the effective stress potential U^*

$$\langle\boldsymbol{\sigma}\otimes\boldsymbol{\sigma}\rangle^{(k)} = \frac{2}{c^{(k)}}\frac{\partial U^*}{\partial\mathbf{M}^{(k)}}, \quad U^* = \frac{1}{2}\langle\boldsymbol{\sigma}\rangle\mathbf{M}^*\langle\boldsymbol{\sigma}\rangle + \boldsymbol{\beta}^*\langle\boldsymbol{\sigma}\rangle - W^* \quad (6.185)$$

obtained in [963].

Invariants $\langle\sigma_0^2\rangle^{(k)}$ and $\langle s_{ij}s_{ij}\rangle^{(k)}$ ($\sigma_0 = \sigma_{ii}/3,\ s_{ij} = \sigma_{ij} - \sigma_0$) are estimated employing the convolutions of Eq. (6.183). For this purpose, we will assume that $\delta\mathbf{M}^{(k)} = 3\delta\bar{p}^{(k)}\mathbf{N}_1 + 2\delta\bar{q}^{(k)}\mathbf{N}_2$, $3\bar{p}^{(k)} = M^{(k)}_{iijj}/3$, $2\bar{q}^{(k)} = (M^{(k)}_{ijij} - 3\bar{p}^{(k)})/5$ ($\mathbf{N}_1 = \boldsymbol{\delta}\otimes\boldsymbol{\delta}/3,\ \mathbf{N}_2 = \mathbf{I} - \mathbf{N}_1$ (see Eq. (2.200)); i.e., we consider a variation $\delta\mathbf{M}^{(k)}$ rather than the tensor $\mathbf{M}^{(k)}$ as such. Taking into account a differential rule of a composite function yields

$$\langle\boldsymbol{\sigma}\rangle\frac{\partial\mathbf{M}^*}{\partial\bar{p}^{(k)}}\langle\boldsymbol{\sigma}\rangle = \langle\boldsymbol{\sigma}\rangle\frac{\partial\mathbf{M}^*}{\partial M^{(k)}_{ijmn}}\langle\boldsymbol{\sigma}\rangle\frac{\partial M^{(k)}_{ijmn}}{\partial\bar{p}^{(k)}}$$

$$= c^{(k)}\langle\sigma_{ij}\sigma_{mn}\rangle^{(k)}3N_{1|ijmn} = 9\langle\sigma_0^2\rangle^{(k)} \quad (6.186)$$

$$\langle\boldsymbol{\sigma}\rangle\frac{\partial\mathbf{M}^*}{\partial\bar{q}^{(k)}}\langle\boldsymbol{\sigma}\rangle = \langle\boldsymbol{\sigma}\rangle\frac{\partial\mathbf{M}^*}{\partial M^{(k)}_{ijmn}}\langle\boldsymbol{\sigma}\rangle\frac{\partial M^{(k)}_{ijmn}}{\partial\bar{q}^{(k)}}$$

$$= c^{(k)}\langle\sigma_{ij}\sigma_{mn}\rangle^{(k)}2N_{2|ijmn} = 2\langle s_{ij}s_{ij}\rangle^{(k)}. \quad (6.187)$$

Then, since the variations $\delta\bar{p}^{(k)}$ and $\delta\bar{q}^{(k)}$ are mutually independent, we can get from (6.183) the representations for the stress invariants averaged over the k's phase

$$\langle\sigma_0^2\rangle^{(k)} = \frac{1}{9c^{(k)}}\frac{\partial\mathbf{M}^*}{\partial\bar{p}^{(k)}}\langle\boldsymbol{\sigma}\rangle\otimes\langle\boldsymbol{\sigma}\rangle - \frac{2}{9c^{(k)}}\frac{\partial W^*}{\partial\bar{p}^{(k)}} + \frac{2}{9c^{(k)}}\frac{\partial\boldsymbol{\beta}^*}{\partial\bar{p}^{(k)}}\langle\boldsymbol{\sigma}\rangle, \quad (6.188)$$

$$\langle s_{ij}s_{ij}\rangle^{(k)} = \frac{1}{2c^{(k)}}\frac{\partial \mathbf{M}^*}{\partial \bar{q}^{(k)}}\langle\boldsymbol{\sigma}\rangle\otimes\langle\boldsymbol{\sigma}\rangle - \frac{1}{c^{(k)}}\frac{\partial W^*}{\partial \bar{q}^{(k)}} + \frac{1}{c^{(k)}}\frac{\partial \boldsymbol{\beta}^*}{\partial \bar{q}^{(k)}}\langle\boldsymbol{\sigma}\rangle, \tag{6.189}$$

where the first, second, and third terms in the right-hand sides of Eqs. (6.188) and (6.189) comprise the appropriate representations for the stress invariants of the corresponding fourth-rank tensors of stresses $\langle\boldsymbol{\sigma}^I\otimes\boldsymbol{\sigma}^I\rangle^{(k)}$, $\langle\boldsymbol{\sigma}^{II}\otimes\boldsymbol{\sigma}^{II}\rangle^{(k)}$, and $\langle\boldsymbol{\sigma}^I\otimes\boldsymbol{\sigma}^{II}\rangle^{(k)}$, respectively, averaged over the volume of the k's phase.

Of course, analogous relations can be derived for the second moment of strains, either through the transformations of Eqs. (6.171)–(6.173) into $\boldsymbol{\varepsilon}$ expressions or by recasting of Eqs. (6.181) and (6.182) in $\boldsymbol{\varepsilon}$ terms by the use of constituent Eqs. (6.39) and (6.40). For example, for pure mechanical stresses $\boldsymbol{\beta}(\mathbf{x})\equiv\mathbf{0}$), we get

$$\langle\boldsymbol{\varepsilon}^I\otimes\boldsymbol{\varepsilon}^I\rangle^{(k)} = \frac{1}{c^{(k)}}\frac{\partial \mathbf{L}^*}{\partial \mathbf{L}^{(k)}}\langle\boldsymbol{\varepsilon}\rangle\otimes\langle\boldsymbol{\varepsilon}\rangle,$$

$$\langle(\varepsilon_0^I)^2\rangle^{(k)} = \frac{1}{9c^{(k)}}\frac{\partial \mathbf{L}^*}{\partial \bar{k}^{(k)}}\langle\boldsymbol{\varepsilon}\rangle\otimes\langle\boldsymbol{\varepsilon}\rangle, \quad \langle\varepsilon_{\rm eq}^2\rangle^{(k)} = \frac{1}{3c^{(k)}}\frac{\partial \mathbf{L}^*}{\partial \bar{\mu}^{(k)}}\langle\boldsymbol{\varepsilon}\rangle\otimes\langle\boldsymbol{\varepsilon}\rangle, \tag{6.190}$$

where $\varepsilon_0 = \varepsilon_{ii}/3$, $\varepsilon_{\rm eq}^2 = 2e_{ij}e_{ij}/3$, $\mathbf{e} = \boldsymbol{\varepsilon} - \varepsilon_0\boldsymbol{\delta}$, and $3\bar{k}^{(k)} = L_{iijj}^{(k)}/3$, $2\bar{\mu}^{(k)} = (L_{ijij}^{(k)} - 3\bar{k}^{(k)})/5$. It should be mentioned that we did not assume isotropy of the tensors $\mathbf{M}^{(k)}$, $\mathbf{L}^{(k)}$, and, because of this, $3\bar{p}^{(k)} \neq (3\bar{k}^{(k)})^{-1}$, $2\bar{q}^{(k)} \neq (2\bar{\mu}^{(k)})^{-1}$ in general. If the elasticity tensor of the k's phase is isotropic $\mathbf{L}^{(k)} = 3k^{(k)}\mathbf{N}_1 + 2\mu^{(k)}\mathbf{N}_2$, then $\bar{p}^{(k)} = p^{(k)}$, $\bar{q}^{(k)} = q^{(k)}$, $\bar{k}^{(k)} = k^{(k)}$, $\bar{\mu}^{(k)} = \mu^{(k)}$, and $3p^{(k)} = (3k^{(k)})^{-1}$, $2q^{(k)} = (2\mu^{(k)})^{-1}$.

Buryachenko [176, 177] and [246, 247] proposed an approximate method of the estimation of the second moment of stresses only inside the matrix from the elastic energy which is exact for the particular case of the composite with incompressible matrix containing rigid inclusions or pores. They used the additional assumption

$$\langle\sigma_0^2\rangle_0 = \langle\sigma_0\rangle_0^2 \tag{6.191}$$

(i.e., perturbed hydrostatic fields in the matrix are neglected). Later, the assumption (6.191) was used [120, 1177, 1259–1261] and equivalent results found by a different method based on the energy approach (see also [1011]), rendering a physical meaning for the adoption of the second stress moment in a homogenization scheme. Without such a connection, the concept of second stress moment would remain purely mathematical. Buryachenko [181] gave estimates of the error caused by the assumptions (6.191) by using Eq. (6.184), which for porous materials gives

$$\langle\sigma_0^2\rangle_0 - \langle\sigma_0\rangle_0^2 = \frac{1}{9(1-c)}\left(\frac{\partial \mathbf{M}^*}{\partial p^{(0)}} - \frac{3\mathbf{N}_1}{1-c}\right)(\langle\sigma\rangle\otimes\langle\sigma\rangle),$$

where the right-hand side equals zero only for an incompressible matrix.

Some particular results of relationships (6.184) and (6.190) were obtained in [1187] (for $\boldsymbol{\beta}\equiv 0$) and [238] (for $\boldsymbol{\sigma}^0\equiv\mathbf{0}$). A more tractable particular case of

Eq. (6.184) is proposed in [661]. Buryachenko and Shermergor [280] considered the generalization of Eq. (6.184) for thermoelastic-electric fields. Relations (6.184) have been obtained for any degree of anisotropy of $\mathbf{M}^*,\ \mathbf{M}^{(i)}\quad (i = 0, 1, \ldots, N)$. For isotropic tensors $\mathbf{M}^{(i)}$, the relations (6.184) reduce to the results [133] and [808] for macroisotropic composites.

In this chapter the exact representation of stress concentrator factors averaged over the volume of the constituents of the two-phase composites and statistically isotropic polycrystalline aggregates composed of crystals of some symmetries was obtained. It was found that exactly analogous results can be obtained in a general case of multiphase composites if the functional dependencies of effective constants on the constituent compliance are known. Indeed, following [808], we will vary a transformation field $\delta\boldsymbol{\beta}(\mathbf{x})$ at the fixed $\mathbf{M}(\mathbf{x})$ instead of variation of $\delta\mathbf{M}(\mathbf{x})$ at the fixed $\boldsymbol{\beta}(\mathbf{x})$. Then varying Eq. (6.173_1) leads to

$$\langle\boldsymbol{\sigma}\rangle\delta\boldsymbol{\beta}^* = \langle\boldsymbol{\sigma}^I\delta\boldsymbol{\beta}\rangle + \langle\delta\boldsymbol{\sigma}^I\boldsymbol{\beta}\rangle = \langle\boldsymbol{\sigma}^I\delta\boldsymbol{\beta}\rangle + \langle\delta\boldsymbol{\sigma}^I\boldsymbol{\varepsilon}^{II}\rangle - \langle\delta\boldsymbol{\sigma}^I(\boldsymbol{\varepsilon}^{II} - \boldsymbol{\beta})\rangle = \langle\boldsymbol{\sigma}^I\delta\boldsymbol{\beta}\rangle$$
$$+ \langle\delta\boldsymbol{\sigma}^I\boldsymbol{\varepsilon}^{II}\rangle - \langle\delta\boldsymbol{\varepsilon}^I\mathbf{L}(\boldsymbol{\varepsilon}^{II} - \boldsymbol{\beta})\rangle = \langle\boldsymbol{\sigma}^I\delta\boldsymbol{\beta}\rangle + \langle\delta\boldsymbol{\sigma}^I\boldsymbol{\varepsilon}^{II}\rangle - \langle\delta\boldsymbol{\varepsilon}^I\boldsymbol{\sigma}^{II}\rangle, \tag{6.192}$$

where the second and the third term in the right-hand side vanish in virtue of (6.171) and (6.174_2) that reduce Eq. (6.192) to $\langle\boldsymbol{\sigma}^I\delta\boldsymbol{\beta}\rangle = \langle\boldsymbol{\sigma}\rangle\delta\boldsymbol{\beta}^*$. In a similar manner, variation of (6.173_2) by varying $\boldsymbol{\beta}(\mathbf{x})$ yields

$$\begin{aligned}
-\delta W^* &= \langle\boldsymbol{\sigma}^{II}\delta\boldsymbol{\beta}\rangle + \langle\delta\boldsymbol{\sigma}^{II}\boldsymbol{\beta}\rangle = \langle\boldsymbol{\sigma}^{II}\delta\boldsymbol{\beta}\rangle + \langle\delta\boldsymbol{\sigma}^{II}\boldsymbol{\varepsilon}^{II}\rangle - \langle\delta\boldsymbol{\sigma}^{II}(\boldsymbol{\varepsilon}^{II} - \boldsymbol{\beta})\rangle\\
&= \langle\boldsymbol{\sigma}^{II}\delta\boldsymbol{\beta}\rangle + \langle\delta\boldsymbol{\sigma}^{II}\boldsymbol{\varepsilon}^{II}\rangle - \langle\delta(\boldsymbol{\varepsilon}^{II} - \boldsymbol{\beta})\boldsymbol{\sigma}^{II}\rangle\\
&= = 2\langle\boldsymbol{\sigma}^{II}\delta\boldsymbol{\beta}\rangle + \langle\delta\boldsymbol{\sigma}^{II}\boldsymbol{\varepsilon}^{II}\rangle - \langle\boldsymbol{\sigma}^{II}\delta\boldsymbol{\varepsilon}^{II}\rangle,
\end{aligned}$$

which can be simplified due to Eqs. (6.171) and (6.182_1) $\langle\boldsymbol{\sigma}^{II}\delta\boldsymbol{\beta}\rangle = -\delta W^*$. A special choice of the varying eigenstrains $(k = 0, 1, \ldots) : \delta\boldsymbol{\beta}(\mathbf{x}) \equiv \delta\boldsymbol{\beta}^{(k)}V^{(k)}(\mathbf{x})$, $\delta\boldsymbol{\beta}^{(k)} = \text{const}\ \ (\mathbf{x} \in v^{(k)})$ analogously to (6.179) allows the explicit representations for the average stresses in the k's phase

$$\langle\boldsymbol{\sigma}^I\rangle^{(k)} = \frac{1}{c^{(k)}}\langle\boldsymbol{\sigma}\rangle\frac{\partial\boldsymbol{\beta}^*}{\partial\boldsymbol{\beta}^{(k)}},\quad \langle\boldsymbol{\sigma}^{II}\rangle^{(k)} = -\frac{1}{c^{(k)}}\frac{\partial W^*}{\partial\boldsymbol{\beta}^{(k)}}, \tag{6.193}$$

under the assumption of fixed compliance $\mathbf{M}(\mathbf{x}),\ \mathbf{x} \in w$.

The preceding considerations show that in order to obtain a quantitative assessment of $\langle\boldsymbol{\sigma}\rangle^{(k)}$ and $\langle\boldsymbol{\sigma}\otimes\boldsymbol{\sigma}\rangle^{(k)}$, it is necessary to know the functional dependence of $\mathbf{M}^*,\ W^*,\ \boldsymbol{\beta}^*$ on both the compliance tensors $\mathbf{M}^{(k)}$ and eigenstrain $\boldsymbol{\beta}_1^{(k)}$. The dependencies can be calculated only approximately, except for some degenerate structures. Therefore, one may think that the relationships are not indeed strictly exact. However, all of them can be measured experimentally, and thus, the exact relationships obtained in this work have an immediate physical sense. Therefore, eventually, the error of numerical results for the first and second field moments is governed by the error of determining the dependencies of $\mathbf{M}^*$, W^*, and $\boldsymbol{\beta}^*$ upon $\mathbf{M}(\mathbf{x})$, and these dependence will be determined below with the help of MEFM, which is physically consistent and requires only weak assumptions.

6.4 Variational Principle of Hashin and Shtrikman

For the first boundary-value problem ($\boldsymbol{\varepsilon}^{\partial w} = \boldsymbol{\varepsilon}^0$, $\partial w_{\mathbf{u}} = \partial w$) (6.4), according to the principle of minimum potential energy (2.105), the total strain enery $W = \langle \boldsymbol{\sigma}\boldsymbol{\varepsilon}\rangle/2$ holds inequality

$$2W \le (\boldsymbol{\varepsilon}^*, \mathbf{L}\boldsymbol{\varepsilon}^*), \tag{6.194}$$

with equality only when a kinematically admissible displacement (2.100) $\mathcal{A}^k \ni \boldsymbol{\varepsilon}^* \equiv \boldsymbol{\varepsilon}^0$. Hereafter, one introduces the scalar production in the space of square integrable functions $(\mathbf{f}, \mathbf{g}) = \bar{w}^{-1}\int_w \mathbf{f}(\mathbf{x})\mathbf{g}(\mathbf{x})\, d\mathbf{x}$. On the other hand, for a statically admissible stress field $\boldsymbol{\sigma}^* \in \mathcal{A}^s$, the prinsiple of complimentary energy (2.107) yields

$$2W \ge 2(\boldsymbol{\sigma}^*, \boldsymbol{\varepsilon}^0) - (\boldsymbol{\sigma}^*, \mathbf{M}\boldsymbol{\sigma}^*) \tag{6.195}$$

with equality only when $\boldsymbol{\sigma}^*$ coincides with the actual stress $\boldsymbol{\sigma}^* \equiv \boldsymbol{\sigma}$.

For the second boundary-value problem ($\boldsymbol{\sigma}^{\partial w} = \boldsymbol{\sigma}^0$, $\partial w_{\sigma} = \partial w$) (6.5), from the potential energy principle (2.105)

$$2W \ge 2(\boldsymbol{\varepsilon}^*, \boldsymbol{\sigma}^0) - (\boldsymbol{\varepsilon}^*, \mathbf{L}\boldsymbol{\varepsilon}^*), \tag{6.196}$$

with equality only when a virtual strain $\boldsymbol{\varepsilon}^*$ coincides with the actual strain $\boldsymbol{\varepsilon}^* \equiv \boldsymbol{\sigma}$. By the complimentary principle (2.107)

$$2W \le (\boldsymbol{\sigma}^*, \mathbf{M}\boldsymbol{\sigma}^*), \tag{6.197}$$

for $\forall\boldsymbol{\sigma}^* \in \mathcal{A}^s$ with equality only when $\boldsymbol{\sigma}^* \equiv \boldsymbol{\sigma}$.

A famous variational principle of Hashin and Shtrikman [618, 619] (see also Hill [641]) is based on both the idea of a homogeneous comparison medium with a stiffness $\mathbf{L}^c$ and compliance $\mathbf{M}^c = (\mathbf{L}^c)^{-1}$ and the use of the notions of the polarization tensors of the stresses $\boldsymbol{\tau}$ and strains $\boldsymbol{\eta}$

$$\boldsymbol{\tau}(\mathbf{x}) = \boldsymbol{\sigma}(\mathbf{x}) - \mathbf{L}^c\boldsymbol{\varepsilon}(\mathbf{x}), \quad \boldsymbol{\sigma}(\mathbf{x}) = \mathbf{L}^c\boldsymbol{\varepsilon}(\mathbf{x}) + \boldsymbol{\tau}(\mathbf{x}) \tag{6.198}$$

$$\boldsymbol{\eta}(\mathbf{x}) = \boldsymbol{\varepsilon}(\mathbf{x}) - \mathbf{M}^c\boldsymbol{\sigma}(\mathbf{x}), \quad \boldsymbol{\varepsilon}(\mathbf{x}) = \mathbf{M}^c\boldsymbol{\sigma}(\mathbf{x}) + \boldsymbol{\eta}(\mathbf{x}) \tag{6.199}$$

instead of the tensors of stresses $\boldsymbol{\sigma}$ and strains $\boldsymbol{\varepsilon}$, respectively. Here, the stress polarization and strain polarization tensors are related by the equation $\boldsymbol{\eta}(\mathbf{x}) = -\mathbf{M}^c\boldsymbol{\tau}(\mathbf{x})$.

For the sake of definiteness, we will consider the first boundary-value problem (6.4) with the prescribed surface displacement $\mathbf{u}(\mathbf{x}) = \boldsymbol{\varepsilon}^0 \cdot \mathbf{x}$ compatible with a uniform strain $\boldsymbol{\varepsilon}^0 \equiv \langle\boldsymbol{\varepsilon}\rangle$ which is consequently the overall average strain $\langle\boldsymbol{\varepsilon}\rangle$ (6.7). The general integral equation (see Chap. 7) for statistically homogeneous media can be presented in the following operator form:

$$\boldsymbol{\varepsilon} = \boldsymbol{\varepsilon}^0 + {}^{\mathcal{P}}\mathbf{U}\boldsymbol{\tau}, \quad {}^{\mathcal{P}}\mathbf{U}\boldsymbol{\tau}(\mathbf{x}) := \int \Big[\mathbf{U}(\mathbf{x}-\mathbf{y})\boldsymbol{\tau}(\mathbf{y}) - \langle\mathbf{U}(\mathbf{x}-\mathbf{y})\boldsymbol{\tau}(\mathbf{y})\rangle\Big]\, d\mathbf{y}, \tag{6.200}$$

where the operator ${}^{\mathcal{P}}\mathbf{U}\boldsymbol{\tau}$ differs from the popular operator $\mathbf{U}^{\mathcal{P}}\boldsymbol{\tau} = \int \mathbf{U}(\mathbf{x} - \mathbf{y})[\boldsymbol{\tau}(\mathbf{y}) - \langle\boldsymbol{\tau}\rangle]\, d\mathbf{y}$ (see for details Sect. 7.2.3) defined by the kernel $\mathbf{U}(\mathbf{x})$ (3.19) and $\mathbf{L}_1(\mathbf{x}) \equiv \mathbf{L}(\mathbf{x}) - \mathbf{L}^c$. We intentionally used the notations ${}^{\mathcal{P}}\mathbf{U} := \mathcal{P}\mathbf{U}$ and $\mathbf{U}^{\mathcal{P}} := \mathbf{U}\mathcal{P} \not\equiv {}^{\mathcal{P}}\mathbf{U}$ associated with a progected operator $\mathcal{P}$ ($\mathcal{P}\mathbf{f} := \mathbf{f} - \langle\mathbf{f}\rangle$) introduced in [106]. Obviously for the arbitrary constant tensors $\mathbf{b}_1, \mathbf{b}_2 = \text{const}$.

$${}^{\mathcal{P}}\mathbf{U}\boldsymbol{\tau} = {}^{\mathcal{P}}\mathbf{U}(\boldsymbol{\tau} + \mathbf{b}_1), \quad (\boldsymbol{\tau}, {}^{\mathcal{P}}\mathbf{U}\boldsymbol{\tau}) = (\boldsymbol{\tau} + \mathbf{b}_2, \mathbf{U}\boldsymbol{\tau}). \tag{6.201}$$

To prove self-adjoint of the operator ${}^{\mathcal{P}}\mathbf{U}$, recall that $(\boldsymbol{\tau}_1, {}^{\mathcal{P}}\mathbf{U}\boldsymbol{\tau}_2) = (\boldsymbol{\tau}_1, \boldsymbol{\varepsilon}_2) = (\boldsymbol{\sigma}_1 - \mathbf{L}^c\boldsymbol{\varepsilon}_1, \boldsymbol{\varepsilon}_2)$, where $\boldsymbol{\sigma}_1$ and $\boldsymbol{\varepsilon}_1$ are the stress and strain associated with arbitrary $\boldsymbol{\tau}_1$ according to Eqs. (6.198_2) and (6.200), respectively, and $\boldsymbol{\varepsilon}_2$ is any strain field similarly consistent with $\boldsymbol{\tau}_2$ and zero displacement boundary conditions (6.4). The virtual principle (2.104)

$$(\boldsymbol{\sigma}_1, \boldsymbol{\varepsilon}_2) = 0 \tag{6.202}$$

and Eq. (6.200) lead to $(\boldsymbol{\tau}_1, \mathbf{U}\boldsymbol{\tau}_2) = -(\mathbf{L}^c\boldsymbol{\varepsilon}_1, \boldsymbol{\varepsilon}_2)$. The operator $\mathbf{U}$ is self-adjoint because the last equality is symmetric with respect to the interchange of indexes due to the symmetry of $\mathbf{L}^c$. It what follows, we see after substitution of $\boldsymbol{\tau}_1 = \boldsymbol{\tau}_2$ into Eq. (6.202) that ${}^{\mathcal{P}}\mathbf{U}$ is negative definite and, therefore, $\mathbf{L}_1^{-1} - {}^{\mathcal{P}}\mathbf{U}$ is positive definite so long as the tensor $\mathbf{L}_1(\mathbf{x})$ is positive definite $\forall \mathbf{x} \in w$ in the sense of a relevant quadratic form. For analysis of a case of negative definiteness of $\mathbf{L}_1$, we will substitute the tensor of polarization stress (6.198) into Eq. (6.200): $-(\boldsymbol{\tau}, {}^{\mathcal{P}}\mathbf{U}\boldsymbol{\tau}) = (\boldsymbol{\eta}, \mathbf{L}^c\boldsymbol{\eta}) - (\boldsymbol{\sigma}, \mathbf{M}^c\boldsymbol{\sigma})$. Summation of the last equality with the equation $(\boldsymbol{\tau}, \mathbf{L}_1^{-1}\boldsymbol{\tau}) = (\mathbf{L}^c\boldsymbol{\eta}, \mathbf{L}_1^{-1}\mathbf{L}^c\boldsymbol{\eta}) = -(\boldsymbol{\eta}, \mathbf{M}_1^{-1}\boldsymbol{\eta}) - (\boldsymbol{\eta}, \mathbf{L}^c\boldsymbol{\eta})$, where the equalities $\boldsymbol{\eta} = -\mathbf{M}^c\boldsymbol{\tau}$ and $\mathbf{L}_1^{-1}\mathbf{L}^c = -\mathbf{M}^c\mathbf{M}_1^{-1} - \mathbf{I}$ were used, leads to the conclusion that the value

$$(\boldsymbol{\tau}, \mathbf{L}_1^{-1}\boldsymbol{\tau}) - (\boldsymbol{\tau}, {}^{\mathcal{P}}\mathbf{U}\boldsymbol{\tau}) = -(\boldsymbol{\eta}, \mathbf{M}_1^{-1}\boldsymbol{\eta}) - (\boldsymbol{\sigma}, \mathbf{M}^c\boldsymbol{\sigma}) \tag{6.203}$$

is negative definite so long as the tensor $\mathbf{M}_1$ is positive definite or, what is the same, $\mathbf{L}_1$ is negative definite. Thus, we have proved the positive (negative) definiteness of the operator $\mathbf{L}_1^{-1} - {}^{\mathcal{P}}\mathbf{U}$ so long as the tensor $\mathbf{L}_1$ is positive (negative) definite at each point $\mathbf{x} \in w$.

From positive (or negative) definiteness of the operator $\mathbf{L}_1^{-1} - \mathbf{U}$, it follows that a functional of an arbitrary "trial" tensor $\boldsymbol{\tau}^*$

$$\mathcal{H}(\boldsymbol{\tau}^*) = (\boldsymbol{\tau}^*, \mathbf{L}_1^{-1}\boldsymbol{\tau}^*) - 2(\boldsymbol{\tau}^*, \boldsymbol{\varepsilon}^0) - (\boldsymbol{\tau}^*, {}^{\mathcal{P}}\mathbf{U}\boldsymbol{\tau}^*) \tag{6.204}$$

attains its extremum $\delta\mathcal{H}(\boldsymbol{\tau}) = 0$ at the real values of the field $\boldsymbol{\tau}$ (6.200). To find the stationary value of the functional (6.204), let $\boldsymbol{\tau}$, $\boldsymbol{\sigma}$, and $\boldsymbol{\varepsilon}$ be the solution of (6.200) and the corresponding stress and strain associated with this $\boldsymbol{\tau}$ according to Eqs. (6.199$_2$) and (6.200), respectively. Then the stationary value of $\mathcal{H}(\boldsymbol{\tau}^*)$ is $-(\boldsymbol{\tau}, \boldsymbol{\varepsilon}^0)$ which is estimated from (6.199$_2$) as

$$-(\boldsymbol{\tau}, \boldsymbol{\varepsilon}^0) = (\mathbf{L}^c\boldsymbol{\varepsilon} - \boldsymbol{\sigma}, \boldsymbol{\varepsilon}^0) = (\boldsymbol{\varepsilon}^0, \mathbf{L}^c\boldsymbol{\varepsilon}^0) - (\langle\boldsymbol{\sigma}\rangle, \boldsymbol{\varepsilon}^0) = 2(W^0 - W), \tag{6.205}$$

where according to Hill's condition (6.16) $W = \langle\boldsymbol{\sigma}\boldsymbol{\varepsilon}\rangle/2$, $W^0 = \boldsymbol{\varepsilon}^0\mathbf{L}^c\boldsymbol{\varepsilon}^0/2$ are the mean elastic energy densities of the composite material and homogeneous comparison medium, respectively. In so doing, the value $\mathcal{H}(\boldsymbol{\tau}^*)$ is minimized (maximized) when the tensor $\mathbf{L}_1(\mathbf{x})$ is positive (negative) definite at each point $\mathbf{x} \in w$. This proved principle is referred to as the Hashin–Shtrikman [618, 619] variational principle which was originally deduced from the field equation rather than directly from Eq. (6.200). Presented derivative differs from the deriavative by Willis [1620] who used the classical GIE with the operator $\mathbf{U}^{\mathcal{P}}\boldsymbol{\tau}(\mathbf{x}) = \int \mathbf{U}(\mathbf{x} - \mathbf{y})[\boldsymbol{\tau}(\mathbf{y}) - \langle\boldsymbol{\tau}\rangle]\, d\mathbf{y}$ instead of ${}^{\mathcal{P}}\mathbf{U}$ in (6.200).

We will establish the connection between the Hashin-Shtrikman variational principle and classical variational principles following [641, 1627]. For this purpose, one will choose as arbitrary trial field $\boldsymbol{\tau}^*$ of the polarization stress tensor that produces the fields of the strain $\boldsymbol{\varepsilon}^*$ and stress $\boldsymbol{\sigma}^*$ according to Eqs. (6.200$_1$) and (6.198$_2$) which, respectively, correspond with the displacement boundary conditions and the equilibrium equation. Then, following the classic minimum potential energy principle (2.106), we will get for the real mean elastic energy density W (2.106)

$$2W \le (\boldsymbol{\varepsilon}^*, \mathbf{L}^c\boldsymbol{\varepsilon}^*) + (\boldsymbol{\varepsilon}^*, \mathbf{L}_1\boldsymbol{\varepsilon}^*). \tag{6.206}$$

The first item in the right-hand side of Eq. (6.206) can be simplified by the use of Eqs. (6.198) and (6.200) and the principle of virtual work (6.202). Then the field $\boldsymbol{\tau}^*$ holds the inequality

$$2(W - W^0) \le -(\boldsymbol{\tau}^*, \mathbf{U}\boldsymbol{\tau}^*) + (\boldsymbol{\varepsilon}^0, \mathbf{L}_1\boldsymbol{\varepsilon}^0) + 2(\boldsymbol{\varepsilon}^0, \mathbf{L}_1^{\mathcal{P}}\mathbf{U}\boldsymbol{\tau}^*) + (\mathbf{U}\boldsymbol{\tau}^*, \mathbf{L}_1\mathbf{U}\boldsymbol{\tau}^*). \tag{6.207}$$

One can recast the last item in Eq. (6.207) in the form

$$({}^{\mathcal{P}}\mathbf{U}\boldsymbol{\tau}^*, \mathbf{L}_1{}^{\mathcal{P}}\mathbf{U}\boldsymbol{\tau}^*) = \Big([\mathbf{L}_1^{-1} - {}^{\mathcal{P}}\mathbf{U}]\boldsymbol{\tau}^* - \boldsymbol{\varepsilon}^0, \mathbf{L}_1\{[\mathbf{L}_1^{-1} - {}^{\mathcal{P}}\mathbf{U}]\boldsymbol{\tau}^* - \boldsymbol{\varepsilon}^0\}\Big) - (\boldsymbol{\tau}^*, \mathbf{L}_1^{-1}\boldsymbol{\tau}^*)$$
$$+2(\boldsymbol{\tau}^*, {}^{\mathcal{P}}\mathbf{U}\boldsymbol{\tau}^*) + 2(\boldsymbol{\varepsilon}^0, \boldsymbol{\tau}^* - \mathbf{L}_1{}^{\mathcal{P}}\mathbf{U}\boldsymbol{\tau}^*) - (\boldsymbol{\varepsilon}^0, \mathbf{L}_1\boldsymbol{\varepsilon}^0). \qquad (6.208)$$

Inasmuch as the real field $\boldsymbol{\tau}$ obeys Eq. (6.200), the first item at the right-hand side of Eq. (6.208) vanishes at $\boldsymbol{\tau}^* = \boldsymbol{\tau}$ and is small if $\boldsymbol{\tau}^*$ is a sufficiently accurate approximation of $\boldsymbol{\tau}$. Then one obtains from (6.207)

$$2(W - W^0) \le -(\boldsymbol{\tau}^*, \mathbf{L}_1^{-1}\boldsymbol{\tau}^*) + (\boldsymbol{\tau}^*, {}^{\mathcal{P}}\mathbf{U}\boldsymbol{\tau}^*) + 2(\boldsymbol{\tau}^*, \boldsymbol{\varepsilon}^0) + (\mathbf{e}^*, \mathbf{L}_1\mathbf{e}^*), \qquad (6.209)$$

where $\mathbf{e}^* \equiv (\mathbf{L}_1^{-1} - {}^{\mathcal{P}}\mathbf{U})\boldsymbol{\tau}^* - \boldsymbol{\varepsilon}^0$ is a residual of the trial field. If $\mathbf{L}^c$ is chosen in this manner that $\mathbf{L}_1$ is negative definite at each point $\mathbf{x} \in w$, then dropping of the last item in the right-hand side of the inequality (6.209) yields

$$2(W - W^0) \le -\mathcal{H}(\boldsymbol{\tau}^*). \qquad (6.210)$$

Thus, the value $2(W - W^0)$ is an extremum of the functional $\mathcal{H}(\boldsymbol{\tau}^*)$. The inequality (6.209) demonstrates that the Hashin–Shtrikman principle is weaker than the corresponding classical variational principle due to the negative definiteness of the item $(\mathbf{e}^*, \mathbf{L}_1\mathbf{e}^*)$.

In a similar manner, the minimum principle of the complementary energy (2.108)

$$2(\boldsymbol{\sigma}^*, \boldsymbol{\varepsilon}^0) - (\boldsymbol{\sigma}^*, \mathbf{M}\boldsymbol{\sigma}^*) \le 2W \qquad (6.211)$$

can be transformed to the form $-\mathcal{H}(\boldsymbol{\tau}^*) - (\mathbf{L}^c\mathbf{e}^*, \mathbf{M}_1^c\mathbf{L}^c\mathbf{e}^*) \le (W - W^0)$ by the use of an equality $\mathbf{L}^c\mathbf{M}_1 + \mathbf{L}_1(\mathbf{M}^c + \mathbf{M}_1) = \mathbf{0}$. By this means if the tensor $\mathbf{M}_1$ is negative definite (or which is the same, $\mathbf{L}_1$ is positive definite), then the inequality (6.211) leads to $-\mathcal{H}(\boldsymbol{\tau}^*) \le 2(W - W^0)$.

We now turn our attention to the homogeneous *second boundary-value* (6.5) problem with prescribed tractions $\mathbf{t}^{\partial w}(\mathbf{x}) \equiv \boldsymbol{\sigma}^0 \cdot \mathbf{n}^{\partial w}(\mathbf{x})$ $(\mathbf{x} \in \partial w)$ compatible with a uniform stress $\boldsymbol{\sigma}^0 = \langle \boldsymbol{\sigma} \rangle$ which is consequently the overall average stress $\langle \boldsymbol{\sigma} \rangle$ (6.10). If the homogeneous boundary conditions produce a homogeneous stress field $\boldsymbol{\sigma}^0$ in the comparison medium, then the operator representations

$$\boldsymbol{\sigma} = \boldsymbol{\sigma}^0 + {}^{\mathcal{P}}\boldsymbol{\Gamma}\boldsymbol{\eta}, \quad {}^{\mathcal{P}}\boldsymbol{\Gamma}\boldsymbol{\eta}(\mathbf{x}) := \int \Big[\boldsymbol{\Gamma}(\mathbf{x} - \mathbf{y})\boldsymbol{\eta}(\mathbf{y}) - \langle \boldsymbol{\Gamma}(\mathbf{x} - \mathbf{y})\boldsymbol{\eta}(\mathbf{y})\rangle\Big]\, d\mathbf{y}, \qquad (6.212)$$

where the operator ${}^{\mathcal{P}}\boldsymbol{\Gamma}\boldsymbol{\eta}$ differs from the popular operator $\boldsymbol{\Gamma}^{\mathcal{P}}\boldsymbol{\eta} = \int \boldsymbol{\Gamma}(\mathbf{x}-\mathbf{y})[\boldsymbol{\eta}(\mathbf{y}) - \langle \boldsymbol{\eta} \rangle]\, d\mathbf{y}$ used in [1620], with the kernel $\boldsymbol{\Gamma}(\mathbf{x}) = -\mathbf{L}^c\boldsymbol{\delta}(\mathbf{x}) - \mathbf{L}^c\mathbf{U}(\mathbf{x})\mathbf{L}^c$ (3.37) The following functional can be associated with Eq. (6.212_2):

$$\mathcal{F}(\boldsymbol{\eta}^*) = (\boldsymbol{\eta}^*, [\mathbf{M}_1^{-1} - {}^{\mathcal{P}}\boldsymbol{\Gamma}]\boldsymbol{\eta}^*) - 2(\boldsymbol{\eta}^*, \boldsymbol{\sigma}^0). \qquad (6.213)$$

The rules for the minimum principles of the elastic energy and complementary energy in the considered case of boundary conditions now change. So, the analog of the principle (6.206), from the complementary energy principle, is

$$2W \leq (\boldsymbol{\sigma}^*, \mathbf{M}^c\boldsymbol{\sigma}^*) + (\boldsymbol{\sigma}^*, \mathbf{M}_1\boldsymbol{\sigma}^*), \tag{6.214}$$

which can be presented in terms of $\boldsymbol{\eta}^*$

$$\begin{aligned} 2(W - W^0) \leq & -(\boldsymbol{\eta}^*, {}^{\mathcal{P}}\boldsymbol{\Gamma}\boldsymbol{\eta}^*) + (\boldsymbol{\sigma}^*, \mathbf{M}_1\boldsymbol{\sigma}^*) \\ & +2(\boldsymbol{\sigma}^0, \mathbf{M}_1^{\mathcal{P}}\boldsymbol{\Gamma}\boldsymbol{\eta}^*) + ({}^{\mathcal{P}}\boldsymbol{\Gamma}\boldsymbol{\eta}^*, \mathbf{M}_1^{\mathcal{P}}\boldsymbol{\Gamma}\boldsymbol{\eta}^*). \end{aligned} \tag{6.215}$$

In the final analysis, we will obtain similarly to the inequality (6.209) $2(W - W^0) \leq -\mathcal{F}(\boldsymbol{\eta}^*) + (\mathbf{L}^c\mathbf{e}^*, \mathbf{M}_1\mathbf{L}^c\mathbf{e}^*)$, and, therefore, $2(W - W^0) \leq -\mathcal{F}(\boldsymbol{\eta}^*)$ at any choice of $\mathbf{M}^c$ with the negative definite $\mathbf{M}_1(\mathbf{x})$ ($\forall\mathbf{x} \in w$) (or, what is the same, $\mathbf{L}_1(\mathbf{x})$ is positive definite). The residual in the inequality (6.214) presented in terms of $\boldsymbol{\eta}^*$ is associated with the residual introduced in the inequality (6.209): $\mathbf{L}^c\mathbf{e}^* = (\mathbf{M}_1^{-1} - {}^{\mathcal{P}}\boldsymbol{\Gamma})\boldsymbol{\eta}^* - \boldsymbol{\sigma}^0$. Furthermore, one can present the functionals $\mathcal{H}$ and $\mathcal{F}$ in the terms $\boldsymbol{\eta}^*$ and $\boldsymbol{\tau}^*$, respectively, where $\boldsymbol{\eta}^*$ and $\boldsymbol{\tau}^*$ are related by Eq. (6.200). Then, it is found to be $\mathcal{H}(\boldsymbol{\tau}^*) = \mathcal{F}(\boldsymbol{\eta}^*)$.

Finally, the following inequalities obtained from the minimum potential energy principle are reciprocal to (6.209) and (6.210) for the second boundary-value problem:

$$-\mathcal{F}(\boldsymbol{\eta}^*) - (\mathbf{e}^*, \mathbf{L}_1\mathbf{e}^*) \leq 2(W - W^0), \quad -\mathcal{F}(\boldsymbol{\eta}^*) \leq 2(W - W^0), \tag{6.216}$$

where $\mathbf{L}_1$ is an arbitrary tensor negatively defined at $\forall\mathbf{x} \in w$.

It should be mentioned that Hill [641] derived the Hashin–Shtrikman principle from the classical energy principles by the use of the field equations without operator-like representation (6.200) (with either operator ${}^{\mathcal{P}}\mathbf{U}$ or $\mathbf{U}^{\mathcal{P}}$). Namely, for the first boundary-value (6.4) problem, the minimum principles (6.194) and (6.195) lead to

$$2(W^0 - W) \geq \ \text{or} \ \leq \mathcal{H}^{\text{HS}} := (\boldsymbol{\tau}^*, \mathbf{L}_1^{-1}\boldsymbol{\tau}^*) - (\boldsymbol{\tau}^*, \boldsymbol{\varepsilon}^0) - (\boldsymbol{\tau}^*, \boldsymbol{\varepsilon}^*) \tag{6.217}$$

as $\mathbf{L}_1$ is negative or positive semi-defined, respectively, throughout w. Instead of "error" $\mathbf{e}^*$ in Eq. (6.209), one uses $\mathbf{e}^* = \mathbf{L}_1^{-1}(\boldsymbol{\sigma}^* - \mathbf{L}\boldsymbol{\varepsilon}^*)$ for obtaining inequality (6.217).

In a similar manner, for the second boundary-value (6.5) problem, the minimum principles (6.196) and (6.197) yield

$$2(W^0 - W) \geq \ \text{or} \ \leq \mathcal{F}^{\text{HS}} := (\boldsymbol{\eta}^*, \mathbf{M}_1^{-1}\boldsymbol{\eta}^*) - (\boldsymbol{\eta}^*, \boldsymbol{\sigma}^0) - (\boldsymbol{\eta}^*, \boldsymbol{\sigma}^*) \tag{6.218}$$

as $\mathbf{L}_1$ is negative or positive semi-defined, respectively.

It is interesting that the classical functional $\mathcal{H}^{\mathrm{HS}}$ and $\mathcal{F}^{\mathrm{HS}}$ are reduced to $\mathcal{H}$ (6.204) and $\mathcal{F}$ (6.213) after the formal substitutions of $\boldsymbol{\varepsilon}^* = \boldsymbol{\varepsilon}^0 +^{\mathcal{P}} \mathbf{U}\boldsymbol{\tau}^*$ (6.200) and $\boldsymbol{\sigma}^* = \boldsymbol{\sigma}^0 +^{\mathcal{P}} \boldsymbol{\Gamma}\boldsymbol{\eta}^*$ (6.212) into the last terms of the inequalities $\mathcal{H}^{\mathrm{HS}}$ (6.217) and $\mathcal{F}^{\mathrm{HS}}$ (6.218), respectively. Thus, following [641, 1575, 1576, 1627], we made sure that the variational principle of Hashin–Shtrikman is inferred from the classical ones. However, estimation of Hashin–Shtrikman's bounds from Hashin–Shtrikman's principle requires acceptence of some additional hypotheses. Because of this, Hashin–Shtrikman's bounds will be considered in Chaps. 8 and 9.

6.5 Hill's Bounds of Effective Elastic Moduli

For definiteness sake, we will consider the first homogeneous boundary-value problem (6.4). Then, the upper bound of the effective modulus $\mathbf{L}^*$ can be estimated by the use of the Hill's condition (6.16) and the classical minimum potential energy principle (6.194)

$$2W \leq \langle \boldsymbol{\varepsilon}^* \mathbf{L} \boldsymbol{\varepsilon}^* \rangle \tag{6.219}$$

where the trial strain field $\boldsymbol{\varepsilon}^* \in \mathcal{A}^k$ is derived from an arbitrary displacement field $\mathbf{u}^*$ consistent with the boundary condition (6.4). In the simplest case of an approximation of $\boldsymbol{\varepsilon}^*$ by a constant $\boldsymbol{\varepsilon}^* = \langle \boldsymbol{\varepsilon} \rangle$, one obtains an inequality

$$\mathbf{L}^* \leq \mathbf{L}^V, \quad \mathbf{L}^V \equiv \langle \mathbf{L} \rangle, \tag{6.220}$$

proposed by Voight [1566]. The inequality (6.220) is understood in the sense of an inequality for the corresponding quadratic form when $a_{ij}(L^V_{ijkl} - L^*_{ijkl})a_{kl} \geq 0$ for any symmetric tensor of the second-order a_{ij}. In so doing, it is possible that for some components of the tensors L_{ijkl} and L^V_{ijkl}, the inequality (6.220$_1$) is violated.

In a similar manner, for the boundary conditions (6.5), the minimum principle of the complementary energy can be written as

$$W \leq (\boldsymbol{\sigma}^*, \mathbf{M}\boldsymbol{\sigma}^*), \tag{6.221}$$

where $\boldsymbol{\sigma}^* \in \mathcal{A}^s$ is a trial stress field obeying the equilibrium equation and boundary conditions (6.5). The simplest approximation of a trial stress field $\boldsymbol{\sigma}^*$ is a constant: $\boldsymbol{\sigma}^* = \text{const}$, which reduce the inequality (6.221) to

$$\mathbf{M}^R \leq \mathbf{M}^*, \tag{6.222}$$

where $\mathbf{M}^R \equiv \langle \mathbf{M} \rangle$ is a definition of a compliance tensor introduced by Reuss [1290]. Then, with inequalities (6.220) and (6.222), one obtains an inequality

$$\mathbf{L}^R \leq \mathbf{L}^* \leq \mathbf{L}^V, \quad \mathbf{M}^V \leq \mathbf{M}^* \leq \mathbf{M}^R \tag{6.223}$$

referred to as the Hill's [639] (see also [1192]) bounds; here, $\mathbf{L}^R := (\mathbf{M}^R)^{-1}$, and $\mathbf{M}^V := (\mathbf{L}^V)^{-1}$. The upper bound (6.223$_1$) is identified as the bound by Voight, while the lower bound is called Reuss's bound. The arithmetic average of the bounds by Voight and Reuss represents Hill's estimation $\mathbf{L}^H = (\mathbf{L}^R + \mathbf{L}^V)/2$.

For statistically isotropic composites, the bounds by Voight and Reuss will be the isotropic tensors which can be presented in Voight notations (2.197), $\mathbf{L}^* = (3k^*, 2\mu^*)$, $\mathbf{L}^V = (3k^V, 2\mu^V)$, and $\mathbf{L}^R = (3k^R, 2\mu^R)$, and the inequality (6.223) acquires the form

$$k^R \leq k^* \leq k^V, \quad \mu^R \leq \mu^* \leq \mu^V. \tag{6.224}$$

In the case of a composite material with isotropic constituents, one obtains

$$k^R = \langle k^{-1}\rangle^{-1}, \quad k^V = \langle k\rangle, \quad \mu^R = \langle \mu^{-1}\rangle^{-1}, \quad \mu^V = \langle \mu\rangle. \tag{6.225}$$

For polycrystals, Hill's bounds (6.223) are estimated as the arithmetic and harmonic means of the local stiffness tensors (5.89)

$$\mathbf{L}^V = \int \mathbf{L}(\mathbf{g}) f(\mathbf{g})\, d\mathbf{g}, \quad \mathbf{M}^R = \int \mathbf{M}(\mathbf{g}) f(\mathbf{g})\, d\mathbf{g}. \tag{6.226}$$

The arithmetic and harmonic means correspond to the assumption of homogeneous strain and stress field, respectively. Note that $\mathbf{L}^V$ and $\mathbf{L}^R$ differing in general give the upper $\mathbf{L}^V$ (Voigt) and lower $\mathbf{L}^R$ (Reuss) bounds for the effective elastic stiffness $\mathbf{L}^*$, the distance between which describes the degree of elastic anisotropy. For instance, in the case of an aggregate of cubic single crystals, this distance can be expressed in terms of the Zener anisotropy ratio Z (2.176): $\parallel \mathbf{L}^V - \mathbf{L}^R \parallel = 6\sqrt{5}(Z-1)^2[5(2Z+3)]^{-1}(L_{1111} - L_{1122})$.

For uniformly distributed grain orientation, tensors $\mathbf{L}^V$ and $\mathbf{L}^R \equiv (\mathbf{M}^R)^{-1}$ are isotropic ones with the bulk k^V, k^R and shear μ^V, μ^R moduli defined by the relations ($J = V, R$)

$$k^J = L^J_{iijj}/9, \quad \mu^J = (L^J_{ijij} - L^J_{iijj}/3)/10, \tag{6.227}$$

which provide the bounds for the effective elastic moduli $\mathbf{L}^* = (3k^*, 2\mu^*)$ (6.224). Due to an equality for the Young modulus $E^{-1} = (3k)^{-1} + (9\mu)^{-1}$, the set of inequalities analogous to (6.224) is valid also for the Young modulus

$$E^R \leq E^* \leq E^V. \tag{6.228}$$

The Hill [639] approach combines the upper and lower bounds as both arithmetic and geometric means of the isotropic bounds by Voigt and Reuss, giving a good approximation for the actual effective elastic constant. $k^H = (k^V + k^R)/2$, $\mu^H = (\mu^V + \mu^H)/2$.

Of course, for the uniformly distributed grain orientation, the Voigt–Reuss estimations (6.226) can be obtained by the direct average of representations (2.169) over all possible orientations **g**. However, the same result can be obtained by the contraction evaluations of (2.170) and (2.171) analogously to (6.227)

$$3k^V = \frac{1}{3}\sum_n(\lambda_n + 6\mu_n + 4\nu_n), \quad 2\mu^V = \frac{2}{15}\sum_n(\lambda_n + 10\nu_n), \tag{6.229}$$

$$(3k^R)^{-1} = \frac{1}{3}\sum_n(p_n + 6q_n + 4r_n), \quad (2\mu^R)^{-1} = \frac{2}{15}\sum_n(p_n + 10r_n). \tag{6.230}$$

Substituting the explicit representations $\lambda_n,\ \mu_n,\ \nu_n$ and analogous representations for $p_n,\ q_n,\ r_n$ for the different symmetric classes (2.170)–(2.177) leads to the representation of the bulk and shear components of the tensors $\mathbf{L}^V$ and $\mathbf{M}^R$:
for orthorhombic symmetry

$$\begin{aligned} 3k^V &= \left[L_{11} + L_{22} + L_{33} + 2(L_{12} + L_{23} + L_{13})\right]/3, \\ 2\mu^V &= \frac{2}{15}\left[L_{11} + L_{22} + L_{33} - L_{12} - L_{23} - L_{13} + 3(L_{44} + L_{55} + L_{66})\right], \end{aligned} \tag{6.231}$$

$$\begin{aligned} (3k^R)^{-1} &= \left[M_{11} + M_{23} + M_{33} + 2(M_{12} + M_{23} + M_{13})\right]/3, \\ (2\mu^R)^{-1} &= \frac{2}{15}\left[M_{11} + M_{22} + M_{33} - M_{12} - M_{23} - M_{13} \right.\\ &\quad \left. + \frac{3}{4}(M_{44} + M_{55} + M_{66})\right], \end{aligned} \tag{6.232}$$

for tetragonal symmetry

$$\begin{aligned} 3k^V &= \left[2L_{11} + L_{33} + 2(L_{12} + 2L_{13})\right]/3, \\ 2\mu^V &= \frac{2}{15}\left[2L_{11} + L_{33} - L_{12} - 2L_{13} + 3(2L_{44} + L_{66})\right], \end{aligned} \tag{6.233}$$

$$\begin{aligned} (3k^R)^{-1} &= \left[2M_{11} + M_{33} + 2(M_{12} + 2M_{13})\right]/3, \\ (2\mu^R)^{-1} &= \frac{2}{15}\left[2M_{11} + M_{33} - M_{12} - 2M_{13} + \frac{3}{4}(2M_{44} + M_{66})\right], \end{aligned} \tag{6.234}$$

for hexagonal symmetry

$$\begin{aligned} 3k^V &= \left[2L_{11} + L_{33} + 2(L_{12} + 2L_{13})\right]/3, \\ 2\mu^V &= \frac{1}{15}\left[7L_{11} + 2L_{33} - 5L_{12} - 4L_{13} + 12L_{44}\right], \end{aligned} \tag{6.235}$$

$$(3k^R)^{-1} = \left[2M_{11} + M_{33} + 2(M_{12} + 2M_{13})\right]/3,$$
$$(2\mu^R)^{-1} = \frac{1}{15}\left[7M_{11} + 2M_{33} - 5M_{12} - 4M_{13} + 3M_{44})\right], \quad (6.236)$$

and for cubic symmetry

$$3k^V = L_{11} + 2L_{12}, \quad 2\mu^V = \frac{2}{5}(L_{11} - L_{12} + 3L_{44}), \quad (6.237)$$

$$(3k^R)^{-1} = M_{11} + 2M_{12}, \quad (2\mu^R)^{-1} = \frac{2}{5}\left(M_{11} - M_{12} + \frac{3}{4}M_{44}\right). \quad (6.238)$$

As expected, $k^V \neq k^R$ and $\mu^V \neq \mu^R$ in general, but for the polycrystals of cubic symmetry $k^V = k^R$ because the local contractions $L_{iijj} =$ const. and $M_{iijj} =$ const. do not vary in the aggregate due to the equality $L_{12} + 2L_{12} = (M_{11} + 2M_{12})^{-1}$. Moreover, the equality $\lambda = k - 2\mu/3$ for the isotropic medium accompanied by the inequality (6.224$_2$) and equality $k^V = k^R$ for the polycrystalline aggregates of cubic crystals leads to $\lambda^V \leq \lambda^* \equiv k^* - 2\mu^* \leq \lambda^R$ [976]. Thus, the upper bound for Lamé constant is the Reuss estimate of this constant, while the lower bound for this constant is given by its Voigt estimate (compare with Eqs. (6.224) and (6.228)).

It should be mentioned that in the contractions (6.227), the tensor containing one of four indexes odd times (e.g., L_{2312}) is absent. Therefore, Eqs. (6.235) and (6.236) obtained for hexagonal crystals are also correct for an aggregate with trigonal crystals, and Eqs. (6.231) and (6.232) describe not only the orthorhombic crystals but also the crystals of monoclinic and triclinic symmetry.

Let us consider the polycrystals of cubic symmetry with the ODF $f(\phi_1, \theta, \phi_2) \equiv f(\theta)$ which implies the transverse isotropy with Ox_3 being the symmetry axis. Then the tensor of the average elastic moduli has five independent constants which can be found by averaging; keeping in mind that $f(\theta)$ does not depend on the angles ϕ_1 and ϕ_2, we can easily estimate the averages

$$\langle c_i^2\rangle = \langle s_i^2\rangle = \frac{1}{2}, \ \langle c_i^4\rangle = \langle s_i^4\rangle = \frac{3}{8}, \ \langle c_i^2 s_i^2\rangle = \frac{1}{8}, \ (i = 1, 2). \quad (6.239)$$

In the case of the series expansion of the ODF $f(\theta) = \sum_{n=0} a_n P_p(\cos\theta)$ in the Legendre functions $P_n(\cos\theta)$, the averages

$$\langle c_1^2\rangle = \frac{1}{3}\left(1 + \frac{2}{5}a_2\right), \ \langle c_1^4\rangle = \frac{1}{5}\left(1 + \frac{4}{7}a_2 + \frac{8}{63}a_4\right),$$
$$\langle s_1^2\rangle = \frac{2}{3}\left(1 - \frac{1}{5}a_2\right), \ \langle s_1^4\rangle = \frac{8}{15}\left(1 - \frac{2}{7}a_2 + \frac{1}{2}a_4\right) \quad (6.240)$$

can be found by the use of an explicit representation of the Legendre polynomials in the operation averages (5.89), (6.226), with taking the orthogonality property

$$\int_{-1}^{1} P_l(x)P_{l'}(x)\,dx = \frac{2}{l+1}\delta_{ll'} \tag{6.241}$$

into account. Substitution of the averages (6.239) and (6.240) into the average Eq. (6.226) leads to

$$\begin{aligned}&\langle L_{11}\rangle = L_{11}^0 + \frac{1}{60}a_4L,\ \langle L_{12}\rangle = L_{12}^0 + \frac{1}{180}a_4L,\ \langle L_{13}\rangle = L_{12}^0 - \frac{1}{45}a_4L,\\ &\langle L_{33}\rangle = L_{11}^0 + \frac{2}{45}a_4L,\ \langle L_{44}\rangle = \frac{1}{2}(L_{11}^0 - L_{12}^0) - \frac{1}{45}a_4L,\end{aligned} \tag{6.242}$$

where $L = L_{11} - L_{12} - 2L_{44}$, and $L_{11}^0 = 1/15(L_{iijj} + 2L_{ijij})$ and $L_{12}^0 = 1/15(2L_{iijj} - L_{ijij})$ are estimated for the uniform random orientation of crystals (6.237) when $a_2 = a_4 = 0$.

The formal generalizations of Hill's [639] estimation $\mathbf{L}^H$ are noteworthy. So an approach [1090] exhibiting a natural reciprocity between stiffness and compliance of effective properties takes into account the recursive arithmetic mean by Voigt and Reuss bounds

$$\begin{aligned}&\mathbf{L}_0^V = \langle \mathbf{L}\rangle, \mathbf{M}_0^R = \langle \mathbf{M}\rangle,\ \mathbf{L}_0^R = \langle \mathbf{L}\rangle^{-1},\ \mathbf{M}_0^V = \langle \mathbf{M}\rangle^{-1},\\ &\mathbf{L}_{i+1}^V = (\mathbf{L}_i^V + \mathbf{R}_i^R)/2,\ \mathbf{M}_{i+1}^R = (\mathbf{M}_i^V + \mathbf{M}_i^R)/2,\\ &\mathbf{L}_{i+1}^R = (\mathbf{M}_{i+1}^R)^{-1},\ \mathbf{M}_{i+1}^V = (\mathbf{L}_{i+1}^V)^{-1}.\end{aligned} \tag{6.243}$$

In the numerical examples being considered, the recursion (6.243) with the initial approximation [639] ($i = 0$) converges in four iterations. This recursive approach has no micromechanical justification other than that effective elastic stiffness and compliance should be reciprocal with one another. A similar level of justification was proposed in a homogenization technique in [22] (see also [1032]) using the geometrical mean of local effective properties

$$\mathbf{L}^A = \exp\Big(\int \ln(\mathbf{L}(\mathbf{g}))f(\mathbf{g})\,d\mathbf{g}\Big),\ \mathbf{M}^A = \exp\Big(\int \ln(\mathbf{M}(\mathbf{g}))f(\mathbf{g})\,d\mathbf{g}\Big) \equiv (\mathbf{L}^A)^{-1}. \tag{6.244}$$

Arithmetic (6.226_1), harmonic (6.226_2), and geometric (6.244) averages were compared by Böhlke and Bertran [135] showing that for cubic crystal aggregates, all of these averages depend on the same irreducible fourth-order tensor, which represents the pure anisotropic portion of the effective elasticity tensor. It should be mentioned that the Voiht–Reuss estimations (6.224) for uniformly distributed grain orientation are based on exploiting two invariants (6.227). Another formal approach

[21] is based on the use of a determinant of the elastic moduli matrix. For example, for the polycrystals of cubic symmetry with the Zener anisotropy ratio Z (2.176), the desired determinant D equals $D = 12KZ^{-1}L_{44}^5$, $K = (L_{11} + 2L_{12})/3$, while for the equivalent isotropic medium with effective elastic moduli $\mathbf{L}^* = (3k^*, 2\mu^*)$, we have $D = 12k^*(\mu^*)^5$. Taking into account an invariantness of K with respect to rotation of the cubic symmetry crystals, we obtain

$$k^* = K, \quad \mu^* = L_{44}Z^{-2/5} \tag{6.245}$$

with identical representation in the approach based on the use of compliance tensor $1/\mu^* = M_{44}Z^{2/5}$. Thus, even in the case of significant differences of μ^V and μ^R (6.224) obtained by the use of linear invariants of $\mathbf{L}$ and $\mathbf{M}$, respectively, the method (6.245) leads to a natural reciprocity of the shear elasticity and compliance.

6.6 Some General Representations for Thermoelectroelasticity of Composites

Piezoelectric materials have the property of converting mechanical energy into electrical energy (direct piezoelectric effect) and vice versa (inverse piezoelectric effect). Piezoelectric materials and especially piezoelectric composites open completely new possibilities in the design of adaptive or smart structure, and very interesting technological applications have been proposed, ranging from aeronautical and automotive structures to miniature positioning devises (see [1115, 1278, 1320]). From another aspect, pronounced anisotropy, texture, and piezoelectric activity are inherent in some polycrystal minerals [1127, 1183, 1424, 1552]. Even a single component polycrystal may be textured, which causes a macroscopic anisotropy [81, 1552, 1553, 1560, 1611]. A natural property of piezoelectric polycrystal is their heterogeneity manifested at multiple length scales [930]. So, piezoelectric ceramics are usually prepared by synthesizing compressed powers of oxide, which produces macroscopically isotropic microporous ceramic [340]. Composite materials such as piezoelectric ceramic fibers embedded in a soft pure elastic matrix are often a better technological solution for a large number of applications such as underwater and medical ultrasonic imaging [456] and transducers for underwater and sonar projector applications. They have superior electromechanical coupling characteristics (better acoustic impedance matching with water and (or) tissue) to conventional piezoelectric ceramic [1415].

6.6.1 Generalized Hill's Conditions and Effective Properties

Let us consider a composite with elastic piezoelectric constituents. All random field under consideration are assumed to be statistically homogeneous and comply with the ergodicity hypothesis (see Sect. 5.2.2). Therefore, averaging over the ensemble of realizations X may be replaced by that over the macrovolume w. For notational convenience, the elastic and electric variable will be treated on equal footing, and with this in mind, we recast the local linear constitutive relations of thermoelectroelasticity for this material (see, e.g., [747, 1033, 1188]) in the notation introduced in [70]). Then, the basic steady-state equations (2.145)–(2.154) can be presented in the form analogous to Eqs. (6.1)–(6.3)

$$\mathcal{D}\boldsymbol{\Sigma} = \mathbf{0}, \ \mathcal{E} = \mathcal{D}\mathcal{U}, \tag{6.246}$$

$$\mathcal{E} = \mathbb{M}\boldsymbol{\Sigma} + \boldsymbol{\Lambda}, \quad \boldsymbol{\Sigma} = \mathbb{L}(\mathcal{E} - \boldsymbol{\Lambda}). \tag{6.247}$$

The general exact relations for the PCM should be mentioned. So, Benveniste [92–94] (see also [280]) obtained the effective eigenstress and stiffness for multiphase composites generalized the relations (6.20), (6.46), (6.77). We will reproduce these and other exact relations following [280]. We extend the Hill's [644] condition (6.16) in the elasticity theory to the piezoelectric composites. For precision, we consider homogeneous boundary conditions in the form

$$\Sigma_{\alpha\beta} n_\beta(\mathbf{x}) = \langle \Sigma_{\alpha\beta} \rangle n_\beta(\mathbf{x}) = t_\alpha(\mathbf{x}), \quad \mathbf{x} \in \partial w, \quad n_4 \equiv \mathbf{0} \tag{6.248}$$

where n_α are the components of an external normal to the domain boundary ∂w, t_i $(i = 1, 2, 3)$ and t_4 are the normal components of the mechanical stress tensor and vector of electric induction. In the case of ergodic fields, one may replace integrating over macrovolume by ensemble averaging. Hereinafter, the Latin indexes range from 1 to 3, and the Greek ones range from 1 to 4. Therefore, in accordance with (6.246), the following equalities hold true:

$$\begin{aligned} \langle \Sigma_{\alpha\beta} w_{\alpha\beta} \rangle &\equiv \bar{w}^{-1} \int \langle \Sigma_{\alpha\beta} w_{\alpha\beta} \rangle \, d\mathbf{x} = \bar{w}^{-1} \Big\langle \int \Sigma_{\alpha\beta} w_{\alpha\beta} \, d\mathbf{x} \Big\rangle \\ &= \bar{w}^{-1} \Big\langle \int t_\alpha U_\alpha \, d\mathbf{s} \Big\rangle = \bar{w}^{-1} \int \langle t_\alpha U_\alpha \rangle \, d\mathbf{s}, \end{aligned} \tag{6.249}$$

where integration in the surface integral is performed over the boundary ∂w, and the Dirac brackets denote ensemble averaging. Separating the displacement field fluctuations $\delta U_\alpha(\mathbf{x}) \equiv U_\alpha(\mathbf{x}) - \langle U_\alpha(\mathbf{x}) \rangle$ in the surface integral, we have

$$\int \langle t_\alpha(\mathbf{x}) U_\alpha(\mathbf{x}) \rangle \, ds = \int \langle t_\alpha(\mathbf{x}) \rangle \langle U_\alpha(\mathbf{x}) \rangle \, d\mathbf{s} + \int \langle t_\alpha(\mathbf{x}) \delta U_\alpha(\mathbf{x}) \rangle \, d\mathbf{s}. \tag{6.250}$$

The last integral in (6.250) is zero because, in view of the boundary conditions (6.248) and the ergodicity, we may replace the averaging over the volume by that over ensemble. Then, we obtain from (6.249) and (6.250) $\langle \Sigma_{\alpha\beta} w_{\alpha\beta}\rangle = \bar{w}^{-1}\int\langle t_\alpha(\mathbf{x})\rangle\langle U_\alpha(\mathbf{x})\rangle\, ds$. But, in accordance with (6.246) and (6.248)

$$\langle \boldsymbol{\mathcal{E}}\rangle = \langle \nabla_{(\alpha} U_{\beta)}\rangle, \quad \langle t_\alpha(\mathbf{x})\rangle = \langle \Sigma_{\alpha\beta}\rangle n_\beta(\mathbf{x}). \tag{6.251}$$

Therefore, using the Gauss theorem, we find from (6.246), (6.250), and (6.251)

$$\langle \Sigma_{\alpha\beta}(\mathbf{x}) w_{\alpha\beta}(\mathbf{x})\rangle = \langle \Sigma_{\alpha\beta}(\mathbf{x})\rangle\langle w_{\alpha\beta}(\mathbf{x})\rangle. \tag{6.252}$$

The equality (6.252) extends the Hill condition to piezoelectric materials and can be obtained if, for the surface ∂w, it is possible to formulate the first boundary problem with given homogeneous fields of generalized strains $\boldsymbol{\mathcal{E}}$. Equation (6.252) was derived without any assumption on the correlation between fields the $\boldsymbol{\sigma}$ and $\boldsymbol{\mathcal{E}}$. The only requirement is that the fields be statistically and kinematically acceptable, i.e., they satisfy conditions (6.246). Formula (6.252) was obtained by Buryachenko and Shermergor [280] and independently in [930].

To derive the macroscopic equation of state, we assume that the local and macroscopic values are related as

$$\boldsymbol{\Sigma}(\mathbf{x}) = \mathbb{B}^*(\mathbf{x})\langle \boldsymbol{\Sigma}\rangle + \mathbb{F}^*(\mathbf{x}) \tag{6.253}$$

Due to the linearity of the problem, tensors $\mathbb{B}^*$ and $\mathbb{F}^*$ exist. Averaging tensor $\mathbb{B}^*$ over the volume of the ith component $\langle \mathbb{B}^*(\mathbf{x})\rangle_i \equiv \mathbb{B}^*_i$, we obtain the tensor of generalized stress concentration. The tensors $\mathbb{B}^*$ and $\mathbb{F}^*$ depend on the physical and geometric properties of the material components and may be found from the solution of the corresponding boundary problem for a microinhomogeneous medium. Averaging of (6.253) at $\theta \equiv 0$ gives an expression determining the effective tensors of generalized compliance $\mathbb{M}^*$ and expansion $\boldsymbol{\Lambda}^*$ in the macroscopic equation of state: $\langle \boldsymbol{\Sigma}\rangle = \mathbb{M}^*\langle \boldsymbol{\sigma}\rangle + \boldsymbol{\Lambda}^*$. At the thermodynamic equilibrium condition stated as $\theta = \langle \theta\rangle$, we find from (6.247) and (6.253): $\mathbb{M}^* = \langle \mathbb{M}\mathbb{B}^*\rangle$, $\boldsymbol{\Lambda}^* = \langle \boldsymbol{\Lambda}\rangle + \langle \mathbb{M}\mathbb{F}^*\rangle$, where $\langle \mathbb{B}^*\rangle = \mathrm{diag}(\mathbf{I}, \boldsymbol{\delta}) \equiv \mathbb{I}$, $\langle \mathbb{F}^*\rangle = \mathbf{0}$.

To find another representation of the tensor $\boldsymbol{\Lambda}^*$, we express the fields of generalized stresses, strains, and thermal expansion in the form (see (6.36) and (6.37))

$$\boldsymbol{\Sigma}(\mathbf{x}) = \boldsymbol{\Sigma}^I(\mathbf{x}) + \boldsymbol{\Sigma}^{II}, \quad \boldsymbol{\mathcal{E}}(\mathbf{x}) = \boldsymbol{\mathcal{E}}^I(\mathbf{x}) + \boldsymbol{\mathcal{E}}^{II}(\mathbf{x}), \quad \boldsymbol{\Lambda}(\mathbf{x}) = \boldsymbol{\Lambda}^I(\mathbf{x}) + \boldsymbol{\Lambda}^{II}(\mathbf{x}), \tag{6.254}$$

where $\boldsymbol{\Sigma}^I$ and $\boldsymbol{\mathcal{E}}^I$ satisfy boundary conditions (6.248) and $\boldsymbol{\Lambda}^I \equiv \mathbf{0}$ and the fields $\boldsymbol{\Sigma}^{II}$ and $\boldsymbol{\mathcal{E}}^{II}$ satisfy zero boundary conditions $t_\alpha \equiv \mathbf{0}$ and $\boldsymbol{\Lambda}^{II} \equiv \boldsymbol{\Lambda}(\mathbf{x})$. In accordance with (6.253), we have $\boldsymbol{\Sigma}^{II}(\mathbf{x}) = \mathbb{F}(\mathbf{x})$. Then for the field under consideration, the local and macroscopic equation of state may be expressed in the form

$$\boldsymbol{\Sigma}^I(\mathbf{x}) = \mathbb{L}(\mathbf{x})\boldsymbol{\mathcal{E}}^I(\mathbf{x}), \quad \boldsymbol{\Sigma}^{II}(\mathbf{x}) = \mathbb{L}(\mathbf{x})[\boldsymbol{\mathcal{E}}^{II}(\mathbf{x}) - \boldsymbol{\Lambda}(\mathbf{x})], \tag{6.255}$$

$$\langle\boldsymbol{\Sigma}^I(\mathbf{x})\rangle = \mathbb{L}^*(\mathbf{x})\langle\boldsymbol{\mathcal{E}}^I(\mathbf{x})\rangle, \quad \langle\boldsymbol{\Sigma}^{II}(\mathbf{x})\rangle = \mathbb{L}^*(\mathbf{x})[\langle\boldsymbol{\mathcal{E}}^{II}(\mathbf{x})\rangle - \boldsymbol{\Lambda}^*(\mathbf{x})]. \tag{6.256}$$

Since $\boldsymbol{\Sigma}$, $\boldsymbol{\Sigma}^I$, $\boldsymbol{\Sigma}^{II}$, $\boldsymbol{\mathcal{E}}$, $\boldsymbol{\mathcal{E}}^I$, and $\boldsymbol{\mathcal{E}}^{II}$ are statically and kinematically admissible fields, then, taking into account Hill's condition (6.252), we have

$$\langle\boldsymbol{\Sigma}^{I\top}\boldsymbol{\mathcal{E}}\rangle = \langle\boldsymbol{\Sigma}^{I\top}\rangle\langle\boldsymbol{\mathcal{E}}\rangle, \quad \langle\boldsymbol{\mathcal{E}}^{I\top}\boldsymbol{\Sigma}\rangle = \langle\boldsymbol{\mathcal{E}}^{I\top}\rangle\langle\boldsymbol{\Sigma}\rangle, \tag{6.257}$$

$$\langle\boldsymbol{\Sigma}^{I\top}\boldsymbol{\mathcal{E}}^I\rangle = \langle\boldsymbol{\Sigma}^{I\top}\rangle\langle\boldsymbol{\mathcal{E}}^I\rangle, \quad \langle\boldsymbol{\Sigma}^{II\top}\boldsymbol{\mathcal{E}}^{II}\rangle = \langle\boldsymbol{\Sigma}^{II\top}\rangle\langle\boldsymbol{\mathcal{E}}^{II}\rangle = 0. \tag{6.258}$$

Subtracting equalities (6.257_1) and (6.257_2) from one another and using (6.255), (6.256), we find $\langle\boldsymbol{\Sigma}^{I\top}\boldsymbol{\Lambda}\rangle = \langle\boldsymbol{\Sigma}^{I\top}\rangle\langle\boldsymbol{\Lambda}^*\rangle$. Substituting (6.253) into the last equation, in view of arbitrariness $\langle\boldsymbol{\Sigma}\rangle$, we have

$$\boldsymbol{\Lambda}^* = \langle\mathbb{B}^{*\top}\boldsymbol{\Lambda}\rangle. \tag{6.259}$$

We see that the matrix of effective generalized thermal expansion $\boldsymbol{\Lambda}^*$ is determined not only by its local value $\boldsymbol{\Lambda}(\mathbf{x})$ but also by the matrix $\mathbb{B}^*(\mathbf{x})$, which relates the local and averaged values of elastic and electric fields. Thus, analogously to the thermoelastic case, the effective generalized thermal expansion $\boldsymbol{\Lambda}^*$ is spatial and orientational averages of local magnetostriction strain in which the generalized stress concentrator factors $\mathbb{B}(\mathbf{x})$ serve as weighting factors. An analogous formula was obtained for the effective magnetostriction of composites and polycrystals with particulate and columnar microstructures in [316].

In some cases, another form of relationship (6.259) is more convenient. To derive it, we use the obvious equalities $\langle\boldsymbol{\mathcal{E}}^{II}\rangle = \langle V\boldsymbol{\mathcal{E}}^{II}\rangle + \langle(1-V)\boldsymbol{\mathcal{E}}^{II}\rangle$, $\langle V\boldsymbol{\Sigma}^{II}\rangle + \langle(1-V)\boldsymbol{\Sigma}^{II}\rangle = 0$. Then, assuming the homogeneity of matrix material, we obtain

$$\boldsymbol{\Lambda}^* = \langle\boldsymbol{\Lambda}\rangle + \langle\mathbb{M}_1\boldsymbol{\Sigma}^{II}V\rangle. \tag{6.260}$$

Formulae (6.259) and (6.260) generalize the similar results (6.46) and (6.48), respectively, of the thermoelasticity theory for piezoelectric composites. These results are commonly obtained by resolving the field $\boldsymbol{\sigma}$ and $\boldsymbol{\varepsilon}$ into components $\boldsymbol{\sigma}^I$ and $\boldsymbol{\varepsilon}^I$ and $\boldsymbol{\sigma}^{II}$ and $\boldsymbol{\varepsilon}^{II}$, similar to (6.254). The same approach was applied to evaluation of the elastic interaction energy of the $\boldsymbol{\sigma}^I$ and $\boldsymbol{\sigma}^{II}$, which will be used in the analysis of inhomogeneous piezoelectric composites.

6.6.2 Effective Energy Functions

Substituting (6.254) in the formula for the energy of generalized strains (compare with (6.61) $\langle W\rangle = \langle\sigma_{ij}(\varepsilon_{ji}-\beta_{ji})\rangle+\langle D_i(E_i-q_i\theta)\rangle \equiv \langle\boldsymbol{\Sigma}^\top(\boldsymbol{\mathcal{E}}-\boldsymbol{\Lambda})\rangle$ as a result, we obtain a decomposition, similar to the thermoelastic case $\langle W\rangle = \langle W^I\rangle + \langle W^{II}\rangle + \langle W^{\text{int}}\rangle$. The terms on the right-hand side of the equality correspond to the energies of the fields $\boldsymbol{\Sigma}^I$ and $\boldsymbol{\Sigma}^{II}$ and to their interaction energy. Averages (6.257) and (6.258) give

$$2\langle W^I\rangle = \langle\boldsymbol{\Sigma}^{I\top}\rangle\langle\boldsymbol{\mathcal{E}}^I\rangle, \quad 2\langle W^{II}\rangle = -\langle\boldsymbol{\Sigma}^{II\top}\boldsymbol{\Lambda}\rangle, \tag{6.261}$$

$$\langle W^{\text{int}}\rangle = \langle\boldsymbol{\Sigma}^{II\top}\rangle\langle\boldsymbol{\mathcal{E}}^I\rangle = \langle\boldsymbol{\Sigma}^{I\top}\rangle\Big(\langle\boldsymbol{\mathcal{E}}^{II}\rangle - \boldsymbol{\Lambda}^*\Big). \tag{6.262}$$

For boundary conditions (6.248) with $\boldsymbol{\Sigma}^{II}(\mathbf{x}) \equiv \mathbf{0}$ and $\mathbf{x} \in \partial w$, we obtain

$$\langle\boldsymbol{\mathcal{E}}^I\rangle = \mathbb{M}^*\langle\boldsymbol{\Sigma}\rangle, \quad \langle\boldsymbol{\Sigma}^{II}\rangle = \mathbf{0}, \quad \langle\boldsymbol{\mathcal{E}}^{II}\rangle = \boldsymbol{\Lambda}^*, \tag{6.263}$$

$$\langle\boldsymbol{\Sigma}^{I\top}\boldsymbol{\mathcal{E}}^I\rangle = \langle\boldsymbol{\Sigma}^\top\rangle\mathbb{M}^*\langle\boldsymbol{\Sigma}\rangle, \quad \langle\boldsymbol{\Sigma}^I\boldsymbol{\mathcal{E}}^{II}\rangle = \langle\boldsymbol{\Sigma}\rangle\boldsymbol{\Lambda}^*, \tag{6.264}$$

$$\langle\boldsymbol{\Sigma}^{II\top}\boldsymbol{\mathcal{E}}^I\rangle = 0, \quad \langle\boldsymbol{\Sigma}^{II\top}\boldsymbol{\mathcal{E}}^{II}\rangle = 0. \tag{6.265}$$

In view of (6.262) and (6.263), the field interaction energy is $\langle W^{\text{int}}\rangle \equiv 0$. Therefore, the total energy can be expressed in the form

$$\langle W\rangle = \langle W^I\rangle + \langle W^{II}\rangle = \frac{1}{2}\Big(\langle\boldsymbol{\Sigma}^\top\rangle\mathbb{M}^*\langle\boldsymbol{\Sigma}\rangle - \langle\boldsymbol{\Sigma}^{II\top}\boldsymbol{\Lambda}\rangle\Big). \tag{6.266}$$

Note that along with the effective tensors $\mathbb{M}^*$ and $\boldsymbol{\Lambda}^*$, energy $\langle W^{II}\rangle \equiv -\langle\mathbb{F}^*\boldsymbol{\Lambda}\rangle/2$ also is an effective function that is determined only by temperature, microstructure geometry, and the material parameters $\mathbb{M}(\mathbf{x})$ and $\boldsymbol{\Lambda}(\mathbf{x})$ and does not depend upon the applied external field $\boldsymbol{\Sigma}^I$. According to the second formulae (6.261), we have $\langle W^{II}\rangle = 0$ at $\boldsymbol{\Lambda}(\mathbf{x}) \equiv$ const. ($\mathbf{x} \in w$), and hence $\langle W^{II}\rangle$ is a measure of the generalized residual microstresses in a microinhomogeneous medium. The value W^{II} can be found both by computation from the definition (6.261) and experimentally. Thus, if the variations of the temperature θ are small as compared with the reference value T_0, then, based on thermodynamic considerations, we obtain (compare with (6.129$_2$)): $2\langle W^{II}\rangle = (C_\sigma^{E*} - \langle C_\sigma^E\rangle)\theta^2/T_0$, where C_σ^{E*} and $\langle C_\sigma^E\rangle$ are the effective and averaged specific heats at constant hydrostatic pressure and the electric field intensity, which can be easily measured experimentally.

6.6.3 Two-Phase Composites

For a two-phase medium with homogeneous components, the problem of calculating the effective generalized tensor of the thermal expansion $\boldsymbol{\Lambda}^*$ reduces to determination of the effective generalized compliance $\mathbb{M}^*$, i.e., $\boldsymbol{\Lambda}^*$ is uniquely expressed through $\mathbb{M}^*$. In fact, assume the homogeneity of the material properties within a component $\mathbf{f}(\mathbf{x}) \equiv$ const, $(\mathbf{x} \in v^{(i)}, i = 0, 1; \mathbf{f} = \mathbb{L}, \mathbb{M}, \boldsymbol{\Lambda})$, and substitute the expressions for stress concentrators of components

$$\langle \mathbb{B}_i^* \rangle = (c^{(i)})^{-1}(\mathbb{M}^{(i)} - \mathbb{M}^{(j)})^{-1}(\mathbb{M}^* - \mathbb{M}^{(j)}) \tag{6.267}$$

$(c^{(i)} = \langle V^{(i)} \rangle,\ i = 0, 1,\ j = 1 - i)$ into the expression for the effective tensor $\boldsymbol{\Lambda}^*$ (6.259). Then, we find

$$\boldsymbol{\Lambda}^* = (\mathbb{M}^* - \mathbb{M}^{(0)})(\mathbb{M}^{(1)} - \mathbb{M}^{(0)})^{-1}\boldsymbol{\Lambda}^{(1)} + (\mathbb{M}^* - \mathbb{M}^{(1)})(\mathbb{M}^{(0)} - \mathbb{M}^{(1)})^{-1}\boldsymbol{\Lambda}^{(0)}. \tag{6.268}$$

Analogous relationships were derived in [92], and [407]. A similar exact link between effective magnetostriction and effective moduli is also established [317]. Formula (6.268) shows that if the electroelastic fields are not coupled within the components ($\mathbf{c} = \mathbf{d} \equiv \mathbf{0}$), then the effective pyroelastic $\mathbf{q}^*$ and thermal expansion $\boldsymbol{\beta}^*$ coefficients are independent of one another in a composite. In this case, any of the coefficients $\boldsymbol{\beta}^*$ or $\mathbf{q}^*$ may be zero. However, this is not the case when the piezoelectric moduli are nonzero at least for one of the components. Then, the tensor $\boldsymbol{\beta}^*$(or $\mathbf{q}^*$) is nonzero even though $\boldsymbol{\beta}(\mathbf{x}) \equiv \mathbf{0}$ (or $\mathbf{q}(\mathbf{x}) \equiv \mathbf{0}$) and $\mathbf{q}(\mathbf{x}) \not\equiv$ const. (or $\boldsymbol{\beta}(\mathbf{x}) \not\equiv$ const).

If the tensor $\boldsymbol{\Lambda}^*$ is known, then one can find the field $\boldsymbol{\Sigma}^{II}$ averaged over the volume of each component i

$$\langle \boldsymbol{\Sigma}^{II} \rangle^{(i)} = (c^{(i)})^{-1}(\mathbb{M}^{(i)} - \mathbb{M}^{(j)})^{-1}(\boldsymbol{\Lambda}^* - \langle \boldsymbol{\Lambda} \rangle). \tag{6.269}$$

From (6.261) and (6.269) we find the exact expression for the averaged stored energy $2\langle W^{II} \rangle = (\boldsymbol{\Lambda}^{(1)} - \boldsymbol{\Lambda}^{(0)})(\mathbb{M}^{(1)} - \mathbb{M}^{(0)})^{-1}(\boldsymbol{\Lambda}^* - \langle \boldsymbol{\Lambda} \rangle)$.

For two-phase composites with anisotropic components, one may obtain not only the averaged fields of generalized thermoelastic stresses but also their exact local values if the local stress concentrators $\mathbb{B}^*(\mathbf{x})$ defined by (6.253) for the corresponding generalized elastic problem are known [94]

$$\mathbb{F}^*(\mathbf{x}) = (\mathbb{B}^*(\mathbf{x}) - \mathbb{I})(\mathbb{M}^{(1)} - \mathbb{M}^{(0)})(\boldsymbol{\Lambda}^{(1)} - \boldsymbol{\Lambda}^{(0)}). \tag{6.270}$$

Since $\langle \boldsymbol{\sigma}^{II} \rangle^{(i)} = \langle \mathbb{F}^* \rangle^{(i)}$, averaging of (6.270) over the volume of a component $v^{(i)}$ $(i = 0, 1)$ yields (6.269).

We note that the results of this section hold true only for two-component composites of matrix or stochastic (mixture) structure. The first type includes, for example, a composite composed of a simply connected homogeneous ambient phase (matrix) and inclusions of similar shape, orientation, and size. In the general case, rocks are textured polycrystals [1552], and Eqs. (6.267)–(6.270) are applicable under certain assumptions. Namely, a rock can be modeled as a composite with inhomogeneities such as cracks or inclusions of a different mineral (e.g., in the form of relatively rigid plates of mica in quartz [1552]). In so doing, the properties of a matrix (actually, a single-phase polycrystal) are estimated via one of the well-known method such as the generalized singular approximation [1379] or the method of effective media (see Chap. 8). The smaller the fluctuations of material properties of a crystal composing a matrix yields the higher accuracy of (6.267)–(6.270).

Exact expressions (6.267)–(6.270) obtained in terms of the theory of two-phase composite can be applied to polycrystals (which comprise the majority of rocks) as follows. Computation of the averaged material fields utilize the discretization of possible orientations of the crystallographic axes, the procedure being traditional in mechanics of polycrystals. Crystallites may be assumed to have two possible orientations only. Then (6.267)–(6.270) may be applied to the two-component medium specified by these orientations.

6.6.4 Discontinuities of Generalized Fields at the Interface Between Components

To determine jumps of generalized stresses and strains at a boundary ∂v_{12} represented by an ideal contact between components $v^{(1)}$ and $v^{(2)}$, we will use projection operators of the second rank $\boldsymbol{\eta}$ and $\boldsymbol{\nu}$, and those of the fourth rank $\mathbf{F}$ and $\mathbf{E}$ introduced in [644] (see Eq. (3.86)) and having "orthogonal" properties (3.87). We construct the following interface projection operators: $\mathbb{H}^{\eta} = \text{diag}(\mathbf{F}, \boldsymbol{\eta})$, $\mathbb{H}^{\boldsymbol{\nu}} = \text{diag}(\mathbf{E}, \boldsymbol{\nu})$ with the properties

$$\mathbb{H}^{\eta} + \mathbb{H}^{\boldsymbol{\nu}} = \text{diag}(\mathbf{I}, \boldsymbol{\delta}) \equiv \mathbb{I}, \quad \mathbb{H}^{\eta}\mathbb{H}^{\eta} = \mathbb{H}^{\eta}, \quad \mathbb{H}^{\boldsymbol{\nu}}\mathbb{H}^{\boldsymbol{\nu}} = \mathbb{H}^{\boldsymbol{\nu}}, \quad \mathbb{H}^{\mathbf{n}}\mathbb{H}^{\boldsymbol{\nu}} = \mathbf{0}. \tag{6.271}$$

Hill [644] and Shermergor and Yakovlev [1379] showed that tensors $\mathbf{F}$, $\mathbf{E}$ and $\mathbb{H}^{\mathbf{n}}$, $\mathbb{H}^{\boldsymbol{\nu}}$, respectively, resolve the tensor g_{ij} and matrix $[g_{ij}, f_k]^{\top}$ into the normal and tangential components, where g_{ij} and f_k are an arbitrary symmetrical second rank tensor and an arbitrary vector, respectively, both specified on the surface ∂v_{12}. The meaning of the resolution is obvious if the unit normal is chosen to be $\mathbf{n} = (0, 0, 1)$. In this case, after the application of the operator $\mathbb{H}^{\eta}$, only the components g_{ij} and f_k that have the index 3 are nonzero.

The operators $\mathbb{H}^{\eta}$ and $\mathbb{H}^{\boldsymbol{\nu}}$ allow the ideal contact condition to be written in the form

$$\mathbb{H}^{\eta}\boldsymbol{\Sigma}^{(1)} = \mathbb{H}^{\eta}\boldsymbol{\Sigma}^{(2)}, \quad \mathbb{H}^{\boldsymbol{\nu}}\boldsymbol{\mathcal{E}}^{(1)} = \mathbb{H}^{\boldsymbol{\nu}}\boldsymbol{\mathcal{E}}^{(2)}. \tag{6.272}$$

Now, we obtain the expressions relating generalized stresses on both sides of the interface ∂v_{12}. According to (6.271) and (6.272$_1$), we have

$$\boldsymbol{\mathcal{E}}^{(2)} = \mathbb{M}^{(2)}\boldsymbol{\Sigma}^{(2)} + \boldsymbol{\Lambda}^{(2)} = \mathbb{M}^{(2)}(\mathbb{H}^{\eta}\boldsymbol{\Sigma}^{(1)} + \mathbb{H}^{\nu}\boldsymbol{\Sigma}^{(2)}) + \boldsymbol{\Lambda}^{(2)}. \tag{6.273}$$

Multiplying both sides of (6.273) by the operator $\mathbb{H}^{\nu}$ and taking into account condition (6.272$_2$), we obtain

$$\boldsymbol{\Sigma}^{(2)} = \mathbb{B}(\mathbf{n})\Big(\boldsymbol{\mathcal{E}}^{(1)} - \mathbb{M}^{(2)}\mathbb{H}^{\eta}\boldsymbol{\Sigma}^{(1)} - \boldsymbol{\Lambda}^{(2)}\Big), \tag{6.274}$$

where $\mathbb{B}(\mathbf{n}) = (\mathbb{H}^{\nu}\mathbb{M}^{(2)}\mathbb{H}^{\nu})^{-1}\mathbb{H}^{\nu}$. Finally, substituting the equation of state (6.247) into the right-hand side of (6.274), we find

$$\boldsymbol{\Sigma}^{(2)} = \Big[\mathbb{I} + \mathbb{B}(\mathbf{n})(\mathbb{M}^{(1)} - \mathbb{M}^{(2)})\Big]\boldsymbol{\Sigma}^{(1)} + \mathbb{B}(\mathbf{n})(\boldsymbol{\Lambda}^{(1)} - \boldsymbol{\Lambda}^{(2)}) \tag{6.275}$$

Similarly, using the conjugation condition for the normal components of the generalized strain tensor of the first material component, we obtain

$$\boldsymbol{\mathcal{E}}^{(2)} = \Big[\mathbb{I} + \mathbb{A}(\mathbf{n})(\mathbb{L}^{(1)} - \mathbb{L}^{(2)})\Big]\boldsymbol{\mathcal{E}}^{(1)} + \mathbb{A}(\mathbf{n})\Big[\mathbb{L}^{(2)}\boldsymbol{\Lambda}^{(2)} - \mathbb{L}^{(1)}\boldsymbol{\Lambda}^{(1)}\Big], \tag{6.276}$$

where $\mathbb{A}(\mathbf{n}) \equiv (\mathbb{H}^{\eta}\mathbb{L}^{(2)}\mathbb{H}^{\eta})^{-1}\mathbb{H}^{\eta}$. Since relationships (6.275) and (6.276) should be equivalent at $\mathbb{M} = \mathbb{L}^{-1}$, the operators $\mathbb{A}(\mathbf{n})$ and $\mathbb{B}(\mathbf{n})$ are related as

$$\mathbb{L}^{(2)}\mathbb{A}(\mathbf{n}) + \mathbb{B}(\mathbf{n})\mathbb{M}^{(2)} = \mathbb{A}(\mathbf{n})\mathbb{L}^{(2)} + \mathbb{M}^{(2)}\mathbb{B}(\mathbf{n}) = \mathbb{I}. \tag{6.277}$$

where operators $\mathbb{B}(\mathbf{n})$ and $\mathbb{A}(\mathbf{n})$, each dependent on the normal vector $\mathbf{n}\perp\partial v_{12}$ and tensor of the properties of a corresponding phase, relate generalized stresses and strains on both sides of the contact. At $\theta = 0$ (6.276), (6.277) obtained in [280] coincide with similar equations obtained earlier in [1379].

6.6.5 *Phase-Averaged First and Second Moments of the Field* $\boldsymbol{\Sigma}$

First, we obtain the averaged single-point second moment of the field $\boldsymbol{\Sigma}$ in terms of the analytical function $\mathbb{M}^* = \mathbb{M}^*(\mathbb{M}^{(i)})$, $(i = 0, 1, \ldots)$ which may be calculated with any desired accuracy or measured experimentally. Averaging is performed over the component $v^{(i)}$. For a given field $\mathbf{t}(\mathbf{x}) \equiv \langle\boldsymbol{\Sigma}\rangle\mathbf{n}(\mathbf{x})$, $\langle\boldsymbol{\Sigma}\rangle = \text{const.}$, $\mathbf{x} \in \partial w$ (see (6.253)), we will vary the local tensor of generalized compliance $\mathbb{M}(\mathbf{x}) \to \mathbb{M}(\mathbf{x}) + \delta\mathbb{M}(\mathbf{x})$, which leads to the variation of the field $\boldsymbol{\Sigma}^I \to \boldsymbol{\Sigma}^I + \delta\boldsymbol{\Sigma}^I$. This technique was applied in the case of a purely elastic field (see Chap. 13). Then, from the relation $2\langle W^I\rangle = \langle \boldsymbol{\Sigma}^{I\top}\mathbb{M}^*\boldsymbol{\Sigma}^I\rangle = \langle \boldsymbol{\Sigma}^{I\top}\rangle\mathbb{M}^*\langle \boldsymbol{\Sigma}^I\rangle$ at $\theta \equiv 0$, we obtain

$$\langle \boldsymbol{\Sigma}^{I\top}\rangle(\delta\mathbb{M}^*)\langle \boldsymbol{\Sigma}^I\rangle = \langle \boldsymbol{\Sigma}^{I\top}(\delta\mathbb{M})\boldsymbol{\Sigma}^I + 2(\delta\boldsymbol{\Sigma}^{I\top})\mathbb{M}\boldsymbol{\Sigma}^I\rangle. \tag{6.278}$$

We transform the second term on the right-hand side of (6.278) by partial integration

$$\langle \delta\boldsymbol{\Sigma}^{I\top}\mathbb{M}\boldsymbol{\Sigma}^I\rangle = (\bar{w})^{-1}\int \delta\boldsymbol{\Sigma}^{I\top}\boldsymbol{\mathcal{U}}\,d\mathbf{s} - \langle(\nabla\delta\boldsymbol{\Sigma}^I)\boldsymbol{\mathcal{U}}^I\rangle \tag{6.279}$$

where $\boldsymbol{\mathcal{U}}$ is the generalized displacement matrix defined by (6.246_2). The surface integral taken over the boundary of domain w is zero at the constant boundary conditions (6.248). The second term on the right-hand side of (6.279) is zero due to static condition (6.246_1). Therefore, we obtain from (6.278) and (6.279)

$$\langle \boldsymbol{\Sigma}^{I\top}\rangle(\delta\mathbb{M}^*)\langle \boldsymbol{\Sigma}^I\rangle = \langle \boldsymbol{\Sigma}^{I\top}(\delta\mathbb{M})\boldsymbol{\Sigma}^I\rangle. \tag{6.280}$$

We assume that the components of the generalized compliance tensor and their variations are homogeneous within each phase, i.e.,

$$\mathbb{M}(\mathbf{x}) \equiv \mathbb{M}^{(i)}V^{(i)}(\mathbf{x}),\ \ \delta\mathbb{M}(\mathbf{x}) \equiv \delta\mathbb{M}^{(i)}V^{(i)}, \tag{6.281}$$

where $\mathbb{M}^{(i)},\ \delta\mathbb{M}^{(i)} = \text{const}$. Then, the value $\delta\mathbb{M}$ in (6.280) may be factored outside the phase averaging, and (6.248) may be written in the following symbolic form:

$$\langle \Sigma^I_{\alpha\beta}\otimes\Sigma^I_{\gamma\delta}\rangle^{(i)} = \frac{1}{c^{(i)}}\frac{\partial\mathbb{M}^*_{\lambda\mu\nu\eta}}{\partial\mathbb{M}^{(i)}_{\alpha\beta\gamma\delta}}\langle\Sigma_{\lambda\mu}\rangle\langle\Sigma_{\nu\eta}\rangle, \tag{6.282}$$

where we took into account that $\langle\boldsymbol{\Sigma}^{II}\rangle \equiv \mathbf{0}$, and hence $\langle\boldsymbol{\Sigma}^I\rangle = \langle\boldsymbol{\Sigma}\rangle$. Expressing (6.282) in terms of generalized strains $\boldsymbol{\mathcal{E}}$, we obtain

$$\langle \mathcal{E}^I_{\alpha\beta}\otimes\mathcal{E}^I_{\gamma\delta}\rangle^{(i)} = \frac{1}{c^{(i)}}\frac{\partial\mathbb{L}^*_{\lambda\mu\nu\eta}}{\partial\mathbb{L}^{(i)}_{\alpha\beta\gamma\delta}}\langle\mathcal{E}_{\lambda\mu}\rangle\mathcal{E}_{\nu\eta}\rangle, \tag{6.283}$$

Here the calculations are carried out at constant values of $\boldsymbol{\Lambda}$. Equations (6.282) and (6.283) were obtained by Buryachenko and Shermergor [280]. Li and Dunn [930] independently obtained the intermediate Eq. (6.280) (without final representation (6.282)) at the condition (6.281). Analogous representations for the field fluctuations describing the transport phenomena (i.e., conductivity) were proposed in [47].

Now, we find the exact relations between the energy $\langle W^{II}\rangle$ and second single point moment of the residual generalized stresses $\langle\boldsymbol{\Sigma}^{II}\otimes\boldsymbol{\Sigma}^{II}\rangle^{(i)}$ $(i = 0, 1, \ldots, N)$ at $\mathbf{t}(\mathbf{x}) \equiv \mathbf{0}$ and $\mathbf{x} \in \partial w$. Let the generalized compliance be varied $\mathbb{M}(\mathbf{x}) \to$

$\mathbb{M}(\mathbf{x}) + \delta\mathbb{M}(\mathbf{x})$ at a given tensor $\boldsymbol{\Lambda}(\mathbf{x})$ and $\mathbf{t}(\mathbf{x}) \equiv \mathbf{0}$, $\mathbf{x} \in \partial w$, as is done solving the thermoelastic problem (see Sect. 6.3.3). Then, we obtain

$$2\langle\delta W^{II}\rangle = \langle\boldsymbol{\Sigma}^{II\top}(\delta\mathbb{M})\boldsymbol{\Sigma}^{II}\rangle - 2\langle(\delta\boldsymbol{\Sigma}^{II\top}\boldsymbol{\Lambda}^{II}\rangle, \quad 2\langle\delta W^{II}\rangle = -\langle(\delta\boldsymbol{\Sigma}^{II})\boldsymbol{\Lambda}^{II}\rangle, \tag{6.284}$$

where we used the condition $\langle(\delta\boldsymbol{\Sigma}^{\top}\boldsymbol{\mathcal{E}}\rangle = 0$. Therefore

$$\langle\boldsymbol{\Sigma}^{II\top}(\delta\mathbb{M})\boldsymbol{\Sigma}^{II}\rangle = -2\langle\delta W^{II}\rangle. \tag{6.285}$$

Here, we took into account that in view of Hill's condition, $\langle(\delta\boldsymbol{\Sigma}^{\top})\boldsymbol{\mathcal{E}}\rangle = 0$. Using the expression for variation of generalized compliance in the form (6.281), we find the final expression

$$\langle\Sigma^{II}\otimes\Sigma^{II}\rangle^{(i)} = -\frac{2}{c^{(i)}}\frac{\partial\langle\, W^{II}\rangle}{\partial\mathbb{M}^{(i)}}, \quad \text{or} \quad \langle\Sigma^{II}_{\alpha\beta}\otimes\Sigma^{II}_{\gamma\delta}\rangle^{(i)} = -\frac{2}{c^{(i)}}\frac{\partial\langle W^{II}\rangle}{\partial\mathbb{M}^{(i)}_{\alpha\beta\gamma\delta}}. \tag{6.286}$$

Equation (6.286) was obtained in [280]. Li and Dunn [930] independently obtained the intermediate Eq. (6.285) (without final representation (6.286)) at the condition (6.281).

If a material is subject not only to temperature fields θ but also to generalized fields $\mathbf{t}$ defined by expression (6.248), then, we obtain resolving the value W into two W^I and W^{II} (6.266) and using (6.282) and (6.286)

$$\langle\Sigma\otimes\Sigma\rangle^{(i)} = \frac{1}{c^{(i)}}\frac{\partial\mathbb{M}^*}{\partial\mathbb{M}^{(i)}}\langle\boldsymbol{\Sigma}\rangle\otimes\langle\boldsymbol{\Sigma}\rangle - \frac{2}{c^{(i)}}\frac{\partial\langle\, W^{II}\rangle}{\partial\mathbb{M}^{(i)}} + \frac{2}{c^{(i)}}\frac{\partial\boldsymbol{\Lambda}^*}{\partial\mathbb{M}^{(i)}}\langle\boldsymbol{\Sigma}\rangle, \tag{6.287}$$

where all partial derivatives are calculated at fixed $\boldsymbol{\Lambda}(\mathbf{x})$. Thus, the fourth rank tensor of the second moment of the generalized stress field $\langle\Sigma\otimes\Sigma\rangle^{(i)}$ averaged over the phase volume $v^{(i)}$ is uniquely determined from the functional relationships $\mathbb{M}^* = \mathbb{M}^*(\mathbb{M}^{(i)})$, $\langle W^{II}\rangle = \langle W^{II}(\mathbb{M}^{(i)})\rangle$, and $\boldsymbol{\Lambda}^* = \boldsymbol{\Lambda}(\mathbb{M}^{(i)})$ in four-dimensional space.

Note that the exact expression for the phase-averaged first moment $\langle\boldsymbol{\Sigma}^I\rangle^{(i)}$ $(i = 0, 1)$ can be found only for a two-phase composite (6.267), which is not the case for general multicomponent media. However, the exact solution for the first moment of the residual field $\langle\boldsymbol{\Sigma}^{II}\rangle^{(i)}$ $(i = 0, 1, \ldots, N)$ can be found if function $\langle W^{II}\rangle = \langle W^{II}(\boldsymbol{\Lambda}^{(i)})\rangle$ is known. For this purpose, let $\boldsymbol{\Lambda}(\mathbf{x}) \to \boldsymbol{\Lambda}(\mathbf{x}) + \delta\boldsymbol{\Lambda}(\mathbf{x})$ be varied in the same manner as in the thermoelastic case, assuming $\mathbb{M}(\mathbf{x})$ to be constant and $\mathbf{t}(\mathbf{x}) \equiv \mathbf{0}$, $\mathbf{x} \in \partial w$. We also assume that the generalized parameter $\boldsymbol{\Lambda}$ and its variation within each phase of the composite are homogeneous $\boldsymbol{\Lambda}(\mathbf{x}) \equiv \boldsymbol{\Lambda}^{(i)}V^{(i)}(\mathbf{x})$, $\delta\boldsymbol{\Lambda}(\mathbf{x}) \equiv \delta\boldsymbol{\Lambda}^{(i)}V^{(i)}(\mathbf{x})$, where $\boldsymbol{\Lambda}^{(i)}, \delta\boldsymbol{\Lambda}^{(i)} = \text{const}$. Hence, we obtain the energy variation in the form

$$-2\langle\delta W^{II}\rangle = \langle\boldsymbol{\Sigma}^{II\top}\delta\boldsymbol{\Lambda}\rangle + \langle(\delta\boldsymbol{\Sigma}^{II\top})\boldsymbol{\Lambda}^I\rangle. \tag{6.288}$$

Let us transform the last term with the use of the equation of state (6.247)

$$\langle(\delta\boldsymbol{\Sigma}^{II\top})\boldsymbol{\Lambda}^{I}\rangle = \langle(\delta\boldsymbol{\Sigma}^{II\top})\boldsymbol{\mathcal{E}}^{II}\rangle\langle\boldsymbol{\Sigma}^{II\top}\delta\boldsymbol{\Lambda}\rangle - \langle\boldsymbol{\Sigma}^{II\top}\delta\boldsymbol{\mathcal{E}}^{II}\rangle. \tag{6.289}$$

Since the admissible fields $\delta\boldsymbol{\Sigma}^{II}$, $\boldsymbol{\mathcal{E}}^{II}$ and $\boldsymbol{\Sigma}^{II}$, $\delta\boldsymbol{\mathcal{E}}^{II}$ obey Hill's condition (6.252), we find substituting (6.289) in (6.288)

$$\langle\boldsymbol{\Sigma}^{II}_{\alpha\beta}\rangle^{(i)} = -\frac{1}{c^{(i)}}\frac{\partial\langle W^{II}\rangle}{\partial\boldsymbol{\Lambda}^{(i)}_{\alpha\beta}} \tag{6.290}$$

($i = 0, 1, \ldots, N$; $k, l = 1, \ldots, 4$). In a manner like (6.193), we can get

$$\langle\boldsymbol{\Sigma}^{I}\rangle^{(i)} = \frac{1}{c^{(i)}}\langle\boldsymbol{\Sigma}\rangle\frac{\partial\boldsymbol{\Lambda}^{*}}{\partial\boldsymbol{\Lambda}^{(i)}}. \tag{6.291}$$

The partial derivatives in Eqs. (6.290) and (6.291) are carried out at fixed values of $\mathbb{M}(\mathbf{x})$.

Relationship (6.287) allows one to determine the fluctuations of the field $\boldsymbol{\Sigma}$ at the point $\mathbf{x} \in v^{(i)}$ ($i = 0, 1, \ldots, N$). For this purpose, it is sufficient to find $\mathbb{M}^*$ at the variation of the generalized compliance in the vicinity of the considered point $\mathbf{x}$. This problem was solved in [280] in the framework of hypothesis (**H**1) only (see Chap. 8) which is the main hypothesis of many micromechanical methods in thermoelasticity. The same is true for the piezoelectric matrix composites. That implies the homogeneity of the stress field single-point moment within the inclusion. This fact enables determination of the value $\langle\boldsymbol{\Sigma}\otimes\boldsymbol{\Sigma}\rangle^{(0)}$ in the matrix material and the second moment of the generalized stress field for the matrix material in the vicinity of a fixed inclusion v_i. Actually, at a point $\mathbf{x}\in\partial v_i$ of the inclusion boundary $\boldsymbol{\Sigma}^{(0)}$ with the external normal $\mathbf{n}$, the outer limiting value of the tensor $\boldsymbol{\Sigma}$ is determined by (6.274). Then, the second moment $\langle\boldsymbol{\Sigma}\otimes\boldsymbol{\Sigma}\rangle_x$ of the field $\boldsymbol{\Sigma}$ for the matrix material in the neighborhood of point $\mathbf{x} \in \partial v_i$ can be found from the expression

$$\begin{aligned}\langle\boldsymbol{\Sigma}\otimes\boldsymbol{\Sigma}\rangle_x = {} & \langle[\mathbb{I}+\mathbb{B}(\mathbf{n})\mathbb{M}_1]\boldsymbol{\Sigma}\otimes[\mathbb{I}+\mathbb{B}(\mathbf{n})\mathbb{M}_1]\boldsymbol{\Sigma}\rangle^{(i)}\\ & + [\mathbb{I}+\mathbb{B}(\mathbf{n})\mathbb{M}_1]\langle\boldsymbol{\Sigma}\rangle^{(i)}\otimes\mathbb{B}(\mathbf{n})\boldsymbol{\Lambda}_1^{(i)}\\ & + \mathbb{B}(\mathbf{n})\boldsymbol{\Lambda}_1^{(i)}\otimes[\mathbb{I}+\mathbb{B}(\mathbf{n})\mathbb{M}_1]\langle\boldsymbol{\Sigma}\rangle^{(i)} + [\mathbb{B}(\mathbf{n})\boldsymbol{\Lambda}_1^{(i)}\otimes\mathbb{B}(\mathbf{n})\boldsymbol{\Lambda}_1^{(i)}].\end{aligned} \tag{6.292}$$

Here, we assumed the field homogeneity within each inclusion v_i. Therefore, (6.292) is approximate, although the exact expressions (6.287) for $\langle\boldsymbol{\Sigma}\otimes\boldsymbol{\Sigma}\rangle^{(i)}$ and (6.290), (6.291) for $\langle\boldsymbol{\Sigma}\rangle^{(i)}$ are used on the right-hand side. Thus, the relationships obtained in this Sect. 6.6 are exact (except (6.292)) and valid for any statistically homogeneous composite with ideal contact between its components and homogeneous boundary conditions. One [262] used the second moment estimations of coupled fields (6.287) and (6.292) for the prediction of the limiting surface (see Sect. 12.1).

Chapter 7
General Integral Equations of Micromechanics of Composite Materials

Abstract The general integral equations (GIE), connecting the stresses and strains in a point being considered and the fields in the surrounding points, are obtained in this chapter for both the random and periodic fields of heterogeneities in the infinite media. The method of obtaining of the GIE is based on a centering procedure of subtraction from both sides of a new initial integral equations their statistical averages obtained without any auxiliary assumptions such as the effective field hypothesis (EFH), which is implicitly exploited in the known centering methods. One proves the absolute convergence of the proposed GIEs which are presented in two equivalent forms for both the stresses and strains. Some particular cases, asymptotic representations (e.g., classical GIE), and simplifications of proposed GIE are presented for either statistically homogeneous, inhomogeneous (graded), or periodic structures of CM.

The need to consider the actual microstructure of composite materials subjected to essentially inhomogeneous mechanical, body force, and temperature loading in micromechanics problems is well-known. Unfortunately, the starting assumptions made in the majority of studies, namely, that the structure of composite media as well as the random fields of stresses is statistically homogeneous and therefore is invariant with respect to the translation, are incorrect. For example, due to some production technologies, the inclusion concentration may be a function of the coordinates [346, 1218, 1219]. The accumulation of damage also occurs locally in stress-concentration regions, for example, at the tip of a macroscopic crack [729]. Furthermore, in layered composite shells, the location of the fibers is random within the periodic layers, and the micromechanics equations are equations with almost periodic coefficients. Finally, *functionally graded materials* (FGMs) also have a statistically inhomogeneous structure which consists of two or more phases which is fabricated with a spatial variation of its composition that may improve the structural response (see Chap. 12). In many problems the material may possess a periodic microstructure formed by the spatial repetition of small microstructures, or unit cells. Such a perfectly regular distribution, of course, does not exist in actual cases, although periodic modeling can be quite useful, since it provides rigorous

V. A. Buryachenko, *Local and Nonlocal Micromechanics of Heterogeneous Materials*, https://doi.org/10.1007/978-3-030-81784-8_7

estimations with a priori prescribed accuracy for various material properties. Another problem is that the development of the theory of homogenization of initially periodic structures with the introduced infinite number of imperfections of unit cells randomly distributed in space has an essentially practical meaning. For example, the accumulation of damage, such as the microcracks in the matrix and debonding of inclusions from the matrix, is essentially random in nature.

The most popular methods of analytical micromechanics are based just on a few basic concepts. The effective field hypothesis (EFH) is apparently the most fundamental and most exploited concept of analytical micromechanics. This concept has directed a development of analytical micromechanics over the last 150 years and made a contribution to their progress incompatible with any another concept. The idea of this concept dates back to Poisson [1225] and Faraday [459] (in the electricity and magnetism context), Mossotti [1097], Clausius [342] (in the dielectric context), Lorenz [968] (in the refractivity context), and Maxwell [1036] (in the conductivity context) who introduced the effective field concept as a local homogeneous field acting on the inclusions and differing from the applied macroscopic one. Landauer [881], Markov [1012], and Scaife [1335] presented comprehensive reviews of the 150-year history of this concept accompanied by some famous formulae with extensive references. Walpole [1575] pioneered the application of the concept to the static of composites under the name uniform image field. Effective field technique was intensively applied in analytical micromechanics of random and periodic structure composites (see for reference, e.g., [200, 747, 1094]) as well as in analytical micromechanics of multiple interacting cracks addressed as traction or pseudo-load [653, 654]. Buryachenko [200] has drawn the conclusion that the effective field concept is used (either explicitly or implicitly) in most popular methods of analytical micromechanics, e.g., the methods of self-consistent fields and effective fields. The concept of the EFH (even if this term is not mentioned) in combination with subsequent assumptions totally dominates (and creates the fundamental limitations) in all four groups of analytical micromechanics in physics and mechanics of heterogeneous media: model methods, perturbation methods, self-consistent methods, and variational ones (see for reference and details [200]).

However, EFH serves the dual function of investigations in analytical micromechanics. On the one hand, the EFH is acting as a background of micromechanical models that is particularly striking in the group of model methods (see, e.g., [643, 812]). On the other hand, the EFH is just a tool exploiting for approximate solution of a so-called general integral equation (GIE). The GIE is the exact integral equation connecting the random fields at the point being considered and the surrounding points. The method of obtaining of this GIE that first comes to mind is the modeling of perturbations produced by the surrounding particles by the Green function technique. However, realization of a direct superposition scheme of summations of these perturbations leads to the integrals which are not absolutely convergent at infinity. There is a very long and dramatic history (which looks as a scientific thriller) of attempts to modify these conditionally convergent integrals resulting from long-range interactions. In particular, an idea of so-called method of renormalization goes back to Lord Rayleigh (1892) [1284] who apparently was the first to encounter the mentioned difficulty in consideration of effective conductivity of periodic system containing a square arrays of cylinders or cubic lattice of spheres. In parallel with the renormalization technique (introduced with the different intuitive levels of rigor), a similar assignment is performed by "noncanonical regularization" proposed by Kröner in 1974 [819] (see also for details [200, 202, 747]) and based on an intuitive introduction of an operation (which was unknown before) of corresponding generalized functions on a constant symmetric tensors. At last almost in 90 years after Lord Rayleigh [1284], several authors like Shermergor [1378], Khoroshun [762], and O'Brian [1143] independently and rigorously proved correctness of GIE previously proposed at the intuitive level. The centering method proposed by Shermergor [1378] for statistically homogeneous media was generalized in [191] to the case of statistically inhomogeneous media. The mentioned proofs were generally recognized as the rigorous ones, and apart from their own significance, they give the answer on the philosophical question what is the primarily either EFH or GIE. Thus, it was proved that the GIE is the primary, while the EFH is just a tool used for an approximate solution of this GIE, and, therefore, GIE can be considered as a background of analytical micromechanics.

Incorrectness of the last statement was established by Buryachenko [201, 202] who proved that so-called exact GIEs are only approximate ones and these approximations are the intrinsic features of these GIEs which were implicitly assumed even before obtaining these GIEs. Moreover, the mentioned assumptions are in fact a some sort of EFH that defines a primogeniture of the EFH with respect to the classical GIEs. However, a primariness of the new GIEs was vindicated very soon by proposal of a new really exact GIE in [201, 202]. The new GIE was obtained by centering procedure of subtraction from both sides of a new initial integral equation their statistical averages obtained without any auxiliary assumptions such as EFH, which is implicitly exploited in the known centering methods. Moreover, one showed that the EFH is a central one, and other concepts play a satellite role providing the conditions for application of the EFH (see for details Chaps. 9–12).

The goal of this chapter is to obtain a generalization of the GIEs (and comparison with classical related equations) for the case of either the statistically inhomogeneous or doubly periodical structures of composite materials subjected to essentially inhomogeneous loading by fields of the stresses, temperature, and body forces. The case of triply periodic structures with random imperfections is also considered.

7.1 General Integral Equations for Matrix Composites of Any Structure

This chapter discusses a certain representative mesodomain w with a characteristic function W and a boundary Γ containing a set $X = (v_i)$ of inclusions v_i with characteristic functions $V_i \quad (i = 1, 2, \ldots)$. At first no restrictions are imposed on the elastic symmetry of the phases or on, for example, geometry of the inclusions. It is assumed that the inclusions can be grouped into components (phases) $v^{(k)}$ $(k = 1, 2, \ldots, N)$ with identical mechanical and geometrical properties (such as the shape, size, orientation, and microstructure of inclusions). We consider the local basic equations of linear thermoelastostatic of composites (6.1)–(6.3). We introduce a "comparison" body, whose mechanical properties $\mathbf{g}^c$ $(\mathbf{g} = \mathbf{L}, \mathbf{M}, \boldsymbol{\alpha}, \boldsymbol{\beta}, \mathbf{f})$ denoted by the upper index c and $\mathbf{L}^c$, $\mathbf{M}^c$ will usually be taken as uniform over w, so that the corresponding boundary-value problem is easier to solve than that for the original body. All tensors $\mathbf{g}$ $(\mathbf{g} = \mathbf{L}, \mathbf{M}, \boldsymbol{\alpha}, \boldsymbol{\beta})$ of material properties are decomposed as $\mathbf{g} \equiv \mathbf{g}^c + \mathbf{g}_1(\mathbf{x}) = \mathbf{g}^c + \mathbf{g}_1^{(m)}(\mathbf{x})$ (3.69). Here and in the following, the upper index $^{(m)}$ indicates the components, and the lower index i indicates the individual inclusions; $v^{(0)} = w \backslash v$, $v \equiv \cup v^{(k)} \equiv \cup v_i$, $V(\mathbf{x}) = \sum V^{(k)} = \sum V_i(\mathbf{x})$, and $V^{(k)}(\mathbf{x})$ are characteristic functions of $v^{(k)}$ that equal to 1 at $\mathbf{x} \in v^{(k)}$ and 0 otherwise, $(m = 0, k;\ k = 1, 2, \ldots, N;\quad i = 1, 2, \ldots)$.

We assume that the phases are perfectly bounded (3.85), so that the displacements and the traction components are continuous across the interphase boundaries, i.e., $[[\boldsymbol{\sigma}\mathbf{n}^{int}]] = \mathbf{0}$ and $[[\mathbf{u}]] = \mathbf{0}$ on the interface boundary Γ^{int} where $\mathbf{n}^{int}$ is the normal vector on Γ^{int} and $[[(.)]]$ is a jump operator. The traction $\mathbf{t}(\mathbf{x}) = \boldsymbol{\sigma}(\mathbf{x})\mathbf{n}(\mathbf{x})$ acting on any plane with the normal $\mathbf{n}(\mathbf{x})$ through the point $\mathbf{x}$ can be represented in terms of displacements $\mathbf{t}(\mathbf{x}) = \hat{\mathbf{t}}(\mathbf{n}, \nabla)\mathbf{u}(\mathbf{x}) + \boldsymbol{\alpha}(\mathbf{x})\mathbf{n}$, where $\hat{t}_{ik}(\mathbf{n}, \nabla) = L_{ijkl}n_j(\mathbf{x})\partial/\partial x_l$. The boundary conditions at the interface boundary are to be considered together with the mixed boundary conditions (4.1) on Γ with the unit outward normal $\mathbf{n}^\Gamma$

$$\mathbf{u}(\mathbf{x}) = \mathbf{u}^\Gamma(\mathbf{x}), \quad \mathbf{x} \in \Gamma_u, \quad \text{and} \quad \boldsymbol{\sigma}(\mathbf{x})\mathbf{n}^\Gamma(\mathbf{x}) = \mathbf{t}^\Gamma(\mathbf{x}). \quad \mathbf{x} \in \Gamma_t, \tag{7.1}$$

where Γ_u and Γ_t are prescribed displacement and traction boundaries such that $\Gamma_u \cup \Gamma_t = \Gamma$, $\Gamma_u \cap \Gamma_t = \emptyset$. $\mathbf{u}^\Gamma(\mathbf{x})$ and $\mathbf{t}^\Gamma(\mathbf{x})$ are, respectively, prescribed displacement on Γ_u and traction on Γ_t; mixed boundary conditions, such as in the case of elastic supports, are possible. Of special practical interest are the homogeneous boundary conditions (6.4) and (6.5)

$$\mathbf{u}^\Gamma(\mathbf{x}) = \boldsymbol{\varepsilon}^\Gamma\mathbf{x}, \quad \boldsymbol{\varepsilon}^\Gamma \equiv \text{const.}, \ \mathbf{x} \in \Gamma, \tag{7.2}$$

$$\mathbf{t}^\Gamma(\mathbf{x}) = \boldsymbol{\sigma}^\Gamma\mathbf{n}^\Gamma(\mathbf{x}), \quad \boldsymbol{\sigma}^\Gamma \equiv \text{const.}, \ \mathbf{x} \in \Gamma, \tag{7.3}$$

where $\boldsymbol{\varepsilon}^\Gamma(\mathbf{x}) = \frac{1}{2}\left[\nabla \otimes \mathbf{u}^\Gamma(\mathbf{x}) + (\nabla \otimes \mathbf{u}^\Gamma(\mathbf{x}))^\top\right]$, $\mathbf{x} \in \Gamma$ and $\boldsymbol{\varepsilon}^\Gamma$ and $\boldsymbol{\sigma}^\Gamma$ are the macroscopic strain and stress tensors, i.e., the given constant symmetric tensors. We will consider the interior problem when the body occupies the interior domain with respect to Γ. Hereafter in analysis of integral equations of microinhomogeneous media, the notations of the domain w and its boundary ∂w are usually replaced by w and Γ, respectively.

In the framework of the traditional scheme, we introduce a homogeneous "comparison" body with homogeneous moduli $\mathbf{L}^c$ and with the inhomogeneous deterministic transformation field $\boldsymbol{\beta}^c(\mathbf{x})$ and body force $\mathbf{f}^c(\mathbf{x})$ (and with solution $\boldsymbol{\sigma}^0$, $\boldsymbol{\varepsilon}^0$, $\mathbf{u}^0$ to the same boundary-value problem). Substituting (6.2) and (6.3) into the equilibrium equation (6.1), we obtain a differential equation with respect to the strain $\boldsymbol{\varepsilon}$

$$\nabla \mathbf{L}^c\big[\boldsymbol{\varepsilon}(\mathbf{x})-\boldsymbol{\beta}^c(\mathbf{x})\big]+\mathbf{f}^c(\mathbf{x}) = -\nabla\big\{\mathbf{L}_1(\mathbf{x})[\boldsymbol{\varepsilon}(\mathbf{x})-\boldsymbol{\beta}(\mathbf{x})]-\mathbf{L}^c\boldsymbol{\beta}_1(\mathbf{x})\big\}-\mathbf{f}_1(\mathbf{x}), \quad (7.4)$$

with a fictitious random "body force" in the right-hand side of the equation, hereafter for all material tensors $\mathbf{g}$ ($\mathbf{L}$, $\mathbf{M}$, $\boldsymbol{\beta}$, $\boldsymbol{\alpha}$, $\mathbf{f}$) the notation $\mathbf{g}_1(\mathbf{x}) \equiv \mathbf{g}(\mathbf{x}) - \mathbf{g}^c$ is used. Then Eq. (7.4) may be reduced to a symmetrized integral form

$$\boldsymbol{\varepsilon}(\mathbf{x}) = \boldsymbol{\varepsilon}^0(\mathbf{x}) + \nabla \int_w \mathbf{G}(\mathbf{x}-\mathbf{y})\nabla\boldsymbol{\tau}(\mathbf{y})\, d\mathbf{y} + \nabla \int_w \mathbf{G}(\mathbf{x}-\mathbf{y})\mathbf{f}_1(\mathbf{y})\, d\mathbf{y}, \quad (7.5)$$

where $\boldsymbol{\tau}(\mathbf{y}) := \mathbf{L}_1(\mathbf{y})[\boldsymbol{\varepsilon}(\mathbf{y}) - \boldsymbol{\beta}(\mathbf{y})] - \mathbf{L}^c\boldsymbol{\beta}_1(\mathbf{y})$ is the stress polarization tensor [see (4.4)] and $\mathbf{G}$ (3.3) (of order $O(\int |x|^{1-d}\, d|\mathbf{x}|$ as $|\mathbf{x}| \to \infty)$ is the infinite homogeneous body Green's function of the Navier equation (3.1) with an elastic modulus tensor $\mathbf{L}^c$. The deterministic function $\boldsymbol{\varepsilon}^0(\mathbf{x})$ is the strain field that would exist in the medium with homogeneous properties $\mathbf{L}^c$ and appropriate boundary conditions.

The strain $\boldsymbol{\varepsilon}^0$ is described by the boundary integral equations (BIE; see, e.g., [60, 156]) on the boundary Γ_0 of the domain w

$$\boldsymbol{\varepsilon}_0(\mathbf{x}) = \int_\Gamma \Big[\nabla\mathbf{G}^{(0)}(\mathbf{x}-\mathbf{s})\underline{\mathbf{t}}_0(\mathbf{s}) - \nabla\mathbf{T}^{(0)\top}(\mathbf{x},\mathbf{s})\mathbf{u}_0(\mathbf{s})\Big]ds$$
$$+\int_w \nabla\mathbf{G}(\mathbf{x}-\mathbf{y})\mathbf{f}^c(\mathbf{y})\, d\mathbf{y}, \quad (7.6)$$

which conforms with the stress field $\boldsymbol{\sigma}^0(\mathbf{x}) = \mathbf{L}^c\boldsymbol{\varepsilon}^0(\mathbf{x}) + \boldsymbol{\alpha}^c(\mathbf{x})$, while $\underline{\mathbf{t}}_0 = \mathbf{t}_0 - \boldsymbol{\alpha}^c \cdot \mathbf{n}^\Gamma(\mathbf{s})$ is a modified traction on the surface Γ with the outer normal $\mathbf{n}^\Gamma(\mathbf{s})$ ($\mathbf{s} \in \Gamma$) The tensor of the "fundamental traction" $\mathbf{T}$ on Γ_0 (called also Kupradze tensor) associated with the tensor of "fundamental displacement" $\mathbf{G}$ is given by Eq. (3.39_2). The representation (7.6) is valid for both the general cases of the first and second boundary-value problems as well as for the mixed boundary-value problem [60, 156, 585]. In particular, for the condition (7.1_1) with $\Gamma_u = \Gamma$ and $\boldsymbol{\beta}$, $\mathbf{f} \equiv \mathbf{0}$ the right-hand side integral over the external surface (7.6) can be considered as a continuation of $\boldsymbol{\varepsilon}_0 = \boldsymbol{\varepsilon}^\Gamma(\mathbf{x})$, $\mathbf{x} \in \Gamma_u$ into w as the strain field that the boundary condition (7.1_1) would generate in the comparison medium with moduli $\mathbf{L}^c$. It should be mentioned that for the constant gravitation loads $f_i^{\rm g} = \rho^c g_i$ and for a centrifugal load $f_i^c = g_{ij}x_j$, the volume integral in Eq. (7.6) can be transformed into a surface integral [156]. Here we introduced a constant mass density ρ^c, a constant gravitation field g_i, as well as the matrix $g_{ij} = $ const.

The Cauchy data $[\mathbf{u}_0(\mathbf{s}), \boldsymbol{\sigma}_0(\mathbf{s})]$ at the smooth surface $\mathbf{s} \in \Gamma$ can be found, e.g., from the conventional BIE taking the limit $\mathbf{x} \to \Gamma_0$

$$\mathbf{k}(\mathbf{s})\mathbf{u}_0(\mathbf{s}) = \int_\Gamma \Big[\mathbf{G}(\mathbf{s}-\boldsymbol{\zeta})\underline{\mathbf{t}}_0(\boldsymbol{\zeta}) - \mathbf{T}^\top(\mathbf{s},\boldsymbol{\zeta})\mathbf{u}_0(\boldsymbol{\zeta})\Big]d\boldsymbol{\zeta}$$

at the boundary conditions (7.1); here $\mathbf{k}(\mathbf{x})$ is the free term coefficient (see, e.g., [354]). Equation (7.6), which is valid for internal points of the body $\mathbf{x} \in w$, is modified in the last equation for the treatment of strong singularities in the surface integrals as $\mathbf{x}$ tends to the boundary Γ. It requires either the definition of singular integrals in the sense of a Cauchy principle value or a regularization procedure in the spirit of a subtraction technique to eliminate or reduce the order of strong singularities [60, 156, 1410, 1411, 1485]. For simplicity we will consider only internal points $\mathbf{x} \in w$ of the microinhomogeneous macrodomain w at sufficient distance from the boundary

$$a \ll |\mathbf{x} - \mathbf{s}|, \ \forall \mathbf{s} \in \Gamma, \tag{7.7}$$

when the validity of Eq. (7.6) takes place except in some "boundary layer" region close to the surface $\mathbf{s} \in \Gamma$ where some boundary data $[\mathbf{u}_0(\mathbf{s}), \mathbf{t}_0(\mathbf{s})]$ (if they are not prescribed by the boundary conditions) will depend on perturbations introduced by all inhomogeneities.

After integration of Eq. (7.5) by parts, it is found that

$$\begin{aligned}\boldsymbol{\varepsilon}(\mathbf{x}) = \boldsymbol{\varepsilon}^0(\mathbf{x}) &+ \int_w \mathbf{U}(\mathbf{x}-\mathbf{y})\boldsymbol{\tau}(\mathbf{y})\, d\mathbf{y} \\ &+ \int_w \nabla \mathbf{G}(\mathbf{x}-\mathbf{y})\mathbf{f}_1(\mathbf{y})\, d + \int_\Gamma \nabla \mathbf{G}(\mathbf{x}-\mathbf{s})\boldsymbol{\tau}(\mathbf{s})\mathbf{n}(\mathbf{s})\, ds,\end{aligned} \tag{7.8}$$

which used that $\nabla_{\mathbf{y}} = -\nabla_{\mathbf{x}}$, and the surface integration is taken over the external surface Γ of the mesodomain w, containing a statistically large number of realizations of inclusions, and the integral operator kernel $\mathbf{U}$ is an even homogeneous generalized function of degree $-d$ defined by Eq. (3.19). For construction of the regularization of a generalized function of the type of derivatives of homogeneous regular function, we will consider a scheme proposed in [537] according to which the tensor $\mathbf{U}(\mathbf{x})$ is split into the singular and regular parts [see Eq. (3.20)]. Both of these terms depend on the shape of an exclusion region prescribed by Eqs. (3.17), while their sum is defined uniquely.

7.2 Random Structure Composites

7.2.1 *General Integral Equation for Random Structure Composites*

Let a full space R^d with a space dimensionality d ($d = 2$ and $d = 3$ for $2D$ and $3D$ problems, respectively) contain a homogeneous matrix $v^{(0)}$ and, in general, a statistically inhomogeneous set $X = (v_i)$ of heterogeneities v_i with indicator functions V_i and bounded by the closed smooth surfaces $\Gamma_i := \partial v_i$

($i = 1, 2, \ldots$) defined by the relations $\Gamma_i(\mathbf{x}) = 0$ ($\mathbf{x} \in \Gamma_i$), $\Gamma_i(\mathbf{x}) > 0$ ($\mathbf{x} \in v_i$), and $\Gamma_i(\mathbf{x}) < 0$ ($\mathbf{x} \notin v_i$). It is assumed that the heterogeneities can be grouped into components (phases) $v^{(q)}$ ($q = 1, 2, \ldots, N$) with identical mechanical and geometrical properties (such as the shape, size, orientation, and microstructure of heterogeneities).

It is assumed that the representative macrodomain w contains a statistically large number of realizations α (providing validity of the standard probability technique) of heterogeneities $v_i \in v^{(k)}$ of the constituent $v^{(k)}$ ($i = 1, 2, \ldots;\ k = 1, 2, \ldots, N$). A random event α belongs to a sample space $\mathcal{A}$, over which a probability density $p(\mathbf{x}, \alpha)$ is defined (see, e.g., [1625]). For any given α, any random function $\mathbf{g}(\mathbf{x}, \alpha)$ (e.g., $\mathbf{g} = V,\ V^{(k)}$) is defined explicitly as one particular member, with label α, of an ensemble realization. Then, the mean, or ensemble average, is defined by the angle brackets enclosing the quantity $\mathbf{g}$

$$\langle \mathbf{g} \rangle(\mathbf{x}) = \int_{\mathcal{A}} \mathbf{g}(\mathbf{x}, \alpha) p(\mathbf{x}, \alpha) d\alpha. \tag{7.9}$$

No confusion will arise below in notation of the random quantity $\mathbf{g}(\mathbf{x}, \alpha)$ if the label α is removed. One treats two material length scales (see, e.g, [1523]): the macroscopic scale L, characterizing the extent of w, and the microscopic scale a, related with the heterogeneities v_i. Moreover, one supposes that applied field varies on a characteristic length scale Λ. The limit of our interests for both the material scales and field one is

$$L \gg \Lambda \geq a \quad \text{or} \quad L \gg \Lambda \gg a, \tag{7.10}$$

where (7.10_2) is called scale separation hypothesis. All the random quantities under discussion are described by statistically inhomogeneous random fields. For the alternative description of the random structure of a composite material, we will use the statistical description presented in Sect. 5.3.3 in terms of the conditional probability densities $\varphi(v_i, \mathbf{x}_i | v_1, \mathbf{x}_1, \ldots, v_n, \mathbf{x}_n)$. We will consider a general case of statistically inhomogeneous media with the homogeneous matrix (e.g., for so-called FGM), when the conditional probability density is not invariant with respect to translation. The average and conditional average taken for the ensemble of a statistically inhomogeneous field $X = (v_i)$ were introduced in Sect. 5.3.3.

Equation (7.8) is centered, that is, from both sides of Eq. (7.8), their statistical averages are subtracted

$$\begin{aligned} \boldsymbol{\varepsilon}(\mathbf{x}) = \langle \boldsymbol{\varepsilon} \rangle(\mathbf{x}) &+ \int_w \langle\langle \mathbf{U}(\mathbf{x} - \mathbf{y}) \boldsymbol{\tau}(\mathbf{y}) \rangle\rangle \, d\mathbf{y} \\ &+ \int_w \langle\langle \nabla \mathbf{G}(\mathbf{x} - \mathbf{y}) \mathbf{f}_1(\mathbf{y}) \rangle\rangle \, d\mathbf{y} + \mathcal{I}^{\Gamma_\epsilon}, \end{aligned} \tag{7.11}$$

where one introduces a centering operation

$$\langle\langle \mathbf{h}(\mathbf{x},\mathbf{y})\mathbf{g}(\mathbf{y})\rangle\rangle := \mathbf{h}(\mathbf{x},\mathbf{y})\mathbf{g}(\mathbf{y}) - \langle \mathbf{h}(\mathbf{x},\mathbf{y})\mathbf{g}(\mathbf{y})\rangle(\mathbf{y}) \tag{7.12}$$

for example, the deterministic $\mathbf{h}$ (e.g., $\mathbf{h} = \mathbf{U}, \nabla\mathbf{G}$) and random $\mathbf{g}$ (e.g., $\mathbf{g} = \boldsymbol{\tau}, \mathbf{f}_1$) functions. The subtraction technique used for the transformation of Eq. (7.8) into Eq. (7.11), also called a centering method, is a major stage of a justification of the so-called locality principle in micromechanics of composite materials (see Sect. 7.5). In so doing, in the right-hand side of Eq. (7.11), the integral over the external surface Γ

$$\mathcal{I}^{\Gamma\epsilon} \equiv \int_{\Gamma} \langle\langle \nabla\mathbf{G}(\mathbf{x}-\mathbf{s})\boldsymbol{\tau}(\mathbf{s})\rangle\rangle \mathbf{n}(\mathbf{s})\, ds \tag{7.13}$$

can be dropped out, because this integral vanishes at sufficient distance $\mathbf{x}$ from the boundary Γ. This means that if $|\mathbf{x}-\mathbf{s}|$ is large enough

$$|\mathbf{x}-\mathbf{s}| \ll a \quad \forall \mathbf{s} \in \Gamma, \tag{7.14}$$

then at the portion of the smooth surface $ds \approx |\mathbf{x}-\mathbf{s}|^{d-1}\, d\boldsymbol{\omega}^s$ with a small solid angle $d\boldsymbol{\omega}^s$, the tensor $\nabla\mathbf{G}(\mathbf{x}-\mathbf{s})|\mathbf{x}-\mathbf{s}|^{d-1}$ depends only on the solid angle $\boldsymbol{\omega}^s$ variables and slowly varies on the portion of the surface $d\mathbf{s}$; in this sense, the tensor $\nabla\mathbf{G}(\mathbf{x}-\mathbf{s})$ is called a "slow" variable of the solid angle $\boldsymbol{\omega}^s$, while the expression in curly brackets on the right-hand-side integral of Eq. (7.13) is a rapidly oscillating function on $d\mathbf{s}$ and is called a "fast" variable. Therefore, we can use a rigorous theory of "separate" integration of "slow" and "fast" variables, according to which (freely speaking) the operation of surface integration may be regarded as averaging; see [472] and its applications [1378]. In consequence of established separation of "slow" and "fast" variables, no confusion arises in understanding of estimations of averaged items $\langle(.)\rangle(\mathbf{S})$ depending on macro-coordinate $\mathbf{S} \in \Gamma_0$, while the corresponding non-averaged items depend on micro-coordinates $\mathbf{s} \in \Gamma_0$. Indeed, the slow variable $\nabla\mathbf{G}(\mathbf{x}-\mathbf{s})$ can be consistent (after the first separation of the fast and slow variables) with the macro-coordinate $\mathbf{S} \in \Gamma$ as $\nabla\mathbf{G}(\mathbf{x}-\mathbf{S})$, while the fast variable $\boldsymbol{\tau}(\mathbf{s},\alpha)$ is in agreement with the micro-coordinate $\mathbf{s} \in \Gamma$ that makes it possible to present the integrand in the integral (7.13) in the formal form $\nabla\mathbf{G}(\mathbf{x}-\mathbf{S})\Big[\boldsymbol{\tau}(\mathbf{s},\alpha) - \langle\boldsymbol{\tau}\rangle(\mathbf{S})\Big]$ leading (after additional separation of the fast and slow variables) to vanishing of integral from the term in the square brackets for statistically large number of random events α. If (as we assume) there is no *long-range* order and the function $\varphi(v_j,\mathbf{x}_j|;v_i,\mathbf{x}_i) - \varphi(v_j,\mathbf{x}_j)$ decays at infinity sufficiently rapidly, then it leads to a degeneration of the surface integral. So, exponential decreasing of this function was obtained in [1621] for spherical inclusions; a more rapidly decreasing function for circle aligned fibers was proposed in [604, 1525].

To express Eq. (7.11) in terms of stresses, we use the identities

$$\mathbf{L}_1(\boldsymbol{\varepsilon}-\boldsymbol{\beta}) = -\mathbf{L}^c\mathbf{M}_1\boldsymbol{\sigma}, \quad \boldsymbol{\varepsilon} = [\mathbf{M}^c\boldsymbol{\sigma}+\boldsymbol{\beta}^c] + [\mathbf{M}_1\boldsymbol{\sigma}+\boldsymbol{\beta}_1]. \tag{7.15}$$

Substituting (7.15) into the right-hand side and the left-hand side of (7.11), respectively, and contracting with the tensor $\mathbf{L}^c$, gives the general integral equation for stresses

$$\boldsymbol{\sigma}(\mathbf{x}) = \langle\boldsymbol{\sigma}\rangle(\mathbf{x}) + \int_w \langle\!\langle \boldsymbol{\Gamma}(\mathbf{x}-\mathbf{y})\boldsymbol{\eta}(\mathbf{y})\rangle\!\rangle\, d\mathbf{y} + \int_w \mathbf{L}^c\langle\!\langle \nabla\mathbf{G}(\mathbf{x}-\mathbf{y})\mathbf{f}_1(\mathbf{y})\rangle\!\rangle\, d\mathbf{y} + \mathcal{I}^{\Gamma\sigma}, \tag{7.16}$$

where we define $\mathcal{I}^{\Gamma\sigma} = -\int_\Gamma \mathbf{L}^c\langle\!\langle \nabla\mathbf{G}(\mathbf{x}-\mathbf{s})\mathbf{L}^c\boldsymbol{\eta}(\mathbf{s})\mathbf{n}(\mathbf{s})\rangle\!\rangle\, d\mathbf{s}$ and the strain polarization tensor $\boldsymbol{\eta}(\mathbf{y}) = \mathbf{M}_1(\mathbf{y})\boldsymbol{\sigma}(\mathbf{y}) + \boldsymbol{\beta}_1(\mathbf{y})$ (4.8). If we assume no long-range order, then the tensor $\mathcal{I}^{\Gamma\sigma}$ degenerates and can be dropped. In (7.16) $\mathbf{M}_1(\mathbf{y})$ and $\boldsymbol{\beta}_1(\mathbf{y})$ are the jumps of the compliance $\mathbf{M}^{(k)}$ and of the eigenstrain $\boldsymbol{\beta}^{(k)}$ inside the component $v^{(k)}$ $(k = 0, \ldots, N)$ with respect to the constant tensors $\mathbf{M}^c$ and $\boldsymbol{\beta}^c$, respectively. The integral operator kernel, $\boldsymbol{\Gamma}(\mathbf{x}-\mathbf{y})$, called the Green stress tensor [820, 822], is defined by Eq. (3.37).

For convenience in the presentation that follows, we will recast Eq. (7.16) in another form, for which we introduce the operation $\boldsymbol{\lambda}^1(\mathbf{x}) = \boldsymbol{\lambda}(\mathbf{x}) - \langle\boldsymbol{\lambda}\rangle_0(\mathbf{x})$ for the random function $\boldsymbol{\lambda}$ (e.g., $\boldsymbol{\lambda} = \boldsymbol{\sigma},\ \boldsymbol{\varepsilon},\ \boldsymbol{\tau},\ \boldsymbol{\eta},\ \mathbf{f}$) with statistical average in the matrix $\langle\boldsymbol{\lambda}\rangle_0(\mathbf{x})$. Then, Eq. (7.16) can be rewritten in the form

$$\boldsymbol{\sigma}(\mathbf{x}) = \langle\boldsymbol{\sigma}\rangle(\mathbf{x}) + \int_w \langle\!\langle \boldsymbol{\Gamma}(\mathbf{x}-\mathbf{y})\boldsymbol{\eta}^1(\mathbf{y})\rangle\!\rangle\, d\mathbf{y} + \int_w \mathbf{L}^c\langle\!\langle \nabla\mathbf{G}(\mathbf{x}-\mathbf{y})\mathbf{f}_1^1(\mathbf{y})\rangle\!\rangle\, d\mathbf{y}. \tag{7.17}$$

The general integral equations (7.12), (7.16), and (7.17) were proposed in [191] for statistically inhomogeneous media for the general case of inhomogeneity of tensors $\boldsymbol{\beta}^c(\mathbf{x}),\ \boldsymbol{\beta}^{(0)}(\mathbf{x}),\ \mathbf{f}^c(\mathbf{x}),\ \mathbf{f}^{(0)}(\mathbf{x})$ for both the general case of the first and second boundary-value problems and for the mixed boundary-value problem.

It should be mentioned that at the infinity $|\mathbf{x}-\mathbf{y}| \to \infty$, the centering operation for $\mathbf{h} = \mathbf{U}, \boldsymbol{\Gamma}$ in Eqs. (7.12) and (7.16) has the following asymptotic representation:

$$\langle\!\langle \mathbf{h}(\mathbf{x},\mathbf{y})\mathbf{g}(\mathbf{y})\rangle\!\rangle := \mathbf{h}(\mathbf{x},\mathbf{y})\mathbf{g}(\mathbf{y}) - \langle\mathbf{h}(\mathbf{x},\mathbf{y})\mathbf{g}(\mathbf{y})\rangle(\mathbf{y}) \to \mathbf{h}(\mathbf{x},\mathbf{y})[\mathbf{g}(\mathbf{y}) - \langle\mathbf{g}(\mathbf{y})\rangle(\mathbf{y})] \tag{7.18}$$

where the expression in square brackets with $\mathbf{g} = \boldsymbol{\tau}, \boldsymbol{\eta}$, respectively, is of the order $O(|\mathbf{x}-\mathbf{y}|^{-d})$ (see Sect. 3.2) as $|\mathbf{x}-\mathbf{y}| \to \infty$, and the integrals in Eqs. (7.12) and (7.16) with the kernels $\mathbf{U}$ and $\boldsymbol{\Gamma}$, respectively, converge absolutely. In a similar manner, the integrals with the body force density converge absolutely. In fact, the kernel $\nabla\mathbf{G}(\mathbf{x}-\mathbf{y})$ is of the order $O(|\mathbf{x}-\mathbf{y}|^{-d+1})$ as $|\mathbf{x}-\mathbf{y}| \to \infty$. In so doing, the term $\mathbf{f}_1^1(\mathbf{y}) - \langle\mathbf{f}_1^1\rangle$, differing from zero only inside inclusions v_j, $(j = 1, 2, \ldots)$, tends to zero with $|\mathbf{x}-\mathbf{y}| \to \infty$ ($\mathbf{x} \in v_i$, $\mathbf{y} \in v_j$) as $\varphi(v_j, \mathbf{x}_j|; v_i, \mathbf{x}_i) - \varphi(v_j, \mathbf{x}_j)$. For no *long-range* order, an absolute convergence of the integrals involved is assured if the function $\varphi(v_j, \mathbf{x}_j|; v_i, \mathbf{x}_i) - \varphi(v_j, \mathbf{x}_j)$ decays at infinity sufficiently rapidly.

Therefore, for $\mathbf{x} \in w$ considered in Eqs. (7.12) and (7.16) and removed far enough from the boundary Γ (7.14), the right-hand-side integrals in (7.12) and (7.16) do not depend on the shape and size of the domain w, and they can be replaced by the integrals over the whole space R^d. With this assumption, we hereafter omit explicitly denoting R^d as the integration domain in the following *exact General Integral Equations* (GIE)

$$\boldsymbol{\varepsilon}(\mathbf{x}) = \langle\boldsymbol{\varepsilon}\rangle(\mathbf{x}) + \int \langle\langle \mathbf{U}(\mathbf{x}-\mathbf{y})\boldsymbol{\tau}^1 \rangle\rangle \, d\mathbf{y} + \int \langle\langle \nabla\mathbf{G}(\mathbf{x}-\mathbf{y})\mathbf{f}_1^1(\mathbf{y}) \rangle\rangle \, d\mathbf{y}, \tag{7.19}$$

$$\boldsymbol{\sigma}(\mathbf{x}) = \langle\boldsymbol{\sigma}\rangle(\mathbf{x}) + \int \langle\langle \boldsymbol{\Gamma}(\mathbf{x}-\mathbf{y})\boldsymbol{\eta}^1(\mathbf{y}) \rangle\rangle \, d\mathbf{y} + \int \mathbf{L}^c \langle\langle \nabla\mathbf{G}(\mathbf{x}-\mathbf{y})\mathbf{f}_1^1(\mathbf{y}) \rangle\rangle \, d\mathbf{y}. \tag{7.20}$$

Thus, there are no difficulties connected with the asymptotic behavior of the generalized functions $\nabla\nabla\mathbf{G}$ and $\boldsymbol{\Gamma}$ at infinity (as $|\mathbf{x}-\mathbf{y}|^{-d}$), and there is no need to postulate either the shape or the size of the integration domain w [943] or to resort to either regularization [822, 836] or renormalization [305, 1040] of integrals that are divergent at infinity [1625], or to consider an auxiliary problem with mixed boundary conditions [158, 159]. The rigorous mathematical analysis of the above-mentioned methods is beyond the purpose of the current presentation. Nevertheless, it should be noted that the disruption of statistical homogeneity of media often leads to additional difficulties, whose resolution by these known mentioned approaches appears to be questionable.

Let the inclusions $v_1, \ldots, v_n$ be fixed, and we define two sorts of effective fields for the strains $\overline{\boldsymbol{\varepsilon}}_i(\mathbf{x})$ and $\widetilde{\boldsymbol{\varepsilon}}_{1,\ldots,n}(\mathbf{x})$ ($i = 1, \ldots, n$; $\mathbf{x} \in v_1, \ldots, v_n$) and the stresses $\overline{\boldsymbol{\sigma}}_i(\mathbf{x})$ and $\widetilde{\boldsymbol{\sigma}}_{1,\ldots,n}(\mathbf{x})$ ($i = 1, \ldots, n$; $\mathbf{x} \in v_1, \ldots, v_n$) by the use of the rearrangement of Eqs. (7.19) and (7.20) in the following form (see the earliest references on related manipulations [238, 268, 270, 746]):

$$\boldsymbol{\vartheta}(\mathbf{x}) = \overline{\boldsymbol{\vartheta}}_i(\mathbf{x}) + \int \mathbf{U}^\theta(\mathbf{x}-\mathbf{y})V_i(\mathbf{y})\boldsymbol{\zeta}^1(\mathbf{y}) \, d\mathbf{y}, \tag{7.21}$$

$$\overline{\boldsymbol{\vartheta}}_i(\mathbf{x}) = \widetilde{\boldsymbol{\varepsilon}}_{1,\ldots,n}(\mathbf{x}) + \sum_{j\neq i} \int \mathbf{U}^\theta(\mathbf{x}-\mathbf{y})V_j(\mathbf{y})\boldsymbol{\zeta}^1(\mathbf{y}) \, d\mathbf{y}, \tag{7.22}$$

$$\begin{aligned}\widetilde{\boldsymbol{\vartheta}}_{1,\ldots,n}(\mathbf{x}) = \langle\boldsymbol{\varepsilon}\rangle(\mathbf{x}) + \int \mathbf{U}^\theta(\mathbf{x}-\mathbf{y})\boldsymbol{\zeta}^1(\mathbf{y})V(\mathbf{y}|; v_1, \mathbf{x}_1; \ldots; v_n, \mathbf{x}_n)\\ -\langle \mathbf{U}^\theta(\mathbf{x}-\mathbf{y})\boldsymbol{\zeta}^1\rangle(\mathbf{y})d\mathbf{y}\end{aligned} \tag{7.23}$$

for $\mathbf{x} \in v_i$, $i = 1, 2, \ldots, n$ and a particular case $\mathbf{f}_1 \equiv \mathbf{0}$. Hereafter for contraction of calculations, we introduce the substitutions

$$(\boldsymbol{\varepsilon}, \boldsymbol{\sigma}) \leftrightarrow \boldsymbol{\vartheta}, \quad (\boldsymbol{\tau}, \boldsymbol{\eta}) \leftrightarrow \boldsymbol{\zeta}, \quad (\boldsymbol{\alpha}, \boldsymbol{\beta}) \leftrightarrow \boldsymbol{\gamma} \tag{7.24}$$

indicating that for each triplet $\boldsymbol{\vartheta} = \boldsymbol{\varepsilon}$, $\boldsymbol{\zeta} = \boldsymbol{\tau}$, $\boldsymbol{\gamma} = \boldsymbol{\alpha}$ ($\mathbf{U}^{\epsilon} = \mathbf{U}$), and $\boldsymbol{\vartheta} = \boldsymbol{\sigma}$, $\boldsymbol{\zeta} = \boldsymbol{\eta}$, $\boldsymbol{\gamma} = \boldsymbol{\beta}$ ($\mathbf{U}^{\sigma} = \boldsymbol{\Gamma}$), Eqs. (7.21)–(7.23) and the subsequent equations with the variables $\boldsymbol{\vartheta}$, $\boldsymbol{\zeta}$ and $\boldsymbol{\gamma}$ are reduced to the corresponding equations for the strain $\boldsymbol{\varepsilon}$ and $\boldsymbol{\sigma}$. Then, considering some conditional statistical averages of the general integral equations (7.19) and (7.20) leads to an infinite system of integral equations that can also be represented in terms of the strain polarization tensor ($n = 1, 2, \ldots$)

$$\langle \boldsymbol{\vartheta} | v_1, \mathbf{x}_1; \ldots; v_n, \mathbf{x}_n \rangle(\mathbf{x}) - \sum_{i=1}^{n} \int \mathbf{U}^{\theta}(\mathbf{x}-\mathbf{y}) \langle V_i(\mathbf{y}) \boldsymbol{\zeta}^1 | v_1, \mathbf{x}_1; \ldots; v_n, \mathbf{x}_n \rangle(\mathbf{y})\, d\mathbf{y}$$
$$= \langle \boldsymbol{\vartheta} \rangle(\mathbf{x}) + \int \mathbf{U}^{\theta}(\mathbf{x}-\mathbf{y}) \langle \boldsymbol{\zeta}^1 |; v_1, \mathbf{x}_1; \ldots; v_n, \mathbf{x}_n \rangle(\mathbf{y}) - \langle \mathbf{U}^{\theta}(\mathbf{x}-\mathbf{y}) \boldsymbol{\zeta}^1 \rangle(\mathbf{y})\, d\mathbf{y}, \tag{7.25}$$

Since $\mathbf{x} \in v_1, \ldots, v_n$ in the nth line of the system can take the values in the inclusions $v_1, \ldots, v_n$, the nth line actually contains n equations.

The definitions of the effective strain (and stress) fields $\overline{\boldsymbol{\varepsilon}}_i(\mathbf{x})$, $\widetilde{\boldsymbol{\varepsilon}}_{1,2,\ldots,n}(\mathbf{x})$ as well as their statistical averages $\langle \overline{\boldsymbol{\varepsilon}}_i \rangle(\mathbf{x})$, $\langle \widetilde{\boldsymbol{\varepsilon}}_{1,2,\ldots,n} \rangle(\mathbf{x})$ are nothing more than notational convenience for different terms of the infinite systems (7.21)–(7.23) and (7.25), respectively. The physical meaning of these fields is the following (graphic illustrations are presented in [268]; see also Fig. 7.1). $\widetilde{\boldsymbol{\varepsilon}}_{1,2,\ldots,n}(\mathbf{x})$ is a strain field in which the chosen fixed inclusions $v_1, \ldots, v_n$ are embedded. This effective field is a random function of all the other positions of the surrounding inhomogeneities, and the average $\langle \widetilde{\boldsymbol{\varepsilon}}_{1,\ldots,n} \rangle(\mathbf{x})$ of $\widetilde{\boldsymbol{\varepsilon}}_{1,\ldots,n}(\mathbf{x})$ over a random realization of these inclusions is equal to the right-hand side of the nth line of the system (7.25). Consequently, each inclusion v_i ($i = 1, \ldots, n$) of the chosen fixed set is in a random (generally speaking, nonhomogeneous) field $\overline{\boldsymbol{\varepsilon}}_i(\mathbf{x})$, ($\mathbf{x} \in v_i$, $i \neq j$, $i, j = 1, 2, \ldots, n$) (7.22), which is the superposition of the effective field $\widetilde{\boldsymbol{\varepsilon}}_{1,\ldots,n}(\mathbf{x})$ and the distribution caused by the other inclusions v_j, ($j \neq i$, $j = 1, \ldots, n$) of the considered set.

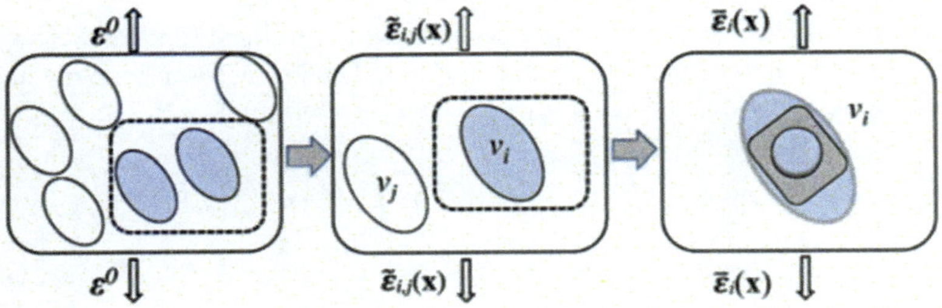

Fig. 7.1 Schematic representation of the effective fields

7.2.2 Some Particular Cases

Thus, the general integral Eqs. (7.12) and (7.16) over the whole space obtained above are exact and do not depend on particular boundary conditions. The subsequent analysis can be done for the comparison medium with any elastic modulus $\mathbf{L}^c$, which necessarily leads to some additional assumptions for the structure of the strain fields in the matrix. Equations (7.12) and (7.16) are much easier to solve when the stress–strain fields are studied only inside the inclusions. Following [257], we will consider two fundamentally different approaches to ensure that the integration on the right-hand sides of Eqs. (7.12), (7.16), and (7.18) extends only over the volume of the inclusion.

In the first one, we postulate

$$\mathbf{L}^c \equiv \mathbf{L}^{(0)}, \quad \boldsymbol{\beta}^c \equiv \boldsymbol{\beta}^{(0)}. \tag{7.26}$$

Then, $\mathbf{L}_1(\mathbf{x}) \equiv \mathbf{0}$ and $\boldsymbol{\beta}_1(\mathbf{x}) \equiv \mathbf{0}$ at $\mathbf{x} \in v^{(0)}$, and the integrands with the arguments $\mathbf{y}$ in both Eqs. (7.18) vanish at $\mathbf{y} \in v^{(0)}$. In the second one, we choose $\mathbf{L}^c$ quite arbitrarily and analyze, e.g., Eq. (7.20). Equation (7.20) being exact for any $\langle \boldsymbol{\eta} \rangle_0(\mathbf{x})$ can be simplified with the additional assumption that the strain polarization tensor in the matrix $\boldsymbol{\eta}(\mathbf{x})$, $(\mathbf{x} \in v^{(0)})$ coincides with its statistical average in the matrix

$$\boldsymbol{\eta}(\mathbf{x}) \equiv \langle \boldsymbol{\eta} \rangle_0(\mathbf{x}), \quad \mathbf{x} \in v^{(0)}. \tag{7.27}$$

In so doing, the assumption (7.26) is more restricted in the sense that the assumption (7.26) yields the assumption (7.27) (the opposite is not true) and, moreover, in such a case the exact equality $\boldsymbol{\eta}(\mathbf{x}) \equiv \langle \boldsymbol{\eta} \rangle_0(\mathbf{x}) \equiv \mathbf{0}$, $\mathbf{x} \in v^{(0)}$ holds. Because of this we will consider only that assumption (7.27) ensured the integration of the integrand in the right-hand side of Eqs. (7.20) only by the volume of inclusions.

We will consider the inclusion field X bounded in one direction such as a laminated structure of some real FGM [1218, 1219]. Then the surface integral (7.8) over a "cylindrical" surface (with the surface area of order $O(\rho^{d-2})$, $\rho = |\mathbf{x} - \mathbf{s}|$) tends to zero with $|\mathbf{x} - \mathbf{s}| \to \infty$ simply because the generalized function $\nabla \mathbf{G}(\mathbf{x} - \mathbf{s})$ is an even homogeneous function of order $-d + 1$. Therefore, for infinite media, the surface integral (7.8) vanishes, and Eq. (7.8) can be rewritten as

$$\boldsymbol{\varepsilon}(\mathbf{x}) = \boldsymbol{\varepsilon}^0(\mathbf{x}) + \int \mathbf{U}(\mathbf{x} - \mathbf{y}) \boldsymbol{\tau}(\mathbf{y}) + \int \nabla \mathbf{G}(\mathbf{x} - \mathbf{y}) \mathbf{f}_1(\mathbf{y})\, d\mathbf{y} \tag{7.28}$$

or, alternatively, in terms of stresses

$$\boldsymbol{\sigma}(\mathbf{x}) = \boldsymbol{\sigma}^0(\mathbf{x}) + \int \boldsymbol{\Gamma}(\mathbf{x} - \mathbf{y}) \boldsymbol{\eta}(\mathbf{y})\, d\mathbf{y} + \int \mathbf{L}^c \nabla \mathbf{G}(\mathbf{x} - \mathbf{y}) \mathbf{f}_1(\mathbf{y})\, d\mathbf{y} \tag{7.29}$$

In so doing the integrals from body forces in Eqs. (7.28) and (7.29) only conditionally converge for a general case of a bounded function $\mathbf{f}_1(\mathbf{y})$; because of this, for the function $\mathbf{f}(\mathbf{y})$, we will assume decay at infinity $|\mathbf{x}-\mathbf{y}| \to \infty$ no less than $O(|\mathbf{x}-\mathbf{y}|^{-\alpha})$, $(\alpha > 0)$ guaranteeing the absolute convergence of body force integrals in Eqs. (7.28) and (7.29). Clearly, in the considered case of X bounded in one direction, Eqs. (7.28) and (7.29) are exact, and the right-hand-side integrals in (7.28) and (7.29) converge absolutely. Equation (7.28) was used in [773, 1521, 1522] for the particular case (7.26) with homogeneous boundary conditions (7.2) and for the inclusion field X with a constant concentration of inclusions within an ellipsoidal domain included in the infinite matrix. Equations (7.12) and (7.16) are more complicated than Eqs. (7.28) and (7.29), nevertheless provide practical advantages because their integrands decay at infinity faster than the integrands involved in Eqs. (7.28) and (7.29).

It should be noted that Eqs. (7.19) and (7.20) exploiting the infinite-homogeneous-body Green functions were obtained from Eqs. (7.11) and (7.16), respectively, at sufficient distance $\mathbf{x}$ from the boundary (7.14) and, therefore, they cannot be used for analysis of boundary layer effects (e.g., free edge effect). For this class of problems, the Green functions for finite domains seem more prospective (see, e.g., [983, 1649] and Sect. 13.5). In such a case, the appropriate general integral equations generalizing Eqs. (7.19) and (7.20) to the finite domains can be obtained in a straightforward manner (see for details Sect. 13.5).

7.2.3 Comparison with Related Equations

We now turn our attention to the discussion of related integral equations and consider the case of a statistically homogeneous medium (with $\mathbf{M}^c = \mathbf{M}^{(0)}$ and $\boldsymbol{\beta} \equiv \mathbf{0}$) and homogeneous boundary conditions either (7.2) or (7.3). Kunin [836] uses the equation over the whole space

$$\boldsymbol{\varepsilon}(\mathbf{x}) = \boldsymbol{\varepsilon}^0 + \int \mathbf{U}(\mathbf{x}-\mathbf{y})\boldsymbol{\tau}(\mathbf{y})\,d\mathbf{y}, \quad \boldsymbol{\sigma}(\mathbf{x}) = \boldsymbol{\sigma}^0 + \int \boldsymbol{\Gamma}(\mathbf{x}-\mathbf{y})\boldsymbol{\eta}(\mathbf{y})\,d\mathbf{y} \tag{7.30}$$

at $\boldsymbol{\varepsilon}^0 \equiv$ const. and $\boldsymbol{\sigma}^0 \equiv$ const., respectively; here $\boldsymbol{\tau} = \mathbf{L}_1\boldsymbol{\varepsilon}$. Equations (7.30_1) and (7.30_2) can be obtained as a straightforward generalization of Eqs. (4.4) and (4.7) (obtained for a finite number of inclusions) to the case of infinite number of inclusions.

Equation (7.30) is correct for a finite array of inclusions, while their use is not justified for the limiting case of a statistically homogeneous field of an infinite number of inhomogeneous in the whole space $w = R^d$. This unjustified generalization leads to well-known convergence difficulties because $\mathbf{U}(\mathbf{x}-\mathbf{y})$, $\boldsymbol{\Gamma}(\mathbf{x}-\mathbf{y})$ are homogeneous generalized functions of degree $-d$ and the integrals in (7.30) are only conditionally convergent, i.e., these integrals depend on the shape of the integration domains. Buryachenko [200–202] analyzed a few known ways how to avoid the difficulties mentioned above and any of them can not be recognized as a mathematically rigorous method.

For statistically homogeneous composites and homogeneous boundary conditions, Lipinski et al. [944] reduced the system (7.6) and (7.8) (at $\mathbf{f},\ \boldsymbol{\beta} \equiv \mathbf{0}$) to an equation analogous to (7.30_1); the subsequent convergence difficulty was overcome by the use of a self-consistent approach which is equivalent in fact to the termination of the constituent $\widetilde{\mathbf{U}}^f(\mathbf{x})$ (3.20) leading to the known convergence problems. Hori and Kubo [652], Ju and Chen [711], and Ju and Tseng [718, 720] (see also [1423]) also used Eq. (7.30_1) and eliminated the difficulties induced by the dependence of a conditionally convergent integral on the shape of integration domain w by the use of an assumption of this shape. Fassi Fehri et al. [462] postulated the size of integration domain in Eq. (7.30_1). Equation (7.30_1) was also used in [713, 722] with the consequent transformation of the known renormalization scheme that is not fully clear to the present author (see also [1037]). The difficulties mentioned above and some others can be avoided easily by using Eqs. (7.12) and (7.16) instead of Eqs. (7.30_1) and (7.30_2), respectively.

Thus, realization of a direct superposition scheme of summations of the perturbations produced by surrounding inclusions leads to the integrals which are not absolutely convergent at infinity. There is a very long and dramatic history (which looks as a scientific thriller) of attempts to modify these conditionally convergent integrals resulted by the long-range interactions. In particular, an idea of so-called method of normalization (or renormalization in analogy to its use in quantum field theory) goes back to Lord Rayleigh [1284] who apparently was the first to encounter the mentioned difficulty in consideration of effective conductivity of periodic system containing a square arrays of cylinders or cubic lattice of spheres. The renormalization procedure consists in subtraction out of the conditionally convergent term by making use of an expression that has the identical convergence properties as the term presenting the difficulty, but which also has a limiting value that is known. The renormalization method was systematically developed for rheological problems [73], for conductivity of random suspensions [696], and for elasticity of composites with spherical particles [305] in terms of perturbations introduced by the dipole strengths of inhomogeneities. However, the correct choice of a renormalizing quantity is not always straightforward that initiated Willis and Acton [1632] (see also [1672]) to propose an alternative method for obtaining a convergent integral expressed through the Green function.

The method is achieved by subtracting from (7.30) the conditionally convergent behavior which is asymptotically closed to $\mathbf{U}(\mathbf{x}-\mathbf{y})\boldsymbol{\tau}(\mathbf{y})$ at $|\mathbf{x}-\mathbf{y}| \to \infty$. For example, the renormalized items

$$[\mathbf{I}+\mathbf{g}(\mathbf{x}-\mathbf{y})]\mathbf{U}(\mathbf{x}-\mathbf{y})\langle\boldsymbol{\tau}\rangle,\quad [\mathbf{I}+\mathbf{g}(\mathbf{x}-\mathbf{y})]\boldsymbol{\Gamma}(\mathbf{x}-\mathbf{y})\langle\boldsymbol{\eta}\rangle \tag{7.31}$$

provide absolute convergence of integrals (7.30_1) and (7.30_2), respectively, for any $\mathbf{g}(\mathbf{x}-\mathbf{y})$ [rather than only for $\mathbf{g}(\mathbf{x}-\mathbf{y}) \equiv \mathbf{0}$] such as $\mathbf{g}(|\mathbf{x}-\mathbf{y}|) \to \mathbf{0}$ at $|\mathbf{x}-\mathbf{y}| \to \infty$

that can lead to the different values of integrals estimated. In particular, comparing non-renormalized (7.30) and renormalized

$$\boldsymbol{\varepsilon}(\mathbf{x}) = \langle\boldsymbol{\varepsilon}\rangle + \int \mathbf{U}(\mathbf{x}-\mathbf{y})[\boldsymbol{\tau}(\mathbf{y}) - \langle\boldsymbol{\tau}\rangle]\, d\mathbf{y}, \tag{7.32}$$

$$\boldsymbol{\sigma}(\mathbf{x}) = \langle\boldsymbol{\sigma}\rangle + \int \boldsymbol{\Gamma}(\mathbf{x}-\mathbf{y})[\boldsymbol{\eta}(\mathbf{y}) - \langle\boldsymbol{\eta}\rangle]\, d\mathbf{y}, \tag{7.33}$$

equations, McCoy [1040, 1041] suggested that one can formally remove the conditionally convergent term appearing in the former by simply setting them equal to zero. There are well-known "noncanonical regularizations" proposed by Kröner [819] (see also [821, 822]) and independently reproposed by Kanaun [736] (see also [747])

$$\int \mathbf{U}(\mathbf{x}-\mathbf{y})\mathbf{h}\, d\mathbf{y} = \mathbf{0}, \quad \text{or} \quad \int \boldsymbol{\Gamma}(\mathbf{x}-\mathbf{y})\mathbf{h}\, d\mathbf{y} = \mathbf{L}^c\mathbf{h}, \tag{7.34}$$

and

$$\int \mathbf{U}(\mathbf{x}-\mathbf{y})\mathbf{h}\, d\mathbf{y} = \mathbf{M}^c\mathbf{h}, \quad \text{or} \quad \int \boldsymbol{\Gamma}(\mathbf{x}-\mathbf{y})\mathbf{h}\, d\mathbf{y} = \mathbf{0}, \tag{7.35}$$

for the first and the second boundary-value problem, respectively; $\mathbf{h}$ is an arbitrary constant symmetric second-order tensor, while, according to (3.37), each of the relations is a consequence of the other from the same pair, either (7.34) or (7.35). In fact, the relations (7.34) and (7.35) introduce an operation of generalized functions $\mathbf{U}$ and $\boldsymbol{\Gamma}$ on a constant symmetric tensor $\mathbf{h} \equiv$ const. (see for comparison [537]). Generalization of the regularization (7.34$_1$) (that is not required in Eq. (7.32)) for any divergence-free second-rank tensor field $\mathbf{h}$ proposed in [393] is also worthy of notice.

For completeness of presentation, we will reproduce the verification of Eqs. (7.34) and (7.35) proposed in Kanaun [736, 737] and reproduced in [836]. Let us assume that $\mathbf{L}_1(\mathbf{x}),\ \mathbf{M}_1(\mathbf{x}) \equiv$ const. throughout the whole space. Analyze first the second boundary-value problem $\boldsymbol{\sigma}^\Gamma \equiv$ const. (7.3) leading to the homogeneous stress state inside w: $\boldsymbol{\sigma}(\mathbf{x}) \equiv \boldsymbol{\sigma}^\Gamma,\ \boldsymbol{\varepsilon}(\mathbf{x}) \equiv \mathbf{M}\boldsymbol{\sigma}^\Gamma$. It is natural to require that Eqs. (7.30$_1$) and (7.30$_2$) give the same result. This is possible only if $\mathbf{U}$ and $\boldsymbol{\Gamma}$ act on constants as in Eqs. (7.35). Analogously, for the first boundary-value problem $\boldsymbol{\varepsilon}^\Gamma \equiv$ const. (7.2), the assumptions (7.34) ensure the homogeneous solution $\boldsymbol{\varepsilon}(\mathbf{x}) \equiv \boldsymbol{\varepsilon}^\Gamma,\ \boldsymbol{\sigma}(\mathbf{x}) = \mathbf{L}\boldsymbol{\varepsilon}^\Gamma$ $(\mathbf{x} \in w)$ by Eq. (7.30$_1$). Thus, the logical scheme of the proof mentioned is as follows: if Eqs. (7.30$_1$) and (7.30$_2$) are valid for a particular case of both the boundary conditions and inhomogeneity of the elastic mismatch ($\mathbf{M}_1(\mathbf{x}) \equiv$ const., $\mathbf{x} \in w$), then either equality (7.34) or (7.35) is valid. The extension of this proof for more practically interesting cases is troublesome because the correctness of Eqs. (7.30$_1$) and (7.30$_2$) is questionable (compare with Eqs. (7.32) and (7.33)).

In the considered conditions, the "noncanonical regularizations" (7.34) and (7.35) provide an insignificant formal simplification of Eqs. (7.11) and (7.15); however, they offer no promise for analysis of statistically inhomogeneous media, where it becomes necessary in general to determine the convolution of $\boldsymbol{\Gamma}$ with an arbitrary piecewise constant function $\langle\boldsymbol{\eta}\rangle(\mathbf{y})$. In so doing, at the obtaining of the new GIE (7.19) and (7.20), no any kind of regularization was required.

At last almost in 90 years after Lord Rayleigh [1284], several authors Shermergor [1378], Khoroshun [762], and O'Brian [1143] independently and rigorously proved correctness of so-called general integral equation (GIE) previously proposed at the intuitive level. Namely, rigorous justification of the renormalization items in Eq. (7.32) and (7.33) $\mathbf{U}(\mathbf{x}-\mathbf{y})\langle\boldsymbol{\tau}\rangle$ [and $\boldsymbol{\Gamma}(\mathbf{x}-\mathbf{y})\langle\boldsymbol{\eta}\rangle$] in the O'Brian [1143] (see also Khoroshun [762]) method was obtained by applying the divergence theorem to the boundary integral in the equation analogous to Eq. (7.5) that leads to Eq. (7.32) with $\langle\boldsymbol{\varepsilon}\rangle \equiv \boldsymbol{\varepsilon}^0$ where the operation of "separate" integration of slow $\nabla\mathbf{G}(\mathbf{x}-\mathbf{s})$ and fast $\boldsymbol{\tau}(\mathbf{s})$ ($\mathbf{s} \in \Gamma$) variables was performed at the intuitive level of justification at the assumption $\boldsymbol{\tau}(\mathbf{s}) \equiv \langle\boldsymbol{\tau}\rangle$. It was explicitly used the macroscopic boundary term at the homogeneous boundary conditions (7.2) that simultaneously defines both the advantage with respect to the classical renormalization method (see e.g., [1632]) and the fundamental limitation of possible generalizations of the proposed approach to both the functional graded materials and inhomogeneous external loading [because $\langle\mathbf{L}_1\boldsymbol{\varepsilon}\rangle$ ≡const. in (7.32)]. An attractive visual interpretation of the obtaining of a version of Eq. (7.32) was proposed in [1300] by the use of consideration of both the direct and reflected interactions between inclusions.

The mentioned deficiency of Eq. (7.32) could be avoided by a centering method systematically developed by Shermergor [1378] also for statistically homogeneous media. This centering method was generalized to the case of statistically inhomogeneous media (i.e., functional graded materials, FGM) by Buryachenko [191] proposed without justification in [258]

$$\boldsymbol{\varepsilon}(\mathbf{x}) = \langle\boldsymbol{\varepsilon}\rangle + \int \mathbf{U}(\mathbf{x}-\mathbf{y})[\boldsymbol{\tau}(\mathbf{y}) - \langle\boldsymbol{\tau}\rangle(\mathbf{y})]\, d\mathbf{y}, \tag{7.36}$$

$$\boldsymbol{\sigma}(\mathbf{x}) = \langle\boldsymbol{\sigma}\rangle + \int \boldsymbol{\Gamma}(\mathbf{x}-\mathbf{y})[\boldsymbol{\eta}(\mathbf{y}) - \langle\boldsymbol{\eta}\rangle(\mathbf{y})]\, d\mathbf{y}. \tag{7.37}$$

It should be mentioned that the particular cases of Eq. (7.36) were also widely used (either explicitly or implicitly) by other authors (see, e.g., [390–392, 983, 984, 1371]). For a purely mechanical loading ($\mathbf{f}$, $\boldsymbol{\beta} \equiv \mathbf{0}$) Eq. (7.8) formally coincides with the analogous relationships obtained in [944] by the use of Green's functions for a bounded domain w when the surface integral in the right-hand side (7.8) vanishes. Its implementation is not trivial because the finite-body Green's function $\mathbf{G}^w$ is generally not known, and replacing $\mathbf{G}^w$ by $\mathbf{G}$ (7.6) if w is large enough leads to well-known convergence difficulties as discussed in detail by Willis [1625]. For homogeneous boundary conditions (7.2) and (7.3) and for $\mathbf{f}$, $\boldsymbol{\beta} \equiv \mathbf{0}$ Eqs. (7.12) and (7.16) and subsequent conversions were considered in [1620, 1625] by a more specific method using essentially the uniformity of the boundary conditions [1143]. It was noted, however, "that the argument given is plausible rather than rigorous and would not apply at all for the boundary conditions that would produce stresses and strains that were not uniform 'on average'" (quoted from [1625]). The method proposed above, (7.5) through (7.16), is devoid of these limitations.

Another widely used simple technique (see [914, 915, 1365, 1628] and the reference in [747]) is worthy of notice. This method is based on a statistical averaging of Eq. (7.30) with subsequent excluding the fields $\boldsymbol{\varepsilon}^0$ and $\boldsymbol{\sigma}^0$ from both the obtained averaged equations and the initial Eq. (7.30). Of course, this technique is exactly the centering method (the initial version of this method was proposed by Shermergor [1378]) exploited for transforming Eq. (7.30) into correct approximate Eqs. (7.32) and (7.33) (compare with the exact Eqs. (7.19) and (7.20), rather than for excluding the deterministic constant known functions $\boldsymbol{\varepsilon}^0$ and $\boldsymbol{\sigma}^0$, respectively. Indeed, denoting Eq. (7.30_1) as $\mathbf{g}(\mathbf{x}) = \mathbf{0}$ leads the mentioned technique to the correct approximate Eq. (7.38_2) (or Eq. (7.32))

$$\mathbf{g}(\mathbf{x}) = \mathbf{0} \;\Rightarrow\; \mathbf{g}(\mathbf{x}) - \langle \mathbf{g} \rangle = \mathbf{0}, \tag{7.38}$$

while the initial Eq. (7.38_1) (or Eq. (7.30)) is questionable. This apparent simplicity of the mentioned technique is explained by the absence of necessity to prove (as in the correct Eqs. (7.11) and (7.38)) vanishing of the surface integrals (7.13) and (7.16) which are lost in Eqs. (7.30). It is interesting that a pioneer in systematic using of the correct centering method for treatment of GIEs mentioned above was apparently Shermergor [1378] who proposed a particular case of these equations for elasticity problem of composites. Presently, this centering method looks very natural and popular (see, e.g., [981, 1625, 1628]).

The related integral Eq. (7.11), averaged over the components and the effective constitutive equations involving inhomogeneous $\langle \boldsymbol{\sigma} \rangle(\mathbf{x})$ and $\langle \boldsymbol{\varepsilon} \rangle(\mathbf{x})$ for statistically homogeneous media, was obtained in [392] by considering an infinite body with only some fictitious distribution of deterministic body-force $\mathbf{f}(\mathbf{x})$ applied that decays sufficiently rapidly at infinity ($|\mathbf{x}| \to \infty$). Then, slightly modifying the approach in [392] (proposed for $\boldsymbol{\beta} \equiv \mathbf{0}$) with the transformation field $\boldsymbol{\beta}_1(\mathbf{x})$ decaying sufficiently rapidly at infinity leads to

$$\nabla \big\{ \mathbf{L}(\mathbf{x})[\boldsymbol{\varepsilon}(\mathbf{x}) - \boldsymbol{\beta}(\mathbf{x})] \big\} + \mathbf{f}(\mathbf{x}) = 0, \tag{7.39}$$

leading to the integral representation over the whole space

$$\boldsymbol{\varepsilon}(\mathbf{x}) = \boldsymbol{\varepsilon}^{0f}(\mathbf{x}) + \int \mathbf{U}(\mathbf{x}-\mathbf{y}) \Big\{ \mathbf{L}_1(\mathbf{y})[\boldsymbol{\varepsilon}(\mathbf{y}) - \boldsymbol{\beta}(\mathbf{y})] - \mathbf{L}^c \boldsymbol{\beta}_1(\mathbf{y}) \Big\} \, d\mathbf{y}, \tag{7.40}$$

rather than (7.11), where

$$\boldsymbol{\varepsilon}^{0f}(\mathbf{x}) = \int \nabla \mathbf{G}(\mathbf{x}-\mathbf{y}) \mathbf{f}(\mathbf{y}) \, d\mathbf{y} \tag{7.41}$$

is a solution for the infinite comparison body with moduli $\mathbf{L}^c$ subjected to the same body-force $\mathbf{f}(\mathbf{x})$. Ensemble averaging of Eq. (7.40) and subtracting the result from the initial Eq. (7.40) leads to Eq. (7.11), which yields the approximate Eq. (19) in [392] for $\boldsymbol{\beta}(\mathbf{x}) \equiv \mathbf{0}$. Kunin [837] noted that the surface traction can be included in the body force as a surface δ-function term, and appropriate integration can be extended to the whole space. Although it has no effect on the correctness of Eqs. (7.15$_1$) and (7.16) obtained by other methods, it should be mentioned that to the author's knowledge, no justification of the introduction of the fictitious body force $\mathbf{f}(\mathbf{x})$ into the present scheme (7.33) through (7.35) has been published before.

Based on the new GIE (7.19) and (7.20), one defined the infinite chains of the effective fields and conditional statistical averages of the strain and stress fields (7.21)–(7.23). In a similar manner, we can construct from the classical GIE (7.37) the corresponding equations for both the effective fields (of strain and stress)

$$\boldsymbol{\vartheta}(\mathbf{x}) = \overline{\boldsymbol{\vartheta}}_i(\mathbf{x}) + \int \mathbf{U}^{\theta}(\mathbf{x}-\mathbf{y})V_i(\mathbf{y})\boldsymbol{\zeta}^1(\mathbf{y})\,d\mathbf{y}, \tag{7.42}$$

$$\overline{\boldsymbol{\vartheta}}_i(\mathbf{x}) = \widetilde{\boldsymbol{\varepsilon}}_{1,\ldots,n}(\mathbf{x}) + \sum_{j\neq i}\int \mathbf{U}^{\theta}(\mathbf{x}-\mathbf{y})V_j(\mathbf{y})\boldsymbol{\zeta}^1(\mathbf{y})\,d\mathbf{y}, \tag{7.43}$$

$$\widetilde{\boldsymbol{\vartheta}}_{1,\ldots,n}(\mathbf{x}) = \langle\boldsymbol{\varepsilon}\rangle(\mathbf{x}) + \int \mathbf{U}^{\theta}(\mathbf{x}-\mathbf{y})\Big[\boldsymbol{\zeta}^1(\mathbf{y})V(\mathbf{y}|;v_1,\mathbf{x}_1;\ldots;v_n,\mathbf{x}_n) - \langle\boldsymbol{\zeta}^1\rangle(\mathbf{y})\Big]d\mathbf{y} \tag{7.44}$$

and the fields

$$\langle\boldsymbol{\vartheta}|v_1,\mathbf{x}_1;\ldots;v_n,\mathbf{x}_n\rangle(\mathbf{x}) - \sum_{i=1}^{n}\int \mathbf{U}^{\theta}(\mathbf{x}-\mathbf{y})\langle V_i(\mathbf{y})\boldsymbol{\zeta}^1|v_1,\mathbf{x}_1;\ldots;v_n,\mathbf{x}_n\rangle(\mathbf{y})\,d\mathbf{y}$$
$$= \langle\boldsymbol{\vartheta}\rangle(\mathbf{x}) + \int \mathbf{U}^{\theta}(\mathbf{x}-\mathbf{y})\Big[\langle\boldsymbol{\zeta}^1|;v_1,\mathbf{x}_1;\ldots;v_n,\mathbf{x}_n\rangle(\mathbf{y}) - \langle\boldsymbol{\zeta}^1\rangle(\mathbf{y})\Big]\,d\mathbf{y}, \tag{7.45}$$

for $\mathbf{x}\in v_i$, $i=1,2,\ldots,n$ and a particular case $\mathbf{f}_1 \equiv \mathbf{0}$. Since $\mathbf{x}\in v_1,\ldots,v_n$ in the nth line of the system can take the values in the inclusions $v_1,\ldots,v_n$, the nth line actually contains n equations.

The original purpose of the renormalizing term was only to provide the absolute convergence of the integral in Eqs. (7.30) that is archived by long-range behavior of the function $\mathbf{U}(\mathbf{x}-\mathbf{y})\langle\mathbf{L}_1\boldsymbol{\varepsilon}\rangle$ at $|\mathbf{x}-\mathbf{y}|\to\infty$. However, the same term is exploited in a short-range domain $|\mathbf{x}-\mathbf{y}|\le 3a$ in the vicinity of the point $\mathbf{x}\in w$. This term is defined only by the average tensor $\langle\mathbf{L}_1\boldsymbol{\varepsilon}\rangle$ and ignores any available microtopological information (possible nonellipsoidal and laminated structure of inhomogeneities, the shape and size of excluded volume, radial distribution function, orientation, etc.). We will demonstrate in the next subsection that Eqs. (7.32) and (7.33) are just approximations of the exact GIEs (7.19) and (7.20) used in the new renormalizing terms uniquely determined from rigorous averaging scheme and directly dependent on the microstructural information.

7.3 Effective Field Hypothesis and Asymptotic Representation of the GIE (7.19) and (7.20)

7.3.1 *Effective Field Hypothesis*

The effective field hypothesis (EFH) is apparently the most fundamental and most exploited concept of analytical micromechanics. The EFH is usually formulated in the following form of a combination of two hypotheses:

Hypothesis (H1a) *Each heterogeneity* v_i *is located in the effective fields*

$$\overline{\boldsymbol{\varepsilon}}_i(\mathbf{y}) \equiv \overline{\boldsymbol{\varepsilon}}(\mathbf{x}_i), \quad \overline{\boldsymbol{\sigma}}_i(\mathbf{y}) \equiv \overline{\boldsymbol{\sigma}}(\mathbf{x}_i) \ (\mathbf{y} \in v_i) \tag{7.46}$$

which are homogeneous over the inclusion v_i.

Hypothesis (H1b) *The perturbation introduced by the heterogeneity* v_i *at the point* $\mathbf{y} \in R^d$ *is defined by the relation*

$$\int \mathbf{U}(\mathbf{y}-\mathbf{x})V_i(\mathbf{x})\boldsymbol{\tau}(\mathbf{x})d\mathbf{x} = \bar{v}_i\mathbf{T}_i^{\epsilon}(\mathbf{y}-\mathbf{x}_i)\boldsymbol{\tau}_{(i)}, \tag{7.47}$$

$$\int \boldsymbol{\Gamma}(\mathbf{y}-\mathbf{x})V_i(\mathbf{x})\boldsymbol{\eta}(\mathbf{x})d\mathbf{x} = \bar{v}_i\mathbf{T}_i^{\sigma}(\mathbf{y}-\mathbf{x}_i)\boldsymbol{\eta}_{(i)}. \tag{7.48}$$

Hereafter

$$\begin{aligned}
&\mathbf{T}_i^{\varepsilon}(\mathbf{y}-\mathbf{x}_i) = \langle \mathbf{U}(\mathbf{y}-\mathbf{x})\rangle_{(i)}, \quad \mathbf{T}_{ji}^{\varepsilon}(\mathbf{x}_j-\mathbf{x}_i) = \langle \mathbf{T}_i^{\varepsilon}(\mathbf{y}-\mathbf{x}_i)\rangle_{(j)},\\
&\mathbf{T}_i^{\sigma}(\mathbf{y}-\mathbf{x}_i) = \langle \boldsymbol{\Gamma}(\mathbf{y}-\mathbf{x})\rangle_{(i)}, \quad \mathbf{T}_{ji}^{\sigma}(\mathbf{x}_j-\mathbf{x}_i) = \langle \mathbf{T}_i^{\sigma}(\mathbf{y}-\mathbf{x}_i)\rangle_{(j)}
\end{aligned} \tag{7.49}$$

are defined by the averages of over the inclusion $\mathbf{y} \in v_j$ and $\mathbf{x} \in v_i$ (either $v_j \neq v_i$ or $v_j = v_i$).

For the ellipsoidal inclusions, the tensors

$$\mathbf{T}_i^{\varepsilon}(\mathbf{x}-\mathbf{x}_i) = \begin{cases} -(\overline{v}_i)^{-1}\mathbf{P}_i(\mathbf{x}) & \text{for } \mathbf{x} \in v_i,\\ (\overline{v}_i)^{-1}\int \mathbf{U}(\mathbf{x}-\mathbf{y})V_i(\mathbf{y})d\mathbf{y} & \text{for } \mathbf{x} \notin v_i, \end{cases}$$

$$\mathbf{T}_i^{\sigma}(\mathbf{x}-\mathbf{x}_i) = \begin{cases} -(\overline{v}_i)^{-1}\mathbf{Q}_i(\mathbf{x}) & \text{for } \mathbf{x} \in v_i,\\ (\overline{v}_i)^{-1}\int \boldsymbol{\Gamma}(\mathbf{x}-\mathbf{y})V_i(\mathbf{y})d\mathbf{y} & \text{for } \mathbf{x} \notin v_i, \end{cases} \tag{7.50}$$

have analytical representations expressed through an isotropic matrix; the tensors $\mathbf{P}_i = \mathbf{P}_i(\mathbf{x}) \equiv -\int \mathbf{U}(\mathbf{x}-\mathbf{y})V_i(\mathbf{y})d\mathbf{y} \equiv \mathbf{S}_i\mathbf{M}^{(0)} \equiv$const. and $\mathbf{Q}_i = \mathbf{Q}_i(\mathbf{x}) \equiv -\int \boldsymbol{\Gamma}(\mathbf{x}-\mathbf{y})V_i(\mathbf{y})d\mathbf{y} \equiv \mathbf{L}^{(0)} - \mathbf{L}^{(0)}\mathbf{P}_i\mathbf{L}^{(0)} \equiv$const. (for $\forall\mathbf{x} \in v_i$) are defined through the Eshelby [449] tensor $\mathbf{S}_i$. For a homogeneous ellipsoidal inclusion v_i, the standard assumption (7.46) (see, e.g., [197]) yields the assumption (7.47) and (7.48); otherwise, the formulae (7.47) and (7.48) define the additional assumptions. In general, the hypothesis **H1b** can be applied to nonellipsoidal inhomogeneous inclusions. However, owing to opportunity that $\boldsymbol{\tau}(\mathbf{x})$ and $\boldsymbol{\eta}(\mathbf{x})$ can vanish at some part of ellipsoidal domain v_i, the choice of ellipsoidal shape of v_i and v_j allows the use of the possible analytical representations (7.50).

It is interesting to compare the assumption (7.47) with more simple representation

$$\int \mathbf{U}(\mathbf{y}-\mathbf{x})V_i(\mathbf{x})\boldsymbol{\tau}(\mathbf{x})d\mathbf{x} = \bar{v}_i\mathbf{U}(\mathbf{x}-\mathbf{x}_i)\boldsymbol{\tau}_{(i)}. \tag{7.51}$$

We will prove that for homogeneous boundary conditions (7.2) and statistically homogeneous media ($n^{(q)}(\mathbf{x}) = n^{(q)} \equiv$const., $q = 1, \ldots, N$), the approximations (7.47) and (7.51) lead to an identical result reducing the new Eq. (7.19) (at $\mathbf{f} \equiv \mathbf{0}$) to the known one (7.32). This statement holds if we prove that contributions made to Eq. (7.19) by the renormalizing terms (7.47) and (7.51) are identical for any macrodomain $\mathbf{x} \in w$ with the internal points at sufficient distances from the boundary

$$\int_w \mathbf{U}(\mathbf{x}-\mathbf{y})\langle\boldsymbol{\tau}\rangle d\mathbf{y} = \sum_{q=1}^{N}\int_w \mathbf{T}_q^{\varepsilon}(\mathbf{x}-\mathbf{x}_q)\overline{v}_q n^{(q)}\langle\boldsymbol{\tau}\rangle_q d\mathbf{x}_q. \tag{7.52}$$

Of course, Eqs. (7.51) and (7.52) are just a shortcut for convenient notations of the renormalizing purts in Eq. (7.32) which have no sense outside of Eq. (7.32). For justification of the equality (7.52), it should be mentioned that for uniform distribution of inclusion centers $\mathbf{x}_q$, all volume of the domain w in the right-hand side of Eq. (7.52) is uniformly covered by the moving ellipsoids v_q. Then any point in the domain $\mathbf{x} \in w$ in the right-hand side integral (7.52) is covered by the same number k of the ellipsoids v_q with homogeneous strain polarization tensor $\boldsymbol{\tau}(\mathbf{y}) \equiv \langle\mathbf{L}_1\boldsymbol{\varepsilon}\rangle_q$ ($\mathbf{y} \in v_q$), and, therefore, the integral over the covered domain w on the right-hand side of Eq. (7.52) is equal (within some probability factor) to k integrals over domain w in the left-hand side. Therefore, in the case $\boldsymbol{\tau}(\mathbf{y}) \equiv$const. inside moving inhomogeneity $\mathbf{y} \in v_q$, both approximation (7.47) and (7.51) reduce Eq. (7.19) to the known one (7.32). However, a condition of homogeneity $\boldsymbol{\tau}(\mathbf{y}) \equiv$const. at $\mathbf{y} \in v_q$ is fulfilled only for homogeneous ellipsoidal inhomogeneities in the framework of an additional hypothesis of effective field homogeneity **H1a** (7.46). Moving away from the effective field hypothesis leads with necessity to inhomogeneity of the stress–strain fields inside the inhomogeneities that tend to result in different predictions of effective moduli based on Eqs. (7.19) and (7.32) even for both the statistically homogeneous media and homogeneous boundary conditions (see for details Chap. 9). This difference is a result of the insensitivity of the renormalizing term $\mathbf{U}(\mathbf{x}-\mathbf{y})\langle\boldsymbol{\tau}\rangle$ [obtained at the approximation (7.47)] in the asymptotic equation (7.32) to the details of heterogeneities of the stress-strain fields inside the inclusions, while a corresponding term $\langle\mathbf{U}(\mathbf{x}-\mathbf{y})\boldsymbol{\tau}\rangle(\mathbf{y})$ [which is exact and obtained without approximations neither (7.47) nor (7.51)] of Eq. (4.8) explicitly depends on the mentioned field inhomogeneity.

7.3.2 *About Renormalized Terms in Eqs. (7.19) and (7.32)*

The introduction of two kinds of conditional averages $\langle\mathbf{g}\rangle^{(q)}(\mathbf{x})$ ($\mathbf{x}\in v^{(q)}\subset w$) and $\langle\mathbf{g}\rangle_q(\mathbf{z})$ ($\mathbf{z}\in v_q$) (5.74) allows us to present the essence of Eq. (7.19) more clearly; in particular, we consider a statistically inhomogeneous function $\mathbf{g}(\mathbf{x},\mathbf{y})=\mathbf{U}(\mathbf{x}-\mathbf{y})\boldsymbol{\tau}(\mathbf{y})$ which is not invariant to translation $\mathbf{x}\to\mathbf{x}+\mathbf{x}'$. We will show that a popular assumption

$$\mathbf{J}_1(\mathbf{x}-\mathbf{y})=\mathbf{J}_2(\mathbf{x}-\mathbf{y}), \tag{7.53}$$

where

$$\mathbf{J}_1(\mathbf{x}-\mathbf{y}):=\langle\mathbf{U}(\mathbf{x}-\mathbf{y})\boldsymbol{\tau}\rangle^{(q)}(\mathbf{y}), \tag{7.54}$$

$$\mathbf{J}_2(\mathbf{x}-\mathbf{y}):=\mathbf{U}(\mathbf{x}-\mathbf{y})\langle\boldsymbol{\tau}\rangle^{(q)}(\mathbf{y}) \tag{7.55}$$

reduces Eq. (7.19) to the classical one (7.32) for statistically homogeneous media [$n^{(q)}(\mathbf{x})=n\equiv$const.]. A popular justification (with an intuitive level of rigor) of Eq. (7.53) is that the deterministic function $\mathbf{U}(\mathbf{x}-\mathbf{y})$ can be always carried out from the brackets $\langle\cdot\rangle(\mathbf{y})$ of statistical average that can be considered as an intuitive generalization of the corresponding basic linear property of an expectation described by an exact equation

$$\langle A\boldsymbol{\tau}\rangle=A\langle\boldsymbol{\tau}\rangle \tag{7.56}$$

for the constant variable $A=$const. and the random one $\boldsymbol{\tau}$. The estimation of $\mathbf{J}_2(\mathbf{x}-\mathbf{y})$ (7.55) with $\langle\boldsymbol{\tau}\rangle(\mathbf{y})^{(q)}=\langle\boldsymbol{\tau}\rangle^{(q)}\equiv$const. is straightforward, while evaluation of $\mathbf{J}_1(\mathbf{x}-\mathbf{y})$ (7.54) is less trivial and should use a two-step scheme of averaging (5.74)

$$\begin{aligned}\mathbf{J}_1(\mathbf{x}-\mathbf{y})&=\Big\langle\int_{v_q^1(\mathbf{y})} n^{(q)}(\mathbf{r})\overline{v}_q^{-1}\int_{v_q}\mathbf{U}(\mathbf{x}-\mathbf{z})\boldsymbol{\tau}(v_q,\mathbf{r}|\mathbf{z}-\mathbf{r})d\mathbf{z}d\mathbf{r}\Big\rangle\\&=\int_{v_q^1(\mathbf{y})} n^{(q)}(\mathbf{r})\overline{v}_q^{-1}\int_{v_q}\mathbf{U}(\mathbf{x}-\mathbf{z})\langle\boldsymbol{\tau}(v_q,\mathbf{r})\rangle(\mathbf{r},\mathbf{z}-\mathbf{r})d\mathbf{z}d\mathbf{r},\end{aligned} \tag{7.57}$$

$$\mathbf{J}_2(\mathbf{x}-\mathbf{y})=\mathbf{U}(\mathbf{x}-\mathbf{y})\int_{v_q^1(\mathbf{y})} n^{(q)}(\mathbf{r})\overline{v}_q^{-1}\int_{v_q}\langle\boldsymbol{\tau}(v_q,\mathbf{r})\rangle(\mathbf{r},\mathbf{z}-\mathbf{r})d\mathbf{z}d\mathbf{r}, \tag{7.58}$$

where v_q is a moving heterogeneity with the center $\mathbf{r}$ and $\boldsymbol{\tau}(v_q, \mathbf{r}|\mathbf{z}-\mathbf{r})$ denotes a random value of a polarization tensor $\boldsymbol{\tau}$ given that the fixed heterogeneity $\mathbf{z} \in v_q$ is centered at $\mathbf{r}$. The second equality in Eq. (7.56) is determined by applying exact Eq. (7.56) to the conditional average of $\boldsymbol{\tau}(v_q, \mathbf{r}|\mathbf{z}-\mathbf{r})$ in a concrete point $\mathbf{z} \in v_q$ where $\mathbf{U}(\mathbf{x}-\mathbf{z})$ is a constant, while $\mathbf{U}(\mathbf{x}-\mathbf{z})$ is a function of $\mathbf{z}$ if $\mathbf{z}$ passes through the cross-section of v_q. Thus, at this stage, the deterministic function $\mathbf{U}(\mathbf{x}-\mathbf{y})$ can be carried out from the brackets of the conditional average at the fixed $\mathbf{z} \in v_q$. The integral in the right-hand side of Eq. (7.58) was obtained as a result of a formal substitution $\mathbf{U}(\mathbf{x}-\mathbf{z}) \equiv \mathbf{I}$ into Eq. (7.57). For statistically homogeneous media ($n^{(q)} \equiv n^{(q)}$ =const.) subjected to the homogeneous boundary conditions (7.3), Eq. (7.58) is reduced to the obvious one $\mathbf{J}_2(\mathbf{x}-\mathbf{y}) = c^{(q)}\mathbf{U}(\mathbf{x}-\mathbf{y})\langle\boldsymbol{\tau}\rangle^{(q)}$. The exact Eqs. (7.57) and (7.58) depend on the variable $\langle\boldsymbol{\tau}(v_q, \mathbf{r})\rangle(\mathbf{r}, \mathbf{z}-\mathbf{r})$ which is deterministic (not random). However, a forthcoming simplification of (7.57$_2$) related with the carrying out of $\mathbf{U}(\mathbf{x}-\mathbf{z})$ from the two integral signs over $\mathbf{r}$ and $\mathbf{z}$ is possible only in the case if $\mathbf{U}(\mathbf{x}-\mathbf{z}) \equiv A$ =const. for both $\forall\mathbf{z} \in v_q$ and $\forall\mathbf{r} \in v_q^1(\mathbf{y})$. This statement finally disproves a previous intuitive declaration mentioned before Eq. (7.56) that the deterministic function $\mathbf{U}(\mathbf{x}-\mathbf{y})$ can always be carried out from the brackets $\langle\cdot\rangle(\mathbf{y})$ of statistical average. Thus, we proved that the representation (7.53) is just an asymptotic approximation (which is fulfilled at $|\mathbf{x}-\mathbf{y}| \to \infty$) of the left-hand side of (7.53) rather than an exact equality, while a very similar Eq. (7.54) (corresponding to Eq. (7.19)) is exact.

7.3.3 *Model Example*

We will consider a model numerical example admitting a simple analytical solution for a quantitative illustration of the assumption (7.46) rather than presenting some concrete micromechanical problem. Namely, one analyzes (see Fig. 7.2) a scalar version of Eq. (7.53) in a $1D$ space of statistically inhomogeneous distribution [$n^{(q)}(\mathbf{y}) \not\equiv$const.] of identical intervals (inclusions $v_i = [-a, a]$) of the length $2a$ with a random polarization tensor $\tau(y) = \tau_0(y)V_q(y)$ described by a stochastic process $\tau_0(y, z) = b\exp(-|y|) + \xi + c\sin(\pi z/a + \zeta)$ where the random variables ζ uniformly distributed on $[-\pi, \pi]$ and ξ have the expectations $M\zeta = 0$ and $M\xi = \tau^0$, respectively. Then in the local coordinate system $\langle\tau\rangle_q(y, z) \equiv \langle\tau\rangle_q(y) = b\exp(-|y|) + \tau^0$ $(z \in [-a, a])$ while

$$\langle\tau\rangle(y) = \int_{y-a}^{y+a} n^{(q)}(r)\frac{1}{2a}\int_{r-a}^{r+a} \langle\tau\rangle_q(r)\, dz\, dr. \tag{7.59}$$

The case $n^{(q)}(\mathbf{y}) \not\equiv$const. describes a functionally graded system, while $b \neq 0$ corresponds to the macroscopically inhomogeneous loading. Then, according to a

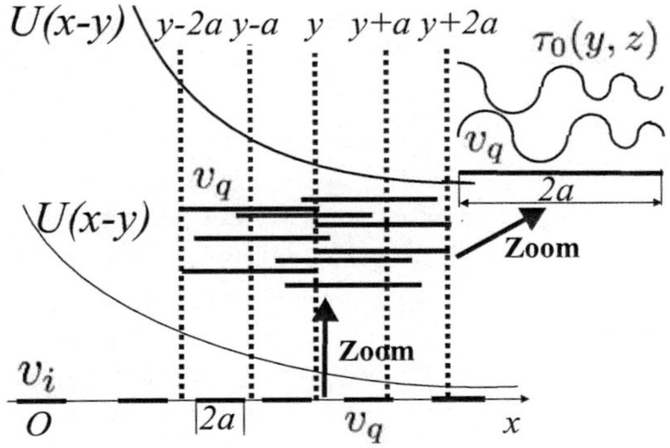

Fig. 7.2 Schematic representation of the functions $U(x-y)$, $\tau_0(y,z)$, and inclusions v_i, v_q

two-step estimation of statistical average (5.74), we get at $|x-y|>2a$ and $U(x-y):=A+B[a/(x-y)]^3$

$$J_1(x-y)=\int_{y-a}^{y+a} n^{(q)}(r)\frac{1}{2a}\int_{r-a}^{r+a} U(x-z)\langle\tau\rangle_q(r)\,dz\,dr, \tag{7.60}$$

$$J_2(x-y)=U(x-y)\int_{y-a}^{y+a} n^{(q)}(r)\langle\tau\rangle_q(r)\,dr. \tag{7.61}$$

The locations of intervals v_q placed in the vicinity of the point y and involved in the integral evaluations (7.60) are schematically depicted in the enlarged part of Fig. 7.1. No confusion will arise due to the overlapping of these intervals v_q because they belong to different infinite realizations of the field X rather than to a single realization X of intervals v_i on the axis Ox.

We see that both $J_1(x-y)$ and $J_2(x-y)$ are sensitive to some nonlocal effects and depend on $\langle\tau\rangle_q(r)$ and $n^{(q)}(r)$ in the vicinity $|r-y|<a$ of the point y rather than only on $\langle\tau\rangle_q(y)$ and $n^{(q)}(y)$ in the point y. For demonstration of the fundamental difference between the averages $J_1(x-y)$ and $J_2(x-y)$, we consider a degenerate case $b=0$ and $n^{(q)}=n=$const. of the statistically homogeneous system when the stochastic process $\tau(y,z)$ is a stationery one (but not ergodic). Then the analytical representations ($g\equiv 2an\tau^0$)

$$J_1(x-y)=gA+gBa^3\{[(x-y)^2-4a^2](x-y)\}^{-1}, \tag{7.62}$$

$$J_2(x-y)=gA+gBa^3(x-y)^{-3}, \tag{7.63}$$

make it possible to estimate a relative error of the approximation (7.53)

$$\Delta\equiv|J_1(x-y)-J_2(x-y)||J_1(x-y)|^{-1}. \tag{7.64}$$

As can be seen, Eq. (7.53) is exactly fulfilled ($\Delta=0$) only for the constant $U(x-y)\equiv A$ ($B=0$). In the other limiting case ($A=0,\ B\neq 0$), the relative error

$\Delta = 4a^2/(x-y)^2$ indicates that the representation (7.53) is just an asymptotic approximation of $\langle U(x-y)\tau\rangle(y)$ az $(x-y)/a \to \infty$; for example, $\Delta = 25\%$ and 4% at $(x-y)/a = 4$ and 10, respectively. The error Δ (7.64) is defined by the inhomogeneity of $U(x-y)$ in the domain $v_y = \{x' : |x'-y| < 2a\}$ (see Fig. 7.2) rather than by the average value of $U(x-y)$ in this domain v_y. We can see in Eq. (7.60) that the two-step average (5.74) and (7.60) is performed at first by the use of the linear property of the expectation (5.74) with the estimation of the statistical average $\langle\tau\rangle_q(y,z)$ in the local coordinate system $z \in [-a,a]$. The second subsequent step of averaging of the weighted polarization amount $\langle\tau\rangle_q(y,z)$ is accomplished over the volumes of the moving inclusions v_q with the centers $y_q \in [y-1, y+a]$, when $U(x-t)$ cannot be moved out of the ensemble average operator as was done in Eq. (7.61). The considered example convincingly confirms the fact that the renormalizing term $\mathbf{U}(\mathbf{x}-\mathbf{y})\langle\boldsymbol{\tau}\rangle(\mathbf{y})$ in the classical GIEl (7.32) is no more than an asymptotic approximation of the exact proposed one $\langle\mathbf{U}(\mathbf{x}-\mathbf{y})\boldsymbol{\tau}\rangle(\mathbf{y})$ (7.19) in a long-range zone as $|\mathbf{x}-\mathbf{y}| \to \infty$. However, the error of the approximation (7.53) is drastically increased in a short-range zone of a few wavelengths (especially if $\mathbf{x}, \mathbf{y} \in v_i$) where interaction of heterogeneities reaches its peak.

Thus, we proved that although the equalities (7.53) intuitively look as Eq. (7.56), the relations (7.53) are just asymptotic approximations of the left-hand sides of (7.53), whereas Eq. (7.56) is exact.

7.4 Periodical Structure Composites

Traditionally, these methods used at the locally elastic analyses of composite materials (CMs) with either the random structure or periodic ones are conceptually different. So, for elastic random structure CMs, a number of micromechanical models, inspired by Eshelby [449], were proposed in the literature for describing the thermoelastic behavior of composites with ellipsoidal inclusions (see for references, e.g., Chaps. 8–14). For homogeneous boundary conditions, periodicity of a composite structure leads to a periodicity of the solution for the stress and strain distributions with the same period as that of the microstructure.

7.4.1 Short Review of Analyses of Periodical Structure Composites

A multi-scale analysis of CMs with fine periodic microstructure assumes that a ratio ε_Ω of the representative size of microstructures (called a unit cell, UC) and the overall dimension L of the structure is negligibly small. In such a case, the term *homogenization* (apparently coined by Babuska [48]; the term *averaging* previously introduced by Bakhvalov is also used [53–55]) describes the asymptotic behavior at

$\varepsilon_\Omega \to 0$ of the differential equations $\mathbf{L}_\varepsilon \mathbf{u}_\varepsilon = \mathbf{0}$ governing the origin boundary-value problem. Loosely speaking, the key of the classical mathematical theory of two-scale asymptotic expansion (see, e.g., [56, 84, 704, 1324], and references therein) is postulating an ansatz for the solution expansion $\mathbf{u}_\varepsilon$ in the form of a power series with the periodic coefficients $\mathbf{u}_i$ of ε_Ω^i, inserting this expansion into $\mathbf{L}_\varepsilon \mathbf{u}_\varepsilon = \mathbf{0}$ and identifying powers of ε_Ω that leads to a cascade of equations for each term $\mathbf{u}_i$. Subsequent average over the unit cell of $\mathbf{u}_0$ gives the asymptotic solution of the effective governing equation $\mathbf{L}^*\mathbf{u} = \mathbf{0}$ for the convergence of the sequence $\mathbf{u}_\varepsilon$ to $\mathbf{u}$. These two-scale expansion methods were incorporated with the finite element analysis (FEA) and extensively infiltrated into engineering research (see, e.g., [477, 731, 732, 1221]). It allows us not only to estimate the macroscopic governing equation but also to evaluate the localized behavior at the fine scale that defines a multi-scale modeling tool in the area of computational mechanics. Generalized mathematical homogenization theory proposed by Fish and coauthors [476, 479] constructs an array of atomistic unit cell problems coupled with a thermo-mechanical continuum problem.

Another alternative version of the homogenization without the two-scale expansions, so-called computational homogenization, was initiated by Suquet [1467]. The key of the computational homogenization is a determination of macroscopic mechanical response (input leading to output, e.g., stress driven by deformation) through the nested solution of a microstructural problem at the microscopic periodic boundary conditions defined from the macroscopic input variables (see, e.g., [798, 1056, 1502]; see also [1030, 1350] and references therein). There are no a priori requirements on the type of the macroscopic constitutive equations (e.g., its nonlinearity) in which the overall stresses and tangent moduli can be defined in terms of discrete forces and displacements on the boundary of unit cell. The most popular type of boundary conditions is periodic displacement and antiperiodic traction. Incorporation of these constrains into the current equilibrium state can be performed by the Lagrange multiplier method, while the solution of the macroscopic nonlinear problems are usually accomplished in an iterative manner by means of rate or incremental form of the governing equations. It should be mentioned that the ansatz for the two-scale expansion method of homogenization is used only to guess the solution form and might not provide an appropriate accuracy especially for the nonlinear problems. In so doing, the methods of the computational homogenization are more general, and its applicability for either the nonlinear, inelastic, or nonlocal problems can be performed in the framework of the same general scheme.

For periodic structure composites, there are also known semi-analytical methods for solving the cell problem. The method of Eshelby transformation strain, taking account of the variability of the field of transformation strain within the inclusion, has been proposed in [1122]. However, because they used very slowly converging series, their approach is hardly applicable. Using the periodic fundamental solution for an isotropic medium [622] (see also [1574]), numerical values of the effective characteristics of dispersionally reinforced composites with isotropic components were obtained in [1326] by multipole expansions. Those authors extended a collocation technique employed in [1141]. Media with arbitrary elastic anisotropy were considered in [868]. Kushch [843, 845] (see also [846]) analyzed the elastic isotropic medium containing several triply periodic lattices of aligned spheroidal isotropic inclusions with different sizes, shapes, and properties (anisotropic cases are considered in [849]).

7.4.2 *Doubly and Triply Periodical Structure Composites*

It is assumed that the representative mesodomain w contains a statistically large number of realizations of ellipsoidal inclusions $v_i \in v^{(1)} \subset R^d$ $(i = 1, 2, \ldots,\ d = 2, 3)$ with identical shape, orientation, and mechanical properties. We now consider a composite medium with particle centers distributed at the nodes of some spatial lattice Λ. Suppose $\mathbf{e}_i$ $(i = 1, 2, 3)$ are linearly independent vectors, so that we can represent any node $\mathbf{m} \in \Lambda$: $\mathbf{x_m} = f_1(m_1)\mathbf{e}_1 + f_2(m_2)\mathbf{e}_2 + f_3(m_3)\mathbf{e}_3$ (5.84), where $\mathbf{m} = (m_1, m_2, m_3)$ are integer-valued coordinates of the node $\mathbf{m}$ in the basis $\mathbf{e}_i$ which are equal in modulus to $|\mathbf{e}_i|$, and $f_i(m_i) - f_i(m_i+1) \not\equiv \text{const.}$, $(i = 1, 2, 3)$. A detailed description of such deterministic structures is presented in Sect. 5.3.4. The representation (5.84) is reduced in some particular cases to the equations describing both the triply (5.85) and doubly (5.86) periodic structures

$$\mathbf{x_m} = m_1\mathbf{e}_1 + m_2\mathbf{e}_2 + m_3\mathbf{e}_3, \tag{7.65}$$

$$\mathbf{x_m} = m_1\mathbf{e}_1 + m_2\mathbf{e}_2 + f_3(m_3)\mathbf{e}_3, \tag{7.66}$$

respectively. The composite material is constructed using building blocks or cells: $w = \cup\Omega_\mathbf{m}$, $v_\mathbf{m} \subset \Omega_\mathbf{m}$. In Sect. 5.3.4, the averages over both the cell Ω_i (5.87) and "moving averaging" cell $\mathcal{V}_\mathbf{x}$ (or moving-window [563]) (5.88) were defined ($\mathbf{x} \in \Omega_i$)

$$\mathbf{g}^\Omega(\mathbf{x}) = \frac{1}{\bar{\Omega}_i}\int_{\Omega_i} \mathbf{g}(\mathbf{y})\, d\mathbf{y}, \quad \langle \mathbf{g}\rangle_\mathbf{x}(\mathbf{x}-\mathbf{y}) = \frac{1}{\bar{\mathcal{V}}_\mathbf{x}}\int_{\mathcal{V}_\mathbf{x}} \mathbf{g}(\mathbf{z}-\mathbf{y})\, d\mathbf{z}. \tag{7.67}$$

By way of illustration, let us consider the case of triply periodic structures (7.65) under the uniform boundary conditions (7.3), and let $\mathbf{g}$ be governed by the boundary condition (7.3) (e.g., $\mathbf{g} \equiv \boldsymbol{\sigma}$). Clearly, for homogeneous boundary conditions (7.3), $\boldsymbol{\sigma}^\Omega(\mathbf{x})$ by (7.67) is an invariant with respect to the cell number i and $\boldsymbol{\sigma}^\Omega(\mathbf{x}) = \langle\boldsymbol{\sigma}\rangle_\mathbf{x}(\mathbf{x}) = \text{const.}$, $\forall\mathbf{x} \in \Omega_i \subset w$ (if $\mathcal{V}_\mathbf{x}$ is a translation of Ω_i). In the general case of inhomogeneous boundary conditions $\boldsymbol{\sigma}^\Gamma(\mathbf{x}) \not\equiv \text{const.}$ (as well as in the case that the condition (7.65) breaks down), $\boldsymbol{\sigma}^\Omega(\mathbf{x})$ is a step function $\boldsymbol{\sigma}^\Omega(\mathbf{x}) \neq \boldsymbol{\sigma}^\Omega(\mathbf{y})$ at $\mathbf{x} \in \Omega_i$ and $\mathbf{y} \in \Omega_j$ $(i \neq j)$ as well as $\boldsymbol{\sigma}^\Omega(\mathbf{x}) \not\equiv \langle\boldsymbol{\sigma}\rangle_\mathbf{x}(\mathbf{x})$ at $\mathbf{x} \in \Omega_i$.

Recall that Eq. (7.8) is valid for concrete realization of the inclusion field X which can be doubly periodical (7.66). In such a case, the function of the operation of a statistical average for random structure composites (see Sect. 7.2) becomes the volume average over the "moving averaging" cell $\mathcal{V}_\mathbf{x}$. By doing so, the transformation of Eq. (7.8) in the framework of the centering method is carried out by subtracting from both sides of Eq. (7.8) their average over the "moving averaging" cell $\mathcal{V}_{\mathbf{x}}$ (7.67$_2$)

$$\boldsymbol{\varepsilon}(\mathbf{x}) = \langle\langle\boldsymbol{\varepsilon}^0\rangle\rangle_{\mathbf{x}}(\mathbf{x}) + \langle\boldsymbol{\varepsilon}\rangle_{\mathbf{x}}(\mathbf{x}) + \int \langle\langle\nabla\mathbf{G}\rangle\rangle_{\mathbf{x}}(\mathbf{x}-\mathbf{y})\mathbf{f}_1(\mathbf{y})\,d\mathbf{y}$$
$$+ \int \langle\langle\mathbf{U}\rangle\rangle_{\mathbf{x}}(\mathbf{x}-\mathbf{y})\boldsymbol{\tau}^1(\mathbf{y})\,d\mathbf{y} + \oint \langle\langle\nabla\mathbf{G}\rangle\rangle_{\mathbf{x}}(\mathbf{x}-\mathbf{s})\boldsymbol{\tau}^1(\mathbf{s})\mathbf{n}(\mathbf{s})\,d\mathbf{s}, \quad (7.68)$$

where $\mathbf{x} \in \Omega_i$, and one introduces a new centering operation over the "moving averaging" cell $\mathbf{x} \in \mathcal{V}_{\mathbf{x}}$ caused by translation of a cell Ω_i (compare with Eq. (7.12))

$$\langle\langle\mathbf{g}\rangle\rangle_{\mathbf{x}}(\mathbf{x}-\mathbf{y}) \equiv \mathbf{g}(\mathbf{x}-\mathbf{y}) - \langle\mathbf{g}\rangle_{\mathbf{x}}(\mathbf{x}-\mathbf{y}). \quad (7.69)$$

For the analysis of integral convergence in Eq. (7.68), we expand $\mathbf{U}(\mathbf{z}-\mathbf{y})$ ($\mathbf{z} \in \mathcal{V}_{\mathbf{x}}$) in a Taylor series about $\mathbf{x}$ and integrate term by term over the cell $\mathcal{V}_{\mathbf{x}}$ with the center $\mathbf{x}$; then

$$\mathbf{U}(\mathbf{z},\mathbf{y}) = \mathbf{U}(\mathbf{x},\mathbf{y}) + (\mathbf{z}-\mathbf{x})\nabla\mathbf{U}(\mathbf{x},\mathbf{y}) + \frac{1}{2}(\mathbf{z}-\mathbf{x})\otimes(\mathbf{z}-\mathbf{x})\nabla\nabla\mathbf{U}(\mathbf{x},\mathbf{y})\ldots,$$
$$\langle\mathbf{U}\rangle_{\mathbf{x}}(\mathbf{x},\mathbf{y}) = \mathbf{U}(\mathbf{x},\mathbf{y}) + \frac{1}{2\bar{V}_{\mathbf{x}}}\int_{\mathcal{V}_{\mathbf{x}}}(\mathbf{z}-\mathbf{x})\otimes(\mathbf{z}-\mathbf{x})\,d\mathbf{z}\nabla\nabla\mathbf{U}(\mathbf{x},\mathbf{y})\ldots.$$

A similar expansion can be performed for the tensor $\nabla\mathbf{G}(\mathbf{x}-\mathbf{y})$ that leads to

$$\langle\langle\mathbf{U}\rangle\rangle_{\mathbf{x}}(\mathbf{x}-\mathbf{y}) = -\frac{1}{2\bar{V}_{\mathbf{x}}}\int_{\mathcal{V}_{\mathbf{x}}}(\mathbf{z}-\mathbf{x})\otimes(\mathbf{z}-\mathbf{x})\,d\mathbf{z}\nabla\nabla\mathbf{U}(\mathbf{x}-\mathbf{y}) + \ldots,$$
$$\langle\langle\nabla\mathbf{G}\rangle\rangle_{\mathbf{x}}(\mathbf{x}-\mathbf{y}) = -\frac{1}{2\bar{V}_{\mathbf{x}}}\int_{\mathcal{V}_{\mathbf{x}}}(\mathbf{z}-\mathbf{x})\otimes(\mathbf{z}-\mathbf{x})\,d\mathbf{z}\nabla\nabla\nabla\mathbf{G}(\mathbf{x}-\mathbf{y}) + \ldots. \quad (7.70)$$

As is evident from Eq. (7.69), the tensor $\langle\langle\mathbf{U}\rangle\rangle_{\mathbf{x}}(\mathbf{x}-\mathbf{y})$ is of the order $O(a_{\mathcal{V}}^{d-1}|\mathbf{x}-\mathbf{y}|^{-d-2})$ with the dropped terms in Eq. (7.70) being of the order $O(a_{\mathcal{V}}^{d+1}|\mathbf{x}-\mathbf{y}|^{-d-4})$ and higher-order terms; $a_{\mathcal{V}} = \sqrt[d]{\bar{V}}$ is a characteristic size of the cell $\mathcal{V}$. Then, the absolute convergence of the volume integral (7.68) is assured because at sufficient distance $\mathbf{x}$ from the boundary Γ and $|\mathbf{x}-\mathbf{y}| \to \infty$, integration over $\mathbf{y}$ can be carried out independently for both $\langle\langle\mathbf{U}\rangle\rangle_{\mathbf{x}}(\mathbf{x}-\mathbf{y})$ (the function of the "slow" variable $\mathbf{x}-\mathbf{y}$) and the expression $\boldsymbol{\tau}^1(\mathbf{y})$ (the function of "fast" variable $\mathbf{y}$), and therefore the volume integral converges absolutely. In a similar manner, the term $\langle\langle\nabla\mathbf{G}\rangle\rangle_{\mathbf{x}}(\mathbf{x}-\mathbf{s})$ in the surface integral (7.68) is of the order $O(a_{\mathcal{V}}^{d-1}|\mathbf{x}-\mathbf{y}|^{-d-1})$, and the surface integral vanishes at $|\mathbf{x}-\mathbf{s}| \to \infty$, $\mathbf{s} \in \Gamma$. For the same reason, the volume integral with the kernel $\langle\langle\nabla\mathbf{G}\rangle\rangle_{\mathbf{x}}(\mathbf{x}-\mathbf{s})$ converges absolutely for bounded functions $\mathbf{f}_1(\mathbf{y})$.

By this means, Eq. (7.68) is reduced to the relation

$$\boldsymbol{\varepsilon}(\mathbf{x}) = \langle\langle \boldsymbol{\varepsilon}^0 \rangle\rangle_{\mathbf{x}}(\mathbf{x}) + \langle \boldsymbol{\varepsilon} \rangle_{\mathbf{x}}(\mathbf{x}) + \int \langle\langle \nabla \mathbf{G} \rangle\rangle_{\mathbf{x}}(\mathbf{x}-\mathbf{y})\mathbf{f}_1(\mathbf{y})\, d\mathbf{y} + \int \langle\langle \mathbf{U} \rangle\rangle_{\mathbf{x}}(\mathbf{x}-\mathbf{y})\boldsymbol{\tau}^1(\mathbf{y})\, d\mathbf{y}, \quad (7.71)$$

where the volume integrals converge absolutely. Expressing Eq. (7.71) in terms of stresses by the use of identities (7.15), the general equation

$$\boldsymbol{\sigma}(\mathbf{x}) = \langle\langle \boldsymbol{\sigma}^0 \rangle\rangle_{\mathbf{x}}(\mathbf{x}) + \langle \boldsymbol{\sigma} \rangle_{\mathbf{x}}(\mathbf{x}) + \int \mathbf{L}^c \langle\langle \nabla \mathbf{G} \rangle\rangle_{\mathbf{x}}(\mathbf{x}-\mathbf{y})\mathbf{f}_1(\mathbf{y}) d\mathbf{y} + \int \langle\langle \boldsymbol{\Gamma} \rangle\rangle_{\mathbf{x}}(\mathbf{x}-\mathbf{y})\boldsymbol{\eta}(\mathbf{y}) d\mathbf{y} \quad (7.72)$$

is obtained.

Evidently, the right-hand-side volume integrals in Eqs. (7.71) and (7.72) converge absolutely, with no restrictions being imposed on the microtopology of the lattice Λ, and Eqs. (7.71) and (7.72) are valid for any deterministic (even nonperiodic) structures. The principal advantages of Eqs. (7.71) and (7.72) as compared with Eq. (7.68) are the lack of the surface integral in Eqs. (7.71) and (7.72) and the local character of Eqs. (7.71) and (7.72). The last-mentioned advantage makes it possible to reduce the analysis of infinite number inclusion problems to the analysis of a finite number of inclusions located in some representative volume element (RVE) [183, 184].

In the interest of obtaining simpler relationships, we will consider the case of "slowly varying" fields $\boldsymbol{\varepsilon}^0(\mathbf{x})$ and $\boldsymbol{\sigma}^0(\mathbf{x})$. Then, expanding $\boldsymbol{\varepsilon}^0(\mathbf{z})$ in a Taylor series about $\mathbf{x}$ gives, by analogy with Eq. (7.70), that $\langle\langle \boldsymbol{\varepsilon}^0 \rangle\rangle_{\mathbf{x}}(\mathbf{x})$ is of order $O(a_{\mathcal{V}}^{d-1} \nabla\nabla \boldsymbol{\varepsilon}^0(\mathbf{x}))$. Therefore, in the interest of obtaining explicit final expressions, we can neglect the term $\langle\langle \boldsymbol{\varepsilon}^0 \rangle\rangle_{\mathbf{x}}(\mathbf{x})$ (7.68) as compared with $\langle \boldsymbol{\varepsilon} \rangle_{\mathbf{x}}(\mathbf{x})$ in the "slowly varying" approximation of $\boldsymbol{\varepsilon}^0(\mathbf{x})$. The same simplification is admissible with respect to the term $\langle\langle \boldsymbol{\sigma}^0 \rangle\rangle_{\mathbf{x}}(\mathbf{x})$ of Eq. (7.72). Then Eqs. (7.71) and (7.72) are reduced to the approximate relationships

$$\boldsymbol{\varepsilon}(\mathbf{x}) = \langle \boldsymbol{\varepsilon} \rangle_{\mathbf{x}}(\mathbf{x}) + \int \langle\langle \nabla \mathbf{G} \rangle\rangle_{\mathbf{x}}(\mathbf{x}-\mathbf{y})\mathbf{f}_1(\mathbf{y})\, d\mathbf{y} + \int \langle\langle \mathbf{U} \rangle\rangle_{\mathbf{x}}(\mathbf{x}-\mathbf{y})\boldsymbol{\tau}(\mathbf{y})\, d\mathbf{y}, \quad (7.73)$$

$$\boldsymbol{\sigma}(\mathbf{x}) = \langle \boldsymbol{\sigma} \rangle_{\mathbf{x}}(\mathbf{x}) + \int \mathbf{L}^c \langle\langle \nabla \mathbf{G} \rangle\rangle_{\mathbf{x}}(\mathbf{x}-\mathbf{y})\mathbf{f}_1(\mathbf{y})\, d\mathbf{y} + \int \langle\langle \boldsymbol{\Gamma} \rangle\rangle_{\mathbf{x}}(\mathbf{x}-\mathbf{y})\boldsymbol{\eta}(\mathbf{y})\, d\mathbf{y}, \quad (7.74)$$

which are exact for linear functions $\boldsymbol{\varepsilon}^0(\mathbf{x})$. Buryachenko [183, 184] has obtained the particular cases of Eq. (7.73) (triply periodical structures with $\mathbf{f}$, $\boldsymbol{\beta} \equiv \mathbf{0}$) and Eq. (7.74) (doubly periodical structures $\mathbf{f} \equiv \mathbf{0}$), respectively.

By way of subsequent simplification, we now consider the case of a triply periodic structure (7.65) under the uniform boundary conditions (7.3). Then, Eqs. (7.71) and (7.72) can be rewritten in the form

$$\boldsymbol{\varepsilon}(\mathbf{x}) = \boldsymbol{\varepsilon}^{\Omega}(\mathbf{x}) + \int \nabla\mathbf{G}(\mathbf{x}-\mathbf{y})\big[\mathbf{f}_1(\mathbf{y}) - \mathbf{f}_1^{\Omega}(\mathbf{y})\big]\,d\mathbf{y}$$
$$+ \int \mathbf{U}(\mathbf{x}-\mathbf{y})\{\boldsymbol{\tau}(\mathbf{y}) - \boldsymbol{\tau}^{\Omega}(\mathbf{y})\}\,d\mathbf{y}, \tag{7.75}$$

$$\boldsymbol{\sigma}(\mathbf{x}) = \boldsymbol{\sigma}^{\Omega}(\mathbf{x}) + \int \mathbf{L}^c\nabla\mathbf{G}(\mathbf{x}-\mathbf{y})\big[\mathbf{f}_1(\mathbf{y}) - \mathbf{f}_1^{\Omega}(\mathbf{x})\big]\,d\mathbf{y}$$
$$+ \int \boldsymbol{\Gamma}(\mathbf{x}-\mathbf{y})\big\{\boldsymbol{\eta}(\mathbf{y}) - \boldsymbol{\eta}^{\Omega}(\mathbf{x})\big\}\,d\mathbf{y}. \tag{7.76}$$

In the pure elastic case with $\mathbf{M}_1^{(0)}(\mathbf{x}), \mathbf{f}(\mathbf{x}), \boldsymbol{\beta}(\mathbf{x}) \equiv \mathbf{0}$ $(\mathbf{x} \in w)$, exact Eq. (7.75) was previously used in [260] for both $\langle \mathbf{L}_1\boldsymbol{\varepsilon}\rangle_{\mathbf{x}}(\mathbf{x}) = (\mathbf{L}_1\boldsymbol{\varepsilon})^{\Omega}(\mathbf{x}) \equiv$ const. and $\langle\boldsymbol{\varepsilon}\rangle_{\mathbf{x}}(\mathbf{x}) = \boldsymbol{\varepsilon}^{\Omega}(\mathbf{x}) \equiv$ const. when in Eq. (7.76) $\langle\boldsymbol{\eta}\rangle_{\mathbf{x}}(\mathbf{x}) = \boldsymbol{\eta}^{\Omega}(\mathbf{x}) \equiv$ const. and $\langle\boldsymbol{\sigma}\rangle_{\mathbf{x}}(\mathbf{x}) = \boldsymbol{\sigma}^{\Omega}(\mathbf{x}) \equiv$ const.. More recently the same problem has also been considered in [1088, 1298] by use of some additional unnecessary assumptions (see for details [200]).

It should be emphasized that for the field X bounded in one direction (e.g., for the field X (7.66)), Eqs. (7.28) and (7.29) are valid as well as for the deterministic field X. Note that for the elastic analysis $(\mathbf{M}_1^{(0)}(\mathbf{x}),\ \mathbf{f}(\mathbf{x}), \boldsymbol{\beta}(\mathbf{x}) \equiv \mathbf{0})$ of triply periodic structures, Fassi Fehri et al. [462] used Eq. (7.28) (for $\boldsymbol{\beta} \equiv \mathbf{0}$) without any renormalization of the right-hand-side integral in Eq. (7.28) which diverges at infinity for the case of triply periodic composites.

7.4.3 *Triply Periodic Particulate Matrix Composites with Imperfect Unit Cells*

We first consider a composite medium with periodic set X of ellipsoidal inclusions with identical shape, orientation, and mechanical properties with particle centers, periodically distributed at the nodes of the same spatial lattice Λ. Suppose $\mathbf{e}_i$ $(i = 1, 2, 3)$ are linearly independent vectors of the principal period of Λ, which determine a unit cell Ω of volume $\overline{\Omega} = |\mathbf{e}_1 \cdot (\mathbf{e}_2 \otimes \mathbf{e}_3)|$, so that we can represent any node $\mathbf{m} \in \Lambda$ in the form (7.65).

Now we will consider a *substitutional* sort of the disordered imperfection of the unit cells which connotes a variability in properties per vertex (see [298, 299, 1165, 1181, 1344, 1406]). In particular, this sort of imperfection describes the effect of randomly missing inclusions, when in some randomly distributed cells, the inclusions are absent and replaced by the matrix. In another sort of imperfection, some inclusions are randomly replaced by one or several kinds of inclusions that, perhaps, are the debonding inclusions in the initially periodic structure. In simplified modeling, the completely debonded inclusions are replaced by the voids of the inclusion size [371]. For a description of the random structure of a composite material, let us introduce a probability density $\varphi(v_i, \mathbf{x}_i)$ of location of the inclusion v_i with the center in the point $\mathbf{x}_i$ and a conditional probability density

$\varphi(v_j, \mathbf{x}_j|v_i, \mathbf{x}_i)$, which is a probability density to find the jth inclusion with the center $\mathbf{x}_j$ in the domain v_j with fixed inclusion v_i with the centers $\mathbf{x}_i$. The notation $\varphi(v_j, \mathbf{x}_j|; v_i, \mathbf{x}_i)$ denotes the case $\mathbf{x}_j \neq \mathbf{x}_i$. Then

$$\varphi(v_i, \mathbf{x}_i) = \sum_{\mathbf{m}} \lambda^{(k)}(\mathbf{x_m})\delta(\mathbf{x}_i - \mathbf{x_m}),\ \varphi(v_j, \mathbf{x}_j|; v_i, \mathbf{x}_i) = \sum_{\mathbf{m}}{}' \lambda^{(l)}(\mathbf{x_m})\delta(\mathbf{x}_j - \mathbf{x_m}), \tag{7.77}$$

where $'$ means that the summation symbol does not contain the argument $\mathbf{x_m} = \mathbf{x}_i$ and $v_i \in v^{(k)}$, $v_j \in v^{(l)}$ $(i, j = 1, 2, \ldots, \ k, l = 1, 2, \ldots, N)$. The statistical averages of random variables $\lambda(\mathbf{x_m})$ $0 \leq \langle \lambda^{(k)}(\mathbf{x_m}) \rangle \equiv \lambda^{(k)} \leq 1$ and $0 \leq \sum_k \lambda^{(k)} \leq 1$ reflect the measure of the imperfection: $\lambda^{(k)} = 0$ means the absence in the lattice Λ of inclusions of the component k; the case $\lambda^{(k)} = 1$ stands for the perfect periodic structure filled by identical inclusions from the component k. In so doing the volume concentration of the k component $c^{(k)}(\lambda^{(k)}) = \lambda^{(k)}\overline{v}^{(k)}/\overline{\Omega}_{\mathbf{m}}$ equals $c^{(k)} \equiv c^{(k)}(1)$ for perfect packing.

In Sects. 7.1 through 7.4, we decomposed the elastic modulus on two summands $\mathbf{L}(\mathbf{x}) = \mathbf{L}^c + (\mathbf{L}(\mathbf{x}) - \mathbf{L}^c)$. The main goal of this decomposition is a subsequent decomposition of a solution found on two summands. One of them specific to $\mathbf{L}^c$ is known, and, moreover, it is a constant for the homogeneous boundary conditions either (7.2) or (7.3) being considered in this section. The second summand corresponding to the perturbation $(\mathbf{L}(\mathbf{x}) - \mathbf{L}^c)$ is estimated by the use of a Green function for the known first problem. The main goal of this section considered as a generalization of the scheme mentioned above is a decomposition of the desired solution on the known solution for the perfect periodic structure and on the perturbation produced by the imperfections in the perfect periodic structure. In an analogy with the notations of Sect. 7.1, we represent the tensor of material properties $\mathbf{f}$ $(\mathbf{f} = \mathbf{L}, \mathbf{M})$ as a decomposition $\mathbf{f}(\mathbf{x}) = \mathbf{f}^c + \mathbf{f}_2(\mathbf{x}) + \mathbf{f}_3(\mathbf{x})$, where $\mathbf{f}(\mathbf{x}) = \mathbf{f}^c$ for $\mathbf{x} \in v^{(0)}$, $\mathbf{f}_2(\mathbf{x}) \equiv \mathbf{f}^{\text{per}}(\mathbf{x}) - \mathbf{f}^c$ is a jump of elastic properties of the perfect imaginary periodic structure with respect to the matrix and $\mathbf{f}_3(\mathbf{x}) \equiv \mathbf{f}(\mathbf{x}) - \mathbf{f}^{\text{per}}(\mathbf{x})$ is a jump of the real elastic properties with respect to the imaginary perfect periodic structure; in the case $\mathbf{f}^c = \mathbf{f}^{(0)} \equiv$ const., obviously, we have $\mathbf{f}_2(\mathbf{x}) = \mathbf{f}_3(\mathbf{x}) \equiv \mathbf{0}$ at $\mathbf{x} \in v^{(0)}$. Usually for the homogeneous matrix $\mathbf{L}^c = \mathbf{L}^{(0)}$; otherwise, it is reasonable to consider the elastic properties of the comparison medium as the Voight (or Reuss) average of elastic properties of the matrix $\mathbf{L}^c = [v^{(0)}]^{-1}\langle \mathbf{L}(\mathbf{x})V^{(0)}(\mathbf{x})\rangle$. The comparison medium with the tensor $\mathbf{f}^c$ will be called by a comparison medium of the first level. In a similar manner, the medium with the material tensor $\mathbf{f}^{\text{per}}(\mathbf{x})$ will be called by a comparison medium of the second level.

Then Eq. (7.4) can be recast in the form $\nabla \mathbf{L}\boldsymbol{\varepsilon}(\mathbf{x}) = -\nabla[\mathbf{L}_2(\mathbf{x})\boldsymbol{\varepsilon}(\mathbf{x})] - \nabla[\mathbf{L}_3(\mathbf{x})\boldsymbol{\varepsilon}(\mathbf{x})]$ (at $\boldsymbol{\beta}(\mathbf{x}), \mathbf{f}(\mathbf{x}) \equiv \mathbf{0}$), leading to the operator representations equivalent to Eq. (7.19)

$$\boldsymbol{\varepsilon}(\mathbf{x}) = \langle \boldsymbol{\varepsilon} \rangle + \mathcal{P}\mathbf{U}\boldsymbol{\eta}_2 + \mathcal{P}\mathbf{U}\boldsymbol{\eta}_3, \tag{7.78}$$

where one has introduced two other strain polarization tensors $\boldsymbol{\eta}_2(\mathbf{y}) = \mathbf{L}_2(\mathbf{y})\boldsymbol{\varepsilon}(\mathbf{y})$ and $\boldsymbol{\eta}_3(\mathbf{y}) = \mathbf{L}_3(\mathbf{y})\boldsymbol{\varepsilon}(\mathbf{y})$ as well as a projector operator $\mathcal{P}$ defined as $(\mathbf{f} = \boldsymbol{\eta}_2, \boldsymbol{\eta}_3)$

$$\mathcal{P}\mathbf{U}\mathbf{f}(\mathbf{x}) := \int \mathbf{U}(\mathbf{x}-\mathbf{y})\mathbf{f}(\mathbf{y}) - \langle \mathbf{U}(\mathbf{x}-\mathbf{y})\mathbf{f}(\mathbf{y})\rangle(\mathbf{y})\, d\mathbf{y}. \tag{7.79}$$

Multiplication of Eq. (7.79) over the periodic tensor $\mathbf{L}_2(\mathbf{x})$ leads to the following equation represented in the operator form $\boldsymbol{\eta}_2 = \mathbf{L}_2\langle\boldsymbol{\varepsilon}\rangle + \mathbf{L}_2\mathcal{P}\mathbf{U}\boldsymbol{\eta}_2 + \mathbf{L}_2\mathcal{P}\mathbf{U}\boldsymbol{\eta}_3$. Following our goal of extraction of the known solution, we introduce the operator $\boldsymbol{\mathcal{A}} \equiv (\mathbf{I} - \mathbf{L}_2\mathcal{P}\mathbf{U})^{-1}$, $\boldsymbol{\mathcal{A}}\mathbf{f} \equiv \int \boldsymbol{\mathcal{A}}(\mathbf{x}-\mathbf{y})\mathbf{f}(\mathbf{y})\, d\mathbf{y}$, reducing to the multiplication over the inhomogeneous tensor $\mathbf{A}^{\text{per}}(\mathbf{x})$ describing the known periodic solution in the case of the action of this operator on the constant tensor $\langle\boldsymbol{\varepsilon}\rangle$

$$\boldsymbol{\mathcal{A}}\langle\boldsymbol{\varepsilon}\rangle = \mathbf{A}^{\text{per}}(\mathbf{x})\langle\boldsymbol{\varepsilon}\rangle. \tag{7.80}$$

Introduction of the operator $\boldsymbol{\mathcal{A}}$ allows one to decompose the desired solution in two parts

$$\boldsymbol{\varepsilon}(\mathbf{x}) = \mathbf{A}^{\text{per}}(\mathbf{x})\langle\boldsymbol{\varepsilon}\rangle + \boldsymbol{\mathcal{A}}\mathbf{L}_2\mathcal{P}\mathbf{U}\boldsymbol{\eta}_3, \tag{7.81}$$

where the first tensorial item in the right-hand side presents the periodic solution and is assumed to be known. The second operator item is produced by the random imperfections contributed in the perfect periodic structure and depends also on the periodic solution described by the operator $\boldsymbol{\mathcal{A}}$.

It should be mentioned that Eqs. (7.78)–(7.81) were obtained as a result of the new GEE (7.19). If as a starting point of the method proposed (7.78)–(7.81) is chosen, the classical GIE (7.32) then a single correction in Eqs. (7.78)–(7.81) will be a replacement of the order of the operators $\mathcal{P}$ and $\mathbf{U}$

$$\mathbf{U}\mathcal{P}\mathbf{f}(\mathbf{x}) := \int \mathbf{U}(\mathbf{x}-\mathbf{y})[\mathbf{f}(\mathbf{y}) - \langle\mathbf{f}\rangle]\, d\mathbf{y}, \tag{7.82}$$

i.e., in the light of Sect. 7.3, the operator $\mathbf{U}\mathcal{P}$ can be considered as an asymptotic approximation of the operator $\mathcal{P}\mathbf{U}$ (although $\mathcal{P}\mathbf{U} \neq \mathbf{U}\mathcal{P}$). Equations (7.81) and (7.82) were solved in [196] (see also [200]) for 3D initially periodic media with comparison medium of either the first level or second one; the relations with the Green's function model [355] and a hierarchical method were established [1508, 1652].

7.5 Concluding Remarks

The final goals of micromechanical research of composites involved in a prediction of both the overall effective properties and statistical moments of stress–strain fields are based on the approximate solution of the exact initial integral equation (7.8). The main difficulty of Eq. (7.8) defined by the surface integral over the boundary Γ is avoided by elimination of this integral by the use of different methods. It leads to the different general integral equations (GIEs) (7.19), (7.32), and (7.36) (and their stress–state analogs) which have different renormalizing terms. Equations either (7.19), (7.32), (7.36), or (7.73) are the theoretical basis of the locality principle by Sokolkin and Tashkinov [1425] which uses the short-range-order effect in the interactions of the periodic or random problem in the boundary-value problem for a domain with a finite number of inhomogeneities. Considering some conditional ensemble averages of the GIEs, either (7.19), (7.32), or (7.36) yields the infinite hierarchy of integral equations (e.g., Eq. (7.25) for GIE (7.19)). These truncated hierarchies of equations are solved as a system of coupled equations. One starts with the last members of these hierarchies, the ones which have the most inclusions held fixed, because these equations do not depend on the others. The fields so obtained give the previous terms in the next equations up the hierarchies. One continues step by step up the hierarchies until the unconditionally averaged fields are finally obtained. However, these standard procedures (differing by both the numbers of coupled equations and assumptions exploited for their solutions; see Chaps. 8–10 for details) have fundamentally diverse backgrounds defined by the features of the renormalizing terms in GIEs either (7.19), (7.32) or (7.36). These features of the initial GIEs are fundamental for subsequently solving the truncated hierarchy involving a rearrangement of each appropriate equation before it is solved. The most successful rearrangements are those which make the right-hand side of the coupled equations reflect the detailed corrections to that basic physics. So, the advantage of Eq. (7.32) with respect to Eq. (7.28) (even for the case when Eq. (7.28) is correct) is explained by faster convergence of the corresponding integrals as $|\mathbf{x}-\mathbf{y}| \to \infty$. The centering method realized by obtaining Eq. (7.32) subtracts the difficult state at infinity from the equation, i.e., roughly speaking, the constant force-dipole density expressed through an alternative technique of the Green's function. This dictates the fundamental limitation of possible generalization of Eq. (7.32) to both the FGMs and inhomogeneous boundary conditions. The mentioned deficiency of Eq. (7.32) was resolved by Eq. (7.36) whose renormalizing term provides an absolute convergence of the integral in Eq. (7.36) at $|\mathbf{x}-\mathbf{y}| \to \infty$ for general cases of the FGMs. However, the same term in Eq. (7.36) is used in a short-range domain $|\mathbf{x}-\mathbf{y}| < 3a$ in the vicinity of the point $\mathbf{x} \in w$. A fundamental deficiency of Eq. (7.36) is a dependence of the renormalizing term $\mathbf{U}(\mathbf{x}-\mathbf{y})\langle\boldsymbol{\tau}(\mathbf{y})\rangle(\mathbf{y})$ [obtained in the framework of the asymptotic approximation (7.51) which is, in fact, the effective field hypothesis **H1b** (7.47)] only on the statistical average $\langle\boldsymbol{\tau}\rangle(\mathbf{y})$ rather than $\langle\mathbf{U}(\mathbf{x}-\mathbf{y})\boldsymbol{\tau}(\mathbf{y})\rangle(\mathbf{y})$. However, the average tensor $\langle\boldsymbol{\tau}\rangle(\mathbf{y})$ ignores any microtopological and material inhomogeneities inside and in vicinity of inclusions (possible nonellipsoidal and laminated structure of inhomogeneities, the shape and size of excluded volume, radial distribution function, orientation, imperfect interface, etc.), while this information directly impacts on the renormalizing term $\langle \mathbf{U}(\mathbf{x}-\mathbf{y})\tau(\mathbf{y})\rangle(\mathbf{y})$ that is critical especially in a short-range domain $|\mathbf{x}-\mathbf{y}| < 3a$. What seems to be only a formal trick [abandoning the use of the approximations of the kind (7.47)] is in reality a new background of analytical micromechanics (7.19) [which does not use an approximation of the kind either (7.46) or regularization (7.34) as in Eqs. (7.32) and (7.30), respectively] yielding the discovery of fundamentally new effects even in the theory of statistically homogeneous media subjected to homogeneous boundary conditions (see for details Chaps. 9–12).

Chapter 8
Multiparticle Effective Field and Related Methods in Micromechanics of Random Structure Composites

Abstract Among the numerous methods in micromechanics mentioned in Chap. 1 (see Willis, J Appl Mech, 50:1202–1209, 1983), in most details, we will consider in this chapter the self-consistent methods based on some mathematical approximations for solving the classical GIE for statistically homogeneous CMs analyzed in Chap. 7. The notion of an effective field in which each particle is located is a basic concept of such powerful methods in micromechanics as the methods of self-consistent fields, effective fields, and its generalization a multiparticle effective field method (MEFM) which includes as the particular cases the methods of the effective medium, Mori-Tanaka, conditional momenta, differential, and Maxwell. Numerical examples for estimations of effective moduli and stress concentration factors are considered for the matrix CMs containing spheroidal inclusions, noncanonical fibers with prescribed random orientations, polifractional spherical inclusions, and clusters of spheroidal inclusions.

The prediction of the behavior composite materials by the use of mechanical properties of constituents and their microstructure is a central problem of micromechanics that is eventually reduced to the estimation of stress fields in constituents. A considerable number of methods are known in the linear theory of composites that yield the effective elastic constants and stress field averages in the components. Appropriate, but by no means exhaustive, references are provided by the reviews [178, 188, 260, 336, 737, 746, 765, 808, 1012, 1075, 1104, 1121, 1378, 1523, 1625, 1628]. It appears today that variants of the effective medium method [643, 812] and the mean field method [89, 1093] are the most popular and widely used methods. The concept of the effective field in combination with subsequent assumptions originated in the physics of multiple scattering of waves (see, e.g., [290, 493, 885]) was intensively applied in micromechanics of random and periodic structure composites (for references see, e.g., [189, 1094] [737, 746]) as well in micromechanics of multiple interacting cracks under the name traction (for reference, see, e.g., [729]) or pseudo-load [654]. The "quasi-crystalline" approximation of Lax [886] is often used for truncation of the hierarchy of integral equations involved in leading to neglect of direct multiparticle interactions of inclusions. The last deficiency was

V. A. Buryachenko, *Local and Nonlocal Micromechanics of Heterogeneous Materials*, https://doi.org/10.1007/978-3-030-81784-8_8

overcome recently by the multiparticle effective field method (MEFM), put forward and developed by the author (references may be found in the survey by Buryachenko [188]). The MEFM is based on the theory of functions of random variables and Green's functions. Within this method a hierarchy of statistical moment equations for conditional averages of the stresses in the inclusions is derived. The hierarchy is established by introducing the notion of an effective field. In this way the interaction of different inclusions is taken directly into account.

8.1 Definitions of Effective Fields and Effective Field Hypotheses

The infinite systems of the integral equations for both the effective fields (7.42)–(7.44) and statistical average fields (7.45) are valid at sufficient distance $\mathbf{x}$ from the boundary ∂w for statistically homogeneous and inhomogeneous media both for the general case of the first and second boundary value problems and for the mixed boundary-value problem in the space dimensionality d ($d = 2, 3$). For the sake of definiteness, in $2D$ case, we will consider a plane-strain problem (see also footnote in Sect. 4.2.1). We will consider the statistical description presented in Sect. 5.3.3 (and used in Chap. 7) in terms of the conditional probability densities $\varphi(v_i, \mathbf{x}_i | v_1, \mathbf{x}_1, \dots, v_n, \mathbf{x}_n)$. The composites being analyzed in this chapter have statistically homogeneous structure (5.20) and are subjected to the homogeneous boundary conditions either (6.4) or (6.5) so that $\langle \boldsymbol{\sigma} \rangle(\mathbf{x})$, $\langle \boldsymbol{\varepsilon} \rangle(\mathbf{x}) \equiv$ const.

Equations (7.45) are much easier to solve when the stress–strain fields are studied only inside the inclusions. There are two fundamentally different assumptions (7.26) and (7.27) to ensuring the integration on the right-hand sides of Eqs. (7.45) extends only over the volume of the inclusion. We will utilize more general assumption (7.27) ensuring that the integration in the right-hand side of Eq. (7.45) is only taken within the volume of inclusions. Here $\boldsymbol{\eta}^1$ is called the modified strain polarization tensor, and one introduced the operation $\boldsymbol{\lambda}^1(\mathbf{x}) \equiv \boldsymbol{\lambda}(\mathbf{x}) - \langle \boldsymbol{\lambda} \rangle^{(0)}$ for the random function $\boldsymbol{\lambda}$ (e.g., $\boldsymbol{\lambda} = \boldsymbol{\sigma}, \boldsymbol{\varepsilon}, \boldsymbol{\eta}$) with statistical average in the matrix $\langle \boldsymbol{\lambda} \rangle^{(0)} \equiv$ const.

8.1.1 *Approximate Effective Field Hypothesis*

To simplify the exact system for the statistical averages of stresses (7.45), we now apply the main hypothesis of many micromechanical methods, the approximation called the effective field hypothesis **H1a** (4.94) and **H1b** (4.95).

According to hypothesis **H1a** (4.94), each inclusion v_i being considered is located in a homogeneous field $\overline{\boldsymbol{\sigma}}(\mathbf{x}_i)$ that significantly simplifies the problem (7.42)

$$\boldsymbol{\sigma}(\mathbf{x}) = \overline{\boldsymbol{\sigma}}_i(\mathbf{x}) + \int \boldsymbol{\Gamma}(\mathbf{x}-\mathbf{y})V_i(\mathbf{y})\big[\mathbf{M}_1(\mathbf{y})\boldsymbol{\sigma}(\mathbf{y}) + \boldsymbol{\beta}_1(\mathbf{y}) - \langle\boldsymbol{\eta}\rangle^{(0)}\big]\,d\mathbf{y}. \tag{8.1}$$

In view of the linearity of the problem, there exist constant fourth- and second-rank tensors $\mathbf{B}^{(i)}(\mathbf{x})$, $\mathbf{R}^{(i)}(\mathbf{x})$, and $\mathbf{C}^{(i)}(\mathbf{x})$, $\mathbf{F}^{(i)}(\mathbf{x})$, such that (see Eqs. (3.175) for comparison)

$$\begin{aligned}\boldsymbol{\sigma}(\mathbf{x}) &= \mathbf{B}^{(i)}(\mathbf{x})\overline{\boldsymbol{\sigma}}(\mathbf{x}_i) + \mathbf{C}^{(i)}(\mathbf{x}),\\ \bar{v}_i\boldsymbol{\eta}^1(\mathbf{x}) &= \mathbf{R}^{(i)}(\mathbf{x})\overline{\boldsymbol{\sigma}}(\mathbf{x}_i) + \mathbf{F}^{(i)}(\mathbf{x}),\end{aligned} \tag{8.2}$$

where $\mathbf{x} \in v_i,\ \ v_i \subset v^{(i)}$ and

$$\begin{aligned}\mathbf{C}^{(i)}(\mathbf{x}) &\equiv \overline{\mathbf{C}}^{(i)}(\mathbf{x}) + \mathbf{B}_i(\mathbf{x})\mathbf{Q}_i\langle\boldsymbol{\eta}\rangle^{(0)}, \quad \mathbf{R}^{(i)}(\mathbf{x}) = \bar{v}_i\mathbf{M}_1^{(i)}(\mathbf{x})\mathbf{B}^{(i)}(\mathbf{x}),\\ \mathbf{F}^{(i)}(\mathbf{x}) &= \bar{v}_i[\mathbf{M}_1^{(i)}(\mathbf{x})\mathbf{C}^{(i)}(\mathbf{x}) + \boldsymbol{\beta}_1^{(i)}(\mathbf{y}) - \langle\boldsymbol{\eta}\rangle^{(0)}],\end{aligned} \tag{8.3}$$

here the tensor $\overline{\mathbf{C}}^{(i)}(\mathbf{x})$ introduced does not depend on $\langle\boldsymbol{\eta}\rangle^{(0)}$. According to Eshelby's theorem ([450]; see Chap. 3), the following relations exist between the averaged tensors (8.2_1) and (8.2_2):

$$\mathbf{R}_i = \overline{v}_i\mathbf{Q}_i^{-1}(\mathbf{I}-\mathbf{B}_i), \quad \mathbf{F}_i = -\overline{v}_i\mathbf{Q}_i^{-1}\mathbf{C}_i \equiv \overline{\mathbf{F}}_i + \widetilde{\mathbf{F}}_i\langle\boldsymbol{\eta}\rangle^{(0)}, \tag{8.4}$$

where $\mathbf{f}_i \equiv \langle\mathbf{f}(\mathbf{x})\rangle_{(i)}$ ($\mathbf{f}$ stands for $\mathbf{B}, \mathbf{C}, \overline{\mathbf{C}}, \mathbf{R}, \mathbf{F}$), the tensor $\mathbf{Q}_i = -\langle\boldsymbol{\Gamma}(\mathbf{x}-\mathbf{y})\rangle_{(i)},\ \mathbf{x},\mathbf{y} \in v_i$ (3.38) is associated with the well-known Eshelby tensor by $\mathbf{S}_i = \mathbf{I} - \mathbf{M}^{(0)}\mathbf{Q}_i$, and the tensors $\overline{\mathbf{F}}_i \equiv -\overline{v}_i\mathbf{Q}_i^{-1}\overline{\mathbf{C}}_i$ and $\widetilde{\mathbf{F}}_i \equiv -\overline{v}_i\mathbf{Q}_i^{-1}\mathbf{B}_i\mathbf{Q}_i$ were introduced for separation of influences of tensors $\boldsymbol{\beta}_1^{(i)}$ and $\langle\boldsymbol{\eta}\rangle^{(0)}$ on the tensor $\mathbf{F}_i$. It should be mentioned that the field $\overline{\boldsymbol{\sigma}}(\mathbf{x}_i)$ can vary with variation of the center $\mathbf{x}_i$ of the inclusion considered, but the field $\overline{\boldsymbol{\sigma}}(\mathbf{y})$ ($\mathbf{y} \in v_i$) is homogeneous over the inclusion v_i. Because of this the application of Eshelby's theorem is correct.

For example, for the homogeneous ellipsoidal domain v_i (3.183), we get

$$\begin{aligned}\mathbf{B}_i &= \left(\mathbf{I} + \mathbf{Q}_i\mathbf{M}_1^{(i)}\right)^{-1}, \quad \mathbf{C}_i = -\mathbf{B}_i\mathbf{Q}_i(\boldsymbol{\beta}_1^{(i)} - \langle\boldsymbol{\eta}\rangle^{(0)}),\\ \mathbf{R}_i &= \bar{v}_i\mathbf{M}_1^{(i)}\mathbf{B}_i, \quad \mathbf{F}_i = \bar{v}_i(\mathbf{I} + \mathbf{M}_1^{(i)}\mathbf{Q}_i)^{-1}(\boldsymbol{\beta}_1^{(i)} - \langle\boldsymbol{\eta}\rangle^{(0)}).\end{aligned} \tag{8.5}$$

In the general case of inclusions v_i, the tensors $\mathbf{B}(\mathbf{x})$ and $\mathbf{C}(\mathbf{x})$ can be found numerically according to the schemes (4.15) and (4.16). For particular cases of coated inclusions, different analytical models were considered in Sect. 3.7.

Averaging (7.45) over the volume of the considered inclusion v_i and using the hypothesis **H1b** (4.95) with (8.2) leads to the infinite system of integral equations

$$\bar{v}_i\langle\boldsymbol{\eta}^1(\mathbf{x})|v_1,\mathbf{x}_1;\ldots;v_n,\mathbf{x}_n\rangle_i - \sum_{j\neq i}^{n}\mathbf{R}_i\mathbf{T}_{ij}(\mathbf{x}_i-\mathbf{x}_j)\langle\boldsymbol{\eta}^1(\mathbf{y})|v_1,\mathbf{x}_1;\ldots;v_n,\mathbf{x}_n\rangle_j$$

$$= \bar{v}_i \boldsymbol{\eta}_i^{10} + \mathbf{R}_i \int \Big[\mathbf{T}_{iq}(\mathbf{x}_i - \mathbf{x}_q) \langle \boldsymbol{\eta}^1(\mathbf{y}) |; v_1, \mathbf{x}_1; \ldots; v_n, \mathbf{x}_n \rangle_q$$

$$\times \varphi(v_q, \mathbf{x}_q |; v_1, \mathbf{x}_1; \ldots; v_n, \mathbf{x}_n) - \mathbf{T}_i(\mathbf{x}_i - \mathbf{x}_q) \langle \boldsymbol{\eta}^1 \rangle n^{(1)} \Big] \, d\mathbf{x}_q, \tag{8.6}$$

where $\bar{v}_i \boldsymbol{\eta}_i^{10} = \mathbf{R}_i \langle \boldsymbol{\sigma} \rangle + \mathbf{F}_i - \bar{v}_i \langle \boldsymbol{\eta} \rangle^{(0)}$ is called the external modified strain polarization tensor in the component $v^{(i)}$.

Using hypothesis **H1**, the system (7.43) for k fixed inclusions with fixed values $\widetilde{\boldsymbol{\sigma}}_{1,\ldots,k}(\mathbf{x})$ $(\mathbf{x} \in v_i,\ i = 1.\ldots, k)$ on the right-hand side of the equations becomes algebraic when the solution (8.2) for one inclusion in the field $\overline{\boldsymbol{\sigma}}(\mathbf{x}_i)$ $(i = 1, \ldots, k)$ is applied

$$\mathbf{R}_i \overline{\boldsymbol{\sigma}}(\mathbf{x}_i) + \mathbf{F}_i = \sum_{j=1}^{k} \mathbf{Z}_{ij} \Big\{ \mathbf{R}_j \widetilde{\boldsymbol{\sigma}}_{1,\ldots,k}(\mathbf{x}_j) + \mathbf{F}_j \Big\}. \tag{8.7}$$

Here the matrix $\mathbf{Z}^{-1}$ has the elements $(\mathbf{Z}^{-1})_{ij} = \mathbf{I}\delta_{ij} - (1 - \delta_{ij}) \mathbf{R}_j \mathbf{T}_{ij}(\mathbf{x}_i - \mathbf{x}_j)$, $(i, j = 1, \ldots, n)$ (4.112). Buryachenko considered [190] (see also Sect. 4.3) the possible refinement of the multiple inclusion interaction problem (7.42) and (7.43) and analyzed the related simplified methods.

8.1.2 Closing Effective Field Hypothesis

For termination of the hierarchy of statistical moment equations (7.45), we will use the closing effective field hypothesis:

Hypothesis (H2a) *For a sufficiently large n, we complete the system (7.45) by the assumption*

$$\langle \widetilde{\boldsymbol{\sigma}}_{1,\ldots,j,\ldots,n+1} \rangle_i(\mathbf{x}) = \langle \widetilde{\boldsymbol{\sigma}}_{1,\ldots,n} \rangle_i(\mathbf{x}) \ \text{ or}$$

$$\langle \widetilde{\boldsymbol{\varepsilon}}_{1,\ldots,j,\ldots,n+1} \rangle_i(\mathbf{x}) = \langle \widetilde{\boldsymbol{\varepsilon}}_{1,\ldots,n} \rangle_i(\mathbf{x}), \tag{8.8}$$

where the right-hand side of the equality does not contain the index $j \neq i$ $(i = 1, \ldots, n;\ j = 1, \ldots, n+1;\ \mathbf{x} \in v_i)$.

In particular, for two fixed inclusions, Eqs. (8.8) are reduced to the following closing assumptions:

$$\langle \widetilde{\boldsymbol{\sigma}}_{i,j} \rangle(\mathbf{x}) = \langle \overline{\boldsymbol{\sigma}}_k \rangle(\mathbf{x}), \ \ \langle \widetilde{\boldsymbol{\varepsilon}}_{i,j} \rangle(\mathbf{x}) = \langle \overline{\boldsymbol{\varepsilon}}_k \rangle(\mathbf{x}), \tag{8.9}$$

at $\mathbf{x} \in v_k$, $k = i, j$.

The hypothesis **H2a**, rewritten in terms of stresses $\boldsymbol{\sigma}(\mathbf{x})$, $(\mathbf{x} \in v_i)$, is a standard closing assumption (see, e.g., [762, 763, 1625]) degenerating to the "quasi-crystalline" approximation by Lax [886] at $n = 1$ (see also [257, 1240] and Sect. 8.3.1).

Hypothesis (H2b) "Quasi-Crystalline" Approximation *It is supposed that the mean value of the effective field (or field) at a point* $\mathbf{x} \in v_i$ *does not depend on the effective field (or field) inside surrounding heterogeneities* $v_j \neq v_i$, $\mathbf{x} \in v_k$, $(k = i, j)$

$$\begin{aligned}
\langle \overline{\boldsymbol{\sigma}} | v_i, \mathbf{x}_i; v_j, \mathbf{x}_j \rangle(\mathbf{x}) &= \langle \overline{\boldsymbol{\sigma}}_k \rangle(\mathbf{x}) \equiv \text{const}, \\
\langle \overline{\boldsymbol{\varepsilon}} | v_i, \mathbf{x}_i; v_j, \mathbf{x}_j \rangle(\mathbf{x}) &= \langle \overline{\boldsymbol{\varepsilon}}_k \rangle(\mathbf{x}) \equiv \text{const}, \quad \text{or} \\
\langle \boldsymbol{\sigma} | v_i, \mathbf{x}_i; v_j, \mathbf{x}_j \rangle(\mathbf{x}) &= \langle \boldsymbol{\sigma}_k \rangle(\mathbf{x}) \equiv \text{const}, \\
\langle \boldsymbol{\varepsilon} | v_i, \mathbf{x}_i; v_j, \mathbf{x}_j \rangle(\mathbf{x}) &= \langle \boldsymbol{\varepsilon}_k \rangle(\mathbf{x}) \equiv \text{const}.
\end{aligned} \tag{8.10}$$

It should be mentioned that the basic closing assumption of the one-particle approximation of the MEFM (called effective field method, EFM) states: "The random variables $\overline{\boldsymbol{\sigma}}_i$ do not depend statistically on the geometric characteristics and elastic constants of the defect v_i" (the citation from Kanaun [738]; see also [746]) that necessarily leads to the "quasi-crystalline" approximation (8.10_1). We will see in Sect. 8.3.1, however, that $\overline{\boldsymbol{\sigma}}_i$ is explicitly dependent on the correlation hole shape v_{ij}^0 which in turn depends on the shape of v_i.

The solution obtained for one inclusion (8.7) and a finite number of inclusions (8.7), which are located in effective fields $\overline{\boldsymbol{\sigma}}(\mathbf{x}_i)$ and $\widetilde{\boldsymbol{\sigma}}_{1,\ldots,n}(\mathbf{x})$, respectively, and the adoption of hypothesis **H2** make it possible to solve the closing system of coupled integral equations [following from Eqs. (7.21)–(7.25) in the framework of the hypothesis **H1b** (4.94) and (4.112)], which is reduced to

$$\begin{aligned}
\langle \widetilde{\boldsymbol{\sigma}}_{1,\ldots,j}(\mathbf{x}) \rangle_i &= \langle \boldsymbol{\sigma} \rangle + \int \Big[\mathbf{T}_{iq}(\mathbf{x}_i - \mathbf{x}_q) \varphi(v_q, \mathbf{x}_q | ; v_1, \mathbf{x}_1; \ldots; v_j, \mathbf{x}_j) \sum_{l=1}^{j+1} \mathbf{Z}_{ql} \\
&\quad \times (\mathbf{R}_l \langle \widetilde{\boldsymbol{\sigma}}_{1,\ldots,j+1}(\mathbf{x}) \rangle_l + \mathbf{F}_l) - \mathbf{T}_i(\mathbf{x}_i - \mathbf{x}_q) \langle \mathbf{R} \overline{\boldsymbol{\sigma}} + \mathbf{F} \rangle n^{(q)} \Big] d\mathbf{x}_q, \\
\langle \widetilde{\boldsymbol{\sigma}}_{1,\ldots,n}(\mathbf{x}) \rangle_i &= \langle \boldsymbol{\sigma} \rangle + \int \Big[\mathbf{T}_{iq}(\mathbf{x}_i - \mathbf{x}_q) \varphi(v_q, \mathbf{x}_q | ; v_1, \mathbf{x}_1; \ldots; v_j, \mathbf{x}_j) \sum_{l=1}^{n} \mathbf{Z}_{ql} \\
&\quad \times (\mathbf{R}_l \langle \widetilde{\boldsymbol{\sigma}}_{1,\ldots,n}(\mathbf{x}) \rangle_l + \mathbf{F}_l) - \mathbf{T}_i(\mathbf{x}_i - \mathbf{x}_q) \langle \mathbf{R} \overline{\boldsymbol{\sigma}} + \mathbf{F} \rangle n^{(q)} \Big] d\mathbf{x}_q.
\end{aligned} \tag{8.11}$$

The tensor $\widetilde{\boldsymbol{\sigma}}_{1,\ldots,n}(\mathbf{x})$ on the right-hand side of the last equation of (8.10) is formed from the tensor $\widetilde{\boldsymbol{\sigma}}_{1,\ldots,n}(\mathbf{x})$ on the left-hand side by replacing one of the indices by q. System (8.11) is linear with respect to $\langle \widetilde{\boldsymbol{\sigma}}_{1,\ldots,j}(\mathbf{x}) \rangle_l$ $(j = 1, \ldots, n; l = 1, \ldots, j)$, and each j-th row with $\langle \widetilde{\boldsymbol{\sigma}}_{1,\ldots,j}(\mathbf{x}) \rangle_l$ contains j equations, since $i = 1, \ldots, j$. The value of $\langle \widetilde{\boldsymbol{\sigma}}_{1,\ldots,n}(\mathbf{x}) \rangle_i$ $(i = 1, \ldots, n)$ can be estimated from the last row in

Eq. (8.11) by the method of successive approximation with all possible positions of the inclusions $v_1, \ldots, v_n$, It is also necessary to take into account the fact that $\langle\widetilde{\boldsymbol{\sigma}}_{1,\ldots,j}(\mathbf{x})\rangle_i \to \langle\overline{\boldsymbol{\sigma}}\rangle_i$ as $|\mathbf{x}_l - \mathbf{x}_i| \to \infty,\ j = 1, \ldots, j;\ l \neq i$. We substitute the value found for $\langle\widetilde{\boldsymbol{\sigma}}_{1,\ldots,n}(\mathbf{x})\rangle_i$ into the right-hand side of the $(n-1)$-th row of system (8.11), determine $\langle\widetilde{\boldsymbol{\sigma}}_{1,\ldots,n-1}(\mathbf{x})\rangle_i$ $(i = 1, \ldots, n-1)$, and so on.

8.1.3 *Effective Field Hypothesis and Composites with One Sort of Inhomogeneities*

We will consider in this subsection the case $\mathbf{L}^c = \mathbf{L}^{(0)}$ when $\langle\boldsymbol{\eta}\rangle^{(0)} \equiv \mathbf{0}$ and the system (8.6) has principally the same structure as the system for the pure elastic problem (with $\mathbf{F} \equiv \mathbf{0}$). Therefore, we can apply the traditional analysis procedure of purely elastic composites and represent $\langle\boldsymbol{\eta}\rangle^{(1)}$ as a linear function of the external field $\boldsymbol{\eta}^0$

$$\langle\boldsymbol{\eta}\rangle^{(1)} = \mathbf{Y}\boldsymbol{\eta}^0 \tag{8.12}$$

and therefore

$$\mathbf{R}\langle\overline{\boldsymbol{\sigma}}\rangle^{(1)} + \mathbf{F} = \mathbf{Y}(\mathbf{R}\boldsymbol{\sigma}^0 + \mathbf{F}). \tag{8.13}$$

Comparison of (8.6) with (8.12) leads to the fact that $\mathbf{Y}$ depends only on the tensors $\mathbf{R}$, $\mathbf{T}_{ij}$, and $\mathbf{T}_j$. The tensor $\mathbf{Y}$ is determined by the purely elastic action (with $\mathbf{F} \equiv \mathbf{0}$) of the surrounding inclusions on the separated one. For a dilute concentration of the inclusions, i.e., $c^{(1)} \to 0$, we have $\mathbf{Y} \to \mathbf{I}$. The actual form of the tensor $\mathbf{Y}$, used in the analysis as an approximation, depends on additional assumptions for closing of the infinite system (8.6). In particular, for purely elastic composites (with $\boldsymbol{\beta}_1 \equiv \mathbf{0}$) with fictitious homogeneous inclusions (3.185), (3.186), such relations will be represented for commonly applied methods of micromechanics, i.e., the effective medium method by Kröner [813] and by Hill [643], Mori–Tanaka [1093] method, MEFM.

Taking into account Eqs. (8.2) and (8.12), from the solution of the purely elastic problem for the composite, we can calculate the average stresses inside the inclusions by

$$\langle\boldsymbol{\sigma}\rangle^{(1)} = \mathbf{B}\mathbf{R}^{-1}\Big[\mathbf{Y}(\mathbf{R}\boldsymbol{\sigma}^0 + \mathbf{F}) - \mathbf{F}\Big] + \mathbf{C}. \tag{8.14}$$

The mean matrix stresses follow simply from the condition $\langle\boldsymbol{\sigma}\rangle = \boldsymbol{\sigma}^0$

$$\langle\boldsymbol{\sigma}\rangle^{(0)} = (c^{(0)})^{-1}\Big[\boldsymbol{\sigma}^0 - c^{(1)}(\mathbf{I}-\mathbf{B})^{-1}\mathbf{C} - c^{(1)}\mathbf{B}\mathbf{R}^{-1}\mathbf{Y}(\mathbf{R}\boldsymbol{\sigma}^0 + \mathbf{F})\Big]. \tag{8.15}$$

The local stresses inside the inclusion, i.e., in the core and in the coating, respectively, are found by

$$\langle\boldsymbol{\sigma}\rangle_i(\mathbf{x}) = \mathbf{C}(\mathbf{x}) + \mathbf{B}(\mathbf{x})\mathbf{R}^{-1}\Big[\mathbf{Y}(\mathbf{R}\boldsymbol{\sigma}^0 + \mathbf{F}) - \mathbf{F}\Big], \tag{8.16}$$

where $\langle\boldsymbol{\sigma}\rangle_i(\mathbf{x})$ is the average of the local stress state at $\mathbf{x} \in v_i \subset v^1$ over an ensemble realization of surrounding inclusions (but not over the volume v_i of a particular inclusion, in contrast to $\langle\boldsymbol{\sigma}\rangle^{(1)}$).

Comparing (6.35) with (8.16) leads to the relation for the average local stress concentration tensor inside the inclusions

$$\langle\boldsymbol{\sigma}\rangle_i(\mathbf{x}) = \langle\mathbf{B}^*\rangle_i(\mathbf{x})\boldsymbol{\sigma}^0 + [\langle\mathbf{B}^*\rangle_i(\mathbf{x}) - \mathbf{B}(\mathbf{x})]\mathbf{R}^{-1}\mathbf{F} + \mathbf{C}(\mathbf{x}), \tag{8.17}$$

where $\langle\mathbf{B}^*\rangle_i(\mathbf{x}) \equiv \mathbf{B}(\mathbf{x})\mathbf{D}$, $(\mathbf{x} \in v_i)$, and the tensor $\mathbf{D} \equiv \mathbf{R}^{-1}\mathbf{Y}\mathbf{R}$ has the simple physical meaning of the action (at $\boldsymbol{\beta}(\mathbf{x}) \equiv \mathbf{0}$) of the surrounding inclusions on the separated one: $\langle\overline{\boldsymbol{\sigma}}\rangle_i = \mathbf{D}\boldsymbol{\sigma}^0$.

Thus, the average of local thermal stresses $\langle\boldsymbol{\sigma}^{II}\rangle_i(\mathbf{x})$ (6.36) over the ensemble is defined by the purely elastic solution for the composite medium (the tensor $\langle\mathbf{B}^*\rangle_i(\mathbf{x})$) as well as by the thermoelastic solution for a single inclusion in an infinite matrix (the tensors $\mathbf{B}(\mathbf{x})$, $\mathbf{C}(\mathbf{x})$, $\mathbf{R}$, $\mathbf{F}$). For two-component composites with identical homogeneous inclusions of any shape, Benveniste and Dvorak [96] obtained an exact relation for nonaveraged local thermal stresses (6.35) and (6.119)

$$\boldsymbol{\sigma}^{II}(\mathbf{x}) = [\mathbf{B}^*(\mathbf{x}) - \mathbf{I}][\mathbf{M}_1^{(1)}]^{-1}\boldsymbol{\beta}_1^{(1)},$$

from which one can derive Eq. (8.17) in the case of ellipsoidal inclusions.

The matrix of stresses in the immediate vicinity of the inclusions v_i, denoted by $\boldsymbol{\sigma}_i^-(\mathbf{n})$, is given by the formula (3.97). Equations (8.16) and (3.97) allow the estimation of the ensemble average of the matrix stresses in the vicinity of the inclusions near a point $\mathbf{x} \in v_i$ with the unit outward normal vector on ∂v_i

$$\begin{aligned}\langle\boldsymbol{\sigma}^-(\mathbf{n})\rangle_x = &\Big[\mathbf{I} + \boldsymbol{\Gamma}^{(0)}(\mathbf{n})\mathbf{M}_1^{(1)}(\mathbf{x})\Big]\Big\{\mathbf{C}(\mathbf{x}) + \mathbf{B}(\mathbf{x})\mathbf{R}^{-1}\Big[\mathbf{Y}(\mathbf{R}\boldsymbol{\sigma}^0 + \mathbf{F}) - \mathbf{F}\Big]\Big\}\\ &+\boldsymbol{\Gamma}(\mathbf{n})\boldsymbol{\beta}_1^{(1)}(\mathbf{x}).\end{aligned} \tag{8.18}$$

For the estimation of the effective compliance, we use relations (6.21) and (8.16) to obtain

$$\mathbf{M}^* = \mathbf{M}^{(0)} + \mathbf{Y}\mathbf{R}n^{(1)}. \tag{8.19}$$

Taking the equality $\boldsymbol{\beta}^* = \boldsymbol{\beta}^{(0)} + \langle\mathbf{M}_1\boldsymbol{\sigma}^{II} + \boldsymbol{\beta}_1\rangle$ (see, e.g., [243] as well as Eq. (6.46)) into account, we find the coefficient of thermal expansion from the relations (8.2) and (8.16): $\boldsymbol{\beta}^* = \boldsymbol{\beta}^{(0)} + \mathbf{Y}\mathbf{F}n^{(1)}$. Equations ($6.79_3$) and (8.16) yield the following relation for the stored energy density:

$$2W^* = -\langle \boldsymbol{\beta}_1(\mathbf{x})\mathbf{C}(\mathbf{x})\rangle - \langle \boldsymbol{\beta}_1(\mathbf{x})\mathbf{B}(\mathbf{x})\rangle \mathbf{R}^{-1}(\mathbf{Y}-\mathbf{I})\mathbf{F}.$$

Interestingly, all three effective quantities $\mathbf{M}^*,\ \boldsymbol{\beta}^*,\ W^*$ can be estimated by the use of unique scheme. For example, for the estimation of the effective compliance $\mathbf{M}^*$, as a first step, it is necessary to solve the elastic problem for a single inclusion in an infinite matrix (e.g., to find the tensor $\mathbf{B}(\mathbf{x})$, see, e.g., Eqs. (4.15) and (4.19) as well as Sect. 3.7.2), and in a second step, the single constant tensor $\mathbf{Y}$ is found from the purely elastic problem ($\boldsymbol{\beta}(\mathbf{x}) \equiv 0$) for the composite with ellipsoidal inclusions with the tensor $\mathbf{R}$ (8.3$_2$). Analogously the problem of evaluating the effective tensors $\boldsymbol{\beta}^*,\ W^*$ and the average stresses inside the components (8.15), (8.17) can be fully solved if one of the two parameters, $\mathbf{Y}$ or $\mathbf{M}^*$, is found (e.g., experimentally). For a proof, assume that $\mathbf{M}^*$ is known. Then (8.19) yields $\mathbf{Y} = (\mathbf{M}^* - \mathbf{M}^{(0)})(\mathbf{R}n^{(1)})^{-1}$, and therefore

$$\boldsymbol{\beta}^* = \boldsymbol{\beta}^{(0)} + (\mathbf{M}^* - \mathbf{M}^{(0)})\mathbf{R}^{-1}\mathbf{F}, \tag{8.20}$$

$$2W^* = -\langle \boldsymbol{\beta}_1(\mathbf{x})\mathbf{B}(\mathbf{x})\rangle \mathbf{R}^{-1}\left[\mathbf{M}^* - \mathbf{M}^{(0)} - \mathbf{R}n^{(1)}\right](\mathbf{R}n^{(1)})^{-1}\mathbf{F} - \langle \boldsymbol{\beta}_1(\mathbf{x})\mathbf{C}(\mathbf{x})\rangle. \tag{8.21}$$

Equation (8.20) can be rewritten in an alternative elegant word formulation:

Theorem *In the framework of the hypothesis* ***H1****, the ratio of the increment of the effective thermal expansion to the increment of the effective compliance is a constant tensor defined by the averaging solution for a single coated inclusion in an infinite matrix (8.4) and does not depend on the concrete statistically homogeneous microstructure of the whole composite material and the detailed microstructure of individual inclusions as well*

$$(\mathbf{M}^* - \mathbf{M}^{(0)})^{-1}(\boldsymbol{\beta}^* - \boldsymbol{\beta}^{(0)}) = \mathbf{R}^{-1}\mathbf{F}. \tag{8.22}$$

In particular, for homogeneous (i.e., noncoated) inclusions, the constant tensors $\mathbf{B},\ \mathbf{C},\ \mathbf{R}$, and $\mathbf{F}$ according to (8.5) are determined by the Eshelby tensor $\mathbf{S}$ and the jumps of the material property tensors $\mathbf{M}_1^{(1)},\ \boldsymbol{\beta}_1^{(1)}$. Then, from (8.20) and (8.21), the classical results (6.106) and (6.108) for two-phase composites are derived. It is not surprising that the exact relations (6.106) and (6.108) are derived from the approximate ones, i.e., (8.20) and (8.21), since the additional assumption **H1** does not expand the class of the considered materials and homogenization methods. The representations (6.106) and (6.108) are formally invariant with respect to the replacement $v^{(1)} \leftrightarrow v^{(0)}$, although this cannot be said about the relations (8.20) and (8.21), obtained for matrix structure composites with ellipsoidal inclusions.

It should be mentioned that all results of this Sect. 8.1.3 were obtained in the framework of hypothesis **H1** only. No restrictions are imposed on the concrete form of the tensor $\mathbf{Y}$ (8.12), on the microtopology of the coated inclusions, or on the inhomogeneity of the stress state in the inclusions (8.16). Moreover, the assumption of an ellipsoidal shape of the inclusion in the hypothesis **H1** was used only in order to obtain analytical solutions (8.4). This is because the tensor $\langle \boldsymbol{\Gamma}(\mathbf{x}-\mathbf{y})\rangle_i$ (3.38) is apparently homogeneous for $\mathbf{x}, \mathbf{y} \in v_i$ for an ellipsoid only (see, e.g., [500, 978]). For nonellipsoidal inclusions, one could assume that in some parts of the region $v_i^c \subset v_i$, the properties $\mathbf{M}_1(\mathbf{x}) \equiv \mathbf{0}$, $\boldsymbol{\beta}_1(\mathbf{x}) \equiv \mathbf{0}$ take place, that is sufficient to include a real nonellipsoidal inclusion $v_i \setminus v_i^c$ into an ellipsoid (with smallest possible volume) and call it the inclusion v_i with a "coating" v_i^c. The further scheme for calculating the tensors $\mathbf{B}$, $\mathbf{C}$ (8.2_1) and overall properties is the same, but the prescribed conditional distributions $\varphi(v_m|\mathbf{x}_m; \mathbf{x}_1, \ldots \mathbf{x}_n)$ will have a larger correlation hole $\cup v_{0i}$, $(i = 1, \ldots n)$ than in the real composite material. This will result in an underestimation of the computed values of $\mathbf{M}^*$ for inclusions that are softer than the matrix and in an overestimation in the opposite case.

It is noteworthy that we do not pursue the goal of presenting in this charter all known methods based on hypothesis **H1**. There are, of course, several other methods that are based on hypothesis **H1**, and no ranking between them is given here. Our main objective is to prove the general relations (8.20) and (8.21) which are valid in the framework of hypothesis **H1** only. No restrictions are imposed on the concrete statistically homogeneous microstructure of the whole composite material with a single sort of coated inclusions being analyzed as well as on the microtopology of coated inclusions or on the inhomogeneity of the stress state in the inclusions.

8.2 Multiparticle Effective Field Method

8.2.1 Average Stresses in the Components

The n-particle approximation solution of problems (8.11) presupposed a solution of systems (8.11) for arbitrary coordinates of the centers of n inclusions and arbitrary orientations of these inclusions. The two-particle approximation and the **H2a** assumption (8.9) (with $i, j = 1, 2$) greatly simplify the problem. This independence of $\langle \widetilde{\boldsymbol{\sigma}}(\mathbf{x})_{1,2}\rangle$ of the spacing between the inclusions v_1 and v_2 (8.9) occurs for the limiting case $|\mathbf{x}_1 - \mathbf{x}_2| \gg \max a_i^k$, where a_i^k $(k = 1, 2, 3;\ i = 1, 2)$ are the semi-axes of the ellipsoidal inclusions v_1 and v_2, respectively. Then the system (8.11) can be solved by analytical methods, so from (8.11), taking (8.7) and (8.9) into account, we get

$$\overline{v}_i\langle \boldsymbol{\eta}_i^1\rangle = \overline{v}_i\boldsymbol{\vartheta}_i^{10} + \mathbf{R}_i\sum_{q=1}^{N}\Big\{\int \mathbf{T}_{iq}(\mathbf{x}_i-\mathbf{x}_q)\mathbf{Z}_{qi}^{21}\varphi(v_q,\mathbf{x}_q|;v_i,\mathbf{x}_i)\,d\mathbf{x}_q\overline{v}_i\langle \boldsymbol{\eta}_i^1\rangle$$
$$= \int\Big[\mathbf{T}_{iq}(\mathbf{x}_i-\mathbf{x}_q)\mathbf{Z}_{qi}^{22}\varphi(v_q,\mathbf{x}_q|;v_i,\mathbf{x}_i) - \mathbf{T}_i(\mathbf{x}_i-\mathbf{x}_q)n^{(q)}\Big]\overline{v}_q\langle \boldsymbol{\eta}_q^1\rangle\,d\mathbf{x}_q\Big\}. \tag{8.23}$$

where the matrix elements $\mathbf{Z}^{21}_{qi}$, $\mathbf{Z}^{22}_{q1}$ are nondiagonal and diagonal elements, respectively, of the binary interaction matrix $\mathbf{Z}_{qi} = (Z^{kl}_{qi})$ $(q, i = 1.\ldots, N,$ $k, l = 1, 2)$ (4.112), (8.7) for the two inclusions v_q and v_i. The algebraic system can be solved for $\langle \boldsymbol{\vartheta}_i \rangle$

$$\overline{v}_i \langle \boldsymbol{\eta}^1_i \rangle = \sum_{j=1}^{N} \mathbf{Y}_{ij} \overline{v}_j \boldsymbol{\eta}^{10}_j, \tag{8.24}$$

where the matrix $\mathbf{Y}^{-1}$ has the following elements $(\mathbf{Y}^{-1})_{ij}$ $(i, j = 1, 2, \ldots, N)$:

$$(\mathbf{Y}^{-1})_{ij} = \delta_{ij} \left[\mathbf{I} - \mathbf{R}_i \sum_{q=1}^{N} \int \mathbf{T}_{iq}(\mathbf{x}_i - \mathbf{x}_q) \mathbf{Z}^{21}_{qi} \varphi(v_q, \mathbf{x}_q |; v_i, \mathbf{x}_i)\, d\mathbf{x}_q \right]$$
$$- \mathbf{R}_i \int \left[\mathbf{T}_{ij}(\mathbf{x}_i - \mathbf{x}_j) \mathbf{Z}^{22}_{jj} \varphi(v_j, \mathbf{x}_j |; v_i, \mathbf{x}_i) - \mathbf{T}_i(\mathbf{x}_i - \mathbf{x}_j) n^{(j)} \right] d\mathbf{x}_j. \tag{8.25}$$

The statistical averages of the effective and stress fields inside the inclusions are obtained from (8.2$_1$) and (8.24)

$$\langle \overline{\boldsymbol{\sigma}} \rangle_i = \mathbf{D}_i \boldsymbol{\sigma}^0 + \mathbf{R}_i^{-1} \sum_{j=1}^{N} (\mathbf{Y}_{ij} - \mathbf{I}\delta_{ij}) \mathbf{F}_j, \quad \mathbf{D}_i = \mathbf{R}_i^{-1} \sum_{j=1}^{N} \mathbf{Y}_{ij} \mathbf{R}_j, \tag{8.26}$$

$$\langle \boldsymbol{\sigma} \rangle_i(\mathbf{x}) = \mathbf{B}_i(\mathbf{x}) \mathbf{R}_i^{-1} \left\{ \sum_{j=1}^{N} \mathbf{Y}_{ij} (\mathbf{R}_j \boldsymbol{\sigma}^0 + \mathbf{F}_j) - \mathbf{F}_i \right\} + \mathbf{C}_i(\mathbf{x}), \tag{8.27}$$

$$\overline{v}_i \langle \boldsymbol{\eta}^1_i \rangle(\mathbf{x}) = \mathbf{R}_i(\mathbf{x}) \mathbf{D}_i \boldsymbol{\sigma}^0 + \mathbf{F}_i(\mathbf{x}) + \mathbf{R}_i(\mathbf{x}) \mathbf{R}_i^{-1} \sum_{j=1}^{N} (\mathbf{Y}_{ij} - \mathbf{I}\delta_{ij}) \mathbf{F}_j, \tag{8.28}$$

$$\langle \boldsymbol{\sigma} \rangle_i = \mathbf{B}_i \left\{ \mathbf{D}_i \boldsymbol{\sigma}^0 + \mathbf{R}_i^{-1} \sum_{j=1}^{N} (\mathbf{Y}_{ij} - \mathbf{I}\delta_{ij}) \mathbf{F}_j \right\} + \mathbf{C}_i, \tag{8.29}$$

where the tensor $\mathbf{D}_i$ $(i = 1, \ldots, N)$ has a simple physical meaning of the action of surrounding inclusions on the separate one: $\langle \overline{\boldsymbol{\sigma}} \rangle_i = \mathbf{D}_i \boldsymbol{\sigma}^0$ for $\boldsymbol{\beta}(\mathbf{x}) \equiv \mathbf{0}$.

The stress in the matrix in the vicinity of the inhomogeneities v_i, $\sigma_i^-(\mathbf{n})$, is given by the formula (3.103) which holds for any shape, anisotropy, and inhomogeneity of the inclusion v_i. Equations (8.27) and (3.103) make it possible to estimate the ensemble average of the stress in the vicinity of the inhomogeneities near the point $\mathbf{x} \in v_i$ with the unit outward normal vector $\mathbf{n} \perp \partial v_i$

$$\langle \boldsymbol{\sigma}^-(\mathbf{n})\rangle_{\mathbf{x}} = \mathbf{B}_i(\mathbf{n})\mathbf{B}_i(\mathbf{x})\mathbf{D}_i\boldsymbol{\sigma}^0 + \mathbf{B}(\mathbf{n})\mathbf{B}_i(\mathbf{x})\mathbf{R}_i^{-1}\sum_{j=1}^{N}(\mathbf{Y}_{ij} - \mathbf{I}\delta_{ij})\mathbf{F}_j$$
$$+ [\mathbf{B}_i(\mathbf{n}) - \mathbf{I}][\mathbf{M}^{(i)}(\mathbf{x}) - \mathbf{M}^{(0)}]^{-1}[\boldsymbol{\beta}^{(i)}(\mathbf{x}) - \boldsymbol{\beta}^{(0)}] + \mathbf{B}_i(\mathbf{n})\mathbf{C}_i(\mathbf{x}), \tag{8.30}$$

where the tensor $\mathbf{B}_i(\mathbf{n})$ (3.103) depends only on the elastic properties of the contacting materials and the direction of the normal $\mathbf{n}$.

It should be mentioned that according to (8.3_1) and (8.4), the right-hand sides of Eqs. (8.27)–(8.29) are the linear functions of tensors $\mathbf{F}_i(\mathbf{x})$, $\mathbf{C}_i(\mathbf{x})$ $(i = 1, \ldots, N)$ depending on $\langle\boldsymbol{\sigma}\rangle^{(0)}$ which can be found by substitution of Eqs. (8.29), (8.3_1), and (8.4) into the obvious equality $\langle\boldsymbol{\sigma}\rangle^{(0)} = (\boldsymbol{\sigma}^0 - \langle\boldsymbol{\sigma}V\rangle)/c^{(0)}$

$$\langle\boldsymbol{\sigma}\rangle^{(0)} = \Big\{c^{(0)}\mathbf{I} + \sum_{i=1}^{N} c^{(i)}\mathbf{B}_i\Big[\mathbf{Q}_i + \mathbf{R}_i^{-1}\sum_{j=1}^{N}[\mathbf{Y}_{ij} - \delta_{ij}\mathbf{I}]\widetilde{\mathbf{F}}_j\Big]\mathbf{M}_1^{(0)}\Big\}^{-1}$$
$$\times\Big\{\boldsymbol{\sigma}^0 - \sum_{i=1}^{N} c^{(i)}\mathbf{B}_i\mathbf{R}_i^{-1}\sum_{j=1}^{N}\mathbf{Y}_{ij}[\mathbf{R}_j\boldsymbol{\sigma}^0 + \overline{\mathbf{F}}_j] - \sum_{i=1}^{N} c^{(i)}(\mathbf{I} - \mathbf{B}_i)^{-1}\overline{\mathbf{C}}_i\Big\}. \tag{8.31}$$

8.2.2 *Effective Properties of the Composite*

After estimating average stresses inside the inclusions (see (8.29)), the problem of calculating effective properties becomes trivial and leads, according to (6.47) and (6.48), to the following representations:

$$\mathbf{M}^* = \mathbf{M}^{(0)} + \sum_{i,j=1}^{N}\mathbf{Y}_{ij}\mathbf{R}_j n^{(i)} + \Big\{\mathbf{I} + \sum_{i,j=1}^{N}\mathbf{Y}_{ij}\widetilde{\mathbf{F}}_j n^{(i)}\Big\}\mathbf{M}_1^{(0)}$$
$$\times\Big\{c^{(0)}\mathbf{I} + \sum_{i=1}^{N} c^{(i)}\mathbf{B}_i\Big[\mathbf{Q}_i + \mathbf{R}_i^{-1}\sum_{j=1}^{N}[\mathbf{Y}_{ij} - \delta_{ij}\mathbf{I}]\widetilde{\mathbf{F}}_j\Big]\mathbf{M}_1^{(0)}\Big\}^{-1}$$
$$\times\Big\{\mathbf{I} - \sum_{i=1}^{N} c^{(i)}\mathbf{B}_i\mathbf{R}_i^{-1}\sum_{j=1}^{N}\mathbf{Y}_{ij}\mathbf{R}_j\Big\}, \tag{8.32}$$

$$\boldsymbol{\beta}^* = \boldsymbol{\beta}^{(0)} + \sum_{i,j=1}^{N}\mathbf{Y}_{ij}\overline{\mathbf{F}}_j n^{(i)} - \Big\{\mathbf{I} + \sum_{i,j=1}^{N}\mathbf{Y}_{ij}\widetilde{\mathbf{F}}_j n^{(i)}\Big\}\mathbf{M}_1^{(0)}$$

$$\times\Big\{c^{(0)}\mathbf{I}+\sum_{i=1}^{N}c^{(i)}\mathbf{B}_i\Big[\mathbf{Q}_i+\mathbf{R}_i^{-1}\sum_{j=1}^{N}[\mathbf{Y}_{ij}-\delta_{ij}\mathbf{I}]\widetilde{\mathbf{F}}_j\Big]\mathbf{M}_1^{(0)}\Big\}^{-1}$$

$$\times\Big\{\sum_{i=1}^{N}c^{(i)}\mathbf{B}_i\mathbf{R}_i^{-1}\sum_{j=1}^{N}\mathbf{Y}_{ij}\overline{\mathbf{F}}_j+\sum_{i=1}^{N}(\mathbf{I}-\mathbf{B}_i)^{-1}\overline{\mathbf{C}}_i\Big\}, \tag{8.33}$$

$$2W^*=-\sum_{i,j=1}^{N}\langle\boldsymbol{\beta}_1^{(i)}(\mathbf{x})\mathbf{B}_i(\mathbf{x})\rangle_i\mathbf{R}_i^{-1}(\mathbf{Y}_{ij}-\mathbf{I}\delta_{ij})\overline{\mathbf{F}}_j c^{(i)}$$

$$-\sum_{i=1}^{N}\langle\boldsymbol{\beta}_1^{(i)}(\mathbf{x})\mathbf{C}_i(\mathbf{x})\rangle_i c^{(i)}-\sum_{i,j=1}^{N}\langle\boldsymbol{\beta}_1^{(i)}(\mathbf{x})\mathbf{B}_i(\mathbf{x})\rangle_i\mathbf{R}_i^{-1}(\mathbf{Y}_{ij}-\mathbf{I}\delta_{ij})\overline{\mathbf{F}}_j$$

$$\times\Big\{c^{(0)}\mathbf{I}+\sum_{i=1}^{N}c^{(i)}\mathbf{B}_i\Big[\mathbf{Q}_i+\mathbf{R}_i^{-1}\sum_{j=1}^{N}[\mathbf{Y}_{ij}-\mathbf{I}\delta_{ij}]\widetilde{\mathbf{F}}_j\Big]\mathbf{M}^{(0)}\Big\}^{-1}$$

$$\times\Big\{\sum_{i=1}^{N}c^{(i)}\mathbf{B}_i\mathbf{R}_i^{-1}\sum_{j=1}^{N}\mathbf{Y}_{ij}\overline{\mathbf{F}}_j+\sum_{i=1}^{N}(\mathbf{I}-\mathbf{B}_i)^{-1}\overline{\mathbf{C}}_i\Big\}. \tag{8.34}$$

In particular, for the case when elastic moduli of comparison medium $\mathbf{M}^c$ coincide with elastic properties of matrix $\mathbf{M}^{(0)}$ (7.26), $\boldsymbol{\eta}(\mathbf{x})\equiv\mathbf{0}$ $(\mathbf{x}\in v^{(0)})$, and there is no need to postulate that the stress field in the matrix is homogeneous. Then formulae (8.32)–(8.34) are simplified substantially

$$\mathbf{M}^*=\mathbf{M}^{(0)}+\sum_{i,j=1}^{N}\mathbf{Y}_{ij}\mathbf{R}_j n^{(i)}, \tag{8.35}$$

$$\boldsymbol{\beta}^*=\boldsymbol{\beta}^{(0)}+\sum_{i,j=1}^{N}\mathbf{Y}_{ij}\mathbf{F}_j n^{(i)}, \tag{8.36}$$

$$2W^*=-\sum_{i,j=1}^{N}\langle\boldsymbol{\beta}_1^{(i)}(\mathbf{x})\mathbf{B}_i(\mathbf{x})\rangle_i\mathbf{R}_i^{-1}(\mathbf{Y}_{ij}-\mathbf{I}\delta_{ij})\mathbf{F}_j c^{(i)}$$

$$-\sum_{i=1}^{N}\langle\boldsymbol{\beta}_1^{(i)}(\mathbf{x})\mathbf{C}_i(\mathbf{x})\rangle_i c^{(i)}. \tag{8.37}$$

Equations (8.35)–(8.37) were obtained in [268] by a different method. For homogeneous inclusions (8.5), the formulae (8.35)–(8.37) are equivalent to the results derived in [238]. For homogeneous ellipsoidal inclusions and some additional assumptions, the relation (8.35) is reduced to the similar representation for effective compliance $\mathbf{M}^*$ obtained by the author [243, 246] from which follows, in turn, a more specific result for the point approximation (4.115) of identical spherical inclusions obtained more recently in [739, 746], and recast in our notation in the following form:

$$\mathbf{M}^* = \mathbf{M}^{(0)} + \mathbf{R}_1 n^{(1)} \left\{ \mathbf{I} - \mathbf{Q}_1 \mathbf{R}_1 n^{(1)} + \bar{\mathbf{J}} \right\}^{-1}, \tag{8.38}$$

$$\bar{\mathbf{J}} \equiv \int \boldsymbol{\Gamma}(\mathbf{x}_i - \mathbf{x}_q)\mathbf{R}_1 \left\{ [\mathbf{I} + \boldsymbol{\Gamma}(\mathbf{x}_i - \mathbf{x}_q)\mathbf{R}_1]^{-1} - \mathbf{I} \right\} \varphi(v_q, \mathbf{x}_q | ; v_i, \mathbf{x}_i)\, d\mathbf{x}_q. \tag{8.39}$$

Further simplifications of the representation of $\mathbf{Y}$ concern the approximate solution of the problem of binary interaction of the inclusions. So the analogous problem for homogeneous ellipsoidal inclusions was solved in [243] by the method of successive approximation (4.118_1) in the framework of point approximation of the inclusions (4.115), when for $\mathbf{Z}_{ji} = \mathbf{I}\delta_{ji} + (1 - \delta_{ji})\mathbf{R}\boldsymbol{\Gamma}(\mathbf{x}_j - \mathbf{x}_i)$, and for

$$\bar{\mathbf{J}} = \int \boldsymbol{\Gamma}(\mathbf{x}_i - \mathbf{x}_q)\mathbf{R}_1 \boldsymbol{\Gamma}(\mathbf{x}_i - \mathbf{x}_q)\mathbf{R}_1 \varphi(v_q, \mathbf{x}_q | ; v_i, \mathbf{x}_i)\, d\mathbf{x}_q,$$

an analytical representation was obtained. The general case of coated inclusions was considered in [269]. Then, for an isotropic matrix $\mathbf{L}^{(0)} = (3k^{(0)}, 2\mu^{(0)})$ and identical spherical inclusions with radius a and isotropic parameters $\mathbf{R}_1 = \frac{4}{3}\pi a^3 (3k^R, 2\mu^R)$, $\varphi(v_q | \mathbf{x}_j; \mathbf{x}_i) = ng(|\mathbf{x}_j - \mathbf{x}_i|) \equiv ng(r)$, one can obtain the following representation for the isotropic tensor:

$$\bar{\mathbf{J}} = 4\pi a^3 c^{(1)} \int_{2a}^{\infty} \mathbf{J}(r) g(r) dr, \ \ \mathbf{J}(r) = (3k^J, 2\mu^J), \tag{8.40}$$

where

$$3k^J = 2\xi^2 3k^R 2\mu^R (3k^{(0)})^2 (2\mu^{(0)})^2 r^{-4},$$

$$2\mu^J = \frac{2}{5r^4}\Big[\xi^2 3k^R 2\mu^R (3k^{(0)})^2 + (2\mu^R)^2 (2\mu^{(0)})^2 (7\gamma^2 - \frac{\eta^2}{4} + 2\xi\eta)\Big](2\mu^{(0)})^2,$$

$$\xi = \frac{1}{3k^{(0)} + 2\mu^{(0)}}, \quad \eta = \frac{1}{3\mu^{(0)}}, \quad \gamma = -\frac{3k^{(0)} + \mu^{(0)}}{3\mu^{(0)}(3k^{(0)} + 4\mu^{(0)})}.$$

8.2.3 Some Related Multiparticle Methods

Another alternative version of the multiparticle method deserves more attention. It is the one proposed and developed by Ju and coauthors [710–713, 722] for a wide class of micromechanical problems. For a clear understanding, we will reproduce the general scheme of their method in our notation that allows us to abandon some unnecessary concepts (such as an eigenstrain and a perturbed strain) and use the rigorous techniques of formal mathematics.

At first, one considers two identical spherical homogeneous inclusions (2D or 3D cases) in an infinite matrix subjected to the uniform remote stress $\boldsymbol{\sigma}^0$, or to put this another way, the system (7.42) and (7.43) is solved in the framework of hypothesis **H1** for $n = 2$, $\widetilde{\boldsymbol{\sigma}}_{1,2}(\mathbf{x}) \equiv \boldsymbol{\sigma}^0$; $\boldsymbol{\beta} \equiv \mathbf{0}$, $\mathbf{M}^c \equiv \mathbf{M}^{(0)}$ and two identical inclusions, with $a_i^1 = a_i^2 = a_i^3$, $i = 1, 2$. The solution

$$\overline{\boldsymbol{\sigma}}(\mathbf{x}_i) = (\mathbf{I} + \mathcal{T}_i(\mathbf{x}_i - \mathbf{x}_j))\boldsymbol{\sigma}^0, \ i = 1, 2, \ j = 3 - i \tag{8.41}$$

was obtained by an approximate method which is equivalent to the inversion of the matrix $(\mathbf{Z}^{-1})_{ij}$ (8.7) by the method of successive approximation [see [190] and Sect. 4.3 for detailed comparative analysis of different popular approximate methods of the solution of the system (7.42) and (7.43)]. Independently of the concrete approximate method, the asymptotic behavior at the infinity of the tensor $\mathcal{T}_i(\mathbf{x}_i - \mathbf{x}_j)$ is well-known: $\mathcal{T}_i(\mathbf{x}_i - \mathbf{x}_j) = O(|\mathbf{x}_i - \mathbf{x}_j|^{-d})$ as $|\mathbf{x}_i - \mathbf{x}_j| \to \infty$.

To obtain the ensemble-average solution of $\langle\boldsymbol{\sigma}\rangle_i$ in the composite within the context of approximate pairwise inclusion interaction, one needs to integrate (8.31) over all possible positions $\mathbf{x}_j$ for a given location $\mathbf{x}_i$ ($i = 1, 2, \ j = 3 - i$)

$$\mathbf{R}_i\langle\overline{\boldsymbol{\sigma}}\rangle(\mathbf{x}_i) = \mathbf{R}_i\Big[\mathbf{I} + \int \mathcal{T}_i(\mathbf{x}_i - \mathbf{x}_j)\varphi(v_j, \mathbf{x}_j|v_i, \mathbf{x}_i)\, d\mathbf{x}_j\Big]\boldsymbol{\sigma}^0. \tag{8.42}$$

According to the assumptions considered in [712, 713, 722], the term in the rectangular brackets in Eq. (8.42) is reduced to the tensor $\boldsymbol{\Gamma}$ defined by Eqs. (4.8) and (36) from [712] and [722], respectively. The passage from Eq. (8.41) to (8.42) as well as the identical passages in related methods is a very popular manipulation in micromechanics (see, e.g., [462, 728, 794, 943, 944, 1692]), which in fact implies the use of one of the equations (7.30). A contradiction of the verification of Eqs. (7.30) was considered in detail in Sect. 7.2. These passages are questionable for the following reasons. In the referenced papers by Ju and coauthors, one uses in actual fact the hypothesis $\langle\widetilde{\boldsymbol{\sigma}}(\mathbf{x})_{i,j}\rangle_k = \langle\overline{\boldsymbol{\sigma}}(\mathbf{x})\rangle_k = \boldsymbol{\sigma}^0$, $(k = i, j, \ \mathbf{x} \in v_k)$ leading necessarily to the consistent perturbation method (considered in Sect. 10.1), differing from Eq. (8.42) by the second summand in the integrands. Then one comes to the known difficulties connected with the absence of absolute convergence of an improper integral in (8.42) since $\mathcal{T}_j(\mathbf{x}_i - \mathbf{x}_j)$ is of the order $|\mathbf{x}_i - \mathbf{x}_j|^{-d}$ (see also Sect. 7.2.2). So, in [712, 713, 718, 720] one eliminated the difficulties produced by the dependence of a conditionally convergent integral on the shape of integration domain w by the use of the postulation of this shape. The size of integration domain in Eq. (8.42) is postulated in [462].

Taking the general Eq. (7.33) into account, for elimination of these errors, Buryachenko [200] corrected Eq. (8.42) by subtraction of the term $\boldsymbol{\Gamma}(\mathbf{x}_j - \mathbf{x}_i)\mathbf{R}_i n^{(i)}$ from the integrand in Eq. (8.42). Similar difficulties with the integral convergence at the estimation of the second moment of stresses were also detected (see for details [200]) in [715, 717, 721–723] where this estimations were applied to a wide class of nonlinear problems such as plasticity and damage phenomena. However, the internal inconsistency of the mentioned source approach of the estimation of both the effective elastic properties and the second moment of stresses will be discussed in detail in Chap. 14, and because of this, the correctness of including these approaches into the nonlinear approaches of both plasticity and damage phenomena is questionable.

8.3 One-Particle Approximation of the MEFM and Mori-Tanaka Approach

8.3.1 One-Particle ("Quasi-Crystalline") Approximation of MEFM

The representations for the effective properties (8.32)–(8.34) were obtained by the MEFM in the framework of the closing hypothesis (8.9) for two interacting inclusions. These representations can be simplified at the replacement of the hypothesis **H2a** by the "quasi-crystalline" approximation **H2b** (8.10). So, applying Eq. (8.10$_1$) immediately for closing of the infinite system (7.45) yields the equation

$$\overline{v}_i\langle\boldsymbol{\vartheta}_i\rangle = \overline{v}_i\boldsymbol{\vartheta}_i^0 + \mathbf{R}_i\sum_{q=1}^{N}\int\Big[\mathbf{T}_{iq}(\mathbf{x}_i-\mathbf{x}_q)\varphi(v_q,\mathbf{x}_q|;v_i,\mathbf{x}_i) \\ -\mathbf{T}_i(\mathbf{x}_i-\mathbf{x}_q)n^{(q)}\Big]\overline{v}_q\langle\boldsymbol{\vartheta}_q\rangle\,d\mathbf{x}_q \tag{8.43}$$

with the solution presented in the form (8.24) where the matrix $\mathbf{Y}^{-1}$ can be reduced to

$$(\mathbf{Y}^{-1})_{ij} = \mathbf{I}\delta_{ij} - \mathbf{R}_i\int[\mathbf{T}_{ij}(\mathbf{x}_i-\mathbf{x}_j)\varphi(v_j,\mathbf{x}_j|;v_i,\mathbf{x}_i) - \mathbf{T}_i(\mathbf{x}_i-\mathbf{x}_j)n^{(j)}]\,d\mathbf{x}_j. \tag{8.44}$$

To make further progress, the hypothesis of *ellipsoidal symmetry* **H3** for the distribution of inclusions, attributed to Willis [1620] (see also [257, 759, 760, 762, 1240]), is widely used:

Hypothesis H3, "Ellipsoidal Symmetry" *The conditional probability density function $\varphi(v_j,\mathbf{x}_j|;v_i,\mathbf{x}_i)$ depends on $\mathbf{x}_j-\mathbf{x}_i$ only through the combination $\rho = |(\mathbf{a}_{ij}^0)^{-1}(\mathbf{x}_j-\mathbf{x}_i)|$*

$$\varphi(v_j,\mathbf{x}_j\ |;v_i,\mathbf{x}_i) = h(\rho), \tag{8.45}$$

where the matrix $(\mathbf{a}_{ij}^0)^{-1}$ (which is symmetric in the indexes i and j, $\mathbf{a}_{ij}^0 = \mathbf{a}_{ji}^0$) defines the ellipsoid excluded volume $v_{ij}^0 = \{\mathbf{x} : |(\mathbf{a}_{ij}^0)^{-1}\mathbf{x}|^2 < 1\}$.

A destination of the hypothesis **H3** is directed toward providing conditions for applying the hypothesis **H1** (see for details Chap. 9) rather than just simplification of Eq. (8.45) (as incorrectly mentioned in [200]). A pair distribution function has "ellipsoidal symmetry" but with an ellipsoid shape differing from the one that defines the inclusion shape. Although the assumed statistics may not be exactly realized in any particular composite, the results of effective moduli estimations are explicit and easily to use. It is crucial for the analyst to be aware of their reasonable choice of the shape of "ellipsoidal" spatial correlation of inclusion location (see Sect. 18.3). For spherical inclusions, the relation (8.45) is realized for a statistical isotropy of the composite structure. It is reasonable to assume that $(\mathbf{a}^0_{ij})^{-1}$ identifies a matrix of affine transformation that transfers the ellipsoid v^0_{ij} being the "excluded volume" ("correlation hole") into a unit sphere, and, therefore, the representation of the matrix $\mathbf{Y}_{ij}$ can be simplified

$$(\mathbf{Y}^{-1})_{ij} = \mathbf{I}\delta_{ij} - \mathbf{R}_i\mathbf{Q}^0_{ij}n^{(j)}, \tag{8.46}$$

where $\mathbf{Q}^0_{ij} \equiv \mathbf{Q}(v^0_{ij})$ (see Eqs. (3.38) and (3.177)) is a constant for the ellipsoidal domain v^0_{ij}. For the sake of simplicity of the subsequent calculation, we will usually assume that the shape of "correlation hole" v^0_{ij} does not depend on the inclusion v_j: $v^0_{ij} = v^0_i$ and $\mathbf{Q}^0_{ij} = \mathbf{Q}^0_i \equiv \mathbf{Q}(v^0_i)$.

Thus, the final results may be significantly simplified under the following additional assumptions:

$$\langle V_j(\mathbf{y})\boldsymbol{\vartheta}_j \,|; v_i, \mathbf{x}_i\rangle = \mathbf{h}_2(\langle\boldsymbol{\vartheta}_j\rangle, \rho), \tag{8.47}$$

where the dependence of the function $\mathbf{h}_2$ on the geometrical parameters of the inclusion v_i is defined by the scalar value ρ (8.45). According to relation (8.47), the conditional averaging properties of the composite have level surfaces which are obtained from the ellipsoidal surfaces ∂v^0_i by the use of a homothetic transformation. Under the assumption of (8.10_1), the equality (8.47) is valid under the simplest conditional probability density (8.45). The found tensor $\mathbf{Y}_{ij}$ (8.46) can be used in all formulae of Sect. 8.2 for estimations of effective thermoelastic properties and stress concentration factors.

According to Eqs. (3.38) and by virtue of the fact that the generalized function $\boldsymbol{\Gamma}(\mathbf{x})$ is a homogeneous function of order $-d$, we obtain—under the assumptions (8.47) and $v^0_{ij} = v^0_i$—the following relation:

$$\sum_{q=1}^{N}\int \boldsymbol{\Gamma}(\mathbf{x}-\mathbf{x}_q)[\langle V_q(\mathbf{x}_q)\boldsymbol{\eta}^1(\mathbf{x}_q)|; v_i, \mathbf{x}_i\rangle - c^{(q)}\langle\boldsymbol{\eta}^1{}_q\rangle]\, d\mathbf{x}_q = \mathbf{Q}^0_i\sum_{q=1}^{N}\langle\boldsymbol{\eta}^1{}_q\rangle c^{(q)} \tag{8.48}$$

Taking the assumptions (8.9), (8.10$_1$), and (8.47) into account, Eq. (8.23) can be combined into a simple equation

$$\overline{v}_i \langle \boldsymbol{\eta}_i^1 \rangle = \overline{v}_i \boldsymbol{\eta}_i^{10} + \mathbf{R}_i \mathbf{Q}_i^0 \sum_{q=1}^{N} \langle \boldsymbol{\eta}_q^1 \rangle c^{(q)}, \tag{8.49}$$

leading to the estimation of average effective fields and stresses inside the inclusions

$$\langle \overline{\boldsymbol{\sigma}} \rangle_i = \boldsymbol{\sigma}^0 + \mathbf{Q}_i^0 \big[\mathbf{I} - \langle \mathbf{R}^v \mathbf{Q}^0 V \rangle\big]^{-1} (\langle \mathbf{R}^v V \rangle \boldsymbol{\sigma}^0 + \langle \mathbf{F}^v V \rangle), \tag{8.50}$$

$$\langle \boldsymbol{\sigma} \rangle_i = \mathbf{B}_i \boldsymbol{\sigma}^0 + \mathbf{C}_i + \mathbf{B}_i \mathbf{Q}_i^0 \big[\mathbf{I} - \langle \mathbf{R}^v \mathbf{Q}^0 V \rangle\big]^{-1} (\langle \mathbf{R}^v V \rangle \boldsymbol{\sigma}^0 + \langle \mathbf{F}^v V \rangle), \tag{8.51}$$

where for convenience one introduces the notations $\mathbf{g}(\mathbf{x}) = \mathbf{g}^{(i)}(\mathbf{x})$, $n(\mathbf{x}) \equiv n^{(i)}$ at $\mathbf{x} \in v^{(i)}$, $(i = 1, \ldots, N;\ \ \mathbf{g} = \mathbf{R}, \mathbf{F}, \mathbf{B}, \mathbf{Q}^0)$ and $\mathbf{f}, \mathbf{g}, n(\mathbf{x}) \equiv 0$ otherwise. Hereafter, the tensors $\mathbf{C}(\mathbf{x}),\ \mathbf{F}(\mathbf{x})$ depend on the average stresses in the matrix

$$\begin{aligned}
\langle \boldsymbol{\sigma} \rangle^{(0)} &= \mathbf{D}^c \Big[(\mathbf{I} - \langle \mathbf{B} V \rangle) \boldsymbol{\sigma}^0 - \langle \overline{\mathbf{C}} V \rangle \\
&\quad - \langle \mathbf{B} \mathbf{Q}^0 V \rangle (\mathbf{I} - \langle \mathbf{R}^v \mathbf{Q}^0 V \rangle)^{-1} (\langle \mathbf{R}^v V \rangle \boldsymbol{\sigma}^0 + \langle \overline{\mathbf{F}}^v V \rangle) \Big], \\
\mathbf{D}^c &= \Big\{ c^{(0)} \mathbf{I} + \langle \mathbf{B} \mathbf{Q} V \rangle \mathbf{M}_1^{(0)} \\
&\quad + \langle \mathbf{B} \mathbf{Q}^0 V \rangle (\mathbf{I} - \langle \mathbf{R}^v \mathbf{Q}^0 V \rangle)^{-1} \langle \mathbf{Q}^{-1} \mathbf{B} \mathbf{Q} V \rangle \mathbf{M}_1^{(0)} \Big\}^{-1},
\end{aligned} \tag{8.52}$$

where the tensor $\mathbf{D}^c \equiv c^{(0)} \mathbf{I}$ if $\mathbf{M}^c = \mathbf{M}^{(0)}$. Equation (8.52) permits obtaining an explicit representation of average stresses inside the inclusions

$$\begin{aligned}
\langle \boldsymbol{\sigma} \rangle_i &= \mathbf{B}_i \boldsymbol{\sigma}^0 + \overline{\mathbf{C}}_i + \mathbf{B}_i \mathbf{Q}_i^0 \big[\mathbf{I} - \langle \mathbf{R}^v \mathbf{Q}^0 V \rangle\big]^{-1} (\langle \mathbf{R}^v V \rangle \boldsymbol{\sigma}^0 + \langle \overline{\mathbf{F}}^v V \rangle) \\
&\quad + \Big[- \mathbf{B}_i \mathbf{Q}_i + \mathbf{B}_i \mathbf{Q}_i^0 (\mathbf{I} - \langle \mathbf{R}^v V \mathbf{Q}^0 \rangle)^{-1} \langle \mathbf{Q}^{-1} \mathbf{B} \mathbf{Q} V \rangle \Big] \mathbf{M}_1^{(0)} \mathbf{D}^c \Big[\boldsymbol{\sigma}^0 - \langle \mathbf{B} V \rangle \boldsymbol{\sigma}^0 \\
&\quad - \langle \overline{\mathbf{C}} V \rangle - \langle \mathbf{B} \mathbf{Q}^0 V \rangle (\mathbf{I} - \langle \mathbf{R}^v \mathbf{Q}^0 V \rangle)^{-1} (\langle \mathbf{R}^v V \rangle \boldsymbol{\sigma}^0 + \langle \overline{\mathbf{F}}^v V \rangle) \Big].
\end{aligned} \tag{8.53}$$

Substitution of Eqs. (8.2$_1$) and (8.53) into (6.79), (6.48) and (6.79$_3$) gives the representation for effective properties

$$\begin{aligned}
\mathbf{M}^* &= \mathbf{M}^c + \langle \mathbf{R}^{0v} V \rangle + \langle \mathbf{R}^{0v} \mathbf{Q}^0 V \rangle \big[\mathbf{I} - \langle \mathbf{R}^v \mathbf{Q}^0 V \rangle\big]^{-1} \langle \mathbf{R}^v V \rangle \\
&\quad + \Big[- \langle \mathbf{R}^{0v} V \rangle + \langle \mathbf{R}^{0v} \mathbf{Q}^0 V \rangle (\mathbf{I} - \langle \mathbf{R}^v V \mathbf{Q}^0 \rangle)^{-1} \langle \mathbf{Q}^{-1} \mathbf{B} \mathbf{Q} V \rangle \Big] \mathbf{M}_1^{(0)} \mathbf{D}^c \\
&\quad \times \Big\{ \mathbf{I} - \langle \mathbf{B} V \rangle - \langle \mathbf{B} \mathbf{Q}^0 V \rangle \big[\mathbf{I} - \langle \mathbf{R}^v \mathbf{Q}^0 V \rangle\big]^{-1} \langle \mathbf{R}^v V \rangle \Big\},
\end{aligned} \tag{8.54}$$

$$\boldsymbol{\beta}^* = \boldsymbol{\beta}^{(0)} + \big[\mathbf{I} - \mathbf{M}_1^{(0)}\langle \mathbf{R}^v\mathbf{Q}^0 V\rangle\big]\big[\mathbf{I} - \langle \mathbf{R}^v\mathbf{Q}^0 V\rangle\big]^{-1}\langle \overline{\mathbf{F}}^v V\rangle$$
$$-\Big[- \langle \mathbf{R}^{0v} V\rangle + \langle \mathbf{R}^{0v}\mathbf{Q}^0 V\rangle(\mathbf{I} - \langle \mathbf{R}^v V\mathbf{Q}^0\rangle)^{-1}\langle \mathbf{Q}^{-1}\mathbf{B}\mathbf{Q}V\rangle\Big]\mathbf{M}_1^{(0)}\mathbf{D}^c$$
$$\times\Big\{\langle \overline{\mathbf{C}}V\rangle + \langle \mathbf{B}\mathbf{Q}^0 V\rangle\big[\mathbf{I} - \langle \mathbf{R}^v\mathbf{Q}^0 V\rangle\big]^{-1}\langle \overline{\mathbf{F}}^v V\rangle\Big\}, \tag{8.55}$$

$$W^* = -\frac{1}{2}\Big\{\langle \boldsymbol{\beta}_1\overline{\mathbf{C}}V\rangle + \langle \boldsymbol{\beta}_1\mathbf{B}\mathbf{Q}^0 V\rangle\big[\mathbf{I} - \langle \mathbf{R}^v\mathbf{Q}^0 V\rangle\big]^{-1}\langle \overline{\mathbf{F}}^v V\rangle\Big\}$$
$$+\frac{1}{2}\Big[- \langle \boldsymbol{\beta}_1\mathbf{B}\mathbf{Q}\rangle + \langle \boldsymbol{\beta}_1\mathbf{B}_i\mathbf{Q}_i^0\rangle(\mathbf{I} - \langle \mathbf{R}^v V\mathbf{Q}^0\rangle)^{-1}\langle \mathbf{Q}^{-1}\mathbf{B}\mathbf{Q}V\rangle\Big]\mathbf{M}_1^{(0)}\mathbf{D}^c$$
$$\times\Big\{\langle \overline{\mathbf{C}}V\rangle + \langle \mathbf{B}\mathbf{Q}^0 V\rangle\big[\mathbf{I} - \langle \mathbf{R}^v\mathbf{Q}^0 V\rangle\big]^{-1}\langle \overline{\mathbf{F}}^v V\rangle\Big\}. \tag{8.56}$$

Equations (8.54)-(8.56) are obtained by one-particle approximation of MEFM (called effective field method, EFM).

Taking into account the assumption of coincidence of elastic moduli of the comparison medium and matrix (7.26) (when $\mathbf{C}_i(\mathbf{x}) = \overline{\mathbf{C}}_i(\mathbf{x}),\ \mathbf{F}_i(\mathbf{x}) = \overline{\mathbf{F}}_i(\mathbf{x})$) leads to simplification of Eqs. (8.51)–(8.56) obtained in [257] by another method

$$\mathbf{M}^* = \mathbf{M}^{(0)} + \Big[\mathbf{I} - \langle \mathbf{R}^v\mathbf{Q}^0 V\rangle\Big]^{-1}\langle \mathbf{R}^v V\rangle, \tag{8.57}$$

$$\boldsymbol{\beta}^* = \boldsymbol{\beta}^{(0)} + (\mathbf{M}^* - \mathbf{M}^{(0)})\langle \mathbf{R}^v V\rangle^{-1}\langle \mathbf{F}^v V\rangle, \tag{8.58}$$

$$W^* = -\frac{1}{2}\Big\{\langle \boldsymbol{\beta}_1\mathbf{C}V\rangle + \langle \boldsymbol{\beta}_1\mathbf{B}\mathbf{Q}^0 V\rangle(\mathbf{M}^* - \mathbf{M}^{(0)})\langle \mathbf{R}^v V\rangle^{-1}\langle \mathbf{F}^v V\rangle\Big\}. \tag{8.59}$$

Equations for the statistical averages of the stresses (8.52) and related fields inside the heterogeneity v_i are also simplified

$$\langle \overline{\boldsymbol{\sigma}}\rangle_i = \boldsymbol{\sigma}^0 + \mathbf{Q}_i^0\big[\mathbf{I} - \langle \mathbf{R}^v\mathbf{Q}^0 V\rangle\big]^{-1}(\langle \mathbf{R}^v V\rangle\boldsymbol{\sigma}^0 + \langle \mathbf{F}^v V\rangle), \tag{8.60}$$

$$\langle \boldsymbol{\sigma}\rangle_i(\mathbf{x}) = \mathbf{B}_i(\mathbf{x})\boldsymbol{\sigma}^0 + \mathbf{C}_i(\mathbf{x})$$
$$+\mathbf{B}_i(\mathbf{x})\mathbf{Q}_i^0\big[\mathbf{I} - \langle \mathbf{R}^v\mathbf{Q}^0 V\rangle\big]^{-1}(\langle \mathbf{R}^v V\rangle\boldsymbol{\sigma}^0 + \langle \mathbf{F}^v V\rangle), \tag{8.61}$$

$$\bar{v}_i\langle \boldsymbol{\eta}_i\rangle(\mathbf{x}) = \mathbf{R}_i(\mathbf{x})\boldsymbol{\sigma}^0 + \mathbf{F}_i(\mathbf{x})$$
$$+\mathbf{R}_i(\mathbf{x})\mathbf{Q}_i^0\big[\mathbf{I} - \langle \mathbf{R}^v\mathbf{Q}^0 V\rangle\big]^{-1}(\langle \mathbf{R}^v V\rangle\boldsymbol{\sigma}^0 + \langle \mathbf{F}^v V\rangle), \tag{8.62}$$

The relations (8.58) and (8.59) do not result as a generalization of Eqs. (8.20) and (8.21) to any number of components, but they were obtained under the additional assumptions (8.43_2) and (8.45). For homogeneous inclusions (3.183), Eqs. (8.57) and (8.58) reduce to some results found in [914, 916] and in [1240] (see also [746, 747, 1438]).

Thus, if the distribution of inclusions is taken to be the same for all inclusions pairs and, therefore, in Eq. (8.48) $\mathbf{Q}_i \equiv \mathbf{Q}(v^{el}) = \text{const.}\ \forall i$ for some ellipsoidal domain v^{el}, then Eq. (8.57) rewritten in terms of elastic moduli coincides with Eq. (8.35) from [1240]. If additionally one assumes that v^{el} has the form a sphere, then Eqs. (8.49) and (8.57) reduce to the results obtained in [916]. For identical homogeneous ellipsoidal inclusions randomly oriented in space, Eq. (8.57) coincides with the result [914]. If one assumes that $\langle \mathbf{R}^v \mathbf{Q}^0 \rangle = \langle \mathbf{R}^v \rangle \langle \mathbf{Q}^0 \rangle$, then Eq. (8.57) reduces to the result [746].

It should be mentioned that for multicomponent media with $N \gg 1$ (e.g., for random orientation of inclusions) in the framework of the hypothesis (8.43_2), either Eqs. (8.51)–(8.56) or Eqs. (8.57)–(8.59) are more convenient from the computational point of view than the direct use of more general Eqs. (8.32)–(8.34) or (8.35)–(8.37), respectively. Indeed, the assumption (8.43_2) yields to the termination of integrals numerically integrable in Eq. (8.27), but the difficulties induced by inversion of multidimensional matrix $\mathbf{Y}^{-1}$ (8.27) remain. In so doing the utilization of Eqs. (8.51)–(8.56) implies only estimation of statistical averages of a few tensors $\langle \mathbf{B}V \rangle$, $\langle \mathbf{C}V \rangle$, $\langle \mathbf{R}^v V \rangle$, $\langle \mathbf{F}^v V \rangle$, $\langle \mathbf{B}\mathbf{Q}^0 V \rangle$, $\langle \boldsymbol{\beta}_1 \mathbf{C} V \rangle$, $\langle \boldsymbol{\beta}_1 \mathbf{B}\mathbf{Q}^0 V \rangle$ that would almost always be preferable for computation.

It should be emphasized that Eqs. (8.58) and (8.59) for one kind of coated inclusions ($N = 1$) formally coincide with analogous equations obtained in [270] only by the use of the hypothesis **H1**. Thus, in the framework of the hypothesis **H1**, the result [270] can be considered as a generalization of classical results by Levin [912] (see also [1304]), which cannot be said about Eqs. (8.58) and (8.59) obtained under the additional assumptions (8.43_2) and (4.7).

Notice that the term one particle for the method considered in this subsection stands for the hypothesis (8.43_2) according to which there is no direct influence of surrounding inclusions v_j on the inclusion v_i being considered. However, such an influence takes place in an average sense via the action of the second summand in the right-hand side of Eq. (8.49). It is interesting that for periodic structure composites considered as a particular realization of a random structure, the hypothesis (8.43_2) is valid exactly. Then the MEFM implies the utilization of the hypothesis **H1** only leading to a high accuracy (in a comparison with exact analytical results) of estimations of both the effective local and nonlocal effective properties (see for details Chap. 13).

A dual structure of effective properties can be derived for a case when the composite is subjected to a boundary displacement conditions corresponding to a uniform strain $\langle \boldsymbol{\varepsilon} \rangle$

$$\mathbf{L}^* = \mathbf{L}^{(0)} + \left[\mathbf{I} - \langle \mathbf{R}^{\epsilon v} \mathbf{P}^0 V \rangle\right]^{-1} \langle \mathbf{R}^{\epsilon v} V \rangle, \tag{8.63}$$

$$\boldsymbol{\alpha}^* = \boldsymbol{\alpha}^{(0)} + (\mathbf{L}^* - \mathbf{L}^{(0)}) \langle \mathbf{R}^{\epsilon v} V \rangle^{-1} \langle \mathbf{F}^{\epsilon v} V \rangle. \tag{8.64}$$

Hereafter $\mathbf{A}$, $\mathbf{C}^{\epsilon}$, $\mathbf{R}^{\epsilon v}$, $\mathbf{F}^{\epsilon v}$ are the concentration factors analogous to $\mathbf{B}$, $\mathbf{C}$, $\mathbf{R}^v$, $\mathbf{F}^v$ (8.2), respectively, $(\mathbf{x} \in v_i \subset v^{(i)})$

$$\boldsymbol{\varepsilon}(\mathbf{x}) = \mathbf{A}^{(i)}(\mathbf{x})\overline{\boldsymbol{\varepsilon}}(\mathbf{x}_i) + \mathbf{C}^{\epsilon(i)}(\mathbf{x}),$$
$$\mathbf{L}_1(\mathbf{x})\boldsymbol{\varepsilon}(\mathbf{x}) = \mathbf{R}^{\epsilon v(i)}(\mathbf{x})\overline{\boldsymbol{\varepsilon}}(\mathbf{x}_i) + \mathbf{F}^{\epsilon v(i)}(\mathbf{x}), \tag{8.65}$$

It is important that the effective moduli $\mathbf{L}^*$ and compliance $\mathbf{M}^*$ under the different boundary conditions are identical. This can be readily proved by invoking the inter-relations (3.116) and (3.117) leading from Eqs. (8.2_1) and (8.65_1) under nonrestrictive equalities $\overline{\boldsymbol{\sigma}} \equiv \boldsymbol{\sigma}^0 = \mathbf{L}^{(0)}\boldsymbol{\varepsilon}^0 \equiv \mathbf{L}^{(0)}\overline{\boldsymbol{\varepsilon}}$ and $\boldsymbol{\sigma}(\mathbf{x}) = \mathbf{L}^{(i)}(\mathbf{x})\boldsymbol{\varepsilon}(\mathbf{x})$ ($\mathbf{x} \in v_i \subset v^{(i)}$). It then follows that $\mathbf{L}^*$ and $\mathbf{M}^*$ can be recast in the form

$$\mathbf{L}^* = \Big[\mathbf{I} - \langle \mathbf{R}^{\epsilon v}\mathbf{P}^0 V\rangle\Big]^{-1}\mathbf{L}^{(0)}\Big[\mathbf{I} - \langle \mathbf{R}^{v}\mathbf{Q}^0 V\rangle\Big],$$
$$\mathbf{M}^* = \Big[\mathbf{I} - \langle \mathbf{R}^{v}\mathbf{Q}^0 V\rangle\Big]^{-1}\mathbf{M}^{(0)}\Big[\mathbf{I} - \langle \mathbf{R}^{\epsilon v}\mathbf{P}^0 V\rangle\Big], \tag{8.66}$$

providing the desirable reciprocal relation between the two. Now that their equivalency has been proved, one may choose either expression for the determination of the effective properties (8.66).

It should be mentioned that in the framework of the "quasi-crystalline" approximation **H2b** (8.10), Eq. (8.43) is equivalent to the equation expressed through the second-order probability function S_{ij}(5.75) (see [1625, 1627]) ($\mathbf{x} \in v^{(i)}$)

$$\boldsymbol{\varepsilon}(\mathbf{x}) = \langle \boldsymbol{\varepsilon}\rangle + \int \mathbf{U}(\mathbf{x}-\mathbf{y})\Big[S_{ij}(\mathbf{x},\mathbf{y})\mathbf{L}_1(\mathbf{y})\boldsymbol{\varepsilon}(\mathbf{y}) - \langle \mathbf{L}_1\boldsymbol{\varepsilon}\rangle\Big]\, d\mathbf{y}, \tag{8.67}$$

which leads to the effective moduli representation (8.63) in the framework of the hypothesis **H3** (8.45) for CMs with the homogeneous ellipsoidal inclusions and $\mathbf{L}^c = \mathbf{L}^{(0)}$.

8.3.2 Mori-Tanaka Approach

As pointed out by Benveniste [89], the essential assumption in the Mori-Tanaka method (MTM) states that each inclusion v_i behaves as an isolated one in the infinite matrix ($\mathbf{M}^c \equiv \mathbf{M}^{(0)}$) and subject to some effective stress field $\overline{\boldsymbol{\sigma}}_i(\mathbf{x})$ (7.42) coinciding with the average stress in the matrix

$$\overline{\boldsymbol{\sigma}}_i \equiv \langle \boldsymbol{\sigma}\rangle^{(0)}. \tag{8.68}$$

The composite is assumed to be subjected to a boundary traction giving rise to a uniform stress $\langle \boldsymbol{\sigma}\rangle$. The assumption (8.68) allows one to define uniquely the effective thermoelastic properties of multicomponent composite materials. Thus, hypothesis (8.68) is more restricted than hypothesis **H1** which gives an opportunity to use the solution (8.1) for the each inclusion v_i and to find the average stress in the matrix by the use of a representation of the average stress in the whole composite over the average stresses in the separate phases

$$\langle\boldsymbol{\sigma}\rangle \equiv c^{(0)}\langle\boldsymbol{\sigma}\rangle^{(0)} + \langle\boldsymbol{\sigma}V\rangle = c^{(0)}\langle\boldsymbol{\sigma}\rangle^{(0)} + \langle\mathbf{B}V\rangle\langle\boldsymbol{\sigma}\rangle^{(0)} + \langle\mathbf{C}V\rangle, \tag{8.69}$$

leading to the representations of statistical average of both the stresses in the matrix and the strain polarization tensor $\boldsymbol{\eta}$ in the inclusions

$$\langle\boldsymbol{\sigma}\rangle^{(0)} = (c^{(0)}\mathbf{I} + \langle\mathbf{B}V\rangle)^{-1}(\langle\boldsymbol{\sigma}\rangle - \langle\mathbf{C}V\rangle), \tag{8.70}$$

$$\langle\boldsymbol{\eta}\rangle_i(\mathbf{x}) = \mathbf{R}^{(i)}(\mathbf{x})(c^{(0)}\mathbf{I} + \langle\mathbf{B}V\rangle)^{-1}(\langle\boldsymbol{\sigma}\rangle - \langle\mathbf{C}V\rangle) + \mathbf{F}^{(i)}(\mathbf{x}). \tag{8.71}$$

Substituting of Eq. (8.71) into the average constitutive equation (6.2) leads to

$$\langle\boldsymbol{\varepsilon}\rangle \equiv \mathbf{M}^{(0)}\langle\boldsymbol{\sigma}\rangle + \boldsymbol{\beta}^{(0)} + \langle\boldsymbol{\eta}V\rangle = \left[\mathbf{M}^{(0)} + \langle\mathbf{R}^{v}V\rangle(c^{(0)}\mathbf{I} + \langle\mathbf{B}V\rangle)^{-1}\right]\langle\boldsymbol{\sigma}\rangle$$
$$+\boldsymbol{\beta}^{(0)} + \langle\mathbf{F}V\rangle - \langle\mathbf{R}^{v}V\rangle(c^{(0)}\mathbf{I} + \langle\mathbf{B}V\rangle)^{-1}\langle\mathbf{C}V\rangle. \tag{8.72}$$

Comparison of the definition of effective parameters $\mathbf{M}^*$ and $\boldsymbol{\beta}^*$ (1.12) with Eq. (8.72) yields to the representation of effective properties and the effective field

$$\mathbf{M}^* = \mathbf{M}^{(0)} + \langle\mathbf{R}^{v}V\rangle(c^{(0)}\mathbf{I} + \langle\mathbf{B}V\rangle)^{-1}, \tag{8.73}$$

$$\boldsymbol{\beta}^* = \boldsymbol{\beta}^{(0)} + \langle\mathbf{F}V\rangle - \langle\mathbf{R}^{v}V\rangle(c^{(0)}\mathbf{I} + \langle\mathbf{B}V\rangle)^{-1}\langle\mathbf{C}V\rangle, \tag{8.74}$$

$$\langle\overline{\boldsymbol{\sigma}}_i\rangle = (c^{(0)}\mathbf{I} + \langle\mathbf{B}V\rangle)^{-1}(\langle\boldsymbol{\sigma}\rangle - \langle\mathbf{C}V\rangle). \tag{8.75}$$

From the definition of the stored energy (6.79_3), in association with (8.75), we obtain

$$2W^* = \langle\boldsymbol{\beta}_1 V\mathbf{B}\rangle(c^{(0)}\mathbf{I} + \langle\mathbf{B}V\rangle)^{-1}\langle\mathbf{C}V\rangle - \langle\boldsymbol{\beta}_1 V\mathbf{C}\rangle. \tag{8.76}$$

Equations (8.73), (8.74), and (8.71) can be recast in an equivalent form convenient for comparison with the analogous representations obtained by the MEFM

$$\mathbf{M}^* = \mathbf{M}^{(0)} + \langle\mathbf{R}^{v}V\rangle(\mathbf{I} - \langle\mathbf{Q}\mathbf{R}^{v}V\rangle)^{-1}, \tag{8.77}$$

$$\boldsymbol{\beta}^* = \boldsymbol{\beta}^{(0)} + \langle\mathbf{F}^{v}V\rangle - \langle\mathbf{R}^{v}V\rangle(\mathbf{I} - \langle\mathbf{Q}\mathbf{R}^{v}V\rangle)^{-1}\langle\mathbf{C}V\rangle, \tag{8.78}$$

$$\langle\boldsymbol{\eta}\rangle_i(\mathbf{x}) = \mathbf{R}_i^{v}(\mathbf{x})(\mathbf{I} - \langle\mathbf{Q}\mathbf{R}^{v}V\rangle)^{-1}\big[\langle\boldsymbol{\sigma}\rangle - \langle\mathbf{C}V\rangle\big] + \mathbf{F}_i^{v}(\mathbf{x}). \tag{8.79}$$

Transformation of Eqs. (8.73) and (8.74) to Eqs. (8.77) and (8.78) was carried out by the use of an equality (8.4) that is valid for ellipsoidal inclusion of any microstructure.

Though once $\mathbf{M}^*$ is found, the displacement at infinity is prescribed to correspond to a uniform strain $\langle\boldsymbol{\varepsilon}\rangle$ that leads to the following representations for the effective thermoelastic properties and effective fields

$$\mathbf{L}^* = \mathbf{L}^{(0)} + \langle\mathbf{R}^{\epsilon v}V\rangle(c^{(0)}\mathbf{I} + \langle\mathbf{A}V\rangle)^{-1}, \tag{8.80}$$

$$\boldsymbol{\alpha}^* = \boldsymbol{\alpha}^{(0)} + \langle \mathbf{F}^{\epsilon v} V\rangle - \langle \mathbf{R}^{\epsilon v} V\rangle (c^{(0)}\mathbf{I} + \langle \mathbf{A} V\rangle)^{-1} \langle \mathbf{C}^{\epsilon} V\rangle, \tag{8.81}$$

$$\langle \overline{\boldsymbol{\varepsilon}}_i\rangle = (c^{(0)}\mathbf{I} + \langle \mathbf{A} V\rangle)^{-1} (\langle \boldsymbol{\varepsilon}\rangle - \langle \mathbf{C}^{\epsilon} V\rangle). \tag{8.82}$$

Taking the equalities (8.72) into account, we can recast the $\mathbf{L}^*$ and $\mathbf{M}^*$ in the form

$$\mathbf{L}^* = [c^{(0)}\mathbf{I} + \langle \mathbf{B} V\rangle]\mathbf{L}^{(0)}[c^{(0)}\mathbf{I} + \langle \mathbf{A} V\rangle]^{-1}, \tag{8.83}$$

$$\mathbf{M}^* = [c^{(0)}\mathbf{I} + \langle \mathbf{A} V\rangle]\mathbf{M}^{(0)}[c^{(0)}\mathbf{I} + \langle \mathbf{B} V\rangle]^{-1}, \tag{8.84}$$

providing the reciprocal relation $(\mathbf{M}^*)^{-1} = \mathbf{L}^*$.

It should be mentioned that the stress concentration factor at the representative inclusion v_i defined by Eq. (8.2_1) depends on two sorts of stress concentration factors. The first one is a "local" stress concentrator factor defined by the solution for one isolated inclusion in the infinite medium that is described by the tensors $\mathbf{B}_i(\mathbf{x})$ and $\mathbf{C}_i(\mathbf{x})$ and which is the same in both the MEFM and MTM. For example, for a pure elastic problem $\boldsymbol{\beta}_1 \equiv \mathbf{0}$, we have (no summation over i)

$$\langle \boldsymbol{\sigma}\rangle_i(\mathbf{x}) = \mathbf{B}_i(\mathbf{x})\mathbf{D}_i\langle \boldsymbol{\sigma}\rangle. \tag{8.85}$$

The second sort of stress concentration factor $\mathbf{D}_i$ is governed by the "global" stress concentration factors at the effective field either (8.60) or (8.75) and differs in the MEFM and the MTM. Moreover, the effective field (8.75) does not depend on the geometrical parameters of the inclusion v_i, while the effective field (8.60) estimated by the MEFM explicitly depends on both the orientation and shape of the inclusion v_i through the tensor $\mathbf{Q}_i^0$. In particular, for 3D uniform random orientation of inclusions, the tensor $\mathbf{D}_i$ estimated by the MTM is an isotropic and does not depend on the orientation and the shape of the inclusion v_i. None of the mentioned properties of the tensor $\mathbf{D}_i$ are fulfilled in the case of evaluation of this tensor by the MTM.

The MTM as defined by Eqs. (8.73) has been called the "direct approach" by Benveniste [89], who showed it is identical for composites with the ellipsoidal homogeneous inclusions to the *equivalent inclusion-average stress* formalism (see, e.g., [1137, 1477, 1489, 1490, 1607, 1608]) which has played a key role in micromechanics and was originally proposed by Eshelby [449, 450] (see Sect. 3.6.1) for one ellipsoidal homogeneous inclusion in the infinite matrix. Namely, the average strains produced by the homogeneous boundary conditions (6.5) in the composite and in the matrix are given respectively by (following [1608])

$$\langle \boldsymbol{\varepsilon}\rangle = \mathbf{M}^*\langle \boldsymbol{\sigma}\rangle, \quad \langle \boldsymbol{\varepsilon}\rangle^{(0)} = \mathbf{M}^{(0)}\langle \boldsymbol{\sigma}\rangle^{(0)}. \tag{8.86}$$

We will also assume that the heterogeneities are subjected to the stress-free strains $\boldsymbol{\beta}_1^{(i)}(\mathbf{x}) \equiv \text{const}$, ($\mathbf{x} \in v^{(i)}$, $\boldsymbol{\beta}^{(0)} \equiv \mathbf{0}$). In each heterogeneity v_i the average strain field and stress fields may be written as the superposition of *mean fields* in the matrix with the additional perturbed strain $\boldsymbol{\varepsilon}_i^{\text{pt}}$ and stress $\boldsymbol{\sigma}_i^{\text{pt}}$

$$\langle\boldsymbol{\varepsilon}\rangle_i = \langle\boldsymbol{\varepsilon}\rangle^{(0)} - \boldsymbol{\beta}_1^{(i)} + \boldsymbol{\varepsilon}_i^{\rm pt}, \quad \langle\boldsymbol{\sigma}\rangle_i = \langle\boldsymbol{\sigma}\rangle^{(0)} + \boldsymbol{\sigma}_i^{\rm pt}, \tag{8.87}$$

which is the essence of *Mori-Tanaka mean field theory*. The argument (which is plausible rather than rigorous) given by Mori-Tanaka is as follows. Since there are many heterogeneities, adding a single heterogeneity into the infinite ensemble will not alter the mean field values $\langle\boldsymbol{\varepsilon}\rangle^{(0)}$ and $\langle\boldsymbol{\sigma}\rangle^{(0)}$, and therefore, the average strains and stresses inside an inclusion can be expressed by (8.87_1) and (8.87_2). The average stress inside the inclusion $\langle\boldsymbol{\sigma}\rangle_i$ then will follow from the stress–strain relation as well as from Eshelby's equivalent-inclusion principle

$$\langle\boldsymbol{\sigma}\rangle_i = \mathbf{L}^{(i)}(\langle\boldsymbol{\varepsilon}\rangle^{(0)} - \boldsymbol{\beta}_1^{(i)} + \boldsymbol{\varepsilon}_i^{\rm pt}) = \mathbf{L}^{(0)}(\langle\boldsymbol{\varepsilon}\rangle^{(0)} - \boldsymbol{\beta}_1^{(i)} + \boldsymbol{\varepsilon}_i^{\rm pt} - \boldsymbol{\varepsilon}_i^{*}), \tag{8.88}$$

differing from the analogous equation for one heterogeneity (3.129) inside the infinite homogeneous matrix by the representation of the average strain inside $v^{(i)}$ in the decomposition form $\langle\boldsymbol{\varepsilon}\rangle_i = \langle\boldsymbol{\varepsilon}\rangle^{(0)} + \boldsymbol{\varepsilon}_i^{\rm pt}$, where the perturbed part $\boldsymbol{\varepsilon}_i^{\rm pt}$ with respect to the strain in the matrix is further related to Eshelby's equivalent "stress-free strain" $\boldsymbol{\varepsilon}_i^{*}$ through Eshelby's tensor as (see Eq. (3.136)): $\boldsymbol{\varepsilon}_i^{\rm pt} = \mathbf{S}_i(\boldsymbol{\varepsilon}_i^{*} + \boldsymbol{\beta}_1^{(i)})$. Equation ($8.87_2$) also provides that the equivalent "eigenstrain" $\boldsymbol{\varepsilon}_i^{*} = \langle\boldsymbol{\varepsilon}\rangle_i - \mathbf{M}^{(0)}\langle\boldsymbol{\sigma}\rangle_i - \boldsymbol{\beta}_1^{(i)}$, or

$$\boldsymbol{\varepsilon}_i^{*} = (\mathbf{M}^{(i)} - \mathbf{M}^{(0)})\langle\boldsymbol{\sigma}\rangle_i.$$

It is interesting that for a pure mechanical case ($\boldsymbol{\beta}(\mathbf{x}) \equiv \mathbf{0}$), the equivalent "eigenstrain" $\boldsymbol{\varepsilon}_i^{*}$ is identical to the average strain polarization tensor $\langle\boldsymbol{\eta}\rangle_i$ inside the inclusion v_i.

We will recast the average stress in the ith phase (8.87_2) by expressing tensors $\mathbf{S}_i$, $\mathbf{L}_1^{(i)}$, and $\mathbf{L}^{(0)}$ through tensors $\mathbf{Q}_i$, $\mathbf{M}_1^{(i)}$, and $\mathbf{M}^{(0)}$ (see Sect. 3.5), respectively, and find

$$\langle\boldsymbol{\sigma}\rangle_i = \mathbf{Q}_i(\boldsymbol{\varepsilon}_i^{*} + \boldsymbol{\beta}^{(i)}) + \langle\boldsymbol{\sigma}\rangle^{(0)} = \langle\boldsymbol{\sigma}\rangle^{(0)} - \mathbf{Q}_i(\mathbf{M}_i^{(i)}\langle\boldsymbol{\sigma}\rangle_i + \boldsymbol{\beta}_1^{(i)}), \tag{8.89}$$

or rearranging

$$\langle\boldsymbol{\sigma}\rangle_i = \mathbf{B}_i\langle\boldsymbol{\sigma}\rangle^{(0)} - \mathbf{B}_i\mathbf{Q}_i\boldsymbol{\beta}_1^{(i)}, \tag{8.90}$$

where $\mathbf{B}_i = (\mathbf{I} + \mathbf{Q}_i\mathbf{M}_1^{(i)})^{-1}$ (3.117) is recognizable as the familiar stress-concentrator tensor of the single inclusion (with $\mathbf{L}^{(i)}$) embedded in the infinite matrix (with $\mathbf{L}^{(0)}$) subjected to the uniform stress field $\langle\boldsymbol{\sigma}\rangle^{(0)}$. For the homogeneous ellipsoidal heterogeneities being assumed in Eqs. (8.87)–(8.89), the basic hypothesis (8.68) of the direct MTM is equivalent to Eq. (8.90) due to the Eqs. (8.2_1) and (8.5). In such a case, reducing Eq. (8.90) to (8.68), repeating subsequent mathematics (8.69)–(8.79) results in the conclusion that Eq. (8.90) leads to the representations for the effective thermoelastic properties (8.77), (8.78) and average strain polarization tensor (8.79) for the composites with homogeneous ellipsoidal heterogeneities.

However, the direct method has a fundamental advantage with respect to the method of equivalent inclusion (8.86)–(8.89) because at the derivative of the initial representation (8.73), no assumptions about both the shape of heterogeneities and their microstructure were used. In so doing, reduction of Eq. (8.73) to Eq. (8.77) does not exclude the possibility that $\mathbf{M}_1(\mathbf{x})$ vanishes at the part of inclusions v_i used for the introduction of the tensors $\mathbf{Q}_i = \mathbf{Q}_i(v_i)$ in Eq. (8.77) (see the comments at the end of Sect. 8.1.3). The direct method is based on a single hypothesis (8.68) which was implicitly used in Eq. (8.87_1) in parallel with the additional assumptions of both the homogeneity of the tensor $\mathbf{L}^{(i)}(\mathbf{x}) = \text{const.}$ ($\mathbf{x} \in v^{(i)}$) and ellipsoidal shape of heterogeneities. Moreover, exploiting a single strong hypothesis (8.68) in the direct method also has a specific merit in shedding new light on the weakness of the direct method that opens up outstanding possibilities to improve the method as was done in the MEFM's hypothesis (8.43_2).

8.3.3 *Effective Properties Estimated via the EFM and MTM at* $\mathbf{Q}_i \equiv \mathbf{Q}_i^0$

At first, we will recast Eqs. (8.57) and (8.58) accompanied with the representation for the average strain polarization tensor $\langle \boldsymbol{\eta}_i \rangle(\mathbf{x})$ (8.49) obtained by the MEFM in an equivalent form convenient for subsequent analysis

$$\mathbf{M}^* = \mathbf{M}^{(0)} + \langle \mathbf{R}^v V \rangle \big[\mathbf{I} - \langle \mathbf{R}^v V \rangle^{-1} \langle \mathbf{R}^v \mathbf{Q}^0 V \rangle \langle \mathbf{R}^v V \rangle\big]^{-1}, \tag{8.91}$$

$$\boldsymbol{\beta}^* = \boldsymbol{\beta}^{(0)} + \langle \mathbf{R}^v V \rangle \big[\mathbf{I} - \langle \mathbf{R}^v V \rangle^{-1} \langle \mathbf{R}^v \mathbf{Q}^0 V \rangle \langle \mathbf{R}^v V \rangle\big]^{-1} \times \langle \mathbf{R}^v V \rangle^{-1} \langle \mathbf{F}^v V \rangle, \tag{8.92}$$

$$\langle \boldsymbol{\eta}_i \rangle(\mathbf{x}) = \mathbf{R}_i^v(\mathbf{x}) \boldsymbol{\sigma}^0 + \mathbf{F}_i^v(\mathbf{x}) + \mathbf{R}_i^v(\mathbf{x}) \mathbf{Q}_i^0 \langle \mathbf{R}^v V \rangle \times \big[\mathbf{I} - \langle \mathbf{R}^v V \rangle^{-1} \langle \mathbf{R}^v \mathbf{Q}^0 V \rangle \langle \mathbf{R}^v V \rangle\big]^{-1} \Big[\boldsymbol{\sigma}^0 + \langle \mathbf{R}^v V \rangle^{-1} \langle \mathbf{F}^v V \rangle\Big]. \tag{8.93}$$

We will consider the equivalence of representations (8.91)–(8.93) and (8.77)–(8.79), respectively, in some particular cases.

In the *first case*

$$\mathbf{Q}_i = \mathbf{Q}_i^0 \equiv \mathbf{Q}(v^{\text{el}}) = \text{const}, \tag{8.94}$$

which takes place for either the aligned or spherical inclusions. In so doing, the anisotropy of inclusions can be arbitrarily changed from one inclusion to another. Buryachenko [200] proved that in the case (8.94)

$$\mathbf{M}^{*\mathrm{EFM}} = \mathbf{M}^{*\mathrm{MTM}}, \ \boldsymbol{\beta}^{*\mathrm{EFM}} = \boldsymbol{\beta}^{*\mathrm{MTM}}, \ \langle \boldsymbol{\eta}_i \rangle^{\mathrm{EFM}}(\mathbf{x}) = \langle \boldsymbol{\eta}_i \rangle^{*\mathrm{MTM}}(\mathbf{x}). \tag{8.95}$$

It should be mentioned that these equivalencies take place in the framework of additional assumptions (8.45) and (8.47) only. No restrictions are imposed on the microtopology of inclusion (the inclusions can be nonellipsoidal and multilayered coated ones) as well as on the stress state in the inclusions.

Equations (8.57) and (8.58) are in rather poor agreement with a multicomponent realization of Mori-Tanaka's scheme (see, e.g., [1203]) the inconsistency of which is well-known: Mori-Tanaka moduli may violate the Hashin–Shtrikman bounds (see [1137, 1256]) and are nonsymmetric for the general biphase composites (see [100, 469]); the convergence difficulties of the Mori-Tanaka scheme are discussed in [1542].

In the *second case*, let all inclusions be isotropic and homogeneous ellipsoids v_i with the shape coinciding with the shape of correlation holes v_i^0 (8.94) and thermoelastic properties

$$\mathbf{M}_1(\mathbf{x}) \equiv \mathbf{M}_1 = \text{const.}, \ \boldsymbol{\beta}_1(\mathbf{x}) \equiv \boldsymbol{\beta}_1 = \text{const.}, \ (\mathbf{x} \in v), \tag{8.96}$$

then the equalities (8.95_1) and (8.95_2) (see for prove [200]) whereas $\langle \boldsymbol{\eta} \rangle_i^{\mathrm{EFM}} = \langle \boldsymbol{\eta} \rangle_i^{\mathrm{MTM}}$ if and only if the ODF satisfies the additional condition of the fiber alignment (8.94); otherwise $\langle \boldsymbol{\eta} \rangle_i^{\mathrm{EFM}} \neq \langle \boldsymbol{\eta} \rangle_i^{\mathrm{MTM}}$. The last inequality immediately yields from a basic assumption according to which the effective field (8.68) acting on the inclusion v_i in the MTM does not depend on the geometrical parameters of the inclusion v_i, in contrast to the effective field estimated by the EFM (8.60), which explicitly depends on the shape of the correlation hole v_i^0 coinciding in the case being considered with the shape of the inclusion v_i. However, angle averaging of total statistical average of strain polarization tensors estimated by both the EFM and MTM coincides $\langle \boldsymbol{\eta} \rangle^{\mathrm{EFM}} = \langle \boldsymbol{\eta} \rangle^{\mathrm{MTM}}$ that in turn leads to the equality of effective thermoelastic properties (8.95_1) and (8.95_2) estimated by these methods.

Only in the particular case of identical aligned inclusions when the "ellipsoidal symmetry" of both a pair distribution function and inclusion coincide, the MT method yields the Hashin–Shtrikman formula (see [1620]). It may be noted there are certain drawbacks of the extension of Mori-Tanaka method to multiphase composites: Mori-Tanaka moduli may violate the Hashin–Shtrikman bounds [1137] and are nonsymmetric for the general biphase composites [100, 469]. However, Nemat-Nasser and Hori [1121] showed that by the energy-based definition, the effective moduli are the symmetric part of the effective elastic moduli defined by the average stress and strain in the composite and, because of this, must always be symmetric. Nevertheless, the mentioned authors proved the symmetry of Mori-Tanaka moduli for biphase composites in two cases. In the first case, the composites are reinforced by the aligned identical anisotropic fibers with the conditions (8.94), for which the equivalence of Mori-Tanaka and the one-particle approximation of the MEFM approaches takes place. The second one is the composite material reinforced by one sort of isotropic fiber with an arbitrary prescribed random orientation.

This case of isotropic spheroidal inclusions with different prescribed random orientations was systematically investigated by Rammerstorfer and coworkers (see, e.g., comprehensive reviews in [136–138, 413, 414, 1203]). In both cases the MT approach is equivalent to the one particle approximation of the MEFM. However, to the author's knowledge, no models exist that satisfy all theoretical criteria mentioned above for arbitrary phase anisotropy and fiber-orientation distributions.

8.3.4 Some Methods Related to the One-Particle Approximation of the MEFM

The conditional-moment method proposed and intensity developed by the Kiev school of micromechanicians (see for references Khoroshun [758, 761–765]) is worthy of critical note. The concrete numerical results were obtained by truncation of the infinite system of integral equations by taking into account only two- and three-point conditional probabilities and by neglecting fluctuations of stresses within the limits of the components. These are equivalent to acceptance of assumption (7.27), (8.43_2) and consideration of homogeneous inclusions. For analytical estimation of integrals involved similar to the integrals in Eq. (8.48), Khoroshun [762, 763] proposed the analytical representation of the two-point density $\langle V_q(\mathbf{x})|; v_i, \mathbf{x}_i\rangle$ conforming with the shape of the inclusion v_i that with necessity leads to acceptance of Eq. (8.48). The concrete numerical results were obtained for aligned homogeneous ellipsoidal inclusions under different choices of comparison media, either

$$\mathbf{M}^c = \langle \mathbf{M}\rangle \quad \text{or} \quad \mathbf{M}^c = \langle \mathbf{M}^{-1}\rangle^{-1} \tag{8.97}$$

if the compliance of the inclusions is less than that of the matrix or otherwise, respectively. Of course, there is no a priori justification for the specific choice of $\mathbf{M}^c$, not counting the condition that the quadratic form $(\mathbf{M}_1\boldsymbol{\sigma})\boldsymbol{\sigma}$, employed in the proof of the Hashin and Shtrikman variational principle, has a constant sign. The only justification for choosing $\mathbf{M}^c$ in the Voight or Reuss estimation of $\mathbf{M}^*$ (see [762, 763]) is the fact that specific experimental data agree with the computing curves. Although the final explicit general representation for effective moduli (similar to (8.51)) was not presented in the conditional-moment method, the equivalence of assumptions admitted leads to the conclusion that the conditional-moment method can be considered as a particular case of the MEFM presented in Sect. 8.2. In addition, in the conditional-moment method, the shape of the inclusions is taken into account via prescribed anisotropy of the conditional probability density $\langle V_q(\mathbf{x})|; v_i, \mathbf{x}_i\rangle$ (or $\varphi(v_q|; v_i, \mathbf{x}_i)$). For equally probable orientation of ellipsoidal inclusions, it is possible to obtain an isotropic function $\varphi(v_q|; v_i, \mathbf{x}_i)$, and the estimation of the effective compliance $\mathbf{M}^*$ will be invariant with respect to the shape of the inclusion. This result can be avoided easily by taking into account directly the shape of the inclusions via the tensor $\mathbf{Q}_q$, as done by Willis [1620, 1627] on the basis of a variational principle. For identical oriented ellipsoidal inclusions, the estimations of effective moduli obtained in [762, 1620] as well as Eq. (8.54) are equivalent.

We note than in order to solve system (7.45), in any case, at least for homogeneous inclusions, it is not necessary to introduce intermediate concepts—effective fields $\overline{\boldsymbol{\sigma}}_i(\mathbf{x})$ and $\widetilde{\boldsymbol{\sigma}}_{1,\ldots,n}(\mathbf{x})$. System (7.45) is linear in the fields $\langle\boldsymbol{\sigma}|v_1,\mathbf{x}_1;\ldots;v_n,\mathbf{x}_n\rangle(\mathbf{x})$, and on closure [762, 763, 1625], analogous to hypothesis **H2**; it becomes finite and can be solved by the methods of linear algebra; this scheme is implemented by the conditional-moment method by Khoroshun [762, 763]. According to the MEFM, information about geometrical and mechanical characteristics is given by the tensors (8.2_1), and the fields $\langle\widetilde{\boldsymbol{\sigma}}_{1,\ldots,n}(\mathbf{x})\rangle$, in contrast to $\langle\boldsymbol{\sigma}|v_1,\mathbf{x}_1;\ldots;v_n,\mathbf{x}_n\rangle(\mathbf{x})$, are weakly nonuniform. This is why, as noted by in [198, 729], even rough assumptions about the structure of the effective fields (8.9) make it possible to obtain correct results. Different versions of closure assumptions in terms of conditional stress fields analogous to hypothesis **H2** for the effective stress fields are known [762, 763, 1625]. The first-order approximations of these similar approaches are given by Eqs. (8.9) and (8.10_1).

Of course, the principal difference between the hypotheses (8.9_1) and (8.10_1) is beyond the scope of direct substitution of the stress field $\boldsymbol{\sigma}(\mathbf{x})$ for the effective field $\overline{\boldsymbol{\sigma}}(\mathbf{x})$, which leads only to quantitative improvement of the estimates (see for details Sect. 8.8). Of more importance is the application of the assumption (8.9_1) after the consideration of multiparticle inclusion interactions (8.7) rather than before as in hypothesis (8.10_1). What seems to be only a formal trick yields to the discovery of fundamentally new nonlocal effects in the theory of functionally graded materials, which will be demonstrated in Chap. 13. However, even for statistically homogeneous composites being considered in this chapter, it will be shown in Sect. 8.8 that the use of assumption (8.9_1) instead of (8.10_1) can lead to a variation of effective elastic moduli by a factor of two or more, a fact that has been confirmed by classical experimental data [803].

In an other popular generalized singular-approximation method, by Shermergor [1378], which is invariant with respect to the shape of the inclusions, the action of the singular integral operator with the kernel $\boldsymbol{\Gamma}$ in the general integral equation (7.32) is replaced by a multiplication on a constant tensor coinciding with the singular component $\boldsymbol{\Gamma}^\delta = -\mathbf{Q}\delta$ for a spherical domain in Eq. (3.38_1). This automatically implies a number of strong assumptions: $\boldsymbol{\Gamma} = \boldsymbol{\Gamma}^\delta$, the field of the stresses in the components are homogeneous, the "quasi-crystalline" approximation (8.43_2) and assumption (8.48) are accepted, and the functions $\varphi(v_q|;v_i,\mathbf{x}_i)$ are isotropic. For this reason, the singular-approximation method is also a consequence of the one-particle approximation of MEFM proposed. The last statements can also be justified by direct reduction of the MEFM to the singular-approximation method if the mentioned assumptions are accepted. Indeed, in the case of coincidence of elastic moduli of the comparison medium and matrix $\mathbf{L}^c = \mathbf{L}^{(0)}$, $\mathbf{Q}_i^0 = \mathbf{Q}_i$, and $c^{(0)} = 0$, Eqs. (8.57) and (8.63) are reduced to

$$\mathbf{M}^* = \langle \mathbf{B}V\rangle^{-1}\langle \mathbf{R}^{v}V\rangle, \quad \mathbf{L}^* = \langle \mathbf{A}V\rangle^{-1}\langle \mathbf{R}^{\epsilon v}V\rangle. \tag{8.98}$$

If in additional $\mathbf{P}(\mathbf{x}),\ \mathbf{Q}(\mathbf{x}) \equiv$ const. (e.g., for the spherical inclusions), then Eqs. (8.98_1) and (8.98_2) can be recast in the following equivalent forms:

$$\begin{aligned} \mathbf{M}^* &= \langle (\mathbf{M}+\mathbf{b}^M)^{-1}\rangle^{-1}\langle \mathbf{M}(\mathbf{M}+\mathbf{b}^M)^{-1}\rangle, \\ (\mathbf{M}^*+\mathbf{b}^M)^{-1} &= \langle (\mathbf{M}+\mathbf{b}^M)^{-1}\rangle, \end{aligned} \tag{8.99}$$

$$\begin{aligned} \mathbf{L}^* &= \langle (\mathbf{L}+\mathbf{b}^L)^{-1}\rangle^{-1}\langle \mathbf{L}(\mathbf{L}+\mathbf{b}^L)^{-1}\rangle, \\ (\mathbf{L}^*+\mathbf{b}^L)^{-1} &= \langle (\mathbf{L}+\mathbf{b}^L)^{-1}\rangle, \end{aligned} \tag{8.100}$$

respectively, where the auxiliary tensors $\mathbf{b}^M$ and $\mathbf{b}^L$ are introduced by the equalities $\mathbf{b}^M = \mathbf{Q}^{-1} - \mathbf{M}^c,\ \mathbf{b}^L = \mathbf{P}^{-1} - \mathbf{L}^c$. The formula (8.100) are exactly the appropriate relations of the generalized singular-approximation method [1378] widely used for analyses of both the composite materials and polycrystals. Other forms of these equations were obtained by the different methods attributed to the terms "absolute disorder" [814], "one-point approximation" [758], "strong isotropy" [142], and "singular approximation" [491]. It was shown in [491, 1333] that if $\mathbf{L}^c = \mathbf{0}$ and $\mathbf{L}^c = \infty$, we obtain the Reuss [1290] and Voight [1567] approximations, respectively. If $\mathbf{L}^c$ is chosen to be equal to the elastic moduli of components with the maximum and minimum rigidities, we get the upper and lower Hashin–Shtrikman [618, 621] bounds (see for details Sect. 8.4). The results of the conditional moment method [758, 762] are obtained if $\mathbf{L}^c$ (or $\mathbf{M}^c$) is taken according to the relations (8.97). Assuming that $\mathbf{L}^c = \mathbf{L}^*$, we arrive at one of the versions of the self-consistent method (see Sect. 9.2).

In the interesting approach by Dvorak and Srinivas [426], a complete formal similarity of the standard techniques for estimating overall elastic and average stress field in the constituents is established for effective medium and Mori-Tanaka methods (see Sect. 8.3.2) and then for Walpole's [1575] formulation of the Hashin–Shtrikman [621] variational bounds for statistically isotropic composites for which the hypotheses **H1** and (8.43_2) are valid. They detected a common analytical regularity of all these methods that allows one to generate any number of related analytical representation for effective properties from other choices of the comparison medium made such that the resulting moduli predictions do not violate the bounds. The generalization of their results for the case of composites with coated inclusions is noteworthy.

8.3.5 Some Analytical Representations for Effective Moduli

Consider first the simple example of a composite consisting of an isotropic matrix containing the identical spherical inclusions of arbitrary structure with uniform random orientation. Then the representations of the effective elastic moduli $\mathbf{L}^* = (3k^*, 2\mu^*)$ obtained by the quasi-crystalline approximation of the MEFM (8.63) and by the MTM (8.80) coincide

$$k^* = k^{(0)} + \frac{c^{(1)} k^{R\epsilon}}{1 - c^{(1)} 3P_1 3k^{R\epsilon}}, \quad \mu^* = \mu^{(0)} + \frac{c^{(1)} \mu^{R\epsilon}}{1 - c^{(1)} 2P_2 2\mu^{R\epsilon}}, \tag{8.101}$$

$$3P_1 = \frac{1}{3k^{(0)} + 4\mu^{(0)}}, \quad 2P_2 = \frac{3(k^{(0)} + 2\mu^{(0)})}{5(3k^{(0)} + 4\mu^{(0)})}, \tag{8.102}$$

where $\mathbf{P}^0 = (3P_1, 2P_2)$ and $(3k^{R\epsilon}, 2\mu^{R\epsilon})$ are the components of the isotropic average polarization tensor $\mathbf{R}^{v\epsilon} = \langle \mathbf{L}_1 \mathbf{A} \rangle$. If the moduli of inclusions are taken as homogeneous and isotropic, then

$$3k^{R\epsilon} = \frac{3k_1^{(1)}}{1 + 3P_1 3k_1^{(1)}}, \quad 2\mu^{R\epsilon} = \frac{2\mu_1^{(1)}}{1 + 2P_2 2\mu^{(1)}}, \tag{8.103}$$

and Eq. (8.101) gives

$$k^* = k^{(0)} + \frac{c^{(1)} k_1^{(1)}}{1 + c^{(0)} 3P_1 3k_1^{(1)}}, \quad \mu^* = \mu^{(0)} + \frac{c^{(1)} \mu_1^{(1)}}{1 + c^{(0)} 2P_2 2\mu_1^{(1)}}. \tag{8.104}$$

The elastic moduli of composites with identical homogeneous aligned ellipsoidal inclusions (with the semi-axes $a_1 = a_2 = a$, a_3, aspect ratio $\gamma = a_3/a$, and a symmetry axis $\mathbf{n}$) also coincide for the MEFM (8.63) and the MTM (8.80) if $\mathbf{P}^0 = \mathbf{P}$

$$\mathbf{L}^* = \mathbf{L}^{(0)} + c^{(1)} \big[\mathbf{I} - c^{(1)} \mathbf{R}^{\epsilon v} \mathbf{P}^0\big]^{-1} \mathbf{R}^{\epsilon v}. \tag{8.105}$$

The tensor $\bar{\mathbf{D}} \equiv \mathbf{I} - n^{(1)} \mathbf{R}^{\epsilon v} \mathbf{P}^0$ presented in Eq. (8.105) can be expressed in the $\mathbf{P}$-basis (21.2.7) $\bar{\mathbf{D}} = (\bar{d}_1, \ldots, \bar{d}_6)$ through the representations $\mathbf{P}^0 = (p_1, \ldots, p_6)$ (3.157) and $\mathbf{R}^{\epsilon v} = (r_1, \ldots, r_6)$ (3.160) as

$$\begin{aligned} &\bar{d}_1 = 1/2 - c^{(1)}(2r_1 p_1 + r_3 p_3), \ \bar{d}_2 = 1 - c^{(1)} r_2 p_2, \\ &\bar{d}_3 = -c^{(1)}(2r_1 p_3 + r_3 p_6), \bar{d}_4 = -c^{(1)}(2r_3 p_1 + r_6 p_3), \\ &\bar{d}_5 = 2 - r_5 p_5/2; \ \bar{d}_6 = 1 - c^{(1)}(2r_3 p_3 + r_6 p_6), \end{aligned} \tag{8.106}$$

which allows the components of the $\mathbf{P}$-basis representation of the effective elastic moduli (see, e.g., [745, 746])

$$\mathbf{L}^* = (k^*, 2m^*, l^*, l^*, 2\mu^*, n^*), \tag{8.107}$$

$$k^* = \frac{\mu^{(0)}}{1 - 2\nu^{(0)}} + \frac{r_1 \bar{d}_6 - r_3 \bar{d}_3}{\Delta}, \ m^* = \mu^{(0)} + \frac{r_2}{2\bar{d}_2}, \ \mu^* = \mu^{(0)} + \frac{r_5}{2\bar{d}_5}, \tag{8.108}$$

$$l^* = \frac{2\mu^{(0)}\nu^{(0)}}{1-2\nu^{(0)}} + \frac{r_3\bar{d}_6 - r_6\bar{d}_3}{\Delta}, \quad n^* = \frac{2\mu^{(0)}(1-\nu^{(0)})}{1-2\nu^{(0)}} + \frac{2(r_6\bar{d}_1 - r_3\bar{d}_4)}{\Delta}, \tag{8.109}$$

where $\Delta = 2(\bar{d}_1\bar{d}_6 - \bar{d}_3\bar{d}_4)$. The case of uniform random orientation of spheroids is easily obtained from (8.105) by replacing the tensors $\mathbf{R}^{\epsilon v}\mathbf{P}^0$ and $\mathbf{R}^{\epsilon v}$ with its isotropic form according to the averaging operations (21.2.15)

$$\begin{aligned}
\mathbf{L}^* &= (3k^*, 2\mu^*), \quad 3k^* = 3k^{(0)} + 3R_1/\tilde{d}_1, \quad 2\mu^* = 2\mu^{(0)} + 2R_2/\tilde{d}_2, \\
3R_1 &= \frac{1}{9}[4(r_1+r_3)+r_6], \quad 2R_2 = \frac{1}{15}[r_1 + 2r_2 - 2(r_3 - 3r_5/4) + r_6], \\
\tilde{d}_1 &= 1 - \frac{1}{3}[8r_1p_1 + 6r_3p_3 + 4(r_1p_3 + r_3p_1) + 2(r_3p_6 + r_6p_3) + r_6p_6], \\
\tilde{d}_2 &= 1 - \frac{2}{15}[2r_1p_1 + 3(r_3p_3 + r_2p_2) - 2(r_1p_3 + r_3p_1) + \frac{3}{4}r_5p_5 \\
&\quad -(r_3p_6 + r_6p_3) + r_6p_6].
\end{aligned} \tag{8.110}$$

In particular, for the limiting case of coin-shaped microcracks (3.158), Eq. (8.110) is reduced to the representation

$$\begin{aligned}
\frac{k^*}{k^{(0)}} &= \left(1 + \frac{19}{9}\frac{1-(\nu^{(0)})^2}{1-2\nu^{(0)}}\epsilon\right)^{-1}, \\
\frac{\mu^*}{\mu^{(0)}} &= \left(1 + \frac{32}{45}\frac{(1-\nu^{(0)})(5-\nu^{(0)})}{2-\nu^{(0)}}\epsilon\right)^{-1}
\end{aligned} \tag{8.111}$$

allowing subsequent simplification in the dilute case when the crack density parameter $\epsilon = n^{(1)}a^3 \to 0$

$$\begin{aligned}
\frac{k^*}{k^{(0)}} &= 1 - \frac{16}{9}\frac{1-(\nu^{(0)})^2}{1-2\nu^{(0)}}\epsilon, \\
\frac{\mu^*}{\mu^{(0)}} &= 1 - \frac{32}{45}\frac{(1-\nu^{(0)})(5-\nu^{(0)})}{2-\nu^{(0)}}\epsilon.
\end{aligned} \tag{8.112}$$

The representations (8.111) were obtained in [89] by the MTM (8.80) (see also a conductivity problem for the 2D case [88]) which is equivalent to the MEFM (8.63) in the case $\mathbf{P}_i^0 = \mathbf{P}_i$ being considered. Formulae (8.112) were proposed in [161, 1581].

In a similar manner, for a composite material made from a transversely isotropic matrix and infinite cylindrical fibers with the axes coinciding with the axis of symmetry of the matrix and fiber, the known components of the **P**-basis (21.2.7) representations of the tensors $\mathbf{L}^{(i)}, \mathbf{P}$ (3.171) and (3.172) allow the expansion coefficients of the effective moduli $\mathbf{L}^*$ (8.107) over the $\mathbf{P}$-basis

$$k^* = k^{(0)} + c^{(1)}k_1^{(1)}\Big[1 + \frac{c^{(0)}k_1^{(1)}}{k^{(0)} + m^{(0)}}\Big]^{-1},$$

$$m^* = m^{(0)} + c^{(1)}m_1^{(1)}\Big[1 + c^{(0)}\frac{m_1^{(1)}(k^{(0)} + 2m^{(0)})}{2m^{(0)}(k^{(0)} + m^{(0)})}\Big]^{-1},$$

$$l^* = l^{(0)} + c^{(1)}l_1^{(1)}\Big[1 + \frac{c^{(0)}k_1^{(1)}}{k^{(0)} + m^{(0)}}\Big]^{-1},$$

$$\mu^* = \mu^{(0)} + c^{(1)}\mu_1^{(1)}\Big[1 + \frac{c^{(0)}\mu^{(1)}}{2\mu^{(0)}}\Big]^{-1},$$

$$n^* = n^{(0)} + c^{(1)}n_1^{(1)} - \frac{c^{(1)}c^{(0)}l_1^{(1)2}}{k^{(0)} + m^{(0)}}\Big[1 + \frac{c^{(0)}k_1^{(1)}}{k^{(0)} + m^{(0)}}\Big]^{-1}. \tag{8.113}$$

In closing, we will consider the analytical representations of elastic moduli obtained by the method of the conditional moments [762, 763, 765]. In particular, for a macroisotropic material of granular-matrix structure consisting of two isotropic phases, the effective moduli $\mathbf{L}^* = (3k^*, 2\mu^*)$ are

$$k^* = \langle k\rangle - \frac{c^{(0)}c^{(1)}(k_1^{(1)})^2}{c^{(0)}k^{(1)} + c^{(1)}k^{(0)} + \tilde{k}}, \quad \mu^* = \langle \mu\rangle - \frac{c^{(0)}c^{(1)}(\mu_1^{(1)})^2}{c^{(0)}\mu^{(1)} + c^{(1)}\mu^{(0)} + \tilde{\mu}},$$
$$\tilde{k} = \frac{4}{3}\mu^c, \ \tilde{\mu} = \frac{\mu^c(9k^c + 8\mu^c)}{6(k^c + 2\mu^c)}, \tag{8.114}$$

where the moduli k^c and μ^c of the comparison medium can be arbitrarily selected, and $k^c = \langle k\rangle$, $\mu^c = \langle \mu\rangle$ and $k^c = \langle k^{-1}\rangle^{-1}$, $\mu^c = \langle \mu^{-1}\rangle^{-1}$ in the cases (8.97_1) and (8.97_2), respectively. For two-component materials of a unidirectionally fibrous structure, whose effective elastic properties are transversally isotropic, the effective constitutive equation can be presented as

$$\langle \sigma_{ij}\rangle = (\lambda_{11}^* - \lambda_{12}^*)\langle \varepsilon_{ij}\rangle + (\lambda_{12}^*\langle \varepsilon_{rr}\rangle + \lambda_{13}^*\langle \varepsilon_{33}\rangle)\delta_{ij},$$
$$\langle \sigma_{33}\rangle = \lambda_{13}^*\langle \varepsilon_{rr}\rangle + \lambda_{33}^*\langle \varepsilon_{33}\rangle, \quad \langle \sigma_{i3}\rangle = 2\lambda_{44}^*\langle \varepsilon_{i3}\rangle,$$

where $i, j, r = 1, 2$, and

$$\frac{1}{2}(\lambda_{11}^* + \lambda_{12}^*) = \langle \lambda + \mu\rangle - \frac{c^{(0)}c^{(1)}(\lambda_1^{(1)} + \mu_1^{(1)})^2}{c^{(0)}(\lambda^{(1)} + \mu^{(1)}) + c^{(1)}(\lambda^{(0)} + \mu^{(0)}) + \mu^c},$$

$$\frac{1}{2}(\lambda_{11}^* - \lambda_{12}^*) = \langle \lambda + \mu\rangle - \frac{c^{(0)}c^{(1)}(\mu_1^{(1)})^2(\lambda^c + 3\mu^c)}{(c^{(0)}\mu^{(1)} + c^{(1)}\mu^{(0)})(\lambda^c + 3\mu^c) + \mu^c(\lambda^c + \mu^c)},$$

$$\lambda_{13}^* = \langle\lambda\rangle - \frac{c^{(0)}c^{(1)}\lambda_1^{(1)}(\lambda_1^{(1)} + \mu_1^{(1)})}{c^{(0)}(\lambda^{(1)} + \mu^{(1)}) + c^{(1)}(\lambda^{(0)} + \mu^{(0)}) + \mu^c},$$

$$\lambda_{44}^* = \langle\mu\rangle - \frac{c^{(0)}c^{(1)}(\mu_1^{(1)})^2}{c^{(0)}\mu^{(1)} + c^{(1)}\mu^{(0)} + \mu^c} \quad (8.115)$$

Here $f_1^{(1)} = f^{(1)} - f^{(0)}$ $(f = \lambda, \mu)$, and $\lambda^c = \langle\lambda\rangle$, $\mu^c = \langle\mu\rangle$ and $\lambda^c = \langle k^{-1}\rangle^{-1} - 2\langle\mu^{-1}\rangle/3$, $\mu^c = \langle\mu^{-1}\rangle^{-1}$.

8.4 Hashin–Shtrikman Bounds Estimations

Variational principles essentially bound the elastic energy of CMs (see Sect. 6.4). Variational bounds are generated via substitution of approximate fields into the strict energy bounds obtained by the volume average of the energy density with the help of lower-order correlation functions. The variational methods represent the most rigorous trend and security aspect of micromechanics of statistically homogeneous media. A critical simplification is a choice of a trial function (e.g., polarization tensor) as a piecewise constant approximation with a constant field inside each phase. After that in Sect. 8.4.1 (following the papers by Willis [1620, 1624] and [1625]), the trial function is substituted into the formula for the strain energy expressed in terms of so-called general integral equation (GIE) (7.32) universally recognized as an exact one (incorrectness of this recognition is explained in Chaps. 7 and 9). However, for obtaining HS bounds in Sect. 8.4.2, we will use a more general trial function in the form of a piecewise constant effective field (rather than the polarization tensor) acting in each phase that gives us an opportunity to take into account a possible inhomogeneity of the polarization tensor inside the noncanonical (i.e., nonellipsoidal) or/and inhomogeneous inclusions.

8.4.1 The Scheme by Willis for Obtaining Hashin–Shtrikman Bounds

In Sect. 6.4 the Hashin–Shtrikman (HS) energy bounds were presented as

$$2(W - W^c) \underset{(\leq)}{\geq} -\mathcal{H}, \quad (8.116)$$

$$\mathcal{H} = (\boldsymbol{\tau}^*, (\mathbf{L}_1^c)^{-1}\boldsymbol{\tau}^*) - 2(\boldsymbol{\tau}^*, \langle\boldsymbol{\varepsilon}\rangle) - (\boldsymbol{\tau}^*, \mathbf{U}\boldsymbol{\tau}^*) \quad (8.117)$$

as $\mathbf{L}_1$ is positive/negative semi-defined. A trial polarization tensor $\boldsymbol{\tau}^*$ extremizes the functional $\mathcal{H}(\boldsymbol{\tau}^*)$ at the real values of the field $\boldsymbol{\tau}(\mathbf{x})$ (Hashin–Shtrikman variational principle)

$$\delta\mathcal{H}(\boldsymbol{\tau}) = 0 \tag{8.118}$$

with the stationary value equals $(W - W^c) = (\boldsymbol{\tau}, \langle\boldsymbol{\varepsilon}\rangle)/2$. We reproduced the functional $\mathcal{H}$ (8.117) with the classical operator $\mathbf{U}\boldsymbol{\tau}(\mathbf{x}) := \int \mathbf{U}(\mathbf{x}-\mathbf{y})[\boldsymbol{\tau}(\mathbf{y})-\langle\boldsymbol{\tau}\rangle]\,d\mathbf{y}$ corresponding to the classical GIE (7.32) rather than with the operator ${}^{\mathcal{P}}\mathbf{U}$ (6.200) as in Eq. (6.204) (this case will be considered in Chap. 9).

In the case of homogeneous inclusions, Willis [1620] proposed a piecewise approximation of a trial field

$$\boldsymbol{\tau}^*(\mathbf{x}, \alpha) = \sum_{i=0}^{1} \boldsymbol{\tau}_i V^{(i)}(\mathbf{x}, \alpha), \ \ \boldsymbol{\tau}_i = \text{const.}, \tag{8.119}$$

where α is a random parameter specifying the concrete realization of the field $V(\mathbf{x})$. We also consider CMs with a piecewise constant approximation of the stiffness tensor

$$\mathbf{L}(\mathbf{x}, \alpha) = \sum_{i=0}^{1} \mathbf{L}^{(i)} V^{(i)}(\mathbf{x}, \alpha), \ \ \mathbf{L}^{(i)} = \text{const.}, \tag{8.120}$$

and ellipsoidal shape of homogeneous identical inclusions v_i (3.183).

For the piecewise constant approximation of the polarization tensor (8.119), the first two summands in the representation $\mathcal{H}$ (8.117) can be estimated immediately

$$(\boldsymbol{\tau}^*, \mathbf{L}_1\boldsymbol{\tau}^*) - 2(\boldsymbol{\tau}^*, \langle\boldsymbol{\varepsilon}\rangle) = \sum_{r=0}^{1} c^{(r)} \Big\{\boldsymbol{\tau}_r [\mathbf{L}_1^{(r)}]^{-1} \boldsymbol{\tau}_r - 2\tau_r \langle\boldsymbol{\varepsilon}\rangle\Big\}. \tag{8.121}$$

Evaluation of the third summand $(\boldsymbol{\tau}^*, \mathbf{U}\boldsymbol{\tau}^*)$ in $\mathcal{H}$ (8.117) is performed by the use of invariantness property of the operator $\mathbf{U}$ (6.201)

$$\begin{aligned}(\boldsymbol{\tau}^*, \mathbf{U}\boldsymbol{\tau}^*) &= n(\boldsymbol{\tau}_1 - \boldsymbol{\tau}_0) \int_{v_i} V_i(\mathbf{x}) \int_{R^d} \mathbf{U}(\mathbf{x}-\mathbf{y}) \\ &\times [V_j(\mathbf{y})\varphi(v_j, \mathbf{x}_j | v_i, \mathbf{x}_i) - c^{(1)}] d\mathbf{y} d\mathbf{x} (\boldsymbol{\tau}_1 - \boldsymbol{\tau}_0) \\ &= c^{(1)}(\boldsymbol{\tau}_1 - \boldsymbol{\tau}_0)\bar{v}_i \mathbf{T}_{ii}(\boldsymbol{\tau}_1 - \boldsymbol{\tau}_0) + n \int_{v_i} V_i(\mathbf{x}) \int_{R^d} \mathbf{U}(\mathbf{x}-\mathbf{y}) \\ &\times [V_j(\mathbf{y})\varphi(v_j, \mathbf{x}_j |; v_i, \mathbf{x}_i) - c^{(1)}] d\mathbf{y} d\mathbf{x} (\boldsymbol{\tau}_1 - \boldsymbol{\tau}_0),\end{aligned} \tag{8.122}$$

where $\mathbf{x}$ belongs to the fixed representative inclusion $\mathbf{x} \in v_i$, while $\mathbf{y}$ pertains to the mowing inclusion v_j. For the homogeneous ellipsoidal inclusions, the tensor $\mathbf{T}_{ii}$ in (8.122$_2$) is expressed through the Eshelby tensor $\mathbf{T}_{ii} = (\bar{v})_i^{-2}\mathbf{P}_i = (\bar{v})_i^{-2}\mathbf{S}_i\mathbf{M}^c$ (4.96) assumed to be below in this subsection. The first equality (8.122$_1$) was obtained due to the fact that integration domains in the integrals $\int_{v_i}(\cdot)d\mathbf{x}$

and $\int_{R^d}(\cdot)d\mathbf{y}$ involving estimation of the inner product $(\boldsymbol{\tau}^*, \mathbf{U}\boldsymbol{\tau}^*)$ are reduced to the integration over the volume of inclusions v_i and v_j because $\boldsymbol{\tau}(\mathbf{x}) - \boldsymbol{\tau}_0 = \boldsymbol{\tau}_1 - \boldsymbol{\tau}_0$ vanishes outside v_i and v_j, and, moreover, $\langle \boldsymbol{\tau} \rangle - \boldsymbol{\tau}_0 = c^{(1)}(\boldsymbol{\tau}_1 - \boldsymbol{\tau}_0)$. The mentioned vanishing of $\boldsymbol{\tau}(\mathbf{x}) - \boldsymbol{\tau}_0$ allows us to use the two-particle distribution function $\varphi(v_j, \mathbf{x}_j | v_i, \mathbf{x}_i)$ (see for details Willis [1621]) instead of the corresponding expressions by Willis [1620] containing two-point correlation functions $\langle V_i(\mathbf{x}) V_j(\mathbf{y}) \rangle$. Furthermore, the second equality (8.122$_2$) was obtained by extraction of the constituent corresponding to the particle $v_i = v_j$ from the internal integral $\int_{R^d}(\cdot)d\mathbf{y}$ (8.122$_1$) that can be formally presented by the equality isolating the term with $\mathbf{x}_i = \mathbf{x}_j$

$$\varphi(v_j, \mathbf{x}_j | v_i, \mathbf{x}_i) = \delta(\mathbf{x}_i - \mathbf{x}_j)\varphi(v_j, \mathbf{x}_j) + \varphi(v_j, \mathbf{x}_j |; v_i, \mathbf{x}_i). \tag{8.123}$$

The next step in rearrangement of Eq. (8.122$_2$) is breaking up the integral with respect to $\mathbf{y}$ over the all space R^d into two parts, the first part being an integral over the correlation hole v_i^0 and the second integral over its compliment ${}^c v_i^0 = R^d \backslash v_i^0$

$$\begin{aligned} n\int_{v_i} V_i(\mathbf{x}) \int_{R^d} &\mathbf{U}(\mathbf{x}-\mathbf{y})[V_j(\mathbf{y})\varphi(v_j, \mathbf{x}_j |; v_i, \mathbf{x}_i) - c^{(1)}]d\mathbf{y}d\mathbf{x} \\ &= -nc^{(1)} \int_{v_i} V_i(\mathbf{x}) \int_{v_i^0} \mathbf{U}(\mathbf{x}-\mathbf{y})d\mathbf{y}d\mathbf{x} + n\int_{v_i} V_i(\mathbf{x}) \\ &\times \int_{{}^c v_i^0} \mathbf{U}(\mathbf{x}-\mathbf{y})[V_j(\mathbf{y})\varphi(v_j, \mathbf{x}_j |; v_i, \mathbf{x}_i) - c^{(1)}]d\mathbf{y}d\mathbf{x} := c^{(1)}\widetilde{\mathbf{P}}_i^0. \end{aligned} \tag{8.124}$$

The representation for the introduced tensor $\widetilde{\mathbf{P}}_i^0$ is simplified

$$\widetilde{\mathbf{P}}_i^0 = \mathbf{P}_i^0 := \mathbf{P}(v_i^0) \tag{8.125}$$

in the case of acceptance of the ellipsoidal symmetry hypothesis **H3** vanishing the second integral in the right-hand side of Eq. (8.124).

We consider the matrix CM with the matrix modulus $\mathbf{L}^{(0)}$ which can be used as a comparison modulus $\mathbf{L}^c = \mathbf{L}^{(0)}$ that leads to identical vanishing of the polarization tensor $\boldsymbol{\tau}_0 = \mathbf{0}$ in the matrix. Then substitution of Eqs. (8.121), (8.122), and (8.124) into Eqs. (8.117) and (8.118) leads to the following linear algebraic equation:

$$[(\mathbf{L}_1^{(1)})^{-1} + \mathbf{P}_i]\boldsymbol{\tau}_1 - c^{(1)}\mathbf{P}_i^0\boldsymbol{\tau}_1 = \langle \boldsymbol{\varepsilon} \rangle. \tag{8.126}$$

which can be immediately solved for $\boldsymbol{\tau}_1$ to obtain a representation for HS bounds

$$\mathbf{L}^{*(\mathrm{HS}[1])} = \mathbf{L}^{(0)} + c^{(1)}[\mathbf{I} - c^{(1)}\mathbf{L}_1^{(1)}\mathbf{A}_i^W\mathbf{P}_i^0]^{-1}\mathbf{L}_1^{(1)}\mathbf{A}_i, \quad \mathbf{A}_i^W := (\mathbf{I} + \mathbf{L}_1^{(1)}\mathbf{P}_i^W)^{-1}, \tag{8.127}$$

where only for the ellipsoidal inclusions $\mathbf{P}^W = \mathbf{P}$. The estimation (8.127) is a lower bound for $\mathbf{L}^*$ if the matrix happens to be the most compliant phase in the CM and a upper bound if the matrix is the stiffest phase.

For CMs containing nonellipsoidal inclusion, the presented scheme is not altered; however, we need to keep in mind that the tensor

$$\mathbf{T}_{ii} := \langle \mathbf{T}_i(\mathbf{x}) \rangle_{(i)}, \quad \mathbf{T}_i(\mathbf{x}) = (\bar{v}_i)^{-1} \int\int V_i(\mathbf{y}) \mathbf{U}(\mathbf{x}-\mathbf{y})\, d\mathbf{y} \tag{8.128}$$

in Eq. (8.122_2) is not currently expressed through the Eshelby tensor via $\mathbf{P}_i$. Because of this, Willis [1620] proposes to replace the tensor $\mathbf{P}_i^W$ in Eqs. (8.127) by the tensor $\bar{v}_i^2 \mathbf{T}_{ii}$ (8.128)

$$\mathbf{P}_i^W \to \bar{v}_i^2 \mathbf{T}_{ii}, \tag{8.129}$$

where $\mathbf{P}_i^W \not\equiv \mathbf{P}_i$ in general. Furthermore, for a noncanonical shape of the inclusion v_i, there is no hope that the correlation hole v_i^0 is an ellipsoid and a microtopology of the CM has "ellipsoidal symmetry" **H3** which servers for vanishing of the second integral in the right-hand side of Eq. (8.124_1). Because of this, the modified Eqs. (8.126) and (8.127) are in fact obtained for CM with a special microtopology (we can call it "modified ellipsoidal symmetry") which corresponds to vanishing of the second integral in the right hand side of Eq. (8.124_1) (otherwise, we need to replace $\mathbf{P}_i^0 \to \widetilde{\mathbf{P}}_i^a$ with the necessary estimation of the second integral in the right-hand side of Eq. (8.124_1)).

It should be mentioned that for the considered CMs with homogeneous ellipsoidal inclusions, a piecewise constant approximation of the polarization tensor (8.119) automatically implies acceptance of the initial hypotheses **H1a**, **H1b** [which is a corollary of **H1a**], and **H2**, while the hypothesis of "ellipsoidal symmetry" **H3** (or "modified ellipsoidal symmetry") is served for vanishing of the second integral in the right-hand side of Eq. (8.124_1).

For comparison of HS bounds $\mathbf{L}^{*\mathrm{HS}[1]}$ (8.127) with the effective moduli estimations, we reproduce the results by EFM (8.63) CM with the identical aligned inhomogeneous inclusions of noncanonical shape and $\mathbf{L}^c = \mathbf{L}^{(0)}$

$$\mathbf{L}^{*[2]} = \mathbf{L}^{(0)} + c^{(1)} \mathbf{R}_i^v (\mathbf{I} - c^{(1)} \mathbf{P}_i^0 \mathbf{R}_i^v)^{-1}, \tag{8.130}$$

$$\mathbf{L}^{*[1]} = \mathbf{L}^{(0)} + c^{(1)} \mathbf{L}_1^{(1)} (\mathbf{I} + \mathbf{P}_i \mathbf{L}_1^{(1)} - c^{(1)} \mathbf{P}_i^0 \mathbf{L}_1^{(1)})^{-1}, \tag{8.131}$$

where the representation (8.130) is reduced to (8.131) for the homogeneous ellipsoidal inclusions when $\mathbf{R}_i = \bar{v}_i \mathbf{L}_1^{(1)} \mathbf{A}_i$. For an arbitrary number of phases and the general case of the reference modulus $\mathbf{L}^c \neq \mathbf{L}^{(0)}$, the representation of $\mathbf{L}^*$ (reducing to Eq. (8.130_1) in the particular case) is presented in Sect. 8.3.1. Furthermore, the representation (8.131) for the homogeneous ellipsoidal inclusions completely coincides with the Hashin–Shtrikman bounds (8.127). The reasons of this coincidence as well as the connections between some different self-consistent approaches were in detail analyzed by Willis [1620, 1621, 1625] (see also [662, 663, 1013, 1240, 1608]); the similar connections for the dynamic case were established in [1610]. Probably the most general reason of the coincidence of Eqs. (8.127) and (8.131) for the CMs containing the ellipsoidal homogeneous aligned inclusions is that in the considered case of CMs, the initial hypotheses **H1a**, **H1b** [which is a corollary of **H1a**], **H2**, and **H3** are uniquely exploited at obtaining both representations (8.127) and (8.131).

8.4.2 *EFH and Hashin–Shtrikman Bounds for CMs with Noncanonical Inhomogeneous Inclusions*

We consider the CM with the identical aligned inhomogeneous inclusions of non-canonical shape. According to hypothesis **H1**, each inclusion v_i being considered is located in a homogeneous field $\overline{\boldsymbol{\varepsilon}}(\mathbf{x}_i)$, while the matrix is subjected to the effective field $\overline{\boldsymbol{\varepsilon}}_0 = (\mathbf{L}_1^{(0)})^{-1}\langle\boldsymbol{\tau}\rangle_0$

$$\boldsymbol{\varepsilon}(\mathbf{x}) = \overline{\boldsymbol{\varepsilon}}_i(\mathbf{x}) + \int \mathbf{U}(\mathbf{x}-\mathbf{y})V_i(\mathbf{y})\big[\mathbf{L}_1(\mathbf{y})\boldsymbol{\varepsilon}(\mathbf{y}) - \langle\boldsymbol{\tau}\rangle_0\big]d\mathbf{y}. \tag{8.132}$$

In view of the linearity of the problem, there exist constant fourth- and second-rank tensors $\mathbf{A}_i(\mathbf{x})$, $\mathbf{R}_i(\mathbf{x})$ and $\mathbf{C}_i(\mathbf{x})$, $\mathbf{F}_i(\mathbf{x})$, such that ($\mathbf{x} \in R^d$)

$$\boldsymbol{\varepsilon}(\mathbf{x}) = \mathbf{A}_i(\mathbf{x})\overline{\boldsymbol{\varepsilon}}(\mathbf{x}_i) + \mathbf{C}_i(\mathbf{x})\overline{\boldsymbol{\varepsilon}}_0, \quad \bar{v}_i\boldsymbol{\tau}(\mathbf{x}) = \mathbf{R}_i(\mathbf{x})\overline{\boldsymbol{\varepsilon}}(\mathbf{x}_i) + \mathbf{F}_i(\mathbf{x})\overline{\boldsymbol{\varepsilon}}_0, \tag{8.133}$$

where $\mathbf{R}_i(\mathbf{x}) = \bar{v}_i\mathbf{L}_1^{(i)}(\mathbf{x})\mathbf{A}_i(\mathbf{x})$, $\mathbf{F}_i(\mathbf{x}) = \bar{v}_i\mathbf{L}_1^{(i)}(\mathbf{x})\mathbf{C}_i(\mathbf{x})$. We will also use the following representations:

$$\bar{v}_i(\boldsymbol{\tau}(\mathbf{x}) - \langle\boldsymbol{\tau}\rangle_0) = \mathbf{R}_i(\mathbf{x})\overline{\boldsymbol{\varepsilon}}(\mathbf{x}_i) + \overline{\mathbf{F}}_i(\mathbf{x})\overline{\boldsymbol{\varepsilon}}_0, \tag{8.134}$$

$$\frac{1}{\bar{v}_i}\int \mathbf{U}(\mathbf{x}-\mathbf{y})V_i(\mathbf{y})\big[\mathbf{R}_i(\mathbf{y})\overline{\boldsymbol{\varepsilon}}(\mathbf{y}_i) + \overline{\mathbf{F}}_i(\mathbf{y})\overline{\boldsymbol{\varepsilon}}_0]d\mathbf{y} = (\mathbf{A}_i(\mathbf{x}) - \mathbf{I})\overline{\boldsymbol{\varepsilon}}_i + \mathbf{C}_i(\mathbf{x})\overline{\boldsymbol{\varepsilon}}_0, \tag{8.135}$$

where Eq. (8.135) was obtained by substitution of Eqs. (8.133_1) and (8.134) into Eq. (8.132). Hereafter

$$\mathbf{C}_i(\mathbf{x}) \equiv \bar{v}_i\mathbf{A}_i(\mathbf{x})\mathbf{T}_i(\mathbf{x}-\mathbf{x}_i)\mathbf{L}_1^{(0)}, \tag{8.136}$$

$$\bar{\mathbf{F}}_i(\mathbf{x}) = \bar{v}_i(\mathbf{L}_1^{(i)}\mathbf{A}_i(\mathbf{x})\mathbf{T}_i(\mathbf{x}-\mathbf{x}_i) - \mathbf{I})\mathbf{L}_1^{(0)}, \tag{8.137}$$

where the tensor $\mathbf{T}_i(\mathbf{x}-\mathbf{x}_i)$ was defined by Eq. (4.131_2).

In opposite to Eq. (8.119), we will use piecewise constant approximation for the effective fields $\overline{\boldsymbol{\varepsilon}}_i$ rather than for the stress polarization tensors $\boldsymbol{\tau}^*(\mathbf{x})$ ($i = 0, 1$)

$$\boldsymbol{\tau}^*(\mathbf{x}, \alpha) = \sum_{i=0}^{1} (\mathbf{R}_i^v(\mathbf{x})\overline{\boldsymbol{\varepsilon}}_i + \mathbf{F}_i^v(\mathbf{x})\overline{\boldsymbol{\varepsilon}}_0) V_i(\mathbf{x}, \alpha), \ \ \overline{\boldsymbol{\varepsilon}}_i \equiv \text{const.}, \tag{8.138}$$

where hereafter $\mathbf{C}_0, \mathbf{F}_0 \equiv \mathbf{0}, \mathbf{B}_0 = \mathbf{I}, \ \mathbf{R}_0 \equiv \mathbf{L}_1^{(0)} \bar{v}_j$ and $\mathbf{f}_j(\mathbf{x}) := \bar{v}_j^{-1} \mathbf{f}_j^v$, $(\mathbf{f} = \mathbf{R}, \mathbf{F})$. Thus, the approximation (8.138) makes it possible a variation of the trial functions $\boldsymbol{\tau}^*(\mathbf{x}, \alpha)$ inside the inclusions, and this variation directly depends (trough the tensors $\mathbf{R}_i^v(\mathbf{x})$ and $\mathbf{F}_i^v(\mathbf{x})$ (8.22)) on both the elastic inhomogeneity and noncanonical shape of inclusions. The simplified piecewise constant approximation of the trial field is obtained by averaging over the phase volumes ($i = 0, \ldots, N$)

$$\boldsymbol{\tau}^*(\mathbf{x}, \alpha) = \sum_{i=0}^{1} (\mathbf{R}_i^v \overline{\boldsymbol{\varepsilon}}_i + \mathbf{F}_i^v \overline{\boldsymbol{\varepsilon}}_0) V_i(\mathbf{x}, \alpha), \ \ \overline{\boldsymbol{\varepsilon}}_i \equiv \text{const.} \tag{8.139}$$

Then from Eq. (8.138), the first two summands in the right-hand side of Eq. (8.139) can be presented in the form

$$(\boldsymbol{\tau}^*, (\mathbf{L}_1)^{-1}\boldsymbol{\tau}^*) - 2(\boldsymbol{\tau}^*, \langle \boldsymbol{\varepsilon} \rangle) = \sum_{i=0}^{1} \Big(\mathbf{R}_i^v(\mathbf{x})\overline{\boldsymbol{\varepsilon}}_i + \mathbf{F}_i^v(\mathbf{x})\overline{\boldsymbol{\varepsilon}}_0, \mathbf{A}_i(\mathbf{x})\overline{\boldsymbol{\varepsilon}}_i + \mathbf{C}_i(\mathbf{x})\overline{\boldsymbol{\varepsilon}}_0 - 2\langle \boldsymbol{\varepsilon} \rangle \Big). \tag{8.140}$$

According to the properties of the operator $\mathbf{U}$ similar to (6.200), we can limit the estimation of integrals in the item $(\boldsymbol{\tau}^*, \mathbf{U}\boldsymbol{\tau}^*)$ only over the volume of inclusions

$$\begin{aligned}(\boldsymbol{\tau}^*, \mathbf{U}\boldsymbol{\tau}^*) = -n \int_{v_i} & V_i(\mathbf{x})\big[\mathbf{R}_i^v(\mathbf{x})\overline{\boldsymbol{\varepsilon}}_i + \overline{\mathbf{F}}_i^v(\mathbf{x})\overline{\boldsymbol{\varepsilon}}_0\big] \\ & \times \int_{R^d} \mathbf{U}(\mathbf{x}-\mathbf{y}) \Big\{ V_j(\mathbf{y}) \Big[\mathbf{R}_j^v(\mathbf{y})\overline{\boldsymbol{\varepsilon}}_j + \overline{\mathbf{F}}_j^v(\mathbf{y})\overline{\boldsymbol{\varepsilon}}_0\Big] \\ & \times \varphi(v_j, \mathbf{x}_j | v_i, \mathbf{x}_i) - c^{(1)} (\mathbf{R}_j^v \overline{\boldsymbol{\varepsilon}}_j + \overline{\mathbf{F}}_j^v \overline{\boldsymbol{\varepsilon}}_0) \Big\} d\mathbf{y} d\mathbf{x}. \end{aligned} \tag{8.141}$$

We rearrange some parts of the Eq. (8.141). Namely, for the inclusion $j = i$, we have

$$\begin{aligned} -n \int_{v_i}\int_{v_i} & V_i(\mathbf{x})\big[\mathbf{R}_i^v(\mathbf{x})\overline{\boldsymbol{\varepsilon}}_i + \overline{\mathbf{F}}_i^v(\mathbf{x})\overline{\boldsymbol{\varepsilon}}_0\big]\mathbf{U}(\mathbf{x}-\mathbf{y})V_i(\mathbf{y})\big[\mathbf{R}_i^v(\mathbf{y})\overline{\boldsymbol{\varepsilon}}_i + \overline{\mathbf{F}}_i^v(\mathbf{y})\overline{\boldsymbol{\varepsilon}}_0\big] d\mathbf{y} d\mathbf{x} \\ = -n \int_{v_i} & V_i(\mathbf{x})\big[\mathbf{R}_i^v(\mathbf{x})\overline{\boldsymbol{\varepsilon}}_i + \overline{\mathbf{F}}_i^v(\mathbf{x})\overline{\boldsymbol{\varepsilon}}_0)\big]\big[(\mathbf{A}_i(\mathbf{x}) - \mathbf{I})\overline{\boldsymbol{\varepsilon}}_i + \mathbf{C}_i(\mathbf{x})\overline{\boldsymbol{\varepsilon}}_0\big] d\mathbf{x} \\ = -\Big(& [\mathbf{R}_i^v(\mathbf{x})\overline{\boldsymbol{\varepsilon}}_i + \overline{\mathbf{F}}_i^v(\mathbf{x})\overline{\boldsymbol{\varepsilon}}_0], [\mathbf{A}_i(\mathbf{x})\overline{\boldsymbol{\varepsilon}}_i + \mathbf{C}_i(\mathbf{x})\overline{\boldsymbol{\varepsilon}}_0]\Big) + c^{(1)}[\mathbf{R}_i^v \overline{\boldsymbol{\varepsilon}}_i + \overline{\mathbf{F}}_i^v \overline{\boldsymbol{\varepsilon}}_0]\overline{\boldsymbol{\varepsilon}}_i. \end{aligned} \tag{8.142}$$

The integral containing the integrand $c^{(1)}(\mathbf{R}_j^v\overline{\boldsymbol{\varepsilon}}_j + \overline{\mathbf{F}}_j\overline{\boldsymbol{\varepsilon}}_0)$ (8.142) can also be reformulated for the excluded volume v_i^0 by the use of self-adjointness of the operator $\mathbf{U}$

$$n\int_{v_i}\int_{^c v_i^0} V_i(\mathbf{x})\big[\mathbf{R}_i^v(\mathbf{x})\overline{\boldsymbol{\varepsilon}}_i + \overline{\mathbf{F}}_i^v(\mathbf{x})\overline{\boldsymbol{\varepsilon}}_0\big]\mathbf{U}(\mathbf{x}-\mathbf{y})\Big[V_i^0(\mathbf{y})c^{(1)}(\mathbf{R}_j^v\overline{\boldsymbol{\varepsilon}}_j + \overline{\mathbf{F}}_j^v\overline{\boldsymbol{\varepsilon}}_0)\Big]d\mathbf{y}d\mathbf{x}$$
$$= -c^{(1)}c^{(1)}(\mathbf{R}_i^v\overline{\boldsymbol{\varepsilon}}_i + \overline{\mathbf{F}}_i^v\overline{\boldsymbol{\varepsilon}}_0)\mathbf{P}_i^0(\mathbf{R}_j^v\overline{\boldsymbol{\varepsilon}}_j + \overline{\mathbf{F}}_j^v\overline{\boldsymbol{\varepsilon}}_0). \tag{8.143}$$

Substitution of Eqs. (8.142) and (8.143) into (8.141) yields

$$(\boldsymbol{\tau}^*, \mathbf{U}\boldsymbol{\tau}^*) = -\Big([\mathbf{R}_i^v(\mathbf{x})\overline{\boldsymbol{\varepsilon}}_i + \overline{\mathbf{F}}_i^v(\mathbf{x})\overline{\boldsymbol{\varepsilon}}_0], [\mathbf{A}_i(\mathbf{x})\overline{\boldsymbol{\varepsilon}}_i + \mathbf{C}_i(\mathbf{x})\overline{\boldsymbol{\varepsilon}}_0]\Big)$$
$$+c^{(1)}[\mathbf{R}_i^v\overline{\boldsymbol{\varepsilon}}_i + \overline{\mathbf{F}}_i^v\overline{\boldsymbol{\varepsilon}}_0][\overline{\boldsymbol{\varepsilon}}_i - c^{(1)}\mathbf{P}_i^0(\mathbf{R}_j^v\overline{\boldsymbol{\varepsilon}}_j + \overline{\mathbf{F}}_j^v\overline{\boldsymbol{\varepsilon}}_0)]$$
$$-n\int_{v_i} V_i(\mathbf{x})\big[\mathbf{R}_i^v(\mathbf{x})\overline{\boldsymbol{\varepsilon}}_i + \overline{\mathbf{F}}_i^v(\mathbf{x})\overline{\boldsymbol{\varepsilon}}_0\big]$$
$$\times\int_{^c v_i} \mathbf{U}(\mathbf{x}-\mathbf{y})\Big[V_j(\mathbf{y})(\mathbf{R}_j^v(\mathbf{y})\overline{\boldsymbol{\varepsilon}}_j + \overline{\mathbf{F}}_j^v(\mathbf{y})\overline{\boldsymbol{\varepsilon}}_0)$$
$$\times\varphi(v_j, \mathbf{x}_j|v_i, \mathbf{x}_i) - c^{(1)}(\mathbf{R}_j^v\overline{\boldsymbol{\varepsilon}}_j + \overline{\mathbf{F}}_j^v\overline{\boldsymbol{\varepsilon}}_0)\Big]d\mathbf{y}d\mathbf{x}. \tag{8.144}$$

Acceptance of the "ellipsoidal symmetry" hypothesis **H3** for the CMs with the homogeneous ellipsoidal inclusions vanished the second integral in Eq. (8.124). In other words, the Hashin–Shtrikman bounds (8.127) were obtained for the CMs with a special "ellipsoidal symmetry" of microtopology which is served just for vanishing of the second integral in Eq. (8.124). In a similar manner, we will estimate the bounds for CMs containing both the noncanonical and inhomogeneous inclusions with "modified ellipsoidal symmetry" of microtopology vanishing the integral in Eq. (8.144) (which is reduced to the second vanishing integral in the right-hand side of Eq. (8.124) for the homogeneous ellipsoidal inclusions).

As in Sect. 8.4.1 we consider the matrix CM with the matrix modulus coinciding with the reference modulus $\mathbf{L}^c = \mathbf{L}^{(0)}$ that also leads to the identical vanishing of polarization tensor $\boldsymbol{\tau}_0 = \mathbf{0}$ in the matrix. Then substitution of Eqs. (8.140) (8.144) into Eqs. (8.117) and (8.118) now gives the following linear algebraic equation:

$$(\mathbf{I} - c^{(1)}\mathbf{P}_i^0\mathbf{R}_i^v)\overline{\boldsymbol{\varepsilon}}_i = \langle\boldsymbol{\varepsilon}\rangle, \tag{8.145}$$

which can be immediately solved for $\boldsymbol{\tau}_1$ to obtain a representation for the effective modulus

$$\mathbf{L}^{*(\mathrm{HS}[2])} = \mathbf{L}^{(0)} + c^{(1)}\mathbf{R}_i^v[\mathbf{I} - c^{(1)}\mathbf{P}_i^0\mathbf{R}_i^v]^{-1}. \tag{8.146}$$

The estimation (8.146) is a lower bound for $\mathbf{L}^*$ if the matrix happens to be the most compliant phase in the CM and a upper bound if the matrix is the stiffest phase.

It is interested that for the CMs with homogeneous ellipsoidal inclusions, $\mathbf{L}^{*(HS)[2]} = \mathbf{L}^{*(HS)[1]}$, and $\mathbf{L}^{*[2]} = \mathbf{L}^{*[1]}$. However, for noncanonical homogeneous inclusions, $\mathbf{L}^{*(HS)[2]} \neq \mathbf{L}^{*(HS)[1]}$ if $\mathbf{L}^{*(HS)[1]}$ is estimated by replacement $\mathbf{P}_i \to \bar{v}_i^2 \mathbf{T}_{ii}$ (8.128), while $\mathbf{L}^{*(HS)[2]}$ takes noncanonical inclusion shape into consideration by the use of the average $\mathbf{R}_i^v := \langle \mathbf{R}_i^v(\mathbf{x}) \rangle_{(i)} \neq (\mathbf{I} + \bar{v}_i^2 \mathbf{L}_1^{(1)} \mathbf{T}_{ii})^{-1}$. In so doing, the representations (8.146) and (8.130_1) coincide with $\mathbf{L}^{*(HS)[2]} = \mathbf{L}^{*[2]}$. It looks like surprise because the sets of initial hypotheses used for obtaining either (8.130_1) or (8.146) distinguish one from another. Indeed, the EFM's estimation (8.130_1) was obtained immediately by solution of the GIE (7.32) in the framework of the hypotheses **H1a**, **H1b**, **H2**, and **H3**. In so doing, HS bounds (8.146) were also obtained from Eq. (7.32) but after its substitution into the Hashin–Shtrikman variational principle (8.117) and (8.118) and subsequent application of the hypotheses **H1a**, **H2**, and **H3**; i.e., the hypothesis **H1b** was not used, and, as we see, the hypothesis **H1b** has no effect on the final result (8.146). Unfortunately, the last statement is incorrect simply because the hypothesis **H1b** from the first is embedded into the GIE (7.32) that was proved in Chap. 7. Nevertheless, we will demonstrate in Chap. 9 that abandonment of the hypothesis **H1b** (and **H3**) is possible.

8.4.3 Interface Boundary Integral Representations

The H-S bounds (8.127) and (8.146) as well as the strain concentration factors are defined by the corresponding tensors $\mathbf{P}_i^{\mathrm{W}}$, $\mathbf{R}_i^v$, and $\mathbf{A}_i(\mathbf{x})$ found for one inclusion inside the infinite matrix. Namely, for this purpose, we consider a model problem with fictitious constitutive equation $\boldsymbol{\sigma}(\mathbf{x}) = \mathbf{L}(\mathbf{x})\boldsymbol{\varepsilon}(\mathbf{x}) + \boldsymbol{\beta}(\mathbf{x})$ with the piecewise constant material parameters $\mathbf{g} = \mathbf{L}, \boldsymbol{\beta}, \boldsymbol{\alpha}$ ($\boldsymbol{\alpha} = -\mathbf{L}\boldsymbol{\beta}$): $\mathbf{g} \equiv \mathbf{g}^{(0)} + \mathbf{g}_1(\mathbf{x}) = \mathbf{g}^{(0)} + \mathbf{g}_1^{(1)}(\mathbf{x})$ where $\mathbf{g}^{(0)}$ =const, $\mathbf{g}(\mathbf{x}) \equiv \mathbf{g}^{(0)}$ at $\mathbf{x} \in v^0$ and $\mathbf{g}_1^{(1)}(\mathbf{x}) \equiv \mathbf{g}_1^{(1)}$ is a homogeneous function of the $\mathbf{x} \in v^{(1)}$. In view of the linearity of the problem, there exist constant fourth- and second-rank tensors $\mathbf{A}_i(\mathbf{x})$, $\mathbf{R}_i(\mathbf{x})$, and $\mathbf{C}_i(\mathbf{x})$, $\mathbf{F}_i(\mathbf{x})$, such that ($\mathbf{x} \in R^d$) [compare with Eqs. (8.133)]

$$\boldsymbol{\varepsilon}(\mathbf{x}) = \mathbf{A}_i(\mathbf{x})\overline{\boldsymbol{\varepsilon}}(\mathbf{x}_i) + \mathbf{C}_i(\mathbf{x})\boldsymbol{\alpha}_1^{(i)}, \quad \bar{v}_i \boldsymbol{\tau}(\mathbf{x}) = \mathbf{R}_i(\mathbf{x})\overline{\boldsymbol{\varepsilon}}(\mathbf{x}_i) + \mathbf{F}_i(\mathbf{x})\boldsymbol{\alpha}_1^{(i)}, \tag{8.147}$$

where

$$\mathbf{R}_i(\mathbf{x}) = \bar{v}_i \mathbf{L}_1^{(i)}(\mathbf{x})\mathbf{A}_i(\mathbf{x}), \tag{8.148}$$

$$\mathbf{F}_i(\mathbf{x}) = \bar{v}_i [\mathbf{L}_1^{(i)}(\mathbf{x})\mathbf{C}_i(\mathbf{x}) + \mathbf{I}], \tag{8.149}$$

$$\mathbf{C}_i(\mathbf{x}) \equiv \bar{v}_i \mathbf{A}_i(\mathbf{x})\mathbf{T}_i(\mathbf{x} - \mathbf{x}_i). \tag{8.150}$$

For the model case of an elastically homogeneous medium $\mathbf{L}_1(\mathbf{x}) \equiv \mathbf{0}$ ($\mathbf{x} \in R^d$), the average strain in the inclusion defines the tensors $\mathbf{T}_{ii}$ (8.128) and $\mathbf{P}_i^{\rm W}$ (8.129) at $\overline{\boldsymbol{\varepsilon}} = \mathbf{0}$

$$\langle \boldsymbol{\varepsilon} \rangle_i \equiv \mathbf{C}_i \boldsymbol{\alpha}_1^{(i)} = \mathbf{T}_{ii} \boldsymbol{\alpha}_1^{(i)}, \quad \mathbf{P}_i^{\rm W} = -\bar{v}_i \mathbf{C}_i. \tag{8.151}$$

Estimations of the mentioned tensors $\mathbf{A}_i(\mathbf{x})$, $\mathbf{R}_i^v$, and $\mathbf{C}_i$ for the noncanonical inclusion shapes were considered by the finite element analysis (FEA) in Sect. 4.2 and in this chapter. However, in the current section, we will use the method of fundamental solutions (MFS; see Sect. 4.3.3) belonging to the class of boundary integral equation methods. Because of this, we will present the mentioned tensors $\mathbf{A}_i(\mathbf{x})$, $\mathbf{R}_i^v$, and $\mathbf{C}_i$ in the form of the boundary integrals over the inclusion interface Γ_i

$$\mathbf{A}_i \overline{\boldsymbol{\varepsilon}}(\mathbf{x}_i) = \bar{v}_i^{-1} \int_{\Gamma_i} \mathbf{u}(\mathbf{s}) \overset{s}{\otimes} \mathbf{n}(\mathbf{s})\, ds, \quad \mathbf{C}_i \boldsymbol{\alpha}_1^{(i)} = \bar{v}_i^{-1} \int_{\Gamma_i} \mathbf{u}(\mathbf{s}) \overset{s}{\otimes} \mathbf{n}(\mathbf{s})\, ds, \tag{8.152}$$

$$\mathbf{R}_i^v \overline{\boldsymbol{\varepsilon}}(\mathbf{x}_i) = \bar{v}_i^{-1} \int_{\Gamma_i} [\mathbf{u}(\mathbf{s}) \overset{s}{\otimes} \mathbf{n}(\mathbf{s}) - \mathbf{M}^{(0)}(\mathbf{s} \otimes \mathbf{t}(\mathbf{s}))]\, ds, \tag{8.153}$$

where the right-hand sides of Eqs. (8.152_1) and (8.153) were estimated at $\overline{\boldsymbol{\varepsilon}}(\mathbf{x}_i) \neq \mathbf{0}$, $\boldsymbol{\alpha}_1^{(i)} = \mathbf{0}$, while the integral in Eq. (8.152_2) was evaluated at $\overline{\boldsymbol{\varepsilon}}(\mathbf{x}_i) = \mathbf{0}$, $\boldsymbol{\alpha}_1^{(i)} \neq \mathbf{0}$.

It is interesting that at the limiting case $c^{(1)} \to 0$, HS bounds (8.127) and (8.146) are reduced to the dilute approximations

$$\mathbf{L}^{*(\mathrm{HS}[1])} = \mathbf{L}^{(0)} + c^{(1)} \mathbf{L}_1^{(i)} \mathbf{A}_i^{\rm W}, \quad \mathbf{L}^{*(\mathrm{HS}[2])} = \mathbf{L}^{(0)} + c^{(1)} \mathbf{L}_1^{(i)} \mathbf{A}_i, \tag{8.154}$$

where a physical meaning of both tensors $\mathbf{A}_i^{\rm W} = (\mathbf{I} + \mathbf{P}_i^{\rm W} \mathbf{L}_1^{(i)})^{-1}$ (8.127) and $\mathbf{A}_i$ (8.152_1) are the average strain concentration factors at the isolated inclusion inside the infinite matrix. Therefore, we should have

$$\mathbf{A}_i = (\mathbf{I} + \mathbf{P}_i^{\rm W} \mathbf{L}_i^{(i)})^{-1}, \tag{8.155}$$

which is not fulfilled in general for a nonellipsoidal inclusion (when $\mathbf{P}_i \neq \mathbf{P}_i^{\rm W}$). Thus, the right-hand side $\mathbf{A}_i^{\rm W} = (\mathbf{I} + \mathbf{P}_i^{\rm W} \mathbf{L}_i^{(i)})^{-1}$ in Eq. (8.155) can be considered as an approximation of the average strain concentration factor $\mathbf{A}_i$ [which can be numerically found with high accuracy; see, e.g., Chaps. 9 and 11]. Namely, the tensors $\mathbf{A}_i^{\rm W}$ and $\mathbf{A}_i = (\mathbf{L}_1^{(i)})^{-1} \mathbf{R}_i^v$ are used in the Hashin–Shtrikman bounds (8.127) and (8.146), respectively. Comparative analyses of the tensors $\mathbf{A}_i^{\rm W}$ and $\mathbf{A}_i$ as well as of the HS bounds $\mathbf{L}^{*(\mathrm{HS}[1])}$ and $\mathbf{L}^{*(\mathrm{HS}[2])}$ will be performed in the next subsection.

8.4.4 Numerical Results

We consider the problem for one inclusion in the infinite homogeneous matrix with large elastic mismatch of the inclusion (the Young modulus $E^{(1)} = 100$ and Poisson ratio $\nu^{(1)} = 0.45$) and the matrix ($E^{(0)} = 1,\ \nu^{(0)} = 0.45$). With the non-essential restriction on space dimensionality d and the shape of inclusions, we will consider $2D$ plane strain problems for statistically homogeneous composites filled by the infinite fibers with the noncircular shape schematically presented in the pictures Fig. 8.1 and described by the curve

$$\begin{cases} (x - R_1 + r_s)^2 + (y - R_2 + r_s)^2 = r_s^2, & \text{for } \{|x| > R_1 - r_s\} \cap \{|y| > R_2 - r_s\}, \\ (x - R_1 + r_s)^2 + (y - R_2 + r_s)^2 = r_s^2, & \text{for } \{|x| > R_1 - r_s\} \cap \{|y| > R_2 - r_s\}, \end{cases} \tag{8.156}$$

which reduces to a circle and a rectangular in the limiting cases $R_1 = R_2 = r_s = a$ and $r_s = 0$, respectively. We will consider the fixed value $R_1 = R_2 = 1$ with the different ratios $\gamma = r_s/R_1$.

We will estimate the tensors (8.152) and (8.153) by the MFS with the number of collocation points $P = 400$ (see Sect. 4.3.3 and [220], where an accuracy of the MFS is estimated; see Fig. 4.5). We analyze the inclusion shape (8.156) with the different aspect ratio $r_s/R_1 = 0.1,\ 0.2,\ 0.4,\ 0.6,\ 0.8,\ 1.0$ and fixed $R_2/R_1 = 1$ which was also considered in [217] for estimation of the average stress concentration tensor. For the current section, we need to estimate a detailed distribution of the strain concentrator factor $\mathbf{A}_i(\mathbf{x})$ ($\mathbf{x} \in v_i$). In Fig. 8.1, one presents the normalized components $f_1(x_1) := A_{i|1111}(x_1)/A^{\text{circle}}_{i|1111}$ in the cross sections $\mathbf{x} = (x_1/R_1, 0)^\top$, where $\mathbf{A}_i^{\text{circle}} \equiv$const ($\mathbf{x} \in v_i$) is a strain concentrator tensor for a circle inclusion. As can be seen from Fig. 8.1, decreasing the ration r_s/R_1 from 1 to 0.1 leads to

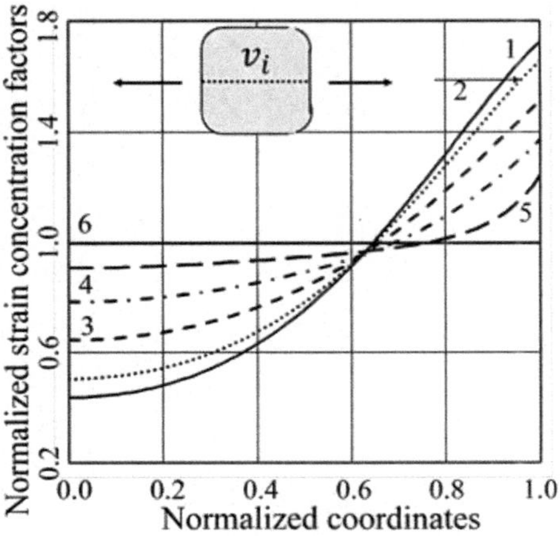

Fig. 8.1 $f_1(x_1)$ vs. x_1/R_1 for $r_s/R_1 = 0.1$ (1), 0.2 (2), 0.4 (3), 0.6 (4), 0.8 (5), and 1 (6)

increasing variation $\rho := f_1(R_1)/f_1(0)$ of the function $f_1(x_1)$ from $\rho = 1$ (at $r_s/R_1 = 1$) to $\rho = 3.94$ (at $r_s/R_1 = 0.1$).

Growing inhomogeneity of the strain distribution $\mathbf{A}_i(\mathbf{x})$ (see transitions from the curve 6 to the curve 1 in Fig. 8.1) inside inclusion leads to increase of the difference between the average strain concentration factor $\mathbf{A}_i$ and its approximation $\mathbf{A}_i^{\mathrm{W}}$ estimated through the average of the transformation concentrator factor $\mathbf{C}_i$ (8.151$_2$) based on assumption of homogeneity of $\mathbf{C}_i(\mathbf{x})$ ($\mathbf{x} \in v_i$). However, the tensors $\mathbf{A}_i(\mathbf{x})$ and $\mathbf{C}_i(\mathbf{x})$ are the nonlinear function of coordinates $\mathbf{x} \in v_i$, and, because of this, $\mathbf{A}_i \neq \mathbf{A}_i^{\mathrm{W}}$ in a general case of the strain inhomogeneity inside inclusion v_i. So, in Fig. 8.2, the curves of the normalized averages $f_2(\gamma) := A_{i|1111}/A_{i|1111}^{\mathrm{circle}}$ and $f_2^{\mathrm{W}}(\gamma) := A_{i|1111}^{\mathrm{W}}/A_{i|1111}^{\mathrm{circle}}$ are presented as the functions of the ratio $\gamma := r_s/R_1$. As can be seen, a difference between the normalized averages $f_2(\gamma)$ and $f_2^{\mathrm{W}}(\gamma)$ is increasing with decrease of the ratio γ.

The found tensors $\mathbf{A}_i^{\mathrm{W}}$ and $\mathbf{A}_i$ allow one to estimate the low Hashin–Shtrikman bounds $\mathbf{L}^{*(\mathrm{HS}[1])}$ (8.127) and $\mathbf{L}^{*(\mathrm{HS}[2])}$ (8.146), respectively. Curves 1 and 2 (see Fig. 8.3) are the new HS bounds for the effective Young modulus $E^*/E^{(0)}$ (8.146) estimated at $\gamma = r_s/R_1 = 0.1$ and $\gamma = 0.2$, respectively, while curves 3 and 4 correspond to the classical formula (8.127) for $\gamma = 0.1$ and $\gamma = 0.2$, respectively. Curve 5 is related with the equality $\mathbf{L}^{*(\mathrm{HS}[1])} = \mathbf{L}^{*(\mathrm{HS}[2])}$ for CMs with the circle inclusions. As can be seen, the classical HS bounds are sensitive with respect to shape of noncanonical inclusions (compare curves 3 and 4 with curve 5).

However, for the new HS bounds (8.146), the mentioned feature of the HS bounds is manifested more significantly. So, for the concentration of inclusions $c^{(1)} = 0.2,\ 0.6,\ 0.8$, the bounds $E^{*(\mathrm{HS}[2])}$ and $E^{*(\mathrm{HS}[1])}$ are 2.53, 5.03, 23.35 and 2.31, 4.12, 11.18 ($\gamma = 0.2$) and 2.61, 5.40, 35.36 and 2.35, 4.25, 12.27 ($\gamma = 0.1$),

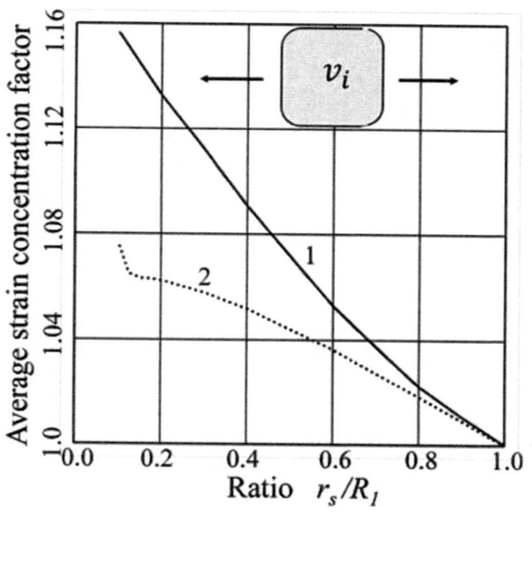

Fig. 8.2 Normalized averages $f_2(\gamma)$ (curve 1) and $f_2^{\mathrm{W}}(\gamma)$ (curve 2) vs. γ

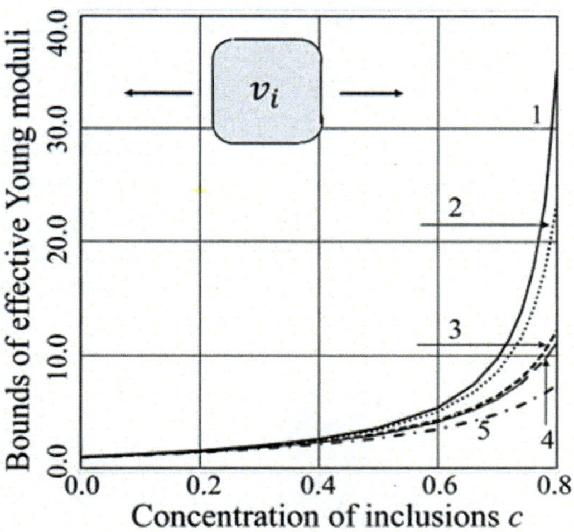

Fig. 8.3 New (1, 2) and classical (3, 4) bounds for $E^*/E^{(0)}$ estimated at $\gamma = 0.1$ (1, 3) and $\gamma = 0.2$ (2, 4), as well as at $\gamma = 1$ (5)

respectively. It means that at $\gamma = 0.1$, $E^{*(\mathrm{HS}[2])}$ and $E^{*(\mathrm{HS}[1])}$ differ by the factors 1.1, 1.3, 2.9 for $c^{(1)} = 0.4,\ 0.6,\ 0.8.$, respectively.

Thus, we qualitatively compared the pairs (HS bounds)-($\mathbf{L}^*$ estimations) one with each other and mentioned that $\mathbf{L}^{*(\mathrm{HS})[1]} = \mathbf{L}^{*[1]}$ and $\mathbf{L}^{*(\mathrm{HS})[2]} = \mathbf{L}^{*[2]}$ in the frameworks of the accepted sets of hypotheses. This coincidence take place only in the framework of the hypothesis **H1b** exploiting (either explicitly or explicitly). From other side, the difference between the HS bounds $\mathbf{L}^{*(\mathrm{HS})[1]}$ and $\mathbf{L}^{*(\mathrm{HS})[2]}$ is explained by the fact that the trial functions (8.139) are more general than the trial functions (8.119). Expansion of the function class involved into an extremizing of the functional $\mathcal{H}$ (8.118) leads to a stiffening (softening) of the bounds of $\mathbf{L}^*$ in the case of stiffer (softer) inclusion with respect to the matrix. Furthermore, the bounds $\mathbf{L}^{*(\mathrm{HS})[2]}$ (8.146) are defined by the polarization concentration factor $\mathbf{R}_i^v(\mathbf{x})$ ($\mathbf{x} \in v_i$) rather than by the homogeneous modulus $\mathbf{L}^{(1)}$ of inclusions v_i [as in (8.127)]. It opens the opportunity for exploiting of Eq. (8.146) for HS bound evaluations for CMs with a continuously varying modulus $\mathbf{L}^{(1)}(\mathbf{x}) \not\equiv$const. ($\mathbf{x} \in v_i$) inside the inclusions. Moreover, an improvement of the HS bounds (8.146) with respect to the bounds (8.127) (compare curves 1 and 2 and 3 and 4, respectively, in Fig. 8.3) was obtained by the use of replacement of the approximation (8.119) by more general approximation (8.139) in combination with the same classical GIE (7.32). Incorporation of the new GIE (7.19) (corresponding to the operator ${}^{\mathcal{P}}\mathbf{U}$ (6.200)) into the HS energy bound (6.208) will be considered in Chap. 9.

8.4.5 Some Representations Related with Hashin–Shtrikman Bounds

In this subsection we consider the case of N phases $v^{(r)}$ $(r = 1, \ldots, N)$ of inhomogeneities where each phase contains identical homogeneous ellipsoidal inclusions and the reference material is the matrix material itself. In the case of acceptance of the "ellipsoidal symmetry" (8.45) hypothesis **H3**, Ponte Castañeda and Willis [1240] (see also [200]) have got the equation for the average polarization tensors $\boldsymbol{\tau}^{(r)}$

$$[(\mathbf{L}_1^{(r)})^{-1} + \mathbf{P}_r]\boldsymbol{\tau}^{(r)} - \sum_{s=1}^{N} c^{(s)}\mathbf{P}^0_{rs}\boldsymbol{\tau}^{(s)} = \langle \boldsymbol{\varepsilon} \rangle \tag{8.157}$$

generalizing Eq. (8.126) to the case of multiphase inclusions where the the symmetric tensors $\mathbf{P}^0_{rs} = \mathbf{P}^0_{sr} = \mathbf{P}(v^0_{rs})$ are associated with the ellipsoidal excluded volume v^0_{rs} (8.126). Equation (8.157) is easily solved

$$\boldsymbol{\tau}^{(r)} = \sum_{s=1}^{N} \mathbf{Y}^{\epsilon}_{rs}\mathbf{R}^{\epsilon v}_r \langle \boldsymbol{\varepsilon} \rangle, \quad [(\mathbf{Y}^{\epsilon})^{-1}]_{rs} = \mathbf{I}\delta_{rs} - c^{(s)}\mathbf{R}^{\epsilon}_r\mathbf{P}^0_{rs}, \tag{8.158}$$

where a concentration factor $\mathbf{Y}^{\epsilon}_{rs}$ is similar to the tensor $\mathbf{Y}_{ij}$ (8.46) introduced for the solution of Eq. (8.43). Analogously to Eq. (8.127), the solution (8.158) is used for obtaining HS estimate for $\mathbf{L}^*$

$$\mathbf{L}^{*(\mathrm{HS}[1])} = \mathbf{L}^{(0)} + \sum_{r=1}^{N} c^{(r)} \Big[\sum_{s=1}^{N} \mathbf{Y}^{\epsilon}_{rs}\mathbf{R}^{\epsilon v}_r \Big], \tag{8.159}$$

which can be simplified when either the shape of excluded volumes v^0_{rs} does not depend on s, i.e., $\mathbf{P}^0_{rs} = \mathbf{P}^0_r$ $(\forall 0 < r, s \leq N)$, or the shape of excluded volumes v^0_{rs} is taken to be the same for all inclusion pairs, $\mathbf{P}^0_{rs} = \mathbf{P}^0$ $(\forall 0 < r, s \leq N)$

$$\mathbf{L}^{*(\mathrm{HS}[1])} = \mathbf{L}^{(0)} + \sum_{r=1}^{N} [\mathbf{I} - \langle \mathbf{R}^{\epsilon v}_r \mathbf{P}^0_r V \rangle]^{-1} \langle \mathbf{R}^{\epsilon v}_r V \rangle, \tag{8.160}$$

$$\mathbf{L}^{*(\mathrm{HS}[1])} = \mathbf{L}^{(0)} + \sum_{r=1}^{N} [\mathbf{I} - \langle \mathbf{R}^{\epsilon v}_r V \rangle \mathbf{P}^0]^{-1} \langle \mathbf{R}^{\epsilon v}_r V \rangle, \tag{8.161}$$

respectively.

The bounds $\mathbf{M}^{*(\mathrm{HS}[1])}$ for the effective compliance $\mathbf{M}^*$ can be obtained by inversion of corresponding $\mathbf{L}^{*(\mathrm{HS}[1])}$. These can also be obtained directly from consideration of the traction boundary value problem (6.5) in a similar manner as the bounds $\mathbf{L}^{*(\mathrm{HS}[1])}$ were found for the displacement boundary conditions (6.4). Extremizing $\mathcal{F}(\boldsymbol{\eta}^*)$ (6.213) is performed analogously to extremizing the functional $\mathcal{H}(\boldsymbol{\tau}^*)$ (6.204) with replacement of $\mathbf{L}_1$ by $\mathbf{M}_1$ and $\mathbf{U}$ by $\boldsymbol{\Gamma}$ taking into account that Eqs. (6.200) and (6.212) must be consistent with $\boldsymbol{\tau} = -\mathbf{L}^c\boldsymbol{\eta}$, $\boldsymbol{\varepsilon}^0 = \langle\boldsymbol{\varepsilon}\rangle$, and $\boldsymbol{\sigma}^0 = \langle\boldsymbol{\sigma}\rangle = \mathbf{L}^c\boldsymbol{\varepsilon}^0 + \langle\boldsymbol{\tau}\rangle$. In such a case, the bounds $\mathbf{M}^{*(\mathrm{HS}[1])}$ for a composite with "ellipsoidal" symmetry (8.45) corresponding to the bounds $\mathbf{L}^{*(\mathrm{HS}[1])}$ (8.160) are given by the relations

$$\mathbf{M}^{*(\mathrm{HS}[1])} = \mathbf{M}^{(0)} + \sum_{r=1}^{N}[\mathbf{I} - \langle\mathbf{R}_r\mathbf{Q}_r^0 V\rangle]^{-1}\langle\mathbf{R}_r V\rangle, \tag{8.162}$$

which are inverse to $\mathbf{L}^{*(\mathrm{HS}[1])}$ (8.160). Here tensors $\mathbf{R}_r$ and $\mathbf{Q}_r^0$ are defined for the homogeneous ellipsoidal inclusions (3.183) by Eqs. (3.184) and (3.106), (3.121), respectively.

The estimation (8.160) was originally proposed in [618, 619, 1575, 1576] (see also [1323]) by another but equivalent method for the particular case of statistically isotropic composites when the tensors $\mathbf{P}_r$ can be assumed to be estimated for the spherical domains v_i^0. A more general case of "ellipsoidal" symmetry was analyzed in [1240, 1620, 1621, 1624, 1627] where Eq. (8.160) was obtained for the particular case $\mathbf{P}_r^0 = \mathbf{P}^0$ $(0 < r \le N)$ with $\mathbf{P}^0$ taking identical "ellipsoidal" form for all phases which might be appropriate to discuss a composite containing aligned ellipsoidal inclusions, as limiting case, thin needles, and flat cracks. Equation (8.160) enables one to evaluate the bounds of effective moduli of anisotropic media [1076]; the case of the transversally isotropic medium was considered in [336, 610].

Further improvements were achieved in higher-order bounds by using three-point correlation functions $S_3^{(1)}(\mathbf{y}_1, \mathbf{y}_2, \mathbf{y}_3)$ (5.75) by Beran and Molyneux [108]; for statistically homogeneous composites, they obtained the bounds on k^* and μ^* which were simplified in [1072] (see also [367, 820, 1039, 1041, 1077, 1520] [1627]). Similar approaches were used for deriving of bounds for the thermal conductivity [104, 620, 1072, 1616] and for the thermal expansion coefficient in the thermoelastic problem (see [549, 1304, 1338, 1378]), whereas the bounds for piezoelectric moduli were obtained in [126, 931]. Simmons and Wang [1407] have summarized in sufficient detail the results of property estimations of cubic symmetry polycrystals. The variation method was also exploited for the analysis of polycrystals with hexagonal, trigonal, tetragonal, and monoclinic symmetry [114, 1595–1598]. Diener et al. [373] have obtained a generalization of Hashin–Shtrikman bounds to the nonlocal media with the effective properties described by the nonlocal effective operator rather than by the tensor $\mathbf{L}$. The bounds of effective properties were estimated by the use of the variational principle generalizing the Hashin–Shtrikman principle to the dynamic case in [1478, 1479].

Let us consider two simplest cases of the choice of $\mathbf{L}^c$. At first, $\mathbf{L}_1(\mathbf{x})$ is positive definite for $\forall\mathbf{x} \in w$ if $\mathbf{L}^c \equiv \mathbf{0}$; then, passing to the limit $\mathbf{L}^c \to \mathbf{0}$ and $\mathbf{P} \to \infty$ in (8.159) leads to the lower bound of the effective moduli $\mathbf{L}^{*(HS[1])-} = \langle\mathbf{M}\rangle^{-1} \equiv \mathbf{L}^R$ which is found to coincide with the lower Reuss bound (6.223). In the opposite case $\mathbf{L}^c \to \infty$, when $\mathbf{P} \to \infty$ and $\mathbf{L}_1(\mathbf{x})$ is negative definite at $\forall \mathbf{x} \in w$. Then the upper bound (8.159) $\mathbf{L}^{*(HS1[1])+} = \langle \mathbf{L} \rangle \equiv \mathbf{L}^V$ coincided with the upper Voight bound (6.223). Thus, the Hashin–Shtrikman bounds (8.159) in the considered limiting cases $\mathbf{L}^c \to \mathbf{0}$ and $\mathbf{L}^c \to \infty$ are reduced to the Hill's bounds (6.223), and, moreover, these limiting cases degenerate the Hashin–Shtrikman variational principle into the minimum principles of the potential and complementary energies.

Following [1575], we will consider a statistically isotropic composite with the isotropic tensors $\mathbf{P}_i = \mathbf{P}^0$ (3.163) and with isotropic comparison medium $\mathbf{L}^c = (3k^c, 2\mu^c)$ and phases $\mathbf{L}^{(r)} = (3k^{(r)}, 2\mu^{(r)})$ $(r = 0, \ldots, N)$. Then Eq. (8.160) can be rearranged

$$\overline{\mathbf{L}} = \Big[\sum_{r=0}^{N} c_r (P_0^{-1} - \mathbf{L}^c + \mathbf{L}^{(r)})^{-1}\Big]^{-1} \sum_{s=0}^{N} c_s (P_0^{-1} - \mathbf{L}^c + \mathbf{L}^{(s)})^{-1} \mathbf{L}^{(s)} \tag{8.163}$$

and presented in Voight symbolic notations

$$\overline{k} = \Big[\sum_{r=0}^{N} c_r (k^{(r)} + \tilde{k})^{-1}\Big]^{-1} - \tilde{k}, \quad \overline{\mu} = \Big[\sum_{r=0}^{N} c_r (\mu^{(r)} + \tilde{\mu})^{-1}\Big]^{-1} - \tilde{\mu}, \tag{8.164}$$

where the components of a symbolic representation of the tensor

$$\mathbf{P}_0^{-1} - \mathbf{L}^c = (3\tilde{k}, 2\tilde{\mu}) = \Big(4\mu^c, \frac{1}{3}\mu^c \frac{9k^c + 8\mu^c}{k^c + 2\mu^c}\Big) \tag{8.165}$$

were introduced by Hill [643]. It can be seen from Eqs. (8.164) that the found values $\overline{k}$ and $\overline{\mu}$ are the monotonically increasing functions of $\tilde{k}$ and $\tilde{\mu}$, which in turn are monotonically increasing functions of k^c and μ^c. Then the lower and upper bounds of the moduli of the comparison medium defining the lower and upper bounds of effective moduli can be taken as $k^{c-} = \min_r(k^{(r)})$, $\mu^{c-} = \min_r(\mu^{(r)})$, and $k^{c+} = \max_r(k^{(r)})$, $\mu^{c+} = \max_r(\mu^{(r)})$, which define the bounds of effective moduli

$$\Big[\sum_{r=0}^{N} c_r (k^{(r)} + \tilde{k}_-)^{-1}\Big]^{-1} - \tilde{k}_- \le k^* \le \Big[\sum_{r=0}^{N} c_r (k^{(r)} + \tilde{k}_+)^{-1}\Big]^{-1} - \tilde{k}_+, \tag{8.166}$$

$$\Big[\sum_{r=0}^{N} c_r (\mu^{(r)} + \tilde{\mu}_-)^{-1}\Big]^{-1} - \tilde{\mu}_- \le \mu^* \le \Big[\sum_{r=0}^{N} c_r (\mu^{(r)} + \tilde{\mu}_+)^{-1}\Big]^{-1} - \tilde{\mu}_+, \tag{8.167}$$

where $\tilde{k}_- = 4\mu^{c-}/3$, $\tilde{k}_+ = 4\mu^{c+}/3$. The values of $\tilde{\mu}_-$ and $\tilde{\mu}_+$ are chosen depending on the relation between the phase elastic moduli. If $(k^{c+} - k^{c-})(\mu^{c+} - \mu^{c-}) > 0$, then

$$\tilde{\mu}_- = \frac{\mu^{c-}}{6}\frac{9k^{c-} + 8\mu^{c-}}{k^{c-} + 2\mu^{c-}}, \quad \tilde{\mu}_+ = \frac{\mu^{c+}}{6}\frac{9k^{c+} + 8\mu^{c+}}{k^{c+} + 2\mu^{c+}}. \tag{8.168}$$

In the opposite case $(k^{c+} - k^{c-})(\mu^{c+} - \mu^{c-}) < 0$, the values

$$\tilde{\mu}_- = \frac{\mu^{c-}}{6}\frac{9k^{c+} + 8\mu^{c-}}{k^{c+} + 2\mu^{c-}}, \quad \tilde{\mu}_+ = \frac{\mu^{c+}}{6}\frac{9k^{c-} + 8\mu^{c+}}{k^{c-} + 2\mu^{c+}}. \tag{8.169}$$

are chosen for the bound estimations (8.167).

It is interesting to mention that for composites (called Hill's [640] medium) with homogeneous shear modulus $\mu(\mathbf{x}) \equiv$ const. when $\mu^{c-} = \mu^{c+} = \mu$, the lower and upper bounds (8.166) coincide, and therefore, the effective bulk and shear moduli are exactly defined

$$k^* = \Big[\sum_{r=0}^{N} c_r\big(k^{(r)} + \frac{4}{3}\mu\big)^{-1}\Big]^{-1} - \frac{4}{3}\mu, \quad \mu^* = \mu. \tag{8.170}$$

Equation (8.160) obtained for the composites can be easily generalized to the case of one-phase polycrystals. In such a polycrystal, the identical grains are randomly located with prescribed orientation of local crystallographic axes $\mathbf{e}_i'$ with respect to the global coordinate system $\mathbf{e}_i$ that is described by the rotation matrix $\mathbf{g}$ (see A1). Then the volume concentration of crystals with orientation enclosed by the interval $(\mathbf{g}, \mathbf{g}+d\mathbf{g})$ equals $f(\mathbf{g})d\mathbf{g}$, and the summation sign in Eq. (8.160) should be replaced by averaging over the possible orientations

$$\mathbf{L}^{*(\mathrm{HS}[1])} = \langle \mathbf{A}(\mathbf{g})\rangle^{-1}\langle \mathbf{A}(\mathbf{g})\mathbf{L}(\mathbf{g})\rangle, \tag{8.171}$$

where $\mathbf{A}(\mathbf{g}) = [\mathbf{I}+\mathbf{L}_1(\mathbf{g})\mathbf{P}(\mathbf{g})]^{-1}$ is a strain concentration factor for one crystal with orientation $\mathbf{g}$ in the infinite matrix with the modulus $\mathbf{L}^c$. In a general case of medium anisotropy, a choice of the comparison medium $\mathbf{L}^{c-} \leq \mathbf{L}^c \leq L^{c+}$ is not trivial. For crystals with a cubic symmetry, the elastic moduli are completely specified by three moduli L_{11}, L_{12}, and L_{44} (2.175) in six-by-six matrix notation. The bulk modulus and two-shear moduli of such a crystal are defined as

$$k = (L_{11} + 2L_{12})/3, \quad \mu_1 = (L_{11} - L_{12})/2, \ \mu_2 = L_{44} \tag{8.172}$$

where $\mu_1 < \mu_2$. Then for statistically isotropic polycrystals [1407], one can set $k^{c-} = k^{c+} = k$, $\mu^{c-} = \mu_1$, $\mu^{c+} = \mu_2$, and (8.171) yields $k^* = k$ and ($\mu_2^1 = \mu_2 - \mu_1$)

$$\bar{\mu}^- = \mu_1 + 3\Big[\frac{12(k + 2\mu_1)}{5\mu_1(3k + 4\mu_1)} + \frac{5}{\mu_2^1}\Big]^{-1}, \quad \bar{\mu}^+ = \mu_2 + 2\Big[\frac{18(k + 2\mu_2)}{5\mu_2(3k + 4\mu_2)} - \frac{5}{\mu_2^1}\Big]. \tag{8.173}$$

8.5 Related Perturbation Methods

8.5.1 Combined MEFM–Perturbation Method

For simplicity, let us consider problems (7.21) and (7.22) for two ellipsoidal inclusions v_i and v_j subjected to a homogeneous field $\widetilde{\boldsymbol{\sigma}}_{1,2}(\mathbf{x})$

$$\overline{\boldsymbol{\sigma}}(\mathbf{x}) = \widetilde{\boldsymbol{\sigma}}_{1,2}(\mathbf{x}) + \mathcal{T}^1_{ij}(\mathbf{x}-\mathbf{x}_j)\widetilde{\boldsymbol{\sigma}}_{1,2}(\mathbf{x}_j) + \mathcal{T}^2_{ij}(\mathbf{x}-\mathbf{x}_j), \tag{8.174}$$

$$\boldsymbol{\sigma}(\mathbf{x}) = \mathbf{B}_i(\mathbf{x})\widetilde{\boldsymbol{\sigma}}_{1,2}(\mathbf{x}) + \mathbf{C}_i(\mathbf{x}) + \mathcal{F}^1_{ij}(\mathbf{x}-\mathbf{x}_j)\widetilde{\boldsymbol{\sigma}}_{1,2}(\mathbf{x}_j) + \mathcal{F}^2_{ij}(\mathbf{x}-\mathbf{x}_j), \tag{8.175}$$

where $\mathbf{x} \in v_i,\ i = 1, 2,\ j = 3 - i$. Reference to some exact and approximate analytical methods of estimation of the tensors $\mathcal{T}^1_i(\mathbf{x}-\mathbf{x}_j)$ and $\mathcal{T}^2_i(\mathbf{x}-\mathbf{x}_j)$ can be found in Chap. 4. Independently on the method of the estimation, the tensors $\mathcal{T}^1, \mathcal{T}^2, \mathcal{F}^1, \mathcal{F}^2$ tend to zero with $|\mathbf{x}-\mathbf{x}_j| \to \infty$ as $|\mathbf{x}-\mathbf{x}_j|^{-d}$. In particular, in the framework of the hypothesis **H1** (4.93) and (4.139), taking Eq. (8.6) into account we get (no sum over $i, j;\ i \neq j$):

$$\mathcal{T}^1_{ij}(\mathbf{x}-\mathbf{x}_j) = \mathbf{R}_i^{-1}[\mathbf{Z}_{ii}\mathbf{R}_i + \mathbf{Z}_{ij}\mathbf{R}_j - \mathbf{R}_i], \tag{8.176}$$

$$\mathcal{T}^2_{ij}(\mathbf{x}-\mathbf{x}_j) = \mathbf{R}_i^{-1}[\mathbf{Z}_{ii}\mathbf{F}_i + \mathbf{Z}_{ij}\mathbf{F}_j - \mathbf{F}_i], \tag{8.177}$$

$$\mathcal{F}^1_{ij}(\mathbf{x}-\mathbf{x}_j) = \mathbf{B}_i(\mathbf{x})\mathbf{R}_i^{-1}[\mathbf{Z}_{ii}\mathbf{R}_i + \mathbf{Z}_{ij}\mathbf{R}_j - \mathbf{R}_i], \tag{8.178}$$

$$\mathcal{F}^2_{ij}(\mathbf{x}-\mathbf{x}_j) = \mathbf{B}_i(\mathbf{x})\mathbf{R}_i^{-1}[\mathbf{Z}_{ii}\mathbf{F}_i + \mathbf{Z}_{ij}\mathbf{F}_j - \mathbf{F}_i]. \tag{8.179}$$

For the sake of simplicity of subsequent calculations we will assume the hypothesis (8.45) where the shape of "excluded volume" v^0_{ij} does not depend on the inclusion v_j: $v^0_{ij} = v^0_i$. Then substituting Eqs. (8.175) into Eq. (8.6) at $n = 1$ leads to

$$\overline{v}_i\langle\boldsymbol{\vartheta}_i\rangle = \overline{v}_i\boldsymbol{\vartheta}^0_i + \mathbf{R}_i\mathbf{Q}^0_i\sum_{j=1}^{N}\langle\boldsymbol{\vartheta}_j\rangle c^{(j)} + \mathcal{J}_i, \tag{8.180}$$

where the notation

$$\mathcal{J}_i = \int \Big\{[\langle\mathbf{R}_i(\mathbf{x})\mathcal{F}^1_{ij}(\mathbf{x}-\mathbf{x}_j)\rangle_i\langle\widetilde{\boldsymbol{\sigma}}_{1,2}\rangle + \langle\mathbf{R}_i(\mathbf{x})\mathcal{F}^2_i(\mathbf{x}-\mathbf{x}_j)\rangle_i]\varphi(v_j,\mathbf{x}_j|v_i,\mathbf{x}_i)$$
$$-\langle\mathbf{R}_i(\mathbf{x})\boldsymbol{\Gamma}(\mathbf{x}-\mathbf{x}_j)\rangle_i\langle\boldsymbol{\vartheta}_j\rangle c^{(j)}[1 - V^0_i(\mathbf{x}_j)]\Big\}\, d\mathbf{x}_j, \tag{8.181}$$

is introduced and the tensor $\mathbf{Q}^0_i$ is estimated by the formula (3.106) with replacement of the volume v_i by the volume v^0_i; in particular, if the shape of "excluded volume" v^0_i can be produced from the surface ∂v_i by a homothetic transformation then

$\mathbf{Q}_i^0 \equiv \mathbf{Q}_i$. At first we assume that the integral (8.181) is known. Then the linear system (8.180) can be solved easily for $\langle \boldsymbol{\sigma} \rangle_i(\mathbf{x})$

$$\langle \boldsymbol{\sigma} \rangle_i(\mathbf{x}) = \mathbf{B}_i(\mathbf{x})\boldsymbol{\sigma}^0 + \mathbf{C}_i(\mathbf{x})$$

$$+ \mathbf{B}_i(\mathbf{x})\mathbf{Q}_i^0[\mathbf{I} - \langle \mathbf{R}^v \mathbf{Q}^0 V \rangle]^{-1} \Big\{ \langle \mathbf{R}^v V \rangle \boldsymbol{\sigma}^0 + \langle \mathbf{F}^v V \rangle + \sum_{i=1}^N c^{(i)} \mathcal{J}_i \Big\} \tag{8.182}$$

where one introduces the notations $\mathbf{f}^v(\mathbf{x}) = \mathbf{f}^{(i)}(\mathbf{x})/\bar{v}^{(i)}$, at $\mathbf{x} \in v^{(i)}$, $(i = 1, \ldots, N;\ \mathbf{f} = \mathbf{R}, \mathbf{F})$. The integrals in Eq. (8.181) are absolutely convergent and do not depend on the shape of the integration region which can be chosen as homothetical to v_i^0. Recall that the stress fields entering in the integrals (8.181) are unknown in actuality, and for solution of Eq. (8.182), we will use the standard assumption of the perturbation method

$$\langle \widetilde{\boldsymbol{\sigma}}(\mathbf{x})_{1,2} \rangle_i = \langle \overline{\boldsymbol{\sigma}}(\mathbf{x}) \rangle_i = \boldsymbol{\sigma}^0, \ i = 1, 2, \ \mathbf{x} \in v_i. \tag{8.183}$$

Then all integrals in Eq. (8.182) valid for arbitrary $\mathbf{M}^c \equiv$ const. are known that makes it possible to estimate the effective properties from Eqs. (6.47) and (6.48) in a similar manner to the analogous process with Eqs. (8.32)–(8.34). Nevertheless, for simplicity of calculations, one assumes $\mathbf{M}^c = \mathbf{M}^{(0)}$ that yields the following *new representations*:

$$\mathbf{M}^* = \mathbf{M}^0 + \Big[\mathbf{I} - \langle \mathbf{R}^v \mathbf{Q}^0 V \rangle\Big]^{-1} \Big\{ \langle \mathbf{R}^v V \rangle + \sum_{i=1}^N c^{(i)} \mathcal{J}_i^{1\mathbf{M}} \Big\}, \tag{8.184}$$

$$\boldsymbol{\beta}^* = \boldsymbol{\beta}^{(0)} + \Big[\mathbf{I} - \langle \mathbf{R}^v \mathbf{Q}^0 V \rangle\Big]^{-1} \Big\{ \langle \mathbf{F}^v V \rangle + \sum_{i=1}^N c^{(i)} \mathcal{J}_i^{2\mathbf{M}} \Big\}, \tag{8.185}$$

$$-2W^* = \langle \boldsymbol{\beta}_1 \mathbf{C} V \rangle + \langle \boldsymbol{\beta}_1 \mathbf{B} \mathbf{Q}^0 V \rangle [\mathbf{I} - \langle \mathbf{R}^v \mathbf{Q}^0 V \rangle]^{-1} \langle \mathbf{F}^v V \rangle$$

$$+ \langle \boldsymbol{\beta}_1 \mathbf{B} \mathbf{Q}^0 V \rangle [\mathbf{I} - \langle \mathbf{R}^v \mathbf{Q}^0 V \rangle]^{-1} \sum_{i=1}^N c^{(i)} \mathcal{J}_i^{2\mathbf{M}} + \sum_{i=1}^N c^{(i)} \mathcal{J}_i^{2\beta}. \tag{8.186}$$

The notation for absolutely convergent integrals is used

$$\mathcal{J}_i^{m\mathbf{M}} = \int \Big\{ \langle \mathbf{R}_i(\mathbf{x}) \mathcal{F}_{ij}^m(\mathbf{x} - \mathbf{x}_j) \rangle_i \varphi(v_j, \mathbf{x}_j | v_i, \mathbf{x}_i)$$

$$- \langle \mathbf{R}_i(\mathbf{x}) \boldsymbol{\Gamma}(\mathbf{x} - \mathbf{x}_j) \rangle_i \mathbf{R}_j n^{(j)} [1 - V_i^0(\mathbf{x}_j)] \Big\} \, d\mathbf{x}_j, \tag{8.187}$$

$$\mathcal{J}_i^{m\beta} = \int \Big\{\langle \boldsymbol{\beta}(\mathbf{x})\mathcal{F}_{ij}^m(\mathbf{x}-\mathbf{x}_j)\rangle_i \varphi(v_j, \mathbf{x}_j|v_i, \mathbf{x}_i)$$

$$-\langle \boldsymbol{\beta}(\mathbf{x})\mathbf{B}_i(\mathbf{x})\boldsymbol{\Gamma}(\mathbf{x}-\mathbf{x}_j)\rangle_i \mathbf{F}_j n^{(j)}[1 - V_i^0(\mathbf{x}_j)]\Big\}\, d\mathbf{x}_j, \qquad (8.188)$$

describing the binary interaction between the inclusions, $m = 1, 2$. If the integration domains of absolutely convergent integrals (8.187) and (8.188) are chosen to be homothetical with the correlation holes v_i^0 $(i = 1, \ldots, N)$, then the second summands in the integrands (8.187) and (8.188) can be dropped out.

It should be mentioned that Eqs. (8.186)–(8.188) can be considered as a first-order approximation of the MEFM (8.35)–(8.37) with the driven term (zero-order approximation) chosen as EFM (8.36)–(8.59); similar ideas were used in [1521] and [189].

8.5.2 *Perturbation Method for Small Concentrations of Inclusions*

In the case of a small concentration of inclusions as well as for a weakly inhomogeneous medium

$$c^{(i)} \ll 1, \quad \text{or} \quad ||\mathbf{M}_1^{(i)}\mathbf{L}^c|| \ll 1 \qquad (8.189)$$

the perturbation method is appropriate. Then instead of hypothesis **H2**, we can apply an assumption (8.183) (in contrast to the combined MEFM-perturbation method), and taking Eq. (8.175) into account, we get

$$\mathbf{M}^* = \mathbf{M}^{(0)} + [\mathbf{I} + \langle \mathbf{R}^v \mathbf{Q}^0 V\rangle]\langle \mathbf{R}^v V\rangle + \sum_{i=1}^N c^{(i)} \mathcal{J}_i^{1\mathbf{M}}, \qquad (8.190)$$

$$\boldsymbol{\beta}^* = \boldsymbol{\beta}^{(0)} + [\mathbf{I} + \langle \mathbf{R}^v \mathbf{Q}^0 V\rangle]\langle \mathbf{F}^v V\rangle + \sum_{i=1}^N c^{(i)} \mathcal{J}_i^{2\mathbf{M}}, \qquad (8.191)$$

$$-2W^* = \langle \boldsymbol{\beta}_1 \mathbf{C} V\rangle + \langle \boldsymbol{\beta}_1 \mathbf{B}\mathbf{Q}^0 V\rangle\langle \mathbf{F}^v V\rangle + \sum_{i=1}^N c^{(i)} \mathcal{J}_i^{2\beta}. \qquad (8.192)$$

differing from the version of the MEFM (8.23) only by the particular solution of the pairwise inclusion interaction problem (8.9) as well as by the additional assumption (8.45) which uniquely reduces the MEFM to the perturbation method (8.190)–(8.192). Equations (8.190)–(8.192) are new. In the case of the exact estimation of the tensor $\mathcal{J}^{1\mathbf{M}}$ (8.188), Eq. (8.190) is reduced to the analogous relation obtained in [305] (see also [75, 777, 1664]) for spherical homogeneous isotropic inclusions of one radius; for an approximate representation of the tensor $\mathcal{J}^{1\mathbf{M}}$ evaluated by the use of Eqs. (4.52), (4.55_1), (8.178), and (8.188), Eq. (8.190) is reduced to the result [267] (see also [1632]). Equations (8.190)–(8.192) can be obtained from Eqs. (8.184)–(8.186) by passage to the limit $c \equiv \langle V \rangle \to 0$ and dropping the terms of order $O(c^3)$ and higher-order terms. Passing to the limit $c \equiv \langle V \rangle \to 0$ in Eqs. (8.57)–(8.59) and dropping the terms of order $O(c^3)$ lead to Eqs. (8.190)–(8.192) with the omitted terms $\mathcal{J}^{1\mathbf{M}}$, $\mathcal{J}^{2\mathbf{M}}$, $\mathcal{J}^{2\beta} \equiv \mathbf{0}$; these representations contain the terms of order $O(c^2)$ although their accuracy has the order $O(c)$. Finally, discarding the terms of order $O(c^2)$ and higher-order terms in Eqs. (8.190)–(8.192) leads to the dilute approximation of effective properties. In the dilute approximation (called also non-interacting approximation), each inhomogeneity interacts only with the matrix, but not with other inhomogeneities (see, e.g., [422]; interaction effect of nanofibers can be significant even for its minor ($c < 0.05$) volume concentration (see, e.g., Sect. 18.3.3 in [200]). Particular cases of this approximation for the effective moduli $\mathbf{L}^*$ of two-phase composites containing homogeneous ellipsoidal inclusions are one of the most rederived formulae for a 40-year period (1957–1999) (for reference, see, e.g., [1012]). Very clear common and concise forms of these representations are attributed to Eshelby [449] and to Krivoglaz and Cherevko [810].

8.5.3 *Perturbation Method for Weakly Inhomogeneous Media*

Another form of the perturbation method is related to another small parameters (8.189$_2$) such as a slight deviation of the moduli $\mathbf{L}(\mathbf{x})$ from a constant value $\mathbf{L}^c \equiv \langle \mathbf{L} \rangle$: $||\mathbf{L}_1\mathbf{M}^c|| \ll 1$. To simplify the subsequent derivations, it is convenient to represent Eq. (7.32) (with $\boldsymbol{\beta}, \mathbf{f} \equiv \mathbf{0}$) in an operator form

$$\mathbf{L}_1\boldsymbol{\varepsilon} = \mathbf{L}_1\langle\boldsymbol{\varepsilon}\rangle + \mathbf{L}_1\boldsymbol{\mathcal{U}}\mathbf{L}_1\boldsymbol{\varepsilon}, \tag{8.193}$$

where $\boldsymbol{\mathcal{U}}$ denotes the linear integral operator with the kernel $\mathbf{U}(\mathbf{x})$: $(\boldsymbol{\mathcal{U}}\boldsymbol{\tau})(\mathbf{x}) = \int \mathbf{U}(\mathbf{x}-\mathbf{y})\big\{\boldsymbol{\tau}(\mathbf{y}) - \langle\boldsymbol{\tau}\rangle(\mathbf{y})\big\}\, d\mathbf{y}$. We construct the solution of Eq. (8.193) by the point Jacobi iterative scheme based on the recursion formula $\mathbf{L}_1\boldsymbol{\varepsilon}_{[n+1]} = \mathbf{L}_1\langle\boldsymbol{\varepsilon}\rangle + \mathbf{L}_1\boldsymbol{\mathcal{U}}\mathbf{L}_1\boldsymbol{\varepsilon}_{[n]}$, where $\{\boldsymbol{\varepsilon}_{(n)}\}$ are treated as an approximation of the solution (8.193), which is presented in the Neumann series form

$$\mathbf{L}_1\boldsymbol{\varepsilon} = \boldsymbol{\mathcal{L}}\mathbf{L}_1\langle\boldsymbol{\varepsilon}\rangle, \ \ \boldsymbol{\mathcal{L}} \equiv \sum_{k=0}^{\infty}\boldsymbol{\mathcal{U}}^k, \tag{8.194}$$

where the power $\boldsymbol{\mathcal{U}}^k$ is defined recursively by the condition $\boldsymbol{\mathcal{U}}^1 = \mathbf{L}_1\boldsymbol{\mathcal{U}}$ and the kernel of $\boldsymbol{\mathcal{U}}^k$ is $\boldsymbol{\mathcal{U}}_k(\mathbf{x},\mathbf{y}) = \mathbf{L}_1(\mathbf{x})\int \mathbf{U}(\mathbf{x}-\mathbf{z})\boldsymbol{\mathcal{U}}_{k-1}(\mathbf{z},\mathbf{y})\, d\mathbf{z}$. Usually the driving term of Eq. (8.193) is used as an initial approximation $\mathbf{L}_1\boldsymbol{\varepsilon}_{[0]} = \mathbf{L}_1\langle\boldsymbol{\varepsilon}\rangle$ which, in combination with Eq. (8.194) and the definition of the effective elastic moduli $\mathbf{L}^*$ (6.21), allows a few iterations of $\mathbf{L}^*$:

$$\mathbf{L}^* = \langle\mathbf{L}\rangle, \quad \mathbf{L}^* = \langle\mathbf{L}\rangle + \langle\mathbf{L}_1\boldsymbol{\mathcal{U}}\mathbf{L}_1\rangle, \tag{8.195}$$

$$\mathbf{L}^* = \langle\mathbf{L}\rangle + \langle\mathbf{L}_1\boldsymbol{\mathcal{U}}\mathbf{L}_1\rangle + \langle\mathbf{L}_1\boldsymbol{\mathcal{U}}\mathbf{L}_1\boldsymbol{\mathcal{U}}\mathbf{L}_1\rangle, \tag{8.196}$$

The representations (8.195) and (8.196) can be obtained from the general Eqs. (7.21)–(7.23) (in the framework of the hypothesis **H1b** (4.94) and (6.21) if the integral system (7.21)–(7.23) is truncated at the $n = 1$ and $n = 2$, and in all integral terms, we assume that $\boldsymbol{\varepsilon}(\mathbf{y}) \equiv \langle\boldsymbol{\varepsilon}\rangle$ (direct interaction between the inclusions is neglected)). Therefore, the perturbation method (8.194) can be considered as a consequence of the MEFM. The perturbation method is attributed to Lifshitz and Rozenzweig [934, 936] and developed for the composite materials and polycrystals by Shermergor and coauthors [359–361, 489–491, 1378] as well as in [103, 106, 814, 1075, 1523]. The sufficient convergence conditions for the Neumann series method (8.194) were obtained in [488] (see also [166]) in the Hilbert space of modified strain fields. To avoid some algebraic manipulation, Buryachenko [200] reproduced the effective moduli representations from these works.

8.6 Effective Medium Methods

8.6.1 Application to Composite Materials

One of the first schemes of the effective medium method (EMM), also called the self-consistent method, was proposed for both the polycrystals [636, 812] and composite materials [167, 643, 1348] and was based on the following hypothesis: each inclusion in the composite material behaves as an isolated one in a homogeneous medium whose properties coincide with the effective properties of the whole composite

$$\mathbf{M}^c \equiv \mathbf{M}^*. \tag{8.197}$$

The field acting on this inclusion $\overline{\boldsymbol{\sigma}}$ coincides with the average field $\overline{\boldsymbol{\sigma}} \equiv \langle\boldsymbol{\sigma}\rangle$ corresponding to the homogeneous traction boundary condition (6.5) for the macrodomain containing a statistically large number of realization of the inclusion field. By doing so, the infinite system (7.21)–(7.22) for the matrix composites being considered is truncated after Eq. (7.27) which provides a way of estimating of effective compliance (6.47) by the use of the solution (8.2) and (8.65) for an isolated inclusion within an infinite effective medium

$$\mathbf{M}^* = \mathbf{M}^{(0)} + \langle\bar{\mathbf{R}}^v V\rangle, \quad \mathbf{M}^* = \mathbf{M}^{(0)} + \langle(\mathbf{M} - \mathbf{M}^{(0)})\mathbf{L}\bar{\mathbf{A}} V\rangle\mathbf{M}^*, \tag{8.198}$$

where the tensors $\bar{\mathbf{R}}^v = (\mathbf{M} - \mathbf{M}^{(0)})\bar{\mathbf{B}}$, $\bar{\mathbf{B}} = \bar{\mathbf{B}}(\mathbf{x})$, and $\bar{\mathbf{A}} = \bar{\mathbf{A}}(\mathbf{x})$ implicitly depend on $\mathbf{M}^*$. Here the bar over the material tensor indicates that it is computed using the fact that the matrix properties coincide with those of the effective medium. It was exploited that $\bar{\mathbf{B}}(\mathbf{x}) = \mathbf{L}(\mathbf{x})\bar{\mathbf{A}}(\mathbf{x})\mathbf{M}^*$ and $\langle\boldsymbol{\varepsilon}\rangle = \mathbf{M}^*\langle\boldsymbol{\sigma}\rangle$.

In a similar manner, for the homogeneous displacement boundary conditions (6.4) accompanied by the EMM assumptions $\mathbf{L}^c = \mathbf{L}^*$, $\overline{\boldsymbol{\varepsilon}} \equiv \langle \boldsymbol{\varepsilon} \rangle$, we get two equivalent representations of effective moduli (6.21)

$$\mathbf{L}^* = \mathbf{L}^{(0)} + \langle (\mathbf{L} - \mathbf{L}^{(0)})\bar{\mathbf{A}} V \rangle, \quad \mathbf{L}^* = \mathbf{L}^{(0)} + \langle (\mathbf{L} - \mathbf{L}^{(0)})\mathbf{M}\bar{\mathbf{B}} V \rangle \mathbf{L}^*. \qquad (8.199)$$

The obtained effective compliances (8.198_1), (8.198_2) and elastic moduli (8.199_1), (8.199_2) are each other's inverses. Indeed, multiplying Eq. (8.199_1) from the right by $(\mathbf{L}^*)^{-1}$ and from the left by $(\mathbf{L}^{(0)})^{-1} = \mathbf{M}^{(0)}$ leads to

$$\mathbf{M}^{(0)} = (\mathbf{L}^*)^{-1} + \langle \mathbf{M}^{(0)}(\mathbf{L} - \mathbf{L}^{(0)})\bar{\mathbf{A}} \rangle (\mathbf{L}^*)^{-1} = (\mathbf{L}^*)^{-1} - \langle (\mathbf{M} - \mathbf{M}^{(0)})\mathbf{L}\bar{\mathbf{A}} \rangle (\mathbf{L}^*)^{-1}. \qquad (8.200)$$

Analogously, multiplying Eq. (8.198_2) by the tensors $(\mathbf{M}^*)^{-1}$ and $(\mathbf{M}^{(0)})^{-1} = \mathbf{L}^{(0)}$ from the right and from the left, respectively, gives

$$\mathbf{L}^{(0)} = (\mathbf{M}^*)^{-1} + \langle \mathbf{L}^{(0)}(\mathbf{M} - \mathbf{M}^{(0)})\bar{\mathbf{B}} \rangle (\mathbf{M}^*)^{-1} = (\mathbf{M}^*)^{-1} - \langle (\mathbf{L} - \mathbf{L}^{(0)})\mathbf{M}\bar{\mathbf{B}} \rangle (\mathbf{M}^*)^{-1}. \qquad (8.201)$$

From comparison of (8.200) with (8.198_1), and (8.201) with (8.198_2), it is seen that the effective compliance $\mathbf{M}^*$ (8.198_1), (8.198_2) and the effective moduli (8.199_1), (8.199_2) are mutually inverse $(\mathbf{L}^*)^{-1} = \mathbf{M}^*$, $(\mathbf{M}^*)^{-1} = \mathbf{L}^*$ for the matrix composite which consists of a homogeneous matrix containing a statistically uniform set of inclusions $v = \cup v^{(i)}$ $(i = 1, \ldots, N)$ (with the indicator function $V(\mathbf{x}) = \sum V^{(i)}$, $i = 1, \ldots, N$) of any shape and microtopology, and with an arbitrary number of phases.

Moreover, we can assume that the composite has no continuous matrix $c^{(0)} \equiv 0$ as it takes place for polycrystals. Then Eqs. (6.21) and (8.199_1) coupled with $\mathbf{L}^{(0)} = \mathbf{L}^*$ imply the equivalent representations for both the strain concentrator factors

$$\langle \bar{\mathbf{A}} V \rangle = \mathbf{I}, \quad \langle (\mathbf{L} - \mathbf{L}^*)\bar{\mathbf{A}} V \rangle = \mathbf{0}, \qquad (8.202)$$

and their counterparts presented in terms of stress concentration factors

$$\langle \bar{\mathbf{B}} V \rangle = \mathbf{I}, \quad \langle (\mathbf{M} - \mathbf{M}^*)\bar{\mathbf{B}} V \rangle = \mathbf{0}. \qquad (8.203)$$

For concrete estimation of the effective properties, we need to know the tensors $\bar{\mathbf{A}}(\mathbf{x})$ $(\mathbf{x} \in v^{(i)}, \ i = 1, \ldots, N)$. In particular, for the homogeneous ellipsoidal inclusions $\bar{\mathbf{A}}(\mathbf{x}) = [\mathbf{I} + \bar{\mathbf{P}}_i(\mathbf{L}^{(i)} - \mathbf{L}^*)]^{-1}$, $\bar{\mathbf{B}}(\mathbf{x}) = [\mathbf{I} + \bar{\mathbf{Q}}_i(\mathbf{M}^{(i)} - \mathbf{M}^*)]^{-1}$, $(\mathbf{x} \in v^{(i)}, \ i = 1, \ldots, N)$ (3.116), (3.117), and $\bar{\mathbf{Q}}_i$ and $\bar{\mathbf{P}}_i$ are defined by Eqs. (3.106) with replacement of $\mathbf{L}^{(0)}$ by $\mathbf{L}^*$. If, however, $\bar{\mathbf{P}}_i = \bar{\mathbf{P}}$ for all i, then Eqs. (8.199_1), (8.202_1), and (8.202_2) are equivalent to the representation

$$\sum_{i=1}^{N} (\mathbf{L}^{(i)} - \mathbf{L}^*)[\mathbf{I} + \bar{\mathbf{P}}(\mathbf{L}^{(i)} - \mathbf{L}^*)]^{-1} = \mathbf{0} \qquad (8.204)$$

allowing subsequent simplified decomposition for composites with the spherical isotropic inclusions in R^d-space ($d = 2, 3$) (see, e.g., [1523])

$$\sum_{i=1}^{n} \frac{c^{(i)}(k^{(i)} - k^*)}{k^{(i)} + 2(d-1)\mu^*/d} = 0, \quad \sum_{i=1}^{n} \frac{c^{(i)}(\mu^{(i)} - \mu^*)}{\mu^{(i)} + g^*} = 0, \tag{8.205}$$

where $g^* = \mu^*[dk^*/2 + (d+1)(d-2)\mu^*/d][k^* + 2\mu^*]^{-1}$, and the obtained representations (8.204), (8.205) are symmetric with respect to the simultaneous interchange $\mathbf{L}^{(i)} \leftrightarrow \mathbf{L}^{(j)}$, $c^{(i)} \leftrightarrow c^{(j)}$ ($i = 0, 1, \ldots, N$.) It is possible to get the simple solutions of (8.205) in the limiting cases of incompressible phase $v^{(0)}$ with the phase $v^{(1)}$ formed by either the rigid inclusions or voids. These formulae predict either infinite or zero effective stiffness if the volume fraction $c^{(1)}$ reaches a so-called percolation threshold $c_p^{(1)}$ [1321]. Such a behavior is a manifestation of the so-called *percolation phenomenon* base on the fact that the grains of the dispersed phase $v^{(1)}$ form a continuous connected path (called an infinite cluster) through the composite if $c^{(1)} \geq c_p^{(1)}$. For example, for the spherical inclusions with $\mu^{(1)}/\mu^{(0)} = \infty$, the percolation threshold $c_p^{(1)} = 2/(d+2)$ which is not sensitive to the microstructure details (e.g., a fractional composition of inclusions and their radial distribution function) and can be quite different from the real percolation threshold of a concrete material.

As another limiting case of the inclusion shape, let us consider an isotropic microcracked medium with a uniform distribution of microcrack orientation. In such a case, we can apply the dilute approximation (8.112) with a replacement of the matrix Poisson ratio $\nu^{(0)}$ by the corresponding ratio of the composite medium ν^*

$$\frac{k^*}{k^{(0)}} = 1 - \frac{16}{9}\frac{1-(\nu^*)^2}{1-2\nu^*}\epsilon, \quad \frac{\mu^*}{\mu^{(0)}} = 1 - \frac{32}{45}\frac{(1-\nu^*)(5-\nu^*)}{2-\nu^*}\epsilon. \tag{8.206}$$

The formulae (8.206) are the solution of the self-consistent approach [169].

It should be mentioned that the representations (8.205) are the particular case of results obtained by the different methods. For example, a choice of the comparison medium as an effective medium (8.197) allows the relations (8.205) by the method of conditional moments (8.114) [762, 1333], by the singular-approximation method (see Sect. 8.3.4 and [1378]), by the method [142] based on the concept of "strong isotropy," as well as by the variation method [1620, 1625]. Obviously, the relations (8.205) always lie between the Hashin–Shtrikman bounds intentionally obtained at the more "soft" or "stiff" values of $\mathbf{L}^c$. Moreover, the EMM estimations (8.205) are realizable in the sense that there is a particular random structure composite whose effective moduli exactly coincide with EM estimations. Such a fractal-like (or self-similar in all-scale) structure justified in [1073, 1074] consists of two-sort granules $v^{(1)}$ and $v^{(2)}$ of dilute concentration randomly distributed in the matrix consisting in turn of much smaller similar grains and so on from a finite size of inclusions to infinitesimally small. Realizability of the EM approach provides a location of the estimations (8.205) within Hashin–Shtrikman bounds.

The obtained representations of effective properties [e.g., (8.199_1), (8.204), (8.205)] are the implicit algebraic systems with respect to yet-unknown effective elastic moduli $\mathbf{L}^*$ providing the effective moduli only implicitly through a straightforward iterative procedure to the fixed point. For example, Eq. (8.199_1) corresponds to the iterative scheme

$$\mathbf{L}^{*[n+1]} = \mathbf{L}^{(0)} + \langle(\mathbf{L} - \mathbf{L}^{(0)})\bar{\mathbf{A}}^{[n]}V\rangle, \tag{8.207}$$

with an initial condition, e.g., either $\mathbf{L}^{*[0]} = \langle\mathbf{L}\rangle$ or $\mathbf{L}^{*[0]} = \langle\mathbf{M}\rangle^{-1}$ if the rigidity of the inclusions is less than that of the matrix or otherwise, respectively. The scheme (8.207) is known to be stable and unique although this statement is plausible rather than rigorous. The fixed point of the mapping (8.207) is nothing more than a solution of the already analyzed Eq. (8.199_1). Therefore, the described scheme (8.207) is actually an iterative procedure of solving Eq. (8.199_1).

In more recent schemes of the effective medium method, the composite material outside a certain layer of matrix surrounding each inclusion is replaced by an effective medium [338, 655, 672, 673, 1413, 1684]. In so doing, Eqs. (8.198_1) and (8.199_1) with obvious corrections of tensors $\bar{\mathbf{B}}(\mathbf{x})$ and $\bar{\mathbf{A}}(\mathbf{x})$ are valid. So, following [338], a composite-sphere model (or three-phase model) considers an RVE as a concentric sphere consisting of a matrix $v^{(0)}$ and a single spherical homogeneous inclusion v_i of radii b and a, respectively, which is inserted in the medium with the yet-unknown elastic moduli $\mathbf{L}^* = (3k^*, 2\mu^*)$. Christensen [337] demonstrated that the composite-sphere model (also called the *generalized self-consistent method*, GSCM) results correlate better with the experimental data on viscosity of suspensions than the MTM (8.80) and the differential scheme considered in Sect. 8.3. Siboni and Benveniste [1390] generalized the (GSCM) to the multiphase composites, while the GSCM was combined with an accumulation of damage in [1339]. The main idea of the EMM lies in the substitution of the strain concentrator factor $\bar{\mathbf{A}}(\mathbf{x})$ into the average scheme of the dilute approximation (8.199_1), although the tensor $\bar{\mathbf{A}}(\mathbf{x})$ can be incorporated into any other average scheme. For example, Luo and Weng [986] integrated $\bar{\mathbf{A}}(\mathbf{x})$ into the Mori-Tanaka scheme and demonstrated in a concrete example that the effective moduli estimated by the modified GSCM lie between the estimations obtained via the MTM and GSCM. The GSCM is based on the assumption that the particles in the three-phase cells are separated by the layer of a pure matrix which is free of the particle phase. However, a variation of properties of this layer along the radius makes it possible to take into account a probability of entrapment of the particle phase in the matrix layer [171, 1512].

To put this another way, the effective medium method (8.198_1) or (8.199_1) is equivalent to the perturbation method of the first order over the concentration of inclusions accompanied by the assumption (8.197). An obvious improvement of the method (8.198_1) is the use of the perturbation method of the second order described in Sect. 8.5.2 accompanied by the assumption (8.197). Moreover, the combination of the MEFM and the effective medium concept (8.197) is equally trivial in a substitution of the equality (8.197) into either Eqs. (8.32) or (8.54).

This direction of a generalization of the EMM is demonstrated in the version of the MEMM proposed in [261] which is based on the solution of Eq. (8.23) where the involved tensors are estimated for $\mathbf{L}^c = \mathbf{L}^*$. It will lead to the representations of effective properties in the form of Eqs. (8.32)–(8.34) where all material tensors are estimated at $\mathbf{L}^c = \mathbf{L}^*$. The mentioned approach is more general than substitution of (8.197) into Eq. (8.54) which, in turn, is more general than the approach [926] (see also [115, 866]) combining the assumption of the Mori-Tanaka (8.68) and effective medium method (8.197) (the last is questionable in the general case of multiphase composites; see Sect. 9.3). The implicit algebraic relations for effective moduli obtained must be solved numerically, in general; see Eq. (8.207).

To avoid confusion, it should be mentioned that the term *self-consistent methods* was originally proposed merely for the methods related to the assumption (8.197) enabling one to get self-consistent implicit algebraic equations for effective elastic moduli, e.g., (8.198_1). We currently use the logical generalization of this term which is widely used to stand for methods based on some closing assumption for effective fields (see, e.g., the hypothesis **H2**) for obtaining a *self-consistent* final system of either integral or differential equations for statistical moments of local fields. Only in the simplified special cases considered in the current work, this system can be solved explicitly in analytical form.

8.6.2 Analysis of Polycrystal Materials

Let us consider (following [1378]) a one-phase polycrystal consisting of homogeneous spherical grains with the same elastic properties and uniform random orientation. Due to the absence of texture, the polycrystal is macroscopically isotropic $\mathbf{L}^* = \mathbf{L}^c = (3k^*, 2\mu^*)$, and the tensor $\mathbf{b}^L$ introduced in Eq. (8.100) is also isotropic

$$\mathbf{b}^L \equiv (3b_k^L, 2b_\mu^L) = \Big(4\mu^*, \frac{\mu^*(9k^*+8\mu^*)}{3(k^*+2\mu^*)}\Big). \qquad (8.208)$$

For averaging in Eq. (8.100_2), we introduce the auxiliary tensors $\mathbf{S}^L$ and $\mathbf{S}^*$

$$(\mathbf{L}+\mathbf{b}^L)\mathbf{S}^L = \mathbf{I}, \quad (\mathbf{L}^*+\mathbf{b}^L)\mathbf{S}^* = \mathbf{I}, \qquad (8.209)$$

allowing us to present Eq. (8.100_2) in the form $\mathbf{S}^* = \langle \mathbf{S}^L \rangle$. Here the components of the isotropic tensor $\mathbf{S}^* \equiv (3S_k^*, 2S_\mu^*) = \big(\frac{1}{3}S^*_{iikk}, \frac{1}{5}(S^*_{ikik} - \frac{1}{3}S^*_{iikk})\big)$ can be found for the nontextured materials from two scalar equations

$$\mathbf{N}_1\mathbf{S}^* = \mathbf{N}_1\mathbf{S}^L, \quad \mathbf{N}_2\mathbf{S}^* = \mathbf{N}_2\mathbf{S}^L. \qquad (8.210)$$

For example, for a cubic symmetry of crystals (2.175), Eqs. (8.210_1) and (8.210_2) combined with Eq. (6.238) give a cubic equation obtained by Hershey [636] and by Kröner [812] via the proposed self-consistent method. Some other examples of a similar approach with applications to the polycrystals for different crystal symmetry can be found, e.g., in [114, 821, 1090, 1091, 1378, 1595].

It should be mentioned that a direct analytical solution of the self-consistent Eqs. (8.199_1) or (8.209) is often cumbersome and hides from view the basic ideas of the method. However, Eq. (8.199_1) and (8.98_2) have an advantage of providing an iterative method with the consecutively repeated scheme, e.g., for the one-particle approximation of the EFM method (8.98_2)

$$\mathbf{L}^{*[n+1]} = \langle \bar{\mathbf{A}}^{[n]} V \rangle^{-1} \langle \bar{\mathbf{R}}^{\epsilon v[n]} V \rangle, \tag{8.211}$$

which is obvious and easily performed (see for details [200]).

A significance of the developed models of polycrystalline materials is beyond the scope of its own field of polycrystals and can be applied to the analysis of some skeletal composites (see Sect. 1.1.1). For example, the open-cell carbon foam microstructure is known to possess a tetrahedral structure of the foam ligaments oriented approximately 109° to each other [541, 1393]. The ligaments of foam materials can be modeled as polycrystalline medium (see, e.g., [598, 1393, 1594]) aggregating from the numerous minute crystals or grains of various sizes and shapes placed in the amorphous matrix and varying along its longitudinal (axial) direction. For example, in approach [1393], because of slenderness, each ligament can be considered as a beam and the tetrahedral cell microstructure with four ligaments as a structure. The subsequent analysis of the random orientation of the unit cells is performed by the use of either Voight or Reuss volume averaging of effective elastic properties estimated for the perfect periodic structures with prescribed different orientation mentioned above. This popular rough approach can be easily improved by the use of the MEFM generalized in this section for the analysis of polycrystalline materials. In so doing, the role of the individual crystal is assumed by the homogeneous effective medium corresponding to the involved periodic structure material with the prescribed orientation of unit cells.

8.7 Differential Methods

In addition to effective field (see Sect. 8.3) and effective medium (see Sect. 9.2.1) methods, the differential effective medium method (EMM) [46, 90, 164, 343, 613, 1135, 1137, 1138, 1302] is used in the mechanics of composite materials. This differential scheme of constructing effective elastic moduli also belongs to the class of the effective medium methods. Differential EMM arose as an alternative to the ambiguous assumptions of EMM, where the value of the effective modulus is used as an estimate of the average strain concentration in an isolated inclusion. Differential EMM is considered as a process of consecutive additions of the inclusion phase in a uniform medium with a modulus equal to the effective modulus of the medium with the previous additions of inclusions to the matrix. This process can take place in two ways. In the first, it is assumed that the inclusion phase consists of an infinite number of fractions with infinite size differences, and successive accommodations of the inclusions in the corresponding uniform medium leads to ascending order of inclusion size. In the second, inclusions of the initial size are added to the medium in infinitesimal amounts to reach the final real concentration. Both methods give equivalent results, since at each iteration step, the same solution for the single-particle problem is used to estimate the effective modulus of the medium with an infinitesimal inclusion in the corresponding uniform medium. Buryachenko and Parton [261] proposed two versions of the multiparticle differential methods for multicomponent composites. In the first one, the tensors analogous to $\bar{\mathbf{R}}_i^{v} \equiv (\mathbf{M}(\mathbf{x}) - \mathbf{M}^{(0)})\bar{\mathbf{B}}_i$ (8.198_1) are determined from the averaged solution of the problem for n inclusions in a uniform medium with tensor properties $\mathbf{M}^*$ and the field $\boldsymbol{\sigma}^0$ given at infinity. In the second one, the tensors $\bar{\mathbf{R}}_i^{v}$ (8.198_1) are estimated at each iteration step based on n-particle MEFM. Thus, both EMM and MEFM are applicable. In this subsection, we do not solve the single-particle problem but instead solve the multiparticle problem at each iteration step of the differential scheme. To do this, we use an algorithm previously proposed in Sect. 9.1. In comparisons of calculations with experimental data, we show the advantage of the multiparticle differential method over the one-particle method.

Following [261], we will estimate effective stiffness

$$\mathbf{L}^* = \mathbf{L}^{(0)} + \sum_{i=1}^{N} \mathbf{R}_i^{\epsilon *} n^{(i)}, \quad \langle \mathbf{L}_1(\mathbf{x})\boldsymbol{\varepsilon}(\mathbf{x}) V^{(i)}(\mathbf{x})\rangle \equiv \mathbf{R}_i^{\epsilon *} n^{(i)} \langle \boldsymbol{\varepsilon}\rangle \tag{8.212}$$

in a pure elastic problem ($\boldsymbol{\beta} \equiv \mathbf{0}$). The first equation in (8.212) can be rewritten in the form

$$\mathbf{L}^* = \mathbf{L}^{(0)} + \sum_{i=1}^{N} \mathbf{R}_i^{\epsilon *} \mathbf{A}_0 n^{(i)} \Big[c^{(0)}\mathbf{I} + \sum_{j=1}^{N} c^{(j)} \mathbf{A}_j^0 \Big]^{-1}, \tag{8.213}$$

$$\langle \boldsymbol{\varepsilon}\rangle_i = \mathbf{A}_i^0 \langle \boldsymbol{\varepsilon}\rangle^{(0)}, \quad \langle \boldsymbol{\varepsilon}\rangle^{(0)} = \mathbf{A}_0 \langle \boldsymbol{\varepsilon}\rangle. \tag{8.214}$$

We will consider a generalization of a differential method scheme [1137, 1138] in the case of a multicomponent inclusion with complex structure. We take the volume w of the composite medium with some finite concentration of inclusions $i = 1, 2, \ldots, N$ (which in general is different from the concentration $c^{(i)}$). The composite medium is replaced by a uniform volume w with tensor properties $\mathbf{L}$, determined from the equation $\langle \boldsymbol{\sigma}\rangle = \mathbf{L}\langle \boldsymbol{\varepsilon}\rangle$. The infinitesimal discrete volume of the uniform medium is removed and replaced by the sum of the components $i = 1, 2, \ldots, N$. That is, the representative volume $dv^{(i)}$ with properties $\mathbf{L}$ is replaced by the same volume $dv^{(i)}$ with properties of the ith component. If $\tilde{\mathbf{R}}_i^{\epsilon v *}$ is the average value of the coefficient of concentration of the polarization tensor ($\tilde{\mathbf{L}}_1(\mathbf{x}) \equiv \mathbf{L}(\mathbf{x}) - \mathbf{L}$)

$$\langle \tilde{\mathbf{L}}_1(\mathbf{x})\boldsymbol{\varepsilon}(\mathbf{x})\rangle_i \equiv \tilde{\mathbf{R}}_i^{\epsilon v*}\langle \boldsymbol{\varepsilon}\rangle \tag{8.215}$$

of element $dv^{(i)}$, then the increment to the tensor of effective properties $\mathbf{L}$ can be found from (8.212) in the form

$$d\mathbf{L} = \sum_{i=1}^{N} \tilde{\mathbf{R}}_i^{\epsilon v*}\frac{dv^{(i)}}{\bar{w}}. \tag{8.216}$$

Since it is convenient to carry out the calculation with the additional volume portions $dc^{(i)}$, it can be shown that (see [1137, 1138]) $dv^{(i)}/\bar{w} = \lambda_i dc/(1-c)$, where $c = \sum_{i=1}^{N} c^{(i)}$, $\lambda_i = c^{(i)}/c$, and consequently

$$d\mathbf{L} = \sum_{i=1}^{N} \tilde{\mathbf{R}}_i^{\epsilon v*}\lambda_i\frac{dc}{1-c}. \tag{8.217}$$

Equation (8.217) is an ordinary differential equation with respect to the unknown tensor $\mathbf{L}$ with initial conditions $\mathbf{L}(0) = \mathbf{L}^{(0)}$, $c^{(i)}(0) = 0$, $c(0) = 0$ $(i = 1, \ldots, N)$ and independent variable c ($\lambda_i = c^{(i)}/c = \text{const}$). The basis for (8.217) for a one-component inclusions is given in [343]. To close (8.217), it is necessary to establish a concrete value for $\tilde{\mathbf{R}}_i^{\epsilon v*}$, which can be determined on the basis of one-particle and multiparticle approaches (see for details [200, 261]). In two limiting cases of inclusion moduli [$\mathbf{L}^{(1)} = (\infty, \infty)$, $\mathbf{L}^{(1)} = (0, 0)$], Zimmerman [1702] obtained a closed analytical representations for the effective moduli for CMs with the rigid spherical inclusions and the pores.

8.8 Numerical Results

8.8.1 Composites with Spheroidal Inhomogeneities

Let us now demonstrate the application of the theoretical results by considering an isotropic composite made of an incompressible isotropic matrix, filled with rigid spherical inclusions of one size ($n = 1$). This example was chosen deliberately because it provides the maximum difference of predictions of effective shear modulus estimated by various methods and was considered by a number of authors. The following alternative radial distribution functions of inclusions (5.97), (5.100), and (5.101). for 3D (5.97) and (5.100) and 2D (5.97) and (5.101) cases. For an isotropic matrix filled by spherical and circle inclusions, the tensors $\mathbf{B}$, $\mathbf{Q}$, and $\mathbf{T}_{ij}(\mathbf{x}_i - \mathbf{x}_j)$ are known (see, e.g., Chap. 4). At first we consider the estimation of the effective shear modulus, giving the analytical representations for a step radial distribution function (5.97) in the 3D case. In so doing, the perturbation methods (8.184) and (8.190) as well as other methods will be rewritten in the terms of elastic moduli. Then we estimate the effective shear modulus by different methods

$$\mu^*/\mu^{(0)} = 1 + 2.5c + 4.844c^2, \tag{8.218}$$

$$\mu^*/\mu^{(0)} = 1 + 2.5c + 5.01c^2, \tag{8.219}$$

$$\mu^*/\mu^{(0)} = 1 + 2.5c/(1-c), \tag{8.220}$$

$$\mu^*/\mu^{(0)} = 1 + 2.5c/(1-31c/16), \tag{8.221}$$

$$\mu^*/\mu^{(0)} = 1 + (1-c)^{-1}(2.5c + 2.344c^2), \tag{8.222}$$

$$\mu^*/\mu^{(0)} = 1 + (1-c)^{-1}(2.5c + 2.51c^2), \tag{8.223}$$

$$\mu^*/\mu^{(0)} = (1-2.5c)^{-1}, \tag{8.224}$$

$$\mu^*/\mu^{(0)} = (1-c)^{-2.5}, \tag{8.225}$$

$$\mu^*/\mu^{(0)} = [1 - 2.5c(1+1.938c)]^{-1}, \tag{8.226}$$

Equations (8.218) and (8.219) are obtained for a small concentration of the inclusions by the perturbation method (4.52), (8.176), (8.190) [with $\mathcal{J}_i^{1\mathbf{M}} = (3J_1^{\mathbf{M}}, 2J_2^{\mathbf{M}})$, $2J_2^{\mathbf{M}} = 31c\mu^{(0)}/16 \approx 1.938c\mu^{(0)}$] and following [305], respectively. In so doing, the isotropic tensors $\mathcal{J}_i^{1\mathbf{M}} = (3J_1^{\mathbf{M}}, 2J_2^{\mathbf{M}})$ used in Eq. (8.218) and (8.219) have the components $2J_2^{\mathbf{M}} = 2.5c \cdot 15/16\mu^{(0)} = 2.344c\mu^{(0)}$ and $2J_2^{\mathbf{M}} = 2.51c\mu^{(0)}$, respectively. Relations (8.220) are calculated via MEFM in the framework of one-particle (or Mori-Tanaka) (8.57) and two-particle (8.38), (4.55), and (4.52) (point approximation) approximations, respectively. Relations (8.222) and (8.223) are estimated by the combined MEFM–perturbation method (8.184) utilizing the tensors $\mathcal{J}_i^{1\mathbf{M}}$ previously evaluated in Eq. (8.218) and (8.219), respectively. The effective medium method (8.198_1) predicts Eq. (8.224), while the differential scheme (see for details [200, 260]) leads to Eq. (8.226); (8.226) is determined using two-particle differential EMM. The differential equations

$$\frac{d^2\mu}{dc^2} = \frac{155}{16}\mu\frac{1}{(1-c)^2}, \quad \mu(0) = \mu^{(0)}, \ \mu^{(1)}(0) = \frac{5}{2}\mu^{(0)} \tag{8.227}$$

and

$$\frac{d^3\mu}{dc^3} = 15\Big(\frac{31}{16}\Big)^2\mu\frac{1}{(1-c)^3}, \quad \mu(0) = \mu^{(0)}, \ \mu^{(1)}(0) = \frac{5}{2}\mu^{(0)}, \ \mu^{(2)}(0) = \frac{155}{16}\mu^{(0)} \tag{8.228}$$

correspond to two-particle differential MEFM of order 2 and 3, respectively, (see for details [200, 260]). The numerical results estimated via Eqs. (8.218)–(8.226) are depicted in Figs. 8.5 and 8.6. As can be seen from Fig. 8.4, the use of Eq. (8.220) leads to underestimating evaluation of the effective shear modulus

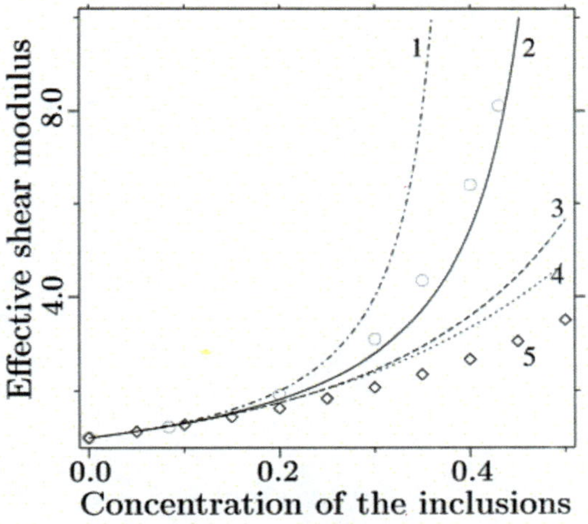

Fig. 8.4 $\mu^*/\mu^{(0)}$ vs c. Experimental data (open circle) and curves calculated by Eqs. (8.224) (1), (8.221) (2), (8.226) (3), (8.222) (4), and (8.220) (diamond symbol, 5)

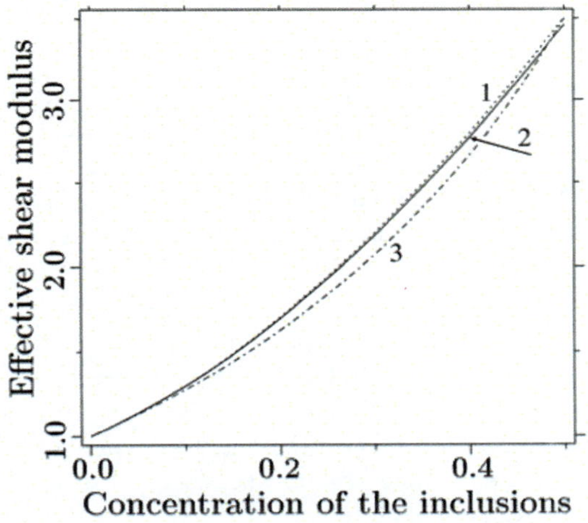

Fig. 8.5 $\mu^*/\mu^{(0)}$ vs c. calculated by Eqs. (8.219) (1), (8.218) (2), and (8.220) (3)

by 2.6 times for $c = 0.43$ compared to the experimental data as well as to the more exact point approximation of weakly interacting inclusions (8.221), which provides the best comparison with experimental data [803] on relative changes of the Newtonian viscosity of a suspension of identical rigid spheres. The results of estimations via Eqs. (8.222) and (8.223) differ from one another by not more than 2%. At the same time, the Mori-Tanaka solution (8.220) differs from the nondilute approximations (8.218) and (8.219) by not more than 5% for the concentration of the inclusions $c \leq 0.5$ (see Fig. 8.5). Therefore, even in the limiting case of an infinite elastic mismatch and $c = 0.5$, the perturbation method (8.174) and (8.190) provides the same accuracy as the Mori-Tanaka method. The relation (8.224) leads to overestimated evaluation of the effective shear modulus.

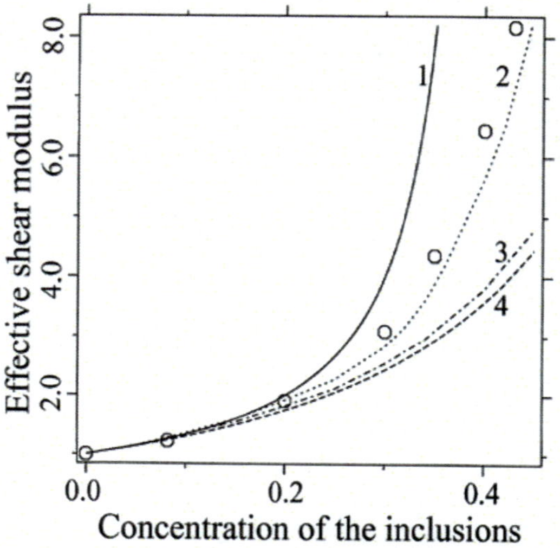

Fig. 8.6 Variation of the relative effective shear modulus $\mu^*/\mu^{(0)}$ calculated by Eqs. (8.224) (1), 8.15.14 with $m - 30$ (2), (8.227) (3), (8.226) (4)

A comparative analysis of differential methods is presented in Fig. 8.6 which shows the experimental data (points) [803] and curves 1–4 computed from Eqs. (8.224), (8.216) with $m = 30$, (8.227), and (8.226), respectively. It is clear that taking the binary interaction of the inclusions into account can increase the accuracy of the differential methods. We now turn to analysis of estimations based on the exact values of the tensors $\mathbf{T}_{ij}(\mathbf{x}_i - \mathbf{x}_j)$ (4.55$_1$). A comparison between the relative shear modulus $\mu^*/\mu^{(0)}$, calculated by the use of the MEFM (8.35), as well as by the point approximation (4.52) of the MEFM (8.38) and (8.39) for different radial functions (5.97) and (5.100), is presented in Fig. 8.7.

Analysis of Figs. 8.5 and 8.8 leads to the conclusion that the high accuracy of Eq. (8.221) exploited the point approximation (4.52), and the step RDF (5.97) is nothing more than fortuitousness because using both the more accurate RDF (5.100) and the solution (4.46) and (4.52) yields to the considerable sacrifice of a prognostic opportunity of the MEFM (see solid curve in Fig. 8.7); it should be added in this connection that the deficiencies of point approximation (4.52) were analyzed by Buryachenko in [183] also for regular structure composites for which an accurate solution is known. In so doing, the MEFM (8.35) exploiting the solution (8.9) and (4.46) of the binary interaction problem admits some improvement based on a more accurate solution of multiple interaction problem [190]. For example, in the combined method motivated by Dyskin and Mühlhaus [428], after estimation of the matrix $\mathbf{Z}_{ij}$ (4.46), we can replace in Eq. (8.27) the tensor $\mathbf{T}_{ij}$ by the tensor $\mathbf{T}_j$ (4.96); that is, the perturbation of surrounding inclusions v_j is replaced with the perturbation at the inclusion center $\mathbf{x}_i$ (see the curve diamond symbol in Fig. 8.7).

We now turn to the analysis of another limiting case of porous materials with an incompressible matrix [181, 244]. The formulae

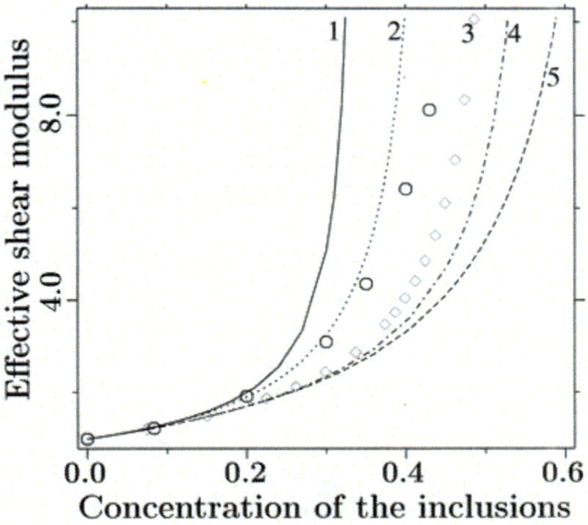

Fig. 8.7 $\mu^*/\mu^{(0)}$ vs c calculated by the MEFM (8.35), (4.52) for RDF (5.100) (1) and (5.97) (2), and by approximation (4.55) for RDF (5.100) (4) and (5.97) (5); combine method (diamond symbol, 3); experimental data (open circle)

$$\mathbf{L}^* = \mu^{(0)}\left(4\left[\frac{1}{c} - 1\right], 2\left[1 - \frac{5}{3}c\frac{1}{1 + \frac{2c}{3}}\right]\right), \tag{8.229}$$

$$\mathbf{L}^* = \mu^{(0)}\left(4\left[\frac{1}{c} - \frac{29}{24}\right], 2\left[1 - \frac{5}{3}c\frac{1}{1 + \frac{5c}{24}}\right]\right), \tag{8.230}$$

$$\mathbf{L}^* = \mu^{(0)}\left(3\frac{1}{c}, 2\left[1 - \frac{16}{15\pi^2}c\frac{1}{1 + \frac{16c}{15\pi^2}}\right]\right), \tag{8.231}$$

$$\mathbf{L}^* = \mu^{(0)}\left(3\left[\frac{1}{c} - \frac{16}{15\pi^2}\right], 2\left[1 - \frac{16}{15\pi^2}c\frac{1}{1 + \frac{16c}{125\pi^2}}\right]\right), \tag{8.232}$$

describe the isotropic effective moduli of composites with identical spherical pores ($a_1 = a_2 = a_3 = a$) (8.229), (8.230), and with identical uniformly oriented coin-shaped cracks $a_1 = a_2 = a \gg a_3$) (8.231), (8.232) estimated by the MT (8.229), (8.231) and by the MEFM (8.230), (8.232); the values $c = \frac{4}{3}\pi a^3 n$ in the expressions presented have different physical meaning for the spherical and coin-shaped pores. In the case of the MEFM estimations, RDF (5.97) and two-particle point approximation (8.38), (8.40) presented in the term of stiffness tensors were used. For reduction of computations, an analysis of porous materials with uniformly oriented coin-shaped cracks was also performed via Eq. (8.38), (8.40) with replacement of the tensors $\mathbf{R}_1^\epsilon, \mathbf{P}_1\mathbf{R}_1^\epsilon n^{(1)}$ by the averaged tensors $\langle\mathbf{R}^\epsilon\rangle$, $(\mathbf{I} - \langle\mathbf{A}\rangle)c$, respectively. We note that for an incompressible matrix and planar spheroidal pores, according to [169], and to [762] $k^* = k^{(0)}$ for any concentration c, which indicates the invalidity of the mentioned theories in the limiting case considered here.

In order to demonstrate the comparison of the available experimental data [901] (see also [1487]) with the predicting capability of the proposed method in the

2D case, Buryachenko [200] (see also [277]) considered the estimation of the effective elastic moduli $\mathbf{L}^*$. Buryachenko and Tandon [281] considered the plane strain state, and first the influence of the noncircular shape of inclusions (4.25) (see Fig. 4.1) on the effective elastic moduli of a composite material with the isotropic components was analyzed. Both the effective moduli (8.57) and stress concentration factors (8.61) were estimated for both aligned and uniformly random orientation of fibers. It was also considered the influence of anisotropy of the matrix.

8.8.2 *Composites with Polifractional Inhomogeneities (Solid Propellants)*

Heterogeneous solid propellants (HSPs) commonly used in aerospace propulsion are likely to play an important role for as long as rockets are built. The typical HSPs consist of multimodal spherical particles of solid oxidizer (ammonium perchlorate, AP) of order 1–100 μm diam embedded in a rubbery fuel binder such as hydroxyl-terminated-polybutadiene (HTPB) or polybutadiene acrylonitrile (PBAN) (see, e.g., [1473]). Designers of rockets are concerned with a number of propellant-related issues, including the burning rate, the thermal and mechanical properties of the propellants, and, for metallized propellants, the behavior of the metal particles at the surface, including agglomeration (see [1003, 1025, 1027, 1028, 1646]).

Both numerical and experimental investigations of random structures are intended for prediction of overall mechanical and physical properties of these media. Suspensions probably are one of the most extensively studied area of highly packed particulate microheterogeneous media. Numerous authors minimize the non-hydrodynamic factors considering the low shear rate Newtonian limit that is most interesting from the point of view of analogy with composite materials. In simplest cases, one assumes that at the plotting of the relative effective viscosity $\eta^*/\eta^{(0)}$ of a suspension versus the adjustable volume fraction c/c_m, viscosity versus concentration plots collapse onto one curve (see for references [30, 1449]): $\eta^*/\eta^{(0)} = [1 - f(c^{(1)}, \ldots, c^{(n)})]^{-Ac_m}$. He and Ekere [629] (see also [333, 372, 1190, 1370]) proposed semi-analytical viscosity model of noncolloidal bimodal suspension where c_m depends upon both the size ratio λ of large and small particles and the volume fraction ξ of large particles. In a limiting case $\lambda \to \infty$, the mentioned models are reduced to a fruitful alternative by Farris [460] (see also [574, 1246, 1668]) based on purely geometric arguments. Assuming that the suspension of the fine particles can be treated as the suspending media for the large particles, [460] simulated the full bimodal mixture of particles by mapping it onto an effective one-modal system which can be simulated more easily. In this model, the fine and coarse fractions are assumed to behave independently of each other, and the viscosity of the mixture of smaller fractions is the medium viscosity for the next largest fraction: $\eta^*/\eta^{(0)} = f(c^{(1)}) \cdot f(c^{(2)}) \cdot \ldots \cdot f(c^{(n)})$ $(a^{(1)} \ll a^{(2)} \ll \ldots \ll a^{(n)})$. A similar approach (under the name a two-step homogenization technique) is well-known (even if the name Farris [460] is not mentioned) in the engineering material science of both the composites with the large size ratio λ of fillers and agglomerated composite materials (for reference, see, e.g., [200, 344, 437]).

Composites with a wide range of bimodal inclusion concentrations are less well understood and call for further investigation. So, a standard so-called multi-particle unit cell method of computational micromechanics is based on periodization of generated random media (see for reference [139, 590, 990]) reducing to the modeling of periodic boundary conditions at the meso cell containing just a few tens of inclusions that eliminates both the boundary and size effects in a windowing approach (see, e.g., [958]). The generation process is based upon Monte Carlo simulation method wherein the particles are generated from a certain aggregate size distribution and then randomly placed into the meso cell in such a way that there is no intersection between the particles. Then, the generated mesostructure models are to be meshed for computational testing of the composite with the finite element method (FEM) in order to estimate the macroscopic response [the use of the boundary element method [320] and multipole method [624, 861] is also known]. In so doing, the pack generations are based on either the hard core model simulation [595, 861, 1351, 1493, 1636, 1640] or the collective rearrangement model (CRM) destined for creation of the close packing [29, 1028, 1458]. Disadvantages of these random generation methods for composites with a wide range of particle concentration were considered by Buryachenko et al. [237, 255].

The advances in imaging techniques [e.g., computed tomography (CT), scanning electron microscopy (SEM), and magnetic resonance imaging (MRI)] require powerful computational methods for numerical analysis of CMs. To simplify the technology of finite element mesh generation for particle reinforced material (e.g., HSP), enrichment techniques are used to account for the material interfaces in the framework of extended FEM (XFEM; see for reference [506, 649]). The geometry of material distribution is described by level set function, which allows one to model the internal boundaries of the microstructure without the adaptation of the mesh [394, 908]. Some alternative approaches based on a real structure can be done mainly in two different ways. The first one can directly import the experimentally obtained real structure into FE software (see for reference [544, 908]). The second approach based on the concept of Statistically Equivalent Periodic Unit Cell (SEPUC) generates the model structures that have similar statistical functions as the experimentally obtained ones with a subsequent substitution of found descriptor into analytical micromechanics models (see, e.g., [145, 893]), [1029, 1241, 1674]. The modeling of composites with bimodal spherical inclusions by the methods of analytical micromechanics (see for reference [200, 1075, 1121], [1523]) has been developed less and necessitates some additional consideration. It is explained by the fact that the most popular methods of analytical micromechanics are essentially one-particle ones (see the current chapter).

So-called discrete structural models alternatively taking binary interactions of inclusions into account are worthy of notice (see, e.g., [530, 531, 1656, 1678]). These methods were initiated Chen and Acrivos [304] who showed that the elastic energy stored in the matrix layer between the spheres is many times greater than that in the rest of the surrounding matrix. This matrix layer can be considered as the elastic rod or spring transferring the interaction force between the spheres that makes it possible to use the principle of physical discretization developed in [3]. A composite material is modeled as an elastic network which seems to be much simpler for mathematical treatment than the original continuum mechanics representation. The fundamental limitations of the method are a simple composite structure and the growing of an approximate error of a spring model with both the growing of inter-particle distances and decreasing of elastic mismatch of constitutives.

Following [207], we will use MEFM (2D case was considered in [195]) which has qualitative benefits following immediately from the consideration of multiparticle interactions and manifesting by a sensitivity of the MEFM's estimations to both the RDF and size distribution of heterogeneities. Namely, we consider the CM with bimodal distribution of the inclusion radii a_i and a_j ($i = 1, 2;\ j = 3 - i$). Then the effective moduli expression (8.35) has the following presentation for the matrix $\mathbf{Y}_{ij}$ ($i = 1, 2;\ j = 3 - i$):

$$(\mathbf{Y}^{-1})_{ij} = \delta_{ij}\left[\mathbf{I} - \mathbf{R}_i \sum_{q=1}^{2} \frac{n^{(i)}}{n^{(q)}} \int \mathbf{T}_{iq}(\mathbf{x}_i - \mathbf{x}_q)\mathbf{Z}^{21}_{qi}\varphi(v_q, \mathbf{x}_q|;\, v_i, \mathbf{x}_i)d\mathbf{x}_q\right]$$
$$- \mathbf{R}_i \frac{n^{(i)}}{n^{(j)}} \int \Big[\mathbf{T}_{ij}(\mathbf{x}_i - \mathbf{x}_j)\mathbf{Z}^{22}_{ji}\varphi(v_q, \mathbf{x}_q|;\, v_i, \mathbf{x}_i) - n^{(j)}\mathbf{T}_i(\mathbf{x}_i - \mathbf{x}_j)\Big]\, d\mathbf{x}_j. \tag{8.233}$$

For comparison with experimental data by Smith [1414], we consider a CM consisting of identical glass beads ($E^{(1)} = 76.0$ GPa, $\nu^{(1)} = 0.23$) embedded in an epoxy matrix ($E^{(0)} = 3.01$ GPa, $\nu^{(0)} = 0.394$). As can be seen from Fig. 8.8, the use of the Mori-Tanaka method (MTM) (8.80) leads to an underestimate of the normalized effective Young modulus $E^*/E^{(0)}$ by 1.5 times for c = 0.5 compared with the experimental data. Much better approximations of $E^*/E^{(0)}$ are given by the MEFM (8.35) with the RDF obtained by the CSM, SCRM, and Willis approximation (WA) (5.100) which are distinguished one from another by less than 0.5%. In so doing, the use of the well-stirred (W-S) RDF (5.97) instead of (5.100) leads to reduction of predicted values $E^*/E^{(0)}$ on 10%.

Let us consider a CM with an incompressible isotropic matrix, filled by rigid identical spherical inclusions. This example was chosen deliberately because it provides the maximum difference between predictions of relative effective shear modulus $\mu^*/\mu^{(0)}$, as estimated by the various methods. In Fig. 8.9 the micromechanical model MEFM (8.35) is analyzed for the effect of choosing different RDFs (see for details Chap. 5). It is interesting that the first peak in Willis approximation (5.100) (see Figs. 5.1 and 5.2) is lower than the first peak in both the collective rearrangement model (CRM) and shaking collective rearrangement model (SCRM) simulation, but the RDF (5.100) decreases slower, and, because of this, the effective shear moduli μ^* for the RDF (5.100) larger than μ^* corresponded to the CRM and SCRM. In contrast to the Willis approximation (5.100), PY

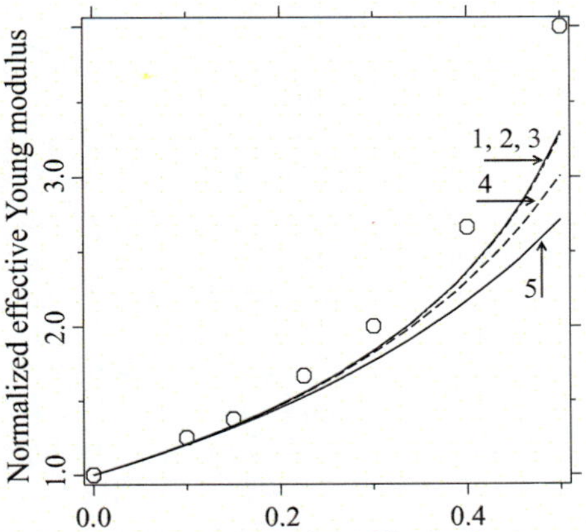

Fig. 8.8 $E^*(c)/E^{(0)}$ vs c estimated by the MEFM with RDF simulated by CRM (1), SCRM (2), WA (3), W-S (4), and MTM (5); open circle—experimental data

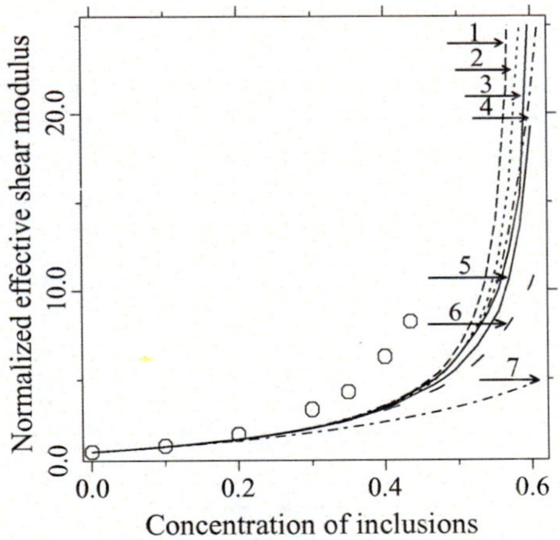

Fig. 8.9 $\mu^*(c)/\mu^{(0)}$ vs c estimated by the MEFM with RDF simulated by WA (1), SCRM [237] (2), CRM (3), SCRM (4), PY (5), W-S (6), and MTM (7); open circle—experimental data [803]

approximation swiftly decreasing after the first peak leads to the values of μ^* which are below corresponding μ^* estimated for both the CRM and SCRM. Just for estimation of the approximation quality SCRM [237], the curve 2 in Fig. 8.9 is evaluated for the RDF predicted from a single SCRM's RDF $g(r, c)$ $(c \equiv 0.5)$ according to Eq. (17) in [237]. As can be seen, curve 2 provides a reasonably accurate approximation of curve 4 corresponding to the SCRM. The effective moduli μ^* as the functions of c $(\mu^* \sim c)$ are very sensitive with respect to the RDF; for example, the use of well-stirred approximation (5.97) leads to reduction of the predicted $\mu^*(c)$ at $c = 0.6$ at least by the factor 2 or 3 with respect to the RDFs estimated by Eq. (5.100) and by CRM and SCRM. Neglecting by the binary interaction of inclusions (8.57) leads to subsequent diminution of $\mu^*(0.6)$ in two times. All RDFs taking a neighboring order into account lead to infinite values of $\mu^*(c)$ for large values of c; however, the limiting values $c \simeq 0.61$ and $c \simeq 0.62$

corresponding the CRM and SCRM, respectively, are better compatible with an experimental value $c_m = 0.637$ than the limiting assessments $c = 0.58$ for the Willis approximation (5.100).

We estimate in Fig. 8.10 the relative effective shear moduli $\mu^*/\mu^{(0)}$ of composites with incompressible matrix containing rigid spherical inclusions with the W-S RDFs (5.97) as well as with the RDF simulated by the SCRM; $\lambda_{21} \equiv a^{(2)}/a^{(1)} = 0.313$ and $\xi_{21} \equiv c^{(2)}/c^{(1)} = 0.333$. As can be seen, the values $\mu^*/\mu^{(0)} \sim c$ estimated by the MEFM for the RDFs corresponding to the SCRM's simulation correlates better with experimental data by Chong et al. [333] than analogous estimations for the W-S RDFs (5.97).

Estimation of these local fields are extremely useful for understanding the evolution of nonlinear phenomena such as plasticity, creep, and damage (see, e.g., [1282, 1483, 1484]). We present Eq. (8.27) in the form $\langle \boldsymbol{\sigma} \rangle_k = \mathbf{B}\mathbf{D}_k \langle \boldsymbol{\sigma} \rangle$, where the tensor $\mathbf{D}_k$ $(k = 1, 2)$ (8.26) (called effective stress concentrator factor) has a simple physical meaning of the action of surrounding inclusions on the separate one $\langle \overline{\boldsymbol{\sigma}} \rangle_k = \mathbf{D}_k \langle \boldsymbol{\sigma} \rangle$. The last equation makes it possible to estimate the ensemble average of the stress (8.30) in the vicinity of the inhomogeneities near the point $\mathbf{x} \in v_i \subset v^{(k)}$ with the unit outward normal vector $\mathbf{n} \perp \partial v_i$ $(k = 1, 2)$ $\langle \boldsymbol{\sigma}^-(\mathbf{n}) \rangle_{\mathbf{x}} = \mathbf{B}(\mathbf{n})\mathbf{B}\mathbf{D}_k \langle \boldsymbol{\sigma} \rangle$, where the tensor $\mathbf{B}(\mathbf{n})$ depends only on the elastic properties of the contacting materials and by the direction of the normal $\mathbf{n}$. Thus, the stress concentrations factors inside and outside inclusions depend on the linear functions of both the elastic solutions $\mathbf{B}$, $\mathbf{B}(\mathbf{n})$ for a single inclusion inside infinite matrix and the concentration factor $\mathbf{D}_k$ defined intrinsically by a micromechanical model. Because of this, we will only analyze a dependence $\mathbf{D}_k \sim c$ $(k = 1, 2)$. As can be seen in Fig. 8.11, the components $D_{k|1111}$ of the stress concentration factors $\mathbf{D}_k$ for both the W-S RDF (5.97) and SCRM's RDF at the large $(k = 1)$ and small $(k = 2)$ inclusions can be distinguished one from another by tens of percent that has a significant practical implementation.

Indeed, for periodic structures, Zhong and Knaus [1687] numerically demonstrated that for a large difference in particle sizes (such as a bimodal distribution), damage occurs at interfaces between the big particles and matrix, while only limited

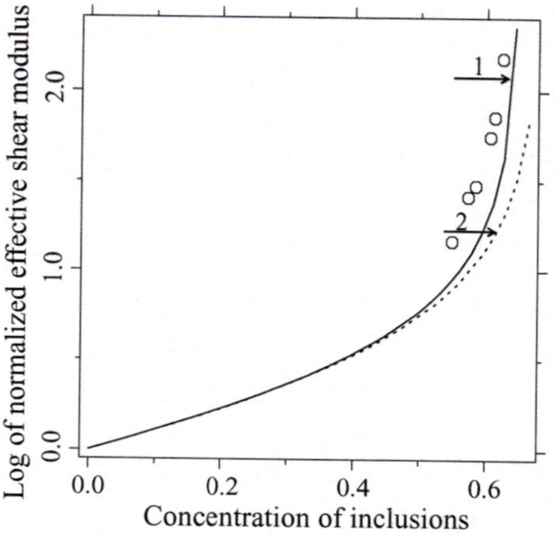

Fig. 8.10 $\text{Log}[\mu^*(c)/\mu^{(0)}]$ vs c estimated by the MEFM with RDFs simulated by the SCRM (curve 1), and Eq. (5.97) (curve 2); open circle—experimental data

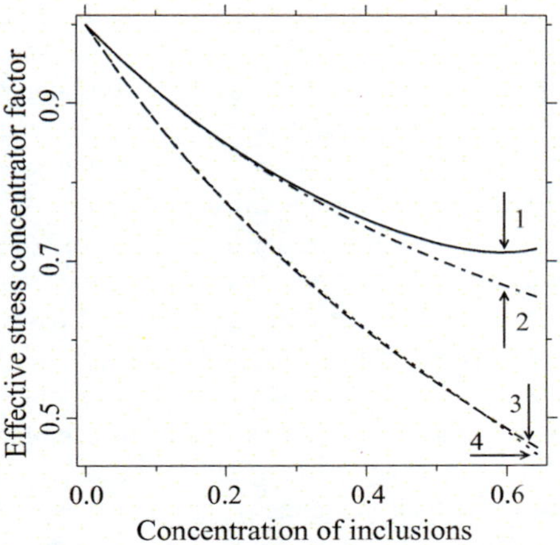

Fig. 8.11 $D_{k|1111}$ vs c for the RDFs simulated by the SCRM (curves 1 and 4), and Eq. (8.233) (curves 2 and 3) at the big inclusions (1,2) and small ones (3,4)

or no damage occurs at interfaces around small ones (that is also experimentally confirmed; see [389, 921]). While these effects of particle interaction and size variation are smoothed out in a large ensemble of particles that are exhibited by the effective elastic moduli, it is foreseeable that they are an important factor in a failure process such as local debonding when the particle size variation leads to a yield-likely response characterized by a nonmonotonic load deformation trace. Such a characteristic may, ultimately, be associated with inhomogeneous deformation such as occurs in shear bands or other strain concentrations, for example. The effect of increased stress concentrator factors at the big particles for random structure bimodal structures pioneered in the current section (see also [207]) is critical for any nonlinear localized phenomena such as onset of yielding, failure initiation, and others. Of prime importance is that the mentioned size distribution effect was discovered for linear elastic composites with ideal contact between the matrix and particles (analysis of non-ideal contact (see, e.g., [1483, 1484]) is not considered now).

It should be mentioned that popular methods using the second-order probability function $S_{ij}(r)$ $(i, j = 0, 1, \ldots, N)$ instead of the $\varphi(v_q, \mathbf{x}_q|v_i, \mathbf{x}_i)$ (see [893, 1625, 1627]). In such a case, Eqs. (8.35) and (8.233) are reduced to

$$\mathbf{M}^* = \mathbf{M}^{(0)} + \sum_{i=1}^{2} c^{(j)}\widehat{\mathbf{Y}}_{ij}\mathbf{M}_1^{(j)},$$

$$(\widehat{\mathbf{Y}}^{-1})_{ij} = \delta_{ij}\mathbf{I} - (c^{(i)})^{-1}\mathbf{M}_1^{(i)}\int \boldsymbol{\Gamma}(\mathbf{x}_i - \mathbf{x}_j)[S_{ij}(\mathbf{x}_i - \mathbf{x}_j) - c^{(i)}c^{(j)}]\,d\mathbf{x}_j \tag{8.234}$$

leading to the final representation $\mathbf{M}^*$ (8.57) in the framework of the hypothesis **H3**. Eqs. (8.234) and (8.57) are invariant with respect to both $S_{ij}(r)$ [at least for statistically isotropic structures] and the size distribution of inclusions. Because of this, numerous estimations of $S_{ij}(r)$ (see, e.g., [831, 893, 907, 952, 1027]) have only a marginal practical use from the perspective of view of micromechanics (see Fig. 8.9). Moreover, a potential usefulness of $S_{ij}(r)$'s estimations is reduced by a simplified assumption of a two-phase approximation ($i, j = 0, 1$) (see, e.g., [688]) of the 2-point probability density $S_{ij}(r)$ even for multimodal distribution ($N > 1$) of particle sizes. $S_{ij}(r)$ is just one of a few possible statistical descriptors which is significantly less sensitive to the structure microtopology than the RDF that provides a smoothness of $S_{ij}(r)$ (in opposite to the RDF's estimations which smoothness can be only achieved by additional sophisticated efforts, e.g., both the parallelization of simulations and the shaking procedure; see Sect. 5.4). Smoothness $S_{ij}(r)$ and its poor sensitivity in comparison with RDF in analysis of random structures provide a possibility for concealment of disadvantages in this simulation. Moreover, coincidence of the $S_{ij}(r)$ in both the real sample and the reconstructed structures generated by the protocol dependent algorithms (see, e.g., [831, 907]) does not assure a coincidence of other more important statistical descriptors (e.g., RDFs). Because of this, comparison of RDFs in both the structure generated by protocol independent algorithm (see Sect. 5.4) and reconstructed structure can be considered as a validation of the protocol dependent algorithm used for configuration reconstruction. Furthermore, geometrical statistical equivalence of the real and reconstructed packs is necessary but not sufficient condition for micromechanical statistical equivalence. In fact, the authors [893] and [907] implicitly introduced (with an intuitive level of rigor) a new concept of Representative Volume Element (RVE) for geometrical statistical equivalence. This concept is not well defined and depends on the statistical descriptors used. However, the concept RVE is well-known in micromechanics for micromechanical statistical equivalence (see this chapter, Chaps. 9–11) which is defined not only by $S_{ij}(\mathbf{x}_i - \mathbf{x}_j)$ (or $\varphi(v_j, \mathbf{x}_j|; v_i, \mathbf{x}_i)$) but also by an elastic particle interaction described by $\varphi(v_j, \mathbf{x}_j|; v_i, \mathbf{x}_i)$ rather than by $S_{ij}(\mathbf{x}_i - \mathbf{x}_j)$.

8.8.3 Nanocomposites

Nanofiber Composites Carbon fibers and plates are currently considered to be most promising. The recent discovery of carbon nanotubes (CNTs) and nanoplates has gained even broader interest due to providing unique properties generated by their structural perfection, small size, low density, high strength, and excellent electronic properties (for reference, see Sect. 1.1.2). Polymeric composites reinforced with nanoelements of nanometer scale recently attracted tremendous attention in material engineering. The general comprehensive reviews of the various types of nanocomposites were given in [513, 676, 709, 1102, 1516]. Due to the high surface area and small interlayer distances, the nanoelements can, in principle, change the properties of polymer matrix due to the changing of polymer morphology and chain conformation. A question of a correct choice of a structural model applicable to nanocomposites presents many challenges for mechanics and will be considered in this chapter from the point of view of allowing for the random orientation and clustering effect of nanoelements in the nanocomposites.

The mechanical and thermal properties of CNTs have not often been measured due to the difficulties of creating homogeneous and uniform samples of nanotubes. As a result, it is desirable to develop the model forming the mechanical properties of CNTs over their unique topology that may overcome the limitations of atomistic simulation concerning the wide range of length scale (10^{-9}–10^{-6}m). Unlike many other fields in science and engineering, the evolution of CNTs to their current level significantly depends on the contributions from modeling and simulation (see, e.g., [507, 882, 1254, 1308, 1465, 1510, 1511, 1675]). On the opposite ends of the length-scale spectrum are computational chemistry and solid mechanics covering quantum mechanical scale, atomistic domain (Monte Carlo and molecular dynamics), microscale methods (e.g., Brownian dynamics (BD), dissipative particle dynamics (DPD), lattice Boltzmann (LB), time-dependent Ginzburg–Lanau method, dynamic density functional theory (DFT) method), and finally macroscopic realm (see Chap. 4). Continuum analysis was linked with atomistic simulation [1308] (see also [1271, 1272]) by incorporating interatomic potential and atomic structures of CNTs directly into the constitutive law through the equalization of the strain energy function on the continuum level to the potential energy stored in the atomic bonds due to an imposed deformation. An equivalent-continuum trussed model by Odegard et al. [1146, 1147] takes into account the discrete nature of the atomic interactions at the nanometer length scale and the interfacial characteristics of the nanotube and the surrounding polymer matrix. With this approach, the mechanical properties of the solid cylinder will be determined by equating the total strain energies of the equivalent-truss and equivalent continuum models under identical loading conditions. Space-frame and truss models [593, 1146], lattice-dynamical (e.g., honeycomb structures), representative volumetric and nanoscale (i.e., molecular Lennard-Jones based) finite elements, and bending simulations have been used to replicate the structural mechanical paradigm at molecular levels.

Since the specific interface area is very large at the nanoscale, it gives rise to substantial property changes in large volume fractions of both matrix and fiber constituents. Some estimates suggest that as much as 30% of the matrix volume may be formed by polymer chains aligned with particle or fiber walls. It was found [562] that the polymer molecules form discrete adsorption layers and their interfacial properties cannot be determined through continuum mechanics. The creation of complete experimental data describing variation of all components of a stiffness tensor in the interphase is coupled with the large difficulties, although these data can, in principle, be obtained by MD simulation by the use of estimation of the effective properties evaluated in the observation windows moving along the cross section of a large sample containing a single nanotube (see, e.g., [1172, 1212]). It was demonstrated [1213] (see also [1147]) by means of atomistic simulation that the anisotropy of a polymer interface of a rigid spherical inclusion is rather weak. The theoretical background of analysis of composites reinforced by the ellipsoidal inclusions coated by anisotropic thin layer with varying along the inclusion surface properties is proposed in [270] (see also [314, 315, 1305]).

In the case of comparable scales of a molecular scale of analyzed structures and the scale of inhomogeneity of internal stresses, for mitigation of the sharp transition from the nonlocal atomistic region to a local continuum, some different coupling methods were proposed (for reference, see [356, 476, 477, 543, 1184, 1273, 1376]). So, the quasi-continuum (QC) method [1065] approximates the global atomic energy functional in the discretized continuum region and the transition region. Picu [1211] has estimated nonlocal elasticity kernels by the use of atomistic simulations based on the atomic scale structure of glasses. In the framework of linear isotropic nonlocal elasticity, it was determined that the kernel for a model material is tensorial, a different function weighting each entry of the stiffness tensor. Different aspects of nonlocal effects (see, e.g., [446, 1232, 1396]) are described in Chaps. 13–19.

Experimental research (see, e.g., [1514]) and molecular dynamic simulation mentioned earlier indicated that nanofibers can be effectively considered in the framework of continuum mechanics as the homogeneous prolate spheroidal anisotropic homogeneous inclusions with a large aspect ratio ($\alpha \cong 1000$). A nanocomposite is modeled as a linearly elastic composite medium subjected to homogeneous remote loading, which consists of a homogeneous matrix containing a statistically homogeneous random field of nanofibers with prescribed random orientation described by the orientation distribution functions (ODF) (5.91)–(5.93). The matrix is described by isotropic elastic properties with Young's modulus $E^{(0)} = 3.01$ GPa, Poisson's ratio $\nu^{(0)} = 0.41$, and coefficient of thermal expansion (CTE) $\boldsymbol{\beta}^{(0)} = \beta_0^{(0)}\boldsymbol{\delta}$, $\beta_0^{(0)} = 10^{-4}(K^{-1})$ (see [1488]). In the parametric analysis, we will consider both isotropic [$E^{(1)} = 240.06$ GPa, $\nu^{(1)} = 0.38$, (see [1514]) $\beta_0^{(1)} = -1.6 \cdot 10^{-6}(K^{-1})$ and transversally anisotropic fibers [$L_{1111}^{(1)} = 457.6$ GPa, $L_{1122}^{(1)} = 8.4$ GPa, $L_{2222}^{(1)} = 14.3$ GPa, $L_{2233}^{(1)} = 5.5$ GPa, $L_{2323}^{(1)} = 27.0$ GPa, see [1146]].

Following [241], we will also consider the modeling of nanocomposites with the cylinder fibers ∂v_i with the smooth ends. The analyses of the effect of fiber and matrix elastic properties on the stress distribution both inside and in the vicinity of a cylindrical fiber embedded in the infinite matrix were performed via the analytical model (see, e.g., [1114, 1269]) and by FEA [555]. The fiber axial ratio, α (varying from 200 until 1000), and the ratio of the Young's modulus of the fiber to that of the matrix were considered in [555] in the analysis of elastic load transfer inside the cylindrical, ellipsoidal, paraboloidal, and conical fibers embedded in the matrix subjected to the axial (only) loading. For example, for a cylindrical fiber, the axial stress distribution along the fiber axis estimated via FEA and approximate analytical models was greatest at the center, decreased steadily over most of the fiber length, and fell rapidly to zero near the fiber end (see [555, 1114, 1269]). In a similar manner the nanocomposites reinforced with the bone-shaped nanofibers synthesized in [1647] can be investigated.

To estimate the tensors $\mathbf{B}(\mathbf{x})$, $\mathbf{C}(\mathbf{x})$, and $\mathbf{Q}_i^0$, it is necessary to solve a series of 3D elasticity theory problems with the only nonzero component of the stress tensor $\overline{\boldsymbol{\sigma}}$

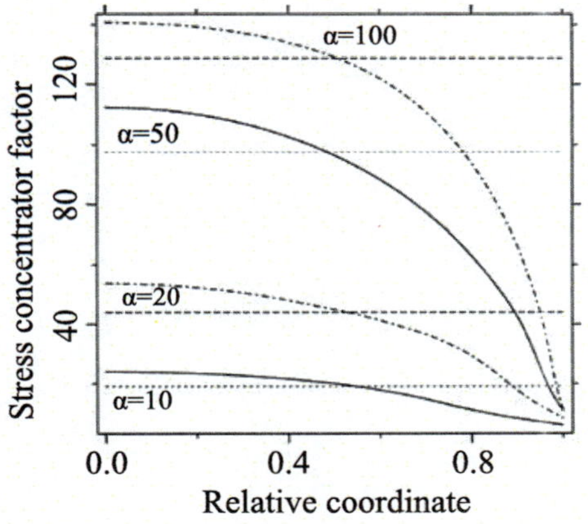

Fig. 8.12 $\sigma_{11}(\mathbf{x})/\langle\sigma_{11}\rangle$ vs x_{r1} for the spheroidal (straight lines) and cylindrical (curved lines) fibers for $\alpha = 100, 50, 20, 10$

or eigenstrain tensor $\boldsymbol{\beta}^{\mathrm{f}}$ (see Chap. 4 and [241]). We analyze stress concentration factors $\sigma_{11}(\mathbf{x})/\langle\sigma_{11}\rangle$ at the axis $\mathbf{x} = (x_1, 0, 0)^\top$ ($x_1 \in [-l - a, l + a]$) in the relative coordinate $x_{r1} = x_1/(l + a) \in [0, 1]$ of one isolated anisotropic fiber v_i in the matrix cube with size $20(l + a)$ subjected to the homogeneous remote loading $\langle\sigma_{ij}\rangle = \delta_{1i}\delta_{1j}$ (see Fig. 8.12). One performed (see [241, 275]) a comparison of experimental data [873] (see also [1516]) with estimations of $\mathbf{L}^*$ obtained by the EFM and MTM for CMs with 1D (5.91) and 3D (5.93) after uniform random orientations of fibers. The difference of the MTM and EFM for 3D (5.93) is more dramatically demonstrated at the analysis of stress concentration factors $\mathbf{D}_i$ (8.85) estimated by these methods for v_i with the axis $\mathbf{x} = (x_1, 0, 0)^\top$. So, the components $\overline{D}_{i|1111}$ and $\overline{D}_{i|1212} = \overline{D}_{i|1313}$ estimated by the EFM vary in the range $0.996 \div 1.0$ that differ from the appropriate tensor components estimated by the MTM by the factors 4.5 and 6.6, respectively.

Now we will consider in more detail the case of 2D uniformly random orientation (5.93) of anisotropic fibers. It was demonstrated [274–276] that if the "correlation hole" v_i^0 is homothetic to spheroidal fiber v_i when $\mathbf{Q}_i^0 \equiv \mathbf{Q}_i$ then the estimations by both the MTM and EFM coincide for E_1^* and E_3^* and differ from one another in the case of cylindrical fibers. In so doing the E_1^* and E_3^* estimated for both the anisotropic and isotropic cylindrical fibers are very close. However, we can observe a significantly more dramatic situation of the loss of symmetry of the matrix of effective elastic moduli estimated by both the EFM and MTM: $L_{1133} \neq L_{3311}^*$ for both spheroidal and cylindrical fibers (see the case for spheroidal fibers in Fig. 8.13). To avoid this contradiction, we will choose another appropriate shape of a correlation hole v_i^0 enclosing the fiber v_i ($i = 1, \ldots$) and prohibited for the centers of surrounding fibers: the domain v_i^0 as an oblate spheroid (with the semiaxes $a_1^0 = a_2^0 = a_1 + a_2$ parallel to the plane Ox_1x_2, and $a_3^0 = 2a_3$) enclosing the prolate spheroid v_i. We get again that the E_1^* and E_3^* estimated by both EFM and MTM for both the anisotropic and isotropic cylindrical fibers are very close. In so doing the matrix L_{ijkl}^* is not symmetric ($L_{1133}^{*MTM} \neq L_{3311}^{*MTM}$) (see Fig. 8.13)

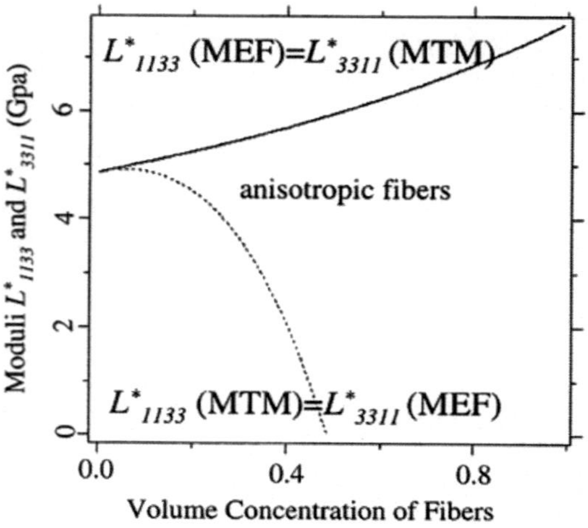

Fig. 8.13 Non-diagonal elements of **L**$*$ vs c

and symmetric ($L^{*EFM}_{1133} = L^{*EFM}_{3311} \approx L^{*MTM}_{3311}$) for the estimations obtained by the MTM and EFM, respectively.

Cilicate Nano Composites The assumption of so-called statistical homogeneity in the case of clay NCs being considered is valid for completely exfoliated nanocomposites which were intensively analyzed in the framework of conventional micromechanics by the MTM (see [1586, 1587]), by Halpin-Tsai (see [600]) equation [165], by the generalized Takayanagi's model method taking the interphase properties into account ([701], see also [796]), as well as by the finite element analysis of multi-inclusion periodic model [589, 591]. Gusev [589] demonstrated a significant difference between the predictions obtained either by the numerical simulation or by the Halpin-Tsai equation for composites by the aligned fibers with the aspect ratio between 10 and 1000.

A particular case of composite materials containing particle-rich and particle-poor regions is the clustered material (see Sect. 5.2.2) where such regions have a concrete shape and random location that does not assume the loss of statistical homogeneity of the composite structure. The concept of clusters is similar to that of "fractal structures" (see, e.g., [457, 1454]), which were successfully applied in [1494] for the statistical analyses of clustered materials by the use of such parameters of statistical descriptors as cluster size ζ, the fractal dimension D, and the radius of gyration R_g^2. Sheng et al. [1374] (see also [165]) have emphasized that the notions of the matrix and inclusions widely used in conventional micromechanics can no longer be directly applied to nanoclay composites due to the hierarchical structure of clusters which, nevertheless, can be treated as a sort of equivalent (or effective) inhomogeneities embedded, in turn, in the bulk matrix material. A similar approach is well-known in the engineering material science of conventional agglomerated composite materials (for reference, see, e.g., [437]) simply because in a polymeric matrix, the filler (even if present in minimum quantities) is always more or less agglomerated. Nielsen [1125] (see also [122, 1124, 1540]) considered agglomerates as a plurality of contacting primary particles behaving like hard effective particles below a certain stress threshold where a high stress should cause the particles constituting an agglomerate to move relative to each other.

A description of the composite cluster structure as well as micromechanical modeling is significantly simplified in the case of clusters of a deterministic structure. For the most part, such clusters (called then the clusters of periodic structure) can be described as "cut-out" from the infinite periodic media. A popular micromechanical modeling can be conditionally described by two limiting kinds of approaches, both of them based on the concept of effective properties of clusters. In the first case, the effective properties of a cluster are estimated for the infinite periodic media (as well as for the infinite statistically homogeneous media) from which a cluster was mentally "cut-out." The effective properties of the clusters were estimated by the Halpin-Tsai equation (see [165, 600]), by MTM (see [119, 1380, 1587]), and by the effective medium method [1618], as well as the investigation of cavity coalescence in inhomogeneously voided ductile solids [891] using the self-consistent method. In the second sort of approaches based on the "cut-out" of a cluster, the effective elastic properties of a cluster extracted from the composite are estimated through the equalization of the strain energy function at the set of inclusions combining the cluster to the potential energy stored in a homogeneous area with effective properties and the sizes of the cluster. It should be mentioned that the choice of the area referred to as a cluster and "cut-out" from the composite for the subsequent analysis by one or another micromechanical method is not unique and defined by the subjective partiality of a researcher (compare [165, 987, 1374]).

The correctness of both abovementioned groups of approaches is questionable in the usual comparability of both the mentioned field and material scales inside the cluster where even the local effective moduli are varying over a cross section of the cluster of either periodic [200] or statistically homogeneous structure (Chap. 13), while a volume concentration of inclusions in either the periodic or random structures is constant. It leads, in turn, to a new class of phenomena of the macroscopic scale based on the use of nonlocal continua (for reference, see Chap. 13 and [253, 799]). One reason for the mentioned incorrectness is explicable on the basis of the fact that the boundary conditions prescribed in both model clusters (see cited papers in [987, 1374, 1586, 1587]) are not compatible with the real contact conditions at the surface of the real considered cluster in the composite material (see for details [200, 749]). In this light, the method of solution of a finite number of interacting spheroidal inclusions inside the infinite matrix developed by Kushch [844, 847, 849] avoids the mentioned discrepancies. The basic idea of the multipole expansion technique proposed consists in expansion of the displacement vector into a series over a set of vectorial functions providing exact satisfaction of the interfacial boundary conditions. Incorporation of the multipole expansion technique into the MEFM in this section will make it possible to abandon the majority of simplified assumptions exploring at the analysis of random structure composites reinforced by the clusters of a deterministic structure.

It is assumed that the clusters can be grouped into components (phases) $\Omega^{(k)}$ $(k = 1, 2, \ldots, N)$ with identical mechanical and geometrical properties (such as the shape, size, orientation, and microstructure of clusters). Each cluster $\Omega_i \subset \Omega^{(k)}$ containing a deterministic field of identical oblate spheroidal inclusions $v_1, \ldots, v_m$ $(m = n^{\Omega(k)})$ with the thermoelastic properties $\mathbf{M}_j^{v(k)}(\mathbf{x})$ and $\boldsymbol{\beta}_j^{v(k)}$ $(j = 1, \ldots, n^{\Omega(k)})$ (see Fig. 18.4). We can define the tensors $\mathbf{B}_i, \mathbf{C}_i, \mathbf{R}_i, \mathbf{F}_i$ for a cluster Ω_i analogously to Eqs. (3.175$_{7,8}$) for the inclusion v_i

$$\mathbf{R}_k = \sum_{j=1}^{n^{\Omega(k)}} \int_{v_j} \mathbf{M}_{j1}^{v(k)}(\mathbf{x})\mathbf{B}_j^{v(k)}(\mathbf{x})\, d\mathbf{x}, \tag{8.235}$$

$$\mathbf{F}_k = \sum_{j=1}^{n^{\Omega(k)}} \int_{v_j} \left[\mathbf{M}_{j1}^{v(k)}(\mathbf{x})\mathbf{C}_j^{v(k)}(\mathbf{x}) + \boldsymbol{\beta}_{j1}^{v(k)}\right] d\mathbf{x}, \tag{8.236}$$

where $\mathbf{f}^{(k)}(\mathbf{x}) = \mathbf{f}_j^{v(k)}(\mathbf{x})$ at $\mathbf{x} \in v_i \subset \Omega^{(k)}$. It should be mentioned that for clustered materials being considered, $\mathbf{R}_k(\mathbf{x}), \mathbf{F}_k(\mathbf{x}) \equiv \mathbf{0}$ for the matrix's part of the cluster $\mathbf{x} \in \Omega^{(k)} \setminus \cup_j v_j$ $(j = 1, \ldots, n^{\Omega(k)};\ k = 1, \ldots, N)$, while the tensors $\mathbf{B}_j^{v(k)}(\mathbf{x})$ and $\mathbf{C}_j^{v(k)}(\mathbf{x})$ are estimated by the method of multipole expansion (see for details [200, 242]). At first, the numerical results are considered for the matrix Epox 862 $(E^{(0)} = 3.01 GPa,\ \nu^{(0)} = 0.41$; see [1488]) with the cluster of nanoinclusions $(E^{(1)} = 300\,\text{GPa},\ \nu^{(1)} = 0.4$; see [1374]) containing n inclusions with $a = 1$, $d = 4b$, and an aspect ratio $\alpha \leq 0.01$, and with the axis of symmetry coinciding with the axis Ox_3 (see Fig. 18.14). We analyze the average stress concentration factors $\langle\sigma_{11}\rangle_{(i)}/\langle\sigma_{11}\rangle$ $(i = 1, \ldots, n,\ \alpha = 0.01$ and $d = 4b,\ a = 1)$ at the inclusions of one isolated cluster Ω_i^h (see Fig. 8.15) in the infinite matrix subjected to the homogeneous loading $\langle\sigma_{ij}\rangle = \delta_{1i}\delta_{1j}$. In Fig. 18.16 it can be seen that the stress concentrator factor $\langle\sigma_{11}\rangle_{(i)}/\langle\sigma_{11}\rangle$ $(i = 1, \ldots, n)$ significantly (almost in three times) decreased from the outer spheroid to the central spheroid of the cluster. Hereafter one depicts the stress concentration factors of inclusions v_i $(i = (n+1)/2, \ldots, n)$ numbered by the subscript $j = 1, \ldots, (n+1)/2$ $(j = i - (n-1)/2$; see Fig. 18.16. Such a behavior is qualitatively confirmed by 2D finite element analysis in [1374] (see also [1695]; Fig. 8.14).

We now turn our attention to the estimation of effective elastic moduli $\mathbf{L}^*$ (8.57) of composites reinforced by randomly located identical aligned clusters. It should be mentioned that in Eqs. (8.235) and (8.236), there is now restriction on the shape of the cluster $\Omega_i \subset \Omega^{(k)}$. However, in the EFM (8.57), the shape of the excluded volume Ω_i^h (correlated with the custer shape: $\Omega_i^h := \{\mathbf{x} = 2\mathbf{y} | \mathbf{y} \in \Omega_i\}$) must be ellipsoidal. Because of this, the ellipsoid Ω_i is chosen proportional to the ellipsoid Ω_i^{ins} inscribed into the stack Ω_i (see Fig. 8.15). The proposed method is logically close to the morphological-pattern approach [148] (see also [1669]) although the concrete estimation method of inclusion interactions inside the cluster (see the

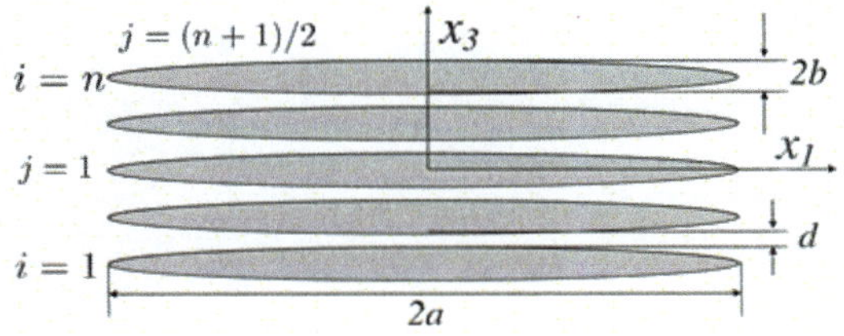

Fig. 8.14 A stack of oblate spheroids

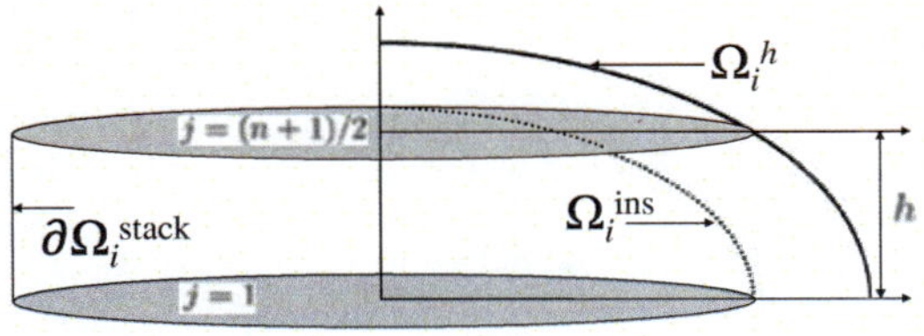

Fig. 8.15 Definition of the shape Ω_i^h

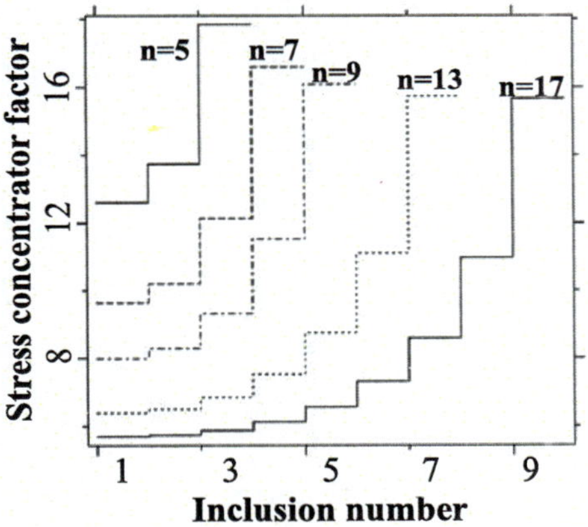

Fig. 8.16 Average stress concentrator factor $\langle\sigma_{11}\rangle_{(j)}/\langle\sigma_{11}\rangle$ in the different spheroids $j = 1, \ldots, (n+1)$ in the clusters containing n inclusions

tensors $\mathbf{R}_j$, $\mathbf{F}_j$) and between the clusters [see the tensors $\mathbf{Y}_{ij}$ (8.44), (8.46)] is different (Fig. 8.16 and 8.17).

As can be seen in Figs. 18.17, increasing the inclusion number in the clusters leads to decreasing the effective Young moduli E_1^* in the longitudinal direction Ox_1 ($\alpha = 0.01$, $d = 4b$, $a = 1$) that quantitatively confirmed intuitive explanations of experimental data [467, 705, 954] and [1382]. We also investigate influence of the inter-layer space $d = 2, 3, 5, 9b$ on the effective Young moduli E_1^* (see Fig. 18.18) at the fixed $\alpha = 0.01$ and $n = 9$. As one would expect, increasingthe interlayer spacing stacking the interaction of the spheroids inside the cluster according to Fig. 18.8 leads to a rise of the effective Young moduli (see experimental data in [793]). From the other side, the viscosity of a system with agglomerates of spherical inclusions is always higher than that of the "well-dispersed" sample (see for reference, e.g., [437, 940, 1113, 1124, 1125, 1679]). In light of this, we analyze in Fig. 18.19 the influence of the inclusion shape in the composite materials reinforced by the clusters with $n = 9$ which are large enough for generating maximum

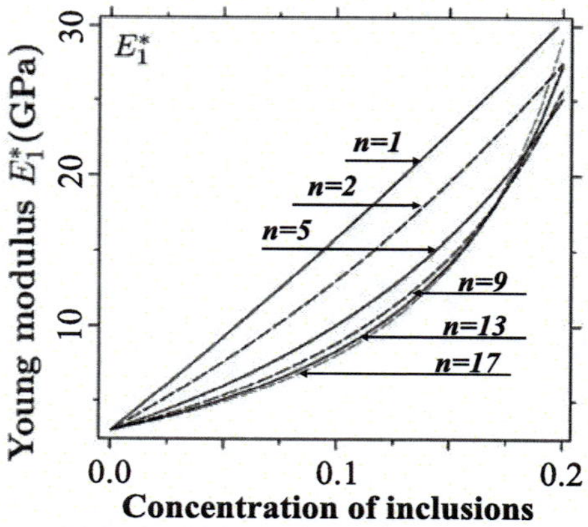

Fig. 8.17 Effective elastic modulus E_1^* vs c for the different inclusion number $n = 1, 2, 5, 13, 17$ in the clusters

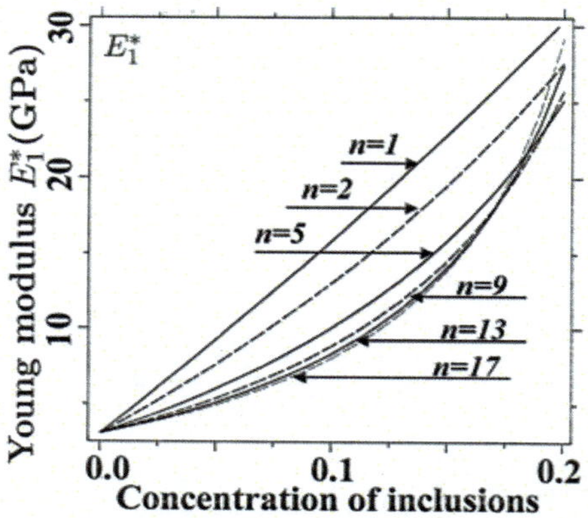

Fig. 8.18 Effective elastic modulus E_1^* vs c for the different interlayer spacing $d = 2, 3, 5, 9b$ ($n = 9$) inclusions

cluster effect. The results of Fig. 18.19 quantitatively confirm intuitively expected decreasing and reverse of the "shielding" effects for the clusters with the less-oblate inclusions. Another limiting case of aligned fibers ($\alpha \gg 1$) arranged into the clusters with the fiber centers placed in the elliptic domains (see [1254]) can also be considered analogously (Figs. 8.18 and 8.19).

8.8.4 Some Prospective Problems in Modeling of Nanocomposites

Although the aligned nanofibers and nanoplates, as was demonstrated in this subsection (see also [1510, 1516, 1645]), are desirable, the real orientation of fibers is usually described by a single parameter of a standard deviation of the Gaussian distribution with the center coinciding with the flow direction (see, e.g., [953]). In the investigation of clay orientation in NCs, Koo et al. [790] have obtained Gaussian

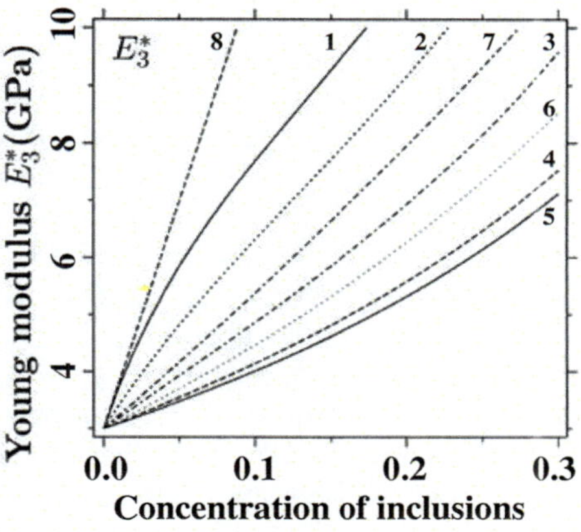

Fig. 8.19 E_3^* vs c for $\alpha = 0.001$ (curve 1); 0.003 (2); 0.01 (3); 0.03 (4), 0.1 (5), 0.3 (6), 1.0 (7), 10.0 (8)

distribution with the center, which indicates that the clay platelets' normals are aligned in the direction perpendicular to the flow direction. They demonstrated quite different shear-induced orientation behaviors with the concentration of silicates. Vaia and Maguire [1551] mentioned that the critical concentration at which packing considerations dictate orientation correlation between nanoelements is 10^{-3} for a plate and 10^{-4} for a rod. These low-volume fractions imply that NCs are not fundamentally isotropic but have a tendency to exhibit anisotropic properties or a domain-like texture that contains nanoelements with a local preferred orientation even though macroscopically the global orientation is isotropic.

We should also mention the special features of load transfer, in tension and in compression and in MWNT-epoxy composites studied in [1336] where it was detected that load transfer in tension was poor in comparison with load transfer in compression, implying that during load transfer to MWNTs, only the outer layers are stressed in tension due to the telescopic inner-wall sliding (reaching at the shear stress 0.5 MPa, see [1666]), whereas all the layers respond in compression. In the framework of continuum mechanics, both the SWNTs [1146] and MWNTs (see [1510]) modeled as the hollow cylinder and multilayer one, respectively, are sometimes replaced by the solid homogeneous cylinder with effective fictitious properties. Pipes and Hubert [1216] investigated the effective mechanical properties of a large array of CNs assembled in helical geometries of circular cross section without the matrix when the nanowall has a cylindrical anisotropy. A cylindrical anisotropy of an SWNT was also estimated [922] (see also [575]) via the molecular structural mechanics method indicating that the radial modulus decreases with increasing tube diameter while the circumferential modulus is insensitive to tube diameter, and roughly equal to the axial modulus. These issues deserve further investigation, for example, by explicitly including graded interface layers (see, e.g., [994, 1613, 1675]). An alternative approach taking into account interfacial effect is based on the concept of surface stress and surface tension which can be clarified by the additional relation between interface/surface stress tensor, the deformation-dependent surface energy, and the deformation-independent surface/interfacial tension (see [679] and Sect. 3.7.3).

One feature characteristic of fiber nanocomposites is that the reinforcing NTs are not straight but rather have curvature or waviness varying through the composites [153, 1178, 1252, 1564]; see also helical NT in [1254, 1380]. This is partially due to the low bending stiffness of nanotubes of the small diameter ($\sim$1 nm) and small elastic Young's modulus in the transversal direction. It is reasonable to assume that the waviness reduces the effectiveness of the NT reinforcement of the polymer at least in the fiber orientation direction. The estimation method of $\mathbf{L}^*$ was proposed in [153] by the use of Mori-Tanaka approach. In the case of isotropic fibers being assumed, the problem was reduced to the estimation of average strain concentration factor for an individual waved fiber in the infinite matrix. This model is invariant with respect to the shape of a correlation hole v_i^0 (simply because $\mathbf{L}^*$ doesn't depend on v_i^0) and cannot be used for analysis of anisotropic fibers even for their similar orientation (because in such a case $\langle \boldsymbol{\eta} \rangle_i \neq \langle \mathbf{L}_1 V_i \rangle_{(i)} \langle \boldsymbol{\varepsilon} \rangle_i$). As shown in Sect. 8.8.3, the effective properties $\mathbf{L}^*$ are defined by an average concentration of stress polarization factor in the fiber rather than average strain concentration factor. The correct method of estimation of the fictitious properties (3.186) and (8.235) is based on the matching of a fictitious inclusion with the same elastic response describe by the tensor $\mathbf{R}^{(k)}$ as a real one. It fundamentally differs from the choice of $\mathbf{L}^{f(k)}$ proposed in [484] and based on the use of the mixture rule only for estimation of the fictitious Young modulus in the direction of the fiber orientation. Moreover, the effective properties estimated for isotropic waved fiber in [153] via Eqs. (3.186) and (8.235) are anisotropic and equivalent. Therefore, a correctness of their exploiting for the estimation of effective properties $\mathbf{L}^*$ of the NCs with an arbitrary prescribed random orientation of waved nanofibers by the MTM is questionable.

Although the FEA approach is quite powerful for the mentioned problems, the analytical models are worthy of more attention. The elegant analytical models [153] and [1380], recently proposed for the fibers of small curvature (both the waved and helical shapes), are based on an assumption that the average stress concentration factor in each section of NT is locally approximated by a straight infinite cylinder approach that of an infinitely long straight NT oriented at the same angle and embedded in an infinite matrix. In this connection, the generalized iteration version of the MEFM is noteworthy. The generalized version takes into account inhomogeneity of the effective fields in the longitudinal fiber direction and its homogeneity over the traversal direction (this idea is initiated by both the numerical results of Fig. 8.12 and by the binary model [1294, 1654]); the last reduces the number of estimated variables. This approach is based on the numerical solution of an appropriate singular integral equation using a new subtraction technique (see for reference on general description, e.g., [370] and Chap. 4) eliminating singularities by exploiting the solutions [153] and [1380]. Substitution of nonlocal solutions for one and two fibers subjected to the inhomogeneous effective fields into the general integral equations for microinhomogeneous media will allow one to obtain implicit representation for the nonlocal integral operators of the effective elastic properties and stress concentrator factor.

It was demonstrated [1446] that the large amount of CNTs is concentrated in agglomerates if the average volume concentration of CNTs in a sample reaches 7.5%. These agglomerates called clusters are usually assumed to be an ellipsoidal domain and cut out from the infinite statistically homogeneous composite sample. Shi et al. [1380] (see also [1178]) used a known approach (called in this study the first sort "cut-out" model) for replacement of these clusters by the fictitious homogeneous inclusions of the same size and the properties estimated as the effective properties for an appropriate infinite statistically homogeneous medium. Such an approach is correct in the limiting case of an infinite difference of scales of both the cluster sizes and the fiber lengths. The reason for an error introduced in the range of comparable mentioned scales was explained in Sect. 13.4. An isolated SWNT bundle without matrix was analyzed [950] via a hybrid atom/continuum model, in which the intratube interaction energy is calculated using the MD-based continuum approach, while the intertube interaction energy is modeled by a usual Lennard-Jones-type potential. Linearized bulk elastic properties of the SWNT bundle with respect to a stable configuration are transversely isotropic and characterized by five independent elastic moduli estimated in this chapter. The prescribed approach of cluster analysis is obviously associated with the second type of "cut-out" modeling of clustered materials considered in Sect. 8.8.3 accompanied with an explanation of their approximative nature. Deterministic cluster structure of nanoropes is best suited for the application of the cluster model considered in Sect. 8.8.3 with obvious replacement of both the arrangement and shape (from oblate spheroidal shape to prolate spheroidal one) of nanoelements in the cluster. An additional challenging issue of nanotechnology considered in [1253] (see also [1675]) is that the nanorope contains not parallel NTs, but rather twisted or woven NTs leading to much better load transfer mechanism in tension than a straight bundle would have. It would be interesting to analyze this problem by the approximate method proposed in the previous paragraph for the investigation of individual waved fiber with incorporation of a solution for a single fiber into the rope problem with subsequent insertion of the solution for an individual rope into the MEFM for the estimation of effective properties of NCs. Another limiting case is CM reinforced by the aligned fibers with the volume fiber concentration described by a step function that is a feature of CM reinforced by the aligned carbon nanotube forests (see [287, 832, 1251]). This case is more simple because all correlation holes v_i^0 are parallel and effective fields $\overline{\boldsymbol{\sigma}}_i(\mathbf{y})$ $(i = 1, 2, \ldots)$ varying along axial fiber direction are identical for all fibers (at least in the case of Lax [886] approximation).

The extraordinary properties of nanoelements are specified by their periodic structure which is not, however, perfect forever and can contain some imperfections (such as defects; see, e.g., [83, 292, 1435]). It was found [1381] that the critical strain of defect nucleation of a CNT is sensitive to its chiral angle but not to its diameter and that with the increase in the Young's modulus of matrix, the critical strain of CNTs decreases. Because of this, the development of the theory of homogenization of initially periodic structures with the introduced infinite number of imperfections of unit cells randomly distributed in space has an essentially practical meaning (see also [196, 200]).

Thus the problems mentioned above have a technical rather than a fundamental character. The MEFM appears to be simple enough in both theoretical and computational aspects and provides the calculation with reasonable accuracy of the effective properties of composites reinforced by either the randomly oriented nanofibers or clustered nanoplates for a whole range of parameters.

8.9 Critical Analysis of Maxwell Homogenization Schemes

At the last decade, an explosive character of development of the Maxwell [1036] method can be observed (see, e.g., comprehensive review [1366]). We will attempt to qualitatively and quantitatively investigate a correspondence between the "Maxwell scheme" and other different basic assumptions and conceptions of micromechanics. A goal of the current study consists in a dissection of fundamental differences between these approaches and their restrictions (see for details [230]).

8.9.1 Classical Maxwell Scheme and Its Comparison with EFM and MTM

At first, for a subsequent comparison, we consider two cases when the EFM's estimations of the effective properties (8.63) and (8.64) are simplified. In the first case, the shapes of excluded volumes v_i^0 coincide with the shapes of inclusions shape $v_i^0 =$ shape v_i $(\forall i)$ that reduces Eqs. (8.63) and (8.64) (see also Eq. (8.160)) to

$$\mathbf{L}^* = \mathbf{L}^{(0)} + \left[c^{(0)}\mathbf{I} + \langle \mathbf{A} V\rangle\right]^{-1} \langle \mathbf{R}^{\epsilon v} V\rangle, \tag{8.237}$$

$$\boldsymbol{\alpha}^* = \boldsymbol{\alpha}^{(0)} + \left[c^{(0)}\mathbf{I} + \langle \mathbf{A} V\rangle\right]^{-1} \langle \mathbf{F}^{\epsilon v} V\rangle, \tag{8.238}$$

$$\langle \boldsymbol{\varepsilon}\rangle_i(\mathbf{x}) = \mathbf{A}_i(\mathbf{x})\mathbf{D}_i\boldsymbol{\varepsilon}^{\Gamma_0}, \quad \mathbf{D}_i = \mathbf{I} + \mathbf{Q}_i^0\left[c^{(0)}\mathbf{I} + \langle \mathbf{A} V\rangle\right]^{-1} \langle \mathbf{R}^{\epsilon v} V\rangle, \tag{8.239}$$

In the second case, all shapes of excluded volumes v_i^0 are identical

$$\text{shape } v_i^0 = \text{ shape } \hat{v}^0 \quad \forall i \tag{8.240}$$

and Eqs. (8.63) and (8.64) (see also Eq. (8.161)) are simplified

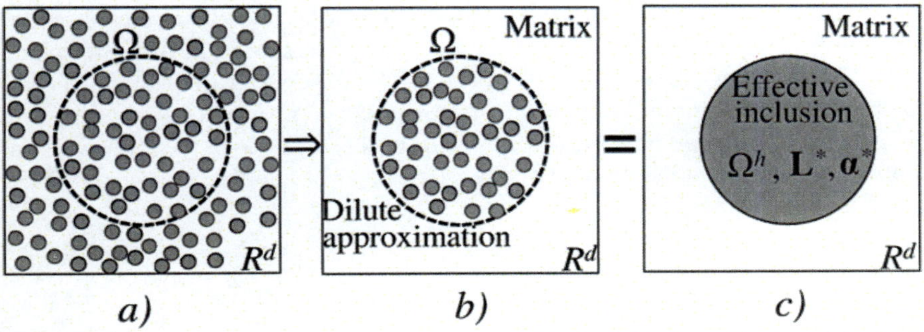

Fig. 8.20 Maxwell scheme

$$\mathbf{L}^* = \mathbf{L}^{(0)} + \left[\langle \mathbf{R}^{\epsilon v} V \rangle^{-1} - \hat{\mathbf{P}}^0\right]^{-1}, \tag{8.241}$$

$$\boldsymbol{\alpha}^* = \boldsymbol{\alpha}^{(0)} + \left[\langle \mathbf{R}^{\epsilon v} V \rangle^{-1} - \hat{\mathbf{P}}^0\right]^{-1} \langle \mathbf{R}^{\epsilon v} V \rangle^{-1} \langle \mathbf{F}^{\epsilon v} V \rangle, \tag{8.242}$$

$$\langle \boldsymbol{\varepsilon} \rangle_i(\mathbf{x}) = \mathbf{B}_i(\mathbf{x}) \hat{\mathbf{D}} \boldsymbol{\varepsilon}^{\Gamma_0}, \quad \hat{\mathbf{D}} = \mathbf{I} + \hat{\mathbf{P}}^0 \left[\langle \mathbf{R}^{\epsilon v} V \rangle^{-1} - \hat{\mathbf{P}}^0\right]^{-1}, \tag{8.243}$$

where $\hat{\mathbf{P}}^0 := \mathbf{P}(\hat{v}^0)$ and $\hat{\mathbf{D}}$ in Eq. (8.243) do not depend on i.

Maxwell's approach to the homogenization problem consists in equating the asymptotic of the disturbance field caused by a cluster Ω (see Fig. 8.20b) of N non-interacting inhomogeneities (perhaps of different sizes and material properties) located inside ellipsoidal domain of actual composite material (see Fig. 8.20a) and equivalent homogeneous ellipsoidal inclusion Ω^h (see Fig. 8.20c) with unknown effective properties. Thus, Maxwell scheme is based on two postulates

P1) dilute approximation of non-interacting inclusions in Ω, (8.244)

P2) ellipsoidal shape of Ω^h. (8.245)

If any of these postulates is violated, the model is not a classical version of Maxwell scheme.

Formalization of this Maxwell scheme is performed by a double using of Eq. (3.79) where instead of the inclusion v_i we consider either the cluster domain Ω containing N non-interacting inhomogeneities or ellipsoidal homogeneous domain Ω^h circumscribing Ω

$$\boldsymbol{\varepsilon}(\mathbf{x}) = \boldsymbol{\varepsilon}^{\Gamma_0} + \sum_{j=1}^{N} \int_{\Omega} \mathbf{U}(\mathbf{x}-\mathbf{y}) \boldsymbol{\tau}(\mathbf{y}) V_j(\mathbf{y}) \, d\mathbf{y}, \tag{8.246}$$

$$\boldsymbol{\varepsilon}(\mathbf{x}) = \boldsymbol{\varepsilon}^{\Gamma_0} + \int_{\Omega^h} \mathbf{U}(\mathbf{x}-\mathbf{z})[(\mathbf{L}^* - \mathbf{L}^{(0)})\boldsymbol{\varepsilon}(\mathbf{z}) + (\boldsymbol{\alpha}^* - \boldsymbol{\alpha}^{(0)})] \, d\mathbf{z}, \tag{8.247}$$

where the homogeneous boundary conditions (6.4) $\boldsymbol{\varepsilon}^{\Gamma_0}$ =const. is prescribed at the infinity. Owing to both homogeneity of the applied strain $\boldsymbol{\varepsilon}^{\Gamma_0}$ and ellipsoidal shape Ω^h, the stress polarization $\boldsymbol{\tau}(\mathbf{y})$ ($\mathbf{y} \in \Omega^h$) is expressed through the Eshelby solution (3.184$_{1,2}$). From the other side, at $|\mathbf{x} - \mathbf{x}_\Omega| \to \infty$ (where $\mathbf{x}_\Omega$ is a center of Ω^h), the strains $\boldsymbol{\varepsilon}(\mathbf{x})$ in both Eqs. (8.246) and (8.247) coincide in the framework of Maxwell scheme, and therefore, the volume averages of integrands in Eqs. (8.246) and (8.247) coincide

$$\sum_{j=1}^{N}[\mathbf{R}_j^{\epsilon}\boldsymbol{\varepsilon}^{\Gamma_0} + \mathbf{F}_j^{\epsilon}] = |\Omega^h|[\mathbf{L}_1^*(\mathbf{I} + \mathbf{P}^{\Omega}\mathbf{L}_1^*)^{-1}\boldsymbol{\varepsilon}^{\Gamma_0} + (\mathbf{I} + \mathbf{L}_1^*\mathbf{P}^{\Omega})^{-1}\boldsymbol{\alpha}_1^*], \tag{8.248}$$

where the right-hand side of Eq. (8.248) was obtained from Eq. (8.246), in fact, in the framework of both the hypothesis **H1b** (4.94) and the dilute approximation solution (3.175$_2$) for the noninteracting inclusions inside the cluster Ω. As can be seen in Eq. (8.248), Maxwell approach is equivalent to equating of the concentration factors of strain polarization tensors inside domains Ω and Ω^h. In so doing, the ellipsoidal shape of the homogeneous domain Ω^h was exploited only in the right-hand side of Eq. (8.248). Equating the corresponding terms in Eq. (8.248) leads to the linear algebraic system for the effective properties

$$\sum_{j=1}^{N}\mathbf{R}_j^{\epsilon} = |\Omega^h|\mathbf{L}_1^*(\mathbf{I} + \mathbf{P}^{\Omega}\mathbf{L}_1^*)^{-1}, \quad \sum_{j=1}^{N}\mathbf{F}_j^{\epsilon} = |\Omega^h|(\mathbf{I} + \mathbf{L}_1^*\mathbf{P}^{\Omega})^{-1}\boldsymbol{\alpha}_1^*, \tag{8.249}$$

which can be easily solved

$$\mathbf{L}^* = \mathbf{L}^{(0)} + \left[\langle \mathbf{R}^{\epsilon v} V\rangle^{-1} - \mathbf{P}^{\Omega}\right]^{-1}, \tag{8.250}$$

$$\boldsymbol{\alpha}^* = \boldsymbol{\alpha}^{(0)} + \left[\langle \mathbf{R}^{\epsilon v} V\rangle^{-1} - \mathbf{P}^{\Omega}\right]^{-1}\langle \mathbf{R}^{\epsilon v} V\rangle^{-1}\langle \mathbf{F}^{\epsilon v} V\rangle, \tag{8.251}$$

$$\langle \boldsymbol{\varepsilon}\rangle_i(\mathbf{x}) = \mathbf{A}_i(\mathbf{x})\mathbf{D}^{\mathrm{MS}}\boldsymbol{\varepsilon}^{\Gamma_0}, \quad \mathbf{D}^{\mathrm{MS}} \equiv \mathbf{I}, \tag{8.252}$$

where the *dilute approximation* (DA) for the stress concentration factor (8.252) exploited by *Maxwell scheme* (MS) is presented for $\boldsymbol{\alpha}_1 \equiv \mathbf{0}$; hereafter $\mathbf{f}_i^v = \mathbf{f}_i/\bar{v}_i$ ($\mathbf{f} = \mathbf{R}^{\epsilon}, \mathbf{F}^{\epsilon}$). It is interesting that representations for $\mathbf{L}^*$ (8.250) and $\boldsymbol{\alpha}^*$ (8.251) obtained by Maxwell scheme completely coincide with the corresponding representations (8.241) and (8.242) obtained by the particular case of EFM if the shape $\hat{v}^0$ (8.240) coincides with the shape Ω. Moreover, if the shape of aligned inclusions coincides with the shape and orientation Ω^h than all three methods EFM and MTM (8.80) and MS (8.250), (8.251) give identical results for the effective properties $\mathbf{L}^*$ and $\boldsymbol{\alpha}^*$ (the case of the strain concentration factors will be considered in the next subsection).

These statements can be symbolically presented in the next schematic consequence (templates)

$$\text{EFM} \Leftrightarrow \begin{bmatrix} \text{INPUT EFM} \\ \text{shape:} \quad \hat{v}^0 \\ \text{DA:} \quad \langle \mathbf{R}_i \rangle \end{bmatrix} \begin{matrix} \Rightarrow \\ \Rightarrow \end{matrix} \begin{bmatrix} \text{INPUT} & & \text{MS} \\ \text{shape } \Omega^h & = & \text{shape } \hat{v}^0 \\ \text{DA:} & & \langle \mathbf{R}_i \rangle \end{bmatrix} \Rightarrow \text{MS} \tag{8.253}$$

$$\text{MTM} \Leftrightarrow \begin{bmatrix} \text{INPUT} & \text{MTM} \\ \text{shape:} & v_1 \\ \text{DA:} & \langle \mathbf{R} \rangle_{(1)}, \langle \mathbf{A} \rangle_{(1)} \end{bmatrix} \begin{matrix} \Rightarrow \\ \Rightarrow \end{matrix} \begin{bmatrix} \text{INPUT} & & \text{MS} \\ \text{shape } \Omega^h & = & \text{shape } v_1 \\ \text{DA:} & & \langle \mathbf{R} \rangle_{(1)} \end{bmatrix} \Rightarrow \text{MS}. \tag{8.254}$$

Namely, let us analyze the two particular cases of EFM and MTM considered. For INPUT boxes for EFM and MTM; at the first step, we need to estimate the averaged solutions for the dilute approximations (DA): $\langle \mathbf{R}^{\epsilon v} V \rangle$ and $\langle \mathbf{A} V \rangle$. After that we need to identify the shapes either $\hat{v}^0$ (8.253) or v_1 (8.254) for EFM or MTM, respectively. At the second step, we create the INPUT boxes for MS by the use of INPUT boxes for either EFM or MTM: assignment of the shape of either $\hat{v}^0$ or v_1 to the shape Ω^h and exploiting of the averaged DA solutions. At the last step, we estimate the effective properties $\mathbf{L}^*$ and $\boldsymbol{\alpha}^*$ by Eqs. (8.250) and (8.251), respectively.

In the author's opinion, most part of the applications of EFM and MTM can be recast in particular cases by the use of templates either (8.253) or (8.254) in the terms of Maxwell scheme (that was demonstrated in comprehensive review by Sevostianov et al. [1366]). Templates (8.253) and (8.254) simplify applications of Maxwell scheme to a wide class of micromechanical problems. From the other side, to the best of the author's knowledge, at the last 30 years, there was not at least one publication where some micromechanical problem analyzed by the Maxwell method was not solved before by either EFM or MTM. Obviously that generalization of Maxwell scheme to the analysis outside the scale separation hypothesis (6.17) is questionable, whereas EFM already demonstrated its efficiency for these problems, e.g., for both the inhomogeneous loading and functional graded materials (see for reference and details Chap. 13).

8.9.2 *Deficiencies of Maxwell Scheme*

In Maxwell scheme, one introduces an unnecessary object called "effective homogeneous ellipsoidal domain" Ω^h which is, perhaps, descriptive but not required and, moreover, presents some obstacles (sometimes irresistible obstacles, some sort of "Procrustean bed") for the generalizations which are already done in the framework of the EFM.

8.9.2.1 Prescribed Random Orientation of Inclusions

The shape Ω^h has no connection with the microstructure of CM (e.g., orientation of inclusions), while $\hat{v}^0$ (8.253) has explicit physical meaning of the excluded volume surrounding the inclusion v_i. Different physical meaning of Ω^h and $\hat{v}^0$ allows us to define a direction which potentially can present the difficulties for application of the Maxwell scheme. So, we can logically present the ellipsoid $\hat{v}^0$ as an ellipsoid of minimal volume circumscribing the union $\cup_i v_i^0$ of all possible orientations of n_i. The trivial geometrical problem of choice of $\hat{v}^0$ in EFM offers a real challenge for MS based on the choice Ω^h; see template (8.253). So, Eq. (8.241) was obtained in [1363] by the Maxwell scheme, although it is immediately follow by the template (8.253). Less trivial case was considered in [1362] (see also [1366]) who proposed to estimate the aspect ratios of Ω^h through the ratios of some convolutions of averaged Hill's tensor $\langle \mathbf{P}_i V \rangle$ for inclusions. It means that geometrical parameters of Ω^h depend on the mechanical properties of the matrix $\mathbf{L}^{(0)}$. Correctness of this approach is questionable twice over. The first is that a necessity of averaging of precisely polarization tensor $\boldsymbol{\tau}(\mathbf{x})$ (but not the Hill's tensor $\mathbf{P}_i$) for estimation of effective properties is dictated by the background concepts of micromechanics; see Eqs. (3.176_1) and (6.55). The second is that the tensor $\mathbf{P}^{\Omega}$ for the effective inclusion Ω^h in the right-hand sides of Eqs. (8.249) is assumed to be dependent on statistical average of Hill's tensors $\langle \mathbf{P} V \rangle$ for the inclusions v_i in the left-hand side of Eqs. (8.249). Furthermore, the Hill's tensors $\mathbf{P}_i$ for the inclusions v_i are absent at the left-hand side of Eqs. (8.249), and, moreover it cannot be uniquely defined for both the nonellipsoidal inclusions and microinhomogeneous (e.g., laminated) ones.

8.9.2.2 Hybrid Composites

The case of CM considered in Sect. 8.9.2.1 may get complicated when we analyze a hybride CM containing two sets of spheroidal inclusions—fibers ($\gamma > 10$) and disks ($\gamma < 0.1$). This problem is already analyzed by EFM with a straightforward generalization of Eqs. (8.63) and (8.64) containing one sort of excluded volume v_i^0 (different by orientation) to the corresponding equations (see for details Eqs. (8.43)–(8.46)) with four different types of excluded volumes v_{ij}^0 (the indexes i, j correlate with both the fibers and discs; see [742]). However, Sevostianov [1362] (see also [1366]) proposed to estimate the shape of Ω^h through the average of both Hill's tensors $\langle \mathbf{P} V \rangle$ and $\langle \mathbf{Q} V \rangle$, and moreover, this shape is also assumed to be dependent on a relative concentration of the inclusion phases with continuous variation of the shape Ω^h between two limiting shapes Ω^h corresponding to the inclusion phase of one type (either only fibers or only discs). This approach is even more questionable than the previous version of MS analyzed in Sect. 8.9.2.1.

8.9.2.3 Prescribed Uniformly Random Orientation of Inclusions

Another potential obstacle for MS is the case where the shape Ω^h is uniquely defined, whereas EFM has physically justified another shape of excluded volumes v_i^0. Indeed, let us consider CM consisting from isotropic matrix reinforced by rigid prolate spheroids with uniformly random distribution (see for details [230]). Then the shape of Ω^f is uniquely spherical, and therefore, the shape of $\hat{v}^0$ is also necessarily spherical. The identical Eqs. (8.241) and (8.249_1) lead to the representation

$$\mathbf{L}^* = \mathbf{L}^{(0)} + [(c^{(1)}\langle\mathbf{P}^{-1}\rangle_\omega)^{-1} - \mathbf{P}^{\Omega}]^{-1}, \qquad (8.255)$$

which was in detail analyzed in [1240]. It was also proved that statistical average over orientations $\langle\mathbf{P}^{-1}\rangle_\omega$ can be presented in the form $\langle\mathbf{P}^{-1}\rangle_\omega = \mathbf{P}^-_\omega\gamma^2$, where $\mathbf{P}^-_\omega$ is some isotropic tensor independent on an aspect ratio $\gamma > 1$ of spheroidal inclusions. Thus, the effective moduli $\mathbf{L}^*$ (8.255) depends on the volume concentration $c^{(1)}$ of inclusions only through the combination $c^{(1)}\gamma^2$. It means that for arbitrary as small as you like $c^{(1)}$ (e.g., $c^{(1)} = 0.01$), there exists the aspect ratio γ^- such that for any $\gamma > \gamma^- \simeq 10$ (see for details [230]), we have $\mathbf{L}^* < \mathbf{0}$ in the sense of quadratic forms. However, it is logical to assume in EFM (8.237) that the excluded volume v_i^0 has the same shape as v_i and has the semiaxes $2a, 2a, 2\gamma a$. In such a case, the identical Eqs. (8.237) and (8.80) for EFM and MTM, respectively, are reduced to the following equation, $\mathbf{L}^* = \mathbf{L}^{(0)} + c^{(1)}(1 - c^{(1)})^{-1}\langle\mathbf{P}^{-1}\rangle_\omega$, which predicts the reasonable values of effective moduli for wider range concentration of inclusion $c^{(1)}$.

Situation is even more dramatic for the hibride composites with uniform orientation of both the rigid fibers $v^{(1)}$ ($\gamma > 10$) and porous discks $v^{(2)}$ ($\gamma < 0.1$). The Maxwell scheme defines a single solution (8.250) for the spherical Ω^h. However, MTM and EFM lead to the different solutions. Moreover, EFM has additional two operating parameters describing the shapes of the excluded volume $v^{(1)0}$ and $v^{(2)0}$. Obviously that an appropriate choice of the shapes and orientations of $v^{(1)0}$ and $v^{(2)0}$ can lead to either the increasing or decreasing of the effective moduli as the function of the inclusion concentration. It means that the variations of effective moduli ($\mathbf{L}^*$ vs c) estimated by the EFM and MS can differ by a sign.

8.9.2.4 Strain Concentration Factors ("Red Flag")

One deliberately considers a simplest case of CM with identical axial homogeneous ellipsoidal inclusions. Buryachenko [230] proved that Mishurova et al. [1082] at the obtaining of an equality $\mathbf{D}^{\mathrm{MS}} = \mathbf{D}^{\mathrm{MTM}}$ failed into the logical trap of mistaking use of background principles of micromechanics even if a necessary condition statistical homogeneity is violated (see Fig. 8.20b). Inconsistency of Maxwell scheme at the exact estimation of the strain concentration factors was also demonstrated by Eqs. (6.98). The identity $\mathbf{D}^{\mathrm{MS}} \equiv \mathbf{I}$ (8.252) is an exact result obtained in the framework of Maxwell scheme for both CMs with inclusions of any shape, microstructure, any prescribed random orientation, and for any shape of effective inclusion Ω^h (therefore, it is not an accidental choice of an inappropriate shape of Ω^h, as mentioned in [1366]). Practical significance of Eq. (8.252) is critically important. It means that for any CMs and any shape of Ω^h, the strain (and stress) concentration factors do not depend on concentration of inclusions (it does not matter that we have either $c^{(1)} = 0.01$ or $c^{(1)} = 0.99$). However, estimation of local fields is critical for CMs with any nonlinear phenomena (see Chap. 12). Therefore, applications of Maxwell schemes for analysis of any nonlinear problem is questionable even if Maxwell schemes sometimes demonstrate "miraculous coincidence" with other classical methods at the estimation of effective moduli (a linear problem).

8.9.3 *Generalized Maxwell Schemes Taking Inclusion Interactions into Account*

As in Fig. 8.20b, let Ω be a cluster mentally "cut-out" from statistically homogeneous or periodic structure CM. We assume that the stress polarization tensor $\boldsymbol{\tau}(\mathbf{x})$ $(\mathbf{x} \in v_i,\ \ \Omega = \cup_i v_i)$

$$\boldsymbol{\tau}(\mathbf{x}) = \mathbf{R}^{\epsilon \mathrm{CL}}(\mathbf{x})\boldsymbol{\varepsilon}^{\Gamma_0} + \mathbf{F}^{\epsilon \mathrm{CL}}(\mathbf{x}), \tag{8.256}$$

is found (see, e.g., Eqs. (8.235) and (8.236)). Repeating Eqs. (8.246)–(8.252), we can get a representation for the effective modulus

$$\mathbf{L}^* = \mathbf{L}^{(0)} + \left\{ \langle \mathbf{R}^{\epsilon \mathrm{CL}}(\mathbf{x}) V(\mathbf{x}) \rangle_{\Omega}^{-1} - \mathbf{P}^{\Omega} \right\}^{(-1)}, \tag{8.257}$$

with the obvious counterparts of Eqs. (8.251) and (8.252); here $\langle (\cdot)(\mathbf{x}) \rangle_{\Omega} = |\Omega^h|^{-1} \int_{\Omega} (\cdot)(\mathbf{x})\ d\mathbf{x}$. Equation (8.257) is general and hold for any shape and microtopology of inclusions as well as for any numerical method used for estimation of the tensors $\mathbf{R}_i^{\epsilon\ \mathrm{CL}}$ and $\mathbf{F}_i^{\epsilon\ \mathrm{CL}}$. In so doing, Mogilevskaya et al. [1084, 1086] independently proposed equivalent particular far-field approach for the cluster with the circular inclusions and microcracks analyzed by complex variables based on technique applicable only to two-dimensional problems (see also [1366]).

In particular, they [1084, 1086] considered triangle packing of identical circle inclusions with expectedly circle effective inclusion (see Fig. 8.21) which is obviously perfectly fitted with the considered both inclusion shape and inclusion packing in the cluster. However, in the framework of the classical Maxwell scheme, we considered in Sect. 8.9.2 a few inconsistencies, and all of them were produced by ambiguous choice of the effective inclusion shape Ω^h. The new opportunities for the potential obstacles for self-consistent applicability of generalized Maxwell scheme are defined by plurality of possible mutual correspondence of both the

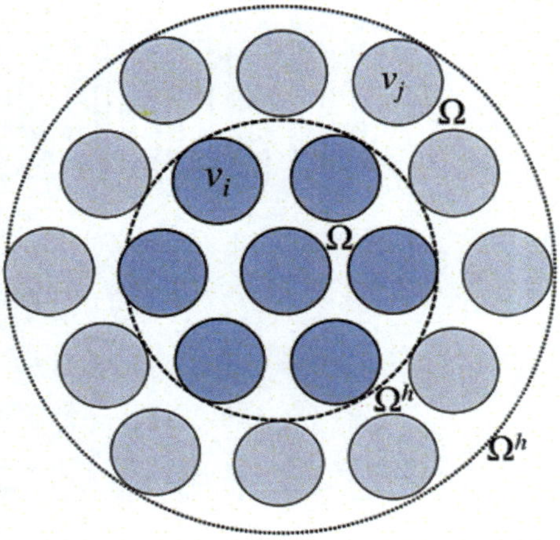

Fig. 8.21 The scheme of clusters with 7 and 19 inclusions

orientations (three angles) and sizes (three lengths) of the inclusions and unit cell. Indeed, let us consider a periodic system of infinite columns containing infinite stacks (as in Fig. 8.14) of oblate spheroids ($\alpha = 0.01$). We can "cut out" the clusters Ω with different numbers n of spheroids v_j by the use of different domains Ω^h. Ambiguous choice of the effective inclusion shape Ω^h leads to both the different averaged distribution of stress concentration factors inside inclusions $v_j \subset \Omega_i$ (see Fig. 8.16) and the different effective Young modulus E_1^* (see Fig. 8.17) for CM with random field of identical stacks with different numbers of inclusions. Especially dramatic inconsistency is verified by Fig. 8.16. Indeed, the stress distribution inside all inclusions in periodic structure must be identical. However, variation of $\langle\sigma_{11}\rangle_{(i)}/\langle\sigma_{11}\rangle$ inside the stack reaches 2.7 times at $n = 17$ (see Fig. 8.16).

The next example that first comes to mind is CM with periodic system of two types of columns containing infinite stacks of either the oblate spheroids or spheres. CM with periodic system of two kinds of layers containing periodic systems of either prolate spheroids or spheres (see Fig. 8.22) is also irresistible obstacle for generalized Maxwell scheme proposed in [1084, 1086].

Another issue is that 40 years of development of homogenization technique (both two-scale asymptotic expansion and computational homogenization; see for reference Sect. 7.4.1) makes computational effectiveness at one UC incomparable with a computational effectiveness of any versions of the Maxwell scheme requiring convergence of results obtained for clusters with growing numbers of UCs. However, for the cubic packing (simple cubic, SC, and body-centered, BCC) of spherical inclusions with $n = 512$ and 559, respectively, in the clusters with a spherical shape Ω^h, the effective properties estimations diverge in the most cases considered for both conductivity [855] and elasticity [853] problems. A reason of divergence is typical for the methods used a direct superposition scheme of summations of the perturbations produced by the inclusions (see for details Sect. 13.4.2). Indeed, the

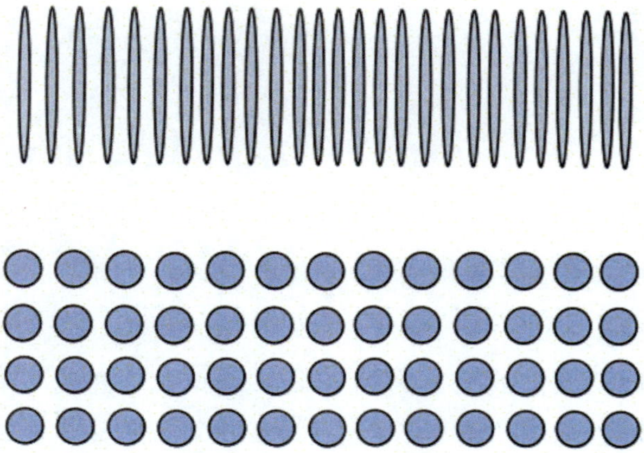

Fig. 8.22 Periodic structure CM with two types of layers

mentioned authors solved the problem (in our notation, $\mathbf{x} \in v_k$, $k = 1, \ldots, N$)

$$\boldsymbol{\varepsilon}(\mathbf{x}) = \boldsymbol{\varepsilon}^{\Gamma_0} + \int \mathbf{U}(\mathbf{x}-\mathbf{y}) \sum_{k=1}^{N} \mathbf{M}_1^{(1)} \boldsymbol{\varepsilon}(\mathbf{y}) V_k(\mathbf{y})\, d\mathbf{y} \tag{8.258}$$

for a finite number of the homogeneous ellipsoidal aligned inclusions v_k ($k = 1, \ldots, N$). It is assumed that the field v_k ($k = 1, \ldots, n$) is cutted out by the domain $v_k \subset \Omega^h$ (with an indicator function $V_\Omega(\mathbf{x})$) from the infinite periodic field of inclsions $v_\mathbf{m}$ with the centers $\mathbf{x_m} = m_1\mathbf{e}_1 + m_2\mathbf{e}_2 + m_3\mathbf{e}_3$ forming the lattice $\mathbf{m} \in \Lambda$. However, a correct counterpart of Eq. (8.259) was obtained at the exploitation of Eq. (7.36) instead of (7.30_1) (see [200]). Namely, owing to its absolute convergence, Eq. (7.36) is equivalent to the following equation:

$$\boldsymbol{\varepsilon}(\mathbf{x}) = \langle \boldsymbol{\varepsilon} \rangle + \int \mathbf{U}(\mathbf{x}-\mathbf{y}) \Big[\sum_{k=1}^{N} \mathbf{M}_1^{(1)} \boldsymbol{\varepsilon}(\mathbf{y}) V_k(\mathbf{y}) - V_\Omega(\mathbf{y}) \langle \mathbf{M}_1^{(1)} \boldsymbol{\varepsilon} \rangle_i \Big]\, d\mathbf{y}, \tag{8.259}$$

whose solution does not depend on the shape of Ω^h if Ω^h is large enough. For definetness, $\mathbf{x} \in v_i$ is a central inclusion of the ellipsoidal domain Ω^h, whereas the strain distributions are identical in all inclsions v_l ($l = 1, \ldots, N$). Equation (8.259) can be solved with any desirable accuracy by the different methods; see, e.g., [200, 230].

Thus, even if the effective properties estimated by Maxwell scheme are appropriate and correlate with some classical micromechanical models (e.g., EFM and MTM) for statistically homogeneous and periodic media in some cases (without warranty of appropriate results in general), its applications for estimation of local stress concentration factors at the inclusions are critically inconsistent. Critical analysis of Maxwell scheme applications [852, 857], and [858] to the modeling of statistically homogeneous CMs is performed in Sect. 13.4.2. We can conclude that a life-asserting optimism of the final slogan in the comprehensive review by Sevostianov et al. [1366] "the Maxwell's methodology of estimating the effective properties is quite alive and well!" is greatly exaggerated.

Chapter 9
Basic Representations of New Background of Analytical Micromechanics

Abstract In this chapter, we consider the solutions of the new GIEs (7.19) and (7.20) by the use of acceptance of different combinations of the classical hypotheses **H1a**, **H1b**, H2, and **H3**. We introduce the concept of *perturbators* which can be found by any available numerical or semianalytical method. This makes it potentially possible to abandon many classical concepts of micromechanics used in most popular methods, namely, effective field hypothesis **H1**, quasi-crystalline approximation **H2**, the hypothesis of "ellipsoidal symmetry" **H3**, and Eshelby tensor. Abandonment of a few different combinations of these hypotheses leads to detection of some new effects that are impossible in the framework of a classical background of micromechanics analyzed in Chap. 8.

The estimations of both the mechanical and transformation effective concentration factors inside a representative heterogeneity $\mathbf{x} \in v_i$

$$\langle \boldsymbol{\varepsilon} \rangle_i(\mathbf{x}) = \mathbf{A}_i^*(\mathbf{x})\langle \boldsymbol{\varepsilon} \rangle + \mathbf{C}_i^{\epsilon *}(\mathbf{x}), \quad \langle \boldsymbol{\sigma} \rangle_i(\mathbf{x}) = \mathbf{B}_i^*(\mathbf{x})\langle \boldsymbol{\varepsilon} \rangle + \mathbf{C}_i^*(\mathbf{x}), \tag{9.1}$$

will be exploited for the following representations of the effective parameters:

$$\mathbf{L}^* = \mathbf{L}^{(0)} + \langle \mathbf{L}_1 \mathbf{A}^* \rangle, \quad \boldsymbol{\alpha}^* = \langle \boldsymbol{\alpha} \rangle + \langle \mathbf{L}_1 \mathbf{C}^{\epsilon *} \rangle, \tag{9.2}$$

$$\mathbf{M}^* = \mathbf{M}^{(0)} + \langle \mathbf{M}_1 \mathbf{B}^* \rangle, \quad \boldsymbol{\beta}^* = \langle \boldsymbol{\beta} \rangle + \langle \mathbf{M}_1 \mathbf{C}^* \rangle, \tag{9.3}$$

$$W^* = -\frac{1}{2}\langle \boldsymbol{\alpha}_1 \mathbf{C}^{\epsilon *} \rangle, \quad W^* = -\frac{1}{2}\langle \boldsymbol{\beta}_1 \mathbf{C}^* \rangle \tag{9.4}$$

of the equivalent double governing equations

$$\langle \boldsymbol{\sigma} \rangle = \mathbf{L}^* \langle \boldsymbol{\varepsilon} \rangle + \boldsymbol{\alpha}^*, \quad \langle \boldsymbol{\varepsilon} \rangle = \mathbf{M}^* \langle \boldsymbol{\sigma} \rangle + \boldsymbol{\beta}^*. \tag{9.5}$$

The stored energies W^* estimated by Eqs. (9.4$_1$) and (9.4$_2$) at $\langle \boldsymbol{\varepsilon} \rangle \equiv \mathbf{0}$ and $\langle \boldsymbol{\sigma} \rangle \equiv \mathbf{0}$, respectively, depend only on the fields $\mathbf{L}(\mathbf{x})$, $\boldsymbol{\alpha}(\mathbf{x})$ and $\mathbf{M}(\mathbf{x})$, $\boldsymbol{\beta}(\mathbf{x})$ (like $\mathbf{L}^*$, $\boldsymbol{\alpha}^*$ and $\mathbf{M}^*$, $\boldsymbol{\beta}^*$, respectively) and act like effective material constants, which can be observed macroscopically (see Sect. 6.2). It should be mentioned that the effective

V. A. Buryachenko, *Local and Nonlocal Micromechanics of Heterogeneous Materials*, https://doi.org/10.1007/978-3-030-81784-8_9

thermoelastic constants $\mathbf{M}^*$, $\boldsymbol{\beta}^*$, W^* play a key role in micromechanics because of their close relationship to the internal stress fluctuations (see Chapter 13 in [200]). Although the GIEs (7.19) and (7.20) are fulfilled for statistically inhomogeneous media, in this chapter, we only analyze statistically homogeneous media subjected to the homogeneous remote loading while the properties of comparison medium $\mathbf{L}^c$, $\boldsymbol{\alpha}^c$ coincide with the properties of matrix $\mathbf{L}^{(0)}$, $\boldsymbol{\alpha}^{(0)}$.

9.1 Stress and Strain Distributions for One Heterogeneity Inside Infinite Matrix

For one inclusion v_i in an infinite matrix with the properties $\mathbf{L}^{(0)}$, $\boldsymbol{\beta}^{(0)}$ $\equiv$const., according to the definitions of the effective strains and stresses, we have the following representations for the perturbations produced by this inclusion v_i in the space $\mathbf{x} \in R^d$

$$\boldsymbol{\varepsilon}(\mathbf{x}) - \overline{\boldsymbol{\varepsilon}}(\mathbf{x}) \equiv \mathcal{L}_i^{\tau}(\mathbf{x}-\mathbf{x}_i, \boldsymbol{\tau}) = \mathcal{L}_i^{\epsilon}(\mathbf{x}-\mathbf{x}_i, \overline{\boldsymbol{\varepsilon}}), \tag{9.6}$$

$$\boldsymbol{\sigma}(\mathbf{x}) - \overline{\boldsymbol{\sigma}}(\mathbf{x}) \equiv \mathcal{L}_i^{\eta}(\mathbf{x}-\mathbf{x}_i, \boldsymbol{\eta}) = \mathcal{L}_i^{\sigma}(\mathbf{x}-\mathbf{x}_i, \overline{\boldsymbol{\sigma}}) \tag{9.7}$$

where

$$\mathcal{L}_i^{\epsilon}(\mathbf{x}-\mathbf{x}_i, \boldsymbol{\tau}) = \int_w \mathbf{U}(\mathbf{x}-\mathbf{y})\boldsymbol{\tau}(\mathbf{y})V_i(\mathbf{y})d\mathbf{y}, \tag{9.8}$$

$$\mathcal{L}_i^{\sigma}(\mathbf{x}-\mathbf{x}_i, \boldsymbol{\eta}) = \int_w \boldsymbol{\Gamma}(\mathbf{x}-\mathbf{y})\boldsymbol{\eta}(\mathbf{y})V_i(\mathbf{y})d\mathbf{y}. \tag{9.9}$$

It should be mentioned that Eqs. (9.6_1) and (9.7_1) with the integral representation of the perturbators $\mathcal{L}_i^{\epsilon}(\mathbf{x}-\mathbf{x}_i, \boldsymbol{\tau})$ (9.8) and $\mathcal{L}_i^{\sigma}(\mathbf{x}-\mathbf{x}_i, \boldsymbol{\eta})$ (9.9) can be obtained one from the other by the use of the identities ($\mathbf{x} \in w$)

$$\boldsymbol{\tau} = -\mathbf{L}^{(0)}\boldsymbol{\eta}, \quad \boldsymbol{\varepsilon} = \mathbf{M}^{(0)}\boldsymbol{\sigma} + \boldsymbol{\beta}^{(0)} + \boldsymbol{\eta}.$$

Indeed, substitutions (9.10_1) and (9.10_2) into the right-hand side and the left-hand side of Eqs. (9.6_1) and (9.8), respectively, lead to (9.7_1), (9.9). In so doing, the strain perturbator $\mathcal{L}_i^{\epsilon}(\mathbf{x}-\mathbf{x}_i, \boldsymbol{\tau})$ and the stress perturbator $\mathcal{L}_i^{\sigma}(\mathbf{x}-\mathbf{x}_i, \boldsymbol{\eta})$ are linked by the relation[1]

$$\mathcal{L}_i^{\sigma}(\mathbf{x}-\mathbf{x}_i, \boldsymbol{\eta}) = \boldsymbol{\tau}(\mathbf{x})V_i(\mathbf{x}) + \mathbf{L}^{(0)}\mathcal{L}_i^{\epsilon}(\mathbf{x}-\mathbf{x}_i, \boldsymbol{\tau}), \tag{9.10}$$

$$\mathcal{L}_i^{\epsilon}(\mathbf{x}-\mathbf{x}_i, \boldsymbol{\tau}) = \boldsymbol{\eta}(\mathbf{x})V_i(\mathbf{x}) + \mathbf{M}^{(0)}\mathcal{L}_i^{\sigma}(\mathbf{x}-\mathbf{x}_i, \boldsymbol{\eta}) \tag{9.11}$$

[1]Formulation of the duality between the stress $\boldsymbol{\tau}$ and strain $\boldsymbol{\eta}$ polarizations (4.2) was actually presented much earlier by Willis [1625].

which can be established by both methods: the substitution of identity (9.10) into Eqs. (9.7$_1$) and the substitution of the constitutive law into Eq. (9.7$_1$). Thus we have proved that Eqs. (9.6$_1$) and (9.7$_1$) are equivalent and the perturbators $\mathcal{L}_i^{\sigma}(\mathbf{x}-\mathbf{x}_i, \boldsymbol{\eta})$ and $\mathcal{L}_i^{\epsilon}(\mathbf{x}-\mathbf{x}_i, \boldsymbol{\tau})$ are related by the equalities (9.10) and (9.11) independently on the concrete representations of these operators [e.g., either Eqs. (9.8) and (9.9) or any other form of Eqs. (9.6$_1$) and (9.7$_1$)].

Strictly speaking, the perturbators $\mathcal{L}_i^{\epsilon}(\mathbf{x}-\mathbf{x}_i, \boldsymbol{\tau})$ (9.6$_1$) and $\mathcal{L}_i^{\sigma}(\mathbf{x}-\mathbf{x}_i, \boldsymbol{\eta})$ (9.7$_1$) and their integral representations (9.8) and (9.9) are just the notations of some problem which is destined to be solved, while the perturbators $\mathcal{L}_i^{\epsilon}(\mathbf{x}-\mathbf{x}_i, \overline{\boldsymbol{\varepsilon}})$ (9.6$_2$) and $\mathcal{L}_i^{\sigma}(\mathbf{x}-\mathbf{x}_i, \overline{\boldsymbol{\sigma}})$ (9.7$_2$) are the solutions of this problem.

We will coincide a construction of the perturbator $\mathcal{L}_i^{\sigma}(\mathbf{x}-\mathbf{x}_i, \overline{\boldsymbol{\sigma}})$ (9.7$_2$) [the perturbator $\mathcal{L}_i^{\epsilon}(\mathbf{x}-\mathbf{x}_i, \overline{\boldsymbol{\varepsilon}})$ (9.6$_2$) can be analyzed in a similar manner]. The FEA is very effective for estimating the perturbator $\mathcal{L}_i^{\sigma}(\mathbf{x}-\mathbf{x}_i, \overline{\boldsymbol{\sigma}})$ (9.7$_2$) [or, that is equivalent, the stress concentration factor $\mathbf{B}_i(\mathbf{x})$ and $\mathbf{C}_i(\mathbf{x})$, see Chap. 4] at the constant effective fields $\overline{\boldsymbol{\sigma}}(\mathbf{x}) \equiv$const., when a prescription of a homogeneous loading at the boundary of a large sample w is obvious. At the same time, analysis of inhomogeneous effective fields is not so straightforward in the FEA. In such a case, the VIE method using a prescription of the effective field $\overline{\boldsymbol{\sigma}}(\mathbf{x})$ only inside the domain $\mathbf{x} \in v_i$ is found to be more effective. A situation is complicated by the fact that the kernel of the operator $\mathcal{L}^{\sigma}(\mathbf{x}-\mathbf{x}_i, \langle\overline{\boldsymbol{\sigma}}_q\rangle)(\mathbf{x}_i)$ is singular at $\mathbf{x}=\mathbf{x}_i$. The mentioned difficulty can be eliminated in the framework of a subtraction technique transforming Eqs. (9.6$_1$) and (9.7$_1$) in the following manner. Namely, let the inclusions v_i be fixed and loaded by the inhomogeneous effective field $\overline{\boldsymbol{\sigma}}_i(\mathbf{x})$

$$\boldsymbol{\sigma}(\mathbf{x}) = \overline{\boldsymbol{\sigma}}_i(\mathbf{x}) + \int \boldsymbol{\Gamma}(\mathbf{x}-\mathbf{y})V_i(\mathbf{y})\boldsymbol{\eta}(\mathbf{y})d\mathbf{y}. \tag{9.12}$$

The difficulties with the troublesome singularities can be avoided if a rearrangement of Eq. (9.12) is performed in the spirit of a subtraction technique used in the modified quadrature method (see, e.g., [370])

$$\boldsymbol{\eta}(\mathbf{x}) = \overline{\boldsymbol{\eta}}_i(\mathbf{x}) + \int \mathbf{K}_i(\mathbf{x},\mathbf{y})\big[\boldsymbol{\eta}(\mathbf{y}) - \boldsymbol{\eta}(\mathbf{x})\big]d\mathbf{y}, \quad \mathbf{x} \in v_i, \tag{9.13}$$

where $\overline{\boldsymbol{\eta}}_i(\mathbf{x}) = \mathbf{E}_i(\mathbf{x})\overline{\boldsymbol{\sigma}}(\mathbf{x}) + \mathbf{H}_1(\mathbf{x})$, $(\mathbf{x} \in v_i)$ is called the effective strain polarization tensor in the inclusion v_i, and (no sum on i)

$$\mathbf{K}_i(\mathbf{x},\mathbf{y}) = \mathbf{E}_i(\mathbf{x})\boldsymbol{\Gamma}(\mathbf{x}-\mathbf{y})V_i(\mathbf{y}), \tag{9.14}$$

$$\mathbf{E}_i(\mathbf{x}) = \mathbf{M}_1(\mathbf{x})[\mathbf{I} + \mathbf{Q}_i(\mathbf{x})\mathbf{M}_1(\mathbf{x})]^{-1}, \tag{9.15}$$

$$\mathbf{H}_i(\mathbf{x}) = [\mathbf{I} + \mathbf{M}_1(\mathbf{x})\mathbf{Q}_i(\mathbf{x})]^{-1}\boldsymbol{\beta}_1(\mathbf{x}), \tag{9.16}$$

where the tensor $\mathbf{Q}_i(\mathbf{x}) \equiv -\int V_i(\mathbf{y})\boldsymbol{\Gamma}(\mathbf{x}-\mathbf{y})d\mathbf{y}$ is found by the FEA (see for details Chap. 4). We rewrite Eq. (9.13) in symbolic form

$$\boldsymbol{\eta} = \overline{\boldsymbol{\eta}}_i + \mathcal{K}_i\boldsymbol{\eta}, \tag{9.17}$$

where $(\mathcal{K}_i\boldsymbol{\eta})(\mathbf{x}) = \int \mathcal{K}_i(\mathbf{x},\mathbf{y})\boldsymbol{\eta}(\mathbf{y})d\mathbf{y}$ defines the integral operator $\mathcal{K}_i$ with the kernel formally represented as $\mathcal{K}_i(\mathbf{x},\mathbf{y}) = \mathbf{K}_i(\mathbf{x},\mathbf{y}) - \delta(\mathbf{x}-\mathbf{y})\int V_i(\mathbf{z})\mathbf{K}_i(\mathbf{x},\mathbf{z})d\mathbf{z}$. We formally write the solution of Eq. (9.17) as

$$\boldsymbol{\eta} = \mathcal{L}_i * \overline{\boldsymbol{\eta}}_i, \tag{9.18}$$

where the inverse operator $\mathcal{L}_i = (\mathbf{I} - \mathcal{K}_i)^{-1}$ will be constructed by the iteration method based on the recursion formula

$$\boldsymbol{\eta}^{[k+1]} = \overline{\boldsymbol{\eta}}_i + \mathcal{K}_i\boldsymbol{\eta}^{[k]} \tag{9.19}$$

to construct a sequence of functions $\{\boldsymbol{\eta}^{[k]}\}$ that can be treated as an approximation of the solution of Eq. (9.17). We prescribed the point Jacobi (called also Richardson and point total-step) iterative scheme for ease of presentation. More advanced algorithm is described in Sect. 4.4.4. Usually, the driving term of this equation is used as an initial approximation

$$\boldsymbol{\eta}^{[0]}(\mathbf{x}) = \overline{\boldsymbol{\eta}}_i(\mathbf{x}), \tag{9.20}$$

which is exact for a homogeneous ellipsoidal inclusion subjected to remote homogeneous stress field $\overline{\boldsymbol{\sigma}}(\mathbf{x}) \equiv \overline{\boldsymbol{\sigma}} =$ const., when $\mathbf{B}_i(\mathbf{x}) = \mathbf{E}_i(\mathbf{x}) \equiv$const. ($\mathbf{x} \in v_i$). The sequence $\{\boldsymbol{\eta}^{[k]}\}$ (9.19) with arbitrary continuous $\boldsymbol{\eta}^{[0]}(\mathbf{x})$ converges to a unique solution $\boldsymbol{\eta}$ if the norm of the integral operator $\mathcal{K}_i$ turns out to be small "enough" (less than 1), and the problem is reduced to the computation of the integrals involved, the density of which is given. The desired for Eq. (9.7$_2$) operator $\mathcal{L}_i^{\sigma}(\mathbf{x}-\mathbf{x}_i,\overline{\boldsymbol{\sigma}})$ can be presented through the found operator $\mathcal{L}_i$

$$\mathcal{L}_i^{\sigma}(\mathbf{x}-\mathbf{x}_i,\overline{\boldsymbol{\sigma}}_i)(\mathbf{x}) = \int \boldsymbol{\Gamma}(\mathbf{x}-\mathbf{y})\mathcal{L}_i * (\mathbf{E}_i\overline{\boldsymbol{\sigma}})(\mathbf{y})V_i(\mathbf{y})d\mathbf{y}. \tag{9.21}$$

A singularity in the right-hand-side integrals in Eqs. (9.9) and (9.21) can be eliminated by way of a subtraction technique

$$\mathcal{L}_i^{g}(\mathbf{x}-\mathbf{x}_i,\mathbf{f}) = \mathbf{Q}_i(\mathbf{x})\mathbf{g}(\mathbf{x}) + \int \boldsymbol{\Gamma}(\mathbf{x}-\mathbf{y})[\mathbf{g}(\mathbf{y})-\mathbf{g}(\mathbf{x})]V_i(\mathbf{y})\,d\mathbf{y}, \tag{9.22}$$

$$\mathcal{L}_i^{g}(\mathbf{x}-\mathbf{x}_i,\mathbf{f}) = \mathbf{Q}_i(\mathbf{x}_m)\mathbf{g}(\mathbf{x}_m) + \int \boldsymbol{\Gamma}(\mathbf{x}-\mathbf{y})[\mathbf{g}(\mathbf{y})-\mathbf{g}(\mathbf{x}_m)]V_i(\mathbf{y})\,d\mathbf{y} \tag{9.23}$$

for $\mathbf{x} \in v_i$ and $\mathbf{x} \notin v_i$, respectively; here $\mathbf{x}_m = \arg\min_{\mathbf{y}} |\mathbf{x}-\mathbf{y}|$ ($\mathbf{y} \in v_i$, $\mathbf{x} \notin v_i$), $\mathcal{L}_i^{g} = \mathcal{L}_i^{\sigma}, \mathcal{L}_i^{\eta}$, $\mathbf{f} = \overline{\boldsymbol{\sigma}}_i, \boldsymbol{\eta}$, and $\mathbf{g}(\mathbf{y}) = \mathcal{L}_i * (\mathbf{E}_i\overline{\boldsymbol{\sigma}}_i)(\mathbf{y}), \boldsymbol{\eta}(\mathbf{y})$, respectively. The subtracted form (9.22) is regular (see for details Chap. 4), and the improper volume integral over the region $v_i\backslash\omega$ absolutely converges and is independent of the shape of an excluded contracted region ω (diam$(\omega) \to 0$). It justifies the applicability of a standard quadrature rule for the numerical estimation of the integral (9.22). Thus, we constructed the solution (9.12) for a perturbation of the stress field inside and outside the inclusion v_i in the operator form obtained by the VIE for an arbitrary effective field $\overline{\boldsymbol\sigma}_i(\mathbf{x})$, $(\mathbf{x} \in v_i)$.

9.2 Some Classical Hypotheses and Approaches Based on the Classical GIE (7.32) and (7.33)

In this subsection, we recast the basic hypotheses and the methods of Chap. 8 in the form adapted for the subsequent presentations.

9.2.1 *Effective Field Hypothesis H1 and Some Perturbator Representations*

The hypothesis **H1a** (6.46) is not altered, while the hypothesis **H1b** (6.47) and (6.48) is reformulated in the term of perturbators.

Hypothesis 1b, H1b *The operators* $\mathcal{L}_k^{\epsilon}(\mathbf{x}-\mathbf{x}_k, \boldsymbol\tau)$ *(9.10₁) and* $\mathcal{L}_k^{\sigma}(\mathbf{x}-\mathbf{x}_k, \boldsymbol\eta)$ *(9.11₁) of perturbations generated by the heterogeneities* v_i *at the point* $\mathbf{x} \notin v_k$ *are reduced to the decoupled tensorial multiplications*

$$\mathcal{L}_i^{\epsilon}(\mathbf{x}-\mathbf{x}_i, \boldsymbol\tau) = \mathbf{L}_k^{\tau}(\mathbf{x}-\mathbf{x}_i)\langle\boldsymbol\tau\rangle_{(i)}, \quad \mathcal{L}_i^{\sigma}(\mathbf{x}-\mathbf{x}_i, \boldsymbol\eta) = \mathbf{L}_i^{\eta}(\mathbf{x}-\mathbf{x}_i)\langle\boldsymbol\eta\rangle_{(i)}. \tag{9.24}$$

In the case of the general definitions of perturbators $\mathcal{L}_i^{\epsilon}(\mathbf{x}-\mathbf{x}_i, \boldsymbol\tau)$ (9.10_1) and $\mathcal{L}_i^{\sigma}(\mathbf{x}-\mathbf{x}_i, \boldsymbol\eta)$ (9.11_1), the assumptions (9.24) are new with the perturbator factors $\mathbf{L}_i^{\tau}(\mathbf{x}-\mathbf{x}_i)$ and $\mathbf{L}_i^{\eta}(\mathbf{x}-\mathbf{x}_i)$ defined by Eqs. (9.24). In the case of representations $\mathcal{L}_i^{\epsilon}(\mathbf{x}-\mathbf{x}_i\boldsymbol\tau)$ (9.10_2) and $\mathcal{L}_i^{\sigma}(\mathbf{x}-\mathbf{x}_i, \boldsymbol\eta)$ (9.11_2), the perturbator factors are reduced to the known ones (7.43) and (7.44) and expressed through the Green's functions $(\mathbf{x} \notin v_i,\ \mathbf{y} \in v_k)$

$$\begin{aligned} \bar v_i^{-1}\mathbf{L}_i^{\tau}(\mathbf{x}-\mathbf{x}_i) &= \langle \mathbf{U}(\mathbf{x}-\mathbf{y})\rangle_{(i)} \equiv \mathbf{T}_i^{\epsilon}(\mathbf{x}-\mathbf{x}_i), \\ \bar v_i^{-1}\mathbf{L}_i^{\eta}(\mathbf{x}-\mathbf{x}_i) &= \langle \boldsymbol\Gamma(\mathbf{x}-\mathbf{y})\rangle_{(i)} \equiv \mathbf{T}_i^{\sigma}(\mathbf{x}-\mathbf{x}_i). \end{aligned} \tag{9.25}$$

According to hypothesis **H1a** and in view of the linearity of the problem, there exist constant fourth $\mathbf{A}_i(\mathbf{x})$, $\mathbf{R}_i^{\epsilon}(\mathbf{x})$, $\mathbf{B}_i(\mathbf{x})$, $\mathbf{R}_i(\mathbf{x})$ and second-rank $\mathbf{C}_i^{\epsilon}(\mathbf{x})$, $\mathbf{C}_i(\mathbf{x})$, $\mathbf{F}_i^{\epsilon}(\mathbf{x})$, $\mathbf{F}_i(\mathbf{x})$ tensors, such that $(\mathbf{x} \in v_i \subset v^{(i)})$

$$\boldsymbol\varepsilon(\mathbf{x}) = \mathbf{A}_i(\mathbf{x})\overline{\boldsymbol\varepsilon}(\mathbf{x}_i) + \mathbf{C}_i^{\epsilon}(\mathbf{x}), \quad \bar v_i\boldsymbol\tau(\mathbf{x}) = \mathbf{R}_i^{\epsilon}(\mathbf{x})\overline{\boldsymbol\varepsilon}(\mathbf{x}_i) + \mathbf{F}_i^{\epsilon}(\mathbf{x}), \tag{9.26}$$

$$\boldsymbol\sigma(\mathbf{x}) = \mathbf{B}_i(\mathbf{x})\overline{\boldsymbol\sigma}(\mathbf{x}_i) + \mathbf{C}_i(\mathbf{x}), \quad \bar v_i\boldsymbol\eta(\mathbf{x}) = \mathbf{R}_i(\mathbf{x})\overline{\boldsymbol\sigma}(\mathbf{x}_i) + \mathbf{F}_i(\mathbf{x}), \tag{9.27}$$

where

$$\mathbf{R}_i^{\epsilon}(\mathbf{x}) = \bar{v}_i \mathbf{L}_1^{(1)}(\mathbf{x})\mathbf{A}_i(\mathbf{x}), \;\; \mathbf{F}_i^{\epsilon}(\mathbf{x}) = \bar{v}_i[\mathbf{L}_1^{(1)}(\mathbf{x})\mathbf{C}_i^{\epsilon} + \boldsymbol{\alpha}_1(\mathbf{x})], \tag{9.28}$$

$$\mathbf{R}_i(\mathbf{x}) = \bar{v}_i \mathbf{M}_1^{(1)}(\mathbf{x})\mathbf{B}_i(\mathbf{x}), \;\; \mathbf{F}_i(\mathbf{x}) = \bar{v}_i[\mathbf{M}_1^{(1)}(\mathbf{x})\mathbf{C}_i + \boldsymbol{\beta}_1(\mathbf{x})]. \tag{9.29}$$

According to Eshelby's [449] theorem, the averaged tensors are linked by relations $\mathbf{R}_i^{\epsilon} = \overline{v}_i\mathbf{P}_i^{-1}(\mathbf{I}-\mathbf{A}_i)$ (9.26), and $\mathbf{R}_i = \overline{v}_i\mathbf{Q}_i^{-1}(\mathbf{I}-\mathbf{B}_i)$ (9.27), where $\mathbf{g}_i \equiv \langle\mathbf{g}(\mathbf{x})\rangle_{(i)} \equiv \bar{v}_i^{-1}\int \mathbf{g}(\mathbf{x})V_i(\mathbf{x})d\mathbf{x}$ ($\mathbf{g}$ stands for $\mathbf{A}_i, \mathbf{B}_i, \mathbf{R}_i^{\epsilon}, \mathbf{R}_i$). In the general case of coated heterogeneities v_i, the tensors $\mathbf{A}_i(\mathbf{x}), \mathbf{B}_i(\mathbf{x}), \mathbf{C}_i^{\epsilon}(\mathbf{x}), \mathbf{C}_i(\mathbf{x})$ can be found by the transformation method by Dvorak and Benveniste [425] (see for reference and details Chap. 4).

Analogously to Eqs. (9.26) and (9.27), due to the linearity of the problem, the perturbators $\boldsymbol{\mathcal{L}}_i^{\epsilon}(\mathbf{x}-\mathbf{x}_i,\boldsymbol{\tau})$ and $\boldsymbol{\mathcal{L}}_i^{\sigma}(\mathbf{x}-\mathbf{x}_i,\boldsymbol{\eta})$ can be decomposed in the framework of the hypothesis **H1a**

$$\boldsymbol{\mathcal{L}}_i^{\epsilon}(\mathbf{x}-\mathbf{x}_i,\boldsymbol{\tau}) = \mathbf{L}_i^{\epsilon}(\mathbf{x}-\mathbf{x}_i)\overline{\boldsymbol{\varepsilon}}(\mathbf{x}_i) + \mathbf{L}_i^{\alpha}(\mathbf{x}-\mathbf{x}_i,\boldsymbol{\alpha}), \tag{9.30}$$

$$\boldsymbol{\mathcal{L}}_i^{\sigma}(\mathbf{x}-\mathbf{x}_i,\boldsymbol{\eta}) = \mathbf{L}_i^{\sigma}(\mathbf{x}-\mathbf{x}_i)\overline{\boldsymbol{\sigma}}(\mathbf{x}_i) + \mathbf{L}_i^{\beta}(\mathbf{x}-\mathbf{x}_i,\boldsymbol{\beta}). \tag{9.31}$$

Comparing Eqs. (9.30), (9.31) with Eqs. (9.26), (9.27), respectively, lead to establishment of the links inside the inclusion $\mathbf{x} \in v_i$

$$\mathbf{L}_i^{\epsilon}(\mathbf{x}-\mathbf{x}_i) = \mathbf{A}_i(\mathbf{x}) - \mathbf{I}, \;\; \mathbf{L}_i^{\alpha}(\mathbf{x}-\mathbf{x}_i,\boldsymbol{\alpha}) = \mathbf{C}_i^{\epsilon}(\mathbf{x}), \tag{9.32}$$

$$\mathbf{L}_i^{\sigma}(\mathbf{x}-\mathbf{x}_i) = \mathbf{B}_i(\mathbf{x}) - \mathbf{I}, \;\; \mathbf{L}_i^{\beta}(\mathbf{x}-\mathbf{x}_i,\boldsymbol{\beta}) = \mathbf{C}_i(\mathbf{x}), \tag{9.33}$$

which are exact in the framework of the hypothesis **H1a** in contrast to the approximations of perturbations (9.24) and (9.25) generated outside the heterogeneity v_i. Here $\mathbf{L}_i^{\alpha}(\mathbf{x}-\mathbf{x}_i,\boldsymbol{\alpha})$ and $\mathbf{L}_i^{\beta}(\mathbf{x}-\mathbf{x}_i,\boldsymbol{\beta})$ are the strain and stress perturbators produced by the eigenstress and eigenstrain, respectively, which can also be decomposed as

$$\mathbf{L}_i^{\alpha}(\mathbf{x}-\mathbf{x}_i,\boldsymbol{\alpha}) = \mathbf{L}_i^{1\alpha}(\mathbf{x}-\mathbf{x}_i,\boldsymbol{\alpha}) + \mathbf{L}_i^{0\alpha}(\mathbf{x}-\mathbf{x}_i,\boldsymbol{\alpha}), \tag{9.34}$$

$$\mathbf{L}_i^{\beta}(\mathbf{x}-\mathbf{x}_i,\boldsymbol{\beta}) = \mathbf{L}_i^{1\beta}(\mathbf{x}-\mathbf{x}_i,\boldsymbol{\beta}) + \mathbf{L}_i^{0\beta}(\mathbf{x}-\mathbf{x}_i,\boldsymbol{\beta}), \tag{9.35}$$

where $\mathbf{L}_i^{1\alpha}(\mathbf{x}-\mathbf{x}_i,\boldsymbol{\alpha})$ and $\mathbf{L}_i^{1\beta}(\mathbf{x}-\mathbf{x}_i,\boldsymbol{\beta})$ vanish at $\mathbf{M}_1 \equiv \mathbf{0}$, while $\mathbf{L}_i^{0\alpha}(\mathbf{x}-\mathbf{x}_i,\boldsymbol{\alpha})$ and $\mathbf{L}_i^{0\beta}(\mathbf{x}-\mathbf{x}_i,\boldsymbol{\beta})$ describe the perturbations introduced in elastically homogeneous media caused by the transformation field $\boldsymbol{\alpha}_1(\mathbf{x})$ (or $\boldsymbol{\beta}_1(\mathbf{x})$, $\mathbf{x} \in R^d$). Analogous to the general relation between the strain and stress perturbators (9.15), the links between the perturbator factors (9.30) and (9.31) can be established by substitution of Eqs. (9.30), (9.31) and the equality $\overline{\boldsymbol{\sigma}}(\mathbf{x}_i) = \mathbf{L}^{(0)}\overline{\boldsymbol{\varepsilon}}(\mathbf{x}_i) + \boldsymbol{\beta}^{(0)}$ into Eq. (9.15).

Equating the concentrating factors related with the effective strain $\overline{\boldsymbol{\varepsilon}}(\mathbf{x}_i)$ and the constant terms in both sides of the obtained equation, we can find the following links:

$$\mathbf{L}_i^{\sigma}(\mathbf{x}-\mathbf{x}_i)\mathbf{L}^{(0)} = \mathbf{R}_i^{\epsilon v}(\mathbf{x})V_i(\mathbf{x}) + \mathbf{L}^{(0)}\mathbf{L}_i^{\epsilon}(\mathbf{x}-\mathbf{x}_i), \quad (9.36)$$

$$\mathbf{L}_i^{\sigma}(\mathbf{x}-\mathbf{x}_i)\boldsymbol{\beta}^{(0)} + \mathbf{L}_i^{\beta}(\mathbf{x}-\mathbf{x}_i,\boldsymbol{\beta}) = \mathbf{F}_i^{\epsilon v}(\mathbf{x})V_i(\mathbf{x}) + \mathbf{L}_i^{(0)}\mathbf{L}_i^{\alpha}(\mathbf{x}-\mathbf{x}_i,\boldsymbol{\alpha}), \quad (9.37)$$

which are not exact ones (as opposite to Eq. (9.15)) and obtained as the initial representations (9.30) and (9.31) in the framework of the hypothesis **H1a**. In a general case of a noncanonical inhomogeneous inclusion v_q, the hypothesis **H1b** (9.24) is not assumed to be valid in the decompositions (9.30) and (9.31). However, due to the definitions (9.30) and (9.31), for the homogeneous effective fields $\overline{\boldsymbol{\varepsilon}}(\mathbf{x}), \overline{\boldsymbol{\sigma}}(\mathbf{x}) \equiv$const., the perturbator factors $\mathbf{L}_i^{\epsilon}(\mathbf{x}-\mathbf{x}_i)$ and $\mathbf{L}_i^{\sigma}(\mathbf{x}-\mathbf{x}_i)$ can be found by any appropriate numerical method (see, e.g., [1694, 1696]). For example, FEA of a large sample with a single small noncanonical heterogeneity v_q $(L = 40a)$ was used for estimation of the tensor $\mathbf{B}_i(\mathbf{x}) \equiv \mathbf{L}_i^{\sigma}(\mathbf{x}-\mathbf{x}_i) + \mathbf{I}$ at $\mathbf{x} \in v_i$ and $\boldsymbol{\beta}(\mathbf{x}) \equiv \mathbf{0}$ (see for details Chap. 4), and a similar scheme was applied for estimations of $\mathbf{L}_i^{0\alpha}(\mathbf{x}-\mathbf{x}_i,\boldsymbol{\alpha})$ and $\mathbf{L}_i^{0\beta}(\mathbf{x}-\mathbf{x}_i,\boldsymbol{\beta})$ $(\mathbf{x} \in R^d)$ in [233]. We use the same approach for evaluation of $\mathbf{L}_i^{\epsilon}(\mathbf{x}-\mathbf{x}_i)$, $\mathbf{L}_i^{\sigma}(\mathbf{x}-\mathbf{x}_q)$ at $\mathbf{x} \notin v_i$.

In the framework of the hypothesis **H1b**, the perturbator factors presented in the forms (9.34) and (9.35) can be expressed through the tensors $\mathbf{T}_i^{\epsilon}(\mathbf{x}-\mathbf{x}_i)$ and $\mathbf{T}_i^{\sigma}(\mathbf{x}-\mathbf{x}_i)$ (9.25), respectively

$$\mathbf{L}_i^{0\alpha}(\mathbf{x}-\mathbf{x}_i,\boldsymbol{\alpha}) = \bar{v}_i\mathbf{T}_i^{\epsilon}(\mathbf{x}-\mathbf{x}_i)\boldsymbol{\alpha}_1^{(i)}, \quad \mathbf{L}_i^{0\beta}(\mathbf{x}-\mathbf{x}_i,\boldsymbol{\alpha}) = \bar{v}_i\mathbf{T}_i^{\sigma}(\mathbf{x}-\mathbf{x}_i)\boldsymbol{\beta}_1^{(i)}, \quad (9.38)$$

$$\mathbf{L}_i^{1\alpha}(\mathbf{x}-\mathbf{x}_i,\boldsymbol{\alpha}) = \bar{v}_i\mathbf{T}_i^{\epsilon}(\mathbf{x}-\mathbf{x}_i)\mathbf{L}_1^{(1)}\mathbf{C}_i^{\epsilon},$$

$$\mathbf{L}_i^{1\beta}(\mathbf{x}-\mathbf{x}_i,\boldsymbol{\alpha}) = \bar{v}_i\mathbf{T}_i^{\sigma}(\mathbf{x}-\mathbf{x}_i)\mathbf{M}_1^{(1)}\mathbf{C}_i, \quad (9.39)$$

$$\mathbf{L}_i^{\epsilon}(\mathbf{x}-\mathbf{x}_i) = \mathbf{T}_i^{\epsilon}(\mathbf{x}-\mathbf{x}_i)\mathbf{R}_i^{\epsilon}, \quad \mathbf{L}_i^{\sigma}(\mathbf{x}-\mathbf{x}_i) = \mathbf{T}_i^{\sigma}(\mathbf{x}-\mathbf{x}_i)\mathbf{R}_i. \quad (9.40)$$

9.2.2 The Hypotheses H2 and H3 and Some Classical Approaches

For termination of the hierarchy of statistical moment equations (7.42)–(7.45) we will use the closing effective field hypothesis called the "quasi-crystalline" approximation by Lax (1952) **H2** (see also Chap. 8). In the framework of the hypothesis **H1** (combining the hypotheses **H1a** and **H1b**), substitution of the solution (9.26) and (9.27) into the first equations of the systems (7.45) at $n = 1$ and at the effective field hypothesis **H2** leads to the solutions $(\mathbf{x} \in v_i,\ i = 1)$ for the statistical averages of strains and stresses fields and for the effective properties (see Chap. 8)

$$\langle \boldsymbol{\varepsilon}\rangle_i(\mathbf{x}) = \mathbf{A}_i(\mathbf{x})(\mathbf{R}_i^{\epsilon})^{-1}\mathbf{Y}^{\epsilon}\mathbf{R}^{\epsilon}{}_i\langle \boldsymbol{\varepsilon}\rangle + n^{(i)}\mathbf{A}_i(\mathbf{x})\mathbf{P}_i^0\mathbf{Y}^{\epsilon}\mathbf{F}_i^{\epsilon} + \mathbf{C}_i^{\epsilon}(\mathbf{x}), \tag{9.41}$$

$$\mathbf{L}^* = \mathbf{L}^{(0)} + n^{(i)}\mathbf{Y}^{\epsilon}\mathbf{R}_i^{\epsilon}, \tag{9.42}$$

$$\boldsymbol{\alpha}^* = \boldsymbol{\alpha}^{(0)} + n^{(i)}\mathbf{Y}^{\epsilon}\mathbf{F}_i^{\epsilon}, \tag{9.43}$$

$$2W^* = -\langle \boldsymbol{\alpha}_1\mathbf{C}_i^{\epsilon}\rangle - \langle \boldsymbol{\alpha}_1\mathbf{A}_i\rangle\mathbf{P}_i^0\mathbf{Y}^{\epsilon}\mathbf{F}_i^{\epsilon}n^{(i)}, \tag{9.44}$$

$$\langle \boldsymbol{\sigma}\rangle_i(\mathbf{x}) = \mathbf{B}_i(\mathbf{x})\mathbf{R}_i^{-1}\mathbf{Y}\mathbf{R}_i\langle \boldsymbol{\sigma}\rangle + n^{(i)}\mathbf{B}_i(\mathbf{x})\mathbf{Q}_i^0\mathbf{Y}\mathbf{F}_i + \mathbf{C}_i(\mathbf{x}), \tag{9.45}$$

$$\mathbf{M}^* = \mathbf{M}^{(0)} + n^{(i)}\mathbf{Y}\mathbf{R}_i, \tag{9.46}$$

$$\boldsymbol{\beta}^* = \boldsymbol{\beta}^{(0)} + n^{(i)}\mathbf{Y}\mathbf{F}_i, \tag{9.47}$$

$$2W^* = -\langle \boldsymbol{\beta}_1\mathbf{C}_i\rangle - \langle \boldsymbol{\beta}_1\mathbf{B}_i\rangle\mathbf{Q}_i^0\mathbf{Y}\mathbf{F}_i n^{(i)}, \tag{9.48}$$

where one sort of identical aligned heterogeneities was considered and the representations (9.44) and (9.48) were estimated at $\langle \boldsymbol{\varepsilon}\rangle \equiv \mathbf{0}$ and $\langle \boldsymbol{\sigma}\rangle \equiv \mathbf{0}$, respectively. Here matrices $\mathbf{Y}^{\epsilon}$ and $\mathbf{Y}$ determine the actions of the surrounding inclusions on the considered one and have the inverse matrices given by

$$(\mathbf{Y}^{\epsilon})^{-1} = \mathbf{I} - \mathbf{R}_i^{\epsilon}\int \Big[\mathbf{T}_{iq}^{\epsilon}(\mathbf{x}_i - \mathbf{x}_q)\varphi(v_q, \mathbf{x}_q|; v_i, \mathbf{x}_i) - \mathbf{T}_i^{\epsilon}(\mathbf{x}_i - \mathbf{x}_q)n^{(1)}\Big]d\mathbf{x}_q, \tag{9.49}$$

$$(\mathbf{Y}^{-1}) = \mathbf{I} - \mathbf{R}_i\int \Big[\mathbf{T}_{iq}(\mathbf{x}_i - \mathbf{x}_q)\varphi(v_q, \mathbf{x}_q|; v_i, \mathbf{x}_i) - \mathbf{T}_i(\mathbf{x}_i - \mathbf{x}_q)n^{(1)}\Big]d\mathbf{x}_q. \tag{9.50}$$

The solutions (9.41)–(9.48) are obtained by the so-called effective field method (EFM). The general case of the closing hypothesis taking n interacting heterogeneities (defining the MEFM) is considered in Chap. 8.

To make further progress, the hypothesis of *ellipsoidal symmetry* (named as "quasi-ellipsoidal") **H3** (see Chap. 7) is now accepted. It is reasonable to assume that $(\mathbf{a}_{ij}^0)^{-1}$ identifies a matrix of affine transformation that transfers the ellipsoid v_{ij}^0 being the "excluded volume" ("correlation hole") into a unit sphere, and, therefore, the representation of the matrixes $\mathbf{Y}^{\epsilon}$ and $\mathbf{Y}$ can be simplified

$$(\mathbf{Y}^{\epsilon})^{-1} = \mathbf{I} - n^{(1)}\mathbf{R}_i^{\epsilon}\mathbf{P}_i^0, \quad \mathbf{Y}^{-1} = \mathbf{I} - n^{(1)}\mathbf{R}_i\mathbf{Q}_i^0, \tag{9.51}$$

where for the sake of simplicity of the subsequent calculation, we will usually assume that the shape of "correlation hole" v_{ij}^0 does not depend on the inclusion v_j: $v_{ij}^0 = v_i^0$ and $\mathbf{P}_{ij}^0 = \mathbf{P}_i^0 \equiv \mathbf{P}(v_i^0)$, $\mathbf{Q}_{ij}^0 = \mathbf{Q}_i^0 \equiv \mathbf{Q}(v_i^0)$.

As pointed out by Benveniste [89], the essential assumption in the Mori-Tanaka [1093] method (MTM) states that each inclusion v_i behaves as an isolated one in the infinite matrix and subjected to some effective field $\overline{\boldsymbol{\varepsilon}}_i$ ($\overline{\boldsymbol{\sigma}}_i$) coinciding with the average strain (stress) in the matrix $\langle\overline{\boldsymbol{\varepsilon}}_i\rangle = \langle\boldsymbol{\varepsilon}\rangle^{(0)}, \ \langle\overline{\boldsymbol{\sigma}}_i\rangle = \langle\boldsymbol{\sigma}\rangle^{(0)}$. Using the last equations as the closing assumption leads to the following representations for the statistical average of strain and stress fields and for the effective properties (see Chap. 8)

$$\langle\boldsymbol{\varepsilon}\rangle_i(\mathbf{x}) = \mathbf{A}_i(\mathbf{x})[c^{(0)}\mathbf{I} + c^{(1)}\mathbf{A}_i]^{-1}(\langle\boldsymbol{\varepsilon}\rangle - \langle\mathbf{C}^{\epsilon}V\rangle) + \mathbf{C}_i^{\epsilon}(\mathbf{x}), \tag{9.52}$$

$$\mathbf{L}^* = \mathbf{L}^{(0)} + n^{(1)}\mathbf{R}_i^{\epsilon}[c^{(0)}\mathbf{I} + c^{(1)}\mathbf{A}_i]^{-1}, \tag{9.53}$$

$$\boldsymbol{\alpha}^* = \boldsymbol{\alpha}^{(0)} + n^{(1)}\mathbf{F}_i^{\epsilon} - n^{(1)}\mathbf{R}_i^{\epsilon}[c^{(0)}\mathbf{I} + c^{(1)}\mathbf{A}_i]^{-1}\mathbf{C}_i^{\epsilon}, \tag{9.54}$$

$$2W^* = -\langle\boldsymbol{\alpha}_1\mathbf{C}^{\epsilon}\rangle - \langle\boldsymbol{\alpha}_1\mathbf{A}_i\rangle[c^{(0)} + c^{(1)}\mathbf{A}_i]^{-1}\langle\mathbf{C}^{\epsilon}V\rangle, \tag{9.55}$$

$$\langle\boldsymbol{\sigma}\rangle_i(\mathbf{x}) = \mathbf{B}_i(\mathbf{x})[c^{(0)}\mathbf{I} + c^{(1)}\mathbf{B}_i]^{-1}(\langle\boldsymbol{\sigma}\rangle - \langle\mathbf{C}V\rangle) + \mathbf{C}_i(\mathbf{x}), \tag{9.56}$$

$$\mathbf{M}^* = \mathbf{M}^{(0)} + n^{(1)}\mathbf{R}_i[c^{(0)}\mathbf{I} + c^{(1)}\mathbf{B}_i]^{-1}, \tag{9.57}$$

$$\boldsymbol{\beta}^* = \boldsymbol{\beta}^{(0)} + n^{(1)}\mathbf{F}_i - n^{(1)}\mathbf{R}_i[c^{(0)}\mathbf{I} + c^{(1)}\mathbf{B}_i]^{-1}\mathbf{C}_i, \tag{9.58}$$

$$2W^* = -\langle\boldsymbol{\beta}_1\mathbf{C}\rangle - \langle\boldsymbol{\beta}_1\mathbf{B}_i\rangle[c^{(0)} + c^{(1)}\mathbf{B}_i]^{-1}\langle\mathbf{C}V\rangle, \tag{9.59}$$

where the representations (9.55) and (9.59) were estimated at $\langle\boldsymbol{\varepsilon}\rangle \equiv \mathbf{0}$ and $\langle\boldsymbol{\sigma}\rangle \equiv \mathbf{0}$, respectively. For the identical ellipsoidal inhomogeneous heterogeneities v_i homothetical to v_i^0, equivalences of Eqs. (9.41)–(9.43), (9.45)–(9.47) and (9.52)–(9.54), (9.56)–(9.58), respectively, are demonstrated in, e.g., [200]. The equivalences of (9.44), (9.48) and (9.55), (9.59) can be established by substitutions of (9.41), (9.45) and (9.52), (9.56) into Eqs. (9.5$_1$) and (9.5$_2$) at $\langle\boldsymbol{\varepsilon}\rangle \equiv \mathbf{0}$ and $\langle\boldsymbol{\sigma}\rangle \equiv \mathbf{0}$, respectively. However, the representations (9.43)–(9.48) do not coincide with (9.52)–(9.59), respectively, even for the identical aligned isotropic heterogeneities if v_i and v_i^0 are not homothetic (particularly, if v_i is not an ellipsoid, see Chap. 8).

Summarizing the classical approaches, we can formulate the following propositions.

Proposition 9.2.1 *For statistically homogeneous media subjected to the homogeneous boundary conditions, linear elastic effective properties* $\mathbf{M}^*$ *and* $\boldsymbol{\beta}^*$ *depend only on stress distributions inside the inhomogeneities* v *but not inside the matrix* $v^{(0)}$.

Proposition 9.2.2 *If the hypotheses* ***H1****,* ***H2****, and* ***H3*** *hold for the statistically homogeneous medium and homogeneous boundary conditions, then the effective properties* $\mathbf{M}^*$ *and* $\boldsymbol{\beta}^*$ *do not depend on the size of the correlation hole* v_i^0 *and the conditional probability density* $\varphi(v_q, \mathbf{x}_q|; v_i, \mathbf{x}_i)$.

9.3 Estimation of Both the Average Fields and Effective Properties Based on the New GIE (7.19) and (7.20)

9.3.1 General Case of Arbitrary Number of Components

In order to simplify the exact systems for both the strains and stresses (7.21)–(7.23), we accept in this subsection the hypotheses **H1a** and **H2**, while the hypotheses **H1b** and **H3** are not used.

Namely, at first we consider conditional averaging of Eqs. (7.22) and (7.25) at $n = 1$ leading to the following representation for the mean of the effective fields in the fixed inclusion $\mathbf{x} \in v_i$:

$$\langle \overline{\boldsymbol{\vartheta}} \rangle_i(\mathbf{x}) = \langle \boldsymbol{\vartheta} \rangle + \int \Big[\boldsymbol{\mathcal{L}}_q^{\vartheta}(\mathbf{x} - \mathbf{x}_q, \langle \overline{\boldsymbol{\vartheta}} |; v_i, \mathbf{x}_i \rangle_q)(\mathbf{x}_q) \varphi(v_q, \mathbf{x}_q |; v_i, \mathbf{x}_i)$$
$$- \boldsymbol{\mathcal{L}}_q^{\vartheta}(\mathbf{x} - \mathbf{x}_q, \langle \overline{\boldsymbol{\vartheta}} \rangle_q)(\mathbf{x}_q) n^{(q)} \Big] d\mathbf{x}_q, \tag{9.60}$$

hereafter the substitutions $(\boldsymbol{\varepsilon}, \boldsymbol{\sigma}) \to \boldsymbol{\vartheta}$ and $(\boldsymbol{\alpha}, \boldsymbol{\beta}) \to \boldsymbol{\gamma}$ indicate that for each pair $\boldsymbol{\vartheta} = \boldsymbol{\varepsilon}, \boldsymbol{\gamma} = \boldsymbol{\alpha}$ and $\boldsymbol{\vartheta} = \boldsymbol{\sigma}, \boldsymbol{\gamma} = \boldsymbol{\beta}$. Closing of Eq. (9.60) is performed by the use of the hypothesis **H2**

$$\langle \overline{\boldsymbol{\vartheta}} \rangle_i(\mathbf{x}) = \langle \boldsymbol{\vartheta} \rangle + \int \boldsymbol{\mathcal{L}}_q^{\vartheta}(\mathbf{x} - \mathbf{x}_q, \langle \overline{\boldsymbol{\vartheta}} \rangle_q)(\mathbf{x}_q) \varphi(v_q, \mathbf{x}_q |; v_i, \mathbf{x}_i) - n^{(q)}] d\mathbf{x}_q, \tag{9.61}$$

In the case of acceptance of the hypothesis **H1a** in Eq. (9.61), the perturbators $\boldsymbol{\mathcal{L}}_q^{\vartheta}(\mathbf{x} - \mathbf{x}_q, \langle \overline{\boldsymbol{\vartheta}} \rangle_q)$ can be split in the forms (9.30) and (9.31) that reduce Eq. (9.61) to

$$\langle \overline{\boldsymbol{\vartheta}} \rangle_i(\mathbf{x}) = \langle \boldsymbol{\vartheta} \rangle + \int [\mathbf{L}_q^{\vartheta}(\mathbf{x} - \mathbf{x}_q) \langle \overline{\boldsymbol{\vartheta}} \rangle_q(\mathbf{x}_q) + \mathbf{L}_q^{\gamma}(\mathbf{x} - \mathbf{x}_q, \boldsymbol{\gamma})]$$
$$\times [\varphi(v_q, \mathbf{x}_q |; v_i, \mathbf{x}_i) - n^{(q)}] d\mathbf{x}_q, \tag{9.62}$$

Equation (9.62) is reduced to the corresponding equation for the mean of both effective strain and effective stress, respectively. Indeed, introduction of notations for the next integrals

$$\mathbf{I}_{iq}^{\vartheta} = \int \langle \mathbf{L}_q^{\vartheta}(\mathbf{x} - \mathbf{x}_q) \rangle_{(i)} [\varphi(v_q, \mathbf{x}_q |; v_i, \mathbf{x}_i) - n^{(q)}] d\mathbf{x}_q,$$
$$\mathbf{I}_{iq}^{\gamma} = \int \langle \mathbf{L}_q^{\gamma}(\mathbf{x} - \mathbf{x}_q, \boldsymbol{\gamma}) \rangle_{(i)} [\varphi(v_q, \mathbf{x}_q |; v_i, \mathbf{x}_i) - n^{(q)}] d\mathbf{x}_q, \tag{9.63}$$

leads to the compact representations of Eq. (9.62) averaged over the volume of the heterogeneity $v_i \in v^{(i)}$ $(i = 1, \ldots, N)$

$$\langle \overline{\boldsymbol{\vartheta}} \rangle_i = \langle \boldsymbol{\vartheta} \rangle + \sum_{q=1}^{N} \Big[\mathbf{I}^{\vartheta}_{iq} \langle \overline{\boldsymbol{\vartheta}} \rangle_q + \mathbf{I}^{\gamma}_{iq} \Big], \tag{9.64}$$

which allows one to obtain the explicit solution

$$\langle \overline{\boldsymbol{\vartheta}} \rangle_i = \sum_{j=1}^{N} \widehat{\mathbf{Y}}^{\vartheta}_{ij} \langle \boldsymbol{\vartheta} \rangle + \sum_{j,q=1}^{N} \widehat{\mathbf{Y}}^{\vartheta}_{ij} \mathbf{I}^{\gamma}_{jq}, \tag{9.65}$$

where the matrix $(\widehat{\mathbf{Y}}^{\vartheta})^{-1}$ has the following four-order tensor elements $(i, j = 1, \ldots, N)$:

$$(\widehat{\mathbf{Y}}^{\vartheta})^{-1}_{ij} = \delta_{ij} \mathbf{I} - \mathbf{I}^{\vartheta}_{ij}. \tag{9.66}$$

The representations for both concentration factors and effective properties result from Eq. (9.65), in which the inverse substitutions $\boldsymbol{\vartheta} \to (\boldsymbol{\varepsilon}, \boldsymbol{\sigma})$, $\gamma \to (\alpha, \beta)$, respectively, should be performed

$$\langle \boldsymbol{\varepsilon} \rangle_i(\mathbf{x}) = \mathbf{L}^{\epsilon}_i(\mathbf{x} - \mathbf{x}_i, \boldsymbol{\alpha}) + [\mathbf{L}^{\epsilon}_i(\mathbf{x} - \mathbf{x}_i) + \mathbf{I}] \Big[\sum_{j=1}^{N} \widehat{\mathbf{Y}}^{\epsilon}_{ij} \langle \boldsymbol{\varepsilon} \rangle + \sum_{j,q=1}^{N} \widehat{\mathbf{Y}}^{\epsilon}_{ij} \mathbf{I}^{\alpha}_{jq} \Big], \tag{9.67}$$

$$\mathbf{L}^* = \mathbf{L}^{(0)} + \sum_{i,j=1}^{N} n^{(i)} \mathbf{R}^{\epsilon}_i \widehat{\mathbf{Y}}^{\epsilon}_{ij}, \tag{9.68}$$

$$\boldsymbol{\alpha}^* = \boldsymbol{\alpha}^{(0)} + \sum_{i=1}^{N} n^{(i)} \mathbf{F}^{\epsilon}_i + \sum_{i,j,q=1}^{N} n^{(i)} \mathbf{R}^{\epsilon}_i \widehat{\mathbf{Y}}^{\epsilon}_{ij} \mathbf{I}^{\alpha}_{jq}, \tag{9.69}$$

$$2W^* = -\langle \boldsymbol{\alpha}_1 \mathbf{C}^{\epsilon} \rangle - \sum_{i=1}^{N} c^{(i)} \langle \boldsymbol{\alpha}_1 \mathbf{A}_i \rangle_{(i)} \sum_{j,q=1}^{N} \widehat{\mathbf{Y}}^{\epsilon}_{ij} \mathbf{I}^{\alpha}_{jq}, \tag{9.70}$$

$$\langle \boldsymbol{\sigma} \rangle_i(\mathbf{x}) = \mathbf{L}^{\sigma}_i(\mathbf{x} - \mathbf{x}_i, \boldsymbol{\beta}) + [\mathbf{L}^{\sigma}_i(\mathbf{x} - \mathbf{x}_i) + \mathbf{I}] \Big[\sum_{j=1}^{N} \widehat{\mathbf{Y}}^{\sigma}_{ij} \langle \boldsymbol{\sigma} \rangle + \sum_{j,q=1}^{N} \widehat{\mathbf{Y}}^{\sigma}_{ij} \mathbf{I}^{\beta}_{jq} \Big], \tag{9.71}$$

$$\mathbf{M}^* = \mathbf{M}^{(0)} + \sum_{i,j=1}^{N} n^{(i)} \mathbf{R}_i \widehat{\mathbf{Y}}^{\sigma}_{ij}, \tag{9.72}$$

$$\boldsymbol{\beta}^* = \boldsymbol{\beta}^{(0)} + \sum_{i=1}^{N} n^{(i)}\mathbf{F}_i + \sum_{i,j,q=1}^{N} n^{(i)}\mathbf{R}_i\widehat{\mathbf{Y}}^{\sigma}_{ij}\mathbf{I}^{\beta}_{jq}, \tag{9.73}$$

$$2W^* = -\langle\boldsymbol{\beta}_1\mathbf{C}\rangle - \sum_{i=1}^{N} c^{(i)}\langle\boldsymbol{\beta}_1\mathbf{B}_i\rangle_{(i)} \sum_{j,q=1}^{N} \widehat{\mathbf{Y}}^{\sigma}_{ij}\mathbf{I}^{\beta}_{jq}, \tag{9.74}$$

where the representations (9.70) and (9.74) were estimated at $\langle\boldsymbol{\varepsilon}\rangle \equiv \mathbf{0}$ and $\langle\boldsymbol{\sigma}\rangle \equiv \mathbf{0}$, respectively. Note that Eqs. (9.67)–(9.74) are new. It should be mentioned that the domains of the tensors $\mathbf{L}^{\vartheta}_i(\mathbf{x}-\mathbf{x}_i)$ and $\mathbf{L}^{\gamma}_i(\mathbf{x}-\mathbf{x}_i,\boldsymbol{\gamma})$ also include a vicinity of the heterogeneity $\mathbf{x} \notin v_i$ rather than only open domain v_i. Because of this, Eqs. (9.67) and (9.71) can also be used for estimation of both the strains $\langle\boldsymbol{\varepsilon}^-(\mathbf{n})\rangle_{\mathbf{y}}$ and stresses $\langle\boldsymbol{\sigma}^-(\mathbf{n})\rangle_{\mathbf{y}}$ in the vicinity of the heterogeneity near the point $\mathbf{y} \in v_i$ with the unit vector $\mathbf{n}$ outward normal to the heterogeneity boundary ∂v_i. In this case, if $\mathbf{x}$ is placed inside the heterogeneity v_i in Eqs. (9.67) and (9.71), then $\mathbf{L}^{\vartheta}_i(\mathbf{x}-\mathbf{x}_i)$, $\mathbf{L}^{\gamma}_i(\mathbf{x}-\mathbf{x}_i,\boldsymbol{\gamma})$ $(\mathbf{x} \in v_i)$ are presented by Eqs. (9.32), (9.33), and, therefore, $\langle\boldsymbol{\varepsilon}^-(\mathbf{n})\rangle_{\mathbf{y}}$, $\langle\boldsymbol{\sigma}^-(\mathbf{n})\rangle_{\mathbf{y}}$ can be found by the use of a technique of interface boundary operators (see for details Sect. 3.4.2).

It will be demonstrated in the next subsection at the example of identical aligned heterogeneities that Eqs. (9.67)–(9.74) can be reduced (at some particular assumptions) to the known ones.

9.3.2 *Some Particular Cases*

For identical aligned heterogeneities, Eqs. (9.67)–(9.74) are reduced to the following ones ($i \equiv 1$):

$$\langle\boldsymbol{\varepsilon}\rangle_1(\mathbf{x}) = \mathbf{L}^{\alpha}_1(\mathbf{x}-\mathbf{x}_1,\boldsymbol{\alpha}) + [\mathbf{L}^{\epsilon}_1(\mathbf{x}-\mathbf{x}_1)+\mathbf{I}]\widehat{\mathbf{Y}}^{\epsilon}_{11}[\langle\boldsymbol{\varepsilon}\rangle + \mathbf{I}^{\alpha}_{11}], \tag{9.75}$$

$$\mathbf{L}^* = \mathbf{L}^{(0)} + n^{(1)}\mathbf{R}^{\epsilon}_1\widehat{\mathbf{Y}}^{\epsilon}_{11}, \tag{9.76}$$

$$\boldsymbol{\alpha}^* = \boldsymbol{\alpha}^{(0)} + n^{(1)}\mathbf{F}^{\epsilon}_1 + n^{(1)}\mathbf{R}^{\epsilon}_1\widehat{\mathbf{Y}}^{\epsilon}_{11}\mathbf{I}^{\alpha}_{11}, \tag{9.77}$$

$$2W^* = -\langle\boldsymbol{\alpha}_1\mathbf{C}^{\epsilon}_1\rangle - \langle\boldsymbol{\alpha}_1\mathbf{A}_1\rangle\widehat{\mathbf{Y}}^{\epsilon}_{11}\mathbf{I}^{\alpha}_{11}, \tag{9.78}$$

$$\langle\boldsymbol{\sigma}\rangle_1(\mathbf{x}) = \mathbf{L}^{\beta}_1(\mathbf{x}-\mathbf{x}_1,\boldsymbol{\beta}) + [\mathbf{L}^{\sigma}_1(\mathbf{x}-\mathbf{x}_1)+\mathbf{I}]\widehat{\mathbf{Y}}^{\sigma}_{11}[\langle\boldsymbol{\sigma}\rangle + \mathbf{I}^{\beta}_{11}], \tag{9.79}$$

$$\mathbf{M}^* = \mathbf{M}^{(0)} + n^{(1)}\mathbf{R}_1\widehat{\mathbf{Y}}^{\sigma}_{11}, \tag{9.80}$$

$$\boldsymbol{\beta}^* = \boldsymbol{\beta}^{(0)} + n^{(1)}\mathbf{F}_1 + n^{(1)}\mathbf{R}_1\widehat{\mathbf{Y}}^{\sigma}_{11}\mathbf{I}^{\beta}_{11}, \tag{9.81}$$

$$2W^* = -\langle\boldsymbol{\beta}_1\mathbf{C}_1\rangle - \langle\boldsymbol{\beta}_1\mathbf{B}_1\rangle\widehat{\mathbf{Y}}^{\sigma}_{11}\mathbf{I}^{\beta}_{11}. \tag{9.82}$$

In the general case of elastic and transformation mismatches ($\mathbf{M}_1(\mathbf{x}), \boldsymbol{\beta}_1(\mathbf{x}) \not\equiv \mathbf{0}$), Eqs. (9.75), (9.77)–(9.79), (9.81), and (9.82) are new ones. Except for notations,

Eq. (9.80) coincides with the corresponding equation obtained in [232] for the case of the pure mechanical loading ($\boldsymbol{\beta}_1(\mathbf{x}) \equiv \mathbf{0}$). In another limiting case for elastically homogeneous media ($\mathbf{M}_1(\mathbf{x}) \equiv \mathbf{0}$) with the random field of inclusions, Eq. (9.79) is reduced to the appropriate equation proposed in [233]. So, for ceramic materials, a common source of internal stresses is the thermal expansion anisotropy (Al_2O_3) or the thermal mismatch (Si–Si_3N_4) in the components as well as phase transformations (ZrO2) that can produce microcracking along grain boundaries or macrocrack propagation [357, 988]. In the mentioned previous publications, it was demonstrated that even in the referred particular cases for the noncanonical and homogeneous heterogeneities, the estimations obtained by the corresponding new Eqs. (9.75)–(9.82), EFM (9.41)–(9.48), and MTM (9.52)–(9.59) quantitatively differ one from another.

In the case of additional acceptance of the hypotheses **H1b** and **H3**, the integrals (9.63) can be simplified by the use of the Eshelby's [449] solution (see for details [202, 203])

$$\mathbf{I}^{\epsilon}_{iq} = n^{(q)}\mathbf{P}^0_i\mathbf{R}^{\epsilon}_q, \quad \mathbf{I}^{\sigma}_{iq} = n^{(q)}\mathbf{Q}^0_i\mathbf{R}_q, \tag{9.83}$$

$$\mathbf{I}^{\alpha}_{iq} = n^{(q)}\mathbf{P}^0_i\mathbf{F}^{\epsilon}_q, \quad \mathbf{I}^{\beta}_{iq} = n^{(q)}\mathbf{Q}^0_i\mathbf{F}_q, \tag{9.84}$$

that can be easily proved. Indeed, analyzing (9.83_2) as an example, we recast Eq. (9.83_2) in the framework of the hypotheses **H1b** and **H3** in the form

$$\mathbf{I}^{\sigma}_{iq} = -n^{(q)} \int_{v^0_i} \langle \mathbf{L}^{\sigma}_q(\mathbf{x}-\mathbf{x}_q)\rangle_{(i)} d\mathbf{x}_q$$
$$+ \int_{R^d \setminus v^0_i} \langle \mathbf{L}^{\sigma}_q(\mathbf{x}-\mathbf{x}_q)\rangle_{(i)} [\varphi(v_q, \mathbf{x}_q|; v_i, \mathbf{x}_i) - n^{(q)}] d\mathbf{x}_q, \tag{9.85}$$

where excluded volume v^0_i is bounded by the surface $\partial v^0_i = \{\mathbf{x}^0_q : |(\mathbf{a}^0_i)^{-1}\mathbf{x}^0_q|^2 = 1\}$, and $\mathbf{L}^{\sigma}_q(\mathbf{x}-\mathbf{x}_q) = \mathbf{T}^{\sigma}_q\mathbf{R}_q$ (9.40). The second integral in Eq. (9.85) can be considered as a summation of integrals over the layered domain $v^{\rm L}_i$ bounded by the homothetic ellipsoidal surfaces $\partial v^{\rm L+}_i = \{\mathbf{x}_q : |(\mathbf{a}^0_i)^{-1}(\mathbf{x}_q/(L+\Delta L))|^2 = 1\}$ and $\partial v^{\rm L-}_i = \{\mathbf{x}_q : |(\mathbf{a}^0_i)^{-1}(\mathbf{x}_q/L)|^2 = 1\}$ with an infinitesimal $0 < \Delta L \ll 1$ and $L > 1$. Each of these layered integral vanishes due to the Eshelby theorem, and, therefore, the second integral in the right-hand side of Eq. (9.85) vanishes. In a similar manner, thinking of the integral representation of $\mathbf{L}^{\sigma}_q(\mathbf{x}-\mathbf{x}_q)$ (9.66) with an integral variable $\mathbf{y} = \mathbf{x}_q + \boldsymbol{\xi}$ [where $\mathbf{x}+\boldsymbol{\xi} \in v^0_i$ (9.85)], we can change the order of integral variables $\mathbf{x}_q$ and $\boldsymbol{\xi}$. Then exploiting Eshelby's theorem reduces Eq. (9.85) to Eq. (9.83_2). The last statement can also be proved analogously to the scheme proposed in [202] (see Sect. 7.3.2) for the uniform distribution of aligned heterogeneities.

Substitution of the equalities (9.83) and (9.84) into Eqs. (9.66) yields

$$(\widehat{\mathbf{Y}}^{\epsilon})^{-1}_{11} = \mathbf{I} - n^{(1)}\mathbf{P}^0_1\mathbf{R}^{\epsilon}_1, \quad (\widehat{\mathbf{Y}}^{\sigma})^{-1}_{11} = \mathbf{I} - n^{(1)}\mathbf{Q}^0_1\mathbf{R}_1 \tag{9.86}$$

which leads to the relations

$$\widehat{\mathbf{Y}}^{\epsilon}_{11} = (\mathbf{R}^{\epsilon}_1)^{-1}\mathbf{Y}^{\epsilon}\mathbf{R}^{\epsilon}_1, \ \ \widehat{\mathbf{Y}}^{\sigma}_{11} = (\mathbf{R}_1)^{-1}\mathbf{Y}\mathbf{R}_1 \tag{9.87}$$

providing the equivalences of Eqs. (9.76), (9.80) and (9.42), (9.46), respectively, e.g.,

$$\mathbf{M}^{*(9.80)} = \mathbf{M}^{(0)} + n^{(1)}\mathbf{R}_1\widehat{\mathbf{Y}}^{\sigma}_{11} = \mathbf{M}^{(0)} + n^{(1)}\mathbf{Y}^{\sigma}_{11}\mathbf{R}_1 = \mathbf{M}^{*(9.46)}. \tag{9.88}$$

The representation (9.81) is reduced to Eq. (9.47) by the use of Eqs. (9.84_2) and (9.87_2)

$$\begin{aligned}\boldsymbol{\beta}^{*(9.81)} &= \boldsymbol{\beta}^{(0)} + n^{(1)}\mathbf{F}_1 + n^{(1)}\mathbf{R}_1\widehat{\mathbf{Y}}^{\sigma}_{11}\mathbf{Q}^0_1\mathbf{F}_1 n^{(1)} = \boldsymbol{\beta}^{(0)} + n^{(1)}\mathbf{F}_1\\ &+ n^{(1)}\mathbf{Y}^{\sigma}_{11}\mathbf{R}_1\mathbf{Q}^0_1\mathbf{F}_1 n^{(1)} = \boldsymbol{\beta}^{(0)} + n^{(1)}\mathbf{Y}^{\sigma}_{11}\mathbf{F}_1 = \boldsymbol{\beta}^{*(9.47)}.\end{aligned} \tag{9.89}$$

Substitution of Eqs. (9.84_2) and (9.87_2) into Eq. (9.79) at $\mathbf{x} \in v_i$ leads to Eq. (9.45)

$$\begin{aligned}\langle\boldsymbol{\sigma}\rangle_i(\mathbf{x})^{(9.79)} &= \mathbf{C}_1(\mathbf{x}) + \mathbf{B}_1(\mathbf{x})\widehat{\mathbf{Y}}^{\sigma}_{11}[\langle\boldsymbol{\sigma}\rangle + n^{(1)}\mathbf{Q}^0_1\mathbf{F}_1]\\ &= \mathbf{C}_1(\mathbf{x}) + \mathbf{B}_1(\mathbf{x})\mathbf{R}_1^{-1}\mathbf{Y}_{11}\mathbf{R}_1\langle\boldsymbol{\sigma}\rangle + \mathbf{B}_1(\mathbf{x})\widehat{\mathbf{Y}}^{\sigma}_{11}\mathbf{Q}^0_1\mathbf{F}_1 n^{(1)}\\ &= \langle\boldsymbol{\sigma}\rangle_i(\mathbf{x})^{(9.45))}\end{aligned} \tag{9.90}$$

In a similar manner, Eq. (9.82) is reduced to Eq. (9.48)

$$\begin{aligned}2W^{*(9.82)} &= -\langle\boldsymbol{\beta}_1\mathbf{C}_1\rangle - \langle\boldsymbol{\beta}_1\mathbf{B}_1\rangle\widehat{\mathbf{Y}}^{\sigma}_{11}\mathbf{Q}^0_1\mathbf{F}_1 n^{(1)}\\ &= -\langle\boldsymbol{\beta}_1\mathbf{C}_1\rangle - \langle\boldsymbol{\beta}_1\mathbf{B}_1\rangle\mathbf{Q}^0_1\mathbf{Y}_{11}\mathbf{F}_1 n^{(1)} = 2W^{*(9.48)}\end{aligned} \tag{9.91}$$

Equivalencies of Eqs. (9.75)–(9.78) and (9.41)–(9.44) are proved analogously to the schemes of Eqs. (9.88)–(9.91).

It should be mentioned that equivalences of Eqs. (9.45), (9.46) and (9.79), (9.80), respectively, in the framework of assumptions **H1b** and **H3** were only approximately proved for some particular composites with inhomogeneous aligned cylindrical inclusions with $\mathbf{M}_1(\mathbf{x}) \neq \mathbf{0}$, $\boldsymbol{\beta}_1(\mathbf{x}) \equiv \mathbf{0}$ [232] and $\mathbf{M}_1(\mathbf{x}) \equiv \mathbf{0}$, $\boldsymbol{\beta}_1(\mathbf{x}) \neq \mathbf{0}$ [233] ($\mathbf{x} \in v^{(1)}$). It was justified by comparison of the numerical estimations obtained for some particular problems with the analytical ones. However, for the considered cases of homogeneous canonical inclusions, acceptance of the hypothesis **H1a** satisfies the hypothesis **H1b**. Now we proved that if the hypotheses **H1**, **H2**, and **H3** hold, then Eqs. (9.41)–(9.48) and (9.75)–(9.82) are equivalent even for the general cases of inhomogeneity of heterogeneities of an arbitrary shape. In contrast to this statement, we will demonstrate in Sect. 9.5 the quantitative differences of estimations (9.41)–(9.48) and (9.75)–(9.82) if the hypothesis **H1b** does not hold for the case of homogeneous noncanonical heterogeneities. A similar difference was qualitatively evaluated for composites with circular inhomogeneous heterogeneities by the VIE method (see Sect. 9.5.2 and [203]).

9.3.3 *Iteration Method and Abandonment of the Hypothesis H1a*

Equation (9.61) containing a generally inhomogeneous effective field $\langle\overline{\boldsymbol{\vartheta}}\rangle_i(\mathbf{x})$ ($\mathbf{x} \in v_i$) can be solved by the iteration method

$$\langle\overline{\boldsymbol{\vartheta}}^{[n+1]}\rangle_i(\mathbf{x}) = \langle\boldsymbol{\vartheta}\rangle + \int \boldsymbol{\mathcal{L}}_q^{\vartheta}(\mathbf{x}-\mathbf{x}_q, \langle\overline{\boldsymbol{\vartheta}}^{[n]}\rangle_q)(\mathbf{x}_q)[\varphi(v_q, \mathbf{x}_q|; v_i, \mathbf{x}_i) - n^{(q)}]d\mathbf{x}_q. \tag{9.92}$$

An initial approximation of the effective field (9.92) can be chosen by different ways, for example, as either the solution (9.65) obtained in the framework of the hypothesis **H1a** or the classical solution (9.45)

$$\begin{aligned} \langle\overline{\boldsymbol{\vartheta}}^{[0]}\rangle_q &= \langle\overline{\boldsymbol{\vartheta}}^{(9.65)}\rangle_q, \\ \langle\overline{\boldsymbol{\sigma}}^{[0]}\rangle_q &= \mathbf{R}_q^{-1}\mathbf{Y}\mathbf{R}_q\langle\boldsymbol{\sigma}\rangle + n^{(q)}\mathbf{Q}_q^0\mathbf{Y}\mathbf{F}_q, \end{aligned} \tag{9.93}$$

respectively. It suggests the Neumann series form the solution $\langle\overline{\boldsymbol{\vartheta}}\rangle_i(\mathbf{x})$ of Eq. (9.92)

$$\langle\overline{\boldsymbol{\vartheta}}\rangle_i(\mathbf{x}) \equiv \lim_{n\to\infty}\langle\overline{\boldsymbol{\vartheta}}^{[n]}\rangle_i(\mathbf{x}) = \mathbf{D}^{\theta}(\mathbf{x})\langle\boldsymbol{\vartheta}\rangle + \mathbf{D}^{\gamma}(\mathbf{x}). \tag{9.94}$$

Subsequent substitution of the effective field representation (9.94) into Eq. (9.18) allows us to get the final representations for the effective properties

$$\mathbf{M}^* = \mathbf{M}^{(0)} + \langle\mathbf{R}^* V\rangle, \quad \boldsymbol{\beta}^* = \boldsymbol{\beta}^{(0)} + \langle\mathbf{F}^* V\rangle, \tag{9.95}$$

where the macroscopic concentration factors $\mathbf{R}^*(\mathbf{x})$ and $\mathbf{F}^*(\mathbf{x})$ are expressed through the found concentration factor of the effective field (9.94) ($\mathbf{x} \in v_i$)

$$\mathbf{R}^*(\mathbf{x}) = \boldsymbol{\mathcal{L}}_i * (\mathbf{E}_i\mathbf{D}^{\sigma})(\mathbf{x}), \quad \mathbf{F}^*(\mathbf{x}) = \boldsymbol{\mathcal{L}}_i * (\mathbf{E}_i\mathbf{D}^{\beta} + \mathbf{H}_i)(\mathbf{x}).$$

Particular cases of the iteration method (9.92)–(9.95) with the corresponding numerical results were considered for 2D CMs reinforced by either the circle inhomogeneous fibers [203] or inclusions of noncanonical shape [235] (see Sect. 9.5.2).

9.3.4 Elastically Homogeneous Medium with Random Residual Stresses

We consider a linearly elastic composite medium, which consists of a homogeneous matrix containing a statistically inhomogeneous random set of noncanonical inclusions. The elastic properties of the matrix and the inclusions are the same ($\mathbf{M}(\mathbf{x}) \equiv \mathbf{M}^{(0)}$), but the stress-free strains are different ($\boldsymbol{\beta}(\mathbf{x}) \not\equiv$const. (see, e.g., [806]). Of course, the first and second moments of residual stresses in composite components can be estimated with the help of passing to the zero limit of the elastic mismatch of different components in the corresponding formulae (e.g., (9.79)). Nevertheless, the desired relationships can be found immediately without any assumptions (see [186]).

Indeed, Eq. (9.60) the perturbators $\mathcal{L}_q^{\vartheta}(\mathbf{x}-\mathbf{x}_q, \langle\overline{\boldsymbol{\vartheta}}|; v_i, \mathbf{x}_i\rangle_q)(\mathbf{x}_q)$ and $\mathcal{L}_q^{\vartheta}(\mathbf{x}-\mathbf{x}_q, \langle\overline{\boldsymbol{\vartheta}}\rangle_q)(\mathbf{x}_q)$ do not depend on the effective field $\langle\overline{\boldsymbol{\vartheta}}|; v_i, \mathbf{x}_i\rangle_q$ and $\langle\overline{\boldsymbol{\vartheta}}\rangle_q$, respectively. Therefore, the splitting Eqs. (9.30) and (9.31) exactly hold (without hypothesis **H1a**) that leads Eq. (9.60) to the exact equation ($\mathbf{x} \in v_i$)

$$\langle\overline{\boldsymbol{\vartheta}}\rangle_i(\mathbf{x}) = \langle\boldsymbol{\vartheta}\rangle(\mathbf{x}) + \int \mathcal{L}_q^{\gamma}(\mathbf{x}-\mathbf{x}_q, \boldsymbol{\gamma})[\varphi(v_q, \mathbf{x}_q|; v_i, \mathbf{x}_i) - \varphi(v_q, \mathbf{x}_q)]\, d\mathbf{x}_q,$$

$$\langle\boldsymbol{\vartheta}\rangle_i(\mathbf{x}) = \langle\overline{\boldsymbol{\vartheta}}\rangle_i(\mathbf{x}) + \mathcal{L}_i^{\gamma}(\mathbf{x}-\mathbf{x}_i, \boldsymbol{\gamma}), \tag{9.96}$$

where, in particular, the perturbators $\mathcal{L}_q^{\gamma}(\mathbf{x}-\mathbf{x}_q, \boldsymbol{\gamma})$ can be expressed through the Green functions

$$\mathcal{L}_q^{\gamma}(\mathbf{x}-\mathbf{x}_q, \boldsymbol{\gamma}) = \int \mathbf{U}^{\gamma}(\mathbf{x}-\mathbf{y})\boldsymbol{\gamma}_1(\mathbf{y})V_q(\mathbf{y})\, d\mathbf{y}, \tag{9.97}$$

with $\mathbf{U}^{\gamma} = \mathbf{U}, \boldsymbol{\Gamma}$ at $\boldsymbol{\gamma} = \boldsymbol{\alpha}, \boldsymbol{\beta}$, respectively.

For an estimation of integrals in Eq. (9.96$_1$) we need to estimate $\mathcal{L}_r^{\gamma r}(\mathbf{x}^r - \mathbf{x}_i, \boldsymbol{\gamma})$ just a single time for a representative inclusion v_i in the representative volume element (RVE) $\mathbf{x} \in D = \{\mathbf{x}^r||\mathbf{x}^r - \mathbf{x}_i| < a_r\}$ of the radius a^r which is defined by a condition that $\varphi(v^r, \mathbf{x}^r|; v_i, \mathbf{x}_i) - \varphi(v^r, \mathbf{x}^r)$ is negligible (in the sense of influence on the local stresses) at $\mathbf{x}^r \notin D$. Indeed, in such a case, any $\mathcal{L}_q^{\gamma}(\mathbf{x}-\mathbf{x}_q, \boldsymbol{\gamma})$ appearing in the integrals in Eq. (9.96$_1$) can be estimated as

$$\mathcal{L}_q^{\gamma}(\mathbf{x}-\mathbf{x}_q, \boldsymbol{\gamma}) = \mathcal{L}_q^{\gamma r}(\mathbf{x}^r - \mathbf{x}_i, \boldsymbol{\gamma}), \tag{9.98}$$

where we need to substitute $\mathbf{x}-\mathbf{x}_q := \mathbf{x}^r - \mathbf{x}_i$.

Thus, we can see from Eqs. (9.96) that the first moment of stresses inside the inclusions is defined by the tensors $\mathcal{L}_q^{\gamma}(\mathbf{x}-\mathbf{x}_q, \boldsymbol{\gamma})$ describing the solution for a single inclusion inside a homogeneous full space R^d as well as by the one-point $\varphi(v_q, \mathbf{x}_q)$ and two-point $\varphi(v_q, \mathbf{x}_q|; v_i, \mathbf{x}_i)$ probability densities. So significant simplicity of the thermoelastic problem defining only by the solution for one isolated inclusion in the whole space R^d is explained by the fact that for elastically homogeneous media, the perturbations produced by the different inclusions do not interact with one another and can be evaluated by a simple summation. However, correctness of the mentioned simple superposition technique for a finite inclusion number is not appropriate for their direct utilization for an infinite number of inclusions. Nevertheless modified superposition scheme used at the obtaining of Eq. (9.96_1) is prevented from this drawback.

Equations (9.96) are a straightforward consequence of the new GIE (7.19) and (7.20). Now we consider the results of the classical GIE (7.33) in a particular case $\mathbf{M}(\mathbf{x}) \equiv \mathbf{M}^{(0)}$, $\boldsymbol{\vartheta} = \boldsymbol{\sigma}, \boldsymbol{\gamma} = \boldsymbol{\beta}$, and the perturbator representation (9.97)

$$\langle \overline{\boldsymbol{\sigma}} \rangle_i(\mathbf{x}) = \langle \boldsymbol{\sigma} \rangle(\mathbf{x}) + \int \Big[\boldsymbol{\mathcal{L}}_q^{\beta}(\mathbf{x} - \mathbf{x}_q, \boldsymbol{\beta}) \varphi(v_q, \mathbf{x}_q | ; v_i, \mathbf{x}_i) - \boldsymbol{\Gamma}(\mathbf{x} - \mathbf{x}_q) \langle \boldsymbol{\beta}_1 \rangle (\mathbf{x}_q) \Big] \, d\mathbf{x}_q,$$

$$\langle \boldsymbol{\vartheta} \rangle_i(\mathbf{x}) = \langle \overline{\boldsymbol{\vartheta}} \rangle_i(\mathbf{x}) + \boldsymbol{\mathcal{L}}_i^{\gamma}(\mathbf{x} - \mathbf{x}_i, \boldsymbol{\gamma}). \tag{9.99}$$

By virtue of the fact that generalized function $\boldsymbol{\Gamma}(\mathbf{x})$ is an even homogeneous function of order $-d$, we have a relation

$$\int \Big[\boldsymbol{\mathcal{L}}_q^{\beta}(\mathbf{x} - \mathbf{x}_q, \boldsymbol{\beta}) \varphi(v_q, \mathbf{x}_q | ; v_i, \mathbf{x}_i) - \boldsymbol{\Gamma}(\mathbf{x} - \mathbf{x}_q) \langle \boldsymbol{\beta}_1 \rangle \Big] \, d\mathbf{x}_q = \mathbf{Q}_i^0 \langle \boldsymbol{\beta} \rangle \tag{9.100}$$

which is valid in the framework of the hypotheses **H1** and **H3**, and $\mathbf{Q}_i^0 = \mathbf{Q}_i^0(v_i^0)$ is associated with the ellipsoidal correlation hole v_i^0. Thus, hypothesis **H2** allows us to obtain from Eq. (9.99) satisfiability of the effective field hypothesis **H1** with the following representation for the effective field (called the effective field method, EFM):

$$\langle \overline{\boldsymbol{\sigma}} \rangle_i = \langle \boldsymbol{\sigma} \rangle + \mathbf{Q}_i^0 \langle \boldsymbol{\beta}_1 \rangle, \tag{9.101}$$

which is exact in the framework of the accepted hypothesis **H3** for statistically homogeneous medium.

Mori-Tanaka [1093] approach does not use the hypothesis **H3**, but an additional closing assumption (8.68) provides a closed representation [called the Mori-Tanaka method (MTM)] ($\mathbf{x} \in v_k \subset v^{(k)}$)

$$\langle \overline{\boldsymbol{\sigma}} \rangle_i = \langle \boldsymbol{\sigma} \rangle - \sum_{k=1}^{N} c^{(k)} \langle \boldsymbol{\mathcal{L}}_k^{\beta}(\mathbf{x} - \mathbf{x}_k, \boldsymbol{\beta}) \rangle_{(k)}. \tag{9.102}$$

For aligned identical homogeneous ellipsoidal inclusions, the relations (9.101) and (9.102) coincide if v_{ij}^0 is homothetic to v_i. However, they are different for either the inhomogeneous ellipsoidal inclusions or noncanonical shape of inclusions. Indeed, the effective fields (9.101) and (9.102) depend on the averaged eigenstrain $\langle \boldsymbol{\beta}_1 \rangle$ and the averaged perturbator $\langle \boldsymbol{\mathcal{L}}_k^{\beta}(\mathbf{x} - \mathbf{x}_k) \rangle_{(k)}$, respectively, which are not reduced one to another for the inhomogeneous inclusions (even in the case of inclusions of ellipsoidal shape).

For accomplishment of this subsection, we estimate the stored energy from $W^* = -\langle \boldsymbol{\beta}_1 \boldsymbol{\sigma}\rangle/2$ (at $\langle \boldsymbol{\sigma}\rangle \equiv \mathbf{0}$) and (9.96)

$$W^* = -\frac{1}{2}\sum_{i=1}^{N}\Big\{c^{(i)}\langle \boldsymbol{\beta}_1(\mathbf{x})\boldsymbol{\mathcal{L}}_i^{\beta}(\mathbf{x}-\mathbf{x}_i,\boldsymbol{\beta})\rangle_i + \int\int V_i(\mathbf{x})\boldsymbol{\beta}_1(\mathbf{x})\boldsymbol{\mathcal{L}}_q^{\beta}(\mathbf{x}-\mathbf{x}_q,\boldsymbol{\beta})$$
$$\times[\varphi(v_q,\mathbf{x}_q|;v_i,\mathbf{x}_i) - \varphi(v_q,\mathbf{x}_q)]\,d\mathbf{x}_q d\mathbf{x}\Big\}. \tag{9.103}$$

Exact Eq. (9.103) was obtained from the GIE (7.20). In the case of homogeneous inclusions ($\boldsymbol{\beta}_1(\mathbf{x}) \equiv$const., $\mathbf{x} \in v^{(i)}$), the hypothesis **H1b** is automatically fulfilled that yields an equivalence of the source GIEs (7.20) and (9.40). This equivalence reduces Eq. (9.103) to the corresponding equation in [186] and [200]. In particular, one may consider a grain structure where the correlation of homogeneous transformations $\boldsymbol{\beta}^{(i)}(\mathbf{x}) \equiv$const. ($\mathbf{x} \in v^{(i)}$) between different grains is lacking [807]. Moreover, we accept the ellipsoidal symmetry hypothesis **H3** (8.45) with Eq. (8.48) that reduces Eq. (9.103) to

$$W^* = \frac{1}{2}\langle(\boldsymbol{\beta} - \langle\boldsymbol{\beta}\rangle)\mathbf{Q}(\boldsymbol{\beta} - \langle\boldsymbol{\beta}\rangle)\rangle \tag{9.104}$$

obtained in [807] (see also [186, 805]).

9.3.5 Comparisons with Some Exact Results

It is interesting to note that along with the classical exact Levin's formula for two phase-composites presented in the form (6.106)

$$\boldsymbol{\beta}^* = \boldsymbol{\beta}^{(0)} + (\mathbf{M}^* - \mathbf{M}^{(0)})(\mathbf{M}_1^{(1)})^{-1}\boldsymbol{\beta}_1^{(1)}, \tag{9.105}$$

it is also known a generalization of Eq. (9.105) to the case of composites with the aligned identical heterogeneities of an arbitrary microstructure

$$\boldsymbol{\beta}^* = \boldsymbol{\beta}^{(0)} + (\mathbf{M}^* - \mathbf{M}^{(0)})\mathbf{R}_1^{-1}\mathbf{F}_1. \tag{9.106}$$

The formula (9.106) was obtained (see Eq. (8.20)) in the framework of hypothesis **H1** (combining the hypotheses **H1a** and **H1b**) only. No restrictions are imposed on the concrete form of the tensor $\widehat{\mathbf{Y}}_{11}$, on the microtopology of the coated inclusions or on the inhomogeneity of the stress state in the heterogeneities. In particular, for homogeneous ellipsoidal inclusions, the tensors $\mathbf{R}_1$ and $\mathbf{F}_1$ are determined by the Eshelby solutions that reduces Eq. (9.106) to Eq. (9.105). In doing so, analog of

Eq. (9.106) can be obtained from the new Eqs. (9.80) and (9.81)

$$\boldsymbol{\beta}^* = \boldsymbol{\beta}^{(0)} + n^{(1)}\mathbf{F}_1 + (\mathbf{M}^* - \mathbf{M}^{(0)})\mathbf{I}^{\beta}_{11}, \tag{9.107}$$

which is reduced to Eq. (9.106) only in the framework of the hypothesis **H1b**.

The formulae (9.105)–(9.107) establish the link between effective transformation parameter $\boldsymbol{\beta}^*$ and effective mechanical parameter $\mathbf{M}^*$. Their specific counterparts are the links between the concentration factors of both the pure mechanical fields $\mathbf{B}^*(\mathbf{x})$ and transformation ones $\mathbf{C}^*(\mathbf{x})$ (see Eq. (9.3)). So, for two-phase composites, the brilliant exact formulae by Benveniste and Dvorak [96] determine the link between the random mechanical and transformation factors in the equation $\boldsymbol{\sigma}(\mathbf{x}) = \mathbf{B}^{*\mathrm{r}}(\mathbf{x})\langle\boldsymbol{\sigma}\rangle + \mathbf{C}^{*\mathrm{r}}(\mathbf{x})$ $(\mathbf{x} \in v^{(1)})$

$$\mathbf{C}^{*\mathrm{r}}(\mathbf{x}) = (\mathbf{B}^{*\mathrm{r}}(\mathbf{x}) - \mathbf{I})(\mathbf{M}_1^{(1)})^{-1}\boldsymbol{\beta}_1^{(1)}, \tag{9.108}$$

rather than the relation between its statistical averages $\mathbf{B}^*(\mathbf{x}) \equiv \langle\mathbf{B}^{*\mathrm{r}}\rangle(\mathbf{x})$ and $\mathbf{C}^*(\mathbf{x}) \equiv \langle\mathbf{C}^{*\mathrm{r}}\rangle(\mathbf{x})$. For composites with identical aligned heterogeneities of an arbitrary microstructure, a counterpart of Eq. (9.106) was also obtained (see Chap. 8) in the framework of hypothesis **H1** only $(\mathbf{x} \in v^{(1)})$

$$\mathbf{C}^*(\mathbf{x}) = \mathbf{C}_i(\mathbf{x}) + [\mathbf{B}^*(\mathbf{x}) - \mathbf{B}_i(\mathbf{x})]\mathbf{R}_i^{-1}\mathbf{F}_i. \tag{9.109}$$

The relation (9.79) yields the equation

$$\mathbf{C}^*(\mathbf{x}) = \mathbf{C}_i(\mathbf{x}) + \mathbf{B}^*(\mathbf{x})\mathbf{I}^{\beta}_{11}, \tag{9.110}$$

where the stress concentration factor $\mathbf{B}^*(\mathbf{x}) = \mathbf{B}_i(\mathbf{x})\widehat{\mathbf{Y}}^{\sigma}_{11}$ $(\mathbf{x} \in v^{(1)})$ was obtained in the framework of the hypotheses **H1a** and **H2**. If in Eq. (9.110) the hypotheses **H1b** and **H3** are additionally accepted, then Eq. (9.110) is reduced to Eq. (9.109) which in turn yields the averaged Eq. (9.108) in the case of homogeneous ellipsoidal heterogeneities.

It should be mentioned that Eqs. (9.105) and (9.108) are exact. Consideration of identical aligned heterogeneities of an arbitrary microstructure in Eqs. (9.106), (9.107), (9.109), and (9.110) requires acceptance of additional assumptions such as **H1** in Eqs. (9.106) and (9.109) as well as the hypotheses **H1a** and **H2** in Eqs. (9.107) and (9.110). Formally we need to know the transformation concentration factors in formulae (9.106), (9.107), (9.109), and (9.110). However, we need to know these functions only inside heterogeneity v_1 in Eqs. (9.106) and (9.109), while in the case of Eq. (9.107) and (9.110), the transformation perturbator should be also known in the vicinity of v_i, and, moreover, the integral $\mathbf{I}^{\beta}_{11}$ (9.107) in (9.110) depends on the conditional probability density $\varphi(v_q, \mathbf{x}_q|; v_1, \mathbf{x}_1)$. It would be interesting to obtain a generalization of Eqs. (9.107) and (9.109) to any closing hypothesis **H2** [as in Eqs. (9.106) and (9.109)] containing any number of interacting heterogeneities. However, more detailed analysis of this problem is beyond the scope of the current presentation.

It is important to stress that if the problem for pure mechanical loading ($\boldsymbol{\beta}_1(\mathbf{x}) \equiv \mathbf{0}$) is exactly solved, and $\mathbf{M}^*$ and $\mathbf{B}^{*\mathrm{r}}(\mathbf{x})$ are exactly found, then the corresponding problem for the residual stresses is also exactly solved, i.e., $\boldsymbol{\beta}^*$ and $\mathbf{C}^{*\mathrm{r}}(\mathbf{x})$ are exactly found. However, the mentioned pure mechanical problem, in general, cannot be exactly solved even for aligned identical ellipsoidal inclusions. Therefore, correctness of the exact representations $\boldsymbol{\beta}^*$ (9.105) and $\mathbf{C}^*(\mathbf{x})$ (9.108) is in general questionable if $\mathbf{M}^*$ and $\mathbf{B}^*(\mathbf{x})$ are evaluated in the framework of additional assumptions. Indeed, the effective properties of residual stresses $\boldsymbol{\beta}^*$ and $\mathbf{C}^*(\mathbf{x})$ should be estimated through the approximation of $\mathbf{M}^*$ and $\mathbf{B}^*(\mathbf{x})$ in the framework of the same assumptions implementing the corresponding representations, namely, Eqs. (9.106), (9.109) or Eqs. (9.107), (9.110).

9.4 Qualitative Analysis of the New GIE, Some Classical Hypotheses, and Propositions

9.4.1 Analysis of Proposition 9.2.1 and Hypothesis H1

In the light of the analyzed approximations (7.47) and (7.48), it is possible to observe an additional unexpected advantage of the operator form of the GIEs (9.60) with respect to its Green's function counterparts (7.20). At first, we should indicate on some ambiguous interpretation related with separation of the microcoordinate $\mathbf{y}$ (7.36) (or $\mathbf{x}_q$ (9.60)) and macrocoordinate (with resolution equal Λ). It is not critical in the case of using rough methods making it possible to estimate only average fields inside inclusions (e.g., $\langle\boldsymbol{\sigma}\rangle_i(\mathbf{y}) = \langle\boldsymbol{\sigma}\rangle^{(k)}$, $\mathbf{y} \in v_i \subset v^{(k)}$). However, more advanced methods (see, e.g., Sect. 9.5) require an advanced technique adapted for separate manipulations with both the micro- and macrocoordinates. It is interesting that the required separation feature is intrinsic in the introduced unique object $\mathcal{L}_q^{\tau}(\mathbf{x}-\mathbf{x}_q, \boldsymbol{\tau})$ (or $\mathcal{L}_i^{\eta}(\mathbf{x}-\mathbf{x}_i, \boldsymbol{\eta})$) because the macrocoordinate $\mathbf{x}$ is presented, (e.g., in Eqs. (9.60), while the microcoordinate $\mathbf{y} \in v_q$ is completely concealed inside $\mathcal{L}_q^{\tau}(\mathbf{x}-\mathbf{x}_q, \boldsymbol{\tau})$ (or $\mathcal{L}_q^{\eta}(\mathbf{x}-\mathbf{x}_q, \boldsymbol{\eta})$). Moreover, the second advantage of Eqs. (9.60) is a psychological one and appearing due to the fact that the concrete forms of operator representations of the perturbators (9.6_1) and (9.7_1) are a priori unknown and may be different from their representations through Green's functions (9.8) and (9.9), and, therefore, there is no temptation to separate the deterministic functions $\mathbf{U}, \boldsymbol{\Gamma}$ and the random ones $\boldsymbol{\tau}$ and $\boldsymbol{\eta}$, respectively, as in the approximations (7.32) and (7.33) ironically recognized as the exact equations.

Moreover, the approximations (7.47) and (7.48) are not just the abstract hypotheses considered for establishment of the difference between the new GIEs (7.19), (7.20) and the classical ones (7.32), (7.33), respectively. Equations (7.47) and (7.48) can be considered as a some sort of a test site for discovery of fundamentally new results in the following sense. If the assumption (7.47) and (7.48) is exactly fulfilled in the framework of both the other hypotheses and methods used, then GIEs (7.19), (7.20) and (7.32), (7.33), respectively, lead to identical estimations of effective material and field parameters. Otherwise, violation of the approximation (7.47) and (7.48) can lead to the large difference between the new (7.19), (7.20) and old (7.32), (7.33) approaches with the possible change of sign of the predicted local statistical average fields $\langle\boldsymbol{\sigma}\rangle_i(\mathbf{x})$ ($\mathbf{x} \in v_i$; see for details Sect. 9.5).

It should be mentioned that the domain of the perturbator operator $\mathcal{L}_q^{\eta}(\mathbf{x} - \mathbf{x}_q, \boldsymbol{\eta})(\mathbf{x})$ (9.7$_1$) is a whole space $\mathbf{x} \in R^d$, and, because of this, some points of the area $\mathbf{x} \in v_i$ in (9.61) can be uncovered by the heterogeneities v_q, and, therefore, the effective stress $\langle\overline{\boldsymbol{\sigma}}\rangle_i(\mathbf{x})$ (9.61) will depend on the stress perturbations $\mathcal{L}_q^{\eta}(\mathbf{x} - \mathbf{x}_q, \boldsymbol{\eta})(\mathbf{x})$ in the vicinity $\mathbf{x} \in v_i \setminus v_q$ of the area v_q rather than only on stress distributions in the "moving" inhomogeneity v_q and $\max_x|\mathbf{x} - \mathbf{x}_q| = 3a$ ($\mathbf{x} \in v_i \setminus v_q$; see Fig. 9.1) for the identical spherical inhomogeneities of the radius a with an isotropic statistically distribution of their centers with the radius $2a$ of the correlation hole v_i^0. Thus, we obtain a fundamental conclusion that effective moduli in general depend not only on the stress distribution inside the inhomogeneities but also on the stresses in the vicinity of inhomogeneities (compare with Proposition 9.2.1). However, if our estimations utilize Eq. (7.33) containing only average strain polarization $\boldsymbol{\eta}_q$ rather than $\langle\boldsymbol{\Gamma}(\mathbf{x} - \mathbf{y})\boldsymbol{\eta}(\mathbf{y})\rangle_q$ in Eq. (7.20) as a renormalizing item, then an influence of stresses in the vicinities of inhomogeneities is degenerated. At the same time, using Eq. (9.61) leads to the necessity of evaluation of stresses in the inhomogeneity vicinities even for a statistically homogeneous field of ellipsoidal homogeneous inclusions (it will be quantitatively demonstrated in Sect. 9.5). Moreover, a fundamental deficiency of Eq. (7.33) is the dependence of the renormalizing item $\boldsymbol{\Gamma}(\mathbf{x} - \mathbf{x}_q)\langle\boldsymbol{\eta}\rangle$ only on the average stress polarization tensor $\langle\boldsymbol{\eta}\rangle$, while a corresponding item $\langle\boldsymbol{\Gamma}(\mathbf{x} - \mathbf{y})\boldsymbol{\eta}(\mathbf{y})\rangle_q(\mathbf{x}_q)$ ($\mathbf{y} \in \mathbf{x}_q$) in the new Eq. (9.61) explicitly depends on details distribution $\boldsymbol{\eta}(\mathbf{y})$ ($\mathbf{y} \in \mathbf{x}_q$). Because of this, the averaging methods used in Eq. (7.33) as a starting element conserve the mentioned deficiency of Eq. (7.33) (at least in some elements of these methods). For example, Chen and Acrivos [305] have estimated the effective elastic moduli through the accurate evaluation of binary interactions of inclusions without the hypothesis **H1**, i.e., in our notations the operator $\mathcal{L}_q^{\sigma}(\langle\overline{\boldsymbol{\sigma}}|; v_i, \mathbf{x}_i\rangle_q)(\mathbf{x})$ (9.61) was estimated at the condition $\langle\widetilde{\boldsymbol{\sigma}}_{i,q}\rangle \equiv \langle\boldsymbol{\sigma}\rangle$ [compare with Eq. (8.43$_2$)]. However, the

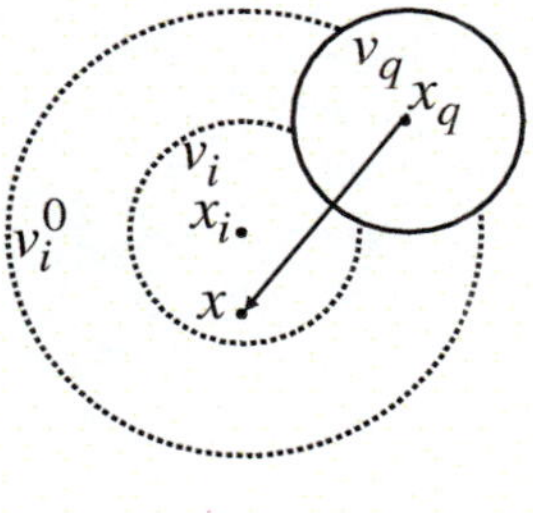

Fig. 9.1 Schematic mutual placements of the domains v_q, v_i and v_i^0

operator $\mathcal{L}_q^{\sigma}(\langle\overline{\boldsymbol{\sigma}}\rangle)(\mathbf{x})$ (9.60) was estimated in [305] at the approximation (9.24) that implicitly implies the use of both the hypothesis **H1b** and Eq. (7.33) (see for details Subsection 10.2.2 in [200]).

From the other hand, although the method (9.45) allows the inhomogeneous statistically averaged tensor $\langle\boldsymbol{\sigma}\rangle_i(\mathbf{y})$ $(\mathbf{y} \in v_i)$, the effective field described by the tensors $\mathbf{Y}$ and $\mathbf{Y}^{\epsilon}$ (9.51) depends only on the average concentration factor $\mathbf{R}$ and $\mathbf{R}^{\epsilon}$, while the effective field in Eq. (9.65) explicitly depends on the inhomogeneous tensors $\mathbf{A}_i(\mathbf{x})$ and $\mathbf{B}_i(\mathbf{x})$ [see Eqs. (9.32) and (3.175)] The detected explicit dependence of the effective properties (9.80) on the detailed concentration factors $\mathbf{A}_i(\mathbf{x})$ and $\mathbf{B}_i(\mathbf{x})$ rather than on their average values allows for an abandonment of the hypothesis **H1b** (9.24) whose accuracy is questionable for the inhomogeneous (e.g., coated) inclusions. In such a case, the statistical average effective field estimated by Eq. (9.95) is found to be inhomogeneous that discards the hypothesis **H1a**. Quantitative estimations of the result of this abandonment of the hypothesis **H1** will be performed in Sect. 9.5 in the framework of the hypothesis **H3** for some particular cases of fiber composites.

9.4.2 *Analysis of Proposition 9.2.2 and Hypotheses H1b and H3*

As was mentioned, addition of the effective filed hypothesis **H1a** by the hypothesis **H1b** (9.24) leads to Eq. (9.51) providing homogeneity of the effective field estimation for the ellipsoidal excluded volume v_i^0. Moreover this estimation of the effective field (and, therefore, of effective moduli) is invariant with respect to the size of the ellipsoidal excluded volume v_i^0. However, the additional hypothesis **H1b** (9.24) is exactly fulfilled only for the homogeneous ellipsoidal inhomogeneity v_i. For either the inhomogeneous inclusion or its nonellipsoidal shape the equality **H1b** (9.24) is just an approximation, and the new general equation (7.20) has an advantage with respect to the popular o (7.33) only based on average strain polarization $\boldsymbol{\eta}_q$ [rather than $\langle\boldsymbol{\Gamma}(\mathbf{x}-\mathbf{y})\boldsymbol{\eta}(\mathbf{y})\rangle_q$]. Then the size of the excluded volume v_i^0 will impact on the effective field (9.65). Indeed, if the radius of the excluded volume v_i^0 in Fig. 9.1 increases from $2a$ to $3a$, then the long distance of influence zone of the inhomogeneity v_q on the effective field $\langle\overline{\boldsymbol{\sigma}}\rangle_i(\mathbf{x})$ will increase from the value $|\mathbf{x}-\mathbf{x}_q| = 3a$ (as in Fig. 9.1) till $|\mathbf{x}-\mathbf{x}_q| = 4a$. This influence of the size of v_i^0 as well as the conditional probability density $\varphi(v_q, \mathbf{x}_q|; v_1, \mathbf{x}_i)$ will be quantitatively estimated in the next subsection, i.e., Proposition 9.2.2 is, in general, incorrect in the case of exploiting of the new GIE (7.20) if either the hypothesis **H1b** or **H3** is abandoned.

A popular explanation of acceptance of the "ellipsoidal symmetry" hypothesis **H3** (see Chap. 8) is that this hypothesis just simplifies Eq. (9.50) reducing this equation to Eq. (9.51) which does not contain the integrals. In a similar manner, a destination of the assumption of the ellipsoidal shape of the excluded volume v_i^0

in the hypothesis **H3** is that this hypothesis just simplifies Eq. (9.100) by the use of analytical known tensor $\mathbf{Q}_i^0$ (expressed through the Eshelby tensor $\mathbf{S}_i^0$) which is exploited instead of a general tensor $\mathbf{Q}_i^0(\mathbf{x})$ found numerically [see, e.g., Sect. 4.4.4]. However, both mentioned assumptions of the hypothesis **H3** have a fundamental conceptual sense rather than only an analytical solution of some particular problem. Exploiting the Eshelby tensor concept in the EFM (9.46) (and in the MEFM) is based on the ellipsoidal shape of the correlation hole v_i^0 rather than on the inclusion shape v_i. An abandonment of either the assumption of the v_i^0's ellipsoidal shape or "ellipsoidal symmetry" hypothesis will necessarily lead to the inhomogeneity of the effective field $\overline{\boldsymbol{\sigma}}_i$ acting on the inclusion v_i that is prohibited for the classical version of the EFM (and MEFM). However, Buryachenko (see Section 9.4 in [200]) proposed a method for a solution based on the general integral equation (7.33) without assumption of the ellipsoidal shape of the excluded volume v_i^0. In such a case, the effective field concentration factor $\mathbf{Y}$ (9.51) has a form

$$\mathbf{Y}^{-1} = \mathbf{I} - \langle \mathbf{R}^{\mathbf{Q}0} V^0 \rangle \tag{9.111}$$

the tensor $\mathbf{R}^{\mathbf{Q}0}(\mathbf{x}) := \langle \mathbf{M}_1(\mathbf{I} - \boldsymbol{\Gamma}\mathbf{M}_1)^{-1}\mathbf{Q}_i^0(\mathbf{x})\rangle_{(i)}$ (see for details Subsection 4.2.3 in [200]) is expressed through the operator $(\mathbf{I} - \boldsymbol{\Gamma}\mathbf{M}_1)^{-1}$. Only for the ellipsoidal correlation hole (exclusive volume) $\mathbf{R}^{\mathbf{Q}0}(\mathbf{x}) = \mathbf{R}_i(\mathbf{x})\mathbf{Q}_i^0$, and, therefore, Eq. (9.111) is reduced to the known representation (9.51) with the constant tensor $\mathbf{Q}_i^0$ depending on the orientation of the correlation hole v_i^0. An advantage of Eqs. (9.80) with respect to Eqs. (9.46) and (9.111) is that Eqs. (9.46) and (9.111) are fundamentally limited by analysis of statistically homogeneous media subjected to the homogeneous boundary conditions, while the system (9.61) can be easily generalized to the statistically inhomogeneous media. Moreover, the method (9.46) and (9.111) conserves a fundamental deficiency of the general integral Eq. (7.33) containing the renormalizing item defining only by the averaged strain polarization tensor $\langle \boldsymbol{\eta} \rangle$. Estimation of the tensor $\mathbf{R}_i^{\mathbf{Q}0}(\mathbf{x})$ in an auxiliary model problem with homogeneous fictitious eigenstrain in v_i^0 implies that influence of surrounding inclusions v_q ($\mathbf{x}_q \in v_i^0$) is defined by the average strain polarization tensors $\langle \boldsymbol{\eta} \rangle_q$ rather than its detailed distribution $\langle \boldsymbol{\eta} \rangle_q(\mathbf{x})$ ($\mathbf{x} \in v_q$). Impact of the last improvement on the estimated effective properties will be considered in Sect. 9.5 for the circle v_i^0 and noncanonical shape of v_i^0 that presents no additional difficulties for the new method (9.80) as opposed to the method (9.46) and (9.111) requiring evaluation of the supplementary tensor $\mathbf{R}_i^{\mathbf{Q}0}(\mathbf{x})$.

In is worthy of note that some popular approaches are related with the GIEs at least in a sense that these approaches constitute the methods of solutions of the GIEs. Indeed, the effective medium method (EMM; see [643, 812] and Chap. 8 for details) are based on the assumptions

$$\mathbf{M}^c = \mathbf{M}^*, \quad \langle \overline{\boldsymbol{\sigma}} \rangle(\mathbf{x}) \equiv \langle \boldsymbol{\sigma} \rangle. \tag{9.112}$$

while the Mori-Tanaka method (MTM; see [1093] and Chap. 8 for details) use the following assumptions:

$$\mathbf{M}^c = \mathbf{M}^{(0)}, \quad \langle\overline{\boldsymbol{\sigma}}\rangle(\mathbf{x}) \equiv \langle\boldsymbol{\sigma}\rangle^{(0)}. \tag{9.113}$$

It is interesting that both sets of assumptions (9.112) and (9.113) in a functional sense fulfil the roles of both the EFH **H1a** [(9.112$_2$) and (9.113$_2$)] and some sort of the closing assumption (instead of **H2**) [(9.112$_1$), (9.112$_2$), and (9.112$_2$)]. These assumptions are enough for estimation of the effective properties (in so doing, the hypotheses **H1b** and **H3** are not required). Thus, we see that acceptance of the assumptions either (9.112) and (9.113) is equivalent to truncation of the GIE (7.25) at $n = 1$ of the integral item depending on the perturbator $\mathcal{L}_j^{\sigma}(\mathbf{x} - \mathbf{x}_j, \overline{\boldsymbol{\sigma}})$ (in the current notations). This truncation is compensated in subsequent manipulations by correction of either $\mathbf{M}^c$ (9.112) or $\langle\overline{\boldsymbol{\sigma}}\rangle(\mathbf{x})$ (9.113). As this takes place, the final representations of the effective properties (9.8) do not depend on a method which can be used for estimations of truncated averaged perturbators in the GIE (7.25). It means that the assumptions (9.112) and (9.113) are so much rough that it does not matter which background of micromechanics (either the old (7.33) or new (7.20) one) was used. Acceptance of the assumptions either (9.112) and (9.113) gives no chance to realize a fundamental advantage of GIEs (7.19), (7.20) over (7.32), (7.33), respectively, lying in the fact that the renormalized items $\langle\mathbf{U}(\mathbf{x} - \mathbf{y})\boldsymbol{\tau}\rangle(\mathbf{y})$ and $\langle\boldsymbol{\Gamma}(\mathbf{x} - \mathbf{y})\boldsymbol{\eta}\rangle(\mathbf{y})$ are directly dependent on the inhomogeneity of field distributions inside inclusions rather than only on its volume averages as in the renormalized items $\mathbf{U}(\mathbf{x} - \mathbf{y})\langle\boldsymbol{\tau}\rangle(\mathbf{y})$ (7.32) and $\boldsymbol{\Gamma}(\mathbf{x} - \mathbf{y})\langle\boldsymbol{\eta}\rangle(\mathbf{y})$ (7.33).

9.5 Numerical Results

9.5.1 Finite Element Analysis in the Framework of the Hypothesis H1a

With the nonessential restriction on space dimensionality d and on the shape of inhomogeneities, we will consider $2D$ plane strain problems for composites reinforced by aligned infinite fibers with noncircular cross section schematically presented in the Fig. 9.2 and described by the curve

$$\begin{cases} (x - R_1 + r_s)^2 + (y - R_2 + r_s)^2 = r_s^2, & \text{for } \{|x| > R_1 - r_s\} \cap \{|y| > R_2 - r_s\}, \\ |x + yR_1/R_2| + |x - yR_1/R_2| = 2R_1, & \text{for } \{|x| < R_1 - r_s\} \cup \{|y| < R_2 - r_s\}, \end{cases} \tag{9.114}$$

which reduces to a circle and a rectangular in the limiting cases $R_1 = R_2 = r_s = a$ and $r_s = 0$, respectively. We will consider the fixed values $R_1 = 1$, $r_s/R_1 = 0.1, 1$

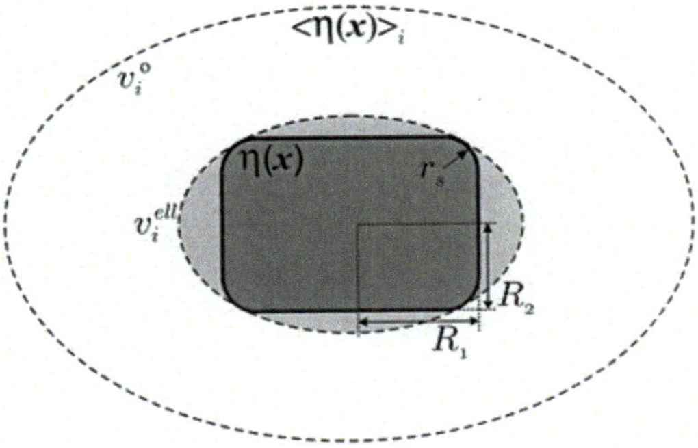

Fig. 9.2 Schematic representation of the EFM

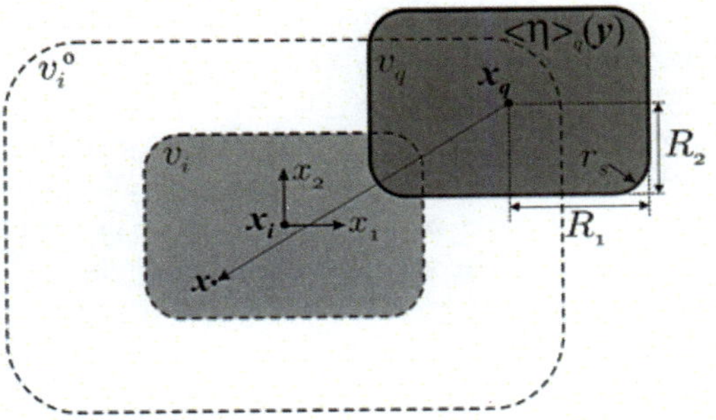

Fig. 9.3 Schematic representation of the new approach

and isotropic constituents with the Young's moduli $E^{(1)} = 100$ and $E^{(0)} = 1$ and Poisson ratio $\nu^{(1)} = \nu^{(0)} = 0.45$.

Figures 9.2 and 9.3 present not only the geometrical parameters of the inhomogeneities v_i and v_q but also a schematic comparison of both the classical approach (EFM; see Fig. 9.2) and the new one (see Fig. 9.3). In the classical approach, the effective field $\langle\overline{\boldsymbol{\sigma}}\rangle_i$ depends only on the volume average of the strain polarization tensor $\langle\boldsymbol{\eta}\rangle_q$ distributed in the ellipsoidal excluded volume v_i^0. Excluded volume v_i^0 in the EFM is chosen to be ellipsoidal one homothetic to the ellipsoid $v_i^{\rm ell}$ enveloping v_i (usually the aspect ratio of $v_i^{\rm ell}$ is taken as R_2/R_1), and, because of this, all estimations by the EFM are invariant with respect to the size of v_i^0. In the new approach, the effective field $\langle\overline{\boldsymbol{\sigma}}\rangle_i(\mathbf{x})$ in the area $\mathbf{x} \in v_i$ occupied by the matrix material (see Fig. 9.3) is generated by moving heterogeneities v_q (see Eq. (9.61)), and the inhomogeneity of the detailed distribution $\langle\boldsymbol{\eta}\rangle_q(\mathbf{y})$ $(y \in v_q)$ affects the effective field $\langle\overline{\boldsymbol{\sigma}}\rangle_i(\mathbf{x})$ $(\mathbf{x} \in v_i)$. Furthermore, the domain of this long-range action is limited not only by $\mathbf{x}_q \in v_i^0$ but by a domain v_i^ϕ for location of inclusion centers $\mathbf{x}_q$ where $\varphi(v_q, \mathbf{x}_r|v_i, \mathbf{x}_i) - n^{(q)}$ is not negligible (in the sense of influence on $\mathbf{M}^*$), and an effect zone of stress perturbations produced by the heterogeneity v_q is limited by the Minkowski addition of domains $\mathbf{x} \in v_{iq}^\oplus \equiv v_i^\phi \oplus v_q$ (see Fig. 9.3) rather than by the domain v_q as in the EFM. Moreover, we note that in classical approaches, there is no a systematic approach to choose the size and shape of the excluded volume v_i^0 (see Fig. 9.3), which impacts on the estimations of effective properties.

A domain, where $\varphi(v_q, \mathbf{x}_q|; v_i, \mathbf{x}_i) - n^{(q)}$ is not negligible, is discretized by the square mesh $\Omega^{\rm sq}$ $(p, k \in Z^1)$

$$\Omega^{\rm sq} := \Big\{(x_1, x_2)^\top \mid (p-1)h < x_1 < ph,\ (k-1)h < x_2 < kh\Big\}, \tag{9.115}$$

that is not optimized (at least for estimation of the perturbator $\mathcal{L}_q^\sigma(\mathbf{x} - \mathbf{x}_q)$), but is efficient. Here h is the discretization step, and x_1, x_2 are local coordinates with origins at the fiber centers, which will be exploited for stress assignment inside and outside the fiber in the Eq. (9.63_1) and estimated in the postprocessing of FEA. In order to compute the integral in Eq. (9.63_1), the discretization (9.115) has been applied considering the simple Simpson numerical integration rule on piecewise constant elements; this choice has been dictated by the simplicity of implementation also for nonregular inclusion shapes even if it does not guarantee a very fast convergence of the results with the discretization step h. It should be mentioned that the main reason of the choice of the square mesh (9.115) is its double usage for two different problems. At first this mesh is exploited for estimation of the effective field $\langle\overline{\boldsymbol{\sigma}}\rangle_i(\mathbf{x})$ (9.65). Second, the same mesh (9.115) is used as the location of the moving inclusion centers $\mathbf{x}_q$ in Eq. (9.63_1). It gives an opportunity to use a solution (9.7) for one heterogeneity in a sample in the nodes of just one realization of the mesh (9.115) which is exploited as an "output" mesh for a solution obtained on a standard inhomogeneous mesh $\Omega^{\rm FEA}$ of the FEA (see, e.g., [478, 1700]). However, in the case of immediate use of the inhomogeneous mesh inside and outside a heterogeneity (which is more effective for estimation of the perturbator factor $\mathcal{L}_q^\sigma(\mathbf{x}-\mathbf{x}_q)$), we will need to estimate the stresses in the nodes of a new mesh generated for each location $\mathbf{x}_q$ of the moving inclusion v_q. Therefore, the square mesh (9.115) is optimal for the current problem.

We are coming now to the analysis of the conditional probability density $\varphi(v_k, \mathbf{x}_k|; v_i, \mathbf{x}_i)$. This function is well investigated only for identical spherical ($3D$ and $2D$ cases) inclusions with a radius a when the pair distribution function $g(\mathbf{x}_i-\mathbf{x}_m) \equiv \varphi(v_i, \mathbf{x}_i|; v_m, \mathbf{x}_m)/n^{(k)}$ depending only on $|\mathbf{x}_m-\mathbf{x}_i|$ is called the radial distribution function (RDF). Two alternative RDFs of inclusion will be examined (see [604, 1525] and Eqs. (5.97), (5.101))

$$g(\mathbf{x}_i - \mathbf{x}_q) \equiv \varphi(v_i, \mathbf{x}_i|; v_q, \mathbf{x}_q)/n^{(i)} = H(r-2a), \tag{9.116}$$

$$g(\mathbf{x}_i - \mathbf{x}_j) = H(r-2a)\Bigg\{1 + \frac{4c}{\pi}H(4a-r) \times \left[\pi - 2\arcsin\left(\frac{r}{4a}\right) - \frac{r}{2a}\sqrt{1-\frac{r^2}{16a^2}}\right]\Bigg\}, \tag{9.117}$$

where H denotes the Heaviside step function, $r \equiv |\mathbf{x}_i - \mathbf{x}_q|$ is the distance between the non-intersecting inclusions v_i and v_q, and c is the volume fraction of fibers of the radius a. The formula (9.116) describes a well-stirred approximation, while Eq. (9.117) takes into account a neighboring order in the distribution of the inclusions. Due to the absence of $\varphi(v_k, \mathbf{x}_k|; v_i, \mathbf{x}_i)$ for non-spherical inclusions v_q, v_i ($\mathbf{x}_i = \mathbf{0}$), we will construct it for identical aligned heterogeneities from the known $g(r/a)$ (9.116) and (9.117) for spherical inclusions in the following manner. Surfaces $\mathbf{s} \in \partial v_i$ and $\mathbf{s}^0 \in \partial v_i^0$ are described in either the polar or spherical coordinate systems by the equations $|\mathbf{s}(\mathbf{n})| \equiv \rho_i(\mathbf{n})$ and $|\mathbf{s}^0(\mathbf{n})| \equiv \rho_i^0(\mathbf{n})$, respectively, and $\rho^0(\mathbf{n}) = 2\rho(\mathbf{n})$, $\mathbf{n} = (\mathbf{x}_k - \mathbf{x}_i)/r$. We consider the alternative function

$$\varphi(v_k, \mathbf{x}_k|; v_i, \mathbf{x}_i) = (1 - V_i^0(\mathbf{x}_k - \mathbf{x}_i))g(r/\rho_i(\mathbf{n}))n^{(k)}, \tag{9.118}$$

where Eq. (9.118) taking into account a neighboring order in the distribution of the inclusions is reduced to the known representation for spherical inclusions $g(r/a)$ (9.117) at $\rho(\mathbf{n}) \equiv a$. According to the best author knowledge, a systematic quantitative investigation of the binary correlation function $\varphi(v_k, \mathbf{x}_k|; v_i, \mathbf{x}_i)$ for the noncanonical shape (and even for the non-spherical one) of inclusions is absent. This issue merits additional detailed consideration which is beyond the scope of the current study.

In order to estimate the approximations introduced by discretizations implemented in the FEA and for the evaluation of the integral equation (9.63_1), we compare numerical results with a reference analytical solution in a plane strain problem. In particular, we consider the representation (9.65) for statistically homogeneous fields with the RDF given in Eq. (9.116) for homogeneous circular heterogeneities of radius R with exact analytical representation of tensors $\mathbf{R}_i$ and $\mathbf{T}_i(\mathbf{x} - \mathbf{x}_i)$ ($\mathbf{x} \in R^2$) as well as analytical estimation of effective stress concentration factor (9.45) of the classical solution. Thus, we compute the effective stress concentration factors $\mathbf{B}_i^*$ in the equation

$$\langle\boldsymbol{\sigma}\rangle_i(\mathbf{x}) = \mathbf{B}_i^*(\mathbf{x})\langle\boldsymbol{\sigma}\rangle, \quad \mathbf{B}_i^*(\mathbf{x}) \equiv \mathbf{B}_i(\mathbf{x})\mathbf{D}_i, \tag{9.119}$$

obtained by substitution of exact analytical representations of the tensors $\mathbf{B}_i$ and $\mathbf{T}_i(\mathbf{x} - \mathbf{x}_i)$ ($\mathbf{x} \in R^2$) into the proposed scheme (9.65) with $h/R = 0.04$ and $h/R = 0.02$. We consider a volume fraction of fibers $c = 0.65$, and we evaluate the stress perturbations $\mathcal{L}_q^\sigma(\mathbf{x} - \mathbf{x}_q)\langle\overline{\boldsymbol{\sigma}}\rangle_q(\mathbf{x}_q)$ (9.7) in the vicinity $\{\mathbf{x} : \max_x|\mathbf{x} - \mathbf{x}_q| = 3a\}$ of the area v_q rather than only in the inhomogeneity $\mathbf{x} \in v_q$ as in the classical approach (9.45). Note that Eq. (9.119) is correct for homogeneous effective field $\overline{\boldsymbol{\sigma}}_i(\mathbf{x})$ when $\mathbf{D}_i$ is constant. In comparison with the exact estimations $D_{1111} = 0.7962$ (9.45), the errors in estimations of D_{1111} resulting from (9.65) due to the discretization Ω^{sq} (9.115) of Eq. (9.65) equal 0.11% and 0.05% for $h = 0.04$ and $h = 0.02$, respectively, while analogous errors for the average component $\overline{B}_{1212}^*$ equal 0.055% and 0.026%, respectively.

The analytical solutions for composites reinforced by the circular heterogeneities make possible error estimation of the FEA subsequently used for noncanonical inclusions (9.114). The commercial finite element code Comsol 3.5 was used for the finite element modeling described in this section. The inclusion and matrix materials were assumed to be isotropic, linear elastic, and modeled with plane strain 3-node triangular elements, while the interface was modeled as perfectly bonded. The infinite dimensions of the matrix were approximated with a length of 40 inclusion diameters: $L = 40R$ (justification of the length $L = 40R$ is performed in [200]). Different regions of discretizations were implemented imposing the smallest element dimensions around to the phase interfaces in order to better capture the high gradients and the jumps in the stress components. Typical meshes contain 18,000÷50,000 elements (approx. 60,000÷200,000 d.o.f.) for a CPU time varying from 40 s to 2 min on a AMD Athlon 64 X2 Dual, 2.2 MHz. The stress perturbator factor $\mathcal{L}_q^\sigma(\mathbf{x}-\mathbf{x}_q)$ $(\mathbf{x} \in R^2)$ (9.63$_1$) estimated by the FEA at the mesh $\Omega^{\rm FEA}$ was used to interpolate $\mathcal{L}_q^\sigma(\mathbf{x}-\mathbf{x}_q)$ at the mesh $\Omega^{\rm sq}$ (9.115) that was exploited for evaluation of the tensor $\mathcal{T}_i(\mathbf{x})$ (9.65) at the mesh $\Omega^{\rm sq}$ (9.115). The mentioned scheme leaded to the errors in estimation of $B^*_{i|1111}(\mathbf{x})$ and $B^*_{i|1212}(\mathbf{x})$ of 0.26% and 0.41%, respectively.

This preliminary computations confirm that sufficient convergence of results is achieved and reveals that the error introduced by the FEA at the mesh $\Omega^{\rm FEA}$ in Eq. (9.7) exceeds the error produced by the numerical computation of the integral equation (9.63) at the mesh $\Omega^{\rm sq}$ (9.115). It is also important to stress that the difference between the new and old approaches does not exceed a computational error, and, therefore, we qualitatively proved that in the considered example of the circle inclusion, both methods, the old (9.87) and new (9.79) ones, which are based on the classical (7.33) and new (7.20) general integral equations, respectively, lead to close numerical results.

It is important to stress that it is incorrect to estimate the influence of the stress pertubator factor $\mathcal{L}_q^\sigma(\mathbf{x}-\mathbf{x}_q)$ (9.63$_1$) only in the inclusion v_q, which means that stress perturbations introduced by the moving inhomogeneity v_q in their vicinity $v^\oplus_{qk} \setminus v_q = \{\mathbf{x}|\, a < |\mathbf{x}-\mathbf{x}_q| < 3a\}$ are neglected. To better show this point, we use the exact analytical representations of tensors $\mathbf{R}_i$ and $\mathbf{T}_i(\mathbf{x}-\mathbf{x}_i)$ $(\mathbf{x} \in R^2)$ substituted into Eq. (9.63$_1$) where for the estimation of $\overline{\boldsymbol{\sigma}}(\mathbf{x})$ $(\mathbf{x} \in v_i)$, only the moving heterogeneities with the centers $|\mathbf{x}_q - \mathbf{x}| < a$ are incorrectly taken into account. Then D_{1111} estimated by the use of incorrect scheme differs from the exact value $D_{1111} = 0.7962$ [corresponding to Eq. (9.45)] by 41%. However, in the case of using the correct new background (7.20), we must estimate the stress perturbation in $\mathbf{x} \in v^\oplus_{qk}$ (that leads to the identical results in both the old approach and the new one) rather than only in $\mathbf{x} \in v_q$.

A large difference of results is obtained in the framework of the backgrounds (7.20) and (7.33) for composites reinforced by nonellipsoidal inclusions demonstrating essentially inhomogeneous stress distribution inside inclusions even in the framework of the hypothesis **H1a**. In more detail, we analyze the inclusion shape (9.114) with the different aspect ratios $R_2/R_1 = 0.1,\ 0.32,\ 0.64,\ 1$, and fixed $r_s/R_1 = 0.1$ (at $c = 0$, see Figs. 9.4 and 9.5). The stress concentration factors for

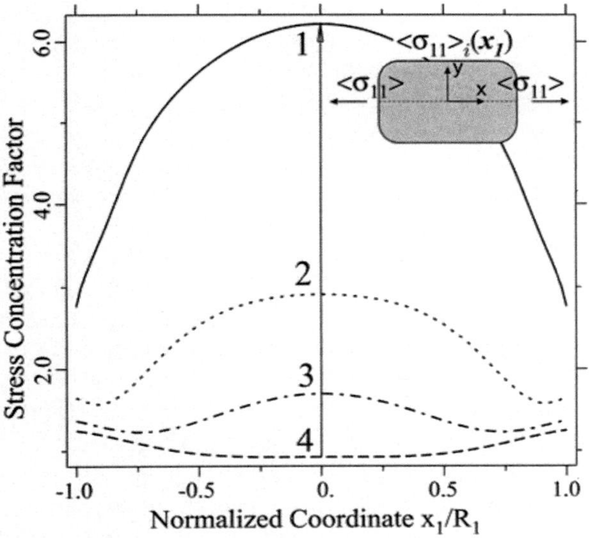

Fig. 9.4 $B_{i|1111}(x_1)$ vs x_1/R_1 for $R_2/R_1 = 0.1$ (1), 0.32 (2), 0.64 (3), and 1.0 (4) and $c = 0$

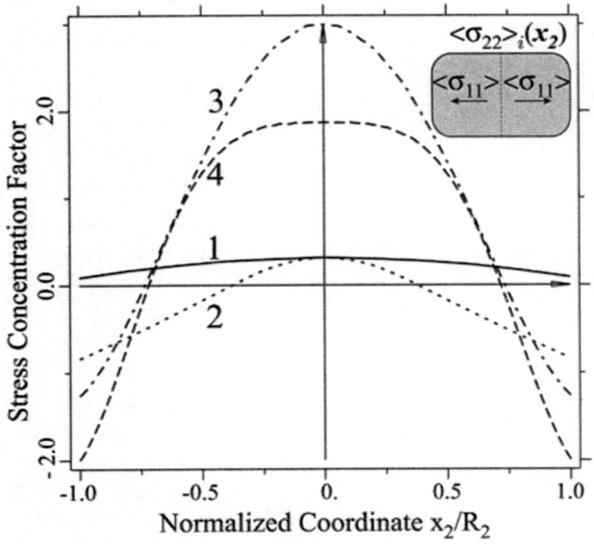

Fig. 9.5 $B_{i|2211}(x_2)$ vs x_2/R_2 for $R_1/R_2 = 0.1$ (1), 0.32 (2), 0.64 (3), and 1.0 (4) and $c = 0$

an isolated inhomogeneity $B_{i|1111}(\mathbf{x})$ in the cross section $\mathbf{x} = (x_1, 0)^\top$ (see Fig. 9.4) grow in both the magnitude and variation with decrease of the ratio R_2/R_1 that is in line with analogous curves obtained in 3D case for the long circular cylinders with the smooth ends (see Chapter 18 in [200]). However, the stress concentration factors $B_{i|2211}(\mathbf{y})$ behave more specifically and demonstrate the change of sign in the cross sections $\mathbf{y} = (0, x_2)^\top$ (see Fig. 9.5). Increase of volume fraction of inclusions from $c = 0$ to $c = 0.7$ leads to progressive variation of curves in Figs. 9.4 and 9.5 (see for details [232]).

Normalized effective Young's moduli $E_1^*/E^{(0)}$ (the case $E_2^*/E^{(0)}$ is considered in [232]) are presented in Fig. 9.6 as the functions of the volume concentration c of inclusions at $R_2/R_1 = 0.32$. Curves 1 and 2 are estimated by the new approach (9.76) for the $\varphi(v_q, \mathbf{x}_q|v_i, \mathbf{x}_i)$ (9.117), (9.118) and (9.116), (9.118), respectively. Curves 3 and 4 are predicted by the EFM and MTM, respectively, which are invariant with respect to the concrete form of $g(r)$, either (9.116) or (9.117), in (9.118). As can be seen, the estimations carried out by the different methods are essentially different at $c > 0.6$. In so doing, the difference between curves 1 and 2 obtained by the new background of micromechanics (NBM) with distinguished functions $\varphi(v_k, \mathbf{x}_k|v_i, \mathbf{x}_i)$ is basically less than the difference between the curves relatives to previous classical methods.

The largest difference between the classical (9.42) and new (9.76) approaches is observed for the component $B^*_{i|2211}(x_2)$ in a cross section $\mathbf{x} = (0, x_2)^\top$ (see Fig. 9.7) where the predicted curves by the new and classical approaches are distinguished by a sign. Curves 4 and 5 obtained by the MTM and EFM, respectively, are invariant with respect to the binary correlation function $\varphi(v_q, \mathbf{x}_q|v_i, \mathbf{x}_i)$ either (9.116) or (9.117). The new approach (9.66) was used for estimation of curves 1–3 where we applied for curves 1 and 2 the excluded volume $v_i^0 = \{\mathbf{x}|\mathbf{x} = 2\mathbf{y}, \mathbf{y} \in v_i\}$ and the binary correlation functions $\varphi(v_q, \mathbf{x}_q|v_i, \mathbf{x}_i)$ (9.117), (9.118) and (9.116), (9.118), respectively. The difference between curves 1 and 2 demonstrates the impact of the $\varphi(v_q, \mathbf{x}_q|v_i, \mathbf{x}_i)$ at the fixed $v_i^0 = \{\mathbf{x}|\mathbf{x} = 2\mathbf{y}, \mathbf{y} \in v_i\}$. In so doing, the dissimilarity of curves 2 and 3 reveals the influence of $v_i^0 = \{\mathbf{x}|\mathbf{x} = 2\mathbf{y}, \mathbf{y} \in v_i\}$ and $v_i^0 =$

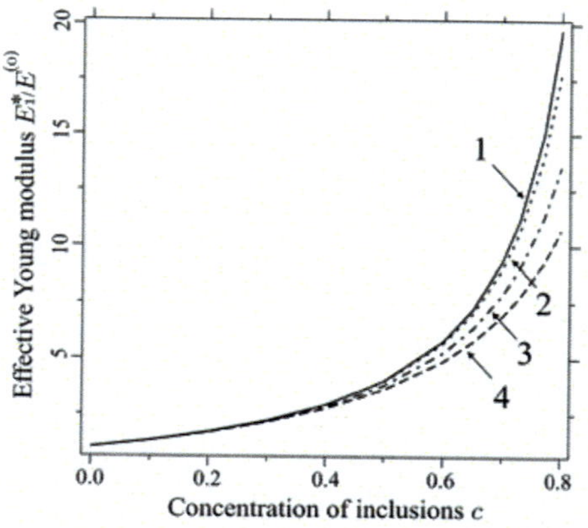

Fig. 9.6 $E_1^*/E^{(0)}$ vs c estimated by the new approach (1 and 2), EFM (3), and Mori-Tanaka scheme (4)

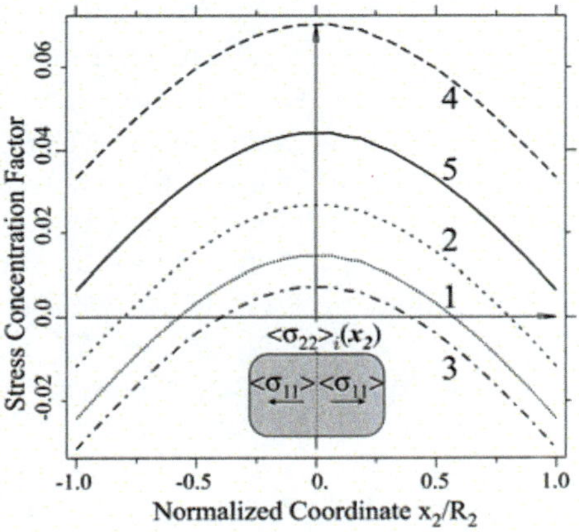

Fig. 9.7 $B^*_{i|2211}(x_2)$ vs x_2/R_2 estimated by the new approach (1, 2, and 3), MTM (4), and EFM (5) for $R_2/R_1 = 0.32$

$\{\mathbf{x}|\mathbf{x} = 3\mathbf{y}, \mathbf{y} \in v_i\}$, respectively, at the fixed $\varphi(v_q, \mathbf{x}_q|v_i, \mathbf{x}_i)$ (9.116), (9.118). It should be mentioned that in Figs. 9.6 and 9.7, the curves estimated for $v_i^0 = \{\mathbf{x}|\mathbf{x} = 2\mathbf{y}, \mathbf{y} \in v_i\}$ by the different methods can be set up in the following order MTM-EFM-[NBM (9.116)]-[NBM (9.117)]. Interestingly, there are known publications (see for reference [200]) which demonstrated equivalence of the MTM and EFM for composites with identical aligned homogeneous ellipsoidal heterogeneities. One [200] obtained analytical representation (without numerical results) of the difference between the effective elastic moduli estimated by either the MTM or EFM in the case of nonellipsoidal inclusions. However, to the best of the author's knowledge, the current section is the first one where the difference of both $\mathbf{L}^*$ (see Fig. 9.6) and $\mathbf{B}^*(\mathbf{x})$ $(\mathbf{x} \in v_i)$ (see Fig. 9.7) estimated by both the MTM and EFM is quantitatively evaluated for the case of nonellipsoidal heterogeneities.

Up to now in this subsection, we only considered the cases of pure mechanical loading ($\boldsymbol{\beta}_1(\mathbf{x}) \equiv \mathbf{0}$, $\mathbf{x} \in R^d$) and compared the corresponding results obtained in the framework of both the classical background (7.34) and new one (7.19). Estimations of the statistical averages of the stresses and strains as well as effective properties were analyzed in [234] for the thermoelastic 2D plane strain problems ($\boldsymbol{\beta}_1(\mathbf{x}) \not\equiv \mathbf{0}$) for CM reinforced by aligned identical homogeneous inclusions of noncanonical shape (9.114). Numerical results $\langle\boldsymbol{\sigma}\rangle(\mathbf{x}) \equiv \mathbf{0}$ $(\mathbf{x} \in R^d)$ were obtained for the average residual stresses $\langle\boldsymbol{\sigma}\rangle_i(\mathbf{y})$ $(\mathbf{y} \in v_i)$ and the effective thermal expansion and stored energy. Some new effects (similar to Figs. 9.4, 9.5, 9.6, and 9.7) were detected that are impossible in the framework of a classical background of micromechanics.

Especially interesting results are obtained if the fluctuation of elastic compliance is negligible ($\mathbf{M}(\mathbf{x})$ =const.) while the stress-free strains fluctuate. Of course, the solution of this problem can be obtained with the help of passing to the zero limit of the elastic mismatch of different components in the corresponding formulae on Sect. 9.3. However, due to their unique properties ($\mathbf{M}(\mathbf{x})$ =const.), one managed to get the exact Eqs. (9.96) which can be solved without any additional hypotheses of micromechanics. A single source of error is the numerical errors related with both the estimation of the perturbator $\mathcal{L}_q^{\beta}(\mathbf{x}-\mathbf{x}_q,\boldsymbol{\beta})$ (expressed, e.g., by Eq. (9.97)) and evaluation of the integral (9.96$_1$) with the found integrand $\mathcal{L}_q^{\beta}(\mathbf{x}-\mathbf{x}_q,\boldsymbol{\beta})$. In so doing, an accuracy of the numerical solution can be provided with any desirable tolerance that can be established by comparison with some particular exact analytical solution (much as Eq. (9.119)). So, for statistically homogeneous CMs with the circle homogeneous inclusion, the exact Eq. (9.98) for the effective residual stresses was compared with the numerical solution (9.96$_1$) obtained by the FEA for $\mathcal{L}_q^{\beta}(\mathbf{x}-\mathbf{x}_q,\boldsymbol{\beta})$ (see for details a case of pure mechanical loading ($\langle\boldsymbol{\sigma}\rangle \neq \mathbf{0}$, $\boldsymbol{\beta}_1(\mathbf{x}) \equiv \mathbf{0}$) and Simpson numerical integration of integral (9.96$_1$) at the mesh (9.115). The numerical errors for the effective residual stresses (9.96$_1$) 1.03% and 0.35% for $h/a = 0.04$ and $h/a = 0.02$, respectively, are comparable with the numerical errors for a pure mechanical loading ($\langle\boldsymbol{\sigma}\rangle \neq \mathbf{0}$, $\boldsymbol{\beta}_1(\mathbf{x}) \equiv \mathbf{0}$) at the beginning of this subsection.

For the noncanonical inclusion shape (9.114), we will evaluate the normalized effective field $\widehat{\sigma}_{kl} = \langle\overline{\sigma}_{kl}\rangle_i/\sigma_{kl}^{\text{ellips}}$ (no sum over $k, l = 1, 2$) in the Eq. (9.97), where $\boldsymbol{\sigma}^{\text{ellips}} \equiv -\mathbf{Q}^{\text{ellips}}\boldsymbol{\beta}_1^{(q)}$ is an analytical representation for the residual stresses in an isolated ellipsoidal inclusion v^{ellips} with an aspect ratio R_2/R_1 (for the circle inclusions $R_1 = R_2$) inside the infinite matrix. The results of normalized effective stresses $\widehat{\boldsymbol{\sigma}}(\mathbf{x})$ are presented in Figs. 9.8 and 9.9 for different methods of estimations and the fixed values $R_2/R_1 = 0.32$ and $c = 0.7$. The largest difference (40%) between the classical (9.98), (9.99) and new (9.97) approaches is observed for the component $\widehat{\sigma}_{22}(\mathbf{x})$ in a cross section $\mathbf{x} = (0, x_2)^{\top}$ (see Fig. 9.9) where curves 4 and 5 were obtained by the MTM and EFM, respectively. It should be mentioned that there are the numerous publications demonstrating equivalence of the MTM's and EFM's estimations for composites with aligned identical ellipsoidal heterogeneities of both the homogeneous and inhomogeneous structures (see for reference Chapter 8 in

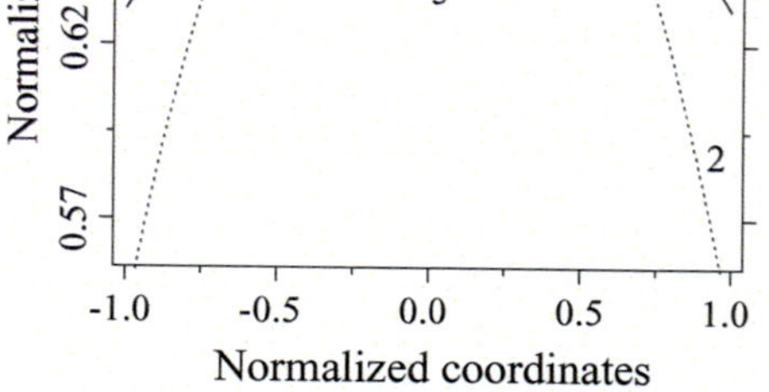

Fig. 9.8 $\widehat{\sigma}_{11}(x_1)$ vs x_1/R_1 estimated for NBM with a step function g (1), NBM with a nonstep function g (2), NBM with the excluded volume v_i^{01} (3), EFM (4), MTM (5)

Fig. 9.9 $\widehat{\sigma}_{22}(x_2)$ vs x_2/R_2 estimated for NBM with a step function g (1), NBM with a nonstep function g (2), NBM with the excluded volume v_i^{01} (3), EFM (4), MTM (5)

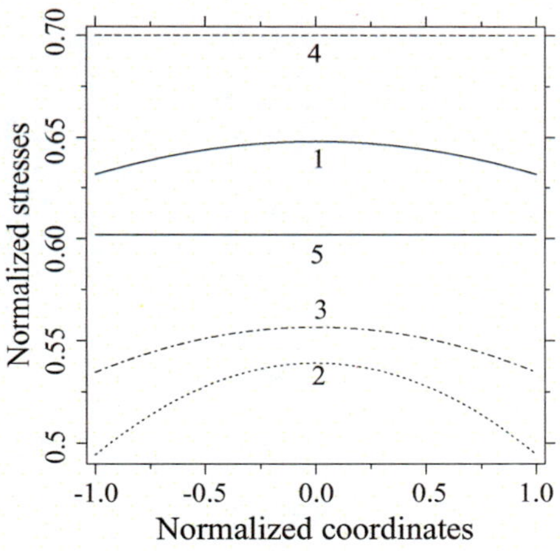

[200]). Figures 9.8 and 9.9 demonstrate an essential difference of these approaches even in a degenerated case of nonellipsoidal homogeneous inclusions with no elastic mismatch. The new approach (9.97) (NBM) was used in Figs. 9.8 and 9.9 for estimating curves 1–3 where we applied for curves 1 and 2 the excluded volume $v_i^0 = \{\mathbf{x}|\mathbf{x} = 2\mathbf{y}, \mathbf{y} \in v_i\}$ and the binary correlation functions $\varphi(v_q, \mathbf{x}_q|v_i, \mathbf{x}_i)$ in the form of either the step function (9.116), (9.118) or the nonstep (9.116), (9.118) one, respectively. The difference between curves 1 and 2 demonstrates the impact of $\varphi(v_q, \mathbf{x}_q|v_i, \mathbf{x}_i)$ at the fixed $v_i^0 = \{\mathbf{x}|\mathbf{x} = 2\mathbf{y}, \mathbf{y} \in v_i\}$. In so doing the dissimilarity of curves 1 and 3 reveals the influence of $v_i^0 = \{\mathbf{x}|\mathbf{x} = 2\mathbf{y}, \mathbf{y} \in v_i\}$ and $v_i^{01} = \{\mathbf{x}|\mathbf{x} = 2.4\mathbf{y}, \mathbf{y} \in v_i\}$, respectively, at the fixed $\varphi(v_q, \mathbf{x}_q|v_i, \mathbf{x}_i)$ (9.116), (9.117). Thus, this new renormalizing in Eq. (9.97) is directly dependent [in opposite to the classical Eq. (7.33)] on the inhomogeneity of stress fields inside the inclusions that lead to the detection of fundamentally new effects in micromechanics such as inhomogeneity of the statistical average of the effective field $\langle\overline{\boldsymbol{\sigma}}\rangle_i(\mathbf{x})$ $(\mathbf{x} \in v_i)$ and their dependence on both the binary correlation functions $\varphi(v_q, \mathbf{x}_q|v_i, \mathbf{x}_i)$ and size of the excluded volume [compare with Eqs. (9.98) and (9.99)].

9.5.2 Iteration Method (Abandonment of the Hypothesis H1a)

All numerical results in Sect. 9.5.1 were obtained in the framework of the Hypothesis **H1a**. Following [235], we consider the numerical results (with an abandonment of the hypotheses **H1a**) obtained by the iteration method (9.92)–(9.95) for CM with the noncanonical shape of homogeneous inclusions (9.114), the square mesh (9.115), the elastic mismatch $E^{(1)}/E^{(0)} = 5$, and $R_2/R_1 = 0.32$; $c = 0.65$. The initial approximation (9.93_1) was estimated by FEA in the framework of the EFH **H1a** (9.95). A fast convergence of the proposed iteration

Fig. 9.10 $B^*_{i|2211}(x_2)$ vs x_2/R_2 estimated by the NBM: initial approximation (1), first iteration (2), fourth iteration (3), eighth iteration (4)

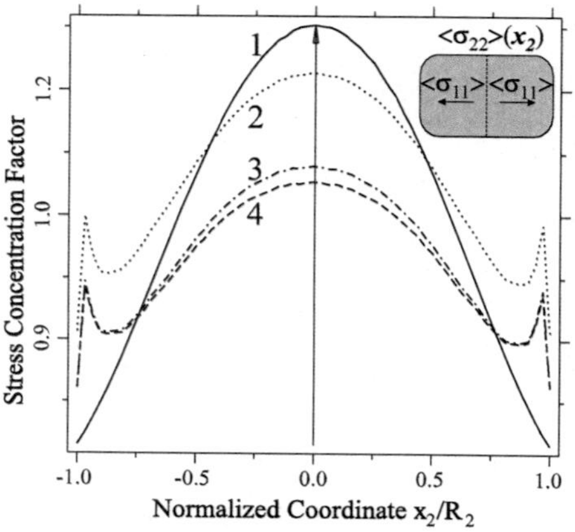

Fig. 9.11 $\overline{B}^{*[n]}_{i|1111}(x_1)$ vs x_1/R_1 estimated by NBM for $n = 0$ (1), $n = 2$ (5), $n = 5$ (4), $n = 6$ (2), and $n = 8$ (3)

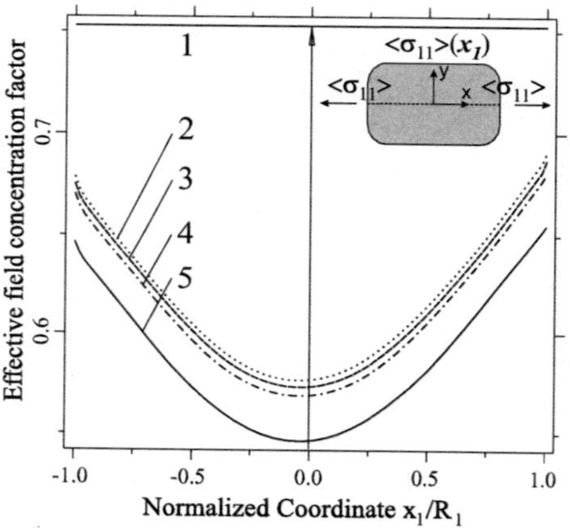

method takes place (see Fig. 9.10); the eighth iteration differs from the sixth, fourth, first, and initial iterations on 5.5%, 20%, 44%, and 101%, respectively. So much significant difference of the values $B^*_{i|2211}(x_2)$ (9.119) in Fig. 9.10 of the initial approximation with the next iterations is understandable if we recall that the initial approximation (9.95) was constructed in the framework of the hypothesis **H1a** when $\langle\overline{\boldsymbol{\sigma}}^{[0]}_1\rangle$ ≡const. (9.95), while in the iterations (9.92), we estimate $\langle\overline{\boldsymbol{\sigma}}^{[n]}_1\rangle(\mathbf{x})$ ≢const. ($\mathbf{x} \in v_1$). The components of the effective stress concentration factors $\overline{B}^{*[n]}_{i|1111}(x_1)$ (in the equation $\langle\overline{\boldsymbol{\sigma}}\rangle_i(\mathbf{x}) = \overline{\mathbf{B}}^*(\mathbf{x})\langle\boldsymbol{\sigma}\rangle$) as the functions of the normalized coordinate x_1/R_1 are presented in Fig. 9.11 for the iterations $n = 0, 2, 4, 8$. As can be seen, a few iterations of Eq. (9.92) produce the inhomogeneous $\overline{B}^{*[n]}_{i|1111}(x_1)$, and this process converges very rapidly. In so doing, a convergence of the effective fields $\overline{B}^{*[n]}_{i|1111}(x_1)$ is faster than convergence of $\mathbf{B}^{*[n]}_i(\mathbf{x})$, while their inhomogeneity is less than inhomogeneity of stress concentration factors.

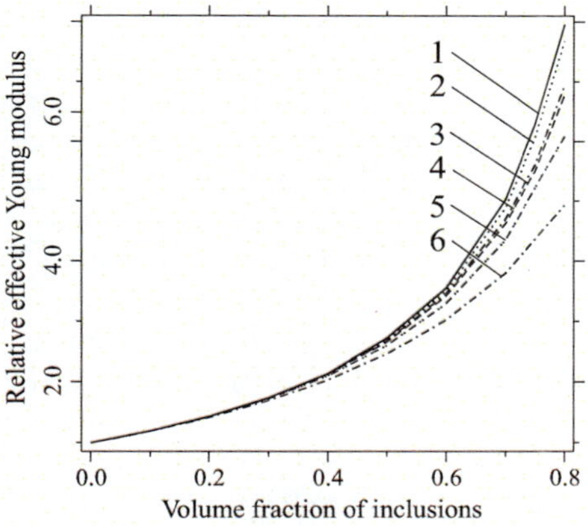

Fig. 9.12 $E_1^*/E^{(0)}$ vs c estimated by the NBM for the RDF (6.4), (6.5) at $n = 0$ (1), $n = 8$ (2), RDF (6.3), (6.5) at $n = 0$ (3), $n = 8$ (4), and by EFM (5), and MTM (6)

The curves of the normalized effective Young's moduli $E_1^*/E^{(0)}$ (the curves $E_2^*/E^{(0)}$ are similar) are presented in Fig. 9.12 as the functions of the volume concentration c. Curves 1–4 are estimated by the new approach (NBM) (9.92), (9.93_1), and (9.95). The RDFs $\varphi(v_q, \mathbf{x}_q|v_i, \mathbf{x}_i)$ (9.117), (9.118) and (9.116), (9.118) were used for estimations of curves 1, 2 and 3, 4, respectively. The initial approximation (9.76) is presented by curves 1 and 3, while curves 2 and 4 illustrate the eighth iteration of the NBM (9.92), (9.93_1), and (9.95). Curves 5 and 6 are predicted by the EFM and MTM, respectively, which are invariant with respect to the concrete form of $g(r)$, either (9.116) or (9.117), in (9.118). As can be seen, the estimations carried out by the different methods are essentially different at $c > 0.6$. In so doing, the difference between curves 1, 2 and 3, 4 obtained by the NBM with distinguished functions $\varphi(v_k, \mathbf{x}_k|v_i, \mathbf{x}_i)$ is basically less than the difference between the curves produced in consequence of the dissimilar methods (compere curves 1–4 with curves 5 and 6). The differences between the initial approximation (1, 3) and corresponding eighth iterations (2, 4) are negligible. Thus, stress concentration factors (Fig. 9.10) are significantly more sensitive values to the choice of the approach than effective elastic moduli (Fig. 9.12).

We demonstrated a quantitative difference between the numerical results obtained by both the new (9.75)–(9.82) and classical (9.41)–(9.48) and (9.52)–(9.59) approaches. These differences are explained by a fundamental reason produced by the distinction of renormalizing terms $\langle \mathbf{U}(\mathbf{x}-\mathbf{y})\boldsymbol{\tau}(\mathbf{y})\rangle$ (7.19) and $\mathbf{U}(\mathbf{x}-\mathbf{y})\langle\boldsymbol{\tau}(\mathbf{y})\rangle$ (7.32) in the new and classical GIEs, respectively. The mentioned difference is only manifested in the case if and only if $\boldsymbol{\tau}(\mathbf{x}) \not\equiv$const. at $\mathbf{x} \in v_i$ for a representative inclusion v_i. The results in Figs. 9.4, 9.5, 9.6, 9.7, 9.8, 9.9, 9.10, 9.11, and 9.12 were obtained for the case when inhomogeneity of stress distributions $\boldsymbol{\sigma}(\mathbf{x})$ inside inclusions $\mathbf{x} \in v_i$ (see Figs. 9.4 and 9.5) is generated as a consequence of noncanonical shape of homogeneous inclusions (9.114). However, the inhomogeneity of stress distributions $\boldsymbol{\sigma}(\mathbf{x})$ inside inclusions $\mathbf{x} \in v_i$ is likewise possible in the ellipsoidal inclusions in the case of their inhomogeneity (see, e.g., [648]). Along this line, one may, for instance, refer to the works [648, 695, 1589, 1609, 1662] concerned with the spatially nonuniform properties of the interphase. Just for concreteness, we assume that circular inclusions contain the cores of the radius $a^c < a$ with the constant moduli $\mathbf{L}^{(1)} \equiv$const, while the moduli $\mathbf{L}^{\text{int}}(\mathbf{x})$ in the interphase with the coating thickness $h = a - a^c$ are taken to vary linearly with the radial distance $r = |\mathbf{x}|$: $\mathbf{L}^{\text{int}}(r) = \mathbf{L}^{(0)} + (\mathbf{L}^{(1)} - \mathbf{L}^{(0)})(a - r)/h$. We assume the matrix is epoxy resin ($\mathbf{L}^{(0)} = (3k^{(0)}, 2\mu^{(0)})$, $k^{(0)} = 3.83$ GPa, and $\mu^{(0)} = 1.27$ GPa) which contains identical circular glass fibers ($\mathbf{L}^{(1)} = (3k^{(1)}, 2\mu^{(1)})$, $k^{(1)} = 34.3$ GPa, and $\mu^{(1)} = 31.3$ GPa). In the iteration method (9.92)–(9.97), the classical solution (9.46) is chosen as an initial approximation (9.93$_2$). In so doing, the perturbator $\boldsymbol{\mathcal{L}}^{\sigma}(\mathbf{x} - \mathbf{x}_q, \overline{\boldsymbol{\sigma}})$ is estimated for arbitrary effective field $\overline{\boldsymbol{\sigma}}$ by the volume integral equation (VIE) method (9.12)–(9.21) with the square mesh Ω^{sq} (9.115), $h/a = 0.03$. As in Sect. 9.5.1, the accuracy of the VIE method was tested for CMs reinforced by the homogeneous circular inclusions in the framework of the hypotheses **H1a**, **H1b**, **H2**, and **H3** when the classical (9.46) and new (9.80) solutions coincide. The numerical error of the VIE method in estimation of stresses inside the inclusions $\langle\boldsymbol{\sigma}\rangle_i(\mathbf{x})$ is 0.09% that is comparable with an application of FEA in Sect. 9.5.1. Going to the iteration method (9.94)–(9.97) in Fig. 9.13 for the coating inclusions $\mathbf{L}^{\text{int}}(r)$, $c = 0.65$ and the step RDF (9.116), we at first estimate the initial approximation $B^{*[0]}_{2211}(\mathbf{x})$ (9.119) corresponding to the classical estimation (9.46) and using the old background (7.33). In Fig. 9.13 the iterations $B^{*[k]}_{2211}(\mathbf{x})$ ($k = 0, 2, 4, 12$) at the axis $\mathbf{x} = (0, x_2)^{\top}$ are presented for $c \equiv \pi a^2 n = 0.65$. The new background (7.19) allows the use of the stress inhomogeneity inside the inclusions for refinement of the renormalizing item in Eq. (9.92) without exploiting hypothesis **H1**. Dramatic difference between the iterations $B^{*[k]}_{2211}(\mathbf{x})$ ($k = 0, 2, 4, 12$) is observed in Fig. 9.13. Indeed, $B^{*[0]}_{1122}(\mathbf{x}) > 0$ at any $\mathbf{x} = (x_1, 0)^{\top}$, while $B^{*[12]}_{1122}(\mathbf{x}) < 0$ at $-0.75a < x_1 < 0.75a$. The proposed

Fig. 9.13 $B^{*[k]}_{2211}(x_2)$ vs x_2/a in inhomogeneous fiber: curves 1, 2, 3, and 4 for $k = 0, 2, 4, 12$, respectively,

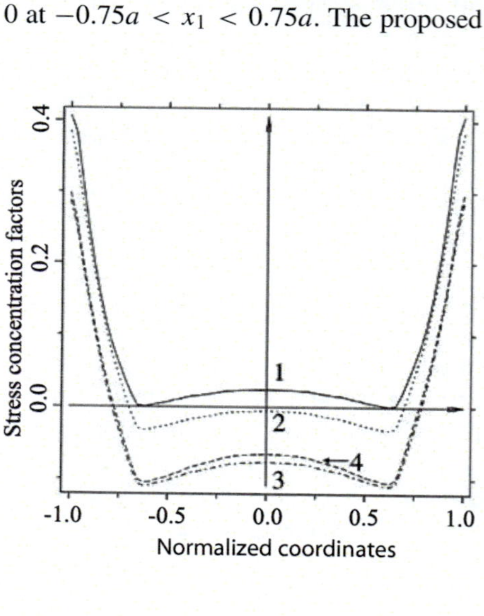

iteration method converges rapidly, and we contend that the 12th iteration provides a difference from the 11th iteration of 0.2%.

9.6 Hashin–Shtrikman Bounds of Effective Moduli Based on the New GIE

In this section we incorporate the new GIE (7.19) (corresponding to the operator ${}^{\mathcal{P}}\mathbf{U}$ (6.200)) into the HS energy bound (6.208). A trial function $\boldsymbol{\tau}^*(\mathbf{x},\alpha)$ is chosen, as before, by Eq. (8.139). Then the first two summands in the functional representation (6.204) are the same as Eq. (8.140), whereas the inner product $(\boldsymbol{\tau}^*, {}^{\mathcal{P}}\mathbf{U}\boldsymbol{\tau})$ is changed with respect to $(\boldsymbol{\tau}^*, \mathbf{U}\boldsymbol{\tau}^*)$ (8.141) $(\mathbf{x}\in v_i,\ \mathbf{y}\in v_j)$

$$
\begin{aligned}
(\boldsymbol{\tau}^*, {}^{\mathcal{P}}\mathbf{U}\boldsymbol{\tau}^*) = &-\int_{v_i} V_i(\mathbf{x})\big[\mathbf{R}_i^v(\mathbf{x})\overline{\boldsymbol{\varepsilon}}_i + \overline{\mathbf{F}}_i^v(\mathbf{x})\overline{\boldsymbol{\varepsilon}}_0\big] \\
&\times \int_{R^d}\Big\{\mathbf{U}(\mathbf{x}-\mathbf{y})V_j(\mathbf{y})(\mathbf{R}_j^v(\mathbf{y})\overline{\boldsymbol{\varepsilon}}_j + \overline{\mathbf{F}}_j^v(\mathbf{y})\overline{\boldsymbol{\varepsilon}}_0)\varphi(v_j,\mathbf{x}_j|v_i,\mathbf{x}_i) \\
&-\langle\mathbf{U}(\mathbf{x}-\mathbf{y})V(\mathbf{y})[\mathbf{R}_j^v(\mathbf{y})\overline{\boldsymbol{\varepsilon}}_j + \overline{\mathbf{F}}_j^v(\mathbf{y})\overline{\boldsymbol{\varepsilon}}_0]\rangle\Big\}d\mathbf{y}d\mathbf{x}, \qquad (9.120)
\end{aligned}
$$

where the integration domains are limited (as in Eq. (8.141)) by the volume of the fixed inclusion v_i and the moving one v_j. By the use of Eq. (8.123), as in Eq. (8.142), we extract from the internal integral $\int_{R^d}(\cdot)d\mathbf{y}$ (9.120) the term corresponding to the inclusion $v_j = v_i$ that completely coincides with the representation (8.142). The difference is only expected at the breaking up of the internal integral $\int_{R^d}(\cdot)d\mathbf{y}$ (9.120), as in Eq. (8.143), into two parts corresponding to the integration domains v_i^0 or ${}^c v_i^0$. However, in a new approach, the mentioned step is completely eliminated, and we have no restrictions on both the shape and the size of the correlation hole v_i^0 as well as on any symmetry (like **H3**) of microtopology described by the two-particle distribution function $\varphi(v_j,\mathbf{x}_j|v_i,\mathbf{x}_i)$.

For demonstration of subsequent arrangement of Eq. (9.120), we introduce, at first, the tensors determined on $\mathbf{x}\in v_i$ (compare with (9.63))

$$
\begin{aligned}
\mathbf{I}_i^{\epsilon}(\mathbf{x}) = \int_{R^d}&\Big\{\mathbf{U}(\mathbf{x}-\mathbf{y})V_j(\mathbf{y})\mathbf{R}_j^v(\mathbf{y})\varphi(v_j,\mathbf{x}_j|;v_i,\mathbf{x}_i) \\
&-\langle\mathbf{U}(\mathbf{x}-\mathbf{y})V(\mathbf{y})\mathbf{R}_j^v(\mathbf{y})\rangle\Big\}d\mathbf{y}d\mathbf{x}, \qquad (9.121)
\end{aligned}
$$

$$
\begin{aligned}
\mathbf{I}_i^{F}(\mathbf{x}) = \int_{R^d}&\Big\{\mathbf{U}(\mathbf{x}-\mathbf{y})V_j(\mathbf{y})\overline{\mathbf{F}}_j^v(\mathbf{y})\varphi(v_j,\mathbf{x}_j|;v_i,\mathbf{x}_i) \\
&-\langle\mathbf{U}(\mathbf{x}-\mathbf{y})V(\mathbf{y})\overline{\mathbf{F}}_j^v(\mathbf{y})\rangle\Big\}d\mathbf{y}d\mathbf{x}. \qquad (9.122)
\end{aligned}
$$

Substitution of the representations (8.142), (9.121) and (9.122) into Eq. (9.120) yields

$$
\begin{aligned}
(\boldsymbol{\tau}^*, {}^{\mathcal{P}}\mathbf{U}\boldsymbol{\tau}^*) = &-\Big([\mathbf{R}_i^v(\mathbf{x})\overline{\boldsymbol{\varepsilon}}_i + \overline{\mathbf{F}}_i^v(\mathbf{x})\overline{\boldsymbol{\varepsilon}}_0], [\mathbf{A}_i(\mathbf{x})\overline{\boldsymbol{\varepsilon}}_i + \mathbf{C}_i(\mathbf{x})\overline{\boldsymbol{\varepsilon}}_0]\Big) \\
&+c^{(1)}[\mathbf{R}_i^v\overline{\boldsymbol{\varepsilon}}_i + \overline{\mathbf{F}}_i^v\overline{\boldsymbol{\varepsilon}}_0]\overline{\boldsymbol{\varepsilon}}_i \\
&-c^{(1)}\langle[\mathbf{R}_i^v(\mathbf{x})\overline{\boldsymbol{\varepsilon}}_i + \overline{\mathbf{F}}_i^v(\mathbf{x})\overline{\boldsymbol{\varepsilon}}_0][\mathbf{J}_i^R(\mathbf{x})\overline{\boldsymbol{\varepsilon}}_j + \mathbf{J}_i^F(\mathbf{x})\overline{\boldsymbol{\varepsilon}}_0]\rangle_{(i)} \qquad (9.123)
\end{aligned}
$$

Furthermore, as in Sect. 4.1, we consider the matrix CM with the matrix modulus coinciding with the reference modulus $\mathbf{L}^c = \mathbf{L}^{0)}$ that also leads to the identical vanishing of the tensors $\boldsymbol{\tau}_0 = \mathbf{0}$ and $\mathbf{C}_i(\mathbf{x})\overline{\boldsymbol{\varepsilon}}_0 = \mathbf{F}_i(\mathbf{x})\overline{\boldsymbol{\varepsilon}}_0 = \overline{\mathbf{F}}_i(\mathbf{x})\overline{\boldsymbol{\varepsilon}}_0 = \mathbf{0}$ in the matrix $\mathbf{x} \in v^{(0)}$. At last, substitution of Eqs. (8.140) and (9.123) into (9.120$_2$) gives the compact expression for the functional $\widetilde{\mathcal{H}}(\boldsymbol{\tau}^*)$

$$
\widetilde{\mathcal{H}}(\boldsymbol{\tau}^*) = -2c^{(1)}(\mathbf{R}_i^v\overline{\boldsymbol{\varepsilon}}_i)\langle\boldsymbol{\varepsilon}\rangle + c^{(1)}(\mathbf{R}_i^v\overline{\boldsymbol{\varepsilon}}_i)\overline{\boldsymbol{\varepsilon}}_i + c^{(1)}\langle[\mathbf{R}_i^v(\mathbf{x})\overline{\boldsymbol{\varepsilon}}_i\big][\mathbf{I}_i^{\epsilon}(\mathbf{x})\overline{\boldsymbol{\varepsilon}}_j]\rangle_{(i)}, \qquad (9.124)
$$

whose extremizing with respect to $\overline{\boldsymbol{\varepsilon}}_i$ leads to the final representations for the effective field $\overline{\boldsymbol{\varepsilon}}_i$, statistical average strain $\langle\boldsymbol{\varepsilon}\rangle(\mathbf{x})$ $(\mathbf{x} \in v_i)$, and the Hashin–Shtrikman bounds $\mathbf{L}^{*(\mathrm{HS})[3]}$ of the effective modulus

$$
\overline{\boldsymbol{\varepsilon}}_i = [\mathbf{I} - (\mathbf{R}_i^v)^{-1}\langle\mathbf{R}_i^v(\mathbf{x})\mathbf{J}_i^{\epsilon}(\mathbf{x})\rangle_{(i)}\Big]^{-1}\langle\boldsymbol{\varepsilon}\rangle, \qquad (9.125)
$$

$$
\langle\boldsymbol{\varepsilon}\rangle(\mathbf{x}) = \mathbf{A}_i(\mathbf{x})\Big[\mathbf{I} - (\mathbf{R}_i^v)^{-1}\langle\mathbf{R}_i^v(\mathbf{x})\mathbf{I}_i^{\epsilon}(\mathbf{x})\rangle_{(i)}\Big]^{-1}, \qquad (9.126)
$$

$$
\mathbf{L}^{*(\mathrm{HS})[3]} = \mathbf{L}^{(0)} + \mathbf{c}^{(1)}\mathbf{R}_i^v\Big[\mathbf{I} - (\mathbf{R}_i^v)^{-1}\langle\mathbf{R}_i^v(\mathbf{x})\mathbf{I}_i^{\epsilon}(\mathbf{x})\rangle_{(i)}\Big]^{-1}, \qquad (9.127)
$$

respectively. $\mathbf{L}^{*(\mathrm{HS})[3]}$ (9.127) is a lower bound for $\mathbf{L}^*$ if $\mathbf{L}^{(0)} < \mathbf{L}^{(1)}$ and a upper bound if $\mathbf{L}^{(0)} > \mathbf{L}^{(1)}$. The tensor $\mathbf{I}_i^{\epsilon}$ (9.121) can be expressed through the perturbator factors (9.63).

The results (9.125)–(9.127) were obtained in the framework of the hypothesis **H1a** (see the approximation (8.138) for the trial polarization tensor $\boldsymbol{\tau}^*(\mathbf{x})$) and **H2** (there is no direct interactions of inclusions v_i and v_j in Eq. (9.120)). In so doing, the hypothesis **H1b** was not used (in contrast to Sect. 8.4) that leads to an inequality $\langle\mathbf{R}_i^v(\mathbf{x})\mathbf{J}_i^R(\mathbf{x})\rangle_{(i)} \neq \mathbf{R}_i^v\langle\mathbf{J}_i^R(\mathbf{x})\rangle_{(i)}$ in (9.125)–(9.127) in a general case of inhomogeneity of $\mathbf{R}_i(\mathbf{x})$ $(\mathbf{x} \in v_i)$ (as it takes place for either the inhomogeneous inclusions or noncanonical ones). In so doing, the hypothesis **H3** and any its versions are not used, and Eqs. (9.125)–(9.128) hold at any two-particle distribution function $\varphi(v_j, \mathbf{x}_j|; v_i, \mathbf{x}_i)$ passing to $\varphi(v_j, \mathbf{x}_j)$ quickly enough at $|\mathbf{x}_i - \mathbf{x}_j| \to \infty$. Moreover, the notion of a correlation hole v_i^0 is absent in the current Sect. 9.6.

However, for ellipsoidal homogeneous (3.183) inclusions, acceptance of the hypothesis **H1a** means automatic acceptance of the hypothesis **H1b** leading to the equalities $\mathbf{R}_i(\mathbf{x}) \equiv$const. $(\mathbf{x} \in v_i)$, $\langle\mathbf{R}_i^v(\mathbf{x})\mathbf{I}_i^{\epsilon}(\mathbf{x})\rangle_{(i)} = \mathbf{R}_i^v\langle\mathbf{J}_i^R(\mathbf{x})\rangle_{(i)}$, and

$\mathbf{L}_i^{\epsilon}(\mathbf{x}-\mathbf{x}_j) = \mathbf{T}_j(\mathbf{x}-\mathbf{x}_j)\mathbf{R}_i$. In the case of additional acceptance of the hypothesis **H3**, the last equality leads to the equality $\langle \mathbf{I}_i^{\epsilon}(\mathbf{x})\rangle_{(i)} = \mathbf{P}_i^0\mathbf{R}_i^v$ (and, therefore, to equivalence of $\mathbf{L}^{*(\mathrm{HS})[1]}$ (8.127) and $\mathbf{L}^{*(\mathrm{HS})[3]}$ (9.127)) that was qualitatively and quantitatively proved in Sects. 9.5.1 and 9.6, respectively.

Just for completeness, we recast Eq. (9.68) for the effective moduli to the case of CMs with the identical inclusions

$$\mathbf{L}^{*[3]} = \mathbf{L}^{(0)} + \mathbf{c}^{(1)}\mathbf{R}_i^v\Big[\mathbf{I} - \langle \mathbf{I}_i^{\epsilon}(\mathbf{x})\rangle_{(i)}\Big]^{-1}. \tag{9.128}$$

For the particular type of CMs with ellipsoidal homogeneous inclusions (3.183), the hypothesis **H1a** assumed to be being accepted yields the hypothesis **H1b**. Additional hypothesis **H3** leads to equivalence of all effective moduli ($\mathbf{L}^{*[1]}, \mathbf{L}^{*[2]}, \mathbf{L}^{*[3]}$) and Hashin–Shtrikman bounds ($\mathbf{L}^{*(\mathrm{HS})[1]}, \mathbf{L}^{*(\mathrm{HS})[2]}, \mathbf{L}^{*(\mathrm{HS})[3]}$). For either noncanonical or inhomogeneous inclusions, the effective moduli representations ($\mathbf{L}^{*[1]}$, $\mathbf{L}^{*[2]}$, $\mathbf{L}^{*[3]}$)) are not reduced to each other in general (the same we can say about HS bounds).

Thus, we qualitatively compared the pairs (HS bounds)-($\mathbf{L}^*$ estimations) one with each other and mentioned that $\mathbf{L}^{*(\mathrm{HS})[1]} = \mathbf{L}^{*[1]}$ and $\mathbf{L}^{*(\mathrm{HS})[2]} = \mathbf{L}^{*[2]}$ in the frameworks of the accepted sets of hypotheses. It is interesting that $\mathbf{L}^{*(\mathrm{HS})[3]} \neq \mathbf{L}^{*[3]}$ in general although both $\mathbf{L}^{*(\mathrm{HS})[3]}$ and $\mathbf{L}^{*[3]}$ were obtained in the framework of the same hypotheses **H1a** and **H2** (without the hypotheses **H1b** and **H3**). A fundamental reason of this difference is that $\mathbf{L}^*$ is estimated through the averaging of a linear function of strains (or polarization tensors), while estimation of $\widetilde{\mathcal{H}}(\boldsymbol{\tau}^*)$ is performed by averaging of square function ($\widetilde{\mathcal{H}}(\boldsymbol{\tau}^*)$ is an energy-like representation) of strains (or polarization tensors). Only in the framework of the additional hypothesis **H1b**, these two different types of averaging coincide (that happened for the pairs $\mathbf{L}^{*(\mathrm{HS})[1]} = \mathbf{L}^{*[1]}$ and $\mathbf{L}^{*(\mathrm{HS})[2]} = \mathbf{L}^{*[2]}$). It is also interesting that new final representation for $\mathbf{L}^{*(\mathrm{HS})[3]}$ (9.127) and $\mathbf{L}^{*[3]}$ (9.128) do not use the classical notions such as Green function, Eshelby tensor, correlation hole, or ellipsoidal symmetry of microtopology of CMs. Instead of these, the new relation of $\mathbf{L}^{*(\mathrm{HS})[3]}$ and $\mathbf{L}^{*[3]}$ explicitly depends just on both an arbitrary two-particle distribution function $\varphi(v_j, \mathbf{x}_j|v_i, \mathbf{x}_i)$ and the perturbation field introduced by a representative isolated inclusion inside and in the vicinity of this inclusion $\mathbf{A}_i(\mathbf{x})$ (9.32) and $\mathbf{L}_i^{\epsilon}(\mathbf{x}-\mathbf{x}_i)$ (9.30) After numerical estimation of some weighted averages of these perturbations, the final analytical representations for both the HS bounds and effective modulus estimations are just as simple as the corresponding classical counterparts (8.127) and (8.130_1) which can also be obtained as a particular cases of the new formulae for CM with ellipsoidal homogeneous inclusions.

9.7 Concluding Remarks

A fundamental deficiency of Eqs. (7.32) and (7.33) is the dependence of the renormalizing terms $\mathbf{U}(\mathbf{x}-\mathbf{y})\langle\boldsymbol{\tau}\rangle(\mathbf{y})$ and $\boldsymbol{\Gamma}(\mathbf{x}-\mathbf{y})\langle\boldsymbol{\eta}\rangle(\mathbf{y})$ [obtained in the framework of the asymptotic approximation of the hypothesis **H1b**] only on the statistical average $\langle\boldsymbol{\tau}\rangle$ and $\langle\boldsymbol{\eta}\rangle$, while the renormalizing terms $\langle\boldsymbol{\mathcal{L}}^{\epsilon}_j(\mathbf{x}-\mathbf{x}_j,\boldsymbol{\tau})\rangle(\mathbf{x}_j)$ and $\langle\boldsymbol{\mathcal{L}}^{\sigma}_j(\mathbf{x}-\mathbf{x}_j,\boldsymbol{\eta})\rangle(\mathbf{x}_j)$ (9.61) explicitly depend on distributions $\langle\boldsymbol{\tau}|v_j,\mathbf{x}_j\rangle(\mathbf{y})$ and $\langle\boldsymbol{\eta}|v_j,\mathbf{x}_j\rangle(\mathbf{y})$ $(\mathbf{y}\in v_j)$, respectively. Because of this, even in the case of statistically homogeneous media subjected to homogeneous boundary conditions, new effects have been found. So, the final classical representations of the effective properties $(\mathbf{M}^*,\boldsymbol{\beta}^*,W^*)$ and effective concentration factors estimated by both the EFM (9.41)–(9.48) and MTM (9.52)–(9.59) depend only on the average concentration factor [e.g., $\mathbf{B}_i,\mathbf{C}_i$ (9.27_1)], while the effective properties and effective concentration factors estimated by the new approach (9.75)–(9.82) implicitly depend on the inhomogeneous concentration tensor [e.g., $\mathbf{B}_i(\mathbf{x}),\mathbf{C}_i(\mathbf{x})$ $(\mathbf{x}\in v_i)$ (9.27_1)]. The detected dependence allows us to abandon the hypothesis **H1b** whose accuracy is questionable for inclusions of noncanonical shape. We obtained a fundamental conclusion that both the effective properties and effective concentration factors in general depend not only on the stress distribution inside the referred heterogeneity [describing by the tensor $\mathbf{B}_i(\mathbf{x}),\mathbf{C}_i(\mathbf{x})$, $\mathbf{x}\in v_i$ (9.27_1)] but also on the stresses in the vicinity of heterogeneity, i.e., extension of $\mathbf{B}_i(\mathbf{x}),\mathbf{C}_i(\mathbf{x})$ (or $\mathbf{L}^{\sigma}_i(\mathbf{x}-\mathbf{x}_i),\mathbf{L}^{\beta}_i(\mathbf{x}-\mathbf{x}_i,\boldsymbol{\beta})),\mathbf{x}\in v^{\oplus}_i$ is necessary. Then the size of the excluded volume v^0_i as well as the binary correlation function $\varphi(v_q,\mathbf{x}_q|v_i,\mathbf{x}_i)$ will impact on both the effective properties and effective concentration factors even in the framework of hypothesis **H2**. In so doing, the concentration factors (see Figs. 9.7 and 9.8) are significantly more sensitive values to the choice of both the approach (either NBM, EFM, or MTM) and the mentioned statistical distributions of heterogeneities [(9.116) or (9.117)] than effective properties (see Fig. 9.6 and [234]).

Thus, we have proposed the new micromechanical model based on the integral equations (7.19) and (7.34) for strain and stress, respectively, and we have introduced the concept of *perturbators* which can be found by any available numerical or semianalytical method. This makes it potentially possible to abandon many classical concepts of micromechanics used in most popular methods, namely, effective field hypothesis **H1**, quasi-crystalline approximation **H2**, the hypothesis of "ellipsoidal symmetry" **H3**, and Eshelby tensor. Abandonment of a few different combinations of these hypotheses leads to detection of some new effects that are impossible in the framework of a classical background of micromechanics. For example, hypotheses **H1a** and **H2** were used, while hypotheses **H1b** and **H3** were not accepted in Sect. 9.5.1 (see also [232, 234]). In Sect. 9.5.2 (see also [203]) has exploited the hypotheses **H2** and **H3**, while the hypotheses **H1a** and **H1b** were not to be accepted for analysis of composites with circular inhomogeneous inclusions. Analysis of composites with the circular homogeneous inclusions (when analytical representation for the perturbator factor (9.31) is known) significantly simplifies the problem that enabled Buryachenko [204] to abandon the hypotheses **H1** and

H2 and use only hypothesis **H3**. Buryachenko [205] also considered the case of circle homogeneous inclusions but with nonlocal constitutive law with the use of the hypotheses **H2** and **H3** and abandonment of the hypothesis **H1** (the list of some other problems where one expects to get fundamentally new results in the case of using of the new background of analytical micromechanics (7.19) is presented in [203]).

A larger difference between the use of the new (7.20) and old (7.34) backgrounds was obtained for composites reinforced by inclusions demonstrating essentially inhomogeneous stress distribution inside isolated inclusions even in the framework of the hypothesis **H1a**. It was quantitatively estimated that the use of the new background (7.20) instead of the old one (7.34) leads to just small corrections of effective properties (effective moduli, coefficient of thermal expansion, and stored energy), while the estimations of the statistical averages of local stresses can be dramatically different with possible change of the sign of predicted local stresses.

Of course, a noncanonical shape of inclusion is dictated by a practical interest rather than by a theoretical exploration. Research shows that composite mechanical properties greatly depend on the fiber shape [1694]. To obtain a better load transfer mechanism and better stress distribution, many different fiber geometries have been experimented and analyzed. Kozaczek et al. [800] demonstrated that the shape of a nonellipsoidal inclusion plays a role in the stress distribution in the grain boundary region; sharp corners raise stress more effectively than rounded edges of oblong-shaped precipitates. CM reinforced by shaped head fibers provide additional mechanical locking in comparison with straight fibers. Zhou [1689] probably first introduced this concept and showed that matrix composites with dumbbell-shaped steel wires have higher strength than those reinforced by straight wires. Tsai et al. [1536] analyzed stress profiles induced during pullout of two chosen shaped head families using a finite element analysis (FEA). Bagwell and Wetherhold [51] (see also [1615]) investigated shaped fiber ends produced by end-impacting and knotting fibers to facilitate anchoring, similar to work with bone-shaped short fibers produced by Zhu and Beyerlein [1696] who evaluated the mechanical behavior of enlarged end short fibers by experimental pullout of aligned fibers from a polyester matrix. Zhou et al. [1694] developed a FEA procedure for inclusion shape optimization maximizing the stiffness of CM and demonstrated that the enlarged end short fiber with many threads is more desirable.

Chapter 10
Inhomogeneity of the First and Second Statistical Moments of Stresses Inside the Heterogeneities of CMs

Abstract Exact second moments of stresses averaged over the phase volumes were obtained in Chap. 6. In the current study, the method of integral equations (see Buryachenko, Micromechanics of Heterogeneous Materials, Springer, New York, 2007) is generalized to the new inhomogeneous effective field concept obtained in the framework of the new background of analytical micromechanics described in Chaps. 7 and 9. This generalization is reduced to the replacement of tensor stress concentration factors by its operator counterparts. In the numerical examples for CMs with the homogeneous circular inclusions, it was demonstrated that even in the framework of the hypothesis **H3**, taking the binary interaction of inclusions into account discovers fundamentally new effects such as a change of a sign in $\langle \boldsymbol{\sigma} \rangle_i(\mathbf{x})$ and strong variation (on 80%) of $\langle \boldsymbol{\sigma} \otimes \boldsymbol{\sigma} \rangle_i(\mathbf{x})$ inside the representative inclusion $\mathbf{x} \in v_i$. In so doing, the average moments $\langle \boldsymbol{\sigma} \rangle_i$ and $\langle \boldsymbol{\sigma} \otimes \boldsymbol{\sigma} \rangle_i$ are only slightly distinguished from the classical results described in Chap. 6.

Although the effective behavior of the composite is traditionally the main focus of micromechanics, it is also essential to supply insight into the statistical description of the local strains and stresses, such as their statistical moments of different order in each phase and at interphase. Estimation of these local fields is extremely useful for understanding the evolution of nonlinear phenomena such as plasticity, creep, and damage. Stress fluctuations in the components of random structure composites represent a measure of inhomogeneity of stress fields in the components; numerical statistical analysis at the inclusion scale level was performed, e.g., by Babuska et al. [49]. The fundamental role of the statistical averages of the second moments of stress concentration factors in nonlinear analysis is explained by the fact that both the yield surface, fiber/matrix interface failure criterion, and the energy release rate are the quadratic functions of the local stress distributions. In so doing, estimation of the effective limiting surfaces separating the linear and nonlinear behavior domains of the mentioned nonlinear phenomena is of profound importance for the practical applications and can be done by the methods describing the linear deformations of composite materials. It should be mentioned that the estimation of the effective elastic moduli is a linear problem, with respect to the stress field analyzed which

V. A. Buryachenko, *Local and Nonlocal Micromechanics of Heterogeneous Materials*, https://doi.org/10.1007/978-3-030-81784-8_10

is less sensitive to the local stress distribution than nonlinear micromechanical problems of elastoplastic deformation, fracture, and fatigue of composite materials depending, at least, on mean-square stress fluctuations in the constituents. The fundamental roles of such inhomogeneities described by the stress fluctuations are discussed in detail in [200, 945, 1239] for a wide class of nonlinear problems of micromechanics such as plasticity, damage, viscosity, or creeping.

Several papers have already been written on the problem of estimation of values of invariants averaged over the volume of the components, which involve particular assumptions or simplifications as, for instance, the two-dimensional model [512, 884] and special correlation function [1161]. A very prospective idea of a perturbation method proposed by Bergman [110] is based on the estimation of the perturbation of an energetic function due to a variation of the material properties, e.g., conductivity and elasticity modulus. This idea was developed for estimation of stress fluctuations in the case of isotropy of materials constants [132, 133, 805] and isotropy of fluctuations [808]. The perturbation method was developed for the exact estimation of all components of the second moment tensor of the pure elastic [1187] and internal residual stresses [280], averaged over the volume of components. It is found that the second moment of the stress field is constant within the inclusions if a homogeneity of some random effective stress fields in the neighborhood of each ellipsoidal inclusion is additionally assumed. Furthermore, an expression for the second moment of the stress tensor in the matrix in the vicinity of the inclusion is derived, and a method for the construction of the correlation function of the internal stresses in separate inclusions is developed. The applicability of the proposed methods for estimating second moments of stresses in the matrix is significantly limited, because they require evaluation of the effective properties of the composite for the general case of matrix anisotropy (even for the case of actually isotropic matrix). This disadvantage was eliminated via the method of integral equations by Buryachenko and Rammerstorfer [265, 268] allowing the estimations of second moments of stresses in the components. Considering both binary and triple interaction of the inclusions, explicit relations for second moments of stresses were obtained. The inhomogeneity of the second moment of stresses in coated inclusions was shown. A comparison between the second invariants of deviator stresses in the matrix estimated via the perturbation method [280] and that estimated via the integral equation method [265, 268] was conducted in [200], Chap. 15. The method proposed was based on the EFH **H1** and the classical GIE (7.33).

In the current study, the approach [265, 268] is generalized (following [204]) to the new inhomogeneous effective field concept obtained in the framework of the new background of analytical micromechanics described in Chaps. 7 and 9. If in some cases this generalization is reduced to the replacement of tensor stress concentration factors by its operator counterparts, then interested readers are referred to the previous publications by Buryachenko [199, 200] cited in the appropriate sections to achieve a deeper understanding of these topics. In the numerical examples for CMs with the homogeneous circular inclusions, it was demonstrated that even in the framework of the hypothesis **H3**, taking the binary interaction of inclusions into account discovers fundamentally new effects such as a change of a sign in $\langle \boldsymbol{\sigma} \rangle_i(\mathbf{x})$

and strong variation (on 80%) of $\langle\boldsymbol{\sigma}\otimes\boldsymbol{\sigma}\rangle_i(\mathbf{x})$ inside the representative inclusion $\mathbf{x}\in v_i$. In so doing, the average moments $\langle\boldsymbol{\sigma}\rangle_i$ and $\langle\boldsymbol{\sigma}\otimes\boldsymbol{\sigma}\rangle_i$ are only slightly distinguished from the classical results described in [200].

10.1 Effective Moduli and the First Statistical Moments Inside Inclusions

10.1.1 General Representations

Fixing the inclusion v_i in the composite material produces the random effective field $\overline{\boldsymbol{\sigma}}_i(\mathbf{x})$ (7.24) which can be recast also in the terms of strain polarization tensors $(\mathbf{x}\in v_i)$

$$\overline{\boldsymbol{\eta}}(\mathbf{x})=\boldsymbol{\eta}_i^{\infty}(\mathbf{x})+\int[\mathbf{K}_i(\mathbf{x},\mathbf{y})\boldsymbol{\eta}(\mathbf{y})V(\mathbf{y}|;v_i,\mathbf{x}_i)-\langle\mathbf{K}_i(\mathbf{x},\mathbf{y})\boldsymbol{\eta}\rangle]d\mathbf{y}, \tag{10.1}$$

where $\boldsymbol{\eta}_i^{\infty}(\mathbf{x})\equiv\mathbf{E}_i(\mathbf{x})\langle\boldsymbol{\sigma}\rangle+\mathbf{H}_i(\mathbf{x})$ is the external strain polarization tensor generated by the homogeneous loading $\langle\boldsymbol{\sigma}\rangle$ (6.5). Acting with the operator $\boldsymbol{\mathcal{L}}_i$ on Eq. (10.1) and taking Eq. (4.131) into account, we obtain

$$\bar{v}_i\boldsymbol{\eta}(\mathbf{x})=\mathbf{R}_i(\mathbf{x})\langle\boldsymbol{\sigma}\rangle+\mathbf{F}_i(\mathbf{x})+\boldsymbol{\mathcal{L}}_i\int[\mathbf{K}_i(\mathbf{z},\mathbf{y})\boldsymbol{\eta}(\mathbf{y})V(\mathbf{y}|;v_i,\mathbf{x}_i)-\langle\mathbf{K}_i(\mathbf{z},\mathbf{y})\boldsymbol{\eta}\rangle]d\mathbf{y}. \tag{10.2}$$

Conditional averaging of Eq. (10.2) with fixed inclusion v_i and exploitation of the solutions (4.124_2), (4.134), and (4.137) in the framework of just binary interacting heterogeneities allow one to reduce Eq. (10.2) to the following:

$$\begin{aligned}\boldsymbol{\mathcal{R}}_i(\langle\overline{\boldsymbol{\sigma}}_i\rangle)(\mathbf{x})=\mathbf{R}_i(\mathbf{x})\langle\boldsymbol{\sigma}\rangle+\int\Big\{&[\boldsymbol{\mathcal{R}}_{i,j}(\langle\widetilde{\boldsymbol{\sigma}}_{i,j}\rangle)(\mathbf{x})+\mathbf{F}_{i,j}(\mathbf{x})]\varphi(v_j,\mathbf{x}_j|;v_i,\mathbf{x}_i)\\&-[\boldsymbol{\mathcal{R}}_{i,j}^{\infty}(\langle\overline{\boldsymbol{\sigma}}_j\rangle)(\mathbf{x})+\mathbf{F}_{i,j}^{\infty}(\mathbf{x})]n^{(j)}\Big\}d\mathbf{x}_j.\end{aligned} \tag{10.3}$$

No confusion will arise hereafter in definition of the operator $\boldsymbol{\mathcal{D}}$ ($\boldsymbol{\mathcal{R}}_{i,j}$, $\boldsymbol{\mathcal{R}}_{i,j}^{\infty}$, $\mathbf{F}_{i,j}$, $\mathbf{F}_{i,j}^{\infty}$) with the kernel $\boldsymbol{\mathcal{D}}(\mathbf{x},\mathbf{y})$ on the inhomogeneous functions $\mathbf{g}(\mathbf{y})$ (e.g., $\mathbf{g}(\mathbf{y})=\langle\widetilde{\boldsymbol{\sigma}}_{i,j}\rangle(\mathbf{y}),\ \langle\overline{\boldsymbol{\sigma}}_j\rangle(\mathbf{y}),\ \mathbf{y}\in V_k;\ V_k=V_i,V_j,V_i+V_j$)

$$\boldsymbol{\mathcal{D}}(\mathbf{g})(\mathbf{x})=\int\boldsymbol{\mathcal{D}}(\mathbf{x},\mathbf{y})\mathbf{g}(\mathbf{y})V_k(\mathbf{y})d\mathbf{y}. \tag{10.4}$$

The operator $\mathcal{D}$ is reduced to the tensor $\mathbf{D}(\mathbf{x})$ on the constant functions $\mathbf{g}(\mathbf{y}) = \mathbf{g} \equiv$const. ($\mathbf{y} \in V_k$)

$$\mathcal{D}(\mathbf{g})(\mathbf{x}) = \mathbf{D}(\mathbf{x})\mathbf{g}, \ \ \mathbf{D}(\mathbf{x}) = \int \mathcal{D}(\mathbf{x}, \mathbf{y}) V_k(\mathbf{y}) d\mathbf{y}. \tag{10.5}$$

No restrictions are imposed on the microtopology of the microstructure and the shape of inclusions as well as on the inhomogeneity of stress field inside the inclusions. However, the main computational advantage of the proposed Eq. (10.3) lies in the fact that such fundamental notions of micromechanics as the Green function is not exploited, and we can analyze any anisotropy of constituents (including the matrix) as well as any shape and any composite structure of inclusions. We constructed Eqs. (10.3) by the use of some building blocks described by the numerical solutions for both one and two inclusions inside the infinite medium subjected to the homogeneous loading at infinity. Just at some additional assumptions, the tensors mentioned above can be expressed through the Green function, Eshelby tensor, and external Eshelby tensor. Equation (10.3) is exact and involves two sorts of statistically averaged effective fields $\langle\overline{\boldsymbol{\sigma}}_i\rangle(\mathbf{x})$ and $\langle\widetilde{\boldsymbol{\sigma}}_{i,j}\rangle(\mathbf{x})$ which can be found from the infinite hierarchy of equations analogous to the system (7.25). These truncated hierarchies of equations will be limited by Eq. (10.3) by the use of the closing assumption considered in the next subsection.

10.1.2 Closing Assumptions

For the closing of Eq. (10.3) we will use the following hypothesis:

H2a *Each pair of inclusions v_j and v_j is subjected to the inhomogeneous field $\widetilde{\boldsymbol{\sigma}}_{i,j}(\mathbf{x})$, and statistical average $\langle\widetilde{\boldsymbol{\sigma}}_{i,j}\rangle(\mathbf{x})$ is defined by the formulae*

$$\langle\widetilde{\boldsymbol{\sigma}}_{i,j}\rangle(\mathbf{x}) = \langle\overline{\boldsymbol{\sigma}}_k\rangle(\mathbf{x}) \tag{10.6}$$

at $\mathbf{x} \in v_k, \ k = i, j.$

The refined version of the closing assumption (10.6) of the MEFM proposed by [199] takes into account both the three-point correlation functions and dependence of the effective fields acting on each pair of heterogeneities v_i and v_j on the distance $|\mathbf{x}_i - \mathbf{x}_j|$.

The adoption of hypothesis **H2a** makes it possible to reduce Eq. (10.3) to the operator one

$$\mathcal{R}_i(\langle\overline{\boldsymbol{\sigma}}_i\rangle)(\mathbf{x}) = \mathbf{R}_i(\mathbf{x})\langle\boldsymbol{\sigma}\rangle + \int \Big\{[\mathcal{R}^I_{i,j}(\langle\overline{\boldsymbol{\sigma}}_i\rangle)(\mathbf{x}) + \mathcal{R}^J_{i,j}(\langle\overline{\boldsymbol{\sigma}}_j\rangle)(\mathbf{x}) + \mathbf{F}_{i,j}(\mathbf{x})]$$
$$\times\varphi(v_j, \mathbf{x}_j|; v_i, \mathbf{x}_i) - [\mathcal{R}^{\infty}_{i,j}(\langle\overline{\boldsymbol{\sigma}}_j\rangle)(\mathbf{x}) + \mathbf{F}^{\infty}_{i,j}(\mathbf{x})]n^{(j)}\Big\} d\mathbf{x}_j. \tag{10.7}$$

which can be solved by the iteration method based on the recursion formula

$$\mathcal{R}_i(\langle\overline{\boldsymbol{\sigma}}_i^{[n+1]}\rangle)(\mathbf{x}) = \mathbf{R}_i(\mathbf{x})\langle\boldsymbol{\sigma}\rangle + \int \Big\{[\mathcal{R}^I_{i,j}(\langle\overline{\boldsymbol{\sigma}}_i^{[n]}\rangle)(\mathbf{x}) + \mathcal{R}^J_{i,j}(\langle\overline{\boldsymbol{\sigma}}_j^{[n]}\rangle)(\mathbf{x}) + \mathbf{F}_{i,j}(\mathbf{x})] \times \varphi(v_j,\mathbf{x}_j|;v_i,\mathbf{x}_i) - [\mathcal{R}^\infty_{i,j}(\langle\overline{\boldsymbol{\sigma}}_j^{[n]}\rangle)(\mathbf{x}) + \mathbf{F}^\infty_{i,j}(\mathbf{x})]n^{(j)}\Big\}d\mathbf{x}_j. \tag{10.8}$$

with an initial approximation considered in Sect. 10.1.3. However, we will use a more convenient method for numerical realization, a modified iteration scheme obtained on the base of perturbations (4.138), (4.139) rather than (4.134), (4.137) ($\mathbf{x}\in v_i$)

$$\langle\overline{\boldsymbol{\sigma}}_i^{[n+1]}\rangle(\mathbf{x}) = \langle\boldsymbol{\sigma}\rangle + \int \Big\{[\mathcal{T}^{I\sigma}_{i,j}(\langle\overline{\boldsymbol{\sigma}}_i^{[n]}\rangle)(\mathbf{x}) + \mathcal{T}^{J\sigma}_{i,j}(\langle\overline{\boldsymbol{\sigma}}_j^{[n]}\rangle)(\mathbf{x}) + \mathbf{T}^{\beta}_{i,j}(\mathbf{x})] \times \varphi(v_j,\mathbf{x}_j|;v_i,\mathbf{x}_i) - [\mathcal{T}^{\sigma\infty}_{i,j}(\langle\overline{\boldsymbol{\sigma}}_j^{[n]}\rangle)(\mathbf{x}) + \mathbf{T}^{\beta\infty}_{i,j}(\mathbf{x})]n^{(j)}\Big\}d\mathbf{x}_j, \tag{10.9}$$

$$\langle\boldsymbol{\sigma}^{[n+1]}\rangle_i(\mathbf{x}) = \mathcal{B}_i(\langle\overline{\boldsymbol{\sigma}}_i^{[n+1]}\rangle)(\mathbf{x}) + \mathbf{C}_i(\mathbf{x}), \tag{10.10}$$

where the decomposition $\mathcal{T}^\sigma_{i,j} = \mathcal{T}^{I\sigma}_{i,j} + \mathcal{T}^{J\sigma}_{i,j}$ was introduced analogously to (4.135).

The solution (10.9) and (10.10) can be formally presented in an operator form $\overline{v}_i\langle\boldsymbol{\eta}^{[n+1]}\rangle_i(\mathbf{x}) = \mathcal{R}_i(\langle\overline{\boldsymbol{\sigma}}_i^{[n+1]}\rangle)(\mathbf{x}) + \mathbf{F}_i(\mathbf{x})$ suggesting the Neumann series form for the solution $\langle\boldsymbol{\eta}\rangle_i(\mathbf{x})$ and $\langle\boldsymbol{\sigma}\rangle_i(\mathbf{x})$ ($\mathbf{x}\in v_i$)

$$\overline{v}_i\langle\boldsymbol{\eta}\rangle_i(\mathbf{x}) \equiv \lim_{n\to\infty}\overline{v}_i\langle\boldsymbol{\eta}^{[n]}\rangle_i(\mathbf{x}) = \mathbf{R}^*_i(\mathbf{x})\langle\boldsymbol{\sigma}\rangle + \mathbf{F}^*_i(\mathbf{x}), \tag{10.11}$$

$$\langle\boldsymbol{\sigma}\rangle_i(\mathbf{x}) \equiv \lim_{n\to\infty}\langle\boldsymbol{\sigma}^{[n]}\rangle_i(\mathbf{x}) = \mathbf{B}^*_i(\mathbf{x})\langle\boldsymbol{\sigma}\rangle + \mathbf{C}^*_i(\mathbf{x}), \tag{10.12}$$

which yields the final representations for the effective properties

$$\mathbf{M}^* = \mathbf{M}^{(0)} + \langle\mathbf{R}^*V\rangle,\quad \boldsymbol{\beta}^* = \boldsymbol{\beta}^{(0)} + \langle\mathbf{F}^*V\rangle,\quad W^* = -\langle\boldsymbol{\beta}_1\mathbf{C}^*V\rangle. \tag{10.13}$$

A convergence of the sequence $\langle\boldsymbol{\eta}^{[n]}\rangle_i(\mathbf{x})$ (10.11) and $\langle\boldsymbol{\sigma}^{[n]}\rangle_i(\mathbf{x})$ (10.12) is analyzed analogously to the sequence (4.98).

The hypothesis **H2a**, rewritten in terms of stresses $\boldsymbol{\sigma}(\mathbf{x})$, ($\mathbf{x}\in v_i$), is a standard closing assumption (see for reference, e.g., Chap. 8) degenerating to the "quasi-crystalline" approximation by Lax [886] ignoring the binary interaction of heterogeneities and assuming homogeneity of the effective fields:

Hypothesis H2b, "Quasi-Crystalline" Approximation *It is supposed that the mean value of the effective field at a point* $\mathbf{x} \in v_i$ *does not depend on the stress field inside surrounding heterogeneities* $v_j \neq v_i$

$$\langle \overline{\boldsymbol{\sigma}} | v_i, \mathbf{x}_i; v_j, \mathbf{x}_j \rangle(\mathbf{x}) = \langle \overline{\boldsymbol{\sigma}}_k \rangle(\mathbf{x}), \quad \langle \overline{\boldsymbol{\sigma}}_k \rangle(\mathbf{x}) \equiv \text{const}, \tag{10.14}$$

at $\mathbf{x} \in v_k, \ (k = i, j)$.

In the framework of the EFH **H1**, the principal difference between the hypotheses **H2a** (10.6) and **H2b** (10.14) was analyzed in Chap. 4. A quantitative difference of results obtained in framework of the hypotheses **H2a** (10.6) and **H2b** (10.14) (without the hypothesis **H1**) will be demonstrated in Sect. 10.3.

It should be mentioned that Buryachenko [203] used the generalized "quasi-crystalline" approximation by Lax [886] (10.14_1) when the assumption (10.14_2) was relaxed: $\langle \overline{\boldsymbol{\sigma}}_k \rangle(\mathbf{x}) \not\equiv \text{const}$ at $\mathbf{x} \in v_k, \ (k = i, j)$. It greatly simplifies the problem (10.7) where in such a case $\mathcal{T}^{\sigma}_{i,j} = \mathcal{T}^{\sigma\infty}_{i,j}$ and $\mathbf{T}^{\beta}_{i,j}(\mathbf{x}) = \mathbf{T}^{\beta\infty}_{i,j}(\mathbf{x})$ ($\mathbf{x} \in v_i$) that reduces Eqs. (10.9) and (10.10) to

$$\langle \overline{\boldsymbol{\sigma}}_i^{[n+1]} \rangle(\mathbf{x}) = \langle \boldsymbol{\sigma} \rangle + \int [\mathcal{T}^{\sigma\infty}_{i,j}(\langle \overline{\boldsymbol{\sigma}}_j^{[n]} \rangle)(\mathbf{x}) + \mathbf{T}^{\beta\infty}_{i,j}(\mathbf{x})]$$

$$\times [\varphi(v_j, \mathbf{x}_j |; v_i, \mathbf{x}_i) - n^{(j)}] d\mathbf{x}_j, \tag{10.15}$$

$$\langle \boldsymbol{\sigma}^{[n+1]} \rangle_i(\mathbf{x}) = \mathcal{B}_i(\langle \overline{\boldsymbol{\sigma}}_i^{[n+1]} \rangle)(\mathbf{x}) + \mathbf{C}_i(\mathbf{x}), \tag{10.16}$$

which are equivalent to the corresponding Eq. (9.92) obtained in [203]. The method uses as a background the new general integral equation (7.20) proposed in [202] and makes it possible to abandon the basic concepts of micromechanics: effective field hypothesis and the hypothesis of "ellipsoidal symmetry." The results of this abandonment were quantitatively estimated for some modeled composite reinforced by aligned continuously inhomogeneous fibers. Some new effects were detected that were impossible in the framework of a classical background (7.33) of micromechanics.

10.1.3 Initial Approximations Based on New Eq. (7.20)

In order to find an initial approximation of Eq. (10.8), we now apply the so-called effective field hypothesis which is the main approximate hypothesis of many micromechanical methods.

Namely, according to the hypothesis **H1a** (4.93), each inclusion v_i is located in the field $\overline{\boldsymbol{\sigma}}_i(\mathbf{y}) \equiv \overline{\boldsymbol{\sigma}}(\mathbf{x}_i)$ ($\mathbf{y} \in v_i$) (7.22) which is homogeneous over the inclusion v_i.

Hypothesis **H1a** allows us to reduce Eq. (10.8) to the algebraic equation

$$\mathbf{R}_i(\mathbf{x})\langle\overline{\boldsymbol{\sigma}}_i\rangle = \mathbf{R}_i(\mathbf{x})\langle\boldsymbol{\sigma}\rangle + \int \Big\{[\mathbf{R}^I_{i,j}(\mathbf{x})\langle\overline{\boldsymbol{\sigma}}_i\rangle + \mathbf{R}^J_{i,j}(\mathbf{x})\langle\overline{\boldsymbol{\sigma}}_j\rangle + \mathbf{F}_{i,j}(\mathbf{x})]$$
$$\times\varphi(v_j,\mathbf{x}_j|;v_i,\mathbf{x}_i) - [\mathbf{R}^\infty_{i,j}(\mathbf{x})\langle\overline{\boldsymbol{\sigma}}_j\rangle + \mathbf{F}^\infty_{i,j}(\mathbf{x})]n^{(j)}\Big\}d\mathbf{x}_j, \quad (10.17)$$

where, in contrast to Eqs. (10.8), (10.15), and (10.16), the effective fields $\langle\overline{\boldsymbol{\sigma}}_k\rangle$ ≡const. at $(\mathbf{x},\mathbf{y} \in v_k,\ k = i,j)$ and the tensors $\mathbf{R}^I_{i,j}(\mathbf{x})$ and $\mathbf{R}^J_{i,j}(\mathbf{x})$ are introduced through the operators $\mathcal{R}^I_{i,j}(\mathbf{x})$ and $\mathcal{R}^J_{i,j}(\mathbf{x})$ (4.133), respectively, according to the definitions (10.4) and (10.15).

For keeping the laying out down, one introduces the notations for the statistical average of all perturbations introduced by surrounding inclusions v_j $(j = 1, 2, \ldots;\ j \neq i)$ into the inclusion v_i

$$\mathbf{R}^{J\,\mathrm{pert}}_{i,j}(\mathbf{x}) = \int \big[\mathbf{R}^J_{i,j}(\mathbf{x})\varphi(v_j,\mathbf{x}_j|;v_i,\mathbf{x}_i) - \mathbf{R}^\infty_{i,j}(\mathbf{x})n^{(j)}\big]d\mathbf{x}_j, \quad (10.18)$$

$$\mathbf{R}^{I\,\mathrm{pert}}_{i,j}(\mathbf{x}) = \int \mathbf{R}^I_{i,j}(\mathbf{x})\varphi(v_j,\mathbf{x}_j|;v_i,\mathbf{x}_i)d\mathbf{x}_j, \quad (10.19)$$

$$\mathbf{F}^{\mathrm{pert}}_i(\mathbf{x}) = \sum_j \int \big[\mathbf{F}_{i,j}(\mathbf{x}))\varphi(v_j,\mathbf{x}_j|;v_i,\mathbf{x}_i) - \mathbf{F}^\infty_{i,j}(\mathbf{x})n^{(j)}\big]d\mathbf{x}_j. \quad (10.20)$$

The volume averages of the tensors $\mathbf{R}^{K\,\mathrm{pert}}_{i,j}(\mathbf{x})$ $(K = I, J)$ and $\mathbf{F}^{\mathrm{pert}}_i(\mathbf{x})$ over the volume of the inclusion v_i are denoted by the $\mathbf{R}^{K\mathrm{pert}}_{i,j}$ $(K = I, J)$ and $\mathbf{F}^{\mathrm{pert}}_i$, respectively. For the small concentration of inclusions $c \equiv \langle V\rangle \ll 1$, the tensors $\mathbf{R}^{K\mathrm{pert}}_{i,j}(\mathbf{x})$ $(K = I, J)$ and $\mathbf{F}^{\mathrm{pert}}_i(\mathbf{x})$ are the linear functions of c. For a pure elastic case ($\boldsymbol{\beta}_1 \equiv \mathbf{0}$), the tensors $\mathbf{R}^{K\mathrm{pert}}_{i,j}(\mathbf{x})$ $(K = I, J)$ have a physical meaning of the perturbations of the statistical average of the strain polarization tensor $\boldsymbol{\eta}$ in the inclusion v_i introduced by the action of surrounding inclusions belonging to the phase $v^{(j)}$.

Volume averaging of Eq. (10.17) leads to the closed algebraic equation

$$\langle\overline{\boldsymbol{\sigma}}_i\rangle = \langle\boldsymbol{\sigma}\rangle + \mathbf{R}_i^{-1}\sum_j(\mathbf{R}^{I\,\mathrm{pert}}_{i,j}\langle\overline{\boldsymbol{\sigma}}_i\rangle + \mathbf{R}^{J\,\mathrm{pert}}_{i,j}\langle\overline{\boldsymbol{\sigma}}_j\rangle + \delta_{ij}\mathbf{F}^{\mathrm{pert}}_j) \quad (10.21)$$

with the solution

$$\langle\overline{\boldsymbol{\sigma}}_i\rangle = \sum_j \mathbf{Y}_{ij}\Big[\langle\boldsymbol{\sigma}\rangle + \mathbf{R}_j^{-1}\mathbf{F}^{\mathrm{pert}}_j\Big], \quad (10.22)$$

$$(\mathbf{Y}^{-1})_{ij} = \delta_{ij}\Big[\mathbf{I} - \mathbf{R}_i^{-1}\sum_k \mathbf{R}^{I\,\mathrm{pert}}_{i,k}\Big] - \mathbf{R}^{J\,\mathrm{pert}}_{i,j}. \quad (10.23)$$

Substituting Eq. (10.22) into the right-hand side of Eq. (10.3) leads to the explicit representation for statistical average of strain polarization tensor in the inclusion v_i

$$\bar{v}_i\langle\boldsymbol{\eta}\rangle_i(\mathbf{x}) = \mathbf{R}_i(\mathbf{x})\langle\boldsymbol{\sigma}\rangle + \mathbf{F}_i(\mathbf{x}) + \mathbf{F}_i^{\text{pert}}(\mathbf{x}) + \sum_{j,m}\big[\mathbf{R}_{i,j}^{I\ \text{pert}}(\mathbf{x})\mathbf{Y}_{im} + \mathbf{R}_{i,j}^{J\ \text{pert}}(\mathbf{x})\mathbf{Y}_{jm}\big]\big[\langle\boldsymbol{\sigma}\rangle + \mathbf{R}_m^{-1}\mathbf{F}_m^{\text{pert}}\big]. \tag{10.24}$$

The mean field of stresses inside the inclusions $\langle\boldsymbol{\sigma}\rangle_i(\mathbf{x})$ is obtained from (4.124), (4.128), and (10.24)

$$\langle\boldsymbol{\sigma}\rangle_i(\mathbf{x}) = \mathbf{B}_i(\mathbf{x})\langle\boldsymbol{\sigma}\rangle + \mathbf{C}_i(\mathbf{x}) + [\bar{v}_i\mathbf{M}_1(\mathbf{x})]^{-1}\mathbf{F}_i^{\text{pert}}(\mathbf{x}) + [\bar{v}_i\mathbf{M}_1(\mathbf{x})]^{-1} \times \sum_{j,m}\big[\mathbf{R}_{i,j}^{I\ \text{pert}}(\mathbf{x})\mathbf{Y}_{im} + \mathbf{R}_{i,j}^{J\ \text{pert}}(\mathbf{x})\mathbf{Y}_{jm}\big]\big[\langle\boldsymbol{\sigma}\rangle + \mathbf{R}_m^{-1}\mathbf{F}_m^{\text{pert}}\big]. \tag{10.25}$$

After estimating average strain polarization tensors inside the inclusions (see Eqs. (10.16) and (10.24)), the problem of calculating effective properties becomes trivial and leads, according to (9.8) and (9.9), to the following *new representations*:

$$\mathbf{M}^* = \mathbf{M}^{(0)} + \sum_{i,j=1}^{N} n^{(i)}\mathbf{R}_i\mathbf{Y}_{ij}, \tag{10.26}$$

$$\boldsymbol{\beta}^* = \boldsymbol{\beta}^{(0)} + \langle\mathbf{F}^v V\rangle + \sum_{i,m=1}^{N} n^{(i)}\mathbf{R}_i\mathbf{Y}_{im}\Big[\langle\mathbf{F}^v V\rangle + \mathbf{R}_m^{-1}\mathbf{F}_m^{\text{pert}}\Big], \tag{10.27}$$

$$\begin{aligned} W^* &= -\frac{1}{2}\langle\boldsymbol{\beta}_1\mathbf{C}\rangle - \frac{1}{2}\langle\boldsymbol{\beta}_1\mathbf{M}_1^{-1}\mathbf{F}^{\text{pert}}\rangle \\ &\quad - \frac{1}{2}\sum_{i,j,m=1}^{N} n^{(i)}\langle\boldsymbol{\beta}_1^{(i)}(\mathbf{x})\mathbf{M}_1^{-1}(\mathbf{x})\mathbf{R}_{i,j}^{I\ \text{pert}}(\mathbf{x})\rangle_{(i)}\mathbf{Y}_{im}\Big[\langle\mathbf{F}^v V\rangle + \mathbf{R}_m^{-1}\mathbf{F}_m^{\text{pert}}\Big] \\ &\quad - \frac{1}{2}\sum_{i,j,m=1}^{N} n^{(i)}\langle\boldsymbol{\beta}_1^{(i)}(\mathbf{x})\mathbf{M}_1^{-1}(\mathbf{x})\mathbf{R}_{i,j}^{J\ \text{pert}}(\mathbf{x})\rangle_{(i)}\mathbf{Y}_{jm}\Big[\langle\mathbf{F}^v V\rangle + \mathbf{R}_m^{-1}\mathbf{F}_m^{\text{pert}}\Big], \end{aligned} \tag{10.28}$$

where $\mathbf{F}_i^v(\mathbf{x}) = \overline{v}_i^{-1}\mathbf{F}_i(\mathbf{x}),\ \mathbf{x} \in v_i \subset v^{(i)}$. It should be mentioned that in Eqs. (10.24) and (10.25), one does not assume that the effective field $\overline{\boldsymbol{\sigma}}_i(\mathbf{x})$ is a homogeneous one that was proposed in the initial version of the MEFM (see Chap. 8 and [199, 200]). Because of this, the tensors $\langle\boldsymbol{\eta}\rangle_i(\mathbf{x}),\ \langle\boldsymbol{\sigma}\rangle_i(\mathbf{x}) \neq$const. due to the inhomogeneity of the tensors $\mathbf{R}_{i,j}^{I\ \text{pert}}(\mathbf{x}),\ \mathbf{R}_{i,j}^{J\ \text{pert}}(\mathbf{x}),\ \mathbf{F}_i^{\text{pert}}(\mathbf{x}) \neq$const. even for the homogeneous ellipsoidal inclusions when $\mathbf{R}_i(\mathbf{x}), \mathbf{F}_i(\mathbf{x}), \mathbf{B}_i(\mathbf{x}), \mathbf{C}_i(\mathbf{x}) =$ const. $(\mathbf{x} \in v_i)$. Therefore, Eqs. (10.24) and (10.25) are in reality the first-order approximation of the stress fields rather than zero one usually used as an initial approximation. The zero-order approximation of $\langle\overline{\boldsymbol{\sigma}}_i\rangle$ and $\langle\overline{\boldsymbol{\eta}}_i\rangle$ could be obtained by direct substitution of Eq. (10.24) into Eqs. (4.124), (4.128)

$$\langle\boldsymbol{\sigma}\rangle_i = \mathbf{C}_i + \mathbf{B}_i \sum_j \mathbf{Y}_{ij}\Big[\langle\boldsymbol{\sigma}\rangle + \mathbf{R}_j^{-1}\mathbf{F}_j^{\rm pert}\Big], \tag{10.29}$$

$$\bar{v}_i\langle\boldsymbol{\eta}\rangle_i = \mathbf{F}_i + \mathbf{R}_i \sum_j \mathbf{Y}_{ij}\Big[\langle\boldsymbol{\sigma}\rangle + \mathbf{R}_j^{-1}\mathbf{F}_j^{\rm pert}\Big]. \tag{10.30}$$

However, the effective properties (10.26) and (10.27) even in such a case will be the same, while the representation (10.28) will be simplified.

10.1.4 Initial Approximations Based on Classical Eq. (7.33)

To make further progress, two next hypotheses are widely used. The first one is a hypothesis **H1b** (4.94) complimenting the basic hypothesis **(H1a)** (4.93). The next hypothesis of *ellipsoidal symmetry* **H3** (8.45) is for the distribution of inclusions $\varphi(v_j, \mathbf{x}_j\ \ |; v_i, \mathbf{x}_i)$. The history and the destination of the hypothesis **H3** were described in Chaps. 8 and 9.

It should be mentioned that Eq. (10.17) used for estimation of the initial approximation (10.29) and (10.30) was obtained from the new GEQ Eq. (7.20). Acceptance of the additional assumption **H1b** in an accompany with the hypothesis of "ellipsoidal symmetry" **H3** allowed Buryachenko [199] to obtain a particular case of the following equation:

$$\mathbf{R}_i(\mathbf{x})\langle\overline{\boldsymbol{\sigma}}_i\rangle = \mathbf{R}_i(\mathbf{x})\langle\boldsymbol{\sigma}\rangle + \mathbf{R}_i(\mathbf{x})\mathbf{Q}_i^0\langle\boldsymbol{\eta}\rangle + \int\Big\{[\mathbf{R}_{i,j}^I(\mathbf{x})\langle\overline{\boldsymbol{\sigma}}_i\rangle + \mathbf{R}_{i,j}^J(\mathbf{x})\langle\overline{\boldsymbol{\sigma}}_j\rangle + \mathbf{F}_{i,j}(\mathbf{x})]$$
$$\times\,\varphi(v_j, \mathbf{x}_j|; v_i, \mathbf{x}_i) - (1 - V_i^0(\mathbf{x}_j))[\mathbf{R}_{i,j}^{\infty}(\mathbf{x})\langle\overline{\boldsymbol{\sigma}}_j\rangle + \mathbf{F}_{i,j}^{\infty}(\mathbf{x})]n^{(j)}\Big\}d\mathbf{x}_j \tag{10.31}$$

instead of Eq. (10.17). The integrals in Eqs. (7.20) and (7.33) (and in all subsequent equations in this study) were absolutely convergent as $|\mathbf{x} - \mathbf{y}| \to \infty$ at $\mathbf{y} \in v^{(j)}$ (see for details Chap. 7). Then, it is tempting to avoid consideration of the infinite range improper integral (10.31) over the whole space by truncating the range at a set of some finite domains, say, the set of the homothetic domains $\omega^{R(k)}$ ($k = 1, \ldots, N$) containing the inscribed spheres with the radius R and with the centers $\mathbf{x}$. In so doing, the mentioned limit of finite range integrals in the right-hand side of Eq. (10.31) exists and does not depend on the shape of the domains $\omega^{R(k)}(\mathbf{x})$. In particular, the shapes of domains $\omega^{R(k)}(\mathbf{x})$ can be chosen coinciding with the shapes of the ellipsoids v_i^0 ($v_i \subset v^{(k)}$). However, the Eshelby tensor $\mathbf{S}_i^0 \equiv \mathbf{S}(\omega^{R(k)}(\mathbf{x}))$, and, because of this, the item containing $(1 - V_i^0(\mathbf{y}))$ can be omitted in the integrals over the mentioned ellipsoidal domains $\omega^{R(k)}(\mathbf{x})$.

In such a case, the tensors (10.18) and (10.20) should be replaced by

$$\bar{\mathbf{R}}_{i,j}^{J\,\mathrm{pert}}(\mathbf{x}) = \mathbf{R}_i\mathbf{Q}_i^0\mathbf{R}_j n^{(j)} + \int \big[\mathbf{R}_{i,j}^J(\mathbf{x})\varphi(v_j,\mathbf{x}_j|;v_i,\mathbf{x}_i) \\ -(1-V_i^0(\mathbf{x}_j))\mathbf{R}_{i,j}^{\infty}(\mathbf{x})n^{(j)}\big]d\mathbf{x}_j, \quad (10.32)$$

$$\bar{\mathbf{F}}_i^{\mathrm{pert}}(\mathbf{x}) = \mathbf{R}_i\mathbf{Q}_i^0\langle\mathbf{F}^v V\rangle + \sum_j \int \big[\mathbf{F}_{i,j}(\mathbf{x}))\varphi(v_j,\mathbf{x}_j|;v_i,\mathbf{x}_i) \\ -(1-V_i^0(\mathbf{x}_j))\mathbf{F}_{i,j}^{\infty}(\mathbf{x})n^{(j)}\big]d\mathbf{x}_j. \quad (10.33)$$

All representations for both the initial approximations and effective properties will be the same as in (10.21)–(10.28) with replacement of the tensors $\mathbf{R}_{i,j}^{J\,\mathrm{pert}}(\mathbf{x})$ (10.18) and $\mathbf{F}_i^{\mathrm{pert}}(\mathbf{x})$ (10.20) by $\bar{\mathbf{R}}_{i,j}^{J\,\mathrm{pert}}(\mathbf{x})$ (10.32) and $\bar{\mathbf{F}}_i^{\mathrm{pert}}(\mathbf{x})$ (10.33), respectively. From these modified representations, we can obtain the results by Buryachenko [199] and [200] obtained at the closing assumptions either $\langle\widetilde{\boldsymbol{\sigma}}_{i,j}\rangle(\mathbf{x}) = \langle\overline{\boldsymbol{\sigma}}_i\rangle$ ≡const. or $\langle\widetilde{\boldsymbol{\sigma}}_{i,j}\rangle(\mathbf{x}) = \langle\overline{\boldsymbol{\sigma}}_j\rangle$ ≡const. instead of (10.6). The tensors $\mathbf{R}_{i,j}^{\infty}(\mathbf{x})$ and $\mathbf{F}_{i,j}^{\infty}(\mathbf{x})$ and the related tensors produced by the constant average strain polarization tensor $\langle\boldsymbol{\eta}\rangle$ in Eqs. (10.31)–(10.33) and in all forthcoming integrals can be omitted if the infinite range improper integrals involved are estimated by the passage to the limit in the set of integrals over the domains $\omega^{R(k)}$ homothetic to v_i^0 that is assumed later on.

It should be mentioned that the initial approximations were obtained in Sects. 10.1.3 and 10.1.4 by the use of the hypothesis **H1a** applied to the different background Eqs. (7.19) and (7.33). Buryachenko [202] proved that an acceptance of the assumptions **H1b** leads to equivalence of Eqs. (7.19) and (7.33). However, for the homogeneous ellipsoidal inclusions, the hypothesis **H1b** is an implication of the hypothesis **H1a**. Therefore, for the homogeneous ellipsoidal inclusions being considered in the numerical examples in Sect. 10.3.1, both the initial approximations obtained in Sects. 10.1.3 and 10.1.4 are identical.

10.1.5 Some Related Integral Equations

Let us ascertain the correlation of Eqs. (10.3) with not numerous known related equations based on the utilization of the solutions for one and two inclusions rather than the direct use of a Green function. In the framework of the standard assumption of a perturbation method

$$\langle\widetilde{\boldsymbol{\sigma}}_{i,j}\rangle \equiv \langle\widetilde{\boldsymbol{\sigma}}_i\rangle \equiv \langle\boldsymbol{\sigma}\rangle, \quad (10.34)$$

Buryachenko and Kushch [239] have obtained (at $\boldsymbol{\beta} \equiv \mathbf{0}$) a cumbersome equation which is equivalent to the following:

$$\bar{v}_i \langle \boldsymbol{\eta} \rangle_i(\mathbf{x}) = \mathbf{R}_i(\mathbf{x})\langle \boldsymbol{\sigma} \rangle + \mathbf{F}_i(\mathbf{x}) + \mathbf{R}_i(\mathbf{x})\mathbf{Q}_i^0 \langle \boldsymbol{\eta} \rangle + \int \Big\{ [\mathbf{R}_{i,j}(\mathbf{x})\langle \boldsymbol{\sigma} \rangle + \mathbf{F}_{i,j}(\mathbf{x})] \\ \times \varphi(v_j, \mathbf{x}_j|; v_i, \mathbf{x}_i) - n^{(j)}[1 - V_i^0(\mathbf{x}_j)][\mathbf{R}_{i,j}^{\infty}(\mathbf{x})\langle \boldsymbol{\sigma} \rangle + \mathbf{F}_{i,j}^{\infty}(\mathbf{x})] \Big\} d\mathbf{x}_j. \tag{10.35}$$

The representation (10.35) can be obtained by direct substitution of the assumption (10.34) into Eq. (10.31). Chen and Acrivos [305] proposed the equation related with Eq. (10.35) for the pure elastic case of the identical homogeneous spherical inclusions ($\boldsymbol{\beta}_1 \equiv \mathbf{0}$, $N = 1$). In our notations this averaged equation can be represented in the form

$$\bar{v}_i \langle \boldsymbol{\eta} \rangle_i = \mathbf{R}_i \langle \boldsymbol{\sigma} \rangle + \mathbf{R}_i \mathbf{Q}_i^0 \langle \boldsymbol{\eta} \rangle + \int \Big[\langle \mathbf{R}_{i,j}(\mathbf{x}) \rangle_{(i)} \langle \boldsymbol{\sigma} \rangle \varphi(v_j, \mathbf{x}_j|; v_i, \mathbf{x}_i) \\ -(1 - V_i^0(\mathbf{x}_j)) n^{(j)} \{\mathbf{X}\boldsymbol{\sigma}'\} \Big] d\mathbf{x}_j, \tag{10.36}$$

where the second summand of the integrand in the brace $\mathbf{X}\boldsymbol{\sigma}'$ is proportional to $\boldsymbol{\sigma}' \equiv \boldsymbol{\sigma}'(\mathbf{x}_i) = \boldsymbol{\sigma}(\mathbf{x}_i) - \langle \boldsymbol{\sigma} \rangle$ which is a perturbation stress at $\mathbf{x}_i$ due to a single inclusion being at $\mathbf{x}_j$ subjected to the applied strain $\langle \boldsymbol{\sigma} \rangle$ at the infinity. The constant tensor $\mathbf{X}$ determined so that the integral will be absolutely convergent has been chosen as $\mathbf{R}_j$. Therefore the item in the brace can be uniquely determined in the term of the solution (4.137) $\mathbf{X}\boldsymbol{\sigma}' = \mathbf{R}_j \langle \mathbf{T}_j^{\sigma\infty}(\mathbf{x}) \rangle_{(i)} \langle \boldsymbol{\sigma} \rangle$. The proposed approximate representation is based on the assumption that the field $\boldsymbol{\sigma}'$ $\equiv$const. in the area v_i. The last limitation can be easily avoided in the framework of the original scheme [305]. Indeed, the item in the braces should be replaced by the relation $\langle \mathbf{R}_{i,j}^{\infty} \rangle_{(i)} \langle \boldsymbol{\sigma} \rangle$, which is the average strain polarization tensor $\langle \boldsymbol{\eta} \rangle_{(i)}$ in the inclusion v_i subjected to both the homogeneous remote field $\langle \boldsymbol{\sigma} \rangle$ and the inhomogeneous field $\mathbf{T}_j^{\sigma\infty}(\mathbf{x})\langle \boldsymbol{\sigma} \rangle$ ($\mathbf{x} \in v_i$). Then Eq. (10.36) has a new representation

$$\langle \boldsymbol{\eta} \rangle_i = \mathbf{R}_i \langle \boldsymbol{\sigma} \rangle + \mathbf{R}_i \mathbf{Q}_i^0 \langle \boldsymbol{\eta} \rangle + \int \Big[\langle \mathbf{R}_{i,j}(\mathbf{x}) \rangle_{(i)} \langle \boldsymbol{\sigma} \rangle \varphi(v_j, \mathbf{x}_j|; v_i, \mathbf{x}_i) \\ -(1 - V_i^0(\mathbf{x}_j)) n^{(j)} \langle \mathbf{R}_{i,j}^{\infty}(\mathbf{x}) \rangle_{(i)} \langle \boldsymbol{\sigma} \rangle \Big] d\mathbf{x}_j, \tag{10.37}$$

which is equivalent to the corresponding relation obtained in [239]. It should be mentioned that the inhomogeneous tensor $\langle \boldsymbol{\sigma} \rangle + \mathbf{T}_j^{\sigma\infty}(\mathbf{x})\langle \boldsymbol{\sigma} \rangle$ ($\mathbf{x} \in v_i$) has a sense of the inhomogeneous effective stress field outside of the inclusion v_j in the area v_i, which can be estimated by different methods not just by the use of the tensors $\mathbf{T}_j^{\sigma\infty}(\mathbf{x})$. A new representation (10.37) coincides with the old one (10.36) at $\mathbf{X}\boldsymbol{\sigma}' = \mathbf{R}_j \langle \mathbf{T}_j^{\sigma\infty}(\mathbf{x}) \rangle_{(i)} \langle \boldsymbol{\sigma} \rangle$ just in the framework of both the case of homogeneous ellipsoidal inclusions and the additional effective field hypothesis **H1**, which is equivalent to the original assumption by Chen and Acrivos [305] $\boldsymbol{\sigma}'(\mathbf{x}) \equiv$const. ($\mathbf{x} \in v_i$). However, Eq. (10.37) is used as a background of the approximate GIE (7.33) which is equivalent to the exact one (7.20) only in the framework of the hypothesis **H1**. Abandonment from this hypothesis leads to a modification of Eq. (10.37)

$$\langle \boldsymbol{\eta} \rangle_i = \mathbf{R}_i \langle \boldsymbol{\sigma} \rangle + \int \Big[\langle \mathbf{R}_{i,j}(\mathbf{x}) \rangle_{(i)} \varphi(v_j, \mathbf{x}_j |; v_i, \mathbf{x}_i) - n^{(j)} \langle \mathbf{R}^{\infty}_{i,j}(\mathbf{x}) \rangle_{(i)} \Big] d\mathbf{x}_j \langle \boldsymbol{\sigma} \rangle, \quad (10.38)$$

which can be obtained from Eq. (10.17) at the condition (10.34). The assumption (10.34) is not used in the current presentation and presented only for exploration of the relation between the proposed Eqs. (10.37) and (10.38) and the known one (10.36).

10.2 Integral Equations for Second Moments of Stresses

10.2.1 Second Moment of Stresses Inside the Heterogeneities

For obtaining an integral representation of the second moments of stresses in the component $v^{(i)}$ of the inclusions ($i = 1, 2, \ldots.$), it is necessary to put the tensor product of (7.22) and (7.23) into $\overline{\boldsymbol{\sigma}}(\mathbf{x})$

$$\begin{aligned}
\overline{\boldsymbol{\sigma}}(\mathbf{x}) \otimes \overline{\boldsymbol{\sigma}}(\mathbf{x}) = {} & \langle \boldsymbol{\sigma} \rangle(\mathbf{x}) \otimes \langle \boldsymbol{\sigma} \rangle(\mathbf{x}) + \langle \boldsymbol{\sigma} \rangle(\mathbf{x}) \otimes \int [\boldsymbol{\Gamma}(\mathbf{x}-\mathbf{x}_p)\boldsymbol{\eta}(\mathbf{x}_p) V_p(\mathbf{x}_p|; v_i, \mathbf{x}_i) \\
& - \langle \boldsymbol{\Gamma}(\mathbf{x}-\mathbf{x}_p)\boldsymbol{\eta} \rangle(\mathbf{x}_p)] d\mathbf{x}_p + \int [\boldsymbol{\Gamma}(\mathbf{x}-\mathbf{x}_q)\boldsymbol{\eta}(\mathbf{x}_q) V_q(\mathbf{x}_q|; v_i, \mathbf{x}_i) \\
& - \langle \boldsymbol{\Gamma}(\mathbf{x}-\mathbf{x}_q)\boldsymbol{\eta} \rangle(\mathbf{x}_q)] d\mathbf{x}_q \otimes \langle \boldsymbol{\sigma} \rangle(\mathbf{x}) + \int \int [\boldsymbol{\Gamma}(\mathbf{x}-\mathbf{x}_p) \\
& \times \boldsymbol{\eta}(\mathbf{x}_p) V_p(\mathbf{x}_p|; v_i, \mathbf{x}_i) - \langle \boldsymbol{\Gamma}(\mathbf{x}-\mathbf{x}_p)\boldsymbol{\eta} \rangle(\mathbf{x}_p)] \\
& \otimes [\boldsymbol{\Gamma}(\mathbf{x}-\mathbf{x}_q)\boldsymbol{\eta}(\mathbf{x}_q) V_q(\mathbf{x}_q|; v_i, \mathbf{x}_i) - \langle \boldsymbol{\Gamma}(\mathbf{x}-\mathbf{x}_q)\boldsymbol{\eta} \rangle(\mathbf{x}_q)] d\mathbf{x}_p d\mathbf{x}_q
\end{aligned} \quad (10.39)$$

with $\mathbf{x} \in v_i$. The right-hand side of Eq. (10.39) is a random function of the arrangements of inclusions $v_p, v_q \neq v_i$ ($p, q = 1, 2, \ldots$). Averaging (10.39) over a realization ensemble leads to

$$\begin{aligned}
\langle \overline{\boldsymbol{\sigma}} \otimes \overline{\boldsymbol{\sigma}} \rangle(\mathbf{x}) = {} & \langle \boldsymbol{\sigma} \rangle(\mathbf{x}) \otimes \langle \boldsymbol{\sigma} \rangle(\mathbf{x}) + \langle \boldsymbol{\sigma} \rangle(\mathbf{x}) \\
& \otimes \int [\langle \boldsymbol{\Gamma}(\mathbf{x}-\mathbf{x}_p)\boldsymbol{\eta}(\mathbf{x}_p) V_p(\mathbf{x}_p) | v_p, \mathbf{x}_p; v_i, \mathbf{x}_i \rangle \varphi(v_p, \mathbf{x}_p|; v_i, \mathbf{x}_i) \\
& - \langle \boldsymbol{\Gamma}(\mathbf{x}-\mathbf{x}_p)\boldsymbol{\eta} \rangle(\mathbf{x}_p)] d\mathbf{x}_p
\end{aligned}$$

$$
\begin{aligned}
&+\int[\langle\boldsymbol{\Gamma}(\mathbf{x}-\mathbf{x}_q)\boldsymbol{\eta}(\mathbf{x}_q)V_q(\mathbf{x}_q)|v_q,\mathbf{x}_q;v_i,\mathbf{x}_i\rangle\varphi(v_q,\mathbf{x}_q|;v_i,\mathbf{x}_i)\\
&-\langle\boldsymbol{\Gamma}(\mathbf{x}-\mathbf{x}_q)\boldsymbol{\eta}\rangle(\mathbf{x}_q)]d\mathbf{x}_q\otimes\langle\boldsymbol{\sigma}\rangle(\mathbf{x})+\iint\Big\{\langle[\boldsymbol{\Gamma}(\mathbf{x}-\mathbf{x}_p)\boldsymbol{\eta}(\mathbf{x}_p)V_p(\mathbf{x}_p)\\
&\otimes[\boldsymbol{\Gamma}(\mathbf{x}-\mathbf{x}_q)\boldsymbol{\eta}(\mathbf{x}_q)V_q(\mathbf{x}_q)]|v_p,\mathbf{x}_p,v_q,\mathbf{x}_q;v_i,\mathbf{x}_i\rangle\\
&\times\varphi(v_p,\mathbf{x}_p,v_q,\mathbf{x}_q\ |;v_i,\mathbf{x}_i)-\langle\boldsymbol{\Gamma}(\mathbf{x}-\mathbf{x}_p)\boldsymbol{\eta}(\mathbf{x}_p)V_p(\mathbf{x}_p)|v_p,\mathbf{x}_p;v_i,\mathbf{x}_i\rangle\\
&\times\varphi(v_p,\mathbf{x}_p|;v_i,\mathbf{x}_i)\otimes\langle\boldsymbol{\Gamma}(\mathbf{x}-\mathbf{x}_q)\boldsymbol{\eta}\rangle(\mathbf{x}_q)-\langle\boldsymbol{\Gamma}(\mathbf{x}-\mathbf{x}_p)\boldsymbol{\eta}\rangle(\mathbf{x}_p)\\
&\otimes\langle\boldsymbol{\Gamma}(\mathbf{x}-\mathbf{x}_q)\boldsymbol{\eta}(\mathbf{x}_q)V_q(\mathbf{x}_q)|v_q,\mathbf{x}_q;v_i,\mathbf{x}_i\rangle\varphi(v_q,\mathbf{x}_q|;v_i,\mathbf{x}_i)\\
&+\langle\boldsymbol{\Gamma}(\mathbf{x}-\mathbf{x}_p)\boldsymbol{\eta}\rangle(\mathbf{x}_p)\otimes\langle\boldsymbol{\Gamma}(\mathbf{x}-\mathbf{x}_q)\boldsymbol{\eta}\rangle(\mathbf{x}_q)\Big\}d\mathbf{x}_p d\mathbf{x}_q. \qquad (10.40)
\end{aligned}
$$

The notations $\langle(.)|v_1,\mathbf{x}_1;\ldots;v_n,\mathbf{x}_n\rangle(\mathbf{x})$ are used for the conditional average taken for the ensemble of a statistically homogeneous field $X=(v_i)$ at the point $\mathbf{x}$, on the condition that there are inclusions at the points $\mathbf{x}_1,\ldots,\mathbf{x}_n$ and $\mathbf{x}_i\neq\mathbf{x}_j$ if $i\neq j$ $(i,j=1,\ldots,n)$. The notations $\langle(.)|;v_1,\mathbf{x}_1;\ldots;v_n,\mathbf{x}_n\rangle(\mathbf{x})$ are used for the case $\mathbf{x}\notin v_1,\ldots,v_n$. The right-hand side of Eq. (10.40) includes double-point $\varphi(v_r,\mathbf{x}_r|;v_i,\mathbf{x}_i)$ $(r=p,q)$ and triple-point $\varphi(v_p,\mathbf{x}_p,v_q,\mathbf{x}_q|;v_i,\mathbf{x}_i)$ conditional probability densities in which the terms with $\mathbf{x}_p=\mathbf{x}_q$ under the condition $\mathbf{x}_p\neq\mathbf{x}_i$, $\mathbf{x}_q\neq\mathbf{x}_i$ may be isolated with the help of the equality

$$
\begin{aligned}
\varphi(v_p,\mathbf{x}_p,v_q,\mathbf{x}_q\ |;v_i,\mathbf{x}_i)&=\delta(\mathbf{x}_p-\mathbf{x}_q)\varphi(v_p,\mathbf{x}_p\ |;v_i,\mathbf{x}_i)\\
&+\varphi(v_p,\mathbf{x}_p\ |;v_i,\mathbf{x}_i)\varphi(v_q,\mathbf{x}_q\ |;v_p,\mathbf{x}_p;v_i,\mathbf{x}_i). \qquad (10.41)
\end{aligned}
$$

Then Eq. (10.40) can be rewritten as

$$
\begin{aligned}
\langle\overline{\boldsymbol{\sigma}}\otimes\overline{\boldsymbol{\sigma}}\rangle_i(\mathbf{x})&=\langle\overline{\boldsymbol{\sigma}}\rangle_i(\mathbf{x})\otimes\langle\overline{\boldsymbol{\sigma}}\rangle_i(\mathbf{x})+\int\langle[\boldsymbol{\Gamma}(\mathbf{x}-\mathbf{x}_p)\boldsymbol{\eta}(\mathbf{x}_p)]\\
&\otimes[\boldsymbol{\Gamma}(\mathbf{x}-\mathbf{x}_p)\boldsymbol{\eta}(\mathbf{x}_p)]|v_p,\mathbf{x}_p;v_i,\mathbf{x}_i\rangle\varphi(v_p,\mathbf{x}_p\ |;v_i,\mathbf{x}_i)d\mathbf{x}\\
&+\int\int\Big\{\langle[\boldsymbol{\Gamma}(\mathbf{x}-\mathbf{x}_p)\boldsymbol{\eta}(\mathbf{x}_p)]\otimes[\boldsymbol{\Gamma}(\mathbf{x}-\mathbf{x}_q)\boldsymbol{\eta}(\mathbf{x}_q)]|v_p,\mathbf{x}_p;v_q,\mathbf{x}_q;v_i,\mathbf{x}_i\rangle\\
&\times\varphi(v_p,\mathbf{x}_p\ |;v_i,\mathbf{x}_i)\varphi(v_q,\mathbf{x}_q\ |;v_p,\mathbf{x}_p;v_i,\mathbf{x}_i)\\
&-[\langle\boldsymbol{\Gamma}(\mathbf{x}-\mathbf{x}_p)\boldsymbol{\eta}(\mathbf{x}_p)|v_p,\mathbf{x}_p;v_i,\mathbf{x}_i\rangle]\otimes[\langle\boldsymbol{\Gamma}(\mathbf{x}-\mathbf{x}_q)\boldsymbol{\eta}(\mathbf{x}_q)|v_q,\mathbf{x}_q;v_i,\mathbf{x}_i\rangle]\\
&\times\varphi(v_p,\mathbf{x}_p\ |;v_i,\mathbf{x}_i)\varphi(v_q,\mathbf{x}_q\ |;v_i,\mathbf{x}_i)\}d\mathbf{x}_q d\mathbf{x}_p \qquad (10.42)
\end{aligned}
$$

with $\mathbf{x}\in v_i$. Here the double integral was transformed in terms of average effective stresses.

From the Eq. (10.42) one can see that the effective stress dispersion

$$
\Delta\overline{\sigma}^{(i)2}_{klmn}(\mathbf{x})\equiv\langle\overline{\sigma}_{kl}\overline{\sigma}_{mn}\rangle_i(\mathbf{x})-\langle\overline{\sigma}_{kl}\rangle_i(\mathbf{x})\langle\overline{\sigma}_{mn}\rangle_i(\mathbf{x}) \qquad (10.43)
$$

presents a determined inhomogeneous function inside the components v_i, $(i = 1, \ldots, N)$. At the same time, the stress dispersion $\Delta\boldsymbol{\sigma}^{(i)2}(\mathbf{x}) \equiv \langle\boldsymbol{\sigma}\otimes\boldsymbol{\sigma}\rangle_i(\mathbf{x}) - \langle\boldsymbol{\sigma}\rangle_i(\mathbf{x})\otimes\langle\boldsymbol{\sigma}\rangle_i(\mathbf{x})$ is connected with the effective stress dispersion (10.43) by

$$\Delta\boldsymbol{\sigma}^{(i)2}(\mathbf{x}) = \boldsymbol{\mathcal{B}}_i * \Delta\overline{\boldsymbol{\sigma}}^{(i)2} * \boldsymbol{\mathcal{B}}_i^{\top}(\mathbf{x}). \tag{10.44}$$

The formula (10.44) allows the estimation of the averaged stress dispersion in the inclusion volume

$$\Delta\boldsymbol{\sigma}^{(i)2} = \bar{v}_i^{-1}\int \Delta\boldsymbol{\sigma}^{(i)2}(\mathbf{x})V_i(\mathbf{x})d\mathbf{x}. \tag{10.45}$$

By the use of the known properties of the second statistical moments, (10.44) and (10.45), we obtain the relations for the dispersion of both the stresses and strain polarizations $\Delta\boldsymbol{\eta}^{(i)2}(\mathbf{x}) := \langle\boldsymbol{\eta}\otimes\boldsymbol{\eta}\rangle_i(\mathbf{x}) = \langle\boldsymbol{\eta}\rangle_i(\mathbf{x})\otimes\langle\boldsymbol{\eta}\rangle_i(\mathbf{x})$ in the component v_i

$$\begin{aligned}\Delta\boldsymbol{\sigma}^{(i)2}(\mathbf{x}) = {} & \langle\boldsymbol{\mathcal{B}}_i * \boldsymbol{\Gamma}(\mathbf{x}-\mathbf{x}_p)\boldsymbol{\eta}(\mathbf{x}_p)\otimes\boldsymbol{\mathcal{B}}_i * \boldsymbol{\Gamma}(\mathbf{x}-\mathbf{x}_p)\boldsymbol{\eta}(\mathbf{x}_p)|v_p,\mathbf{x}_p;v_i,\mathbf{x}_i\rangle \\ & \times\varphi(v_p,\mathbf{x}_p\,|;v_i,\mathbf{x}_i)d\mathbf{x}_p + \int\!\!\int\Big\{\langle\boldsymbol{\mathcal{B}}_i * \boldsymbol{\Gamma}(\mathbf{x}-\mathbf{x}_p)\boldsymbol{\eta}(\mathbf{x}_p) \\ & \otimes\boldsymbol{\mathcal{B}}_i * \boldsymbol{\Gamma}(\mathbf{x}-\mathbf{x}_q)\boldsymbol{\eta}(\mathbf{x}_q)|v_p,\mathbf{x}_p;v_q,\mathbf{x}_q;v_i,\mathbf{x}_i\rangle \\ & \times\varphi(v_p,\mathbf{x}_p\,|;v_i,\mathbf{x}_i)\varphi(v_q,\mathbf{x}_q\,|;v_p,\mathbf{x}_p;v_i,\mathbf{x}_i) \\ & - \langle\boldsymbol{\mathcal{B}}_i * \boldsymbol{\Gamma}(\mathbf{x}-\mathbf{x}_p)\boldsymbol{\eta}(\mathbf{x}_p)|v_p,\mathbf{x}_p;v_i,\mathbf{x}_i\rangle\varphi(v_p,\mathbf{x}_p\,|;v_i,\mathbf{x}_i) \\ & \otimes\langle\boldsymbol{\mathcal{B}}_i * \boldsymbol{\Gamma}(\mathbf{x}-\mathbf{x}_q)\boldsymbol{\eta}(\mathbf{x}_q)|v_q,\mathbf{x}_q;v_i,\mathbf{x}_i\rangle\varphi(v_q,\mathbf{x}_q\,|;v_i,\mathbf{x}_i)\Big\}d\mathbf{x}_q d\mathbf{x}_p,\end{aligned} \tag{10.46}$$

$$\begin{aligned}\Delta\boldsymbol{\eta}^{(i)2}(\mathbf{x}) = {} & \int\langle\boldsymbol{\mathcal{R}}_i^v * \boldsymbol{\Gamma}(\mathbf{x}-\mathbf{x}_p)\boldsymbol{\eta}(\mathbf{x}_p)\otimes\boldsymbol{\mathcal{R}}_i^v * \boldsymbol{\Gamma}(\mathbf{x}-\mathbf{x}_p)\boldsymbol{\eta}(\mathbf{x}_p)|v_p,\mathbf{x}_p;v_i,\mathbf{x}_i\rangle \\ & \times\varphi(v_p,\mathbf{x}_p\,|;v_i,\mathbf{x}_i)d\mathbf{x}_p + \int\!\!\int\Big\{\langle\boldsymbol{\mathcal{R}}_i^v * \boldsymbol{\Gamma}(\mathbf{x}-\mathbf{x}_p)\boldsymbol{\eta}(\mathbf{x}_p) \\ & \otimes\boldsymbol{\mathcal{R}}_i^v * \boldsymbol{\Gamma}(\mathbf{x}-\mathbf{x}_q)\boldsymbol{\eta}(\mathbf{x}_q)|v_p,\mathbf{x}_p;v_q,\mathbf{x}_q;v_i,\mathbf{x}_i\rangle \\ & \times\varphi(v_p,\mathbf{x}_p\,|;v_i,\mathbf{x}_i)\varphi(v_q,\mathbf{x}_q\,|;v_p,\mathbf{x}_p;v_i,\mathbf{x}_i) \\ & - \langle\boldsymbol{\mathcal{R}}_i^v * \boldsymbol{\Gamma}(\mathbf{x}-\mathbf{x}_p)\boldsymbol{\eta}(\mathbf{x}_p)|v_p,\mathbf{x}_p;v_i,\mathbf{x}_i\rangle\varphi(v_p,\mathbf{x}_p\,|;v_i,\mathbf{x}_i) \\ & \otimes\langle\boldsymbol{\mathcal{R}}_i^v * \boldsymbol{\Gamma}(\mathbf{x}-\mathbf{x}_q)\boldsymbol{\eta}(\mathbf{x}_q)|v_q,\mathbf{x}_q;v_i,\mathbf{x}_i\rangle\varphi(v_q,\mathbf{x}_q\,|;v_i,\mathbf{x}_i)\Big\}d\mathbf{x}_q d\mathbf{x}_p,\end{aligned} \tag{10.47}$$

respectively, and $\boldsymbol{\mathcal{R}}_i^v \equiv \bar{v}_i^{-1}\boldsymbol{\mathcal{R}}_i$ The relations (10.46) and (10.47) are derived by the use of both the triple and double interaction of the heterogeneities. As may be seen from Eqs. (7.8) and (7.9), the neglect of binary interaction is tantamount to assuming homogeneity of the effective stresses inside the component v_i

$$\langle \overline{\boldsymbol{\sigma}} \otimes \overline{\boldsymbol{\sigma}} \rangle_i(\mathbf{x}) = \langle \overline{\boldsymbol{\sigma}} \rangle_i(\mathbf{x}) \otimes \langle \overline{\boldsymbol{\sigma}} \rangle_i(\mathbf{x}), \tag{10.48}$$

$$\langle \boldsymbol{\sigma} \otimes \boldsymbol{\sigma} \rangle_i(\mathbf{x}) = \langle \boldsymbol{\sigma} \rangle_i(\mathbf{x}) \otimes \langle \boldsymbol{\sigma} \rangle_i(\mathbf{x}), \tag{10.49}$$

$$\langle \boldsymbol{\eta} \otimes \boldsymbol{\eta} \rangle_i(\mathbf{x}) = \langle \boldsymbol{\eta} \rangle_i(\mathbf{x}) \otimes \langle \boldsymbol{\eta} \rangle_i(\mathbf{x}). \tag{10.50}$$

However, substitution of the equalities (10.48), (10.49), and (10.50) into the right-hand sides of Eqs. (10.42), (10.46), and (10.47), respectively, does not mean that Eq. (10.48), (10.49), and (10.50) hold for the left-hand sides of Eqs. (10.42), (10.46), and (10.47). The following new approximations for the dispersion can be obtained by taking only binary interaction of the inclusions into account:

$$\Delta \boldsymbol{\sigma}^{(i)2}(\mathbf{x}) = \int \langle \boldsymbol{\mathcal{B}}_i * \boldsymbol{\Gamma}(\mathbf{x} - \mathbf{x}_p) \boldsymbol{\eta}(\mathbf{x}_p) \\ \otimes \boldsymbol{\mathcal{B}}_i * \boldsymbol{\Gamma}(\mathbf{x} - \mathbf{x}_p) \boldsymbol{\eta}(\mathbf{x}_p) | v_p, \mathbf{x}_p; v_i, \mathbf{x}_i \rangle \varphi(v_p, \mathbf{x}_p \,|; v_i, \mathbf{x}_i) d\mathbf{x}_p, \tag{10.51}$$

$$\Delta \boldsymbol{\eta}^{(i)2}(\mathbf{x}) = \int \langle \boldsymbol{\mathcal{R}}_i^v * \boldsymbol{\Gamma}(\mathbf{x} - \mathbf{x}_p) \boldsymbol{\eta}(\mathbf{x}_p) \\ \otimes \boldsymbol{\mathcal{R}}_i^v * \boldsymbol{\Gamma}(\mathbf{x} - \mathbf{x}_p) \boldsymbol{\eta}(\mathbf{x}_p) | v_p, \mathbf{x}_p; v_i, \mathbf{x}_i \rangle \varphi(v_p, \mathbf{x}_p \,|; v_i, \mathbf{x}_i) d\mathbf{x}_p, \tag{10.52}$$

Ignoring effective stress fluctuations in the inclusions v_p in the integral terms (10.46) and (10.47) leads to other new representations

$$\Delta \boldsymbol{\sigma}^{(i)2}(\mathbf{x}) = \int [\boldsymbol{\mathcal{B}}_i * \mathbf{T}_p(\mathbf{x} - \mathbf{x}_p) \langle \boldsymbol{\eta} \rangle_p] \\ \otimes [\boldsymbol{\mathcal{B}}_i * \mathbf{T}_p(\mathbf{x} - \mathbf{x}_p) \langle \boldsymbol{\eta} \rangle_p] \varphi(v_p, \mathbf{x}_p |; v_i, \mathbf{x}_i) \bar{v}_p^2 d\mathbf{x}_p \\ + \int \int [\boldsymbol{\mathcal{B}}_i * \mathbf{T}_p(\mathbf{x} - \mathbf{x}_p) \langle \boldsymbol{\eta} \rangle_p] \otimes [\boldsymbol{\mathcal{B}}_i * \mathbf{T}_q(\mathbf{x} - \mathbf{x}_q) \langle \boldsymbol{\eta} \rangle_q] \\ \otimes \widetilde{\varphi}(v_q, \mathbf{x}_q \,|; v_p, \mathbf{x}_p; v_i, \mathbf{x}_i) \varphi(v_p, \mathbf{x}_p \,|; v_i, \mathbf{x}_i) \bar{v}_p \bar{v}_q d\mathbf{x}_p d\mathbf{x}_q, \tag{10.53}$$

$$\Delta \boldsymbol{\eta}^{(i)2}(\mathbf{x}) = \int [\boldsymbol{\mathcal{R}}_i^v * \mathbf{T}_p(\mathbf{x} - \mathbf{x}_p) \langle \boldsymbol{\eta} \rangle_p] \\ \otimes [\boldsymbol{\mathcal{R}}_i^v * \mathbf{T}_p(\mathbf{x} - \mathbf{x}_p) \langle \boldsymbol{\eta} \rangle_p] \varphi(v_p, \mathbf{x}_p |; v_i, \mathbf{x}_i) \bar{v}_p^2 d\mathbf{x}_p \\ + \int \int [\boldsymbol{\mathcal{R}}_i^v * \mathbf{T}_p(\mathbf{x} - \mathbf{x}_p) \langle \boldsymbol{\eta} \rangle_p] \otimes [\boldsymbol{\mathcal{R}}_i^v * \mathbf{T}_q(\mathbf{x} - \mathbf{x}_q) \langle \boldsymbol{\eta} \rangle_q] \\ \times \widetilde{\varphi}(v_q, \mathbf{x}_q \,|; v_p, \mathbf{x}_p; v_i, \mathbf{x}_i) \varphi(v_p, \mathbf{x}_p \,|; v_i, \mathbf{x}_i) \bar{v}_p \bar{v}_q d\mathbf{x}_p d\mathbf{x}_q, \tag{10.54}$$

where $\widetilde{\varphi}(v_q, \mathbf{x}_q \,|; v_p, \mathbf{x}_p; v_i, \mathbf{x}_i) \equiv \varphi(v_q, \mathbf{x}_q \,|; v_p, \mathbf{x}_p; v_i, \mathbf{x}_i) - \varphi(v_q, \mathbf{x}_q \,|; v_i, \mathbf{x}_i)$. The relations (10.53) and (10.54) can be called as "far-field approximation."

10.2.2 Stress Fluctuations Inside the Matrix

In a similar manner, it is possible to obtain the estimation of the second stress moment averaged over a volume of the matrix. At first we define the conditional probability density $\varphi(v_p, \mathbf{x}_p \;|; v_q, \mathbf{x}_q; \ldots; v_0, \mathbf{x}_0)$ under the condition that the inclusions $v_q, \ldots$ are located at points $\mathbf{x}_q, \ldots$, whereas the matrix material appears at the point $\mathbf{x}_0$. In analogy to the above considerations, we obtain representation for a stress dispersion in the matrix $\Delta\boldsymbol{\sigma}^{(0)2}(\mathbf{x}) := \langle\boldsymbol{\sigma}\otimes\boldsymbol{\sigma}\rangle_0 - \langle\boldsymbol{\sigma}\rangle_0 \otimes \langle\boldsymbol{\sigma}\rangle_0$

$$
\begin{aligned}
\Delta\boldsymbol{\sigma}^{(0)2}(\mathbf{x}) &= \int \langle[\boldsymbol{\Gamma}(\mathbf{x}_0-\mathbf{x}_p)\boldsymbol{\eta}(\mathbf{x}_p)]\otimes[\boldsymbol{\Gamma}(\mathbf{x}_0-\mathbf{x}_p)\boldsymbol{\eta}(\mathbf{x}_p)]|v_p,\mathbf{x}_p; v_0,\mathbf{x}_0\rangle \\
&\times\varphi(v_p,\mathbf{x}_p \;|; v_0,\mathbf{x}_0)d\mathbf{x}_p \\
&+\int\int\Big\{\langle[\boldsymbol{\Gamma}(\mathbf{x}_0-\mathbf{x}_p)\boldsymbol{\eta}(\mathbf{x}_p)]\otimes[\boldsymbol{\Gamma}(\mathbf{x}_0-\mathbf{x}_q)\boldsymbol{\eta}(\mathbf{x}_q)]|v_p,\mathbf{x}_p; v_q,\mathbf{x}_q; v_0,\mathbf{x}_0\rangle \\
&\times\varphi(v_p,\mathbf{x}_p \;|; v_0,\mathbf{x}_0)\varphi(v_q,\mathbf{x}_q \;|; v_p,\mathbf{x}_p; v_0,\mathbf{x}_0) \\
&-[\langle\boldsymbol{\Gamma}(\mathbf{x}_0-\mathbf{x}_p)\boldsymbol{\eta}(\mathbf{x}_p)|v_p,\mathbf{x}_p; v_i,\mathbf{x}_i\rangle\varphi(v_p,\mathbf{x}_p \;|; v_0,\mathbf{x}_0)] \\
&\otimes[\langle\boldsymbol{\Gamma}(\mathbf{x}_0-\mathbf{x}_q)\boldsymbol{\eta}(\mathbf{x}_q)|v_q,\mathbf{x}_q; v_i,\mathbf{x}_i\rangle\varphi(v_q,\mathbf{x}_q \;|; v_0,\mathbf{x}_0)]\Big\}d\mathbf{x}_q d\mathbf{x}_p.
\end{aligned}
\tag{10.55}
$$

In contrast to (10.46), the second stress moment in the matrix, (10.55), does not depend on $\mathbf{x}_0 \in v_0$. A more approximate estimation of the second moment $\langle\boldsymbol{\sigma}\otimes\boldsymbol{\sigma}\rangle_0$ can be obtained in direct analogy to (10.51), (10.53) by replacing $v_i,\ \mathbf{x}_i$ by $v_0,\ \mathbf{x}_0$. So, in the framework of the effective field hypothesis **H1**, we have

$$
\begin{aligned}
\Delta\boldsymbol{\sigma}^{(0)2}(\mathbf{x}) &= \int \langle[\mathbf{T}_p(\mathbf{x}_0-\mathbf{x}_p)\boldsymbol{\eta}_p\bar{v}_p]\otimes[\mathbf{T}_p(\mathbf{x}_0-\mathbf{x}_p)\boldsymbol{\eta}_p\bar{v}_p]|v_p,\mathbf{x}_p; v_0,\mathbf{x}_0\rangle \\
&\times\varphi(v_p,\mathbf{x}_p \;|; v_0,\mathbf{x}_0)d\mathbf{x}_p \\
&+\int\int\Big\{\langle[\mathbf{T}_p(\mathbf{x}_0-\mathbf{x}_p)\boldsymbol{\eta}_p]\otimes[\mathbf{T}_q(\mathbf{x}_0-\mathbf{x}_q)\boldsymbol{\eta}_q]|v_p,\mathbf{x}_p; v_q,\mathbf{x}_q; v_0,\mathbf{x}_0\rangle \\
&\times\varphi(v_p,\mathbf{x}_p \;|; v_0,\mathbf{x}_0)\varphi(v_q,\mathbf{x}_q \;|; v_p,\mathbf{x}_p; v_0,\mathbf{x}_0) \\
&-[\mathbf{T}_p(\mathbf{x}_0-\mathbf{x}_p)\langle\boldsymbol{\eta}_p|v_p,\mathbf{x}_p; v_i,\mathbf{x}_i; v_0,\mathbf{x}_0\rangle\varphi(v_p,\mathbf{x}_p \;|; v_0,\mathbf{x}_0)] \\
&\times[\mathbf{T}_q(\mathbf{x}_0-\mathbf{x}_q)\langle\boldsymbol{\eta}_q|v_q,\mathbf{x}_q; v_i,\mathbf{x}_i; v_0,\mathbf{x}_0\rangle\varphi(v_q,\mathbf{x}_q \;|; v_0,\mathbf{x}_0)]\Big\}\bar{v}_p\bar{v}_q d\mathbf{x}_q d\mathbf{x}_p.
\end{aligned}
\tag{10.56}
$$

"Far-field approximation" leads to the following simplification of Eq. (10.56):

$$\Delta\boldsymbol{\sigma}^{(0)2}(\mathbf{x}) = \int [\mathbf{T}_p(\mathbf{x}_0 - \mathbf{x}_p)\langle\boldsymbol{\eta}\rangle_p \bar{v}_p] \otimes [\mathbf{T}_p(\mathbf{x}_0 - \mathbf{x}_p)\langle\boldsymbol{\eta}\rangle_p \bar{v}_p] \varphi(v_p, \mathbf{x}_p \,|; v_0, \mathbf{x}_0) d\mathbf{x}_p$$
$$+ \int\int [\mathbf{T}_p(\mathbf{x}_0 - \mathbf{x}_p)\langle\boldsymbol{\eta}\rangle_p] \otimes [\mathbf{T}_q(\mathbf{x}_0 - \mathbf{x}_q)\langle\boldsymbol{\eta}\rangle_q] \bar{v}_p \bar{v}_q$$
$$\times [\varphi(v_q, \mathbf{x}_q \,|; v_p, \mathbf{x}_p; v_0, \mathbf{x}_0) - \varphi(v_q, \mathbf{x}_q \,|; v_0, \mathbf{x}_0)] \varphi(v_p, \mathbf{x}_p \,|; v_0, \mathbf{x}_0) d\mathbf{x}_q d\mathbf{x}_p. \tag{10.57}$$

The new representations (10.46) and (10.56) of the stress dispersion are exact and depend on both the triple and double interactions of heterogeneities. The Eqs. (10.46) and (10.56) do not depend on which basic integral equations either (7.20) or (7.33) they were obtained. Their most strong simplifications were obtained in the framework of far-field approximation (10.53) and (10.57) which depend only on the average polarization tensors inside heterogeneities $\langle\boldsymbol{\eta}\rangle_p$. More accurate simplifications of the right-hand sides of Eqs. (10.46) and (10.56) will be established in the next section in the framework of the effective field hypothesis.

10.2.3 Second Moment of Stresses at the Interface

The stress $\sigma_i^-(\mathbf{n})$ in the matrix in the vicinity of the inhomogeneities v_i near the point $\mathbf{x} \in v_i$ with the unit outward normal vector $\mathbf{n} \perp \partial v_i$ is given by the formula

$$\boldsymbol{\sigma}^-(\mathbf{n}) = \mathbf{B}^-(\mathbf{n})\boldsymbol{\sigma}^+(\mathbf{x}) + (\mathbf{B}^-(\mathbf{n}) - \mathbf{I})(\mathbf{M}_1(\mathbf{x}))^{-1}\boldsymbol{\beta}_1(\mathbf{x}), \tag{10.58}$$

which holds for any shape, anisotropy, and inhomogeneity of the inclusion v_i, and the representation of the interface tensor $\mathbf{B}(\mathbf{n})$ is presented in Sect. 3.4.2. Equations (10.58), (10.7), and (10.8) make it possible to estimate the ensemble average of the stress $\langle\boldsymbol{\sigma}^-(\mathbf{n})\rangle_{\mathbf{x}}$ in the vicinity of the inhomogeneity near the point $\mathbf{x} \in v_i$. The second moment of stresses in the vicinity of the heterogeneity can be found from (10.58) and (10.46)

$$\langle\boldsymbol{\sigma}^-(\mathbf{n}) \otimes \boldsymbol{\sigma}^-(\mathbf{n})\rangle_{\mathbf{x}} = \langle[\mathbf{B}(\mathbf{n})\boldsymbol{\sigma}] \otimes [\mathbf{B}(\mathbf{n})\boldsymbol{\sigma}]\rangle_i(\mathbf{x})$$
$$+ \left\{[\mathbf{B}(\mathbf{n}) - \mathbf{I}][\mathbf{M}_1^{(i)}(\mathbf{x})]^{-1}\boldsymbol{\beta}_1^{(i)}(\mathbf{x})\right\} \otimes [\mathbf{B}(\mathbf{n})\langle\boldsymbol{\sigma}\rangle_i(\mathbf{x})]$$
$$+ \langle\boldsymbol{\sigma}^-(\mathbf{n})\rangle_{\mathbf{x}} \otimes \left\{[\mathbf{B}(\mathbf{n}) - \mathbf{I}][\mathbf{M}_1^{(i)}(\mathbf{x})]^{-1}\boldsymbol{\beta}_1^{(i)}(\mathbf{x})\right\}, \tag{10.59}$$

where the first term on the right-hand side can be obtained by using Eq. (10.46). Expressions for $\langle\boldsymbol{\sigma}\rangle_i(\mathbf{x})$ and $\langle\boldsymbol{\sigma}^-(\mathbf{n})\rangle_{\mathbf{x}}$ are given by formulae (10.8) and (10.58), respectively. Relation (10.59) was proposed in [199] by a less general and more cumbersome method for the homogeneous stress moments $\langle\boldsymbol{\sigma} \otimes \boldsymbol{\sigma}\rangle_i$ and $\langle\boldsymbol{\sigma}\rangle_i$ in the right-hand side in (10.59). It should be mentioned that for the calculation of the invariants $\langle\boldsymbol{\sigma}^-(\mathbf{n})\mathbf{N}_1\boldsymbol{\sigma}^-(\mathbf{n})\rangle_{\mathbf{x}}$ and $\langle\boldsymbol{\sigma}^-(\mathbf{n})\mathbf{N}_2\boldsymbol{\sigma}^-(\mathbf{n})\rangle_{\mathbf{x}}$, it is necessary to estimate all components of the averaged second moment of the stresses inside the inclusion v_i by Eq. (10.46) [in contrast to Eqs. (10.41)].

10.3 Second Moments of Stresses Inside the Components

10.3.1 Initial Approximation

Although Eqs. (10.46) and (10.56) are exact, they are implicit ones with respect to the conditional second moments of the strain polarization tensor $\boldsymbol{\eta}$. Therefore, the solutions of Eqs. (10.46) and (10.56) are reduced to the analysis of the infinite systems of coupled integral equations. The closing of this infinite systems of integral equations in the two-particle approximation can be performed with the help of the following assumptions ($i = 1, 2;\quad j = 3 - i$):

$$\langle \widetilde{\boldsymbol{\sigma}}(\mathbf{x}_i)_{i,j} \otimes \widetilde{\boldsymbol{\sigma}}(\mathbf{x}_j)_{i,j}\rangle = \langle \overline{\boldsymbol{\sigma}}\rangle_i \otimes \langle \overline{\boldsymbol{\sigma}}\rangle_j,\ \langle \widetilde{\boldsymbol{\sigma}}(\mathbf{x}_i)_{i,j} \otimes \widetilde{\boldsymbol{\sigma}}(\mathbf{x}_i)_{i,j}\rangle = \langle \overline{\boldsymbol{\sigma}} \otimes \overline{\boldsymbol{\sigma}}\rangle_i. \quad (10.60)$$

The assumption of no correlation between the values of the field $\widetilde{\boldsymbol{\sigma}}(\mathbf{x})_{i,j}$ at different points (10.60) does not mean the lack of correlation between the stresses $\boldsymbol{\sigma}(\mathbf{x}_i)$ and $\boldsymbol{\sigma}(\mathbf{x}_j)$. The statistical average stresses $\langle \boldsymbol{\sigma}\rangle_i(\mathbf{x})$ ($\mathbf{x} \in v_i$) may be obtained by using formulae (10.7) and (10.8). For the calculation of a second one-point effective stress moment averaged over the heterogeneity volume $\langle \boldsymbol{\sigma}\otimes\boldsymbol{\sigma}\rangle_i$, Buryachenko [175] applied a cumbersome approximation method. For estimation of $\langle \boldsymbol{\sigma} \otimes \boldsymbol{\sigma}\rangle_i(\mathbf{x})$, we will consider a modification of the approach presented in [268] (see also Chapter 13 in [200]) used for evaluation of $\langle \boldsymbol{\sigma} \otimes \boldsymbol{\sigma}\rangle_i$ and based on simplification of Eq. (10.46) obtained in the framework of the effective field hypothesis **H1** of the binary interaction of the inclusions.

Using hypothesis **H1** (combining the hypotheses **H1a** and **H1b**), the system (7.23) for the pair fixed inclusions with fixed values $\widetilde{\boldsymbol{\sigma}}_{i,j}(\mathbf{x})$ ($\mathbf{x} \in v_i,\ i, j = 1, 2,\ j = 4.4.i$) on the right-hand side of the equations becomes algebraic when the solution (4.127), (4.128) for one inclusion in the field $\overline{\boldsymbol{\sigma}}_i(\mathbf{x}_i)$ is applied

$$\overline{\boldsymbol{\sigma}}_i(\mathbf{x}_i) = \mathbf{R}_i^{-1}\Big\{\sum_{k=1}^{2} \mathbf{Z}_{ij}^{1k}\big[\mathbf{R}_k\widetilde{\boldsymbol{\sigma}}_{i,j}(\mathbf{x}_k) + \mathbf{F}_k\big] - \mathbf{F}_i\Big\}, \quad (10.61)$$

where the matrix $\mathbf{Z}$ was defined by Eq. (4.143). For two fixed inhomogeneities $v_i,\ v_j$, we have the second statistical moment of the effective stresses inside the inhomogeneity v_i obtained by taking the tensor product of (7.22) ($k = 2$)

$$\langle \overline{\boldsymbol{\sigma}} \otimes \overline{\boldsymbol{\sigma}}\rangle_i = \Big\langle\Big\{\mathbf{R}_i^{-1}\Big[\sum_{k=1}^{2} \mathbf{Z}_{ij}^{1k}(\mathbf{R}_k\widetilde{\boldsymbol{\sigma}}_{i,j}(\mathbf{x}_k) + \mathbf{F}_k) - \mathbf{F}_i\Big]\Big\}$$
$$\otimes\Big\{\mathbf{R}_i^{-1}\Big[\sum_{k=1}^{2} \mathbf{Z}_{ij}^{1k}(\mathbf{R}_k\widetilde{\boldsymbol{\sigma}}_{i,j}(\mathbf{x}_k) + \mathbf{F}_k) - \mathbf{F}_i\Big]\Big\}\Big\rangle \quad (10.62)$$

with $(i, j = 1, 2)$. Here $\overline{\boldsymbol{\sigma}}(\mathbf{x}_i)$ and $\widetilde{\boldsymbol{\sigma}}_{i,j}(\mathbf{x}_k)$ $(\mathbf{x}_i \in v_i,\ \mathbf{x}_j \in v_j,\ k = i, j,\ i, j = 1, 2)$ are assumed to be homogeneous random fields in the inclusions $v_i,\ v_j$. According to (10.60) we use the following assumption to modify (10.46) in the framework of the effective field hypothesis:

$$\langle \boldsymbol{\eta}_i \otimes \boldsymbol{\eta}_j \rangle = \langle \boldsymbol{\eta}_i \rangle \otimes \langle \boldsymbol{\eta}_j \rangle, \quad i \neq j, \tag{10.63}$$

$$\langle \boldsymbol{\eta}_i \otimes \boldsymbol{\eta}_i \rangle = \left\langle \left\{ \boldsymbol{\Gamma}_i + \sum_{k=1}^{2} \mathbf{S}_{ij}^{1k} \overline{\boldsymbol{\sigma}}_k \right\} \otimes \left\{ \boldsymbol{\Gamma}_i + \sum_{k=1}^{2} \mathbf{S}_{ij}^{1k} \overline{\boldsymbol{\sigma}}_k \right\} \right\rangle, \tag{10.64}$$

where $(k = 1, 2)$

$$\mathbf{S}_{ij}^{1k} = \mathbf{Z}_{ij}^{1k} \mathbf{R}_k, \quad \boldsymbol{\Gamma}_i = \sum_{k=1}^{2} \mathbf{Z}_{ij}^{1k} \mathbf{F}_k. \tag{10.65}$$

The equations (10.63) and (10.64) are exact for the limiting case $|\mathbf{x} - \mathbf{y}| \gg \max a_m^l$, where $\mathbf{x} \in v_i,\ \mathbf{y} \in v_j\ (l = 1, \ldots, d;\ m = i, j)$.

Then from (10.43) with (10.63) and (10.64), we find the relation for second moments of stresses in the inclusions

$$\langle \overline{\boldsymbol{\sigma}} \otimes \overline{\boldsymbol{\sigma}} \rangle_i = \sum_{j=1}^{N} \mathbf{X}_{ij} \sum_{p=1}^{N} \{ (\delta_{pj} + \mathbf{H}_{jp}^3) \langle \overline{\boldsymbol{\sigma}} \rangle_p \otimes \langle \overline{\boldsymbol{\sigma}} \rangle_j$$
$$+ \mathbf{H}_{jp}^4 \langle \overline{\boldsymbol{\sigma}} \rangle_j \otimes \langle \overline{\boldsymbol{\sigma}} \rangle_p + \mathbf{P}_{jp}^1 \langle \overline{\boldsymbol{\sigma}} \rangle_p + \mathbf{P}_{jp}^2 \langle \overline{\boldsymbol{\sigma}} \rangle_j + \mathbf{P}_{jp}^3 \}, \tag{10.66}$$

where the tensors $\mathbf{H}_{ip}^k,\ \mathbf{P}_{ip}^l\ (k = 1, 2, 3, 4;\ l = 1, 2, 3)$ are represented in Appendix 21.3.2. The following tensor product notation has been applied:

$$[(B_1 \otimes B_2)(b_1 \otimes b_2)]_{pqrs} = B_{1pqkl} b_{1kl} B_{2rsmn} b_{2mn}, \tag{10.67}$$

$$[(B_1 \otimes b_1) b_2)]_{pqrs} = B_{1pqkl} b_{2kl} b_{1rs},\ [(b_1 \otimes B_1) b_2)]_{pqrs} = b_{1pq} B_{1rsmn} b_{2mn}, \tag{10.68}$$

where $\mathbf{B}_1,\ \mathbf{B}_2$ and $\mathbf{b}_1,\ \mathbf{b}_2$ stand for any fourth- and second-order tensor, respectively. Here the inverse matrix $\mathbf{X}^{-1}$ of $\mathbf{X}$ has the following elements: $(\mathbf{X}^{-1})_{ij} = \delta_{ij} \left\{ \mathbf{I} \otimes \mathbf{I} - \sum_{q=1}^{N} \mathbf{H}_{iq}^1 \right\} - \mathbf{H}_{ij}^2$, $(i, j = 1, \ldots, N)$. The inversion of the eighth-order tensor $\mathbf{X}$ was carried out by the use of the tensor analogy of the Taylor expansion $(1 - x)^{-1} = \sum x^n$, $(n = 0, 1, \ldots)$; in the considered examples of composites, ten terms of the expansion provide an error which is less than 1 percent. On the right-hand side of Eq. (10.66), the values $\langle \overline{\boldsymbol{\sigma}} \rangle_i$ $(i = 1, \ldots, N)$ are obtained from (10.22). A simplification of Eq. (10.66) can be achieved by accepting the quasi-crystalline approximation (10.14), when the equality (10.48) holds in the right-hand side of Eq. (10.66) that reduces Eq. (10.66) to

$$\langle \overline{\sigma} \otimes \overline{\sigma} \rangle_i = \langle \overline{\sigma} \rangle_i \otimes \langle \overline{\sigma} \rangle_i + \sum_{p=1}^{N} \Big\{ \mathbf{H}^1_{ip} \langle \overline{\sigma} \rangle_i \otimes \langle \overline{\sigma} \rangle_i + \mathbf{H}^2_{ip} \langle \overline{\sigma} \rangle_p \otimes \langle \overline{\sigma} \rangle_p + \mathbf{H}^3_{ip} \langle \overline{\sigma} \rangle_p$$

$$\otimes \langle \overline{\sigma} \rangle_i + \mathbf{H}^4_{ip} \langle \overline{\sigma} \rangle_i \otimes \langle \overline{\sigma} \rangle_p + \mathbf{P}^1_{ip} \langle \overline{\sigma} \rangle_p + \mathbf{P}^2_{ip} \langle \overline{\sigma} \rangle_i + \mathbf{P}^3_{ip} \Big\}, \qquad (10.69)$$

An estimation of the average stress dispersion (10.45) can be obtained from Eq. (10.44) in the framework of the effective field hypothesis by the use of Eq. (10.66). In a similar manner, the average second moment of the strain polarization tensor $\langle \eta \otimes \eta \rangle_i$ can be obtained

$$\langle \eta \otimes \eta \rangle_i = \mathbf{R}_i \langle \overline{\sigma} \rangle_i \otimes \mathbf{F}_i + \mathbf{F}_i \otimes \mathbf{R}_i \langle \overline{\sigma} \rangle_i + \mathbf{F}_i \otimes \mathbf{F}_i$$
$$+ (\mathbf{R}_i \otimes \mathbf{R}_i) \sum_{j=1}^{N} \mathbf{X}_{ij} \sum_{p=1}^{N} \{ (\delta_{pj} + \mathbf{H}^3_{jp}) \langle \overline{\sigma} \rangle_p \otimes \langle \overline{\sigma} \rangle_j$$
$$+ \mathbf{H}^4_{jp} \langle \overline{\sigma} \rangle_j \otimes \langle \overline{\sigma} \rangle_p + \mathbf{P}^1_{jp} \langle \overline{\sigma} \rangle_p + \mathbf{P}^2_{jp} \langle \overline{\sigma} \rangle_j + \mathbf{P}^3_{jp} \}, \qquad (10.70)$$

where an analog of the tensor product notation (10.67) and (10.68) was used. The representation (10.70) can be exploited in the right-hand side of Eq. (10.51) for estimation of the inhomogeneous second moment of stresses $\langle \sigma \otimes \sigma \rangle_i(\mathbf{x})$ $(\mathbf{x} \in v_i)$ (10.51).

An estimation of the second stress moment averaged over the matrix can be found in analogy to the derivation of (10.45) and (10.66). If we restrict our consideration to binary interaction of the inclusions and use the assumption $\langle \overline{\sigma} \otimes \overline{\sigma} | v_i, \mathbf{x}_i; v_0, \mathbf{x}_0 \rangle_i \equiv \langle \overline{\sigma} \otimes \overline{\sigma} \rangle_i$, we obtain

$$\langle \sigma \otimes \sigma \rangle_0 = \langle \sigma \rangle_0 \langle \sigma \rangle_0 + \sum_{p=1}^{N} \{ \mathbf{H}^{10}_p \langle \overline{\sigma} \otimes \overline{\sigma} \rangle_p + \mathbf{P}^{10}_p \langle \overline{\sigma} \rangle_p + \mathbf{P}^{30}_p \}, \qquad (10.71)$$

where the tensors $\mathbf{H}^{10}_p$, $\mathbf{P}^{10}_p$, $\mathbf{P}^{30}_p$ are represented in Appendix 21.3.2. The quantities $\langle \overline{\sigma} \otimes \overline{\sigma} \rangle_p$ and $\langle \overline{\sigma} \rangle_p$ can be estimated by the use of Eqs. (10.66) and (10.22), respectively. The new Eqs. (10.66)–(10.71) formally coincide with the corresponding equations by Buryachenko [200]. A difference is defined by distinction of the tensors $\mathbf{H}^k_{ip}$, $\mathbf{P}^l_{ip}$ $(k = 1, 2, 3, 4;\ l = 1, 2, 3$ which are reduced to the corresponding tensors in [200] by replacements $\mathbf{T}_p(\mathbf{x} - \mathbf{x}_p) \to \mathbf{T}_{ip}(\mathbf{x}_i - \mathbf{x}_p)$.

10.3.2 High-Order Approximation

The initial approximation of the second moment of the effective stresses $\langle (\overline{\sigma} \otimes \overline{\sigma})^{[0]} \rangle_i$ was found in Sect. 10.1 in the framework of the effective field hypothesis **H1a** and **H1b**. Substitution of the found $\langle (\overline{\sigma} \otimes \overline{\sigma})^{[0]} \rangle_i$ into the right-hand side of Eq. (10.42) allows us to obtain a recursive formula for estimation of the inhomogeneous effective stress dispersion (10.43) ($\mathbf{x} \in v_i$) and the related quantities

$$\Delta\overline{\sigma}^{(i)2[k+1]}_{klmn}(\mathbf{x}) = \sum_{p=1}^{N} \Big\{ \mathbf{H}^1_{ip}(\mathbf{x})\langle(\boldsymbol{\eta}\otimes\boldsymbol{\eta})^{[k]}\rangle_i + \mathbf{H}^2_{ip}(\mathbf{x})\langle(\boldsymbol{\eta}\otimes\boldsymbol{\eta})^{[k]}\rangle_p + \mathbf{P}^3_{ip}(\mathbf{x})$$

$$+ \mathbf{H}^3_{ip}(\mathbf{x})\langle\boldsymbol{\eta}\rangle_p \otimes \langle\boldsymbol{\eta}\rangle_i + \mathbf{H}^4_{ip}(\mathbf{x})\langle\boldsymbol{\eta}\rangle_i \otimes \langle\boldsymbol{\eta}\rangle_p + \mathbf{P}^1_{ip}(\mathbf{x})\langle\boldsymbol{\eta}\rangle_p + \mathbf{P}^2_{ip}(\mathbf{x})\langle\boldsymbol{\eta}\rangle_i \Big\}, \quad (10.72)$$

$$\langle(\boldsymbol{\eta}\otimes\boldsymbol{\eta})^{[q+1]}\rangle_r(\mathbf{x}) = \langle\boldsymbol{\eta}\rangle_r(\mathbf{x}) \otimes \langle\boldsymbol{\eta}\rangle_r(\mathbf{x}) + \mathcal{R}_r \Delta\overline{\sigma}^{(i)2[k+1]}(\mathbf{x})\mathcal{R}_r^\top, \quad (10.73)$$

$$\langle(\boldsymbol{\sigma}\otimes\boldsymbol{\sigma})^{[q+1]}\rangle_r(\mathbf{x}) = \langle\boldsymbol{\sigma}\rangle_r(\mathbf{x}) \otimes \langle\boldsymbol{\sigma}\rangle_r(\mathbf{x}) + \mathcal{B}_r \Delta\overline{\sigma}^{(i)2[k+1]}(\mathbf{x})\mathcal{B}_r^\top, \quad (10.74)$$

where the tensors $\mathbf{H}^l_{ip}(\mathbf{x}), \mathbf{P}^l_{ip}(\mathbf{x})$ ($l = 1, 2, 3$) are defined in Appendix 21.3.2 and the statistical averages $\langle\overline{\boldsymbol{\sigma}}^{[k]}\rangle_i(\mathbf{x})$ and $\langle\boldsymbol{\eta}\rangle_r \equiv \lim_{k\to\infty}\langle\boldsymbol{\eta}^{[k]}\rangle_r$ ($r = i, p$) are found from the independent recursive relations (10.9) and (10.10), respectively.

It should be mentioned that the summands under the sum sign in the right-hand side of Eq. (10.72) depend on the volume averages $\langle(\boldsymbol{\eta}\otimes\boldsymbol{\eta})^{[k]}\rangle_i$ and $\langle\boldsymbol{\eta}\rangle_i$ that makes it possible to reduce the integral equation (10.42) to the algebraic one (10.72) where the tensors $\mathbf{H}^l_{ip}(\mathbf{x}), \mathbf{P}^l_{ip}(\mathbf{x})$ ($l = 1, 2, 3$) are estimated only one time. Exploiting instead of $\langle(\boldsymbol{\eta}\otimes\boldsymbol{\eta})^{[k]}\rangle_i$ and $\langle\boldsymbol{\eta}\rangle_i$ their non-averaged counterparts $\langle(\boldsymbol{\eta}\otimes\boldsymbol{\eta})^{[k]}\rangle_i(\mathbf{x})$ and $\langle\boldsymbol{\eta}\rangle_i(\mathbf{x})$ is obvious; however, this generalization will take essentially longer computing time. For example, the summands containing the tensors $\mathbf{H}^3_{ip}(\mathbf{x})$ and $\mathbf{H}^4_{ip}(\mathbf{x})$ should be replaced by the operators ($\mathbf{x} \in v_i$, no sum on i, p)

$$\int \mathcal{T}^{P\eta}_{i,p}(\langle\boldsymbol{\eta}\rangle_p)(\mathbf{x}) \otimes \mathcal{T}^{I\eta}_{i,p}(\langle\boldsymbol{\eta}\rangle_i)(\mathbf{x})\varphi(v_p, \mathbf{x}_p|; v_i, \mathbf{x}_i)d\mathbf{x}_p, \quad (10.75)$$

$$\int \mathcal{T}^{I\eta}_{i,p}(\langle\boldsymbol{\eta}\rangle_i)(\mathbf{x}) \otimes \mathcal{T}^{P\eta}_{i,p}(\langle\boldsymbol{\eta}\rangle_p)(\mathbf{x})\varphi(v_p, \mathbf{x}_p|; v_i, \mathbf{x}_i)d\mathbf{x}_p, \quad (10.76)$$

respectively, where the inhomogeneous $\langle\boldsymbol{\eta}\rangle_r(\mathbf{x})$ ($\mathbf{x} \in v_r$, $r = i, p$) are found from the independent recursive relation (10.8). The operator $\mathcal{T}^{\eta}_{i,p} = \mathcal{T}^{I\eta}_{i,p} + \mathcal{T}^{P\eta}_{i,p}$ determined analogously to Eq. (4.139) is reduced in the framework of the hypothesis **H1** to the tensors $\mathbf{T}^{I\eta}_{i,p}$ and $\mathbf{T}^{P\eta}_{i,p}$ (4.147) presented through the matrix $\mathbf{Z}_{ip}$ by Eq. (4.146$_{1,2}$) with the replacement $\mathbf{R}_i \equiv \mathbf{I}$. A simplified construction of the operator $\mathcal{T}^{\eta}_{i,p}$ (4.148) allows us to take the inhomogeneities of $\langle\boldsymbol{\eta}\rangle_i(\mathbf{y})$ and $\langle\boldsymbol{\eta}\rangle_p(\mathbf{z})$ ($\mathbf{y} \in v_i$, $\mathbf{z} \in v_p$) into account.

10.4 Elastically Homogeneous Composites with Randomly Distributed Residual Microstresses

In this section specializations of the general formulae given in the preceding sections are derived. Especially interesting results are obtained if the mismatch in the elastic compliance is negligible, i.e., $\mathbf{M}(\mathbf{x}) \equiv \mathbf{M}^{(0)} = \text{const}$, and the stress-free strains $\boldsymbol{\beta}$ fluctuate. Some relations represented below can be found immediately without the assumptions of MEFM.

The relations for the stored energy of elastically homogeneous composites (9.103) and (9.104) were obtained previously. Utilizing Eq. (6.177) with a special variation of the compliance tensor $\mathbf{M} \to \mathbf{M}^{(0)} + \delta\mathbf{M}$ ($\mathbf{M}, \delta\mathbf{M} \equiv \text{const.}$) leads to the exact representation for the second moment of residual stresses averaged over both the whole composite and each phase (6.185)

$$\langle \boldsymbol{\sigma} \otimes \boldsymbol{\sigma} \rangle = -2\frac{\partial W^*}{\partial \mathbf{M}^{(0)}}, \quad \langle \boldsymbol{\sigma} \otimes \boldsymbol{\sigma} \rangle_i = -\frac{2}{c^{(i)}}\frac{\partial W^*}{\partial \mathbf{M}^{(i)}}\bigg|_{\boldsymbol{\beta}}, \tag{10.77}$$

respectively. Unfortunately, Eq. (10.77_2) is less simple and requires a consideration of multiparticle interaction of the inclusions. This can be done by the use of MEFM. Nevertheless, relation (10.77_1) is useful for testing of the exactness of the integral equation method proposed in the present chapter for the estimation of second moments of stresses in the components.

For elastically homogeneous medium with $\mathbf{M}(\mathbf{x}) \equiv \mathbf{M}^{(0)}$ and $\boldsymbol{\eta}(\mathbf{x}) = \boldsymbol{\beta}_1(\mathbf{x})$, Eq. (10.46) is simplified

$$\begin{aligned}\Delta\boldsymbol{\sigma}^{(i)2}(\mathbf{x}) = &\int \boldsymbol{\Gamma}(\mathbf{x}-\mathbf{x}_p)\boldsymbol{\beta}_1(\mathbf{x}_p) \otimes \boldsymbol{\Gamma}(\mathbf{x}-\mathbf{x}_p)\boldsymbol{\beta}_1(\mathbf{x}_p)\varphi(v_p,\mathbf{x}_p \,|;\, v_i,\mathbf{x}_i)d\mathbf{x}_p \\ &+\int\!\!\int \boldsymbol{\Gamma}(\mathbf{x}-\mathbf{x}_p)\boldsymbol{\beta}_1(\mathbf{x}_p) \otimes \boldsymbol{\Gamma}(\mathbf{x}-\mathbf{x}_q)\boldsymbol{\beta}_1(\mathbf{x}_q) \\ &\times\varphi(v_p,\mathbf{x}_p \,|;\, v_i,\mathbf{x}_i)\widetilde{\varphi}(v_q,\mathbf{x}_q \,|;\, v_p,\mathbf{x}_p;\, v_i,\mathbf{x}_i)d\mathbf{x}_q d\mathbf{x}_p,\end{aligned} \tag{10.78}$$

where $\widetilde{\varphi}(v_q,\mathbf{x}_q \,|;\, v_p,\mathbf{x}_p;\, v_i,\mathbf{x}_i) \equiv \varphi(v_q,\mathbf{x}_q \,|;\, v_p,\mathbf{x}_p;\, v_i,\mathbf{x}_i) - \varphi(v_q,\mathbf{x}_q \,|;\, v_i,\mathbf{x}_i)$. The following new approximations for the second dispersion can be obtained by taking only binary interaction of the inclusions into account

$$\Delta\boldsymbol{\sigma}^{(i)2}(\mathbf{x}) = \int \boldsymbol{\Gamma}(\mathbf{x}-\mathbf{x}_p)\boldsymbol{\beta}_1(\mathbf{x}_p) \otimes \boldsymbol{\Gamma}(\mathbf{x}-\mathbf{x}_p)\boldsymbol{\beta}_1(\mathbf{x}_p)\varphi(v_p,\mathbf{x}_p \,|;\, v_i,\mathbf{x}_i)d\mathbf{x}_p. \tag{10.79}$$

Ignoring effective stress fluctuations in the inclusions v_p in the integral terms (10.78) leads to other new representations

$$\Delta\boldsymbol{\sigma}^{(i)2}(\mathbf{x}) = \int \mathcal{L}_p^{\beta}(\mathbf{x}-\mathbf{x}_p,\beta)\otimes\mathcal{L}_p^{\beta}(\mathbf{x}-\mathbf{x}_p,\beta)\varphi(v_p,\mathbf{x}_p\,|;v_i,\mathbf{x}_i)d\mathbf{x}_p$$
$$+\int\int \mathcal{L}_p^{\beta}(\mathbf{x}-\mathbf{x}_p,\beta)\otimes\mathcal{L}_q^{\beta}(\mathbf{x}-\mathbf{x}_q,\beta)$$
$$\times\varphi(v_p,\mathbf{x}_p\,|;v_i,\mathbf{x}_i)\widetilde{\varphi}(v_q,\mathbf{x}_q\,|;v_p,\mathbf{x}_p;v_i,\mathbf{x}_i)d\mathbf{x}_q d\mathbf{x}_p, \quad (10.80)$$

The relation (10.80) can be called as "far-field approximation." In a similar manner, the stress dispersion on the matrix can be found

$$\Delta\boldsymbol{\sigma}^{(0)2} = \int \mathcal{L}_p^{\beta}(\mathbf{x}_0-\mathbf{x}_p,\beta)\otimes\mathcal{L}_p^{\beta}(\mathbf{x}_0-\mathbf{x}_p,\beta)\varphi(v_p,\mathbf{x}_p\,|;v_0,\mathbf{x}_0)d\mathbf{x}_p$$
$$+\int\int \mathcal{L}_p^{\beta}(\mathbf{x}_0-\mathbf{x}_p,\beta)\otimes\mathcal{L}_q^{\beta}(\mathbf{x}_0-\mathbf{x}_q,\beta)$$
$$\times\varphi(v_p,\mathbf{x}_p\,|;v_0,\mathbf{x}_0)\widetilde{\varphi}(v_q,\mathbf{x}_q\,|;v_p,\mathbf{x}_p;v_0,\mathbf{x}_0)d\mathbf{x}_q d\mathbf{x}_p, \quad (10.81)$$

The representations (10.80) and (10.81) were obtained in [186].

Therefore, the exact estimation of the stress dispersion $\Delta\boldsymbol{\sigma}^2 \equiv \langle\boldsymbol{\sigma}\otimes\boldsymbol{\sigma}\rangle$ at $\langle\boldsymbol{\sigma}\rangle \equiv \mathbf{0}$ (10.77_1) offers a test of the accuracy of evaluation (10.80) and (10.81) which uses dissimilar approximate conditional probability densities. By virtue of the fact that the average stresses in the ellipsoidal inclusions are calculated exactly by Eqs. (9.101_2) and (9.103), the difference $\overline{\Delta}\boldsymbol{\sigma}^2 = \langle\boldsymbol{\sigma}\otimes\boldsymbol{\sigma}\rangle - \sum_{j=0}^{N} c_j\langle\boldsymbol{\sigma}\rangle_j\otimes\langle\boldsymbol{\sigma}\rangle_j$ can be used for the estimation of the accuracy of calculated stress fluctuations inside the components (10.80) and (10.81). Buryachenko [186] (see also [200]) demonstrated that for the CMs with spherical particles, the approximate formulas (10.80) and (10.81) under the binary step function (5.97) can be used for the estimation of stress fluctuations $\langle\Delta\boldsymbol{\sigma}^2\rangle_i$ with sufficient accuracy. It was also estimated inhomogeneity of stress dispersions inside the inclusions $\Delta\boldsymbol{\sigma}^{(i)2}(\mathbf{x})$ ($\mathbf{x}\in v^{(i)}$) for the different RDFs (5.97) and (5.100). It should be mentioned that for strength prediction, precisely stress dispersions $\Delta\boldsymbol{\sigma}^{(i)2}$ ($\imath = 0, 1$) should be used rather than the exact stress fluctuation in the whole composites $\Delta\boldsymbol{\sigma}^2 \equiv \langle\boldsymbol{\sigma}\otimes\boldsymbol{\sigma}\rangle - \langle\boldsymbol{\sigma}\rangle\otimes\langle\boldsymbol{\sigma}\rangle$ (as in an approach [1161]); it was demonstrated in [186] that $\Delta\boldsymbol{\sigma}^{(i)2}$ ($\imath = 0, 1$) is less than $\Delta\boldsymbol{\sigma}^2$ by a factor 5.

10.5 Related Method of Estimations of the Second Moments of Stresses

It should be mentioned that for the pure elastic case ($\boldsymbol{\beta}_1(\mathbf{x}) \equiv \mathbf{0}$), an alternative formula for the estimation of the second moment $\langle\mathbf{ss}\rangle_0$ was proposed in [714] and [721] without any justification

$$\langle \mathbf{ss} \rangle_0 = H^0 + \int \big[\langle \mathbf{ss} | v_p, \mathbf{x}_p; v^{(0)}, \mathbf{x}_0 \rangle - H^0\big]\varphi(v_p, \mathbf{x}_p |; v^{(0)}, \mathbf{x}_0)\, d\mathbf{x}_p, \qquad (10.82)$$

where $H^0 = (\mathbf{B}_1\boldsymbol{\sigma}^0)(\mathbf{B}_1\boldsymbol{\sigma}^0)$, whereas the estimation of the second conditional moment $\langle \mathbf{ss} | v_p, \mathbf{x}_p; v^{(0)}, \mathbf{x}_0 \rangle$ was fulfilled on the basis of some additional assumptions which are similar (but not equivalent) to the "far-field approximation" (10.57)

$$\langle \mathbf{ss} | v_p, \mathbf{x}_p; v^{(0)}, \mathbf{x}_0 \rangle = (\mathbf{B}_1\boldsymbol{\sigma}^0 + \mathbf{T}_p(\mathbf{x}_0, \mathbf{x}_p)\mathbf{B}_2\boldsymbol{\sigma}^0)(\mathbf{B}_1\boldsymbol{\sigma}^0 + \mathbf{T}_p(\mathbf{x}_0, \mathbf{x}_p)\mathbf{B}_2\boldsymbol{\sigma}^0). \qquad (10.83)$$

Here the tensors $\mathbf{B}_1, \mathbf{B}_2 = \text{const.}$ are defined again by an average scheme.

More correct estimations of second conditional moment of stresses were proposed by Eq. (10.57) (see also [265, 268]). It is well-known that integrals such as in (10.82) with (10.83) are not absolutely convergent simply because $\mathbf{T}_p(\mathbf{x}_0 - \mathbf{x}_p)$ is of order $|\mathbf{x}_0 - \mathbf{x}_p|^{-d}$. In dealing with this difficulty, one [714, 721] reduced a volume improper integral to a particular case of a triple integral, a procedure that is questionable. Moreover, formulae (10.82) and (10.83) were used by the afore-mentioned authors for the estimation of second moments of stresses $\langle \mathbf{ss} \rangle_0$ in the elastoplastic analysis, when $\boldsymbol{\beta}_1(\mathbf{x}) \not\equiv \mathbf{0}$, or, in other words, the authors mentioned that the second $\langle \mathbf{ss} \rangle_0$ and first $\langle \mathbf{s} \rangle_0$ moments are invariant with respect to the transformation strain tensor (e.g., $\boldsymbol{\beta}_1$) (a mathematically consistent consideration of this problem that is more exact is presented in [238, 265]).

10.6 Numerical Results

10.6.1 Effective Moduli and the First Statistical Moments of Stresses Inside the Inclusions

For the sake of definiteness, we will consider a plane strain $2D$ problem for the random structure composites containing the circle inclusions (the fibers). The different radial distribution functions (RDF) for the inclusions will be examined at the estimation of effective parameters (10.1)–(10.13). We consider a pure mechanical problem ($\boldsymbol{\beta} \equiv \mathbf{0}$) and assume the matrix is epoxy resin [$\mathbf{L}^{(0)} = (3k^{(0)}, 2\mu^{(0)})$, $k^{(0)} = 3.24$ GPa and $\mu^{(0)} = 1.16$ GPa] which contains identical circular SCS-0 fibers [$\mathbf{L}^{(1)} = (3k^{(1)}, 2\mu^{(1)})$, $k^{(1)} = 190.47$ GPa and $\mu^{(1)} = 173.91$ GPa]. Two alternative radial distribution functions (RDF) $g(r)$ of inclusions will be examined, the well-stirred approximation (9.116) and the model (9.117) (see [604, 1525]), taking into account a neighboring order in the distribution of the inclusions, where $r \equiv |\mathbf{x}_i - \mathbf{x}_q|$ is the distance between the nonintersecting inclusions v_i and v_q and c is the volume fraction of fibers of the radius a.

Figures 10.1 and 10.2 present a schematic comparison of both the classical approach (EFM; see Fig. 10.1) and the new one (see Fig. 10.2) in the framework of hypotheses **H2b** and **H3** for the circle inclusions $v_k = \{\mathbf{x} \in R^2 |\ |\mathbf{x}| \le a\}$

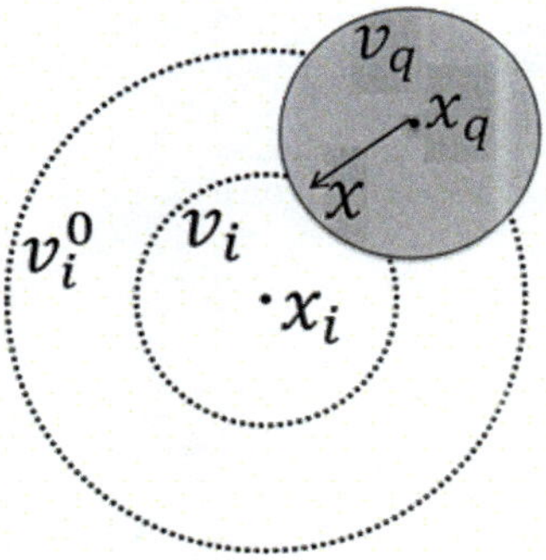

Fig. 10.1 Schematic representation of the EFM

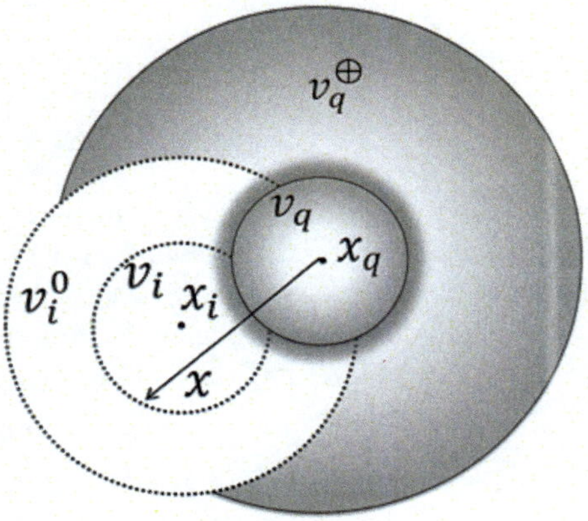

Fig. 10.2 Schematic representation of the new approach

($k = i, q$) with the RDF (5.101) (although a generalization to a convex symmetric inclusions can also be presented; see Figs. 9.2 and 9.3). In the classical approach, the effective field $\langle\overline{\boldsymbol{\sigma}}\rangle_i$ depends only on the volume average of the strain polarization tensor $\langle\boldsymbol{\eta}\rangle_q(\mathbf{x}) \equiv \langle\boldsymbol{\eta}\rangle_q$ =const. ($\mathbf{x} \in v_q$, marked by a homogeneous shadow; see Fig. 10.1) inside the moving inclusions v_q with the centers $\mathbf{x}_q$ distributed in the ellipsoidal excluded volume $v_i^0 \equiv v_i \oplus v_i = 2v_i$, where the left-hand side of the last equality is the Minkowski sum and the right-hand side the enlargement by a factor of 2, and all estimations by the EFM are invariant with respect to the size of v_i^0. In the new approach, the effective field $\langle\overline{\boldsymbol{\sigma}}\rangle_i(\mathbf{x})$ in the area $\mathbf{x} \in v_i$ occupied by the matrix material (see Fig. 10.2) is generated by the moving heterogeneities v_q (see Eqs. (7.15) and (7.16)) with the inhomogeneous strain polarization tensor $\langle\boldsymbol{\eta}\rangle_q(\mathbf{y})$ ($y \in v_q$, marked by inhomogeneous shadow). Furthermore, the domain of this long-range action is limited not only by $\mathbf{x}_q \in v_i^0$ but by a domain v_i^{ϕ} for location of inclusion centers $\mathbf{x}_q$ where $\varphi(v_q, \mathbf{x}_r|v_i, \mathbf{x}_i) - n^{(q)}$ is not negligible (in the sense of influence on $\mathbf{M}^*$) and an effect zone of stress perturbations produced by the heterogeneity v_q is limited by the Minkowski addition of domains $\mathbf{x} \in v_{iq}^{\oplus} \equiv v_i^{\phi} \oplus v_q$ (see Fig. 10.2) rather than by the domain v_q as in the EFM. Because of this, the estimations of effective moduli by the new approach in general depend on the size

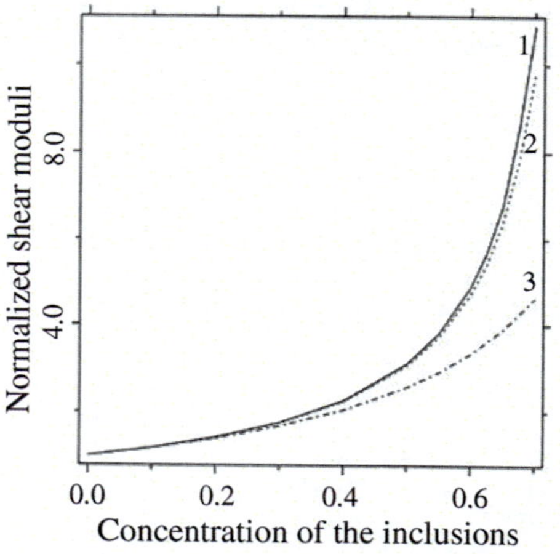

Fig. 10.3 Normalized shear moduli $\mu^*/\mu^{(0)}$ estimated by the use of the new method (1), MEFM (2), and Mori-Tanaka approach (3)

of excluded volume v_i^0. However, the new approach (7.15) and (7.16) is reduced to the classical one in the framework of the hypothesis **H1** when the homogeneity of the strain polarization tensor $\langle \boldsymbol{\eta} \rangle_q(\mathbf{x}) \equiv \langle \boldsymbol{\eta} \rangle_q$ =const. ($\mathbf{x} \in v_q$) is assumed.

We begin our numerical analysis from the evaluations of effective elastic moduli $\mathbf{L}^* = 2k^*_{[2]}\mathbf{N}_1 + 2\mu^*_{[2]}\mathbf{N}_2$ ($\mathbf{N}_1 = \boldsymbol{\delta} \otimes \boldsymbol{\delta}/2,\ \mathbf{N}_2 = \mathbf{I} - \mathbf{N}_1$) for the RDF (7.16). For the fiber composites, it is the plane-strain bulk modulus $k^{(0)}_{[2]}$ (and $k^*_{[2]}$)—instead of the 3D bulk modulus $k^{(0)}_{[3]}$—that plays the significant role: $k^{(0)}_{[2]} = k^{(0)}_{[3]} + \mu^{(0)}_{[3]}/3$, $\mu^{(0)}_{[2]} = \mu^{(0)}_{[3]}$. The normalized shear moduli $\mu^*/\mu^{(0)}$ (the bulk moduli $k^*/k^{(0)}$ are presented in [204]) are presented in Fig. 10.3, for both the classical approach [corresponding to the initial approximation obtained by both two particles (the curve 2) and one particle (the curve 3) MEFM] and new one [corresponding to the eighth iteration (10.11) and (10.13)], the curve 1. As can be seen, the distinctions between curves 1 and 2 equal to 6.4% and 10.2% at $c = 0.70$ for $\mu^*/\mu^{(0)}$ and $k^*/k^{(0)}$, respectively. In so doing, curves 2 and 3 can differ on 16% and 81% for the bulk and shear moduli, respectively. Thus, even in the framework of the hypothesis **H1** (initial approximation), the two-particle approximation of the MEFM (curve 2) leads to significant correction of $\mathbf{L}^*$ estimated by the one-particle approximation of the MEFM (which is equivalent to the Mori-Tanaka approach in the considered case, curve 3). In so doing, the use of the advanced new approach (curve 1) leads to mere small corrections of $\mathbf{L}^*$ with respect to the estimation obtained by the two-particle approximation of the MEFM.

However, it was demonstrated in Chap. 9 (see also [203]) that stress concentration factors are significantly more sensitive values to the choice of the approach than effective elastic moduli. Now we will estimate the means of stress concentration factors ($\boldsymbol{\beta} \equiv \mathbf{0}$ $\langle \boldsymbol{\sigma} \rangle_i(\mathbf{x}) = \mathbf{B}^*(\mathbf{x})\langle \boldsymbol{\sigma} \rangle$, defined analogously to Eq. (10.11). For the sake of definiteness, the composite plate is loaded by the tension along the axis Ox: $\langle \sigma_{ij} \rangle = \delta_{i1}\delta_{j1}$. Then the stress concentration factor $B^*_{kl11}(\mathbf{x})$ defines the stress

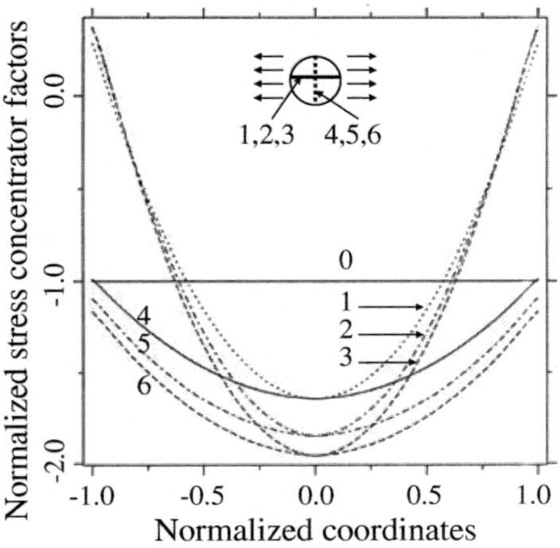

Fig. 10.4 $\langle\sigma_{22}^{\text{nor}[k]}\rangle(x_1)$ (curves 1, 2, 3) and $\langle\sigma_{22}^{\text{nor}[k]}\rangle(x_2)$ (curves 4, 5, 6) vs x_j/a ($j = 1, 2$): curves 0, 1, 2, 3 and 0,4, 5, 6 for $k = 0, 1, 2, 3$, respectively

distribution inside the heterogeneity v_i $\langle\sigma_{kl}\rangle_i(\mathbf{x}) = B^*_{kl11}(\mathbf{x})\langle\sigma_{11}\rangle_i$ that allows one to define a normalized stress concentration factor $\langle\sigma^{\text{nor}}_{kl}\rangle_i(\mathbf{x}) \equiv B^*_{kl11}(\mathbf{x})/|B^{MEFM*}_{kl11}|$ (no sum on k, l) where $\mathbf{B}^{MEFM*}$ is the initial stress concentration factor estimated by the classical version of the MEFM $\langle\boldsymbol{\sigma}\rangle_i(\mathbf{x}) = \mathbf{B}^{MEFM*}\langle\boldsymbol{\sigma}\rangle_i$ (10.25). In Fig. 10.4 the components $\langle\sigma^{\text{nor}}_{22}\rangle_i(\mathbf{x})$ demonstrating maximum dependence on $\mathbf{x} = (x_1, 0)^\top$ (curves 1, 2, and 3) and $\mathbf{x} = (0, x_2)^\top$ (curves 4, 5, and 6) are presented at $c = 0.60$ for the initial $\langle\sigma_{22}^{\text{nor}[0]}\rangle_i(\mathbf{x}) \equiv -1$ (curve 0), first $\langle\sigma_{22}^{\text{nor}[1]}\rangle_i(\mathbf{x})$ (curves 1, 4), second $\langle\sigma_{22}^{\text{nor}[2]}\rangle_i(\mathbf{x})$ (curves 2, 5), and third $\langle\sigma_{22}^{\text{nor}[3]}\rangle_i(\mathbf{x})$ (curves 3, 6) iterations of normalized stress concentration factor. A fast convergence of the proposed iteration method is detected: the third iteration $\langle\sigma_{22}^{\text{nor}[3]}\rangle_i(\mathbf{0})$ differs from the second $\langle\sigma_{22}^{\text{nor}[2]}\rangle_i(\mathbf{0})$, fourth, fifth, and eighth by 11.1%, 5.8%, 5.3%, and 5.4 %, respectively, while the difference between the eighth iteration and the seventh, sixth, and fifth iterations equal to 0.015%, 0.51%, and 0.027%, respectively. It should be mentioned that $\langle\sigma_{22}^{\text{nor}[8]}\rangle_i(\mathbf{0})$ differs from the initial approximation $\langle\sigma_{22}^{\text{nor}[0]}\rangle_i \equiv -1$ by a factor of 1.97, while $\langle\sigma_{22}^{\text{nor}[8]}\rangle_i(\mathbf{x})$ ($\mathbf{x} = (1, 0)^\top$) and $\langle\sigma_{22}^{\text{nor}[0]}\rangle_i \equiv -1$ distinguish from one another by a sign. Nevertheless, the volume average $\langle\sigma_{22}^{\text{nor}[8]}\rangle_i = -0.9909$ is distinguished from $\langle\sigma_{22}^{\text{nor}[0]}\rangle_i \equiv -1$ by nothing more than 0.9% that explains a small dissimilarity of effective moduli $\mathbf{L}^*$ estimated by the new approach and MEFM (curves 1 and 2, respectively) demonstrated in Fig. 10.3. Such prominent differences of the old and new approaches are based on the abandonment of the effective field hypothesis **H1** in the new approach (see for details [204]).

The normalized stress concentration factors $\langle\sigma_{11}^{\text{nor}[k]}\rangle(\mathbf{x})$ for $\mathbf{x} = (x_1, 0)^\top$ and $\mathbf{x} = (0, x_2)^\top$ are presented in Figs. 10.5 and 10.6, respectively, at $c =$ 0.0, 0.1, 0.2, 0.3, 0.4, 0.5, 0.6, and 0.7 (curves 0, 1, 2, 3, 4, 5, 6, and 7, respectively). The most significant variation of $\langle\sigma_{11}^{\text{nor}[k]}\rangle(\mathbf{x})$ takes place along the axis Ox_2 where $\langle\sigma_{11}^{\text{nor}[k]}\rangle(x_2)$ varies on 14% at $c = 70\%$. However, we can observe in Figs. 10.7 and 10.8 significantly more dramatic situation with the component

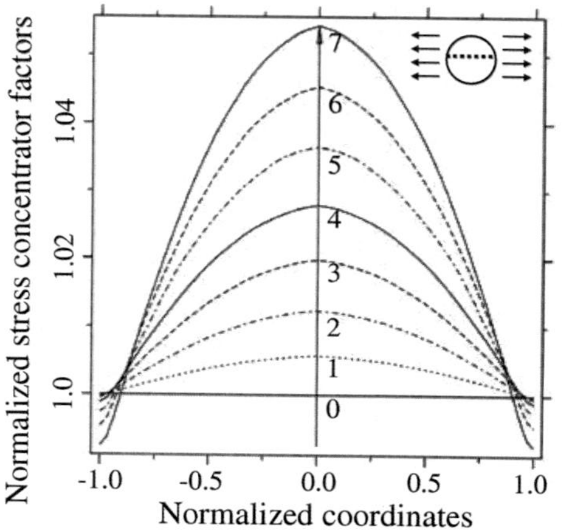

Fig. 10.5 $\langle\sigma_{11}^{\mathrm{nor}[8]}\rangle(x_1)$ vs x_1/a: curves 0, 1, 2, 3, 4, 5, 6, and 7 for $c =$ 0.0, 0.1, 0.2, 0.3, 0.4, 0.5, 0.6, and 0.7, respectively

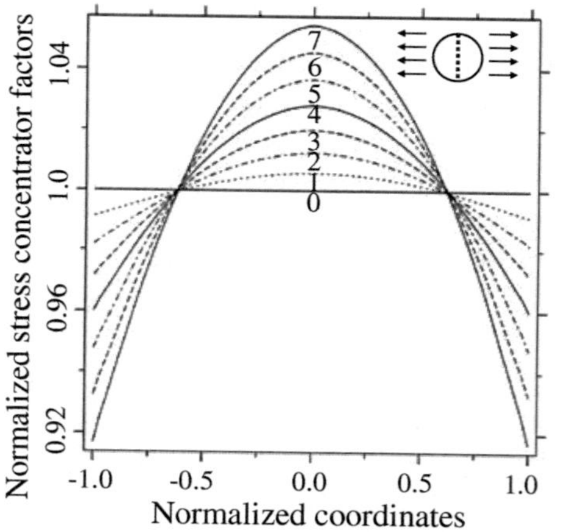

Fig. 10.6 $\langle\sigma_{11}^{\mathrm{nor}[8]}\rangle(x_2)$ vs x_2/a: curves 0, 1, 2, 3, 4, 5, 6, and 7 for $c =$ 0.0, 0.1, 0.2, 0.3, 0.4, 0.5, 0.6, and 0.7, respectively

$\langle\sigma_{22}^{\mathrm{nor}[8]}\rangle(\mathbf{x})$ for $\mathbf{x} = (x_1, 0)^\top$ (Fig. 10.7) and $\mathbf{x} = (0, x_2)^\top$ (Fig. 10.9), where at $c = 0.70$ $\langle\sigma_{22}^{\mathrm{nor}[k]}\rangle(x_1)$ changes a sign and varies in the range from 0.94 ($x_1/a = \pm 1$) to -2.34 ($x_1 = 0$). The component $\langle\sigma_{22}^{\mathrm{nor}[8]}\rangle(x_2)$ in a transversal cross section varies from -1.23 at $x_2/a = \pm 1$ till -2.34 at $x_2 = 0$. The difference between the curves $\langle\sigma_{kl}^{\mathrm{nor}[8]}\rangle(\mathbf{x})$ and $\langle\sigma_{kl}^{\mathrm{nor}[0]}\rangle_i$ increase with the growing of the concentration of fibers c and using the advanced proposed method leads to significant refinement of the result obtained at the local level inside the fibers. However, the surprising thing is that so much dramatic variation of local stresses in the fibers $\langle\sigma_{22}^{\mathrm{nor}[8]}\rangle(\mathbf{x})$ reflects almost not at all on their conditional volume averages: $\langle\sigma_{22}^{\mathrm{nor}[8]}\rangle_i = -0.995$ and $\langle\sigma_{22}^{\mathrm{nor}[0]}\rangle_i \equiv -1.0$ (the sign "−" is explained by the definition $\langle\sigma_{kl}^{\mathrm{nor}}\rangle_i(\mathbf{x}) \equiv B^*_{kl11}(\mathbf{x})/|B^{MEFM*}_{kl11}|$ where $B^{MEFM*}_{2211} = -0.058$ in the case being considered).

Fig. 10.7 $\langle\sigma_{22}^{\mathrm{nor}[8]}\rangle(x_1)$ vs x_1/a: curves 0, 1, 2, 3, 4, 5, 6, and 7 for $c =$ 0.0, 0.1, 0.2, 0.3, 0.4, 0.5, 0.6, and 0.7, respectively

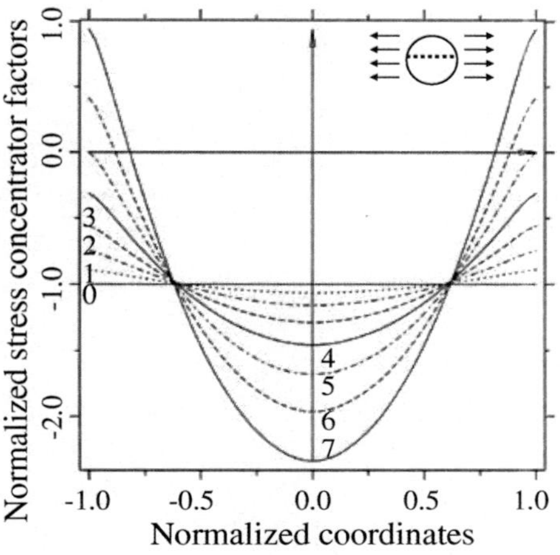

Fig. 10.8 $\langle\sigma_{22}^{\mathrm{nor}[8]}\rangle(x_2)$ vs x_2/a: curves 0, 1, 2, 3, 4, 5, 6, and 7 for $c =$ 0.0, 0.1, 0.2, 0.3, 0.4, 0.5, 0.6, and 0.7, respectively

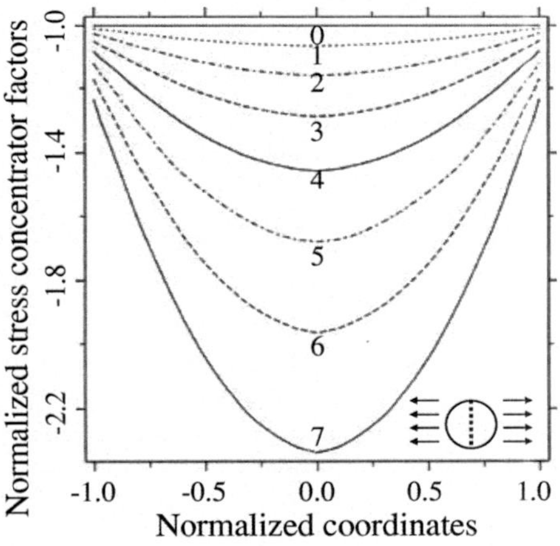

The interaction effects generating the local stresses produced by inclusions are highly sensitive to their locations describing by the RDF. Influence of both the RDF (9.116) and (9.117) on stress concentration factors $B^{*[k]}_{1111}(x_2)$ and $B^{*[k]}_{2211}(x_1)$ are presented in Figs. 10.9 and 10.10, respectively, for $k = 0, 8$ and $c = 0.7$. As can be seen in Fig. 10.9, the difference of $B^{*[0]}_{1111}$ estimated by the use of the RDF (9.116) and (9.117) equals 4.4% that is compatible with the distinction of $B^{*[8]}_{1111}(0)$ equals 4.3%. However, a situation is drastically changing with the replacement of the RDF (9.116) and (9.117) for the components $B^{*[k]}_{2211}(x_1)$ when the difference between $B^{*[0]}_{2211}$ evaluated for the RDF (9.116) and (9.117) equals 8.4% is increasing for eighth iteration till the distinction 26.2% at $x_1 = 0$ while $B^{*[k]}_{2211}(\pm 1)$ differs from one another by a factor 5.7. Thus, the local stress concentration factors $\mathbf{B}^{*[8]}(\mathbf{x})$ are

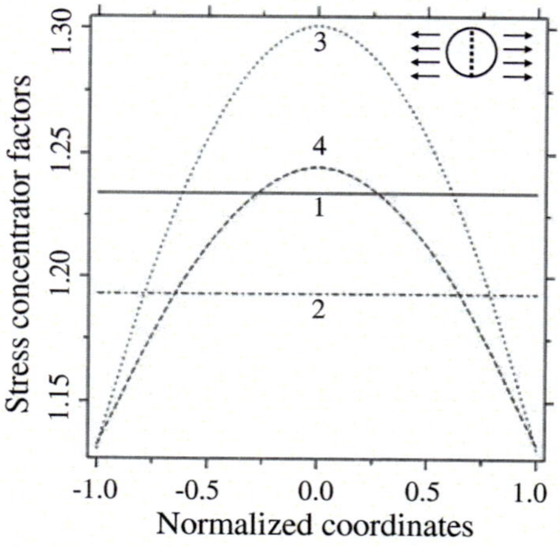

Fig. 10.9 $B^{*[k]}_{1111}(x_2)$ vs x_2/a for $k = 0$ and 8 (curves 1, 2 and 3, 4) estimated for the RDF (9.117) (curves 1, 3) and (9.116) (curves 2, 4, respectively)

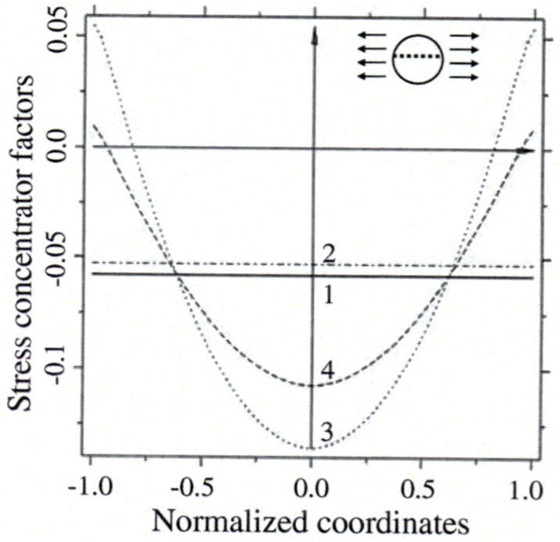

Fig. 10.10 $B^{*[k]}_{2211}(x_1)$ vs x_1/a for $k = 0$ and 8 (curves 1, 2 and 3,4) estimated for the RDF (9.117) (curves 1, 3) and (9.116) (curves 2, 4, respectively)

significantly more sensitive to the choice of the RDF than both the effective elastic moduli and average stress concentration factors $\mathbf{B}^{*[0]}$.

10.6.2 Stress Fluctuations Inside the Inclusions

We begin our numerical analysis of the second moment of stresses from the evaluations of the stress fluctuations averaged over the inclusion volume $\Delta\boldsymbol{\sigma}^{(i)2}$ (10.45) and estimated by both the differential method (see Chap. 6) and the initial approximation of the integral equations method (10.70). Both the effective field hypothesis **H1** and the classical version of the MEFM are used in both methods for estimation of the effective compliance $\mathbf{M}^*$ (10.26), (4.144) and the averaged effective fields $\langle\overline{\boldsymbol{\sigma}}\rangle_i$ (10.22), (4.144), respectively. For definiteness sake,

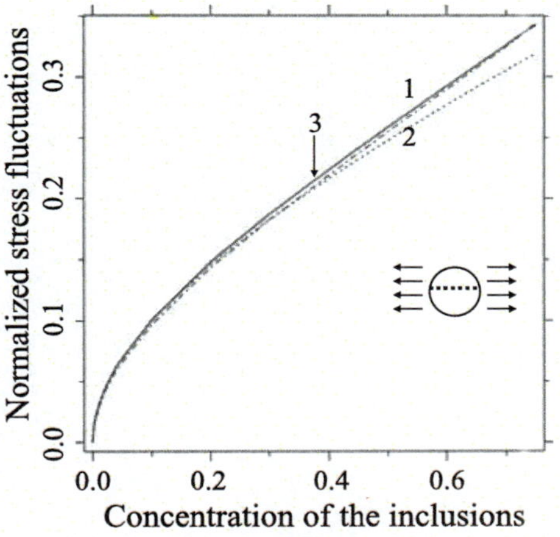

Fig. 10.11 $\bar{\Delta}\sigma^{(i)}_{1111}$ vs c estimated by the methods (6.185), (10.60), and (10.74) (curves 1, 2, and 3, respectively)

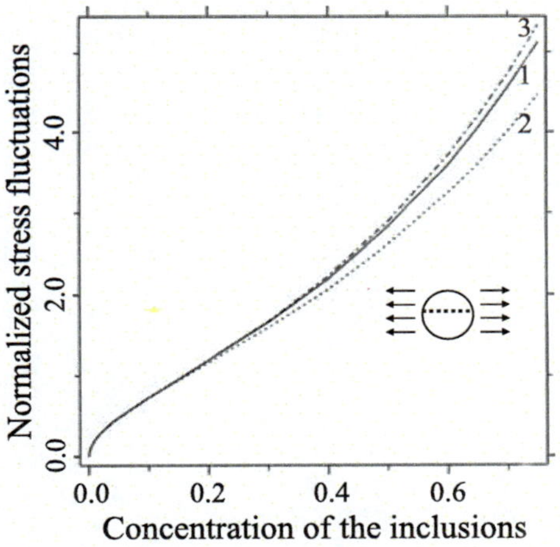

Fig. 10.12 $\bar{\Delta}\sigma^{(i)}_{2222}$ vs c estimated by the methods (6.185), (10.60), and (10.74) (curves 1, 2, and 3, respectively)

the composite plate with the structure described by the RDF (9.117) is loaded by the tension along the axis Ox: $\langle\sigma_{ij}\rangle = \delta_{i1}\delta_{j1}$. The initial stress concentration factor $\mathbf{B}^{*MEFM}_i$ estimated by the classical version of the MEFM $\langle\boldsymbol{\sigma}\rangle_i = \mathbf{B}^{*MEFM}_i\langle\boldsymbol{\sigma}\rangle$ (10.29), (4.144) allows us to estimate the normalized stress fluctuations $\bar{\Delta}\sigma^{(i)}_{klmn} \equiv (\Delta\sigma^{(i)2}_{klmn}/B^{*MEFM}_{i|kl11}B^{*MEFM}_{i|mn11})^{1/2}$ (no sum on k, l, m, n). The components $\bar{\Delta}\sigma^{(i)}_{1111}$ and $\bar{\Delta}\sigma^{(i)}_{2222}$ as the functions of the volume concentration of fibers $c = \pi a^2 n$ are presented in Figs. 10.11 and 10.12, respectively. Even for high-value $c = 0.75$, the differences between the estimations obtained by two different methods (curves 1 and 2) do not exceed 13 % and 10% for the components $\bar{\Delta}\sigma^{(i)}_{2222}$ and $\bar{\Delta}\sigma^{(i)}_{1111}$, respectively. Curves 3 estimated by Eqs. (10.74) and (10.45) will be analyzed later. It should be mentioned that acceptance of the quasi-crystalline approximation (10.14) automatically neglecting the binary interaction of inclusions

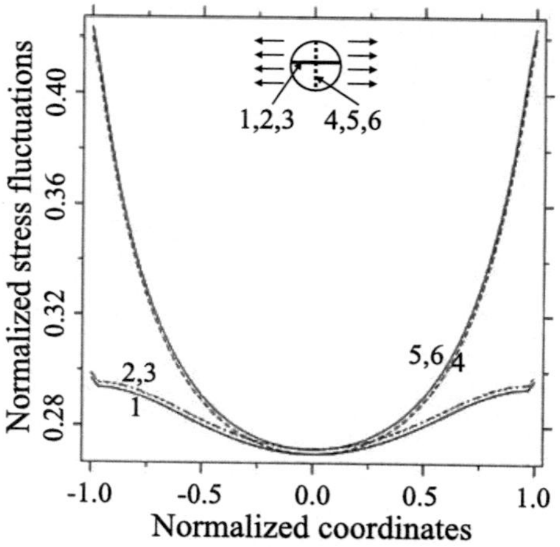

Fig. 10.13 $\bar{\Delta\sigma}^{(i)[k]}_{1111}(x_1)$ (curves 1, 2, 3) and $\bar{\Delta\sigma}^{(i)[k]}_{1111}(x_2)$ (curves 4, 5, 6) vs x_1 and x_2, respectively, $k = 1$ (1, 4), $k = 2$ (2, 5), $k = 3$ (3,6)

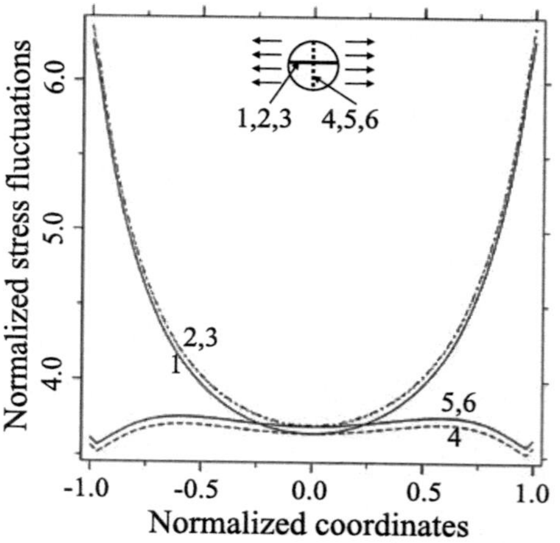

Fig. 10.14 $\bar{\Delta\sigma}^{(i)[k]}_{2222}(x_1)$ (curves 1, 2, 3) and $\bar{\Delta\sigma}^{(i)[k]}_{2222}(x_2)$ (curves 4, 5, 6) vs x_1 and x_2, respectively, for $k = 1$ (1, 4), $k = 2$ (2, 5), $k = 3$ (3,6)

leads to a degenerated result $\Delta\boldsymbol{\sigma}^{(i)2} \equiv \mathbf{0}$ in the differential method (see Chap. 6). In so doing $\Delta\boldsymbol{\sigma}^{(i)2} \equiv \mathbf{B}_i \Delta\overline{\boldsymbol{\sigma}}^{(i)2}\mathbf{B}_i^\top \neq \mathbf{0}$ in Eq. (10.69); for example, $\Delta\overline{\boldsymbol{\sigma}}^{(i)}_{2222} = 4.67$ (10.69) at $c = 0.75$, while $\Delta\overline{\boldsymbol{\sigma}}^{(i)}_{2222} = 5.34$ in Eq. (10.72).

We now turn our attention to the analysis of the inhomogeneous normalized stress fluctuations $\bar{\Delta\sigma}^{(i)}_{klmn}(\mathbf{x}) \equiv (\Delta\sigma^{(i)2}_{klmn}(\mathbf{x})/B^{*MEFM}_{i|kl11} B^{*MEFM}_{i|mn11})^{1/2}$ (10.72), (10.74) along the cross sections $\mathbf{x} = (x_1, 0)^\top$ and $\mathbf{x} = (0, x_2)^\top$. The components $\bar{\Delta\sigma}^{(i)[k]}_{1111}(\mathbf{x})$ and $\bar{\Delta\sigma}^{(i)[k]}_{2222}(\mathbf{x})$ of the k's iterations are presented in Figs. 10.13 and 10.14 for $k = 1, 2, 3$ and $c = 0.65$. A fast convergence of the proposed iteration method is detected: the fifth iteration $\bar{\Delta\sigma}^{(i)[5]}_{2222}(\mathbf{x})$ differs from the third, second, and first iterations by 0.01%, 0.15%, and 1.4%, respectively. As can be seen in Figs. 10.11, 10.12, 10.13, and 10.14, the components $\bar{\Delta\sigma}^{(i)[k]}_{2222}$ and $\bar{\Delta\sigma}^{(i)[k]}_{2222}(\mathbf{x})$ of the averaged and inhomogeneous stress fluctuations, respectively, far exceed

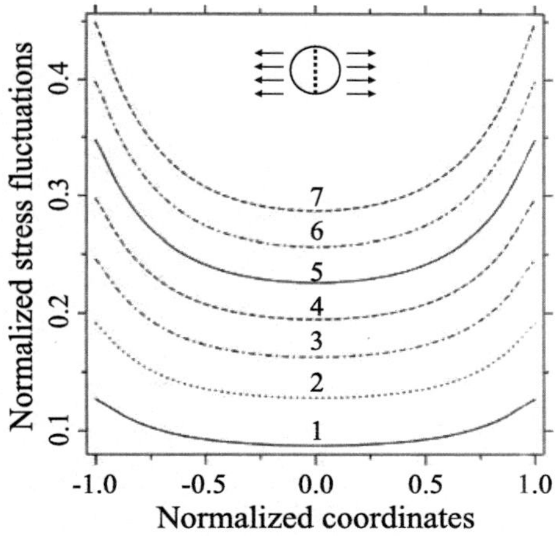

Fig. 10.15 $\bar{\Delta\sigma}^{(i)[5]}_{1111}(x_2)$ vs x_2 at $c =$ 0.1, 0.2, 0.3, 0.4, 0.5, 0.6, 07 (curves 1, 2, 3, 4, 5, 6, 7)

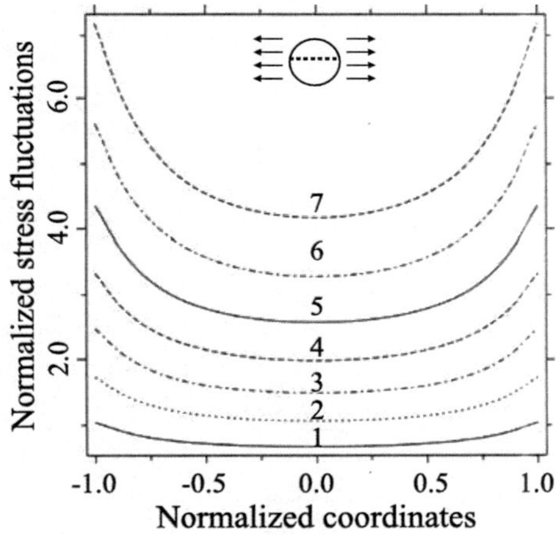

Fig. 10.16 $\bar{\Delta\sigma}^{(i)[5]}_{2222}(x_1)$ vs x_1 at $c =$ 0.1, 0.2, 0.3, 0.4, 0.5, 0.6, 0.7 (curves 1, 2, 3, 4, 5, 6, 7)

the components $\bar{\Delta\sigma}^{(i)[k]}_{1111}$ and $\bar{\Delta\sigma}^{(i)[k]}_{1111}(\mathbf{x})$, respectively, along the cross sections $\mathbf{x} = (x_1, 0)^\top$ and $\mathbf{x} = (0, x_2)^\top$.

The normalized stress fluctuations $\bar{\Delta\sigma}^{(i)[5]}_{1111}(\mathbf{x})$ along the cross sections $\mathbf{x} = (0, x_2)^\top$ are presented in Fig. 10.15 at $c = 0, 0.1, 0.2, 0.3, 0.4, 0.5, 0.6$, and 0.7 (curves 1, 2, 3, 4, 5, 6, and 7, respectively); $\bar{\Delta\sigma}^{(i)[5]}_{1111}(x_2)$ varies on 56% at $c = 0.7$. However, we can observe in Fig. 10.16 more significant variation of the component $\bar{\Delta\sigma}^{(i)[5]}_{2222}(\mathbf{x})$ for $\mathbf{x} = (x_1, 0)^\top$, where at $c = 0.70$ $\bar{\Delta\sigma}^{(i)[5]}_{2222}(\mathbf{x})$ varies on 79% in the range from 7.17 ($x_1/a = \pm 1$, $x_2 = 0$) to 4.02 at the axis $x_1 = 0$. In light of so much strong inhomogeneity of stress fluctuations $\bar{\Delta\sigma}^{(i)[5]}_{klmn}(\mathbf{x})$, of distinctive interest is a comparison in Figs. 10.11 and 10.12 of averaged stress fluctuations $\bar{\Delta\sigma}^{(i)[5]}_{klmn}$ (curves 3) with the corresponding values $\bar{\Delta\sigma}^{(i)}_{klmn}$ estimated in the framework of the effective field hypothesis **H1** by both the differential method (6.185) (curves 2) and the initial approximation of the integral equations

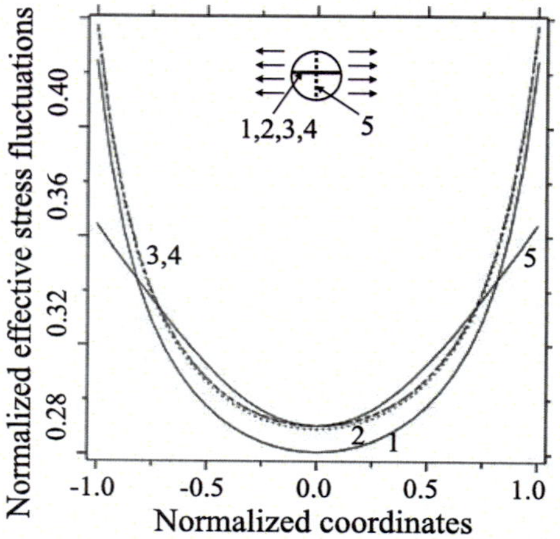

Fig. 10.17 $\bar{\Delta\overline{\sigma}}^{(i)[k]}_{2222}(\mathbf{x})$ vs x_1 (curves 1, 2, 3, 4) and x_2 (curve 5) at $k = 1(1), 2(2), 3(3), 4(4,5)$

method (10.70) (curves 1). It can be seen that the maximum deviation between the curves is reached for the component $\bar{\Delta}\sigma^{(i)}_{2222}$. However, the deviation between the curves 1 and 3 estimated by the method of integral equations with the use of the hypothesis **H1** (10.70) and without one (10.72), (10.74), respectively, does not exceed 4.4.% at $c = 0.75$. Thus, the eventual abandonment of the hypothesis **H1** leads to only slight correction of the averaged stress fluctuations $\bar{\Delta}\sigma^{(i)[5]}_{klmn}$, while the local values $\bar{\Delta}\sigma^{(i)[5]}_{2222}(\mathbf{x})$ can vary along the cross-section of fibers on 79%. So much prominent and systematic differences of the old (10.70) and new (10.72) approaches are based on the abandonment from the effective field hypothesis **H1** in the new approach. We will estimate the normalized effective stress fluctuations $\bar{\Delta}\overline{\sigma}^{(i)[k]}_{pqrs}(\mathbf{x}) \equiv (\Delta\overline{\sigma}^{(i)2}_{pqrs}(\mathbf{x})/\overline{B}^{*MEFM}_{i|pq11}\,\overline{B}^{*MEFM}_{i|rs11})^{1/2}$, where $\Delta\overline{\boldsymbol{\sigma}}^{(i)2}(\mathbf{x})$ is estimated by Eqs. (10.72), whereas $\overline{B}^{*MEFM}_{i}(\mathbf{x})$ is evaluated by Eq. (10.25) in the framework of the EFH **H1**. Normalized effective stress fluctuations $\bar{\Delta}\overline{\sigma}^{(i)[k]}_{2222}(\mathbf{0})$ equal to 4.607, 4.690, 7.801, 4.702, and 4.702 at k =1, 2, 3, 4, 7 (see Fig. 10.17) that demonstrates their fast convergence. $\bar{\Delta}\overline{\sigma}^{(i)[4]}_{2222}(\mathbf{x})$ is lowered on 24% as the coordinate x_1 varies from the fiber boundary to their center, whereas the volume averages $\bar{\Delta}\overline{\sigma}^{(i)[4]}_{2222} = 4.23$ and $\bar{\Delta}\overline{\sigma}^{(i)[0]}_{2222} = 4.05$ differ from one another by only 3.6%.

The local stresses generating by the interaction effects of heterogeneities are highly sensitive to their locations describing by the RDF. Influence of both the RDF (9.116) and (9.117) on stress concentration factors $\mathbf{B}^*(\mathbf{x})$ was estimated in Sect. 10.1, while their influences on the normalized stress fluctuations $\bar{\Delta}\sigma^{(i)[5]}_{2222}(x_1)$ are presented in Fig. 10.18, for k =0, 5 and $c = 0.7$. As can be seen in Fig. 10.18, the difference of $\bar{\Delta}\sigma^{(i)[5]}_{2222}(x_1)$ estimated by the use of the RDF (9.116) and (9.117) equals 25%, while the distinction of their averaged values $\bar{\Delta}\sigma^{(i)[5]}_{2222}$ does not exceed 19%. Thus, the local stress fluctuations $\bar{\Delta}\boldsymbol{\sigma}^{(i)[5]}(\mathbf{x})$ are expectedly more sensitive to the choice of the RDF than their averaged values $\bar{\Delta}\boldsymbol{\sigma}^{(i)[5]}$.

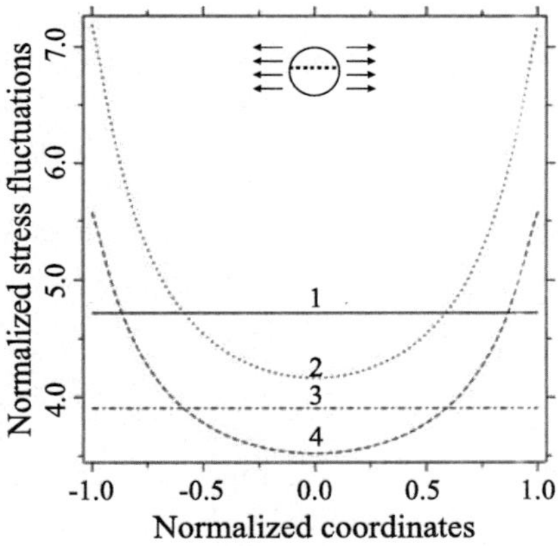

Fig. 10.18 $\bar{\Delta}\sigma^{(i)[k]}_{2222}(x_1)$ vs x_1 for RDF (9.117) (curves 1, 2) and (9.116) (curves 3, 4) at $k = 0$ (1,3) and 5 (2,4)

10.7 Conclusion

In the numerical examples for CMs with the homogeneous circular inclusions, it was demonstrated that even in the framework of the hypothesis **H3**, taking the binary interaction of inclusions into account discovers a fundamentally new effects such as a change of a sign in $\langle\boldsymbol{\sigma}\rangle_i(\mathbf{x})$ (see Figs. 10.4 and 10.7) and strong variation (on 80%) of $\langle\boldsymbol{\sigma}\otimes\boldsymbol{\sigma}\rangle_i(\mathbf{x})$ (see Fig. 10.16) inside the representative inclusion $\mathbf{x}\in v_i$. In so doing, the average moments $\langle\boldsymbol{\sigma}\rangle_i$ and $\langle\boldsymbol{\sigma}\otimes\boldsymbol{\sigma}\rangle_i$ (see Fig. 10.3, 10.9, and 10.10) are only slightly distinguished from the classical results described in [200]. The estimated second moment of stresses $\langle\boldsymbol{\sigma}\otimes\boldsymbol{\sigma}\rangle_i(\mathbf{x})$ ($\mathbf{x}\in v_i$) allows the representation for the second moment of stresses in the vicinity of the heterogeneity near the point $\mathbf{z}\in v_i$: $\langle\boldsymbol{\sigma}^-(\mathbf{n})\otimes\boldsymbol{\sigma}^-(\mathbf{n})\rangle_{\mathbf{z}}$ (10.59). The last second moment was used in Section 13 in [200] for evaluation of the effective failure envelopes of composites reinforced by aligned infinite fibers at the assumption $\langle\boldsymbol{\sigma}\otimes\boldsymbol{\sigma}\rangle_i(\mathbf{z}) = \langle\boldsymbol{\sigma}\otimes\boldsymbol{\sigma}\rangle_i$. However, we can see in Figs. 10.13, 10.14, 10.15, 10.16, 10.17, and 10.18 that maximums of the different components of stress fluctuations $\bar{\Delta}\sigma^{(i)}_{klmn}(\mathbf{x})$ are reached precisely in the vicinity $\mathbf{z}\in v_i$ of the fiber interface where the difference between $\bar{\Delta}\sigma^{(i)}_{klmn}(\mathbf{z})$ and $\bar{\Delta}\sigma^{(i)}_{klmn}$ can archive to a few tens percent (compare Figs. 10.9 and 10.16). This estimation refinement of the local stress fluctuations $\bar{\Delta}\sigma^{(i)}_{klmn}(\mathbf{z})$ with respect to their averaged values $\bar{\Delta}\sigma^{(i)}_{klmn}$ has immediate practical interest at the evaluation of the failure initiation envelope (or onset of yielding) of the interface in the composite materials.

An alternative analysis method of the failure phenomena proposed by Lipton [945, 946] examines the distribution of extreme values for the stress in the linear elastic regime. The focus here is to assess the size and location of the region over which the magnitude of the stress (or equivalent positive-definite function of stress σ^{eq}) exceeds a nominal value t. The stress distribution function $\lambda_i(t)$ gives the volume of the over-stressed zone $s^{(i)}\subset v^{i)}$ where σ^{eq} exceeds the value

$t > 0$. However, the rigorous bounds on $\lambda_i(t)$ were obtained by the use of some statistical moments of stresses averaged over the volume v^i rather than the local values of these moments in each point $\mathbf{x} \in v^{(i)}$. Subdivision of heterogeneities $v^{(i)} = \cup v_j^{(i)}$ allows us to obtain a generalization of the mentioned results to the case of inhomogeneous statistical moments of stresses inside heterogeneities considered in the current chapter. However, more detailed analysis of the failure phenomena is beyond the scope of the current study.

Finally, a necessity of verifying the new approach should be mentioned. Unfortunately, validating the numerical results obtained is currently beyond the capabilities of standard experimental techniques based on the measurement of effective elastic moduli $\mathbf{L}^*$ because the difference of $\mathbf{L}^*$ estimated by the new and classical approaches is negligible (see Figs. 10.3), whereas the local stress concentration factors can be distinguished by a sign (see Figs. 10.9 and 10.10). Therefore, there is an increasing need to provide very detailed experimental information about the local stress concentration factors of numerically investigated materials at the microscale. The availability of 3D imaging tools such as computed X-ray tomography and microtomography, confocal optical microscopy, or magnetic resonance imaging, combined with the increased power of modern computers allowing to process large quantities of data at reasonable costs, has made it recently possible to use these opportunities for verification of the new approach. So, the new volumetric digital image correlation (VDIC) relying on the X-ray tomographic imaging of naturally occurring material texture within samples requires an ability to reproduce the position of image points with "high precision" because the determination of relative deformations requires two images for comparison purposes to extract strain measures. Full-field measurements of strains by VDIC (and by other imaging tools) are potentially well suited to analyze the specific mechanical properties of composite materials (see, e.g., [565, 1474]). However, the mentioned opportunities of experimental methods are not yet realized at the level required for verification of the new approach even for the composites with the constituents described by local constitutive laws considered in the current chapter. The case of composites with nonlocal constitutive laws of components (see [205, 206]) calls for further investigations.

The second direction of a possible verification is a comparison of results obtained by both the new approach and the computational micromechanics one based on periodization [1316] reducing to the modeling of periodic boundary conditions at the unit cell containing just a few hundred n inclusions with the centers $\mathbf{x}_{\alpha\beta}$ in β's Monte Carlo realization ($\alpha = 1, \ldots n, \quad \beta = 1, \ldots N$) (see, e.g., [136, 137, 589, 591, 735, 849, 960, 1342, 1351, 1706]). These approaches are reduced to evaluation of the volume stress average $\langle \boldsymbol{\sigma} \rangle_1$ through the estimation of average stresses over both n inclusions in a unite cell and N random generations of sets of n inclusions by the Monte Carlo simulations: $\langle \boldsymbol{\sigma} \rangle_1 \approx (nN)^{-1} \sum_\alpha \sum_\beta \langle \boldsymbol{\sigma}(\mathbf{x}_{\alpha\beta}, \mathbf{x}) \rangle_{(i_{\alpha\beta})}$. where $\boldsymbol{\sigma}(\mathbf{x}_{\alpha\beta}, \mathbf{x})$ is a stress distribution in the local coordinate system in the inclusion $\mathbf{x} \in v_{i_{\alpha\beta}}$ with the center $\mathbf{x}_{\alpha\beta}$. Obviously, an estimation of the desired statistical stress averages [similar to Eq. (10.25)] can be obtained by more simple formula

$\langle\boldsymbol{\sigma}\rangle_1(\mathbf{x}) \approx (nN)^{-1}\sum_\alpha\sum_\beta \boldsymbol{\sigma}(\mathbf{x}_{\alpha\beta},\mathbf{x})$ (see, e.g., [148]), which was not, nevertheless, investigated in detail (although the histograms of stress distributions in the matrix were found [1533]; see also cumulative probability function of maximum traction at the interface [861, 864]).

At last, the next opportunity of uncompleted verification is a consideration of particular materials for which a solution obtained by the new approach is exact. So, Buryachenko and Brun [233] deliberately considered a simplified example of composites with zero elastic mismatch and nonzero thermoelastic mismatch when it is not necessary to use one of the closing assumptions (e.g., **H2a** or **H2b**) required at the consideration of composites with elastic mismatch. It was qualitatively and quantitatively proved [at $\mathbf{M}_1(\mathbf{x}) \equiv \mathbf{0}$, $\boldsymbol{\beta}_1(\mathbf{x}) \not\equiv \mathbf{0}$, $\mathbf{x} \in R^d$] that the classical approach based on the general asymptotic representation (7.33) and the hypotheses **H1** and **H3** can lead to infinite error (i.e. correction of sign) the estimation of the average effective fields $\langle\overline{\boldsymbol{\sigma}}\rangle_i$ in comparison with exact estimations of $\langle\overline{\boldsymbol{\sigma}}\rangle_i$ $(\mathbf{x} \in v_i)$ obtained by the use of the new background (7.20) without any assumptions.

Chapter 11
Interface Integral Technique in Thermoelasticity of Random Structure Matrix CMs

Abstract The representations of the effective properties (effective moduli, thermal expansion, and stored energy) are expressed through the statistical averages of the interface polarization tensors introduced apparently for the first time. New GIE is obtained in terms of a perturbators introduced by the boundary interface integrals estimated by the method of fundamental solution for a single inclusion inside the infinite matrix. This enables us to reconsider basic concepts of micromechanics such as effective field hypothesis, quasi-crystalline approximation, and the hypothesis of "ellipsoidal symmetry" in terms of the interface polarization tensors. Some new effects are detected that are impossible in the framework of a classical background of micromechanics. GIEs of Stokes flow in porous media are proposed.

11.1 Interface Polarization Tensors and Effective Properties

The concept of the polarization tensors (4.2) and (4.3) is an important ingredient of the effective moduli theory based on the estimations of the polarization tensors $\boldsymbol{\tau}$ (4.2$_1$) and $\boldsymbol{\eta}$ (4.2$_2$) averaged over the inclusion volume v. However, the representations $\boldsymbol{\tau}$ (4.3$_1$) and $\boldsymbol{\eta}$ (4.3$_2$) allow one to use only surface averages of some transformations of these tensors. So, we introduce the new notions of the stress $\boldsymbol{\tau}^s(\mathbf{s})$ and strain $\boldsymbol{\eta}^s(\mathbf{s})$ *interface polarization tensors* (4.35)

$$\boldsymbol{\tau}^s(\mathbf{s}) = \mathbf{t}(\mathbf{s}) \overset{S}{\otimes} \mathbf{s} - \mathbf{L}^{(0)}[\mathbf{u}(\mathbf{s}) \otimes \mathbf{n}(\mathbf{s})] - [\boldsymbol{\alpha}^{(0)}\mathbf{n}(\mathbf{s})] \overset{S}{\otimes} \mathbf{s}, \tag{11.1}$$

$$\boldsymbol{\eta}^s(\mathbf{s}) = \mathbf{u}(\mathbf{s}) \overset{S}{\otimes} \mathbf{n}(\mathbf{s}) - \mathbf{M}^{(0)}[\mathbf{s} \otimes \mathbf{t}(\mathbf{s})] - [\boldsymbol{\beta}^{(0)}\mathbf{n}(\mathbf{s})] \overset{S}{\otimes} \mathbf{s}, \tag{11.2}$$

defined at the inclusion surface $\mathbf{s} \in \Gamma$.

V. A. Buryachenko, *Local and Nonlocal Micromechanics of Heterogeneous Materials*, https://doi.org/10.1007/978-3-030-81784-8_11

The volume averages of the strains and stresses inside the inclusions can be expressed through the averages over the inclusion boundaries by the use of the Gauss's theorem

$$\langle \boldsymbol{\varepsilon} V\rangle = \frac{1}{\bar{v}}\int_{\Gamma} \mathbf{u}(\mathbf{s}) \overset{S}{\otimes} \mathbf{n}(\mathbf{s})\, ds, \tag{11.3}$$

$$\langle \boldsymbol{\sigma} V\rangle = \frac{1}{\bar{v}}\int_{\Gamma} \mathbf{t}(\mathbf{s}) \overset{S}{\otimes} \mathbf{s}\, ds, \tag{11.4}$$

$$\langle \boldsymbol{\alpha}^{(0)} V\rangle = \frac{1}{\bar{v}}\int_{\Gamma} [\boldsymbol{\alpha}^{(0)}\mathbf{n}(\mathbf{s})] \overset{S}{\otimes} \mathbf{s}\, ds, \tag{11.5}$$

where in Eqs. (11.4) and (11.5) one used the equalities $\nabla(\mathbf{s}\otimes\boldsymbol{\sigma}) = \boldsymbol{\sigma}$ and $\nabla(\mathbf{s}\otimes\boldsymbol{\alpha}^{(0)}) = \boldsymbol{\alpha}^{(0)}$, respectively. Then the effective properties (6.47), (6.48), and (6.55) can be expressed through the boundary interface integrals

$$\mathbf{L}^* = \mathbf{L}^{(0)} + \mathbf{S}^{\epsilon *}, \quad \mathbf{S}^{\epsilon *}\langle\boldsymbol{\varepsilon}\rangle := \frac{1}{\bar{v}}\int_{\Gamma} \boldsymbol{\tau}^{sI}(\mathbf{s})ds, \tag{11.6}$$

$$\boldsymbol{\alpha}^* = \boldsymbol{\alpha}^{(0)} + \mathbf{S}^{\alpha *}, \quad \mathbf{S}^{\alpha *}(\mathbf{x}) := \frac{1}{\bar{v}}\int_{\Gamma} \boldsymbol{\tau}^{sII}(\mathbf{s})ds, \tag{11.7}$$

$$W^* = -\frac{1}{2\bar{v}}\int_{\Gamma} \boldsymbol{\alpha}_1(\mathbf{s})[\mathbf{u}(\mathbf{s}) \overset{S}{\otimes} \mathbf{n}(\mathbf{s})]ds, \tag{11.8}$$

$$\mathbf{M}^* = \mathbf{M}^{(0)} + \mathbf{S}^{\sigma *}, \quad \mathbf{S}^{\sigma *}\langle\boldsymbol{\sigma}\rangle := \frac{1}{\bar{v}}\int_{\Gamma} \boldsymbol{\eta}^{sI}(\mathbf{s})ds, \tag{11.9}$$

$$\boldsymbol{\beta}^* = \boldsymbol{\beta}^{(0)} + \mathbf{S}^{\beta *}, \quad \mathbf{S}^{\beta *}(\mathbf{x}) := \frac{1}{\bar{v}}\int_{\Gamma} \boldsymbol{\eta}^{sII}(\mathbf{s})ds, \tag{11.10}$$

$$W^* = -\frac{1}{2\bar{v}}\int_{\Gamma} \boldsymbol{\beta}_1(\mathbf{s})[\mathbf{u}(\mathbf{s}) \overset{S}{\otimes} \mathbf{n}(\mathbf{s})]ds, \tag{11.11}$$

for the boundary conditions (6.4) and (6.5), respectively. Hereafter $\boldsymbol{\zeta}^{sI}$ and $\boldsymbol{\zeta}^{sII}$ ($\boldsymbol{\zeta} = \boldsymbol{\eta}, \boldsymbol{\tau}$) correspond to the decompositions (6.36) and (6.37). The stored energies W^* are estimated by Eqs. (11.8) and (11.11) at $\langle\boldsymbol{\varepsilon}\rangle \equiv \mathbf{0}$ and $\langle\boldsymbol{\sigma}\rangle \equiv \mathbf{0}$, respectively. The representation (11.6) for a particular case of the isotropic matrix and $\boldsymbol{\alpha}(\mathbf{x}) \equiv \mathbf{0}$ is reduced to the corresponding equation proposed in [304, 1310].

A comparison of Eqs. (6.47), (6.48), and (6.55) and Eqs. (11.6)–(11.10), respectively, yields a conclusion that the averages of the volume polarization tensors $\boldsymbol{\tau}(\mathbf{x})$ (4.3_1) and $\boldsymbol{\eta}(\mathbf{x})$ (4.3_2), and the averages of the surface polarization tensors $\boldsymbol{\tau}^s(\mathbf{s})$ (11.1) and $\boldsymbol{\eta}^s(\mathbf{s})$ (11.2), respectively, are coincident with each other

$$\mathbf{R}^{\epsilon *} = \mathbf{S}^{\epsilon *}, \quad \mathbf{R}^{\alpha *} = \mathbf{S}^{\alpha *}, \quad \mathbf{R}^{\sigma *} = \mathbf{S}^{\sigma *}, \quad \mathbf{R}^{\beta *} = \mathbf{S}^{\beta *}, \tag{11.12}$$

and

$$2W^* = -\langle \boldsymbol{\alpha}_1 \mathbf{C}^{\epsilon *} \rangle = -\left\langle \boldsymbol{\alpha}_1 [\mathbf{C}^{u*}(\mathbf{s}) \overset{S}{\otimes} \mathbf{n}(\mathbf{s})] \right\rangle, \tag{11.13}$$

$$2W^* = -\langle \boldsymbol{\beta}_1 \mathbf{C}^{*} \rangle = -\left\langle \boldsymbol{\beta}_1 [\mathbf{C}^{t*}(\mathbf{s}) \overset{S}{\otimes} \mathbf{s}] \right\rangle, \tag{11.14}$$

where one used the interface counterparts of Eqs. (9.1) for a representative heterogeneity $\mathbf{s} \in \Gamma_i$

$$\langle \mathbf{u} \rangle_i(\mathbf{s}) = \mathbf{A}_i^{u*}(\mathbf{s})\langle \boldsymbol{\varepsilon} \rangle + \mathbf{C}_i^{u*}(\mathbf{s}), \quad \langle \boldsymbol{\sigma} \rangle_i(\mathbf{s}) = \mathbf{B}_i^{t*}(\mathbf{s})\langle \boldsymbol{\sigma} \rangle + \mathbf{C}_i^{t*}(\mathbf{s}), \tag{11.15}$$

respectively.

Moreover, there are identical interrelations inside the pairs of the polarization tensors $\boldsymbol{\eta}, \boldsymbol{\tau}$ and $\boldsymbol{\eta}^s, \boldsymbol{\tau}^s$ $(\mathbf{x} \in R^d,\ \mathbf{s} \in \Gamma)$

$$\boldsymbol{\tau}(\mathbf{x}) = -\mathbf{L}^{(0)}\boldsymbol{\eta}(\mathbf{x}), \quad \boldsymbol{\tau}^s(\mathbf{s}) = -\mathbf{L}^{(0)}\boldsymbol{\eta}^s(\mathbf{s}), \tag{11.16}$$

although $\boldsymbol{\tau}(\mathbf{x})$, $\boldsymbol{\eta}(\mathbf{x})$, $\boldsymbol{\tau}^s(\mathbf{s})$, and $\boldsymbol{\eta}^s(\mathbf{s})$, respectively, cannot be compared between the pairs directly because of the different domains of definition. It is interesting that due to the continuity of both the traction $\mathbf{t}(\mathbf{s})$ and displacement $\mathbf{u}(\mathbf{s})$ in a vicinity of the interface $\mathbf{s} \in \Gamma_i$, we have the following equalities

$$\boldsymbol{\tau}^s(\mathbf{s}^+) = \boldsymbol{\tau}^s(\mathbf{s}^-), \quad \boldsymbol{\eta}^s(\mathbf{s}^+) = \boldsymbol{\eta}^s(\mathbf{s}^-) \tag{11.17}$$

where $\mathbf{s} = \mathbf{s}^{\pm}$ and $\mathbf{s}^+ \in \Gamma_i^+, \mathbf{s}^- \in \Gamma_i^-$ are placed from the side of the inhomogeneity v_i (Γ_i^+) and the matrix $v^{(0)}$ (Γ_i^-), respectively. In so doing, the polarization tensors $\boldsymbol{\tau}(\mathbf{s})$ and $\boldsymbol{\eta}(\mathbf{s})$ are discontinuous at the interface

$$\boldsymbol{\tau}(\mathbf{s}^+) \neq \boldsymbol{\tau}(\mathbf{s}^-), \quad \boldsymbol{\eta}(\mathbf{s}^+) \neq \boldsymbol{\eta}(\mathbf{s}^-) \tag{11.18}$$

for either (or both) elastic $[\mathbf{M}(\mathbf{s}^+) \neq \mathbf{M}(\mathbf{s}^-)]$ or (and) thermal $[\boldsymbol{\alpha}(\mathbf{s}^+) \neq \boldsymbol{\alpha}(\mathbf{s}^-)]$ mismatch. Equalities (11.16) and (11.17) define a physical meaning and injectivity of the notions of the interface polarization tensors $\boldsymbol{\tau}^s(\mathbf{s})$ (11.1), and $\boldsymbol{\eta}^s(\mathbf{s})$ (11.2).

11.2 Interface Integral Technique

11.2.1 General Integral Equation for Microinhomogeneous Infinite Medium

Hereafter for contraction of calculations, we introduce the substitutions

$$(\boldsymbol{\varepsilon}, \boldsymbol{\sigma}) \leftrightarrow \boldsymbol{\vartheta}, \quad (\boldsymbol{\tau}, \boldsymbol{\eta}) \leftrightarrow \boldsymbol{\zeta}, \quad (\mathbf{u}, \mathbf{t}) \leftrightarrow \boldsymbol{\chi}, \tag{11.19}$$

indicating that for each duplet $\boldsymbol{\vartheta} = \boldsymbol{\varepsilon}, \boldsymbol{\zeta} = \boldsymbol{\tau}$ and $\boldsymbol{\vartheta} = \boldsymbol{\sigma}, \boldsymbol{\zeta} = \boldsymbol{\eta}$, whereas in the duplet $\boldsymbol{\xi}$ both variables $\mathbf{u}$ and $\mathbf{t}$ are used simultaneously. In so doing, the perturbators $\mathcal{L}_k^{\vartheta}(\mathbf{x} - \mathbf{x}_k, \boldsymbol{\zeta})$ (4.44) and $\mathcal{L}_k^{\chi}(\mathbf{x} - \mathbf{x}_k, \boldsymbol{\chi})$ (4.45) are defined by the volume $\boldsymbol{\zeta}(\mathbf{x})$ $(\mathbf{x} \in v_i)$ and the boundary $\boldsymbol{\chi}(\mathbf{s})$ $(\mathbf{s} \in \Gamma_i)$ values of the corresponding fields.

Let us consider an arbitrary random realization α of inclusions in the domain w described by an analog of Eq. (4.44) generalized to any number of inhomogeneities. Then the centering method (see for details Chap. 9) subtracting from both sides of the mentioned equation their statistical averages leads to the *general integral equation* (GIE) $(\mathbf{x} \in v_i;\ \boldsymbol{\vartheta} = \boldsymbol{\varepsilon}, \boldsymbol{\sigma};\ \boldsymbol{\zeta} = \boldsymbol{\tau}, \boldsymbol{\eta})$

$$\boldsymbol{\vartheta}(\mathbf{x}, \alpha) = \langle \boldsymbol{\vartheta} \rangle(\mathbf{x}) + \int_w [\mathcal{L}_k^{\vartheta}(\mathbf{x} - \mathbf{x}_k, \boldsymbol{\zeta}) V_k^{\delta}(\mathbf{x}_k, \alpha) - \langle \mathcal{L}_k^{\vartheta}(\mathbf{x} - \mathbf{x}_k, \boldsymbol{\zeta}) \rangle(\mathbf{x}_k)] d\mathbf{x}_k \tag{11.20}$$

expressed through the volume perturbators $\mathcal{L}_k^{\vartheta}(\mathbf{x} - \mathbf{x}_k, \boldsymbol{\zeta})$ (4.44). Equation (11.20) is only obtained at the internal points $\mathbf{x} \in w$ of the macrodomain w at sufficient distance from the boundary $a \ll |\mathbf{x} - \mathbf{s}|$, $\forall \mathbf{s} \in \Gamma_0$ that guarantees an absolute convergence of the integral involved. Therefore, for $\mathbf{x} \in w$ far enough from the boundary Γ_0 (11.20), the right-hand side integral in (11.20) does not depend on the shape and size of the domain w, and it can be replaced by the integrals over the whole space R^d; the domain integration R^d, as well as the label α, will be omitted hereafter for simplicity of notations.

In a similar manner, the GIE expressed through the perturbator $\mathcal{L}_k^{\chi}(\mathbf{x} - \mathbf{x}_k, \boldsymbol{\chi})$ (4.45) can be obtained $(\mathbf{x} \in \Gamma_i;\ \boldsymbol{\chi} = (\mathbf{u}, \mathbf{t}))$

$$\boldsymbol{\chi}(\mathbf{x}) = \langle \boldsymbol{\chi} \rangle(\mathbf{x}) + \int [\mathcal{L}_k^{\chi}(\mathbf{x} - \mathbf{x}_k, \boldsymbol{\chi}) V_k^{\delta}(\mathbf{x}_k) - \langle \mathcal{L}_k^{\chi}(\mathbf{x} - \mathbf{x}_k, \boldsymbol{\chi}) \rangle(\mathbf{x}_k)] d\mathbf{x}_k. \tag{11.21}$$

A particular case of the Eq. (11.21) was proposed in [214] in the problem of slow flow in incompressible fluid through assemblages of a random field of fixed solid particles. Other particular cases of Eq. (11.21) were proved for both the elastic CMs $(\boldsymbol{\alpha}(\mathbf{x}) \equiv \mathbf{0}$; see [218] and [219] and the elastically homogeneous media with random field of noncanonical inclusions ($\mathbf{M}(\mathbf{x}) \equiv$const., $\boldsymbol{\alpha}(\mathbf{x}) \not\equiv$const.; see [217].

The term "general" is used with respect to Eqs. (11.20) and (11.21) in the sense that we did not accept any assumptions related with either estimation of perturbators $\mathcal{L}_k^{\vartheta}(\mathbf{x} - \mathbf{x}_k, \boldsymbol{\zeta})$ (4.44) and $\mathcal{L}_k^{\chi}(\mathbf{x} - \mathbf{x}_k, \boldsymbol{\chi})$ (4.45) or microstructure of heterogeneous medium (e.g., statistical homogeneity). In so doing, we only consider the physical phenomena described in Chap. 9.

Let the inclusions $v_1, \ldots, v_n$ be fixed, and we define two sorts of effective fields $\overline{\boldsymbol{\chi}}_i(\mathbf{x})$ and $\widetilde{\boldsymbol{\chi}}_{1,\ldots,n}(\mathbf{x})$ $(i = 1, \ldots, n;\ \mathbf{x} \in v_1, \ldots, v_n)$ by the use of the rearrangement of Eq. (11.21) in the following operator forms (analogous case of these manipulation approach for the perturbator $k(\mathbf{x} - \mathbf{x}_k, \boldsymbol{\zeta})$ (4.44) is given in Chap. 8) $(\mathbf{x} \in \Gamma_i)$

$$\boldsymbol{\chi}(\mathbf{x}) = \overline{\boldsymbol{\chi}}_i(\mathbf{x}) + \mathcal{L}_i^{\chi}(\mathbf{x} - \mathbf{x}_i, \boldsymbol{\chi}),$$

$$\overline{\boldsymbol{\chi}}_i(\mathbf{x}) = \widetilde{\boldsymbol{\chi}}_{1,\ldots,n}(\mathbf{x}) + \sum_{j \neq i}^{n} \mathcal{L}_j^{\chi}(\mathbf{x} - \mathbf{x}_j, \boldsymbol{\chi}),$$

$$\widetilde{\chi}_{1,\ldots,n}(\mathbf{x}) = \langle\chi\rangle(\mathbf{x}) + \int [\mathcal{L}_k^{\chi}(\mathbf{x}-\mathbf{x}_k,\chi)V^{\delta}(\mathbf{x}_k|;v_1,\mathbf{x}_1;\ldots;v_n,\mathbf{x}_n)$$
$$-\langle\mathcal{L}_k^{\chi}(\mathbf{x}-\mathbf{x}_k,\chi)\rangle(\mathbf{x}_k)]d\mathbf{x}_k, \qquad (11.22)$$

respectively, for $\mathbf{x} \in \Gamma_i,\ i = 1,2,\ldots,n$; here $V^{\delta}(\mathbf{x}_k|;v_1,\mathbf{x}_1;\ldots;v_n,\mathbf{x}_n) = \sum_m \delta(\mathbf{x}_k - \mathbf{x}_m) - \sum_{i=1}^{n} \delta(\mathbf{x}_k - \mathbf{x}_i)$ is a random delta function of heterogeneity centers $\mathbf{x}_m$ $(m = 1,2,\ldots)$ under the condition that $\mathbf{x}_k \neq \mathbf{x}_i,\ \mathbf{x}_i \neq \mathbf{x}_j$ if $i \neq j$ $(i,j = 1,\ldots,n)$. The definitions of the effective fields $\overline{\chi}_i(\mathbf{x})$, $\widetilde{\chi}_{1,2,\ldots,n}(\mathbf{x})$ as well as their statistical averages $\langle\overline{\chi}_i\rangle(\mathbf{x})$, $\langle\widetilde{\chi}_{1,2,\ldots,n}\rangle(\mathbf{x})$ are nothing more than a notation convenience for different terms of the infinite system (11.22).

Then, considering some conditional statistical averages of the GIE (11.21) leads to an infinite system of integral equations $(\mathbf{x} \in \Gamma_j,\ j = 1,2\ldots n)$

$$\langle\chi|v_1,\mathbf{x}_1;\ldots;v_n,\mathbf{x}_n\rangle(\mathbf{x}) - \sum_{i=1}^{n}\langle\mathcal{L}_i^{\chi}(\mathbf{x}-\mathbf{x}_i,\chi)|v_1,\mathbf{x}_1;\ldots;v_n,\mathbf{x}_n\rangle_i$$
$$= \langle\chi\rangle(\mathbf{x}) + \int \big\{\langle\mathcal{L}_j^{\chi}(\mathbf{x}-\mathbf{x}_j,\chi|;v_1,\mathbf{x}_1;\ldots;v_n,\mathbf{x}_n\rangle_j\varphi(v_j,\mathbf{x}_j|;v_1,\mathbf{x}_1,\ldots,v_n,\mathbf{x}_n)$$
$$-\langle\mathcal{L}_j^{\chi}(\mathbf{x}-\mathbf{x}_j,\chi)\rangle(\mathbf{x}_j)\big\}d\mathbf{x}_j. \qquad (11.23)$$

Since $\mathbf{x} \in \Gamma_1,\ldots,\Gamma_n$ in the n-th line of the system can take the values inside the inclusion surfaces $\Gamma_1,\ldots,\Gamma_n$, the n-th line actually contains n equations. Statistical averaging $\langle(\cdot)\rangle_i$ stands for the averaging over the all surrounding heterogeneities at the fixed v_i, while the average $\langle(\cdot)\rangle(\mathbf{x}_j)$ implies the averaging overall possible location of v_i with possible dependence of this average on the macro-coordinate $\mathbf{x}_i$ as for FGMs.

Introduction of the effective field (11.22) makes it possible to define another sort of the perturbators (4.45_2). In such a case, the GIEs (11.23) can be transformed to the new GIEs in terms of effective fields $(\mathbf{x} \in \Gamma_k,\ j = 1,2,\ldots,n)$

$$\langle\overline{\chi}|v_1,\mathbf{x}_1;\ldots;v_n,\mathbf{x}_n\rangle(\mathbf{x}) - \sum_{i\neq k}^{n}\langle\mathcal{L}_i^{\chi}(\mathbf{x}-\mathbf{x}_i,\overline{\chi})|v_1,\mathbf{x}_1;\ldots;v_n,\mathbf{x}_n\rangle_i$$
$$= \langle\chi\rangle(\mathbf{x}) + \int \big\{\langle\mathcal{L}_j^{\chi}(\mathbf{x}-\mathbf{x}_j,\overline{\chi})|;v_1,\mathbf{x}_1;\ldots;v_n,\mathbf{x}_n\rangle_j\varphi(v_j,\mathbf{x}_j|;v_1,\mathbf{x}_1,\ldots,v_n,\mathbf{x}_n)$$
$$-\langle\mathcal{L}_j^{\chi}(\mathbf{x}-\mathbf{x}_j,\overline{\chi})\rangle(\mathbf{x}_j)\big\}d\mathbf{x}_j. \qquad (11.24)$$

The GIEs (11.22), (11.23), and (11.24) can be considered as the general operator equations where the operators $\mathcal{L}_j^{\chi}(\mathbf{x}-\mathbf{x}_j,\chi)$ and $\mathcal{L}_j^{\chi}(\mathbf{x}-\mathbf{x}_j,\overline{\chi})$ can be found by any available numerical method. It is implied that the variable χ denotes the duplet $\chi = (\mathbf{u},\mathbf{t})^{\top}$ (11.19_3) of the displacement and traction used in the BIEs. In Chap. 9 we proved absolute convergence of GIE (11.20) for the volume inclusion perturbators (4.44) that holds also for the surface inclusion perturbators (4.45) in

Eq. (11.21). Indeed, for the porous materials ($\boldsymbol{\chi}(\mathbf{s}) = \mathbf{u}(\mathbf{s}),\ \mathbf{s} \in \partial v$), the integrand of Eq. (11.24) is of order $O(|\mathbf{x}-\mathbf{x}_j|^{2d-1})$ as $|\mathbf{x}-\mathbf{x}_j| \to \infty$, while for CMs reinforced by the rigid inclusions ($\boldsymbol{\chi}(\mathbf{s}) = \mathbf{t}(\mathbf{s}),\ \mathbf{s} \in \partial v$), the integrand of Eq. (11.24) tends to zero with $|\mathbf{x}-\mathbf{x}_j| \to \infty$ as $O(|\mathbf{x}-\mathbf{x}_j|^{2d})$.

11.2.2 Some Classical Hypotheses and Approaches

The basic hypotheses of micromechanics were analyzed in Chaps. 9 and 10 in the form adopted to the use of the volume inclusion perturbators $\boldsymbol{\mathcal{L}}^{\vartheta}(\mathbf{x}-\mathbf{x}_k, \overline{\boldsymbol{\vartheta}})$. We present in this subsection the same hypotheses in the form which are most adopted for the subsequent application of the boundary interface inclusion perturbators $\boldsymbol{\mathcal{L}}^{\chi}(\mathbf{x}-\mathbf{x}_k, \overline{\boldsymbol{\chi}})$.

In order to approximately solve the exact system (11.24), we now apply the so-called effective field hypothesis (EFH) which is the main approximate hypothesis of many micromechanical methods (see for references and details Chaps. 8 and 9)

Hypothesis H1a *Each heterogeneity v_i is located in the field ($\boldsymbol{\vartheta} = \boldsymbol{\varepsilon}, \boldsymbol{\sigma}$)*

$$\overline{\boldsymbol{\vartheta}}_i(\mathbf{y}) \equiv \overline{\boldsymbol{\vartheta}}(\mathbf{x}_i)\ (\mathbf{y} \in v_i), \tag{11.25}$$

which is homogeneous over the heterogeneity v_i.

For the boundary variables ($\mathbf{s} \in \Gamma_i$), the hypothesis **H1** can be presented in the local coordinate systems connected with v_i

$$\overline{\mathbf{u}}_i(\mathbf{s}) \equiv \overline{\boldsymbol{\varepsilon}}(\mathbf{x}_i)(\mathbf{s}-\mathbf{x}_i) \quad \overline{\mathbf{t}}_i(\mathbf{s}) \equiv \overline{\boldsymbol{\sigma}}(\mathbf{x}_i)\mathbf{n}(\mathbf{s}), \tag{11.26}$$

where $\mathbf{n}^{\Gamma_i}(\mathbf{s})$ is unit outward normal at $\mathbf{s} \in \Gamma_i$.

In some methods (e.g., the MEFM), this basic hypothesis **H1a** is complemented by a satellite hypothesis presented in the form of the perturbator rather than the Green's function:

Hypothesis H1b *The volume operator $\boldsymbol{\mathcal{L}}_k^{\vartheta}(\mathbf{x}-\mathbf{x}_k, \boldsymbol{\zeta})$(4.44) of a perturbation generated by the heterogeneity v_i at the point $\mathbf{x} \notin v_k$ is reduced to the decoupled tensorial multiplications ($\boldsymbol{\zeta} = \boldsymbol{\tau}, \boldsymbol{\eta};\ \ \boldsymbol{\vartheta} = \boldsymbol{\sigma}, \boldsymbol{\varepsilon}$)*

$$\boldsymbol{\mathcal{L}}_k^{\vartheta}(\mathbf{x}-\mathbf{x}_k, \boldsymbol{\zeta}) = \mathbf{L}_k^{\zeta}(\mathbf{x}-\mathbf{x}_k)\langle\boldsymbol{\zeta}\rangle_{(k)}. \tag{11.27}$$

It should be mentioned that the popular formulation of the EFH (hypothesis **H1**) is a combination of the hypotheses **H1a** and **H1b**.

For the perturbator $\boldsymbol{\mathcal{L}}_k^{\vartheta}(\mathbf{x}-\mathbf{x}_k, \boldsymbol{\zeta})$ expressed through the Green's functions (4.48), the assumptions (11.27) are reduced to the known ones (see Sect. 4.3.3) with the perturbator factors (4.54) presented in terms of both the internal $\mathbf{S}_i$ and external

$\mathbf{S}_i(\mathbf{x})$ Eshelby [449] tensors. In such a case, the new GIE (11.20) is reduced to the classical one (see also for references and details Chap. 8) ($\boldsymbol{\vartheta} = \boldsymbol{\sigma},\ \boldsymbol{\zeta} = \boldsymbol{\eta}$)

$$\boldsymbol{\sigma}(\mathbf{x}, \alpha) = \langle \boldsymbol{\sigma} \rangle(\mathbf{x}) + \int \boldsymbol{\Gamma}^{(0)}(\mathbf{x} - \mathbf{x}_k)[\boldsymbol{\eta}(\mathbf{x}_k) V_k^{\delta}(\mathbf{x}_k, \alpha) - \langle \boldsymbol{\eta} \rangle(\mathbf{x}_k)] d\mathbf{x}_k, \tag{11.28}$$

where the renormalizing term $\boldsymbol{\Gamma}^{(0)}(\mathbf{x} - \mathbf{x}_k)\langle \boldsymbol{\eta} \rangle(\mathbf{x}_k)$ is defined by the volume average $\langle \boldsymbol{\eta} \rangle(\mathbf{x}_k)$ of the volume polarization tensor rather than the detailed distribution $\boldsymbol{\eta}(\mathbf{x})$ ($\mathbf{x} \in v_k$) as in Eq. (11.20). In Chap. 8 we presented the history of obtaining of the GIE (11.28) (which is dating back to Lord Rayleigh [1284]) and proved that (11.28) is, in fact, just approximation of the really GIE (11.20) obtained in the framework of the hypothesis **H1b**.

It should be also mentioned that the significant mentioned achievements in obtaining of GIEs for linear elastic CMs were formally generalized in [412] and [155] to the problems for the slow flow of incompressible fluid through assemblages of a random field of fixed rigid particles. They introduced a filtration counterpart of the hypothesis **H1b** based on estimation of a drag force at the representative particle v_i which does not depend on the details of both the particle shape and traction distribution at the particle surface (that is intrinsic to $\langle \boldsymbol{\vartheta} \rangle_{(k)}$ in (11.27)). This approximation of the renormalyzing term by the drag force is widely used in order to avoid the expensive construction of the far-field resistance matrix by the uses of direct calculation of the far-field hydrodynamic force (see, e.g., [155, 1391]). At last, a filtration counterpart of Eq. (11.21) with analyses of its approximations is considered in Sect. 11.4.

Different methods can be employed to truncate the hierarchies (11.23) and (11.24) (see for references, e.g., [646] and Chaps. 8–10). Probably the most formal closing assumption is the following (the corresponding assumption for the fields $\boldsymbol{\vartheta}$ and $\overline{\boldsymbol{\vartheta}}$ are formulated analogously).

Hypothesis H2a *For a sufficiently large n, the systems (11.23) and (11.24) are completed by the equations*

$$\begin{aligned}
&\langle \boldsymbol{\mathcal{L}}_j^{\chi}(\mathbf{x} - \mathbf{x}_j, \boldsymbol{\chi}) \mid ; v_1, \mathbf{x}_1; \ldots; v_q, \mathbf{x}_q; \ldots, v_{n+1}, \mathbf{x}_{n+1} \rangle_j \\
&\qquad = \langle \boldsymbol{\mathcal{L}}_j^{\chi}(\mathbf{x} - \mathbf{x}_j, \boldsymbol{\chi}) |; v_1, \mathbf{x}_1; \ldots; v_n, \mathbf{x}_n \rangle_j, \\
&\langle \boldsymbol{\mathcal{L}}_j^{\chi}(\mathbf{x} - \mathbf{x}_j, \overline{\boldsymbol{\chi}}) \mid ; v_1, \mathbf{x}_1; \ldots; v_q, \mathbf{x}_q; \ldots; v_{n+1}, \mathbf{x}_{n+1} \rangle_j \\
&\qquad = \langle \boldsymbol{\mathcal{L}}_j^{\chi}(\mathbf{x} - \mathbf{x}_j, \overline{\boldsymbol{\chi}}) |; v_1, \mathbf{x}_1; \ldots; v_n, \mathbf{x}_n \rangle_j,
\end{aligned} \tag{11.29}$$

respectively, where the right-hand sides of these equalities don't contain the index $q \neq j$ ($j = 1, \ldots, n;\ q = 1, \ldots, n+1;\ \mathbf{x} \in \Gamma_j$).

In opposite to Eq. (11.23), the sum in the left-hand side of Eq. (11.24) does not contain a summand with the index k. The truncated hierarchy of Eqs. (11.23) and (11.24) are solved as the systems of coupled equations. One starts with the last equation of the hierarchy which has the most inclusions held fixed, because this equation does not depend on the previous ones. The obtained field gives the forcing term in the previous equation of the hierarchy. One continues step by step up the hierarchy until the first equations in the systems (11.23) and (11.24).

In the simple case $n = 1$, the closing effective field hypothesis is called "quasi-crystalline approximation" by Lax [886] which neglects direct interaction between each pair of heterogeneities although such an interaction takes place through the effective field generated by all surrounding inclusions.

Hypothesis H2b, "Quasi-Crystalline" Approximation *It is supposed that the mean value of the effective field at a point* $\mathbf{x} \in v_i$ *does not depend on the stress field inside surrounding heterogeneities* $v_j \neq v_i$ $(\mathbf{x} \in v_i,\ \mathbf{s} \in \Gamma_i)$

$$\langle \overline{\boldsymbol{\vartheta}}_i(\mathbf{x})|; v_j, \mathbf{x}_j\rangle = \langle \overline{\boldsymbol{\vartheta}}_i\rangle, \quad \langle \overline{\boldsymbol{\chi}}_i(\mathbf{s})|; v_j, \mathbf{x}_j\rangle = \langle \overline{\boldsymbol{\chi}}_i\rangle. \tag{11.30}$$

To make further progress, the hypothesis of "ellipsoidal symmetry" **H3** (8.45) for the distribution of inclusions attributed to Willis [1620] is widely used.

11.2.3 Solution of GIEss

Solution of the GIEs in terms of the volume inclusion perturbators $\mathcal{L}_k^{\vartheta}(\mathbf{x}-\mathbf{x}_k, \boldsymbol{\zeta})$ was performed by Buryachenko [211] (see Chap. 9) for the different partial acceptance of the hypotheses **H1–H3**. At first we recast these approaches in terms of both the perturbators presented by the boundary integral operators and the volume effective fields. After that we direct our attention on the use of the boundary inclusion perturbators $\mathcal{L}_k^{\chi}(\mathbf{x} - \mathbf{x}_k, \boldsymbol{\chi})$ and its modifications for the interface effective field $\overline{\boldsymbol{\chi}}(\mathbf{s})$. Hypotheses **H1–H3** allow one to obtain from Eq. (11.28) the explicit solution by the well-known of the effective field method (EFM; see for references and details Chap. 8) for the stress distribution (the strain field can be considered in a similar manner) and the effective properties for identical aligned heterogeneities $(v_i = v_j = v_1,\ \mathbf{x} \in v_1)$.

It should be mentioned that the effective compliances (9.46) and (9.57) are traditionally expressed through the averages of the tensors $\mathbf{B}_i(\mathbf{x})$ and $\mathbf{R}_i(\mathbf{x})$ over the inclusion volumes v_i. However, these tensors can be also estimated through the averages over the inclusion interface $\mathbf{s} \in \Gamma_i$ of the fields $\boldsymbol{\vartheta}(\mathbf{s})$ $(\boldsymbol{\vartheta} = \boldsymbol{\varepsilon}, \boldsymbol{\sigma})$ corresponding to the homogeneous loading $\overline{\boldsymbol{\sigma}} \equiv$const., $\boldsymbol{\beta}_1(\mathbf{x}) \equiv \mathbf{0}$ $(c \to 0)$ of a single inclusion v_i

$$\mathbf{B}_i\overline{\boldsymbol{\sigma}} = \frac{1}{\overline{v}_i}\int_{\Gamma_i} \mathbf{t}(\mathbf{s}) \overset{s}{\otimes} \mathbf{s}\, ds, \quad \mathbf{R}_i\overline{\boldsymbol{\sigma}} = \frac{1}{\overline{v}_i}\int_{\Gamma_i} \boldsymbol{\eta}^s(\mathbf{s}) ds. \tag{11.31}$$

For the CMs with noncanonical shape of inclusions, we have performed in Chap. 9 (ar $\boldsymbol{\beta}_1(\mathbf{x}) \equiv \mathbf{0}$) a comparative analysis of the EFM (9.41)–(9.48) and Mori–Tanaka approach (9.52)–(9.59), respectively, with the tensors $\mathbf{B}_i$ and $\mathbf{R}_i$ represented by the interface boundary integrals (11.31).

The tensors $\mathbf{C}_i$ and $\mathbf{F}_i$ [used in (9.47), (9.48) and (9.58), (9.59)] are estimated at $\overline{\boldsymbol{\sigma}} \equiv \mathbf{0}$ and $\boldsymbol{\beta}_1(\mathbf{x}) \not\equiv \mathbf{0}$ by the representation

$$\mathbf{C}_i = \frac{1}{\bar{v}_i} \int_{\Gamma_i} \mathbf{t}(\mathbf{s}) \overset{S}{\otimes} \mathbf{s}\, ds, \quad \mathbf{F}_i = \frac{1}{\bar{v}_i} \int_{\Gamma_i} \boldsymbol{\eta}^s(\mathbf{s}) ds. \tag{11.32}$$

The displacement $\mathbf{u}(\mathbf{s})$ and the traction $\mathbf{t}(\mathbf{s})$ ($\mathbf{s} \in \Gamma_i$) are found by the MFS (see Section 5). The effective properties (9.41)–(9.48) and (9.52)–(9.59) are estimated through the boundary integrals as in Eqs. (11.31) and (11.32). In so doing, the stress concentration tensor $\mathbf{B}_i(\mathbf{x})$ in Eqs. (3.42) is estimated by Eq. (4.88) at $i = 1$.

In order to simplify a counterpart of the exact systems (11.24) for the volume effective field $\overline{\boldsymbol{\sigma}}(\mathbf{x})$, we accept the hypotheses **H1a** and **H2b** while the hypotheses **H1b** and **H3** are not used. At first, for the new GIE (11.20), this leads to the representation for the mean of the effective fields (9.62) in the fixed inhomogeneity $\mathbf{x} \in v_i$. The average stress concentration factor (9.79) and the effective compliance (9.80) are expressed through the volume perturbator $\mathbf{L}_q^{\sigma}(\mathbf{x} - \mathbf{x}_q)$ (11.27) which was estimated by the VIE method (see [203]) and by the FEA (see for references [211]).

Now we consider the GIE (11.21) for the surface variables $\boldsymbol{\chi}(\mathbf{s})$ ($\mathbf{s} \in \Gamma$) instead of the volume ones $\boldsymbol{\zeta}(\mathbf{x})$ ($\mathbf{x} \in v$) (11.20). In order to simplify the exact systems for the surface effective fields $\overline{\boldsymbol{\chi}}(\mathbf{s})$ ($\mathbf{s} \in \Gamma_i$) (11.24), we accept the hypotheses **H1a** and **H2b** while the hypotheses **H1b** and **H3** are not used. This leads to the following representation for the mean of the surface effective fields at the boundary $\mathbf{s} \in \Gamma_i$ of the fixed inhomogeneity $\mathbf{s} \in v_i$

$$\langle \overline{\boldsymbol{\chi}} | v_i, \mathbf{x}_i \rangle(\mathbf{s}) = \langle \boldsymbol{\chi} \rangle(\mathbf{s}) + \int \langle \boldsymbol{\mathcal{L}}_j^{\chi}(\mathbf{s} - \mathbf{x}_j, \overline{\boldsymbol{\chi}}) \rangle [\varphi(v_j, \mathbf{x}_j | ; v_i, \mathbf{x}_i) - n] d\mathbf{x}_j. \tag{11.33}$$

Equation (11.33) holds true for the general cases of statistically inhomogeneous CM loaded by an arbitrary inhomogeneous remote loading. However, for simplicity of results obtained, one only considers hereafter the statistically homogeneous media loaded by the homogeneous boundary conditions (6.5) (the conditions (6.4) can be analyzed in a similar manner). Even in such a case, the statistical average $\langle \boldsymbol{\chi} \rangle(\mathbf{s})$ ($\mathbf{s} \in \Gamma_i$) is inhomogeneous one

$$\langle \boldsymbol{\chi} \rangle(\mathbf{s}) = ([\mathbf{M}^* \boldsymbol{\sigma}^{\Gamma}](\mathbf{s} - \mathbf{x}_i), \boldsymbol{\sigma}^{\Gamma} \mathbf{n}^{\Gamma_i}(\mathbf{s}))^{\top}, \tag{11.34}$$

which explicitly depends on both the placement of $\mathbf{s}$ at the interface Γ_i and the effective compliance $\mathbf{M}^*$.

Due to acceptance of the hypothesis **H1a** (11.26_2), a tensorial product of Eq. (11.33) on $\mathbf{s}$ ($\mathbf{s} \in \Gamma_i$) and their subsequent averaging over the interface Γ_i at $\boldsymbol{\chi} = \mathbf{t}$ leads to ($\mathbf{s} \in \Gamma_i$)

$$\langle \overline{\boldsymbol{\sigma}}_i \rangle = \langle \boldsymbol{\sigma} \rangle + \int \Big[\langle \mathbf{s} \overset{S}{\otimes} \mathbf{L}_q^{t\sigma}(\mathbf{s} - \mathbf{x}_q) \rangle_{(i)} \langle \overline{\boldsymbol{\sigma}}_i \rangle + \langle \mathbf{s} \overset{S}{\otimes} \mathbf{C}_q^{t}(\mathbf{s} - \mathbf{x}_q) \rangle_{(i)} \Big]$$

$$\times[\varphi(v_q,\mathbf{x}_q|;v_i,\mathbf{x}_i)-n(\mathbf{x}_q)]d\mathbf{x}_q, \tag{11.35}$$

where one used the equalities (11.4) and (4.61) and $\langle \mathbf{n}^{\Gamma_i}(\mathbf{s}) \overset{S}{\otimes} \mathbf{s}\rangle_{(i)} = \mathbf{I}$. Equation (11.35) allows one to obtain the explicit solution for identical aligned heterogeneities ($v_i = v_q = v_1,\ \mathbf{x}\in v_1$)

$$\langle\overline{\boldsymbol{\sigma}}\rangle_1(\mathbf{x}) = \check{\mathbf{Y}}[\langle\boldsymbol{\sigma}\rangle + \check{\mathbf{I}}^{\beta}], \tag{11.36}$$

$$\langle\boldsymbol{\sigma}\rangle_1(\mathbf{x}) = \mathbf{B}_1(\mathbf{x})\check{\mathbf{Y}}[\langle\boldsymbol{\sigma}\rangle + \check{\mathbf{I}}^{\beta}] + \mathbf{C}_1(\mathbf{x}), \tag{11.37}$$

$$\mathbf{M}^* = \mathbf{M}^{(0)} + n^{(1)}\langle \mathbf{R}_1^{t\sigma}(\mathbf{s})\rangle_1\check{\mathbf{Y}}, \tag{11.38}$$

$$\boldsymbol{\beta}^* = \boldsymbol{\beta}^{(0)} + n^{(1)}\langle \mathbf{F}_1^{t}(\mathbf{s})\rangle_1 + n^{(1)}\langle \mathbf{R}_1^{t\sigma}(\mathbf{s})\rangle_1\check{\mathbf{Y}}\check{\mathbf{I}}^{\beta}, \tag{11.39}$$

$$2W^* = -\langle\boldsymbol{\beta}_1[\mathbf{C}_1^t(\mathbf{s})\overset{S}{\otimes}\mathbf{s}]\rangle_1 - n^{(1)}\langle\boldsymbol{\beta}_1[\mathbf{B}_1^{t\sigma}(\mathbf{s})\overset{S}{\otimes}\mathbf{s}]\rangle_1\check{\mathbf{Y}}\check{\mathbf{I}}^{\beta}, \tag{11.40}$$

$$\check{\mathbf{Y}}^{-1} = \mathbf{I} - \int \big\langle \mathbf{s}\overset{S}{\otimes}\mathbf{L}_q^{t\sigma}(\mathbf{s}-\mathbf{x}_q)\big\rangle_{(i)}[\varphi(v_q,\mathbf{x}_q|;v_i,\mathbf{x}_i)-n^{(q)}]d\mathbf{x}_q, \tag{11.41}$$

$$\check{\mathbf{I}}^{\beta} = \int \big\langle \mathbf{s}\overset{S}{\otimes}\mathbf{C}_q^{t}(\mathbf{s}-\mathbf{x}_q)\big\rangle_{(i)}[\varphi(v_q,\mathbf{x}_q|;v_i,\mathbf{x}_i)-n^{(q)}]d\mathbf{x}_q. \tag{11.42}$$

The representations for both the average stress concentration factor and the effective properties in Eqs. (11.36)–(11.42) are similar to Eqs. (9.75)–(11.33) with the replacements $\langle \mathbf{L}_q^{\sigma}(\mathbf{x}-\mathbf{x}_q)\rangle_{(i)} \to \langle \mathbf{s}\overset{S}{\otimes}\mathbf{L}_q^{t\sigma}(\mathbf{s}-\mathbf{x}_q)\rangle_{(i)}$, $\langle \mathbf{C}_q(\mathbf{x}-\mathbf{x}_q)\rangle_{(i)} \to \langle \mathbf{s}\overset{S}{\otimes}\mathbf{C}_q^{t}(\mathbf{s}-\mathbf{x}_q)\rangle_{(i)}$, $\langle\boldsymbol{\beta}_1\mathbf{C}_1\rangle \to \langle\boldsymbol{\beta}_1[\mathbf{C}_1^t(\mathbf{s})\overset{S}{\otimes}\mathbf{s}]\rangle_1$, and $\langle\boldsymbol{\beta}_1\mathbf{B}_1\rangle \to \langle\boldsymbol{\beta}_1[\mathbf{B}_1^{t\sigma}(\mathbf{s})\overset{S}{\otimes}\mathbf{s}]\rangle_1$ used.

All equations in Sects. 9.3.2 and 11.2.3 were derived for the perturbators expressed in both the general operator form and the boundary integral equations. However, the final results (9.75)–(9.82) and (11.36)–(11.42), respectively, are equivalent for both perturbator representations because both approaches use the same selection of hypotheses **H1a** and **H2b**, while the hypotheses **H1b** and **H3** are not exploited. Furthermore, the BIE methods (including MFS) have the well-known advantage of reduction of dimension by one that can be crucial for the analyst. From another side, the representations (9.75)–(9.82) can be easily applied to CMs with inhomogeneous inclusions ($\mathbf{M}(\mathbf{x}) \not\equiv$const., $\mathbf{x}\in v^{(1)}$) that present a real challenge to the BIE methods (see, e.g. (11.36)–(11.42)). Thus, each method [either FEA or BIE used for the particular problems (9.75)–(9.82) and (11.36)–(11.42), respectively] has advantages and disadvantages, and it is crucial for the analyst to be aware of their range of applications. It should be mentioned that the final representations (11.36)–(11.42) are reduced to the corresponding equations for the composites with either zero eigenstrain mismatch ($\mathbf{M}_1(\mathbf{x}) \not\equiv \mathbf{0}$, $\boldsymbol{\beta}_1(\mathbf{x}) \equiv \mathbf{0}$; see [218, 219]), zero elastic mismatch ($\mathbf{M}_1(\mathbf{x}) \equiv \mathbf{0}$, $\boldsymbol{\beta}_1(\mathbf{x}) \not\equiv \mathbf{0}$; see[221]), or nonzero elastic and eigenstrain mismatch ($\mathbf{M}_1(\mathbf{x}) \not\equiv \mathbf{0}$, $\boldsymbol{\beta}_1(\mathbf{x}) \not\equiv \mathbf{0}$; see[224]).

11.2.4 Iteration Method in Terms of the Interface Perturbator Factor

The representations (11.36)–(11.42) can be considered as an initial approximation of the solution of Eq. (11.33). Equation (11.33), in its turn, can be solved by both the successive approximations and the quadrature method leading to full linear systems. So, the counterpart of Eq. (11.33) in terms of the fields $\boldsymbol{\vartheta}$ (11.19) and (11.20) contains a homogeneous statistical average $\langle\boldsymbol{\vartheta}\rangle(\mathbf{x})$ ≡const. that gave an opportunity to us (see Sect. 9.3.3) to use for solution of Eq. (11.20) the iteration method accompanied by the volume integral equation method which is a rather time-consuming procedure. However, inhomogeneity of the field $\langle\boldsymbol{\chi}\rangle(\mathbf{s})$ makes the use of the same scheme for Eq. (11.33) more cumbersome. The known disadvantage of the iteration methods such as a necessity of a proof of its convergence is eliminated in the quadrature method which is very popular although it requires solving linear systems of high order. However, the last difficulties are excluded by the BIE method with the well-known advantages of reduction of dimension by one that is realized at the solution of Eq. (11.33) by the MFS. Namely, we present a perturbation $\overline{\mathcal{L}}_i^{\chi} * \overline{\boldsymbol{\chi}}(\mathbf{x})$ introduced by all surrounding inclusions into the effective field $\langle\overline{\boldsymbol{\chi}}|v_i,\mathbf{x}_i\rangle(\mathbf{s})$ $(\mathbf{s}\in\Gamma_i)$

$$\overline{\mathcal{L}}_i^{\chi} * \langle\overline{\boldsymbol{\chi}}\rangle(\mathbf{s}) := \int \langle\boldsymbol{\mathcal{L}}_q^{\chi}(\mathbf{s}-\mathbf{x}_q,\overline{\boldsymbol{\chi}})\rangle[\varphi(v_q,\mathbf{x}_q|;v_i,\mathbf{x}_i)-n]d\mathbf{x}_q, \tag{11.43}$$

in the matrix form that leads Eq. (11.33) to the matrix equation

$$\left(\mathbf{I}-\overline{\mathcal{L}}_i^{\chi}\right) * \langle\overline{\boldsymbol{\chi}}\rangle(\mathbf{s}) = \langle\boldsymbol{\chi}\rangle(\mathbf{s}). \tag{11.44}$$

The matrix Eq. (11.44) is explicitly solved $(\mathbf{s}\in\Gamma_i)$

$$\langle\overline{\boldsymbol{\chi}}\rangle(\mathbf{s}) = \boldsymbol{\mathcal{Y}}_i * \langle\boldsymbol{\chi}\rangle(\mathbf{s}), \quad \boldsymbol{\mathcal{Y}}_i = (\mathbf{I}-\overline{\mathcal{L}}_i^{\chi})^{-1}. \tag{11.45}$$

where the $2dS\times 2dS$ matrix $\boldsymbol{\mathcal{Y}}_i$ (11.45) (rather than the $3(d-1)\times 3(d-1)$ tensor $\mathbf{Y}$ (9.51)) determining the action of the surrounding inclusions is estimated according to the following two steps scheme. At first, we estimate the perturbator $\boldsymbol{\mathcal{L}}_i^{\chi}(\mathbf{s}-\mathbf{x}_i,\overline{\boldsymbol{\chi}})$ $(\mathbf{s}\in\Gamma_i)$ in the form of the $2dS\times 2dS$ matrix by the MFS (see Sect. 4.3) for an infinite matrix with a single small noncanonical heterogeneity v_i. The perturbation outside of inclusion v_q $\langle\boldsymbol{\mathcal{L}}_q^{\chi}(\mathbf{s}-\mathbf{x}_q,\overline{\boldsymbol{\chi}})\rangle_q$ $(\mathbf{s}\in\Gamma_i)$ is estimated as $\langle\boldsymbol{\mathcal{L}}_q^{\chi}(\mathbf{s}-\mathbf{x}_q,\langle\overline{\boldsymbol{\chi}}|;v_i,\mathbf{x}_i\rangle)(\mathbf{s})+[\mathbf{M}^{(0)}\langle\overline{\boldsymbol{\sigma}}\rangle_i](\mathbf{x}_i-\mathbf{x}_q))\rangle_q$ where $\langle\overline{\boldsymbol{\sigma}}\rangle_i$ is the statistical average of the effective field $\overline{\boldsymbol{\sigma}}$ (11.26_2). In the second step, we estimate the integral (11.43). In so doing, the sets of both the source and field points and the mesh in Eq. (11.43) are different (see for details Section 6). The representations (11.43)–(11.45) expressed though the boundary-type perturbator factors $\boldsymbol{\mathcal{L}}_i^{\chi}$ (11.43) coincide, except for notations, with the corresponding equations expressed through the volume perturbator factors (which were found by either the FEA or the VIE method, see Chap. 9).

In the NBM the effective field either $\langle\overline{\boldsymbol{\sigma}}\rangle_i(\mathbf{x})$ (9.64) or $\langle\overline{\boldsymbol{\chi}}\rangle(\mathbf{s})$ (11.33) in the area either $\mathbf{x}\in v_i$ or $\mathbf{s}\in\Gamma_i$, respectively, occupied by the matrix material is generated by a moving heterogeneities v_q (see Eqs. (9.64) and (11.33)), and the inhomogeneity of the detailed distribution either $\mathbf{L}_q^{\sigma}(\mathbf{x}-\mathbf{x}_q)$ $(x\in v_q)$ or $\mathcal{L}_q^{\chi}(\mathbf{s}-\mathbf{x}_q,\overline{\boldsymbol{\chi}})$ $(\mathbf{s}\in\Gamma_i)$ affects the effective fields either $\langle\overline{\boldsymbol{\sigma}}\rangle_i(\mathbf{x})$ $(\mathbf{x}\in v_i)$ or $\langle\overline{\boldsymbol{\chi}}\rangle(\mathbf{s})$ $(\mathbf{s}\in\Gamma_i)$, respectively. Furthermore, the domain of this long-range action is limited not only by $\mathbf{x}_q\in v_i^0$ but by a domain v_i^{ϕ} for location of inclusion centers $\mathbf{x}_q$ where $\varphi(v_q,\mathbf{x}_r|v_i,\mathbf{x}_i)-n^{(q)}$ is not negligible (in the sense of influence on the effective fields), and an effect zone of stress perturbations produced by the heterogeneity v_q is limited by the Minkowski addition of domains $\mathbf{x}\in v_{iq}^{\oplus}\equiv v_i^{\phi}\oplus v_q$ rather than by the domain v_q as in the EFM. Moreover, we note that in classical approaches, there is no systematic approach to choose the size and shape of the excluded volume v_i^0, which impacts on the estimations of effective properties.

The found statistical average of the effective field $\overline{\boldsymbol{\chi}}(\mathbf{s})$ allows one to get a statistical average of the interface counterparts of both the concentration and polarization tensors (4.60) $(\mathbf{s}\in v_i)$

$$\langle\boldsymbol{\chi}\rangle(\mathbf{s})=\mathcal{B}_i^{\chi}*\mathcal{Y}_i*\langle\boldsymbol{\chi}\rangle(\mathbf{s}), \tag{11.46}$$

$$\bar{v}_i\langle\boldsymbol{\eta}^s\rangle(\mathbf{s})=\mathcal{R}_i^{\chi}*\mathcal{Y}_i*\langle\boldsymbol{\chi}\rangle(\mathbf{s}). \tag{11.47}$$

Splitting of the operator $\mathcal{Y}_i$ (11.46) and (11.47) in a spirit of Eq. (4.61)

$$\mathcal{Y}_i=\mathcal{Y}_i^u+\mathcal{Y}_i^t \tag{11.48}$$

on two summands $\mathcal{Y}^u$ and $\mathcal{Y}^t$ acting on the displacement and traction, respectively, at the boundary Γ_i, and substitution of Eq. (11.46) into the effective constitutive Eq. (11.9) yields

$$\langle\boldsymbol{\chi}\rangle(\mathbf{s})=\langle\mathcal{B}_i^{\chi}*\mathcal{Y}_i^u*\mathbf{M}^*[\langle\boldsymbol{\sigma}\rangle(\mathbf{s}-\mathbf{x}_i)]\rangle_i+\langle\mathcal{B}_i^{\chi}*\mathcal{Y}_i^t*[\langle\boldsymbol{\sigma}\rangle\mathbf{n}(\mathbf{s})]\rangle_i, \tag{11.49}$$

$$\begin{aligned}\mathbf{M}^*\langle\boldsymbol{\sigma}\rangle&=\mathbf{M}^{(0)}\langle\boldsymbol{\sigma}\rangle+n^{(1)}\langle\mathcal{R}_i^{\chi}*\mathcal{Y}_i^u*\mathbf{M}^*[\langle\boldsymbol{\sigma}\rangle(\mathbf{s}-\mathbf{x}_i)]\rangle_i\\&\quad+n^{(1)}\langle\mathcal{R}_i^{\chi}*\mathcal{Y}_i^t*[\langle\boldsymbol{\sigma}\rangle\mathbf{n}(\mathbf{s})]\rangle_i,\end{aligned} \tag{11.50}$$

where the representation (11.34) for the statistical average field $\langle\boldsymbol{\chi}\rangle(\mathbf{s})$ was used. The Eqs. (11.49) and (11.50) can be solved by the method of successive approximations

$$\langle\boldsymbol{\chi}_{[n+1]}\rangle(\mathbf{s})=\langle\mathcal{B}_i^{\chi}*\mathcal{Y}_i^u*[\mathbf{M}_{[n]}^*\langle\boldsymbol{\sigma}\rangle](\mathbf{s}-\mathbf{x}_i)\rangle_i+\langle\mathcal{B}_i^{\chi}*\mathcal{Y}_i^t*[\langle\boldsymbol{\sigma}\rangle\mathbf{n}(\mathbf{s})]\rangle_i, \tag{11.51}$$

$$\begin{aligned}\mathbf{M}_{[n+1]}^*\langle\boldsymbol{\sigma}\rangle&=\mathbf{M}^{(0)}\langle\boldsymbol{\sigma}\rangle+n^{(1)}\langle\mathcal{R}_i^{\chi}*\mathcal{Y}_i^u*[\mathbf{M}_{[n]}^*\langle\boldsymbol{\sigma}\rangle](\mathbf{s}-\mathbf{x}_i)\rangle_i\\&\quad+n^{(1)}\langle\mathcal{R}_i^{\chi}*\mathcal{Y}_i^t*[\langle\boldsymbol{\sigma}\rangle\mathbf{n}(\mathbf{s})]\rangle_i,\end{aligned} \tag{11.52}$$

for $3(d-1)$ basic loading ($k, l, i, j = 1, \ldots, d;\ i \neq j$, no sum on i)

$$\langle\sigma_{kl}\rangle = \delta_{ki}\delta_{li}, \quad \langle\sigma_{kl}\rangle = (\delta_{ki}\delta_{lj} + \delta_{li}\delta_{kj})/2 \tag{11.53}$$

with an initial approximation $\mathbf{M}^*_{[0]} = \mathbf{M}^{*(9.80)}$ defined by Eqs. (9.80) and (11.31) for the volume homogeneous effective field $\overline{\boldsymbol{\vartheta}} \equiv$const.

11.2.5 Numerical Results

We will consider $2D$ plane–strain problems for composites reinforced by aligned infinite fibers with noncircular cross section described by the curve (9.114) (see Fig. 9.2), which reduces to a circle and a rectangular in the limiting cases $R_1 = R_2 = r_s = a$ and $r_s = 0$, respectively. We will consider isotropic constituents with the Young's moduli $E^{(1)} = 100$, $E^{(0)} = 1$, Poisson ratios $\nu^{(1)} = \nu^{(0)} = 0.45$, and the eigenstrains $\boldsymbol{\beta}_1(\mathbf{x}) = \beta_1^0 \boldsymbol{\delta} V_q(\mathbf{x})$ ($v_q \in v^{(1)}$) as well as zero average stresses $\langle\boldsymbol{\sigma}\rangle \equiv \mathbf{0}$.

In order to estimate the approximations introduced by discretizations implemented in the MFS in Sect. 4.3.3 and for the evaluation of the solution (11.31), we compare numerical results with a reference Eshelby analytical solution for a single circle inclusion inside the infinite matrix in a plane–strain problem subjected to the uniform remote loading $\overline{\boldsymbol{\sigma}} \equiv$const. Buryachenko [215] (see Sect. 4.3.4) demonstrated that for the discretization parameters $S = 400$ (4.67), $\lambda = 1.0E-5$ (4.81), and $D = 0.9R$ (4.71) (which also will be used in a subsequent estimations), the numerical error is $1.6E{-}5\%$. The corresponding errors for residual stresses ($\mathbf{M}_1(\mathbf{x}) \equiv \mathbf{0}, \boldsymbol{\beta}_1 \not\equiv \mathbf{0}$) [217] and ($\mathbf{M}_1(\mathbf{x}) \not\equiv \mathbf{0}, \boldsymbol{\beta}_1 \not\equiv \mathbf{0}$) [224] were $1.2E{-}6\%$ and $3.2E-4\%$, respectively.

A domain, where $\varphi(v_q, \mathbf{x}_q|; v_i, \mathbf{x}_i) - n^{(q)}$ is not negligible, is discretized by the square mesh $\Omega^{\rm sq}$ ($p, k \in Z^1$) (9.115) that is not optimized (at least for estimation of the perturbator $\boldsymbol{\mathcal{L}}^{\xi}_q(\mathbf{x} - \mathbf{x}_q)$) but is efficient. The main reason of the choice of the square mesh (9.115) is described in Sect. 9.6. Estimation of the perturbator $\boldsymbol{\mathcal{L}}^{\chi}_j(\mathbf{x} - \mathbf{x}_j, \overline{\boldsymbol{\chi}})$ (4.45) makes it possible to turn our attention to the analysis of CM by substitution of $\mathbf{L}^{t\sigma}_j(\mathbf{s} - \mathbf{x}_j)$ found into Eq. (11.41). Before proceeding to a CM's analysis, at first we consider some intermediate problems providing an exact analytical solutions. Namely, in Sect. 9.3.2 it was proved that acceptance of the hypotheses **H1a, H2a**, and **H3** for the CM with statistically homogeneous field of identical homogeneous ellipsoidal inclusions reduces with necessity Eqs. (9.75)–(9.82) to the EFM's analytical representations (9.41)–(9.48). This equivalence proved for pure mechanical loading ($\boldsymbol{\beta}_1 \equiv \mathbf{0}$) can be generalized to the thermoelastic problem when

$$n^{(1)}\mathbf{Q}_i^0\mathbf{R}_i = \int \langle \mathbf{L}_q^{\sigma}(\mathbf{x}-\mathbf{x}_q)\rangle_{(i)}[\varphi(v_q,\mathbf{x}_q|;v_i,\mathbf{x}_i)-n^{(q)}]d\mathbf{x}_q, \qquad (11.54)$$

$$n^{(1)}\mathbf{Q}_i^0\mathbf{F}_i = \int \langle \mathbf{C}_q^{\sigma}(\mathbf{x}-\mathbf{x}_q)\rangle_{(i)}[\varphi(v_q,\mathbf{x}_q|;v_i,\mathbf{x}_i)-n^{(q)}]d\mathbf{x}_q. \qquad (11.55)$$

Except for notations, Eqs. (11.54) and (11.55) can be reformulated in terms of the interface perturbator factor $\mathbf{L}_q^{t\sigma}(\mathbf{s}-\mathbf{x}_q)$ and $\mathbf{C}_q^{t}(\mathbf{s}-\mathbf{x}_q)$ $(\mathbf{s}\in v_i)$

$$n^{(1)}\mathbf{Q}_i^0\mathbf{R}_i = \int \big\langle \mathbf{s}\overset{S}{\otimes}\mathbf{L}_q^{t\sigma}(\mathbf{s}-\mathbf{x}_q)\big\rangle_{(i)}[\varphi(v_q,\mathbf{x}_q|;v_i,\mathbf{x}_i)-n^{(q)}]d\mathbf{x}_q, \qquad (11.56)$$

$$n^{(1)}\mathbf{Q}_i^0\mathbf{F}_i = \int \big\langle \mathbf{s}\overset{S}{\otimes}\mathbf{C}_q^{t}(\mathbf{s}-\mathbf{x}_q)\big\rangle_{(i)}[\varphi(v_q,\mathbf{x}_q|;v_i,\mathbf{x}_i)-n^{(q)}]d\mathbf{x}_q. \qquad (11.57)$$

Hereafter the conditional probability density $\varphi(v_k,\mathbf{x}_k|;v_i,\mathbf{x}_i)$ is described by a the step function of the well-stirred approximation (9.116).

An accuracy assessment of both the perturbator estimations and the integral evaluations (9.63), (9.66), (11.41), and (11.42) at the mesh $\Omega^{\rm sq}$ (9.115) can be performed by comparison of the approximate right-hand sides of Eqs. (11.54)–(11.57) with the corresponding exact analytical representations in the left-hand sides of Eqs. (11.54)–(11.57). It was achieved in the case of the FEA's representations of the perturbator (4.44) for the pure mechanical loading (11.54) ($\boldsymbol{\beta}_1\equiv\mathbf{0}$ [232]) and for elastically homogeneous medium with the random thermal mismatch (11.55) ($\mathbf{M}_1\equiv\mathbf{0}$ [233]). A similar numerical analysis of Eq. (11.56) for the MFS's representation of the interface perturbator factor $\mathbf{L}_q^{t\sigma}(\mathbf{x}-\mathbf{x}_q)$ (4.45) was performed in [218] for the pure mechanical loading ($\boldsymbol{\beta}_1\equiv\mathbf{0}$). One [217] estimated by the MFS the statistical averages of residual stresses in elastically homogeneous medium with random field of noncanonical inclusions ($\mathbf{M}_1\equiv\mathbf{0},\boldsymbol{\beta}_1\not\equiv\mathbf{0}$). A general case of both the elastic and thermal mismatch ($\boldsymbol{\beta}_1,\mathbf{M}_1\not\equiv\mathbf{0}$) was tested in [223]. In particular, it was estimated the errors of the right-hand side of Eq. (11.57) 0.93%, 0.42%, and 0.22% at the discretization step of the mesh $\Omega^{\rm sq}$ (9.115) $h=0.06R$; $0.03R$, and $h=0.015R$, respectively.

These preliminary computations confirm that sufficient convergence of results is achieved and reveals that the error introduced by the numerical computation of the integrals in Eqs. (11.56) (see [63]) and (11.57) at the mesh $\Omega^{\rm sq}$ (9.115) exceeds the error of the perturbator (4.45) estimations by the MFS (see Sect. 4.3.3). It is also important to stress that the difference between the new and old approaches does not exceed a computational error, and, therefore, we qualitatively proved that in the considered example of the circle inclusion both methods the old (9.41)–(9.48) and new [(9.52)–(9.59) and (11.36)–(11.42)] ones which are based on the classical (11.28) and new [(11.20) and (11.20), respectively] general integral equations, respectively, lead to close numerical results.

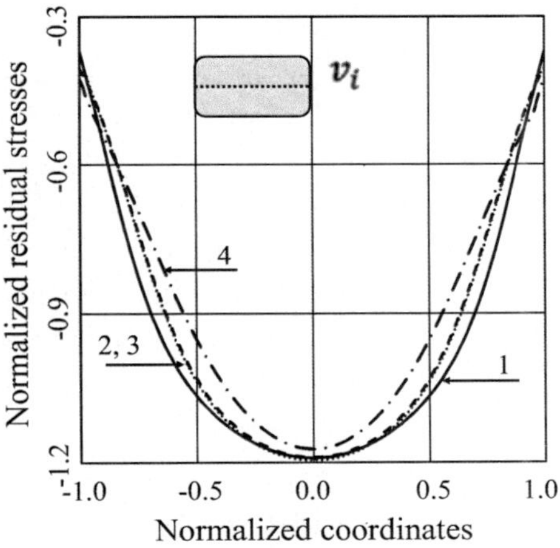

Fig. 11.1 $\widehat{\sigma}_{i|11}(x_1)$ vs. x_1/R_1 for $R_2/R_1 = 0.2$ (1), 0.32 (2, MFS), 0.32 (3, FEA), and 0.6 (4)

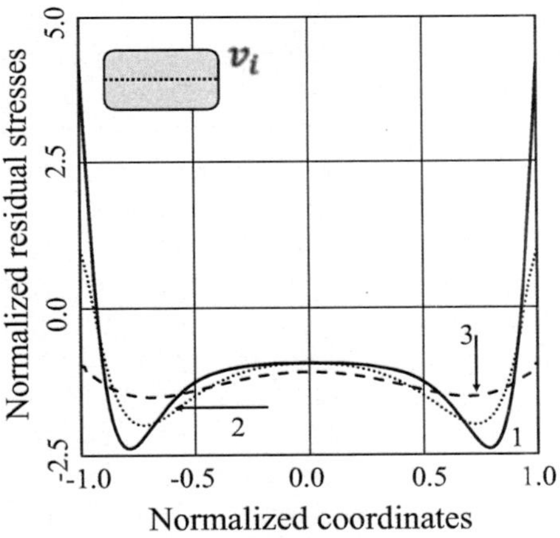

Fig. 11.2 $\widehat{\sigma}_{i|22}(x_1)$ vs. x_1/R_1 for $R_1/R_2 = 0.2$ (1), 0.32 (2), and 0.6 (3)

In the framework of the backgrounds (11.21) for composites with nonellipsoidal inclusions v_q and the homogeneous eigenstrain $\boldsymbol{\beta}_1(\mathbf{y}) \equiv$const ($\mathbf{y} \in v_q$), it is expected that the greater inhomogeneity of the concentration factors (11.37) or an isolated heterogeneity v_q ($c = 0$) inside infinite matrix leads to the greater difference between the old (11.28) and new (11.21) approaches. In more details we will analyze the inclusion shape (9.114) with the different aspect ratios $R_2/R_1 = 0.2,\ 0.32, and\ 0.60$ and fixed $r_s/R_1 = 0.1$ (see Figs. 11.1 and 11.2). The normalized residual stresses $\widehat{\sigma}_{i|kk}(\mathbf{x})$ ($\mathbf{x} \in v_i$) are estimated by the MFS (enriched by the TR procedure) (4.81) for a single inclusion in the infinite matrix (i.e. $c = 0$), and the components $\widehat{\sigma}_{i|11}(\mathbf{x})$ and $\widehat{\sigma}_{i|22}(\mathbf{x})$ in the cross section $\mathbf{x} = (x_1, 0)^\top$ are presented in Figs. 11.1 and 11.2, respectively. The values $\widehat{\sigma}_{i|11}(\mathbf{x})$ (see Fig. 11.1) at $R_2/R_1 = 0.32$ were compared with the corresponding FEA performed in [233]) for the size of a sample $L/R_1 = 40$. The maximum difference between the

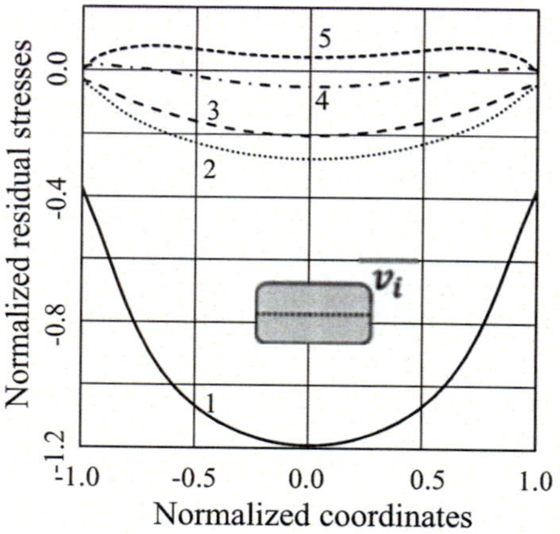

Fig. 11.3 $\langle\widehat{\sigma}_{11}\rangle(x_1)$ vs. x_1/R_1 estimated by the NBM (11.37) (curves 1, 3, 5) and EFM (9.45) (curves 1, 2, 4)

curves 2 (dotted curve, MFS) and 3 (dashed curve, FEA) in Fig. 11.1 is not more than 0.5% at $x_1 = 0$. As can be seen in Fig. 11.2, the variations of $\widehat{\sigma}_{i|22}(\mathbf{x})$ are significantly growing with decreasing of the ratio R_2/R_1, while the component $\widehat{\sigma}_{i|11}(\mathbf{x})$ is considerably less sensitive values to the choice of the ratio R_2/R_1.

In greater detail we present the normalized stress $\widehat{\boldsymbol{\sigma}}_i^*(x_1)$ (Fig. 11.3) for the composite materials filled by inclusions with $R_2/R_1 = 0.2$ providing the most inhomogeneous stress distribution $\widehat{\sigma}_{i|22}(x_1)$ in the isolated inclusion v_i (4.56) (see Fig. 11.2). The normalized stress distributions $\langle\widehat{\sigma}\rangle_i(x_1)$ are estimated in the frameworks of both the new background $\langle\widehat{\sigma}\rangle_i^{\rm NBM}(x_1)$ (11.37) (curves 1, 3, 5) and the classical one (the EFM) $\langle\widehat{\sigma}\rangle_i^{\rm EFM}(x_1)$ (9.45) (curves 1, 2, 4) for a step conditional probability density $\varphi(v_q, \mathbf{x}_q|v_i, \mathbf{x}_i)$ (9.116) and the volume fractions $c = 0.0,\ 0.5,\ 0.8$, respectively. Increasing of volume fraction of inclusions from $c = 0$ (curve 1) to $c = 0.5$ (curve 2 and 3) leads to reduction of residual stresses in the cross section $\mathbf{x} = (x_1, 0)^\top$. Subsequent increasing of volume fraction of inclusions from $c = 0.5$ to $c = 0.8$ (curve 4 and 5) can lead to change of sign of residual stresses along the mentioned cross section. So, for $c = 0.5$ the normalized stresses $\langle\widehat{\sigma}_{11}\rangle^{\rm EFM}(0)$ and $\langle\widehat{\sigma}_{11}\rangle^{\rm NBM}(0)$ in the center of inclusions differ from one another by 26% while at the ends of inclusions $\langle\widehat{\sigma}_{11}\rangle^{\rm EFM}(\pm R_1) \approx \langle\sigma_{11}\rangle^{\rm NBM}(\pm R_1)$. More dramatic difference between the estimations of normalized stresses obtained by both the NBM and the EFM is observed for the inclusion concentration $c = 0.8$. So, $\langle\widehat{\sigma}_{11}\rangle(x_1) < 0$ at $|x_1|/R_1 \leq 0.67$ for the estimations obtained by the EFM (curve 4) while $\langle\widehat{\sigma}_{11}\rangle(x_1) > 0$ at $\forall|x_1| \leq R_1$ for the NBM's estimations (curve 5), i.e., estimations of $\langle\widehat{\sigma}_{11}\rangle(x_1)$ obtained at $|x_1|/R_1 \leq 0.67$ by both the NBM (11.47) and the EFM (9.45) differ by a sign. Comparative analysis of the EFM's (9.41)–(9.48) and MTM's (9.52)–(9.59) estimations for the noncanonical heterogeneity shapes was performed in [232–234] and [235] (see Chap. 9) by the FEA, and it is not considered here.

The effective elastic moduli $\mathbf{L}^*$ were estimated in terms of both the volume perturbator factors (9.46) [232]) and the interface perturbator factor (11.38), while

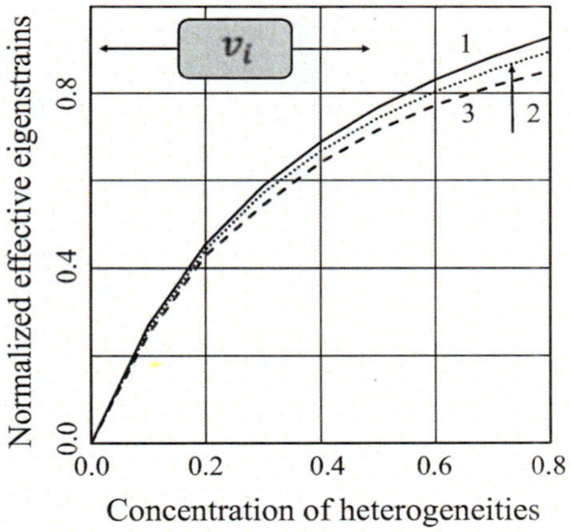

Fig. 11.4 $\widehat{\beta}^*_{11}$ vs. c for elliptical (1) and noncanonical (2, 3) heterogeneities

the stored energy (9.48) was evaluated in [232]; these problems are not considered in more details here. Normalized effective eigenstrains $\widehat{\beta}^*_{pq} \equiv (\beta^*_{pq} - \beta^{(0)}_{pq})/\beta^{(1)}_{pq}$ (no sum on $p = q = 1, 2$) as a function of the volume fraction of heterogeneities c at $R_2/R_1 = 0.2$ are presented in Figs. 11.4 for $p = q = 1$. Curve 1 in Fig. 11.4 is estimated by the EFM (9.47) for CM reinforced by the canonical elliptical heterogeneities with aspect ratio $R_2/R_1 = 0.2$, while the curve 2 is evaluated by the same EFM (9.47) for CM with the noncanonical heterogeneities (9.114) $R_2/R_1 = 0.2$ and $r_s/R_1 = 0.1$. In so doing, the curve 3 corresponds to the NBM's method (11.39). As can be seen, the estimations of the effective eigenstrains $\widehat{\boldsymbol{\beta}}^*$ carried out by the different methods are closely aligned in all range of the volume fraction c.

A large difference of results is obtained in the framework of the backgrounds (9.45)–(9.48) and (11.36)–(11.39) for composites reinforced by nonellipsoidal inclusions demonstrating essentially inhomogeneous stress distribution inside inclusions even in the framework of the hypothesis **H1a**. In more details we analyze the inclusion shape (9.114) with the aspect ratios $R_2/R_1 = 0.2$ and fixed $r_s/R_1 = 0.1$ (at $c = 0$, see Fig. 11.5). The stress concentration factors for an isolated inhomogeneity $B_{i|1111}(x_1)$ and $B_{i|2211}(x_2)$ in the cross sections $\mathbf{x} = (x_1/R_1, 0)^\top$ and $\mathbf{x} = (0, x_2/R_2)^\top$ (see Fig. 11.5) grows in both the magnitude and variation with decreasing of the ratio R_2/R_1. For CMs, the solution for a single inclusion (see Sect. 4.3 and Fig. 11.5) obtained by the MFS was used in both the old (9.45)–(9.48) and new (11.36)–(11.39) approaches. However, in the new approach (NA) (11.36)–(11.39), the hypothesis **H1b** was not used while this hypothesis was exploited in the classical EFM (9.45)–(9.48) with the aspect ratio v_i^0 equals to R_2/R_1. The largest difference between the classical (9.45)–(9.48) and new (11.36)–(11.39) approaches are observed for the component $B^*_{i|2211}(x_2)$ ($c = 0.7$) in a cross section $\mathbf{x} = (0, x_2/R_2)^\top$ (see Fig. 11.6), where the curves predicted by the new and classical approaches are

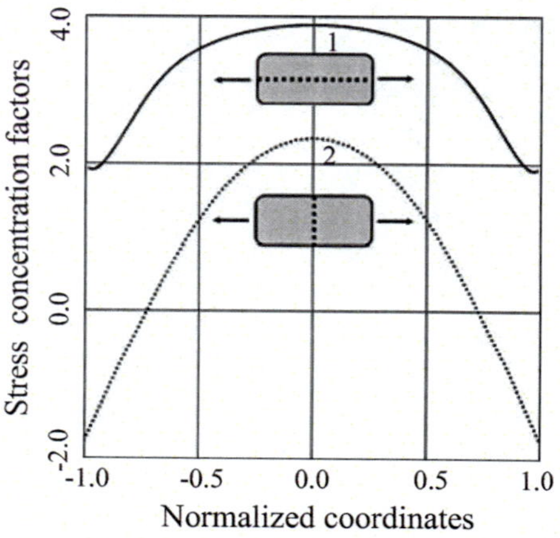

Fig. 11.5 $B_{i|1111}(x_1)$ (1) and $100 \cdot B_{i|2211}(x_2)$ (2) vs. x_1/R_1 and x_2/R_2, respectively

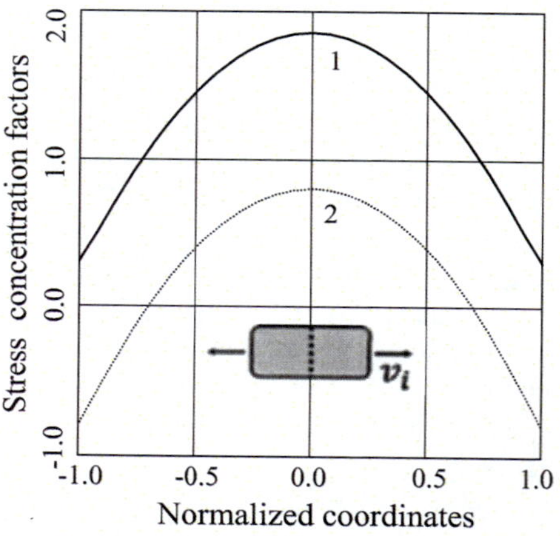

Fig. 11.6 $100 \cdot B^*_{i|2211}(x_2)$ vs. x_2/R_2 estimated by the EFM (9.45)–(9.48) (1) and NA (11.37)–(11.42) (2)

distinguished by a sign. Normalized effective Young's moduli $E_1^*/E^{(0)}$ and $E_2^*/E^{(0)}$ are presented in Figs. 11.7 and 11.8, respectively, as the functions of the volume concentration c of inclusions at $R_2/R_1 = 0.2$. Curves 1 and 2 are estimated by the new approach (11.36)–(11.39) and classical one (9.45)–(9.48), respectively, for the $\varphi(v_q, \mathbf{x}_q|v_i, \mathbf{x}_i)$ (9.116), respectively, in both Figs. 11.6 and 11.7. The curves 3 are predicted by the EFM for the homogeneous ellipsoidal inclusion v_i and v_i^0 with the aspect ratio equals R_2/R_1.

Thus, we have proposed the new background of micromechanics based on the new general integral equation (11.21) which makes it potentially possible to abandon many classical concepts of micromechanics used in most popular methods, namely: effective field hypothesis **H1**, quasi-crystalline approximation **H2**, the hypothesis of "ellipsoidal symmetry" **H3**, and Eshelby tensor (see for details [203] and Chap. 9). In the present chapter, numerical results were obtained for composites with statistically homogeneous field of homogeneous aligned prolate identical

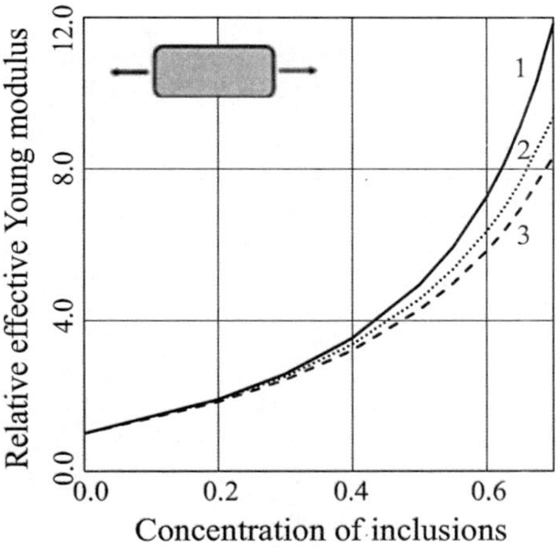

Fig. 11.7 $E_1^*/E^{(0)}$ vs. c estimated by the NA (1), EFM for noncanonical inclusions (2), and by the EFM for canonical inclusions

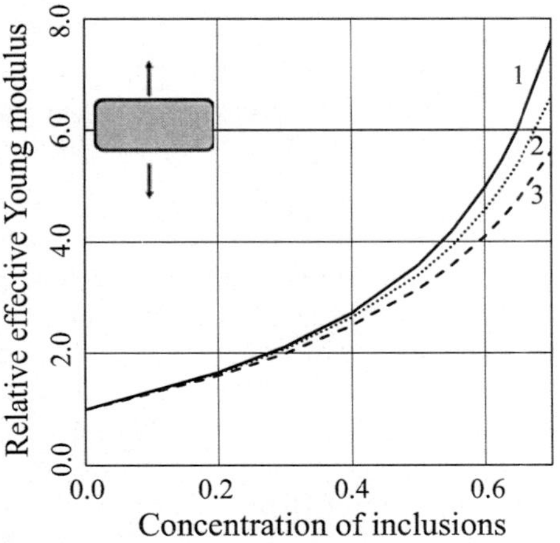

Fig. 11.8 $E_2^*/E^{(0)}$ vs. c estimated by the NA (1), EFM for noncanonical inclusions (2), and by the EFM for canonical inclusions

heterogeneities of noncanonical shape. Hypotheses **H1a** and **H2** were used, while the hypotheses **H1b** and **H3** were not to be accepted. A fundamental deficiency of Eq. (11.28) is a dependence of the renormalizing term $\boldsymbol{\Gamma}(\mathbf{x}-\mathbf{y})\langle\boldsymbol{\eta}\rangle(\mathbf{y})$ [obtained in the framework of the asymptotic approximation of the hypothesis **H1b**] only on the statistical average $\langle\boldsymbol{\eta}\rangle(\mathbf{y})$, while the renormalizing term either $\langle\boldsymbol{\mathcal{L}}_k^{\vartheta}(\mathbf{x}-\mathbf{x}_k,\boldsymbol{\zeta})\rangle(\mathbf{x}_k)$ in Eq. (11.20) or $\langle\boldsymbol{\mathcal{L}}_k^{\chi}(\mathbf{x}-\mathbf{x}_k,\boldsymbol{\chi})\rangle(\mathbf{x}_k)$ (11.21) explicitly depends on details distribution $\langle\boldsymbol{\eta}|v_i,\mathbf{x}_i\rangle(\mathbf{y})$ $(\mathbf{y}\in v_i)$ or $\langle\boldsymbol{\chi}|v_i,\mathbf{x}_i\rangle(\mathbf{s})$ $(\mathbf{s}\in\Gamma_i)$, respectively.

The detected dependence allows us to abandon the hypothesis **H1b** whose accuracy is questionable for inclusions of noncanonical shape. We obtained a fundamental conclusion that both the effective properties and effective concentration factors in general depend not only on the stress distribution inside the referred heterogeneity [describing by the tensor $\mathbf{B}_i(\mathbf{x}),\mathbf{C}_i(\mathbf{x})$, $\mathbf{x}\in v_i$ but also on the stresses in the vicinity of heterogeneity, i.e., extension of $\boldsymbol{\mathcal{L}}_k^{\vartheta}(\mathbf{x}-\mathbf{x}_k,\boldsymbol{\zeta})$ (11.20) or

$\mathcal{L}_k^{\chi}(\mathbf{x}-\mathbf{x}_k, \chi)$ (11.21) for $\mathbf{x} \in v_i^{\oplus}$ is necessary. Then the size of the excluded volume v_i^0 as well as the binary correlation function $\varphi(v_q, \mathbf{x}_q | v_i, \mathbf{x}_i)$ will impact on both the effective properties and effective concentration factors even in the framework of hypothesis **H2**. In so doing, the concentration factors (see Fig. 11.3) are significantly more sensitive values to the choice of the approach (either NBM or EFM) than effective properties (see Figs. 11.4, 11.7, and 11.8).

Thus, we have proposed the new micromechanical model based on the integral equation (11.21) for the interface displacements and tractions presented in the form of the *interface perturbators*, while the effective properties (11.6)–(11.11) are expressed through the new notions of the *interface polarization tensors*. It opens up in micromechanics of random structure CMs the strong opportunities for systematic exploitation of the BIE methods (and, in particular, the MFS) which have the well-known advantage of reduction of dimension by one that can be crucial for the analyst.

11.3 General Integral Equations in Micromechanics of CMs with Imperfectly Bonded Interfaces

Interfacial bonding conditions play an important role that control key properties of the composite materials (CM) such as stiffness, strength, and fracture behavior. In the literature, three kinds of model are often used to simulate the properties of interface regions in CM. The modeling of all three kind of imperfect interface is described in Sect. 3.7.3.

The popular approaches of analysis of CMs with imperfect interface properties are usually based on the coupling of two problems. At first the problem for one heterogeneity inside an infinite matrix subjected to the homogeneous remote stresses is solved for the different interface bonded laws of both the heterogeneity and matrix (see, e.g., the references above). The second problem is incorporating the mentioned solution into one of the known micromechanical schemes, e.g., in the MTM which is totally dominating in micromechanics of CMs with the accepted imperfectly bonded interfaces (see, e.g., [314, 402, 1373, 1483, 1484, 1651] and other references mentioned above). However, all mentioned methods of micromechanics are based on the effective field hypothesis **H1** (see Chaps. 7 and 8) critically analyzed in Chaps. 9–11. Moreover, a concrete average scheme of micromechanics is initially determined by both the EFH and additional ones used for the approximate solution of exact initial integral equations connecting the random stress fields at the point being considered and the surrounding points. The main goal of the current section is obtaining such general integral equation taking into account the actual microstructure and possible imperfectly bonded interfaces of CMs subjected to essentially inhomogeneous mechanical, body force, and temperature loading.

11.3.1 Stress and Strain Distributions for One Heterogeneity Inside Macrodomain w

Let us assume that the domain w contains one heterogeneity $v_k \subset w$. Substituting (6.2) and (6.3_1) into the equilibrium equation (6.1) leads to a differential equation with respect to the displacement $\mathbf{u}$

$$\nabla \cdot [\mathbf{L}^c(\mathbf{x})(\nabla \otimes \mathbf{u}(\mathbf{x})) + \boldsymbol{\alpha}^c(\mathbf{x})] + \mathbf{f}^c(\mathbf{x}) = -\nabla \cdot [\mathbf{L}_1(\mathbf{x})(\nabla \otimes \mathbf{u}(\mathbf{x})) + \boldsymbol{\alpha}_1(\mathbf{x})] - \mathbf{f}_1(\mathbf{x}) \tag{11.58}$$

with a fictitious random "body-force" in the right-hand side of the equation. Then Eq. (11.58) can be reduced to a symmetrized integral form

$$\boldsymbol{\varepsilon}(\mathbf{x}) = \boldsymbol{\varepsilon}^0(\mathbf{x}) + \nabla \overset{S}{\otimes} \int_w \mathbf{G}(\mathbf{x}-\mathbf{y})\Big\{\nabla \cdot [\mathbf{L}_1(\mathbf{y})(\nabla \otimes \mathbf{u}(\mathbf{y})) + \boldsymbol{\alpha}_1(\mathbf{y})] + \mathbf{f}_1(\mathbf{y})\Big\} V_k(\mathbf{y}) d\mathbf{y}, \tag{11.59}$$

where $\mathbf{G}$ is the infinite body Green's function of the Navier equation with homogeneous elastic modulus tensor $\mathbf{L}^c$, defined by

$$\nabla \left\{\mathbf{L}^c[\nabla \otimes \mathbf{G}(\mathbf{x})]\right\} = -\boldsymbol{\delta}\delta(\mathbf{x}), \tag{11.60}$$

of order $O\big(\int |\mathbf{x}|^{1-d} d|\mathbf{x}|\big)$ as $|\mathbf{x}| \to \infty$; the symbols $\overset{S}{\otimes}$ are replaced by $\otimes$ in Eqs. (11.58)–(11.60) due to the symmetry of $\mathbf{L}^c$ and $\mathbf{L}_1$. $\boldsymbol{\varepsilon}_0(\mathbf{x})$ (11.59) is the strain corresponding to a homogeneous domain w with modulus tensor $\mathbf{L}^c$ and which is described by the boundary integral equation (BIE, see, e.g., [59, 156]) on the boundary Γ_0 of the domain w

$$\begin{aligned} \boldsymbol{\varepsilon}^0(\mathbf{x}) = & \int_{\Gamma_0} \Big[\nabla \overset{S}{\otimes} \mathbf{G}(\mathbf{x}-\mathbf{s})\underline{\mathbf{t}}_0(\mathbf{s}) - \nabla \overset{S}{\otimes} \mathbf{T}^\top(\mathbf{x},\mathbf{s})\mathbf{u}_0(\mathbf{s})\Big] ds \\ & + \int_w \nabla \overset{S}{\otimes} \mathbf{G}(\mathbf{x}-\mathbf{s})\mathbf{f}^c(\mathbf{y}) d\mathbf{y}, \end{aligned} \tag{11.61}$$

which confirms with the stress field $\boldsymbol{\sigma}^0(\mathbf{x}) = \mathbf{L}^c\boldsymbol{\varepsilon}^0(\mathbf{x}) + \boldsymbol{\alpha}^c(\mathbf{x})$; $\underline{\mathbf{t}}_0(\mathbf{s}) \equiv \mathbf{t}_0(\mathbf{s}) - \boldsymbol{\alpha}^{(0)}\mathbf{n}^{\Gamma_0}(\mathbf{s})$ is the modified traction on the surface Γ_0 with the outer normal $\mathbf{n}^{\Gamma_0}(\mathbf{s})$ ($\mathbf{s} \in \Gamma_0$). The tensor of the "fundamental traction" $\mathbf{T}$ on Γ_0 (called also Kupradze tensor) associated with the tensor of "fundamental displacement" $\mathbf{G}$ (3.39) is given by Eq. (3.39_2).

It should be mentioned that Eqs. (11.58) and (11.59) are well-known for the ideal interface Γ_k (3.196) and presented for an illustrative demonstration of corrections of these equations for the discontinuity of the displacement or/and stress fields at the interface Γ_k. In the case of discontinuities of $\mathbf{u}(x)$ and $\boldsymbol{\sigma}(\mathbf{x})$ at the surfaces Γ_k^u and Γ_k^σ ($\Gamma_k = \Gamma_k^u \cup \Gamma_k^\sigma$, $\Gamma_k^u \cap \Gamma_k^\sigma = \emptyset$), respectively, Eq. (11.58) should be considered in the class of distributions with additional Dirac distribution concentrated on the surfaces (3.178) of each jump discontinuity Γ_k^u and Γ_k^σ defined by the equations $\Gamma_k^u(\mathbf{x})=0$ and $\Gamma_k^\sigma(\mathbf{x})=0$, respectively

$$\nabla[\mathbf{L}^c(\mathbf{x})(\nabla\otimes\mathbf{u}(\mathbf{x}))\boldsymbol{\alpha}^c(\mathbf{x})]+\mathbf{f}^c(\mathbf{x})=-\nabla\cdot[\mathbf{L}_1(\mathbf{x})(\nabla\otimes\mathbf{u}(\mathbf{x}))+\boldsymbol{\alpha}_1(\mathbf{x})]$$
$$-[[\boldsymbol{\sigma}]]\cdot\mathbf{n}(\mathbf{x})\delta_{\Gamma_k^\sigma}+\nabla\cdot\{\mathbf{L}_1^{(0)}(\mathbf{x})([[\mathbf{u}]]\otimes\mathbf{n}(\mathbf{x}))\}\delta_{\Gamma_k^u}-\mathbf{f}_1(\mathbf{x}), \quad (11.62)$$

where one usually assumes that either

$$\Gamma_k^u=\Gamma_k,\ \Gamma_k^\sigma=\emptyset \quad \text{or} \quad \Gamma_k^\sigma=\Gamma_k,\ \Gamma_k^u=\emptyset. \quad (11.63)$$

By rearranging Eq. (11.59) into an integral form and applying Gauss' theorem, we obtain

$$\boldsymbol{\varepsilon}(\mathbf{x})=\boldsymbol{\varepsilon}^0(\mathbf{x})+\int_w \mathbf{U}(\mathbf{x}-\mathbf{y})\boldsymbol{\tau}(\mathbf{y})d\mathbf{y}+\int_w \nabla\overset{S}{\otimes}\mathbf{G}(\mathbf{x}-\mathbf{y})\mathbf{f}_1(\mathbf{y})d\mathbf{y}$$
$$-\int_{\Gamma_k^u}\nabla\overset{S}{\otimes}\mathbf{T}^\top(\mathbf{x}-\mathbf{s})[[\mathbf{u}(\mathbf{s})]]d\mathbf{s}-\int_{\Gamma_k^\sigma}\nabla\overset{S}{\otimes}\mathbf{G}(\mathbf{x}-\mathbf{s})[[\boldsymbol{\sigma}(\mathbf{s})]]\cdot\mathbf{n}(\mathbf{s})d\mathbf{s}, \quad (11.64)$$

where we have also used the rule $\nabla_\mathbf{x}\mathbf{G}(\mathbf{x}-\mathbf{y})=-\nabla_\mathbf{y}\mathbf{G}(\mathbf{x}-\mathbf{y})$; the volume integral kernel $U_{kijl}(\mathbf{x})=[\nabla_k\nabla_l G_{ij}(\mathbf{x})]_{(ki)(jl)}$ is an even homogeneous generalized function of degree $-d$. The surface integrals over Γ_k^u and Γ_k^σ (11.58) coincide with the corresponding integrals considered by Asaro [34] and Sharma and Ganti [1368], respectively.

To express Eq. (11.64) in terms of stresses, we use the identities

$$\mathbf{L}_1(\boldsymbol{\varepsilon}-\boldsymbol{\beta})=-\mathbf{L}^c\mathbf{M}_1\boldsymbol{\sigma}, \quad \boldsymbol{\varepsilon}=[\mathbf{M}^c\boldsymbol{\sigma}+\boldsymbol{\beta}^c]+[\mathbf{M}_1\boldsymbol{\sigma}+\boldsymbol{\beta}_1]. \quad (11.65)$$

Substituting (11.65) into the right-hand side and the left-hand side of (11.64), respectively, and contracting with the tensor $\mathbf{L}^c$ give the integral equation for stresses

$$\boldsymbol{\sigma}(\mathbf{x})=\boldsymbol{\sigma}^0(\mathbf{x})+\int_w \boldsymbol{\Gamma}(\mathbf{x}-\mathbf{y})\boldsymbol{\eta}(\mathbf{y})d\mathbf{y}+\int_w \mathbf{L}^c\nabla\mathbf{G}(\mathbf{x}-\mathbf{y})\mathbf{f}_1(\mathbf{y})\mathbf{y}$$
$$-\int_{\Gamma_k^u}\mathbf{L}^c\nabla\mathbf{T}^\top(\mathbf{x}-\mathbf{s})[[\mathbf{u}(\mathbf{s})]]d\mathbf{s}-\int_{\Gamma_k^\sigma}\mathbf{L}^c\nabla\mathbf{G}(\mathbf{x}-\mathbf{s})[[\boldsymbol{\sigma}(\mathbf{s})]]\cdot\mathbf{n}(\mathbf{s})d\mathbf{s}, \quad (11.66)$$

where the integral operator kernel, $\boldsymbol{\Gamma}(\mathbf{x}-\mathbf{y})\equiv-\mathbf{L}^c[\mathbf{I}\delta(\mathbf{x}-\mathbf{y})+\mathbf{U}(\mathbf{x}-\mathbf{y})\mathbf{L}^c]$, is called the Green stress tensor, see Kröner [820, 822].

Analysis of Eqs. (11.64) and (11.66) generates an introduction of general notions of the *perturbators* of the strain $\mathcal{L}_k^\epsilon(\mathbf{x}-\mathbf{x}_k,\widehat{\boldsymbol{\tau}})$, the stress $\mathcal{L}_k^\sigma(\mathbf{x}-\mathbf{x}_k,\widehat{\boldsymbol{\eta}})$, and the displacement $\mathcal{L}_k^u(\mathbf{x}-\mathbf{x}_k,\widehat{\boldsymbol{\tau}})$

$$\mathcal{L}_k^{\epsilon}(\mathbf{x}-\mathbf{x}_k,\widehat{\boldsymbol{\tau}}) \equiv \boldsymbol{\varepsilon}(\mathbf{x})-\boldsymbol{\varepsilon}^0(\mathbf{x}), \tag{11.67}$$

$$\mathcal{L}_k^{\sigma}(\mathbf{x}-\mathbf{x}_k,\widehat{\boldsymbol{\eta}}) \equiv \sigma(\mathbf{x})-\sigma^0(\mathbf{x}), \tag{11.68}$$

$$\mathcal{L}_k^{u}(\mathbf{x}-\mathbf{x}_k,\widehat{\boldsymbol{\tau}}) \equiv \mathbf{u}(\mathbf{x})-\mathbf{u}^0(\mathbf{x}), \tag{11.69}$$

which produce the perturbation of the fields $\boldsymbol{\varepsilon}(\mathbf{x})$, $\sigma(\mathbf{x})$, and $\mathbf{u}(\mathbf{x})$, respectively, in the point $\mathbf{x}$ due to the insertion of the heterogeneity center into the point $\mathbf{x}_k$. $\widehat{\boldsymbol{\tau}} and\ \widehat{\boldsymbol{\eta}}$ are the symbolic notations of dependance of $\mathcal{L}_k^{\epsilon}(\mathbf{x}-\mathbf{x}_k,\widehat{\boldsymbol{\tau}})$, $\mathcal{L}_k^{\sigma}(\mathbf{x}-\mathbf{x}_k,\widehat{\boldsymbol{\eta}})$, and $\mathcal{L}_k^{u}(\mathbf{x}-\mathbf{x}_k,\widehat{\boldsymbol{\tau}})$ on $\boldsymbol{\tau}(\mathbf{x})$, $\boldsymbol{\eta}(\mathbf{x})$ $(\mathbf{x}\in v_k)$, respectively, $\mathbf{f}_1(\mathbf{x})$ $(\mathbf{x}\in v_k)$, $[[\mathbf{u}(\mathbf{x})]]$ $(\mathbf{x}\in\Gamma_k^u)$, and $[[\boldsymbol{\sigma}(\mathbf{x})]]$ $(\mathbf{x}\in\Gamma_k^{\sigma})$. Obviously, the perturbators $\mathcal{L}_k^{\epsilon}(\mathbf{x}-\mathbf{x}_k,\widehat{\boldsymbol{\tau}})$, $\mathcal{L}_k^{\sigma}(\mathbf{x}-\mathbf{x}_k,\widehat{\boldsymbol{\eta}})$, and $\mathcal{L}_k^{u}(\mathbf{x}-\mathbf{x}_k,\widehat{\boldsymbol{\tau}})$ are generalizations of the relevant multiplied Green's functions with some polarizations of a point defect to the impacts of heterogeneity v_k of the finite size with possible imperfection of bounding conditions. The mentioned perturbators can be found by any available numerical method, such as the volume integral equation (VIE), boundary element method (BEM), FEM, hybrid FEM–BEM, multipole expansion method, and complex potential method among others (see for references [200, 544, 959]). Each method has advantages and disadvantages, and it is crucial for the analyst to be aware of their range of applications.

We obtained representation of the perturbators by the hybrid VIE–BIE (11.64) and (11.66)

$$\begin{aligned}\mathcal{L}_k^{\epsilon}(\mathbf{x}-\mathbf{x}_k,\widehat{\boldsymbol{\tau}}) = &\int_w \mathbf{U}(\mathbf{x}-\mathbf{y})\boldsymbol{\tau}(\mathbf{y})d\mathbf{y} + \int_w \nabla\overset{S}{\otimes}\mathbf{G}(\mathbf{x}-\mathbf{y})\mathbf{f}_1(\mathbf{y})dy\\ &-\int_{\Gamma_k^u} \nabla\overset{S}{\otimes}\mathbf{T}^{\top}(\mathbf{x}-\mathbf{s})[[\mathbf{u}(\mathbf{s})]]ds\\ &-\int_{\Gamma_k^{\sigma}} \nabla\overset{S}{\otimes}\mathbf{G}(\mathbf{x}-\mathbf{s})[[\boldsymbol{\sigma}(\mathbf{s})]]\cdot\mathbf{n}(\mathbf{s})ds,\end{aligned} \tag{11.70}$$

$$\begin{aligned}\mathcal{L}_k^{\sigma}(\mathbf{x}-\mathbf{x}_k,\widehat{\boldsymbol{\eta}}) = &\int_w \boldsymbol{\Gamma}(\mathbf{x}-\mathbf{y})\boldsymbol{\eta}(\mathbf{y})d\mathbf{y} + \int_w \mathbf{L}^c\nabla\mathbf{G}(\mathbf{x}-\mathbf{y})\mathbf{f}_1(\mathbf{y})\mathbf{y}\\ &-\int_{\Gamma_k^u} \mathbf{L}^c\nabla\mathbf{T}^{\top}(\mathbf{x}-\mathbf{s})[[\mathbf{u}(\mathbf{s})]]ds\\ &-\int_{\Gamma_k^{\sigma}} \mathbf{L}^c\nabla\mathbf{G}(\mathbf{x}-\mathbf{s})[[\boldsymbol{\sigma}(\mathbf{s})]]\cdot\mathbf{n}(\mathbf{s})ds,\end{aligned} \tag{11.71}$$

$$\begin{aligned}\mathcal{L}_k^{u}(\mathbf{x}-\mathbf{x}_k,\widehat{\boldsymbol{\tau}}) = &\int_w \nabla\mathbf{G}(\mathbf{x}-\mathbf{y})\boldsymbol{\tau}(\mathbf{y})dy + \int_w \mathbf{G}(\mathbf{x}-\mathbf{y})\mathbf{f}_1(\mathbf{y})dy\\ &-\int_{\Gamma_k^u} \mathbf{T}^{\top}(\mathbf{x}-\mathbf{s})[[\mathbf{u}(\mathbf{s})]]ds\\ &-\int_{\Gamma_k^{\sigma}} \mathbf{G}(\mathbf{x}-\mathbf{s})[[\boldsymbol{\sigma}(\mathbf{s})]]\cdot\mathbf{n}(\mathbf{s})ds,\end{aligned} \tag{11.72}$$

The perturbators $\mathcal{L}_k^{\epsilon}(\mathbf{x}-\mathbf{x}_k,\widehat{\boldsymbol{\tau}})$, $\mathcal{L}_k^{\sigma}(\mathbf{x}-\mathbf{x}_k,\widehat{\boldsymbol{\eta}})$, and $\mathcal{L}_k^{u}(\mathbf{x}-\mathbf{x}_k,\widehat{\boldsymbol{\tau}})$ can be found from joint solutions of Eqs. (11.67)–(11.72) and the contact conditions at the interfaces

Γ_k^u and Γ_k^σ. These contact conditions can be described by the limited cases (3.200)–(3.206) where only either the displacement jump or traction jump is presented. However, a generalization of these conditions to a formalism which is able to take into account both previous schemes at the same time is of strong practical importance [296, 473, 474, 578, 579, 1193]). In any way, the perturbators in the form defined by Eqs. (11.67)–(11.72) are also valid for subsequent applications.

For simplicity we will consider only internal points $\mathbf{x} \in w$ of the macrodomain w at sufficient distance from the boundary $\mathbf{S} \in \Gamma$ (7.7). The Cauchy data $[\mathbf{u}_0(\mathbf{s}), \boldsymbol{\sigma}_0(\mathbf{s})]$ at the smooth surface $\mathbf{s} \in \Gamma_0$ can be found from the conventional BIE taking the limit $\mathbf{x} \to \Gamma_0$

$$\mathbf{k}(\mathbf{s})\mathbf{u}_0(\mathbf{s}) = \int_{\Gamma_0} \Big[\mathbf{G}(\mathbf{s}-\boldsymbol{\zeta})\underline{\mathbf{t}}_0(\boldsymbol{\zeta}) - \mathbf{T}^\top(\mathbf{s}, \boldsymbol{\zeta})\mathbf{u}_0(\boldsymbol{\zeta})\Big] d\boldsymbol{\zeta} + \boldsymbol{\mathcal{L}}_k^u(\mathbf{s}-\mathbf{x}_k, \widehat{\boldsymbol{\tau}}) \tag{11.73}$$

at the boundary conditions (6.4) and (6.5); here $\mathbf{k}(\mathbf{x})$ is the free term coefficient (see, e.g., [354]).

It should be mentioned that Eqs. (11.67) and (11.68) with the integral representation of the perturbators $\boldsymbol{\mathcal{L}}_k^\epsilon(\mathbf{x}-\mathbf{x}_k, \widehat{\boldsymbol{\tau}})$ (11.70) and $\boldsymbol{\mathcal{L}}_k^\sigma(\mathbf{x}-\mathbf{x}_k, \widehat{\boldsymbol{\eta}})$ (11.71) can be obtained one from the another by the use of identities ($\mathbf{x} \in w$)

$$\boldsymbol{\varepsilon} = \mathbf{M}^c\boldsymbol{\sigma} + \boldsymbol{\beta}^c + \boldsymbol{\eta}, \quad \boldsymbol{\tau} = -\mathbf{L}^c\boldsymbol{\eta}. \tag{11.74}$$

Indeed, substitutions (11.74_1) and (11.74_2) into Eqs. (11.68) and (11.70), respectively, lead to (11.67) and (11.71). In so doing, the strain perturbator $\boldsymbol{\mathcal{L}}_k^\epsilon(\mathbf{x}-\mathbf{x}_k, \widehat{\boldsymbol{\tau}})$ and the stress perturbator $\boldsymbol{\mathcal{L}}_k^\sigma(\mathbf{x}-\mathbf{x}_k, \widehat{\boldsymbol{\eta}})$ are linked by the relation

$$\boldsymbol{\mathcal{L}}_k^\sigma(\mathbf{x}-\mathbf{x}_k, \widehat{\boldsymbol{\eta}}) = \boldsymbol{\tau}(\mathbf{x})V_k(\mathbf{x}) + \mathbf{L}^c\boldsymbol{\mathcal{L}}_k^\epsilon(\mathbf{x}-\mathbf{x}_k, \widehat{\boldsymbol{\tau}}), \tag{11.75}$$

which can be established by both methods: the substitution of identity (11.74_2) into Eq. (11.71) and the substitution of the constitutive law (6.2_1) into Eq. (11.68). Thus we have proved that Eqs. (11.67) and (11.68) are equivalent and the perturbators $\boldsymbol{\mathcal{L}}_k^\sigma(\mathbf{x}-\mathbf{x}_k, \widehat{\boldsymbol{\eta}})$ and $\boldsymbol{\mathcal{L}}_k^\epsilon(\mathbf{x}-\mathbf{x}_k, \widehat{\boldsymbol{\tau}})$ are related by the equality (11.75) independently on the concrete representations of these operators [e.g., either Eqs. (11.70), (11.71) or any other form of Eqs. (11.67) and (11.68)].

11.3.2 General Integral Equations

Let us consider an arbitrary random realization α of inclusions in the domain w

$$\begin{aligned}\boldsymbol{\varepsilon}(\mathbf{x}, \alpha) = \boldsymbol{\varepsilon}_0(\mathbf{x}, \alpha) &+ \int_w \boldsymbol{\mathcal{L}}_k^\epsilon(\mathbf{x}-\mathbf{x}_k, \widehat{\boldsymbol{\tau}})V_k^\delta(\mathbf{x}_k, \alpha)d\mathbf{x}_k \\ &+ \int_{\Gamma_0} \nabla\overset{S}{\otimes}\mathbf{G}(\mathbf{x}-\mathbf{s})\boldsymbol{\tau}^{[[u]]}(\mathbf{s}, \alpha)\mathbf{n}(\mathbf{s})d\mathbf{s},\end{aligned} \tag{11.76}$$

$$\boldsymbol{\sigma}(\mathbf{x}, \alpha) = \boldsymbol{\sigma}_0(\mathbf{x}, \alpha) + \int_w \mathcal{L}_k^{\sigma}(\mathbf{x} - \mathbf{x}_k, \widehat{\boldsymbol{\eta}}) V_k^{\delta}(\mathbf{x}_k, \alpha) d\mathbf{x}_k$$
$$- \int_{\Gamma_0} \mathbf{L}^c \nabla \mathbf{G}(\mathbf{x} - \mathbf{s}) \mathbf{L}^c \boldsymbol{\eta}^{[[u]]}(\mathbf{s}, \alpha) \mathbf{n}(\mathbf{s}) ds, \qquad (11.77)$$

where $\boldsymbol{\varepsilon}_0(\mathbf{x}, \alpha)$ and $\boldsymbol{\sigma}_0(\mathbf{x}, \alpha)$ can be defined through Eq. (11.61) in which the Cauchy data $[\mathbf{u}_0(\mathbf{s}, \alpha), \boldsymbol{\sigma}_0(\mathbf{s}, \alpha)]$ at $\mathbf{s} \in \Gamma_0$ depend on the total impacts of all perturbators $\mathcal{L}_k^u(\mathbf{s} - \mathbf{x}_k, \widehat{\boldsymbol{\tau}})$ [analogously to Eq. (11.70)] presented in the form of the volume integrals (11.64). The parameters

$$\boldsymbol{\tau}^{[[u]]}(\mathbf{s}, \alpha) \equiv \boldsymbol{\tau}(\mathbf{s}, \alpha) + \mathbf{L}_1(\mathbf{x})([[\mathbf{u}(\mathbf{x}, \alpha)]] \otimes \mathbf{n}(\mathbf{x}))\}\delta_{\Gamma_k^u}, \qquad (11.78)$$

$$\boldsymbol{\eta}^{[[u]]}(\mathbf{s}, \alpha) \equiv \boldsymbol{\eta}(\mathbf{s}, \alpha) - \mathbf{M}^c \mathbf{L}_1(\mathbf{x})([[\mathbf{u}(\mathbf{x}, \alpha)]] \otimes \mathbf{n}(\mathbf{x}))\}\delta_{\Gamma_k^u} \qquad (11.79)$$

are introduced in the surface integrals in (11.76) and (11.77) by using Gauss' theorem in Eq. (3.8) and taking into account the possibility of intersection of some part of heterogeneities v_k with the macroboundary Γ_0 where $\boldsymbol{\tau}^{[[u]]}(\mathbf{s}, \alpha)$, $\boldsymbol{\eta}^{[[u]]}(\mathbf{s}, \alpha)$ $\not\equiv \mathbf{0}$ $(\mathbf{s} \in \Gamma_0)$. In a particular case of floating heterogeneities with $\boldsymbol{\tau}(\mathbf{s}, \alpha), \boldsymbol{\eta}(\mathbf{s}, \alpha) \equiv \mathbf{0}$ $(\mathbf{s} \in \Gamma_0)$, the boundary integrals in Eq. (11.73) vanish.

It should be mentioned that the equations for both the strains (11.76) and stresses (11.77) are equivalent and can be obtained one from another by use of identities (11.74) and (11.75). This equivalence leads to the equivalence of the dual effective laws with reciprocally inverse effective properties for stiffness and compliance.

Now, we apply the centering method initially proposed in a particular form by Shermergor [1378] for statistically homogeneous media subjected to homogeneous boundary conditions. However, even in the mentioned simplified cases, the centering method has an advantage with respect to the method by O'Brian [1143] using the Gauss theorem applied to the asymptotic representation $(\boldsymbol{\tau}(\mathbf{s}) \equiv \langle\boldsymbol{\tau}\rangle)$ of the boundary integral corresponding to the boundary integral in Eq. (11.76). This centering procedure subtracting from both sides of the mentioned integral equation their statistical averages obtained without any auxiliary assumptions (e.g., EFH) leads to the following new equation for statistically inhomogeneous media

$$\boldsymbol{\varepsilon}(\mathbf{x}, \alpha) = \langle\boldsymbol{\varepsilon}\rangle(\mathbf{x}) + \int_w [\mathcal{L}_k^{\epsilon}(\mathbf{x} - \mathbf{x}_k, \widehat{\boldsymbol{\tau}}) V_k^{\delta}(\mathbf{x}_k, \alpha) - \langle\mathcal{L}_k^{\epsilon}(\mathbf{x} - \mathbf{x}_k, \widehat{\boldsymbol{\tau}})\rangle(\mathbf{x}_k)] d\mathbf{x}_k$$
$$+ \mathcal{L}_{\epsilon}^{\Gamma}, \qquad (11.80)$$

$$\boldsymbol{\sigma}(\mathbf{x}, \alpha) = \langle\boldsymbol{\sigma}\rangle(\mathbf{x}) + \int_w [\mathcal{L}_k^{\sigma}(\mathbf{x} - \mathbf{x}_k, \widehat{\boldsymbol{\eta}}) V_k^{\delta}(\mathbf{x}_k, \alpha) - \langle\mathcal{L}_k^{\sigma}(\mathbf{x} - \mathbf{x}_k, \widehat{\boldsymbol{\eta}})\rangle(\mathbf{x}_k)] d\mathbf{x}_k$$
$$+ \mathcal{L}_{\sigma}^{\Gamma}. \qquad (11.81)$$

In the right-hand side of Eqs. (11.80) and (11.81), assuming traction boundary conditions ($\Gamma_0 \equiv \Gamma_t$) (6.5), the integrals over the external surface Γ_0

$$\mathcal{L}^{\Gamma}_{\epsilon} \equiv \int_{\Gamma_0} \Big[\nabla \overset{S}{\otimes} \mathbf{G}(\mathbf{x}-\mathbf{s})\boldsymbol{\tau}^{[[u]]}(\mathbf{s},\alpha) - \big\langle \nabla \overset{S}{\otimes} \mathbf{G}(\mathbf{x}-\mathbf{s})\boldsymbol{\tau}^{[[u]]}\big\rangle(\mathbf{s})\Big]\mathbf{n}(\mathbf{s})ds$$
$$-\int_{\Gamma_0} \Big[\nabla \overset{S}{\otimes} \mathbf{T}^{\top}(\mathbf{x}-\mathbf{s})\mathbf{u}_0(\mathbf{s},\alpha) - \big\langle \nabla \overset{S}{\otimes} \mathbf{T}^{\top}(\mathbf{x}-\mathbf{s})\mathbf{u}_0\big\rangle(\mathbf{s})\Big]ds, \qquad (11.82)$$

$$\mathcal{L}^{\Gamma}_{\sigma} \equiv -\mathbf{L}^c \int_{\Gamma_0} \Big[\nabla \mathbf{G}(\mathbf{x}-\mathbf{s})\mathbf{L}^c \boldsymbol{\eta}^{[[u]]}(\mathbf{s},\alpha) - \langle \nabla \mathbf{G}(\mathbf{x}-\mathbf{s})\mathbf{L}^c \boldsymbol{\eta}^{[[u]]}\rangle(\mathbf{s})\Big]\mathbf{n}(\mathbf{s})ds$$
$$-\mathbf{L}^{(0)} \int_{\Gamma_0} \Big[\nabla \mathbf{T}^{\top}(\mathbf{x}-\mathbf{s})\mathbf{u}_0(\mathbf{s},\alpha) - \langle \nabla \mathbf{T}^{\top}(\mathbf{x}-\mathbf{s})\mathbf{u}_0\rangle(\mathbf{s})\Big]ds \qquad (11.83)$$

vanish at the sufficient distance $\mathbf{x}$ (assumed hereafter) from the boundary Γ_0 (7.7), when the validity of separation of length scale Eq. (7.10) holds. The first integrals in Eqs. (11.82) and (11.83) are the centered boundary integral terms in Eqs. (11.76) and (11.77), respectively. The second boundary integrals in Eqs. (11.82) and (11.83) are in fact the centered items $\boldsymbol{\varepsilon}^0(\mathbf{x},\alpha) - \langle \boldsymbol{\varepsilon}^0\rangle(\mathbf{x})$ and $\boldsymbol{\sigma}^0(\mathbf{x},\alpha) - \langle \boldsymbol{\sigma}^0\rangle(\mathbf{x})$ (see Eq. (11.61)), respectively, which can be different from zero in a "boundary layer" region close to the surface $\mathbf{s} \in \Gamma_0$ where boundary data $[\mathbf{u}_0(\mathbf{s}), \mathbf{t}_0(\mathbf{s})]$ not prescribed by the boundary conditions depend on perturbations introduced by all inhomogeneities, and, therefore $\boldsymbol{\varepsilon}_0(\mathbf{x}) = \boldsymbol{\varepsilon}_0(\mathbf{x},\alpha)$, $\boldsymbol{\sigma}_0(\mathbf{x}) = \boldsymbol{\sigma}_0(\mathbf{x},\alpha)$. Vanishing of all integrals in Eqs. (11.82) and (11.83) can be proved analogously to analysis in [233].

Thus, the last items $\mathcal{L}^{\Gamma}_{\epsilon}$ and $\mathcal{L}^{\Gamma}_{\sigma}$ in Eqs. (11.80) and (11.81), respectively, can be omitted. However, the remaining integrals in Eqs. (11.80) and (11.81) depend only on the perturbators $\mathcal{L}^{\epsilon}_{k}(\mathbf{x}-\mathbf{x}_k, \widehat{\boldsymbol{\tau}})$ and $\mathcal{L}^{\sigma}_{k}(\mathbf{x}-\mathbf{x}_k, \widehat{\boldsymbol{\eta}})$ presenting notations of the numerical solutions (11.67) and (11.68) which are currently general and not related with a concrete numerical method. We only need to know that $\mathcal{L}^{\epsilon}_{k}(\mathbf{x}-\mathbf{x}_k, \widehat{\boldsymbol{\tau}})$ and $\mathcal{L}^{\sigma}_{k}(\mathbf{x}-\mathbf{x}_k, \widehat{\boldsymbol{\eta}})$ behave as $O(|\mathbf{x}-\mathbf{x}_k|^{-d})$ at the infinity $|\mathbf{x}-\mathbf{x}_k| \to \infty$. The volume integrals in (11.80) and (11.81) converge absolutely for both the statistically homogeneous and inhomogeneous random set X of heterogeneities. Indeed, even for the FGMs, the integrands in the square brackets in Eqs. (11.80) and (11.81) are of order $O(|\mathbf{x}-\mathbf{x}_k|^{-2d})$ as $|\mathbf{x}-\mathbf{x}_k| \to \infty$, and the integrals in Eqs. (11.80) and (11.81) converge absolutely. For no *long-range* order assumed, the function $\varphi(v_j, \mathbf{x}_j|; v_i, \mathbf{x}_i) - \varphi(v_j, \mathbf{x}_j)$ decays at infinity sufficiently rapidly and guarantees an absolute convergence of the integrals involved. Therefore, for $\mathbf{x} \in w$ far enough from the boundary Γ_0 (7.7) (that vanishes the integrals in Eqs. (11.76) and (11.77)), the right-hand side integrals in (11.80) and (11.81) do not depend on the shape and size of the domain w, and they can be replaced by the integrals over the whole space R^d. With this assumption Eqs. (11.80) and (11.81) take the form

$$\boldsymbol{\varepsilon}(\mathbf{x},\alpha) = \langle \boldsymbol{\varepsilon}\rangle(\mathbf{x}) + \int [\mathcal{L}^{\epsilon}_{k}(\mathbf{x}-\mathbf{x}_k, \widehat{\boldsymbol{\tau}})V^{\delta}_{k}(\mathbf{x}_k,\alpha)$$
$$-\langle \mathcal{L}^{\epsilon}_{k}(\mathbf{x}-\mathbf{x}_k, \widehat{\boldsymbol{\tau}})\rangle(\mathbf{x}_k)]d\mathbf{x}_k, \qquad (11.84)$$

$$\boldsymbol{\sigma}(\mathbf{x},\alpha) = \langle\boldsymbol{\sigma}\rangle(\mathbf{x}) + \int [\mathcal{L}_k^\sigma(\mathbf{x}-\mathbf{x}_k,\widehat{\boldsymbol{\eta}})V_k^\delta(\mathbf{x}_k,\alpha)$$
$$-\langle\mathcal{L}_k^\sigma(\mathbf{x}-\mathbf{x}_k,\widehat{\boldsymbol{\eta}})\rangle(\mathbf{x}_k)]d\mathbf{x}_k, \qquad (11.85)$$

where the domain integration R^d will be omitted hereafter for simplicity of notation.

11.3.3 Simple Example of Residual Stresses

We consider residual stresses (called also the internal microstructural stresses) in elastically homogeneous matrix ($\mathbf{L}(\mathbf{x}) = \mathbf{L} \equiv$const., $\mathbf{x} \in R^d$) random structure composites with statistically homogeneous field of spherical identical inclusions v_i ($i = 1, 2, \ldots$) and $\langle\boldsymbol{\sigma}\rangle \equiv \mathbf{0}$. A simplified problem with zero elastic mismatch is deliberately chosen because in such a case it is not necessary to use one of the closing assumptions (e.g. "quasi-crystalline" approximation **H2b**) required at the consideration of composites with elastic mismatch. We analyze a simplest case of imperfect interface called dislocation-like model (see Sect. 3.7.3) when the local eigenstrains $\boldsymbol{\beta}_1(\mathbf{y})$ are presented by the Dirac distribution δ_{Γ_i} on the interfaces of inclusions $\mathbf{y} \in \Gamma_i$

$$\boldsymbol{\beta}_1(\mathbf{y}) = \mathbf{f}(\mathbf{n}_i(\mathbf{y})) \otimes \mathbf{n}_i(\mathbf{y})\delta_{\Gamma_i}, \qquad (11.86)$$

where $\mathbf{n}_i(\mathbf{y})$ is the normal vector on Γ_i and $\mathbf{f} = (f_1, \ldots, f_d)^\top$. The case (4.21) is a limiting asymptotic approach of the model with varying eigenstrains inside the inclusions (see for details [234]). In particular, for

$$\beta_{1|kl}(\mathbf{y}) = a[n_{i|k}(\mathbf{y})n_{i|l}(\mathbf{y}) - \frac{1}{3}\delta_{kl}n_{i|k}(\mathbf{y})n_{i|k}(\mathbf{y})]\delta_{\Gamma_i},$$

both the volume and surface averages over the inclusion v_i vanish for any a =const. (11.86).

$$\langle\boldsymbol{\beta}_1\rangle_i \equiv \langle\boldsymbol{\beta}_1\rangle_i^\Gamma \equiv \mathbf{0}, \qquad (11.87)$$

where $\langle(\cdot)\rangle_i^\Gamma = (\text{mes}\Gamma_i)^{-1}\int_{\Gamma_i}(\cdot)d\mathbf{y}$. In the framework of the hypotheses **H1** and **H3** (classical approach), the estimation of the average effective stresses $\langle\overline{\boldsymbol{\sigma}}\rangle_i(\mathbf{x})$ ($\mathbf{x} \in v_i$) are well-known (see for references [200])

$$\langle\overline{\boldsymbol{\sigma}}\rangle_i(\mathbf{x}) = \mathbf{Q}_i\langle\boldsymbol{\beta}_1\rangle \equiv \mathbf{0}, \qquad (11.88)$$

where $\mathbf{Q}_i = \mathbf{L}(\mathbf{I} - \mathbf{P}_i\mathbf{L})$ (11.86). However, the exact new approach estimation is

$$\langle\overline{\boldsymbol{\sigma}}\rangle_i(\mathbf{x}) = \int \mathcal{L}^\sigma(\mathbf{x}-\mathbf{x}_q,\boldsymbol{\beta}_1)[\varphi(v_q,\mathbf{x}_q|;v_i,\mathbf{x}_i) - \varphi(v_q,\mathbf{x}_q|;v_i,\mathbf{x}_i)]d\mathbf{x}_q, \qquad (11.89)$$

$$\mathcal{L}^\sigma(\mathbf{x}-\mathbf{x}_q,\boldsymbol{\beta}_1) = \int_{\Gamma_q} \boldsymbol{\Gamma}(\mathbf{x}-\mathbf{y})[\boldsymbol{\beta}_1^\delta(\mathbf{y}) - \boldsymbol{\beta}_1^\delta(\mathbf{x})]d\mathbf{y}, \qquad (11.90)$$

where $\boldsymbol{\beta}_1(\mathbf{y}) = \boldsymbol{\beta}_1^{\delta}(\mathbf{y})\delta_{\Gamma_q}$ $(\mathbf{y} \in v_q)$ and the integrals (4.33) and (4.34) converge absolutely. The estimation (11.89) presents a nonvanished effective stress $\langle\overline{\boldsymbol{\sigma}}\rangle_i(\mathbf{x}) \not\equiv \mathbf{0}$ and $\langle\overline{\boldsymbol{\sigma}}\rangle_i \neq \mathbf{0}$ that points to an infinite error of the prediction by the EFM $\langle\overline{\boldsymbol{\sigma}}\rangle_i = \mathbf{0}$ (11.88. Moreover, changing of a sign in the eigenstrain mismatch $\boldsymbol{\beta}_1(\mathbf{y}) \not\equiv \mathbf{0}$ $(\mathbf{y} \in v_q)$ remains fixed $\langle\overline{\boldsymbol{\sigma}}\rangle_i \equiv \mathbf{0}$ (11.88) and changes the signs in the estimations $\langle\overline{\boldsymbol{\sigma}}\rangle_i(\mathbf{x}) \not\equiv \mathbf{0}$ and $\langle\overline{\boldsymbol{\sigma}}\rangle_i \neq \mathbf{0}$ (11.89). Furthermore, a small perturbation of $\boldsymbol{\beta}_1(\mathbf{y}) \not\equiv \mathbf{0}$ $(\mathbf{y} \in v_q)$ can lead to any desirable sign of $\langle\overline{\boldsymbol{\sigma}}\rangle_i$ (11.88) while the signs of $\langle\overline{\boldsymbol{\sigma}}\rangle_i(\mathbf{x})$ and $\langle\overline{\boldsymbol{\sigma}}\rangle_i$ (11.89) are unaltered, and, therefore, the estimations (11.88) and (11.89) can have the different signers.

11.4 General Boundary Integral Equations of Stokes Flow Inporous Media

The problem of slow flow through assemblages of random field of fixed solid particles is a basic one for processes involving flow through porous media. There exists a large body of theoretical and experimental studies of Stokes flows in porous media, i.e., flows for which the Reynolds number is negligibly small. One of the general methods for solutions of such a class of problems is the boundary integral equation (BIE) technique, where, singular forces distributed, e.g., on the surface of particles, depend on the velocity of an ambient flow, thus yielding an integral equation for the singularity strengths. However, although the formulation of BIEs describing the flow has been known for many years [872], their effective solution was proposed later (see, e.g., [947, 1060]). There are also known numerical solutions for the many-body problem with the hydrodynamic interactions among particles of general shape (see, e.g. [412, 466, 871, 1060]) requiring development of advanced numerical techniques when two particles approach each other [777, 1391, 1449]. For dilute systems, two-body interactions may suffice, but at even relatively low-volume fractions of solids, many-body hydrodynamic interactions of infinite number of particles become important for both quantitative and qualitative behavior prediction.

Moreover, determining the interactions among an infinite number of hydrodynamically interacting particles is additionally complicated by the long-range nature of the interactions. The fluid velocity disturbance caused by a particle incorporated into the an ambient flow decays as $1/r$, where r is the distance from the particle. Therefore, realization of a direct superposition scheme of summations of these disturbances leads to the integrals which are not absolutely convergent at infinity, and therefore the hydrodynamic interactions cannot be truncated and no simple pairwise additive approximation can be proposed. Significance of these convergence difficulties is the generally recognized, and several procedures have been devised for overcoming them, resulting in well-posed, absolutely convergent expressions for the effective material and field parameters in mechanics and physics in the so-called general integral equation (GIE) connecting the random fields at the point being considered and the surrounding points. In particular, probably the most attractive method by O'Brian [1143] is very convenient for adaptation to simulation, and Brady and coauthors (see, e.g. [154, 155]) have been advocating its use as an exact procedure for treating the long-range hydrodynamic interactions. Diverse suspension properties as sedimentation, permeability, and rheology can all be treated with the same framework for any arrangement of particles. Brady et al. [155] have proposed generalization of the O'Brian [1143] method and proved the absolute convergence of the integral involved (in contrast with the methods by [24, 332, 504] and [871], requiring a special treatment of corresponding integrals).

Unfortunately, the known GIEs obtained by the classical methods [1143] and [1378] and recognized by Brady and coauthors (see, e.g. [154, 155]) as the exact ones are only approximate ones, and these approximations are the intrinsic feature of these GIEs based on asymptotic representation for the fluid velocity disturbances produced by the remote particles. The goal of the current section is obtaining of exact general boundary integral equations of Stokes flow in porous media.

11.4.1 Stokes Equations and Fundamental Solutions

Let a porous medium occupy an open bounded domain $w \subset R^d$ with a smooth boundary Γ^0 and with an indicator function W and space dimensionality d ($d = 2$ and $d = 3$ for $2D$ and $3D$ problems, respectively) contain a matrix (porous) $v^{(0)}$ and, in general, a statistically inhomogeneous set $X = (v_i)$ of fixed rigid heterogeneities v_i with indicator functions V_i and bounded by the closed smooth surfaces $\Gamma_i := \partial v_i$ ($i = 1, 2, \ldots$) defined by the relations $\Gamma_i(\mathbf{x}) = 0$ ($\mathbf{x} \in \Gamma_i$), $\Gamma_i(\mathbf{x}) > 0$ ($\mathbf{x} \in v_i$), and $\Gamma_i(\mathbf{x}) < 0$ ($\mathbf{x} \notin v_i$). It is assumed that the heterogeneities can be grouped into components (phases) $v^{(q)}$ $(q = 1, 2, \ldots, N)$ with identical mechanical and geometrical properties (such as the shape, size, orientation, and microstructure of heterogeneities). The particular cases of the mentioned structure for both the periodic and statistically homogeneous fields X were considered by [45, 154, 155, 332, 622, 1208, 1209, 1604]).

We first consider the local basic equations of Stokesian fluid dynamics. The velocity field $\mathbf{u}(\mathbf{x})$ at any point $\mathbf{x}$ defines the rate-of-strain tensor $\boldsymbol{\varepsilon}(\mathbf{x}) = [\nabla \otimes \mathbf{u} + (\nabla \otimes \mathbf{u})^\top)]/2$. The fluid stress tensor $\boldsymbol{\sigma}(\mathbf{x})$ in an isotropic Newtonian fluid with dynamic viscosity μ

$$\boldsymbol{\sigma}(\mathbf{x}) = \boldsymbol{\sigma}(\mathbf{u}, p)(\mathbf{x}) := -p(\mathbf{x})\boldsymbol{\delta} + 2\mu\boldsymbol{\varepsilon}(\mathbf{x}) \tag{11.91}$$

is expressed through the pressure $p(\mathbf{x})$ and $\boldsymbol{\varepsilon}(\mathbf{x})$; $\boldsymbol{\delta}$ is the unit second-order tensor. The governing equations for incompressible Stokes flow in a Newtonian fluid are

$$\nabla p = \mu \nabla^2 \mathbf{u}, \quad \nabla \cdot \mathbf{u} = 0, \tag{11.92}$$

where Eq. (11.91_1) derives from the Stokes-Navier equations by neglecting the convective acceleration terms in comparison to the viscous terms, $\mu\nabla^2\mathbf{u}$. This neglect is provided vanishingly by small particle Reynolds number, Re:= $\rho \mathbf{u}a/\mu \ll 1$, considered in the case of translating bodies or streaming flows; ρ is a fluid density. The continuity equation (11.91$_2$) expresses conservation of mass, so everywhere the velocity satisfies the solenoidal constraint. The interfaces between the constituent phases of CM are classically assumed to be perfect, i.e., the potential $\mathbf{u}$ and the traction $\mathbf{t}(\mathbf{x}) \equiv \boldsymbol{\sigma}(\mathbf{x}) \cdot \mathbf{n}(\mathbf{x})$ components are continuous across the interphase boundaries: $[[\mathbf{u}(\mathbf{x})]] = \mathbf{0}$, $[[\boldsymbol{\sigma}(\mathbf{x})]] \cdot \mathbf{n}(\mathbf{x}) = \mathbf{0}$. In particular, for the rigid inclusions being assumed, $\mathbf{u}(\mathbf{x}) = \mathbf{0}$ $(\mathbf{x} \in \Gamma^{(1)})$.

Consider the solution in the homogeneous fluid domain w without inclusions enclosed by the surface Γ^0. The Stokes equations (11.92$_1$) can be recast in integral form expressed through the fundamental solutions for the velocity and pressure (see [872, 1243, 1244]), satisfying

$$\mu \nabla^2 \mathcal{J}(\mathbf{x}, \mathbf{y}) - \nabla \mathcal{P}(\mathbf{x}, \mathbf{y}) = -2\omega_d \delta(\mathbf{x} - \mathbf{y})\boldsymbol{\delta}, \quad \text{div} \mathcal{J} = 0. \tag{11.93}$$

Here the compact representations of the fundamental solutions for both 2D and 3D cases are

$$\mathcal{J}(\mathbf{r}) = \gamma(r)\boldsymbol{\delta} + \frac{\mathbf{r}\mathbf{r}}{r^d}, \quad \mathcal{P}(\mathbf{r}) = -2\mu \nabla \gamma(r) = 2\mu \frac{\mathbf{r}}{r^d}, \tag{11.94}$$

where $\gamma(r) = -\ln r$ and $\gamma(r) = r^{-1}$ for $d = 2$ and $d = 3$, respectively, and ω_d is the surface area of the unit sphere in R^d; $r = |\mathbf{r}|$, $\mathbf{r} = \mathbf{x} - \mathbf{y}$. Here $\mathcal{J}$ is the free-space Green function or propagator for Stokes flow, known alternatively as the Stokeslet or Oseen tensor. In some literature, one encounters the mentioned terminology for the combination $\mathcal{J}/(2\omega_d \mu)$; see, e.g., [125, 657].

The boundary conditions at the perfect interface boundary are considered together with the mixed boundary conditions on the external boundary of w: $\Gamma^0 := \partial w$ with the unit outward normal $\mathbf{n}^\Gamma$

$$\mathbf{u}(\mathbf{x}) = \mathbf{u}^\Gamma(\mathbf{x}), \quad \mathbf{x} \in \Gamma_u, \tag{11.95}$$

$$\boldsymbol{\sigma}(\mathbf{x})\mathbf{n}^\Gamma(\mathbf{x}) = \mathbf{t}^\Gamma(\mathbf{x}), \quad \mathbf{x} \in \Gamma_t, \tag{11.96}$$

where Γ_u and Γ_t are prescribed displacement and traction boundaries such that $\Gamma_u \cup \Gamma_t = \Gamma^0$, $\Gamma_u \cap \Gamma_t = \emptyset$. $\mathbf{u}^\Gamma(\mathbf{x})$ and $\mathbf{t}^\Gamma(\mathbf{x})$ are, respectively, prescribed displacement on Γ_u and traction on Γ_t. Of special practical interest are the homogeneous boundary conditions (6.4) and (6.5).

The resulting boundary integral equations involving the boundary values of the velocity and traction can be expressed through the surface integrals over the enclosing surface Γ^0

$$\mathbf{c}(\mathbf{x})\mathbf{u}(\mathbf{x}) = \frac{1}{2\omega_d \mu} \int_{\Gamma^0} \Big[- \mathcal{J}(\mathbf{x} - \mathbf{s})\boldsymbol{\sigma}(\mathbf{u}, p)(\mathbf{s})\mathbf{n}(\mathbf{s}) + \mu \Big(\boldsymbol{\Sigma}(\mathbf{x} - \mathbf{s})\mathbf{n}(\mathbf{s}) \Big)^\top \mathbf{u}(\mathbf{s}) \Big] ds,$$

$$c_p(\mathbf{x}) p(\mathbf{x}) = \frac{1}{2\omega_d} \int_{\Gamma^0} [-\mathcal{P}(\mathbf{x} - \mathbf{s})\boldsymbol{\sigma}(\mathbf{u}, p) + 2\mu \boldsymbol{\Pi}(\mathbf{x} - \mathbf{s})\mathbf{u}(\mathbf{s})] \cdot \mathbf{n}(\mathbf{s}) ds, \tag{11.97}$$

where $\mathbf{s}$ is a point on the surface Γ^0 and $\mathbf{n}(\mathbf{s})$ is the surface normal vector pointing into the fluid. For $\mathbf{x} \in w\backslash\Gamma^0$ the coefficients $\mathbf{c}(\mathbf{x}) \equiv \boldsymbol{\delta}$, $c_p(\mathbf{x}) \equiv 1$, while for the boundary $\mathbf{x} \in \Gamma^0$, these coefficients are determined from the problem geometry; in particular, if $\mathbf{x}$ is on a smooth portion of the boundary, then $\mathbf{c}(\mathbf{x}) = \boldsymbol{\delta}/2$, $c_p(\mathbf{x}) = 1/2$. It is understood that the representation of p is unique only up to an additive constant. The triadic stress field $\boldsymbol{\Sigma} = \boldsymbol{\sigma}(\boldsymbol{\mathcal{J}}, -\boldsymbol{\mathcal{P}})$ of the Oseen tensor is directly expressed through the pressure field of the Oseen tensor $\boldsymbol{\mathcal{P}}$ and $\nabla\boldsymbol{\mathcal{J}}$ as

$$\Sigma_{ijk} = -\mathcal{P}_j\delta_{ik} + \mu(\mathcal{J}_{ij,k} + \mathcal{J}_{kj,i}) = -2d\mu\frac{r_ir_jr_k}{r^{d+2}} \tag{11.98}$$

which holds the equation

$$\frac{\partial}{\partial x_k}\Sigma_{ijk}(\mathbf{x}-\mathbf{s}) = -2\omega_d\delta_{ij}\delta(\mathbf{x}-\mathbf{s}). \tag{11.99}$$

The tensor $\boldsymbol{\Pi}(\mathbf{x}-\mathbf{s}) = \partial\boldsymbol{\mathcal{P}}(\mathbf{x}-\mathbf{s})/\partial\mathbf{n}(\mathbf{s})$ has the components

$$\Pi_{ij}(\mathbf{r}) = 2(d-1)\mu\Big[\frac{\delta_{ij}}{r^d} + d\frac{r_ir_j}{r^{d+2}}\Big]. \tag{11.100}$$

The solution of Eqs. (11.97) can be represented in an operator form (in an analogy with a grand mobility matrix; see [412] and [154])

$$\begin{pmatrix}\mathbf{u}\\ p\end{pmatrix} = \begin{pmatrix}\boldsymbol{\mathcal{L}}_0^{ut} & \boldsymbol{\mathcal{L}}_0^{uu}\\ \boldsymbol{\mathcal{L}}_0^{pt} & \boldsymbol{\mathcal{L}}_0^{pu}\end{pmatrix}\begin{pmatrix}\mathbf{t}^{\Gamma}\\ \mathbf{u}^{\Gamma}\end{pmatrix}, \tag{11.101}$$

where the pairs $(\boldsymbol{\mathcal{L}}_0^{ut}, \boldsymbol{\mathcal{L}}_0^{pt})$ and $(\boldsymbol{\mathcal{L}}_0^{uu}, \boldsymbol{\mathcal{L}}_0^{pu})$ are the respective single- and double-layer hydrodynamic potentials defined by $(\mathbf{x} \in w \cup \Gamma^0)$

$$\boldsymbol{\mathcal{L}}_0^{ut}(\cdot)(\mathbf{x}) := -\frac{1}{2\omega_d\mu\mathbf{c}(\mathbf{x})}\int_{\Gamma^0}\boldsymbol{\mathcal{J}}(\mathbf{x}-\mathbf{s})(\cdot)ds, \tag{11.102}$$

$$\boldsymbol{\mathcal{L}}_0^{pt}(\cdot)(\mathbf{x}) := -\frac{1}{2\omega_d c_p(\mathbf{x})}\int_{\Gamma^0}\boldsymbol{\mathcal{P}}(\mathbf{x}-\mathbf{s})(\cdot)ds, \tag{11.103}$$

$$\boldsymbol{\mathcal{L}}_0^{uu}(\cdot)(\mathbf{x}) := \frac{1}{2\omega_d\mu\mathbf{c}(\mathbf{x})}\int_{\Gamma^0}\Big(\boldsymbol{\Sigma}(\mathbf{x}-\mathbf{s})\mathbf{n}(\mathbf{s})\Big)^{\top}(\cdot)ds, \tag{11.104}$$

$$\boldsymbol{\mathcal{L}}_0^{pu}(\cdot)(\mathbf{x}) := \frac{\mu}{\omega_d c_p(\mathbf{x})}\int_{\Gamma^0}\Big(\boldsymbol{\Pi}(\mathbf{x}-\mathbf{s})(\cdot)\Big)\mathbf{n}(\mathbf{s})ds. \tag{11.105}$$

In (11.101) the boundary conditions are the Cauchy data $\mathbf{u}(\mathbf{s}) = \mathbf{u}^{\Gamma}(\mathbf{s})$ (11.95) and $\mathbf{t}(\mathbf{s}) := \boldsymbol{\sigma}(\mathbf{u}, p)(\mathbf{s}) \cdot \mathbf{n}(\mathbf{s}) = \mathbf{t}^{\Gamma}(\mathbf{s})$ (11.96) $(\mathbf{s} \in \Gamma^0)$ of the solution to the Stokes equations (11.92).

The representation (11.101) is valid for both the general cases of the first and second boundary-value problems as well as for the mixed boundary-value problem (11.95) and (11.96). In particular, for the conditions (11.95) and (11.96), the first row in the right-hand of Eq. (11.101) can be considered as a continuation of $\boldsymbol{\varepsilon}^{\Gamma}(\mathbf{x})$, $\mathbf{x} \in \Gamma_u \equiv \Gamma$, i.e., (11.95), into w as the velocity field that the boundary condition (11.95) and (11.96) would generate in the fluid medium with homogeneous viscosity μ.

11.4.2 Single Rigid Inclusion in a Macrodomain

For a single rigid inclusion v_i placed at sufficient distance $\mathbf{x}$ from the macroboundary Γ^0 $a \ll |\mathbf{x}-\mathbf{s}|$ and $\forall \mathbf{s} \in \Gamma$ (7.7), Eq. (11.101) can be recast in the form (see [776, 947])

$$\mathbf{u}(\mathbf{x}) - \mathbf{u}^{\infty}(\mathbf{x}) := \mathcal{L}_i^{ut}\mathbf{t}^{\Gamma_i} = \mathcal{L}_i^{ut}(\mathbf{x}-\mathbf{x}_i, \mathbf{t}^{\Gamma}), \tag{11.106}$$

$$p(\mathbf{x}) - p^{\infty}(\mathbf{x}) := \mathcal{L}_i^{pt}\mathbf{t}^{\Gamma_i} = \mathcal{L}_i^{pt}(\mathbf{x}-\mathbf{x}_i, \mathbf{t}^{\Gamma}), \tag{11.107}$$

which demonstrates that the flow past a rigid particle is represented with just single-layer potential. Here a pair ($\mathbf{u}^{\infty}(\mathbf{x})$ and $p^{\infty}(\mathbf{x})$) of an ambient flow field is presented by the solution of Eq. (11.101) for the homogeneous macrodomain w in the absence of the particle. In fact, the right-hand sides of Eqs. (11.106) and (11.107) define the *perturbators* (or disturbances) $\mathcal{L}_i^{ut}(\mathbf{x}-\mathbf{x}_i, \mathbf{t}^{\Gamma})$ and $\mathcal{L}_i^{pt}(\mathbf{x}-\mathbf{x}_i, \mathbf{t}^{\Gamma})$ which produce the perturbations of the fields $\mathbf{u}(\mathbf{x})$ and $p(\mathbf{x})$, respectively, in the point $\mathbf{x}$ due to insertion of the heterogeneity center into the point $\mathbf{x}_i$. The submatrices and boundary charge with the index Γ_i for the inclusion surfaces in Eqs. (11.106) and (11.107) are defined analogously to the corresponding values with the index Γ^0 by the use of the replacement $\Gamma^0 \to \Gamma_i$ in the consistent representations for the macrodomain boundary Γ^0 (see, e.g., Eqs. (11.102) and (11.103)). Because the particles are rigid, there is no velocities within the particle, and, hence, the double-layer hydrodynamic potentials vanish for rigid particles v_i in Eqs. (11.106) and (11.107); $\mathbf{t}^{\Gamma_i} = \boldsymbol{\sigma}(\mathbf{s})\mathbf{n}(\mathbf{s})$ ($\mathbf{s} \in \Gamma_i$) is the force density at the point $\mathbf{s}$ on the surface of the particle v_i and $\mathbf{n}(\mathbf{s})$ is the surface normal vector pointing into the fluid. Thus, the disturbance fields (11.106_1) and (11.107_1) are represented by the surface distributions of point forces whose strength densities are the boundary traction.

11.4.3 General Integral Equation for Microinhomogeneous Infinite Medium

For the domain w containing the rigid stationary particles v_j ($j = 1, 2, \ldots$), Eqs. (11.106) and (11.107) can be transformed to

$$\begin{pmatrix} \mathbf{u} \\ p \end{pmatrix} = \sum_{i=1} \begin{pmatrix} \mathcal{L}_j^{ut}\mathbf{t}^{\Gamma_j} \\ \mathcal{L}_j^{pt}\mathbf{t}^{\Gamma_j} \end{pmatrix} + \begin{pmatrix} \mathcal{L}_0^{ut} & \mathcal{L}_0^{uu} \\ \mathcal{L}_0^{pt} & \mathcal{L}_0^{pu} \end{pmatrix} \begin{pmatrix} \mathbf{t}^{\Gamma} \\ \mathbf{u}^{\Gamma} \end{pmatrix}, \tag{11.108}$$

where the summation indicates that the integration is to be performed overall particle surfaces within w.

We replace the summation in Eq. (11.108) by the volume integral over the increasing sequence w_p at an arbitrary random realization α of inclusions ($\mathbf{x} \in v^{(0)} \cup \Gamma^{(1)}$)

$$\begin{pmatrix} \mathbf{u}(\mathbf{x}) \\ p(\mathbf{x}) \end{pmatrix} = \begin{pmatrix} \int_{w_p} \mathcal{L}_j^{ut}(\mathbf{x}-\mathbf{x}_j, \mathbf{t}) V_j^{\delta}(\mathbf{x}_j, \alpha) d\mathbf{x}_j \\ \int_{w_p} \mathcal{L}_j^{pt}(\mathbf{x}-\mathbf{x}_j, \mathbf{t}) V_j^{\delta}(\mathbf{x}_j, \alpha) d\mathbf{x}_j \end{pmatrix} + \begin{pmatrix} \mathcal{L}_0^{ut} & \mathcal{L}_0^{uu} \\ \mathcal{L}_0^{pt} & \mathcal{L}_0^{pu} \end{pmatrix} \begin{pmatrix} \mathbf{t}^{\Gamma} \\ \mathbf{u}^{\Gamma} \end{pmatrix}, \tag{11.109}$$

Now we center Eq. (11.109) by the use of statistical averages, i.e., from both sides of Eq. (11.109) their statistical averages are subtracted

$$\mathbf{u}(\mathbf{x}, \alpha) = \langle \mathbf{u} \rangle(\mathbf{x}) + \lim_{w_p \to R^d} \int_{w_p} [\mathcal{L}_j^{ut}(\mathbf{x}-\mathbf{x}_j, \mathbf{t}) V_j^{\delta}(\mathbf{x}_j, \alpha) - \langle \mathcal{L}_j^{ut}(\mathbf{x}-\mathbf{x}_j, \mathbf{t}) \rangle (\mathbf{x}_j)] d\mathbf{x}_j + \mathcal{L}_0^{u}(\mathbf{x}), \tag{11.110}$$

$$p(\mathbf{x}, \alpha) = \langle p \rangle(\mathbf{x}) + \lim_{w_p \to R^d} \int_{w_p} [\mathcal{L}_j^{pt}(\mathbf{x}-\mathbf{x}_j, \mathbf{t}) V_j^{\delta}(\mathbf{x}_j, \alpha) - \langle \mathcal{L}_j^{ut}(\mathbf{x}-\mathbf{x}_j, \mathbf{t}) \rangle (\mathbf{x}_j)] d\mathbf{x}_j + \mathcal{L}_0^{p}(\mathbf{x}). \tag{11.111}$$

No divergences occur in (11.110) and (11.111) for a finite region w_p bounded by the surface Γ_p^0. This is an arbitrary surface immersed in an unbounded medium either statistically homogeneous or inhomogeneous, i.e., the field X continues outside of Γ_p^0. Without loss of generality, we consider a case of floating particle field X in w_p. The domain w in Eqs. (11.110) and (11.111) was replaced on w_p for definition of the improper integral over the whole space R^d (11.106) as a limit ($p \to \infty$) of definite integrals over an increasing sequence w_p filling R^d. We prove that these limits exist independently on the particular choice of closed bounded regions w_p in R^d such that any w_p is contained in every w_{p_1} for which $p_1 > p$ and for any bounded region w' in R^d, there is p_2 such that $w' \subset w_{p_2}$.

Justification of absolute convergence of the integral in Eq. (11.110) (the integral in Eq. (11.111) is considered analogously) at the condition (11.110) was performed in [214] similarly to Sect. 7.2.1.

We now turn our attention to the analysis of the term $\mathcal{L}_0^{u} = \mathcal{L}_0^{ut}\mathbf{t}^{\Gamma} + \mathcal{L}_0^{uu}\mathbf{u}^{\Gamma}$ (11.110) ($\mathcal{L}_0^{p}$ (11.111) can be considered analogously) defined by the submatrices $\mathcal{L}_0^{ut}$ and $\mathcal{L}_0^{uu}$. In particular, for the summand $\mathcal{L}_0^{ut}\mathbf{t}^{\Gamma}$ (the summand $\mathcal{L}_0^{uu}\mathbf{u}^{\Gamma}$ can be analyzed in a similar manner), we have the following surface centered integral

$$\begin{aligned}\mathcal{L}_0^{ut}\mathbf{t}^{\Gamma} = & \lim_{w_p\to R^d}\int_{\Gamma_p^0}\langle\!\langle \mathcal{J}(\mathbf{x}-\mathbf{s})\boldsymbol{\sigma}(\mathbf{u},p)(\mathbf{s})\mathbf{n}(\mathbf{s})\rangle\!\rangle ds \\ = & \lim_{w_p\to R^d}\int_{\Gamma_{up}^0}\langle\!\langle \mathcal{J}(\mathbf{x}-\mathbf{s})\boldsymbol{\sigma}(\mathbf{u},p)(\mathbf{s})\mathbf{n}(\mathbf{s})\rangle\!\rangle ds \\ & + \lim_{w_p\to R^d}\int_{\Gamma_{tp}^0}\langle\!\langle \mathcal{J}(\mathbf{x}-\mathbf{s})\boldsymbol{\sigma}(\mathbf{u},p)(\mathbf{s})\mathbf{n}(\mathbf{s})\rangle\!\rangle ds, \end{aligned} \tag{11.112}$$

where one introduces a centering operation $\langle\!\langle(\cdot)\rangle\!\rangle(\mathbf{y})$ (7.12) and the boundary integral in the left-hand side of Eq. (11.112) over the external surface Γ_p^0 is subdivided over the integrals over Γ_{up}^0 and Γ_{tp}^0 ($\Gamma_p^0 = \Gamma_{up}^0 \cup \Gamma_{tp}^0$) with prescribed boundary conditions for $\mathbf{u}(\mathbf{s})$ and $\mathbf{t}(\mathbf{s})$ at Γ_{up}^0 and Γ_{tp}^0, respectively. The second integral in right-hand side of Eq. (11.112) equals zero independently on the size of w_p because $\mathbf{t}(\mathbf{s})$ is a deterministic function at $\mathbf{s} \in \Gamma_{tp}^0$, and, therefore, $\mathbf{t}(\mathbf{s}) \equiv \langle\mathbf{t}\rangle(\mathbf{s})$ ($\mathbf{s} \in \Gamma_{tp}^0$). The first integral in the right-hand side of Eq. (11.112) can be also dropped out, because this integral vanishes only asymptotically at both $w_p \to \infty$ and sufficient distance $\mathbf{x}$ from the boundary Γ_p^0 (7.7) (it was proved in [214] analogously to Sect. 7.2.1).

Thus, Eqs. (11.110) and (11.111) can be reduced to the new *general integral equations* (GIE) ($\mathbf{x} \in \Gamma_i$)

$$\boldsymbol{\vartheta}(\mathbf{x}) = \langle\boldsymbol{\vartheta}\rangle(\mathbf{x}) + \int \big[\mathcal{L}_j^{\theta t}(\mathbf{x}-\mathbf{x}_j,\mathbf{t})V_j^{\delta}(\mathbf{x}_j) - \langle\mathcal{L}_j^{\theta t}(\mathbf{x}-\mathbf{x}_j,\mathbf{t})\rangle(\mathbf{x}_j)\big]d\mathbf{x}_j \tag{11.113}$$

where the integration domain R^d and the label α are omitted hereafter for simplicity of notations. For contraction of calculations, we introduce the substitution

$$(\mathbf{u}, p) \leftrightarrow \boldsymbol{\vartheta} \tag{11.114}$$

indicating that for each $\boldsymbol{\vartheta} = \mathbf{u}$ and $\boldsymbol{\vartheta} = p$, Eq. (11.113) and the subsequent equations with the variable $\boldsymbol{\vartheta}$ are reduced to the corresponding equations for $\mathbf{u}$ and p. Obviously, that differentiation and symmetrization of Eq. (11.110) allow to recast Eqs. (11.110) and (11.111) in the form (11.113) where the substitution

$$(\boldsymbol{\varepsilon}, p) \leftrightarrow \boldsymbol{\vartheta} \tag{11.115}$$

is used.

Representation of Eq. (11.113) in the terms $(\boldsymbol{\varepsilon}, p)$ allows one to recast Eq. (11.113) for the stresses $\boldsymbol{\sigma}(\mathbf{u}, p)$ (11.91) and the traction $\mathbf{t}(\mathbf{x}) = \boldsymbol{\sigma}(\mathbf{u}, p)\mathbf{n}(\mathbf{x})$ ($\mathbf{x} \in \Gamma_i$)

$$\mathbf{t} \leftrightarrow \boldsymbol{\vartheta}, \tag{11.116}$$

where the perturbator ($\mathbf{x} \in \Gamma_i$)

$$\mathbf{t}(\mathbf{x}) - \mathbf{t}^{\infty}(\mathbf{x}) := \mathcal{L}_j^{tt}(\mathbf{x}-\mathbf{x}_j, \mathbf{t}) = \left(\nabla \mathcal{L}_j^{ut}(\mathbf{x}-\mathbf{x}_j, \mathbf{t}) - \mathcal{L}_j^{pt}(\mathbf{x}-\mathbf{x}_j, \mathbf{t})\right)^{\top} \mathbf{n}(\mathbf{x}) \tag{11.117}$$

is expressed through the perturbators $\mathcal{L}_j^{ut}(\mathbf{x}-\mathbf{x}_j, \mathbf{t})$ and $\mathcal{L}_j^{pt}(\mathbf{x}-\mathbf{x}_j, \mathbf{t})$, $\mathbf{n}(\mathbf{x})$ is the outward normal vector on $\mathbf{x} \in \Gamma_i$ from v_i to $v^{(0)}$, and $\mathbf{t}^{\infty}(\mathbf{x}) := -p^{\infty}(\mathbf{x})\boldsymbol{\delta} + \mu[\nabla \otimes \mathbf{u}^{\infty}(\mathbf{x}) + (\nabla \otimes \mathbf{u}^{\infty}(\mathbf{x}))^{\top}]$. In the case of the representations of $\mathcal{L}_j^{ut}(\mathbf{x}-\mathbf{x}_j, \mathbf{t})$ and $\mathcal{L}_j^{pt}(\mathbf{x}-\mathbf{x}_j, \mathbf{t})$ through the Green functions (11.106) and (11.107), respectively, the perturbator $\mathcal{L}_j^{tt}(\mathbf{x}-\mathbf{x}_j, \mathbf{t})$ is also expressed through the Green functions. It should be mentioned that Eq. (11.113) for $\boldsymbol{\vartheta} = \mathbf{t}$ (11.116) is the Fredholm integral equation of the second kind in terms of the unknown surface traction, rather than the ill-posed velocity formulation of the Fredholm integral equation of the first kind (11.113) for $\boldsymbol{\vartheta} = \mathbf{u}$ (11.114).

The coupled infinite system of the new GIEs for corresponding effective fields was obtained in [214] analogously to Eqs. (11.22)–(11.24). In particular, introduction of the effective field makes it possible to define another sort of the perturbators

$$\mathcal{L}_j^{\theta\theta}(\mathbf{x}-\mathbf{x}_j, \overline{\boldsymbol{\vartheta}}_j) \equiv \boldsymbol{\vartheta}(\mathbf{x}) - \overline{\boldsymbol{\vartheta}}_j(\mathbf{x}). \tag{11.118}$$

In such a case, the GIEs (11.113) can be transformed to the new GIEs in terms of effective fields ($\mathbf{x} \in \Gamma_k,\ k = 1, 2, \ldots, n$)

$$\begin{aligned} &\langle \overline{\boldsymbol{\vartheta}} | v_1, \mathbf{x}_1; \ldots; v_n, \mathbf{x}_n\rangle(\mathbf{x}) - \sum_{j\neq i}^{n} \langle \mathcal{L}_j^{\theta\theta}(\mathbf{x}-\mathbf{x}_j, \overline{\boldsymbol{\vartheta}}) | v_1, \mathbf{x}_1; \ldots; v_n, \mathbf{x}_n\rangle_j \\ &= \langle \boldsymbol{\vartheta}\rangle(\mathbf{x}) + \int \{\langle \mathcal{L}_j^{\theta\theta}(\mathbf{x}-\mathbf{x}_j, \overline{\boldsymbol{\vartheta}})|; v_1, \mathbf{x}_1; \ldots; v_n, \mathbf{x}_n\rangle_j \varphi(v_j, \mathbf{x}_j|; v_1, \mathbf{x}_1, \ldots, v_n, \mathbf{x}_n) \\ &\quad - \langle \mathcal{L}_j^{\theta\theta}(\mathbf{x}-\mathbf{x}_j, \overline{\boldsymbol{\vartheta}})\rangle(\mathbf{x}_j)\} d\mathbf{x}_j. \end{aligned} \tag{11.119}$$

Strictly speaking, the perturbators $\mathcal{L}_j^{\theta t}(\mathbf{x}-\mathbf{x}_j, \mathbf{t})$ (11.106) and (11.107) and their integral representations (11.102) and (11.103), respectively, are just the notations of some problem which is destined to be solved, while the perturbator $\mathcal{L}_j^{\theta\theta}(\mathbf{x} - \mathbf{x}_j, \overline{\boldsymbol{\vartheta}}_j)$ (11.118) is the solutions of this problem. Analytical representations for the operator $\mathcal{L}_j^{\theta\theta}(\mathbf{x}-\mathbf{x}_j, \overline{\boldsymbol{\vartheta}}_j)$ are only known for the particular canonical shapes of particle and specific effective fields. So, for a homogeneous flow $\overline{\mathbf{u}}(\mathbf{x})$ and a spherical (and cylindrical) single inclusion v_j of radius a, the operators $\mathcal{L}_j^{\theta\theta}(\mathbf{x} - \mathbf{x}_j, \overline{\boldsymbol{\vartheta}}_j)$ are reduced to the tensorial products $\mathcal{L}_j^{\theta\theta}(\mathbf{x}-\mathbf{x}_j, \overline{\boldsymbol{\vartheta}}_j) = \mathcal{L}_j^{\theta\theta}(\mathbf{x}-\mathbf{x}_j)\overline{\mathbf{u}}(\mathbf{x}_j)$ where the tensors $\mathcal{L}_j^{\theta\theta}(\mathbf{x}-\mathbf{x}_j)$ are expressed through the Oseen tensor, and it's the second derivative (see for details [24, 877, 1341]).

Furthermore, although the perturbator $\mathcal{L}_j^{\theta t}(\mathbf{x} - \mathbf{x}_j, \mathbf{t})$ is defined through the Green's functions (see Eqs. (11.102) and (11.103)), it can be formally found as a numerical representation for the perturbation (see Eqs. (11.106) and (11.107)) introduced in the point $\mathbf{x}$ due to insertion of the heterogeneity center into the point $\mathbf{x}_j$. The mentioned perturbators (11.106_1) and (11.107_1) can be found by any available numerical method, such as boundary element method (BEM), finite element method (FEM), hybrid FEM–BEM, multipole expansion method, complex potential method, and others (see for references [45, 1244, 1604]). Each method has advantages and disadvantages, and it is crucial for the analyst to be aware of their range of applications. A distinct advantage of the BEMs compared to the FEMs methods is that the BEMs require meshing only the boundary surface of computational domain as opposed to the entire 3D domain. From other side, the FEM is supported by well-developed commercial softwares and gives strong advantages in terms of CPU time. The FEA is especially effective for estimations of perturbators at the constant fields.

Thus, the GIEs (11.113) and (11.119) can be considered as the general operator equations where the operators $\boldsymbol{\mathcal{L}}_j^{\theta t}(\mathbf{x}-\mathbf{x}_j,\mathbf{t})$ and $\boldsymbol{\mathcal{L}}_j^{\theta\theta}(\mathbf{x}-\mathbf{x}_j,\overline{\boldsymbol{\vartheta}})$ can be found by any available numerical method.

11.4.4 Known GIEs as the Approximations of the New Exact GIEs

We proved in Chap. 7 that the classical GIEs (7.19) and (7.20) are just approximation of the new exact GIEs (7.19) and (7.20) obtained in the framework of the EFH **H1b** (7.47) and (7.48). We will show in this subsection that the new exact GIEs (11.113) and (11.119) in filtration problems are asymptotically reduced to the known approximative GIEs proposed by Brady and coauthors ([155, 1391], see also [294], [458]). Namely, a filtration counterpart of the EFH **H1b** (7.47) and (7.48) in elasticity of CMs can be presented as a simplification of the single-layer potential (11.113) at $\theta=\mathbf{u}$

$$\langle\boldsymbol{\mathcal{L}}_j^{ut}(\mathbf{x}-\mathbf{x}_j,\mathbf{t})\rangle(\mathbf{x}_j)=\frac{n}{2\omega_d\mu}\boldsymbol{\mathcal{J}}(\mathbf{x}-\mathbf{x}_j)\langle\mathbf{F}^{\Gamma}\rangle(\mathbf{x}_j), \tag{11.120}$$

where the drag force

$$\langle\mathbf{F}^{\Gamma}\rangle(\mathbf{x}_j)=\int_{\Gamma_j}\langle\boldsymbol{\sigma}\rangle_j(\mathbf{s})\mathbf{n}(\mathbf{s})ds \tag{11.121}$$

at the representative particle v_j (with the center $\mathbf{x}_j$) does not depend on the details of both the particle shape and traction distribution at the particle surface. The approximations similar to (11.120) are widely used in order to avoid the expensive construction of the far-field resistance matrix by the uses of direct calculation of the far-field hydrodynamic force (see, e.g. [155, 1391]). Substitution of the approximation (11.120) into Eq. (11.113) at $\theta=\mathbf{u}$ yields

$$\mathbf{u}(\mathbf{x}) = \langle \mathbf{u} \rangle(\mathbf{x}) + \int [\mathcal{L}_j^{ut}(\mathbf{x}-\mathbf{x}_j, \mathbf{t}) V_j^\delta(\mathbf{x}_j) - \frac{n}{2\omega_d \mu} \mathcal{J}(\mathbf{x}-\mathbf{x}_j) \langle \mathbf{F}^\Gamma \rangle (\mathbf{x}_j)] d\mathbf{x}_j. \tag{11.122}$$

An approximation of the first summand in the integrand of Eq. (11.122) in a spirit of Eq. (11.120) defines the next simplification of Eq. (11.122)

$$\mathbf{u}(\mathbf{x}) = \langle \mathbf{u} \rangle(\mathbf{x}) + \frac{1}{2\omega_d \mu} \int \mathcal{J}(\mathbf{x}-\mathbf{x}_j) [\mathbf{F}^\Gamma(\mathbf{x}_j) V_j^\delta(\mathbf{x}_j) - n \langle \mathbf{F}^\Gamma \rangle (\mathbf{x}_j)] d\mathbf{x}_j. \tag{11.123}$$

Here $\mathbf{F}^\Gamma(\mathbf{x}_j)$ is a random drag force at the particle v_j estimated by Eq. (11.121) without operation of statistical average $\langle(\cdot)\rangle(\mathbf{x}_j)$. It should be mentioned that Eq. (11.123) was obtained by the use of more restricted assumptions than their elastic counterpart (7.47) and (7.48) proposed for the asymptotic approximation (7.51) only for the renormalyzing term. Equation (11.121) for statistically homogeneous media subjected to homogeneous loading (when $\langle \mathbf{F}^\Gamma \rangle(\mathbf{x}_j) = \langle \mathbf{F}^\Gamma \rangle \equiv$const.) coincides with the equations obtained in [155] and [154] by the use of generalization of the O'Brian's [1143] method. Thus, we established that the exact GIE (11.113) proposed is reduced to the known approximate equation (7.32) at some additional assumptions (see, e.g. (11.120)) for statistically homogeneous media subjected to homogeneous loading.

We may also obtain higher-order corrections of the renormalizing term $\langle \mathcal{L}_k^{\theta t}(\mathbf{x} - \mathbf{x}_k, \mathbf{t}) \rangle (\mathbf{x}_k)$ in which the coefficients are moments of the surface traction $\mathbf{t}^{\Gamma_k}$ and thus obtain the multipole expansion. Formally, we set $|\mathbf{x}-\mathbf{x}_k| \gg |\mathbf{s}|$ and take the Taylor series of $\mathcal{J}(\mathbf{x}-\mathbf{x}_k-\mathbf{s})$ in $\mathbf{s}$ about $\mathbf{s} = \mathbf{0}$

$$\langle \mathcal{L}_{k|i}^{\theta t}(\mathbf{x}-\mathbf{x}_k, \mathbf{t}) \rangle (\mathbf{x}_k) = \frac{1}{2\omega_d \mu} \sum_{n=0}^{\infty} \int_{\Gamma_k} \langle t_j s_{k_1} \ldots s_{k_n} \rangle (\mathbf{s}) d\mathbf{s} \mathcal{J}_{ij,k_1\ldots k_n}(\mathbf{x}-\mathbf{x}_k). \tag{11.124}$$

For the deterministic problem (without of statistical average operation $\langle(\cdot)\rangle(\mathbf{s})$), Eq. (11.124) provides a representation for the field perturbation (or disturbance) $\boldsymbol{\vartheta}(\mathbf{x}) - \overline{\boldsymbol{\vartheta}}_k(\mathbf{x})$ at any point in the fluid in terms of a multipole expansion. The leading term in the right-hand side of Eq. (11.124), or monopole, is the zero moment with the vector coefficient $\mathbf{F}^\Gamma$ describing the total force and corresponding to an integral of the Stokeslet distribution. The first moment or dipole, a second-order tensor, contains antisymmetric and symmetric parts, which are identified as the total torque and the stresslet. These lowest-order multipole moments were then used in [412] (see also [294], where bimodal suspensions are analyzed) to describe the velocity disturbance produced by each sphere and to form with the aid of Faxen's formulas the grand mobility matrix for the N-sphere system. In fact the grand mobility matrix has been truncated at the level of force dipoles or stresslets. This is a far-field approximation to the interaction among particles including terms up to $O(r^{-5})$, where r is a characteristic antiparticle spacing. For isolated fixed inclusion v_k, the decomposition (11.124) makes it possible to derive an accurate approximation of the left-hand side of Eq. (11.124) by retaining of a large enough number of decomposition terms (see, e.g. [776]). However, for the random fields, no constructive and efficient way of estimation of statistical averages $\langle t_j s_{k_1} \ldots s_{k_n} \rangle(\mathbf{s})$ of the solution was found, and, therefore, a practical effectiveness of this scheme in Eq. (11.119) is questionable.

Thus, a construction of the renormalization terms in the filtration counterpart of Eqs. (7.32) based on the transformation of the macrosurface integral in the method by O'Brian [1143] (see [155, 412]) is only asymptotically fulfilled. These asymptotic representations (e.g., (7.51) and (11.120)) provide both the appropriate accuracy and absolute convergence of integrals involved at $|\mathbf{x}-\mathbf{x}_k| \to \infty$. However, the same renormalization term (e.g., (11.120)) is used in a short-range domain $|\mathbf{x} - \mathbf{x}_k| < 3a$ in the vicinity of the point $\mathbf{x} \in w$. A fundamental deficiency of the mentioned method is a dependence of the renormalizing term only on the statistical average of traction $\mathbf{t}^{\Gamma_k}(\mathbf{s})$ ($\mathbf{s} \in \Gamma_k$) with some weighting functions rather than $\langle \boldsymbol{\mathcal{L}}_k^{\theta t}(\mathbf{x} - \mathbf{x}_k, \mathbf{t}) \rangle(\mathbf{x}_k)$ (11.113).

The proposed renormalizing term $\langle \boldsymbol{\mathcal{L}}_k^{\theta t}(\mathbf{x} - \mathbf{x}_k, \mathbf{t}) \rangle(\mathbf{x}_k)$ can be estimated for any inhomogeneous effective field $\overline{\boldsymbol{\vartheta}}_k(\mathbf{x})$ (11.118) evaluated in a solution procedure of the system (11.119). In so doing, the renormalizing term in the known approach (see [155, 412]) is based on the solution for a single remote inclusion v_k ($|\mathbf{x}-\mathbf{x}_k| \to \infty$) averaged with some weighting functions and subjected to the homogeneous effective field. Therefore, the GIEs based on the asymptotic approximations (11.120) and (11.124) of a renormalizing term are not really general (in contrast with Eq. (11.113) and (11.119)).

It should be emphasized that the chapter is dedicated to obtaining of the new exact GIEs (11.113) and (11.119) rather than to solution of these equations. This solution can be performed at some additional assumptions similar to Sect. 11.2.3.

Chapter 12
Nonlinear Problems in Micromechanics of CMs

Abstract For linearizing nonlinear functions (such as the yield or strength criterion, dissipative function) in this chapter, physically consistent assumptions for the dependence of these functions on the second moment of stresses (see Chaps. 6 and 10) are employed that explained by the fact that both the yield surface, fiber/matrix interface failure criterion, and the energy release rate are the quadratic functions of the local stress distributions. The method of integral equations (Chap. 10) easily allows one to estimate the inhomogeneous second moment of stresses in the inclusions and at each point on the interface between the matrix and inclusions. A local theory of elastoplastic deformation is proposed where homogeneity of the plastic strains is assumed inside the matrix and in individual subdomains of the coatings, which are considered as the individual components.

In contrast to the extensive work on the elastic behavior of composites, only a limited number of studies for the nonlinear range have been published (for references see [421, 1239]). Standard methods of successive approximation may be used in nonlinear problems of composites with random structure, linearized problems being solved at each stage. The well-known concept of secant moduli has been combined with the hypothesis of the homogeneity of the increments of the plastic or creep strains [117, 248, 249, 349, 420, 1491, 1573]. Since the widely used method of average strains is capable of estimating only the average stresses in the components, its use for the linearization of functions describing nonlinear effects, e.g., strength [33, 1287], yielding [938], damage accumulation ([556, 801]; see also the theories combining the damage accumulation and life-fraction rule [236, 1448]), hardening [1258], creep [1697], and dynamic viscoplasticity [1699] may be problematical. Due to the significant inhomogeneity of the stress fields in the components (especially in the matrix), such linearization entails physical inconsistencies, which was discussed in detail in [251, 266] within the context of predicting the flow behavior of porous media. Alternatively, for linearizing nonlinear functions (such as the yield or strength criterion, dissipative function), physically consistent assumptions for the dependence of these functions on the second moment of stresses are employed. In this way, the results of Chap. 10 for

V. A. Buryachenko, *Local and Nonlocal Micromechanics of Heterogeneous Materials*, https://doi.org/10.1007/978-3-030-81784-8_12

the volume averages of the stress second moments will be used. The fundamental role of the statistical averages of the second moments of stress concentration factors in nonlinear analysis is explained by the fact that both the yield surface, fiber/matrix interface failure criterion, and the energy release rate are the quadratic functions of the local stress distributions. In so doing, estimation of the effective limiting surfaces separating the linear and nonlinear behavior domains of the mentioned nonlinear phenomena is of profound importance for the practical applications and can be done by the methods describing the linear deformations of composite materials. It should be mentioned that the estimation of the effective elastic moduli is a linear problem, with respect to the stress field analyzed which is less sensitive to the local stress distribution than nonlinear micromechanical problems of elastoplastic deformation, fracture, and fatigue of composite materials depending, at least, on mean-square stress fluctuations in the constituents (see, e.g., [188, 945, 1239]). The estimations of second moment of stresses are defined by both the random stress fluctuations in the components and the inhomogeneity of the statistical average of stress fields in the constituents; the separation of these stresses presents a real challenge in the framework of the perturbation method considered in Sect. 6.3.3. However, the method of integral equations easily allows one to estimate the inhomogeneous second moment of stresses in the inclusions and at each point on the interface between the matrix and inclusions. The dispersion of these interface stresses, defined only by stress fluctuations, will be used in this chapter for the prediction of the failure initiation. The failure initiation is dependent on the size and the volume fraction of inclusions, their surface treatment, matrix, and inclusion properties. A change of the adhesion properties of the interface, defined by the surface treatment, has a smaller effect on modulus than on strength. Indeed, even poor adhesion between the constituents does not appear to be an important factor as long as the frictional forces between the phases are not exceeded by the interface stress. Because of this, the estimation of the failure initiation envelope of the interface is of practical interest.

12.1 Effective Limiting Surface

12.1.1 Limiting Surface for Bulk Stresses

The failure analysis of composite materials considers the initiation and accumulation of damage occurring in each phase of the material and involves several types of local degradation processes including matrix microcracking (type I), interfacial debonding (type II), fiber breakages (type III), etc. Generally, these failure mechanisms may initiate concurrently in an early loading stage and progressively accumulate inside the materials (see for references and detail [371, 452, 453, 757, 867, 1050, 1285, 1545]). The first type of damage processes are those that relate to matrix degradation. They include matrix microcracking and pseudo-delamination.

The second type of degradation models describes interfacial decohesion and related mechanisms, such as fiber matrix friction and fiber pull-out processes. Let us assume that the well-known tensor-polynomial strength criterion (see, e.g., [1538]) describes the initiation of failure mechanisms of the types I and III for each component, i.e., the equivalent stress is given by

$$\boldsymbol{\Pi}^{(i)}(\boldsymbol{\sigma}) = \boldsymbol{\Pi}^{1(i)}\boldsymbol{\sigma} + \boldsymbol{\Pi}^{2(i)}(\boldsymbol{\sigma}\otimes\boldsymbol{\sigma}) + \boldsymbol{\Pi}^{3(i)}(\boldsymbol{\sigma}\otimes\boldsymbol{\sigma}\otimes\boldsymbol{\sigma}) + \ldots = 1, \qquad (12.1)$$

where $i = 0, 1, \ldots$, and the second-, fourth-, and sixth-rank tensors of strength $\boldsymbol{\Pi}^1$, $\boldsymbol{\Pi}^2$, $\boldsymbol{\Pi}^3$ are expressed through technical strength parameters for different classes of material symmetry [1506, 1686]. It should be mentioned that in Eq. (12.1), the tensor $\boldsymbol{\sigma} = \boldsymbol{\sigma}(\mathbf{x})$ stands the local stresses in the composite material. Because of this, the criterion (12.1) drown his name Tsai-Wu criterion used macroscopic stresses $\langle\boldsymbol{\sigma}\rangle$ from just a notation convenience.

Similarly to (12.1) for onset of yielding in the case of elastic–plastic composites, a quadratic yield condition ($i = 0, 1, \ldots, N$)

$$\boldsymbol{\Pi}^{(i)}(\sigma) \equiv \Pi^{2(i)}_{ijmn}\sigma_{ij}\sigma_{mn} = 1, \quad \frac{3}{2}\mathbf{s}\mathbf{s} = \tau_i^2 \qquad (12.2)$$

is used [109]. Here Eq. (12.2_1) generalizes some classical criteria, as, for example, the von Mises equivalent stress criterion (12.22) for which we have $\boldsymbol{\Pi}^{2(i)} = 3\mathbf{N}_2/(2\tau_i^2)$ and $\boldsymbol{\Pi}^{2(i)}(\boldsymbol{\sigma}) = \frac{1}{2\tau_i^2}\{(\sigma_{11}-\sigma_{22})^2 + (\sigma_{22}-\sigma_{33})^2 + (\sigma_{11}-\sigma_{33})^2 + 3[(\sigma_{12})^2 + (\sigma_{13})^2 + (\sigma_{23})^2]\} = 1$, where τ_i stands for the yield stress of the component $v^{(i)}$.

The limiting surfaces $\boldsymbol{\Pi}^{(i)}(\boldsymbol{\sigma}) = 1$ (12.1) and (12.2) represent nonlinear functions of $\boldsymbol{\sigma}$. Usually in the treatment of random structure composite homogeneity of the stress or strain fields are used even for these nonlinear functions. Thus, a common way to produce an effective failure envelope for the composite materials is substitution of the component average stress values into the formula (12.1) [33, 420, 770, 1287] ($i = 0, 1, \ldots$)

$$\boldsymbol{\Pi}^*(\boldsymbol{\sigma}) \equiv \max_i \Big[\boldsymbol{\Pi}^{1(i)}\langle\boldsymbol{\sigma}\rangle^{(i)} + \boldsymbol{\Pi}^{2(i)}(\langle\boldsymbol{\sigma}\rangle^{(i)}\otimes\langle\boldsymbol{\sigma}\rangle^{(i)}) + \boldsymbol{\Pi}^{3(i)}(\langle\boldsymbol{\sigma}\rangle^{(i)}\otimes\langle\boldsymbol{\sigma}\rangle^{(i)}\otimes\langle\boldsymbol{\sigma}\rangle^{(i)}) + \ldots\Big] = 1. \qquad (12.3)$$

This hypothesis is, however, in some sense inconsistent. For example, let us consider an isotropic porous material with matrix yielding properties described by the von Mises criterion (12.22). In this case, under hydrostatic loading the condition $\langle\sigma_{ij}\rangle = \langle\sigma_0\rangle\delta_{ij}$ leads to $\langle\mathbf{s}\rangle^{(0)} \equiv \mathbf{0}$ irrespective of the microstructure of the pores and the method of calculation of $\langle\boldsymbol{\sigma}\rangle^{(0)}$. This would mean that such a porous material has an infinite yield stress under hydrostatic compression as mentioned, e.g., in [251, 1258]. It is believed that the following definition of effective limiting surface proposed in [246, 247] and based on fewer assumptions is more correct (see, e.g., [188, 1239])

$$\widetilde{\mathbf{\Pi}}^*(\boldsymbol{\sigma}) \equiv \max_i \Big[\mathbf{\Pi}^{1(i)}\langle\boldsymbol{\sigma}\rangle^{(i)} + \mathbf{\Pi}^{2(i)}\langle\boldsymbol{\sigma}\otimes\boldsymbol{\sigma}\rangle^{(i)} + \mathbf{\Pi}^{3(i)}\langle\boldsymbol{\sigma}\otimes\boldsymbol{\sigma}\otimes\boldsymbol{\sigma}\rangle^{(i)} + \ldots\Big] = 1,$$

$$\widetilde{\mathbf{\Pi}}^*(\boldsymbol{\sigma}) \equiv \max_i \Big\{\frac{3}{2\tau_i^2}\mathbf{N}_2(\langle\sigma\otimes\sigma\rangle^{(i)})\Big\} = 1, \qquad (12.4)$$

where the estimations of average stress moments of different orders $\langle\boldsymbol{\sigma}\rangle^{(i)}$, $\langle\boldsymbol{\sigma}\otimes\boldsymbol{\sigma}\rangle^{(i)}$, and $\langle\boldsymbol{\sigma}\otimes\boldsymbol{\sigma}\otimes\boldsymbol{\sigma}\rangle^{(i)}$ $(i = 0, 1, \ldots)$ can be found by the use of the relevant formulae of Chap. 10. In particular, for local limiting surfaces of von Mises's type (12.2) and according to Holder's inequality $\langle\mathbf{fg}\rangle \leq \langle\mathbf{f}^p\rangle^{1/p}\langle\mathbf{g}^q\rangle^{1/q}$, $1/p+1/q = 1$, $p, q = \text{const.}$, we have

$$\widetilde{\mathbf{\Pi}}^*(\langle\boldsymbol{\sigma}\rangle) \geq \mathbf{\Pi}^*(\langle\boldsymbol{\sigma}\rangle). \qquad (12.5)$$

Here, the equalities are obtained only for homogeneous stress fields in the components, which holds for the mixture theory models with either parallel or series arrangements of the components (for instance, the schemes of Voight and Reuss).

Let us show the physical consistency of the effective strength criterion (12.4_1) (in contrast to (12.3)) subjected to the thermal loading $(\langle\boldsymbol{\sigma}\rangle \equiv \mathbf{0})$. In fact, let us consider a two-component isotropic composite with isotropic phases. In this case one may observe that symmetry requires that average stresses inside both components will be hydrostator, one $\langle\sigma_{kl}\rangle^{(1)} \equiv \langle\sigma_{kl}\rangle^{(0)}(c-1)/c \equiv \sigma_{11}^0\delta_{kl}$; in so doing the microstructure and the method of calculation of average stresses inside the components (for example, (8.29) or any other formula) influence the value of scalar σ_{11}^0 but have no effect on the tensor structure of the fields $\langle\boldsymbol{\sigma}\rangle^{(0)}$ and $\langle\boldsymbol{\sigma}\rangle^{(1)}$. Then the composite strength is dictated by the strength of the component which is to be found under conditions of hydrostatic tension and is not determined by the strength of second component. If the strength of the second component falls far short of the strength of the first one, we will obtain an improper prediction of composite strength. In fact, according to Chaps. 6 and 10, the average values of second deviator invariant inside each component $\langle\mathbf{ss}\rangle^{(i)} \neq 0$ $(s_{kl} \equiv \sigma_{kl} - \sigma_{nn}\delta_{kl}/3;\ i = 0, 1, \ldots)$. Therefore the composite strength is defined by the strength of second more weak component at the cost of the fluctuations of the stress deviator.

From the above discussion, one should not conclude that for all real composites, the criterion (12.4_1) provides a qualitatively better approximation for the actual onset of yielding at any applied stress state. The particular case of criterion (12.3) was applied in [1491] to determinate the tensile behavior of a silica/epoxy composite and found good agreement with experimental results. Such an approach is reasonably acceptable for a particle-reinforced solid under a stress state with only low triaxiality (e.g., pure tension or shear; see [120, 1258]. However, for the criterion Eq. (12.3) under a pure hydrostatic loading, the macroisotropic composite would never yield. To remove this undesired feature, the different versions of criterion (12.4_1) can be used.

12.1.2 Limiting Surface for Interface Stresses

The modeling of global composite behavior necessarily requires the consideration of interface degradation in addition to the matrix and fiber phase degradation (12.1). This requires not only the calculation of the stress at the interface but equally the identification and the application of a local failure criterion. In a similar manner with Eq. (12.1), we can present a tensor-polynomial failure criterion for the interface failure initiation

$$\boldsymbol{\Pi}_a^{(i)}(\mathbf{n},\boldsymbol{\sigma}) = \boldsymbol{\Pi}_a^{1(i)}(\mathbf{n})\boldsymbol{\sigma}^-(\mathbf{n}) + \boldsymbol{\Pi}_a^{2(i)}(\mathbf{n})[\boldsymbol{\sigma}^-(\mathbf{n})\otimes\boldsymbol{\sigma}^-(\mathbf{n})] + \ldots = 1, \tag{12.6}$$

where $\boldsymbol{\sigma}^-(\mathbf{n})$ is the limiting stress within the matrix near the inclusion boundary $\mathbf{x} \in \partial v_i$ with the unit outward normal vector $\mathbf{n}$. Generally speaking, adhesion strength parameters $\boldsymbol{\Pi}_a^{1(i)}(\mathbf{n})$, $\boldsymbol{\Pi}_a^{2(i)}(\mathbf{n})$, $\boldsymbol{\Pi}_a^{3(i)}(\mathbf{n})$, which convey the normal and the shear debonding at the interface between the matrix and the fibers, differ from $\boldsymbol{\Pi}^{1(i)}$, $\boldsymbol{\Pi}^{2(i)}$, and $\boldsymbol{\Pi}^{3(i)}$.

We will present now popular local criteria that convey the normal and the shear debonding at the interface between the matrix and the fibers. The interface stresses $\boldsymbol{\sigma}_{\mathbf{n}}^- \equiv \boldsymbol{\sigma}^-(\mathbf{n})\mathbf{n}$ can be partitioned as $\boldsymbol{\sigma}_{\mathbf{n}}^- = \mathbf{N}^{\mathbf{n}}\boldsymbol{\sigma}^-(\mathbf{n}) + \mathbf{T}^{\mathbf{n}}\boldsymbol{\sigma}^-(\mathbf{n})$, where $\mathbf{N}^{\mathbf{n}}$ and $\mathbf{T}^{\mathbf{n}}$ are the three-rank functions of the normal $\mathbf{n}$ such that $N_{ikl}^{\mathbf{n}} = n_i n_k n_l$, $T_{ikl}^{\mathbf{n}} = (\delta_{ik}n_l + \delta_{il}n_k)/2 - n_i n_k n_l$, where the tensors $N_{ikl}^{\mathbf{n}}$ and $T_{ikl}^{\mathbf{n}}$ symmetrical under the interchanges $k \leftrightarrow l$ generate the normal $\boldsymbol{\sigma}_n^- = \mathbf{N}^{\mathbf{n}}\boldsymbol{\sigma}^-(\mathbf{n})$ and tangential components $\boldsymbol{\sigma}_\tau^- = \mathbf{T}^{\mathbf{n}}\boldsymbol{\sigma}^-(\mathbf{n})$ of the traction $\boldsymbol{\sigma}^-\mathbf{n}$ with the magnitudes $\sigma_n \equiv \|\boldsymbol{\sigma}_n^-\| = \sigma_{kl}^- n_k n_l$ and $\sigma_\tau \equiv \|\boldsymbol{\sigma}_\tau^-\| = \sqrt{\sigma_{kl}^-\sigma_{kl}^- - (\sigma_n)^2}$, respectively.

In a simple maximum stress criterion, the normal σ_n and tangential σ_τ components are compared to maximum values $\sigma_n^{\max}$ and $\sigma_\tau^{\max}$ characterizing the interface, and the failure tensors $\boldsymbol{\Pi}_a^{2(i)}(\mathbf{n},\boldsymbol{\sigma})$ have a form

$$\bar{\boldsymbol{\Pi}}_a^{2(i)}(\mathbf{n},\boldsymbol{\sigma}) = \max\left[\frac{\|\mathbf{N}^{\mathbf{n}}\boldsymbol{\sigma}^-(\mathbf{n})\|}{\sigma_n^{\max}}, \frac{\|\mathbf{T}^{\mathbf{n}}\boldsymbol{\sigma}^-(\mathbf{n})\|}{\sigma_\tau^{\max}}\right] = \max\left[\frac{\sigma_n}{\sigma_n^{\max}}, \frac{\sigma_\tau}{\sigma_\tau^{\max}}\right] = 1. \tag{12.7}$$

Other type of criteria taking into account the friction problem were considered by a number of authors (for references see [1107]). The Coulomb form [32] permits the introduction of the friction coefficient at the interface by the use of a linear combination between the normal and the shear interface stresses

$$\check{\boldsymbol{\Pi}}_a^{2(i)}(\mathbf{n},\boldsymbol{\sigma}) = \frac{\|\mathbf{N}^{\mathbf{n}}\boldsymbol{\sigma}^-(\mathbf{n})\|}{\sigma_n^{\max}} + \frac{\|\mathbf{T}^{\mathbf{n}}\boldsymbol{\sigma}^-(\mathbf{n})\|}{\sigma_\tau^{\max}} = 1. \tag{12.8}$$

Logical generalization of tensor-polynomial criteria (12.1) to the interface failure initiation was proposed in [1464] (see also [870]) for the cylindrical fibers, which in our more general notations has the form

$$\hat{\mathbf{\Pi}}_a^{2(i)}(\mathbf{n}, \boldsymbol{\sigma}) = \frac{(\mathbf{N}^{\mathbf{n}}\boldsymbol{\sigma}^-(\mathbf{n}))(\mathbf{N}^{\mathbf{n}}\boldsymbol{\sigma}^-(\mathbf{n}))}{(\sigma_n^{\max})^2} + \frac{(\mathbf{T}^{\mathbf{n}}\boldsymbol{\sigma}^-(\mathbf{n}))(\mathbf{T}^{\mathbf{n}}\boldsymbol{\sigma}^-(\mathbf{n}))}{(\sigma_\tau^{\max})^2} = 1. \tag{12.9}$$

The criterion (12.9) can be recast in the tensor-polynomial form (12.6) with the tensors $\mathbf{\Pi}_a^{1(i)}(\mathbf{n}, \boldsymbol{\sigma}) \equiv \mathbf{0}$ and $\Pi_{a|ijkl}^{2(i)}(\mathbf{n}, \boldsymbol{\sigma}) = \left[(\sigma_n^{\max})^{-2} - (\sigma_\tau^{\max})^{-2}\right] n_i n_j n_k n_l + (2\sigma_\tau^{\max})^{-2}(\delta_{ik} n_{jl} + \delta_{il} n_{jk} + \delta_{jl} n_{ik} + \delta_{jk} n_{il})$.

Because the criteria (12.7)–(12.9) should predict the identical stresses of the failure initiation at the pure local normal and shear stresses, the empirical interfacial strengths corresponding to tension and shear should be the same in criteria (12.7)–(12.9), and, therefore, the failure envelope (12.9) is inserted between the failure surfaces (12.7) and (12.8) $\check{\mathbf{\Pi}}_a^{2(i)}(\mathbf{n}, \boldsymbol{\sigma}) < \hat{\mathbf{\Pi}}_a^{2(i)}(\mathbf{n}, \boldsymbol{\sigma}) < \bar{\mathbf{\Pi}}_a^{2(i)}(\mathbf{n}, \boldsymbol{\sigma})$, where the equalities

$$\check{\mathbf{\Pi}}_a^{2(i)}(\mathbf{n}, \boldsymbol{\sigma}) = \hat{\mathbf{\Pi}}_a^{2(i)}(\mathbf{n}, \boldsymbol{\sigma}) = \bar{\mathbf{\Pi}}_a^{2(i)}(\mathbf{n}, \boldsymbol{\sigma}) \tag{12.10}$$

in general hold just for the normal $\mathbf{n}$ with either the pure normal or pure tangential local loading $\boldsymbol{\sigma}^-(\mathbf{n}) \equiv \mathbf{N}^{\mathbf{n}}\boldsymbol{\sigma}^-(\mathbf{n})$, or $\boldsymbol{\sigma}^-(\mathbf{n}) \equiv \mathbf{T}^{\mathbf{n}}\boldsymbol{\sigma}^-(\mathbf{n})$. The equalities also hold in some particular cases of correlations between the elastic and strength properties of constituents. For example, the equalities (12.10) are valid for any $\mathbf{n}$ for the limiting type of soft fibers (hole, $\mathbf{L}^{(1)} \equiv \mathbf{0}$) when $\boldsymbol{\sigma}_{\mathbf{n}}^- \equiv \mathbf{0}$ and the interface failure is degenerated into the failure of the matrix in the vicinity of the interface. In another limiting case of perfect sliding $\sigma_\tau^{\max} = 0$ (for references see [1107]), the equality is also valid for any $\mathbf{n}$. Moreover, the last statement also holds if under the failure initiation one understands the normal debonding $(\mathbf{u}^+ - \mathbf{u}^-)\mathbf{n} > 0$.

It should be mentioned that the transverse strength of the reinforced fiber composites is usually significantly higher than that of the matrix. The popular assumption is that the strengths of the interface and the matrix are equal (i.e., the bonding between fiber and matrix is assumed to be perfect; see, e.g., [542]). As an approximation, $\sigma_\tau^{\max}$ is also taken to be a half of $\sigma_n^{\max}$ as usually assumed in a homogeneous isotropic material (see, e.g., [870]). In light of the heuristic level of justification, the importance of the fundamental experimental work [1487] can scarcely be exaggerated. If the possibility of interfacial fracture is taken into account, the macrostrength criterion can be expressed in the following forms, analogous to Eqs. (12.3) and (12.4$_1$) ($i = 0, 1, \ldots$)

$$\mathbf{\Pi}_a^*(\boldsymbol{\sigma}) = \max\left\{\mathbf{\Pi}^*(\langle\boldsymbol{\sigma}\rangle),\ \max_i \max_{\mathbf{n}}\left[\mathbf{\Pi}_a^{1(i)}(\mathbf{n})\langle\boldsymbol{\sigma}^-(\mathbf{n})\rangle_x\right.\right.$$
$$\left.\left. +\mathbf{\Pi}_a^{2(i)}(\mathbf{n})\langle\boldsymbol{\sigma}^-(\mathbf{n})\rangle_x \otimes \langle\boldsymbol{\sigma}^-(\mathbf{n})\rangle_x + \ldots\right]\right\} = 1, \tag{12.11}$$

$$\widetilde{\mathbf{\Pi}}_a^*(\boldsymbol{\sigma}) = \max\left\{\widetilde{\mathbf{\Pi}}^*(\langle\boldsymbol{\sigma}\rangle),\ \max_i \max_{\mathbf{n}}\left[\mathbf{\Pi}_a^{1(i)}(\mathbf{n})\langle\boldsymbol{\sigma}^-(\mathbf{n})\rangle_x\right.\right.$$

$$+\boldsymbol{\Pi}_a^{2(i)}(\mathbf{n})\langle\boldsymbol{\sigma}^-(\mathbf{n})\otimes\boldsymbol{\sigma}^-(\mathbf{n})\rangle_x+\ldots\Big]\Big\}=1, \qquad (12.12)$$

respectively, where $\langle\boldsymbol{\sigma}^-(\mathbf{n})\rangle_x$ and $\langle\boldsymbol{\sigma}^-(\mathbf{n})\otimes\boldsymbol{\sigma}^-(\mathbf{n})\rangle_x$ are the statistical moments of limiting stresses within the matrix near the inclusion boundary $\mathbf{x}\in\partial v_i$ with the unit outward normal vector $\mathbf{n}$; see, e.g., (8.30) and (10.59). These criteria can be viewed as intermediate statements between the local and averaged (12.3) and (12.4$_1$) criteria.

It should be mentioned that special so-called localized models of plasticity of particulate composites have been developed. Herve and Zaoui [638] proposed a generalization of the elastic three-phase spherical model by Hashin [607, 611] toward the elastoplastic case in the framework of a secant modulus concept. We will not discuss here in general the known advantages of flow theory over deformation theory of plasticity in the case of nonradial loading that takes place usually at the local level near the inclusions even if the overall stress path is radial. We just look at the method of calculation of onset of yielding in the layer model [638], which used the criterion (in our notations) $\sqrt{1.5\langle\boldsymbol{\sigma}^-(\mathbf{n})\rangle_{s_i}:\mathbf{N}_2:\langle\boldsymbol{\sigma}^-(\mathbf{n})\rangle_{s_i}}-\tau_0^{(0)}=0$, where $\langle\boldsymbol{\sigma}^-(\mathbf{n})\rangle_{s_i}$ is the surface average of the stresses $\boldsymbol{\sigma}^-(\mathbf{n})$ over the surface of the inclusion s_i. As a consequence of the properties of the interface operator $\boldsymbol{\Gamma}(\boldsymbol{n})$ (3.97), this average is given by $\langle\boldsymbol{\sigma}^-(\mathbf{n})\rangle_{s_i}\equiv\langle\boldsymbol{\sigma}\rangle^{(1)}$ (at least for an ellipsoidal inclusion in the framework of the hypothesis **H1**; see Chap. 8), and, therefore, for porous materials one would obtain an infinite overall initial yield stress for pure hydrostatic loading σ_0^0. More recently a yield criterion [147] was proposed $\sqrt{1.5\langle\{\langle\boldsymbol{\sigma}^-(\mathbf{n})\rangle_x:\mathbf{N}_2:\langle\boldsymbol{\sigma}^-(\mathbf{n})\rangle_x\}\rangle_{s_i}}-\tau_0^{(0)}=0$, which contains inside the generalized yield surface (12.12) in a dimensionless coordinate system $X_\tau=\sigma_0^0/\tau_0^{(0)},\ Y_\tau=(1.5\mathbf{s}^0:\mathbf{s}^0)^{1/2}/\tau_0^{(0)}$, where $\sigma_0^0\equiv\boldsymbol{\delta}:\boldsymbol{\sigma}^0/3,\ \mathbf{s}^0\equiv\mathbf{N}_2\boldsymbol{\sigma}^0$.

12.1.3 Effective Fracture Surface for an Isolated Crack

Let us consider a coin-shaped crack of radius R^c with center $\mathbf{x}^c=(0,0,0)$ and unit normal $\mathbf{n}=(1,0,0)$ to the crack surface S in an infinite elastic homogeneous medium. The energy release rate may be defined by means of $\mathbf{K}\equiv(K_1,K_2,K_3)^\top$ as $\mathcal{J}=\boldsymbol{\Lambda}(\mathbf{K}\otimes\mathbf{K})$, where for the general anisotropic material the matrix Λ_{ij} $(i,j=1,2,3)$ is symmetric [69, 1293]. For an isotropic material, the matrix Λ_{ij} is diagonal: $\boldsymbol{\Lambda}=(2\mu)^{-1}\mathrm{diag}(1-\nu,1-\nu,1)$. The energy release rate $\mathcal{J}$ could alternatively be expressed as a path-independent line J^r Rice's integral which is invariant with respect to the integration along an arbitrary path encircling the crack tip (see, e.g., [1635]). The fracture criterion that will be used for the remainder of this work therefore based on the equality $\mathcal{J}=\mathcal{J}^c\equiv 2\gamma$, where γ called fracture surface energy. For homogeneous external loading, the equation $\mathcal{J}=2\gamma$ defines the surface of second order in six-dimensional space of stresses $\boldsymbol{\sigma}$. In particular, for coin-shaped crack with radius R^c, the last formula turns into

$$\frac{2(1-\nu)R^c}{\pi\mu}\Big(\sigma_{11}^2+\Big[\frac{2}{2-\nu}\Big]^2\sigma_{12}^2+\Big[\frac{2(1-\nu)^2}{2-\nu}\Big]\sigma_{13}^2\Big)=2\gamma, \tag{12.13}$$

(see, e.g., [1108]). Thus we have obtained the strength criterion (12.13) in terms of (12.2) in which the strength tensor $\mathbf{\Pi}^4$ is a function of the size and orientation of the crack. This analogy permits the use of strength calculation method of composites developed in the next section.

We will shortly reproduce an estimation scheme (see for details [187]) of effective fracture surface for an isolated crack in the elastically homogeneous medium with random residual microstresses. Namely, we can define $\mathbf{K}(\mathbf{z})$ and $\mathbf{J}(\theta,\mathbf{z})$ by the integrals (see [524, 1293])

$$\mathbf{K}(\mathbf{z})=\int_S \mathbf{k}(\mathbf{z},\mathbf{s})\mathbf{t}(\mathbf{s})\,ds,\quad \mathbf{J}(\theta,\mathbf{z})=\int_S \mathbf{g}(\theta,\mathbf{z},\mathbf{s})\mathbf{t}(\mathbf{s})\,ds, \tag{12.14}$$

respectively, over surface $\mathbf{s}\in S$ from both the crack-face weight functions $\mathbf{k}(\mathbf{z},\mathbf{s})$ and tensor crack-face weight vector function $\mathbf{g}(\theta,\mathbf{z},\mathbf{s})$; the "traction" $\mathbf{t}(\mathbf{s}) = -\mathbf{n}(\mathbf{s})\boldsymbol{\sigma}(\mathbf{s})$ ($\mathbf{s}\in S$) at the crack faces corresponds to the remote loading $\boldsymbol{\sigma}(\mathbf{x})$. For a coin-shaped crack in an isotropic medium, the relations of $\mathbf{K}(\mathbf{z})$ (12.14_1) can be significantly simplified. We will use the solution [454] according to which the representation (12.14_1) can actually be expressed in the convenient complex form. Kachanov and Laures [730] eliminated the singularities in proposed integrals by way of transfer to the new coordinate system; Buryachenko [200] corrected some misprints made in [730] and [729] in these representations.

At first, the problem for macrocrack S interacting with a single inclusion was considered (see also [1103, 1291, 1292]) by the use of Eq. (12.14_1) utilizing weight functions and exactly defined tractions $\mathbf{t}(\mathbf{s})$ ($\mathbf{s}\in S$) induced by the external loading and residual stresses. After that, let us consider a crack within elastically homogeneous medium containing a statistically homogeneous inclusion field with the eigenstrains. The expected value of the tensor $\mathbf{K}$ is defined by relation (12.14_1): $\langle\mathbf{K}(\mathbf{z})\rangle=\int\mathbf{k}(\mathbf{z},\mathbf{s})\langle\mathbf{t}(\mathbf{s})\rangle\,ds$, where the expected value of the traction $\langle\mathbf{t}(\mathbf{s})\rangle$ may be exactly found as $\langle\mathbf{t}(\mathbf{s})\rangle=-\mathbf{n}(\mathbf{s})\boldsymbol{\sigma}^0$. Thus the expectation value $\langle\mathbf{K}(\mathbf{z})\rangle$ and $\langle\mathbf{J}(\theta,\mathbf{z})\rangle$ coincide with the corresponding values $\mathbf{K}^0(\mathbf{z})$ and $\mathbf{J}^0(\theta,\mathbf{z})$, respectively, for an isolated crack inside the infinite medium without residual stresses. For a general case of statistically inhomogeneous field of inclusions, one can found in a similar manner the conditional averages $\langle\mathbf{K}(\mathbf{z})|v_i,\mathbf{x}_i)\rangle$ and $\langle\mathbf{J}(\theta,\mathbf{z}|v_i,\mathbf{x}_i)\rangle$ by the use of estimation of conditional tractions $\langle\mathbf{t}(\mathbf{s})|v_i,\mathbf{x}_i)\rangle$ evaluated by Eqs. (8.79) and (8.80). At last, the second moment of stresses in the vicinity of the crack tip (having a singularity ρ^{-1} are expressed by the representation which are very similar to Eqs. (10.80) and (10.80) where the perturbators $\mathcal{L}_q^{\beta}(\mathbf{x}-\mathbf{x}_q,\beta)$ and $\mathcal{L}_q^{\beta}(\mathbf{x}_0-\mathbf{x}_q,\beta)$ should be replaced by $\mathbf{J}(\theta,\mathbf{z}|v_i,\mathbf{x}_i)$ and $\mathbf{J}(\theta,\mathbf{z}|v_0,\mathbf{x}_0)$, respectively. After that, the effective energy release rate $\mathcal{J}^*(\mathbf{z})$ can be estimated by different combination of $\langle\mathcal{J}(\mathbf{z})|v_k,\mathbf{x}_k\rangle := \mathbf{\Lambda}\langle\mathbf{K}(\mathbf{z})|v_k,\mathbf{x}_k)\otimes\mathbf{K}(\mathbf{z})|v_k,\mathbf{x}_k)\rangle$ ($k=0,1,2,\ldots$) (see for details [187]).

It should be mentioned that the described criteria $\mathcal{J}^*(\mathbf{z})$ proposed for a special case of a single microcrack loaded by the random residual stress field can be easily generalized to the cases of both the composites and microcracked media. Indeed, e.g., a fracture criterion for media with identical aligned coin-shaped cracks is based on the estimation of $\mathcal{J}^*(\mathbf{z}) = \boldsymbol{\Lambda}(\langle \mathbf{K}(\mathbf{z}) \otimes \mathbf{K}(\mathbf{z}) \rangle_i)$ for a representative crack v_i. In the framework of the effective field hypothesis **(H1)**, $\mathbf{K}(\mathbf{z}) = \mathbf{K}_i^{\sigma}(\mathbf{z})\overline{\boldsymbol{\sigma}}_i$ where $\mathbf{K}_i^{\sigma} \equiv$ const. is assumed to be known (see, e.g., [454, 1108]). Then $\mathcal{J}^*(\mathbf{z}) = \boldsymbol{\Lambda}(\mathbf{K}_i^{\sigma}(\mathbf{z}) \otimes \mathbf{K}_i^{\sigma}(\mathbf{z}))(\langle \overline{\boldsymbol{\sigma}} \otimes \overline{\boldsymbol{\sigma}} \rangle_i)$ are found after estimation of the second moment of the effective field $\langle \overline{\boldsymbol{\sigma}} \otimes \overline{\boldsymbol{\sigma}} \rangle_i$ obtained, e.g., by Eq. (10.42). The proposed criterion is reduced to the simplified formulae obtained under the additional assumptions such as the dilute approximation $\langle \overline{\boldsymbol{\sigma}} \otimes \overline{\boldsymbol{\sigma}} \rangle_i \equiv \langle \boldsymbol{\sigma} \rangle \otimes \langle \boldsymbol{\sigma} \rangle$ [374, 895, 1711] and the effective medium approach $\langle \overline{\boldsymbol{\sigma}} \otimes \overline{\boldsymbol{\sigma}} \rangle_i \equiv \langle \boldsymbol{\sigma} \rangle \otimes \langle \boldsymbol{\sigma} \rangle$ and $\boldsymbol{\Lambda} = \boldsymbol{\Lambda}(\mathbf{L}^*)$, $\mathbf{K}_i^{\sigma} = \mathbf{K}_i^{\sigma}(\mathbf{L}^*)$ [1277].

The proposed strength and fracture criteria are based on the determination of conditional averages either $\langle \boldsymbol{\Pi}^{(k)}(\boldsymbol{\sigma}) \rangle$ or $\langle \mathcal{J}(\mathbf{z}) | v_J, \mathbf{x}_J \rangle$ $(k = 0, 1, \ldots, N;\ J = 0, i;\ i = 0, 1, \ldots)$. In fracture mechanics for random loading, another approach is known. One of these methods is based on the calculation of distribution $F_Y(y)$ of some fracture parameter Y [141]. Such a parameter can be identified with different nonlinear functions of local stresses. For example, for $Y = \boldsymbol{\Pi}^{(k)}(\boldsymbol{\sigma})$ (12.1), we have a critical value $y^{\text{crit}} = 1$; in a similar manner in fracture mechanics for $Y = \mathcal{J}(\mathbf{K})$, we have $y^{\text{crit}} = 2\gamma$. Thereafter the first-order estimation of the fraction of fractured component f (or grain faces) is calculated as $f = 1 - F_Y(y)$. Usually the approach assumes that the damage density is small enough so that the interaction of fractured elements is negligible small.

Then pertinent damages (or fracture probability of the component) can be defined by the relations [141] $(k = 0, 1, \ldots N)$

$$f^{(k)} = 1 - \int d\Phi^{(k)}(\boldsymbol{\sigma}), \text{ or } f^{(k)}(\mathbf{z}|v_k, \mathbf{x}_k) = 1 - \int d\Phi^{(k)}(\mathbf{K}|v_k, \mathbf{x}_k), \tag{12.15}$$

where $\Phi^{(k)}(\boldsymbol{\sigma})$ and $\Phi^{(k)}(\mathbf{K}|v_k, \mathbf{x}_k)$ are the Gaussian distributions with the expectation values $\langle \boldsymbol{\sigma} \rangle^{(k)}$ and $\langle \mathbf{K}(\mathbf{z}) | v_k, \mathbf{x}_k \rangle\rangle$ and the covariance matrices $\Delta\boldsymbol{\sigma}^{(k)2}$ and $\Delta\mathbf{K}^2(\mathbf{z}|v_k, \mathbf{x}_k)$, respectively. In the right-hand side integrals of Eqs. (12.15), the integral domains are determined by one of two inequalities $(k = 0, 1, \ldots, N)$ $\boldsymbol{\Pi}^{(k)}(\boldsymbol{\sigma}) \leq 1$ for Eq. (12.15_1) and $\boldsymbol{\Lambda}^{(k)}(\mathbf{K} \otimes \mathbf{K}) \leq 2\gamma^{(k)}$ or $K_1 < 0$ for Eq. (12.15_2). The most simple phenomenological ways of fracture probability calculation for composites [1422] are based either on the total probability formula or extreme value distribution: $f = \sum_{k=0}^{N} c^{(k)} f^{(k)}$, $\quad f = 1 - \prod_{k=0}^{N}(1 - f^{(k)})$, respectively.

Ortiz and Molinari [1161] were the first who considered the particular case of this scheme with an application to random structure composites (see also [941, 988, 1164]). Because the residual stresses are self-equilibrating, we have $\delta\boldsymbol{\sigma} \equiv \boldsymbol{\sigma} - \langle \boldsymbol{\sigma} \rangle = \boldsymbol{\sigma}$. Thereupon Ortiz and Molinari [1161] (see also [995]) have assumed that the residual stresses are normally distributed with a zero expectation in each point of the composite and the covariance matrix $\mathbf{K}^{\sigma} = \langle \delta\boldsymbol{\sigma} \otimes \delta\boldsymbol{\sigma} \rangle$. The indicated approximation of the real process $\delta\boldsymbol{\sigma}$ by Gaussian distributions with zero average stress $\langle \boldsymbol{\sigma} \rangle \equiv \mathbf{0}$ can lead to too crude estimations if conditional average stresses $\langle\boldsymbol{\sigma}\rangle^{(i)} \neq \mathbf{0}$ $(i = 0, \ldots, N)$ exactly estimated by Eq. (9.103) for CMs with spherical inclusions. Therefore, the definition of the conditional covariance $\mathbf{K}^{\sigma(k)} = \langle\boldsymbol{\sigma}\otimes\boldsymbol{\sigma}\rangle^{(k)} - \langle\boldsymbol{\sigma}\rangle^{(k)}\otimes\langle\boldsymbol{\sigma}\rangle^{(k)}$ seems to be more correct (see for details [187] and [200]). It was demonstrated that the random residual stress fluctuations inside the components are not in excess of 10% of the average values $|\langle\sigma_{11}\rangle^{(i)}|$. Thus, the probability that the stresses inside the component $v^{(i)}$ exceed the value $1.3|\langle\sigma_{11}\rangle^{(i)}|$ is much smaller than predicted in [1161].

Numerous numerical results were presented in [186, 187, 192, 200] for both statistically homogeneous and inhomogeneous random fields of ellipsoidal inclusions (with $\mathbf{M}(\mathbf{x})$ ≡const). In particular, it was estimated $\langle\mathbf{K}(\mathbf{z})|v_k, \mathbf{x}_k)\rangle$, $\langle\mathbf{K}(\mathbf{z})|v_k, \mathbf{x}_k)\otimes\mathbf{K}(\mathbf{z})|v_k, \mathbf{x}_k)\rangle$, and $\langle\mathcal{J}(\mathbf{z})|v_k, \mathbf{x}_k\rangle := \mathbf{\Lambda}\langle\mathbf{K}(\mathbf{z})|v_k, \mathbf{x}_k)\otimes\mathbf{K}(\mathbf{z})|v_k, \mathbf{x}_k)\rangle$, and $\delta\mathcal{J}*(\mathbf{z}) := \langle\mathcal{J}(\mathbf{z})\rangle - \mathcal{J}(\langle\mathbf{K}(\mathbf{z})\rangle)$. It is interesting (see for details [187]) that for sufficiently strong material, the fracture probability of the matrix is more than fracture probability of the inclusion notwithstanding the fact that $\langle K_I(\mathbf{z})|v_1, \mathbf{x}_1\rangle > 0 > \langle K_I(\mathbf{z})|v_0, \mathbf{x}_0\rangle$. Such a result stands clear when one takes into account the greater level of SIF fluctuation inside the matrix. The fracture probability will be determined fundamentally by the maximum SIF fluctuations (other than conditional average SIF) that arouses considerable interest in actual practice. The popular problem of a macrocrack in the cluster (or the cloud) of randomly distributed inclusions with eigenstrains and the estimations of statistical averages of the stresses $\langle\boldsymbol{\sigma}\rangle(\mathbf{x})$ (see [200]) are exploited for the estimation of statistical averages of stress intensity factors at the crack tip (see, e.g., [152, 168, 186, 878, 1045, 1462]). More accurate analysis is performed in the framework of the conditional averages of both the SIF $\mathbf{K}$ and $\mathcal{J}$ in [187] and [200].

12.1.4 Numerical Results for the Effective Limiting Surfaces

At first let us consider as an example a composite material consisting of an incompressible matrix satisfying the von Mises yield criterion (12.22) and containing either the rigid inclusions or voids. Then the onset of yielding of the composite is defined by

$$\frac{1}{2c^{(0)}}\frac{\partial M^*_{mnpq}}{\partial q^{(0)}}\langle\sigma_{mn}\rangle\langle\sigma_{pq}\rangle = \frac{2}{3}\tau_0^2. \tag{12.16}$$

For a macroisotropic composite, we will obtain an elliptical criterion for the onset of yielding

$$\frac{X_\tau^2}{a_\tau^2} + \frac{Y_\tau^2}{b_\tau^2} = 1, \tag{12.17}$$

in the coordinate system of normalized overall stresses: $X_\tau = \langle\sigma_0\rangle/\tau_0$, $Y_\tau = \langle\tau\rangle/\tau_0$, $\langle\tau\rangle = (3\langle\mathbf{s}\rangle\langle\mathbf{s}\rangle/2)^{1/2}$. The effective moduli (8.221), (8.230) and (8.232) estimated by the MEFM are consistent with the yields

$$a_\tau^2 = 0, \quad b_\tau^2 = (1-c)\left(1 + \frac{5}{2}c\frac{1}{1-31c/16}\right), \tag{12.18}$$

$$a_\tau^2 = \frac{4}{9c}\left(1 - \frac{29}{24}c\right), \quad b_\tau^2 = (1-c)\left(1 - \frac{5}{3}c\frac{1}{1+5c/24}\right), \tag{12.19}$$

$$a_\tau^2 = \frac{1}{3c}\left(1 - \frac{16}{15\pi^2}c\right), \quad b_\tau^2 = \left(1 - \frac{16}{15\pi^2}c\frac{1}{1-16c/125\pi^2}\right), \tag{12.20}$$

for the composites with identical rigid spherical inclusions ($a_1 = a_2 = a_3 = a$) (12.18), with identical spherical pores ($a_1 = a_2 = a_3 = a$) (12.19), and with identical uniformly oriented coin-shaped cracks ($a_1 = a_2 = a \gg a_3$) (12.20); the values $c = \frac{4}{3}\pi a^3 n$ (where n in a number density) in the expressions presented have different physical meaning for the spherical and coin-shaped pours. The relations (12.19) and (12.20) show that yielding will also take place under purely hydrostatic loading, which is in contradiction to the results for onset of yielding obtained by conventional yielding analysis of composites based on the assumption of homogeneity of the microstress fields in the matrix, Eq. (12.6).

For the "quasi-crystalline" approximation (8.43), (8.57), Eqs. (12.16) and (12.17) give

$$a_\tau^2 = 0, \quad b_\tau^2 = (1-c)\left(1 + \frac{5}{2}c\frac{1}{1-c}\right), \tag{12.21}$$

$$a_\tau^2 = \frac{4}{9c}(1-c), \quad b_\tau^2 = (1-c)\left(1 - \frac{5}{3}c\frac{1}{1+2c/3}\right), \tag{12.22}$$

$$a_\tau^2 = \frac{1}{3c}, \quad b_\tau^2 = \left(1 - \frac{16}{15\pi^2}c\frac{1}{1+16c/15\pi^2}\right). \tag{12.23}$$

The representations (12.18), (12.19), (12.21), and (12.22) were obtained in [246] for the case of perfect plasticity of composite materials. More recently the relations (12.21)–(12.23) were also obtained in [1052, 1235, 1236] using variational methods; the relations (12.19) and (12.20) were obtained in [180] for linear elastic composites.

Figure 12.1 shows a comparison of the above surfaces in terms of normalized overall stresses X_τ and Y_τ for a chosen spherical void volume fraction of $c = 0.25$. As expected, the yield surface (12.19) is enclosed inside the yield surface (12.22) because, according to [1235, 1236, 1633], the yield surface (12.22) is a rigorous upper bound for all other yield surfaces.

For comparison with yield surfaces (12.19) and (12.22), we now consider the application of the integral equation method (IEM) for estimating the second moment of stresses $\langle\boldsymbol{\sigma}\otimes\boldsymbol{\sigma}\rangle_0$ by the integral equation method (10.71) (with an

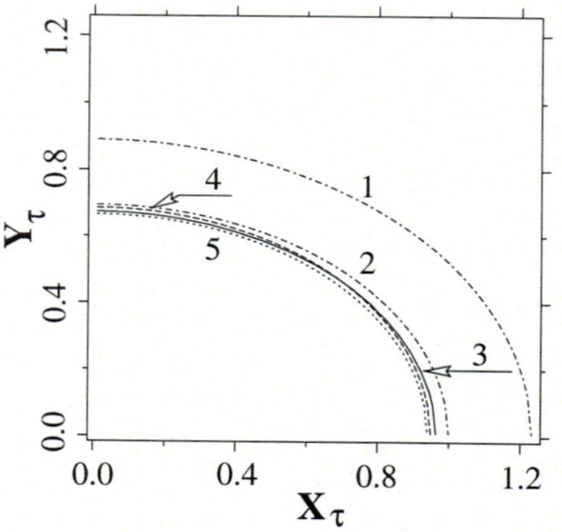

Fig. 12.1 Plots of the yield surface calculated via the method [714] (line 1), via the EFM (12.22) (line 2), via the MEFM (12.18) (solid line 3), via the IEM (10.57) (dashed line 4), via the refined IEM (10.71)

additional hypothesis **H1b**; see [265, 268]) and by the relation (10.57) ignoring the stress fluctuations in the inclusions with only the first integral in (10.57). In Fig. 12.1 we see that in both cases the yield surfaces are located inside the rigorous upper bound (12.22). It should be noted that Ju and coauthors [714, 719, 721] have used the estimation of the averaged second moment of the stress deviator $\langle \mathbf{ss} \rangle^{(0)}$ for the determination of yield surfaces. For this purpose they proposed an approximate integral equation method ignoring the stress fluctuations inside the inclusions; they suggested to evaluate the effective yield surface by the formula $(1-c)^2 \langle \mathbf{ss} \rangle^{(0)} = 2\tau_0^2/3$, where a factor $(1-c)^2$ is added without any justification. However, their results differ from Eqs. (10.57) and do not fall within the rigorous upper bound (12.22); see Fig. 12.1 and [266, 268].

For regular porous structures, the criteria of Gurson [584] ($\beta = 1$) as well as Koplik and Needleman [791] ($\beta = 1.25$)

$$Y_\tau^2 + 2c\beta \cosh\left[3X_\tau/(2\tau_0)\right] = 1 - \beta^2 c^2 \tag{12.24}$$

(see also [143, 1546]) construct the limiting surface by the numerical averaging of the dissipative function inside the matrix. Analogously to [499] it is interesting to note that for sufficiently small values X_τ/τ_0, we find from (12.24) by using a Taylor expansion of $\cosh(3X_\tau/2\tau_0)$ that for second-degree surfaces for the onset of yielding, Eq. (12.17) has parameters

$$a_\tau^2 = \frac{4}{9c}(1-\beta c)^2, \quad b_\tau^2 = (1-\beta c)^2. \tag{12.25}$$

The predictions of the MEFM, Eq. (12.19), and the results of Koplik and Needleman [791], Eq. (12.25); of Ponte Castañeda, Eq. (12.22); and of Gurson, Eq. (12.25), respectively, are related by the inequalities $b_\tau^{\text{MEFM}} < b_\tau^{\text{KN}} < b_\tau^{\text{P-C}} < b_\tau^{\text{G}}$ for $c > 0.2$. The difference between b_τ^{MEFM} and b_τ^{KN} is less than 5% for $c < 0.5$. The approximity of estimates by the methods mentioned was recognized [181, 278]. Buryachenko and Lipanov [246] showed that the two-particle approximation of the MEFM is superior to the one-particle approximation in representing the experimental data for uniaxial compressive loading (pressure p) of porous sintered electrolytic nickel subjected to uniaxial compression inside a die mold. The predicted dependences of the pressure leading to onset of yielding for $c \to 0$ are asymptotically similar to the calculated curves.

Let us analyze an influence of residual stresses on the limiting surface estimated. First we will consider elastically homogeneous isotropic medium containing statistically homogeneous of identical homogeneous spherical inclusions with radius a and eigenstrains $\boldsymbol{\beta}_1(\mathbf{x}) \equiv \beta_1^{(1)}\boldsymbol{\delta} = \text{const.}$ $(\mathbf{x} \in v^{(1)})$. The step function (5.97) is applied for the calculation of the stress fluctuations inside the inclusions and in the matrix. For the representation of numerical results in dimensionless form. we define a slight modification of a normalizing coefficient $\eta \equiv -3Q^k\beta_{10}c^{1/2}$, where $3Q^k$ represents the bulk component of the tensor $\mathbf{Q}_1 = (3Q^k, 2Q^\mu)$ and $\boldsymbol{\beta}(\mathbf{x}) \equiv \beta_{10}\boldsymbol{\delta}$ $(\mathbf{x} \in v^{(1)})$. The physical meaning of $\eta c^{-1/2}$ is defined as a hydrostatic stress inside a single isolated inclusion inside infinite homogeneous matrix (at $\mathbf{M}(\mathbf{x}) \equiv$const.). In addition we define the dimensionless variables $Y^{(i)} \equiv (3\langle \mathbf{s}\rangle : \langle \mathbf{s}\rangle/2)^{1/2}/\tau_0^{(i)}$, $X^{(i)} \equiv \eta/\tau_0^{(i)}$ for the component $v^{(i)}$ $(i = 0, 1)$. Then the criterion for onset of yielding (12.7) of the component $v^{(i)}$ is circumscribed by an ellipsoidal curve (do not sum to i)

$$\frac{(X^{(i)})^2}{(a_\tau^{(i)})^2} + \frac{(Y^{(i)})^2}{(b_\tau^{(i)})^2} = 1 \tag{12.26}$$

in the coordinates $X^{(i)}$*and* $Y^{(i)}$. The semiaxis $b^{(i)} = 1$ does not depend on the phase transformation strains $\boldsymbol{\beta}_1(1)$ and on the microtopology of the composite. The semiaxes

$$a_\tau^{(0)} = \left[3\Delta\sigma^{(0)2}/(2\eta^2)\right]^{-1/2}, \quad a_\tau^{(1)} = \left[3\Delta\sigma^{(1)2}/(2\eta^2)\right]^{-1/2} \tag{12.27}$$

where the stress fluctuations $\Delta\sigma^{(0)2}$ and $\Delta\sigma^{(1)2}$ were estimated in [185] by the simplified Eqs. (10.81) and (10.80), respectively, containing only the first integrals in the right-hand sides of Eqs. (10.81) and (10.80). The value of $a_\tau^{(i)}$ does not depend on the inclusion concentration c if the step radius distribution function (5.97) is assumed. The obtained criterion (12.26) is invariant with respect to the hydrostatic portion of the external stresses. For the mean field criterion (12.6), we obtain a degenerated criterion for onset of yielding of the component $v^{(i)}$ that is invariant with respect to transformation strain $\boldsymbol{\beta}_1$, when $b_\tau^{(i)} = 1$, $a_\tau^{(i)} = \infty$ in (12.26): $(Y^{(i)})^2 = 1$.

Figure 12.2 shows the curves 1, 2, and 3 calculated by using the criteria (12.27$_1$), (12.27$_2$), and (12.6), respectively, for the matrix and the inclusions ($\nu = 0.3$, the range $0.1 \le \nu \le 0.499$ was considered in [185]). The curve of yielding of the matrix (12.27$_1$) is enclosed by the curve for yielding of the inclusions (12.27$_2$). Hence, if both components have the same initial yield stress,

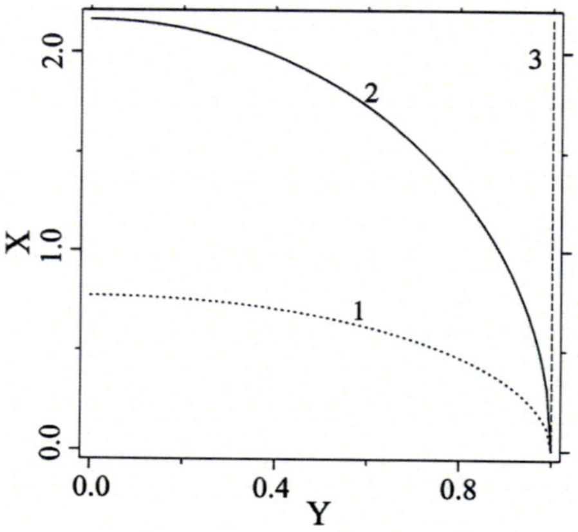

Fig. 12.2 $X^{(i)}$ vs. $Y^{(i)}$. The curves are estimated for the matrix [(12.27$_1$), (1)] and the inclusions [(12.27$_2$), (2)] and for Eq. (12.6), (3)

$\tau_0^{(0)} = \tau_0^{(1)}$, the yielding of the composite due to a stress-free strain $\boldsymbol{\beta}_1 \not\equiv 0$ (e.g., caused by a thermal mismatch) will begin from the matrix.

The authors [268] considered a numerical example for a composite consisting of isotropic homogeneous components with both the thermal and elastic mismatch and having identical spherical inclusions with yield properties according to (12.2). The effective limited surface is in general an ellipsoidal with an additional axes $Z_\tau^{(i)} \equiv \langle\sigma_0\rangle/\tau^{(i)}$ in Eq. (12.2). Moreover, for the spherical inclusion, this ellipsoid is degenerated into an elliptical cylinder with a symmetry axis which does not depend on the concentration of the inclusions. The pictures were presented for the cross sections $X_\tau = 0$ estimated by the both the MEFM and IEM for the different c.

Let us compare the estimates obtained by local averaging criteria (12.11) and (12.12) involving the interface stresses, with Eq. (12.4) based on utilizing of bulk stresses inside the phases. In the general case of loading we plot in Fig. 12.3 the effective yield surface described by Eq. (12.11) (the surface (12.12) is very close) of the porous material considered in Fig. 12.1. Just for comparison, we reproduce with the same notations the yield surfaces estimated via Eqs. (12.3), (12.22), and (12.18). As can be seen, the yield surface corresponding to the yield condition (12.11) accompanied by Eq. (10.59) is not elliptic (see also [181, 266]) and plotted for the particular case of external loading $\langle\sigma_{ij}\rangle = \alpha\delta_{i3}\delta_{j3} + \beta\delta_{ij}$, where α and β are arbitrary variables with $\alpha > 0$. As indicated in Fig. 12.3, the criterion (12.11) as well as (12.22) and (12.18) shows that yielding will also take place under purely hydrostatic loading.

The authors [277] quantitatively investigated the performance of the approaches in Sect. 12.1.2 to the failure analysis of fiber composites. Nonelliptical effective failure envelopes were estimated by both the MEFM and EFM for the RDF (5.97) and (5.101) and the different failure initiation criteria at the interface.

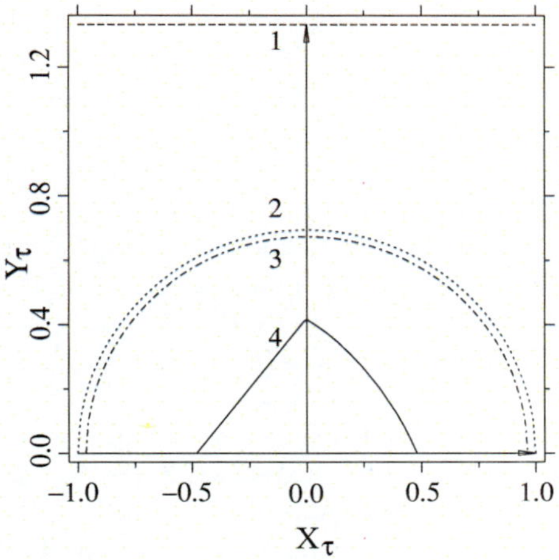

Fig. 12.3 $Y^{(i)}$ vs. $X^{(i)}$ for the matrix, calculated by Eqs. (12.3) and (8.31) (1), by Eq. (12.22) (2), by (12.18) (3), and by Eqs. (12.4) and (10.59) (4)

12.1.5 Possible Generalizations

The scheme presented in Sects. 10.4 and 12.1.3 can be easily generalized for the case of the elastic mismatch of inclusions in the framework of the method of successive approximations with the realization of the MEFM at each iteration. So, the initial problem for multiple interacting inclusions (4.106), analogous to the macrocrack–microcrack interaction problem considered in [339, 729], should be accompanied by additional equation for the effective field for the crack $\overline{\boldsymbol{\sigma}}^{\text{crack}}(\mathbf{s})$, or, what is the same—for the traction $\mathbf{t}(\mathbf{s})$, coupled with the SIF by the use of explicit integral representation of the tensor $\boldsymbol{\mathcal{L}}^{\text{crack}}$ (4.133). The further realization of the nonlocal version of the MEFM method (13.2) should be accompanied by as additional equation for an isolated single crack subjected to an unknown effective field $\overline{\boldsymbol{\sigma}}^{\text{crack}}(\mathbf{s})$ that makes the problem essentially nonlocal one even for statistically homogeneous inclusion field.

In so doing, the application of the iteration method that is simplified in the framework of "quasi-crystalline" approximation (8.11) is preferable.

A problem of damage processes in brittle matrix composites under tension parallel to the fiber direction [62, 291, 775, 1038, 1099, 1100, 1174, 1199, 1358] is of considerable interest in practice. After crack nucleation, two models of damage evaluation are possible: multiple cracking and crack expansion. The first model can be described in a spirit of the standard models for multicomponent composites. The second model is determination of an appropriate force-separation law that characterized the restraining traction provided by the fibers. The popular simplified models for overall responses of composites with bridged cracks have the following drawbacks: two-dimensional bridging analysis leading to the neglect of the spatial variation of crack tip behavior, replacement of surrounding fiber-reinforced composites by a homogeneous isotropic material with effective properties (see, e.g., [291, 483]), and a continuous bridging modeling replacing the action of the fibers by a smeared bridging stress distribution [1100]. The exact and approximate solution for the one and deterministic field of fibers bridging the crack [291, 470, 483] is known. The stochastic aspects of damage processes are valid ones for further investigation.

It should be mentioned that there is some similarity between Eq. (13.2) for microinhomogeneous medium and equations involved for bridged crack model that coincide between one to other except the notations in the framework of simplified fiber-bridged crack model [291] assuming a coincidence of elastic properties of the matrix and fibers as well as a small variation of bridging force inside each fiber (analog of hypothesis **H1**). Then in Eq. (13.2) the integral over the whole space should be replaced by the integral over the crack surface $\mathbf{s} \in S$. The tensor $\mathbf{B}_i(\mathbf{s}_i)$ (8.2) is estimated from the solution for a single fiber bridging a crack, and the matrix of binary interactions $\mathbf{Z}_{ij}(\mathbf{s}_i, \mathbf{s}_j)$ is also known. The knowledge of tensors mentioned makes it possible to utilize the scheme of a two-particle approximation of the MEFM for finding of the nonlocal dependence of bridging forces $\boldsymbol{\sigma}(\mathbf{s})$ on the applied stress field $\boldsymbol{\sigma}^0(\mathbf{s})$. The subsequent estimation of effective parameters $\langle \mathbf{K}(\mathbf{z})\rangle$ and $\langle \mathcal{J}(\mathbf{z})\rangle$ ($\mathbf{z} \in \Gamma^{\text{crack}}$) is obvious. The scheme presented was in actuality realized in [1099] in the framework of the Mori–Tanaka method for the homogeneous $\boldsymbol{\sigma}^0 \equiv$ const. and a limiting case of a completely failed matrix material (i.e., two half-spaces connected by fibers are considered). These assumptions allow one to eliminate from consideration all nonlocal effects induced in the general case by the nonhomogeneity of the effective field $\overline{\boldsymbol{\sigma}}(\mathbf{x})$.

Evolution of nonlinear phenomena such as viscoplasticity, creep, and damage is a new challenge for designers of rockets. When one tries to estimate the equivalent stress in the strength theories as well as in nonlinear creep theory, or when the yield function in plasticity theory is considered, squares of the first invariant or the second invariant of the deviator of local stresses are frequently used. For unimodal distribution of particles, estimations of the required second statistical moments of local stresses can be performed by both the perturbation method and the method of integral equations (see for details Chap. 13 in [200]). Generalization of the mentioned methods to the case of the multimodal distribution of particles (which is inherent in the HSP) is straightforward. It opens a new avenue of questions on some specific nonlinear problems such as ignition, hot spot analysis, and detonation. So the authors [374, 471, 576, 1026, 1483] indicated the following mechanisms of energy concentrations in locations called "hot spot" causing localized melting, ignition, and fast burn of the reactive material. All of these problems were partially solved for HSPs and explosives by simplified micromechanical models such as the dilute approximation ($\langle\overline{\boldsymbol{\sigma}}\rangle_i = \langle\boldsymbol{\sigma}\rangle$; see [374, 1711]), the effective medium approximation considering each defect inside the effective medium ($\mathbf{L}^{(0)} = \mathbf{L}^*$; see [1277]), and the mixture role of damage mechanics. It can also be mentioned that substitution of estimations of both the first and second statistical moments of local stresses performed by the proposed approach (see Sects 12.3) into the mentioned nonlinear problems is straightforward.

12.2 Nonlinear Composites

Although significant advances have been achieved with respect to disorder and nonlinearity micromechanics separately, the situation becomes quite complex and interesting when both are important. When plastic deformations occur, the homogeneity of the mechanical properties of the components is lost, and the local properties of the phases become position-dependent. Exact solutions via analytical or numerical methods can be obtained only for model composites having deterministic phase arrangements, such as composites with regular structure (see, e.g., [4, 140, 420, 1117, 1119, 1120, 1472, 1605]). Alternatively, nonlinear multiphase materials may be described by self-consistent schemes. For example, a special case of infinitely small concentrations of heterogeneities in nonlinear matrix composites was studied [416, 892], and then differential self-consistent schemes were used to obtain approximate constitutive relations applicable for a large range of inclusion concentrations. In addition, effective medium methods were employed for carrying out analysis of elastoplastic polycrystals [677, 1123]. The three-phase spherical model [611] was used in [1698] to predict the creep behavior of metal matrix composites (see also [147, 638]) and in [1159] for elastoplastic analysis of thermal cycling of particulate composites. A principal role of the concentration of plastic strains in the vicinity of the inclusions in the overall deformation properties of composite materials was shown [635, 1475]. The method [110] was extended in [1676] for linear problem and made a representation of the effective nonlinear dielectric constant and cubic susceptibility which effectively used a perturbation solution. Agarwal and Dutta Gupta [10] proposed a single-grain T-matrix approach for the calculation of the third- and fifth-order nonlinear susceptibilities of a heterogeneous medium. A determination of the third-order elastic constant by using the perturbation approximation was obtained in [58].

Other noteworthy achievements in the theory of nonlinear composite are connected with generalizations of the Hashin–Shtrikman [621] variational principles to nonlinear materials, as proposed in [1480] (see also [1158, 1239]). Bounds for the effective properties of a broad class of nonlinear composites were determined in [1052, 1630, 1633]. A new variational procedure that can be used to obtain general types of bounds as well as estimates for the effective properties of nonlinear composites was developed [1233–1235].

Standard methods of successive approximation may be used in nonlinear problems of composites with random structure, linearized problems being solved at each stage. So, in a classical method of successive approximation, the secant tensor is defined by the average strains in the constituent (see, e.g., [58, 248, 249, 769]). The well-known concept of secant moduli has been combined with the hypothesis of the homogeneity of the increments of the plastic or creep strains [117, 249, 420, 1491]. Since the widely used method of average strains is capable of estimating only the average stresses in the components, its use for the linearization of functions describing nonlinear effects, e.g., strength ([33, 1287], yielding [938], hardening [1258], creep [1698], and dynamic viscoplasticity [1699] may be problematical.

Owing to the significant inhomogeneity of the stress fields in the components (especially in the matrix), such linearization entails physical inconsistencies, which will be discussed within the context of predicting the flow behavior of porous media. When the investigations are conducted with the aim of determining effective nonlinear elastic parameters, a concept of secant modulus is used. But at each step the linearized problem is solved with the help of the MEFM which has none of the disadvantages of Mori–Tanaka's method (for details see Chap. 8). In the modified approach, the nonlinear dependence of elastic properties on the strain field is taken into account, and one assumes that the effective strain for the phase being considered equals the second moment of strains averaged over this phase [179, 180, 1469]. In doing so the exact estimates [808, 1187] (as distinct from those contained in [117, 1258, 1491]) in relation to the average in volume of anisotropic components single-point second moments of the stress fields are used. A similar approach was employed previously in [179, 180] for estimating the overall moduli of nonlinear elastic composites. Finally, the application of the proposed theory to the case of different problems will be considered (following Buryachenko [181]): the effective moduli, the yield surface, the power-law creep response, and the elastoplastic deformation of composites with compressible and incompressible matrix are predicted.

12.2.1 Linearization Schemes

We consider statistically homogeneous matrix composites consisting of a homogeneous matrix containing a statistically uniform set of inclusions. The local strain and stress tensors satisfy the linearized strain–displacement (6.3) relations and the equilibrium equations (6.1) (with no body force acting), respectively. We consider a mesodomain w containing a statistically large number of realization of inclusions and subjected to the homogeneous boundary conditions (6.5) generating a homogeneous macroscopic stress $\boldsymbol{\sigma}^0 = \langle \boldsymbol{\sigma} \rangle$; the phases are assumed to be perfectly connected.

We will consider a constitutive equation

$$\boldsymbol{\sigma}(\mathbf{x}) = \mathbf{L}(\boldsymbol{\varepsilon}, \mathbf{x})\boldsymbol{\varepsilon} + \boldsymbol{\mathcal{L}}(\boldsymbol{\varepsilon}, \mathbf{x})\boldsymbol{\varepsilon} \otimes \boldsymbol{\varepsilon}, \tag{12.28}$$

where $\otimes$ stands for a tensor product. $\mathbf{L}(\boldsymbol{\varepsilon}, \mathbf{x})$ and $\boldsymbol{\mathcal{L}}(\boldsymbol{\varepsilon}, \mathbf{x})$ are the second-order elastic compliance and the third-order elastic stiffness, respectively. Hereafter we assume that $\mathbf{L}(\boldsymbol{\varepsilon}, \mathbf{x})$ and $\boldsymbol{\mathcal{L}}(\boldsymbol{\varepsilon}, \mathbf{x})$ depend on $\boldsymbol{\varepsilon}$ only through the equivalent strains $\varepsilon^e_k(\mathbf{x})$: $\mathbf{L}(\boldsymbol{\varepsilon}, \mathbf{x}) \equiv \mathbf{L}(\varepsilon^e_1, \mathbf{x})$, $\boldsymbol{\mathcal{L}}(\boldsymbol{\varepsilon}, \mathbf{x}) \equiv \boldsymbol{\mathcal{L}}(\varepsilon^e_2, \mathbf{x})$ $(k = 1, 2)$, and: $\varepsilon^e_k = \mathbf{E}_k\boldsymbol{\varepsilon} + \mathcal{E}_k\boldsymbol{\varepsilon} \otimes \boldsymbol{\varepsilon}$, where $\mathbf{E}_k and\ \mathcal{E}_k$ $(k = 1, 2)$ are the some second-order and fourth-order tensors, respectively. In particular, for the Murnaghan stain potential w (2.78), we obtain from expression $\boldsymbol{\sigma} = \partial w/\partial \boldsymbol{\varepsilon}$

$$L_{ijkl}(\boldsymbol{\varepsilon}) = (3k, 2\mu) \equiv 3k\mathbf{N}_1 + 2\mu\mathbf{N}_2, \quad \mathbf{E}_1 = \mathbf{E}_2 = \mathbf{0}, \quad \mathcal{E}_1 = \mathcal{E}_2 = \mathbf{0},$$

$$\mathcal{L}_{ijklmn} = a\delta_{ij}\delta_{kl}\delta_{mn} + b(\delta_{ij}I_{klmn} + \delta_{mn}I_{ijkl} + \delta_{kl}I_{mnij}) + dJ_{ijklmn}, \quad (12.29)$$

where $J_{ijklmn} = (I_{ipmn}I_{jpkl} + I_{ipkl}I_{jpmn})/2$. For the Kauderer strain potential w (2.83), we have

$$\mathbf{L}(\boldsymbol{\varepsilon}) = \mathbf{L}^{[0]} + \boldsymbol{\psi}(\boldsymbol{\varepsilon}), \;\; \boldsymbol{\mathcal{L}} \equiv \mathbf{0}, \;\; \mathbf{L}^{[0]} = (3k, 2\mu), \;\; \boldsymbol{\psi}(\boldsymbol{\varepsilon}) = 2\gamma\varepsilon_{\rm eq}^2\mathbf{N}_2, \quad (12.30)$$

where $\varepsilon_{\rm eq}^2 \equiv 2\mathbf{e} : \mathbf{e}/3$, $\mathbf{e} \equiv \boldsymbol{\varepsilon} - \varepsilon_0\boldsymbol{\delta}$, $\varepsilon_0 \equiv \varepsilon_{ii}/3$, and γ is a nonlinear parameter. The tensor of material properties $\mathbf{f} \equiv \mathbf{f}^{(0)} + \mathbf{f}_1(\mathbf{x})$ ($\mathbf{f} = \mathbf{L}, \mathbf{M}, \boldsymbol{\mathcal{L}}, \mathbf{M} \equiv \mathbf{L}^{-1}$) is assumed to be constant in the matrix $v^{(0)} = w\backslash v$ ($v \equiv \cup v_i$, $i = 1, 2, \ldots$) $\mathbf{f}(\mathbf{x}) \equiv \mathbf{f}^{(0)}$ and in each inclusion phase v_i ($i = 1, 2, \ldots$) $\mathbf{f}(\mathbf{x}) = \mathbf{f}^{(0)} + \mathbf{f}_1^{(i)}$; hereinafter the upper index of materials properties tensor, given in the parentheses, shows the number of the respective components. The rule of calculation of these piecewise constant tensors is described below. We recast the constitutive equation in the form

$$\boldsymbol{\sigma}(\mathbf{x}) = \mathbf{L}(\boldsymbol{\varepsilon}, \mathbf{x})\boldsymbol{\varepsilon}(\mathbf{x}) + \boldsymbol{\alpha}(\boldsymbol{\varepsilon}, \mathbf{x}), \;\; \boldsymbol{\alpha}(\boldsymbol{\varepsilon}, \mathbf{x}) = \boldsymbol{\mathcal{L}}(\boldsymbol{\varepsilon}, \mathbf{x})\boldsymbol{\varepsilon} \otimes \boldsymbol{\varepsilon}, \quad (12.31)$$

where $k = 0, 1, \ldots, N$. In such a case, Eq. (12.31) formally coincides with an appropriate equation of thermoelasticity (6.2). The effective moduli of the second and third orders in the relation $\langle\boldsymbol{\sigma}\rangle = \mathbf{L}^*\langle\boldsymbol{\varepsilon}\rangle + \boldsymbol{\mathcal{L}}^*\langle\boldsymbol{\varepsilon}\rangle\otimes\langle\boldsymbol{\varepsilon}\rangle$ can be found by averaging local equation (12.31)

$$\mathbf{L}^* = \mathbf{L}^{(0)} + \langle\mathbf{L}_1\mathbf{A}^*\rangle, \quad \boldsymbol{\mathcal{L}}^* = \sum_{i=1}^{N}\langle\mathbf{L}_1^{(i)}\boldsymbol{\mathcal{F}}_1^{(i)}\rangle^{(i)} + \sum_{k=0}^{N}\langle\boldsymbol{\mathcal{L}}^{(k)}\rangle\boldsymbol{\mathcal{F}}_2^{(k)}, \quad (12.32)$$

where the tensors of the fourth-rank $\mathbf{A}^*$, the sixth-rank $\boldsymbol{\mathcal{F}}_1$, and the eighth-rank $\boldsymbol{\mathcal{F}}_2$ define the average strain concentrations in a component $\mathbf{x} \in v^{(k)}$ ($k = 0, 1, \ldots, N$) $\langle\boldsymbol{\varepsilon}\rangle^{(k)} = \mathbf{A}^*(\mathbf{x}) + \boldsymbol{\mathcal{F}}_1(\mathbf{x})\langle\boldsymbol{\varepsilon}\rangle \otimes \langle\boldsymbol{\varepsilon}\rangle$, $\langle\boldsymbol{\varepsilon} \otimes \boldsymbol{\varepsilon}\rangle^{(k)} = \boldsymbol{\mathcal{F}}_2(\mathbf{x})\langle\boldsymbol{\varepsilon}\rangle \otimes \langle\boldsymbol{\varepsilon}\rangle$. To obtain final results that can be visualized, we adopt linearization of the Eqs. (12.28) and (12.31) and subsequent equations, which assumes the homogeneity of $\boldsymbol{\varepsilon}(\mathbf{x}) \otimes \boldsymbol{\varepsilon}(\mathbf{x})$, $\mathbf{L}(\boldsymbol{\varepsilon}, \mathbf{x})$, and $\boldsymbol{\mathcal{L}}(\boldsymbol{\varepsilon}, \mathbf{x})$ within the phase $v^{(k)}$ ($k = 0, 1, \ldots, N$): $\boldsymbol{\varepsilon}(\mathbf{x})\otimes\boldsymbol{\varepsilon}(\mathbf{x}) \equiv \langle\boldsymbol{\varepsilon}(\mathbf{x})\otimes\boldsymbol{\varepsilon}(\mathbf{x})\rangle^{(k)}$, $\mathbf{L}(\boldsymbol{\varepsilon}, \mathbf{x}) = \mathbf{L}(\langle\varepsilon_1^e\rangle^{(k)}, \mathbf{x})$, and $\boldsymbol{\mathcal{L}}(\boldsymbol{\varepsilon}, \mathbf{x}) \equiv \boldsymbol{\mathcal{L}}(\langle\varepsilon_2^e\rangle^{(k)}, \mathbf{x})$ at $\mathbf{x} \in v^{(k)}$. This method of linearization assumes homogeneity of material parameters within the phases $\mathbf{x} \in v^{(k)}$, $k = 0, 1, \ldots, N$):

$$\mathbf{L}(\boldsymbol{\varepsilon}, \mathbf{x}) \equiv \mathbf{L}^{s(k)} = \mathbf{L}(\langle\varepsilon_1^e\rangle^{(k)}), \;\; \boldsymbol{\mathcal{L}}(\boldsymbol{\varepsilon}, \mathbf{x}) \equiv \boldsymbol{\mathcal{L}}^{s(i)} = \boldsymbol{\mathcal{L}}(\langle\varepsilon_2^e\rangle^{(k)}). \quad (12.33)$$

Hereafter the superscript ^{s} indicates the calculation of the values with the help of L^s. The method of estimation of secant modulus $\mathbf{L}^{s(k)}$, $\boldsymbol{\mathcal{L}}^{s(k)}$, ($k = 0, 1, \ldots$) is based on application of a method of successive approximation for the solution of nonlinear problems (see, e.g., [566, 1152]). The constitutive equation (12.31) then yields

$$\boldsymbol{\sigma}^{[1]} = \mathbf{L}^{s[1]}\boldsymbol{\varepsilon}^{[1]}, \tag{12.34}$$

$$\boldsymbol{\sigma}^{[2]} = \mathbf{L}^{s[2]}\boldsymbol{\varepsilon}^{[2]} + \boldsymbol{\alpha}^{[2]}, \ \boldsymbol{\alpha}^{[2]} \equiv \boldsymbol{\mathcal{L}}^{s[2]}\boldsymbol{\varepsilon}^{[1]} \otimes \boldsymbol{\varepsilon}^{[1]}, \ldots \tag{12.35}$$

$$\boldsymbol{\sigma}^{[n+1]} = \mathbf{L}^{s[n+1]}\boldsymbol{\varepsilon}^{[n+1]} + \boldsymbol{\alpha}^{[n+1]}, \ \boldsymbol{\alpha}^{[n+1]} \equiv \boldsymbol{\mathcal{L}}^{s[n+1]}\boldsymbol{\varepsilon}^{[n]} \otimes \boldsymbol{\varepsilon}^{[n]}, \tag{12.36}$$

where the upper index in parentheses indicates the iteration number of the tensors.

The tensors $\mathbf{L}^s$ and $\boldsymbol{\mathcal{L}}^s$ are not known in general spiking. The problem of their estimation even at a local level is a substantially nonlinear one because of nonlinear dependence between strains and stresses. Since the secant moduli $\mathbf{L}^{s(i)}$ and $\boldsymbol{\mathcal{L}}^{s(i)}$ (and consequently $\mathbf{L}^{s*}$ and $\boldsymbol{\mathcal{L}}^{s*}$) are not known a priori, an iterative procedure is usually required. We may start by assuming ($\mathbf{x} \in v^{(k)}, \ k = 0, 1, \ldots$): $\mathbf{L}^{s(k)[0]}(\mathbf{x}) \equiv \mathbf{L}^{(k)}(\mathbf{0}, \mathbf{x}), \ \boldsymbol{\mathcal{L}}^{s(k)[0]}(\mathbf{x}) \equiv \boldsymbol{\mathcal{L}}^{(k)}(\mathbf{0}, \mathbf{x}), \ \boldsymbol{\alpha}^{[0]} = \mathbf{0}$, and calculate the corresponding $\mathbf{L}^{*[0]}$ and $\boldsymbol{\mathcal{L}}^{*[0]}$ by the appropriate equations of Chap. 8. The other secant modulus then follows from equations ($\mathbf{x} \in v^{(k)}; \ k = 0, 1, \ldots; \ n = 0, 1, \ldots$)

$$\mathbf{L}^{s(k)[n+1]}(\mathbf{x}) = \mathbf{L}\Big(\langle \varepsilon_1^{e[n]} \rangle^{(i)}, \mathbf{x}\Big), \ \boldsymbol{\mathcal{L}}^{s(k)[n+1]}(\mathbf{x}) = \boldsymbol{\mathcal{L}}\Big(\langle \varepsilon_2^{e[n]} \rangle^{(i)}, \mathbf{x}\Big), \tag{12.37}$$

$$\boldsymbol{\varepsilon}^{(k)[n+1]}(\mathbf{x}) \otimes \boldsymbol{\varepsilon}^{(k)[n+1]}(\mathbf{x}) = \langle \boldsymbol{\varepsilon} \otimes \boldsymbol{\varepsilon} \rangle^{(k)[n]} \tag{12.38}$$

where $[n]$ is the number of an iteration; $\langle \varepsilon_1^{e[n]} \rangle^{(k)} and \ \langle \varepsilon_2^{e[n]} \rangle^{(k)}$ are defined by $\mathbf{L}^{s*} = \mathbf{L}^{s*[n]}, and \ \boldsymbol{\mathcal{L}}^{s(k)} = \boldsymbol{\mathcal{L}}^{s(k)[n]}$. If $||\mathbf{I} - \mathbf{L}^{s*[n]}/\mathbf{L}^{s*[n]}|| \equiv \xi \ll 1$, then the solution is found; usually $n \leq 5$ at $\xi = 10^{-2}$.

A widely distributed method of linearization in nonlinear elasticity problems assumes homogeneity the field of stresses or strains (contrary to Eq. (12.33) when computing material tensors, which describe nonlinear effects. One usually equates to terms ($\mathbf{x} \in v^{(k)}; \ k = 0, 1, \ldots$)

$$\mathbf{L}(\mathbf{x}) = \mathbf{L}^{s(k)} \equiv \mathbf{L}(\mathbf{E}_1 \langle \boldsymbol{\varepsilon} \rangle^{(k)} + \mathcal{E}_1 \langle \boldsymbol{\varepsilon} \rangle^{(k)} \otimes \langle \boldsymbol{\varepsilon} \rangle^{(k)}), \tag{12.39}$$

$$\boldsymbol{\mathcal{L}}(\mathbf{x}) = \boldsymbol{\mathcal{L}}^{s(k)}(\mathbf{x}) \equiv \boldsymbol{\mathcal{L}}(\mathbf{E}_2 \langle \boldsymbol{\varepsilon} \rangle^{(k)} + \mathcal{E}_2 \langle \boldsymbol{\varepsilon} \rangle^{(k)} \otimes \langle \boldsymbol{\varepsilon} \rangle^{(k)}). \tag{12.40}$$

Then for the Murnaghan strain potential w (2.78) with the elastic moduli (12.29), we get for the second-order approximation (12.35) $\boldsymbol{\mathcal{L}}^* = \sum_{k=0}^{N} c_i \boldsymbol{\mathcal{L}}^{(k)}(\mathbf{A}_k^* \otimes \mathbf{A}_k^* \otimes \mathbf{A}_k^*)$, where the strain concentrator factors $\mathbf{A}_k^*$ are found for the first-order approximation (12.34) by some averaging method prescribed in Chap. 8. In particular for the isotropic composites with the spherical inclusions and the components described by elastic potential (2.78), the analytical representations for the tensor $\boldsymbol{\mathcal{L}}^*$ were obtained in [248] (see also [200]), where the numerical results are presented (and compared with the method [1021, 1024]) for the porous steel. A porous material with the matrix described by the Kauderer potential (12.30) in the framework of the linearization scheme (12.39) was also considered with a comparative (with [1021]) numerical analysis.

Hypothesis (12.39) is inconsistent for potential (2.78). In fact, let us consider an isotropical porous material with a matrix, the elastic properties of which are described by the stress potential (2.78). In such a case $\langle \varepsilon_{ij} \rangle = \langle \varepsilon_0 \rangle \delta_{ij}$ at the hydrostatic loading condition irrespective of microstructure of pores and the method of calculation of $\langle \boldsymbol{\varepsilon} \rangle^{(0)}$ (e.g., by the formula (8.31) or by any other formula), we obtain that $\langle \mathbf{e} \rangle^{(0)} \equiv \mathbf{0}$ ($\mathbf{e} \equiv \boldsymbol{\varepsilon} - \varepsilon_0 \boldsymbol{\delta}$). It means that this porous material with incompressible nonlinear elastic modulus has only linear elastic deformations at hydrostatic compression.

12.2.2 Modified Linearization Scheme

The abovementioned inconsistency can be easily avoided if the linearizations (12.39) and (12.40) are replaced by the linearizations

$$\mathbf{L}(\mathbf{x}) = \mathbf{L}^{s(k)} \equiv \mathbf{L}\big(\mathrm{E}_1 \langle \boldsymbol{\varepsilon} \rangle^{(k)} + \mathcal{E}_1 \langle \boldsymbol{\varepsilon} \otimes \boldsymbol{\varepsilon} \rangle^{(k)}\big), \tag{12.41}$$

$$\boldsymbol{\mathcal{L}}(\mathbf{x}) = \boldsymbol{\mathcal{L}}^{s(k)}(\mathbf{x}) \equiv \boldsymbol{\mathcal{L}}\big(\mathrm{E}_2 \langle \boldsymbol{\varepsilon} \rangle^{(k)} + \mathcal{E}_2 \langle \boldsymbol{\varepsilon} \otimes \boldsymbol{\varepsilon} \rangle^{(k)}\big). \tag{12.42}$$

where $k = 0, 1, \ldots, N$, and the second moment of strains $\langle \boldsymbol{\varepsilon} \otimes \boldsymbol{\varepsilon} \rangle^{(k)}$ are estimated by one of the methods considered in Chap. 11. In particular, at the second-order approximation (12.35), this value can be estimated by the perturbation method (6.190) for the purely elastic case ($\boldsymbol{\alpha} \equiv \mathbf{0}$)

$$\langle \boldsymbol{\varepsilon} \otimes \boldsymbol{\varepsilon} \rangle^{(k)[1]} = \frac{1}{c^{(k)}} \frac{\partial \mathbf{L}^{s*[0]}}{\partial \mathbf{L}^{s(k)[0]}} \langle \boldsymbol{\varepsilon} \rangle \otimes \langle \boldsymbol{\varepsilon} \rangle, \tag{12.43}$$

while the subsequent approximations $\langle \boldsymbol{\varepsilon} \otimes \boldsymbol{\varepsilon} \rangle^{(k)[n]}$ should be estimated through consideration of an appropriate thermoelastic problem (e.g., like Eqs. (6.190) and (8.35)).

Let us consider, as an example, a porous material containing the spherical identical voids in the incompressible matrix with the elastic properties described by the Kauderer potential (12.30). Then we have an initial approximation of the effective moduli $\mathbf{L}^{*[0]} = \mu^{(0)}(4f_1, 2f_2)$, where

$$f_1 = \frac{1}{c} - \frac{29}{24}, \quad f_2 = \Big(1 - \frac{35c}{24}\Big)\Big(1 + \frac{5c}{24}\Big)^{-1}, \tag{12.44}$$

$$f_1 = \frac{1}{3}\Big(\frac{1}{c} - \frac{16}{15\pi^2}\Big), \quad f_2 = \Big(1 - \frac{448c}{375\pi}\Big)\Big(1 - \frac{848c}{375\pi^2}\Big)^{-1}, \tag{12.45}$$

for the spherical pores ($a_1 = a_2 = a_3 = a,\ c = 4/3\pi a^3 n$), and coin-shaped cracks ($a_1 = a_2 = a \gg a_3,\ c = 4/3\pi a^3 n$), respectively, are found by the point approximation of the MEFM (8.230) and (8.232). Then the effective isotropic tensors $\boldsymbol{\psi}^*$ (12.30) obtained from either linearization conditions (12.39) or nonlinear ones (12.41) and (12.43) are given in [180] $\boldsymbol{\psi}^* = \gamma^{(0)}(0,\ f_2^2 \langle \varepsilon \rangle_{\text{eq}}^2)$, and $\boldsymbol{\psi}^* = \gamma^{(0)}(1-c)^{-1}(16 f_1^2 \langle \varepsilon_0 \rangle^2 + 2 f_1 f_2 \langle \varepsilon \rangle_{\text{eq}}^2,\ 8 f_1 f_2 \langle \varepsilon_0 \rangle^2 + f_2^2 \langle \varepsilon \rangle_{\text{eq}}^2)$, respectively, where $\langle \varepsilon \rangle_{\text{eq}}^2 = 2\langle \mathbf{e} \rangle \langle \mathbf{e} \rangle / 3$, $\varepsilon_0 = \langle \varepsilon_{ii} \rangle / 3$, $\mathbf{e} = \boldsymbol{\varepsilon} - \varepsilon_0 \boldsymbol{\delta}$. Due to the incompressibility of the matrix, the obtained representations are consistent with an infinite number of iterations (12.37).

Now let us consider a composite material with only one phase demonstrating nonlinear properties $\boldsymbol{\sigma}(\mathbf{x}) = \mathbf{L}^{s(i_1)}\boldsymbol{\varepsilon}(\mathbf{x})$, $\mathbf{L}^{s(i_1)} = \mathbf{L}(\mathbf{x}, \varepsilon_1^e)$, $\mathbf{x} \in v^{(i_1)}$, while the linear law $\boldsymbol{\sigma}(\mathbf{x}) = \mathbf{L}^{s(j)[0]}\boldsymbol{\varepsilon}(\mathbf{x})$, $\mathbf{x} \in v^{(j)}$, $(j = 1, \ldots, N;\ j \neq i_1)$ holds good for other phases. In such a case, the general iteration scheme (12.37) can be simplified. Indeed, we assume that a monotonically increasing proportional loading

$$\langle \varepsilon_{ij}(t) \rangle = \xi_{ij}\varepsilon^t(t) \tag{12.46}$$

takes place, where ξ_{ij} is a desired proportional constant and $\varepsilon^t(t)$ is a monotonically increasing scalar function depending on a time parameter t. At the first step, we will give a small increment of the second strain invariant $\langle \varepsilon \rangle_{\text{eq}}^{(i_1)2}$ in the phase $v^{(i_1)}$. Thereafter we will estimate the secant modulus values according to Eqs. (12.41) and (12.43) and find $\mathbf{L}^{s*}$ by the use of the determined $\mathbf{L}^{s(i_1)}$. For the desired proportional loading (12.46), the value $\varepsilon^t(t)$ is given by

$$\varepsilon^t(t) = \left[\frac{1}{3c^{(i_1)}}\frac{\partial \mathbf{L}^{s*}}{\partial \mu^{s(i_1)}}\langle \boldsymbol{\xi} \rangle \otimes \langle \boldsymbol{\xi} \rangle\right]^{-1/2}, \tag{12.47}$$

which provides a way of estimating the components $\langle \varepsilon_{ij} \rangle$ from (12.46). The average stresses $\langle \boldsymbol{\sigma} \rangle$ can be found from the equation $\langle \boldsymbol{\sigma} \rangle = \mathbf{L}^{s*}\langle \boldsymbol{\varepsilon} \rangle$ through the known tensors $\mathbf{L}^{s*}$ and $\langle \boldsymbol{\varepsilon} \rangle$. The second step of deformations at the stress–strain curve can be performed by increasing the values $\langle \varepsilon \rangle_{\text{eq}}^{(i_1)2}$. This procedure continues up to a value of $\varepsilon^t(t) = 1$; see for details [180]. Similar modified secant moduli theory based on the variational principles ([1236] and [1470]) was independently proposed by Suquet [1469].

It should be mentioned that in contrast, the problem involving geometrically nonlinear materials has hardly been investigated, which is understandable given the inherent mathematical difficulties that lead to some simplifications of the models proposed. So, Hashin [612] used the composite spheres assemblage model, taking the large strains of the spherical element into account. To derive the second-order elasticity equations, one uses the method of successive approximation [118] with a power series expansion of the displacements, stresses, and their gradients in a certain small parameter. In so doing, the case of dilute concentration of spherical inclusions was considered in [684, 980], and [1152]. The case of a finite concentration of ellipsoidal inclusions was analyzed in [1022, 1023] (see also [767]) by the method of conditional moments (linear versions of this method were described in Chap. 8) at each step of successive approximation; the average strains in the components estimated at the first step were used at the second step, or, in other words, one used the linearization (12.36). A similar approach with the construction of infinite number of recursions in the scheme of successive approximation method was realized in [823] by the perturbation method combined with the assumption $\boldsymbol{\Gamma}(\mathbf{x}) = -\mathbf{Q}\delta(\mathbf{x})$. In the model [245] (see also [200]), the authors generalize the Hashin's [612] model which presumes equivalent strain problems for a porous medium and a thick-walled spherical shell and equality of the ratio of the volumes of the pore and spherical element to the porosity of the composite medium being modeled. The model [245] uses the advantage of the MEFM and places the thick-walled spherical elements $v^{\rm sp}$ (with the gas pressure inside the hole) in the matrix with prescribed effective stress field at infinity. It was demonstrated (with graphical representations of numerical results) that for small porosity, the deformation behaviors at the positive and negative hydrostatic loading are fundamentally different owing to nonlinear effects in the thick-walled spherical elements $v^{\rm sp}$. Since a material always contains a certain number of pores, it follows that perfectly isotropic inextensible materials do not exist. Moreover, the parameters of different elastic potentials of isotropically deformable materials are generally determined in hydrostatic compression (see [329, 330]). In light of the above analysis, the use of these parameters in the region of large hydrostatic tension would appear to be incorrect.

Other methods of analysis of nonlinear properties are based on the *incremental method* or *tangent modulus concept* (for details see [1239]). Constitutive law is written in the incremental form $\dot{\boldsymbol\sigma} = \mathbf{L}_t(x, \boldsymbol\varepsilon)\dot{\boldsymbol\varepsilon}$ where the tangent tensors $\mathbf{L}_t(\mathbf{x}, \boldsymbol\varepsilon)$ are replaced by tensors $\mathbf{L}_t^{(i)}(\varepsilon_1^e)$, which are constant in each phase. The most obvious choice for the effective strain of phase $v^{(i)}$ is the average strain over this phase $\varepsilon_1^e(\mathbf{x}) = \langle\boldsymbol\varepsilon\rangle^{(i)}$ $(\mathbf{x} \in v^{(i)})$ which can be modified taking the second moment of strains into account. $\mathbf{L}_t^{(i)}$ is generally anisotropic (even for isotropic phases) that yields to known difficulties in the estimation of the effective tangent stiffness $\mathbf{L}^*$.

12.2.3 Deformation Plasticity Theory of Composite Materials

It is assumed that the rheological properties of composite media with isotropic components are described by the theory of small elastoplastic strains under monotonic, proportional loading. We will not discuss here in general the known advantages of flow theory over deformation theory of plasticity in the case of nonradial loading, which takes place usually at the local level even if the overall stress path is radial.

Specifically, the total strain $\boldsymbol\varepsilon$ is written as the sum of elastic $\boldsymbol\varepsilon^e$ and plastic $\boldsymbol\varepsilon^p$ contributions $\boldsymbol\varepsilon = \boldsymbol\varepsilon^e + \boldsymbol\varepsilon^p$ (2.226). The local equation for the elastic material state, which relates the stress tensor $\boldsymbol\sigma(x)$ and the elastic strain tensor $\boldsymbol\varepsilon^e(\mathbf{x})$, is given in the form (2.197)

$$\boldsymbol\sigma(\mathbf{x}) = \mathbf{L}(\mathbf{x})\boldsymbol\varepsilon^e(\mathbf{x}),\ \mathbf{L}(\mathbf{x}) = \big(3k^{(j)}, 2\mu^{(j)}\big), \tag{12.48}$$

where $\mathbf{x} \in v^{(j)}$, $j = 0, 1, \ldots, N$. The relation between flow stress and plastic strains is represented by the expression (2.227): $\tau = \tau_0 + f(\varepsilon_{\rm eq}^p)$, $f(0) = 0$ in terms of the von Mises effective stress τ and the effective strain $\epsilon_{\rm eq}^p$ (2.228); τ_0 is the initial yield stress, and f is a nonlinear function (2.229) describing the material's hardening behavior.

One may determine for each phase j ($j = 0, 1, \dots, N$) the Young's secant modulus $E^{s(j)}$ (2.232) as well as (due to the plastic incompressibility) the secant Poisson's ratio $\nu^{s(j)}$ and shear modulus $\mu^{s(j)}$ (2.233). Hereafter the superscript ^{s} indicates the calculation of the parameter under consideration with the help of the secant modulus E^s. The effective compliances $\mathbf{M}^{s*}$ and $\mathbf{M}^*$ in the macroscopic equations of state,

$$\mathbf{M}^{s*}\langle\boldsymbol{\sigma}\rangle = \langle\boldsymbol{\varepsilon}\rangle, \quad \mathbf{M}^*\langle\boldsymbol{\sigma}\rangle = \langle\boldsymbol{\varepsilon}^e\rangle, \tag{12.49}$$

are defined by the relations $\mathbf{M}^{s*} = \langle\mathbf{M}^s\mathbf{B}^{s*}\rangle$, $\mathbf{M}^* = \langle\mathbf{M}\mathbf{B}^*\rangle$, where $\mathbf{M}(\mathbf{x}) = \mathbf{L}(\mathbf{x})^{-1}$ is the position-dependent compliance tensor. The tensors $\mathbf{B}^{s*}$ and $\mathbf{B}^*$ are found from the solutions of the elastoplastic problem $\boldsymbol{\sigma}(x) = \mathbf{B}^{s*}(\mathbf{x})\langle\boldsymbol{\sigma}\rangle$ and the elastic problem $\boldsymbol{\sigma}(x) = \mathbf{B}^*(\mathbf{x})\langle\boldsymbol{\sigma}\rangle$ (at $E^s = E$), respectively. Once $\mathbf{M}^{s*}$ and $\mathbf{M}^*$ have been found, the plastic strain of the composite can be determined from the unloading (elastic) process, i.e.,

$$\epsilon_*^p = (\mathbf{M}^{s*} - \mathbf{M}^*)\langle\boldsymbol{\sigma}\rangle. \tag{12.50}$$

The composite begins its plastic deformation, when

$$\max_i \frac{\widetilde{\tau}^{(i)}}{\tau_0^{(i)}} = 1, \quad \widetilde{\tau}^{(i)} \equiv \sqrt{\frac{3}{2}\langle s_{kl}s_{kl}\rangle^{(i)}} \tag{12.51}$$

($i = 0, 1, \dots, N$) is fulfilled. Up to the moment when yielding begins, the problem of estimating $\widetilde{\tau}^{(i)}$ is linear, which allows the use of closed relations for calculating the yield surface ($i = 0, 1, \dots, N$)

$$F_y(\langle\boldsymbol{\sigma}\rangle) \equiv \max_i \frac{1}{2\tau_0^{(i)}}\left[\frac{3}{c^{(i)}}\frac{\partial\mathbf{M}^*}{\partial q^{(i)}}\langle\boldsymbol{\sigma}\rangle \otimes \langle\boldsymbol{\sigma}\rangle\right]^{1/2} = 1. \tag{12.52}$$

Equation (12.52) determines the second-order surface $F_y(\langle\boldsymbol{\sigma}\rangle) = 1$ for the onset of yielding within the six-dimensional stress space which depends on $\langle\sigma_0\rangle$. The yield surfaces of the components $\widetilde{\tau}^{(i)} = \tau_0^{(i)}$ may be embedded within each other, or they may intersect in the space of macrostresses $\langle\boldsymbol{\sigma}\rangle$.

Now we shall deal with the elastoplastic state, when for the sake of definiteness, the surface of i_1 is embedded within the others, or we may examine those loading regimes for which the yield surface i_1 is reached first. In this case the strain law (2.227) is true for a component i_1, whereas Eq. (12.48) holds for the other components. Then the macroscopic equation of state relating the tensors $\boldsymbol{\sigma}$ and $\boldsymbol{\varepsilon}$ may be found by the following simple algorithm (see [1258]).

In the first step we add a small increment to the effective plastic strain $\epsilon_{\rm eq}^{p(i_1)}$ in the component i_1. Then we will estimate $\widetilde{\tau}^{(i_1)}$ and the values of the secant modulus in accordance with the formulae (2.227) and (2.232)

$$\widetilde{\tau}^{(i_1)} = \tau_0^{(i_1)} + f\left(\varepsilon_{\rm eq}^{p(i_1)}\right), \quad E^{s(i_1)} = \left[\frac{1}{E} + \frac{\varepsilon_{\rm eq}^{p(i_1)}}{\widetilde{\tau}^{(i_1)}}\right]^{-1}. \tag{12.53}$$

Using E^s we obtain $\mathbf{M}^{s*}$ from Eq. (8.35). For a desired proportional loading $\langle\boldsymbol{\sigma}\rangle = \boldsymbol{\xi}\sigma^t(t)$ ($\sigma^t(t)$ is a monotonically increasing scalar function) the value of $\sigma^t(t)$ is given by

$$\sigma^t(t) = 2\widetilde{\tau}^{(i_1)} \left[\frac{3}{c^{(i_1)}} \frac{\partial \mathbf{M}^{s*}}{\partial q^{s(i_1)}} \boldsymbol{\xi} \otimes \boldsymbol{\xi}\right]^{-1/2}, \tag{12.54}$$

which allows the components σ_{ij}^0 to be determined. By using $\mathbf{M}^{s*}$, $\boldsymbol{\sigma}^0$, and Eqs. (12.49) and (12.50), we find the total strain $\langle\boldsymbol{\varepsilon}\rangle$ and the overall plastic strain ϵ_*^p. By increasing the value $\langle\varepsilon_{\rm eq}^{p(i_1)}\rangle$, the stress–strain curves of the composite during the second step of deformation can be obtained.

This procedure is continued up to the yielding of the next component i_2 (where applicable). This occurs when condition $\widetilde{\tau}^{(i)} = \tau_0^{(i)}$ is met for $i = i_2$

$$\sigma^t(t) = 2\widetilde{\tau}^{(i_2)} \left[\frac{3}{c^{(i_2)}} \frac{\partial \mathbf{M}^{s*}}{\partial q^{s(i_2)}} \boldsymbol{\xi} \otimes \boldsymbol{\xi}\right]^{-1/2}. \tag{12.55}$$

Let us assume that at the moment of interest, the components $i_1, \ldots, i_m$ ($m > 1$) are in the plastic state. In that event the problem of estimating the macro parameters $\mathbf{M}^{s*}$ and ϵ_*^p on the basis of the given values $\varepsilon_{\rm eq}^{p(i)}$ ($i = i_1, \ldots, i_m$) becomes more difficult and is equivalent to a search for a minimum of the m-dimensional function with the components $|\widetilde{\tau}^{(j)} - f(\epsilon_{\rm eq}^{p(i)})|$ ($j = i_1, \ldots, i_m$), which depends on m arguments of $\varepsilon_{\rm eq}^{p(j)}$ ($i = i_1, \ldots, i_m$), where $\widetilde{\tau}^{(j)}$ is defined by Eq. $\widetilde{\tau}^{(i)} = \tau_0^{(i)}$. With small increments of the loading parameter at each iteration, this problem can be solved by standard gradient methods.

Note that, as a consequence of the heterogeneity of the field of stresses in each component, the value $\varepsilon_{\rm eq}^{p(j)}$ ($j = i_1, \ldots, i_m$) depends in a complex manner on the local values of the effective plastic strain $\epsilon_{\rm eq}^p(\mathbf{x})$ ($\mathbf{x} \in v_j$) and the average plastic deformations $\langle\epsilon^p\rangle_j$ in the component j. The values $\epsilon^p(\mathbf{x})$ and $\langle\epsilon^p\rangle_j$ ($\mathbf{x} \in v_j$; $j = i_1, \ldots, i_m$) cannot be found from the macroscopic deformation parameters because the problem is itself nonlinear. With this general principle in mind, we will estimate the components of an average plastic deformation on the basis of the secant modulus concept $\epsilon^{p(j)} \equiv [\mathbf{M}^{s(j)} - \mathbf{M}^{(j)}]\langle\boldsymbol{\sigma}\rangle^{(j)} \neq \langle\epsilon^p\rangle^{(j)}$, where the average stresses $\langle\boldsymbol{\sigma}\rangle_j$ in a component are calculated from Eq. (8.29). It is worth noting that

$$\varepsilon_{\rm eq}^{p(j)} \neq \left[\frac{2}{3}\epsilon_{kl}^{p(j)}\epsilon_{kl}^{p(j)}\right]^{1/2}, \quad \varepsilon_{\rm eq}^{p(j)} \neq \left[\frac{2}{3}\langle\epsilon_{kl}^p\rangle^{(j)}\langle\epsilon_{kl}^p\rangle^{(j)}\right]^{1/2} \tag{12.56}$$

and, in consequence, the flow law (2.230) written in terms of $\epsilon^{p(j)}$ and $\epsilon^{p*(j)}$ and $\langle\boldsymbol{\sigma}\rangle^{(j)}$, $\langle\widetilde{\tau}\rangle^{(j)}$ is not satisfied.

Finally, it should be noted that generally the local stress state is not monotonical and proportional even with monotonical and proportional external loading. However, using the linearization $\widetilde{\tau}^{(i)} = \tau_0^{(i)}$, one can see from Eqs. (12.53) and (12.54) that the effective stress and the plastic strain are homogeneous, monotonical, and proportional functions inside each component for at least a particular case of the composites consisting of an incompressible matrix containing rigid inclusions or pores, and, therefore, the above secant modulus concept can be used.

Now we consider the general case of an incompressible matrix with finite values of the parameters $\mu^{(0)} < \infty$, $\tau_0 > 0$. In this case the representation for the secant Young's modulus of the matrix must take the general form (2.232)

$$E^{s(0)} = \left(\frac{1}{E^{(0)}} + \frac{\varepsilon_{\rm eq}^{p(0)}}{\widetilde{\tau}^{(0)}}\right)^{-1}, \tag{12.57}$$

where the ratio $\varepsilon_{\rm eq}^{p(0)}/\widetilde{\tau}^{(0)}$ can be expressed in terms of the macrostresses $\langle\boldsymbol{\sigma}\rangle$ as

$$\frac{\epsilon_{\rm eq}^{p(0)}}{\widetilde{\tau}^{(0)}} = \frac{g(\widetilde{\tau}^{(0)} - \tau_0^{(0)})}{\widetilde{\tau}^{(0)}}, \quad \widetilde{\tau}^{(0)} = \left[\frac{3}{2}\mathbf{F}(\langle\boldsymbol{\sigma}\rangle \otimes \langle\boldsymbol{\sigma}\rangle)\right]^{1/2}, \quad \mathbf{F} = \frac{1}{2(1-c)}\frac{\partial \mathbf{M}^{s*}}{\partial q^{s(0)}}. \tag{12.58}$$

Here the function g is introduced, which is the inverse of the nonlinear hardening function (2.227). It is assumed that $g \equiv 0$ for $\widetilde{\tau}^{(0)} < \tau_0^{(0)}$.

For either the pores or rigid inclusions, taking into account the relations (12.58) and $\mathbf{F} = 2\mu^{s(0)}(1-c)^{-1}\mathbf{M}^{s*}$, we obtain the following expressions for the overall stress potential

$$W^{c*} = \frac{3}{4}(1-c)\left[\frac{1}{E^{(0)}} + \frac{g(\widetilde{\tau}^{(0)} - \tau_0^{(0)})}{\widetilde{\tau}^{(0)}}\right]\mathbf{F}(\langle\boldsymbol{\sigma}\rangle \otimes \langle\boldsymbol{\sigma}\rangle) \tag{12.59}$$

and for the overall plastic strain $\epsilon_*^p = 1.5(1-c)g(\widetilde{\tau}^{(0)} - \tau_0^{(0)})[\widetilde{\tau}^{(0)}]^{-1}\mathbf{F}\langle\boldsymbol{\sigma}\rangle$. It is interesting that the overall plastic deformation of the composite ϵ_*^p is independent of the initial modulus of the matrix $\mu^{(0)}$.

At present, it is difficult to construct sufficiently simple analytical solutions, and we will use numerical calculations on the basis of Eqs. (12.50), (12.52), (8.31), and (8.35). The authors [181, 231] considered a material containing spherical pores with numerical results for the elastic and plastic properties of an aluminum matrix given in [1258]. It was estimated the plastic deformations $\epsilon_{*ij}^p = \epsilon_{*0}^p\delta_{ij}$ for different values of porosity c in dependence of the hydrostatic stress $\sigma_{ij}^0 = \sigma^0\delta_{ij}$, taking into account the solution of problem (12.50). As expected, there is only a moderate influence of the matrix compressibility on the elastoplastic deformation of porous materials over a wide range of values of $\nu^{(0)}$ in comparison with Eq. (8.230). Accordingly, the assumption of an incompressible matrix $\nu^{(0)} = 0.5$ leads to an increase of 1–2% of the calculated plastic deformations. Note that according to criterion (12.3), we will obtain $\epsilon^p_{*0} \equiv 0$ for any σ^0_0, or, in other words, the porous material has infinite values of initial yield stress under hydrostatic compression in the case of the classical model.

The modified secant modulus model of plasticity described above was proposed in [181, 231] and is equivalent to the model proposed independently in [658–660]. The related approach [120, 1177, 1260, 1261] is based on the approximative method of estimation of second moments of stresses (as opposed to (6.185)) and used the assumption (10.83) which was utilized also in the flow model of plasticity in [177, 247]. The deformation theory of plasticity in the case of nonradial loading, which takes place usually at the local level, even if the overall stress path is radial, has the known difficulties avoided easily by an incremental plasticity model described below in Sects. 12.2.4 and 12.2.5.

12.2.4 Power-Law Creep

In the case of a matrix with zero initial yield stress,

$$\mu^{(0)}, \frac{k^{(0)}}{\mu^{(0)}} \to \infty, \quad \tau_0^{(0)} = 0 \tag{12.60}$$

the relation between flow stress and plastic strain, Eq. (12.53), takes the form ($\varepsilon^p_{ii} \equiv 0$)

$$\tau = h\left(\varepsilon^p_{\text{eq}}\right)^n, \quad \sigma_{ij} = \frac{2}{3} h\left(\varepsilon^p_{\text{eq}}\right)^{n-1} \varepsilon^p_{ij}. \tag{12.61}$$

The local stress potential corresponding to the constitutive relation (12.61) can be expressed as

$$w^c = \frac{n}{n+1} h^{-\frac{1}{n}} \left[\frac{3}{2}\mathbf{N}_2(\boldsymbol{\sigma} \otimes \boldsymbol{\sigma})\right]^{\frac{1}{2}+\frac{1}{2n}} = \frac{n}{n+1} h^{-\frac{1}{n}} \tau^{1+\frac{1}{n}}, \tag{12.62}$$

where the tensor $\mathbf{N}_2 = (0, 1)$ is isotropic and contains no hydrostatic contributions. If ε^p is considered as a rate of deformation (strain rate) tensor, Eqs. (12.61) and (12.62) describe a power-law creep of the matrix. In the theory of non-Newtonian fluids, it is customary to express the second Eq. (12.61) in the form

$$s_{ij} = 2\eta^{(0)} \varepsilon^p_{ij}, \quad 2\eta^{(0)} \equiv \frac{2}{3} h \left(\varepsilon^p_{\text{eq}}\right)^{n-1}, \tag{12.63}$$

where $\eta^{(0)}$ is a nonlinear viscosity of the matrix.

In order to estimate the second moment of the stress inside the matrix of the composites with the incompressible matrix containing either the pores or rigid inclusions, we will make use of Eqs. (6.185) (8.221), (8.230), and (8.232) with consequent passage to the limit (12.60). Insofar our concern is only with the plastic deformation ε^p of a single component, it is clearly advantageous to use the iterative scheme (12.53) and (12.54). For the initial increment $\varepsilon_{\rm eq}^{p(0)}$, we find $\widetilde{\tau}^{(0)}$ according to Eq. (12.53$_1$), the secant modulus of the matrix takes the form $E^{s(0)} = \widetilde{\tau}^{(0)}/\varepsilon_{\rm eq}^{p(0)}$, and the second moment of the stress inside the matrix is obtained as (6.185)

$$\langle s_{ij}s_{ij}\rangle^{(0)} = F_{mnpq}\langle\sigma_{mn}\rangle\langle\sigma_{pq}\rangle. \tag{12.64}$$

For pores or rigid inclusions, the tensor

$$\mathbf{F} \equiv \frac{1}{2(1-c)}\frac{\partial \mathbf{M}^{s*}}{\partial q^{s(0)}} \tag{12.65}$$

depends only on the microtopology of the composite and is independent of the mechanical properties of the matrix. From the relations (8.221), (8.230), and (8.232) it can be shown that

$$\mathbf{F} \equiv \frac{1}{(1-c)}2\mu^{s(0)}\mathbf{M}^{s*} \tag{12.66}$$

for the considered configurations. It can be proved that the identity (12.66) is valid either for rigid inclusions or for pores in an incompressible matrix. Actually, according to Hill's condition (see Sect. 6.1), we obtain for rigid inclusions or for pores $\mathbf{M}^{s*}(\langle\boldsymbol{\sigma}\rangle\otimes\langle\boldsymbol{\sigma}\rangle) = (1-c)\mathbf{M}^{s(0)}\langle\boldsymbol{\sigma}\otimes\boldsymbol{\sigma}\rangle^{(0)}$, from which Eq. (12.66) follows under the assumption of an incompressible matrix. Therefore, the actual value of the effective stress in the matrix $\widetilde{\tau}^{(0)}$ can be obtained directly from (12.64) without employing the iterative process (12.53) and (12.54).

From (12.64) and (12.66) together with the principle of minimum complementary energy, the expression for the overall stress potential is found as

$$w^{c*} = (1-c)\langle w^c\rangle^{(0)} = \frac{n}{n+1}(1-c)h^{-\frac{1}{n}}\left[\frac{3}{2}\mathbf{F}(\langle\boldsymbol{\sigma}\rangle\otimes\langle\boldsymbol{\sigma}\rangle)\right]^{\frac{1}{2}+\frac{1}{2n}} \tag{12.67}$$

and the overall plastic strain is obtained as

$$\varepsilon_{\rm eq}^{p} = \frac{3}{2}(1-c)h^{-\frac{1}{n}}\left[\frac{3}{2}\mathbf{F}(\langle\boldsymbol{\sigma}\rangle\otimes\langle\boldsymbol{\sigma}\rangle)\right]^{\frac{1}{2n}-\frac{1}{2}}. \tag{12.68}$$

Rather than employing relations (12.67) and (12.68), we will present our result in a more standard manner making use of a "reference tensor"

$$\widetilde{\mathbf{F}} = (1-c)^n \mathbf{F}^{\frac{1}{2}+\frac{n}{2}}. \tag{12.69}$$

As can be seen below, the tensor $\widetilde{\mathbf{F}}$ shows the effect of the inclusions on the variation of the rheological properties of the composite. Obviously, we have $\widetilde{\mathbf{F}} \to \mathbf{N}_2$ for $c \to 0$. For $n = 1$ the tensor $\widetilde{\mathbf{F}} \equiv 2\mu^{s(0)}\mathbf{M}^{s*}$ agrees with the well-known linear results on the relative change of the effective compliance with respect to the matrix properties.

The expression for the tensor $\widetilde{\mathbf{F}}$ for the two-particle approximation of the MEFM was obtained in [247] by a special method for spherical inclusions and by the use of the representations (8.221) and (8.230). General Eq. (12.67) for rigid inclusion and its generalization for the viscoplastic matrix was obtained in [176, 279] in the framework of the same method of the estimation of the second moment of stresses in the incompressible matrix used the assumption (10.83). From relation (12.67), which was obtained using the one-particle approximation of the MEFM, the analogous results [1052, 1234, 1235, 1480] follow, which were obtained by variational methods. It was shown that the upper bound for linear elastic comparison composite $\mathbf{L}^*$ (e.g., for porous materials) generates an upper bound for w^{c*}. However, linear estimated lower bounds (e.g., for composites with rigid inclusions) cannot be used, in general, for generation of lower bounds for the nonlinear composite by the variational procedure proposed which can be used nevertheless for obtaining estimations for nonlinear composites by application of a fairly accurate estimation for linear comparison composites described in Chap. 8.

Let us now consider in more detail the case of rigid inclusions (8.221) [porous materials (8.230) and (8.232) can be investigated in the same manner] by the use of the terminology of the theory of non-Newtonian suspensions in a fluid matrix (12.63). Then the tensor $\mathbf{F}$, Eq. (12.64), has the structure of the tensor $\mathbf{N}_2$, with zero hydrostatic contributions, and for a suspension of the particles, we have a flow law related to (12.63)

$$\langle s_{ij}\rangle = 2\eta^*\varepsilon^p_{ij}, \quad 2\eta^* = \frac{2}{3}h(\varepsilon^p_{\rm eq})^{n-1}, \tag{12.70}$$

where $\varepsilon^p_{\rm eq} = (2\varepsilon^p_{ij}\varepsilon^p_{ij}/3)^{1/2}$ is an effective overall plastic strain and $\eta^* = \eta^{(0)}$ for $c = 0$. Therefore, for the same rate of shear strain ϵ^p_*, we find from (12.63) and (12.68)–(12.70) the relative change of the effective viscosity as $\eta^*/\eta^{(0)} = (2\widetilde{f}^{\mu})^{-1}$*and* $\eta^{(0)} = h(\varepsilon^p_{\rm eq})^{n-1}/3$, where $2\widetilde{f}^{\mu}$ is the shear component of the Voight representation of the tensor $\widetilde{\mathbf{F}} = (3\widetilde{f}^k, 2\widetilde{f}^{\mu})$.

For the two-particle and one-particle approximations of the MEFM, we find from Eqs. (8.221) and (12.69) and Eqs. (8.220) and (12.69), respectively,

$$\frac{\eta^*}{\eta^{(0)}} = \left[1 + \frac{5}{2}c\left(1 - \frac{31}{16}c\right)^{-1}\right]^{(n+1)/2}(1-c)^{(1-n)/2}, \tag{12.71}$$

$$\frac{\eta^*}{\eta^{(0)}} = \left(1 + \frac{3}{2}c\right)^{(n+1)/2} (1-c)^{-n}. \tag{12.72}$$

Relations (12.71) and (12.72) were obtained in [247] via the MEFM. An expression analogous to Eq. (12.72) was subsequently proposed and analyzed in detail in [1234, 1235] via a variational method. For dilute concentrations of the inclusions ($c \to 0$), the limits of Eqs. (12.71) and (12.72) take the form

$$\eta^*/\eta^{(0)} = 1 + \gamma c, \quad \gamma = (3 + 7n)/4, \tag{12.73}$$

which agree with the Einstein linear result $\gamma = 5/2$ for $n = 1$. In Fig. 12.4 the experimental data [809] are compared with the results of Eq. (12.73) and with predictions obtained in [63] and [892] from numerical studies of a sphere in an infinite matrix of a power-law material (dilute approximation). For nondilute concentrations of the inclusions, Fig. 12.5 shows a comparison of the experimental data [601] ($n = 0.41$) with the results of formulae (12.71) and (12.72) and the dilute approximation (of first degree with respect to c) [892].

12.2.5 *Elastic–Plastic Behavior of Elastically Homogeneous Materials with Random Residual Microstresses*

We assume that the rheological properties of the composite medium with, generally speaking, anisotropic components are described by the theory of small elastic–plastic strains under arbitrarily varying external loading. Additive decomposition of the incremental of total strain tensor $d\boldsymbol{\varepsilon}$ is used in each component of the composite (2.213), i.e., $\boldsymbol{\varepsilon} = \boldsymbol{\varepsilon}^e + \boldsymbol{\varepsilon}^t + \boldsymbol{\varepsilon}^p$, with the elastic strains $\boldsymbol{\varepsilon}^e$, the so-called transformation

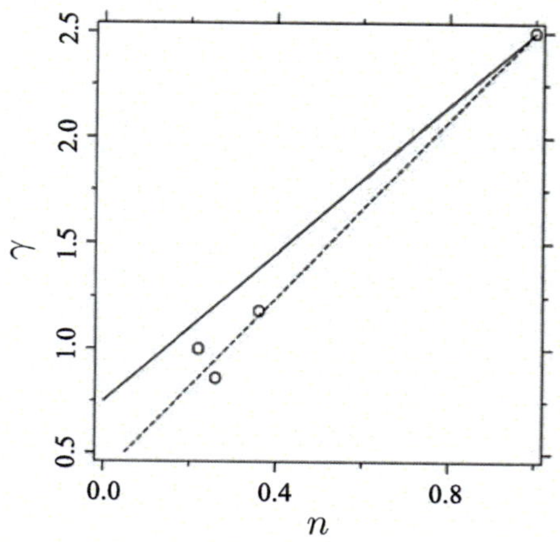

Fig. 12.4 Parameter γ (12.73) vs. n. Experimental data (circle) and calculations via the MEFM (solid line) and dilute approximation [892] (dotted line)

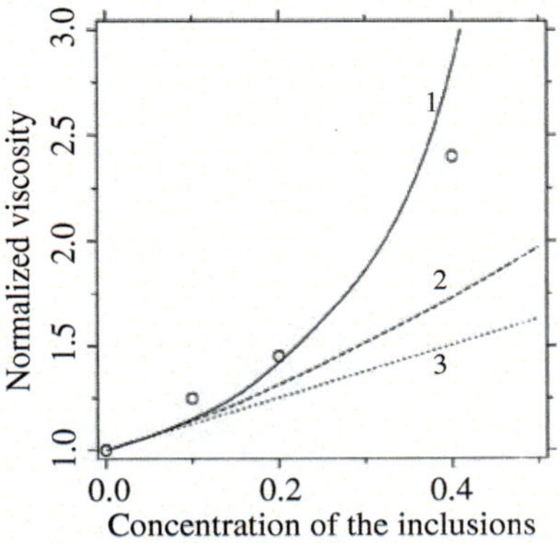

Fig. 12.5 $\eta^*/\eta^{(0)}$ vs. c. Experimental data (circle) and predictions from Eqs. (12.71) (1) and (12.72) (2) and from the dilute approximation (3)

strains $\boldsymbol{\varepsilon}^t$ including thermal strains, too, and plastic strains $\boldsymbol{\varepsilon}^p$; $\boldsymbol{\beta} \equiv \boldsymbol{\varepsilon}^t + \boldsymbol{\varepsilon}^p$. The local basic Eqs. (6.1)–(6.3) are assumed to be accomplished (with no body forces acting). The uniform traction boundary condition at the surface ∂w of the sample domain w (6.5) generating a homogeneous macroscopic stress $\boldsymbol{\sigma}^0$ is given, and the phases are assumed to be perfectly connected.

For the description of the behavior of the matrix and the elastic–plastic coating, the so-called J_2-flow theory with combined isotropic–kinematic hardening is used; see Eqs. (2.215)–(2.217) for details. $\mathbf{s}^a$ is the deviator of the active stresses; $\mathbf{a}^p$ is a symmetric second-order tensor corresponding to the "back stress" defining the location of the center of the yield surface in the deviatoric stress space. For the evaluation of the back stress $\mathbf{a}^p$, Ziegler's hardening rule (2.218) is used. The material behaves elastically if $f < 0$ or if $f = 0$ and $(\partial f/\partial \boldsymbol{\sigma}) : d\boldsymbol{\sigma} \leq 0$; elastic–plastic deformations take place under active loading, when $f = 0$ and $(\partial f/\partial \boldsymbol{\sigma}) : d\boldsymbol{\sigma} > 0$; see Eq. (2.220).

With nonessential restriction on the number of phases, we will consider (following [185, 264]) two-component statistically homogeneous composites containing identical aligned inclusions. The elastic properties of the matrix and the inclusions are the same, but the so-called "stress-free strains," i.e., the strain contributions due to temperature loading, phase transformations, and the plastic strains, fluctuate.

At first we will recast the representations obtained in Chaps. 8 and 13 for the statistical moments of residual stresses in the phases in the form convenience for subsequent calculations. So, for two-component composites the relation (9.103) for the average stresses inside inclusions can be represented in unit form for the inclusions ($i = 1$) and the matrix ($i = 0$)

$$\langle \boldsymbol{\sigma} \rangle^{(i)} = \boldsymbol{\sigma}^0 + \mathbf{C}^{(i)} \boldsymbol{\beta}_1^{(1)}, \quad \mathbf{C}^{(i)} = (-1)^i (1 - c^{(i)}) \mathbf{Q}_i, \tag{12.74}$$

where $\boldsymbol{\sigma}^0 = \langle \boldsymbol{\sigma} \rangle$. The representations for the second moments of stresses inside the phases (10.79) and (10.81) taking Eq. (12.74) and only binary interaction of inclusions into account can be recast in terms of the average second invariant of deviator stress stresses inside the inclusions (do not sum i)

$$\langle\boldsymbol{\sigma}\mathbf{N}_2\boldsymbol{\sigma}\rangle^{(i)} - \langle\boldsymbol{\sigma}\rangle^{(i)}\mathbf{N}_2\langle\boldsymbol{\sigma}\rangle^{(i)} = \widetilde{\mathbf{B}}^{(i)}\big(\langle\boldsymbol{\sigma}\rangle^{(i)} - \boldsymbol{\sigma}^0\big) \otimes \big(\langle\boldsymbol{\sigma}\rangle^{(i)} - \boldsymbol{\sigma}^0\big), \tag{12.75}$$

where

$$\widetilde{\mathbf{B}}^{(i)} = \mathbf{N}_2 : \widetilde{\mathbf{C}}^{(i)}\Big[\big(\mathbf{C}^{(i)}\big)^{-1} \otimes \big(\mathbf{C}^{(i)}\big)^{-1}\Big],$$

$$\widetilde{\mathbf{C}}^{(0)} = \int \big[\mathbf{T}_p(\mathbf{x}_0 - \mathbf{x}_p)\overline{v}_p\big] \otimes \big[\mathbf{T}_p(\mathbf{x}_0 - \mathbf{x}_p)\overline{v}_p\big]\,\varphi(v_p|\mathbf{x}_p;\mathbf{x}_0)\,d\mathbf{x}_p,$$

$$\widetilde{\mathbf{C}}^{(1)} = \overline{v}_1^{-1}\int\!\!\int \big[\mathbf{T}_p(\mathbf{x} - \mathbf{x}_p)\overline{v}_p\big] \otimes \big[\mathbf{T}_p(\mathbf{x} - \mathbf{x}_p)\overline{v}_p\big]\,V_1(\mathbf{x})\varphi(v_p|\mathbf{x}_p;\mathbf{x}_1)\,d\mathbf{x}\,d\mathbf{x}_p.$$

Since our main interest here is the consideration of the elastic–plastic deformations of the whole of the components $v^{(i)}$ $(i = 0, 1)$ and for the sake of simplicity, we are using the yield criterion (12.4), instead of (12.8). In Sect. 12.1.1 we demonstrated that this assumption is physically consistent, in contrast with (12.3).

The composite starts yielding when

$$f^* \equiv \max_i[\widetilde{\tau}^{(i)}/\tau_0^{(i)}] - 1 = 0, \quad (i = 0, 1). \tag{12.76}$$

Of course, if only the first onset of yielding is concerned, the problem of estimating $\widetilde{\tau}^{(i)}$ is an elastic one.

At first let us consider situations in which at zero time $\boldsymbol{\sigma}^0 = \mathbf{0}$ and $f^* < 0$. Under monotonically increased external active loading, elastic–plastic deformations will start to appear, when $f^* = 0$, according to (12.76), and $[(\partial f^*/\partial\boldsymbol{\sigma}^0) : d\boldsymbol{\sigma}^0 + (\partial f^*/\partial\theta)d\theta] > 0$. The initial yield surfaces of the components (see (12.4)) may be embedded in each other or they may intersect in the space spanned by the macrostresses $\boldsymbol{\sigma}^0$ and temperature θ. We define the state of the composite as being elastic–plastic if at least one component $v^{(s)}$ is in an elastic–plastic state. In this case the macroscopic constitutive law can be found as follows.

Taking into account the estimation for the yield surface (12.4_2) and (12.76), the increments of homogeneous plastic strains in the component $v^{(s)}$,

$$d\boldsymbol{\varepsilon}^{p(s)} = d\lambda^{(s)}\frac{\partial \widetilde{f}^{(s)}}{\partial\langle\boldsymbol{\sigma}\rangle^{(s)}}, \tag{12.77}$$

are defined by the use of the functions $\widetilde{f}^{(s)} \equiv \widetilde{\tau}^{(s)} - F^{(s)}(\gamma^{(s)},\theta) = 0$, $\widetilde{\tau}^{(s)} = \big(1.5\langle\mathbf{s}^a : \mathbf{s}^a\rangle^{(s)}\big)^{1/2}$. At each incremental step of the external stresses and of the temperature, respectively, we assume homogeneous plastic strains $\boldsymbol{\varepsilon}^{p(s)}$ and increments of the hardening parameters inside the component $v^{(s)}$. If Ziegler's rule (2.218) is used, we have

$$d\mathbf{a}^{p(s)} = d\gamma^{(s)}A\big(\gamma^{(s)}\big)\langle\mathbf{s}^{a(s)}\rangle^{(s)}, \quad A\big(\gamma^{(s)}\big) = H^{(s)}/F^{(s)}\big(\gamma^{(s)},\theta\big), \tag{12.78}$$

where $H^{(s)}$ is the current plastic tangent modulus of the uniaxial stress–plastic strain curve:

$$H^{(s)} = \frac{\partial \widetilde{\tau}^{a(s)}}{\partial \gamma^{(s)}}, \quad d\widetilde{\tau}^{a(s)} \equiv \Big(\frac{3}{2}\langle d\mathbf{s}^a : d\mathbf{s}^a\rangle^{(s)}\Big)^{1/2}, \quad d\gamma^{(s)} = \Big(\frac{2}{3}\, d\boldsymbol{\varepsilon}^{p(s)} : d\boldsymbol{\varepsilon}^{p(s)}\Big)^{1/2}.$$

Here the estimations (13.20) and (13.23) can be used for the evaluation of the second moment $\langle d\mathbf{s}^a : d\mathbf{s}^a\rangle^{(s)} \equiv \langle d\mathbf{s} : d\mathbf{s}\rangle^{(s)} - 2\langle d\mathbf{s}\rangle^{(s)} : \mathbf{a}^{(s)} + \mathbf{a}^{p(s)} : \mathbf{a}^{p(s)}$.

It should be noted that the assumption of homogeneity of the stress and strain fields inside each component, as used in (12.3) for mean field theory, leads to the following relation for the increments of the plastic strains inside the component $v^{(s)}$ [420, 423]:

$$d\boldsymbol{\varepsilon}^{p(s)} = d\lambda^{(s)} \frac{\partial f^{(s)}}{\partial \langle \boldsymbol{\sigma}\rangle^{(s)}}, \tag{12.79}$$

where the yield function, $f^{(s)} \equiv \tau^{(s)} - F^{(s)}(\gamma^{(s)}, \theta) = 0$, is defined by the average stresses inside the component $\tau^{(s)} = \big(1.5\langle \mathbf{s}^a\rangle^{(s)} : \langle \mathbf{s}^a\rangle^{(s)}\big)^{1/2}$.

Of course, if the formulation (12.79) (based on estimation of $\tau^{(s)}$) is used, it is not necessary to estimate the second stress moments inside the component $v^{(s)}$. In reality for statistically isotropic composites containing spherical inclusions, we see from (12.74) that the deviator of the active stresses inside the component $\langle \mathbf{s}^a\rangle^{(s)} \equiv \mathbf{N}_2\big(\boldsymbol{\sigma}^0 - \mathbf{a}^{p(s)}\big)$ does not depend on the stress-free strain field $\boldsymbol{\beta}$ and the internal microtopology of the composite, and, therefore, residual stresses does not affect the elastic–plastic strains of the composite under consideration. However, in subsequent discussion it will be shown that residual stresses are of fundamental importance.

We now turn our attention to the discussion of the yield condition (12.79_1) under consideration of residual stresses. In view of some further simplifications of the relations (12.77)–(12.78_1), we restrict ourselves to two-phase materials. Taking Eq. (12.75) into account we obtain from (12.79_2)

$$\widetilde{\tau}^{(s)} = \Big[\frac{3}{2}\langle \mathbf{s}^a\rangle^{(s)} : \langle \mathbf{s}^a\rangle^{(s)} + \frac{3}{2}\widetilde{\mathbf{B}}^{(s)}\Big((\langle \boldsymbol{\sigma}\rangle^{(s)} - \boldsymbol{\sigma}^0) \otimes (\langle \boldsymbol{\sigma}\rangle^{(s)} - \boldsymbol{\sigma}^0)\Big)\Big]^{1/2}. \tag{12.80}$$

For consistency of the plastic deformation process the following holds in analogy to (2.221):

$$d\widetilde{f}^{(s)} \equiv \frac{\partial \widetilde{f}^{(s)}}{\partial \langle \boldsymbol{\sigma}\rangle^{(s)}} : d\langle \boldsymbol{\sigma}\rangle^{(s)} + \frac{\partial \widetilde{f}^{(s)}}{\partial \boldsymbol{\sigma}^0} : d\boldsymbol{\sigma}^0 + \frac{\partial \widetilde{f}^{(s)}}{\partial \mathbf{a}^{p(s)}} : d\mathbf{a}^{p(s)} + \frac{\partial \widetilde{f}^{(s)}}{\partial \gamma^{(s)}} d\gamma^{(s)} + \frac{\partial \widetilde{f}^{(s)}}{\partial \theta} d\theta = 0. \tag{12.81}$$

At first we will calculate some partial derivatives in (12.81) by using (no s sum)

$$\frac{\partial \widetilde{f}^{(s)}}{\partial \langle \boldsymbol{\sigma}\rangle^{(s)}} = \frac{3}{2F^{(s)}}\Big[\langle \mathbf{s}^a\rangle^{(s)} + \widetilde{\mathbf{B}}^{(s)}\big(\langle \boldsymbol{\sigma}\rangle^{(s)} - \boldsymbol{\sigma}^0\big)\Big], \tag{12.82}$$

$$\frac{\partial \widetilde{f}^{(s)}}{\partial \boldsymbol{\sigma}^0} = -\frac{3}{2F^{(s)}}\widetilde{\mathbf{B}}^{(s)}(\langle \boldsymbol{\sigma}\rangle^{(s)} - \boldsymbol{\sigma}^0), \tag{12.83}$$

$$\frac{\partial \widetilde{f}^{(s)}}{\partial \mathbf{a}^{p(s)}} = -\frac{3}{2F^{(s)}}\langle \mathbf{s}^a\rangle^{(s)}, \quad \frac{\partial \widetilde{f}^{(s)}}{\partial \gamma^{(s)}} = -\frac{\partial F^{(s)}}{\partial \gamma^{(s)}}, \quad \frac{\partial \widetilde{f}^{(s)}}{\partial \theta} = -\frac{\partial F^{(s)}}{\partial \theta}, \tag{12.84}$$

From (12.74), (12.77), and (12.78$_1$), the relations for the differential of the average stresses and hardening parameters inside component $v^{(s)}$ are obtained

$$d\langle \boldsymbol{\sigma}\rangle^{(s)} = d\boldsymbol{\sigma}^0 + \mathbf{C}^{(s)}\Big(d\boldsymbol{\varepsilon}_1^{t(s)} + (-1)^{(s+1)} d\lambda^{(s)} \frac{\partial \widetilde{f}^{(s)}}{\partial \langle \boldsymbol{\sigma}\rangle^{(s)}}\Big), \tag{12.85}$$

$$d\gamma^{(s)} = d\lambda^{(s)}\Big(\frac{2}{3}\frac{\partial \widetilde{f}^{s}}{\partial \langle \boldsymbol{\sigma}\rangle^{(s)}} : \frac{\partial \widetilde{f}^{(s)}}{\partial \langle \boldsymbol{\sigma}\rangle^{(s)}}\Big)^{1/2}, \tag{12.86}$$

$$d\mathbf{a}^{p(s)} = d\lambda^{(s)} A(\gamma^{(s)})\langle \mathbf{s}^a\rangle^{(s)}\left(\frac{2}{3}\frac{\partial \widetilde{f}^{s}}{\partial \langle \boldsymbol{\sigma}\rangle^{(s)}} : \frac{\partial \widetilde{f}^{(s)}}{\partial \langle \boldsymbol{\sigma}\rangle^{(s)}}\right)^{1/2}. \tag{12.87}$$

Substitution of the Eqs. (12.82)–(12.87) in (12.81) leads to the relation of proportionality factor (do not sum to s, $s = 0, 1$): $d\lambda^{(s)} = b_s/\boldsymbol{\beta}_{ss}$, where

$$b_s = -\Big(\frac{3}{2}\langle \mathbf{s}^a\rangle^{(s)} : d\boldsymbol{\sigma}^0 + \frac{\partial \widetilde{f}^{(s)}}{\partial \langle \boldsymbol{\sigma}\rangle^{(s)}}\mathbf{C}^{(s)}\, d\boldsymbol{\varepsilon}_1^{t(1)} - \frac{\partial F^{(s)}}{\partial \theta} d\theta\Big), \tag{12.88}$$

$$\begin{aligned}\boldsymbol{\beta}_{ss} &= (-1)^s \frac{\partial \widetilde{f}^{(s)}}{\partial \langle \boldsymbol{\sigma}\rangle^{(s)}}\mathbf{C}^{(s)}\frac{\partial \widetilde{f}^{(s)}}{\partial \langle \boldsymbol{\sigma}\rangle^{(s)}} \\ &\quad -\Big[\Big(\frac{\widetilde{\tau}^{(s)}}{F^{(s)}}\Big)^2 H^{(s)} + \frac{\partial F^{(s)}}{\partial \gamma^{(s)}}\Big]\Big[\frac{2}{3}\frac{\partial \widetilde{f}^{(s)}}{\partial \langle \boldsymbol{\sigma}\rangle^{(s)}}\mathbf{N}_2\frac{\partial \widetilde{f}^{(s)}}{\partial \langle \boldsymbol{\sigma}\rangle^{(s)}}\Big]^{1/2}.\end{aligned} \tag{12.89}$$

Let us consider the process of the deformation, when for each component $v^{(i)}$ $(i = 0, 1)$ the yield conditions are fulfilled

$$\widetilde{f}^{(i)} = 0, \quad \frac{\partial \widetilde{f}^{(i)}}{\partial \langle \boldsymbol{\sigma}\rangle^{(i)}} : d\langle \boldsymbol{\sigma}\rangle^{(i)} + \frac{\partial \widetilde{f}^{(i)}}{\partial \theta}\, d\theta > 0. \tag{12.90}$$

Then from reduced formulae (12.81)–(12.87), only Eq. (12.85) will be changed

$$d\langle \boldsymbol{\sigma}\rangle^{(s)} = d\boldsymbol{\sigma}^0 + \mathbf{C}^{(s)}\Big(d\boldsymbol{\varepsilon}_1^{t(1)} + d\lambda^{(1)}\frac{\partial \widetilde{f}^{(1)}}{\partial \langle \boldsymbol{\sigma}\rangle^{(1)}} - d\lambda^{(0)}\frac{\partial \widetilde{f}^{(0)}}{\partial \langle \boldsymbol{\sigma}\rangle^{(0)}}\Big). \tag{12.91}$$

By substituting the relations (12.82)–(12.84), (12.86), (12.87), and (12.91) in (12.81), we will obtain the system of two linear equations for the proportionality factors $d\lambda^{(i)}$ $(i = 0, 1)$: $\sum_{j=0}^{1} \boldsymbol{\beta}_{ij} d\lambda^{(j)} = b_i$, where b_i and $\boldsymbol{\beta}_{ii}$ $(i = 0, 1)$ are determined by (12.88) and (12.89) and the off-diagonal matrix elements

$\boldsymbol{\beta}_{ij}$ $(i = 0, 1;\ j = 2 - i)$ are defined by the relations

$$\boldsymbol{\beta}_{ij} = (-1)^j \frac{\partial \widetilde{f}^{(i)}}{\partial \langle \boldsymbol{\sigma} \rangle^{(i)}} \mathbf{C}^{(i)} \frac{\partial \widetilde{f}^{(j)}}{\partial \langle \boldsymbol{\sigma} \rangle^{(j)}}. \tag{12.92}$$

Then we get $d\lambda^{(i)} = \sum_{j=0}^{1} h_{ij} b_j$, where h_{ij} can be considered functions of hardening $h_{00} = \boldsymbol{\beta}_{11}/\Delta$, $h_{01} = -\boldsymbol{\beta}_{01}/\Delta$, $h_{10} = -\boldsymbol{\beta}_{10}/\Delta$, $h_{11} = \boldsymbol{\beta}_{00}/\Delta$, *and* $\Delta = \boldsymbol{\beta}_{00}\boldsymbol{\beta}_{11} - \boldsymbol{\beta}_{10}\boldsymbol{\beta}_{01}$.

This way for prescribed paths of loading $\boldsymbol{\sigma}(t)$ and $\theta(t)$ (where t is a monotonically varying parameter), the system of Eqs. (12.79) becomes a system of 6 (or 12, respectively, for both phases yielding) ordinary differential equations for the determination of all components of $d\boldsymbol{\varepsilon}^{p(s)}$. In so doing only five or ten, respectively, these differential equations from ones are independent of each other by virtue of the yield condition (12.79$_1$) or (12.90), respectively. After integration of this system, we will find the plastic strains inside each component and find the total strains: $\langle \boldsymbol{\varepsilon} \rangle^{(s)} = \mathbf{M}\langle \boldsymbol{\sigma} \rangle^{(s)} - \boldsymbol{\varepsilon}^{t(s)} - \boldsymbol{\varepsilon}^{p(s)}$. The overall plastic strains $\boldsymbol{\varepsilon}^{po}$ and the average stresses are defined by the relations

$$\boldsymbol{\varepsilon}^{po} = \langle \boldsymbol{\varepsilon}^{p} \rangle, \quad \langle \boldsymbol{\varepsilon} \rangle = \mathbf{M}\boldsymbol{\sigma}^0 + \langle \boldsymbol{\varepsilon}^{t} \rangle + \boldsymbol{\varepsilon}^{po}. \tag{12.93}$$

The known numerical schemes [423, 875, 1182] are modified due to the additional concentration factors $\widetilde{\mathbf{B}}^{(i)}$ $(i = 0, 1)$ (12.75) which are calculated only once from the elastic problem. Here the backward difference method with a variable integration step is used for integrating Eq. (12.77). A Newton-type iterative procedure is used, starting with the elastic predictor which provides a first estimate of the stress increment in each component.

As a numerical example, we consider a composite with ferritic matrix containing identical spherical austenite inclusions ($c = 0.5$, $\tau_0 = \tau_0(°\mathrm{C})$). The thermoelastoplastic constants as usually found in the literature (see [1612]) are displayed in Table 12.1.

Here the thermal expansion coefficient $\mathbf{m} \equiv m_0\boldsymbol{\delta}$ defines the transformation strains (cf. Eq. (2.213)) inside the components $\boldsymbol{\varepsilon}^t_{10} = (m_0^{(1)} - m_0^{(0)})\theta = (m_0^{(1)} - m_0^{(0)})(T - T^{\max})$ due to a temperature difference $(T - T^{\max})$ between the stress-free reference temperature $T^{\max} = 900\,°\mathrm{C}$ (from which the cooldown starts) and a current temperature T. Just for the sake of theoretical interest, let us consider the components with temperature-independent parameters $\tau_0, h, n =$ const and small kinematic hardening $H^{(i)} = 0.1\tau^{(i)}$ and $h^{(i)} = 0$ $(i = 0, 1)$. Let the composite being loaded by a constant uniaxial external stress $\sigma^0_{ij} = \sigma^0_1\delta_{i1}\delta_{j1}$ and by temperature cycles with one cycle described by $T = T^{\min} + (T^{\max} - T^{\min})(t - 1)[H(t-1) - H(1-t)]$, $\quad t \in (0, 2)$, where H is the Heaviside step function and $T^{\min}$ is a minimum temperature.

In Fig. 12.6 typical calculated curves for the variation of the first component ε^{po}_{11} of the overall plastic strain tensor (12.93$_1$) under the action of different external stresses $\sigma^0_1 = 0,\ 0.125,$ *and* 0.135 GPa and four temperature cycles $900\,°\mathrm{C} \rightarrow$

Table 12.1 Thermoelastoplastic constants

Phase	τ_0 [GPa]	$m_0 \cdot 10^6$ [°C^{-1}]	μ [GPa]	ν
Ferrite	0.32	12.5	76.9	0.3
Austenite	0.20	17.1	76.9	0.3

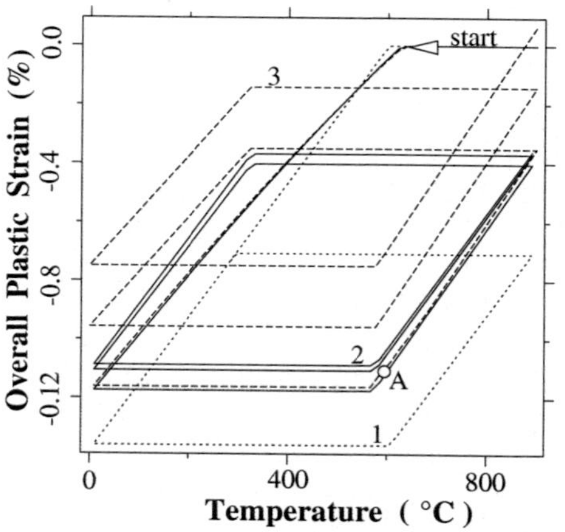

Fig. 12.6 ε_{11}^{po} vs. T at fixed external loading $\sigma_1^0 = 0$ (1), 0.125 GPa (2), 0.135 GPa (3)

0 °C → 900 °C are shown. Analogous results are presented in Fig. 12.7 for 400 temperature cycles. The curves show the growth of the longitudinal overall plastic strain ε_{11}^{po} accumulated from cycle to cycle. This means that only the plastic strain after completion of each cycle is represented in Fig. 12.7 as a function of cycle number $l = 0, \ldots, 400$. From Figs. 12.6 and 12.7, we see that for $\sigma_1^0 \leq 0.125$ GPa, the overall plastic strain after a few temperature cycles $l \leq 4$ converges to a steady-state response (or steady cycle), i.e., a closed loop is received. In structural analysis such behavior is specified as plastic shakedown (or cyclic collapse) (see references [1155, 1227, 1279, 1439]). A further increase of the parameters c or /and $(T^{\max} - T^{\min})/2$ or /and σ_1^0 leads to plastic shakedown (curves 1 and 2 in Fig. 12.6 and curve 1 Fig. 12.7), and, finally, ratchetting (or incremental collapse) appears—see curve 3 in Fig. 12.6 and curves 2 and 3 in Fig. 12.7. (see for details [185, 200]); curves 1, 2, and 3 are obtained under the assumption of kinematic hardening, and curve 4 is for mixed kinematic–isotropic hardening. Similar observations were made in experimental investigations with austenitic–ferritic duplex steel specimens under pure thermal cycles [76, 1612].

12.2.6 A Local Theory of Elastoplastic Deformations of Metal Matrix Composites

It should be mentioned that the results presented in Sect. 12.2.5 were obtained by the use of assumption (12.77) which might be too approximate in particular cases. When plastic deformations occur, the homogeneity of the mechanical properties

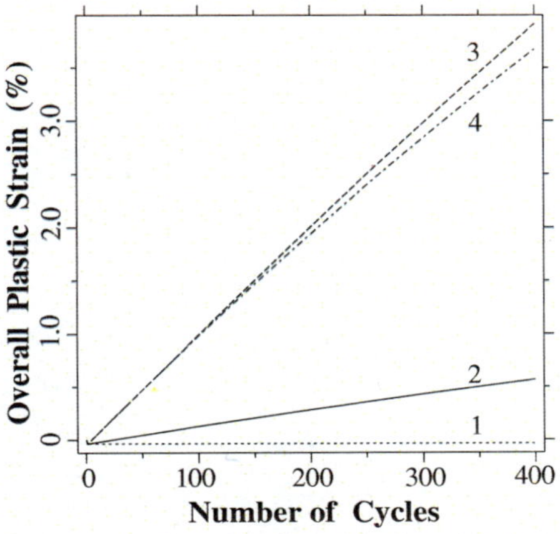

Fig. 12.7 ε_{11}^{po} at σ_1^0= 0.125 GPa (1), 0.127 GPa (2), 0.135 GPa (3 and 4)

of the components is lost, and the local properties of the phases become position-dependent. Appropriate, but by no means exhaustive, references about particulate MMCs are provided in the review [680] (see also [1079, 1472]). We will consider (following [272, 273], see also [200]) a local theory of elastoplastic deformation of random structure composites in the framework of flow theory and small elastoplastic strains. In this model each inclusion consists of an elastic core and a layer of thin coatings. The mechanical properties of the coatings are the same as that of the matrix. Homogeneity of the plastic strains is assumed inside the matrix and in individual subdomains of the coatings, which are considered as the individual components. By this means the proposed method can be considered as a logical extension of the transformation field analysis (TFA) by Dvorak [419] to random arrangement, when the phases with inhomogeneous stress states are subdivided into a finite number of subdomains with homogeneous stress states and plastic strains. At each step of the incremental macroscopic stresses, changes of the average stresses within the components are estimated with the help of the MEFM considered in Chap. 8 for linear problems. For a single inclusion, the micromechanical approach is based on the Green function technique as well as on the Hill's interfacial operators; the thin-layer hypothesis and the assumption of a homogeneous stress state in the core are used (see Chap. 4). For a dilute concentration of the inclusions, the proposed model is tested by the use of finite element analysis. The employment of the proposed theory for predicting the elastoplastic deformation behavior of a model material with a finite concentration of inclusions is shown.

12.2.6.1 Geometrical Structure of the Components

At first, due to the analyzed model, we consider a slight modification of a statistical description of statistically homogeneous composites presented in Sect. 5.3.1. Namely, we discuss a mesodomain w with a characteristic function W containing

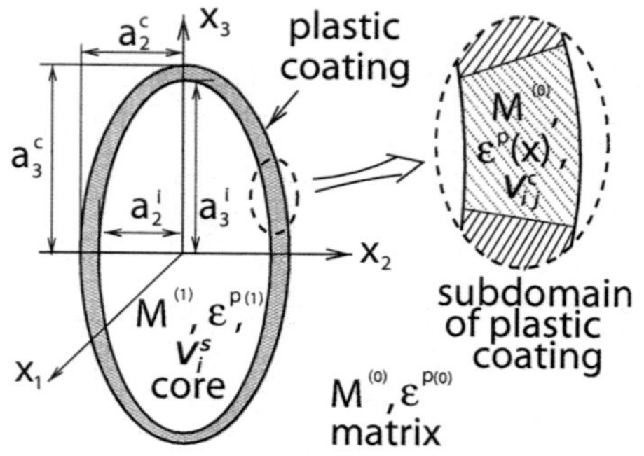

Fig. 12.8 Local model of elastoplastic deformation

a set $X = (V_i, \mathbf{x}_i, \omega_i)$, $(i = 1, 2, \ldots)$ of coated ellipsoidal inclusions v_i with characteristic functions V_i, centers x_i (that forms a Poisson set), semiaxes a_j^c $(j = 1, 2, 3)$, and aggregate of Euler angles ω. It is assumed that the inclusions v_i have identical mechanical and geometrical properties. Each inclusion consists of an elastic ellipsoidal core $v_i^s \subset v_i$ with semiaxes a_j^s $(j = 1, 2, 3)$, characteristic function $V_i^s(\mathbf{x})$, and a thin coating $v_i^c \equiv v_i \setminus v_i^s$ bounded by a homothetic ellipsoidal surface ∂v_i^c with semiaxes $a_j^c = a_j^s(1 + \xi)$ $(j = 1, 2, 3,\ 0 < \xi < 1)$ and a characteristic function $V_i^c(\mathbf{x}) \equiv V_i(\mathbf{x}) - V_i^s$; here ξ is the relative thickness of the coating (see Fig. 12.8). The mechanical properties $\mathbf{g}(\mathbf{x}) = \mathbf{g}^{(0)}$ $(\mathbf{g} = \mathbf{L}, \tau^0, h, n)$ are the same for both the matrix $v^{(0)} \equiv w \setminus \cup v_i$ and the coating v_i^c. The plastic strains are constant in the matrix and are an inhomogeneous function along the surface ∂v_i^s. The volume v_i^c is subdivided along ∂v_i^s into several local volumes v_{ij}^c $(mes\ v_{ij}^c \ll mes\ v_i^c,\ j = 1, \ldots n^c)$, such that the plastic strains $\boldsymbol{\varepsilon}^p(\mathbf{x}),\ (\mathbf{x} \in v_{ij}^c)$ are constant inside the each individual subdomain v_{ij}^c. In the core of the inclusions, we assume $\mathbf{g}(\mathbf{x}) = \mathbf{g}^{(0)} + \mathbf{g}_1(\mathbf{x}) = \mathbf{g}^{(0)} + \mathbf{g}_1; \mathbf{g}_1(\mathbf{x}) = \text{const}$ for $\mathbf{x} \in v^{(1)}$, and any elastic mismatch of the matrix and inclusion core can take place. The upper index of the material properties tensor, written in parentheses, shows the number of the respective component: $^{(0)}$ correspond to the matrix, $^{(1)}$ to the core of the inclusions, and $^{(k)}$ $(k = 2, n^c + 1)$ to the coating.

It is assumed that the representative domain w contains a statistically large number of realization of the inclusion field v_k; all random quantities under discussion are described by statistically homogeneous ergodic random fields, and, hence, the ensemble averaging could be replaced by volume averaging $\langle(.)\rangle$ and $\langle(.)\rangle^{(k)}$ $(k = 0, \ldots, n^c + 1)$ for the overall average and for phase averages, respectively. The bar appearing above the region represents its measure, $\overline{v} \equiv \text{mes}\ v$. $V^{(k)}$ is the characteristic function of $v^{(k)}$. The average over an individual inclusion $v_i \in v^{(1)}$ $(i = 1, 2, \ldots)$ is defined by $\langle(.)\rangle^{(i)} = \langle(.)\rangle^{(1)}$ if $v_i \in v^{(1)}$. $c^{(k)}$ is the concentration, i e., volume fraction, of component $v^{(k)} : c^{(k)} = \langle V^{(k)}\rangle;\ c^{(1)} = \overline{v}_m^s n^{(1)},\ c^{(0)} = 1 - \sum c^{(l)}$ $(k = 0, \ldots, n^c + 1;\ l = 1, \ldots, n^c + 1,\ m = 1, 2, \ldots)$. In the following the notation, $\langle(.)(\mathbf{x})|\mathbf{x}_1; \ldots; \mathbf{x}_m\rangle$ is used for the conditional average taken for the ensemble of a statistically homogeneous ergodic field $X = (v_i)$, on the condition that there are inclusions at the points $\mathbf{x}_1, \ldots, \mathbf{x}_m$ and $\mathbf{x}_1 \neq \ldots \neq \mathbf{x}_m$. The notation $\langle (.)(\mathbf{y})|; \mathbf{x}_1; \ldots; \mathbf{x}_m\rangle$ means the case $\mathbf{y} \notin v_1, \ldots, v_m$.

12.2.6.2 Average Stresses Inside the Components and Overall Elastic Moduli

The average stresses within the components are estimated for linear intermediate problems with the help of the MEFM considered in Chap. 8. In view of the linearity of the problem, there exist constant fourth- and second-rank tensors $\mathbf{B}^{(j)}$, and $\mathbf{C}^{(jk)}$, respectively, connecting the stresses inside the component $\mathbf{x} \in v^{(j)}$ $(j = 1, \ldots n^c + 1)$ and effective field acting on the inclusion v_i being considered

$$\boldsymbol{\sigma}(\mathbf{x}) = \mathbf{B}^{(j)}\overline{\boldsymbol{\sigma}}_i + \sum_{k=1}^{n^c+1} \mathbf{C}^{(jk)}\boldsymbol{\varepsilon}_1^{p(k)}, \tag{12.94}$$

$(\mathbf{x} \in v^{(j)})$, where $\mathbf{C}^{(j1)} \equiv \mathbf{0}$. The tensors $\mathbf{B}^{(j)}$ and $\mathbf{C}^{(jk)}$ can be found in the framework of the thin-layer hypothesis considered in Sect. 3.7.2. Namely, one obtains from Eq. (3.193) an approximate solution for a single ellipsoidal inclusion with thin coating under the approximate assumption of a homogeneous stress state in the core, $\boldsymbol{\sigma}^i$, and at the infinity $\overline{\boldsymbol{\sigma}}_i$: $\boldsymbol{\sigma}(\mathbf{x}) \equiv \boldsymbol{\sigma}^i = \text{const}, \quad \mathbf{x} \in v_i^s$. Then the stresses in the coating $\boldsymbol{\sigma}^c(\mathbf{s})$ are found by the relation $\boldsymbol{\sigma}^c(\mathbf{s}) = \boldsymbol{\sigma}^i + \boldsymbol{\Gamma}(\mathbf{n})\left[\mathbf{M}_1^{(1)}(\mathbf{s})\boldsymbol{\sigma}^i - \boldsymbol{\varepsilon}^p(\mathbf{s})\right]$, where the interface operator $\boldsymbol{\Gamma}(\mathbf{n})$ (3.97) is defined by Eq. (3.90); $\mathbf{n}$ is the unit outward normal vector on s_-^i in the point $\mathbf{s} \in s_-^i$, s_-^i is the outer surface of the boundary ∂v_i^s. Let the volume v_i^c be subdivided into one layer of n^c individual local volumes v_{ij}^c (mes $v_{ij}^c \ll$ mes v_i^c, $j = 1, \ldots n^c$), such that the plastic strain $\boldsymbol{\varepsilon}^p(\mathbf{x})$ and $\boldsymbol{\sigma}^c(\mathbf{x})$ $(\mathbf{x} \in v_{ij}^c)$ are constant inside each subdomain v_{ij}^c. For example, in a spherical coordinate system (θ, φ, r) coinciding with the semiaxes a_j^s, this subdivision can be done by the use of cutting the coating v_i^c along surfaces $\theta = \text{const}$, $\varphi = \text{const}$. Estimations for the tensors $\mathbf{B}^{(j)}$, $\mathbf{C}^{(jk)}$ $(j, k = 1, \ldots, n^c{+}1)$ can be obtained

$$\begin{aligned}
\mathbf{B}^{(1)} &= (\mathbf{I} + \mathbf{Q}\mathbf{M}_1^{(1)})^{-1}, \ \mathbf{B}^{(l)} = \left[\mathbf{I} + \tilde{\boldsymbol{\Gamma}}(n_l)\mathbf{M}_1^{(1)}\right]\mathbf{B}^{(1)}, \\
\mathbf{C}^{(1l)} &= -\mathbf{B}^{(1)}\mathbf{Q}\left[\delta_{1l} + \frac{\overline{v}_{il}}{\overline{v}_i}(1 - \delta_{1l})\right] + \frac{\overline{v}_{il}}{\overline{v}_i}(1 - \delta_{1l})\mathbf{B}^{(1)}\boldsymbol{\Gamma}(\mathbf{x}_i - \mathbf{y}_{il}), \\
\mathbf{C}^{(ml)} &= \left[\mathbf{I} + \tilde{\boldsymbol{\Gamma}}(n_m)\mathbf{M}_1^{(1)}\right]\mathbf{C}^{(1l)} + \tilde{\boldsymbol{\Gamma}}(n_m)(\delta_{1l} - \delta_{ml}),
\end{aligned} \tag{12.95}$$

The representations for both the averaged stresses and strain polarization tensors can be found from (12.94) and (12.95)

$$\langle \boldsymbol{\sigma}\rangle^{(i)} = \mathbf{B}\langle \overline{\boldsymbol{\sigma}}(x)\rangle^{(i)} + \mathbf{C}, \quad \bar{v}_1\langle \boldsymbol{\eta}\rangle^{(i)} = \mathbf{R}\langle \overline{\boldsymbol{\sigma}}\rangle^{(i)} + \mathbf{F},$$

$$\mathbf{B} \equiv \frac{1}{1-c^{(0)}} \sum_{j=1}^{n^c+1} c^{(j)} \mathbf{B}^{(j)} = \mathbf{I} + \frac{\overline{v}_i^s}{\overline{v}_i} (\mathbf{B}^{(1)} - \mathbf{I}),$$

$$\mathbf{C} = \frac{1}{1-c^{(0)}} \sum_{j,k=1}^{n^c+1} c^{(j)} \mathbf{C}^{(jk)} \boldsymbol{\varepsilon}_1^{p(k)}, \tag{12.96}$$

where $l, m = 2, \dots, n^c + 1$; $\mathbf{x}_i$ and $\mathbf{y}_{il}$ are the centers of the domains v_i and v_{il}^c, respectively; and $\mathbf{n}_l$ is the unit outward normal vector on s_-^i in the point $\mathbf{y}_{il}$. The tensors $\mathbf{R}$ and $\mathbf{F}$ are found by the use of the Eshelby theorem $\mathbf{R} = \overline{v}_i \mathbf{Q}^{-1}(\mathbf{I} - \mathbf{B})$, $\mathbf{F} = -\overline{v}_i \mathbf{Q}^{-1} \mathbf{C}$ (see Eq. (3.181)); the tensor $\mathbf{Q}$ is associated with the well-known Eshelby tensor $\mathbf{S}$ by $\mathbf{S} = \mathbf{I} - \mathbf{M}^{(0)} \mathbf{Q}$ (3.132).

In Sect. 8.1.3 we proved that in the framework of the effective field hypothesis, **(H1)** only the effective parameters $\mathbf{M}^*$ and $\boldsymbol{\varepsilon}_*^p$ governing the overall constitutive relation $\langle \boldsymbol{\varepsilon} \rangle = \mathbf{M}^* \langle \boldsymbol{\sigma} \rangle + \boldsymbol{\varepsilon}_*^p$ as well as statistical average of the local stresses inside the inclusions have the general representation. In particular, from Eq. (12.94) we get

$$\mathbf{M}^* = \mathbf{M}^{(0)} + \mathbf{Y} \mathbf{R} n^{(1)}, \ \boldsymbol{\varepsilon}_*^p = \boldsymbol{\varepsilon}^{p(0)} + \mathbf{Y} \mathbf{F} n^{(1)}, \tag{12.97}$$

$$\langle \boldsymbol{\sigma} \rangle^{(j)} = \mathbf{B}^{*(j)} \boldsymbol{\sigma}^0 + \sum_{k=0}^{n^c+1} \mathbf{C}^{*(jk)} \boldsymbol{\varepsilon}_1^{p(k)}, \tag{12.98}$$

where $\mathbf{C}^{*(j0)}, \mathbf{C}^{*(j1)} \equiv 0$, $(j = 0, \dots, n^c + 1)$ and $(l = 1, \dots n^c + 1)$:

$$\mathbf{B}^{*(l)} = \mathbf{B}^{(l)} \mathbf{R}^{-1} \mathbf{Y} \mathbf{R}, \quad \mathbf{B}^{*(0)} = \frac{\mathbf{I}}{c^{(0)}} - \sum_{l=1}^{n^c+1} \frac{c^{(l)}}{c^{(0)}} \mathbf{B}^{*(l)}, \quad \mathbf{C}^{*(0l)} = -\sum_{k=1}^{n^c+1} \frac{c^{(k)}}{c^{(0)}} \mathbf{C}^{*(kl)},$$

$$\mathbf{C}^{*(jl)} = \mathbf{C}^{(jl)} + \mathbf{B}^{(j)} \mathbf{R}^{-1} (\mathbf{Y} - \mathbf{I}) \left[\overline{v}_i^s \mathbf{M}_1^{(1)} \mathbf{C}^{(1l)} + \frac{c^{(l)} \mathbf{I}}{n^{(1)}} \right], \tag{12.99}$$

The tensor $\mathbf{Y}$ is determined by the purely elastic action (with $\mathbf{F} \equiv \mathbf{0}$) of the surrounding inclusions on the separated one. For a dilute concentration of the inclusions, i.e., $c^{(1)} \to 0$, we have $\mathbf{Y} \to \mathbf{I}$. The actual form of the tensor $\mathbf{Y}$, used in the analysis as an approximation, depends on additional assumptions for closing of the infinite system (8.6). For the purely elastic case (with $\mathbf{F} \equiv 0$), such relations are represented in Chap. 8. Concrete numerical analysis in Sect. 12.2.6.4 will be performed for $\mathbf{Y}$ obtained by the MEFM (8.35).

12.2.6.3 Elastoplastic Deformation

In our case for varied loading, onset of yielding appears in the individual subdomain of the coating v_{ij}^c or in the remaining matrix $v^{(0)}$, respectively, if the corresponding yield criterion

$$\widetilde{\tau}^{(j)} = \tau_0^{(0)}, \quad \widetilde{\tau}^{(j)} \equiv \sqrt{1.5\langle\boldsymbol{\sigma}\rangle^{(j)} : \mathbf{N}_2 : \langle\boldsymbol{\sigma}\rangle^{(j)}}, \tag{12.100}$$

$(j = 0, 1, \ldots, n^c + 1)$ is used. The composite starts to deform plastically, when $\max_j \widetilde{\tau}^{(j)} - \tau_0^{(0)} = 0$ $(j = 0, \ldots, n^c + 1)$ is fulfilled. Up to the moment when yielding starts, the problem of estimating $\widetilde{\tau}^{(j)}$ is linear, which allows the use of the relations (12.98) for calculating the yield surface

$$\widetilde{f}(\langle\boldsymbol{\sigma}\rangle) = \max_j \sqrt{1.5(\mathbf{B}^{*(j)}\boldsymbol{\sigma}^0)^\top : \mathbf{N}_2 : (\mathbf{B}^{*(j)}\boldsymbol{\sigma}^0)} - \tau_0^{(0)} = 0, \tag{12.101}$$

$(j = 0, \ldots, n^c + 1)$. The initial yield surfaces of the components (i.e., the individual subdomains) (12.100) may be embedded within each other, or they may intersect in the space of macrostresses $\langle\boldsymbol{\sigma}\rangle$. It should be mentioned that Zhu and Weng [1698] used a criterion, equivalent to (12.100), in that approach at each increment of external loading the increment of homogeneous inelastic strains in the matrix was assumed to be the average of the local inelastic strain increments over the matrix volume. This, however, would lead to zero increments of inelastic strains in the case of hydrostatic loading of macroisotropic media.

Now we shall deal with the elastoplastic state, when under successive external loading elastic–plastic deformations will take place for the components $i_1, \ldots, i_r$

$$\widetilde{f}^{(i_q)} = 0, \quad \frac{\partial \widetilde{f}^{(i_q)}}{\partial \langle\boldsymbol{\sigma}\rangle^{(i_q)}} : d\langle\boldsymbol{\sigma}\rangle^{(i_q)} > 0, \tag{12.102}$$

where the index i_q, $(q = 1, \ldots, r)$ passes through the number $(i_1, \ldots, i_r)$ of plastically deformed components $v^{(i_q)}$, $(0 \le i_q \le n^c + 1,\ i_q \ne 1)$ and where a homogeneous yield criterion inside the volume $v^{(i_q)}$ is assumed

$$\widetilde{f}^{(i_q)} \equiv \widetilde{\tau}^{(i_q)} - F^{(i_q)}\big(\gamma^{(i_q)}\big) = 0, \quad \widetilde{\tau}^{(i_q)} = \sqrt{1.5\langle\mathbf{s}^a\rangle^{(i_q)} : \langle\mathbf{s}^a\rangle^{(i_q)}}, \tag{12.103}$$

where $\mathbf{s}^a = \mathbf{N}_2(\boldsymbol{\sigma} - \mathbf{a}^p)$ and $\mathbf{a}^p$ is a back stresses. By using the associated flow rule, the yield function $\widetilde{f}^{(i_q)}$ is taken as plastic potential function of the matrix from which the incremental plastic strains inside the matrix can be determined as

$$d\boldsymbol{\varepsilon}^{p(i_q)} = d\lambda^{(i_q)} \frac{\partial \widetilde{f}^{(i_q)}}{\partial \langle\boldsymbol{\sigma}\rangle^{(i_q)}}. \tag{12.104}$$

At each incremental step of the external stresses, homogeneous plastic strains $\boldsymbol{\varepsilon}^{p(i_q)}$ and increments of hardening parameters are assumed within each subdomain. With Ziegler's rule (2.218) we obtain

$$d\mathbf{a}^{p(i_q)} = d\gamma^{(i_q)} A \langle \mathbf{s}^a \rangle^{(i_q)}, \quad A = \frac{H^{(i_q)}}{F^{(i_q)}(\gamma^{(i_q)})}, \quad H^{(i_q)} = \frac{\partial \widetilde{\tau}^{a(i_q)}}{\partial \gamma^{(i_q)}}, \tag{12.105}$$

$$d\gamma^{(i_q)} = \sqrt{2\, d\boldsymbol{\varepsilon}^{p(i_q)} : d\boldsymbol{\varepsilon}^{p(i_q)}/3}, \quad d\widetilde{\tau}^{a(i_q)} \equiv \sqrt{1.5 \langle d\mathbf{s}^a \rangle^{(i_q)} : \langle d\mathbf{s}^a \rangle^{(i_q)}}. \tag{12.106}$$

where $H^{(i_q)}$ is the plastic tangent modulus, derived from the uniaxial stress–plastic strain curve of the matrix material. From the requirement of consistency of the plastic deformation process, we come up with

$$\frac{\partial \widetilde{f}^{(i_q)}}{\partial \langle \boldsymbol{\sigma} \rangle^{(i_q)}} : d\langle \boldsymbol{\sigma} \rangle^{(i_q)} + \frac{\partial \widetilde{f}^{(i_q)}}{\partial \mathbf{a}^{p(i_q)}} : d\mathbf{a}^{p(i_q)} + \frac{\partial \widetilde{f}^{(i_q)}}{\partial \gamma^{(i_q)}} d\gamma^{(i_q)} = 0. \tag{12.107}$$

The partial derivatives in (12.107) are found under the following assumptions: $\boldsymbol{\varepsilon}^{p(i_q)}$ and $\langle \boldsymbol{\sigma} \rangle^{p(i_q)} = \text{const}$ $(q = 1, \ldots, r)$. From (12.98), (12.104), (12.105), and (12.106), the relations for the differential of average stresses and hardening parameters inside the component $v^{(q)}$ can be derived as

$$d\langle \boldsymbol{\sigma} \rangle^{(i_q)} = \mathbf{B}^{*(i_q)}\, d\boldsymbol{\sigma}^0 - \sum_{k=1}^{r} \mathbf{C}^{*(i_q i_k)} \left[\lambda^{(i_k)} \frac{\partial \widetilde{f}^{(i_k)}}{\partial \langle \boldsymbol{\sigma} \rangle^{(i_k)}} - \lambda^{(0)} \frac{\partial \widetilde{f}^{(0)}}{\partial \langle \boldsymbol{\sigma} \rangle^{(0)}} \right], \tag{12.108}$$

$$d\gamma^{(i_q)} = d\lambda^{(i_q)}, \qquad d\mathbf{a}^{p(i_q)} = d\lambda^{(i_q)} A\big(\gamma^{(i_q)}\big) \langle \mathbf{s}^a \rangle^{(i_q)}. \tag{12.109}$$

In Eq. (12.109) $d\lambda^{(0)} \equiv 0$ and $\partial \widetilde{f}^{(0)}/\partial \langle \boldsymbol{\sigma} \rangle^{(0)} \equiv 0$ if the conditions (12.102) are not met for the component $v^{(0)}$. Substitution of Eqs. (12.108) and (12.109) in (12.107) leads to the following relation for the proportionality factor

$$d\lambda^{(i_q)} = \sum_{s=1}^{r} \beta_{qs} b_s, \quad b_s = \frac{\partial \widetilde{f}^{(i_s)}}{\partial \langle \boldsymbol{\sigma} \rangle^{(i_s)}} : \mathbf{B}^{*(i_s)} : d\boldsymbol{\sigma}^0. \tag{12.110}$$

The elements of the inverse of the matrix $(\boldsymbol{\beta})$ are given by

$$\boldsymbol{\beta}_{qs}^{-1} = \frac{\partial \widetilde{f}^{(i_q)}}{\partial \langle \boldsymbol{\sigma} \rangle^{(i_q)}} \left[\mathbf{C}^{*(i_q i_s)} (\delta_{i_s 0} - 1) + \delta_{i_s 0} \sum_{n=1}^{r} \mathbf{C}^{*(i_q i_n)} \right] \frac{\partial \widetilde{f}^{(i_s)}}{\partial \langle \boldsymbol{\sigma} \rangle^{(i_s)}} + 2 \frac{\partial F^{(i_q)}}{\partial \gamma^{(i_q)}} \delta_{qs}. \tag{12.111}$$

Plastic deformations in the component $v^{(i_q)}$ take place if the condition $d\lambda^{(i_q)} > 0$ is satisfied together with the local yield criterion (12.102) and (12.103_1). By this means for a prescribed loading path $\boldsymbol{\sigma}^0(t)$ (where t is a monotonically varying parameter), the system of Eqs. (12.108) becomes a system of six ordinary differential equations for the determination of all components of the tensor $\langle \boldsymbol{\sigma} \rangle^{(q)}$. In so doing only five of these six equations are independent by virtue of the yield condition (12.102). After integration of the system (12.108) the averaged plastic strains inside each subdomain are found, and the averaged strains are given by

$\langle\boldsymbol{\varepsilon}\rangle^{(i_q)} = \mathbf{M}^{(0)}\langle\boldsymbol{\sigma}\rangle^{(i_q)} + \boldsymbol{\varepsilon}^{p(i_q)}$. The overall plastic strains $\boldsymbol{\varepsilon}^p_*$ and the total overall strains are defined by the relations $\boldsymbol{\varepsilon}^p_* = \sum_j (\mathbf{B}^{*(j)})^\top \boldsymbol{\varepsilon}^{p(j)} c^{(j)}$, $(j = 0, \ldots, n^c + 1)$, $\langle\boldsymbol{\varepsilon}\rangle = \mathbf{M}^*\boldsymbol{\sigma}^0 + \boldsymbol{\varepsilon}^p_*$. The numerical integration of Eq. (12.108) under the restriction (12.103$_1$) can be carried out by different integration schemes analogous to the treatment of homogeneous materials with vertex yield surfaces as, for example, described in [1121, 1163, 1182, 1283]. The same methodology usually is applied in particular versions of mean field methods in the mechanics of composite materials [423, 686, 687, 875]. In this work a backward difference method with a variable integration step has been used as the integration scheme for Eq. (12.104). An iterative procedure is used, starting with the elastic predictor (12.110) which provides a first estimate of the stress increment in the component. This means that at each step an Euler-backward scheme is used in combination with a Newton method, with the initial guess being determined by the elastic predictor.

12.2.6.4 Numerical Results

We consider an isotropic composite consisting of a steel matrix and identical spherical carbidic inclusions with elastoplastic parameters (Table 12.2).

At first we consider the case of a dilute concentration of the inclusions $c^{(1)} \ll 1$. This leads to $\mathbf{Y} \equiv \mathbf{I}$ (see (12.97)), and we have the possibility to check the quality of our local plasticity model by the use of finite element analysis (FEA). In the local plasticity theory, the subdivision of the coating v^c_i into one layer of individual subdomains v^c_{ij} $(j = 1, \ldots, n^c)$ is employed by the use of cross sections $\theta = n\pi/n^c$ in a spherical coordinate system (θ, ϕ, r), the origin of which coincides with the center of the inclusion. Hereafter we will restrict our problems to axisymmetric loading, and, therefore, the plastic strains in the ribbon $a < r < a(1+\xi)$, $k\pi/n^c < \phi < (k+1)\pi/n^c$ $(k = 0, \ldots, n^c - 1)$ can be obtained by the rotation around the axis $\theta = 0$. It was obtained that both subdivisions of the coating ($n^c = 11$ and $n^c = 33$) with $\xi \le 0.1$ lead to similar results for the local values of the von Mises effective plastic strains $\gamma^{(i)}$ $(i = 2, \ldots, n^c + 1)$, and, as a consequence, from here on, we will consider the thickness $\xi = 0.1$. For uniaxial loading $\sigma^0_{ij} = \sigma^0_{33}\delta_{i3}\delta_{j3}$, $\sigma^0_{33} \sim t$, the influence of the mesh width of the uniform subdivision (π/n^c) of the coating on the local values of the von Mises effective plastic strains $\gamma^{(i)}$ $(i = 2, \ldots, n^c + 1)$ (12.106) is studied; here t is the time, i.e., a monotonically increasing parameter. The results shown in Fig. 12.9 were calculated for $\xi = 0.1$ and for both $n^c = 11$ and for $n^c = 33$. In Fig. 12.9 the values $\gamma(\mathbf{x})$ as a function of the polar angle $0 < \theta < 90°$ are shown for $\sigma^0_{33} = 2.8$ GPa, whereas onset of yielding takes place at $\sigma^0_{33} = 2.2$ GPa. As can be seen for the considered case, both

Table 12.2 Elastoplastic constants

Phase	k [GPa]	ν	τ_0 [GPa]	h [GPa]	n
Matrix	175	0.3	2.75	1.5	0.5
Inclusions	300	0.25	∞	–	–

subdivisions of the coating ($n^c = 11$ and $n^c = 33$) lead to similar results for the local values of the von Mises effective plastic strains $\gamma^{(i)}$ ($i = 2, \ldots, n^c + 1$) (12.106). The increase of the degree of subdivision, $n^c = 33$, leads to an increase in the calculated values $\boldsymbol{\varepsilon}_*^{pn}$ which is, however, smaller than 0.5%.

Let us now study the finite concentration of carbidic inclusions ($c^{(1)} = 0.25$), for which the FEA–unit cell analysis is not able to capture real random arrangements of the inclusions. Whereas the solution of the corresponding linear problem can be found by any known method (see Chaps. 8 and 9), in the current presentation we will use only the solution of the linear elastic problem (12.97) and (12.98) obtained by the two-particle approximation of MEFM. Let us assume uniaxial tensional loading $\sigma_{ij}^0 = \sigma_{33}^0\delta_{i3}\delta_{j3}$, $\sigma_{33}^0 \sim t$. It was performed the comparable estimates (see for details [200, 273]) obtained by the proposed model and by other popular methods. There are different versions of mean field methods employing flow theory [420] and secant concept method (see, e.g., [1258] and also Sect. 12.2.3). In Fig. 12.10 the curves 4 and 5 are calculated in the framework of the mean field method (at $\xi = 0$) by the secant concept method and by the flow theory, respectively. Modified improved methods utilize the estimations of statistical averages in each component of either the stress potential or the second moment of stresses (12.4) that can be found by perturbation methods and by the method of integral equations (see Chap. 10) as well as by variational methods (for references see [1239]). For our elastoplastic analysis in the framework of homogeneity of plastic strains in the matrix ($\xi = 0$), we will use the exact perturbation method in combination with the secant concept method [181] as well as the method of integral equations in combination with flow theory. The corresponding curves 2 and 3, respectively, are presented in Fig. 12.10. Finally, the proposed local plasticity model takes into account the inhomogeneity of plastic strains in some matrix layer around the inclusions ($\xi = 0.1$) as well as the inhomogeneity of the first-order moment of the stresses in the coating (curve 1). At the same time in the considered case ($\sigma_{33}^0 = 3\,\text{GPa}$), the modified secant approach (curve 2) predicts values of the overall plastic strain ε_{*33}^p which are considerably smaller than those calculated via the use of both FEA–unit cell model and the local model of plasticity for random packing of the inclusions (curve 1). As can be seen in Fig. 12.10 (see for details [200, 273]) the better discretization of the stress–plastic strain state in the proposed model (curve 1) leads to significantly improved predictions of the overall plastic strains (with respect to the comparison with the FEA–unit cell model for face-centered cubic (FCC) packing of inclusion, curve 6) in comparison with all mentioned models based on the assumption of plastic strain homogeneity in the components.

12.3 Concluding Remarks

Other important achievements of the theory of nonlinear composite are connected with extensions of the Hashin–Shtrikman [621] variational principles for nonlinear composites, as proposed in pioneering work by Talbot and Willis [1480] utilizing a

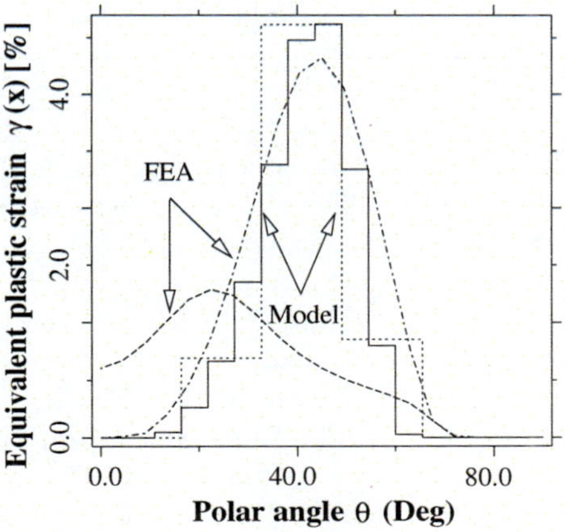

Fig. 12.9 $\gamma(\mathbf{x})$ vs. θ, calculated via FEA (dot–dashed curve for $|\mathbf{x}| = a + 0$, dotted curve for $|\mathbf{x}| = a(1+\xi) - 0$) and by the proposed model (dashed curve for $n^c = 11$, solid curve for $n^c = 33$)

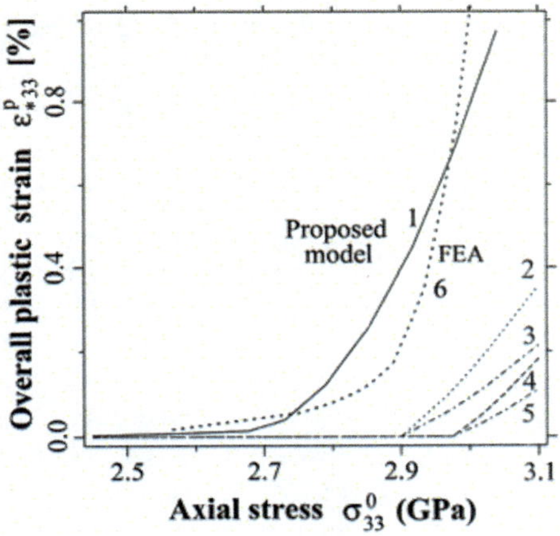

Fig. 12.10 ε^p_{*33} ε^p_{*33} vs. σ^0_{33} calculated via the proposed model (12.100) (1), via a modified approach (2, 3) , via the MFM (4, 5), and via FEA (6)

linear homogeneous reference material. Their work is followed by the introduction of several new variational principles (pioneered in [1234, 1235], and [1468]; for references see a review by Ponte Castañeda and Suquet [1239]) making use of appropriately chosen *linear heterogeneous reference material*, which allow the determination of Hashin–Shtrikman and more general bounds and estimates (e.g., self-consistent estimates and three-point bounds), directly from the corresponding bounds and estimates for linear composites with the same microstructural distribution of phases. These include the variational principles for general classes of nonlinear composites [681, 1236, 1469, 1481, 1633], for power-law composites [1468], for perfectly plastic composites [1158], for viscoplastic polycrystals [1429], and for elasto-viscoplastic composites [874]. Ponte Castañeda [1237, 1238] has shown that taking into account both the first and the second moment per phase in the linear comparison composite instead of just one of these quantities leads to more accurate predictions of overall behavior (see also [874]).

In so doing the Hashin–Shtrikman [621] variational principle and bounds for the linear comparison composites is used in their classical version (see for details Sect. 8.4.1); it would be interesting to use the generalized Hashin–Shtrikman bounds (considered in Sects. 8.4.2 and 9.7) which can lead to significant corrections of classical results for noncanonical shape of inclusions. From other side, Markov [1013] justified that the estimations of the effective moduli by the effective field method (EFM, see for details Sect. 8.3) coincide with one of the Hashin–Shtrikman bounds. It is interesting that $\langle s_{ij}s_{ij}\rangle^{(1)}$ is defined by the exact equations for W^* (10.77$_2$) and (6.116) and for $\boldsymbol{\beta}^*$ (6.113). We see that W^* and $\boldsymbol{\beta}^*$ depend on the effective bulk modulus k^* and do not depend on the effective shear modulus μ^*. In so doing k^* can be estimated by one or another micromechanical model, e.g., EFM or MEFM. In particular, for the CM with the isotropic constituents and identical spherical inclusions, the EFM (ignoring direct interaction of inclusions) leads to the simple representation (8.104) which depends on $k^{(1)}$, $k^{(0)}$, and $\mu^{(0)}$ rather than $\mu^{(1)}$. If k^* does not depend on $\mu^{(1)}$, then we get an exact chain of equalities $0 = \partial k^*/\partial\mu^{(1)} = \partial\beta^*/\partial\mu^{(1)} = \partial W^*/\partial\mu^{(1)}$. It means that $\langle s_{ij}s_{ij}\rangle^{(1)} \equiv 0$, and, therefore, a parameter $a_\tau^{(1)} \equiv \infty$ (12.26) if we use EFM instead of either MEFM or IEM; in so doing the elliptical curve 2 in Fig. 12.2 is degenerated into the straight line 3.

In general, the variational approaches in nonlinear micromechanics of CMs are based on consideration of some nonlinear functions of the fields (e.g., either the second moment of the fields or the field potentials) averaged over the volumes of constituents. However, Fig. 12.10 demonstrates drastically improved accuracy of the overall plastic strains estimated by the local model proposed in comparison with the models based on the evaluation of the average second moments of stresses in the constituents. In so doing, the local model at each step of loading is conceptually linear when the concentration factors of both the stress and eigenstress in Eqs. (12.97)–(12.99) are estimated a single time by the linear MEFM for all history of any loading. In fact, this is a version of the transformation field analysis by Dvorak [419] assuming the uniform plastic strain fields inside the subdomains; this assumption was generalized in a model-reduction technique, called the nonuniform transformation field analysis, based on a decomposition of the local fields of internal variables on a reduced basis of modes, to analyze the effective response of composite materials (see for references and details [1053–1055]). In its turn, nonlinear variational approach [1239] makes effective use of the various estimates available for linear composites, allowing the determination of corresponding estimates for nonlinear composites. It would be interesting to incorporate the mentioned linear estimations (12.97)–(12.99) into the nonlinear variational approach [1239]. Furthermore, the local model proposed can be applied to a wide class of nonlinear problems for which the local properties of the components become location-dependent: nonlinear elasticity and conductivity, viscosity and creeping, and viscoplasticity. It can be generalized easily to any number of thin coating layers of nonellipsoidal inclusions (see [270]) with a distribution of size, shape, orientation, and properties, to a consideration of statistical averages of the second moments of stresses in both the matrix and the coating. It is only important that response of a "coated" inclusion is defined by the Eqs. (12.94) and (12.96_1), notwithstanding the inclusion can be considered as some sort of a "black box." In particular, a coin-shaped cohesive microcrack (Barenblatt-Dugdale-type) [932] can be analyzed as such a "black box." The principal limitation of this approach is due to the assumption of statistical homogeneity of the composite microstructure. Nevertheless, in the light of the researches on functionally graded composites performed by the MEFM (see Chap. 13), obviously, consideration of the local model of elastoplastic deformation of graded materials is also possible.

It should be mentioned about machine learning (ML) predicting the behavior of general conventional heterogeneous materials under irreversible processes such as inelastic deformation when the offline stage (estimated a single time) only uses the linear elastic properties of each material phase, making it efficient. In so doing, physics-based models and simulations provide the only avenues for building models with the potential for high-fidelity extrapolation. It is interesting that the local model of elastoplastic deformation of CMs proposed in Sect. 12.2.6 can be easily recast in terms of machine learning and data mining (see, e.g., [134, 583]), e.g., such as a mechanistic data-driven two-scale approach (see [924, 964]). Indeed, the proposed approach includes two major innovations (in terms of [964]): (1) dataset generation and compression algorithm (12.97)–(12.99), k-means clustering (i.e., subdividing over the subdomains v_i^s, v_{ij}^c, $v^{(0)}$; see Fig. 12.8) during the offline stage of the method to homogenize the linear local features of the material microstructure into a group of clusters (that is performed a single time), and (2) a new method (see Sect. 12.2.6.4) called in [964] self-consistent clustering analysis used in the online stage for outperforming the training data in the design space that is valid for any local plasticity laws of each material phase without the need for additional calibration. Possible generalizations of the local model proposed are described in the previous paragraph. Of course, a major challenge is to balance the expense of data generation and the resulting performance of the applied ML model. The proposed model (see Sect. 12.2.6) provides a high-accuracy prediction (see Figs. 12.9 and 12.10) because it is based on physically justified advanced micromechanical modeling. From other side, a dataset generation is computationally effective in the sense of computing time but it is pretty expensive in the sense of the theory's complexity (compare Eqs. (12.97)–(12.99) with [924, 964]).

Chapter 13
Nonlocal Effects in Micromechanics of Locally Elastic CMs

Abstract For either a statistically inhomogeneous medium or inhomogeneous remote boundary conditions, the eventual abandonment of the hypothesis of statistically homogeneous fields leads to a *nonlocal* coupling between statistical averages of the stress and strain tensors. Nonlocal effective integral operator estimated by the MEFM in this chapter is reduced to the differential one for the smoothed applied fields. Statistically homogeneous CMs are analyzed by either the Fourier transform method, quadrature method, or the iteration one. The statistically inhomogeneous (graded) CMs are analyzed by either the quadrature method or the iteration one. Solution for a microinhomogeneous half-space is presented through the operator form of perturbators in the framework of the generalized quasi-crystalline approximation.

When the statistical average stress in a composite medium varies sufficiently slowly compared to the size scale of the microstructure (i.e., roughly speaking the separation of the external and internal scales take place), the material macroscopically behaves as a homogeneous body with some effective moduli which can be estimated by different methods referred to earlier. If the condition of the separation of scales summarized earlier is not valid, the material's response cannot be adequately described in the framework of the local theory of elasticity for the homogeneous medium.

For analysis of the widely separated scales (but not "too widely"), a number of *phenomenological* approaches have been proposed to enhance the continuum model by nonlocal terms, introduced either through an integral equation or through an additional gradient equation including one or several intrinsic length scales and based on the assumption that the forces between material points can be long range in character (see [11, 78, 445, 485, 523, 1301] and further references therein). The integral formulation may be reduced to a gradient form by truncating the series expansion of the nonlocal kernel. Although an approximation, the gradient representation usually leads to problems that are more tractable than those obtained within the integral formulation. The different modifications of these phenomenological constitutive relations were used for the analyses of various size effects such as strain

V. A. Buryachenko, *Local and Nonlocal Micromechanics of Heterogeneous Materials*, https://doi.org/10.1007/978-3-030-81784-8_13

gradient hardening, size effect in torsion and indentation, strain gradient plasticity, and nonlocal damage mechanics.

Micromechanical models are used to answer a fundamental question of how length scales in the effective constitutive equations could be directly derived from the geometrical and mechanical properties of constituent phases. The mechanics of generalized continua has actively developed from the pioneer works [441–443, 445, 815, 818, 825, 836] establishing nonlocal elastic theories based on the lattice theory, on generalized Cosserat continua approximately considering independent microdisplacements and a scale parameter in a medium, as well as on idea that cohesive forces have a long-range and are in fact nonlocal. Kröner [818] showed that the eventual abandonment of the hypothesis of statistically homogeneous fields leads to a *nonlocal* coupling between statistical averages of the stress and strain tensors when the statistical average stress is given by an integral of the field quantity weighted by some tensorial kernel, i.e., the *nonlocal* effective elastic operator $\mathcal{L}^*$: $\langle\boldsymbol{\sigma}\rangle(\mathbf{x}) = \int \mathcal{L}^*(\mathbf{x}, \mathbf{y})\langle\boldsymbol{\varepsilon}\rangle(\mathbf{y})d\mathbf{y}$. It leads to the violation of Hill's condition, and a meaningfulness of the RVE notion is lowered. For clarification of this violation, we introduce two material length scales (see, e.g., [1523]): the macroscopic scale L, characterizing the extent of w, and the microscopic scale a, related with the heterogeneities v_i. Moreover, one supposes that applied field varies on a characteristic length scale Λ. The limit of our interests for both the material scales and field one are either

$$L \gg \Lambda \geq a \quad \text{or} \quad L \gg \Lambda \gg a, \tag{13.1}$$

where the inequalities (13.1$_2$) are called a *scale separation hypothesis*. The reason for this violation is that in micromechanics of random structure composites, the assumption (see Chap. 6) is usually made that all characteristic lengths associated with the spatial variations of the statistical mean field quantities are significantly large (13.1$_2$) compared to all characteristic lengths associated with the spatial variations in the material's properties (such as the size of the inclusions, the mean distance between them, and the characteristic length of the variation of the inclusion number density). If this assumption is valid, the governing equations for the ensemble-averaged fields are identical in form with the usual local elastic equations for homogeneous media. Otherwise, the effective constitutive law is nonlocal. This is especially true if the inclusion number density varies over distances that are comparable to the particle size. Thus, even if RVE is introduced as a minimum size of a microinhomogeneous medium providing the same (in some sense) effective elastic response as an infinite microinhomogeneous medium, this notion cannot be consistent. Indeed, the size of such as RVE will depend on the kind of effective elastic parameters (both local and nonlocal) involved in a priori unknown type of effective constitutive equation, and, moreover, it will depend on the method used for estimation of this effective constitutive equation (for details see Chaps. 9 and 11). Indeed, estimation of an effective elastic moduli by the one-particle effective field method (EFM) (see [257, 738]) does not depend on the size of RVE, in contrast with the evaluation of $\mathbf{L}^*$ by the MEFM (see Chap. 9).

If the field $\langle \boldsymbol{\varepsilon} \rangle(\mathbf{y})$ (see, e.g., [399]) is varying sufficiently slowly in the neighborhood of an arbitrary point $\mathbf{y}$, then it can be expanded about $\mathbf{x}$ in a Taylor series, and, therefore, the integral operator of nonlocal effective elastic properties can be considered as a differential one; see Beran and McCoy [106], and Levin [913] considered a weakly inhomogeneous medium. Analysis of highly contrasted statistically homogeneous media is simplified for sufficiently smooth external loading permitting the use of the Taylor expansion for the statistical average of a stress field and the application of a Fourier transform method. The scheme summarized informally above is usually based on the hypothesis of the homogeneity of the effective field in which each particle is located. The "quasi-crystalline" approximation by Lax [886] is often used for truncation of the hierarchy of integral equations involved leading to neglect of direct multiparticle interactions of inclusions. This reduces the analysis to the use of statistical information of up to two-point correlation functions and allows one to derive explicit relations for the nonlocal overall differential operator by the different methods: via the effective field method [746] or via the method of conditional moments [764]. Diener et al. [373] derived bounds on the Fourier transform of the nonlocal overall operator in the special case of "cell-structure" models. The nonlocal influence of density variations in an otherwise homogeneous medium was analyzed in [1629] assuming an exponential form for the two-point correlation function. An advantage of the rigorous approach [390–392] is that it is based on variational principles, providing the bounds of approximations; Drugan and Willis obtained an elegant explicit representation of the nonlocal effective differential operators of the second- and fourth-order and systematically estimated long-range action of a nonlocal effect and the necessary sizes of representative volume elements providing *a'priori* prescribed accuracy of the evaluation of nonlocal effects. A similar Hashin–Shtrikman formalism was exploited in [981, 982] for the analysis of nonlocal effects produced by random fluctuations in the applied body force. Besides the usual (nonlocal) effective modulus operator, Luciano and Willis [981, 982, 984] estimated another operator that provides a contribution to the mean stress field with itself.

The main disadvantage of "quasi-crystalline" approximation consisting of neglecting of direct multiparticle interactions of inclusions was overcome recently by the multiparticle effective field method (MEFM) (references may be found in Chap. 8), where the interaction of different inclusions is taken directly into account in the framework of the homogeneity hypothesis of the effective field. Combining the MEFM with the standard scheme of the Fourier transform method usually used for the analysis of the statistically inhomogeneous stress fields allowed [182, 250, 267] to obtain the explicit representation of a nonlocal overall operator in the form of the second-order differential operator. Buryachenko [184] estimated the nonlocal integral operator of the overall constitutive equation for periodic structure composites by the use of the first iteration of the iteration method and showed the reduction of the integral operator obtained to the differential operator involved derived by the Fourier transform method.

More accurate analysis of the periodic structures takes into account higher-order terms ε, when one can see that the stresses in the effective medium depend on higher-order derivatives of displacements with respect to the coordinates considered, in particular, in couple-stress elasticity [150, 496, 522]. The variational asymptotic approach proposed in [1418] may provide a better approximation to the real solution than the "purely" asymptotic approach in the sequential powers of the small parameter ε. The reason for this is that the variational construction is intrinsically such that it tends to minimize the error in a certain variational sense for the final values ε, while the asymptotic approach simply constructs the perturbations assuming that ε is very small. The effective material parameters were determined from the response of a unit cell under either displacement, displacement periodic, or traction boundary conditions [151]. It was found that the three boundary conditions result in hierarchies of both couple-stress moduli and characteristic length. Smyshlyaev and Cherednichenko [1418] (see also [151, 1176]) proposed a combination of the effective medium and micromechanical theories leading to reasonably accurate interface stresses. Vanin [1554] obtained an analytical representation for couple-stress effective elastic properties of doubly periodic fiber-reinforced medium; in the framework of moment elasticity for the effective medium, he estimated the stress concentration around a circular hole in a composite plate.

Up to now we considered statistically homogeneous media. However, it is well-known that the possible starting assumptions may be wrong. For example, due to some production technologies, the inclusion concentration may be a function of the coordinates (see [346, 923, 1218, 1219] and also Fig. 13.1). The accumulation of damage also occurs locally in stress concentration regions [891], for example, at the tip of a macroscopic crack [729]. Furthermore, in layered composite shells, the location of the fibers is random within the periodic layers (see Fig. 13.2), and the micromechanics equations are equations with almost periodic coefficients. Finally, *functionally graded materials* (FGMs) have been the subject of intense research efforts from mid-1980s when this term was originated in Japan in the framework of a national project to develop heat-shielding structural materials for the future Japanese space program. FGMs are composites consisting of two or more phases that are fabricated with a spatial variation of composition which may improve the structural response (see, e.g., [1015, 1095, 1130, 1471]) In particular cases of the

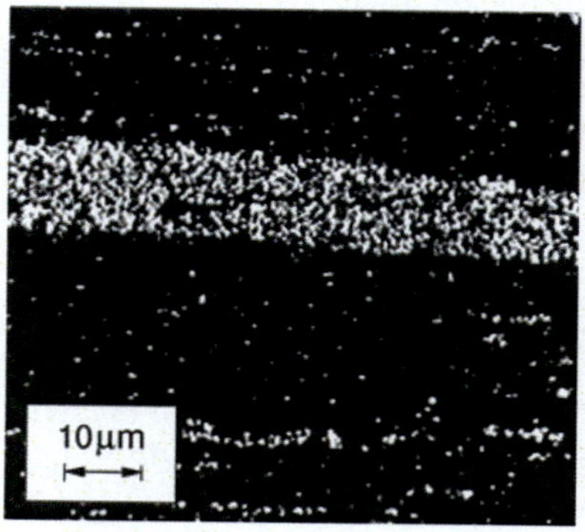

Fig. 13.1 The metallograph of high speed steel reinforced by carbide inclusion sand produced by electro-slag remelting by [1218, 1219])

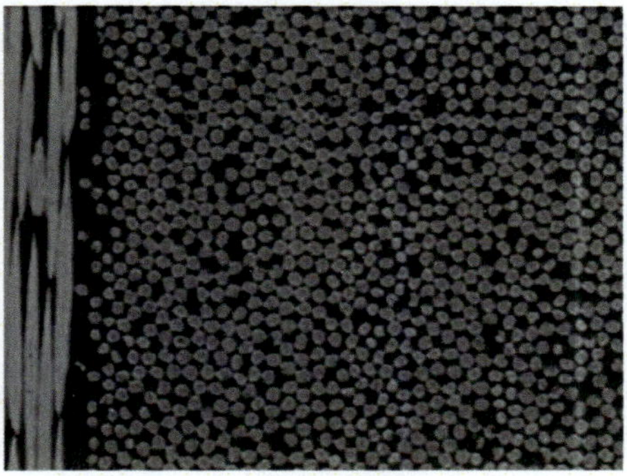

Fig. 13.2 Fragment of the original micrograph of fiber composite material microstructure near the boundary between two plies with the different orientation of aligned fibers

above problems, it is possible to use different generalizations of the known methods for random structures and periodic ones.

FGMs of deterministic structure (e.g., doubly periodic composites) were considered in [184, 200] (see also [1117, 1214, 1215, 1606]). Analysis of random structure composites is a significantly more complicated problem. A considerable number of methods are known in the linear theory of statistically homogeneous composites, which yield the effective thermoelastic constants and stress field averages in the components. Classification and analysis of these methods were considered in Chap. 8. The most popular so-called one-particle methods (such as, e.g., effective medium and Mori–Tanaka methods) provide relatively simple analytical estimations. However, they have a principal limitation following immediately from the basic hypothesis of these methods about statistical homogeneity of the composite structure. In so doing, the MEFM does not make use of a number of hypotheses that form the basis of the traditional one-particle methods, and, moreover, the MEFM includes as particular cases the aforementioned methods. However, MEFM has not only quantitative advantages (in comparison with experimental data and exact analytical solutions for some regular structures) but also qualitative benefits following immediately from the consideration of multiparticle interactions. From such considerations it can be concluded that the final relations for effective properties depend explicitly not only on the local concentration of the inclusions but also on at least binary correlation functions of the inclusions that will also be considered in this chapter. Therefore, effective properties as well as local statistical average stresses in the components are nonlocal functions of the inclusion concentration; this was shown in [258] under some additional assumptions for pure mechanical loading of FGM with homogeneous inclusions.

13.1 The Nonlocal Integral Equation

Equation (10.9) obtained for the homogeneous $\langle\boldsymbol{\sigma}\rangle$ can be generalized to the case of statistically inhomogeneous problem (when $\langle\boldsymbol{\sigma}\rangle(\mathbf{x}) \not\equiv$const., $\mathbf{x} \in v_i$)

$$\langle\overline{\boldsymbol{\sigma}}_i\rangle(\mathbf{x}) = \langle\boldsymbol{\sigma}\rangle(\mathbf{x}) + \sum_{j=1}^{N} \int \Big\{[\mathcal{T}_{i,j}^{I\sigma}(\langle\overline{\boldsymbol{\sigma}}_i\rangle)(\mathbf{x}) + \mathcal{T}_{i,j}^{J\sigma}(\langle\overline{\boldsymbol{\sigma}}_j\rangle)(\mathbf{x}) + \mathbf{T}_{i,j}^{\beta}(\mathbf{x})]$$
$$\times\varphi(v_j, \mathbf{x}_j|; v_i, \mathbf{x}_i) - [\mathcal{T}_{i,j}^{\sigma\infty}(\langle\overline{\boldsymbol{\sigma}}_j\rangle)(\mathbf{x}) + \mathbf{T}_{i,j}^{\beta\infty}(\mathbf{x})]n^{(j)}\Big\}d\mathbf{x}_j, \tag{13.2}$$

Solution of Eq. (13.2) will be considered by two method. The first method is based on generalized "quasi-crystalline" approximation when $\langle\overline{\boldsymbol{\sigma}}_k\rangle(\mathbf{x}) \not\equiv$ const at $\mathbf{x} \in v_k$, $(k = i, j)$. The second method taking the binary interaction of inclusions uses the hypothesis **H1a**.

In the framework of the hypothesis **H1a**, the linear operators with respect to the effective fields in Eq. (13.2) are reduced to factoring. We introduce the notation for the factors averaged over the inclusion volumes v_i and v_j

$$\mathbf{T}_{ij}^{\sigma}(\mathbf{x}_i, \mathbf{x}_j) = \big\langle\mathbf{T}_{i,j}^{I\sigma}(\mathbf{x})\big\rangle_{(i)(j)}\varphi(v_j, \mathbf{x}_j|; v_i, \mathbf{x}_i), \tag{13.3}$$

$$\mathbf{F}_{ij}^{\sigma}(\mathbf{x}_i, \mathbf{x}_j) = \big\langle\mathbf{T}_{i,j}^{J\sigma}(\mathbf{x})\big\rangle_{(i)(j)}\varphi(v_j, \mathbf{x}_j|; v_i, \mathbf{x}_i) - \big\langle\mathbf{T}_{i,j}^{\sigma\infty}(\mathbf{x})\big\rangle_{(i)(j)}n^{(j)}, \tag{13.4}$$

$$\mathbf{F}_{ij}^{\beta}(\mathbf{x}_i, \mathbf{x}_j) = \big\langle\mathbf{T}_{i,j}^{\beta}(\mathbf{x})\big\rangle_{(i)(j)}\varphi(v_j, \mathbf{x}_j|; v_i, \mathbf{x}_i) - \langle\mathbf{T}_{i,j}^{\beta\infty}(\mathbf{x})\rangle_{(i)(j)}n^{(j)}. \tag{13.5}$$

In particular, for statistically homogeneous media (5.20), when the probability densities $\varphi(v_i, \mathbf{x}_i)$ and $\varphi(v_q, \mathbf{x}_q|v_i, \mathbf{x}_i)$ are insensitive to translations

$$\varphi(v_i, \mathbf{x}_i) = n^{(i)} \equiv \text{const.}, \ \ \varphi(v_q, \mathbf{x}_q + \mathbf{x}|v_i, \mathbf{x}_i + \mathbf{x}) \equiv \varphi(v_q, \mathbf{x}_q|v_i, \mathbf{x}_i) \tag{13.6}$$

for any $\mathbf{x}$, the integral kernels (13.3)–(13.5) are the convolution ones: $\mathbf{T}_{iq}^{\sigma}(\mathbf{x}_i, \mathbf{x}_q) = \mathbf{T}_{iq}^{\sigma}(\mathbf{x}_i - \mathbf{x}_q)$, $\mathbf{F}_{iq}^{\sigma}(\mathbf{x}_i, \mathbf{x}_q) = \mathbf{F}_{iq}^{\sigma}(\mathbf{x}_i - \mathbf{x}_q)$, and $\mathbf{F}_{iq}^{\beta}(\mathbf{x}_i, \mathbf{x}_q) = \mathbf{F}_{iq}^{\beta}(\mathbf{x}_i - \mathbf{x}_q)$.

Then Eq. (13.2) is reduced to the following one

$$\langle\overline{\boldsymbol{\sigma}}_i\rangle(\mathbf{x}_i) = \langle\boldsymbol{\sigma}\rangle_{(i)}(\mathbf{x}_i) + \sum_{j=1}^{N} \int \Big[\mathbf{T}_{ij}^{\sigma}(\mathbf{x}_i, \mathbf{x}_j) + \mathbf{F}_{ij}^{\sigma}(\mathbf{x}_i, \mathbf{x}_j)\Big]d\mathbf{x}_j\langle\overline{\boldsymbol{\sigma}}_i\rangle(\mathbf{x}_i)$$
$$+\sum_{j=1}^{N} \int \mathbf{F}_{ij}^{\sigma}(\mathbf{x}_i, \mathbf{x}_j)\Big[\langle\overline{\boldsymbol{\sigma}}_j(\mathbf{x}_j)\rangle - \langle\overline{\boldsymbol{\sigma}}_i\rangle(\mathbf{x}_i)\Big]d\mathbf{x}_j + \sum_{j=1}^{N} \int \mathbf{F}_{ij}^{\beta}(\mathbf{x}_i, \mathbf{x}_j)d\mathbf{x}_j, \tag{13.7}$$

and, therefore,

$$\langle\overline{\boldsymbol{\sigma}}_i\rangle(\mathbf{x}_i) = \sum_{j=1}^{N} \mathbf{Y}_{ij}\Big\{\langle\boldsymbol{\sigma}\rangle_{(j)}(\mathbf{x}_j) + \mathbf{F}_j^{\beta}(\mathbf{x}_j)$$
$$+\sum_{q=1}^{N}\int \mathbf{F}_{jq}^{\sigma}(\mathbf{x}_j,\mathbf{x}_q)\Big[\langle\overline{\boldsymbol{\sigma}}_q(\mathbf{x}_q)\rangle - \langle\overline{\boldsymbol{\sigma}}_j\rangle(\mathbf{x}_q)\Big]d\mathbf{x}_j\Big\}, \tag{13.8}$$

where $\mathbf{F}_j^{\beta}(\mathbf{x}_j) = \sum_{q=1}^{N}\int \mathbf{F}_{jq}^{\beta}(\mathbf{x}_j,\mathbf{x}_q)d\mathbf{x}_q$, whereas the matrix $\mathbf{Y}^{-1}$ has the following elements

$$(\mathbf{Y}^{-1})_{ij} = \delta_{ij}\Big[\mathbf{I} - \sum_{q=1}^{N}\int \mathbf{T}_{iq}^{\sigma}(\mathbf{x}_i,\mathbf{x}_q)d\mathbf{x}_q\Big] - \int \mathbf{F}_{ij}^{\sigma}(\mathbf{x}_i,\mathbf{x}_j)d\mathbf{x}_j \tag{13.9}$$

for $i, j = 1, 2, \ldots, N$. This matrix $\mathbf{Y}$ determines the action of the surrounding inclusions on the isolated one (in the case of both the statistically homogeneous structure and homogeneous loading) and is defined simply by the solution of the problem for purely mechanical loading (with $\boldsymbol{\beta} \equiv 0$).

Then Eq. (13.8) can be rewritten in compact form

$$\mathcal{E} = \mathbf{Y}(\mathcal{E}^{av} + \mathcal{E}^{\beta}) + \mathcal{K}\mathcal{E}, \quad \mathbf{x} \in v_i, \tag{13.10}$$

where $\mathcal{E}(\mathbf{x}) \equiv (\langle\overline{\boldsymbol{\sigma}}_1\rangle(\mathbf{x}_1), \ldots, \langle\overline{\boldsymbol{\sigma}}_N\rangle(\mathbf{x}_N))^{\top}$, $\mathcal{E}^{\beta}(\mathbf{x}) \equiv (\mathbf{F}_1^{\beta}(\mathbf{x}_1), \ldots, \mathbf{F}_N^{\beta}(\mathbf{x}_N))^{\top}$, $\mathcal{E}^{av}(\mathbf{x}) \equiv (\langle\boldsymbol{\sigma}_{(1)}\rangle(\mathbf{x}_1), \ldots, \langle\boldsymbol{\sigma}_{(N)}\rangle(\mathbf{x}_N))^{\top}$, and $(\mathcal{K}\mathcal{E})(\mathbf{x}) = \int \mathbf{K}(\mathbf{x},\mathbf{y})[\mathcal{E}(\mathbf{y}) - \mathcal{E}(\mathbf{x})]\,d\mathbf{y}$ defines the integral operator $\mathcal{K}$ with the matrix kernel $\mathbf{K} \equiv \mathbf{Y}\mathbf{F}^{\sigma}(\mathbf{x}_j,\mathbf{x}_q)$ with elements formally represented as

$$\mathbf{K}_{iq}(\mathbf{x}_i,\mathbf{x}_q) = \sum_{j=1}^{N}\mathbf{Y}_{ij}\mathbf{F}_{jq}^{\sigma}(\mathbf{x}_j,\mathbf{x}_q). \tag{13.11}$$

Thus we reduced the integral equation (13.7) to the standard form of the operator equation (13.9) with the regular integral kernel of convolution type that can be solved by such known methods as the method of mechanical quadratures, successive approximations, and Fourier transform methods. We formally write the solution of Eq. (13.9) as $\mathcal{E} = \mathcal{L}(\mathcal{E}^{av} + \mathcal{E}^{\beta})$, where the inverse operator $\mathcal{L} = (\mathbf{I} - \mathcal{K})^{-1}$ will be constructed via three methods mentioned in the next section.

13.2 Methods for the Solution of the Nonlocal Integral Equations

13.2.1 *Direct Quadrature Method*

Application of the method of mechanical quadratures (called also Nystrom method) is very popular [899, 900] although it is necessary to solve linear systems of high order even for smoothly varying load. In effect the Nystrom method transforms the integral equation problem (13.9) with regular kernel into the linear algebra problem in any case $\mathcal{E}_k = \sum_l w_l \mathbf{Y}_{kl}\mathcal{E}^{av}_l + \sum_l w_l \mathcal{K}_{kl}\mathcal{E}_l$, where $\mathcal{E}_k = \mathcal{E}(\mathbf{x}_k)$, $\mathcal{E}_l = \mathcal{E}(\mathbf{x}_l)$, $\mathcal{E}^{av}_k = \mathcal{E}^{av}(\mathbf{x}_k)$, and w_l are the weights $(k, l = 1, 2, \ldots)$ of some known quadrature rule [370]. For periodic function $\mathcal{E}^{av}$, the integral (13.11) with the infinite range is in fact reduced to the integral over the unit cell. Since the kernel $\mathcal{K}(\mathbf{x}, \mathbf{y})$ decays at infinity fast enough (as $O(|\mathbf{x}-\mathbf{y}|^{-2d})$, then for $\mathcal{E}^{av}$ stabilizing at infinity ($\mathcal{E}^{av}(\mathbf{x}) \to$ const. as $|\mathbf{x}| \to \infty$), it is tempting to avoid consideration of the infinite range problem by truncating the range at some finite domain, say, the sphere with the radius R and the characteristic function describing by the Heaviside step function $H(R-|\mathbf{x}-\mathbf{y}|)$: $\int \mathbf{K}(\mathbf{x}, \mathbf{y})[\mathcal{E}(\mathbf{y})-\mathcal{E}(\mathbf{x})]\, d\mathbf{y} = \lim_{R\to\infty} \int \mathbf{K}(\mathbf{x}, \mathbf{y})[\mathcal{E}(\mathbf{y})-\mathcal{E}(\mathbf{x})] H(R-|\mathbf{x}-\mathbf{y}|)\, d\mathbf{y}$. The effectiveness of this scheme in some concrete numerical examples will be demonstrated in Sect. 13.4.4. Analysis of more general cases of the behavior of the function $\mathcal{E}^{av}$ based on the use of a rule developed directly for the infinite range (such as Gauss–Laguerre and Gauss–Hermite rules [370]) is performed in a straightforward manner and will not be considered in more detail in the current presentation.

13.2.2 *The Iteration Method*

Although the direct quadrature method usually causes no problems of accuracy, for a large number of unknown variables, N its $O(N^3)$ cost dependence can lead to surprisingly long computing time. The obvious way of reducing this cost is to construct an iterative scheme for Eq. (13.9). Because of this, at first we will solve Eq. (13.9) by the method of successive approximations, which is also called the Neumann series method. The connection with the Fourier transform method will be considered in Sect. 13.3.3.

For simplicity we will consider the point Jacoby iterative method based on the recursion formula

$$\mathcal{E}^{[k+1]} = \mathbf{Y}(\mathcal{E}^{av} + \mathcal{E}^{\beta}) + \mathcal{K}\mathcal{E}^{[k]} \tag{13.12}$$

to construct a sequence of functions $\left\{\mathcal{E}^{[k]}\right\}$ that can be treated as an approximation of the solution of Eq. (13.9). We will not analyze other methods, for example, the accelerated Liebmann method (called also extrapolated Gauss–Seidel method) which is usually "faster" than the point Jacobi method and has the computational advantage that it does not require simultaneous storage of the two iterations $\mathcal{E}^{[k+1]}$ and $\mathcal{E}^{[k]}$ [1558].

Usually the driving term of Eq. (13.12) is used as an initial approximation

$$\mathcal{E}^{[0]}(\mathbf{x}) = \mathbf{Y}(\mathcal{E}^{av}(\mathbf{x}) + \mathcal{E}^{\beta}(\mathbf{x})), \quad \langle\overline{\boldsymbol{\sigma}}_i\rangle(\mathbf{x}_i) = \sum_{j=1}^{N} \mathbf{Y}_{ij}\big[\langle\boldsymbol{\sigma}\rangle_{(j)}(\mathbf{x}_j) + \mathbf{F}_j^{\beta}(\mathbf{x}_j)\big], \tag{13.13}$$

which is exact (in the framework of hypotheses **H1** and **H2**) for a statistically homogeneous media subjected to a homogeneous boundary conditions either (7.2) or (7.3). The matrix $\mathbf{Y}$ determines the "local" action of the surrounding inclusions on the one v_i, while the integral operator kernel $\mathcal{K}$ describes a "nonlocal" action of these inclusions. The next two iterations have the form

$$\mathcal{E}^{[1]}(\mathbf{x}) = \mathbf{Y}\mathcal{E}^{av}(\mathbf{x}) + \int \mathbf{K}(\mathbf{x},\mathbf{y})\mathbf{Y}[\mathcal{E}^{av}(\mathbf{y}) + \mathcal{E}^{\beta}(\mathbf{y}) - \mathcal{E}^{av}(\mathbf{x}) - \mathcal{E}^{\beta}(\mathbf{x})]\,d\mathbf{y}, \tag{13.14}$$

$$\begin{aligned}\mathcal{E}^{[2]}(\mathbf{x}) = {} & \mathbf{Y}\mathcal{E}^{av}(\mathbf{x}) + \int \mathbf{K}(\mathbf{x},\mathbf{y})\mathbf{Y}[\mathcal{E}^{av}(\mathbf{y}) + \mathcal{E}^{\beta}(\mathbf{y}) - \mathcal{E}^{av}(\mathbf{x}) - \mathcal{E}^{\beta}(\mathbf{x})]\,d\mathbf{y} \\ & + \int \mathbf{K}(\mathbf{x},\mathbf{y})\mathbf{Y}\Big\{\int \mathbf{K}(\mathbf{y},\mathbf{z})\mathbf{Y}\big[\mathcal{E}^{av}(\mathbf{z}) + \mathcal{E}^{\beta}(\mathbf{z}) - \mathcal{E}^{av}(\mathbf{y}) - \mathcal{E}^{\beta}(\mathbf{y})\big]\,d\mathbf{z} \\ & - \int \mathbf{K}(\mathbf{x},\mathbf{z})\mathbf{Y}\big[\mathcal{E}^{av}(\mathbf{z}) + \mathcal{E}^{\beta}(\mathbf{z}) - \mathcal{E}^{av}(\mathbf{x}) - \mathcal{E}^{\beta}(\mathbf{x})\big]\,d\mathbf{z}\Big\}\,d\mathbf{y},\end{aligned} \tag{13.15}$$

and again proceeding formally; it suggests the Neumann series form for the solution $\mathcal{E}$ of (13.9)

$$\mathcal{E} = \mathcal{L}(\mathcal{E}^{av} + \mathcal{E}^{\beta}), \quad \mathcal{L} \equiv \sum_{k=0}^{\infty} \mathcal{K}^k, \tag{13.16}$$

where $\mathcal{L}$ is the inverse integral operator with the kernel $\mathcal{L}(\mathbf{x},\mathbf{y})$, and the power $\mathcal{K}^k$ is defined recursively by the condition $\mathcal{K}^1 = \mathcal{K}$, and the kernel of $\mathcal{K}^k$ is (see, e.g., [1217]) $\mathcal{K}_k(\mathbf{x},\mathbf{y}) = \int \mathcal{K}(\mathbf{x},\mathbf{z})\mathcal{K}_{k-1}(\mathbf{z},\mathbf{y})\,d\mathbf{z}$. In effect the iteration method (13.12) transforms the integral equation problem (13.9) into the linear algebra problem (13.14) and (13.16) in any case. The sequence $\{\mathcal{E}^{[k]}\}$ (13.12) with arbitrary continuous $\mathcal{E}^{av}(\mathbf{x}) + \mathcal{E}^{\beta}$ converges to a unique solution $\mathcal{E}$ (13.9) for the kernel of $\mathcal{K}$ "small" enough, which is to say that

$$\|\mathcal{K}\|_{\infty,v_i} \equiv \max_{\mathbf{x}\in v_i} \int |\mathcal{K}(\mathbf{x},\mathbf{y})|\,d\mathbf{y} < 1. \tag{13.17}$$

As will be shown, only a few iterations of Eq. (13.12) are necessary; these iterations prove very much faster than a direct inversion of the operator $\mathbf{I} - \mathcal{K}$ by the quadrature method. In so doing, condition (13.17) is fulfilled in all examples considered.

13.2.3 *The Fourier Transform Method for Statistically Homogeneous Media*

One assumes that the statistical average of strain polarization tensor $\langle \boldsymbol{\eta}_q \rangle(\mathbf{x})_q$ belongs to the class of m-times continuously differential functions $C^m(w)$. Since we desire an explicit representation for the stress concentration factor, we will approximate $\langle \boldsymbol{\eta}_q \rangle(\mathbf{x})_q$ by the first terms of its Taylor expansion (see for details [451])

$$\langle \overline{\boldsymbol{\sigma}}_q \rangle(\mathbf{x}_q) \approx \sum_{|\alpha|=0}^{m} \frac{1}{\alpha!}[\otimes(\mathbf{x}_q - \mathbf{x}_i)]^\alpha \nabla^\alpha \langle \overline{\boldsymbol{\sigma}}_q \rangle(\mathbf{x}_i), \tag{13.18}$$

where for the multi-indices of nonnegative integers $\alpha = (\alpha_1, \ldots, \alpha_d) \in Z_+^d$ and $\beta = (\beta_1, \ldots, \beta_d) \in Z_+^d$ $(d = 2, 3)$, the following notations were used:

$$\alpha + \beta = (\alpha_1 + \beta_1, \ldots, \alpha_d + \beta_d), \quad |\alpha| \equiv \alpha_1 + \ldots + \alpha_d,$$
$$\nabla^\alpha = \frac{\partial^{\alpha_1}}{\partial x_1^{\alpha_1}} \cdots \frac{\partial^{\alpha_d}}{\partial x_d^{\alpha_d}}, \quad \sum_{|\alpha|=0}^{m} = \sum_{|\alpha|=0}^{m} \sum_{\alpha_1+\ldots+\alpha_d=|\alpha|},$$
$$(\otimes \mathbf{x})^\alpha = (x_1)^{\alpha_1} \ldots (x_d)^{\alpha_d}, \quad \alpha! = \alpha_1! \ldots \alpha_d!. \tag{13.19}$$

Taking the expansions (13.18) into account, Eq. (13.5) can be rewritten in compact form

$$\mathcal{P}(\mathbf{x}_i, \nabla)\mathcal{E}(\mathbf{x}_i) = \mathcal{E}^{av}(\mathbf{x}_i) + \mathcal{E}^\beta(\mathbf{x}_i), \tag{13.20}$$

where the matrix of the linear partial differential operator $\mathcal{P}(\mathbf{x}_i, \nabla) = [\mathcal{P}(\mathbf{x}_i, \nabla)_{ij}]$ $(i, j = 1, \ldots, N)$ has the operator elements

$$\mathcal{P}(\mathbf{x}_i,\nabla)_{ij} = \delta_{ij}\Big[\mathbf{I} - \sum_{q=1}^{N} \int \mathbf{T}_{iq}^\sigma(\mathbf{x}_i, \mathbf{x}_q) d\mathbf{x}_q\Big]$$
$$- \sum_{|\alpha|=0}^{m} \frac{1}{\alpha!} \int \mathbf{F}_{ij}^\sigma(\mathbf{x}_i, \mathbf{x}_j) \otimes (\mathbf{x}_i - \mathbf{x}_j)^\alpha d\mathbf{x}_j \cdot \nabla^\alpha. \tag{13.21}$$

It should be mentioned that Eq. (13.20) and the results of Sects. 13.2.1 and 13.2.2 were obtained for the general cases of statistically inhomogeneous media. The analytical solution of Eq. (13.20) can be obtained with the additional assumption of statistical homogeneity (13.8) of random structures being considered. In such a case, the kernels (13.5) and (13.6) of involved integral operators are the convolution ones. Then for statistically homogeneous media (13.8), the differential operator $\mathcal{P}(\mathbf{x}_i, \nabla)$ (13.20) is a linear partial differential operator with constant coefficients: $\mathcal{P}(\mathbf{x}_i, \nabla) \equiv \mathcal{P}(\nabla)$.

Considering that Eq. (13.20) is a differential equation with constant coefficients, the method of solution that first comes to mind is using the Fourier transform method to transform the differential problem of solving (13.20) into the division problem of solving the multiplicative equation [1385, 1532]. The Fourier transformation $\widetilde{\mathbf{g}}(\boldsymbol{\xi})$ of a function $\mathbf{g}(\mathbf{x})$ and its inverse are defined by the formulae (3.4) provided, of course, that the integrals on the right-hand sides of Eqs. (3.4) are convergent. Using the properties of Fourier transforms of fundamental functions $F(\nabla^\alpha \mathbf{g}(\mathbf{x})) = (i\boldsymbol{\xi})^\alpha \widetilde{\mathbf{g}}(\boldsymbol{\xi})$ and $F^{-1}((i\boldsymbol{\xi})^\alpha \widetilde{\mathbf{g}}(\boldsymbol{\xi})) = \nabla^\alpha \mathbf{g}(\mathbf{x})$ enables one to transform the linear differential equation with the constant coefficient (13.20) to the algebraic multiplicative equation $\mathcal{P}(i\boldsymbol{\xi})\widetilde{\mathcal{E}}(\boldsymbol{\xi}) = \widetilde{\mathcal{E}^{av}}(\boldsymbol{\xi})$, where the symbol $\mathcal{P}(i\boldsymbol{\xi})$ of the operator $\mathcal{P}(\nabla)$ is a polynomial with complex coefficients in R^d of real transform variable $\boldsymbol{\xi} = (\xi_1, \ldots, \xi_d)^\top$. Taking advantage of (3.4$_2$) into account, we can write

$$\mathcal{E}(\mathbf{x}) = (2\pi)^{-d} \int e^{i\xi\cdot\mathbf{x}} \mathcal{P}^{-1}(i\boldsymbol{\xi}) \widetilde{\mathcal{E}^{av}}(\boldsymbol{\xi})\, d\boldsymbol{\xi} \tag{13.22}$$

which should be a solution of (13.20) in view of (3.4). Equation (13.22) provides a convenient way of calculating the average polarization tensor $\mathcal{E}(\mathbf{x})$ for given average $\widetilde{\mathcal{E}^{av}}(\mathbf{x})$ if we know the transformed properties $\mathcal{P}^{-1}(i\boldsymbol{\xi})$. To facilitate explicit results, we restrict attention to "long-wave" approximations and approximate $\mathcal{P}^{-1}(i\boldsymbol{\xi})$ by its Taylor expansion about $\boldsymbol{\xi} = \mathbf{0}$

$$\mathcal{P}^{-1}(i\boldsymbol{\xi}) = \sum_{k=0}^{m} \mathcal{Y}_{(k)}(i\boldsymbol{\xi}), \tag{13.23}$$

where the functions $\mathcal{Y}_{(k)} = [\mathcal{Y}_{(k)ij}] = \sum_{|\alpha|=k} \mathcal{Y}^\alpha(i\boldsymbol{\xi})$ are constructed by the functions $\mathcal{Y}^\alpha(i\boldsymbol{\xi})$ proportional to $(i\boldsymbol{\xi})^\alpha$. Then (13.22) and (13.23) lead to the representation of concentration of strain polarization by differential operator

$$\mathcal{E}(\mathbf{x}) = \sum_{k=0}^{m} \mathcal{Y}_{(k)}(\nabla)\big[\mathcal{E}^{av}(\mathbf{x}) + \mathcal{E}^{\beta}(\mathbf{x})\big]. \tag{13.24}$$

The local part of the differential operator (13.24) coincides with the local part of the integral operator (13.13): $\mathcal{Y}_{(0)}(\nabla) \equiv \mathbf{Y} = \big[\mathbf{Y}_{ij}\big]$ and $\mathcal{E}^{\beta}(\mathbf{x}) \equiv$const. Truncating

Eq. (13.24) after the first four sets of terms, making use of Eq. (13.23), gives the following equations on $\mathcal{Y}_{(m)}(\nabla)$ $(m = 2, 4)$

$$\mathcal{Y}_{(2)}(\nabla) = -\mathbf{Y}\mathcal{B}_2(\nabla)\mathbf{Y}, \tag{13.25}$$

$$\mathcal{Y}_{(4)}(\nabla) = -\mathbf{Y}\mathcal{B}_4(\nabla)\mathbf{Y} + 6\mathbf{Y}\mathcal{B}_2(\nabla)\mathbf{Y}\mathcal{B}_2(\nabla)\mathbf{Y}, \tag{13.26}$$

where

$$\mathcal{B}_k(\nabla) = \sum_{|\alpha|=k} \mathcal{B}^{\alpha}(\nabla), \quad \mathcal{B}^{\alpha}(\nabla) = [\mathcal{B}^{\alpha}_{ij}(\nabla)], \tag{13.27}$$

$$\mathcal{B}^{\alpha}_{ij}(\nabla) = \frac{1}{\alpha!}\int \mathbf{F}^{\sigma}_{ij}(\mathbf{x}_i - \mathbf{x}_j)\big[\otimes(\mathbf{x}_j - \mathbf{x}_i)^{\alpha}\big]\,d\mathbf{x}_j \cdot \nabla^{\alpha}, \tag{13.28}$$

$\alpha \in Z^d_+,\ k, k_1, \ldots, k_n \in Z^1_+,\ n = 1, \ldots, m$. For statistically isotropic composites, the differential operators $\mathcal{B}_k(\nabla)$ (and $\mathcal{Y}_{(k)}(\nabla)$) of odd order $k = 2n-1$ $(n = 1, \ldots)$ equal zero identically, simply because, for considered composites, the generalized functions $\mathcal{F}_{ij}(\mathbf{x}_i - \mathbf{x}_j)$ will be even homogeneous functions, and, because of this, the terms with the odd order $\mathcal{B}_k(\nabla)$ are absent in the expansions (13.25) and (13.26).

13.3 Average Stresses in the Components and Effective Properties for Statistically Homogeneous Media

13.3.1 Differential Representations

The mean effective fields of stresses inside the inclusions $\langle\overline{\sigma}\rangle_i(\mathbf{x}_i)$ is obtained from and (13.24):

$$\langle\overline{\sigma}_i\rangle_i(\mathbf{x}_i) = \sum_{j=1}^{N} \mathbf{Y}_{ij}\big[\langle\boldsymbol{\sigma}\rangle_{(j)}(\mathbf{x}_j) + \mathbf{F}^{\beta}_j\big] + \sum_{j=1}^{N}\sum_{k=1}^{m} \mathcal{Y}_{(k)ij}(\nabla)\langle\boldsymbol{\sigma}\rangle_{(j)}(\mathbf{x}_i), \tag{13.29}$$

and therefore

$$\bar{v}_i\langle\boldsymbol{\eta}\rangle_i(\mathbf{x}_i) = \mathbf{R}_i \sum_{j=1}^{N} \mathbf{Y}_{ij}\big[\langle\boldsymbol{\sigma}\rangle_{(j)}(\mathbf{x}_j) + \mathbf{F}^{\beta}_j(\mathbf{x}_j)\big] + \mathbf{F}_i$$
$$+\mathbf{R}_i \sum_{j=1}^{N}\sum_{k=1}^{m} \mathcal{Y}_{(k)ij}(\nabla)\langle\boldsymbol{\sigma}\rangle_{(j)}(\mathbf{x}_i). \tag{13.30}$$

Taking the ensemble average of a local constitutive equation combined with Eq. (13.24) gives a macroscopic constitutive equation that relates $\langle \boldsymbol{\varepsilon} \rangle(\mathbf{x})$ and $\langle \boldsymbol{\sigma} \rangle(\mathbf{x})$

$$\langle \boldsymbol{\varepsilon} \rangle(\mathbf{x}) = \mathbf{M}^* \langle \boldsymbol{\sigma} \rangle(\mathbf{x}) + \boldsymbol{\beta}^* + \mathcal{M}^*(\langle \boldsymbol{\sigma} \rangle)(\mathbf{x}), \tag{13.31}$$

$$\mathbf{M}^* = \mathbf{M}^{(0)} + \sum_{i,j=1}^{N} \mathbf{R}_i \mathbf{Y}_{ij} n^{(i)}, \tag{13.32}$$

$$\boldsymbol{\beta}^* = \boldsymbol{\beta}^{(0)} + \sum_{j=1}^{N} \mathbf{F}_j n^{(j)} + \sum_{i,j=1}^{N} n^{(i)} \mathbf{R}_i \mathbf{Y}_{ij} \mathbf{F}_j^{\beta}, \tag{13.33}$$

$$\mathcal{M}^*(\langle \boldsymbol{\sigma} \rangle)(\mathbf{x}) = \sum_{i,j=1}^{N} \sum_{k=1}^{m} n^{(i)} \mathbf{R}_i \mathcal{Y}_{(k)ij}(\nabla) \langle \boldsymbol{\sigma} \rangle(\mathbf{x}). \tag{13.34}$$

The differential operator $\mathcal{M}^*(\langle \boldsymbol{\sigma} \rangle)$ of the second order ($m = 2$) (13.34) is reduced to the analogous relation proposed in [182] for the identical homogeneous inclusions (3.183) ($N = 1$).

13.3.2 The Reduction of Integral Overall Constitutive Equations to Differential Ones

If $\mathcal{E}^{av}(\mathbf{x})$ belongs to $C^m(w)$, then substituting its Taylor expansion analogous to (13.18) into the first-order iteration for the average strain polarization tensor $\mathcal{E}_{(1)}(\mathbf{x})$ (13.14) reduces this integral equation to the differential one

$$\mathcal{E}_{(1)}(\mathbf{x}) = [\mathcal{Z}_{(0)}(\nabla) + \mathcal{Z}_{(1)}(\nabla)] \mathcal{E}^{av}(\mathbf{x}), \tag{13.35}$$

where $\mathcal{Z}_{(0)}(\nabla) \equiv \mathbf{Y}$ and the differential operators with the constant coefficients

$$\mathcal{Z}_{(1)}(\nabla) = \sum_{k=1}^{m} \mathcal{Z}_{(1)}^{k}(\nabla),$$

$$\mathcal{Z}_{(1)}^{k}(\nabla) = \sum_{|\alpha|=k} \frac{1}{\alpha!} \int \mathbf{K}(\mathbf{x}, \mathbf{y}) \mathbf{Y} [\otimes (\mathbf{y} - \mathbf{x})^{\alpha}] \, d\mathbf{y} \nabla^{\alpha}, \tag{13.36}$$

can be recast, according to the notations (13.11) and (13.27), in the form $\mathcal{Z}_{(1)}^{k}(\nabla) = \mathbf{Y} \mathcal{B}_k(\nabla) \mathbf{Y}$. For statistically isotropic composites, all odd operators $\mathcal{B}^{\alpha}(\nabla) \equiv \mathbf{0}$ at $|\alpha| = 2n - 1, \ n = 1, \ldots$.

In a similar manner, substitution of the Taylor expansion $\mathcal{E}^{av}(\mathbf{x})$ for (13.18) into the representation for the second iteration (13.15) leads to the differential equation

$$\mathcal{E}_{(2)}(\mathbf{x}) = [\mathcal{Z}_{(0)}(\nabla) + \mathcal{Z}_{(1)}(\nabla) + \mathcal{Z}_{(2)}(\nabla)]\mathcal{E}^{av}(\mathbf{x}), \tag{13.37}$$

where for the representation of $\mathcal{Z}_{(2)}(\nabla)$, we transform the iterated integral in Eq. (13.15) to the following form

$$\int \mathbf{K}(\mathbf{x},\mathbf{y})\mathbf{Y}\Big\{\int \mathbf{K}(\mathbf{y},\mathbf{z})\mathbf{Y}\big[\mathcal{E}^{av}(\mathbf{z}) - \mathcal{E}^{av}(\mathbf{y})\big]\,d\mathbf{z}$$
$$- \int \mathbf{K}(\mathbf{x},\mathbf{z})\mathbf{Y}\big[\mathcal{E}^{av}(\mathbf{z}) - \mathcal{E}^{av}(\mathbf{x})\big]\,d\mathbf{z}\Big\}\,d\mathbf{y}$$
$$= \int \mathbf{Y}\mathcal{F}(\mathbf{x}-\mathbf{y})\mathbf{Y}\mathcal{B}^{\alpha}(\nabla)\mathbf{Y}[\mathcal{E}^{av}(\mathbf{y}) - \mathcal{E}^{av}(\mathbf{x})]\,d\mathbf{y}. \tag{13.38}$$

Then the repeated use of the Taylor expansion for the function $\mathcal{E}^{av}(\mathbf{y})$ leads to the final representation $\mathcal{Z}_{(2)}(\nabla) = \sum_{k=2}^{m} \mathcal{Z}^{k}_{(2)}(\nabla)$; hereafter $\mathcal{Z}^{k}_{(m)}(\nabla) = \sum_{|\alpha^1+\ldots\alpha^m|=k} \mathbf{Y}\mathcal{B}^{\alpha^1}(\nabla) \ldots \mathbf{Y}\mathcal{B}^{\alpha^m}(\nabla)\mathbf{Y}$ and $\alpha^i = (\alpha^i_1,\ldots,\alpha^i_d) \in Z^d_+$; $i = 1,\ldots,m$; $k = m, m+1,\ldots$; $k,m \in Z^1_+$. The solutions (13.35) and (13.37) coincide with the first- (13.14) and the second- (13.15) order approximation, respectively, only for both $\mathcal{E}^{av}(\mathbf{x}) \in C^{\infty}(w)$ and taking into account infinite number of terms of the Taylor expansion of $\mathcal{E}^{av}(\mathbf{x})$, $m = \infty$ (13.18). The construction of the following differential analogs $\mathcal{Z}_{(n)}$, $(n = 3, 4, \ldots)$ of integral iterations $\mathcal{K}^n$ (13.16) is obvious. In so doing, the differential operators $\mathcal{Y}_{(n)}$ $(n = 0, 1, \ldots)$ (13.24) can be obtained by truncations of differential operators $\mathcal{Z}_{(m)}$ $(m = 0, \ldots, n)$

$$\mathcal{Y}_{(1)}(\nabla) = \mathcal{Z}^1_{(1)}, \quad \mathcal{Y}_{(2)}(\nabla) = \mathcal{Z}^2_{(1)} + \mathcal{Z}^2_{(2)}, \tag{13.39}$$

$$\mathcal{Y}_{(3)}(\nabla) = \mathcal{Z}^3_{(1)} + \mathcal{Z}^3_{(2)} + \mathcal{Z}^3_{(3)}, \tag{13.40}$$

$$\mathcal{Y}_{(4)}(\nabla) = \mathcal{Z}^4_{(1)} + \mathcal{Z}^4_{(2)} + \mathcal{Z}^4_{(3)} + \mathcal{Z}^4_{(4)}, \tag{13.41}$$

and so on. For instance, for the estimation of the operator $\mathcal{Y}_{(2)}(\nabla)$, it is enough to apply the first-order approximation (13.14) to the quadratic polynomial approximation of the Taylor expansion (13.18) of the function $\mathcal{E}^{av}(\mathbf{x})$. The operator $\mathcal{Y}_{(m)}$ can be estimated by application of the $(m-1)$ order approximation of integral operator $\mathcal{K}^{(m-1)}$ (13.16) to the m's polynomial approximation of the Taylor expansion (13.18) of the function $\mathcal{E}^{av}(\mathbf{x})$. Again, all operators $\mathcal{Z}^{2n-1}_{(m)}$ $(n = 1, 2, \ldots,\ m = 1, \ldots, 2n-1)$ as well as the operators $\mathcal{Z}^{2n}_{(m)}$ $(n = 1, 2, \ldots,\ m = 1, \ldots, 2n)$ constructed by the operators $\mathcal{B}_{2k-1}(\nabla)$ $(k < n)$ of odd order vanish for statistically isotropic media.

13.3.3 "Quasi–Crystalline" Approximation

Buryachenko [203] used the generalized "quasi-crystalline" approximation by Lax [886] (10.14$_1$) when the assumption (10.14$_2$) was relaxed: $\langle\overline{\boldsymbol\sigma}_k\rangle(\mathbf{x}) \not\equiv$ const at $\mathbf{x} \in v_k$, $(k = i, j)$. It greatly simplifies the problem (13.2) where in such a case $\mathcal{T}^{\sigma}_{i,j} = \mathcal{T}^{\sigma\infty}_{i,j}$ and $\mathbf{T}^{\beta}_{i,j}(\mathbf{x}) = \mathbf{T}^{\beta\infty}_{i,j}(\mathbf{x})$ $(\mathbf{x} \in v_i)$ that reduces Eq. (13.2) to

$$\langle\overline{\boldsymbol\sigma}_i\rangle(\mathbf{x}) = \langle\boldsymbol\sigma\rangle(\mathbf{x}) + \int [\mathcal{T}^{\sigma\infty}_{i,j}(\langle\overline{\boldsymbol\sigma}_j\rangle)(\mathbf{x}) + \mathbf{T}^{\beta\infty}_{i,j}(\mathbf{x})] \times[\varphi(v_j, \mathbf{x}_j|; v_i, \mathbf{x}_i) - n^{(j)}(\mathbf{x}_q)]d\mathbf{x}_j, \tag{13.42}$$

which is equivalent to generalization of Eq. (9.92) to statistically inhomogeneous problems

$$\langle\overline{\boldsymbol\sigma}\rangle_i(\mathbf{x}) = \langle\boldsymbol\vartheta\rangle(\mathbf{x}) + \int \mathcal{L}^{\sigma}_q(\mathbf{x}-\mathbf{x}_q, \langle\overline{\boldsymbol\sigma}\rangle_q)(\mathbf{x}_q)[\varphi(v_q, \mathbf{x}_q|; v_i, \mathbf{x}_i) - n^{(q)}(\mathbf{x}_q)]d\mathbf{x}_q. \tag{13.43}$$

The method uses as a background the new general integral equation (7.20) proposed in [202] and makes it possible to abandon the basic concepts of micromechanics: effective field hypothesis and the hypothesis of "ellipsoidal symmetry". The results of this abandonment were quantitatively estimated for some modeled composite reinforced by aligned continuously inhomogeneous fibers. Some new effects were detected that were impossible in the framework of a classical background (7.33) of micromechanics.

Of course, Eqs. (13.42) and (13.43) can be solved by either the quadrature method (see Sect. 13.2.1) or iteration method (see Sect. 13.2.2). However, for statistically homogeneous structure considered, we will only apply the Fourier transform method. Namely, for identical aligned inclusions ($N = 1$), we introduce $(\mathbf{x} \in v_i,\ \ \mathbf{y} \in v_j)$

$$\mathcal{T}^{\sigma\infty}_{ij}(\mathbf{x}, \mathbf{y}) = \mathcal{T}^{\sigma\infty}_{i,j}(\mathbf{x})[\varphi(v_j, \mathbf{x}_j|; v_i, \mathbf{x}_i) - n^{(j)}], \tag{13.44}$$

$$\mathbf{F}^{\beta}_{ij}(\mathbf{x}, \mathbf{x}_j) = \langle\mathbf{T}^{\beta\infty}_{i,j}(\mathbf{x})\rangle_{(j)}[\varphi(v_j, \mathbf{x}_j|; v_i, \mathbf{x}_i) - n^{(j)}]. \tag{13.45}$$

Then Eq. (13.20) is reduced to $(\mathbf{x} \in v_i)$ form

$$\mathcal{P}^{Q}(\mathbf{x}, \nabla)\mathcal{E}(\mathbf{x}) = \mathcal{E}^{av}(\mathbf{x}) + \mathcal{E}^{\beta}(\mathbf{x}), \tag{13.46}$$

where $\mathcal{P}^{Q}(\mathbf{x}, \nabla) = [\mathcal{P}^{Q}(\mathbf{x}, \nabla)_{ij}]$ $(i, j = 1, \ldots, N)$ has the operator elements $(\mathbf{y} \in v_j)$

$$\mathcal{P}^{Q}(\mathbf{x},\nabla)_{ij} = \delta_{ij}\mathbf{I} - \sum_{|\alpha|=0}^{m} \frac{1}{\alpha!}\int \langle\mathcal{T}^{\sigma\infty}_{ij}(\mathbf{x}, \mathbf{y}) \otimes (\mathbf{x}-\mathbf{y})^{\alpha}\rangle_{(j)}d\mathbf{x}_j \cdot \nabla^{\alpha}. \tag{13.47}$$

Then Eqs. (13.25)–(13.27) and (13.31)–(13.34) are hold with the altering

$$\mathbf{Y}_{ij}(\nabla) \to \mathbf{Y}^{Q}_{ij}(\mathbf{x}) = \delta_{ij}\mathbf{I} - \int \langle \mathcal{T}^{\sigma\infty}_{ij}(\mathbf{x}-\mathbf{y})\rangle_{(j)}\, d\mathbf{x}_j, \tag{13.48}$$

$$\mathcal{B}^{\alpha}_{ij}(\nabla) \to \mathcal{B}^{Q\alpha}_{ij}(\nabla) := \frac{1}{\alpha!}\int \langle \mathcal{T}^{\sigma\infty}_{ij}(\mathbf{x}-\mathbf{y}) \otimes (\mathbf{x}-\mathbf{y})^{\alpha}\rangle_{(j)}\, d\mathbf{x}_j \cdot \nabla^{\alpha}, \tag{13.49}$$

It should be mentioned that nonlocal effects in Sects. 13.1 and 13.2 and in Sects. 13.3.1 and 13.3.2 were estimated taking binary interaction of inclusions (hypothesis **H2a**) into account in the framework of the hypothesis **H1a** (hypothesis **H1b** was not used). In so doing, in quasi-crystalline approximation in the current Subsection, the hypotheses **H1a** and **H1b** were not exploited, whereas the hypothesis **H2b** was accepted. Generalization of the mentioned methods to the case of acceptance of the hypothesis **H2a** (without the hypotheses **H1a** and **H1b**) is straightforward.

13.3.4 Approximations in the Framework of the Hypothesis H1b

Fundamental difference of the methods in this chapter up to now from the corresponding methods in [200] is using of the new GIE (7.20) (without the hypothesis **H1b**) instead of the classical GIE (7.33) (obtained in the framework of the hypothesis **H1b**). Nevertheless, just for completeness, we shortly present in the current subsection the asymptotic representations (in the framework of the hypothesis **H1b**) of the mentioned methods.

Indeed, binary interaction of inclusions in the framework of the effective field hypotheses (4.93) and (4.94) is described by the matrix $\mathbf{Z}$ (4.143) ($\mathbf{x} \in v_i$, no sum on i, j)

$$(\mathbf{Z}^{-1})_{ij} = \mathbf{I}\delta_{ij} - (1-\delta_{ij})\mathbf{R}_j\mathbf{T}_{ij}(\mathbf{x}_i - \mathbf{x}_j), \tag{13.50}$$

where for simplicity $i = 1, 2,\ \ j = 3 - i$ are taken and the matrix $\mathbf{Z}^{21}_{ij}$ and $\mathbf{Z}^{22}_{ij}$ are nondiagonal and diagonal elements, respectively, of the matrix $\mathbf{Z}_{ij} = (\mathbf{Z}^{lm}_{ij})$ ($l, m = 1, 2$). Then the tensors (13.3)–(13.5) are represented in the forms

$$\mathbf{T}^{\sigma}_{ij}(\mathbf{x}_i, \mathbf{x}_j) = \mathbf{T}_{ij}(\mathbf{x}_i - \mathbf{x}_j)\mathbf{Z}^{21}_{ji}\mathbf{R}_i\varphi(v_j, \mathbf{x}_j|; v_i, \mathbf{x}_i), \tag{13.51}$$

$$\mathbf{F}^{\sigma}_{ij}(\mathbf{x}_i, \mathbf{x}_j) = \mathbf{T}_{ij}(\mathbf{x}_i - \mathbf{x}_j)[\mathbf{Z}^{22}_{ji}\varphi(v_j, \mathbf{x}_j|; v_i, \mathbf{x}_i) - n^{(j)}]\mathbf{R}_j, \tag{13.52}$$

$$\mathbf{F}^{\beta}_{ij}(\mathbf{x}_i, \mathbf{x}_j) = \mathbf{T}_{ij}(\mathbf{x}_i - \mathbf{x}_j)\Big\{[\mathbf{Z}^{22}_{ji}\mathbf{F}_j + \mathbf{Z}^{21}_{ji}\mathbf{F}_i]\varphi(v_j, \mathbf{x}_j|; v_i, \mathbf{x}_i) - n^{(j)}\mathbf{F}_j\Big\}. \tag{13.53}$$

Equations (13.31)–(13.34) hold unaltered at the replacement of Eqs. (13.3)–(13.5) by approximate representations (13.51)–(13.53).

In so doing, application of the hypothesis **H1b** (without hypothesis **H1a**) yields more significant simplifications in the quasi-crystalline approximation (13.43)–(13.48). So, the representations (13.43) and (13.45) are approximated by ($\mathbf{x} \in v_i$, $\mathbf{y} \in v_j$)

$$\mathcal{T}^{\sigma\infty}_{ij}(\mathbf{x},\mathbf{y}) = [\mathbf{T}_{i,j}(\mathbf{x}_i - \mathbf{x}_j)\varphi(v_j, \mathbf{x}_j|; v_i, \mathbf{x}_i) - \mathbf{T}_i(\mathbf{x}_i - \mathbf{x}_j)n^{(j)}]\mathbf{R}_j,$$

$$\mathbf{F}^{\beta}_{ij}(\mathbf{x},\mathbf{y}) = \mathbf{T}_{ij}(\mathbf{x}_i - \mathbf{x}_j)\mathbf{F}_j[\varphi(v_j, \mathbf{x}_j|; v_i, \mathbf{x}_i) - n^{(j)}]$$

and Eqs. (13.46)–(13.49) are simplified.

A "quasi-crystalline" approximation, which is equivalent to the assumption $\mathbf{Z}_{ij} = \mathbf{I}\delta_{ij}$, was used in [390, 392, 764, 1421, 1710], where the authors obtained the differential operator $\mathcal{M}^*$ (13.34) of the second and fourth orders for an arbitrary comparison moduli and for another conditional probability density $\langle V^{(i)}(\mathbf{x})V^{(q)}(\mathbf{y})\rangle$ describing the random structure of composites instead of $\varphi(v_q, \mathbf{x}_q|v_i, \mathbf{x}_i)$.

13.3.5 Numerical Analysis of Nonlocal Effects for Statistically Homogeneous Composites

We will estimate effective nonlocal properties of statistically homogeneous composite material with an isotropic incompressible matrix containing spherical identical rigid inclusions [182]. This example with the infinite contrast between the two phases was chosen deliberately because it provides the maximum difference of predictions of effective elastic response estimated by various methods and was considered by a number of authors. We will consider a well-stirred RDF $g(r) \equiv \varphi(v_i, \mathbf{x}_i|; v_q, \mathbf{x}_q)/n^{(1)} = H(r - 2a)$ where H denotes the Heaviside step function, $r \equiv |\mathbf{x}_i - \mathbf{x}_q|$ is the distance between the nonintersecting inclusions v_i and v_q, and $c(\mathbf{x}) \equiv 0.4$. Let $\mathcal{L}^*$ be the nonlocal effective elastic operator in the governing equation $\langle \boldsymbol{\varepsilon}\rangle(\mathbf{x}) = \mathbf{L}^*(\mathbf{x})\langle \boldsymbol{\varepsilon}\rangle(\mathbf{x}) + \mathcal{L}^*(\langle \boldsymbol{\varepsilon}\rangle)(\mathbf{x})$, where

$$\mathcal{L}^*(\langle \boldsymbol{\varepsilon}\rangle)(\mathbf{x}) = b_1\delta_{kl}\Delta\langle\varepsilon_0\rangle(\mathbf{x}) + b_2\partial_{kl}\langle\varepsilon_0\rangle(\mathbf{x}) + b_3\Delta\langle \mathrm{e}_{kl}\rangle(\mathbf{x}), \tag{13.54}$$

and $\varepsilon_0 \equiv \boldsymbol{\varepsilon}\boldsymbol{\delta}/3$, $\mathbf{e} \equiv \boldsymbol{\varepsilon} - \varepsilon_0\boldsymbol{\delta}$, $\Delta \equiv \partial_i\partial_i$. In Fig. 13.3 a comparison between the normalized parameters $b_3^{\mu} \equiv b_3(2\mu^{(0)}a)^{-2}$ calculated by the use of far-field $\mathbf{Z}_{ij} = \mathbf{I}\delta_{ij} + (1 - \delta_{ij})\mathbf{U}(\mathbf{x}_i - \mathbf{x}_j)\mathbf{R}$ and the "quasi-crystalline" $\mathbf{Z}_{ij} = \mathbf{I}\delta_{ij}$ approximations is presented. For sufficiently large concentrations of the inclusions, the values b_3^{μ} (solid and dotted curves) differ by a factor of six or even more.

Now we will consider the 2D case (plane–strain problem) for the two-phase composites consisting of an isotropic incompressible matrix reinforced by a statistically homogeneous field of rigid identical circle fibers of radius a. We will examine two alternative RDFs of inclusions (9.114) and (9.116). Estimation of

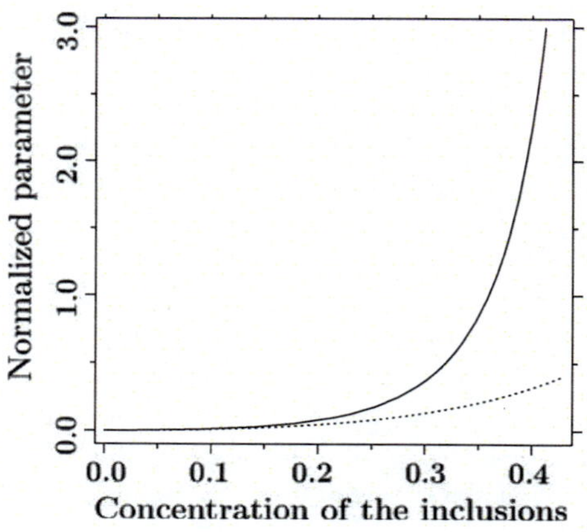

Fig. 13.3 Normalized parameters b_3^{μ} vs. c: the far-field (solid curve) and "quasi-crystalline" (doted curve) approximation

effective modulus $\mathbf{L}^*$ of composites with RDFs (9.114) and (9.116) was already considered in Chap. 8. Our approach to addressing this in Sect. 13.4.4 will be to employ the nonlocal equations for stress concentration tensors we derived in the previous sections by the iteration and Fourier transform methods. We will consider ensemble-averaged stress $\langle\boldsymbol{\sigma}\rangle(\mathbf{x})$ and determine the stress concentration factors $\langle\boldsymbol{\sigma}\rangle_i(\mathbf{x})$ at the inclusions by the quadrature method (QM), by the iteration method (IM) and, in some particular cases, by the Fourier transform methods (FTM). The quantitative results we obtain will be for the 2D case (plane–strain problem) for the two-phase composites consisting of an isotropic matrix reinforced by a random dispersion of isotropic identical circle particles.

Let us now demonstrate the application of the theoretical results by considering an isotropic highly filled composite ($c = 0.6$) made of an incompressible isotropic matrix $\mathbf{L}^{(0)} = (\infty, 2\mu^{(0)})$, filled with rigid circle inclusions $\mathbf{L}^{(1)} = (\infty, \infty)$ of one size $a = 1$, $N = 1$. Again, infinite elastic mismatch of constituents was chosen deliberately. We shall consider the response for a normal loading that varies with position in its loading direction

$$\langle\sigma_{ij}\rangle(\mathbf{x}) = f(x_1)\delta_{1i}\delta_{1j}. \tag{13.55}$$

with four specific cases of the functions

$$f_1(x_1) = \sin\left(\frac{\pi}{4}x_1\right), \quad f_2(x_1) = 0.6579|x_1|^{2.001}e^{-0.2422x_1^2}, \tag{13.56}$$

$$f_3(x_1) = 0.6584|x_1|^{1.999}e^{-0.2422x_1^2}, \quad f_4(x_1) = 0.6580x_1^2e^{-0.2420x_1^2}. \tag{13.57}$$

The loading (13.55) and (13.56_1) with $f_1(x_1) \in C^{\infty}(R)$ was analyzed by the FTM in detail in [390, 392], for the different arguments and the different concentrations of the spherical inclusions. We can only define the mth derivatives of the functions $f_j(0)$ ($j = 2, 3$) if $f_j^{(m)}(-0) = f_j^{(m)}(+0)$. Now $f_2 \in C^2$ and $f_3 \in C^1$ and

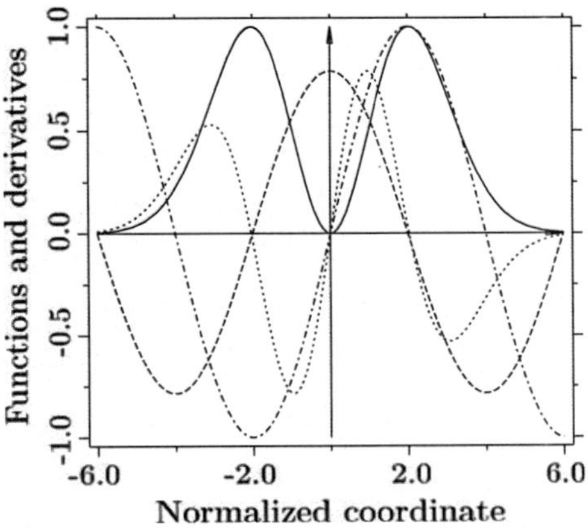

Fig. 13.4 The functions and their first derivatives vs. argument x_1: $f_2(x_1)$ (solid line), $f_1(x_1)$ (dot-dashed line), $f_1'(x_1)$ (dashed line), $f_2'(x_1)$ (dotted line)

Table 13.1 Comparative analysis of different functions (13.56) and (13.57)

	$\max\langle\sigma_{11}\rangle_{i(0)}$	$\langle\sigma_{11}\rangle_{i(7)}(\mathbf{0})$	$\max\langle\sigma_{11}\rangle_{i(7)}$	$\Delta_1(\%)$	$\Delta_2(\%)$
f_1	1.2446	0.0	1.3106	0	5.3
f_2	1.2439	−0.4295	1.5664	34.5	25.9
f_3	1.2430	−0.4300	1.5580	34.6	25.3
f_4	1.2441	−0.4296	1.5592	35.0	26.2

the third and the second derivatives do not exist for the functions f_2 and f_3, respectively. The functions f_j $(j = 2, 3, 4)$ have approximately the same max and the same max of their first derivatives as the function f_1: $|\max f_j(x_1) - 1| < 10^{-4}$, $|\max f_j'(x_1) - \max f_2'(y_1)| < 10^{-4}$, $(j = 2, 3, 4;\ x_1, y_1 \in R)$ (see Fig. 13.4). In so doing $f_2''(0) = 0$ and, therefore, $\mathcal{Y}_{(2)}\langle\boldsymbol{\sigma}\rangle(\mathbf{0}) = \mathbf{0}$ and the actions of the next order differential operators, (13.26) $\mathcal{Y}_{(2)}\langle\boldsymbol{\sigma}\rangle(\mathbf{x})$ cannot be defined for the function $f_2(x_1)$ due to non-differentiability of the function $f_2''(x_1)$. Thus, independently from the concrete micromechanical average scheme, the FTM will predict zero nonlocal effects at the point $\mathbf{0}$ for the function $f_2(x_1)$. Analogous analysis leads to a more dramatic conclusion for the function $f_3(x_1)$, that is, the FTM cannot be applied in principle to the field (13.55) and (13.57$_1$) because the differentiable function $f_3(x_1)$ which is very close to $f_2(x_1) \in C^2$ and $f_4(x_1) \in C^\infty$ is not twice differentiable at $x_1 = 0$.

The comparative results estimated by the MEFM for the radial distribution function (9.116) $(c = 0.6)$ and for different functions $f_j(x_1)$ $(j = 1, \ldots, 4)$ for 0th and 7th iterations are presented in Table 13.1; the results for functions f_2, f_3, f_4 differ from one another by less than 1%. Table 13.1 also gives the quantitative analysis of nonlocal effects presented in the terms of the stress variations $\Delta_1(\%) \equiv \langle\sigma_{11}\rangle_{i(7)}(\mathbf{0})/\max\langle\sigma_{11}\rangle_{i(0)}$; $\Delta_2(\%) \equiv \langle\max\sigma_{11}\rangle_{i(7)}/\max\langle\sigma_{11}\rangle_{i(0)}$. Table 13.2 presents nth iterations of the statistical average of stresses at the origin $\mathbf{0}$ estimated by the iteration method $\langle\sigma_{11}\rangle_{i(n)}(\mathbf{0})$ $(n = 0, 7, 20)$ and by the quadrature method (13.18) $\langle\sigma_{11}^{QM}\rangle_i(\mathbf{0})$. The 7th and 20th iterations differ from one other

Table 13.2 Comparative analysis of estimations by the IM and QM

	$\langle\sigma_{11}\rangle_{i(0)}(\mathbf{0})$	$\langle\sigma_{11}\rangle_{i(7)}(\mathbf{0})$	$\langle\sigma_{11}\rangle_{i(20)}(\mathbf{0})$	$\langle\sigma_{11}^{QM}\rangle_i(\mathbf{0})$
f_1	0.0	0.0	0.0	0.0
f_2	0.0	−0.4295	−0.4419	−0.4426
f_3	0.0	−0.4300	−0.4433	−0.4440
f_4	0.0	−0.4296	−0.4429	−0.4427

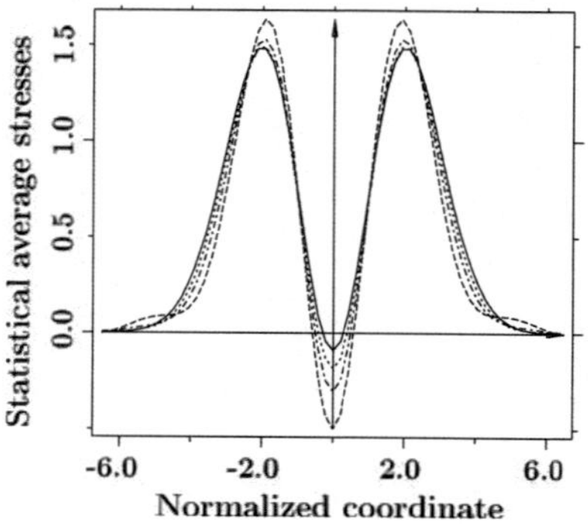

Fig. 13.5 The statistical averages $\langle\sigma_{11}^{QM}\rangle_i(x_1)$ vs. x_1 estimated for the functions $f_2(x_i)$ (13.56$_2$) and $g_2(r)$ (9.116) by the MEFM: $c = 0.15$ (solid line), 0.30 (dotted line), 0.45 (dot-dashed line), 0.60 (dashed line)

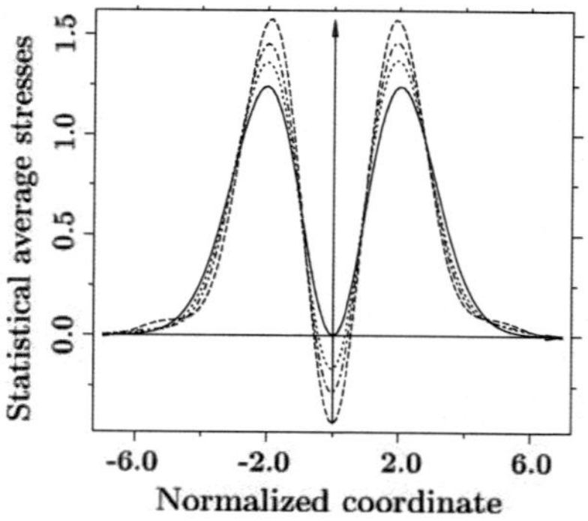

Fig. 13.6 $\langle\sigma_{11}\rangle_i(x_1)$ vs. x_1 estimated for $c = 0.6$, the functions $f_2(x_i)$ (13.56$_2$) and $g_2(r)$ (9.116) via the MEFM. Zero-order (solid line), first-order (dotted line), second-order (dot-dashed line), seventh-order (dashed line) approximations

by 3%; the difference of the estimations by the quadrature method and the 20th iterations is not over 0.2%. In Fig. 13.5 one presents the stresses $\langle\sigma_{11}^{QM}\rangle_i(x_1)$ for the function f_2 (13.56$_2$), radial distribution function $g(\mathbf{x}_i - \mathbf{x}_j)$ (9.116), and $c = 0.15,\ 0.3,\ 0.45,\ 0.6$ estimated by the QM which differ from the estimations of 20th iterations by the IM less than 0.2%. Comparison of Figs. 9.2 with 13.3 and 13.5 leads to the conclusion that the values of both the nonlocal stresses $\langle\sigma_{11}^{QM}\rangle_i(x_1) - \langle\sigma_{11}\rangle_{i(0)}(x_1)$ and nonlocal differential operator $\boldsymbol{\mathcal{L}}^*$ are more sensitive than the local effective modulus $\mathbf{L}^*$ to the value of the volume fiber concentration c. The first few iterations of estimations of stresses in the fibers $\langle\sigma_{11}\rangle_i(x_1)$ for the function f_2 are presented in Fig. 13.6. The first-order approximation of the IM leads to $\langle\sigma_{11}\rangle_{i(1)}(\mathbf{0})/\ \max\langle\sigma_{11}\rangle_{i(0)}(x_1) = 0.2134$ provided, according to Eqs. (13.35) and (13.36), by the derivatives $f_2^{(m)}$, $m = 4, 6, \ldots$ (if they exist),

Table 13.3 Comparative analysis of different methods and the functions $g(r)$ (9.114)–(9.116)

Method	$g(r)$	$\max\langle\sigma_{11}\rangle_{i(0)}$	$\langle\sigma_{11}\rangle_{i(7)}(\mathbf{0})$	$\max\langle\sigma_{11}\rangle_{i(7)}$	$\Delta_1(\%)$	$\Delta_2(\%)$
MEFM	(9.114)	1.2146	−0.3564	1.4938	29.3	23.0
MEFM	(9.116)	1.2439	−0.4295	1.5664	34.5	25.9
MT	(9.114)	1.1247	−0.2311	1.2955	20.5	15.2
MT	(9.116)	1.1247	−0.2412	1.3163	21.4	17.0
PM	(9.114)	1.1631	−0.1542	1.2830	13.3	10.3
PM	(9.116)	1.1336	−0.1995	1.2976	17.6	14.7

whereas the second-order approximation of the FTM leads to the degenerate result $\mathcal{Y}_{(2)}(\nabla)\langle\boldsymbol{\sigma}\rangle_i(\mathbf{0}) = \mathbf{0}$ (nonlocal effect is absent at the point $\mathbf{x} = \mathbf{0}$).

In Table 13.3 only the function f_2 is analyzed in the framework of the iteration scheme. A comparative analysis of the MEFM (13.12), Mori–Tanaka method (MT) (13.43), and the perturbation method (PM) (see [253] and Sect. 13.4.3) for the different radial correlation functions (9.114) and (9.116) ($c = 0.6$) is presented. As can be seen, the MEFM is most sensitive to the choice of radial distribution function and leads to the maximum nonlocal effects predicted. It is interesting that nonlocal response, in contrast to the local one, estimated by the Mori–Tanaka method, depends on the radial distribution function $g(r)$. The difference of estimations for the different average methods analyzed in Table 13.3 by the quadrature method (13.18) and 7th iteration is usually not over 3% as in Table 13.1.

Thus, in some specific numerical examples, the feasibility, efficiency, and accuracy of all three numerical methods of solution of nonlocal integral equation (13.11), quadrature method (QM), iteration method (IM), and Fourier transform method (FTM) were emphatically demonstrated. Their qualitative comparative analyses will be presented in Sect. 13.6.

13.4 The Effective Properties of Statistically Inhomogeneous Media

13.4.1 *Local Effective Properties of FGMs*

Equation (13.15) was obtained for a statistically inhomogeneous field of inclusions as well. In particular, we will consider a finite ellipsoidal cluster of inclusions w^0 with semiaxes a_w^k, $(k = 1, 2, 3;\ a_w^1 \geq a_w^2 \geq a_w^3)$ with a homogeneous distribution of identical homogeneous ellipsoidal inclusions within the cluster (see Fig. 13.7a)

$$\varphi(v_i, \mathbf{x}_i) = \begin{cases} n & \text{for } \mathbf{x} \in w^0, \\ 0 & \text{for } \mathbf{x} \notin w^0, \end{cases} \quad \varphi(v_q, \mathbf{x}_q | v_i, \mathbf{x}_i) = \begin{cases} n & \text{for } \mathbf{x} \in w^0 \setminus v_{qi}^0, \\ 0 & \text{for } \mathbf{x} \in (w \setminus w^0) \cup v_{qi}^0. \end{cases} \tag{13.58}$$

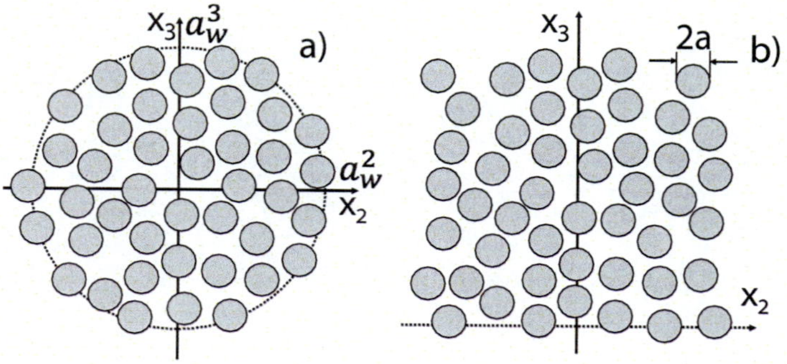

Fig. 13.7 Schematic representations of the inclusion clusters: (**a**) the ellipsoidal cluster, (**b**) the cluster of the form of an upper half-space

However, obtaining the analytical solution (13.24) in the traditional framework of Fourier transform method poses major problems because $\mathcal{P}(\mathbf{x}, \nabla)$ in Eq. (13.15) has variable coefficients. Since we desire an explicit representation for $\langle \boldsymbol{\sigma} \rangle(\mathbf{x}_i)$, we will omit the terms $\mathcal{Y}_{(k)}(\nabla)$ $(k \geq 1)$ in in Eq. (13.24): $\mathcal{Y}^{\alpha}_{jq} \equiv \mathbf{0}$ $(\forall\alpha > \mathbf{0})$. This means that all subsequent estimations of both average effective fields and local effective properties should be considered as zero-order approximations of appropriate nonlocal operators. For the zero-order approximations, we have the ordinary theory of thermoelasticity at the macrolevel; for the following approximations, a weak dispersion theory takes place.

The conditional statistical average of the stresses $\langle \boldsymbol{\sigma} | v_i, \mathbf{x}_i \rangle(\mathbf{x})$ inside the inclusions $\mathbf{x} \in v_i$ is obtained from (13.30)

$$\langle \boldsymbol{\sigma} | v_i, \mathbf{x}_i \rangle(\mathbf{x}) = \langle \mathbf{C}_i \rangle(\mathbf{x}) + \mathbf{B}_i(\mathbf{x}) \sum_{j=1}^{N} \mathbf{Y}_{ij} \big[\langle \boldsymbol{\sigma} \rangle(\mathbf{x}) + \mathbf{F}^{\beta}_j \big]. \tag{13.59}$$

In (13.59) it is necessary to distinguish between the conditional average $\langle \boldsymbol{\sigma} | v_i, \mathbf{x}_i \rangle(\mathbf{x})$ at the point $\mathbf{x}_i \in v_i$ and the statistical average $\langle \boldsymbol{\sigma} \rangle(\mathbf{x})$. The statistical average of the stresses in the matrix $\mathbf{x} \in v^{(0)}$ is found simply from the relations $\langle \boldsymbol{\sigma} | v^{(0)}, \mathbf{x} \rangle(\mathbf{x}) = [\langle \boldsymbol{\sigma} \rangle(\mathbf{x}) - \langle \boldsymbol{\sigma} V | v^{\text{incl}}, \mathbf{x} \rangle(\mathbf{x})]/c^{(0)}$. On the right side of the last equation, the statistical average $\langle \boldsymbol{\sigma} \rangle(\mathbf{x})$ and the conditional statistical average $\langle \boldsymbol{\sigma} V | v^{\text{incl}}, \mathbf{x} \rangle(\mathbf{x})$ (under the condition that there is an inclusion at the point $\mathbf{x} \in v^{\text{incl}}$, $v^{\text{incl}} \equiv \cup v^{(k)}$, $k = 1, \ldots, N$) are used.

From (13.59) the representations for the estimation of the local effective parameters follow

$$\mathbf{M}^*(\mathbf{x}) = \mathbf{M}^{(0)} + \sum_{i,j=1}^{N} \langle \mathbf{R}_i \rangle(\mathbf{x}) \mathbf{Y}_{ij}(\mathbf{x}) \mathbf{R}_j n^{(i)}(\mathbf{x}), \tag{13.60}$$

$$\boldsymbol{\beta}^*(\mathbf{x}) = \boldsymbol{\beta}^{(0)} + \sum_{j=1}^{N} \langle \mathbf{F}_j \rangle(\mathbf{x}) n^{(j)} + \sum_{i,j=1}^{N} \langle \mathbf{R}_i \rangle(\mathbf{x}) \mathbf{Y}_{ij}(\mathbf{x}) \mathbf{F}_j^{\beta}, \tag{13.61}$$

which governs the statistically averaged constitutive relation $\langle \boldsymbol{\varepsilon} \rangle(\mathbf{x}) = \mathbf{M}^*(\mathbf{x}) \langle \boldsymbol{\sigma} \rangle(\mathbf{x}) + \boldsymbol{\beta}^*(\mathbf{x})$. Only for a statistically homogeneous ergodic field X and uniform boundary conditions, the local effective parameters coincide with the global ones: $\mathbf{M}^*(\mathbf{x}),\ \boldsymbol{\beta}^*(\mathbf{x}) \equiv$ const. It should be mentioned that for the homogeneous ellipsoidal inclusions $\mathbf{R}_i(\mathbf{x}), \mathbf{F}_i(\mathbf{x}) \equiv$const. at $\mathbf{x} \in v_i$, but the statistical averages $\langle \mathbf{R}_i \rangle(\mathbf{x}), \langle \mathbf{F}_i \rangle(\mathbf{x}) \not\equiv$const. at the macropoint $\mathbf{x} \in (w^0 \oplus v_i) \setminus w^0$.

Only at first glance are the relations (13.60) and (13.61) equivalent to (13.32) and (13.33), respectively, for global effective properties. The main difference is that $\mathbf{M}^*(\mathbf{x})$ and $\boldsymbol{\beta}^*(\mathbf{x})$ depend on the parameters of the inclusion distribution not only at the point $\mathbf{x}$ but also in a certain neighborhood of that point leading to a so-called nonlocal effect, though, of course, the effective parameter $\mathbf{M}^*(\mathbf{x})$ (13.60) is the local one in the sense of "nonlocal elasticity" theory.

The diameter of this region mentioned above is estimated as three times the characteristic dimension of the inclusions (see also [182, 184, 267]). As a result, a statistically inhomogeneous composite medium behaves like a macroscopically inhomogeneous medium with local effective parameter $\mathbf{M}^*(\mathbf{x})$ determined for a non-local distribution of the inclusions in a certain neighborhood of the point considered. Relations (13.60) and (13.61) can be used to find the stress and strain distributions $\langle \boldsymbol{\sigma} \rangle(\mathbf{x}) and\ \langle \boldsymbol{\varepsilon} \rangle(\mathbf{x})$ in the mesodomain w for prescribed boundary conditions, and a conditional statistical average stress field inside the inclusions (13.59) can be estimated. Therefore, the present approach explicitly couples the microstructural details with the global analysis.

A popular macroscopic approach is the modeling of FGMs as macroscopically elastic materials, in which the material properties are graded but continuous and are described by a local effective constitutive equation (see, e.g., [438, 1245]). Obviously, for the one-particle approximation of MEFM (13.48), we obtain a trivial dependence of $\mathbf{M}^*(\mathbf{x})$ on the point $\mathbf{x}$, from which estimations by the Mori–Tanaka method can be derived for identical aligned inclusions. Mori–Tanaka methods are widely used in the analysis of graded materials [1112, 1573, 1709], although in the case of statistically inhomogeneous materials, the Mori–Tanaka hypothesis is particularly questionable (see also [1659] analyzed in Subsection 10.5.3). We can say the same about other one-particle methods considered in [1709] (see also [1288, 1289, 1365]) based on the hypothesis of a statistically homogeneous structure of the composites. Nevertheless, these theories provide an acceptable accuracy if the gradients of statistical averages of stress fields are small enough.

13.4.2 Numerical Results of Estimation of Effective Properties of FGMs

It should be mentioned that the final explicit representation for local thermoelastic properties (13.60) and (13.61) were obtained for unrestricted probability densities $\varphi(v_i, \mathbf{x}_i)$ and $\varphi(v_q, \mathbf{x}_q|v_i, \mathbf{x}_i)$. For statistically homogeneous composites, the conditional probability density $\varphi(v_q, \mathbf{x}_q|v_i, \mathbf{x}_i)$ was investigated in detail by theoretical and experimental methods (at least for spherical inclusions). However, for graded composites ergodicity is lost, i.e., one cannot equal ensemble and volume average, and a rigorous theoretical and experimental justification of analytical representations for the functions $\varphi(v_q, \mathbf{x}_q|v_i, \mathbf{x}_i)$ is required. For this purpose we choose simple step functions (13.58_1) and (13.59_2) in order to reduce the influence of undesirable side effects on the interesting particular results obtained. The step functions are the limiting cases of real distribution functions of the concentration of inclusions in FGM described in [1267] in the framework of the theory of a general Poisson processes.

If $a_w^3 \gg a_i^k$, $(k = 1, 2, 3)$, the local effective compliance of the cluster (13.60) is equal to the global effective compliance $\mathbf{M}^*$ of the corresponding statistically homogeneous composite except, in a certain boundary layer, the thickness of which will be evaluated later. Then the average stresses in the interior of the cluster are determined by the relations (8.2) and (8.6) $\langle\boldsymbol{\sigma}\rangle(\mathbf{x}) = \mathbf{B}(w^0)\boldsymbol{\sigma}^0$, where for the calculation of the tensor $\mathbf{B}(w^0)$ by the formula (8.6), the tensor $\mathbf{M}^{(i)}$ must be replaced by the global effective compliance $\mathbf{M}^*$. Then the average value of the stresses in the inclusions can be found from Eqs. (13.59) and (13.61): $\langle\boldsymbol{\sigma}|v_i, \mathbf{x}_i\rangle = (c^{(1)}\mathbf{M}_1^{(1)})^{-1}(\mathbf{M}^* - \mathbf{M}^{(0)})\mathbf{B}(w^0)\boldsymbol{\sigma}^0$, which differs from the expression for the average stresses in the inclusions of a statistically homogeneous composite, $\langle\boldsymbol{\sigma}\rangle_i = (c^{(1)}\mathbf{M}_1^{(1)})^{-1}(\mathbf{M}^* - \mathbf{M}^{(0)})\boldsymbol{\sigma}^0$ by the factor $\mathbf{B}(w^0) \neq \mathbf{I}$. This difference depends on the shape but not on the size of cloud w^0.

Therefore, the use of any large but finite number of inclusions for estimating the statistical average stresses in the inclusions is generally illegitimate. However, according to (13.54) a finite number of inclusions can be used for the estimation of the local effective compliance (13.54) which is equal to the global effective compliance of the corresponding statistically homogeneous composite except in a certain boundary layer ∂w^0. For evaluation of the thickness of this layer, let us calculate the local effective modulus $\mathbf{L}^*(\mathbf{x}) = \mathbf{L}^*(x_3)$ (13.54) as a function of the distance of the point $\mathbf{x} = (x_1, x_2, x_3)$ from the flat boundary $x_3 = 0$ of an inclusion cluster of the form of an upper half-space (see Fig. 13.7b) simulating the real structure of some graded composites (Fig. 13.1). Estimations of the normalized effective moduli $\mathbf{L}^{\mu*}(x_3) = \mathbf{L}^{\mu*}(x_3, c^{(1)}) = \mathbf{L}^*(x_3, c^{1)})/2\mu^{(0)}$ by Eq. (13.54) are presented in Fig. 13.8 for the concentration of the rigid spherical inclusions (with radius a) inside the cloud $c^{(1)} \equiv \overline{v}_i n = 0.4$; the matrix is assumed to be incompressible. In the cluster's interior (i.e., when $x_3 \to \infty$) $\mathbf{L}^*(\infty)$ coincides with the isotropic effective moduli $\mathbf{L}^*$ for the statistically homogeneous medium. Near the cluster boundary, the tensor of the effective moduli $\mathbf{L}^*(x_3)$ is transversally

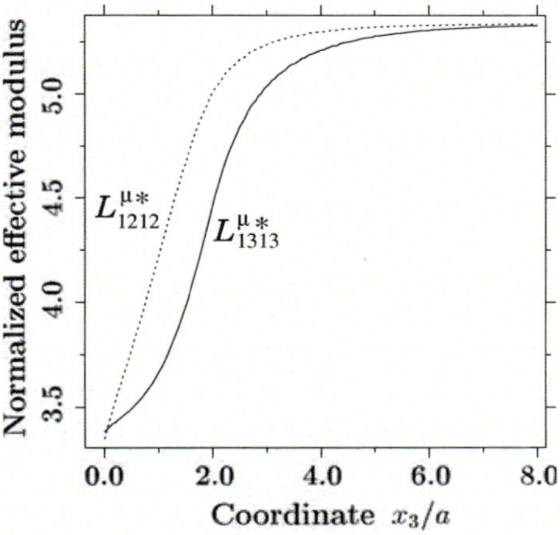

Fig. 13.8 Normalized local effective moduli $\mathbf{L}^{\mu*}(x_3)$ as functions of the distance x_3/a from the cloud boundary for $c^{(1)} = 0.4$: $L^{\mu*}_{1212}(x_3)$ (dotted curve) and $L^{\mu*}_{1313}(x_3)$ (solid curve)

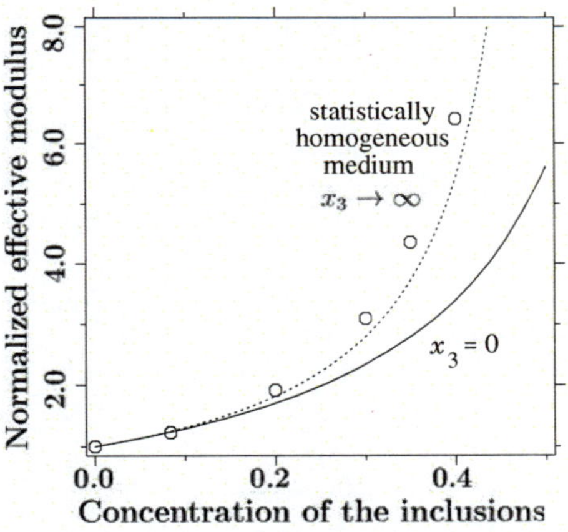

Fig. 13.9 Normalized effective shear moduli $L^{\mu*}_{1212}(x_3, c)$ vs. $c^{(1)}$, calculated for $x_3 = 0$ (solid curve) and $x_3 \to \infty$ (dotted curve) as well as experimental data by [803] ((circle), $x_3 = \infty$)

isotropic. As can be seen from Fig. 13.8, the thickness of the boundary layer with significant variation of the effective elastic moduli is not in excess of two times the size (diameter) of the inclusions. It is obvious that in the layer $-0.5a < x_3 < 0$ the number density $\varphi(v_i, \mathbf{x}) \equiv 0$, and $0 < \langle c^{(1)} \rangle(\mathbf{x}) < c^{(1)}$, and, therefore, $L^{\mu(0)}_{3333} < L^{\mu*}_{3333}(x_3, c^{(1)}) < L^{\mu*}_{3333}(0, c^{(1)})$. For simplicity, in the evaluation of $L^{*}_{3333}(x_3, c^{(1)})$, one can assume that $\mathbf{Y}(x_3) = \mathbf{Y}(0)$ for $-0.5a < x_3 < 0$. For the two positions: $x_3 = 0$ and ∞, numerical estimations for $L^{\mu*}_{3333}(x_3, c^{(1)})$ as well as experimental data [803] for statistically homogeneous media (Newtonian viscosity of a suspension of identical rigid spheres) are presented in Fig. 13.9 for the range of concentration $0 < c^{(1)} < 0.5$. As can be seen, for sufficiently large concentrations of the inclusions, the values of the local effective moduli L^{*}_{3333} near the boundary ($x_3 = 0$) and in the interior of the inclusion cluster ($x_3 \to \infty$) differ by a factor of two or even more.

In an opposite case, let us consider the estimation of the conditional average of the residual stresses ($\boldsymbol{\sigma}^0 \equiv 0$) in spherical clouds w^0 with spherical inclusions (with the radius a), when the fluctuations of the elastic moduli vanish, i.e., $\mathbf{L}(\mathbf{x}) \equiv \mathbf{L}^{(0)} = \text{const}$, and the stress-free strains $\boldsymbol{\beta}$ fluctuate. In such a case, the centering operation used at the obtaining of both Eqs. (9.96) and (9.101) is not necessary and these equation can be reduced to the same one

$$\langle\overline{\boldsymbol{\sigma}}\rangle_i(\mathbf{x}) = \boldsymbol{\sigma}^0(\mathbf{x}) + \int \boldsymbol{\mathcal{L}}_j^{\beta}(\mathbf{x}-\mathbf{x}_j, \boldsymbol{\beta})\varphi(v_j, \mathbf{x}_q|; v_i, \mathbf{x}_i)\, d\mathbf{x}_j, \tag{13.62}$$

where $\boldsymbol{\mathcal{L}}_j^{\beta}(\mathbf{x}-\mathbf{x}_j, \boldsymbol{\beta})$ is expressed by Eq. (9.99) and $\boldsymbol{\mathcal{L}}_j^{\beta}(\mathbf{x}-\mathbf{x}_j, \boldsymbol{\beta}) = \mathbf{T}_{i,j}^{\beta\infty}(\mathbf{x})$. For the sake of definiteness, we will analyze a composite with isotropic components $\mathbf{L}^{(0)} = (3k^{(0)}, 2\mu^{(0)})$, $\boldsymbol{\beta}_1^{(1)} = \beta_{01}\boldsymbol{\delta}$ and homogeneous distributions of identical spherical inclusions (13.58_1) within spherical clouds with the center $\mathbf{0}$ and with the radius a_w (see Fig. 13.7a). For the concentration of the inclusions inside the cloud $c(\mathbf{x}) = 0.5$ ($\mathbf{x} \in w^0$) and Poisson's ratio $\nu^{(0)} = 0.3$, conditional normalized statistical averages $\langle\boldsymbol{\sigma}^{\mu\beta}|v_i, \mathbf{x}_i\rangle(\mathbf{x}_i)$ ($\boldsymbol{\sigma}^{\mu\beta} := \boldsymbol{\sigma}/(\mu^{(0)}\beta_{01})$ (13.62) and the isotropic average $\langle\boldsymbol{\sigma}^{\mu\beta}\rangle_i$ (9.101) of the corresponding statistically homogeneous composite as well as the isotropic average $\langle\boldsymbol{\sigma}^{\mu\beta}\rangle(\mathbf{x}_i)$ are presented in Fig. 13.10. From this figure it can be seen that these stresses at the boundary of the cluster and at its center differ by a factor of more than two; $\mathbf{x}_i = (0, 0, x_3)$ and $a_w = 10a$. The conditional average inside the cluster $\langle\boldsymbol{\sigma}_{33}|v_i, \mathbf{x}_i\rangle(\mathbf{x}_i)$ changes more than 25% in the boundary layer with thickness $2a$. As this takes place in the boundary layer, the tensor $\langle\boldsymbol{\sigma}|v_i, \mathbf{x}_i\rangle(\mathbf{x}_i)$ is anisotropic, and the absolute magnitude of its third component is larger than that of the first one: $|\langle\sigma_{33}|v_i, \mathbf{x}_i\rangle(\mathbf{x}_i)| > |\langle\sigma_{11}|v_i, \mathbf{x}_i\rangle(\mathbf{x}_i)|$.

A similar spherical cloud w^0 of the rigid spherical inclusions in the incompressible matrix was analyzed in [267]. It was assumed that $n(\mathbf{x}) = 0.4/\bar{v}_i$ at $\mathbf{x} \in w^0$ and $n(\mathbf{x}) \equiv 0$ at $\mathbf{x} \notin w^0$. In Fig. 13.11 the estimations of normalized effective moduli $\mathbf{L}^{\mu*} \equiv \mathbf{L}^*(\mathbf{x})/(2\mu^{(0)})$, $\mathbf{L}^*(\mathbf{x}) \equiv [\mathbf{M}^*(\mathbf{x})]^{-1}$ as functions of the

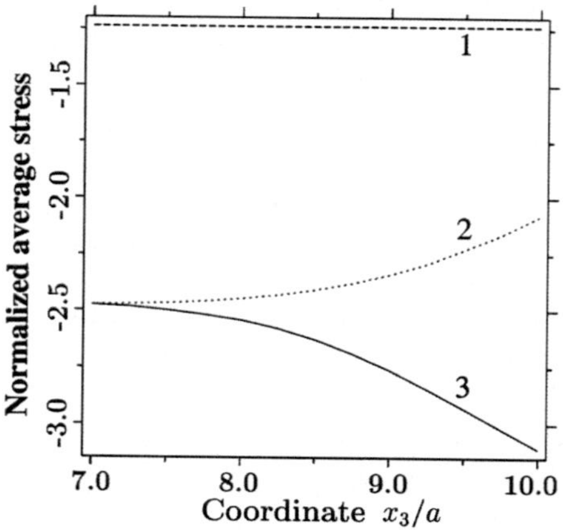

Fig. 13.10 Normalized average stresses vs. x_3/a for $c^{(1)}$=0.5: $\langle\sigma_{33}^{\mu\beta}|v_i, \mathbf{x}_i\rangle(\mathbf{x}_i)$ (3), $\langle\sigma_{11}^{\mu\beta}|v_i, \mathbf{x}_i\rangle(\mathbf{x}_i)$ (2), $\langle\sigma_{11}^{\mu\beta}\rangle_i \equiv \langle\sigma_{11}^{\mu\beta}\rangle(\mathbf{x}_i)$ (1)

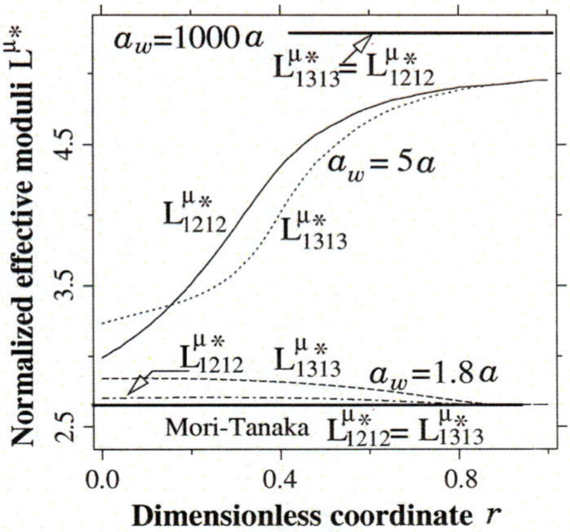

Fig. 13.11 Normalized effective shear moduli $L^{\mu*}_{1212}(r)$ and $L^{\mu*}_{1313}(r)$ vs. r at $c^{(1)} = 0.4$ for the different sizes of clusters

dimensionless coordinate r are represented for $\mathbf{x} = (0, 0, a_w(1 - r))$, where r represents the relative distance from the cloud boundary. In the interior of a large cloud ($a_w \to \infty$), sufficiently far away from its boundary, $\mathbf{L}^*(\infty)$ coincides with the isotropic effective moduli $\mathbf{L}^*$ for the statistically homogeneous medium. Near the boundary of the cloud, the tensors of the effective moduli $\mathbf{L}^{\mu*}(\mathbf{x})$ are transversally isotropic and vary significantly within the boundary layer $a_w^2 \geq \mathbf{x}^2 > (a_w - 4a)^2$ (nonlocal *boundary layer effect*). The character of the dependence $\mathbf{L}^*(\mathbf{x}) = \mathbf{L}^*(r)$ varies (increases or decreases monotonically or nonmonotonically with $r = |\mathbf{x}|$) with the variation of the cloud size a_w (*scale effect*). For $a_w = 2.2a$ the curves $L^{\mu*}_{1212}(r)$ and $L^{\mu*}_{1313}(r)$ are nonmonotonical functions of r (see Fig. 13.11 and [194]). For $a_w < 2a$ the values $\mathbf{L}^*(\mathbf{0})$ (13.54) coincide with the Mori–Tanaka approach with degeneration of binary interaction effect [89]. It is interesting that the scale effect analogous to the one described above for elastic graded composite was estimated in [1206] numerically for effective conductivity over the cross section of the composite layer of finite thickness between two homogeneous half-spaces.

Thus, the effective moduli can be obtained after "removing" of the boundary layers of the cloud; this procedure is conceptually similar to the minus-sampling method (see Chap. 5) of statistical analysis. Possible generalization of the FGM described is obvious and related with using of GIE Eq. (7.19) instead of (7.32). It offers the opportunities to abandon accepted hypotheses **H1**a, **H1b**, and **H3** (see for references and details Chap. 9) with statistical average in the point (5.74) instead of (5.69). Moreover, estimation of $\mathbf{L}^*(\mathbf{x})$ is just a local approximation of *nonlocal effective operator* which can be estimated in the framework of nonlocal theory presented in Sects. 13.1 and 13.2.

Of course, the mentioned ellipsoidal cluster can be directly extracted from the image (e.g., micro-CT) of composite sample MVE. In such a case the effective moduli can be estimated as a proportionality factor between the volume average (over the cloud with "removed" boundary layer) strain and the volume average stress, or, that is the same, as a fictitious effective ellipsoidal inclusion, see Figs. 13.8 and 13.11. Estimation of the stress–strain fields can be estimated by different methods which are most adapted for analysis of the inclusions inside the infinite matrix (see for references Chap. 4 and [849]). This scheme was effectively realized in the generalized Maxwell scheme proposed in [852, 857], and [858] and called "a rigorous and exact method of micromechanics." The mentioned straightforward scheme does not require intermedium steps such as estimations statistical descriptors and subsequent statistical simulation (see for references Chap. 5).

Obvious achievements in [852, 857] and [858] did not protect the authors from the acceptance of the most fundamental and, simultaneously, most famous incorrectness (or mistake) of analytical micromechanics, which has a long and dramatic history pioneered by Rayleigh [1284]. So, Rayleigh [1284] apparently was the first to encounter the convergence difficulty of Eq. (8.256) in consideration of effective conductivity of periodic system. Dramatic history of development of GIE from (7.30) to (7.36) and (7.19) was described in Chap. 7. Of course, Eq. (8.248) (or equivalent equations from [857, 858]) for the finite cluster of either the noninteractive or interacting N inclusions is correct and equivalent to Eq. (7.30_1). However, it is also well-known that the integral in (7.30_1) is not absolutely convergent at infinity meaning that this integral depends on both the shape and size of the integration domain (see for detals Chap. 7). This birthmark of Eq. (7.30_1) insistingly manifested in Eq. (8.248) (with interacting inclusions) where increasing size of cluster $v_i \subset \Omega$ $(i = 1, \ldots, N)$ with convergence of the corresponding results like (8.257) is a requirement of Maxwell scheme (for either the noninteractive or interacting N inclusions). Direct superposition scheme of summations of the inclusion perturbations proposed in [852, 857] and [858] is a "red flag" (see for references and details Chaps. 7 and 9) signalizing that the integrals exploited are not absolutely convergent at infinity. Thus, there is no warranty that in all cases "effective properties converge to their exact values with increasing RVE size" (see [857]). The mentioned conclusion is referred to any superposition method of estimation of inclusion interactions (e.g., either the complex potential or multipole expansion), of any accuracy as high as you like and any physical phenomena (e.g., either elasticity or conductivity). Probably, this fundamental inconsistency of the theory is a reason of the absence of convergence for the spherical Ω^h and the cubic packing of spherical inclusions in [855] and [853].

The mentioned difficulties with convergence is overcome in this section (see Figs. 13.8, 13.9, 13.10, and 13.11) in the framework of the functionally graded CMs based on the GIE (7.36). Although both Figs. 8.20b and 13.7a identically represent the circular clusters of the circular inclusions, it does not need to emphasize that FGM's theory proposed has nothing common with Maxwell scheme. Even if we forget (for the time being) about inconsistency of the current level of Maxwell scheme, there is conceptual difference between FGM theory and Maxwell scheme. The last is based on the *volume average* $\{\boldsymbol{\tau}\}_{\Omega^h}$ (meaning a summation) over Ω^h of all local polarization tensors $\boldsymbol{\tau}(\mathbf{x})$, whereas FGM theory is based on extraction from all local polarization tensors $\boldsymbol{\tau}(\mathbf{x})$ $(\mathbf{x} \in \Omega)$ their *statistical average* $\langle\boldsymbol{\tau}\rangle(\mathbf{X})$ in each macropoint $\mathbf{X} \in \Omega^h$. However, "Procrustean bed" of "rigorous" Maxwell homogenization scheme (see [852, 857], and [858]) appeared even more eloquently and spectacularly. Namely, one suggests to use all available detailed information for local polarization tensors $\boldsymbol{\tau}(\mathbf{x})\ \forall\mathbf{x}\in\Omega$ (or, more exactly, interface polarization tensor; see Chap. 11) for all random realizations just for obtaining of one constant $\mathbf{L}^*$ (some sort of average temperature in a hospital). Obtaining of so much detailed local fields estimations (which is critical for any nonlinear problem; see Chap. 12) is problematic for other competitive methods. Combining of the FGM theory with the mentioned Maxwell scheme is proposed by Buryachenko [230].

13.4.3 Perturbation Method

In the case of a small concentration of the inclusions $c \ll 1$ (9.20_1) as well as for weakly inhomogeneous medium or $||\mathbf{M}_1^{(i)}\mathbf{L}^{(0)}|| \ll 1$ (9.20_2), the perturbation method is appropriate. Then instead of hypothesis **H2** (8.10), we can use the assumption $\langle\widetilde{\boldsymbol{\sigma}}_{i,q}(\mathbf{x})\rangle_j = \langle\boldsymbol{\sigma}\rangle(\mathbf{x}_j)$ $(j=i,q)$, and Eq. (13.7) is reduced to

$$\langle\overline{\boldsymbol{\sigma}}_i\rangle(\mathbf{x}_i) = \langle\boldsymbol{\sigma}\rangle_{(i)}(\mathbf{x}_i) + \sum_{j=1}^N \int \Big[\mathbf{T}^{\sigma}_{ij}(\mathbf{x}_i,\mathbf{x}_j) + \mathbf{F}^{\sigma}_{ij}(\mathbf{x}_i,\mathbf{x}_j)\Big] d\mathbf{x}_j \langle\boldsymbol{\sigma}\rangle_{(i)}(\mathbf{x}_i)$$
$$+ \sum_{j=1}^N \int \mathbf{F}^{\sigma}_{ij}(\mathbf{x}_i,\mathbf{x}_j)\Big[\langle\boldsymbol{\sigma}\rangle_{(j)}(\mathbf{x}_j) - \langle\boldsymbol{\sigma}\rangle_{(i)}(\mathbf{x}_i)\Big] d\mathbf{x}_j$$
$$+ \sum_{j=1}^N \int \mathbf{F}^{\beta}_{ij}(\mathbf{x}_i,\mathbf{x}_j) d\mathbf{x}_j, \qquad (13.63)$$

leading to the overall constitutive equations (13.31) with effective properties

$$\mathbf{M}^{*\mathrm{per}} = \mathbf{M}^{(0)} + \sum_{i=1}^N n^{(i)}\mathbf{R}_i \Big\{\mathbf{I} + \sum_{j=1}^N \int \Big[\mathbf{T}^{\sigma}_{ij}(\mathbf{x}_i,\mathbf{x}_j) + \mathbf{F}^{\sigma}_{ij}(\mathbf{x}_i,\mathbf{x}_j)\Big] d\mathbf{x}_j\Big\}, \qquad (13.64)$$

$$\boldsymbol{\beta}^{*\mathrm{per}} = \boldsymbol{\beta}^{(0)} + \sum_{i=1}^N n^{(i)}\Big[\mathbf{F}_i + \mathbf{R}_i\mathbf{F}^{\beta}_i\Big], \qquad (13.65)$$

$$\mathcal{M}^{*\mathrm{per}}(\langle\boldsymbol{\sigma}\rangle)(\mathbf{x}) = \sum_{i,j=1}^N n^{(i)}\mathbf{R}_i \int \mathbf{F}^{\sigma}_{ij}(\mathbf{x}_i,\mathbf{x}_j)\Big[\langle\boldsymbol{\sigma}\rangle_{(j)}(\mathbf{x}_j) - \langle\boldsymbol{\sigma}\rangle_{(i)}(\mathbf{x}_i)\Big] d\mathbf{x}_j. \qquad (13.66)$$

For representation of the integral operator (13.66) in the differential form, we expand $\langle\boldsymbol{\sigma}\rangle(\mathbf{x}_q)$ about $\mathbf{x}_i$ in a Taylor series and integrate term by term over the whole space

$$\mathcal{M}^{*\mathrm{per}}(\langle\boldsymbol{\sigma}\rangle)(\mathbf{x}) = \sum_{i,q=1}^N \sum_{|\alpha|=1}^{\infty} n^{(i)}\mathcal{B}^{\alpha}_{iq}(\nabla)\mathbf{R}_q\langle\boldsymbol{\sigma}\rangle(\mathbf{x}). \qquad (13.67)$$

Equations (13.64)–(13.67) at the assumptions (13.54)–(13.56) and $\mathbf{Z}_{\text{ij}} = \mathbf{I}\delta_{ij}$ are reduced to the analogous relations proposed in [182] (see also [106]) for identical homogeneous inclusions (3.183) ($N = 1$); a simplified approach [267] used a point approximation $\mathbf{T}_{ij}(\mathbf{x}_i - \mathbf{x}_j) = \boldsymbol{\Gamma}(\mathbf{x}_i - \mathbf{x}_j)$ of binary interactions of the inclusions described by the matrix $\mathbf{Z}_{ij}$ (4.118) obtained from (4.112) by a few iterations of a successive approximation method.

It should be mentioned that effective properties can be estimated by the use of the evaluation of average strains in the components. In so doing, utilization of the perturbation method as well as the combined MEFM–perturbation method reduced to an inequality $\mathbf{L}^* \neq (\mathbf{M}^*)^{-1}$ in contrast to the MEFM (13.32). Because of this Eqs. (13.65)–(13.67) will be employed if $\mathbf{M}_1(\mathbf{x})$ is positive definite; otherwise a dual scheme based on the estimation of average strains $\langle \boldsymbol{\varepsilon} \rangle_i$ and effective moduli $\mathbf{L}^*$ will be used.

13.4.4 Combined MEFM–Perturbation Method

Instead of a standard perturbation method considered in Sect. 13.4.3, we will analyze one possible version of the combined MEFM–perturbation method described in Subsection 9.1.1 and will apply the hypotheses $\langle \widetilde{\boldsymbol{\sigma}}(\mathbf{x})_{1,2} \rangle_i = \langle \overline{\boldsymbol{\sigma}}(\mathbf{x}) \rangle_i = \langle \boldsymbol{\sigma} \rangle(\mathbf{x}),\ i = 1, 2,\ \mathbf{x} \in v_i$ (9.14) to Eq. (13.7) only for estimation of the second integral on the right-hand side of Eq. (13.7) rather than to the first integral in Eq. (13.7). Then from (13.7) we get the overall constitutive equation: $\langle \boldsymbol{\varepsilon} \rangle(\mathbf{x}) = \mathbf{M}^*(\mathbf{x})\langle \boldsymbol{\sigma} \rangle(\mathbf{x}) + \boldsymbol{\beta}^*(\mathbf{x}) + \boldsymbol{\mathcal{M}}^{*\text{per}}(\langle \boldsymbol{\sigma} \rangle)(\mathbf{x})$, where the local thermoelastic parameters described by the tensors $\mathbf{M}^*(\mathbf{x}_i),\ \boldsymbol{\beta}^*(\mathbf{x}_i)$ coincide with (13.60) and (13.61) and the nonlocal part of effective compliance is defined by

$$\boldsymbol{\mathcal{M}}^{*\text{per}}(\langle \boldsymbol{\sigma} \rangle)(\mathbf{x}) = \sum_{i,j=1}^{N} n^{(i)} \mathbf{R}_i \mathbf{Y}_{ij} \int \mathbf{F}^{\sigma}_{jq}(\mathbf{x}_j, \mathbf{x}_q) \Big[\langle \boldsymbol{\sigma} \rangle_{(q)}(\mathbf{x}_q) - \langle \boldsymbol{\sigma} \rangle_{(j)}(\mathbf{x}_j) \Big] d\mathbf{x}_q. \tag{13.68}$$

For representation of the integral operator (13.68) in differential form, we expand $\langle \boldsymbol{\sigma} \rangle(\mathbf{x}_q)$ about $\mathbf{x}_i$ in a Taylor series and integrate term by term over the whole space

$$\boldsymbol{\mathcal{M}}^{*\text{per}}(\langle \boldsymbol{\sigma} \rangle)(\mathbf{x}_i) = \sum_{|\alpha|=1}^{\infty} \boldsymbol{\mathcal{B}}^{\text{per}}_{\alpha}(\mathbf{x}_i) \nabla^{\alpha} \langle \boldsymbol{\sigma} \rangle(\mathbf{x}_i), \tag{13.69}$$

$$\boldsymbol{\mathcal{B}}^{\text{per}}_{\alpha}(\mathbf{x}_i) \equiv \frac{1}{\boldsymbol{\alpha}!} \sum_{i,j,q=1}^{N} n^{(i)} \mathbf{R}_i \mathbf{Y}_{ij}(\mathbf{x}_i) \int \mathbf{F}^{\sigma}_{jq}(\mathbf{x}_j, \mathbf{x}_q)[\otimes(\mathbf{x}_q - \mathbf{x}_j)]^{\alpha} d\mathbf{x}_q. \tag{13.70}$$

At the approximations (13.54)–(13.56), Eq. (13.70) is reduced at to the corresponding representation in [200].

13.5 Random Structure Matrix Composites in a Half-Space

The investigation of the free-edge effect in micromechanics of composites has a long history initiated in pioneer papers by Pagano (see [1175, 1270, 1285, 1312]). Significant progress in this field was achieved for the composites of regular structures (for references see [2, 1176]). If the material is locally nonperiodic, such as for the case of periodic media in the vicinity of the free edge, the asymptotic expansion technique generally leads to the poor approximation of the local fields. These difficulties have been partially overcome by employing an asymptotic expansion of a different nature in the boundary layer region and matching the solutions in the outer and inner regions using one of the matching rules [1325]. The theoretical foundation of this boundary layer and edge effects in homogenization theory were developed in [1325]; the appropriate analyses were carried out for laminated composites [72, 406, 1173], as well as for composites reinforced by fibers [482] and by particulates [1148]. Pagano and Yuan [1176] (where additional references can be found) proposed a combination of the effective medium and micromechanical theories leading to reasonably accurate interface stresses. The scheme of matching rules for the analysis of a rectilinear crack in a composite material of doubly periodic structure [733] (see also [731]) was developed. However, the analogous results for the random structure composites are significantly less developed. The perturbation method for weakly inhomogeneous media was proposed for the free-edge analysis of microinhomogeneous half-space by Podalkov and Romanov [1222] (see also [966, 967]), who constructed dispersion of the strains by this method in a first-order approximation. Luciano and Willis [983] proposed a methodology for obtaining a nonlocal effective constitutive operator in the vicinity of the free edge of composites with any elastic contrast between the constituent properties; they numerically analyzed a thermal conductivity problem for a random 2D model structure specified by a two-point exponential correlation function. However, to the author's knowledge, the analogous elastic problem for an arbitrary elastic contrast between the constituent properties is solved only for simplest random structures, such as a dilute concentration of spherical inclusions [863]. The last classical problem for the 2D case will be considered in this section.

Sillling [1400] (see also [1648, 1660]) proposed a data-driven technique to learn nonlocal constitutive laws for stress wave propagation models. The method is an optimization-based technique in which the nonlocal kernel function is derived so that when used in a nonlocal solver, it generates solutions that closely match high-fidelity data. The optimal kernel therefore acts as a homogenized nonlocal continuum model that accurately reproduces wave motion in a smaller-scale, more detailed model that can include multiple materials.

It should be mentioned that self-assembled quantum dots (QDs) (and quantum wires, QWs) have attached substantial attention in recent years [123, 606]. A semiconductor QD is a small volume of one semiconductor material, with dimensions of usually a few nanometers, buried within a second semiconductor material called a matrix. QD and QW structures are fabricated by spontaneous growth of small islands (including zero-dimensional dots and one-dimensional wires) from a wetting layer due to its mismatch strain to the substrate according to the Stranski–Krastanov (SK) mechanism. A common simplification assumed by many researchers is that the thermal or lattice mismatch plays the dominant role and an elastic mismatch can be neglected. In such a case among the most prominent continuum techniques for stress analysis (the finite elements, Green's function, and the Fourier series; see, e.g., [433, 1104, 1170, 1171]), the Green's function method acquires additional benefits. It is explained by the fact that the stress fields acting on the considered inclusion from the surrounding ones does not depend on the stress distribution inside this inclusion and can be obtained by direct superposition.

Independent of practical applications, the general inclusion problem (with zero elastic mismatch) initiated in a pioneer paper by Eshelby [449] was extensively developed (see for references, e.g., [254, 592, 1170, 1171, 1691]). In particular, the inhomogeneous inclusions in [1690] are first broken up into small cuboidal elements, which each are then treated as cuboidal homogeneous inclusions with initial eigenstrains plus unknown equivalent eigenstrains using Eshelby's equivalent inclusion method; the explicit closed-form solutions for a cuboidal inclusion with uniform eigenstrains were given in [952] (see also [1591]). Expanding the eigenstrain in a polynomial form and exploiting the Eshelby's equivalent inclusion method, one [962] calculated the eigenstrain and the elastic field by the use of the fundamental solution for Mindlin's problem.

Along with 3D problems, the solutions of 2D problems for the inclusions of arbitrary shape in a two-phase plane were also derived [1306, 1466] using the conformal mapping approach. By means of analytical continuation and conformal mapping, Ru et al. [1306] (see also [1590]) derived analytic solutions for the Eshelby's problem of a two-dimensional inclusion of arbitrary shape with uniform eigenstrains in a plane, half-plane, and two bonded half-planes. In such a case, Kushch et al. [863] combined the Muskhelishvili [1110] method of complex potentials with the Fourier integral transform technique reducing a primary boundary-value elastostatics problem for a half-plane containing elliptic inclusions to an ordinary well-posed set of linear algebraic equations. A similar problem was also solved by the boundary integral equations method [689] as well as by the volume integral equation method (see [898] and Chap. 4). Buryachenko and Pagano [254] (see also [898]) considered the hybrid BIE and VIE method for multiscale analysis of multiple interacting inhomogeneous inclusions of essentially different scale with its distinctive in 10^7 times that models a half-space with inclusions. Meshing is only performed over both the inclusion volumes and the surface of the macroinclusion.

The related phenomena of a size effect in microinhomogeneous media considered in [628, 674, 1166, 1195] (see for references [1318]), should be mentioned. The authors detected the dependence of effective elastic properties obtained by the volume average of both the local strains and stresses in the mesosample on the size of the samples being analyzed as well as on the boundary conditions (Dirichlet- or Neumann-type) acting either on this sample or on the subdomain in the domain decomposition method [1705]. In view of the aforementioned, these effects can be explained by the multiple interactions between inclusions as well as by the interactions of inclusions with the boundary of the sample. Because of this the effective constitutive equation produced via the ensemble averaging procedure will be a nonlocal one, and both local and nonlocal parts of the effective operator will depend on the coordinate of the macropoint being considered.

We will consider a half-space $w = R_0^- = \{\mathbf{x}|x_3 \leq 0\}$ containing one inclusion (the case of the finite number of inclusions is analogous). For one inclusion v_i in a half-space matrix with the properties $\mathbf{L}^{(0)}$ and $\boldsymbol{\beta}^{(0)}$ $\equiv$const., according to the definitions of the effective strains and stresses, we have the following representations for the perturbations produced by this inclusion v_i in the half-spaces $\mathbf{x} \in R^-$ and $\boldsymbol{\zeta} \in \partial R^-$

$$\boldsymbol{\varepsilon}(\mathbf{x}) - \overline{\boldsymbol{\varepsilon}}(\mathbf{x}) \equiv \mathcal{L}_i^{\epsilon\tau}(\mathbf{x}-\mathbf{x}_i, \boldsymbol{\tau}) + \mathcal{L}^{\epsilon u}(\mathbf{x}) = \mathcal{L}_i^{\epsilon\epsilon}(\mathbf{x}-\mathbf{x}_i, \overline{\boldsymbol{\varepsilon}}) + \mathcal{L}_i^{\epsilon u}(\mathbf{x}), \quad (13.71)$$

$$\mathbf{u}(\boldsymbol{\zeta}) - \overline{\mathbf{u}}(\boldsymbol{\zeta}) \equiv \mathcal{L}_i^{u\tau}(\boldsymbol{\zeta}-\mathbf{x}_i, \boldsymbol{\tau}) + \mathcal{L}^{uu}(\boldsymbol{\zeta}) = \mathcal{L}_i^{u\epsilon}(\boldsymbol{\zeta}-\mathbf{x}_i, \overline{\boldsymbol{\varepsilon}}) + \mathcal{L}_i^{uu}(\boldsymbol{\zeta}) \quad (13.72)$$

where $\overline{\boldsymbol{\varepsilon}}_i(\mathbf{x})$ and $\overline{\mathbf{u}}_i(\boldsymbol{\zeta})$ are the effective fields existing in the homogeneous half-space without the inclusion and corresponding to the nonrandom body force $\mathbf{f}$ and to boundary conditions that can be on the displacements or the tractions or mix. In the case of the hybrid BIE–VIE method considered in [254], the perturbators in Eqs. (13.71) and (13.72) are expressed through the Green function for entire space R^d

$$\mathcal{L}_i^{\epsilon\tau}(\mathbf{x}-\mathbf{x}_i, \boldsymbol{\tau}) = \int V_i(\mathbf{y})\mathbf{U}(\mathbf{x}-\mathbf{y})\boldsymbol{\tau}(\mathbf{y})\, d\mathbf{y},$$

$$\mathcal{L}^{\epsilon u}(\mathbf{x}) = \int_{\partial w} \nabla_{\mathbf{x}}\mathbf{T}(\mathbf{x}-\mathbf{s})\mathbf{u}(\mathbf{s})\, d\mathbf{s}, \quad (13.73)$$

$$\mathcal{L}_i^{u\tau}(\boldsymbol{\zeta}-\mathbf{x}_i, \boldsymbol{\tau}) = \int V_i(\mathbf{y})\nabla\mathbf{G}(\boldsymbol{\zeta}-\mathbf{y})\boldsymbol{\tau}(\mathbf{y})\, d\mathbf{y},$$

$$\mathcal{L}^{uu}(\boldsymbol{\zeta}) = \int_{\partial w} \mathbf{T}(\boldsymbol{\zeta}-\mathbf{s})\mathbf{u}(\mathbf{s})\, d\mathbf{s}, \quad (13.74)$$

where the singular integrals (13.73_2) and (13.74_2) can be simplified by the use of the physical regularization (4.34).

Strictly speaking, the perturbators $\mathcal{L}_i^{\epsilon\tau}(\mathbf{x}-\mathbf{x}_i, \boldsymbol{\tau})$, $\mathcal{L}^{\epsilon u}(\mathbf{x})$ (13.71_1), and $\mathcal{L}_i^{u\tau}(\mathbf{x}-\mathbf{x}_i, \boldsymbol{\tau})$, $\mathcal{L}^{uu}(\boldsymbol{\zeta})$ (13.72_1) and their integral representations (13.73) and (13.74), respectively, are just the notations of some problem which is destined to be solved, while the perturbators $\mathcal{L}_i^{\epsilon\epsilon}(\mathbf{x}-\mathbf{x}_i, \overline{\boldsymbol{\varepsilon}})$, $\mathcal{L}_i^{\epsilon u}(\mathbf{x})$ (13.71_2), $\mathcal{L}_i^{u\epsilon}(\mathbf{x}-\mathbf{x}_i, \overline{\boldsymbol{\varepsilon}})$, and $\mathcal{L}_i^{uu}(\boldsymbol{\zeta})$ (13.72_2) are the solutions of this problem. It should be mentioned that $\mathcal{L}^{\epsilon u}(\mathbf{x})$ (13.73_2) and $\mathcal{L}^{uu}(\boldsymbol{\zeta})$ (13.74_2) do not explicitly depend on the inclusion

v_i; however, dependence of the solutions $\mathcal{L}_i^{\epsilon u}(\mathbf{x})$ and $\mathcal{L}_i^{uu}(\boldsymbol{\zeta})$ on inclusion v_i is indicated by the index i.

For infinite realization of the inclusion field, the centering operation analogously to Eq. (9.60) leads to the following equation for the fixed inclusion $\mathbf{x} \in v_i$, $\boldsymbol{\zeta} \in R^-$

$$\langle \overline{\boldsymbol{\varepsilon}} \rangle_i(\mathbf{x}) = \langle \boldsymbol{\varepsilon} \rangle(\mathbf{x}) + \int \Big[\{\mathcal{L}_q^{\epsilon\epsilon}(\mathbf{x} - \mathbf{x}_q, \langle \overline{\boldsymbol{\varepsilon}} |; v_i, \mathbf{x}_i \rangle_q)(\mathbf{x}_q) + \mathcal{L}_q^{\epsilon u}(\mathbf{x})\} \varphi(v_q, \mathbf{x}_q |; v_i, \mathbf{x}_i)$$

$$- \{\mathcal{L}_q^{\epsilon\epsilon}(\mathbf{x} - \mathbf{x}_q, \langle \overline{\boldsymbol{\varepsilon}} \rangle_q)(\mathbf{x}_q) + \mathcal{L}_q^{\epsilon u}(\mathbf{x})\} n^{(q)} \Big] V(R_a^-)(\mathbf{x}_q)\, d\mathbf{x}_q, \qquad (13.75)$$

$$\langle \overline{\mathbf{u}} \rangle_i(\boldsymbol{\zeta}) = \langle \mathbf{u} \rangle(\boldsymbol{\zeta}) + \int \Big[\{\mathcal{L}_q^{u\epsilon}(\boldsymbol{\zeta} - \mathbf{x}_q, \langle \overline{\boldsymbol{\varepsilon}} |; v_i, \mathbf{x}_i \rangle_q)(\mathbf{x}_q) + \mathcal{L}_q^{\epsilon u}(\boldsymbol{\zeta})\} \varphi(v_q, \mathbf{x}_q |; v_i, \mathbf{x}_i)$$

$$- \{\mathcal{L}_q^{u\epsilon}(\boldsymbol{\zeta} - \mathbf{x}_q, \langle \overline{\boldsymbol{\varepsilon}} \rangle_q)(\mathbf{x}_q) + \mathcal{L}_q^{uu}(\boldsymbol{\zeta})\} n^{(q)} \Big] V(R_a^-)(\mathbf{x}_q)\, d\mathbf{x}_q, \qquad (13.76)$$

where $V(R_a^-) := \{\mathbf{x} |\ x_3 \leq -a\}$ is the indicator function of a domain for possible location of the inclusion centers. Closing of Eqs. (13.75)–(13.76) is performed by the use of the hypothesis **H2b**

$$\langle \overline{\boldsymbol{\varepsilon}} \rangle_i(\mathbf{x}) = \langle \boldsymbol{\varepsilon} \rangle(\mathbf{x}) + \int \{\mathcal{L}_q^{\epsilon\epsilon}(\mathbf{x} - \mathbf{x}_q, \langle \overline{\boldsymbol{\varepsilon}} \rangle_q)(\mathbf{x}_q) + \mathcal{L}_q^{\epsilon u}(\mathbf{x})\}$$

$$\times [\varphi(v_q, \mathbf{x}_q |; v_i, \mathbf{x}_i) - n^{(q)}] V(R_a^-)(\mathbf{x}_q)\, d\mathbf{x}_q, \qquad (13.77)$$

$$\langle \overline{\mathbf{u}} \rangle_i(\boldsymbol{\zeta}) = \langle \mathbf{u} \rangle(\boldsymbol{\zeta}) + \int \{\mathcal{L}_q^{u\epsilon}(\boldsymbol{\zeta} - \mathbf{x}_q, \langle \overline{\boldsymbol{\varepsilon}} \rangle_q)(\mathbf{x}_q) + \mathcal{L}_q^{\epsilon u}(\boldsymbol{\zeta})\}$$

$$\times [\varphi(v_q, \mathbf{x}_q |; v_i, \mathbf{x}_i) - n^{(q)}] V(R_a^-)(\mathbf{x}_q)\, d\mathbf{x}_q. \qquad (13.78)$$

Equations (13.77) and (13.78) can be solved by the iteration method analogously to the iterations (9.92). Owing to inhomogeneity of $\langle \boldsymbol{\varepsilon} \rangle(\mathbf{x})$ in general (as in Eq. (13.7)), the solution of Eqs. (13.77) and (13.78) will be presented by the nonlocal operators with subsequent representation of the effective properties also by the nonlocal operators.

Equations (13.77)–(13.78) can be simplified for the elastically homogeneous materials with the random field of inclusions ($\mathbf{L}_1(\mathbf{x}) \equiv \mathbf{0}$, $\boldsymbol{\alpha}_1(\mathbf{x}) \not\equiv \mathbf{0}$)

$$\langle \overline{\boldsymbol{\varepsilon}} \rangle_i(\mathbf{x}) = \langle \boldsymbol{\varepsilon} \rangle(\mathbf{x}) + \int \{\mathbf{L}_q^{\epsilon\alpha}(\mathbf{x} - \mathbf{x}_q, \boldsymbol{\alpha}) + \mathcal{L}_q^{\epsilon u}(\mathbf{x})\}$$

$$\times [\varphi(v_q, \mathbf{x}_q |; v_i, \mathbf{x}_i) - n^{(q)}] V(R_a^-)(\mathbf{x}_q)\, d\mathbf{x}_q, \qquad (13.79)$$

$$\langle \overline{\mathbf{u}} \rangle_i(\boldsymbol{\zeta}) = \langle \mathbf{u} \rangle(\boldsymbol{\zeta}) + \int \{\mathbf{L}_q^{u\alpha}(\mathbf{x} - \mathbf{x}_q, \boldsymbol{\alpha}) + \mathcal{L}_q^{\epsilon u}(\boldsymbol{\zeta})\}$$

$$\times [\varphi(v_q, \mathbf{x}_q |; v_i, \mathbf{x}_i) - n^{(q)}] V(R_a^-)(\mathbf{x}_q)\, d\mathbf{x}_q, \qquad (13.80)$$

where $\mathbf{L}_q^{\epsilon\alpha}(\mathbf{x}-\mathbf{x}_q,\boldsymbol{\alpha}) = \nabla\mathbf{L}_q^{u\alpha}(\mathbf{x}-\mathbf{x}_q,\boldsymbol{\alpha})$. As can be seen, the statistical average of the effective field $\langle\overline{\boldsymbol{\varepsilon}}\rangle_i(\mathbf{x})$ is defined by summation of all perturbations $\mathbf{L}_q^{\epsilon\alpha}(\mathbf{x}-\mathbf{x}_q,\boldsymbol{\alpha}) + \boldsymbol{\mathcal{L}}_q^{\epsilon u}(\mathbf{x})$ (13.71_1) produced by surrounding inclusions v_q with the centers in the domain where the function $[\varphi(v_q,\mathbf{x}_q|;v_i,\mathbf{x}_i) - n^{(q)}]V(R_a^-)(\mathbf{x}_q)$ does not vanish. Of course, the mentioned perturbators can be estimated by any available numerical method rather than the hybrid BIE–VIE method (13.73). So, the solution of the intersecting elliptical inclusions in a half-plane was obtained in [863] by a combination of the Muskhelishvili's method of complex potentials with the Fourier integral transform technique. After that the averaging of both the residual strains (or stress) and effective eigenstrain are performed by a modified average operation (5.74)

$$c^{(q)}(\mathbf{x})\langle\mathbf{g}\rangle^{(q)}(\mathbf{x}) = \int_{v_q^1(\mathbf{x})\cap R_a^-} n^{(q)}(\mathbf{y})\langle\mathbf{g}|v_q,\mathbf{y}\rangle(\mathbf{x},\mathbf{y}-\mathbf{x})\,d\mathbf{y}. \tag{13.81}$$

In particular, for estimation of the effective eigenstress $\boldsymbol{\alpha}^*(\mathbf{x}) = \boldsymbol{\alpha}^{(0)} + \langle\boldsymbol{\alpha}_1\rangle(\mathbf{x})$ (eigenstress $\boldsymbol{\beta}^*$ can be estimated in a similar manner), we need to substitute in Eq. (13.81) $\mathbf{g}(v_q,\mathbf{y}-\mathbf{x}) = \boldsymbol{\alpha}_1^{(1)}V(v_q,\mathbf{y})(\mathbf{x})$, where $V(v_q,\mathbf{y})(\mathbf{x})$ is an indicator function of the inclusion v_q with the moving center $\mathbf{y}$. At the estimation of the effective residual strains inside the inclusions $\langle\boldsymbol{\varepsilon}^{II}\rangle^{(1)}(\mathbf{x})$ in Eq. (13.81), we need to substitute

$$\mathbf{g}(v_q,\mathbf{y}-\mathbf{x}) = \mathbf{L}_q^{\epsilon\alpha}(\mathbf{x},\boldsymbol{\alpha}) + \boldsymbol{\mathcal{L}}_q^{\epsilon u}(\mathbf{x}) + \langle\overline{\boldsymbol{\varepsilon}}\rangle_q(\mathbf{x}). \tag{13.82}$$

for the moving inclusion v_q with the center $\mathbf{y}$; here a summand $\langle\overline{\boldsymbol{\varepsilon}}\rangle_q(\mathbf{x})$ is estimated by Eq. (13.79) (see for details [240]).

The scheme (13.79)–(13.82) holds for any statistically inhomogeneous fields of random eigenstresses in elastically homogeneous half-space. In the case of elastic mismatch $\mathbf{L}_1(\mathbf{x}), \boldsymbol{\alpha}_1(\mathbf{x}) \not\equiv \mathbf{0}$, Eqs. (13.77) and (13.78) can be simplified in the framework of a dilute approximation, when $\langle\overline{\boldsymbol{\varepsilon}}\rangle_i(\mathbf{x}) = \langle\boldsymbol{\varepsilon}\rangle \equiv$const. and, therefore, owing to the splitting of the operators (13.71_2)

$$\boldsymbol{\mathcal{L}}_q^{\epsilon\epsilon}(\mathbf{x}-\mathbf{x}_q,\langle\overline{\boldsymbol{\varepsilon}}\rangle_q)(\mathbf{x}_q) + \boldsymbol{\mathcal{L}}_q^{\epsilon u}(\mathbf{x}) = \mathbf{L}_q^{\epsilon\epsilon}(\mathbf{x}-\mathbf{x}_q)\langle\boldsymbol{\varepsilon}\rangle + \mathbf{L}_q^{\epsilon\alpha}(\mathbf{x}-\mathbf{x}_q). \tag{13.83}$$

Substitution of Eq. (13.83) into Eqs. (13.82), (6.54_1), and (6.54_2) enables one to obtain the final representations for both the statistical average fields $\langle\boldsymbol{\varepsilon}^{II}\rangle^{(1)}(\mathbf{x})$ and the effective properties $\mathbf{L}^*(\mathbf{x})$ and $\boldsymbol{\alpha}^*(\mathbf{x})$ of the local approximation of the effective constitutive equation $\langle\boldsymbol{\sigma}\rangle(\mathbf{x}) = \mathbf{L}^*(\mathbf{x})\langle\boldsymbol{\varepsilon}\rangle + \boldsymbol{\alpha}^*(\mathbf{x})$. This scheme was realized for 2D half-plane with elliptical inclusions in [863], where the operators (13.83), (13.71_2), and (13.72_2) were estimated by a combination of the Muskhelishvili's method of complex potentials with the Fourier integral transform technique (see also [200]). It was demonstrated that the free-edge effects for both $\mathbf{M}^*(\mathbf{x})$ and $\langle\boldsymbol{\sigma}^I\rangle^{(1)}(\mathbf{x})$ are manifested in a wide strip $-5a \le x_3 \le 0$ that reflects inhomogeneity of the stresses inside the inclusions (distinct from the Eshelby solution) in the vicinity of the free edge.

It should be mentioned that both the local and nonlocal effective operators (see, e.g., Eqs. (13.34) and (13.60)) considered in the current Chapter are associated with the statistical averaging of stresses and strains in a macropoint $\mathbf{x} \in w$ (and in its vicinity) according to the operation (5.74) rather than with the averaging over the sample w (as in the popular approaches in the scale effect estimations, see, e.g., [628, 674, 1166, 1195, 1318]).

13.6 Concluding Remarks

Let us discuss the main scheme as well as short sketch of limitations and of possible generalization and application of the methods proposed.

The obtained relations depend on the values associated with the mean distance between inclusions and do not depend on the other characteristic size, i.e., the mean inclusion diameter. This fact may be explained by the initial use of the hypothesis **H1** dealing with the homogeneity of the field $\overline{\boldsymbol{\sigma}}(\mathbf{x})$ inside each inclusion. In the case of a variable representation of $\overline{\boldsymbol{\sigma}}(\mathbf{x})$ $(\mathbf{x} \in v_i)$, for instance, in polynomial form, the mean size of the inclusions will be contained in the nonlocal dependence of microstresses on the average stress $\langle \boldsymbol{\sigma} \rangle(\mathbf{x})$. Such an improvement based on the abandonment from hypothesis **H1a** and **H1b** was considered in Sect. 13.3.3.

It should be mentioned that QM, IM, and FTM have a series of advantages and disadvantages, and it is crucial for the analyst to be aware of their range of application. The QM caused no problems of accuracy but can become less effective in the CPU time sense than the IM if the number of variables increases. The IM (13.16) has two known drawbacks. First, the Neumann series ensures the existence of solutions to integral equations of the second kind only for sufficiently small kernels (13.17), and, second, in general, it cannot be summed in closed form. Of course Eq. (13.16) can be solved directly by the quadrature method even if the condition (13.17) is not valid. However, strongly inhomogeneous problems may lead to much larger numbers of quadrature points, making iteration potentially worthwhile. Moreover, increasing the problem dimensionality (from 2D to 3D) raises the number of nodes to the dimensional power, and the situation changes radically. As was shown, only a few iterations of Eq. (13.16) are necessary; these iterations prove very much faster than a direct inversion of the operator $\mathbf{I} - \mathcal{K}$ by the quadrature method. In so doing, the condition (13.17) is fulfilled in all examples considered for the highly filled composites with an infinite contrast in moduli between the phases.

The reduction of integral operators to the differential ones allows an understanding of drawbacks of the FTM. The first one is that for obtaining an m order differential operator, it is necessary that $\mathcal{E}^{av}$ belongs to $C^m(w)$. In so doing, the IM providing the accuracy of differential operator of an infinite order does not have even continuity of $\mathcal{E}^{av}$ since integration is a smoothing operation and the right-hand side integral (13.11) is likely to be a rather smooth function even when $\mathcal{E}^{av}(\mathbf{x})$ is very jagged. But even the m'th approximation of the IM contains the differential operators of infinite orders (although not all; see Eq. (13.36)) that are lost in the FTM. The question of the convenience of using one method over another is solved also in favor of the IM because in the FTM, it is necessary to calculate the cumbersome tensors $\mathcal{Y}_{(m)}$ (13.24) and $\mathcal{B}_m$ (13.28), completeness of estimation of which increase dramatically with m, while in the IM it is enough to estimate a single tensor $\mathbf{K}(\mathbf{x}_i - \mathbf{x}_q)$ (13.11) and consecutively to apply the recursion scheme (13.16) the completeness of which does not depend on the iteration number. Thus, the advantage of the FTM comprised of obtaining analytical explicit relations dwindles in light of the disadvantages mentioned above such as a requirement of smoothness of $\mathcal{E}^{av} \in C^m(w)$ and an intricacy of analytical calculations. Nevertheless, there is some advantage of the FTM naturally connected with the explicit representation of the differential operators when tensors $\mathcal{Y}_{(m)}$ (13.24) and $\mathcal{B}_m$ (13.28) need to be estimated just one time and can be used for subsequent analysis of any $\mathcal{E}^{av} \in C^m(w)$ smooth enough (although, of course, the question of which order of differential operator (13.25) provides a priori prescribed accuracy of the solution of Eq. (13.11) and how it is connected with the smoothness of the function $\mathcal{E}^{av}$ remains valid).

We indicated only mathematical and computational difficulties in the use of the FTM which can be solved, at least in principle, at the cost of great effort if an analytical solution is necessary. However, it should be mentioned that there is an extremely important class of micromechanical problems for statistically inhomogeneous media (such as functionally graded and clustered materials considered in the current chapter, analysis of which by FTM is questionable). The breakdown of the assumption of statistical homogeneity leads to the inequality $\mathbf{Y}(\mathbf{x}_i) \not\equiv$ const. Then the average stresses $\langle \boldsymbol{\sigma} \rangle(\mathbf{x}) \not\equiv$ const. and $\mathcal{K}\mathcal{E} \not\equiv \mathbf{0}$ even at the homogeneous boundary conditions either (7.2) or (7.3). However, even this simplest case of homogeneous boundary conditions leads to a fundamental prohibitive obstacle against the use of the FTM. Indeed, the inhomogeneity $\langle \boldsymbol{\sigma} \rangle(\mathbf{x}) \not\equiv$ const. yields the inequality $\mathcal{K}(\mathbf{x}_i, \mathbf{x}_q) \neq \mathcal{K}(\mathbf{x}_i - \mathbf{x}_q)$, and, therefore, the linear differential operator (13.20) has variable coefficients. Then the transfer from Eqs. (13.20) to Eq. (13.25), respectively, based on the properties of the Fourier transform, is difficult, and applicability of the FTM is questionable. In so doing, the use of the IM for analysis of statistically inhomogeneous media (see Subsection 13.2.2) inserts requires a slight modification of the scheme presented for application of the iteration method for research of periodic graded composites (see Sect. 11.3).

Chapter 14
Micromechanics of Strongly Nonlocal Strain-Type and Weakly Nonlocal CMs

Abstract One considers statistically homogeneous CMs subjected to homogeneous remote boundary conditions. However, unlike Chaps. 6–13, the constitutive law at the microlevel is assumed to be nonlocal. Different nonlocal models are considered: strongly nonlocal (strain-type) and slightly nonlocal (strain gradient theories, stress-gradient theories). The nonlocal counterparts are obtained as the generalization of classical general results for locally elastic CMs (see Chap. 6): Hill–Mandel condition, effective moduli definition, and exact second moment of stresses average over the phases. It is demonstrated that effective constitutive equations are formally described by the locally elastic equations. GIEs for locally elastic CMs (see Chap. 7) are generalized to CMs with nonlocal properties of the phases. In so doing, the solution scheme of the GIE presented in Chap. 9 is generalized to their nonlocal counterparts.

Usually the general works in thermostatics of composites abound with studies of constitutive models for the mechanical behavior of solids. These works fall within the category of simple nonpolar materials [1136], for which the stress at a given point uniquely depends on the deformation and temperature at that point only. However, such a consideration implicitly implies that the material can be treated as a continuum at an arbitrarily small scale. It means that the properties and behaviors of materials captured by the classical continuum mechanics are invariant with respect to time and length scales, and more notably size effects cannot be captured by this classical mechanics [678]. Real materials, however, have complicated internal structures with length scales ranging over many orders of magnitude. Simultaneous modeling over the entire range of length scales is prohibitively expensive for practical applications since it requires coupling of continuum mechanics approaches with both molecular mechanics and quantum mechanics models. Affordable material models should select some resolution level below which the microstructural details are not explicitly "visible" to the model and need to be taken into account approximately and implicitly, by an appropriate definition of "effective" material properties (see, e.g., [78]). Unlike many other fields in material science, the evolution of nanocomposites (NC) to its

V. A. Buryachenko, *Local and Nonlocal Micromechanics of Heterogeneous Materials*, https://doi.org/10.1007/978-3-030-81784-8_14

anticipated level of importance in general significantly depends on the contributions from modeling and simulation. Computational approaches, based on the Newton's equations of motion at the atomistic level [heretofore referred to as the molecular dynamics (MD) approach], generate the question: "Should we altogether abandon classical field theory and appeal to atomic theory only?" (see Eringen [444]). The answer depends on the ratio of external and internal scales. The MD approaches are currently limited to the nanoscale (where all dimensions are at the nanoscale) and cannot deal with the micro-length scales (which are traditionally analyzed by a continuum mechanics approach).

A gap between the continuum mechanics and the discrete point-mass models of molecular dynamics (or interatomic potentials based on quantum mechanics) is bridged by a special mechanism using various forms of generalized (so-called nonlocal) continuum formulations, dealing with materials that are nonsimple or polar, or both, where the nonlocal variables are simply volume averages of the corresponding local variables over some nonlocal representative volume element (RVE) around the material point being considered. In the intermediate state with comparable scales, linear theory of nonlocal elasticity proposed independently by various authors Kröner, Eringen, Kunin, Bažant, and others (see, e.g., [440, 446, 815, 826, 835, 836], as well as [12, 78, 79, 283, 445, 888, 1034, 1568] see also comprehensive reviews [1369, 1530, 1534]) incorporates important features of lattice theory in an accompany with classical elasticity theory in the long-wave length limit. Following definitions by [1035], it is possible to classify "generalized continuum mechanics" all what is not covered in the restricted framework of the Cauchy model exposed in the prerequisite Chap. 2. In a non-exhaustive manner, generalized continua can be roughly subdivided on the weakly nonlocal theories and strongly nonlocal approaches. *Weakly nonlocal* theories cover higher-order models considering material points with additional degrees of freedom, besides classical translation (e.g., rotation [350], stretch or expand [445], see also [9, 447, 568]) and higher-grade models including higher-order derivatives of the displacement into the classical elastic Lagrangian; see [567, 568, 1067, 1068] which were essentially developed later (see, e.g., the comprehensive reviews [12, 78, 1231]). So, the Cosserat theory [350] was developed assuming that the material is a composition of rigid particles, and the particles exhibit independent rotations. In the context of the micromorphic theory [439], the material is a composition of a large number of material particles, and each one of these particles can move, deform, and rotate. The micromorphic theory [441] is a generalized model of continua, which can be simplified to the micropolar/Cosserat theory and microstretch theory.

Sources of the nonlocality "implicitly" introduced in weakly nonlocal theories are the gradients of the micro-deformations and the kinematical variables of the coupling between the different degrees of freedom. However, *strongly nonlocal* (commonly also named in the literature "nonlocal theories") elasticity theory represents the nonlocal fields using convolutions of the kinetical and kinematical variables that "explicitly" model the long-range effects of the material's kinematical degrees of freedom. Such a theory takes into account long-range (nonlocal) interactions which implies that the quantity as stress is integral functional of strain with the kernel which weights the elastic parameters of the material. True development of this modeling took place in the 1960s with the works of Kröner and Datta [816, 824] and Kunin [835] generalized in [446, 836, 1301]. Strongly nonlocal methods, which commonly used the convolution kernel of material properties, can be roughly subdivided on the strain-type and the displacement-type. In the *strain-type* a convolution kernel is convolved with the strain (in the constitutive equation; see [446, 816, 824] and Sects. 14.1–14.9), whereas in the *displacement-type* a convolution kernel is convolved with the displacement (in the motion equation, see [834, 835, 839, 1301]); the displacement-type strongly nonlocal model is reduced to the strain-type one [837]. Under certain conditions, an approximation to the integral nonlocal theory can be mimicked by the so-called strain gradient elasticity. Subsequently, the nonlocal microcontinuum mechanics was developed as a more general version of nonlocal mechanics [445, 446]. A unified approach of nonlocal field theories for elastic solids, viscous fluids, electromagnetic solids and fluids, memory-dependent elastic solids, and media with microstructure was developed and presented in a book [446]. At last, *peridynamics* by Silling [1396], according to [1035], is special version of the nonlocal theory (displacement-type) of continua (see for references and much more details Chaps. 16–19). One of the main feature of peridynamics is absent of any derivatives of displacements that gives the great opportunities for generality and exploration of model elastic continua with nonsmooth deformation phenomena such as, e.g., a fracture.

At the present time, a great variety of nonlocal models, involving either spatial integrals or gradients of strain or internal variables, have been developed for microscopically homogeneous materials although relatively little work has gone with respect to composite materials. The popular approaches of analysis of composites with nonlocal constitutive properties are usually based on coupling of two problems. At first the problem for one heterogeneity (either the particle, flake, or fiber) inside an infinite matrix subjected to the homogeneous remote stresses is solved for the different constitutive laws of both the heterogeneity and matrix (see, e.g., a comprehensive review [592] as well as [158, 1530]). The second problem is incorporating the mentioned solution into one of the known micromechanical schemes (e.g., Mori–Tanaka method; see [89, 1093]). As far as inclusion problems are concerned, one [829] has obtained analytical representation for stress distribution around a circular inhomogeneity in an infinite plane subjected to a uniform compressed remote loading for the case when nonlocal effects described by integral operators are the same for both the matrix and inhomogeneity. Size dependence of both the inside and outside micropolar Eshelby tensor for the spherical and cylindrical inclusions (same properties as matrix) in an infinite Cosserat medium is established by Cheng and He [324, 325] (see also [303, 1006, 1685]). For the second gradient elasticity with couple stresses, Zhang and Sharma [1682] evaluated the complete three-dimensional (strain gradient) Eshelby tensor for spherical inclusion in closed form (Reid and Gooding [1286] considered a two-dimensional problem with the eigenstrain restricted to a pure dilatation) although the corresponding inhomogeneity problem was analyzed by an approximate perturbation method. They have postulated that this equivalent inclusion method can be applied in an average sense for a micropolar material that is usually utilized for evaluating the average stress for nonellipsoidal heterogeneity in classical micromechanics. The exact average stress in fibers for a 2D micropolar coated fiber–matrix composite was estimated in [1650] who incorporated the obtained solution into the extending Mori–Tanaka's method to a micropolar composite (see also [955, 1372]).

It should be mentioned some correlations between the concepts of homogenization and generalized continua. The first case corresponds to locally elastic heterogeneous materials which, owing to homogenization (see the conditions described in Chap. 13), should be considered as effective generalized continua. The second case corresponds to generalized statistically homogeneous CMs subjected to homogeneous remote loading that leads to locally elastic effective behavior of homogenized medium. This case is considered in Chaps. 14 and 16–19. The last case corresponds to violation of at least one condition of homogeneity (statistical homogeneity of the structure or/and homogeneity of remote loading) for generalized heterogeneous material that leads to description of its effective behavior in the framework of generalized continua. We will demonstrate the scheme how for the considered models of generalized continua at the local level, the third case can be formally presented as an extension of the second case to the conditions of the first case.

14.1 Nonlocal Constitutive Equation

Comparing molecular scale of analyzed structures and the scale of inhomogeneity of internal stresses by necessity yields nonlocal character of the constitutive law in the area of coupling of the mentioned scales. Because of this, the popular basic locality assumption of constitutive law acting as a bridge between nano- and micromechanics can be considered as an approximation of the real nonlocal constitutive law. The present analysis should make it clear that micromechanical modeling based on continuum mechanics must be used with caution for nanocomposites.

Following Eringen [446], the stress–strain relationship in nonlocal linear thermoelasticity is given by the integral ("strongly nonlocal") equation

$$\boldsymbol{\sigma}(\mathbf{x}) = \boldsymbol{\mathcal{L}} * \boldsymbol{\varepsilon}(\mathbf{x}) + \boldsymbol{\alpha}(\mathbf{x}), \quad \boldsymbol{\mathcal{L}} * (.)(\mathbf{x}) = \int_w \boldsymbol{\mathcal{L}}(\mathbf{x}, \mathbf{y})(.)(\mathbf{y})d\mathbf{y},$$

$$\boldsymbol{\alpha}(\mathbf{x}) = \int_w \mathbf{m}(\mathbf{x}, \mathbf{y})\theta(\mathbf{y})d\mathbf{y}, \tag{14.1}$$

where the elastic integral operator $\boldsymbol{\mathcal{L}}$ is defined by the kernel $\boldsymbol{\mathcal{L}}(\mathbf{x}, \mathbf{y})$ in a certain representative mesodomain w with a characteristic function W and space dimensionality d ($d = 2$ and $d = 3$ for $2D$ and $3D$ problems, respectively). It is assumed that the difference $\theta(\mathbf{y}) \equiv \theta = (T - T^0)$ between the reference temperature T^0 =const. and a current temperature T is uniform in space. Because of this, the action of the kernel of nonlocal thermal expansion coefficient $\mathbf{m}(\mathbf{x}, \mathbf{y})$ is reduced to the transformation strain tensor $\boldsymbol{\alpha}(\mathbf{x})$. The reciprocal constitutive equation is given by a similar representation

$$\boldsymbol{\varepsilon}(\mathbf{x}) = \boldsymbol{\mathcal{M}}*\boldsymbol{\sigma}(\mathbf{x}) + \boldsymbol{\beta}(\mathbf{x}) \tag{14.2}$$

with obvious correlations between the parameters involved into Eqs. (14.1) and (14.2)

$$\boldsymbol{\mathcal{L}}*\boldsymbol{\mathcal{M}} = \boldsymbol{\mathcal{M}}*\boldsymbol{\mathcal{L}} = \mathbf{I}, \quad \boldsymbol{\mathcal{M}}*\boldsymbol{\alpha} = -\boldsymbol{\beta}, \quad \boldsymbol{\mathcal{L}}*\boldsymbol{\beta} = -\boldsymbol{\alpha}, \tag{14.3}$$

where $\mathbf{I}$ is the unit tensor. The material parameters are assumed to be subjected to the following symmetry regulations (see, e.g., [817, 825, 835] $\boldsymbol{\mathcal{K}} = \boldsymbol{\mathcal{L}}, \boldsymbol{\mathcal{M}}$; $\boldsymbol{\gamma} = \boldsymbol{\alpha}, \boldsymbol{\beta}$]

$$\mathcal{K}_{ijkl} = \mathcal{K}_{ijlk} = \mathcal{K}_{jikl}, \tag{14.4}$$

$$\gamma_{ij} = \gamma_{ji}, \quad \mathcal{K}_{ijkl}(\mathbf{x}, \mathbf{y}) = \mathcal{K}_{klij}(\mathbf{y}, \mathbf{x}). \tag{14.5}$$

For the homogeneous media, $\boldsymbol{\mathcal{K}} = \boldsymbol{\mathcal{L}}, \boldsymbol{\mathcal{M}}$ (and $\mathbf{m}$) are even functions of $\mathbf{x} - \mathbf{y}$

$$\boldsymbol{\mathcal{K}}(\mathbf{x}, \mathbf{y}) = \boldsymbol{\mathcal{K}}(\mathbf{x} - \mathbf{y}) = \boldsymbol{\mathcal{K}}(\mathbf{y} - \mathbf{x}). \tag{14.6}$$

Since real materials have strong short-range forces between neighboring inclusions, it seems physically reasonable (see [817]) to split the kernel $\boldsymbol{\mathcal{K}}(\mathbf{x}, \mathbf{y})$ into a short-range (described by the Dirac's δ delta function) and long-range part by setting ($\boldsymbol{\mathcal{K}} = \boldsymbol{\mathcal{L}}, \boldsymbol{\mathcal{M}}, \ \mathbf{K} = \mathbf{L}, \mathbf{M}$)

$$\boldsymbol{\mathcal{K}}(\mathbf{x}, \mathbf{y}) = \mathbf{K}(\mathbf{x})\delta(\mathbf{x} - \mathbf{y}) + \boldsymbol{\mathcal{K}}'(\mathbf{x}, \mathbf{y}). \tag{14.7}$$

Indeed, the action of the nonlocal operator $\boldsymbol{\mathcal{K}}$ on an piecewise continuous bounded function $\mathbf{f}$ (e.g., $\mathbf{f} = \boldsymbol{\sigma}, \boldsymbol{\varepsilon}$) can be presented as

$$\boldsymbol{\mathcal{K}} * \mathbf{f}(\mathbf{x}) = \mathbf{K}(\mathbf{x})\mathbf{f}(\mathbf{x}) + \boldsymbol{\mathcal{K}}' * \mathbf{f}(\mathbf{x}), \quad \mathbf{K}(\mathbf{x}) = \int_w \boldsymbol{\mathcal{K}}(\mathbf{x}, \mathbf{y})d\mathbf{y},$$

$$\boldsymbol{\mathcal{K}}' * \mathbf{f}(\mathbf{x}) = \int_w \boldsymbol{\mathcal{K}}(\mathbf{x}, \mathbf{y})(\mathbf{f}(\mathbf{y}) - \mathbf{f}(\mathbf{x}))d\mathbf{y}, \tag{14.8}$$

and, therefore, $\mathbf{K}(\mathbf{x})\delta(\mathbf{x} - \mathbf{y})$ can be interpreted as a classical local operator (e.g., elastic modulus $\mathbf{L}(\mathbf{x})$ or stiffness tensor $\mathbf{M}(\mathbf{x})$), while $\mathbf{I}\delta(\mathbf{x} - \mathbf{y})$ is considered as the identity operator. Regularized operator $\boldsymbol{\mathcal{K}}'$ vanishes identically for any uniform field $\mathbf{f}$ in w. In particular, for the local properties

$$\boldsymbol{\mathcal{L}}(\mathbf{x}, \mathbf{y}) = \mathbf{L}(\mathbf{x})\delta(\mathbf{x} - \mathbf{y}), \quad \boldsymbol{\mathcal{M}}(\mathbf{x}) = \mathbf{M}(\mathbf{x})\delta(\mathbf{x} - \mathbf{y}), \tag{14.9}$$

the operator relations (14.3) for the materials parameters are reduced to the algebraic ones

$$\mathbf{L} = \mathbf{M}^{-1}, \quad \mathbf{M}\boldsymbol{\alpha} = -\boldsymbol{\beta}, \quad \mathbf{L}\boldsymbol{\beta} = -\boldsymbol{\alpha}.$$

In particular, for isotropic constituents the local stiffness tensor $\mathbf{L}(\mathbf{x})$ (14.9) is given in terms of the local bulk modulus $k(\mathbf{x})$ and the local shear modulus $\mu(\mathbf{x})$, and the local eigenstrain $\boldsymbol{\beta}(\mathbf{x})$ is given in terms of the bulk component $\beta_0(\mathbf{x})$ by the relations

$$\mathbf{L}(\mathbf{x}) = (dk, 2\mu) \equiv dk(\mathbf{x})\mathbf{N}_1 + 2\mu(\mathbf{x})\mathbf{N}_2, \quad \boldsymbol{\beta}(\mathbf{x}) = \beta_0(\mathbf{x})\boldsymbol{\delta}, \tag{14.10}$$

$\mathbf{N}_1 = \boldsymbol{\delta} \otimes \boldsymbol{\delta}/d$, $\mathbf{N}_2 = \mathbf{I} - \mathbf{N}_1$ ($d = 2$ or 3); $\boldsymbol{\delta}$ and $\mathbf{I}$ are the unit second-order and fourth-order tensors. For the fiber composites, it is the plane–strain bulk modulus $k_{[2]}$—instead of the 3D bulk modulus $k_{[3]}$—that plays the significant role: $k_{[2]} = k_{[3]} + \mu_{[3]}/3$, $\mu_{[2]} = \mu_{[3]}$.

It should be mentioned that the decomposition of $\mathcal{K}(\mathbf{x}, \mathbf{y})$ (2.7) on the formal $\mathcal{K}'(\mathbf{x}, \mathbf{y})$ and singular $\mathbf{K}(\mathbf{x})\delta(\mathbf{x} - \mathbf{y})$ components is not unique for a general form of the generalized function $\mathcal{K}$. So, the Green strain tensor is split into two parts with a singular constituent associated with some infinitely small exclusion region, while the sum of these parts, being the left-hand side of (2.7), is defined uniquely (see, e.g., Section 3.2 in [200]). However, for the operator $\mathcal{K}$ describing elastic properties, the decomposition (2.7) is assumed to be unique (see, e.g., [1232]).

The local versions of the nonlocal operators can be also obtained as a limiting case of the following representation

$$\mathcal{K}(\mathbf{x}, \mathbf{y}) = \mathbf{K}(\mathbf{x})\lambda(\mathbf{x}, \mathbf{y}), \tag{14.11}$$

where the scalar attenuation function $\lambda(\mathbf{x}, \mathbf{y})$ (called also the nonlocal weight function or the nonlocal averaging function) for a homogeneous full space $w = R^d$ ($d = 2, 3$) depends only on the distance $r = |\mathbf{x} - \mathbf{y}|$ and can be expressed as $\lambda(\mathbf{x}, \mathbf{y}) = \lambda_\infty(|\mathbf{x} - \mathbf{y}|)$, where λ_∞ is an even function. To remove ambiguity, the normalizing condition

$$\int_w \lambda(\mathbf{x}, \mathbf{y}) d\mathbf{y} = 1, \quad \forall \mathbf{x} \in w = R^d \tag{14.12}$$

is assumed to be fixed and gives to $\mathbf{K}$ the physical meaning of elastic stiffness and compliance under uniform straining and stressing, respectively. For a finite domain w, the weight function is usually adjusted (see, e.g., [78, 706]) such that the nonlocal field (e.g., $\boldsymbol{\sigma}$) corresponding to a constant local field [e.g., $\boldsymbol{\varepsilon} \equiv$const in Eq. (14.1)] remains constant even in the vicinity of a boundary of w. This is guaranteed if the weight function satisfies the normalizing condition (14.12) even for $w \neq R^d$ that provides a constancy of the nonlocal field (e.g., $\boldsymbol{\sigma}$ in (14.1)) if

$$\lambda(\mathbf{x},\mathbf{y}) = \lambda_\infty(\mathbf{x}-\mathbf{y})\Big[\int_w \lambda_\infty(\mathbf{x}-\mathbf{y})d\mathbf{y}\Big]^{-1}. \tag{14.13}$$

In particular, widely used nonlocal weight functions in R^d $(d = 1, 2, 3)$ (see, e.g., [1228]) are the Gaussian function

$$\lambda_\infty^{\text{Gauss}}(r) = \frac{1}{(2\pi a_\lambda)^{d/2}} \exp\Big(-\frac{r^2}{2a_\lambda^2}\Big), \tag{14.14}$$

the bell polynomial-shaped function

$$\lambda_\infty^{\text{bell}}(r) = \begin{cases} c_\text{b}^\lambda\Big(1-\frac{r^2}{a_\lambda^2}\Big)^2, & r \le a_\lambda, \\ 0, & r \ge a_\lambda, \end{cases} \tag{14.15}$$

and the conical function (used in lattice dynamic)

$$\lambda_\infty^{\text{con}}(r) = \begin{cases} c_\text{c}^\lambda\Big(1-\frac{r}{a_\lambda}\Big), & r \le a_\lambda, \\ 0, & r \ge a_\lambda, \end{cases} \tag{14.16}$$

where c_b^λ and c_c^λ for $d = 1, 2, 3$ equal $15/(16a_\lambda), 3/(\pi a_\lambda^2), 105/(32\pi a_\lambda^3)$, $1/a_\lambda, 3/(\pi a_\lambda^2), and 3/(\pi a_\lambda^3)$, respectively. It was proved in [888] that the anisotropy of $\mathbf{K}$ (14.11) need not to be the same as those of the nonlocal kernel function $\alpha(\mathbf{x},\mathbf{y})$ (14.11), thus permitting the fact that a solid might possess different symmetry properties on the bulk than on the crystallographic scale (Angström-scale). The attenuation functions (14.15) and (14.16) with a finite support contain a parameter with the dimension of length a_λ playing the role of the interaction radius, because $\lambda_\infty(r)$ vanishes for $r > a_\lambda$. Then in the limiting case $\lambda_\infty(\mathbf{x}) \to \delta(\mathbf{x})$ [if e.g., $a_\lambda \to 0$ in Eqs. (14.14)–(14.16)], a nonlocal operator $\mathcal{K}$ is reduced to the local one $\mathcal{K}(\mathbf{x},\mathbf{y}) \to \mathbf{K}(\mathbf{x})\delta(\mathbf{x}-\mathbf{y})$.

It should be mentioned that nonlocal operators can be described not only by the integral operators but also by the differential ones, and the strain gradient theory can be considered as successive approximations to the nonlocal integral theory (see, e.g., [816]). The differential operator can be obtained by truncation of the corresponding Taylor approximation of the integral one, and, because of this, the form of the differential operator is less general. The strain (or stress) gradients appear after performing the integrations when $\boldsymbol{\varepsilon}(\mathbf{y})$ (2.1) [or $\boldsymbol{\sigma}(\mathbf{y})$ (2.2)] is written as a Taylor series around point $\mathbf{x}$. Indeed, the integral operator $\mathcal{K}$ at the m-times continuously differentiable functions $\mathbf{f} \in C^m(w)$ can be reduced to the differential one if the function $\mathbf{f}$ is approximated by the first terms of its Taylor expansion

$$\mathbf{f}(\mathbf{x}) \approx \sum_{|\alpha|=0}^{m} \frac{1}{\alpha!}[\otimes(\mathbf{x}-\mathbf{y})]^{\alpha}\nabla^{\alpha}\mathbf{f}(\mathbf{y}), \tag{14.17}$$

where for the multi-indices of nonnegative integers $\alpha = (\alpha_1, \ldots, \alpha_d) \in Z^d_+$, $(d = 2, 3)$, the following notations were used

$$\nabla^{\alpha} = \frac{\partial^{\alpha_1}}{\partial x_1^{\alpha_1}} \cdots \frac{\partial^{\alpha_d}}{\partial x_d^{\alpha_d}}, \quad \sum_{|\alpha|=0}^{m} = \sum_{|\alpha|=0}^{m} \sum_{\alpha_1+\cdots+\alpha_d=|\alpha|},$$
$$(\otimes\mathbf{x})^{\alpha} = (x_1)^{\alpha_1}\ldots(x_d)^{\alpha_d}, \quad \alpha! = \alpha_1!\ldots\alpha_d!, \quad |\alpha| \equiv \alpha_1 + \cdots + \alpha_d. \tag{14.18}$$

Taking the expansions (14.17) into account, integral representation (14.1_2) can be rewritten in the differential ("weakly nonlocal") form ($\mathcal{K} = \mathcal{L}, \mathcal{M}$)

$$\mathcal{K}*\mathbf{f}(\mathbf{x}) = \mathcal{P}(\mathbf{x}, \nabla)\mathbf{f}(\mathbf{x}), \tag{14.19}$$

where the linear partial differential operator

$$\mathcal{P}(\mathbf{x},\nabla) = \sum_{|\alpha|=0}^{m} \mathbf{p}_{\alpha}(\mathbf{x})\nabla^{\alpha}, \quad \mathbf{p}_{\alpha}(\mathbf{x}) = \frac{1}{\alpha!}\int \mathcal{K}(\mathbf{x},\mathbf{y}) \otimes (\mathbf{x}-\mathbf{y})^{\alpha}\, d\mathbf{y} \tag{14.20}$$

has the constant coefficients $\mathbf{p}_{\alpha}(\mathbf{x}) \equiv$const. for homogeneous infinite media $w = R^d$; these constant coefficients of the odd orders ($|\alpha| = 2k-1,\ k = 1, \ldots$) vanish. However, in the vicinity of the interface surface between two homogeneous media, the coefficients $\mathbf{p}_{\alpha}(\mathbf{x})$ (14.20) are in general inhomogeneous inside each constituent. Popular strain gradient (see, e.g., [1067, 1499]; see for details Sections 14.10 and 14.11) theory is based on representation of some energy function (e.g., strain energy function W or Helmholtz free energy density F) in the form of a quadratic function of the strain and gradient of strains, e.g., $W = \frac{1}{2}(\boldsymbol{\varepsilon}\mathbf{L}\boldsymbol{\varepsilon} + \alpha\nabla\boldsymbol{\varepsilon}\mathbf{L}\nabla\boldsymbol{\varepsilon})$, where α is some scale constant. Such representation for W leads to the constitutive gradient law with constant coefficients, which are usually used also in the vicinity of the interface surface between two homogeneous media (see, e.g., [1286, 1682]). The mentioned representation for W was generalized to $W = \frac{1}{2}[\boldsymbol{\varepsilon}\mathbf{L}\boldsymbol{\varepsilon} + \alpha\mathcal{K}'(\boldsymbol{\varepsilon})\mathbf{L}\mathcal{K}'(\boldsymbol{\varepsilon})]$ (see [1232]), where the argument $\mathcal{K}'$ (14.8) acts in a way similar to the strain gradient for strain gradient-dependent materials; in particular, like the strain gradient, $\mathcal{K}'$ vanishes identically for any uniform strain field in w. The differential form of the nonlocal operator (2.19) and (2.20) has obvious advantages due to the more readily realized finite element analysis (see, e.g., [388, 1570]) with respect to the integral form (see, e.g., [1228]). However, due to generality of integral representations, we will use the nonlocal integral models because the gradient models are usually derived as approximations of corresponding integral models in the immediate (infinitely closed) vicinity of the point being considered.

14.2 Effective Elastic Moduli

14.2.1 *Influence Functions and Effective Stiffness and Compliance*

In the framework of strongly nonlocal (integral type) elasticity [431, 815], Eringen and coworkers (see for references [445, 446]), we consider a simplified theory for linear (macroscopically) homogeneous isotropic elasticity, which differs from the classical one in the stress–strain constitutive relation only, whereas the equilibrium and compatibility equations remain unaltered. Because of this, we will accompany the constitutive equations (14.1) and (14.2) by the remaining basic equations of thermoelasticity of composites with zero body-force

$$\nabla\boldsymbol{\sigma}(\mathbf{x}) = \mathbf{0}, \tag{14.21}$$

$$\boldsymbol{\varepsilon}(\mathbf{x}) = \text{Def}\mathbf{u}(\mathbf{x}), \quad \text{Inc}\boldsymbol{\varepsilon}(\mathbf{x}) = \mathbf{0}, \tag{14.22}$$

and to one of two homogeneous boundary conditions

$$\mathbf{u}(\mathbf{x}) = \mathbf{u}^{\partial w}(\mathbf{x}) = \boldsymbol{\varepsilon}^{\partial w}\cdot\mathbf{x}, \ \boldsymbol{\varepsilon}^{\partial w} \equiv \text{const.}, \ \mathbf{x}\in\partial w_{\mathbf{u}}, \tag{14.23}$$

$$\boldsymbol{\sigma}(\mathbf{x})\cdot\mathbf{n}^{\partial w}(\mathbf{x}) = \mathbf{t}^{\partial w}(\mathbf{x}) \equiv \boldsymbol{\sigma}^{\partial w}\cdot\mathbf{n}^{\partial w}(\mathbf{x}), \ \ \boldsymbol{\sigma}^{\partial w} = \text{const.}, \ \mathbf{x}\in\partial w_{\sigma}. \tag{14.24}$$

Here $(\text{Def})_{ijk} = (\delta_{ik}\nabla_j + \delta_{jk}\nabla_i)/2$ and $(\text{Inc})_{ikmn} = \epsilon_{ipm}\epsilon_{kqn}\nabla_p\nabla_q$ are the operator of deformation and incompatibility, respectively, and ϵ_{ipm} is the antisymmetric Levi–Civita tensor; both $\boldsymbol{\varepsilon}^{\partial w}$ and $\boldsymbol{\sigma}^{\partial w}$ are symmetric constant given tensors, while $\mathbf{n}^{\partial w}$ is the outside normal to ∂w. For materials with long-range elastic forces (2.11), uniform straining of the finite body w with an indicator function W^w does not generate a uniform stress. To restore equilibrium, additional forces are needed in the entire boundary layer of thickness a. Because of this, we will study a stress state generated in w by the boundary conditions either (14.23) or (14.24) outside this boundary layer. In the case described by Eq. (14.23), the boundary conditions produce uniform strain $\boldsymbol{\varepsilon}^{\partial w}$ in a homogeneous material. The surface constraints in the case (14.24) are provided by the surface traction of the kind that would produce uniform internal stress in a homogeneous material. We assumed that only one of the following two representations of ∂w takes place: $\partial w = \partial w_{\mathbf{u}}$ $(\partial w_\sigma \equiv 0)$ or $\partial w = \partial w_\sigma$ $(\partial w_{\mathbf{u}} \equiv 0)$. One treats two material length scales: the macroscopic scale Λ, characterizing the extent of w, and the microscopic scale a, related with the heterogeneities v_i, which can contain a few phases $V^{(k)}$. The limit of our interests for the material scales is $\Lambda \gg a \simeq a_\lambda$. All the random quantities under discussion are described by statistically homogeneous random fields (see for details Sect. 5.3.1) satisfying the ergodicity condition when the ensemble averaging $\langle(\cdot)\rangle$ could be replaced by volume averaging $\{(\cdot)\}$.

Next we introduce an inner product for two arbitrary integrable symmetric random tensor functions $\mathbf{f}(\mathbf{x})$ and $\mathbf{g}(\mathbf{x})$ $(\mathbf{x}\in w)$ as $(\mathbf{f},\mathbf{g}) \equiv \{\mathbf{f}:\mathbf{g}\} = \langle\mathbf{f}:\mathbf{g}\rangle =$

$\langle f_{ij} g_{ji} \rangle$ due to both the ergodicity and statistical homogeneity of the fields $\mathbf{f}(\mathbf{x})$ and $\mathbf{g}(\mathbf{x})$. Then we can define

$$(\mathbf{f}, (\mathcal{K}\mathbf{g})) \equiv \langle \mathbf{f}(\mathcal{K}*\mathbf{g}) \rangle = \frac{1}{\bar{w}} \sum_{k=0}^{N} \int_{v^{(k)}} \mathbf{f}(\mathbf{x}) \sum_{l=0}^{N} \int_{v^{(l)}} \mathcal{K}(\mathbf{x}, \mathbf{y}) \mathbf{g}(\mathbf{y}) d\mathbf{y} d\mathbf{x} \tag{14.25}$$

for the symmetric tensor kernel $\mathcal{K}(\mathbf{x}, \mathbf{y})$ $(\mathbf{x}, \mathbf{y} \in w)$ (14.4) and (14.5). It is assumed in the work that for any symmetric tensor functions $\mathbf{f}, \mathbf{g}(\mathbf{x}) \in C(v^{(k)})$ $(k = 1, \ldots, N)$ of the second order and symmetric operator $\mathcal{K}$ (14.4) and (14.5) described above (e.g., $\mathbf{f}, \mathbf{g} = \boldsymbol{\varepsilon}, \boldsymbol{\sigma}$; $\mathcal{K} = \mathcal{L}, \mathcal{M}$), the relation

$$\langle \mathbf{f}(\mathcal{K}*\mathbf{g}) \rangle = \sum_{k=0}^{N} c^{(k)} \langle \mathbf{f}(\mathcal{K}*\mathbf{g}V^{(k)}) \rangle^{(k)} \equiv \frac{1}{\bar{w}} \sum_{k=0}^{N} \int_{v^{(k)}} \mathbf{f}(\mathbf{x}) \int_{v^{(k)}} \mathcal{K}(\mathbf{x}, \mathbf{y}) \mathbf{g}(\mathbf{y}) d\mathbf{y} d\mathbf{x} \tag{14.26}$$

is realized. The assumption (14.26) is quite a strong one, postulated that the nonlocal response of a specific phase $v^{(k)}$ depends only on the strain in this phase, i.e., nonlocal interactions between two different phases is excluded. However, for the operator $\mathcal{K}$ with the symmetry properties (14.4) and (14.5), the integration order in the relation (14.26) can be changed and (14.26) yields

$$(\mathbf{f}, (\mathcal{K}*\mathbf{g})) = ((\mathcal{K}*\mathbf{f}), \mathbf{g}), \tag{14.27}$$

i.e., the operator $\mathcal{K}$ is self-adjoint. Thus, the operator $\mathcal{K}$ is self-adjoint (14.27) if the assumption (14.26) holds, which takes place, e.g., if only one phase has nonlocal properties. The equality (14.27) can also be proven if a few phases (e.g., the inclusions with different orientations) have identical homogeneous nonlocal mechanical properties. Self-adjointness of the operator $\mathcal{K}$ (14.27) was also proven by [1016] if the attenuation functions $\lambda(\mathbf{x}, \mathbf{y})$ are identical for all phases with nonlocal properties. Thus, each from a few sufficient conditions provides a self-adjointness of the operator $\mathcal{K}$. However, we pursue the goal of describing of neither all more general sufficient conditions providing a self-adjointness of the operator $\mathcal{K}$ nor basic classes of models, from the most general to the most specific, and identifying which results are valid for each class. Our main objective is to establish that for some classes of materials, the operator $\mathcal{K}$ is self-adjoint that we will assume [along with the symmetry conditions (14.24) and (14.5)] in the subsequent analyses. A special kernel in the form (14.11) considered as an example is not assumed in the current presentation, although Buryachenko [206] precisely analyzed the form (14.11).

We will reproduce the classical representations of the stress and strain averages through the corresponding boundary conditions as well as the Hill's [640]) condition (6.10)

$$\{\boldsymbol{\sigma} : \boldsymbol{\varepsilon}\} = \{\boldsymbol{\sigma}\} : \{\boldsymbol{\varepsilon}\}, \quad \langle \boldsymbol{\sigma} : \boldsymbol{\varepsilon} \rangle = \langle \boldsymbol{\sigma} \rangle : \langle \boldsymbol{\varepsilon} \rangle \tag{14.28}$$

which holds for any admissible $\mathbf{u}$ ($\boldsymbol{\varepsilon} = \mathbf{Def}\mathbf{u}$) and self-equilibrium $\boldsymbol{\sigma}$ compatible with the homogeneous boundary conditions (14.23) and (14.24), respectively. We note that in Eq. (14.28) $\boldsymbol{\sigma}$ and $\boldsymbol{\varepsilon}$ need not be connected by a specific stress–strain relation (local or nonlocal).

We now turn our attention to the effective constitutive law for purely mechanical loading ($\boldsymbol{\alpha} \equiv \boldsymbol{\beta} \equiv \mathbf{0}$). Independently on both local and nonlocal nature of the constitutive laws in the micro- (or nano-) points, the field macrovariables $\langle\boldsymbol{\sigma}\rangle$ and $\langle\boldsymbol{\varepsilon}\rangle$ depend only on the homogeneous boundary conditions, either (14.23) or (14.24), and are related by a local material effective tensor of either the elasticity or compliance

$$\langle\boldsymbol{\sigma}\rangle = \mathbf{L}^*\langle\boldsymbol{\varepsilon}\rangle, \quad \text{or} \quad \langle\boldsymbol{\varepsilon}\rangle = \mathbf{M}^*\langle\boldsymbol{\sigma}\rangle, \tag{14.29}$$

respectively. Establishment of the constitutive linkage between the suitably chosen micro- (or nano-) and macrofield is reduced to the estimation of the tensor *influence functions* $\mathbf{A}^*_\varepsilon(\mathbf{x})$, $\mathbf{B}^*_\varepsilon(\mathbf{x})$, and $\mathbf{A}^*_\sigma(\mathbf{x})$, $\mathbf{B}^*_\sigma(\mathbf{x})$ for the displacement (14.23) and traction (14.24) homogeneous boundary conditions

$$\boldsymbol{\varepsilon}(\mathbf{x}) = \mathbf{A}^*_\varepsilon(\mathbf{x})\{\boldsymbol{\varepsilon}\}, \quad \boldsymbol{\sigma}(\mathbf{x}) = \mathbf{B}^*_\varepsilon(\mathbf{x})\{\boldsymbol{\varepsilon}\}, \tag{14.30}$$

$$\boldsymbol{\varepsilon}(\mathbf{x}) = \mathbf{A}^*_\sigma(\mathbf{x})\{\boldsymbol{\sigma}\}, \quad \boldsymbol{\sigma}(\mathbf{x}) = \mathbf{B}^*_\sigma(\mathbf{x})\{\boldsymbol{\sigma}\}, \tag{14.31}$$

respectively. Averaging of Eqs. (14.30_1) and (14.31_2) leads to the conclusion that

$$\{\mathbf{A}^*_\varepsilon(\mathbf{x})\} = \{\mathbf{B}^*_\sigma(\mathbf{x})\} = \mathbf{I}. \tag{14.32}$$

The local volume averages of the influence functions over $v^{(r)}$ may be referred to as the *mechanical concentrator factors* (e.g., $\mathbf{A}^{*(r)}_\varepsilon$, $r = 0, 1, \ldots, N$). Due to linearity of the problem, these influencing functions exist (but are generally unknown) and related by the obvious equations

$$\begin{aligned}
\mathbf{B}^*_\varepsilon(\mathbf{x}) &= \boldsymbol{\mathcal{L}}*\mathbf{A}^*_\varepsilon(\mathbf{x}), \quad \mathbf{A}^*_\varepsilon(\mathbf{x}) = \boldsymbol{\mathcal{M}}*\mathbf{B}^*_\varepsilon(\mathbf{x}),\\
\mathbf{A}^*_\varepsilon(\mathbf{x}) &= \mathbf{A}^*_\sigma(\mathbf{x})\mathbf{L}^*, \quad \mathbf{B}^*_\varepsilon(\mathbf{x}) = \mathbf{B}^*_\sigma(\mathbf{x})\mathbf{L}^*,
\end{aligned} \tag{14.33}$$

$$\begin{aligned}
\mathbf{B}^*_\sigma(\mathbf{x}) &= \boldsymbol{\mathcal{L}}*\mathbf{A}^*_\sigma(\mathbf{x}), \quad \mathbf{A}^*_\sigma(\mathbf{x}) = \boldsymbol{\mathcal{M}}*\mathbf{B}^*_\sigma(\mathbf{x}),\\
\mathbf{A}^*_\sigma(\mathbf{x}) &= \mathbf{A}^*_\varepsilon(\mathbf{x})\mathbf{M}^*, \quad \mathbf{B}^*_\sigma(\mathbf{x}) = \mathbf{B}^*_\varepsilon(\mathbf{x})\mathbf{M}^*.
\end{aligned} \tag{14.34}$$

In deriving the equalities (14.33_3), (14.33_4), and (14.34_3), (14.34_4), we assumed that the so-called Hill's weak assumption (see for details Sect. 6.1.2 and [353]) is fulfilled hereafter. If Hill's weak assumption is satisfied and if the average strains $\{\boldsymbol{\varepsilon}\}$ coincide for both homogeneous boundary conditions (14.23) and (14.24), according to the hypothesis made, then the average stresses $\{\boldsymbol{\sigma}\}$ also coincide for both homogeneous boundary conditions (14.23) and (14.24). In such a case, the equivalence of the dual effective material laws $\mathbf{L}^* = (\mathbf{M}^*)^{-1}$ takes place.

The influence functions (14.30) and (14.31) play a fundamental role in obtaining the macroscopic elastic response. Indeed, substitutions of the averaged multiscale linkages $\langle\boldsymbol{\sigma}\rangle = \langle\mathbf{B}^*_\varepsilon\rangle\langle\boldsymbol{\varepsilon}\rangle$ (14.30$_2$) and $\langle\boldsymbol{\varepsilon}\rangle = \langle\mathbf{A}^*_\sigma\rangle\langle\boldsymbol{\sigma}\rangle$ (14.31$_1$) into the averaged constitutive equations (14.29$_1$) and (14.29$_2$) yield estimations of the effective stiffness and compliance

$$\mathbf{L}^* = \{\mathbf{B}^*_\varepsilon(\mathbf{x})\},\quad \mathbf{M}^* = \{\mathbf{A}^*_\sigma(\mathbf{x})\}, \tag{14.35}$$

respectively. The representations (14.35) are reduced to the classical relations [640]

$$\mathbf{L}^* = \{\mathbf{L}(\mathbf{x})\mathbf{A}^*_\varepsilon(\mathbf{x})\},\quad \mathbf{M}^* = \{\mathbf{M}(\mathbf{x})\mathbf{B}^*_\sigma(\mathbf{x})\},$$

respectively, only for the local constitutive laws at the fine scale (14.9).

Similarly to the locally elastic case (4.2) and (4.3), we introduce the stress $\boldsymbol{\tau}$ and strain $\boldsymbol{\eta}$ polarization tensors ($\mathbf{x} \in R^d$)

$$\boldsymbol{\tau}(\mathbf{x}) = \boldsymbol{\sigma}(\mathbf{x}) - [\boldsymbol{\mathcal{L}}^{(0)}*\boldsymbol{\varepsilon}(\mathbf{x}) + \boldsymbol{\alpha}^{(0)}],\ \ \boldsymbol{\eta}(\mathbf{x}) = \boldsymbol{\varepsilon}(\mathbf{x}) - [\boldsymbol{\mathcal{M}}^{(0)}*\boldsymbol{\sigma}(\mathbf{x}) + \boldsymbol{\beta}^{(0)}], \tag{14.36}$$

where $\boldsymbol{\tau} = -\mathbf{L}^{(0)}\boldsymbol{\eta}$. Substituting Eqs. (14.32$_1$), (14.33$_1$), and (14.32$_2$), (14.34$_2$) into Eqs. (14.35$_1$) and (14.35$_2$), respectively, leads to the representations of the effective properties

$$\begin{aligned}\mathbf{L}^* &= \mathbf{L}^{(0)} + \langle\boldsymbol{\tau}\rangle,\quad \mathbf{L}^* = \mathbf{L}^{(0)} + \{[\mathbf{B}^*_\varepsilon(\mathbf{x}) - \boldsymbol{\mathcal{L}}^{(0)}*\mathbf{A}^*_\varepsilon(\mathbf{x})]V(\mathbf{x})\},\\ \mathbf{M}^* &= \mathbf{M}^{(0)} + \langle\boldsymbol{\eta}\rangle,\quad \mathbf{M}^* = \mathbf{M}^{(0)} + \{[\mathbf{A}^*_\sigma(\mathbf{x}) - \boldsymbol{\mathcal{M}}^{(0)}*\mathbf{B}^*_\sigma(\mathbf{x})]V(\mathbf{x})\}\end{aligned} \tag{14.37}$$

expressed only through the mechanical concentration factors inside the inclusions for a particular case of local elastic properties of the matrix $\boldsymbol{\mathcal{L}}^{(0)}(\mathbf{x},\mathbf{y}) \equiv \mathbf{L}^{(0)}V^{(0)}(\mathbf{x})\delta(\mathbf{x}-\mathbf{y})$ (14.9). The relations (14.37) have the following local counterparts (for $\boldsymbol{\mathcal{L}}(\mathbf{x},\mathbf{y}) = \mathbf{L}(\mathbf{x})\delta(\mathbf{x}-\mathbf{y})$)

$$\mathbf{L}^* = \mathbf{L}^{(0)} + \{[\mathbf{L}(\mathbf{x}) - \mathbf{L}^{(0)}]\mathbf{A}^*_\varepsilon(\mathbf{x})V(\mathbf{x})\},\ \mathbf{M}^* = \mathbf{M}^{(0)} + \{[\mathbf{M}(\mathbf{x}) - \mathbf{M}^{(0)}]\mathbf{B}^*_\sigma(\mathbf{x})V(\mathbf{x})\} \tag{14.38}$$

The difference between Eqs. (14.37) and (14.38) looks very small. However, what seems to be only a formal trick leads to a fundamental new result: in the framework of effective field hypothesis of micromechanics, we don't need to know for constitutive phases a concrete constitutive law (local or nonlocal, see for comparison [1145, 1147]) [1211] which is just a loss of information contained in the mentioned concentrator factors $\mathbf{A}_\varepsilon(\mathbf{x})$, $\mathbf{B}_\varepsilon(\mathbf{x})$ (or $\mathbf{A}_\sigma(\mathbf{x})$, $\mathbf{B}_\sigma(\mathbf{x})$). We only need to know that these laws are linear, and we need to know both concentration factors $\mathbf{A}_\varepsilon(\mathbf{x})$, $\mathbf{B}_\varepsilon(\mathbf{x})$ (or $\mathbf{A}_\sigma(\mathbf{x})$, $\mathbf{B}_\sigma(\mathbf{x})$) for single isolated representative heterogeneity $\mathbf{x} \in v_i$ inside an infinite matrix. The used micromechanical model (see for details [206]) defines a concrete representation of the effective properties through these concentrator factors (for single isolated heterogeneity), which can be found directly from molecular dynamics (MD) simulation. However, MD modeling is beyond the scope of the current study.

14.2.2 *Energetic Definition of Effective Elasticity*

Thus, the influence functions (14.30) and (14.31) act as weighting factors in averages giving the effective stiffness (14.35_1) and compliance (14.35_2). We will consider the properties of these functions in more detail. From the symmetry of the tensors $\{\boldsymbol{\varepsilon}\}$ and $\{\boldsymbol{\sigma}\}$, it follows that all influence tensors are symmetric inside the index pairs ($\mathbf{C} = \mathbf{A}^*_\varepsilon, \mathbf{B}^*_\varepsilon, \mathbf{A}^*_\sigma, \mathbf{B}^*_\sigma$): $C_{ijkl} = C_{jikl} = C_{ijlk}$. From (14.35) and from the symmetry properties of the values at the fine scale $\boldsymbol{\mathcal{L}}(\mathbf{x})$, $\boldsymbol{\mathcal{M}}(\mathbf{x})$ and $\mathbf{A}^*(\mathbf{x})$, $\mathbf{B}^*(\mathbf{x})$, the symmetry of macroproperties $\mathbf{L}^*$ and $\mathbf{M}^*$ inside the index pairs (minor symmetry) follows. The symmetry of $\mathbf{L}^*$ and $\mathbf{M}^*$ between the index pairs (major symmetry, e.g., $L^*_{ijkl} = L^*_{klij}$) can be obtained from energetic analysis. Indeed, for the homogeneous displacement boundary conditions (14.24), substitution of Eqs. (14.30) into Hill's condition (14.28) leads to

$$\{\boldsymbol{\varepsilon} : \boldsymbol{\sigma}\} = \Big\{(\mathbf{A}^*_\varepsilon\{\boldsymbol{\varepsilon}\}) : (\mathbf{B}^*_\varepsilon\{\boldsymbol{\varepsilon}\})\Big\} = \Big\{(\{\boldsymbol{\varepsilon}\}\mathbf{A}^{*\top}_\varepsilon) : (\mathbf{B}^*_\varepsilon\{\boldsymbol{\varepsilon}\})\Big\}, \tag{14.39}$$

where $\mathbf{h}^\top$ (e.g., $\mathbf{h} = \mathbf{A}^*_\varepsilon$) denotes the transposition of the tensor $\mathbf{h}$. Since the volume average $\{\boldsymbol{\varepsilon}\}$ is a constant tensor, Eq. (14.39) can be recast as

$$\{\boldsymbol{\varepsilon} : \boldsymbol{\sigma}\} = \{\boldsymbol{\varepsilon}\} : \{\mathbf{A}^{*\top}_\varepsilon \mathbf{B}^*_\varepsilon\}\{\boldsymbol{\varepsilon}\}, \tag{14.40}$$

and compared with the result of substitution of the effective constitutive equation (14.29_1) into Hill's condition (14.28)

$$\{\boldsymbol{\varepsilon} : \boldsymbol{\sigma}\} = \{\boldsymbol{\varepsilon}\} : \mathbf{L}^*\{\boldsymbol{\varepsilon}\}, \tag{14.41}$$

which can be considered as a *new energetic definition* of the effective elasticity $\mathbf{L}^*$. The two definitions (14.35_1) and (14.41) are not in general equivalent to one another for arbitrary inhomogeneous media, and the extent to which they agree in their computation of $\mathbf{L}^*$ provides a partial check on the validity of the "overall modulus" concept.

Since Eqs. (14.40) and (14.41) coincide for any constant symmetric field $\{\boldsymbol{\varepsilon}\}$, the effective modulus $\mathbf{L}^*$ can be expressed as

$$\mathbf{L}^* = \{\mathbf{A}^{*\top}_\varepsilon \mathbf{B}^*_\varepsilon\}. \tag{14.42}$$

For the proof of a symmetry $\mathbf{L}^* = (\mathbf{L}^*)^\top$, we will substitute Eq. (14.33_1) into Eq. (14.42) and find, according to (14.26)

$$\mathbf{L}^* = (\bar{w})^{-1}\sum_{k=0}^{N}\int_{v^{(k)}}\mathbf{A}_\varepsilon^{*\top}(\mathbf{x})\int_{v^{(k)}}\boldsymbol{\mathcal{L}}(\mathbf{x},\mathbf{y})\mathbf{A}_\varepsilon^*(\mathbf{y})d\mathbf{y}d\mathbf{x}$$

$$= (\bar{w})^{-1}\sum_{k=0}^{N}\int_{v^{(k)}}\int_{v^{(k)}}\mathbf{A}_\varepsilon^{*\top}(\mathbf{x})\boldsymbol{\mathcal{L}}(\mathbf{y},\mathbf{x})d\mathbf{x}\mathbf{A}_\varepsilon^*(\mathbf{y})d\mathbf{y}, \tag{14.43}$$

$$\mathbf{L}^{*\top} = (\bar{w})^{-1}\sum_{k=0}^{N}\int_{v^{(k)}}\int_{v^{(k)}}\mathbf{A}_\varepsilon^{*\top}(\mathbf{y})\boldsymbol{\mathcal{L}}^\top(\mathbf{x},\mathbf{y})d\mathbf{y}\mathbf{A}_\varepsilon^*(\mathbf{x})d\mathbf{x}. \tag{14.44}$$

Owing to the self-adjointness of the operator of $\boldsymbol{\mathcal{L}}$ (14.27) and exchanging $\mathbf{x} \leftrightarrow \mathbf{y}$, comparison of the right-hand sides of Eqs. (14.43) and (14.44) yields the symmetry of $\mathbf{L}^*$, i.e., $\mathbf{L}^{*\top} = \mathbf{L}^*$. Thus, if the nonlocal elastic operator at the fine scale is self-adjoint (14.27), then the effective elastic operator (which is the tensor $\mathbf{L}^*$ in our case) is also self-adjoint.

In a similar manner, for homogeneous traction boundary conditions (14.24), we obtain an analogous energetic definition of the effective compliance and their symmetric expression

$$\{\boldsymbol{\varepsilon}\boldsymbol{\sigma}\} = \{\boldsymbol{\sigma}\}\mathbf{M}^*\{\boldsymbol{\sigma}\}, \tag{14.45}$$

$$\mathbf{M}^* = \{\mathbf{B}_\sigma^{*\top}\mathbf{A}_\sigma^*\}, \quad \mathbf{M}^{*\top} = \mathbf{M}^*, \tag{14.46}$$

respectively.

14.3 Effective Eigenfields end Overall Elastic and Stored Energies

14.3.1 Field Decomposition and Effective Eigenstrain

We are coming now to consideration of the general constitutive equations of thermoelasticity (14.1) and (14.2) accompanied by both the basic Eqs. (14.21) and (14.22) and the boundary conditions (14.23) and (14.24). In such a case, the overall constitutive relations of the representative volume are given in the terms of the volume averages (or statistical averages)

$$\{\boldsymbol{\sigma}\} = \mathbf{L}^*\{\boldsymbol{\varepsilon}\} + \boldsymbol{\alpha}^*, \quad \{\boldsymbol{\varepsilon}\} = \mathbf{M}^*\{\boldsymbol{\sigma}\} + \boldsymbol{\beta}^*, \tag{14.47}$$

where $\boldsymbol{\alpha}^* = -\mathbf{L}^*\boldsymbol{\beta}^*, \ \boldsymbol{\beta}^* = -\mathbf{M}^*\boldsymbol{\alpha}^*$.

For the estimation of effective transformation $\boldsymbol{\alpha}^*$ and $\boldsymbol{\beta}^*$, we decompose the overall field as

$$\boldsymbol{\sigma}(\mathbf{x}) = \boldsymbol{\sigma}^I(\mathbf{x}) + \boldsymbol{\sigma}^{II}(\mathbf{x}), \quad \boldsymbol{\varepsilon}(\mathbf{x}) = \boldsymbol{\varepsilon}^I(\mathbf{x}) + \boldsymbol{\varepsilon}^{II}(\mathbf{x}), \tag{14.48}$$

with the sources

$$\boldsymbol{\alpha}^I(\mathbf{x}) = \boldsymbol{\beta}^I(\mathbf{x}) = 0, \quad \text{and} \quad \boldsymbol{\alpha}^{II}(\mathbf{x}) = \boldsymbol{\alpha}(\mathbf{x}), \quad \boldsymbol{\beta}^{II}(\mathbf{x}) = \boldsymbol{\beta}(\mathbf{x}). \tag{14.49}$$

We assume homogeneous traction boundary conditions

$$\mathbf{t}^I(\mathbf{x}) = \mathbf{t}^{\partial w_\sigma}(\mathbf{x}) = \langle \boldsymbol{\sigma} \rangle \cdot \mathbf{n}(\mathbf{x}), \ \ \mathbf{t}^{II}(\mathbf{x}) = \mathbf{0}, \tag{14.50}$$

$\mathbf{x} \in \partial w$. The field I is solely created at $\boldsymbol{\beta}(\mathbf{x}) \equiv \mathbf{0}$ by the applied traction compatible with a homogeneous stress field $\langle \boldsymbol{\sigma} \rangle$. The field II is generated by the random stress-free strains $\boldsymbol{\beta}(\mathbf{x})$ in the absence of boundary traction. Because of this, we may call $\boldsymbol{\sigma}^I$ the load stress field and $\boldsymbol{\sigma}^{II}$ the residual stress field. Consequently, the constitutive equations decompose into two kinds of equations at the fine level

$$\boldsymbol{\sigma}^I = \boldsymbol{\mathcal{L}}*\boldsymbol{\varepsilon}^I, \ \text{or} \ \boldsymbol{\varepsilon}^I = \boldsymbol{\mathcal{M}}*\boldsymbol{\sigma}^I, \tag{14.51}$$

$$\boldsymbol{\sigma}^{II} = \boldsymbol{\mathcal{L}}*\boldsymbol{\varepsilon}^{II} + \boldsymbol{\alpha}, \ \text{or} \ \boldsymbol{\varepsilon}^{II} = \boldsymbol{\mathcal{M}}*\boldsymbol{\sigma}^{II} + \boldsymbol{\beta} \tag{14.52}$$

and the macroscopic level

$$\langle \boldsymbol{\sigma}^I \rangle = \mathbf{L}^* \langle \boldsymbol{\varepsilon}^I \rangle, \ \text{or} \ \langle \boldsymbol{\varepsilon}^I \rangle = \mathbf{M}^* \langle \boldsymbol{\sigma}^I \rangle, \tag{14.53}$$

$$\langle \boldsymbol{\sigma}^{II} \rangle = \mathbf{L}^* \langle \boldsymbol{\varepsilon}^{II} \rangle + \boldsymbol{\alpha}^*, \ \text{or} \ \langle \boldsymbol{\varepsilon}^{II} \rangle = \mathbf{M}^* \langle \boldsymbol{\sigma}^{II} \rangle + \boldsymbol{\beta}^*, \tag{14.54}$$

respectively.

Taking into account the definitions of the two fields I and II and the boundary conditions, we find the following tensor equations for the average fields

$$\langle \boldsymbol{\sigma}^I \rangle = \langle \boldsymbol{\sigma} \rangle, \quad \langle \boldsymbol{\varepsilon}^I \rangle = \mathbf{M}^* \langle \boldsymbol{\sigma} \rangle, \quad \langle \boldsymbol{\sigma}^{II} \rangle = \mathbf{0}, \quad \langle \boldsymbol{\varepsilon}^{II} \rangle = \boldsymbol{\beta}^*. \tag{14.55}$$

Since $\boldsymbol{\sigma}^I$ and $\boldsymbol{\sigma}^{II}$ are statically admissible stress fields, while $\boldsymbol{\varepsilon}^I$ and $\boldsymbol{\varepsilon}^{II}$ are kinematically admissible strain fields, then, according to the Hill's condition, we obtain

$$\langle \boldsymbol{\sigma}^I \boldsymbol{\varepsilon}^I \rangle = \langle \boldsymbol{\sigma} \rangle \mathbf{M}^* \langle \boldsymbol{\sigma} \rangle, \quad \langle \boldsymbol{\varepsilon}^I \boldsymbol{\sigma}^{II} \rangle = 0, \quad \langle \boldsymbol{\varepsilon}^{II} \boldsymbol{\sigma}^I \rangle = \boldsymbol{\beta}^* \langle \boldsymbol{\sigma} \rangle, \quad \langle \boldsymbol{\varepsilon}^{II} \boldsymbol{\sigma}^{II} \rangle = 0. \tag{14.56}$$

We can recast Eq. (14.56_3) by the use of Eqs. (14.47) and (14.52)

$$\begin{aligned} \boldsymbol{\beta}^* \langle \boldsymbol{\sigma} \rangle &= \langle \boldsymbol{\varepsilon}^{II} (\boldsymbol{\mathcal{L}}*\boldsymbol{\varepsilon}^I) \rangle = \langle (\boldsymbol{\varepsilon}^{II} - \boldsymbol{\beta})(\boldsymbol{\mathcal{L}}*\boldsymbol{\varepsilon}^I) \rangle + \langle \boldsymbol{\beta} (\boldsymbol{\mathcal{L}}*\boldsymbol{\varepsilon}^I) \rangle \\ &= \langle \boldsymbol{\varepsilon}^I \boldsymbol{\sigma}^{II} \rangle + \langle \boldsymbol{\beta} \boldsymbol{\sigma}^I \rangle, \end{aligned} \tag{14.57}$$

where the self-adjointness of the operator $\boldsymbol{\mathcal{L}}$ (14.27) and an equivalence of the averaging operations $\{(\cdot)\}$ and $\langle(\cdot)\rangle$ were exploited. Substitution of Eqs. (14.56_2)

and (14.31_2) into Eq. (14.57) yields the representation of the effective eigenstrain through the mechanical influence function of the pure mechanical loading problem (14.31_2)

$$\boldsymbol{\beta}^* = \{\mathbf{B}_\sigma^{*\top}\boldsymbol{\beta}\}, \tag{14.58}$$

which is a generalization of the classical representation by Levin (1967) $\boldsymbol{\beta}^* = \langle\mathbf{B}_\sigma^{*\top}\boldsymbol{\beta}\rangle$ proposed for the composites with the local constitutive laws at the fine scale (14.9). The effective eigenstrain $\boldsymbol{\beta}^*$ (14.58) can also be presented in the form

$$\boldsymbol{\beta}^* = -\{\mathbf{A}_\sigma^{*\top}\boldsymbol{\alpha}\}, \tag{14.59}$$

by substitutions of Eqs. (14.34_1) and (14.3_3) into Eq. (14.58). Thus the effective residual strain $\boldsymbol{\beta}^*$ (14.58), (14.59) is defined by the averaging of eigenstress $\boldsymbol{\beta}(\mathbf{x})$ [or eigenstrain $\boldsymbol{\alpha}(\mathbf{x})$], with the weight $\mathbf{B}_\sigma^{*\top}(\mathbf{x})$ [or $\mathbf{A}_\sigma^{*\top}(\mathbf{x})$] depending only on the elastic properties of the medium.

It is interesting to mention that the fundamental representation (14.59) can also be obtained by an independent procedure that uses the elastic reciprocal theorem in the form proposed by Dvorak and Benveniste [418]

$$\int_{\partial w} t_i u_i' d\mathbf{s} + \int_w (b_i u_i' - \alpha_{ij}\varepsilon_{ij}')d\mathbf{x} = \int_{\partial w} t_i' u_i d\mathbf{s} + \int_w (b_i' u_i - \alpha_{ij}'\varepsilon_{ij})d\mathbf{x}, \tag{14.60}$$

where u_i and ε_{ij} are the displacement and strain caused by the system t_i, b_i, α_{ij}, whereas u_i' and ε_{ij}' are the displacement and strain caused by the system t_i', b_i', α_{ij}'. Recasting of Eq. (14.60) in terms of the fields I and II leads to the following work equation

$$\int_w \mathbf{t}^I(\mathbf{s})\mathbf{u}^{II}(\mathbf{s})d\mathbf{s} = -\int_w \boldsymbol{\alpha}^{II}(\mathbf{x})\boldsymbol{\varepsilon}^I(\mathbf{x})d\mathbf{x}. \tag{14.61}$$

Since $\mathbf{t}^I = \langle\boldsymbol{\sigma}\rangle \cdot \mathbf{n}(\mathbf{x})$ (14.50), one can rewrite Eq. (14.61) in the indicial notations

$$\begin{aligned}\langle\sigma_{ij}^I\rangle \int_{\partial w} \frac{1}{2}(u_i^{II} n_j + u_j^{II} n_i)d\mathbf{s} &= -\int_w \alpha_{kl}^{II}(\mathbf{x}) A_{klrs}^{\sigma *}(\mathbf{x})\langle\sigma_{rs}\rangle d\mathbf{x} \\ &= \langle\boldsymbol{\sigma}_{rs}^I\rangle \int_w (A_{rskl}^{\sigma *})^\top \alpha_{kl}^{II} d\mathbf{x}.\end{aligned} \tag{14.62}$$

Note first that the integral on the left-hand side of Eq. (14.62) is [by definition (14.23), (14.28)] $\langle\boldsymbol{\varepsilon}^{II}\rangle$, which is equal to the overall eigenstrain $\boldsymbol{\beta}^*$ (14.54_4). Exchanging of the subscripts $rs \leftrightarrow ij$ in (14.62) leads to the solution for the overall eigenstrain (14.59) which is thus recovered.

However, another form of Eq. (14.59) is possible for the matrix composites with the local homogeneous properties (14.18) of the matrix $\mathbf{x} \in v^{(0)} \subset w$ with the characteristic function $V^{(0)}$ $(\mathbf{x} \in v^{(0)})$

$$\mathbf{M}(\mathbf{x}) \equiv \mathbf{M}^{(0)} = \text{const.}, \quad \boldsymbol{\beta}(\mathbf{x}) = \boldsymbol{\beta}^{(0)} = \text{const.} \tag{14.63}$$

To obtain it, we resort to the obvious equalities

$$\langle \boldsymbol{\varepsilon}^{II} \rangle = \langle V(\mathbf{x})\boldsymbol{\varepsilon}^{II}(\mathbf{x}) \rangle + \langle V^{(0)}(\mathbf{x})\boldsymbol{\varepsilon}^{II}(\mathbf{x}) \rangle, \quad \mathbf{0} = \langle \boldsymbol{\sigma}^{II} \rangle. \tag{14.64}$$

Then averaging of the constitutive equation at the fine scale (14.2) over the volume of the matrix and inclusions $v \subset w$ with the characteristic function $V(\mathbf{x})$ ($V^{(0)}(\mathbf{x}) + V(\mathbf{x}) \equiv 1$)

$$\langle V^{(0)}(\mathbf{x})\boldsymbol{\varepsilon}^{II}(\mathbf{x}) \rangle = \langle V^{(0)}(\mathbf{x})\mathbf{M}^{(0)}\boldsymbol{\sigma}^{II}(\mathbf{x}) \rangle + \langle V^{(0)}(\mathbf{x})\boldsymbol{\beta}(\mathbf{x}) \rangle, \tag{14.65}$$

$$\langle V(\mathbf{x})\boldsymbol{\varepsilon}^{II}(\mathbf{x}) \rangle = \langle V(\mathbf{x})\boldsymbol{\mathcal{M}}*\boldsymbol{\sigma}^{II}(\mathbf{x}) \rangle + \langle V(\mathbf{x})\boldsymbol{\beta}(\mathbf{x}) \rangle \tag{14.66}$$

leads to the representation

$$\boldsymbol{\beta}^* = c^{(0)}\boldsymbol{\beta}^{(0)} + \langle V(\mathbf{x})(\boldsymbol{\varepsilon}^{II}(\mathbf{x}) - \mathbf{M}^{(0)}\boldsymbol{\sigma}^{II}(\mathbf{x})) \rangle \tag{14.67}$$

at the assumption of homogeneity of matrix elastic properties (14.63); here $c^{(0)} = \langle V^{(0)} \rangle$ is a volume fraction of the matrix.

The representation (14.67) has some advantages over Eq. (14.59). First of all, we did not use the volume average (4.16), nor did we use the assumptions of homogeneity of boundary conditions and statistical homogeneity of composite microstructure. Because of this, Eq. (14.67) can be easily generalized to the case of statistically inhomogeneous media and recast in terms of statistical averages

$$\boldsymbol{\beta}^*(\mathbf{x}) = \boldsymbol{\beta}^{(0)}\langle V^{(0)} \rangle(\mathbf{x}) + \langle V(\boldsymbol{\varepsilon}^{II} - \mathbf{M}^{(0)}\boldsymbol{\sigma}^{II}) \rangle(\mathbf{x}), \tag{14.68}$$

where the argument $\mathbf{x}$ reflects potential variation of the effective material and field characteristics within the space of statistically inhomogeneous media.

In a similar manner, the first homogeneous boundary-value problem (14.23) can be analyzed when the following decomposition of the boundary displacement takes place

$$\mathbf{u}^I(\mathbf{x}) = \boldsymbol{\varepsilon}^{\partial w_\mathbf{u}} \cdot \mathbf{x}, \quad \boldsymbol{\varepsilon}^{\partial w_\mathbf{u}} \equiv \text{const.}, \quad \mathbf{u}^{II}(\mathbf{x}) \equiv \mathbf{0}, \quad \mathbf{x} \in \partial w_\mathbf{u} \equiv \partial w, \tag{14.69}$$

which leads to the average fields replacing the averages (14.51)

$$\langle \boldsymbol{\varepsilon}^I \rangle = \langle \boldsymbol{\varepsilon} \rangle, \quad \langle \boldsymbol{\sigma}^I \rangle = \mathbf{L}^*\langle \boldsymbol{\varepsilon} \rangle, \quad \langle \boldsymbol{\varepsilon}^{II} \rangle = \mathbf{0}, \quad \langle \boldsymbol{\sigma}^{II} \rangle = \boldsymbol{\alpha}^*. \tag{14.70}$$

$$\langle \boldsymbol{\sigma}^I \boldsymbol{\varepsilon}^I \rangle = \langle \boldsymbol{\varepsilon} \rangle \mathbf{L}^* \langle \boldsymbol{\varepsilon} \rangle, \quad \langle \boldsymbol{\varepsilon}^I \boldsymbol{\sigma}^{II} \rangle = \boldsymbol{\alpha}^* \langle \boldsymbol{\varepsilon} \rangle, \quad \langle \boldsymbol{\varepsilon}^{II} \boldsymbol{\sigma}^I \rangle = 0, \quad \langle \boldsymbol{\varepsilon}^{II} \rangle \langle \boldsymbol{\sigma}^{II} \rangle = 0. \tag{14.71}$$

Consider the loading case $\langle \boldsymbol{\varepsilon} \rangle \equiv \mathbf{0}$, and note from (14.47) there is $\boldsymbol{\alpha}^* = \langle \boldsymbol{\mathcal{L}}*\boldsymbol{\varepsilon}(\mathbf{x}) + \boldsymbol{\alpha}(\mathbf{x}) \rangle$. Then we will get the representations of the effective eigenstress analogous to the representations for the effective eigenstrain (14.59) and (14.67)

$$\boldsymbol{\alpha}^* = -\langle \mathbf{B}_\varepsilon^{*\top} \boldsymbol{\beta} \rangle, \quad \boldsymbol{\alpha}^* = c^{(0)} \boldsymbol{\alpha}^{(0)} + \langle V(\boldsymbol{\sigma}^{II} - \mathbf{L}^{(0)} \boldsymbol{\varepsilon}^{II}) \rangle. \tag{14.72}$$

Substitution of Eqs. (14.3_2) and (14.33_2) into Eq. (14.72_1) leads to

$$\boldsymbol{\alpha}^* = \langle \mathbf{A}_\varepsilon^{*\top} \boldsymbol{\alpha} \rangle, \tag{14.73}$$

which can be considered as a generalization of the corresponding Levin's (1967) representation to composites with nonlocal properties of constituents.

14.3.2 Transformation Influence Functions

The transformation fields $\boldsymbol{\alpha}$ (14.1) and $\boldsymbol{\beta}$ (14.2) cause residual fields to develop presenting in w even in the case of the vanished applied strain (14.3) or stress (14.4); their evaluation has been analyzed in [425] for the local properties (14.9) of constitutives. In particular, the residual local fields were presented in terms of certain transformation influence functions with establishment of their relation to the analogous mechanical influence functions (14.30) and (14.31) for the special case of piecewise uniform eigenfields. For the use of an analogy between the mentioned theory and their nonlocal version, we will assume that the nonlocal operators $\mathcal{L}$ and $\mathcal{M}$ are presented in the form (14.11) and for the phase $\mathbf{x} \in v^{(r)}$ $(r = 0, 1, \ldots, N)$

$$\boldsymbol{\alpha}(\mathbf{x}) = \bar{\boldsymbol{\alpha}}^{(r)} f_\alpha^{(r)}(\mathbf{x}), \quad \boldsymbol{\beta}(\mathbf{x}) = \bar{\boldsymbol{\beta}}^{(r)} f_\beta^{(r)}(\mathbf{x}), \tag{14.74}$$

where $f_\alpha^{(r)}(\mathbf{x})$ and $f_\alpha^{(r)}(\mathbf{x})$ are the scale functions and the modified eigenfields $\bar{\boldsymbol{\alpha}}^{(r)}$ and $\bar{\boldsymbol{\beta}}^{(r)}$ are the constant tensors. Then the total strain or stress caused in $v^{(s)}$ by the uniform fields $\{\boldsymbol{\varepsilon}\} = \langle \boldsymbol{\varepsilon} \rangle$ or $\{\boldsymbol{\sigma}\} = \langle \boldsymbol{\sigma} \rangle$, and the transformation fields (14.74), is represented in the following form that extends (14.30) and (14.31) as ($\mathbf{x} \in v^{(s)}$)

$$\boldsymbol{\varepsilon}(\mathbf{x}) = \mathbf{A}_\varepsilon^*(\mathbf{x}) \langle \boldsymbol{\varepsilon} \rangle + \sum_{r=0}^{N} \mathbf{D}_\varepsilon^{sr}(\mathbf{x}) \bar{\boldsymbol{\beta}}^{(r)}, \quad \boldsymbol{\sigma}(\mathbf{x}) = \mathbf{B}_\varepsilon^*(\mathbf{x}) \langle \boldsymbol{\varepsilon} \rangle + \sum_{r=0}^{N} \mathbf{F}_\varepsilon^{sr}(\mathbf{x}) \bar{\boldsymbol{\alpha}}^{(r)}, \tag{14.75}$$

$$\boldsymbol{\varepsilon}(\mathbf{x}) = \mathbf{A}_\sigma^*(\mathbf{x}) \langle \boldsymbol{\sigma} \rangle + \sum_{r=0}^{N} \mathbf{D}_\sigma^{sr}(\mathbf{x}) \bar{\boldsymbol{\beta}}^{(r)}, \quad \boldsymbol{\sigma}(\mathbf{x}) = \mathbf{B}_\sigma^*(\mathbf{x}) \langle \boldsymbol{\sigma} \rangle + \sum_{r=0}^{N} \mathbf{F}_\sigma^{sr}(\mathbf{x}) \bar{\boldsymbol{\alpha}}^{(r)}. \tag{14.76}$$

Here the $\mathbf{D}_\varepsilon^{sr}(\mathbf{x})$ and $\mathbf{D}_\sigma^{sr}(\mathbf{x})$ are the *eigenstrain influence functions*, while the $\mathbf{F}_\varepsilon^{sr}(\mathbf{x})$ and $\mathbf{F}_\sigma^{sr}(\mathbf{x})$ are the corresponding *eigenstress influence functions*; their local volume averages over $v^{(s)}$ may be referred to as the *transformation concentrator factors* $\mathbf{D}_\varepsilon^{sr}, \mathbf{D}_\sigma^{sr}, \mathbf{F}_\varepsilon^{sr}, and \mathbf{F}_\sigma^{sr}$.

We will establish now some general relations between the transformation influence functions. Substituting Eq. (14.75_1) at $\langle \boldsymbol{\varepsilon} \rangle = \mathbf{0}$ into the constitutive Eq. (14.1_1) and comparing the result with (14.75_2) leads to ($\mathbf{x} \in v^{(s)}$, $s = 0, 1, \ldots, N$)

$$\boldsymbol{\alpha}(\mathbf{x})+\sum_{r=0}^{N}\boldsymbol{\mathcal{L}}*\mathbf{D}_{\varepsilon}^{sr}(\mathbf{x})\bar{\boldsymbol{\beta}}^{(r)}=\sum_{r=0}^{N}\mathbf{F}_{\varepsilon}^{sr}(\mathbf{x})\bar{\boldsymbol{\alpha}}^{(r)}. \tag{14.77}$$

Since Eq. (14.77) should hold for any chosen $\bar{\boldsymbol{\beta}}^{(r)}$, it follows that for any phase $\mathbf{x}\in v^{(s)}$ and $v^{(r)}$

$$(1-\delta_{rs})\boldsymbol{\alpha}(\mathbf{x})+\boldsymbol{\mathcal{L}}*\mathbf{D}_{\varepsilon}^{sr}(\mathbf{x})\bar{\boldsymbol{\beta}}^{(r)}=\mathbf{F}_{\varepsilon}^{sr}(\mathbf{x})\bar{\boldsymbol{\alpha}}^{(r)}. \tag{14.78}$$

In an analogous way, one can obtain the following relations ($\mathbf{x}\in v^{(s)}$)

$$(1-\delta_{rs})\boldsymbol{\beta}(\mathbf{x})+\boldsymbol{\mathcal{M}}*\mathbf{F}_{\varepsilon}^{sr}(\mathbf{x})\bar{\boldsymbol{\alpha}}^{(r)}=\mathbf{D}_{\varepsilon}^{sr}(\mathbf{x})\bar{\boldsymbol{\beta}}^{(r)}, \tag{14.79}$$

$$(1-\delta_{rs})\boldsymbol{\alpha}(\mathbf{x})+\boldsymbol{\mathcal{L}}*\mathbf{D}_{\sigma}^{sr}(\mathbf{x})\bar{\boldsymbol{\beta}}^{(r)}=\mathbf{F}_{\sigma}^{sr}(\mathbf{x})\bar{\boldsymbol{\alpha}}^{(r)}, \tag{14.80}$$

$$(1-\delta_{rs})\boldsymbol{\beta}(\mathbf{x})+\boldsymbol{\mathcal{M}}*\mathbf{F}_{\sigma}^{sr}(\mathbf{x})\bar{\boldsymbol{\alpha}}^{(r)}=\mathbf{D}_{\sigma}^{sr}(\mathbf{x})\bar{\boldsymbol{\beta}}^{(r)}. \tag{14.81}$$

The definition of the transformation concentration factors makes it possible to obtain another pair of representations for $\boldsymbol{\beta}^*$ and $\boldsymbol{\alpha}^*$ additionally to (14.59) and (14.72). Indeed, considering the homogeneous traction boundary conditions (14.50) and taking (14.55$_4$) into account yields the expression

$$\boldsymbol{\beta}^*=\sum_{s,r=0}^{N}c^{(s)}\mathbf{D}_{\sigma}^{sr}\bar{\boldsymbol{\beta}}^{(r)}. \tag{14.82}$$

An analogous derivation leads to

$$\boldsymbol{\alpha}^*=\sum_{s,r=0}^{N}c^{(s)}\mathbf{F}_{\varepsilon}^{sr}\bar{\boldsymbol{\alpha}}^{(r)}. \tag{14.83}$$

The conditions providing consistency of these results with the previous ones (14.59) and (14.72) can be found by simple comparison of the mentioned corresponding equations. So, equating the right-hand sides of Eqs. (4.13) and (5.9) and assigning the eigenfields $\boldsymbol{\alpha}(\mathbf{x})$ and $\boldsymbol{\beta}(\mathbf{x})$ ($\mathbf{x}\in v^{(r)}$) while all others eigenfields vanish, one can obtain

$$\sum_{s=0}^{N}c^{(s)}\mathbf{D}_{\sigma}^{sr}\bar{\boldsymbol{\beta}}^{(r)}=-c^{(r)}\langle\mathbf{A}_{\sigma}^{*\top}\boldsymbol{\alpha}\rangle_r,\quad \sum_{s=0}^{N}c^{(s)}\mathbf{F}_{\varepsilon}^{sr}\bar{\boldsymbol{\alpha}}^{(r)}=-c^{(r)}\langle\mathbf{B}_{\varepsilon}^{*\top}\boldsymbol{\beta}\rangle_r, \tag{14.84}$$

where a representation (14.84$_2$) was obtained analogously to (14.84$_1$). For both the local homogeneous properties (14.9) of the volume $v^{(r)}$ and the possible nonlocal properties of all other phases, the links (14.84) are simplified

$$\sum_{s=0}^{N} c^{(s)} \mathbf{D}_{\sigma}^{sr} \mathbf{M}^{(r)} = c^{(r)} \langle \mathbf{A}_{\sigma}^{*\top} \rangle_r, \quad \sum_{s=0}^{N} c^{(s)} \mathbf{F}_{\varepsilon}^{sr} \mathbf{L}^{(r)} = c^{(r)} \langle \mathbf{B}_{\varepsilon}^{*\top} \rangle_r. \tag{14.85}$$

Other relations for the transformation concentration factors can be obtained by averaging of Eqs. (14.75$_1$) and (14.76$_2$) over the volume w taking Eq. (14.32) into account

$$\sum_{s=0}^{N} c^{(s)} \sum_{r=0}^{N} \mathbf{D}_{\varepsilon}^{sr} \bar{\boldsymbol{\beta}}^{(r)} = \mathbf{0}, \quad \sum_{s=0}^{N} c^{(s)} \sum_{r=0}^{N} \mathbf{F}_{\sigma}^{sr} \bar{\boldsymbol{\alpha}}^{(r)} = \mathbf{0}. \tag{14.86}$$

The representations (14.86) can be simplified due to the possible independent selection of the modified eigenfields $\bar{\boldsymbol{\alpha}}^{(r)}$ and $\bar{\boldsymbol{\beta}}^{(r)}$

$$\sum_{s=0}^{N} c^{(s)} \mathbf{D}_{\varepsilon}^{sr} = \mathbf{0}, \quad \sum_{s=0}^{N} c^{(s)} \mathbf{F}_{\sigma}^{sr} = \mathbf{0}. \tag{14.87}$$

14.3.3 Overall Elastic Energy and Stored Energy

We will show that the effective strain energy W and potential energy Π can be expressed in terms of only the average values of the state variables and the composite effective properties. For this purpose these average energy function densities are presented in terms of the constitutive properties (see [78, 835, 1228])

$$W = \frac{1}{2} \int_w \boldsymbol{\sigma}(\boldsymbol{\varepsilon} - \boldsymbol{\beta})\, d\mathbf{y} = \frac{1}{2} \int_w \int_w (\boldsymbol{\varepsilon}(\mathbf{x}) - \boldsymbol{\beta}(\mathbf{x}))^\top$$
$$\times \boldsymbol{\mathcal{L}}(\mathbf{x}, \mathbf{y})(\boldsymbol{\varepsilon}(\mathbf{y}) - \boldsymbol{\beta}(\mathbf{y}))\, d\mathbf{x} d\mathbf{y}, \tag{14.88}$$

$$\Pi = W - \int_w \rho \mathbf{b} \mathbf{u}\, d\mathbf{x} - \int_{\partial w_\sigma} \mathbf{t}^{\partial w_\sigma} \mathbf{u}\, ds. \tag{14.89}$$

Hereafter, we assume without loss of generality that $\bar{w} = 1$. For the homogeneous boundary conditions (14.23) and (14.24), due to the assumed statistical homogeneity of the random function involved and applicability of the ergodicity hypothesis, both the volume and surface averages in Eqs. (14.88) and (14.89) can be replaced by the appropriate statistical averages. Then the average elastic energy W for the homogeneous boundary conditions (14.50) and (14.69) at $\mathbf{b} \equiv \mathbf{0}$ can be decomposed for the fields I and II (14.48) and (14.49)

$$W \equiv \frac{1}{2} \langle (\boldsymbol{\varepsilon} - \boldsymbol{\beta}) \boldsymbol{\sigma} \rangle = \frac{1}{2} \langle \boldsymbol{\varepsilon}^I \boldsymbol{\sigma}^I \rangle + \frac{1}{2} \langle \boldsymbol{\varepsilon}^I \boldsymbol{\sigma}^{II} \rangle + \frac{1}{2} \langle (\boldsymbol{\varepsilon}^{II} - \boldsymbol{\beta}) \boldsymbol{\sigma}^I \rangle$$

$$+\frac{1}{2}\langle(\boldsymbol{\varepsilon}^{II}-\boldsymbol{\beta})\boldsymbol{\sigma}^{II}\rangle = W^I + W^{II} + W^{\mathrm{Int}}, \quad W^I = \frac{1}{2}\langle\boldsymbol{\varepsilon}^I\boldsymbol{\sigma}^I\rangle, \tag{14.90}$$

$$W^{II} = \frac{1}{2}\langle(\boldsymbol{\varepsilon}^{II}-\boldsymbol{\beta})\boldsymbol{\sigma}^{II}\rangle, \quad W^{\mathrm{Int}} = \frac{1}{2}\langle\boldsymbol{\varepsilon}^I\boldsymbol{\sigma}^{II}\rangle + \frac{1}{2}\langle(\boldsymbol{\varepsilon}^{II}-\boldsymbol{\beta})\boldsymbol{\sigma}^I\rangle, \tag{14.91}$$

where the superscripts I, II, and Int stand for energetic values caused by the fields $\boldsymbol{\sigma}^I$ and $\boldsymbol{\sigma}^{II}$ as well as by the interaction of $\boldsymbol{\sigma}^I$ and $\boldsymbol{\sigma}^{II}$, respectively. We will also introduce the average value of the strain energy density stored in the residual stress field (shortly called stored energy)

$$W^* \equiv \frac{1}{2}\langle\boldsymbol{\sigma}(\boldsymbol{\varepsilon}-\boldsymbol{\beta})\rangle = -\frac{1}{2}\langle\boldsymbol{\sigma}\boldsymbol{\beta}\rangle, \tag{14.92}$$

where the Hill's condition (14.28) was used together with the assumed condition $\langle\boldsymbol{\sigma}\rangle = \mathbf{0}$. As we will see, the stored energy W^* depends only on $\boldsymbol{\beta}(\mathbf{s})$ and transformation influence functions and is thus as effective constant of a composite material in parallel with $\mathbf{L}^*$ and $\boldsymbol{\beta}^*$.

For the traction boundary conditions (14.50), according to the average representations ($14.56_{2,3}$) and (14.58), we can recast (14.90) as

$$W^I = \frac{1}{2}\langle\boldsymbol{\sigma}^I\rangle\langle\boldsymbol{\varepsilon}^I\rangle, \quad W^{II} = -\frac{1}{2}\langle\boldsymbol{\beta}\boldsymbol{\sigma}^{II}\rangle, \quad W^{\mathrm{Int}} = 0 \tag{14.93}$$

Thus the overall strain energy density W is the sum of the strain energy density W^I due to the applied load and $W^{II} = W^*$ stored by the transformation stress field

$$W = W^I + W^{II}, \quad W^I = \frac{1}{2}\langle\boldsymbol{\sigma}\rangle\mathbf{M}^*\langle\boldsymbol{\sigma}\rangle, \quad W^{II} = W^* = -\frac{1}{2}\langle\boldsymbol{\beta}\boldsymbol{\sigma}^{II}\rangle. \tag{14.94}$$

The straightforward substitution of the decompositions (14.92) and (14.93) into the representation for the potential energy density (14.89) at $\mathbf{b} \equiv \mathbf{0}$ and applying the divergence theorem to the last integral in Eq. (14.89) lead to the decomposition of the potential energy at the traction boundary conditions

$$\Pi = \Pi^I + \Pi^{II} + \Pi^{\mathrm{Int}}, \tag{14.95}$$

$$\Pi^I = W^I - \int_{\partial\mathcal{E}} \mathbf{t}^{\partial\mathcal{E}}\langle\mathbf{u}^I\rangle\, ds = -W^I, \tag{14.96}$$

$$\Pi^{II} = W^{II}, \quad \Pi^{\mathrm{Int}} = -\boldsymbol{\beta}^*\langle\boldsymbol{\sigma}^I\rangle. \tag{14.97}$$

Thus, the strain energy W and potential energy Π can be expressed through the average stresses and the effective parameters

$$W = \frac{1}{2}\langle\boldsymbol{\sigma}\rangle\mathbf{M}^*\langle\boldsymbol{\sigma}\rangle + W^*, \quad \Pi = -\frac{1}{2}\langle\boldsymbol{\sigma}\rangle\mathbf{M}^*\langle\boldsymbol{\sigma}\rangle - \boldsymbol{\beta}^*\langle\boldsymbol{\sigma}\rangle + W^*. \tag{14.98}$$

For the displacement boundary conditions (14.69), the decomposition of the strain energy W presented in the form

$$W \equiv \frac{1}{2}\langle(\sigma-\alpha)\varepsilon\rangle = \frac{1}{2}\langle\sigma^I\varepsilon^I\rangle + \frac{1}{2}\langle\sigma^I\varepsilon^{II}\rangle + \frac{1}{2}\langle(\sigma^{II}-\alpha)\varepsilon^I\rangle$$
$$+\frac{1}{2}\langle(\sigma^{II}-\alpha)\varepsilon^{II}\rangle = W^I + W^{II} + W^{\mathrm{Int}}, \quad W^I = \frac{1}{2}\langle\sigma^I\rangle\langle\varepsilon^I\rangle,$$
$$W^{II} = \frac{1}{2}\langle(\sigma^{II}-\alpha)\varepsilon^{II}\rangle, \quad W^{\mathrm{Int}} \equiv \frac{1}{2}\langle\sigma^I\varepsilon^{II}\rangle + \frac{1}{2}\langle(\sigma^{II}-\alpha)\varepsilon^I\rangle \tag{14.99}$$

can be simplified due to the averages (14.70), (14.71), and (14.72_1). Therefore the decompositions (14.93) and (14.95) should be replaced by

$$W^I = \Pi^I = \frac{1}{2}\langle\boldsymbol{\varepsilon}\rangle\mathbf{L}^*\langle\boldsymbol{\varepsilon}\rangle, \quad W^{II} = \Pi^{II} = -\frac{1}{2}\langle\boldsymbol{\alpha}\boldsymbol{\varepsilon}^{II}\rangle, \quad W^{\mathrm{Int}} = \Pi^{\mathrm{Int}} = 0. \tag{14.100}$$

Therefore, the total strain energy and potential energy density become

$$W = \Pi = W^I + W^* = \frac{1}{2}\langle\boldsymbol{\varepsilon}\rangle\mathbf{L}^*\langle\boldsymbol{\varepsilon}\rangle - \frac{1}{2}\langle\boldsymbol{\alpha}\boldsymbol{\varepsilon}^{II}\rangle. \tag{14.101}$$

Thus, the average energy densities (14.88) and (14.89) can be expressed by the effective thermoelastic constants $\mathbf{L}^*$, $\boldsymbol{\beta}^*$, and W^* for the homogeneous boundary conditions (14.23) and (14.24) . The stored energies W^*, like $\boldsymbol{\alpha}^*$ and $\boldsymbol{\beta}^*$, depend only on the fields $\boldsymbol{\alpha}(\mathbf{x})$ and $\boldsymbol{\beta}(\mathbf{x})$ and the influence transformation functions and act like effective material constants, which can be observed macroscopically (see, e.g., Chap. 6). It should be mentioned that the effective eigenfields $\boldsymbol{\alpha}^*$ and $\boldsymbol{\beta}^*$ and stored energy W^* depend not only on the material properties but also on its "loading" by temperature changes θ =const. or similar sources of eigenfields. So the effective material parameters are not $\boldsymbol{\alpha}^*$, $\boldsymbol{\beta}^*$, and W^* itself but something that transforms the temperature change into $\boldsymbol{\alpha}^*$, $\boldsymbol{\beta}^*$, and W^*. Because of this, strictly speaking, the real effective parameters are $\boldsymbol{\alpha}^*/\theta$, $\boldsymbol{\beta}^*/\theta$, and W^*/θ^2 (which do not depend on θ) rather than $\boldsymbol{\alpha}^*$, $\boldsymbol{\beta}^*$, and W^*, respectively.

The stored energy W^* used in Eqs. (14.98) and (14.101) can be expressed through the influence transformation functions as ($\mathbf{x} \in v^{(s)}$)

$$W^* = -\frac{1}{2}\sum_{s,r=0}^{N} c^{(s)}\langle\boldsymbol{\beta}(\mathbf{x})\mathbf{F}_{\sigma}^{sr}(\mathbf{x})\bar{\boldsymbol{\alpha}}^{(r)}\rangle^{(s)},$$
$$W^* = -\frac{1}{2}\sum_{s,r=0}^{N} c^{(s)}\langle\boldsymbol{\alpha}(\mathbf{x})\mathbf{D}_{\varepsilon}^{sr}(\mathbf{x})\bar{\boldsymbol{\beta}}^{(r)}\rangle^{(s)}, \tag{14.102}$$

for the homogeneous traction (14.50) and displacement (14.49) boundary conditions, respectively. It should be mentioned that the effective thermoelastic constants

$\mathbf{L}^*, \boldsymbol{\beta}^*, and W^*$ play a key role in micromechanics because of their close relationship to both the internal stress fluctuations and average stresses in the constituents.

14.4 Perturbation Method for Estimation of Statistical Moments of Fields

14.4.1 Stress Fluctuations

In this section, we derive exact representations for all components for second-order moments of the stress and strain fields averaged over the volume of a component by utilizing the functional dependence of effective properties on the thermoelastic constants of the component being considered (see for references Chap. 6). This functional dependence can be obtained only by employing a special theoretical micromechanical model rather than from an experimental measurement of effective properties.

For concreteness, we will consider homogeneous traction boundary conditions (14.50) and decompose the overall field as $\boldsymbol{\sigma}(\mathbf{x}) = \boldsymbol{\sigma}^I(\mathbf{x}) + \boldsymbol{\sigma}^{II}(\mathbf{x})$, with the sources (14.48) and (14.49) $\boldsymbol{\beta}^I(\mathbf{x}) = \mathbf{0}, \quad \mathbf{t}^I(\mathbf{y}) = \mathbf{t}^{\partial w_\sigma}(\mathbf{y})$ and $\boldsymbol{\beta}^{II}(\mathbf{x}) = \boldsymbol{\beta}(\mathbf{x}), \quad \mathbf{t}^{II}(\mathbf{y}) = \mathbf{0}$ $(\mathbf{x} \in w, \quad \mathbf{y} \in \partial w_\sigma)$ for the field $\boldsymbol{\sigma}^I$ and $\boldsymbol{\sigma}^{II}$, respectively, where $\mathbf{t}^{\partial w_\sigma}$ is a prescribed traction in the boundary conditions (14.50). For fixed boundary conditions (14.50), we first consider a small variation of the compliance kernel $\delta\boldsymbol{\mathcal{M}}(\mathbf{x}, \mathbf{y})$ as compared to $\boldsymbol{\mathcal{M}}(\mathbf{x}, \mathbf{y})$: $\boldsymbol{\mathcal{M}}(\mathbf{x}, \mathbf{y}) \to \boldsymbol{\mathcal{M}}(\mathbf{x}, \mathbf{y}) + \delta\boldsymbol{\mathcal{M}}(\mathbf{x}, \mathbf{y})$ implying corresponding variations of the stress and strain fields: $\boldsymbol{\sigma}(\mathbf{x}) \to \boldsymbol{\sigma}(\mathbf{x}) + \delta\boldsymbol{\sigma}(\mathbf{x})$ and $\boldsymbol{\varepsilon}(\mathbf{x}) \to \boldsymbol{\varepsilon}(\mathbf{x}) + \delta\boldsymbol{\varepsilon}(\mathbf{x})$, respectively. Since both $\boldsymbol{\sigma}(\mathbf{x})$ and $\boldsymbol{\sigma}(\mathbf{x}) + \delta\boldsymbol{\sigma}(\mathbf{x})$ satisfy the same boundary condition, we obtain from (14.55$_1$) $\langle\delta\boldsymbol{\sigma}(\mathbf{x})\rangle = \mathbf{0}$. Moreover, $\delta\boldsymbol{\sigma}(\mathbf{x})$ and $\delta\boldsymbol{\varepsilon}(\mathbf{x})$ accomplish the requirements of the Hill's condition (14.28) leading to the scale equalities

$$\langle\delta\boldsymbol{\sigma}\boldsymbol{\varepsilon}\rangle = \langle\delta\boldsymbol{\sigma}\rangle\langle\boldsymbol{\varepsilon}\rangle = 0, \quad \langle\boldsymbol{\sigma}\delta\boldsymbol{\varepsilon}\rangle = \langle\boldsymbol{\sigma}\rangle\langle\delta\boldsymbol{\varepsilon}\rangle \tag{14.103}$$

which are fulfilled for both $\boldsymbol{\varepsilon}^I$ and $\boldsymbol{\varepsilon}^{II}$ combined with both fields $\boldsymbol{\sigma}^I$ and $\boldsymbol{\sigma}^{II}$.

Varying (14.56$_1$) in combination with Eqs. (14.103$_1$) and (14.51) leads to

$$\langle\boldsymbol{\sigma}\rangle\delta\mathbf{M}^*\langle\boldsymbol{\sigma}\rangle = \langle\delta\boldsymbol{\sigma}^I(\boldsymbol{\mathcal{M}} * \boldsymbol{\sigma}^I)\rangle + \langle\boldsymbol{\sigma}^I(\delta\boldsymbol{\mathcal{M}} * \boldsymbol{\sigma}^I)\rangle + \langle\boldsymbol{\sigma}^I(\boldsymbol{\mathcal{M}} * \delta\boldsymbol{\sigma}^I)\rangle = \langle\boldsymbol{\sigma}^I\delta\mathbf{M}\boldsymbol{\sigma}^I\rangle, \tag{14.104}$$

where the self-adjointness of the operator $\boldsymbol{\mathcal{M}}$ (14.27) was also used. In a similar manner, variation of Eq. (14.94$_3$) by varying of $\mathbf{M}(\mathbf{x})$ at the fixed $\boldsymbol{\beta}(\mathbf{x})$ yields

$$\begin{aligned} 2\delta W^* &= \langle\delta\boldsymbol{\sigma}^{II}(\boldsymbol{\mathcal{M}} * \boldsymbol{\sigma}^{II})\rangle + \langle\boldsymbol{\sigma}^{II}(\delta\boldsymbol{\mathcal{M}} * \boldsymbol{\sigma}^{II})\rangle + \langle\boldsymbol{\sigma}^{II}(\boldsymbol{\mathcal{M}} * \delta\boldsymbol{\sigma}^{II})\rangle \\ &= \langle\delta\boldsymbol{\sigma}^{II}\boldsymbol{\varepsilon}^{II}\rangle + \langle\boldsymbol{\sigma}^{II}(\delta\boldsymbol{\mathcal{M}} * \boldsymbol{\sigma}^{II})\rangle + 2\langle\boldsymbol{\varepsilon}^{II}\delta\boldsymbol{\sigma}^{II}\rangle - 2\langle\boldsymbol{\beta}\delta\boldsymbol{\sigma}^{II}\rangle \end{aligned} \tag{14.105}$$

where the first and the third terms in the right-hand side of Eq. (14.105) vanish by virtue of (14.103$_1$). Combining Eqs. (14.92) and (14.105) leads to

$$\langle \boldsymbol{\sigma}^{II}(\delta\boldsymbol{\mathcal{M}}*\boldsymbol{\sigma}^{II})\rangle = -2\delta W^*. \tag{14.106}$$

Finally, variations of the equalities $\langle \boldsymbol{\varepsilon}^{I}\boldsymbol{\sigma}^{II}\rangle = 0$ (14.56$_2$) and $\langle \boldsymbol{\sigma}^{II}\boldsymbol{\varepsilon}^{I}\rangle = 0$ with variation of the self-adjoint operator $\mathbf{M}(\mathbf{x})$ (14.27) and fixed $\boldsymbol{\beta}(\mathbf{x})$ give

$$\begin{aligned} 0 &= \langle \delta\boldsymbol{\sigma}^{I}(\boldsymbol{\mathcal{M}}*\boldsymbol{\sigma}^{II})\rangle + \langle \boldsymbol{\sigma}^{I}(\delta\boldsymbol{\mathcal{M}}*\boldsymbol{\sigma}^{II})\rangle + \langle \boldsymbol{\sigma}^{I}(\boldsymbol{\mathcal{M}}*\delta\boldsymbol{\sigma}^{II})\rangle \\ &= \langle \delta\boldsymbol{\sigma}^{I}\boldsymbol{\varepsilon}^{II}\rangle - \langle \delta\boldsymbol{\sigma}^{I}\boldsymbol{\beta}\rangle + \langle \boldsymbol{\sigma}^{I}(\delta\boldsymbol{\mathcal{M}}*\boldsymbol{\sigma}^{II})\rangle + \langle \boldsymbol{\varepsilon}^{I}\delta\boldsymbol{\sigma}^{II}\rangle \\ &= \langle \boldsymbol{\sigma}^{I}(\delta\boldsymbol{\mathcal{M}}*\boldsymbol{\sigma}^{II})\rangle - \delta\boldsymbol{\beta}^*\langle\boldsymbol{\sigma}\rangle, \end{aligned} \tag{14.107}$$

$$\begin{aligned} 0 &= \langle \delta\boldsymbol{\sigma}^{II}(\boldsymbol{\mathcal{M}}*\boldsymbol{\sigma}^{I})\rangle + \langle \boldsymbol{\sigma}^{II}(\delta\boldsymbol{\mathcal{M}}*\boldsymbol{\sigma}^{I})\rangle + \langle \boldsymbol{\sigma}^{II}(\boldsymbol{\mathcal{M}}*\delta\boldsymbol{\sigma}^{I})\rangle \\ &= \langle \delta\boldsymbol{\sigma}^{II}\boldsymbol{\varepsilon}^{I}\rangle + \langle \boldsymbol{\sigma}^{II}(\delta\boldsymbol{\mathcal{M}}*\boldsymbol{\sigma}^{I}\rangle + \langle \boldsymbol{\varepsilon}^{II}\delta\boldsymbol{\sigma}^{I}\rangle - \langle \boldsymbol{\beta}\delta\boldsymbol{\sigma}^{I}\rangle \\ &= \langle \boldsymbol{\sigma}^{II}\delta(\boldsymbol{\mathcal{M}}*\boldsymbol{\sigma}^{I})\rangle - \delta\boldsymbol{\beta}^*\langle\boldsymbol{\sigma}\rangle, \end{aligned} \tag{14.108}$$

where the last equalities in (14.107) and (14.108) were obtained utilizing Eqs. (14.56$_2$), (14.57), (14.103$_1$), and (14.103$_2$) as well as a chain of equalities $\langle \delta\boldsymbol{\sigma}^{I}\boldsymbol{\beta}\rangle = \langle \delta(\mathbf{B}^*_\sigma\langle\boldsymbol{\sigma}\rangle)\boldsymbol{\beta}\rangle = \langle\boldsymbol{\sigma}\rangle\langle\delta\mathbf{B}^{*\top}_\sigma\boldsymbol{\beta}\rangle = \langle\boldsymbol{\sigma}\rangle\delta\boldsymbol{\beta}^* = \delta\boldsymbol{\beta}^*\langle\boldsymbol{\sigma}\rangle$.

Here one may use a special variation of the kernel of the compliance tensor operator ($k = 0, 1, \ldots$)

$$\delta\boldsymbol{\mathcal{M}}(\mathbf{x},\mathbf{y}) \equiv \delta\mathbf{M}^{(k)}V^{(k)}(\mathbf{x})\delta(\mathbf{x}-\mathbf{y}),\ \ \delta\mathbf{M}^{(k)} = \text{const},\quad \mathbf{x},\mathbf{y}\in w. \tag{14.109}$$

It should be mentioned that $\boldsymbol{\mathcal{M}}(\mathbf{x},\mathbf{y})$ is a nonlocal operator, whereas $\delta\mathbf{M}^{(k)}$ is a constant symmetric (minor and major) positive define tensor. In this case $\delta\mathbf{M}$ may be taken outside the averaging sign over the volume of component $v^{(k)}$ $(k = 0, 1, \ldots, N)$. Furthermore we have

$$\delta\mathbf{M}^* = \frac{\partial\mathbf{M}^*}{\partial\mathbf{M}^{(k)}}\delta\mathbf{M}^{(k)},\quad \delta W^* = \frac{\partial W^*}{\partial\mathbf{M}^{(k)}}\delta\mathbf{M}^{(k)},\quad \delta\boldsymbol{\beta}^* = \frac{\partial\boldsymbol{\beta}^*}{\partial\mathbf{M}^{(k)}}\delta\mathbf{M}^{(k)} \tag{14.110}$$

which together with Eqs. (14.104), (14.106), (14.107), and (14.109) lead to the final expressions for the second moments of stresses

$$\langle\boldsymbol{\sigma}^{I}\otimes\boldsymbol{\sigma}^{I}\rangle^{(k)} = \frac{1}{c^{(k)}}\frac{\partial\mathbf{M}^*}{\partial\mathbf{M}^{(k)}}\langle\boldsymbol{\sigma}\rangle\otimes\langle\boldsymbol{\sigma}\rangle, \tag{14.111}$$

$$\langle\boldsymbol{\sigma}^{II}\otimes\boldsymbol{\sigma}^{II}\rangle^{(k)} = -\frac{2}{c^{(k)}}\frac{\partial W^*}{\partial\mathbf{M}^{(k)}},\quad \langle\boldsymbol{\sigma}^{I}\otimes\boldsymbol{\sigma}^{II}\rangle^{(k)} = \frac{1}{c^{(k)}}\frac{\partial\boldsymbol{\beta}^*}{\partial\mathbf{M}^{(k)}}\langle\boldsymbol{\sigma}\rangle. \tag{14.112}$$

In most cases, the correct type of index operations can be figured out from the structure of the equation, but we will also present an index form if it is not obvious [e.g., in Eqs. (7.9) and (7.10)]

$$\langle\sigma^I_{pq}\sigma^I_{rs}\rangle^{(k)} = \frac{1}{c^{(k)}}\frac{\partial M^*_{ijmn}}{\partial M^{(k)}_{pqrs}}\langle\sigma_{ij}\rangle\langle\sigma_{mn}\rangle, \tag{14.113}$$

$$\langle\sigma^{II}_{pq}\sigma^{II}_{rs}\rangle^{(k)} = -\frac{2}{c^{(k)}}\frac{\partial W^*}{\partial M^{(k)}_{pqrs}}, \quad \langle\sigma^I_{pq}\sigma^{II}_{rs}\rangle^{(k)} = \frac{1}{c^{(k)}}\frac{\partial \beta^*_{ij}}{\partial M^{(k)}_{pqrs}}\langle\sigma_{ij}\rangle. \tag{14.114}$$

Here the partial derivatives in Eqs. (14.111)–(14.114) are estimated as a directional derivative of the function $\mathbf{f}$ ($\mathbf{f} = \mathbf{M}^*,\ W^*,\ \boldsymbol{\beta}^*$) in the direction $M^{(k)}_{ijkl}$

$$\frac{\partial \mathbf{f}}{\partial M^{(k)}_{ijpq}} = \lim_{\xi\to 0}\frac{\mathbf{f}(\mathcal{M}(\mathbf{x},\mathbf{y}) + \xi M^{(k)}_{ijkl}V^{(k)}(\mathbf{x})\delta(\mathbf{x}-\mathbf{y}),\boldsymbol{\beta}(\mathbf{x})) - \mathbf{f}(\mathcal{M}(\mathbf{x},\mathbf{y}),\boldsymbol{\beta}(\mathbf{x}))}{\xi M^{(k)}_{ijkl}}, \tag{14.115}$$

where according to (14.109) $M^{(k)}_{ijkl}$ is an arbitrary constant symmetric tensor and ξ is chosen from real numbers; $\mathbf{f}(\mathcal{M}(\mathbf{x},\mathbf{y}),\boldsymbol{\beta}(\mathbf{x}))$ is the effective parameter at the real (unperturbed) values $\mathcal{M}(\mathbf{x},\mathbf{y})$ and $\boldsymbol{\beta}(\mathbf{x})$ $(\mathbf{x},\mathbf{y} \in w)$. Although Eqs. (14.111)–(14.114) look analogous to the corresponding equations for the composites with local properties of constituents (14.9), they are essentially different because the tensor $\mathbf{M}^{(k)}$ in Eqs. (14.111)–(14.114) is related with a special variation (14.109) of nonlocal compliance in Eq. (14.115) rather than with the real values of this compliance describing by the nonlocal operator $\mathcal{M}$ (14.2). Because of this, one uses in Eqs. (7.9)–(7.13) the variation $\delta\mathbf{M}^{(k)}$ rather than the tensor $\mathbf{M}^{(k)}$ as such.

Summation of Eqs. (14.111) and (14.112) yields the representation for the second moment of the overall stresses $\boldsymbol{\sigma} = \boldsymbol{\sigma}^I + \boldsymbol{\sigma}^{II}$

$$\langle\boldsymbol{\sigma}\otimes\boldsymbol{\sigma}\rangle^{(k)} = \frac{1}{c^{(k)}}\frac{\partial\mathbf{M}^*}{\partial\mathbf{M}^{(k)}}\langle\boldsymbol{\sigma}\rangle\otimes\langle\boldsymbol{\sigma}\rangle - \frac{2}{c^{(k)}}\frac{\partial W^*}{\partial\mathbf{M}^{(k)}} + \frac{2}{c^{(k)}}\frac{\partial\boldsymbol{\beta}^*}{\partial\mathbf{M}^{(k)}}\langle\boldsymbol{\sigma}\rangle, \tag{14.116}$$

where the partial derivatives are calculated under the assumption of fixed transformation fields $\boldsymbol{\beta}(\mathbf{x})$, $\mathbf{x} \in w$. The left-hand side of Eq. (7.14) holds the symmetry conditions (minor and major) due to the symmetry of the tensor $\delta\mathbf{M}^{(k)}$.

Invariants $\langle\sigma_0^2\rangle^{(k)}$ and $\langle s_{ij}s_{ij}\rangle^{(k)}$ $(\sigma_0 = \sigma_{ii}/3,\ s_{ij} = \sigma_{ij} - \sigma_0)$ are estimated employing the convolutions of Eqs. (14.113) and (14.114). For this purpose, we will assume that in (14.109) $\delta\mathbf{M}^{(k)} = 3\delta\bar{p}^{(k)}\mathbf{N}_1 + 2\delta\bar{q}^{(k)}\mathbf{N}_2$ $(\mathbf{N}_1 = \boldsymbol{\delta}\otimes\boldsymbol{\delta}/3,\ \mathbf{N}_2 = \mathbf{I}-\mathbf{N}_1)$, where $\delta\bar{p}^{(k)}$ and $\delta\bar{q}^{(k)}$ are some infinitesimal increments of independent variables $\bar{p}^{(k)}$ and $\bar{q}^{(k)}$ which have no connections with the compliance operator $\mathcal{M}(\mathbf{x},\mathbf{y})$. Taking into account a differential rule of a composite function, we get

$$\langle\boldsymbol{\sigma}\rangle\frac{\partial\mathbf{M}^*}{\partial\bar{p}^{(k)}}\langle\boldsymbol{\sigma}\rangle = \langle\boldsymbol{\sigma}\rangle\frac{\partial\mathbf{M}^*}{\partial M^{(k)}_{ijmn}}\langle\boldsymbol{\sigma}\rangle\frac{\partial M^{(k)}_{ijmn}}{\partial\bar{p}^{(k)}}$$
$$= c^{(k)}\langle\sigma_{ij}\sigma_{mn}\rangle^{(k)}3N_{1|ijmn} = 9\langle\sigma_0^2\rangle^{(k)}, \tag{14.117}$$

$$\langle\boldsymbol{\sigma}\rangle\frac{\partial\mathbf{M}^*}{\partial\bar{q}^{(k)}}\langle\boldsymbol{\sigma}\rangle = \langle\boldsymbol{\sigma}\rangle\frac{\partial\mathbf{M}^*}{\partial M^{(k)}_{ijmn}}\langle\boldsymbol{\sigma}\rangle\frac{\partial M^{(k)}_{ijmn}}{\partial\bar{q}^{(k)}}$$

$$= c^{(k)}\langle \sigma_{ij}\sigma_{mn}\rangle^{(k)} 2N_{2|ijmn} = 2\langle s_{ij}s_{ij}\rangle^{(k)}. \qquad (14.118)$$

Then, since the variations $\delta\bar{p}^{(k)}$ and $\delta\bar{q}^{(k)}$ are mutually independent, we can get from (14.113) and (14.114) the representations for the stress invariants averaged over the k's phase

$$\langle\sigma_0^2\rangle^{(k)} = \frac{1}{9c^{(k)}}\frac{\partial \mathbf{M}^*}{\partial \bar{p}^{(k)}}\langle\boldsymbol{\sigma}\rangle\otimes\langle\boldsymbol{\sigma}\rangle - \frac{2}{9c^{(k)}}\frac{\partial W^*}{\partial \bar{p}^{(k)}} + \frac{2}{9c^{(k)}}\frac{\partial \boldsymbol{\beta}^*}{\partial \bar{p}^{(k)}}\langle\boldsymbol{\sigma}\rangle, \qquad (14.119)$$

$$\langle s_{ij}s_{ij}\rangle^{(k)} = \frac{1}{2c^{(k)}}\frac{\partial \mathbf{M}^*}{\partial \bar{q}^{(k)}}\langle\boldsymbol{\sigma}\rangle\otimes\langle\boldsymbol{\sigma}\rangle - \frac{1}{c^{(k)}}\frac{\partial W^*}{\partial \bar{q}^{(k)}} + \frac{1}{c^{(k)}}\frac{\partial \boldsymbol{\beta}^*}{\partial \bar{q}^{(k)}}\langle\boldsymbol{\sigma}\rangle, \qquad (14.120)$$

where the first, second, and third terms in the right-hand sides of Eqs. (14.119) and (14.120) comprise the appropriate representations for the stress invariants of the corresponding fourth-rank tensors of stresses $\langle\boldsymbol{\sigma}^I\otimes\boldsymbol{\sigma}^I\rangle^{(k)}$, $\langle\boldsymbol{\sigma}^{II}\otimes\boldsymbol{\sigma}^{II}\rangle^{(k)}$, and $2\langle\boldsymbol{\sigma}^I\otimes\boldsymbol{\sigma}^{II}\rangle^{(k)}$, respectively, averaged over the volume of the k's phase.

Of course, analogous relations can be derived for the second moment of strains, either through the transformations of Eqs. (14.55)–(14.57) and (14.94_3) into $\boldsymbol{\varepsilon}$-expressions or by recasting of Eqs. (14.111) and (14.112) in $\boldsymbol{\varepsilon}$-terms by the use of the constituent equations (14.51) and (14.52). For example, for pure mechanical stresses ($\boldsymbol{\beta}(\mathbf{x}) \equiv \mathbf{0}$), we get

$$\langle\boldsymbol{\varepsilon}^I\otimes\boldsymbol{\varepsilon}^I\rangle^{(k)} = \frac{1}{c^{(k)}}\frac{\partial \mathbf{L}^*}{\partial \mathbf{L}^{(k)}}\langle\boldsymbol{\varepsilon}\rangle\otimes\langle\boldsymbol{\varepsilon}\rangle, \qquad (14.121)$$

$$\langle(\varepsilon_0^I)^2\rangle^{(k)} = \frac{1}{9c^{(k)}}\frac{\partial \mathbf{L}^*}{\partial \bar{k}^{(k)}}\langle\boldsymbol{\varepsilon}\rangle\otimes\langle\boldsymbol{\varepsilon}\rangle, \quad \langle\varepsilon_{\rm eq}^2\rangle^{(k)} = \frac{1}{3c^{(k)}}\frac{\partial \mathbf{L}^*}{\partial \bar{\mu}^{(k)}}\langle\boldsymbol{\varepsilon}\rangle\otimes\langle\boldsymbol{\varepsilon}\rangle, \qquad (14.122)$$

where $\varepsilon_0 = \varepsilon_{ii}/3$, $\varepsilon_{\rm eq}^2 = 2e_{ij}e_{ij}/3$, $\mathbf{e} = \boldsymbol{\varepsilon} - \varepsilon_0\boldsymbol{\delta}$, $\delta\mathbf{L}^{(k)} = 3\delta\bar{k}^{(k)}\mathbf{N}_1 + 2\delta\bar{\mu}^{(k)}\mathbf{N}_2$, and all partial derivatives in Eqs. (14.121) and (14.122) are calculated analogously to the relation (14.115). It should be mentioned that we did not assume isotropy of the tensors $\mathbf{M}^{(k)}$*and* $\mathbf{L}^{(k)}$ used in (14.116) and (14.121), respectively, simply because we are dealing with the infinitesimal isotropic increments $\delta\mathbf{M}^{(k)}$ and $\delta\mathbf{L}^{(k)}$ rather than with the tensors $\mathbf{M}^{(k)}$ and $\mathbf{L}^{(k)}$ as such and $\delta\mathbf{M}^{(k)} \not\equiv (\delta\mathbf{L}^{(k)})^{-1}$ in general.

Some particular results of relationships (14.116) and (14.121) for the local properties at the fine scale (14.9) were obtained in [1187] (for $\boldsymbol{\beta} \equiv 0$) and [238] (for $\boldsymbol{\sigma}^0 \equiv \mathbf{0}$). A more tractable particular case of Eq. (14.116) is proposed in [661]. It was considered [280] the generalization of Eqs. (14.116) for thermo–elastic–electric fields related by the local constitutive laws at the fine scale. Relations (14.116) have been obtained for any degree of anisotropy of $\mathbf{M}^*$, $\mathbf{M}^{(i)}$ ($i = 0, 1, \ldots, N$). For isotropic tensors $\mathbf{M}^{(i)}$, the relations (14.116) reduce to the results in [133] and [808] for macroisotropic composites containing only constituents with the local properties. A perturbation method for a micropolar composite analogous to Eq. (14.116) was used in [307] to estimate the average second-order stress/couple-stress moment in the phases.

14.4.2 Local Fluctuation of Stresses

From the relation (14.116), one may generally estimate the stress fluctuations at any point $\mathbf{x} \in v_i \subset v^{(k)}$ $(i = 1, 2, \ldots;\ k = 1, \ldots, N)$. For this purpose it is sufficient to choose an auxiliary domain $v_i' \subset v_i$ $(\mathbf{x} \in v_i',\ \ i = 1, 2, \ldots)$ with a characteristic function $V_i'(\mathbf{x})$ in such a manner that $\overline{v}_i' \to 0$ and the variation $\delta\mathbf{M}$ (7.7) differ from zero only in the v_i' region. So it is necessary to evaluate $\mathbf{f}$ $(\mathbf{f} = \mathbf{M}^*, W^*, \boldsymbol{\beta}^*)$ with the operator $\boldsymbol{\mathcal{M}}(\mathbf{x}, \mathbf{y})$ being changed into $\boldsymbol{\mathcal{M}}(\mathbf{x}, \mathbf{y}) \to \boldsymbol{\mathcal{M}}(\mathbf{x}, \mathbf{y}) + \delta\boldsymbol{\mathcal{M}}(\mathbf{x}, \mathbf{y})$

$$\delta\boldsymbol{\mathcal{M}}(\mathbf{x}, \mathbf{y}) \equiv \delta\mathbf{M}^{(k)} V_i'(\mathbf{x})\delta(\mathbf{x} - \mathbf{y}), \ \ \delta\mathbf{M}^{(k)} = \text{const}, \ \ \mathbf{x} \in v_i', \ \ \mathbf{y} \in w, \tag{14.123}$$

where, for definiteness sake, the domain v_i' is a sphere with the center $\mathbf{x}'$ and radius $a_i' \to 0$.

Then Eq. (7.14) can be recast in the local representation for any point $\mathbf{x}' \in v_i$ in the heterogeneity $v_i \subset v^{(k)}$

$$\langle \boldsymbol{\sigma} \otimes \boldsymbol{\sigma} \rangle_i(\mathbf{x}') = \frac{\partial \mathbf{M}^*}{\partial \mathbf{M}^{(k)}(\mathbf{x}')} \langle \boldsymbol{\sigma} \rangle \otimes \langle \boldsymbol{\sigma} \rangle - 2\frac{\partial W^*}{\partial \mathbf{M}^{(k)}(\mathbf{x}')} + 2\frac{\partial \boldsymbol{\beta}^*}{\partial \mathbf{M}^{(k)}(\mathbf{x}')} \langle \boldsymbol{\sigma} \rangle, \tag{14.124}$$

where the partial derivatives $(\mathbf{f} = \mathbf{M}^*, W^*, \boldsymbol{\beta}^*)$

$$\frac{\partial \mathbf{f}}{\partial M_{ijkl}^{(k)}(\mathbf{x}')} = \lim_{\xi, a_i' \to 0} \frac{\overline{v}_i}{c^{(k)} \overline{v}_i' \xi M_{ijkl}^{(k)}} \times \Big[\mathbf{f}(\boldsymbol{\mathcal{M}}(\mathbf{x}, \mathbf{y}) + \xi M_{ijkl}^{(k)} V_i'(\mathbf{x})\delta(\mathbf{x} - \mathbf{y}), \boldsymbol{\beta}(\mathbf{x})) - \mathbf{f}(\boldsymbol{\mathcal{M}}(\mathbf{x}, \mathbf{y}), \boldsymbol{\beta}(\mathbf{x}))\Big] \tag{14.125}$$

are calculated analogously to Eq. (7.13) under the assumption of fixed transformation fields $\boldsymbol{\beta}(\mathbf{x}),\ \mathbf{x} \in w$.

Equations (7.17)–(7.20) can be generalized to the estimations of the statistical second moments of stresses in the point $\mathbf{x}' \in v_i \subset v^{(k)}$ analogously to Eq. (7.22).

14.4.3 Average Stresses in the Constituents

Well-known are the exact representations of stress concentration factors averaged over the volume of the constituents of the two-phase classic composites and statistically isotropic polycrystalline aggregates composed of crystals of some symmetries (see for references Chap. 6). It will be now demonstrated that analogous results can be obtained in a general case of multiphase composites with the nonlocal properties of constituents if the functional dependencies of effective constants on the constituent eigenstress are known. Indeed, following a scheme of the local constitutive law (see [808]), we will vary a transformation field $\delta\boldsymbol{\beta}(\mathbf{x})$ at the fixed $\boldsymbol{\mathcal{M}}(\mathbf{x}, \mathbf{y})$ instead of variation of $\delta\boldsymbol{\mathcal{M}}(\mathbf{x}, \mathbf{x})$ (14.109) at the fixed $\boldsymbol{\beta}(\mathbf{x})$. Then varying

Eq. (14.57) and taking (14.56_2) into account lead to

$$\begin{aligned}\langle\boldsymbol{\sigma}\rangle\delta\boldsymbol{\beta}^* &= \langle\delta\boldsymbol{\sigma}^I\boldsymbol{\beta}\rangle + \langle\boldsymbol{\sigma}^I\delta\boldsymbol{\beta}\rangle = \langle\delta\boldsymbol{\sigma}^I\boldsymbol{\varepsilon}^{II}\rangle - \langle\delta\boldsymbol{\sigma}^I(\boldsymbol{\varepsilon}^{II}-\boldsymbol{\beta})\rangle + \langle\boldsymbol{\sigma}^I\delta\boldsymbol{\beta}\rangle\\ &= \langle\delta\boldsymbol{\sigma}^I\boldsymbol{\varepsilon}^{II}\rangle - \langle\delta\boldsymbol{\varepsilon}^I(\boldsymbol{\mathcal{L}}*(\boldsymbol{\varepsilon}^{II}-\boldsymbol{\beta}))\rangle + \langle\boldsymbol{\sigma}^I\delta\boldsymbol{\beta}\rangle\\ &= \langle\delta\boldsymbol{\sigma}^I\boldsymbol{\varepsilon}^{II}\rangle - \langle\delta\boldsymbol{\varepsilon}^I\boldsymbol{\sigma}^{II}\rangle + \langle\boldsymbol{\sigma}^I\delta\boldsymbol{\beta}\rangle, \end{aligned} \tag{14.126}$$

where the self-adjointness of the operator $\boldsymbol{\mathcal{L}}$ (14.27) has been used, and the first and the second term in the right-hand side vanish by virtue of (14.55) and (14.106_2) that reduce Eq. (14.126) to $\langle\boldsymbol{\sigma}^I\delta\boldsymbol{\beta}\rangle = \langle\boldsymbol{\sigma}\rangle\delta\boldsymbol{\beta}^*$. In a similar manner, variation of (14.92) by varying $\boldsymbol{\beta}(\mathbf{x})$ yields

$$\begin{aligned} -\delta W^* &= \langle\delta\boldsymbol{\sigma}^{II}\boldsymbol{\beta}\rangle + \langle\boldsymbol{\sigma}^{II}\delta\boldsymbol{\beta}\rangle = \langle\delta\boldsymbol{\sigma}^{II}\boldsymbol{\varepsilon}^{II}\rangle - \langle\delta\boldsymbol{\sigma}^{II}(\boldsymbol{\varepsilon}^{II}-\boldsymbol{\beta})\rangle + \langle\boldsymbol{\sigma}^{II}\delta\boldsymbol{\beta}\rangle\\ &= \langle\delta\boldsymbol{\sigma}^{II}\boldsymbol{\varepsilon}^{II}\rangle - \langle\delta(\boldsymbol{\varepsilon}^{II}-\boldsymbol{\beta})\boldsymbol{\sigma}^{II}\rangle + \langle\boldsymbol{\sigma}^{II}\delta\boldsymbol{\beta}\rangle\\ &= 2\langle\boldsymbol{\sigma}^{II}\delta\boldsymbol{\beta}\rangle + \langle\delta\boldsymbol{\sigma}^{II}\boldsymbol{\varepsilon}^{II}\rangle - \langle\boldsymbol{\sigma}^{II}\delta\boldsymbol{\varepsilon}^{II}\rangle, \end{aligned} \tag{14.127}$$

which can be simplified due to Eqs. (14.55) and (14.112_1) as $\langle\boldsymbol{\sigma}^{II}\delta\boldsymbol{\beta}\rangle = -\delta W^*$. A special choice of the varying eigenstrains $(k = 0, 1, \ldots) : \delta\boldsymbol{\beta}(\mathbf{x}) \equiv \delta\boldsymbol{\beta}^{(k)}V^{(k)}(\mathbf{x})$, $\delta\boldsymbol{\beta}^{(k)} = \text{const}\ \ (\mathbf{x}\in v^{(k)})$, analogously to (14.109), allows the explicit representations for the average stresses in the k's phase

$$\langle\boldsymbol{\sigma}^I\rangle = \frac{1}{c^{(k)}}\langle\boldsymbol{\sigma}\rangle\frac{\partial\boldsymbol{\beta}^*}{\partial\boldsymbol{\beta}^{(k)}},\quad \langle\boldsymbol{\sigma}^{II}\rangle = -\frac{1}{c^{(k)}}\frac{\partial W^*}{\partial\boldsymbol{\beta}^{(k)}}, \tag{14.128}$$

where the partial derivatives are calculated similarly to the representation (14.115) for an arbitrary constant symmetric tensor $\beta_{ij}^{(k)}$ $(\mathbf{f} = W^*, \boldsymbol{\beta}^*)$

$$\frac{\partial\mathbf{f}}{\partial\beta_{ij}^{(k)}} = \lim_{\xi\to 0}\frac{\mathbf{f}(\boldsymbol{\mathcal{M}}(\mathbf{x},\mathbf{y}),\boldsymbol{\beta}(\mathbf{x}) + \xi\beta_{ij}^{(k)}V^{(k)}(\mathbf{x})) - \mathbf{f}(\boldsymbol{\mathcal{M}}(\mathbf{x},\mathbf{y}),\boldsymbol{\beta}(\mathbf{x}))}{\xi\beta_{ij}^{(k)}}, \tag{14.129}$$

under the assumption of fixed operator compliance $\boldsymbol{\mathcal{M}}(\mathbf{x},\mathbf{y})$ $(\mathbf{x},\mathbf{y}\in w)$. There is no restriction on homogeneity of $\boldsymbol{\beta}(\mathbf{x})$ $(\mathbf{x}\in v^{(k)})$ in the relations (14.128) (in contrast with the known corresponding representations for the composites with the local properties of constitutives, see Chap. 6).

The preceding considerations show that in order to obtain a quantitative assessment of $\langle\boldsymbol{\sigma}\rangle^{(k)}$ and $\langle\boldsymbol{\sigma}\otimes\boldsymbol{\sigma}\rangle^{(k)}$, it is necessary to know the functional dependence of $\mathbf{M}^*$, W^*, and $\boldsymbol{\beta}^*$ on both the compliance operator $\boldsymbol{\mathcal{M}}^{(k)}$ and eigenstrain $\boldsymbol{\beta}^{(k)}$. These dependencies can be calculated only approximately, except for some degenerate structures. Therefore, one may think that the relationships are not indeed strictly exact. However, all of them can be measured experimentally, and thus the exact relationships obtained in this work have an immediate physical sense. Therefore, the error in numerical results for the first and second field moments is governed by the error in determining the dependencies of $\mathbf{M}^*$, $\boldsymbol{\beta}^*$, and W^* upon $\boldsymbol{\mathcal{M}}(\mathbf{x})$ and $\boldsymbol{\beta}(\mathbf{x})$.

These dependencies will be determined in subsequent publications with the help of various micromechanical models.

14.5 General Integral Equations and Effective Fields

Substituting the constitutive equation and the Cauchy equation into the equilibrium equation, we obtain a differential equation with respect to the strain $\boldsymbol{\varepsilon}$

$$\nabla[\mathcal{L}*\boldsymbol{\varepsilon}(\mathbf{x})+\boldsymbol{\alpha}(\mathbf{x})]+\mathbf{f}(\mathbf{x})=\mathbf{0}. \tag{14.130}$$

We assume that the properties of the comparison medium and the matrix coincide. The operator moduli of the matrix is $\boldsymbol{\mathcal{L}}^{(0)}$, whereas $\boldsymbol{\alpha}^c(\mathbf{x}), \boldsymbol{\beta}^c(\mathbf{x})$ and body force $\mathbf{f}^c(\mathbf{x})$ are inhomogeneous deterministic transformation fields (and with solution $\boldsymbol{\sigma}^0$, $\boldsymbol{\varepsilon}^0$, $\mathbf{u}^0$ to the same boundary-value problem). So, Eq. (14.130) transforms to the Navier-like governing equation with respect to the displacement $\mathbf{u}$

$$\nabla[\boldsymbol{\mathcal{L}}^{(0)}*\boldsymbol{\varepsilon}(\mathbf{x})+\boldsymbol{\alpha}^{(0)}(\mathbf{x})]+\mathbf{f}^c(\mathbf{x})=-\nabla\boldsymbol{\tau}(\mathbf{x})-\mathbf{f}_1(\mathbf{x}), \tag{14.131}$$

with a fictitious random "body-force" in the right-hand side of the equation. Hereafter, we introduce the stress $\boldsymbol{\tau}$ and strain $\boldsymbol{\eta}$ polarization tensors by two equivalent ways ($\mathbf{x}\in R^d$) (compare with Eqs. (4.2) and (4.2))

$$\boldsymbol{\tau}(\mathbf{x})=(\boldsymbol{\mathcal{L}}_1*\boldsymbol{\varepsilon})(\mathbf{x})+\boldsymbol{\alpha}(\mathbf{x}),\ \boldsymbol{\eta}(\mathbf{x})=(\boldsymbol{\mathcal{M}}_1*\boldsymbol{\sigma})(\mathbf{x})+\boldsymbol{\beta}_1(\mathbf{x}),\ \boldsymbol{\tau}=-\boldsymbol{\mathcal{L}}^{(0)}*\boldsymbol{\eta}, \tag{14.132}$$

$$\boldsymbol{\tau}(\mathbf{x})=\boldsymbol{\sigma}(\mathbf{x})-[(\boldsymbol{\mathcal{L}}^{(0)}*\boldsymbol{\varepsilon})(\mathbf{x})+\boldsymbol{\alpha}^{(0)}],\ \ \boldsymbol{\eta}(\mathbf{x})=\boldsymbol{\varepsilon}(\mathbf{x})-[(\boldsymbol{\mathcal{M}}^{(0)}*\boldsymbol{\sigma})(\mathbf{x})+\boldsymbol{\beta}^c], \tag{14.133}$$

which are simply a notational convenience. In (14.132) and (14.133), $\boldsymbol{\mathcal{N}}_1(\mathbf{y})$ and $\boldsymbol{\gamma}_1(\mathbf{y})$ are the jumps of the operator $\boldsymbol{\mathcal{N}}^{(k)}$ and eigenfield $\boldsymbol{\gamma}^{(k)}$ ($\boldsymbol{\mathcal{N}}=\boldsymbol{\mathcal{L}},\boldsymbol{\mathcal{M}};\ \boldsymbol{\gamma}=\boldsymbol{\alpha},\boldsymbol{\beta}$) inside the component $v^{(k)}$ ($k=0,\ldots,N$) with respect to the operator $\boldsymbol{\mathcal{N}}^{(0)}$ and $\boldsymbol{\gamma}^{(0)}$, respectively.

Then Eq. (14.131) may be reduced to a symmetrized integral form

$$\boldsymbol{\varepsilon}(\mathbf{x})=\boldsymbol{\varepsilon}^0(\mathbf{x})+\nabla\int_w \mathbf{G}(\mathbf{x}-\mathbf{y})\Big\{\nabla\boldsymbol{\tau}(\mathbf{y})+\mathbf{f}_1(\mathbf{y})\Big\}\,d\mathbf{y}, \tag{14.134}$$

where $\mathbf{G}$ is the infinite homogeneous body Green's function of the Navier equation with an elastic operator $\mathcal{L}^{(0)}$. The Green's function $\mathbf{G}$ is assumed to be known for any operator $\boldsymbol{\mathcal{L}}^c$ being considered. The deterministic function $\boldsymbol{\varepsilon}^0(\mathbf{x})$ is the strain field that would exist in the medium with homogeneous properties $\mathcal{L}^c$ and appropriate boundary conditions at $\Gamma=\partial w$. Different nonstandard boundary conditions for gradient elasticity problems were considered in [1228, 1231]. We will not specify these boundary conditions because the final integral representation obtained in this study does not contain the strain $\boldsymbol{\varepsilon}^0(\mathbf{x})$ and holds for any appropriate boundary conditions.

After integration of Eq. (14.134) by parts and centering of the obtained equation by the use of statistical averaging, it is found (in a similar manner as in Chap. 7) that for infinite number of inclusions

$$\boldsymbol{\varepsilon}(\mathbf{x}) = \langle\boldsymbol{\varepsilon}\rangle(\mathbf{x}) + \int \langle\langle \mathbf{U}(\mathbf{x}-\mathbf{y})\boldsymbol{\tau}(\mathbf{y})\rangle\rangle(\mathbf{y}) + \int \langle\langle \nabla\mathbf{G}(\mathbf{x}-\mathbf{y})\mathbf{f}_1\rangle\rangle(\mathbf{y})d\mathbf{y}d\mathbf{y}, \qquad (14.135)$$

where one introduces a centering operation

$$\langle\langle \mathbf{U}(\mathbf{x}-\mathbf{y})\mathbf{g}\rangle\rangle(\mathbf{y}) \equiv \mathbf{U}(\mathbf{x}-\mathbf{y})\mathbf{g}(\mathbf{y}) - \langle \mathbf{U}(\mathbf{x}-\mathbf{y})\mathbf{g}\rangle(\mathbf{y}). \qquad (14.136)$$

To express Eq. (14.135) in terms of stresses, we use the identities

$$\boldsymbol{\tau} = -\mathcal{L}^{(0)} * \boldsymbol{\eta}, \quad \boldsymbol{\varepsilon} = [\boldsymbol{\mathcal{M}}^{(0)} * \boldsymbol{\sigma} + \boldsymbol{\beta}^{(0)}] + \boldsymbol{\eta} \qquad (14.137)$$

Substituting (14.137_1) and (14.137_2) into the right-hand side and the left-hand side of (14.135), respectively, and contracting with the operator $\mathcal{L}^{(0)}$ gives the general integral equation for stresses

$$\boldsymbol{\sigma}(\mathbf{x}) = \langle\boldsymbol{\sigma}\rangle(\mathbf{x}) + \int \langle\langle \boldsymbol{\Gamma}(\mathbf{x}-\mathbf{y}) * \boldsymbol{\eta}(\mathbf{y})\rangle\rangle(\mathbf{y})\, d\mathbf{y} + \int \boldsymbol{\mathcal{L}}^c * \langle\langle \nabla\mathbf{G}(\mathbf{x}-\mathbf{y})\mathbf{f}_1(\mathbf{y})\rangle\rangle(\mathbf{y})\, d\mathbf{y}. \qquad (14.138)$$

The integral operator kernel $\boldsymbol{\Gamma}(\mathbf{x}-\mathbf{y}) = -\boldsymbol{\mathcal{L}}^{(0)} * (\mathbf{I}\delta(\mathbf{x}-\mathbf{y}) + \mathbf{U}\boldsymbol{\mathcal{L}}^{(0)} *)$ is called the Green stress tensor (see [822]). Convergence of the integrals (14.135) and (14.138) can be proved analogously to Chap. 7.

We introduce two sorts of effective fields based on the Eq. (14.135). Namely, let the inclusions $v_1, \ldots, v_n$ be fixed, and we can define two sorts of effective fields $\overline{\boldsymbol{\varepsilon}}_i(\mathbf{x})$ and $\widetilde{\boldsymbol{\varepsilon}}_{1,\ldots,n}(\mathbf{x})$ $(i = 1, \ldots, n;\ \mathbf{x} \in v_1, \ldots, v_n)$ by the use of the rearrangement of Eq. (14.135) in the following form (see for similar manipulations Eqs. (7.35)–(7.37))

$$\boldsymbol{\varepsilon}(\mathbf{x}) = \overline{\boldsymbol{\varepsilon}}_i(\mathbf{x}) + \int \mathbf{U}(\mathbf{x}-\mathbf{y})V_i^{\lambda}(\mathbf{y})\boldsymbol{\tau}(\mathbf{y})d\mathbf{y}, \qquad (14.139)$$

$$\overline{\boldsymbol{\varepsilon}}_i(\mathbf{x}) = \widetilde{\boldsymbol{\varepsilon}}_{1,\ldots,n}(\mathbf{x}) + \sum_{j\neq i} \int \mathbf{U}(\mathbf{x}-\mathbf{y})V_j^{\lambda}(\mathbf{y})\boldsymbol{\tau}(\mathbf{y})d\mathbf{y}, \qquad (14.140)$$

$$\widetilde{\boldsymbol{\varepsilon}}_{1,\ldots,n}(\mathbf{x}) = \langle\boldsymbol{\varepsilon}\rangle(\mathbf{x}) + \int \big\{\mathbf{U}(\mathbf{x}-\mathbf{y})\boldsymbol{\tau}(\mathbf{y})V^{\lambda}(\mathbf{y}|; v_1, \mathbf{x}_1; \ldots; v_n, \mathbf{x}_n) - \langle \mathbf{U}(\mathbf{x}-\mathbf{y})\boldsymbol{\tau}\rangle(\mathbf{y})\big\}d\mathbf{y}, \qquad (14.141)$$

for $\mathbf{x} \in v_i,\ i = 1, 2, \ldots, n$. In Eqs. (14.139)–(14.141) the indicator function $V_i^\lambda(\mathbf{y})$ of the domain $v_i^\lambda = v_i \oplus b(\mathbf{0}, b_\lambda)$ is used instead of $V_i(\mathbf{y})$ (7.35)–(7.37) because $\boldsymbol{\tau}(\mathbf{y})$ only vanishes outside the domain v_i^λ (rather than v_i in locally elastic case in Eqs. (7.35)–(7.37)). Here $\oplus$ is a Minkovski dilation (5.18) with the ball $b(\mathbf{0}, b_\lambda)$ of radius a_λ centered at the origin $\mathbf{0}$.

Considering some conditional statistical averages of the general integral equation (14.135) leads to an infinite system of *new integral equations* ($n = 1, 2, \ldots$) (compare with (7.25))

$$\langle \boldsymbol{\varepsilon}|v_1, \mathbf{x}_1; \ldots; v_n, \mathbf{x}_n\rangle(\mathbf{x}) - \sum_{i=1}^{n} \int \mathbf{U}(\mathbf{x}-\mathbf{y})\langle V_i^\lambda(\mathbf{y})\boldsymbol{\tau}|v_1, \mathbf{x}_1; \ldots; v_n, \mathbf{x}_n\rangle(\mathbf{y})d\mathbf{y}$$

$$= \langle \boldsymbol{\varepsilon}\rangle(\mathbf{x}) + \int \big\{\langle \mathbf{U}(\mathbf{x}-\mathbf{y})\boldsymbol{\tau}|; v_1, \mathbf{x}_1; \ldots; v_n, \mathbf{x}_n\rangle(\mathbf{y}) - \langle \mathbf{U}(\mathbf{x}-\mathbf{y})\boldsymbol{\tau}\rangle(\mathbf{y})\big\}d\mathbf{y}. \quad (14.142)$$

Since $\mathbf{x} \in v_1, \ldots, v_n$ in the n-th line of the system can take the values in the inclusions $v_1, \ldots, v_n$, the n-th line actually contains n equations.

It should be mentioned that acceptance of the Hypothesis **H1b** (7.47) reduces Eqs. (14.135) to the following one (at $\mathbf{f}_1 \equiv \mathbf{0}$)

$$\boldsymbol{\varepsilon}(\mathbf{x}) = \langle \boldsymbol{\varepsilon}\rangle(\mathbf{x}) + \int \Big[\mathbf{U}(\mathbf{x}-\mathbf{y})\boldsymbol{\tau}(\mathbf{y}) - \mathbf{U}(\mathbf{x}-\mathbf{y})\langle \boldsymbol{\tau}\rangle(\mathbf{y})\Big]d\,\mathbf{y}, \quad (14.143)$$

which is well-known in locally elastic micromechanics (see Chap. 7).

14.6 One Heterogeneity Inside Infinite Matrix

Equations (14.139)–(14.142) were obtained for statistically inhomogeneous inclusion fields. However, in subsequent s ections the statistically homogeneous media subjected to homogeneous boundary conditions (14.23) are considered. Moreover, according to hypothesis **H1a** being accepted and in view of the linearity of the problem, there exist constant fourth- and second-rank tensors $\mathbf{A}_i^\varepsilon(\mathbf{x}),\ \mathbf{B}_i^\varepsilon(\mathbf{x}),\ \mathbf{R}_i^\varepsilon(\mathbf{x})$ and $\mathbf{C}_i^\varepsilon(\mathbf{x}),\ \mathbf{C}_i^\sigma(\mathbf{x}),\ \mathbf{F}_i^\varepsilon(\mathbf{x})$, respectively, such that

$$\boldsymbol{\varepsilon}(\mathbf{x}) = \mathbf{A}_i^\varepsilon(\mathbf{x})\overline{\boldsymbol{\varepsilon}}(\mathbf{x}_i) + \mathbf{C}_i^\varepsilon(\mathbf{x}), \quad (14.144)$$

$$\boldsymbol{\sigma}(\mathbf{x}) = \mathbf{B}_i^\varepsilon(\mathbf{x})\overline{\boldsymbol{\varepsilon}}(\mathbf{x}_i) + \mathbf{C}_i^\sigma(\mathbf{x}), \quad (14.145)$$

$$\overline{v}_i\boldsymbol{\tau}(\mathbf{x}) = \mathbf{R}_i^\varepsilon(\mathbf{x})\overline{\boldsymbol{\varepsilon}}(\mathbf{x}_i) + \mathbf{F}_i^\varepsilon(\mathbf{x}), \quad \mathbf{x} \in v_i, \quad (14.146)$$

where $v_i \subset v^{(i)}$ and

$$\mathbf{R}_i^\varepsilon(\mathbf{x}) = \bar{v}_i[\mathbf{B}_i^\varepsilon(\mathbf{x}) - \boldsymbol{\mathcal{L}}^{(0)} * \mathbf{A}_i^\varepsilon(\mathbf{x})], \quad (14.147)$$

$$\mathbf{F}_i^{\varepsilon}(\mathbf{x}) = \bar{v}_i[\mathbf{C}_i^{\sigma}(\mathbf{x}) - \mathcal{L}^{(0)}*\mathbf{C}_i^{\varepsilon}(\mathbf{x}) + \boldsymbol{\alpha}_1(\mathbf{x})]. \tag{14.148}$$

For both the local properties of constituents and homogeneous ellipsoidal heterogeneity $\mathbf{x} \in v_i$

$$\mathcal{L}(\mathbf{x}, \mathbf{y}) \equiv [\mathbf{L}^{(0)} + \mathbf{L}_1^{(i)} V_i(\mathbf{x})]\delta(\mathbf{x} - \mathbf{y}), \quad \boldsymbol{\alpha}(\mathbf{x}) \equiv \boldsymbol{\alpha}^{(0)} + \boldsymbol{\alpha}_1^{(i)} V_i(\mathbf{x}), \tag{14.149}$$

the tensors (14.144)–(14.146) are constant and have the analytical Eshelby representations (see Chap. 4).

For the parametric analysis of nonlocal effect influence, we will estimate the mechanical influence functions $\mathbf{A}_i^{\varepsilon}(\mathbf{x})$, $\mathbf{B}_i^{\varepsilon}(\mathbf{x})$, $\mathbf{A}_i^{\sigma}(\mathbf{x})$, and $\mathbf{B}_i^{\sigma}(\mathbf{x})$ for the isolated ellipsoidal inclusions v_i with the elastic properties described by the nonlocal operator $\mathcal{L}_i$ $(\equiv \mathcal{M}_i^{-1})$ and the infinite matrix with the operator $\mathcal{L}^{(0)}$. The solution can be presented in the form of an integral equation

$$\boldsymbol{\varepsilon}(\mathbf{x}) = \overline{\boldsymbol{\varepsilon}}(\mathbf{x}) + \int \mathbf{U}(\mathbf{x} - \mathbf{y})[\mathcal{L}_i*\boldsymbol{\varepsilon}(\mathbf{y}) - \mathcal{L}^{(0)}*\boldsymbol{\varepsilon}(\mathbf{y}) + \boldsymbol{\alpha}_1(\mathbf{y})]V_i(\mathbf{y})d\mathbf{y} \tag{14.150}$$

Equation (14.150) is suitable for solution by the iteration method (see, e.g., Chap. 4) although it can be easily solved by the quadrature method as well. With the aim of using a subtraction technique, we decompose the integrand of Eq. (14.150) in the following manner ($\mathbf{x}, \mathbf{y} \in v_i$)

$$\begin{aligned}
\mathcal{L}_i*\boldsymbol{\varepsilon}(\mathbf{y}) - \mathcal{L}^{(0)}*\boldsymbol{\varepsilon}(\mathbf{y}) &= \Big\{[\mathcal{L}_i*\boldsymbol{\varepsilon}(\mathbf{y}) - \overline{\mathbf{L}}(\mathbf{y})\boldsymbol{\varepsilon}(\mathbf{y})] - [\mathcal{L}_i*\boldsymbol{\varepsilon}(\mathbf{x}) - \overline{\mathbf{L}}(\mathbf{x})\boldsymbol{\varepsilon}(\mathbf{x})]\Big\} \\
&+ \Big\{[\overline{\mathbf{L}}(\mathbf{y})\boldsymbol{\varepsilon}(\mathbf{y}) - \overline{\mathbf{L}}^{(0)}(\mathbf{y})\boldsymbol{\varepsilon}(\mathbf{y})] - [\overline{\mathbf{L}}\boldsymbol{\varepsilon}(\mathbf{x}) - \overline{\mathbf{L}}^{(0)}(\mathbf{x})\boldsymbol{\varepsilon}(\mathbf{x})]\Big\} \\
&+ \Big\{[\mathcal{L}_i*\boldsymbol{\varepsilon}(\mathbf{x}) - \overline{\mathbf{L}}(\mathbf{x})\boldsymbol{\varepsilon}(\mathbf{y})] + [\overline{\mathbf{L}}\boldsymbol{\varepsilon}(\mathbf{x}) - \overline{\mathbf{L}}^{(0)}(\mathbf{x})\boldsymbol{\varepsilon}(\mathbf{x})]\Big\} \\
&= \Big\{\overline{\mathcal{L}}_1*\boldsymbol{\varepsilon}(\mathbf{y}) - \overline{\mathcal{L}}_1*\boldsymbol{\varepsilon}(\mathbf{x})\Big\} + \Big\{\overline{\mathbf{L}}_1(\mathbf{y})\boldsymbol{\varepsilon}(\mathbf{y}) - \overline{\mathbf{L}}_1(\mathbf{x})\boldsymbol{\varepsilon}(\mathbf{x})\Big\} \\
&+ \overline{\mathcal{L}}_1*\boldsymbol{\varepsilon}(\mathbf{x}) + \overline{\mathbf{L}}_1(\mathbf{x})\boldsymbol{\varepsilon}(\mathbf{x}),
\end{aligned} \tag{14.151}$$

where

$$\overline{\mathcal{L}}_1*\boldsymbol{\varepsilon}(\mathbf{y}) \equiv \mathcal{L}_i*\boldsymbol{\varepsilon}(\mathbf{y}) - \overline{\mathbf{L}}(\mathbf{y})\boldsymbol{\varepsilon}(\mathbf{y}), \quad \overline{\mathbf{L}}_1(\mathbf{y}) \equiv \overline{\mathbf{L}}(\mathbf{y}) - \overline{\mathbf{L}}^{(0)}(\mathbf{y}), \tag{14.152}$$

$$\overline{\mathbf{L}}(\mathbf{y}) \equiv \int \mathcal{L}(\mathbf{y}, \mathbf{z})V_i(\mathbf{z})d\mathbf{z}, \quad \overline{\mathbf{L}}^{(0)}(\mathbf{y}) \equiv \int \mathcal{L}^{(0)}(\mathbf{y}, \mathbf{z})V_i(\mathbf{z})d\mathbf{z}, \tag{14.153}$$

and $\mathcal{L}*\boldsymbol{\varepsilon}(\mathbf{y}) \equiv \int \mathcal{L}(\mathbf{y}, \mathbf{z})\boldsymbol{\varepsilon}(\mathbf{z})V_i(\mathbf{z})d\mathbf{z}$. In such a case Eq. is reduced to a nonsingular equation

$$\boldsymbol{\varepsilon}(\mathbf{x}) = \overline{\boldsymbol{\varepsilon}}(\mathbf{x}) - \mathbf{P}_i(\mathbf{x})\overline{\mathbf{L}}_1(\mathbf{x})\boldsymbol{\varepsilon}(\mathbf{x}) - \mathbf{P}_i(\mathbf{x})\overline{\mathcal{L}}_1*\boldsymbol{\varepsilon}(\mathbf{x}) - \mathbf{P}_i(\mathbf{x})\boldsymbol{\alpha}_1(\mathbf{x})$$

$$+\int \mathbf{U}(\mathbf{x}-\mathbf{y})[\overline{\mathcal{L}}_1 * \boldsymbol{\varepsilon}(\mathbf{y}) - \overline{\mathcal{L}}_1 * \boldsymbol{\varepsilon}(\mathbf{x})]V_i(\mathbf{y})d\mathbf{y}$$

$$+\int \mathbf{U}(\mathbf{x}-\mathbf{y})[\overline{\mathbf{L}}_1(\mathbf{y})\boldsymbol{\varepsilon}(\mathbf{y}) - \overline{\mathbf{L}}_1(\mathbf{x})\boldsymbol{\varepsilon}(\mathbf{x}) + \boldsymbol{\alpha}_1(\mathbf{y}) - \boldsymbol{\alpha}_1(\mathbf{x})]V_i(\mathbf{y})d\mathbf{y}, \quad (14.154)$$

where $\mathbf{P}_i(\mathbf{x}) := \int \mathbf{U}(\mathbf{x}-\mathbf{y})V_i(\mathbf{y})d\mathbf{y}$. Nonsingularity of Eq. 14.154) can be proved analogously to justification of nonsingularity of the corresponding local counterpart suggested considered in Chap. 4. Equation (14.154) can be easily transformed to the form most suitable for the iteration solution

$$\begin{aligned}\boldsymbol{\varepsilon}^{[n+1]}(\mathbf{x}) &= \mathbf{A}_i^{\epsilon}(\mathbf{x})\overline{\boldsymbol{\varepsilon}}(\mathbf{x}) - \mathbf{A}_i^{\epsilon}(\mathbf{x})\mathbf{P}_i(\mathbf{x})\boldsymbol{\alpha}_1(\mathbf{x}) - \mathbf{A}_i^{\epsilon}(\mathbf{x})\mathbf{P}_i(\mathbf{x})\overline{\mathcal{L}}_1 * \boldsymbol{\varepsilon}^{[n]}(\mathbf{x})\\ &+\mathbf{A}_i^{\epsilon}(\mathbf{x})\int \mathbf{U}(\mathbf{x}-\mathbf{y})[\overline{\mathcal{L}}_1 * \boldsymbol{\varepsilon}^{[n]}(\mathbf{y}) - \overline{\mathcal{L}}_1 * \boldsymbol{\varepsilon}^{[n]}(\mathbf{x})]V_i(\mathbf{y})d\mathbf{y}\\ &+\mathbf{A}_i^{\epsilon}(\mathbf{x})\int \mathbf{U}(\mathbf{x}-\mathbf{y})[\overline{\mathbf{L}}_1(\mathbf{y})\boldsymbol{\varepsilon}^{[n]}(\mathbf{y})\\ &-\overline{\mathbf{L}}_1(\mathbf{x})\boldsymbol{\varepsilon}^{[n]}(\mathbf{x}) + \boldsymbol{\alpha}_1(\mathbf{y}) - \boldsymbol{\alpha}_1(\mathbf{x})]V_i(\mathbf{y})d\mathbf{y} \end{aligned} \quad (14.155)$$

with the driving term usually chosen as an initial approximation

$$\boldsymbol{\varepsilon}^{[0]}(\mathbf{x}) = \mathbf{A}_i^{\epsilon}(\mathbf{x})\overline{\boldsymbol{\varepsilon}}(\mathbf{x}) - \mathbf{A}_i^{\epsilon}(\mathbf{x})\mathbf{P}_i(\mathbf{x})\boldsymbol{\alpha}_1(\mathbf{x}), \ \ \mathbf{A}_i^{\epsilon}(\mathbf{x}) := (\mathbf{I} + \mathbf{P}_i(\mathbf{x})\overline{\mathbf{L}}_1(\mathbf{x}))^{-1}, \quad (14.156)$$

which is exact for both the local properties of constituents and homogeneous ellipsoidal heterogeneity v_i subjected to remote homogeneous effective field

$$\overline{\boldsymbol{\varepsilon}}(\mathbf{x}) \equiv \text{const.}\ (\mathbf{x} \in v_i). \quad (14.157)$$

If the $[n]$th iteration of the strain $\boldsymbol{\varepsilon}^{[n]}(\mathbf{x})$ is estimated inside the heterogeneity $\mathbf{x} \in v_i$, it can be used for evaluation of the strain outside v_i $(\mathbf{x} \notin v_i)$

$$\begin{aligned}\boldsymbol{\varepsilon}^{[n+1]}(\mathbf{x}) &= \overline{\boldsymbol{\varepsilon}}(\mathbf{x}) - \mathbf{P}_i(\mathbf{x}_m)\overline{\mathbf{L}}_1(\mathbf{x}_m)\boldsymbol{\varepsilon}^{[n]}(\mathbf{x}_m) - \mathbf{P}_i(\mathbf{x}_m)\overline{\mathcal{L}}_1 * \boldsymbol{\varepsilon}^{[n]}(\mathbf{x}_m)\\ &-\mathbf{P}_i(\mathbf{x}_m)\boldsymbol{\alpha}_1(\mathbf{x}_m) + \int \mathbf{U}(\mathbf{x}-\mathbf{y})[\overline{\mathcal{L}}_1 * \boldsymbol{\varepsilon}^{[n]}(\mathbf{y}) - \overline{\mathcal{L}}_1 * \boldsymbol{\varepsilon}^{[n]}(\mathbf{x}_m)]V_i(\mathbf{y})d\mathbf{y}\\ &+\int \mathbf{U}(\mathbf{x}-\mathbf{y})[\overline{\mathbf{L}}_1(\mathbf{y})\boldsymbol{\varepsilon}^{[n]}(\mathbf{y}) - \overline{\mathbf{L}}_1(\mathbf{x}_m)\boldsymbol{\varepsilon}^{[n]}(\mathbf{x}_m)\\ &+\boldsymbol{\alpha}_1(\mathbf{y}) - \boldsymbol{\alpha}_1(\mathbf{x}_m)]V_i(\mathbf{y})d\mathbf{y}, \end{aligned} \quad (14.158)$$

where $\mathbf{x}_m = \arg\min_{\mathbf{z}} |\mathbf{x}-\mathbf{z}|$ $(\mathbf{z} \in v_i)$ is chosen for elimination of possible singularities in the integrands (14.158) at $|\mathbf{x}-\mathbf{y}| \to 0$ $(\mathbf{y} \in v_i,\ \mathbf{x} \notin v_i)$.

The scheme of the method of successive approximation (14.155) and (14.158) can be presented in symbolic form $\mathbf{x} \in R^d$)

$$\boldsymbol{\varepsilon}(\mathbf{x}) = \boldsymbol{\varepsilon}^{[0]}(\mathbf{x}) + \mathcal{K}^{\varepsilon} * \boldsymbol{\varepsilon}(\mathbf{x}), \quad (14.159)$$

where nonsingular operator $\mathcal{K}^\varepsilon$ has obvious integral representation according to Eqs. (14.155) and (14.158). Equation (14.158) suggests the Neumann series form for the solution $\boldsymbol{\varepsilon}$ of (14.150) $\boldsymbol{\varepsilon} = (\sum_{k=0}^{\infty} \mathcal{K}^k)\boldsymbol{\varepsilon}^{[0]}$, where the power $\mathcal{K}^k$ is defined recursively by the conditions $\mathcal{K}^0 = \mathbf{I}$ and $\mathcal{K}^1 = \mathcal{K}$, and the kernel of $\mathcal{K}^k$ is $\mathcal{K}_k(\mathbf{x}, \mathbf{y}) = \int \mathcal{K}(\mathbf{x}, \mathbf{z})\mathcal{K}_{k-1}(\mathbf{z}, \mathbf{y})d\mathbf{z}$ (see, e.g., Chap. 4). The sequence $\{\boldsymbol{\varepsilon}^{[k]}\}$ (14.155) and (14.158) with arbitrary continuous $\boldsymbol{\varepsilon}^{[0]}(\mathbf{x})$ converges to a unique solution $\boldsymbol{\varepsilon}$ if the norm of the integral operator $\mathcal{K}$ turns out to be small "enough" (less than 1) $\|\mathcal{K}\|_{\infty, v_i} \equiv \max_{\mathbf{x}\in v_i} \int |\mathcal{K}(\mathbf{x}, \mathbf{y})|\, d\mathbf{y} < 1$ and the problem is reduced to computation of the integrals involved, the density of which is given. In effect the iteration method (14.155) and (14.158) transforms the integral equation problem (14.150) into a linear algebra problem in any case. Of course Eq. (14.159) can be solved directly by the quadrature method even if this condition of smallness is not valid.

It is interesting that in the case of local elasticity $\mathcal{L}(\mathbf{x}, \mathbf{y}) = \mathbf{L}(\mathbf{x})\delta(\mathbf{x} - \mathbf{y})$ (14.9), we need to estimate one integral over the volume of the inclusion, while in the case of nonlocal elasticity, we need to estimate just three similar integrals. It is well-known in the context of micropolar elasticity that the strains are nonuniform even for the homogeneous elastic properties if the inclusion. Therefore, the popular assumption of homogeneity of strain (or stress) field inside the ellipsoid is just the first term in a perturbation series expansion involving the difference in the moduli of the inhomogeneity matrix (see, e.g., [1682]). In contrast to the mentioned popular approximation, our approach provides any desirable accuracy to the solution (14.150). After finding the strain field $\boldsymbol{\varepsilon}(\mathbf{x})$ (14.155) and (14.158), the stress field $\boldsymbol{\sigma}(\mathbf{x})$ can be found from the constitutive equation (14.1). Therefore, the operator representations of the stress–strain state ($\mathbf{x} \in R^d$)

$$\boldsymbol{\varepsilon}(\mathbf{x}) = \mathcal{A}_i^\varepsilon * \overline{\boldsymbol{\varepsilon}}(\mathbf{x}) + \mathbf{C}_i^\varepsilon(\mathbf{x}), \quad \boldsymbol{\sigma}(\mathbf{x}) = \mathcal{B}_i^\varepsilon * \overline{\boldsymbol{\varepsilon}}(\mathbf{x}) + \mathbf{C}_i^\sigma(\mathbf{x}), \tag{14.160}$$

can be considered as found for any inhomogeneous effective field $\overline{\boldsymbol{\varepsilon}}(\mathbf{x})$. For the homogeneous effective field (14.157), the operators (14.160) are reduced to the concentration factors (14.144) and (14.145): $\mathcal{A}_i^\varepsilon * \overline{\boldsymbol{\varepsilon}}(\mathbf{x}) = \mathbf{A}_i^\varepsilon(\mathbf{x})\overline{\boldsymbol{\varepsilon}}, \mathcal{B}_i^\varepsilon * \overline{\boldsymbol{\varepsilon}}(\mathbf{x}) = \mathbf{B}_i^\varepsilon(\mathbf{x})\overline{\boldsymbol{\varepsilon}}$.

After obtaining of the solution , we will introduce the operators A^ε and A^τ describing a perturbation of the strain field inside and outside the inclusion v_i ($\mathbf{x} \in R^d$)

$$\int \mathbf{U}(\mathbf{x}-\mathbf{y})\boldsymbol{\tau}(\mathbf{y})V_i(\mathbf{y})d\mathbf{y} = \boldsymbol{\varepsilon}(\mathbf{x}) - \overline{\boldsymbol{\varepsilon}}_i(\mathbf{x}) \equiv \mathsf{A}_i^\varepsilon(\overline{\boldsymbol{\varepsilon}}_i)(\mathbf{x}) \equiv \mathsf{A}_i^\tau(\boldsymbol{\tau})(\mathbf{x}), \tag{14.161}$$

$$\mathsf{A}_i^\varepsilon(\overline{\boldsymbol{\varepsilon}}_i)(\mathbf{x}) = \int \mathbf{U}(\mathbf{x} - \mathbf{y})[\mathcal{L}_1 * \mathcal{A}_i^\varepsilon * \overline{\boldsymbol{\varepsilon}}(\mathbf{y}) + \mathcal{L}_1 * \mathbf{C}_i^\varepsilon(\mathbf{y}) + \boldsymbol{\alpha}_1(\mathbf{y})]V_i(\mathbf{y})d\mathbf{y}, \tag{14.162}$$

$$\mathsf{A}_i^\tau(\boldsymbol{\tau})(\mathbf{x}) = \int \mathbf{U}(\mathbf{x} - \mathbf{y})\boldsymbol{\tau}(\mathbf{y})V_i(\mathbf{y})d\mathbf{y}. \tag{14.163}$$

The right-hand sides of Eqs. (14.162) and (14.163) can be also estimated in the manner of the subtraction technique inside (14.154) and outside (14.158) the inclusion v_i ($\mathbf{x} \in R^d$).

It should be mentioned that nonlocal elastic properties can be described not only by the integral operators but also by the differential ones. The differential operator can be obtained by truncation of the corresponding Tailor approximation of the integral one (14.36), and, because of this, the form of the differential operator is less general. However, the differential form of the nonlocal operator (14.160) and (14.161) has obvious advantage due to the more readily realized FEA (see, e.g., [388, 1570]) with respect to the integral form (see, e.g., [1228]). Therefore, for estimations of the tensors $\mathbf{A}_i^{\varepsilon}(\mathbf{x})$ and $\mathbf{B}_i^{\varepsilon}(\mathbf{x})$ (rather than the operators (14.160)) by the use of the method considered in Chap. 4 for the local problems, this approach can be more effective than the volume integral equation method (14.150).

14.7 Effective Elastic Properties

For termination of the hierarchy of statistical moment Eqs. (14.142), we will use in this section the closing effective field hypothesis, quasi-crystalline approximation **H2b**. To make further progress, the hypothesis of "ellipsoidal symmetry" **H3** for the distribution of inclusions is exploited (see for details Chap. 8).

Statistical averaging of Eqs. (14.141) at $n = 1$ leads to the following representation for the mean of the effective field in the heterogeneity v_i

$$\langle \overline{\boldsymbol{\varepsilon}} \rangle_i(\mathbf{x}) = \langle \boldsymbol{\varepsilon} \rangle(\mathbf{x}) + \int [\mathbf{A}_q^{\varepsilon}(\langle \overline{\boldsymbol{\varepsilon}} |; v_i, \mathbf{x}_i \rangle_q)(\mathbf{x})\varphi(v_q, \mathbf{x}_q |; v_i, \mathbf{x}_i) - \mathbf{A}_q^{\varepsilon}(\langle \overline{\boldsymbol{\varepsilon}} \rangle_q)(\mathbf{x})n^q(\mathbf{x}_q)]d\mathbf{x}_q, \tag{14.164}$$

where $\langle \overline{\boldsymbol{\varepsilon}} |; v_i, \mathbf{x}_i \rangle_q \equiv \langle \overline{\boldsymbol{\varepsilon}} |; v_i, \mathbf{x}_i \rangle_q(\mathbf{y})$ is a conditional statistical average of $\overline{\boldsymbol{\varepsilon}}(\mathbf{y})$ varying along the fixed heterogeneity $\mathbf{y} \in v_q$ at the fixed v_i, while $\langle \overline{\boldsymbol{\varepsilon}} \rangle_q \equiv \langle \overline{\boldsymbol{\varepsilon}} \rangle_q(\mathbf{y})$ is a statistical average $\overline{\boldsymbol{\varepsilon}}(\mathbf{y})$ inside the heterogeneity $\mathbf{y} \in v_q$.

In quasi-crystalline approximation **H2b**

$$\langle \boldsymbol{\varepsilon} |; v_i, \mathbf{x}_i \rangle_q(\mathbf{y}) = \langle \boldsymbol{\varepsilon} \rangle_q(\mathbf{y}), \quad \langle \overline{\boldsymbol{\sigma}} |; v_i, \mathbf{x}_i \rangle_q(\mathbf{y}) = \langle \overline{\boldsymbol{\sigma}} \rangle_q(\mathbf{y}), \quad (\mathbf{y} \in v_q) \tag{14.165}$$

and Eq. (14.164) is simplified

$$\langle \overline{\boldsymbol{\varepsilon}} \rangle_i(\mathbf{x}) = \langle \boldsymbol{\varepsilon} \rangle(\mathbf{x}) + \int \mathbf{A}_q^{\varepsilon}(\langle \overline{\boldsymbol{\varepsilon}} \rangle_q)(\mathbf{x})[\varphi(v_q, \mathbf{x}_q |; v_i, \mathbf{x}_i) - n^q(\mathbf{x}_q)]d\mathbf{x}_q, \tag{14.166}$$

For statistically homogeneous media when $n(\mathbf{x}_q) = n^{(q)} \equiv$const., it is logical to assume the acceptance of the additional hypothesis of "ellipsoidal symmetry" **H3** leading to the simplification

$$\int \mathbf{A}_q^{\varepsilon}(\langle\overline{\boldsymbol{\sigma}}\rangle_q)(\mathbf{x})[\varphi(v_q,\mathbf{x}_q|;v_i,\mathbf{x}_i)-n^{(q)}(\mathbf{x}_q)][1-V_{iq}^0(\mathbf{x}_q)]d\mathbf{x}_q=\mathbf{0}, \tag{14.167}$$

which can be presented in an equivalent form by using Green's function

$$\int \Big\{\langle \mathbf{U}(\mathbf{x}-\mathbf{y})\boldsymbol{\mathcal{L}}_1{*}\boldsymbol{\varepsilon}\rangle_q(\mathbf{y})+\langle \mathbf{U}(\mathbf{x}-\mathbf{y})\boldsymbol{\alpha}_1(\mathbf{y})\rangle_q\Big\}$$
$$\times\,[\varphi(v_q,\mathbf{x}_q|;v_i,\mathbf{x}_i)-n^{(q)}]V_q(\mathbf{y})[1-V_{iq}^0(\mathbf{y})]d\mathbf{y}=\mathbf{0}. \tag{14.168}$$

Then Eq. (14.166) leads to

$$\langle\overline{\boldsymbol{\varepsilon}}\rangle_i(\mathbf{x})=\langle\boldsymbol{\varepsilon}\rangle(\mathbf{x})-\int n^{(q)}\mathbf{A}_q^{\varepsilon}(\langle\overline{\boldsymbol{\varepsilon}}\rangle_q)(\mathbf{x})V_i^0(\mathbf{x}_q)d\mathbf{x}_q. \tag{14.169}$$

Accuracies of the assumptions (14.165) and (14.167) will be estimated in Sect. 14.9. However, we will perform subsequent solution of Eq. (14.166) rather than Eq. (14.169) keeping in mind that the hypothesis **H3** is accepted.

The regular integral equations (14.166) and (14.169) have no singularities and can be solved by a direct quadrature method which usually causes no problems of accuracy. However, for a large number of unknown variables N, its $O(N^3)$ cost dependence can lead to surprisingly long computing time. The obvious way of reducing this cost is to construct an iterative scheme which will be considered now; namely, we will use the iteration method when the initial constant effective field $\langle\overline{\boldsymbol{\varepsilon}}\rangle_q(\mathbf{x})=\langle\overline{\boldsymbol{\varepsilon}}\rangle_q\equiv$const. is estimated from the classical approach (14.144) and (8.25) (or (8.44)). Indeed, for $\langle\overline{\boldsymbol{\varepsilon}}\rangle_q\equiv$const., the statistical average of strains inside the heterogeneity v_q is found to be inhomogeneous [see Eqs. (14.144) and (14.145)]

$$\langle\boldsymbol{\varepsilon}\rangle_q(\mathbf{x})=\mathbf{A}_q^{\varepsilon}(\mathbf{x})\langle\overline{\boldsymbol{\varepsilon}}\rangle_q+\mathbf{C}_q^{\varepsilon}(\mathbf{x}),\quad \langle\boldsymbol{\tau}\rangle_q(\mathbf{x})=\mathbf{R}_q^{v\varepsilon}(\mathbf{x})\langle\overline{\boldsymbol{\varepsilon}}\rangle_q+\mathbf{F}_q^{v\varepsilon}(\mathbf{x}), \tag{14.170}$$

in a general case of the heterogeneity v_q, and $\mathbf{R}_q^{v\varepsilon}=\bar{v}_q^{-1}\mathbf{R}_q^{\varepsilon}$, $\mathbf{F}_q^{v\varepsilon}=\bar{v}_q^{-1}\mathbf{F}_q^{\varepsilon}$. In such a case, the right-hand side of Eq. (14.166) generates inhomogeneous field $\langle\overline{\boldsymbol{\varepsilon}}\rangle_i(\mathbf{x})$. For elimination of this difficulty, we will use the additional condition of the effective field hypothesis **H1b** (see Chap. 7), when a perturbation introduced by the inhomogeneity v_q is defined by the stress polarization tensor $\langle\boldsymbol{\tau}\rangle_q$ averaged over the volume v_q ($\mathbf{x}\in R^d$)

$$\mathbf{A}_q^{\varepsilon}(\langle\overline{\boldsymbol{\varepsilon}}\rangle_q)(\mathbf{x})=\bar{v}_q\mathbf{T}_q^{\varepsilon}(\mathbf{x}-\mathbf{x}_q)\langle\boldsymbol{\tau}\rangle_{(q)}, \tag{14.171}$$

Equation (14.171) comprises the approximation **H1b** providing the equivalence of representations (14.138) and (14.143). Then acceptance of the hypothesis **H3** reduces Eq. (14.166) to the classical representation for the effective field $\langle\overline{\boldsymbol{\varepsilon}}\rangle_i(\mathbf{x})$ (see Sect. 8.3)

$$\langle\overline{\boldsymbol{\varepsilon}}\rangle_i(\mathbf{x}) = \langle\boldsymbol{\varepsilon}\rangle + \mathbf{P}_i^0(\mathbf{x})\sum_{q=1}^{n} c^{(q)}\langle\boldsymbol{\tau}\rangle_{(q)}, \tag{14.172}$$

The tensor $\mathbf{P}_i^0(\mathbf{x})$ (14.172) (and, therefore, the effective field $\langle\overline{\boldsymbol{\varepsilon}}\rangle_i(\mathbf{x})$) is inhomogeneous in a general case of nonlocal properties $\boldsymbol{\mathcal{L}}^{(0)}$ even for an ellipsoidal shape of the excluded volume v_i^0. Combining the averaged Eqs. (14.170) and (14.172) leads to the final representations for the statistical averaged tensors of both the effective field and stress polarization tensor

$$\langle\overline{\boldsymbol{\varepsilon}}\rangle_i^{[0]}(\mathbf{x}) = \langle\boldsymbol{\varepsilon}\rangle + \mathbf{P}_i^0[\mathbf{I} - \langle\mathbf{R}^{v\varepsilon}\mathbf{P}^0 V\rangle]^{-1}(\langle\mathbf{R}^{v\varepsilon}V\rangle\langle\boldsymbol{\varepsilon}\rangle + \langle\mathbf{F}^{v\varepsilon}V\rangle), \tag{14.173}$$

$$\begin{aligned}\langle\boldsymbol{\tau}\rangle_i^{[0]}(\mathbf{x}) &= \mathbf{R}_i^{v\varepsilon}(\mathbf{x})\langle\boldsymbol{\varepsilon}\rangle + \mathbf{F}_i^{v\varepsilon}(\mathbf{x}) + \mathbf{R}_i^{v\varepsilon}(\mathbf{x})\mathbf{P}_i^0[\mathbf{I} - \langle\mathbf{R}^{v\varepsilon}\mathbf{P}^0 V v\rangle]^{-1} \\ &\quad\times(\langle\mathbf{R}^{v\varepsilon}V\rangle\langle\boldsymbol{\varepsilon}\rangle + \langle\mathbf{F}^{v\varepsilon}V\rangle),\end{aligned} \tag{14.174}$$

where the tensor $\mathbf{P}_i^0$ is taken to be constant after averaging Eq. (14.172). The representations (14.173) and (14.174) will be considered as the initial approximation in the iteration scheme constructing the inhomogeneous effective field ($\mathbf{x} \in v_i$, $\mathbf{y} \in v_q$)

$$\langle\overline{\boldsymbol{\varepsilon}}\rangle_i^{[n+1]}(\mathbf{x}) = \langle\boldsymbol{\varepsilon}\rangle + \int \mathbf{A}_q^{\tau}(\langle\boldsymbol{\tau}\rangle_q^{[n]})(\mathbf{x})[\varphi(v_q, \mathbf{x}_q|; v_i, \mathbf{x}_i) - n^q(\mathbf{x}_q)]d\mathbf{x}_q, \tag{14.175}$$

$$\langle\boldsymbol{\tau}\rangle_q^{[n+1]}(\mathbf{y}) = \boldsymbol{\mathcal{R}}_q^{\varepsilon} * \langle\overline{\boldsymbol{\varepsilon}}\rangle_q^{[n+1]}(\mathbf{y}) + \mathbf{F}_q^{\varepsilon}(\mathbf{y}), \tag{14.176}$$

where $\boldsymbol{\mathcal{R}}_q^{\varepsilon} = \boldsymbol{\mathcal{L}}_1 * (\mathbf{I} + \mathbf{A}_q^{\varepsilon})$ and Eq. (14.175) in the case of the assumption (14.167) is reduced to the following one

$$\langle\overline{\boldsymbol{\varepsilon}}\rangle_i^{[n+1]}(\mathbf{x}) = \langle\boldsymbol{\varepsilon}\rangle - \int \mathbf{A}_q^{\tau}(\langle\boldsymbol{\tau}\rangle_q^{[n]})(\mathbf{x})n^{(q)}(\mathbf{x}_q)V_i^0(\mathbf{x}_q)d\mathbf{x}_q. \tag{14.177}$$

The scheme (14.175) and (14.176) forms the Neumann series for the solution $\boldsymbol{\tau}$

$$\langle\boldsymbol{\tau}\rangle_i(\mathbf{x}) \equiv \lim_{n\to\infty}\langle\boldsymbol{\tau}^{[n]}\rangle_i(\mathbf{x}) = \mathbf{R}_i^{v\varepsilon *}(\mathbf{x})\langle\boldsymbol{\varepsilon}\rangle + \mathbf{F}_i^{v\varepsilon *}(\mathbf{x}), \tag{14.178}$$

which yields the final representations for the effective properties

$$\mathbf{L}^* = \mathbf{L}^{(0)} + \langle\mathbf{R}^{v\varepsilon *}V\rangle, \quad \boldsymbol{\alpha}^* = \boldsymbol{\alpha}^{(0)} + \langle\mathbf{F}^{v\varepsilon *}V\rangle. \tag{14.179}$$

It should be mentioned that Eq. (14.166) with an accuracy to notations coincides with Eqs. (13.42) and (13.43) for locally elastic CMs. Therefore, Eq. (14.166) can be solved by either the quadrature method (see Sect. 13.2.1) or iteration method (see Sect. 13.2.2), or the Fourier transform method (as it was mentioned in Chap. 13 with respect to Eqs. (13.42) and (13.43). In so doing, Eq. (14.166) holds for any inhomogeneity of both the statistical average field $\langle\boldsymbol{\varepsilon}\rangle(\mathbf{x}) \not\equiv$const. and concentration of inclusions $n^{(k)}$ $\not\equiv$cons. In the case of the iteration method, the solution of Eq. (14.166) will be obtained in the form of a nonlocal operator (13.10). Thus, the third case of correspondence between the concepts of homogenization and generalized continua (see Introduction in this Chapter) can be considered in a straightforward manner as an application of the developed methods of Chap. 13 to Eq. (14.166). Of course, we cannot say anything about the kernel of the effective nonlocal operator which is, in fact, an algorithm of estimation of the effective output response $\langle\boldsymbol{\sigma}\rangle(\mathbf{x})$ as a nonlocal function of statistically inhomogeneous input $\langle\boldsymbol{\varepsilon}\rangle(\mathbf{x})$.

14.8 Numerical Results

With the nonessential restriction on space dimensionality d and the shape of inhomogeneities, we will consider $2D$ problems for composites reinforced by cylindrical infinite fibers. The domains of inclusions v_i in the matrix are discretized by a square meshes (9.115) (see for details [206]). Two alternative radial distribution functions (RDF) of inclusion (9.116) (well-stirred approximation) and (9.117) will be examined. We will consider a pure mechanical problem ($\boldsymbol{\beta} \equiv \mathbf{0}$) for the composite with the isotropic constituents (14.9) and assume the matrix is epoxy resin with the locally elastic properties ($\boldsymbol{\mathcal{L}}^{(0)}(\mathbf{x},\mathbf{y}) \equiv \mathbf{L}^{(0)}\delta(\mathbf{x}-\mathbf{y})$, $\mathbf{L}^{(0)}$ $\equiv$const., $k^{(0)} = 3.83$ GPa, $\mu^{(0)} = 1.27$ GPa), which contains identical circular glass fibers ($k^{(1)} = 34.3$ GPa and $\mu^{(1)} = 31.3$ GPa) (14.10) with nonlocal stiffness (14.11) described by a polynomial bell-shaped attenuation function (14.15) $\lambda_\infty^{\text{bell}}(r) = 3(\pi a_\lambda^2)\{(1-(r/a_\lambda)^2)^2\}$, where the Macaulay brackets $\{\cdot\}$ denote the positive part, defined as $\{x\} = \max(0,x)$ (see, e.g., [706]). Thus, the assumption (14.26$_1$) holds, and the operator $\boldsymbol{\mathcal{L}}$ is self-adjoint (14.27). In most details we will consider the case $a_\lambda/a = 0.5$ although other ratios a_λ/a will be also analyzed in a few comparative examples.

Due to better accessibility, we will estimate the influence functions for the stresses rather than for the strains (related with the corresponding stresses by the constitutive law)

$$\langle\boldsymbol{\sigma}\rangle_i(\mathbf{x}) = \mathbf{B}^*(\mathbf{x})\langle\boldsymbol{\sigma}\rangle,\quad \langle\overline{\boldsymbol{\sigma}}\rangle_i(\mathbf{x}) = \overline{\mathbf{B}}^*(\mathbf{x})\langle\boldsymbol{\sigma}\rangle,\quad (\mathbf{x}\in v_i\subset v^{(1)}). \tag{14.180}$$

At first, we will analyze the results obtained in the framework the hypotheses **H2b** and **H3** for the well-stirred RDF (9.116). In Fig. 14.1 the iterations $B_{1111}^{*[k]}(\mathbf{x})$ $(k = 0, 2, 4, 14)$ at the axis $\mathbf{x} = (x_1, 0)^\top$ are presented for $c \equiv \pi a^2 n = 0.65$. The initial approximation $B_{1111}^{*[0]}(\mathbf{x})$ corresponding to the classical estimation (14.172) and using the old background (14.143) reveals their essential inhomogeneity (8.2%) even in the framework of the effective field hypothesis **H1**. The new background (14.135) allows the use of this inhomogeneity for refinement of the renormalizing item in Eq. (14.143) without exploiting of the hypothesis **H1**. The twelfth iteration $B_{1111}^{*[1111}(\mathbf{x})$ differs from the initial approximation $B_{1111}^{*[0]}(\mathbf{x})$ on 7.87%, while the

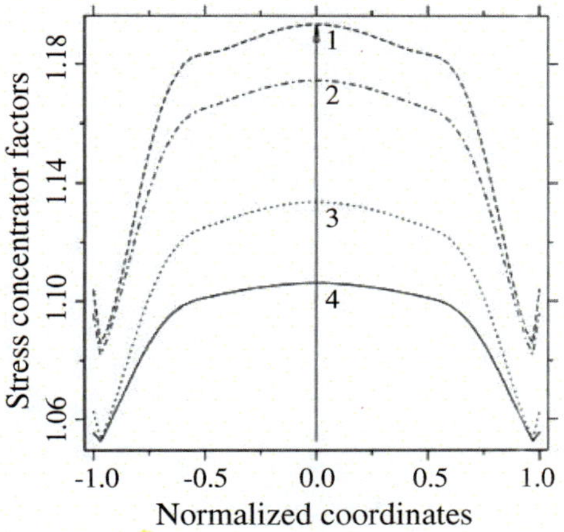

Fig. 14.1 Stress concentration factors $B^{*[k]}_{1111}(x_1)$ vs. normalized coordinate x_1/a in inhomogeneous fiber: curves 1, 2, 3, 4 for $k = 14, 4, 2, 0$, respectively

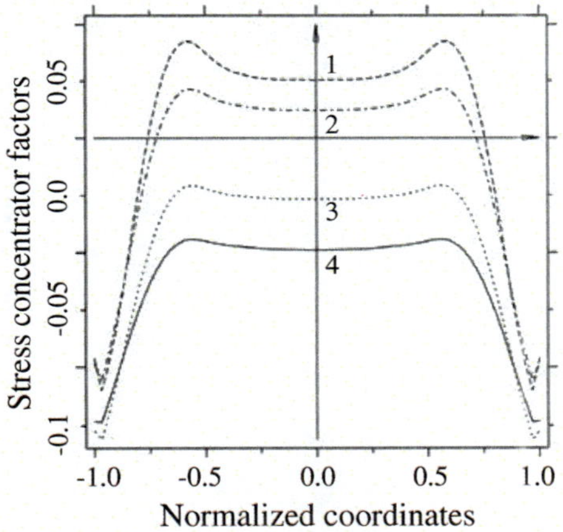

Fig. 14.2 Stress concentration factors $B^{*[k]}_{1122}(x_1)$ vs. normalized coordinate x_1/a in inhomogeneous fiber: curves 1, 2, 3, 4 for $k = 14, 4, 2, 0$, respectively

14th and 13th iterations are distinguished from one another by 0.04%. Of even greater difference, the results obtained for the backgrounds (14.135) and (14.143) are observed for the component $B^{*[k]}_{1122}(\mathbf{x})$ $(k = 0, 2, 4, 14)$ at $\mathbf{x} = (x_1, 0)^\top$ in Fig. 14.2. Indeed, $B^{*[14}_{1122}(\mathbf{x}) > 0$ at $-.75a < x_1 < 0.75a$, while $B^{*[0]}_{1122}(\mathbf{x}) < 0$ at any $\mathbf{x} = (x_1, 0)^\top$. Again, the proposed iteration method converges rapidly, and we contend that 14th iteration provides a difference from 13th iteration on 0.06% at $|x_1| = a$.

So much prominent and systematic difference of the old and new approaches is based on the abandonment from effective field hypothesis **H1** in the new approach. In particular, a tensor of effective stress concentration factor $\langle\overline{\boldsymbol{\sigma}}\rangle_i(\mathbf{x})$ (14.180) was estimated in [206]. It was obtained that $\overline{B}^{*[14]}_{1122}(\mathbf{x})$ differs from both $\overline{B}^{*[0]}_{1122}(\mathbf{x})$ and $\overline{B}^{*[13]}_{1111}(\mathbf{x})$ on 65.9% and 0.1%, respectively. Thus, violation of the hypothesis **H1** is quantitatively estimated.

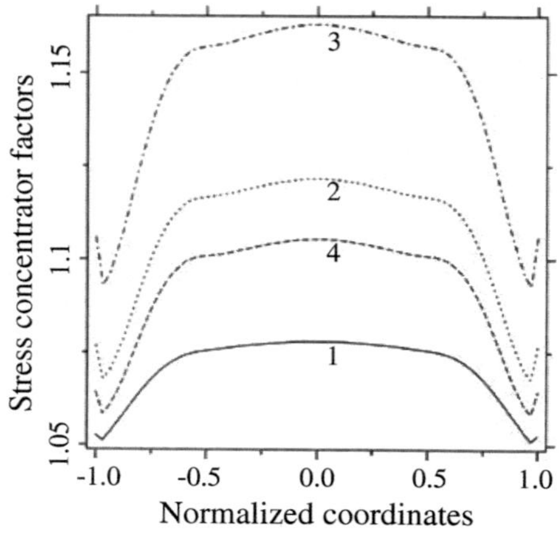

Fig. 14.3 $B^{*[k]}_{1111}(x_1)$ vs. x_1/a for the different a^0 and RDF: curves 1 ($k=0$), 2 [RDF (9.117), $a^0=2a,\ k=14$], 3 [RDF (9.117), $a^0=3a,\ k=14$], 4 [RDF (9.118), $a^0=2a,\ k=14$]

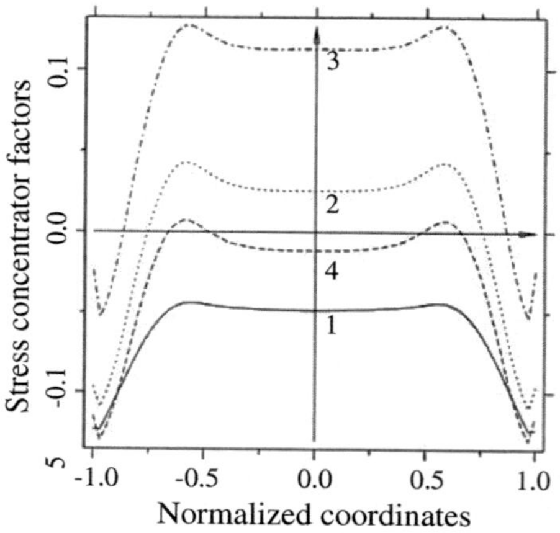

Fig. 14.4 $B^{*[k]}_{1122}(x_1)$ vs. x_1/a for the different a^0 and RDF: curves 1 ($k=0$), 2 [RDF (9.117), $a^0=2a,\ k=14$], 3 [RDF (9.117), $a^0=3a,\ k=14$], 4 [RDF (9.118), $a^0=2a,\ k=14$]

We now turn our attention to the analysis of the size of the circle excluded volume v_i^0 with the radius a^0 on the stress concentration factor $\mathbf{B}^{*[k]}(\mathbf{x})$ also for the RDF (9.116) and Eqs. (14.175)–(14.177). We will compare the estimation of $\mathbf{B}^{*[k]}(\mathbf{x})$ carried out for $a^0=3a$. with previously obtained results for $a^0=2a$ (see Figs. 14.1 and 14.2). The components $B^{*[k]}_{1111}(\mathbf{x})$ and $B^{*[k]}_{1122}(\mathbf{x})$ ($\mathbf{x}=(x_1,0)^\top$) are presented in Figs. 14.3 and 14.4, respectively, for $k=0$ (curves 1) and $k=14$ for both $a^0=2a$ (curves 2) and $a^0=14.8$. (curves 3). The RDF (9.116) provides a fulfillment of the "ellipsoidal symmetry" hypothesis **H3**, and, because of this, the classical representations for $\mathbf{B}^{*[0]}(\mathbf{x})$ are invariant to the size of v_i^0, while $B^{*[14]}_{1111}(\mathbf{x})$ and $B^{*[14]}_{1122}(\mathbf{x})$ estimated by the new approach (14.177) for $a^0=2a$ and $a^0=3a$. differ at $x_1=0$ one from another by 7.0% and by a factor of 2.7, respectively. This detected difference between $\mathbf{B}^{*[0]}(\mathbf{x})$ and $\mathbf{B}^{*[14]}(\mathbf{x})$ is explained by a fundamental deficiency of the classical background (14.143) where the renormalizing item $\mathbf{U}(\mathbf{x}-\mathbf{x}_q)\langle\boldsymbol{\tau}\rangle$ depends only on the average strain polarization tensor $\langle\boldsymbol{\tau}\rangle$, while a

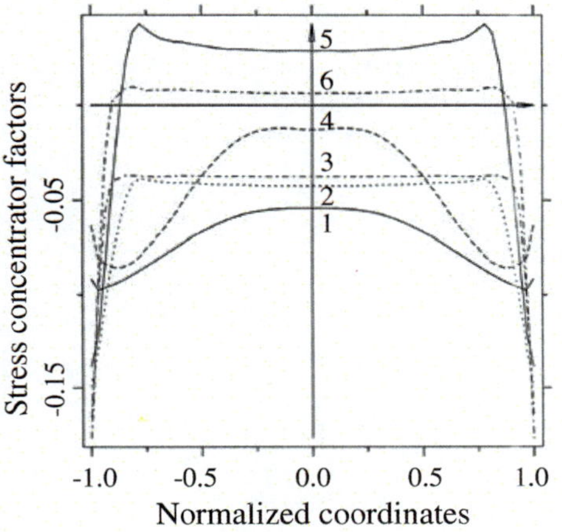

Fig. 14.5 $B^{*[k]}_{1122}(x_1)$ vs. x_1/a for $k = 0$: 1 ($a_\lambda/a = 1.0$), 2 ($a_\lambda/a = 0.25$), 3 ($a_\lambda/a = 0.1$) and for $k = 14$: 4($a_\lambda/a = 1.0$), 5 ($a_\lambda/a = 0.25$), 6 ($a_\lambda/a = 0.1$)

corresponding item $\langle \mathbf{U}(\mathbf{x}-\mathbf{y})\boldsymbol{\tau}(\mathbf{y})\rangle_q(\mathbf{x}_q)$ $(\mathbf{y} \in \mathbf{x}_q)$ in the new Eq. (14.135) explicitly depends on details distribution $\eta(\mathbf{y})$ $(\mathbf{y} \in \mathbf{x}_q)$. The last feature of Eq. (14.135) also leads to dependence of $B^{*[14]}(\mathbf{x})$ on the stress perturbations $\mathbf{A}^{\tau}_q(\langle\boldsymbol{\tau}\rangle^{[n]}_q)(\mathbf{x})$ (14.175) in the vicinity $\mathbf{x} \in v_i \setminus v_q$ rather than only on stress distributions in v_q and $\max_x|\mathbf{x}-\mathbf{x}_q| = (l+1)a$ $(\mathbf{x} \in v_i \setminus v_q)$ for the RDF (14.175) with the radius of excluded volume $a^0 = la$ $(l = 2, \ldots)$.

Finally, we compare the influence of the RDF (9.116) and (9.117) at $a^0 = 2a$ on estimation of $\mathbf{B}^{*[k]}(\mathbf{x})$ in Figs. 14.3 and 14.4. Needless to mention that $\mathbf{B}^{*[0]}(\mathbf{x})$ estimated in (14.174) and (14.180) is invariant with respect to the RDF in the framework of the hypotheses **H1**, **H2b**, and **H3** while $B^{*[14]}_{1111}(\mathbf{x})$ and $B^{*[14]}_{1122}(\mathbf{x})$ estimated for the RDF (9.116) (curves 2) and (9.117) (curves 3) (when only hypothesis **H1** is violated) distinguished one from another by 2.8% ($B^{*[14}_{1111}(\mathbf{x})$) and 54% ($B^{*[14]}_{1122}(\mathbf{x})$), respectively. Thus, the hypothesis **H3** applied to the new background Eq. (14.135) even under conditions of "quasi-crystalline" approximation **H2b** does not lead to a simplification (14.167) which takes place for the classical background (14.143) and (14.135) (14.171). The indicated differences demonstrating fundamentally new effects inherent in the new approach (14.175)–(14.179) far exceed the iteration error between 13th and 14th iterations of corresponded components which are less than 0.04% and 0.06% for $B^{*[k]}_{1111}(\mathbf{x})$ and $B^{*[14]}_{1122}(\mathbf{x})$, respectively, $(k = 13, 14)$.

Just for completeness, we will estimate an influence of the interaction radius a_λ on the stress concentrator factor $\mathbf{B}^{*[0]}(\mathbf{x})$ and $\mathbf{B}^{*[14]}(\mathbf{x})$ $(\mathbf{x} = (x_1, 0)^\top)$ for the RDF (9.116) with $a^0 = 2a$. In addition to Figs. 14.1 and 14.2 displaying the results for $a_\lambda/a = 0.5$, we are demonstrating the similar estimations $B^{*[k]}_{1122}(\mathbf{x})$ $(k = 0, 14)$ for $a_\lambda/a = 0.1,\ 0.25$, and $a_\lambda/a = 1$ in Fig. 14.5. As can be seen, $B^{*[0]}_{1112}(\mathbf{x})$ and $B^{*[14]}_{1112}(\mathbf{x})$ differ one from another by a sign for $a_\lambda/a = 0.1,\ 0.25$ at $|\mathbf{x}| < 0.84a$, while for $a_\lambda/a = 1$, $B^{*[0]}_{1122}(\mathbf{0})$ and $B^{*[14]}_{1122}(\mathbf{0})$ differ by a factor of three. The similar

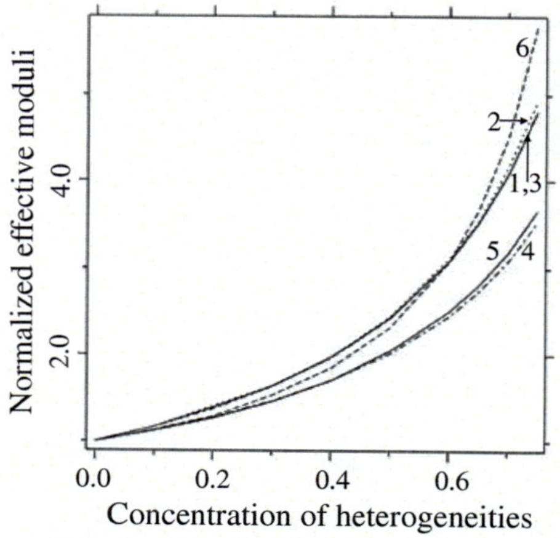

Fig. 14.6 Normalized effective moduli $\mu^*/\mu^{(0)}$ (1, 2 , 3) and $k^*/k^{(0)}$ (4, 5, 6) estimated for the local elasticity of fibers (1,4) and nonlocal ones

differences for the components $B^{*[0]}_{1111}(\mathbf{0})$ and $B^{*[14]}_{1111}(\mathbf{0})$ are 3.3%, 6.18%, and 4.3% for $a_\lambda/a = 0.1,\ 0.25$ and $a_\lambda/a = 1$, respectively.

We complete our numerical analysis by estimating the isotropic effective moduli $\mathbf{L}^* = 2k^*_{[2]}\mathbf{N}_1 + 2\mu^*_{[2]}\mathbf{N}_2$ ($\mathbf{N}_1 = \boldsymbol{\delta}\otimes\boldsymbol{\delta}/2,\ \mathbf{N}_2 = \mathbf{I}-\mathbf{N}_1$). For the fiber composites, it is the plane–strain bulk modulus $k^{(0)}_{[2]}$ (and $k^*_{[2]}$)—instead of the 3D bulk modulus $k^{(0)}_{[3]}$—that plays the significant role: $k^{(0)}_{[2]} = k^{(0)}_{[3]} + \mu^{(0)}_{[3]}/3,\ \mu^{(0)}_{[2]} = \mu^{(0)}_{[3]}$. The normalized bulk and shear moduli $k^*/k^{(0)}$ and $\mu^*/\mu^{(0)}$ are presented in Fig. 14.6 for both the local properties of the constituents (14.148) and nonlocal ones when the effective moduli are estimated by both the classical approach [corresponding to the stress concentration factors $\mathbf{B}^{*[0]}(\mathbf{x})$] and new one [corresponding to the 14's iteration $\mathbf{B}^{*[14]}(\mathbf{x})$]. As can be seen, the nonlocality of constituent properties (at $k = 14$) leads to an increasing of both the effective shear $\mu^*/\mu^{(0)}$ and bulk $k^*/k^{(0)}$ properties estimated by the classical approach ($k = [0]$) just on 0.2% and 0.3%, respectively. It should be mentioned that for the micropolar constituents (see [1650]), the effective shear modulus increases with decreasing fiber diameter at the same fiber volume fraction, while the effective in plane bulk modulus predicted in micropolar theory is identical to that predicted by Cauchy theory, since in micropolar theory, only rigid rotation is considered for the microstructure (the dilatation effect of the microstructure is not included). Turning back to Fig. 14.6, we observe that the new approach ($k = 14$) does not lead to a pronounced variation of the effective shear modulus $\mu^*/\mu^{(0)}$. However, the distinctions of the estimations $k^*/k^{(0)}$ between two approaches ($k = [0]$ and $k = [14]$) reach 41% for $c = 0.7$. In so doing, the stress concentration factors $\mathbf{B}^{*[0]}_{1122}(\mathbf{x})$ and $\mathbf{B}^{*[14]}_{1122}(\mathbf{x})$ in these approaches at the points $\mathbf{x} = (x_1, 0)^\top$ of fibers can differ by a sign at $|x_1| < 0.75.a$ (see Fig. 14.2). Thus, stress concentration factors are significantly more sensitive values to the choice of the approach than effective elastic moduli.

Thus, we established that a nonlocal constitutive law either inside or outside the heterogeneities leads to a stress field inhomogeneity in a homogeneous ellipsoidal inclusion even in the framework of hypothesis **H1a**. In its turn, this inhomogeneity can produce a correction of a sign of the effective local stress concentration factor $\mathbf{B}^*(\mathbf{x})$ estimated in the framework of new (14.135) and classical (14.143) backgrounds (see Fig. 14.5), whereas the corresponding effective moduli are not too different from one another (see Fig. 14.6). In the current chapter, the simplest case of a nonlocal constitutive law inside the circle heterogeneities was intentionally considered because it enables one to eliminate the impacts of a lot of other effects: nonelliptical shape of inhomogeneous heterogeneities (as in Chap. 9) with, perhaps, nonideal interface (including sliding, debonding, van der Waals and electrostatic effects, cohesive phenomena, as well as surface stress and surface tension ones, as in Chap. 11), binary interaction effects (as in Chap. 10), and nonlocal effects related with the special features of both the microstructure and applied loading, as in Chap. 12 (inhomogeneous remote loading, functionally graded materials, clustered materials, bounded media, contact of microinhomogeneous media, macro-inhomogeneity inside microinhomogeneous medium). All mentioned problems deserve further investigations that are, however, beyond the scope of the current publication.

14.9 Strain Gradient Theories

14.9.1 First Strain Gradient Medium

In the linearized theory of Mindlin's form II first strain gradient elasticity ([1067, 1068], see also for details and references [888, 889, 1220, 1427, 1693]), strain energy density of an homogeneous centrosymmetric material is a quadratic form of both the strain $\boldsymbol{\varepsilon}$ and its gradient $\boldsymbol{\kappa} = \nabla\boldsymbol{\varepsilon}$

$$W(\boldsymbol{\varepsilon}, \nabla\boldsymbol{\varepsilon}) = \frac{1}{2}\boldsymbol{\varepsilon} : \mathbf{L} : \boldsymbol{\varepsilon} + \frac{1}{2}\boldsymbol{\kappa} \,\therefore\, \mathbf{C} \,\therefore\, \boldsymbol{\kappa} = \frac{1}{2}L_{ijkl}\varepsilon_{ij}\varepsilon_{kl} + \frac{1}{2}C_{ijmkln}\partial_m\varepsilon_{ij}\partial_n\varepsilon_{kl}, \tag{14.181}$$

where $\therefore$ denotes the contraction over the last three indices of the left operand and the first three indices of the right operand. In form I, the strain energy density is assumed to be a quadratic form of the strain and the second-order derivatives of the displacement. Form III postulates that the strain energy density is a function of the strain, the rotation gradient, and the symmetric second gradient of the displacement. For the forms I and III, as well as the second strain gradient $W(\boldsymbol{\varepsilon}, \nabla\boldsymbol{\varepsilon}, \nabla\nabla\boldsymbol{\varepsilon})$ theories, the elastic energy can be decomposed in a similar manner. Usually in the literature dedicated to strain gradient modeling, one uses the form II (14.181) which leads to a symmetric total stress tensor and referred to as the first strain gradient model [39]. The tensor $\mathbf{L}$, the standard rank-four tensor of elastic constants, and $\mathbf{C}$, rank-six tensor of strain gradient elastic constants, have the symmetries

$$L_{ijkl} = L_{jikl} = L_{ijlk} = L_{klij},$$
$$C_{ijmkln} = C_{jimkln} = C_{ijmlkn} = C_{klnijm}. \tag{14.182}$$

The Cauchy stress $\boldsymbol{\sigma}$ and double stress $\boldsymbol{\mu}$ tensors are derived from the strain energy (14.181) as

$$\sigma_{ij} = \frac{\partial W}{\partial \varepsilon_{ij}} = L_{ijkl}\varepsilon_{kl}, \quad \mu_{ijm} = \frac{\partial W}{\partial(\partial_m \varepsilon_{ij})} = C_{ijmkln}\varepsilon_{kl,n}, \tag{14.183}$$

respectively.

From the energy density expression (14.181), via appeal to the Euler–Lagrange equations, the Navier-like governing static has a form

$$\partial_j(\sigma_{ij} - \partial_m \mu_{ijm}) = -f_i, \tag{14.184}$$

which can be recast in the following equation for the displacements

$$\mathbb{L}^{\mathrm{M}}\mathbf{u} = -\mathbf{f}, \;\; \mathbb{L}^{\mathrm{M}} = \mathbb{L} - \mathbb{C}, \;\; \mathbb{L}_{ik} = L_{ijkl}\partial_j\partial_l, \;\; \mathbb{C}_{ik} = C_{ijmklm}\partial_j\partial_l\partial_m\partial_n. \tag{14.185}$$

Here the Mindlin operator $\mathbb{L}^{\mathrm{M}}$ is expressed through the classical Navier operator $\mathbb{L}$ and fourth-order operator $\mathbb{C}$ due to the higher-order gradients of displacements in Eq. (14.181). On the other hand, the total stress tensor $\hat{\boldsymbol{\sigma}}$ defined as a variational derivative

$$\hat{\sigma}_{ij} := \frac{\delta W}{\delta \varepsilon_{ij}} = \frac{\partial W}{\partial \varepsilon_{ij}} - \partial_k \frac{\partial W}{\partial \partial_k \varepsilon_{ij}} \tag{14.186}$$

and reading as

$$\hat{\sigma}_{ij} := \sigma_{ij} - \partial_m \mu_{ijm} \tag{14.187}$$

reduce Eq. (14.184) to the simple form

$$\partial_j(\hat{\sigma}_{ij}) = -f_i. \tag{14.188}$$

In the case of the finite domain w, the boundary conditions on the boundary $\Gamma = \partial w$ have the form

$$\begin{cases} \mathbf{t} = \mathbf{t}^{\Gamma}, & \text{or } \mathbf{u} = u^{\Gamma} \\ \mathbf{m} = \mathbf{m}_i^{\Gamma}, & \text{or } \partial_{\mathbf{n}}\mathbf{u} = \mathbf{g}^{\Gamma}, \end{cases} \tag{14.189}$$

where $\mathbf{t}^{\Gamma}$, $\mathbf{m}^{\Gamma}$, $\mathbf{u}^{\Gamma}$, and $\mathbf{g}^{\Gamma}$ are the traction, double traction, displacement vector, and normal gradient of displacement, respectively, prescribed on $\Gamma = \partial w$ assumed to be smooth; traction vector **t** and double traction vector **m** that persist in (14.189) are

$$\mathbf{t} = \mathbf{n} \cdot \boldsymbol{\sigma} - \nabla_S \cdot (\mathbf{n} \cdot \boldsymbol{\mu}) - H\mathbf{nn} : \boldsymbol{\mu}, \quad \mathbf{m} = \mathbf{nn} : \boldsymbol{\mu}, \tag{14.190}$$

where **n** is the external unite normal to Γ; $H = -\nabla_S \cdot \mathbf{n}$ is the mean curvature of Γ and $\nabla_S = \nabla - \mathbf{n}\partial_n$ is the surface gradient operator. In particular, for the first and the second homogeneous boundary value, the conditions on $\mathbf{x} \in \Gamma$ are the following

$$\begin{aligned} \mathbf{u} = \boldsymbol{\varepsilon}^{\Gamma} \cdot \mathbf{x}, \quad \partial_{\mathbf{n}}\mathbf{u} = \mathbf{0}, \\ \mathbf{t} = \boldsymbol{\sigma}^{\Gamma} \cdot \mathbf{n}, \quad \mathbf{m}^{\Gamma} = \mathbf{0}, \end{aligned} \tag{14.191}$$

respectively, where $\boldsymbol{\varepsilon}^{\Gamma}(\mathbf{x}) \equiv$const. and $\boldsymbol{\sigma}^{\Gamma}(\mathbf{x}) \equiv$const. ($\mathbf{x} \in \Gamma$) are the prescribed constant symmetric tensors.

For the isotropic materials

$$L_{ijkl} = \lambda\delta_{ij}\delta_{kl} + \mu(\delta_{ik}\delta_{jl} + \delta_{il}\delta_{jk}), \tag{14.192}$$

$$\begin{aligned} C_{ijklmn} &= c_1/2(\delta_{il}\delta_{jk}\delta_{mn} + \delta_{im}\delta_{jk}\delta_{ln} + \delta_{ik}\delta_{jm}\delta_{ln} + \delta_{ik}\delta_{jl}\delta_{mn}) \\ &+c_2/2(\delta_{ij}\delta_{lk}\delta_{mn} + \delta_{ij}\delta_{mk}\delta_{ln} + \delta_{in}\delta_{jk}\delta_{lm} + \delta_{ik}\delta_{jn}\delta_{ml}) \\ &+2c_3\delta_{ij}\delta_{kn}\delta_{lm} + c_4(\delta_{im}\delta_{jl}\delta_{kn} + \delta_{il}\delta_{jm}\delta_{kn}) \\ &+c_5/2(\delta_{in}\delta_{jm}\delta_{kl} + \delta_{in}\delta_{jl}\delta_{km} + \delta_{im}\delta_{jn}\delta_{lk} + \delta_{il}\delta_{jn}\delta_{mk}), \end{aligned} \tag{14.193}$$

where λ and μ are the Lame's constants, whereas c_1, c_2, c_3, and c_4, c_5 are the five strain gradient parameters. Some well-known simplified strain gradient theories, such as couple-stress theory, Aifantis, Kleinert, and Wei and Hutchinson's models, have similar forms of governing equation [527, 993, 1682, 1685] as that of the general isotropic second gradient theory, it is therefore interesting to explore their connection. In the following, we focus on Eshelby tensor **S** for simplicity, and the other Eshelby tensors can be analyzed in the same way.

In particular, for a simplified model by Altan and Aifantis [23] (see also [39] as well as a fractional generalization [13]), the strain energy density W (14.181) reads

$$W(\boldsymbol{\varepsilon}, \nabla\boldsymbol{\varepsilon}) = \lambda(\boldsymbol{\delta} : \boldsymbol{\varepsilon})^2 + \mu\boldsymbol{\varepsilon} : \boldsymbol{\varepsilon} + l_{\mathrm{A}}^2[\lambda(\boldsymbol{\delta} : \nabla\boldsymbol{\varepsilon})\cdot(\boldsymbol{\delta} : \nabla\boldsymbol{\varepsilon}) + \mu\nabla\boldsymbol{\varepsilon} \therefore \nabla\boldsymbol{\varepsilon}], \tag{14.194}$$

that corresponds to the gradient parameters (14.193)

$$c_1 = \lambda l_{\mathrm{A}}^2/2, \quad c_2 = c_3 = 0, \qquad c_4 = c_5 = \mu l_{\mathrm{A}}^2/2, \tag{14.195}$$

with the length scale parameter l_{A} and the double stress tensor

$$\mu_{ijk} = \mu_{jik} = l_{\mathrm{A}}^2\sigma_{ij,k} = \lambda l_{\mathrm{A}}^2\delta_{ij}\kappa_{nnk} + 2\mu l_{\mathrm{A}}^2\kappa_{ijk}. \tag{14.196}$$

The isotropic Mindlin operator of first strain gradient elasticity

$$\mathbb{L}^{\mathrm{M}}_{ik} = (\lambda + 2\mu)[1 - l_1^2 \Delta]\partial_i \partial_k + \mu[1 - l_2^2 \Delta](\delta_{ik}\Delta - \partial_i \partial_k) \tag{14.197}$$

has the two characteristic lengths of Mindlin's isotropic first strain gradient elasticity of form II [1067]

$$l_1^2 = 2(c_1 + c_2 + c_3 + c_4 + c_5)/(\lambda + 2\mu), \quad l_2^2 = 2(a_3 + a_4)/\mu \tag{14.198}$$

connecting with l_{A} (14.193) as $l_1 = l_2 = l_{\mathrm{A}}$, i.e., Aifantis theory has two equal characteristic lengths.

Couple-stress (CS) theory is the form III in Mindlin [1067] theory with only rotation gradient for micro-deformation gradient is obtained for

$$c_1 = c_2 = c_3 = 0, \quad c_4 = \mu l_{\mathrm{CS}}^2, \quad c_5 = -\mu l_{\mathrm{CS}}^2 \tag{14.199}$$

connecting with the two characteristic lengths in second gradient theory (14.198) as $l_1 = 0,\ l_2 = l_{\mathrm{CS}}$.

According to Kleinert strain gradient (KSG) theory (see, e.g., [783, 1682]), we set

$$c_1 = \frac{1 + 2\mu}{2} l_{1\mathrm{KSG}}^2 - \frac{1}{2}\mu l_{2\mathrm{KSG}}^2, \quad c_3 = \frac{1}{2}\mu l_{2\mathrm{KSG}}^2, \quad c_2 = c_4 = c_5 = 0, \tag{14.200}$$

i.e., $l_1 = l_{1\mathrm{KSG}}$ and $l_2 = l_{1\mathrm{KSG}}$.

At last, according to the strain energy density function of Wei–Hutchinson (W–H) strain gradient theory, we set

$$c_1 = c_2 = c_3 = c_5 = 0, \quad c_4 = E_{\mathrm{W-H}}^2, \tag{14.201}$$

that corresponds to the equalities $l_1 = l_{\mathrm{W-H}}\sqrt{2E/(\lambda + 2\mu)}$ and $l_2 = l_{\mathrm{W-H}}\sqrt{2E/\mu}$, where E is Young's modulus. The number of strain gradient elastic constants reduces to three for linear, isotropic, strain gradient models proposed in [876] and [1693].

We define the Green tensor of the Mindlin operator $\mathbb{L}^{\mathrm{M}}$ (14.185) for the infinite homogeneous medium with the constants $\mathbf{L}$ and $\mathbf{C}$ (14.181)

$$\mathbb{L}^{\mathrm{M}}_{ik} G^{\mathrm{M}}_{kj}(\mathbf{x}) = -\delta_{ij}\delta(\mathbf{x}) \tag{14.202}$$

vanishing at the infinite and allowing to obtain a solution of Eq. (14.185) in the convolution form

$$u_k = G^{\mathrm{M}}_{kj} * f_j. \tag{14.203}$$

Let us consider an inhomogeneity v_i with the centers $\mathbf{x}_i$ and the material parameters $\mathbf{L}^{(1)}(\mathbf{x})$ and $\mathbf{C}^{(1)}(\mathbf{x})$ (14.181) in the infinite homogeneous matrix w with the material constants $\mathbf{L}^{(1)}$ and $\mathbf{C}^{(1)}$ (14.181). It is assumed that some remote boundary conditions (14.184) of the first boundary-value problem produces the fields $\overline{\boldsymbol{\varepsilon}}(\mathbf{x})$ and $\overline{\boldsymbol{\kappa}}(\mathbf{x})$ $(\mathbf{x} \in w)$ in the homogeneous matrix without inhomogeneity v. Then the equilibrium Eq. (14.186) with zero body-force can be presented in the form

$$\mathbb{L}^{\mathrm{M}\,(0)}\mathbf{u} = -\mathbb{L}_1^{\mathrm{M}}\mathbf{u}, \tag{14.204}$$

where a fictitious body-force in the right-hand side of Eq. (14.203) is defined by the material parameters $\mathbf{L}_1(\mathbf{x}) := \mathbf{L}(\mathbf{x}) - \mathbf{L}^{(0)}$ and $\mathbf{C}_1(\mathbf{x}) := \mathbf{C}(\mathbf{x}) - \mathbf{C}^{(0)}$ vanishing in the matrix $\mathbf{x} \in w \setminus v$.

Analogously to [1419], we introduce the stress and double stress polarization tensors

$$\boldsymbol{\tau}(\mathbf{x}) = \boldsymbol{\sigma}(\mathbf{x}) - \mathbf{L}^{(0)} : \boldsymbol{\varepsilon}(\mathbf{x}), \quad \boldsymbol{\pi}(\mathbf{x}) = \boldsymbol{\mu}(\mathbf{x}) - \mathbf{C}^{(0)} \therefore \boldsymbol{\kappa}(\mathbf{x}), \tag{14.205}$$

vanishing in the matrix $\mathbf{x} \in w \setminus v_i$. Substituting (14.203) into Eq. (14.204) leads to

$$\mathbb{L}^{\mathrm{M}\,(0)}\mathbf{u} = -\nabla[\boldsymbol{\tau} - \nabla\boldsymbol{\kappa}]. \tag{14.206}$$

Just for subsequent notational convenience associated with the locally elastic case (4.9), we introduce the pairs

$$\hat{\boldsymbol{\tau}} = \begin{pmatrix} \boldsymbol{\tau} \\ \boldsymbol{\pi} \end{pmatrix}, \quad \hat{\boldsymbol{\varepsilon}} = \begin{pmatrix} \boldsymbol{\varepsilon} \\ \boldsymbol{\kappa} \end{pmatrix}, \quad \hat{\overline{\boldsymbol{\varepsilon}}} = \begin{pmatrix} \overline{\boldsymbol{\varepsilon}} \\ \overline{\boldsymbol{\kappa}} \end{pmatrix}, \quad \hat{\mathbf{U}} = \begin{pmatrix} \mathbf{U}^{\epsilon\tau} & \mathbf{U}^{\epsilon\pi} \\ \mathbf{U}^{\kappa\tau} & \mathbf{U}^{\kappa\pi} \end{pmatrix}, \tag{14.207}$$

where $\hat{\overline{\boldsymbol{\varepsilon}}} = (\overline{\boldsymbol{\varepsilon}}, \overline{\boldsymbol{\kappa}})$ is a field that would be exist in the medium with homogeneous properties $\mathbf{L}^{(0)}$ and $\mathbf{C}^{(0)}$ and appropriate remote boundary condition; $\mathbf{U}^{\epsilon\tau} = \nabla\nabla\mathbf{G}^{\mathrm{M}}$, $\mathbf{U}^{\epsilon\pi} = -\nabla\nabla\nabla\mathbf{G}^{\mathrm{M}}$, $\mathbf{U}^{\kappa\tau} = \nabla\nabla\nabla\mathbf{G}^{\mathrm{M}}$, and $\mathbf{U}^{\kappa\tau} = -\nabla\nabla\nabla\nabla\mathbf{G}^{\mathrm{M}}$; and the Green functions and their nonsingular first and second derivatives for the isotropic and anisotropic media are proposed in [889] and [1220]. Particularly for spherical, cylindrical, and ellipsoidal inclusions, Eshelby tensors and their volume averages are obtained in an analytical form in [993]. It is found that the Eshelby tensors are not uniform inside the inclusion even for a spherical inclusion and their variations depend on the two characteristic lengths l_1 and l_2 of second gradient theory (14.197). It is also demonstrated that the existing Green's functions and Eshelby tensors of couple-stress theory (14.199) (see [1685]), Aifantis (14.195) (see [527]), Kleinert (14.200) (see [1682]), and Wei–Hutchinson (14.201) special strain gradient theories could be recovered as special cases.

Transforming the equilibrium Eq. (14.205) by the use of the Green function $\mathbf{G}^{\mathrm{M}}$ (14.202) and integration by parts, it is found a convolution form of the following integral equation in the new variables (14.207)

$$\hat{\boldsymbol{\varepsilon}} = \hat{\overline{\boldsymbol{\varepsilon}}} + \hat{\mathbf{U}} * \hat{\boldsymbol{\tau}}, \tag{14.208}$$

which is coincide with Eq. (4.9$_1$) with an accuracy to notations. In particular, for the homogeneous remote boundary conditions (14.189$_2$), $\hat{\overline{\boldsymbol{\varepsilon}}} = (\boldsymbol{\varepsilon}^{\Gamma}, \mathbf{0})^{\top}$.

The integral Eq. (14.208) with nonsingular convolution kernel $\hat{\mathbf{U}}$ and their solution can be presented in an operator form (9.6)

$$\hat{\boldsymbol{\varepsilon}}(\mathbf{x}) - \hat{\overline{\boldsymbol{\varepsilon}}}(\mathbf{x}) := \mathcal{L}_i^{\epsilon}(\mathbf{x} - \mathbf{x}_i, \hat{\boldsymbol{\tau}}) = \mathcal{L}_i^{\epsilon}(\mathbf{x} - \mathbf{x}_i, \hat{\overline{\boldsymbol{\varepsilon}}}),$$
$$\bar{v}_i \hat{\boldsymbol{\tau}}(\mathbf{x}) = \mathcal{R}_i^{\epsilon}(\mathbf{x} - \mathbf{x}_i, \hat{\overline{\boldsymbol{\varepsilon}}}), \tag{14.209}$$

where the solutions $\mathcal{L}_i^{\epsilon}(\mathbf{x} - \mathbf{x}_i, \hat{\overline{\boldsymbol{\varepsilon}}})$ and $\mathcal{R}_i^{\epsilon}(\mathbf{x} - \mathbf{x}_i, \hat{\overline{\boldsymbol{\varepsilon}}})$ (14.209) can be found, e.g., either by the volume integral equation method (see Sect. 4.4) or by finite element analysis (see, e.g., [39, 388, 1571] and Sects. 4.2 and 4.4).

14.9.2 Homogenization of the First Strain Gradient CMs

This subsection discusses as in Chaps. 7–9 a certain representative mesodomain w with an indicator function W and a boundary Γ containing a set $X = (v_i) \subset v$ of inclusions v_i with indicator functions $V_i \quad (i = 1, 2, \ldots)$. At first no restrictions are imposed on the elastic symmetry of the phases or on, for example, geometry of the inclusions. All tensors $\mathbf{g}$ ($\mathbf{g} = \mathbf{L}, \mathbf{C}$) (14.181) of material properties are decomposed as $\mathbf{g} \equiv \mathbf{g}^{(0)} + \mathbf{g}_1(\mathbf{x}) = \mathbf{g}^{(0)} + \mathbf{g}_1^{(m)}(\mathbf{x})$ (3.69), where $\mathbf{L}^{(0)}$ and $\mathbf{C}^{(0)}$ (14.181) are the constants of the homogeneous matrix $v^{(0)} = w \setminus v$.

Hill's [640] condition (6.10) is a milestone in micromechanics. It was generalized to different CMs in the framework of generalized continuum mechanics: to Cosserat (micropolar) [495, 933, 951, 1167], simplified strain gradient continua [525], couple-stress continua [518], and strongly nonlocal media (see Sects. 14.3.1 and 17.2.1). Following Gao [525], we consider the case of simplified strain gradient elasticity theory (14.194). Namely, the volume average W_w of the strain average (14.180) over the macrovolume $\mathbf{x} \in w$ can be presented in terms of field quantities on the boundary $\mathbf{s} \in \Gamma = \partial w$ as

$$W_w = \frac{1}{2\bar{w}} \int_w (\sigma_{ij}\varepsilon_{ij} + \mu_{ijk}\kappa_{ijk}) d\mathbf{x} = \frac{1}{2\bar{w}} \int_{\Gamma} (\sigma_{ij}u_i n_j + \mu_{ijk}\varepsilon_{ij}n_k) d\mathbf{s}. \tag{14.210}$$

Then a generalized Hill's lemma for the model (14.194) can be obtained (see for details [525]) for the volume averages

$$\{\sigma_{ij}\varepsilon_{ij} + \mu_{ijk}\kappa_{ijk}\} - \{\sigma_{ij}\}\{\varepsilon_{ij}\} - \{\mu_{ijk}\}\{\kappa_{ijk}\}$$
$$= \frac{1}{2\bar{w}} \int_{\Gamma} (\sigma_{ik}n_k - \{\sigma_{ik}\}n_k)(u_i - \{\varepsilon_{ij}\}x_j) d\mathbf{s}$$

$$+\frac{1}{2\bar{w}}\int_{\Gamma}(\mu_{ijk}n_k - \{\mu_{ijk}\}n_k)(\varepsilon_{ij} - \{\varepsilon_{ij}\})d\mathbf{s}. \tag{14.211}$$

Furthermore, Hill's condition, following from (14.211)

$$\{\sigma_{ij}\varepsilon_{ij} + \mu_{ijk}\kappa_{ijk}\} = \{\sigma_{ij}\}\{\varepsilon_{ij}\} + \{\mu_{ijk}\}\{\kappa_{ijk}\}, \tag{14.212}$$

holds for any from the following four admissible boundary conditions on $\mathbf{s} \in \Gamma$ vanishing the integrands in the right-hand side of Eq. (14.212): the traction-type boundary conditions

$$\sigma_{ik}n_k = \{\sigma_{ik}\}n_k, \quad \mu_{ijk}n_k = \{\mu_{ijk}\}n_k, \tag{14.213}$$

the displacement-type boundary conditions

$$u_i = \{\varepsilon_{ij}\}x_j, \quad \varepsilon_{ij} = \{\varepsilon_{ij}\}, \tag{14.214}$$

the mixed traction and strain boundary conditions

$$\sigma_{ik}n_k = \{\sigma_{ik}\}n_k, \quad \varepsilon_{ij} = \{\varepsilon_{ij}\}, \tag{14.215}$$

and the mixed double stress traction and displacement boundary conditions

$$\mu_{ijk}n_k = \{\mu_{ijk}\}n_k, \quad u_i = \{\varepsilon_{ij}\}x_j. \tag{14.216}$$

At the vanishing scale length $l_{\text{A}} = 0$ (14.194), when $\boldsymbol{\mu} = \mathbf{0}$ and $\hat{\boldsymbol{\sigma}} = \boldsymbol{\sigma}$, the Hill's lemma (14.211) and Hill's conditions (14.212) degenerate to their locally elastic classical counterparts (see Sect. 6.1.1). For generalized continua with $l_{\text{A}} \neq 0$, it is required that both the integrals on the right-hand side of Eq. (14.211) vanish in order to satisfy the generalized Hill's condition (14.212). This leads to the additional constraints indicated in Eqs. (14.213_2), (14.214_2), (14.215_2), and (14.216_2) in prescribing the boundary conditions on a homogeneous comparison material.

Another opportunity of exploiting of boundary integral representation for the strain energy W_w (14.210) is related with deriding of generalization of Eshelby [448] formula (6.74) that allows to estimate the difference between strain energy of heterogeneous material W_w (consisting of matrix and inclusions) and corresponding strain energy W_0 for pure matrix under the same loading conditions

$$W_w - W_0 = \frac{1}{2\bar{w}}\int_{\Gamma}(\mathbf{n}\cdot\boldsymbol{\sigma}\cdot\mathbf{u} - \mathbf{n}\cdot\boldsymbol{\sigma}_0\cdot\mathbf{u}_0 + \mathbf{n}\cdot\boldsymbol{\mu} : \nabla\mathbf{u} - \mathbf{n}\cdot\boldsymbol{\mu}_0 : \nabla\mathbf{u}_0)d\mathbf{s}, \tag{14.217}$$

$$W_0 = \frac{1}{2\bar{w}}\int_{\Gamma}(\mathbf{t}_0\cdot\mathbf{u}_0 + \mathbf{m}_0\cdot\partial_{\mathbf{n}}\mathbf{u}_0)d\mathbf{s}, \tag{14.218}$$

Following [1427], Eq. (14.217) can be simplified

$$W_w = W_0 + \frac{1}{2\bar{w}}\int_{\Gamma}(\mathbf{t}\cdot\mathbf{u} - \mathbf{t}_0\cdot\mathbf{u}_0 + \mathbf{m}\cdot\partial_{\mathbf{n}}\mathbf{u} - \mathbf{m}_0\cdot\partial_{\mathbf{n}}\mathbf{u}_0)ds, \tag{14.219}$$

where the definitions of tractions $\mathbf{t}$ and $\mathbf{t}_0$ and double tractions $\mathbf{m}$ and $\mathbf{m}_0$ are presented by Eqs. (14.190). Finally, Eq. (14.219) is reduced to the generalized Eshelby formula

$$W_w = W_0 + \frac{1}{2\bar{w}}\int_{\partial v^{(1)}}(\mathbf{t}\cdot\mathbf{u}_0 - \mathbf{t}_0\cdot\mathbf{u} + \mathbf{m}_0\cdot\partial_{\mathbf{n}}\mathbf{u} - \mathbf{m}\cdot\partial_{\mathbf{n}}\mathbf{u}_0)ds, \tag{14.220}$$

with the surface integration over the inclusion boundaries $\partial v^{(1)}$ and the prescribed the traction-type boundary conditions (14.213) on the macroboundary Γ; Eq. (14.218) should be modified for another boundary conditions (14.214)–(14.216) (see for details [1427]). At the vanishing scale length $\mathbf{C} \equiv \mathbf{0}$ (14.181), when $\boldsymbol{\mu} = \mathbf{0}$ and $\hat{\boldsymbol{\sigma}} = \boldsymbol{\sigma}$, we have classical definitions of tractions $\mathbf{t} = \mathbf{n}\cdot\boldsymbol{\sigma}$, and $\mathbf{t}_0 = \mathbf{n}\cdot\boldsymbol{\sigma}_0$ and Eq. (14.218) is reduced to the classical Eshelby formula (6.74).

Let us consider the displacement-type homogeneous boundary conditions (14.191_1) leading to the uniform stress field $\langle\boldsymbol{\sigma}\rangle = \boldsymbol{\sigma}_0 = \mathbf{L}^{(0)}\boldsymbol{\varepsilon}^{\Gamma}$, $\boldsymbol{\mu} \equiv \mathbf{0}$ [992] inside the homogeneous media with the material tensors $\mathbf{L}^{(0)}$ and $\mathbf{C}^{(0)}$ (14.181). Moreover, for the composite materials, $\{\boldsymbol{\kappa}\} = \mathbf{0}$ due to the boundary condition (14.191_1) and the right-hand side of Eq. (14.210) is expressed through the classical effective moduli $\mathbf{L}^*$ due to linearity of the problem

$$\{\boldsymbol{\varepsilon}\}\mathbf{L}^*\{\boldsymbol{\varepsilon}\} = 2W_w. \tag{14.221}$$

Substitution of Eqs. (14.220) and (14.218) into the right-hand side of Eq. (14.221) yields

$$\{\boldsymbol{\varepsilon}\}\mathbf{L}^*\{\boldsymbol{\varepsilon}\} = \{\boldsymbol{\varepsilon}\}\mathbf{L}^{(0)}\{\boldsymbol{\varepsilon}\} + \frac{1}{\bar{w}}\int_{v^{(1)}}(\boldsymbol{\sigma} : \boldsymbol{\varepsilon}^{\Gamma} - \boldsymbol{\sigma}_0 : \boldsymbol{\varepsilon})ds, \tag{14.222}$$

where the Gauss's theorem was used. Owing to the constitutive law for the homogeneous comparison medium $\boldsymbol{\sigma}_0 = \mathbf{L}^{(0)}\boldsymbol{\varepsilon}^{\Gamma}$ and arbitrariness of the constant field $\{\boldsymbol{\varepsilon}\} = \boldsymbol{\varepsilon}^{\Gamma}$, Eq. (14.218) is reduced to the final representation for the effective modulus

$$\mathbf{L}^* = \mathbf{L}^{(0)} + c^{(1)}\langle\boldsymbol{\tau}\rangle^{(1)}, \tag{14.223}$$

where the stress polarization tensor $\boldsymbol{\tau}$ (14.205) is exploited and volume average over the inclusion volume v_1 is replaced by a statistical average inside inclusions $v^{(1)}$. It is interesting that the new Eq. (14.223) is formally coincided with the classical representation (6.55) for locally elastic CM when the comparison medium coincides with the matrix (compare also with Eqs. (14.37) and (6.235) for strongly nonlocal continua micromechanics). In both Eqs. (6.55) and (14.223), one extracts the classical tensor $\mathbf{L}^{(0)}$ of local elasticity although the matrix $v^{(0)}$ in Eq. (14.223) demonstrates the gradient effects ($\mathbf{C} \not\equiv \mathbf{0}$). It looks that the matrix with the different scale effects in all three cases of (14.37), (6.235), and (14.223) is deformed identically to the locally elastic matrix (with the modulus $\mathbf{L}^{(0)}$). However, it does not mean that at the estimation of the effective moduli (14.223), the matrix can be replaced by a locally elastic one because the term $\langle \boldsymbol{\tau} \rangle^{(1)}$ depends on both the elastic properties (either local or gradient) of the matrix and micromechanical model of homogenization (see for details next subsection).

14.9.3 GIE for the First Strain Gradient CMs and Effective Properties

Let us consider an arbitrary random realization α of inclusions in the domain w described by Eq. (14.209$_1$) generalized to any number of inhomogeneities. Then the centering method (see for details Chap. 9) subtracting from both sides of the mentioned equation their statistical averages leads to the *general integral equation* (GIE) ($\mathbf{x} \in v_i$)

$$\hat{\boldsymbol{\varepsilon}}(\mathbf{x}, \alpha) = \langle \hat{\boldsymbol{\varepsilon}} \rangle(\mathbf{x}) + \int [\boldsymbol{\mathcal{L}}_k^{\epsilon}(\mathbf{x} - \mathbf{x}_k, \hat{\boldsymbol{\tau}}) V_k^{\delta}(\mathbf{x}_k, \alpha) - \langle \boldsymbol{\mathcal{L}}_k^{\epsilon}(\mathbf{x} - \mathbf{x}_k, \hat{\boldsymbol{\tau}}) \rangle(\mathbf{x}_k)] d\mathbf{x}_k \tag{14.224}$$

expressed through the volume perturbators $\boldsymbol{\mathcal{L}}_k^{\epsilon}(\mathbf{x} - \mathbf{x}_k, \hat{\boldsymbol{\tau}})$ (14.209$_1$). Equation (14.224) is only obtained at the internal points $\mathbf{x} \in w$ of the macrodomain w at sufficient distance from the boundary $a \ll |\mathbf{x} - \mathbf{s}|$, $\forall \mathbf{s} \in \Gamma$. Owing to absolute convergence of the integral (14.224), the domain w can be replaced by the integrals over the whole space R^d; the domain integration R^d, as well as the label α, will be omitted hereafter for simplicity of notations. Equation (14.224) is obtained for any inhomogeneous remote loading (14.189) generating a general case of inhomogeneous $\langle \hat{\boldsymbol{\varepsilon}} \rangle(\mathbf{x})$ and any statistically homogeneous (5.62$_2$) or inhomogeneous (5.62$_1$) structures described by the conditional probability density functions $\varphi(v_i, \mathbf{x}_i | v_1, \mathbf{x}_1, \ldots, v_n, \mathbf{x}_n)$.

Equation (14.224) with an accuracy to notations coincides with Eq. (11.20) and can be solved in the framework of different sets of accepted hypotheses either **H1a**, **H1b**, **H2a**, **H2b**, or **H3** (see Chap. 8 for details). In particular, if we accept the hypothesis **H2b** whereas the hypotheses **H1a**, **H1b**, and **H3** are not used, then Eq. (14.224) at $n = 1$ is reduced to the following equation for the effective field

$$\langle \overset{\wedge}{\boldsymbol{\varepsilon}}{}^{[n+1]} \rangle(\mathbf{x}) = \langle \hat{\boldsymbol{\varepsilon}} \rangle(\mathbf{x}) + \int \boldsymbol{\mathcal{L}}_k^{\epsilon}(\mathbf{x} - \mathbf{x}_k, \overset{\wedge}{\boldsymbol{\varepsilon}}{}^{[n]}) [\varphi(v_q, \mathbf{x}_q |; v_i, \mathbf{x}_i) - n^{(q)}(\mathbf{x}_q)] d\mathbf{x}_q \tag{14.225}$$

formally similar to Eq. (9.92) with the difference of the additional opportunities $\langle \hat{\boldsymbol{\varepsilon}} \rangle(\mathbf{x}) \not\equiv$const. and $n^{(q)}(\mathbf{x}_q) \not\equiv$const. in Eq. (14.225).

Equation (14.225) can be solved by iteration method analogously to Eq. (9.92). For both the statistically homogeneous CMs (5.62_2) and homogeneous boundary conditions (14.191_1), an initial approximation of the effective field (14.225) can be chosen by different ways, for example, as the solution (14.225) obtained in the framework of the hypothesis **H1a** (compare with Eq. (9.65))

$$\langle \hat{\overline{\boldsymbol{\varepsilon}}}^{[0]} \rangle(\mathbf{x}) = \hat{\mathbf{Y}} \langle \hat{\boldsymbol{\varepsilon}} \rangle, \quad (\hat{\mathbf{Y}})^{-1} = \mathbf{I} - \int \langle \mathbf{L}_k^{\epsilon}(\mathbf{x} - \mathbf{x}_k) \rangle_{(i)} [\varphi(v_q, \mathbf{x}_q | ; v_i, \mathbf{x}_i) - n^{(q)}(\mathbf{x}_q)] d\mathbf{x}_q, \tag{14.226}$$

where the operator $\mathcal{L}_k^{\epsilon}(\mathbf{x} - \mathbf{x}_k, \hat{\overline{\boldsymbol{\varepsilon}}}) = \mathbf{L}_k^{\epsilon}(\mathbf{x} - \mathbf{x}_k) \hat{\overline{\boldsymbol{\varepsilon}}}$ is factored in the framework of the hypothesis **H1a**. If the hypotheses **H1b** and **H3** are additionally accepted, then the initial approximation is simplified

$$\langle \hat{\overline{\boldsymbol{\varepsilon}}}^{[0]} \rangle(\mathbf{x}) = \mathbf{Y} \langle \hat{\boldsymbol{\varepsilon}} \rangle, \quad (\mathbf{Y})^{-1} = \mathbf{I} - n^{(1)} \mathbf{R}_i^{\epsilon} \mathbf{P}_i^0, \tag{14.227}$$

where the operator $\mathcal{R}_i^{\epsilon}(\mathbf{x} - \mathbf{x}_i, \hat{\overline{\boldsymbol{\varepsilon}}}) = \mathbf{R}_i^{\epsilon}(\mathbf{x} - \mathbf{x}_i) \hat{\overline{\boldsymbol{\varepsilon}}}$ is factored owing to the hypothesis **H1a** and the tensor $\mathbf{P}_i^0 = \mathbf{S}_i^0 \mathbf{M}^{(0)}$ for the excluded volume v_i^0 is expressed through the averaged Eshelby tensor $\mathbf{S}_i^0 = \langle \mathbf{S}_i^0(\mathbf{x}) \rangle_{(i)}$, where the inhomogeneous Eshelby tensor $\mathbf{S}_i^0(\mathbf{x})$ ($\mathbf{x} \in v_i$) was obtained in [1682] (see also [527, 1357, 1372]).

It suggests the Neumann series form the solution $\langle \overline{\boldsymbol{\vartheta}} \rangle_i(\mathbf{x})$ of Eq. (14.225)

$$\langle \hat{\overline{\boldsymbol{\varepsilon}}} \rangle(\mathbf{x}) = \lim_{n \to \infty} \langle \hat{\overline{\boldsymbol{\varepsilon}}}^{[n]} \rangle(\mathbf{x}) \equiv \mathbf{D}^{\epsilon}(\mathbf{x}) \langle \hat{\boldsymbol{\varepsilon}} \rangle, \tag{14.228}$$

otherwise, we get

$$\mathbf{D}^{\epsilon} = \mathbf{D}^{\epsilon}(\mathbf{x}) = \hat{\mathbf{Y}}, \quad \mathbf{D}^{\epsilon} = \mathbf{D}^{\epsilon}(\mathbf{x}) = \mathbf{Y}, \quad \mathbf{D}^{\epsilon} = \left[\mathbf{I} + c^{(1)} \langle \mathbf{L}_i^{\epsilon}(\mathbf{x} - \mathbf{x}_i \rangle_{(i)}\right]^{-1} \tag{14.229}$$

for Eqs. (14.226), (14.227) , and Mori–Tanaka approach (MTA) [89], respectively.

Thus, for both the statistically homogeneous CMs (5.62_2) and homogeneous boundary conditions (14.191_1), the effective moduli (14.223) for the first strain gradient CMs are defined by

$$\mathbf{L}^* = \mathbf{L}^{(0)} + n^{(1)} \langle \mathbf{R}_i^{\epsilon}(\mathbf{x} - \mathbf{x}_i, \mathbf{D}^{\epsilon}) \rangle_{(i)}, \tag{14.230}$$

reducing to the algebraic representations for the particular cases of Eqs. (14.229_1), (14.229_2), and (14.229_3)

$$\mathbf{L}^* = \mathbf{L}^{(0)} + n^{(1)} \mathbf{R}_i^{\epsilon} \mathbf{D}^{\epsilon} \tag{14.231}$$

with the constant tensors $\mathbf{R}_i^{\epsilon}$, $\mathbf{D}^{\epsilon}$. It should be mentioned that the effective moduli $\mathbf{L}^*$ (14.231) estimated by the new EFM (14.229_1), classical EFM (14.229_2), and MTA (14.229_3) only coincide for both the homogeneous ellipsoidal aligned inhomogeneities v_i and vanishing graded effects $\mathbf{C}(\mathbf{x}) \equiv \mathbf{0}$. If any of the mentioned conditions is violated, all three estimations of $\mathbf{D}^{\epsilon}$ (and, therefore, $\mathbf{L}^*$) (14.229$_1$), (14.229$_2$) and (14.229$_3$) differ from each other.

It should be also noted that in micromechanics of CM with gradient properties of constituents, simplified versions of the MTA (corresponding to the locally elastic CMs with aligned ellipsoidal homogeneous inhomogeneities) are usually used. So, the authors [991, 992, 1372, 1419, 1682] assumed homogeneity of the polarization tensor $\hat{\boldsymbol{\tau}}(\mathbf{x})$ (14.207) inside inhomogeneity $\mathbf{x} \in v_i$ (see for details the locally elastic case in Sect. 8.3.2) with an approximation $\mathbf{S}_i = \langle \mathbf{S}_i(\mathbf{x}) \rangle_{(i)}$. From other side, Tran [1530] (see also [1531]) used the formula (8.73) for locally elastic CM, where the tensor $\mathbf{B}_i = \langle \mathbf{B}_i(\mathbf{x}) \rangle_{(i)}$ was accurately estimated, whereas for $\mathbf{R}_i$ one used the equality $\mathbf{R}_i = (\mathbf{M}^{(i)} - \mathbf{M}^{(0)})\mathbf{B}_i$ which is an approximation in a general case of inhomogeneity v_i. In so doing, in the proposed generalized MTA (14.231) and (14.229$_3$), one only exploits the effective field hypothesis (8.68) without any other approximation. However, the tensor $\mathbf{S}_i^0 = \langle \mathbf{S}_i^0(\mathbf{x}) \rangle_{(i)}$ was used in another EFM (14.227) for the excluded volume v_i^0 (rather than for inclusion v_i) where the mention equality is appropriate (see for details Sect. 8.3.1). Moreover, the mentioned assumption of homogeneity of the polarization tensor $\hat{\boldsymbol{\tau}}(\mathbf{x})$ (14.207) inside inhomogeneity $\mathbf{x} \in v_i$ was not exploited in the proposed Eqs. (14.229)–(14.231).

14.9.4 Strain Gradient Plasticity Model of CMs and Polycrystals

In strain gradient plasticity model, we assume an additive decomposition of the total strain $\boldsymbol{\varepsilon}$ and strain gradient $\boldsymbol{\kappa}$ tensors on the elastic and plastic contributions tensor

$$\boldsymbol{\varepsilon} = \boldsymbol{\varepsilon}^e + \boldsymbol{\varepsilon}^p, \quad \boldsymbol{\kappa} = \boldsymbol{\kappa}^e + \boldsymbol{\kappa}^p. \tag{14.232}$$

The von Mises plasticity model (2.215), Mises effective stress τ, and plastic strain $\varepsilon^p_{\rm eq}$ (2.216) should be generalized as

$$\widetilde{f} := \Gamma_e - \Gamma_{e0} - F(\mathcal{E}_p) = 0, \tag{14.233}$$

$$\Gamma_e := \Big(\frac{3}{2} s_{ij} s_{ij} + \sum_{q=1}^{3} [l_q^{-2} \mathbb{S}_{ijk}^{(q)} \mathbb{S}_{ijk}^{(q)}]\Big)^{1/2}, \tag{14.234}$$

$$\mathcal{E}_p := \Big(\frac{3}{2} \varepsilon_{ij}^p \varepsilon_{ij}^p + \sum_{p=1}^{3} [l_p^2 \kappa_{ijk}^{p(q)} \kappa_{ijk}^{p(q)}]\Big)^{1/2}. \tag{14.235}$$

Here $\mathbb{S}_{ijk}^{(q)}$ and $\kappa_{ijk}^{p(q)}$ $(q = 1, 2, 3)$ are orthogonal decompositions of the deviatoric part $\mathbb{S}$ of the double stress tensor $\boldsymbol{\mu}$ and $\boldsymbol{\kappa}^p$, respectively (see for details [485, 1572]).

For the local elasticity of CMs, estimation of the effective field surface (see Eqs. (12.51), (12.52), and (12.58) is based on the evaluation of the second moment of stresses $\langle\boldsymbol{\sigma}\otimes\boldsymbol{\sigma}\rangle^{(r)}$ (6.185) in each phase $r=0,\dots,N$. For the strain gradient plasticity, the estimations $\langle\boldsymbol{\sigma}\otimes\boldsymbol{\sigma}\rangle^{(r)}$ should be complimented by corresponding second moment of the deviatoric part $\mathbb{S}$ of the double stress tensor

$$\langle\mathbb{S}\otimes\mathbb{S}\rangle^{(r)}=\frac{1}{1-c^{(r)}}\frac{\partial\mathbf{M}^{s*}}{\partial\mathbf{C}^{(r)}}\langle\boldsymbol{\sigma}\rangle\otimes\langle\boldsymbol{\sigma}\rangle, \tag{14.236}$$

which can be obtained in a similar manner as Eq. (6.185). After that, the scheme (12.53)–(12.58) of the deformation theory of plasticity can be formally recast for the strain gradient model.

For flow theory of plasticity, following the associated flow rule, the plastic strain rate tensor in the phase $v^{(0)}$ is normal to the yield surface, and a linear relation generalized of a flow theory is

$$\dot{\mathcal{E}}_i^{p(r)}=\frac{1}{h(\Gamma_e^{(r)})}\frac{\partial\widetilde{f}^{(r)}}{\partial\Lambda_i^{(r)}}\dot{\Gamma}_e^{(r)}, \tag{14.237}$$

where the vectors $\mathcal{E}_i^{p(r)}$ and Λ_i $(i=1,\dots,23)$ of the 23 components including the components of $\boldsymbol{\varepsilon}^{p(r)}$, $\boldsymbol{\kappa}^{p(r)}$ and $\mathbf{s}$, $\mathbb{S}$: $\mathcal{E}^{p(r)}=(\boldsymbol{\varepsilon}^{p(r)},\boldsymbol{\kappa}^{p(r)})^\top$, $\boldsymbol{\Lambda}^{(r)}=(\mathbf{s}^{(r)},\mathbb{S}^{(r)})^\top$, respectively; and the hardening rate h are chosen such that the uniaxial tensile response is reproduced (see for details [485, 1572]).

It should be mentioned that for the polycrystal plasticity model, Smyshlyaev and Fleck [1421] formally presented a stress potential for the linear crystal as a quadratic form of $\boldsymbol{\sigma}$ and $\boldsymbol{\mu}$ (14.183) similar to the strain energy (14.181). Introduction of the field pairs and the Green functions as in [1419] and Eqs. (14.205) provides opportunity to use a generalized Hashin–Shtrikman principle where, as for locally elastic case (see Sect. 6.4), the generalized strain polarization tensor is a constant inside the phases. This is a popular assumption in the strain gradient elasticity of CMs (see for references Sect. 14.10.3) which was effectively avoided in the proposed Eqs. (14.229)–(14.231). The same scheme of Sects. 14.10.1–14.10.3 can be recast for the polycrystal plasticity model [1421].

Thus, the main improvement of strain gradient plasticity models for both the CMs [485] and polycrystals [1421] is defined by improvement of micromechanical modeling of CMs with strain gradient properties of constituents. Namely, the authors [485, 992, 1372, 1421, 1682] assumed homogeneity of eigenstrain inside inhomogeneity v_i with an approximation $\mathbf{S}_i=\langle\mathbf{S}_i(\mathbf{x})\rangle_{(i)}$, whereas in the approach proposed in Sect. 14.10.3, one only exploits the effective field hypothesis **H1** without any other approximation.

14.10 Stress-Gradient Theories

14.10.1 Stress-Gradient Elasticity Model

In this subsection we shortly reproduce the scheme of the stress-gradient elasticity (proposed in [497] and developed in [158, 1315, 1531]) adapted for application of the EFM proposed before for locally elastic CMs (see Chap. 9). In the strain gradient model (14.181), the strain-energy density W depends on the strain $\boldsymbol{\varepsilon}$ and its first gradient $\nabla\boldsymbol{\varepsilon}$. Likewise, in the stress-gradient model of Forest and Sab [497] (see for references and details), the complementary strain energy density W^c depends on the stress $\boldsymbol{\sigma}$ and its first gradient $\nabla\boldsymbol{\sigma}$. More precisely, the stress-gradient $\nabla\boldsymbol{\sigma}$ is decomposed as the orthogonal sum of a trace-free part $\mathbf{R}$ and its complement that is fully defined by the body forces $\mathbf{f}$

$$\nabla\sigma = \mathbf{R} - \frac{2}{d+1}\mathbf{I}\cdot\mathbf{f}, \quad \mathbf{R}:\boldsymbol{\delta} = \mathbf{0}, \quad \mathbf{R}\therefore\mathbf{I}\cdot\mathbf{f} = 0. \tag{14.238}$$

The trace-free part $\mathbf{R}$ of the stress-gradient $\nabla\boldsymbol{\sigma}$ is the orthogonal projection of $\nabla\boldsymbol{\sigma}$ onto the space of third-rank, trace-free tensors, and we write $\mathbf{R} = \mathbf{I}'_6 \therefore \nabla\boldsymbol{\sigma}$, where the sixth-rank tensor $I'_{6|ijkpqr} = I_{ijpq}\delta_{kr} - (I_{pq(ir}\delta_{j)k})/2$ [1531] is defined as the orthogonal projection (in the sense of the $\therefore$ scalar product). Being a projector, $\mathbf{I}'_p$ enjoys the classical property $\mathbf{I}'_p \therefore \mathbf{I}'_p = \mathbf{I}'_p$.

The complementary strain energy density W^c of an linear homogeneous centrosymmetric material is a quadratic form of both $\boldsymbol{\sigma}$ and $\mathbf{R}$

$$W^c(\boldsymbol{\sigma},\mathbf{R}) = \frac{1}{2}\boldsymbol{\sigma}:\mathbf{M}:\boldsymbol{\sigma} + \frac{1}{2}\mathbf{R}\therefore\mathbf{S}\therefore\mathbf{R}, \tag{14.239}$$

where $\mathbf{M}$ and $\mathbf{S}$ are the classical and generalized compliances which are tensors with minor and major symmetries. Besides, owing to the fact that $\mathbf{S}$ operates on the space of trace-free, third-rank tensors, we must have $\mathbf{I}'_6 \therefore \mathbf{S} \therefore \mathbf{I}'_6 = \mathbf{S}$. In particular, a simplified model for isotropic, linearly elastic stress-gradient materials, the complementary strain energy density was proposed by Brisard [158] (see also [1230]) in much the same manner as (14.194)

$$W^c(\boldsymbol{\sigma},\mathbf{R}) = \frac{1}{4\mu}\Big[\boldsymbol{\sigma}:\boldsymbol{\sigma} - \frac{\nu}{1+\nu}(\boldsymbol{\delta}:\boldsymbol{\sigma})^2\Big] + \frac{l^2}{4\mu}\Big[\mathbf{R}\therefore\mathbf{R} - \frac{\nu}{1+\nu}(\boldsymbol{\delta}:\mathbf{R})\cdot(\boldsymbol{\delta}:\mathbf{R})\Big], \tag{14.240}$$

where l is a stress-gradient material length.

The local equations are found from the minimization of the complementary strain energy W^c under the constraint that the local stress $\boldsymbol{\sigma}$ be in equilibrium equation with the body forces $\mathbf{f}$

$$\mathbf{div}\sigma = -\mathbf{f}.$$

The strain measures, the energy-conjugate to the stress $\boldsymbol{\sigma}$ and trace-free variable $\mathbf{R}$, are

$$\mathbf{e} = \mathbf{M} : \boldsymbol{\sigma} = \frac{\partial W^c}{\partial \boldsymbol{\sigma}}, \quad \boldsymbol{\phi} = \mathbf{S} \therefore \mathbf{R} = \frac{\partial W^c}{\partial \mathbf{R}}, \tag{14.241}$$

where the total strain $\mathbf{e}$ is not necessarily the symmetric gradient of the displacement $\mathbf{u}$. A deviatoric third-order tensor $\boldsymbol{\phi}$ is called microdisplacements in [497]. It was found that the field $\mathbf{e}$ can be written in terms of $\boldsymbol{\phi}$; the flux $\boldsymbol{\psi} := \boldsymbol{\phi} + \mathbf{I} \cdot \mathbf{u}$ and of a displacement field $\mathbf{u}$ by a compatibility condition

$$\mathbf{e} = \boldsymbol{\psi} \cdot \nabla = \mathbf{u} \overset{S}{\otimes} \nabla + \boldsymbol{\phi} \cdot \nabla, \quad \mathbf{u} = \boldsymbol{\psi} : \boldsymbol{\delta}/2, \quad \boldsymbol{\phi} = \mathbf{I}'_6 \therefore \boldsymbol{\psi}. \tag{14.242}$$

Equations (14.239) can be considered as an inversion of Eq. (14.183)

$$\boldsymbol{\sigma} = \mathbf{L} : \mathbf{e}, \quad \mathbf{R} = \mathbf{C} \therefore \boldsymbol{\phi}, \tag{14.243}$$

where $\mathbf{L} = (\mathbf{M})^{-1}$ is a classical stiffness, whereas the generalized compliance $\mathbf{S}$ and stiffness $\mathbf{C}$ are inverse to each other $\mathbf{C} \therefore \mathbf{S} = \mathbf{S} \therefore \mathbf{C} = \mathbf{I}'_6$.

Besides, across any discontinuity surface, the flux $\boldsymbol{\psi}$ and the full stress tensor $\boldsymbol{\sigma}$ are continuous

$$[[\boldsymbol{\phi} \cdot \mathbf{n} + \mathbf{sym}(\mathbf{u} \otimes \mathbf{u})]] = \mathbf{0}, \quad [[\boldsymbol{\sigma}]] = \mathbf{0}, \tag{14.244}$$

where $\mathbf{n}$ denotes the normal to the interface, while $[[(\cdot)]]$ denotes the jump of $(\cdot)$ across the interface.

Different boundary conditions are considered in [1531]. In particular, the following either the full stress or mixed boundary conditions generalize the classical homogeneous traction boundary conditions on the boundary $\mathbf{x} \in \Gamma$

$$\begin{aligned} &\boldsymbol{\sigma}(\mathbf{x}) = \boldsymbol{\sigma}^{\Gamma}, \\ &\boldsymbol{\sigma}(\mathbf{x}) \cdot \mathbf{n}(\mathbf{x}) = \boldsymbol{\sigma}^{\Gamma} \cdot \mathbf{n}(\mathbf{x}), \quad \boldsymbol{\nu}(\mathbf{x}) \cdot (\boldsymbol{\psi}(\mathbf{x}) \cdot \mathbf{n}(\mathbf{x})) \cdot \boldsymbol{\nu}(\mathbf{x}) = \mathbf{0}, \end{aligned} \tag{14.245}$$

respectively, for the divergence-free $\boldsymbol{\sigma}$ and symmetric constant prescribed stress $\boldsymbol{\sigma}^{\Gamma}(\mathbf{x})$ =const.; $\boldsymbol{\nu} := \boldsymbol{\delta} - \mathbf{n} \otimes \mathbf{n}$ is the projection onto the target plane to the boundary. In the homogeneous displacement-type boundary conditions, it is assumed that $(\mathbf{x} \in \Gamma)$

$$\mathbf{u}(\mathbf{x}) = \boldsymbol{\varepsilon}^{\Gamma} \cdot \mathbf{x}, \quad \boldsymbol{\phi}(\mathbf{x}) \cdot \mathbf{n}(\mathbf{x}) = \mathbf{0}, \tag{14.246}$$

where $\boldsymbol{\varepsilon}^{\Gamma}$ is a constant tensor.

Analogously to (14.205), we introduce the stress and stress-gradient polarization tensors

$$\boldsymbol{\tau}(\mathbf{x}) = \boldsymbol{\sigma}(\mathbf{x}) - \mathbf{L}^{(0)} : \mathbf{e}(\mathbf{x}), \quad \boldsymbol{\pi}(\mathbf{x}) = \mathbf{R}(\mathbf{x}) - \mathbf{C}^{(0)} \therefore \boldsymbol{\phi}(\mathbf{x}), \tag{14.247}$$

vanishing in the matrix $\mathbf{x} \in w \setminus v_i$. Similarly to Eq. (14.208) (see also [1531]), we get an equation for one inhomogeneity in the infinite matrix

$$\hat{\mathbf{e}} = \hat{\overline{\mathbf{e}}} + {}^{\mathbf{e}}\hat{\mathbf{U}} * \hat{\boldsymbol{\tau}}, \tag{14.248}$$

where

$$\hat{\boldsymbol{\tau}} = \begin{pmatrix} \boldsymbol{\tau} \\ \boldsymbol{\pi} \end{pmatrix}, \quad \hat{\mathbf{e}} = \begin{pmatrix} \mathbf{e} \\ \boldsymbol{\phi} \end{pmatrix}, \quad \hat{\overline{\mathbf{e}}} = \begin{pmatrix} \overline{\mathbf{e}} \\ \overline{\boldsymbol{\phi}} \end{pmatrix}, \quad {}^{\mathbf{e}}\hat{\mathbf{U}} = \begin{pmatrix} {}^{\mathbf{e}}\mathbf{U}^{e\tau} & {}^{\mathbf{e}}\mathbf{U}^{e\pi} \\ {}^{\mathbf{e}}\mathbf{U}^{\phi\tau} & {}^{\mathbf{e}}\mathbf{U}^{\phi\pi} \end{pmatrix}, \tag{14.249}$$

where $\hat{\overline{\mathbf{e}}} = (\overline{\mathbf{e}}, \overline{\boldsymbol{\pi}})$ is a field that would exist in the medium with homogeneous properties $\mathbf{M}^{(0)}$ and $\mathbf{S}^{(0)}$ and appropriate remote boundary condition; all Green operators ${}^{\mathbf{e}}\mathbf{U}^{e\pi}$, ${}^{\mathbf{e}}\mathbf{U}^{\phi\tau}$, and ${}^{\mathbf{e}}\mathbf{U}^{\phi\pi}$ can be deduced from the sole operator ${}^{\mathbf{e}}\mathbf{U}^{e\tau}$, obtained by Tran [1530] for some isotropic simplified model.

The integral Eq. (14.248) with nonsingular convolution kernel $\hat{\mathbf{U}}$ and their solution can be presented in an operator form (9.6)

$$\hat{\mathbf{e}}(\mathbf{x}) - \hat{\overline{\mathbf{e}}}(\mathbf{x}) := {}^{\mathbf{e}}\mathcal{L}_i^{\tau}(\mathbf{x} - \mathbf{x}_i, \hat{\boldsymbol{\tau}}) = {}^{\mathbf{e}}\mathcal{L}_i^{e}(\mathbf{x} - \mathbf{x}_i, \hat{\overline{\mathbf{e}}}),$$
$$\bar{v}_i \hat{\boldsymbol{\tau}}(\mathbf{x}) = {}^{\mathbf{e}}\mathcal{R}_i^{e}(\mathbf{x} - \mathbf{x}_i, \hat{\overline{\mathbf{e}}}), \tag{14.250}$$

where the solutions ${}^{\mathbf{e}}\mathcal{L}_i^{e}(\mathbf{x} - \mathbf{x}_i, \hat{\overline{\mathbf{e}}})$ and ${}^{\mathbf{e}}\mathcal{R}_i^{e}(\mathbf{x} - \mathbf{x}_i, \hat{\overline{\mathbf{e}}})$ can be found, e.g., either by the volume integral equation method (see Sect. 4.4) by finite element analysis (see, e.g., [39, 388, 1571] and Sects. 4.2. and 4.4)

14.10.2 *Homogenization of Stress-Gradient CMs*

In this subsection we consider (following [1530, 1531]) the basic principles of homogenization of micromechanics of stress-gradient random structure composites. We explore the conditions on a divergence-free stress tensor $\boldsymbol{\sigma}_1$, a displacement vector $\mathbf{u}_2$, a microdisplacement tensor ϕ_2, and total strain $\mathbf{e}_2$ defined by Eq. (14.242_1) under which the micro- and macroscopic internal works coincide (Hill's lemma) for no body forces, that is

$$\langle \boldsymbol{\sigma}_1 : \mathbf{e}_2 + (\boldsymbol{\sigma}_1 \otimes \nabla) \therefore \phi_2 \rangle = \langle \boldsymbol{\sigma}_1 \rangle : \langle \mathbf{e}_2 \rangle, \tag{14.251}$$

where the macroscopic $\langle \boldsymbol{\sigma}_1 \otimes \nabla \rangle \therefore \langle \phi_2 \rangle$ has been discarded, because the homogenized stress-gradient material is expected to behave as a Cauchy material (no macroscopic gradient effect). It was proved that the left-hand side can be presented in terms of fields at the boundary $\mathbf{s} \in \Gamma = \partial w$ of macrodomain

$$\langle \boldsymbol{\sigma}_1 : \mathbf{e}_2 + (\boldsymbol{\sigma}_1 \otimes \nabla) \therefore \boldsymbol{\phi}_2 \rangle = \frac{1}{\bar{w}} \int_\Gamma \boldsymbol{\sigma}_1 : \big(\mathbf{u}_2 \overset{S}{\otimes} \mathbf{n} + \boldsymbol{\phi}_2 \cdot \mathbf{n}\big) ds, \quad (14.252)$$

where the surface integral representation was obtained with no assumptions on the values of $\boldsymbol{\sigma}_1, \mathbf{u}_2$, and $\boldsymbol{\phi}_2$ at the boundary Γ with the outer normal $\mathbf{n}$. As a result, plugging $\boldsymbol{\sigma}_1 \equiv$const. into Eq. (14.252), we find the following asymptotic expression of the macroscopic strain (see for details [1531])

$$\langle \mathbf{e}_2 \rangle = \big\langle \nabla \overset{S}{\otimes} \mathbf{u} \big\rangle \quad (14.253)$$

for macroscopically large domain w with statistically homogeneous inclusion field. Therefore, Hill–Mandel lemma holds, provided that

$$\frac{1}{\bar{w}} \int_\Gamma \boldsymbol{\sigma}_1 : \big(\mathbf{u}_2 \overset{S}{\otimes} \mathbf{n} + \boldsymbol{\phi}_2 \cdot \mathbf{n}\big) ds = \langle \boldsymbol{\sigma}_1 \rangle : \langle \mathbf{e}_2 \rangle, \quad (14.254)$$

It is first assumed the homogeneous displacement-type boundary condition (14.246) corresponding to $\mathbf{u}_2 = \boldsymbol{\varepsilon}^\Gamma \cdot \mathbf{x}$ and $\boldsymbol{\phi}_2(\mathbf{x}) \cdot \mathbf{n}(\mathbf{x}) = \mathbf{0}$ $(\mathbf{x} \in \Gamma)$. Substitution of $\boldsymbol{\sigma}_1, \mathbf{u}_2, \boldsymbol{\phi}_2$, and $\boldsymbol{\sigma}_1, \boldsymbol{\varepsilon}^\Gamma \cdot \mathbf{x}, \mathbf{0}$ into the left-hand side and right-hand side of Eq. (14.252), respectively, leads to

$$\langle \boldsymbol{\sigma}_1 : \mathbf{e}_2 + \big(\boldsymbol{\sigma}_1 \otimes \nabla\big) \therefore \boldsymbol{\phi}_2 \rangle = \langle \boldsymbol{\sigma}_1 \rangle : \boldsymbol{\varepsilon}^\Gamma, \quad (14.255)$$

which further implies (taking $\boldsymbol{\sigma}_1 =$ const. $= \langle \boldsymbol{\sigma}_1 \rangle$) that $\boldsymbol{\varepsilon}^\Gamma = \langle \mathbf{e}_2 \rangle$, and Hill–Mandel lemma (14.251) holds. Thus, we can consider $\boldsymbol{\varepsilon}^\Gamma$ as a loading parameter, whereas the macroscopic stress is then computed *a posteriori* as the volume average $\langle \boldsymbol{\sigma} \rangle$ of the local stress $\boldsymbol{\sigma}$. It allows to determine the apparent stiffness $\mathbf{L}^*$ as the linear operator which maps $\langle \boldsymbol{\varepsilon} \rangle = \boldsymbol{\varepsilon}^\Gamma$ to $\langle \boldsymbol{\sigma} \rangle$

$$\langle \boldsymbol{\sigma} \rangle = \mathbf{L}^* : \langle \mathbf{e} \rangle. \quad (14.256)$$

In a similar manner, we can prove that Hill–Mandel lemma (14.251) holds (see [1530]) at the homogeneous traction-type boundary conditions (14.245) that allows us to determine the apparent compliance

$$\langle \boldsymbol{\varepsilon} \rangle = \mathbf{M}^* : \langle \boldsymbol{\sigma} \rangle. \quad (14.257)$$

14.10.3 GIE of Micromechanics of Stress-Gradient CMs and Effective Moduli

For one inclusion inside infinite matrix, Eq. (14.250) with an accuracy to notations coincides with Eq. (14.209). Then, in a similar manner with Eq. (14.224), we can obtain the GIE with subsequent its solution in the framework of the hypothesis **H2b** (without the hypotheses **H1a**, **H1b** and **H3**)

$$\langle \hat{\mathbf{e}}^{[n+1]} \rangle(\mathbf{x}) = \langle \hat{\mathbf{e}} \rangle(\mathbf{x}) + \int {}^{\mathbf{e}}\boldsymbol{\mathcal{L}}_k^e(\mathbf{x} - \mathbf{x}_k, \hat{\mathbf{e}}^{[n]})[\varphi(v_q, \mathbf{x}_q | ; v_i, \mathbf{x}_i) - n^{(q)}(\mathbf{x}_q)] d\mathbf{x}_q \qquad (14.258)$$

which is hold for any possible inhomogeneity either $\langle \hat{\boldsymbol{\varepsilon}} \rangle(\mathbf{x}) \neq$const. or $n^{(q)}(\mathbf{x}_q) \neq$const.

For both the statistically homogeneous CMs (5.62$_2$) and homogeneous boundary conditions (14.191$_1$), an initial approximation of the effective field (14.225) can be chosen by different ways, for example, as the solution (14.225) obtained in the framework of the hypothesis **H1a** (compare with Eq. (9.65))

$$\langle \hat{\mathbf{e}}^{[0]} \rangle(\mathbf{x}) = \tilde{\mathbf{Y}} \langle \hat{\mathbf{e}} \rangle, \quad (\tilde{\mathbf{Y}})^{-1} = \mathbf{I} - \int \langle {}^{\mathbf{e}}\mathbf{L}_k^{\epsilon}(\mathbf{x} - \mathbf{x}_k) \rangle_{(i)} [\varphi(v_q, \mathbf{x}_q | ; v_i, \mathbf{x}_i) - n^{(q)}(\mathbf{x}_q)] d\mathbf{x}_q, \qquad (14.259)$$

where the operator ${}^{\mathbf{e}}\boldsymbol{\mathcal{L}}_k^e(\mathbf{x} - \mathbf{x}_k, \hat{\overline{\boldsymbol{\varepsilon}}}) = {}^{\mathbf{e}}\mathbf{L}_k^e(\mathbf{x} - \mathbf{x}_k)\hat{\overline{\boldsymbol{\varepsilon}}}$ is factored in the framework of the hypothesis **H1a**. If the hypotheses **H1b** and **H3** are additionally accepted, then the initial approximation is simplified

$$\langle \hat{\mathbf{e}}^{[0]} \rangle(\mathbf{x}) = \mathbf{Y} \langle \hat{\mathbf{e}} \rangle, \quad (\mathbf{Y})^{-1} = \mathbf{I} - n^{(1)} \mathbf{R}_i^e \mathbf{P}_i^0, \qquad (14.260)$$

where the operator ${}^{\mathbf{e}}\boldsymbol{\mathcal{R}}_i^e(\mathbf{x} - \mathbf{x}_i, \hat{\overline{\mathbf{e}}}) = {}^{\mathbf{e}}\mathbf{R}_i^e(\mathbf{x} - \mathbf{x}_i)\hat{\overline{\mathbf{e}}}$ is factored owing to the hypothesis **H1a** and the tensor $\mathbf{P}_i^0 = \mathbf{S}_i^0 \mathbf{M}^{(0)}$ for the excluded volume v_i^0 is expressed through the Eshelby $\mathbf{S}_i^0 = \langle \mathbf{S}_i^0(\mathbf{x}) \rangle_{(i)}$ found in [1530].

It suggests the Neumann series form the solution $\langle \overline{\mathbf{e}} \rangle_i(\mathbf{x})$ of Eq. (14.258)

$$\langle \hat{\overline{\mathbf{e}}} \rangle(\mathbf{x}) = \lim_{n \to \infty} \langle \hat{\mathbf{e}}^{[n]} \rangle(\mathbf{x}) \equiv {}^{\mathbf{e}}\mathbf{D}(\mathbf{x}) \langle \hat{\mathbf{e}} \rangle, \qquad (14.261)$$

otherwise, we get

$$^{\mathbf{e}}\mathbf{D} = {}^{\mathbf{e}}\mathbf{D}(\mathbf{x}) = \tilde{\mathbf{Y}}, \quad {}^{\mathbf{e}}\mathbf{D} = {}^{\mathbf{e}}\mathbf{D}(\mathbf{x}) = \mathbf{Y}, \qquad (14.262)$$

$$^{\mathbf{e}}\mathbf{D} = \left[\mathbf{I} + c^{(1)} \langle {}^{\mathbf{e}}\mathbf{L}_i^e(\mathbf{x} - \mathbf{x}_i) \rangle_{(i)}\right]^{-1} \qquad (14.263)$$

for Eqs. (14.259), (14.260) and Mori–Tanaka approach (MTA) [89], respectively. It should be mentioned that the comments in the last paragraph of Subsection 14.10.3 with respect to micromechanics of strain gradient CMs are also hold for the stress-gradient CMs currently considered.

Thus for both the statistically homogeneous CMs (5.62$_2$) and homogeneous boundary conditions (14.191$_1$), the effective moduli for the stress-gradient CMs are defined similarly to (14.230)

$$\mathbf{L}^* = \mathbf{L}^{(0)} + n^{(1)} \langle {}^{\mathbf{e}}\boldsymbol{\mathcal{R}}_i^e(\mathbf{x} - \mathbf{x}_i, {}^{\mathbf{e}}\mathbf{D}) \rangle_{(i)}, \qquad (14.264)$$

reducing to the algebraic representations for the particular cases of Eqs. (14.262_1), (14.262_2) and (14.230)

$$\mathbf{L}^* = \mathbf{L}^{(0)} + n^{(1)} {}^{\mathbf{e}}\mathbf{R}_i^e {}^{\mathbf{e}}\mathbf{D} \tag{14.265}$$

with the constant tensors ${}^{\mathbf{e}}\mathbf{R}_i^e$, ${}^{\mathbf{e}}\mathbf{D}$. It should be mentioned that the effective moduli $\mathbf{L}^*$ (14.264) estimated by the new EFM (14.262_1), classical EFM (14.262_2), and MTA (14.263) only coincide for both the homogeneous ellipsoidal aligned inhomogeneities v_i and vanishing graded effects $\mathbf{C}(\mathbf{x}) \equiv \mathbf{0}$. If any of the mentioned conditions is violated, all three estimations of $\mathbf{L}^*$ (14.262_1), (14.229_2), and (14.263) differ from each other.

14.11 Short Review of Higher-Order Micromechanical Modeling of CMs and Concluding Remarks

In summarizing, it should be noted a unified scheme used in this chapter for micromechanical analyses of CMs with the constituent properties described by different types of generalized continuum mechanics. Namely, homogenization scheme is based on the use of generalized Hill–Mandel lemma (14.28), (14.212) and (14.251) with both the direct and energetic definitions of effective moduli. The next step is obtaining decomposition of material property tensors with extraction of the properties of a comparison medium (corresponding to the properties of the matrix) and solution of the problem [see Eqs. (14.160), (14.209) and (14.248)] for one inclusion inside the infinite homogeneous matrix subjected to homogeneous remote fields. The third step is a construction of the GIE in terms of the operator form [see Eqs. (14.164), (14.224) and (14.258)] for the solutions obtained at the second step. The last fourth step is a solution of the obtained equations [see Eqs. (14.177), (14.228), and (14.261)] in the framework of different sets of hypotheses (described in Chap. 9) with subsequent substitution the effective moduli representations obtained at the first step. The most effective method of solution is the iteration method using as an initial approximation the generalized versions of the classical either EFM and MTA (in so doing the average Eshelby tensor $\langle \mathbf{S}_i(\mathbf{x}) \rangle_{(i)}$ is not enough). Of course, the mention scheme allows a straightforward generalization to either (or both) the binary interaction of inclusions (as in Chap. 10), inhomogeneous loading or statistically inhomogeneous structure (as in Chap. 13).

In summary we present a short review of higher-order models (adapted to the unified averaging scheme used in this chapter) generalizing classical elasticity by allowing material points the additional degrees of freedom, besides classical translation [e.g., rotation, (Cosserat media [350]), stretch or expand (e.g., microstretch media [445]), or micro-deformation (e.g., micromorphic media [447, 568, 1067]), see also for references [1530, 1534]]. In particular, the motion of a linear *Cosserat medium* [350] is described by a displacement field, $\mathbf{u}$, and a micro-rotation field, $\boldsymbol{\vartheta}$, attributed to each material point. The associated deformation $\mathbf{e}$ and torsion-curvature tensors $\boldsymbol{\kappa}$ and their dual quantities, the force stress $\boldsymbol{\sigma}$ and the couple-stress tensor $\boldsymbol{\mu}$, are

$$\mathbf{e} = \mathbf{u} \otimes \nabla + \epsilon \boldsymbol{\vartheta}, \quad \kappa = \boldsymbol{\vartheta} \otimes \nabla, \tag{14.266}$$

$$\boldsymbol{\sigma} = \partial W(\mathbf{e}, \boldsymbol{\kappa})/\partial \mathbf{e}, \quad \boldsymbol{\mu} = \partial W(\mathbf{e}, \boldsymbol{\kappa})/\partial \boldsymbol{\kappa}, \tag{14.267}$$

respectively, where ϵ denotes the permutation tensor. Here the tensors $\boldsymbol{\sigma}$ and $\boldsymbol{\mu}$ are not necessarily symmetric and the strain energy density function $W(\mathbf{e}, \boldsymbol{\kappa})$ for a linear, isotropic, centrosymmetric Cosserat medium reads (compare with (14.181))

$$W(\mathbf{e}, \boldsymbol{\kappa}) = \frac{1}{2}\mathbf{e} : \mathbf{L} : \mathbf{e} + \frac{1}{2}\boldsymbol{\kappa} : \mathbf{C} : \boldsymbol{\kappa}, \tag{14.268}$$

where the elasticity tensors $\mathbf{L}$ and $\mathbf{C}$ obey the major symmetry, i.e., $L_{ijkl} = L_{klij}$, $C_{ijkl} = C_{klij}$ but not minor symmetries due to the asymmetry of the strain measures. Besides two classical Lamé constants λ and μ [Pa], the Cosserat model (14.268) introduces four elastic constants c_1, c_2, c_3, and c_4 [N]. The elastic equilibrium of a Cosserat medium loaded at its boundary ($\mathbf{x} \in \Gamma$, $\mathbf{n}(\mathbf{x}) \perp \Gamma$) only

$$\boldsymbol{\sigma} \cdot \nabla = \mathbf{0}, \quad \mu \cdot \nabla - \epsilon \sigma = \mathbf{0}, \tag{14.269}$$

are added by the displacement-type or traction-type boundary conditions (compare with (14.191))

$$\mathbf{u} = \mathbf{u}^{\Gamma}, \quad \boldsymbol{\vartheta} = \boldsymbol{\vartheta}^{\Gamma} \quad \text{or} \tag{14.270}$$

$$\mathbf{t} := \boldsymbol{\sigma}(\mathbf{x}) \cdot \mathbf{n} = \mathbf{t}^{\Gamma}, \quad \mathbf{m} := \boldsymbol{\mu}(\mathbf{x}) \cdot \mathbf{n}(\mathbf{x}) = \mathbf{m}^{\Gamma}, \tag{14.271}$$

respectively. For the homogeneous boundary conditions $\boldsymbol{\vartheta}^{\Gamma}, \mathbf{m}^{\Gamma} = \mathbf{0}$, and $\mathbf{u}^{\Gamma} = \boldsymbol{\varepsilon}^{\Gamma} \cdot \mathbf{x}$, $\mathbf{t}^{\Gamma} = \boldsymbol{\sigma}^{\Gamma} \cdot \mathbf{n}(\mathbf{x})$, where the tensors $\boldsymbol{\varepsilon}^{\Gamma}$ and $\boldsymbol{\sigma}^{\Gamma}$ are given and constant.

If the micro-rotation is constrained to be equal to the macrorotation vector, $\boldsymbol{\vartheta} = \frac{1}{2}\,\mathrm{curl}\mathbf{u}$, the Cosserat model (14.266)–(14.268) is reduced to the so-called *couple-stress model* [1066] (see for references [1530, 1534]). It follows directly that the curvature tensor, $\boldsymbol{\kappa} = \nabla\boldsymbol{\vartheta}$, is deviatoric $\zeta_{ii} = 0$, and the couple-stress model corresponding to (14.268) has two additional scale parameters (instead of four as in Eq. (14.268)). In modified couple-stress model [1653], one postulated additional equilibrium equations for moments of couples that force the couple-stress tensor to be symmetric and reduce the number of scale parameters in $\mathbf{C}$ (14.268) to one.

The motion of a *micromorphic body* is described by a classical translation and an additional second-order micro-deformation tensor $\boldsymbol{\vartheta}$ [447, 1067] that was generalized by Germain [538] introduced additional third- and higher-order micro-deformation tensor. The micromorphic theory can be simplified to the micropolar/Cosserat theory and microstretch theory (see [441] and [1369]) by observing some changes over the kinematical variables (see [441] and [1369]). The associated deformation fields are the classical strain tensor $\boldsymbol{\varepsilon}$, the relative deformation $\mathbf{e}$, and the micro-deformation gradient tensor $\boldsymbol{\zeta}$, whereas their energy-conjugate quantities are the symmetric force stress tensor $\boldsymbol{\sigma}$, the generally nonsymmetric relative force stress tensor $\mathbf{s}$ and third-rank stress tensor $\boldsymbol{\mu}$:

$$\boldsymbol{\varepsilon} = \mathbf{u} \overset{S}{\otimes} \nabla, \quad \mathbf{e} = \mathbf{u} \otimes \nabla - \boldsymbol{\vartheta}, \quad \boldsymbol{\zeta} = \boldsymbol{\vartheta} \otimes \nabla, \tag{14.272}$$

$$\boldsymbol{\sigma} = \partial W / \partial \boldsymbol{\varepsilon}, \quad \mathbf{s} = \partial W / \partial \mathbf{e}, \quad \boldsymbol{\mu} = \partial W / \partial \boldsymbol{\zeta}. \tag{14.273}$$

The constitutive equations for linear elastic centrosymmetric micromorphic materials read (compare with (14.183))

$$\boldsymbol{\sigma} = \mathbf{L} : \boldsymbol{\varepsilon}, \quad \mathbf{s} = \mathbf{C}' : \mathbf{e}, \quad \boldsymbol{\mu} = \mathbf{C} \therefore \boldsymbol{\zeta}, \tag{14.274}$$

where the elasticity tensors $\mathbf{L}$, $\mathbf{C}'$, and $\mathbf{C}$ display the major symmetries and $\mathbf{L}$ has also the usual minor symmetries. The equilibrium of a micromorphic body loaded at its boundary Γ only (with neither body simple nor double forces) reads

$$(\boldsymbol{\sigma} + \mathbf{s}) \cdot \nabla = \mathbf{0}, \quad \boldsymbol{\mu} \cdot \nabla + \mathbf{s} = \mathbf{0}. \tag{14.275}$$

with either the displacement-type or traction-type boundary conditions prescribed on the boundary by the tensors $\mathbf{u}^{\Gamma}$, $\boldsymbol{\vartheta}^{\Gamma}$ or $\mathbf{t}^{\Gamma} = \boldsymbol{\sigma} \cdot \mathbf{n}$, $\mathbf{m}^{\Gamma} = \boldsymbol{\mu} \cdot \mathbf{n}$, respectively (similarly to Eqs. (14.189)). In nonlocal micromorphic model by Eringen [445, 446], the strain energy density function depends on the kinematical variables at two different points of the micromorphic continuum that provides a convolution type of the constitutive equations (14.274).

The first step of micromechanical modeling of CM is a homogenization scheme based on the Hill–Mandel conditions generalized to Cosserat (micropolar) medium in [518, 933, 951, 1167]) (compare with (14.212))

$$\{\sigma_{ij}\varepsilon_{ij} + \mu_{ij}\kappa_{ij}\} = \{\sigma_{ij}\}\{\varepsilon_{ij}\} + \{\mu_{ij}\}\{\kappa_{ij}\}, \tag{14.276}$$

It was explored the preconditions under which Hill–Mandel condition (14.276) for Cosserat continuum holds. Application of the homogeneous either displacement-type ($\boldsymbol{\varepsilon}^{\Gamma}$ =const, $\boldsymbol{\vartheta}^{\Gamma} = \mathbf{0}$) (14.270) or traction-type ($\boldsymbol{\sigma}^{\Gamma}$ =const, $\mathbf{m}^{\Gamma} = \mathbf{0}$) (14.271) boundary condition allows us to define the effective stiffness (or compliance) (compare with (14.219))

$$\{\boldsymbol{\varepsilon}\} : \mathbf{L}^* : \{\boldsymbol{\varepsilon}\} = 2W_w = \{\boldsymbol{\sigma} : \boldsymbol{\varepsilon} + \boldsymbol{\mu} : \boldsymbol{\kappa}\}. \tag{14.277}$$

For obtaining of a solution for one inclusion inside homogeneous infinite medium [second step, as Eqs. (14.160), (14.209) and (14.248)], we define the polarization tensors

$$\boldsymbol{\tau}(\mathbf{x}) = \boldsymbol{\sigma}(\mathbf{x}) - \mathbf{L}^{(0)} : \boldsymbol{\varepsilon}(\mathbf{x}), \quad \boldsymbol{\pi}(\mathbf{x}) = \boldsymbol{\mu}(\mathbf{x}) - \mathbf{C}^{(0)} : \boldsymbol{\kappa}(\mathbf{x}), \tag{14.278}$$

vanishing in the matrix $\mathbf{x} \in w \setminus v_i$ and coinciding with Eqs. (14.205) with an accuracy of notations. Introduction of the field pairs and the Green functions as in Eqs. (14.205) provides a solution of the problem for one inclusion inside the infinite homogeneous matrix [similar to Eqs. (14.160), (14.209), and (14.248)] subjected to homogeneous remote fields $\hat{\bar{\boldsymbol{\varepsilon}}} = (\bar{\boldsymbol{\varepsilon}}, \bar{\boldsymbol{\vartheta}})^\top$. This problem is solved in [1419] for a spherical inclusion in an incompressible couple-stress infinite media. The authors [1650] solved directly Eshelby's inhomogeneity (not inclusion) problem for a Cosserat matrix reinforced by circular Cauchy fibers. The strain localization tensor was then obtained in closed form, with no need to resort to the equivalent inclusion approximation (as in [1372]). For subsequent estimations of the effective moduli by the classical Mori–Tanaka approach (in the version [662, 663]), the authors estimated the average stresses inside the inclusions. However, the initial basic solution (like $\mathcal{L}_i^e(\mathbf{x} - \mathbf{x}_i, \hat{\bar{\boldsymbol{\varepsilon}}}$ (14.225); see also [1531]) for one inclusion comprises a special interest for incorporation into the advanced micromechanical method (like (14.225)) requiring the field estimations $(\mathbf{e}(\mathbf{x}), \boldsymbol{\kappa}(\mathbf{x}))^\top$ in the vicinity of a representative inclusion v_i $|\mathbf{x} - \mathbf{x}_i| \leq 3a$ rather than only inside the inclusion $\mathbf{x} - \mathbf{x}_i| \leq a$ (see for details Chap. 9).

The third step is a construction of the GIE in terms of the operator form [as Eqs. (14.164), (14.224), and (14.258)] for the solutions (like $\mathcal{L}_i^\epsilon(\mathbf{x} - \mathbf{x}_i, \hat{\bar{\boldsymbol{\varepsilon}}})$ (14.225). At last, the last fourth step is a solution of the obtained equations [similarly to Eqs. (14.177), (14.228), and (14.261)] in the framework of different sets of hypotheses (described in Chap. 9) with subsequent substitution the effective moduli representations obtained at the first step. The most effective method of solution is the iteration method using as an initial approximation the generalized versions of the classical either EFM or MTA. Of course, the mentioned scheme allows a straightforward generalization to either (or both) the binary interaction of inclusions (as in Chap. 10), inhomogeneous loading, or statistically inhomogeneous structure (as in Chap. 13).

Other direction of exploration of generalized continuum micromechanics is worthy of notice. This direction presented by variational methods generalizing the classical Hashin–Shtrikman, [618, 619], variational principle , and bounds was pioneered by Smyshlyaev and Fleck [1419] who considered linear, isotropic, incompressible couple-stress CMs. Modifications of this approach were applied late for bound estimations of plasticity strain gradient materials [1420], polycrystals materials [1421] and incompressible strain gradient plasticity for two-phase materials [486], as well as fundamentally different stress-gradient CMs (see [160, 1530]). Extensions of the classical Hashin–Shtrikman bounds were obtained by substitution of two types of polarization tensors [like (14.205) or (14.247)] and related operator-like representations [like (14.208) or (14.248)] into the corresponding variational principle. In all cases the version of the classical GIE (7.36) was used as a source key for subsequent generalization of Hashin–Shtrikman variational principle and bounds. The new GIE (7.19) with (6.200) leading to the generalized Hashin–Shtrikman variational principle (6.210) and (6.216) and the bounds (9.127) for locally elastic CMs may be of some interest. Replacement of the classical GIE (7.36) by the new GIE (7.19) for exploration of generalized continuum micromechanics looks especially promising owing to availability of nonlocal effects producing, in general, a field inhomogeneity inside inclusions (even in the case of spherical inclusions) that provides a fundamental difference between the results obtained at the base at either the classical GIE (7.36) or the new GIE (7.19) (see for details Sect. 14.9 and Chap. 9).

Chapter 15
Subsequent Generalizations of Theory and Related Problems

Abstract In this chapter we obtain a unified operator form of the GIE for the general cases of local and nonlocal problems, static, and wave motion phenomena for composite materials with periodic and random structures containing inclusions with perfect and imperfect interfaces and subjected to any number of coupled or uncoupled, homogeneous, or inhomogeneous external fields of different physical nature. Estimations of the effective properties and both the first and second statistical moments of fields in the constituents of CMs are presented in a general form of perturbations introduced by the heterogeneities and taking into account a possible imperfection of interface conditions. Some particular cases, asymptotic representations, and simplifications are presented for linear thermoelastic cases, conductivity problem, problems for piezoelectric and other coupled phenomena, composites with nonlocal elastic properties of constituents, and the wave propagation in composites with electromagnetic, optic, and mechanical responses.

15.1 Basic Constitutive Equations

Following Rogula [1301], we consider a linear response problem, or a basic constitutive equation

$$\boldsymbol{\sigma}(\mathbf{x}) = (\boldsymbol{\mathcal{L}} * \boldsymbol{\varepsilon})(\mathbf{x}) + \boldsymbol{\alpha}(\mathbf{x}), \tag{15.1}$$

which physical meaning need not be prescribed here and the current notations (15.1) usually used in elasticity theory can have a different sense. In particular, we can consider elasticity with $\boldsymbol{\varepsilon}$-displacement, and $\boldsymbol{\sigma}$-external forces or with $\boldsymbol{\varepsilon}$-strains, $\boldsymbol{\sigma}$-stresses, $\boldsymbol{\alpha}$-eigenstresses, or electrostatics with $\boldsymbol{\varepsilon}$-electric field, $\boldsymbol{\sigma}$-free charge, or $\boldsymbol{\varepsilon}$-electric field, $\boldsymbol{\sigma}$-electric induction, etc. In the case of a strongly nonlocal theory (strain type, see for references Chap. 14), the operator $\boldsymbol{\mathcal{L}}$ is represented by the integral operator with the kernel $\boldsymbol{\mathcal{L}}(\mathbf{x}, \mathbf{y})$

$$\boldsymbol{\mathcal{L}} * (\cdot) = \int \boldsymbol{\mathcal{L}}(\mathbf{x}, \mathbf{y})(\cdot)(\mathbf{y})d\mathbf{y}, \quad \boldsymbol{\alpha}(\mathbf{x}) = \int \mathbf{m}(\mathbf{x}, \mathbf{y})\Delta T(\mathbf{y})d\mathbf{y}, \tag{15.2}$$

V. A. Buryachenko, *Local and Nonlocal Micromechanics of Heterogeneous Materials*, https://doi.org/10.1007/978-3-030-81784-8_15

while the transformation field is described by the function $\boldsymbol{\alpha}(\mathbf{x})$ corresponding to an assumption (in the case of thermoelasticity) that a current temperature T is uniform in the space and the action of the nonlocal thermal expansion coefficient is reduced to the transformation strain tensor $\boldsymbol{\alpha}(\mathbf{x})$. In the case of the local theory, the operator $\boldsymbol{\mathcal{L}}$ is reduced to the tensor

$$\boldsymbol{\mathcal{L}}(\mathbf{x}, \mathbf{y}) = \mathbf{L}(\mathbf{x})\delta(\mathbf{x} - \mathbf{y}), \;\; \mathbf{m}(\mathbf{x}, \mathbf{y}) = \mathbf{m}(\mathbf{x})\delta(\mathbf{x} - \mathbf{y}), \tag{15.3}$$

where $\delta(\mathbf{x} - \mathbf{y})$ is the delta function. In the framework of strongly nonlocal (strain type) elasticity [431, 815], we follow a simplified theory for linear (macroscopically) homogeneous isotropic elasticity (see for references [78, 445, 446]), which differs from the classical one in the stress–strain constitutive relation (15.1) and (15.2) only, whereas the equilibrium and compatibility equations remain unaltered. The current interest in nonlocal simulation is driven largely by a practical need in the design, fabrication, and characterization of nanocomposites containing heterogeneities with at least one size of nano-length scale.

In the event of the coupled multifield linear response, the application of driving fields $\boldsymbol{\varepsilon} \equiv (\boldsymbol{\varepsilon}^1, \ldots, \boldsymbol{\varepsilon}^M)$ to the material induces in it "fluxes" $\boldsymbol{\sigma} \equiv (\boldsymbol{\sigma}^1, \ldots, \boldsymbol{\sigma}^M)$ that are related linearly to the driving fields, the latter being derivable from potentials $\mathbf{u} = (\mathbf{u}^1, \ldots, \mathbf{u}^M)$ of different tensorial ranks $\boldsymbol{\varepsilon}^\alpha = \nabla\mathbf{u}^\alpha$ ($\alpha = 1, \ldots, M$), e.g., a vector potential in elasticity and scalar potential in conductivity ($i, j = 1, \ldots, d$)

$$\varepsilon_{ij} = \frac{1}{2}\Big(\frac{\partial u_i}{\partial x_j} + \frac{\partial u_j}{\partial x_i}\Big), \quad \varepsilon_i = \frac{\partial u}{\partial x_i}, \tag{15.4}$$

respectively. A symmetry and positive definiteness of the $M \times M$ response matrix $\boldsymbol{\mathcal{L}} = (\mathcal{L}_{\alpha\beta})$ of submatrices $\mathcal{L}_{\alpha\beta} = (\mathcal{L}_{i\alpha j\beta})$ were considered in [1062] and [1063] by the use of analysis of some energy function and Onsager reciprocity relations. For the sake of definiteness, in the 2D case we will consider a plane driving field problem. At first no restrictions are imposed on the elastic symmetry of the phases or on the geometry of heterogeneities.[1]

The reciprocal constitutive equation is given by a similar representation

$$\boldsymbol{\varepsilon}(\mathbf{x}) = \boldsymbol{\mathcal{M}}{*}\boldsymbol{\sigma}(\mathbf{x}) + \boldsymbol{\beta}(\mathbf{x}) \tag{15.5}$$

with obvious correlations between the parameters involved into Eqs. (15.1) and (15.5)

$$\boldsymbol{\mathcal{L}}{*}\boldsymbol{\mathcal{M}} = \boldsymbol{\mathcal{M}}{*}\boldsymbol{\mathcal{L}} = \mathbf{I}, \quad \boldsymbol{\mathcal{M}}{*}\boldsymbol{\alpha} = -\boldsymbol{\beta}, \quad \boldsymbol{\mathcal{L}}{*}\boldsymbol{\beta} = -\boldsymbol{\alpha}, \tag{15.6}$$

[1] For example, it is known that for 2D elastic problems the plane–strain state is only possible for material symmetry no lower than orthotropic (see e.g. [910] that will be assumed hereafter in 2D case.

which are reduced to the algebraic relations for the local properties

$$\mathbf{L}{*}\mathbf{M} = \mathbf{M}{*}\mathbf{L} = \mathbf{I}, \quad \mathbf{M}{*}\boldsymbol{\alpha} = -\boldsymbol{\beta}, \quad \mathbf{L}{*}\boldsymbol{\beta} = -\boldsymbol{\alpha}, \tag{15.7}$$

where

$$\mathbf{I} = \mathrm{diag}(\mathbf{I}^1, \ldots, \mathbf{I}^M) \tag{15.8}$$

is the unit tensor combined from the unit $\mathbf{I}^\alpha$ ($\alpha = 1, \ldots, M$) fourth-order and second-order tensors for the vector (15.4_1) and scalar (15.4_2) potentials, respectively.

We introduce a *comparison body*, whose mechanical properties $\mathbf{g}^c$ ($\mathbf{g} = \boldsymbol{\mathcal{L}}, \boldsymbol{\mathcal{M}}, \boldsymbol{\alpha}, \boldsymbol{\beta}$) denoted by the upper index c and $\mathbf{g}^c$ will usually be taken as uniform over w, so that the corresponding boundary-value problem is easier to solve than that for the original body. All tensors $\mathbf{g}$ ($\mathbf{g} = \boldsymbol{\mathcal{L}}, \boldsymbol{\mathcal{M}}, \boldsymbol{\alpha}, \boldsymbol{\beta}$) of material properties are decomposed as $\mathbf{g} \equiv \mathbf{g}^c + \mathbf{g}_1(\mathbf{x}) = \mathbf{g}^c + \mathbf{g}_1^{(m)}(\mathbf{x})$. The upper index $^{(m)}$ indicates the components, and the lower index i indicates the individual inclusions; $v^{(0)} = w \backslash v$, $v \equiv \cup v^{(k)} \equiv \cup v_i$, $V(\mathbf{x}) = \sum V^{(k)}(\mathbf{x}) = \sum V_i(\mathbf{x})$, and $V^{(k)}(\mathbf{x})$ is an indicator function of $v^{(k)}$ equals 1 at $\mathbf{x} \in v^{(k)}$ and 0; otherwise, ($m = 0, k$; $k = 1, 2, \ldots, N$; $i = 1, 2, \ldots$). $V^\delta(\mathbf{x}) \equiv \sum_i \delta(\mathbf{x} - \mathbf{x}_i)$ is the delta function of random set of heterogeneities centers (called also density field, see [1240, 1456]). The introduction of jumps of material properties allows one to define the flux $\boldsymbol{\tau}$ and driving $\boldsymbol{\eta}$ polarization tensors by two equivalent ways ($\mathbf{x} \in R^d$)

$$\boldsymbol{\tau}(\mathbf{x}) = \boldsymbol{\mathcal{L}}_1 * \boldsymbol{\varepsilon}(\mathbf{x}) + \boldsymbol{\alpha}_1(\mathbf{x}), \ \boldsymbol{\eta}(\mathbf{x}) = \boldsymbol{\mathcal{M}}_1 * \boldsymbol{\sigma}(\mathbf{x}) + \boldsymbol{\beta}_1(\mathbf{x}), \ \boldsymbol{\tau} = -\boldsymbol{\mathcal{L}}^c * \boldsymbol{\eta}, \tag{15.9}$$

$$\boldsymbol{\tau}(\mathbf{x}) = \boldsymbol{\sigma}(\mathbf{x}) - [\boldsymbol{\mathcal{L}}^c * \boldsymbol{\varepsilon}(\mathbf{x}) + \boldsymbol{\alpha}^c(\mathbf{x})], \ \boldsymbol{\eta}(\mathbf{x}) = \boldsymbol{\varepsilon}(\mathbf{x}) - [\boldsymbol{\mathcal{M}}^c * \boldsymbol{\sigma}(\mathbf{x}) + \boldsymbol{\beta}^c(\mathbf{x})], \tag{15.10}$$

which are simply a notational convenience and vanish inside the matrix $\boldsymbol{\tau}(\mathbf{x}) \equiv \boldsymbol{\eta}(\mathbf{x}) \equiv \mathbf{0}$ ($\mathbf{x} \in v^{(0)}$) if

$$\boldsymbol{\mathcal{L}}^c = \boldsymbol{\mathcal{L}}^{(0)}, \quad \boldsymbol{\alpha}^c = \boldsymbol{\alpha}^{(0)}. \tag{15.11}$$

Hereafter $\boldsymbol{\mathcal{M}}_1(\mathbf{y})$ and $\boldsymbol{\beta}_1(\mathbf{y})$ are the jumps of the operator $\boldsymbol{\mathcal{M}}^{(k)}$ and of the eigenstrain $\boldsymbol{\beta}^{(k)}$ inside the component $v^{(k)}$ ($k = 0, \ldots, N$) with respect to the operators $\boldsymbol{\mathcal{M}}^c$ and $\boldsymbol{\beta}^c$. It is interesting, that for the local elasticity (15.3) with $\mathbf{L}^c = \mathbf{L}^{(0)}$, $\boldsymbol{\alpha}(\mathbf{x}) \equiv \mathbf{0}$, the popular polarization tensors (15.9) (see, e.g., [1625]) are attributed to Hashin and Shtrikman [618] and Hill [641] who introduced, in fact, these notions in the equivalent form (15.10). For both local and nonlocal elasticity theory , Eqs. (15.9) and (15.10) are also equivalent; however, the form (15.10) is preferable for subsequent manipulation because Eq. (15.10) does not explicitly use the constitutive equation (15.1) for the heterogeneities $\mathbf{x} \in v$.

Let the fields be governed by the linear elliptic partial differential operator equation ($\mathbf{x} \in R^d$)

$$\widehat{\mathcal{L}}\mathbf{u}(\mathbf{x}) = -\mathbf{f}(\mathbf{x}), \quad \text{e.g. } \widehat{\mathcal{L}} = \nabla\boldsymbol{\mathcal{L}} * \nabla + \mathbf{b}, \tag{15.12}$$

with the functions $\mathbf{b}(\mathbf{x})$ and $\mathbf{f}(\mathbf{x})$. The fundamental solution $\mathbf{G}(\mathbf{x}, \mathbf{y})$ for the whole homogeneous space R^d defined through the operator $\widehat{\mathcal{L}}^c = \nabla\boldsymbol{\mathcal{L}}^c * \nabla + \mathbf{b}^c$ by the equality

$$\widehat{\mathcal{L}}^c\mathbf{G}(\mathbf{x}, \mathbf{y}) = -\boldsymbol{\delta}\delta(\mathbf{x} - \mathbf{y}) \tag{15.13}$$

has the order $O(\int |\mathbf{x}|^{1-d} d|\mathbf{x}|)$ as $|\mathbf{x}| \to \infty$; here $\boldsymbol{\delta} = \text{diag}(\boldsymbol{\delta}^1, \ldots, \boldsymbol{\delta}^M)$ is the unit tensor combined from the unit $\boldsymbol{\delta}^\alpha$ $(\alpha = 1, \ldots, M)$ second-order tensors δ_{ij} and scalar 1 for the vector (15.4_1) and scalar (15.4_2) potentials, respectively. We additionally introduce (similarly to the momentum polarization tensor [1622]) a polarization tensor

$$\boldsymbol{\pi}(\mathbf{x}) = \mathbf{b}_1(\mathbf{x})\mathbf{u}(\mathbf{x}) + \mathbf{f}_1(\mathbf{x}). \tag{15.14}$$

All polarization tensors vanish inside the matrix

$$\boldsymbol{\tau}(\mathbf{x}), \boldsymbol{\eta}(\mathbf{x}), \boldsymbol{\pi}(\mathbf{x}) \equiv \mathbf{0}, \quad \mathbf{x} \in v^{(0)} \tag{15.15}$$

if the properties of the comparison medium and the matrix coincide

$$\boldsymbol{\mathcal{L}}^c \equiv \boldsymbol{\mathcal{L}}^{(0)}, \ \ \boldsymbol{\beta}^c \equiv \boldsymbol{\beta}^{(0)}, \ \ \mathbf{b}^c \equiv \mathbf{b}^{(0)}, \ \ \mathbf{f}^c \equiv \mathbf{f}^{(0)}. \tag{15.16}$$

The interfaces between the constituent phases of CM are classically assumed to be perfect. If the phases are perfectly bonded, the potential $\mathbf{u}$ and the traction $\mathbf{t}(\mathbf{x}) \equiv \boldsymbol{\sigma}(\mathbf{x}) \cdot \mathbf{n}(\mathbf{x})$ components are continuous across the interphase boundaries, i.e.,

$$[[\mathbf{u}(\mathbf{x})]] = \mathbf{0}, \quad [[\boldsymbol{\sigma}(\mathbf{x})]] \cdot \mathbf{n}(\mathbf{x}) = \mathbf{0}, \tag{15.17}$$

on the interface boundary $\mathbf{x} \in \Gamma = \cup\Gamma_i$ $(i = 1, \ldots)$ (assumed to be sufficiently smooth) where $\mathbf{n}$ is the outward normal vector on Γ from v to $v^{(0)}$ and $[[(.)]]$ =(out)−(in) is the jump operator. We will also consider a case of an imperfect interface

$$[[\mathbf{u}(\mathbf{x})]] \neq \mathbf{0}, \quad \mathbf{x} \in \Gamma_i^u, \tag{15.18}$$

$$[[\boldsymbol{\sigma}(\mathbf{x})]] \cdot \mathbf{n}(\mathbf{x}) \neq \mathbf{0}, \quad \mathbf{x} \in \Gamma_i^\sigma, \tag{15.19}$$

with the jumps (15.18) and (15.19) at the Γ_i^u and Γ_i^σ $(\Gamma_i^u, \Gamma_i^\sigma \subset \Gamma_i)$, respectively, described by the different models of interface imperfections. One usually assumes that either

$$\Gamma_i^u = \Gamma_i, \ \Gamma_i^\sigma = \emptyset \ \text{ or } \ \Gamma_i^\sigma = \Gamma_i, \ \Gamma_i^u = \emptyset. \tag{15.20}$$

It is assumed, that the boundary conditions prescribed at the infinity ($|\mathbf{x}| \to \infty$) generate the field ($\mathbf{x} \in R^d$)

$$\boldsymbol{\sigma}^0(\mathbf{x}) = (\boldsymbol{\mathcal{L}}^c * \boldsymbol{\varepsilon}^0)(\mathbf{x}) + \boldsymbol{\alpha}^c \tag{15.21}$$

in a homogeneous medium

$$\boldsymbol{\mathcal{L}}^c(\mathbf{y}+\mathbf{x}, \mathbf{z}+\mathbf{x}) = \boldsymbol{\mathcal{L}}^c(\mathbf{y}, \mathbf{z}), \quad \boldsymbol{\alpha}^c(\mathbf{x}) \equiv \boldsymbol{\alpha}^c = \text{const.},$$
$$\mathbf{b}^c(\mathbf{x}) \equiv \mathbf{b}^c = \text{const.}, \quad \mathbf{f}^c(\mathbf{x}) \equiv \mathbf{f}^c = \text{const.} \tag{15.22}$$

without heterogeneities ($V(\mathbf{x}) \equiv 0, \ \forall \mathbf{x}, \mathbf{y}, \mathbf{z} \in R^d$).

For the homogeneous loading, Eq. (15.21) is presented by the constant parameters

$$\boldsymbol{\sigma}^0 = \mathbf{L}^c \boldsymbol{\varepsilon}^0 + \boldsymbol{\alpha}^c, \quad \boldsymbol{\sigma}^0, \boldsymbol{\varepsilon}^0 \equiv \text{const.}, \tag{15.23}$$

where $\mathbf{L}^c(\mathbf{x}) \equiv \int \boldsymbol{\mathcal{L}}^c(\mathbf{x}, \mathbf{y})d\mathbf{y}$ =const. In such a case, independently on both the local and nonlocal nature of the constitutive law (15.1) in the micro- (nano-) points of a statistically homogeneous medium (5.61_2), the field macrovariables $\langle \boldsymbol{\sigma} \rangle$ and $\langle \boldsymbol{\varepsilon} \rangle$ are related by the local material effective constant tensors

$$\langle \boldsymbol{\sigma} \rangle = \mathbf{L}^* \langle \boldsymbol{\varepsilon} \rangle + \boldsymbol{\alpha}^*, \quad \langle \boldsymbol{\varepsilon} \rangle = \mathbf{M}^* \langle \boldsymbol{\sigma} \rangle + \boldsymbol{\beta}^*, \tag{15.24}$$

where $\mathbf{L}^*$ and $\mathbf{M}^*$ are expressed through the field concentration factors

$$\boldsymbol{\varepsilon}(\mathbf{x}) = \mathbf{A}^*_{\epsilon}(\mathbf{x})\langle \boldsymbol{\varepsilon} \rangle + \mathbf{C}^*_{\epsilon\epsilon}(\mathbf{x}), \quad \boldsymbol{\sigma}(\mathbf{x}) = \mathbf{B}^*_{\epsilon}(\mathbf{x})\langle \boldsymbol{\varepsilon} \rangle + \mathbf{C}^*_{\sigma\epsilon}(\mathbf{x}), \tag{15.25}$$
$$\boldsymbol{\varepsilon}(\mathbf{x}) = \mathbf{A}^*_{\sigma}(\mathbf{x})\langle \boldsymbol{\sigma} \rangle + \mathbf{C}^*_{\epsilon\sigma}(\mathbf{x}), \quad \boldsymbol{\sigma}(\mathbf{x}) = \mathbf{B}^*_{\sigma}(\mathbf{x})\langle \boldsymbol{\sigma} \rangle + \mathbf{C}^*_{\sigma\sigma}(\mathbf{x}), \tag{15.26}$$

for the prescribed $\langle \boldsymbol{\varepsilon} \rangle$ (15.25) and $\langle \boldsymbol{\sigma} \rangle$ (15.26), respectively. Then a straight forward generalization of the corresponding results for the nonlocal elasticity (see Chap. 14) leads to

$$\mathbf{L}^* = \langle \mathbf{B}^{*\top}_{\epsilon} \rangle, \quad \mathbf{M}^* = \langle \mathbf{A}^{*\top}_{\sigma} \rangle, \tag{15.27}$$
$$\boldsymbol{\alpha}^* = \langle \mathbf{A}^{*\top}_{\epsilon} \boldsymbol{\alpha} \rangle = -\langle \mathbf{B}^{*\top}_{\epsilon} \boldsymbol{\beta} \rangle, \tag{15.28}$$
$$\boldsymbol{\beta}^* = -\langle \mathbf{A}^{*\top}_{\sigma} \boldsymbol{\alpha} \rangle = \langle \mathbf{B}^{*\top}_{\sigma} \boldsymbol{\beta} \rangle, \tag{15.29}$$
$$W^* = -\frac{1}{2}\langle \boldsymbol{\alpha}_1 \mathbf{C}^*_{\epsilon\epsilon} \rangle = -\frac{1}{2}\langle \boldsymbol{\beta}_1 \mathbf{C}^*_{\sigma\sigma} \rangle, \tag{15.30}$$

which are reduced to the classical results by [640] and [912] for the local thermoelasticity. The stored energy $W^* := \langle \boldsymbol{\alpha}\boldsymbol{\varepsilon} \rangle$ and $W^* := \langle \boldsymbol{\beta}\boldsymbol{\sigma} \rangle$ (at $\mathbf{f}(\mathbf{x}), \mathbf{b}(\mathbf{x}) \equiv \mathbf{0}$) corresponding to $\langle \boldsymbol{\varepsilon} \rangle \equiv \mathbf{0}$ and $\langle \boldsymbol{\sigma} \rangle \equiv \mathbf{0}$ are expressed by Eqs. (15.30_1) and (15.30_2), respectively.

The stored energies W^*, like $\boldsymbol{\alpha}^*$ and $\boldsymbol{\beta}^*$, depend only on the fields $\boldsymbol{\alpha}(\mathbf{x}) and \boldsymbol{\beta}(\mathbf{x})$ and the transformation functions and act like effective material constants, which can be observed macroscopically (see, e.g., Chap. 6). It should be mentioned that the effective eigenfields $\boldsymbol{\alpha}^*$, $\boldsymbol{\beta}^*$, and stored energy W^* depend not only on the material properties but also on its "loading" by temperature changes ΔT =const. or similar sources of eigenfields. So the effective material parameters can be obtained from $\boldsymbol{\alpha}^*$, $\boldsymbol{\beta}^*$, and W^* by its normalization with respect to ΔT. Because of this, strictly speaking, the real effective parameters are $\boldsymbol{\alpha}^*/\Delta T$, $\boldsymbol{\beta}^*/\Delta T$, and $W^*/(\Delta T)^2$ (which do not depend on ΔT) rather than $\boldsymbol{\alpha}^*$, $\boldsymbol{\beta}^*$, and W^*, respectively.

15.2 General Integral Equations (GIEs) for CMs

15.2.1 General Integral Equation in Terms of the Green's Functions

Substituting Eq. (15.4) and the constitutive equation (15.1) into the governing equation (15.12), we obtain a differential equation with respect to the potential $\mathbf{u}$

$$\nabla[(\mathcal{L}*\nabla\mathbf{u})(\mathbf{x})+\boldsymbol{\alpha}(\mathbf{x})]+\mathbf{b}(\mathbf{x})\mathbf{u}(\mathbf{x})+\mathbf{f}(\mathbf{x})=\mathbf{0}. \tag{15.31}$$

In the framework of the traditional scheme, we introduce a homogeneous "comparison" body (15.20) with the solution $\boldsymbol{\sigma}^0(\mathbf{x})$, $\boldsymbol{\varepsilon}^0(\mathbf{x}) = \nabla\mathbf{u}^0(\mathbf{x})$ corresponding to the same boundary-value problem. So Eq. (15.31) can be transformed to the Navier-like governing equation with respect to the potential field $\mathbf{u}$

$$\nabla[(\mathcal{L}^c*\nabla\mathbf{u})(\mathbf{x})+\boldsymbol{\alpha}^c]+\mathbf{b}^c\mathbf{u}(\mathbf{x})+\mathbf{f}^c=-\nabla\boldsymbol{\tau}(\mathbf{x})-\boldsymbol{\pi}(\mathbf{x}), \tag{15.32}$$

with a fictitious random "body force" in the right-hand side of the equation, where $\boldsymbol{\tau}(\mathbf{x})$ and $\boldsymbol{\pi}(\mathbf{x})$ are defined by Eqs. (15.9_1), (15.10_1), and (15.14), respectively. Then Eq. (15.32) can be reduced to a symmetrized integral form:

$$\boldsymbol{\varepsilon}(\mathbf{x})=\boldsymbol{\varepsilon}^0(\mathbf{x})+\nabla\int\mathbf{G}(\mathbf{x}-\mathbf{y})\Big\{\nabla\boldsymbol{\tau}(\mathbf{y})+\boldsymbol{\pi}(\mathbf{y})\Big\}\,d\mathbf{y}, \tag{15.33}$$

where $\mathbf{G}$ is the infinite homogeneous body Green's function of the Navier-like equation with a response operator $\widehat{\mathcal{L}}^c$ (15.12). The Green's function $\mathbf{G}$ is assumed to be known for any operator $\widehat{\mathcal{L}}^c$ (15.12) being considered. The deterministic function $\boldsymbol{\varepsilon}^0(\mathbf{x})$ is the driving field (15.19) that would exist in the medium with homogeneous properties (15.20) and appropriate boundary conditions at the infinity $|\mathbf{x}| \to \infty$.

The improper integral over the whole space R^d (15.33) is defined as a limit ($p \to \infty$) of definite integrals over an increasing sequence w_p filling R^d

$$\boldsymbol{\varepsilon}(\mathbf{x}) = \boldsymbol{\varepsilon}^0(\mathbf{x}) + \lim_{w_p \to R^d} \int_{w_p} \nabla_{\mathbf{x}} \mathbf{G}(\mathbf{x}-\mathbf{y}) \Big\{ \nabla_{\mathbf{y}} \boldsymbol{\tau}(\mathbf{y}) + \boldsymbol{\pi}(\mathbf{y}) \Big\} \, d\mathbf{y}, \tag{15.34}$$

where the limit exist independently on the particular choice of closed bounded regions w_p in R^d such that any w_p is contained in every w_{p_1} for which $p_1 > p$ and for any bounded region w' in R^d, there is p_2 such that $w' \subset w_{p_2}$. After integration of Eq. (15.34) by parts, it is found that

$$\begin{aligned} \boldsymbol{\varepsilon}(\mathbf{x}) = \boldsymbol{\varepsilon}^0(\mathbf{x}) + \lim_{w_p \to R^d} \Bigg\{ \int_{w_p} \mathbf{U}(\mathbf{x}-\mathbf{y})\boldsymbol{\tau}(\mathbf{y}) \, d\mathbf{y} + \int_{w_p} \nabla \mathbf{G}(\mathbf{x}-\mathbf{y})\boldsymbol{\pi}(\mathbf{y}) \, d\mathbf{y} \\ + \int_{\Gamma_p^0} \nabla \mathbf{G}(\mathbf{x}-\mathbf{s})\boldsymbol{\tau}(\mathbf{s})\mathbf{n}(\mathbf{s}) \, d\mathbf{s} \Bigg\}, \end{aligned} \tag{15.35}$$

which used that $\nabla_{\mathbf{y}} = -\nabla_{\mathbf{x}}$, $\nabla = \nabla_{\mathbf{x}}$, and the surface integration is taken over the external surface Γ_p^0 of the mesodomain w_p containing a statistically large number of inclusion realizations, and the integral operator kernel $\mathbf{U}$ is an even homogeneous generalized function of degree $-d$ defined as

$$U_{kljl}(\mathbf{x}) = [\nabla_k \nabla_l G_{ij}(\mathbf{x})]_{(ki)(jl)}, \quad U_{kl}(\mathbf{x}) = \nabla_k \nabla_l G(\mathbf{x}) \tag{15.36}$$

for the vector (15.4_1) and scalar (15.4_2) potentials, respectively. For construction of the regularization of a generalized function of the type of derivatives of homogeneous regular function, we use a scheme proposed in Gelfand and Shilov [537] according to which the tensor $\mathbf{U}(\mathbf{x})$ is split into the singular $\mathbf{U}^s(\mathbf{x})$ and formal $\mathbf{U}^f(\mathbf{x})$ parts

$$\mathbf{U}(\mathbf{x}) = \mathbf{U}^s(\mathbf{x}) + \mathbf{U}^f(\mathbf{x}), \tag{15.37}$$

where $\mathbf{U}^s(\mathbf{x}) = \delta(\mathbf{x})\widetilde{\mathbf{U}}^s$ ($\widetilde{\mathbf{U}}^s \equiv$const.) is a singular function associated with some infinitely small exclusion region and $\mathbf{U}^f(\mathbf{x}) \equiv 1/r^{-d}\widetilde{\mathbf{U}}^f(\mathbf{n})$ ($\mathbf{x} = |\mathbf{x}|\mathbf{n}$) is a formal function. Both of these terms depend on the shape of an exclusion region being prescribed, while their sum, being the left-hand side of (15.37), is defined uniquely.

Now we center Eq. (15.35) by the use of statistical averages, i.e., from both sides of Eq. (15.35), their statistical averages are subtracted

$$\begin{aligned} \boldsymbol{\varepsilon}(\mathbf{x}) = \langle \boldsymbol{\varepsilon} \rangle(\mathbf{x}) + \lim_{w_p \to R^d} \Bigg\{ \int_{w_p} \langle\!\langle \mathbf{U}(\mathbf{x}-\mathbf{y})\boldsymbol{\tau} \rangle\!\rangle(\mathbf{y}) d\mathbf{y} \\ + \int_{w_p} \langle\!\langle \nabla \mathbf{G}(\mathbf{x}-\mathbf{y})\boldsymbol{\pi}(\mathbf{y}) \rangle\!\rangle(\mathbf{y}) d\mathbf{y} \Bigg\} + \mathcal{I}^{\Gamma_\epsilon}, \end{aligned} \tag{15.38}$$

where one introduces a centering operation

$$\langle\!\langle \mathbf{h}(\mathbf{x},\mathbf{y})\mathbf{g}(\mathbf{y}) \rangle\!\rangle(\mathbf{y}) \equiv \mathbf{h}(\mathbf{x},\mathbf{y})\mathbf{g}(\mathbf{y}) - \langle \mathbf{h}(\mathbf{x},\mathbf{y})\mathbf{g}(\mathbf{y}) \rangle(\mathbf{y}), \tag{15.39}$$

and in the right-hand side of Eq. (15.38), the integral over the external surface $\Gamma_p^0 := \partial w_p$

$$\mathcal{I}^{\Gamma\epsilon} \equiv \lim_{w_p \to R^d} \int_{\Gamma_p^0} \langle\langle \nabla \mathbf{G}(\mathbf{x}-\mathbf{s})\boldsymbol{\tau}(\mathbf{x})\rangle\rangle(\mathbf{s})\mathbf{n}(\mathbf{s})d\mathbf{s}, \tag{15.40}$$

can be dropped out, because this integral vanishes at sufficient distance $\mathbf{x}$ from the boundary Γ_p^0:

$$a \ll |\mathbf{x}-\mathbf{s}|, \ \forall \mathbf{s} \in \Gamma_p^0. \tag{15.41}$$

This means that if $|\mathbf{x}-\mathbf{s}|$ is large enough for $\forall \mathbf{s} \in \Gamma_p^0$, then at the portion of the smooth surface $d\mathbf{s} \approx |\mathbf{x}-\mathbf{s}|^{d-1} d\boldsymbol{\omega}^s$ with a small solid angle $d\boldsymbol{\omega}^s$, the tensor $\nabla \mathbf{G}(\mathbf{x}-\mathbf{s})|\mathbf{x}-\mathbf{s}|^{d-1}$ depends only on the solid angle $\boldsymbol{\omega}^s$ variables and slowly varies on the portion of the surface $d\mathbf{s}$; in this sense the tensor $\nabla \mathbf{G}(\mathbf{x}-\mathbf{s})$ is called a "slow" variable of the solid angle $\boldsymbol{\omega}^s$, while the expression $\boldsymbol{\tau}(\mathbf{s})$ is a rapidly oscillating function on $d\mathbf{s}$ and is called a "fast" variable. Therefore we can use a rigorous theory of "separate" integration of "slow" and "fast" variables, according to which (freely speaking) the operation of surface integration may be regarded as averaging [see for details, e.g., [472] and its applications in elasticity [1378]]. In consequence of established separation of "slow" and "fast" variables, no confusion is possible in understanding of estimations of averaged items $\langle(.)\rangle(\mathbf{S})$ depending on macro-coordinate $\mathbf{S} \in \Gamma_p^0$, while the corresponding unaveraged items depend on micro-coordinates $\mathbf{s} \in \Gamma_p^0$. Indeed, the slow variable $\nabla \mathbf{G}(\mathbf{x}-\mathbf{s})$ can be consistent (after the first separation of the fast and slow variables) with the macro-coordinate $\mathbf{S} \in \Gamma_p^0$ as $\nabla \mathbf{G}(\mathbf{x}-\mathbf{S})$, while the fast variable $\boldsymbol{\tau}(\mathbf{s})$ is in agreement with the micro-coordinate $\mathbf{s} \in \Gamma_p^0$ that makes it possible to present the integrand in the integral (15.40) in the formal form $\nabla \mathbf{G}(\mathbf{x}-\mathbf{S})\Big[\boldsymbol{\tau}(\mathbf{s}) - \langle\boldsymbol{\tau}\rangle(\mathbf{S})\Big]$ leading (after additional separation of the fast and slow variables) to vanishing of integral from the term in the square brackets for statistically large number of random events α. If (as we assume) there is no *long-range* order and the function $\varphi(v_j, \mathbf{x}_j|; v_i, \mathbf{x}_i) - \varphi(v_j, \mathbf{x}_j)$ decays at infinity sufficiently rapidly (see [604, 1525, 1621]), then it leads to a degeneration of the surface integral.

To express Eq. (15.38) in terms of fluxes, we use the identities

$$\mathcal{L}_1 * (\boldsymbol{\varepsilon} - \boldsymbol{\beta}) = -\mathcal{L}^c * \big[\boldsymbol{\mathcal{M}}_1 * \boldsymbol{\sigma}\big], \tag{15.42}$$

$$\boldsymbol{\varepsilon} = [\boldsymbol{\mathcal{M}}^c * \boldsymbol{\sigma} + \boldsymbol{\beta}^c] + [\boldsymbol{\mathcal{M}}_1 * \boldsymbol{\sigma} + \boldsymbol{\beta}_1]. \tag{15.43}$$

Substituting (15.42) and (15.43) into the right-hand side and the left-hand side of (15.38), respectively, and contracting with the operator $\mathcal{L}^c$ give the general integral equation for the stresses

$$\boldsymbol{\sigma}(\mathbf{x}) = \langle\boldsymbol{\sigma}\rangle(\mathbf{x}) + \lim_{w_p\to R^d}\left\{\int_{w_p}\langle\langle\boldsymbol{\Gamma}(\mathbf{x}-\mathbf{y})*\boldsymbol{\eta}(\mathbf{y})\rangle\rangle(\mathbf{y})\,d\mathbf{y}\right. \tag{15.44}$$

$$\left. + \int_{w_p}\boldsymbol{\mathcal{L}}^c*\langle\langle\nabla\mathbf{G}(\mathbf{x}-\mathbf{y})\boldsymbol{\pi}(\mathbf{y})\rangle\rangle(\mathbf{y})\,d\mathbf{y}\right\} + \boldsymbol{\mathcal{I}}^{\Gamma\sigma}.$$

where we define

$$\boldsymbol{\mathcal{I}}^{\Gamma\sigma} = -\lim_{w_p\to R^d}\int_{\Gamma_p^0}\boldsymbol{\mathcal{L}}^c*\langle\langle\nabla\mathbf{G}(\mathbf{x}-\mathbf{s})\boldsymbol{\mathcal{L}}^c*\boldsymbol{\eta}(\mathbf{s})\rangle\rangle(\mathbf{s})\mathbf{n}(\mathbf{s})\,d\mathbf{s}. \tag{15.45}$$

If we assume no long-range order, then the tensor $\boldsymbol{\mathcal{I}}^{\Gamma\sigma}$ degenerates analogously to the tensor $\boldsymbol{\mathcal{I}}^{\Gamma\epsilon}$ (15.39) and can be dropped. The integral operator kernel

$$\boldsymbol{\Gamma}(\mathbf{x}-\mathbf{y}) = -\boldsymbol{\mathcal{L}}^c*(\mathbf{I}\delta(\mathbf{x}-\mathbf{y}) + \mathbf{U}\boldsymbol{\mathcal{L}}^c*) \tag{15.46}$$

is called the Green flux tensor (see the local elasticity case (15.3) in [820, 822]).

For a statistically inhomogeneous field X, the dependence of the statistical averages $\langle(.)\rangle(\mathbf{y})$ (15.39) in Eqs. (15.38) and (15.44) on the current coordinate $\mathbf{y}$ is of fundamental importance. But even in this case, the expressions in the averaging brackets $\langle\langle(\cdot)\rangle\rangle(\mathbf{y})$ in Eqs. (15.38) and (15.44) are of the order $O(|\mathbf{x}-\mathbf{y}|^{-2d})$ as $|\mathbf{x}-\mathbf{y}|\to\infty$, and the integrals in Eqs. (15.38) and (15.44) with the kernels $\mathbf{U}$ and $\boldsymbol{\Gamma}$, respectively, converge absolutely. In a similar manner, the integrals with the body-force density converge absolutely. In fact, $\langle\langle\nabla\mathbf{G}(\mathbf{x}-\mathbf{y})\boldsymbol{\pi}(\mathbf{y})\rangle\rangle(\mathbf{y})$ tends to zero with $|\mathbf{x}-\mathbf{y}|\to\infty$ ($\mathbf{x}\in v_i$, $\mathbf{y}\in v_j$) as $O(|\mathbf{x}-\mathbf{y}|^{-d+1})\big[\varphi(v_j,\mathbf{x}_j|;v_i,\mathbf{x}_i)-\varphi(v_j,\mathbf{x}_j)\big]$. For no *long-range* order, an absolute convergence of the integrals involved is assured if the function $\varphi(v_j,\mathbf{x}_j|;v_i,\mathbf{x}_i)-\varphi(v_j,\mathbf{x}_j)$ decays at infinity sufficiently rapidly. Therefore, for $\mathbf{x}\in w_p$ considered in Eqs. (15.38), (15.44) and removed far enough from the boundary Γ_p^0, the right-hand side integrals in (15.38) and (15.44) do not depend on the shape and size of the domain w_p, and they can be replaced by the integrals over the whole space R^d that defines the general integral equations (GIEs)

$$\boldsymbol{\varepsilon}(\mathbf{x}) = \langle\boldsymbol{\varepsilon}\rangle(\mathbf{x}) + \int\langle\langle\mathbf{U}(\mathbf{x}-\mathbf{y})\boldsymbol{\tau}(\mathbf{y})\rangle\rangle(\mathbf{y})d\mathbf{y}$$

$$+ \int\langle\langle\nabla\mathbf{G}(\mathbf{x}-\mathbf{y})\boldsymbol{\pi}(\mathbf{y})\rangle\rangle(\mathbf{y})d\mathbf{y}, \tag{15.47}$$

$$\boldsymbol{\sigma}(\mathbf{x}) = \langle\boldsymbol{\sigma}\rangle(\mathbf{x}) + \int\langle\langle\boldsymbol{\Gamma}(\mathbf{x}-\mathbf{y})*\boldsymbol{\eta}(\mathbf{y})\rangle\rangle(\mathbf{y})\,d\mathbf{y}$$

$$+ \int\boldsymbol{\mathcal{L}}^c*\langle\langle\nabla\mathbf{G}(\mathbf{x}-\mathbf{y})\boldsymbol{\pi}(\mathbf{y})\rangle\rangle(\mathbf{y})\,d\mathbf{y}. \tag{15.48}$$

Thus, there are no difficulties connected with the asymptotic behavior of the functions $\mathbf{U}$, $\boldsymbol{\Gamma}$, and $\nabla\mathbf{G}$ at infinity (as $|\mathbf{x}-\mathbf{y}|^{-d}$ and $|\mathbf{x}-\mathbf{y}|^{-d+1}$, respectively).

It should be mentioned that particular linear elastic cases (for both local and nonlocal constitutive Eq. (15.1)) of Eqs. (15.47) and (15.48) were also proved in Chap. 7 for a bounded domain w with prescribed boundary conditions at the boundary $\Gamma^0 := \partial w$ which is far removed from $\mathbf{x} \in w$ as in Eq. (15.41). However, the method used is more cumbersome than proposed in this chapter because it involves the additional "fast" variables at Γ^0 where boundary data $[\mathbf{u}^0(\mathbf{s}), \mathbf{t}^0(\mathbf{s})]$ not prescribed by the boundary conditions depend on perturbations introduced by all inhomogeneities, and, therefore, $\boldsymbol{\varepsilon}^0(\mathbf{x}) = \boldsymbol{\varepsilon}^0(\mathbf{x}, \alpha)$, $\boldsymbol{\sigma}^0(\mathbf{x}) = \boldsymbol{\sigma}^0(\mathbf{x}, \alpha)$. The boundary problem becomes even more complicated for nonlocal constitutive Eq. (15.1) corresponding to different nonstandard boundary conditions (see, e.g., [1229]). The surface integrals over Γ^0 related with the boundary conditions on Γ^0 vanish similar to the integrals (15.40) and (15.45), but the proof of this vanishing takes some additional efforts which are absent in the method proposed in the current chapter and based on the analysis of integrals over the increasing sequence w_p filling R^d (15.35).

15.2.2 *One Inclusion in the Matrix*

For one heterogeneity v_k in the infinite matrix, one introduces the general notions of the *perturbators* of the driving field $\mathcal{L}_k^{\tau}(\mathbf{x}-\mathbf{x}_k, \widehat{\boldsymbol{\tau}})$, the flux $\mathcal{L}_k^{\eta}(\mathbf{x}-\mathbf{x}_k, \widehat{\boldsymbol{\eta}})$, and the potential $\mathcal{L}_k^{u}(\mathbf{x}-\mathbf{x}_k, \widehat{\boldsymbol{\tau}})$

$$\mathcal{L}_k^{\tau}(\mathbf{x}-\mathbf{x}_k, \widehat{\boldsymbol{\tau}}) \equiv \boldsymbol{\varepsilon}(\mathbf{x}) - \boldsymbol{\varepsilon}^0(\mathbf{x}), \tag{15.49}$$

$$\mathcal{L}_k^{\eta}(\mathbf{x}-\mathbf{x}_k, \widehat{\boldsymbol{\eta}}) \equiv \boldsymbol{\sigma}(\mathbf{x}) - \boldsymbol{\sigma}^0(\mathbf{x}), \tag{15.50}$$

$$\mathcal{L}_k^{u\tau}(\mathbf{x}-\mathbf{x}_k, \widehat{\boldsymbol{\tau}}) \equiv \mathbf{u}(\mathbf{x}) - \mathbf{u}^0(\mathbf{x}), \tag{15.51}$$

which produce the perturbations of the fields $\boldsymbol{\varepsilon}(\mathbf{x})$, $\sigma(\mathbf{x})$, and $\mathbf{u}(\mathbf{x})$, respectively, in the point $\mathbf{x}$ due to the insertion of the heterogeneity center into the point $\mathbf{x}_k$. $\widehat{\boldsymbol{\tau}} and \widehat{\boldsymbol{\eta}}$ are the symbolic notations of dependance of $\mathcal{L}_k^{\tau}(\mathbf{x}-\mathbf{x}_k, \widehat{\boldsymbol{\tau}})$, $\mathcal{L}_k^{\eta}(\mathbf{x}-\mathbf{x}_k, \widehat{\boldsymbol{\eta}})$, and $\mathcal{L}_k^{u\tau}(\mathbf{x}-\mathbf{x}_k, \widehat{\boldsymbol{\tau}})$ on $\boldsymbol{\tau}(\mathbf{x})$, $\eta(\mathbf{x})$ $(\mathbf{x} \in v_k)$, respectively, and $\mathbf{f}_1(\mathbf{x})$ $(\mathbf{x} \in v_k)$, $[[\mathbf{u}(\mathbf{x})]]$ $(\mathbf{x} \in \Gamma_k^u)$, and $[[\boldsymbol{\sigma}(\mathbf{x})]] \cdot \mathbf{n}$ $(\mathbf{x} \in \Gamma_k^{\sigma})$. Obviously, the perturbators $\mathcal{L}_k^{\tau}(\mathbf{x}-\mathbf{x}_k, \widehat{\boldsymbol{\tau}})$, $\mathcal{L}_k^{\eta}(\mathbf{x}-\mathbf{x}_k, \widehat{\boldsymbol{\eta}})$, and $\mathcal{L}_k^{u\tau}(\mathbf{x}-\mathbf{x}_k, \widehat{\boldsymbol{\tau}})$ are the generalizations of multiplications of the corresponding Green's functions and some polarizations of a point defect to the impacts of heterogeneity v_k of the finite size with the possible imperfection of boundary conditions. In particular, the perturbators can be presented through the Green's functions for the perfect interface conditions (see Eq. (15.17)) and can be modified for the case of imperfect interface (see Eqs. (15.18) and (15.19)) in the following manner (see for details [208])

$$\mathcal{L}_k^{\tau}(\mathbf{x}-\mathbf{x}_k, \widehat{\boldsymbol{\tau}}) = \int_{v_k} \mathbf{U}(\mathbf{x}-\mathbf{y})\boldsymbol{\tau}(\mathbf{y})d\mathbf{y} + \int_{v_k} \nabla\mathbf{G}(\mathbf{x}-\mathbf{y})\boldsymbol{\pi}(\mathbf{y})d\mathbf{y}$$

$$-\int_{\Gamma_k^u} \nabla\mathbf{T}^\top(\mathbf{x}-\mathbf{s})[[\mathbf{u}(\mathbf{s})]]ds - \int_{\Gamma_k^\sigma} \nabla\mathbf{G}(\mathbf{x}-\mathbf{s})[[\boldsymbol{\sigma}(\mathbf{s})]]\cdot\mathbf{n}(\mathbf{s})ds, \tag{15.52}$$

$$\begin{aligned}\mathcal{L}_k^\eta(\mathbf{x}-\mathbf{x}_k,\widehat{\boldsymbol{\eta}}) = &\int_{v_k} \boldsymbol{\Gamma}(\mathbf{x}-\mathbf{y})*\boldsymbol{\eta}(\mathbf{y})d\mathbf{y} + \int_{v_k} \mathcal{L}^c*\nabla\mathbf{G}(\mathbf{x}-\mathbf{y})\boldsymbol{\pi}(\mathbf{y})d\mathbf{y}\\ &-\int_{\Gamma_k^u} \mathcal{L}^c*\nabla\mathbf{T}^\top(\mathbf{x}-\mathbf{s})[[\mathbf{u}(\mathbf{s})]]ds\\ &-\int_{\Gamma_k^\sigma} \mathcal{L}^c*\nabla\mathbf{G}(\mathbf{x}-\mathbf{s})[[\boldsymbol{\sigma}(\mathbf{s})]]\cdot\mathbf{n}(\mathbf{s})ds,\end{aligned} \tag{15.53}$$

$$\begin{aligned}\mathcal{L}_k^{u\tau}(\mathbf{x}-\mathbf{x}_k,\widehat{\boldsymbol{\tau}}) = &\int_{v_k} \nabla\mathbf{G}(\mathbf{x}-\mathbf{y})\boldsymbol{\tau}(\mathbf{y})d\mathbf{y} + \int_{v_k} \mathbf{G}(\mathbf{x}-\mathbf{y})\boldsymbol{\pi}(\mathbf{y})d\mathbf{y}\\ &-\int_{\Gamma_k^u} \mathbf{T}^\top(\mathbf{x}-\mathbf{s})[[\mathbf{u}(\mathbf{s})]]ds - \int_{\Gamma_k^\sigma} \mathbf{G}(\mathbf{x}-\mathbf{s})[[\boldsymbol{\sigma}(\mathbf{s})]]\cdot\mathbf{n}(\mathbf{s})ds,\end{aligned} \tag{15.54}$$

The perturbators $\mathcal{L}_k^\tau(\mathbf{x}-\mathbf{x}_k,\widehat{\boldsymbol{\tau}})$, $\mathcal{L}_k^\eta(\mathbf{x}-\mathbf{x}_k,\widehat{\boldsymbol{\eta}})$, and $\mathcal{L}_k^{u\tau}(\mathbf{x}-\mathbf{x}_k,\widehat{\boldsymbol{\tau}})$ can be found from joint solutions of Eqs. (15.49)–(15.51) and (15.52)–(15.54), respectively, and the contact conditions at the interfaces Γ_k^u (15.18) and Γ_k^σ (15.19). The volume integrals in Eqs. (15.52)–(15.54) are estimated over the open domains where the boundaries Γ_i^u and Γ_i^σ are excluded. The integral operator kernel $\boldsymbol{\Gamma}(\mathbf{x}-\mathbf{y}) = -\mathcal{L}^c*(\mathbf{I}\delta(\mathbf{x}-\mathbf{y})+\mathbf{U}\mathcal{L}^c*)$ is called the Green flux tensor. For the local elasticity case (15.3), $\boldsymbol{\Gamma}(\mathbf{x}-\mathbf{y})$ is the Green stress tensor (see, e.g., [820]), while the tensor of the "fundamental traction" $\mathbf{T}$ on Γ_k^u (called also Kupradze tensor; see, e.g., [59]) associated with the tensor of "fundamental displacement" $\mathbf{G}$ is given by Eq. (3.39_2).

It should be mentioned that the driving field perturbator $\mathcal{L}_k^\tau(\mathbf{x}-\mathbf{x}_k,\widehat{\boldsymbol{\tau}})$ and the flux perturbator $\mathcal{L}_k^\eta(\mathbf{x}-\mathbf{x}_k,\widehat{\boldsymbol{\eta}})$ are linked by the relation

$$\mathcal{L}_k^\eta(\mathbf{x}-\mathbf{x}_k,\widehat{\boldsymbol{\eta}}) = \boldsymbol{\tau}(\mathbf{x})V_k(\mathbf{x}) + \mathcal{L}^c*\mathcal{L}_k^\tau(\mathbf{x}-\mathbf{x}_k,\widehat{\boldsymbol{\tau}}), \tag{15.55}$$

which can be established by the substitution of both the constitutive law (15.1) and the identities

$$\mathcal{L}_1*(\boldsymbol{\varepsilon}-\boldsymbol{\beta}) = -\mathcal{L}^c*\big[\boldsymbol{\mathcal{M}}_1*\boldsymbol{\sigma}\big], \tag{15.56}$$

$$\boldsymbol{\varepsilon} = [\boldsymbol{\mathcal{M}}^c*\boldsymbol{\sigma}+\boldsymbol{\beta}^c] + [\boldsymbol{\mathcal{M}}_1*\boldsymbol{\sigma}+\boldsymbol{\beta}_1] \tag{15.57}$$

into Eq. (15.50). Here $\boldsymbol{\mathcal{M}}_1(\mathbf{y})$ and $\boldsymbol{\beta}_1(\mathbf{y})$ are the jumps of the operator $\boldsymbol{\mathcal{M}}^{(k)}$ and the eigenstrain $\boldsymbol{\beta}^{(k)}$ inside the component $v^{(k)}$ $(k = 0,\ldots,N)$ with respect to the operators $\boldsymbol{\mathcal{M}}^c$ and $\boldsymbol{\beta}^c$, respectively. Thus, we have proved that Eqs. (15.49) and (15.50) are equivalent and the perturbators $\mathcal{L}_k^\eta(\mathbf{x}-\mathbf{x}_k,\widehat{\boldsymbol{\eta}})$ and $\mathcal{L}_k^\tau(\mathbf{x}-\mathbf{x}_k,\widehat{\boldsymbol{\tau}})$

are related by the equality (15.55) independently on the concrete representations of these operators.

We introduce the so-called effective fields $\overline{\boldsymbol{\varepsilon}}(\mathbf{x})$, $\overline{\boldsymbol{\sigma}}(\mathbf{x})$, and $\overline{\mathbf{u}}(\mathbf{x})$ as the local fields acting on the heterogeneity $\mathbf{x} \in v_k$, which are coincide with the fields $\boldsymbol{\varepsilon}^0(\mathbf{x})$, $\boldsymbol{\sigma}^0(\mathbf{x})$, and $\mathbf{u}^0(\mathbf{x})$, respectively, for the current problem for a single inclusion. Introduction of the effective fields makes it possible to define another sort of the perturbators

$$\mathcal{L}_k^{\epsilon}(\mathbf{x} - \mathbf{x}_k, \overline{\boldsymbol{\varepsilon}}) \equiv \boldsymbol{\varepsilon}(\mathbf{x}) - \overline{\boldsymbol{\varepsilon}}(\mathbf{x}), \tag{15.58}$$

$$\mathcal{L}_k^{\sigma}(\mathbf{x} - \mathbf{x}_k, \overline{\boldsymbol{\sigma}}) \equiv \boldsymbol{\sigma}(\mathbf{x}) - \overline{\boldsymbol{\sigma}}(\mathbf{x}), \tag{15.59}$$

$$\mathcal{L}_k^{u}(\mathbf{x} - \mathbf{x}_k, \overline{\mathbf{u}}) \equiv \mathbf{u}(\mathbf{x}) - \overline{\mathbf{u}}(\mathbf{x}). \tag{15.60}$$

Strictly speaking, the perturbators $\mathcal{L}_k^{\tau}(\mathbf{x} - \mathbf{x}_k, \widehat{\boldsymbol{\tau}})$ (15.49), $\mathcal{L}_k^{\eta}(\mathbf{x} - \mathbf{x}_k, \widehat{\boldsymbol{\eta}})$ (15.50), and $\mathcal{L}_k^{u\tau}(\mathbf{x} - \mathbf{x}_k, \widehat{\boldsymbol{\tau}})$ (15.51) and their integral representations (15.52)–(15.54) are just the notations of some problem which is destined to solve, while the perturbators $\mathcal{L}_k^{\epsilon}(\mathbf{x} - \mathbf{x}_k, \overline{\boldsymbol{\varepsilon}})$ (15.58), $\mathcal{L}_k^{\sigma}(\mathbf{x} - \mathbf{x}_k, \overline{\boldsymbol{\sigma}})$ (15.59), and $\mathcal{L}_k^{u}(\mathbf{x} - \mathbf{x}_k, \overline{\mathbf{u}})$ (15.60) are the solutions of this problem. In the case of the integral representation (15.52)–(15.54), this solutions should be found from the joint solutions of Eqs. (15.49)–(15.51) and (15.52)–(15.54), respectively, and the contact conditions at the interfaces Γ_k^u and Γ_k^{σ} (linear elastic case is considered in Sect. 6.2 where the other methods are presented if the integral representations (15.52)–(15.54) are not used).

The mentioned perturbators (15.49)–(15.51) and (15.58)–(15.60) can be found by any available numerical method, such as the volume integral equation (VIE), boundary element method (BEM), FEM, hybrid FEM–BEM, multipole expansion method, complex potential method, and other (see for references Chap. 4 and [544, 849, 959, 1353]). Each method has advantages and disadvantages, and it is crucial for the analyst to be aware of their range of applications. In particular, the VIE method for the ideal contact conditions enables one to restrict discretization to the inclusions only (in contrast to the FEA), and an inhomogeneous structure of inclusions (see, e.g., [313, 695, 1662]) presents no problem in the framework of the same numerical scheme (compared to the standard BEM). The VIE method (see Chap. 4) has well-developed routines for the solution of integral equations (such as the iteration method and the quadrature schemes) and allows to analyze arbitrary inhomogeneous fields $\boldsymbol{\varepsilon}^0(\mathbf{x})$ and $\boldsymbol{\sigma}^0(\mathbf{x})$. However, the VIE method is quite time-consuming, and no optimized commercial softwares exist for its application. From other side, the FEM is supported by well-developed commercial softwares and gives strong advantages in terms of CPU time. The FEA is especially effective for estimations of perturbators at the constant fields $\boldsymbol{\varepsilon}^0, \boldsymbol{\sigma}^0$ =const. (see, e.g., Chap. 9).

Equations (15.58) and (15.59) can be recast in other operator forms

$$\boldsymbol{\varepsilon}(\mathbf{x}) = \mathcal{B}_i^{\epsilon}(\overline{\boldsymbol{\varepsilon}})(\mathbf{x}) + \mathbf{C}_i^{\epsilon}(\mathbf{x}), \quad \bar{v}_i \boldsymbol{\tau}(\mathbf{x}) = \mathcal{R}_i^{\epsilon}(\overline{\boldsymbol{\varepsilon}})(\mathbf{x}) + \mathbf{F}_i^{\epsilon}(\mathbf{x}), \tag{15.61}$$

$$\boldsymbol{\sigma}(\mathbf{x}) = \mathcal{B}_i^{\sigma}(\overline{\boldsymbol{\sigma}})(\mathbf{x}) + \mathbf{C}_i^{\sigma}(\mathbf{x}), \quad \bar{v}_i \boldsymbol{\eta}(\mathbf{x}) = \mathcal{R}_i^{\sigma}(\overline{\boldsymbol{\sigma}})(\mathbf{x}) + \mathbf{F}_i^{\sigma}(\mathbf{x}), \tag{15.62}$$

where

$$\mathcal{B}_i^{\epsilon}(\overline{\boldsymbol{\varepsilon}})(\mathbf{x}) = \mathcal{L}_i^{\epsilon}(\mathbf{x} - \mathbf{x}_i, \overline{\boldsymbol{\varepsilon}}) + \mathbf{I}\overline{\boldsymbol{\varepsilon}}(\mathbf{x}) - \mathcal{L}_i^{\epsilon}(\mathbf{x} - \mathbf{x}_i, \mathbf{0}),$$
$$\mathbf{C}_i^{\epsilon}(\mathbf{x}) = \mathcal{L}_i^{\epsilon}(\mathbf{x} - \mathbf{x}_i, \mathbf{0}), \tag{15.63}$$
$$\mathcal{B}_i^{\sigma}(\overline{\boldsymbol{\sigma}})(\mathbf{x}) = \mathcal{L}_i^{\sigma}(\mathbf{x} - \mathbf{x}_i, \overline{\boldsymbol{\sigma}}) + \mathbf{I}\overline{\boldsymbol{\sigma}}(\mathbf{x}) - \mathcal{L}_i^{\sigma}(\mathbf{x} - \mathbf{x}_i, \mathbf{0}),$$
$$\mathbf{C}_i^{\sigma}(\mathbf{x}) = \mathcal{L}_i^{\sigma}(\mathbf{x} - \mathbf{x}_i, \mathbf{0}), \tag{15.64}$$
$$\mathcal{R}_i^{\epsilon}(\overline{\boldsymbol{\varepsilon}})(\mathbf{x}) = \bar{v}_i \mathcal{L}_1^{(1)} * \mathcal{B}_i^{\epsilon}(\overline{\boldsymbol{\varepsilon}})(\mathbf{x}), \quad \mathbf{F}_i^{\epsilon}(\mathbf{x}) = \bar{v}_i\big[\mathcal{L}_1^{(1)} * \mathbf{C}_i^{\epsilon}(\mathbf{x}) + \boldsymbol{\alpha}_1(\mathbf{x})\big], \tag{15.65}$$
$$\mathcal{R}_i^{\sigma}(\overline{\boldsymbol{\sigma}})(\mathbf{x}) = \bar{v}_i \mathcal{M}_1^{(1)} * \mathcal{B}_i^{\sigma}(\overline{\boldsymbol{\sigma}})(\mathbf{x}), \quad \mathbf{F}_i^{\sigma}(\mathbf{x}) = \bar{v}_i\big[\mathcal{M}_1^{(1)} * \mathbf{C}_i^{\sigma}(\mathbf{x}) + \boldsymbol{\beta}_1(\mathbf{x})\big]. \tag{15.66}$$

15.2.3 Operator Form of General Integral Equations

Let us consider an arbitrary random realization α of inclusions in the increasing sequence w_p

$$\boldsymbol{\varepsilon}(\mathbf{x}, \alpha) = \boldsymbol{\varepsilon}^0(\mathbf{x}, \alpha) + \lim_{w_p \to R^d} \bigg\{ \int_{w_p} \mathcal{L}_k^{\tau}(\mathbf{x} - \mathbf{x}_k, \widehat{\boldsymbol{\tau}}) V_k^{\delta}(\mathbf{x}_k, \alpha) d\mathbf{x}_k$$
$$+ \int_{\Gamma_p^0} \nabla\mathbf{G}(\mathbf{x} - \mathbf{s})\boldsymbol{\tau}^{[[u]]}(\mathbf{s}, \alpha)\mathbf{n}(\mathbf{s}) ds \bigg\}, \tag{15.67}$$

$$\boldsymbol{\sigma}(\mathbf{x}, \alpha) = \boldsymbol{\sigma}^0(\mathbf{x}, \alpha) + \lim_{w_p \to R^d} \bigg\{ \int_{w_p} \mathcal{L}_k^{\eta}(\mathbf{x} - \mathbf{x}_k, \widehat{\boldsymbol{\eta}}) V_k^{\delta}(\mathbf{x}_k, \alpha) d\mathbf{x}_k$$
$$- \int_{\Gamma_p^0} \mathbf{L}^c * \nabla\mathbf{G}(\mathbf{x} - \mathbf{s})\mathbf{L}^c * \boldsymbol{\eta}^{[[u]]}(\mathbf{s}, \alpha)\mathbf{n}(\mathbf{s}) ds \bigg\}, \tag{15.68}$$

where the parameters $\boldsymbol{\tau}^{[[u]]}(\mathbf{s}, \alpha)$ and $\boldsymbol{\eta}^{[[u]]}(\mathbf{s}, \alpha)$ are introduced in the surface integrals in (15.67) and (15.68) appearing to the use of the Gauss theorem analogously to Eq. (15.53) and taking into account the possibility of intersection of some part of heterogeneities v_i with the macroboundary Γ_p^0 where $\boldsymbol{\tau}^{[[u]]}(\mathbf{s}, \alpha), \boldsymbol{\eta}^{[[u]]}(\mathbf{s}, \alpha) \not\equiv \mathbf{0}$ ($\mathbf{s} \in \Gamma_p^0$). The parameters $\boldsymbol{\tau}^{[[u]]}(\mathbf{s}, \alpha)$ and $\boldsymbol{\eta}^{[[u]]}(\mathbf{s}, \alpha)$ are the summations of the known eigenfields $\boldsymbol{\tau}(\mathbf{s}, \alpha)$ and $\boldsymbol{\eta}(\mathbf{s}, \alpha)$ ($\mathbf{s} \in v_i$) as in Eqs. (15.58) and (15.63), respectively, and some additional items corresponding, e.g., to a possible imperfection of interfaces Γ_i (see for details a particular elastic case in [208]). However, the surface integrals in Eqs. (15.67) and (15.68) will be eliminated in subsequent transformations of Eqs. (15.67) and (15.68), and, because of this, the parameters $\boldsymbol{\tau}^{[[u]]}(\mathbf{s}, \alpha)$ and $\boldsymbol{\eta}^{[[u]]}(\mathbf{s}, \alpha)$ are not considered here in more details.

Now, we apply the centering method of subtraction from both sides of Eqs. (15.67) and (15.68) their statistical averages obtained without any auxiliary assumptions (such as, e.g., EFH). It leads to the following new equation for statistically inhomogeneous media

$$\boldsymbol{\varepsilon}(\mathbf{x},\alpha) = \langle\boldsymbol{\varepsilon}\rangle(\mathbf{x}) + \lim_{w_p\to R^d}\int_{w_p}[\boldsymbol{\mathcal{L}}_k^{\tau}(\mathbf{x}-\mathbf{x}_k,\widehat{\boldsymbol{\tau}})V_k^{\delta}(\mathbf{x}_k,\alpha)$$
$$-\langle\boldsymbol{\mathcal{L}}_k^{\tau}(\mathbf{x}-\mathbf{x}_k,\widehat{\boldsymbol{\tau}})\rangle(\mathbf{x}_k)]d\mathbf{x}_k + \boldsymbol{\mathcal{L}}_{\epsilon}^{\Gamma}, \tag{15.69}$$

$$\boldsymbol{\sigma}(\mathbf{x},\alpha) = \langle\boldsymbol{\sigma}\rangle(\mathbf{x}) + \lim_{w_p\to R^d}\int_{w_p}[\boldsymbol{\mathcal{L}}_k^{\eta}(\mathbf{x}-\mathbf{x}_k,\widehat{\boldsymbol{\eta}})V_k^{\delta}(\mathbf{x}_k,\alpha)$$
$$-\langle\boldsymbol{\mathcal{L}}_k^{\eta}(\mathbf{x}-\mathbf{x}_k,\widehat{\boldsymbol{\eta}})\rangle(\mathbf{x}_k)]d\mathbf{x}_k + \boldsymbol{\mathcal{L}}_{\sigma}^{\Gamma}. \tag{15.70}$$

In the right-hand side of Eqs. (15.69) and (15.70), one introduced the integrals over the external surface Γ_p^0

$$\boldsymbol{\mathcal{L}}_{\epsilon}^{\Gamma} \equiv \lim_{w_p\to R^d}\int_{\Gamma_p^0}\Big[\nabla\mathbf{G}(\mathbf{x}-\mathbf{s})\boldsymbol{\tau}^{[[u]]}(\mathbf{s},\alpha) - \langle\nabla\mathbf{G}(\mathbf{x}-\mathbf{s})\boldsymbol{\tau}^{[[u]]}\rangle(\mathbf{s})\Big]\mathbf{n}(\mathbf{s})ds, \tag{15.71}$$

$$\boldsymbol{\mathcal{L}}_{\sigma}^{\Gamma} \equiv -\lim_{w_p\to R^d}\mathbf{L}^{(0)}\int_{\Gamma_p^0}\Big[\nabla\mathbf{G}(\mathbf{x}-\mathbf{s})\mathbf{L}^{(0)}\boldsymbol{\eta}^{[[u]]}(\mathbf{s},\alpha)$$
$$-\langle\nabla\mathbf{G}(\mathbf{x}-\mathbf{s})\mathbf{L}^{(0)}\boldsymbol{\eta}^{[[u]]}\rangle(\mathbf{s})\Big]\mathbf{n}(\mathbf{s})ds, \tag{15.72}$$

which vanish at the sufficient distance $\mathbf{x}$ (assumed hereafter) from the boundary Γ_p^0 (15.59), when the validity of separation of length scale (13.1$_2$) holds. Vanishing of the integrals (15.71) and (15.72) can be proved analogously to the analyses of Eqs. (15.58) and (15.63).

Thus, the remaining integrals in Eqs. (15.69) and (15.70) depend only on the perturbators $\boldsymbol{\mathcal{L}}_k^{\tau}(\mathbf{x}-\mathbf{x}_k,\widehat{\boldsymbol{\tau}})$ and $\boldsymbol{\mathcal{L}}_k^{\eta}(\mathbf{x}-\mathbf{x}_k,\widehat{\boldsymbol{\eta}})$ presenting notations of the numerical solutions (15.49) and (15.50) which are quite general and not related with a concrete numerical method. We only need to know that $\boldsymbol{\mathcal{L}}_k^{\tau}(\mathbf{x}-\mathbf{x}_k,\widehat{\boldsymbol{\tau}})$ and $\boldsymbol{\mathcal{L}}_k^{\eta}(\mathbf{x}-\mathbf{x}_k,\widehat{\boldsymbol{\eta}})$ behave as $O(|\mathbf{x}-\mathbf{x}_k|^{-d})$ at the infinity $|\mathbf{x}-\mathbf{x}_k|\to\infty$. The volume integrals in (15.69) and (15.70) converge absolutely for both the statistically homogeneous and inhomogeneous random set X of heterogeneities. Indeed, even for the FGMs, the integrands in the square brackets in Eqs. (15.69) and (15.70) are of order $O(|\mathbf{x}-\mathbf{y}|^{-2d})$ as $|\mathbf{x}-\mathbf{y}|\to\infty$, and the integrals in Eqs. (15.69) and (15.70) converge absolutely. Therefore, Eqs. (15.69) and (15.70) can be reduced [analogously to Eqs. (15.65) and (15.66)] to the new GIEs

$$\boldsymbol{\varepsilon}(\mathbf{x},\alpha)=\langle\boldsymbol{\varepsilon}\rangle(\mathbf{x})+\int[\boldsymbol{\mathcal{L}}_k^{\tau}(\mathbf{x}-\mathbf{x}_k,\widehat{\boldsymbol{\tau}})V_k^{\delta}(\mathbf{x}_k,\alpha)-\langle\boldsymbol{\mathcal{L}}_k^{\tau}(\mathbf{x}-\mathbf{x}_k,\widehat{\boldsymbol{\tau}})\rangle(\mathbf{x}_k)]d\mathbf{x}_k, \tag{15.73}$$

$$\boldsymbol{\sigma}(\mathbf{x},\alpha)=\langle\boldsymbol{\sigma}\rangle(\mathbf{x})+\int[\boldsymbol{\mathcal{L}}_k^{\eta}(\mathbf{x}-\mathbf{x}_k,\widehat{\boldsymbol{\eta}})V_k^{\delta}(\mathbf{x}_k,\alpha)-\langle\boldsymbol{\mathcal{L}}_k^{\eta}(\mathbf{x}-\mathbf{x}_k,\widehat{\boldsymbol{\eta}})\rangle(\mathbf{x}_k)]d\mathbf{x}_k, \tag{15.74}$$

where the integration domain R^d will be omitted hereafter for simplicity of notations.

It should be mentioned that Eqs. (15.73) and (15.74) are equivalent and can be obtained one from the other by the use of the identities (15.55)–(15.57). This equivalence leads to the equivalence of the dual effective laws with reciprocally inverse effective properties. Furthermore, although Eqs. (15.73) and (15.74) depend only on the perturbators (15.58) and (15.59), their derivation was performed by the use of the Green function [see Eqs. (15.67) and (15.68)], and, therefore, the linear response (15.1) for the matrix was assumed. However, the linear response (2.8) was not presumed for the heterogeneities v_i and their interfaces Γ_i $(i = 1, 2, \ldots)$. Therefore, Eqs. (15.73) and (15.74) can be applied to a lot of problems with nonlinear properties of heterogeneities, such as accumulation of plastic deformations in the coating surrounding heterogeneities (see the localized model of plasticity of composites, Chapter 16 in [200]) and cohesive zone model (CZM, see, e.g., [1483, 1484]).

Let the inclusions $v_1, \ldots, v_n$ be fixed, and we define two sorts of effective fields for the driving fields $\overline{\boldsymbol{\varepsilon}}_i(\mathbf{x}), \widetilde{\boldsymbol{\varepsilon}}_{1,\ldots,n}(\mathbf{x})$ and fluxes $\overline{\boldsymbol{\sigma}}_i(\mathbf{x}), \widetilde{\boldsymbol{\sigma}}_{1,\ldots,n}(\mathbf{x})$ $(i = 1, \ldots, n;\ \mathbf{x} \in v_1, \ldots, v_n)$ by the use of rearrangements of Eqs. (15.72) and (15.73) in the following operator forms (an analogous particular case of this manipulation approach for the linear elastic problem with the perfect interface is given in Chap. 9; see also [202, 203])

$$\boldsymbol{\vartheta}(\mathbf{x}) = \overline{\boldsymbol{\vartheta}}_i(\mathbf{x}) + \boldsymbol{\mathcal{L}}_i^{\zeta}(\mathbf{x} - \mathbf{x}_i, \widehat{\boldsymbol{\zeta}}), \tag{15.75}$$

$$\overline{\boldsymbol{\vartheta}}_i(\mathbf{x}) = \widetilde{\boldsymbol{\vartheta}}_{1,\ldots,n}(\mathbf{x}) + \sum_{j\neq i}^{n} \boldsymbol{\mathcal{L}}_j^{\zeta}(\mathbf{x} - \mathbf{x}_j, \widehat{\boldsymbol{\zeta}}), \tag{15.76}$$

$$\widetilde{\boldsymbol{\vartheta}}_{1,\ldots,n}(\mathbf{x}) = \langle \boldsymbol{\vartheta} \rangle(\mathbf{x}) + \int [\boldsymbol{\mathcal{L}}_k^{\zeta}(\mathbf{x} - \mathbf{x}_k, \widehat{\boldsymbol{\zeta}}) V^{\delta}(\mathbf{x}_k|; v_1, \mathbf{x}_1; \ldots; v_n, \mathbf{x}_n) \\ - \langle \boldsymbol{\mathcal{L}}_k^{\zeta}(\mathbf{x} - \mathbf{x}_k, \widehat{\boldsymbol{\zeta}}) \rangle (\mathbf{x}_k)] d\mathbf{x}_k \tag{15.77}$$

for $\mathbf{x} \in v_i$, $i = 1, 2, \ldots, n$; here $V^{\delta}(\mathbf{x}_k|; v_1, \mathbf{x}_1; \ldots; v_n, \mathbf{x}_n) = \sum_m \delta(\mathbf{x}_k - \mathbf{x}_m) - \sum_{i=1}^{n} \delta(\mathbf{x}_k - \mathbf{x}_i)$ is a random delta function of heterogeneity centers $\mathbf{x}_m$ $(m = 1, 2, \ldots)$ under the condition that $\mathbf{x}_k \neq \mathbf{x}_i$, $\mathbf{x}_i \neq \mathbf{x}_j$ if $i \neq j$ $(i, j = 1, \ldots, n)$. Hereafter for contraction of calculations, we introduce the substitutions

$$(\boldsymbol{\varepsilon}, \boldsymbol{\sigma}) \leftrightarrow \boldsymbol{\vartheta}, \quad (\boldsymbol{\tau}, \boldsymbol{\eta}) \leftrightarrow \boldsymbol{\zeta}, \quad (\boldsymbol{\alpha}, \boldsymbol{\beta}) \leftrightarrow \boldsymbol{\gamma} \tag{15.78}$$

indicating that for each triplet $\boldsymbol{\vartheta} = \boldsymbol{\varepsilon}$, $\boldsymbol{\zeta} = \boldsymbol{\tau}$, $\boldsymbol{\gamma} = \boldsymbol{\alpha}$ and $\boldsymbol{\vartheta} = \boldsymbol{\sigma}$, $\boldsymbol{\zeta} = \boldsymbol{\eta}$, $\boldsymbol{\gamma} = \boldsymbol{\beta}$, Eqs. (15.75)–(15.77) and the subsequent equations with the variables $\boldsymbol{\vartheta}$, $\boldsymbol{\zeta}$, and $\boldsymbol{\gamma}$ are reduced to the corresponding equations for the driving field $\boldsymbol{\varepsilon}$ and flux $\boldsymbol{\sigma}$. The definitions of the effective fields $\overline{\boldsymbol{\vartheta}}_i(\mathbf{x})$, $\widetilde{\boldsymbol{\vartheta}}_{1,2,\ldots,n}(\mathbf{x})$ as well as their statistical averages $\langle \overline{\boldsymbol{\vartheta}}_i \rangle(\mathbf{x})$, $\langle \widetilde{\boldsymbol{\vartheta}}_{1,2,\ldots,n} \rangle(\mathbf{x})$ are nothing more than a notation convenience for the different terms of the infinite system (15.75)–(15.77), respectively.

Then, considering some conditional statistical averages of the general integral equations (15.72) and (15.73) leads to an infinite system of *new integral equations* $(n = 1, 2, \ldots)$

$$\langle \boldsymbol{\vartheta} | v_1, \mathbf{x}_1; \ldots; v_n, \mathbf{x}_n \rangle(\mathbf{x}) - \sum_{i=1}^{n} \langle \boldsymbol{\mathcal{L}}_i^{\zeta}(\mathbf{x} - \mathbf{x}_i, \widehat{\boldsymbol{\zeta}}) | v_1, \mathbf{x}_1; \ldots; v_n, \mathbf{x}_n \rangle_i$$

$$= \langle \boldsymbol{\vartheta} \rangle(\mathbf{x}) + \int \big\{ \langle \boldsymbol{\mathcal{L}}_j^{\zeta}(\mathbf{x} - \mathbf{x}_j, \widehat{\boldsymbol{\zeta}}) |; v_1, \mathbf{x}_1; \ldots; v_n, \mathbf{x}_n \rangle_j \varphi(v_j, \mathbf{x}_j |; v_1, \mathbf{x}_1, \ldots, v_n, \mathbf{x}_n)$$

$$- \langle \boldsymbol{\mathcal{L}}_j^{\zeta}(\mathbf{x} - \mathbf{x}_j, \widehat{\boldsymbol{\zeta}}) \rangle(\mathbf{x}_j) \big\} d\mathbf{x}_j, \qquad (15.79)$$

Since $\mathbf{x} \in v_1, \ldots, v_n$ in the n-th line of the system can take the values inside the inclusions $v_1, \ldots, v_n$, the n-th line actually contains n equations. Statistical averaging $\langle (\cdot) \rangle_i$ stands for the averaging over all surrounding heterogeneities at the fixed v_i, while the average $\langle (\cdot) \rangle(\mathbf{x}_j)$ implies the averaging over all possible location of v_i with possible dependence of this average on the macro-coordinate $\mathbf{x}_i$ as for FGMs.

Introduction of both the effective fields (15.75) and another sort of the perturbators (15.58) and (15.59) makes it possible to recast the GIE (15.79) in terms of the effective fields ($\mathbf{x} \in v_k,\ k = 1, 2, \ldots$)

$$\langle \overline{\boldsymbol{\vartheta}} | v_1, \mathbf{x}_1; \ldots; v_n, \mathbf{x}_n \rangle(\mathbf{x}) - \sum_{i \neq k}^{n} \langle \boldsymbol{\mathcal{L}}_i^{\theta}(\mathbf{x} - \mathbf{x}_i, \overline{\boldsymbol{\vartheta}}) | v_1, \mathbf{x}_1; \ldots; v_n, \mathbf{x}_n \rangle_i$$

$$= \langle \boldsymbol{\vartheta} \rangle(\mathbf{x}) + \int \big\{ \langle \boldsymbol{\mathcal{L}}_j^{\theta}(\mathbf{x} - \mathbf{x}_j, \overline{\boldsymbol{\vartheta}}) |; v_1, \mathbf{x}_1; \ldots; v_n, \mathbf{x}_n \rangle_j \varphi(v_j, \mathbf{x}_j |; v_1, \mathbf{x}_1, \ldots, v_n, \mathbf{x}_n)$$

$$- \langle \boldsymbol{\mathcal{L}}_j^{\theta}(\mathbf{x} - \mathbf{x}_j, \overline{\boldsymbol{\vartheta}}) \rangle(\mathbf{x}_j) \big\} d\mathbf{x}_j. \qquad (15.80)$$

15.2.4 Some Particular Cases and Approximate GIEs

It should be mentioned that for the perfect interface (15.17) and the integral representations of the perturbators (15.76) and (15.77), the new GIEs (15.72) and (15.73) can be presented through the Green functions

$$\boldsymbol{\varepsilon}(\mathbf{x}) = \langle \boldsymbol{\varepsilon} \rangle(\mathbf{x}) + \int [\mathbf{U}(\mathbf{x} - \mathbf{y}) \boldsymbol{\tau}(\mathbf{y}) - \langle \mathbf{U}(\mathbf{x} - \mathbf{y}) \boldsymbol{\tau} \rangle(\mathbf{y})] d\mathbf{y}$$

$$+ \int [\nabla \mathbf{G}(\mathbf{x} - \mathbf{y}) \boldsymbol{\pi}(\mathbf{y}) - \langle \nabla \mathbf{G} \boldsymbol{\pi} \rangle(\mathbf{y})] d\mathbf{y}, \qquad (15.81)$$

$$\boldsymbol{\sigma}(\mathbf{x}) = \langle \boldsymbol{\sigma} \rangle(\mathbf{x}) + \int [\boldsymbol{\Gamma}(\mathbf{x} - \mathbf{y}) * \boldsymbol{\eta}(\mathbf{y}) - \langle \boldsymbol{\Gamma} \boldsymbol{\eta} \rangle(\mathbf{y})] d\mathbf{y}$$

$$+ \boldsymbol{\mathcal{L}}^{(0)} * \int [\nabla \mathbf{G}(\mathbf{x} - \mathbf{y}) \boldsymbol{\pi}(\mathbf{y}) - \langle \nabla \mathbf{G}(\mathbf{x} - \mathbf{y}) \boldsymbol{\pi} \rangle(\mathbf{y})] d\mathbf{y}. \qquad (15.82)$$

Contrary to the new GIEs having both the operator form (15.72) and (15.73) and the Green's function representation (15.81)–(15.82), the classical GIEs are only known in the form expressed through the Green's functions (the case of the local elasticity (15.3) and $\boldsymbol{\pi}, \mathbf{b}_1, \mathbf{f}_1 \equiv \mathbf{0}$ was considered in Chap. 7)

$$\boldsymbol{\varepsilon}(\mathbf{x}) = \langle\boldsymbol{\varepsilon}\rangle(\mathbf{x}) + \int \mathbf{U}(\mathbf{x}-\mathbf{y})[\boldsymbol{\tau}(\mathbf{y}) - \langle\boldsymbol{\tau}\rangle(\mathbf{y})]d\mathbf{y}$$
$$+ \int \nabla\mathbf{G}(\mathbf{x}-\mathbf{y})[\boldsymbol{\pi}(\mathbf{y}) - \langle\boldsymbol{\pi}\rangle(\mathbf{y})]d\mathbf{y}, \tag{15.83}$$

$$\boldsymbol{\sigma}(\mathbf{x}) = \langle\boldsymbol{\sigma}\rangle(\mathbf{x}) + \int \boldsymbol{\Gamma}(\mathbf{x}-\mathbf{y})*[\boldsymbol{\eta}(\mathbf{y}) - \langle\boldsymbol{\eta}\rangle(\mathbf{y})]d\mathbf{y}$$
$$+\boldsymbol{\mathcal{L}}^{(0)}*\int \nabla\mathbf{G}(\mathbf{x}-\mathbf{y})[\boldsymbol{\pi}(\mathbf{y}) - \langle\boldsymbol{\pi}\rangle(\mathbf{y})]d\mathbf{y}, \tag{15.84}$$

where one used the point approximations

$$\langle\mathbf{U}(\mathbf{x}-\mathbf{y})\boldsymbol{\tau}\rangle(\mathbf{y}) = \mathbf{U}(\mathbf{x}-\mathbf{y})\langle\boldsymbol{\tau}\rangle(\mathbf{y}), \tag{15.85}$$
$$\langle\boldsymbol{\Gamma}(\mathbf{x}-\mathbf{y})*\boldsymbol{\eta}\rangle(\mathbf{y}) = \boldsymbol{\Gamma}(\mathbf{x}-\mathbf{y})*\langle\boldsymbol{\eta}\rangle(\mathbf{y}), \tag{15.86}$$
$$\langle\nabla\mathbf{G}(\mathbf{x}-\mathbf{y})\boldsymbol{\pi}\rangle(\mathbf{y}) = \nabla\mathbf{G}(\mathbf{x}-\mathbf{y})\langle\boldsymbol{\pi}\rangle(\mathbf{y}). \tag{15.87}$$

Buryachenko [203, 208] has estimated the error of the approximations (15.85) and (15.86) in some model examples and proved an incorrectness of a popular justification (with an intuitive level of rigor) of Eqs. (15.85) and (15.86) asserting that the deterministic functions $\mathbf{U}(\mathbf{x}-\mathbf{y})$ and $\boldsymbol{\Gamma}(\mathbf{x}-\mathbf{y})$ can be always carried out from the brackets $\langle(\cdot)\rangle(\mathbf{y})$ of statistical average in Eqs. (15.81) and (15.82). Equations (15.83) and (15.84) were proposed by Buryachenko [271] for the FGMs. Additional assumption of the statistical homogeneity of media (5.61_2) subjected to the homogeneous boundary conditions (15.23) reduces Eqs. (15.83) and (15.84) to the known ones

$$\boldsymbol{\varepsilon}(\mathbf{x}) = \langle\boldsymbol{\varepsilon}\rangle + \int \mathbf{U}(\mathbf{x}-\mathbf{y})[\boldsymbol{\tau}(\mathbf{y}) - \langle\boldsymbol{\tau}\rangle]d\mathbf{y}$$
$$+ \int \nabla\mathbf{G}(\mathbf{x}-\mathbf{y})[\boldsymbol{\pi}(\mathbf{y}) - \langle\boldsymbol{\pi}\rangle]d\mathbf{y}, \tag{15.88}$$

$$\boldsymbol{\sigma}(\mathbf{x}) = \langle\boldsymbol{\sigma}\rangle + \int \boldsymbol{\Gamma}(\mathbf{x}-\mathbf{y})*[\boldsymbol{\eta}(\mathbf{y}) - \langle\boldsymbol{\eta}\rangle]d\mathbf{y}$$
$$+\boldsymbol{\mathcal{L}}^{(0)}*\int \nabla\mathbf{G}(\mathbf{x}-\mathbf{y})[\boldsymbol{\pi}(\mathbf{y}) - \langle\boldsymbol{\pi}\rangle]d\mathbf{y}. \tag{15.89}$$

Moreover, for a particular case of the field X bounded in one direction (such as a laminated structure of some real FGM, see, e.g., [1218] and Chapter 12 in [200]), Eqs. (15.88) and (15.89) are reduced to the equations

$$\boldsymbol{\varepsilon}(\mathbf{x}) = \boldsymbol{\varepsilon}^0 + \int \mathbf{U}(\mathbf{x}-\mathbf{y})\boldsymbol{\tau}(\mathbf{y})d\mathbf{y} + \int \nabla\mathbf{G}(\mathbf{x}-\mathbf{y})\boldsymbol{\pi}(\mathbf{y})d\mathbf{y}, \tag{15.90}$$

$$\boldsymbol{\sigma}(\mathbf{x}) = \boldsymbol{\sigma}^0 + \int \boldsymbol{\Gamma}(\mathbf{x}-\mathbf{y}) * \boldsymbol{\eta}(\mathbf{y})d\mathbf{y} + \boldsymbol{\mathcal{L}}^{(0)} * \int \nabla\mathbf{G}(\mathbf{x}-\mathbf{y})\boldsymbol{\pi}(\mathbf{y})d\mathbf{y}. \tag{15.91}$$

which were also widely used for statistically homogeneous media subjected to the homogeneous boundary conditions. In such a case, the improper integrals in Eqs. (15.90) and (15.91) do not absolutely convergent at infinity where the integrands behave as $O((|\mathbf{x}-\mathbf{y}|^{-d})$.

Following Beran and McCoy [106, 107], we introduce a projector operator $\mathcal{P}$ called the deviation operator which projects out the deviation part of the random function standing on its right side in the form of a single or product expressions: $\mathcal{P}\mathbf{f} \equiv \mathbf{f} - \langle \mathbf{f}\rangle$, $\mathcal{P}\mathbf{fg} \equiv \mathbf{fg} - \langle \mathbf{fg}\rangle$, and $\mathcal{P}^2 = \mathcal{P}$. Equations (15.72) and (15.88) can be presented in the operator forms

$$\boldsymbol{\varepsilon}(\mathbf{x}) = \langle\boldsymbol{\varepsilon}\rangle(\mathbf{x}) + (\mathcal{P}\mathbf{U} * \boldsymbol{\tau})(\mathbf{x}) + (\mathcal{P}\nabla\mathbf{G} * \boldsymbol{\pi})(\mathbf{x}), \tag{15.92}$$

$$\boldsymbol{\varepsilon}(\mathbf{x}) = \langle\boldsymbol{\varepsilon}\rangle(\mathbf{x}) + (\mathbf{U} * \mathcal{P}\boldsymbol{\tau})(\mathbf{x}) + (\nabla\mathbf{G} * \mathcal{P}\boldsymbol{\pi})(\mathbf{x}), \tag{15.93}$$

with the analogous recasting of Eqs. (15.66) and (15.89), respectively. Thus, the operators $\mathcal{P}$ and $\mathbf{U} and \nabla\mathbf{G}$ are not commutative: $\mathcal{P}\mathbf{U} \neq \mathbf{U}\mathcal{P} and\ \mathcal{P}\nabla\mathbf{G} \neq \nabla\mathbf{G}\mathcal{P}$. It is obvious that the projector operator $\mathcal{P}$ is a particular case of the centering operation $\mathcal{P}\mathbf{f} \equiv \langle\langle\mathbf{f}\rangle\rangle$ for statistically homogeneous media (5.61_2). In a similar manner, Eqs. (15.92) and (15.93) are the particular case of Eqs. (15.81) and (15.84), respectively, presented here just for its recasting in terms of the popular projector operator $\mathcal{P}$.

15.2.5 Composites of Periodical Structure

For the periodic structures mentioned above, we considered in Sect. 7.4 the general integral equations (7.73) for the particular static problem ($\mathbf{b} \equiv \mathbf{0}$) with uncoupled ($M = 1$) local constitutive equation. Straightforward generalization of these equations to the wave motion problem for the composites with the coupled nonlocal properties of constituents leads to the following new GIEs (see for details [208])

$$\boldsymbol{\varepsilon}(\mathbf{x}) = \langle\boldsymbol{\varepsilon}\rangle_{\mathbf{x}}(\mathbf{x}) + \int \langle\langle \boldsymbol{\mathcal{L}}_k^{\tau}(\mathbf{x}-\mathbf{x}_k, \widehat{\boldsymbol{\tau}})\rangle\rangle_{\mathbf{x}} d\mathbf{x}_k, \tag{15.94}$$

$$\boldsymbol{\sigma}(\mathbf{x}) = \langle\boldsymbol{\sigma}\rangle_{\mathbf{x}}(\mathbf{x}) + \int \langle\langle \boldsymbol{\mathcal{L}}_k^{\eta}(\mathbf{x}-\mathbf{x}_k, \widehat{\boldsymbol{\eta}})\rangle\rangle_{\mathbf{x}} d\mathbf{x}_k, \tag{15.95}$$

which are exact for linear functions $\boldsymbol{\varepsilon}^0(\mathbf{x})$ and $\boldsymbol{\sigma}^0(\mathbf{x})$, whereas the operation $\langle\!\langle(\cdot)\rangle\!\rangle_{\mathbf{x}}$ is defined by Eq. (7.69).

In particular, for the ideal interface (15.17) and integral representations for the perturbators (15.52) and (15.53), Eqs. (15.93) and (15.94) are reduced to

$$\boldsymbol{\varepsilon}(\mathbf{x})=\langle\boldsymbol{\varepsilon}\rangle_{\mathbf{x}}(\mathbf{x})+\int[\langle\!\langle\mathbf{U}(\mathbf{x}-\mathbf{y})\boldsymbol{\tau}(\mathbf{y})\rangle\!\rangle_{\mathbf{x}}+\langle\!\langle\nabla\mathbf{G}(\mathbf{x}-\mathbf{y})\boldsymbol{\pi}(\mathbf{y})\rangle\!\rangle_{\mathbf{x}}]d\mathbf{y}, \tag{15.96}$$

$$\boldsymbol{\sigma}(\mathbf{x})=\langle\boldsymbol{\sigma}\rangle_{\mathbf{x}}(\mathbf{x})+\int[\langle\!\langle\boldsymbol{\Gamma}(\mathbf{x}-\mathbf{y})\boldsymbol{\eta}(\mathbf{y})\rangle\!\rangle_{\mathbf{x}}+\boldsymbol{\mathcal{L}}^{(0)}\langle\!\langle\nabla\mathbf{G}(\mathbf{x}-\mathbf{y})\boldsymbol{\pi}(\mathbf{y})\rangle\!\rangle_{\mathbf{x}}]d\mathbf{y}. \tag{15.97}$$

In the case of asymptotic approximations analogous to Eqs. (7.47) and (7.48), Eqs. (15.96) and (15.97) are reduced to the known ones

$$\boldsymbol{\varepsilon}(\mathbf{x})=\langle\boldsymbol{\varepsilon}\rangle_{\mathbf{x}}(\mathbf{x})+\int[\langle\!\langle\mathbf{U}(\mathbf{x}-\mathbf{y})\rangle\!\rangle_{\mathbf{x}}\boldsymbol{\tau}(\mathbf{y})+\langle\!\langle\nabla\mathbf{G}(\mathbf{x}-\mathbf{y})\rangle\!\rangle_{\mathbf{x}}\boldsymbol{\pi}(\mathbf{y})d\mathbf{y}, \tag{15.98}$$

$$\boldsymbol{\sigma}(\mathbf{x})=\langle\boldsymbol{\sigma}\rangle_{\mathbf{x}}(\mathbf{x})+\int[\langle\!\langle\boldsymbol{\Gamma}(\mathbf{x}-\mathbf{y})\rangle\!\rangle_{\mathbf{x}}\boldsymbol{\eta}(\mathbf{y})+\boldsymbol{\mathcal{L}}^{(0)}\langle\!\langle\nabla\mathbf{G}(\mathbf{x}-\mathbf{y})\rangle\!\rangle_{\mathbf{x}}\boldsymbol{\pi}(\mathbf{y})d\mathbf{y}, \tag{15.99}$$

which were analyzed in details and compared with other known equations by Buryachenko [200] (see also [747]) for the particular static problem ($\mathbf{b}\equiv\mathbf{0}$) with the uncoupled [$M=1$; see (15.8)] local constitutive equation (15.1) and (15.3).

15.3 General Representations for Solution

15.3.1 Two Heterogeneities in the Infinite Matrix

Let us assume that two inclusions v_i and v_j are placed in an infinite homogeneous matrix subjected to the inhomogeneous field $\widetilde{\boldsymbol{\vartheta}}_{i,j}(\mathbf{x})$ ($\boldsymbol{\vartheta}=\boldsymbol{\varepsilon},\boldsymbol{\sigma};\ \mathbf{x}\in R^d$). For $\mathbf{x}\in v_i$, Eqs. (15.75) and (15.76) can be recast in the form

$$\boldsymbol{\vartheta}(\mathbf{x})=\overline{\boldsymbol{\vartheta}}_i(\mathbf{x})+\boldsymbol{\mathcal{L}}_i^{\theta}(\mathbf{x}-\mathbf{x}_i,\widetilde{\boldsymbol{\vartheta}}), \tag{15.100}$$

$$\overline{\boldsymbol{\vartheta}}_i(\mathbf{x})=\boldsymbol{\mathcal{L}}_j^{\theta}(\mathbf{x}-\mathbf{x}_j,\widetilde{\boldsymbol{\vartheta}}). \tag{15.101}$$

We can transform Eqs. (15.100) and (15.101) into the following ones ($\mathbf{x}\in v_i$)

$$\boldsymbol{\vartheta}(\mathbf{x})-\widetilde{\boldsymbol{\vartheta}}_{i,j}(\mathbf{x})-\boldsymbol{\mathcal{L}}_i^{\theta}(\mathbf{x}-\mathbf{x}_i,\widetilde{\boldsymbol{\vartheta}}_{i,j})=\boldsymbol{\mathcal{L}}_{i,j}^{\theta}(\mathbf{x}-\mathbf{x}_j,\widetilde{\boldsymbol{\vartheta}}_{i,j}) \tag{15.102}$$

defining the perturbator $\boldsymbol{\mathcal{L}}_{i,j}^{\theta}(\mathbf{x}-\mathbf{x}_j,\widetilde{\boldsymbol{\vartheta}}_{i,j})$ which can be found by any numerical method (e.g., such as BIE and FEA, complex potential method, and others) analogously to the operator $\boldsymbol{\mathcal{L}}_i^{\theta}(\mathbf{x}-\mathbf{x}_i,\overline{\boldsymbol{\vartheta}}_i)$. It should be mentioned that the operators $\boldsymbol{\mathcal{L}}_i^{\theta}(\mathbf{x}-\mathbf{x}_i,\widetilde{\boldsymbol{\vartheta}}_{i,j})$ and $\boldsymbol{\mathcal{L}}_{i,j}^{\theta}(\mathbf{x}-\mathbf{x}_j,\widetilde{\boldsymbol{\vartheta}}_{i,j})$ (15.102) act on the effective fields $\widetilde{\boldsymbol{\vartheta}}_{i,j}(\mathbf{x})$

at $\mathbf{x} \in v_i$ and $\mathbf{x} \in v_i, v_j$, respectively, and the kernel of the operator $\boldsymbol{\mathcal{L}}^{\theta}_{i,j}$ can be decomposed ($K = I, J$)

$$\boldsymbol{\mathcal{L}}^{\theta}_{i,j}(\mathbf{x},\mathbf{y}) = \boldsymbol{\mathcal{L}}^{\theta I}_{i,j}(\mathbf{x},\mathbf{y}) + \boldsymbol{\mathcal{L}}^{\theta J}_{i,j}(\mathbf{x},\mathbf{y}), \quad \boldsymbol{\mathcal{L}}^{\theta K}_{i,j}(\mathbf{x},\mathbf{y}) = \boldsymbol{\mathcal{L}}^{\theta}_{i,j}(\mathbf{x},\mathbf{y})V_k(\mathbf{y}), \tag{15.103}$$

where one follows Mura's [1104] tensorial indicia notation, i.e., repeated lowercase indices are summed up from 1 to d, while uppercase indices always take on the same numbers as the corresponding lowercase ones but are not summed up.

The fields inside the heterogeneity v_i can be estimated from Eq. (15.102)

$$\boldsymbol{\vartheta}(\mathbf{x}) - \boldsymbol{\mathcal{B}}^{\theta}_{i}(\widetilde{\boldsymbol{\vartheta}}_{i,j})(\mathbf{x}) - \mathbf{C}^{\theta}_{i}(\mathbf{x}) = \boldsymbol{\mathcal{B}}^{\theta}_{i,j}(\widetilde{\boldsymbol{\vartheta}}_{i,j})(\mathbf{x}) + \mathbf{C}^{\theta}_{i,j}(\mathbf{x}), \tag{15.104}$$

where the operator concentrations and transformation concentration factors are defined through the corresponding perturbators analogously to Eqs. (15.63) and (15.64)

$$\boldsymbol{\mathcal{L}}^{\theta}_{i,j}(\mathbf{x}-\mathbf{x}_j, \widetilde{\boldsymbol{\vartheta}}_{i,j}) = \boldsymbol{\mathcal{B}}^{\theta}_{i,j}(\widetilde{\boldsymbol{\vartheta}}_{i,j})(\mathbf{x}) + \mathbf{C}^{\theta}_{i,j}(\mathbf{x}). \tag{15.105}$$

Let us replace the inhomogeneity v_j by the fictitious inclusion $\mathcal{L}_1(\mathbf{y}) \equiv \mathbf{0}$ ($\mathbf{y} \in v_i$) with the matrix properties $\boldsymbol{\mathcal{L}}^{(0)}$ and the fictitious eigenfield ($\boldsymbol{\gamma} = \boldsymbol{\alpha}, \boldsymbol{\beta}$ and $\boldsymbol{\vartheta} = \boldsymbol{\varepsilon}, \boldsymbol{\sigma}$, respectively)

$$\boldsymbol{\gamma}^{\text{fict}}_1(\mathbf{x}) = \boldsymbol{\mathcal{R}}^{\theta}_{i}(\widetilde{\boldsymbol{\vartheta}}_{i,j})(\mathbf{x}) - \mathbf{F}^{\theta}_{i}(\mathbf{x}), \tag{15.106}$$

so that the constitutive equation in the inclusion v_j is $\boldsymbol{\sigma}(\mathbf{y}) = \boldsymbol{\mathcal{L}}^{(0)} * \boldsymbol{\varepsilon}(\mathbf{y}) + \boldsymbol{\beta}^{\text{fict}}_1(\mathbf{y})$. Then the polarization tensors $\boldsymbol{\zeta}(\mathbf{y}) = \boldsymbol{\gamma}^{\text{fict}}_1(\mathbf{y})$ ($\mathbf{y} \in v_j$) can be assumed to be known, and Eqs. (15.102) and (15.104) can be reduced to

$$\boldsymbol{\vartheta}(\mathbf{x}) - \widetilde{\boldsymbol{\vartheta}}_{i,j}(\mathbf{x}) - \boldsymbol{\mathcal{L}}^{\theta}_{i}(\mathbf{x}-\mathbf{x}_i, \widetilde{\boldsymbol{\vartheta}}_{i,j}) = \boldsymbol{\mathcal{L}}^{\theta\infty}_{i,j}(\mathbf{x}-\mathbf{x}_j, \widetilde{\boldsymbol{\vartheta}}_{i,j}), \tag{15.107}$$

$$\boldsymbol{\vartheta}(\mathbf{x}) - \boldsymbol{\mathcal{B}}^{\theta}_{i}(\widetilde{\boldsymbol{\vartheta}}_{i,j})(\mathbf{x}) - \mathbf{C}^{\theta}_{i}(\mathbf{x}) = \boldsymbol{\mathcal{B}}^{\theta\infty}_{i,j}(\widetilde{\boldsymbol{\vartheta}}_{i,j})(\mathbf{x}) + \mathbf{C}^{\theta\infty}_{i,j}(\mathbf{x}), \tag{15.108}$$

respectively, defining the new perturbator $\boldsymbol{\mathcal{L}}^{\theta\infty}_{i,j}(\mathbf{x}-\mathbf{x}_j, \widetilde{\boldsymbol{\vartheta}}_{i,j})$ ($\mathbf{x} \in v_i$) and the concentrations $\boldsymbol{\mathcal{B}}^{\theta\infty}_{i,j}(\widetilde{\boldsymbol{\vartheta}}_{i,j})(\mathbf{x})$, $\mathbf{C}^{\theta\infty}_{i,j}(\mathbf{x})$ which can be estimated analogously to $\boldsymbol{\mathcal{L}}^{\theta}_{i,j}(\mathbf{x}-\mathbf{x}_j, \widetilde{\boldsymbol{\vartheta}}_{i,j})$ (15.107), and $\boldsymbol{\mathcal{B}}^{\theta\infty}_{i,j}(\widetilde{\boldsymbol{\vartheta}}_{i,j})(\mathbf{x})$, $\mathbf{C}^{\theta\infty}_{i,j}(\mathbf{x})$ (15.108), respectively.

Thus, the tensors $\boldsymbol{\mathcal{B}}^{\theta}_{i}(\overline{\boldsymbol{\vartheta}}_{i})(\mathbf{x})$ and $\mathbf{C}^{\theta}_{i}(\mathbf{x})$ ($\mathbf{x}, \mathbf{y} \in v_i$) describe the field $\boldsymbol{\vartheta}(\mathbf{x})$ in the isolated inclusion v_i in the infinite matrix subjected to the loading $\overline{\boldsymbol{\vartheta}}_i(\mathbf{x})$ ($\mathbf{x} \in v_i$). In so doing, the tensors $\boldsymbol{\mathcal{B}}^{\theta}_{i,j}(\widetilde{\boldsymbol{\vartheta}}_{i,j}(\mathbf{x}))$ and $\mathbf{C}^{\theta}_{i,j}(\mathbf{x})$ ($\mathbf{x} \in v_i$) define the perturbation of the field $\boldsymbol{\vartheta}(\mathbf{x})$ introduced by the placement of the inclusion v_j interacting with the inclusion v_i. The tensors $\boldsymbol{\mathcal{B}}^{\theta\infty}_{i,j}(\widetilde{\boldsymbol{\vartheta}}_{i,j})(\mathbf{x})$ and $\mathbf{C}^{\theta\infty}_{i,j}(\mathbf{x})$ describe another perturbation of the field $\boldsymbol{\vartheta}(\mathbf{x})$ at $\mathbf{x} \in v_i$ produced by the inclusion v_j interacting just with the external loading $\widetilde{\boldsymbol{\vartheta}}_{i,j}(\mathbf{y})$ ($\mathbf{y} \in v_j$) but not with the inclusion v_i.

In so doing, the tensors $\mathcal{B}^{\theta\infty}_{i,j}(\widetilde{\boldsymbol{\vartheta}})(\mathbf{x})$ and $\mathbf{C}^{\theta\infty}_{i,j}(\mathbf{x})$ are defined on the whole space $\mathbf{x} \in R^d$, while $\mathcal{B}^{\theta}_{i,j}(\widetilde{\boldsymbol{\vartheta}})(\mathbf{x})$ and $\mathbf{C}^{\theta}_{i,j}(\mathbf{x})$ are defined only on $\mathbf{x} \in R^d \setminus v_j$. The tensors $\mathcal{B}^{\theta\infty}_{i,j}(\widetilde{\boldsymbol{\vartheta}})(\mathbf{x})$ and $\mathbf{C}^{\theta\infty}_{i,j}(\mathbf{x})$ can be considered as the first-order approximation of the tensors $\mathcal{B}^{\infty}_{i,j}(\widetilde{\boldsymbol{\vartheta}})(\mathbf{x})$ and $\mathbf{C}^{\infty}_{i,j}(\mathbf{x})$ by the iteration method.

In a similar manner, we can define the effective field perturbators $\mathcal{J}^{\theta}_{i,j}$, $\mathbf{T}^{\gamma}_{i,j}(\mathbf{x})$, and $\mathcal{J}^{\theta\infty}_{i,j}$, $\mathbf{T}^{\gamma\infty}_{i,j}(\mathbf{x})$ describing the perturbation of the effective field $\overline{\boldsymbol{\vartheta}}_i(\mathbf{x}) - \widetilde{\boldsymbol{\vartheta}}_{i,j}(\mathbf{x})$ introduced by both the heterogeneity v_j and the fictitious inclusion with the response operator $\mathcal{L}^{(0)}$ and eigenfield $\boldsymbol{\gamma}_1^{\text{fict}}(\mathbf{y})$ ($\mathbf{y} \in v_j$, $\mathbf{x} \in v_i$, $\mathbf{z} \in R^d$) (15.106) and (15.107), respectively,

$$\overline{\boldsymbol{\vartheta}}_i(\mathbf{x}) - \widetilde{\boldsymbol{\vartheta}}_{i,j}(\mathbf{x}) = \mathcal{J}^{\theta}_{i,j}(\widetilde{\boldsymbol{\vartheta}}_{i,j})(\mathbf{x}) + \mathbf{T}^{\gamma}_{i,j}(\mathbf{x}), \tag{15.109}$$

$$\overline{\boldsymbol{\vartheta}}_i(\mathbf{z}) - \widetilde{\boldsymbol{\vartheta}}_{i,j}(\mathbf{z}) = \mathcal{J}^{\theta\infty}_{i,j}(\widetilde{\boldsymbol{\vartheta}}_{i,j})(\mathbf{x}) + \mathbf{T}^{\gamma\infty}_{i,j}(\mathbf{x}), \tag{15.110}$$

where the operators $\mathcal{J}^{\theta}_{i,j}$ and $\mathcal{J}^{\theta\infty}_{i,j}$ on the constant $\widetilde{\boldsymbol{\vartheta}}_{i,j}(\mathbf{y})$ =const. are reduced to the tensors with the corresponding contraction of the kernels

$$\mathbf{J}^{\theta}_{i,j}(\mathbf{x}) = \langle \mathcal{J}^{\theta}_{i,j}(\mathbf{x}, \mathbf{y}) \rangle_{(j)}, \quad \mathbf{J}^{\theta\infty}_{j}(\mathbf{z}) = \langle \mathcal{J}^{\theta\infty}_{j}(\mathbf{z}, \mathbf{y}) \rangle_{(j)} \tag{15.111}$$

($\mathbf{y} \in v_j$, $\mathbf{x} \in v_i$, $\mathbf{z} \in R^d$), respectively. No confusion will arise hereafter in definition of the operator $\mathcal{D}$ ($\mathcal{B}^{\theta}_{i}$, $\mathcal{B}^{\theta}_{i,j}$, $\mathcal{J}^{\theta}_{i,j}$, $\mathcal{J}^{\theta\infty}_{i,j}$) with the kernel $\mathcal{D}(\mathbf{x}, \mathbf{y})$ on the inhomogeneous functions $\mathbf{g}(\mathbf{y})$ (e.g., $\mathbf{g}(\mathbf{y}) = \langle \widetilde{\boldsymbol{\vartheta}}_{i,j} \rangle(\mathbf{y})$, $\langle \overline{\boldsymbol{\vartheta}}_j \rangle(\mathbf{y})$, $\mathbf{y} \in V_k$; $V_k = V_i, V_j, V_i + V_j$)

$$\mathcal{D}(\mathbf{g})(\mathbf{x}) = \int \mathcal{D}(\mathbf{x}, \mathbf{y}) \mathbf{g}(\mathbf{y}) V_k(\mathbf{y}) d\mathbf{y}. \tag{15.112}$$

The operator $\mathcal{D}$ is reduced to the tensor $\mathbf{D}(\mathbf{x})$ on the constant functions $\mathbf{g}(\mathbf{y}) = \mathbf{g} \equiv$const. ($\mathbf{y} \in V_k$)

$$\mathcal{D}(\mathbf{g})(\mathbf{x}) = \mathbf{D}(\mathbf{x})\mathbf{g}, \quad \mathbf{D}(\mathbf{x}) = \int \mathcal{D}(\mathbf{x}, \mathbf{y}) V_k(\mathbf{y}) d\mathbf{y}. \tag{15.113}$$

For ascertaining the connections between $\mathcal{B}^{\theta}_{i,j}(\mathbf{x})$, $\mathbf{C}^{\theta}_{i,j}(\mathbf{x})$ and $\mathcal{J}^{\theta}_{i,j}(\widetilde{\boldsymbol{\vartheta}}_{i,j})(\mathbf{x})$, we recast Eq. (15.105) in terms of the effective fields

$$\mathcal{J}^{\theta}_{i,j}(\widetilde{\boldsymbol{\sigma}}_{i,j})(\mathbf{x}) = (\mathcal{B}^{\theta}_{i})^{-1}(\mathcal{B}^{\theta}_{i,j}(\widetilde{\boldsymbol{\sigma}}_{i,j}))(\mathbf{x}), \quad \mathbf{T}^{\gamma}_{i,j}(\mathbf{x}) = (\mathcal{B}^{\theta}_{i})^{-1}(\mathbf{C}^{\theta}_{i,j})(\mathbf{x}). \tag{15.114}$$

15.3.2 General Representations for the Effective Fields

Fixing the inclusion v_i in the composite material produces the random effective field $\overline{\boldsymbol{\vartheta}}_i(\mathbf{x})$ (15.80) which can be recast for $n = 1$ as ($\mathbf{x} \in v_i$)

$$\langle\overline{\boldsymbol{\vartheta}}_i\rangle(\mathbf{x}) = \langle\boldsymbol{\vartheta}\rangle(\mathbf{x}) + \int \big\{\langle\boldsymbol{\mathcal{L}}_j^{\theta}(\mathbf{x}-\mathbf{x}_j, \overline{\boldsymbol{\vartheta}})|; v_i, \mathbf{x}_i\rangle_j \varphi(v_j, \mathbf{x}_j|; v_i, \mathbf{x}_i)$$
$$-\langle\boldsymbol{\mathcal{L}}_j^{\theta}(\mathbf{x}-\mathbf{x}_j, \overline{\boldsymbol{\vartheta}}_j)\rangle(\mathbf{x}_j)\big\} d\mathbf{x}_j. \qquad (15.115)$$

In Eq. (15.115) the conditional perturbator $\langle\boldsymbol{\mathcal{L}}_j^{\theta}(\mathbf{x}-\mathbf{x}_j, \overline{\boldsymbol{\vartheta}})|; v_i, \mathbf{x}_i\rangle_j$ can be expressed through the explicit perturbator for two interacting heterogeneities subjected to the field $\widetilde{\boldsymbol{\vartheta}}_{i,j}$

$$\langle\overline{\boldsymbol{\vartheta}}_i\rangle(\mathbf{x}) = \langle\boldsymbol{\vartheta}\rangle(\mathbf{x}) + \int \big\{[\boldsymbol{\mathcal{J}}_{i,j}^{\theta}(\langle\widetilde{\boldsymbol{\vartheta}}_{i,j}\rangle)(\mathbf{x}) + \mathbf{T}_{i,j}^{\gamma}(\mathbf{x})]\varphi(v_j, \mathbf{x}_j|; v_i, \mathbf{x}_i)$$
$$-[\boldsymbol{\mathcal{J}}_{i,j}^{\theta\infty}(\langle\widetilde{\boldsymbol{\vartheta}}_{i,j}\rangle)(\mathbf{x}) + \mathbf{T}_{i,j}^{\gamma\infty}(\mathbf{x})]\big\} d\mathbf{x}_j. \qquad (15.116)$$

No restrictions are imposed on the microtopology of the microstructure and the shape of inclusions as well as on the field inhomogeneity inside the inclusions. However, the main computational advantage of the proposed Eq. (15.116) lies in the fact that such fundamental notions of micromechanics as the Green's function are not exploited, and we can analyze any anisotropy of constituents (including the matrix) as well as any shape and any composite structure of inclusions. We constructed Eqs. (15.116) by the use of some building blocks described by the numerical solutions for both one and two inclusions inside the infinite medium subjected to the homogeneous loading at infinity. Just at some additional assumptions, the tensors mentioned above can be expressed through the Green's function, Eshelby [450] tensor, and external Eshelby tensor. Equation (15.116) is exact and involves two sorts of statistically averaged effective fields $\langle\overline{\boldsymbol{\vartheta}}_i\rangle(\mathbf{x})$ and $\langle\widetilde{\boldsymbol{\vartheta}}_{i,j}\rangle(\mathbf{x})$ which can be found from the infinite hierarchy of equations analogous to the system (15.80). These truncated hierarchies of equations will be limited by Eq. (15.116) and analyzed for both the random and periodic structures by the use of the closing assumption considered in the next subsection.

15.4 Some Particular Cases of the Perturbators and Corresponding Tensors for Some Constitutive Laws

Although the equations in Sections 15.2 and 15.3 are fulfilled for statistically inhomogeneous media subjected to inhomogeneous loading, in the next sections, the GIEs presented will be only applied to statistically homogeneous media subjected to the homogeneous remote loading for estimations of both the concentration inside a representative heterogeneity and effective properties.

15.4.1 Closing Assumption

For the closing of Eq. (15.116), we will use the following hypothesis:

(H2a) *Each pair of inclusions* v_i *and* v_j *is subjected to the inhomogeneous field* $\widetilde{\boldsymbol{\vartheta}}_{i,j}(\mathbf{x})$, *and statistical average* $\langle\widetilde{\boldsymbol{\vartheta}}_{i,j}\rangle(\mathbf{x})$ *is defined by the formulae*

$$\langle\widetilde{\boldsymbol{\vartheta}}_{i,j}\rangle(\mathbf{x}) = \langle\overline{\boldsymbol{\vartheta}}_k\rangle(\mathbf{x}) \qquad (15.117)$$

at $\mathbf{x} \in v_k,\ k = i, j$.

The refined version of the closing assumption (15.117) of the MEFM proposed in [199] takes into account both the three-point correlation functions and dependence of the effective fields acting on each pair of heterogeneities v_i and v_j on the distance $|\mathbf{x}_i - \mathbf{x}_j|$.

The adoption of hypothesis **H2a** makes it possible to reduce Eq. (15.116) to the operator one

$$\langle\overline{\boldsymbol{\vartheta}}_i\rangle(\mathbf{x}) = \langle\boldsymbol{\vartheta}\rangle(\mathbf{x}) + \int \Big\{\Big[\mathcal{J}^{I\theta}_{i,j}(\mathbf{x}-\mathbf{x}_j, \langle\overline{\boldsymbol{\vartheta}}_i\rangle) + \mathcal{J}^{J\theta}_{i,j}(\mathbf{x}-\mathbf{x}_j, \langle\overline{\boldsymbol{\vartheta}}_j\rangle)\Big]$$
$$\times\varphi(v_j, \mathbf{x}_j|; v_i, \mathbf{x}_i) - \mathcal{J}^{\theta\infty}_{i,j}(\mathbf{x}-\mathbf{x}_j, \langle\overline{\boldsymbol{\vartheta}}_j\rangle)\Big\}d\mathbf{x}_j, \qquad (15.118)$$

where a decomposition $\mathcal{J}^{\theta}_{i,j} = \mathcal{J}^{I\theta}_{i,j} + \mathcal{J}^{J\theta}_{i,j}$ was introduced analogously to Eq. (15.103). An integral Eq. (15.118) can be solved by the iteration method based on the recursion formula

$$\langle\overline{\boldsymbol{\vartheta}}_i^{[n+1]}\rangle(\mathbf{x}) = \langle\boldsymbol{\vartheta}\rangle(\mathbf{x}) + \int \{[\mathcal{J}^{I\theta}_{i,j}(\langle\overline{\boldsymbol{\vartheta}}_i^{[n]}\rangle)(\mathbf{x}) + \mathcal{J}^{J\theta}_{i,j}(\langle\overline{\boldsymbol{\vartheta}}_j^{[n]}\rangle)(\mathbf{x}) + \mathbf{T}^{\gamma}_{i,j}(\mathbf{x})]$$
$$\times\varphi(v_j, \mathbf{x}_j|; v_i, \mathbf{x}_i) - [\mathcal{J}^{\theta\infty}_{i,j}(\langle\overline{\boldsymbol{\vartheta}}_j^{[n]}\rangle)(\mathbf{x}) + \mathbf{T}^{\gamma\infty}_{i,j}(\mathbf{x})]\}d\mathbf{x}_j, \qquad (15.119)$$

$$\langle\boldsymbol{\vartheta}^{[n+1]}\rangle_i(\mathbf{x}) = \mathcal{B}^{\theta}_i(\langle\overline{\boldsymbol{\vartheta}}_i^{[n+1]}\rangle)(\mathbf{x}) + \mathbf{C}^{\theta}_i(\mathbf{x}), \qquad (15.120)$$

$$\bar{v}_i\langle\boldsymbol{\zeta}^{[n+1]}\rangle_i(\mathbf{x}) = \mathcal{R}^{\theta}_i(\langle\overline{\boldsymbol{\vartheta}}_i^{[n+1]}\rangle)(\mathbf{x}) + \mathbf{F}^{\theta}_i(\mathbf{x}), \qquad (15.121)$$

with an initial approximation considered in Sects. 5.2 and 5.3. The solution (15.119)–(15.121) can be formally presented in the Neumann series form for the solution $\langle\boldsymbol{\zeta}\rangle_i(\mathbf{x})$ and $\langle\boldsymbol{\vartheta}\rangle_i(\mathbf{x})$ $(\mathbf{x} \in v_i)$

$$\bar{v}_i\langle\boldsymbol{\zeta}\rangle_i(\mathbf{x}) \equiv \bar{v}_i \lim_{n\to\infty}\langle\boldsymbol{\zeta}^{[n]}\rangle_i(\mathbf{x}) = \mathbf{R}^{\theta*}_i(\mathbf{x})\langle\boldsymbol{\sigma}\rangle + \mathbf{F}^{\theta*}_i(\mathbf{x}),$$

$$\langle \boldsymbol{\vartheta} \rangle_i(\mathbf{x}) \equiv \lim_{n\to\infty} \langle \boldsymbol{\vartheta}^{[n]} \rangle_i(\mathbf{x}) = \mathbf{B}_i^{\theta *}(\mathbf{x})\langle \boldsymbol{\sigma} \rangle + \mathbf{C}_i^{\theta *}(\mathbf{x}), \tag{15.122}$$

which yields the final representations for the effective properties

$$\mathbf{K}^* = \mathbf{K}^{(0)} + \langle \mathbf{R}^{\theta *} V \rangle, \quad \boldsymbol{\gamma}^* = \boldsymbol{\gamma}^{(0)} + \langle \mathbf{F}^{\theta *} V \rangle, \quad W^* = -\langle \boldsymbol{\gamma}_1 \mathbf{C}^{\theta *} V \rangle \tag{15.123}$$

for $\mathbf{K} = \mathbf{L}, \mathbf{M}$, $\boldsymbol{\vartheta} = \boldsymbol{\varepsilon}, \boldsymbol{\sigma}$, and $\boldsymbol{\gamma} = \boldsymbol{\alpha}, \boldsymbol{\beta}$, respectively. A convergence of the sequences $\langle \boldsymbol{\zeta}^{[n]} \rangle_i(\mathbf{x})$ (15.122$_1$) and $\langle \boldsymbol{\vartheta}^{[n]} \rangle_i(\mathbf{x})$ (15.118$_2$) can be proved analogously to the locally elastic case [203].

The hypothesis **H2a**, rewritten in terms of the fields $\boldsymbol{\vartheta}(\mathbf{x})$, $(\mathbf{x} \in v_i)$, is a standard closing assumption (see for references, e.g., [200, 747, 1625]) degenerating to the "quasi-crystalline" approximation by Lax [886] which ignores the binary interaction of heterogeneities and assumes homogeneity of the effective fields:

Hypothesis H2b, "Quasi-Crystalline" Approximation *It is supposed that the mean value of the effective field at a point* $\mathbf{x} \in v_i$ *does not depend on the field inside surrounding heterogeneities* $v_j \neq v_i, \mathbf{x} \in v_k, \ (k = i, j)$

$$\langle \overline{\boldsymbol{\vartheta}} | v_i, \mathbf{x}_i; v_j, \mathbf{x}_j \rangle(\mathbf{x}) = \langle \overline{\boldsymbol{\vartheta}}_k \rangle(\mathbf{x})$$
$$\langle \overline{\boldsymbol{\vartheta}}_k \rangle(\mathbf{x}) \equiv \text{const.} \tag{15.124}$$

In the framework of the EFH **H1** (see for details Sect. 5.2), the principal difference between the hypotheses **H2a** (15.117) and **H2b** (15.124) (see for details Chapter 13 in [200]) is beyond the scope of the direct substitution of the field $\boldsymbol{\vartheta}(\mathbf{x})$ for the effective field $\overline{\boldsymbol{\vartheta}}(\mathbf{x})$. Of more importance is an acceptance of the assumption (15.117) after the consideration of multiparticle inclusion interactions (15.109) rather than before as in hypothesis (15.124). What seems to be only a formal trick yields to the discovery of fundamentally new nonlocal effects in the theory of functionally graded materials, which was demonstrated in Chapter 13 in [200] for the locally elastic case. However, even for statistically homogeneous composites, it was shown that the use of the assumption (15.117) instead of (15.124) can lead to a variation of the concentration factors by a factor of two or more.

It is possible a straightforward generalization of "quasi-crystalline" approximation by Lax [886] (15.124$_1$) when the assumption (15.124$_2$) is relaxed: $\langle \overline{\boldsymbol{\vartheta}}_k \rangle(\mathbf{x}) \not\equiv$ const at $\mathbf{x} \in v_k$, $(k = i, j$; a case of the local elasticity was considered Chap. 9; see also [203]) that is equivalent to the equalities

$$\langle \mathcal{L}_j^{\tau}(\mathbf{x} - \mathbf{x}_j, \widehat{\boldsymbol{\tau}}_{(j)}) |; v_1, \mathbf{x}_1 \rangle = \langle \mathcal{L}_j^{\tau}(\mathbf{x} - \mathbf{x}_j, \widehat{\boldsymbol{\tau}}_{(j)}) \rangle,$$
$$\langle \mathcal{L}_j^{\eta}(\mathbf{x} - \mathbf{x}_j, \widehat{\boldsymbol{\eta}}_{(j)}) |; v_1, \mathbf{x}_1 \rangle = \langle \mathcal{L}_j^{\eta}(\mathbf{x} - \mathbf{x}_j, \widehat{\boldsymbol{\eta}}_{(j)}) \rangle. \tag{15.125}$$

It greatly simplifies the problem (15.119) where in such a case $\mathcal{J}_{i,j}^{\theta} = \mathcal{J}_{i,j}^{\theta\infty}$ $(\mathbf{x} \in v_i)$ reducing Eqs. (15.119) to

$$\langle\overline{\boldsymbol{\vartheta}}_i^{[n+1]}\rangle(\mathbf{x}) = \langle\boldsymbol{\vartheta}\rangle(\mathbf{x}) + \int [\mathcal{J}_{i,j}^{\theta\infty}(\langle\overline{\boldsymbol{\vartheta}}_j^{[n]}\rangle) + \mathbf{T}_{i.j}^{\gamma\infty}(\mathbf{x})]$$
$$\times[\varphi(v_j, \mathbf{x}_j|; v_i, \mathbf{x}_i) - n^{(j)}(\mathbf{x}_j)]d\mathbf{x}_j. \qquad (15.126)$$

For local elasticity this method uses as a background the new GIE (15.81) and (15.82) proposed by Buryachenko [202] and makes it possible to abandon the basic concepts of micromechanics: effective field hypothesis and the hypothesis of "ellipsoidal symmetry" (see Sect. 15.4.4). The results of this abandonment for local elasticity were quantitatively estimated for some modeled composites reinforced by aligned continuously inhomogeneous fibers. Some new effects were detected that were impossible in the framework of the classical background (15.83) and (15.84) of micromechanics.

15.4.2 Effective Field Hypothesis H1a and Initial Approximation

In order to find an initial approximation of Eqs. (15.119) and (15.126), we now apply the so-called effective field hypothesis which is the main approximate hypothesis of many micromechanical methods (see for references and details Chap. 9 and [211]):

Hypothesis (H1a) *Each inclusion v_i is located in the field (15.76)*

$$\overline{\boldsymbol{\vartheta}}_i(\mathbf{y}) \equiv \overline{\boldsymbol{\vartheta}}(\mathbf{x}_i) \ (\mathbf{y} \in v_i) \qquad (15.127)$$

which is homogeneous over the inclusion v_i.

Hypothesis **(H1a)** allows us to reduce Eq. (15.119) to the algebraic equation

$$\langle\overline{\boldsymbol{\vartheta}}_i\rangle(\mathbf{x}) = \langle\boldsymbol{\vartheta}\rangle(\mathbf{x}) + \int \big\{[\mathbf{J}_{i,j}^{I\theta}(\mathbf{x}-\mathbf{x}_j)\langle\overline{\boldsymbol{\vartheta}}_i\rangle(\mathbf{x}_i) + \mathbf{J}_{i,j}^{J\theta}(\mathbf{x}-\mathbf{x}_j)\langle\overline{\boldsymbol{\vartheta}}_j\rangle(\mathbf{x}_j)$$
$$+\mathbf{J}_{i,j}^{\gamma}(\mathbf{x}-\mathbf{x}_j)]\varphi(v_j, \mathbf{x}_j|; v_i, \mathbf{x}_i)$$
$$-[\mathbf{J}_{i,j}^{\theta\infty}(\mathbf{x}-\mathbf{x}_j)\langle\overline{\boldsymbol{\vartheta}}_j\rangle(\mathbf{x}_j) + \mathbf{J}_{i,j}^{\gamma\infty}(\mathbf{x}-\mathbf{x}_j)]\big\}d\mathbf{x}_j, \qquad (15.128)$$

where the linear operators $\mathcal{J}_{i,j}^{\theta}$ and $\mathcal{J}_{i,j}^{\theta\infty}$ (15.119) are decomposed and reduced to the tensors (15.113) at the applying to the constant effective fields (15.127)

$$\mathcal{J}_{i,j}^{I\theta}(\mathbf{x}{-}\mathbf{x}_j, \langle\overline{\boldsymbol{\vartheta}}_i\rangle) + \mathcal{J}_{i,j}^{J\theta}(\mathbf{x}-\mathbf{x}_j, \langle\overline{\boldsymbol{\vartheta}}_j\rangle)$$
$$= \mathbf{J}_{i,j}^{I\theta}(\mathbf{x}-\mathbf{x}_j)\langle\overline{\boldsymbol{\vartheta}}_i\rangle(\mathbf{x}_i) + \mathbf{J}_{i,j}^{J\theta}(\mathbf{x}-\mathbf{x}_j)\langle\overline{\boldsymbol{\vartheta}}_j\rangle(\mathbf{x}_j) + \mathbf{T}_{i,j}^{\gamma}(\mathbf{x}), \qquad (15.129)$$

$$\mathcal{J}_{i,j}^{\theta\infty}(\mathbf{x}{-}\mathbf{x}_j, \langle\overline{\boldsymbol{\vartheta}}_j\rangle) = \mathbf{J}_{i,j}^{\theta\infty}(\mathbf{x}-\mathbf{x}_j)\langle\overline{\boldsymbol{\vartheta}}_j\rangle(\mathbf{x}_j) + \mathbf{T}_{i,j}^{\gamma\infty}(\mathbf{x}). \qquad (15.130)$$

For keeping the laying out down, one introduces the notations for the statistical average of all perturbations introduced by surrounding inclusions v_j ($j = 1, 2, \ldots;\ j \neq i$) into the inclusion v_i

$$\mathbf{J}_{i,j}^{J\theta\,\text{pert}}(\mathbf{x}-\mathbf{x}_j) = \int [\mathbf{J}_{i,j}^{J\theta}(\mathbf{x}-\mathbf{x}_j)\varphi(v_j, \mathbf{x}_j|; v_i, \mathbf{x}_i) - \mathbf{J}_{i,j}^{\theta\infty}(\mathbf{x}-\mathbf{x}_j)n^{(j)}]d\mathbf{x}_j, \tag{15.131}$$

$$\mathbf{J}_{i,j}^{I\theta\,\text{pert}}(\mathbf{x}-\mathbf{x}_j) = \int \mathbf{J}_{i,j}^{I\theta}(\mathbf{x}-\mathbf{x}_j)\varphi(v_j, \mathbf{x}_j|; v_i, \mathbf{x}_i)d\mathbf{x}_j, \tag{15.132}$$

$$\mathbf{J}_i^{\gamma\,\text{pert}}(\mathbf{x}-\mathbf{x}_j) = \sum_j^N \int [\mathbf{T}_{i,j}^{\gamma}(\mathbf{x})\varphi(v_j, \mathbf{x}_j|; v_i, \mathbf{x}_i) - \mathbf{T}_{i,j}^{\gamma\infty}(\mathbf{x})n^{(j)}]d\mathbf{x}_j. \tag{15.133}$$

The volume averages over $\mathbf{x} \in v_i$ of the tensors $\mathbf{J}_{i,j}^{K\theta\,\text{pert}}(\mathbf{x})$ $(K = I, J)$ and $\mathbf{J}_i^{\gamma\text{pert}}(\mathbf{x})$ over the volume of the inclusion v_i are denoted by the $\mathbf{J}_{i,j}^{K\theta\,\text{pert}}$ $(K = I, J)$ and $\mathbf{J}_i^{\gamma\text{pert}}$, respectively. For the small concentration of inclusions $c \equiv \langle V \rangle \ll 1$, the tensors $\mathbf{J}_{i,j}^{K\theta\text{pert}}(\mathbf{x})$ $(K = I, J)$ and $\mathbf{J}_i^{\gamma\,\text{pert}}(\mathbf{x})$ are the linear functions of c. For a pure elastic case ($\boldsymbol{\beta}_1 \equiv \mathbf{0}$), the tensors $\mathbf{J}_{i,j}^{K\theta\,\text{pert}}(\mathbf{x})$ $(K = I, J)$ have a physical meaning of the perturbations of the statistical average of the effective fields $\langle \overline{\boldsymbol{\vartheta}}_i \rangle(\mathbf{x})$ in the inclusion v_i introduced by the action of surrounding inclusions belonging to the phase $v^{(j)}$.

Volume averaging over $\mathbf{x} \in v_i$ of Eq. (15.128) leads to the closed integral equation with respect to an inhomogeneous over macro-coordinate $\mathbf{x}_i$ of the effective fields $\langle \overline{\boldsymbol{\vartheta}}_i \rangle(\mathbf{x}_i)$. For the case (considered hereafter) of statistically homogeneous media (5.61$_2$) subjected to the homogeneous boundary conditions (15.21) and (15.23) when $\langle \overline{\boldsymbol{\vartheta}}_i \rangle(\mathbf{x}_i) \equiv \langle \overline{\boldsymbol{\vartheta}}_i \rangle \equiv$const., this integral equation is reduced to the algebraic one

$$\langle \overline{\boldsymbol{\vartheta}}_i \rangle = \langle \boldsymbol{\vartheta} \rangle + \sum_{j=1}^N (\mathbf{J}_{i,j}^{I\theta\,\text{pert}} \langle \overline{\boldsymbol{\vartheta}}_i \rangle + \mathbf{J}_{i,j}^{J\theta\,\text{pert}} \langle \overline{\boldsymbol{\vartheta}}_j \rangle + \delta_{ij}\mathbf{J}_j^{\gamma\,\text{pert}}) \tag{15.134}$$

with the solution

$$\langle \overline{\boldsymbol{\vartheta}}_i \rangle = \sum_{j=1}^N \mathbf{Y}_{ij}\Big[\langle \boldsymbol{\vartheta} \rangle + \mathbf{J}_j^{\gamma\,\text{pert}}\Big], \tag{15.135}$$

$$(\mathbf{Y}^{-1})_{ij} = \delta_{ij}\Big[\mathbf{I} - \sum_{k=1}^N \mathbf{J}_{i,k}^{I\theta\,\text{pert}}\Big] - (1-\delta_{ij})\mathbf{J}_{i,j}^{J\theta\,\text{pert}}. \tag{15.136}$$

Substituting Eq. (15.135) into the right-hand side of Eq. (15.128) leads to the explicit representation for the statistical averages of effective fields in the inclusion v_i

$$\langle\overline{\boldsymbol{\vartheta}}\rangle_i(\mathbf{x}) = \langle\boldsymbol{\vartheta}\rangle + \mathbf{J}_i^{\gamma\ \mathrm{pert}}(\mathbf{x})$$
$$+ \sum_{j,m=1}^{N} \big[\mathbf{J}_{i,j}^{I\theta\ \mathrm{pert}}(\mathbf{x})\mathbf{Y}_{jm} + \mathbf{J}_{i,j}^{J\theta\ \mathrm{pert}}(\mathbf{x})\mathbf{Y}_{jm}\big]\big[\langle\boldsymbol{\vartheta}\rangle + \mathbf{J}_m^{\gamma\ \mathrm{pert}}\big]. \quad (15.137)$$

After estimating the average polarization tensors inside the inclusions (see Eqs. (15.61), (15.62), (15.113), and (15.137)) the problem of calculating the effective properties becomes trivial and leads, according to (15.123), to the following *new representations*

$$\langle\boldsymbol{\vartheta}\rangle_i(\mathbf{x}) = \mathbf{B}^\theta\langle\boldsymbol{\vartheta}\rangle + \mathbf{C}^\theta + \boldsymbol{\mathcal{B}}_i^\theta(\mathbf{J}_i^{\gamma\ \mathrm{pert}})(\mathbf{x}) + \sum_{j,m=1}^{N} \big[\boldsymbol{\mathcal{B}}_i^\theta(\mathbf{J}_{i,j}^{I\theta\ \mathrm{pert}})(\mathbf{x})\mathbf{Y}_{im}$$
$$+ \boldsymbol{\mathcal{B}}_i^\theta(\mathbf{J}_{i,j}^{J\theta\ \mathrm{pert}})(\mathbf{x})\mathbf{Y}_{jm}\big]\big[\langle\boldsymbol{\vartheta}\rangle + \mathbf{J}_m^{\gamma\ \mathrm{pert}}\big], \quad (15.138)$$

$$\mathbf{K}^* = \mathbf{K}^{(0)} + \langle\mathbf{R}^{\theta v}V\rangle + \sum_{i,j,m=1}^{N} n^{(i)}\big[\langle\boldsymbol{\mathcal{R}}_i^\theta(\mathbf{J}_{i,j}^{I\theta\ \mathrm{pert}})\rangle_{(i)}\mathbf{Y}_{im}$$
$$+ \langle\boldsymbol{\mathcal{R}}_i^\theta(\mathbf{J}_{i,j}^{J\theta\ \mathrm{pert}})\rangle_{(i)}\mathbf{Y}_{jm}\big], \quad (15.139)$$

$$\boldsymbol{\gamma}^* = \langle\boldsymbol{\gamma}\rangle + \langle\mathbf{F}^{\theta v}V\rangle + \sum_{i=1}^{N} n^{(i)}\langle\boldsymbol{\mathcal{R}}_i^\theta(\mathbf{J}_{i,j}^{\gamma\ \mathrm{pert}})\rangle_{(i)} + \sum_{i,j,m=1}^{N} n^{(i)}$$
$$\times \big[\langle\boldsymbol{\mathcal{R}}_i^\theta(\mathbf{J}_{i,j}^{I\theta\ \mathrm{pert}})\rangle_{(i)}\mathbf{Y}_{im} + \langle\boldsymbol{\mathcal{R}}_i^\theta(\mathbf{J}_{i,j}^{J\theta\ \mathrm{pert}})\rangle_{(i)}\mathbf{Y}_{jm}\big]\mathbf{J}_m^{\gamma\ \mathrm{pert}}, \quad (15.140)$$

$$W^* = -\frac{1}{2}\langle\boldsymbol{\gamma}_1\mathbf{C}^\theta\rangle - \frac{1}{2}\sum_{i=1}^{N} n^{(i)}\langle\boldsymbol{\gamma}_1^{(i)}\boldsymbol{\mathcal{B}}_i^\theta(\mathbf{J}_i^{\beta\ \mathrm{pert}})\rangle_{(i)}$$
$$-\frac{1}{2}\sum_{i,j,m=1}^{N} n^{(i)}\langle\boldsymbol{\gamma}_1^{(i)}\boldsymbol{\mathcal{B}}_i^\theta(\mathbf{J}_{i,j}^{I\theta\ \mathrm{pert}})\rangle_{(i)}\mathbf{Y}_{im}\mathbf{J}_m^{\theta\ \mathrm{pert}}$$
$$-\frac{1}{2}\sum_{i,j,m=1}^{N} n^{(i)}\langle\boldsymbol{\gamma}_1^{(i)}\boldsymbol{\mathcal{B}}_i^\theta(\mathbf{J}_{i,j}^{J\theta\ \mathrm{pert}})\rangle_{(i)}\mathbf{Y}_{jm}\mathbf{J}_m^{\gamma\ \mathrm{pert}}, \quad (15.141)$$

for $\mathbf{K} = \mathbf{L}, \mathbf{M}$, $\boldsymbol{\vartheta} = \boldsymbol{\varepsilon}, \boldsymbol{\sigma}$, and $\boldsymbol{\gamma} = \boldsymbol{\alpha}, \boldsymbol{\beta}$, respectively; here $\boldsymbol{\mathcal{F}}_i^v(\mathbf{x}) := \overline{v}_i^{-1}\boldsymbol{\mathcal{F}}_i(\mathbf{x})$, $\mathbf{x} \in v_i \subset v^{(i)}$ ($\boldsymbol{\mathcal{F}}_i = \mathbf{F}_i, \mathbf{R}_i$). It should be mentioned that the effective field $\langle\overline{\boldsymbol{\vartheta}}\rangle_i(\mathbf{x})$ (15.137) is an inhomogeneous one, while this field was assumed to be homogeneous in the initial version of the MEFM (see [199] and [200]). Because of this the tensors $\langle\boldsymbol{\zeta}\rangle_i(\mathbf{x})$, $\langle\boldsymbol{\vartheta}\rangle_i(\mathbf{x}) \neq$const. even for the linear elastic problem for the homogeneous ellipsoidal inclusions when $\mathbf{R}_i^\theta(\mathbf{x}), \mathbf{F}_i^\theta(\mathbf{x}), \mathbf{B}_i^\theta(\mathbf{x}), \mathbf{C}_i^\theta(\mathbf{x}) = \mathrm{const.}$ ($\mathbf{x} \in v_i$). Therefore, Eq. (15.137) is in reality the first-order approximation of the effective fields rather than a zero one usually used as an initial approximation (see

Eq. (15.119). The zero-order approximation of $\langle \overline{\boldsymbol{\vartheta}}_i \rangle$ (15.135) reduces Eqs. (15.138)–(15.141) to the simplified representations

$$\langle \boldsymbol{\vartheta} \rangle_i(\mathbf{x}) = \mathbf{B}^\theta(\mathbf{x})\langle \boldsymbol{\vartheta} \rangle + \mathbf{C}_i^\theta(\mathbf{x}) + \mathbf{B}_i^\theta(\mathbf{x}) \sum_{j=1}^N \mathbf{Y}_{ij}\big[\langle \boldsymbol{\vartheta} \rangle + \mathbf{J}_j^{\gamma\ \mathrm{pert}}\big], \tag{15.142}$$

$$\mathbf{K}^* = \mathbf{K}^{(0)} + \sum_{i,j=1}^N n^{(i)} \mathbf{R}_i^\theta \mathbf{Y}_{ij}, \tag{15.143}$$

$$\boldsymbol{\gamma}^* = \boldsymbol{\gamma}^{(0)} + \langle \mathbf{F}^{\theta v} V \rangle + \sum_{i,j=1}^N n^{(i)} \mathbf{R}_i^\theta \mathbf{Y}_{ij} \mathbf{J}_j^{\gamma\ \mathrm{pert}}, \tag{15.144}$$

$$W^* = -\frac{1}{2}\langle \boldsymbol{\gamma}_1 \mathbf{C}^\theta \rangle - \frac{1}{2} \sum_{i,j}^N n^{(i)} \langle \boldsymbol{\gamma}_1^{(i)} \mathbf{B}_i^\theta \rangle_{(i)} \mathbf{Y}_{ij} \mathbf{J}_{i,j}^{\gamma\ \mathrm{pert}}. \tag{15.145}$$

Acceptance of the hypothesis **H2b** (15.124) instead of **H2a** (15.117) reduces the iteration Eqs. (15.119)–(15.126), while the initial approximations (15.138)–(15.145) can be also simplified by the substitution

$$\mathbf{J}_{i,j}^{I\theta}(\mathbf{x} - \mathbf{x}_i) = \mathbf{0}, \tag{15.146}$$

$$\mathbf{J}_{i,j}^{J\theta}(\mathbf{x} - \mathbf{x}_i) = \mathbf{J}_{i,j}^{J\theta\infty}(\mathbf{x} - \mathbf{x}_i), \tag{15.147}$$

$$\mathbf{J}_{i,j}^{\gamma}(\mathbf{x} - \mathbf{x}_i) = \mathbf{J}_{i,j}^{\gamma\infty}(\mathbf{x} - \mathbf{x}_i). \tag{15.148}$$

15.4.3 Some Additional Hypotheses and Initial Approximations

To make further progress, two next hypotheses are widely used. The first one is a hypothesis complimenting the basic hypothesis (**H1a**):

Hypothesis 1b, H1b *The operators* $\mathcal{L}_k^\zeta(\mathbf{x} - \mathbf{x}_i, \widehat{\boldsymbol{\zeta}})$ *of perturbations generated by the heterogeneities* v_i *at the point* $\mathbf{x} \notin v_i$ *are defined by the relations*

$$\mathcal{L}_i^\zeta(\mathbf{x} - \mathbf{x}_i, \widehat{\boldsymbol{\zeta}}) = \mathcal{L}_i^\zeta(\mathbf{x} - \mathbf{x}_i, \widehat{\boldsymbol{\zeta}}_{(i)}). \tag{15.149}$$

Here the volume average of $\widehat{\boldsymbol{\zeta}}(\mathbf{y})$ ($\mathbf{y} \in v_i$) over the volume of v_i denoted as $\widehat{\boldsymbol{\zeta}}_{(i)}$ is considered below. It is interesting that hypothesis **H1a** is equivalent to

$$\mathcal{L}_i^\theta(\mathbf{x} - \mathbf{x}_i, \overline{\boldsymbol{\vartheta}}) = \mathcal{L}_i^\theta(\mathbf{x} - \mathbf{x}_i, \overline{\boldsymbol{\vartheta}}_{(i)}). \tag{15.150}$$

Only for a homogeneous ellipsoidal heterogeneity v_i with $\mathbf{b}(\mathbf{x}), \mathbf{f}(\mathbf{x}) \equiv \mathbf{0}$ ($\mathbf{x} \in R^d$) and the perfect interface (15.15), the assumption (15.150) yields the assumption (15.149); otherwise Eq. (15.149) defines the additional assumption. For the case of the local elasticity (15.3) and $\mathbf{b}_1, \mathbf{f}_1 \equiv \mathbf{0}$, Buryachenko [202] (see Chaps. 7 and 9), proved that the acceptance of the assumptions **H1b** leads to equivalence of Eqs. (15.81), (15.82), and (15.83), (15.84), respectively. Moreover, all methods used as a background Eqs. (15.83) and (15.84) implicitly exploit the Hypothesis **H1b**. It should be mentioned that the popular formulation of the EFH (hypothesis **H1**) is a combination of the hypotheses **H1a** and **H1b**.

We decompose $\widehat{\boldsymbol{\vartheta}}_{(i)}$ in such a manner that the perturbators (15.149) of the volume average of polarization tensors are described as

$$\mathcal{L}_i^{\tau}(\mathbf{x}-\mathbf{x}_i, \widehat{\boldsymbol{\tau}}_{(i)}) = \int_{v_i} \mathbf{U}(\mathbf{x}-\mathbf{y})d\mathbf{y}\widehat{\boldsymbol{\tau}}^{\mathbf{f}}_{(i)} + \int_{v_i} \nabla\mathbf{G}(\mathbf{x}-\mathbf{y})d\mathbf{y}\boldsymbol{\pi}_{(i)}, \tag{15.151}$$

$$\mathcal{L}_i^{\eta}(\mathbf{x}-\mathbf{x}_i, \widehat{\boldsymbol{\eta}}_{(i)}) = \int_{v_i} \boldsymbol{\Gamma}(\mathbf{x}-\mathbf{y}) * d\mathbf{y}\widehat{\boldsymbol{\eta}}^{\mathbf{f}}_{(i)} + \int_{v_i} \mathcal{L}^c * \nabla\mathbf{G}(\mathbf{x}-\mathbf{y})d\mathbf{y}\boldsymbol{\pi}_{(i)}, \tag{15.152}$$

$$\mathcal{L}_i^{u\tau}(\mathbf{x}-\mathbf{x}_i, \widehat{\boldsymbol{\tau}}_{(i)}) = \int_{v_i} \nabla\mathbf{G}(\mathbf{x}-\mathbf{y})d\mathbf{y}\widehat{\boldsymbol{\tau}}^{\mathbf{f}}_{(i)} + \int_{v_i} \mathbf{G}(\mathbf{x}-\mathbf{y})d\mathbf{y}\boldsymbol{\pi}_{(i)}, \tag{15.153}$$

where $\widehat{\boldsymbol{\tau}}^{\mathbf{f}}_{(i)}$ and $\widehat{\boldsymbol{\eta}}^{\mathbf{f}}_{(i)}$ do not depend on $\mathbf{b}_1(\mathbf{x})$ and $\mathbf{f}_1(\mathbf{x})$ and are defined by the averages of Eq. (15.10) over the closed domain $v_i \cup \Gamma_i$ taking into account the jumps (15.18) and (15.19) at the interface Γ_i

$$\begin{aligned}\widehat{\boldsymbol{\tau}}^{\mathbf{f}}_{(i)} &= \boldsymbol{\sigma}_{(i)} - [\mathcal{L}^c * \boldsymbol{\varepsilon}_{(i)} + \boldsymbol{\alpha}^c] + \frac{1}{\bar{v}_i}\int_{\Gamma_i^{\sigma}} ([[\boldsymbol{\sigma}]]\cdot\mathbf{n})\otimes\mathbf{s}ds \\ &\quad - \frac{1}{2\bar{v}_i}\mathcal{L}^c * \int_{\Gamma_i^{\epsilon}} ([[\mathbf{u}]]\otimes\mathbf{n} + \mathbf{n}\otimes[[\mathbf{u}]])ds,\end{aligned} \tag{15.154}$$

$$\begin{aligned}\widehat{\boldsymbol{\eta}}^{\mathbf{f}}_{(i)} &= \boldsymbol{\varepsilon}_{(i)} - [\mathcal{M}^c * \boldsymbol{\sigma}_{(i)} + \boldsymbol{\beta}^c] - \frac{1}{\bar{v}_i}\mathcal{M}^c * \int_{\Gamma_i^{\sigma}} ([[\boldsymbol{\sigma}]]\cdot\mathbf{n})\otimes\mathbf{s}ds \\ &\quad + \frac{1}{2\bar{v}_i}\int_{\Gamma_i^{\epsilon}} ([[\mathbf{u}]]\otimes\mathbf{n} + \mathbf{n}\otimes[[\mathbf{u}]])ds,\end{aligned} \tag{15.155}$$

while the averages $\boldsymbol{\varepsilon}_{(i)}$ and $\boldsymbol{\sigma}_{(i)}$ in (15.154) and (15.155) are estimated over the open region v_i (see, e.g., [400, 614]).

For $\mathbf{b}_1(\mathbf{x}), \mathbf{f}_1(\mathbf{x}) \equiv \mathbf{0}$, according to the hypothesis **H1a** and in view of the linearity of the problem, there exist the tensors $\mathbf{R}_i^{\epsilon}(\mathbf{x}), \mathbf{R}_i^{\sigma}(\mathbf{x}), \mathbf{F}_i^{\epsilon}(\mathbf{x}), \mathbf{F}_i^{\sigma}(\mathbf{x})$, such as ($\mathbf{x} \in R^d$)

$$\bar{v}_i\widehat{\boldsymbol{\tau}}^{\mathbf{f}}(\mathbf{x}) = \mathbf{R}_i^{\epsilon}(\mathbf{x})\overline{\boldsymbol{\varepsilon}}(\mathbf{x}_i) + \mathbf{F}_i^{\epsilon}(\mathbf{x}), \tag{15.156}$$

$$\bar{v}_i\widehat{\boldsymbol{\eta}}^{\mathbf{f}}(\mathbf{x}) = \mathbf{R}_i^{\sigma}(\mathbf{x})\overline{\boldsymbol{\sigma}}(\mathbf{x}_i) + \mathbf{F}_i^{\sigma}(\mathbf{x}). \tag{15.157}$$

Representations (15.151)–(15.153) can be recast in the new notations ($\mathbf{x} \in R^d$)

$$\mathcal{L}_i^{\tau}(\mathbf{x}-\mathbf{x}_i, \widehat{\boldsymbol{\tau}}_{(i)}) = \bar{v}_i \mathbf{T}^{\epsilon}(\mathbf{x}-\mathbf{x}_i)\widehat{\boldsymbol{\tau}}^{\mathbf{f}}_{(i)} + \bar{v}_i \mathbf{T}^{\nabla G}(\mathbf{x}-\mathbf{x}_i)\boldsymbol{\pi}_{(i)}, \tag{15.158}$$

$$\mathcal{L}_i^{\eta}(\mathbf{x}-\mathbf{x}_i, \widehat{\boldsymbol{\eta}}_{(i)}) = \bar{v}_i \mathbf{T}^{\sigma}(\mathbf{x}-\mathbf{x}_i)\widehat{\boldsymbol{\eta}}^{\mathbf{f}}_{(i)} + \bar{v}_i \mathcal{L}^c * \mathbf{T}^{\nabla G}(\mathbf{x}-\mathbf{x}_i)\boldsymbol{\pi}_{(i)}, \tag{15.159}$$

$$\mathcal{L}_i^{u\tau}(\mathbf{x}-\mathbf{x}_i, \widehat{\boldsymbol{\tau}}_{(i)}) = \bar{v}_i \mathbf{T}^{\nabla G}(\mathbf{x}-\mathbf{x}_i)\widehat{\boldsymbol{\tau}}^{\mathbf{f}}_{(i)} + \bar{v}_i \mathbf{T}^{G}(\mathbf{x}-\mathbf{x}_i)\boldsymbol{\pi}_{(i)}, \tag{15.160}$$

where one introduces the perturbator factors

$$\mathbf{T}_i^{\varepsilon}(\mathbf{x}–\mathbf{x}_i) = \begin{cases} -(\overline{v}_i)^{-1}\mathbf{P}_i(\mathbf{x}) & \text{for}\mathbf{x} \in v_i, \\ (\overline{v}_i)^{-1} \int \mathbf{U}(\mathbf{x}-\mathbf{y})V_i(\mathbf{y})d\mathbf{y} & \text{for}\mathbf{x} \notin v_i, \end{cases} \tag{15.161}$$

$$\mathbf{T}_i^{\sigma}(\mathbf{x}–\mathbf{x}_i) = \begin{cases} -(\overline{v}_i)^{-1}\mathbf{Q}_i(\mathbf{x}) & \text{for}\mathbf{x} \in v_i, \\ (\overline{v}_i)^{-1} \int \boldsymbol{\Gamma}(\mathbf{x}-\mathbf{y})V_i(\mathbf{y})d\mathbf{y} & \text{for}\mathbf{x} \notin v_i, \end{cases} \tag{15.162}$$

$$\mathbf{T}_i^{\nabla G}(\mathbf{x}–\mathbf{x}_i) = \begin{cases} -(\overline{v}_i)^{-1}\mathbf{P}_i^{\nabla G}(\mathbf{x}) & \text{for}\mathbf{x} \in v_i, \\ (\overline{v}_i)^{-1} \int \nabla\mathbf{G}(\mathbf{x}-\mathbf{y})V_i(\mathbf{y})d\mathbf{y} & \text{for}\mathbf{x} \notin v_i, \end{cases} \tag{15.163}$$

$$\mathbf{T}_i^{G}(\mathbf{x}–\mathbf{x}_i) = \begin{cases} -(\overline{v}_i)^{-1}\mathbf{P}_i^{G}(\mathbf{x}) & \text{for}\mathbf{x} \in v_i, \\ (\overline{v}_i)^{-1} \int \mathbf{G}(\mathbf{x}-\mathbf{y})V_i(\mathbf{y})d\mathbf{y} & \text{for}\mathbf{x} \notin v_i, \end{cases} \tag{15.164}$$

with the known point approximations

$$\mathbf{T}_i^{\varepsilon}(\mathbf{x}-\mathbf{x}_i) = \mathbf{U}(\mathbf{x}-\mathbf{x}_i), \quad \mathbf{T}_i^{\sigma}(\mathbf{x}-\mathbf{x}_i) = \boldsymbol{\Gamma}(\mathbf{x}-\mathbf{x}_i), \tag{15.165}$$

$$\mathbf{T}_i^{\nabla G}(\mathbf{x}-\mathbf{x}_i) = \nabla\mathbf{G}(\mathbf{x}-\mathbf{x}_i), \quad \mathbf{T}_i^{G}(\mathbf{x}-\mathbf{x}_i) = \mathbf{G}(\mathbf{x}-\mathbf{x}_i). \tag{15.166}$$

Here $\mathbf{T}_i^{\varepsilon}(\mathbf{x}–\mathbf{x}_i)$ (15.161) and $\mathbf{T}_i^{\sigma}(\mathbf{x}–\mathbf{x}_i)$ (15.162) have analytical representations inside ($\mathbf{x} \in v_i$) and outside ($\mathbf{x} \notin v_i$) of an ellipsoidal inclusion v_i presented through the Eshelby [449] tensor $\mathbf{S}_i$ and considered in [200] (where additional references can be found) for some particular problems in more detail. Obviously the equalities (15.158)–(15.160) are only asymptotically fulfilled at $|\mathbf{x}-\mathbf{x}_i| \to \infty$, and it is possible to propose a formal counterexample where an error of any approximations (15.158)–(15.160) equals infinity, e.g., if $\mathbf{g}(\mathbf{x}) \not\equiv \mathbf{0}$ and $\langle \mathbf{g} \rangle_{(i)} = \mathbf{0}$ ($\mathbf{g} = \boldsymbol{\tau}, \boldsymbol{\eta}, \mathbf{x} \in v_i$). The most popular point approximations (15.165) and (15.166) (which are simultaneously the most crude) were implicitly used by many authors for some particular problems (see for early references, e.g., [106]) including some sort of centering (see [1672]). A quantitative analysis of results obtained by the use of some representations (15.158)–(15.160) was performed by Buryachenko [203] for the local uncoupled elasticity.

15.4.4 Approximate Representation of Some Tensors and Operators. Some Asymptotic Representations

Effective field hypothesis **H1a** means that for the σ's components (ϵ's components can be presented analogously) $\mathbf{J}_j^{\sigma\infty}(\mathbf{x}) = \mathbf{T}_j^{\sigma}(\mathbf{x}-\mathbf{x}_j)\mathbf{R}_j^{\sigma}$, $\mathbf{J}_j^{\beta\infty}(\mathbf{x}) = \mathbf{T}_j^{\sigma}(\mathbf{x}-\mathbf{x}_j)\mathbf{F}_j^{\sigma}$, where hereafter the tensors $\mathbf{T}_j^{\sigma}(\mathbf{x}-\mathbf{x}_j) \equiv \langle\boldsymbol{\Gamma}(\mathbf{x}-\mathbf{y})\rangle_{(j)}$, $\mathbf{T}_{ij}^{\sigma}(\mathbf{x}_i-\mathbf{x}_j) \equiv \langle\mathbf{T}_j^{\sigma}(\mathbf{x}-\mathbf{x}_j)\rangle_{(i)}$ are expressed (for the local elastic constitutive law) through the external Eshelby tensor $\mathbf{S}_j(\mathbf{x}-\mathbf{x}_j) \equiv -\bar{v}_j\langle\mathbf{U}(\mathbf{x}-\mathbf{y})\rangle_{(j)}\mathbf{L}^{(0)}$ ($\mathbf{y} \in v_j,\ \mathbf{x} \in R^d$): $\mathbf{T}_j^{\sigma}(\mathbf{x}) = \mathbf{L}^{(0)}\big[\mathbf{S}_j(\mathbf{x}-\mathbf{x}_j) - \mathbf{I}V_j(\mathbf{x})\big]$. The acceptance of the effective field hypothesis **H1** for both interacting inclusions v_i and v_i reduces the integral Eqs. (15.109) describing the solution for the binary interacting inclusions to the algebraic ones. This allows us to significantly simplify the estimation of the tensors $\mathbf{J}_{i,j}^{\theta}(\mathbf{x}-\mathbf{x}_j)$ and $\mathbf{J}_{i,j}^{\gamma}(\mathbf{x}-\mathbf{x}_j)$ in the matrix representation of the solution of Eq. (15.109) ($\mathbf{x} \in v_i$, no sum on i, j)

$$\overline{\boldsymbol{\vartheta}}_i(\mathbf{x}) - \widetilde{\boldsymbol{\vartheta}}_{i,j}(\mathbf{x}) = \mathbf{J}_{i,j}^{I\theta}(\mathbf{x}-\mathbf{x}_j)(\widetilde{\boldsymbol{\vartheta}}_{i,j})_{(i)} + \mathbf{J}_{i,j}^{J\theta}(\mathbf{x}-\mathbf{x}_j)(\widetilde{\boldsymbol{\vartheta}}_{i,j})_{(j)} + \mathbf{J}_{i,j}^{\gamma}(\mathbf{x}-\mathbf{x}_j), \tag{15.167}$$

where

$$\mathbf{J}_{i,j}^{I\theta}(\mathbf{x}-\mathbf{x}_j) = \mathbf{T}_j^{\theta}(\mathbf{x}-\mathbf{x}_j)\mathbf{R}_j^{\theta}\mathbf{Z}_{ij}^{21}, \tag{15.168}$$

$$\mathbf{J}_{i,j}^{J\theta}(\mathbf{x}-\mathbf{x}_j) = \mathbf{T}_j^{\theta}(\mathbf{x}-\mathbf{x}_j)\mathbf{R}_j^{\theta}\mathbf{Z}_{ij}^{22}, \tag{15.169}$$

$$\mathbf{J}_{i,j}^{\gamma}(\mathbf{x}-\mathbf{x}_j) = \mathbf{T}_j^{\theta}(\mathbf{x}-\mathbf{x}_j)\mathbf{F}_j^{\theta} + \mathbf{T}_j^{\theta}(\mathbf{x}-\mathbf{x}_j)\mathbf{R}_j^{\theta} \times[\mathbf{Z}_{ij}^{12}\mathbf{T}_{ij}^{\theta}(\mathbf{x}_i-\mathbf{x}_j)\mathbf{F}_j^{\theta} + \mathbf{Z}_{ij}^{22}\mathbf{T}_{ij}^{\theta}(\mathbf{x}_i-\mathbf{x}_j)\mathbf{F}_i^{\theta}], \tag{15.170}$$

and the matrix $\mathbf{Z}^{-1}$ has the elements

$$(\mathbf{Z}^{-1})_{ij} = \mathbf{I}\delta_{ij} - (1-\delta_{ij})\mathbf{T}_{ij}^{\theta}(\mathbf{x}_i-\mathbf{x}_j)\mathbf{R}_j^{\theta}, \tag{15.171}$$

expressed through the submatrices $\mathbf{Z}_{ij}^{21}$ and $\mathbf{Z}_{ij}^{22}$ which are nondiagonal and diagonal elements, respectively, of the matrix $\mathbf{Z}_{ij} = (\mathbf{Z}_{ij}^{lm})$ ($l, m = 1, 2$).

To make further progress, the hypothesis of "ellipsoidal symmetry" for the distribution of inclusions attributed to Willis [1620] (see also [759, 762, 1240]) is widely used:

Hypothesis H3, "Ellipsoidal Symmetry" *The conditional probability density function* $\varphi(v_j, \mathbf{x}_j|; v_i, \mathbf{x}_i)$ *depends on* $\mathbf{x}_j - \mathbf{x}_i$ *only through the combination* $\rho = |(\mathbf{a}_{ij}^0)^{-1}(\mathbf{x}_j-\mathbf{x}_i)|$

$$\varphi(v_j, \mathbf{x}_j\ |; v_i, \mathbf{x}_i) = h(\rho), \tag{15.172}$$

where the matrix $(\mathbf{a}_{ij}^0)^{-1}$ *(which is symmetric in the indexes* i *and* j, $\mathbf{a}_{ij}^0 = \mathbf{a}_{ji}^0$*) defines the ellipsoid excluded volume* $v_{ij}^0 = \{\mathbf{x} : |(\mathbf{a}_{ij}^0)^{-1}\mathbf{x}|^2 < 1\}$.

A destination of the hypothesis **H3** is directed toward providing conditions for applying the hypothesis **H1** (see for details [232, 233]). The use of the satellite hypothesis **H3** has no sense without the hypothesis **H1**. If the hypotheses **H1**, **H2b**, and **H3** hold for the statistically homogeneous material and the field problems, then the effective properties $\mathbf{L}^*$ and $\boldsymbol{\alpha}^*$ do not depend on the size of the correlation hole v_i^0 and the binary correlation function $\varphi(v_j, \mathbf{x}_j \mid; v_i, \mathbf{x}_i)$.

Thus, acceptance of the hypotheses **H1a**, **H1b**, **H2a**, and **H3** reduces Eqs. (15.131)–(15.133) to the same equations with the following corrections of the tensors involved ($K = I, J;\ \mathbf{z} \in v_i$)

$$\mathbf{J}_{i,j}^{K\theta}(\mathbf{z} - \mathbf{x}_j) := \langle \mathbf{J}_{i,j}^{K\theta}(\mathbf{x} - \mathbf{x}_j)\rangle_{(i)},\ \mathbf{J}_{i,j}^{\theta\infty}(\mathbf{z} - \mathbf{x}_j) := \langle \mathbf{J}_{i,j}^{\theta\infty}(\mathbf{x} - \mathbf{x}_j)\rangle_{(i)}, \quad (15.173)$$

$$\mathbf{J}_{i,j}^{\gamma}(\mathbf{z} - \mathbf{x}_j) := \langle \mathbf{J}_{i,j}^{\gamma}(\mathbf{x} - \mathbf{x}_j)\rangle_{(i)},\ \mathbf{J}_{i,j}^{\gamma\infty}(\mathbf{z} - \mathbf{x}_j) := \langle \mathbf{J}_{i,j}^{\gamma\infty}(\mathbf{x} - \mathbf{x}_j)\rangle_{(i)}. \quad (15.174)$$

Acceptance of the hypothesis **H2b** instead of **H2a** leads to the following simplification ($k, m = 1, 2$)

$$\mathbf{Z}_{ij}^{km} = \mathbf{I}\delta_{km},\quad \mathbf{J}_{ij}^{I\theta\ \mathrm{pert}}(\mathbf{x}) \equiv \mathbf{0},\ \ \mathbf{J}_{ij}^{J\theta\ \mathrm{pert}}(\mathbf{x}) \equiv \mathbf{I} - n^{(i)}\mathbf{P}_i^{\theta 0}\mathbf{R}_j^{\theta}, \quad (15.175)$$

where $\mathbf{P}_i^{\theta 0} = \mathbf{P}_i^0, \mathbf{Q}_i^0$ for $\boldsymbol{\vartheta} = \boldsymbol{\varepsilon}, \boldsymbol{\sigma}$, respectively.

In particular, for one sort of identical aligned heterogeneities ($i = N = 1$), Eqs. (15.142)–(15.145) and (15.173)–(15.175) for the fields $\boldsymbol{\vartheta}$ are reduced to the representation obtained by the effective field method (EFM, called also one-particle approximation of the MEFM; see the case of the local elasticity (15.3) and $\mathbf{b}_1, \mathbf{f}_1 \equiv \mathbf{0}$ in [200])

$$\langle \boldsymbol{\vartheta}\rangle_i(\mathbf{x}) = \mathbf{B}_i^{\theta}(\mathbf{x})\Big[(\mathbf{R}_i^{\theta})^{-1}\mathbf{Y}\mathbf{R}_i^{\theta}\langle \boldsymbol{\vartheta}\rangle + n^{(i)}\mathbf{P}_i^{\theta 0}\mathbf{Y}\mathbf{F}_i^{\theta}\Big] + \mathbf{C}_i^{\theta}(\mathbf{x}), \quad (15.176)$$

$$\mathbf{K}^* = \mathbf{K}^{(0)} + n^{(i)}\mathbf{Y}\mathbf{R}_i^{\theta}, \quad (15.177)$$

$$\boldsymbol{\gamma}^* = \boldsymbol{\gamma}^{(0)} + n^{(i)}\mathbf{Y}\mathbf{F}_i^{\theta}, \quad (15.178)$$

$$2W^* = -\langle \boldsymbol{\gamma}_1\mathbf{C}_i^{\theta}\rangle - \langle \boldsymbol{\gamma}_1\mathbf{B}_i^{\theta}\rangle\mathbf{P}_i^{\theta 0}\mathbf{Y}\mathbf{F}_i^{\theta}n^{(i)} \quad (15.179)$$

for $\mathbf{K} = \mathbf{L}, \mathbf{M}$; $\boldsymbol{\vartheta} = \boldsymbol{\varepsilon}, \boldsymbol{\sigma}$, and $\boldsymbol{\gamma} = \boldsymbol{\alpha}, \boldsymbol{\beta}$, respectively. Here $\mathbf{Y}^{-1} = \mathbf{I} - n^{(1)}\mathbf{P}_i^{\theta 0}\mathbf{R}_i^{\theta}$ and Eq. (15.179) was estimated at $\langle \boldsymbol{\vartheta}\rangle \equiv \mathbf{0}$.

As pointed out by Benveniste [89] (see also [1608]) for the case of the local elasticity (15.3) and $\mathbf{b}_1, \mathbf{f}_1 \equiv \mathbf{0}$, the essential assumption in the Mori–Tanaka [1093]) method (MTM) states that each inclusion v_i behaves as an isolated one in the infinite matrix and subjected to some effective field $\overline{\boldsymbol{\vartheta}}_i$ coinciding with the average tensor $\langle \boldsymbol{\vartheta}\rangle^{(0)}$ in the matrix ($\boldsymbol{\vartheta} = \boldsymbol{\varepsilon}, \boldsymbol{\sigma}$)

$$\langle\overline{\boldsymbol{\varepsilon}}_i\rangle = \langle\boldsymbol{\varepsilon}\rangle^{(0)}, \quad \langle\overline{\boldsymbol{\sigma}}_i\rangle = \langle\boldsymbol{\sigma}\rangle^{(0)}. \tag{15.180}$$

Using Eq. (15.180) as the closing assumption leads to the following representations for both the statistical averages of the fields $\langle\boldsymbol{\vartheta}\rangle_i(\mathbf{x})$ and the effective properties

$$\langle\boldsymbol{\vartheta}\rangle_i(\mathbf{x}) = \mathbf{B}_i(\mathbf{x})[c^{(0)}\mathbf{I} + c^{(1)}\mathbf{B}_i^{\theta}]^{-1}(\langle\boldsymbol{\vartheta}\rangle - \langle\mathbf{C}^{\theta}V\rangle) + \mathbf{C}_i^{\theta}(\mathbf{x}), \tag{15.181}$$

$$\mathbf{K}^* = \mathbf{K}^{(0)} + n^{(1)}\mathbf{R}_i^{\theta}[c^{(0)}\mathbf{I} + c^{(1)}\mathbf{B}_i^{\theta}]^{-1}, \tag{15.182}$$

$$\boldsymbol{\gamma}^* = \boldsymbol{\gamma}^{(0)} + n^{(1)}\mathbf{F}_i^{\theta} - n^{(1)}\mathbf{R}_i^{\theta}[c^{(0)}\mathbf{I} + c^{(1)}\mathbf{B}_i^{\theta}]^{-1}\mathbf{C}_i^{\theta}, \tag{15.183}$$

$$2W^* = -\langle\boldsymbol{\gamma}_1\mathbf{C}^{\theta}\rangle - \langle\boldsymbol{\gamma}_1\mathbf{B}_i^{\theta}\rangle[c^{(0)} + c^{(1)}\mathbf{B}_i^{\theta}]^{-1}\langle\mathbf{C}^{\theta}V\rangle, \tag{15.184}$$

for $\mathbf{K} = \mathbf{L}, \mathbf{M}$; $\boldsymbol{\vartheta} = \boldsymbol{\varepsilon}, \boldsymbol{\sigma}$, and $\boldsymbol{\gamma} = \boldsymbol{\alpha}, \boldsymbol{\beta}$, respectively; here the representations (15.184) were estimated at $\langle\boldsymbol{\vartheta}\rangle \equiv \mathbf{0}$. For the identical ellipsoidal inhomogeneous heterogeneities v_i homothetical to v_i^0, equivalences of Eqs. (15.177)–(15.179) and (15.181)–(15.184), respectively, are demonstrated in, e.g., [199] for the local elasticity (15.3). However, the representations (15.177)–(15.179) do not coincide with (15.181)–(15.184), respectively, even for the identical aligned isotropic heterogeneities v_i if v_i and v_i^0 are not homothetic (in particular, if v_i is not an ellipsoid, see Chap. 9).

At last, the effective medium method (EMM; see a linear elastic counterpart proposed by Kröner [812] and Hill [643]), also called the self-consistent method, is based on the following hypothesis: each inclusion $\mathbf{x} \in v_i$ in the composite material behaves as an isolated one in a homogeneous medium whose properties coincide with the effective properties of the whole composite

$$\mathbf{K}^c = \mathbf{K}^*, \quad \langle\overline{\boldsymbol{\vartheta}}\rangle(\mathbf{x}) \equiv \langle\boldsymbol{\vartheta}\rangle \tag{15.185}$$

for $\mathbf{K} = \mathbf{L}, \mathbf{M}$, and $\boldsymbol{\vartheta} = \boldsymbol{\varepsilon}, \boldsymbol{\sigma}$, respectively. Thus, we see that acceptance of the assumptions either (15.180) or (15.185) is equivalent to truncation in the GIE (15.80) at $n = 1$ of the integral item depending on the perturbator $\boldsymbol{\mathcal{L}}_j^{\sigma}(\mathbf{x} - \mathbf{x}_j, \overline{\boldsymbol{\sigma}})$. This truncation is compensated in subsequent manipulations by correction of either $\langle\overline{\boldsymbol{\sigma}}\rangle(\mathbf{x})$ (15.180_2) or $\mathbf{M}^c$ (15.185_1). In so doing, the final representations of the effective properties (15.24) do not depend on the method which can be used for estimations of truncated averaged perturbators in the GIE (15.80). It means that the assumptions (15.180) and (15.185) are so much rough that it does not matter which background of micromechanics (either the old (15.83) or new (15.81) one) was used.

15.5 Second Moment of Fields

15.5.1 General Representations of Second Moment of Fields

For obtaining an integral representation of the second moments of fields in the component $v^{(i)}$ of the inclusions ($i = 1, 2, \ldots.$), we estimate the second moment of effective fields as a tensor product of (15.76) and (15.77) (at $n = 1$) into $\overline{\boldsymbol{\vartheta}}(\mathbf{x})$

$$\overline{\boldsymbol{\vartheta}}(\mathbf{x})\otimes\overline{\boldsymbol{\vartheta}}(\mathbf{x}) = \langle\boldsymbol{\vartheta}\rangle(\mathbf{x}) \otimes \langle\boldsymbol{\vartheta}\rangle(\mathbf{x}) + \langle\boldsymbol{\vartheta}\rangle(\mathbf{x}) \otimes \int [\boldsymbol{\mathcal{L}}_p^{\zeta}(\mathbf{x}-\mathbf{x}_p, \widehat{\boldsymbol{\zeta}})$$
$$\times V_p^{\delta}(\mathbf{x}_p|; v_i, \mathbf{x}_i) - \langle\boldsymbol{\mathcal{L}}_p^{\zeta}(\mathbf{x}-\mathbf{x}_p, \widehat{\boldsymbol{\zeta}})\rangle(\mathbf{x}_p)]d\mathbf{x}_p$$
$$+ \int [\boldsymbol{\mathcal{L}}_q^{\zeta}(\mathbf{x}-\mathbf{x}_q, \widehat{\boldsymbol{\zeta}})V_q^{\delta}(\mathbf{x}_q|; v_i, \mathbf{x}_i) - \langle\boldsymbol{\mathcal{L}}_q^{\zeta}(\mathbf{x}-\mathbf{x}_q, \widehat{\boldsymbol{\zeta}})\rangle(\mathbf{x}_q)]d\mathbf{x}_q \otimes \langle\boldsymbol{\vartheta}\rangle(\mathbf{x})$$
$$+ \int\int [\boldsymbol{\mathcal{L}}_p^{\zeta}(\mathbf{x}-\mathbf{x}_p, \widehat{\boldsymbol{\zeta}})V_p^{\delta}(\mathbf{x}_p|; v_i, \mathbf{x}_i) - \langle\boldsymbol{\mathcal{L}}_p^{\zeta}(\mathbf{x}-\mathbf{x}_p, \widehat{\boldsymbol{\zeta}})\rangle(\mathbf{x}_p)]$$
$$\otimes [\boldsymbol{\mathcal{L}}_q^{\zeta}(\mathbf{x}-\mathbf{x}_q, \widehat{\boldsymbol{\zeta}})V_q^{\delta}(\mathbf{x}_q|; v_i, \mathbf{x}_i) - \langle\boldsymbol{\mathcal{L}}_q^{\zeta}(\mathbf{x}-\mathbf{x}_q, \widehat{\boldsymbol{\zeta}})\rangle(\mathbf{x}_q)]d\mathbf{x}_p d\mathbf{x}_q \quad (15.186)$$

with $\mathbf{x} \in v_i$. The right-hand side of Eq. (15.186) is a random function of the arrangements of inclusions $v_p, v_q \neq v_i$ $(p, q = 1, 2, \ldots)$. Averaging (15.186) over a realization ensemble leads to ($\mathbf{x} \in v_i$)

$$\langle\overline{\boldsymbol{\vartheta}} \otimes \overline{\boldsymbol{\vartheta}}\rangle(\mathbf{x}) = \langle\boldsymbol{\vartheta}\rangle(\mathbf{x}) \otimes \langle\boldsymbol{\vartheta}\rangle(\mathbf{x})$$
$$+ \langle\boldsymbol{\vartheta}\rangle(\mathbf{x}) \otimes \int \big[\langle\boldsymbol{\mathcal{L}}_p^{\zeta}(\mathbf{x}-\mathbf{x}_p, \widehat{\boldsymbol{\zeta}})|; v_i, \mathbf{x}_i\rangle_p \varphi(v_p, \mathbf{x}_p|; v_i, \mathbf{x}_i)$$
$$- \langle\boldsymbol{\mathcal{L}}_p^{\zeta}(\mathbf{x}-\mathbf{x}_p, \widehat{\boldsymbol{\zeta}})\rangle(\mathbf{x}_p)\big]d\mathbf{x}_p$$
$$+ \int \big[\boldsymbol{\mathcal{L}}_q^{\zeta}(\mathbf{x}-\mathbf{x}_q, \widehat{\boldsymbol{\zeta}})|; v_i, \mathbf{x}_i\rangle_q \varphi(v_q, \mathbf{x}_q|; v_i, \mathbf{x}_i)$$
$$- \langle\boldsymbol{\mathcal{L}}_p^{\zeta}(\mathbf{x}-\mathbf{x}_q, \widehat{\boldsymbol{\zeta}})\rangle(\mathbf{x}_q)\big]d\mathbf{x}_q \otimes \langle\boldsymbol{\vartheta}\rangle(\mathbf{x})$$
$$+ \int\int \Big\{\langle[\boldsymbol{\mathcal{L}}_p^{\zeta}(\mathbf{x}-\mathbf{x}_p, \widehat{\boldsymbol{\zeta}}) \otimes \boldsymbol{\mathcal{L}}_p^{\zeta}(\mathbf{x}-\mathbf{x}_p, \widehat{\boldsymbol{\zeta}})|v_p, \mathbf{x}_p, v_q, \mathbf{x}_q; v_i, \mathbf{x}_i\rangle_p$$
$$\times \varphi(v_p, \mathbf{x}_p, v_q, \mathbf{x}_q \;|; v_i, \mathbf{x}_i)$$
$$- \langle\boldsymbol{\mathcal{L}}_p^{\zeta}(\mathbf{x}-\mathbf{x}_p, \widehat{\boldsymbol{\zeta}})|; v_i, \mathbf{x}_i\rangle_p \varphi(v_p, \mathbf{x}_p|; v_i, \mathbf{x}_i) \otimes \langle\boldsymbol{\mathcal{L}}_q^{\zeta}(\mathbf{x}-\mathbf{x}_q, \widehat{\boldsymbol{\zeta}})\rangle(\mathbf{x}_q)$$
$$- \langle\boldsymbol{\mathcal{L}}_p^{\zeta}(\mathbf{x}-\mathbf{x}_p, \widehat{\boldsymbol{\zeta}})\rangle(\mathbf{x}_p) \otimes \langle\boldsymbol{\mathcal{L}}_q^{\zeta}(\mathbf{x}-\mathbf{x}_q, \widehat{\boldsymbol{\zeta}})|; v_i, \mathbf{x}_i\rangle_q \varphi(v_q, \mathbf{x}_q|; v_i, \mathbf{x}_i)$$
$$+ \langle\boldsymbol{\mathcal{L}}_p^{\zeta}(\mathbf{x}-\mathbf{x}_p, \widehat{\boldsymbol{\zeta}})\rangle(\mathbf{x}_p) \otimes \langle\boldsymbol{\mathcal{L}}_q^{\zeta}(\mathbf{x}-\mathbf{x}_q, \widehat{\boldsymbol{\zeta}})\rangle(\mathbf{x}_q)\Big\}d\mathbf{x}_p d\mathbf{x}_q. \quad (15.187)$$

The right-hand side of Eq. (15.187) includes double-point $\varphi(v_r, \mathbf{x}_r|; v_i, \mathbf{x}_i)$ $(r = p, q)$ and triple-point $\varphi(v_p, \mathbf{x}_p, v_q, \mathbf{x}_q|; v_i, \mathbf{x}_i)$ conditional probability densities in which the terms with $\mathbf{x}_p = \mathbf{x}_q$ under the condition $\mathbf{x}_p \neq \mathbf{x}_i,\ \mathbf{x}_q \neq \mathbf{x}_i$ can be isolated with the help of the equality $\varphi(v_p, \mathbf{x}_p, v_q, \mathbf{x}_q\ |; v_i, \mathbf{x}_i) = \delta(\mathbf{x}_p - \mathbf{x}_q)\varphi(v_p, \mathbf{x}_p\ |; v_i, \mathbf{x}_i) + \varphi(v_p, \mathbf{x}_p\ |; v_i, \mathbf{x}_i)\varphi(v_q, \mathbf{x}_q\ |; v_p, \mathbf{x}_p; v_i, \mathbf{x}_i)$. Then Eq. (15.187) can be rewritten as

$$\begin{aligned}
\langle \overline{\boldsymbol{\vartheta}} \otimes \overline{\boldsymbol{\vartheta}} \rangle_i(\mathbf{x}) &= \langle \overline{\boldsymbol{\vartheta}} \rangle_i(\mathbf{x}) \otimes \langle \overline{\boldsymbol{\vartheta}} \rangle_i(\mathbf{x}) \\
&\quad + \int \langle \boldsymbol{\mathcal{L}}_p^{\zeta}(\mathbf{x}-\mathbf{x}_p, \widehat{\boldsymbol{\zeta}}) \otimes \boldsymbol{\mathcal{L}}_p^{\zeta}(\mathbf{x}-\mathbf{x}_p, \widehat{\boldsymbol{\zeta}})|; v_i, \mathbf{x}_i \rangle_p \varphi(v_p, \mathbf{x}_p\ |; v_i, \mathbf{x}_i) d\mathbf{x}_p \\
&\quad + \int\int \Big\{ \langle \boldsymbol{\mathcal{L}}_p^{\zeta}(\mathbf{x}-\mathbf{x}_p, \widehat{\boldsymbol{\zeta}}) \otimes \boldsymbol{\mathcal{L}}_q^{\zeta}(\mathbf{x}-\mathbf{x}_q, \widehat{\boldsymbol{\zeta}})|; v_i, \mathbf{x}_i \rangle_{pq} \\
&\quad \times \varphi(v_p, \mathbf{x}_p\ |; v_i, \mathbf{x}_i)\varphi(v_q, \mathbf{x}_q\ |; v_p, \mathbf{x}_p; v_i, \mathbf{x}_i) \\
&\quad - \langle \boldsymbol{\mathcal{L}}_p^{\zeta}(\mathbf{x}-\mathbf{x}_p, \widehat{\boldsymbol{\zeta}})|; v_i, \mathbf{x}_i \rangle_p \otimes \langle \boldsymbol{\mathcal{L}}_q^{\zeta}(\mathbf{x}-\mathbf{x}_q, \widehat{\boldsymbol{\zeta}})|; v_i, \mathbf{x}_i \rangle_q \\
&\quad \times \varphi(v_p, \mathbf{x}_p\ |; v_i, \mathbf{x}_i)\varphi(v_q, \mathbf{x}_q\ |; v_i, \mathbf{x}_i) \Big\} d\mathbf{x}_q d\mathbf{x}_p
\end{aligned} \tag{15.188}$$

with $\mathbf{x} \in v_i$. Here the double integral was transformed in terms of average effective fields.

From Eq. (15.188) one can see that the effective field dispersion

$$\Delta\overline{\theta}_{klmn}^{(i)2}(\mathbf{x}) := \langle \overline{\theta}_{kl}\overline{\theta}_{mn} \rangle_i(\mathbf{x}) - \langle \overline{\theta}_{kl} \rangle_i(\mathbf{x}) \langle \overline{\theta}_{mn} \rangle_i(\mathbf{x}) \tag{15.189}$$

presents a determined inhomogeneous function inside the components v_i $(i = 1, \ldots, N)$. At the same time, the field dispersion $\Delta\boldsymbol{\vartheta}^{(i)2}(\mathbf{x}) := \langle \boldsymbol{\vartheta} \otimes \boldsymbol{\vartheta} \rangle_i(\mathbf{x}) - \langle \boldsymbol{\vartheta} \rangle_i(\mathbf{x}) \otimes \langle \boldsymbol{\vartheta} \rangle_i(\mathbf{x})$ is connected with the effective field dispersion (15.189) by

$$\Delta\boldsymbol{\vartheta}^{(i)2}(\mathbf{x}) = \boldsymbol{\mathcal{B}}_i^{\theta} * \Delta\overline{\boldsymbol{\vartheta}}^{(i)2} * \boldsymbol{\mathcal{B}}_i^{\theta\top}(\mathbf{x}). \tag{15.190}$$

The formula (15.190) allows the estimation of the averaged field dispersion in the inclusion volume

$$\Delta\boldsymbol{\vartheta}^{(i)2} = \bar{v}_i^{-1} \int \Delta\boldsymbol{\vartheta}^{(i)2}(\mathbf{x}) V_i(\mathbf{x}) d\mathbf{x}. \tag{15.191}$$

By the use of the known properties of the second statistical moments, (15.190) and (15.189), we obtain the relations for the average second moments of the fields $\langle \boldsymbol{\vartheta} \otimes \boldsymbol{\vartheta} \rangle_i(\mathbf{x})$ in the component v_i

$$\begin{aligned}
\langle \boldsymbol{\vartheta} \otimes \boldsymbol{\vartheta} \rangle_i(\mathbf{x}) &= \langle \boldsymbol{\vartheta} \rangle_i(\mathbf{x}) \otimes \langle \boldsymbol{\vartheta} \rangle_i(\mathbf{x}) \\
&\quad + \int \langle \boldsymbol{\mathcal{B}}_i * \boldsymbol{\mathcal{L}}_p^{\zeta}(\mathbf{x}-\mathbf{x}_p, \widehat{\boldsymbol{\zeta}}) \otimes \boldsymbol{\mathcal{B}}_i * \boldsymbol{\mathcal{L}}_p^{\zeta}(\mathbf{x}-\mathbf{x}_p, \widehat{\boldsymbol{\zeta}})|v_p, \mathbf{x}_p; v_i, \mathbf{x}_i \rangle
\end{aligned}$$

$$
\begin{aligned}
&\times \varphi(v_p, \mathbf{x}_p \mid; v_i, \mathbf{x}_i) d\mathbf{x}_p + \int\int \Big\{ \langle \boldsymbol{\mathcal{B}}_i * \boldsymbol{\mathcal{L}}_p^{\zeta}(\mathbf{x} - \mathbf{x}_p, \widehat{\boldsymbol{\zeta}}) \\
&\otimes \boldsymbol{\mathcal{B}}_i * \boldsymbol{\mathcal{L}}_q^{\zeta}(\mathbf{x} - \mathbf{x}_q, \widehat{\boldsymbol{\zeta}}) | v_p, \mathbf{x}_p; v_q, \mathbf{x}_q; v_i, \mathbf{x}_i \rangle \\
&\times \varphi(v_p, \mathbf{x}_p \mid; v_i, \mathbf{x}_i) \varphi(v_q, \mathbf{x}_q \mid; v_p, \mathbf{x}_p; v_i, \mathbf{x}_i) \\
&- \langle \boldsymbol{\mathcal{B}}_i * \boldsymbol{\mathcal{L}}_p^{\zeta}(\mathbf{x} - \mathbf{x}_p, \widehat{\boldsymbol{\zeta}}) | v_p, \mathbf{x}_p; v_i, \mathbf{x}_i \rangle \varphi(v_p, \mathbf{x}_p \mid; v_i, \mathbf{x}_i) \\
&\otimes \langle \boldsymbol{\mathcal{B}}_i * \boldsymbol{\mathcal{L}}_q^{\zeta}(\mathbf{x} - \mathbf{x}_q, \widehat{\boldsymbol{\zeta}}) | v_q, \mathbf{x}_q; v_i, \mathbf{x}_i \rangle \varphi(v_q, \mathbf{x}_q \mid; v_i, \mathbf{x}_i) \Big\} d\mathbf{x}_q d\mathbf{x}_p.
\end{aligned} \tag{15.192}
$$

The relation (15.192) is derived by the use of both the triple and double interaction of the heterogeneities. As may be seen from Eq. (15.192), the neglecting of binary interactions is tantamount to assuming homogeneity of the effective fields inside the component v_i

$$
\langle \overline{\boldsymbol{\vartheta}} \otimes \overline{\boldsymbol{\vartheta}} \rangle_i(\mathbf{x}) = \langle \overline{\boldsymbol{\vartheta}} \rangle_i(\mathbf{x}) \otimes \langle \overline{\boldsymbol{\vartheta}} \rangle_i(\mathbf{x}), \tag{15.193}
$$

$$
\langle \boldsymbol{\vartheta} \otimes \boldsymbol{\vartheta} \rangle_i(\mathbf{x}) = \langle \boldsymbol{\vartheta} \rangle_i(\mathbf{x}) \otimes \langle \boldsymbol{\vartheta} \rangle_i(\mathbf{x}). \tag{15.194}
$$

However, substitution of the equalities (15.193) and (15.194) into the right-hand sides of Eqs. (15.188) and (15.192) does not mean that Eqs. (15.193) and (15.194) hold for the left-hand sides of Eqs. (15.188) and (15.192). The following new approximations for the second moments can be obtained by taking only binary interaction of inclusions into account

$$
\begin{aligned}
\langle \boldsymbol{\vartheta} \otimes \boldsymbol{\vartheta} \rangle_i(\mathbf{x}) =& \langle \boldsymbol{\vartheta} \rangle_i(\mathbf{x}) \otimes \langle \boldsymbol{\vartheta} \rangle_i(\mathbf{x}) + \int \langle \boldsymbol{\mathcal{B}}_i * \boldsymbol{\mathcal{L}}_p^{\zeta}(\mathbf{x} - \mathbf{x}_p, \widehat{\boldsymbol{\zeta}}) |; v_i, \mathbf{x}_i \rangle \\
&\otimes \langle \boldsymbol{\mathcal{B}}_i * \boldsymbol{\mathcal{L}}_p^{\zeta}(\mathbf{x} - \mathbf{x}_p, \widehat{\boldsymbol{\zeta}}) |; v_i, \mathbf{x}_i \rangle \varphi(v_p, \mathbf{x}_p \mid; v_i, \mathbf{x}_i) d\mathbf{x}_p,
\end{aligned} \tag{15.195}
$$

Ignoring effective field fluctuations in the inclusions v_p in the integral terms (15.192) (see (15.119) and (15.120)) leads to other new representations

$$
\begin{aligned}
\langle \boldsymbol{\vartheta} \otimes \boldsymbol{\vartheta} \rangle_i(\mathbf{x}) =& \langle \boldsymbol{\vartheta} \rangle_i(\mathbf{x}) \otimes \langle \boldsymbol{\vartheta} \rangle_i(\mathbf{x}) + \int [\boldsymbol{\mathcal{B}}_i * \boldsymbol{\mathcal{L}}_p^{\zeta}(\mathbf{x} - \mathbf{x}_p, \langle \widehat{\boldsymbol{\zeta}}_{(p)} \rangle)] \\
&\otimes [\boldsymbol{\mathcal{B}}_i * \boldsymbol{\mathcal{L}}_p^{\zeta}(\mathbf{x} - \mathbf{x}_p, \langle \widehat{\boldsymbol{\zeta}}_{(p)} \rangle)] \varphi(v_p, \mathbf{x}_p |; v_i, \mathbf{x}_i) d\mathbf{x}_p \\
&+ \int\int [\boldsymbol{\mathcal{B}}_i * \boldsymbol{\mathcal{L}}_p^{\zeta}(\mathbf{x} - \mathbf{x}_p, \langle \widehat{\boldsymbol{\zeta}}_{(p)} \rangle)] \otimes [\boldsymbol{\mathcal{B}}_i * \boldsymbol{\mathcal{L}}_q^{\zeta}(\mathbf{x} - \mathbf{x}_q, \langle \widehat{\boldsymbol{\zeta}}_{(q)} \rangle)] \\
&\times \widetilde{\varphi}(v_q, \mathbf{x}_q \mid; v_p, \mathbf{x}_p; v_i, \mathbf{x}_i) \varphi(v_p, \mathbf{x}_p \mid; v_i, \mathbf{x}_i) d\mathbf{x}_p d\mathbf{x}_q,
\end{aligned} \tag{15.196}
$$

where $\widetilde{\varphi}(v_q, \mathbf{x}_q \mid; v_p, \mathbf{x}_p; v_i, \mathbf{x}_i) \equiv \varphi(v_q, \mathbf{x}_q \mid; v_p, \mathbf{x}_p; v_i, \mathbf{x}_i) - \varphi(v_q, \mathbf{x}_q \mid; v_i, \mathbf{x}_i)$. The relation (15.196) can be called as "far-field approximation." Locally elastic counterparts of Eqs. (15.192)–(15.196) were considered in [200].

15.5.2 Field Fluctuations Inside the Matrix

In a similar manner, it is possible to obtain the estimation of the second moment of the field $\boldsymbol{\vartheta}$ averaged over a volume of the matrix. At first we define the conditional probability density $\varphi(v_p, \mathbf{x}_p \;|; v_q, \mathbf{x}_q; \ldots; v_0, \mathbf{x}_0)$ under the condition that the inclusion centers of $v_q, \ldots$ are located in the points $\mathbf{x}_q, \ldots$, whereas the matrix material appears at the point $\mathbf{x}_0$. In analogy to the above considerations, we obtain

$$\begin{aligned}
\langle \boldsymbol{\vartheta} \otimes \boldsymbol{\vartheta} \rangle_0 = {} & \langle \boldsymbol{\vartheta} \rangle_0 \otimes \langle \boldsymbol{\vartheta} \rangle_0 \\
& + \int \langle \boldsymbol{\mathcal{L}}_p^{\xi}(\mathbf{x}_0 - \mathbf{x}_p, \widehat{\boldsymbol{\zeta}}) \otimes \boldsymbol{\mathcal{L}}_p^{\xi}(\mathbf{x}_0 - \mathbf{x}_p, \widehat{\boldsymbol{\zeta}}) |; v_0, \mathbf{x}_0 \rangle \varphi(v_p, \mathbf{x}_p \;|; v_0, \mathbf{x}_0) d\mathbf{x}_p \\
& + \int \int \Big\{ \langle \boldsymbol{\mathcal{L}}_p^{\xi}(\mathbf{x}_0 - \mathbf{x}_p, \widehat{\boldsymbol{\zeta}}) \otimes \boldsymbol{\mathcal{L}}_q^{\xi}(\mathbf{x}_0 - \mathbf{x}_q, \widehat{\boldsymbol{\zeta}}) |; v_q, \mathbf{x}_q; v_0, \mathbf{x}_0 \rangle \\
& \times \varphi(v_p, \mathbf{x}_p \;|; v_0, \mathbf{x}_0) \varphi(v_q, \mathbf{x}_q \;|; v_p, \mathbf{x}_p; v_0, \mathbf{x}_0) \\
& - \langle \boldsymbol{\mathcal{L}}_p^{\xi}(\mathbf{x}_0 - \mathbf{x}_p, \widehat{\boldsymbol{\zeta}}) |; v_0, \mathbf{x}_0 \rangle \varphi(v_p, \mathbf{x}_p \;|; v_0, \mathbf{x}_0) \\
& \otimes \langle \boldsymbol{\mathcal{L}}_q^{\xi}(\mathbf{x}_0 - \mathbf{x}_q, \widehat{\boldsymbol{\zeta}}) |; v_0, \mathbf{x}_0 \rangle \varphi(v_q, \mathbf{x}_q \;|; v_0, \mathbf{x}_0) \Big\} d\mathbf{x}_q d\mathbf{x}_p. \qquad (15.197)
\end{aligned}$$

In contrast to (15.192), the second field moment in the matrix, (15.197), does not depend on $\mathbf{x}_0 \in v_0$. A more approximative estimation of the second moment $\langle \boldsymbol{\vartheta} \otimes \boldsymbol{\vartheta} \rangle_0$ can be obtained in direct analogy to (15.195) and (15.196) by replacing v_i, $\mathbf{x}_i$ by v_0, $\mathbf{x}_0$. So, in the framework of the effective field hypothesis **H1** (see Eqs. (15.119) and (15.120)), we have

$$\begin{aligned}
\langle \boldsymbol{\vartheta} \otimes \boldsymbol{\vartheta} \rangle_0 = {} & \langle \boldsymbol{\vartheta} \rangle_0 \otimes \langle \boldsymbol{\vartheta} \rangle_0 \\
& + \int \langle \boldsymbol{\mathcal{L}}_p^{\xi}(\mathbf{x} - \mathbf{x}_p, \widehat{\boldsymbol{\zeta}}_{(p)}) \\
& \otimes \boldsymbol{\mathcal{L}}_p^{\xi}(\mathbf{x} - \mathbf{x}_p, \widehat{\boldsymbol{\zeta}}_{(p)}) |; v_0, \mathbf{x}_0 \rangle \varphi(v_p, \mathbf{x}_p \;|; v_0, \mathbf{x}_0) d\mathbf{x}_p \\
& + \int \int \Big\{ \langle \boldsymbol{\mathcal{L}}_p^{\xi}(\mathbf{x} - \mathbf{x}_p, \widehat{\boldsymbol{\zeta}}_{(p)}) \otimes \boldsymbol{\mathcal{L}}_q^{\xi}(\mathbf{x} - \mathbf{x}_q, \widehat{\boldsymbol{\zeta}}_{(q)}) | v_p, \mathbf{x}_p; v_q, \mathbf{x}_q; v_0, \mathbf{x}_0 \rangle \\
& \times \varphi(v_p, \mathbf{x}_p \;|; v_0, \mathbf{x}_0) \varphi(v_q, \mathbf{x}_q \;|; v_p, \mathbf{x}_p; v_0, \mathbf{x}_0) \\
& - \langle \boldsymbol{\mathcal{L}}_p^{\xi}(\mathbf{x} - \mathbf{x}_p, \widehat{\boldsymbol{\zeta}}_{(p)}) | v_p, \mathbf{x}_p; v_0, \mathbf{x}_0 \rangle \otimes \langle \boldsymbol{\mathcal{L}}_q^{\xi}(\mathbf{x} - \mathbf{x}_q, \widehat{\boldsymbol{\zeta}}_{(q)}) |; v_0, \mathbf{x}_0 \rangle \\
& \times \varphi(v_p, \mathbf{x}_p \;|; v_0, \mathbf{x}_0) \varphi(v_q, \mathbf{x}_q \;|; v_0, \mathbf{x}_0) \Big\} d\mathbf{x}_q d\mathbf{x}_p. \qquad (15.198)
\end{aligned}$$

"Far-field approximation" leads to the following simplification of Eq. (15.198)

$$
\begin{aligned}
\langle \boldsymbol{\vartheta} \otimes \boldsymbol{\vartheta} \rangle_0 &= \langle \boldsymbol{\vartheta} \rangle_0 \otimes \langle \boldsymbol{\vartheta} \rangle_0 \\
&\quad + \int \mathcal{L}_p^{\zeta}(\mathbf{x} - \mathbf{x}_p, \langle \widehat{\boldsymbol{\zeta}}_{(p)} \rangle) \otimes \mathcal{L}_p^{\zeta}(\mathbf{x} - \mathbf{x}_p, \langle \widehat{\boldsymbol{\zeta}}_{(p)} \rangle) \varphi(v_p, \mathbf{x}_p \,|; v_0, \mathbf{x}_0) d\mathbf{x}_p \\
&\quad + \int \int \Big\{ \mathcal{L}_p^{\zeta}(\mathbf{x} - \mathbf{x}_p, \langle \widehat{\boldsymbol{\zeta}}_{(p)} \rangle) \\
&\quad \otimes \mathcal{L}_q^{\zeta}(\mathbf{x} - \mathbf{x}_q, \langle \widehat{\boldsymbol{\zeta}}_{(q)} \rangle) [\varphi(v_q, \mathbf{x}_q \,|; v_p, \mathbf{x}_p; v_0, \mathbf{x}_0) \\
&\quad - \varphi(v_q, \mathbf{x}_q \,|; v_0, \mathbf{x}_0)] \varphi(v_p, \mathbf{x}_p \,|; v_0, \mathbf{x}_0) \Big\} d\mathbf{x}_q d\mathbf{x}_p. \qquad (15.199)
\end{aligned}
$$

The new representations (15.192) and (15.198) of the field second moments are exact and depend on both the triple and double interactions of heterogeneities. Equations (15.192) and (15.198) do not depend on which basic integral equations either (15.82) or (15.84) they were obtained although the values $\widehat{\boldsymbol{\zeta}}(\mathbf{x} \in v_p, v_q)$ in the right-hand side of the Eqs. (15.192) and (15.198) depend on the estimations used either Eq. (15.82) or Eq. (15.84). Their most strong simplifications were obtained in the framework of far-field approximation (15.196) and (15.199) which depend only on the average polarization tensors inside heterogeneities $\langle \widehat{\boldsymbol{\zeta}}_{(p)} \rangle$. More accurate simplifications of the right-hand sides of Eqs. (15.192) and (15.196) can be established in the framework the scheme proposed by Buryachenko [204] (see Chap. 10) for the linear elastic composites with the local constitutive laws.

15.6 Particular Cases

We will consider in this section some particular schemes of generations of the perturbators and their analytical asymptotic representations for some constitutive laws and heterogeneities shapes. Some of these particular schemes were proposed by the author before and shortly summarized in this section for clarification of unified presentations of particular problems.

15.6.1 Thermoelasticity of CMs

We will consider the local basic equations of thermoelastostatics ($\mathbf{b}(\mathbf{x}) \equiv \mathbf{0}$) of composites with no body forces acting ($\mathbf{f}(\mathbf{x}) \equiv \mathbf{0}$) and the perfect interface conditions (15.17). It is assumed that the properties of both the comparison medium and the matrix coincide (15.16). In such a case, only the first integrals in Eqs. (15.52)–(15.54) do not vanish and

$$
\begin{aligned}
&\mathcal{L}_k^{\tau}(\mathbf{x} - \mathbf{x}_k, \widehat{\boldsymbol{\tau}}) = \mathcal{L}_k^{\tau}(\mathbf{x} - \mathbf{x}_k, \boldsymbol{\tau}), \;\; \mathcal{L}_k^{\eta}(\mathbf{x} - \mathbf{x}_k, \widehat{\boldsymbol{\eta}}) = \mathcal{L}_k^{\eta}(\mathbf{x} - \mathbf{x}_k, \boldsymbol{\eta}), \\
&\mathcal{L}_k^{u\tau}(\mathbf{x} - \mathbf{x}_k, \widehat{\boldsymbol{\tau}}) = \mathcal{L}_k^{u\tau}(\mathbf{x} - \mathbf{x}_k, \boldsymbol{\tau}). \qquad (15.200)
\end{aligned}
$$

For the ellipsoidal heterogeneities (4.55)

$$x_m(a^{-2})_{ml}x_l \leq 1, \quad a_{kl} := a_k\delta_{kl}, \tag{15.201}$$

the tensors $\mathbf{P}_i(\mathbf{x})$ and $\mathbf{Q}_i(\mathbf{x})$ are constant inside the domain (4.55) $\mathbf{x} \in v_i$ and represented through the Eshelby [449] tensor $\mathbf{S}_i$

$$\mathbf{P}_i(\mathbf{x}) \equiv \mathbf{P}_i = \mathbf{S}_i\mathbf{M}^{(0)}, \quad \mathbf{Q}_i(\mathbf{x}) \equiv \mathbf{Q}_i = \mathbf{L}^{(0)}(\mathbf{I} - \mathbf{S}_i). \tag{15.202}$$

The tensors $\mathbf{T}_i^{\varepsilon}(\mathbf{x}–\mathbf{x}_i)$ (15.161) and $\mathbf{T}_i^{\sigma}(\mathbf{x}–\mathbf{x}_i)$ (15.162) outside the domain $\mathbf{x} \notin v_i$ are also expressed through the external Eshelby tensor $\mathbf{S}_i(\mathbf{x})$ analogously to Eq. (15.202) (see for references and details Chap. 3). The tensors $\mathbf{T}_{ij}^{\theta}(\mathbf{x}_i - \mathbf{x}_j)$ (15.171) have the analytical representations for the ellipsoidal inclusions of different sizes and orientations (see for references and details Chap. 4).

Moreover, for the homogeneous ellipsoidal domain $\mathbf{x} \in v_i$ (4.55), in the framework of the hypothesis **H1a**, the kernels of the operators $\mathcal{B}_i^{\theta}$ and the tensors $\mathbf{C}^{\theta}$ $(\theta = \epsilon, \sigma)$ are expressed through the tensors (15.202) $(\mathbf{x}, \mathbf{y} \in v_i)$

$$\mathcal{B}^{\epsilon}(\mathbf{x}, \mathbf{y}) = \mathbf{B}_i^{\epsilon}(\mathbf{x})\delta(\mathbf{x} - \mathbf{y}), \ \mathbf{B}_i^{\epsilon}(\mathbf{x}) = (\mathbf{I} + \mathbf{P}_i\mathbf{L}_1^{(i)})^{-1}, \ \mathbf{C}_i^{\epsilon} = \mathbf{B}_i^{\epsilon}\mathbf{P}_i\boldsymbol{\alpha}_1^{(i)}, \tag{15.203}$$

$$\mathcal{B}^{\sigma}(\mathbf{x}, \mathbf{y}) = \mathbf{B}_i^{\sigma}(\mathbf{x})\delta(\mathbf{x} - \mathbf{y}), \ \mathbf{B}_i^{\sigma}(\mathbf{x}) = (\mathbf{I} + \mathbf{Q}_i\mathbf{M}_1^{(i)})^{-1}, \ \mathbf{C}_i^{\sigma} = \mathbf{B}_i^{\sigma}\mathbf{Q}_i\boldsymbol{\beta}_1^{(i)}. \tag{15.204}$$

It should be mentioned that the operator representation for effective material and field parameters was obtained for any shape, microstructure, and any linear imperfect interface properties. In particular, for the ellipsoidal and spherical inclusions, the operators (perturbators) $\mathcal{L}_i^{\theta}(\mathbf{x} - \mathbf{x}_i, \overline{\boldsymbol{\vartheta}})$, $\mathcal{L}_i^{\theta}(\mathbf{x} - \mathbf{x}_i, \widetilde{\boldsymbol{\vartheta}}_{i,j})$ $\mathcal{L}_{i,j}^{\theta}(\mathbf{x} - \mathbf{x}_j, \widetilde{\boldsymbol{\vartheta}}_{i,j})$ (15.102) or $\mathcal{L}_i^{\chi}(\mathbf{s} - \mathbf{x}_i, \overline{\boldsymbol{\chi}})$ (4.45) averaged over the volume v_i or the interface Γ_i, respectively, were estimated for the spring model (3.201) (see [424], [614–616, 850]) and Gurtin–Murdoch interface [587] model (3.205) (see [314, 315, 848, 1368, 1373]). The spherical and circular inclusions with Steigmann–Ogden [1440] curvature-dependent interfacial model were analyzed in [526, 1085].

For demonstration of potency of our efforts in the direction of procurement of general equations, we will shortly reproduce some particular results obtained (or expected) for thermoelastic problems considered in this Subsection 15.7.1.

Result 15.1.1 For homogeneous loading (15.21) of statistically homogeneous media (5.61_2) with aligned identical homogeneous ellipsoidal heterogeneities, hypotheses **H2b** and **H3** lead to equivalence of the new Eqs. (15.121) and (15.122) and classic Eqs. (15.125) and (15.126), respectively (see Chap. 9), with the identical effective material (15.22) and field [$\langle\boldsymbol{\varepsilon}\rangle_i$, $\langle\boldsymbol{\sigma}\rangle_i$, see (15.176), (15.177)] properties.

Result 15.1.2 The effective material and field properties estimated by the use of the new Eqs. (15.143) and (15.144) depend on the field distribution in the vicinity of heterogeneities [even for the conditions of *Result 15.1.1*, in contrast with Eqs. (15.88) and (15.89)] in the basic auxiliary one-particle problem (15.49), (15.50) as well as on the size of the excluded volume v_i^0 and the RDF even in the framework of hypothesis **H2b** and **H3** (see for details Chap. 9 and [203, 232, 233]).

Result 15.1.3 For both the homogeneous nonellipsoidal heterogeneities (Chap. 9 and [232, 233]) and circle inclusions with continuously inhomogeneous coating (Chap. 9 and [203]), inhomogeneity of field distributions inside heterogeneity in the auxiliary one-particle problem (15.49), (15.50) for $\boldsymbol{\varepsilon}^0(\mathbf{x}), \boldsymbol{\sigma}^0(\mathbf{x})$ ≡const. can lead to the different signs of local stresses $\langle\boldsymbol{\sigma}\rangle_i(\mathbf{z})$ ($\mathbf{z} \in v_i$) estimated by the use of Eqs. (15.81) and (15.82) and Eqs. (15.83) and (15.84), respectively, while its corresponding effective material properties (15.143) and (15.177) are not too different from one another.

Result 15.1.4 Taking into account binary interaction of spherical heterogeneities in CM satisfying conditions of *Result 15.1.1* (with replacement of hypothesis **H2b** on hypothesis **H2a** at $n = 2$) can lead to the different signs of local stresses $\langle\boldsymbol{\sigma}\rangle_i(\mathbf{z})$ ($\mathbf{z} \in v_i$) and average particle ones $\langle\boldsymbol{\sigma}\rangle_i$ estimated by the use of Eqs. (15.82) and (15.89), respectively, while its corresponding effective material properties (15.23) are not too different from one another (see Chap. 10 and [204]).

It was mentioned in Chaps. 9 and 11 (and in [208, 232, 233]) that it is expected to get a larger difference (which can reach infinity with the change of the sign of predicted local statistical average fields) between the results obtaining the use of either Eqs. (15.81), (15.82), or (15.88), (15.89) for composites reinforced by heterogeneities demonstrating greater inhomogeneity of field distributions inside heterogeneities. The last inhomogeneity can be produced by any one (or a few) reason including such as nonellipsoidal shape of heterogeneities, inhomogeneity of its properties (e.g., laminated structure or imperfect interface), binary interaction of heterogeneities, inhomogeneity of external loading (15.20), statistical inhomogeneity of composite structure (5.61), nonlocal constitutive laws (15.1), dynamic effects, and effect of size boundness of CM. In so doing, a difference of effective properties estimated by the use of either Eqs. (15.81), (15.82), or (15.88), (15.89) is not significant.

We mentioned a few meaning of the term generality in the context of evaluation of significance of the new approach. The first meaning is that the operator form of GIE (15.72) and (15.73) generalizes the corresponding GIEs for the different particular cases of constituent equations considered in Chaps. 9–11 and 14. The second meaning of generality of the new approach is demonstrated even for both the particular constitutive equations and partial perturbator representations. For example, for the locally elastic phenomena (15.3) with Green's function representations (15.52)–(15.54) of the perturbators, the new GIEs (15.81) and (15.82) are more universal than the old ones (15.88)–(15.89) obtained at the assumptions **H1b** (15.85) and (15.86) (see Result 15.1.3). Moreover, the Green function technique for the mentioned case is just a particular tool for representation of a solution for the inclusion interaction problem. In so doing, a more general expression for this solution in terms of perturbators (15.49)–(15.51) with a subsequent use in Eqs. (15.72) and (15.73) has the additional interesting advantage with respect to Eqs. (15.88)–

(15.89) because there is no reason to the questionable caring out from the statistical brackets of some deterministic function as in Eqs. (15.85) and (15.86). However, both the generality and novelty of GIEs obtained [see Eqs. (15.72), (15.73), (15.79) and (15.80)] provide not only the theoretical interest but also the critical practical benefits dictated by drastically improved accuracy of local fields.

Inhomogeneity of the mechanical properties inside ellipsoidal aligned identical inclusions (4.55$_2$) $\mathbf{L}_1(\mathbf{x})$ and $\boldsymbol{\alpha}_1(\mathbf{x})$ $\not\equiv$const.($\mathbf{x} \in v_i$) also leads to violation of the hypothesis **H1b** and, as a consequence, to discrepancy of estimations (15.142)–(15.145), (15.176)–(15.179), and (15.181)–(15.184), respectively, in the framework of the hypotheses **H2b** and **H3**. In so doing, the estimations (15.176)–(15.179) and (15.181)–(15.184) coincide for the identical shape of v_i and v_i^0, while the evaluations of the field concentration factors (15.142) and (15.176) can differ by a sign (see for details [203]). One proposed iteration method of volume integral equations for estimation of the operator perturbators (15.200) for the single isolated inhomogeneity v_i subjected to an inhomogeneous effective field with subsequent substitution of Eqs. (15.200) into a particular case of iteration scheme (15.126).

Another origin of the field inhomogeneity inside inclusions $\boldsymbol{\vartheta}(\mathbf{x})$ $\not\equiv$const. ($\mathbf{x} \in v_i$) is a binary interaction between heterogeneities. Buryachenko [204] analyzed a particular case of the iteration scheme (15.119)–(15.121) with the initial approximation (15.176) and acceptance of the hypotheses **H1b**, **H2a**, and **H3**. The numerical results were obtained for composites with identical homogeneous circle inclusions with the analytical approximative representations (15.168)–(15.170) for the perturbator factors (15.167) for the binary interacting inclusions. It was demonstrated the possible different signs for the local field concentration factors estimated with [(15.172), (15.173)] and without hypothesis **H1a**, while a variation of a field dispersion (15.190), (15.196) along a cross section of the inclusion $\mathbf{x} \in v_i$ can reach 80%. In so doing, the corresponding effective material properties are not too different from one another. The mentioned scheme can be directly applied for the composites with nonellipsoidal inclusion shape and the abandoned hypothesis **H3**, while rejection of the hypothesis **H1b** will lead to significant increasing of a computing performance.

It should be mentioned the fundamental role of the statistical averages of the second moments of stress concentration factors which are critical in the popular nonlinear analyses due to the fact that both the yield surface, fiber/matrix interface failure criterion, and the energy release rate are the quadratic functions of the local stress distributions. The perturbation method was developed for the exact estimation of all components of the second moment tensor $\langle \boldsymbol{\vartheta} \otimes \boldsymbol{\vartheta} \rangle_i$ averaged over the volume of components $v_i \subset v^{(i)}$ (for the elastic case, see for references and details [200]). It is found that the second moment of the field is constant within the inclusions if a homogeneity of some random effective fields $\overline{\boldsymbol{\vartheta}}(\mathbf{x})$ $\equiv$const. ($\mathbf{x} \in v_i$) in the neighborhood of each ellipsoidal inclusion was additionally assumed. Exploiting of the method of integral equations (15.196) makes it possible to estimate inhomogeneous second field moment $\langle \boldsymbol{\vartheta} \otimes \boldsymbol{\vartheta} \rangle_i(\mathbf{x})$ ($\mathbf{x} \in v_i$) inside inclusions. Equations (15.195), (15.196), and (15.198), (15.199) are the operator generalization of corresponding representations (see [206]) proposed for the integral representations of corresponding perturbators (15.52), (15.53). It was demonstrated that an effective field dispersion (15.189) inside the circle identical inclusions at $c = 0.7$ is reached 80%. It is possible generalization of the mentioned results [206] to the case of nonellipsoidal shape of inclusions with imperfect interface.

Exploiting of the new background (15.80) is even more attractive because this background [in opposite to the classical one (15.83), (15.84)] is sensitive to the field inhomogeneity inside the particles on the side generated, in general, by the imperfect interface conditions. In so doing, it does not invite further investigation additionally to research already performed for one-particle solution (15.49) and (15.50) with either the LSM, CZM, or ISM (see Sects. 3.7.3 and 15.6.1) interface. Indeed, the use of averaged tensors $\mathbf{R}_i^\theta$ (15.156), (15.157) implies that the tensors $\mathbf{R}_i^\theta(\mathbf{x})$ are previously estimated that in its turn means that the tensors $\mathcal{L}_i^\theta(\mathbf{x} - \mathbf{x}_i, \overline{\boldsymbol{\vartheta}})$ (15.49) and (15.50) are found. Therefore, we can immediately exploit either Eq. (15.126), (used hypotheses **H1a** and **H2b**) or Eq. (15.119)–(15.121) (used hypotheses **H1a** and **H1b** at $n = 2$). Subsequent improvement of the mentioned methods is related with abandonment of hypothesis **H1a** when variation of the effective fields $\langle\overline{\boldsymbol{\vartheta}}\rangle_i(\mathbf{y})$ $(\mathbf{y} \in v_i)$ can be estimated by the iteration method analogously to the counterpart problem for CM with a perfect interface (see for details [203, 208]).

Imperfection of interface conditions (with either LSM, CZM, or ISM, see Sect. 3.7.3) leads to additional opportunity for stress inhomogeneity inside the particles, and, therefore, after addition of imperfect interface conditions into the Results 15.1.2–15.1.4, it can be considered as expected results for CM with imperfect interface.

15.6.2 *Thermoconductivity of CMs*

It is well-known that the equations describing the quasistatic conditions of such processes as heat and mass transfer, electric conductivity and permittivity, and filtration of a Newtonian liquid in undeformable cracked-porous media (see, e.g., [74, 516, 599, 881, 1389]) are mathematically equivalent. For reasons of mathematical analogy, we consider the basic equations of steady-state transfer process

$$\nabla \mathbf{q} = 0, \quad \mathbf{q} = -\boldsymbol{\kappa}\nabla T, \tag{15.205}$$

where T is a potential (e.g., temperature), $\boldsymbol{\kappa}$ is a conductivity tensor, $\mathbf{q}$ is a flux vector, and the first Eq. (15.205$_1$) is a transfer equation (Fourier, Ficks, Ohm, Darcy, etc.). Equations (15.205) are simpler field equations than Eqs. (6.1)–(6.3) that simplifies its analysis.

The immediate consequence of Eqs. (15.65), (15.89), (15.94), and (15.96) are

$$\nabla T(\mathbf{x}) = \langle\nabla T\rangle(\mathbf{x}) + \int [\mathbf{U}(\mathbf{x}-\mathbf{y})\boldsymbol{\kappa}_1(\mathbf{x})\nabla T(\mathbf{y}) - \langle\mathbf{U}(\mathbf{x}-\mathbf{y})\boldsymbol{\kappa}_1\nabla T\rangle(\mathbf{y})]d\mathbf{y}, \tag{15.206}$$

$$\nabla T(\mathbf{x}) = \langle \nabla T \rangle(\mathbf{x}) + \int \mathbf{U}(\mathbf{x}-\mathbf{y})[\kappa_1(\mathbf{y})\nabla T(\mathbf{y}) - \langle \kappa_1 \nabla T \rangle(\mathbf{y})]d\mathbf{y}, \qquad (15.207)$$

$$\nabla T(\mathbf{x}) = \langle \nabla T \rangle + \int \mathbf{U}(\mathbf{x}-\mathbf{y})[\kappa_1(\mathbf{y})\nabla T(\mathbf{y}) - \langle \kappa_1 \nabla T \rangle]d\mathbf{y}, \qquad (15.208)$$

$$\nabla T(\mathbf{x}) = \nabla T^0 + \int \mathbf{U}(\mathbf{x}-\mathbf{y})\kappa_1(\mathbf{y})\nabla T(\mathbf{y})d\mathbf{y} \qquad (15.209)$$

where $\mathbf{U}(\mathbf{x}-\mathbf{y})$ is defined by Eq. (15.54_2) and the interface conditions (15.15) are assumed to be perfect. The total majority of methods investigating conductivity problem of CMs (see for references and details [197]) use (either implicitly or explicitly) EFH and, therefore, imply Eqs. (15.207) and (15.208). The basic historical steps of obtaining of (15.207) and (15.208) (see e.g., [200, 747]) in fact duplicate the corresponding development stages of Eqs. (15.81) and (15.88) analyzed in Sect. 15.2.4. In the analogous ways, an incorrect use of Eq. (15.209) for both statistically homogeneous [1207] and inhomogeneous [1659] composites was performed with the same questionable justifications critically analyzed in Sect. 6.1 for elastic problem (see also Chapter 2 in [200]).

Depending on the numerous ways of closing the integral equations involved (15.206)–(15.209), the methods in this case are classified in an analogy with Chap. 8. Four groups of methods for determining the effective coefficients are known. The first group is that of *model treatments*, replacing the real stochastic structure of composites by specific ones that have intuitive physical interpretation and lead to particular exact results. Probably the most famous special models are the fractal-type microinhomogeneous media composed from composite Hashin's spherical or cylindrical coated inclusions [609, 620] (see also [102, 1075]) of all possible sizes, from finite down to the infinitesimally small, as well as Dykhne's [427] two-dimensional, two-phase composites with statistically equivalent phases. Another sort of a special structure is the regular ones [404, 756, 1043, 1046] or periodic structures containing randomly dispersed inhomogeneities in the unit cell [591, 849, 1328]; percolation models, in particular, belong to this group [781, 1321, 1387].

The perturbation methods of the second group give correct results in the case of the availability of a small parameter such as either the small differences of the conduction coefficients [1345, 1346, 1378] of the components or the small concentration of inhomogeneities as in the perturbation method of the second order (9.22) [331, 696, 971, 972]. For example, let us consider the weakly inhomogeneous macroscopically isotropic media with the local isotropic conductivity $\kappa(\mathbf{x}) = \varkappa(\mathbf{x})\delta$ represented in the form $\kappa(\mathbf{x}) = \bar{\kappa} + \kappa'(\mathbf{x})$ with a small fluctuation of $\kappa'(\mathbf{x})$ about its mean value. Then, following the scheme of the correlation approximation method [1378], we get

$$\varkappa^*/\bar{\varkappa} = 1 - (\langle \varkappa^2 \rangle - \bar{\varkappa}^2)/(3\bar{\varkappa}^2) \qquad (15.210)$$

proposed in [163] and rederived by many authors (see, e.g., [879]). Continuation of the described scheme of the successive approximation method and holding of subsequent terms allows the expansion of the effective conductivity to be found in powers of the difference in the phase conductivity; see [1359] and [1523]. Let us consider another sort of perturbation methods when a concentration of inhomogeneities is a small parameter in the expansion of the effective conductivity of matrix composites

$$\boldsymbol{\kappa}^* = \boldsymbol{\kappa}^{(0)} + \sum\nolimits_j \mathbf{b}_j c^j + o(c^m), \tag{15.211}$$

where the coefficients $\mathbf{b}_j$ $(j = 1, \ldots, m)$ are independent of the volume concentration of inclusions $c = c^{(1)}$ but depend on the conductivities of the phases. At the dilute concentration of inhomogeneities when an interaction between the inclusions is negligible and an effective field acting on each inclusion coincides with an applied remote one $\overline{\nabla T} = \mathbf{G}^{\Gamma_0}$, we get

$$\boldsymbol{\kappa}^* = \boldsymbol{\kappa}^{(0)} + n\mathbf{R}^t + o(c^2), \tag{15.212}$$

where n is a numerical density of inhomogeneities: $n = c/\bar{v}_i$, and the tensor $\mathbf{R}^t$ is defined through the averaged polarization tensor $\bar{v}_i \boldsymbol{\kappa}_1 \nabla T(\mathbf{x}) = \mathbf{R}_i^t(\mathbf{x})\mathbf{G}^{\partial w}$, $\mathbf{R}^t \equiv \langle \mathbf{R}_i^t \rangle$ analogous to the tensor $\mathbf{R}_i^\epsilon$ (3.175). The polarizability tensor for the homogeneous ellipsoidal inclusion $\mathbf{R}_i^t(\mathbf{x}) \equiv \mathbf{R}_i^t =$ const. is analytically expressed through the *depolarization* tensor $\mathbf{S}_i$ (3.220) similar to the Eshelby tensor (3.131)

$$\mathbf{R}_i^t = \bar{v}_i \boldsymbol{\kappa}_1^{(i)} \mathbf{A}_i^t, \quad \mathbf{A}_i^t = (\boldsymbol{\delta} + \mathbf{S}_i \mathbf{k}^{(0)} \boldsymbol{\kappa}_1^{(i)})^{-1}. \tag{15.213}$$

Estimation of the higher-order terms in the expansion (15.216), starting with $o((c^{(1)})^2)$, requires an analysis of m multiparticle interaction. So, Jeffrey [696] solved exactly the problem of two identical spheres in a uniform field $\mathbf{Q}$ prescribed at infinity by using bispherical coordinates and Legendre polynomials. He obtained an explicit representation for the c^2 coefficient of the expansion (15.216) in terms of the radial distribution function $g(r)$, which was generalized in [335] (see also [380, 650]) to the d-dimensional case. The problem for two spherical isotropic inclusions with different radii and properties was solved in [1450]. Pair interactions of coated or debonded spheres were estimated in [971] by using a twin spherical expansion technique. The same problem for two aligned spheroidal inclusions was solved in [972] via the boundary collocation method; Chen and Yang [318] used a generalization of the method [1096] of polynomial representations of the effective fields and stress fields inside the inclusions; consequently, the case the matrix anisotropy was analyzed via the use of an appropriate coordinate transform reducing the initial anisotropic problem to to an analogous problem but with an isotropic matrix. The fast multipole method was used in [572] for the interaction analysis of numerous inclusions of arbitrary shape. A complete full-field solution is obtained in [860] by multipole expansion method for the multiparticle models comprising inhomogeneities of diverse shape, size, orientation, and properties of anisotropic ellipsoidal inclusions in anisotropic matrix. Thorpe [1509] proposed conformal mapping to study a medium containing the nonconductive or superconducted inclusions of polygonal shape. Buryachenko and Murov [252] used the solution [696] in the form of a slowly convergent series to demonstrate the high accuracy of the solution obtained in the framework of the hypothesis **(H1)**; the exact expression of the order c^2 (15.216) was obtained.

Variational methods providing two bounds of effective properties for the strong-contrast composites are presented by two-point bounds by Hashin and Shtrikman [620] (see also [609]) and by three-point bounds by Beran [104] (see also [1071, 1072, 1519], as well as Sect. 8.4.

The fourth group of *self-consistent methods* is based on expressing the solution of the steady-state transfer equation with random rapidly oscillating coefficients in terms of the Green's function of the analogous equation for a homogeneous medium. Depending on the numerous ways of closing the integral equations obtained, the methods in this case are classified as the method of effective medium [170, 608, 786], generalized self-consistent method [1069], differential method [1074, 1205, 1703], Mori–Tanaka–Eshelby method of the average fields [88, 300, 625, 626, 1600], the singular approximation method [1389], the method of conditional moments [769], and the MEFM method [252] (where one presents comparison of the different method estimations with experimental data [1051]). Below we briefly summarize mentioned results keeping in mind that they were usually obtained by the methods analogous to ones considered in Chap. 8 (see also [162, 881, 1012, 1075, 1378, 1523]). So, in the effective medium method (EMM), the effective conductivity $\boldsymbol{\kappa}^*$ can be found from Eq. (15.217) where the concentration factor $\mathbf{A}_i^t$ (15.213) of the phase $v^{(i)}$ $(i = 1, \ldots, N)$ is estimated by the formula (15.217$_2$) by the use of replacement $\boldsymbol{\kappa}^{(0)} \to \boldsymbol{\kappa}^*$

$$\boldsymbol{\kappa}^* = \boldsymbol{\kappa}^{(0)} + \sum_{i=1}^{N} c^{(i)}(\boldsymbol{\kappa}^{(i)} - \boldsymbol{\kappa}^{(0)})\mathbf{A}_i^t(\boldsymbol{\kappa}^*), \tag{15.214}$$

where $\mathbf{A}_i^t(\boldsymbol{\kappa}^*) = [\boldsymbol{\delta} + \mathbf{S}_i(\boldsymbol{\kappa}^*)(\boldsymbol{\kappa}^*)^{-1}(\boldsymbol{\kappa}^{(i)} - \boldsymbol{\kappa}^*)]^{-1}$. Particular cases for spherical inclusions are considered in [164] for two-phase composites (see also [880, 1701]) and in [1523] for dD case. A fractal-like (or self-similar in all -scale) generalization of the EMM was proposed in [1074]. Torquato [1523] demonstrated that the EMM approach can violate three-point bounds taking more realistic geometrical model of composites into account. Another inconsistency is related with the assumption that each inclusion is incorporated into the homogeneous medium with effective properties of the composite that completely ignores the availability of some matrix gaps surrounding the test inclusion. Acrivos and Chang [6] eliminated this disadvantage by covering the test inclusion by a layer of the matrix with subsequent incorporation of the composed inhomogeneity into the effective medium. A popular choice justification of a thickness h of the matrix layer surrounding inclusion with a radius a as $a^3/(a+h)^3 = c^{(1)}$ cannot be recognized as a rigorous.

Equation (15.214) was obtained by the averaging of the flux $\mathbf{q}(\mathbf{x}) = \boldsymbol{\kappa}(\mathbf{x})\nabla T(\mathbf{x})$ over the macrodomain w. The average field approximation [1226] uses the averaging operations of the different fields such as a temperature gradient $\nabla T(\mathbf{x}) = \mathbf{k}(\mathbf{x})\mathbf{q}(\mathbf{x})$ leading to the different representations of effective properties for a general case of homogeneous ellipsoidal inclusions which coincide with (15.214) for the isotropic composites containing homogeneous isotropic spherical inclusions (see [1132]). The implicit Eq. (15.214) can in general be solved only numerically via, e.g., the iteration scheme $\boldsymbol{\kappa}^{*[n+1]} = \mathbf{f}(\boldsymbol{\kappa}^{*[n]})$ with an initial approximation $\boldsymbol{\kappa}^{*[0]} = \boldsymbol{\kappa}^{[0]}$. However, for two-phase composites, Eq. (15.214) is a quadratic equation for the effective conductivity with explicit solution which (as well as Eq. (15.214)) is simplifying in two limiting cases of either the perfectly conducting particles ($\varkappa^{(1)}/\varkappa^{(0)} = \infty$) or ideally insulating inclusions ($\varkappa^{(1)}/\varkappa^{(0)} = 0$) (see for details [200]). It should be mentioned that the formulae obtained predict either infinite or zero effective conductivity if the volume fraction $c^{(1)}$ reaches so-called percolation threshold $c_p^{(1)}$ with asymptotic behavior of $\boldsymbol{\kappa}^*$ in the vicinity of $c^{(1)} = c_p^{(1)}$ considered in [352, 781, 1321].

Another extension of the EM approach is the differential effective medium (DEM) method which is also implicit but not symmetric with respect to the interchange $\varkappa^{(i)} \leftrightarrow \varkappa^{(j)}$, $c^{(i)} \leftrightarrow c^{(j)}$ $(i = 0, 1, \ldots, N)$. The DEM approach was extensively described for the elastic problems in Sect. 8.7. A conductivity analog of one version of the DEM method can be recast for a two-phase isotropic composites with spherical isotropic inclusions in the following form

$$\frac{\partial \varkappa^*}{\partial c} = \frac{d\varkappa^*}{1-c} \frac{\varkappa^{(1)} - \varkappa^*}{\varkappa^{(1)} + (d-1)\varkappa^*}, \tag{15.215}$$

with the initial condition $\varkappa^*(c = 0) = \varkappa^{(0)}$ and $d = 2, 3$. Equation (15.215) can be analytically integrated and simplified for the limiting cases $\varkappa^{(1)}/\varkappa^{(0)} = \infty$ or $\varkappa^{(1)}/\varkappa^{(0)} = 0$. It is interesting that Sen et al. [1360] got excellent agreement of the predicted conductivity $\varkappa^* = (1 - c)^{3/2}\varkappa^{(0)}$ for ideally insulating inclusions $\varkappa^{(1)}/\varkappa^{(0)} = 0$. The DEM approach (15.215) was generalized in [1138] to the composites with a variety of shape and compositions of inclusions. The DEM scheme was also considered for randomly oriented ellipsoidal inclusions [554] and for cylindrical fibers [1044]. Avelaneda [46] considered realizability of the DEM scheme for multiphase composites containing anisotropic inclusions of arbitrary shape.

Thin interphase layer between two media is modeled in terms of imperfect interface conditions by [101], while [1012] and [1526] treat nonideal contact of a spherical inclusion of the radius a in two limiting case of both the large (3.225) and small (3.226) conductivity $\boldsymbol{\kappa}^{\text{int}} = \kappa_0^{\text{int}}\boldsymbol{\delta}$ of interphase with a small thickness $\delta \ll a$ (the cases of an ellipsoidal heterogeneity with these nonideal contacts were analyzed in [101, 851, 1070]). For the conditions of "superconducting" interface (3.225), the temperature field is continuous at the interface, but the heat flux suffers a jump that corresponds to the ISM (see Sect. 3.7.3), whereas for "resisting" conditions (3.226), the heat flux remains continuous, but the temperature suffers a jump on the interface that parallels to the CZM (see Sect. 3.7.3). Pavanello et al. [1193] (see also [580]) introduce a generalized formalism, which is able to take into account the properties of both previous schemes, and, at the same time, it implements more complex behaviors, already observed in recent investigations. The next step is an estimation of the tensor $\mathbf{R}^t$ (like (3.227)) in a solution for one isolate particle in the infinite matrix $\bar{v}_i \langle \kappa_1 \nabla T \rangle_i = \mathbf{R}^t \overline{\nabla T}$ corresponding to the general problem (4.124) and subsequent substitution into (15.206) in the framework of some versions of hypotheses **H1** and **H2**.

The nonlinear and nonlocal problems of transport phenomena are worthy of more attention. It is interesting that the scheme of the perturbation method of the estimation of the second moment of fields (6.185) was proposed precisely in thermal conductivity theory [110]. The variance of the fields within each phase was derived in [47, 105] and [347]. In so doing Cheng and Torquato [322] detected a double-peak character of probability density functions of fields for two-dimensional dispersions, and, therefore, the second moment of the field is a crude characteristic of the field fluctuation in the composite; they used Monte Carlo simulation techniques for occupancy of 64 inclusions in a periodic unit cell with subsequent numerical solution of integral equations involved in the fast multipole method (for references see [571, 573]). Knowledge of field distribution is important in understanding many crucial materials properties such as the breakdown phenomenon and the nonlinear behavior of composites [1321, 1322, 1330]. So, Stroud and Hui [1461] obtained an exact expression for the effective nonlinear conductivity $\boldsymbol{\kappa}^*$ for composites with a dilute concentration of nonlinear inclusions in a linear host.

Zeng et al. [1676] proposed an approximate method for the estimation of $\boldsymbol{\kappa}^*$ for composites with a weakly nonlinear constitutive relation when the linearization (6.185) and (12.36) was used to accompany the initial approximation (12.34) of $\boldsymbol{\kappa}(\mathbf{x})$. The same approximation was used in [911, 1460] to accompany the effective medium method. The scheme mentioned was generalized more recently in [180] (see also Chap. 12) in the framework of the MEFM method. Variational method of estimation of the effective energy and conductivity was used in [550] through the estimation of the effective conductivity for the linear comparison composite.

Progress in research of nonlocal effects is more moderate and was achieved under special assumptions. So, Beran and McCoy [107] used a perturbation method for the analysis of statistically homogeneous composite materials with small conductivity mismatch implying the application of assumptions for the flux $\mathbf{Q}$ analogous to Eq. (8.183); for the estimation of nonlocal effective properties, they used Fourier transform method described in Sect. 13.2.3. Also by the use of the perturbation method [516], the variation of the local effective conductivity was detected near a wall having unsymmetric distribution of a constituent with respect to the boundary (see for comparison Fig. 14.3 in [200]). Stagfen [1436] considered the case of very long and thin highly conducting fibers of so-called semidilute concentration; via the use of statistical averaged equations and slender body theory, the perturbation created by the surrounding fibers is demonstrated to be an integral of the temperature gradient about any point weighted by a function that is the appropriate nonlocal conductivity.

Among the papers dedicated to graded materials, the following articles are noteworthy. In the analysis of the low volume fraction composites, the pair-interacting inclusions with a special renormalization were considered [755] (see also [1207]) to avoid divergent integrals similar to [696, 697] for statistically homogeneous suspensions that allow one to obtain concrete numerical results only for a step function $c(\mathbf{x}) = c^{\infty}H(x_1)$; a boundary layer effect similar to the elastic case depicted in Fig. 13.11 was detected. Buevich and Ustinov [172] presented a more heuristic analysis based on the effective medium method; they explicitly assumed that the mean temperatures of different components are the same and obtained effective constitutive equation containing only the first derivative of the potential [compare with Eqs. (13.63) and (13.67)]. A more general numerical model of a nonuniform composite with spherical inclusions was considered [1007] taking into account a difference between the average temperatures in the phases. It is interesting that scaler effect analogous to the one detected in [267] (see also Chap. 13 and Fig. 13.11) for elastic graded composite was estimated in [1206] (see also [1658] numerically for effective conductivity over the cross section of the composite layer of finite thickness between two homogeneous half-spaces.

Thus, the conductivity problems of CMs with both perfect and imperfect interface coincides with the elasticity problems considered in Subsections 15.7.1 with a precision of notations. In a similar manner, we can obtain the integral equations and its solutions corresponding to the abandonment from either any one or a few hypotheses **H1a**, **H1b**, **H2b**, and **H3** in a general integral Eq. (15.206). Therefore, after the replacements of the word "elasticity" on the word "conductivity," the *Results 15.1.1–15.1.4* (see Subsection 15.7.1) for elasticity problems can be considered as the expected results for conductivity ones.

15.6.3 Coupled Problems in Micromechanics of CMs

We will consider the general integral equations for the coupled problems at the example of thermoelectroelasticity equations for CMs. For rotational convenience the elastic and electric variable will be treated on equal footing, and with this in mind, we recast the local linear constitutive relations of thermoelectroelasticity for this material (see, e.g., [747, 1033, 1188]) in the notation introduced in [70]). Then the basic steady-state equations (2.145)–(2.154) can be presented in the form analogous to Eqs. (6.1)–(6.3)

$$\mathcal{D}\boldsymbol{\Sigma} = \mathbf{0}, \ \mathcal{E} = \mathcal{D}\mathcal{U}, \tag{15.216}$$

$$\mathcal{E} = \mathbb{M}\boldsymbol{\Sigma} + \boldsymbol{\Lambda}, \ \ \boldsymbol{\Sigma} = \mathbb{L}(\mathcal{E} - \boldsymbol{\Lambda}), \tag{15.217}$$

where the notations are described by (2.145), (2.146), and (2.148). It is assumed that the properties of both the comparison medium and the matrix coincide.

We suppose that the contact between components is ideal, and the normal components of the stress tensor and the electric induction vector are continuous, as well as the tangential components of the strain tensor and electrical field intensity. Owing to similarity of Eqs. (6.1)–(6.3) and (15.216) and (15.217), the theory of piezoelectric CM (PCM) retraces at a particular instant the path of development of the theory of microinhomogeneous elastic media, exhibiting substantial progress (see for references, e.g., [200] and [869, 1667] for random and periodic structure, respectively). In light of the analogy mentioned in our brief survey, we will not consider in detail the GIE and average schemes of PCM; one may refer instead to the appropriate schemes of Subsection 15.7.1 (see also [200, 747]).

For the basic equations (15.216) and (15.217), the GIEs (15.73), (15.88), and (15.90) are reduced to the following equations for $\boldsymbol{\Sigma}$ (corresponding equations for $\mathcal{E}$ can be presented in a similar manner)

$$\boldsymbol{\Sigma}(\mathbf{x}) = \langle\boldsymbol{\Sigma}\rangle(\mathbf{x}) + \int [\boldsymbol{\Gamma}(\mathbf{x}-\mathbf{y})\boldsymbol{\eta}(\mathbf{y}) - \langle\boldsymbol{\Gamma}(\mathbf{x}-\mathbf{y})\boldsymbol{\eta}\rangle(\mathbf{y})]\, d\mathbf{y}, \tag{15.218}$$

$$\boldsymbol{\Sigma}(\mathbf{x}) = \langle\boldsymbol{\Sigma}\rangle(\mathbf{x}) + \int \boldsymbol{\Gamma}(\mathbf{x}-\mathbf{y})[\boldsymbol{\eta}(\mathbf{y}) - \langle\boldsymbol{\eta}\rangle]\, d\mathbf{y}, \tag{15.219}$$

$$\boldsymbol{\Sigma}(\mathbf{x}) = \boldsymbol{\Sigma}^0 + \int \boldsymbol{\Gamma}(\mathbf{x}-\mathbf{y})\boldsymbol{\eta}(\mathbf{y})\, d\mathbf{y}, \tag{15.220}$$

respectively, where the polarization tensor $\boldsymbol{\eta}(\mathbf{x}) = \mathbb{M}_1(\mathbf{x})\boldsymbol{\Sigma}(\mathbf{x}) + \boldsymbol{\Lambda}_1(\mathbf{x})$, and the coupled flux Green's tensor $\boldsymbol{\Gamma}(\mathbf{x})$ is defined by Eqs. (15.64), (15.54), and (15.12). Here Eq. (15.219) holds only for statistically homogeneous material (5.62_2) and field (15.21) problems, while Eq. (15.220) holds true only for the fields X bounded in one direction (7.28).

A comprehensive review with numerous references related with universal relations in piezoelectric (and thermo-magneto-electro-elastic) composites can be found in Chapter 17 in [200] (see also [1255]). So, based on the contour integral representation of Green's function derived in [368], Dunn and Taya [410] obtained explicit expressions for a set of four tensors corresponding to Eshelby's elastic tensors for ellipsoidal inclusion and incorporated this solution into several generalized micromechanical models such as Mori and Tanaka's approach, dilute concentration approach, the differential method of successive smoothing, and self-consistent model (see also [407–409, 411, 927]). It cannot be doubted that application of the most popular generalized Mori–Tanaka approach (see, e.g., [311, 669, 670, 840]) for the investigation of PCMs with a general case of random orientation of piezoelectric fibers must lead the incorrect results (see, e.g., Sect. 8.3.2). Any average scheme of effective properties estimations of the PCM uses a solution of a single inclusion in an infinite piezoelectric solid (see for references Sect. 3.8.3). The known micromechanical approaches considered in Chap. 8 were generalized in a straightforward manner for the analysis of PCM; these are the multi-inclusion model [928, 1121], conditional moment method [766, 768], generalized singular approximation [1379], effective medium method [929], one-particle approximation of the MEFM [919], self-consistent and differential schemes [1592], and the MEFM [259, 262]. As for elastic case (see [95, 616]), the imperfect interface models obtained through the asymptotic methods can be regarded as micromechanical models linking the interface response to the behavior of the material layer constituting the bonding. According to these techniques, the thickness of the layer is considered as a small parameter ϵ, whereas soft interface model and rigid interface model correspond to the constitutive coefficients depending on ϵ and $1/\epsilon$, respectively. Multiphysics interface material behavior, spanning from uncoupled phenomena, such as conduction and elasticity, was generalized in [405, 1361] to coupled multifield and multiphysics theories, such as piezoelectricity and continua with microstructure. The elastic model for an ellipsoidal inclusion with thin coating [1262] (see also Sect. 3.7.2) was generalized in [378, 379] to a linear spring interface model for piezoelectric CM with ellipsoidal inclusions. Fibrous piezoelectric CMs with general imperfect interfaces (with the jumps of driving fields and fluxes) was considered in [580].

Of course, the absence of coupled field effect for the matrix essentially simplifies the micromechanical problems because in such a case the Green's function $\mathcal{G}$ has a block-diagonal structure defined by the Green's functions of uncoupled fields. Then the generalized Eshelby tensor and the tensors (3.127) have the same structure and, as it is well-known, lead to nothing more than a slight modification of softwares supporting the approaches described in Chap. 8 for elastic problems; this simplification was used, e.g., in [262, 702]. Usually a comparison medium without coupled effects is used (see [768, 1116]), whereas Buryachenko and Parton [259] introduced for the matrix the new field variables without coupling effects. The field coupling in the matrix complicates essentially a micromechanical analysis that extends the importance of a fundamental trend in micromechanics of coupled fields based on the diagonalization of response matrixes $\mathbb{M}$ by a matrix decomposition technique proposed and taking the beginning from the work by Straley [1455], who showed that by the use of a linear transform of the real fluxes, the problem of evaluation effective thermoelectric properties for a two-phase composite can be reduced to the conductivity problem of a composite. This method was generalized in [1063] (see also [98, 310, 312]) (where more references may be found) to the analysis of any number of coupled fields with scalar potentials; then for the field being considered, any known results existing in the single-field effect can be transformed into coupled-field responses.

These and other methods can be classified with respect to the GIE as two groups of methods. The methods of the first one initially accept very rough hypotheses [as in EMM (9.112) and MTM (9.113)] which allow us to obtain the final results without using of GIE. Analogously to the locally elastic composites considered in Sect. 15.6.1, the coupled counterparts of both the EMM (9.112) and MTM (9.113) (see, e.g. [377, 920, 973] and referenced in [200]) are, in fact, the approximate solutions of the GIE (15.215)–(15.220). In so doing, these solutions are so much rough that it does not enable one to use all opportunities of the different GIEs (see the comments after Eq. (9.113)). The methods of the second group use one or another GIE, but in many cases a choice of this GIE as well as its subsequent solution is performed at some additional weakly justified assumptions. For example, for a statistically homogeneous infinite inclusion field, popular versions of the EFM are based on the initial incorrect exploiting of Eq. (15.220) which is transformed to the correct Eq. (15.219) by the use of the scheme such as (7.34) (see, e.g., [308, 917]).

Thus, to the best of the author's knowledge, the mentioned approaches and the most part of other ones either explicitly or implicitly use either Eq. (15.219) or Eq. (15.220), with the different level or rigor. In so doing, the new exact GIE (15.218) holds true for any statistically inhomogeneities of material (5.62_1) and field (15.21) parameters and directly takes into account dependence of the renormalized item $\langle \mathbf{\Gamma}(\mathbf{x}-\mathbf{y})\boldsymbol{\eta}\rangle(\mathbf{y})$ on the inhomogeneity of the polarization tensor $\boldsymbol{\eta}(\mathbf{y})$ ($\mathbf{y} \in v$) inside heterogeneities. Because of this, the Results 15.1.1–15.1.4 can be considered as the expected results for the coupled problems.

15.6.4 Wave Propagation in CMs

We study a monochromatic elastic wave of frequency ω that propagates in a composite medium with the local elastic properties $\mathbf{L}(\mathbf{x})$ and $\boldsymbol{\alpha}(\mathbf{x})$ and the mass density $\rho(\mathbf{x})$ (15.1) and (15.3) while $\mathbf{f}(\mathbf{x})$, $[[\mathbf{u}(\mathbf{x})]]$, $[[\boldsymbol{\sigma}(\mathbf{x})]]\cdot\mathbf{n}(\mathbf{x}) \equiv \mathbf{0}$. If the dependence of time t is defined by the factor $\exp(i\omega t)$, the displacement field u_i in the medium has the form $u_i(\mathbf{x},t) = u_i(\mathbf{x})exp(i\omega t)$, and amplitude $u_i(\mathbf{x})$ of this field satisfies the following wave motion equation (see, e.g., [1622])

$$\boldsymbol{\mathcal{L}}^{\omega}(\mathbf{x})\mathbf{u}(\mathbf{x}) = \mathbf{0}, \quad \mathcal{L}^{\omega}_{ik}(\mathbf{x}) = \nabla_j L_{ijkl}(\mathbf{x})\nabla_l + \delta_{ik}\rho(\mathbf{x})\omega^2 \tag{15.221}$$

with the corresponding time-reduced wave Green's tensor satisfying the reduced wave equation for an infinite homogeneous comparison body

$$\boldsymbol{\mathcal{L}}^{\omega c}(\mathbf{x})\mathbf{g}(\mathbf{x},\mathbf{y}) = -\boldsymbol{\delta}\delta(\mathbf{x}-\mathbf{y}), \quad \boldsymbol{\mathcal{L}}^{\omega c} = \nabla\mathbf{L}^c\nabla + \boldsymbol{\delta}\rho^c\omega^2, \tag{15.222}$$

that is, one assumes that the coefficient $\mathbf{b}(\mathbf{x}) = \boldsymbol{\delta}\rho(\mathbf{x})\omega^2$ (15.13).

The perturbators $\boldsymbol{\mathcal{L}}^{\tau}_k(\mathbf{x}-\mathbf{x}_k,\widehat{\boldsymbol{\tau}})$ and $\boldsymbol{\mathcal{L}}^{u\tau}_k(\mathbf{x}-\mathbf{x}_k,\widehat{\boldsymbol{\tau}})$ are presented by Eqs. (15.52) and (15.54), respectively, where the surface integrals are omitted. We consider $\mathbf{u}$ and $\boldsymbol{\varepsilon}$ as independent variables and introduce a symbolic notation $\boldsymbol{\mathcal{E}} = (\mathbf{u},\boldsymbol{\varepsilon})^\top$ for which the coupled forms of the GIEs lead from the GIEs (15.81), (15.88), and (15.90)

$$\boldsymbol{\mathcal{E}}(\mathbf{x}) = \langle\boldsymbol{\mathcal{E}}\rangle(\mathbf{x}) + \int [\mathbf{U}^{\omega}(\mathbf{x}-\mathbf{y})\boldsymbol{\mathcal{T}}(\mathbf{y}) - \langle\mathbf{U}^{\omega}(\mathbf{x}-\mathbf{y})\boldsymbol{\mathcal{T}}\rangle(\mathbf{y})]\,d\mathbf{y}, \tag{15.223}$$

$$\boldsymbol{\mathcal{E}}(\mathbf{x}) = \langle\boldsymbol{\mathcal{E}}\rangle + \int \mathbf{U}^{\omega}(\mathbf{x}-\mathbf{y})[\boldsymbol{\mathcal{T}}(\mathbf{y}) - \langle\boldsymbol{\mathcal{T}}\rangle]\,d\mathbf{y}, \tag{15.224}$$

$$\boldsymbol{\mathcal{E}}(\mathbf{x}) = \boldsymbol{\mathcal{E}}^0 + \int \mathbf{U}^{\omega}(\mathbf{x}-\mathbf{y})\boldsymbol{\mathcal{T}}(\mathbf{y})\,d\mathbf{y}, \tag{15.225}$$

respectively, where

$$\boldsymbol{\mathcal{E}} = \begin{pmatrix} \mathbf{u} \\ \boldsymbol{\varepsilon} \end{pmatrix}, \quad \boldsymbol{\mathcal{T}} = \begin{pmatrix} \pi \\ \tau \end{pmatrix}, \quad \mathbf{U}^{\omega} = \begin{pmatrix} \mathbf{g} & \nabla\mathbf{g} \\ \nabla\mathbf{g} & \nabla\nabla\mathbf{g} \end{pmatrix}. \tag{15.226}$$

Here Eqs. (15.224) (see [1622], see also for references [200, 747]) and (15.225) hold only for statistically homogeneous material (5.62_2) and field (15.21) problems and for the fields X bounded in one direction (7.28), respectively, while the new exact GIE (15.223) is valid for any statistically inhomogeneous medium (5.62_1). Fish et al. [480] proposed a dispersive nonlinear theory for periodic media and estimated the dispersion and micro-inertia effects in either explicit or implicit integration scheme.

After that the scheme of derivation of effective properties is quite similar to the one given in Chap. 8 for the static case. So, for the case in which the mechanical properties of the components differ slightly, the perturbation method of the first order of an accuracy with respect to the concentration of inclusions $\overline{\boldsymbol{\varepsilon}}(\mathbf{x}_i) \equiv \langle \boldsymbol{\varepsilon} \rangle (\mathbf{x}_i)$ has been used in [582, 1641] to estimate the effective wave propagation velocities and the attenuation factors. One [1378], discarding a density fluctuation ($\rho_1 \equiv 0$), used a similar method of the correlation approximation for obtaining of the long- and short-wave asymptotic solutions of the dispersion equation. The effective medium approach widely employed (see, e.g., [362, 510, 748, 1144, 1317]) is based on solution of the problem for one ellipsoidal inclusion (see, e.g., [511, 1622]) in the matrix with the effective properties of the medium and with specified load at infinity. Sato and Shindo [1331] considered scattering of elastic waves combining a dilute approximation approach with a numerical solution obtained by the boundary element method for a single partially debonded elliptical inclusion. Variational principles for dynamic problems [1626, 1631] were used in [1478, 1479] for estimation of the dispersion and attenuation of elastic waves in random composites with the ellipsoidal inclusions. It should be mentioned that the T-matrix approximation method [1004, 1556] and polarization approximation method [1623], as well as the one-particle variant of the effective field method [297, 743, 746, 747], consider an isolated inclusion in the matrix with a self-consistent field acting at infinity, whose parameters are independent of the properties of the inclusion in question ("quasi-crystalline" approximation). Use of this "quasi-crystalline" approximation reduces the problem to a single-particle one, thus entailing substantial error in estimating the coefficient in the second degree with respect to the concentration of inclusions c, in the relationship between the effective parameters of the medium and small c value. A general approach for setting up an effective wave operator proposed in [256] and reproduced in [200] is based on the generalization of the static version of the MEFM.

Fundamentally new effects of using of the static counterparts of GIEs (15.223) and (15.224) were detected (see Sects. 6.1 and 6.2) for any problem and method providing estimation of inhomogeneous stress distributions inside inclusions (such as, e.g., nonellipsoidal shape of inclusions, its binary interactions corresponding to the hypothesis **H2a**, imperfect interface conditions, violation of the hypothesis **H3**). All these sources of the stress inhomogeneities and their impacts on the estimated effective material and field parameters are also valid for GIEs (15.223) and (15.224). However, there is an additional reason of stress inhomogeneity inside inclusions which is absent in all static problems and intrinsic in basic wave motion phenomena (15.221) and (15.222) even for a single ($c = 0$) homogeneous ellipsoidal inclusion $\mathbf{x} \in v_i$ subjected to the homogeneous effective field $\overline{\boldsymbol{\varepsilon}}(\mathbf{x}) \equiv$const. ($\mathbf{x} \in v_i$). Namely, even in such a case, the fields of the strain $\boldsymbol{\varepsilon}(\mathbf{x})$ and displacement $\mathbf{u}(\mathbf{x})$ are inhomogeneous inside the inclusion (see for details, [747]), and, therefore, for the composite materials considered, the hypothesis **H1b** is not valid in general even in the framework of the hypotheses **H1a**, **H2b**, and **H3**. Thus, a wave motion counterpart of *Result* considered in Sect. 15.6.1 does not hold in general.

A popular transformation of analogs of Eq. (15.225) (widely used also for statistically homogeneous media) to the correct approximate Eq. (15.224) is performed in a spirit of the scheme (7.38) (see, e.g., [308, 492, 747, 1622, 1623]). A wave motion counterpart of the EMM (9.112) (see for references [747], where the different versions of the EMM can be also found) along with its static case (9.112) can also be considered as a truncation of the corresponding GIE (15.225) (or (15.224)). Furthermore, a dealing only with one type of scalar waves (e.g., electric waves) propagating significantly simplifies analysis of corresponding equations which coincide within notations with Eqs. (15.221), (15.222) where the tensors of the forth ($\mathbf{L}$) and first ($\mathbf{u}$, $\boldsymbol{\delta}$) orders should be formally replaced by the tensors of the second order and the scalar, respectively (see for references [747]). All comments mentioned with respect to Eqs. (15.223)–(15.225) for elastic waves propagation also remain in force for the scalar wave problems. Generalization of GIEs (15.223)–(15.225) to the consideration of coupled effects (e.g., electromagnetic waves, electromagnetic and elastic waves in thermo-magneto-electro-elastic composites; see, e.g., [308, 920, 973] is straightforward: we only need to extend the block's structures (15.226) by some further blocks corresponding to the appropriate coupled phenomena (15.213) and (15.214).

An upsurge in interest in wave propagation phenomena has been prompted by the recent proliferation of metamaterials (in both the theoretical and practical senses) which are no more than the composites exhibiting exceptional frequency electromagnetic, optic, and mechanical responses not readily observed in nature. Metamaterials are artificially engineered periodic or random microstructures displaying strong resonance behavior at frequencies at which the inclusions themselves and the distances between them are small which can possess simultaneously negative dielectric constant and magnetic permeability in the context of electromagnetics and negative effective mass density and negative elastic constants in the case of elasticity. The original design of metamaterials is the combination of metallic split ring resonators and metallic rods, which realize the negative permeability and the negative permittivity, respectively, while in succeeding years it was performed experimental or/and computational investigations of different inclusion shapes (e.g., silver dendritic cells, U-shaped nanostructures, sphere—rod structures, split hollow spheres) and its microstructure (e.g., coated structures); see [376, 561, 675, 902, 949, 1555]. The most popular methods of effective properties estimations are numerical homogenization of periodic structures and EMM (see

Sect. 15.4.4 and (9.112)) of random structures of spherical coated nanoparticles (e.g., [561, 865, 1593]). However, application of the new GIE (15.223) instead of a classical one (15.224) opens the new avenue of attack on the metamaterial's investigations. Namely, metamaterials exhibit unique opportunity for simultaneous display of a few sources of field inhomogeneities inside the inclusions (coated structure, noncanonical shape, nanoscale size, imperfect interface, wave motion phenomena) violating the hypothesis **H1b** even in the framework of the hypothesis **H1a**. In such a case, all results mentioned in the preceding Sects. 15.6.1 and 15.6.2 are expected to be manifested for the metamaterials even more dramatically than for the conventional composites considered before.

Chapter 16
Bond-Based Peridynamics

Abstract In this chapter, we consider a nonlocal peridynamic theory proposed by Silling (J Mech Phys Solids 48:175–209, 2000) where the equilibrium of a material point is achieved by a summation of internal forces produced by surrounding points over finite distance (called a horizon). A simplified version being derived from this approach is the so-called bond-based approach, in which interactions only occur between pairs of material points within a horizon. One defines linear admissibility condition and angular admissibility condition, isotropy property. Definitions of the microelastic material, pairwise potential function, and peridynamic stress are presented. Volumetric boundary conditions are defined, and the linearized microelastic model is considered. Damage is introduced by permitting damaged bonds to break irreversibly by bond removing if the bond stretch exceeds a critical stretch.

Proposal of peridynamics by Silling [1396] (see also [129, 694, 1000, 1404]), a nonlocal theory of solid mechanics, initiated the explosive character of the progress in different physical phenomena based on the replacement of the classical partial differential equation of balance of linear momentum by the integral equation which is free of any spatial derivatives of displacement. In the nonlocal peridynamic theory, the equilibrium of a material point is achieved by a summation of internal forces produced by surrounding points over finite distance (called a horizon), while in the classical theory, such interactions are only exerted by adjacent points through the contact forces. In classical continuum mechanics, the partial differential equations fail to be applicable on deformation singularities (e.g., at the microcracks and "sharp corners" at the polycrystal) where the necessary derivatives are not defined. when undefined derivatives created at these regions require additional modeling techniques such as the extended finite element method. The mentioned difficulties are completely avoided in peridynamics where any spatial derivatives of displacement are absent and the governing equations of peridynamics are not altered anywhere in the domain of interest, even at material discontinuities. Moreover, due to its nonlocality and similarity to molecular dynamics, the peridynamics can naturally handle the long-range forces (e.g., van der Waals forces), that is critical in nanome-

V. A. Buryachenko, *Local and Nonlocal Micromechanics of Heterogeneous Materials*, https://doi.org/10.1007/978-3-030-81784-8_16

chanics. The effectiveness of peridynamic models has already been demonstrated in several sophisticated applications, including damage accumulation, the fracture and failure of composites of deterministic structure, crack instability, the fracture of polycrystals, phase transition, and nanofiber networks (see, e.g., [37, 38, 365, 1000]). While the similar problems were solved in the framework of local theory, their peridynamic formulation needs considerably higher computational costs simply because each material point requires connection with multiple neighbors within a horizon through the integral equation, while in the local theory each point interacts with the points at the infinitesimally small distance through the differential operator. However, the mentioned drawback of peridynamics is, with usury, compensated by the opportunities offered by peridynamics in solving problems with discontinuities. Generally, in peridynamics, the state-based approach permits the response of a material at a point to depend collectively on the deformation of all bonds connected to the point within its finite radius horizon [1402, 1404] via a response function that completely describes the interaction. It means that the forces between two peridynamic nodes depend also on deformations of other bonds surrounding these nodes within the horizons. The horizon can encompass discontinuities or different materials. It should also be mentioned that a computationally efficient, staggered, nonlocal multiscale model by Fish et al. [481] (allowing effective implementation of the nonlocal and peridynamics type of methods) employs current information for the point under consideration $\mathbf{x}$ and past information from its local neighborhood from the source point $\hat{\mathbf{x}}$ taken from the previous load increment (or time step).

A simplified version derived from this approach is the so-called bond-based approach, in which interactions only occur between pairs of material points within a horizon. As is well known, a direct consequence of this assumption is that Poisson's ratio for isotropic linear materials is fixed at $\nu = 1/4$ (3D and 2D plane strain) or $\nu = 1/3$ (2D plane stress) (see [539, 1396]). The major advantages of the state-based approach include a material response depending on collective quantities (like volume change or shear angle), which allows constitutive models from the conventional theory of solid mechanics to be incorporated directly within the peridynamic approach (see, e.g., [14, 1402]). However, this book will use the bond-based approach as it is most suitable to the chosen implementation.

In this chapter, we summarize the bond-based peridynamic model introduced by Silling [1396] (see also [909, 1401]).

16.1 Basic Formulation

16.1.1 Peridynamic Equilibrium Equation

Let $w \subset R^d$ ($d = 1, 2, 3$) be the reference configuration of a solid body at time $t = 0$ with d being the dimension of the Euclidean space. The body is composed of material points located at generic positions $\mathbf{x}$ to which infinitesimal volumes $dV_{\mathbf{x}}$

are associated. Consider two points $\mathbf{x}$ and $\hat{\mathbf{x}}$ with the relative position of these two material points in the reference configuration (i.e., at $t = 0$)

$$\boldsymbol{\xi} = \hat{\mathbf{x}} - \mathbf{x} \tag{16.1}$$

and their relative displacement at time $t \geq 0$

$$\boldsymbol{\eta} = \mathbf{u}(\hat{\mathbf{x}}) - \mathbf{u}(\mathbf{x}) = \hat{\mathbf{u}} - \mathbf{u}. \tag{16.2}$$

Thus, $\boldsymbol{\xi} + \boldsymbol{\eta}$ is a relative position of the points in the deformed configuration (see Fig. 16.1).

It is assumed that each pair $\mathbf{x}$ and $\hat{\mathbf{x}}$ interact through a vector-valued pairforce (called also a *bond force*) whose value is the force vector that the point located at $\mathbf{x}_j$ (in the reference configuration) exerts on the point located at $\mathbf{x}$ (also in the reference configuration). Then the *peridynamic equation of motion* is given by

$$\rho\ddot{\mathbf{u}}(\mathbf{x}, t) = \hat{\mathcal{L}}(\mathbf{u})(\mathbf{x}, t) + \mathbf{b}(\mathbf{x}, t), \tag{16.3}$$

$$\hat{\mathcal{L}}(\mathbf{u})(\mathbf{x}, t) := \int \mathbf{f}(\mathbf{u}(\hat{\mathbf{x}}, t) - \mathbf{u}(\mathbf{x}, t), \hat{\mathbf{x}} - \mathbf{x}, \hat{\mathbf{x}})d\hat{\mathbf{x}}, \tag{16.4}$$

where $\rho(\mathbf{x})$ is the mass density, $\ddot{\mathbf{u}}(x)$ is the acceleration of the point $\mathbf{x}$, and $\mathbf{b}(\mathbf{x}, t)$ is a given body force density, which represents the external force per unit reference volume. The third argument $\mathbf{x}$ of $\mathbf{f}$ (16.3) can be dropped for the homogeneous media $\mathbf{f}(\mathbf{u}(\hat{\mathbf{x}}) - \mathbf{u}(\mathbf{x}), \hat{\mathbf{x}} - \mathbf{x}, \hat{\mathbf{x}}) = \mathbf{f}(\mathbf{u}(\hat{\mathbf{x}}) - \mathbf{u}(\mathbf{x}), \hat{\mathbf{x}} - \mathbf{x})$. The dimension of $\mathbf{f}$ is $[\mathbf{f}] = F/L^{2d}$, where L and F denote the physical dimensions of length and force, respectively. All constitutive behaviors are described by the function $\mathbf{f}$ which is assumed to be Riemann integrable without any other restrictions on their smoothness with respect to the arguments. The bond force $\mathbf{f}$ does not contain history-dependent variable, i.e., we consider a peridynamic material without memory.

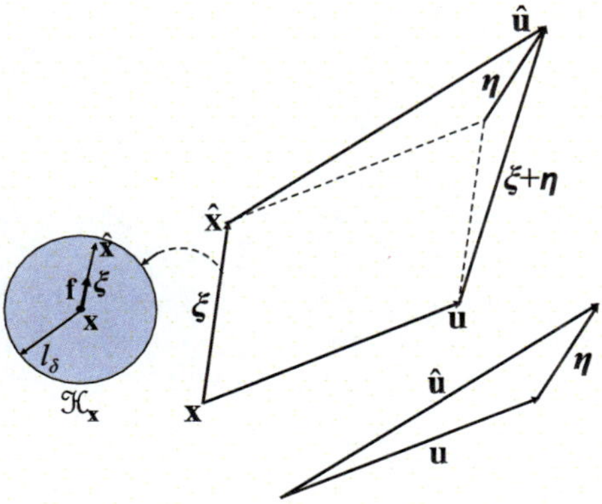

Fig. 16.1 Relative vectors $\boldsymbol{\xi}$, $\boldsymbol{\eta}$, and the family of $\mathbf{x}$: $\mathcal{H}_\mathbf{x}$

The *peridynamic equilibrium equation* is given by

$$\hat{\mathcal{L}}(\mathbf{u})(\mathbf{x}) + \mathbf{b}(\mathbf{x}) = \mathbf{0}, \quad \hat{\mathcal{L}}(\mathbf{u})(\mathbf{x}, t) = \int \mathbf{f}(\mathbf{u}(\hat{\mathbf{x}}) - \mathbf{u}(\mathbf{x}), \hat{\mathbf{x}} - \mathbf{x}, \hat{\mathbf{x}}) d\hat{\mathbf{x}}, \tag{16.5}$$

which states that the total force on the point $\mathbf{x}$ is zero. If this equation is satisfied with a zero displacement and no loading force, the reference configuration will be called equilibrated. The reference configuration is called pairwise equilibrated if $\mathbf{f}(\mathbf{0}, \boldsymbol{\xi}) = \mathbf{0}, \quad \forall \boldsymbol{\xi} \neq \mathbf{0}$. Just for subsequent convenience, we will use a locally elastic counterpart (see Eqs. (6.1)–(6.3)) of Eq. (16.5) in the form

$${}^L\hat{\mathcal{L}}(\mathbf{u})(\mathbf{x}) + \mathbf{b}(\mathbf{x}) = \mathbf{0}, \quad {}^L\hat{\mathcal{L}}(\mathbf{u})(\mathbf{x}) := \nabla[\mathbf{L}\nabla\mathbf{u}], \tag{16.6}$$

where ${}^L\hat{\mathcal{L}}(\mathbf{u})(\mathbf{x})$ is an elliptic differential operator of the second order. The superscript ${}^L(\cdot)$ will be corresponded to the local case.

The further points are apart, the weaker their interaction. Only points $\hat{\mathbf{x}}$ inside some neighborhood $\mathcal{H}_{\mathbf{x}}$ (called the family of $\mathbf{x}$) of $\mathbf{x}$ interact with $\mathbf{x}$:

$$\mathbf{f}(\boldsymbol{\eta}, \boldsymbol{\xi}, \mathbf{x}) \equiv \mathbf{0} \quad \forall \hat{\mathbf{x}} \notin \mathcal{H}_{\mathbf{x}}, \tag{16.7}$$

which means that the point $\mathbf{x}$ cannot "see" outside $\mathcal{H}_{\mathbf{x}}$. The vector $\boldsymbol{\xi} = \hat{\mathbf{x}} - \mathbf{x}$ ($\hat{\mathbf{x}} \in \mathcal{H}_{\mathbf{x}}$) is called a *bond* to $\mathbf{x}$, and the collection of all bonds to $\mathbf{x}$ forms the domain $\mathcal{H}_{\mathbf{x}}$. Without loss of generality, it is assumed that a shape of $\mathcal{H}_{\mathbf{x}}$ is spherical:

$$\mathcal{H}_{\mathbf{x}} = \{\hat{\mathbf{x}} : \ |\hat{\mathbf{x}} - \mathbf{x}| \leq l_c\} \tag{16.8}$$

and a positive number l_c, called the *horizon*, does not depend on $\mathbf{x}$ (although there is no restriction on the shape $\mathcal{H}_{\mathbf{x}}$). The pairwise force function $\mathbf{f}$ is required to have the following properties:

$$\mathbf{f}(-\boldsymbol{\eta}, -\boldsymbol{\xi}, \mathbf{x} + \boldsymbol{\xi}) \equiv -\mathbf{f}(\boldsymbol{\eta}, \boldsymbol{\xi}, \mathbf{x}) \quad \forall \boldsymbol{\eta}, \boldsymbol{\xi}, \quad \forall \mathbf{x} \in w \tag{16.9}$$

meaning assurance of conservation of linear momentum, and

$$(\boldsymbol{\eta} + \boldsymbol{\xi}) \times \mathbf{f}(\boldsymbol{\eta}, \boldsymbol{\xi}, \mathbf{x}) \equiv \mathbf{0} \quad \forall \boldsymbol{\eta}, \boldsymbol{\xi}, \tag{16.10}$$

which assures conservation of angular momentum and means that the force vector between these particles $\hat{\mathbf{x}}$ and $\mathbf{x}$ is parallel to their current relative position. Equations (16.9) and (16.10) are called the *linear admissibility condition* and *angular admissibility condition*, respectively. The conditions (16.9) and (16.10) are equivalent to representation of $\mathbf{f}$ through the most general scalar-valued function F

$$\mathbf{f}(\boldsymbol{\eta}, \boldsymbol{\xi}) = F(\boldsymbol{\eta}, \boldsymbol{\xi})(\boldsymbol{\eta} + \boldsymbol{\xi}), \tag{16.11}$$

$$F(-\boldsymbol{\eta}, -\boldsymbol{\xi}) = F(\boldsymbol{\eta}, \boldsymbol{\xi}). \tag{16.12}$$

A peridynamic material is said to be *isotropic* if and only if the response of this material is independent of their orientation, i.e.,

$$F(\mathbf{Q}\boldsymbol{\eta}, \mathbf{Q}\boldsymbol{\xi}) = F(\boldsymbol{\eta}, \boldsymbol{\xi}), \quad \forall\boldsymbol{\eta}, \boldsymbol{\xi}, \quad \forall\mathbf{Q} \in \mathcal{O}^+, \tag{16.13}$$

where $\mathcal{O}^+$ denotes the set of all proper orthogonal tensors. Equation (16.13) means invariantness with respect to rotations of the triangle with two sides of lengths, $|\boldsymbol{\eta}|$ and $|\boldsymbol{\xi}|$, and the angle between these sides, $\cos^{-1}(\boldsymbol{\xi} \cdot \boldsymbol{\eta})/(|\boldsymbol{\xi}|\,|\boldsymbol{\eta}|)$. The mentioned triangle is equivalent to the triangle with the sides $|\boldsymbol{\eta} + \boldsymbol{\xi}|$ and $|\boldsymbol{\xi}|$ and the same angle. Thus, the most general form of **f** for an isotropic material is

$$\mathbf{f}(\boldsymbol{\eta}, \boldsymbol{\xi}) = I(p, q, r)(\boldsymbol{\eta} + \boldsymbol{\xi}), \quad \forall\boldsymbol{\xi}, \boldsymbol{\eta}, \tag{16.14}$$

where

$$p = |\boldsymbol{\xi} + \boldsymbol{\eta}|, \quad q = \boldsymbol{\xi} \cdot \boldsymbol{\eta}, \quad r = |\boldsymbol{\xi}| \tag{16.15}$$

are invariant with respect to rigid rotations.

16.1.2 Elasticity

A peridynamic material is called *microelastic* if

$$\int_\Gamma \mathbf{f}(\boldsymbol{\eta}, \boldsymbol{\xi}) \cdot d\boldsymbol{\eta} = 0 \tag{16.16}$$

for any $\boldsymbol{\xi} \neq \mathbf{0}$ and any closed curve Γ with differential vector path $d\boldsymbol{\eta}$ along Γ. It means that the line integral is path independent so that the net work done by the response force along any closed curve is zero, similar to elasticity in the classical theory. According to Stokes' theorem, a necessary and sufficient condition for (16.16) with **f** continuously differentiable in $\boldsymbol{\eta}$ is

$$\nabla_\eta \times \mathbf{f}(\boldsymbol{\eta}, \boldsymbol{\xi}) = \mathbf{0}, \tag{16.17}$$

where the vector curl operator $\nabla_\eta\times$ is evaluated with respect to the coordinates of $\boldsymbol{\eta}$. Applying (16.11)–(16.17) yields the equality

$$\frac{\partial F}{\partial \boldsymbol{\eta}}(\boldsymbol{\eta}, \boldsymbol{\xi}) \times (\boldsymbol{\eta} + \boldsymbol{\xi}) = \mathbf{0} \quad \forall\boldsymbol{\xi} \neq \mathbf{0}, \tag{16.18}$$

which means that the derivative of F with respect to $\boldsymbol{\eta}$ must be in the direction of $\boldsymbol{\eta} + \boldsymbol{\xi}$. Subsequent integration of this derivative leads to the conclusion that F

depends only on both $\boldsymbol{\xi}$ and the length of $\boldsymbol{\eta}+\boldsymbol{\xi}$. Therefore, the most general form of $\mathbf{f}$ for a microelastic material is given by

$$\mathbf{f}(\boldsymbol{\eta}, \boldsymbol{\xi}) = H(p, \boldsymbol{\xi})(\boldsymbol{\xi}+\boldsymbol{\eta}), \tag{16.19}$$

$$p = |\boldsymbol{\xi}+\boldsymbol{\eta}|, \quad \forall \boldsymbol{\xi}, \boldsymbol{\eta}, \tag{16.20}$$

where $H(p, \boldsymbol{\xi})$ is another scalar-valued even function (differing from F by the relative distance $|\boldsymbol{\xi}+\boldsymbol{\eta}|$). Thus, the bonds between points in a microelastic material can be interpreted as being a (possibly nonlinear) spring, because the force between every pair of points $\mathbf{x}_i$ and $\mathbf{x}_j$ is only dependent on the length of $\boldsymbol{\xi}+\boldsymbol{\eta}$ in the deformed configuration.

Another consequence of Stokes' theorem is the equivalence of (16.13) to the existence of a differentiable and scalar-valued function (called the *pairwise potential function* w) such that the pairwise force function can be represented as

$$\mathbf{f}(\boldsymbol{\eta}, \boldsymbol{\xi}) = \frac{\partial \widetilde{w}}{\partial \boldsymbol{\eta}}(\boldsymbol{\eta}, \boldsymbol{\xi}). \tag{16.21}$$

Furthermore, it can be shown that the micropotential depends on the relative displacement vector $\boldsymbol{\eta}$ only through the scalar distance between the deformed points and, therefore,

$$\widetilde{w}(\boldsymbol{\eta}, \boldsymbol{\xi}) = \widetilde{w}(|\boldsymbol{\eta}+\boldsymbol{\xi}|, \boldsymbol{\xi}). \tag{16.22}$$

Combining (16.21) and (16.22) and differentiating the latter with respect to the components of $\boldsymbol{\eta}$ yields

$$\mathbf{f}(\boldsymbol{\eta}, \boldsymbol{\xi}) = f(|\boldsymbol{\eta}+\boldsymbol{\xi}|, \boldsymbol{\xi})\frac{\boldsymbol{\eta}+\boldsymbol{\xi}}{|\boldsymbol{\eta}+\boldsymbol{\xi}|} = f(|\boldsymbol{\eta}+\boldsymbol{\xi}|, \boldsymbol{\xi})\mathbf{e} \tag{16.23}$$

with the scale-valued function

$$f(|\boldsymbol{\eta}+\boldsymbol{\xi}|, \boldsymbol{\xi}) = \frac{\partial \widetilde{w}(p, \boldsymbol{\xi})}{\partial p} \quad \forall \boldsymbol{\xi} \text{ and } \forall p := |\boldsymbol{\eta}+\boldsymbol{\xi}| \tag{16.24}$$

and the unit vector

$$\mathbf{e} := \frac{\boldsymbol{\eta}+\boldsymbol{\xi}}{|\boldsymbol{\eta}+\boldsymbol{\xi}|} \tag{16.25}$$

along the direction of the bond between $\mathbf{x}_i$ and $\mathbf{x}_j$ in the deformed configuration $\boldsymbol{\eta}+\boldsymbol{\xi}$.

If a microelastic material is isotropic, then comparison of (16.19) with (16.11) demonstrates that H depends on its second argument $\boldsymbol{\xi}$ only through $|\boldsymbol{\xi}|$ and I does not depend on $q = \boldsymbol{\xi} \cdot \boldsymbol{\eta}$. Therefore, the most general form of w for a microelastic isotropic material is

$$\widetilde{w}(\eta, \xi) = \widetilde{w}(|\xi + \eta|, |\xi|), \tag{16.26}$$

where $\hat{w}$ is a scalar-valued function.

Transformation from microelasticity to macroelasticity is performed by the introduction of the local strain energy density (the energy per unit volume in the body at a given point)

$$\mathcal{W}(\mathbf{x}_i) = \frac{1}{2}\int_{\mathcal{H}_{\mathbf{x}_i}} \widetilde{w}(\eta, \xi)d\mathbf{x}_j, \tag{16.27}$$

where the factor of 1/2 appears because each endpoint of a bond "owns" only half the energy in the bond and the forces in the prestressed reference configuration are assumed to be vanished since otherwise these forces can be incorporated into the body force **b**.

One of the simplest nonlinear models that has been suggested as a particular case of the general representation (16.19) is the proportional microelastic material model with

$$\mathbf{f}(\eta, \xi) = f(|\eta + \xi|, \xi)\mathbf{e}, \quad f(|\eta + \xi|, \xi) = c(\xi)s \tag{16.28}$$

or

$$H(|\eta + \xi|, \xi) = \frac{c(\xi)s}{|\eta + \xi|}, \tag{16.29}$$

i.e., peridynamic force is a linear function of the bond stiffness, also called *spring stiffness*, c, though, generally, it can be a varying function of the relative position. A scale variable s is the current bond stretch, which is defined as the change of length of the bond ξ per its unit length

$$s = \frac{|\eta + \mathbf{x}| - |\xi|}{|\eta + \xi|}. \tag{16.30}$$

Equation (16.29) was initially proposed by Silling and Askari [1401] as the prototype micro-brittle model with damage (see Sect. 16.5).

16.1.3 Peridynamic Stress

The peridynamic theory is traditionally based on the use of the displacement field $\mathbf{u}(\mathbf{x})$ rather than either the stress $\boldsymbol{\sigma}(\mathbf{x})$ or strain $\boldsymbol{\varepsilon}(\mathbf{x})$ fields which are not conceptually necessary. However, the introduction of the notion of stress is helpful, as one can use it to formulate stress-strain relations for exploiting the well-developed tool of the classical elasticity theory in the subsequent application of the present theory for heterogeneous materials. Moreover, it is too restrictive to assume pairwise equilibration when the force between any two particles vanishes, and because of this, a so-called unstressed configuration will be defined from the following explanation.

When a homogeneous, microelastic body $w = w^+ \cup w^-$ is divided into two parts by a plane $\mathbf{x} \in \mathcal{P}$, then one of these parts, w^+, exerts a force on the other part, w^-, formally described by

$$w^+ = \{\mathbf{x}' \in w : (\mathbf{x}' - \mathbf{x}) \cdot \mathbf{n} \geq 0\}, \quad w^- = \{\mathbf{x}' \in w : (\mathbf{x}' - \mathbf{x}) \cdot \mathbf{n} \leq 0\}, \tag{16.31}$$

where $\mathbf{x}$ is a point in $\mathcal{P}$ and $\mathbf{n}$ is a unit vector normal to the dividing plane $\mathcal{P}$ (see Fig. 16.3). The load density $\mathbf{L}(\hat{\mathbf{x}})$ at a point in $\hat{\mathbf{x}} \in w^-$ is calculated by integrating the pairwise force function over all of the points interacting with in w^+

$$\mathbf{L}(\hat{\mathbf{x}}) = \int_{w^+} \mathbf{f}(\mathbf{u}' - \hat{\mathbf{u}}, \mathbf{x}' - \hat{\mathbf{x}}) dv_{\mathbf{x}'}. \tag{16.32}$$

The force acting on the domain w^+ can then be calculated by integrating the load density over the volume of every point contained within w^- as

$$\mathbf{F}^-(\mathbf{x}) = \int_{w^-} \mathbf{L}(\hat{\mathbf{x}}) dV_{\hat{\mathbf{x}}} = \int_{w^-} \int_{w^+} \mathbf{f}(\mathbf{u}' - \hat{\mathbf{u}}, \mathbf{x}' - \hat{\mathbf{x}}) dv_{\mathbf{x}'} v_{\hat{\mathbf{x}}} \tag{16.33}$$

that defines the force acting on region w^-. This force is then divided by the cross-sectional area $\mathcal{A}^{\mathcal{P}} = \mathcal{H}_{\mathbf{x}} \cap \mathcal{P}$ formed by the separating plane yielding (see [366])

$$\mathbf{F}^- / \text{mes}\mathcal{A}^{\mathcal{P}} = \boldsymbol{\sigma}\mathbf{n}, \tag{16.34}$$

where, as before, $\mathbf{n}$ is the unit vector perpendicular to the plane $\mathbf{x} \in \mathcal{P}$ of separation and $\boldsymbol{\sigma}$ is the Piola-Kirchhoff stress tensor in the reference configuration.

The integration over w^- can be split up by an integration over the area $\mathcal{P}$ and the set of colinear points $\mathcal{L}$:

$$\mathbf{F}^-(\mathbf{x}) = \int_{\mathcal{P}} \int_{\mathcal{L}} \int_{w^+} \mathbf{f}(\mathbf{u}' - \hat{\mathbf{u}}, \mathbf{x}' - \hat{\mathbf{x}}) dv_{\mathbf{x}'} d\hat{l} dA, \tag{16.35}$$

where

$$\mathcal{L} = \{\hat{\mathbf{x}} \in w^- : \hat{\mathbf{x}} = \mathbf{x} - s\mathbf{n}, \ 0 \leq s < \infty\} \tag{16.36}$$

and $d\hat{l}$ represents the differential path length over $\mathcal{L}$. Dividing the total force by the area of the plane separating the parts, the *areal force density* $\boldsymbol{\tau}(\mathbf{x}, \mathbf{n})$ (see [1396]) is defined as

$$\boldsymbol{\tau}(\mathbf{x}, \mathbf{n}) = \int_{\mathcal{L}} \int_{w^+} \mathbf{f}(\mathbf{u}' - \hat{\mathbf{u}}, \mathbf{x}' - \hat{\mathbf{x}}) dv_{\mathbf{x}'} d\hat{l}. \tag{16.37}$$

The areal force density in the peridynamic theory is close to the concept of traction in the conventional theory. Now, a configuration is called unstressed if

$$\boldsymbol{\tau}(\mathbf{x}, \mathbf{n}) = \mathbf{0}, \quad \forall \mathbf{x} \in w, \quad \forall \mathbf{n}. \tag{16.38}$$

Integration over the sets w^+ and $\mathcal{L}$ is also useful when calculating the total work done by the bonds between particles.

As mentioned by Silling [1396], this definition of $\boldsymbol{\tau}(\mathbf{x}, \mathbf{n})$ is most useful in the case of a homogeneous deformation (i.e., a deformation whose gradient is a constant matrix) when

$$\boldsymbol{\tau}(\mathbf{x}, \mathbf{n}) = \boldsymbol{\sigma}\mathbf{n} \quad \forall \mathbf{x} \in w, \quad \forall \mathbf{n}, \tag{16.39}$$

where $\boldsymbol{\sigma}$ is really a Piola-Kirchhoff stress tensor, because $\boldsymbol{\tau}$ represents the force per unit area in the reference configuration. Moreover, its linear dependence on $\mathbf{n}$ is not obvious, and it provides no explicit formula for the first Piola-Kirchhoff stress tensor in terms of the pairwise force function $\mathbf{f}$. This issue was overcome later in [1403]) where the peridynamic stress tensor $\boldsymbol{\sigma}$ was defined through the function $\mathbf{f}$. Namely, for subsequent convenience, one introduces a vector-valued function $\hat{\mathbf{f}}$: $R^d \times R^d \to R^d$ by

$$\hat{\mathbf{f}}(\mathbf{p}, \mathbf{q}) = \begin{cases} \mathbf{f}(\mathbf{u}(\mathbf{p}) - \mathbf{u}(\mathbf{q}), \mathbf{p} - \mathbf{q}, \mathbf{q}), & \text{if } \mathbf{p}, \mathbf{q} \in w, \\ \mathbf{0}, & \text{otherwise.} \end{cases} \tag{16.40}$$

It is assumed that $\hat{\mathbf{f}}(\hat{\mathbf{x}}, \mathbf{x})$ is Riemann integrable that does not imply the bounding of $\hat{\mathbf{f}}(\hat{\mathbf{x}}, \mathbf{x})$ as $|\hat{\mathbf{x}} - \mathbf{x}| \to 0$. Then by adapting Cauchy's notion of stress in a crystal, one can define the "peridynamic stress" $\boldsymbol{\sigma}(\mathbf{z})$ at the point $\mathbf{z}$ to be the total force that all material points $\hat{\mathbf{x}}$ to the right of $\mathbf{z}$ exert on all material particles to its left (see, e.g., [1405, 1601]). In particular, for 1D case,

$$\sigma(x) = \int_{\mathcal{H}^-_{x-l_c}} \left\{ \int_{\mathcal{H}^+_{r+l_c}} \hat{f}(r, s) ds \right\} dr, \tag{16.41}$$

where $\mathcal{H}^-_{x-l_c} = \{r : x - l_c < r < x\}$ and $\mathcal{H}^+_{r+l_c} = \{s : x < s < r + l_c\}$. For 2D and 3D cases ($d = 2, 3$),

$$\boldsymbol{\sigma}(\mathbf{x}) = \boldsymbol{\mathcal{L}}^\sigma(\mathbf{u}), \tag{16.42}$$

$$\boldsymbol{\mathcal{L}}^\sigma(\mathbf{u}) := \frac{1}{2} \int_S \int_0^\infty \int_0^\infty (y + z)^{d-1} \hat{\mathbf{f}}(\mathbf{x} + y\mathbf{m}, \mathbf{x} - z\mathbf{m}) \otimes \mathbf{m} dz dy d\Omega_{\mathbf{m}}, \tag{16.43}$$

where S stands for the unit sphere and $d\Omega_{\mathbf{m}}$ denotes a differential solid angle on S in the direction of any unit vector $\mathbf{m}$. Equations (16.42) and (16.43) were proved in Lehoucq and Silling [909] for 3D case, while the case $d = 2$ can be justified in a similar manner (see [216]). Equations (16.42) and (16.43) at $d = 1$ can also be reduced to Eq. (16.41) by the variable exchange $x - z \mapsto r, x + y \mapsto s$. Indeed, the origin-centered unit 1D "sphere" is the set {-1 , 1}, which has a measure of 2. Then the integrand of Eq. (16.42) is reduced to the integrand of Eq. (16.41) by the use of the equality (16.9).

As can be seen, no connection to areal force density was provided in Eq. (16.43). Note that through the function **f**, the dependence of $\boldsymbol{\sigma}$ on the deformation is nonlocal. Lehoucq and Silling [909] proved that the spatial divergence of this tensor is equal to the density of internal forces in the reference configuration, i.e.,

$$\mathrm{div}\boldsymbol{\sigma}(\mathbf{x}) = \int_w \mathbf{f}(\mathbf{x}', \mathbf{x}) dv_{\mathbf{x}'} \quad \forall \mathbf{x} \in R^d - \partial w, \tag{16.44}$$

if **f** is continuously differentiable on $(R^d - \partial w) \times (R^d - \partial w) - \mathcal{I}$, where $\mathcal{I} = \{(\mathbf{p}, \mathbf{q}) \in R^d \times R^d \mid \mathbf{p} = \mathbf{q}\}$. So the equation of motion in peridynamics can be rewritten to the form

$$\rho \ddot{\mathbf{u}}(\mathbf{x}) = \mathrm{div}\boldsymbol{\sigma}(\mathbf{x}) + \mathbf{b}(\mathbf{x}), \tag{16.45}$$

which is formally similar to the equation of motion in the conventional locally elastic theory. The stress tensor $\boldsymbol{\sigma}$ is the analog of the first Piola-Kirchhoff stress tensor in the classical elasticity (see for details [909, 1197, 1404]).

16.2 Boundary Volumetric Conditions

Due to nonlocality, the equilibrium equation (16.2) should be accompanied by a "boundary" condition, imposed as a volumetric constraint in the so-called interaction domain w_Γ (opposite to the local elasticity case where the boundary conditions are imposed precisely at the bounding surface $\Gamma^{(0)}$) (see for details [771, 1396]); i.e., the nonlocal boundary w_Γ is a d-dimensional region unlike its $(d-1)$-dimensional counterpart Γ^0 in local problems. The interaction domain w_Γ contains points **y** not in w interacting with points $\mathbf{x} \in w$. A variety of choices for the domain w_Γ having positive volume are possible (see for details [397]); $\overline{\overline{w}} = w \cup w_\Gamma$ stands for the *nonlocal closure* of w. The most popular shape for w_Γ with prescribed forces and displacements is a boundary layer of thickness given by the horizon l_c (see [996]): $w_\Gamma = \{w \oplus \mathcal{H}_0\} \backslash w$, where $w \oplus \mathcal{H}_0$ is a Minkowski outer parallel (dilation) of w ($\mathcal{A} \oplus \mathcal{B} := \cup_{\mathbf{x} \in \mathcal{A}, \mathbf{y} \in \mathcal{B}} \{\mathbf{x} + \mathbf{y}\}$); in such a case, w is the internal region of $\overline{\overline{w}}$ (see [1396]). We further assume that $w_\Gamma = w_{\Gamma u} \cup w_{\Gamma \sigma}$, where $w_{\Gamma u}$ and $w_{\Gamma \sigma}$ are domains over which "Dirichlet" and "Neumann" volume constraints are to be applied and $w_{\Gamma u} \cap w_{\Gamma \sigma} = \emptyset$. Points in w interact with points in w_Γ through the pairwise force function **f**, and, therefore, these external forces must be supplied through the fictitious force density **b**. In particular, the nonlocal "Dirichlet" volume-constrained problem is given as (see [397])

$$\begin{cases} -\widehat{\mathcal{L}}(\mathbf{u}) = \mathbf{b} & \text{on } w, \\ \mathbf{u} = \mathbf{h}, & \text{on } w_{\Gamma u} = w_{\Gamma}, \end{cases} \tag{16.46}$$

while the nonlocal "Neumann" volume-constrained problem is presented by

$$\begin{cases} -\widehat{\mathcal{L}}(\mathbf{u}) = \mathbf{b} & \text{on } w, \\ -\widehat{\mathcal{L}}(\mathbf{u}) = \mathbf{g}, & \text{on } w_{\Gamma\sigma} = w_{\Gamma}. \end{cases} \tag{16.47}$$

The conditions (16.46_2) and (16.47_2) are called *displacement conditions* and *force loading conditions*, respectively. Relabeling **g** as **b** over the interaction domain w_Γ, the representation (16.2) has the equivalent form

$$-\widehat{\mathcal{L}}(\mathbf{u}) = \mathbf{b} \text{ on } w \cup w_\Gamma. \tag{16.48}$$

In a similar manner, if Dirichlet-type volumetric boundary conditions (these conditions will be usually used in the subsequent presentation; see Fig. 16.2) are prescribed, the functional $\widehat{\mathcal{L}}(\mathbf{u})(\mathbf{x})$ (16.2) should be modified as follows:

$$\widehat{\mathcal{L}}(\mathbf{u})(\mathbf{x}) \longmapsto \widehat{\mathcal{L}}(\mathbf{u})(\mathbf{x}) + \widehat{\mathcal{L}}^{\Gamma u}(\mathbf{h})(\mathbf{x}), \tag{16.49}$$

$$\widehat{\mathcal{L}}^{\Gamma u}(\mathbf{h})(\mathbf{x}) := \int_{w_{\Gamma u}} \mathbf{f}(\mathbf{h} - \mathbf{u}, \mathbf{y} - \mathbf{x}, \mathbf{x}) d\mathbf{y}, \tag{16.50}$$

where the shorthand $\mathbf{h} = \mathbf{u}(\mathbf{y})$ has been used. Thus, both the volumetric boundary conditions (16.46) and (16.47) can be incorporated into one equilibrium equation:

$$\breve{\mathcal{L}}(\mathbf{u})(\mathbf{x}) = \breve{\mathbf{b}}(\mathbf{x}), \quad \mathbf{x} \in w \cup w_\Gamma \tag{16.51}$$

which does require any additional supplements.

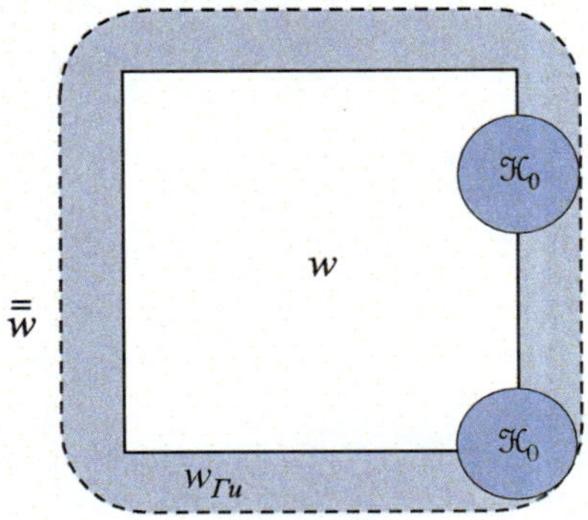

Fig. 16.2 Domain w with the nonlocal boundary $w_{\Gamma u}$

It is assumed that $\overline{\overline{w}}$ is obtained by cutting out of a domain from a medium filling the entire space R^d. The interaction domain w_Γ is covered by the surfaces $\Gamma_{\rm r} := \{\mathbf{y}: \ \text{dist}(\mathbf{y}, w) = \text{r}\}$ $(0 \le \text{r} \le l_c)$ so that for $\forall \mathbf{y} \in w_\Gamma$ $(\mathbf{y} \in \Gamma_{\rm r})$ one can attach a normal vector $\mathbf{n}(\mathbf{y}) \perp \Gamma_{\rm r}$; i.e., $\Gamma^{(0)}$ and Γ_{l_c} are the boundaries of w and w_Γ, respectively. In such a case, the Dirichlet (16.46_2) or Neumann (16.47_2) volumetric boundary conditions are called homogeneous loading conditions if there exist some symmetric constant tensors $\boldsymbol{\varepsilon}^{w_\Gamma}$ or $\boldsymbol{\sigma}^{w_\Gamma}$ such that

$$\mathbf{h}(\mathbf{y}) = \boldsymbol{\varepsilon}^{w_\Gamma}\mathbf{y}, \ \forall \mathbf{y} \in w_{\Gamma u} = w_\Gamma, \tag{16.52}$$

$$\mathbf{g}(\mathbf{y}) = \boldsymbol{\sigma}^{w_\Gamma}\mathbf{n}(\mathbf{y}), \ \forall \mathbf{y} \in w_{\Gamma\sigma} = w_\Gamma, \tag{16.53}$$

respectively. There are no specialized restrictions on the shape and smoothness of Γ_0 which is defined just by convenience of representation. For example, it is convenient for visibility, although not an essential feature of the theory, to assume that Γ^0 has a circle (or oval) shape ($d = 2$). It is possible to perform a domain decomposition of w_Γ over nonoverlapping subdomains w^i_Γ $(w_\Gamma = \cup w^i_\Gamma, w^i_\Gamma \cap w^j_\Gamma = \emptyset, \ \forall i \neq j)$ such that mes w^i_Γ/mes $w \ll 1$ and $\mathbf{n}(\mathbf{y}) \simeq \mathbf{n}(\mathbf{y}_i)$ for $\forall \mathbf{y} \in w^i_\Gamma$, where $\mathbf{y}_i$ is a center of Γ_{l_c} at the boundary of w^i_Γ.

16.3 Linearized Microelastic Model

A linearization of the vector force function can be introduced by solving the equation in an incremental approach dealing with small displacement that means (see Silling [1397])

$$\sup_{|\boldsymbol{\xi}| \le l_\delta} |\boldsymbol{\eta}| \ll l_\delta \quad \forall \mathbf{x} \in w, \tag{16.54}$$

which is a weaker condition than the corresponding requirement in the classical local elasticity $|$ grad$\mathbf{u}| \ll 1$ because it permits us to consider jump discontinuities in $\mathbf{u}$. Then the vector force function in a neighbor of the point $(\mathbf{0}, \boldsymbol{\xi})$ can be represented by a Taylor series expansion of the first order

$$\mathbf{f}(\boldsymbol{\eta}, \boldsymbol{\xi}) = \mathbf{C}(\boldsymbol{\xi})\boldsymbol{\eta} + \mathbf{f}(\mathbf{0}, \boldsymbol{\xi}) \tag{16.55}$$

with the corresponding micropotential given by

$$\widetilde{w}(\boldsymbol{\eta}, \boldsymbol{\xi}, \mathbf{x}) = \frac{1}{2}\boldsymbol{\eta}^\top \cdot \mathbf{C}(\boldsymbol{\xi}, \mathbf{x})\boldsymbol{\eta}, \tag{16.56}$$

where $\mathbf{C}(\boldsymbol{\xi})$ is a second-order tensor, called *micromodulus*, which in general is expressed through the vector force as

$$\mathbf{C}(\boldsymbol{\xi}) = \frac{\partial \mathbf{f}}{\partial \boldsymbol{\eta}}(\boldsymbol{\eta}, \boldsymbol{\xi}) = \boldsymbol{\xi} \otimes \frac{\partial F}{\partial \boldsymbol{\eta}}(\mathbf{0}, \boldsymbol{\xi}) + F(\mathbf{0}, \boldsymbol{\xi})\mathbf{I} \ \forall \boldsymbol{\xi}, \tag{16.57}$$

where $\otimes$ is the dyadic product and Eq. (16.11) was used. For consistency with Newton's third law (also following the requirement from (16.9)), the micromodulus function $\mathbf{C}$ for the homogeneous materials must be symmetric with respect to its argument as well as with respect to its tensor structure (due to Eq. (16.17)):

$$\mathbf{C}(-\boldsymbol{\xi}) = \mathbf{C}(\boldsymbol{\xi}) = \mathbf{C}^\top(\boldsymbol{\xi}) \quad \forall \boldsymbol{\xi}. \tag{16.58}$$

Substitution of Eq. (16.55) into Eq. (16.9) leads to the following property of $\mathbf{C}(\boldsymbol{\xi}, \mathbf{x})$:

$$\mathbf{C}(\boldsymbol{\xi}, \mathbf{x}) = \mathbf{C}(-\boldsymbol{\xi}, \mathbf{x} + \boldsymbol{\xi}). \tag{16.59}$$

For example, for the micromodulus functions with the step-function and triangular profiles,

$$\mathbf{C}(\boldsymbol{\xi}) = \mathbf{C}V(\mathcal{H}_{\mathbf{x}}), \ \ \mathbf{C}(\boldsymbol{\xi}) = \mathbf{C}(1 - |\boldsymbol{\xi}|/l_c)V(\mathcal{H}_{\mathbf{x}}), \tag{16.60}$$

respectively, where $V(\mathcal{H}_{\mathbf{x}})$ is the indicator function of $\mathcal{H}_{\mathbf{x}}$. The peridynamic solution of Eq. (16.3) with $\mathbf{C}$ described by Eq. (16.60) is investigated in detail by both numerical and analytical methods in 1D case (see [131, 1058, 1405, 1601]) and 2D case (see [666, 667]).

As mentioned in [1404], formally, the linearized equation in the peridynamic model (16.3), (16.4), and (16.55) is the same as the proposed by Kunin [834–836]. However, in the peridynamic equation, the kernel is derived by linearization of a material model in the nonlinear theory and therefore allows the micromodulus function in (16.55) to be obtained in an unambiguous way.

A necessary and sufficient condition for the tensorial symmetry of $\mathbf{C}(\boldsymbol{\xi})$ (16.58_2) is the existence of a scalar-valued function λ such that

$$\mathbf{C}(\boldsymbol{\xi}) = \lambda(\boldsymbol{\xi})\boldsymbol{\xi} \otimes \boldsymbol{\xi} + F_0(\boldsymbol{\xi})\mathbf{I} \quad \forall \boldsymbol{\xi}, \tag{16.61}$$

where

$$\lambda(\boldsymbol{\xi}) = \frac{1}{|\boldsymbol{\xi}|^2}\boldsymbol{\xi} \cdot \frac{\partial F}{\partial \boldsymbol{\eta}}(\mathbf{0}, \boldsymbol{\xi}), \quad F_0(\boldsymbol{\xi}) := F(\mathbf{0}, \boldsymbol{\xi}) \ \forall \boldsymbol{\xi} \tag{16.62}$$

are also symmetric due to Eqs. (16.56) and (16.12), respectively. It should be also mentioned that for pairwise equilibrated reference configuration assumed

$$F_0(\boldsymbol{\xi}) \equiv 0, \ \ \text{and} \ \mathbf{f}(\mathbf{0}, \boldsymbol{\xi}) \equiv \mathbf{0} \tag{16.63}$$

in Eqs. (16.38) and (16.50), respectively. If a linear microelastic material is additionally isotropic, then $\lambda(\boldsymbol{\xi})$ and $F_0(\boldsymbol{\xi})$ depend on $\boldsymbol{\xi}$ only through $|\boldsymbol{\xi}|$.

For the special case of a linear microelastic material,

$$\mathbf{C}(\boldsymbol{\xi}) = c\frac{\boldsymbol{\xi} \otimes \boldsymbol{\xi}}{|\boldsymbol{\xi}|^3} V(\mathcal{H}_{\mathbf{x}}), \quad \text{i.e.,} \quad C_{ij}(\boldsymbol{\xi}) = c\frac{\xi_i \xi_j}{(\xi_k \xi_k)^{3/2}} V(\mathcal{H}_{\mathbf{x}}), \tag{16.64}$$

where a constant c depends again on the material but also on the dimension d of the problem. The proportionality factor c (16.64) (or $\mathbf{C}$ (16.58)) is to be determined in such a way that the deformation energy density (or the constitutive equation) of a homogeneous body under homogeneous loading arising from the peridynamic model coincides with the corresponding value from the classical linear elasticity theory (see, e.g., [434, 436, 1396]).

In particular, for the proportional microelastic model (16.29), we get

$$\frac{\partial F}{\partial \boldsymbol{\eta}}(\boldsymbol{\eta}, \boldsymbol{\xi}) = \frac{c(\boldsymbol{\xi})}{|\boldsymbol{\xi}|}\frac{\partial}{\partial \boldsymbol{\eta}}\left(1 - \frac{|\boldsymbol{\xi}|}{|\boldsymbol{\eta} + \boldsymbol{\xi}|}\right) = \frac{c(\boldsymbol{\xi})}{|\boldsymbol{\eta} + \boldsymbol{\xi}|^3}(\boldsymbol{\eta} + \boldsymbol{\xi}) \tag{16.65}$$

with its evaluation at $(\mathbf{0}, \boldsymbol{\xi})$

$$\frac{\partial F}{\partial \boldsymbol{\eta}}(\mathbf{0}, \boldsymbol{\xi}) = \frac{c(\boldsymbol{\xi})\boldsymbol{\xi}}{|\boldsymbol{\xi}|^3}. \tag{16.66}$$

Substituting Eq. (16.65) into Eq. (16.57) gives exactly the special case of a linear elastic material (16.64).

16.4 Homogeneous Deformations

For an isotropic microelastic body $w = w^+ \cap w^-$ divided into two parts, w^+ and w^-, by a plane P (see Fig. 16.3), $\boldsymbol{\tau}(\mathbf{x}, \mathbf{n})$ can be estimated as follows (see [1402]). Let an orthonormal basis $\{\mathbf{e}_1, \mathbf{e}_2, \mathbf{e}_3\}$ be given and $\mathbf{x}$ is taken as the origin with $\mathbf{n}$ pointing in the direction of $\mathbf{e}_1$. Let w be a body subjected to the homogeneous deformation given by $u_1 = \varepsilon_{11}x_1$, $u_2 \equiv u_3 \equiv 0$, where ε_{11} is a constant and $|\varepsilon_{11}| \ll 1$. Then $\eta_1 = \varepsilon_{11}\xi_1$, $\eta_2 \equiv \eta_3 \equiv 0$, and, due to Eqs. (16.56) and (16.61),

$$f_1 = \varepsilon_{11}\lambda(r)\xi_1^3, \quad f_2 = \varepsilon_{11}\lambda(r)\xi_1^2\xi_2, \quad f_3 = \varepsilon_{11}\lambda(r)\xi_1^2\xi_3, \tag{16.67}$$

where $r = |\xi|$ and one assumed $F_0 = 0$ for simplicity. We now estimate the stress component $\sigma_{11} = \tau_1(\mathbf{e}_1)$, where $\boldsymbol{\tau}$ defined by Eq. (16.37) can be recast in the form

$$\boldsymbol{\tau}(0, \mathbf{e}_1) = \int_0^\infty \int_{w^+} \mathbf{F}(0, |\mathbf{x}' + s\mathbf{n}|)(\mathbf{x}' + s\mathbf{n})dv_{\mathbf{x}'}ds, \tag{16.68}$$

where $\mathbf{x}' - \hat{\mathbf{x}} = \mathbf{x}' + s\mathbf{n}$ and $d\hat{l} = ds$ have been used for the integral over $\mathcal{L}$. Setting $\boldsymbol{\xi} = \mathbf{x}' + s\mathbf{e}_1$ with a spherical coordinate (see Fig. 16.3)

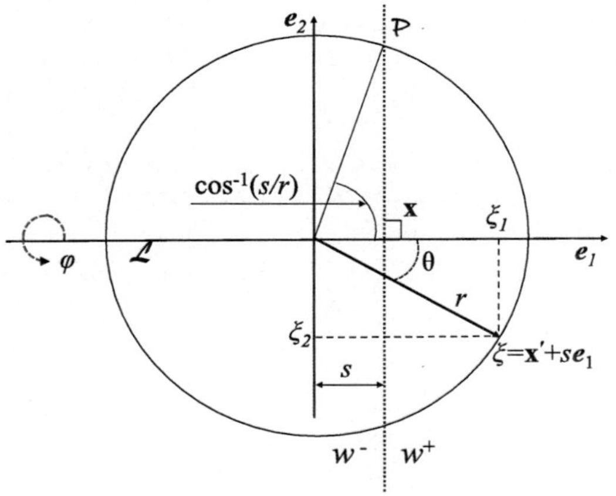

Fig. 16.3 A circular body w divided into w^+ and w^- with the spherical coordinates

$$\xi_1 = r\cos\theta,\ \ \xi_2 = r\sin\theta\cos\phi,\ \ \xi_3 = r\sin\theta\sin\phi,\ \ r = |\boldsymbol{\xi}|, \tag{16.69}$$

and $dv_{\mathbf{x}'} = r^2 \sin\theta d\phi d\theta dr$, Eq. (16.60) leads to

$$\boldsymbol{\tau}(0, \mathbf{e}_1) = \varepsilon \int_0^\infty \int_0^r \int_0^{\cos^{-1}(s/r)} \int_0^{2\pi} \lambda(r)(r\cos\theta)^3 r^2 \sin\theta d\phi d\theta ds dr := \Lambda, \tag{16.70}$$

where

$$\Lambda = \frac{2\pi}{5}\int_0^\infty \lambda(r) r^6 dr. \tag{16.71}$$

Similar calculation can be performed for the other components of $\boldsymbol{\sigma}$. To exemplify, for the components in the direction of $\mathbf{e}_2$, the variables in Eq. (16.69) should be changed to $\xi_1 = r\sin\theta\cos\phi,\ \xi_2 = \cos\theta,\ \xi_3 = r\sin\theta\sin\phi$. Comparing the results with the stresses for corresponding to their locally elastic counterparts, we find

$$[\sigma] = \varepsilon \begin{bmatrix} \Lambda & 0 & 0 \\ 0 & \Lambda/3 & 0 \\ & 0 & \Lambda/3 \end{bmatrix} = \varepsilon \begin{bmatrix} l+2m & 0 & 0 \\ 0 & l & 0 \\ 0 & 0 & l, \end{bmatrix} \tag{16.72}$$

where l and μ are the Lamé moduli. From this, we can express the peridynamic parameter Λ (which is not directly measurable) through the easily measurable quantities l, μ, E, ν, k

$$l = \mu = \frac{\Lambda}{3},\ \ \nu = \frac{1}{4},\ \ E = \frac{3\Lambda}{6},\ \ k = \frac{5\Lambda}{9}, \tag{16.73}$$

where ν is Poisson's ratio and E and k are Young's and bulk moduli, respectively.

Correspondence of the parameters of the peridynamic model to their locally elastic counterpart can be also determined in such a way that the deformation energy of a homogeneous body under homogeneous loading arising from the peridynamic model coincides with the energy known from the classical theory. More precisely, consider a large homogeneous body under isotropic extension in the proportional microelastic model (16.28) (see [1401]), i.e., s is constant for all $\boldsymbol{\xi}$, and $\boldsymbol{\eta} = s\boldsymbol{\xi}$. Defining $\xi = |\boldsymbol{\xi}|$ and $\eta = |\boldsymbol{\eta}|$, we have $\eta = s\xi$. Substituting scale function $f = cs = c\eta/\xi$ into the definition of the micropotential (16.23) leads to $\tilde{w} = c\eta^2/(2\xi) = cs^2\xi/2$. Then the local strain energy density (16.28) can be calculated as

$$\mathcal{W}(\mathbf{x}) = \frac{1}{2}\int_{\mathcal{H}_{\mathbf{x}}} \tilde{w} dv_{\mathbf{x}} = \frac{cs^2}{2}\int_0^{l_\delta}\int_0^{2\pi}\int_0^{\pi} r^3 \sin\theta d\theta d\phi dr = \frac{\pi c s^2 l_\delta^4}{4}. \tag{16.74}$$

In the classical linear elasticity theory, we have equal the strain energy density

$$\mathcal{W} = \frac{9ks^2}{2}, \tag{16.75}$$

where k denotes the bulk modulus. Following the equating of Eqs. (16.74) and (16.75), the representation for the spring constant c in the model (16.28) can be defined as

$$c = \frac{18k}{\pi l_\delta^4} = \frac{12E}{\pi l_\delta^4} \tag{16.76}$$

with $\nu = 1/4$. In a similar manner, we find a spring constant in the two-dimensional case (see [539])

$$c = \frac{12E}{\pi t(1+\nu) l_\delta^3} \tag{16.77}$$

with $\nu = 1/3$ for plane stress and $\nu = 1/4$ for plane strain; t is a plane structure thickness. For the one-dimensional case, we get

$$c = \frac{18k}{5\pi l_\delta^2}. \tag{16.78}$$

A linearization of the pairwise force function (16.28) of the proportional microelastic model (see Eqs. (16.65) and (16.66)) gives exactly the linear model (16.64) with the constant of proportionality c (depending on the dimension d of w) as above (16.76)–(16.78).

Following the same procedure performed to calculate the micromodulus (16.77), we obtain the constant (16.79_1) and conical (16.79_3) micromodulus functions $c(\boldsymbol{\xi})$ in 2D (16.28) plane stress conditions (see [131, 594])

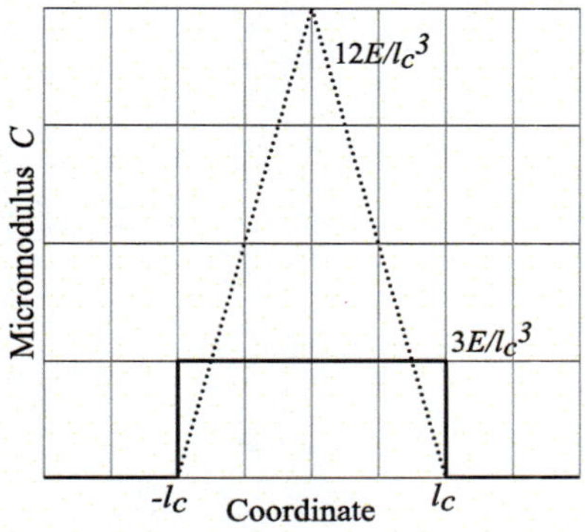

Fig. 16.4 Constant and triangle profiles

$$c(\boldsymbol{\xi}) = c_0, \quad c_0 = \frac{6E}{\pi(1-\nu)l_\delta^3} \text{ and } c(\boldsymbol{\xi}) = c_1\Big(1 - \frac{|\boldsymbol{\xi}|}{l_\delta}\Big), \quad c_1 = \frac{24E}{\pi(1-\nu)l_\delta^3}, \tag{16.79}$$

as well as in 1D case

$$c(\boldsymbol{\xi}) = c_0, \quad c_0 = \frac{2E}{\pi l_\delta^2 A} \text{ and } c(\boldsymbol{\xi}) = c_1\Big(1 - \frac{|\boldsymbol{\xi}|}{l_\delta}\Big), \quad c_1 = \frac{6E}{\pi l_\delta^2 A}, \tag{16.80}$$

respectively; here, A is a constant cross-sectional area of a bar.

The corresponding elastic moduli of the locally elastic medium can be also found by equating the stress fields in both the locally elastic and peridynamic media in the special case of a homogeneous deformation of the form $\mathbf{u}(\mathbf{x}) = \boldsymbol{\varepsilon}\mathbf{x}$ for some constant $\boldsymbol{\varepsilon}$. So, for 1D cases, we obtain for the constant (16.60_1) and triangle (16.60_2) micromodulus functions (see [1405]) at $A = 1$

$$C(\xi) = C_0, \quad c_0 = \frac{3E}{l_\delta^3} \text{ and } C(\xi) = C_1\Big(1 - \frac{|\xi|}{l_\delta}\Big), \quad c_1 = \frac{12E}{l_\delta^3}, \tag{16.81}$$

respectively (see Fig. 16.4).

16.5 Damage and Failure Models

Damage is a basic concept in engineering design exploited for predictions about the initiation, propagation, and fracture of materials without molecular scale which can be too complex for practical engineering analysis (see the pioneering works by Kachanov [725] and Rabotnov [1268] as well as [727] and [801, 1569]). Damage mechanics uses state variables (which are defined by an indication method) representing their effects on the exploitation properties of the material such as the stiffness and remaining life. Popular state variables can be measured and used

Fig. 16.5 Schematic diagram of the peridynamic force

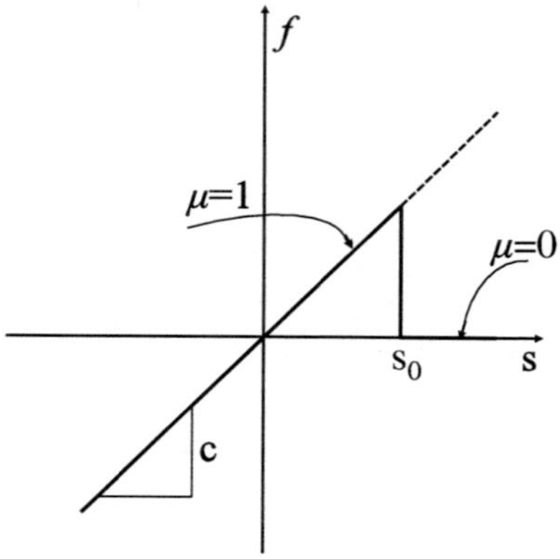

for proper definition of the damage variable through density change, resistivity change, acoustic emission, change of the fatigue limit, change of stiffness, and microcrack density used in micromechanical damage formulations for prediction of both damage initiation and evolution. Physically justified damage theory was proposed in peridynamics in [1401], where damage is introduced by permitting damaged bonds to break irreversibly, i.e., damaging of the bond is considered as bond removing if the bond stretch exceeds a critical stretch for some time). Damage in a bond is treated as bond breakage where some scalar function called damage jumps discontinuously from 0 to 1 if the bond strain exceeds some critical bond strain.

In the prototype microelastic brittle (PMB) constitutive model by Silling and Askari [1401], the scalar-valued function f of the proportional model (16.28) is defined as

$$f(|\boldsymbol{\eta}+\boldsymbol{\xi}|, \boldsymbol{\xi}) = \mu(\boldsymbol{\xi})c(\boldsymbol{\xi})s, \tag{16.82}$$

where $\mu(\boldsymbol{\xi})$ is the history-dependent scalar-valued function which takes into consideration bond failure state and it is expressed as (see Fig. 16.5)

$$\mu(\boldsymbol{\xi}) = \begin{cases} 1 & s < s_0 \\ 0 & s \geq s_0. \end{cases} \tag{16.83}$$

A failure criterion is based on a maximum time-history stretch ($\max(s(t)) \geq s_0$, for $t > 0$) that a bond can withstand during its loading. As soon as the stretch overcomes this limit, the bond is irreversibly broken and it is not considered anymore in the computation (no stretch or other variables are computed for the next steps). This means that if the current stretch is less than the limit (i.e., it is still computed), all the stretch values of the previous time steps satisfy this condition too.

A schematic diagram of the peridynamic force versus the bond stretch for a PMB material is illustrated in Fig. 16.5. To determine s_0, we follow the procedure introduced in [1401]. The critical bond stretch s_0 can somehow be linked to known macroscopic quantities such as the critical energy release rate of the material G_0 when the homogeneous structure is subjected to isotropic extension. This relation is established under the assumptions of complete separation of the fracture surfaces and of the absence of other dissipative mechanisms near the crack tip. The work required to break a single bond for the PMB model is equal to

$$w_0(\boldsymbol{\xi}) = \int_0^{s_0} cs(\xi ds) = \frac{cs_0^2\xi}{2}, \tag{16.84}$$

where the relation $d\eta = \xi ds$ has been used. All the bonds crossing the mentioned separating surface have to be broken, and the energy per unit surface area required to break them all is equal to the critical energy release rate G_0 of the classical continuum theory

$$G_0 = \int_0^{2\pi} \int_0^{l_\delta} \int_z^{l_\delta} \int_0^{\cos^{-1}(z/\xi)} \frac{cs_0^2\xi}{2} \xi^2 \sin\varphi d\varphi d\xi d\theta dz. \tag{16.85}$$

A virtual explanation of how to compute this equation is presented in Fig. 16.6 in the spherical coordinate system centered at A. Along the dashed line with $0 \leq z \leq l_\delta$, the bond connecting point A beneath the fracture surface with point B located above the surface will be broken. We then sum the energy contribution of all its bonds. After the evaluation of these integrals (16.85) and substitution of Eq. (16.76), we find

$$s_0 = \sqrt{\frac{10G_0}{\pi c l_\delta^5}} = \sqrt{\frac{5G_0}{9kl_\delta}} \tag{16.86}$$

for 3D cases.

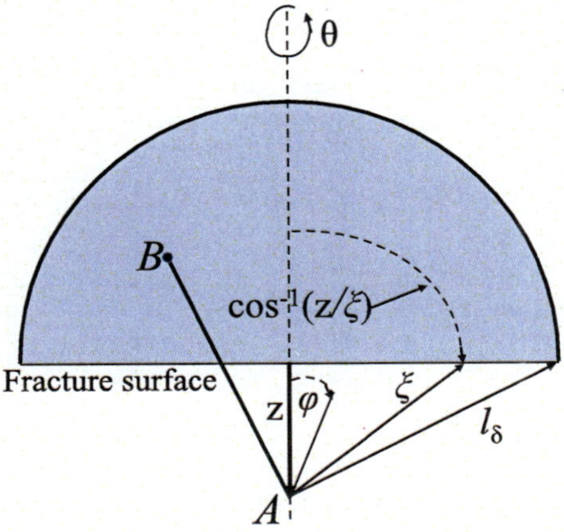

Fig. 16.6 Schematic representation of calculation of Eq. (16.85)

For a 2D case, Ha and Bobaru [594] evaluate the work G_0 required to break all bonds per unit fracture area as

$$G_0 = 2 \int_0^{\delta} \int_z^{\delta} \int_0^{\cos^{-1}(z/\xi)} \frac{c(\xi) s_0^2 \xi}{2} \xi d\varphi d\xi dz. \tag{16.87}$$

Substituting the constant (16.79) and conical (16.81) micromodulus functions and rearranging for s_0, we can rewrite this equation to obtain

$$s_0 = \sqrt{\frac{4\pi G_0}{9El_\delta}} \quad \text{and} \quad s_0 = \sqrt{\frac{5\pi G_0}{9El_\delta}}, \tag{16.88}$$

respectively.

The local damage at a point $\mathbf{x}$ is defined as an average damage function within the horizon

$$\phi(\mathbf{x}) = 1 - \frac{\int_{\mathcal{H}_{\mathbf{x}}} \mu(\mathbf{x}, \boldsymbol{\xi}) d\boldsymbol{\xi}}{\int_{\mathcal{H}_{\mathbf{x}}} d\boldsymbol{\xi}}. \tag{16.89}$$

Therefore, it is the ratio of the sum of broken interactions and all the initial interactions of the points. A point with no broken bonds will have a damage index $\phi = 0$ and a point with all of its bonds broken will have a damage index $\phi = 1$. Thus, the local damage is not sensitive to the detailed distribution of bond breakages described by the damage function $\mu(\mathbf{x}, \boldsymbol{\xi})$ (16.83).

16.6 Thermo-mechanical Peridynamics

Up to now, the case of pure mechanical loading of peridynamic medium was only considered. The basic equations of thermoperidynamics can be obtained from the corresponding peridynamic counterparts (16.43), (16.55), and (16.56) and the volumetric constraint problems (16.46) and (16.47) by generalization of the replacements $\boldsymbol{\eta}$ to the case of availability of microeigenstrains $\widetilde{\boldsymbol{\beta}}^x(\hat{\mathbf{x}}, \mathbf{x})$ as

$$\boldsymbol{\eta}(\hat{\mathbf{x}}, \mathbf{x}) \mapsto \boldsymbol{\eta}^t(\hat{\mathbf{x}}, \mathbf{x}) = \boldsymbol{\eta}(\hat{\mathbf{x}}, \mathbf{x}) - \tilde{\boldsymbol{\beta}}(\hat{\mathbf{x}}, \mathbf{x}). \tag{16.90}$$

So for the linear thermoperidynamic media usually considered in the subsequent presentation, the bond force (16.55) can be generalized (following [80, 771, 772]) as

$$\mathbf{f}^t(\boldsymbol{\eta}, \boldsymbol{\xi}) = \mathbf{C}(\boldsymbol{\xi})\boldsymbol{\eta}^t, \quad \boldsymbol{\eta}^t(\hat{\mathbf{x}}, \mathbf{x}) = \boldsymbol{\eta}(\hat{\mathbf{x}}, \mathbf{x}) - \widetilde{\boldsymbol{\beta}}^x(\hat{\mathbf{x}}, \mathbf{x}) \ \ \forall \boldsymbol{\eta}, \boldsymbol{\xi} \tag{16.91}$$

with the micropotential $\widetilde{w}$ given (up to some additive constant) by

$$\widetilde{w}(\boldsymbol{\eta}, \boldsymbol{\xi}) = \boldsymbol{\eta}^{\top} \cdot \mathbf{C}(\boldsymbol{\xi})(\boldsymbol{\eta}/2 - \widetilde{\boldsymbol{\beta}}^{x}(\boldsymbol{\xi})). \tag{16.92}$$

It is assumed that for the homogeneous materials

$$\widetilde{\boldsymbol{\beta}}^{x}(\hat{\mathbf{x}}, \mathbf{x}) = \widetilde{\boldsymbol{\beta}} \cdot (\hat{\mathbf{x}} - \mathbf{x}), \tag{16.93}$$

that corresponds to the constant temperature in the case of a thermoelastic problem. In particular, under hydrostatic loading $\boldsymbol{\eta} = s\boldsymbol{\xi}$ of isotropic materials with the thermal expansion $\boldsymbol{\beta}^t$ (i.e., $\boldsymbol{\beta}^t = \beta^t\theta\boldsymbol{\delta}$), the elastic energy of local elasticity

$${}^{L}\mathcal{W} = 9k(s^2/2 - \beta^t\theta s), \tag{16.94}$$

where $\theta = T - T_0$ denotes the temperature changes with respect to a reference temperature T_0 and β^tT is a thermal expansion. For the same loading of an isotropic peridynamic material (16.91) and (16.55), (16.64) ($\widetilde{\boldsymbol{\beta}} = \widetilde{\beta}\boldsymbol{\delta}$),

$$\mathbf{f}(\boldsymbol{\eta}, \boldsymbol{\xi}) = \frac{c}{|\boldsymbol{\xi}|}(\eta - \widetilde{\boldsymbol{\beta}}\boldsymbol{\xi}) = c(s - \widetilde{\beta}), \quad \widetilde{w}(\boldsymbol{\eta}, \boldsymbol{\xi}) = c|\boldsymbol{\xi}|\Big(\frac{s^2}{2} - \widetilde{\beta}s\Big). \tag{16.95}$$

From (16.27) and (16.95_2), we can find the peridynamic strain energy density in the three-dimensional case

$$\mathcal{W} = \frac{c(s^2/2 - \widetilde{\beta}s)}{2}\int_0^{l_c}\int_0^{2\pi}\int_0^{\pi} r^3 \sin\psi d\psi d\varphi dr = \frac{c(s^2/2 - \widetilde{\beta}s)l_c^4\pi}{2}. \tag{16.96}$$

Equating Eqs. (16.96) and (16.94) leads to the equality (16.76) and the conclusion that the microeigenstrain $\widetilde{\boldsymbol{\beta}}^{x} = \widetilde{\beta}\boldsymbol{\xi}$ is related with the thermal expansion β^t as

$$\widetilde{\beta} = \frac{18k\beta^t\theta}{\pi l_\delta^4}, \quad \frac{12E\beta^t\theta}{\pi t(1+\nu)l_\delta^3}, \quad \frac{18k\beta^t\theta}{5\pi l_\delta^2} \tag{16.97}$$

for 3D, 2D, and 1D cases, respectively. The representations (16.97_2) and (16.97_3) were obtained through Eqs. (16.77) and (16.78), respectively, analogous to the representation (16.97_1).

The *areal force density* $\boldsymbol{\tau}(\mathbf{x}, \mathbf{n})$ (see [1396]) is obtained by modification of Eq. (16.37)

$$\boldsymbol{\tau}(\mathbf{x}, \mathbf{n}) = \int_{\mathcal{L}}\int_{w^+} \mathbf{f}\Big(\mathbf{u}' - \hat{\mathbf{u}} - \widetilde{\boldsymbol{\beta}}(\mathbf{x}' - \hat{\mathbf{x}}), \mathbf{x}' - \hat{\mathbf{x}}\Big) dv_{\mathbf{x}'} d\hat{l}, \tag{16.98}$$

while the stress field for linear thermoperidynamic follows from Eqs. (16.43) and (16.90)

$$\boldsymbol{\sigma}(\mathbf{x}) = \boldsymbol{\mathcal{L}}^{\sigma t}(\mathbf{u}), \tag{16.99}$$

$$\mathcal{L}^{\sigma t}(\mathbf{u}) := \frac{1}{2}\int_S \int_0^\infty \int_0^\infty (y+z)^{d-1}\hat{\mathbf{f}}^t(\mathbf{x}+y\mathbf{m}, \mathbf{x}-z\mathbf{m}) \otimes \mathbf{m} dz dy d\Omega_{\mathbf{m}}, \tag{16.100}$$

where

$$\hat{\mathbf{f}}^t(\mathbf{p},\mathbf{q}) = \begin{cases} \mathbf{f}\Big(\mathbf{u}(\mathbf{p}) - \mathbf{u}(\mathbf{q}) - \widetilde{\boldsymbol{\beta}}(\mathbf{p}-\mathbf{q}), \mathbf{p}-\mathbf{q}, \mathbf{q}\Big), & \text{if } \mathbf{p}, \mathbf{q} \in w, \\ \mathbf{0}, & \text{otherwise.} \end{cases} \tag{16.101}$$

Due to nonlocality, the thermoperidynamic equilibrium equation

$$\widehat{\mathcal{L}}(\mathbf{u}^t)(\mathbf{x}) + \mathbf{b}(\mathbf{x}) = \mathbf{0}, \tag{16.102}$$

$$\widehat{\mathcal{L}}(\mathbf{u}^t)(\mathbf{x}) = \int_{\mathcal{H}_{\mathbf{x}}} \mathbf{C}(\mathbf{x},\hat{\mathbf{x}})[\boldsymbol{\eta}(\hat{\mathbf{x}},\mathbf{x}) - \widetilde{\boldsymbol{\beta}}^x(\hat{\mathbf{x}},\mathbf{x})]d\hat{\mathbf{x}} \tag{16.103}$$

should be accompanied by a “boundary” condition, imposed as a volumetric constraint in the so-called interaction domain w_Γ (see for details [771, 1396]). In particular, if Dirichlet-type volumetric boundary conditions are presented, the functional $\widehat{\mathcal{L}}(\mathbf{u}-\widetilde{\boldsymbol{\beta}})(\mathbf{x})$ (16.100) should be modified as follows:

$$\widehat{\mathcal{L}}(\mathbf{u})(\mathbf{x}) \longmapsto \widehat{\mathcal{L}}(\mathbf{u}-\widetilde{\boldsymbol{\beta}})(\mathbf{x}) + \widehat{\mathcal{L}}^{\Gamma ut}(\mathbf{u}^*)(\mathbf{x}), \tag{16.104}$$

$$\widehat{\mathcal{L}}^{\Gamma ut}(\mathbf{u}^*)(\mathbf{x}) := \int_{w_{\Gamma u}} \mathbf{f}\Big(\mathbf{u}^* - \mathbf{u} - \widetilde{\boldsymbol{\beta}}(\mathbf{x}^*-\mathbf{x}), \mathbf{x}^*-\mathbf{x}\Big)d\mathbf{x}^*, \tag{16.105}$$

where the shorthand $\mathbf{u}^* = \mathbf{u}^*(\mathbf{x}^*)$ has been used.

For the homogeneous domain w with the linear properties $\mathbf{C}^{(0)}(\mathbf{x},\hat{\mathbf{x}})$, $\widetilde{\boldsymbol{\beta}}^{x(0)}(\mathbf{x},\hat{\mathbf{x}})$ (16.90) subjected to the homogeneous boundary conditions, Eq. (16.100) is reduced to the constitutive equation of local elasticity with the coefficients $\mathbf{L}^{(0)}$, $\boldsymbol{\alpha}^{(0)}$ =const. expressed through $\mathbf{C}^{(0)}(\mathbf{x},\hat{\mathbf{x}})$, $\widetilde{\boldsymbol{\beta}}^{x(0)}(\mathbf{x},\hat{\mathbf{x}})$. It enables one to define the action of the operator $\mathcal{L}^{\sigma t}$ on the linear functions $\mathbf{u}(\mathbf{x}) = \boldsymbol{\varepsilon}\mathbf{x}$ and $\widetilde{\boldsymbol{\beta}}^{(0)}\mathbf{x}$

$$\mathcal{L}^{\sigma t}(\boldsymbol{\varepsilon}\boldsymbol{\xi} - \widetilde{\boldsymbol{\beta}}^{(0)}\boldsymbol{\xi}) = \mathbf{L}^{(0)}\boldsymbol{\varepsilon} + \boldsymbol{\alpha}^{(0)}, \tag{16.106}$$

where $\boldsymbol{\alpha}^{(0)} = -\mathbf{L}^{(0)}\boldsymbol{\beta}^{(0)}$, while $\widetilde{\boldsymbol{\beta}}^{(0)}$ and $\boldsymbol{\beta}^{t(0)}$ for isotropic medium are related by representation (16.97). Equation (16.106) recovers and generalizes the result by Silling [1396] which states that for any microelastic material with microeigenstrains, the corresponding conventional thermoelastic material (in the sense of homogeneous deformations) can be found. However, the converse is not in general true.

Chapter 17
Background of Peridynamic Micromechanics

Abstract In this chapter, some background principles of peridynamic micromechanics are presented. The classical representations of effective elastic moduli through the mechanical influence functions for elastic CMs are generalized to the case of peridynamics, and the energetic definition of effective elastic moduli is proposed. Effective properties of thermoperidynamic CM are expressed through the introduced local stress polarization tensor averaged over the extended inclusion phase. Generalized Hill's condition is proved; the admissible displacement and force fields are defined. The theorem of work and energy, Betti's reciprocal theorem, and the theorem of virtual work are proved. The variational principles and generalized Hill's bounds in micromechanics of linear peridynamic random structure composites are established. The GIEs of peridynamic micromechanics are proposed.

Inspired by Eshelby [449], a number of micromechanical models were proposed in the literature for describing the thermoelastic behavior of composites with ellipsoidal inclusions (see for references, e.g., Chap. 8 and [1523]). Apart from these models exploiting microtopological information of the composite structure (e.g., the shape of heterogeneities, their orientation and binary correlation functions), the general results establishing the links between the effective properties (effective elastic moduli, effective thermal expansion, and effective specific heat) and the corresponding mechanical and transformation influence functions are worthy of notice. The works of this general direction were inspired by Hill [640] and Levin [912], where one obtained the classical representations of effective elastic moduli and thermal expansion through the mechanical influence functions. Development of the mentioned methods (see [883, 1304, 1338]) achieved apparently the most generality in [424], where one obtained some exact results and developed the uniform field theory to their current level. Some basic representations analogous to the mentioned above were generalized in [189] (see Chap. 14) to the nonlocal thermoelasticity of CMs with differential equilibrium equation. In opposite to micromechanics of thermoelastic CM, the general theory of peridynamic CM is not nearly developed. In particular, the theory of peridynamic CM was only considered

V. A. Buryachenko, *Local and Nonlocal Micromechanics of Heterogeneous Materials*, https://doi.org/10.1007/978-3-030-81784-8_17

in the framework of assumption of replacement of the random structure by a deterministic one (e.g., mixture theory and laminate theory (see [38, 665, 771]) as well as periodic structures (see [18])). Moreover, historically, micromechanicians usually work with the differential equation of motion rather than with the integral one (as in peridynamics). Because of this, even the background concepts of micromechanics (see for details, e.g., Chap. 6) such as the effective moduli and effective eigenfields were not yet defined to the recent time in the theory of random structure peridynamic CMs. This chapter (see also [209] and [216]) is dedicated to closing this gap between micromechanical approaches and peridynamic models and to generalization of the mentioned general results for thermoelastic CMs to its peridynamic counterparts.

The variational methods represent the most rigorous trend and security aspect of micromechanics of statistically homogeneous media. Variational bounds of the effective moduli are generated via substitution of approximate fields into the strict energy bounds obtained by the volume average of the energy density with the help of lower-order correlation functions. The chapter will be concerned with the determination of variational bounds on the effective moduli of CMs, in particular Hill [640] and Hashin-Shtrikman [621] bounds obtained and developed to the modern level of generality in the classical papers by Willis (see [1625, 1627, 1628]). Approximate character of the variational bounds and their restricted opportunities (for the formula manipulations) of the variational methods in comparison with the different self-consistent methods is explained by a fundamental reason. Indeed, the self-consistent methods in linear statics of mechanics of CMs were exploited in a wide class of micromechanical problems for CMs with nonellipsoidal, coated, and continuously inhomogeneous heterogeneities with possible nonideal interface as well as for different nonlocal problems (inhomogeneous remote loading, functionally graded materials, clustered materials, bounded media, nanocomposites, nonlocal constitutive law, e.g., peridynamics) (see for references, e.g., [212]). Analyses of so much diverse (in the sense of microtopology and constitutive laws) problems are essentially simplified due to the linear dependence of the effective properties on average fields (strains and stresses) inside the inclusions. In so doing, Hashin-Shtrikman (HS) bounds are based on the estimation of energy-like quadratic function averaged over the volumes of the phases that hopelessly complicate the problem due to necessity to take into account the randomness of coupled multiparticle interactions. Generalized Hill's [640] bounds of the effective moduli of linear peridynamic random structure composites considered in this chapter (see also [227]) do not require analysis of multiparticle interactions (which are taken into account at the estimation of effective moduli; see, e.g., [224, 225]).

At last, the general integral equations (GIEs) connecting the displacement fields in the point (in the referenced inclusion) being considered and the surrounding points (of other inclusions) are proposed. These new GIEs are obtained in this chapter (see also [209, 227]) by a centering procedure of subtraction from both sides of a new initial integral equation their statistical averages obtained without any auxiliary assumptions which are implicitly exploited in the known centering methods.

17.1 Decomposition of Material Properties and Some Averages

For two-phase composite medium containing any two points $\mathbf{x}$ and $\hat{\mathbf{x}}$ in R^d, the micromodulus $\mathbf{C}(\boldsymbol{\xi}) = \mathbf{C}(\mathbf{x}, \hat{\mathbf{x}})$ (16.59) ($\boldsymbol{\xi} = \hat{\mathbf{x}} - \mathbf{x}$) is given by the formula

$$\mathbf{C}(\mathbf{x}, \hat{\mathbf{x}}) = \begin{cases} \mathbf{C}^{(1)}(\mathbf{x}, \hat{\mathbf{x}}), & \text{for } \mathbf{x}, \hat{\mathbf{x}} \in v, \\ \mathbf{C}^{(0)}(\mathbf{x}, \hat{\mathbf{x}}), & \text{for } \mathbf{x}, \hat{\mathbf{x}} \in v^{(0)}, \\ \mathbf{C}^{i}(\mathbf{x}, \hat{\mathbf{x}}), & \text{for } \mathbf{x} \in v, \hat{\mathbf{x}} \in v^{(0)} \text{or } \ \mathbf{x} \in v^{(0)}, \hat{\mathbf{x}} \in v, \end{cases} \tag{17.1}$$

which can also be presented in the form

$$\mathbf{C}(\mathbf{x}, \hat{\mathbf{x}}) = \mathbf{C}^{(1)}(\mathbf{x}, \hat{\mathbf{x}})V^{(1)}(\mathbf{x})V^{(1)}(\hat{\mathbf{x}}) + \mathbf{C}^{(0)}(\mathbf{x}, \hat{\mathbf{x}})V^{(0)}(\mathbf{x})V^{(0)}(\hat{\mathbf{x}})$$
$$+\mathbf{C}^{i}(\mathbf{x}, \hat{\mathbf{x}})[V^{(1)}(\mathbf{x})V^{(0)}(\hat{\mathbf{x}}) + V^{(0)}(\mathbf{x})V^{(1)}(\hat{\mathbf{x}})]. \tag{17.2}$$

The material parameters $\mathbf{C}^{(1)}$ and $\mathbf{C}^{(0)}$ are intrinsic to each phase and can be determined through experiments. Bonds connecting points in the different materials are characterized by micromodulus $\mathbf{C}^{i}$, which can be chosen such that, e.g.,

$$\mathbf{C}^{i}(\mathbf{x}, \hat{\mathbf{x}}) = (\mathbf{C}^{(0)}(\mathbf{x}, \hat{\mathbf{x}}) + \mathbf{C}^{(1)}(\mathbf{x}, \hat{\mathbf{x}}))/2 \tag{17.3}$$

(see [17, 18, 1401]), where $\mathbf{x}, \hat{\mathbf{x}} \in v_{\Gamma}$ is called the interaction interface (see for details [1354]). Although there are no technical issues with correction of discretized peridynamic equation in the extended interface $\mathbf{x} \in v_{\Gamma}$, it is helpful to assume for simplicity only that the peridynamic operator in the extended interface (fuzzy interface, i.e., not sharp as it would be in a local formulation) is described by Eq. (16.5) with the constant horizon and micromodulus (17.1) determined as an average value of the micromoduli in the matrix and inclusion (17.3) (see also [998], where one proposed a variation of the extended interface micromodulus $\mathbf{C}^{i}(\mathbf{x}, \hat{\mathbf{x}})$ in a spirit of the functionally graded materials theory described in, e.g., [200]). Adaptive grid refinement technique was proposed by [130] for the analysis of peridynamic problems in the vicinity of interface that involves a variable horizon size.

In a similar manner, for two-phase composite medium containing any two points $\mathbf{x}$ and $\hat{\mathbf{x}}$ in R^d, the microeigenstrain $\widetilde{\boldsymbol{\beta}}^{x}(\mathbf{x}, \hat{\mathbf{x}})$ (16.94) is presented in the form

$$\widetilde{\boldsymbol{\beta}}^{x}(\hat{\mathbf{x}}, \mathbf{x}) = [\widetilde{\boldsymbol{\beta}}^{(1)}\hat{\mathbf{x}} - \widetilde{\boldsymbol{\beta}}^{(0)}\mathbf{x}]V^{(1)}(\hat{\mathbf{x}})V^{(0)}(\mathbf{x}) + [\widetilde{\boldsymbol{\beta}}^{(0)}\hat{\mathbf{x}} - \widetilde{\boldsymbol{\beta}}^{(1)}\mathbf{x}]V^{(0)}(\hat{\mathbf{x}})V^{(1)}(\mathbf{x}).$$
$$+\Big\{\widetilde{\boldsymbol{\beta}}^{(1)}V^{(1)}(\mathbf{x})V^{(1)}(\hat{\mathbf{x}}) + \widetilde{\boldsymbol{\beta}}^{(0)}V^{(0)}(\mathbf{x})V^{(0)}(\hat{\mathbf{x}})\Big\}\boldsymbol{\xi}. \tag{17.4}$$

One of the basic initial manipulations in locally elastic micromechanics is a decomposition of material property tensors $\mathbf{g}(\mathbf{x})$ ($\mathbf{g} = \mathbf{L}, \mathbf{M}, \boldsymbol{\alpha}, \boldsymbol{\beta}$) as

$$\mathbf{g}(\mathbf{x}) = \mathbf{g}^{(0)} + \mathbf{g}_1(\mathbf{x}), \tag{17.5}$$

where $\mathbf{g}_1(\mathbf{x}) \equiv \mathbf{0}$ in the homogeneous matrix $v^{(0)}$ with the properties $\mathbf{g}^{(0)} \equiv$const. The peridynamic counterpart of the decomposition (17.5) is

$$\mathbf{C}(\mathbf{x}, \hat{\mathbf{x}}) = \mathbf{C}^{(0)}(\mathbf{x}, \hat{\mathbf{x}}) + \mathbf{C}_1(\mathbf{x}, \hat{\mathbf{x}}),$$
$$\widetilde{\boldsymbol{\beta}}^x(\hat{\mathbf{x}}, \mathbf{x}) = \widetilde{\boldsymbol{\beta}}^{x(0)}(\hat{\mathbf{x}}, \mathbf{x}) + \widetilde{\boldsymbol{\beta}}_1^x(\hat{\mathbf{x}}, \mathbf{x}), \tag{17.6}$$

where the properties $\mathbf{C}^{(0)}(\mathbf{x}, \hat{\mathbf{x}})$ and $\widetilde{\boldsymbol{\beta}}^{x(0)}(\hat{\mathbf{x}}, \mathbf{x})$ of a homogeneous matrix $v^{(0)}$ are invariant with respect to translation ($\mathbf{x} \to \mathbf{x} + \mathbf{x}'$, $\hat{\mathbf{x}} \to \hat{\mathbf{x}} + \mathbf{x}'$). In so doing, the jump of the properties $\mathbf{C}_1(\mathbf{x}, \hat{\mathbf{x}})$ and $\widetilde{\boldsymbol{\beta}}_1^x(\hat{\mathbf{x}}, \mathbf{x})$ vanishes inside the truncated matrix $v^{l(0)} := w \subset v^{l(1)}$ (see Eq. (17.1)) rather than inside the matrix $v^{(0)}$ (as in locally elastic micromechanics (17.5)). Hereafter, $v^{l(1)} = v^{(1)} \oplus \mathcal{H}_0$ is an extended inclusion phase obtained by Minkowski outer parallel (dilation) of the inclusion phase $v^{(1)}$.

It is assumed that the representative macrodomain w contains a statistically large number of realizations α (providing validity of the standard probability technique) of heterogeneities $v_i \in v^{(k)}$ of the constituent $v^{(k)}$ ($i = 1, 2, \ldots; \ k = 1, 2, \ldots, N$). A random event α belongs to a sample space $\mathcal{A}$, over which a probability density $p(\mathbf{x}, \alpha)$ is defined (see, e.g., [1625]). For any given α, any random function $\mathbf{g}(\mathbf{x}, \alpha)$ (e.g., $\mathbf{g} = V, V^{(k)}, \boldsymbol{\sigma}, \boldsymbol{\varepsilon}$) is defined explicitly as one particular member, with label α, of an ensemble realization. Then, the mean or ensemble average is defined by the angle brackets enclosing the quantity $\mathbf{g}$

$$\langle \mathbf{g} \rangle(\mathbf{x}) = \int_{\mathcal{A}} \mathbf{g}(\mathbf{x}, \alpha) p(\mathbf{x}, \alpha) d\alpha. \tag{17.7}$$

No confusion will arise below in notation of the random quantity $\mathbf{g}(\mathbf{x}, \alpha)$ if the label α is removed. One treats two material length scales (see, e.g., [1523]): the macroscopic scale L, characterizing the extent of w, and the microscopic scale a, related with the heterogeneities v_i. Moreover, one supposes that applied field varies on a characteristic length scale Λ. We consider the scale separation of both the material scales and field ones either

$$L \gg \Lambda \gg a \geq l_\delta \quad \text{or} \quad L \gg \Lambda \geq a \geq l_\delta \tag{17.8}$$

for statistically homogeneous or inhomogeneous fields, respectively.

The material and field decompositions usually used in locally elastic micromechanics correspond to the volume partition of a full space $w = R^d$ over the volumes of the matrix and inclusions

$$1 = c^{(0)} + c^{(1)}, \quad w = v^{(0)} + v^{(1)}, \quad W(\mathbf{x}) = V^{(0)}(\mathbf{x}) + V^{(1)}(\mathbf{x}).$$

Then for statistically homogeneous random function $\mathbf{f}(\mathbf{x})$ (e.g., $\mathbf{f} = \boldsymbol{\sigma}, \boldsymbol{\varepsilon}$) satisfying the ergodicity condition assumed, the spatial average estimated over one sufficiently large sample and statistical mean coincide for both the whole volume w and the individual constituent $v^{(k)}$ $(k = 0, 1)$:

$$\langle \mathbf{f} \rangle = \{\mathbf{f}\} \equiv \lim_{w \uparrow R^d} |w|^{-1} \int_w \mathbf{f}(\mathbf{x})(\mathbf{x}) d\mathbf{x},$$

$$\langle \mathbf{f} \rangle^{(k)} = \{\mathbf{f}\}^{(k)} \equiv \lim_{w \uparrow R^d} |w|^{-1} \int_w \mathbf{f}(\mathbf{x})(\mathbf{x}) V^{(k)} d\mathbf{x}, \tag{17.9}$$

where $|w| = \text{mes}\, w$. Hereafter, $n^{(k)}$ is a number density of the component $v^{(1)} \ni v_i$ and $c^{(1)}$ is the concentration, i.e., volume fraction, of the component $v_i \in v^{(1)}$: $c^{(1)} = \langle V^{(1)} \rangle = \overline{v}_i n^{(1)}$, $\overline{v}_i = \text{mes} v_i$ $(i = 1, 2, \ldots)$, $c^{(0)} = 1 - \langle V \rangle = 1 - \langle V^{(1)} \rangle$.

However, in peridynamics, the material and field decompositions correspond to another partition of the entire space

$$1 = c^{l(0)} + c^{l(1)}, \quad w = v^{l(0)} + v^{l(1)}, \quad W(\mathbf{x}) = V^{l(0)}(\mathbf{x}) + V^{l(1)}(\mathbf{x}), \tag{17.10}$$

over the extended inclusions $v^{l(1)} := v^{(1)} \oplus \mathcal{H}_0$ and the truncated matrix $v^{l(0)} := w \backslash v^{l(1)}$ with the indicator functions $V^{l(1)}(\mathbf{x})$ and $V^{l(0)}(\mathbf{x})$, respectively. Hereafter, we use the volume averages (e.g., $\mathbf{f} = \boldsymbol{\sigma}, \boldsymbol{\varepsilon}$)

$$\langle \mathbf{f} \rangle^{l(k)} = \{\mathbf{f}\}^{l(k)} \equiv \lim_{w \uparrow R^d} |w|^{-1} \int_w \mathbf{f}(\mathbf{x})(\mathbf{x}) V^{l(k)} d\mathbf{x},$$

$$\langle \mathbf{f} \rangle = \{\mathbf{f}\} = c^{l(0)} \{\mathbf{f}\}^{l(0)} + c^{l(1)} \{\mathbf{f}\}^{l(1)}, \tag{17.11}$$

over the extended inclusions $v^{l(1)}$ $(k = 1)$ and the truncated matrix $v^{l(0)}$ $(k = 0)$, respectively. It should be mentioned that the inclusions v_i and v_j cannot overlap while overlapping of the corresponding extended inclusions v_i^l and v_j^l is possible. However, for simplicity purposes, we will also assume nonoverlapping of the extended inclusions v_i^l and v_j^l.

The volume averages of the strains and stresses inside the extended inclusion v_i^l can be expressed through the averages over the external inclusion boundary Γ_i^- by the use of Gauss's theorem

$$\boldsymbol{\varepsilon}^{l(1)} = \boldsymbol{\varepsilon}^{l\omega(1)} := \frac{1}{\bar{w}} \sum_i \int_{\Gamma_i^-} \mathbf{u}(\mathbf{s}) \overset{s}{\otimes} \mathbf{n}(\mathbf{s})\, ds, \tag{17.12}$$

$$\boldsymbol{\sigma}^{l(1)} = \boldsymbol{\sigma}^{l\omega(1)} := \frac{1}{\bar{w}} \sum_i \int_{\Gamma_i^-} \mathbf{t}(\mathbf{s}) \overset{s}{\otimes} \mathbf{s}\, ds, \tag{17.13}$$

where in Eq. (17.13) one used the equality $\nabla(\mathbf{s} \otimes \boldsymbol{\sigma}) = \boldsymbol{\sigma}$ and $\mathbf{a} \overset{s}{\otimes} \mathbf{b}$ represents the symmetric part of the second-order tensor, e.g., $\mathbf{a} \overset{s}{\otimes} \mathbf{b} = (\mathbf{a} \otimes \mathbf{b} + \mathbf{b} \otimes \mathbf{a})/2$. The peridynamic theory is traditionally based on the use of the displacement field $\mathbf{u}(\mathbf{x})$ rather than either the stress $\boldsymbol{\sigma}(\mathbf{x})$ or strain $\boldsymbol{\varepsilon}(\mathbf{x})$ (obtained by numerical differentiation of the displacement field $\mathbf{u}(\mathbf{x})$) fields which are not conceptually necessary. The absence of numerical differentiation defines an advantage of the surface average (17.12) with respect to the volume average (17.11). The surface average of stresses (17.13) has the well-known advantage of the reduction of dimension (see Eq. (17.11)) by one that can be crucial for the analyst.

17.2 General Representations for Effective Moduli

17.2.1 Admissible Fields and Generalized Hill's Equality

It is assumed that $\overline{\overline{w}}$ is obtained by cutting out of a domain from statistically homogeneous heterogeneous medium filling the entire space R^d. $\overline{\overline{w}} = w \cup w_\Gamma$ stands for the *nonlocal closure* of w (see Sect. 16.2), whereas the nonlocal boundary w_Γ is a d-dimensional region unlike its $(d-1)$-dimensional counterpart Γ^0 in local problems. The volumetric homogeneous boundary conditions either (16.46) or (16.47) are prescribed at the interaction macrodomain W_Γ. There are no specialized restrictions on the shape and smoothness of Γ^0 (and w_Γ) which is defined just by convenience of representation. For example, it is convenient for visibility, although not an essential feature of the theory, to assume that Γ^0 has a circle (or oval) shape ($d = 2$). The actual displacement, stress, and force fields produced in the domain w by the applied forces and boundary conditions are not assumed to be known now. Instead, for some general descriptions of volumetric boundary value problems (16.46) and (16.47), we introduce the notion of admissible fields corresponding to the homogeneous volumetric boundary conditions (16.52) and (16.53), respectively, which are defined by complying the following admissible conditions (locally elastic counterparts were considered, e.g., in [353, 1263]). A sufficiently smooth (of class C^0) displacement field $\mathbf{u}$ with the displacement loading conditions (16.46$_2$) ($\mathbf{x} \in \overline{\overline{w}},\ \mathbf{y} \in w_{\Gamma u}$)

$$\mathcal{A}^k(\mathbf{u}) = \{\mathbf{u}|\mathbf{u}(\mathbf{x}) \in C^0(\overline{\overline{w}}),\ \mathbf{u}(\mathbf{y}) = \mathbf{h}(\mathbf{y}),\ \mathbf{y} \in w_{\Gamma u}\} \tag{17.14}$$

is called a *kinematically admissible displacement field*. $\mathbf{u} \in \mathcal{A}^k(\mathbf{u})$ (17.14) generates a symmetric compatible field $\boldsymbol{\varepsilon}$, while a stress field may not satisfy the equilibrium equation (16.45).

A sufficiently smooth (of class C^1) force field satisfying the equilibrium equation (16.5) and the force loading condition (16.47) ($\mathbf{x} \in \overline{\overline{w}}$)

$$\mathcal{A}^s(\mathbf{f}) = \{\mathbf{f}|\mathbf{f}(\mathbf{x}) \in C^1(\overline{\overline{w}}),\ \widehat{\mathcal{L}}(\mathbf{f}) = -\breve{\mathbf{b}}\} \tag{17.15}$$

is referred to as a *statically admissible force field*. In so doing, a tensor $\boldsymbol{\varepsilon}$ is not necessarily compatible.

The goal of this chapter is to obtain some asymptotic representations for the volume averages {.} at the condition $L/a \to \infty$. We define the *mean value* (or volume average) of an integrable tensor field $\boldsymbol{\zeta}$ (e.g., $\boldsymbol{\zeta} = \boldsymbol{\varepsilon}, \boldsymbol{\sigma}$) given on the macrodomain $\overline{\overline{w}} \subset R^d$ $(d = 2, 3)$ as

$$\{\boldsymbol{\zeta}\} = |\overline{\overline{w}}|^{-1} \int_{\overline{\overline{w}}} \boldsymbol{\zeta}(\mathbf{x})\, d\mathbf{x}, \quad |\overline{\overline{w}}| := \text{mes}\, \overline{\overline{w}}, \tag{17.16}$$

i.e., the domain w in Eq. (17.9) is replaced by $\overline{\overline{w}}$. Then, according to the Gauss theorem and Cauchy relations, we can express the volume averages of stress and strain by integrals over the respective boundary values $(\mathbf{b}(\mathbf{y}) \equiv \mathbf{0}, \mathbf{y} \in \overline{\overline{w}})$

$$\{\sigma_{ij}\} = |\overline{\overline{w}}|^{-1} \int_{\overline{\overline{w}}} \sigma_{ik} x_{k,j}\, d\mathbf{x} = |\overline{\overline{w}}|^{-1} \int_{\Gamma_{lc}} t_{(i}(\mathbf{s}) s_{j)}\, d\mathbf{s}, \tag{17.17}$$

$$\{\varepsilon_{ij}\} = |\overline{\overline{w}}|^{-1} \int_{\Gamma_{lc}} u_{(i}(\mathbf{s}) n_{j)}(\mathbf{s})\, d\mathbf{s}. \tag{17.18}$$

In the particular form of the homogeneous boundary conditions (16.52), Eq. (17.18) with $\nabla\mathbf{x} = \mathbf{x}\nabla = \boldsymbol{\delta}$ leads to the conclusion that for any kinematically admissible strain field corresponding to the homogeneous displacement loading conditions (16.52) $\boldsymbol{\varepsilon}^{w\Gamma} \equiv$ const., the mean value $\{\boldsymbol{\varepsilon}\}$ of $\boldsymbol{\varepsilon}$ is just $\boldsymbol{\varepsilon}^{w\Gamma}$:

$$\{\varepsilon_{ij}\} = |\overline{\overline{w}}|^{-1} \int_{\Gamma_{lc}} \varepsilon^{w\Gamma}_{(ik} s_k n_{j)}(\mathbf{s})\, d\mathbf{s} = \varepsilon^{w\Gamma}_{ij} = \text{const.} \tag{17.19}$$

because in Eq. (17.18) $\mathbf{u}(\mathbf{x}) = \boldsymbol{\varepsilon}^{w\Gamma}\mathbf{x}$ $(\mathbf{x} \in \Gamma_{l_c} \subset w_\Gamma)$. In a similar manner, the homogeneous force loading condition (16.53) incorporated into Eq. (17.17) leads to the equality for the statically admissible stress field corresponding to $\boldsymbol{\sigma}^{w\Gamma}(\mathbf{x})$:

$$\{\sigma_{ij}\} = |\overline{\overline{w}}|^{-1} \int_{\Gamma_{lc}} \sigma^{w\Gamma}_{(ik} n_k(\mathbf{s}) s_{j)}\, d\mathbf{s} = \sigma^{w\Gamma}_{ij} = \text{const.} \tag{17.20}$$

Thus, the mean values of the strain $\{\boldsymbol{\varepsilon}\}$ and stress $\{\boldsymbol{\sigma}\}$ coincide with $\boldsymbol{\varepsilon}^{w\Gamma}$ and $\boldsymbol{\sigma}^{w\Gamma}$ defined by the homogeneous volumetric boundary conditions (16.52) and (16.53), respectively.

Now we will consider the generalization of Hill's [640] condition which enables one to split a volume average of an energy-like representation into a product of volume averages of stress and strain fields:

$$\{\boldsymbol{\sigma} : \boldsymbol{\varepsilon}\} = \{\boldsymbol{\sigma}\} : \{\boldsymbol{\varepsilon}\} \tag{17.21}$$

for any admissible displacement $\mathbf{u}$ and equilibrium stress $\boldsymbol{\sigma}$ compatible with the homogeneous volumetric boundary conditions (16.52) and (16.53), respectively.

Indeed, the next obvious chain of equalities for the displacement loading conditions (16.52) ($w_\Gamma = w_{\Gamma u}$)

$$\{\boldsymbol{\sigma} : \boldsymbol{\varepsilon}\} = |\overline{\overline{w}}|^{-1} \int_{\Gamma_{lc}} t_i u_i \, d\mathbf{s} = |\overline{\overline{w}}|^{-1} \{\varepsilon_{ik}\} \int_{\Gamma_{lc}} t_i x_k \, d\mathbf{s} = \{\boldsymbol{\sigma}\} : \{\boldsymbol{\varepsilon}\} \quad (17.22)$$

proves Hill's condition (17.21); here, the second equality is defined by Eq. (16.52). In a similar manner, Hill's condition (17.21) can be obtained for the homogeneous force loading condition (16.53). It should be emphasized that $\boldsymbol{\sigma}$ and $\boldsymbol{\varepsilon}$ need not be connected by a specific stress-strain relation (e.g., either (16.5) or (16.6)).

For statistically homogeneous random function $\mathbf{f}(\mathbf{x})$ satisfying the ergodicity condition (17.11) assumed, Hill's condition can be presented in two equivalent forms involving either the volume averages (17.21) or statistical averages

$$\langle \boldsymbol{\sigma} : \boldsymbol{\varepsilon} \rangle = \langle \boldsymbol{\sigma} \rangle : \langle \boldsymbol{\varepsilon} \rangle. \quad (17.23)$$

In so doing, the displacement field $\mathbf{u}(\mathbf{x})$ compatible with the homogeneous volumetric loading either (16.52) or (16.52)

$$\langle \mathbf{u} \rangle (\mathbf{x}) = \boldsymbol{\varepsilon}^{w_\Gamma} \mathbf{x} \neq \{\mathbf{u}\}. \quad (17.24)$$

Hill's condition holds for any compatible strain field $\boldsymbol{\varepsilon}(\mathbf{x})$ and equilibrium stress field $\boldsymbol{\sigma}(\mathbf{x})$ not necessarily related to each other by a specific stress-strain relation. The result (17.23) does not demand statistical independence of stress and strain fields because it is only a scalar condition.

The basic subject of the theory of statistically homogeneous linear peridynamic composites is the characterization of the *overall* or *equivalent* elastic behavior of composites. It means that one is asking not for the detailed stress-strain state in each point, but only for the (volume or statistical) averages of the stress state. This involves finding the *overall* or *effective* moduli as the moduli of some fictitious homogeneous medium with the same average stress-strain state as for the microinhomogeneous one being considered. It means that there exists a *representative volume element* Ω (RVE) of the body under consideration that (see [640]) "(a) is structurally entirely typical of the whole mixture on average, and (b) contains a sufficient number of inclusions for the apparent overall moduli to be effectively independent of the values" of force loading conditions (16.53) and displacement ones (16.52), so long as these values are "macroscopically uniform" (this description is a slight modification of the corresponding definition for locally elastic composites by Hill [640] accentuated by the quotes). These apparent or *effective* moduli fluctuate about a mean value with a wavelength smaller than the dimension l of the sample, which is, however, much larger than the characteristic size of the body inhomogeneity a and much smaller than the characteristic dimension L of the body itself: $a \ll l \ll L$. The effect of such fluctuations becomes insignificant within a few wavelengths from the interaction domain w_Γ. The contribution of this boundary layer effect to any volume averages becomes negligible if the RVE becomes large enough.

17.2.2 *Effective Elastic Moduli*

Thus, a notion of RVE is critical to the concrete transition from micro- to macrolevels. This transition involves two main problems: (a) to define field and material macrovariables of an effective constitutive equation and (b) to determine how the macrovariables are related to the microvariables of the local constitutive equations as well as to the microtopology of a composite material. The term RVE was originally proposed in the analysis of statistically homogeneous linear elastic media when the answer to the problem (a) is obvious. Indeed, independently on both local and nonlocal nature of the constitutive laws in the micro (or nano) points, the field macrovariables $\langle \boldsymbol{\sigma} \rangle$ and $\langle \boldsymbol{\varepsilon} \rangle$ depend only on the homogeneous nonlocal boundary conditions, either (16.52) or (16.53), and are related by a local material effective tensor of either the elasticity or compliance:

$$\{\boldsymbol{\sigma}\} = \mathbf{L}^*\{\boldsymbol{\varepsilon}\}, \quad \{\boldsymbol{\varepsilon}\} = \mathbf{M}^*\{\boldsymbol{\sigma}\}, \tag{17.25}$$

respectively. Establishment of the constitutive linkage between the suitably chosen micro- (or nano-) and macrofield is reduced to the estimation of the tensor *influence functions* $\mathbf{A}^*_\varepsilon(\mathbf{x})$, $\mathbf{B}^*_\varepsilon(\mathbf{x})$ and $\mathbf{A}^*_\sigma(\mathbf{x})$, $\mathbf{B}^*_\sigma(\mathbf{x})$ for the nonlocal homogeneous boundary conditions (16.52) or (16.52):

$$\mathbf{u}(\mathbf{x}) = \mathbf{A}^*_u(\mathbf{x})\{\boldsymbol{\varepsilon}\}, \quad \boldsymbol{\varepsilon}(\mathbf{x}) = \mathbf{A}^*_\epsilon(\mathbf{x})\{\boldsymbol{\varepsilon}\}, \quad \boldsymbol{\sigma}(\mathbf{x}) = \mathbf{B}^*_\epsilon(\mathbf{x})\{\boldsymbol{\varepsilon}\}, \tag{17.26}$$

$$\mathbf{u}(\mathbf{x}) = \mathbf{B}^*_u(\mathbf{x})\{\boldsymbol{\sigma}\}, \quad \boldsymbol{\varepsilon}(\mathbf{x}) = \mathbf{A}^*_\sigma(\mathbf{x})\{\boldsymbol{\sigma}\}, \quad \boldsymbol{\sigma}(\mathbf{x}) = \mathbf{B}^*_\sigma(\mathbf{x})\{\boldsymbol{\sigma}\}, \tag{17.27}$$

respectively. Averaging Eqs. (17.26_2) and (17.27_3) leads to the conclusion that

$$\{\mathbf{A}^*_\epsilon(\mathbf{x})\} = \{\mathbf{B}^*_\sigma(\mathbf{x})\} = \mathbf{I}. \tag{17.28}$$

The local volume averages of the influence functions over $v^{(r)}$ may be referred to as the *mechanical concentration factors* (e.g., $\mathbf{A}^{*(r)}_\varepsilon$, $r = 0, 1, \ldots, N$). Due to the linearity of the problem, these influencing functions exist (but are generally unknown) and related by the obvious equations

$$\mathbf{B}^*_\varepsilon(\mathbf{x}) = \boldsymbol{\mathcal{L}}^\sigma * \mathbf{A}^*_u(\mathbf{x}), \quad \mathbf{A}^*_\varepsilon(\mathbf{x}) = \nabla \overset{S}{\otimes} \mathbf{A}^*_u(\mathbf{x}), \tag{17.29}$$

$$\mathbf{A}^*_\varepsilon(\mathbf{x}) = \mathbf{A}^*_\sigma(\mathbf{x})\mathbf{L}^*, \quad \mathbf{B}^*_\varepsilon(\mathbf{x}) = \mathbf{B}^*_\sigma(\mathbf{x})\mathbf{L}^*, \tag{17.30}$$

$$\mathbf{B}^*_\sigma(\mathbf{x}) = \boldsymbol{\mathcal{L}}^\sigma * \mathbf{B}^*_u(\mathbf{x}), \quad \mathbf{A}^*_\sigma(\mathbf{x}) = \nabla \overset{S}{\otimes} \mathbf{B}^*_u(\mathbf{x}), \tag{17.31}$$

$$\mathbf{A}^*_\sigma(\mathbf{x}) = \mathbf{A}^*_\varepsilon(\mathbf{x})\mathbf{M}^*, \quad \mathbf{B}^*_\sigma(\mathbf{x}) = \mathbf{B}^*_\varepsilon(\mathbf{x})\mathbf{M}^*, \tag{17.32}$$

where $\overset{s}{\otimes}$ is the operation of the symmetrized gradient. In deriving the equalities (17.30$_1$), (17.30$_2$) and (17.32$_1$), (17.32$_2$), we assumed that the so-called Hill's weak assumption (see for details [353] and Chap. 6) is fulfilled hereafter. If Hill's weak assumption is satisfied, and if the average strains $\{\boldsymbol{\varepsilon}\}$ coincide for both homogeneous boundary conditions (16.52) and (16.53), according to the hypothesis made, then the average stresses $\{\boldsymbol{\sigma}\}$ also coincide for both nonlocal homogeneous boundary conditions (16.52) and (16.53). In such a case, the equivalence of the dual effective material laws $\mathbf{L}^* = (\mathbf{M}^*)^{-1}$ takes place.

The influence functions (17.26) and (17.27) play a fundamental role in obtaining the macroscopic elastic response. Indeed, substitutions of the averaged multiscale linkages $\langle\boldsymbol{\sigma}\rangle = \langle\mathbf{B}^*_\varepsilon\rangle\langle\boldsymbol{\varepsilon}\rangle$ (17.26$_3$) and $\langle\boldsymbol{\varepsilon}\rangle = \langle\mathbf{A}^*_\sigma\rangle\langle\boldsymbol{\sigma}\rangle$ (17.27$_2$) into the averaged constitutive equations (17.25$_1$) and (17.25$_2$) yield estimations of the effective stiffness and compliance:

$$\mathbf{L}^* = \{\mathbf{B}^*_\varepsilon(\mathbf{x})\}, \quad \mathbf{M}^* = \{\mathbf{A}^*_\sigma(\mathbf{x})\}, \tag{17.33}$$

respectively, while substitution of the representation (17.29$_1$) into Eq. (17.33$_1$) leads to another form of the effective moduli

$$\mathbf{L}^* = \{\boldsymbol{\mathcal{L}}^\sigma(\mathbf{A}^*_u)\}. \tag{17.34}$$

It should be mentioned that a similar formula for $\mathbf{M}^*$ (17.33$_2$) depends on $\mathbf{B}^*_u(\mathbf{x})$ for the force loading conditions (16.53) defined by an inverse operator $(\boldsymbol{\mathcal{L}}^\sigma)^{-1}$; this operator is in general unknown. Because of this, the force loading conditions (16.53) are considered in less detail than the displacement loading conditions (16.52). The representations (17.33) are reduced to the classical relations (see [640])

$$\mathbf{L}^* = \{\mathbf{L}(\mathbf{x})\mathbf{A}^*_\varepsilon(\mathbf{x})\}, \quad \mathbf{M}^* = \{\mathbf{M}(\mathbf{x})\mathbf{B}^*_\sigma(\mathbf{x})\},$$

respectively, only for the local constitutive laws at the fine scale (6.62).

We consider now the properties of the stiffness (17.33$_1$) and compliance (17.33$_2$) in more detail. From the symmetry of the tensors $\{\boldsymbol{\varepsilon}\}$ and $\{\boldsymbol{\sigma}\}$, it follows that all influence tensors are symmetric inside the index pairs ($\mathbf{C} = \mathbf{A}^*_\varepsilon, \mathbf{B}^*_\varepsilon, \mathbf{A}^*_\sigma, \mathbf{B}^*_\sigma$): $C_{ijkl} = C_{jikl} = C_{ijlk}$. From (17.23) and from the symmetry properties of the values at the fine scale $\boldsymbol{\mathcal{L}}^c(\mathbf{x})$ and $\mathbf{A}^*_u(\mathbf{x})$, the symmetry of macroproperties $\mathbf{L}^*$ inside the index pairs (minor symmetry) follows. The symmetry of $\mathbf{L}^*$ between the index pairs (major symmetry, e.g., $L^*_{ijkl} = L^*_{klij}$) will be considered in the next subsection from energetic analysis.

17.2.3 Energetic Definition of Effective Elastic Moduli

At first, we introduce an inner product for two arbitrary integrable random vector functions $\mathbf{f}(\mathbf{x})$ and $\mathbf{g}(\mathbf{x})$ ($\mathbf{x} \in \overline{\overline{w}}$) as $(\mathbf{f}, \mathbf{g}) \equiv \{\mathbf{f} : \mathbf{g}\} = \langle \mathbf{f} : \mathbf{g} \rangle = \langle f_{ij} g_{ji} \rangle$ due to both the ergodicity and statistical homogeneity of the fields $\mathbf{f}(\mathbf{x})$ and $\mathbf{g}(\mathbf{x})$. We consider the operator $\hat{\mathcal{L}}$ of the equilibrium equation ($\boldsymbol{\xi} = \mathbf{y} - \mathbf{x}$)

$$\widehat{\mathcal{L}}(\mathbf{u})(\mathbf{x}) + \mathbf{b}(\mathbf{x}) = \mathbf{0}, \quad \widehat{\mathcal{L}}(\mathbf{u})(\mathbf{x}) = \int \mathbf{C}(\boldsymbol{\xi}, \mathbf{x})[\mathbf{u}(\mathbf{y}) - \mathbf{u}(\mathbf{x})]\mathbf{y}, \tag{17.35}$$

and prove that this operator is self-adjoint ($\mathbf{u}, \mathbf{v} \in L_2(\overline{\overline{w}})$)

$$(\mathbf{v}, \widehat{\mathcal{L}}(\mathbf{u})) = (\widehat{\mathcal{L}}(\mathbf{v}), \mathbf{u}) \tag{17.36}$$

by the use of the next obvious chain of equalities ($\boldsymbol{\xi} = \mathbf{y} - \mathbf{x}$)

$$\begin{aligned}(\mathbf{v}, \widehat{\mathcal{L}}(\mathbf{u})) &= \{\mathbf{v}(\widehat{\mathcal{L}} * \mathbf{u})\} = |\overline{\overline{w}}|^{-1} \int_{\overline{\overline{w}}} \mathbf{v}(\mathbf{x}) \int_{\overline{\overline{w}}} \mathbf{C}(\boldsymbol{\xi}, \mathbf{x})[\mathbf{u}(\mathbf{y}) - \mathbf{u}(\mathbf{x})]dyd\mathbf{x} \\ &= -\frac{1}{2}|\overline{\overline{w}}|^{-1} \int_{\overline{\overline{w}}} \Big\{ \int_{\overline{\overline{w}}} [\mathbf{v}(\mathbf{y}) - \mathbf{v}(\mathbf{x})]\mathbf{C}(\boldsymbol{\xi}, \mathbf{x})[\mathbf{u}(\mathbf{y}) - \mathbf{u}(\mathbf{x})]d\mathbf{y} \Big\} d\mathbf{x} \\ &= (\widehat{\mathcal{L}}(\mathbf{v}), \mathbf{u}),\end{aligned} \tag{17.37}$$

where the third equality was obtained using the property of $\mathbf{C}(\boldsymbol{\xi}, \mathbf{x})$ (16.59) combining with Fubini's theorem for the iterated double integral (see also [16]).

A new *energetic definition* of the effective elastic moduli $\mathbf{L}^*$ can be obtained from a representation of the total strain energy through the influence functions ($\boldsymbol{\xi} = \mathbf{y} - \mathbf{x}$)

$$\begin{aligned}W = \{W(\mathbf{x})\} &= \frac{1}{4}\Big\{\int_{\mathcal{H}_x} \boldsymbol{\eta}^\top \mathbf{C}(\boldsymbol{\xi}, \mathbf{x})\boldsymbol{\eta} d\mathbf{y}\Big\} \\ &= \frac{1}{4}\Big\{\int_{\mathcal{H}_x} \{\boldsymbol{\varepsilon}\}\mathbf{A}_u^{*\eta\top} \mathbf{C}(\boldsymbol{\xi}, \mathbf{x})\mathbf{A}_u^{*\eta}\{\boldsymbol{\varepsilon}\}d\mathbf{y}\Big\},\end{aligned} \tag{17.38}$$

where $\mathbf{A}_u^{*\eta}(\boldsymbol{\xi}) := \mathbf{A}_u^*(\mathbf{y}) - \mathbf{A}_u^*(\mathbf{x})$. From other side, W can be represented as a total strain energy of homogeneous effective medium subjected to nonlocal homogeneous boundary conditions (16.46_2) and (16.52)

$$W = \frac{1}{2}\{\boldsymbol{\varepsilon}\}\mathbf{L}^*\{\boldsymbol{\varepsilon}\}. \tag{17.39}$$

Equating Eqs. (17.38) and (17.39) for any constant symmetric fields $\{\boldsymbol{\varepsilon}\}$ yields the new representation for the *energetic definition* of the effective elastic modulus

$$\mathbf{L}^* = \frac{1}{2}\left\{\int_{\mathcal{H}_x} \mathbf{A}_u^{*\eta\top}(\boldsymbol{\xi})\mathbf{C}(\boldsymbol{\xi},\mathbf{x})\mathbf{A}_u^{*\eta}(\boldsymbol{\xi})d\mathbf{y}\right\}. \tag{17.40}$$

For the proof of a symmetry $\mathbf{L}^* = (\mathbf{L}^*)^\top$, we recast Eq. (17.40), according to (17.37), as

$$\mathbf{L}^* = \frac{1}{2}|\overline{\overline{w}}|^{-1}\int_{\overline{\overline{w}}} \mathbf{A}_u^{*\eta\top}(\boldsymbol{\xi})\int_{\overline{\overline{w}}} \mathbf{C}(\boldsymbol{\xi},\mathbf{x})\mathbf{A}_u^{*\eta}(\boldsymbol{\xi})d\mathbf{y}d\mathbf{x}.$$

$$= \frac{1}{2}|\overline{\overline{w}}|^{-1}\int_{\overline{\overline{w}}}\int_{\overline{\overline{w}}} \mathbf{A}_u^{*\eta\top}(\boldsymbol{\xi})\mathbf{C}(\boldsymbol{\xi},\mathbf{y})d\mathbf{x}\mathbf{A}_u^{*\eta}(\boldsymbol{\xi})d\mathbf{y}, \tag{17.41}$$

$$\mathbf{L}^{*\top} = \frac{1}{2}|\overline{\overline{w}}|^{-1}\int_{\overline{\overline{w}}}\int_{\overline{\overline{w}}} \mathbf{A}_u^{*\eta\top}(\boldsymbol{\xi})\mathbf{C}(\boldsymbol{\xi},\mathbf{y})d\mathbf{y}\mathbf{A}_u^{*\eta}(\boldsymbol{\xi})d\mathbf{x}. \tag{17.42}$$

Owing to the self-adjointness of the operator of $\widehat{\boldsymbol{\mathcal{L}}}$ (17.37) and exchanging $\mathbf{x} \leftrightarrow \mathbf{y}$, comparison of the right-hand sides of Eqs. (17.41) and (17.42) yields the symmetry of $\mathbf{L}^*$, i.e., $\mathbf{L}^{*\top} = \mathbf{L}^*$. Thus, the self-adjointness of the operator of $\widehat{\boldsymbol{\mathcal{L}}}$ (17.37) at the fine scale yields self-adjointness of the effective elastic operator (which is the tensor $\mathbf{L}^*$ in our case).

It is possible to obtain another form for energetic definition of the effective elastic elasticity $\mathbf{L}^*$. Indeed, for the homogeneous displacement loading conditions (16.52), substitution of Eqs. (17.26) into Hill's condition (17.21) leads to

$$\{\boldsymbol{\varepsilon\sigma}\} = \Big\{(\mathbf{A}_\varepsilon^*\{\boldsymbol{\varepsilon}\})(\mathbf{B}_\varepsilon^*\{\boldsymbol{\varepsilon}\})\Big\} = \Big\{(\{\boldsymbol{\varepsilon}\}\mathbf{A}_\varepsilon^{*\top})\big(\mathbf{B}_\varepsilon^*\{\boldsymbol{\varepsilon}\}\big)\Big\}. \tag{17.43}$$

Since the volume average $\{\boldsymbol{\varepsilon}\}$ is a constant tensor, Eq. (17.43) can be recast as

$$\{\boldsymbol{\varepsilon\sigma}\} = \{\boldsymbol{\varepsilon}\}\{\mathbf{A}_\varepsilon^{*\top}\mathbf{B}_\varepsilon^*\}\{\boldsymbol{\varepsilon}\}, \tag{17.44}$$

and compared with the result of substitution of the effective constitutive equation (17.25$_1$) into Hill's condition (17.21):

$$\{\boldsymbol{\varepsilon\sigma}\} = \{\boldsymbol{\varepsilon}\}\mathbf{L}^*\{\boldsymbol{\varepsilon}\}, \tag{17.45}$$

which leads to another *energetic definition* of the effective elasticity $\mathbf{L}^*$

$$\mathbf{L}^* = \{\mathbf{A}_\varepsilon^{*\top}\boldsymbol{\mathcal{L}}^\sigma(\mathbf{A}_u^*)\}. \tag{17.46}$$

The three definitions (17.33), (17.39), and (17.46) are not in general equivalent to one another for arbitrary inhomogeneous media, and the extent to which they agree in their computation of $\mathbf{L}^*$ provides a partial check on the validity of the "overall modulus" concept.

17.3 Thermoperidynamic Composites

17.3.1 Basic Equations of Thermoperidynamics

Up to now, the case of pure mechanical loading was only considered. For the local thermoelastic problem, the constitutive equation (17.2) should be modified:

$$\boldsymbol{\sigma}(\mathbf{x}) = \mathbf{L}(\mathbf{x})\boldsymbol{\varepsilon}(\mathbf{x}) + \boldsymbol{\alpha}(\mathbf{x}), \quad \text{or} \quad \boldsymbol{\varepsilon}(\mathbf{x}) = \mathbf{M}(\mathbf{x})\boldsymbol{\sigma}(\mathbf{x}) + \boldsymbol{\beta}(\mathbf{x}), \tag{17.47}$$

where $\boldsymbol{\beta}(\mathbf{x})$ and $\boldsymbol{\alpha}(\mathbf{x}) \equiv -\mathbf{L}(\mathbf{x})\boldsymbol{\beta}(\mathbf{x})$ are second-order tensors of local eigenstrains and eigenstresses. In particular, for isotropic constituents, $\boldsymbol{\beta}(\mathbf{x}) = \beta^{\mathrm{T}}(\mathbf{x})\theta\boldsymbol{\delta}$, where $\theta = T - T_0$ denotes the temperature changes with respect to a reference temperature T_0 and β^{T} is a thermal expansion.

For thermoperidynamics media, the bond force (16.55) can be generalized (following [80, 771, 772, 1000]) to the case of availability of microeigenstrains $\widetilde{\boldsymbol{\beta}}^x(\hat{\mathbf{x}}, \mathbf{x})$

$$\mathbf{f}^t(\boldsymbol{\eta}, \boldsymbol{\xi}) = \mathbf{C}(\boldsymbol{\xi})\boldsymbol{\eta}^t, \quad \boldsymbol{\eta}^t(\hat{\mathbf{x}}, \mathbf{x}) = \boldsymbol{\eta}(\hat{\mathbf{x}}, \mathbf{x}) - \widetilde{\boldsymbol{\beta}}^x(\hat{\mathbf{x}}, \mathbf{x}) \;\; \forall \boldsymbol{\eta}, \boldsymbol{\xi} \tag{17.48}$$

with the micropotential $\widetilde{w}$ given (up to some additive constant) by

$$\widetilde{w}(\boldsymbol{\eta}, \boldsymbol{\xi}) = \boldsymbol{\eta}^\top \cdot \mathbf{C}(\boldsymbol{\xi})(\boldsymbol{\eta}/2 - \widetilde{\boldsymbol{\beta}}^x). \tag{17.49}$$

It is assumed that for the homogeneous materials

$$\widetilde{\boldsymbol{\beta}}^x(\hat{\mathbf{x}}, \mathbf{x}) = \widetilde{\boldsymbol{\beta}} \cdot (\hat{\mathbf{x}} - \mathbf{x}), \tag{17.50}$$

that corresponds to the constant temperature in the case of a thermoelastic problem. For two-phase composite medium containing any two points $\mathbf{x}$ and $\hat{\mathbf{x}}$ in R^d, $\widetilde{\boldsymbol{\beta}}^x(\mathbf{x}, \hat{\mathbf{x}})$ is presented in the form (17.4).

The basic equations of thermoperidynamics can be obtained from the corresponding peridynamic counterparts (16.55), (16.42), and (16.43) and the volumetric constraint problems (16.46) and (16.47) by the replacements

$$\boldsymbol{\eta}(\hat{\mathbf{x}}, \mathbf{x}) \mapsto \boldsymbol{\eta}^t(\hat{\mathbf{x}}, \mathbf{x}) := \boldsymbol{\eta}(\hat{\mathbf{x}}, \mathbf{x}) - \widetilde{\boldsymbol{\beta}}^x(\hat{\mathbf{x}}, \mathbf{x}), \tag{17.51}$$

$$\mathbf{u}(\mathbf{x}) \mapsto \mathbf{u}^t(\mathbf{x}) := \mathbf{u}(\mathbf{x}) - \widetilde{\boldsymbol{\beta}}(\mathbf{x}) \cdot \mathbf{x}. \tag{17.52}$$

For example, the integral equation (16.43) can be recast for the thermoperidynamic case as

$$\boldsymbol{\sigma}(\mathbf{x}) = \boldsymbol{\mathcal{L}}^{\sigma}(\mathbf{u}^t)(\mathbf{x}), \quad \mathbf{u}^t(\mathbf{x}) = \mathbf{u}(\mathbf{x}) - \widetilde{\boldsymbol{\beta}}(\mathbf{x}) \cdot \mathbf{x}. \tag{17.53}$$

In particular, for the homogeneous domain w with the properties $\mathbf{C}^{(0)}(\mathbf{x}, \hat{\mathbf{x}})$, $\widetilde{\boldsymbol{\beta}}^{x(0)}(\mathbf{x}, \hat{\mathbf{x}})$ subjected to the homogeneous volumetric boundary conditions (16.52), Eq. (17.53) is reduced to the constitutive equation of local elasticity (17.47) with the coefficients $\mathbf{L}^{(0)}$, $\boldsymbol{\alpha}^{(0)}$ =const. expressed through $\mathbf{C}^{(0)}(\mathbf{x}, \hat{\mathbf{x}})$, $\widetilde{\boldsymbol{\beta}}^{x(0)}(\mathbf{x}, \hat{\mathbf{x}})$. It enables one to establish the action of the operator $\boldsymbol{\mathcal{L}}^{\sigma(0)}$ (defined by the bond force $\mathbf{f}^{(0)}(\boldsymbol{\eta}, \boldsymbol{\xi})$ with, e.g., the parameters $\mathbf{C}^{(0)}(\mathbf{x}, \hat{\mathbf{x}})$, $\widetilde{\boldsymbol{\beta}}^{x(0)}(\mathbf{x}, \hat{\mathbf{x}})$) on the linear functions $\mathbf{u}(\mathbf{x}) = \boldsymbol{\varepsilon}\mathbf{x}$ and $\widetilde{\boldsymbol{\beta}}^{(0)}\mathbf{x}$

$$\boldsymbol{\mathcal{L}}^{\sigma(0)}(\boldsymbol{\varepsilon}\boldsymbol{\xi} - \widetilde{\boldsymbol{\beta}}^{(0)}\boldsymbol{\xi}) = \mathbf{L}^{(0)}\boldsymbol{\varepsilon} + \boldsymbol{\alpha}^{(0)}, \tag{17.54}$$

where $\boldsymbol{\alpha}^{(0)} = -\mathbf{L}^{(0)}\boldsymbol{\beta}^{(0)}$, $\boldsymbol{\beta}^{(0)} := \widetilde{\boldsymbol{\beta}}^{(0)}$. Equation (17.54) recovers and generalizes the result by Silling [1396] which states that for any microelastic material with microeigenstrains, the corresponding conventional thermoelastic material (in the sense of homogeneous deformations) can be found. However, the converse is not in general true.

Just for completeness, we mention about another sort of nonlocal models of thermoelasticity which are based on the differential equilibrium equation (17.1) and Cauchy one (17.3), while the algebraic constitutive equation (17.2) is replaced by the integral (called *strongly nonlocal*, strain type) representation (see Chap. 14)

$$\boldsymbol{\sigma}(\mathbf{x}) = \boldsymbol{\mathcal{L}}{*}\boldsymbol{\varepsilon}(\mathbf{x}) + \boldsymbol{\alpha}(\mathbf{x}), \tag{17.55}$$

$$\boldsymbol{\mathcal{L}}{*}(\cdot)(\mathbf{x}) = \int_w \boldsymbol{\mathcal{L}}(\mathbf{x}, \mathbf{y})(\cdot)d\mathbf{y}, \quad \boldsymbol{\alpha}(\mathbf{x}) = \int_w \mathbf{m}(\mathbf{x}, \mathbf{y})\Delta T(\mathbf{y})d\mathbf{y}, \tag{17.56}$$

where the elastic integral operator $\boldsymbol{\mathcal{L}}$ is defined by the kernel $\boldsymbol{\mathcal{L}}(\mathbf{x}, \mathbf{y})$ in $w \subset R^d$. It is assumed that the difference $\Delta T(\mathbf{y}) = (T - T^0)$ between the reference temperature T^0 =const. and the current temperature T is uniform in the space. Because of this, the action of the kernel of nonlocal thermal expansion coefficient $\mathbf{m}(\mathbf{x}, \mathbf{y})$ is reduced to the transformation strain tensor $\boldsymbol{\alpha}(\mathbf{x})$.

17.3.2 *Effective Eigenfields*

We are coming now to consideration of statistically homogeneous thermoperidynamic macrodomain w subjected to the homogeneous Dirichlet-type volumetric boundary conditions (16.52). In such a case, the overall constitutive relations of the representative volume are given in terms of the volume averages (or statistical averages)

$$\langle \boldsymbol{\sigma} \rangle = \mathbf{L}^* \langle \boldsymbol{\varepsilon} \rangle + \boldsymbol{\alpha}^*, \quad \langle = \mathbf{M}^* \langle \boldsymbol{\sigma} \rangle + \boldsymbol{\beta}^*, \tag{17.57}$$

where $\boldsymbol{\alpha}^* = -\mathbf{L}^* \boldsymbol{\beta}^*, \ \ \boldsymbol{\beta}^* = -\mathbf{M}^* \boldsymbol{\alpha}^*$.

For the estimation of effective transformation fields $\boldsymbol{\alpha}^*$ and $\boldsymbol{\beta}^*$, we decompose the overall fields as

$$\boldsymbol{\sigma}(\mathbf{x}) = \boldsymbol{\sigma}^I(\mathbf{x}) + \boldsymbol{\sigma}^{II}(\mathbf{x}), \quad \boldsymbol{\varepsilon}(\mathbf{x}) = \boldsymbol{\varepsilon}^I(\mathbf{x}) + \boldsymbol{\varepsilon}^{II}(\mathbf{x}), \tag{17.58}$$

with the sources

$$\widetilde{\boldsymbol{\alpha}}^{xI}(\hat{\mathbf{x}}, \mathbf{x}) = \widetilde{\boldsymbol{\beta}}^{xI}(\hat{\mathbf{x}}, \mathbf{x}) = 0, \text{ and } \widetilde{\boldsymbol{\alpha}}^{xII}(\hat{\mathbf{x}}, \mathbf{x}) = \widetilde{\boldsymbol{\alpha}}^{x}(\hat{\mathbf{x}}, \mathbf{x}), \ \ \widetilde{\boldsymbol{\beta}}^{xII}(\hat{\mathbf{x}}, \mathbf{x}) = \widetilde{\boldsymbol{\beta}}^{x}(\hat{\mathbf{x}}, \mathbf{x}). \tag{17.59}$$

We assume homogeneous displacement loading conditions:

$$\mathbf{u}^I(\mathbf{x}) = \boldsymbol{\varepsilon}^{w_\Gamma} \cdot \mathbf{x}, \ \ \mathbf{u}^{II}(\mathbf{x}) = \mathbf{0}, \tag{17.60}$$

$\mathbf{x} \in w_\Gamma$, $\boldsymbol{\varepsilon}^{w_\Gamma}$ =const. The field I is solely created at $\widetilde{\boldsymbol{\beta}}^{x}(\hat{\mathbf{x}}, \mathbf{x}) \equiv \mathbf{0}$ by the applied displacement loading compatible with a homogeneous strain field $\langle \boldsymbol{\varepsilon} \rangle$. The field II is generated by the random stress-free strains $\widetilde{\boldsymbol{\beta}}^{x}(\mathbf{x}, \hat{\mathbf{x}})$ with the zero displacement loading (16.52). Because of this, we may call $\boldsymbol{\sigma}^I$ the load stress field and $\boldsymbol{\sigma}^{II}$ the residual stress field. Consequently, the constitutive equations decompose into two kinds of equations at the fine level

$$\boldsymbol{\sigma}^I = \boldsymbol{\mathcal{L}}^\sigma(\mathbf{u}^I), \ \ \boldsymbol{\sigma}^{II} = \boldsymbol{\mathcal{L}}^\sigma(\mathbf{u}^{II} - \widetilde{\boldsymbol{\beta}}^{x}), \tag{17.61}$$

and the macroscopic level

$$\langle \boldsymbol{\sigma}^I \rangle = \mathbf{L}^* \langle \boldsymbol{\varepsilon}^I \rangle, \ \ \langle \boldsymbol{\sigma}^{II} \rangle = \mathbf{L}^* \langle \boldsymbol{\varepsilon}^{II} \rangle + \boldsymbol{\alpha}^*, \tag{17.62}$$

respectively. Taking into account the definitions of the two fields I and II and the volumetric boundary conditions (17.60), we find the following tensor equations for the average fields

$$\langle \boldsymbol{\varepsilon}^I \rangle = \langle \boldsymbol{\varepsilon} \rangle, \quad \langle \boldsymbol{\sigma}^I \rangle = \mathbf{L}^* \langle \boldsymbol{\varepsilon} \rangle, \quad \langle \boldsymbol{\varepsilon}^{II} \rangle = \mathbf{0}, \quad \langle \boldsymbol{\sigma}^{II} \rangle = \boldsymbol{\alpha}^*. \tag{17.63}$$

Since $\boldsymbol{\sigma}^I$ and $\boldsymbol{\sigma}^{II}$ are statically admissible stress fields, while $\boldsymbol{\varepsilon}^I$ and $\boldsymbol{\varepsilon}^{II}$ are kinematically admissible strain fields, then, according to Hill's condition, we obtain

$$\langle \boldsymbol{\sigma}^I \boldsymbol{\varepsilon}^I \rangle = \langle \boldsymbol{\varepsilon} \rangle \mathbf{L}^* \langle \boldsymbol{\varepsilon} \rangle, \ \ \langle \boldsymbol{\varepsilon}^I \boldsymbol{\sigma}^{II} \rangle = \boldsymbol{\alpha}^* \langle \boldsymbol{\varepsilon} \rangle, \ \ \langle \boldsymbol{\varepsilon}^{II} \boldsymbol{\sigma}^I \rangle = 0, \ \ \langle \boldsymbol{\varepsilon}^{II} \rangle \langle \boldsymbol{\sigma}^{II} \rangle = 0. \tag{17.64}$$

In a similar manner, the homogeneous force loading value problem (16.52) can be analyzed when the following decomposition of the force loading takes place ($\boldsymbol{\sigma}^{w\Gamma}$ =const., $\mathbf{x} \in w_\Gamma$):

$$-\widehat{\mathcal{L}}(\mathbf{u}^I)(\mathbf{x}) = \boldsymbol{\sigma}^{w\Gamma}\mathbf{n}(\mathbf{x}), \quad \widehat{\mathcal{L}}(\mathbf{u}^{II})(\mathbf{x}) \equiv \mathbf{0}, \tag{17.65}$$

which leads to the average fields replacing the averages (17.63) and (17.64)

$$\langle \boldsymbol{\sigma}^I \rangle = \langle \boldsymbol{\sigma} \rangle, \quad \langle \boldsymbol{\varepsilon}^I \rangle = \mathbf{M}^* \langle \boldsymbol{\sigma} \rangle, \quad \langle \boldsymbol{\sigma}^{II} \rangle = \mathbf{0}, \quad \langle \boldsymbol{\varepsilon}^{II} \rangle = \boldsymbol{\beta}^*, \tag{17.66}$$

$$\langle \boldsymbol{\sigma}^I \boldsymbol{\varepsilon}^I \rangle = \langle \boldsymbol{\sigma} \rangle \mathbf{M}^* \langle \boldsymbol{\sigma} \rangle, \quad \langle \boldsymbol{\varepsilon}^I \boldsymbol{\sigma}^{II} \rangle = 0,$$
$$\langle \boldsymbol{\varepsilon}^{II} \boldsymbol{\sigma}^I \rangle = \boldsymbol{\beta}^* \langle \boldsymbol{\sigma} \rangle, \quad \langle \boldsymbol{\varepsilon}^{II} \boldsymbol{\sigma}^{II} \rangle = 0, \tag{17.67}$$

respectively.

17.3.3 Transformation Influence Functions

The transformation fields $\widetilde{\boldsymbol{\beta}}$ (17.50) and $\widetilde{\boldsymbol{\alpha}}$ (17.59) cause residual fields presenting in w even in the case of the vanished loading conditions (16.46) or (16.47); their evaluation has been analyzed in [425] for the local properties (17.47) of constituents, while one Buryachenko [205] and [206] considered a similar problem for strongly nonlocal elasticity (17.55). In particular, the residual local fields were presented in terms of certain transformation influence functions with the establishment of their relation to the analogous mechanical influence functions (17.26) and (17.27) for the special case of piecewise uniform eigenfields. For the use of an analogy between the mentioned theories and their peridynamic version, we will assume that the nonlocal operator $\widehat{\mathcal{L}}$ is presented in the form (17.6) and (17.48), and for the phase $\mathbf{x} \in v^{(r)}$ $(r = 0, 1, \ldots, N)$

$$\widetilde{\boldsymbol{\alpha}}(\mathbf{x}) = \bar{\boldsymbol{\alpha}}^{(r)} V_\alpha^{(r)}(\mathbf{x}), \quad \widetilde{\boldsymbol{\beta}}(\mathbf{x}) = \bar{\boldsymbol{\beta}}^{(r)} V_\beta^{(r)}(\mathbf{x}), \tag{17.68}$$

where $V_\alpha^{(r)}(\mathbf{x})$ and $V_\alpha^{(r)}(\mathbf{x})$ are the scale indicator functions and the modified eigenfields $\bar{\boldsymbol{\alpha}}^{(r)}$ and $\bar{\boldsymbol{\beta}}^{(r)}$ are the constant tensors. Then the total fields (displacement, strain, or stress) caused in $v^{(s)}$ by the uniform loading fields $\boldsymbol{\varepsilon}^{w\Gamma} = \langle \boldsymbol{\varepsilon} \rangle$ (16.52) and $\boldsymbol{\sigma}^{w\Gamma} = \langle \boldsymbol{\sigma} \rangle$ (16.53) and the transformation fields (17.68) are represented in the following form that extends (17.26) and (17.27) as ($\mathbf{x} \in v^{(s)}$)

$$\mathbf{u}(\mathbf{x}) = \mathbf{A}^*_{u\varepsilon}(\mathbf{x})\langle \boldsymbol{\varepsilon} \rangle + \sum_{r=0}^{N} \mathbf{D}^{sr}_{u\varepsilon}(\mathbf{x})\bar{\boldsymbol{\beta}}^{(r)}, \tag{17.69}$$

$$\boldsymbol{\varepsilon}(\mathbf{x}) = \mathbf{A}^*_{\varepsilon}(\mathbf{x})\langle \boldsymbol{\varepsilon} \rangle + \sum_{r=0}^{N} \mathbf{D}^{sr}_{\varepsilon}(\mathbf{x})\bar{\boldsymbol{\beta}}^{(r)}, \tag{17.70}$$

$$\boldsymbol{\sigma}(\mathbf{x}) = \mathbf{B}^*_{\varepsilon}(\mathbf{x})\langle\boldsymbol{\varepsilon}\rangle + \sum_{r=0}^{N} \mathbf{F}^{sr}_{\varepsilon}(\mathbf{x})\bar{\boldsymbol{\alpha}}^{(r)}, \tag{17.71}$$

$$\mathbf{u}(\mathbf{x}) = \mathbf{A}^*_{u\sigma}(\mathbf{x})\langle\boldsymbol{\sigma}\rangle + \sum_{r=0}^{N} \mathbf{D}^{sr}_{u\sigma}(\mathbf{x})\bar{\boldsymbol{\beta}}^{(r)}, \tag{17.72}$$

$$\boldsymbol{\varepsilon}(\mathbf{x}) = \mathbf{A}^*_{\sigma}(\mathbf{x})\langle\boldsymbol{\sigma}\rangle + \sum_{r=0}^{N} \mathbf{D}^{sr}_{\sigma}(\mathbf{x})\bar{\boldsymbol{\beta}}^{(r)}, \tag{17.73}$$

$$\boldsymbol{\sigma}(\mathbf{x}) = \mathbf{B}^*_{\sigma}(\mathbf{x})\langle\boldsymbol{\sigma}\rangle + \sum_{r=0}^{N} \mathbf{F}^{sr}_{\sigma}(\mathbf{x})\bar{\boldsymbol{\alpha}}^{(r)}, \tag{17.74}$$

respectively. Here, $\mathbf{D}^{sr}_{\varepsilon}(\mathbf{x})$ and $\mathbf{D}^{sr}_{\sigma}(\mathbf{x})$ are the *eigenstrain influence functions*, while $\mathbf{F}^{sr}_{\varepsilon}(\mathbf{x})$ and $\mathbf{F}^{sr}_{\sigma}(\mathbf{x})$ are the corresponding *eigenstress influence functions*; their local volume averages over $v^{(s)}$ may be referred to as the *transformation concentration factors* $\mathbf{D}^{sr}_{\varepsilon}$, $\mathbf{D}^{sr}_{\sigma}$, $\mathbf{D}^{sr}_{u\varepsilon}$, $\mathbf{D}^{sr}_{u\sigma}$, and $\mathbf{F}^{sr}_{\varepsilon}$, $\mathbf{F}^{sr}_{\sigma}$.

We will establish now some general relations between the transformation influence functions. Substituting Eq. (17.69) at $\langle\boldsymbol{\varepsilon}\rangle = \mathbf{0}$ into the constitutive equation (17.53) and comparing the result with (17.71) leads to the equation ($\mathbf{x} \in v^{(s)}, s = 0, 1, \ldots, N$)

$$\boldsymbol{\alpha}(\mathbf{x}) + \sum_{r=0}^{N} \boldsymbol{\mathcal{L}}^{\sigma(s)} * \mathbf{D}^{sr}_{\varepsilon}(\mathbf{x})\bar{\boldsymbol{\beta}}^{(r)} = \sum_{r=0}^{N} \mathbf{F}^{sr}_{u\varepsilon}(\mathbf{x})\bar{\boldsymbol{\alpha}}^{(r)}, \tag{17.75}$$

obtained by the use of Eqs. (17.54) and (17.68); here, $\boldsymbol{\mathcal{L}}^{\sigma(s)}$ is defined by the bond force $\mathbf{f}^{(s)}(\boldsymbol{\eta}, \boldsymbol{\xi})$ with, e.g., the parameters $\mathbf{C}^{(s)}(\mathbf{x}, \hat{\mathbf{x}})$, $\widetilde{\boldsymbol{\beta}}^{x(s)}(\mathbf{x}, \hat{\mathbf{x}})$. Since Eq. (17.75) should hold for any chosen $\bar{\boldsymbol{\beta}}^{(r)}$, it follows that for any phase $\mathbf{x} \in v^{(s)}$ and $v^{(r)}$

$$\delta_{rs}\bar{\boldsymbol{\alpha}}^{(r)} + \boldsymbol{\mathcal{L}}^{\sigma(s)} * \mathbf{D}^{sr}_{u\varepsilon}(\mathbf{x})\bar{\boldsymbol{\beta}}^{(r)} = \mathbf{F}^{sr}_{\varepsilon}(\mathbf{x})\bar{\boldsymbol{\alpha}}^{(r)}. \tag{17.76}$$

The representations analogous to Eq. (17.76) for the force loading $\boldsymbol{\sigma}^{w\Gamma} = \langle\boldsymbol{\sigma}\rangle$ (16.53) and the transformation fields (17.68) can be presented in a like manner; however, we do not consider it due to the comments after Eq. (17.34).

The definition of the transformation concentration factors and Eqs. (17.66) make it possible to obtain another pair of representations for $\boldsymbol{\beta}^*$ and $\boldsymbol{\alpha}^*$ additionally to (17.63$_4$). Indeed, considering the homogeneous force loading (16.53) and taking (17.66$_4$) into account yields the expression

$$\boldsymbol{\beta}^* = \sum_{s,r=0}^{N} c^{(s)}\mathbf{D}^{sr}_{\sigma}\bar{\boldsymbol{\beta}}^{(r)}. \tag{17.77}$$

An analogous derivation leads to

$$\boldsymbol{\alpha}^* = \sum_{s,r=0}^{N} c^{(s)} \mathbf{F}_{\varepsilon}^{sr} \bar{\boldsymbol{\alpha}}^{(r)}. \tag{17.78}$$

The conditions providing consistency of these results with the previous ones (17.28) can be found by averaging Eqs. (17.70) and (17.74) over the volume w and taking Eq. (17.28) into account

$$\sum_{s=0}^{N} c^{(s)} \sum_{r=0}^{N} \mathbf{D}_{\varepsilon}^{sr} \bar{\boldsymbol{\beta}}^{(r)} = \mathbf{0}, \quad \sum_{s=0}^{N} c^{(s)} \sum_{r=0}^{N} \mathbf{F}_{\sigma}^{sr} \bar{\boldsymbol{\alpha}}^{(r)} = \mathbf{0}. \tag{17.79}$$

The representations (17.79) can be simplified due to the possible independent selection of the modified eigenfields $\bar{\boldsymbol{\alpha}}^{(r)}$ and $\bar{\boldsymbol{\beta}}^{(r)}$ $(r = 0, \ldots, N)$:

$$\sum_{s=0}^{N} c^{(s)} \mathbf{D}_{\varepsilon}^{sr} = \mathbf{0}, \quad \sum_{s=0}^{N} c^{(s)} \mathbf{F}_{\sigma}^{sr} = \mathbf{0}. \tag{17.80}$$

At last, the next relations between the eigenfield influence functions can be established analogous to Eqs. (17.29_2) and (17.31_2):

$$\mathbf{D}_{\varepsilon}^{sr}(\mathbf{x}) = \nabla \overset{S}{\otimes} \mathbf{D}_{u\varepsilon}^{sr}(\mathbf{x}), \quad \mathbf{D}_{\sigma}^{sr}(\mathbf{x}) = \nabla \overset{S}{\otimes} \mathbf{D}_{u\sigma}^{sr}(\mathbf{x}). \tag{17.81}$$

17.4 Effective Thermoelastic Properties for the Matrix Composites

17.4.1 *Effective Moduli and Effective Eigenstrains*

For the homogeneous displacement loading (16.52), the representations for the effective moduli $\mathbf{L}^*$ (17.33_1) and effective eigenstress $\boldsymbol{\alpha}^*$ (17.63_4) are obtained for both the local (conventional thermoelasticity (17.47)) and nonlocal (strongly nonlocal thermoelasticity (17.55) and thermoperidynamic (17.6)) problems. However, a direct practical use of Eqs. (17.33_1) and (17.63_4) involves difficulties because it requests the estimation of average fields inside each phase of both the load stresses $\langle \boldsymbol{\sigma}^I \rangle^{(k)}$ and residual ones $\langle \boldsymbol{\sigma}^{II} \rangle^{(k)}$ $(k = 0, 1, \ldots, N)$. The mentioned equations are significantly simplified for the conventional thermoelasticity (17.47) of matrix composites with the homogeneous matrix when the decompositions

$$\mathbf{L}(\mathbf{x}) = \mathbf{L}^{(0)} + \mathbf{L}_1(\mathbf{x}), \quad \boldsymbol{\alpha}(\mathbf{x}) = \boldsymbol{\alpha}^{(0)} + \boldsymbol{\alpha}_1(\mathbf{x}) \tag{17.82}$$

(superscript ${}^L(.)$ is omitted for convenience) hold and the jumps of the properties $\mathbf{L}_1(\mathbf{x})$ and $\boldsymbol{\alpha}_1(\mathbf{x})$ vanish inside the matrix $\mathbf{x} \in v^{(0)}$. Averaging the constitutive equation at the fine scale (17.47$_1$) yields

$$\langle \boldsymbol{\sigma} \rangle = \mathbf{L}^{(0)} \langle \boldsymbol{\varepsilon} \rangle + \boldsymbol{\alpha}^{(0)} + \langle {}^L\boldsymbol{\tau} \rangle, \tag{17.83}$$

where

$${}^L\boldsymbol{\tau}(\mathbf{x}) = \mathbf{L}_1(\mathbf{x})\boldsymbol{\varepsilon}(\mathbf{x}) + \boldsymbol{\alpha}_1(\mathbf{x}) \tag{17.84}$$

denotes the stress polarization tensor which is simply a rotational convenience and vanishes inside the matrix: ${}^L\boldsymbol{\tau}(\mathbf{x}) \equiv \mathbf{0}$ ($\mathbf{x} \in v^{(0)}$). Then

$$\mathbf{L}^* = \mathbf{L}^{(0)} + \mathbf{R}^*, \quad \boldsymbol{\alpha}^* = \boldsymbol{\alpha}^{(0)} + \mathbf{F}^*, \quad \langle {}^L\boldsymbol{\tau} \rangle := \mathbf{R}^* \langle \boldsymbol{\varepsilon} \rangle + \mathbf{F}^*. \tag{17.85}$$

The representation (17.63$_4$) can also be simplified by the use of auxiliary obvious equations

$$\langle \boldsymbol{\sigma}^{II} \rangle = \langle V(\mathbf{x})\boldsymbol{\sigma}^{II}(\mathbf{x}) \rangle + \langle V^{(0)}(\mathbf{x})\boldsymbol{\sigma}^{II}(\mathbf{x}) \rangle, \quad \mathbf{0} = \langle \boldsymbol{\varepsilon}^{II} \rangle. \tag{17.86}$$

Then averaging the constitutive equation at the fine scale (17.47$_1$) over the volume of the matrix with the indicator function $V^{(0)}(\mathbf{x})$ ($V^{(0)}(\mathbf{x}) \equiv 1 - V(\mathbf{x})$)

$$\langle V^{(0)}(\mathbf{x})\boldsymbol{\sigma}^{II}(\mathbf{x}) \rangle = \langle V^{(0)}(\mathbf{x})\mathbf{L}^{(0)}\boldsymbol{\varepsilon}^{II}(\mathbf{x}) \rangle + \langle V^{(0)}(\mathbf{x})\boldsymbol{\alpha}(\mathbf{x}) \rangle \tag{17.87}$$

leads to the representation

$$\boldsymbol{\alpha}^* = \boldsymbol{\alpha}^{(0)} + \langle {}^L\boldsymbol{\tau}^{II} \rangle \tag{17.88}$$

at the assumption of homogeneity of matrix thermoelastic properties. Hereafter, $c^{(0)} = \langle V^{(0)} \rangle$ is a volume fraction of the matrix, and one used the residual stress polarization tensor

$${}^L\boldsymbol{\tau}^{II} = V(\mathbf{x})\{\boldsymbol{\sigma}^{II}(\mathbf{x}) - [\mathbf{L}^{(0)}\boldsymbol{\varepsilon}^{II}(\mathbf{x}) + \boldsymbol{\alpha}^{(0)}]\}. \tag{17.89}$$

A fundamental advantage of the representations (17.85) and (17.88) with respect to Eqs. (17.33$_1$) and (17.63$_4$), respectively, is a necessity to estimate the stress and strain fields only inside the inclusions ($\mathbf{x} \in v$) rather than inside the entire composite ($\mathbf{x} \in w$). It is found to be possible due to the fact that a constant tensor can be always carried out from the brackets $\langle(\cdot)\rangle$ of statistical averages, e.g.,

$$\langle \mathbf{L}^{(0)}\boldsymbol{\varepsilon} \rangle = \mathbf{L}^{(0)} \langle \boldsymbol{\varepsilon} \rangle, \quad \langle \mathbf{L}^{(0)}\boldsymbol{\varepsilon}^{II} \rangle = \mathbf{L}^{(0)} \langle \boldsymbol{\varepsilon}^{II} \rangle. \tag{17.90}$$

We will propose the generalization of Eqs. (17.85) and (17.88) to the thermoperidynamic composites by the use of generalizations of some basic auxiliary representations for micromechanics of conventional thermoelasticity. Namely, an analog of the tensorial decompositions (17.82) can be presented in the next form for the operator (17.53_1)

$$\mathcal{L}^{\sigma}(\mathbf{u}^t)(\mathbf{x}) = \mathcal{L}^{\sigma(0)}(\mathbf{u}^t)(\mathbf{x}) + \mathcal{L}_1^{\sigma}(\mathbf{u}^t)(\mathbf{x}), \tag{17.91}$$

where $\mathcal{L}^{\sigma(0)}$ denotes an action of the operator $\mathcal{L}^{\sigma}$ on the medium with the material properties of the matrix defined by the bond force $\mathbf{f}^{t(0)}$ (see Eq. (17.54); e.g., for the bond force (17.48_1), we have $\mathbf{C}(\boldsymbol{\xi}, \mathbf{x}) \equiv \mathbf{C}^{(0)}(\boldsymbol{\xi})$, $\widetilde{\boldsymbol{\beta}}^x(\mathbf{y}, \mathbf{x}) \equiv \widetilde{\boldsymbol{\beta}}^{x(0)}(\mathbf{y}, \mathbf{x})$) and the displacement fields $\mathbf{u}(\mathbf{y})$, $\mathbf{u}(\mathbf{x})$ of the real composite material. The jump operator $\mathcal{L}_1^{\sigma}(\mathbf{u}^t)(\mathbf{x})$ defined by Eq. (17.91) is called the *local stress polarization tensor* (compare with Eq. (17.84)) and represented as

$$\boldsymbol{\tau}(\mathbf{x}) = \mathcal{L}_1^{\sigma}(\mathbf{u}^t)(\mathbf{x}) = \frac{1}{2}\int_S \int_0^{\infty} \int_0^{\infty} (y+z)^{d-1} \hat{\mathbf{f}}_1^t(\mathbf{x} + y\mathbf{m}, \mathbf{x} - z\mathbf{m}) \otimes \mathbf{m}\, dz\, dy\, d\Omega_{\mathbf{m}}, \tag{17.92}$$

where $\hat{\mathbf{f}}_1^t(\mathbf{p}, \mathbf{q}) := \hat{\mathbf{f}}^t(\mathbf{p}, \mathbf{q}) - \hat{\mathbf{f}}^{t(0)}(\mathbf{p}, \mathbf{q})$. In particular, for the bond force (17.48_1), the bond force jump $\hat{\mathbf{f}}_1^t(\mathbf{x} + y\mathbf{m}, \mathbf{x} - z\mathbf{m})$ is called the *micropolarization* tensor

$$\widetilde{\boldsymbol{\tau}}(\mathbf{x}, \hat{\mathbf{x}}) := \hat{\mathbf{f}}_1^t(\mathbf{x}, \hat{\mathbf{x}}) = \mathbf{C}_1(\mathbf{x}, \hat{\mathbf{x}})\boldsymbol{\eta}(\hat{\mathbf{x}}, \mathbf{x}) + \widetilde{\boldsymbol{\alpha}}_1^x(\hat{\mathbf{x}}, \mathbf{x}) \tag{17.93}$$

and

$$\begin{aligned} \boldsymbol{\sigma}(\mathbf{x}) = \mathcal{L}^{\sigma}(\mathbf{u}^t)(\mathbf{x}) &= \frac{1}{2}\int_S \int_0^{\infty} \int_0^{\infty} (y+z)^{d-1} \Big\{\mathbf{C}(\mathbf{x} + y\mathbf{m}, \mathbf{x} - z\mathbf{m}) \\ &\times [\mathbf{u}(\mathbf{x} + y\mathbf{m}) - \mathbf{u}(\mathbf{x} - z\mathbf{m})] + \widetilde{\boldsymbol{\alpha}}^x(\mathbf{x} + y\mathbf{m}, \mathbf{x} - z\mathbf{m})\Big\} \otimes \mathbf{m}\, dz\, dy\, d\Omega_{\mathbf{m}}, \\ \boldsymbol{\tau}(\mathbf{x}) = \mathcal{L}_1^{\sigma}(\mathbf{u}^t)(\mathbf{x}) &= \frac{1}{2}\int_S \int_0^{\infty} \int_0^{\infty} (y+z)^{d-1} \Big\{\mathbf{C}_1(\mathbf{x} + y\mathbf{m}, \mathbf{x} - z\mathbf{m}) \\ &\times [\mathbf{u}(\mathbf{x} + y\mathbf{m}) - \mathbf{u}(\mathbf{x} - z\mathbf{m})] + \widetilde{\boldsymbol{\alpha}}_1^x(\mathbf{x} + y\mathbf{m}, \mathbf{x} - z\mathbf{m}) \otimes \mathbf{m}\, dz\, dy\, d\Omega_{\mathbf{m}}. \end{aligned} \tag{17.94}$$

Here, the jumps of material properties are introduced:

$$\begin{aligned} \mathbf{C}(\mathbf{x}, \hat{\mathbf{x}}) &:= \mathbf{C}^{(0)}(\mathbf{x}, \hat{\mathbf{x}}) + \mathbf{C}_1(\mathbf{x}, \hat{\mathbf{x}})V(\hat{\mathbf{x}}), \\ \widetilde{\boldsymbol{\alpha}}^x(\hat{\mathbf{x}}, \mathbf{x}) &:= -\mathbf{C}(\mathbf{x}, \hat{\mathbf{x}})\widetilde{\boldsymbol{\beta}}^x(\hat{\mathbf{x}}, \mathbf{x}) = \widetilde{\boldsymbol{\alpha}}_0^x(\hat{\mathbf{x}}, \mathbf{x}) + \widetilde{\boldsymbol{\alpha}}_1^x(\hat{\mathbf{x}}, \mathbf{x})V(\hat{\mathbf{x}}), \end{aligned} \tag{17.95}$$

where

$$\begin{aligned} \widetilde{\boldsymbol{\alpha}}_0^x(\mathbf{x}, \hat{\mathbf{x}}) &:= -\mathbf{C}^{(0)}(\mathbf{x}, \hat{\mathbf{x}})\widetilde{\boldsymbol{\beta}}^{x(0)}(\mathbf{x}, \hat{\mathbf{x}}), \\ \widetilde{\boldsymbol{\alpha}}_1^x(\mathbf{x}, \hat{\mathbf{x}}) &:= \mathbf{C}^{(0)}(\mathbf{x}, \hat{\mathbf{x}})\widetilde{\boldsymbol{\beta}}^{x(0)}(\mathbf{x}, \hat{\mathbf{x}}) - \mathbf{C}(\mathbf{x}, \hat{\mathbf{x}})\widetilde{\boldsymbol{\beta}}^x(\mathbf{x}, \hat{\mathbf{x}}). \end{aligned} \tag{17.96}$$

Thus, the operator equation (17.53) for the thermoperidynamic composite material can be decomposed as

$$\boldsymbol{\sigma}(\mathbf{x}) = \boldsymbol{\mathcal{L}}^{\sigma(0)}(\mathbf{u}^t)(\mathbf{x}) + \boldsymbol{\tau}(\mathbf{x}). \tag{17.97}$$

For subsequent averaging of Eq. (17.97), we perform a peridynamic analog of the thermoelastic counterpart operation (17.90)

$$\begin{aligned}\langle \boldsymbol{\mathcal{L}}^{\sigma(0)}(\mathbf{u}^t)\rangle(\mathbf{x}) &= \frac{1}{2}\Big\langle \int_S \int_0^\infty \int_0^\infty (y+z)^{d-1} \hat{\mathbf{f}}^{t(0)}(\mathbf{x}+y\mathbf{m}, \mathbf{x}-z\mathbf{m}) \otimes \mathbf{m} dz dy d\Omega_{\mathbf{m}} \Big\rangle \\ &= \frac{1}{2}\int_S \int_0^\infty \int_0^\infty (y+z)^{d-1} \mathbf{f}^{t(0)}(\langle \mathbf{u}^t \rangle \mathbf{u}^t\rangle(\mathbf{x}+y\mathbf{m}) - \langle \mathbf{u}^t\rangle(\mathbf{x}-z\mathbf{m}), (y+z)\mathbf{m}) \\ &\otimes \mathbf{m} dz dy d\Omega_{\mathbf{m}} = \mathbf{L}^{(0)}\langle \boldsymbol{\varepsilon}\rangle + \boldsymbol{\alpha}^{(0)}. \end{aligned} \tag{17.98}$$

Here, the equality (17.98_2) was obtained due to the linearity of $\mathbf{f}^{t(0)}$ from the random variable $\mathbf{u}$, while all remaining functions (e.g., $\mathbf{C}^{(0)}(\mathbf{x}, \hat{\mathbf{x}})$ and $\widetilde{\boldsymbol{\beta}}(\hat{\mathbf{x}}, \mathbf{x})$, and the integration limits) are deterministic at the fixed $\mathbf{x}$ and $\hat{\mathbf{x}}$. Because of this, analogous to the averaged equation (17.83), a deterministic function describing the matrix properties (e.g., $\mathbf{C}^{(0)}(\mathbf{x}, \hat{\mathbf{x}})$) can be carried out (together with the corresponding integral operations) from the averaging operation $\langle(\cdot)\rangle$ at each $\mathbf{x}$ and $\hat{\mathbf{x}}$. In turn, the equality (17.98_3) follows from Eqs. (17.23) and (17.54).

In fact, the modulus $\mathbf{L}^{(0)}$ is defined through the influence functions leading to $\mathbf{L}^{(0)} = \boldsymbol{\mathcal{L}}^{\sigma(0)}(\mathbf{I}\boldsymbol{\xi})$, where $\boldsymbol{\xi} = (y+z)\mathbf{m}$ and $\mathbf{L}^{(0)}$ is determined by Eq. (17.98). From other side, the modulus $\mathbf{L}^{(0)}$ can be equivalently determined through the energetic definition

$$\mathbf{L}^{(0)} := \frac{1}{2}\Big\langle \int_{\mathcal{H}_{\mathbf{x}}} \boldsymbol{\xi}^\top \mathbf{C}^{(0)}(\boldsymbol{\xi}, \mathbf{x})\boldsymbol{\xi} d\boldsymbol{\xi} \Big\rangle = \frac{1}{2}\int_{\mathcal{H}_{\mathbf{x}}} \boldsymbol{\xi}^\top \mathbf{C}^{(0)}(\boldsymbol{\xi}, \mathbf{x})\boldsymbol{\xi} d\boldsymbol{\xi} \tag{17.99}$$

(do not mix up the modulus $\mathbf{L}^{(0)}$ with the modulus ${}^L\mathbf{L}^{(0)}$ of the locally elastic matrix in Eq. (17.83)). The equality (17.100_2) holds due to the matrix homogeneity when the micromodulus $\mathbf{C}^{(0)}(\boldsymbol{\xi}, \mathbf{x})$ is invariant with respect to the parallel translation $\mathbf{C}^{(0)}(\boldsymbol{\xi}, \mathbf{x}) = \mathbf{C}^{(0)}(\boldsymbol{\xi} + \mathbf{y}, \mathbf{x} + \mathbf{y})\ \forall \mathbf{y} \in R^d$.

Equation (17.98) has a very attractive physical meaning. Namely, we formally replaced $\mathbf{C}(\mathbf{x}, \hat{\mathbf{x}})$ in the external inclusion phase v^l by the micromodulus of the matrix $\mathbf{C}^{(0)}(\mathbf{x}, \hat{\mathbf{x}})$, while a real random strongly inhomogeneous displacement field $\mathbf{u}(\mathbf{x})$ ($\mathbf{x} \in w$) is fixed. Nevertheless, the average of the stress operator $\langle \boldsymbol{\mathcal{L}}^{\sigma}(\tilde{\boldsymbol{\tau}}^{(0)})\rangle(\mathbf{x})$ defined at this displacement field exactly coincides with the term $\mathbf{L}^{(0)}\langle \boldsymbol{\varepsilon}\rangle$ for the corresponding locally elastic composite material. It means that in the average sense, the peridynamic truncated matrix (with the micromodulus $\mathbf{C}^{(0)}(\mathbf{x}, \hat{\mathbf{x}})$) is deformed identically to the locally elastic matrix (with the modulus $\mathbf{L}^{(0)}$) deformation. In so doing, the modulus $\mathbf{L}^{(0)}$ is uniquely expressed through $\mathbf{C}^{(0)}(\mathbf{x}, \hat{\mathbf{x}})$ as an effective modulus of the peridynamic CM with zero concentration of inclusions (17.99).

No assumptions about $\mathbf{C}^{(0)}(\mathbf{x}, \hat{\mathbf{x}})$ (e.g., either a spherical shape of $\mathcal{H}_{\mathbf{x}}$ or isotropy $\lambda(\boldsymbol{\xi}) = \lambda(|\boldsymbol{\xi}|)$ of a micromodulus (16.64)) were used at the derivative of Eq. (17.97).

Therefore, the representation (17.98) leads to simplification of the averaged equation (17.97) (compare with Eq. (17.83)):

$$\langle \boldsymbol{\sigma} \rangle = \mathbf{L}^{(0)} \langle \boldsymbol{\varepsilon} \rangle + \boldsymbol{\alpha}^{(0)} + \langle \boldsymbol{\tau} \rangle, \tag{17.100}$$

that yields the representation for the effective elastic modulus analogous to Eq. (17.85)

$$\mathbf{L}^* = \mathbf{L}^{(0)} + \mathbf{R}^{\epsilon *}, \quad \langle \boldsymbol{\tau}^I \rangle = \mathbf{R}^{\epsilon *} \langle \boldsymbol{\varepsilon} \rangle, \tag{17.101}$$

where $\boldsymbol{\tau}^I(\mathbf{x})$ is defined by $\boldsymbol{\tau}(\mathbf{x})$ (see, e.g., Eq. (17.92)) with the replacement $\mathbf{u} \to \mathbf{u}^I$ and $\widetilde{\boldsymbol{\beta}}^x(\hat{\mathbf{x}}, \mathbf{x}) \equiv \mathbf{0}$.

In a similar manner, combining Eq. (17.100) with Eq. (17.101) at $\langle \boldsymbol{\varepsilon} \rangle = \mathbf{0}$ leads to the effective eigenstress representation

$$\boldsymbol{\alpha}^* = \boldsymbol{\alpha}^{(0)} + \mathbf{R}^{\alpha *}, \quad \mathbf{R}^{\alpha *} = \langle \boldsymbol{\tau}^{II} \rangle, \tag{17.102}$$

where $\boldsymbol{\tau}^{II}(\mathbf{x})$ is defined by $\boldsymbol{\tau}(\mathbf{x})$ (see, e.g., Eq. (17.92)) with the replacement $\mathbf{u} \to \mathbf{u}^{II}$. Equations (17.101) and (17.102) were used in [209] and [210] for the estimation of both the effective stiffness and effective eigenstrain, respectively, of a simplified 1D random structure (a randomly heterogeneous infinite bar).

It should be clarified that in the average sense, the peridynamic truncated matrix is deformed identically to the locally elastic matrix (with the modulus $\mathbf{L}^{(0)}$) (16.26). However, it does not mean that at the estimation of the effective moduli (16.28), the matrix can be replaced by a locally elastic one because the term $\langle \boldsymbol{\tau} \rangle$ depends on the elastic properties (either local or nonlocal) of the matrix.

17.4.2 *Violation of the Generalized Levin's Representation*

Formally, the averaged equations (17.57) for the thermoelastic (17.47) and thermoperidynamic (17.53) composites are identical (compare Eqs. (17.85) and (17.88) with Eqs. (17.101) and (17.102), respectively), although the polarization tensors ${}^L\boldsymbol{\tau}(\mathbf{x})$ (17.84), ${}^L\boldsymbol{\tau}^{II}(\mathbf{x})$ (17.89) and micropolarization tensors $\boldsymbol{\tau}(\mathbf{x})$ (17.92), $\boldsymbol{\tau}^{II}(\mathbf{x})$, respectively, are conceptually different. However, a common feature of $\boldsymbol{\tau}(\mathbf{x})$ (17.92), $\boldsymbol{\tau}^{II}(\mathbf{x})$, and ${}^L\boldsymbol{\tau}(\mathbf{x})$ (17.84), ${}^L\boldsymbol{\tau}^{II}(\mathbf{x})$ (17.89), respectively, is that all these tensors vanish inside the modified matrix ($\mathbf{x} \in v^{(0)l} := w \backslash v^l, \quad v^l := v \oplus \mathcal{H}_{\mathbf{0}}$) and expressed through the jumps of material properties of both the inclusions and matrix (described by the operators and tensors, respectively); for the local thermoelastic problem, $v^{(0)l} \equiv v^{(0)}$ and $\mathcal{H}_{\mathbf{0}} = \emptyset$. Moreover, one [205] (see Chap. 14) proved that for strongly nonlocal (strain type) thermoelasticity described by the constitutive equation (17.55) (instead of Eq. (17.47)), the effective properties representations (17.33), (17.101), and (17.102) also hold true.

Unfortunately, the effective properties representations for CM with both the local (17.47) and strongly nonlocal (17.55) thermoelastic and thermoperidynamic (17.53) properties do not always coincide, but do so often enough that these formal coincidences can lead to some confusion in our expectations. Namely, for CM with the local constitutive laws at the fine scale (17.47), there is a well-known Levin's [912] sort representation of the effective eigenfields through the mechanical influence functions of the pure mechanical loading problem

$$\boldsymbol{\beta}^* = \langle \mathbf{B}_\sigma^* \boldsymbol{\beta} \rangle, \quad \boldsymbol{\alpha}^* = \langle \mathbf{A}_\varepsilon^* \boldsymbol{\alpha} \rangle. \tag{17.103}$$

It was proved in Chap. 14 that for strongly nonlocal (strain type) thermoelasticity described by the constitutive equation (17.55) (instead of Eq. (17.47)), the effective eigenfield representations (17.103) also hold true. So, correctness of Eq. (17.103_1) was established (see [24]) by the use in Hill's equality (17.67_3) and the self-adjointness of the operator $\boldsymbol{\mathcal{L}}$ (17.55) defined on the symmetric second-order tensors (such as $\boldsymbol{\varepsilon}$). Analogous manipulations for the peridynamic composites are awkwardly expressed for the operator $\widehat{\boldsymbol{\mathcal{L}}}$ (17.37) (which is a peridynamic counterpart of $\boldsymbol{\mathcal{L}}$ (17.55)) is self-adjoint on the vector functions (such as $\mathbf{u}$). Because of this, despite the expectation analogy, Eqs. (17.103) do not in general hold for thermoperidynamic composites (17.53). Indeed, as a particular case of Eqs. (17.103), let us consider the CM with random fluctuation of eigenfields $\boldsymbol{\beta}(\mathbf{x}), \boldsymbol{\alpha}(\mathbf{x}) \neq$const and the homogeneous properties of pure mechanical parameters (e.g., either $\mathbf{L}(\mathbf{x}) \equiv \mathbf{L}^{(0)}$ =const., $\boldsymbol{\mathcal{L}} \equiv \boldsymbol{\mathcal{L}}^{(0)}$, or $\mathbf{C}(\boldsymbol{\xi}, \mathbf{x}) \equiv \mathbf{C}^{(0)}(\boldsymbol{\xi})$). Then in all three cases of constitutive laws at the fine scale mentioned above, the mechanical influence functions $\mathbf{B}_\sigma^*(\mathbf{x}) = \mathbf{A}_\varepsilon^*(\mathbf{x}) \equiv \mathbf{I}$ $\forall \mathbf{x} \in w$, and Eqs. (17.103) are reduced to the effective eigenfield representations

$$\boldsymbol{\beta}^* = \langle \boldsymbol{\beta} \rangle, \quad \boldsymbol{\alpha}^* = \langle \boldsymbol{\alpha} \rangle, \tag{17.104}$$

which do not depend on the remaining material parameters (e.g., either $\mathbf{L}^{(0)}$, $\boldsymbol{\mathcal{L}}^{(0)}$, or $\mathbf{C}^{(0)}(\boldsymbol{\xi})$). However, Eqs. (17.104) hold for both the local (17.47) and strongly nonlocal (17.55) thermoelastic CMs, while for the thermoperidynamic CMs, $\boldsymbol{\beta}^*$, in general, depends on $\mathbf{C}^{(0)}(\boldsymbol{\xi})$, and Eqs. (17.104) (and, therefore, Eqs. (17.103)), in general, only hold in the limit of the horizon going to zero $l_c \to 0$ (see a numerical example for 1D case in at the end of this section).

A $1D$ example of an infinite heterogeneous peridynamic bar $x \in R$ with vanishing force loading conditions ($\langle \sigma \rangle = 0$) is examined for the next geometrical and mechanical properties of constituents. The infinitely long bar of constant cross section contains the randomly distributed identical inclusions of the size $2a$. The micromodulus functions with the step-function profile (16.60_1) with the same horizon $l_c/a = 0.02, \ 0.1, \ 0.25$ are considered for both the inclusions and matrix with the ratios of the corresponding thermoelastic counterparts $\beta^{(1)}/\beta^{(0)} = 5$ (17.54) and $E(x) \equiv E^{(0)}$ =const. Here, $E^{(0)}$ is Young's modulus related with the micromodulus $C(x) = C^{(0)} V(\mathcal{H}_x)$ by the equation $C^{(0)} = 12E^{(0)}/l_c^3$ (see [1601]).

The midpoint quadrature scheme is realized for the uniform grid with a constant step size $\Delta x = a/1000$ (corresponding to $l_c/\Delta x = 20, 100, 250$, respectively); an accuracy of the numerical solution for a single inclusion in the infinite homogeneous bar is analyzed by comparison of results obtained for $\Delta x = l_c/250,\ l_c/100$ and $l_c/20$ (see for details [209, 210]).

At first, a single inclusion $v_i(0)$ in the infinite homogeneous bar is considered for $g(y) = 0$ ($y \in w_{\Gamma\sigma}$, $|y| \gg a$) (16.53) and $\Delta x = a/1000$. For simplicity of subsequent evaluation of the integrals, one accepts a dilute approximation ignoring interaction between inclusions v_i and v_j ($i \neq j,\ i, j \in [1, 2, \ldots)$), impossibility of overlapping of domains $v_i^l \cap v_j^l = \emptyset$ ($v_k^l = \{x : |x - x_k| \leq a + l_c\},\ k = i, j$), and assuming that each inclusion v_i behaves as an isolated one inside a homogeneous bar subjected to some effective homogeneous loading. The infinitely long bar containing the randomly distributed identical inclusions with both the elastic ($E^{(1)} \neq E^{(0)}$) and thermal ($\beta^{(1)} \neq \beta^{(0)}$) mismatches was analyzed in [213]. In particular, for this peridynamically homogeneous medium with random eigenstrains, we get $E^* \equiv E^{(0)}$. One also obtained an exact (with a numerical accuracy of integral evaluations, see for details [213]) representation for α^* in the framework of the assumption mentioned above

$$\alpha^* = \alpha^{(0)} + \frac{n}{2(a + l_c)} \int_{z-a}^{z+a} \int_{r-a-l_c}^{r+a+l_c} F(s)dsdr. \tag{17.105}$$

Here, the function $F(s)$ represents a solution for a single inclusion $v_i(0)$ inside infinite homogeneous bar with vanishing force loading conditions at the infinity

$$\tau(s) = F(s), \tag{17.106}$$

vanishing outside v_i^l rather than $v_i \subset v_i^l$ as in the local case ${}^L\tau(s)$ (17.84).

In Fig. 17.1, one presents the normalized stress fields $\sigma^{II}(x)/(\beta^{(1)} - \beta^{(0)})$ demonstrating strongly inhomogeneous behavior in the area $|x/a - 1| \leq 2l_c/a$. A smooth variation of the residual stresses $\sigma^{II}(x)$ in comparison with the local elastic limit ${}^L\sigma^{II}(x) \equiv 0$ reflects a nonlocal nature of peridynamic phenomena. As can be seen in Fig. 17.2, $\beta^*/\beta^{(0)}$ essentially dependent on l_c and do not depend on $E^{(0)}$ varying in the interval $E^{(0)} \in [0.001, 1000]$. Thus, the results obtained show that the thermoperidynamic theory predicts some features in an apparently simple one-dimensional equilibrium problem that would not be presented in the classical solution (17.104). So, the effective eigenstrain β^* exactly predicted in the classical local theory (17.104) does not depend on the elastic properties ($E^{(0)}$ and $E^{(1)}$) of constituents, while β^* evaluated in the thermoperidynamic theory depends on the micromoduli $C^{(0)}(x, \hat{x})$ and $C^{(1)}(x, \hat{x})$ (compare the curves 1 and 2–4 in Fig. 17.2); see also [210], where the case $E(x) \not\equiv$const. was considered.

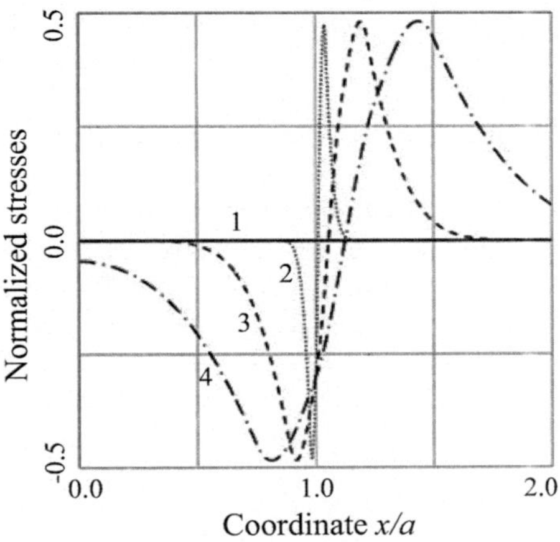

Fig. 17.1 Normalized stresses $\sigma^{II}(x)/(\beta^{(1)} - \beta^{(0)})$ vs. x/a estimated for $l_c/a = 0$ (1), $l_c/a = 0.02$ (2), $l_c/a = 0.1$ (3), and $l_c/a = 0.25$ (4)

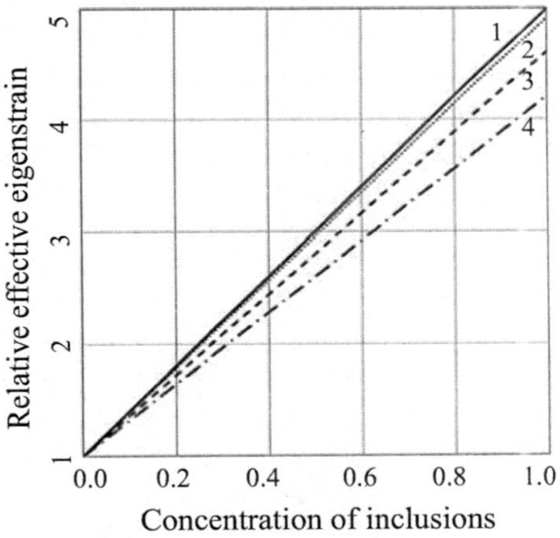

Fig. 17.2 Relative eigenstrains $\beta^*/\beta^{(0)}$ vs. x/a estimated for $l_c/a = 0$ (1), $l_c/a = 0.02$ (2), $l_c/a = 0.1$ (3), and $l_c/a = 0.25$ (4)

17.4.3 Interface Integral Representations of Effective Properties

For obtaining the peridynamic counterpart of the averages of interface polarization tensors (11.1) and (11.2) introduced in Chap. 11, we mention at first that the volume average (17.100) of the local polarization tensor (17.92) is performed over the extended inclusion phase v^l. Then the volume average of the peridynamic analog of the decomposition (17.84) can be presented in the form

$$\langle \boldsymbol{\tau} \rangle_i^l = \langle \boldsymbol{\sigma} \rangle_i^l - \frac{1}{2}\Big\langle \int_S \int_0^\infty \int_0^\infty (y+z)^{(d-1)} \Big\{ \mathbf{C}^{(0)}(\mathbf{x} + y\mathbf{m}, \mathbf{x} - z\mathbf{m}) \times [\mathbf{u}(\mathbf{x} + y\mathbf{m}) - \mathbf{u}(\mathbf{x} - z\mathbf{m})] \Big\} \otimes \mathbf{m} dz dy d\Omega_{\mathbf{m}} \Big\rangle_i^l. \quad (17.107)$$

Equation (17.107) has a very attractive physical meaning. Namely, we formally replaced $\mathbf{C}_1(\mathbf{x}, \hat{\mathbf{x}})$ (17.92) in the extended inclusion phase v^l by the micromodulus of the matrix $\mathbf{C}^{(0)}(\mathbf{x}, \hat{\mathbf{x}})$, while a real random strongly inhomogeneous displacement field $\mathbf{u}(\mathbf{x})$ $(\mathbf{x} \in v_i^l)$ in CM is fixed. Then in the averaging operator of the integral in Eq. (17.107), $\mathbf{u}(\mathbf{x} + y\mathbf{m})$ and $\mathbf{u}(\mathbf{x} - z\mathbf{m})$ are only random, while the rest functions (such as $\mathbf{C}^{(0)}$ and the integration limits) are deterministic at the fixed $(\mathbf{x} + y\mathbf{m})$ and $(\mathbf{x} - z\mathbf{m})$. Because of this, analogous to Eq. (17.98), a deterministic function $\mathbf{C}^{(0)}$ can be carried out (together with the corresponding integral operations) from the averaging operation $\langle(.)\rangle_i^l$ at each $(\mathbf{x}+y\mathbf{m})$ and $(\mathbf{x}-z\mathbf{m})$: $\langle\mathbf{C}^{(0)}(\mathbf{x}+y\mathbf{m}, \mathbf{x}-z\mathbf{m})\mathbf{u}(\mathbf{x}+y\mathbf{m})\rangle_i^l = \mathbf{C}^{(0)}(\mathbf{x}+y\mathbf{m}, \mathbf{x}-z\mathbf{m})\langle\mathbf{u}(\mathbf{x}+y\mathbf{m})\rangle_i^l = \mathbf{C}^{(0)}(\mathbf{x}+y\mathbf{m}, \mathbf{x}-z\mathbf{m})\langle\boldsymbol{\varepsilon}\rangle_i^l(\mathbf{x}+y\mathbf{m})$ that reduce Eq. (17.107) to

$$\langle\boldsymbol{\tau}\rangle_i^l = \left\langle\mathbf{t}(\mathbf{s}) \overset{S}{\otimes} \mathbf{s}\right\rangle_i^{l\omega} - \mathbf{L}^{(0)}\left\langle\mathbf{u}(\mathbf{s}) \overset{S}{\otimes} \mathbf{n}(\mathbf{s})\right\rangle_i^{l\omega} - \boldsymbol{\alpha}^{(0)}\left\langle\mathbf{n}(\mathbf{s}) \overset{S}{\otimes} \mathbf{s}\right\rangle_i^{l\omega}, \quad (17.108)$$

where Eqs. (17.12) and (17.13) were used. At obtaining Eq. (17.108) (as well as Eq. (17.98)), one essentially exploited the statistical homogeneity of microstructure and homogeneity of external loading that yield a linearity of statistical average of the displacement field $\langle\mathbf{u}\rangle(\mathbf{x}) = \langle\boldsymbol{\varepsilon}\rangle\mathbf{x}$.

It is interesting to compare Eq. (17.108) and the locally elastic counterparts (11.6) and (11.7) in Chap. 11. The limit of vanishing length scale

$$l_\delta/a \to 0 \quad (17.109)$$

reduces Eq. (17.108) to Eq. (11.6). At the left-hand sides of these equations, the averages of the polarization tensor ${}^L\boldsymbol{\tau}$ and the local polarization tensor $\boldsymbol{\tau}$ are presented. At the right-hand side of Eq. (11.6), we see the average $\langle{}^L\boldsymbol{\tau}^s\rangle^\omega$ of the interface polarization tensor ${}^L\boldsymbol{\tau}^s$. Although the averages in the right-hand sides of Eqs. (17.108) and (11.6)) are very similar, but the author is faced with difficulties to introduce the peridynamic counterpart of the interface polarization tensor ${}^L\boldsymbol{\tau}^s$ because the interface traction $\mathbf{t}(\mathbf{s})$ and displacement $\mathbf{u}(\mathbf{s})$ are appeared in Eq. (17.108) after the volume average operation (17.107) rather than before as in Eq. (11.1). Thus, a peridynamic analog of Eqs. (11.6) and (11.7) is

$$\mathbf{L}^* = \mathbf{L}^{(0)} + \mathbf{S}^{\epsilon*}, \quad \boldsymbol{\alpha}^* = \boldsymbol{\alpha}^{(0)} + \mathbf{S}^{\alpha*}, \quad (17.110)$$

where

$$\mathbf{S}^{\epsilon*}\langle\boldsymbol{\varepsilon}\rangle = \langle\mathbf{t}^I(\mathbf{s}) \overset{S}{\otimes} \mathbf{s}\rangle_i^{l\omega} - \mathbf{L}^{(0)}\langle\mathbf{u}^I(\mathbf{s}) \overset{S}{\otimes} \mathbf{n}(\mathbf{s})\rangle_i^{l\omega}, \quad (17.111)$$

$$\mathbf{S}^{\alpha*}\langle\boldsymbol{\varepsilon}\rangle = \left\langle\mathbf{t}^{II}(\mathbf{s}) \overset{S}{\otimes} \mathbf{s}\right\rangle_i^{l\omega} - \mathbf{L}^{(0)}\left\langle\mathbf{u}^{II}(\mathbf{s}) \overset{S}{\otimes} \mathbf{n}(\mathbf{s})\right\rangle_i^{l\omega} - \boldsymbol{\alpha}^{(0)}\left\langle\mathbf{n}(\mathbf{s}) \overset{S}{\otimes} \mathbf{s}\right\rangle_i^{l\omega} \quad (17.112)$$

and

$$\mathbf{S}^{\epsilon*} = \mathbf{R}^{\epsilon*}, \quad \mathbf{R}^{\alpha*} = \mathbf{S}^{\alpha*}, \quad (17.113)$$

where it was used a decomposition of the fields $\mathbf{f} = \boldsymbol{\sigma}, \boldsymbol{\varepsilon}$ analogous to Eq. (17.58). Also in an analogy with (11.6), $\mathbf{S}^{\epsilon *}\langle \boldsymbol{\varepsilon} \rangle$ is called average local interface field, although we keep in mind that local interface polarization tensor is not defined. It can be seen a similarity between ${}^L\mathbf{S}^{\epsilon *}$ (11.6) and $\mathbf{S}^{\epsilon *}$ (17.111) which is more manifested than similarity between ${}^L\mathbf{R}^*$ (see Eq. 6.55) and $\mathbf{R}^*$ (see Eq. (6.55). In both cases, ${}^L\mathbf{S}^{\epsilon *}$ (11.6) and $\mathbf{S}^{\epsilon *}$ (17.111), we need to estimate the surface averages of formally the same function $\mathbf{t}(\mathbf{s}) \overset{S}{\otimes} \mathbf{s} - \mathbf{L}^{(0)}[\mathbf{u}(\mathbf{s}) \otimes \mathbf{n}(\mathbf{s})]$ over the surfaces Γ_i and Γ_i^-, respectively, although the local modulus $\mathbf{L}^{(0)}$ (determined by Eq. (17.99)) is not defined at the external interface $\mathbf{s} \in \Gamma_i^-$.

Thus, for statistically homogeneous thermoperidynamic CMs subjected to homogeneous volumetric boundary loading, one proved that the effective behavior of these CMs is governed by the conventional effective constitutive equations (17.57) as for the local thermoelasticity theory. It was made by the most exploitation of the popular tools and concepts used in conventional local thermoelasticity of CMs and adapted to thermoperidynamic. These are decomposition of real fields into two constituents of the load and residual fields (17.58) as well as extraction from the material properties a constituent of the matrix properties (17.100). The general results establishing the links between the effective properties (effective elastic moduli, effective thermal expansion) and the corresponding mechanical and transformation influence functions are obtained by the use of decomposition of local fields into the load and residual fields. The effective properties $\mathbf{L}^*$ and $\boldsymbol{\alpha}^*$ of thermoperidynamic CM are expressed through the introduced local stress polarization tensor (17.92) averaged over the extended inclusion phase $\langle \boldsymbol{\tau} \rangle(\mathbf{x})$ ($\mathbf{x} \in v^l$) rather than in an entire space $\langle \boldsymbol{\tau} \rangle(\mathbf{x})$ ($\mathbf{x} \in R^d$) (compare Eqs. (17.100)–(17.102) with the corresponding Eqs. (17.85) for the local elasticity). A detected similarity of results for both the peridynamic and locally elastic composites is explained by the fundamental reasons because the methods used for obtaining the mentioned results widely exploit Hill's [640] (17.22) condition and the self-adjointness of the stress operator $\widehat{\mathcal{L}}$ (17.36). This similarity opens a way for straightforward expansion of analytical micromechanics tools (developed for the locally elastic CMs; see Chaps. 8–13 to the new area of peridynamic micromechanics of random structure CMs. However, the effective properties representations for composites with both the local thermoelastic and nonlocal thermoperidynamic properties do not always coincide. So, the representations of the effective eigenfields through the mechanical influence functions generalizing Levin's [912] representation (17.103) do not in general hold for thermoperidynamic CMs that are demonstrated for $1D$ numerical example (see Fig. 17.2).

17.5 Constitutive Model and Strain Energy

For the local elasticity, the relations between the stresses $\boldsymbol{\sigma}$ and strains $\boldsymbol{\varepsilon}$ are a constitutive model prescribed by the tensors of either the stiffness $\mathbf{L}$ (6.2_1) or compliance $\mathbf{M} = \mathbf{L}^{-1}$ (6.2$_2$). A peridynamic counterpart of the constitutive equation (6.2$_1$) with the stiffness tensor $\mathbf{L}$ is represented by Eq. (16.42):

$$\boldsymbol{\sigma}(\mathbf{x}) = \mathcal{L}^{\eta}(\boldsymbol{\eta})(\mathbf{x}), \tag{17.114}$$

where the linear operator $\mathcal{L}^{\eta}(\boldsymbol{\eta}) \equiv \mathcal{L}^{\sigma}(\mathbf{u})$ (16.42) is introduced for notation convenience. Our goal is a construction of the inversion of the linear operator $\mathcal{L}^{\eta}(\boldsymbol{\eta})$

$$\boldsymbol{\eta}(\mathbf{x}) = \mathcal{M}^{\sigma}(\boldsymbol{\sigma}), \tag{17.115}$$

which is a nonlocal counterpart of Eq. (6.2$_2$) mapping a vector space $\boldsymbol{\sigma} \in R^n$ into $\boldsymbol{\eta} \in R^n$. This construction will be performed by a slight modification of the iteration method proposed in [223].

For the construction of the inversion of the linear operator $\mathcal{L}^{\eta}(\boldsymbol{\eta})$, let us consider the governing equation for an infinite R^d $(d = 1, 2, 3)$ homogeneous peridynamic medium (16.5)

$$\widehat{\mathcal{L}}^{(0)}(\mathbf{u}^0)(\mathbf{x}) + \mathbf{b}(\mathbf{x}) = \mathbf{0} \tag{17.116}$$

with the micromodulus $\mathbf{C}^{(0)}$. The self-equilibrated body force density $\mathbf{b}(\mathbf{x})$ (17.114) produces a displacement corresponding to the homogeneous stress

$$\mathbf{u}^0(\mathbf{x}) \equiv \boldsymbol{\varepsilon}^0\mathbf{x}, \quad \boldsymbol{\varepsilon}^0 = \mathbf{M}^{(0)}\{\boldsymbol{\sigma}\} \equiv \text{const}, \tag{17.117}$$

where the compliance $\mathbf{M}^{(0)}$ of the homogeneous matrix will be defined later. We are coming now to consideration of a single fixed inclusion $v_i(\mathbf{0})$ with the micromodulus $\mathbf{C}^{(1)}(\boldsymbol{\xi})$ in the infinite homogeneous medium with the properties $\mathbf{C}^{(0)}(\boldsymbol{\xi})$. Then the solution $\mathbf{u}(\mathbf{x})$ can be split into two ones as

$$\mathbf{u}(\mathbf{x}) = \mathbf{u}^0(\mathbf{x}) + \mathbf{u}_1(\mathbf{x})$$

where

$$\widehat{\mathcal{L}}^{(0)}(\mathbf{u}^0)(\mathbf{x}) + \mathbf{b}(\mathbf{x}) = \mathbf{0},$$
$$\widehat{\mathcal{L}}(\mathbf{u}_1)(\mathbf{x}) + \widehat{\mathcal{L}}_1(\mathbf{u}^0)(\mathbf{x}) = \mathbf{0}, \tag{17.118}$$

and the operator $\widehat{\mathcal{L}}$ is decomposed as $\widehat{\mathcal{L}} \equiv \widehat{\mathcal{L}}^{(0)} + \widehat{\mathcal{L}}_1$ (17.114), where $\widehat{\mathcal{L}}_1$ is defined by Eq. (16.5) with the replacements $\mathbf{C}(\mathbf{x}, \hat{\mathbf{x}}) \to \mathbf{C}_1(\mathbf{x}, \hat{\mathbf{x}}) := \mathbf{C}(\mathbf{x}, \hat{\mathbf{x}}) - \mathbf{C}^{(0)}(\mathbf{x}, \hat{\mathbf{x}})$; the jump of the micromodulus $\mathbf{C}_1(\mathbf{x}, \hat{\mathbf{x}})$ vanishes inside the truncated matrix $\mathbf{x}, \hat{\mathbf{x}} \in v^{l(0)}$. The decomposition $\widehat{\mathcal{L}} \equiv \widehat{\mathcal{L}}^{(0)} + \widehat{\mathcal{L}}_1$ is supported by a single assumption of the matrix homogeneity meaning invariantness of the tensor $\mathbf{C}^{(0)}(\mathbf{x}, \hat{\mathbf{x}})$ with respect to parallel transaction. However, the main advantage of the decompositions (17.118) is elimination in (17.118) of the nonlocal volumetric boundary conditions prescribed over a nonzero volume (see, e.g., (16.47) and [396, 397, 771, 996, 1396]) of Eq. (17.118$_2$), while the solution of Eq. (17.118$_1$) is assumed to be known. Indeed, Eq. (17.118) can be recast in the form (16.49) ($p, q = \ldots, -2, -1, 0, 1, 2, \ldots$)

$$\sum_q \mathbb{K}_{pq} \mathbf{u}_{1|q} = \tilde{\mathbf{b}}_p, \tag{17.119}$$

where the infinite banded matrix $\mathbb{K}_{pq}$ is defined by Eq. (16.50) while a fictitious body force has the following representation:

$$\tilde{\mathbf{b}}_p = -\sum_{r \in \mathcal{H}_p} \mathbf{C}_1(\mathbf{x}_p, \mathbf{x}_r)(\mathbf{M}^{(0)}\{\boldsymbol{\sigma}\})(\mathbf{x}_r - \mathbf{x}_p)\bar{v}_{pr}. \tag{17.120}$$

The solution of Eqs. (17.119) and (17.120) by the iteration method is described in detail by the iteration method in Subsection 19.1.3. Thus, the basic problem is assumed to be solved, and, therefore, the inverse linear operator $\boldsymbol{\mathcal{M}}^{\sigma}$ (17.115) (found from Eq. (17.119)) also has a matrix representation

$$\boldsymbol{\eta}(\mathbf{x}) = \mathbf{M}^{\sigma}\boldsymbol{\sigma}, \quad \mathbf{M}^{\sigma} = (\mathbf{L}^{\eta})^{-1}. \tag{17.121}$$

For the actual displacements $\mathbf{u} = \mathbf{v}$, the inner product $(\widehat{\mathcal{L}}(\mathbf{u}), \mathbf{u})$ defines the total strain energy ($\boldsymbol{\xi} = \mathbf{y} - \mathbf{x}$)

$$W^u(\mathbf{u}) := W = \frac{1}{2}(\widehat{\mathcal{L}}(\mathbf{u}), \mathbf{u}) = \frac{1}{4}\Big\{\int_{\mathcal{H}_x} \boldsymbol{\eta}^{\top}\mathbf{C}(\boldsymbol{\xi}, \mathbf{x})\boldsymbol{\eta}\, d\mathbf{y}\Big\}, \tag{17.122}$$

where for a notational convenience, the total strain energy is denoted by $W^u(\mathbf{u})$ rather than W (16.27).

We can also recast the strain energy W as a function of the stress fields by substitution of the constitutive equation (17.121) into Eq. (17.122):

$$W^{\sigma}(\boldsymbol{\sigma}) := W = \frac{1}{4}\Big\{\int_{\mathcal{H}_x} \boldsymbol{\sigma}^{\top}\breve{\mathbf{M}}\boldsymbol{\sigma}\, d\mathbf{y}\Big\}, \quad \breve{\mathbf{M}} := (\mathbf{M}^{\sigma})^{\top}\mathbf{C}(\boldsymbol{\xi}, \mathbf{x})\mathbf{M}^{\sigma}. \tag{17.123}$$

The limit of vanishing length scale $l_{\delta}/a \to 0$ reduces the peridynamic representations (17.122) and (17.123) to their locally elastic counterparts for the strain energy

$${}^{L}W^{\epsilon}(\boldsymbol{\varepsilon}) = \frac{1}{2}\{\boldsymbol{\varepsilon}^{\top}\mathbf{L}\boldsymbol{\varepsilon}\}, \tag{17.124}$$

$${}^{L}W^{\sigma}(\boldsymbol{\sigma}) = \frac{1}{2}\{\boldsymbol{\sigma}^{\top}\mathbf{M}\boldsymbol{\sigma}\}, \tag{17.125}$$

respectively, and ${}^{L}W^{\epsilon}(\boldsymbol{\varepsilon}) = {}^{L}W^{\sigma}(\boldsymbol{\sigma})$ because $\mathbf{M}^{\top}\mathbf{L}\mathbf{M} = \mathbf{M}$.

A formal similarity of the strain energy representations for both the local elasticity (17.124), (17.125) and peridynamics (17.122), (17.123), respectively, opens an opportunity for a straightforward modification of the corresponding variational principles of local elasticity (see [353, 422, 639, 640, 1263]).

17.6 Minimum Principles of Linear Peridynamics

17.6.1 Some Properties of Strain Energy

For consideration of some properties of strain energy, we introduce the following notation for an *admissible set*, $\mathbf{s} = [\mathbf{u}, \mathbf{f}] \in \mathcal{A}^{ks}(\mathbf{u}, \mathbf{f})$, where $\mathbf{u} \in \mathcal{A}^{k}(\mathbf{u})$ is an admissible displacement field and $\mathbf{f} \in \mathcal{A}^{s}(\mathbf{f})$ is an admissible force field. It should be emphasized that $\mathbf{u}$ and $\mathbf{f}$ need not be related by a specific constitutive model, e.g., (16.5) and (17.114). The pair $[\mathbf{b}, \mathbf{g}]$ is called the *external force system*.

Lemma *For two displacements* $\mathbf{u}(\mathbf{x})$ *and* $\tilde{\mathbf{u}}(\mathbf{x})$ $(\mathbf{x} \in \overline{\overline{w}})$, *respectively, the following relation between the strain energies takes place:*

$$W^u(\mathbf{u} + \tilde{\mathbf{u}}) = W^u(\mathbf{u}) + W^u(\tilde{\mathbf{u}}) - (\widehat{\mathcal{L}}(\mathbf{u}), \tilde{\mathbf{u}}). \tag{17.126}$$

Indeed, due to the self-adjointness of the peridynamic operator $\widehat{\mathcal{L}}$ (17.36)

$$(\widehat{\mathcal{L}}(\mathbf{u} + \tilde{\mathbf{u}}), \mathbf{u} + \tilde{\mathbf{u}}) = (\widehat{\mathcal{L}}(\mathbf{u}), \mathbf{u}) + (\widehat{\mathcal{L}}(\tilde{\mathbf{u}}), \tilde{\mathbf{u}}) + 2(\widehat{\mathcal{L}}(\mathbf{u}), \tilde{\mathbf{u}}), \tag{17.127}$$

that implies the relation (17.126).

We consider a peridynamic medium under an external body force density $\mathbf{b}(\mathbf{x})$ $(\mathbf{x} \in w)$ and external "traction" $\mathbf{g}(\mathbf{x})$ $(\mathbf{x} \in w_{\Gamma\sigma})$. Effecting the inner product of both sides of Eqs. (16.47) with kinematically admissible displacement $\mathbf{u} \in \mathcal{A}^{k}(\mathbf{u})$ and taking Eq. (17.122) into account, we prove

Theorem of Work and Energy

$$(\mathbf{b}, \mathbf{u}) + (\mathbf{g}, \mathbf{u}) = -(\widehat{\mathcal{L}}(\mathbf{u}), \mathbf{u}) = 2W^u(\mathbf{u}), \tag{17.128}$$

which states that the work done by the external force system $\breve{\mathbf{b}} = [\mathbf{b}, \mathbf{g}]$ is equal to twice the total strain energy $2W^u(\mathbf{u})$ stored in $\overline{\overline{w}}$ when the body is in equilibrium.

Further, due to the positive definiteness of $\mathbf{C}$, the following theorem takes place:

Theorem of Positive Work *For any admissible set* $[\mathbf{u}, \mathbf{f}] \in \mathcal{A}^{ks}$ *corresponding to the external force system* $[\mathbf{b}, \mathbf{g}]$, *the work produced by* $[\mathbf{b}, \mathbf{g}]$ *is nonnegative*

$$W^u(\mathbf{u}) \geq 0, \tag{17.129}$$

and vanishes only for the rigid displacement $\mathbf{u}$.

Indeed, due to the positive definiteness of $\mathbf{C}$, the strain energy $W^u(\mathbf{u}) \geq 0$ (17.122), and the work produced by $[\mathbf{b}, \mathbf{g}]$ (17.128) is nonnegative. If $W^u(\mathbf{u}) = 0$, then $\boldsymbol{\eta}^\top \mathbf{C}(\boldsymbol{\xi}, \mathbf{x})\boldsymbol{\eta}$ and $\boldsymbol{\eta}$ must vanish on w. Then $\mathbf{u}$ must be a rigid displacement. Conversely, if $\mathbf{u}$ is a rigid displacement, the relative displacement $\boldsymbol{\eta}$ vanishes with the strain energy $W^u(\mathbf{u})$ (17.122), and, therefore, the work produced by the external system $[\mathbf{b}, \mathbf{g}]$ (17.128) is zero.

Due to the linearity of the peridynamic operator $\widehat{\mathcal{L}}$, we have the following:

Principle of Superposition for Admissible Sets *If* $\mathbf{s} = [\mathbf{u}, \mathbf{f}] \in \mathcal{A}^{ks}$ *and* $\widetilde{\mathbf{s}} = [\widetilde{\mathbf{u}}, \widetilde{\mathbf{f}}] \in \mathcal{A}^{ks}$ *are the admissible sets corresponding to the external force systems* $[\mathbf{b}, \mathbf{g}]$ *and* $[\widetilde{\mathbf{b}}, \widetilde{\mathbf{g}}]$, *respectively, then for any* $\forall \alpha, \widetilde{\alpha} \in R$, *a linear combination* $\alpha\mathbf{s} + \widetilde{\alpha}\widetilde{\mathbf{s}} \in A^{ks}$ *is an admissible set corresponding to the external force system* $\alpha[\mathbf{b}, \mathbf{g}] + \widetilde{\alpha}[\widetilde{\mathbf{b}}, \widetilde{\mathbf{g}}]$ *and*

$$\begin{aligned} \alpha[\mathbf{u}, \mathbf{f}] + \widetilde{\alpha}[\widetilde{\mathbf{u}}, \widetilde{\mathbf{f}}] &= [\alpha\mathbf{u} + \widetilde{\alpha}\widetilde{\mathbf{u}}, \alpha\mathbf{f} + \widetilde{\alpha}\widetilde{\mathbf{f}}], \\ \alpha[\mathbf{b}, \mathbf{g}] + \widetilde{\alpha}[\widetilde{\mathbf{b}}, \widetilde{\mathbf{g}}] &= [\alpha\mathbf{b} + \widetilde{\alpha}\widetilde{\mathbf{b}}, \alpha\mathbf{g} + \widetilde{\alpha}\widetilde{\mathbf{g}}]. \end{aligned} \tag{17.130}$$

Additionally, combining the theorem of work and energy (17.128) with the properties of self-adjointness of the peridynamic operator $\widehat{\mathcal{L}}$ (17.122) leads to

Betti's Reciprocal Theorem Let $[\mathbf{u}, \mathbf{f}] \in \mathcal{A}^{ks}$ and $[\widetilde{\mathbf{u}}, \widetilde{\mathbf{f}}] \in \mathcal{A}^{ks}$ be two independent admissible sets corresponding to the external force systems $[\mathbf{b}, \mathbf{g}]$ and $[\widetilde{\mathbf{b}}, \widetilde{\mathbf{g}}]$, respectively. Then

$$(\mathbf{b}, \widetilde{\mathbf{u}}) + (\mathbf{g}, \widetilde{\mathbf{u}}) = (\widetilde{\mathbf{b}}, \mathbf{u}) + (\widetilde{\mathbf{g}}, \mathbf{u}) = -(\widehat{\mathcal{L}}(\mathbf{u}), \widetilde{\mathbf{u}}) = (\widehat{\mathcal{L}}(\widetilde{\mathbf{u}}), \mathbf{u}). \tag{17.131}$$

Betti's theorem states that the work of the first external system $[\mathbf{b}, \mathbf{g}]$ on the displacement $\widetilde{\mathbf{u}}$ produced by the second system $[\widetilde{\mathbf{b}}, \widetilde{\mathbf{g}}]$ equals to the work done by the second external system $[\widetilde{\mathbf{b}}, \widetilde{\mathbf{g}}]$ over the displacement $\mathbf{u}$ caused by the first system $[\mathbf{b}, \mathbf{g}]$.

At last, we turn our attention to one of the most general principles in mechanics, *principle of virtual work*. We introduce a kinematically admissible virtual displacement $\delta\mathbf{u} \in \mathcal{A}^k(\mathbf{u})$ consisting of the prescribed volumetric boundary conditions (16.46), i.e., $\delta\mathbf{u}(\mathbf{x}) = \mathbf{0}$ $(\mathbf{x} \in w_{\Gamma u})$. The admissible forces $\mathbf{f}$ and the external force system $[\mathbf{b}, \mathbf{g}]$ (which may but do not need to be related with $\delta\mathbf{u}$ by a constitutive model) remain constant along the virtual deformation path $\delta\mathbf{u}$. Reconstruction of the derivative of Eq. (17.128) with virtual term $\delta\mathbf{u}$ instead of $\mathbf{u}$ implies the

Virtual Work Theorem

$$(\mathbf{b}, \delta\mathbf{u}) + (\mathbf{g}, \delta\mathbf{u}) = -(\widehat{\mathcal{L}}, \delta\mathbf{u}) = \frac{1}{2}\Big\{\int_{\mathcal{H}_x} \boldsymbol{\eta}^\top \mathbf{C}(\boldsymbol{\xi}, \mathbf{x})\delta\boldsymbol{\eta} d\mathbf{y}\Big\}, \tag{17.132}$$

where $\delta\eta := \delta\mathbf{u}(\mathbf{y}) - \delta\mathbf{u}(\mathbf{x})$ and the summand $(\mathbf{g}, \delta\mathbf{u})$ is estimated only over $w_{\Gamma\sigma}$ because $\delta\mathbf{u}(\mathbf{x}) = \mathbf{0}$ on $\mathbf{x} \in w_{\Gamma u}$.

17.6.2 Minimum Principles

We now turn to consideration of the principle of minimum potential energy and the minimum complementary energy characterizing the solution of the mixed problem (16.51).

Principle of Minimum Potential Energy *Let the functional*

$$\Pi(\mathbf{s}) = \Pi(\mathbf{u}, \mathbf{b}, \mathbf{g}) := W^u(\mathbf{u}) - \int_w \mathbf{b} \cdot \mathbf{u} d\mathbf{x} - \int_{w_{\Gamma\sigma}} \mathbf{g} \cdot \mathbf{u} d\mathbf{x} \tag{17.133}$$

be defined on the space of admissible sets $\mathbf{s} = [\mathbf{u}, \mathbf{f}] \in \mathcal{A}^{ks}$. *If* $\mathbf{s}$ *is a solution of the mixed problem (16.51), then*

$$\Pi(\mathbf{s}) \leq \Pi(\widetilde{\mathbf{s}}) \tag{17.134}$$

for every $\widetilde{\mathbf{s}} \in \mathcal{A}^{ks}$, *and the equality is reached only if* $\widetilde{\mathbf{s}} = \mathbf{s}$ *modulo a rigid displacement.*

Here, the *potential energy* $\Pi(\mathbf{s})$ (17.133) corresponding to the given external force system $[\mathbf{b}, \mathbf{g}]$ equals to a summation of the strain energy $W^u(\mathbf{u})$ given by (17.122) and the potential of the external force system $[\mathbf{b}, \mathbf{g}]$. The last integral in Eq. (17.133) is estimated over the part of the interaction boundary $w_{\Gamma\sigma}$ where the force loading $\mathbf{g}$ is prescribed.

In order to prove this principle, let $\mathbf{s}$ be a solution of the mixed problem (16.51) and $\widetilde{\mathbf{s}} \in \mathcal{A}^{ks}$ be an arbitrary admissible set. Then $\mathbf{s}' := \widetilde{\mathbf{s}} - \mathbf{s} = [\mathbf{u}', \mathbf{f}']$ is also admissible (see Eqs. (17.130)) and, according to (17.126),

$$W^u(\widetilde{\mathbf{u}}) - W^u(\mathbf{u}) = W^u(\mathbf{u}') - (\widehat{\mathcal{L}}(\mathbf{u}), \mathbf{u}'). \tag{17.135}$$

The last term in Eq. (17.135) can be transformed, due to the theorem of work and energy (17.128), as

$$-(\widehat{\mathcal{L}}(\mathbf{u}), \mathbf{u}') = (\mathbf{b}, \mathbf{u}') + (\mathbf{g}, \mathbf{u}'). \tag{17.136}$$

Then Eqs. (17.135) and (17.136) imply, due to definition (17.133),

$$\Pi(\widetilde{\mathbf{s}}) - \Pi(\mathbf{s}) = W^u(\mathbf{u}'). \tag{17.137}$$

Due to the positive definiteness of **C**, $W^u(\mathbf{u}') \geq 0$ (17.129) and

$$\Pi(\mathbf{s}) \leq \Pi(\widetilde{\mathbf{s}}) \quad \text{on } \mathcal{A}^{ks}. \tag{17.138}$$

In so doing, $\Pi(\mathbf{s}) = \Pi(\widetilde{\mathbf{s}})$ if and only if $\mathbf{s}' = \widetilde{\mathbf{s}} - \mathbf{s} = \mathbf{0}$, i.e., $\widetilde{\mathbf{s}} = \mathbf{s}$ modulo a rigid displacement and the proof of the principle of minimum potential energy is complete.

The minimum potential energy theorem asserts that when potential energy of linear peridynamic body has a minimum, the displacement field is that of actual equilibrium state and it satisfies volumetric boundary conditions prescribed at $w_{\Gamma u}$.

Next, we recast the strain energy W of a peridynamic medium in equilibrium as the function of peridynamic stress field $W = W^\sigma(\boldsymbol{\sigma})$ (17.123). We also consider a statically admissible force $\mathbf{f} \in \mathcal{A}^s$ defining the peridynamic stresses $\boldsymbol{\sigma} \in \mathcal{A}^s$ (16.42) and satisfying the equilibrium and the force boundary loading (16.47). Then the following holds:

Principle of Minimum Complementary Energy *Let $\Pi^*(\widetilde{\boldsymbol{\sigma}})$ (called complementary energy) be a functional on $\mathcal{A}^s$*

$$\Pi^*(\widetilde{\boldsymbol{\sigma}}) = W^\sigma(\widetilde{\boldsymbol{\sigma}}) - \int_{w_{\Gamma u}} \widehat{\mathcal{L}}(\widetilde{\mathbf{f}}) \cdot \mathbf{h} dx \tag{17.139}$$

for every $\widetilde{\mathbf{f}} \in \mathcal{A}^s$. Then for the solution $\mathbf{u}, \boldsymbol{\sigma}$ of Eq. (16.47) with prescribed displacement loading $\mathbf{u}(\mathbf{x}) = \mathbf{h}(\mathbf{x})$ $(\mathbf{x} \in w_{\Gamma u})$

$$\Pi^*(\boldsymbol{\sigma}) \leq \Pi^*(\widetilde{\boldsymbol{\sigma}}) \tag{17.140}$$

for any $\widetilde{\boldsymbol{\sigma}} \in \mathcal{A}^s$, and equality holds if and only if $\widetilde{\boldsymbol{\sigma}} = \boldsymbol{\sigma}$.

Indeed, let $\mathbf{u}, \boldsymbol{\sigma}$ be a solution of the mixed problem (16.51) and $\widetilde{\boldsymbol{\sigma}} \in \mathcal{A}^s$ be a statically admissible stress field. We define $\mathbf{f}' = \widetilde{\mathbf{f}} - \mathbf{f}$ and $\boldsymbol{\sigma}' = \widetilde{\boldsymbol{\sigma}} - \boldsymbol{\sigma}$. Then $\mathbf{f}'$ satisfies the equilibrium equation (16.47$_1$) with the external force system $\check{\mathbf{b}}'(\mathbf{x}) = \mathbf{0}$ $(\mathbf{x} \in w \cup w_{\Gamma\sigma})$. Due to a self-adjointness of the operator $\widehat{\mathcal{L}}$ (17.37), Eq. (17.126) can be recast as

$$W^\sigma(\widetilde{\boldsymbol{\sigma}}) - W^\sigma(\boldsymbol{\sigma}) = W^\sigma(\boldsymbol{\sigma}') - (\widehat{\mathcal{L}}(\mathbf{f}'), \mathbf{u}), \tag{17.141}$$

that leads to

$$\Pi^*(\widetilde{\boldsymbol{\sigma}}) - \Pi^*(\boldsymbol{\sigma}) = W^\sigma(\boldsymbol{\sigma}'). \tag{17.142}$$

At obtaining (17.142), the item $(\widehat{\mathcal{L}}(\mathbf{f}'), \mathbf{u})$ (17.141) was transformed by the use of the decomposition $\overline{\overline{w}} = (w \cup w_{\Gamma\sigma}) \cup w_{\Gamma u}$ of the integration domain $\overline{\overline{w}}$. Since $W^\sigma(\boldsymbol{\sigma}')$ is nonnegative, we finally obtain $\Pi^*(\boldsymbol{\sigma}) \leq \Pi^*(\widetilde{\boldsymbol{\sigma}})$ and $\Pi^*(\boldsymbol{\sigma}) = \Pi^*(\widetilde{\boldsymbol{\sigma}})$ if and only if $\boldsymbol{\sigma}' = \widetilde{\boldsymbol{\sigma}} - \boldsymbol{\sigma} = \mathbf{0}$.

17.6.3 Upper and Lower Bounds for the Strain Energy

A particular interest in subsequent estimation of bounds on overall properties of CMs is based on estimations of bounds for both the potential energy and compliance energy in the special cases of vanishing body force $\mathbf{b}(\mathbf{x}) \equiv \mathbf{0}$ ($\mathbf{x} \in w$) and the homogeneous loading either (16.52) or (16.53) at the volumetric boundary w_Γ. For the first volumetric boundary value problem with prescribed homogeneous displacement loading (16.52) $\boldsymbol{\varepsilon}^{w_\Gamma}(\mathbf{x})$ =const. ($\mathbf{x} \in w_{\Gamma u} = w_\Gamma, \quad w_{\Gamma\sigma} = \emptyset$), it follows from the definition (17.133) that

$$W^u(\mathbf{u}) = \Pi(\mathbf{u}) \leq \Pi(\widetilde{\mathbf{u}}) = W^u(\widetilde{\mathbf{u}}) \tag{17.143}$$

for any kinematically admissible displacement $\widetilde{\mathbf{u}} \in \mathcal{A}^k$ and the actual displacement $\mathbf{u}$. On the other hand, due to $w_{\Gamma u} = w_\Gamma$, it follows from the theorem of work and energy (17.128) that the last term in the expression for $\Pi^*(\sigma)$ in Eq. (17.139) equals to $-2W$. Therefore, $\Pi^*(\boldsymbol{\sigma}) = -W$, and Eq. (17.140) implies

$$W^\sigma(\boldsymbol{\sigma}) \geq \langle\widetilde{\boldsymbol{\sigma}}\rangle\boldsymbol{\varepsilon}^{w_\Gamma} - W^\sigma(\widetilde{\boldsymbol{\sigma}}) \tag{17.144}$$

for any statically admissible $\widetilde{\boldsymbol{\sigma}} \in \mathcal{A}^s$ and actual $\boldsymbol{\sigma}$ fields.

For the second volumetric boundary value problem, where the volumetric boundary forces $\mathbf{g}(\mathbf{x})$ with the corresponding homogeneous stresses $\boldsymbol{\sigma}(\mathbf{x}) = \boldsymbol{\sigma}^{w_\Gamma}$ =const. ($\mathbf{x} \in w_{\Gamma\sigma} = w_\Gamma, \quad w_{\Gamma u} = \emptyset$) are applied, we conclude from the theorem of work and energy that the last term in Eq. (17.133) equals to $-2W^\sigma(\mathbf{s})$ and, therefore, $\Pi(\mathbf{s}) = -W^u(\mathbf{u})$ (17.133). Then Eqs. (17.134) and (17.139) imply

$$W^u(\mathbf{u}) \geq \boldsymbol{\sigma}^{w_{\Gamma\sigma}}\langle\widetilde{\boldsymbol{\varepsilon}}\rangle - W^u(\widetilde{\mathbf{u}}), \tag{17.145}$$

$$W^\sigma(\boldsymbol{\sigma}) \leq W^\sigma(\widetilde{\boldsymbol{\sigma}}) \tag{17.146}$$

for any statically admissible $\widetilde{\mathbf{g}}$ generating $\widetilde{\boldsymbol{\sigma}}$ and actual $\mathbf{u}$, $\boldsymbol{\sigma}$ fields.

17.7 Generalized Hill's Bounds

One has demonstrated in Sect. 5.3 that the special choices of the volumetric homogeneous boundary conditions either (16.52) or (16.53) make it possible to estimate the upper and lower bonds of the strain energy. Subsequent bounds estimations of effective properties are defined by a special choice of homogeneous either the kinematic or static admissible fields satisfying to the prescribed volumetric boundary condition either (16.52) or (16.53), respectively. To demonstrate these estimations, we at first present some extraction from the previously introduced definitions of effective moduli and compliances (see Chap. 6).

17.7.1 *Effective Elastic Moduli*

The basic subject of the theory of statistically homogeneous linear peridynamic composites is the characterization of the *overall* or *equivalent* elastic behavior of composites. This involves finding the *overall* or *effective* moduli as the moduli of some fictitious homogeneous medium with the same average stress-strain state as for the microinhomogeneous one being considered. It means that there exists a *representative volume element* Ω (RVE) of the body under consideration that (see [640]) "(a) is structurally entirely typical of the whole mixture on average, and (b) contains a sufficient number of inclusions for the apparent overall moduli to be effectively independent of the values" of force loading conditions (16.53) and displacement ones (16.52), so long as these values are "macroscopically uniform" (this description is a slight modification of the corresponding definition for locally elastic composites by Hill [640] accentuated by the quotes) in the framework of scale separation (7.10_2).

Due to the RVE concept, independently on the nature of the constitutive laws (either local or nonlocal) in the micro (or nano) points, the field macrovariables $\langle\boldsymbol{\sigma}\rangle$ and $\langle\boldsymbol{\varepsilon}\rangle$ depend only on the homogeneous nonlocal boundary conditions, either (16.52) or (16.53), and are related by a local material effective tensor of either the elasticity or compliance:

$$\{\boldsymbol{\sigma}\} = \mathbf{L}^*\{\boldsymbol{\varepsilon}\}, \quad \{\boldsymbol{\varepsilon}\} = \mathbf{M}^*\{\boldsymbol{\sigma}\}, \tag{17.147}$$

respectively. Owing to the scale separation hypothesis (7.10_2), statistical homogeneities of both the material and field parameters, and homogeneity of the boundary loading, the nonlocal boundary conditions do not affect the statistical average of the constitutive equation (17.147).

Establishment of the constitutive linkage between the suitably chosen micro- (or nano-) and macrofield is reduced to the estimation of the tensor *influence functions* $\mathbf{A}^*_\varepsilon(\mathbf{x})$, $\mathbf{B}^*_\varepsilon(\mathbf{x})$ and $\mathbf{A}^*_\sigma(\mathbf{x})$, $\mathbf{B}^*_\sigma(\mathbf{x})$ (do not miss with the influence functions in peridynamics, see [1356, 1405]) for the nonlocal homogeneous boundary conditions (16.52) or (16.53):

$$\mathbf{u}(\mathbf{x}) = \mathbf{A}^*_u(\mathbf{x})\{\boldsymbol{\varepsilon}\}, \quad \boldsymbol{\varepsilon}(\mathbf{x}) = \mathbf{A}^*_\epsilon(\mathbf{x})\{\boldsymbol{\varepsilon}\}, \quad \boldsymbol{\sigma}(\mathbf{x}) = \mathbf{B}^*_\epsilon(\mathbf{x})\{\boldsymbol{\varepsilon}\}, \tag{17.148}$$

$$\mathbf{u}(\mathbf{x}) = \mathbf{B}^*_u(\mathbf{x})\{\boldsymbol{\sigma}\}, \quad \boldsymbol{\varepsilon}(\mathbf{x}) = \mathbf{A}^*_\sigma(\mathbf{x})\{\boldsymbol{\sigma}\}, \quad \boldsymbol{\sigma}(\mathbf{x}) = \mathbf{B}^*_\sigma(\mathbf{x})\{\boldsymbol{\sigma}\}, \tag{17.149}$$

respectively. Averaging Eqs. (17.148_2) and (17.149_3) leads to the conclusion that

$$\{\mathbf{A}^*_\epsilon(\mathbf{x})\} = \{\mathbf{B}^*_\sigma(\mathbf{x})\} = \mathbf{I}. \tag{17.150}$$

The relationships between the influence function are established and the effective stiffness and compliance are expressed through these influence functions

$$\mathbf{L}^* = \{\mathbf{B}^*_\varepsilon(\mathbf{x})\}, \quad \mathbf{M}^* = \{\mathbf{A}^*_\sigma(\mathbf{x})\}, \quad \mathbf{L}^* = \{\boldsymbol{\mathcal{L}}^\sigma(\mathbf{A}^*_u)\}. \tag{17.151}$$

One proved the equivalence of the dual effective material laws $\mathbf{L}^* = (\mathbf{M}^*)^{-1}$ as well as the symmetry of macroproperties $\mathbf{L}^*$ inside the index pairs (minor symmetry) and between the index pairs (major symmetry, e.g., $L^*_{ijkl} = L^*_{klij}$).

An *energetic definition* of the effective elastic elasticity $\mathbf{L}^*$ can be obtained from a representation of the total strain energy through the influence functions ($\boldsymbol{\xi} = \mathbf{y}-\mathbf{x}$)

$$W^u(\mathbf{u}) = \{W(\mathbf{x})\} = \frac{1}{4}\Big\{\int_{\mathcal{H}_x} \boldsymbol{\eta}^\top \mathbf{C}(\boldsymbol{\xi},\mathbf{x})d\mathbf{y}\boldsymbol{\eta}\Big\} = \frac{1}{4}\Big\{\int_{\mathcal{H}_x} \{\boldsymbol{\varepsilon}\}\mathbf{A}_u^{*\eta\top}\mathbf{C}(\boldsymbol{\xi},\mathbf{x})\mathbf{A}_u^{*\eta}d\mathbf{y}\{\boldsymbol{\varepsilon}\}\Big\}, \tag{17.152}$$

where $\mathbf{A}_u^{*\eta}(\boldsymbol{\xi}) := \mathbf{A}_u^*(\mathbf{y}) - \mathbf{A}_u^*(\mathbf{x})$. From other side, W can be represented as a total strain energy of homogeneous effective medium subjected to nonlocal homogeneous boundary conditions (16.53)

$$W = \frac{1}{2}\{\boldsymbol{\varepsilon}\}\mathbf{L}^*\{\boldsymbol{\varepsilon}\}. \tag{17.153}$$

Equating Eqs. (17.152) and (17.153) for any constant symmetric fields $\{\boldsymbol{\varepsilon}\}$ yields the representation for the *energetic definition* of the effective elastic modulus

$$\mathbf{L}^* = \frac{1}{2}\Big\{\int_{\mathcal{H}_x} \mathbf{A}_u^{*\eta\top}(\boldsymbol{\xi})\mathbf{C}(\boldsymbol{\xi},\mathbf{x})\mathbf{A}_u^{*\eta}(\mathbf{x})d\mathbf{y}\Big\}. \tag{17.154}$$

The representations (17.151) and (17.154) for the effective elastic moduli are very similar to the corresponding representations for CMs with nonlocally elastic (in the sense of Eringen [446]) properties of constituents (see Chap. 14).

From other side, substitution of the influence function $\mathbf{B}_\sigma^*(\mathbf{x})$ (17.149$_3$) into the strain energy representation (17.123) expressed through the stress field $\boldsymbol{\sigma}(\mathbf{x})$

$$W^\sigma(\boldsymbol{\sigma}) = \frac{1}{2}\Big\{\int_{\mathcal{H}_x} \{\boldsymbol{\sigma}\}\mathbf{B}_\sigma^{*\top}\breve{\mathbf{M}}\mathbf{B}_\sigma^* d\mathbf{y}\{\boldsymbol{\sigma}\}\Big\}, \tag{17.155}$$

and subsequent comparison with the corresponding representation for the homogeneous effective medium, leads to the following new expression for the effective compliance

$$\mathbf{M}^* = \frac{1}{2}\Big\{\int_{\mathcal{H}_x} \mathbf{B}_\sigma^{*\top}\breve{\mathbf{M}}\mathbf{B}_\sigma^* d\mathbf{y}\Big\}. \tag{17.156}$$

17.7.2 *Generalized Hill's Bounds on the Effective Stiffness and Compliance*

The historically significant bounds by Hill [639] and [640] (see also [1192]) assert that, in fact, the simple estimations by Voigt [1566] and Reuss [1290] determine the upper and lower bounds, respectively, on the effective linear elastic properties of CMs. These bounds were obtained by substitution of specially chosen admissible fields into the corresponding representations of strain energy bounds (17.143) and (17.146). We modify this classical scheme in the framework of peridynamics.

Namely, the generalized Voigt bound adopts the same linear applied displacement field as a local displacement field $\widetilde{\mathbf{u}} = \widetilde{\boldsymbol{\varepsilon}} \cdot \mathbf{x}$ in all phases $\mathbf{x} \in w$. This field is kinematically admissible $\widetilde{\mathbf{u}} \in \mathcal{A}^k(\mathbf{u})$, since it satisfied to the prescribed homogeneous displacement loading $\widetilde{\boldsymbol{\varepsilon}} = \boldsymbol{\varepsilon}^{w\Gamma}$ (16.52). Then, according to (17.148$_1$), the concentration factor $\mathbf{A}^*_u(\mathbf{x}) \equiv \mathbf{x} \cdot \mathbf{I}$, and the actual value $W^u(\mathbf{u})$ of the strain energy (17.152) of CM does not exceed (see (17.143)) the approximation $W^u(\widetilde{\mathbf{u}})$ ($\boldsymbol{\xi} = \mathbf{y} - \mathbf{x}$)

$$W^u(\mathbf{u}) = \frac{1}{2}\{\boldsymbol{\varepsilon}\}\mathbf{L}^*\{\boldsymbol{\varepsilon}\} \leq \frac{1}{4}\Big\{\int_{\mathcal{H}_x} \{\boldsymbol{\varepsilon}\}\boldsymbol{\xi}\mathbf{C}(\boldsymbol{\xi}, \mathbf{x})\boldsymbol{\xi}d\mathbf{y}\{\boldsymbol{\varepsilon}\}\Big\} := W^V = \frac{1}{2}\{\boldsymbol{\varepsilon}\}\mathbf{L}_V\{\boldsymbol{\varepsilon}\}, \tag{17.157}$$

for the admissible field

$$\widetilde{\mathbf{u}}(\mathbf{x}) = \{\boldsymbol{\varepsilon}\} \cdot \mathbf{x} \tag{17.158}$$

that suggests that the generalized Voigt estimation of effective moduli is

$$\mathbf{L}_V = \frac{1}{2}\Big\{\int_{\mathcal{H}_x} \boldsymbol{\xi}\mathbf{C}(\boldsymbol{\xi}, \mathbf{x})\boldsymbol{\xi}d\mathbf{y}\Big\} \tag{17.159}$$

and $\mathbf{L}_V - \mathbf{L}^*$ is positive semidefinite:

$$\mathbf{L}^* \leq \mathbf{L}_V. \tag{17.160}$$

The inequality (17.160) is understood in the sense of an inequality for the corresponding quadratic form when $a_{ij}(L^V_{ijkl} - L^*_{ijkl})a_{kl} \geq 0$ for any symmetric tensor of the second order a_{ij}. In so doing, it is possible that for some components of the tensors L^V_{ijkl} and L^*_{ijkl}, the inequality (17.160) is violated.

We transform the representation (17.159) by the use of the micromodulus and locally elastic counterpart decompositions

$$\mathbf{C}(\boldsymbol{\xi}, \mathbf{x}) = \mathbf{C}^{(0)}(\boldsymbol{\xi}, \mathbf{x}) + \mathbf{C}_1(\boldsymbol{\xi}, \mathbf{x}), \tag{17.161}$$

$${}^L\mathbf{L}(\mathbf{x}) = {}^L\mathbf{L}^{(0)} + {}^L\mathbf{L}_1(\mathbf{x}), \tag{17.162}$$

respectively. The decomposition (17.162) is very popular in classical micromechanics (see, e.g., [14, 16, 46]). The decomposition (17.161) enables one to get the following Voigt estimates:

$$\mathbf{L}_V = \mathbf{L}^{(0)} + \frac{c^{l(1)}}{2\bar{v}_i^l} \int_{v_i^l} \int_{\mathcal{H}_x} \boldsymbol{\xi}^\top \mathbf{C}_1(\boldsymbol{\xi}, \mathbf{x})\boldsymbol{\xi} d\mathbf{y} d\mathbf{x}, \tag{17.163}$$

$${}^L\mathbf{L}_V = \langle {}^L\mathbf{L} \rangle = {}^L\mathbf{L}^{(0)} + \langle {}^L\mathbf{L}_1 \rangle, \tag{17.164}$$

where in Eq. (17.163) one used the definition of the average (17.11)

$$\{\mathbf{f}(\mathbf{x})V_i^l(\mathbf{x})\} = c^{(1)} \frac{\bar{v}_i^l}{\bar{v}_i} \{\mathbf{f}(\mathbf{x})V_i^l(\mathbf{x})\}_i^l. \tag{17.165}$$

A fundamental advantage of the representation (17.163) is a necessity to estimate the corresponding integrals only over the extended inclusion ($\mathbf{x} \in v_i^{l(1)}$) rather than over the entire composite (17.159) ($\mathbf{x} \in w$). It is found to be possible due to the fact that the tensor $\mathbf{C}_1(\mathbf{x}, \hat{\mathbf{x}})$ vanishes inside the truncated matrix $\mathbf{x}$ or $\hat{\mathbf{x}} \in v^{l(0)}$.

The decomposition (17.161) allows us to obtain a decomposition of the Voigt approximation W^V (17.157) of the elastic energy ($\boldsymbol{\xi} = \mathbf{y} - \mathbf{x}$)

$$W^V = W^{\epsilon(0)} + c^{l(1)} \{W_1^V(\mathbf{x})\}_i^l, \quad W_1^V(\mathbf{x}) = \frac{1}{2}\{\boldsymbol{\varepsilon}\} \int_{\mathcal{H}_x} \boldsymbol{\xi} \mathbf{C}_1(\boldsymbol{\xi}, \mathbf{x})\boldsymbol{\xi} d\mathbf{y} \{\boldsymbol{\varepsilon}\}, \tag{17.166}$$

where $W^{\epsilon(0)} := \{\boldsymbol{\varepsilon}\}\mathbf{L}^{(0)}\{\boldsymbol{\varepsilon}\}/2$ is the strain energy of the homogeneous matrix medium (at $\mathbf{C}_1(\boldsymbol{\xi}, \mathbf{x}) \equiv \mathbf{0}$) with the stiffness

$$\mathbf{L}^{(0)} = \frac{1}{2} \int_{\mathcal{H}_x} \boldsymbol{\xi} \mathbf{C}^{(0)}(\boldsymbol{\xi}, \mathbf{x})\boldsymbol{\xi} d\mathbf{y}, \tag{17.167}$$

and $W_1^V(\mathbf{x})$ is a jump of the local strain energy density inside the extended inclusions v_i^l with respect to the matrix.

The generalized Reuss bound is based on the approximation of statically admissible field $\widetilde{\boldsymbol{\sigma}}(\mathbf{x})$ by a constant coinciding with the average stress field $\{\boldsymbol{\sigma}\}$ applied to RVI, or, in other words, concentration factors $\mathbf{B}_\sigma^*(\mathbf{x}) \equiv \mathbf{I}$ and ($\mathbf{x} \in w$)

$$\widetilde{\boldsymbol{\sigma}}(\mathbf{x}) = \{\boldsymbol{\sigma}\}. \tag{17.168}$$

The actual complementary energy $\Pi^*(\boldsymbol{\sigma})$ (17.155) and its approximation $\Pi^*(\widetilde{\boldsymbol{\sigma}})$ must satisfy the inequality (17.140) reducing to (17.146)

$$W^\sigma(\boldsymbol{\sigma}) = \frac{1}{2}\{\boldsymbol{\sigma}\}\mathbf{M}^*\{\boldsymbol{\sigma}\} \leq \frac{1}{4}\left\{ \int_{\mathcal{H}_x} \{\boldsymbol{\sigma}\}\breve{\mathbf{M}}(\boldsymbol{\xi}, \mathbf{x}) d\mathbf{y} \{\boldsymbol{\sigma}\} \right\} := W^R = \frac{1}{2}\{\boldsymbol{\sigma}\}\mathbf{M}_R\{\boldsymbol{\sigma}\}. \tag{17.169}$$

This asserts that the generalized Reuss estimate of effective compliance is

$$\mathbf{M}_R = \frac{1}{2}\Big\{ \int_{\mathcal{H}_x} \breve{\mathbf{M}}(\boldsymbol{\xi}, \mathbf{x}) d\mathbf{y} \Big\} \tag{17.170}$$

and $\mathbf{M}_R - \mathbf{M}^*$ is positive semidefinite

$$\mathbf{M}_R \geq \mathbf{M}^*. \tag{17.171}$$

As in Eqs. (17.161) and (17.162), the decompositions

$$\breve{\mathbf{M}}(\boldsymbol{\xi}, \mathbf{x}) = \breve{\mathbf{M}}^{(0)}(\boldsymbol{\xi}, \mathbf{x}) + \breve{\mathbf{M}}_1(\boldsymbol{\xi}, \mathbf{x}), \tag{17.172}$$

$$\mathbf{M}(\mathbf{x}) = \mathbf{M}^{(0)} + \mathbf{M}_1(\mathbf{x}) \tag{17.173}$$

provide a way to obtain the following Reuss estimates:

$$\mathbf{M}_R = \mathbf{M}^{(0)} + \frac{c^{l(1)}}{2v_i^l} \int_{v_i^l} \int_{\mathcal{H}_x} \breve{\mathbf{M}}_1(\boldsymbol{\xi}, \mathbf{x}) d\mathbf{y} d\mathbf{x}, \tag{17.174}$$

$${}^L\mathbf{M}_R = \langle \mathbf{M} \rangle = \mathbf{M}^{(0)} + \langle \mathbf{M}_1 \rangle, \tag{17.175}$$

where $\mathbf{M}^{(0)} = (\mathbf{L}^{(0)})^{-1}$ can be also obtained directly from Eq. (17.170).

The Reuss approximation of the strain energy W^R can be split by the use of the decomposition (17.172)

$$W^R = W^{\sigma(0)} + c^{l(1)} \{ W_1^R(\mathbf{x}) \}_i^l, \quad W_1^R(\mathbf{x}) = \frac{1}{2} \{\boldsymbol{\sigma}\} \int_{\mathcal{H}_x} \breve{\mathbf{M}}_1(\boldsymbol{\xi}, \mathbf{x}) \boldsymbol{\xi} d\mathbf{y} \{\boldsymbol{\sigma}\}, \tag{17.176}$$

where $W^{\sigma(0)} := \{\boldsymbol{\sigma}\}\mathbf{M}^{(0)}\{\boldsymbol{\sigma}\}/2$ is the strain energy of the homogeneous matrix medium (at $\mathbf{C}_1(\boldsymbol{\xi}, \mathbf{x}) \equiv \mathbf{0}$) with the compliance $\mathbf{M}^{(0)}$.

Thus, we obtained the upper and lower bounds for the effective stiffness $\mathbf{L}^*$ and compliance $\mathbf{M}^* = (\mathbf{L}^*)^{-1}$ for both the peridynamic and locally elastic

$$\mathbf{L}_R \leq \mathbf{L}^* \leq \mathbf{L}_V, \quad \mathbf{M}_V \leq \mathbf{M}^* \leq \mathbf{M}_R, \tag{17.177}$$

$${}^L\mathbf{L}_R \leq {}^L\mathbf{L}^* \leq^L \mathbf{L}_V, \quad {}^L\mathbf{M}_V \leq^L \mathbf{M}^* \leq^L \mathbf{M}_R, \tag{17.178}$$

respectively, where the notations $\mathbf{M}_V = (\mathbf{L}_V)^{-1}$, ${}^L\mathbf{M}_V = ({}^L\mathbf{L}_V)^{-1}$ and $\mathbf{L}_R = (\mathbf{M}_R)^{-1}$, ${}^L\mathbf{L}_R = ({}^L\mathbf{M}_R)^{-1}$ were introduced.

Hill's [639] estimations combine the upper and lower bounds as both arithmetic and harmonic means

$$\mathbf{L}_H = (\mathbf{L}_R + \mathbf{L}_V)/2, \quad \mathbf{M}_H = (\mathbf{M}_R + \mathbf{M}_V)/2, \tag{17.179}$$

$${}^L\mathbf{L}_H = ({}^L\mathbf{L}_R + {}^L\mathbf{L}_V)/2, \quad {}^L\mathbf{M}_H = ({}^L\mathbf{M}_R + {}^L\mathbf{M}_V)/2. \tag{17.180}$$

It should be mentioned that due to the scale separation hypotheses (7.10$_2$), the classical Hill's bounds (17.178) are scale invariant, depend only upon the moduli and the phase concentrations, and do not depend on the size and shape of inclusions as well as on their relative distributions. The representations (17.164) and (17.170) are equally simple and do not require additional efforts with respect to each other. In so doing, in the generalized Hill's bounds (17.163) and (17.174), comparable scales of the inclusion size and horizon are taken into account $a \geq l_\delta$ that leads to the dependence of the bounds on both the size and shape of the inclusions. The averaging for the estimation of $\mathbf{L}_V$ (17.163) and $\mathbf{M}_R$ (17.174) was performed over the randomly inhomogeneous macrodomain w that presents a considerable challenge. As this takes place, estimations of the generalized Hill's bounds (17.163) and (17.148) are reduced to the integration of the integrands $\boldsymbol{\xi}\mathbf{C}_1\boldsymbol{\xi}$ and $\breve{\mathbf{M}}_1$, respectively, only over one representative extended inclusion v_i^l. Thus, the generalized Hill's bounds (17.177) represent averaging over the extended inclusions v_i^l of some sort of the global stiffness and compliance matrices, respectively, rather than the solutions of the corresponding boundary value problems used for estimations of effective moduli (see for details [224, 225]).

17.7.3 Numerical Examples

We consider a heterogeneous peridynamic bar with the triangle-function profiles (16.79$_2$) (see [131]) ($k = 0, 1$; $E^{(1)}/E^{(0)} = 0.5$)

$$C(\xi) = \begin{cases} C^{(k)}(1 - |\xi|/l_\delta)/|\xi|, & \text{for } |\xi| < l_\delta, \\ 0, & \text{for } |\xi| > l_\delta, \end{cases}, \quad C^{(k)} = 6E^{(k)}/l_\delta^2, \tag{17.181}$$

as well as a heterogeneous peridynamic bar with the step-function profiles (16.79$_1$) ($k = 0, 1$)

$$C(\xi) = \begin{cases} C^{(k)}, & \text{for } |\xi| < l_\delta, \\ 0, & \text{for } |\xi| > l_\delta, \end{cases}, \quad C^{(k)} = 3E^{(k)}/l_\delta^3. \tag{17.182}$$

The midpoint quadrature scheme (similar to [210, 212]) is realized for the uniform grid $x_i = h(0, \pm 1, \pm 2, \ldots)$; $2N^a + 1$ and n_l stand for the numbers of the material points in the inclusion v_i and the horizon n_l, respectively. In all cases, $N^a = 1000$ was used; a discretization $n_l = 250$ provides an accuracy of 0.02% at the estimation of $\mathbf{L}^{(0)}$ (17.167).

In Figs. 17.3 and 17.4, one presents the generalized Hill's bounds for the micromoduli (17.181) and (17.182), respectively. The curves 1–4 correspond to the upper bounds L_V estimated for $l_\delta/a = n_l/N^a = 0,\ 0.1,\ 0.25$, and $n_l/N^a = 0.5$, respectively. The curves 5, 6, 7, and 8 were estimated for the lower bounds L_R at $n_l/N^a = 0,\ 0.1,\ 0.25$, and $n_l/N^a = 0.5$, respectively. As can be seen, L_R depends

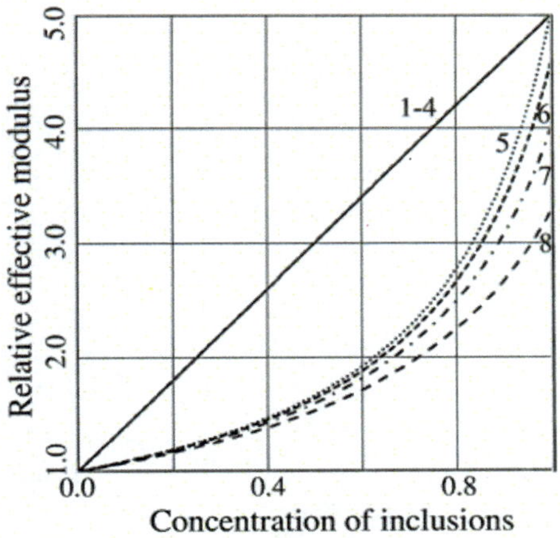

Fig. 17.3 Relative bounds of effective moduli $E_V/E^{(0)}$ (curves 1–4) and $E_R/E^{(0)}$ (curves 5–8) vs. c for $l_\delta/a =$ 0(1, 5), 0.1(2, 6), 0.25(3, 7), 0.5(4, 8)

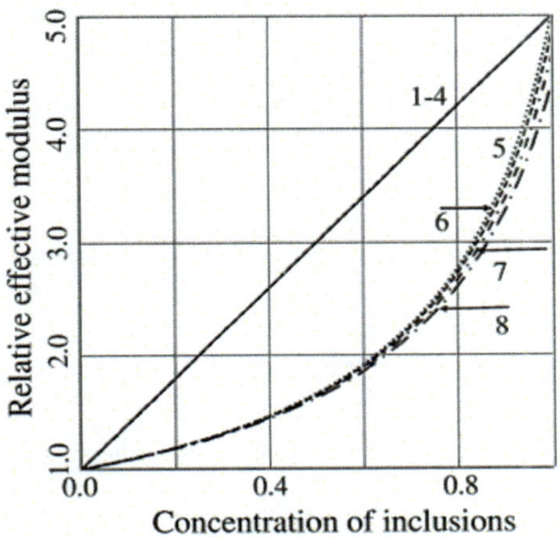

Fig. 17.4 Relative bounds of effective moduli $E_V/E^{(0)}$ (curves 1–4) and $E_R/E^{(0)}$ (curves 5–8) vs. c for $l_\delta/a =$ 0(1, 5), 0.1(2, 6), 0.25(3, 7), 0.5(4, 8)

on both the ratio l_δ/a and the particular choice of the micromoduli (either (17.181) or (17.182)). The limit of vanishing length scale $l_\delta \to 0$ reduces the peridynamic representations L_R (17.174) to their locally elastic counterpart ${}^L L_R$ (17.175).

It is interesting that L_V for both profiles (17.181) and (17.182) is not sensitive to the nonlocal effects. So, the curves 1–4 are distinguished one from another not more than 0.04%. For explanation of this fact, one presents the normalized jump of the local strain energy density $W_1^V(x)$ (at $\{\boldsymbol{\varepsilon}\} = 1$) as a function of coordinates x/a for the micromoduli (17.181) and (17.182); see Figs. 17.5 and 17.6, respectively. The step functions 1 correspond to the locally elastic solution, while the curves 2 and 3 were estimated for $l_\delta/a = 0.1$ and $l_\delta/a = 0.5$, respectively. It should be mentioned that the areas of the domains $B_1C_1D_1$ and $A_1C_1D_1$ (for $l_\delta/a = 0.1$ and $l_\delta/a = 0.5$, respectively) present the values of the elastic energies reducing the elastic energy inside a representative inclusion for the locally elastic case. In a similar manner, the areas of the domains $B_2C_2D_2$ and $A_2C_2D_2$ (for $l_\delta/a = 0.1$ and $l_\delta/a = 0.5$,

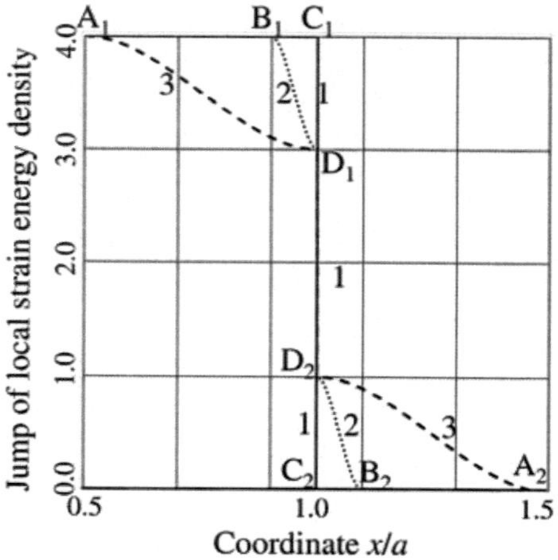

Fig. 17.5 $W_1^V(x)$ (at $\{\boldsymbol{\varepsilon}\} = 1$) vs. x/a for $l_\delta/a =$ 0.0 (1), 0.1 (2), and 0.5(3)

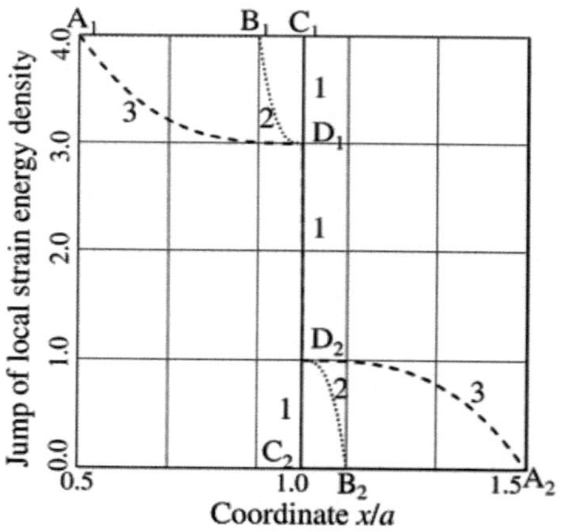

Fig. 17.6 $W_1^V(x)$ (at $\{\boldsymbol{\varepsilon}\} = 1$) vs. x/a for l_δ/a =0.0 (1), 0.1 (2), and 0.5(3)

respectively) present the values of the elastic energies increasing the elastic energy in the matrix in the vicinity of the representative inclusion for the locally elastic case. In so doing, the areas of $B_1C_1D_1$ and $A_1C_1D_1$ equal to the areas $B_2C_2D_2$ and $A_2C_2D_2$, respectively, with an accuracy of 0.52%. Thus, the Voigt approximation W^V (17.166) of the total elastic energy and Voigt estimate L_V (17.163) are invariant (with the mentioned accuracy) with respect to the nonlocal effects in 1D case for the analyzed profiles (17.181) and (17.182). Analysis of 2D and 3D cases requires additional consideration.

It should be mentioned that a synergism effect between the elastic mismatch $E^{(1)}/E^{(0)}$ and the nonlocal scale l_δ/a is manifested in the curves $E^*/E^{(0)} \sim c$. In parallel with Hill's bounds, $E^*/E^{(0)}$ were also estimated by the basic generalized micromechanical models, such as the Mori-Tanaka approach (MTA) and effective field method (EFM; see Chap. 19). In so doing, for 1D cases, the nonlocal effects depending on l_δ/a is more manifested in both MTA and EFM rather than in Hill's bound estimates.

Thus, the admissible displacement and force fields are defined and the strain energy is expressed through either the displacement or the stress fields. The theorem of work and energy, Betti's reciprocal theorem, and the theorem of virtual work are proved. Principles of minimum of both potential energy and complementary energy are generalized. The strain energy bounds are estimated for both the displacement and force homogeneous volumetric boundary conditions. The classical representations of effective elastic moduli through the mechanical influence functions for elastic CM are generalized to the case of peridynamics, and the energetic definition of effective elastic moduli is proposed. Generalized Hill's bounds on the effective elastic moduli of peridynamic random structure composites are obtained. A fundamental restriction of the variational principles and generalized Hill's bounds proposed is a linearity of the problem (16.55) (corresponding to small relative displacements). The proposed generalized Hill's bounds initiate investigations in the area of variational methods representing the most rigorous trend and security aspects of micromechanics. In contrast to the classical Hill's bounds, in the new generalized Hill's bounds, comparable scales of the inclusion size and horizon are taken into account that leads to the dependence of the bounds on both the size and shape of the inclusions.

17.8 General Integral Equations

In this section, one considers peridynamic statistically homogeneous composite materials occupying the full space R^d $(d = 1, 2, 3)$. It is well known that the background of micromechanics of media governed by conventional differential equations of the local balance of linear momentum (see, e.g., (16.6)) is formed by the so-called general integral equation (GIE). The GIE is the exact integral equation connecting the random fields at the point being considered and all surrounding points. There is a very long and dramatic history of attempts to obtain the exact version of this equation (see Chap. 7) for locally elastic CMs. The new GIE was obtained by a centering procedure of subtraction from both sides of a new initial integral equation their statistical averages obtained without any auxiliary assumptions, which are implicitly exploited in the known centering methods. Generalization of the GIE to peridynamic composites will be proposed in this section as a revision of representations obtained by Buryachenko [209].

17.8.1 General Integral Equation in the Operator Form

At first, we consider a single fixed inclusion $v_i(\mathbf{0})$ with the micromodulus $\mathbf{C}^{(1)}(\boldsymbol{\xi})$ in the infinite homogeneous medium with the properties $\mathbf{C}^{(0)}(\boldsymbol{\xi})$. Then the solution

$\mathbf{u}(\mathbf{x})$ can be split into two ones as $\mathbf{u} = \mathbf{u}^0 + \mathbf{u}_1$, where the solution of Eq. (17.118$_1$) is assumed to be known

$$\mathbf{u}(\mathbf{x}) - \mathbf{u}^0(\mathbf{x}) \equiv \mathcal{L}^{u\pi}(\mathbf{x} - \mathbf{x}_i, \boldsymbol{\pi}) = \int_{v_i^l} \mathbf{G}(\mathbf{x} - \mathbf{y})\boldsymbol{\pi}(\mathbf{y})d\mathbf{y}, \tag{17.183}$$

where

$$\boldsymbol{\pi}(\mathbf{y}) := \widehat{\mathcal{L}}_1(\mathbf{u}^t)(\mathbf{y}) = \int_{\mathcal{H}_{\mathbf{y}}} \widetilde{\boldsymbol{\tau}}(\mathbf{y}, \hat{\mathbf{x}})d\hat{\mathbf{x}} \tag{17.184}$$

is a local polarization tensor obtained by averaging the micropolarization tensor $\widetilde{\boldsymbol{\tau}}(\mathbf{x}, \hat{\mathbf{x}})$ (17.93) over $\mathcal{H}_{\mathbf{x}}$. The numerical solution of Eq. (17.183) is presented in Chap. 19, whereas Eq. (17.183$_2$) represents the solution $\mathbf{u}^t(\mathbf{x}) - \mathbf{u}^{t0}(\mathbf{x})$ expressed through the Green function $\mathbf{G}$ for the infinite homogeneous matrix $\mathcal{L}^{(0)}$ (see [1588, 1602]).

Let us consider an arbitrary random realization α of inclusions in the increasing sequence w_p

$$\mathbf{u}(\mathbf{x}, \alpha) = \mathbf{u}^0(\mathbf{x}) + \lim_{w_p \to R^d} \int_{w_p} \mathcal{L}_k^{u\pi}(\mathbf{x} - \mathbf{x}_k, \boldsymbol{\pi}) V_k^\delta(\mathbf{x}_k, \alpha)d\mathbf{x}_k, \tag{17.185}$$

where $\mathbf{u}^0(\mathbf{x})$ is an auxiliary displacement field produced by a fictitious deterministic body force $\mathbf{b}(\mathbf{x})$ with the compact support. The improper integral over the whole space R^d is defined as a limit ($p \to \infty$) (17.185) of definite integrals over an increasing sequence w_p filling R^d where the limit exists independently on the particular choice of closed bounded regions w_p in R^d such that any w_p is contained in every w_{p_1} for which $p_1 > p$ and for any bounded region w' in R^d there is p_2 such that $w' \subset w_{p_2}$.

For elimination of $\mathbf{u}^0(\mathbf{x})$, we apply the centering method of subtraction from both sides of Eq. (17.185) their statistical averages obtained without any auxiliary assumptions that leads to the following new equation for statistically inhomogeneous media:

$$\mathbf{u}^t(\mathbf{x}, \alpha) = \langle \mathbf{u}^t \rangle(\mathbf{x}) + \lim_{w_p \to R^d} \int_{w_p} \big[\mathcal{L}_k^{u\pi}(\mathbf{x} - \mathbf{x}_k, \boldsymbol{\pi}) V_k^\delta(\mathbf{x}_k, \alpha) - \langle \mathcal{L}_k^{u\pi}(\mathbf{x} - \mathbf{x}_k, \boldsymbol{\pi}) \rangle(\mathbf{x}_k)\big]d\mathbf{x}_k. \tag{17.186}$$

For a statistically inhomogeneous field X, the dependence of the statistical averages $\langle(.)\rangle(\mathbf{x}_k)$ in Eq. (17.186) on the current coordinate $\mathbf{x}_k$ is of fundamental importance. But even in this case, the expression in the averaging brackets $\langle\langle(\cdot)\rangle\rangle(\mathbf{x}_k)$ in Eq. (17.186) is of the order $O(|\mathbf{x} - \mathbf{x}_k|^{-2d})$ as $|\mathbf{x} - \mathbf{x}_k| \to \infty$, and the integral in Eq. (17.186) converges absolutely. Therefore, for $\mathbf{x} \in w_p$ considered in Eq. (17.186) and removed far enough from the boundary ∂w_p, the right-hand-side integral in (17.186) does not depend on the shape and size of the domain w_p, and it can be replaced by the integral over the whole space R^d. Thus, the general integral equations (GIEs) for the displacement and strains are

$$\mathbf{u}(\mathbf{x}) = \langle \mathbf{u}\rangle(\mathbf{x}) + \int \big[\mathcal{L}_k^{u\pi}(\mathbf{x}-\mathbf{x}_k,\boldsymbol{\pi})V_k^{\delta}(\mathbf{x}_k) - \langle \mathcal{L}_k^{u\pi}(\mathbf{x}-\mathbf{x}_k,\boldsymbol{\pi})\rangle(\mathbf{x}_k)\big]d\mathbf{x}_k, \tag{17.187}$$

$$\boldsymbol{\varepsilon}(\mathbf{x}) = \langle \boldsymbol{\varepsilon}\rangle(\mathbf{x}) + \int \big[\mathcal{L}_k^{\varepsilon\pi}(\mathbf{x}-\mathbf{x}_k,\boldsymbol{\pi})V_k^{\delta}(\mathbf{x}_k) - \langle \mathcal{L}_k^{\varepsilon\pi}(\mathbf{x}-\mathbf{x}_k,\boldsymbol{\pi})\rangle(\mathbf{x}_k)\big]d\mathbf{x}_k, \tag{17.188}$$

respectively, where $\mathcal{L}_k^{\varepsilon\pi} := \nabla\mathcal{L}_k^{u\pi}$, and the domain integration R^d and the label α are omitted hereafter for simplicity of notation.

In so doing, the counterparts of Eqs. (17.187) and (17.187) for locally elastic composites have a form

$$\mathbf{u}(\mathbf{x}) = \langle \mathbf{u}\rangle(\mathbf{x}) + \int \big[{}^L\mathcal{L}_k^{u\tau}(\mathbf{x}-\mathbf{x}_k,{}^L\boldsymbol{\tau})V_k^{\delta}(\mathbf{x}_k) - \langle {}^L\mathcal{L}_k^{u\tau}(\mathbf{x}-\mathbf{x}_k,{}^L\boldsymbol{\tau})\rangle(\mathbf{x}_k)\big]d\mathbf{x}_k, \tag{17.189}$$

$$\boldsymbol{\varepsilon}(\mathbf{x}) = \langle \boldsymbol{\varepsilon}\rangle(\mathbf{x}) + \int \big[{}^L\mathcal{L}_k^{\varepsilon\tau}(\mathbf{x}-\mathbf{x}_k,{}^L\boldsymbol{\tau})V_k^{\delta}(\mathbf{x}_k) - \langle {}^L\mathcal{L}_k^{\varepsilon\tau}(\mathbf{x}-\mathbf{x}_k,{}^L\boldsymbol{\tau})\rangle(\mathbf{x}_k)\big]d\mathbf{x}_k. \tag{17.190}$$

If the perturbators $\mathcal{L}_k^{u\pi}(\mathbf{x}-\mathbf{x}_k,\boldsymbol{\pi})$ and ${}^L\mathcal{L}_k^{u\tau}(\mathbf{x}-\mathbf{x}_k,\boldsymbol{\tau})$ are expressed through the corresponding Green functions $\mathbf{G}$ and ${}^L\mathbf{G}$, respectively, Eqs. (17.187), (17.187) and (17.189), (17.190) can be recast as

$$\mathbf{u}(\mathbf{x}) = \langle \mathbf{u}\rangle(\mathbf{x}) + \int \big[\mathbf{G}(\mathbf{x}-\mathbf{y})\boldsymbol{\pi}(\mathbf{y})V(\mathbf{y}) - \langle \mathbf{G}(\mathbf{x}-\mathbf{y})\boldsymbol{\pi}(\mathbf{y})\rangle(\mathbf{y})\big]d\mathbf{y}, \tag{17.191}$$

$$\boldsymbol{\varepsilon}(\mathbf{x}) = \langle \mathbf{u}\rangle(\mathbf{x}) + \int \big[\nabla\mathbf{G}(\mathbf{x}-\mathbf{y})\boldsymbol{\pi}(\mathbf{y})V(\mathbf{y}) - \langle \nabla\mathbf{G}(\mathbf{x}-\mathbf{y})\boldsymbol{\pi}(\mathbf{y})\rangle(\mathbf{y})\big]d\mathbf{y}, \tag{17.192}$$

$$\mathbf{u}(\mathbf{x}) = \langle \mathbf{u}\rangle(\mathbf{x}) + \int \big[\nabla^L\mathbf{G}(\mathbf{x}-\mathbf{y})^L\boldsymbol{\tau}(\mathbf{y})V(\mathbf{y}) - \langle \nabla^L\mathbf{G}(\mathbf{x}-\mathbf{y})^L\boldsymbol{\tau}(\mathbf{y})\rangle(\mathbf{y})\big]d\mathbf{y}, \tag{17.193}$$

$$\begin{aligned}\boldsymbol{\varepsilon}(\mathbf{x}) = \langle \boldsymbol{\varepsilon}\rangle(\mathbf{x}) + \int &\big[\nabla\nabla^L\mathbf{G}(\mathbf{x}-\mathbf{y})^L\boldsymbol{\tau}(\mathbf{y})V(\mathbf{y}) \\ &-\langle \nabla\nabla^L\mathbf{G}(\mathbf{x}-\mathbf{y})^L\boldsymbol{\tau}(\mathbf{y})\rangle(\mathbf{y})\big]d\mathbf{y},\end{aligned} \tag{17.194}$$

respectively. For the simplified approximations

$$\langle \mathbf{G}(\mathbf{x}-\mathbf{y})\boldsymbol{\pi}(\mathbf{y})\rangle(\mathbf{y}) = \mathbf{G}(\mathbf{x}-\mathbf{y})\langle \boldsymbol{\pi}(\mathbf{y})\rangle(\mathbf{y}), \tag{17.195}$$

$$\langle \nabla^L\mathbf{G}(\mathbf{x}-\mathbf{y})^L\boldsymbol{\tau}(\mathbf{y})\rangle(\mathbf{y}) = \nabla^L\mathbf{G}(\mathbf{x}-\mathbf{y})\langle {}^L\boldsymbol{\tau}(\mathbf{y})\rangle(\mathbf{y}) \tag{17.196}$$

(which are only asymptotically valid at $|\mathbf{x}-\mathbf{y}| \to \infty$), Eqs. (17.191)–(17.194) are reduced to

$$\mathbf{u}(\mathbf{x}) = \langle \boldsymbol{\varepsilon} \rangle(\mathbf{x}) + \int \mathbf{G}(\mathbf{x}-\mathbf{y})\big[\boldsymbol{\pi}(\mathbf{y})V(\mathbf{y}) - \langle \boldsymbol{\pi}(\mathbf{y})\rangle(\mathbf{y})\big]d\mathbf{y}, \qquad (17.197)$$

$$\boldsymbol{\varepsilon}(\mathbf{x}) = \langle \boldsymbol{\varepsilon} \rangle(\mathbf{x}) + \int \nabla\mathbf{G}(\mathbf{x}-\mathbf{y})\big[\boldsymbol{\pi}(\mathbf{y})V(\mathbf{y}) - \langle \boldsymbol{\pi}(\mathbf{y})\rangle(\mathbf{y})\big]d\mathbf{y}, \qquad (17.198)$$

$$\mathbf{u}(\mathbf{x}) = \langle \mathbf{u} \rangle(\mathbf{x}) + \int \nabla^L\mathbf{G}(\mathbf{x}-\mathbf{y})\big[{}^L\boldsymbol{\tau}(\mathbf{y})V(\mathbf{y}) - \langle {}^L\boldsymbol{\tau}(\mathbf{y})\rangle(\mathbf{y})\big]d\mathbf{y}, \quad (17.199)$$

$$\boldsymbol{\varepsilon}(\mathbf{x}) = \langle \boldsymbol{\varepsilon} \rangle(\mathbf{x}) + \int \nabla\nabla^L\mathbf{G}(\mathbf{x}-\mathbf{y})\big[{}^L\boldsymbol{\tau}(\mathbf{y})V(\mathbf{y}) - \langle {}^L\boldsymbol{\tau}(\mathbf{y})\rangle(\mathbf{y})\big]d\mathbf{y}, \ (17.200)$$

respectively. Both the qualitative and quantitative estimations of errors produced by the popular simplified approximations of Eq. (17.200) (based on the differential equilibrium equation (16.6)) were considered in Chaps. 7 and 9.

17.8.2 Infinite Coupled System of General Integral Equations

Let the inclusions $v_1, \ldots, v_n$ be fixed, and we define two sorts of effective fields for the displacement field $\overline{\mathbf{u}}_i(\mathbf{x})$, $\widetilde{\mathbf{u}}_{1,\ldots,n}(\mathbf{x})$ ($i = 1, \ldots, n$; $\mathbf{x} \in v_1, \ldots, v_n$) by the use of the rearrangement of Eqs. (17.187) and (17.187) in the following operator forms (an analogous particular case of these manipulation approaches for local elasticity problem is given in Chap. 7):

$$\mathbf{u}(\mathbf{x}) = \overline{\mathbf{u}}_i(\mathbf{x}) + \mathcal{L}_i^{u\pi}(\mathbf{x}-\mathbf{x}_i, \boldsymbol{\pi}),$$

$$\overline{\mathbf{u}}_i(\mathbf{x}) = \widetilde{\mathbf{u}}_{1,\ldots,n}(\mathbf{x}) + \sum_{j\neq i}^{n} \mathcal{L}_j^{u\pi}(\mathbf{x}-\mathbf{x}_j, \boldsymbol{\pi}),$$

$$\widetilde{\mathbf{u}}_{1,\ldots,n}(\mathbf{x}) = \langle \mathbf{u} \rangle(\mathbf{x}) + \int [\mathcal{L}_k^{u\pi}(\mathbf{x}-\mathbf{x}_k, \boldsymbol{\pi})V^{\delta}(\mathbf{x}_k|; v_1, \mathbf{x}_1; \ldots; v_n, \mathbf{x}_n)$$
$$-\langle \mathcal{L}_k^{u\pi}(\mathbf{x}-\mathbf{x}_k, \boldsymbol{\pi})\rangle(\mathbf{x}_k)]d\mathbf{x}_k, \qquad (17.201)$$

$$\boldsymbol{\varepsilon}(\mathbf{x}) = \overline{\boldsymbol{\varepsilon}}_i(\mathbf{x}) + \mathcal{L}_i^{\varepsilon\pi}(\mathbf{x}-\mathbf{x}_i, \boldsymbol{\pi}),$$

$$\overline{\boldsymbol{\varepsilon}}_i(\mathbf{x}) = \widetilde{\boldsymbol{\varepsilon}}_{1,\ldots,n}(\mathbf{x}) + \sum_{j\neq i}^{n} \mathcal{L}_j^{\varepsilon\pi}(\mathbf{x}-\mathbf{x}_j, \boldsymbol{\pi}),$$

$$\widetilde{\boldsymbol{\varepsilon}}_{1,\ldots,n}(\mathbf{x}) = \langle \boldsymbol{\varepsilon} \rangle(\mathbf{x}) + \int [\mathcal{L}_k^{\varepsilon\pi}(\mathbf{x}-\mathbf{x}_k, \boldsymbol{\pi})V^{\delta}(\mathbf{x}_k|; v_1, \mathbf{x}_1; \ldots; v_n, \mathbf{x}_n)$$
$$-\langle \mathcal{L}_k^{\varepsilon\pi}(\mathbf{x}-\mathbf{x}_k, \boldsymbol{\pi})\rangle(\mathbf{x}_k)]d\mathbf{x}_k \qquad (17.202)$$

for $\mathbf{x} \in v_i,\ i = 1, 2, \ldots, n$; here, $V^{\delta}(\mathbf{x}_k|; v_1, \mathbf{x}_1; \ldots; v_n, \mathbf{x}_n) = \sum_m \delta(\mathbf{x}_k - \mathbf{x}_m) - \sum_{i=1}^{n} \delta(\mathbf{x}_k - \mathbf{x}_i)$ is a random delta function of heterogeneity centers $\mathbf{x}_m$ $(m = 1, 2, \ldots)$ under the condition that $\mathbf{x}_k \neq \mathbf{x}_i,\ \mathbf{x}_i \neq \mathbf{x}_j$ if $i \neq j$ $(i, j = 1, \ldots, n)$. The definitions of the effective fields $\overline{\mathbf{u}}_i(\mathbf{x})$, $\widetilde{\mathbf{u}}_{1,2,\ldots,n}(\mathbf{x})$ as well as their statistical averages $\langle \overline{\mathbf{u}}_i \rangle(\mathbf{x})$, $\langle \widetilde{\mathbf{u}}_{1,2,\ldots,n} \rangle(\mathbf{x})$ are nothing more than a notation convenience for different terms of the infinite systems (17.201) and (17.202).

Considering some conditional statistical averages of GIE (17.187) and (17.187) leads to an infinite system of *new integral equations* $(n = 1, 2, \ldots)$

$$\langle \mathbf{u} \mid v_1, \mathbf{x}_1; \ldots; v_n, \mathbf{x}_n \rangle(\mathbf{x}) - \sum_{i=1}^{n} \langle \mathcal{L}_i^{u\pi}(\mathbf{x} - \mathbf{x}_i, \boldsymbol{\pi}) | v_1, \mathbf{x}_1; \ldots; v_n, \mathbf{x}_n \rangle_i$$
$$= \langle \mathbf{u} \rangle(\mathbf{x}) + \int \big\{ \langle \mathcal{L}_j^{u\pi}(\mathbf{x} - \mathbf{x}_j, \boldsymbol{\pi}) |; v_1, \mathbf{x}_1; \ldots; v_n, \mathbf{x}_n \rangle_j$$
$$\times \langle \varphi(v_j, \mathbf{x}_j |; v_1, \mathbf{x}_1, \ldots, v_n, \mathbf{x}_n) - \mathcal{L}_j^{u\pi}(\mathbf{x} - \mathbf{x}_j, \boldsymbol{\pi}) \rangle(\mathbf{x}_j) \big\} d\mathbf{x}_j, \quad (17.203)$$

$$\langle \boldsymbol{\varepsilon} \mid v_1, \mathbf{x}_1; \ldots; v_n, \mathbf{x}_n \rangle(\mathbf{x}) - \sum_{i=1}^{n} \langle \mathcal{L}_i^{\varepsilon\pi}(\mathbf{x} - \mathbf{x}_i, \boldsymbol{\pi}) | v_1, \mathbf{x}_1; \ldots; v_n, \mathbf{x}_n \rangle_i$$
$$= \langle \boldsymbol{\varepsilon} \rangle(\mathbf{x}) + \int \big\{ \langle \mathcal{L}_j^{\varepsilon\pi}(\mathbf{x} - \mathbf{x}_j, \boldsymbol{\pi}) |; v_1, \mathbf{x}_1; \ldots; v_n, \mathbf{x}_n \rangle_j$$
$$\times \langle \varphi(v_j, \mathbf{x}_j |; v_1, \mathbf{x}_1, \ldots, v_n, \mathbf{x}_n) - \mathcal{L}_j^{u\pi}(\mathbf{x} - \mathbf{x}_j, \boldsymbol{\pi}) \rangle(\mathbf{x}_j) \big\} d\mathbf{x}_j. \quad (17.204)$$

Since $\mathbf{x} \in v_1, \ldots, v_n$ in the nth line of the systems can take the values inside the inclusions $v_1, \ldots, v_n$, the nth line actually contains n equations. Statistical averaging $\langle(\cdot)\rangle_i$ stands for the averaging over all surrounding heterogeneities at the fixed v_i, while the average $\langle(\cdot)\rangle(\mathbf{x}_j)$ implies the averaging over all possible locations of v_i with possible dependence of this average on the macrocoordinate $\mathbf{x}_i$ as for FGMs.

Equation (17.183) shows that the operator $\mathcal{L}_k^{u\pi}(\mathbf{x} - \mathbf{x}_k, \boldsymbol{\pi})$ (and analogously $\mathcal{L}_k^{\varepsilon\pi}(\mathbf{x} - \mathbf{x}_k, \boldsymbol{\pi})$) has the physical interpretation of perturbations introduced by a single heterogeneity v_k in the infinite homogeneous matrix subjected by the effective fields $\overline{\mathbf{u}}_i(\mathbf{x})$ and $\overline{\mathbf{e}}_i(\mathbf{x})$, respectively, where at first no restrictions are imposed on the inhomogeneities of the effective fields $\overline{\mathbf{u}}_i(\mathbf{x})$ and $\overline{\mathbf{e}}_i(\mathbf{x})$.

Introduction of the effective fields (17.201_2) and (17.202_2) makes it possible to define another sort of the perturbators $\mathcal{L}_k^{uu}(\mathbf{x} - \mathbf{x}_k, \overline{\mathbf{u}})$ and $\mathcal{L}_k^{\varepsilon\varepsilon}(\mathbf{x} - \mathbf{x}_k, \overline{\boldsymbol{\varepsilon}})$ which are, in fact, the solutions of the basic problem for one inclusion inside the infinite peridynamic matrix. In such a case, GIEs (17.203) and (17.204) can be recast in terms of effective fields $(\mathbf{x} \in v_k,\ k = 1, 2, \ldots)$

$$\langle \overline{\mathbf{u}} \mid v_1, \mathbf{x}_1; \ldots; v_n, \mathbf{x}_n \rangle(\mathbf{x}) - \sum_{i \neq k}^{n} \langle \mathcal{L}_i^{uu}(\mathbf{x} - \mathbf{x}_i, \overline{\mathbf{u}}) | v_1, \mathbf{x}_1; \ldots; v_n, \mathbf{x}_n \rangle_i$$
$$= \langle \mathbf{u} \rangle(\mathbf{x}) + \int \big\{ \langle \mathcal{L}_j^{uu}(\mathbf{x} - \mathbf{x}_j, \overline{\mathbf{u}}) |; v_1, \mathbf{x}_1; \ldots; v_n, \mathbf{x}_n \rangle_j$$
$$\times \langle \varphi(v_j, \mathbf{x}_j |; v_1, \mathbf{x}_1, \ldots, v_n, \mathbf{x}_n) - \langle \mathcal{L}_j^{uu}(\mathbf{x} - \mathbf{x}_j, \overline{\mathbf{u}}) \rangle(\mathbf{x}_j) \big\} d\mathbf{x}_j \quad (17.205)$$

$$
\begin{aligned}
&\langle\overline{\boldsymbol{\varepsilon}}\mid v_1,\mathbf{x}_1;\ldots;v_n,\mathbf{x}_n\rangle(\mathbf{x})-\sum_{i\neq k}^{n}\langle\boldsymbol{\mathcal{L}}_i^{\varepsilon\varepsilon}(\mathbf{x}-\mathbf{x}_i,\overline{\boldsymbol{\varepsilon}})|v_1,\mathbf{x}_1;\ldots;v_n,\mathbf{x}_n\rangle_i\\
&=\langle\mathbf{u}\rangle(\mathbf{x})+\int\Big\{\langle\boldsymbol{\mathcal{L}}_j^{\varepsilon\varepsilon}(\mathbf{x}-\mathbf{x}_j,\overline{\boldsymbol{\varepsilon}})|;v_1,\mathbf{x}_1;\ldots;v_n,\mathbf{x}_n\rangle_j\\
&\quad\times\langle\varphi(v_j,\mathbf{x}_j|;v_1,\mathbf{x}_1,\ldots,v_n,\mathbf{x}_n)-\boldsymbol{\mathcal{L}}_j^{\varepsilon\varepsilon}(\mathbf{x}-\mathbf{x}_j,\overline{\boldsymbol{\varepsilon}})\rangle(\mathbf{x}_j)\Big\}d\mathbf{x}_j. \quad (17.206)
\end{aligned}
$$

There are a couple fundamental advantages of Eqs. (17.205) and (17.206) with respect to Eqs. (17.191)–(17.200). So, Eqs. (17.205) and (17.206) have no Green functions, and, because of this, analyses of Eqs. (17.205) and (17.206) in a full space $\mathbf{x}\in R^d$ are not a restriction of the theory proposed (we can consider a half space $\{\mathbf{x}=(x_1,\ldots,x_d)^\top|x_1\geq 0\}$). However, the most intriguing feature of Eqs. (17.205) and (17.206) is the absence of a constitutive law in these equations. As can be seen, Eq. (17.206) completely coincides with the corresponding equation for the locally elastic CMs (see [213]). It means that the same equation can be used for both the peridynamic CMs and locally elastic CMs. The GIEs (17.205) and (17.206) proposed are adapted to the straightforward generalizations of corresponding methods of local thermoelastic micromechanics (see, e.g., [211, 213]).

Chapter 18
Peridynamic Micromechanics of Random Structure Composites

Abstract In this chapter, we consider the solution methods of the GIE of peridynamic micromechanics. These methods are based on extraction from the material properties a constituent of the matrix properties. Effective moduli are expressed through the average local interface polarization tensor over the surface of the extended inclusion phase rather than over an entire space. Any spatial derivatives of displacement fields are not required. The basic hypotheses of locally elastic micromechanics are generalized to their peridynamic counterparts. In particular, in the generalized effective field method (EFM) proposed, the classical effective field hypothesis is relaxed, and the hypothesis of the ellipsoidal symmetry of the random structure of CMs is not used. One demonstrates some similarity and difference with respect to other methods (the dilute approximation and Mori-Tanaka approach) of micromechanics of peridynamic CMs.

Estimation of macroscopic effective response of heterogeneous media with random structures in an averaged (or homogenized) meaning in terms of the mechanical and geometrical properties of constituents is a central focus of micromechanics denoted as *micro-to-macro modeling*. The general results establishing the links between the effective properties and the corresponding mechanical and transformation influence functions were inspired by Hill [640] for locally elastic composites. Some basic representations analogous to the mentioned above were generalized in Chap. 17 to the thermoperidynamics of CMs. The displacement field estimations in the constituents, in turn, are based on a substitution into the one or another micromechanical scheme of a solution (called basic problem) for one inclusion inside the infinite matrix subjected to some effective field. So, for locally elastic random structure CMs, a number of micromechanical models inspired by Eshelby [449] (see Chap. 3) were proposed in the literature for describing the thermoelastic behavior of composites with ellipsoidal inclusions (see Chaps. 8–12). Numerical solutions for the basic problem are considered in Sect. 18.1, whereas the different micromechanical models are generalized to their peridynamic counterparts in Sects. 18.2–18.4.

V. A. Buryachenko, *Local and Nonlocal Micromechanics of Heterogeneous Materials*, https://doi.org/10.1007/978-3-030-81784-8_18

18.1 Modeling of One Inclusion in the Infinite Matrix

At first, the scheme by the FEA for the locally elastic basic problem is generalized to the peridynamic sample of increasing size subjected to the volumetric homogeneous displacement loading. While numerically solving a problem initially formulated on the unbounded domain, one typically truncates this domain. The artificial volumetric boundary conditions (AVBC) are set up at the external boundary of a final domain obtained from the original infinite domain by truncation (see [1539]). In the simplest case of a homogeneous remote loading at infinity, the corresponding artificial boundary conditions are also assumed to be homogeneous. A necessary requirement of AVBC is that the solution found inside the truncated domain is close to the corresponding solution in the origin infinite domain. The infinite medium is *modeled* by an increased size sample with controlled stabilization of the solution for this sample.

The theory of multiscale analysis of locally elastic composites of periodic structure is well developed (see for references, e.g., Chap. 19). It is usually implied that a ratio of the representative size of microstructure and the overall dimension of the structure is negligibly small. An alternative sort of truncation method in peridynamics of CM is periodization method when a unit cell (UC; see [223, 998]) size is increased, while the inclusion size is fixed or, in other words, dilute concentration of inclusion is analyzed, i.e., the infinite medium with one inclusion is modeled as a periodic structure CM with dilute concentrations of inclusions. For the peridynamic periodic structure CM, one introduces new volumetric periodic boundary conditions (PBC) at the interaction boundary of a representative unit cell (UC). Introduction of the volumetric periodic boundary conditions allows us to generalize the classical computational homogenization approach to their peridynamic version.

In the next approach, the displacement field is decomposed as linear displacement corresponding to the homogeneous loading of the infinite homogeneous medium and a perturbation field introduced by one inclusion. This perturbation field is found by the iteration method for entirely infinite sample analogous to a solution for the nonlocally elastic (in the sense of Eringen [446]) basic problem considered in Chap. 14. The discretization of the nonlocal linear equations of peridynamics is reduced to the linear matrix equations. The direct method of triangulation of the corresponding $N \times N$ matrix A becomes prohibitive in terms of computer time and storage if the matrix is quite large for either (or both) 2D and 3D dimensions or comparable scale of nonlocal effects (defined by the horizon) and the field and material scales. Furthermore, most large problems are sparse, and the sparsity is lost to a considerable extent during the triangulation procedure. For dealing with these very large matrices with many nonzero entries, it is advisable to use a class of iterative methods which never alter the matrix A and require the storage of a few vectors of length N at a time.

In many physics and engineering areas, nature may be best represented or modeled by unbounded or semi-unbounded domains. While the existence and uniqueness of solutions of finite systems have been completely resolved, much remains unknown for infinite systems corresponding to the problems for one inclusion in the infinite matrix. An excellent detailed account of the colorful history of infinite matrices is given in [111]. Establishment of a sufficient condition guarantying the existence and uniqueness of infinite matrix equations is usually restricted by the matrices of special structures (see [1386]) like diagonal dominance, tridiagonal, sign distributions, etc. which are frequently nonsingular. The purpose of this section is to propose a method for solution of infinite matrix equations corresponding to the problems for one inclusion in the infinite peristaltic matrix. Establishment of the right-hand side of this equation as a discretization of a fictitious body force with compact support has a beneficial effect on the development of an iteration solution method which falls out as a bonus.

Immediate consideration of the infinite medium can be performed by the Green function (see [1588, 1602]) technique which is reduced to the volume integral equation method (VIEM) well developed in the theory of locally elastic CMs (see Chaps. 3 and 4). Introduction of the micropolarization tensor by Buryachenko ([210] and [216]) (generalizing the notion of polarization tensor in local elasticity; see, e.g., Chaps. 3 and 4) provides a great scope for generalization of the VIEM to its peridynamic counterpart that can be easily performed as a straightforward exploitation of the method in Chap. 14 proposed for solution of a corresponding basic problem with nonlocally elastic (in the sense of Eringen [446]) properties of constituents.

18.1.1 Truncation Method

We consider a domain w with an inclusion $v = v^{(1)} \subset w$ and the interaction domain w_Γ forming a *nonlocal closure* of w: $\overline{\overline{w}} = w \cup w_\Gamma$. A nonlocal "Dirichlet" volume-constrained problem is given by Eq. (16.46) with prescribed homogeneous volumetric boundary conditions (16.52). In numerical implementation described by Silling and Askari [1401], the region w is discretized into a set of nodes p, each with a finite known volume (called *full volume*) $\bar{v}_p = h^d$ defined by the size h. Taken together, the nodes p form a square grid $\mathbf{x}_p$ (square mesh, $\Omega^{\rm sq}$), $i = 1, \ldots, d$, $\mathbf{x}_p = (x_1, \ldots, x_d)^\top$, $\mathbf{p} = (p_1, \ldots, p_d) \in Z^d$): $\Omega^{\rm sq} = \{(x_1, \ldots, x_d)^\top \,|\, x_i = hp_i\}$. The total number of nodes covering the total macrovolume w is $N^{\max}$. The spatially discretized form of the equilibrium equations (16.5) and (16.55) replaces the integral by the finite sum for each node p

$$\sum_{q \in \mathcal{H}_p} \mathbf{C}(\mathbf{x}_p, \mathbf{x}_q)(\mathbf{u}_q - \mathbf{u}_p)\bar{v}_{pq} = -\mathbf{b}_p, \tag{18.1}$$

where a midpoint-type integration scheme is used with subscripts denoting the node number $\mathbf{u}_p = \mathbf{u}(\mathbf{x}_p)$, $\mathbf{b}_p = \mathbf{b}(\mathbf{x}_p)$. For every interaction between the nodes q and

p, only the volume fraction $\bar{v}_{pq}$ of the volume $\bar{v}_q$ of the node q inside the cutoff distance l_δ of the node p is counted. Estimation of the partial volumes $\bar{v}_{pq}$ (partial volume algorithm) for $d = 1, 2, 3$ was proposed in [1185] (see also [1355]). In modified algorithm [664] (with an accuracy estimation [1355]), the family of the node p may include nodes q outside the horizon $|\mathbf{x}_q - \mathbf{x}_p| > l_\delta$ that allows for a nonvanishing force state for this pair of nodes with subsequent correction for partial volumes. The mentioned algorithms were proposed for the family points completely belonging to the same phase (either $\mathcal{H}_p \subset v^{(0)}$ or $\mathcal{H}_p \subset v^{(1)}$). If the line segment $[\mathbf{x}_p, \mathbf{x}_q]$ intersects the geometrical interface Γ separating the phases $v^{(0)}$ and $v^{(1)}$, then the locally elastic model requires separate interface conditions defined on Γ_i. In contrast to local elasticity, in peridynamics, one introduces a d-dimensional set $\mathbf{x} \in v_\Gamma$ (called extended interface or interaction interface) of thickness $2l_\delta$ (including Γ as a central $d-1$-dimensional surface Γ) where a new peridynamic operator is defined on displacements $\mathbf{u}(\mathbf{x})$ (see for details [1354, 1356]). It was shown that nonlocal interface problems converge to their classical local counterparts in the limit of the horizon going to zero $l_\delta/a \to 0$. Although there are no technical issues with correction of discretized peridynamic equation in the extended interface $\mathbf{x} \in v_\Gamma$, it is helpful to assume for simplicity only that the peridynamic operator in the extended interface is described by Eq. (17.1) with the constant horizon and micromodulus determined as an average value of the micromoduli in the matrix and inclusion (17.3), see [18] [and also [998]) where one proposed a variation of the extended interface micromodulus $\mathbf{C}^i(\mathbf{x}, \hat{\mathbf{x}})$ in a spirit of the functionally graded materials theory described in, e.g., [200]). The equilibrium equation (18.1) can be presented in a standard matrix form ($\mathbf{x}_p \in w$)

$$\sum_q \mathbb{K}_{pq}\mathbf{u}_q = \mathbf{b}_p, \tag{18.2}$$

$$\mathbb{K}_{pq} = \mathbf{C}(\mathbf{x}_p, \mathbf{x}_q)\bar{v}_{pq} - \delta_{pq} \sum_{r \in \mathcal{H}_p} \mathbf{C}(\mathbf{x}_p, \mathbf{x}_r)\bar{v}_{pr}, \tag{18.3}$$

where $\mathbb{K}_{pq}$ is a rectangular matrix $\big(dN^{\max} \times d(N^{\max} + N^{w\Gamma})\big)$ while a displacement $\mathbf{u}_q$ is presented by a vector $\big(d(N^{\max} + N^{w\Gamma}) \times 1\big)^\top$. Equation (18.2) should be solved in accompany with the corresponding homogeneous volumetric boundary conditions (e.g., Dirichlet-type volumetric boundary condition (16.46)). It means inversion of a global square stiffness matrix $\hat{\mathbb{K}}$ $(dN^{\max} \times dN^{\max})$ for all nodes $p \in [1, dN^{\max}]$ in Eq. (18.4) where a displacement $\mathbf{u}$ and a fictitious body force vector $\hat{\mathbf{b}}$ (defined by the volumetric boundary conditions (16.46) and (16.52)) are presented by the vectors $(dN^{\max} \times 1)^\top$

$$\sum_q \hat{\mathbb{K}}_{pq}\mathbf{u}_q = \hat{\mathbf{b}}_p, \tag{18.4}$$

$$\hat{\mathbb{K}}_{pq} = \mathbb{K}_{pq} W(\mathbf{x}_q), \tag{18.5}$$

$$\hat{\mathbf{b}}_p = -\sum_{r \in \mathcal{H}_p} \mathbf{C}(\mathbf{x}_p, \mathbf{x}_r)\boldsymbol{\varepsilon}^{w\Gamma}(\mathbf{x}_r - \mathbf{x}_p)\bar{v}_{pr} W^{w\Gamma}(\mathbf{x}_r) \tag{18.6}$$

where $\mathbf{h} \equiv \boldsymbol{\varepsilon}^{w\Gamma}\mathbf{x}$ ($\boldsymbol{\varepsilon}^{w\Gamma} \equiv$const., $\mathbf{x} \in w_\Gamma$) (16.52), and the indicator function $W^{\overline{\overline{w}}}$ of the nonlocal closure $\overline{\overline{w}} = w \cup w_\Gamma$ is split as

$$W^{\overline{\overline{w}}}(\mathbf{x}) = W(\mathbf{x}) + W^{w\Gamma}(\mathbf{x}). \tag{18.7}$$

Thus, the solution $\mathbf{u}(\mathbf{x})$ ($\mathbf{x} \in w$) based on inverting a truncated matrix $\hat{\mathbb{K}}_{pq}$ ($p, q = 1, \ldots, dN^{\max}$) is assumed to be found. In truncation domain method, an infinite integration domain R^d is replaced by corresponding finite domain $w \subset R^d$ with control of the error. For estimation of the size of the trancated domain, let us the geometrical boundary Γ of the domain w in 2D is prescribed in the polar coordinate system $\Gamma(\mathbf{x}) = \{\mathbf{x} : x_1 = r(\phi)\cos(\phi), x_2 = r(\phi)\sin(\phi)\}$. The positions of the points $\mathbf{y} \in \Gamma^\rho$ of the surface Γ^ρ are expressed through the points $\mathbf{x} \in \Gamma$ by a homothetic transformation:

$$\mathbf{y} = \mathbf{x}_i + \rho(\mathbf{x} - \mathbf{x}_i), \tag{18.8}$$

where $\mathbf{x}_i = \mathbf{0}$ is a homothetic center coinciding with the particle center. Once the parameter $1 < \rho$ is chosen, the distribution of the points $\mathbf{y} \in \Gamma^\rho$ is determined. For a sequence $\rho_k = k$, we estimate a sequence $\mathbf{u}^\rho$ in the domain w^ρ with the boundary Γ^ρ and analyze a convergence of $\mathbf{u}^\rho$ by estimation of tolerance defined by the relative difference of $\mathbf{u}^\rho$ and $\mathbf{u}^{\rho+1}$:

$$\Delta = ||\mathbf{u}^{\rho+1} - \mathbf{u}^\rho||_{L_2}/||\mathbf{u}^{\rho+1}||_{L_2}, \quad \Delta \le 0.1\%$$
$$||(\cdot)||_{L_2} = \Big((\text{mes}\, v_\Gamma)^{-1}\int_{v_\Gamma}(\cdot)^2(\mathbf{y})d\mathbf{y}\Big)^{1/2}, \tag{18.9}$$

where the L_2 norm is evaluated over the domain v_Γ. The simulation is stopped when the tolerance Δ reaches 0.1%. We control the variations of $\mathbf{u}^\rho$ only over the domain $v^l := v \cup v_\Gamma$ because according to [210] and [217], the effective moduli depend on the strain distribution $\mathbf{u}(\mathbf{x})$ only in the domain $\mathbf{x} \in v^l$.

The found solution $\mathbf{u}(\mathbf{x})$ ($\mathbf{x} \in w$) allows us to finalize the solution of the basic problem by the following representations for the displacements and strains:

$$\mathbf{u}(\mathbf{x}_q) = \mathbf{A}^u(\mathbf{x}_q)\boldsymbol{\varepsilon}^{w\Gamma}, \quad \boldsymbol{\varepsilon}(\mathbf{x}_q) = \mathbf{A}^\epsilon(\mathbf{x}_q)\boldsymbol{\varepsilon}^{w\Gamma}, \quad (\mathbf{x}_q \in v^l), \tag{18.10}$$

respectively, where $\mathbf{A}^u(\mathbf{x}_q)$ is the displacement influence function while the strain influence function $\mathbf{A}^\epsilon(\mathbf{x}_q)$ is obtained from $\mathbf{A}^u(\mathbf{x}_q)$ by numerical differentiation. Hereafter, we introduced the notion of the field (displacement, strain, and stress) influence functions widely used in micromechanics (see Chaps. 6 and 17) (no confusion with the term "influence (or weighting) function" exploited in peridynamics;

see, e.g., [1356] and [1404]). It should be mentioned that the strain influence function $\mathbf{A}^{\epsilon}(\mathbf{x}_q)$ averaged over the inclusion v is used in the different methods of micromechanics of locally elastic CMs (see Chaps. 6 and 8). However, the peridynamic theory is traditionally based on the use of the displacement fields $\mathbf{u}(\mathbf{x})$ rather than the strain $\boldsymbol{\varepsilon}(\mathbf{x})$ fields which are not conceptually necessary. Moreover, the fictitious body forces (18.6) with a discontinuity at $\mathbf{x} \in \Gamma$ also produce a discontinuity of $\mathbf{u}(\mathbf{x})$ at $\mathbf{x} \in \Gamma$ because in general the displacement field $\mathbf{u}(\mathbf{x})$ has the same smoothness as the body force field (see [1405]). For elimination of this difficulty, Buryachenko [223] proposed a micromechanical model of peridynamic CMs based on the averaging of weighted displacement influence function $\mathbf{A}^{u}(\mathbf{s})$ over the surface $\mathbf{s} \in \partial v^l$. However, just for better visualization (e.g., for comparison with a locally elastic solution) of comparative numerical 1D analyses of different methods (see Section 6), we will estimate the strain influence function $\mathbf{A}^{\epsilon}(\mathbf{x}_q)$.

Of course, a classical problem for a finite peridynamic sample (with or without inclusions) was intensively analyzed in many publications even for more general constitutive laws (such as nonlinear and state-based approaches [1185, 1355, 1356, 1401, 1404]) and more advanced numerical methods. So, in mesh adaptivity [130, 131] and [633], a given quadrature model is refined where additional resolution is needed (e.g., in the vicinity of interface). In the coarsening approach by Silling [1398], the detailed geometry of a body is replaced by a succession of different geometries providing essentially the same result as in the original detailed problem. Macek and Silling [996] (see also [999]) implemented the peridynamic approach into the ABAQUS commercial FE code by the use of truss elements as mesh with appropriate stiffness properties which represents the peridynamic bonds. Thus, there is a room for improvement of the quadrature method considered (18.2)–(18.7) and its generalization to more general constitutive laws. The goal of the current study is to formulate a systematic scheme for obtaining the basic problem solutions (18.10) having control of the error (18.9). The form of this solution is adapted for subsequent incorporation into one or another micromechanical method (e.g., the generalized Mori-Tanaka approach; see for details [18]) of random structure peridynamic CMs. In general, the displacement (18.10_1) and strain (18.10_2) influence functions depend on the size of the domain w. However, the satisfaction of the tolerance requirements (18.9) provides a stabilization of the solutions $\mathbf{A}^{u}(\mathbf{x})$ and $\mathbf{A}^{\epsilon}(\mathbf{x})$ in the extended inclusion $\mathbf{x} \in v^l$. It should be mentioned that the estimation of the tolerance (18.9) in the same domain $\mathbf{x} \in v_{\Gamma}$ for two solutions $\mathbf{u}^{\rho+1}$ and $\mathbf{u}^{\rho}$ with the different domains of definition is more convenient than control of the limit

$$a/L \to 0. \tag{18.11}$$

18.1.2 Periodization Method

An alternative sort of truncation method is periodization method where the problem for one inclusion inside the infinite matrix is replaced by a periodic structure with an increasing ratio $a/l^{\Omega} \to 0$ of a unit cell (UC) size l^{Ω} and the inclusion size a. So, the domain w is constructed from a periodic field of the cubic unit cells (UCs) Ω_α: $w = \cup\Omega_\alpha$, where each UC with the side length $2l^{\Omega}$ is labeled by the multiplet of integer numbers $\boldsymbol{\alpha} = (\alpha_1, \dots, \alpha_d) \in Z^+$; here, $\alpha_i = 0, \pm 1$ ($i = 1, \dots, d$). The centers of UCs are presented as $\mathbf{x}_\alpha^{\Omega} = 2l^{\Omega}\alpha_1\mathbf{e}_1 + \dots + 2l^{\Omega}\alpha_d\mathbf{e}_d$ in the Cartesian coordinate frame $\mathbf{e}_1, \dots, \mathbf{e}_d$ (see for details Chap. 19). Determination of the microstructural displacements $\mathbf{u}(\mathbf{x})$ ($\mathbf{x} \in \Omega_{00}$) in accompany with the volumetric periodic boundary conditions (PBC) at the dilute concentration of inclusions ($a/l^{\Omega} \to 0$; compare with Eq. (18.11)) allows to finalize the solution of the basic problem with estimation of the displacement $\mathbf{A}^u(\mathbf{x}_q)$ (18.10_1) and strain $\mathbf{A}^{\epsilon}(\mathbf{x}_q)$ (18.10_2) influence functions. Control of accuracy is executed by the tolerance requirement (18.9). It should be mentioned that the theory of computational homogenization of periodic structure peridynamic CMs was developed in [223] (see Chap. 19). However, a dilute approximation ($a/l^{\Omega} \ll 1$) with prescribed tolerance requirement (18.9) was not considered before.

18.1.3 Iterative Methods

One of the main features of linear peridynamics (16.5) and (16.55) is that each unknown $\mathbf{u}_p$ appears in only a few equations $2n^l + 1$ (where $2l_\delta = 2n^l h$) of the linear system in a "neighborhood" of the pth equation. In such a case, for the full space $w = R^d$, the infinite coefficient matrix $\mathbb{K}$ is banded with some bandwidth k^b (so that $\mathbb{K}_{pq} = 0$ at $|p - q| > k^b$) defined by both the numeration of nodes p and n^l. In particular, the bandwidth for 1D problem equals n_l.

Let us consider the governing equation for an infinite R^d ($d = 1, 2, 3$) homogeneous peridynamic medium (16.5)

$$\widehat{\mathcal{L}}^{(0)}(\mathbf{u}^0)(\mathbf{x}) + \mathbf{b}(\mathbf{x}) = \mathbf{0} \tag{18.12}$$

with the micromodulus $\mathbf{C}^{(0)}$. In particular, we consider a case when the self-equilibrated body force density $\mathbf{b}(\mathbf{x})$ (18.12) produces a displacement corresponding to the homogeneous strain

$$\mathbf{u}^0(\mathbf{x}) \equiv \boldsymbol{\varepsilon}^0\mathbf{x}, \quad \boldsymbol{\varepsilon}^0 = \text{const.} \tag{18.13}$$

We are coming now to consideration of a single fixed inclusion $v_i(\mathbf{0})$ with the micromodulus $\mathbf{C}^{(1)}(\boldsymbol{\xi})$ in the infinite homogeneous medium with the properties $\mathbf{C}^{(0)}(\boldsymbol{\xi})$. Then the solution $\mathbf{u}(\mathbf{x})$ can be split into two ones as

$$\mathbf{u}(\mathbf{x}) = \mathbf{u}^0(\mathbf{x}) + \mathbf{u}_1(\mathbf{x}), \tag{18.14}$$

where

$$\widehat{\mathcal{L}}^{(0)}(\mathbf{u}^0)(\mathbf{x})+\mathbf{b}(\mathbf{x})=\mathbf{0}, \tag{18.15}$$

$$\widehat{\mathcal{L}}(\mathbf{u}_1)(\mathbf{x})+\widehat{\mathcal{L}}_1(\mathbf{u}^0)(\mathbf{x})=\mathbf{0}, \tag{18.16}$$

and the operator $\widehat{\mathcal{L}}$ is decomposed as $\widehat{\mathcal{L}} \equiv \widehat{\mathcal{L}}^{(0)} + \widehat{\mathcal{L}}_1$ (18.12), where $\widehat{\mathcal{L}}_1$ is defined by Eq. (18.12) with the replacements $\mathbf{C}^{(0)} \to \mathbf{C}_1$. In so doing, Eq. (18.16) is more preferable to solve than Eqs. (16.5) and (16.55) (see, e.g., [1603]) because a fictitious body force density of Eq. (18.16) has a compact support $\widehat{\mathcal{L}}_1(\mathbf{u}^0)(\mathbf{x}) \equiv \mathbf{0}$ $(\mathbf{x} \notin v_i^l$, $v_i^l = \{\mathbf{x} : |\mathbf{x}| \leq a + l_\delta\})$. Hence, $\mathbf{u}_1(\mathbf{x})$ passes to the constant $\mathbf{u}_1(\mathbf{x}) \to$const. at the removal of $\mathbf{x}$ from v_i^l: $|\mathbf{x}-\mathbf{y}| \to \infty$ $(\forall \mathbf{y} \in v_i^l)$, and, therefore, for numerical solution of Eq. (18.16), only some finite domain $\mathbf{x} \in v_i^{l0} := \{\mathbf{x} : \ |\mathbf{x}| < a^{l0}\}$ should be discretized. However, the main advantage of the decompositions (18.15) and (18.16) is elimination in (18.16) of the nonlocal volumetric boundary conditions prescribed over a nonzero volume (see, e.g., (16.46) and [397, 771, 996, 1396]) of Eq. (18.16), while the solution of Eq. (18.15) is assumed to be known.

Indeed, Eq. (18.16) can be recast in the form (18.2) $(p, q = \ldots, -2, -1, 0, 1, 2\ldots)$

$$\sum_q \mathbb{K}_{pq}\mathbf{u}_{1|q} = \tilde{\mathbf{b}}_p, \tag{18.17}$$

where the infinite banded matrix $\mathbb{K}_{pq}$ is defined by Eq. (18.3) while a fictitious body force has the following representation:

$$\tilde{\mathbf{b}}_p = -\sum_{r\in\mathcal{H}_p} \mathbf{C}_1(\mathbf{x}_p, \mathbf{x}_r)\boldsymbol{\varepsilon}^0(\mathbf{x}_r - \mathbf{x}_p)\bar{v}_{pr}. \tag{18.18}$$

It should be mentioned that the indices $p, q = \ldots, -2, -1, 0, 1, 2, \ldots$ in Eq. (18.17) are renumbered with respect to the indices $p, q = (1, dN^{\max})$ (18.4). In general, the matrix $\mathbb{K}$ (18.17) is an infinite matrix; however, in our current practice, the matrix $\mathbb{K}$ (18.17) is, in fact, a truncated submatrix $(\mathbb{K}_{pq})$ $(p, q = [N^{\text{finite}}, N^{\text{finite}}]$ with the size $2N^{\text{finite}} \times 2N^{\text{finite}}$ described below. The vector $\tilde{\mathbf{b}}_p$ in the right-hand side of Eq. (18.17) has a compact support $(\tilde{\mathbf{b}}_p \equiv \mathbf{0},\ \forall \mathbf{x}_p \in R^d \setminus v^l)$ because $\mathbf{C}_1(\mathbf{x}_p, \mathbf{x}_q)$ vanishes in the domain $R^d \setminus v^l$. It means that nonzero elements of the vector $\tilde{\mathbf{b}}_p$ are limited by the indices $|p| \leq d(N^v + N^{v\Gamma})$, where N^v and $N^{v\Gamma}$ are the numbers of nodes in the inclusion v and the external part of the interaction interface $v^l \setminus v$, respectively.

The last established feature of the right-hand side of Eq. (18.17) allows us to exploit the iterative methods which are especially effective for analyses of infinite banded matrices. The popular iterative methods pass from one iteration to the next by modifying one or a few components of an approximate vector solution at a time. Improvement of iterations is controlled by the residual vector $\tilde{\mathbf{b}} - \mathbb{K}\mathbf{u}$. Namely, Eq. (18.17) can be written in the form $(k = 1, 2, \ldots)$

$$\mathbb{D}\mathbf{u}_1^{(k)} = \mathbb{F}\mathbf{u}_1^{(k-1)} + \tilde{\mathbf{b}}_p, \tag{18.19}$$

where $\mathbb{K} = \mathbb{D} - \mathbb{F}$ is a splitting and $\mathbf{u}_1^{(0)}$ is a starting vector. For the iteration to be practical, it must be easy to solve the system (18.19). In the simplest case, this is the Jacobi method where $\mathbb{D} = \mathrm{diag}(\ldots, \mathbb{K}_{-1-1}, \mathbb{K}_{00}, \mathbb{K}_{11}, \ldots)$ is a diagonal matrix. The diagonal entries of $\mathbb{D}$ are all nonzero that enables the inverse infinite matrix $\mathbb{D}^{-1} = \mathrm{diag}(\ldots, \mathbb{K}_{-1-1}^{-1}, \mathbb{K}_{00}^{-1}, \mathbb{K}_{11}^{-1}, \ldots)$ to be found. The inversion $\mathbb{D}^{-1}$ of the diagonal matrix makes it possible to reduce Eq. (18.19) to an explicit recursive relation

$$\mathbf{u}_1^{(k)} = \mathbb{B}\mathbf{u}_1^{(k-1)} + \tilde{\tilde{\mathbf{b}}}, \quad \mathbb{B} := \mathbb{D}^{-1}\mathbb{F}, \quad \tilde{\tilde{\mathbf{b}}} := \mathbb{D}^{-1}\tilde{\mathbf{b}} \tag{18.20}$$

with the driving term usually chosen as an initial approximation

$$\mathbf{u}_1^{(0)} = \tilde{\tilde{\mathbf{b}}}. \tag{18.21}$$

It is also a possible choice of the initial approximation $\mathbf{u}_1^{(0)}$ in the form of the locally elastic solution (16.6)

$$\mathbf{u}_1^{(0)} = {}^L\mathbf{u}_1. \tag{18.22}$$

It should be mentioned that the use of the initial approximation (18.22) implies, in fact, the correction of the $\tilde{\mathbf{b}}(\mathbf{x})$ at $\mathbf{x} \in w_\Gamma$ analogous to Eq. (18.6) with the replacement $\boldsymbol{\varepsilon}^{w_\Gamma} \to {}^L\boldsymbol{\varepsilon}_1$.

Thus, the indices p of nonzero elements of the vectors $\tilde{\mathbf{b}}_p$ and $\mathbf{u}_{1|p}^{(0)}$ are bounded by the inequality $|p| \le d(N^v + N^{v\Gamma})$. The iteration matrix $\mathbb{B}$ (completely determining the convergence (or not) of an iterative method (18.19)) is the infinite banded matrix (as $\mathbb{K}$) with the same bandwidth k^b. Then multiplication of the matrix $\mathbb{B}$ and the vector $\mathbf{u}_{1|p}^{(0)}$ leads to the vector $\mathbf{u}_{1|p}^{(1)}$ with the indices of nonzero elements bounded by an inequality $|p| \le d(N^v + N^{v\Gamma} + k^b)$. In a similar manner, the k's iteration $\mathbf{u}_{1|p}^{(k)}$ has a compact support with the indices of nonzero elements $|p| \le N^{(k)} := d(N^v + N^{v\Gamma} + k^b \times k)$ that defines the size of the domain v_i^{l0}. It should be mentioned that the matrix $\mathbb{B}$ (18.20$_1$) is an infinite matrix. However, owing to the mentioned compact support of $\mathbf{u}_1^{(k-1)}$, only the submatrix $(\mathbb{B})_{pq}$, $p, q = -[d(N^v + N^{v\Gamma} + k^b \times (k-1)), d(N^v + N^{v\Gamma} + k^b \times (k-1))]$, is used. It means that in the iteration scheme (18.20$_1$), one exploits the finite matrix $(\mathbb{B})_{pq}$, $p, q = -[d(N^v + N^{v\Gamma} + k^b \times (k-1)), d(N^v + N^{v\Gamma} + k^b \times (k-1))]$, with the size depending on the iteration number k.

In the Jacobi iteration, each component of the next iteration $\mathbf{u}_1^{(k+1)}$ is expressed through all components of the previous step $\mathbf{u}_1^{(k)}$. However, in the *Gauss-Seidel* iteration, the newly computed component $u_{1|p}^{(k)}$ is changed within a working vector that is redefined at each step ($|p| \le N^{(k)}$)

$$u_{1|p}^{(k+1)} = (\mathbb{D}_{pp})^{-1} \left[\sum_{q=-N^{(k)}}^{p-1} \mathbb{F}_{pq} u_{1|q}^{(k+1)} + \sum_{p+1}^{q=N^{(k)}} \mathbb{F}_{pq} u_{1|q}^{(k)} + \tilde{b}_p \right]. \quad (18.23)$$

The *successive overrelaxation* method (SOR) corrects the pth component following the relaxation sequence ($0 < \omega \leq 1$)

$$u_{1|p}^{(k+1)} = \omega u_{1|p}^{GS} + (1-\omega) u_{1|p}^{(k)}, \quad (18.24)$$

in which $u_{1|p}^{GS}$ is defined by the expression on the right-hand side of (18.23); note that if $\omega = 1$, then this is just the Gauss-Seidel method.

The sequence $\mathbf{u}_1^{(0)}, \mathbf{u}_1^{(1)}, \mathbf{u}_1^{(2)}, \ldots$ converges to a unique solution $\mathbf{u} = \mathbf{u}_1$ if the norm of the iteration matrix $\mathbb{B}$ turns out to be small "enough" (less than 1) $\|\mathbb{B}\| < 1$ where $\|\mathbb{B}\|$ denotes the Frobenius norm of $\mathbb{B}$. A convergence of $\mathbf{u}^{(k)}$ is controlled by estimation of tolerance defined by the relative difference of $\mathbf{u}^{(k+1)}$ and $\mathbf{u}^{(k)}$ and the residual vector

$$\Delta_2 = ||\mathbf{u}^{(k+1)} - \mathbf{u}^{(k)}||_{L_2} / ||\mathbf{u}^{(k+1)}||_{L_2}, \quad \Delta_3 = ||\tilde{\mathbf{b}} - \mathbb{K}\mathbf{u}^{(k+1)}||_{L_2} / ||\mathbf{u}^{(k+1)}||_{L_2}, \quad (18.25)$$

where the L_2 norm is evaluated over the domain v^l (as in Eq. (18.9$_2$)). The simulation is stopped when the tolerance $\max(\Delta_2, \Delta_3)$ reaches 0.1%.

It may appear that the ill-conditioning issue (see, e.g., [131]) that occurs with the use of some micromodulus function prevents computing the numerical solution to the static problem (18.4) or (18.17). A measure of the ill-posedness of the mentioned equations can be performed by using the condition number $\text{cond}\mathbb{K} = \|\mathbb{K}\|\|\mathbb{K}^{-1}\|$. If $\mathbb{K}$ is highly ill-conditioned, one can use a regularization method to solve this system. So, the known Tikhonov[1507] regularization equation is

$$\tilde{\mathbb{K}}\mathbf{u}_1 = \mathbf{b}^K, \quad \tilde{\mathbb{K}} := (\mathbb{K}^\top \mathbb{K} + \lambda \mathbf{I}), \quad (18.26)$$

where $\mathbf{b}^K := \mathbb{K}^\top \tilde{\mathbf{b}}$, $\mathbf{I}$ is the unit infinite matrix, and the superscript $^\top$ signifies the transpose. The optimal value λ is chosen according to the method described in, e.g., [603, 937] and [1008].

The methods for inversion of the nonsingular finite matrix $\tilde{\mathbb{K}}$ (18.4) are well developed (see, e.g., [559, 1313]). We modify the Tikhonov regularization method [1507] for the infinite matrix $\tilde{\mathbb{K}}$ by using the preconditioner and iteration for solving an ill-posed linear infinite system (18.17). So, we begin with the following preconditioner:

$$\mathbf{P} = \mathbb{K}^\top + \lambda \mathbb{K}^{-1}. \quad (18.27)$$

Applying the operator $\mathbf{P}$ (18.27) to Eq. (18.17), we obtain the following equation:

$$(\mathbb{K}^\top + \lambda\mathbb{K}^{-1})\mathbb{K}\mathbf{u} = (\mathbb{K}^\top + \lambda\mathbb{K}^{-1})\tilde{\mathbf{b}}, \tag{18.28}$$

which is reduced to

$$\tilde{\mathbb{K}}\mathbf{u} = \mathbf{b}^K + \lambda\mathbf{u}. \tag{18.29}$$

It is interesting that Eq. (18.29) is similar to the Tikhonov regularization equation (18.26) with additional term $\lambda\mathbf{u}$ at the right-hand side in Eq. (18.29). It gives an opportunity to form an iterative process to find the solution of $\mathbf{u}$ by

$$\tilde{\mathbb{K}}\mathbf{u}_1^{[n+1]} = \mathbf{b}^K + \lambda\mathbf{u}_1^{[n]}. \tag{18.30}$$

Starting from $\mathbf{u}_1^{[n]} = \mathbf{0}$ $(n = 0)$, we can apply the SOR to solve the above linear system and generate a sequence of vectors $\mathbf{u}_1^{(k)[n]}$ $(k = 1, \ldots)$. In each inner iteration, we apply the SOR to solve the linear system (18.26) under a convergence criterion specified in Eq. (18.26). When the following convergence criterion is satisfied, we also stop the outer iterative sequence, and then the solution of $\mathbf{u}_1$ is obtained. Thus, the basic problem is assumed to be solved and the displacement $\mathbf{A}^u(\mathbf{x}_q)$ (18.10$_1$) and strain $\mathbf{A}^\epsilon(\mathbf{x}_q)$ (18.10$_2$) influence functions are found with the required tolerance (18.25). It should be mentioned that the solution of the basic problem in [209] is, in fact, a simplified first-order approximation of the iteration method (18.19) proposed before at the intuitional level of rigorousness for exclusively 1D case without any tolerance requirements.

18.1.4 Green Function Method

Let us consider an infinite homogeneous matrix with the micromodulus $\mathbf{C}^{(0)}$ subjected to some body force density $\mathbf{b}(\mathbf{x})$ corresponding to the prescribed displacement $\overline{\mathbf{u}}(\mathbf{x})$ (called the effective field). In particular, for the homogeneous loading

$$\overline{\mathbf{u}}(\mathbf{x}) \equiv \overline{\boldsymbol{\varepsilon}}\mathbf{x}, \quad \overline{\boldsymbol{\varepsilon}} = \text{const}. \tag{18.31}$$

A solution of the basic problem can be split into two summands as $\mathbf{u}(\mathbf{x}) = \overline{\mathbf{u}}(\mathbf{x}) + \mathbf{u}_1(\mathbf{x})$ by the use of a replacement $\mathbf{u}^0 \to \overline{\mathbf{u}}$ in Eq. (18.21). Then Eq. (18.23) can be easily converted

$$\widehat{\mathcal{L}}^{(0)}(\mathbf{u}_1)(\mathbf{x}) = -\widehat{\mathcal{L}}_1(\overline{\mathbf{u}})(\mathbf{x}) \tag{18.32}$$

for applying the Green function ($\mathbf{G}$; see [1588] and [1602]) technique

$$\mathbf{u}(\mathbf{x}) - \overline{\mathbf{u}}(\mathbf{x}) \equiv \mathcal{L}^{u\pi}(\mathbf{x} - \mathbf{x}_i, \overline{\pi}) = \int_{v_i^l} \mathbf{G}(\mathbf{x} - \mathbf{y})\overline{\pi}(\mathbf{y})d\mathbf{y}, \tag{18.33}$$

$$\overline{\pi}(\mathbf{y}) := \widehat{\mathcal{L}}_1(\mathbf{u})(\mathbf{y}) = \int_{\mathcal{H}_\mathbf{y}} \widetilde{\boldsymbol{\tau}}(\mathbf{y}, \hat{\mathbf{x}})d\hat{\mathbf{x}}, \tag{18.34}$$

where $\overline{\pi}(\mathbf{y})$ is a local average polarization tensor obtained by averaging the micropolarization tensor (17.94) over $\mathcal{H}_\mathbf{y}$. Obviously, $\overline{\pi}(\mathbf{y})$ is some sort of a fictitious body force in Eq. (18.32) vanishing outside the extended inclusion (Minkowski dilation) $v_i^l := v_i \oplus \mathcal{H}_\mathbf{0}$. The method (18.33) and (18.34) is also called the volume integral equation (VIE) method as the corresponding method in local elasticity

$$\mathbf{u}(\mathbf{x}) - \overline{\mathbf{u}}(\mathbf{x}) \equiv^L \mathcal{L}^{u\tau}(\mathbf{x} - \mathbf{x}_i, \boldsymbol{\tau}) = \int_{v_i} \nabla^L \mathbf{G}(\mathbf{x} - \mathbf{y})^L \boldsymbol{\tau}(\mathbf{y}) d\mathbf{y}. \tag{18.35}$$

Although Eq. (18.35_2) is quite general, its disadvantage is defined by the limited opportunities of the Green function construction for a general case of a medium anisotropy, while the evaluation of the operator form (18.35_1) by the numerical quadrature methods is well developed.

The discretized form of Eq. (18.33) ($\mathbf{x}_r, \mathbf{x}_p \in R^d; \mathbf{x}_q \in \mathcal{H}_{\mathbf{x}_p}$)

$$\mathbf{u}(\mathbf{x}_r) - \overline{\mathbf{u}}(\mathbf{x}_r) = \sum_r \mathbf{G}(\mathbf{x}_r - \mathbf{x}_p)\bar{v}_{rp} \sum_q \mathbb{K}^1_{pq}\mathbf{u}_q \tag{18.36}$$

contains the matrix $\mathbb{K}^1_{pq}$ defined analogously to the matrix $\mathbb{K}_{pq}$ (18.3) with a replacement $\mathbf{C}(\mathbf{x}, \hat{\mathbf{x}}) \to \mathbf{C}_1(\mathbf{x}, \hat{\mathbf{x}})$. It should be mentioned that the micromodulus jump $\mathbf{C}_1(\mathbf{x}, \hat{\mathbf{x}})$ vanishes outside the extended inclusion v_i^l, and, because of this, $\mathbf{x}_r, \mathbf{x}_p, \mathbf{x}_q \in v_i^l$. It means that the matrix equation (18.36) is, in fact, finite and its numerical solution should present no problem that does not consider in more detail in the current presentation.

18.1.5 Numerical Results for 1D Case

At first, we consider the truncation method for 1D peridynamic bar $\bar{\bar{w}}$ of the length $2L + 2l_\delta$ and cross-sectional area $A = 1$ containing one inclusion $v = [-a, a]$ and the volumetric boundary w_Γ: $\bar{\bar{w}} = w_\Gamma \cup w$, $w_\Gamma = [-L - l_\delta, -L] \cup [L, L + l_\delta]$. The midpoint quadrature scheme (similar to [131, 369, 664]) is realized for the uniform grid $x_i = (0, \pm 1, \pm 2, \ldots, \pm(N^{\max} + N^{w\Gamma}))$ with the numbers of material points $2N^{\max} + 1$ and $2N^{w\Gamma} + 1$ inside both w and the interaction boundary w_Γ, respectively, and a constant step size $h = L/N^{\max}$.

We consider a heterogeneous peridynamic bar with the step-function profiles (16.82_1) ($k = 0, 1$, $E^{(1)}/E^{(0)} = 5$); the triangle profile (16.82_2) (and the micromodulus (16.64)) can be considered in a similar manner.

For the periodic structure CMs, the rule of the interaction boundary w_Γ is performed by the interaction UC's boundary Ω_Γ. In Fig. 18.1, one presents the

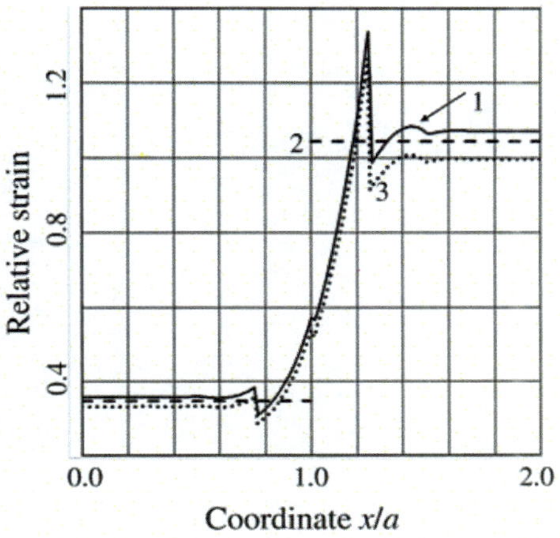

Fig. 18.1 Relative strains $\tilde{\varepsilon}(x)$ (3) and $\tilde{\varepsilon}^{\Omega}(x)$ (1) vs. x/a. Locally elastic solution ${}^{L}\tilde{\varepsilon}^{\Omega}(x)$ (2)

estimation of the relative strains $\tilde{\varepsilon}(x) := \varepsilon(x)/\varepsilon^0$ and $\tilde{\varepsilon}^{\Omega}(x) := \varepsilon(x)/\langle\varepsilon\rangle$ in the peridynamic composite bar by the truncation method (curve 3) and periodization one (curve 1), respectively, at $a/L = a/l^{\Omega} = 0.06$ and $N^a = 200$. In fact, the relative strains $\tilde{\varepsilon}(x)$ and $\tilde{\varepsilon}^{\Omega}(x)$ are the strain influence functions (see Eq. (17.26)). The curves of these strain influence functions vs. x/a behave according to the general features of peridynamic solutions for a homogeneous peridynamic bar with a fictitious body force considered in [1405] and [131]. Indeed, the fictitious body forces (18.6) with a discontinuity at $x = \pm a$ also produce a discontinuity of $u(x)$ at $x = \pm a$ (it is not demonstrated in Fig. 18.1) (see [1405]). In addition, the discontinuity in the micromodulus $C(\xi)$ (16.82) at $|\xi| = l_\delta$ has a further effect on the smoothness of $u(x)$ with a corresponding discontinuity of the derivatives $u^{(k)}$ of the order $k = 1, 2, \ldots$ at $x = \pm(a + kl_\delta)$ (it is well observed for curves 1 and 3 in Fig. 18.1 in the form of peaks of $\varepsilon(x)$ $(k = 0)$ at $x = a \pm l_\delta,\ l_\delta/a0.2)$. The locally elastic solution for the periodic bar ${}^{L}\tilde{\varepsilon}^{\Omega}(x)$ is shown by curve 2. The solution by the truncation method (curve 3) is close to the corresponding locally elastic solution $\tilde{\varepsilon}(x) = [1 + 2H(|x|/a - 1)]/3$ at $|x| - a > 2l_\delta$, where $H(x)$ is the Heaviside function.

For the iterative method with the initial approximation (18.22), we estimate the influence of the ratio $N^a/N^{\max}$ with the fixed $N^{\max} = 2500$, $l_\delta/a = 0.25$, and the iteration number $k = 100$. In Fig. 18.2, one presents the curves $\varepsilon(x) \sim x/a$ with the different ratios $N^a/N^{\max} = 1/25, 1/10, 1/2$ corresponding to $\max_x \varepsilon(x)/\varepsilon^0 \equiv \varepsilon(1.25)/\varepsilon^0 = 1.333,\ 1.347,\ 1.357$, respectively. As can be seen, a change of the ratio $N^a/N^{\max}$ in 12 times yields a negligible distinction of curves 1 and 3 that indicates on the effectiveness and robustness of the iterative method proposed (so $\Delta_2 = 3.4E-4,\ \Delta_3 = 6.3E-9$) in 1D case.

We analyze the influence of iteration number on the distribution of $\varepsilon(x)/\varepsilon^0 \sim x/a$ for the fixed $N^a = 600$ and $l_\delta/a = 0.25$. As can be seen in Fig. 18.3, increasing the iteration number k from 3 to 10 and 100 (see curves 1, 2, and 3, respectively, in Fig. 18.3) leads to variation of the pick of strains $\max \varepsilon(x)/\varepsilon^0 = \max \varepsilon(1.25)/\varepsilon^0$

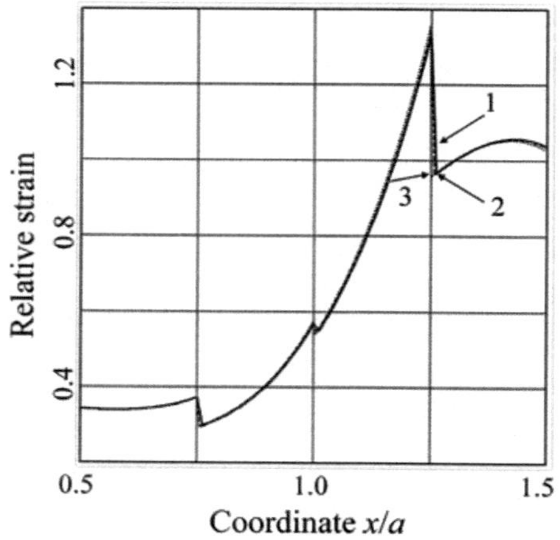

Fig. 18.2 Relative strain $\tilde{\varepsilon}(x)$ vs. x/a for $N^a/N^{\max} =$ 1/2 (1), 1/10 (2), 1/25 (3)

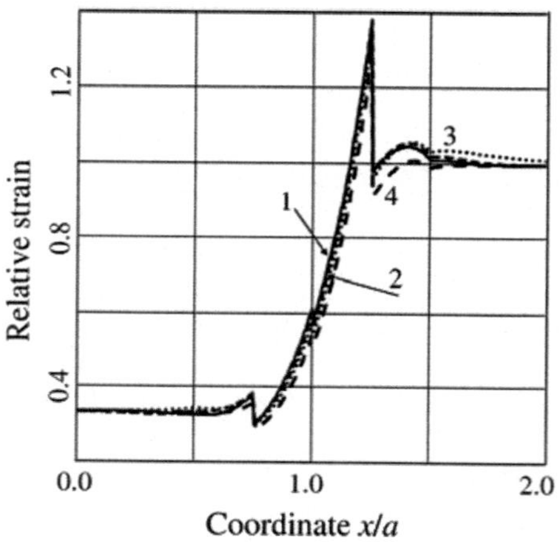

Fig. 18.3 Relative strains $\tilde{\varepsilon}(x)$ vs. x/a estimated by the iterative method $k = 3$ (1), 10 (2), and 100 (3) and truncation method (4)

from 1.381 to 1.379 and 1.348, respectively. It means that ten iterations provide sufficient accuracy. Curve 4 in Fig. 18.3 estimated by the truncation method (as in Fig. 18.1) has the pick $\max \varepsilon(x)/\varepsilon^0 = \varepsilon(1.25)/\varepsilon^0 = 1.31$. In so doing, the compute times spent for estimation of curves 2 and 4 differ by two decimal orders.

In Fig. 18.4, the fixed $N^a = 600$ is considered with the different scale ratios $l^c/a = 0.05, 0.25, 0.5, 1.0$ (curves 1–4, respectively) corresponding to the $\max_x \varepsilon(x) = 1.344, 1.356, 1.357, 1.364$ and $\min_x \varepsilon(x) = 0.297, 0.296, 0.300, 0.333$, respectively. As it is expected, $\varepsilon(x) \to^L \varepsilon(x)$ in the limit of the horizon going to zero $l_\delta/a \to 0$ at some distance of x from the inclusion boundary $x = a$. The peridynamic curves 1–4 estimated for $N^a = 600$ and $N^a = 200$ (only the results for $N^a = 600$ are depicted in Fig. 18.4), respectively, are pretty close to each other with the exception of the point $x = a$ where the curve undergoes unexpectedly large break. Due to this reason, the estimations are performed for $N^a = 600$.

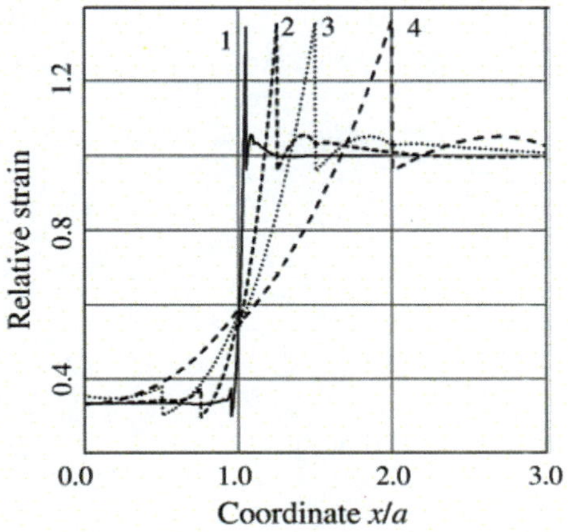

Fig. 18.4 Relative strain vs. x/a for $l^c/a = 0.05$, $0.25,\ 0.5,\ 1.0$ (curves 1–4, respectively)

Thus, one presented four different methods for the modeling of the basic problem for one inclusion inside the infinite matrix subjected to some homogeneous effective remote field. The truncation method combined with commercial software FEA (e.g., ABAQUS) is apparently most popular in engineering practice for locally elastic media. The mentioned features of the different methods are analyzed in Chap. 4 in the context of locally elastic problems. However, in the case of peridynamics, computational advantage of the truncation method realized by the FEA [996] is leveled off (with respect to the VIE method) due to nonlocality of the constitutive law. Nevertheless, universality of commercial software FEA provides a long-term advantage of the truncation method in engineering practice. In general, for dealing with the infinite banded matrices (18.17), it is advisable to use a class of iterative methods proposed which never alter the infinite matrix $\mathbb{K}$ and require the storage of a few vectors of finite length at a time. One demonstrated the effectiveness and robustness of the iterative method proposed in 1D case.

18.2 Basic Hypotheses of Peridynamic Micromechanics

We generalize the basic hypothesis of local elasticity of CMs to their peridynamic counterparts. To simplify the exact GIEs (17.205) and (17.206), we now apply the main hypothesis of many micromechanical methods called the effective field hypothesis (EFH):

(H1a) *Each inclusion v_i is located in the homogeneous strain field $\overline{\boldsymbol{\varepsilon}}_i(\mathbf{y})$ corresponding to the linear displacement $\overline{\mathbf{u}}(\mathbf{y})$ $(\mathbf{y} \in v_i^l)$:*

$$\overline{\boldsymbol{\varepsilon}}_i(\mathbf{y}) \equiv \overline{\boldsymbol{\varepsilon}}(\mathbf{x}_i), \quad \overline{\mathbf{u}}(\mathbf{y}) = \overline{\boldsymbol{\varepsilon}}(\mathbf{x}_i)(\mathbf{y} - \mathbf{x}_i). \tag{18.37}$$

In some methods (e.g., the multiparticle EFM (MEFM); see Chap. 8), this basic hypothesis **H1a** is complemented by a satellite hypothesis, which can be presented in peridynamics as

(H1b) *The perturbation introduced by the inclusion* v_i *at the point* $\mathbf{y} \notin v_i$ *is defined by the relation*

$$\int \mathbf{G}(\mathbf{y}-\mathbf{x})\overline{\boldsymbol{\pi}}(\mathbf{x})V_i^l(\mathbf{x}) = \bar{v}_i^l \mathbf{T}_i^{\varepsilon}(\mathbf{y}-\mathbf{x}_i)\overline{\boldsymbol{\pi}}_i, \tag{18.38}$$

$$\mathcal{L}^{\varepsilon\tau}(\mathbf{y}-\mathbf{x}_i, \overline{\boldsymbol{\tau}}) = \mathbf{L}^{\varepsilon\tau}(\mathbf{y}-\mathbf{x}_i)\overline{\boldsymbol{\tau}}_i. \tag{18.39}$$

Hereafter, $\overline{\boldsymbol{\tau}}_i \equiv \langle \overline{\boldsymbol{\tau}}(\mathbf{x})V_i(\mathbf{x})\rangle_{(i)}$ is an average over the volume of the inclusion v_i^l (but not over the ensemble), $\langle(.)\rangle_i \equiv \langle\langle(.)\rangle_{(i)}\rangle$, and the tensor

$$\mathbf{T}_i^{\varepsilon}(\mathbf{y}-\mathbf{x}_i) = \langle \mathbf{G}(y-x)\rangle_i. \tag{18.40}$$

The tensor $\mathbf{T}_i^{\varepsilon}(\mathbf{y}-\mathbf{x}_i)$ for locally elastic problem has the analytical Eshelby [449] representations for ellipsoidal homogeneous inclusion in an isotropic matrix with local properties (see Chap. 3). In the last case, the standard assumption (18.37) (see Chap. 4) yields the assumptions (18.38) and (18.39); otherwise, the formulae (18.38) and (18.39) define an additional assumption. The assumption **H1b** reduces the exact equations (17.187) and (17.187) to their approximations (17.197) and (17.198), respectively.

For termination of the hierarchy of the statistical moment equations (17.205) and (17.206) and their subsequent simplification, we will analyze the applicability of the hypotheses **H2b** and **H3** formulated in Chap. 8.

Thus, the solutions of Eqs. (18.26) and (18.36) for the displacements and strains can be considered in the form ($\mathbf{x} \in R^d$)

$$\mathbf{u}(\mathbf{x}) - \overline{\mathbf{u}}(\mathbf{x}) = \mathcal{L}_i^{uu}(\mathbf{x}-\mathbf{x}_i, \overline{\mathbf{u}}), \quad \boldsymbol{\varepsilon}(\mathbf{x}) - \overline{\boldsymbol{\varepsilon}}(\mathbf{x}) = \mathcal{L}_i^{\varepsilon\varepsilon}(\mathbf{x}-\mathbf{x}_i, \overline{\boldsymbol{\varepsilon}}), \tag{18.41}$$

respectively, and the strain perturbator $\mathcal{L}_i^{\varepsilon\varepsilon}(\mathbf{x}-\mathbf{x}_i, \overline{\mathbf{u}})$ is obtained from the displacement perturbator $\mathcal{L}_i^{uu}(\mathbf{x}-\mathbf{x}_i, \overline{\mathbf{u}})$ by numerical differentiation. For the homogeneous effective strain $\overline{\boldsymbol{\varepsilon}}$ ≡const. (18.37), the linear integral operators $\mathcal{L}_i^{uu}(\mathbf{x}-\mathbf{x}_i, \overline{\mathbf{u}})$ and $\mathcal{L}_i^{\varepsilon\varepsilon}(\mathbf{x}-\mathbf{x}_i, \overline{\mathbf{u}})$ ($\mathbf{x} \in R^d$) are decomposed as

$$\mathcal{L}_i^{uu}(\mathbf{x}-\mathbf{x}_i, \overline{\mathbf{u}}) = \mathbf{L}_i^{u\epsilon}(\mathbf{x}-\mathbf{x}_i)\overline{\boldsymbol{\varepsilon}}, \quad \mathcal{L}_i^{\varepsilon\varepsilon}(\mathbf{x}-\mathbf{x}_i, \overline{\boldsymbol{\varepsilon}}) = \mathbf{L}_i^{\varepsilon\varepsilon}(\mathbf{x}-\mathbf{x}_i)\overline{\boldsymbol{\varepsilon}}. \tag{18.42}$$

18.3 Micromechanical Modeling

Generalization of the classical micromechanical models of locally elastic CMs to their peridynamic counterpart is performed by the popular tools and concepts used in conventional elasticity of CMs and adapted to peridynamics. Effective moduli are expressed through the introduced local stress polarization tensor averaged over the extended inclusion phase or the external interaction interface of inclusions rather than in an entire space. In the last case, any spatial derivatives of displacement fields are not required. One proposes a generalization of a dilute approximation method to their peridynamic counterpart in the sense that the volume fraction of the particles is small and their mutual interaction can be neglected. As in a classical approach, the essential assumption in the generalized Mori-Tanaka method (MTM) states that each extended inclusion behaves as an isolated one in the infinite matrix and subject to some effective strain field coinciding with the average strains in the truncated matrix.

18.3.1 Dilute Approximation Method (DAM)

Let the dispersion contains identical particles, of a fixed shape vi, randomly aligned throughout the matrix. We assume that the dispersion is dilute, in the sense that the volume fraction, $c^{(1)}$, of the particles is small and their mutual interaction can be neglected, i.e., each particle can be thus imagined as single, immersed into an unbounded matrix

$$c^{(1)} \ll 1, \quad \overline{\boldsymbol{\varepsilon}} = \langle \boldsymbol{\varepsilon} \rangle. \tag{18.43}$$

Then substitution of Eq. (18.43) into Eqs. (17.92) and (18.42) leads to

$$\mathbf{L}^{*}{=}\mathbf{L}^{(0)} + c^{l(1)}\mathbf{R}_i^l, \tag{18.44}$$

where ($\mathbf{x}, \mathbf{z}, \mathbf{y} \in v_i^l$)

$$\begin{aligned} \mathbf{R}_i^l &{=}\langle \mathbf{R}_i(\mathbf{x})\rangle_i^l, \quad \mathbf{R}_i(\mathbf{x})\langle\overline{\boldsymbol{\varepsilon}}\rangle = \mathcal{L}^{\sigma}(\mathbf{C}_1\boldsymbol{\eta})(\mathbf{x}), \\ \boldsymbol{\eta}(\mathbf{z}, \mathbf{y}) &{=}\Big[(\mathbf{z}-\mathbf{y})^{\top} + \mathbf{L}^{u\epsilon}(\mathbf{z}-\mathbf{x}_i) - \mathbf{L}^{u\epsilon}(\mathbf{y}-\mathbf{x}_i)\Big]\overline{\boldsymbol{\varepsilon}}. \end{aligned} \tag{18.45}$$

Estimation of the volume average of local polarization tensor $\langle \boldsymbol{\tau}\rangle_i^l = \mathbf{R}_i^l\langle\boldsymbol{\varepsilon}\rangle_i$ (18.44) can be equivalently replaced by the average of the local interface fields (17.108)

$$\mathbf{R}_i^l\langle\boldsymbol{\varepsilon}\rangle = \mathbf{R}_i^{l\omega}\langle\boldsymbol{\varepsilon}\rangle := \langle\big[\mathcal{L}^{\sigma}(\mathbf{C}\boldsymbol{\eta})(\mathbf{s})\mathbf{n}(\mathbf{s})\big]\otimes\mathbf{s}\rangle_i^{l\omega} - \mathbf{L}^{(0)}\Big[\mathbf{I} + \langle\mathbf{L}^{u\epsilon}(\mathbf{s}-\mathbf{x}_i)\otimes\mathbf{n}(\mathbf{s})\rangle_i^{l\omega}\Big]\langle\overline{\boldsymbol{\varepsilon}}\rangle, \tag{18.46}$$

where $\boldsymbol{\eta}$ is defined in Eq. (18.45).

At the limit of vanishing length scale $ł_\delta/a \to 0$, the tensor ${}^{L}\mathbf{R}(\mathbf{x})$ is found from the equation ($\mathbf{x} \in v_i$)

$${}^{L}\boldsymbol{\tau}(\mathbf{x}) ={}^{L}\mathbf{R}_i(\mathbf{x})\langle\boldsymbol{\varepsilon}\rangle. \tag{18.47}$$

In particular, for the homogeneous ellipsoidal inclusion (3.70), ${}^L\mathbf{R}_i(\mathbf{x})$ is expressed through the Eshelby tensor $\mathbf{S}_i$

$$^L\mathbf{R}_i(\mathbf{x}) = ^L\mathbf{L}_1(\mathbf{I} + \mathbf{S}_i^L\mathbf{M}^{(0)L}\mathbf{L}_1)^{-1}.$$

It should be mentioned that the effective moduli can be estimated by the use of the evaluation of average stresses in the components ($\mathbf{M}^{(0)} := (\mathbf{L}^{(0)})^{-1}$)

$$c^{(1)} \ll 1, \quad \overline{\boldsymbol{\varepsilon}} = \mathbf{M}^{(0)}\langle\boldsymbol{\sigma}\rangle, \tag{18.48}$$

that leads to the following representation:

$$\mathbf{M}^* = \mathbf{M}^{(0)}\Big[\mathbf{I} - c^{l(1)}\mathbf{R}_i^l\Big]. \tag{18.49}$$

In so doing, the utilization of the dilute approximation method (DAM) reduces to an inequality $\mathbf{L}^* \neq (\mathbf{M}^*)^{-1}$, although two versions (18.44) and (18.49) coincide in the asymptotic $c^{(1)} \to 0$. The estimations (18.44) and (18.49) are consistent only up to the first order $O(c^{(1)})$ of the volume fraction of the inhomogeneities, providing however in this case an exact expression (since the particle interactions are totally neglected). Because of this, Eq. (18.44) will be exploited if $\mathbf{C}(\mathbf{x}, \hat{\mathbf{x}}) - \mathbf{C}^{(0)}(\mathbf{x}, \hat{\mathbf{x}})$ is positive definite; otherwise, a dual scheme (18.49) based on the estimation of average stresses and effective compliance will be used. The dilute expressions are strictly valid only for noninteracting inclusions. Equations (18.44) and (18.49) were obtained in the framework of dilute approximation when the inclusions do not "feel" any influence of surrounding neighbors, i.e., they are loaded by the unperturbed applied remote field. For locally elastic CM with ellipsoidal homogeneous inclusions, the representations (18.44) and (18.49) are reduced to the corresponding equations obtained by the Eshelby method. Because of this, the approaches (18.44) and (18.49) are also called the generalized Eshelby method.

18.3.2 Mori-Tanaka Method (MTM)

Modeling of the overall behavior of composites with nondilute inclusion volume fraction must (explicitly or implicitly) account for the interaction between inclusions, i.e., for the effects of the surrounding inclusions on the stress and strain fields in the vicinity of a given inclusion. Now consider a model of CM where for a typical inhomogeneity v_i, the effects (or the existence) of other inhomogeneities are communicated to it through the displacement, strain, and stress fields in its surrounding matrix material. When an infinite number of inhomogeneities randomly distributed in the matrix are analyzed, the average field in the matrix can be considered as good approximations of the actual fields in the matrix surrounding each inhomogeneity. As pointed out by Benveniste [89], the essential assumption in the Mori-Tanaka method (MTM) in locally elastic micromechanics states that each inclusion v_i behaves as an isolated one in the infinite matrix and subject to some effective stress field $\overline{\boldsymbol{\varepsilon}}(\mathbf{x})$ coinciding with the average stress in the matrix:

$$\overline{\boldsymbol{\varepsilon}}(\mathbf{x}) = \langle \boldsymbol{\varepsilon} \rangle^{(0)}. \tag{18.50}$$

The assumption (18.50) makes it possible to determine uniquely the effective elastic properties of multicolored composite materials. Thus, the hypothesis (18.50) is more restricted than the hypothesis **H1** which gives an opportunity to use the solution for each inclusion v_i and to find the average stress in the matrix by the use of a representation of the average strain in the whole composite over the average strain in the separate phases:

$$\langle \boldsymbol{\varepsilon} \rangle = c^{l(0)} \langle \boldsymbol{\varepsilon} \rangle^{l(0)} + \langle \boldsymbol{\varepsilon} V^l \rangle = c^{l(0)} \langle \boldsymbol{\varepsilon} \rangle^{l(0)} + \langle \mathbf{A}_i(\mathbf{x}) V^l(\mathbf{x}) \rangle \langle \boldsymbol{\varepsilon} \rangle^{l(0)}, \tag{18.51}$$

hereafter ($\mathbf{x} \in v_i^l$)

$$\begin{aligned} \mathbf{A}_i(\mathbf{x}) &:= \mathbf{L}_i^{\varepsilon\varepsilon}(\mathbf{x} - \mathbf{x}_i) + \mathbf{I}, \\ \mathbf{A}_i^l &= |v_i^l|^{-1} \langle \mathbf{A}_i(\mathbf{x}) V_i^l(\mathbf{x}) \rangle = \mathbf{I} + \langle \mathbf{L}_i^{\varepsilon\varepsilon}(\mathbf{x} - \mathbf{x}_i) \rangle_i^l. \end{aligned} \tag{18.52}$$

The right-hand side of Eq. (18.51) can be also found through the displacements at the external interface Γ_i^+. Indeed, by the use of the Gauss-Ostrogradsky theorem, we get the average strain at the extended inclusion v_i^l

$$\langle \boldsymbol{\varepsilon} \rangle_i^l =: \mathbf{A}_i^l \langle \boldsymbol{\varepsilon} \rangle^{l(0)} = \langle \mathbf{n} \overset{S}{\otimes} \mathbf{u} \rangle_i^{l\omega} := |v_i^l|^{-1} \int_{\Gamma_i^-} \mathbf{n}(\mathbf{s}) \overset{S}{\otimes} \mathbf{u}(\mathbf{s}) ds, \tag{18.53}$$

where $\mathbf{u}(\mathbf{s})$ is a displacement on the interface boundary $\mathbf{s} \in \Gamma_i^-$ with the outward normal unit vector $\mathbf{n}(\mathbf{s})$ on Γ_i^-. Thus, the volume average strain concentration factor $\mathbf{A}_i^l$ can be also presented through the interface weighted average of the perturbator $\mathbf{L}_i^{u\epsilon}(\mathbf{s} - \mathbf{x}_i)$

$$\mathbf{A}_i^l = \mathbf{A}_i^{l\omega} := \mathbf{I} + \langle \mathbf{n}(\mathbf{s}) \overset{S}{\otimes} \mathbf{L}_i^{u\epsilon}(\mathbf{s} - \mathbf{x}_i) \rangle_i^{l\omega}. \tag{18.54}$$

Among the formally equivalent concentration factors $\mathbf{A}_i^l$ and $\mathbf{A}_i^{l\omega}$, the first one $\mathbf{A}_i^l$ has additional numerical error produced by numerical differentiation of $\mathbf{L}^{u\epsilon}(\mathbf{x} - \mathbf{x}_i)$ (which is not differentiable) at the estimation of $\mathbf{L}_i^{\epsilon\epsilon}(\mathbf{x} - \mathbf{x}_i)$ (compare Eqs. (18.52_3) and (18.54_2)). The use of $\mathbf{A}_i^{l\omega}$ (18.54_2) instead of $\mathbf{A}_i^l$ (18.52_3) has a fundamental advantage because an evaluation of the strain concentration factor $\mathbf{A}_i(\mathbf{x})$ ($\mathbf{x} \in v_i^l$) is not required.

Equations (18.50) and (18.53) lead to the representations of statistical average of both the strains in the matrix and the stress polarization tensor $\boldsymbol{\tau}$ in the inclusions $\mathbf{x} \in v_i^l$ (see for details [226]). Then the averaging of the constitutive equation (4.18) yields

$$\langle \boldsymbol{\sigma} \rangle = \Big[\mathbf{L}^{(0)} + \mathbf{R}_i^l \big(c^{l(0)} + c^{l(1)} \mathbf{A}_i^l\big)^{-1}\Big] \langle \boldsymbol{\varepsilon} \rangle.$$

$$\langle \boldsymbol{\sigma} \rangle = \Big[\mathbf{L}^{(0)} + c^{l(1)} \mathbf{R}_i^{l\omega} \big(c^{l(0)} + c^{l(1)} \mathbf{A}_i^{l\omega}\big)^{-1}\Big] \langle \boldsymbol{\varepsilon} \rangle. \tag{18.55}$$

Comparison of the definition of effective moduli $\mathbf{L}^*$ with Eq. (18.55) leads to the representation of effective properties and effective fields:

$$\mathbf{L}^{*\mathrm{MTM}} = \mathbf{L}^{(0)} + c^{l(1)} \mathbf{R}_i^l \mathbf{D}_i, \ \ \langle \overline{\boldsymbol{\varepsilon}} \rangle = \mathbf{D}_i^{\mathrm{MTM}} \langle \boldsymbol{\varepsilon} \rangle \tag{18.56}$$

$$\mathbf{D}_i^{\mathrm{MTM}} = \big[c^{l(0)} + c^{l(1)} \mathbf{A}_i^l\big]^{-1}, \tag{18.57}$$

$$\mathbf{L}^{*\omega\,\mathrm{MTM}} = \mathbf{L}^{(0)} + c^{l(1)} \mathbf{R}_i^{l\omega} \mathbf{D}^{\omega\,\mathrm{MTM}}, \ \ \langle \overline{\boldsymbol{\varepsilon}} \rangle = \mathbf{D}^{\omega\mathrm{MTM}} \langle \boldsymbol{\varepsilon} \rangle, \tag{18.58}$$

$$\mathbf{D}^{\omega\mathrm{MTM}} = \big[c^{l(0)} + c^{l(1)} \mathbf{A}_i^{l\omega}\big]^{-1}. \tag{18.59}$$

As can be seen from comparing Eqs. (18.43$_2$) and (18.50), the only difference between the dilute approximation and the Mori-Tanaka methods is the choice of the effective field $\overline{\boldsymbol{\varepsilon}}$ representing how the effects of other inclusions are taken into account. So, the dilute approximation implies no interaction between the inhomogeneities and, therefore, the strain in the matrix coincides with the remote strain $\langle \boldsymbol{\varepsilon} \rangle$. The Mori-Tanaka method also assumes the absence of all inclusions; however, the influence of these removed inclusions is described by the average strain $\langle \boldsymbol{\varepsilon} \rangle^{l(0)}$ in the truncated matrix (instead of $\langle \boldsymbol{\varepsilon} \rangle$). Because of this, both the MTM and DAM are invariant with respect to the possible overlapping of the extended inclusions v_i^l and v_j^l mentioned after Eq. (18.3). As it is expected, the Mori-Tanaka estimations (18.56)–(18.59) are reduced to the dilute approximation (18.44) in the limiting case $c^{l(1)} \to 0$. It demonstrates a role of the DAM as a test-drive for another micromechanical method which should be reduced to the DAM in the limiting case $c^{l(1)} \to 0$.

Equations (18.56) and (18.57) coincide within notations with the corresponding MTM for the locally elastic CMs

$${}^L\mathbf{L}^{*\mathrm{MTM}} = \mathbf{L}^{(0)} + {}^L\mathbf{R}_i \, {}^L\mathbf{D}_i^{\mathrm{MTM}}, \quad {}^L\mathbf{D}_i^{\mathrm{MTM}} = [c^{(0)} + c^{(1)} \, {}^L\mathbf{A}_i^l]^{-1}, \tag{18.60}$$

with conceptually different tensors ${}^L\mathbf{R}_i$ and ${}^L\mathbf{A}_i$ considered as locally elastic counterparts of the tensors $\mathbf{R}_i^l \to {}^L\mathbf{R}_i$ and $\mathbf{A}_i^l \to {}^L\mathbf{A}_i$ as $l_\delta \to 0$.

For locally elastic CMs with the ellipsoidal homogeneous inclusions, the MTM defined by Eq. (18.60) and called the "direct approach" is identical (see [89]) to the "equivalent inclusion-average stress" formalism (see [1093, 1608]). No assumptions about both the shape of heterogeneities and their microstructure were used at the derivative of Eqs. (18.44), (18.49), and (18.58). The direct method based on a single hypothesis (18.50) also has a specific merit in shedding new light on the weakness of the direct method that opens up outstanding possibilities to improve the method as it is done in the effective field method (EFM) generalizing its locally elastic counterpart (see Chaps. 8–10) to the peridynamic analog in the next subsection.

18.3.3 Effective Field Method (Abandonment of the Hypotheses H1b and H3)

Statistical averaging of Eqs. (18.35) and (18.36) at $n = 1$ leads to the following representation for the mean of the effective field in the heterogeneity v_i:

$$\langle\overline{\mathbf{u}}\rangle_i = \langle\mathbf{u}\rangle(\mathbf{x}) + \int \Big\{\langle\mathcal{L}_q^{uu}(\mathbf{x}-\mathbf{x}_q,\overline{\mathbf{u}})|;v_i,\mathbf{x}_i\rangle_q\varphi(v_q,\mathbf{x}_q|;v_i,\mathbf{x}_i) -n^{(j)}\langle\mathcal{L}_q^{uu}(\mathbf{x}-\mathbf{x}_q,\overline{\mathbf{u}})\rangle(\mathbf{x}_q)\Big\}d\mathbf{x}_q \tag{18.61}$$

$$\langle\overline{\boldsymbol{\varepsilon}}\rangle_i = \langle\boldsymbol{\varepsilon}\rangle(\mathbf{x}) + \int \Big\{\langle\mathcal{L}_q^{\varepsilon\varepsilon}(\mathbf{x}-\mathbf{x}_q,\overline{\boldsymbol{\varepsilon}})|;v_i,\mathbf{x}_i\rangle_q\varphi(v_q,\mathbf{x}_q|;v_i,\mathbf{x}_i) -n^{(j)}\langle\mathcal{L}_q^{\varepsilon}(\mathbf{x}-\mathbf{x}_q,\overline{\boldsymbol{\varepsilon}})\rangle(\mathbf{x}_q)\Big\}d\mathbf{x}_q. \tag{18.62}$$

Hereafter, $\langle\overline{\boldsymbol{\varepsilon}}|;v_i,\mathbf{x}_i\rangle_q \equiv \langle\overline{\boldsymbol{\varepsilon}}|;v_i,\mathbf{x}_i\rangle_q(\mathbf{y})$ is a conditional statistical average of $\overline{\boldsymbol{\varepsilon}}(\mathbf{y})$ varying along the fixed heterogeneity $\mathbf{y}\in v_q$ at the fixed v_i, while $\langle\overline{\boldsymbol{\varepsilon}}\rangle_q \equiv \langle\overline{\boldsymbol{\varepsilon}}\rangle_q(\mathbf{y})$ is a statistical average $\overline{\boldsymbol{\varepsilon}}(\mathbf{y})$ inside the heterogeneity $\mathbf{y}\in v_q$.

In the framework of the quasicrystalline approximation **H2b** (8.10), Eqs. (18.61) and (18.62) are simplified:

$$\langle\overline{\mathbf{u}}\rangle_i(\mathbf{x}) = \langle\mathbf{u}\rangle(\mathbf{x}) + \int \mathcal{L}_q^{uu}(\mathbf{x}-\mathbf{x}_q,\langle\overline{\mathbf{u}}\rangle_q)[\varphi(v_q,\mathbf{x}_q|;v_i,\mathbf{x}_i)-n^{(j)}]d\mathbf{x}_q, \tag{18.63}$$

$$\langle\overline{\boldsymbol{\varepsilon}}\rangle_i(\mathbf{x}) = \langle\boldsymbol{\varepsilon}\rangle(\mathbf{x}) + \int \mathcal{L}_q^{\varepsilon\varepsilon}(\mathbf{x}-\mathbf{x}_q,\langle\overline{\boldsymbol{\varepsilon}}\rangle_q)[\varphi(v_q,\mathbf{x}_q|;v_i,\mathbf{x}_i)-n^{(j)}]d\mathbf{x}_q. \tag{18.64}$$

Acceptance of the EFH **H1a** (18.37) allows one to get from (18.64) an explicit representation of the average effective strain

$$\langle\overline{\boldsymbol{\varepsilon}}\rangle_i = \mathbf{D}_i\langle\boldsymbol{\varepsilon}\rangle,\quad \mathbf{D}_i = \Big\{\mathbf{I} - \int \mathbf{L}_{qi}^{\varepsilon\varepsilon}(\mathbf{x}_i-\mathbf{x}_q)[\varphi(v_q,\mathbf{x}_q|;v_i,\mathbf{x}_i)-n^{(j)}]d\mathbf{x}_q\Big\}^{-1}, \tag{18.65}$$

where $\mathbf{L}_{qi}^{\varepsilon\varepsilon}(\mathbf{x}_i-\mathbf{x}_q) := \langle\mathbf{L}_{jq}^{\varepsilon\varepsilon}(\mathbf{x}-\mathbf{x}_q)\rangle_{(i)}$.

The limit of vanishing length scale $l_\delta \to 0$ reduces Eq. (18.65) to the corresponding representation obtained in the framework of a new background of micromechanics by Buryachenko [211]. Acceptance of additional hypotheses **H1b** and **H3** yields the classical formulae of the effective field method (EFM; see Chapter 8)

$$ {}^L\mathbf{L}^{*\mathrm{EFM}} = {}^L\mathbf{L}^{(0)} + n^{(1)}\,{}^L\mathbf{R}_i\,{}^L\mathbf{D}_i^{\mathrm{EFM}}, \tag{18.66}$$

$$ {}^L\mathbf{D}_i^{\mathrm{EFM}} = \big(\mathbf{I} - \mathbf{S}_i^0\,{}^L\mathbf{M}^{(0)}\,{}^L\mathbf{R}_i\big)^{-1}. \tag{18.67}$$

For the homogeneous ellipsoidal inclusions (4.92), the EFM's estimations (18.66) and (18.67) coincide with the Mori-Tanaka representations (18.60). However, it may be noted that there are certain drawbacks of the extension of the Mori-Tanaka method to multiphase composites: Mori-Tanaka moduli may violate the Hashin-Shtrikman bounds [621] and are nonsymmetric for the general biphase composites (see for references, e.g., Chap. 8). In contrast with the EFM, these deficiencies are persisted for generalization of the MTM to the peridynamic CMs. It should be mentioned that Eqs. (18.66) and (18.67) were obtained from the classical equation (17.197), while Eq. (18.65) has been gotten from the new operator GIE (18.65) which is formally identical to the corresponding GIE for locally elastic CMs (see [212]). It should be mentioned that we did not suppose in Eqs. (18.66) and (18.67) that inclusions v_i are ellipsoidal homogeneous ones, and, therefore, the tensor ${}^L\mathbf{R}_i$ is assumed to be found numerically without exploiting the Eshelby [449] theorem. However, the Eshelby [449] theorem is used for the homogeneous ellipsoidal domain v_i^0 that manifested by the tensor $\mathbf{S}_i^0 = \mathbf{S}(v_i^0)$ in Eq. (18.67).

The deficiency of the solution (18.65) is the evaluation of the operator $\mathcal{L}_q^{\varepsilon\varepsilon}(\mathbf{x} - \mathbf{x}_q, \langle\overline{\boldsymbol{\varepsilon}}\rangle_q)$ through the operator $\mathcal{L}_q^{uu}(\mathbf{x}-\mathbf{x}_q, \langle\overline{\mathbf{u}}\rangle_q)$ by numerical differentiation of the displacement field (18.41). However, the displacement field $\mathbf{u}(\mathbf{x})$ has a discontinuity at the interface $\mathbf{x} \in \Gamma_i$ (see [1405]). For elimination of numerical error due to the numerical differentiation, one considers an explicit solution of Eq. (18.63) in the framework of the hypothesis **H1a** when ($\mathbf{x} \in v_q$)

$$ \overline{\boldsymbol{\varepsilon}}(\mathbf{x}) = \overline{\boldsymbol{\varepsilon}}(\mathbf{x}_q), \tag{18.68}$$

and, therefore,

$$ \langle\overline{\mathbf{u}}\rangle_i(\mathbf{x}) = \langle\mathbf{u}\rangle(\mathbf{x}) + \int \mathbf{L}_{qi}^{u\epsilon}(\mathbf{x} - \mathbf{x}_q, \langle\overline{\boldsymbol{\varepsilon}}\rangle_q)[\varphi(v_q, \mathbf{x}_q|; v_i, \mathbf{x}_i) - n^{(q)}]d\mathbf{x}_q. \tag{18.69}$$

Indeed, similar to (18.53), the average effective strain at the extended inclusion v_i^l can be represented as

$$ \langle\overline{\boldsymbol{\varepsilon}}\rangle_i^l = \langle\mathbf{n} \overset{S}{\otimes} \overline{\mathbf{u}}\rangle_i^{l\omega} := |v_i^l|^{-1}\int_{\Gamma_i^-} \mathbf{n}(\mathbf{s}) \overset{S}{\otimes} \overline{\mathbf{u}}(\mathbf{s})ds. \tag{18.70}$$

Then a tensor product of Eq. (18.69) over the outward normal $\mathbf{n}(\mathbf{s})\perp\Gamma_i^-$ and its subsequent surface averaging over the interface boundary Γ_i^- yield ($\mathbf{s}\in\Gamma_i^-$)

$$\langle\overline{\boldsymbol{\varepsilon}}\rangle_i = \mathbf{D}_i^{\omega\mathrm{EFM}}\langle\boldsymbol{\varepsilon}\rangle,$$
$$\mathbf{D}_i^{\omega\mathrm{EFM}} = \Big\{\mathbf{I} - \int \langle\mathbf{n}(\mathbf{s})\overset{S}{\otimes}\mathbf{L}_q^{u\epsilon}(\mathbf{s}-\mathbf{x}_q)\rangle_i^{l\omega}[\varphi(v_q,\mathbf{x}_q|;v_i,\mathbf{x}_i) - n^{(q)}]d\mathbf{x}_q\Big\}^{-1}, \tag{18.71}$$

where one uses the equality $\langle\mathbf{n}(\mathbf{s})\overset{S}{\otimes}\langle\mathbf{u}\rangle(\mathbf{s})\rangle_i^{l\omega} = \langle\boldsymbol{\varepsilon}\rangle$ ($\mathbf{s}\in\Gamma_i^-$).

Thus, both Eqs. (18.65_1) and (18.71_1) are presented in an identical form leading to the effective modulus representation

$$\langle\overline{\boldsymbol{\varepsilon}}\rangle_i = \mathbf{D}_i\langle\boldsymbol{\varepsilon}\rangle, \quad \mathbf{L}^* = \mathbf{L}^{(0)} + \mathbf{R}_i^l\mathbf{D}_i. \tag{18.72}$$

The constant tensor $\mathbf{D}_i$ [$\mathbf{D}_i = \mathbf{D}_i^{\mathrm{MTM}}, \mathbf{D}_i^{\omega\,\mathrm{MTM}}, \mathbf{D}_i^{\mathrm{EFM}}, \mathbf{D}_i^{\omega\mathrm{EFM}}$ are different in Eqs. (18.57), (18.59), (18.65), and (18.71)] has a physical meaning of effective strain concentration factor at the representative inclusion v_i in the CM; for the dilute concentration of inclusions ($c^{(1)}\to 0$), we have $\mathbf{D}_i\to\mathbf{I}$, and Eqs. (18.58) and ($18.72_2$) are reduced to Eq. (18.44). It should be mentioned that the solutions (18.65) and (18.71) were obtained in the framework of the hypotheses **H1a** and **H2**, while the hypotheses **H1b** and **H3** were not used.

18.3.4 Effective Field Method (Abandonment of the Hypotheses H1a, H1b, and H3)

An abandonment of the effective field hypothesis **H1a** assumed now gives no way of decomposing the operators $\mathcal{L}_q^{uu}(\mathbf{x}-\mathbf{x}_q,\overline{\mathbf{u}})$ and $\mathcal{L}_q^{\varepsilon\varepsilon}(\mathbf{x}-\mathbf{x}_q,\overline{\boldsymbol{\varepsilon}})$ (as in Eqs. (18.65) and (18.71)). In such a case, the first that comes to mind is a solution of Eq. (18.63) by the iteration method

$$\langle\overline{\mathbf{u}}^{[n+1]}\rangle_i(\mathbf{x}) = \langle\mathbf{u}\rangle(\mathbf{x}) + \int \mathcal{L}_q^{uu}(\mathbf{x}-\mathbf{x}_q,\langle\overline{\mathbf{u}}^{[n]}\rangle_q)[\varphi(v_q,\mathbf{x}_q|;v_i,\mathbf{x}_i) - n^{(j)}]d\mathbf{x}_q \tag{18.73}$$

with the initial approximation $\langle\overline{\mathbf{u}}^{[0]}\rangle_i(\mathbf{x}) = \langle\overline{\boldsymbol{\varepsilon}}^{(18.71)}\rangle_i\mathbf{x}$ ($\mathbf{x}\in v_i^l$) defined by the solution (18.71). The scheme (18.73) forms the Neumann series for the solution ($\mathbf{x}\in v_i^l$)

$$\langle\overline{\mathbf{u}}\rangle_i(\mathbf{x}) := \langle\lim_{n\to\infty}\overline{\mathbf{u}}^{[n+1]}\rangle_i(\mathbf{x}) = \mathbf{D}_i^u(\mathbf{x})\langle\boldsymbol{\varepsilon}\rangle, \tag{18.74}$$

which yields the final representations for the effective properties

$$\mathbf{L}^* = \mathbf{L}^{(0)} + \langle \mathbf{R}_i^D V^l \rangle, \quad \mathbf{R}_i^D(\mathbf{x})\langle \boldsymbol{\varepsilon} \rangle = \mathcal{L}^\sigma(\mathbf{C}_1 \boldsymbol{\eta}^D), \tag{18.75}$$

$$\boldsymbol{\eta}^D(\mathbf{x},\mathbf{y}) = \Big[\mathbf{D}_i^u(\mathbf{x}) + \mathbf{L}_i^{uu}(\mathbf{x}-\mathbf{x}_i, \mathbf{D}_i^u(\mathbf{x}-\mathbf{x}_i)) \\ - \mathbf{D}_i^u(\mathbf{y}) - \mathbf{L}_i^{uu}(\mathbf{y}-\mathbf{x}_i, \mathbf{D}_i^u(\mathbf{y}-\mathbf{x}_i))\Big]\langle \boldsymbol{\varepsilon} \rangle, \tag{18.76}$$

where one introduced $\mathcal{L}_i^{uu}(\mathbf{x}-\mathbf{x}_i, \mathbf{D}_i^u(\mathbf{x}-\mathbf{x}_i)) := \mathbf{L}_i^{uu}(\mathbf{x}-\mathbf{x}_i, \mathbf{D}_i^u(\mathbf{x}-\mathbf{x}_i))\langle \boldsymbol{\varepsilon} \rangle$. It should be mentioned that the tensor $\mathbf{D}^u(\mathbf{x})$ (18.74) is an inhomogeneous function of coordinates $\mathbf{x} \in v_i^l$, while $\mathbf{D}_i \equiv \text{const.}$ (18.65) and (18.71). Furthermore, the solutions (18.75) and (18.76) were obtained in the framework of the hypothesis **H2**, while the hypotheses **H1a**, **H1b**, and **H3** were not used (compare with the assumptions of obtaining $\mathbf{D}_i^{\omega\text{EFM}} \equiv \text{const.}$).

The meaning of acceptance or abandonment of one or another hypothesis (either **H1a**, **H1b**, **H2**, or **H3**) is explained in locally elastic micromechanics by Buryachenko [211, 212]. The destination of the hypothesis **H3** is directed toward providing conditions for applying the hypothesis **H1** (see for details [232, 234]). If the hypotheses **H1**, **H2**, and **H3** hold for the statistically homogeneous material and field problem in local elasticity, then the effective moduli $\mathbf{L}^*$ do not depend on the size of the correlation hole v_i^0 and the binary correlation function (that is not true for peridynamic problem). Simultaneous acceptance of the hypotheses **H1**, **H2**, and **H3** yields a significant simplification of the effective modulus representation in locally elastic micromechanics (18.66) and (18.67) due to an opportunity of exploiting the Eshelby (1957) theorem (even for the CMs with nonellipsoidal inclusions v_i; see Eq. (18.67)). However, implication of a simplified technique based on the Eshelby [449] theorem is questionable in peridynamic micromechanics (see for details Sect. 17.1), and, because of this, the proposed generality (abandonment of the hypotheses **H1a**, **H1b**, and **H3**) is a forced challenging achievement rather than just pursuing an abstruse theoretical exercise.

18.3.5 The Scheme of Estimation of the Effective Moduli

Independently on the averaging method, the effective modulus estimation is performed according to the following two-step scheme. At first, we estimate the perturbator factors $\mathbf{L}_i^{u\epsilon}(\mathbf{x}-\mathbf{x}_i)$ (18.42$_1$) or $\mathbf{L}_i^{\epsilon\epsilon}(\mathbf{x}-\mathbf{x}_i)$ (18.42$_2$) by either the truncation method or the Green function method (see Sect. 7.1). In the second step for both the dilute approximation and the MTM, we estimate the weighted averages of the perturbator factors over either the volume of the extended reference inclusion $v_i^l = v_i \cup v_{\Gamma i}^-$ or the external interaction interface Γ_i^- (see Fig. 18.5 and Sect. 16.4). The scheme of the generalized EFM is more sophisticated and based on the additional estimation of the integrals, either (18.65) or (18.71), with the integrands produced by the perturbations of "moving" inclusions v_q. An action of the moving perturbator ($\mathbf{L}_i^{u\epsilon}(\mathbf{x}-\mathbf{x}_i)$ (18.42$_1$) or $\mathbf{L}_i^{\epsilon\epsilon}(\mathbf{x}-\mathbf{x}_i)$ (18.42$_2$)) is averaged

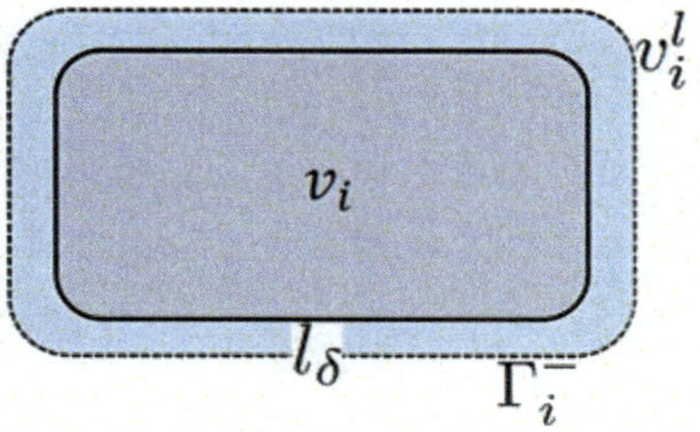

Fig. 18.5 Extended inclusion $v_i^l = v_i \cup v_{\Gamma i}^-$ with the external interaction interface Γ_i^-

by the use of the conditional probability $\varphi(v_q, \mathbf{x}_q|; v_i, \mathbf{x}_i)$. In so doing, the meshes used in both the one inclusion perturbation and its averaging specified by one or another micromechanical model can be different.

In more detail, we consider the EFM because it has both the same averages $\mathbf{R}_i^l$ (or $\mathbf{R}_i^{l\omega}$) (as the DA and MTM) and the additional averages (18.65) or (18.71). In the generalized classical approach (compare (17.195) and (17.196)), the effective fields (18.35) or (18.36) depend only on the volume average of the local polarization tensor $\langle\overline{\boldsymbol{\pi}}\rangle_q$ distributed in the ellipsoidal excluded volume v_i^0. Because of this, all estimations (18.66) and (18.67) by the classical EFM are invariant with respect to detailed field distribution inside v_i^l. In the new approach, the effective field $\langle\overline{\mathbf{u}}\rangle_i(\mathbf{x})$ in the area $\mathbf{x} \in v_i^l$ (imaginary inclusion) occupied by the matrix material is generated by a moving heterogeneity v_q (see Eqs. (18.65) and (18.71)), and the inhomogeneity of the detailed distribution $\langle\overline{\boldsymbol{\pi}}\rangle_q(\mathbf{y})$ $(y \in v_q)$ affects the effective field $\langle\overline{\mathbf{u}}\rangle_i(\mathbf{x})$ $(\mathbf{x} \in v_i^l)$. Furthermore, the domain of this long-range action is limited not only by $\mathbf{x}_q \in v_i^0$ but also by a domain v_i^ϕ for the location of inclusion centers $\mathbf{x}_q$ where $\varphi(v_q, \mathbf{x}_q|v_i, \mathbf{x}_i) - n^{(q)}$ is not negligible (in the sense of influence on $\mathbf{L}^*$). An effect zone of field perturbations produced by the heterogeneity v_q^l is limited by the Minkowski addition of domains $\mathbf{x} \in v_q^\oplus \equiv v_q^\phi \oplus v_q^l$ (see Fig. 18.5) rather than by the domain v_q^l as in the classical EFM. It should be mentioned that for the dilute approximation $\varphi(v_q, \mathbf{x}_q|v_i, \mathbf{x}_i) = n^{(q)}$ at $\mathbf{x}_q \in R^d \backslash v_i^0$ and the domain v_i^ϕ is reduced to v_i^0. In particular, for the spherical inclusions $v_i = \{\mathbf{x}|\ \mathbf{x}| \le a\}$, there is known excluded volume $v_i^0 = \{\mathbf{x}||\mathbf{x}| \le 2a\}$ and $v_i^\oplus = \{\mathbf{x}||\mathbf{x}| \le 3a\}$. However, for the ellipsoidal inclusion, v_i^0 is not ellipsoid in general. Nevertheless, we will use a popular hypothesis of ellipsoidal shape of the excluded volume $v_i^0 = \{\mathbf{x}|\ \mathbf{x}/2 \in v_i\}$ of the ellipsoidal inclusion v_i. Moreover, we will use a similar assumption for the nonellipsoidal inclusion v_i: $v_i^0 = \{\mathbf{x}|\ \mathbf{x}/2 \in v_i\}$ and $v_i^\phi = \{\mathbf{x}|\ \mathbf{x}/3 \in v_i\}$ (see Fig. 18.6). In so doing, an ellipsoidal shape of v_i^0 (as in the classical EFM (18.66) and (18.67)) is not required that was explained in the locally elastic case in [234].

We note that in classical methods, there is no systematic approach to choose the size and shape of the excluded volume v_i^0. A domain, where $\varphi(v_q, \mathbf{x}_q|; v_i, \mathbf{x}_i) - n^{(q)}$ is not negligible, is discretized by the square mesh Ω^{sq} (18.12), which impacts on the estimations of effective properties. The square mesh Ω^{sq} will be exploited for field assignment inside and outside the inclusion in Eqs. (18.65) and (18.71)

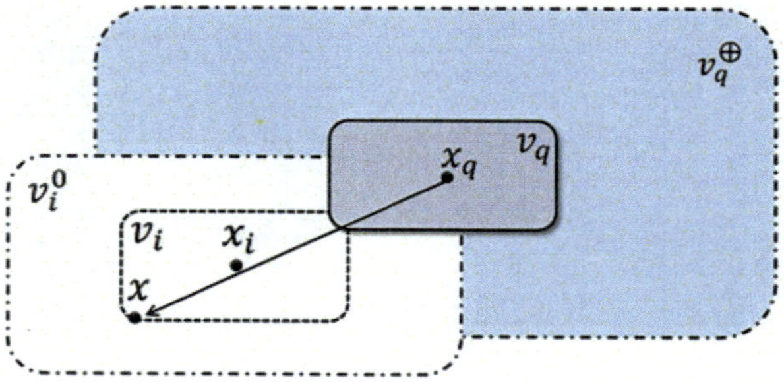

Fig. 18.6 Fixed imaginary inclusion v_i inside excluded volume v_i^0 and the moving inclusion v_q inside domain $v_q^{\oplus}$

estimated in the postprocessing of the solution in Sect. 17.1. In order to compute the integral in Eq. (18.65) (and (18.71)), the discretization (18.12) is applied considering the simple Simpson numerical integration rule on piecewise constant elements; this choice is dictated by the simplicity of implementation also for nonregular inclusion shapes even if it does not guarantee a very fast convergence of the results with the discretization step h. It should be mentioned that the main reason of the choice of the square mesh (18.12) is its double use for two different problems. At first, this mesh is exploited for estimation of the effective field perturbator factors $\mathbf{L}_i^{u\epsilon}(\mathbf{x}-\mathbf{x}_i)$ (18.42_1) or $\mathbf{L}_i^{\epsilon\epsilon}(\mathbf{x}-\mathbf{x}_i)$ (18.42_2). At the second, the same mesh (18.12) is used as the location of the moving inclusion centers $\mathbf{x}_q$ in Eq. (18.65) (and (18.71)). It gives an opportunity to use a solution $\mathbf{L}_i^{u\epsilon}(\mathbf{x}-\mathbf{x}_i)$ (18.42_1) (or $\mathbf{L}_i^{\epsilon\epsilon}(\mathbf{x}-\mathbf{x}_i)$ (18.42_2)) for one heterogeneity in a sample in the nodes of just one realization of the mesh (18.12) which is exploited as an "output" mesh for a solution obtained on some inhomogeneous mesh Ω^{inhom} used, e.g., in the FEA (see, e.g., [478, 1700]). The values of the variables $\mathbf{u}$ at any point $\mathbf{x}_i \in v_i^l$ can be calculated as a function of the values of these variables at the surrounding nodes $\mathbf{x}_i \in \omega_i \subset \Omega^{\text{inhom}}$ using the following expression (see, e.g., [1700]):

$$\mathbf{u}(\mathbf{x}_i) = \sum_q N^q(\mathbf{x}_i)\mathbf{u}^q, \tag{18.77}$$

where N^q are element shape functions, $\mathbf{u}^q$ are nodal values of $\mathbf{u}$, and the sum ranges over the number of nodes associated with an element $\mathbf{x}_i \in \omega_i \subset \Omega^{\text{inhom}}$.

However, in the case of immediate use of the inhomogeneous mesh inside and outside a heterogeneity (which is, perhaps, more effective for estimation of the perturbator factors), we will need to estimate the fields in the nodes of a new mesh generated for each location $\mathbf{x}_q$ of the moving inclusion v_q. Therefore, the square mesh (18.12) is optimal for the current problem.

It should be mentioned that the conditional probability density $\varphi(v_k, \mathbf{x}_k|; v_i, \mathbf{x}_i)$ is well investigated only for identical spherical ($3D$ and $2D$ cases) inclusions with a radius a when the pair distribution function $g(\mathbf{x}_i - \mathbf{x}_m) \equiv \varphi(v_i, \mathbf{x}_i|; v_m, \mathbf{x}_m)/n^{(k)}$ depending only on $|\mathbf{x}_m - \mathbf{x}_i|$ is called the radial distribution function (RDF; see for references and details, e.g., [237]). However, even for spherical inclusions and acceptance of the hypothesis **H2**, the effective moduli depend on the RDF (in contrast with the locally elastic EFM (18.66) and (18.67)).

18.3.6 Nonlocal Effective Operator

The iteration scheme (18.73) was proposed for the linear average strain $\langle \mathbf{u}\rangle(\mathbf{x})$. In a general case of $\langle \mathbf{u}\rangle(\mathbf{x})$, we recast Eq. (18.73) in the form ($\mathbf{y} \in v_q^l;\ \mathbf{x}, \mathbf{z} \in v_i^l$)

$$\langle \overline{\mathbf{u}}\rangle_i(\mathbf{x}) = \langle \mathbf{u}\rangle(\mathbf{x}) + \int \mathcal{L}_q^{uu}(\mathbf{x} - \mathbf{x}_q, \langle \overline{\mathbf{u}}\rangle_q(\mathbf{z}))[\varphi(v_q, \mathbf{x}_q|; v_i, \mathbf{x}_i) - n^{(j)}]d\mathbf{x}_q$$
$$+ \int \mathcal{L}_q^{uu}[\mathbf{x} - \mathbf{x}_q, \langle \overline{\mathbf{u}}\rangle_q(\mathbf{y}) - \langle \overline{\mathbf{u}}\rangle_q(\mathbf{z})][\varphi(v_q, \mathbf{x}_q|; v_i, \mathbf{x}_i) - n^{(j)}]d\mathbf{x}_q. \tag{18.78}$$

From the first line in Eq. (18.78), we form an equation in the domain v_i^l and solve it by the iteration method

$$\langle \overline{\mathbf{u}}\rangle_i(\mathbf{x}) = \langle \mathbf{u}\rangle(\mathbf{x}) + \int \mathcal{L}_q^{uu}(\mathbf{x} - \mathbf{x}_q, \langle \overline{\mathbf{u}}\rangle_q(\mathbf{z}))[\varphi(v_q, \mathbf{x}_q|; v_i, \mathbf{x}_i) - n^{(j)}]d\mathbf{x}_q \tag{18.79}$$

that allows one to get the iteration solution of Eq. (18.78)

$$\langle \overline{\mathbf{u}}^{[n+1]}\rangle_i(\mathbf{x}) = \langle \mathbf{u}\rangle(\mathbf{x}) + \int \mathcal{L}_q^{uu}(\mathbf{x} - \mathbf{x}_q, \langle \overline{\mathbf{u}}^{[n]}\rangle_q(\mathbf{z}))[\varphi(v_q, \mathbf{x}_q|; v_i, \mathbf{x}_i) - n^{(j)}]d\mathbf{x}_q$$
$$+ \int \mathcal{L}_q^{uu}[\mathbf{x} - \mathbf{x}_q, \langle \overline{\mathbf{u}}^{[n]}\rangle_q(\mathbf{y})$$
$$- \langle \overline{\mathbf{u}}^{[n]}\rangle_q(\mathbf{z})][\varphi(v_q, \mathbf{x}_q|; v_i, \mathbf{x}_i) - n^{(j)}]d\mathbf{x}_q. \tag{18.80}$$

The scheme (18.80) forms the Neumann series for the solution ($\mathbf{x} \in v_i^l$)

$$\langle \overline{\mathbf{u}}\rangle_i(\mathbf{x}) := \big\langle \lim_{n\to\infty} \overline{\mathbf{u}}^{[n+1]}\big\rangle_i(\mathbf{x}) = \mathcal{D}_i^u(\langle \mathbf{u}\rangle)(\mathbf{x}), \tag{18.81}$$

which yields the final representations for the effective nonlocal operator

$$\langle \boldsymbol{\sigma}\rangle(\mathbf{x}) = \mathcal{L}^{\sigma}(\langle \widetilde{\boldsymbol{\tau}}^{(0)}\rangle) + \langle \boldsymbol{\tau}\rangle(\mathbf{x}),\ \ \langle \widetilde{\boldsymbol{\tau}}^{(0)}\rangle(\mathbf{x}, \hat{\mathbf{x}}) := \mathbf{C}^{(0)}(\mathbf{x}, \hat{\mathbf{x}})(\langle \mathbf{u}\rangle(\hat{\mathbf{x}}) - \langle \mathbf{u}\rangle(\mathbf{x})) \tag{18.82}$$

where the statistical average of the local stress micropolarization tensor ($\mathbf{x}, \mathbf{z}, \mathbf{y} \in v_i^l$)

$$\langle \boldsymbol{\tau} \rangle(\mathbf{x}) = \int_{v_i^{l1}(\mathbf{x})} n^{(1)} \boldsymbol{\mathcal{L}}^{\sigma}(\mathbf{C}_1 \boldsymbol{\eta}^D)(\mathbf{x}-\mathbf{y}) d\mathbf{y}, \tag{18.83}$$

$$\boldsymbol{\eta}^D(\mathbf{z}, \mathbf{t}) = \Big[\boldsymbol{\mathcal{D}}^u(\langle \mathbf{u} \rangle)(\mathbf{z}) - \boldsymbol{\mathcal{D}}^u(\langle \mathbf{u} \rangle)(\mathbf{t}) + \boldsymbol{\mathcal{L}}^{uu}(\mathbf{z} - \mathbf{x}_i, \boldsymbol{\mathcal{D}}^u(\langle \mathbf{u} \rangle)) - \boldsymbol{\mathcal{L}}^{uu}(\mathbf{t} - \mathbf{x}_i, \boldsymbol{\mathcal{D}}^u(\langle \mathbf{u} \rangle)) \Big] \tag{18.84}$$

is reduced to the dilute approximation at $\boldsymbol{\mathcal{D}}^u \to \mathbf{I}$ (see for details Buryachenko, 2020c).

For the homogeneous loading of statistically homogeneous medium (when $\boldsymbol{\varepsilon}\rangle(\mathbf{x}) \equiv$ const.), the operator $\boldsymbol{\mathcal{D}}^u(\langle \mathbf{u} \rangle)$ is reduced to a linear algebraic representation

$$\boldsymbol{\mathcal{D}}^u(\langle \mathbf{u} \rangle))(\mathbf{x}) = \mathbf{D}^u(\mathbf{x}) \langle \boldsymbol{\varepsilon} \rangle \tag{18.85}$$

that reduces Eqs. (18.82)–(18.84) to (18.75), (18.76).

18.3.7 Numerical Schemes and Results for 1D Case

At first, we consider the truncation method for 1D peridynamic bar $\bar{\bar{w}}$ of the length $2L + 2l_\delta$ and cross-sectional area $A = 1$ containing one inclusion $v_i = [-a, a]$ and the volumetric boundary w_Γ: $\bar{\bar{w}} = w_\Gamma \cup w$, $w_\Gamma = [-L - l_\delta, -L] \cup [L, L + l_\delta]$. The midpoint quadrature scheme (similar to [131, 369, 664] and Sect. 18.1.5) is realized for the uniform grid $x_i = h(0, \pm 1, \pm 2, \ldots, \pm(N^{\max} + N^{w\Gamma}))$ with the numbers of material points $2N^{\max} + 1$ and $2N^{w\Gamma} + 1$ inside both w and the interaction boundary w_Γ, respectively, and a constant step size $h = L/N^{\max}$; $2N^a + 1$ and n_l stand for the numbers of the material points in the inclusion v_i and the horizon n_l, respectively.

Following [228], we consider a heterogeneous peridynamic bar with the triangle-function profiles (16.64) and (16.64) (see [131]) ($k = 0, 1$, $E^{(1)}/E^{(0)} = 5.0$). Given a node distribution $x_p = 0, \pm 1, \pm 2, \ldots, \pm N^{\max}$ over the length of v_i^{l0}, we discretize Eq. (18.15) at the nodes x_p using the composite midpoint integration method

$$\sum_{q=p-n_l}^{q=p+n_l} C(x_p, x_q)(\tilde{u}_q - u_p)\bar{v}_{pq} = -\sum_{q=p-n_l}^{q=p+n_l} C(x_p, x_q)\varepsilon^{w\Gamma} x_q \bar{v}_{pq} W^{w\Gamma}(x_q), \tag{18.86}$$

where $\bar{v}_{pj}$ is the portion of node q "volume" covered by the horizon of node p and a constant step size $h = l_\delta/n_l = L/N^{\max}$ is assumed; $\tilde{u}_q = u_q$ if $|q| \leq N^{\max}$ and $\tilde{u}_q = 0$ if $|q| > N^{\max}$. The error in the coordinate difference formula (18.88) is well known to be $O(h^2)$ (see [1401]). The effectiveness of the quadrature scheme (18.86) for the homogeneous infinite bar was demonstrated in [131] who also proved uniform convergence of the peridynamic solutions to the classical solutions of static and dynamic elasticity problems in 1D case in the limit of the horizon going to zero.

After estimation of $u(x_p)$ ($|x_p| \leq L$) by the quadrature method (18.86), the peridynamic stress distribution and the micropolarization tensor (17.92) can be presented in a similar form:

$$\sigma(x_p) = \sum_{q=j-n_l}^{j} \sum_{r=j}^{q+n_l} C(x_q, x_r)[u(x_r) - u(x_q)]\bar{v}_{qp}\bar{v}_{rp},$$

$$\tau(x_p) = \sum_{q=j-n_l}^{j} \sum_{r=j}^{q+n_l} C_1(x_q, x_r)[u(x_r) - u(x_q)]\bar{v}_{qp}\bar{v}_{rp}. \tag{18.87}$$

Equation (18.87$_2$) allows us to estimate the average micropolarization tensor

$$R_i^l \varepsilon^{w\Gamma} = \frac{2}{N^{\max} + N^{w\Gamma}} \times \sum_{p=-N^{\max}-N^{w\Gamma}}^{N^{\max}+N^{w\Gamma}} \sum_{q=j-n_l}^{j} \sum_{r=j}^{q+n_l} C_1(x_q, x_r)[u(x_r) - u(x_q)]\bar{v}_{qp}\bar{v}_{rp}, \tag{18.88}$$

while the average strain concentration factor for the isolated inclusion $A_i^l \varepsilon^{w\Gamma}$ can be found from the solution (18.86)

$$A_i^l \varepsilon^{w\Gamma} = (a + l_\delta)^{-1}(u_{N^a+n^l} - u_{-N^a-n^l})/2. \tag{18.89}$$

The external interface average of the strain concentration factor $A^{l\omega}$ and the expression for the average of the local interface fields $R_i^{l\omega}$ can be presented in a similar form:

$$A_i^{l\omega} = 1 + (a + l_\delta)^{-1}(u_{N^a+n^l} - u_{-N^a-n^l})/(2\varepsilon^{w\Gamma}), \tag{18.90}$$

$$R_i^{l\omega} = \frac{1}{2\varepsilon^{w\Gamma}} \sum_{k=1}^{2} \sum_{q=J-n_l}^{J} \sum_{r=J}^{q+n_l} C(x_q, x_r)(u_r - u_q)\bar{v}_{qJ}\bar{v}_{rJ} - E^{(0)} A_i^{l\omega}. \tag{18.91}$$

Following Sects. 18.3.2 and 18.3.3, the effective field concentration factors $D_i^{\omega MTM}$ (18.58) and $D_i^{\omega EFM}$ (18.71) can be found ($J = (-1)^k(N^a + n_l)$):

$$(D_i^{\omega MTM})^{-1} = 1 - c^{(1)}\left(1 + \frac{l_\delta}{a}\right) + \frac{c^{(1)}}{2a\varepsilon^{w\Gamma}}(u_{N^a+n^l} - u_{-N^a-n^l}), \tag{18.92}$$

$$\left(D_i^{\omega EFM}\right)^{-1} = 1 + \frac{c^{(1)}}{4a(a+l_\delta)\varepsilon^{w\Gamma}} \sum_{k=1}^{2} (-1)^k \times \sum_{j=-2N^a}^{2N^a} \Big[u_{J-j} - \varepsilon^{w\Gamma} h(J-j)\Big] \bar{V}_{ji}, \tag{18.93}$$

where the well-stirred approximation (5.97) was used with $n^{(1)} = c^{(1)}/2a$.

Substitutions of Eqs. (18.90), (18.91), and (18.92) into Eqs. (18.44), (18.49), (18.59), and (18.71) lead to the effective Young modulus representations:

$$E^* = E^{(0)} + c^{l(1)} R_i^{l\omega}, \tag{18.94}$$

$$E^* = E^{(0)}\big[1 - c^{l(1)} R_i^{l\omega}\big]^{-1}, \tag{18.95}$$

$$E^* = E^{(0)} + c^{l(1)} R_i^{l\omega} D_i^{\omega MTM}, \tag{18.96}$$

$$E^* = E^{(0)} + c^{l(1)} R_i^{l\omega} D_i^{\omega EFM}, \tag{18.97}$$

for the generalized dilute approximations (18.94) and (18.95), the generalized MTM (18.96), and the generalized EFM (18.97), respectively.

It should be mentioned that the effective modulus estimations (18.94)–(18.96) are expressed through the representations (18.90) and (18.91) which are valid for both the uniform and nonuniform grids. In so doing, the generalized EFM estimations (18.93) and (18.97) were obtained at the implying of coincidence of the grids exploited for the evaluation of both the perturbator factor $\mathbf{L}_i^{u\epsilon}(\mathbf{x} - \mathbf{x}_i)$ (18.42_1) and the location of the moving inclusions $\mathbf{x}_q$ (18.71). However, in the case of nonuniform refinement, application of Eq. (18.93) requests a preliminary interpolation at the grid Ω^{sq} (18.12) of the perturbator factor $\mathbf{L}_i^{u\epsilon}(\mathbf{x} - \mathbf{x}_i)$ (18.42) estimated at the nonuniform refined grid Ω^{inhom}. It will be obtained through the expression (18.45) with the linear shape function.

Estimation of the effective Young modulus can also be obtained from Eqs. (18.94)–(18.97) in the limit of vanishing length scale $l_\delta/a \to 0$. Indeed, substitutions of the local limit ($A_i^l \to A_i^{(1)}$ and $R_i^l \to R_i$)

$$A_i^{(1)} = E^{(0)} \hat{E}^{(1)}, \quad R_i = (E^{(1)} - E^{(0)}) E^{(0)} \hat{E}^{(1)} \tag{18.98}$$

into Eqs. (18.93)–(18.96) yield

$$E^* = E^{(0)}(1 + c^{(1)}) - c^{(1)}(E^{(0)})^2 \hat{E}^{(1)}, \tag{18.99}$$

$$\hat{E}^* = c^{(0)} \hat{E}^{(0)} + c^{(1)} \hat{E}^{(1)}, \tag{18.100}$$

$$E^* = \Big[c^{(0)} \hat{E}^{(0)} + c^{(1)} \hat{E}^{(1)}\Big]^{-1}, \tag{18.101}$$

$$E^* = \left[c^{(0)}\hat{E}^{(0)} + c^{(1)}\hat{E}^{(1)}\right]^{-1}, \tag{18.102}$$

for the dilute approximations (18.99) and (18.100), the MTM (18.101), and the EFM (18.102), respectively. It is interesting that the representations (18.100)–(18.102) coincide with the exact one and differ from Eq. (18.99). The mentioned coincidence is explained by a fundamental reason described, e.g., in [228].

A 1D example of an infinite heterogeneous peridynamic bar is analyzed for the next geometrical and mechanical properties of constituents. An infinitely long bar of a constant cross section $A = 1$ contains the randomly distributed identical inclusions of the size $2a$. At first, the micromodulus functions with the step-function profile (18.86) with the same horizon $l_\delta/a = 0.05,\ 0.25$, and 0.5 are considered for both the inclusions and matrix with the ratios of the corresponding elastic counterparts $E^{(1)}/E^{(0)} = 5$; the ratio of the inclusion size and the sample size is fixed at $a/L = 0.2$. The midpoint quadrature scheme (18.88) is realized for the uniform grid with a constant step size $h = L/4000$ (corresponding to $l_\delta/h = 20, 100, 200$, respectively); an accuracy of the numerical solution for a single inclusion in the infinite homogeneous bar is analyzed by comparison of results obtained for $h = l_\delta/200,\ l_\delta/100$, and $l_\delta/20$.

The curves $u(x)$ vs. x/a behave according to the general features of peridynamic solutions for a homogeneous peridynamic bar subjected to the fictitious body force mentioned by Silling et al. [1405] (see also [131]). So, $u_1(x)$ also has a discontinuity at $x = \pm a$ because in general the displacement field $u(x)$ has the same smoothness as the body force field (see [1405]). The displacements $u(x - x_i)$ $(x \in v_i^l)$ as the functions of the relative coordinates x/a are presented in Fig. 18.7 for $l_\delta/a = 0, 0.05,\ 0.25$, and 0.5.

Fig. 18.7 Displacement $u(x)$ vs. x/a: $l_\delta/a = 0(1)$, $0.1(2)$, $0.25(3)$, $0.5(4)$

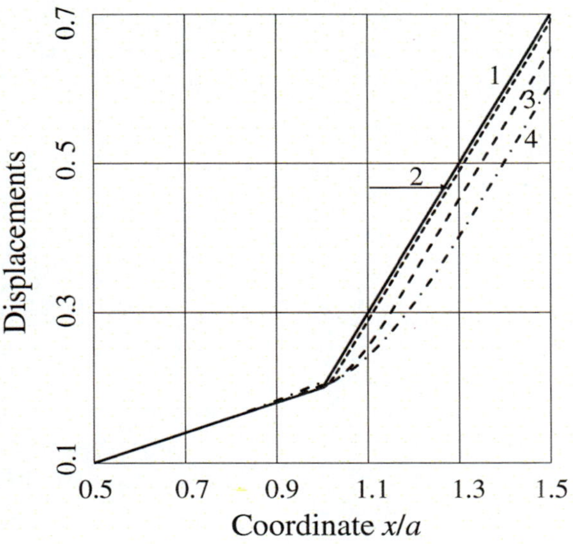

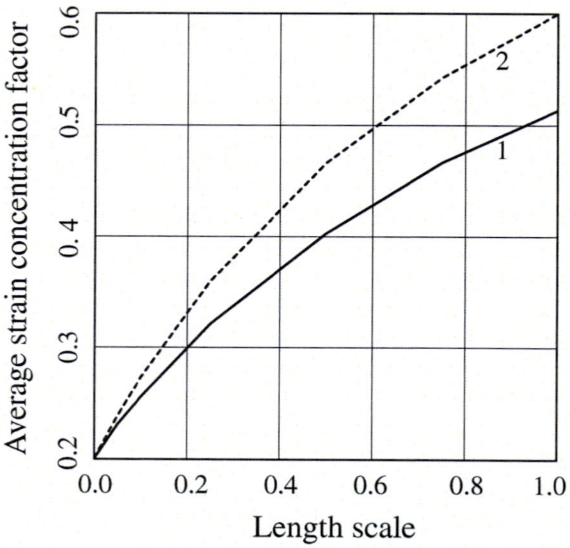

Fig. 18.8 Average strain concentration factor vs. l_δ/a: $A_i^{l\omega}$ (1) and $A_{\text{loc}}^{l\omega}$ (2)

It should be mentioned that the volume average of the strain concentration factor $\mathbf{A}_i^l$ (18.52) coincides with the interface average of the strain concentration factor $\mathbf{A}_i^{l\omega}$ (18.54) based on the average of the interface displacements $\mathbf{u}(\mathbf{s})$ ($\mathbf{s} \in \Gamma_i^-$) rather than the strain $\boldsymbol{\varepsilon}(\mathbf{x})$ ($\mathbf{x} \in v_i^l$), i.e., estimation of the strain fields $\boldsymbol{\varepsilon}(\mathbf{x})$ ($\mathbf{x} \in vi^l$) is not required. The dependence $A_i^{l\omega}$ on the ratio l_δ/a is presented in Fig. 18.8 (curve 1). As can be seen, in the limit of vanishing length scale l_δ/a, $A_i^{l\omega}$ is reduced to the locally elastic solution for the strain concentration factor inside the inclusion $A_i^{(1)} = 0.2$ (18.98). It should be mentioned that the growing function $A_i^{l\omega} \sim l_\delta/a$ was obtained by averaging over the growing area $\bar{v}_i^l = 2(a + l_\delta)$. In such a case, for extraction of the nonlocal effects, we need to compare $A_i^{l\omega}$ with the formal average $A_{\text{loc}}^{l\omega} = (A_i^{(1)} + l_\delta/a)/(1 + l_\delta/a)$ (see curve 2 in Fig. 18.8) of the local concentration factor

$$A_i(x) = \begin{cases} A_i^{(1)}, & \text{for } |a| < a, \\ 1, & \text{for } a \le |x| \le a + l_\delta \end{cases} \tag{18.103}$$

averaged over the extended inclusion $v_i = \{x||x| \le a + l_\delta\}$. The difference between curves 1 and 2 in Fig. 18.8 defined by the nonlocal effects leads to the additional rigidity of the inclusion.

An estimation of $A_i^{l\omega}$ makes it possible to turn our attention to the analysis of CM by the substitution of $A_i^{l\omega}$ into the representation for D^{MTM}. The effective field concentration factor D^{EFM} is defined by Eq. (18.97) through the displacement perturbator factor $L^{u\epsilon}(x - x_i)$. We estimate now an influence of refinement on the estimation of both the interface average of the strain concentration factor $\mathbf{A}_i^{l\omega}$ (18.54) and the effective moduli $E^{*\,\mathrm{MTM}}$ (18.96) $E^{*\mathrm{EFM}}$ (18.97). Namely, following [131], we consider two types of grid refinement: uniform and nonuniform refinement. In uniform refinement, we estimate the displacements at the uniform but denser grids (see Fig. 18.9), while for nonuniform, selected regions are refined, while other are not. We only consider $m-$type convergence corresponding to the cases of increasing density of the nodes with the fixed horizon l_δ =const. So the uniform grid (a) is refined with respect to the grid (b). Following the "visibility criterion" (see [131]), the refinement areas for nonuniform grid cover the loaded zones. If a refined solution is sought at a point x, then the points of applications of the loads are not 'seen' by any nodes of the course region. However, the role of "loading" in the current case is performed by the fictitious body force $\tilde{\mathbf{b}}(\mathbf{x})$ (see Eqs. (18.18) and (18.19)) vanishing outside of the extended inclusion v_i^l.

The numerical results for the strain concentration factor $A^{l\omega}$ and the relative effective moduli $\tilde{E}^{*\,\mathrm{MTM}} = E^{*\,\mathrm{MTM}}/E^{(0)}$ (18.96) and $\tilde{E}^{*\mathrm{EFM}} = E^{*\mathrm{EFM}}/E^{(0)}$ (18.97) are presented in Table 18.1. Rows (a), (b), and (c) were estimated for the uniform grid, while the results in rows (d)–(g) were obtained for the nonuniform grids. The results of row (a) were obtained for uniformly refined grid at $N^{\max} = 4000$, rows (b), (d), (f), and (g) correspond to $N^{\max} = 2000$, while row (c) was estimated for the course uniform grid with $N^{\max} = 1000$ (see Fig. 18.10). In all cases, $a/L^{\max} = 5$, $l_\delta/a = 0.25$, and $c^{(1)} = 0.8$. The columns CV(%) indicate the coefficients of variation (CV(%)) of the values in the corresponding row and the previous column with respect to row (a); for example, 1.89% at the intersection of row (c) with the second column CV(%) means CV(%) of the number 3.170 $\tilde{E}^{*\,\mathrm{MTM}}$ with respect to 3.111 $\tilde{E}^{*\,\mathrm{MTM}}$. As can be seen, reduction of nodes to $N^{\max} = 1000$ (row (c)) significantly decreases an accuracy of estimations of both the strain concentration factor and effective moduli. Because of this, we analyze now the sensitivity of solutions in rows (d)–(g) (at $N^{\max} = 2000$) to the location of the refinement region with respect to the loaded area $v_i^l = \{x| -a - l_\delta \le x \le a + l_\delta\}$.

Case (d) (See Fig. 18.10) The refined zone $[-2l_p l_\delta, 2l_p l_\delta]$ $(0 < l_p < 2)$ is completely covered by the inclusion $v_i = [a, a]$. The ratio h_r of the step sizes of the refined region and the unrefined area is $h_r = h^{(1)}/h^{(0)}$; $4l_r l_\delta$ is the size of the refined region. This ratio h_r defines the node numbers in the refined zone $n^{(0)} = 2[L^{\max} - 2l_r l_\delta - 2l_r l_\delta/h^{(1)}]/N^{\max}$ and the course zone $n^{(1)} = 4l_r l_\delta/h^{(1)}$. Thus, the refinement process is defined by the ratio parameters l_p and h_p and the location of the refined region. The pair of rows (d_1) and (d_2) (as well as the pairs (e_1) and (e_2) and (f_1) and (f_2)) correspond to two densities of the refined region $h_p = 0.5$ and $h_p = 0.2$, respectively. We will also consider $l_p = 1.5$ in all cases (d)–(g) (exemptions are (g_4) and (g_5)).

Case (e) (See Fig. 18.10e) The refined areas $[-a - l_p l_\delta, -a + l_p l_\delta] \cup [a - l_p l_\delta, a + l_p l_\delta]$ cover the displacement discontinuity $x \pm a$ and the nodes in the unrefined regions do not cover $-a$ and a with their horizon. The ratios h_r and l_p completely

Fig. 18.9 Schematic plot of uniform mesh: (**a**) refined mesh, (**b**) coarse mesh

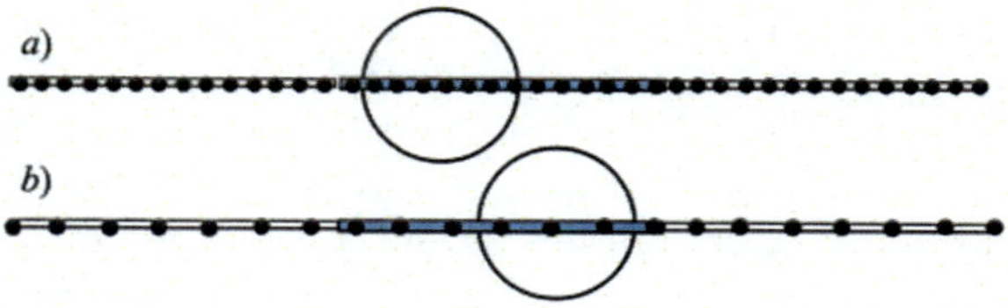

Table 18.1 The estimations of $A^{l\omega}$, $\tilde{E}^{*\,\mathrm{MTM}}$, and $\tilde{E}^{*\mathrm{EFM}}$ for the different cases of the grid refinements

Cases	$A_i^{l\omega}$	CV(%)	$\tilde{E}^{*\,\mathrm{MTM}}$	CV(%)	$\tilde{E}^{*\mathrm{EFM}}$	CV(%)
a	0.3214	–	3.111	–	3.032	–
b	0.3221	0.21	3.123	0.38	3.040	0.29
c	0.3136	2.45	3.170	1.89	3.054	0.73
d_1	0.3219	0.15	3.122	0.35	3.043	0.36
d_2	0.3205	0.28	3.134	0.73	3.051	0.62
e_1	0.3220	0.18	3.101	0.32	3.018	0.46
e_2	0.3218	0.12	3.102	0.28	3.021	0.36
f_1	0.3225	0.34	3.109	0.06	3.027	0.16
f_2	0.3199	0.46	3.125	0.45	3.057	0.82
g_1	0.3225	0.31	3.109	0.06	3.035	0.16
g_2	0.3213	0.03	3.112	0.03	3.033	0.03
g_3	0.3189	0.77	3.108	0.09	3.066	1.12
g_4	0.3173	1.28	3.144	0.06	3.030	0.06
g_3	0.3231	0.52	3.099	0.38	3.031	0.03

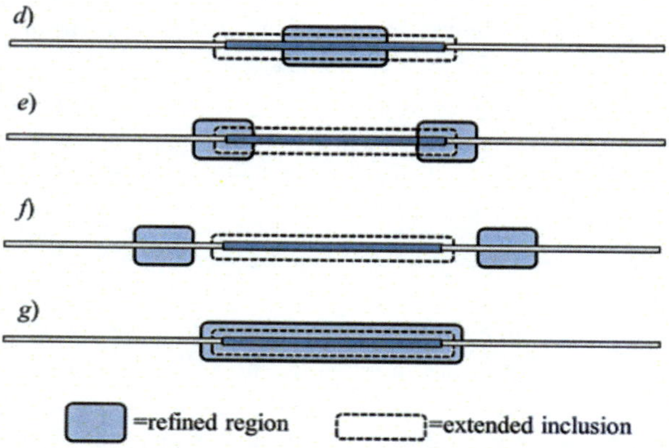

Fig. 18.10 Four cases for location of refinement region with respect to the extended inclusion

define (analogous to the case (c)) the step sizes $h^{(1)}$ and $h^{(0)}$ and the node numbers $n^{(1)}$ and $n^{(0)}$ inside the refined and unrefined regions, respectively.

Case (f) (See Fig. 18.10f) The refined regions $[-a-2l_pl_\delta, -a-l_pl_\delta]\cup[a+l_pl_\delta, a+2l_pl_\delta]$ have no intersection with the "loading" area represented by the extended inclusion $v_i^l = [-a-l_\delta, a+l_\delta]$.

Case (g) (See Fig. 18.10g) The refined regions $[-a-l_pl_\delta, a+l_pl_\delta]$ completely cover the extended inclusion $v_i^l = [-a-l_\delta, a+l_\delta]$ so that the point of application of the fictitious body force $\tilde{\mathbf{b}}(\mathbf{x})$ ($\mathbf{x} \in v_i^l$) is not "seen" by the nodes in the unrefined area. Rows (g_1), (g_2), and (g_3) correspond to three densities of the refined region with $h_p = 0.5$, $h_p = 0.4$, and $h_p = 0.2$, respectively.

Rows (d)–(f) were obtained for nonuniform refinement with the same ratio $4l_pl_\delta/L^{\max} = 0.075$ of the lengths of refined regions $4l_pl_\delta$ and a hole sample $2L^{\max}$. Due to the fixed total number $N^{\max} = 2000$ of nodes in rows (d)–(g), a significant node densification in the refined zone (as $h_p = 0.2$) can lead to reduction of grid density in the unrefined region with possible decreasing accuracy in this region. The worst accuracy is detected for the case (f); it means that the refinement in the matrix outside of the extended inclusion leads to decreasing accuracy. Most likely, nonuniform refinement in the loading region improves the accuracy in the solutions compared with the solutions exploited the same number of nodes in the sample (compare rows (b), (d_1), (d_2), (e_1), (e_2), (g_1), (g_2)). From other side, the lengths of the refined regions $4l_pl_\delta$ occupy 0.075 (cases (d)–(f)) of the total length $2L^{\max}$, and, because of this, decreasing h_p from 1.0 to 0.5 and 0.2 leads to increasing of the distances between the nodes in the rest area from $5.27E-3$ to $6.06E-3$ and $7.40E-3$, respectively. In so doing, for the case (g), decreasing h_p from 1.0 to 0.5, 0.4, and 0.2 leads to the increasing of distances between the nodes in the unrefined region from $5.27E-3$ to $6.583E-3$, $7.24E-3$, and $1.05E-2$, respectively (see rows (g_1), (g_2), (g_3)). As can be seen (compare the results for the uniform and nonuniform refinement in rows (b) and (g_3), respectively), improvement of accuracy in the refined region $[-a-l_pl_\delta, a+l_pl_\delta]$ does not compensate the loss of accuracy in the coarsened regions. Rows (g_4) and (g_5) were estimated for $h_r = 0.4$ (as row (g_2)) with $l_p = 0$ and $l_p = 1$, respectively, while row (g_2) was evaluated at $l_p = 1.5$. The worst accuracy at $l_p = 0$ corresponds to the case (g_4) when the loading zone $[-a-l_pl_\delta, a+l_pl_\delta]$ is not refined at the most critical area $[-a-l_pl_\delta, a]\cup[a, a+l_pl_\delta]$.

It should be mentioned that the methods of adaptive meshing and coarsening (see [130, 131] and [1398]) have been studied for the homogeneous peridynamic media and periodic structures. However, adaptive grid refinement algorithms for the heterogeneous peridynamic materials (basic problem; see "Introduction") with automatic refining and coarsening and its subsequent implementation into the micromechanical models for random structure composites are a subject of further study.

Now we compare the estimations of the effective Young modulus by the dilute approximations (18.94) and (18.95), the generalized Mori-Tanaka approach (18.96), and the generalized EFM (18.97) (see Figs. 18.7, 18.8, and 18.9, corresponding to the scales $l_\delta/a = 0.1$, 0.25, and 0.5, respectively). For the dilute approximations of the locally elastic case ($l_\delta/a = 0$), curves 1 and 2 in Figs. 17.11–17.13 were estimated by the exact equations (18.100) and (18.99), respectively. The dilute approximations of the peridynamic composite bars are described by curves 3 and 4 (see Figs. 18.11–18.13) estimated by Eqs. (18.95) and (18.94), respectively. Curve 5 was plotted by the generalized Mori-Tanaka approach (18.96), while the generalized EFM estimations are plotted by curve 6. As can be seen, the nonlocal effects have a weak action on the dilute approximation (18.94) (compare curves 2 and 4 in Figs. 18.11–18.13), while the effective moduli E^* estimated by the dilute approximation (18.95) can differ by a factor of 2 for the different scales l_δ/a (see Fig. 18.11). In so doing, both the generalized Mori-Tanaka approach (18.96) and the generalized dilute approximation (18.95) lead to closed results (compare curves 3 and 5 in Figs. 18.11–18.13) with softer (on 0.7%; see Fig. 18.13) results corresponding to Eq. (18.96). Nonlocal effects for the generalized EFM (18.97) are less manifested than for both the generalized dilute approximation (18.95) and the Mori-Tanaka approach (18.96) (compare curve 6 with curves 3 and 5). It is interesting to compare the influence of the different micromoduli (18.86) (this chapter) and (18.87) (see [226]) with the same local limit $E^{(1)}$ and $E^{(0)}$. It can be seen that the last two methods (18.95) and (18.96) exhibit greater dependency on the ratio of nonlocality l_δ/a (Figs. 18.12 and 18.13).

In Fig. 18.14, we only present the estimations $E^*/E^{(0)} \sim c$ obtained by the generalized EFM for the scales $l_\delta/a = 0, 0.05, 0.25, 0.5$, and $l_\delta/a = 0.75$. As can be seen, increasing the nonlocality parameter l_δ/a from 0 to 0.75 leads to the stiffening of the effective modulus $E/E^{(0)}$ on 40% at $c = 1$. It should be clarified that all results and descriptions in Sects. 17.3.1–17.3.5 were presented for dD ($d = 1, 2, 3$) cases. However, the numerical results in Subsection 17.3.6 are only obtained for a simplified one-dimensional (1D) structure because it is possible to dispense with some assumptions (widely used in two-dimensional (2D) and three-dimensional (3D) cases; see the locally elastic counterparts in Sections 8–12) and to focus our attention to the direct use of a large body of analytical and numerical results obtained for a 1D homogeneous and inhomogeneous peridynamic bar of

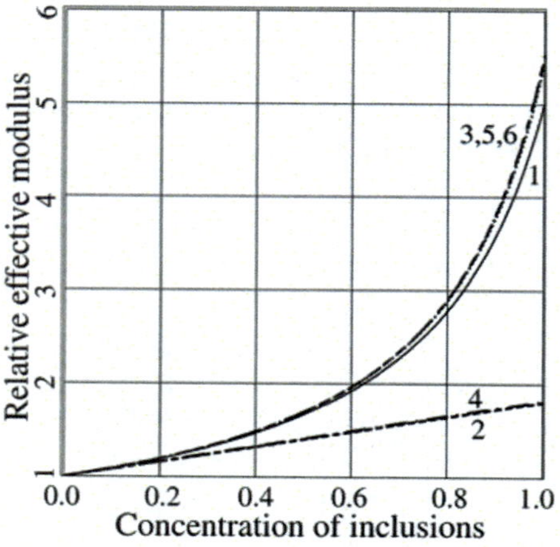

Fig. 18.11 $E^*/E^{(0)}$ vs. c for $l_\delta = 0.1$ estimated by Eqs. (18.100)(1), (18.99)(2), (18.95)(3), (18.94)(4), (18.96)(5), (18.97)(6)

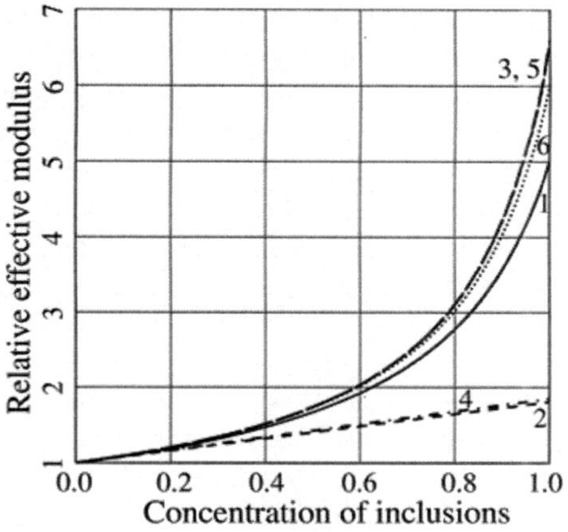

Fig. 18.12 $E^*/E^{(0)}$ vs. c for $l_\delta = 0.25$ estimated by Eqs. (18.100)(1), (18.99)(2), (18.95)(3), (18.94)(4), (18.96)(5), (18.97)(6)

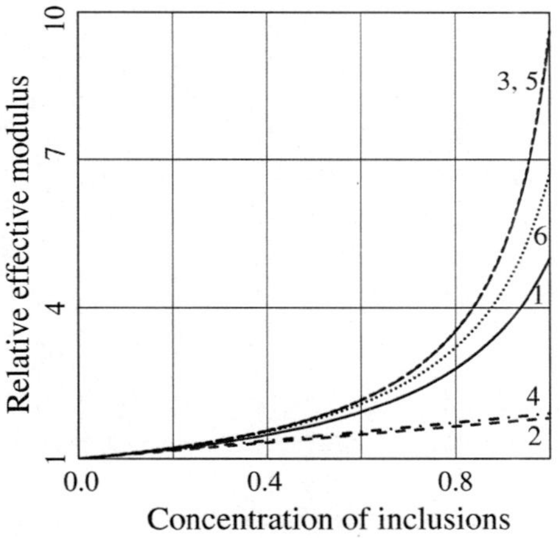

Fig. 18.13 $E^*/E^{(0)}$ vs. c for $l_\delta = 0.5$ estimated by Eqs. (18.100)(1), (18.99)(2), (18.95)(3), (18.94)(4), (18.96)(5), (18.97)(6)

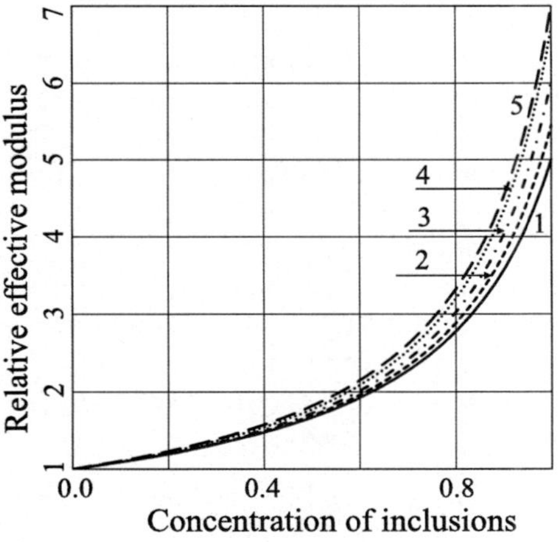

Fig. 18.14 $E^*/E^{(0)}$ vs. c estimated by Eq. (18.97) for $l_\delta = 0.0$ (1), 0.1 (2), 0.25 (3), 0.5 (4), and 0.75(5)

deterministic structure (see, e.g.,[131, 435, 1058, 1401, 1405, 1601]). Indeed, for locally elastic micromechanics, $1D$ solution has known exact analytical representations (18.98) and (18.99) coinciding with the classical versions of the MTM (18.101) and the EFM (18.102). It gave us a unique opportunity to be focused on the analyses of nonlocal effects rather than on the estimation of the inclusion interactions (locally elastic counterparts of these interactions were considered in, e.g., [211]).

Thus, one proposes a generalization of a dilute approximation method to their peridynamic counterpart in the sense that the volume fraction of the particles is small and their mutual interaction can be neglected. As in a classical approach, the essential assumption in the generalized Mori-Tanaka method (MTM) states that each extended inclusion behaves as an isolated one in the infinite matrix and subject to some effective strain field coinciding with the average strains in the truncated matrix. The direct Mori-Tanaka method based on a single hypothesis (18.56)–(18.59) also has a specific merit in shedding new light on the weakness of the direct method that opens up outstanding possibilities to improve the method as was done in the effective field method (EFM) generalizing its locally elastic counterpart (see for references [200]). In the generalized EFM proposed, the effective field is evaluated from self-consistent estimations by the use of closing corresponding new general integral equations (18.35) and (18.36) in the framework of the quasicrystalline approximation. One demonstrates some similarity and difference between the mentioned methods. It should be mentioned that the effective modulus estimations in all three methods are obtained by the use of averaging of either strain concentration factors (e.g., $\mathbf{A}_i^l$ (18.52)) or the displacement concentration factors (e.g., $\mathbf{A}_i^{l\omega}$ (18.54)) over the extended inclusions or the external interaction interface of inclusions, respectively. New external interaction interface averaging operation (18.54) and (18.71) was introduced as a straightforward generalization of interface integral technique proposed in [224] (see Chap. 12 in locally elastic micromechanics. In such a case, any spatial derivatives of displacement fields are not required. Elimination of the requirement mentioned is critical because the displacement fields $\mathbf{u}(\mathbf{x})$ ($\mathbf{x} \in v^l$) are not differentiable inside the extended inclusions.

It should be mentioned that homogeneity hypothesis **H1a** (18.37) (used in Sects. 17.3.1–17.3.3) does not assume any restrictions on both the material micro-topology of inclusions and the field inhomogeneity of $\mathbf{u}(\mathbf{x})$, $\boldsymbol{\varepsilon}(\mathbf{x})$, $\boldsymbol{\sigma}(\mathbf{x})$ ($\mathbf{x} \in R^d$) (e.g., either singularity, discontinuity, or nonlocality). Finally, the methods proposed in this chapter allow us to completely abandon the hypotheses (as in [211]) **H1a**, **H1b**, and **H3**, while the hypothesis **H2** can be used for multiparticle generality. The significance of the results obtained is beyond the scope of just a proposal of the generalized EFM. Of more importance is a development of a tool which can be easily modified and implemented for a wide class of problems of peridynamic micromechanics, e.g., effective deformation of heterogeneous media subjected to inhomogeneous body forces (see Sect. 17.4), state-based peridynamics, coupled problems, and nonlinear problems. However, consideration of these problems goes far beyond the scope of the current study.

18.4 Effective Deformation of Peridynamic Random Structure Bar Subjected to Inhomogeneous Body Force

It should be mentioned that the results indicated above for the random structure peridynamic composites can be considered as a straightforward generalization of the corresponding classical results for locally elastic composites which were obtained for both statistically homogeneous composites and homogeneous remote loading. In such a case, the effective properties of both the peridynamic composites and locally elastic ones are described by a constant tensor of local effective moduli. However, even for locally elastic composites subjected to inhomogeneous loading, the effective deformations are described by a nonlocal (either the differential or integral) operator (see for references, e.g., [200]) relating a statistical average of stresses in the point being considered with a statistical average of strains in the vicinity of this point. On the other hand, to the best of the authors' knowledge, there are no academic publications analyzing the mentioned effective deformations of locally elastic composites due to prescribed inhomogeneous body forces. Thus, this chapter dedicated to the estimation of effective deformations of 1D statistically homogeneous peridynamic composites for the prescribed self-equilibrated body forces has no known locally elastic counterpart. A simplified one-dimensional (1D) structure is analyzed because it is possible to dispense with some assumptions (widely used in two-dimensional (2D) and three-dimensional (3D) cases) and to focus our attention to the direct use of a large body of analytical and numerical results obtained for a 1D homogeneous and inhomogeneous peridynamic bar of deterministic structure (see, e.g., [1401, 1405] and also [131, 435, 1058, 1601, 1603]). A subsequent simplification is described by the dilute approximation method assuming the coincidence of the effective field with the applied field when the interaction of inclusions is negligible.

18.4.1 Locally Elastic Solution for the Infinite Bar with Inclusion

For the local elasticity theory, the constitutive law has the form

$$\sigma(x) = E(x)\varepsilon(x), \ \varepsilon(x) = \hat{E}(x)\sigma(x), \tag{18.104}$$

where $\hat{E}(x) := E^{-1}(x)$. Let us a peridynamic bar $w = R$ of the infinite length and cross section $A = 1$ contains the inclusion $v = \{x| \, |x| \leq a\}$ and the corresponding

Young moduli are $E^{(0)}$ and $E^{(1)}$, respectively. A self-equilibrated body force with the compact support $w^b = \{x|\ |x| \leq a^b\}$ is described by the equation

$$b(x) = \begin{cases} 2B - 2B|x - c^b - a^b/2|/a^b, & \text{for} 0 \leq x - c^b \leq a^b, \\ 2B|x - c^b + a^b/2|/a^b - 2B, & \text{for } -a^b \leq x - c^b \leq 0, \end{cases} \tag{18.105}$$

with the symmetry center c^b. The bar with the inclusion and the body force are presented in Fig. 18.15 for the case $c^b = 0$ and $a^b = a$. In the coordinate system with the origin coinciding with the symmetry center $x = c^b = 0$ of the body force density (18.105), the equilibrium equation

$$AE(x)\frac{d^2u}{dx^2} = -b(x) \tag{18.106}$$

has the boundary conditions

$$u(x)\big|_{x=0} = 0, \quad \frac{du}{dx}\bigg|_{x=a^b} = 0. \tag{18.107}$$

The solution of the displacement fields (18.106) and (18.107) is described by the integral representation ($x \geq 0$)

$$u(x) = \frac{1}{A}\Big[\int_0^x \frac{1}{E(y)}dy \int_0^{a^b} b(z)dz - \int_0^x \frac{1}{E(y)} \int_0^y b(z)dzdy\Big] \tag{18.108}$$

corresponding to the stress distribution

$$\sigma(x) = -\int_x^{a^b} b(x)dx. \tag{18.109}$$

The solutions for $x \leq 0$ and an arbitrary $c^b \neq 0$ can be obtained in a similar manner. So, the displacement distributions $u(x) \sim x$ are plotted in Fig. 18.16 for $c^b/a^b = -1.5,\ -1.0,\ 0.0,\ 1.0,\ 1.5$ (curves 1–5, respectively) and the fixed $a^b = a = 1.0$, $B = 1.0$, $E^{(1)} = 5$, $E^{(0)} = 1$. As can be seen, the displacement fields are the constants outside the domain $w^b = \{x|\ |x| \leq a^b\}$:

$$u(x) = \begin{cases} u(c^b + a^b) =^L u_j^b, & \text{for } x \geq c^b + a^b, \\ u(c^b - a^b) =^L u_j^b, & \text{for } x \leq c^b - a^b, \end{cases} \tag{18.110}$$

where $j = \text{sign}(x - c^b)$ and $u(x) = 0$ at $x = c^b$. The curves ${}^Lu_+^b \sim x/a$ and ${}^Lu_-^b \sim x/a$ are plotted in [229].

If for two body forces $b_1(x)$ and $b_2(x)$ with the corresponding symmetry centers $c_1 = c^b$ and $c_2 = c^b$, respectively, the domains w^{b_1} and w^{b_2} do not intersect the

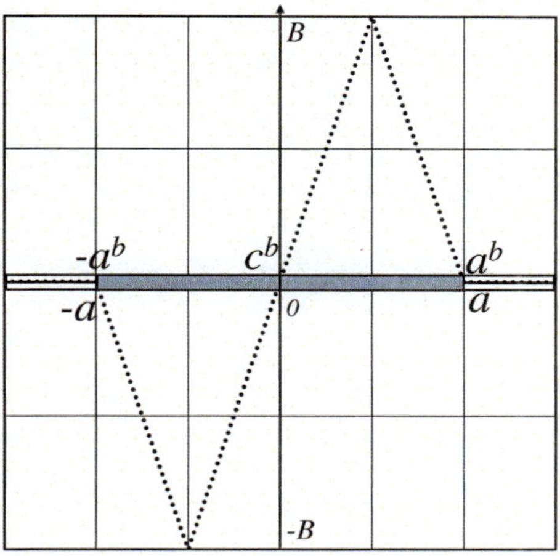

Fig. 18.15 Body force $b(x)$ (18.105) vs. coordinates x/a

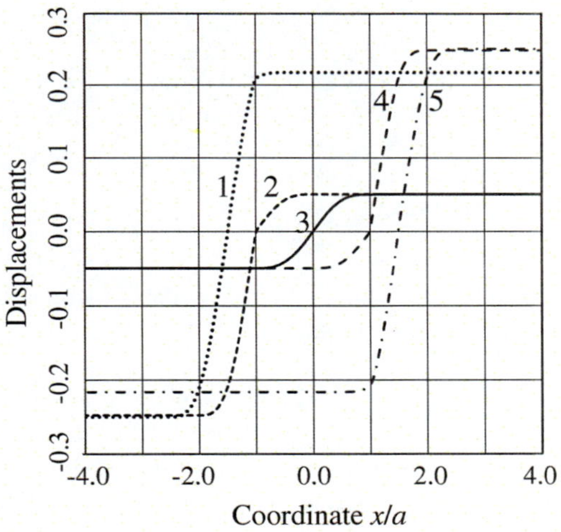

Fig. 18.16 $u(x)$ (18.108) vs. x/a for $c^b/a^b = -1.5$ (1), -1.0 (2), 0.0 (3), 1.0 (4), 1.5 (5)

domain v, then the corresponding displacement distributions $u_1(x)$ and $u_2(x)$ can be obtained by a parallel transfer: $u_1(x) = u_2(x + c_1 - c_2)$. For the homogeneous bar ($E(x) \equiv E^{(0)}$), the solution $u(x)$ can be found from Eq. (18.106) by substitution $E(x) \equiv E^{(0)}$ with the following correction of Eq. (18.110):

$$u(x) = \begin{cases} u(c^b + a^b) = ^L u_j^{b(0)}, & \text{for } x \geq c^b + a^b, \\ u(c^b - a^b) = ^L u_j^{b(0)}, & \text{for } x \leq c^b - a^b. \end{cases} \tag{18.111}$$

18.4.2 Single Inclusion in the Infinite Homogeneous Medium: Decomposition and Truncation Methods

Decomposition and truncation methods were proposed in Subsection 17.1.1 as applied to the prescribed homogeneous volumetric BCs (16.46_2) and (16.52). We modify these methods as applied to the prescribed inhomogeneous body force $b(x)$ (16.46_1). Namely, we decompose the material parameters and operators analogous to Eq. (18.16):

$$C(x, \hat{x}) = C^{(0)}(x, \hat{x}) + C_1(x, \hat{x}), \tag{18.112}$$

$$\widehat{\mathcal{L}} = \widehat{\mathcal{L}}^{(0)} + \mathcal{L}_1, \tag{18.113}$$

$$\mathcal{L}^{\sigma} = \mathcal{L}^{\sigma(0)} + \mathcal{L}_1^{\sigma}, \tag{18.114}$$

where the stress operator $\mathcal{L}^{\sigma}$ defines a *peridynamic stress polarization* parameter $\boldsymbol{\tau} = \mathcal{L}_1^{\sigma} * u(s)$ (18.87_2) generalizing the classical definition of the local elasticity ${}^L\tau(s) := (E^{(1)}(s) - E^{(0)}(s))\varepsilon(s)$.

We also decompose a displacement field $u(x)$ as $u(x) = \overline{u}(x) + u_1(x)$, where $\overline{u}(x)$ is some displacement field inside homogeneous medium (i.e., if $C(x, \hat{x}) \equiv C^{(0)}(x, \hat{x})$) generated by a self-equilibrium body force density $b(x)$:

$$\widehat{\mathcal{L}}^{(0)}(\overline{u})(x) + b(x) = 0, \quad \overline{u}(x) = -\big(\widehat{\mathcal{L}}^{(0)}\big)^{-1} b(x). \tag{18.115}$$

If the body force $b(x)$ considered has a compact support w^b ($b(x) \equiv 0$ at $x \notin w^b$), a produced displacement field is a constant in the remote volumetric boundary

$$\overline{u}(y) \equiv \overline{u}^{b(0)} \equiv \text{const.}, \quad (y \in w_{\Gamma u}) \tag{18.116}$$

and coincides with the constant locally elastic counterpart ${}^L\overline{u}^{b(0)} := \overline{u}(y) = \overline{u}^{b(0)} \equiv$ const. ($y \in w_{\Gamma u}$), see Eq. (18.110).

We are coming now to consideration of a single fixed inclusion $v_i(0)$ with the micromodulus $C^{(1)}(\xi)$ in the infinite homogeneous medium with the properties $C^{(0)}(\xi)$. Then the solution $u(x)$ split into the constituents $\overline{u}(x)$ and $u_1(x)$ is described by Eqs. (18.116) and

$$\widehat{\mathcal{L}}(u_1)(x) + \widehat{\mathcal{L}}_1(\overline{u})(x) = 0, \tag{18.117}$$

respectively. In so doing, Eq. (18.117) is more preferable to solve than Eq. (16.5) (see, e.g., [1603]) because a fictitious body force density in Eq. (18.117) has a compact support $\mathcal{L}_1(\overline{u})(x) \equiv 0$ ($x \notin v_i^l$, $v_i^l = \{x : |x| \leq a + l_\delta\}$). However, the main advantage of the decompositions (18.115) and (18.117) is elimination in (18.117) of the nonlocal volumetric boundary conditions prescribed over a nonzero volume (see, e.g., [396, 771, 996, 1396]), while the solution of Eq. (18.115) can be found more easily.

Thus, the solutions of Eq. (18.117) for the displacements can be considered in the form ($x \in R^d, \quad y \in v_i^l$)

$$u_1(x) := u(x) - \overline{u}(x) = \widehat{\mathcal{L}}_i^{uu}(x - x_i, \overline{u}(y)), \tag{18.118}$$

which can be expressed through the operators $\widehat{\mathcal{L}}$ and $\widehat{\mathcal{L}}_1$:

$$\widehat{\mathcal{L}}_i^{uu}(x - x_i, \overline{u}(y)) = -(\widehat{\mathcal{L}})^{-1}\widehat{\mathcal{L}}_1(\overline{u}(y)). \tag{18.119}$$

Representation (18.118) can be also presented for the compact support $b(x)$ as

$$u_1(x) := u(x) - \overline{u}(x) = \widehat{\mathcal{L}}_i^{ub}(x - x_i, b) := (\widehat{\mathcal{L}}_i)^{-1}\widehat{\mathcal{L}}_{1i}\big(\widehat{\mathcal{L}}^{(0)}\big)^{-1}b(x), \tag{18.120}$$

where $\overline{u}(x)$ is defined by Eq. (18.115$_2$).

Equations (18.115) and (18.117) are presented for the infinite peridynamic bar $w = R$. However, the numerical quadrature methods of this solution imply a finite size of the bar. Let us consider 1D peridynamic bar w of the length $2L + 2l_\delta$ and cross-sectional area $A = 1$ subjected to the self-equilibrium body force with the volumetric boundary condition (18.116) in the interaction remote boundary $w_\Gamma = [-L - l_\delta, -L] \cup [L, L + l_\delta]$. In numerical implementation described in [1401], one considers a discretization into a set of nodes p, each with a finite known volume (called *full volume*) $\bar{v}_p = h$ defined by the size h. For a given distribution x_j, $j = 0, \pm 1, \pm 2, \ldots, \pm(N^L + n_c)$ over the length of $[-L - l_\delta, L + l_\delta]$, we discretize Eq. (16.5) and the boundary conditions (18.116) at the nodes $x_j = jh$ that leads to the linear algebraic system ($i, j \in [-N^L, N^L]$)

$$\widehat{\mathcal{L}}_{ij}^{(0)}\overline{u}_j = -\hat{b}_i \tag{18.121}$$

with the modified body force vector described below. Hereafter, $u_j = u(x_j)$ denotes the nodal displacements, V_{pj} is the portion of node p "volume" covered by the horizon of node j, and a constant step size $h = l_\delta/n_l = L/N^L$ is assumed in the midpoint integration method used. Nonzero elements of the matrix L_{ij} are defined by the micromodulus $C(x_p, x_j)$ (17.1) ($i \in [-N^L, N^L]$, $i + p \in [-N^L, N^L]$)

$$\widehat{\mathcal{L}}_{i,i+p}^{(0)} = C^{(0)}(x_p, x_i)V_{pi} - \sum_{k=i-n_c}^{i+n_c} C^{(0)}(x_k, x_i)\bar{v}_{ki}, \tag{18.122}$$

while the modified body force $\hat{b}_i = \hat{b}(x_i)$ is defined as ($i \in [-N^L, N^L]$)

$$-\hat{b}_i = -b_i + \sum_{k=i-n_c}^{i+n_c} C^{(0)}(x_k, x_i)\overline{u}_k^{b(0)}\bar{v}_{ki}\Big(1 - \chi^{[-L,L]}(x_k)\Big), \tag{18.123}$$

where $\chi^S(x)$ is the indicator function of a segment S (e.g., $S = [-L, L]$). The second summand in the right-hand side of Eq. (18.123) vanishes outside the boundary layer $[-L, -L + l_\delta] \cup [L - l_\delta, L]$ and loading region w^b.

For one inhomogeneity $v = [-a, a]$ in the homogeneous matrix $w \subset R^d$, the corresponding matrix equations can be obtained from Eqs. (18.121) and (18.122) by substitution $\widehat{\mathcal{L}}^{(0)} \to \widehat{\mathcal{L}}, \overline{u} \to u_1$ and $C^{(0)} \to C$, respectively,

$$\sum_j \widehat{\mathcal{L}}_{pq} u_{1|j} = -\breve{b}_i. \tag{18.124}$$

In so doing, the modified body force $\hat{b}$ should be replaced by the fictitious body force $\breve{b}$ $(i, q \in [-N^L, N^L])$

$$-\breve{b}_i = -\sum_{k=i-n_c}^{k=i+n_c} C_1(x_i, x_k)\overline{u}_k \overline{V}_{ik} - \sum_{k=i-n_c}^{k=i+n_c} C^{(0)}(x_i, x_k)$$
$$\times \big({}^L\overline{u}_j^b - {}^L\overline{u}_j^{b(0)}\big)\big(1 - \chi^{[-L,L]}(x_k)\big)\bar{v}_{ik}, \tag{18.125}$$

where $j = \text{sign}(k - i)$; $\overline{u}$ (18.122) is found from Eq. (18.121), while ${}^L\overline{u}_j^b(x)$ (18.110) and ${}^L\overline{u}_j^{b(0)}(x)$ $\equiv$const. (18.110) $(x \in w_{\Gamma u})$ are defined as the locally elastic solutions for the domain w with the inclusion v $({}^L L(x) \not\equiv {}^L L^{(0)}(x))$ analogous to (18.116) and the homogeneous domain $({}^L L(x) \equiv^L L^{(0)}(x))$, respectively.

After estimation of $u(x_i)$ $(-L < x_i < L)$ by the quadrature method (18.124) and (18.125), the peridynamic stress distribution (18.87) can be presented in a similar form $(i \in [-N^L, N^L], k \in [-N^L - n_l, N^L + n_l])$:

$$\sigma(x_i) = \sum_{p=i-n_l}^{i} \sum_{q=i}^{p+n_l} C(x_p, x_q)[\hat{u}(x_q) - \hat{u}(x_p)]\bar{v}_{qi}\bar{v}_{pi}, \tag{18.126}$$

$$\hat{u}(x_k) = \chi^{[-L,L]}(x_k)u(x_k) + (1 - \chi^{[-L,L]}(x_k))\overline{u}(x_k). \tag{18.127}$$

Thus, the stress field inside the segment $x \in w = [-L, L]$ is estimated through both the displacement found in Eqs. (18.121) and(18.124) in this segment $x \in w = [-L, L]$ and the prescribed displacement (18.116) on the volumetric boundary $x \in w_{\Gamma u}$.

The error in the coordinate difference formulae (18.121)–(18.127) is well known to be $O(h^2)$ (see [1401]). The effectiveness of the quadrature scheme (18.121)–(18.127) for the homogeneous infinite bar was demonstrated in by [131] who also proved uniform convergence of the peridynamic solutions to the classical solutions of static and dynamic elasticity problems in 1D case in the limit of the horizon going to zero. A similar accuracy is expected to be accomplished for the numerical schemes considered in Section 18.4.2.

18.4.3 Micromechanical Modeling

The perturbator $\widehat{\mathcal{L}}_k^{ub}(x-x_k, b)$ (18.120) is, in fact, the solutions of the basic problem for one inclusion inside the infinite peridynamic matrix subjected to the body force $b(x)$. Let the dispersion contains identical particles, of a fixed shape v_i, randomly aligned throughout the matrix. We assume that the dispersion is dilute, in the sense that the volume fraction, $c^{(1)}$, of the particles is small and their mutual interaction can be neglected, i.e., each particle can be thus imagined as single, immersed into an unbounded matrix

$$c^{(1)} \ll 1, \quad \overline{u}(x) = \langle u \rangle(x). \tag{18.128}$$

In such a case, the conditional statistical averages of the general integral equation GIE can be recast in terms of the perturbators $\widehat{\mathcal{L}}_j^{ub}(x-x_j, b)$ (18.120)

$$\langle u \rangle(x) = u^{b(0)}(x) + \int \widehat{\mathcal{L}}_j^{ub}(x-x_j, b)\varphi(v_j, x_j)dx_j, \tag{18.129}$$

where the displacement field $u^{b(0)}(x)$ in the infinite homogeneous matrix (with $v = \emptyset$) is

$$u^{b(0)}(x) = -(\widehat{\mathcal{L}}^{(0)})^{-1}b(x). \tag{18.130}$$

Thus, the statistical average of the displacements $\langle u \rangle(x)$ explicitly depends on both the probability density $\varphi(v_j, x_j)$ of the locations of inclusions v_j and the perturbators $\widehat{\mathcal{L}}_j^{ub}(x-x_j, b)$. We recast Eq. (18.129) in the form

$$\langle u \rangle(x) = u^{b(0)}(x) + \Delta\langle u \rangle(x), \tag{18.131}$$

where the summand $\Delta\langle u \rangle(x)$ describes the increment of the statistical average of displacement due to the influence of inclusions.

For the composite material, the body force $b(x)$ produces a stress field

$$\langle \sigma \rangle(x) = \mathcal{L}^{\sigma}(\widehat{\mathcal{L}}^{(0)}\eta^{(0)})(x) + \langle \tau \rangle(x), \tag{18.132}$$

where

$$\eta^{(0)}(z,t) = (\widehat{\mathcal{L}}^{(0)})^{-1}b(z) - (\widehat{\mathcal{L}}^{(0)})^{-1}b(t), \tag{18.133}$$

$$\langle \tau \rangle(x) = \int_{v_i^{l1}(x)} n^{(1)} \int_{\Gamma_q^-} \left[\mathcal{L}^{\sigma}(\langle \eta \rangle)(s)n(s)\right] \otimes s ds dy$$

$$- L^{(0)} \int_{v_i^{l1}(x)} n^{(1)} \int_{\Gamma_q^-} \left\{ [\widehat{\mathcal{L}}^{-1} \hat{\mathcal{L}}_1 - 1](\widehat{\mathcal{L}}^{(0)})^{-1} b(s) \right\} \otimes n(s) ds dy, \tag{18.134}$$

$$\langle \eta \rangle (z,t) = \widehat{\mathcal{L}}^{-1} \widehat{\mathcal{L}}_1 (\hat{\mathcal{L}}^{(0)})^{-1} b(z) - \widehat{\mathcal{L}}^{-1} \widehat{\mathcal{L}}_1 (\hat{\mathcal{L}}^{(0)})^{-1} b(t). \tag{18.135}$$

The first summand in the right-hand side of Eq. (18.133) describes the stress distribution inside a homogeneous bar. The second summand is defined by an influence of the inclusion and vanished at the vanishing of elastic mismatch $\widehat{\mathcal{L}}_1 \to 0$. Analogously Eqs. (18.131) and (18.132) can be recast in the form

$$\langle \sigma \rangle (x) = \sigma^{b(0)} + \Delta \langle \sigma \rangle (x), \tag{18.136}$$

where $\sigma^{b(0)}$ and $\Delta\langle \sigma \rangle (x)$ corresponding to the first and second summands in the right-hand side of Eq. (18.132), respectively, describe both the stress distribution inside homogeneous matrix bar and the stress increment produced by noninteracting inclusions.

18.4.4 Numerical Results

At first, we consider the truncation method for 1D peridynamic bar $\bar{\bar{w}}$ of the length $2L + 2l_c$ and cross-sectional area $A = 1$ containing one inclusion $v_0 = [-a, a]$ and the volumetric boundary w_Γ: $\bar{\bar{w}} = w_\Gamma \cup w$, $w_\Gamma = [-L - l_c, -L] \cup [L, L + l_c]$. The midpoint quadrature scheme (similar to [211] and [212]) is realized for the uniform grid $x_i = (0, \pm 1, \pm 2, \ldots, \pm(N^{\max} + N^{w\Gamma}))$ with the numbers of material points $2N^{\max} + 1$ and $2N^{w\Gamma} + 1$ inside both w and the interaction boundary w_Γ, respectively, and a constant step size $h = L/N^{\max} = a/N^a$, where $2N^a$ is the number of material points in the inclusion. We consider a heterogeneous peridynamic bar with the step-function profiles (16.82_1) ($k = 0, 1$, $E^{(1)}/E^{(0)} = 5$.

At first, we analyze a homogeneous bar with the properties of the matrix $C^{(0)}(\xi)$ subjected to the body force (18.105) with $B = 1$ and $a^b = a$. The displacements $\overline{u}(x)$ (18.121) in the matrix vs. x/a are presented in Fig. 18.17 for $l_\delta/a = 0(1)$, $0.2(2)$, $0.75(3)$, and $1(4)$ and $h = a/200$, $L/a = 10$. It is interesting that estimations of the displacement fields for the cases $h/a = 100$ and $L/a = 4$ lead to correction of the results on 0.1% and 0.2%, respectively. It demonstrated the effectiveness and robustness of the quadrature truncation method proposed in 1D case. The subsequent estimations are performed for $N^a = 200$ and $L/a = 10$. The curves of displacement functions $u(x)$ vs. x/a behave according to the general features of peridynamic solutions for a homogeneous peridynamic bar with a fictitious body force considered in [131] and [1405].

In Fig. 18.18, the displacements $u(x)$ as the functions of the coordinates x/a are presented for the different scale ratios $l_\delta/a = 0,\ 0.5,\ 0.75, 1$ for the inclusion $v_0 = [-a, a]$ with the center coinciding with the origin $x = 0$. As it is expected,

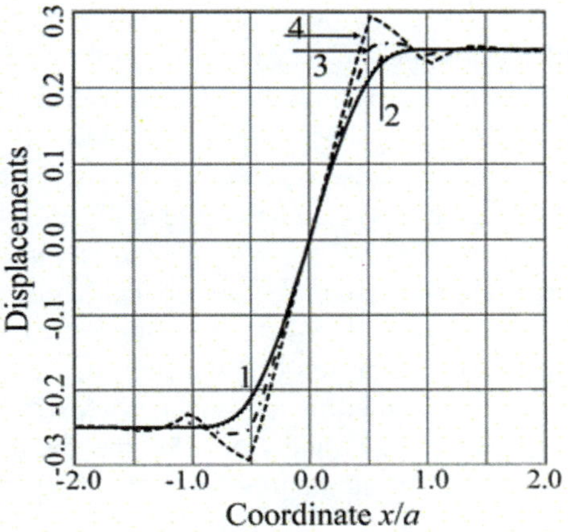

Fig. 18.17 Displacements $\overline{u}(x)$ (18.121) in the matrix vs. x/a for $l_\delta/a = 0(1)$, $0.2(2)$, $0.75(3)$, and $1(4)$

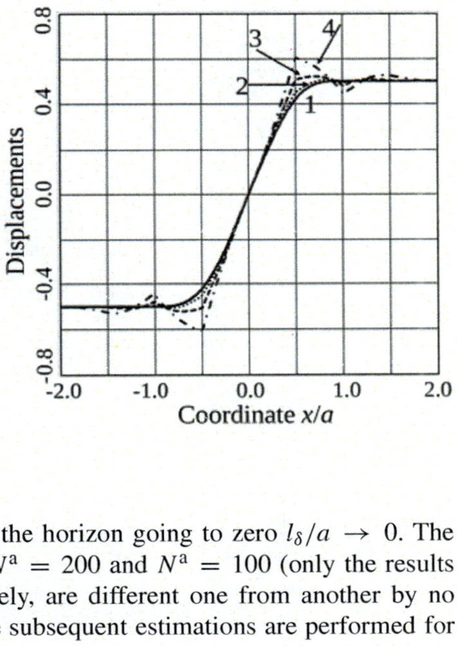

Fig. 18.18 Displacements $u(x)$ vs. x/a for $l_\delta/a = 0(1)$, $0.5(2)$, $0.75(3)$, and $1(4)$

$u(x) \to^L u(x)$ (18.108) in the limit of the horizon going to zero $l_\delta/a \to 0$. The peridynamic curves 1–4 estimated for $N^{\rm a} = 200$ and $N^{\rm a} = 100$ (only the results for $N^{\rm a} = 200$ are depicted), respectively, are different one from another by no more than 0.1%. Due to this reason, the subsequent estimations are performed for $N^a = 300$.

Estimation of the statistical averages of the displacement field $\langle u\rangle(x)$ (18.129) is explicitly expressed through the perturbators $\widehat{\mathcal{L}}_j^{ub}(x - x_j, b)$ (18.120) produced by the inclusion with the center x_i under the action of the body force $b(x)$ with the symmetry center $c^b = 0$. It means a necessity of the estimation of the values $(\widehat{\mathcal{L}}_j)^{-1}\widehat{\mathcal{L}}_{1j}(\widehat{\mathcal{L}}^{(0)})^{-1}$ for all possible locations of the inclusions v_j in Eq. (18.129). However, we will use an equivalent simplification reducing estimations of all sets $(\widehat{\mathcal{L}}_j)^{-1}\widehat{\mathcal{L}}_{1j}(\widehat{\mathcal{L}}^{(0)})^{-1}$ to an evaluation $(\widehat{\mathcal{L}}_j)^{-1}\widehat{\mathcal{L}}_{1j}(\widehat{\mathcal{L}}^{(0)})^{-1}$ for a single inclusion $v_j = v_0 = [-a, a]$ under the effects of a set of the body force $b(x)$ with the moving symmetry centers $c^b = -x_j$. In so doing, the perturbators $\widehat{\mathcal{L}}_j^{ub}(x - x_j, b)$ (18.120) are defined for the fictitious body force (18.125) expressed through both the displacement $\overline{u}$ (18.121) in the homogeneous peridynamic matrix and the locally elastic solutions ${}^L\overline{u}_j^b(x)$ (18.109) and ${}^L\overline{u}_j^{b(0)}(x)$ ≡const. (18.110) in the volumetric boundary $x \in w_{\Gamma u}$ for the domain w with the inclusion v. The curves $u(x)$ vs. x/a are presented in Fig. 18.19 for the fixed inclusion $v_0 = [-a, a]$ with the different symmetry centers $c^b = -3/2a, \ -a, \ 0, \ a, \ 3/2a$ of the body forces $b(x)$ (18.105). As it was mentioned, discontinuity of the elastic properties produces the discontinuities of the displacement fields $u(x)$ at $x = \pm a$ for $c^b/a = -1.5, \ -1.0, 0.0, \ 1.0, \ 1.5$, while the discontinuities of $u(x)$ for $c^b = 0$ are less manifested because in such a case, the body force $b(x)$ vanishes at $|x| \geq a$.

After estimation of $u(x_i)$ $(-N^{\max} < i < N^{\max})$ by the quadrature method (18.124) and (18.125), the peridynamic stress distribution $\sigma(x_i)$ can be found in a similar manner (18.126). The stresses $\sigma(x)$ as the functions of coordinates x/a are presented in Fig. 18.20 for the different scale ratios $l_\delta/a = 0, \ 0.2, \ 0.75, 1$

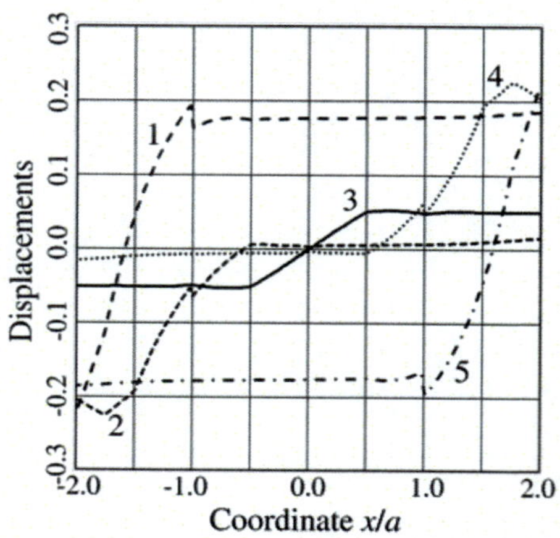

Fig. 18.19 Displacements $u(x)$ vs. x/a for $c^b/a = -1.5$ (1), -1.0 (2), 0.0 (3), 1.0 (4), 1.5 (5)

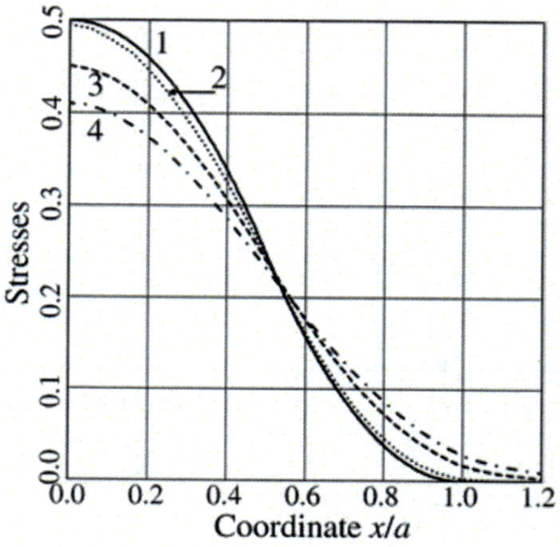

Fig. 18.20 The stresses $\sigma(x)$ vs. x/a for $l_\delta/a = 0$ (1), 0.5 (2), 0.75 (3), and 1 (4)

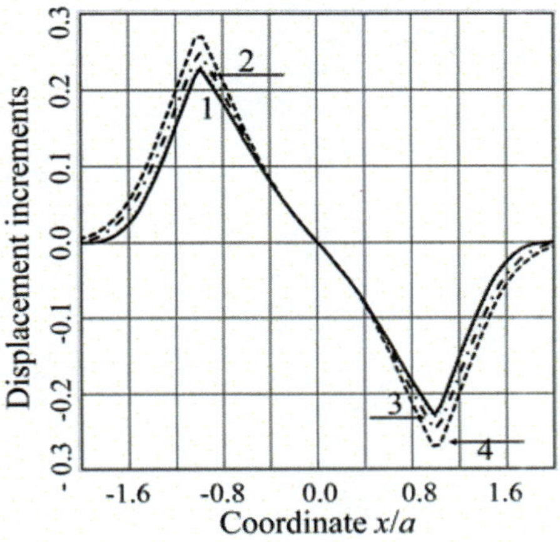

Fig. 18.21 Displacement increments $\Delta u(x)$ (18.131) vs. x/a for $l_\delta/a = 0(1), 0.2(2), 0.75(3), 1(4)$

for the inclusion $v_0 = [-a, a]$ and the body force $b(x)$ with the symmetry center $c^b = 0$. As can be seen, increasing the nonlocality scale l_δ leads to the smoothing of stress distribution fields with increasing stresses outside the inclusion $|x| \geq a$ where the body force vanishes ($b(x) \equiv 0$ at $|x| \geq 0$).

Thus, we obtained all required preliminary solutions for both the homogeneous bar [$u^{b(0)}$ (18.130) and $\sigma^{b(0)}$] and one inclusion in the homogeneous bar (18.124) and (18.126) for the different possible mutual locations of the centers of both the inclusion x_i and body force support domain c^b. It is enough for estimation of the increments of the statistical averages of both the displacement $\Delta\langle u\rangle(x)$ (18.131) and stress $\Delta\langle\sigma\rangle(x)$ (18.136). As can be seen in Fig. 18.21, increasing the nonlocal effects described by the ratio l_δ/a yields an effective stiffening of the composite bar. In so doing, the displacement increments $\Delta\langle u\rangle(x)$ (18.131) vary inside the domain with vanishing body force ($|x| \geq a$) even for locally elastic bar ($l_\delta = 0$; see curve 1 in Fig. 18.21). It means that even for statistically homogeneous locally elastic bar loaded by the inhomogeneous body force, the effective deformation demonstrates the nonlocal effects (as it is expected due to the general theory of micromechanics; see, e.g., [200]). Increasing the nonlocal parameter l_δ/a leads to the extension of domains with nonvanished both the displacement and stress increments (see Figs. 18.21 and 18.22, respectively).

It is known that for both statistically homogeneous composites and homogeneous remote loading, the effective properties of both the peridynamic composites and locally elastic ones are described by constant tensor of local effective moduli. If any from two mentioned conditions is violated, the effective deformations are described by a nonlocal (either the differential or integral) operator depending on the prescribed inhomogeneous body forces. This chapter dedicated to the estimation of effective deformations of 1D statistically homogeneous peridynamic composite bar for the prescribed self-equilibrated body forces has no known locally elastic counterpart. We demonstrated computationally, through 1D static examples, the analytical results presented in this chapter. In particular, we showed numerically

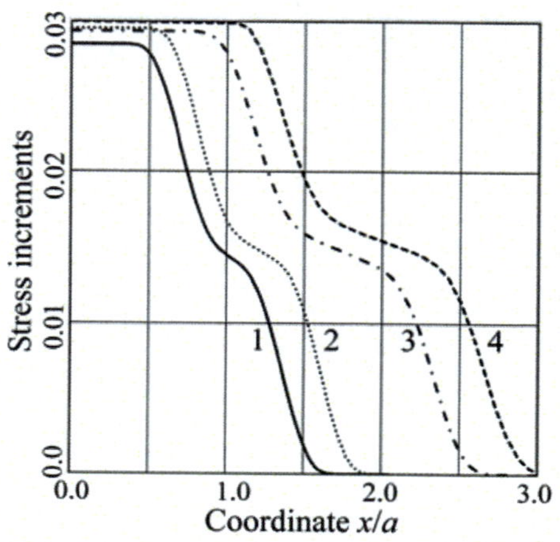

Fig. 18.22 Stress increments $\Delta\sigma(x)$ (18.136) vs. x/a for $l_\delta/a = 0(1),\ 0.2(2),\ 0.75(3),$ and $1(4)$

the convergence of peridynamic models to locally elastic models for deterministic structures. We demonstrated that, in contrast to the locally elastic model, a solution discontinuity may appear at the material interface separating the inclusion and the matrix. A subsequent simplification is described by the dilute approximation method assuming the coincidence of the effective field with the applied field when the interaction of inclusions is negligible. The method is based on the estimation of a perturbator introduced by one inclusion that is, in fact, the solutions of the basic problem for one inclusion inside the infinite peridynamic matrix subjected to the body force. The statistical averages of both the displacement and stresses are estimated by summation of these perturbators for all possible locations of inclusions.

It should be mentioned that the dilute approximation method used in this chapter can be considered as a first-order approximation of the generalized effective field method (EFM) which was already proposed in peridynamic micromechanics for the homogeneous loading (see Sect. 18.3.4). In so doing, a zero-order approximation is exploited for the effective field $\overline{u}(x)$ (see Eq. (18.128)). However, for self-equilibrated body force with a compact support (as, e.g., in Eq. (18.105)), more accurate self-consistent estimation of the effective fields $\overline{u}(x)$ acting on the inclusions should lead to the stabilization of nonvanished displacement increments at the infinity (see, e.g., Figs. 18.17 and 18.18) for the CMs with either locally elastic or peridynamic properties of constituents. A generalization of the EFM to arbitrary inhomogeneous body forces is far beyond the scope of the current study and and can be the subject of further research in this area, but the proposal of the generalized dilute approximation method is a critical first step in understanding the nature of nonlocal effects for CMs.

18.5 The Effective Displacements and Stresses for CMs with Prescribed Body Force

In this section, we consider CM in R^d subjected to self-equilibrated body force density $\mathbf{b}(\mathbf{x})$ with compact support $w^b = \{\mathbf{x} \mid |\mathbf{x}| \leq a^b\}$. Then Eq. (17.185) for the compact support of the body force density can be recast in terms of the effective field analogous to Eq. (18.73) ($\mathbf{x} \in v_i^l$, $\mathbf{y} \in v_q^l$)

$$\langle \overline{\mathbf{u}}^{[n+1]} \rangle_i(\mathbf{x}) = \mathbf{u}^{b(0)}(\mathbf{x}) + \int \mathcal{L}_q^{uu}(\mathbf{x} - \mathbf{x}_q, \langle \overline{\mathbf{u}}^{[n]} \rangle_q(\mathbf{y}))\varphi(v_q, \mathbf{x}_q|; v_i, \mathbf{x}_i)d\mathbf{x}_q, \quad (18.137)$$

with the initial approximation $\langle \overline{\mathbf{u}}^{[0]} \rangle_i(\mathbf{x}) = \mathbf{u}^{b(0)}(\mathbf{x})$ ($\mathbf{x} \in v_i^l$, $\mathbf{x}_i \in R^d$), i.e., the initial approximation coincides with the method of the dilute approximation (similar to 1D Eq. (18.129)). $\mathbf{u}^{b(0)}(\mathbf{x})$ is a displacement field corresponding to the body force density analogous to 1D case (18.130).

The recursive scheme (18.137) forms the Neumann series for the solution ($\mathbf{x} \in v_i^l$)

$$\langle \overline{\mathbf{u}} \rangle_i(\mathbf{x}) := \big\langle \lim_{n\to\infty} \overline{\mathbf{u}}^{[n+1]} \big\rangle_i(\mathbf{x}) = \mathcal{D}_i^u(\mathbf{u}^{b(0)})(\mathbf{x}), \quad (18.138)$$

with the nonlocal operator $\mathcal{D}_i^u$ in the right-hand side.

Equation (18.138) allows us to estimate a statistical average of the displacement field in the macropoint $\mathbf{X}$

$$\langle \mathbf{u} \rangle_i(\mathbf{X}) := c^{l(0)} \langle \mathbf{u} \rangle^{l(0)}(\mathbf{X}) + c^{l(1)} \langle \mathbf{u} \rangle^{l(1)}(\mathbf{X}) \quad (18.139)$$

expressed through the statistical averages of displacements in the point $\mathbf{X}$ in the matrix $\langle \mathbf{u} \rangle^{l(0)}(\mathbf{X})$ and inclusions $\langle \mathbf{u} \rangle^{l(1)}(\mathbf{X})$. Indeed, at first, we built some auxiliary set $v_i^1(\mathbf{X})$ with the indicator function $V_i^1(\mathbf{X})$ and the boundary $\partial v_i^1(\mathbf{X})$ formed by the centers of translated ellipsoids $v_k(\mathbf{0})$ around the fixed point $\mathbf{X}$. We construct $v_i^1(\mathbf{X})$ as a limit $v_{ki}^0 \to v_i^1(\mathbf{X})$ if a fixed ellipsoid v_k is shrinking to the point $\mathbf{X}$. Then

$$\begin{aligned}\langle \mathbf{u} \rangle^{l(0)}(\mathbf{X}) &= \mathbf{u}^{b(0)}(\mathbf{X}) + \int \mathcal{L}_q^{uu}(\mathbf{X} - \mathbf{x}_q, \langle \overline{\mathbf{u}} \rangle_q(\mathbf{y}))\varphi(v_q, \mathbf{x}_q|; \mathbf{X})d\mathbf{x}_q \\ &= \mathbf{u}^{b(0)}(\mathbf{X}) + \int \mathcal{L}_q^{uu}(\mathbf{X} - \mathbf{x}_q, \mathcal{D}_q^u(\mathbf{u}^{b(0)}))\varphi(v_q, \mathbf{x}_q|; \mathbf{X})d\mathbf{x}_q. \quad (18.140)\end{aligned}$$

In a similar manner, $\langle \mathbf{u} \rangle^{l(1)}(\mathbf{X})$ can be estimated as

$$c^{l(1)} \langle \mathbf{u} \rangle^{l(1)}(\mathbf{X}) = c^{l(1)} \mathbf{u}^{b(0)}(\mathbf{X}) + \int_{v_i^1(\mathbf{X})} n^{(1)}(\mathbf{x}_q)\mathcal{L}_q^{uu}(\mathbf{X} - \mathbf{x}_q, \mathcal{D}_q^u(\mathbf{u}^{b(0)}))d\mathbf{x}_q, \quad (18.141)$$

and, therefore, the statistical average of displacements (18.139) is expressed through the body force density

$$\langle \mathbf{u}\rangle(\mathbf{X}) = \mathbf{u}^{b(0)}(\mathbf{X}) + \int \mathcal{L}_q^{uu}(\mathbf{X}-\mathbf{x}_q, \mathcal{D}_q^u(\mathbf{u}^{b(0)}))\varphi(v_q, \mathbf{x}_q|; \mathbf{X})d\mathbf{x}_q$$
$$+ \int_{v_i^1(\mathbf{X})} n^{(1)}(\mathbf{x}_q)\mathcal{L}_q^{uu}(\mathbf{X}-\mathbf{x}_q, \mathcal{D}_q^u(\mathbf{u}^{b(0)}))d\mathbf{x}_q. \tag{18.142}$$

The statistical average of the constitutive law leads to

$$\langle \boldsymbol{\sigma}\rangle(\mathbf{X}) = \mathcal{L}^{\sigma}((\widetilde{\boldsymbol{\tau}}^{(0)}))(\mathbf{X}) + \langle \boldsymbol{\tau}\rangle(\mathbf{X}). \tag{18.143}$$

Here, the statistical average of $\widetilde{\boldsymbol{\tau}}^{(0)}(\mathbf{x}, \hat{\mathbf{x}}) = \mathbf{C}^{(0)}(\mathbf{x}, \hat{\mathbf{x}})\boldsymbol{\eta}(\mathbf{x}, \hat{\mathbf{x}})$ is expressed through the statistical average $\langle \boldsymbol{\eta}(\mathbf{x}, \hat{\mathbf{x}})\rangle = \langle \mathbf{u}\rangle(\hat{\mathbf{x}}) - \langle \mathbf{u}\rangle(\mathbf{x})$ (18.142). The statistical average of the local polarization tensor $\langle \boldsymbol{\tau}\rangle(\mathbf{X})$ is also obtained by averaging over v_i^1

$$\langle \boldsymbol{\tau}\rangle(\mathbf{X}) = \int_{v_i^1(\mathbf{X})} n^{(1)}\mathcal{L}^{\sigma}(\mathbf{C}_1\boldsymbol{\eta}^D)(\mathbf{y}-\mathbf{X})d\mathbf{y}, \tag{18.144}$$

$$\boldsymbol{\eta}^D(\mathbf{x}, \mathbf{y}) = \Big[\mathcal{D}_i^u(\mathbf{u}^{b(0)})(\mathbf{x}) + \mathcal{L}_i^{uu}(\mathbf{x}-\mathbf{x}_i, \mathcal{D}_i^u(\mathbf{u}^{b(0)})(\mathbf{x}-\mathbf{x}_i))$$
$$-\mathcal{D}_i^u(\mathbf{u}^{b(0)})(\mathbf{y}) - \mathcal{L}_i^{uu}(\mathbf{y}-\mathbf{x}_i, \mathcal{D}_i^u(\mathbf{u}^{b(0)})(\mathbf{y}-\mathbf{x}_i))\Big]. \tag{18.145}$$

18.6 Concluding Remarks

For the media subjected to remote homogeneous volumetric boundary loading, one proved that the effective behavior of this media is governed by conventional effective constitutive equation as for the local elasticity. It was made by the most exploitation of the popular tools and concepts used in conventional elasticity of CMs and adapted to peridynamics. Effective moduli are expressed through the introduced local stress polarization tensor averaged over the extended inclusion phase rather than in an entire space. Classical methods of locally elastic micromechanics: dilute approximation, Mori-Tanaka approach (MTA), and the effective field method (EFM). In the generalized EFM proposed, the effective field is evaluated from self-consistent estimations by the use of closing a corresponding new general integral equation (17.205) in the framework of the quasicrystalline approximation. One demonstrates some similarity and difference between the mentioned methods. It should be mentioned that the effective modulus estimations in all three methods are obtained by the use of averaging of the displacement concentration factors (e.g., $\mathbf{A}_i^{1\omega}$) (18.54) over the external interaction interface of inclusions, respectively. New external interaction interface averaging operation (18.54) and (18.71) was introduced as a straightforward generalization of interface integral technique proposed in Chap. 11) in locally elastic micromechanics. In such a case, any spatial derivatives of displacement fields are not required. Elimination of the requirement mentioned is critical because the displacement fields $\mathbf{u}(\mathbf{x})$ ($\mathbf{x} \in v^l$) are not differentiable inside the extended inclusions.

It was established formal similarity of the operator forms of GIEs for both locally elastic CMs and nonlocal ones that opens the opportunities to straightforward generalization of their solutions for locally elastic CMs (see Chap. 9) to their nonlocal counterparts. The methods proposed in this chapter allow us to completely abandon the hypotheses (as in Chap. 9) **H1a**, **H1b**, and **H3**, while the hypothesis **H2** can be used for multiparticle generality. The significance of the current study is beyond the scope of just a proposal of the generalized EFM. Of more importance is a development of a tool which can be easily modified and implemented for a wide class of problems of peridynamic micromechanics, e.g., effective deformation of heterogeneous media subjected to inhomogeneous body forces (see Sect. 18.4), state-based peridynamics, coupled problems, and nonlinear problems. However, consideration of these problems goes far beyond the scope of the current publication.

Chapter 19
Computational Homogenization in Linear Peridynamic Micromechanics of Periodic Structure CMs

Abstract In this chapter, new volumetric periodic boundary conditions (PBCs) are introduced at the interaction boundary of a representative unit cell (UC). A generalization of Hill's equality to peridynamic composites is proved. The discretization of the equilibrium equation acts as a macro-to-micro transition of the deformation-driven type, where the overall deformation is controlled. Determination of the microstructural displacements allows one to estimate the peridynamic traction at the geometric UC's boundary, which is exploited for the estimation of the macroscopic stresses and the effective moduli. Effective moduli of peridynamic medium with periodically distributed damage are estimated. The volumetric Bloch-Floquet periodic boundary condition is proposed. The scheme of data-driven learning of nonlocal models is presented.

Estimation of the macroscopic effective response of heterogeneous media (with either random or periodic structures) in an averaged (or homogenized) meaning in terms of the mechanical and geometrical properties of constituents is a central focus of micromechanics denoted as *micro-to-macro modeling*. Traditionally, these methods used at the locally elastic analyses of composite materials (CMs) with either the random structure or periodic ones are conceptually different. So, for locally elastic random structure CMs, a number of micromechanical models, inspired by Eshelby [449], were proposed in the literature for describing the thermoelastic behavior of composites with ellipsoidal inclusions (see for references, e.g., Chaps. 8–14). The peridynamic counterpart of these researches is presented in Chaps. 17 and 18. The locally elastic micromechanics of CMs with fine periodic microstructure was considered in Sect. 7.4. The research in homogenization of peridynamic periodic structure CMs is less well developed, and just a few discussions in this area are published. So, Alali and Lipton [18] constructed two-scale solution expansion (with neither numerical results nor definition of the effective moduli) splitting into a microscopic component tracking the dynamics at the length scale of the heterogeneities and a macroscopic component tracking the volume averaged (homogenized) dynamics. By considering the micromechanics of a layered composite under uniaxial stress, Silling [1399] demonstrated that nonuniformity

V. A. Buryachenko, *Local and Nonlocal Micromechanics of Heterogeneous Materials*, https://doi.org/10.1007/978-3-030-81784-8_19

of the macroscopic strain field leads to nonlocality in a homogenized model. The peridynamic counterpart of the computational homogenization direction is started by Madenci with coauthors [998] (see also [381, 382, 1644]) who proposed the peridynamic unit cell model (subjected to the classical periodic boundary conditions) for prediction of the effective properties. The most important feature of the computational homogenization is a description of the periodic boundary conditions at the unit cell. The current work is dedicated to systematic generalization of classical locally elastic version of the *computational homogenization* to their peridynamic counterpart. We refer to the boundary of the UC separating the adjoining UCs one from another as the geometric boundary of the UC. For the locally elastic case, the governing equation exploiting partial differential form is added by specific boundary conditions at the geometric boundary of the UC such as periodic displacement and antiperiodic traction. For the peridynamic linear bond-based model being considered and utilizing the integral governing equation, the material points in the vicinity of the geometric boundary of the UC directly interact within the horizon. It defines a volumetric region (called an interaction boundary of the UC) surrounding the geometric boundary and limited by the surfaces at the horizon distance from the geometric boundary of the UC. Introduction of the volumetric periodic boundary conditions allows us to generalize the classical computational homogenization approach to their peridynamic version. We focus on the "deformation driven" procedure formulated at the macroscopic level as follows: given a macroscopic strain, determine the macroscopic stress and the constitutive tangent, based on the response of the underlying microstructure. A "stress driven" model (given a macroscopic stress, estimate the strain) is also possible. However, such a procedure does not directly fit into the standard displacement-based finite element framework usually employed in both local elasticity and peridynamics.

19.1 Periodic and Volumetric Periodic Boundary Conditions

Let a linear elastic body occupy an open simply connected bounded domain $w \subset R^d$ with a smooth boundary Γ_w and with an indicator function W and space dimensionality d ($d = 1$, $d = 2$, and $d = 3$ for $1D$, $2D$, and $3D$ problems, respectively). The domain w is constructed from a periodic field of the cubic unit cells (UCs) Ω_α: $w = \cup\Omega_\alpha$, where each UC with the side length $2l^\Omega$ is labeled by the multiplet of integer numbers $\boldsymbol{\alpha} = (\alpha_1, \ldots, \alpha_d) \in Z^+$, here $\alpha_i = 0, \pm 1$ ($i = 1, \ldots, d$). The centers of UCs are presented as $\mathbf{x}_\alpha^\Omega = 2l^\Omega\alpha_1\mathbf{e}_1 + \ldots + 2l^\Omega\alpha_d\mathbf{e}_d$ in the Cartesian coordinate frame $\mathbf{e}_1, \ldots, \mathbf{e}_d$. Each identical unit cell contains a homogeneous matrix $v^{(0)}$ and field $X = (v_i)$ of heterogeneities v_i with indicator functions V_i and bounded by the closed smooth surfaces Γ_i ($i = 1, 2, \ldots$) defined by the relations $\Gamma_i(\mathbf{x}) = 0$ ($\mathbf{x} \in \Gamma_i$), $\Gamma_i(\mathbf{x}) > 0$ ($\mathbf{x} \in v_i$), and $\Gamma_i(\mathbf{x}) < 0$ ($\mathbf{x} \notin v_i$). In the context of the principle of separation of scales assumed, the microscopic length scale $|\Omega_\alpha|$ is much smaller than the characteristic size of the macroscopic sample $|w|$.

19.1.1 Locally Elastic Unit Cell with Periodic Boundary Condition

We first consider the local basic equations of elastostatics of composites for the stresses $\boldsymbol{\sigma}$, strains $\boldsymbol{\varepsilon}$, and displacements $\mathbf{u}$ (6.1)–(6.3). Substitution of Eqs. (6.2) and (6.3_1) into Eq. (6.1) yields a representation of the equilibrium equation (6.1) in the form

$$ {}^L\widehat{\mathcal{L}}(\mathbf{u})(\mathbf{x}) + \mathbf{b}(\mathbf{x}) = \mathbf{0}, \quad {}^L\widehat{\mathcal{L}}(\mathbf{u})(\mathbf{x}) := \nabla[\mathbf{L}\nabla\mathbf{u}(\mathbf{x})], \tag{19.1} $$

where ${}^L\widehat{\mathcal{L}}(\mathbf{u})(\mathbf{x})$ is an elliptic differential operator of the second order. The ideal boundary conditions at the interface boundaries will be considered together with the homogeneous boundary conditions on Γ_w with the unit outward normal $\mathbf{n}^{\Gamma_w}$

$$ \mathbf{u}^{w\Gamma}(\mathbf{x}) = \boldsymbol{\varepsilon}^{w\Gamma}\mathbf{x}, \ \boldsymbol{\varepsilon}^{w\Gamma} \equiv \text{const.}, \ \mathbf{x} \in \Gamma_w, \tag{19.2} $$

$$ \mathbf{t}^{w\Gamma}(\mathbf{x}) = \boldsymbol{\sigma}^{w\Gamma}\mathbf{n}^{w\Gamma}(\mathbf{x}), \ \ \boldsymbol{\sigma}^{w\Gamma} = \text{const.}, \ \mathbf{x} \in \Gamma_w, \tag{19.3} $$

where $\boldsymbol{\varepsilon}^{w\Gamma}(\mathbf{x}) = \frac{1}{2}\big[\nabla\mathbf{u}^{w\Gamma}(\mathbf{x}) + (\nabla\mathbf{u}^{w\Gamma}(\mathbf{x}))^\top\big]$, $\mathbf{x} \in \Gamma_w$, and $\boldsymbol{\sigma}^{w\Gamma}$ are the given constant symmetric tensors of the macroscopic strain and stress, respectively. We will consider the interior problem when the body occupies the interior domain with respect to the boundary Γ_w. However, the traction boundary conditions (19.3) are not appropriate in the deformation-driven procedure analyzed in this chapter, and they will not be considered in more detail.

For simplicity of notations, we will consider 2D cases $w = \cup\Omega_{ij}$ $(i, j = 0, \pm1, \pm2, \ldots)$ with the square unit cells Ω_{ij}. Let a representative unit cell Ω_{00} with the corner points $\mathbf{x}^c_{ij}$ $(i, j = \pm1)$ has the boundary $\Gamma^0 = \cup\Gamma^0_{ij}$ where the boundary partition $\mathbf{x}^0_{ij} \in \Gamma^0_{ij}$ separates the UCs Ω_{00} and Ω_{ij} $(i = 0, \pm1,\ j = \pm(1 - |i|)$; see Fig. 19.1). A representative volume element (or the unit cell (UC)) Ω_{00} is deformed in a repetitive way identical to its neighbors. Compatibility of deformation of two adjacent UCs demands compatibility of two pairs of the opposite boundaries implying both identical shape of both boundaries and opposite sign of tractions acting on those boundaries. For visualization of the corresponding periodic boundary conditions (PBCs), an alternative *toroidal edge correction* (see Sect. 5.4.2 and [1297]) is often used in which each 2D square box Ω_{00} can be regarded as a torus, so that points on opposite edges are considered to be closed (a 3D case is analyzed in a similar manner). Then Ω_{00} can be considered to be part of a grid of identical boxes, forming a border around the pattern inside Ω_{00}. Deformation of the macrodomain w subjected to the macroscopic homogeneous boundary conditions (19.2) is reduced to the deformation of the unit cell Ω_{00} subjected to the PBC

$$ \mathbf{u}(\mathbf{x}^0_{ij}) = \mathbf{u}(\mathbf{x}^0_{kl}) + \boldsymbol{\varepsilon}^{w\Gamma}(\mathbf{x}^0_{ij} - \mathbf{x}^0_{kl}), \ \ \mathbf{t}(\mathbf{x}^0_{ij}) = -\mathbf{t}(\mathbf{x}^0_{kl}) \tag{19.4} $$

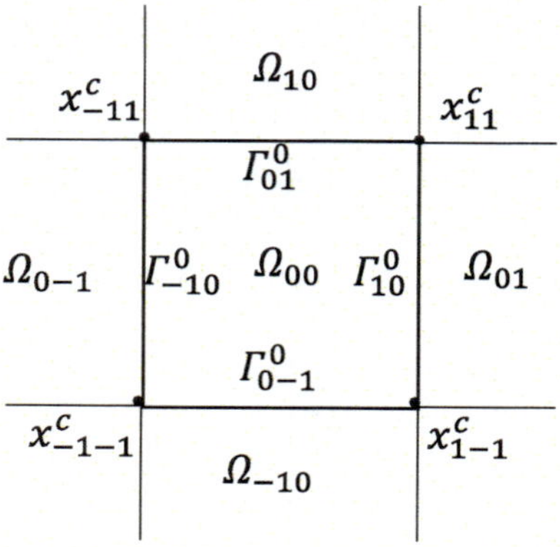

Fig. 19.1 A unit cell Ω_{00} with the boundary Γ^0_{ij} and the surrounding UCs Ω_{ij}

representing periodic displacement (19.4$_1$) and antiperiodic tractions $\mathbf{t}(\mathbf{x}^0_{mn}) = \boldsymbol{\sigma}(\mathbf{x}^0_{mn})\mathbf{n}(\mathbf{x}^0_{mn})$ (either $(mn) = (ij)$ or $(mn) = (kl)$ (19.4$_2$), respectively, on the boundaries $\mathbf{x}^0_{ij} \in \Gamma^0_{ij}$ and $\mathbf{x}^0_{kl} \in \Gamma^0_{kl}$ $(i = 0, \pm 1,\ j = \pm(1 - |i|),\ k = -i,\ l = -j)$ with the outward normal unit vectors $\mathbf{n}(\mathbf{x}^0_{ij}) = -\mathbf{n}(\mathbf{x}^0_{kl})$, respectively. The position vectors $\mathbf{x}^0_{ij} \in \Gamma^0_{ij}$ or $\mathbf{x}^0_{kl} \in \Gamma^0_{kl}$ are expressed through the position vectors of the corner points $\mathbf{x}^c_{mn}$ $(i = 0, \pm 1,\ j = \pm(1 - |i|),\ k = -i,\ l = -j,\ m, n = \pm 1)$

$$\mathbf{x}^0_{ij} = \mathbf{x}^0_{kl} + \mathbf{x}^c_{1,-1} - \mathbf{x}^c_{-1,-1}, \quad \mathbf{x}^0_{ij} = \mathbf{x}^0_{kl} + \mathbf{x}^c_{-1,1} - \mathbf{x}^c_{-1,-1} \tag{19.5}$$

for the $i = 1,\ j = 0$ and $i = 0,\ j = 1$, respectively, of the boundary Γ^0.

19.1.2 *Volumetric Boundary Conditions in Peridynamic Micromechanics*

One treats three material length scales (see, e.g., [1523]): the macroscopic scale L, characterizing the extent of w, the microscopic scale l^{Ω}, related with a size of a unit cell Ω^{ϵ}, and the horizon l_{δ}. Moreover, one supposes that applied field varies on a characteristic length scale Λ. The limit of our interests for both the material scales and field ones is

$$L \gg \Lambda \gg l^{\Omega} \geq l_{\delta}. \tag{19.6}$$

In locally elastic micromechanics, the scale separation hypothesis (13.1) is the basis hypothesis of the first-order computational homogenization technique initially restricted to small deformations and/or linear material behavior, although these methods are fully general and have been elaborated over time for large deformations with arbitrary nonlinear material behavior at the fine scale (see for references and details [535, 798, 1030, 1056, 1467, 1502]). The current work is dedicated to generalization of the mentioned first-order computational homogenization technique to their peridynamic counterpart.

Due to nonlocality, the equilibrium equation (16.5) should be accompanied by a "boundary" condition, imposed as a volumetric constraint in the so-called interaction domain w_Γ (opposite to the local elasticity case where the boundary conditions are imposed precisely at the bounding surface $\Gamma^{(0)}$) (see for details [1396]); i.e., the nonlocal boundary w_Γ is a d-dimensional region unlike its $(d-1)$-dimensional counterpart Γ^0 in local problems. We consider homogeneous volumetric boundary conditions (16.52) or (16.53). As for local elasticity, it is assumed that the representative unit cell $\mathbf{x} \in \Omega_{00}$ inside the macrodomain w subjected to the homogeneous displacement loading at $w_\Gamma(\mathbf{y})$ (19.6) ($|\mathbf{x}-\mathbf{y}| \gg l^\Omega$) deforms in a repetitive way identical to its neighbors. Pure geometrical part of this problem is similar to the problem of random packing of spheres in the periodic unit cells that has been studied very extensively due to their technological importance, and their capability of modeling the simple liquid, concentrated suspensions, amorphous and powder materials (see for references and details Sect. 5.4 and [237]). Deformation of a UC is equivalent to the deformation of an isolated unit cell subjected to some volumetric periodic boundary condition (PBC) following from field compatibility demands with respect to the opposite edges that will be performed by the *volumetric toroidal edge correction* (see Fig. 19.2). Namely, if for the field point $\mathbf{x} \in \Omega_{00}$ some source point $\mathbf{y} = \mathbf{x} + \xi$ ($|\xi| \leq l_\delta$) leaves the central simulation box Ω_{00}, then one of its image particles enters the central box Ω_{00} from the opposite direction. Each of the image particles in the neighboring boxes (the numbers of these neighboring boxes equal to 8 and 26 in $2D$ and $3D$ cases, respectively) moves in exactly the same way. Precisely this method will be explored hereafter in this chapter for the elimination of the boundary effect by the use of volumetric boundary conditions. So, we start with a cube $\Omega_{00} = \{[-l^\Omega, l^\Omega]^d\}$ with the center $\mathbf{0} \in R^d$ and its periodically located neighbors of Ω_{00} labeled by the multiplet of integer numbers $\boldsymbol{\alpha} = (\alpha_1, \ldots, \alpha_d) \in Z^+$; here, $\alpha_i = 0, \pm 1$ ($i = 1, \ldots, d$). N field points $\mathbf{x}_i$ ($i = 1, \ldots, N$) in the central cube Ω_{00} are periodically reflected as $\mathbf{x}_i^\alpha$ into the neighboring cubes Ω_α. If the source point $\mathbf{x}+\xi \in \Omega_\alpha$, then the image point (interacting with the point $\mathbf{x}_i$) is $\mathbf{x}+\xi-2\boldsymbol{\alpha}^l \in \Omega_{00}$, where $\boldsymbol{\alpha}^l := l^\Omega \boldsymbol{\alpha}$. Volumetric toroidal edge correction allows us to analyze all N particles in the central box Ω_{00} without using the source points in the neighboring cubes Ω_α ($|\boldsymbol{\alpha}| \neq 0$).

In analogy with the volumetric boundary domain w_Γ, we introduce a volumetric unit cell boundary $\Omega_\Gamma = \Gamma^+ \cup \Gamma^-$, where Γ^+ and Γ^- are the boundary layers (internal and external, respectively) separated by the geometric boundary Γ^0 and having a thickness given by the horizon l_δ (the UC Ω_{00} with some parts of the internal and external volumetric boundaries is presented in Fig. 19.2). For simplicity, we construct Γ^+ and Γ^- for 2D case for the square Ω_{00} by the straight line segments (or the plates for 3D) which are parallel to the corresponding side of Ω_{00}. Moreover, in analogy with the parts Γ_α^0 of the UC boundary Γ^0, we introduce the partitions of both the internal Γ^+ and external Γ^- volumetric boundaries. The

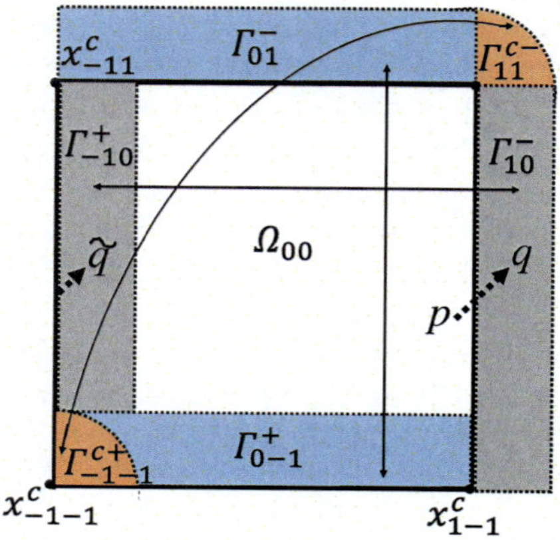

Fig. 19.2 Unit cell Ω_{00} with the volumetric periodic boundaries Γ^+_{ij} and Γ^-_{ij}

external volumetric boundary is decomposed as $\Gamma^- = \cup_\alpha \Gamma^-_\alpha \cup_\alpha \Gamma^{c-}_\alpha$, where Γ^-_α is rectangular with the sides Γ^0_α and l_δ, while Γ^{c-}_α are the quarters of the circle attached to the corner points $\mathbf{x}^c_\alpha$. For a general form of the UC Ω_{00}, the external volumetric boundary Γ^- can be expressed through the Minkowski dilation $\Gamma^- = \{\Omega_{00} \oplus \mathcal{H}_\mathbf{0}\} \backslash \Omega_{00}$ with the partitions $\Gamma^-_\alpha = \Gamma^- \cap \Omega_\alpha$ (if $|\alpha_1|+\ldots+|\alpha_d| = d-1$) and $\Gamma^{c-}_\alpha = \Gamma^- \cap \Omega_\alpha$ (if $|\alpha_1| + \ldots + |\alpha_d| = d$); for $d = 3$ and $|\alpha_1| + |\alpha_2| + |\alpha_3| = 1$, the case $\Gamma^- \cap \Omega_\alpha$ corresponds to the edges. The internal volumetric boundary is decomposed as $\Gamma^+ = \cup_\alpha \Gamma^+_\alpha$, where Γ^+_α is rectangular with the sides Γ^0_α and l_δ. Intersections of Γ^+_α with each other include the quarters of a circle Γ^{c+}_β attached to the corner points $\mathbf{x}^c_\beta$ and obtained by parallel transition of the corresponding parts Γ^{c-}_α ($\boldsymbol{\beta} = -\boldsymbol{\alpha}$). The parts Γ^+_{01}, Γ^-_{10}, and Γ^{c-}_{11} and their imaged parts Γ^-_{0-1}, Γ^+_{-10}, and Γ^{c+}_{-1-1} are shaded in Fig. 19.2. According to the equilibrium equation (16.5), the displacement field $\mathbf{u}(\mathbf{x})$ ($\mathbf{x} \in \Omega_{00}$) depends only on the fields $\mathbf{u}(\hat{\mathbf{x}})$ of the family $\hat{\mathbf{x}} \in \mathcal{H}_\mathbf{x}$. For $\mathbf{x} \in \Omega_{00}$ and $\mathcal{H}_\mathbf{x} \subset \Omega_{00}$ (i.e., $\mathbf{x} \notin \Gamma^+$), we do not need to take interaction of the opposite volumetric boundaries into account. However, for $\mathbf{x} \in \Gamma^+$, some points of the family $\mathbf{y} \in \mathcal{H}_\mathbf{x}$ do not fall into the UC $\mathbf{y} \notin \Omega_{00}$ that required the establishment of the tying constraints linking both the position points and the fields of the opposite volumetric boundaries Γ^+ and Γ^-. Let us for definiteness, the collocation point $\mathbf{x}_p$ belongs to the right internal volumetric boundary $\mathbf{x}_p \in \Gamma^+_{10}$ (for 2D case, see Fig. 19.2) and the source point $\mathbf{x}_q = \mathbf{x}_p + \boldsymbol{\xi} \in \Gamma^-_{10}$ with $x_{p|1} < x_{q|1} < l^\Omega + l_\delta$ and $-l^\Omega + l_\delta < x_{q|2} < l^\Omega - l_\delta$. Then the imaging point $\mathbf{x}_{\tilde{q}} = (\mathbf{x}_{\tilde{q}|1}, \mathbf{x}_{\tilde{q}|2})$ of the node q has the coordinates $x_{\tilde{q}|1} = x_{q|1} - 2l^\Omega$ and $x_{\tilde{q}|2} = x_{q|2}$. If the source point $\mathbf{x}_q$ belongs to the external volumetric boundary $\mathbf{x} \in \Gamma^{c-}_{11}$ (a quarter of circle), then the imaged point $\mathbf{x}_{\tilde{q}} \in \Gamma^{c+}_{-1-1}$ has the coordinates $x_{\tilde{q}|1} = x_{q|1} - 2l^\Omega$ and $x_{\tilde{q}|2} = x_{q|2} - 2l^\Omega$. Other correspondence between the material points in Γ^+ and Γ^- can be considered in a similar manner. In general, the nodes q of the same family $q \in \mathcal{H}_p$ of p can belong to the different adjusted parts of the external volumetric boundary (e.g., $\mathbf{x}_q \in \Gamma^-_\alpha$, Γ^{c-}_α), while the imaged points

$\mathbf{x}_{\tilde{q}} \in \Gamma_{\beta}^{+}$, Γ_{β}^{c+} ($\boldsymbol{\beta} := -\boldsymbol{\alpha}$) can belong to the different nonoverlapping families. Each source point $\mathbf{x}_q$ belonging to the external volumetric boundary $\mathbf{x}_q \in \Gamma^-$ has only one imaged point $\mathbf{x}_{\tilde{q}} \in \Gamma^+$, while some imaged points $\mathbf{x}_{\tilde{q}} \in \Gamma^+$ can have either $1, \ldots, d$ or $d+1$ corresponding source points $\mathbf{x}_q$; for example, for $d = 2$ (see Fig. 19.2), any imaged point $\mathbf{x}_{\tilde{q}} \in \Gamma_{-1-1}^{c+}$ corresponds to three source points $\mathbf{x}_q$ belonging to Γ_{11}^{c-}, Γ_{10}^{-}, and Γ_{01}^{-}. In so doing, the micromodulus $\mathbf{C}(\mathbf{x}_p, \mathbf{x}_q)$ is estimated at the initial collocation p and source $q \in \mathcal{H}_p$ nodes even for the points $\mathbf{x}_q \in \Gamma^-$.

Thus, if the source point $\mathbf{x}_p + \boldsymbol{\xi} \in \Omega_\alpha$ ($\boldsymbol{\xi} \in \mathcal{H}_p$), then the peridynamic counterpart of the locally elastic PBC (19.4) called *new volumetric periodic boundary conditions* (PBCs) represents periodic displacements and antiperiodic tractions

$$\mathbf{u}(\mathbf{x}_p + \boldsymbol{\xi}) = \mathbf{u}(\mathbf{x}_p + \boldsymbol{\xi} - 2\boldsymbol{\alpha}^l) + 2\boldsymbol{\varepsilon}^{w\Gamma}\boldsymbol{\alpha}^l, \tag{19.7}$$

$$\mathbf{t}(\mathbf{x}_\alpha^0) = -\mathbf{t}(\mathbf{x}_{\boldsymbol{\beta}}^0), \tag{19.8}$$

respectively, where $\boldsymbol{\beta} = -\boldsymbol{\alpha}$ and $\mathbf{x}_{\boldsymbol{\alpha}}^0$ are defined analogously to Eq. (19.5).

For the homogeneous displacement loading conditions (16.5), the spatial average estimated over both the sufficiently large macrosample w and the individual constitutive $v^{(k)} \subset w$ ($k = 0, 1, \ldots, N$; $\mathbf{f} = \boldsymbol{\sigma}, \boldsymbol{\varepsilon}$, $|w| = \text{mes}\, w$)

$$\{\mathbf{f}\} = \lim_{w \uparrow R^d} |w|^{-1} \int_w \mathbf{f}(\mathbf{x}) d\mathbf{x}, \quad \{\mathbf{f}\}^{(k)} = \lim_{w \uparrow R^d} |v^{(k)}|^{-1} \int_w \mathbf{f}(\mathbf{x}) V^{(k)} d\mathbf{x}, \tag{19.9}$$

coincides with the corresponding averages over the UC

$$\{\mathbf{f}\} = \langle \mathbf{f} \rangle^{\Omega} := |\Omega_{00}|^{-1} \int_{\Omega_{00}} \mathbf{f}(\mathbf{x}) d\mathbf{x},$$

$$\{\mathbf{f}\}^{(k)} = \langle \mathbf{f} \rangle^{\Omega(k)} := |v^{(k)}|^{-1} \int_{\Omega_{00}} \mathbf{f}(\mathbf{x}) V^{(k)} d\mathbf{x}. \tag{19.10}$$

Hereafter, $n^{(k)}$ is a number density of component $v^{(k)} \ni v_i$ and $c^{(k)}$ is the concentration, i.e., volume fraction, of the component $v_i \in v^{(k)}$: $c^{(k)} = \{V^{(k)}\} = \overline{v}_i n^{(k)}$, $\overline{v}_i = \text{mes} v_i$ ($k = 1, 2, \ldots, N$; $i = 1, 2, \ldots$), $c^{(0)} = 1 - \{V\}$.

19.2 General Representations for Effective Moduli

19.2.1 Generalization of Hill's Equality

The goal of the current consideration is to obtain some asymptotic representations for the volume averages $\{.\}$ at the condition $L/a \to \infty$. For the volumetric homogeneous boundary conditions either (16.4) or (16.5), we define the *mean value* (or volume average) of an integrable tensor field $\boldsymbol{\zeta}$ (e.g., $\boldsymbol{\zeta} = \boldsymbol{\varepsilon}, \boldsymbol{\sigma}, \boldsymbol{\sigma} : \boldsymbol{\varepsilon}$) given on the macrodomain $\overline{\overline{w}} \subset R^d$ $(d = 2, 3)$ as the averages over both the domain $\overline{\overline{w}}$ and the UC Ω_{00}

$$\{\boldsymbol{\zeta}\} = |\overline{\overline{w}}|^{-1} \int_{\overline{\overline{w}}} \boldsymbol{\zeta}(\mathbf{x})\, d\mathbf{x} = |\Omega_{00}|^{-1} \int_{\Omega_{00}} \boldsymbol{\zeta}(\mathbf{x})\, d\mathbf{x} := \langle \boldsymbol{\zeta} \rangle^{\Omega}, \tag{19.11}$$

where $|\overline{\overline{w}}| := \text{mes}\, \overline{\overline{w}}$ and the domain w in Eq. (19.9) is replaced by $\overline{\overline{w}}$. Then, according to the Gauss theorem and Cauchy relations, we can express the volume averages of stress and strain by integrals over the respective boundary values on w_Γ and Γ^0 $(\mathbf{b}(\mathbf{y}) \equiv \mathbf{0}, \mathbf{y} \in \overline{\overline{w}})$

$$\{\sigma_{ij}\} = |\overline{\overline{w}}|^{-1} \int_{\overline{\overline{w}}} \sigma_{ik} x_{k,j}\, d\mathbf{x} = |\overline{\overline{w}}|^{-1} \int_{\Gamma_{l_\delta}} t_{(i}(\mathbf{s}) s_{j)}\, d\mathbf{s},$$

$$\{\varepsilon_{ij}\} = |\overline{\overline{w}}|^{-1} \int_{\Gamma_{l_\delta}} u_{(i}(\mathbf{s}) n_{j)}(\mathbf{s})\, d\mathbf{s}, \tag{19.12}$$

$$\{\sigma_{ij}\} = \langle \sigma_{ij} \rangle^{\Omega} = |\Omega_{00}|^{-1} \int_{\Gamma^0} t_{(i}(\mathbf{s}) s_{j)} d\mathbf{s},$$

$$\{\varepsilon_{ij}\} = \langle \varepsilon_{ij} \rangle^{\Omega} = |\Omega_{00}|^{-1} \int_{\Gamma^0} u_{(i}(\mathbf{s}) n_{j)}(\mathbf{s}) d\mathbf{s}, \tag{19.13}$$

respectively.

In the particular form of the homogeneous boundary conditions (16.52), Eq. (19.13) with $\nabla \mathbf{x} = \mathbf{x}\nabla = \boldsymbol{\delta}$ leads to the conclusion that for any kinematically admissible strain field corresponding to the homogeneous displacement loading conditions (16.52) $\boldsymbol{\varepsilon}^{w_\Gamma} \equiv$ const., the mean value $\{\boldsymbol{\varepsilon}\}$ of $\boldsymbol{\varepsilon}$ is just $\boldsymbol{\varepsilon}^{w_\Gamma}$:

$$\{\varepsilon_{ij}\} = |\overline{\overline{w}}|^{-1} \int_{\Gamma_{l_\delta}} \varepsilon^{w_\Gamma}_{(ik} s_k n_{j)}(\mathbf{s})\, d\mathbf{s} = \varepsilon^{w_\Gamma}_{ij} = \text{const.} \tag{19.14}$$

because in Eq. (19.13) $\mathbf{u}(\mathbf{x}) = \boldsymbol{\varepsilon}^{w_\Gamma}\mathbf{x}$ $(\mathbf{x} \in \Gamma_{l_\delta} \subset w_\Gamma)$. In a similar manner, the homogeneous force loading condition (16.53) incorporated into Eq. (19.12) leads to the equality for the statically admissible stress field corresponding to $\boldsymbol{\sigma}^{w_\Gamma}(\mathbf{x})$:

$$\{\sigma_{ij}\} = |\overline{\overline{w}}|^{-1} \int_{\Gamma_{l_\delta}} \sigma^{w_\Gamma}_{(ik} n_k(\mathbf{s}) s_{j)}\, d\mathbf{s} = \sigma^{w_\Gamma}_{ij} = \text{const.} \tag{19.15}$$

Thus, the mean values of the strain $\{\boldsymbol{\varepsilon}\}$ and stress $\{\boldsymbol{\sigma}\}$ coincide with $\boldsymbol{\varepsilon}^{w_\Gamma}$ and $\boldsymbol{\sigma}^{w_\Gamma}$ defined by the homogeneous volumetric boundary conditions (16.52) and (16.53), respectively.

Generalization of Hill's [640] condition (see for details [223]) enables one to split the volume averages (over $\overline{\overline{w}}$ and Γ^0) of an energy-like representation into a product of volume averages of stress and strain fields:

$$\{\boldsymbol{\sigma} : \boldsymbol{\varepsilon}\} = \{\boldsymbol{\sigma}\} : \{\boldsymbol{\varepsilon}\}, \quad \langle \boldsymbol{\sigma} : \boldsymbol{\varepsilon}\rangle^{\Omega} = \langle \boldsymbol{\sigma}\rangle^{\Omega} : \langle \boldsymbol{\varepsilon}\rangle^{\Omega} \tag{19.16}$$

for any admissible displacement $\mathbf{u}$ and equilibrium stress $\boldsymbol{\sigma}$ compatible with the homogeneous volumetric boundary conditions (16.52) and (16.53), respectively.

19.2.2 Effective Elastic Moduli

Thus, if the volumetric "macroscopically uniform" boundary condition either (16.52) or (16.53) is prescribed and the scale separation hypothesis (19.6) holds, then Hill's equality (19.16) is valid and the operation of the volume macroscopic average $\{\mathbf{f}\}$ ($\mathbf{f} = \boldsymbol{\varepsilon}, \boldsymbol{\sigma}, \boldsymbol{\sigma} : \boldsymbol{\varepsilon}$) coincides with the operation over the unit cell $\langle \mathbf{f}\rangle^{\Omega}$. Moreover, for statistically homogeneous media, the volume macroscopic average $\{\mathbf{f}\}$ also coincides with the statistical average $\langle \mathbf{f}\rangle$. The mentioned equivalence makes it possible to recast some basic properties of the effective elastic moduli obtained for statistically homogeneous media in terms of the volume averages $\{\mathbf{f}\}$ to the corresponding representations for the periodic media in terms of the UC averages $\langle \mathbf{f}\rangle^{\Omega}$.

So, independently on either the local or nonlocal nature of the constitutive laws in the micro points, the field macrovariables $\langle\langle \boldsymbol{\sigma}\rangle\rangle$ and $\langle\langle \boldsymbol{\varepsilon}\rangle\rangle$ depend only on the homogeneous nonlocal boundary conditions, either (16.52) or (16.53), and are related by a local material effective tensor of either the elasticity or compliance:

$$\langle\langle \boldsymbol{\sigma}\rangle\rangle = \mathbf{L}^{*}\langle\langle \boldsymbol{\varepsilon}\rangle\rangle, \quad \langle\langle \boldsymbol{\varepsilon}\rangle\rangle = \mathbf{M}^{*}\langle\langle \boldsymbol{\sigma}\rangle\rangle, \tag{19.17}$$

respectively, where the symbol $\langle\langle \mathbf{f}\rangle\rangle$ ($\mathbf{f} = \boldsymbol{\sigma}, \boldsymbol{\varepsilon}$) stands for either the volume macroscopic average or the UC average

$$\langle\langle \mathbf{f}\rangle\rangle = \{\mathbf{f}\} \text{ or } \langle\langle \mathbf{f}\rangle\rangle = \langle \mathbf{f}\rangle^{\Omega}, \tag{19.18}$$

respectively. In so doing, $\{\boldsymbol{\varepsilon}\}$ and $\{\boldsymbol{\sigma}\}$ are estimated by Eqs. (19.14) and (19.15) through the macroscopic volumetric homogeneous boundary conditions (16.52) and (16.53), respectively; the UC averages $\langle \boldsymbol{\varepsilon}\rangle^{\Omega}$ and $\langle \boldsymbol{\sigma}\rangle^{\Omega}$ are evaluated by Eqs. (19.13_2) and (19.13_1) through the displacements and traction at the geometric UC boundary Γ^0.

Establishment of the constitutive linkage between the suitably chosen micro- (or nano-) and macrofield is reduced to the estimation of the tensor *influence functions* $\mathbf{A}^{*}_{\varepsilon}(\mathbf{x})$, $\mathbf{B}^{*}_{\varepsilon}(\mathbf{x})$ and $\mathbf{A}^{*}_{\sigma}(\mathbf{x})$, $\mathbf{B}^{*}_{\sigma}(\mathbf{x})$ (do not miss with the influence functions in peridynamics, see [1356, 1405]) for the nonlocal homogeneous boundary conditions (16.52) or (16.53):

$$\mathbf{u}(\mathbf{x}) = \mathbf{A}^{*}_{u}(\mathbf{x})\langle\langle \boldsymbol{\varepsilon}\rangle\rangle, \quad \boldsymbol{\varepsilon}(\mathbf{x}) = \mathbf{A}^{*}_{\epsilon}(\mathbf{x})\langle\langle \boldsymbol{\varepsilon}\rangle\rangle, \quad \boldsymbol{\sigma}(\mathbf{x}) = \mathbf{B}^{*}_{\epsilon}(\mathbf{x})\langle\langle \boldsymbol{\varepsilon}\rangle\rangle, \tag{19.19}$$

$$\mathbf{u}(\mathbf{x}) = \mathbf{B}^{*}_{u}(\mathbf{x})\langle\langle \boldsymbol{\sigma}\rangle\rangle, \quad \boldsymbol{\varepsilon}(\mathbf{x}) = \mathbf{A}^{*}_{\sigma}(\mathbf{x})\langle\langle \boldsymbol{\sigma}\rangle\rangle, \quad \boldsymbol{\sigma}(\mathbf{x}) = \mathbf{B}^{*}_{\sigma}(\mathbf{x})\langle\langle \boldsymbol{\sigma}\rangle\rangle, \tag{19.20}$$

respectively. Averaging Eqs. (19.19$_2$) and (19.20$_3$) leads to the conclusion that

$$\langle\langle \mathbf{A}^*_\epsilon(\mathbf{x})\rangle\rangle = \langle\langle \mathbf{B}^*_\sigma(\mathbf{x})\rangle\rangle = \mathbf{I}. \tag{19.21}$$

The influence functions (19.19) and (19.20) play a fundamental role in obtaining the macroscopic elastic response. Indeed, substitutions of the averaged multiscale linkages $\langle\langle \boldsymbol{\sigma}\rangle\rangle = \langle\langle \mathbf{B}^*_\varepsilon\rangle\rangle\langle\langle \boldsymbol{\varepsilon}\rangle\rangle$ (19.19$_3$) and $\langle\langle \boldsymbol{\varepsilon}\rangle\rangle = \langle\langle \mathbf{A}^*_\sigma\rangle\rangle\langle\langle \boldsymbol{\sigma}\rangle\rangle$ (19.20$_2$) into the averaged constitutive equations (19.17$_1$) and (19.17$_2$) yield estimations of the effective stiffness and compliance:

$$\mathbf{L}^* = \langle\langle \mathbf{B}^*_\varepsilon(\mathbf{x})\rangle\rangle, \quad \mathbf{M}^* = \langle\langle \mathbf{A}^*_\sigma(\mathbf{x})\rangle\rangle, \tag{19.22}$$

respectively, while substitution of the representation (19.19$_1$) into Eq. (19.22$_1$) leads to another form of the effective moduli

$$\mathbf{L}^* = \langle\langle \boldsymbol{\mathcal{L}}^\sigma(\mathbf{A}^*_u)\rangle\rangle. \tag{19.23}$$

It should be mentioned that a similar formula for $\mathbf{M}^*$ (19.22$_2$) depends on $\mathbf{B}^*_u(\mathbf{x})$ for the force loading conditions (16.53) defined by an inverse operator $(\boldsymbol{\mathcal{L}}^\sigma)^{-1}$; this operator is in general unknown. Because of this, the force loading conditions (16.53) are considered in less detail than the displacement loading conditions (16.52). The representations (19.22) are reduced to the classical relations (see [640])

$$\mathbf{L}^* = \langle\langle \mathbf{L}(\mathbf{x})\mathbf{A}^*_\varepsilon(\mathbf{x})\rangle\rangle, \quad \mathbf{M}^* = \langle\langle \mathbf{M}(\mathbf{x})\mathbf{B}^*_\sigma(\mathbf{x})\rangle\rangle, \tag{19.24}$$

respectively, only for the local constitutive laws at the fine scale (6.2).

We consider now the properties of the stiffness (19.22$_1$) and compliance (19.22$_2$) in more detail. From the symmetry of the tensors $\langle\langle \boldsymbol{\varepsilon}\rangle\rangle$ and $\langle\langle \boldsymbol{\sigma}\rangle\rangle$, it follows that all influence tensors are symmetric inside the index pairs ($\mathbf{C} = \mathbf{A}^*_\varepsilon, \mathbf{B}^*_\varepsilon, \mathbf{A}^*_\sigma, \mathbf{B}^*_\sigma$): $C_{ijkl} = C_{jikl} = C_{ijlk}$. From (19.16) and from the symmetry properties of the values at the fine scale $\boldsymbol{\mathcal{L}}^c(\mathbf{x})$ and $\mathbf{A}^*_u(\mathbf{x})$, the symmetry of macroproperties $\mathbf{L}^*$ inside the index pairs (minor symmetry) follows. The symmetry of $\mathbf{L}^*$ between the index pairs (major symmetry, e.g., $L^*_{ijkl} = L^*_{klij}$) is proved analogously to [216] by the use of both Hill's equality (19.16) and self-adjointness of the peridynamic operator $\widehat{\mathcal{L}}$ (16.5) at the fine scale (see [16]).

A new *energetic definition* of the effective elastic elasticity $\mathbf{L}^*$ can be obtained from both the average of the strain energy $W = \langle\langle W(x)\rangle\rangle$ expressed through the influence functions $\mathbf{A}^{*\eta}_u(\boldsymbol{\xi}) := \mathbf{A}^*_u(\mathbf{y}) - \mathbf{A}^*_u(\mathbf{x})$ and as a total strain energy of homogeneous effective medium subjected to nonlocal homogeneous boundary conditions (16.52) and (16.53):

$$W = \frac{1}{2}\langle\langle \boldsymbol{\varepsilon}\rangle\rangle \mathbf{L}^* \langle\langle \boldsymbol{\varepsilon}\rangle\rangle. \tag{19.25}$$

Then two alternative energetic definitions of the effective moduli can be obtained:

$$\mathbf{L}^* = \frac{1}{2}\left\langle\!\!\left\langle \int_{\mathcal{H}_x} \mathbf{A}_u^{*\eta\top}(\boldsymbol{\xi})\mathbf{C}(\boldsymbol{\xi},\mathbf{x})\mathbf{A}_u^{*\eta}(\boldsymbol{\xi})d\mathbf{y} \right\rangle\!\!\right\rangle, \tag{19.26}$$

$$\mathbf{L}^* = \langle\!\langle \mathbf{A}_\varepsilon^{*\top} \boldsymbol{\mathcal{L}}^\sigma(\mathbf{A}_u^*)\rangle\!\rangle, \tag{19.27}$$

where the last equality (19.27) was obtained at additional use of Hill's equality (19.16). The three definitions (19.22), (19.26), and (19.27) are not in general equivalent to one another for arbitrary inhomogeneous media, and the extent to which they agree in their computation of $\mathbf{L}^*$ provides a partial check on the validity of the "overall modulus" concept.

It should be mentioned that all three definitions (19.22), (19.26), and (19.27) of the effective modulus $\mathbf{L}^*$ are based on estimations of averaging the different values over the volume of the unit cell Ω_{00}. In so doing, Eqs. (19.17) have a dual destiny which is not only a definition of the effective modulus $\mathbf{L}^*$ as a proportionality factor of both the macroscopic stresses $\{\boldsymbol{\sigma}\}$ and strains $\{\boldsymbol{\varepsilon}\}$ but also it presents (in light of Eqs. (19.13)) a key idea for the micro-to-macro modeling. Indeed, Eqs. (19.17) hold for the UC averages $\langle\mathbf{f}\rangle^{\Omega}$ (rather than only for the volume macroscopic average $\{\mathbf{f}\}$) which can be expressed through the averages over the UC geometric boundary Γ^0 (19.13). The last averages formally coincide with the corresponding averages for locally elastic CMs. However, the formal similarity of averages $\langle\mathbf{f}\rangle^{\Omega}$ does not protect from conceptional differences of these notions in either the local elasticity or peridynamics. Indeed, in the locally elastic CMs, the averages $\langle\mathbf{f}\rangle^{\Omega}$ (19.13) are performed on a class of functions periodical on the geometric boundary Γ^0 (19.4), while in the peridynamics CMs, the same averages are estimated in the class of functions with the volumetric periodicity (19.7) and (19.8). In its turn, the volumetric periodicity (19.7) allows us to use only both the displacements $\mathbf{u}(\mathbf{x})$ inside the internal volumetric boundary $\mathbf{x} \in \Gamma^+$ and the tractions $\mathbf{t}(\mathbf{s})$ on the geometric boundary $\mathbf{s} \in \Gamma^0$. Thus, the key feature of computational homogenization is not altered in either the locally elastic CMs or the peridynamic ones: a determination of macroscopic mechanical response (input leading to output, e.g., stress driven by deformation) through the nested solution of a microstructural problem at the microscopic periodic boundary conditions (geometric or volumetric) defined from the macroscopic input variables.

19.3 Decomposition of the Material and Field Parameters and Effective Modulus Representations

For the homogeneous displacement loading ((19.2) or (16.52)), many relationships between the averages (for both the statistical averages and UC averages) are formally similar. For emphasizing of this similarity, we will analyze the corresponding representation for both structures. No confusion will arise from our simultaneous consideration of statistically homogeneous and periodic media which use the different averaging operations $\langle(\cdot)\rangle$ and $\langle(\cdot)\rangle^{\Omega}$, respectively. The variables corresponding to the locally elastic medium are labeled by a superindex ^{L}.

For the matrix composites with homogeneous matrix being considered, we decompose the material parameters (the elastic modulus $\mathbf{L}$, micromodulus $\mathbf{C}$, peridynamic operator $\widehat{\mathcal{L}}$ (16.5), and peridynamic stress operator $\hat{\mathcal{L}}^{\sigma}$ (16.43)) as

$$^L\mathbf{L}(\mathbf{x}) = {}^L\mathbf{L}^{(0)} + \mathbf{L}_1(\mathbf{x}), \tag{19.28}$$

$$\mathbf{C}(\mathbf{x}, \hat{\mathbf{x}}) = \mathbf{C}^{(0)}(\mathbf{x}, \hat{\mathbf{x}}) + \mathbf{C}_1(\mathbf{x}, \hat{\mathbf{x}}), \tag{19.29}$$

$$\widehat{\mathcal{L}} = \widehat{\mathcal{L}}^{(0)} + \widehat{\mathcal{L}}_1, \tag{19.30}$$

$$\mathcal{L}^{\sigma} = \mathcal{L}^{\sigma(0)} + \mathcal{L}_1^{\sigma}, \tag{19.31}$$

where ${}^L\mathbf{L}_1(\mathbf{x}) := {}^L\mathbf{L}(\mathbf{x}) - {}^L\mathbf{L}^{(0)}$ and $\mathbf{C}_1(\mathbf{x}, \hat{\mathbf{x}}) := \mathbf{C}(\mathbf{x}, \hat{\mathbf{x}}) - \mathbf{C}^{(0)}(\mathbf{x}, \hat{\mathbf{x}})$ are the jumps of elastic modulus and micromodulus, respectively, with respect to the corresponding properties of the matrix. Obviously, ${}^{\mathbf{L}}{}_1(\mathbf{x}) \equiv \mathbf{0}$ vanishes outside the inclusion phase $\mathbf{x} \notin v$, while $\mathbf{C}_1(\mathbf{x}, \hat{\mathbf{x}}) \equiv \mathbf{0}$ outside the extended inclusion phase $v^l := v \oplus \mathcal{H}_0$ (obtained by a Minkowski dilation of v when either $\mathbf{x} \notin v^l$ or $\hat{\mathbf{x}} \notin v^l$). In a similar manner, the operators $\widehat{\mathcal{L}}^{(0)}$ and $\widehat{\mathcal{L}_1}$ [and $\mathcal{L}^{\sigma(0)}$ and $\mathcal{L}_1^{\sigma}$] are expressed by the representations (16.5$_2$), (16.55) [and (16.43)] at $\mathbf{C}(\mathbf{x}, \hat{\mathbf{x}}) = \mathbf{C}^{0)}(\mathbf{x}, \hat{\mathbf{x}})$ and $\mathbf{C}(\mathbf{x}, \hat{\mathbf{x}}) = \mathbf{C}_1(\mathbf{x}, \hat{\mathbf{x}})$, respectively. In so doing, $\widehat{\mathcal{L}}_1 * \mathbf{u}(\mathbf{x})$ and $\mathcal{L}_1^{\sigma} * \mathbf{u}(\mathbf{x})$ vanish inside the modified matrix $\mathbf{x} \in v^{(0)l} := w \backslash v^l$. The tensor ${}^{\mathbf{L}}{}_1$ in local elasticity and the operator $\mathcal{L}_1^{\sigma}$ define the so-called stress polarization and *peridynamic stress polarization*

$$^L\boldsymbol{\tau}(x) := {}^L\mathbf{L}_1(\mathbf{x})\boldsymbol{\varepsilon} \quad \text{and} \quad \boldsymbol{\tau}(\mathbf{x}) := \mathcal{L}_1^{\sigma} * \mathbf{u}(\mathbf{x}), \tag{19.32}$$

which vanish at $\mathbf{x} \in v^{(0)}$ and $\mathbf{x} \in v^{(0)l}$, respectively.

Thus, the decompositions of material properties (19.28)–(19.31) are equivalent to "extraction" of the material properties of the matrix (which are invariant to translations), while the jumps of the corresponding material properties of either the statistically homogeneous or periodic variables vanish inside the modified matrix $v^{(0)l}$. The same idea is used for the field decomposition based on "extraction" of a field corresponding to the homogeneous displacement loading ((19.2) or (16.52))

$$\boldsymbol{\varepsilon}(\mathbf{x}) = \boldsymbol{\varepsilon}_0(\mathbf{x}) + \boldsymbol{\varepsilon}_1(\mathbf{x}), \quad \boldsymbol{\varepsilon}_0(\mathbf{x}) := \boldsymbol{\varepsilon}^{w\Gamma} \tag{19.33}$$

$$\mathbf{u}(\mathbf{x}) = \mathbf{u}_0(\mathbf{x}) + \mathbf{u}_1(\mathbf{x}), \quad \mathbf{u}_0(\mathbf{x}) := \boldsymbol{\varepsilon}^{w\Gamma}\mathbf{x}. \tag{19.34}$$

The statistical averages $\langle(\cdot)\rangle$ and UC averages $\langle(\cdot)\rangle^{\Omega}$ of the fluctuation fields $\boldsymbol{\varepsilon}_1$ and $\mathbf{u}_1$ vanish:

$$\langle\boldsymbol{\varepsilon}_1\rangle = \langle\boldsymbol{\varepsilon}_1\rangle^{\Omega} = \mathbf{0}, \quad \langle\mathbf{u}_1\rangle^{\Omega} = \mathbf{0}, \tag{19.35}$$

for statistically homogeneous and periodic structures, respectively.

The material operators of the matrix act on a linear function as

$$\widehat{\mathcal{L}}^{(0)} * \mathbf{u}_0(\mathbf{x}) \equiv \mathbf{0}, \quad \mathcal{L}^{\sigma(0)} * \mathbf{u}_0(\mathbf{x}) \equiv \mathbf{L}^{(0)}\boldsymbol{\varepsilon}_0, \tag{19.36}$$

(where $\mathbf{L}^{(0)}$ is determined by Eq. (17.99)) that allows us to decompose both the equilibrium equation (16.5) and the constitutive equation (16.43) as

$$\widehat{\mathcal{L}} * \mathbf{u}_1(\mathbf{x}) = -\widehat{\mathcal{L}}_1 * \mathbf{u}_0(\mathbf{x}), \tag{19.37}$$

$$\mathcal{L}^{\sigma} * \mathbf{u}(\mathbf{x}) = \mathbf{L}^{(0)}\boldsymbol{\varepsilon}_0 + \mathcal{L}^{\sigma(0)} * \mathbf{u}_1(\mathbf{x}) + \mathcal{L}_1^{\sigma} * \mathbf{u}(\mathbf{x}), \tag{19.38}$$

which hold for both the random and periodic fields $\mathbf{u}(\mathbf{x})$. The decompositions (19.37) and (19.38) are accompanied by the decomposed volumetric boundary conditions (19.7) and (19.8) for $\mathbf{u}_0(\mathbf{x})$ and $\mathbf{u}_1(\mathbf{x})$ ($\mathbf{x}+\boldsymbol{\xi}-2\boldsymbol{\alpha}^l \in \Gamma^+$ and $\mathbf{x}+\boldsymbol{\xi} \in \Omega_\alpha$, $|\boldsymbol{\alpha}| \neq 0$):

$$\mathbf{u}_0(\mathbf{x}+\boldsymbol{\xi}) = \mathbf{u}_0(\mathbf{x}+\boldsymbol{\xi}-2\boldsymbol{\alpha}^l) + 2\boldsymbol{\varepsilon}^{wr}\boldsymbol{\alpha}^l \equiv \boldsymbol{\varepsilon}^{wr}(\mathbf{x}+\boldsymbol{\xi}), \tag{19.39}$$

$$\mathbf{t}_0(\mathbf{x}_{\boldsymbol{\alpha}}^0) = -\mathbf{t}_0(\mathbf{x}_{\boldsymbol{\beta}}^0) \equiv \mathbf{L}^{(0)}\boldsymbol{\varepsilon}^{wr}\mathbf{n}(\mathbf{x}_{\boldsymbol{\alpha}}^0), \tag{19.40}$$

$$\mathbf{u}_1(\mathbf{x}+\boldsymbol{\xi}) = \mathbf{u}_1(\mathbf{x}+\boldsymbol{\xi}-2\boldsymbol{\alpha}^l), \tag{19.41}$$

$$\mathbf{t}_1(\mathbf{x}_{\boldsymbol{\alpha}}^0) = -\mathbf{t}_1(\mathbf{x}_{\boldsymbol{\beta}}^0), \tag{19.42}$$

where $\boldsymbol{\beta} = -\boldsymbol{\alpha}$ and $\mathbf{x}_{\boldsymbol{\alpha}}^0$ are defined analogously to Eq. (19.5).

The governing effective equations

$$\langle \boldsymbol{\sigma} \rangle = \mathbf{L}^* \langle \boldsymbol{\varepsilon} \rangle, \quad \langle \boldsymbol{\sigma} \rangle^{\Omega} = \mathbf{L}^* \langle \boldsymbol{\varepsilon} \rangle^{\Omega}, \tag{19.43}$$

are obtained by averaging Eq. (19.37) that leads to the effective modulus representation

$$\mathbf{L}^* = \mathbf{L}^{(0)} + \mathbf{R}^*, \tag{19.44}$$

with the tensor $\mathbf{R}^*$ defined either for statistically homogeneous or periodic media as

$$\langle \boldsymbol{\tau} \rangle = \mathbf{R}^* \boldsymbol{\varepsilon}_0, \tag{19.45}$$

$$\langle \boldsymbol{\tau} \rangle^{\Omega} = \mathbf{R}^* \boldsymbol{\varepsilon}_0, \tag{19.46}$$

respectively. For statistically homogeneous media, one [216] proved Eqs. (19.44) and (19.45) (that means vanishing the average $\langle \mathcal{L}^{\sigma(0)}(\mathbf{u}_1) \rangle = \mathbf{0}$) by the use of linearity property of $\hat{\mathbf{f}}^{(0)} \sim \mathbf{u}$ combined with Fubini's theorem for the iterated double integral. In a similar manner, for the periodic medium

$$\langle \mathcal{L}^{\sigma(0)}(\mathbf{u}_1) \rangle^{\Omega} = \frac{1}{2} \Big\langle \int_S \int_0^\infty \int_0^\infty (y+z)^{d-1} \hat{\mathbf{f}}^{(0)}(\mathbf{x}+y\mathbf{m}, \mathbf{x}-z\mathbf{m}) \otimes \mathbf{m}\, dz\, dy\, d\Omega_{\mathbf{m}} \Big\rangle^{\Omega}$$

$$= \frac{1}{2}\int_S\int_0^\infty\int_0^\infty (y+z)^{d-1}\mathbf{f}^{(0)}(\langle \mathbf{u}_1(\mathbf{x}+y\mathbf{m})\rangle^{\Omega}$$

$$- \langle \mathbf{u}_1(\mathbf{x}-z\mathbf{m})\rangle^{\Omega}, (y+z)\mathbf{m}) \otimes \mathbf{m} dz dy d\Omega_{\mathbf{m}} = \mathbf{0}, \qquad (19.47)$$

where the last equality is obtained by substitution of Eq. (19.35$_3$).

Thus, we obtained formally similar representations for effective elastic moduli for statistically homogeneous (19.44), (19.45) and periodic structure (19.44), (19.46) peridynamic CMs. These equations are reduced to their counterparts for locally elastic CMs in the limit of the horizon going to zero $l_\delta/a \to 0$ and $\boldsymbol{\tau}(\mathbf{x}) \to^{L} \boldsymbol{\tau}(\mathbf{x})$ (19.32).

19.4 Numerical Schemes

19.4.1 Bounded Domain Subjected to the Volumetric Displacement Homogeneous Boundary Condition

In Chap. 18, we considered the discretized form (18.4)–(18.6) of the equilibrium equation (17.35) with prescribed homogeneous volumetric boundary conditions (16.52). Now we consider these equations on the extended central UC $\mathbf{x}_p, \mathbf{x}_q \in \Omega_{00} \cup \Gamma^- \subset w$

$$\sum_q \hat{\mathbb{K}}_{pq}\mathbf{u}_q = \hat{\mathbf{b}}_p, \qquad (19.48)$$

$$\hat{\mathbb{K}}_{pq} = \mathbb{K}_{pq} V^{\Omega_{00}}(\mathbf{x}_q), \;\; \mathbb{K}_{pq} = \mathbf{C}(\mathbf{x}_p, \mathbf{x}_q)\bar{v}_{pq} - \delta_{pq}\sum_{r\in\mathcal{H}_p}\mathbf{C}(\mathbf{x}_p, \mathbf{x}_r)\bar{v}_{pr}, \qquad (19.49)$$

$$\hat{\mathbf{b}}_p = -\sum_{r\in\mathcal{H}_p}\mathbf{C}(\mathbf{x}_p, \mathbf{x}_r)\boldsymbol{\varepsilon}^{w\Gamma}(\mathbf{x}_r - \mathbf{x}_p)\bar{v}_{pr} V^{\Omega_{\Gamma 0}^-}(\mathbf{x}_r), \qquad (19.50)$$

where one introduced the indicator functions $V^{\Omega_{00}}$ and $V^{\Omega_{\Gamma 0}^-}(\mathbf{x})$ of the domains Ω_{00} and Γ^-, respectively. Our goal is a reformulation of Eqs. (19.48)–(19.50) by the use of the volumetric PBC (19.7) and (19.8) so that a new equation will be defined on the UC Ω_{00} rather than on the extended UC $\Omega_{00} \cup \Gamma^-$.

19.4.2 Direct Approach for the Infinite Periodic Medium

However, in the framework of both the scale separation hypothesis (19.6) and the volumetric homogeneous boundary conditions (16.52), the problem on the macrodomain $\mathbf{x} \in w$ is reduced to the microproblem on the unit cell $\mathbf{x} \in \Omega_{00}$

with the volumetric PBC (19.7) and (19.8) found from the macroscopic input variable. The volumetric PBC (16.52) are defined on the volumetric boundary $\Omega_{\Gamma 0}$ (also called the interaction UC boundary in analogy with the interaction interface introduced in [1354]) surrounding the geometric boundary Γ^0 and limited by the surfaces at the horizon distance from the geometric boundary Γ^0 of the UC. In the simplest case, the material properties (such as both the micromodulus and horizon) are identical for any points in the interaction boundary $\Omega_{\Gamma 0}$, while the displacements of directly interacting points from the different sides of the geometric boundary are different. Otherwise, it is possible overlapping of the interaction boundary $\Omega_{\Gamma 0}$ with the interaction interface v_Γ of the inclusion $v_i \subset \Omega_{00}$ embedded into the UC Ω_{00} (see for details [1354]). It should be additionally mentioned that the different portions of the interaction boundary $\Omega_{\Gamma 0}$ of the UC Ω_{00} belong to 3, 9, and 27 different UCs for $d = 1, 2$, and 3, respectively. The condition of the displacement periodicity (19.4) at the geometric boundary Γ^0 is extended to the volumetric periodicity of displacements for all points inside the interaction boundary of the UC (19.7). We will demonstrate that due to the volumetric PBC (19.7), the reformulated equation (19.48) depends only on the displacements $\mathbf{u}(\mathbf{x})$ inside the unit cell $\mathbf{x} \in \Omega_{00}$ rather than on the displacements $\mathbf{u}(\mathbf{y})$ inside the extended unit cell $\mathbf{y} \in \Omega_{00} \cup \Gamma^-$. Due to the volumetric displacement periodicity (19.7), the special antiperiodic traction boundary condition (19.8) is not required for the establishment of the micro-to-macro displacement relationship, although this classical traction condition at the geometric boundary Γ^0 of the UC is needed for estimations of both the overall stresses and effective moduli.

Indeed, we recast (19.48) at $\mathbf{b} \equiv \mathbf{0}$ assumed in the form ($\mathbf{x}_p \in \Omega_{00}$)

$$\sum_{q \in \mathcal{H}_p} \mathbf{C}(\mathbf{x}_p, \mathbf{x}_q)(\widetilde{\mathbf{u}}_q - \mathbf{u}_p)\bar{v}_{pq} = -\mathbf{b}_p^{\Omega}. \tag{19.51}$$

If the collocation point $\mathbf{x}_p \in \Omega_{00}$ does not fall into the internal volumetric boundary $\mathbf{x} \notin \Gamma^+$, then $\mathcal{H}_p \subset \Omega_{00}$ and Eq. (19.51) (at $\widetilde{\mathbf{u}}_q = \mathbf{u}_q$, $\mathbf{b}^{\Omega} = \mathbf{0}$) coincides with Eq. (19.48). However, if $\mathbf{x}_p \in \Omega_{00}$ falls into $\Gamma^+ \subset \Omega_{00}$ so that some source point $\mathbf{x}_q = \mathbf{x}_p + \boldsymbol{\xi}$ belongs to some part of the external volumetric boundary $\mathbf{x}_q \in \Gamma^- \cap \Omega_\alpha$, then the volumetric PBC (19.7) and (19.8) should be taken into account that leads to the following images of both the node $\widetilde{\mathbf{q}}$, node coordinates $\widetilde{\mathbf{x}}_q$, node displacements $\widetilde{\mathbf{u}}_q$, and $\mathbf{b}_p^{\Omega}$:

$$\widetilde{\mathbf{q}} = \begin{cases} \mathbf{q}, & \text{for } \mathbf{x}_q \in \Omega_{00}, \\ \mathbf{q} - (2N^{\Omega} + 1)\boldsymbol{\alpha}, & \text{for } \mathbf{x}_q \in \Gamma^- \cap \Omega_\alpha, \end{cases} \tag{19.52}$$

$$\widetilde{\mathbf{x}}_{\widetilde{q}} = \begin{cases} \mathbf{x}_q, & \text{for } \mathbf{x}_q \in \Omega_{00}, \\ \mathbf{x}_q - 2\boldsymbol{\alpha}^l, & \text{for } \mathbf{x}_q \in \Gamma^- \cap \Omega_\alpha, \end{cases} \tag{19.53}$$

$$\widetilde{\mathbf{u}}_{\widetilde{q}} = \begin{cases} \mathbf{u}(\mathbf{x}_q), & \text{for } \mathbf{x}_q \in \Omega_{00}, \\ \mathbf{u}(\mathbf{x}_q - 2\boldsymbol{\alpha}^l), & \text{for } \mathbf{x}_q \in \Gamma^- \cap \Omega_\alpha, \end{cases} \tag{19.54}$$

$$\mathbf{b}_p^\Omega = -2 \sum_{r \in \mathcal{H}_p} (1 - V^{\Omega_{00}}(\mathbf{x}_r))\mathbf{C}(\mathbf{x}_p, \mathbf{x}_r)\boldsymbol{\varepsilon}^{w\Gamma}\boldsymbol{\alpha}^l, \tag{19.55}$$

with $V^{\Omega_{00}}$ being the indicator function of the UC Ω_{00}. Here, for each $\mathbf{x}_q \notin \Omega_{00}$ falling in some part $\Gamma^- \cap \Omega_\alpha$ (that uniquely defines $\boldsymbol{\alpha}$), one uses precisely this $\boldsymbol{\alpha}$ at the estimation of all right-hand sides of Eqs. (19.52)–(19.55). The imaged index $\widetilde{\mathbf{q}}$ (used for estimations of the imaged variables $\widetilde{\mathbf{x}}_{\widetilde{\mathbf{q}}}$ and $\widetilde{\mathbf{u}}_{\widetilde{\mathbf{q}}}$) is produced from the index $\mathbf{q}$ if $\mathbf{x}_\mathbf{q} \in \Gamma^- \cap \Omega_\alpha$. In so doing, the micromodulus $\mathbf{C}(\mathbf{x}_p, \mathbf{x}_\mathbf{q})$ is estimated at the initial arguments $\mathbf{x}_p$ and $\mathbf{x}_\mathbf{q}$ (rather than $\mathbf{x}_p$ and $\widetilde{\mathbf{x}}_{\widetilde{\mathbf{q}}}$). In the case of notational convenience, we use equivalent multiplets of integer numbers $\mathbf{r} = (r_1, \ldots, r_d) \in [-N^\Omega, N^\Omega]^d$ instead of index r $(r = p, q)$. The collocation node p $(\mathbf{x}_p \in \Omega_{00})$ with the source node q $(\mathbf{x}_q \in \Gamma_{10}^-)$ and the imaged node $\widetilde{q}$ $(\mathbf{x}_{\widetilde{q}} \in \Gamma_{-10}^+)$ are presented in Fig. 19.2.

Equation (19.51) can be also presented in the form (19.48) (with the corresponding multiplets $\mathbf{p}, \mathbf{q} \in [-N^\Omega, N^\Omega]^d$)

$$\sum_q \mathbb{K}_{pq}^\Omega \mathbf{u}_q = \mathbf{b}_p^\Omega. \tag{19.56}$$

All nonzero elements of the matrix $\mathbb{K}_{pq}$ (19.48) coincide with the corresponding nonzero elements of the $\mathbb{K}_{pq}^\Omega$ (19.56)

$$\mathbb{K}_{pq}^\Omega = \mathbb{K}_{pq} \tag{19.57}$$

with the multiplets $\mathbf{p}, \mathbf{q} \in [-N^\Omega, N^\Omega]^d$. However, the matrix $\mathbb{K}_{pq}^\Omega$ (19.56) has additional nonzero elements if $\mathbf{x}_q$ in Eq. (19.51) falls into the volumetric external boundary $\mathbf{x}_q \in \Gamma_{\boldsymbol{\alpha}}^-$

$$\mathbb{K}_{p\widetilde{\mathbf{q}}}^\Omega = \mathbf{C}(x_p, x_q)\bar{v}_{pq}, \tag{19.58}$$

where $\mathbf{x}_q \in \Gamma_{\boldsymbol{\alpha}}^-$ in the right-hand side of Eq. (19.52), while $\widetilde{\mathbf{q}} \in [-N^\Omega, N^\Omega]^d$ defined by Eq. (19.52) belongs to the UC nodes: $\mathbf{x}_p, \mathbf{x}_{\widetilde{\mathbf{q}}} \in \Omega_{00}$. The emergence of these additional nonzero elements (19.57) is explained by periodical involvement of the displacements $\mathbf{u}(\mathbf{x}+\boldsymbol{\xi})$ and $\mathbf{u}(\mathbf{x}+\boldsymbol{\xi}-2\boldsymbol{\alpha}^l)$ (19.7) at the opposite sides $\mathbf{x}+\boldsymbol{\xi} \in \Gamma^-$ and $\mathbf{x}+\boldsymbol{\xi}-2\boldsymbol{\alpha}^l \in \Gamma^+$ of the volumetric bounds of the unit cell w_Γ^Ω. In so doing, a link between $\mathbf{u}(\mathbf{x}+\boldsymbol{\xi})$ and $\mathbf{u}(\mathbf{x}+\boldsymbol{\xi}-2\boldsymbol{\alpha}^l)$ (19.7) described by a parallel transfer $2\boldsymbol{\alpha}^l$ forms the given vector (fictitious body force) $\mathbf{b}_p^\Omega$ (19.56) defined by Eq. (19.55).

After the estimation of $\mathbf{u}(\mathbf{x}_p)$ $(\mathbf{x}_p \in \Omega_{00})$ in Eq. (19.56), the peridynamic stress distribution can be found by Eq. (16.43):

$$\sigma(\mathbf{x}) = \frac{1}{2}\int_S \int_0^\infty \int_0^\infty (y+z)^{d-1}\mathbf{C}((y+z)\mathbf{m}, \mathbf{x}-z\mathbf{m})$$
$$\times[\hat{\mathbf{u}}(\mathbf{x}+y\mathbf{m}) - \hat{\mathbf{u}}(\mathbf{x}-z\mathbf{m})] \otimes \mathbf{m}dzdyd\Omega_{\mathbf{m}} \qquad (19.59)$$

with the modified displacements $\hat{\mathbf{u}}(\mathbf{x}_{\widetilde{q}})$ (at $\mathbf{x}_q \in \Omega_\alpha$) which this bout is replaced by the following discretized representations (with either $\mathbf{x}_q = \mathbf{x}+y\mathbf{m}$ or $\mathbf{x}_q = \mathbf{x}-z\mathbf{m}$):

$$\hat{\mathbf{u}}_{\widetilde{q}} = \begin{cases} \mathbf{u}(\mathbf{x}_q), & \text{for } \mathbf{x}_q \in \Omega_{00}, \\ \mathbf{u}(\mathbf{x}_q - 2\alpha^l) + 2\boldsymbol{\varepsilon}^{w\Gamma}\alpha^l, & \text{for } \mathbf{x}_q \in \Gamma_\alpha^-, \end{cases} \qquad (19.60)$$

where the imaged node $\widetilde{\mathbf{q}} \in [N^\Omega, N^\Omega]^d$ is estimated by Eq. (19.56). Thus, the stress field inside the UC $\mathbf{x} \in \Omega_{00}$ is estimated through both the displacements found in Eq. (19.57) in this UC $\mathbf{x} \in \Omega_{00}$ and the coupled displacement (19.7) on the volumetric unit cell boundary $\mathbf{x} \in \Gamma^+$. It should be mentioned that the left-hand side of Eq. (19.59) (and, therefore, the term in the square brackets in Eq. (19.59)), in fact, depends only on the displacements inside the unit cell $\mathbf{x}_{\widetilde{q}} \in \Omega_{00}$, while the estimation of $\mathbf{C}((y+z)\mathbf{m}, \mathbf{x}-z\mathbf{m})$ in Eq. (19.59) is directly performed according to Eq. (16.64) or (16.79) with the possible locations of their arguments inside the volumetric external boundary $(y+z)\mathbf{m}, \mathbf{x}-z\mathbf{m} \in \Gamma^-$. Moreover, at the estimation of the stresses $\sigma(\mathbf{x})$ inside the UC $\mathbf{x} \in \Omega_{00}$ (19.59), we used the volumetric displacement PBC (19.7) but not the antiperiodic traction PBC (19.8). However, due to the substitution of the volumetric PBC (19.7) into the equilibrium equation (19.51), the antiperiodic traction PBC (19.8) is automatically fulfilled.

19.4.3 Decomposition Approach for the Infinite Periodic Medium

In the decomposition approach for the periodic peridynamic CMs, the solution of Eq. (16.5) is decomposed as in Eq. (19.34) ($\mathbf{x} \in \Omega_{00}$)

$$\mathbf{u}(\mathbf{x}) = \mathbf{u}_0(\mathbf{x}) + \mathbf{u}_1(\mathbf{x}), \quad \mathbf{u}_0(\mathbf{x}) := \boldsymbol{\varepsilon}^{\Gamma_w}\mathbf{x}, \qquad (19.61)$$

where $\mathbf{u}_0(\mathbf{x}) := \boldsymbol{\varepsilon}^{\Gamma_w}\mathbf{x}$ is a linear constituent of the displacement corresponding to the macrostrain $\boldsymbol{\varepsilon}^{\Gamma_w}$ and $\mathbf{u}_1(\mathbf{x})$ stands for a fluctuation part of the displacement with a vanishing average over the unit cell (19.35_3). In so doing, the decomposed peridynamic operator $\widehat{\mathcal{L}} = \widehat{\mathcal{L}}^{(0)} + \widehat{\mathcal{L}}_1$ (19.37) of the equilibrium equation (16.5) is considered with the the decomposed volumetric boundary conditions (19.42)–(19.45) for $\mathbf{u}_0(\mathbf{x})$ and $\mathbf{u}_1(\mathbf{x})$.

Substitution of the volumetric PBC (19.39)–(19.42) and the field decomposition (19.61) into the equilibrium equation (19.37) with subsequent discretization allows one to get the following linear algebraic system analogous to Eq. (19.56) (with the corresponding multiplets $\mathbf{p}, \mathbf{q} \in [-N^\Omega, N^\Omega]^d$):

$$\sum_q \mathbb{K}^{\Omega}_{pq} \mathbf{u}_{1|q} = -\mathbf{b}^{\Omega 0}_p, \tag{19.62}$$

where the matrix $\mathbb{K}^{\Omega}_{pq}$ (19.62) coincides with the matrix $\mathbb{K}^{\Omega}_{pq}$ (19.56), while a fictitious body force $\mathbf{b}^{\Omega 0}_p$ is defined by the following equation:

$$\mathbf{b}^{\Omega 0}_p = \sum_{r \in \mathcal{H}_p} \mathbf{C}_1(\mathbf{x}_p, \mathbf{x}_r) \boldsymbol{\varepsilon}^{w\Gamma} (\mathbf{x}_r - \mathbf{x}_p) \bar{v}_{pr}, \tag{19.63}$$

which is simpler than the fictitious body force (19.55) formed due to the second term of the right-hand side of the volumetric PBC (19.7), while the vector (19.63) depending on the volumetric macroscopic BC (19.2) has no relations with the periodicity conditions.

After the estimation of $\mathbf{u}_1(\mathbf{x})$ ($\mathbf{x} \in \Omega_{00}$) by the quadrature method (19.62), the peridynamic stress distribution (16.43) can be found by the decomposition (19.38) of Eq. (19.59):

$$\begin{aligned}\boldsymbol{\sigma}(\mathbf{x}) = & \frac{1}{2} \int_S \int_0^\infty \int_0^\infty (y+z)^{d-1} \mathbf{C}^{(0)}((y+z)\mathbf{m}, \mathbf{x} - z\mathbf{m}) \\ & \times [\hat{\mathbf{u}}_1(\mathbf{x} + y\mathbf{m}) - \hat{\mathbf{u}}_1(\mathbf{x} - z\mathbf{m})] \otimes \mathbf{m} dz dy d\Omega_{\mathbf{m}} \\ & + \frac{1}{2} \int_S \int_0^\infty \int_0^\infty (y+z)^{d-1} \mathbf{C}_1((y+z)\mathbf{m}, \mathbf{x} - z\mathbf{m}) \\ & \times [\hat{\mathbf{u}}(\mathbf{x} + y\mathbf{m}) - \hat{\mathbf{u}}(\mathbf{x} - z\mathbf{m})] \otimes \mathbf{m} dz dy d\Omega_{\mathbf{m}} + \mathbf{L}^{(0)} \boldsymbol{\varepsilon}^{w\Gamma}\end{aligned} \tag{19.64}$$

with the modified displacements $\hat{\mathbf{u}}_1(\mathbf{x}_{\tilde{q}})$ and $\hat{\mathbf{u}}(\mathbf{x}_{\tilde{q}}) = \hat{\mathbf{u}}_0(\mathbf{x}_{\tilde{q}}) + \hat{\mathbf{u}}_1(\mathbf{x}_{\tilde{q}})$ (at $\mathbf{x}_q \in \Omega_\alpha$) which this bout is replaced by the following discretized representations (with either $\mathbf{x}_q = \mathbf{x} + y\mathbf{m}$ or $\mathbf{x}_q = \mathbf{x} - z\mathbf{m}$):

$$\hat{\mathbf{u}}_1(\mathbf{x}_{\tilde{q}}) = \begin{cases} \mathbf{u}_1(\mathbf{x}_q), & \text{for } \mathbf{x}_q \in \Omega_{00}, \\ \mathbf{u}_1(\mathbf{x}_q - 2\boldsymbol{\alpha}^l), & \text{for } \mathbf{x}_q \in \Gamma^- \cap \Omega_\alpha, \end{cases} \tag{19.65}$$

$$\hat{\mathbf{u}}(\mathbf{x}_{\tilde{q}}) = \begin{cases} \mathbf{u}(\mathbf{x}_q), & \text{for } \mathbf{x}_q \in \Omega_{00}, \\ \mathbf{u}(\mathbf{x}_q - 2\boldsymbol{\alpha}^l) + 2\boldsymbol{\varepsilon}^{w\Gamma}\boldsymbol{\alpha}^l, & \text{for } \mathbf{x}_q \in \Gamma^- \cap \Omega_\alpha, \end{cases} \tag{19.66}$$

where the imaged node $\tilde{\mathbf{q}} \in [N^\Omega, N^\Omega]^d$ is estimated by Eq. (19.56).

Analogous to Eq. (19.59), it can be demonstrated that the stress field inside the UC $\mathbf{x} \in \Omega_{00}$ only depends on the the displacements in this UC $\mathbf{x} \in \Omega_{00}$ found in Eq. (19.62). Due to the substitution of the volumetric PBC (19.65) and (19.66) into the equilibrium equation, the antiperiodic traction boundary condition (19.42) (which is not used at the estimation of $\mathbf{u}_1(\mathbf{x})$ (19.62)) is automatically fulfilled. Although the condition (19.35$_3$) is not directly used at the displacement field estimation $\mathbf{u}_1(\mathbf{x})$, it can be exploited as a measure of accuracy of the corresponding quadrature method (either the direct method (19.56) or the decomposition method (19.62)).

Of course, the problems of both the direct approach (19.56) and the decomposition one (19.62) are formally equivalent. Indeed, the equilibrium equations (16.5) and (19.37) coincide, and the volumetric PBC (19.7), (19.8) and (19.39)–(19.42) are also identical. We proposed both methods (19.56) and (19.62) because their locally elastic counterparts are equally used in the classical computational homogenization.

19.4.4 Estimation of Effective Moduli

By the use of the Gauss-Ostrogradsky theorem, we define the overall macrostress $\{\boldsymbol{\sigma}\} = \langle\boldsymbol{\sigma}\rangle^{\Omega}$ and the overall macrostrain $\{\boldsymbol{\varepsilon}\} = \langle\boldsymbol{\varepsilon}\rangle^{\Omega}$ of the UC Ω_{00}:

$$\langle\boldsymbol{\sigma}\rangle^{\Omega} := |\Omega_{00}|^{-1}\int_{\Omega_{00}} \boldsymbol{\sigma}(\mathbf{x})d\mathbf{x} = |\Omega_{00}|^{-1}\int_{\Gamma^0} \mathbf{t}(\mathbf{s}) \overset{S}{\otimes} \mathbf{s}ds, \tag{19.67}$$

$$\langle\boldsymbol{\varepsilon}\rangle^{\Omega} := |\Omega_{00}|^{-1}\int_{\Omega_{00}} \boldsymbol{\varepsilon}(\mathbf{x})d\mathbf{x} = |\Omega_{00}|^{-1}\int_{\Gamma^0} \mathbf{u}(\mathbf{s}) \overset{S}{\otimes} \mathbf{n}(\mathbf{s})ds, \tag{19.68}$$

in terms of the traction $\mathbf{t}(\mathbf{s}) := \boldsymbol{\sigma}(\mathbf{s})\mathbf{n}(\mathbf{s})$ and the displacement $\mathbf{u}(\mathbf{s})$ on the geometric boundary of the UC $\mathbf{s} \in \Gamma^0$ with the outward normal unit vectors $\mathbf{n}(\mathbf{s})$ on Γ^0; here, $\overset{S}{\otimes}$ stands for a symmetrized tensorial product $(\mathbf{a} \overset{S}{\otimes} \mathbf{b})_{[ij]} := (a_i b_j + a_j b_i)/2$. Both pairs of Eqs. ($19.67_1$), ($19.68_1$) and ($19.67_2$), ($19.68_2$) are equivalent, but the pair of Eqs. (19.67_2) and (19.68_2) is preferable because it only requires the estimation of the traction $\mathbf{t}(\mathbf{s}) := \boldsymbol{\sigma}(\mathbf{s})\mathbf{n}(\mathbf{s})$ and the displacement $\mathbf{u}(\mathbf{s})$ at the geometric UC boundary $\mathbf{s} \in \Gamma^0$. Moreover, the use of the definition (19.68_1) is even more restrictive because (19.68_1) exploits the strain field $\varepsilon(x)$ usually found by numerical differentiation of $\mathbf{u}(\mathbf{x})$ ($\mathbf{x} \in \Omega_{00}$), while $\mathbf{u}(\mathbf{x})$ is defined on the geometric boundary Γ^0.

For the periodic peridynamic CMs subjected to the homogeneous volumetric boundary conditions either (16.52) or (16.53), the effective stiffness $\mathbf{L}^*$ is estimated as a proportionality factor between the UC's averages of the stresses $\langle\boldsymbol{\sigma}\rangle^{\Omega}$ and strains $\langle\boldsymbol{\varepsilon}\rangle^{\Omega}$:

$$\langle\boldsymbol{\sigma}\rangle^{\Omega} = \mathbf{L}^*\langle\boldsymbol{\varepsilon}\rangle^{\Omega}. \tag{19.69}$$

Estimation of the involved macrovariables is performed by the following *micro-to-macro* transition. Namely, we showed that due to the volumetric PBC (19.7) and (19.8), the governing equation depends only on the displacement inside the UC Ω_{00} rather than on the displacement inside the extended UC $\Omega_{00} \oplus \mathcal{H}_0$. In fact, the volumetric PBC (19.7) incorporated into the equilibrium equation (16.5) produces the new closed integral equation which does not require any additional boundary conditions. The discretization of this equation in the form of either (19.56) or (19.62) acts as a macro-to-micro transition of the *deformation-driven* type, where the overall deformation $\boldsymbol{\varepsilon}^{w\Gamma} = \langle \boldsymbol{\varepsilon} \rangle^{\Omega}$ is controlled. Determination of the microstructural displacements $\mathbf{u}(\mathbf{x})$ ($\mathbf{x} \in \Omega_{00}$) in accompany with the volumetric PBC either (19.60) or (19.65) allows one to estimate the peridynamic traction $\mathbf{t}(\mathbf{s}) := \boldsymbol{\sigma}(\mathbf{s})\mathbf{n}(\mathbf{s})$ at the geometric UC's boundary $\mathbf{s} \in \Gamma^0$. In so doing, due to the incorporated volumetric periodicity (19.7) of the displacement $\mathbf{u}(\mathbf{x})$ ($\mathbf{x} \in w_{\Gamma}^{\Omega}$) into Eq. (19.56) or (19.62), the condition of the traction antiperiodicity (19.8) is not used and automatically fulfilled. In its turn, the found traction $\mathbf{t}(\mathbf{s})$ ($\mathbf{s} \in \Gamma^0$) is exploited for the estimation of the macroscopic stresses $\langle \boldsymbol{\sigma} \rangle^{\Omega}$ by the use of Eq. (19.67_1) with subsequent evaluation of the effective moduli $\mathbf{L}^*$ by Eq. (19.69).

From other side, the linear algebraic system (19.56) or (19.62) is solved at the boundary conditions (19.68_2) imposed on the collocation points of the geometric boundary Γ^0 of the UC Ω_{00} and served as a constraint linear equation. These constraints can be incorporated by means of Lagrange multipliers in a straightforward manner (see, e.g., [997]) which can be also extended to consider the nonlinear problems. However, a solution of the current linear problem can be also performed in a standard iterative manner due to the linearity of the fictitious body force (19.63) (or (19.55)) $\mathbf{b}^{\Omega 0} := \underline{\mathbf{b}}^{\Omega 0} \langle \boldsymbol{\varepsilon} \rangle^{\Omega}$ with respect to $\langle \boldsymbol{\varepsilon} \rangle^{\Omega}$. Indeed, at the initial macroscopic strain $\langle \boldsymbol{\varepsilon} \rangle^{\Omega}_{[0]}$, we found the microdisplacements $\mathbf{u}_{[n]}(\mathbf{x})$ ($n = 1$) (19.62) and the next step of $\langle \boldsymbol{\varepsilon} \rangle^{\Omega}_{[n]}$ (19.68_2) which is used for the estimation of the next step of the right-hand side of Eq. (19.62) $\mathbf{b}^{\Omega 0}_{[n+1]} = \underline{\mathbf{b}}^{\Omega 0} \langle \boldsymbol{\varepsilon} \rangle^{\Omega}_{[n]}$. In the considered numerical examples for 1D problems, five iterations described provide an accuracy of $1.0E - 14\%$ for the constraint equation (19.68_2).

19.4.5 *Choice of the Unit Cell*

The unit cell of a periodic area is not always uniquely determined. The choice of the most appropriate unit cell describing a periodic structure is often based on convenience consideration. So, the same periodic structure can be interpreted as the repetition of more than one UC with different geometry and level of complexity. For example, Fig. 19.3a demonstrates an infinite composite plate with circular inclusions. Both repetitive volume elements Ω_a and Ω_b depicted in Fig. 19.3a are suitable elementary UCs. However, sometimes one from the possible UCs is better than the other due to the possible simplification of the mathematical analysis. So, in the UC Ω_b, the intersections inside both pairs of geometric boundaries of the inclusion Γ_i and the UC Γ^0 and the interaction interface v_Γ (17.3) and the volumetric UC boundary Ω_Γ take place for any scale parameter l_δ/a. Handling of the interaction interfaces (or the volumetric boundaries) requires additional mathematical refinements (see, e.g., [130, 1354]) which becomes even more entangled in the case of intersection of these domains. The mentioned difficulties are avoided for the UC Ω_a (see Fig. 19.3a) at least if the scale parameter l_δ/a is small enough.

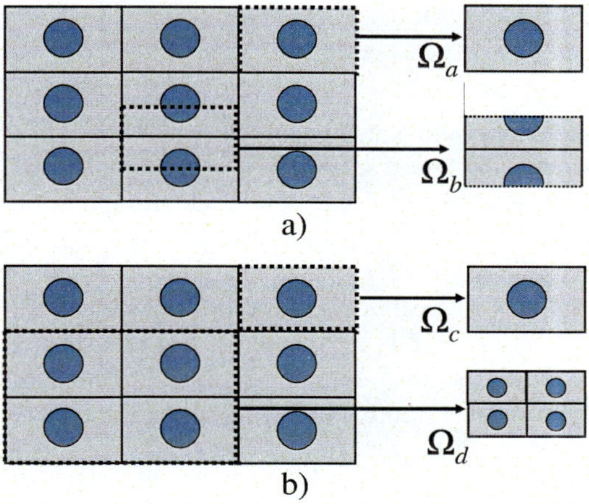

Fig. 19.3 (**a**) Elementary unit cell Ω_a and Ω_b. (**b**) Irreducible UC Ω_c and reducible UC Ω_d

Thus, the geometric selection of the best UC is dictated by the level of mathematical simplicity of the peridynamic problem solution at the selected UC.

On the other hand, it is important to make sure that the chosen UC is irreducible, i.e., that the selected UC is identifiable in the composite domain carrying all mechanical and geometric features necessary to a suitable micro-to-macro modeling. For example, both UCs in Fig. 19.3b can be interpreted as repetitive volume elements, but only element Ω_c is irreducible in the sense that its sizes cannot be reduced with the conservation of the same effective fields and material parameters of the CM. In such a case, comparison of the local fields estimated by the use of both the irreducible and irreducible UCs can be considered as a test of the method correctness. So, for the locally elastic 2D CM with the circle fibers arranged into an equilateral triangle array, Jiang et al. [64] estimated the effective moduli $\mathbf{L}^*$ by the volume average of the local fields (19.67_1) and (19.68_1) in the UCs with the sizes $k\delta_0$ ($k = 1, 2, 4$), where δ_0 is the size of an irreducible UC. In the case of homogeneous either the displacement (19.2) or traction (19.3) boundary conditions on the UC boundary, the effective moduli $\mathbf{L}^*$ depend on both the size of the UC (either $k = 1, 2$ or $k = 3$) and the homogeneous boundary conditions (either the displacement (19.2) or traction (19.3). In so doing, the effective moduli $\mathbf{L}^*$ of locally elastic CMs estimated for the UCs subjected to the PBC (19.4) are invariant to the size of the UC. A similar invariantness of the effective moduli $\mathbf{L}^*$ holds for periodic peridynamic CM that will be demonstrated in Section 6 for 1D periodic heterogeneous bar.

19.5 Numerical Schemes for 1D Case

A $1D$ example of an infinite periodic peridynamic bar is analyzed for the next geometric and mechanical properties of constituents. An infinitely long bar of a constant cross section $A = 1$ contains the periodically distributed identical inclusions of the size $2a = 2$. The midpoint quadrature scheme (similar to [209]) is realized for the uniform grid $x_i = (0, \pm 1, \pm 2, \ldots, \pm N^\Omega)$ with the numbers of material points $2N^\Omega + 1$ and $2n^a + 1$ inside both the UC Ω_0 and the inclusion $v_0 \subset \Omega_0$, respectively, and a constant step size $h = l^\Omega / N^\Omega$.

Section 19.5 is presented as a 1D illustrative example of a direct consequence of representations obtained in Sects. 19.1–19.4. Some results of this section were proposed by more entangled and less well-founded method in [221] initially analyzing 1D case.

19.5.1 Locally Elastic Heterogeneous Bar of Periodic Structure

Let us consider a locally elastic heterogeneous bar with the equilibrium equation

$$d(E(x)du/dx)/dx = -b(x). \tag{19.70}$$

For a piecewise constant Young modulus $E(x) = E^{(k)} V^{(k)}(x)$ $(x \in v^{(k)},\ \hat{E}^{(k)} \equiv (E^{(k)})^{-1},\ k = 0, 1)$, the effective constitutive equation

$$\langle\!\langle \sigma \rangle\!\rangle = E^* \langle\!\langle \varepsilon \rangle\!\rangle \tag{19.71}$$

is defined by the effective modulus E^* and the homogeneous strains inside the phases $V^{(k)}$ $(k = 0, 1)$ $(\sigma(x) \equiv \langle\!\langle \sigma \rangle\!\rangle \equiv$const.)

$$\begin{aligned} &E^* = [c^{(1)} \hat{E}^{(1)} + c^{(0)} \hat{E}^{(0)}]^{-1}, \quad \langle\!\langle \varepsilon \rangle\!\rangle = c^{(0)} {}^{L}\varepsilon^{0} + c^{(1)} {}^{L}\varepsilon^{1}, \\ &{}^{L}\varepsilon^{0} \equiv E^* \hat{E}^{(0)} \langle\!\langle \varepsilon \rangle\!\rangle, \quad {}^{L}\varepsilon^{1} \equiv E^* \hat{E}^{(1)} \langle\!\langle \varepsilon \rangle\!\rangle. \end{aligned} \tag{19.72}$$

Equations (19.71) and (19.72_1) hold for the heterogeneous bar of both the statistically homogeneous structure and the periodic one.

We consider an infinite heterogeneous bar $w = (-\infty, \infty)$ constructed from the infinite number of unit cells containing identical inclusions v_k

$$w = \cup \Omega_k, \quad v_k \subset \Omega_k \quad (k = 0, \pm 1, \pm 2, \ldots), \tag{19.73}$$

e.g., for the representative unit cell $\Omega_0 = [-l^\Omega, l^\Omega]$ and $v_0 = [-a, a]$ $(a < L^\Omega)$. The UC Ω_0 is subjected to both the periodic displacement and antiperiodic traction (see Eq. (19.4))

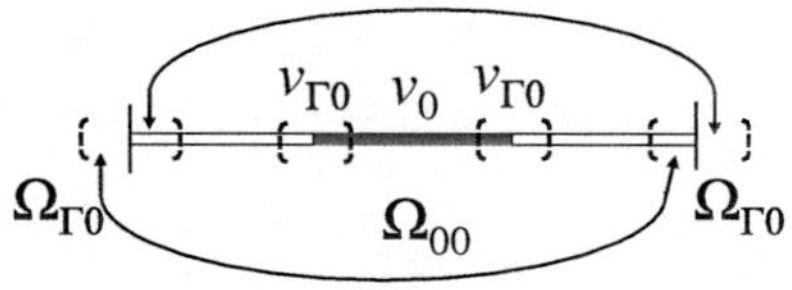

Fig. 19.4 1D UC Ω_{00} with both the interaction interface $v_{\Gamma 0}$ and the interaction boundary Ω_Γ of the UC

$$u(-l^\Omega) = u(l^\Omega) - 2l^\Omega \varepsilon^\Gamma, \tag{19.74}$$

$$t(-l^\Omega) = -t(l^\Omega) \tag{19.75}$$

boundary conditions, respectively, on the boundary $\Gamma_0 = -l^\Omega \cup l^\Omega$ of the representative unit cell (UC) Ω^0.

19.5.2 Peridynamic Heterogeneous Bar of Periodic Structure

We consider a heterogeneous peridynamic bar with the step-function profiles $C^{(k)}(\xi)$ $(k = 0, 1)$ (16.82). The triangle profile (16.83) can be considered in a similar manner. The UC Ω_{00}, the inclusion $v_0 \in \Omega_{00}$, the extended interface Γ_i^v, and the volumetric UC boundary w_Γ^Ω are reduced in the 1D case to $\Omega_{00} = [-l^\Omega, l^\Omega]$, $v_0 = [-a, a]$, $v_{\Gamma 0} = [-a - l_\delta, -a + l_\delta] \cup [a - l_\delta, a + l_\delta]$, $\Omega_{\Gamma 0} = [-l^\Omega - l_\delta, l^\Omega + l_\delta] \cup [l^\Omega - l_\delta, l^\Omega + l_\delta]$, $\Gamma_{-1}^- = [-l^\Omega - l_\delta, -l^\Omega]$, $\Gamma_1^- = [l^\Omega, l^\Omega + l_\delta]$, $\Gamma_{-1}^+ = [-l^\Omega, -l^\Omega + l_\delta]$, and $\Gamma_1^+ = [l^\Omega - l_\delta, l^\Omega]$ (see Fig. 19.4).

In the decomposition approach (the direct approach was considered in [221] for 1D case) for the periodic peridynamic bar, the solution of Eq. (16.5) is decomposed as ($x \in [l^\Omega, l^\Omega]$) (19.61)

$$u(x) = u_0(x) + u_1(x), \quad u_0(x) := \langle\varepsilon\rangle^\Omega x, \quad \langle\varepsilon_1\rangle^\Omega = 0, \tag{19.76}$$

where $u_0(x)$ is a linear constituent of the displacement corresponding to the macrostrain, while the fluctuation part $u_1(x)$ has a vanishing average over the UC (19.35$_3$). We also decompose the peridynamic operator $\mathcal{L} = \mathcal{L}^{(0)} + \mathcal{L}_1$ of the equilibrium equation (19.37)

$$\mathcal{L} * u_1(x) + \mathcal{L}_1 * u^{(0)} = 0 \tag{19.77}$$

with the decomposed volumetric PBC (19.65) and (19.66) for $u_0(x)$ and $u_1(x)$ ($x \in [l^\Omega, l^\Omega]$, $x + \xi \in \Gamma^-$, $x + \zeta := x + \xi - 2\ \text{sign}(x) l^\Omega \in \Gamma^+$)

$$u_0(x + \xi) = u_0(x + \xi - 2\ \text{sign}(x) l^\Omega) + 2\langle\varepsilon\rangle^\Omega\ \text{sign}(x) l^\Omega \equiv \langle\varepsilon\rangle^\Omega (x + \xi), \tag{19.78}$$

$$t_0(l^\Omega) = -t_0(-l^\Omega) \equiv E^{(0)} \langle\varepsilon\rangle^\Omega n(l^\Omega), \tag{19.79}$$

$$u_1(x+\xi) = u_1(x+\xi-2\ \mathrm{sign}(x)l^{\Omega}), \tag{19.80}$$

$$t_1(l^{\Omega}) = -t_1(-l^{\Omega}). \tag{19.81}$$

We refer to the boundary (e.g., $\Gamma_0 = -l^{\Omega} \cup l^{\Omega}$) of the UC separating the adjoining UCs (e.g., Ω_0 and Ω_1) one from another as the geometric boundary of the UC. For the locally elastic case, the governing equation exploiting partial differential form is added by specific boundary conditions at the geometric boundary Γ_0 of the UC such as periodic displacement (19.79) and antiperiodic traction (19.81). For the peridynamic bond-based model being considered and utilizing the integral governing equation (19.77), the material points in the vicinity of the geometric boundary Γ_0 of the UC directly interact within the horizon. It defines a volumetric boundary (called the interaction boundary of the UC; see Fig. 19.4) $\Omega_{\Gamma 0} := [-l^{\Omega} - l_c, -l^{\Omega} + l_c] \cup [l^{\Omega} - l_c, l^{\Omega} + l_c]$ surrounding the geometric boundary Γ_0 and limited by the surfaces at the horizon distance from the geometric boundary of the UC. There are no restrictions on the possible intersections of the interaction domains $v_{\Gamma 0}$ and $\Omega_{\Gamma 0}$. In the simplest case, the material properties (such as both the micromodulus and horizon) are identical for any points in the interaction boundary, while the displacements of directly interacting points from the different sides of the geometric boundary are different. Otherwise, it is possible overlapping of the interaction boundary $\Omega_{\Gamma 0}$ with the interaction interface $v_{\Gamma 0} := [-a - l_c, -a + l_c] \cup [a - l_c, a + l_c]$ of the inclusion $v_0 \subset \Omega_0$ embedded into the UC Ω_0 (see for details [1354]). It should be additionally mentioned that the different portions of the interaction boundary $\Omega_{\Gamma 0}$ of the UC Ω_0 belong to three different UCs (Ω_0 and $\Omega_{\pm 1}$) for 1D case and to nine UCs for 2D case. The condition of displacement periodicity at the geometric boundary is extended to the volumetric periodicity of displacements for all points inside the interaction boundary of the UC. We will demonstrate that due to the volumetric PBC (19.78) and (19.79), Eq. (19.77) depends only on the displacement $u(x)$ inside the unit cell $x \in \Omega_0 = [-l^{\Omega}, l^{\Omega}]$ rather than on the displacement $u(y)$ inside the extended unit cell $y \in [-l^{\Omega} - l_\delta, l^{\Omega} + l_\delta]$. Due to the volumetric displacement periodicity (19.78) and (19.79), the special traction boundary conditions (19.80) and (19.81) are not required for the establishment of the micro-to-macro displacement relationship, although these classical traction conditions (19.80) and (19.81) at the geometric boundary of the UC are needed for estimations of both the overall stresses and effective moduli. Of course, a formal simple setting $u_1 \equiv 0$ holds Eq. (19.79), while the equilibrium equation (19.76) is not valid. It means that $u_1 \not\equiv 0$ in general and $u(x)$ is not linear.

The 1D counterpart of Eq. (19.62) is ($i, j \in [-N^{\Omega}, N^{\Omega}]$)

$$\sum_j \mathbb{K}^{\Omega}_{ij} u_{1|j} = -b^{\Omega 0}_i. \tag{19.82}$$

Some nonzero elements of the matrix $\mathbb{K}^{\Omega}_{ij}$ are defined by the micromodulus $C(x_p, x_j)$ ($i \in [-N^{\Omega}, N^{\Omega}], i + p \in [-N^L, N^L]$; see Eq. (19.49))

$$\mathbb{K}^{\Omega}_{i,i+p} = C(x_p, x_i)V_{pi} - \sum_{k=i-n_l}^{i+n_l} C(x_k, x_i)\bar{v}_{ki}, \tag{19.83}$$

while the case $i + p \notin [-N^{\Omega}, N^{\Omega}]$ corresponds to the additional nonzero elements $(i \in [-N^{\Omega}, N^{\Omega}],\ p \in [1, n_l];$ see Eq. (19.58))

$$\mathbb{K}^{\Omega}_{i,i-p+2N^{\Omega}+1} = C(x_p, x_i)\bar{v}_{pi}, \quad (i - p < -N^{\Omega}), \tag{19.84}$$

$$\mathbb{K}^{\Omega}_{i,i+p-2N^{\Omega}-1} = C(x_p, x_i)\bar{v}_{pi}, \quad (i + p > N^{\Omega}). \tag{19.85}$$

A fictitious body force $b^{\Omega 0}_i = b^{\Omega 0}(x_i)$ is defined by the following equation (see Eq. (19.63)):

$$b^{\Omega 0}_i = \langle \varepsilon \rangle^{\Omega} \sum_{k=i-n_l}^{i+n_l} C_1(x_k, x_i)\bar{v}_{ki}(x_k - x_i). \tag{19.86}$$

It is interesting to compare the structures of the stiffness matrices $\mathbb{K}^{\Omega}_{i,i+p}$ corresponding to the 1D cases for the locally elastic (16.6) and peridynamic (19.82) equilibrium equations. So, the known elements $\mathbb{K}^{\Omega}_{i,i+p}$ (19.83) of the discretized peridynamic equation (19.77) form a band with the bandwidth equal to n_l ($n_l = n_c$, see symbolic drawing in Fig. 19.5a). The nonzero elements (19.84) and (19.85) appearing due to the volumetric PBC (19.78) and (19.80) correspond to upper triangular submatrix and low triangular one, respectively, with n_l rows and columns (see Fig. 19.5a). In so doing, a locally elastic counterpart of the band (19.83) is presented by three diagonals (see Fig. 19.5b), while the PBC (19.74) produces two nonzero elements $\mathbb{K}^{\Omega}_{-N^{\Omega},N^{\Omega}}$ and $\mathbb{K}^{\Omega}_{N^{\Omega},-N^{\Omega}}$ (see two crosses in Fig. 19.5b). Thus, in the limit of the horizon going to zero $l_{\delta}/a \to 0$ reducing the peridynamic heterogeneous bar to their locally elastic counterpart, two triangular submatrices corresponding to the volumetric PBC (19.78) and (19.80) in Fig. 19.5a are shrunk into two nonzero elements corresponding to the PBC (19.74) which are marked by the crosses in Fig. 19.5b.

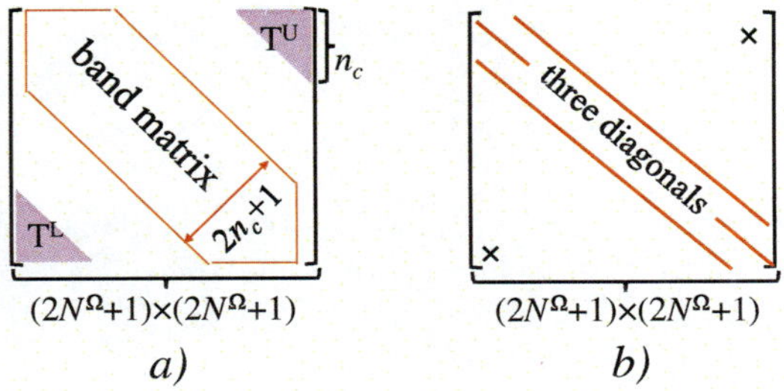

Fig. 19.5 Symbolic representation of the matrix $\mathbb{K}^{\Omega}_{i,j}$ for the peridynamic bar (**a**) and locally elastic bar (**b**)

After the estimation of $u_1(x_i)$ $(-N^\Omega < i < N^\Omega)$ by the quadrature method (19.82) and (19.86), the peridynamic stress distribution (18.87) can be found by the decomposition (19.38)

$$\sigma(x_i) = \sum_{p=i-n_l}^{i} \sum_{q=i}^{p+n_l} C^{(0)}(x_p, x_q)[\hat{u}_1(x_q) - \hat{u}_1(x_p)]\bar{v}_{qi}\bar{v}_{pi} + E^{(0)}\langle\varepsilon\rangle^\Omega$$

$$+ \sum_{p=i-n_l}^{i} \sum_{q=i}^{p+n_l} C_1(x_p, x_q)[\hat{u}(x_q) - \hat{u}(x_p)]\bar{v}_{qi}\bar{v}_{pi} \tag{19.87}$$

with the modified displacements $\hat{u}_1(x_k)$ and $\hat{u}(x_k) = u_0(x_k) + \hat{u}_1(x_k)$ which this bout is replaced by the following representations:

$$\hat{u}_1(x_k) = u_1(x_k), \quad \hat{u}(x_k) = u(x_k), \tag{19.88}$$

$$\hat{u}_1(x_m) = u_1(x_m - (2N^\Omega + 1)h\ \mathrm{sgn}(x_m)), \tag{19.89}$$

$$\hat{u}(x_m) = u(x_m - (2N^\Omega + 1)h\ \mathrm{sgn}(x_m)) + \langle\varepsilon\rangle^\Omega(2N^\Omega + 1)h\ \mathrm{sgn}(x_m) \tag{19.90}$$

at $k \in [-N^\Omega, N^\Omega]$ and $m \in [-N^\Omega - n_l, -N^\Omega - 1] \cup [N^\Omega + 1, N^\Omega + n_l]$, respectively, and $\mathrm{sgn}(x)$ is the sign function: $\mathrm{sgn}(x) = x/|x|$ $(x \neq 0)$ and $\mathrm{sgn}(x) = 0$ $(x = 0)$. Thus, the stress field inside the segment $x_k \in w = [-l^\Omega, l^\Omega]$ only depends on the the displacements found in Eqs. (19.83)–(19.87) in this segment $x_j \in w = [-l^\Omega, l^\Omega]$. Owing to the substitution of the volumetric PBC (19.89) and (19.92) into the constitutive equation (19.87), the antiperiodic traction boundary condition (19.81) (which is not used at the estimation of $u_1(x)$ (19.83)–(19.87)) is automatically fulfilled. Although the condition (19.89) is not directly used at the displacement field estimation $u_1(x_k)$, it can be exploited as a measure of accuracy of the corresponding quadrature method (either the direct method or the decomposition method (19.83)–(19.7.18)).

The discretized equations (19.83)–(19.87) were obtained for one inclusion in the UC. However, these equations can be easily modified for a few inclusions n^{inc} in the UC when the indicator function $V^{(1)}(x)$ is presented as

$$V^{(1)}(x) = \sum_{j=1}^{n^{\mathrm{inc}}}[H(x - x_j^a + a) - H(x - x_j^a - a)], \quad V^{(0)}(x) = 1 - V^{(1)}(x), \tag{19.91}$$

where x_j^a $(j = 1, \ldots n^{\mathrm{inc}})$ is the coordinates of the inclusion centers with an identical size $2a$ and $H(x)$ is the Heaviside function. The micromodulus in the equilibrium equation (17.35) is estimated by Eq. (17.2) with the indicator functions described by Eq. (19.91).

The dependence of the macroscopic stress $\langle\sigma\rangle^{\Omega}$ on the overall macrostrain $\langle\varepsilon\rangle^{\Omega}$ is determined by the effective moduli E^* (see Eq. (19.69))

$$\langle\sigma\rangle^{\Omega} = E^*\langle\varepsilon\rangle^{\Omega}, \tag{19.92}$$

where the overall macrostress $\langle\sigma\rangle^{\Omega}$ and the overall macrostrain $\langle\varepsilon\rangle^{\Omega}$ of the UC Ω_0 presented as the particular cases of Eqs. (19.67) and (19.68)

$$\langle\sigma\rangle^{\Omega} := |\Omega_0|^{-1}\int_{\Omega_0}\sigma(x)dx = |\Omega_0|^{-1}\int_{\partial\Omega_0} t(s)sds, \tag{19.93}$$

$$\langle\varepsilon\rangle^{\Omega} := |\Omega_0|^{-1}\int_{\Omega_0}\varepsilon(x)dx = |\Omega_0|^{-1}\int_{\partial\Omega_0} u(s)n(s)ds \tag{19.94}$$

are obtained in terms of the traction $t(s) := \sigma(s)n(s)$ and the displacement $u(s)$ on the geometric boundary of the unit cell $s \in \partial\Omega_0 = -l^{\Omega} \cup l^{\Omega}$ with the outward normal unit vectors $n(s) = n(\pm l^{\Omega}) = \pm 1$, respectively. Both pairs of Eqs. (19.93_1), (19.94_1) and (19.93_2), (19.94_2) are equivalent, but the pair of Eqs. (19.93_2) and (19.94_2) is preferable because it only requires the estimation of the traction $t(s) := \sigma(s)n(s)$ and the displacement $u(s)$ in two points of the geometric UC boundary $\pm l^{\Omega}$. Moreover, the use of the definition (19.94_1) is even more restrictive because (19.94_1) exploits the strain $\varepsilon(x)$ usually found by numerical differentiation of $u(x)$ ($x \in [-l^{\Omega}, l^{\Omega}]$), while $u(x)$ has the discontinuities at $x = \pm a$ (see for details [131]). Furthermore, at the estimation of the stresses $\sigma(x)$ inside the unit cell $x_k \in [-l^{\Omega}, l^{\Omega}]$ (19.87), we used the volumetric displacement PBC (19.88)–(19.88) but not the antiperiodic traction (19.81). However, owing to the substitution of the volumetric PBC (19.81) into the equilibrium equation (19.77), the antiperiodic PBC (19.81) is automatically fulfilled.

It should be mentioned that the different aspects of "homogenization" with respect to the peridynamic CMs were also considered before in [18, 130, 998, 1399]. However, the definition of the effective elastic modulus (19.92) through the averaged tractions (19.93_2) and displacement function (19.94_2) is first proposed for peridynamic periodic CMs subjected to the new volumetric PBC (19.78)–(19.81) introduced in the current study. For example, Bobaru and Ha [130] (the papers [18, 998] and [1399] were discussed at the beginning of this chapter) have proposed a hybrid method for the analysis of peridynamic functionally graded materials (FGMs). At first, in the framework of classical micromechanics, a popular locally homogenized model is proposed for the estimation of local effective modulus by the mixture rule that means acceptance of the scale separation hypothesis (quantitative restrictions of this hypothesis for the analysis of FGM were considered in [200]). This hypothesis was also implicitly applied at the second step, when the peridynamic micromoduli of the homogenized FGM are assumed to be estimated through the local effective moduli evaluated at the first step. In this chapter, we do not use this very popular engineering hypothesis of scale separation; the numerical examples for estimations of effective moduli of peridynamic CMs with comparable scales of both the horizon l_δ and the unit cell size l^Ω will be considered in the next section.

19.5.3 Numerical Examples

The micromodulus functions with the step-function profile (19.76$_1$) with the same horizon $l_\delta/a = 0.05,\ 0.25,\ 0.5,\ 1.0$ are considered for both the inclusions and matrix with the ratio of the corresponding elastic counterparts (19.76$_2$) $E^{(1)}/E^{(0)} = 3$. It should be mentioned that the variation of the smoothed displacement curves $u(x) \sim x$ (as well as the stress ones) is visually unclearly manifested. Because of this, we will pay our attention to analyses of strain field distributions found by numerical differentiation of $u(x)$: $\varepsilon = \partial u(x)/\partial x$. Buryachenko [221] has analyzed the CMs with the concentration of inclusions $c = 0.1$ for both the locally elastic solution ${}^L\varepsilon(x) \sim x$ (19.72) and the peridynamic ones (19.77)–(19.86) at $l_\delta/a = 0.25$, and $N^\Omega = 500$ and $N^\Omega = 2000$. The curves of relative strains $\tilde{\varepsilon} := \varepsilon(x)/\langle\varepsilon\rangle^\Omega$ vs. x/a behave according to the general features of peridynamic solutions for a homogeneous peridynamic bar with a fictitious body force mentioned in [131] and [1405]. As it is expected, $\varepsilon(x) \to^L \varepsilon(x)$ in the limit of the horizon going to zero $l_\delta/a \to 0$ at some distance of x from the inclusion boundary $x = a$. The peridynamic curves estimated for $N^\Omega = 500$ and $N^\Omega = 2000$, respectively, are pretty close to each other with the exception of the point $x = a$ where the curve estimated for $N^\Omega = 500$ undergoes unexpectedly large break. Owing to this reason, the subsequent estimations are performed for $N^\Omega = 2000$. In particular, in Fig. 19.6, we fix both the inclusion concentration $c = 0.4$ and the material points $2N^\Omega + 1 = 4001$ in the UC, while the horizon ratio l_δ/a varies: $l_\delta/a = 0,\ 0.05,\ 0.25,\ 0.5$ for the curves of the relative strains $\tilde{\varepsilon} := \varepsilon(x)/\langle\varepsilon\rangle^\Omega \sim x/a$ (curves 1, 2, 3, 4, respectively). As can be seen, the first peaks at curves 2, 3, 4 at $x/a = 1 + l_\delta/a$ are lightly growing from 1.732 to 1.842, while the peaks at $x/a = 1 - l_\delta/a$ are growing from 0.490 to 0.523. A behavior of curves 2, 3, and 4 is qualitatively consistent with the corresponding curves $du(x)/dx$ for the peridynamic diffusion (see [1354]) in the vicinity of the interaction interface $[a - l_\delta, a + l_\delta]$. As it is expected, $\varepsilon(x) \to {}^L\varepsilon(x)$ in the limit of the horizon going to zero $l_c/a \to 0$. An adaptive grid refinement technique was proposed in [326] for the analysis of peridynamic problems in the vicinity of an interface that involves a variable horizon size. However, the analysis of possible numerical approaches improving accuracy of the particular problem (such as in Fig. 19.6) is beyond the scope of the current publication focused on the proposal of computational homogenization technique in the theory of peridynamic periodic CMs.

The next step of parametric analysis is the estimation of effective elastic properties E^* for the periodic peridynamic bar (19.92) which are directly defined by the average stress $\langle\sigma\rangle^\Omega$ (19.93) and average strain $\langle\varepsilon\rangle^\Omega$ (19.94) rather than the average polarization tensor $\langle\tau\rangle$ (17.94) for the random medium. In Table 19.1,

Table 19.1 Relative effective Young modulus $E^*/E^{(0)} \sim c$

Method	$c = 0.1$	$c = 0.2$	$c = 0.3$	$c = 0.4$	$c = 0.5$	$c = 0.6$
Local	1.071	1.154	1.250	1.364	1.500	1.667
Random	1.062	1.133	1.214	1.307	1.416	1.544
Period_1	1.161	1.165	1.263	1.388	1.591	1.729
Period_2	1.068	1.162	1.268	1.391	1.592	1.731
Period_3	1.073	1.149	1.247	1.362	1.501	1.669

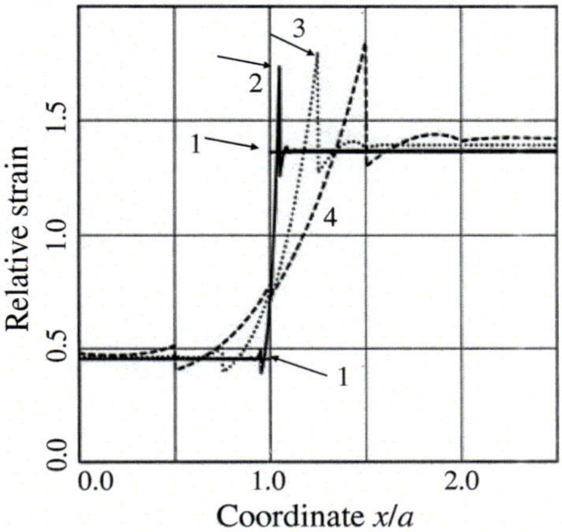

Fig. 19.6 Relative strains $\widetilde{\varepsilon}(x)$ vs. x/a for $c = 0.4$ at $l_\delta/a = 0.0(1)$, $0.05(2)$, $0.25(3)$, and $0.5(4)$

the first line denoted as *Local* contains the exact locally elastic representations of the effective Young modulus $E^*/E^{(0)}$ (19.72) (which is invariant with respect to either the statistically homogeneous random structure or the periodic one) for the different concentrations of inclusions $c \in [0, 0.6]$. The second line *Random* presents estimations of $E^*/E^{(0)}$ (18.94) for the random structure peridynamic bar with the horizon $l_\delta/a = 0.25$ and the step size $h = l_\delta/200$. The third and fourth lines denoted as *Period*_1 and *Period*_2 correspond to the estimation of $E^*/E^{(0)}$ (19.92) by the direct approach (19.51)–(19.55) and the decomposition approach (19.76)–(19.81), respectively, for the periodic structure peridynamic bars with the horizons $l_\delta/a = 0.25$, while the row *Period*_3 is estimated by Eqs. (19.51)–(19.55) for $l_\delta/a = 0.05$; other parameters $N^\Omega = 2000$, $a = 1$ are the same.

The limit of vanishing length scale $l_\delta/a \to 0$ reduces the peridynamic solution $E^* \sim c$ (19.92)–(19.94) to the classical exact result ${}^LE^* \sim c$ (19.72) (compare lines 1, 4, and 5). It is interesting that taking the nonlocal peridynamic properties of constituents into account leads to the softening and stiffening of E^* in comparison with ${}^LE^*$ for the composite bar with the random structure and periodic one, respectively. In so doing, the estimation of E^* (18.94) for the random structures is entirely approximate and obtained in the framework of dilute approximation, while the methods proposed for the analysis of periodic structures (19.51)–(19.55) and the decomposition approach (19.76)–(19.81) do not exploit any additional assumption

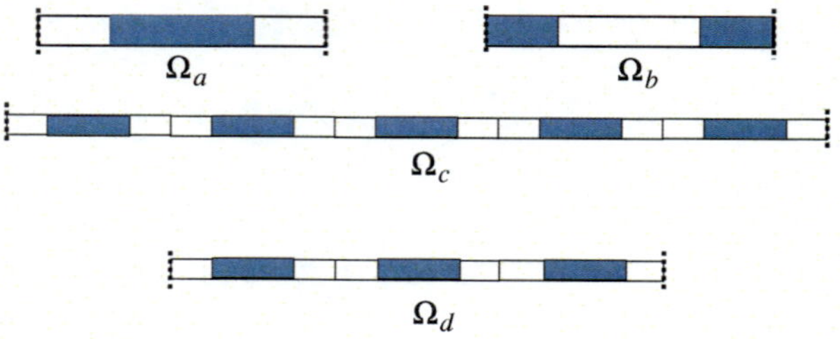

Fig. 19.7 Unit cells Ω_a, Ω_b, Ω_c, and Ω_d

and have a controlled numerical error defined by a discretization scale h. However, at the dilute concentration of inclusions $c = 0.1$ (and $l_\delta/a = 0.25$) when the influence of a structure arrangement (either the random or periodic) is negligible, the estimations E^* for either the random or periodic structures are close to each other (1.062 and 1.068, respectively) and softer than the locally elastic exact value ${}^LE^* = 1.071$.

Of course, a partition of the representative UC Ω_{00} on both the matrix $v^{(0)}$ and the inclusion v_1 is not more than convention. A single restriction is a fixed size of the inclusion $2a$, while the matrix size $2l^\Omega - 2a$ is varying. This is due to the fact (which we are going to prove) that the estimated local fields are invariant with respect to the UC shape (Ω_a and Ω_b (see Fig. 19.7), which are the 1D of the counterparts of UCs Ω_a and Ω_b from Fig. 19.3a). Indeed, a case $a = l^\Omega/2$ (i.e., $c^{(1)} = 0.5$) with $v_0 = [-a, a]$ is symmetric, and we can compare the results $\tilde{\varepsilon}(x) \sim x/a$ ($x \in \Omega_{00}$) obtained for either the stiff inclusions in the soft matrix $E^{(1)} = 3E^{(0)}$ (19.76_2) or the soft inclusions in the rigid matrix $E^{(0)} = 3E^{(1)}$ (19.76_2) (see curves 1 for UC Ω_a, $\tilde{\varepsilon}^{[1]}(x) \sim x/a$, and 2 for UC Ω_b, $\tilde{\varepsilon}^{[2]}(x) \sim x/a$, respectively, in Fig. 19.7 for $l_\delta/a = 0.5$ and $N^\Omega = 2000$). In so doing, for the locally elastic composite bar, the relative strains ${}^L\tilde{\varepsilon} = 0.5$ and ${}^L\tilde{\varepsilon} = 2.0$ for the stiff and soft segments of the bar, respectively, see curve 3 for UC Ω_a in Fig. 19.8. Owing to the symmetry of the problem, we must have $\tilde{\varepsilon}^{[1]}(a - y) \equiv \tilde{\varepsilon}^{[2]}(a + y)$ and $\tilde{\varepsilon}^{[1]}(a + y) \equiv \tilde{\varepsilon}^{[2]}(a - y)$ for $\forall y \in [0, a]$. Indeed, this equality is numerically fulfilled with the numerical error not exceeding 0.55% and 0.32% for $N^\Omega = 1000$ and $N^\Omega = 2000$, respectively.

Up to now, we considered the cases for one inclusion in the representative UC Ω_{00}. However, a prescription of the indicator function $V^{(1)}(x)$ in the form (19.87) allows us to analyze any number of inclusions n^{inc} with any their arrangement in the UC Ω_{00}. In particular, we compare the results $\tilde{\varepsilon}(x) \sim x/a$ at $c^{(1)} = 0.5$ estimated for three different periodic distributions of identical inclusions (see Fig. 19.7) $x_i^a = 0, \pm\Delta, \pm 2\Delta$ ($n^{\mathrm{inc}} = 5$, UC Ω_c), $x_i^a = 0, \pm\Delta$ ($n^{\mathrm{inc}} = 3$, UV Ω_d), and $x_i^a = 0$ ($n^{\mathrm{inc}} = 1$,UC Ω_a), where Δ stands for a distance between the centers of inclusions $2a < \Delta = 2l^\Omega/n^{\mathrm{inc}}$. In Fig. 19.9, a curve $\tilde{\varepsilon}(x) \sim x/a$ is presented for five inclusions $n^{\mathrm{inc}} = 5$ at $l_\delta/a = 0.5$ and $N^\Omega = 2000$. Due to the periodicity of all examples considered, the values $\tilde{\varepsilon}(x) \sim x/a$ at $0 < x < \Delta/2$ for $n^{\mathrm{inc}} = 1,\ 3,\ 5$

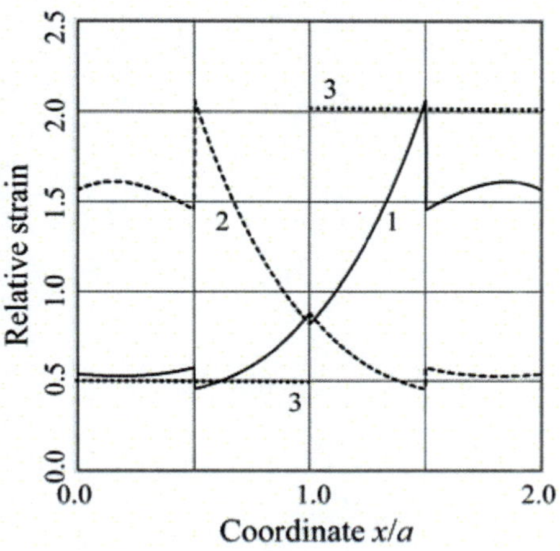

Fig. 19.8 Relative strains $\tilde{\varepsilon}^{[1]}(x)(1)$ and $\tilde{\varepsilon}^{[2]}(x)(1)$ vs. x/a for $c^{(1)} = 0.5$: $l_c/a = 0.5$ (curves 1 and 2)

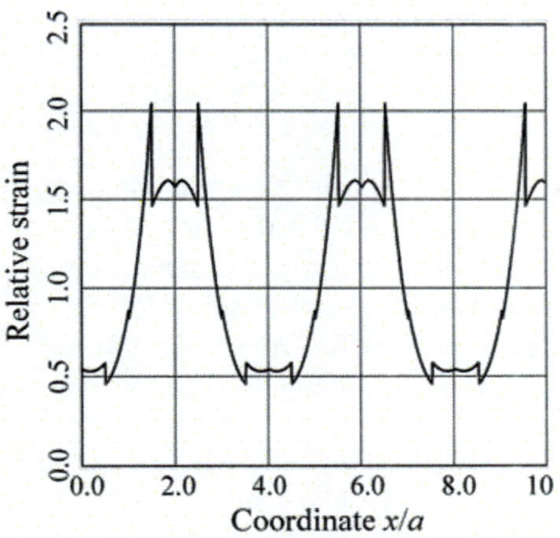

Fig. 19.9 Relative strain $\tilde{\varepsilon}^{[1]}(x)$ vs. x/a for five inclusions n^{inc} in UC

should coincide that is confirmed by numerical difference between these values which is not exceeding 0.36%.

The classical problem of micromechanics is the estimation of effective elastic properties E^* for the periodic peridynamic bar (19.53) which are directly defined by the average stress $\langle\sigma\rangle^{\Omega}$ (19.89) and average strain $\langle\varepsilon\rangle^{\Omega}$ (19.90). In Fig. 19.10, the curves $E^*/E^{(0)} \sim c^{(1)}$ 1–4 are estimated for $l_\delta/a = 0,\ 0.1,\ 0.25, 0.5$, respectively, at $N^{\Omega} = 2000$. The limit of vanishing length scale $l_\delta/a \to 0$ reduces the peridynamic solution $E^* \sim c^{(1)}$ (19.88) to the classical exact result ${}^{L}E^* \sim c$ (19.71$_1$) (compare curve 1 with curves 2, 3, and 4). It is interesting that taking the nonlocal peridynamic properties of constituents into account leads to the stiffening of E^* in comparison with ${}^{L}E^*$ for the composite bar of the periodic structure.

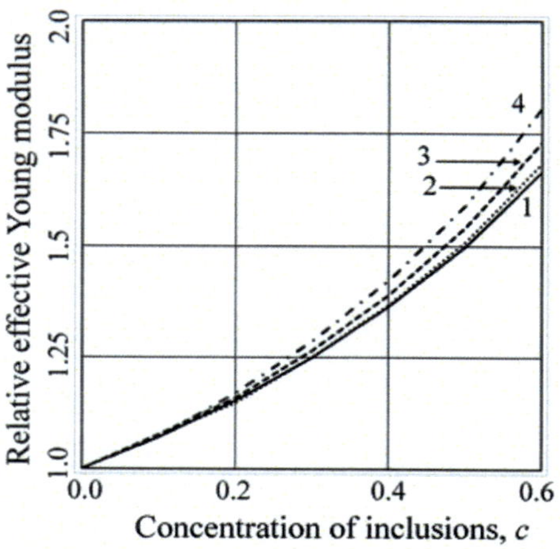

Fig. 19.10 Relative effective Young modulus $E^*/E^{(0)}$ estimated for $l_c/a = 0.0$ (1), 0.1 (2), 0.25 (3), and 0.5 (4)

19.6 Effective Moduli of Peridynamic Medium with Periodically Distributed Damage

Damage is a basic concept in engineering design exploited for predictions about the initiation, propagation, and fracture of materials without molecular scale which can be too complex for practical engineering analysis (see the pioneering works by Kachanov [725] and Rabotnov [1268] as well as [727, 801]). Damage mechanics uses state variables (which are defined by an indication method) representing their effects on the exploitation properties of the material, e.g., the stiffness and remaining life. Popular state variables can be measured and used for proper definition of the damage variable through density change, resistivity change, acoustic emission, change of the fatigue limit, change of stiffness, and microcrack density used in micromechanical damage formulations for prediction of both damage initiation and evolution. Physically justified damage theory was proposed in peridynamics (see [1396, 1401]), where damage is introduced by permitting damaged bonds to break irreversibly, i.e., damaging of the bond is considered as its removing if the bond stretch exceeds a critical stretch for some time. Damage in a bond is treated as bond breakage where some scalar function called damage jumps discontinuously from 0 to 1 if the bond strain exceeds some critical bond strain. In other words, damaged medium is considered as a composite material in which some two-point micromodulus functions vanish. The section is dedicated to the estimation of the effective moduli of damaged medium (initially homogeneous) as the functions of damage, damage functions, and micromodulus profile. One also analyzes either the translation-invariant damage functions or non-translation-invariant ones. In so doing, all mentioned parameters of damage are fixed and assumed to be known, i.e., we do not consider accumulation of damage and their dependence on the applied loading as well as crack initiation and propagation (see, e.g., [15, 594]).

The term "composite materials" is widely used in the current section due to an intensive exploitation of the micromechanical tools previously developed for exploration of CMs (of either periodic or random structure) with either locally elastic or peridynamic constituents. In so doing, we keep in mind the application of some adopted versions of the micromechanical approaches mentioned above to the particular cases of matrix CMs consisting of a homogeneous matrix filled by either the microcracks and voids (in the locally elastic case) or the bond breakage (in the peridynamic case).

19.6.1 Decomposition of the Material and Field Parameters

According to [1401], a damage is considered as missing of the corresponding bonds. Then for an undamaged (pristine) material $C(x,\hat{x}) \equiv C^{(0)}(x,\hat{x})$, while the micromodulus of damaged medium can be presented in the form

$$C(x,\hat{x}) = \mu(x,\hat{x})C^{(0)}(x,\hat{x}), \tag{19.95}$$

where $\mu(x,\hat{x})$ is a damage function that takes on a value of 0 or 1 ($\xi = x - \hat{x}$)

$$\mu(x,\xi) = \begin{cases} 0, & \text{the bond } \xi \text{ is missing,} \\ 1, & \text{the bond } \xi \text{ is acting.} \end{cases} \tag{19.96}$$

The local damage at a point x is defined as an average damage function within horizon

$$\varphi(x) = 1 - \int_{\mathcal{H}_x} \mu(x,\xi)d\xi \Big/ \int_{\mathcal{H}_x} d\xi. \tag{19.97}$$

Thus, the local damage is not sensitive to the detailed distribution of bond breakages described by the damage function $\mu(x,\xi)$ (19.96).

We decompose the material parameter (the micromodulus) (19.95) (see for details [222]) as

$$C(x,\hat{x}) = C^{(0)}(x,\hat{x}) + C_1(x,\hat{x}), \tag{19.98}$$

where $C_1(x,\hat{x}) := C(x,\hat{x}) - C^{(0)}(x,\hat{x}) = (\mu(x,\xi) - 1)C^{(0)}(x,\hat{x})$ is a jump of micromodulus with respect to the micromodulus of the matrix (undamaged medium). The decomposition (19.98) leads to the decomposition of both the peridynamic operator (19.30) and the stress one (19.31)

$$\mathcal{L} = \mathcal{L}^{(0)} + \mathcal{L}_1, \quad \mathcal{L}^{\sigma} = \mathcal{L}^{\sigma(0)} + \mathcal{L}_1^{\sigma}, \tag{19.99}$$

where the operators $\mathcal{L}^{(0)}$ and $\mathcal{L}_1$ [and $\mathcal{L}^{\sigma(0)}$ and $\mathcal{L}_1^{\sigma}$] are expressed by the representations (17.35) [and (17.94)] at $C(x,\hat{x}) = C^{(0)}(x,\hat{x})$ and $C(x,\hat{x}) = C_1(x,\hat{x})$, respectively.

We also use the displacement field decomposition (19.76) with the volumetric PBC (19.78)–(19.81). Substitution of the volumetric PBC (19.80) and the field decomposition (19.76) into the equilibrium equation (19.77) with subsequent discretization allows one to get the following linear algebraic system ($i, j \in [-N^{\Omega}, N^{\Omega}]$):

$$\sum_j \mathbb{K}_{ij}^{\Omega\mu} u_{1|j} = -b_i^{\Omega 0\mu}, \tag{19.100}$$

where the matrix $\mathbb{K}_{ij}^{\Omega\mu}$ and coefficient vector $b^{\Omega 0\mu}$ are expressed through $\mathbb{K}_{ij}^{\Omega\mu}$ (19.82) and $b^{\Omega 0\mu}$ (19.86) determined before

$$\mathbb{K}_{ij}^{\Omega\mu} = \mu_{ij}\mathbb{K}_{ij}^{\Omega}, \quad b_i^{\Omega 0\mu} = -b_i^{\Omega 0} + \langle\varepsilon\rangle^{\Omega} \sum_{k=i-n_c}^{i+n_c} \mu_{ik} C^{(0)}(x_k, x_i)\bar{v}_{ki}(x_k - x_i). \tag{19.101}$$

After the estimation of $u_1(x_i)$ ($-N^{\Omega} < i < N^{\Omega}$) by the quadrature method (19.100) and (19.78)–(19.81), the peridynamic stress distribution (17.94) can be found by the use of the decomposition (19.38)

$$\begin{aligned}\sigma(x_i) = &\sum_{p=i-n_l}^{i} \sum_{q=i}^{p+n_l} C^{(0)}(x_p, x_q)[\hat{u}_1(x_q) - \hat{u}_1(x_p)]\bar{v}_{qi}\bar{v}_{pi} + E^{(0)}\langle\varepsilon\rangle^{\Omega} \\ &+ \sum_{p=i-n_l}^{i} \sum_{q=i}^{p+n_l} (\mu_{qp} - 1) C^{(0)}(x_p, x_q)[\hat{u}(x_q) - \hat{u}(x_p)]\bar{v}_{qi}\bar{v}_{pi}\end{aligned} \tag{19.102}$$

with the modified displacements $\hat{u}_1(x_k)$ and $\hat{u}(x_k) = u_0(x_k) + \hat{u}_1(x_k)$ defined by Eqs. (19.88)–(19.90).

At last, the average fields $\langle\sigma\rangle^{\Omega}$ and $\langle\varepsilon\rangle^{\Omega}$ and the effective Young modulus are found by Eqs. (19.93), (19.94), and (19.92), respectively.

19.6.2 Numerical Results

A $1D$ example of an infinite periodically damaged peridynamic bar is analyzed for the next geometric and mechanical properties of constituents. We consider an infinitely long bar of a constant cross section $A = 1$ with the micromodulus functions described by both the step-function (SF) profile (16.82_1) and the triangle-

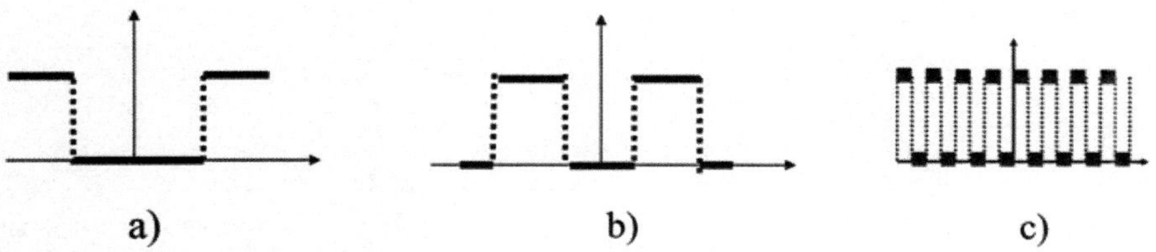

Fig. 19.11 Sketches of the damage functions (**a**), (**b**), and (**c**)

function (TF) profile (16.82_2); the same values of the undamaged Young modulus $E^{(0)} = 1$ and the horizon l_c are analyzed in both cases of the profiles.

The midpoint quadrature scheme (similar to [213]) is realized for the uniform grid $x_i = (0, \pm 1, \pm 2, \ldots, \pm N^{\Omega})$ with the numbers of material points $2N^{\Omega} + 1$ and $2n^l + 1$ inside both the UC Ω_0 and the horizon l_c, respectively, and a constant step size $h = l^{\Omega}/N^{\Omega}$. For both cases of the micromoduli (16.82_1) and (16.82_2) (see Fig. 16.4), the following damage functions are tested ($i \in [-N^{\Omega}, N^{\Omega}]$, $p \in [-n^l, n^l]$):

$$\mu_{ip} = \begin{cases} 0, & |p| \leq 0.5n^l, \\ 1, & \text{otherwise}, \end{cases} \quad \mu_{ip} = \begin{cases} 0, & 0.25n^l \leq |p| \leq 0.75n^l, \\ 1, & \text{otherwise}, \end{cases} \tag{19.103}$$

$$\mu_{ip} = \begin{cases} 0, & p \text{ is odd}, \\ 1, & p \text{ is even}. \end{cases} \tag{19.104}$$

Thus, all three damage functions (19.103_1), (19.103_2), and (19.104) (damage functions a, b, and c, respectively; see Fig. 19.11) correspond to the homogeneous local damage $\varphi(x) \equiv 0.5$ ($x \in \Omega_0$) (19.97). E^* of the damaged peridynamic bar are estimated for the different combinations of the micromodulus profiles SF and TF and the various damage functions (19.103_1) (a), (19.103_2) (b), and (19.104) (c). The accuracies of these numerical estimations are analyzed by comparison of results obtained for the following step sizes and scale parameters: $h = l^{\Omega}/400$, $n^l/n^{\Omega} = 0.1$ (case 1), $h = l^{\Omega}/400$, $n^l/n^{\Omega} = 1.0$ (case 2), $h = l^{\Omega}/2000$, $n^l/n^{\Omega} = 0.1$ (case 3), $h = l^{\Omega}/2000$, $n^l/n^{\Omega} = 1.0$ (case 4), respectively.

As can be seen in Table 19.2, the effective moduli E^* differ one from another for the different micromodulus profiles (either TF or SF) and the same damage function (either (a), (b), or (c)). Moreover, we get the different E^* at the fixed damage function (either (a), (b), or (c)) and the different micromodulus profiles (either SF or TF). It is also interesting that the effective moduli E^* estimated for both SF and TF and the damage functions (c) are close to the popular engineering approximation (see, e.g., [727])

$$E^* = \big[1 - \varphi(x)\big] E^{(0)}. \tag{19.105}$$

Table 19.2 Relative effective Young modulus $E^*/E^{(0)}$

Profile	SF	TF	SF	TF	SF	TF
Dam. func.	(a)	(b)	(c)	(a)	(b)	(c)
Case 1	0.8654	0.4014	0.5181	0.6686	0.6596	0.5012
Case 2	0.8740	0.4058	0.5019	0.6856	0.6847	0.5000
Case 3	0.8740	0.4109	0.5019	0.6835	0.6875	0.5000
Case 4	0.8746	0.4066	0.5007	0.6867	0.6875	0.5000

It is not an accidental coincidence. Indeed, for the homogeneous loading ($\mathbf{u}(x) = \varepsilon x$), it can be seen from (19.97)–(19.97 and) the following constitutive equations (see [1405]):

$$\sigma = E^{(0)}\varepsilon, \;\; E^{(0)} = \int_0^\infty r^2 C^{(0)}(r)\, dr \tag{19.106}$$

that Eq. (19.105) is asymptotically fulfilled for any dimension ($d = 1, 2, 3$) and any continuous undamaged micromodulus profile $C^{(0)}$ (e.g., either SF or TF) in the case of the homogeneous damage distribution (like (19.104)) if the node number n^l within horizon is large enough (compare the rows corresponding to the case pair 1 and 3 as well as the case pair 2 and 4 in Table 19.2). Let us compare the effective moduli E^* estimated in cases 3 and 4 (with the same h and the scale parameters $n^l/n^\Omega = 0.1$ and 1, respectively) which have a locally elastic counterpart problem for periodic CMs containing either one or ten voids in the unit cell. The close values of the rows corresponding to cases 3 and 4 indicate on correctness of the averaging scheme proposed and based on the volumetric PBCs (19.78)–(19.81). By this means, the theory of averaging of periodic peridynamic media is verified (at least for particular cases of the damage model for 1D case).

The results of case 2 have a comparable accuracy with the results corresponding to case 4, although their step sizes $h = l^\Omega/400$ and $h = l^\Omega/2000$ differ by a factor of 5. Comparison of columns 1–3 of the micromodulus SF (16.82_1) with the same local damage $\varphi(x) \equiv 0.5$ leads to a conclusion that removing the remote bounds (as in damage function (19.103_2)) leads to more significant reduction of the stiffness E^* than removing the bonds placed near the center of the family (as in damage function (19.103_1)). The case of the homogeneous distribution of removed bonds (19.104) occupies an intermediate position in effect of the damage function on the effective modulus E^*. The mentioned feature of the damage functions is less manifested for the triangle profile of the micromodulus TF (16.82_2) where the remote bounds (at $|\xi| \geq 0.75$) of the family have a reduced rigidity in comparison with the central bonds of the family.

It should be mentioned that the concept of peridynamic damage (19.95) (initiated by Silling [1396]; see also [1401]) expressed through the damage function (19.96) is universally recognized by peridynamic society and widely used for analyses of the finite size samples of either initially homogeneous media or CMs. However, the infinite damaged media is of special interest just because the classical continuum damage (CD) theory (see the books [727, 801, 1569]) was initially conceptually introduced for statistically homogeneous media as a state variable affecting on the exploitation properties of this infinite media. In other words, the so-called scale separation hypothesis is implicitly assumed to be valid that means that both the sample scale and the applied field scale infinitely exceed a damage scale. Moreover, it is also interesting to establish a relationship between the CD theories and the peridynamic damage.

The popular idea for the incorporation of existing classical CD models into the peridynamic damage theory is an approximation of some peridynamic parameters by a function depending on a single parameter φ, which is commonly the case in some of the most popular damage models used in practice. So, one [1543] proposed to describe an influence function of damaged medium to be proportional to the conventional (radial) influence function (for a pristine medium) with the proportionality factor $(1 - \varphi)$ depending upon the state of damage between two points defining a bond. Han et al. [602] used an approximation for the damaged stiffness tensor (called the effective stiffness in our notation) defined by the micromodulus which, in its turn, depends on the critical local damage rather than on the damage function. Thus, the known incorporations of CD concept into peridynamics are performed by the presentations of one or another peridynamic parameter (e.g., either influence function, micropotential, or micromodulus) as the functions of a state of damage (which is commonly the case in some of the most popular damage models used in practice). In the fundamental work by Littlewood et al. [948], the damage state (which is conceptually more general than the local damage φ) is introduced within the most general thermodynamic framework with the strain energy density function presented as a dot product of micropotential of undamaged bonds and the damage state.

However, according to Table 19.2 (where each value $E^*/E^{(0)}$ is estimated for the same local damage $\varphi = 0.5$ (19.97) and the same $E^{(0)}$), the effective modulus E^* has significantly more sensitive dependence on details of both the damage parameters (such as the damage function (19.96)) and the peridynamic properties (such as micromodulus profile either SF (16.82_1) or (SF (16.82_2)) rather than on only one parameter of the local damage (19.97)). Only in a limiting case of both the homogeneous distribution of damaged bonds within horizon (like (19.104)) and the large node number n^l, the peridynamic periodic damage model is reduced to the classical CD model (19.105). In some sense, a goal of the proposed approach can be considered as inversion with respect to the goals of the known approaches (see [602, 1543], and the most general version of peridynamic damage theory proposed in [948]). We do not try to incorporate the concept of the CD theory into peridynamics. Rather the opposite, we extracted, in particular, apparently most popular equation of CD theory (19.105) as a particular case of the most simple peridynamic damage theory (19.95)–(19.97).

It should be mentioned that in engineering applications, the inhomogeneous in the space effective modulus (19.105) (and their 2D and 3D counterparts) is very popular. This inhomogeneity means (from a micromechanical point of view, see Chap. 12 in [200]) a consideration of the CM as a functionally graded material, when the effective properties are described, in general, by the effective integral operator rather than by the effective modulus. In such a case, an algebraic form of the effective constitutive equation (either (19.92) or (19.105)) with $E^*(x) \not\equiv$const. has a practical meaning in the framework of the scale separation hypothesis. It means that one treats three material length scales (see, e.g., [1523]): the macroscopic scale L, characterizing the extent of w, the microscopic scale l^Ω, related with a size of a unit cell Ω_0, and the horizon l_δ. Moreover, one supposes that applied field varies on a characteristic length scale Λ. The limit of our interests for both the material scales and field ones is $L \gg \Lambda \gg l^\Omega \geq l_\delta$.

We have considered the most simple cases with the fixed both the undamaged micromodulus $C^{(0)}(x, \xi)$ and the translation-invariant damage functions $\mu(x, \xi)$ (i.e., $\mu(x, \xi) = \mu(x + y, \xi)$, $\forall x, y \in \Omega_0$) inside the unit cell $x \in \Omega_0$ of the periodic medium subjected to the homogeneous volumetric PBC (19.78)–(19.81) with $\langle \varepsilon \rangle^\Omega \equiv$const. In terms of micromechanics of locally elastic CMs, it means an estimation of effective moduli of periodic medium subjected to the homogeneous PBC. The next step in the development of the approach proposed is a consideration of translation non-invariantness of a damage function, i.e., $\mu(x, \xi) \not\equiv \mu(x - y, \xi)$.

Namely, we analyze a damaged domain $v^\mu := [-a^\mu, a^\mu] \subset \Omega_0$ with the indicator function $V^\mu(x)$ ($x \in \Omega_0$) so that a modified damage function $\mu^{\rm m}_{ip}$ is produced from the damage functions (19.103) and (19.104) as

$$\mu^{\rm m}_{ip} = \begin{cases} \mu_{ip}, & x_i \in v^\mu, \\ 1, & x_i \notin v^\mu, \end{cases} \tag{19.107}$$

which can be considered as a model for the peridynamic generalization of the classical feature of damage localization. For definiteness, we will use in Eq. (19.107) the damage function $\mu(x, \xi)$ (19.104) providing the most significant reduction of stiffness of the peridynamic medium with the step-function profile SF (16.82_1).

At first, we consider the popular engineering approach based on the so-called scale separation hypothesis. Namely, we assume that the effective modulus $E^{*[1]}$ is estimated for the damaged domain v^μ by the use of Table 19.2. Then the locally elastic approximation of the Young modulus inside the unit cell

$$E(x) = E^{*[1]} V^\mu(x) + E^{(0)}[1 - V^\mu(x)] \tag{19.108}$$

is assumed to be asymptotically fulfilled at

$$l^c / a^\mu \ll 1 \tag{19.109}$$

that leads to the exact known representations for both the total effective modulus and the strain concentration factors of the periodic damaged bar

$$\langle \sigma \rangle^\Omega = E^{*[2]} \langle \varepsilon \rangle^\Omega, \quad E^{*[2]} = [c^{(1)} / E^{*[1]} + c^{(0)} / E^{(0)}]^{-1},$$

$$\varepsilon(x) = \begin{cases} E^{*[2]}/E^{(0)}\langle\varepsilon\rangle^{\Omega}, & x \in \Omega_0 \backslash v^{\mu}, \\ E^{*[2]}/E^{*[1]}\langle\varepsilon\rangle^{\Omega}, & x \in v^{\mu}, \end{cases} \tag{19.110}$$

where the assumptions $c^{(1)} = \langle V^{\mu}(x)\rangle^{\Omega}$ and $c^{(0)} = 1 - c^{(1)}$ are admitted.

In Fig. 19.12, one presents the effective elastic modulus $E^{*[2]} \sim c^{(1)}$ (curve 1) estimated by the engineering approach (19.110_2) obtained in the framework of the scale separation hypothesis (19.109). One also presents the results of the direct peridynamic estimation of E^* (19.92) obtained without additional assumption (like (19.109)) but taking into account the translation non-invariantness of a damage function (19.104) and (19.107). Curves 2 and 3 of $E^* \sim c^{(1)}$ are plotted for the different scale ratios $n^l/N^{\Omega} = 0.2$ and $n^l/N^{\Omega} = 0.5$, respectively. As can be seen, $E^{*[2]} = E^*$ at $c^{(1)} = 0$ and $c^{(1)} = 1$, and, moreover, the intermediate values $E^{*[2]} \sim c^{(1)}$ and $E^* \sim c^{(1)}$ are pretty close one to another at $0 < c^{(1)} < 1$. Thus, the known engineering approach (19.110_2) provides an admissible accuracy for the effective modulus estimation E^* (if we use, of course, the peridynamic estimations of $E^{*[1]}$ for the damaged cloud v^{μ}). However, this illusion of an engineering accuracy is dissipated if we compare the strain concentration factors obtained by either the engineering approach (19.110_3) or the corresponding values $\varepsilon(x)$ evaluated by Eqs. (19.100) and (19.101) with the non-translation-invariant damage function (19.103_2) and (19.107).

Indeed, in Fig. 19.13, one presents the strain concentration factors $A^*(x)$: $\varepsilon(x) = A^*(x)\langle\varepsilon\rangle^{\Omega}$ ($\langle\varepsilon\rangle^{\Omega} = 1$) estimated for the fixed number of nodes in both the UC $N^{\Omega} = 2000$ and the damaged domain $n^{\mu} = 400$ ($n^{\mu} := (a^{\mu}/l^{\Omega})N^{\Omega}$). Curve 1 corresponds to the engineering approach (19.110_3), while curves 2, 3, and 4 are estimated by Eqs. (19.100) and (19.101) with the non-translation-invariant damage function (19.104), (19.107) with the scale ratios $l_c/a^{\mu} = 0.5,\ 1$, and 2.5, respectively. It can be seen that increasing the nonlocal effects (manifested by growing scale ratios l_c/a^{μ}) leads to significant redistribution of strains $\varepsilon(x)$. So, the strains inside the damaged domain v^{μ} decrease in average, while the strain concentration factors in the undamaged matrix $\Omega_0\backslash v^{\mu}$ significantly increase in some points. The strain distribution inside the undamaged medium $\Omega\backslash v^{\mu}$ is strongly inhomogeneous described by a factor of 3 with the maximum $A^*(x)$ in the vicinity of the damaged domain v^{μ}. The detected strain redistributions in damaged material quantitatively explain a qualitative feature of damaging proposed in [36] where the breakage of individual bonds more likely generates its neighbors break, because more load is shifted to their neighbors. As more and more bonds break, a progressive process of damage accumulation occurs in the material. So, the detected feature of redistribution of the strain concentration factors due to nonlocal effects is critical for subsequent prediction of damage evolution (that is beyond the scope of the current study.

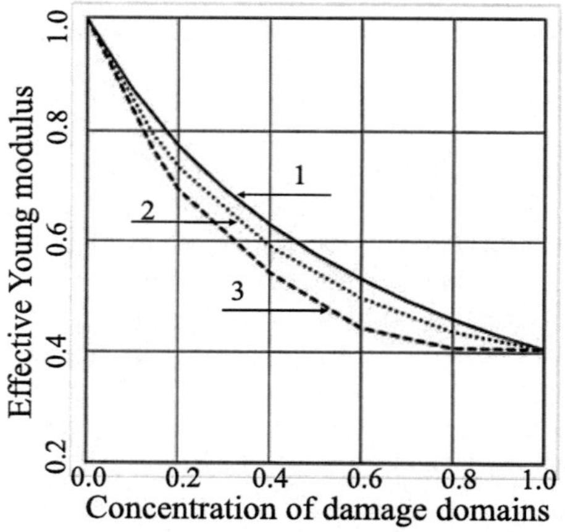

Fig. 19.12 Relative effective Young modulus $E^* \sim c^{(1)}$ estimated by either Eq. (19.110$_2$) (curve 1) or Eq. (19.92) at $n^l/N^\Omega = 0.2$ and $n^l/N^\Omega = 0.5$ (curves 2 and 3, respectively)

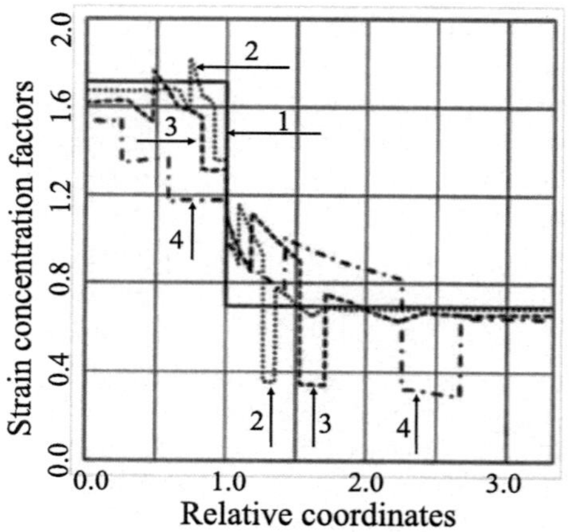

Fig. 19.13 Strain concentration factors $A^*(x)$ vs. relative coordinates x/a^μ estimated either by Eqs. (19.110$_3$) (curve 1) or Eqs. (19.100) and (19.101) at $l_c/a^\mu = 0.5,\ 1$, and 2.5 (curves 2, 3, 4)

19.7 Volumetric Bloch-Floquet Periodic Boundary Condition

For the locally elastic periodic CMs, logical generalization of the PBC for the static problems is the Bloch periodic boundary conditions (BPBCs) for the wave propagation problems. One of the popular wave-based methods is a wave finite element method applying directly the Bloch [128] periodic boundary conditions (BPBCs) to the unit cell mass and stiffness matrices with proper defined Bloch operators effectively limiting the computation to a single unit cell. Proposal of the new volumetric BPBCs opens the opportunities for analyses of a wide class of wave propagation in periodic peridynamic composites (e.g., metamaterials).

19.7.1 Discretized Peridynamic Equation in Frequency Domain

We consider a harmonic loading when the field quantities $\hat{f}$ are harmonic at frequency ω, that are (for 1D case),

$$\check{f}(x,t) = f(x)\exp(-i\omega t), \tag{19.111}$$

where $\check{f}(x,t)$ represent the field variables (e.g., force density $\check{b}$, stress $\check{\sigma}$, displacement $\check{u}$, acceleration $\ddot{\check{u}}$), whereas f represents their periodic parts (e.g., $f = b, \sigma, u$) which are generally the complex coefficients carrying frequency-dependent magnitude and phase information. Substituting Eq. (19.111) into Eq. (19.103) transforms Eq. (19.103) into the frequency domain

$$\mathcal{L}^{\omega} * u(x) + b(x) = 0, \quad \mathcal{L}^{\omega} * u(x) := \rho\omega^2 u(x) + \int_{-\infty}^{\infty} C(x,\hat{x})[u(\hat{x}) - u(x)]d\hat{x}. \tag{19.112}$$

Thus, the integrodifferential equation (19.103) is reduced to the integral equation (19.112) with the time variation removed and instead frequency is introduced as a parameter. The locally elastic counterpart ${}^{L}\mathcal{L}^{\omega}$ of the operator $\mathcal{L}^{\omega}$ has a representation

$${}^{L}\mathcal{L}^{\omega} * u(x) + b(x) = 0, \quad {}^{L}\mathcal{L}^{\omega} * u(x) := \rho\omega^2 u(x) + E\frac{\partial^2 u(x)}{\partial^2 x} \tag{19.113}$$

with the Young modulus E.

Let us consider an infinite heterogeneous bar $w = (-\infty, \infty)$ constructed from the infinite number of unit cells containing identical inclusions v_k (19.73), e.g., for the representative unit cell $\Omega_0 = [-l^{\Omega}, l^{\Omega}]$ with the boundary $\Gamma^0 = \Gamma^0_{-1} \cup \Gamma^0_1 := -l^{\Omega} \cup l^{\Omega}$ and $v_0 = [-a, a]$ $(a < L^{\Omega})$. In view of the periodicity of the bar, we have

$$C(x,\hat{x}) = C(x+ml^{\Omega}, \hat{x}+ml^{\Omega}), \;\; \rho(x) = \rho(x+ml^{\Omega}), \;\; V^{(k)}(x) = V^{(k)}(x+ml^{\Omega}), \tag{19.114}$$

where $m \in Z$ is an integer.

In numerical implementation described by Silling and Askay [1401], the region w is discretized into a set of nodes p, each with a finite known volume (called full volume) defined by the size h. Taken together, the nodes form a grid $x_j = jh$ $(j = 0, \pm1, \pm2, \ldots)$ covering the total bar w. The spatially discretized form of Eq. (19.113) replaces the integral by the finite sum for each node j. For a given distribution x_j, $j = 0, \pm1, \pm2, \ldots$, it leads to the infinite linear algebraic system $(i, j = 0, \pm1, \pm2, \ldots)$

$$K_{ij}u_j - \omega^2 M u_i = -b_i \tag{19.115}$$

with the given vector b_i corresponding to the known fictitious body force density. Hereafter, $u_j = u(x_j)$ ($\mathbf{u} = (\ldots, -u_2, -u_1, u_0, u_1, u_2, \ldots)^\top$) denotes the nodal displacements, $\bar{v}_{pj}$ is the portion of node p "volume" covered by the horizon of node j, and a constant step size $h = l_\delta / n_l$ is assumed in the midpoint integration method used. We designedly use the notations popular in a well-developed area of structural dynamics, where **K** and **M** are called the stiffness matrix and mass matrix, respectively, which transform the nodal displacement into the internal nodal force vector. The mass matrix is a scalar matrix, but for notional convenience, it has been transformed to a matrix form by multiplication with an identity tensor $\mathbf{M} = \rho\mathbf{I}$:

$$\mathbf{Ku} - \omega^2\mathbf{Mu} = -\mathbf{b}. \tag{19.116}$$

The nonzero elements of the band matrix K_{ij} (with the bandwidth equal to n_l) are defined by the micromodulus $C(x_p, x_j)$ (19.107) ($i = 0, \pm 1, \ldots, p \in [-n_l, n_l]$; $i, p \in Z$)

$$K_{i,i+p} = C(x_p, x_i)\bar{v}_{pi} - \sum_{k=i-n_c}^{i+n_c} C(x_k, x_i)\bar{v}_{ki}, \tag{19.117}$$

while for the locally elastic counterpart, the matrix K_{ij} is a tridiagonal one ($i = 0, \pm 1, \ldots, p \in [-1, 1]$)

$$K_{i,i+p} = \frac{E}{h^2}(3|p| - 2). \tag{19.118}$$

19.7.2 Bloch-Floquet Condition for Locally Elastic CMs

For the solution of Eq. (19.116), we need to complement it by the boundary conditions reflecting a periodic structure of a bar (19.114). In Section 19.7.2, we consider a locally elastic composite bar with the stiffness matrix (19.118). Strictly, when the segment $\Omega_0 = [-l^\Omega, l^\Omega]$ is repeated, there will be overlapping of the segments Ω_k at the boundaries. However, the Bloch-Floquet theorem can be used to remove the redundant displacement fields **u** at the segment boundaries $\pm l^\Omega(1+2m)$ (m is an integer) so that the model effectively limits the computation to the unit cell. We reproduce the Bloch-Floquet condition for locally elastic CMs in a form adopted for subsequent representation of the analysis of peridynamic CMs.

The Bloch-Floquet theorem relates the time harmonic response at a given point in a unit cell to the corresponding point in an adjacent unit cell. The displacements and tractions at the unit cell boundaries Γ_α^0 and Γ_β^0 ($\alpha = -\beta = \pm 1$) are related by Bloch [128] conditions (the one-dimensional temporal version of the Bloch theory is known as the Floquet theory; see [487])

$$u(x_\alpha) = \lambda_\beta u(x_\beta), \quad t(x_\alpha) = -\lambda_\alpha t(x_\beta), \tag{19.119}$$

where

$$\lambda_\beta = \lambda_\alpha^{-1} = e^{-2ikl^\Omega \operatorname{sign}(\beta)}, \tag{19.120}$$

k is the wave vector, $i = \sqrt{-1}$, and $t(\pm l^\Omega) = n(\pm l^\Omega)\sigma(\pm l^\Omega)$ denotes the traction defined by the stresses σ on the boundaries $\pm l^\Omega$ with the outward normal unit vectors $n(\pm l^\Omega) = \pm 1$, respectively.

For the line $i \in [N^\Omega, N^\Omega]$ of Eqs. (19.116) and (19.118), $\mathbf{u}$ holds two grid points assigned to neighboring cells $\mathbf{u}^- = (u_{N^\Omega-1}, u_{N^\Omega+1})^\top$ which can be removed by the Bloch conditions. So, according to Eq. (19.119_1), the displacements of the reduced model $\bar{\mathbf{u}} = ({}^0\mathbf{u}, \mathbf{u}^+, \mathbf{u}^-)^\top$ can be related to $\hat{\mathbf{u}} = ({}^0\mathbf{u}, \mathbf{u}^+)^\top$ using the transformation matrix

$$\begin{bmatrix} {}^0\mathbf{u} \\ \mathbf{u}^+ \\ \mathbf{u}^- \end{bmatrix} = \mathbf{P} \begin{bmatrix} {}^0\mathbf{u} \\ \mathbf{u}^+ \end{bmatrix}, \quad \mathbf{P} = \begin{bmatrix} {}^0\mathbf{I} & 0 \\ \mathbf{0} & {}^2\mathbf{I} \\ \mathbf{0} & {}^2\mathbf{I}^\lambda \end{bmatrix}, \tag{19.121}$$

where the matrix ${}^0\mathbf{I}$ in $\mathbf{P}$ represents the identity matrix of size $2N^\Omega - 1$ equal to the number of the internal nodes $x_j \in (l^\Omega, l^\Omega)$ in the unit cell, while ${}^2\mathbf{I} = \operatorname{diag}(1, 1)$ and ${}^2\mathbf{I}^\lambda = \operatorname{diag}(\lambda_{-1}, \lambda_1)$ are the diagonal matrices 2×2.

$$\begin{aligned} {}^0\mathbf{u} &= (u_{-N^\Omega+1}, \ldots, u_0, \ldots, u_{N^\Omega-1})^\top, \\ \mathbf{u}^+ &= (u_{-N^\Omega}, u_{N^\Omega})^\top, \\ \mathbf{u}^- &= (u_{N^\Omega+1}, u_{-N^\Omega-1})^\top \end{aligned} \tag{19.122}$$

are the displacement sets at the internal, boundary, and the external boundary grid points, respectively. The Bloch boundary transformation matrix, $\mathbf{P}$, links the mixed vector $\bar{\mathbf{u}} = ({}^0\mathbf{u}, \mathbf{u}^+, \mathbf{u}^-)^\top$ to the so-called Bloch mode synthesis vector $\hat{\mathbf{u}} = ({}^0\mathbf{u}, \mathbf{u}^+)^\top$. The main idea with Bloch mode synthesis (BMS [675]) method is to describe a free unit cell model with a set of Ritz mode shapes, and then couple these mode shapes, in essence to themselves, via Bloch boundary conditions in order to create a reduced order model of the periodic medium.

19.7.3 Volumetric Bloch-Floquet Periodic Boundary Condition

For the peridynamic composite bar (19.116) and (19.117), we consider the arbitrary row i corresponding to the node $x_i \in \Omega_0 = [-l^\Omega, l^\Omega]$ inside a representative unit cell Ω_0. The nonzero elements of the row i are placed in the columns $j = i+p$ ($p \in [-n_l, n_l]$) that means that unit cell degrees of freedom (DOF) in $x_i \in [-l^\Omega, l^\Omega]$ are defined by DOF in the extended unit cell $y_j \in [-l^\Omega - l_c, l^\Omega + l_c]$. Indeed, we extract a submatrix form of the matrix equations (19.116) and (19.117)

$$\sum_j (K_{ij} - \omega^2 M_{ij}) u_j = b_i, \tag{19.123}$$

which is defined on the extended central UC $x_j \in \Omega_0 \cup \Gamma_0^- \subset w$, i.e., UC is considered as an isolated UC with the volumetric boundary $\Omega_{\Gamma 0} = \Gamma_0^+ \cup \Gamma_0^-$ presented as a union of both the internal boundary $\Gamma_0^+ = [-l^\Omega, -l^\Omega + l_c] \cup [l^\Omega - l_c, l^\Omega]$ and the external one $\Gamma_0^- := [-l^\Omega - l_c, -l^\Omega) \cup (l^\Omega, l^\Omega + l_c]$. The sizes of the matrices K_{ij} and M_{ij} (19.123) are $(2N^\Omega + 1) \times (2N^\Omega + 2n_c + 1)$. Obviously, that substitution of a part $u(\mathbf{x}_j)$ ($\mathbf{x}_j \in \Omega_0 \cup \Gamma^-$) of the solution (19.116) vanishes Eq. (19.123). Our goal is a reformulation of Eq. (19.123) by the use of the volumetric Block condition so that a new equation will be defined on the UC Ω_0 rather than on the extended UC $\Omega_0 \cup \Gamma_0^-$.

Straightforward generalization of the Bloch-Floquet conditions to the peridynamic media forms the new *volumetric Bloch-Floquet conditions* representing periodic displacements ($x \in [-l^\Omega, l^\Omega]$, $\xi \in [0, l_c]$, $\alpha = -\beta = \pm 1$)

$$u^\alpha(x + \operatorname{sign}(x)\xi) = \lambda_\beta u^\beta(x + \operatorname{sign}(x)[\xi - 2l^\Omega]), \tag{19.124}$$

and antiperiodic tractions

$$t(x_\alpha) = -\lambda_\alpha t(x_\beta), \tag{19.125}$$

where $t(\pm l^\Omega) = n(\pm l^\Omega)\sigma(\pm l^\Omega)$ denotes the traction defined by the stresses σ (19.111) on the boundaries $\pm l^\Omega$ with the outward normal unit vectors $n(\pm l^\Omega) = \pm 1$, respectively. In Eq. (19.124), position x belongs to one of the internal volumetric boundaries $x \in [-l^\Omega, -l^\Omega + l_c]$ or $x \in [l^\Omega - l_c, l^\Omega]$, while $x +$ signξ ($|\xi| \le l_c$) falls into the external volumetric boundary either $x + \operatorname{sign}\xi \in [-l^\Omega - l_c, -l^\Omega]$ or $x + \operatorname{sign}\xi \in [l^\Omega, l^\Omega + l_c]$. If both mentioned inequality constrains on x and $x + \xi$ are fulfilled, then $u(x + \xi)$ in the left-hand side of Eq. (19.124) is estimated by the right-hand side of Eq. (19.124). Equation (19.124) established the links between the points $x + \operatorname{sign}\xi$ and $x + \operatorname{sign}[\xi - 2l^\Omega]$ in the neighbor unit cells. In the case $\xi = 0$ and $x \in \Gamma_0 = -l^\Omega \cup l^\Omega$, the volumetric Bloch-Floquet conditions (19.124) and (19.125) are reduced to their locally elastic counterparts (19.119). Therefore, the volumetric Bloch-Floquet conditions (19.124) and (19.125) serve as a peridynamic generalization of the classical Bloch-Floquet conditions (19.119).

We refer to the boundary (e.g., $\Gamma_0 = -l^\Omega \cup l^\Omega$) of the UC separating the adjoining UCs (e.g., Ω_0 and Ω_1) one from another as the geometric boundary of the UC. For the locally elastic case, the governing equation exploiting partial differential form is added by specific boundary conditions at the geometric boundary Γ_0 of the UC such as periodic displacement and antiperiodic traction (19.119). For the peridynamic bond-based model being considered and utilizing the integral governing equation (19.112), the material points in the vicinity of the geometric boundary Γ_0 of the UC directly interact within the horizon. It defines a volumetric boundary (called the interaction boundary of the UC; see Fig. 16.2) $\Omega_{\Gamma 0}$: $\Gamma_0^+ \cup \Gamma_0^- = [-l^{\Omega} - l_c, -l^{\Omega} + l_c] \cup [l^{\Omega} - l_c, l^{\Omega} + l_c]$ surrounding the geometric boundary Γ_0 and limited by the surfaces at the horizon distance from the geometric boundary of the UC.

Following Eq. (19.124), the displacements at the extended UC $[-l^{\Omega} - l_c, l^{\Omega} + l_c]$ of the reduced model $({}^0\mathbf{u}, \mathbf{u}^+, \mathbf{u}^-)^\top$ are formed from the UC displacements $({}^0\mathbf{u}, \mathbf{u}^+)^\top$ by the use of the transformation matrix

$$\begin{bmatrix} {}^0\mathbf{u} \\ \mathbf{u}^+ \\ \mathbf{u}^- \end{bmatrix} = \hat{\mathbf{P}} \begin{bmatrix} {}^0\mathbf{u} \\ \mathbf{u}^+ \end{bmatrix}, \quad \hat{\mathbf{P}} = \begin{bmatrix} {}^{0\mathrm{p}}\mathbf{I} & \mathbf{0} \\ \mathbf{0} & {}^{\mathrm{p}}\mathbf{I} \\ \mathbf{0} & {}^{\mathrm{p}}\mathbf{I}^{\lambda} \end{bmatrix}, \tag{19.126}$$

where the matrices ${}^{0\mathrm{p}}\mathbf{I}$ and ${}^{\mathrm{p}}\mathbf{I}$ in $\mathbf{P}$ represent the identity matrices of sizes $2N^{\Omega} - 2n_c - 1$ and $2n_c + 2$, respectively, while ${}^{\mathrm{p}}\mathbf{I}^{\lambda} = \mathrm{diag}(\overbrace{\lambda_{-1}, \ldots, \lambda_{-1}}^{n_c+1}, \overbrace{\lambda_1, \ldots, \lambda_1}^{n_c+1})$ is a diagonal matrix of the size $2n_c + 2$. The displacement sets (compare with (19.122))

$$\begin{aligned} {}^0\mathbf{u} &= [u_{-N^{\Omega}+n_c+1}, \ldots, u_0, \ldots, u_{N^{\Omega}-n_c-1}]^\top, \\ \mathbf{u}^+ &= [u_{-N^{\Omega}}, \ldots, u_{-N^{\Omega}+n_c}]^\top \cup [u_{N^{\Omega}-n_c}, \ldots, u_{N^{\Omega}}]^\top, \\ \mathbf{u}^- &= [u_{N^{\Omega}}, \ldots, u_{N^{\Omega}+n_c}]^\top \cup [u_{-N^{\Omega}-n_c}, \ldots, u_{-N^{\Omega}}]^\top \end{aligned} \tag{19.127}$$

are defined at the internal, internal boundary, and the external boundary grid points, respectively. The new volumetric Bloch boundary transformation matrix, $\hat{\mathbf{P}}$, links the union of sets of interior, internal, and external boundary layers DOF $\overline{\mathbf{u}} = ({}^0\mathbf{u}, \mathbf{u}^+, \mathbf{u}^-)^\top$ to the so-called volumetric Bloch mode synthesis vector $\hat{\mathbf{u}} = ({}^0\mathbf{u}, \mathbf{u}^+)^\top$ corresponding to the points $x_i \in \Omega_0$, $i \in [-N^{\Omega}, N^{\Omega}]$, of the periodic unit cell Ω_0.

Substituting Eq. (19.124) into Eq. (19.116) leads to the time-reduced wave equation in the form

$$(\mathbf{K} - \omega^2\mathbf{M})\hat{\mathbf{P}}\hat{\mathbf{u}} = -\mathbf{b}. \tag{19.128}$$

Premultiplying the equation of motion (19.124) by the complex conjugate matrix transform $\hat{\mathbf{P}}^\dagger$,

$$\hat{\mathbf{P}}^\dagger = \begin{bmatrix} {}^{0\mathrm{p}}\mathbf{I} & \mathbf{0} & \mathbf{0} \\ \mathbf{0} & {}^{\mathrm{p}}\mathbf{I} & {}^{\mathrm{p}}\mathbf{I}^{\lambda\dagger} \end{bmatrix}, \tag{19.129}$$

of $\hat{\mathbf{P}}$ forms the final equations of motion

$$(\hat{\mathbf{K}} - \omega^2\hat{\mathbf{M}})\hat{\mathbf{u}} = -\hat{\mathbf{b}}, \tag{19.130}$$

where the volumetric Bloch reduction mass and stiffness square matrices

$$\hat{\mathbf{M}} = \hat{\mathbf{P}}^{\dagger}\mathbf{M}\hat{\mathbf{P}}, \quad \hat{\mathbf{K}} = \hat{\mathbf{P}}^{\dagger}\mathbf{K}\hat{\mathbf{P}}, \quad \hat{\mathbf{b}} = \hat{\mathbf{P}}^{\dagger}\mathbf{b} \tag{19.131}$$

have the size $2N^{\Omega}+1$, and ${}^{\mathrm{p}}\mathbf{I}^{\lambda\dagger} = \mathrm{diag}(\overbrace{\lambda_1,\ldots,\lambda_1}^{n_c+1},\overbrace{\lambda_{-1},\ldots,\lambda_{-1}}^{n_c+1})$. Equation (19.130) is the reduction of Eq. (19.116) because Eq. (19.130) is defined on the periodic UC $\Omega_0 = [-l^{\Omega}, l^{\Omega}]$ rather than the extended UC $\Omega_0 \cup \Gamma_0^- = [-l^{\Omega} - l_c, l^{\Omega} + l_c]$ (as Eq. (19.116)).

It should be mentioned that we intentionally have used the presentation of the volumetric Bloch transformation (19.126) in the same form as adopted version of the known Bloch transformation (19.121), although the sizes of both all submatrices and the displacement vectors in Eqs. (19.121) and (19.126) are different due to the nonlocal nature of Eq. (19.126). The reason of exploiting of this similarity of Eqs. (19.121) and (19.126) is obtaining of the final discretized equation (19.130) proposed for UC in isolation from other UCs. Equation (19.130) and the sizes $2N^{\Omega}+1$ of its matrices $\hat{\mathbf{M}}$ and $\hat{\mathbf{K}}$ formally coincide with their well-known locally elastic counterpart (see, e.g., [534, 675, 802]) that opens up wide opportunities for a straightforward exploitation of numerical methods developed for analyses of eigenvalue problems for periodic locally elastic CMs. An upsurge in interest in wave propagation phenomena has been prompted by the recent proliferation of metamaterials (in both the theoretical and practical senses) displaying strong resonance behavior at frequencies at which the inclusions themselves and the distances between them are small that can possess simultaneously negative dielectric constant and magnetic permeability in the context of electromagnetics, and negative effective mass density and negative elastic constants in the case of elasticity (see for references [211]).

Of course, the scheme of the volumetric BPBC proposed in Sect. 19.7.3 can be generalized to dD ($d = 2, 3$) case in a straightforward manner. The next step is a generalization of the theory proposed for bond-based case to the state-based peridynamic CMs.

19.8 Scheme of Data-Driven Learning of Nonlocal Models

We now consider a composite medium with inclusion centers distributed at the nodes of some spatial lattice $\mathbf{m} \in \Lambda$ (5.85). We introduce the high-fidelity (HF) model representing the system at microscopic scale

$$\mathcal{L}_{\mathrm{HF}}[\mathbf{u}](\mathbf{x}) = -\mathbf{b}(\mathbf{x}), \tag{19.132}$$

where self-equilibrated body force has a compact support: $\mathbf{b}(\mathbf{x}) \equiv \mathbf{0}$ at $|\mathbf{x}| \geq a^b$. The operator $\mathcal{L}_{\mathrm{HF}}$ can correspond to the different constitutive models

$$\mathcal{L}_{\rm HF}[\mathbf{u}](\mathbf{x}) = \nabla\mathbf{L}(\mathbf{x})\nabla\mathbf{u}(\mathbf{x}),\ \mathcal{L}_{\rm HF}[\mathbf{u}](\mathbf{x}) = \nabla\mathbf{L}(\mathbf{x})\nabla\mathbf{u}(\mathbf{x}) + \nabla\nabla\mathbf{C}(\mathbf{x})\nabla\nabla\mathbf{u}(\mathbf{x}), \quad (19.133)$$

$$\mathcal{L}_{\rm HF}[\mathbf{u}](\mathbf{x}) = \nabla\int\alpha(\mathbf{x},\mathbf{y})\nabla\mathbf{u}\,dy,\ \mathcal{L}_{\rm HF}[\mathbf{u}](\mathbf{x}) = \int\mathbf{K}(\mathbf{x},\mathbf{y})[\mathbf{u}(\mathbf{x}) - \mathbf{u}(\mathbf{y})]\,dy, \quad (19.134)$$

of local elasticity, strain gradient model (14.184), strongly nonlocal strain-type model (14.1) and (14.21), and peridynamics, respectively. We assume that the body force is fixed, whereas the CM defined by the grid $\mathbf{x_m}$ is moving. Just for concreteness, we consider the locally elastic model (19.133_1), and we designate some configuration $\mathbf{x_m}$ as an initial and mark them by the lower index $_0$ with the corresponding stiffness $\mathbf{L}_0(\mathbf{x})$ (19.133_1) and solution at the microlevel $\mathbf{u}_0(\mathbf{x})$ (19.132). At the parallel transition of the grid $\mathbf{x_m}$ over the vector χ, the corresponding stiffness $\mathbf{L}(\mathbf{x},\chi)$ (19.133_1) and the solution $\mathbf{u}(\mathbf{x},\chi)$ (19.132) will also change

$$\mathbf{L}(\mathbf{x},\chi) = \mathbf{L}_0(\mathbf{x}-\chi),\quad \mathbf{b}(\mathbf{x},\chi) = \mathbf{b}_0(\mathbf{x}),\quad \mathbf{u}(\mathbf{x},\chi) \not\equiv \mathbf{u}_0(\mathbf{x}-\chi), \quad (19.135)$$

where the inequality (19.135_3) holds because $\mathbf{b}(\mathbf{x})$ is fixed. Of course, the same solution $\mathbf{u}(\mathbf{x},\chi)$ can be obtained in an alternate case when the field $\mathbf{L}(\mathbf{x})$ is fixed, whereas the body force $\mathbf{b}(\mathbf{x})$ is moving

$$\mathbf{L}(\mathbf{x},\chi) = \mathbf{L}_0(\mathbf{x}),\quad \mathbf{b}(\mathbf{x},\chi) = \mathbf{b}_0(\mathbf{x}+\chi),\quad \mathbf{u}(\mathbf{x},\chi) \not\equiv \mathbf{u}_0(\mathbf{x}+\chi). \quad (19.136)$$

Thus, for each $\chi \in \mathcal{V}_\mathbf{x}$, we get the solution $\mathbf{u}(\mathbf{x},\chi)$ obtained in the framework of either Eq. (19.135) or (19.136) that makes it possible to define an effective (or macroscopic) displacement

$$\langle\mathbf{u}\rangle(\mathbf{x}) = \frac{1}{\bar{\mathcal{V}}_\mathbf{x}}\int_{\mathcal{V}_\mathbf{x}}\mathbf{u}(\mathbf{x},\chi)\,d\chi,\qquad \mathbf{x}\in w. \quad (19.137)$$

It is interesting that Eq. (19.137) formally looks as averaging over the moving window $\mathcal{V}_\mathbf{x}$ (5.88), although the operation (19.137) is conceptually different and obtained by averaging over the number of the displacement realizations $\mathbf{u}(\mathbf{x},\chi)$ produced by a parallel transform of either $\mathbf{L}(\mathbf{x},\chi)$ (19.135) or $\mathbf{b}(\mathbf{x},\chi)$ (19.136) rather than by averaging of one realization $\mathbf{u}_0(\mathbf{x})$ over the moving window $\mathcal{V}_\mathbf{x}$ (5.88) (see Eq. (19.135) or (19.136), respectively).

Our goal is a construction of provably robust data-driven surrogate equation of the effective nonlocal model with the same $\mathbf{b}(\mathbf{x})$ as in (19.132)

$$\mathcal{L}_{\rm K}[\widetilde{\mathbf{u}}](\mathbf{x}) = -\mathbf{b}(\mathbf{x}),\quad \mathcal{L}_{\rm K}[\widetilde{\mathbf{u}}](\mathbf{x}) := \int\mathbf{K}(|\mathbf{x}-\mathbf{y}|)(\widetilde{\mathbf{u}}(\mathbf{y}) - \widetilde{\mathbf{u}}(\mathbf{x}))\,dy, \quad (19.138)$$

which solutions $\widetilde{\mathbf{u}}_k(\mathbf{x})$ corresponding to a set $\mathbf{b}_k(\mathbf{x})$ are close to the collection of effective displacements $\langle\mathbf{u}_k\rangle(\mathbf{x})$ corresponding to the same set $\mathbf{b}_k(\mathbf{x})$ ($k = 1,\ldots,N$). Obtaining an optimal surrogate model for the kernel function $\mathbf{K}^*$ is achieved by solving an optimization problem of the following form:

$$\mathbf{K}^* = \underset{\mathbf{K}}{\operatorname{argmin}} \sum_{k=1}^{N} ||\widetilde{\mathbf{u}}_k(\mathbf{b}_k, \mathbf{x}) - \langle \mathbf{u}_k \rangle(\mathbf{b}_k, \mathbf{x})||_{l_2}^2 + \mathcal{R}(\mathbf{K}), \tag{19.139}$$

where the addition argument $\mathbf{b}_k$ is introduced for emphasizing the explicit dependence of the functions on $\mathbf{b}_k$, whereas $\langle \mathbf{u} \rangle_k(\mathbf{b}_k, \mathbf{x})$ and $\widetilde{\mathbf{u}}_k(\mathbf{b}_k, \mathbf{x})$ belong to the collections with the same sets $\mathbf{b}_k(\mathbf{x})$

$$\mathcal{D} = \{\langle \mathbf{u}_k \rangle(\mathbf{b}_k, \mathbf{x}), \mathbf{b}_k(\mathbf{x})\}_{k=1}^N, \quad \widetilde{\mathcal{D}} = \{\widetilde{\mathbf{u}}_k(\mathbf{b}_k, \mathbf{x}), \mathbf{b}_k(\mathbf{x})\}_{k=1}^N, \tag{19.140}$$

respectively. Here, the l_2 norm is taken over $\mathbf{x} \in w$, and $\mathcal{R}(\mathbf{K})$ is a regularization term on the coefficients that improves the conditioning of the optimization problem (e.g., Tikhonov regularization), see [1507]. Note that the above training procedure only makes use of the macroscale data $\mathcal{D}$ and $\widetilde{\mathcal{D}}$ which can only describe the deformation of the effective medium under certain loads $\{\mathbf{b}_k(\mathbf{x})\}_{k=1}^N$.

Thus, a surrogate equation (19.138) describing the effective behavior of the locally elastic CM (19.133_1) follows. At the first step, for each $\mathbf{b}_k(\mathbf{x})$ ($k = 1, \ldots, N$; see (19.140_1)), we estimate $\mathbf{u}(\mathbf{x}, \chi)$ in the microscopic problem (19.132) for each $\chi \in \mathcal{V}_\mathbf{x}$ by the use of parallel transition $\mathbf{L}(\mathbf{x}, \chi)$ (19.135) or $\mathbf{b}(\mathbf{x}, \chi)$ (19.136). At the second step, the effective displacements $\langle \mathbf{u} \rangle(\mathbf{b}, \mathbf{x})$ corresponding to $\mathbf{b}_k(\mathbf{x})$ ($k = 1, \ldots, N$) are estimated by Eq. (19.137). At the third step, the collection $\mathcal{D}$ (19.140_1) is formed. At last, the effective model (19.138) approximating the obtained effective deformation behavior $\mathcal{D}$ of CM is estimated by solving the numerical optimization problem (19.139) by the use, e.g., Adam optimizer [778] after each step of gradient descent (see for details [1660] and [1661]), where the kernel $\mathbf{K}$ is presented as a linear combination of Bernstein basis polynomials. The main difficulties at the first and second steps are related with the estimation of nonperiodic solution $\mathbf{u}(\mathbf{x}, \chi)$ on the infinite periodic medium $\mathbf{L}(\mathbf{x})$. However, due to a compact support of the body forces $\{\mathbf{b}_k(\mathbf{x})\}_{k=1}^N$ being considered, the problem on R^d is reduced, in fact, to the analysis on the finite size domain that significantly simplifies the problem.

It should be mentioned that Silling [1400] (see also [1648, 1660, 1661]) proposed a data-driven technique to learn nonlocal constitutive laws for stress wave propagation models. A difference between (19.138) and static counterpart of You et al. [1660]

$$\mathbf{K}^* = \underset{\mathbf{K}}{\operatorname{argmin}} \sum_{k=1}^{N} ||\widetilde{\mathbf{u}}_k(\mathbf{x}) - \mathbf{u}_k(\mathbf{x})||_{l_2}^2 + \mathcal{R}(K) \tag{19.141}$$

is that in a similar equation (19.141), one used the high-fidelity solution of Eq. (19.132) instead of the averaged HF solution $\langle \mathbf{u} \rangle(\mathbf{b}, \mathbf{x})$ (19.137). Exploiting the averaged HF solution $\langle \mathbf{u} \rangle(\mathbf{x})$ (19.137) leads to obtaining optimization problems (19.139) containing only the macro-coordinated $x \in w$ which are not retailed with local coordinates in one or another unit cell. You et al. [1660] also investigated a dynamic case with oscillating source $\mathbf{b}(\mathbf{x}, t)$ when we can avoid the difficulties related with boundary conditions and edge effects. In such a case, instead of moving $\mathbf{b}(\mathbf{x}, \boldsymbol{\chi})$ ($\boldsymbol{\chi} \in \mathcal{V}_{\mathbf{x}}$), we will need to use $\mathbf{b}(\mathbf{x}, \boldsymbol{\chi}, t)$ with the subsequent estimation of $\mathbf{u}(\mathbf{x}, \boldsymbol{\xi}, t)$ and $\langle \mathbf{u} \rangle(\mathbf{b}, \mathbf{x}, t)$. Therefore, we can modify Eq. (7) in [1660] in the same manner as we modified Eq. (19.141) in the form of Eq. (19.139). Loading by dynamic body force $\mathbf{b}(x, t)$ is necessary for a surrogate model with nonlocal dynamics properties (see, e.g., [39]): $\rho[\ddot{\widetilde{\mathbf{u}}})(\mathbf{x}) - l_d^2 \nabla^2 \ddot{\widetilde{\mathbf{u}}})(\mathbf{x})] = \mathcal{L}_{\mathrm{K}}[\widetilde{\mathbf{u}}](\mathbf{x}) + \mathbf{b}(\mathbf{x})$, where ρ is the mass density, whereas l_d denotes an internal dynamic scale parameter.

From other side, we do not perform any subsequent constructions (as in [1648]) for a surrogate model (19.138) describing the mechanics of the material at the macroscale. So, the authors Xu et al. [1648] averaged Eqs. (19.132) and (19.133_1) over the unit cells (UCs) Ω_i with the fixed grid $\mathbf{x}_i$ (5.85) and substituted the averaged displacement field $\langle \widetilde{\mathbf{u}} \rangle_i^{\Omega}(\mathbf{x}) = |\Omega_i|^{-1} \int_{\Omega_i} \widetilde{\mathbf{u}}(\boldsymbol{\chi})\, d\boldsymbol{\chi}$ into the analog of Eq. (19.141) instead of $\widetilde{\mathbf{u}}_k(\mathbf{x})$. It should be mentioned that averaging $\langle \widetilde{\mathbf{u}} \rangle_i^{\Omega}(\mathbf{x})$ over the UCs leads to piecewise constant functions as opposite to the continuous ones obtained by averaging over the "moving averaging" cell $\mathcal{V}_{\mathbf{x}}(\boldsymbol{\chi}) := \{\boldsymbol{\chi} \mid |\mathbf{x}-\boldsymbol{\chi}| \leq a^p\}$ (or moving window (5.88)). In so doing, the kernel $\mathbf{K}(|\mathbf{x}|)$ is found as a discrete function defined on the grid points (5.85) (see for details [1648]).

It should be also mentioned that a choice of a linearized bond-based peridynamic model (19.138) for approximating the effective model is not the only possible. Other models, e.g., the state-based peridynamics, the strain gradient elasticity (19.133_2), strongly nonlocal elasticity by Eringen (19.134_1), and some others, can be tested as the different hypotheses. In such a case, it appears the problem of *discrimination of hypotheses* with designing an additional discrimination experiment $\mathcal{D}$. It is a well-developed area of statistically designed experiment. Furthermore, the optimization problem (19.139) is formulated for the effective strain fields $\langle \mathbf{u}_k \rangle(\mathbf{b}_k, \mathbf{x})$ of the homogenized medium for the set $\mathbf{b}_k(\mathbf{x})$ (19.140). It does not explicitly depend on both the microstructure and strain microfields. Moreover, the optimization problem (19.139) also holds for random structure CM when $\langle \mathbf{u}_k \rangle(\mathbf{b}_k, \mathbf{x})$ is the effective strain field estimated, e.g., by Eq. (18.139). In so doing, the effective nonlocal operator for random structure CMs was estimated before in Chap. 13 and Sect. 18.3.6 for the locally elastic and peridynamic CMs at the microscale, respectively. These operators were presented, in fact, in the form of algorithms mapping the statistical average stress field on the statistical average strain field by the use of the solution of microscopic problem at each inhomogeneous statistical average stress field. Obviously, the data-driven surrogate equation (19.138) at the macroscale is preferable.

Thus, the data-driven learning of the nonlocal model (19.139) is universal and can be used for both the different types of material models at the microstructure and the different kinds of effective material models for CM of either the periodic or random structure for both the static and dynamic problems.

19.9 Concluding Remarks

In this chapter, we extended the classical computational homogenization approach of locally elastic periodic media to the area of peridynamic CMs of periodic structure. In peridynamic periodic structure CMs, we need to distinguish a geometric boundary of UC and an interaction boundary. The former is the boundary separating the adjoining UCs one from another which is also served for the description of the PBC in the classical computational homogenization approach. In so doing, the latter is a volumetric region surrounding the geometric boundary that contains all long-range inter-material interactions including some material points in the adjoining UCs. These two boundaries coincide in classical local elasticity in the limit of vanishing nonlocality or local limit. The condition of displacement periodicity at the geometric boundary in local elasticity is extended to the volumetric periodicity of displacements for all points inside the interaction boundary of the UC. We showed that due to the volumetric PBC, the governing equation depends only on the displacements inside the UC rather than on the displacement inside the extended UC. The discretization of this equation in the form of either (19.56) or (19.62) acts as a macro-to-micro transition of the *deformation-driven* type, where the overall deformation $\boldsymbol{\varepsilon}^{w\Gamma}$ is controlled. Determination of the microstructural displacements $\mathbf{u}(\mathbf{x})$ ($\mathbf{x} \in \Omega_{00}$) in accompany with the volumetric PBC either (19.60) or (19.65) allows one to estimate the peridynamic traction $\mathbf{t}(\mathbf{s}) := \boldsymbol{\sigma}(\mathbf{s})\mathbf{n}(\mathbf{s})$ at the geometric UC's boundary $\mathbf{s} \in \Gamma^0$ which is exploited for the estimation of the macroscopic stresses $\langle \boldsymbol{\sigma} \rangle^{\Omega}$ by the use of Eq. (19.67_1) with subsequent evaluation of the effective moduli $\mathbf{L}^*$ by Eq. (19.69).

It should be mentioned that the volumetric boundary conditions (16.46) and (16.47) were proposed in [1396] and [996]. In so doing, these volumetric constraints gave birth to consideration of many mathematical problems, e.g., well-posedness results for the bond-based and state-based peridynamic models, and formulation of a variational principle (see for details [397, 1049]). Analyses of such a class of problems in the context of the new volumetric PBC are beyond the scope of the current study and deserve further investigations. From another side, generalized mathematical homogenization theory proposed by Fish and coauthors [476, 479] constructs an array of atomistic unit cell problems coupled with a thermomechanical continuum problem. The coarse scale problem is derived directly from atomistic simulations with long-range interaction forces without making any *a priori* assumption about its mathematical structure. However, this sort of nonlocal problems belonging to the class of asymptotic expansion approaches is also beyond the scope of the current study.

The main achievements of this chapter are an introduction of the new volumetric PBC and development of general computational homogenization theory for peridynamic periodic structure composites. It opens wide opportunities for a straightforward generalization of well-developed local theory of computational homogenization ([798, 1056, 1502]; see also [1030]) to their peridynamic counterparts including different physical phenomena (e.g., state-based models, diffusion, viscosity, thermoelastoplasticity, damage accumulation, wave propagation) for CMs with different periodic structures (e.g., polycrystals, fiber network, hybrid structures, foam materials). Moreover, one obtained formally similar representations for effective elastic moduli for statistically homogeneous CMs and periodic structure peridynamic CMs (see Sect. 19.2) for the homogeneous boundary conditions (19.2) and (19.3) and the volumetric PBC (19.7) and (19.8). A detected similarity of results for both the periodic peridynamic CMs (see Sects. 19.2 and 19.3) and random structure CMs (see Sects. 17.2 and 17.3) is explained by a fundamental reason. Indeed, the methods used for obtaining these results widely exploit the analogy between the averaged Hill's conditions (17.2) and (17.3) and the self-adjointness of both the elasticity and peridynamic operators. The mentioned similarity makes it possible to recast some basic relationships of the effective elastic moduli obtained for statistically homogeneous elastic media in terms of $\langle \mathbf{f} \rangle$ by the corresponding representations for the periodic peridynamic media in terms of the UC averages $\langle \mathbf{f} \rangle^{\Omega}$ (see Sects. 19.2 and 19.3).

Chapter 20
Conclusion

> Tis strange,- but true, for Truth is always strange- Stranger than fiction
>
> —Lord Byron. *Don Juan.*

> If you can look into the seeds of time, And say which grain will grow and which will not, Speak then to me ...
>
> —William Shakespeare. *Macbeth.*

> And no, we don't know where it will lead. We just know there's something much bigger than any of us here.
>
> —Steve Jobs.

It should be mentioned that the sketch of micromechanics of random structure composites can be subdivided into two schematic blocks (this classification as any other classification cannot be perfect and reflects just one from possible points of view). The first one contains both the analytical and numerical solutions for one and a finite number of heterogeneities in the infinite homogeneous matrix. This first block (called the basis) is usually associated with computational mechanics becoming more and more ambitious because of the advances in modern computer software and hardware. On one hand, some models have been developed with the goal of minimizing empirical elements and assumptions (e.g., for nonlocal and nonlinear phenomena). In many cases, the resolution of microscopic phenomena has led to improved accuracy and offers the possibility of solving previously intractable problems. On the other hand, we are ambitious to attack increasingly large systems. Such methods are based on the wide exploitation of Monte Carlo simulation with forthcoming numerical analysis for each random realization of multiparticle interactions of microinhomogeneities taking into account both the complicated constitutive properties (coupled, nonlocal, and nonlinear), imperfect interfaces, and large number of interacting heterogeneities. However, at the present level of computer hardware and software, they are practical only for realizations containing no more than a few thousand inhomogeneities.

V. A. Buryachenko, *Local and Nonlocal Micromechanics of Heterogeneous Materials*, https://doi.org/10.1007/978-3-030-81784-8_20

In parallel with computational micromechanics, the classical analytical micromechanics (the second block called a superstructure) represented by the model, perturbation, variational, and self-consistent methods presents, first of all, a set of both the hypotheses and tools for interactions of these hypotheses with the numerical results of the first block. We mentioned about the total domination in the second block of the EFH defining, in actual truth, the essence of classical analytical micromechanics. The EFH played a crucial progressive and encouraged role in the history of micromechanics and provided its current level of development. So, the basic hypotheses of the second block are EFH and quasicrystalline approximation **H2b** (8.10) were proposed 160 and 70 years ago, respectively, while the basic result of the fist block was obtained only in 1957, although the Eshelby [449] theorem has a fundamental conceptual sense rather than only an analytical solution of some particular problem for the ellipsoidal homogeneous inclusion. The basic hypotheses **H1**, **H2**, and **H3** (defining EFM) were so much general that made it possible to realize a constantly increasing achievement of the first block. However, explosive character of the progress in the first block (especially in front of nanotechnology challenges) has led to the conflict with successively increasing gap between the high level of presented possibilities of the first block and the poor consumer's opportunities of the second block rigidly bounded by the classical (first) background of analytical micromechanics (based on the EFH proposed by Poisson [1534], Faraday [1274], Mossotti [1494], Clausius [1234], Lorenz [1444], Maxwell [1475], 1824–1880) (it looks like a conflict in the social study between production relations (superstructure) and the increasing development of productive forces (basis); see [1020]). As an attractive example of such a gap, we can mention a conventional incorporation of molecular dynamic simulation of nanofiber elastic moduli (implicitly assuming a local nature of continuum mechanics law at the nanoscale that is questionable) into Mori-Tanaka scheme based on even more strong hypothesis than **H1** (see for detailed discussion in [205, 206] and Chap. 8). This conflict is presently overcome in a conceptual sense very effectively by the new GIEs (7.19) and (7.20) forming a new (just a second) background of analytical micromechanics that allow one to completely abandon the hypotheses **H1** and **H3**, while the hypothesis **H2** can be used for multiparticle generality. The researches can now forget about the basic restrictions of analytical micromechanics (e.g., the Eshelby tensor and the hypotheses **H1** and **H3**) and use the numerical solutions for one and a few heterogeneities obtained by any available method of computational micromechanics.

In such a case, the renormalizing terms $\langle \mathcal{L}_j^{\tau}(\mathbf{x}-\mathbf{x}_j,\boldsymbol{\tau})\rangle(\mathbf{x}_j)$ and $\langle \mathcal{L}_j^{\eta}(\mathbf{x}-\mathbf{x}_j,\boldsymbol{\eta})\rangle(\mathbf{x}_j)$ (see Eqs. (15.73), (15.74), (15.96), (15.97)) explicitly depend on the distributions $\langle\boldsymbol{\tau}|v_j,\mathbf{x}_j\rangle(\mathbf{y})$ and $\langle\boldsymbol{\eta}|v_j,\mathbf{x}_j\rangle(\mathbf{y})$ $(\mathbf{y}\in v_j^{\oplus})$, respectively, while a fundamental deficiency of Eqs. (7.36), (7.37) and (15.98), (15.99) is defined by the dependence of the renormalizing terms $\mathbf{U}(\mathbf{x}-\mathbf{y})\langle\boldsymbol{\tau}\rangle(\mathbf{y})$ and $\boldsymbol{\Gamma}(\mathbf{x}-\mathbf{y})\langle\boldsymbol{\eta}\rangle(\mathbf{y})$ (obtained in the framework of the asymptotic approximation of the hypothesis **H1b**) only on the statistical averages $\langle\boldsymbol{\tau}\rangle$ and $\langle\boldsymbol{\eta}\rangle$. It is expected that the greater inhomogeneity of the field concentration factor for a single heterogeneity inside infinite matrix leads to the greater difference between the new and classical approaches (based on generalizations of either Eq. (7.19) or (7.36), respectively) with the possible change of sign of predicted local fields (see Chaps. 9 and 11). Buryachenko [203, 204, 206, 207] and Buryachenko and Brun [232–235] (see Chap. 9) have performed some particular comparative analysis of numerical results obtained by the use of either Eq. (7.19) or (7.36) for the perfect interface (3.196). Imperfection of interface conditions (3.200)–(3.206) leads in general to an additional opportunity for the stress inhomogeneity inside the heterogeneities, and, therefore, a fundamental difference of results obtained in the framework of the new background (7.19) and classical one (7.36) is expected. Moreover, the basic models of imperfect interface can be considered as the limiting asymptotic approaches of interphase models (see for references Sect. 3.7.3). More general cases of inhomogeneities of either the elastic moduli $\mathbf{L}(\mathbf{x}) \not\equiv$ const. or eigenstrains $\boldsymbol{\beta}(\mathbf{x}) \not\equiv$ const. ($\mathbf{x} \in v_i$; $i = 1, 2 \ldots$) were considered in Chap. 9 with demonstration of a significant difference between the new (7.19) and old (7.36) approaches at the estimation of the local stresses $\langle \boldsymbol{\sigma} \rangle_i(\mathbf{x})$ ($\mathbf{x} \in v_i$).

A progress of the EFM was defined, first of all, by generalization of the hypotheses exploited in the method and, in a second, by generalization of both the constitutive laws and microtopology of constituent phases. In particular, a multiparticle EFM (MEFM) (see Chap. 8) is based on the hypotheses **H1a**, **H1b**, multiparticle generalization of **H2a**, and **H3**. The next fundamental step is a proposal of a new general integral equation (GIE) (see Chaps. 7 and 15) of microinhomogeneous media forming the second background of micromechanics (the first one is based on the EFH; see above). The critical features of the new GIE are the absence of a direct dependence of GIE on both the Green function and the constitutive law (either local or nonlocal) without restrictions of the conventional micromechanics (e.g., acceptance of the hypotheses **H1** and **H3**) that offers opportunities for a fundamental jump in multiscale and multiphysics researches with drastically improved accuracy of local field estimations (even to the point of correction of a sign; see Sect. 9.6) that is critical for advanced material development (especially for nonlinear modeling). It was established formal similarity of the operator forms of GIEs for both locally elastic CMs and nonlocal ones (strongly nonlocal and weakly nonlocal; see Chaps. 14 and 17) that opens the opportunities to straightforward generalization of their solutions for locally elastic CMs to their nonlocal counterparts. For both the local and nonlocal micromechanics, the general results establishing the links between the effective properties (effective elastic moduli, effective thermal expansion) and the corresponding mechanical and transformation influence functions (do not miss with the influence functions in peridynamics) are obtained by the use of both the decomposition of local fields into the load and residual fields as well as extraction from the material properties a constituent of the matrix properties. Effective moduli are expressed through the introduced generalized local polarization tensors (which are different for the different cases of either strongly nonlocal or weakly nonlocal constitutive equations; see Chaps. 14 and 17) averaged over the extended inclusion phase or the external interaction interface of inclusions rather than in an entire space. In the last case, any spatial derivatives of displacement fields are not required; elimination of this requirement is critical in peridynamic micromechanics because the displacement fields are not differentiable inside the extended inclusions. It should be mentioned that the meaning of acceptance or abandonment of one or another hypothesis (either **H1a**, **H1b**, **H2**, or **H3**) is explained in locally elastic micromechanics in Chaps. 8 and 9. In particular, the destination of the hypothesis **H3** is directed toward providing conditions for applying the hypothesis **H1**. If the hypotheses **H1**, **H2**, and **H3** hold for the statistically homogeneous material and field problem in local elasticity, then the effective moduli $\mathbf{L}^*$ do not depend on the size of the correlation hole v_i^0 and the binary correlation function (that is not true for nonlocal material properties). Simultaneous acceptance of the hypotheses **H1**, **H2**, and **H3** yields a significant simplification of the effective modulus representation in locally elastic micromechanics (see Chaps. 8 and 9) due to an opportunity of exploiting the Eshelby [449] theorem (even for the CMs with nonellipsoidal inclusions v_i). However, implication of a simplified technique based on the Eshelby [449] theorem is questionable in nonlocal micromechanics (see for details Chaps. 14 and 18).

Thus, indeed, the EFM is sufficiently flexible and based on physically clear hypotheses which can be modified and improved if necessary (up to abandonment of these hypotheses) in the framework of a unique scheme for analyses of a wide class of static and dynamic, local and nonlocal (strongly nonlocal and weakly nonlocal), linear and nonlinear multiscale problems of composite materials with deterministic (periodic and nonperiodic), random (statistically homogeneous and inhomogeneous, so-called graded), and mixed (periodic structures with random imperfections) structures in bounded and unbounded domains, containing coated or uncoated inclusions of any shape and orientation and subjected to coupled or uncoupled, homogeneous or inhomogeneous external fields of different physical natures. The proposed generality (abandonment of the hypotheses **H1a**, **H1b**, **H2b**, and **H3**) is forced by practice challenging achievement rather than just pursuing an abstruse theoretical exercise.

Chapter 21
Appendix

21.1 Parametric Representation of Rotation Matrix

Three parametric representations of the rotation tensor $\mathbf{g} \in SO(3)$ (2.6) have been widely used for the description of the local system. In the first case, let the matrix $\mathbf{g} = \|g_{ij}\|$, $(i, j = 1, \ldots, 3)$ be the rotation matrix for coordinates $\mathbf{x}'$ of an arbitrary point in the reference frame orthonormalized $\mathbf{e}'$ $(\mathbf{e}'_i \cdot \mathbf{e}'_j = \delta_{ij})$ expressed by means of the coordinates $\mathbf{x}$ in the crystal orthonormalized coordinate system $\mathbf{e}$ $(\mathbf{e}_i \cdot \mathbf{e}_j = \delta_{ij})$ with the same origin of coordinates:

$$\mathbf{x}' = \mathbf{g}\mathbf{x}. \tag{21.1.1}$$

We require that the orthonormalized basis is right-handedness $\mathbf{e}_1 \cdot \mathbf{e}_2 \otimes \mathbf{e}_3 = 1$ and $\mathbf{e}_i \cdot \mathbf{e}_i = \delta_{ij}$ $(\mathbf{e}_i \cdot \mathbf{e}_i = \delta_{ij})$, where $\mathbf{e}_i \cdot \mathbf{e}_j$ $(\mathbf{e}_i \cdot \mathbf{e}_j)$ denotes the projection of the ith basic vector on the jth one (or vice versa). The expansion of the new basis $\mathbf{e}_1, \mathbf{e}_2, \mathbf{e}_3$ into the vectors of the old one $\mathbf{e}'_{i'} = g_{i'j}\mathbf{e}_j$ is defined by coefficients $g_{i'j}$ called a *cosine matrix*, because each element $g_{i'j}$ is the cosine of the angle between two corresponding axes $g_{i'j} = \mathbf{e}'_i \cdot \mathbf{e}_j = \cos\alpha_{i'j}$. Since the rotation does not alter the length or angles, it leaves the scalar production of any two vectors invariant. That is, if $\mathbf{x}' = \mathbf{g}\mathbf{x}$ and $\mathbf{y}' = \mathbf{g}\mathbf{x}$, then the substitution of Eq. (21.1.1) into the left-hand side of the equality $(\mathbf{x}', \mathbf{y}') = (\mathbf{x}, \mathbf{y})$ leads to $\sum_{i,k,l} g_{ik}g_{il}x_kx_l = \sum_k x_ky_k$. Equating the coefficient of x_ky_l yields the equality $\mathbf{g}^\top\mathbf{g} = \boldsymbol{\delta}$, meaning the *orthogonality* of the matrix $\mathbf{g}$: $\mathbf{g}^\top = \mathbf{g}^{-1}$. Thus, the matrix $\|g_{ji'}\|$ is simultaneously both the inverse of the initial matrix $\|g_{i'j}\|$ and its *transpose*. The relation $\mathbf{g}^\top\mathbf{g} = \boldsymbol{\delta}$ provides six independent constant conditions on the components of $\mathbf{g}$ such that only three components of $\mathbf{g}$ are independent. The set of all second-order orthogonal tensors forms a group under the inner product called the three-dimensional orthogonal group $O(3)$. The orthogonal transformation can be divided into proper rotation (if $\Delta \equiv \det\|g_{i'j}\| = 1$, which forms the special orthogonal group $SO(3) \subset O(3)$) and

V. A. Buryachenko, *Local and Nonlocal Micromechanics of Heterogeneous Materials*, https://doi.org/10.1007/978-3-030-81784-8_21

improper rotation, or rotation with reflection (if $\Delta = -1$). The unity element of the group $SO(3)$ in the dyadic representation is $\boldsymbol{\delta} = \delta_{ij}\mathbf{e}_i \otimes \mathbf{e}_j$.

The second one is the axis-angle parameterization of the rotation tensor $\mathbf{g} \in SO(3)$ specified by an axis of rotation $\mathbf{n}$ ($|\mathbf{n}| = 1$) and the angle of rotation, ω, taken in the right-handed sense about $\mathbf{n}$ (the interested reader is referred to [1431]). Then the components of $\mathbf{g}$ are completely described by the vector $\omega\mathbf{n}$:

$$g_{ij}(\omega\mathbf{n}) = \delta_{ij}\cos\omega - \varepsilon_{ijk}n_k \sin\omega + (1-\cos\omega)n_i n_j, \tag{21.1.2}$$

where ε_{ijk} are the components of a permutation tensor. A particularly useful class of such axis-angle parameterizations, the neo-Eulerian (see, e.g., [830]), is obtained by scaling the axis of rotation by a function of the angle $\omega\mathbf{n} \to f(\omega)\mathbf{n}$ which includes the Rodrigues vector $\boldsymbol{\rho} = \tan(\omega/2)\mathbf{n}$ with the property that rotations about a fixed axis trace straight lines in the parametric space. In the case of all possible rotations, the direction of $\mathbf{n}$ is restricted to the surface of unit hemisphere centered on the origin of $\mathbf{n}$: $-\pi < \omega \leq \pi$. Points lying in this "ball" of parameters (called π sphere) specify all physically possible $\omega\mathbf{n}$ rotations. In so doing, the rotation vectors $\pm\pi\mathbf{n}$ have identical physical meaning. If θ and ψ are the ordinary spherical coordinates of the vector $\mathbf{n}$, then $\mathbf{n} = \sin\theta\cos\psi\mathbf{e}_1 + \sin\theta\sin\psi\mathbf{e}_2 + \cos\theta\mathbf{e}_3$, with the range of allowed variables $0 \leq \theta \leq \pi$ and $0 \leq \psi < 2\pi$, and

$$\mathbf{g} = \begin{pmatrix} mpr - ns & ms + npr & qr \\ -mps - nr & mr - nps & -qs \\ -mq & -nq & p \end{pmatrix}, \tag{21.1.3}$$

where $m = \cos\theta$, $n = \sin\theta$, $p = \cos\psi$, $q = \sin\psi$, $r = \cos\omega$, and $s = \sin\omega$. The orthogonal tensor $\mathbf{g}(\omega\mathbf{n})$ can also be represented by the formula (see, e.g., [586, 1048]) $\mathbf{g}(\omega\mathbf{n}) = \boldsymbol{\delta} + \sin\omega\mathbf{N} + (1-\cos\omega)\mathbf{N}^2 = e^{\omega\mathbf{P}}$, where the three-dimensional skew-symmetric tensor $\mathbf{N}$ with components $N_{ij} = -\epsilon_{ijk}n_k$ is introduced to represent the unit vector $\mathbf{n}$, and the exponential representation $e^{\omega\mathbf{N}} \equiv \sum_{k=1}^{\infty}\mathbf{N}^k/k!$ is widely used in computational solid mechanics (see [1048]).

However, we will usually use the third parameterization for the rotation matrix $\mathbf{g} \in SO(3)$ exploring the Euler angles ϕ_1, θ, ϕ_2 (see, e.g., [174]) with the range of allowed parameters constituting a rectangular box with two of its bounded surfaces excluded $0 \leq \phi_1 < 2\pi$, $0 \leq \phi_2 < 2\pi$, $0 \leq \theta \leq \pi$. Let the coordinate axes (Ox, Oy, Oz) be rotated into the axes (Ox', Oy', Oz') (initially coinciding one with another) using the matrix $\mathbf{g}(\phi_1, \theta, \phi_2)$, which can be presented in the form of a product of three successive rotations: the rotation $\mathbf{g}(\phi_1, 0, 0)$ through the angle ϕ_1 about the axis Oz (the axes (Ox, Oy, Oz) transform into the axes $(O\bar{x}, O\bar{y}, Oz)$), the rotation $g(0, \theta, 0)$ to the angle θ around the axis $O\bar{x}$ (the axes $(O\bar{x}, O\bar{y}, Oz)$ transform into the axes $(O\bar{x}, Oy'', Oz')$), and the rotation $\mathbf{g}(0, 0, \phi_2)$ to the angle ϕ_2 ones again about the axis Oz' (the axes $(O\bar{x}, Oy'', Oz')$ transform into the axes (Ox', Oy', Oz')):

$$\mathbf{g}(\phi_1,0,0)=\begin{pmatrix} c_2 & s_2 & 0\\ -s_2 & c_2 & 0\\ 0 & 0 & 1\end{pmatrix},\quad \mathbf{g}(0,\theta,0)=\begin{pmatrix} 1 & 0 & 0\\ 0 & c_1 & s_1\\ 0 & -s_1 & c_1\end{pmatrix},\quad \mathbf{g}(0,0,\phi_2)=\begin{pmatrix} c_3 & s_3 & 0\\ -s_3 & c_3 & 0\\ 0 & 0 & 1\end{pmatrix},$$

where $c_1 = \cos(\theta)$, $s_1 = \sin(\theta)$, $c_2 = \cos(\phi_1)$, $s_2 = \sin(\phi_1), c_3 = \cos(\phi_2), \mathit{and}\ s_3 = \sin(\phi_2)$. Then an arbitrary matrix $\mathbf{g}(\phi_1,\theta,\phi_2)$ can be expressed as a multiplication of the matrices corresponding to the Euler rotations with the angles ϕ_1,θ,ϕ_2: $\mathbf{g}(\phi_1,\theta,\phi_2)=\mathbf{g}(0,0,\phi_2)\mathbf{g}(0,\theta,0)\mathbf{g}(\phi_1,0,0)$, or, in more detail,

$$\mathbf{g}=\begin{pmatrix} c_2c_3-c_1s_2s_3 & s_2c_3+c_1c_2s_3 & s_1s_3\\ -c_2s_3-c_1s_2c_3 & -s_2s_3+c_1c_2c_3 & s_1c_3\\ s_1s_2 & -s_1c_2 & c_1\end{pmatrix}. \tag{21.1.4}$$

The matrix $\mathbf{g}(\phi_1,\theta,\phi_2)$ can be also determined in the term of the matrices (21.1.2): $\mathbf{g}(\phi_1,\theta,\phi_2) = g(\phi_1\mathbf{e}_1)g(\theta\mathbf{e}_3)g(\phi_2\mathbf{e}_1)$ when the unit vector $\mathbf{e}_3 = \mathbf{e}_3(\theta,\psi)$, $\psi = \pi/2-\phi_1$, specifies points on the rotation unit sphere S^2. If we substitute $\phi_1 \to \pi-\phi_2$ and $\phi_2 \to \pi-\phi_1$ in the matrix $\mathbf{g}$ (21.1.4), we obtain its transpose, which coincides with the inverse matrix according to (21.1.4). Therefore, the rotation matrix $\mathbf{g}(\phi_1,\theta,\phi_2)$ has the inverse matrix $\mathbf{g}(\pi-\phi_2,\theta,\pi-\phi_1)$.

Thus, if the angle between the new and old axes is known, Eq. (21.1.1) is used; if the axis and the angle of rotation are known, Eq. (21.1.2) is exploited; and if the Eulerian angles are known, Eq. (21.1.4) is used.

21.2 Second- and Fourth-Order Tensors of Special Structures

In a number of micromechanical problems, one uses the tensors of a special kind depending on the Kronecker delta and on a unit vector. The expansion of these tensors in a special basis allows us to simplify the algebraic operations with ones such as a multiplication and an inversion (see [745, 836, 1364, 1579]).

21.2.1 E-Basis

Let us consider the fourth-order tensors made up of the unit vector $\mathbf{n}$ and the Kronecker deltas $\boldsymbol{\delta}$. These tensors are assumed to be symmetric in the first and the second pairs of indices but, in general, nonsymmetric with respect to permutation of the pairs. All tensors of this structure can be represented as a linear combination of six linearly independent basis tensors:

$$E^{(1)}_{ijkl} = \delta_{i(k}\delta_{l)j}, \quad E^{(2)}_{ijkl} = \delta_{ij}\delta_{kl}, \quad E^{(3)}_{ijkl} = \delta_{ij}n_k n_l,$$
$$E^{(4)}_{ijkl} = n_i n_j \delta_{kl}, \quad E^{(5)}_{ijkl} = n_{i)}n_{(l}\delta_{k)(j}, \quad E^{(6)}_{ijkl} = n_i n_j n_k n_l. \tag{21.2.1}$$

The tensors $\mathbf{E}^{(k)}$ can be represented in the matrix form

$$\mathbf{E}^{(1)} = \begin{pmatrix} 1 & 0 & 0 & 0 & 0 & 0 \\ 0 & 1 & 0 & 0 & 0 & 0 \\ 0 & 0 & 1 & 0 & 0 & 0 \\ 0 & 0 & 0 & \frac{1}{2} & 0 & 0 \\ 0 & 0 & 0 & 0 & \frac{1}{2} & 0 \\ 0 & 0 & 0 & 0 & 0 & \frac{1}{2} \end{pmatrix}, \quad \mathbf{E}^{(2)} = \begin{pmatrix} 1 & 1 & 1 & 0 & 0 & 0 \\ 1 & 1 & 1 & 0 & 0 & 0 \\ 1 & 1 & 1 & 0 & 0 & 0 \\ 0 & 0 & 0 & 0 & 0 & 0 \\ 0 & 0 & 0 & 0 & 0 & 0 \\ 0 & 0 & 0 & 0 & 0 & 0 \end{pmatrix},$$

$$\mathbf{E}^{(3)} = \begin{pmatrix} n_1^2 & n_2^2 & n_3^2 & n_1n_2 & n_1n_3 & n_2n_3 \\ n_1^2 & n_2^2 & n_3^2 & n_1n_2 & n_1n_3 & n_2n_3 \\ n_1^2 & n_2^2 & n_3^2 & n_1n_2 & n_1n_3 & n_2n_3 \\ 0 & 0 & 0 & 0 & 0 & 0 \\ 0 & 0 & 0 & 0 & 0 & 0 \\ 0 & 0 & 0 & 0 & 0 & 0 \end{pmatrix}, \mathbf{E}^{(4)} = \begin{pmatrix} n_1^2 & n_2^2 & n_3^2 & 0 & 0 & 0 \\ n_1^2 & n_2^2 & n_3^2 & 0 & 0 & 0 \\ n_1^2 & n_2^2 & n_3^2 & 0 & 0 & 0 \\ n_1n_2 & n_1n_2 & n_1n_2 & 0 & 0 & 0 \\ n_1n_3 & n_1n_3 & n_1n_3 & 0 & 0 & 0 \\ n_2n_3 & n_2n_3 & n_2n_3 & 0 & 0 & 0 \end{pmatrix},$$

$$\mathbf{E}^{(5)} = \frac{1}{4}\begin{pmatrix} 4n_1^2 & 0 & 0 & 2n_1n_2 & 2n_1n_3 & 0 \\ 0 & 4n_2^2 & 0 & 2n_1n_2 & 0 & 2n_2n_3 \\ 0 & 0 & 4n_3^2 & 0 & 2n_1n_3 & 2n_2n_3 \\ 2n_1n_2 & 2n_1n_2 & 0 & n_1^2+n_2^2 & n_2n_3 & n_1n_3 \\ 2n_1n_3 & 0 & 2n_1n_3 & n_2n_3 & n_1^2+n_3^2 & n_1n_2 \\ 0 & 2n_2n_3 & 2n_2n_3 & n_1n_3 & n_1n_2 & n_2^2+n_2^3 \end{pmatrix},$$

$$\mathbf{E}^{(6)} = \begin{pmatrix} n_1^4 & n_1^2n_2^2 & n_1^2n_3^2 & n_1^3n_2 & n_1^3n_3 & n_1^2n_2n_3 \\ n_1^2n_2^2 & n_2^4 & n_2^2n_3^2 & n_1n_2^3 & n_1n_2^2n_3 & n_2^3n_3 \\ n_1^2n_3^2 & n_2^2n_3^2 & n_3^4 & n_1n_2n_3^2 & n_1n_3^3 & n_2n_3^3 \\ n_1^3n_2 & n_1n_3^3 & n_1n_2n_3^2 & n_1^2n_2^2 & n_1^2n_2n_3 & n_1n_2^2n_3 \\ n_1^3n_3 & n_1n_2^2n_3 & n_1^3n_3 & n_1^2n_2n_3 & n_1^2n_3^2 & n_1n_2n_3^2 \\ n_1^2n_2n_3 & n_2^2n_3 & n_2n_3^3 & n_1n_2^2n_3 & n_1n_2n_3^2 & n_2^2n_3^2 \end{pmatrix}. \tag{21.2.2}$$

In particular, the linear space of the isotropic tensors symmetric with respect to permutation of the first and the second pairs is defined only by the two tensors $\mathbf{E}^{(1)}$ and $\mathbf{E}^{(2)}$. The basis tensors $\mathbf{E}^{(1)}$ and $\mathbf{E}^{(2)}$ form the orthogonal basis $\mathbf{N}_1 = \mathbf{E}_2/3$, and $\mathbf{N}_2 = \mathbf{E}_1 - \mathbf{E}_2/3$ (2.17) in which the arbitrary isotropic tensor of the fourth order can be decomposed as in Eq. (2.197).

Let us define the operation of multiplication (contraction over two indices):

$$(\mathbf{E}^{(\alpha)} : \mathbf{E}^{(\beta)})_{ijkl} \equiv \mathbf{E}^{(\alpha)}_{ijpq}E^{(\beta)}_{pqkl}. \tag{21.2.3}$$

Table 21.1 "Multiplication" matrix $a_\gamma^{\alpha\beta}$ of the basic tensors $\mathbf{E}^{(k)}$ $(k = 1, \dots, 6)$

	$\mathbf{E}^{(1)}$	$\mathbf{E}^{(2)}$	$\mathbf{E}^{(3)}$	$\mathbf{E}^{(4)}$	$\mathbf{E}^{(5)}$	$\mathbf{E}^{(6)}$
$\mathbf{E}^{(1)}$	$\mathbf{E}^{(1)}$	$\mathbf{E}^{(2)}$	$\mathbf{E}^{(3)}$	$\mathbf{E}^{(4)}$	$\mathbf{E}^{(5)}$	$\mathbf{E}^{(6)}$
$\mathbf{E}^{(2)}$	$\mathbf{E}^{(2)}$	$3\mathbf{E}^{(2)}$	$3\mathbf{E}^{(2)}$	$\mathbf{E}^{(2)}$	$\mathbf{E}^{(3)}$	$\mathbf{E}^{(3)}$
$\mathbf{E}^{(3)}$	$\mathbf{E}^{(3)}$	$\mathbf{E}^{(2)}$	$\mathbf{E}^{(3)}$	$\mathbf{E}^{(2)}$	$\mathbf{E}^{(3)}$	$\mathbf{E}^{(3)}$
$\mathbf{E}^{(4)}$	$\mathbf{E}^{(4)}$	$3\mathbf{E}^{(4)}$	$3\mathbf{E}^{(6)}$	$\mathbf{E}^{(4)}$	$\mathbf{E}^{(6)}$	$\mathbf{E}^{(6)}$
$\mathbf{E}^{(5)}$	$\mathbf{E}^{(5)}$	$\mathbf{E}^{(4)}$	$\mathbf{E}^{(6)}$	$\mathbf{E}^{(4)}$	$\frac{1}{2}(\mathbf{E}^{(5)} + \mathbf{E}^{(6)})$	$\mathbf{E}^{(6)}$
$\mathbf{E}^{(6)}$	$\mathbf{E}^{(6)}$	$\mathbf{E}^{(4)}$	$\mathbf{E}^{(6)}$	$\mathbf{E}^{(4)}$	$\mathbf{E}^{(6)}$	$\mathbf{E}^{(6)}$

The linear tensor space, spanned by the basis $\mathbf{E}^\alpha$ $(\alpha = 1, \dots, 6)$, is closed with respect to the operation of multiplication (21.2.3) and therefore forms a closed algebra where the inner product (21.2.3)

$$(\mathbf{E}^{(\alpha)} : \mathbf{E}^{(\beta)})_{ijkl} = a_\gamma^{\alpha\beta} E_{ijkl}^{(\gamma)} \tag{21.2.4}$$

is defined by the matrix $a_\gamma^{\alpha\beta}$ (the left column represents the left factor in the left-hand side of Eq. (21.2.4) see Table 21.1).

21.2.2 P-Basis

The analysis of the general transversally isotropic body with symmetry axis defined by the unit vector $\mathbf{n}$, we will begin from the second-order tensors. We introduce the basis containing two tensors $\eta_{ij} = n_i n_j$ and $\nu_{ij} \equiv \delta_{ij} - \eta_{ij}$

$$\nu_{ik}\nu_{kj} = \nu_{ij}, \quad \eta_{ik}\eta_{kj} = \eta_{ik}, \quad \nu_{ik}\eta_{kj} = \nu_{jk}\eta_{ki} = 0 \tag{21.2.5}$$

decomposing the Kronecker delta as $\delta_{ij} = \nu_{ij} + \eta_{ij}$. The most general second-order tensor $\mathbf{A}$ reflecting the transverse isotropy is defined by the construction $A_{ij} = a_1\nu_{ij} + a_2\eta_{ij}$. By the use of a straightforward checking, we can confirm that the inverse $\mathbf{A}^{-1}$ of the tensor $\mathbf{A}$ $(\mathbf{A}^{-1}\mathbf{A} = \boldsymbol{\delta})$ as well as the product $\mathbf{AC}$ with the tensor $C_{ij} = c_1\nu_{ij} + c_2\eta_{ij}$ can be found in the closed form

$$(\mathbf{A}^{-1})_{ij} = a_1^{-1}\nu_{ij} + a_2^{-1}\eta_{ij}, \quad (\mathbf{AC})_{ij} = a_1c_1\nu_{ij} + a_2c_2\eta_{ij}. \tag{21.2.6}$$

Let us consider now the tensors of the fourth-order tensors constructed from Kronecker deltas and the unit vectors $\mathbf{n}$, which are symmetric in the first and the second pairs of indices, but not symmetric with respect to permutation of the pairs. The tensors with such a structure can be constructed as the various independent "outer" products of ν_{ij} and n_{kl} and can be represented as a linear combination of the following six linearly independent tensors, which, thus, play the role of a basis:

$$P^{(1)}_{ijkl} = \nu_{ij}\nu_{kl}, \quad P^{(2)}_{ijkl} = \nu_{i(k}\nu_{l)j} - \nu_{ij}\nu_{kl}/2, \quad P^{(3)}_{ijkl} = \nu_{ij}n_k n_l,$$
$$P^{(4)}_{ijkl} = n_i n_j \nu_{kl}, \quad P^{(5)}_{ijkl} = n_{i)}n_{(k}\nu_{l)(j}, \quad P^{(6)}_{ijkl} = n_i n_j n_k n_l. \tag{21.2.7}$$

The basis tensors $\mathbf{E}^{(k)}$ and $\mathbf{P}^{(k)}$ are related one with another by the relations $\mathbf{P}^{(i)} = a^{ij}\mathbf{E}^{(j)}$, $\mathbf{E}^{(i)} = (a^{-1})^{ij}\mathbf{P}^{(j)}$, with the matrices

$$a^{ij} = \begin{pmatrix} 0 & 1 & -1 & -1 & 0 & 1 \\ 1 & -\frac{1}{2} & \frac{1}{2} & \frac{1}{2} & -2 & \frac{1}{2} \\ 0 & 0 & 1 & 0 & 0 & -1 \\ 0 & 0 & 0 & 1 & 0 & -1 \\ 0 & 0 & 0 & 0 & 1 & -1 \\ 0 & 0 & 0 & 0 & 0 & 1 \end{pmatrix}, \quad (a^{-1})^{ij} = \begin{pmatrix} \frac{1}{2} & 1 & 0 & 0 & 2 & 1 \\ 1 & 0 & 1 & 1 & 0 & 1 \\ 0 & 0 & 1 & 0 & 0 & 1 \\ 0 & 0 & 0 & 1 & 0 & 1 \\ 0 & 0 & 0 & 0 & 1 & 1 \\ 0 & 0 & 0 & 0 & 0 & 1 \end{pmatrix}. \tag{21.2.8}$$

Six basis tensors (21.2.7) form also a closed algebra with respect to the operation of multiplication described by the matrix $a^{\alpha\beta}_{\gamma}$ (the left column represents the left factor in the left-hand side of Eq. (21.2.8)).
It should be mentioned that this production operation is not commutative, so $\mathbf{P}^{(3)}\mathbf{P}^{(4)} = \mathbf{P}^{(1)}$ but $\mathbf{P}^{(4)}\mathbf{P}^{(3)} = 2\mathbf{P}^{(6)}$.

The general fourth-order tensor for a transverse isotropy may be constructed as an arbitrary linear combination of the six elementary tensors

$$\mathbf{X} = \sum_{k=1}^{6} X_k \mathbf{P}^{(k)} \equiv (X_1,\ X_2,\ X_3,\ X_4,\ X_5,\ X_6), \tag{21.2.9}$$

where the second equality uses the briefer symbolic notation. Any inner product $(\mathbf{XY})_{ijkl} = X_{ijpq}Y_{pqkl}$ of two tensors $\mathbf{X} = \sum_{k=1}^{6} X_k\mathbf{P}^{(k)}$ and $\mathbf{Y} = \sum_{k=1}^{6} Y_k\mathbf{P}^{(k)}$, which is generally noncommutative, can be calculated readily means of the multiplication Table 21.2 for the elementary tensors:

$$\mathbf{XY} = (2X_1Y_1 + X_3Y_4,\ X_2Y_2,\ 2X_1Y_3 + X_3Y_6,\ 2X_4Y_1 + X_6Y_4,\ X_5Y_5/2,\ X_6Y_6 + 2X_4Y_3). \tag{21.2.10}$$

By reducing this product to the unit tensor $\mathbf{XY} = \mathbf{I}$, we may calculate the inverse of $\mathbf{X}$ as

Table 21.2 "Multiplication" matrix $a^{\alpha\beta}_{\gamma}$ of the basic tensors $\mathbf{P}^{(k)}$ $(k = 1, \ldots, 6)$

	$\mathbf{P}^{(1)}$	$\mathbf{P}^{(2)}$	$\mathbf{P}^{(3)}$	$\mathbf{P}^{(4)}$	$\mathbf{P}^{(5)}$	$\mathbf{P}^{(6)}$
$\mathbf{P}^{(1)}$	$2\mathbf{P}^{(1)}$	0	$2\mathbf{P}^{(3)}$	0	0	0
$\mathbf{P}^{(2)}$	0	$\mathbf{P}^{(2)}$	0	0	0	0
$\mathbf{P}^{(3)}$	0	0	0	$\mathbf{P}^{(1)}$	0	$\mathbf{P}^{(3)}$
$\mathbf{P}^{(4)}$	$2\mathbf{P}^{(4)}$	0	$2\mathbf{P}^{(6)}$	0	0	0
$\mathbf{P}^{(5)}$	0	0	0	0	$0.5\mathbf{P}^5$	0
$\mathbf{P}^{(6)}$	0	0	0	$\mathbf{P}^{(4)}$	0	$\mathbf{P}^{(6)}$

$$\mathbf{X}^{-1} = \left(\frac{X_6}{2\Delta}, \frac{1}{X_2}, -\frac{X_3}{\Delta}, -\frac{X_4}{\Delta}, \frac{4}{X_5}, \frac{2X_1}{\Delta}\right), \tag{21.2.11}$$

where $\Delta = 2(X_1X_6 - X_3X_4)$. In the case when the tensor X_{ijkl} is symmetric with respect to permutation of the pairs ij and kl, one can show that the inverse tensor $(\mathbf{X}^{-1})_{ijkl}$ possesses the same symmetry. However, this property is generally lost in an inner product of the tensors with such symmetry (Table 21.2).

In particular, if Ox_3 is the axis of transverse symmetry and, therefore, $\mathbf{n} = (0, 0, 1)^\top$, the tensors $\mathbf{P}^{(k)}$ $(k = 1, \ldots, 6)$ (21.2.7) have the following nonzero tensor components:

$$\begin{aligned}
&P^{(1)}_{1111} = P^{(2)}_{2222} = P^{(1)}_{1122} = P^{(1)}_{2211} = 1, \quad P^{(6)}_{3333} = 1,\\
&P^{(2)}_{1212} = P^{(2)}_{2121} = P^{(2)}_{1221} = P^{(2)}_{2112} = P^{(2)}_{1111} = P^{(2)}_{2222} = \frac{1}{2},\\
&P^{(2)}_{1122} = P^{(2)}_{2211} = -\frac{1}{2}, \quad P^{(3)}_{1133} = P^{(3)}_{2233} = 1, \quad P^{(4)}_{3311} = P^{(4)}_{3322} = 1,\\
&P^{(5)}_{1313} = P^{(5)}_{2323} = P^{(5)}_{1331} = P^{(5)}_{2332} = P^{(5)}_{3131} = P^{(5)}_{3232} = P^{(5)}_{3131} = P^{(5)}_{3232} = \frac{1}{4}.
\end{aligned} \tag{21.12}$$

In particular, a general form of transversally isotropic fourth-order tensor $\mathbf{A} = (a_1, \ldots, a_6)$ has the following components:

$$\mathbf{A} = ((A_{1111} + A_{1122})/2,\ 2A_{1212},\ A_{1133},\ A_{3311},\ 4A_{1313},\ A_{3333}). \tag{21.2.13}$$

The isotropic tensors $\mathbf{L} = (L_1, \ldots, L_6)$ and $\mathbf{M} = (M_1, \ldots, M_6)$ of elastic moduli and stiffness have the components $\mathbf{L} = (\lambda + \mu, 2\mu, \lambda, \lambda, 4\mu, \lambda + 2\mu)$, and

$$\mathbf{M} = \left(\frac{(1-\nu)}{4\mu(1+\nu)}, \frac{1}{2\mu}, \frac{-\nu}{2\mu(1+\nu)}, \frac{-\nu}{2\mu(1+\nu)}, \frac{1}{2\mu}, \frac{1}{2\mu(1+\nu)}\right), \tag{21.2.14}$$

respectively, where $\lambda = 2\mu\nu/(1 - 2\nu)$. The unit fourth rank tensor has the representation $\mathbf{I} = (1/2, 1, 0, 0, 2, 1)$.

If we have a uniform distribution of orientations (statistical isotropy) of inclusions with the properties represented through the tensors $\mathbf{P}^{(i)}$ (21.2.7) and $\boldsymbol{\eta}$, $\boldsymbol{\nu}$ (21.2.5), it can be useful the averaged tensors given by

$$\begin{aligned}
&\langle \mathbf{P}^{(1)} \rangle = \frac{2}{15}(10\mathbf{N}_1 + \mathbf{N}_2), \quad \langle \mathbf{P}^{(2)} \rangle = \frac{4}{15}(5\mathbf{N}_1 + 2\mathbf{N}_2), \quad \langle \mathbf{P}^{(5)} \rangle = \frac{1}{5}\mathbf{N}_2,\\
&\langle \mathbf{P}^{(3)} \rangle = \langle \mathbf{P}^{(4)} \rangle = \frac{2}{15}(5\mathbf{N}_1 - \mathbf{N}_2), \quad \langle \mathbf{P}^{(6)} \rangle = \frac{1}{15}(5\mathbf{N}_1 + 2\mathbf{N}_2),
\end{aligned} \tag{21.2.15}$$

$$\langle \boldsymbol{\eta} \rangle = \frac{1}{3}\boldsymbol{\delta}, \quad \langle \boldsymbol{\nu} \rangle = \frac{2}{3}\boldsymbol{\delta}. \tag{21.2.16}$$

21.2.3 B-Basis

It turns out that a symbolic representation by Walpole [1579] (see also [465, 642, 1577]) generalizing the form (2.197) for the isotropic tensors holds for any fourth-order tensors with transverse isotropy with respect to the axis **n**. The basis (called Walpole's basis) in the space of fourth-order pair-symmetric tensors is given by the tensors

$$B^{(1)}_{ijkl} = \nu_{ij}\nu_{kl}/2, \quad B^{(2)}_{ijkl} = \eta_{ij}\eta_{kl}, \quad B^{(3)}_{ijkl} = \nu_{i(k}\nu_{l)j} - \nu_{ij}\nu_{kl}/2,$$
$$B^{(4)}_{ijkl} = 2\eta_{i)(k}\nu_{l)(j}, \quad B^{(5)}_{ijkl} = \eta_{ij}\nu_{kl}, \quad B^{(6)}_{ijkl} = \nu_{ij}\eta_{kl}, \tag{21.2.17}$$

related with another basis tensor $\mathbf{P}^{(i)}$ $(i = 1, \ldots, 6)$ (21.2.7) according to the following rule:

$$\mathbf{B}^{(1)} = \mathbf{P}^{(1)}/2, \quad \mathbf{B}^{(2)} = \mathbf{P}^{(6)}, \quad \mathbf{B}^{(3)} = \mathbf{P}^{(2)},$$
$$\mathbf{B}^{(4)} = 2\mathbf{P}^{(5)}, \quad \mathbf{B}^{(5)} = \mathbf{P}^{(4)}, \quad \mathbf{B}^{(6)} = \mathbf{P}^{(3)}. \tag{21.2.18}$$

By testing we immediately obtain a multiplication-convolution of two tensors

$$\mathbf{B}^{(\alpha)} \cdot \mathbf{B}^{(\beta)} = a^{\alpha\beta}_{\gamma}\mathbf{B}^{(\gamma)}, \tag{21.2.19}$$

where the matrix $a^{\alpha\beta}_{\gamma}$ is prescribed by Table 21.3 (the left column of the table represents the left factor). It should be mentioned that while $\mathbf{B}^{(1)}$, $\mathbf{B}^{(2)}$, $\mathbf{B}^{(3)}$, and $\mathbf{B}^{(4)}$ are full symmetric and idempotent, $\mathbf{B}^{(5)}$ and $\mathbf{B}^{(6)}$ are only pair symmetric, and because of this, the production operation (21.2.19) is not commutative, so $\mathbf{B}^{(5)}\mathbf{B}^{(6)} = 2\mathbf{B}^{(2)}$ but $\mathbf{B}^{(6)}\mathbf{B}^{(5)} = 2\mathbf{B}^{(1)}$. The general fourth-order tensor $\mathbf{X}$ for a transverse isotropy can be decomposed in the basis $\mathbf{B}^{(\alpha)}$ $(\alpha = 1, \ldots, 6)$

$$\mathbf{X} = \sum_{\alpha=1}^{6} X_\alpha \mathbf{B}^{(\alpha)} \equiv (X_1, X_2, X_3, X_4, X_5, X_6)^W, \tag{21.2.20}$$

where the second equality introduces Walpole's 6×1 vector of symbolic notation. The superscript ^{W} (21.2.20) is introduced for indication of dissimilarity from the analogous tensor decomposition in the tensor basis $\mathbf{P}^{\alpha}$ $(\alpha = 1, \ldots, 6)$ (see Eq. (21.2.18)). Walpole's components X_α $(\alpha = 1, \ldots 6)$ of the tensor $\mathbf{X}$ are found through a suitable contraction of the indices of $\mathbf{X}$:

$$X_1 = \frac{1}{2}B^{(1)}_{mmij}X_{ijkl}B^{(1)}_{klnn}, \quad X_2 = B^{(2)}_{mmij}X_{ijkl}B^{(1)}_{klnn},$$
$$X_3 = \frac{1}{2}B^{(3)}_{mnij}X_{ijkl}B^{(3)}_{klmn}, \quad X_4 = \frac{1}{2}B^{(4)}_{mnij}X_{ijkl}B^{(4)}_{klmn},$$
$$X_5 = \frac{1}{2}B^{(2)}_{mmij}X_{ijkl}B^{(1)}_{klnn}, \quad X_1 = \frac{1}{2}B^{(1)}_{mmij}X_{ijkl}B^{(2)}_{klnn}. \tag{21.2.21}$$

Table 21.3 "Multiplication" matrix $a_\gamma^{\alpha\beta}$ of the basic tensors $\mathbf{B}^{(k)}$ $(k = 1, \ldots, 6)$

	$\mathbf{B}^{(1)}$	$\mathbf{B}^{(2)}$	$\mathbf{B}^{(3)}$	$\mathbf{B}^{(4)}$	$\mathbf{B}^{(5)}$	$\mathbf{B}^{(6)}$
$\mathbf{B}^{(1)}$	$\mathbf{B}^{(1)}$	0	0	0	0	$\mathbf{B}^{(6)}$
$\mathbf{B}^{(2)}$	0	$\mathbf{B}^{(2)}$	0	0	$\mathbf{B}^{(5)}$	0
$\mathbf{B}^{(3)}$	0	0	$\mathbf{B}^{(3)}$	0	0	0
$\mathbf{B}^{(4)}$	0	0	0	$\mathbf{B}^{(4)}$	0	0
$\mathbf{B}^{(5)}$	$\mathbf{B}^{(5)}$	0	0	0	0	$2\mathbf{B}^{(2)}$
$\mathbf{B}^{(6)}$	0	$\mathbf{B}^{(6)}$	0	0	$2\mathbf{B}^{(1)}$	0

It should be mentioned that if the first four components of the decomposition (21.2.20) are positive, the tensor $\mathbf{X}$ is positively defined. If $X_5 = X_6$, then the tensor $\mathbf{X}$ is diagonally symmetric (i.e., symmetric with respect to permutation of the pairs ij and kl) and Walpole's components (21.2.20) are connected with five familiar independent elastic moduli:

$$X_1 = 2k, \quad X_2 - 2X_5^2/X_1 = E_3, \quad X_5/X_1 = \nu_3, \quad X_3 = 2\mu_{12}, \quad X_4 = 2\mu_{13}, \tag{21.2.22}$$

namely, with the plane-strain bulk modulus k, the transverse shear modulus μ_{12}, the axial shear modulus μ_{13}, and the axial Young modulus and Poisson's ratio E_3 and ν_3, respectively.

The space of pair-symmetric transversely isotropic tensors, spanned by the basis $\mathbf{B}^{(\alpha)}$, is closed with respect to sum $\mathbf{X} + \mathbf{Y}$ and product $\mathbf{XY}$ operations between two tensors and hence forms an algebra. The sum of Walpole's associated vectors $\mathbf{X} = (X_1, \ldots, X_6)^W$ and $\mathbf{Y} = (\mathbf{Y}_1, \ldots, Y_6)^W$ is defined as the sum of the homologous components

$$\mathbf{X} + \mathbf{Y} = (X_1 + Y_1, \ldots, X_6 + Y_6)^W. \tag{21.2.23}$$

The inner product, which is generally noncommutative, can be calculated readily means of the multiplication Table 21.3 for the elementary tensors

$$\mathbf{XY} = (X_1Y_1+2X_6Y_5,\ X_2Y_2+2X_5Y_6,\ X_3Y_3,\ X_4Y_4,\ X_5Y_1+X_2Y_5,\ X_6Y_2+X_1Y_6)^W. \tag{21.2.24}$$

By reducing this product to the unit tensor $\mathbf{XY} = \mathbf{I}$, we may calculate the inverse of $\mathbf{X}$ as

$$\mathbf{X}^{-1} = \left(\frac{X_2}{\Delta}, \frac{X_1}{\Delta}, \frac{1}{X_3}, \frac{1}{X_4}, -\frac{X_5}{\Delta}, -\frac{X_6}{\Delta}\right)^W, \tag{21.2.25}$$

where $\Delta = X_1X_2 - 2X_5X_6$ and Walpole's associated vector of the identity tensor $\mathbf{I} = (1, 1, 1, 1, 0, 0)^W$. In the case when the tensor X_{ijkl} is symmetric with respect to permutation of the pairs ij and kl, one can show that the inverse tensor $(\mathbf{X}^{-1})_{ijkl}$ possesses the same symmetry. However, this property is generally lost in an inner product of the tensors with such symmetry.

21.3 Analytical Representation of Some Tensors

21.3.1 Exterior-Point Eshelby Tensor

Let us rewrite the tensor $\mathbf{T}_{ij}(\mathbf{x}_i - \mathbf{x}_j)$ (4.95) in the form $\mathbf{T}_{ij}(\mathbf{x}_i - \mathbf{x}_j) = \mathbf{L}^{(0)}\mathbf{T}^S_{ij}(\mathbf{x}_i - \mathbf{x}_j)$ where the dimensionless tensor $\mathbf{T}^S_{ij}(\mathbf{x}_i - \mathbf{x}_j)$ for the isotropic medium $\mathbf{L}^{(0)}$ with two d-dimensional ($d = 2, 3$) spherical inclusions is known:

$$\mathbf{T}^S_{ij}(\mathbf{x}_i - \mathbf{x}_j) = \sum_{p=1}^{6} b^S_p \mathbf{E}^{(p)}, \qquad (21.3.1)$$

where the basis tensors $\mathbf{E}^{(p)}$ were introduced by Eq. (21.2.1) and the values b^S_p $(p = 1, \ldots, 6)$ are defined by the equations

$$b^S_1 = \frac{1}{\omega_d(1-\nu^{(0)})r^d}(1 - 2\nu^{(0)} + \frac{d}{d+2}\rho^2),$$

$$b^S_2 = \frac{1}{2\omega_d(1-\nu^{(0)})r^d}(-1 + 2\nu^{(0)} + \frac{d}{d+2}\rho^2),$$

$$b^S_3 = \frac{d}{2\omega_d(1-\nu^{(0)})r^d}(1 - \nu^{(0)}\rho^2), \quad b^S_4 = \frac{d}{2\omega_d(1-\nu)r^d}(1 - \nu^{(0)} - \nu^{(0)}\rho^2),$$

$$b^S_5 = \frac{2d\nu^{(0)}}{\omega_d(1-\nu^{(0)})r^d}(1 - \rho^2), \quad b^S_6 = \frac{d+2}{2\omega_d(1-\nu^{(0)})r^d}(-d + (d+4)\rho^2), \qquad (21.3.2)$$

and $\nu^{(0)} \equiv (3k^{(0)} - 2\mu^{(0)})/(6k^{(0)} + 2\mu^{(0)})$ is Poisson's ratio of the matrix, $r \equiv |\mathbf{x}_i - \mathbf{x}_j|$, $\rho^2 \equiv [(a_i)^2 + (a_j)^2]/r^2$. For two identical inclusions, the representations (21.3.1) and (21.3.2) are reduced to the equivalent relations obtained in [1632], and in [712] for the 3D problem as well as in [722] for the 2D problem. The case of 3D spheres of different sizes is considered in [265].

We will reproduce an analytical representation for the external Eshelby tensor $\mathbf{S}(\mathbf{x})$ ($\mathbf{x} \notin v_i$) following [716]. Let us consider an ellipsoidal inclusion v_i: $x_i x_i / a_I^2 \le 1$ with the semiaxes a_i and the center placed at the original of the Cartesian coordinate system $\mathbf{x} = (x_1, x_2, x_3)^\top$. Herein the tensorial notations by Mura [1104] are used: repeated lowercase indices are summed up from 1 to 3, while uppercase indices which take on the same numbers as the lowercase ones are not summed up. For any point $\mathbf{x} = (x_1, x_2, x_3)^\top$ located outside the ellipsoid v_i, one can define an imaginary ellipsoidal v_i^{image} $x_i x_i (a_I^2 + \lambda)^{-1} = 1$ (λ =const.> 0), with the unit outward normal vector $n_i = x_i\left[(a_I^2 + \lambda)\sqrt{\theta(\lambda)}\right]^{-1}$, at the point $\mathbf{x} \in v_i^{\text{image}}$ placed at the surface of the v_i^{image}, where $\theta(\lambda) = \theta_i(\lambda)\theta_i(\lambda)$ and $\theta_i(\lambda) = x_i/(a_I^2 + \lambda)$. Then the external-point Eshelby tensor was found as

$$
\begin{aligned}
S_{ijkl} = {} & S_{IK}^{(1)}(\lambda)\delta_{ij}\delta_{kl} + S_{IJ}^{(2)}(\lambda)(\delta_{ik}\delta_{jl} + \delta_{il}\delta_{jk}) + S_I^{(3)}(\lambda)(\lambda)\delta_{ij}n_k n_l \\
& + S_K^{(4)}\delta_{kl}n_i n_j + S_I^{(5)}(\lambda)(\delta_{ik}n_j n_l + \delta_{il}n_j n_k) \\
& + S_J^{(6)}(\lambda)(\delta_{jk}n_i n_l + \delta_{jl}n_i n_k) + S_{IJKL}^{(7)}(\lambda)n_i n_j n_k n_l
\end{aligned} \tag{21.3.3}
$$

with the explicit compact representation for the tensors

$$
\begin{aligned}
S_I^{(3)} &= \frac{\rho^3(\lambda)}{2(1-\nu)}[1-\rho_I^2(\lambda)], \quad S_I^{(4)} = \frac{\rho^3(\lambda)}{2(1-\nu)}[1-2\nu-\rho_I^2(\lambda)], \\
S_I^{(5)} &= \frac{\rho^3(\lambda)}{2(1-\nu)}[\nu-\rho_I^2(\lambda)], \quad S_I^{(6)} = \frac{\rho^3(\lambda)}{2(1-\nu)}[\nu-\rho_I^2(\lambda)], \\
S_{IJKL}^{(7)} &= \frac{\rho^3(\lambda)}{2(1-\nu)}\Big\{2[\rho_I^2(\lambda)+\rho_J^2(\lambda)+\rho_K^2(\lambda)+\rho_L^2(\lambda)] \\
& \quad +\rho_m(\lambda)\rho_m(\lambda) - \frac{4\rho_M^2(\lambda)\theta_m(\lambda)\theta_m(\lambda)}{\theta(\lambda)} - 5\Big\},
\end{aligned} \tag{21.3.4}
$$

where $\rho_I(\lambda) = a_I(a_I^2+\lambda)^{-1/2}$, $\rho(\lambda) = [\rho_1(\lambda)\rho_2(\lambda)\rho_3(\lambda)]^{1/3}$. The tensors $\mathbf{S}^{(1)}$ and $\mathbf{S}^{(2)}$ are represented by the standard elliptic integrals, which can be integrated explicitly for the spheroidal inclusions ($a_1 \neq a_2 = a_3$) with the aspect ratio $\alpha = a_1/a_2$, and the following representations take place:

$$
\begin{aligned}
4(1-\nu)S_{11}^{(1)}(\lambda) &= -\Big[4\nu + \frac{2}{\alpha^2-1}\Big]g(\lambda) - \frac{2\rho_1^3(\lambda)}{3(\alpha^2-1)} \\
& \quad + \Big[4\nu + \frac{2}{\alpha^2-1}\Big]\rho_1(\lambda)\rho_2^2(\lambda), \\
4(1-\nu)S_{12}^{(1)}(\lambda) &= -\Big[4\nu - \frac{2\alpha^2+1}{\alpha^2-1}\Big]g(\lambda) + \Big[4\nu - \frac{2\alpha^2}{\alpha^2-1}\Big]\rho_1(\lambda)\rho_2^2(\lambda), \\
4(1-\nu)S_{21}^{(1)}(\lambda) &= -\Big[2\nu + \frac{2\alpha^2+1}{\alpha^2-1}\Big]g(\lambda) - \frac{2\alpha^2}{\alpha^2-1}\rho_1(\lambda)\rho_2^2(\lambda), \\
4(1-\nu)S_{22}^{(1)}(\lambda) &= -\Big[2\nu - \frac{4\alpha^2-1}{4(\alpha^2-1)}\Big]g(\lambda) + \frac{\alpha^2\rho_2^4(\lambda)}{2(\alpha^2-1)\rho_1(\lambda)}, \\
4(1-\nu)S_{11}^{(2)}(\lambda) &= -\Big[4\nu - \frac{4\alpha^2-2}{\alpha^2-1}\Big]g(\lambda) - \frac{2\rho_1^3(\lambda)}{3(\alpha^2-1)} \\
& \quad - \Big[4\nu - \frac{4\alpha^2-2}{\alpha^2-1}\Big]\rho_1(\lambda)\rho_2^2(\lambda),
\end{aligned}
$$

$$4(1-\nu)S_{12}^{(2)}(\lambda) = -\Big[\nu + \frac{\alpha^2+2}{\alpha^2-1}\Big]g(\lambda) - \Big[2\nu + \frac{2}{\alpha^2-1}\Big]\rho_1(\lambda)\rho_2^2(\lambda),$$

$$4(1-\nu)S_{22}^{(2)}(\lambda) = \Big[2\nu - \frac{4\alpha^2-7}{4(\alpha^2-1)}\Big]g(\lambda) + \frac{\alpha^2\rho_2^4(\lambda)}{2(\alpha^2-1)\rho_1(\lambda)}, \qquad (21.3.5)$$

where $S_{13}^{(1)} = S_{12}^{(1)}$, $S_{31}^{(1)} = S_{21}^{(1)}$, $S_{23}^{(1)} = S_{32}^{(1)} = S_{33}^{(1)} = S_{22}^{(1)}$, and $S_{13}^{(2)} = S_{21}^{(2)} = S_{31}^{(2)} = S_{12}^{(2)}$, $S_{23}^{(2)} = S_{32}^{(2)} = S_{33}^{(2)} = S_{22}^{(2)}$. Here,

$$g(\lambda) = \begin{cases} -\frac{\alpha^2}{\alpha^2-1}\frac{\rho_2^2(\lambda)}{\rho_1(\lambda)} + \frac{\alpha}{(\alpha^2-1)^{3/2}}\ln\Big[(\alpha^2-1)^{1/2}\rho_2(\lambda) + \frac{\alpha\rho_2(\lambda)}{\rho_1(\lambda)}\Big], & \text{for} \alpha > 1, \\ -\frac{\alpha^2}{\alpha^2-1}\frac{\rho_2^2(\lambda)}{\rho_1(\lambda)} + \frac{\alpha}{(1-\alpha^2)^{3/2}}\tan^{-1}\frac{\alpha}{(1-\alpha^2)^{1/2}\rho_1(\lambda)}, & \text{for} \alpha < 1, \end{cases}$$

$$\lambda = 0.5\Big[r^2 - a_1^2 - a_2^2 + \sqrt{(r^2+a_1^2-a_2^2)^2 - 4(a_1^2-a_2^2)x_1^2}\Big], \; r = |\mathbf{x}|. (21.3.6)$$

In particular, we have for a circle cylinder ($\alpha \to \infty$, $x_1 = 0$, $a = a_2 = a_3$, $j = 2, 3$) $g(\alpha) = -a_2^2 r^{-2}$, $\lambda = r^2 - a_2^2$, $\rho_1 = 1, \rho_j = a_j/r, \theta_1 = n_1 = 0$, $\theta_j = n_j = x_j/r, \theta = 1$, and

$$S_{11}^{(1)} = 8\nu\rho_2^2, \quad S_{12}^{(1)} = 2(2\nu-1)\rho_2^2, \quad S_{21}^{(1)} = 2\rho_2^2, \quad S_{22}^{(1)} = (2\nu-1)\rho_2^2 + \rho_2^4/2,$$

$$S_{11}^{(2)} = 8\rho_2^2, \quad S_{12}^{(2)} = (1-\nu)\rho_2^2, \quad S_{22}^{(2)} = (1-2\nu)\rho_2^2 + \rho_2^4/2,$$

$$S_{ijkl}^{(7)} = \frac{\rho_2^2}{(1-\nu)}\Big\{\rho_i^2(\lambda) + \rho_j^2(\lambda) + \rho_k^2(\lambda) + \rho_l^2(\lambda) - 2\rho_2^2 - 2\Big\}. \qquad (21.3.7)$$

For a disc ($\alpha \to 0$)

$$g(\alpha) = 0, \quad \lambda = 0.5\Big[r^2 - a_2^2 + \sqrt{(r^2-a_2^2)^2 + 4a_2^2x_1^2}\Big],$$

$$\rho_1 = 0, \; \rho_j = 2a_j\Big[r^2 + a_j^2 + \sqrt{(r^2-a_j^2)^2 + 4a_j^2x_1^2}\Big]^{-1/2},$$

$$\theta_1 = x_1/\lambda, \; \theta_j = 2x_j\Big[r^2 + a_j^2 + \sqrt{(r^2-a_j^2)^2 + 4a_j^2x_1^2}\Big]^{-1/2}, \quad (21.3.8)$$

and, in particular, for $r = x_1$: $\lambda = r^2$, $\rho_1 = 0$, $\rho_j = a_j(r^2+a_j^2)^{-1/2}$, $\theta_1 = r^{-1}$, $\theta_j = 0, \theta = r^{-2}, n_1 = 1, n_j = 0$. For spherical inclusions ($\alpha = 1$):

$$S_{ijkl} = \frac{\rho^3}{30(1-\nu)}\Big[(3\rho^2+10\nu-5)\delta_{ij}\delta_{kl} + (3\rho^2-10\nu+5)(\delta_{ik}\delta_{jl}+\delta_{il}\delta_{jk})$$
$$+15(1-\rho^2)\delta_{ij}n_kn_l + 15(1-2\nu-\rho^2)\delta_{kl}n_in_j + 15(7\rho^2-5)n_in_jn_kn_l$$
$$+15(\nu-\rho^2)(\delta_{ik}n_jn_l + \delta_{il}n_jn_k + \delta_{jk}n_in_l + \delta_{jl}n_in_k)\Big], \qquad (21.3.9)$$

where $n_i = x_i/r$ and $\rho = \rho_i = a/r$ $(i = 1, 2, 3)$. In the case of the spherical inclusions v_i and $\mathbf{x} \in v_j$, we have $\mathbf{T}^S_{ij}(\mathbf{x}_i - \mathbf{x}_j) \equiv \langle \mathbf{S}(\mathbf{x}) \rangle_j$.

21.3.2 Some Tensors Describing Fluctuations of Residual Stresses

The notations of Eq. (10.66):

$$
\begin{aligned}
\mathbf{H}^1_{ip}(\mathbf{x}) &= \int [\mathbf{T}_p(\mathbf{x}-\mathbf{x}_p)\mathbf{S}^{21}_{ip} \otimes \mathbf{T}_p(\mathbf{x}-\mathbf{x}_p)\mathbf{S}^{21}_{ip}\varphi(v_p, \mathbf{x}_p|; v_i, \mathbf{x}_i)d\mathbf{x}_p,\\
\mathbf{H}^2_{ip}(\mathbf{x}) &= \int [\mathbf{T}_p(\mathbf{x}-\mathbf{x}_p)\mathbf{S}^{22}_{ip} \otimes \mathbf{T}_p(\mathbf{x}-\mathbf{x}_p)\mathbf{S}^{22}_{ip}]\varphi(v_p, \mathbf{x}_p|; v_i, \mathbf{x}_i)d\mathbf{x}_p,\\
\mathbf{H}^3_{ip}(\mathbf{x}) &= \int [\mathbf{T}_p(\mathbf{x}-\mathbf{x}_p)\mathbf{S}^{22}_{ip} \otimes \mathbf{T}_p(\mathbf{x}-\mathbf{x}_p)\mathbf{S}^{21}_{ip}]\varphi(v_p, \mathbf{x}_p|; v_i, \mathbf{x}_i)d\mathbf{x}_p,\\
\mathbf{H}^4_{ip}(\mathbf{x}) &= \int [\mathbf{T}_p(\mathbf{x}-\mathbf{x}_p)\mathbf{S}^{21}_{ip} \otimes \mathbf{T}_p(\mathbf{x}-\mathbf{x}_p)\mathbf{S}^{22}_{ip}]\varphi(v_p, \mathbf{x}_p|; v_i, \mathbf{x}_i)d\mathbf{x}_p,\\
\mathbf{P}^1_{ip}(\mathbf{x}) &= \int [\mathbf{T}_p(\mathbf{x}-\mathbf{x}_p)\mathbf{S}^{22}_{ip} \otimes \mathbf{T}_p(\mathbf{x}-\mathbf{x}_p)\boldsymbol{\gamma}_p\\
&\quad +\mathbf{T}_p(\mathbf{x}-\mathbf{x}_p)\boldsymbol{\gamma}_p \otimes \mathbf{T}_p(\mathbf{x}-\mathbf{x}_p)\mathbf{S}^{22}_{ip}]\varphi(v_p, \mathbf{x}_p|; v_i, \mathbf{x}_i)d\mathbf{x}_p,\\
\mathbf{P}^2_{ip}(\mathbf{x}) &= \int [\mathbf{T}_p(\mathbf{x}-\mathbf{x}_p)\mathbf{S}^{21}_{ip} \otimes \mathbf{T}_p(\mathbf{x}-\mathbf{x}_p)\boldsymbol{\gamma}_p\\
&\quad +\mathbf{T}_p(\mathbf{x}-\mathbf{x}_p)\boldsymbol{\gamma}_p \otimes \mathbf{T}_p(\mathbf{x}-\mathbf{x}_p)\mathbf{S}^{21}_{ip}]\varphi(v_p, \mathbf{x}_p|; v_i, \mathbf{x}_i)d\mathbf{x}_p,\\
\mathbf{P}^3_{ip}(\mathbf{x}) &= \int [\mathbf{T}_p(\mathbf{x}-\mathbf{x}_p)\boldsymbol{\gamma}_p \otimes \mathbf{T}_p(\mathbf{x}-\mathbf{x}_p)\boldsymbol{\gamma}_p]\varphi(v_p, \mathbf{x}_p|; v_i, \mathbf{x}_i)d\mathbf{x}_p, \qquad (21.3.10)
\end{aligned}
$$

where $\mathbf{S}^{21}_{ip} = \mathbf{Z}^{21}_{ip}\mathbf{R}_i$, $\mathbf{S}^{22}_{ip} = \mathbf{Z}^{22}_{ip}\mathbf{R}_p$, $\boldsymbol{\gamma}_p = \mathbf{Z}^{21}_{ip}\mathbf{F}_i + \mathbf{Z}^{22}_{ip}\mathbf{R}_p$, and the volume averages $\mathbf{f} \equiv \langle \mathbf{f}(\mathbf{x}) \rangle_{(i)}$, $(\mathbf{f} = \mathbf{H}^k_{ip}, \mathbf{P}^l_{ip})$; $k = 1, 2, 4.4.4.4.$ $l = 1, 2, 4.4.$ differ from the corresponding tensors introduced in Section 13 by Buryachenko [200]. The notations of Eq. (10.71):

$$
\begin{aligned}
\mathbf{H}^{10}_p &= \int [\mathbf{T}_p(\mathbf{x}_0-\mathbf{x}_p)\mathbf{R}_p \otimes \mathbf{T}_p(\mathbf{x}_0-\mathbf{x}_p)\mathbf{R}_p]\varphi(v_p, \mathbf{x}_p|; v_0, \mathbf{x}_0)d\mathbf{x}_p,\\
\mathbf{P}^{10}_p &= \int [\mathbf{T}_p(\mathbf{x}_0-\mathbf{x}_p)\mathbf{R}_p \otimes \mathbf{T}_p(\mathbf{x}_0-\mathbf{x}_p)\mathbf{F}_p\\
&\quad +\mathbf{T}_p(\mathbf{x}_0-\mathbf{x}_p)\mathbf{F}_p \otimes \mathbf{T}_p(\mathbf{x}_0-\mathbf{x}_p)\mathbf{R}_p]\varphi(v_p, \mathbf{x}_p|; v_0, \mathbf{x}_0)d\mathbf{x}_p,\\
\mathbf{P}^{30}_p &= \int [\mathbf{T}_p(\mathbf{x}_0-\mathbf{x}_p)\mathbf{F}_p \otimes \mathbf{T}_p(\mathbf{x}_0-\mathbf{x}_p)\mathbf{F}_p]\varphi(v_p, \mathbf{x}_p|; v_0, \mathbf{x}_0)d\mathbf{x}_p. \qquad (21.3.11)
\end{aligned}
$$

References

1. ABAQUS *Theory Manual for Version 6.2-1*, Pawtucket, RI: Hibbitt, Karlsson, and Sorenson, Inc., 2001.
2. Aboudi J, Arnold SM, Bednarcyk BA (2013) *Micromechanics of Composite Materials: A Generalized Multiscale Analysis Approach* Butterworth-Heinemann, Oxford, UK
3. Absi E, Prager W (1975) A comparison of equivalencies and finite element methods. *Comput. Meth. Appl. Mech. Eng.*, **6**:15–99
4. Accorsi ML, Nemat-Nasser S (1986) Bounds on the overall elastic and instantaneous elastoplastic moduli of periodic composites. *Mech Mater* **5**:209–220
5. Achenbach JD, Zhu H (1990) Effect of interfaces on micro and macromechanical behavior of hexagonal-array fiber-composites. *J Appl Mech,* **57**:956–963
6. Acrivos A, Chang E (1986) A model for estimating transport quantities in two-phase materials. *Phys Fluids,* **29**:3–4
7. Adams BL, Morris PR, Wang TT, Willden KS, Wright SI (1987) Description of orientation coherence in polycrystalline materials. *Acta Metall,* **35**:2935–2946
8. Adams BL, Olson T (1998) The metho structure – properties linkage in polycrystals. *Prog Mater Sci,* **43**:1–88
9. Aero EL, Kuvshinskii EV (1960) Fundamental equations of the theory of elastic media with rotationally interacting particles. *Fizika Tverdogo Tela,* **2**:1399. (In Russian. Engl. Transl. *Sov. Phys., Solid State* (1961) **2**:1272–1281
10. Agarwal GS, Dutta Gupta S (1988) T-matrix approach to the nonlinearb susceptibilities of heterogeneous media. *Phys Rev,* **A38**:5678–5687
11. Aifantis EC (1999) Gradient deformation models at nano, micro, and macro scales. *J Engng Mater Technol,* **121**:189–202
12. Aifantis EC (2003) Update on a class of gradient theories. *Mechanics of Materials*, **35**:259–280
13. Aifantis EC (2003) Fractional generalizations of gradient mechanics. In: Tarasov, V.E., Ed. De Gruyter (Eds): *Handbook of Fractional Calculus with Applications*, Berlin, Germany, **4**:241–262
14. Aguiar AR, Fosdick R, (2014) A constitutive model for a linearly elastic peridynamic body *Mathematics and Mechanics of Solids*, **19**:502–523
15. Agwai A, Guven I, Madenci E (2011) Predicting crack propagation with peridynamics: a comparative study *Int. J. Fracture*, **171**:65-78
16. Aksoylu B, Parks ML (2011) Variational theory and domain decomposition for nonlocal problems *Applied Mathematics and Computation*, **217**:6498–6515
17. Alali B, Gunzburger M (2015) Peridynamics and material interfaces *J. Elast.*, **120**:225–248

V. A. Buryachenko, *Local and Nonlocal Micromechanics of Heterogeneous Materials*, https://doi.org/10.1007/978-3-030-81784-8

18. Alali B, Lipton R, (2012) Multiscale dynamics of heterogeneous media in the peridynamic formulation. *J. Elast.*, **106**:71–103
19. Alcaraz AN, Duhau RS, Fernández JR, Harrowell P, Miracle DB (2008) Dense amorphous packing of binary hard sphere mixtures with chemical order: The stability of a solute ordered approximation. *J. Non-Crystalline Solids*, **54**:3171–3178
20. Allen MP, Tildesley DJ (1987) *Computer Simulations of Liquids*, Oxford Science Publications, Oxford
21. Alexandrov K (1965) Average values of tensorial variables. *DAN SSSR (Soviet Phys-Docl)*, **164**:800–802 (In Russian)
22. Alexandrov K, Aisenberg L (1966) A method of calculating the physical constants of polycrystalline materials. *DAN SSSR*, **167**:1028 (In Russian. English translation: *Soviet Phys-Docl*, **11**:323-325)
23. Altan B, Aifantis E (1997) On some aspects in the special theory of gradient elasticity". *J. Mech.l Behavior of Materials*, **8**:231–282
24. Altenberger AR, Dahler JD, Tirrell MV (1985) A mean-field theory of suspension viscosity. *Macromolecules*. **18**:2752–2755
25. Alves CJS (2009) On the choice of source points in the method of fundamental solutions. *Engineering Analysis with Boundary Elements*, **33**:1348–1361
26. Alves CJS, Antunes PRS (2005) The method of fundamental solutions applied to the calculation of eigenfrequencies and eigenmodes of 2D simplyconnected shapes. *Comput Mater Continua*, **2**:251–66
27. Alves CJS, Silvestre AL (2004) Density results using Stokesltes and a method of fundamental solutions for the Stokes equations. *Engineering Analysis with Boundary Elements*, **28**:1245–1252
28. Amparano, FE, Xib Y, Roh Y-S (2000) Experimental study on the effect of aggregate content on fracture behavior of concrete. *Engineering Fracture Mechanics*, **67**:65–84
29. Annapragada SR, Sun D, Garimella SV (2007) Prediction of effective thermo-mechanical properties of particulate composites. *Computational Materials Science*, **40**:255–266
30. Arefinia R, Shojaei A (2006) On the viscosity of composite suspensions of aluminum and ammonium perchlorate particles dispersed in hydroxyl terminated polybutadiene - New empirical model. *J. Colloid and Interface Sci.*, **299**:962–971
31. Armington M (1991) Limit distributions of the states and inhomogenization in random media. *Acta Mechan*, **88**:27–59
32. Arnould JF (1982) Etude de l'endommagement des materiaux composites par decohesion fibre-matrice. In: Corvino A et al. (eds), *Comptes Rendus des Troisiemes Journees Nationales sur les Composites (JNC-3)*. AMAS, Paris, 159–169
33. Arsenault RJ, Taya M (1987) Thermal residual stress in metal matrix composite. *Acta Metall*, **35**:651–659
34. Asaro RJ, (1975) Somigliana dislocations and internal stresses; with application to second phase hardening. *Int. J. Engng Sci.*, **13**:271–286
35. Asaro RJ, Barnett DM (1975) The non-uniform transformation strain problem for an anisotropic ellipsoidal inclusions. *J Mech Phys Solids*, **23**: 77–83
36. Askari A, Azdoud Y, Han F, Lubineau G, Silling S (2015) Peridynamics for analysis of failure in advanced composite materials *Numerical Modelling of Failure in Advanced Composite Materials*, Woodhead Publishing Series in Composites Science and Engineering, 331–350
37. Askari E, Bobaru F, Lehoucq RB, Parks ML, Silling SA, Weckner O (2009) Peridynamics for multiscale materials modeling. *Journal of Physics: Conference Series*, **125**:012078
38. Askari E, Xu J, Silling SA (2006) Peridynamic analysis of damage and failure in composites. *44th AIAA Aerospace Sciences Meeting and Exhibition*, **AIAA 2006–88**, Reno, NV, 1–12
39. Askes H, Aifantis EC (2011) Gradient elasticity in statics and dynamics: An overview of formulations, length scale identification procedures, finite element implementations and new results. *Int. J. Solids Structures* **48**:1962–1990

40. Aste T, Saadatfar M, Sakellariou A, Senden TJ (2004) Investigating the geometrical structure of disordered sphere packings. *Physica*, **A339**:16–23
41. Aste T, Saadatfar M, Senden TJ (2005) Geometrical structure of disordered sphere packings. *Phys. Rev.*, **E71**:061302
42. Atkin RJ, Fox N (1990) *An Introduction to the Theory of Elasticity*. Longman, London
43. Atluri SN (2004) *The meshless method (MLPG) for domain & BIE discretizations*. Tech Science Press.
44. Attard, P (1989) Spherically inhomogeneous fluids. II. Hard-sphere solute in a hard-sphere solvent. *J. Chem. Phys.*, **91**:3083–3089
45. Auriault JL, Boutin C, Geindreau C (2011) *Homogenization of coupled phenomena in heterogenous media*. Wiley-ISTE, Hoboken,NY
46. Avellaneda M (1987) Iterated homogenization, differential effective medium theory and applications. *Commun Pure Appl Math*, **40**: 527–554
47. Axel F (1992) Bounds for field fluctuations in two-phase materials. *J Appl Phys*, **72**:1217–1220
48. Babuska I (1976) Homogenization and its application. Mathematical and computational problems. In Lions, J.-L, Glowinski, R. (Eds.) *Numerical Solution of Partial Differential Equations*. III. Academic Press, New York, 89–116
49. Babuska I, Anderson B, Smith PJ, Levin K (1999) Damage analysis of fiber composites. Part I: Statistical analysis on fiber scale. *Comput Methods Appl Mech Engng*, **172**:27–77
50. Baer MR, Hall CA, Gustavsen RL, Hooks DE, Sheffield SA (2007) Isentropic loading experiments of a plastic bonded explosive and constituents. *J. Appl. Phys.*, **101**:034906
51. Bagwell RM, Wetherhold RC (2005) End-shaped copper fibers in an epoxy matrix–predicted versus actual fracture toughening *Theoretical and Applied Fracture Mechanics*, **43**:181–188
52. Bahei-El-Din YA, Khire R, Hajela P (1010) Multiscale transformation field analysis of progressive damage in fibrous laminates. *Int. J. Multiscale Comput. Eng.*, **8**:69–80
53. Bakhvalov NS (1974) Averaged characteristics of bodies with periodic structure. *Docl. Akad. Nauk SSSR*, **218**:1046–1048
54. Bakhvalov NS (1975) Averaging of partial differential equations with rapidly oscillating coefficients. *Docl. Akad. Nauk SSSR*, **221**:516–519
55. Bakhvalov NS (1975) Averaging of nonlinear partial differential equations with rapidly oscillating coefficients. *Docl. Akad. Nauk SSSR*, **225**:249–252
56. Bakhvalov NG, Panasenko G (1989) *Homogenization: Averaging Processes in Periodic Media*. Kluwer, Dordrecht
57. Ball JM (1977) Convexity conditions and existence theorems in nonlinear elasticity. *Arch Rational Mech Anal*, **63**:337-403
58. Ballabh TK, Paul M, Middya TR, Basu AN (1992) Theoretical multiple-scattering calculation of nonlinear elastic constants of disordered solids. *Phys Rev*, **B45**:2761–2771
59. Ballas J, Sladek J, Sladek V (1989) *Stress Analysis by Boundary Element Methods*. Elsevier, Amsterdam
60. Banerjee PK (1994) *The Boundary Element Methods in Engineering*. McGraw-Hill, London, New York
61. Banerjee B, Adams DO (2004) On predicting the effective elastic properties of polymer bonded explosives using the recursive cell method *Int. J. Solids Structures*, **41**:481–509
62. Bao G, Sao Z (1992) Remarks on crack-bridging concepts. *Appl Mech Rev*, **45**:355–366
63. Bao G, Hutchinson JW, McMeeking RM (1991) Particle reinforcement of ductile matrices against plastic flow and creep. *Acta Metall*, **39**:1871–1880
64. Barbero EJ (2017) *Introduction to Composite Materials Design*. Taylor & Francis, CRC Press, Boca Raton
65. Barenblatt GI (1962) The mathematical theory of equilibrium of cracks in brittle fracture. *Advances in Applied Mechanics*. **7**:55–129.
66. Barker GC (1993) Computer simulation of granular materials. In: Mehta A (ed) *Granular Matter - An Interdisciplinary Approach*. Springer Verlag, Berlin

67. Bargmann S, Klusemann B, Markmann J, Schnabel JE, Schneider K, Soyarslan C, Wilmers J (2018) Generation of 3D representative volume elements for heterogeneous materials: A review. *Progress in Materials Science*, **96**:322–384
68. Barnett DM (1972) The precise evaluation of derivatives of the anisotropic elastic Green's functions. *Phys Stat Sol (b)*, **49**:741-748
69. Barnett DM, Asaro RJ (1972) The fracture mechanics of slit like cracks in anisotropic elastic media. *J Mech Phys Solids*, **20**:353–366
70. Barnett DM, Lothe J (1975) Dislocation and line charges in anisotropic piezoelectric insulators. *Phys Stat Solids (b)*, **67**:105–111
71. Barrioa C, Solana JR (2005) Mapping a hard-sphere fluid mixture onto a single component hard-sphere fluid. *Physica*, **A 351**:387–403
72. Bar-yoseph P, Avrashi J (1986) New variational-asymptotic formulations in elastic composite materials. *J Appl Math Phys*, **37**:305–321
73. Batchelor GK (1972) Sedimentation in a dilute dispersion of spheres. *J Fluid Mech*, **52**:245–268
74. Batchelor GK (1974) Transport properties of two-phase materials with random structure. *Ann-Rev Fluid Mech*, **6**:227–255
75. Batchelor GK, Green JT (1972) The determination of the bulk stress in a suspension of spherical particles to order c^2. *J Fluid Mech*, **56**:401–427
76. Bauer F, Böhm HJ, Fischer FD, Rammerstorfer FG (1987) Finite element analysis of drastic shape changes of cylinders due to thermal cycling. *Comput Struct*, **26**: 263–274.
77. Baus M, Colot JL (1987) Thermodynamics and structure of a fluid of hard rods, disks, spheres, of hyperspheres from rescaled virial expansion. *Phys Rev*, **A36**:3912–3925
78. Bazănt ZP, Jirásek M (2002) Nonlocal integral formulations of plasticity and damage Survey of progress. *J. Engineering Mechanics*, **128**:1119–1149.
79. Bazănt ZP, Luo W, Chau VT, Bessa MA (2016) Wave dispersion and basic concepts of peridynamics compared to classical nonlocal damage models. *J. Appl. Mech.*, **83**:111004
80. Beckmann R, Mella R, Wenman MR (2013) Mesh and timestep sensitivity of fracture from thermal strains using peridynamics implemented in Abaqus. *Comput. Methods Appl. Mech. Engrg.*, **263**:71–
81. Belikov VP, Aleksandrov KS, Ryzhova TV (1970) *Elastic Properties of Rock-Forming Minerals and Rocks*. Nauka, Moscow (In Russian)
82. Belytschko T, Krongauz Y, Organ D, Fleming M, Krysl P (1996) Meshless method An overview and recent developments. *Comput Methods Applied Mech Eng*. **139**:3–47
83. Belytschko T, Xiao SP, Schatz GC, Ruoff RS (2002) Atomistic simulation of nanotube fracture. *Phys Rev*, **B65**:235430-1-8
84. Bensoussan A, Lions JL, Papanicolaou G (1978) *Asymptotic Analysis for Periodic Structures*. North-Holland, Amsterdam, New York
85. Bennet CH (1972) Serially deposited amorphous aggregates of hard spheres. *J Appl Phys*, **43**:2727–2734
86. Bennett KC, Luscher DJ (2019) Effective thermoelasticity of polymer-bonded particle composites with imperfect interfaces and thermally expansive interphases. *J. of Elasticity* **136**:55-85.
87. Benveniste Y (1985) The effective mechanical behavior of composite materials with imperfect contact between the constituents. *Mech Mater*, **4**:197–208
88. Benveniste Y (1986) On the effective thermal conductivity of multiphase composites. *J Appl Math Phys (ZAMP)*, **37**:696–713
89. Benveniste Y (1987a) A new approach to application of Mori-Tanaka's theory in composite materials. *Mech Mater*, **6**:147–157
90. Benveniste Y (1987b) A differential effective medium theory with a composite sphere embedding. *J Appl Mech*, **54**:466–468
91. Benveniste Y (1992) The determination of the elastic and electric fields in piezoelectric inhomogeneity. *J Appl Phys*, **72**:1086–1095

92. Benveniste Y (1993a) Universal relations in piezoelectric composites with eigenstress and polarization fields I, binary media: local fields and effective behavior. *J Appl Mech,* **60**: 265–269
93. Benveniste Y (1993b) Universal relations in piezoelectric composites with eigenstress and polarization fields II, multiphase media: effective behavior. *J Appl Mech,* **60**: 270–275
94. Benveniste Y (1994) Exact results concerning the local fields and effective properties in piezoelectric composites. *J Eng Mater Technol,* **116**: 260-267
95. Benveniste Y (2006) A general interface model for a three-dimensional curved thin anisotropic interphase between two anisotropic media. *J Mech Phys Solids,* **54**:708–734
96. Benveniste Y, Dvorak GJ (1990a) On a correspondence between mechanical and thermal effects in two–phase composites. In: Weng GJ, Taya M, Abe H (eds) *Micromechanics and Inhomogeneity, The Toshio Mura 65th Anniversary Volume.* Springer-Verlag, New York, 65–81
97. Benveniste Y, Dvorak GJ (1990b) On a correspondence between mechanical and thermal effects in composites with slipping interfaces. In: Dvorak GJ (ed), *Inelastic Deformation of Composite Materials.* IUTAM Symposium, Springer-Verlag, Berlin, 77–98
98. Benveniste Y, Dvorak GJ (1992) On uniform fields and universal relations in piezoelectric composites *J Mech Phys Solids,* **40**:1295–1312
99. Benveniste Y, Dvorak GJ, Chen T (1989) Stress fields in composites with coated inclusions. *Mech Mater,* **7**:305–317
100. Benveniste Y, Dvorak GJ, Chen,T (1991) On diagonal and elastic symmetry of the approximate effective stiffness tensor of heterogeneous media. *J Mech Phys Solids,* **39**:929–946
101. Benveniste Y, Miloh T (1986) The effective conductivity of composites with imperfect contact at constituent interfaces. *Int J Eng Sci,* **24**:1537–1552
102. Benveniste Y, Milton GW(2003) New exact results for the e'ective electric, elastic, piezoelectric and other properties of composite ellipsoid assemblages. *J Mech Phys Solids,* **51**:1773 –1813
103. Beran MJ (1968) *Statistical Continuum Theories.* John Wiley & Sons, New York
104. Beran M (1974) Application of statistical theories for the determination of thermal, electrical and magnetic properties of heterogeneous materials. In: Sendeckyj GP (ed), *Mechanics of Composite Materials.* Academic Press, New York, **2**:209–249
105. Beran M (1980) Field fluctuations in a two-phase random medium. *J Math Phys,* **21**:2583–2585
106. Beran MJ, McCoy JJ (1970a) Mean field variations in a statistical sample of heterogeneous linearly elastic solids. *Int J Solid Struct,* **6**:1035–1054
107. Beran MJ, McCoy JJ (1970b) Mean field variation in random media. *Quart Appl Math,* **28**:245–257
108. Beran M, Molyneux J (1966) Use of classical variational principles for the effective bulk modulus in heterogeneous media. *Quart Appl Math,* **24**:107-125
109. Bergander H (1995) Finite plastic constitutive laws for finite deformations. *Acta Mechan,* **109**:79–99
110. Bergman DJ (1978) The dielectric constant of a composite material — a problem of classical physics. *Phys Rep,* **43C**:377–407
111. Bernkopf M (1968) A history of infinite matrices. *Arch. Hist. Exact Sci.,* **4**:308–358
112. Berryman JG (1983) Random close packing of hard spheres and disks. *Phys Rev,* **A 27**:1053–1061
113. Berryman JG (1985) Measurement of spatial correlation functions using image processing techniques. *J Appl Phys,* **57**:2374–2384
114. Berryman JG (2005) Bounds and self-consistent estimates for elastic constants of random polycrystals with hexagonal, trigonal, and tetragonal symmetries. *Int J Solids Struct,* **53**:2141–2173
115. Berryman JG, Berge PA (1996) Critique of two explicit schemes for estimating elastic properties of multiphase composites. *Mech Mater* **22**:149–164

116. Berveiller M, Fassi-Fenri O, Hihi A (1987) The problem of two plastic and heterogeneous inclusions in an anisotropic medium. *Int J Engng Sci,* **25**:691–709
117. Berveiller M, Zaoui A (1979) An extension of the self-consistent scheme to plastically flowing polycrystals. *J Mech Phys Solids,* **26**:325–344
118. Bharata S, Levinson M (1978) Signorini's perturbation scheme for a general reference configuration in finite electrostatics. *Arch Rat Mech Anal,* **67**:365–394
119. Bhattacharyya A, Lagoudas DC (2000) Effective elastic moduli of two-phase transversely isotropic composites with aligned clustered fibers. *Acta Mechan,* **145**:65–93
120. Bhattacharyya A, Weng GJ (1994) The elastoplastic behavior of a class of two–phase composites containing rigid inclusions. *Appl Mech Rev,* **47**(Part 1):S45–S65
121. Bildstein B, Kahl G (1994) Triplet correlation functions for hard-spheres: computer simulation results. *J Chem Phys,* **100**:5882–5893
122. Bilger N, Auslender F, Bornert M, Jean-Claude Michel J-C, Moulinec H, Suquet P, Zaoui A (2005) Effect of a nonuniform distribution of voids on the plastic response of voided materials: a computational and statistical analysis. *Int J Solids Struct,* **42**:517–538
123. Bimberg D, Grundmann M, Ledentsov NN (1998) *Quantum Dot Heterostructures.* John Wiley and Sons, New York
124. Binder K, Heerman DW (1997) *Monte Carlo Simulation in Statistical Physics: an Introduction.* Springer, Berlin NY
125. Bird RB, Hassaren O, Amstrong RC, Curtiss C. (1987)*Dynamics of Polymeric Liquids*, 2nd edition, Miley, NY
126. Bisegna P, Luciano R (1996) Variational bounds for the overall properties of piezoelectric composites *J. Mech. Physics Solids*, **44**:583-602.
127. van Blaaderen A, Wiltzius P (1995) Real-space structure of colloidal hard-sphere glasses. *Science*, **270**(5239), 1177–1179
128. Bloch F (1928) U ber die Quantenmechanik der Electron in Kristallgittern, *Zeifschrift fur Physik* **52**:555–600.
129. Bobaru F, Foster J, Geubelle P, Silling S. (Editors) (2016) *Handbook of Peridynamic Modeling*, CRC Press, Boca Raton, FL.
130. Bobaru F, Ha YD (2011) Adaptive refinement and multiscale modeling in 2D peridynamics. *Int. J. Multiscale Comput. Eng.*, **9**:635–659
131. Bobaru F, Yang M, Alves LF, Silling SA, Askari A, Xu J (2009) Convergence, adaptive refinement, and scaling in 1d peridynamics. *Int. J. Numerical Methods Engng*, **77**:852–877.
132. Bobeth M, Diener G (1986) Field fluctuations in multicomponents mixtures. *J Mech Phys Solids,* **34**:1–17
133. Bobeth M, Diener G (1987) Static elastic and thermoelastic field fluctuations in multiphase composites. *J Mech Phys Solids 35*:137–145
134. Bock FE, Aydin RC, Cyron C J, Huber N, Kalidindi SR, Klusemann B. (2019) A review of the application of machine learning and data mining approaches in continuum materials mechanics. *Frontiers in Materials*, **6**, 110 (23 pages)
135. Böhlke T, Bertran A (2001) The evolution of Hooke's law due to texture development in FCC polycrystals. *Int J Solids Struct,* **38**:9437–9459
136. Böhm H (2004) Continuum models for the thermomechanical behavior of discontinuously reinforced materials. *Adv Engng Mater* **6**:626–633
137. Böhm HJ (2021) *A Short Introduction to Basic Aspects of Continuum Micromechanics.* ILSB Report / ILSB-Arbeitsbericht 206, Vienna University of Technology, Austria
138. Böhm HJ, Eckschlager A, Han W (2002) Multi-inclusion unit cell models for metal matrix composites with randomly oriented discontinuous reinforcements. *Comput Mater Sci,* **25**:42–53
139. Böhm HJ, Han W, Eckschlager A (2006) Multi-inclusion unit cell studies of reinforcement stresses and particle failure in discontinuously reinforced ductile matrix composites. *CMES Computer Modeling in Engineering & Sciences*, **5**:5–20
140. Böhm HJ, Rammerstorfer FG, Weissenbek E (1993) Some simple models for micromechanical investigations of fiber arrangement effect in MMCs. *Comput Mater Sci,* **1**:177–194

141. Bolotin VV (1993) Random initial defects and fatigue life prediction. In: Sobczuk K (ed), *Stochastic Approach to Fatigue: Experiments, Modeling and Reliability Estimation.* Springer-Verlag, Vienna, 121–163
142. Bolotin VV, Moskalenko VN (1968) Determination of the elastic constants of a microinhomogeneous medium. *Zh Priklad Mekh Tekhn Fiz (J Appl Mech Tech Phys)* N1:66–72 (In Russian)
143. Bompard P, Dan W, Guennounl F, Francois D (1987) Mechanics and fracture behavior of porous materials. *Engng Fract Mech,* **28**:627–642
144. Bonfoh N, Dreistadt C, Sabar H (2017) Micromechanical modeling of the anisotropic thermal conductivity of ellipsoidal inclusion-reinforced composite materials with weakly conducting interfaces. *Int. J. Heat Mass Transfer*, **108**:1727–1739
145. Borbély A, Kenesei P, Biermann H (2006) Estimation of the effective properties of particle-reinforced metal matrix composites from microtomographic reconstructions. *Acta Materialia*, **54**:2735–2744
146. Borisenko AI, Tarapov IE (1968) *Vector and Tensor Analysis with Applications.* Prentice-Hall, Englewood Cliffs, NY
147. Bornert M, Hervé E, Stolz C, Zaoui A (1994) Self-consistent approaches and strain heterogeneities in two-phase elastoplastic materials. *Appl Mech Rev,* **47**(1, Part 2):S66–S76
148. Bornert M, Stolz C, Zaoui A (1996) Morphologically representative pattern-based bounding in elasticity. *J Mech Phys Solids,* **44**:307-331
149. Boudreaux DS, Gregor JM (1977) Structure simulation of transition-metal-metalloid glasses. *J Appl Phys,* **48**:152–158
150. Boutin C (1996) Microstructural effects in elastic composites. *Int J Solids Struct,* **33**:1023–1051
151. Bouyge F, Jasiuk I, Ostoja-Starzewski M (2001) A micromechanically based couple-stress model of an elastic two-phase composite. *Int J Solids Struct,* **38**:1721–1735
152. Bover AF, Ortiz W (1993) The influence of grain size on the toughness of monolithic ceramic. *J Engng Mater Technol,* **115**:228–236
153. Bradshaw RD, Fisher FT, Brinson LC (2003) Fiber waviness in nanotube-reinforced polymer composites: II. Modeling via numerical approximation of the dilute strain concentration tensor. *Compos Sci Technol,* **63**:1705–1722
154. Brady JF, Bossis G (1988) Stokesian dynamics. *Ann. Rev. Fluid Mech.* **20**:111–157
155. Brady JF, Phillips RJ, Lester JC, Bossis G (1988) Dynamic simulation of hydrodynamically interacting suspensions. *J. Fluid Mech,* **195**:257–280
156. Brebbia CA, Telles JCF, Wrobel LC (1984) *Boundary Element Techniques.* Springer-Verlag, Berlin
157. Brechet YJM (1994) Clusters, plasticity and damage: a missing link? *Mater Sci Engng,* **A175**:63–69
158. Brisard S (2017) *Mémoire d'Habilitation à Diriger des Recherches From microstructure to macroscopic properties Some applications of the Hashin–Shtrikman principle.* Mechanics of materials [physics.class-ph]. Université Paris-Est, Paris
159. Brisard S, Sab K, Dormieux L (2013) New boundary conditions for the computation of the apparent stiffness of statistical volume elements. *J. Mech. Physics Solids*, **61**:2638–2658
160. Brisard S, Tran VP, Sab K (2021) Morphology-dependent Hashin–Shtrikman bounds on the effective properties of stress-gradient materials *European Journal of Mechanics-A/Solids* , **85**:104072
161. Bristow JR (1960) Microcracks, and the static and dynamic elastic constants of annealed heavily coldworked metals. *Br J Appl Phys,* **11**:81–85
162. Brosseau C (2006) Modelling and simulation of dielectric heterostructures: a physical survey from an historical perspective. *J Phys D: Appl Phys*, **39**:1277–1294
163. Brown WF (1955) Solid mixture permittivities. *J Phys Chem,* **23**:1514–1517
164. Bruggeman DAG (1935) Berechnung verschiedener physikalischer Konstante von heterogenete substanze I: Dielektrizitätskonstanten und leitfähigkeiten der misckörper aus isotropen substanzen. *Annal Physik,* **24**:636–679

165. Brune DA, Bicerano J (2002) Micromechanics of nanocomposites: comparison of tensile and compressive elastic moduli, and prediction of effects of incomplete exfoliation and imperfect alignment on modulus. *Polymer,* **43**:369–387
166. Bruno OP (1991) Taylor expansions of bounds for the effective conductivity and the effective elastic moduli of multicomponent composites and polycrystals. *Asympt Anal,* **4**:339–365
167. Budiansky Y (1965) On the elastic moduli of some heterogeneous material. *J Mech Phys Solids,* **13**:223–227
168. Budiansky B, Hutchinson JW, Lambropoulos JC (1983) Continuum theory of dilatant transformation toughening in ceramics. *Int J Solids Struct,* **19**:337–355
169. Budiansky B, O'Connel RJ (1976) Elastic moduli of cracked solids. *Int J Solids Struct,* **12**:81–91
170. Buevich YA (1992) Heat and mass transfer in disperse media. I. Average field equations. *Int J Heat Mass Transfer,* **35**:2445–2452
171. Buevich YuA, Shelchkova, IN (1978) Flow of dense suspensions. *Prog Aerospace Sci,* **18**:121–150
172. Buevich YA, Ustinov VA (1995) Effective conductivity of a macroscopically inhomogeneous dispersions. *Int J Heat Mass Transfer,* **38**:381–389
173. Bui HD (1978) Some remarks about the formulation of three-dimensional thermoelastic problems by integral equations. *Int J Solid Struct,* **14**:935–939
174. Bunge HJ (1982) *Texture Analysis in Mater Science.* Butterworths, Boston
175. Buryachenko VA (1987) Correlation function of stress field in matrix composites. *Mekhanika Tverdogo Tela* $N^0$3, 69–76 (In Russian. Engl Transl. *Mech Solids,* **22**: 66–73)
176. Buryachenko VA (1990a) Effective viscoplasticity parameters of suspension. *Inzhenerno Fiz Zhurnal,* **58**:452–456 (In Russian. Engl Transl. *J Engng Phys,* **58(3)**:331–334)
177. Buryachenko VA (1990b) Prediction of macroproperties of composite medium by the elastoplastic effective field method. *Probl Prochnosti* (11), 57–76 (In Russian. Engl Transl. *Strength Mater,* **22**:1645–1650)
178. Buryachenko VA (1993a) *Effective Physicomechanical Properties of Random Structure Composites.* D. Sc. Thesis, 397p., SP Timoshenko Institute of Mechanics of NAS of Ukraine, Kiev (In Russian)
179. Buryachenko VA (1993b) Mechanics of random structure composites. In: Miravete A (ed) *Proc. of 9th Int. Conf. on Composites.* Univ. Zaragoza, Spain **3**:398-405
180. Buryachenko VA (1993c) Effective strength properties of elastic physically nonlinear composites. In: Marigo JJ, Rousselier G (eds), *Proc. of the MECAMAT Conf. Micromechanics of Materials.* Editions Eyrolles, Paris, 567–578
181. Buryachenko VA (1996) The overall elastoplastic behavior of multiphase materials. *Acta Mechan,* **119**:93–117
182. Buryachenko VA (1998) Some nonlocal effects in graded random structure matrix composites. *Mech Res Commun,* **25**:117–122
183. Buryachenko VA (1999a) Triply periodical particulate matrix composites in varying external stress fields. *Int J Solids Struct,* **36**:3837–3859
184. Buryachenko VA (1999b) Effective thermoelastic properties of graded doubly periodic particulate composites in varying external stress fields. *Int J Solids Struct,* **36**:3861–3885
185. Buryachenko VA (1999c) Thermo-elastoplastic deformation of elastically homogeneous materials with a random field of inclusions. *Int J Plast,* **15**:687–720
186. Buryachenko VA (2000a) Internal residual stresses in elastically homogeneous solids: I. Statistically homogeneous stress fluctuations. *Int J Solids Struct,* **37**:4185–4210
187. Buryachenko VA (2000b) Internal residual stresses in elastically homogeneous solids: II. Stress fluctuations near a crack tip and effective energy release rate. *Int J Solids Struct,* **37**:4211–4238
188. Buryachenko VA (2001a) Multiparticle effective field and related methods in micromechanics of composite materials. *Appl Mech Rev,* **54**:1–47
189. Buryachenko VA (2001b) Multiparticle effective field and related methods in micromechanics of random structure composites. *Math Mech Solids,* **6**:577–612

190. Buryachenko VA (2001c) A simple method of multiple inclusion interaction problem. *Int J Comput Civil Struct Engng,* **1**:7–25
191. Buryachenko VA (2001d) Locality principle and general integral equations of micromechanics of composite materials. *Math Mech Solids,* **6**:299–321
192. Buryachenko VA (2002) Stress fluctuations near a crack tip in ceramic materials and effective energy release rate. In: Brand RC, Sakai M, Munz D, Shevchenko VJ, White KW (eds) *Fracture Mechanics of Ceramics, V.13: Crack-Microstructure interection, R-Curve Behavior, Environmental Effects in Fracture, and Standardtization.* Kluver, Norwell MA, 47–62
193. Buryachenko VA (2004a) Foreword to Special Issue on Recent Advances in Micromechanics of Composite Materials. *Int J Multiscale Comput Engng,* **2**:vii–viii
194. Buryachenko VA (2004b) Multiscalar mechanics of nonlocal effects in heterogeneous materials. *Int J Multiscale Comput Engng,* **2**:1-14
195. Buryachenko VA (2005a) Effective elastic moduli and stress concentrator factors in random structure aligned fiber composites. *ZAMP,* **56**:1107–1115
196. Buryachenko VA (2005b) Effective elastic moduli of triply periodic particulate matrix composites with imperfect unit cells. *Int J Solids Struct,* **42**:4811–4832
197. Buryachenko VA (2007) Generalization of the multiparticle effective field method in static of random structure matrix composites. *Acta Mechan,* **188**:167–208
198. Buryachenko VA, Bechel VT (2000) A volume integral equation method for multiple inclusion interaction problems. *Compos Sci Technol,* **60**:2465–2469
199. Buryachenko VA (2007a) Generalization of the multiparticle effective field method in static of random structure matrix composites. *Acta Mechanica,* **188**:167–208
200. Buryachenko VA (2007b) *Micromechanics of Heterogeneous Materials.* Springer, NY
201. Buryachenko VA (2010a) On some background of micromechanics of random structure matrix composites. *Int. J. Pure and Applied Mathematics.* **59**:163–179.
202. Buryachenko VA (2010b) On the thermo-elastostatics of heterogeneous materials. I. General integral equation. *Acta Mech,* **213**:359–374.
203. Buryachenko VA (2010c) On the thermo-elastostatics of heterogeneous materials. II. Analyze and generalization of some basic hypotheses and propositions. *Acta Mech,* **213**:375-398.
204. Buryachenko VA (2011a) Inhomogeneity of the first and second statistical moments of stresses inside the heterogeneities of random structure matrix composites. *Int. J. Solids and Structures,* **48**:1665–1687.
205. Buryachenko VA (2011b) On thermoelastostatics of composites with nonlocal properties of constituents. I. General representations for effective material and field parameters. *Int. J. Solids and Structures,* **48**:1818–1828.
206. Buryachenko VA (2011c) On thermoelastostatics of composites with nonlocal properties of constituents. II. Estimation of effective material and field parameters. *Int. J. Solids and Structures,* **48**:1829–1845.
207. Buryachenko V (2012) Modeling of random bimodal structures of composites (application to solid propellant) II. Estimation of effective elastic moduli. *Comput. Model. Engng & Sciences (CMES),* **85**(5). 417–446.
208. Buryachenko VA (2013) General integral equations of micromechanics of composite materials with imperfectly bonded interfaces. *Int. J. Solids and Structures,* **50**:3190–3206.
209. Buryachenko V (2014a) Effective elastic modulus of heterogeneous peristatic bar of random structure. *J. Solids and Structures,* **51**:2940-2948
210. Buryachenko VA (2014b) Some general representations in thermoperistatics of random structure composites. *Int. J. Multiscale Comput. Enging.,* **12**:331–350
211. Buryachenko V (2014c) Solution of general integral equations of micromechanics of heterogeneous materials. *J. Solids and Structures,* **51**:3823–3843 (130 refs)
212. Buryachenko V (2015a) General integral equations of micromechanics of heterogeneous materials. *Int. J. Multiscale Comput. Enging.,* **13**:11-53 (170 refs.)

213. Buryachenko VA (2015b) Effective thermoelastic properties of heterogeneous thermoperistatic bar of random structure. *Int. J. Multiscale Comput. Enging.*, **13**:55-71
214. Buryachenko VA (2015c) General integral equations of Stokes flow through the random structure porous media. *Int. J. Multiscale Comput. Enging.*, **13**:375–392
215. Buryachenko VA (2016) Estimation of effective elastic moduli of random structure composites by the method of fundamental solutions. *Engineering Analysis with Boundary Elements* **62**:13–21
216. Buryachenko VA (2017a) Effective properties of thermoperistatic random structure composites: some background principles. *Math. Mech. of Solids.*, **22**:366–1386
217. Buryachenko V.A (2017b) Statistical average of residual stresses in elastically homogeneous medium with random field of noncanonical inclusions. *Computers and Structures*, **187**:24–34
218. Buryachenko V (2017c) Method of fundamental solutions in micromechanics of elastic random structure composites. *Int. J. Solids and Structures*, **124**:135-150
219. Buryachenko VA (2018a) General interface integral equations in elasticity of random structure composites. Eds. S. A. Meguid, G. J. Weng. *Micromechanics and Nanomechanics of Composite Solids*, Springer, NY, 469-506
220. Buryachenko V (2018b) Effective field hypothesis in Hashin-Shtrikman bounds estimations on effective moduli of composites with noncanonical inhomogeneous inclusions. *Mechanics of Materials*, **119**:16-24
221. Buryachenko V (2018c) Effective elastic modulus of heterogeneous peristatic bar of periodic structure. *Computers and Structures*, **202**:129-139
222. Buryachenko V (2018d) Effective elastic modulus of damaged peristatic bar of periodic structure. *J. Multiscale Comput. Enging*, **16**:101–118
223. Buryachenko V (2019a) Computational homogenization in linear elasticity of peristatic periodic structure composites. *Math. Mech. of Solids.*, **24**:2497-2525
224. Buryachenko V (2019b) Interface integral technique in thermoelasticity of random structure matrix composites. *Math. Mech. of Solids*, **24**:2785–2813
225. Buryachenko V (2019c) Modeling of one inclusion in the infinite peristatic matrix. *J. Peridynamics and Nonlocal Modeling Published*, **1**:75–87
226. Buryachenko V (2020a) Generalized Mori-Tanaka approach in micromechanics of peristatic random structure composites. *J. Peridynamics and Nonlocal Modeling* , **2** :26–49
227. Buryachenko V (2020b) Variational principals and generalized Hill's bounds in micromechanics of peristatic random structure composites. *Math. Mech. of Solids*, **25**:682–704
228. Buryachenko V (2020c) Generalized effective field method in peridynamic micromechanics of random structure composites. *Int. J. Solid Structure* **202**:765-786
229. Buryachenko V (2020d) Effective deformation of peridynamic random structure bar subjected to inhomogeneous body-force. *J. Multiscale Comput. Enging*, **18**:569-585
230. Buryachenko V (2021) Critical analysis of generalized Maxwell homogenization schemes and related prospective problems. *Math. Mech. of Solids*, **26**: (submitted)
231. Buryachenko VA, Böhm HJ, Rammerstorfer FG (1996) Modelling of the overall elastoplastic behavior of multiphase materials by the effective field method. In: Pineau A, Zaoui Z (eds) *IUTAM Symp. Micromech. of Plasticity and Damage of multiphase materials*. Kluwer, Dordrecht, 35–42
232. Buryachenko VA, Brun M (2011) FEA in elasticity of random structure composites reinforced by heterogeneities of noncanonical shape. *Int. J. Solid Struct.* **48**:719–728.
233. Buryachenko VA, Brun M (2012a) Random residual stresses in elasticity homogeneous medium with inclusions of noncanonical shape. *Int. J. Multiscale Comput. Enging.* **10**:261-279.
234. Buryachenko VA, Brun M (2012b) Thermoelastic effective properties and stress concentrator factors of composites reinforced by heterogeneities of noncanonical shape. *Mechanics of Materials*, **53**:91–110.

235. Buryachenko VA, Brun M (2013) Iteration method in linear elasticity of random structure composites containing heterogeneities of noncanonical shape. *Int. J. Solids and Structures*, **50**:1130–1140.
236. Buryachenko VA, Goikhman BD (1985) Integrodifferential criterion of the strength of aging polymers. *Problemy Prochnosti*, **17**(8):88–91 (In Russian. Engl Transl *Strength of Mater* **17**:1138–1141)
237. Buryachenko V, Jackson T, Amadio G (2012) Modeling of random bimodal structures of composites (application to solid propellant) I. Simulation of random packs. *Comput. Model. Engng & Sciences (CMES)*, **85**(5), 379–416.
238. Buryachenko VA, Kreher WS (1995) Internal residual stresses in heterogeneous solids — a statistical theory for particulate composites. *J Mech Phys Solids*, **43**:1105–1125
239. Buryachenko VA, Kushch VI (2006) Effective transverse elastic moduli of composites at non-dilute concentration of a random field of aligned fibers. *ZAMP* , **57**:491–505
240. Buryachenko VA, Kushch VI (2007) Statistical properties of local residual micro stresses in elastically homogeneous half-space. *Int J Multiscale Comput Engng*, **4**, 733–754.
241. Buryachenko VA, Kushch VI, Dutka VA, Roy A (2008) Effective elastic properties of nanocomposites reinforced by cylindrical nanofibers. *Acta Mechanica*, **98**, 129–146
242. Buryachenko VA, Kushch VI, Roy A (2007) Effective thermoelastic properties of random structure composites reinforced by the clusters of deterministic structure (application to clay nanocomposites). *Acta Mechanica*, **192**, 135–167
243. Buryachenko VA, Lipanov AM (1986a) Stress concentration ellipsoidal inclusions and effective thermoelastic properties of composite materials. *Priklad Mekh*, (11):105–111 (In Russian. Engl Transl. *Soviet Appl Mech*, **22**(11):1103–1109)
244. Buryachenko VA, Lipanov AM (1986b) Equations of mechanics for gas-saturated porous media. *Priklad Mekh Tekhn Fiz*, (4):106–109 (In Russian. Engl Transl. *J Appl Mech Tech Phys*, **27**:577–581)
245. Buryachenko VA, Lipanov AM (1988) Equation of isotropic deformation for gas-saturated materials with allowance for large strains of spherical pores. *Priklad Mekh Tekhn Fiz*, (4), 120–124 (In Russian. Engl Transl. *J Appl Mech Tech Phys*, **29**:565–569)
246. Buryachenko VA, Lipanov AM (1989a) Effective field method in the theory of perfect plasticity of composite materials. *Priklad Mekh Tekhn Fiz*, (3):149–155 (In Russian. Engl Transl. *J Appl Mech Tech Phys*, **30**:482–487)
247. Buryachenko VA, Lipanov AM (1989b) Predicting the parameters of a nonlinear flow of multicomponent mixtures. *Priklad Mekh Tekhn Fiz*, (4):53–57 (In Russian. Engl Transl. *J Appl Mech Tech Phys*, **30**:558–562)
248. Buryachenko VA, Lipanov AM (1990a) Effective moduli of elasticity of composites of third order. *Priklad Mekh Tekhn Fiz*, **31**(6):118–123 (In Russian. Engl Transl. *J Appl Mech Tech Phys*, **31**:909–913)
249. Buryachenko VA, Lipanov AM (1990b) Effective characteristics of elastic physically nonlinear composites. *Prikladnaya Mekhanika*, **26**(1):12–16 (In Russian. Engl Transl. *Sov Appl Mech*, **26**:9–13)
250. Buryachenko VA, Lipanov AM (1992) Thermoelastic stress concentration at ellipsoidal inclusions in matrix composites in the region of strongly varying external stress and temperature fields. In: Naimark OB, Evlampieva SE (eds) *Deformation and Fracture of Structural-Inhomogeneous Materials*. AN SSSR, Sverdlovsk, 12–19 (In Russian)
251. Buryachenko VA, Lipanov AM, Kozhevnikova YG (1991) Effective properties of elasto-plastic porous materials. *Problemy Prochnosti*, (1):36–39 (In Russian. Engl Transl. *Strength Mater*, **23**:41–45)
252. Buryachenko VA, Murov V A (1991) Effective conductivity of matrix composites. *Inzhenerno Fiz Zhurnal*, **61**(2):305–312 (In Russian. Engl Transl. *J Engng Phys*, **61**:1041–1047)
253. Buryachenko VA, Pagano NJ (2003) Nonlocal models of stress concentrations and effective thermoelastic properties of random structure composites. *Math Mech of Solids*, **8**:403–433
254. Buryachenko VA, Pagano NJ (2005) Multiscale analysis of multiple interacting inclusions problem: finite number of interacting inclusions. *Math Mech Solids*, **10**:25–62

255. Buryachenko VA, Pagano NJ, Kim RY, Spowart JE (2003) Quantitative description of random microstructures of composites and their effective elastic moduli. *Int J Solids Struct,* **40**:47–72
256. Buryachenko VA, Parton VZ (1990a) Effective Helmholtz operator for matrix composites. *Izv AN SSSR, Mekh Tverd Tela* (3):55–63 (In Russian. Engl Transl. *Mech Solids,* **25**:60–69
257. Buryachenko VA, Parton VZ (1990b) One-particle approximation of the effective field method in the statics of composites. *Mekh Kompoz Mater,* (3):420–425 (In Russian. Engl Transl. *Mech Compos Mater,* **26**(3):304–309)
258. Buryachenko VA, Parton VZ (1990c) Effective parameters of statistically inhomogeneous matrix composites. *Izv AN SSSR, Mekh Tverd Tela,* (6):24–29 (In Russian. Engl Transl. *Mech Solids,* **25**:22–28)
259. Buryachenko VA, Parton VZ (1991) Effective parameters of static conjugating physical-mechanical fields in matrix composites. *Fiziko-Khimichescaja Mech Mater,* **27**(4):105–111 (In Russian. Engl Transl. *Sov Mater Sci,* **27**:428–433)
260. Buryachenko VA, Parton VZ (1992a) Effective field method in the statics of composites. *Priklad Mekh Tekhn Fiz,* (5):129–140 (In Russian. Engl Transl. *J Appl Mech Tech Phys,* **33**:735–745)
261. Buryachenko VA, Parton VZ (1992b) Multi-particle differential methods in the statics of composites. *Priklad Mekh Tekhn Fiz,* (3):148–156 (In Russian. Engl Transl. *J Appl Mech Tech Phys,* **33**:455–462)
262. Buryachenko VA, Parton VZ (1992c) Effective strength parameters of composites in coupled physicomechanical fields. *Priklad Mekh Tekhn Fiz,* (4), 124–130 (In Russian. Engl Transl. *J Appl Mech Tech Phys,* **33**:589–593)
263. Buryachenko VA, Rammerstorfer FG (1996a) Thinly coated inclusion with stress free strains in an elastic medium. *Mech Res Commun,* **23**:505–509
264. Buryachenko VA, Rammerstorfer FG (1996b) Elastoplastic behavior of elastically homogeneous materials with a random field of inclusions. In: Markov K (ed), *Continuum Models and Discrete Systems.* World Scientific, Singapore, 140–147
265. Buryachenko VA, Rammerstorfer FG (1997a) Elastic stress fluctuations in random structure particulate composites. *Eur J Mech A/Solids,* **16**:79–102
266. Buryachenko VA, Rammerstorfer FG (1997b) A local theory of elastoplastic deformations of random structure composites. *Z Angew Math Mech,* **77**(S1):S61–S62
267. Buryachenko VA, Rammerstorfer FG (1998a) Micromechanics and nonlocal effects in graded random structure matrix composites. In: Bahei-El-Din YA, Dvorak GJ (eds) *IUTAM Symp. on Transformation Problems in Composite and Active Materials.* Kluwer, Dordrecht, 197–206
268. Buryachenko VA, Rammerstorfer FG (1998b) Thermoelastic stress fluctuations in random structure coated particulate composites. *Eur J Mechanics A/Solids,* **17**:763–788
269. Buryachenko VA, Rammerstorfer FG (1999) On the thermoelasticity of random structure particulate composites. *Z Angew Math Phys,* **50**:934–947
270. Buryachenko VA, Rammerstorfer FG (2000) On the thermostatics of composites with coated inclusions. *Int J Solids Struct,* **37**:3177–3200
271. Buryachenko VA, Rammerstorfer FG (2001) Local effective thermoelastic properties of graded random structure composites. *Arch Appl Mech,* **71**:249–272
272. Buryachenko VA, Rammerstorfer FG, Plankensteiner AF (1997) A local theory of elastoplastic deformations of random structure composites. *Z Angew Math Mech,* **7**(S1):S61–S62
273. Buryachenko VA, Rammerstorfer FG, Plankensteiner AF (2002) A local theory of elastoplastic deformation of two-phase metal matrix random structure composites. *J Appl Mech,* **69**:489–496
274. Buryachenko VA, Roy A (2005a) Effective elastic moduli of nanocomposites with prescribed random orientation of nanofibers. *Composites B,* **36**:405–416
275. Buryachenko VA, Roy A (2005b) Effective thermoelastic moduli and stress concentrator factors of nanocomposites. *Acta Mechan,* **177**:149–169

276. Buryachenko VA, Roy A, Lafdi K, Anderson KL, Chellapilla S (2005) Multi-scale mechanics of nanocomposites including interface: experimental and numerical investigation. *Comp Sci Technol,* **65**:2435–2465
277. Buryachenko VA, Schoeppner G (2004) Effective elastic and failure properties of fiber aligned composites. *Int J Solids Struct,* **41**:4827–4844
278. Buryachenko VA, Scorbov YS, Gunin SV (1991a) Effective parameters of the strength of matrix composites. *Probl Prochnosti* (12):47–51 (In Russian. Engl Transl. *Strength Mater,* **23**(12))
279. Buryachenko VA, Scorbov YS, Gunin SV, Lysov VV (1991b) Effective rheological parameters of suspensions with polyfractional filler. *Inzhenerno Fiz Zhurnal,* **61**:928–933 (In Russian. Engl Transl. *J Engng Phys,* **61**:1478–1482)
280. Buryachenko VA, Shermergor TD (1995) Material and field characteristics of piezoelectric rocks. Some exact results. *Fiz Zemli,* (8):32–42 (In Russian. Engl Transl. *Phys of the Solid Earth,* (1996) **31**:665–672)
281. Buryachenko VA, Tandon GP (2004) Estimation of effective elastic properties of random structure composites for arbitrary inclusion shape and anisotropy of components using finite element analysis. *Int J Multiscale Comput Engng,* **2**:29–45
282. Butrak O., Kil'nitskaya TI, Mikhas'kiv VV (2012) The scattering of an harmonic elastic wave by a volume inclusion with a thin interlayer *J. Applied Math. Mechanics*, **76**, 342–347
283. Capriz G (1989) *Continua with Microstructure*. Springer, Berlin Heidelberg New York.
284. Cargill III GS (1994) Random packing for amorphous binary alloys. *J Phys Chem Solids,* **55**:1375–1380
285. Carlson DE (1972) Linear thermoelasticity. In: Truesdell C (ed) *Flügge's Handbuch der Physik VI a/2*. Springer-Verlag, Berlin-Heidelberg-New York, 297–346
286. Carrado K (2003) Polymer-clay nanocomposites. In: Shonaike GO, Advani SA (eds), *Advance Polymer Materials*. CRC Press, Boca Raton, London. NY, 349–437
287. Cebeci H, de Villoria RG, Hart AJ, Wardle BL (2009) Multifunctional properties of high volume fraction aligned carbon nanotube polymer composites with controlled morphology *Compos. Science Technology*, **69**:2649–2656
288. Cemlins A (1988) Representation of two-phase flows by volume averaging. *Int J Multiphase Flow,* **14**:81-91
289. Cesarano III J, McEuen MJ, Swiler T (1995) Computer simulation of particle packing. *Intern SAMPE Technical Conf,* **27**:658–665
290. Chaban IA (1965) Self-consistent field approach to calculation of the effective parameters of microinhomogeneous media. *Akust Zhurn,* **10**:351–358 (In Russian. Engl Transl. *Soviet Physics-Acoustics,* **10**:298–302)
291. Chandra A, Huang Y, Hu RX (1997) Crack-size dependence of overall responses of fiber-reinforced composites with matrix cracking. *Int J Solids Struct,* **34**:3837–3857
292. Chandra N, Namilae S, Shet C (2004) Local elastic properties of carbon nanotubes in the presence of Stone-Wales defects. *Phys Rev,* **B69**:094101
293. Chang JS, Cheng CH (1992) Thermoelastic properties of composites with short coated fibers. *Int J Solids Struct,* **29**:2259–2279
294. Chang C, Powell RL (1993) Dynamic simulation of bimodal suspensions of hydrodynamically interacting spherical particles. *J. Fluid Mech.*, **253**:1–25
295. Chang T, Huajian Gao H (2003) Size-dependent elastic properties of a single-walled carbon nanotube via a molecular mechanics model. *J Mech Phys Solids,* **51**:1059 – 1074
296. Chatzigeorgiou G, Javili A, Meraghni F (2019) Micromechanical method for effective piezoelectric properties and electromechanical fields in multi-coated long fiber composites. *Int. J. Solids Structures*, **159**:21–39
297. Chekin BC (1970) Effective parameters of elastic medium with randomly distributed cracks. *Izv AN SSSR, Fiz Zemli,* N10:13–21 (In Russian. Engl Transl. *Phys Solid Earth,* **5**)
298. Chen C, Lu TJ, Fleck NA (1999) Effect of imperfections on the yielding of two-dimensional foams. *J Mech Phys Solids,* **47**:2235–2272

299. Chen C, Fleck NA (2002) Size effects in the constrained deformation of metallic foams. *J Mech Phys Solids*, **50**:955–977
300. Chen CH, Wang YC (1996) Effective thermal conductivity of misoriented short-fiber reinforced composites. *Mech Mater*, **23**:217–228
301. Chen FG, Young K (1977) Inclusions of arbitrary shape in an elastic medium. *J Math Phys*, **18**:1412–1416
302. Chen G, Zhou J (1992) *Boundary Element Methods*. Academic Press, London
303. Chen H, Liu X, Hu G (2008) Overall plasticity of micropolar composites with interface effect. *Mechanics of Materials*, **40**:721–728
304. Chen HS, Acrivos A (1978a) The solution of the equations of linear elasticity for an infinite region containing two spherical inclusions. *Int J Solids and Struct*, **14**:331–348
305. Chen HS, Acrivos A (1978b) The effective elastic moduli of composite materials containing spherical inclusions at non-dilute concentrations. *Int J Solids Struct*, **14**:349–364
306. Chen JT, Hong H-K (1999) Review of dual boundary element methods with emphasis on hypersingular integrals and divergent series. *Appl. Mech. Rev.*, **52**:17–33.
307. Chen KH, Chen JT, Kao JH (2008) Regularized meshless method for antiplane shear problems with multiple inclusions. *Int. J. Numer. Meth. Engng*, **73**:1251–1273.
308. Chen P, Shen Y (2007) Propagation of axial shear magneto—electro–elastic waves in piezoelectric—piezomagnetic composites with randomly distributed cylindrical inhomogeneities. *Int. J. Solids and Structures*, **44**:1511–1532.
309. Chen T (1993a) Thermoelastic properties and conductivity of composites reinforced by spherically anisotropic inclusions, *Mechan of Mater* **14**:257–268
310. Chen T (1993b) Piezoelectric properties of multiphase fibrous composites: some theoretical results. *J Mech Phys Solids*, **41**:1781–1794
311. Chen TY (1994) Micromechanical estimates of the overall thermoelectroelastic moduli of multiphase fibrous composites. *Int J Solids Struct*, **31**:3099–3111
312. Chen T (1999) Exact moduli and bounds of two-phase composites with coupled multifield linear responses *J Mech Phys Solids*, **45**:385–398
313. Chen T, Dvorak GJ, Benveniste Y (1990) Stress fields in composites reinforced by coated cylindrically orthotropic fibers *Mechanics of Materials*, **9**:17–32
314. Chen T, Dvorak GJ, Yu CC (2007) Size-dependent elastic properties of unidirectional nanocomposites with interface stresses. *Acta Mechan*, **188**:39–54
315. Chen T, Dvorak GJ, Yu CC (2007) Solids containing spherical nano-inclusions with interface stresses effective properties and thermal-mechanical connections. *Int. J. Solids and Structures*, **44**:941–955.
316. Chen T, Nan C-W, Weng GJ, Chen G-X (2003a) Unified approach for the estimate of effective magnetostriction of composites and polycrystals with particulate and columnar microstructures. *Phys Rev*, **B68**:224406
317. Chen T, Nan C-W, Weng GJ (2003b) Exact connections between effective magnetostriction and effective elastic moduli of fibrous composites and polycrystals. J Appl Phys, **94**:491–495
318. Chen T, Yang S-H (1995) The problem of thermal conductivity for two ellipsoidal inhomogeneities in an anisotropic medium and its relevance to composite materials. *Acta Mechan*, **111**:41–58
319. Chen W, Tanaka M (2002) A meshless, integration-free, and boundary-only RBF technique *Computers & Mathematics with Applications*, **43**: 379–391
320. Chen X, Papathanasiou TD (2004) Interface stress distributions in transversely loaded continuous fiber composites: parallel computation in multi-fiber RVEs using the boundary element method. *Comp Sci Technol*, **64**:1101–1114
321. Cheng AH-D, Chen CS, Golberg MA, Rashed YF (2001) BEM for thermoelasticity and elasticity with body force – a revisit. *Engineering Analysis with Boundary Elements*, **25**:377–387.
322. Cheng H, Torquato S (1997) Electric-field fluctuations in random dielectric composites. *Phys Rev*, **B56**:8060–8068

323. Cheng J, Jordan EH, Walker KP (1997) Gauss integration applied to a Green's function formulation for cylindrical fiber composites. *Mechan Mater* **26**:247–267
324. Cheng ZQ, He LH (1995) Micropolar elastic fields due to a spherical inclusion. *Int. J. Eng. Sci.*, **33**:389–397
325. Cheng ZQ, He LH (1997) Micropolar elastic fields due to a circular cylindrical inclusion. *Int. J. Eng. Sci.*, **35**:659–686
326. Cheng Z, Zhang G, Wang Y, Bobaru F (2015) A peridynamic model for dynamic fracture in functionally graded materials *Composite Structures*, **133**:529–546
327. Cheng YF, Guo SJ, Lay HY (2000) Dynamic simulation of random packing of spherical particles. *Powder Technol,* **107**:123–130
328. Cherkaoui M, Sabar H, Berveiller M (1995) Elastic composites with coated reinforcements: a micromechanical approach for non homothetic topology. *Int J Engng Sci,* **33**:829–843
329. Chernikh KF (1986) *Nonlinear Theory of Elasticity in Machine-Building Calculations.* Mashinostroenie, Leningrad (In Russian)
330. Chernikh KF, Shubina IM (1978) Allowance for the compressibility of rubber. Mekhanika Elastomerov. *Kuban. Univ, Krasnodar,* **263**:56–62 (In Russian)
331. Chiew Y-C, Glandt ED (1987) Effective conductivity of dispersion: the effect of resistance at the particle surfaces. *Chem Engng Sci,* **42**:2677–2685
332. Childress S (1972) Viscous flow past a random array of spheres. *J. Chem. Phys.*, **56**:2527–2539
333. Chong JS, Christiansen EB, Baer AD (1971) Rheology of concentrated suspensions. *J. Applied Polymer Science*, **15**:2007–2021
334. Chou T-W (1992) *Microstructural Design of Fiber Composites.* Cambridge Solid State Science Series. Cambridge University Press, Cambridge, UK, 99–168
335. Choy TA, Alexandropoulos A, Thorpe MF (1998) Dielectric function for a material containing hyperspherical inclusions in $O(c^2)$: I. Multipole expansions; II. Method of images. *Proc Roy Soc Lond Ser,* **A454**(1975):1973–1992, 1993–2013
336. Christensen RM (1979) *Mechan of Composite Materials.* Wiley Interscience, New York
337. Christensen RM (1990) A critical evaluation for a class of micromechanics models. *J Mech Phys Solids,* **38**:379–404
338. Christensen RM, Lo KH (1979) Solutions for effective shear properties in three phase sphere and cylinder models. *J Mech Phys Solids,* **27**:315–330
339. Chudnovsky A, Wu S (1993) Evaluation of energy release rate in the crack-microcrack interaction problem. *Int J Solids Structure,* **29**:1699-1709
340. Chueng HT, Kim HG (1987) Characteristics of domain in tetragonal phase PZT ceramics. *Ferroelectrics,* **76**:327-333
341. Clarke AS, Willey JD (1987) Numerical simulation of the dense random packing of a binary mixture of hard spheres: amorphous metals. *Phys Rev,* **B35**:7350–7356
342. Clausius R (1879) *Die Mechanische Behandlung der Electricität.* Vieweg, Braunshweig
343. Cleary MP, Chen LW, Lee SM (1980) Self-consistent techniques for heterogeneous solids. *J Engng Mech,* **106**(5):861–871
344. Clements BE, Mas EM (2004) A theory for plastic-bonded materials with a bimodal size distribution of filler particles *Modelling Simul. Mater. Sci. Eng.*, **12**:407–421
345. Glüge R, Weber M, Bertram A (2012) Comparison of spherical and cubical statistical volume elements with respect to convergence, anisotropy, and localization behavior. *Comput. Mater. Sci.*, **63**:91–104
346. Conlon KT, Wilkinson DS (1996) Microstructural inhomogeneity and the strength of particulate metal matrix composites. In: Pineau A, Zaoui A (eds) *IUTAM Symp. on Micromechanics of Plasticity and Damage of Multiphase Materials,* Kluwer, Dordrecht, 347–354
347. Corcolle R, Préault V, Dani L (2012) Romain Corcolle, Valentin Préault, and Laurent Dani Second order moments in linear smart material composites. *IEEE Transactions on Magnetics*, **48**:663-666

348. Corson PB (1974) Correlation function for predicting properties of heterogeneous materials. I. Experimental measurement of spatial correlation functions in multiphase solids. *J Appl Phys,* **45**:3159–3164
349. Corvasce F, Lipinski P, Berveiller M (1990) The effects of thermal, plastic and elastic stress concentration on the overall behavior of metal matrix composites. In: Dvorak GJ (ed), *Inelastic Deformation of Composite Materials.* Springer-Verlag, New York, 389–408
350. Cosserat E, Cosserat F (1909) *Théorie des corps déformables.* A. Hermann et Fils, Paris.
351. Cox DR, Isham V (1980) *Point Processes.* Chapman and Hall, London and NY
352. Creswik RJ, Farah HA, Poole CP (1998) *Introduction to Renormalization Group Methods Physics.* John Wiley & Sons, New York
353. Cristescu ND, Craciun E-M, Soós E (2003) *Mechanics of Elastic Composites.* Chapman & Hall/CRC, Boca Raton, FL
354. Cruse TA (1974) An improved boundary-integral equation method for three dimensional elastic stress analysis. *Comput. Struct.* **4**:741–754
355. Curtin WA (1999) Stochastic damage evaluation and failure in fiber-reinforced composites. Hutchinson JW, Wu TJ (eds), *Adv Appl Mech,* **36**:163–253
356. Curtin WA, Miller RE (2003) Atomistic/continuum coupling in computational materials science. *Modelling Simul Mater Sci Engng,* **11**:R33–R68
357. Cutler RA, Vircar AV (1985) The effect of binder thickness and residual stresses on the fracture toughness of cement carbide. *J Mater Sci,* **20**:3557–3573
358. Dao M, Gu P, Maewal A, Asaro RJ (1997) A micromechanical study of residual stresses in functionally graded materials. *Acta Mater,* **45**:3265–3276
359. Darinskii BM, Fokin AG, Shermergor TD (1967) The calculation of elastic moduli of polycrystalline agragates. *Zh Prikl Mekh Thekhn Fiziki,* **6**(5):123 (In Russian. Engl Transl. *J Appl Mech Tech Phys,* **8**:79–82)
360. Darinskii BM, Shermergor TD (1964) Temperature relaxation in polycrystals of cubic structure. *Fizika Metallov i Metallovedenie,* **18**:645 (In Russian)
361. Darinskii BM, Shermergor TD (1965) Elastic moduli of cubic polycrystals. *Zh Prikl Mekh Thekhn Fiziki,* **6**(4):79–82 (In Russian)
362. Datta SK (1977) A self-consistent approach to multiple scattering by elastic ellipsoidal inclusions. *J Appl Mech,* **44**:657–661
363. Davies GJ (1973) *Solidification and Casting.* Appl Sci Publication, Oxford
364. Davis IL, Carter RG (1989) Random particle packing by reduced dimension algorithms. *J Appl Phys,* **67**:1022–1029
365. Dayal K, Bhattacharya K (2006) Kinetics of phase transformations in the peridynamic formulation of continuum mechanics. *J. Mech. Physics Solids*, **54**:1811–1842
366. Decklever J, Spanos P (2016) Nanocomposite material properties estimation and fracture analysis via peridynamics and Monte Carlo simulation. *Probabilistic Engineering Mechanics* **44** , 77–88
367. Dederichs P H, Zeller R (1973) Variational treatment of the elastic constants of disordered materials. *Z Physik,* **259**:103–116
368. Deeg WF (1980) *The Analysis of Dislocation, Crack and Inclusion Problems in Piezoelectric Solids.* PhD Thesis, Stanford University, Stanford, CA
369. D'Elia M, Du Q, Glusa C, Gunzburger M, Tian X, Zhou Z (2020) Numerical methods for nonlocal and fractional models. *Acta Numerica*, **29**, 1–124
370. Delves LM, Mohamed JL (1985) *Computational Methods for Integral Equations.* Cambridge University Press, Cambridge, UK
371. Desrumaux F, Meraghi F, Benzeggagh ML (2001) Generalized Mori-Tanaka scheme to model anisotropic damage using numerical Eshelby tensor. *J Compos Mater* **35**:603–624
372. D'Haene P, Mewis J (1994) Rheological characterization of bimodal colloidal dispersions. *Rheol Acta*, **33**:165–174
373. Diener G, Hurrich A, Weissbarth J (1984) Bounds on the non–local effective elastic properties of composites. *J Mech Phys Solids,* **32**:21–39

374. Dienes JK, Zuo QH, Kershner JD (2006) Impact initiation of explosives and propellants via statistical crack mechanics. *J Mech Phys Solids*, **54**, 1237–1275
375. Diggle PJ (2003) *Statistical Analysis of Spatial Point Patterns* (2nd edn). Academic Press, New York
376. Ding C-L, Zhao X-P (2011) Multi-band and broadband acoustic metamaterial with resonant structures *J. Phys. D Appl. Phys.*, **44**:215402 (8pp.)
377. Dinzart F, Sabar H (2011) Magneto-electro-elastic coated inclusion problem and its application to magnetic-piezoelectric composite materials. *Int. J. Solids and Structures*, **48**:2393–2401.
378. Dinzart F, Sabar H (2017) New micromechanical modeling of the elastic behavior of composite materials with ellipsoidal reinforcements and imperfect interfaces. *Int.l J. of Solids and Structures*, **108**:254-262
379. Dinzart F, Sabar H. (2019) Electroelastic ellipsoidal inclusion with imperfect interface and its application to piezoelectric composite materials *Int. J. Solids Structures*, **136–137**, 241–249
380. Djiordjevic BR, Hetherington JH, Thorpe, MF (1996) Spectral function for a conducting sheet containing circular inclusions. *Phys Rev*, **B35**:14862–14871
381. Diyaroglu C, Madenci E, Phan N (2019) Peridynamic homogenization of microstructures with orthotropic constituents in a finite element framework *Composite Structures*, **227**:111334
382. Diyaroglu C, Madenci E, Stewart RJ, Zobi SS (2019) Combined peridynamic and finite element analyses for failure prediction in periodic and partially periodic perforated structures *Composite Structures*, **227**:111481
383. Döge G (2000) Grand canonical simulation of hard-disc systems by simulated tempering. In: Mecke KR, Stoyan D (eds) *Statistical Physics and Spatial Statistics: The Art of Analyzing and Modeling Spatial Structures and Pattern Formation*. Lecture Nnotes in Physics, Vol. 554, Berlin
384. Donev A, Torquato S, Stillinger FH (2005) Pair correlation function characteristics of nearly jammed disordered and ordered hard-sphere packings. *J. of Computational Physics*, **202**:737–764
385. Dong CY, Bonnet M (2002) An integral formulation for steady-state elastoplastic contact over a coated half-plane. *Comput Mech*, **28**:105–121
386. Dong CY, Lo, SH, Cheung YK (2002) Application of the boundary-domain integral equation in elastic inclusion problems. *Engng Anal Bound Elements*, **26**:471–477
387. Dong, C.Y, Lo, S.H, Cheung, Y.K (2003) Stress analysis of inclusion problems of various shapes in an infinite anisotropic elastic medium. *Comput. Methods Appl. Mech. Engrg.* **192**:683-696
388. Dorgan RJ, Voyiadjis GZ (2006) A mixed finite element implementation of a gradient-enhanced coupled damage-plasticity model. *Int. J. Damage Mechanics*, **15**:201-235
389. Draughn RA (1981) Effects of temperature on mechanical properties of composite dental restorative materials. *J. Biomedical Mater. Research* **15**:489–495
390. Drugan WJ (2000) Micromechanics-based variational estimations for a higher-order nonlocal constitutive equation and optimal choice of effective moduli for elastic composites. *J Mech Phys Solids*, **48**:1359–1387
391. Drugan WJ (2003) Two exact micromechanics-based nonlocal constitutive equations for random linear elastic composite materials. *J Mech Phys Solids*, **51**:1745–1772
392. Drugan WJ, Willis JR (1996) A micromechanics-based nonlocal constitutive equation and estimates of representative volume elements for elastic composites. *J Mech Phys Solids*, **44**:497–524
393. Drugan WJ, Willis JR (2016) A micromechanics-based nonlocal constitutive equation incorporating three-point statistics for random linear elastic composite materials. *J Mech Phys Solids*, **91**:291–1052
394. Du C, Ying Z, Jiang S (2010) Extended finite element method and its application in heterogeneous materials with inclusions. *IOP Conf. Series Materials Science and Engineering.*, **10**:012083

395. Du JK, Shen YP, Ye DY, Yue FR (2004) Scattering of anti-plane shear waves by a partially debonded magneto-electro-elastic circular cylindrical inhomogeneity. *Int J Engin Sci,* **42**:887–913
396. Du Q, Gunzburger M, Lehoucq RB, Zhou K (2012) Analysis and approximation of nonlocal diffusion problems with volume constraints. *SIAM Review,* **54**:667–696.
397. Du Q, Gunzburger M, Lehoucq RB, Zhou K (2013). Analysis of the volume-constrained peridynamic Navier equation of linear elasticity. *J. Elast.*, **113**:193–217.
398. Du X, Ostoja-Starzewski M (2006) On the scaling from statistical to representative volume element in thermoelasticity of random materials. *Networks Heterogeneous Media*, **1**:259–274
399. Du ZZ, McMeeking R M, Schmauder S (1995) Transverse yielding and matrix flow past the fibers in metal matrix composites. *Mech Mater* **21**:159–167
400. Duan HL, Karihaloo BL (2007) Thermo-elastic properties of heterogeneous materials with imperfect interfaces Generalized Levin's formula and Hill's connections. *J. Mechanics Physics Solids*, **55**:1037–1052.
401. Duan HL, Wang J, Huang ZP, Karihaloo BL (2005) Eshelby formalism for nano-inhomogeneities. *Proc Roy Soc Lond Ser,* **A461**:3335–3353
402. Duan HL, Yi X, Huang ZP, Wang J (2007) A unified scheme for prediction of effective moduli of multiphase composites with interface effects Part II—Application and scaling laws. *Mech. Materials*, **39**:94–103
403. Duan ZP, Kienzler R, Herrmann G (1986) An integral equation method and its application to defect mechanics. *J Mechan Phys Solids,* **34**:539–561
404. Dul'nev GH, Malarev BI (1990) Percolation theory in the conductivity theory of inhomogeneous media. *Inzhenerno Fiz Zhurnal,* **39**:522–539 (In Russian)
405. Dumont , Serpilli M, Rizzoni R, Lebon FC (2020) Numerical validation of multiphysic imperfect interfaces models. *Frontiers in Materials*, **2020**, **7**:158 (14 pages)
406. Dumonted H (1986) Study of a boundary layer problem in elastic composite materials. *Math Model Numer Anal,* **20**:265–286
407. Dunn ML(1993) Exact relations between the thermoelectroelastic moduli of heterogeneous materials. *Proc Roy Soc Lond,* **A441**:549–557
408. Dunn ML (1994a) Electroelastic Green's functions for transversely isotropic piezoelectric media and their applications to the solutions of inclusion and inhomogeneity problems. *Int J Engrg Sci,* **32**:119–131
409. Dunn ML (1994b) Thermally induced Relds in electroelastic composite materials: average Relds and effective behavior. *J Engng Mat Technol,* **116**:200–207
410. Dunn ML, Taya M (1993a) Micromechanics predictions of the effective electroelastic moduli of piezoelectric composites. *Int J Solids Struct,* **30**:161-175
411. Dunn ML, Taya M (1993b) An analysis of piezoelectric composite materials containing ellipsoidal inhomogeneities. *Proc R Soc Lond: A***443**: 265–287
412. Durlofsky L, Brady JF, Bossis G (1987) Dynamic simulation of hydrodynamically interacting particles. *J. Fluid Mech,* **180**:21–49
413. Duschlbauer D, Böhm HJ, Petterman HE (2006) Computational simulation of composites reinforced by planar random fibers homogenization and localization by unit cell and mean field approaches. *J. Composite Materials.*, **40**:2217–2234
414. Duschlbauer D, Pettermann HE, Böhm HJ (2003) Mori Tanaka based evaluation of inclusion stresses in composites with nonaligned reinforcements. *Scripta Materialia,* **48**:223–228
415. Dutta A, Mashelkar RA (1989) Thermal conductivity of structured liquids. In: Hartnet JP, Irvin TF (eds), *Advances in Heat Transfer,* Academic Press, New York, **18**:161–239
416. Duva JM (1988) A constitutive description of nonlinear materials containing voids. *Mech Mater* **5**:137–144
417. Dvorak GJ (1990) On uniform fields in heterogeneous media. *Proc Roy Soc Lond,* **A431**:89–110
418. Dvorak GJ (1992a) On some exact results in thermoelasticity of composite materials. *J Thermal Stresses,* **15**:211–228

419. Dvorak GJ (1992b) Transformation field analysis of inelastic composite materials. *Proc R Soc Lond,* **A437**:311–327
420. Dvorak GJ (1993) Nadai lecture–micromechanics of inelastic composite materials: theory and experiment. *J Engng Mater Tech,* **115**:327–338
421. Dvorak GJ (2000) Composite materials: inelastic behavior, damage, fatigue, and fracture. *Int J Solids Struct,* **37**:155–170
422. Dvorak GJ (2013) *Micromechanics of Composite Materials.* Springer, Dordrecht
423. Dvorak GJ, Bahei-El-Din YA, Wafa AM (1994) The modeling of inelastic composite materials with the transformation field analysis. *Modelling Simul Mater Sci Engng,* **2**:571–585
424. Dvorak GJ, Benveniste Y (1992a) On the thermomechanics of composites with imperfectly bonded interfaces and damage. *Int J Solids Struct,* **29**:2907–2919
425. Dvorak GJ, Benveniste Y (1992b) On transformation strains and uniform fields in multiphase elastic media. *Proc Roy Soc Lond,* **A437**:291–310
426. Dvorak GJ, Srinivas MV (1999) New estimations of overall properties of heterogeneous solids. *J Mech Phys Solids,* **47**:899–920
427. Dykhne AM (1970) Conductivity of a two-dimensional two-phase system. *J Experiment Theor Phys (JETP)* 59:110–116 (In Russian. Engl Trunsl. *Soviet Phys*, (1971) **32**:63–65)
428. Dyskin AV, Mühlhaus HB (1995) Equilibrium bifurcations in dipole asymptotics model of periodic crack arrays. In: Mühlhaus HBM (ed) *Continuum Models for Materwith Microstructure.* John Wiley & Sons, New York, 69–104
429. Ebbesen TW (1997) *Carbon Nanotubes: Preparation and Properties.* CRC Press, Boca Raton, FL
430. Echlin MP, Lenthe WC, Pollock TM (2014) Three-dimensional sampling of material structure for property modeling and design. *Integ.Mater. Manuf. Innov.*, **3**:1–14
431. Edelen DGB, Laws N (1971) On the thermodynamics of systems with nonlocality. *Arch. Ration. Mech. Anal.*, **43**:36–44
432. Eischen JW, Torquato S (1993) Determining elastic behavior of composites by the boundary element method. *J Appl Phys,* **74**:159–170
433. Ellaway SW, Fauxa DA (2002) Effective elastic stiffnesses of InAs under uniform strain. *J Appl Phys,* **92**:3027–3033
434. Emmrich E, Weckner O (2006) The peridynamic equation of motion in non-local elasticity theory. In C. A. Mota Soares *et al.* (eds.), *III European Conference on Computational Mechanics. Solids, Structures and Coupled Problems in Engineering*. Springer, Dordrecht
435. Emmrich E, Weckner O (2007a) Analysis and numerical approximation of an integro-differential equation modeling non-local effects in linear elasticity. *Math. Mech. Solids,* **12**:363–384
436. Emmrich E, Weckner O (2007b) On the well-posedness of the linear peridynamic model and its convergence towards the Navier equation of linear elasticity. *Commun. Math. Sci.* **5**:851–864
437. Enikolopyan NS, Fridman ML, Stalnova IO, Popov VL (1990) Filled polymers: mechanical properties and processability. *Adv Polym Sci,* **96**:1–67
438. Erdogan F (1995) Fracture mechanics of functionally graded materials. *Compos Engng,* **5**:753–770
439. Eringen AC (1964) Simple microfluids. *Int. J. Eng. Sci.*, **2**:205–217
440. Eringen AC (1966) A unified theory of thermomechanical materials. *Int. J. Eng. Sci.*, **4**:179–202
441. Eringen AC (1968) Mechanics of micromorphic continua. In: Kröner E (ed), *Mechan of Generalized Continua.* Springer-Verlag, Berlin, 18–35
442. Eringen AC (1976) Nonlocal polar field theories. In: Eringen AC (ed), *Continuum Phys. Vol. IV. Polar and Nonlocal Field Theories.* Academic Press, New York, 205–267
443. Eringen AC (1978) Nonlocal continuum mechanics and some applications. In: Barut AO (ed) *Nonlinear Equations in Phys and Mathematics.* Reidel, Dordrecht, 271–318

444. Eringen AC (1987) Theory of nonlocal elasticity and some application. *Res Mechanica*, **21**:313–342
445. Eringen AC (1999) *Microcontinuum Field Theories I. Foundations and Solids*. Springer-Verlag, Berlin
446. Eringen AC (2002) *Nonlocal Continuum Field Theories*. Springer-Verlag, New York
447. Eringen AC, Suhubi ES (1964) Nonlinear theory of simple micro-elastic solids. *Int. J. Eng. Sci.* **2**, 189–203
448. Eshelby JD (1956) The continuum theory of lattice defects. In: Seitz F, Turnbull D (eds), *Solid State Phys*. Academic Press, New York, **3**:79–144
449. Eshelby JD (1957) The determination of the elastic field of an ellipsoidal inclusion, and related problems. *Proc Roy Soc Lond*, **A241**:376–396
450. Eshelby JD (1961) Elastic inclusion and inhomogeneities. In: Sneddon IN, Hill R (eds), *Prog in Solid Mechan.* North-Holland, Amsterdam, **2**:89–140
451. Eskin GI (1981) *Boundary Value Problems for Elliptic Pseudodifferential Equations*. American Mathematical Society, Providence, RI
452. Evans AG (1987) Microfracture from thermal expansion anisotropy; I. Single phase systems. *Acta Metall*, **26**:1845–1853
453. Evans AG (1989) The new high-toughness ceramics. In: Wei RP, Gangloff RP (eds), *Fracture Mechanics: Perspectives and Directions (Twentieth Symposium)* Amer Soc Testing and Mater, 267–291
454. Fabrikant VI (1989) Complete solutions to some mixed boundary value problems in elasticity. In: Hutchinson JW, Wu TJ (eds) *Advances in Applied Mechanics* Academic Press, New York, 153–223
455. Fairweather G, Karageorghis A (1998) The method of fundamental solutions for elliptic boundary value problems. *Advances in Computational Mathematics*, **9** 69–95
456. Fakri N, Azrar LL, Bakkali LE (2003) Electroelastic behavior modeling of piezoelectric composite materials containing spatially oriented reinforcements. *Int J Solids Struct*, **40**:361–384
457. Falconer K (2014) *Fractal Geometry Mathematical Foundations and Applications*. Chichester, John Wiley
458. Fan X-J, Phan-Thien N, Zheng R (1998) Complemented double layer boundary element method for periodic suspensions. *Z Angew Math Phys*, **49**:167–193
459. Faraday M (1838) Experimental researches on electricity. *Philos Trans R Soc Lond*, II Ser. 1ff
460. Farris RJ (1968) Prediction of the viscosity of multimodal suspensions from unimodal viscosity data. *Trans. Society of Rheology*, **12**:281–301
461. Fasshauer GE (2016) Meshfree methods, *Handbook of Theoretical and Computational Nanotechnology*, M. Rieth and W. Schommers (eds.), American Scientific Publishers, **2** 33–97
462. Fassi–Fehri O, Hihi A, Berveiller M (1989) Multiple site self consistent scheme. *Int J Engng Sci*, **27**:495–502
463. Feder J (1980) Random sequential adsorption *J Theor Biol*, **87**:237–254
464. Fedoryuk MV (1988) Diffraction of acoustic waves on a triaxial ellipsoid. *J Acoustics*, **34**:160–164 (In Russian)
465. Federico S, Grillo A, Herzog W (2004) A transversally isotropic composite with a statistical distribution of spheroidal inclusions: a geometrical approach to overall properties. *J Mech Phys Solids*, **52**:2309–2327
466. Felderhof BU (1989) Hydrodynamic interactions in suspensions with periodic boundary conditions. *Physica*, **A 159**:1-18
467. Feng X (2001) Effective elastic moduli of polymer-layered silicate nanocomposites. *Chin Sci Bull*, **46**:1130–1133
468. Ferrante FJ, Arwade SR, Graham-Brady LL (2005) A translation model for non-stationary, non-Gaussian random processes. *Probabilistic Engin Mech*, **20**:215–228

469. Ferrari M (1991) Asymmetry and the high concentration limit of the Mori–Tanaka effective medium theory. *Mech Mater,* **11**:251–256
470. Fett T, Munz D (1993) Influence of bridging interactions on the lifetime behavior of coarse-grained Al_2O_3. *J Eur Ceram Soc,* **12**:131–138
471. Field JE, Bourne NK, Palmer SJP, Walley SM (1992) Hot-spot ignition mechanisms for explosives and propellants. *Philos. Trans. R. Soc.*, **A339**:269–283
472. Filatov AN, Sharov LV (1979) *Integral Inequalities and the Theory of Nonlinear Oscillations.* Nauka, Moscow (In Russian)
473. Firooz S, Chatzigeorgiou G, l Meraghni F, Javili A (2020) Bounds on size effects in composites via homogenization accounting for general interfaces *Continuum Mech. Thermodynamics*, **32**:173–206
474. Firooz S, Javili A (2019) Understanding the role of general interfaces in the overall behavior of composites and size effects. *Comput. Materials Science*, **162**:245–254
475. Fish J (1992) The s-version of the finite element method. *Comput Struct,* **43**:539–547
476. Fish J (2006) Bridging the scales in nano engineering and science *J Nanoparticle Res,* **8**:577–594
477. Fish J (2014) *Practical Multiscaling*. John Wiley & Sons, Chichester, UK
478. Fish J, Belytschko T (2007) *A First Course in Finite Elements,* Chichester, England; Hoboken, NJ John Wiley & Sons.
479. Fish J, Chen W, Li R (2007) Generalized mathematical homogenization of atomistic media at finite temperatures in three dimensions. *Comput. Methods Appl. Mech. Engrg.*, **196**:908–922
480. Fish J, Filonova V, Kuznetsov S (2012b) Micro-inertia effects in nonlinear heterogeneous media *Int. J. Numer. Meth. Engng*, **91**:1406–1426
481. Fish J, Jiang T, Yuan Z (2012a) A staggered nonlocal multiscale model for a heterogeneous medium. *Int. J. Numer. Meth. Engng*, **91**:142–157
482. Fish J, Wagiman A (1993) Multiscale finite element method for a locally nonperiodic heterogeneous medium. *Comput Mech,* **12**:164–180
483. Fisher FD, Mayrtofer K (1997) The crack-closure effect for a penny-shaped crack bridged by an arbitrary located single fiber. *Composites,* **B28**:425–432
484. Fisher FT, Bradshaw RD, Brinson LC (2003) Fiber waviness in nanotube-reinforced polymer composites: I. Modulus predictions using effective nanotube properties. *Composites Sci and Technol,* **63**:1689–1703
485. Fleck NA, Hutchinson JW (1997) Strain gradient plasticity. In: Hutchinson JW, Wu TY (eds), *Advances in Appl Mech,* Academic Press, New York, **33**:295–361
486. Fleck NA, Willis JR (2004) Bounds and estimates for the effect of strain gradients upon the effective plastic properties of an isotropic two-phase composite. *J. Mech. Phys. Solids.* **52**:1855–1888
487. Floquet G (1883) Sur les equations differentielles linearies a coefficients periodiques. *Ann. de l'Ecole Normale Supérieur*, **12**:47–88
488. Fokin AG (1984) A method of solving problems of the linear theory of elasticity. *Prikl Math Mech,* **48**:436–446 (In Russian. Engl Transl. *J Appl Math Mech,* **48**:315–323)
489. Fokin AG, Shermergor TD (1968a) The boundaries of the effective elastic moduli for inhomogeneous solids. *Zhurnal Prikl Mekh Tekhnic Fiziki,* **9**(4):39–46 (In Russian. Engl Transl. *J Appl Mech Tech Phys,* **9**:381–388)
490. Fokin AG, Shermergor TD (1968b) Calculation of elastic moduli of inhomogeneous materials. *Mekhanika Polymerov,* **4**:624–630 (In Russian. Engl Transl. *Mech Comp Mater* **4**:481–486)
491. Fokin AG, Shermergor TD (1969) Calculation of the effective elastic moduli of composite materials with multiphase interactions taken into consideration. *Zhurnal Prikl Mekh Tekhnic Fiziki,* **10**:51–57 (In Russian. Engl Transl. *J Appl Mech Tech Phys,* **10**:48–54)
492. Fokin AG, Shermergor TD (1989) Theory of propagation of elastic waves in nonhomogeneous media. *Mekhan. Kompoz. Materialov*, **25**(5), 821–832 (In Russian. Engl Transl. *Mech. Compos. Materials*, (1990) **25**:600–609.)

493. Foldy LL (1945) The multiple scattering of waves. I. General theory of isotropic scattering by randomly distributed scatters. *Phys Rev,* **67**:107–117
494. Fond C, Riccardi A, Schirrer R, Montheillet F (2001) Mechanical interaction between spherical inhomogeneities: an assessment of a method based on the equivalent inclusion. *Eur J Mech A/Solids,* **20**:59–75
495. Forest S, Dendievel R, Canova GR (1999) Estimating the overall properties of heterogeneous Cosserat materials. *Model. Simul. Mater. Sci. Eng.*, **7**:829–840
496. Forest S, Sab K (1998) Cosserat overall modeling of heterogeneous materials *Mech Res Commun,* **25**:449–454
497. Forest S, Sab K (2012) Stress gradient continuum theory. *Mechanics Research Communications*, **40**:16–25
498. Fornes TD, Paul DR (2003) Modeling properties of nylon 6/clay nanocomposites using composite theories *Polymers,* **44**:4993–5013
499. Fotiu P, Irschik H, Ziegler F (1991) Micromechanical foundations of dynamic plasticity with applications to damaging structures In: Brüller O, Mannl V, Najar J (eds), *Advances in Continuum Mechan*. Springer-Verlag, Berlin, 338–349
500. Franciosi P (2005) On the modified Green operator integral for polygonal, polyhedral and other non-ellipsoidal inclusions. *Int J Solids Struct,* **42**:3509–3531
501. Franciosi P (2010) The boundary-due terms in the Green operator of inclusion patterns from distant to contact and to connected situations using radon transforms Illustration for spheroid alignments in isotropic media. *Int. J. Solids Struct.* **47**:304–319.
502. Francqueville F, Gilormini P, Diani J (2019) Representative volume elements for the simulation of isotropic composites highly lled with monosized spheres - *Int. J. Solids and Structures*, **158**:277–286
503. Franciosi P, Lebail H (2004) Anisotropy features of phase and particle spatial pair distributions in various matrix/inclusions structures. *Acta Mater,* **52**:3161–3172
504. Freed KF, Muthukumar M (1978) On the Stokes problem for a suspension of spheres at finite concentrations. *J. Chem. Phys.*, **68**:2088–2096
505. Freidin AB, Kucher VA (2016) Solvability of the equivalent inclusion problem for an ellipsoidal inhomogeneity. *Math. Mech. Solids*, **21**:255–262
506. Fries TP, Belytschko T (2010) The extended/generalized finite element method an overview of the method and its applications. *Int. J. Numer. Meth. Engng,* **84**:253–304
507. Friesecke G, James RD (2000) A scheme for the passage from atomic to continuum theory for thin films, nanotubes and nanorods. *J Mech Phys Solids,* **48**:1519–1540
508. Frisch HL (1965) Statistics of random media. *Trans Soc Rheol,* **9**:293–312
509. Fu LS (1982) Mechanical aspects of NDE by sound and ultrasound. *Appl Mech Rev,* **35**:1047–1057
510. Fu LS (1987) Dynamic moduli and located damage in composites. In: Vary A (ed), *Material Analysis by Ultrasonics: Metals, Ceramics, Composites.* Noyes Data Corp, New York, 225–248.
511. Fu LS, Mura T (1983) The determination of elastodynamic fields of an ellipsoidal inhomogeneity. *J Appl Mech,* **50**:390–396
512. Fu Y, Evans AG (1985) Some effects of microcracks on the mechanical properties of brittle solids; I. Stress-strain relations. *Acta Metall,* **33**:1515–1523
513. Fu S-Y, Feng X-Q, Lauke B, Mai Y-W (2008) Effects of particle size, particle/matrix interface adhesion and particle loading on mechanical properties of particulate–polymer composites. *Composites*, **B 39**:933–961
514. Fu Y, Klimkowski KJ, Rodin GJ, Berger E, Browne JC, Singer JK, Van De Geijn, RA, Vemaganti KS (1998) A fast solution method for three-dimensional many-particle problems of linear elasticity. *Int J Numer Methods Engng,* **42**:1215–1229
515. Fung YC (1965) *Foundations of Solid Mechanics.* Prentice-Hall, Englewood Cliffs, NJ
516. Furmañski P (1997) Head conduction in composites: homogenization and macroscopic behavior. *Appl Mech Rev,* **50**:327–356

517. Furukawa K, Imai K, Kurashige M (2000) Simulated effect of box size and wall on porosity of random packing of spherical particles. *Acta Mechan,* **140**:219–231
518. Gad AI, Gao XL (2020) Extended Hill's lemma for non-Cauchy continua based on a modified couple stress theory. *Acta Mech,* **231**:977–997 (2020)
519. Gajdošík J, Zeman J, Sejnoha M (2006) Qualitative analysis of fiber composite microstructure Influence of boundary conditions. *Probabilistic Engineering Mechanics,* **21**:317–329
520. Galli M, Cugnoni J, Botsis J (2012) Numerical and statistical estimates of the representative volume element of elastoplastic random composites. *Eur. J. Mech. A/Solids,* **33**:31–38
521. Gallier S (2009) A stochastic pocket model for aluminum agglomeration in solid propellants. *Propellants Explos. Pyrotech.,* **34**:97–105
522. Gambin B, Kröner E (1989) High order terms in the homogenized stress-strain relation of periodic elastic media. *Phys Stat Sol,* **151**:513–519
523. Ganghoffer JF, de Borst R (2000) A new framework in nonlocal mechanics. *Int J Engng Sci,* **38**:453–486
524. Gao H, (1988) Nearly circular shear mode cracks. *Int J Solids Struct,* **24**:177–193
525. Gao XL (2016) Extended Hill's lemma for non-Cauchy continua based on the simplified strain gradient elasticity theory. *J. Micromech. Mol. Phys.,* **3**:1640004-1–1640004-13
526. Gao X, Huang Z, Fang D (2017) Curvature-dependent interfacial energy and its effects on the elastic properties of nanomaterials. *Int. J. Solids Structures,* **113–114**:100–107
527. Gao XL, Ma HM (2010) Strain gradient solution for Eshelby's ellipsoidal inclusion problem. *Proc. R. Soc.* **A466**:2425–2446
528. Garboczi EJ, Bentz DP (1997) Analytical formulas for interfacial transition zone properties. *Advanced Cement Based Materials,* **6**:99-108
529. Garboczi EJ, Day (1995) An algorithm for computing the effective linear properties of heterogeneous materials. *J Mechan Phys Solids,* **43**:1349–1362
530. Garishin OC (1997) Physical discretezation and its application in the structural model of composite. In: Moshev, V.V, Svistkov, A.L, Garishin, D.C, Evlampieva, S.E, Rogovoy, A.A, Kovrov, V.N, Komar, L.A, Golotina, L.A, Kozhevnikov A.L. *Structural mechanisms of the mechanical properties of particulate polymer composites,* Ural Branch of RAS, Perm, 80–168 (In Russian)
531. Garishin OC, Moshev VV (2002) Damage model of elastic rubber particulate composites. *Theoretical Applied Fracture Mechanics.* **38**:63–69
532. Gavazzi AC, Lagoudas DC (1990) On the evaluation of Eshelby's tensor and its application to the elastoplastic fibrous composites. *Comput Mech,* **7**:13–19
533. Gavrikov VL, Stoyan D (1995) The use of marked point processes in ecological and environmental forest studies. *Environmental and Ecological Statistics,* **2**:331–344
534. Gazalet J, Duponta S, Kastelik JC, Rolland Q, Djafari-Rouhani B (2013) A tutorial survey on waves propagating in periodic media: Electronic, photonic and phononic crystals. Perception of the Bloch theorem in both real and Fourier domains *Wave Motion,* **50**, 619–654
535. Geers MGD, Kouznetsova VG, Matouš K, Yvonnet J (2017) Homogenization methods and multiscale modeling Nonlinear problems. In Stein E, Borst E, Hughes TJR (eds). *Encyclopedia of Computational Mechanics,* John Wiley & Sons, Ltd.
536. Gel'fand IM, Milos RA, Shapiro SY (1963) *Representations of the Rotation and Lorentz Groups and Their Applications.* Pergamon, Oxford
537. Gel'fand IA, Shilov G (1964) *Generalized Functions.* Academic Press, **1**, New York
538. Germain P (1973) The method of virtual power in continuum mechanics. part 2 Microstructure. *SIAM J. Appl. Math.,* **25**:556–575
539. Gerstle W, Sau N, Silling SA (2005) Peridynamic modeling of plain and reinforced concrete structures. In. *18th Intern. Confer. on Structural Mechanics in Reactor Technology (SMiRT 18),* Beijing, China, number SMIRT18-B01-2, 54–68
540. Giannelis EP, Krishnamoorti R, Manias E (1999) Polymer-silicate nanocomposites: model systems for confined polymers and polymer brushes. *Adv Polym Sci,* **138**:107–147
541. Gibson LJ, Ashby MF (1997) *Cellular Solids: Structure and Properties,* (2nd edn), Cambridge University Press, Cambridge, UK

542. Ghassemieh E, Nassehi V (2001) Prediction of failure and fracture mechanisms of polymeric composites using finite element analysis. Part 2: Fibre reinforced composites. *Polym Compos,* **22**:542-554
543. Ghoneim NM, Busso EP, Kioussis N, Huang H (2003) Multiscale modeling of nanomechanics and micromechanics: an overview. *Philos Mag,* **83**:3475–3528
544. Ghosh S (2011) *Micromechanical Analysis and Multi-Scale Modeling Using the Voronoi Cell Finite Element Method (Computational Mechanics and Applied Analysis).* CRC Press, Boca Raton
545. Ghosh S, Kubair DV (2017) Exterior statistics based boundary conditions for representative volume elements of elastic composites. *J. Mech. Physics Solids,* **95**:1–24
546. Ghosh S, Moorthy S (2004) Three dimensional Voronoi cell finite element model for microstructures with ellipsoidal heterogeneties. *Comput Mech,* **34**:510–531
547. Ghosh S, Mukhopadhyay S N (1991) A two-dimensional automatic mesh generator for finite element analysis for random composites. *Compos & Struct,* **41**:245–256
548. Ghosh S, Nowak Z, Lee K (1997) Quantitative characterization and modeling of composite microstructures by Voronoi cells. *Acta Mater,* **45**:2215–2234
549. Gibiansky LV, Torquato S (1997) Thermal expansion of isotropic multiphase composites and polycrystals. *J. Mech. Phys. Solids,* **45**:1223–1252
550. Gibiansky L, Torquato S (1998) New approximation for the effective energy of nonlinear conducting composites. *J Appl Phys,* **84**:301–305
551. Gibson L, Ashby M (1998) *Cellular Solids.* Cambridge University Press, Cambridge, UK
552. Gidaspow D, Huilin L (1998) Equation of state and radial distribution functions of FCC particles in a CFB. *AICHE J.,* **44**:279–293
553. Gilormini P, Brechet Y (1999) Syntheses: mechanical properties of heterogeneous media: Which material for which model? Which model for which material? *Modelling Simul Mater Sci Engng,* **7**:805–816
554. Giordano S (2005) Order and disorder in heterogeneous material microstructure: electric and elastic characterisation of dispersion of pseudo-oriented spheroids. *Int J Engng Sci,* **43**:1033–1058
555. Goh KL, Aspden RM, Mathias KJ, Hukins DW (2004) Finite-element analysis of the effect of material properties and the fiber shape on stresses in an elastic fiber embedded in an elastic matrix in a fiber-composite material. *Pro Roy Soc London A,* **460**:2339-2352
556. Goikhman BD, Buryachenko VA (1980) One variant of the topokinetic model of fracture of solids. *Problemy Prochnosti,* (2):28–31 (In Russian. Engl Trans. *Strength Mater* **12**:152–157
557. Goldberg MA, Chen CS (1999) The method of fundamental solutions for potential, Helmholtz and diffu- sion problems. In M. A. Goldberg, editor, *Boundary Integral Methods Numerical and Mathematical Aspects,* Pineridge Press, Southampton, Boston, 103-176
558. Golovchan VT (1974) The solution of static boundary-value problems for the elastic body constrained by spherical surfaces. *Dokl AN Ukr SSSR* (1):61–64 (In Ukrainian)
559. Golub GH, Van Loan CF. (2012) *Matrix Computations.* Johns Hopkins University Press, Baltimore, MD, USA
560. Gonzalez A, Roman FL, White JA (1999) A test-particle method for the calculation of the three-particle distribution function of the hard-sphere fluid: density functional theory and simulation. *J Phys: Condens Matter,* **11**:3789–3998
561. Gordon JA, Ziolkowski RW (2008) CNP optical metamaterials. *Optics Express,* **16**:6692–6716
562. Gou J, Minaie B, Wang B, Liang Z, Zhang C (2004) Computational and experimental study of interfacial bonding of single-walled nanotube reinforced composites. *Comput Mater Sci,* **31**:225–236
563. Graham-Brady LL, Siragy EF, Baxter SC (2003) Analysis of heterogeneous composites based on moving-window techniques. *J Engng Mech,* **129**:1054–1064
564. Gray LJ, Ghosh D, Kaplan T (1996) Evaluation of the anisotropic Green's function in three dimensional elasticity. *Comput Mech,* **17**:255-261

565. Grédiac M (2004) The use of full-field measurement methods in composite material characterization interest and limitations. *Composites* **A35**:751–761
566. Green AE, Adkins JE (1970) *Large Elastic Deformations*. Clarendon Press, Oxford
567. Green AE, Rivlin R S (1964a) Simple force and stress multipoles. *Arch. Ration. Mech. Anal.*, **16**:325–353
568. Green AE, Rivlin R S (1964b) Multipolar continuum mechanics. *Arch. Ration. Mech. Anal.*, **17**:113–147
569. Green AE, Zerna W (1968) *Theoretical Elasticity* (2nd edn). Oxford University Press, Oxford
570. Green PI, Sibson R (1977) Computing Dirichlet tesselations in the plane. *Computer J*, **21**:168–173
571. Greengard L, Helsing J (1998) On the numerical evaluation of elastostatic fields in locally isotropic two-dimensional composites. *J Mechan Phys of Solids*, **46**:1441–1462.
572. Greengard L, Moura M (1994) On the numerical evaluation of electrostatic fields in composite materials. *Acta Numerica*, **3**:379–410
573. Greengard L, Rokhlin V (1997) A new version of the fast multipole method for the Laplace equation in three dimensions. *Acta Numerica*, **6**:229–270
574. Greenwood R, Luckham PF, Gregory T (1998) Minimising the viscosity of concentrated dispersions by using bimodal particle size distributions. *Colloids and Surfaces A Physicochemical and Engineering Aspects*, **144**:139–147
575. Griebel M, Hamaekers J (2004) Molecular dynamics simulations of the elastic moduli of polymer-carbon nanotube composites. *Computer Meth Appl Mech Engng*, **193**:1773–1788
576. Grinfeld MA (2009) Voids as triggers of hot spots in solids. *Shock Compression of Condensed Matter*, edited by M. L. Elert, W. T. Buttler, M. D. Furnish, W. W. Anderson, and W. G. Proud. 201–204
577. Grujicic M, Zhag Y (1998) Determination of effective elastic properties of functionally graded materials using Voronoi cell finite element method. *Mater Sci Engng*, **A251**: 64–76
578. Gu ST, He QC, Pensée V (2015) Homogenization of fibrous piezoelectric composites with general imperfect interfaces under anti-plane mechanical and in-plane electrical loadings. *Mechanics Materials*, **88**:12–29
579. Gu ST, Liu JT, He QC (2014) Size-dependent effective elastic moduli of particulate composites with interfacial displacement and traction discontinuities. *Int. J. Solids Struct.*, **51**, 2283–2296
580. Gu ST, Wang AL, Xu Y, He QC (2015) Closed-form estimates for the effective conductivity of isotropic composites with spherical particles and general imperfect interfaces. *Int. J. Heat Mass Tran.* **83**, 317–326
581. Gubernatis IE (1979) Long-wave approximations for the scattering of elastic waves from flaws with applications to ellipsoidal voids and inclusions. *J Appl Phys*, **50**:4046–4058
582. Gubernatis IE, Domany E, Krymhansl IA, Huberman M (1977) The Born approximation in the theory of the scattering of elastic waves by flows. *J Appl Phys*, **48**:2812–2819
583. Guo K, Yang Z, Yu HC, Buehler MJ (2021) Artificial intelligence and machine learning in design of mechanical materials. *Materials Horizons*, **202**, DOI: 10.1039/D0MH01451F
584. Gurson AL (1977) Continuum theory of ductile rupture by void nucleation and growth: Part I–Yield criteria and flow rules for porous ductile media. (Ed. C Truesdell), *J Eng Mater Technol*, **99**:2–15
585. Gurtin ME (1972) The linear theory of elasticity. In: Truesdell C (ed) *Flügge's Handbuch der Physik VI a/2*. Springer-Verlag, Berlin, 1–295
586. Gurtin ME (1976) *An Introduction to Continuum Mechanics*. Academic Press, New York
587. Gurtin ME, Murdoch AI (1975) A continuum theory of elastic material surfaces.*Archive for Rational Mechanics and Analysis*, **59**:291–323.
588. Gurtin ME, Weissmuller J, Larche F (1998) The general theory of curved deformable interfaces in solids at equilibrium. *Philos. Mag.* **A 78**:1093–1109
589. Gusev AA (2001) Numerical identification of the potential of whisker-filled polymers. *Macromolecules*, **34**:3081–3093

590. Gusev A, Lusti HR, Hine PJ (2002) Stiffness and thermal expansion of short fiber composites with fully aligned fibers. *Adv. Eng. Mater.*, **4**:927–931
591. Guseva O, Lusti HR, Gusev AA (2004) Matching thermal expension of mica-polymer nanocomposites and metals. *Model Simul Mater Sci Engng* , **12**:S101–S105
592. Gutkin MY (2006) Elastic behavior of defects in nanomaterials I. Models for infinite and semi-infinite media. *Rev. Adv. Mater. Sci.* **13**:125-161.
593. Guz IA, Rodger AA, Guz AN, Rushchitsky JJ (2007) Developing the mechanical models for nanomaterials *Composites Part A*, **38**:1234–1250
594. Ha YD, Bobaru F (2010) Studies of dynamic crack propagation and crack branching with peridynamics. *Int. J. Fract*, **162**:229–244
595. Häfner S, Eckardt S, Luther T, Könke C (2006) Mesoscale modeling of concrete Geometry and numerics. *Computers Structures*, **84**:450–461
596. Hahn UA, Micheletti R, Pohlink R, Stoyan D, Wendrock H (1999) Stereological analysis and modelling of gradient structures. *J Microsc*, **195**:113–124
597. Hall P (1988) *Introduction to the Theory of Coverage Processes.* John Willey & Sons, NY
598. Hall R (1993) Effective moduli of cellular materials. *J Reinf Plast Comp*, **12**:186-197
599. Halle DK (1976) The physical properties of composite materials *J Mater Sci*, **11**:2105–2141
600. Halpin JC, Kardos JL (1976) The Halpin-Tsai equations: a review. *Polym Engng Sci*, **16**:344–352
601. Han CD (1974) Rheological properties of calcium carbonate filled polypropylene melts. *J Appl Polym Sci*, **18**:821–829
602. Han F, Lubineau G, Azdoud Y (2016) Adaptive coupling between damage mechanics and peridynamics a route for objective simulation of material degradation up to complete failure. *J. Mechanics Physics of Solids*, **94**:453–472
603. Hansen PC (1998) *Rank-Deficient and Discrete Ill-Posed Problems Numerical Aspects of Linear Inversion*, Philadelphia, PA SIAM
604. Hansen JP, McDonald IR (1986) *Theory of Simple Liquids.* Academic Press, New York
605. Harper LT, Qian C, Turner TA, Li S, Warrior NA (2012) Representative volume elements for discontinuous carbon fibre composites—Part 2 Determining the critical size. *Compos. Sci. Technol.*, **72**:204–210
606. Harrison P (2000) *Quantum Wells, Wires and Dots: Theoretical and Computational Phys.* John Wiley & Sons, New York
607. Hashin Z (1962) The elastic moduli of heterogeneous materials. *J Appl Mech*, **29**:143–150
608. Hashin Z (1968) Assessment of the self consistent scheme approximation: conductivity of particulate composites. *J Compos Mater*, **2**:284–300
609. Hashin Z (1972) *Theory of Fiber Reinforced Materials.* NASA Contractor report CR-1974, NASA, Washington, DC
610. Hashin Z (1979) Analysis of properties of fiber composites with anisotropic constituents. *J Appl Mech*, **46**:543–550
611. Hashin Z (1983) Analysis of composite materials–a survey. *J Appl Mech*, **50**:481–505
612. Hashin Z (1985) Large isotropic elastic deformation of composites and porous media. *J Mech Phys Solids*, **21**:711–720
613. Hashin Z (1988) The differential scheme and its application to cracked materials. *J Mech Phys Sol*, **36**:719–733
614. Hashin Z (1991a) Thermoelastic properties of particular composites with imperfect interface. *J Mech Phys Solids*, **39**:745–762
615. Hashin Z (1991b) The spherical inclusion with imperfect interface. *J. Appl. Mech.*, **58**:444–449
616. Hashin Z (2002) Thin interphase imperfect interface in elasticity with application to coated fiber composites. *J. Mech. Phys. Solids.*, **50**:2509–2537
617. Hashin Z, Rosen BW (1964) The elastic moduli of fiber-reinforced materials. *J Appl Mech*, **31**:223–232
618. Hashin Z, Shtrikman S (1962a) On some variational principles in anisotropic and nonhomogeneous elasticity. *J Mech Phys Solids*, **10**:335–342

619. Hashin Z, Shtrikman S (1962b) A variational approach to the theory of the elastic behavior of polycrystals. *J Mech Phys Solids*, **10**:343–352
620. Hashin Z, Shtrikman S (1962c) A variational approach to the theory of the effective magnetic permeability of multiphase materials. *J Appl Phys*, **35**:3125–3131
621. Hashin Z, Shtrikman S (1963) A variational approach to the theory of the behavior of multiphase materials. *J Mech Phys Solids*, **11**: 127–140
622. Hasimoto H (1959) On the periodic fundamental solutions of the Stokes equations and their application to viscous flow past a cubic array of spheres. *J. Fluid Mech*. **5**:317–328
623. Hatami-Marbini H, Shodja HM (2007) Thermoelastic fields of a functionally graded coated inhomogeneity with sliding/perfect interfaces. *J. Appl. Mech.*, **74**:389–398
624. Hatch RL, Davis IL (2006) Mechanical properties for an arbitrary arrangement of rigid spherical particles embedded in an elastic matrix. *(Preprint)*. ATK Launch Systems Group, Brigham City, UT
625. Hatta H, Taya M (1985) Effective thermal conductivity of a misoriented short fiber composite. *J Appl Phys*, **58**:2478–2486
626. Hatta H, Taya M (1986) Equivalent inclusion method for steady state heat conduction in composites. *Int J Engng Sci*, **24**:1159–1172
627. Hatta Y, Taya M (1987) Thermal stress in a coated short fiber composite. *J Engng Mater*, **109**:59–63
628. Hazanov S (1999) On micromechanics of imperfect interfaces in heterogeneous bodies smaller than the representative volume. *Int J Engng Sci*, **37**:847–861
629. He D, Ekere NN (2001) Structure simulation of concentrated suspensions of hard spherical particles *AIChE J*, **47**:53–59
630. He D, Ekere NN, Cai L (1999) Computer simulation of random packing of unequal particles. *Phys Rev*, **E60**:7098
631. He LH, Li ZR (2006) Impact of surface stress on stress concentration *Int. J. Solids Struct.*, **43**:6208–6219
632. Henderson D, Plischke M (1987) Sum rules for the pair-correlation function of inhomogeneous fluid: results for the hard-sphere-hard-wall system. *Proc Roy Soc Lond*, **A410**:409–420
633. Henke SF, Shanbhag S (2014) Mesh sensitivity in peridynamic simulations. *Computer Physics Communications*, **185**:181–193
634. Herczynski R, Pienkowska I (1980) Toward a statistical theory of suspension. *Annu Rev Fluid Mech*, **12**:137–269
635. Herrmann KP, Mihovsky IM (1994) On the modeling of the inelastic thermomechanical behavior and the failure of fibre-reinforced composites–a unified approach. In: Markov KZ (ed) *Advances in Math Modelling of Composite Materials*. World Scientific, Singapore, 141–191
636. Hershey V (1954) The elasticity of an isotropic aggregate of anisotropic cubic crystals. *J Appl Mech*, **21**:236–241
637. Hervé E (2002) Thermal and thermoelastic behaviour of multiply coated inclusion-reinforced composites. *Int J Solids Struct*, **39**:1041-1058
638. Hervé E, Zaoui A (1993) N-layered inclusion-based micromechanical modeling. *Int J Engng Sci*, **31**:1–10
639. Hill R (1952) The elastic behavior of a crystalline aggregate. *Proc Phys Soc*, **A65**:349–354
640. Hill R (1963a) Elastic properties of reinforced solids: some theoretical principles. *J Mech Phys Solids*, **11**:357–372
641. Hill R (1963b) New derivations of some elastic extremum principles. *Prog in Appl Mechanics*. The Prager Anniversary Volume. Macmillan, New York, 99–106
642. Hill R (1964) Theory of mechanical properties of fiber-strength materials: I. Elastic behavior. *J Mech Phys Solids*, **12**:199–212
643. Hill R (1965) A self-consistent mechanics of composite materials. *J Mech Phys Solids*, **13**:212–222

644. Hill R (1983) Interfacial operators in the mechanics of composite media. *J Mech Phys Solids,* **31**:347–357
645. Hill R (1984) On macroscopic effects of heterogeneity in elastoplastic media at finite strain. *Math. Proc. Camb. Phil. Soc.*, **95**:481–494
646. Hinch EJ (1977) An averaged-equation approach to particle interactions in a fluid suspension. *J Fluid Mech,* **83**:695–720
647. Hinrichsen E L, Feder J, Jossang T (1986) Geometry of random sequential adsorption. *J Statist Phys,* **44**:793–827
648. Hirai T, Sasaki M, Niino M (1987) CVD in-situ ceramic composites. *J. Mater. Sci. Soc. Japan* **36**:1205–1211
649. Hiriyur B, Waisman H, Deodatis G (2011) Uncertainty quantification in homogenization of heterogeneous microstructures modeled by XFEM. *Int. J. Numer. Meth. Engng.* , vol. 88, 257–278
650. Honein E, Honein T, Herrmann G (1990) On two circular inclusions in harmonic problems. *Quart Appl Math,* **50**:479–499
651. Honein T, Herrmann G (1990) On bounded inclusions with circular or straight boundaries in plane elastostatic *J Appl Phys,* **57**:850–856
652. Hori M, Kubo J (1998) Analysis of probabilistic distribution and range of average stress in each phase of heterogeneous materials. *J Mech Phys Solids,* **46**:537–556
653. Hori H, Nemat-Nasser S (1985) Elastic field of interacting inhomogeneities. *Int J Solids Struct,* **21**:731–745
654. Hori M, Nemat-Nasser S (1987) Interacting microcracks near the tip in the process zone of a macrocrack. *J Mech Phys Solids,* **35**:601-629
655. Hori M, Nemat-Nasser S (1993) Double-inclusion model and overall moduli of multi-phase composites. *Mech Mater,* **14**:189–206
656. Hsiao GC, Steinbach O, Wendland WL (2002) Domain decomposition methods via boundary integral equations. *J. Computational and Applied Mathematics* **125**:521–537
657. Hsiao GC, Wendland WL (2008) *Boundary Integral Equations*. Springer-Verlag, Berlin
658. Hu G (1996) A method of plasticity for general aligned spheroidal void or fiber-reinforced composites. *Int J Plast,* **12**:439–449
659. Hu G (1997) Composite plasticity based on matrix average second order stress moment. *Int J Solids Struct,* **34**:1007–1015
660. Hu G, Tian F (1997) Micromechanical analysis of fatigue properties of metal-matrix composites. *Mech Res Commun,* **24**:65–68
661. Hu GK, Weng GJ (1998) Influence of thermal residual stresses on the composite macroscopic behavior. *Mech Mater* **27**:229–240
662. Hu GK , Weng GJ (2000) The connections between the double inclusion model and the Ponte Castañeda—Willis, Mori–Tanaka, and Kuster—Toksoz models. *Mech. Mater.* **32**:495–503
663. Hu GK, Weng GJ (2000) Some reflections on the Mori–Tanaka and Ponte Castañeda—Willis methods with randomly oriented ellipsoidal inclusions. *Acta Mech.* **140**:31–40
664. Hu W, Ha YD, Bobaru F (2010) *Numerical integration in peridynamics*, Tech. rep, University of Nebraska-Lincoln
665. Hu W, Ha YD, Bobaru F (2012a) Peridynamic model for dynamic fracture in unidirectional fiber-reinforced composites. *Comput. Methods Appl. Mech. Engrg.*, **217–220**:247–261
666. Hu W, Ha YD, Bobaru F (2012b) Peridynamic model for dynamic fracture in unidirectional fiber-reinforced composites. *Comput. Methods Appl. Mech. Engrg.*, **217–220**:247–261
667. Hu W, Ha YD, Bobaru F, Silling SA (2012c) The formulation and computation of the nonlocal J-integral in bond-based peridynamics. *Int. J. Fract.*, **176**:195–206
668. Huang JH, Furuhashi R, Mura T (1993) Frictional sliding inclusions. *J Mech Phys Solids,* **41**:247–265
669. Huang JH, Kuo W-S (1997) The analysis of piezoelectric/piezomagnetic composite materials containing ellipsoidal inclusions *J. Applied Physics*, **81**:1378–1386
670. Huang JH, Liu H-K, Dai W-L (2000) The optimized fiber volume fraction for magnetoelectric coupling effect in piezoelectric-piezomagnetic continuous fiber reinforced composites. *Int J Engng Sci,* **38**:1207–1217

671. Huang JH, Yu JS (1994) Electroelastic Eshelby tensors for an ellipsoidal piezoelectric inclusion. *Compos Engng,* **4**:1169–1182
672. Huang Y, Hu KX, Chandra AA (1994) A generalized self-consistent mechanics method for composite materials with multiphase inclusions. *J Mech Phys Solids,* **94**:491–502
673. Huang Y, Hu KX (1995) A generalized self-consistent mechanics method for solids containing elliptical inclusions. *J Appl Mech,* **62**:566–572
674. Huet C (1990) Application of variational concepts to size effects in elastic heterogeneous bodies. *J Mech Phys Solids,* **38**:813–841
675. Hussein MI, Leamy MJ, Ruzzene M (2014) Dynamics of phononic materials and structures: Historical origins, recent progress, and future outlook. *Applied Mechanics Reviews,* **66**:040802 (52 pages)
676. Hussain F, Hojjati M, Okamoto M, Gorga RE (2006) Review article: Polymer-matrix nanocomposites, processing, manufacturing, and application: An overview. *J Compos Mater,* **40**:1511–1575
677. Hutchinson JW (1976) Bounds and self–consistent estimates for creep of polycrystalline materials. *Proc Roy Soc Lond Series* **A348**:101–127
678. Hütter G (2017) Homogenization of a Cauchy continuum towards a micromorphic continuum. *J. Mech. Phys. Solid.* **99**:394–408
679. Ibach H (1997) The role of surface stress in reconstruction, epitaxial growth and stabilization of mesoscopic structures. *Surf Sci Rep,* **29**(5-6):193–263
680. Ibrahim IA, Mohammed FA, Lavernia EJ (1991), Particulate reinforced metal matrix composites–a review. *J Mater Sci,* **26**:1137–1156
681. Idiart M, Ponte Castañeda P (2007) Variational linear comparison bounds for nonlinear composites with anisotropic phases. I. General results. *Proc. Royal Society A.*, **463**:907–924
682. Iijima S (1991) Helical microtubes of graphitic carbon. *Nature,* **354**:56–58
683. Illian J, Penttinen A, Stoyan H, Stoyan D (2008) *Statistical Analysis and Modeling of Spatial Point Patterns.* Willey & Sons, Chichester.
684. Imam A, Johnson GC, Ferrari M (1995) Determination of the overall moduli in second order incompressible elasticity. *J Mech Phys Solids,* **43**:1087–1104
685. Ingber MS, Papathanasiou TD (1997) A parallel-supercomputing investigation of the stiffness of aligned, short-fiber-reinforced composites using the boundary element method. *Int J Numer Methods Engng,* **40**:3477–3491
686. Isupov LP (1996a) Constitutive equations of plastic anisotropic composite medium. In: Pineau A, Zaoui Z (eds), *IUTAM Symp Micromech of Plasticity and Damage of multiphase materials,* Kluwer, Dordrecht, 91–98
687. Isupov LP (1996b) A variant of the theory of plasticity of two-phase composite media. *Acta Mechan,* **119**:65–78
688. Jackson TL, Hooks DE, Buckmaster J (2011) Modeling the microstructure of energetic materials with realistic constituent morphology. *Propellants Explos. Pyrotech.*, **36**:252–258
689. Jammes M, Mogilevskaya SG, Crouch SL (2009) Multiple circular nano-inhomogeneities and/or nano-pores in one of two joined isotropic elastic half-planes *Engineering Analysis with Boundary Elements*, **33**:233-248
690. Jasiuk I, Kouider MW (1993) The effect of an inhomogeneous interphase on the elastic constants of transversely isotropic composites. *Mech Mater,* **15**:53–63
691. Jasiuk I, Mura T, Tsuchida E (1988) Thermal stresses and thermal expansion coefficient of short fiber composites with sliding interfaces. *J Engng Mater Tech,* **110**:96–100
692. Jasiuk I, Sheng PY, Tsuchida E (1997) A spherical inclusion in an elastic half-space under shear. *J Appl Mech,* **64**:471–479
693. Jasiuk I, Tsuchida E, Mura T (1987) The sliding inclusion under shear. *Int. J. Solids Struct.*, **23**:1373–1385
694. Javili A, Morasata R, Oterkus E (2019) Peridynamics review. *Mathematics and Mechanics of Solids*, **24**:3714–3739
695. Jayaraman K, Reifsnider KL (1992) Residual stresses in a composite with continuously varying Young's modulus in the fiber/matrix interphase. *J. Comp. Mater.* **26**:770–791

696. Jeffrey DJ (1973) Conduction through a random suspension of spheres. *Proc Roy Soc Lond,* **A335**:355–367
697. Jeffrey DJ (1974) Group expansion for the bulk properties of a statistically homogeneous, random suspension. *Proc Roy Soc London,* **A338**:505–516
698. Jensen EBV (1998) *Local Stereology.* World Science, Singapore
699. Jerier J-F, Richefeu V, Imbault D, Donzé FV (2010) Packing spherical discrete elements for large scale simulations. *Comput. Methods Applied Mech. Engng,* **199**:1668–1676
700. Jeulin D (2001) Random structure models for homogenization and fracture satistics. In: Jeulin D, Ostoja-Starzewski M (eds), *Mechanics of random and multiscale microstructures,* Springer, Wien-New York, 33–91
701. Ji XL, Jing JK, Jiang W, Jiang BZ (2002) Tensile modulus of polymer nanocomposites. *Polym Eng Sci,* **42**:983–993
702. Jiang B, Fang D-N, Hwang K-C (1999) A unified model for piezocomposites with non-piezoelectric matrix and piezoelectric inclusions. *Int J Solids Structure,* **37**:2707–2733
703. Jiang M, Jasiuk I, Ostoja-Starzewski M (2002) Apparent elastic and elastoplastic behavior of periodic composites. *Int J Solids and Struct,* **439**:199–212
704. Jikov VV, Kozlov SM, Oleinik OA (1994) *Homogenization of Differential Operators and Integral Functionals.* Springer-Verlag, Berlin.
705. Jimenez G, Ogata N, Kawai H, Ogihara T (1997) Structure and thermal/mechanical properties of poly (-caprolactone)-clay blend. *J Appl Polymer Sci,* **64**:2211–2220
706. Jirasek M, Rolshoven S (2003) Comparison of integral-type nonlocal plasticity models for strain-softening materials. *Int. Engng Science* **41**:1553–1602.
707. Jodrey WS, Tory M (1985) Computer simulation of close random packing of equal spheres. *Phys Rev,* **A32**:2347
708. Johnson WC, Earmme YY, Lee JK (1980) Application of the strain field associated with an inhomogeneous precipitate. I: Theory. *J Appl Mech,* **47**:775–780
709. Jordan J, Jacob KI, Tannenbaum R, Sharaf MA, Jasiuk I (2005) Experimental trends in polymer nanocomposites–a review. *Materials Science Engineering* **A 393**:1–11
710. Ju JW, Chen TM (1992) Micromechanics and effective moduli of elastic composites with randomly dispersed inhomogeneities. *Macroscopic Behavior of Heterogeneous Materials from the Microstructure.* ASME, ADM 147 NY, 95–109
711. Ju JW, Chen TM (1994a) Effective elastic moduli of two-dimensional brittle solids with interacting microcracks, I: Basic formulations. *J Appl Mech,* **61**:349–357
712. Ju JW, Chen TM (1994b) Micromechanics and effective moduli of elastic composites containing randomly dispersed ellipsoidal inhomogeneities. *Acta Mechan,* **103**:103–121
713. Ju JW Chen TM (1994c) Effective elastic moduli of two-phase composites containing randomly dispersed spherical inhomogeneities. *Acta Mechan,* **103**:123–144
714. Ju JW, Chen TM (1994e) Micromechanics and effective elastoplastic behavior of two-phase metal matrix composites. *J Engng Mater Tech,* **116**:310–318
715. Ju JW, Lee HK (2001) A micromechanical damage model for effective elastoplastic behavior of partially debonded ductile matrix composites. *Int J Solids Struct,* **38**:6307–6332
716. Ju JW, Sun, LZ (1999) A novel formulation for the exterior point Eshelby's tensor of an ellipsoidal inclusion. *J Appl Mech,* **66**:570–572
717. Ju JW, Sun LZ (2001) Effective elastoplastic behavior of metal matrix composites containing randomly located aligned spheroidal inhomogeneities. Part I: micromechanics. *Int J Solids Struct,* **38**:183–201
718. Ju JW, Tseng KH (1992) A three-dimensional micromechenical theory for brittle solids with interacting microcracks. *Int J Damage Mech,* **1**:102–131
719. Ju JW, Tseng KN (1994) Effective elastoplastic behavior of two-phase metal matrix composites: micromechanics and computational algorithms. In: Voyiadjis GZ, Ju JW (eds), *Inelasticity and Micromechanics of Metal Matrix Composites.* Elsevier, Amsterdam, 121–141
720. Ju JW, Tseng KH (1995) Improved two-dimensional micromechanical theory for brittle solids with randomly located interacting microcracks. *Int. J. Damage Mechanics* **4**:23–57

721. Ju JW, Tseng KH (1996) Effective elastoplastic behavior of two-phase ductile matrix composites: a micromechanical framework. *Int J Solids Struct,* **33**:4327–4291
722. Ju JW, Zhang XD (1998) Micromechanics and effective transverse elastic moduli of composites with randomly located aligned circular fibers. *Int J Solids Struct,* **35**:941–960
723. Ju JW, Zhang XD (2001) Effective elastoplastic behaviour of ductile matrix composites containing randomly located aligned circular fibers. *Int J Solids Struct,* **38**:4045–4069
724. Jung M, Steinbach O (2002) A finite element—boundary element algorithm for inhomogeneous boundary value problems. Computing **68**:1–17
725. Kachanov LM (1958) On the time to rupture under creep conditions. *Izv. AN SSSR, Ofd. Tekhn. Nauk.* vol. **8**:26–31 (in Russian) [Also available in Kachanov LM (1999) Rupture time under creep conditions. *Int. J. Fracture* vol. **97**:1–18]
726. Kachanov LM (1971) *Foundation of Theory of Plasticity.* North-Holland, Amsterdam
727. Kachanov LM (1986) *Introduction to Continuum Damage Mechanics.* Netherlands Maritinus Nijhoff Publishers, Springer
728. Kachanov M (1987) Elastic solids with many cracks: a simple method of analysis *Int J Solids Struct,* **23**:23–43
729. Kachanov M (1993) Elastic solids with many cracks and related problems In: Hutchinson JW, Wu TJ (eds), *Adv Appl Mechan.* Academic Press, New York, **30**:259–445
730. Kachanov M, Laures J-P (1989) Three-dimensional problem of strongly interacting arbitrarily located penny-shaped cracks. *Int J Fracture,* **41**:289–313
731. Kalamkarov AL (1992) *Composite and Reinforced Elements of Construction.* John Wiley & Sons, New York
732. Kalamkarov AL, Kolpakov AG (1997) *Analysis, Design and Optimization of Composite Structures.* John Wiley & Sons, New York
733. Kalamkarov AL, Kudriavtsev BA, Parton VZ (1990) The boundary-layer method in the fracture mechanics of composites of periodic structure. *Prikl Matem Mech,* **54**:322–328 (In Russian, Engl Transl. *J Appl Math Mech,* **54**:266–271)
734. Kaloerov SA, Goryanskaya ES (1995) Two-dimensional stress state of multiply-connected anisotropic solid with holes and cracks. *Theor Appl Mech,* **25**:45–56
735. Kaminski MM (2005) *Computational Mechanics of Composite Materials: Sensitivity, Randomness, and Multiscale Behaviour.* Springer-Verlag, London
736. Kanaun SK (1977) Self-consistent field approximation for an elastic composite medium. *Zhurnal Prikladnoi i Tehknich Fiziki,* **18**(2):160–169 (In Russian. Engl Transl. *J Appl Mech Techn Phys,* **18**:274–282)
737. Kanaun SK (1982) The effective field method in linear problems of statics of composite media. *Prikl. Matem. Mechanika,* **46**:655–665. (In Russian. Engl. Transl. *J Appl Math Mech,* **46**:520–528)
738. Kanaun SK (1983) Elastic medium with random fields of inhomogeneities. In: Kunin IA *Elastic Media with Microstructure.* Springer–Verlag, Berlin, **2**:165–228
739. Kanaun SK (1990) Self-consistent averaging schemes in the mechanics of matrix composite materials. *Mekhanika Kompozitnikh Materialov,* **26**:702–711 (In Russian. Engl Transl. *Mech Compos Mater,* **26**:984–992)
740. Kanaun S (2016) Efficient homogenization techniques for elastic composites Maxwell scheme vs. effective field method. *Int. J. Engng Science,* **103**:19–34
741. Kanaun SK (2020) *Heterogeneous Media: Local Fields, Effective Properties, and Wave Propagation.* Elsevier, Oxford, UK
742. Kanaun SK, Jeulin D (2001) Elastic properties of hybrid composites by the effective field approach. *J. Mech. Physics of Solids,* **49**:2339–2367
743. Kanaun SK, Levin VM (1984) Development of effective wave operator for medium with isolated inhomogeneities. *Mech Solids,* (5):67–76 (In Russian)
744. Kanaun SK, Levin VM (1986) Propagation of elastic waves through media with thin crack-like inclusions. *Prikladnaya Matematika i Mekhanika,* **50**:309-319 (In Russian. Engl Transl. *J Appl Math Mech,* **50**:231–239

745. Kanaun SK, Levin VM (1993) *Effective Field Method in Mechanics of Composite Materials.* University of Petrozavodsk, Petrozavodsk (In Russian)
746. Kanaun SK, Levin VM (1994) Effective field method on mechanics of matrix composite materials. In: Markov KZ (ed), *Advances in Math Modelling of Composite Materials.* World Scientific, Singapore, 1–58
747. Kanaun KK, Levin VM (2008) *Self-Consistent Methods for Composites.* Vol. **1**, **2**, Springer, Dordrecht
748. Kanaun SK, Levin VM, Sabina FJ (2004) Propagation of elastic waves in composites with random set of spherical inclusions (effective medium approach). *Wave Motion*, **40**:69–88
749. Kanit T, Forest S, Galliet I, Mounoury V, Jeulin D (2003) Determination of the size of the representative volume element for random composites: statistical and numerical approach. *Int J Solids Struct*, **40**:3647–3679
750. Kansal AR, Truskett TM, Torquato S (2000) Nonequilibrium hard-disk packing with controlled orientational order. *J Chem Phys*, **113**:4844–4851
751. Kanuparthi S, Subbarayan G, Siegmund T, Sammakia B (2009) The effect of polydispersivity on the thermal conductivity of particulate thermal interface materials. *IEEE Trans. Components Packaging Technologies*, **32**:424–434
752. Karageorghis A, Smyrlis Y-S (2007) Matrix decomposition MFS algorithms for elasticity and thermo-elasticity problems in axisymmetric domains. *J. Computat. Applied Mathematics*, **206** 774–795
753. Karlsson LM, Liljeborg A (1994) Second-order stereology for pores in translucent alumina studied by confocal scanning laser microscopy. *J Microsc,* **175**:186–194
754. Keer LM, Lin W (1990) Analysis of cracks in transversally isotropic media. In: Weng GJ, Taya M, Abe H (eds), *Micromechanics and Inhomogeneity. The Toshio Mura 65the Anniversary Volume.* Springer-Verlag, New York, 187–195
755. Keiller RA, Feuillebous F (1993) Head conduction through an inhomogeneous suspension. *Proc Roy Soc Lond*, **A440**:717–726
756. Keller JB (1963) Conductivity of a medium containing a dense array of perfectly conducting spheres or cylinders or nonconducting cylinders. *J Appl Phys,* **34**:991–993
757. Kerans RJ, Hay RS, Parthasaraty TA, Cinibulk MK (2002) Interface design for oxidation-resistant ceramic composites. *J Am Ceram Soc*, **85**:2599–2632
758. Khoroshun LP (1967) The theory of isotropic deformation of elastic bodies with random inhomogeneities. *Priklad Mech,* **3**:12–19 (In Russian)
759. Khoroshun LP (1972) Elastic properties of materials reinforced by uni-directional short fibers. *Prikladnaya Mekhanika*, **8**(12), 86–92 (In Russian. Engl. Transl. *Soviet Appl. Mech.* **8**:1358–1363)
760. Khoroshun LP (1974) Prediction of thermoelastic properties of materials strengthened by unidirectional discrete fibers. *Prikladnaya Mekhanika*, **10**(12), 23–30 (In Russian. Engl. Transl. *Soviet Appl. Mech.* **10**:1288–1293)
761. Khoroshun LP (1977) About heat conductivity equations of composites. *Docladu Acad Nauk Ukraine.* **A**(7):630–634 (In Russian)
762. Khoroshun LP (1978) Random functions theory in problems on the macroscopic characteristics of microinhomogeneous media. *Priklad Mekh,* **14**(2):3–17 (In Russian. Engl Transl. *Soviet Appl Mech,* **14**:113–124)
763. Khoroshun LP (1987) Conditional-moment method in problems of the mechanics of composites. *Priklad Mekh,* **23**(10):100–108 (In Russian. Engl Transl. *Soviet Appl Mech,* **23**:989–998)
764. Khoroshun L (1996) On a mathematical model for inhomogeneous deformation of composites. *Priklad Mekh,* **32**(5):22–29 (In Russian. Engl Transl. *Int Appl Mech,* **32**:341–348)
765. Khoroshun LP (2000) Mathematical models and methods of the mechanics of stochastic composites. *Prikl Mekh,* **30**(10):30–62 (In Russian. Engl Transl. *Int Appl Mech,* **30**:1284–1316
766. Khoroshun LP, Dorodnykh TI (2004) The effective piezoelectric properties of polycrystals with the trigonal symmetry. *Acta Mechan,* **169**:203–219

767. Khoroshun LP, Maslov BP (1993) *Nonlinear Properties of Composite Materials of Stochastic Structure.* Naukova Dumka, Kiev (In Russian)
768. Khoroshun LP, Maslov BP, Leshchenko PV (1989) *Prediction of Effective Properties of Piezoactive Composites.* Naukova Dumka, Kiev (In Russian)
769. Khoroshun LP, Maslov BP, Shikula EN, Nazarenko LV (1993) *Statistical Mechanics and Effective Properties of Materials.* Naukova Dumka, Kiev (In Russian)
770. Khoroshun LP, Vesalo YA (1987) Theory of the effective properties of ideally plastic composites. *Priklad Mekh,* **23**(1):86–90 (In Russian. Engl Transl. *Int Appl Mech,* **23**:76–79)
771. Kilic B (2008) *Peridynamic theory for progressive failure prediction in homogeneous and heterogeneous materials. PhD Thesis*, Dep. Mechan. Engng, The University of Arisona, 1–262
772. Kilic B, Madenci E (2009) Structural stability and failure analysis using peridynamic theory. *Int. J. Non-Linear Mechanics*, **44**:8
773. Kim J, Torquato S (2020) Effective elastic wave characteristics of composite media. *New J. Phys.*, **22**, 123050 (20 pages)
774. Kim JC, Auh KH, Martin DM (2000) Multi-level particle packing model of ceramic agglomerates. *Model Simul Mater Sci Engng,* **8**:159–168
775. Kim RY, Pagano NJ (1991) Crack initiation in unidirectional brittle-material composite. *J Am Ceram Soc,* **74**:1082–1090
776. Kim S, Karrila SJ (1991) *Microhydrodynamics.* Butterworth–Heinemann, Oxford
777. Kim S, Mifflin RT (1985) The resistance and mobility functions of two equal spheres in low-Reynolds-number flow. *Phys Fluid,* **28**:2033–2045
778. Kingma DP, Ba J (2014) Adam: A method for stochastic optimization, *arXiv:1412.6980*
779. Kinoshita N, Mura T (1971) Elastic fields of inclusions in anisotropic media. *Phys Stat Sol,* **(a)5**:759–768
780. Kiriakie K, Polyzos D, Valavanides M (1997) Low-frequency scattering of coated spherical obstacles. *J Engng Math,* **31**:379–395
781. Kirkpatrick S (1973) Percolation and conductivity. *Rev Mod Phys,* **45**:574–588
782. Kirkwood JG (1935) Statistical mechanics of fluid mixtures. *J Chem Phys,* **3**:300–313
783. Kleinert H (1989) *Gauge Fields in Condensed Matter*, vol. 2. World Scientific, ISBN 9971-50-210-0
784. Knott GM, Jackson TL, Buckmaster J (2001) Random packing of heterogeneous propellants. *AIAA J,* **39**:678–686
785. Kochevets S, Buckmaster J, Jackson TL, Hegab A (2001) Random propellant packs and the flames they support. *AIAA J Propul Power*, vol. 17:883–891
786. Koelman JM, Kuijper A (1997) An effective medium model for the effective conductivity of N-component anisotropic and percolating mixture. *Physica,* **A247**:10–22
787. Koiter WT (1960) General theorems for elastic-plastic solids. In: Sneddon IN, Hill IR (eds), *Progress in Solid Mechan.* North-Holland, Amsterdam, 165–221
788. Kondrachuk AV, Shapovalov GG, Kartuzov VV (1997) Simulation modeling of the randomly nonuniform structure of powders. Two-dimensional formulation of the problem. *Poroshkovaya Metallurgiya,* (1-2):111-118 (In Russian. Engl Translation. *Powder Metall Metal Ceram,* **36**:101–106)
789. König D, Carvajal-Gonzalz S, Downs AM, Vassy J, Rigaut JP (1991) Modelling and analysis of 3-D arrangements of particles by point process with examples of application to biological data obtained by confocal scanning light microscopy. *J Microscopy,* **161**:405–433
790. Koo CM, Kim SO, Chung IE (2003) Study on morphology evolution, orientational behavior, and anisotropic phase formation of highly filled polymer-layered silicate nanocomposites. *Macromolecules,* **36**:2748–2757
791. Koplik J, Needleman A (1988) Void growth and coalescence in porous plastic solids. *Int J Solids Struct,* **24**:835–853
792. Korneev VA, Petrashen GI (1987) Calculation of diffraction fields from an elastic cylinder. *Prob Dyn Theory Seismic Wave Propaga,* **27**:45–69

793. Kornmann X, Thomann R, Mülhaupt R, Finter J, Berglund L (2002) High performance epoxy-layered silicate nanocomposites. *Polym Engng Sci,* **42**:1815–1826
794. Kosheleva AA (1983) Method of multipolar expansion in the mechanics of matrix composites. *Mekhanika Kompozititnykh Materialov,* **19**(3):416–422 (In Russian. Engl. Transl. *Mech Compos Mater* **19**:301–307)
795. Kosmodamiansky AS (1972) *Stress Distribution in the Isotropic Multiply-Connected Solids.* University Publ, Donetsk (In Russian)
796. Kotal M, Bhowmick AK (2015) Polymer nanocomposites from modified clays Recent advances and challenges. *Progress in Polymer Science,* **51**:127–187
797. Kotlarchyk M, Chen S-H (1983) Analysis of small angle neutron scattering spectra from polydisperse interacting colloids. *J. Chem. Phys.*, vol. 79, 2461–2469
798. Kouznetsova VG, Brekelmans WAM, Baaijens FPT. (2001) An approach to micro—macro modeling of heterogeneous materials, *Comput. Mech.* **27**:37–48
799. Kouznetsova VG, Geers MGD, Brekelmans WAM (2004) Multi-scale second-order computational homogenization of multi-phase materials: a nested finite element solution strategy. *Comput Methods Appl Mech Engng,* **193**:5525–5550
800. Kozaczek KJ, Sinharoy A, Ruud CO, McIlree AR (1995) Micromechanical modeling of microstress fields around carbide precipitates in alloy 600. *Model Simul Mater Sci Engng,* **3**:829–843
801. Krajcinovic D (1996) *Damage Mechan.* Elsevier, Amsterdam
802. Krattiger D, Hussein MI (2014) Bloch mode synthesis Ultrafast methodology for elastic band-structure calculations -*Physical Review E,* **90**:063306
803. Kreger IW (1972) Rheology of monodisperse lattices. *Adv Colloid and Interface Sci,* **3**:111–136
804. Kreher W (1988) Internal stresses and relations between effective thermoelastic properties of stochastic solids–some exact solutions. *Z Angew Math Mech,* **68**:147–154
805. Kreher W (1990) Residual stresses and stored elastic energy of composites and polycrystals. *J Mech Phys Solids,* **38**: 115–128
806. Kreher W, Janssen R (1992) On microstructural residual stresses in particle reinforced ceramics. *J Eur Ceram Soc,* **10**:167-173
807. Kreher W, Molinari A (1993) Residual stresses in polycrystals as influenced by grain shape and texture. *J Mech Phys Solids,* **41**:1955-1977
808. Kreher W, Pompe W (1989) *Internal Stresses in Heterogeneous Solids.* Akademie-Verlag, Berlin
809. Kremesec VJ, Slattery JC (1977) The apparent stress-deformation of spheres in power-model fluid. *J Rheol,* **21**:459–461
810. Krivoglaz M, Cherevko A (1959) On the elastic moduli of a two-phase solid. *Phiz Metallov Metallovedenie,* **8**(2):161–168 (In Russian. Engl Transl. *Phys Metals Metall,* **8**:1-4)
811. Kröner E (1953) Das Fundamentalintegral der anisotropen elastischen Differentialgleichungen. *Z Physik,* **136**:402-410
812. Kröner E (1958) Berechnung der elastischen Konstanten des Vielkristalls aus den Konstanstanten des Einkristalls. *Z Physik,* **151**:504–518
813. Kröner E (1961) Zur plastischen Verformung des Vielkristalls, *Acta Metall,* **9**:155–161
814. Kröner E (1967a) Elastic moduli of perfectly disodered composite material. J *Mech Phys Solids,* **15**:319–329
815. Kröner E (1967b) Elasticity theory of materials with long range cohesive forces. *Int J Solid Struct,* **3**:731–742
816. Kröner E (1968) Interrelations between various branches of continuum mechanics. In Kröner, E. (Ed.) *Mechanics of generalized continua.* Springer-Verlag, Berlin/Heidelberg, 330–340
817. Kröner E (1970) The problem of non-locality in the mechanics of solids Review of present status. In Simmons J, de Wit R, Bullough R (Eds.), *Fundamental Aspects of Dislocation Theory.* Nat. Bur. Stand. (US), Washington, **II**:729–736
818. Kröner E (1972) *Statistical Continuum Mechanics.* Springer-Verlag, Vienna–New York

819. Kröner E (1974) On the physics and mathematics of self-stresses. In Zeman, J.L. and Ziegler, F. *Topics in Applied Continuum Mechanics*. Springer-Verlag, Wien, 22–38
820. Kröner E (1977) Bounds for effective moduli of disordered materials. *J Mech Phys Solids*, **25**:137–155
821. Kröner E (1986) Statistical modeling. In: Gittus J, Zarka J (eds), *Modeling Small Deformations of Polycrystals*. Elsevier, London/NY, 229–291.
822. Kröner E (1990) Modified Green function in the theory of heterogeneous and/or anisotropic linearly elastic media. In: Weng GJ, Taya M, Abe H (eds) *Micromechanics and Inhomogeneity. The Toshio Mura 65th Anniversary Volume.* Springer–Verlag, New York, 197–211
823. Kröner E (1994) Nonlinear elastic properties of micro-heterogeneous media. *J Engng Mater Technol,* **116**:325–330
824. Kröner E, Datta BK (1966) Nichtlokal Elastostatik Ableitung aus der Gittertheorie. *Z. Phys.*, **196**:203–211
825. Kröner E, Datta BK (1970) Non-local theory of elasticity for a finite inhomogeneous medium–a derivation from lattice theory. In: Simmons J, de Wit R, Bullough R (eds), *Fundamental Aspects of Dislocation Theory.* Nat Bur Stand (US), Washington, 737–746
826. Krumhansl JA (1965) Generalized continuum field representation for lattice vibrations. In Wallis, R. F. (Ed.) *Lattice dynamics*. Pergamon, London, 627–634
827. Kubair DV, Ghosh S (2017) Exterior statistics based boundary conditions for establishing statistically equivalent representative volume elements of statistically nonhomogeneous elastic microstructures. *Int. J. Silids Structures*, bf 112:106-121
828. Kukarni R, Ochoa O (2006) Transverse and longitudinal CTE measurements of carbon fibers and their impact on interfacial residual stresses in composites. *J Compos Mater*, **40**, 734–754
829. Kumasaka H, Hirashima K-I (1996) Stress distributions around circular inclusion in infinite plane for nonlocal elasticity (Matrix and circular inclusion have the same nonlocal coefficients). *JSME Int. Journal*, **A39**:192–196
830. Kumar A, Dawson PR (1998) Modeling crystallographic texture evolution with finite elements over neo-Eulerian orientation spaces. *Comput Methods Appl Mech Engng*, **153**:259–302
831. Kumar NC, Matouš K, Geubelle PH (2008) Reconstruction of periodic unit cells of multimodal random particulate composites using genetic algorithms. *Computational Materials Science*, **42**:352–367
832. Kundalwal SI, Ray MC (2012) Effective properties of a novel composite reinforced with short carbon fibers and radially aligned carbon nanotubes *Mechanics Materials*, **53**:47–60
833. Kunin IA (1963) Theory of dislocations. In: Shouten AY. (ed), *Tensorial Analysis for Physicists*. Nauka, Moskow, 373–450 (In Russian)
834. Kunin IA (1966) Theory of elasticity with spatial dispersion. One-dimensional complex structure. *Prikl. Mat. Mekh.*, **30**:866–874 (In Russian. Engl Trunsl. *J. Appli. Mathematics and Mechanics* **30**:1025–1034)
835. Kunin IA (1967) Inhomogeneous elastic medium with nonlocal interaction. *Zhurn. Prikl. Mekhan. Tekhn. Fiziki*, **8**:60–66 (In Russian. Engl Transl. *J. Appl. Mechan. Techn. Phys.*, **8**:41–44)
836. Kunin IA (1983) *Elastic Media with Microstructure*. Springer-Verlag, Berlin, **2**
837. Kunin IA (1984) On foundations of the theory of elastic media with microstructure *Int J Solids Struct,* **22**:969–978
838. Kunin IA, Sosnina EG (1971) Ellipsoidal inhomogeneity in the elastic medium. *Dokladi AN SSSR,* **37**:306-315 (In Russian. Engl Transl. *Sov Phys Dokl,* **16**:571–575)
839. Kunin IA, Waisman AM (1970) On problems of the non-local theory of elasticity. In: Simmons J, de Wit R, Bullough R (eds) *Fundamental Aspects of Dislocation Theory.* Nat Bur Stand (US), Washington, 747–757
840. Kuo WS, Huang JH (1997) On the effective electroelastic properties of piezoelectric composites containing spatially oriented inclusions. *Int J Solids Struct,* **34**:2445-2461
841. Kupradze VD, Aleksidze MA (1964) The method of functional equations for the approximate solution of certain boundary value problems. *Z. Vychisl Matimat Fiz.* , **4**:683–715

842. Kurita R, Weeks ER (2010) Experimental study of random-close-packed colloidal particles. *Physical Rev.*, **82**:011403
843. Kushch VI (1987) Computation of the effective elastic moduli of a granular composite material of regular structure. *Priklad Mekh,* (4):57–61 (In Russian. Engl Transl. *Soviet Appl Mech,* **23**:362–364)
844. Kushch VI (1996) Elastic equilibrium of a medium containing finite number of aligned spheroidal inclusions. *Int J Solids Struct,* **33**:1175–1189
845. Kushch VI (1997a) Microstresses and effective elastic moduli of a solid reinforced by periodically distributed spheroidal particles. *Int J Solids Struct,* **34**:1353–1366
846. Kushch VI (1997b) Conductivity of a periodic particle composite with transversely isotropic phases. *Proc Roy Soc Lond,* **A453**:65–76
847. Kushch VI (1998a) Elastic equilibrium of a medium containing a finite number of arbitrarily oriented spheroidal inclusions. *Int J Solids Struct,* **35**:1187–1198
848. Kushch (2018) Stress field and effective elastic moduli of periodic spheroidal particle composite with Gurtin–Murdoch interface. *Int. J. Eng. Sci.*, **132**, 79–96
849. Kushch V (2020) *Micromechanics of Composites Multipole Expansion Approach.* Amsterdam Butterworth-Heinemann.
850. Kushch (2020b) Elastic fields and effective stiffness of ellipsoidal particle composite using the representative unit cell model and multipole expansion method. *Int. J. Engineering Science*, **154**:103336
851. Kushch (2020c) Numerical algorithm of multipole expansion method for conductivity of ellipsoidal particle composite *J. Computational Physics*, **418**: 109642
852. Kushch VI, Knyazeva AG (2016). Finite cluster model and effective conductivity of a composite with non-randomly oriented elliptic inclusions. *Acta Mechanica*, **227**:113–126
853. Kushch VI, Mogilevskaya SG (2016) On convergence of the generalized Maxwell scheme conductivity of composites containing cubic arrays of spherical particles. *Philosophical Magazine Letters*, **96**:392–401
854. Kushch VI, Mogilevskaya SG, Stolarski HK, Crouch SL (2011) Elastic interaction of spherical nanoinhomogeneities with Gurtin–Murdoch type interfaces. *J. Mech. Phys. Solids*, **59**:1702–1716
855. Kushch VI, Mogilevskaya SG, Stolarski HK, Crouch SL (2013) Evaluation of the effective elastic moduli of particulate composites based on Maxwell's concept of equivalent inhomogeneity microstructure-induced anisotropy, *J. Mech. Mater. Struct*, **8**:283–303
856. Kushch VI, Sevostianov I (2004) Effective elastic properties of the particulate composite with transversely isotropic phases. *Int J Solids Struct,* **41**:885–906
857. Kushch VI, Sevostianov I (2016a) The "rigorous "Maxwell homogenization scheme in 2d elasticity Effective stiffness tensor of composite with elliptic inhomogeneities. *Mechanics of Materials*, **103**:44–54
858. Kushch VI, Sevostianov I (2016b) Maxwell homogenization scheme as a rigorous method of micromechanics Application to effective conductivity of a composite with spheroidal particles. *Int. J. Engineering Science*, **98**:36–50
859. Kushch VI, Sevostianov I (2020) Ellipsoidal inhomogeneity in elliptically orthotropic elastic solid. *Int. J. Solids Structures*, **206**:282–291
860. Kushch VI, Sevostianov I, Giraud A (2017) Local fields and effective conductivity tensor of ellipsoidal particle composite with anisotropic constituents. *Proc. R. Soc. A* , **473**:20170472.
861. Kushch VI, Sevostianov I, Mishnaevsky L (2008) Stress concentration and effective stiffness of aligned fiber reinforced composite with anisotropic constituents. *Int. J. Solids and Structures*, **45**:5103–5117
862. Kushch VI, Shmegera SV, Buryachenko VA (2005) Interacting elliptic inclusions by the method of complex potentials. *Int J Solids Struct,* **42**:5491–5512
863. Kushch VI, Shmegera SV, Buryachenko VA (2006) Elastic equilibrium of a half plane containing a finite array of elliptic inclusions. *Int J Solids Struct,* **43**:3459–3483
864. Kushch VI, Shmegera SV, Mishnaevsky L (2008) Meso cell model of fiber reinforced composite: interface stress statistics and debonding paths. *Int. J. Solids Structures*, **45**:2758–2784

865. Kussow A-G, Akyurtlu A, Angkawisittpan N (2008) Optically isotropic negative index of refraction metamaterial *Phys. Stat. Sol.*, **b245**:992–997
866. Kuster GT, Toksöz MN (1974) Velocity and attenuation of seismic waves in two-phase media: I. Theoretical formulation. *Geophysics*, **39**:587–606
867. Kutlu Z, Chang FK (1995) Composite panels containing multiple through-the-width delaminations and subjected to compression. Part I: Analysis. *Compos Struct*, **31**:273–296
868. Kuznetsov SV (1991) Microstructural stress in porous media. *Priklad Mech*, **27**(11):23–28 (In Russian. Engl Transl. *Soviet Appl Mech*, **27**:750–755)
869. Kuznetsov S, Fish J (2012) Mathematical homogenization theory for electroactive continuum. *Int. J. Numer. Meth. Engng*, **91**:1199–1226
870. Kwon YW, Eren H (2000) Micromechanical study of interface stresses and failure in fibrous composites using boundary element method. *Polym Polym Compos*, **8**:369–386
871. Ladd AJC (1990) Hydrodynamic transport coefficients of random dispersions of hard spheres. *J. Chem. Phys.*, **93**:3484–3494
872. Ladyzhenskaya OA (1963) *The Mathematical Theory of Viscous Incompressible Flow*. Gordon Breach, NY
873. Lafdi K, Matzek M (2003) Carbon nanofibers as a nano-reinforcement for polymeric nanocomposites. *The 35th Int SAMPE Technical Conference*, Dayton, Ohio
874. Lahellec N, Suquet P (2013). Effective response and field statistics in elasto-plastic and elastoviscoplastic composites under radial and non-radial loadings. *Int. J. Plasticity*, **42**:1–30
875. Lagoudas DS, Gavazzi AC, Nigam H (1991) Elastoplastic behavior of metal matrix composites based on incremental plasticity and the Mori–Tanaka averaging scheme. *Comput Mech*, **8**:193–203
876. Lam D, Yang F, Chong A, Wang J, Tong P (2003). Experiments and theory in strain gradient elasticity. *J. Mech. Phys. Solids.*, **51**:1477–1508
877. Lamb H (1951) *Hydrodynamics*. Dover, New York
878. Lambropolous JC (1986) Shear, shape and orientation effects in transformation toughening. *Int J Solids Struct*, **22**:1083–1106
879. Landau LD, Lifshitz EM (1960) *Electrodynamics of Continuum Media*. Pergamon Press, Oxford
880. Landauer R (1952) The electrical resistance of binary metallic mixtures. *J Appl Phys*, **23**:779–784
881. Landauer R (1978) Electric conductivity in inhomogeneous media. In: Garland JC, Tanner DB (eds) *Electric, Transport and Optical Properties of Inhomogeneous Media*. American Institute of Physics, New York, 2–43
882. Lau K-T, Gu C, Hui D (2006) A critical review on nanotube and nanotube/nanoclay related polymer composite materials *Composites*, **B37**:425–436
883. Laws N (1973) On the thermostatics of composite materials. *J Mech Phys Solids*, **21**:9–17
884. Laws N, Lee JC (1989) Microcracking in polycrystalline ceramics: elastic isotropy and thermal anisotropy. *J Mech Phys Solids*, **37**:603–618
885. Lax M (1951) Multiple scattering of waves. *Rev Modern Phys*, **23**:287–310
886. Lax M (1952) Multiple scattering of waves II. The effective fields dense systems. *Phys Rev*, **85**:621–629
887. Lay PT (1967) Potentiels élastiques, tenseurs de Green et de Neumann. *J Mećanique*, **6**:212–242
888. Lazar M, Agiasofitou E, Po G (2020) Three-dimensional nonlocal anisotropic elasticity a generalized continuum theory of Ångström-mechanics *Acta Mechanica*, **231**:743–781
889. Lazar M, Po G (2018) On Mindlin's isotropic strain gradient elasticity Green tensors, regularization, and operator-split. *J. Micromech. Molecular Physics*, **3**:1840008 (39 pages)
890. LeBaron PC, Wang Z, Pinnavaia TJ (1999) Polymer-layered silicate nanocomposites: an overview. *Appl Clay Sci*, **15**:11–29
891. Leblond JD, Perrin G (1999) A self-consistent approach to coalescence of cavities in inhomogeneously voided ductile solids. *J Mech Phys Solids*, **47**:1823–1841

892. Lee BJ, Mear ME (1991) Effect of inclusion shape on the stiffness of non-linear two-phase composites. *J Mech Phys Solids*, **39**:627–649
893. Lee CH, Gillman AS, Matouš K (2011) Computing overall elastic constants of polydisperse particulate composites from microtomographic data. *J. Mech. Phys. Solids*, **59**:1838–1857
894. Lee HK, Simonovic S (2001) A damage constitutive model of progressive debonding in aligned discontinuous fiber composites. *Int J Solids Struct*, **38**:875–895
895. Lee HK, Simonovic S, Shin DK (2004) A computational approach for prediction of the damage evolution and crushing behavior of chopped random fiber composites. *Comput Mater Sci*, **29**:459–474
896. Lee J, Han M (2020) Three-dimensional volume integral equation method for solving isotropic/anisotropic inhomogeneity problems *Mathematics*, **8**:1866 (26 pages)
897. Lee J, Ku D, Mal A (2011) Elastic analysis of a half-plane with multiple inclusions using volume integral equation method *Eng. Analysis with Boundary Elements*, **35**, 564-574
898. Lee, J.K, Mal, A (2011) Elastic analysis of a half-plane containing an inclusion and a void using a mixed volume and boundary integral equation method *Enging Analysis Boundary Elements*, **35**:915–924
899. Lee J, Mal A (1997) A volume integral equation technique for multiple inclusion and crack interaction problems. *J Appl Mech*, **64**:23–31
900. Lee J, Mal A (1998) Characterization of matrix damage in metal matrix composites under transverse loads. *Comput Mech*, **21**:339–346
901. Lee JA, Mykkanen DL (1987) *Metal and Polymer Matrix Composites.* Noyes Data Corporation, New York
902. Lee JH, Singer JP, Thomas EL (2012) Micro-Ř/nanostructured mechanical metamaterials *Adv. Mater.*, **24**:4782–4810
903. Lee JK, Johnson WC (1984) Elastic interaction and elastoplastic deformation of inhomogeneities. In: Bilby BA, Miller KJ, Willis JR (eds), *Fundamentals of deformation and Fracture. Eshelby Memorial Symposium.* Cambridge University Press, Cambridge, UK, 145–162
904. Lee M, Jasiuk I, Tsuchida E (1991) The sliding circular inclusion in an elastic half-plane. *J Appl Mech*, **59**:S57–S64
905. Lee S, Kim Y (2019) Applicability of the interface spring model for micromechanical analyses with interfacial imperfections to predict the modified exterior Eshelby tensor and effective modulus. *Math. Mech. Solids*, **24**, 2944–2960
906. Lee S, Lee J, Ryu S (2019). Modified Eshelby tensor for an anisotropic matrix with interfacial damage. *Math. Mech. Solids*, **24**, 1749–1762
907. Lee Y, Fang C, Tsou Y-R, Lu L-S, Yang C-T (2009) A packing algorithm for three-dimensional convex particles *Granular Matter*, **11**:307–315
908. Legrain G, Cartraud P, Perreard I, Moës N (2011) An X-FEM and level set computational approach for image-based modelling Application to homogenization *Int. J. Numerical Methods in Engineering*, **86**:915–934
909. Lehoucq RB, Silling SA (2008) Force flux and the peridynamic stress tensor. *J. Mech. Phys. Solids*, **56**:1566–1577
910. Lekhnitskii AG (1963) *Theory of Elasticity of an Anisotropic Elastic Body.* Holder Day, San Francisco
911. Levi O, Bergman DJ (1994) Critical behavior of the weakly nonlinear conductivity and flicker noise of two-component composites. *Phys Rev*, **B50**:3652–3660
912. Levin VM (1967) Thermal expansion coefficient of heterogeneous materials. *Izv AN SSSR, Mekh Tverd Tela*, (2):88–94 (In Russian. Engl Transl. *Mech Solids*, **2**(2):58–61)
913. Levin VM (1971) The relation between mathematical expectations of stress and strain tensors in elastic microheterogeneous media. *Prikladnaya Matematika i Mekhanika* (In Russian. Engl Transl. *J Appl Math Mech*, **35**:694–701)
914. Levin VM (1975) Determination of effective elastic moduli of composite materials. *Docl Akad Nauk SSSR*, **220**:1042–1045 (In Russian. Engl Transl. *Sov Phys Docl*, **20**:147–148)

915. Levin VM (1976) Determination of composite material elastic and thermoelastic constants. *Izv AN SSSR, Mech Tverd Tela,* (6):137–145 (In Russian. Engl Transl. *Mech Solids,* **11**(6):119–126)
916. Levin VM (1977) On the stress concentration in inclusions in composite materials. *Prikl Matem Mekh,* **41**:735–743 (In Russian. Engl Transl. *J Appl Mathem Mech,* **41**:735–743)
917. Levin VM, Michelitsch TM, Gao H (2002) Propagation of electroacoustic waves in the transversely isotropic piezoelectric medium reinforced by randomly distributed cylindrical inhomogeneities. *Int. J. Solids and Structures,* **39**:5013–5051
918. Levin VM, Michelitsch T, Sevostianov I (2000) Spheroidal inhomogeneity in a transversely isotropic piezoelectric medium *Archive of Applied Mechanics,* **70**:673–693
919. Levin VM, Rakovskaja M. I, Kreher W. S (1999) The effective thermoelectroelastic properties of microinhomogeneous materials. *Int J Solids Struct,* **36**:2683–2705
920. Levin VM, Valdiviezo-Mijangos O, Sabina GJ (2011) Propagation of electroacoustic axial shear waves in a piezoelectric medium reinforced by continuous fibers. *Int. J. Engng Science,* **49**:1232–1243.
921. Lewandowski JJ, Liu C, Hunt Jr. WH (1989) Effects of matrix microstructure and particle distribution on fracture of an aluminum metal matrix composites. *Mater. Sci. Engineering,* **A 107**:241–255
922. Li C, Chou TW (2004) Elastic properties of single-walled carbon nanotubes in transverse directions *Phys Rev,* **B69**:073401
923. Li C, Ellyin F (1998) A mesomechanical approach to inhomogeneous particulate composites undergoing localized damage: part I–a mesodomain simulation. *Int J Solids Struct,* **36**:5529–5544
924. Li H, Kafka OL, Gao J, Yu C, Nie Y. *et al.* (2019): Clustering discretization methods for generation of material performance databases in machine learning and design optimization. *Comput. Mechanics,* **64**, 281–305
925. Li H, Zhong W-F, Li G-F (1985) On the method of equivalent inclusion in elastodynamics and scattering fields of ellipsoidal inhomogeneities. *Appl Math Mech,* **6**:489–498
926. Li JY (1999) On micromechanics approximation for the effective thermoelastic moduli of multi-phase composite materials. *Mech Mater* **31**:149–159
927. Li JY (2000a) The effective electroelastic moduli of textured piezoelectric polycrystalline aggregates. *J Mech Phys Solids,* **48**:529–552
928. Li JY (2000b) Magnetoelectroelastic multi-inclusion and inhomogeneity problems and their applications in composite materials. *Int J Engng Sci,* **38**:1993–2011
929. Li JY (2004) The effective pyroelectric and thermal expansion coefficients of ferroelectric ceramics. *Mechan Mater* **36**:949-958
930. Li JY, Dunn ML (1999) Analysis of microstructural fields in heterogeneous piezoelectric solids. *Int J Engng Sci,* **37**:665–685
931. Li JY, Dunn ML (2001) Variational bounds for the effective moduli of heterogeneous piezoelectric solids. *Philosophical Magazine A,* **81**:903–926
932. Li S, Wang G (2008) *Introduction to micromechanics and nanomechanics.* World Scientific Publishing Co. Pte. Ltd, Singapore
933. Li X, Zhang J, Zhang X (2011) Micro–macro homogenization of gradient-enhanced Cosserat media. *Eur. J. Mech. A Solids,* **30**:362–372)
934. Lifshitz IM, Rozenzweig LN (1946) Theory of elastic properties of polycristals. *Zh Eksp Teor Fiz,* **16**:967–980 (In Russian)
935. Lifshitz IM, Rozenzweig LN (1947) On the construction of the Green's tensor for the basic equation of the theory of elasticity of an anisotropic infinite medium. *Zh Eksp Teor Fiz,* **17**:783–791 (In Russian)
936. Lifshitz IM, Rozenzweig LN (1951) Corrections of the paper "Properties of polycrystals". *Zh Eksp Teor Fiz,* **21**:1184 (In Russian)
937. Lin J, Chen W, Wang F (2011) A new investigation into regularization techniques for the method of fundamental solutions. *Mathematics and Computers in Simulation,* **81**:1144–1152

938. Lin SC, Yang C, Mura T, Iwakuma T (1992) Average elastic-plastic behavior of composite materials. *Int J Solids Struct George Herrmann 70th Anniversary Issue,* **28**:1859–1872
939. Lingois P, Berglund L (2002) Modeling elastic properties and volume change in dental composites. *J. Mater. Sci.*, **37**:4573–4579
940. Lipatov YS (1995) *Polymer Reinforcement. ChemTech Publishing.* Toronto, Ontario, Canada
941. Lipetzky P, Kreher W (1994) Statistical analysis of crack advance in ceramic composites. *Mech Mater,* **20**:225-240
942. Lipinski P, Barhdadi EH, Cherkaoui M (2006) Micromechanical modelling of an arbitrary ellipsoidal multi-coated inclusion. *Philos Mag,* **86**:1305–1326
943. Lipinski P, Berveiller M (1989) Elastoplasticity of micro-inhomogeneous metals at large strains. *Int J Plast,* **5**:149–172
944. Lipinski P, Berveiller, M, Reubrez E, Morreale J (1995) Transition theories of elastic-plastic deformation of metallic polycrystals. *Arch Appl Mech,* **65**:295-311
945. Lipton R (2003) Assessment of the local stress state through macroscopic variables. *Philos Trans Roy Soc Lond,* **361**:921-946
946. Lipton R (2004) Bounds on the distribution of extreme values for the stress in composite materials. *J. Mech. Phys. Solids*, **52**:1053-1069
947. Liron N, Batra E (1992) Motion of a rigid particle in Stokes flow a new second-kind boundary-integral equation formulation *J . Fluid Mec.*, **238**:579–598
948. Littlewood DJ, Silling SA, Mitchell JA, Seleson PD, Bond SD, Parks ML, Turner DZ, Burnett DJ, Ostien J, Gunzburger M (2015) *Strong Local-Nonlocal Coupling for Integrated Fracture Modeling*, Tech. Rep. SAND2015-7998, Sandia National Laboratories, Albuquerque, NM, 2015
949. Liu B, Zhao X, Zhu W, Luo W, Cheng X (2008) Multiple pass-band optical left-Handed metamaterials based on random dendritic cells. *Adv. Funct. Mater.*, **18**:3523–3528
950. Liu JZ, Zheng Q-Z, Wang L-F, Jiang Q (2005) Mechanical properties of single-walled carbon nanotube bundles as bulk materials *J Mech Phys Solids,* **53**:123–142
951. Liu Q (2015) A new version of Hill's lemma for Cosserat continuum. *Arch. Appl. Mech.*, **85**:761–773
952. Liu SB, Jin XQ, Wang ZJ, Keer LM, Wang Q (2012) Analytical solution for elastic fields caused by eigenstrains in a half-space and numerical implementation based on FFT. *Int. J. Plast*. **35**:135–154
953. Liu T, Kumar S (2003) Quantitative characterization of SWNT orientation by polarized Raman spectroscopy. *Chem Phys Lett,* **378**:257–262
954. Liu W, Hoa SV, Pugh M (2005) Organoclay-modified high performance epoxy nanocomposites. *Compos Sci Technol,* **65**:307–316
955. Liu X, Hu G (2005) A continuum micromechanical theory of overall plasticity for particulate composites including particle size effect. *Int. J. Plasticity*, **21**:777–799
956. Liu YJ (2000) On the simple-solution method and non-singular nature of the BIE/BEM – review and some new results. *Engng Anal Bound Elements* **24**:789–795
957. Liu YJ (2009) *Fast multipole boundary element method theory and applications in engineering* Cambridge University Press. Cambridge.
958. Liu Y, Greene MS, Chen W, Dikin DA, Liu WK (2013) Computational microstructure characterization and reconstruction for stochastic multiscale material design. *Computer-Aided Design*. **45**:65–76
959. Liu YJ, Mukherjee S, Nishimura N, Schanz M, Ye W, Sutradhar A, Pan E, Dumont NA, Frangi A, Saez A (2011) Recent advances and emerging applications of the boundary element method. *Applied Mechanics Reviews*, **64**:031001 (38 pages)
960. Liu YJ, Nishimura N, Otani Y (2005) Large-scale modeling of carbon-nanotube composites by a fast multipole boundary element method. *Comput Mater Sci,* **34**:173–187
961. Liu YJ, Rudolphi TJ (1999) New identities for fundamental solutions and their applications to non-singular boundary element formulations. *Comput. Mechanics* **24**:286–292
962. Liu YJ, Song G, Yin HM (2015) Boundary effect on the elastic field of a semi-infinite solid containing inhomogeneities. *Proc. R. Soc*, **A471**:20150174

963. Liu Y, Ponte Castañeda P (2004) Second-order theory for the effective behavior and field fluctuations in viscoplastic polycrystals. *J Mech Phys Solids*,467–495
964. Liu Z, Bessa MA, Liu WK (2016) Self-consistent clustering analysis: An efficient multi-scale scheme for inelastic heterogeneous materials. *Comput. Methods Appl. Mech. Engng*, **306**, 319–341
965. Lochmann K, Oger L, Stoyan D (2006) Statistical analysis of random sphere packings with variable radius distribution *Solid State Sciences*, **8**:1397–1413
966. Lomakin VA (1970) *Statistical Problems of the Mechanics of Solid Deformable Bodies*. Nauka, Moscow (In Russian)
967. Lomakin VA, Sheinin VI (1974) Stress concentration at the boundary of a randomly inhomogeneous elastic body. *Mekh Tverdogo Tela*, **9**(2):65–70 (In Russian Engl Transl *Mech Solids*, **9**(2):58–63)
968. Lorenz L (1880) Uber die Refraktionskonstante, *Ann. Phys. Chem.*, **11**, 70ff.
969. Lotwick HW (1982) Simulations on some spatial hard core models, and the complete packing problem *J Statist Comp Simul*, **15**:295–314
970. Lu GQ, Ti LB, Ishizaki K (1994) A new algorithm for simulating the random packing of monosized powder in CIP processes. *Mater Manufact Processes*, **9**:601–621
971. Lu SY, Lin HC (1996a) Effective conductivity of composites with spherical inclusions: effect of coating and detachment *J Appl Phys*, **79**:609–618
972. Lu SY, Lin HC (1996b) Effective conductivity of composites containing aligned spheroidal inclusions of finite conductivity *J Appl Phys*, **79**:6761–6769
973. Lu XY, Li H, Wang B, (2011) Theoretical analysis of electric, magnetic and magnetoelectric properties of nano-structured multiferroic composites. *J. Mech. Phys. Solids*, **59**:1966–1977
974. Lubachevsky BD, Stillinger FH (1990) Geometric properties of random disk packing. *J Statist Phys*, **60**:561–583
975. Lubachevsky BD, Stillinger FH, Pinson EN (1991) Disks vs spheres: contrasting properties of random packing. *J Statist Phys*, **64**:501–524
976. Lubarda VA (1998) A note on the effective Lamé constants of polycrystalline aggregate of cubic crystals. *J Appl Mech*, **65**:769–770
977. Lubarda VA (2002) *Elastoplasticity Theory*. CRC Press, Boca Raton, FL
978. Lubarda VA, Markenscoff X (1998a) On the absence of Eshelby properties for non-ellipsoidal inclusions. *Int J Solids Struct*, **35**:3405–3411
979. Lubarda VA, Markenscoff X (1998b) On the stress field in sliding ellipsoidal inclusions with shear eigenstrain *J Appl Mech*, **65**:858–862
980. Lubarda VA, Richmond O (1999) Second-order analysis of dilute distribution of spherical inclusions. *Mech Mater* **31**:1–8
981. Luciano R, Willis JR (2000) Bounds of nonlocal effective relations for random composites loaded by configuration-dependent body force. *J Mech Phys Solids*, **48**:1827–1849
982. Luciano R, Willis JR (2001) Non-local effective relations for fibre-reinforced composites loaded by configuration-dependent body forces. *J Mech Phys Solids*, **49**:2705–2717
983. Luciano R, Willis JR (2003) Boundary-layer correlations for stress and strain field in randomly heterogeneous materials. *J Mech Phys Solids*, **51**:1075–1088
984. Luciano R, Willis JR (2004) Non-local constitutive equations for functionally graded materials. *Mech Mater*, **36**:1195–1206
985. Luijten E (2006) Introduction to cluster Monte Carlo algorithms. *Lecture Notes in Physics*, **703**:13-38
986. Luo HA, Weng GJ (1987) On Eshelby's inclusion problem in a three-phase spherically concentric solid, and a modification of Mori–Tanaka's method. *Mech Mater*, **6**:347–361.
987. Luo J-J, Daniel IM (2003) Characterization and modeling of mechanical behavior of polymer/clay nanocomposites. *Compos Sci Technol*, **63**:1607–1616
988. Luo J, Stevens R (1993) Residual stress and microcracking in Sic-MgO composites. *J. European Ceramic Soc.* **12**, 369-375
989. Lurie AI (1990) *Nonlinear Theory of Elasticity*. North-Holland, Amsterdam, The Netherlands

990. Lusti HR, Gusev AA (2004) Finite element predictions for the thermoelastic properties of nanotube reinforced polymers. *Modelling Simul Mater Sci Engng*, **12**:107–119
991. Ma H (2010) *Solutions of Eshelby-type inclusion problems and a related homogenization method based on a simplified strain gradient elasticity theory*. PhD Thesis, Texas A&M University, USA
992. Ma HM, Gao XL (2010) A new homogenization method based on a simplified strain gradient elasticity theory. *Acta Mech*, **225**:1075–1091
993. Ma H, Hu G, Wei Y, Liang L (2018) Inclusion problem in second gradient elasticity. *Int. J. Engineering Science*, **132**:60–78
994. Ma PG, Siddiqui NA, Marom G, Kim JK (2010) Dispersion and functionalization of carbon nanotubes for polymer-based nanocomposites: A review *Composites Part A: Applied Science Manufacturing*, **41**:1345–1367
995. Ma Q, Clarke DR (1994) Piezospectroscopic determination of residual stresses in polycrystalline alumina. *J Am Ceram Soc*, **77**:298–302
996. Macek RW, Silling SA (2007) Peridynamics via finite element analysis *Finite Elements in Analysis and Design*, **43**:1169–1178
997. Madenci E, Barut A, Futch M (2016) Peridynamic differential operator and its applications. *Comput. Methods Appl. Mech. Engrg*, **304**:408–451
998. Madenci E, Barut A, Phan ND (2017) Peridynamic unit cell homogenization, *58th AIAA/ASCE/AHS/ASC Structures, Structural Dynamics, and Materials Conference*, AIAA SciTech Forum, (AIAA 2017-1138)
999. Madenci E, Dorduncua N, Barut A, Phanb N (2018) A state-based peridynamic analysis in a finite element framework. *Engineering Fracture Mechanics*, **195**:104–128
1000. Madenci E, Oterkus E (2014) *Peridynamic Theory and Its Applications.*. Springer, NY
1001. Maekava ZI, Hamada H Yokoyama A (1989) Lamination theory of composite material with complex fiber orientation. *Proc ICCS*, **5**:701–714
1002. Maggi F, Bandera A, Galfetti L, De Luca LT, Jackson TL (2010) Efficient solid rocket propulsion for access to space. *Acta Astronautica*, **66**:1563–1573
1003. Maggi F, Stafford S, Jackson TL, Buckmaster J (2008) Nature of packs used in propellant modeling. *Physical Review*, **E77**:046107
1004. Mal AK, Knopoff L (1967) Elastic wave velocities in two-component systems. *I. Inst Math Appl*, **3**:376–387
1005. Mandel J (1972) Plasticité Classique, Viscoplasticité (CISM Courses and Lectures), **97**, Springer-Verlag, New York
1006. Maranganti R, Sharma P (2007) Strain field calculation in embedded quantum dots and wires. *J. Comput. Theoret. Nanoscience*, **4**:715–738
1007. Marchioro M, Prosperitti A (1999) Heat conduction in a non-uniform composite with spherical inclusions. *Proc Roy Soc Lond*, **A455**:1483–1508
1008. Marin L (2005) A meshless method for solving the Cauchy problem in three-dimensional elastostatics. *Computers and Mathematics with Applications*, **50**:73–92
1009. Marin L, Karageorghis A (2013) The MFS for the Cauchy problem in two-dimensional steady-state linear thermoelasticity. *Int. J. Solids Structures*, **50**:3387–3398.
1010. Marin L, Karageorghis A (2013) The MFS–MPS for two-dimensional steady-state thermoelasticity problems. *Engineering Analysis with Boundary Elements*, **37**:1004–1020.
1011. Markov KZ (1981) "One-particle" approximation in mechanics of composite materials. In: Brulin O, Hsieh RKT (eds), *Continuum Models of Discrete Systems*. North-Holland, Amsterdam, 441–448
1012. Markov KZ (2000) Elementary micromechanics of heterogeneous media. In: Markov K, Preziosi L (eds), *Heterogeneous Media. Micromechanics, Modelling, Methods, and Simulations*. Birkhäuser, Boston, 1–162
1013. Markov KZ (2001) Justification of an effective field method in elasto-statics of heterogeneous solids. *J Mech Phys Solids*, **49**:2621–2634
1014. Markov KZ, Willis JR (1998) On the two-point correlation function for dispersions of nonoverlapping spheres. *Math Models Methods Appl Sci*, **8**:359–377

1015. Markworth AJ, Ramesh KS, Parks WP (1995) Review. Modeling studies applied to functionally graded materials. *J Mater Sci,* **30**:2183–2192
1016. Marotti de Sciarra F (2008) Variational formulations and a consistent finite-element procedure for a class of nonlocal elastic continua*Int. J. Solids Struct.*, **45**:4184–4202
1017. Martin JB (1975) *Plasticity: Fundamentals and General Results.* MIT Press, Cambridge, MA
1018. Martins NFM, Rebelo M (2013) A meshfree method for elasticity problems with interfaces *Applied Mathematics and Computation*, **219**:10732–10745
1019. Maruyama B, Alam K (2002) Carbon nanotubes and nanofibers in composite materials. *SAMPE J,* **38**:60–69
1020. Marx K (1867) *Das Kapital: Kritik der politischen Oekonomie*. Volume **1-3**: Der Produktionsprozess des Kapitals (1 ed.). Hamburg: Verlag von Otto Meissner. https://doi.org/10.3931/e-rara-25773
1021. Maslov BP (1979) Macroscopic third-order elastic moduli. *Prikl Mekhanika,* **15**:57-61 (In Russian)
1022. Maslov BP (1981) Effective constants of the theory of geometrically nonlinear solids. *Prikl Mekh,* **17**(5):45–50 (In Russian. Engl Transl. *Soviet Appl Mech,* **17**:439–444)
1023. Maslov BP (2000) Stress concentration in incompressible multicomponent materials. *Prikl Mekh,* **36**(3):118-114 (In Russian. Engl Transl. *Int Appl Mech,* **36**:384–390)
1024. Maslov BP, Khoroshun LP (1972) Effective characteristics of elastic, physically nonlinear inhomogeneous media. *Izvestia AN SSSR Mech Tverd Tela,* (2):149-153 (In Russian)
1025. Massa L, Jackson TL, Short M (2003) Numerical solution of three-dimensional heterogeneous solid propellants. *Combust. Theory Modeling*, **7**:579–602
1026. Massoni J, Saurel R, Baudin G, Demol G (1999) A mechanistic model for shock initiation of solid explosives. *Physics of Fluids*, **11**:710–736
1027. Matouš K, Geubelle PH (2006) Multiscale modelling of particle debonding in reinforced elastomers subjected to finite deformations. *Int. J. Numerical Methods in Engineering*, **65**:190–223
1028. Matouš, K, Inglis H.M, Gu X, Rypl, D, Jackson, T.L, Geubelle, H.P (2007) Multiscale modeling of solid propellants: From particle packing to failure. *Composites Science and Technology*, **67**:1694–1708
1029. Matouš K, Lepš M, Zeman J, Šejnoha M (2000) Applying genetic algorithms to selected topics commonly encountered in engineering practice. *Computer Methods in Applied Mechanics and Engineering*, **190**:1629–1650
1030. Matouš K, Geers MGD, Kouznetsova VG, Gillman A. (2017) A review of predictive nonlinear theories for multiscale modeling of heterogeneous materials *J. Comput. Physics*, **330**:192–220
1031. Materon G (1989) *Random Set and Integral Geometry.* John Wiley & Sons, New York
1032. Matthes S, Humbert M (1995) On the principle of a geometrical mean of even-rank tensors for textured polycrystals. *J Appl Cryst,* **28**:254–266
1033. Maugin GA (1988) *Continuum Mechanics of Electromagnetic Solids.* North-Holland, Amsterdam
1034. Maugin GA (1998) On the structure of the theory of polar elasticity. *Philos. Trans. R. Soc. Lond.*, **A356**:1367–1395
1035. Maugin GA (2017) *Non-Classical Continuum Mechanics. A Dictionary.* Springer Nature, Singapore
1036. Maxwell JC (1873) A Treatise on Electricity and Magnetism, Dover, New York (1954). (Republication of 3rd edition of 1892.)
1037. Mazilu P, Ju JW (1996) Comments and author's reply on "Micromechanics and effective moduli of elastic composites containing randomly dispersed ellipsoidal inclusions" by Ju JW, Chen TM, *Acta Mechan 1994* **103**:103-121. *Acta Mechan,* **114**:235–239.
1038. McCartney LN (1987) Mechanics of matrix cracking in brittle-matrix fiber-reinforced composites. *Proc Roy Soc Lond,* **A409**:329–350

1039. McCoy JJ (1970) On the displacement field in an elastic medium with random variation of material properties. *Rec Adv Engng Sci*, **5**, Gordon and Breach, New York
1040. McCoy JJ (1979) On the calculation of bulk properties of heterogeneous materials. *Q Appl Math*, **36**:137–149
1041. McCoy JJ (1981) Macroscopic response of continue with random microstructure. In: Nemat-Nasser S (ed) *Mechanics Today*. Pergamon Press, Oxford, **6**:1–40
1042. McGeary RK (1961) Mechanical packing of spherical particles. *J. Am. Ceram. Soc.*, **44**:513–522
1043. McKenzie DR, McPedran RC, Derrick GH (1978) The conductivity of lattices of spheres. II. The body centered and face centered cubic lattices. *Proc Roy Soc Lond*, **A362**:211–232
1044. McLaughlin R (1977) A study of the differential scheme for composite materials. *Int J Engrg Sc*, **15**:237–244
1045. McMeeking RM, Evans AC (1982) Mechanics of transformation toughening in brittle materials. *J Amer Ceram Soc*, **69**:242–246
1046. McPedran RC, McKenzie DR (1978) The conductivity of lattices of spheres. I. The simple cubic lattice. *Proc Roy Soc Lond*, **A359**:45–63
1047. McPedran RC, Movchan AB (1994) The Rayleigh multiple method for linear elasticity. *J Mechan Phys Solids*, **42**:711–727
1048. Mehrabadi MA, Cowin SC, Jaric J (1995) Six-dimensional orthogonal tensor representation of the rotation about an axis in three dimensions. *Int J Solids Struct*, **32**:439–449
1049. Mengesha T, Du Q (2014) The bond-based peridynamic system with Dirichlet-type volume constraint. *Proc. R. Soc. Edinburgh*, **A 144**:161–186
1050. Meraghni F, Blaman CJ, Benzeggagh ML (1996) Effect of interfacial decohesion on stiffness reduction in a random discontinuous-fibre composite containing matrix microcracks. *Compos Sci Technol*, **56**:541–555
1051. Meridith RE, Tobias CW (1960) Resistance to potential flow through a cubical array of spheres. *J Appl Phys*, **31**:1270–1274
1052. Michel JC, Suquet P (1992) The constitutive law of nonlinear viscous and porous materials. *J Mech Phys Solids*, **40**:783–812
1053. Michel JC, Moulinec H, Suquet P (1999) Effective properties of composite materials with periodic microstructure: a computational approach *Comput. Meth. Appl. Mech.Engng.* **172**:109–143
1054. Michel JC, Moulinec H, Suquet P (2001) A computational method for linear and nonlinear composites with arbitrary phase contrast. *Int . J. Numer. Meth. Engng.* **52**:139–160
1055. Michel JC, Suquet P (2016) A model-reduction approach in micromechanics of materials preserving the variational structure of constitutive relations. *J. Mech. Physics Solids*, **90**:254–285
1056. Miehe C, Koch A (2002) Computational micro-to-macro transition of discretized microstructures undergoing small strain. *Arch. Appl. Mech.* **72**:300–317
1057. Mikata Y (2001) Explicit determination of piezoelectric Eshelby tensors for a spheroidal inclusion. *Int J Solids Struct*, **38**:7045–7063
1058. Mikata Y (2012) Analytical solutions of peristatic and peridynamic problems for a 1D infinite rod *Int. J. Solids and Structures*, **49**:2887–2897
1059. Mikata Y, Taya M (1986) Thermal stress in a coated short fiber composite. *J Appl Mech*, **53**:681–689
1060. Mikhaylenko MA, Povitsky A (2014) Optimal allocation of boundary singularities for stokes flows about pairs of particles *Engineering Analysis with Boundary Elements*, **41**:122-138
1061. Mikhlin SG, Proössdorf S (1980) *Singular Integral Operators*. Springer-Verlag, Berlin, New York
1062. Milgrom M (1990) Linear response of general composite systems to many coupled fields. *Phys. Rev.* **B 41**:12484–12494
1063. Milgrom M, Shtrikman S (1989) Linear response of two-phase composites with cross moduli: exact universal relations. *Phys Rev*, **A40**:1568–1575

1064. Miller RE, Shenoy VB (2000) Size-dependent elastic properties of nanosized structural elements. *Nanotechnology*, **11**:139–147
1065. Miller RE, Tadmor EB (2002) The quasicontinuum method Overview, applications, and current directions. *J. Computer Aided Mater. Des.*, **9**:203–239
1066. Mindlin RD, Tiersten HF (1962) Effects of couple-stresses in linear elasticity. *Arch. Ration. Mech. Anal.*, **11**:415–448
1067. Mindlin RD (1964) Micro-structure in linear elasticity. *Arch. Rational Mech. Anal.*, **16**:51–78
1068. Mindlin RD, Eshel NN (1968) On the first strain gradient theories in linear elasticity. *Int. J. Solids Structures*, **4**:109–124
1069. Miloh T, Benveniste Y (1988) A generalized self-consistent method for the effective conductivity of composites with ellipsoidal inclusions and cracked bodies *J Appl Phys*, **63**:789–7796
1070. Miloh T, Benveniste Y (1999) On the effective conductivity of composites with ellipsoidal inhomogeneities and highly conducting interfaces. *Proc Roy Soc Lond*, **A455**:2687–2706
1071. Milton GW (1981) Bounds on the transport and optical properties of a two-component composite material. *J Appl Phys*, **52**:5294–5304
1072. Milton GW (1982) Bounds on the elastic and transport properties of two-component composites. *J Mech Phys Solids*, **30**:177–191
1073. Milton GW (1984) Correlation of the electromagnetic and elastic properties of composites and microgeometries corresponding with effective medium approximations. In: Johnson DL, Sen PN (eds), *Physics and Chemistry of Porous Media*. American Institute of Physics, New York, 66–77
1074. Milton GW (1985) The coherent potential approximation is a realizable effective medium scheme. *Commun Math Phys*, **99**:463–500
1075. Milton GW (2002) *The Theory of Composites*. Cambridge University Press, Cambridge, UK
1076. Milton GW, Kohn RV (1988) Variational bounds on the effective elastic moduli of anisotropic composites. *J Mech Phys Solids*, **36**:597–629
1077. Milton GW, Phan-Tien N (1982) New bounds on effective elastic moduli of two-component materials. *Proc Roy Soc Lond*, **A380**:305–331
1078. Mindlin RD (1948) Stress distribution around a hole near the edge of a plate under tension. *Proc Soc Exp Stress Anal*, **5**:56–68
1079. Miracle DB (2005) Metal matrix composites–From science to technological significance. *Comp Sci Technol*, **65**:2526–2540
1080. Mishchenko MI. (2020) Comprehensive the matic T-matrix reference database: a 2017–2019 update. *J Quant Spectrosc Radiat Transfer*, **242**:106692
1081. Mishnaevsky LL (2006) Functionally gradient metal matrix composites: Numerical analysis of the microstructure-strength relationships. *Comp Sci Technol*, **66**:1873–1887
1082. Mishurova T, Cabeza S, Bruno G, Sevostianov I (2016) Average phase stress concentrations in multiphase metal matrix composites under compressive loading. *Int. J. Engineering Science*, **106**:245–261
1083. Mogilevskaya SG, Crouch S (2013) Combining Maxwell's methodology with the BEM for evaluating the two-dimensional effective properties of composite and micro-cracked materials. *Comput Mech*, **51**:377–389
1084. Mogilevskaya SG, Crouch SL, Stolarski HK, Benusiglio A (2010) Equivalent inhomogeneity method for evaluating the effective elastic properties of unidirectional multi-phase composites with surface/interface effects. *Int. J. Solids Structures*, **47**:407–418
1085. Mogilevskaya SG, Kushch VI, Zemlyanova AY (2019) Displacements representations for the problems with spherical and circular material surfaces. *The Quarterly J. Mechanics App. Mathematics*, **72**, 449–471
1086. Mogilevskaya SG, Stolarski HK, Crouch SL (2012) On Maxwell's concept of equivalent inhomogeneity When do the interactions matter? *J. Mech. Physics Solids*, **60**:391–417
1087. Molchanov IS (1997) *Statistics of the Boolean Model Model for Practitioners and Mathematicans*. John Wiley & Sons, New York

1088. Molinari A, El Mouden M (1996) The problem of elastic inclusions at finite concentration. *Int J Solid Struct,* **33**:3131–3150
1089. Monetto I, Drugan WJ (2004) A micromechanics-based nonlocal constitutive equation for elastic composites containing randomly oriented spheroidal heterogeneities *J Mech Phys Solids,* **52**:359–393
1090. Morawiec A (1994) Review of deterministic methods of calculations of physical elastic constants. *Textures Microstruct,* **22**:139–167
1091. Morawiec A (1996) The effective elastic constants of quasi-isotropic polycrystalline materials composed of cubic phase. *Phys Stat Sol,* **A155**:353–364
1092. Morawiec A, Field DP (1996) Rodrigues parameterization for orientation and misorientation distributions. *Philos Mag,* **A73**:1111–1128
1093. Mori T, Tanaka K (1973) Average stress in matrix and average elastic energy of materials with misfitting inclusions. *Acta Metall,* **21**:571–574
1094. Morse PM, Feshbach H (1953) *Methods of Theoretical Physics.* Parts I and II. McGraw-Hill, Maidenhead
1095. Mortensen A, Suresh S (1995) Functionally graded metals and metal–ceramic composites Part 1. Processing. *Int. Mater. Reviews,* **40**:239–265
1096. Moschovidis ZA, Mura T (1975) Two-ellipsoidal inhomogeneities by the equivalent inclusion method. *J Appl Mech,* **42**:847–852
1097. Mossotti OF (1850) Discussione analitica sul'influenza che l'azione di un mezzo dielettrico ha sulla distribuzione dell'electricitá alla superficie di piú corpi elettrici disseminati in eso. *Mem Mat Fis della Soc Ital di Sci in Modena,* **24**:49–74
1098. Moumen AE, Kanit T, Imad A (2021) Numerical evaluation of the representative volume element for random composites. *European Journal of Mechanics / A Solids*, **86**:104181
1099. Movchan NV, Willis JR (1997) Influence of spatial correlations on crack bridging by frictional fibers. *Engng Fracture Mech,* **58**:571–579
1100. Movchan NV, Willis JR (1998) Penny-shaped crack bridged by fibres. *Quart Appl Math,* **56**:327–340
1101. Mukherjee S, Liu Y (2013) The boundary element method *Int. J. Computational Meth.*, **10**:1350037 (91 pages).
1102. Müller K, Bugnicourt E, Latorre M, Jorda M, Sanz YE, Lagaron JM, Miesbauer O, Bianchin A, Hankin S, Bölz U, *et al.* (2017) Review on the processing and properties of polymer nanocomposites and nanocoatings and their applications in the packaging, automotive and solar energy fields. *Nanomaterials* **7**:74.
1103. Müller WH, Gao H, Chiu C-H, Schmauder S (1996) A semi-infinite crack in front of a circular thermally mismatched heterogeneity. *Int J Solids Struct,* **33**:731–746
1104. Mura T (1987) *Micromechanics of Defects in Solids.* Martinus Nijhoff, Dordrecht
1105. Mura T, Furuhashi R (1984) The elastic inclusion with a sliding interface. *J. Appl. Mech.*, **51**:308–310.
1106. Mura T, Jasiuk I, Tuschida B (1985) The stress field of a sliding inclusion. *Int. J. Solids Struct.*, **21**:1165–1179
1107. Mura T, Shodia HM, Hirose Y (1996) Inclusions problems. *Appl Mechan Rev,* **49**(10), Part 2):S118–127
1108. Murakami Y (1987) *Stress Intensity Factors Handbook.* Pergamon Press, Oxford, **2**
1109. Murata I, Mori T, Nakagawa M Continuous energy Monte Carlo calculations of randomly distributed spherical fuels in high-temperature gas-cooled reactor based on a statistical geometry model. *Nuclear Sci Engng,* **123**:96–109
1110. Muskhelishvili NI (1953) *Some Basic Problems of the Mathematical Theory of Elasticity.* P. Noordhoff, Groningen, XXXI
1111. Mustoe GGW (1984) Advanced integration schemes over boundary elements and volume cells for two and three dimensional nonlinear analysis. In. Banerjee, P.K, Mukherjee, S. *Development in Boundary Element Methods* **3**:213-270
1112. Nadeau JC, Ferrari M (1999) Microstructural optimization of a functionally graded transversely isotropic layer. *Mechan of Mater* **31**:637–651

1113. Nagarajan B, Schoen MAW, Trudel S, Qureshi AJ, Mertiny P (2020) Rheology-assisted microstructure control for printing magnetic composites–material and process development. *Polymers*, **12**:2143 (19 pages)
1114. Nairn JA (1997) On the use of shear-lag methods for analysis of stress transfer in unidirectional composites. *Mech Mater*, **26**:63–80
1115. Nan C-W, Bichurin MI, Dong S, Viehland D, Srinivasan G (2008) Multiferroic magnetoelectric composites Historical perspective, status, and future directions. *J. Appl. Phys*, **103**:031101 (39 pages)
1116. Nan C-W, Clarke DR (1997) Effective properties of ferroelectric and/or ferromagnetic composites: a unified approach and its application. *J Am Ceram Soc*, **60**:1333–1340
1117. Nakamura T, Suresh S (1993) Effects of thermal residual stresses and fiber packing on deformation of metal-matrix composites *Acta Metall Mater*, **41**:1665–1681
1118. Needleman A (1990) An analysis of decohesion along an imperfect interface. *Int. J. Fract.*, **42**:21–40
1119. Needleman A (2000) Computational mechanics at the mesoscale. *Acta mater*, **48**:105–124
1120. Needleman A, Tvergaard V (1993) Comparison of crystal plasticity and isotropic hardening predictions for metal-matrix composites. *J Appl Mech*, **60**:70–76
1121. Nemat-Nasser S, Hori M (1993) *Micromechanics: Overall Properties of Heterogeneous Materials*. Elsevier, North-Holland.
1122. Nemat–Nasser S, Iwakuma T, Hejazi M (1982), On composites with periodic structure. *Mech Mater* **1**:239–267
1123. Nemat-Nasser S, Obata M (1986) Rate–dependent, finite deformation of polycrystals. *Proc Roy Soc Lond*, **A407**:377–404
1124. Nielsen LE (1974) *Mechanical properties of polymers and composites*. Marcel Dekker, Inc, NY
1125. Nielsen LE (1979) Dynamic mechanical properties of polymers filled with aggregated particles. *J Polym Sci: Polym Phys*, **17**:1897–1901
1126. Nigmatulin RI (1979) Spatial averaging in the mechanics of heterogeneous and dispersed systems. *Int J Multiphase Flow*, **5**:353–385
1127. Nikitin AN, Rusakova EI, Ivankina TI (1989) On the theory of origination of piezoelectric textures in rocks. *Izv Acad Sci USSR Phys Solid Earth*, (6):49–60 (In Russian)
1128. Nishimura N (2002) Fast multipole accelerated boundary integral equation methods. *Appl Mech Rev* **55**:299–324
1129. Noda N-A, Matsumo T (1998) Singular integral equation method for interaction between elliptic inclusions. *J Appl Mech*, **65**:310–319
1130. Noda N, Nakai S, Tsuji T (1998) Thermal stresses in functionally graded materials of particle-reinforced composite. *JSME Int J*, **A41**:178–184
1131. Nogales S, Böhm HJ (2008) Modeling of the thermal conductivity and thermomechanical behavior of diamond reinforced Composites. *Int. J. Engng. Sci.* **46**:606–619
1132. Noh TW, Song PH, Sievers AJ (1991) Self-consistency conditions for the effective medium approximation in composite materials. *Phys Rev*, **B44**:5459–5464
1133. Nolan GT, Kavanagh PE (1992) Computer simulation of random packing of hard spheres. *Powder Technol*, **72**:149–155
1134. Nolan GT, Kavanagh PE (1995) Octahedral configurations of random close packing. *Powder Technology* **83**:253–258
1135. Norris AN (1985) A differential scheme for the effective moduli of composites. *Mech Mater*, **4**:1–16
1136. Noll W (1972) A new mathematical theory of simple materials. *Arch. Ration. Mech. Anal.*, **48**:1–50
1137. Norris AN (1989) An examination of the Mori-Tanaka effective medium approximation for multiphase composites. *J Appl Mech*, **56**:83–88
1138. Norris AN, Callegari AJ, Sheng PA (1985) A generalized differential effective medium theory. *J Mech Phys Solids*, **33**:525–543

1139. Nowak ER, Knight JB, Povinelli ML, Jaeger HM, Nagel SR (1997) Reversibility and irreversibility in the packing of vibrated granular material. *Powder Technology*, **94**:79–83
1140. Nozaki H, Shindo Y (1998) Effect of interface layers on elastic wave propagation in a fiber-reinforced metal-matrix composite. *Int J Engin Sci,* **36**:383–394
1141. Nunan KC, Keller JB (1984) Effective elasticity tensor of a periodic composite. *J Mech Phys Sol,* **32**:259–280
1142. Nye JF (1957) *Physical Properties of Crystals.* Oxford University Press, Oxford
1143. O'Brian RW (1979) A method for the calculation of the effective transport properties of suspensions of interacting particles. *J Fluid Mech,* **91**:17–39
1144. O'Connel RJ, Budiansky B (1974) Seismic velocities in dry and saturated cracked solids. *J Geophys Res,* **79**:5412–5426
1145. Odegard GM, Gates TS, Nicholson LM, Wise KE (2002) Equivalent-continuum modeling of nano-structured materials. *Composites Sci. Technology,* **62**:1869–1880
1146. Odegard GM, Gates TS, Wise KE, Park C, Siochi EJ (2003) Constitutive modeling of nanotube-reinforced polymer composites. *Compos Sci Technol,* **63**:1671–1687
1147. Odegard GM, Clancy TC, Gates TS (2005) Modeling of the mechanical properties of nanoparticle/polymer composites. *Polymers,* **46**:553–562
1148. Oden JT, Zohdi TI (1997) Analysis and adaptive modeling of highly heterogeneous elastic structures. *Comput Methods Appl Mechan Engng,* **148**:367–391
1149. Ogata Y, Tanemura M (1981) Estimation of interaction potentials of spatial point-patterns through the maximum-likelihood procedure. *Ann Inst Statist Math,* **33**:315–338
1150. Ogata Y, Tanemura M (1984) Likelihood analysis of spatial point-patterns. *J Roy Statist Soc,* **B46**:496–518
1151. Ogden RW (1972) Large deformation isotropic elasticity: on the correlation of theory and experiment for compressible rubberlike solids. *Proc Roy Soc Lond,* **A328**:567–583
1152. Ogden RW (1974) On the overall moduli of non-linear elastic composite materials *J Mech Phys Solids,* **22**:541–554
1153. Ogden RW (1997) *Non-linear Elastic Deformation.* Dover, Mineola, New York
1154. Oger L, Troadec JP, Gervois A, Medvedev N (1998) Computer simulation and tessellations of granular materials. In: Rivier N, Sadoc JF (eds), *Foams and Emulsions.* Kluver, Dordrecht, 527–545
1155. Ohno N, Wang J-D (1994) Kinematic hardening rules for simulation of ratchetting behavior. *Eur J Mech A/Solids,* **13**:519–531
1156. Ohser J, Mücklich F (2000) *Statistical Analysis of Microstructures in Material Science.* John Wiley & Sons, Chichester
1157. Okabe A, Boots B, Sugihara K (1992) *Spatial Tessellations.* John Wiley & Sons, New York
1158. Olson T (1994) Improvements on a Taylor's upper bound for rigid-plastic composites. *Mater Sci Eng A,* **175**:15–19
1159. Olsson M, Giannakopoulos AE, Suresh S (1995) Elastoplastic analysis of thermal cycling: ceramic particles in a metallic matrix. *J Mech Phys Solids,* **43**:1639–1671
1160. Ornstein LS, Zernike F (1914) Accidental deviation of density and opalescence at the critical points of a single substance. *Proc Acad Sci (Amsterdam),* **17**:793–806.
1161. Ortiz M, Molinari A (1988) Microstructural thermal stresses in ceramic materials. *J Mech Phys Solids,* **36**:385–400
1162. Ortiz M, Pandolfi A (1999) Finite-deformation irreversible cohesive element for three-dimensional crack-propagation analysis. *Int. J. Numer. Methods Eng.,* **44**:1267–1282
1163. Ortiz M, Popov EP (1985) Accuracy and stability of integration algorithms for elastoplastic constitutive relations. *Int J Numer Methods Engng,* **21**:1561–1576
1164. Ortiz M, Suresh S (1993) Statistical properties of residual stresses and intergranular fracture in ceramic materials. *J Appl Mech,* **60**:77-84
1165. Ostoja-Starzewski M (2002) Lattice models in micromechanics. *Appl Mech Rev 55*:35–60
1166. Ostoja-Starzewski M (2007) *Microstructural Randomness and Scaling in Mechanics of Materials.* Chapman & Hall/CRC, Boca Raton

1167. Ostoja-Starzewski M (2011) Macrohomogeneity condition in dynamics of micropolar media. *Arch. Appl. Mech.*, **81**:899–906
1168. Ostoja-Starzewski M, Kale S, Karimi P, Malyarenko A, Raghavan B, Ranganathan SI, Zhang J (2016) Scaling to RVE in random media. *Adv. Appl. Mech.*, **49**:111–211
1169. Othmani Y, Delannay L, Doghri I (2011) Equivalent inclusion solution adapted to particle debonding with a non-linear cohesive law. *Int. J. Solids Struct.*, **48**:3326–3335
1170. Ovid'ko IA, Sheinerman AG (2005) Elastic fields of inclusions in nanocomposite solids. *Rev Adv Mater Sci*, **9**:17–33
1171. Ovidk̀o IA, Sheinerman AG (2010) Elastic fields of nanoscopic inclusions in nanocomposites *Materials Physics and Mechanics*, **10**:1–29
1172. Ozmusul MS, Picu RC (2002) Elastic moduli of particulate composites with graded filler-matrix interfaces. *Polym Compos*, **23**:110–119
1173. Pagano NJ (1987) Free-edge stress fields in composite laminates. *Int J Solids Struct*, **14**:401–406
1174. Pagano NJ, Kim RY (1994) Progressive microcracking in unidirectional brittle matrix composites. *Mech Compos Mater Struct*, **1**:3–29
1175. Pagano NJ, Rybicki EF (1974) On the significance of effective modulus solution for fibrous composites. *J Compos Mater* **8**:214–228
1176. Pagano NJ, Yuan FG (2000) Significance of effective modulus theory (homogenization) in composite laminate mechanics. *Compos Sci Technol*, **60**:2471–2488
1177. Pan HH, Weng GJ (1993) Thermal stress relief by plastic deformation in aligned two-phase composites. *Compos Engng*, **3**:219–234
1178. Pan J, Bian L, Zhao H, Zhao Y (2016) A new micromechanics model and effective elastic modulus of nanotube reinforced composites *Computational Materials Science*, **113**:21–26
1179. Pan L, Adams DO, Rizzo FJ (1999) Boundary element analysis for composite materials and a library of Green's functions. *Composites & Struct*, **66**:685–693
1180. Pan YC, Chou TW (1976) Point force solution for an infinite transversely isotropic solid. *J Appl Mech*, **43**:608-612
1181. Panasenko GP (1983) Averaging processes in frame constructions with random properties. *Zhurnal Vychislitel'noi Matematiki i Matematicheskoi Fiziki*, **23**:1098-1109 (In Russian. Engl Transl. *USSR Comput Maths Math Phys*, **23**(5):48–55)
1182. Papadopoulos P, Taylor RL (1994) On the application of multi-step integration methods to infinitesimal elastoplasticity. *Int J Numer Methods Engng*, **37**:3169–3184
1183. Parkhomenko EI (1968) *Electrization in Rocks*. Nauka, Moscow (In Russian)
1184. Parks ML, Bochev PB, Lehoucq RB (2008) Connecting atomistic-to-continuum coupling and domain decomposition. *Multiscale Modeling & Simulation*, **7**:362–380
1185. Parks ML, Seleson P, Plimpton SJ, Silling SA, Lehoucq RB (2011) *Peridynamics with LAMMPS A user guide v0.3 beta, SAND Report 2011–8523*, Sandia National Laboratories, Albuquerque, NM and Livermore, CA
1186. Parnell WJ (2016) The Eshelby, Hill, moment and concentration tensors for ellipsoidal inhomogeneities in the Newtonian potential problem and linear elastostatics. *J Elasticity*, **125**:231–294
1187. Parton VZ, Buryachenko VA (1990) Stress fluctuation in elastic composites. *Dokladi AN SSSR*, **310**:1075–1078 (In Russian. Engl Transl. *Sov Phys Docl*, **35**:191–193)
1188. Parton VZ, Kudryavtsev BA (1988) *Electromagnetic Elasticity of Piezoelectric and Electrically Conductive Bodies*. Nauka, Moscow
1189. Parton VZ, Perlin PI (1982) *Integral Equation Method in Elasticity*. MIR, Moscow
1190. Patlazhan SA (1993) Effective viscosity theory of a random concentrated suspension of polydisperse hard spheres. *Physica* vol. A198:385–400
1191. Partridge PW, Brebbia CA, Wrobel LC (1992) *The dual reciprocity boundary element method*. Computational Mechanics Publications, Southampton
1192. Paul B (1960) Prediction of the elastic constants of multiphase materials. *Trans Am Inst Min Metail Pet Engng*, **218**:36–41

1193. Pavanello F, Manca F, Palla PL, Giordano S (2012) Generalized interface models for transport phenomena Unusual scale effects in composite nanomaterials. *J. Appl. Phys.*, **112**,084306
1194. Paxton JP, Mowles ED, Lukehart CM, Witzig AJ (2001) Polymer material property enhancement through nanocomposite technology. In: Hyer MW (ed), *Proc of the American Society for Composites. Sixteenth Technical Conference.* Loos AG, Blacksburg, VA
1195. Pecullan S, Gibianski LV, Torquato S (1999) Scale effects on the elastic behavior of periodic behavior of periodic and hierarchical two-dimensional composites. *J Mech Phys Solids,* **47**:1509–1542
1196. Peirce AP, Napier JAL (1995) A spectral multipole method for effecient solution of large-scale boundary element models in elastostatics. *Int J Numer Methods Engng,* **38**:4009–4039
1197. Pelech P (2019) The peridynamic stress tensors and the non-local to local passage, *J. of Applied Math. and Mechanics*, **99**(6)::e201800010
1198. Percus JK, Yevick GJ (1958) Analysis of classical statistical mechanics by means of collective coordinates. *Phys Rev,* **110**:1–13
1199. Perelmuter M (2013) Boundary element analysis of structures with bridged interfacial cracks. *Computational Mechanics*, **51**:523–534
1200. Perez MM, Wrobel LC (1996) An integral-equation formulation for anisotropic elastostatics. *J Appl Mech,* **63**:891–902
1201. Perlin PI (1976) Application of the regular representation of singular integrals to the solution of the second fundamental problem of the theory of elasticity. *Prikl Metem Mekhan,* **40**:366–371 (In Russian. Engl Transl. *J Appl Math Mech,* **40**:342–347)
1202. Peterson RE (1974) *Stress Concentration Factors.* John Wiley & Sons, New York
1203. Pettermann HE, Bohm HJ, Rammerstorfer FG (1997) Some direction dependent properties of matrix–inclusion type composites with given reinforcement orientation distributions. *Composites,* **B28**:253–265.
1204. Phan-Thien N, Kim S (1994) *Microstructures in Elastic Media: Principles and Computational Methods.* Oxford University Press, New York
1205. Phan-Thien N, Pham DC (2000) Differential multiphase models for polydispersed spheroidal inclusions: thermal conductivity and effective viscosity. *Int J Engng Sci,* **38**:73–88
1206. Phelan PE, Niemann RC (1998) Effective thermal conductivity of a thin, randomly oriented composite material. *J Heat Transfer,* **120**:971–976
1207. Pfeil K, Klingenberga DJ (2004) Nonlocal electrostatics in heterogeneous suspensions using a point-dipole model *J. Appl. Phys.*, **96**:5341–5348
1208. Phillips RJ (1989) *Multi-sphere Hydrodynamic Models of Suspensions and Porous Media.* PhD Thesis, Dep. of Chemical Engineering, U. California, Davis
1209. Phillips RJ, Brady JF, Bossis G (1988) Hydrodynamic transport properties of hard-sphere dispersions. II. Porous media *Phys. Fluids* **31**:3473–3479
1210. Picka J (2012) Statistical inference for disordered sphere packings. *Statist. Surv.*, **6**:74–112
1211. Picu RC (2002) Non-local elasticity kernels extracted from atomistic simulations. In: Zavaliangos A, Tikare V, Olevsky EA (eds), *Modelling and Numerical Simulation of Materials Behavior and Evolution.* Mater Res Soc , Warrendale, PA, **731**:71–76
1212. Picu RC, Ozmusul MS (2003) Structure of linear polymeric chains confined between impenetrable spherical walls. *J Chem Phys,* **118**:11239–11248
1213. Picu RC, Sarvestani A, Ozmusul MS (2004) Elastic moduli of polymer nanocomposites derived from their molecular structure. In: Harik VM (ed), *Trends in Nanoscale Mechanics: Analysis of Nanostructured Materials and Multiscale Modeling.* Kluwer, Dordrecht, 61–88
1214. Pindera M-J, Aboudi J, Arnold SM (1995) Limitations of the uncoupled, RVE–based micromechanical approach in the analysis of functionally graded composites. *Mech Mater* **20**:77–94
1215. Pindera M-J, Dunn P (1997) Evaluation of the higher-order theory for functionally graded materials via the finite-element method. *Compos Engng,* **7**:109–119

1216. Pipes RB, Hubert P (2003) Helical carbon nanotube arrays: thermal expansion. *Compos Sci Technol*, **63**:1571–1579
1217. Pipkin AC (1991) *A Course on Integral Equations*. Springer-Verlag, New York
1218. Plankensteiner AF, Böhm HJ, Rammerstorfer FG, Buryachenko VA (1996) Hierarchical modeling of the mechanical behavior of high speed steels as layer–structured particulate MMCs. *J Physique IV*, **6**:C6-395–C6-402
1219. Plankensteiner AF, Böhm HJ, Rammerstorfer FG, Buryachenko VA, Hackl G (1997) Modeling of layer–structured high speed tool steel. *Acta Metall Mater*, **45**:1875–1887
1220. Po G, Admal NG, Lazar M (2019) The Green tensor of Mindlin's anisotropic first strain gradient elasticity *Materials Theory*, **3** 3 (16 pages)
1221. Pobedrya BY (1984) *Mechan of Composite Materials*. MGU, Moscow (In Russian)
1222. Podalkov VV, Romanov VA (1984) Deformation of an elastic anisotropic micro-inhomogeneous half-space. *J Appl Math Mech*, **47**:383–388
1223. Podil'chuk YuN (1984) *The Boundary Value Problems of Statics of Elastic Body*. Naukova Dumka, Kiev (In Russian)
1224. Podil'chuk YuN (2001) Exact analytical solutions of three-dimensional static thermoelastic problems for a transversely isotropic body in curvilinear coordinate systems. *Int Appl Mech*, **37**:728–761
1225. Poisson S (1824) Mémoire sur la théorie du magnétisme. *Mem de l' Acad R de France* V 247–338.
1226. Polder D, Van Santen JH (1946) The effective permeability of mixtures of solids. *Physica*, **XII**:257–271
1227. Polizzotto C (1994) On elastic plastic structure under cyclic load. *Eur J Mech A/Solids*, **13**(4-Suppl):149–173
1228. Polizzotto C (2001) Nonlocal elasticity and related variational principles. *Int. J. Solids Structures*, **38**:7359–7380
1229. Polizzotto C (2003) Gradient elasticity and nonstandard boundary conditions. *Int. J. Solids Structures*, **40**:7399–7423
1230. Polizzotto C (2014) Stress gradient versus strain gradient constitutive models within elasticity. *Int. J. Solids Structures*, **51**:1809–1818
1231. Polizzotto C (2015) A unifying variational framework for stress gradient and strain gradient elasticity theories. *European Journal of Mechanics, A/Solids*, **49**:430–440
1232. Polizzotto C, Fuschi P, Pisano AA (2006) A nonhomogeneous nonlocal elasticity model. *Eur J Mech A/Solids*, **25**, 308–333
1233. Ponte Castañeda P (1988), New variational principles in plasticity and their application to composite materials. *J Mech Phys Solids*, **40**: 1757–1788
1234. Ponte Castañeda P (1991a) The effective mechanical properties of nonlinear isotropic composites. *J Mech Phys Solids*, **39**:45–71
1235. Ponte Castañeda P (1991b) Effective properties in power-law creep. In: Cocks ALF, Ponter ARS (eds) *Micromechanics of Creep of Brittle Materials*. Elsevier, London, **2**:218–229
1236. Ponte Castañeda P (1992) New variational principles in plasticity and their application to composite materials. *J Mech Phys Solids*, **40**:1757–1788
1237. Ponte Castañeda P (2002) Second-order homogenization estimates for nonlinear composites incorporating field fluctuations I-Theory, *J. Mech. Phys. Solids*, **50**:737–757
1238. Ponte Castañeda P (2002) Second-order homogenization estimates for nonlinear composites incorporating field fluctuations II-Application, *J. Mech. Phys. Solids*, **50**:759–782
1239. Ponte Castañeda P, Suquet P (1998) Nonlinear composites. In: Hutchinson JW, Wu TJ (eds), *Adv Appl Mech* **34**:171–302
1240. Ponte Castañeda P, Willis JR (1995) The effect of spatial distribution on the effective behavior of composite materials and cracked media. *J Mech Phys Solids*, **43**:1919–1951
1241. Povirk GL (1995) Incorporation of microstructural information into models of two-phase materials, *Acta Metallurgica et Materialia*, **43**:3199–3206
1242. Povstenko YZ (1993) Theoretical investigation of phenomena caused by heterogeneous surface tension in solids. *J. Mech. Phys. Solids*, **41**:1499–1514

1243. Pozrikidis C (1992) *Boundary Integral and Singularity Methods for Linearized Viscous Flow.* Cambridge University Press, NY
1244. Pozrikidis C (2011) *Introduction to Theoretical and Computational Fluid Dynamics.* NY Oxford University Press.
1245. Praveen GN, Reddy JN (1998) Nonlinear transient thermoelastic analysis of functionally graded ceramics-metal plates. *Int J Solids Struct,* **35**:4437–4476
1246. Probstein RF. Sengun MZ. Tseng T-C (1994) Bimodal model of concentrated suspension viscosity for distributed particle sizes. *J. Rheol.*, **38**:811–829
1247. Prosperetti A (1998) Ensemble averaging techniques for disperse flows. In: Drew DA, Joseph DD, Passma SL (eds) *Particulate Flows Processing and Rheology.* Springer-Verlag, New York 99–136
1248. Pyrz R (1994) Quantitative description of the microstructure of composites. Part I: Morphology of unidirectional composite systems. *Compos Sci Technol,* **50**:197–208
1249. Pyrz R (2004) Microstructural description of composites–statistical methods. In: Böhm H (ed), *CISM Courses and Lectures.* Springer, Udine, **464**:173–233
1250. Pyrz R, Bochenek B (1998) Topological disorder of microstructure and its relation to the stress field. *Int J Solids Struct,* **35**:2413–2427
1251. Qi HJ, Teo KBK, Lau KKS, Boyce MC, Milne WI, Robertson J, Gleason KK (2003) Determination of mechanical properties of carbon nanotubes and vertically aligned carbon nanotube forests using nanoindentation. *J Mech Phys Solids,* **51**:2213–2237
1252. Qian D, Dickey EC, Andrews R, Rantell T (2000) Load transfer and deformation mechanisms in carbon nanotube- polystyrene composites *Appl Phys Lett,* **76**: 2868–2870
1253. Qian D, Liu WK, Ruoff RS (2003) Load transfer mechanism in carbon nanotube ropes. *Compos Sci Technol,* **63**:1561–1569
1254. Qian D, Wagner GJ, Liu WK, Yu M-F, Ruoff RS (2002) Mechanics of carbon nanotubes. *Appl Mech Rev,* **55**:495–533
1255. Qin Q, Yang Q-S (2008) *Macro-Micro Theory on Multifield Coupling Behavior of Heterogeneous Materials.* Springer, Berlin
1256. Qiu YP, Weng GJ (1990) On the application of Mori-Tanaka's theory involving transversely isotropic spheroidal inclusions. *J Engng Sci,* **28**:1121–1137
1257. Qiu YP, Weng GJ (1991a) Elastic moduli of thickly coated particle and fiber–reinforced composites. *J Appl Mech,* **58**:388–398
1258. Qiu YP, Weng GJ (1991b) The influence of estimation shape on the overall elastoplastic behavior of a two-phase isotropic composite. *Int J Solids Struct,* **27**:1537–1550
1259. Qiu YP, Weng GJ (1992) A theory of plasticity for porous materials and particle–reinforced composites. *J Appl Mech,* **59**:261-268
1260. Qiu YP, Weng GJ (1993) Plastic potential and yield function of porous materials with aligned and randomly oriented spheroidal voids. *Int J Plast,* **9**:271–290
1261. Qiu YP, Weng GJ (1995) An energy approach to the plasticity of a two-phase composite containing aligned inclusions. *J Appl Mech,* **62**:1039–1046
1262. Qu J (1993) The effect of slightly weakened interfaces on the overall elastic properties of composite materials. *Mech Mater,* **14**:269–281
1263. Qu J, Cherkaoui M (2006)*Fundamentals of Micromechanics of Solids.* Hoboken, NJ John Wiley & Sons, Inc.
1264. Quang HL, He Q-C (2008) Variational principles and bounds for elastic inhomogeneous materials with coherent imperfect interfaces *Mechanics of Materials*, **40**:865–884
1265. Quang HL, Pham DC, Bonnet G, He Q-C (2013) Estimations of the effective conductivity of anisotropic multiph
1266. Quintanilla J (1999) Microstructure and properties of random heterogeneous materials: a review of theoretical results. *Polym Engng Sci,* **39**:559–585
1267. Quintanilla J, Torquato S (1997) Microstructure functions for a model of statistically inhomogeneous random media. *Phys Rev,* **E55**:1558–1565
1268. Rabotnov YN (1959) A mechanism of a long time failure. In *Creep problems in structural members*, *AN SSSR*, 5–7, (In Russian)

1269. Räisänen VI, Herrmann HJ (1999) Stress transfer in dilute short-fiber reinforced composites. *J Mater Sci*, **34**:897–904
1270. Raghavan P, Moorthy S, Ghosh S, Pagano NJ (2001) Revisiting the composite laminate problem with an adaptive multi-level computational model. *Composite Sci Technol*, **61**:1017–1040
1271. Rafiee R, Moghadam RM (2014) On the modeling of carbon nanotubes a critical review *Composites Part B Engineering*, **56**:435–449
1272. Rafii-Tabar H (2004) Computational modelling of thermo-mechanical and transport properties of carbon nanotubes *Physics Reports*, **390**:235–452
1273. Rafii-Tabar H, Ghavanloob E, Fazelzadeh SA (2016) Nonlocal continuum-based modeling of mechanical characteristics of nanoscopic structures. *Physics Reports*, **638**, 1–97
1274. Rajabi M, Hasheminejad SM (2009) Acoustic resonance scattering from a multilayered cylindrical shell with imperfect bonding. *Ultrasonics*, **49**:682–695
1275. Raghavan P, Moorthy S, Ghosh S, Pagano NJ (2001) Revisiting the composite laminate problem with an adaptive multi-level computational model. *Compos Sci Technol*, **61**:1017–1040
1276. Rajapakse RKND, Shah AH (1988) Hybrid modeling of semi-infinite media. *Int J Solids Struct*, **24**:1205–1224
1277. Rajendran AM, Grove DJ (1996) Modeling the shock response of silicon carbide, boron carbide, and titanium diboride. *Int J Impact Eng*, **18**:611–631
1278. Ramadan KS, Sameoto D, Evoy S (2014) A review of piezoelectric polymers as functional materials for electromechanical transducers *Smart Mater. Struct.*, **23**:033001
1279. Rammerstorfer FG (1992) *Leichtbau–Repetitorium*. Oldenbourg-Verlag, Wien-München
1280. Rank E, Krause R (1997) A multiscale finite-element method. *Computers Struct*, **64**:139–144
1281. Rankenburg IC, Zieve RJ (2001) Influence of shape on ordering of granular systems in two dimensions. *Phys Rev*, **E63**:61303-1–61303-9
1282. Ravichandran G, Liu CT (1995) Modeling constitutive behavior of particulate composites undergoing damage. *Int. J. Solids and Structures*, **32**:979–990
1283. Ray SK, Utku S (1989), A numerical model for the thermo-elasto-plastic behavior of a material. *Int J Numer Methods Engng*, **28**:1103–1114
1284. Rayleigh L (1892) On the influence of obstacles arranged in rectangular order upon the properties of a medium. *Philosophical Magazine*, **34**:481–502
1285. Reddy JN (Ed) (1994) *Mechan of Composite Materials. Selected works of N. J. Pagano*. Kluwer, Dordrecht
1286. Reid A. C. E., Gooding, R. J. (1992) Inclusion problem in a two-dimensional nonlocal elastic solid. *Physical Review B*, **46**, 6045–6049
1287. Reifsnider KL, Gao Z (1991) A micromechanics models for composites under fatigue loading. *Int J Fatigue*, **13**:149–156
1288. Reiter T, Dvorak GJ (1998) Micromechanical models of functionally graded composite materials. In: Bahei-El-Din YA, Dvorak GJ (eds), *IUTAM Symp. on Transformation Problems in Composite and Active Materials*. Kluwer, Dordrecht, 173–184
1289. Reiter T, Dvorak GJ, Tvergaard V (1997) Micromechanical models for graded composite materials. *J Mech Phys Solids*, **45**:1281–1302
1290. Reuss A (1929) Berechnung der Fliessgrenze von Mischkristallen auf Grund der Plastizitatsbedingung fur Einkristalle. *Z Angew Math Mech*, **9**:49–58
1291. Rice RW, Pohanka RC (1979) The grain size dependence of spontaneous-cracking in ceramics. *J Am Ceram Soc*, **62**:559-563
1292. Rice JR (1985) Three-dimensional elastic crack tip interaction with transformation strains and dislocations. *Int J Solids Struct*, **21**:781–791
1293. Rice JR (1989) Weight function theory for three-dimensional elastic crack analysis. In: Wei RP, Gangloff RP (eds), *Fracture Mechanics: Perspectives and Directions (Twentieth Symp)*. Amer Soc Test Mater, Philadelphia, 29–57

1294. Rinaldi RG, Blacklock M, Bale H, Begley MR, Cox BN (2012) Generating virtual textile composite specimens using statistical data from micro-computed tomography: 3D tow representations *J. Mechanics Physics Solids*, **60**:1561–1581
1295. Ripley BD (1977) Modeling spatial patterns. *J Roy Statist Soc,* **B39**:172–212
1296. Ripley BD (1979) Tests of "randomness" for spatial point patterns. *J. R. Statist. Soc.*, **B 41**:368–374
1297. Ripley BD (1981) *Spatial Statistic.* John Wiley & Sons, New York
1298. Rodin GJ (1993) The overall elastic response of materials containing spherical inhomogeneities *Int J Solids Struct,* **30**:1849–1863
1299. Rodin GJ, Hwang YL (1991) On the problem of linear elasticity for an infinite region containing a finite number of spherical inclusions. *Int J Solids Struct,* **27**:145–159
1300. Rodin G, Weng J (2014) On reflectedinteractionsinelasticsolids containing inhomogeneities. *J. Mechanics Physics of Solids*, **68**:197–209
1301. Rogula D (1982) Nonlocal theory of material media. *CISM Courses and Lectures,* **268**. Springer-Verlag, Vienna, New York
1302. Roscoe R (1952) The viscosity of a suspension of rigid spheres. *Br J Appl Phys,* **3**:267–268
1303. Rosen BW (1970) Thermoelastic energy functions and minimum energy principles for composite materials. *Int J Engng Sci,* **8**:5–18
1304. Rosen BW, Hashin Z (1970) Effective thermal expansion coefficients and specific heat of composite materials. *Int J Engng Sci,* **8**:157–173
1305. Ru CQ, Schiavone P (1997) A circular inclusion with circumferentially inhomogeneous interface in antiplane shear. *Proc Roy Soc Lond,* **A453**:2551-2572
1306. Ru CQ, Schiavone P, Mioduchowski A (2001) Elastic fields in two jointed half-planes with an inclusion of arbitrary shape. *Z Angew Math Phys,* **52**:18–32
1307. Rudge JF, Holness MB, Smith GC (2008) Quantitative textural analysis of packings of elongate crystals *Contrib Mineral Petrol*, **156**:413–429
1308. Ruoff RS, Qian D, Liu WK (2003) Mechanical properties of carbon nanotubes theoretical predictions and experimental measurements *C. R. Physique*, **4**:993–100
1309. Russ JC (2002) *The Image Processing Handbook.* CRC Press, Boca Raton, FL
1310. Russel W B, Acrivos A (1972) On the effective moduli of composite materials slender rigid inclusions at dilute concentrations. *J Appl Math Phys (ZAMP)*, **23**:434–464
1311. Russel WB, Saville DA, Schowalter WR (1989) *Colloidal Dispersions.* Cambridge University Press, Cambridge, UK
1312. Rybicki EF, Pagano NJ (1976) A study of the influence of microstructure on the modified effective modulus approach for composite laminates. *Proc 1975 Int Conf Composite Mater* **2**:198–207
1313. Saad Y (1996) *Iterative Methods for Sparse Linear Systems*, PWS Pub. Co, Boston
1314. Saadatfar M (2009) Computer simulation of granular materials. *Computing in Science & Engineering*, **11**:66–74
1315. Sab K, Legoll F, Forest S (2016). Stress gradient elasticity theory existence and uniqueness of solution. *J. Elast.*, **123**:179–201
1316. Sab K, Nedjar B (2005) Periodization of random media and representative volume element size for linear composites. *C R Mecanique*, **333**:187–195
1317. Sabina FJ, Smyshlyaev VP, Willis JR (1993) Self-consistent analysis of waves in a matrix-inclusion composite.–I. Aligned spheroidal inclusions. *J Mech Phys Solids,* **41**:1573–1588
1318. Saeb S, Steinmann P, Javili A (2016) Aspects of computational homogenization at finite deformations a unifying review from Reuss' to Voigt's bound *Appl. Mech. Rev.*, **68**,(5), 050801 (33 pages)
1319. Safadi AY (1996) *A Numerical Technique for Determining the Elastic Field of an Arbitrary Shaped Inhomogeneity.* PhD. Thesis, Northwestern University, Evanston
1320. Safari A, Akdogan EK (2008) *Piezoelectric and Acoustic Materials for Transducer Applications*. Springer, United Kingdom
1321. Sahimi M (1998) Non-linear and non-local transport processes in heterogeneous media: from long-range correlated percolation to fracture and materials breakdown. *Phys Rep,* **306**:213–395

1322. Sahimi M (2003) *Heterogeneous Materials II Nonlinear and Breakdown Properties*, Springer-Verlag, Berlin
1323. Salerno GM, Watt JP (1986) Walpole bounds on the effective elastic moduli of isotropic multicomponent composites. *J Appl Phys*, **60**:1618–1624
1324. Sanchez-Palencia E (1980) *Homogenization Techniques and Vibration Theory.* Lecture Notes in Physics, No. 127, Springer–Verlag, Berlin
1325. Sanchez-Palencia E (1987) Boundary layers and edge effects in composites. In: Sanchez-Palencia E, Zaoui A (eds), *Homogenization Techniques for Composite Media.* Lecture Notes in Physics, Springer-Verlag, Berlin **272**:121–147
1326. Sangani A, Lu W (1987) Elastic coefficients of composites containing spherical inclusions in a periodic array. *J Mech Phys Sol*, **35**:1–21
1327. Sangani AS, Mo G (1997) Elastic interactions in particulate composites with perfect as well as imperfect interfaces. *J Mech Phys Solids*, **45**:2001–2031
1328. Sangani AS, Yao C (1997) Transport processes in random array of cylinders. I. Thermal conduction. *Phys Fluids*, **31**:2426–2434
1329. Saroukhani S, Vafadari R, Andersson R, Larsson F, Runesson K (2015) On statistical strain and stress energy bounds from homogenization and virtual testing. *Europ. J. Mechanics A/Solids*, **51**:77–95
1330. Sarychev AK, Shalaev VM (20000) Electromagnetic field fluctuations and optical nonlinearities in metal-dielectric composites. *Physics Reports*, **335**:275–371
1331. Sato H, Shindo Y (2002) Influence of microstructure on scattering of plane elastic waves by a distribution of partially debonded elliptical inclusions. *Mech Mater* **34**:401–409
1332. Savin GN (1961) *Stress Concentration Around Holes.* Pergamon Press, New York
1333. Savin GN, Khoroshun LP (1972) Problem of elastic constants of randomly reinforsed materials. *Mechanics of Composite Media and Related Problems of Analysis.* Nauka, Moscow, 437–444 (In Russian)
1334. Sawant S, Muliana A (2009) A multiscale framework for analyzing thermo-viscoelastic behavior of fiber metal laminates. *Int. J. Multiscale Comput. Enging.* **7**:351–370
1335. Scaife BKP (1989) *Principle of Dielectrics.* Oxford University Press, Oxford, UK
1336. Schadler LS, Giannaris SC, Ajayan PM (1998) Load transfer in carbon nanotube epoxy composites. *Appl Phys Lett*, **73**:3842–3844
1337. Schaertl W, Sillescu H (1994) Brownian dynamics of polydisperse colloidal hard spheres Equilibrium structures and random close packings. *J. Statistical Phys.*, **77**:1007-1025
1338. Schapery RA (1968) Thermal expansion coefficients of composite materials based on energy principles. *J Compos Mater*, **2**:380–404
1339. Schapery RA (1986) A micromechanical model for non-linear viscoelastic behavior of particle-reinforced rubber with distributed damage. *Eng Fract Mech*, **25**:845–867
1340. Schclar NA (1994) *Anisotropic Analysis Using Boundary Elements.* Comput Mech Publ, Southampton and Boston
1341. Schlichting H (1979) *Boundary-layer Theory*. McGraw Hill, NY.
1342. Schmauder S (2002) Computational mechanics. *Annu Rev Mater Res*, **32**, 437–465
1343. Schnack E, Szikrai S, Türke K (1998) Local effects in engineering with macro-elements. *Comput Methods Appl Mech Engng*, **157**:299–309
1344. Schraad MW, Triantafyllidis N (1997) Scale effects in media with periodic and nearly periodic microstructures. I. Macroscopic properties. *J Appl Mech*, **64**:751–62
1345. Schulgasser K (1976a) Relationship between single-crystal and polycrystal electrical conductivity. *J Appl Phys*, **47**:1880–1886
1346. Schulgasser K (1976b) On the conductivity of fiber-reinforced materials. *J Math Phys*, **17**:382–387
1347. Schulgasser K (1987) Thermal expansion of polycrystalline aggregates with texture. *J Mech Phys Solids*, **35**:35-42
1348. Scorohod VV (1961) Calculation of the effective isotropic moduli of disperse solid systems. *Poroshkovaya Metallurgiya (Powder Metall)*, (1):50–51 (In Russian)

1349. Sedov LL (1966) *Foundations of the Non-Linear Mechanics of Continua.* Pergamon Press, Oxford
1350. Segurado J, Lebensohn RA, LLorca J (2018) Computational homogenization of polycrystals. *Advances Applied Mechanics* **51**:1–114
1351. Segurado J, Llorca J (2002) A numerical approximation to the elastic properties of sphere-reinforced composites. *J. Mech. Phys. Solids*, **50**:2107–2121
1352. Segurado J, Gonzallez C, Llorca J (2003) A numerical investigation of the effect of particle clustering on the mechanical properties of composites. *Acta Mater,* **51**:2355–2369
1353. Sejnoha M, Zeman J (2013) *Micromechanics in Practice.* WIT Press, Southampton, UK
1354. Seleson, P, Gunzburger, M, Parks, M.L, 2013. Interface problems in nonlocal diffusion and sharp transitions between local and nonlocal domains. *Comput. Methods Appl. Mech. Engrg.*, **266**:185-204
1355. Seleson P, Littlewood DJ (2016) Convergence studies in meshfree peridynamic simulations. *Comput. Mathematics with Applications*, :2432–2448
1356. Seleson P, Parks ML (2011) On the role of the influence function in the peridynamic theory. *Int. J. Multiscale Comput. Eng.*, **9**:689–706
1357. Selmi A (2019) Green's function and Eshelby's tensor based on Mindlin's 2nd gradient model an explicit study of cylindrical inclusion case. *J. Multiscale Modelling*, **10**:1850007
1358. Selvadurai APS (2015) Bridged defects in continuously and discretely reinforced solidsCrack-bridging in a unidirectionally fibre-reinforced plate *J Eng Math*, **95**:359–380
1359. Sen AK, Torquato S (1989) Effective conductivity of anisotropic two-phase composite medium. *Phys Rev,* **B39**:4504–4515
1360. Sen P, Scala C, Cohen MH (1981) A self-similar model for sedimentary rocks with application to the dielectric constant of fused glass beards. *Geophysics,* **46**:781–795
1361. Serpilli M, Rizzoni R, Lebon F, Dumont S (2019). An asymptotic derivation of a general imperfect interface law for linear multiphysics composites. *Int. J. Solids Struct.*, **180–181**: 97–107
1362. Sevostianov I (2014) On the shape of efective inclusion in the maxwell homogenization scheme for anisotropic elastic composites. *Mech. Mater.* **75**:45–59
1363. Sevostianov I, Giraud A (2013) Generalization of maxwell homogenization scheme for elastic material containing inhomogeneities of diverse shape. *Int. J. Engineering Science*, **64**:23–36
1364. Sevostianov I, Kachanov M (2002) Explicit cross-property correlation for anisotropic two-phase composite materials. *J Mech Phys Solids,* **30**:252–282
1365. Sevostianov IB, Levin VM, Pompe W (1998) Evolution of the mechanical properties of ceramics during drying. *Phys. Stat. Sol.* **(a) 166**:817–828
1366. Sevostianov I, Mogilevskaya SC, Kushch VI (2019) Maxwell's methodology of estimating effective properties Alive and well. *Int. J. Engineering Science*, **140**:35–88
1367. Shan Z Gokhale AM (2002) Representative volume element for non-uniform micro-structure. *Comput Mater Sci,* **24**:361–379
1368. Sharma P, Ganti S (2004) Size-dependent Eshelby-s tensor for embedded nano-inclusions incorporating surface/interface energies. *J Appl Mech,* **71**:663–671
1369. Shaat M, Ghavanloob E, Fazelzadehb SA (2020) Review on nonlocal continuum mechanics Physics, material applicability, and mathematics. *Mechanics of Materials*, **150**:103587
1370. Shapiro AP, Probstein RF (1992) Random packings of spheres and fluidity limits of monodisperse and bidisperse suspensions. *Phys. Rev. Lett.*, **68**:1422–1425
1371. Sharif-Khodaei Z, Zeman J (2008) Microstructure-based modeling of elastic functionally graded materials: One dimensional case. *J. Mechanics of Materials and Structures*, **3**:1773–1796
1372. Sharma P, Dasgupta A (2002) Average elastic field and scale-dependent overall properties of heterogeneous micropolar materials containing spherical and cylindrical inhomogeneities. *Phys. Rev.*, **B 66**:224110, 1–10
1373. Sharma, P. and Wheeler, L.T, (2007) Size-dependent elastic state of ellipsoidal nano-inclusions incorporating surface/interface tension. *J. Applied Mechanics*, **74**:447–454

1374. Sheng N, Boyce MC, Parks DM, Rutledge GC, Abes JI, Cohen RE (2004) Multiscale micromechanical modeling of polymer/clay nanocomposites and the effective clay particle. *Polymers,* **45**:487-506
1375. Shenoy VB (2002) Size-dependent rigidities of nanosized torsional elements. *Int. J. Solids Struct.*, **39**:4039–4052
1376. Shenoy VB, Miller R, Tadmor EB, Rodney D, Phillipsa R, Ortiz M (1999) An adaptive finite element approach to atomic-scale mechanics–the quasicontinuum method. *J Mech Phys Solids,* **36**:500–531
1377. Sherman DI (1959) On the problem of plane strain in non-homogeneous media. In: Olszak W (ed), *Nonhomogeneity in Elasticity and Plasticity.* Pergamon Press, 3–20
1378. Shermergor TD (1977) *The Theory of Elasticity of Microinhomogeneous Media.* Nauka, Moscow (In Russian)
1379. Shermergor TD, Yakovlev VB (1993) Concentration of coupled electrical mechanical fields on a crystallite surface in textured quartz. *Izv Acad Sci Russ Phys Solid Earth,* **32**:89–94 (In Russian)
1380. Shi D-L, Feng X-Q, Huang YY, Hwang K-C, Gao H (2004) The effect of nanotube waviness and agglomeration on the nanotube-reinforced composites. *J Engng Mater Technol,* **126**:250–257
1381. Shi DL, Feng XQ, Jiang HQ, Huang Y, Huang KS (2005) Multiscale analysis of fracture of carbon nanotubes embedded in composites. *Int J Fracture*, **134**:369–386
1382. Shia D, Hui CY, Burnside SD, Giannelis EP (1998) An interface model for the prediction of Young's modulus of layered silicate-elastomer nanocomposites. *Polym Compos,* **19**:608–617
1383. Shibata S, Jasiuk I, Mori T, Mura T (1990) Successive iteration method applied to composites containing sliding inclusions: effective modulus and elasticity. *Mech Mater* **9**:229–243
1384. Shidlovskii AK, Glushkov EH, Reztsov VF, Snarskaya GI (1989) Generalized form of Eshelby principle in electodynamic of inhomogeneous media and some its application. *Dokladi AN Ukraine SSR,* **A**(3):82–86
1385. Shilov GE (1968) *Generalized Functions and Partial Differential Equations.* Gordon & Breach, New York
1386. Shivakumar KC, Sivakumar KC (2009) A review of infinite matrices and their applications. *Linear Algebra and its Applications*, **430**:976–998
1387. Shklovskii BI, Efros AL (1979) *Electronic Properties of Doped Semiconductors.* Nauka, Moscow (In Russian)
1388. Shubin AB (1995) On maximum density of random packing of the identical solid spheres. *Rasplavy* (1):92–97 (In Russian)
1389. Shvidler MI (1985) *Statistical Hydrodynamics of Porous Media.* Nauka, Moscow. (In Russian)
1390. Siboni G, Benveniste Y (1991) A micromechanical model for the effective thermomechanical behaviour of multiphase composite media. *Mech Mater* **11**:107–122
1391. Sierou A, Brady JF (2001) Accelerated Stokesian dynamics simulations. *J. Fluid Mech,* **448**:115–146
1392. Sierou A, Brady JF (2002) Rheology and microstructure in concentrated noncolloidal suspensions. *J. Rheology*, **46**:1031–1056
1393. Sihn S, Roy AK (2004) Modeling and prediction of bulk properties of open-cell carbon foam. *J Mech Phys Solids,* **52**:167–191
1394. Silberschmidt V (2008) Account for random microstructure in multiscale models. In Kwon, Y.W, Allen, D.H, Talreja, R.R. (eds) *Multiscale Modeling and Simulation of Composite Materials and Structures*. Springer, NY. 1–35
1395. Silbert LE, Ertas D, Grest GS, Halsey TC, Levine D (2002) Geometry of frictionless and frictional sphere packings. *Physical Review*, **E65**:031304
1396. Silling S (2000) Reformulation of elasticity theory for discontinuities and long-range forces. *J. Mech. Physics of Solids* **48**:175–209

1397. Silling S (2010) Linearized theory of peridynamic states. *Journal of Elasticity*, **99**:85–111
1398. Silling S (2011) A coarsening method for linear peridynamics *Int. J. Multiscale Computational Engng*, **9**:609–622
1399. Silling SA (2014) Origin and effect of nonlocality in a composite. *J. Mechanics of Materials and Structures*, **9**:245–258
1400. Silling S (2020) Propagation of a stress pulse in a heterogeneous elastic bar. *Sandia Report SAND2020-8197*, Sandia National Laboratories.
1401. Silling SA, Askari E (2005) A meshfree method based on the peridynamic model of solid mechanics. *Comput. Struct.* **83**:1526–1535
1402. Silling SA, Epton M, Weckner O, Xu J, Askari E (2007) Peridynamic states and constitutive modeling. *J. Elasticity*, **88**:151–184
1403. Silling SA, Lehoucq RB (2008) Convergence of peridynamics to classical elasticity theory, *J. Elasticity*, **93**:13–37
1404. Silling SA, Lehoucq RB (2010) Peridynamic theory of solid mechanics. *Adv. Appl.Mech.*, **44**:73–168
1405. Silling SA, Zimmermann M, Abeyaratne R (2003) Deformation of a peridynamic bar. *J. Elasticity*, **73**:173–190
1406. Silva MJ, Hayes WC, Gibson LJ (1995) The effect of non-periodic microstructure on the elastic properties of two-dimensional cellular solids. *Int J Mech Sci,* **37**:1161–1177
1407. Simmons G, Wang H (1971) *Single Crystal Elastic Constants and Calculates Aggregate Properties.* MIT Press, Cambridge/London
1408. Sinelnikov NN, Mazo MA, Berlin AA (1997) Dense packing of random binary assemblies of disks. *J Phys I France,* **7**:247–254
1409. Sirotin YuI, Shaskolskaya MP (1982) *Fundamentals of Crystal Physics.* Mir Publishers, Moscow
1410. Sladek V, Sladek J (1998a) Some computational aspects associated with singular kernels. *Singular Integrals in Boundary Element Methods.* Eds. V. Sladek and J. Sladek. CMP, Southampton
1411. Sladek V, Sladek J (1998b) Singular integrals and boundary elements. *Computer Methods in Applied Mechanics and Engineering* **157**:251–266
1412. Sloan CS, Cowell MD, Lehnhoff TF (1999) The effect of a large hole on the stress concentration factor of a satellite hole in a tension field. *J Pressure Vessel Piping,* **121**:252–256
1413. Smith JC (1974) Correction and extension of van der Poel-s method for calculating the shear modulus of a particulate composite. *J Res Natl Bur Stand Sect,* **A78**:355–361
1414. Smith JC (1976) Experimental values for the elastic constants of a particulate-filled glassy polymer. *J Res Nat Bur Stand US,* **80A**:45–49
1415. Smith WA (1989) The role of piezocomposites in ultrasonic transducers. *Proceedings of the IEEE 1989 Ultrasonic Symposium,* 755–766
1416. Smith P, Torquato S (1988) Computer simulation results for the two-point probability functions of composite media. *J Comput Phys,* **76**:176–191
1417. Smyrlis Y-S (2009) Mathematical foundation of the MFS for certain elliptic systems in linear elasticity. *Numer. Math.*, **112**:319–340
1418. Smyshlyaev VP, Cherednichenko KD (2000) A rigorous derivation of strain gradient effects in the overall behavior of periodic heterogeneous media. *J Mech Phys Solids,* **48**:1325–1357
1419. Smyshlyaev VP, Fleck NA (1994) Bounds and estimates for linear composites with strain gradient effects. *J Mech Phys Solids,* **42** 1851–1882
1420. Smyshlyaev VP, Fleck NA (1995) Bounds and estimates for the overall plastic behaviour of composites with strain gradient effects. *Proc. R. Soc. Lond, A*, **451**:795–810
1421. Smyshlyaev VP, Fleck NA (1996) The role of strain gradients in the grain size effects for polycrystals. *J Mech Phys Solids,* **44**:465–495
1422. Sobczuk K, Spencer BF (1991) *Random Fatigue: From Data to Theory.* Academic Press, New York

1423. Sobczyk K, Trebicki J, Movchan AB (2007) Characterization of random microstructural stresses and fracture estimation *European Journal of Mechanics-A/Solids*, **26**:573–591
1424. Sobolev GA, Demin VN (1980) *Mechanoelectric Phenomena in the Earth.* Nauka, Moscow (In Russian)
1425. Sokolkin YV, Tashkinov AA (1984) *Deformation and Fracture Mechanics of Structurally Inhomogeneous Bodies.* Nauka, Moscow (In Russian)
1426. Sokolnikov IS (1983) *Mathematical Theory of Elasticity* (2nd edn). Robert E. Krieger, Melbourne
1427. Solyaev Y, Lurie S, Ustenko A (2020) On the relations between direct and energy based homogenization approaches in second gradient elasticity. In Abali BE, Giorgio I (Eds.) *Developments and Novel Approaches in Biomechanics and Metamaterials*, Springer Nature, Switzerland AG, 443–457
1428. Somigliana C (1886) Sopra l'equilibrio di un corpo elastico isotropo. *Il Nuovo Cimento*, **19**:84–90
1429. Song D, Ponte Castañeda P (2018) A multi-scale homogenization model for one-grained porous viscoplastic polycrystals I Finite-strain theory. *J. Mech. Physics of Solids*, **115**:102–122
1430. Sotiropolous DA, Achenbach JD, Zhu H (1987) An inverse scattering method to characterize inhomogeneities in elastic solids. *J Appl Phys*, **62**:2771–2777
1431. Spenser AJM (1980) *Continuum Mechanics.* John Wiley & Sons, New York
1432. Spitalsky Z, Tasis D, Papagelis K, Galiotis C (2010) Carbon nanotube–polymer composites: chemistry, processing, mechanical and electrical properties *Progress in polymer science*, **35**:357–401
1433. Spitzig WA, Kelly JF, Richmond O (1985) Quantitative characterization of second-phase populations. *Metallography*, **18**:235–261
1434. Spowart JE, Maruyama B, Miracle DB (2001) Multi-scale characterization of spatially heterogeneous systems: implications for discontinuously reinforced metal-matrix composite microstructures. *Mater Sci Engng*, **A307**:51–66
1435. Srivastava D, Wei C, Cho K (2003) Nanomechanics of carbon nanotubes and composites. *Appl Mech Rev*, **56**:215–230
1436. Stagfen ESG (1988) A nonlocal theory for the heat transport in composites containing highly conducting fibrous inclusions. *Phys Fluids*, **31**:2405–2425
1437. Stafford DS, Jackson TL (2010) Using level sets for creating virtual random packs of non-spherical convex shapes. *J. Comput. Physics*. **229**:3295–3315
1438. Stang H (1986) Strength of composite materials with small cracks in the matrix. *Int J Solids Struct*, **22**:1259–1277
1439. Stein E, Zhang G (1992) Theoretical and numerical shakedown analysis for kinematic hardening materials. In: Owen DRJ, Oñate E, Hinton E (eds), *Computational Plasticity. Fundamental and Applications.* Pineridge Press, Swansea 1:1–25
1440. Steigmann DJ, Ogden RW (1999) Elastic surface-substrate interactions. *Proc. R. Soc. Lond.*, **A455**, 437–474
1441. Steinbach O (2000) On a hybrid boundary element method. *Numer. Math.* **84**:679–695
1442. Steinbach O (2008) *Numerical Approximation Methods for Elliptic Boundary Value Problems. Finite and Boundary Elements.* Springer, NY
1443. Steinhauser OM, Hiermaier S (2009) A review of computational methods in materials science examples from shock-wave and polymer physics. *Int. J. Mol. Sci.*, **10**:5135–5216
1444. Stell G (1991) Statistical mechanics applied to random-media problem. In: Owen DRJ, Oñate E, Hinton E (eds) *Mathematics of Random Media.* Lectures in Applied Mathematics **27**:109–127
1445. Stell G, Rirvold PA (1987) Polydispersity in fluid, dispersions, and composites: some theoretical results. *Chem Engng Commun*, **51**:233–260
1446. Stephan C, Nguen TP, Chapelle ML, Lefrant S (2000) Characterization of single-walled carbon nanotubes-PMMA composites. *Synth Methods*, **108**:139–149

1447. Sternberg E, Sadowsky MA (1952) On the axisymmetric problem of the elasticity for an infinite region containing two spherical inclusions. *J Appl Mech,* **19**:19–27
1448. Stigh U (2006) Continuum damage mechanics and the life-fraction rule. *J Appl Mech,* **73**:702–704
1449. Stickel JJ, Powell RL (2005) Fluid mechanics and rheology of dense suspensions *Annu. Rev. Fluid Mech.,* **37**:129–149
1450. Stoy RD (1989) Solution procedure for the Laplace equation in bispherical coordinates for two sphere in uniform external field: parallel orientation. *J Appl Phys,* **65**:2611–2615
1451. Stoyan D (1998) Random sets: models and statistics. *Int Statistical Rev,* **66**:1-27
1452. Stoyan D (2000) Basic ideas of spatial statistics. In: Mecke KR, Stoyan D (eds) *Statistical Physics and Spatial Statistics: The Art of Analyzing and Modeling Saptial Structures and Pattern Formation.* Lecture Notes in Physics, Springer, Berlin, 554
1453. Stoyan D, Kendall WS, Mecke J (1995) *Stochastic Geometry and Its Applications.* John Wiley & Sons, Chichester
1454. Stoyan D, Stoyan H (1994) *Fractals, Random Shapes and Point Fields. Methods of Geometric Statistics.* J Wiley & Sons, Chichester
1455. Straley JP (1981) Thermoelectric properties of inhomogeneous materials. *J Phys D: Appl Phys,* **14**:2101–2105
1456. Stratonovich RL (1963) *Topics in the Theory of Random Noise.* Gordon and Breach, New York
1457. Stratton JA (1941) *Electromagnetic Theory.* McGraw-Hill, New York
1458. Stroeven M, Askes H, Sluys LJ (2004) Numerical determination of representative volumes for granular materials. *Comput. Methods Appl. Mech. Engrg.* **193**:3221–3238
1459. Strouboulis T, Copps K, Babuska I (2001) The generalized finite element method. *Computer Meth Appl Mech Engng,* **190**:4081–4193
1460. Stroud D (1998) The effective medium approximation: some recent development. *Superlatt Microstruct,* **23**:567–573
1461. Stroud D, Hui PM (1988) Nonlinear succeptibilities of granular materials. *Phys Rev,* **B37**:8719–8724
1462. Stump DM, Budiansky B (1989) Crack-growth resistance in transformation-toughened ceramics. *Int J Solids Struct,* **25**:635-646
1463. Sukumar N, Chopp DL, Moes N, Belytschko T (2001) Modeling holes and inclusions by level sets in the extended finite-element method *Computer Meth Appl Mech Engng,* **190**:6183–6200
1464. Sun CT, Zhou SG (1988) Failure of quasi-isotropic composite laminates with free edges. *J Reinf Plast Comp,* **7**:515
1465. Sun L, Gibson RF, Gordaninejad F, Suhr J (2009) Energy absorption capability of nanocomposites a review. *Composites Science Technology,* **69**:2392–2409
1466. Sun YF, Peng YZ (2003) Analytic solutions for the problems of an inclusion of arbitrary shape embedded in a half-plane. *Appl Math Comput,* **140**:105–113
1467. Suquet P-M (1985) Local and global aspects in the mathematical theory of plasticity. In Sawczuk, A, *et al.* (Eds.), *Plasticity Today.* Elsevier, London, New York, 279–309
1468. Suquet P (1993) Overall potentials and extremal surfaces of power law or ideally plastic composites. *J Mech Phys Solids,* **41**:981–1002
1469. Suquet P (1995) Overall properties of nonlinear composites: A modified secant moduli theory and its link with Ponte Castañeda's nonlinear variational procedure. *C R Acad Sci Paris, Série IIb,* **320**:563–571
1470. Suquet P, Ponte Castañeda P (1993) Small-contrast perturbation expansions for the effective properties of nonlinear composites. *C. R. Acad. Sci. Paris II,* **317**:1515–522
1471. Suresh S, Mortensen A (1998) *Fundamentals of Functionally Graded Materials : Processing and Thermomechanical Behaviour of Graded Metals and Metal-Ceramic Composites.* IOM Communications, London
1472. Suresh S, Mortensen A, Needlman A (1993) *Fundamentals of Metal-Matrix Composites.* Butterworth-Heinemann, Boston

1473. Sutton GP, Biblarz O (2003) *Rocket Propulsion Elements*. John Wiley & Sons, NY
1474. Sutton MA, Orteu JJ, Schreier HW (2009) *Image Correlation for Shape, Motion and Deformation Measurements*. NY, Springer
1475. Taggart DG, Bassani JL (1991) Elastic–plastic behavior of particle reinforced composites—influence of residual stresses. *Mechan of Mater* **12**:63–80
1476. Takano N, Okuno Y (2004) Three-scale finite element analysis of heterogeneous media by asymptotic homogenisation and mesh superposition method. *Int J Solids Struct,* **41**:4121–4135
1477. Takao Y, Taya M (1985) Thermal expansion coefficients and thermal stresses in an aligned short fiber composite with application to a short carbon fiber/aluminum. *J Appl Mech,* **52**:806–810
1478. Talbot DR, Willis JR (1982a) Variational estimates for dispersion and attenuation of waves in random composites. I. General theory. *Int I Solids Struct,* **18**:673–683
1479. Talbot DRS, Willis JR (1982b) Variational estimates for dispersion and attenuation of waves in random composites. II. Isotropic composites. *Int I Solids Struct,* **18**:685–698
1480. Talbot DRS, Willis JR (1985) Variational principles for nonlinear inhomogeneous media. *IMA J Appl Math,* **35**:39–54
1481. Talbot DRS, Willis JR (1992) Some explicit bounds for the overall behaviour of nonlinear composites. *Int J Solids Struct,* **29**:1981–1987
1482. Talpaert YR (2002) *Tensor Analysis and Continuum Mechanics*. Kluwer, Dordrecht
1483. Tan H, Huang Y, Liu C, Inglis HM, Ravichandran G, Geubelle PH (2007) The uniaxial tension of particle-reinforced composite materials with nonlinear interface debonding. *Int. J. Solids Struct.* **44**:1809–1822
1484. Tan H, Huang Y, Liu C, Ravichandran G, Paulino GH (2007) Constitutive behaviors of composites with interface debonding the extended Mori–Tanaka method for uniaxial tension. *Int. J. Fract.*, **146**:139–148
1485. Tanaka M, Sladek V, Sladek J (1994) Regularization techniques applied to boundary element methods. *Applied Mechanics Reviews* **47**, 457–499
1486. Tanimoto Y, Nishiwaki T, Nemoto K, Ben G (2004) Effect of filler content on bending properties of dental composites Numerical simulation with the use of the finite-element method. *J. Biomedical Materials Research,* **71B**:188-195
1487. Tandon GP, Kim RY, Bechel VT (2004) Construction of the fiber-matrix interfacial failure in a polymer matrix composites. *Int J Multiscale Comput Engng,* **2**:101-114
1488. Tandon GP, Kim RY, Rice BP (2002) Influence of vapor-grown carbon nanocomposites on thermomechanical properties of graphite-epoxy composites. *Proc. American Society for Composites 17th Technical Conference.* Purdue University, West Lafayette, IN, Paper 2039
1489. Tandon GP, Weng GJ (1984) The effect of aspect ratio of inclusions on the elastic properties of unidirectionally aligned composites. *Polym Compos,* **5**:327–333
1490. Tandon GP, Weng GJ (1986) Average stress in the matrix and effective moduli of randomly oriented composites. *Compos Sci Technol,* **27**:111–132
1491. Tandon GP, Weng GJ (1988) A theory of particle–reinforced plasticity. *J Appl Mech,* **55**:126–135
1492. Tanemura M (1979) On random complete packing by discs. *Ann Inst Statist Math,* **31**:351–365
1493. Tawerghi E, Yi Y-B (2009) A computational study on the effective properties of heterogeneous random media containing particulate inclusions. *J. Physics. D. Applied Physics,* **42**:175409
1494. Taya M (1990) Some thoughts on inhomogeneous distribution of fillers in composites. In: Weng GJ, Taya M, Abe H (eds), *Micromechanics and Inhomogeneity, The Toshio Mura 65th Anniversary Volume.* Springer-Verlag, New York, 433–447
1495. Taya M, Arsenault RJ (1989) *Metal Matrix Composites. Thermomechanical Behavior.* Pergamon Press, Oxford
1496. Taya M, Hayashi S, Kobayashi AS, Yoon HS (1990) Toughening of a particulate-reinforced ceramic-matrix composite by thermal residual stress. *J Amer Ceram Soc,* **73**:1382-1391

1497. Telles JCF, Brebbia CA (1991) Boundary element solution for half-plane problems. *Int J Solids Struct,* **17**:1149–1158
1498. Temizer I, Wu T, Wriggers P (2013) On the optimality of the window method in computational homogenization. *Int. J. Eng. Sci.,* **64**:66–73
1499. Tenek LT, Aifantis EC (2001) On some applications of gradient elasticity to composite materials *Composite Structures*, **53**:189–197
1500. Teodosiu C (1982) *Elastic Models of Crystal Defects.* Academei, Bucuresty, Springer, Berlin
1501. Terada K, Hori M, Kyoya T, Kikuchi N (2000) Simulation of the multi-scale convergence in computational homogenization approaches. *Int. J. Solids Struct.*, **37**:2285–2311
1502. Terada K, Kikuchi N (2001) A class of general algorithms for multi-scale analyses of heterogeneous media, *Comput. Methods Appl. Mech. Eng.* **190**:5247–5464
1503. Terrones M (2003) Science technology of the twenty-first century: synthesis, properties, and applications of carbon nanotubes. *Annu Rev Mater Res*, **33**:419–501
1504. Tewari A, Gokhale AM, Spowart JE, Miracle DB (2004) Quantitative characterization of spatial clustering in three-dimensional microstructures using two-point correlation functions. *Acta Mater,* **52**:307–319
1505. Theocaris PS (1987) *The Concept of Mesophase in Composites.* Springer-Verlag, Berlin
1506. Theocaris PS (1991) The elliptic paraboloidal failure criterion for cellular solids and brittle forms. *Acta Mechan,* **89**:93–121
1507. Tikhonov AN, Arsenin VY (1986) *Methods for Solving Ill-Posed Problems,* Moscow Nauka
1508. Thomson R, Zhou SJ, Carlsson AE, Tewary VK (1992) Lattice imperfections studied by use of lattice Green's functions. *Phys Rev,* **B17**:10613-10622
1509. Thorpe MF (1992) The conductivity of a sheet containing a few polygonal holes and/or superconducting inclusions *Proc Roy Soc Lond,* **A437**:215–227
1510. Thostenson ET, Li CY, Chou TW (2005) Nanocomposites in context. *Compos Sci Technol,* **65**:491–516
1511. Thostenson ET, Ren ZF, Chou T-W (2001) Advances in the science and technology of carbon nanotubes and their composites: a review. *Compos Sci and Technol,* **61**:1899–1912
1512. Thovert JE, Acrivos A (1989) The effective thermal conductivity of a random polydispersed suspension of spheres to order c^2. *Chem Engng Commun,* **82**:177–191
1513. Throop GJ, Bearman RJ (1965) Numerical solution of the Percus-Yervick equation for the hard-sphere potential. *J Chem Phys,* **42**:2408–2411
1514. Tibbetts GG, McHugh JJ (1999) Mechanical properties of vapor-grown carbon fiber composites with thermoplastic matrices. *J Mater Res,* **14**:2871–2880
1515. Ting TCT (1996) *Anisotropic Elasticity. Theory and Applications.* Oxford University Press, New York, Oxford
1516. Tjong SC (2006) Structural and mechanical properties of polymer nanocomposites. *Mater Sci Engng,* **R53**:73–197
1517. Tobochnik J, Chapin PM (1988) Monte Carlo simulation of hard spheres near random closest packing using sphrical boundary conditions. *J Chem Phys,* **88**:5824–5830
1518. Toebes ML, van Heeswijk JMP, Bitter JH, van Dillen AJ, de Jong KP (2004) The influence of oxidation on the texture and the number of oxygen-containing surface groups of carbon nanofibers. *Carbon,* **42**:307–315
1519. Torquato S (1980) *Microscopic Approach to Transport in Two-Phase Random Media.* PhD Thesis, State University of New York at Stony Brook
1520. Torquato S (1991) Random heterogeneous media: microstructure and improved bounds on effective properties. *Appl Mech Rev,* **44**(2):37–75
1521. Torquato S (1997) Effective stiffness tensor of composite media – I. Exact series expansion. *J Mech Phys Solids,* **45**:1421–1448
1522. Torquato S (1998) Effective stiffness tensor of composite media: II. Application to isotropic dispersions. *J Mech Phys Solids,* **45**:1421–1448
1523. Torquato S (2002a) *Random Heterogeneous Materials: Microstucture and Macroscopic Properties.* Springer-Verlag, New York, Berlin

1524. Torquato S (2002b) Statistical description of microstructures. *Annu Rev Mater Res*, **32**:77–111
1525. Torquato S, Lado F (1992) Improved bounds on the effective elastic moduli of random arrays of cylinders. *J Appl Mech*, **59**:1–6
1526. Torquato S, Rintoul MD (1995) Effect of the interface on the properties of composite media. *Phys Rev Lett*, **75**:4067–4070
1527. Torquato S, Stell G (1985) Microstructure of two-phase random media. *J Chem Phys*, **82**:980–987
1528. Torquato S, Truskett TM, Debenetti PG (2000) Is random close packing of spheres well defined? *Phys Rev Lett*, **84**:2064-2067
1529. Toselli A, Widlund O (2005) *Domain Decomposition Methods. Algorithms and Theory*. Berlin Springer.
1530. Tran VP (2016) *Modélisation à plusieurs échelles d'un milieu continu hétérogène aléatoire*. PhD Thesis. Université Paris-Est, Paris
1531. Tran VP, Brisard S, Guilleminot J, Sab, K (2018) Mori–Tanaka estimates of the effective elastic properties of stress-gradient composites *Int. J. Solids Structures*, **146**:55–68
1532. Treves F (1980) *Introduction to Pseudodifferential and Fourier Integral Operators*. Plenum Press, New York, **1**
1533. Trias D, Costa J, Mayugo JA, J.E. Hurtado JE (2006) Random models versus periodic models for fibre reinforced composites. *Comput Mater Sci*, **38**:316–324
1534. Trovalusci P (2014) Molecular approaches for multifield continua origins and current developments. In: T. Sadowsky and P. Trovalusci, editors, *Multiscale Modeling of Complex Materials phenomenological, theoretical and computational aspects*, **556** CISMCourses and Lectures, 211–278. Springer, Berlin
1535. Truesdell CA (1991) *A First Course in Rational Continuum Mechanics*. (2nd edn). Academic Press, New York, **1**
1536. Tsai JH, Patra A, Wetherhold R (2005) Finite element simulation of shaped ductile fiber pullout using a mixed cohesive zone/friction interface model *Composites*, **A36**:827–838
1537. Tsai JH (2009) The method of fundamental solutions with dual reciprocity for three-dimensional thermoelasticity under arbitrary body forces. *Engineering Computations International Journal for Computer-Aided Engineering and Software*, **26**:229-244
1538. Tsai SW, Wu EM (1971) A general theory of strength for anisotropic materials. *J Compos Mater*, **5**:58–80
1539. Tsynkov SV (1998) Numerical solution of problems on unbounded domains. A review *Applied Numerical Mathematics*, **27**:465–532
1540. Tszeng TC (1998) The effect of particle clustering on the mechanical behavior of particle reinforced composites. *Composites* **29B**:299–308
1541. Turnbull D, Cormia RL (1960) A dynamic hard sphere model. *J Appl Phys*, **31**:674–678
1542. Turner PA, Signorelli JW, Bertinetti MA, Bolmaro RE (1999) Explicit method for calculating the effective properties and micromechanical stresses: an application to an alumina-SiC composites. *Philos Mag*, **79**:1379–1394
1543. Tupek MR, Rimoli JJ, Radovitzky R (2013) An approach for incorporating classical continuum damage models in state-based peridynamics. *Comput. Methods Appl. Mechanics and Enging*, *263*:20–26, 2013
1544. Tvergaard, V, 1990. Effect of fiber debonding in a whisker-reinforced metal. *Mater. Sci. Eng.*, **125**:203–213
1545. Tvergaard V, Hutchinson JW (1988) Microcracking in ceramics by thermal expansion or elastic anisotropy. *J Am Ceram Soc*, **71**:157-166
1546. Tvergaard V, Needleman A (1997) Nonlocal effects on localization in a void-sheet. *Int J Solids Struct*, **34**:2221–2238
1547. Twersky V (1978) Acoustic bulk parameters in distribution of pair correlated scatterers. *J Acoust Soc Am*, **64**:1710–1719
1548. Umetani S-I (1988) *Adaptive Boundary Element Methods in Elastostatics*. Computational Mech Publ, Southampton, Boston

1549. Usuki A, Kojima Y, Kawasumi M, Okada A, Fukushima Y, Kurauchi T et al (1993) Synthesis of nylon 6-clay hybrid. *J Mater Res,* **8**:1179–1184
1550. Vaia RA, Giannelis EP (2001) Polymer nanocomposites: status and opportunities. *MRS Bull,* 394–401
1551. Vaia RA, Maguire JF (2007) Polymer Nanocomposites with Prescribed Morphology: Going beyond Nanoparticle-Filled Polymers *Chem. Mater.*, **19**:2736–2751
1552. Valter K, Kurtasov SF, Nikitin AN, Torina EG (1993a) Modeling of deformation textures in high temperature quartz. *Izv Acad Sci Russ Phys Solid Earth,* (6):25–48 (In Russian)
1553. Valter K, Nikitin AN, Shermergor TD, Yakovlev VD (1993b) Determination of effective electroelastic constants of polycrystalline rocks. *Izv Acad Sci Russ Phys Solid Earth* (6):83-88 (In Russian)
1554. Vanin GA (1996) Plane strain gradient theory of multilevel media. *Mekh Tverdogo Tela,* (3):5–15 (In Russian. Engl Transl. *Mech Solids,* **31**(3):2–11)
1555. Varadan VV, Kim K (2012) Fabrication of 3-D metamaterials using LTCC techniques for high-frequency applications *IEEE Trans. Components, Packaging Manufac. Technol.*, **2**:410–417
1556. Varadan VK, Ma Y, Varadan VV (1985) A multiple scattering theory for elastic wave propagation in discrete random media. *J Acoust Soc Amer,* **77**:375–385
1557. Varadan VK, Varadan VV (Editors) (1980) *Acoustic Electomagnetic and Elastic Wave-Scattering Focus on the T-Matrix Approach.* Pergamon, New York
1558. Varga RS (2000) *Matrix Iterative Analysis.* Springer, Berlin
1559. Vasiliev VV (2017) *Mechanics of Composites Structures.* Taylor & Francis, Washington
1560. Venk GP (1993) Development of deformation textures in rocks. *Izv Acad Sci Russ Phys Solid Earth,* (6):5-36 (In Russian)
1561. Verlet L, Weis JJ (1972) Perturbation theory for the thermodynamic properties of simple liquids. *Mol Phys,* **24**:1013–1024
1562. Verruijt A (1997) A complex variable solution for a deforming circular tunnel in an elastic half-plane. *Int J Numer Analyt Methods Geomechan,* **21**:77–89
1563. Verruijt A (1998) Deformations of an elastic half plane with a circular cavity. *Int J Solid Struct,* **35**:2795–2804
1564. Vigolo B, Penicaud AP, Couloun C, Sauder S, Pailler R, Journet C, Bernien P, Poilin P (2000) Macroscopic fibers and ribbons of oriented carbon nanotubes. *Science,* **290**:1331–1334
1565. Vinson JR, Sierakowski RL (2003) *The Behavior of Structures Composed of Composites Materials.* Kluwer, Dordrecht, Boston
1566. Voight W (1889) Uber die Beziehung zwischen den beiden Elastizitatskon-stanten isotroper Korper. *Wied Ann,* **38**:573–587
1567. Voight W (1910) *Lehrbuch der Kristallphysik.* B.G. Teubner, Leipzig und Berlin
1568. Voyiadjis GZ (Ed.) (2019) *Handbook of Nonlocal Continuum Mechanics for Materials and Structures.* Springer, NY
1569. Voyiadjis GZ, Kattan PI (2012) *Advances in Damage Mechanics: Metals and Metal Matrix Composites.* Elsevier Science, Oxford, UK
1570. Voyiadjis GZ, Song CR (2006) *The Coupled Theory of Mixtures in Geomechanics with Applications.* Springer-Verlag, Berlin Heidelberg
1571. Voyiadjis GZ, Song CR (2020) *Gradient-enhanced Continuum Plasticity.* Elsevier, Amsterdam
1572. Voyiadjis GZ, Yaghoobi M (2019) *Size Effects in Plasticity: From Macro to Nano.* Academic Press, London
1573. Wakashima K, Tsukamoto H (1991) Mean-field micromechanics model and its application to the analysis of thermomechanical behavior of composite materials. *Mater Sci Engng,* **A32**:883–892
1574. Walker KP, Jordan EH, Freed AD (1990) Equivalence of Green's function and the Fourier series representation of composites with periodic microstructure. In: Weng GJ, Taya M, Abe H (eds) *Micromechanics and Inhomogeneity.* Springer-Verlag, New York, 535–558

1575. Walpole LJ (1966a) On the bounds for the overall elastic moduli of inhomogeneous system. I. *J Mech Phys Solids,* **14**:151–162
1576. Walpole LJ (1966b) On the bounds for the overall elastic moduli of inhomogeneous system. II. *J Mech Phys Solids,* **14**:289–301
1577. Walpole LJ (1969) On the overall elastic moduli of composite materials. *J Mech Phys Solids,* **17**:235–251
1578. Walpole LJ (1978) A coated inclusion in an elastic medium. *Math Proc Camb,* **83**:495–506
1579. Walpole LJ (1981) Elastic behavior of composite materials: theoretical foundations. *Adv Appl Mech,* **21**:169–242
1580. Walpole LJ (1997) An inclusion in one of two joined isotropic elastic half-spaces. *IMA J Appl Math,* **59**:193–209
1581. Walsh JB (1965) The effect of cracks on the compressibility of rocks. *J Geophys Res,* **70**:381–389
1582. Wang B (1994) Effective behaviour of piezoelectric composites. In: Ostoja-Starzevski M, Jasiuk I (eds), *Micromechanics of Random Media. Appl Mech Rev,* **47**:112–121
1583. Wang B, Liu Y (1990) The average field in piezoelectric media with randomly distributed inclusions. In: Hsieh KTN (ed), *Mechanical Modeling of New electromagnetic Materials.* Elsevier, Amsterdam, 313–318
1584. Wang H, Yao Z (2005) A new fast boundary element method for large scale analysis of mechanical propertias in 3D particle-reinforced composites. *Computer Modeling in Engineering & Sciences,* **4**:85-95
1585. Wang J, Huang Z, Duan H, Yu S, Feng X, Wang G, Zhang W, Wang T (2011) Surface stress effect of nanostructured materials. *Acta Mechanica Solida Sinica,* **24**:53–82
1586. Wang J, Pyrz R (2004a) Prediction of the overall moduli of layered silicate-reinforced nanocomposites–part I: basic theory and formulas. *Composites Sci Technol,* **64**:925–934
1587. Wang J, Pyrz R (2004b) Prediction of the overall moduli of layered silicate-reinforced nanocomposites–part II: analyses. *Compos Sci Technol,* **64**:935–944
1588. Wang L, Xu J, Wang J (2017) Static and dynamic Green's functions in peridynamics *Journal of Elasticity,* **126**:95–125
1589. Wang W, Jasiuk I (1998) Effective elastic constants of particulate composites with inhomogeneous interphases. *J. Comp. Mater.,* **32**:1391–424
1590. Wang X, Zhou K (2014) An inclusion of arbitrary shape in an infinite or semi-infinite isotropic multilayered plate *Int. J. Applied Mechanics,* **6**:1450001 (21 pages)
1591. Wang Z, Yu H, Wang Q (2016) Analytical solutions for elastic fields caused by eigenstrains in two joined and perfectly bonded half-spaces and related problems *Int. J. Plasticity,* **76**:128
1592. Wang Z, Zhu J, Jin XY, Chen WQ, Zhan C (2014) Effective moduli of ellipsoidal particle reinforced piezoelectric composites with imperfect interfaces *J. Mechanics Physics Solids,* **65**: 138–156
1593. Wani SN, Sangani AS, Sureshkumar R (2012) Effective permittivity of dense random particulate plasmonic composites *J. Opt. Soc. Am.,* **B 29**:1443–1455.
1594. Warren WE, Kraynik AM (1997) Linear elastic behavior of a low-density Kelvin foam with open cells. *J Appl Mech,* **64**:787-794
1595. Watt JP (1976) The elastic properties of composite materials. *Rev Geophys Res,* **14**:541–563
1596. Watt JP (1979) Hashin-Shtrikman bounds of the effective elastic moduli of polycrystals with orthorhombic symmetry. *J Appl Phys,* **50**:6290–6295
1597. Watt JP (1980) Hashin-Shtrikman bounds of the effective elastic moduli of polycristals with monoclinic symmetry. *J Appl Phys,* **51**:1520–1524
1598. Watt JP, Peselnic L (1980) Clarification of the Hashin-Shtrikman bounds of the effective elastic moduli of polycristals with hexagonal, trigonal, and tetragonal symmetries. *J Appl Phys,* **51**:1525–1531
1599. Webb MD, Davis IL (2006) Random particle packing with large particle size variations using reduced-dimension algorithms. *Powder technology*, vol. 167, 10-19
1600. Weber L, Fischer C, Mortensen A (2003) On the influence of the shape of randomly oriented, non-conducting inclusions in a conducting matrix on the effective electrical conductivity. *Acta Mater,* **51**:495–505

1601. Weckner O, Abeyaratne R (2005) The effect of long-range forces on the dynamics of a bar. *J. Mech. Physics of Solids* **53**:705–728
1602. Weckner O, Brunk G, Epton MA, Silling SA, Askari E. (2009) Green's functions in non-local three-dimensional linear elasticity. *Proc. R. Soc.*, **A 465**:3463–3487
1603. Weckner O, Emmrich E (2005) Numerical simulation of the dynamics of a nonlocal, inhomogeneous, infinite bar. *J. Comp. Appl. Mech.*, **6**:311–319
1604. Weinbaum S, Ganatos P, Yan Z (1990) Numerical multipole and boundary integral equation techniques in Stokes flow *Annual Review of Fluid Mechanics*, **22**:275–316
1605. Weissenbek E, Böhm HJ, Rammerstorfer FG (1994) Micromechanical investigations of arrangement effects in particle reinforced metal matrix composites. *Comput Mater Sci*, **3**:263–278
1606. Weissenbek E, Pettermann HE, Suresh S (1997) Numerical simulation of plastic deformation in compositionally graded metal-ceramic structures. *Acta Mater,* **45**:3401–3417
1607. Weng GJ (1984) Some elastic properties of reinforced solids with special reference to isotropic ones containing spherical inclusions. *Int J Engng Sci*, **22**:845–856
1608. Weng GJ (1990) The theoretical connection between Mori–Tanaka's theory and the Hashin–Shtrikman–Walpole bounds. *Int J Engng Sci*, **28**:1111–1120
1609. Weng GJ (2003) Effective bulk moduli of two functionally graded composites. *Acta Mechanica*, **166**:57—67
1610. Weng GJ (2010) A dynamical theory for the Mori–Tanaka and Ponte Castañeda–Willis estimates. *Mech. Mater*, **42**:886–893
1611. Wenk HR, Van Houtte P (2004) Texture and anisotropy. *Rep Prog Phys*, **67**:1367–1428
1612. Werner E, Siegmund T, Fischer FD (1994) A computer study of the thermomechanical deformation behavior of a duplex steel. *Comput Mater Sci*, **3**:279–285
1613. Wernik JM, Meguid SA (2011) Multiscale modeling of the nonlinear response of nano-reinforced polymers *Acta Mechanica*, **217**:1–16
1614. Wertheim MS (1963) Exact solution of the Percus-Yevick integral equation for hard spheres. *Phys Rev Lett*, **10**:321–323
1615. Wetherhold RC, Lee FK (2001) Shaped ductile fibers to improve the toughness of epoxyma-trix composites. *Compos. Sci. Technol.*, **61**:517–553
1616. Wiener O (1912) Die theorie des mischkörpers fr̈ das feld des stationar̈en strm̈ung. Erste abhandlung die mttelswertsïze fr̈ kraft, polarisation und energie. *Abt Math-Physichen Klasse Königl Säcsh Gessel Wissen*, **36**(6):509–604
1617. Widom W (1966) Random sequential addition of hard spheres to a volume. *J Chem Phys*, **44**:3888–3894
1618. Wilkinson DS, Pompe W, Oeschener M (2001) Modeling the mechanical behavior of heterogeneous multi-phase materials. *Prog in Mater Sci*, **46**:379–405
1619. Willis JR (1965) The elastic interaction energy of dislocation loops in anisotropic media. *Q J Mech Appl Math*, **18**:419–433
1620. Willis JR (1977) Variational and related methods for the overall properties and self-consistent estimates for the overall properties. *J Mech Phys Solids*, **25**:185–202
1621. Willis JR (1978) Variational principles and bounds for the overall properties of composites. In: Provan JW (ed), *Continuum Models of Disordered Systems*. University of Waterloo Press, Waterloo 185–215
1622. Willis JR (1980a) A polarization approach to the scattering of elastic waves I. Scattering by a single inclusion. *J Mech Phys Solids*, **28**:287–305
1623. Willis JR (1980b) A polarization approach to the scattering of elastic waves. II: Multiple scattering from inclusions. *J Mech Phys Solids*, **28**:307–326
1624. Willis JR (1980c) Relationships between derivatives of the overall properties of composites by perturbation expansion and variational principles. In: Nemat-Nasser S (ed), *Variational Methods in Mechanics of Solids*. Pergamon Press, New York, 59–66
1625. Willis JR (1981a) Variational and related methods for the overall properties of composites. *Adv Appl Mech*. **21**:1–78

1626. Willis IR (1981b) Variational principles for dynamic problems for inhomogeneous elastic media. *Wave Motion,* **3**:1–11
1627. Willis JR (1982) Elasticity theory of composites. In: Hopkins HA, Sewell MI (eds), *Mechanics of Solids, The Rodney Hill 60th Anniversary Volume.* Pergamon Press, Oxford, 653–686
1628. Willis JR (1983) The overall elastic response of composite materials. *J Appl Mech,* **50**:1202–1209
1629. Willis JR (1985) The nonlocal influence of density variations in a composite. *Int J Solids Struct,* **21**:805–817
1630. Willis JR (1991) On methods for bounding the overall properties of nonlinear composites. *J. Mechanics Physics Solids,* **39**:73–86
1631. Willis JR (1997) Dynamics of composites. In *Continuum Micromechanics,* CISM Lecture Notes. Springer, New York, NY, 265–290
1632. Willis JR, Acton JR (1976) The overall elastic moduli of a dilute suspension of spheres. *Q J Mechan Appl Math,* **29**:163–177
1633. Willis JR, Talbot DRS (1990) Variational methods in the theory of random composite materials. In: Maugin GA (ed) *Continuum Models and Discrete systems.* Longman Scientific & Technical, Harlow, UK, **1**:113–131
1634. Wilson RB, Cruse TA (1978) Efficient implementation of anisotropic three dimensional boundary-integral equation stress analysis. *Int J Numer Meth Engng,* **12**:1283–1397
1635. Wilson WK, Yu LW (1979) The use of the J-integral in thermal stress crack problems. *Int J Fracture,* **15**:377–387
1636. Wissler M, Lusti HR, Oberson C, Widmann-Schupak AH, Zappini G, Gusev AA (2003) Non-additive effects in the elastic behavior of dental composites. *Advanced Engineering Materials,* **5**:113–116
1637. Withers PJ (1989) The determination of the elastic field of an ellipsoidal inclusion in a transversally isotropic medium, and its relevance to composite materials. *Philos Magazine,* **A59**:750–781
1638. Wong E, Seehan PE, Lieber CM (1997) Nanobeam mechanics elasticity, strength, and toughness of nanorods and nanotubes. *Science,* **277**:1971–1975
1639. Wong SC, Mai YW (2003) Performance synergism in polymer-based hybrid materials. In: Shonaike GO, Advani SG (eds), *Adv Polymeric Materials. Structure Property Relationships.* CRC Press, Boca Raton, FL, 439–477.
1640. Wriggers P, Moftah SO (2006) Mesoscale models for concrete Homogenisation and damage behavior. *Finite Elements in Analysis and Design,* **42**:623–636
1641. Wu RS, Aki K (1985) Elastic wave scattering by a random medium and small-scale inhomogeneities in the lithosphere. *J Geoph Res,* **B90**:10261–10273
1642. Wu TT (1966). The effect of inclusion shape on the elastic moduli of a two-phase material. *Int. J. Solids Structures,* **2**:1–8
1643. Xia M, Hamada H, Maekawa Z (1995) Flexural stiffness of injection molded glass fiber reinforced thermoplastics. *Int Polym Process,* **10**:74-81
1644. Xia W, Oterkus E, Oterkus S (2020) Peridynamic modelling of periodic microstructured materials *Procedia Structural Integrity,* **28**, 820–828
1645. Xie XL, Mai YW, Zhou XP (2005) Dispersion and alignment of carbon nanotubes in polymer matrix: a review *Materials science engineering: R: Reports,* **49**:89–112
1646. Xu F, Aravas N, Sofronis P (2008) Constitutive modeling of solid propellant materials with evolving microstructural damage. *J. Mechanics Physics Solids,* **56**:2050–2073
1647. Xu TT, Fisher FT, Brinson LC, Ruoff RS (2003) Bone-shape nanomaterials for nanocomposite application. *Nano Lett,* **3**:1135–1139
1648. Xu X, D'Elia M, Foster JT (2021) A machine-learning framework for peridynamic material models with physical constraints *arXiv preprint arXiv:2101.01095*
1649. Xu XF (2009) Generalized variational principles for uncertainty quantification of boundary value problems of random heterogeneous materials. *J. Eng. Mech.,* **135**:1180-1188

1650. Xun F, Hu G, Huang Z (2004) Effective in plane moduli of composites with a micropolar matrix and coated fibers *Int. J. Solids Structures*, **41**:247–265
1651. Yanase K, Ju JW (2012) Effective elastic moduli of spherical particle reinforced composites containing imperfect interfaces. *Int. J. Damage Mech*, **21**:97–127
1652. Yang B, Tewary VK (2005) Green's function-based multiscale modeling of defects in a semi-infinite silicon substrate. *Int J Solids Struct*, **42**:4722–4737
1653. Yang F, Chong A, Lam D, Tong P (2002) Couple stress based strain gradient theory for elasticity. *Int. J. Solids Struct.*, **39**:2731–2743
1654. Yang Q, Cox B (2003) Spatially averaged local strains in textile composites via the binary model formulation. *J Engng Mater Technol*, **125**:418–425
1655. Yang S, Gokhale AM, Shan Z (2000) Utility of microstructure modeling for simulation of micro-mechanical response of composites containing non-uniformly distributed fibers. *Acta Mater*, **48**:2307–2322
1656. Yanovsky VE, Zgaevskii VE (2004) Mechanical properties of high elastic polymer matrix composites filled with rigid particles Nanoscale consideration of the interfacial problem *Composite Interfaces*, **11**:245–261
1657. Yao Z, Kong F, Wang H, Wang P (2004) 2D Simulation of composite materials using BEM. *Engng Anal Bound Elements***28**:927–935
1658. Yin HM, Paulino GH, Buttlar WG, Sun LZ (2005) Effective thermal conductivity of two-phase functionally graded particulate composites. *J Appl Phys*, **98**:063704
1659. Yin HM, Paulino GH, Buttlar WG, Sun LZ (2007) Micromechanics-based thermoelastic model for functionally graded particulate materials with particle interactions. *J Mech Phys Solids*, **55**:132–160
1660. You H, Yu Y, Silling S, D'Elia M (2020) Data-driven learning of nonlocal models: from high-fidelity simulations to constitutive laws. *arXiv:2012.04157*
1661. You H, Yu Y, Traskb N, Gulianb M, D'Elia M (2020) Data-driven learning of robust nonlocal physics from high- delity synthetic data. *arXiv:2005.10076*
1662. You LH, You XY, Zheng ZY (2006) Thermomechanical analysis of elastic–plastic fibrous composites comprising an inhomogeneous interphase *Computational Materials Science*, **36**:440–450.
1663. Young DL, Jane SJ, Fan CM, Murugesan K, Tsai CC (2006) The method of fundamental solutions for 2D and 3D Stokes flows. *J. Comput. Physics*, **211** 1–8
1664. Yoon BJ, Kim S (1987) Note on the direct calculation of mobility functions for two equal-size spheres in Stokes flow. *J Fluid Mech*, **185**:437–446
1665. Yu HY, Sanday SC, Rath BB, Chang CI (1995) Elastic field due to defects in transversely isotropic bimaterials. *Proc Roy Soc Lond*, **A449**:11–30
1666. Yu M-F, Yakobson BI, Ruo RS (2000) Controlled sliding and pullout of nested shells in individual multiwalled nanotubes. *J Phys Chem*, **B104**:8764–8767
1667. Zabihyan R, Mergheima J, Javili A, Steinmanna P (2018) Aspects of computational homogenization in magneto-mechanics Boundary conditions, RVE size and microstructure composition *Int. J. Solids Structures*, **130–131**:105–121
1668. Zaman AA, Moudgil BM (1998) Rheology of bidisperse aqueous silica suspensions A new scaling method for the bidisperse viscosity *J. Rheol.*, **42**:21–39
1669. Zaoui A (2002) Continuum micromechanics: Survey. *J Engng Mech*, **128**:808–816
1670. Zapparov KI, Perlin PI (1976) Numerical solution of plane elasticity theory problems for regions of complicated configuration. *Prikl Mekhan*, **12**(5):103–108 (in Russian)
1671. Zecevic M, Bennett KS, Luscher DJ, Lebensohn RA (2021) New self-consistent homogenization for thermo-elastic polycrystals with imperfect interfaces. *Mechanics Materials*, **155**, 103651
1672. Zeller R, Dederichs PH (1973) Elastic constants of polycrystals. *Phys. Stat. Sol.*, **a55**:831–842
1673. Zeman J (2003) *Analysis of Composite Materials with Random Microstructures*. PhD Thesis. Czech TU in Prague

1674. Zeman J, Šejnoha M (2001) Numerical evaluation of effective elastic properties of graphite fiber tow impregnated by polymer matrix. *J. Mech. Physics of Solids*, **49**:69–90
1675. Zeng QH, Yua AB, Lu GQ (2008) Multiscale modeling and simulation of polymer nanocomposites. *Prog. Polym. Sci.*, *33*:191–269
1676. Zeng X, Bergman DJ, Hui PM, Stroud D (1988) Effective medium theory for weakly nonlinear composites. *Phys Rev*, **B38**:10970–10973
1677. Zerda AS, Lesser AJ (2001) Intercalated clay nanocomposites: morphology, mechanics, and fracture behavior. *J Polym Sci Part B: Polym Phys*, **39**:1137–1146
1678. Zgaevsky VE (1977) Elastic and viscoelastic properties of polymers filled with solid particles. *Int. J. Polym. Mater.*, **6**:109–124
1679. Zhang H, Lib F, Wang X (2020) The effective elastic properties of chain-like structural particle-reinforced composites *Mechanics of Materials*, **144**:103344
1680. Zhang J, Tanaka M (2007) Systematic study of thermal properties of CNT composites by the fast multipole hybrid boundary node method. *Engng Anal Bound Elements* **31**:388–401
1681. Zhang P, Huang Y, Geubelle PH, Klein PA, Hwang KC (2002) The elastic modulus of single-wall carbon nanotubes: a continuum analysis incorporating interatomic potential. *Int J Solids Struct*, **39**:3893–3906
1682. Zhang X, Sharma P (2005) Inclusions and inhomogeneities in strain gradient elasticity with couple stresses and related problems. *Int. J. Solids Struct.* **42**:3833–3851
1683. Zhao YH, Weng GJ (2002) The effect of debonding angle on the reduction of effective moduli of particle and fiber-reinforced composites. *J Appl Mech*, **69**, 292–302
1684. Zheng Q-S, Du D-X (2001) An explicit and universally applicable estimate for the effective properties of multiphase composites which accounts for inclusion distribution. *J Mech Phys Solids*, **49**:2765–2788
1685. Zheng Q-S, Zhao Z-H (2004) Green's function and Eshelby's fields in couple-stress elasticity *Int. J. Multiscale Comput. Engng.* **2**:15–27
1686. Zhiging J, Tennysin RC (1989) Closure of cubic tensor polynomial failure surface. *J Comp Mech*, **23**:208–231
1687. Zhong XA, Knaus WG (2000) Effects of particle interaction and size variation on damage evolution in filled elastomers. *Mechanics Compos. Materials Structures*, **7**:35–53
1688. Zhong Z, Meguid SA (1997) On the elastic field of a spherical inhomogeneity with an imperfectly bonded interface. *J. Elast.*, **46**:91–113
1689. Zhou BL (1994) The biomimetic study of composite materials. *J Met*, **46**(2), 57–62
1690. Zhou K, Keer LM, Wang X, Ai X, Sawamiphakdi K, Glaws P, Paire M, Che F (2012) Interaction of multiple inhomogeneous inclusions beneath a surface *Comput. Methods Appl. Mech. Engineering*, **217–220**:25–33
1691. Zhou K, Hoh HJ, Wang X, Keer LM, Pang JHL, Song B (2013) A review of recent works on inclusions. *Mech. Mater*, **60**:144–158
1692. Zhou SA, Hsieh RK (1986) Statistical theory of elastic materials with micro-defects. *Int J Engng Sci*, **24**:1195–1206
1693. Zhou S, Li A, Wang B (2016) A reformulation of constitutive relations in the strain gradient elasticity theory for isotropic materials. *Int. J. Solids Struct.*, **80**:28–37
1694. Zhou Y, Li C, Mason JJ (2005) Shape optimization of randomly oriented short fibers for bone cement reinforcements. *Materials Science Engineering* **A393**:374–381
1695. Zhu LJ, Narh KA (2004) Numerical simulation of the tensile modulus of nanoclay-filled polymer composites. *J Polym Sci Part B: Polym Phys*, **42**:2391–2406
1696. Zhu YT, Beyerlein IJ (2002) Bone-shaped short fiber composites-an overview *Materials Science and Engineering*, **A326**:208–227
1697. Zhu ZG, Weng GJ (1989) Creep deformation of particle-strengthened metal-matrix composites. *J Engng Mater Technol*, **111**:99–105
1698. Zhu ZG, Weng GJ (1990) A local theory for the calculation of overall creep strain of particle–reinforced composites. *Int J Plast*, **6**:449–469
1699. Ziegler F (1992) Developments in structural dynamic viscoplasticity including ductile damage. *Z Angew Math Mech*, **72**(4):T5–T15

1700. Zienkiewicz OZ, Taylor RL (2005) *The Finite Element Method for Solid and Structural Mechanics*, Amsterdam, Elsevier
1701. Ziman JM (1979) *Models of Disorder*. Cambridge University Press, New York
1702. Zimmerman RW (1991) Elastic moduli of a solid containing spherical inclusions. *Mech Mater,* **12**:17-24
1703. Zimmerman RW (1996) Effective conductivity of a two-dimensional medium containing elliptical inhomogeneities *Proc Roy Soc Lond,* **A452**:1713–1727
1704. Zinchenko AZ (1994) Algorithm for random close packing of spheres with periodic boundary conditions. *J Comput Phys,* **114**: 298–307
1705. Zohdi T, Wriggers P (1999) A domain decomposition method for bodies with heterogeneous microstructure based on material regularization. *Int J Solids Struct,* **36**:2507–2525
1706. Zohdi TI, Wriggers P (2005) *Introduction to Computational Micromechanics.* Springer, Berlin
1707. Zou RP, Xu JQ, Feng CL, Yu AB, Johnston S, Standish N (2003) Packing of multi-sized mixtures of wet coarse spheres. *Powder Technol.*, **130**:77–83
1708. Zozulya VV, Gonzalez-Chi PI (1999) Weakly singular, singular and hypersingular integrals in 3-D elasticity and fracture mechanics *J. Chinese Institute of Engineering*, **22**:763-775
1709. Zuiker JR (1995) Functionally graded materials: choice of micromechanics model and limitations in property variation. *Compos Engng,* **5**:807–819
1710. Zuiker JR, Dvorak GJ (1994) The effective properties of functionally graded composites–I. Extension of the Mori–Tanaka method to linearly varying fields. *Compos Engng,* **4**:19–35
1711. Zuo QH, Addessio FL, Dienes JK, Lewis MW (2006) A rate-dependent damage model for brittle materials based on the dominant crack. *Int J Solids Struct,* **43**:3350–3380

Index

A

Admissibility condition in peridynamics, 728
Admissible fields
- condition, 728
- kinematically, 34, 220, 255, 621, 752, 753, 777, 783
- set, 776–778
- statically, 34, 220, 268, 621, 752, 753, 789, 856

Agglomerates, 4, 375, 384, 385, 387, 391
Aggregates, 4, 46, 263, 633
Anisotropic media, 38, 46, 62, 78, 100, 105, 114, 193, 247, 270, 355, 382, 527, 543, 653, 714
Approximation
- asymptotic, 520, 689
- Born, 106
- correlation, 713
- dilute, 350, 361, 364, 394, 529, 603, 770, 811, 844
- effective field, 295, 308, 327, 452
- far field, 461, 468
- general integral equation, 262, 265, 275, 294, 298, 308
- high order, 464
- initial, 146, 148, 149, 417, 453, 462, 577, 644, 695
- interacting inclusions, 115, 144, 149
- iteration method, 145, 155, 361, 406, 540, 576, 602
- local, 599
- long wave limit, 107, 579, 608
- method of fundamental solutions, 134
- point, 115, 150, 372, 598, 687
- quasi-crystalline (*see* Quasi-crystalline approximation)
- singular-approximation method, 719
- successive (*see* Iterations)
- T-matrix, 105, 722
- volume integral equation method, 142
- widely space inclusions, 150

Arbitrary shape inclusion, 72, 92, 600, 716
Average
- conditional, 186, 188, 297, 326, 412, 447, 527, 594
- constitutive equations, 223, 225, 494, 560, 581, 599, 617, 640, 756
- elastic properties, 219–223, 331, 409
- energy functions, 36, 233, 269, 614
- ensemble, 181, 187, 188, 220, 266, 378, 457, 581
- orientation of, 183, 197, 248, 340, 366, 383, 396
- polarization tensors, 92, 103, 106, 129, 255, 396, 485, 635, 662, 673, 719, 768, 773, 806, 822
- statistical, 11, 188, 229, 287, 320, 486, 590, 685, 815
- strains, 12, 332, 616, 759
- stresses, 236, 245, 319, 328, 524, 561, 580, 634, 756
- stress intensity tensor, 530
- volume, 167, 181, 191, 219, 302, 364, 596, 617, 681, 751, 763, 772, 855

B

Back-stresses, 55, 551
Basis tensors, 86–90, 908

V. A. Buryachenko, *Local and Nonlocal Micromechanics of Heterogeneous Materials*, https://doi.org/10.1007/978-3-030-81784-8

Body-force, 118, 289, 503, 635, 676, 801, 847
Boolean model, 172
Born approximation, 106
Boundary condition, 29, 680
  displacement, 33, 117, 216, 234, 615
  homogeneous, 33, 219, 242, 291
  ideal, 75, 105, 243, 273
  imperfect, 96–100, 244, 502, 689, 710
  periodic, 851
  traction, 33, 34, 220, 236, 280, 615
  volumetric, 761, 798
  volumetric Bloch-Floquet periodic, 888
Boundary integral equation method, 125–142, 1113
Boundary layer effect, 223, 228, 289, 593, 718
Bounds
  complementary energy, 35, 259, 548
  conductivity, 273
  elastic energy, 259, 754
  Hashin-Shtrikman, 342, 439, 715
  higher order, 355
  Hill, 260, 780
  Reuss, 261
  Voigt, 261
Bridging mechanism, 536
Bulk modulus, 32, 53, 247, 612, 740

C
Carbon fiber composites, 6
Carbon nanotubes, 7, 380
Cauchy
  data, 127, 281, 506
  elastic materials, 30, 666
  principle value, 65
  relation, 42, 220, 753
  stress tensor, 26, 651
Centering method, 277, 279, 293, 486, 657, 683, 789
Ceramic matrix composites, 5, 177, 265, 414
Cherry-pit potential, 179
Clustered composites
  definition of, 4
  of deterministic structure, 385, 386, 391
  Matern field, 4, 174
  Neyman-Scott field, 174
  radius, 175
  of random structure, 198, 228, 384, 599–604
  statistical description, 174–175, 211, 384
  strip model, 177
Coating inclusion, 91, 319, 559–560
Collective rearrangement model, 200, 207, 375
Comparison medium, 12, 255, 281, 306, 322, 328, 635
Compatibility condition, 22, 25, 160, 242, 615, 662, 853
Complex potential method, 50, 111
Compliance, 9, 38, 42, 47, 95, 223, 254, 354, 433, 474, 592, 755, 774, 780
Composite material
  classification of manufacturing, 8
  definition of, 1
  geometrical classication, 2
  mechanical properties classication, 5
Computational homogenization, 301, 796
Conduction
  depolarization tensor, 103
  effective properties, 713, 715–717
  ellipsoidal coated inclusion, 103, 104
  general integral equation, 718
  Green's function, 101
  symmetry, 101
Conservation laws
  angular momentum, 27, 728
  continuity, 28
  energy, 28
  linear momentum, 27, 728
Constitutive equation. *see* Nonlinear elasticity and; Nonlolal elasticity and; Peridynamics
  associated law, 32, 56
  effective, 230, 617
  hyperelasticity, 30
  at interface, 99
  J2-flow theory, 55, 551, 564
  linear thermoelasticity, 39, 43, 109
  power-law creep, 547
  secant modulus concept, 58, 539, 540, 542
  thermoelectroelasticity, 42, 265, 718
  yield vertex, 55, 522
Correlation hole. *see* Excluded volume
Coulomb form, 525
Coupling between elastic field and
  bulk and surface tensors, 99
  electric, 40–43, 109, 266–268
  electromagnetic, 723
  temperature, 36
Covariance matrix, 161, 529
Cracked medium, 340, 364, 742
Cracks
  energy release rate, 522, 711
  Eshelby solution, 95, 347
  fracture surface energy, 527
  interacting, 115, 151
  stress intensity factors, 530
Cubic packing

body-centered, 195, 399
face-centered, 195, 210, 564
Cubic symmetry, 48, 263, 265, 355, 366
Cyclic loading, 552
Cylinder, 86, 103, 383, 922

D

Damage, 365, 522, 535, 570, 726, 731, 741–744
See also Cracked medium
Data-driven learning in
elastoplasticity of composites, 796
nonlocal micromechanics, 599, 894
Decomposition of
boundary conditions, 228, 623, 865
concentration factor, 326, 500
constitutive equations, 229
elastic energy, 228, 233, 626, 788
elastic moduli, 52, 615
generalized fields, 620, 860, 882
material properties, 72, 116, 654, 673, 749, 783
operators, 154, 689, 698, 769, 776, 817, 861, 881
second derivatives of Green's functions, 63–67
strain fields, 77, 117, 229, 332, 550, 659, 761
stress field, 117, 229, 333, 656
tensors, 95, 98, 154, 908
Density function of
energy, 652, 667
probability, 160, 176, 197, 326, 702
Depolarization tensor, 103, 714
Dielectric constant, 41
Differential methods, 367
Diffusion. *see* Conduction
Dilute approximation, 350, 361, 364, 394, 529, 550, 811
Dirac's delta function, 60, 97, 503, 509, 611
Dirichlet tessellation, 173
Discontinuity, 23, 504, 662, 800, 807, 827
Dispersion
colloidal, 179
matrix, 161
nonoverlapping spheres, 171, 179, 194, 204
radial distribution function, 170, 179, 187, 194, 428
stresses, 459, 464, 705
Distribution
bimodal, 213, 378
binomial, 162, 172
"ellipsoidal symmetry," 326, 410, 490, 701
function, *n* particle, 177, 185, 193, 326, 344, 439
function, orientation, 197–199, 248, 383
function, radial, 170, 186, 187, 200, 428
generalized function, 63, 67, 97
Gibbs, 178
inclusion size, 100, 197, 214, 375, 786
normal, 163, 529
Poisson, 162, 172
probability, 160, 172, 183
Drucker's postulate, 56
Duhamel-Newmann law, 38

E

Edge
correction, 169, 171, 205, 853
effect, 114, 167, 601, 897
Effective
compliance, 224, 318
conductivity, 714
eigenstrains, 230, 232, 247, 499, 603, 622
eigenstresses, 229, 240, 603
elastic moduli, 9, 223, 226, 231, 240, 260, 361, 386, 423, 615, 741, 747, 755, 757, 781, 857
energy functions, 232, 238, 269, 672
energy release rate, 529
fields, 135, 144, 147, 149, 277, 287, 295, 316, 464, 487, 580, 685, 692, 693, 793, 814
fracture surface, 526
Gibbs energy, 238
interface fields, 491
limiting surfaces, 530
medium methods, 362, 365, 425, 703
nonlocal operator, 10
plastic strain increment, 561
properties, 172, 191, 266, 322, 335, 363, 382, 394, 410, 415, 424, 484, 581, 589, 592, 598, 618, 643, 694, 717, 775, 779, 843, 891
specific heat, 42, 238, 269
Effective conductivity
bounds, 354
differential method, 715
dilute approximation, 722, 770
effective medium method, 713
MEFM, 715
model methods, 713
Mori-Tanaka method, 331
nonlinear properties, 717

Effective conductivity (*cont.*)
nonlocal properties, 616, 717
percolation threshold, 713, 716
perturbation methods, 713
Effective elastic moduli
analytical representations, 338, 369, 429, 430, 433
apparent bounds, 222
clustered composites, 386
combined MEFM-perturbation method, 598
composites with imperfect interface, 90, 246, 502, 680
definition of, 17, 223, 225, 481
differential methods, 367, 372
doubly periodical composites, 195, 302
effective field method (EFM), 296
effective medium approach, 362, 364, 425, 703
functionally graded materials, 187, 572, 595
generalized self-consistent method, 365
half-space, 599
Hashin-Shtrikman bounds, 346, 354, 439
Hill's bounds, 260, 780
interface definition, 483
Maxwell schemes, 394, 597
MEFM, generalized, 410, 452
MEFM, one-particle approximation, 325
MEFM, standard, 325
periodic system with random imperfections, 307
perturbation method, 358–362
polifractional composites, 374
polycrystals, 248, 263, 367
self-consistent method, 362, 366
triply periodical composites, 268, 302, 305, 706, 708
Effective field hypotheses
approximate, 144, 391, 488, 496
closing, 315, 409, 449, 489
eigenstrains, 38, 81, 141, 229, 231, 240, 499, 528, 530, 600, 764
eigenstresses, 38, 229, 240, 759
link to Mori-Tanaka approach, 325
link to quasi-crystalline approximation, 315, 325, 450, 490
Ellipsoidal inclusion
coated, 94
conductivity problem, 100–105
elastic problem, 90
Eshelby tensor, 81
fictitious, 81
imaginary, 119, 820, 914
imperfect interface, 95
piezoelectric problem, 109
scattering of elastic waves, 106
Energy
complementary, 34, 40, 661
conservation equation, 28
decomposition, 236
effective functions, 232, 269
Gibbs, 38, 233
Helmholtz free-energy, 36, 233
interaction, 236
internal, 28, 36
potential, 35, 40, 233, 780
release rate, 527, 743
stored, 241, 317, 676
strain, 30–31, 233, 626, 649, 651, 731, 773, 782
stress, 32
Entropy, 29, 36, 237
Equivalent inclusions method, 80, 118, 336
Ergodicity condition, 166, 176, 189, 615
Eshelby formula, 235, 658
Eshelby tensor
conductivity, 103
elasticity, 81–90, 569, 596, 619, 782
piezoelectricity, 109
related tensors, 81
symmetry properties, 85
Excluded volume, 185, 202, 294, 309, 326, 348, 354, 358, 396, 410, 425, 428, 432, 435, 470, 494, 643, 647, 659, 702, 710, 819, 820

F

Failure, 379, 480, 522–523, 525, 741
Fast and slow variables, 284, 678
Fiber-reinforced composites, 3, 535, 572
Finite element analysis, 119, 426, 430, 443, 497, 564
Flow law, 56, 546, 549
Fluctuations
of SIF, 530
of stresses, 474, 633
Flux, 101, 672, 673
Foams, 3, 367
Fourier
transform, 60, 68, 76, 106, 180
transform method, 571, 578
Fractal, 365, 384, 713
Fracture
criterion, 527
mechanics, 89

probability, 529
surface, 527, 743
toughness, 5
Functionally graded materials
boundary layer effect, 227, 289, 510, 592, 718
definition of, 277, 575
deterministic structures, 222
elastically homogeneous composites, 603
free edge effect, 289, 599
local effective elastic moduli, 592
modeling, 591, 740
nonlocal integral equation, 574
statistical description, 176

G
Gaussian distribution, 162, 388, 529
Gauss's divergence theorem, 22, 222, 235, 245, 484, 507, 656, 683, 751, 753, 856
Gauss-Seidel method, 804
General integral equations
coupled phenomena, 671, 688, 718
interface —, 485–488, 492
local elasticity, 290, 293
peridynamics, 796, 798–847
periodic structures, 305–307, 688
Stokes flow, 510–520
strain gradient, 660
stress gradient, 664
strongly nonlocal, strain-type, 635
wave propagation, 721
Generalized self-consistent method, 338
Gibbs
energy, 38, 41, 232
point field, 177
Green's function
for conductivity problem, 100
coupled phenomena, 600, 718
for displacements, 59
dynamic, 67
dynamic, steady-state, 71, 722
isotropic medium, 62, 83
peridynamics, 789, 846
for piezoelectric problem, 109
second derivative of, 63–67
Stokes flow, 512
strain gradient, 654
for stresses, 67, 504, 636, 678
stress gradient, 665
strongly nonlocal, strain-type, 630
Gurson criterion, 533

H
Half-space, 536
Hard core model, 180, 202, 204
Hardening, 58, 546, 551, 554, 562, 660
Heaviside step function, 70, 429, 576, 807, 874
Helmholtz
equation, 71
free-energy, 57, 233
Hexagonal symmetry, 50, 61, 247, 262
Hierarchy of equations, 311, 314, 409, 489
Hill's
bounds, local elasticily, 219–223
bounds, peridynamics, 783
condition for
Cosserat medium, 597
electroelasticity, 266
local elasticity, 219
peridynamics, 781
strain gradient medium, 649
stress gradient medium, 662
strongly nonlocal, strain-type, 615
estimation, 264
interface operator, 75, 110, 272
lemma, 221, 654
medium, 357
tensors, 79
weak assumption, 225
Homogeneity
of comparison medium, 9, 12, 255, 257, 258, 281, 404, 656
of effective fields, 127, 133, 144, 150, 316, 391, 405, 571, 639, 694
of eigenstrain, 82, 533, 660
statistical, 165, 168
Homogeneous boundary conditions, 34, 190, 220, 221, 223, 225, 226, 243, 258, 266, 276, 289, 292–298, 302, 577, 605, 615, 617, 626, 666, 696, 752, 756, 851, 857, 862, 899
Homogenization. *see* Effective elastic effective elastic

I
Imperfect interface model
cohesive zone, 99
general integral equations, 502, 676
Gurtin-Murdoch, 99
linear spring, 98, 265
sliding, 90
Steigmann and Ogden, 99
thermal contact, 104

Imperfections in unite cells, 391
Inclusions
arbitrary shape, 72, 111
coated ellipsoidal, 90, 94
conductivity problem for ellipsoidal–, 100
cylindrical fiber in the transversally isotropic medium, 89
ellipsoidal (*see* Oblate spheroids, prolate spheroids)
elliptical, 103, 112, 142, 534
fictitious ellipsoidal–, 91
–inside imaginary ellipsoid, 119
interaction (*see* Interaction of inclusions)
jump of stresses at the interface, 77
piezoelectric problem for ellipsoidal–, 109
simulation of geometrical distribudistributions of–(*see* Monte-Carlo simulation)
wave problem for ellipsoidal–, 105
Incremental method, 54, 543, 547
Increments
of elastic moduli, 615
of strains, 55, 543, 553
Indicator function, 92, 164, 169, 187, 191, 282, 400, 511, 615, 654, 673, 737, 751, 762, 765, 799, 838, 845, 850, 862, 864, 874, 878, 886
Intensity, 26, 165–167
Interaction
energy, 235, 268
of inclusions, 109
approximative methods, 94, 141, 391
boundary integral equations, BIE, 113
complex potentials method, 111
hybrid VEE and BIE method, 114, 492
multipole expansions, 112, 114, 375
volume integral equation, VIE, method, 142
Interface
boundary conditions
ideal, 75, 276
effective limiting surface, 522
imperfect, *see* Imperfect interface model
integral representations of effective moduli, 483, 758
integral technique, 485
operator
thermoelastic problem, 91, 98
thermo piezoelectric problem, 110, 270
polarisation tensor, 129
stresses, second moment, 462, 464
Inversion of tensors, 50, 78, 463, 803, 907
Isotropic
elastic constants, 31, 49, 51, 316, 339, 366, 594, 649
medium, 31, 62, 65, 67, 70, 78, 89, 106, 107, 263, 355, 511, 528, 533, 653, 716, 744
statistically–, 166, 178, 192, 194, 357, 580
transversally–, 46, 62, 89, 241, 247, 382, 595, 911
Iteration method
deformation plasticity theory, 543
doubly periodical structures, 195
generalizations of Hill's estimation, 264
half-space composite medium, 599
interacting inclusions, 150, 156
interface integral technique, 485
interface perturbator factors, 493
nonlinear elastic composites, 538
nonlocal effects, 577, 638, 802
perturbation method, 360
self-consistent estimations, 364, 715
statistically homogeneous structures, 641, 692
volume integral equation method, 142, 406, 493

J
J2-flow theory, 55
jamming limit, 201, 205, 209

K
Kirkwood approximation, 187

L
Lame's elastic constants, 31, 51
Laminated structures, 3, 176, 288, 396, 599, 688
Lax approximation. *see* Quasi-crystalline approximation
Lennard-Jones potential, 179, 381, 391
Levin formula, 241, 420, 622, 768
Ligament, 367
Limiting surface
failure, 523, 526
fracture, 527, 529
interface strength, 525, 5430
yielding, 523, 533
Linearization, 521, 540, 737
Local
constitutive law, 223, 230, 581
effective modulus, 588, 592
theory of elastoplastic deformations, 556

Locality principle, 284, 308
Long-range order, 185, 192, 284, 508, 678
Long-wave approximations, 106, 579

M
Material symmetry. *see* Symmetry material
Matrix representation of symmetric tensors, 43
Maxwell schemes, 392–401
Metal matrix composites, 5, 556
Micromechanics, 8–13
Minimum principle of
  complementary energy, 35, 40, 258
  peridynamics, 776–780
  potential energy, 35, 40, 255
Minkowski addition and subtraction, 164, 427, 494, 734, 750, 806
Mises yield criterion, 56, 523, 543, 659
Misorientation distribution function, 197
Molecular dynamic simulation, 201, 382, 608, 725
Monte-Carlo simulations
  in conductivity problems, 713
  in elastic problems, 481
  of geometrical structures, 191, 205
Mori-Tanaka mean field theory, 331, 432, 470, 589, 812
Moving averaging cell, 196, 302, 897
Moving domain, 202, 204, 296, 428, 469, 603, 818
Multi-index, 578, 614

N
Nanofiber composites
  modeling, 382, 391, 903
  properties, 7, 380
Nanofibers, 7, 380, 382, 388
Nanoplates. *see* Silicate clay nanocomposites
Newtonian viscosity of suspension, 214, 371, 593
No long-range order, 185, 193, 284, 508, 678
Nonlinear conductivity, 717
Nonlinear elasticity, 31, 540
Nonlocal
  constitutive low
    higher order models, 666
    peridynamics, 726
    strain gradient, 650
    stress gradient, 660
    strongly nonlocal, strain-type, 610
  effective elastic operator
    differential, 585, 598
    in a half-space, 601
    integral, 820, 892
  effects, 10, 569, 585, 607
Nonlocal integral equation, solution
  combined MEFM–perturbation method, 598
  direct quadrature method, 576
  fourier transform method, 578
  iteration method, 576
Non-overlapping model, 172, 736, 751
Number density, 165, 177, 185, 196, 213, 531, 592, 751, 855

O
Oblate spheroid, 87, 383, 387, 399
Oblate spheroids, prolate pheroids, 91
Orientation distribution function
  definition of, 197
  series expansion, 199
  uniform, 198, 247, 338, 383
Orthotropic symmetry, 46, 53, 101, 672

P
Particle approximation
  n–, 115, 319
  one–, 315, 325, 336
  two–, 319
Particle reinforced composites, 3, 171, 176, 196, 227, 375, 385, 426, 499, 524
Percolation threshold
  conductivity, 714
  elasticity, 382
Peridynamic micromechanics
  decomposition of material and field parameters, 749, 760, 781, 859
  dilute approximation method, 811
  effective eigenstrain, 768
  effective field method, 814
  effective moduli, 754, 756, 764, 768
  extended inclusion, 750, 751, 772, 813
  Hill's bounds, 783–786
  minimum principles, 776–780
  Mori–Tanaka method, 812
  virtual work, 777
Peridynamics
  bond-based, 726
  bond force, 727
  damage, 741
  horizon, 726, 728
  force density, 732
  microelastic, 729
  micromodulus, 736
  potential function, 730

Peridynamics (*cont.*)
spring stiffness, 731, 740
state based, 726
stress, 732, 733
Periodic structures, viii, 183, 302, 305, 399, 688, 829, 849
Permutation tensor, 20, 667
Perturbation methods
combined method, 358, 598
conductivity, 714
small concentrations of inclusions, 360, 370, 451, 528, 597
stress fluctuations, 629
weakly inhomogeneous media, 360, 599
Perturbator, 129–134, 405, 407, 486, 498, 504, 514, 601, 681, 689, 721, 791, 810, 844
Plane-strain and plane-stress problems, 54, 116, 470, 648, 726, 913
Plasticity
back-stresses, 55
deformation theory, 57, 543
effective limiting surface, 523, 527, 530
flow theory, 55, 551, 557
flow theory of elastically homogeneous composites, 550
localized models, 685
plastic coating, 558
ratchetting, 556
shakedown, 556
Point fields
Gibbs field, 178
Poisson field, 170
Poisson's ratio, 51
Polarization tensors
coupled, 719
double stress, 653
interface, 129, 483, 506, 771
micro, 766
momentum, 106, 674, 725
stress gradient, 663
volume, 92, 118, 154, 254, 306, 406, 465, 487, 619, 633, 681, 699
Polycrystal
bounds of effective conductivity, 715
bounds of effective moduli, 260, 261, 355
definition of, 8, 46
effective constants, 12, 250, 361
piezoelectric, 268
plasticity model, 659
Polymeric matrixes composites, 4, 6, 7, 200, 380, 385
Polynomial conservation theorem, 78
Porous materials, effective properties
conductivity, 712
elastic moduli, 374, 391
elastoplastic deformation, 546, 556
filtration, 100
limiting surface, 530–535
nonlinear elastic properties, 540, 541
stress fluctuations, 266, 459
Positive definite, 40, 72, 76, 101, 256, 355, 598, 672, 776, 779, 913
Prager's rule, 56
Probability
distribution, 160
measure, 160
Palm distribution, 167
space, 160
Probability density
conditional, 162, 183, 186, 411, 428, 457, 496, 701, 819
definition of, 160, 161, 166, 325
of indicator function, 191, 363
normal, 163
normalized condition, 183
n-particle, 178, 184–188
n-point, 191, 193
number, 177, 185, 187, 196
orientation, 197
Prolate spheroid, 91, 382, 397
Pyroelectric coefficients, 42

Q

Quadrature method, 61, 116, 142, 146, 405, 493, 572, 576, 640, 682, 770, 806, 837
Quantum dot and wire, 600
Quasi-crystalline approximation, 314, 325, 338, 449, 475, 490, 531, 571, 583, 641, 694, 722, 815

R

Radial distribution function
analytical representations, 202, 429
definition of, 168, 170, 186, 187
link to 2-point probability density, 193, 195
Percus-Yevick approximation, 180, 202, 214
simulation, 209, 212
well-stirred approximation, 185, 202
Willis approximation, 203
Random
marked process, 183
packing, 200–202, 204–208, 210–212, 853

point fields, 164, 167, 176
variables, 10, 160, 171
Regularization, 63, 79, 128, 139, 291, 677, 804, 896
Representative volume element, 223, 226, 304, 380, 418, 781, 851
Reuss approximation, 10, 260, 264, 338, 524, 785
Rodrigues vector, 198, 906

S

Scattering of elastic waves, 105, 722
Self-consistent method, 13, 338, 346, 362–367, 715
Series expansion
in combined MEFM-perturbation method, 580, 598
Green function, 106
iteration method, 145, 406, 435, 449, 493, 576, 640
Neumann, 145, 449, 577, 640, 693, 817, 845
in perturbation method, 597
in spherical harmonics, 112, 199, 264
of strain energy, 32
Taylor, 303, 519, 597, 613, 736
Shear modulus, 51, 53, 261, 357, 470, 593
Silicate clay nanocomposites
exfoliated, 8, 384
intercalated, 8
modeling, 384, 388
properties, 8
Single-crystal
cubic symmetry, 48, 50, 261
elastic constants, 46–53
orientation, 197, 198
piezoelectric coefficients, 45
Singularity, 63, 128, 406, 528
Skeletization, 173
Slipping interfaces, 243, 247
Somigliana identity, 118
Specific heat at
constant strain, 37, 42, 237
constant stress, 38, 238
constant stress and electric field, 269
Statistical
average, 188, 190
basic descriptors, 167, 203
dependence, 162, 183
description of indicator functions, 191
ergodicity, 166
homogeneity, 165, 185, 186, 211, 220, 241
homogeneity of clustered fields, 174
inhomogeneity, 176, 188, 191, 283
isotropy, 166, 247, 261
uniformity, 166
Statistical moment of stresses
elastically homogeneous composites, 466
estimation by integral equations, 461, 465, 474
estimation by perturbation method, 253, 631, 660
first order, 320, 328, 383, 404, 419, 468, 494, 644
second order, 330
using in nonlinear problems, 523, 539, 544, 547, 553
Statistical second-order descriptors
autocorrelation function, 192
intensity function, 168
L-function, 170
nearest neighbor distribution, 167
radial distribution function, 170
two-particle probability density, 183, 186
Stereological technique, 192
Stokes flow, 510
Stokes theorem, 23
Stored energy, 236, 240, 249, 270, 317, 331, 466, 627, 675, 775, 780
Strain
energy, 30–34, 40, 233, 626–629, 649, 651, 655, 660, 731, 740, 757
invariants, 25, 26
polarization (*see* Tensor polarization volume)
stress-free, 75, 82, 230, 418, 551, 594, 621, 761
Strain energy forms
Hook, 38
Kauderer, 31, 538, 540
Mooney-Rivlind, 30
Murnaghan, 31, 538, 540
neo-Hookean, 31
series expansion, 32
Stress
concentration, 95, 121, 267, 317, 332, 351, 374, 386, 394, 405, 429, 436, 471, 491, 578, 645
intensity, 530
invariants, 26, 27, 58, 253, 264, 265, 461, 524, 551, 632
polarization (*see* Tensor polarization volume)
residual, 220, 230, 236, 249, 273, 418, 466, 528, 603, 622, 627, 762, 770, 917

Structures of composite materials
deterministic, 3, 12, 166, 195, 227, 278, 300–307, 385, 399, 688, 795, 806(*See also* Periodic and)
foam (cellular), 2, 3, 367
laminates, 2, 4, 176, 288, 294, 396, 599, 688, 747
matrix, 3, 5, 177, 230, 279, 305, 362, 535, 557, 599, 623, 714, 764, 860
polycrystalline, 3 (*See also* Polycrystal)
random, 3, 12, 141, 167, 181, 191, 202, 282, 311, 364, 379, 483, 523, 557, 573, 796, 877 (*See also* Random and)
Summation convention, 21, 44
Superposition techniques, 118, 147, 278, 290, 399, 419, 510, 596
Symmetry
conductivity, 100
diagonal, 32
effective modulus, 224, 248, 262, 339, 619, 756, 781, 858
Eshelby tensor, 85, 86
material, 32, 46, 666, 672
elastic tensors, 32, 46–50, 60
orientation distribution function, 197
pair, 32
structures of ellipsoidal, 325, 335, 345, 355, 410, 424, 453, 583, 641, 695, 701

T

Tensor
algebra, 19–25
basis–(fourth order), 86–90, 339, 907
basis–(second order), 905
fundamental displacement (Green's), 59, 67, 117
fundamental stress, 66
fundamental traction, 67, 117
matrix representation, 43
strain, 24, 29, 44
stress, 26, 27, 31
symmetric, 21
unit, 22, 51
Tessellation
Delaunay, 206
Dirichlet, 173
Voronoi, 172
Thermal
boundary conditions, 39
conductivity (*see* Conductivity)
conductivity equation, 100
contact, 104
effective thermal expansion, 242, 247, 268, 318, 773
expansion, 38, 42, 55, 96, 382, 415, 610, 745, 759
loading, 524, 555
mismatch, 414, 485
strain, 49
stresses, 316
Thermoelasticity, 36, 39, 45, 708
Thermoelectroelasticity, 42, 44, 265, 718
Thermoperistatics, 759, 769
Thin-layer hypothesis, 91, 95, 98, 557, 559
Threshold. *see* Percolation
Traction, 26, 30, 33, 39, 96, 97, 116, 126, 130, 136, 230, 250, 362, 516, 525, 650, 655, 667, 674, 733, 776, 851, 867, 890
Tsai-Wu criterion, 523
Two-phase composites
effective constants, 238, 716
elastic concentrator factors, 239, 243
statistical description, 191
thermal stress-concentrator factors, 240, 242, 420, 553
thermoelectroelastic concentrator, 270

U

Uniform
boundary conditions (*see* Homogeneous boundary conditions)
distribution, 162, 163, 166, 171, 174, 178, 192, 196, 208
field theory, 242
macroscopically, 280, 366
piecewise–local properties, 232, 762
random orientation, 196, 199, 247, 264, 338, 383, 911
strain, 220, 255, 329, 624
stress, 220, 258, 615, 656
Unit cell, 12, 195, 226, 301, 305, 367, 375, 399, 480, 564, 572, 576, 713, 796, 800, 850–865, 888

V

Variational principle, 255, 342, 564, 722
Virtual work, 35, 233, 246, 257, 777
Voight approximation, 10, 261, 306, 338, 356, 549
Volume concentration, 165, 174, 177, 191, 201, 306, 357, 361, 392, 397, 714
Volume integral equation method, 142